Genes VII

Benjamin Lewin

OXFORD
UNIVERSITY PRESS

OXFORD
UNIVERSITY PRESS

Great Clarendon Street, Oxford OX2 6DP

Oxford University Press is a department of the University of Oxford.
It furthers the University's objective of excellence in research, scholarship,
and education by publishing worldwide in

Oxford New York

Athens Auckland Bangkok Bogotá Buenos Aires Calcutta
Cape Town Chennai Dar es Salaam Delhi Florence Hong Kong Istanbul
Karachi Kuala Lumpur Madrid Melbourne Mexico City Mumbai
Nairobi Paris São Paulo Singapore Taipei Tokyo Toronto Warsaw
with associated companies in Berlin Ibadan

Published in the United States
by Oxford University Press Inc., New York

First published 2000

A catalogue record for this book is available from the British Library

Library of Congress Cataloging in Publication Data
(Data applied for)

ISBN 0–19–879277–8 (Pbk)
0–19–879276–X (Hbk)

Typeset by J&L Composition Ltd, Filey, North Yorkshire
Printed in The United States of America

Preface

This edition of GENES marks the most significant reorganization for several editions. It is time to acknowledge that the power of direct analysis of the genome is making a significant difference to our approach. This is now recognized in the first part of GENES, which starts with the concept of the gene as a segment of DNA coding for protein and then proceeds directly to the characterization of the genome in terms of its content of genes.

The rapidity of our advance in analyzing the genome has led to some confusion about means and ends. A genome is no more than a sum of the genes it contains, although the number and complexity of these genes gives us important insights into the nature of their connection into regulatory networks. The focus of this book, as its name suggests, therefore remains firmly on the gene.

The thesis of GENES is that only by understanding the structure and function of the gene itself will we be able in turn to understand the operation of the genome as a whole. Although the emphasis has shifted to the characterization of eukaryotic genes, and therefore to their analysis by the direct techniques of molecular biology rather than the subtlety of genetics, the classical approach remains intellectually penetrating. It remains an aim of this book to integrate both approaches in the context of a unified approach to prokaryotes and eukaryotes.

We consider the gene from all aspects: the basic forms that it takes, the numbers and relationships among genes in a genome, their packaging into chromosomes, the process of gene expression from transcription through translation, the reproduction and safeguarding of the structure of the gene, and finally some aspects of the overall circuitry through which genotype determines phenotype.

Almost two decades after the first edition of GENES, this latest edition is a celebration of the remarkable progress during that time, and, I hope, an indication of the future to come.

B. L.
Cambridge, Massachussetts
January 2000

Outline

Contents

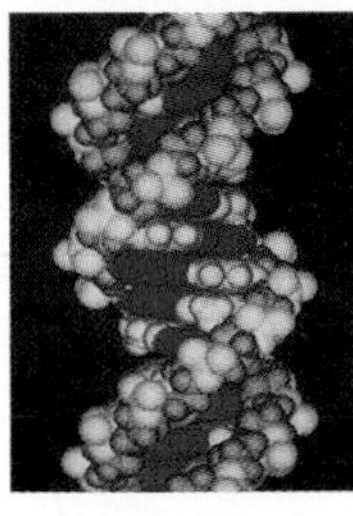

Part 1 **Genes** 1

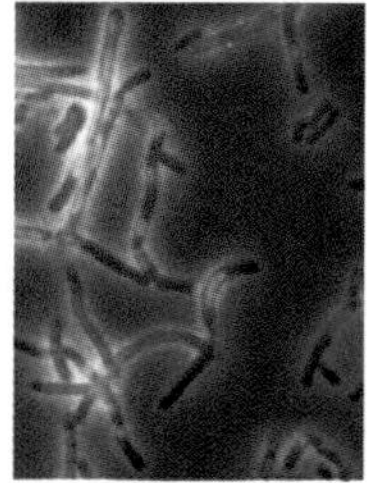

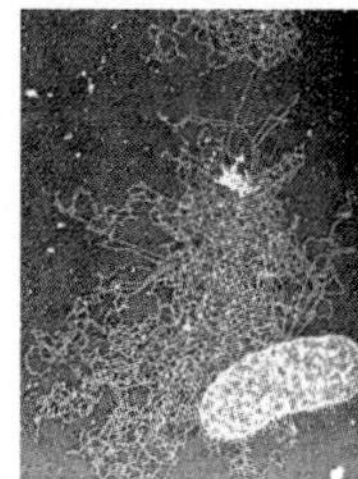

20 **Initiation of transcription** 617

21 **Regulation of transcription** 649

22 **Nuclear splicing** 685

23 **Catalytic RNA** 719

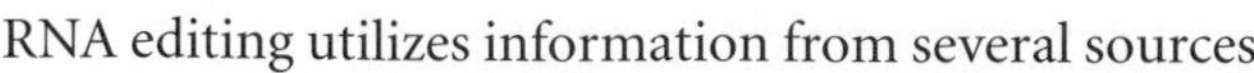

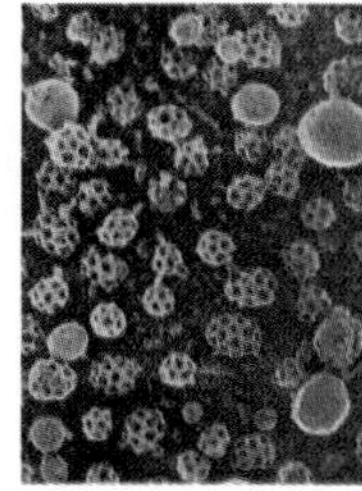

Part 1

Genes

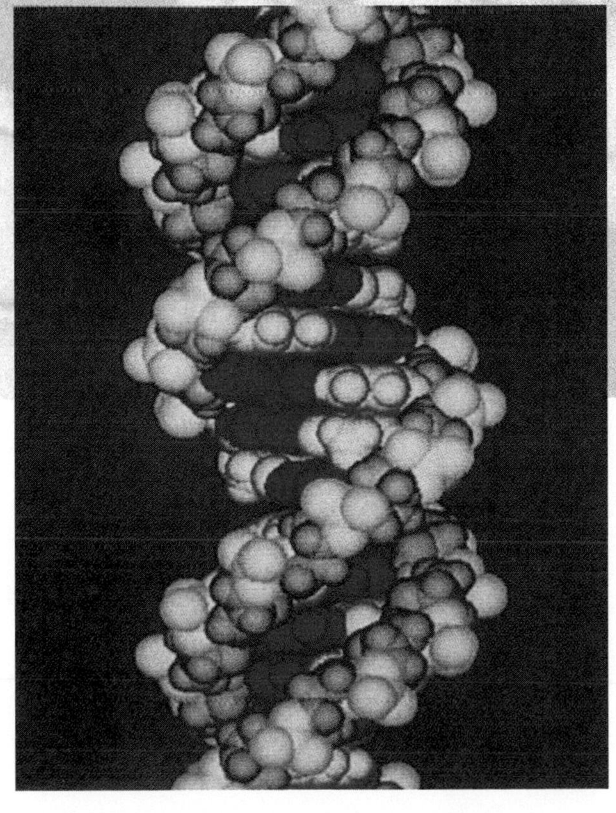

Chapter 1

Genes are DNA

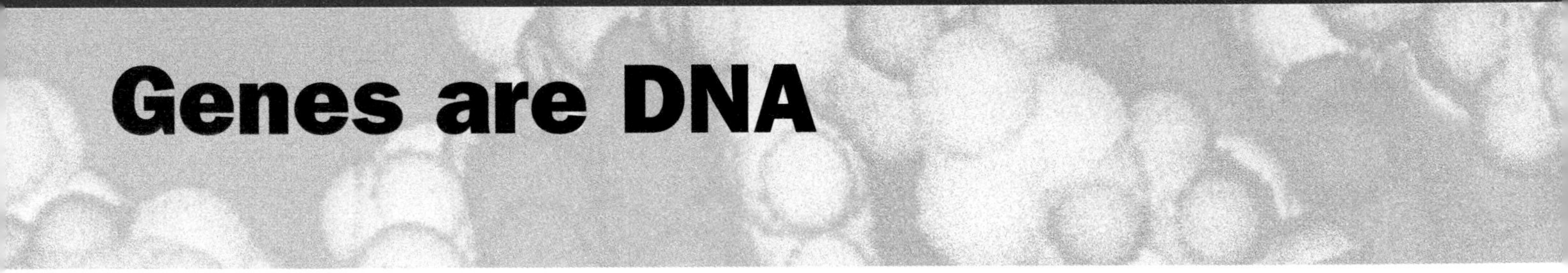

THE basic nature of the gene was defined by Mendel more than a century ago. Summarized in his two laws, the gene was recognized as a "particulate factor" that passes unchanged from parent to progeny. A gene may exist in alternative forms (alleles).

In diploid organisms, which have two sets of chromosomes, one copy of each chromosome is inherited from each parent. This is the same behavior that is displayed by genes. The equivalence led to the discovery that chromosomes in fact carry the genes.

The next step was the demonstration that each chromosome consists of a linear array of genes. Mendel's laws predict that genes carried on different chromosomes will segregate independently. However, genes that are on the same chromosome show linked inheritance. The basic observation is that genes on different chromosomes recombine at random from one generation to the next, whereas genes that are linked show a reduction in recombination, that is, they tend to stay together.

Genetic analysis allows the construction of a linkage map that connects all the genes carried by one chromosome. The genetic map of a linkage group corresponds to the physical existence of the chromosome.

On the genetic maps of higher organisms established during the first half of this century, the genes are arranged like beads on a string. They occur in a fixed order, and genetic recombination involves transfer of corresponding portions of the string between homologous chromosomes. The gene is to all intents and purposes a mysterious object (the bead), whose relationship to its surroundings (the string) is unclear.

The resolution of the recombination map of a higher eukaryote is restricted by the small number of progeny that can be obtained from each mating.

Figure 1.1 A brief history of genetics.

1865	Genes are particulate factors
1903	Chromosomes are hereditary units
1910	Genes lie on chromosomes
1913	Chromosomes contain linear arrays of genes
1927	Mutations are physical changes in genes
1931	Recombination is caused by crossing over
1944	DNA is the genetic material
1945	A gene codes for a protein
1953	DNA is a double helix
1958	DNA replicates semiconservatively
1961	Genetic code is triplet
1977	DNA can be sequenced
1997	Genomes can be sequenced

Recombination occurs so infrequently between nearby points that it is rarely observed between different mutations in the same gene. By moving to a microbial system in which a very large number of progeny can be obtained from each genetic cross, it became possible to demonstrate that recombination occurs within genes. It follows the same rules that were previously deduced for recombination between genes.

Mutations within a gene can be arranged into a linear order, showing that the gene itself has the same linear construction as the array of genes on a chromosome. So the genetic map is linear within as well as between loci: it consists of an unbroken sequence within which the genes reside. This conclusion segues naturally into the modern view that the genetic material of a chromosome consists of an uninterrupted length of DNA that represents many genes.

A genome consists of the entire set of chromosomes for any particular organism, and therefore comprises a series of DNA molecules, each of which contains a series of many genes. The ultimate definition of a genome is to determine the sequence of the DNA of each chromosome.

Figure 1.1 summarizes the stages in the transition from the historical concept of the gene to the modern definition of the genome. In this chapter, we analyze the properties of the gene in terms of its basic molecular construction. From the demonstration that a gene consists of DNA, and that a chromosome consists of a long stretch of DNA representing many genes, we move to the physical basis for replication and recombination of genes, and for their expression as protein products. In Chapter 2, we take up in more detail the organization of the gene and its representation of proteins. In Chapter 3, we consider the total number of genes, and in Chapter 4 we discuss other components of the genome and the maintenance of its organization.

DNA is the genetic material

THE idea that genetic material is nucleic acid had its roots in the discovery of **transformation** in 1928. The bacterium *Pneumococcus* kills mice by causing pneumonia. The virulence of the bacterium is determined by its **capsular polysaccharide.** This is a component of the surface that allows the bacterium to escape destruction by the host. Several **types** (I, II, III) of pneumococcus have different capsular polysaccharides. They have a **smooth** (S) appearance.

Each of the smooth pneumococcal types can give rise to variants that fail to produce the capsular polysaccharide. These bacteria have a **rough** (R) surface (consisting of the material that was beneath the capsular polysaccharide). They are **avirulent.** They do not kill the mice, because the absence of the polysaccharide allows the animal to destroy the bacteria.

When smooth bacteria are killed by heat treatment, they lose their ability to harm the animal. But inactive, heat-killed S bacteria and the ineffectual variant R bacteria together have a quite different effect from either bacterium by itself. **Figure 1.2** shows that when they are injected together into an animal, the mouse dies as the result of a pneumococcal infection. Virulent S bacteria can be recovered from the mouse postmortem.

In this experiment, the dead S bacteria were of type III. The live R bacteria had been derived from type II. The virulent bacteria recovered from the mixed infection had the smooth coat of type III. So some property of the dead type III S bacteria can **transform** the live R bacteria so that they make the type III capsular polysaccharide, and as a result become virulent.

The component of the dead bacteria responsible for transformation was called the **transforming principle.** It was purified by developing a cell-free system, in which extracts of the dead S bacteria could be added to the live R bacteria before injection into the animal. The classic studies of Avery showed chemically in 1944 that the isolated transforming principle is **deoxyribonucleic acid (DNA)**.

The surprise of this result is indicated by the fact that, at this time, DNA was not even known to be a component of pneumococcus, although of course it had been recognized for many decades as a major component of eukaryotic chromosomes. In showing that the genetic material of a prokaryote is DNA, this result therefore offered a unifying view for the basis of heredity in bacteria and higher organisms.

The implications of the result were captured by the

Figure 1.2 The transforming principle is DNA.

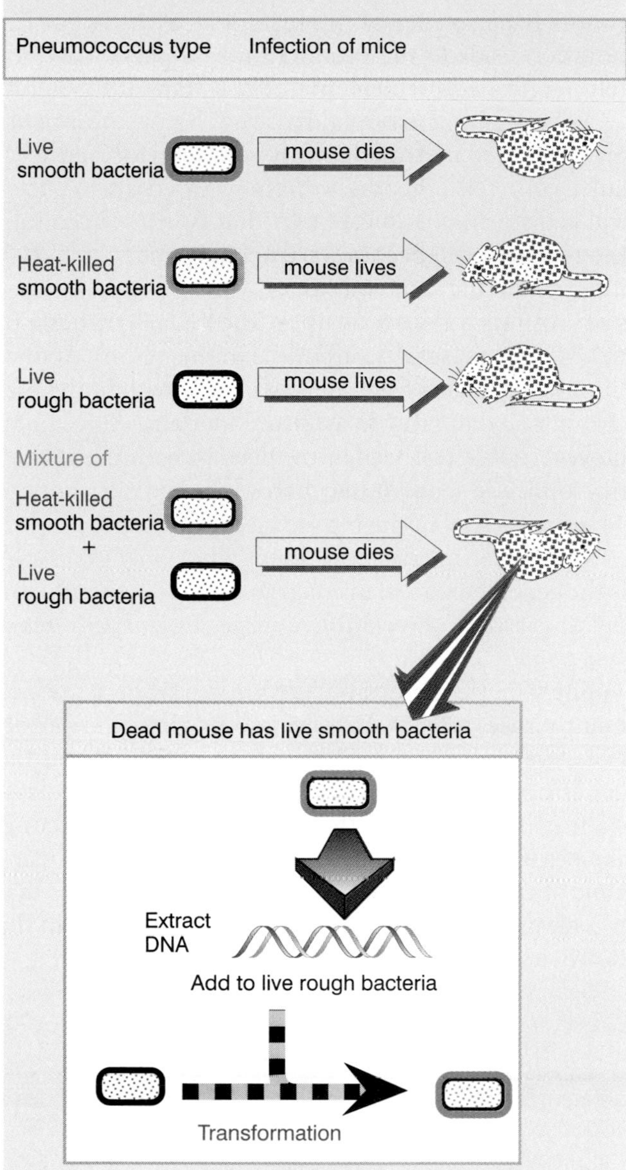

Figure 1.3 The genetic material of phage T2 is DNA.

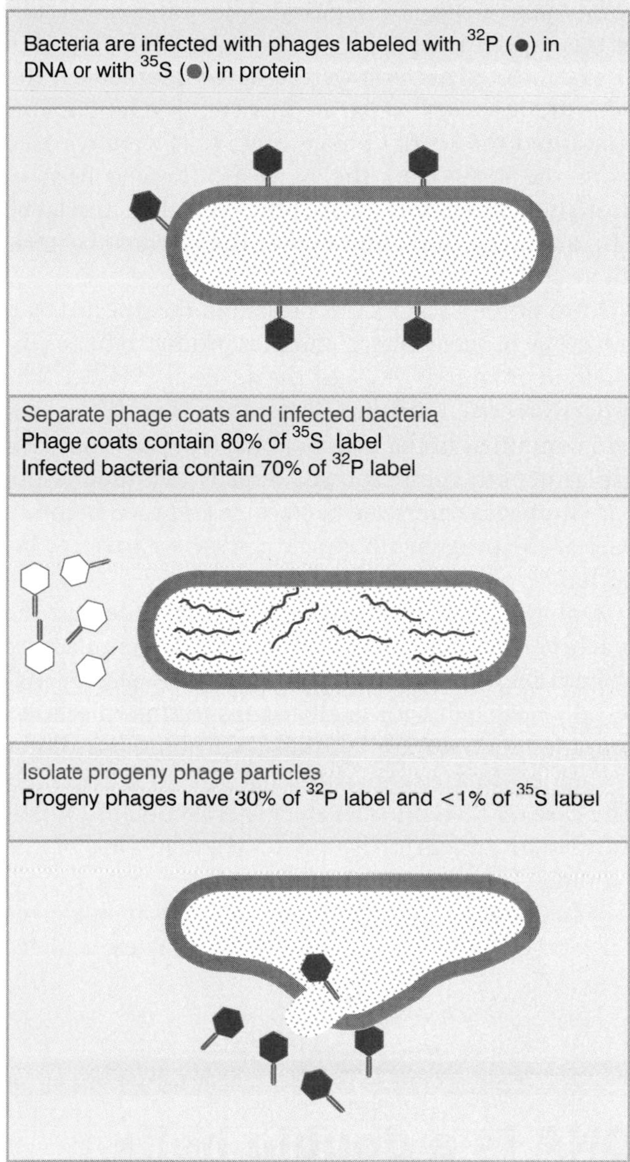

original paper. "The inducing substance, on the basis of its chemical and physical properties, appears to be a highly polymerized and viscous form of DNA. On the other hand, the type III capsular polysaccharide, the synthesis of which is evoked by this transforming agent, consists chiefly of a nonnitrogenous polysaccharide... So it is evident that the inducing substance and the substance produced in turn are chemically distinct and biologically specific in their action and that both are requisite in determining the type specificity of the cells of which they form a part."

This discussion marked the introduction of a distinction between the genetic material and the products of its expression, a view that became the implicit basis for subsequent studies.

After the transforming principle had been shown to consist of DNA, the next step was to demonstrate that DNA provides the genetic material in a quite different system. Phage T2 is a virus that infects the bacterium *E. coli*. When phage particles are added to bacteria, they adsorb to the outside surface, some material enters the bacterium, and then ~20 minutes later each bacterium bursts open (lyses) to release a large number of progeny phage.

In 1952, Hershey and Chase infected bacteria with T2 phages that had been radioactively labeled *either* in their DNA component (with ^{32}P) *or* in their protein component (with ^{35}S). **Figure 1.3** illustrates the results of this experiment.

The infected bacteria were agitated in a blender, and two fractions were separated by centrifugation. One contained the empty phage coats that were released from the surface of the bacteria; these consist of protein and therefore carried the ^{35}S radioactive label. The other fraction consisted of the infected bacteria themselves.

Most of the ^{32}P label was present in the infected bacteria. The progeny phage particles produced by the infection contained ~30% of the original ^{32}P label. The progeny received very little—less than 1%—of the protein contained in the original phage population. This experiment therefore showed directly that the DNA of parent phages enters the bacterium and then becomes part of the progeny phages, exactly the pattern of inheritance expected of genetic material.

A phage (virus) reproduces by commandeering the machinery of an infected host cell to manufacture more copies of itself. The phage possesses genetic material whose behavior is analogous to that of cellular genomes: its traits are faithfully reproduced, and they are subject to the same rules that govern inheritance. The case of T2 reinforces the general conclusion that the genetic material is DNA, whether part of the genome of a cell or virus.

When DNA is added to populations of single eukaryotic cells growing in culture, the nucleic acid enters the cells, and in some of them results in the production of new proteins. At first performed with DNA extracted *en masse*, these experiments now can be routinely performed with purified DNA whose incorporation leads to the production of a particular protein. **Figure 1.4** depicts one of the standard systems.

Although for historical reasons these experiments are described as **transfection** when performed with eukaryotic cells, they are a direct counterpart to bacterial transformation. The DNA that is introduced into the recipient cell becomes part of its genetic material, inherited in the same way as any other part. Its expression confers a new trait upon the cells (synthesis of thymidine kinase in the example in the figure). At first, these experiments were successful only with individual cells adapted to grow in a culture medium. Since then, however, DNA has been introduced into mouse eggs by microinjection; and it may become a stable part of the genetic material of the mouse (see Chapter 17).

Such experiments show directly not only that DNA is the genetic material in eukaryotes, but also that *it can be transferred between different species and yet remain functional.*

The genetic material of all known organisms and many viruses is DNA. However, some viruses use an alternative nucleic acid, **ribonucleic acid** (**RNA**), as the genetic material. Although its chemical formula is slightly different from that of DNA, in these circumstances RNA exercises the same role. The general principle of the nature of the genetic material, then, is that it is always nucleic acid; in fact, it is DNA except in the RNA viruses.

DNA is a double helix

A nucleic acid consists of a polynucleotide chain. **Figure 1.5** shows that the backbone of the chain consists of an alternating series of pentose (sugar) and phosphate residues. A purine or pyrimidine **nitrogenous base** is linked to the sugar. DNA takes its name from its sugar (2-deoxyribose); RNA is named for its sugar (ribose). The difference is that the sugar in RNA has an OH group at the 2 position of the pentose ring.

Each nucleic acid contains four types of base. The same two purines, adenine and guanine, are present in both DNA and RNA. The two pyrimidines in DNA are cytosine and thymine; in RNA uracil is found instead of thymine. The only difference between uracil and thymine is the presence of a methyl substituent at position C_5. The bases are usually referred to by their initial letters. DNA contains A, G, C, T, while RNA contains A, G, C, U.

The nitrogenous base is linked to position 1 on the pentose ring by a glycosidic bond from N_1 of pyrimidines or N_9 of purines. To avoid ambiguity between the numbering systems of the heterocyclic rings and the sugar, positions on the pentose are given a prime (′).

The polynucleotide chain is constructed by linking

Figure 1.4 Eukaryotic cells can acquire a new phenotype as the result of transfection by added DNA.

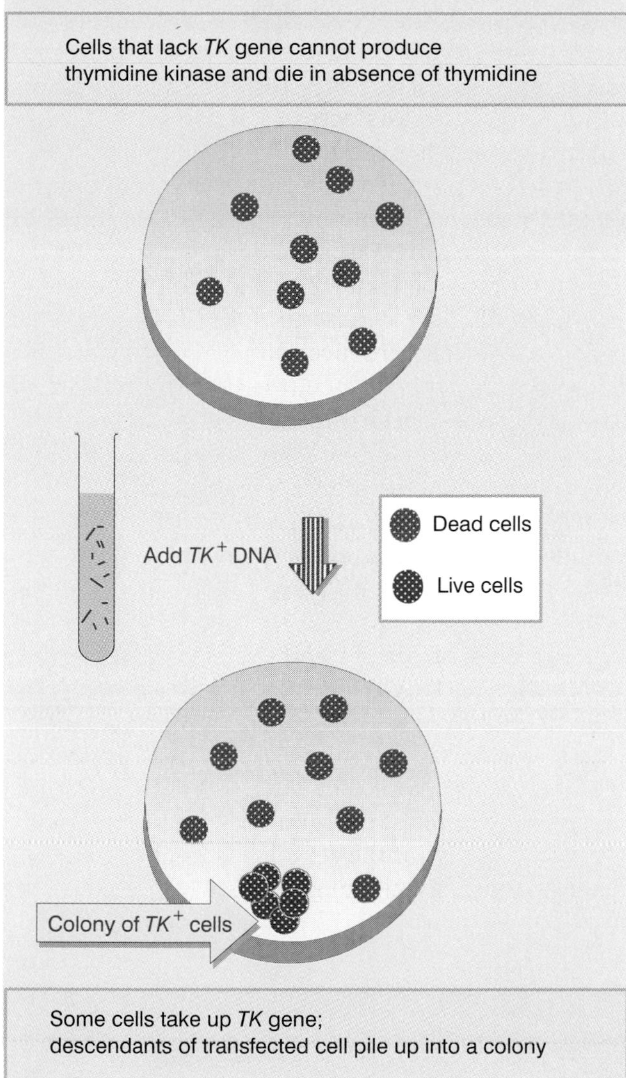

Figure 1.5 A polynucleotide chain consists of a series of 5'-3' sugar-phosphate links that form a backbone from which the bases protrude.

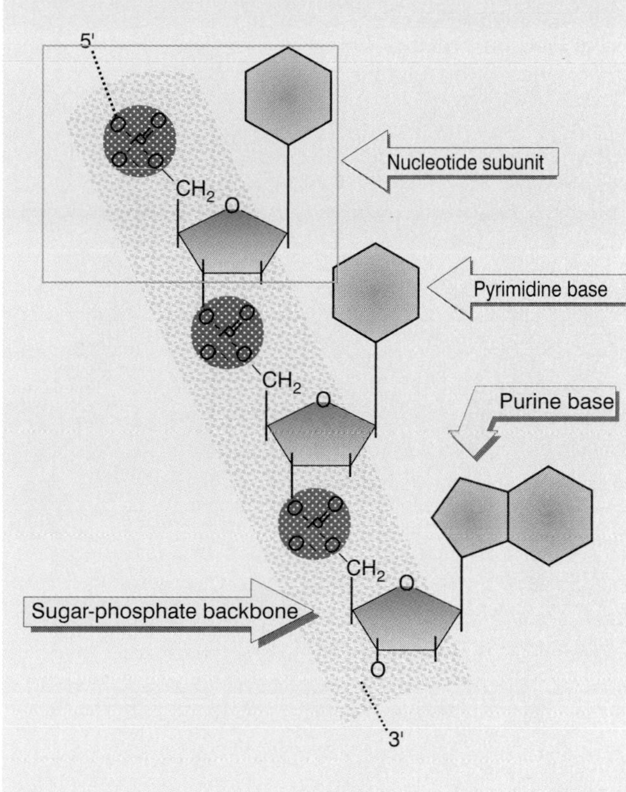

the 5′ position of one pentose ring to the 3′ position of the next pentose ring via a phosphate group. So the sugar-phosphate backbone is said to consist of 5′–3′ phosphodiester linkages. The nitrogenous bases "stick out" from the backbone.

The terminal nucleotide at one end of the chain has a free 5′ group; the terminal nucleotide at the other end has a free 3′ group. It is conventional to write nucleic acid sequences in the 5′→3′ direction—that is, from the 5′ terminus at the left to the 3′ terminus at the right.

The observation that the bases are present in different amounts in the DNAs of different species led to the concept that the *sequence of bases is the form in which genetic information is carried.* By the 1950s, the concept of genetic information was common: the twin problems it posed were working out the structure of the nucleic acid, and explaining how a sequence of bases in DNA could represent the sequence of amino acids in a protein.

Three notions converged in the construction of the double helix model for DNA by Watson and Crick in 1953:

- X-ray diffraction data showed that DNA has the form of a regular helix, making a complete turn every 34 Å (3.4 nm), with a diameter of ~20 Å (2 nm). Since the distance between adjacent nucleotides is 3.4 Å, there must be 10 nucleotides per turn.
- The density of DNA suggests that the helix must contain two polynucleotide chains. The constant diameter of the helix can be explained if the bases in each chain face inward and are restricted so that a purine is always opposite a pyrimidine, avoiding partnerships of purine-purine (too thick) or pyrimidine-pyrimidine (too thin).

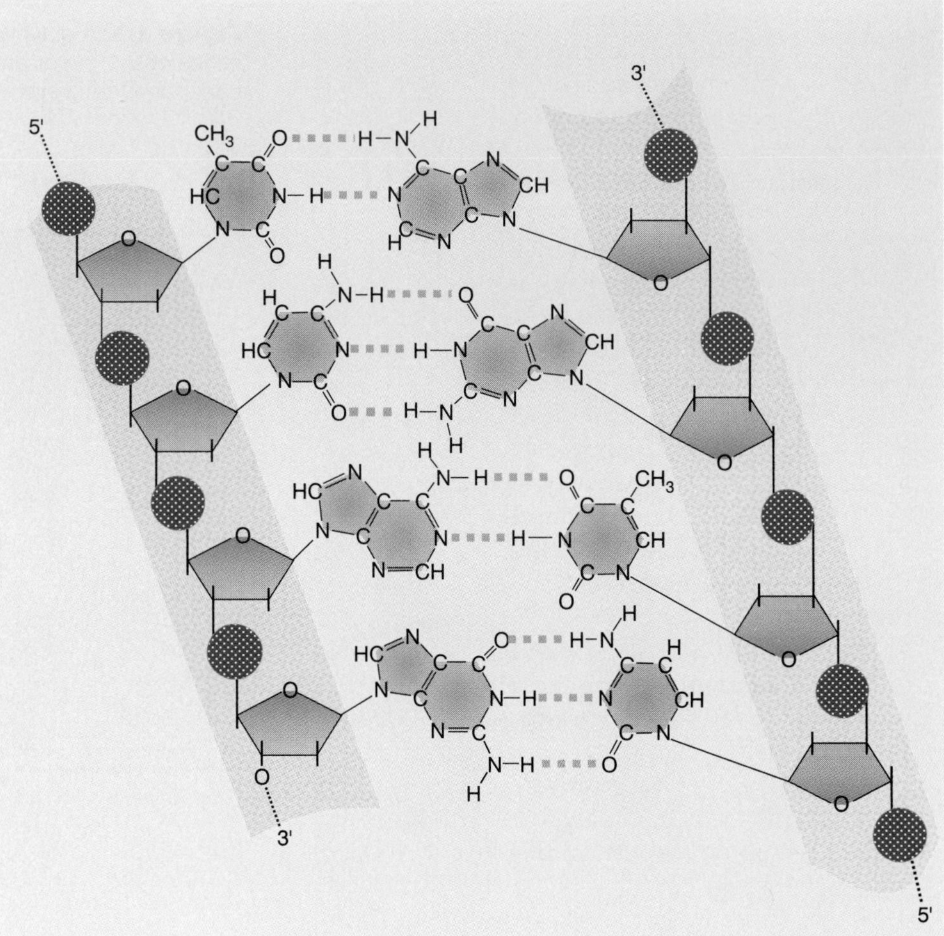

Figure 1.6 The double helix maintains a constant width because purines always face pyrimidines in the complementary A–T and G–C base pairs. The sequence in the figure is
T–A
C–G
A–T
G–C

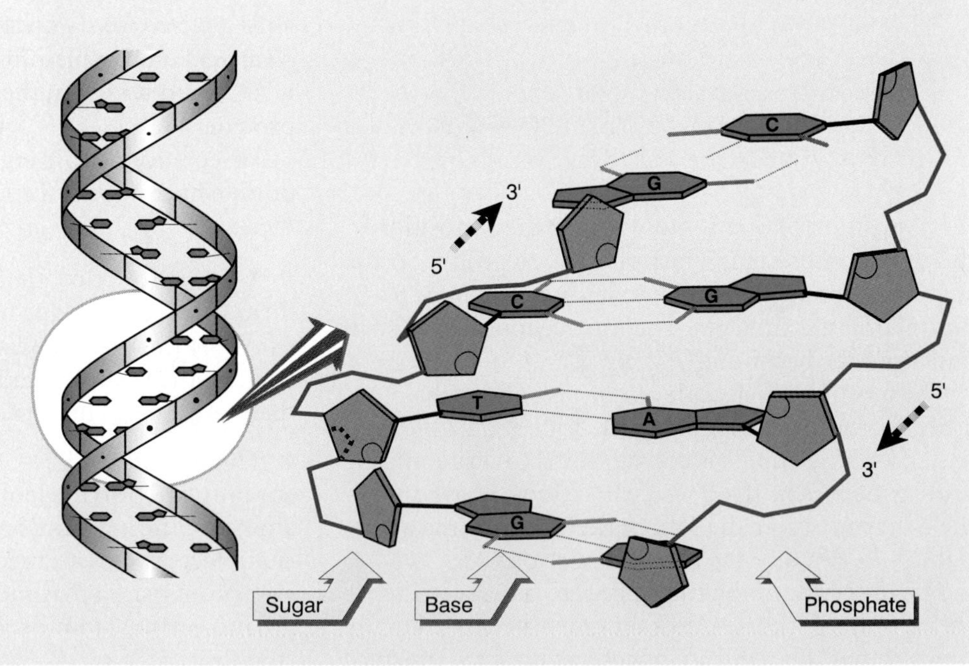

Figure 1.7 Flat base pairs lie perpendicular to the sugar-phosphate backbone.

■ Irrespective of the actual amounts of each base, the proportion of G is always the same as the proportion of C in DNA, and the proportion of A is always the same as that of T. So the composition of any DNA can be described by the proportion of its bases, that is G + C. This ranges from 26% to 74% for different species.

Watson and Crick proposed that the two polynucleotide chains in the double helix associate by *hydrogen bonding between the nitrogenous bases.* G can hydrogen bond specifically only with C, while A can bond specifically only with T. These reactions are described as **base pairing,** and the paired bases (G with C, or A with T) are said to be **complementary.**

The model requires the two polynucleotide chains to run in opposite directions (**antiparallel**), as illustrated in **Figure 1.6.** Looking along the helix, therefore, one strand runs in the 5′→3′ direction, while its partner runs 3′→5′.

The sugar-phosphate backbone is on the outside and carries negative charges on the phosphate groups. When DNA is in solution *in vitro*, the charges are neutralized by the binding of metal ions, typically by Na^+. In the natural state *in vivo*, positively charged proteins provide some of the neutralizing force. These proteins play an important role in determining the organization of DNA in the cell.

The bases lie on the inside. They are flat structures, lying in pairs perpendicular to the axis of the helix. Consider the double helix in terms of a spiral staircase: the base pairs form the treads, as illustrated schematically in **Figure 1.7.** Proceeding along the helix, bases are stacked above one another, in a sense like a pile of plates.

Each base pair is rotated ~36° around the axis of the helix relative to the next base pair. So ~10 base pairs make a complete turn of 360°. The twisting of the two strands around one another forms a double helix with a **narrow groove** (~12 Å across) and a **wide groove** (~22 Å across), as can be seen from the scale model in **Figure 1.8.** The double helix is **right-handed;** the turns run clockwise looking along the helical axis. These features represent the accepted model for what is known as the **B-form** of DNA.

Figure 1.8 The two strands of DNA form a double helix.

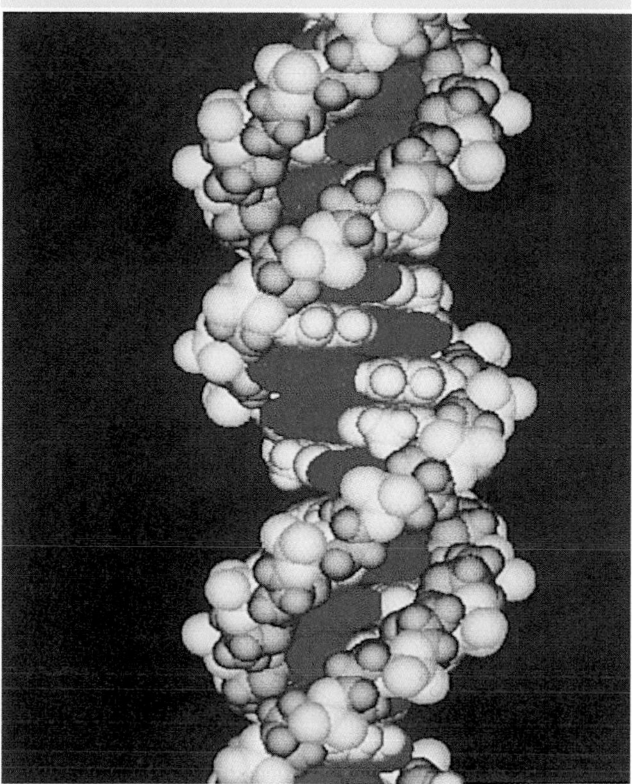

DNA replication is semiconservative

It is crucial that the genetic material is reproduced accurately. Because the two polynucleotide strands are joined only by hydrogen bonds, they are able to separate without requiring breakage of covalent bonds. The specificity of base pairing suggests that each of the separated **parental strands** could act as a **template** for the synthesis of a complementary **daughter strand,** as depicted in **Figure 1.9.** The principle is that a new daughter strand is assembled on each parental strand. The sequence of the daughter strand is dictated by the

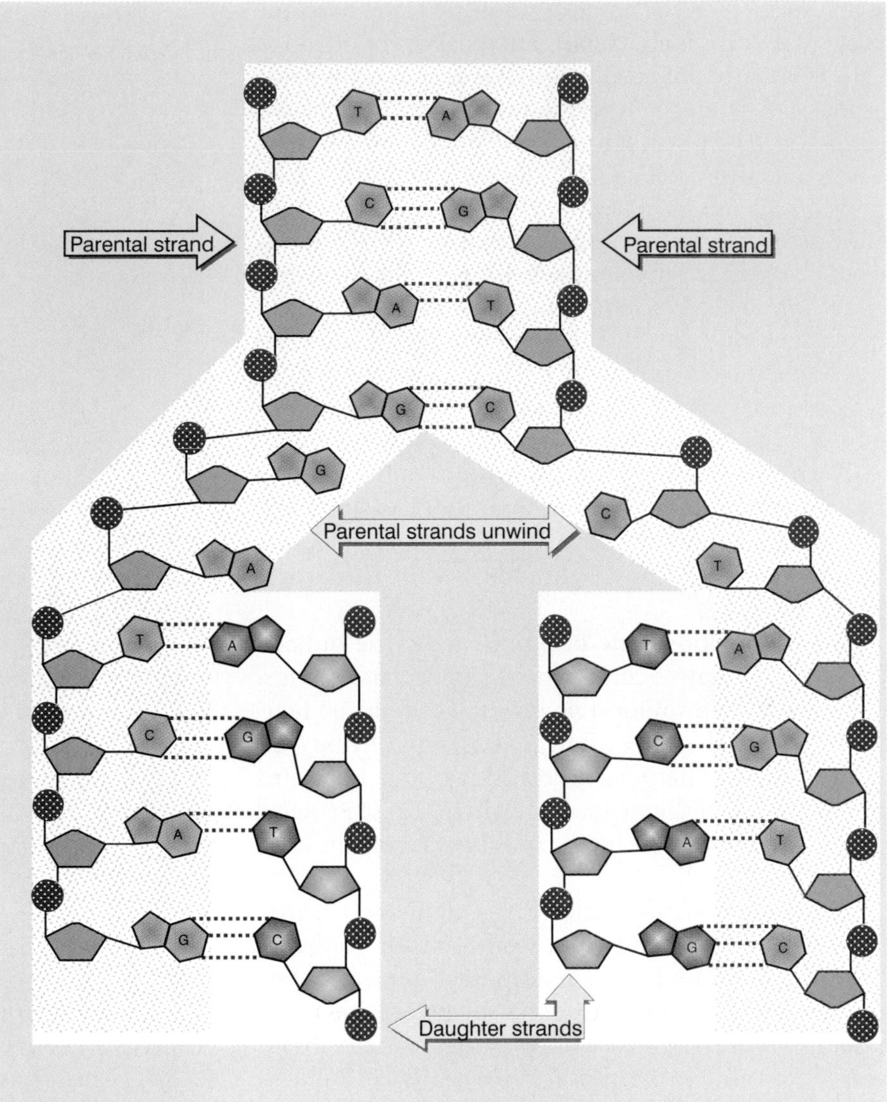

Figure 1.9 Base pairing provides the mechanism for replicating DNA.

parental strand; an *A* in the parental strand causes a *T* to be placed in the daughter strand, a parental *G* directs incorporation of a daughter *C*, and so on.

The top part of the figure shows a **parental duplex** and the lower part shows the two **daughter duplexes** that are being produced by complementary base pairing. The (unreplicated) parental duplex consists of the original two parental strands. Replication requires the two parental strands to be separated so that each can be used as a template for synthesis of a complement. Each of the daughter duplexes is identical in sequence with the original parent, and contains one parental strand and one newly synthesized strand. *The structure of DNA carries the information needed to perpetuate its sequence.*

The consequences of this mode of replication are illustrated for the DNA molecule as a whole in **Figure 1.10.** The parental duplex is replicated to form two daughter duplexes, each of which consists of one parental strand and one (newly synthesized) daughter strand. *The unit conserved from one generation to the next is one of the two individual strands comprising the parental duplex.* This behavior is called **semiconservative replication.**

The figure illustrates a prediction of this model. If the parental DNA carries a "heavy" density label because the organism has been grown in medium containing a suitable isotope (such as ^{15}N), its strands can be distinguished from those that are synthesized when the organism is transferred to a medium containing normal "light" isotopes.

The parental DNA consists of a duplex of two heavy strands (red). After one generation of growth in light medium, the duplex DNA is "hybrid" in density—it consists of one heavy parental strand (red) and one light daughter strand (blue). After a second generation, the two strands of each hybrid duplex have separated; each gains a light partner, so that now half of the duplex DNA remains hybrid while half is entirely light (both strands are blue).

The individual strands of these duplexes all are entirely heavy or entirely light. This pattern was confirmed experimentally in the Meselson-Stahl experiment of 1958, which followed the semiconservative replication of DNA through three generations of growth of *E. coli.* When DNA was extracted from bacteria and its density measured by centrifugation, the DNA formed bands corresponding to its density—heavy for parental, hybrid for the first generation, and half hybrid and half light in the second generation.

Figure 1.10 Replication of DNA is semiconservative.

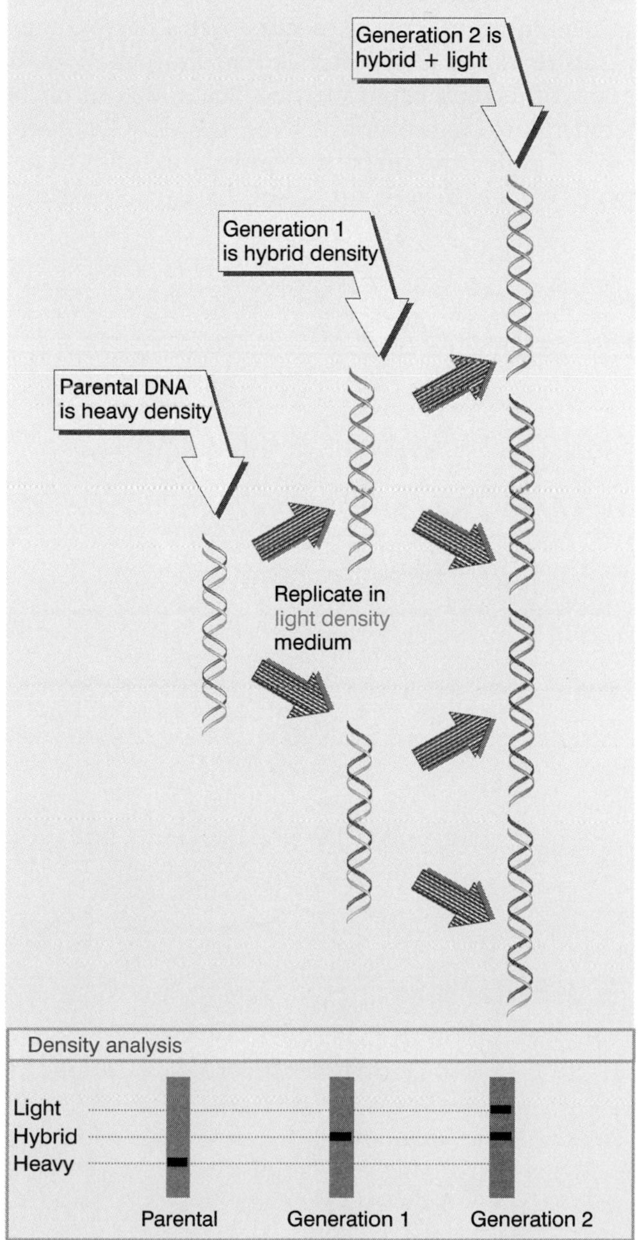

Replication involves a major disruption of the structure of DNA. However, although the two strands of the parental duplex must separate, they do not exist as single strands. The disruption of structure is only transient and is reversed as the daughter duplex is formed. So only a small part of the DNA loses the duplex structure at any moment.

The helical structure of a molecule of DNA engaged in replication is illustrated in **Figure 1.11.** The nonreplicated region consists of the parental duplex, opening into the replicated region where the two daughter duplexes have formed. The double helical structure is disrupted at the junction between the two regions, called the **replication fork.** Replication involves movement of the replication fork along the parental DNA, so there is a continuous unwinding of the parental strands and rewinding into daughter duplexes.

The synthesis of nucleic acids is catalyzed by specific enzymes, which recognize the template and undertake the task of catalyzing the addition of subunits to the polynucleotide chain that is being synthesized. The enzymes are named according to the type of chain that is synthesized: **DNA polymerases** synthesize DNA, and **RNA polymerases** synthesize RNA.

Degradation of nucleic acids also requires specific enzymes: **deoxyribonucleases (DNAases)** degrade

Figure 1.11 The replication fork is the region of DNA in which there is a transition from the unwound parental duplex to the newly replicated daughter duplexes.

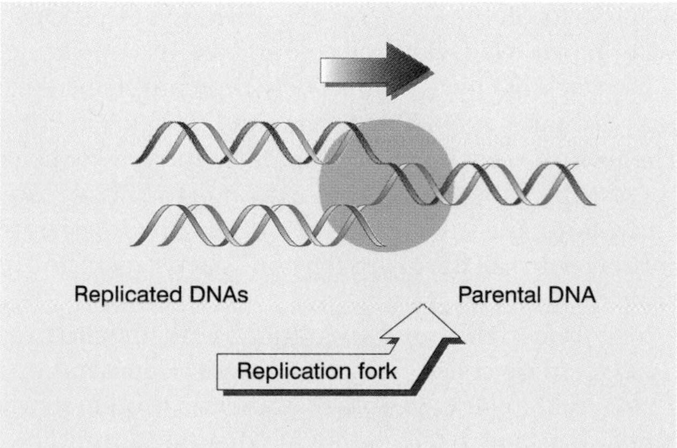

DNA, and **ribonucleases** (**RNAases**) degrade RNA. The nucleases fall into the general classes of **exonucleases** and **endonucleases.** Endonucleases cut individual bonds *within* RNA or DNA molecules, generating discrete fragments. They are involved in cutting reactions. Exonucleases remove residues one at a time from the end of the molecule, generating mononucleotides. They are involved in trimming reactions.

Nucleic acids hybridize by base pairing

A crucial property of the double helix is the ability to separate the two strands without disrupting covalent bonds. This makes it possible for the strands to separate and reform under physiological conditions at the (very rapid) rates needed to sustain genetic functions. The specificity of the process is determined by complementary base pairing.

The concept of base pairing is central to all processes involving nucleic acids. Disruption of the base pairs is a crucial aspect of the function of a double-stranded molecule, while the ability to form base pairs is essential for the activity of a single-stranded nucleic acid. **Figure 1.12** shows that the same set of base-pairing interactions that maintain the double helix, or that are used during replication to assemble duplicates, enable complementary single-stranded nucleic acids to form a duplex structure. An intramolecular duplex region can form by base pairing between two complementary sequences that are part of a single-stranded molecule. Or a single-stranded molecule may base pair with an independent, complementary single-stranded molecule to form an intermolecular duplex. Formation of duplex regions from single-stranded nucleic acids is most important for RNA, but single-stranded DNA also exists (in the form of viral genomes). Base pairing between independent complementary single strands is not restricted to DNA–DNA or RNA–RNA, but can also occur between a DNA molecule and an RNA molecule.

The lack of covalent links between complementary strands makes it possible to manipulate DNA *in vitro.* The noncovalent forces that stabilize the double helix are disrupted by heating or by exposure to low salt concentration. The two strands of a double helix separate entirely when all the hydrogen bonds between them are broken.

The process of strand separation is called **denaturation** or (more colloquially) **melting.** ("Denaturation" is also used to describe loss of authentic protein structure; it is a general term implying that the natural conformation of a macromolecule has been converted to some other form.)

Denaturation of DNA occurs over a narrow temperature range and results in striking changes in many of its physical properties. The midpoint of the temperature range over which the strands of DNA separate is called the **melting temperature** (T_m). It depends on the proportion of G•C base pairs. Because

Figure 1.12 Base pairing occurs in duplex DNA and also in intra- and inter-molecular interactions in single-stranded RNA (or DNA).

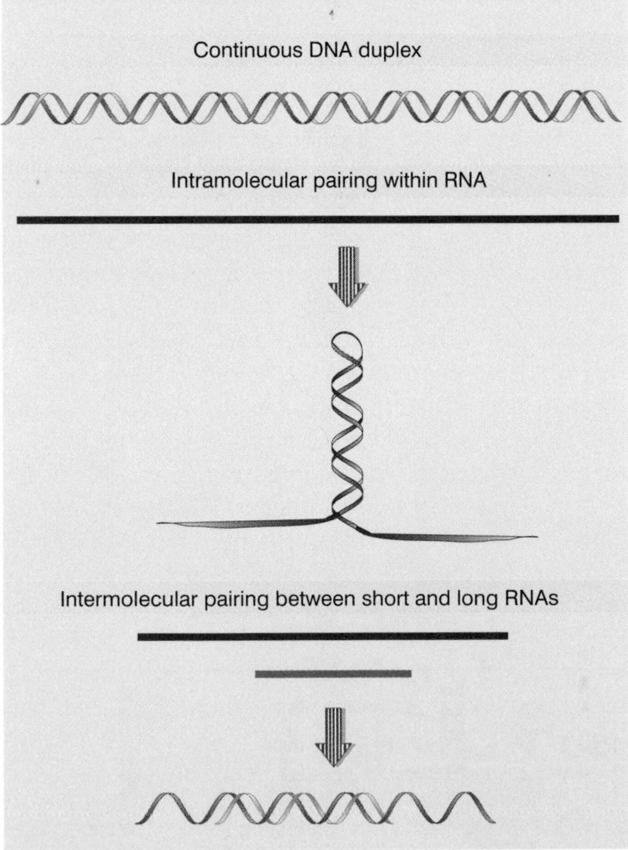

each G•C base pair has three hydrogen bonds, it is more stable than an A•T base pair, which has only two hydrogen bonds. The more G•C base pairs are contained in a DNA, the greater the energy that is needed to separate the two strands. In solution under approximately physiological conditions, a DNA that is 40% G•C—a value typical of mammalian genomes—denatures with a T_m of about 87°C. So without intervention from cellular systems, duplex DNA is stable at the temperature prevailing in the cell.

Nucleic acid sequences can be compared in terms of either similarity or complementarity:

- *Similarity* between two sequences is given in principle by the proportion of bases (for single-stranded sequences) or base pairs (for double-stranded sequences) that is identical. This is the measure that is usually used when (for example) comparing genes, but it requires direct determination of the sequence.
- *Complementarity* is determined by the rules for base pairing between A•T and G•C. In a perfect duplex of DNA, the strands are precisely complementary. If we compare two different but related double-stranded molecules, each strand of the first molecule will be similar to one strand of the second molecule and will be (partly) complementary to the other strand of the second molecule. *Complementarity can be measured directly by the ability of two single-stranded nucleic acids to base pair with each another.* If double-stranded molecules are denatured into single strands, the complementarity between the single strands can be used to indicate the similarity between the original duplex molecules.

It is possible to measure complementarity because the denaturation of DNA is reversible under appropriate conditions. The ability of the two separated complementary strands to reform into a double helix is called **renaturation.** It is illustrated in **Figure 1.13.**

Renaturation depends on specific base pairing between the complementary strands. The reaction takes place in two stages. First, single strands of DNA in the solution encounter one another by chance; if their sequences are complementary, the two strands base pair to generate a short double-helical region. Then the region of base pairing extends along the molecule by a zipper-like effect to form a lengthy duplex molecule. Renaturation of the double helix restores the original properties that were lost when the DNA was denatured.

Renaturation describes the reaction between two complementary sequences that were separated by denaturation. However, the technique can be extended to allow any two complementary nucleic acid sequences to **anneal** with each other to form a duplex structure. The reaction is generally described as **hybridization** when nucleic acids from different sources are involved, as in the case when one preparation consists of DNA and the other consists of RNA. *The ability of two nucleic acid preparations to hybridize constitutes a precise test for their complementarity since* only *complementary sequences can form a duplex structure.*

Figure 1.13 Denatured single strands of DNA can renature to give the duplex form.

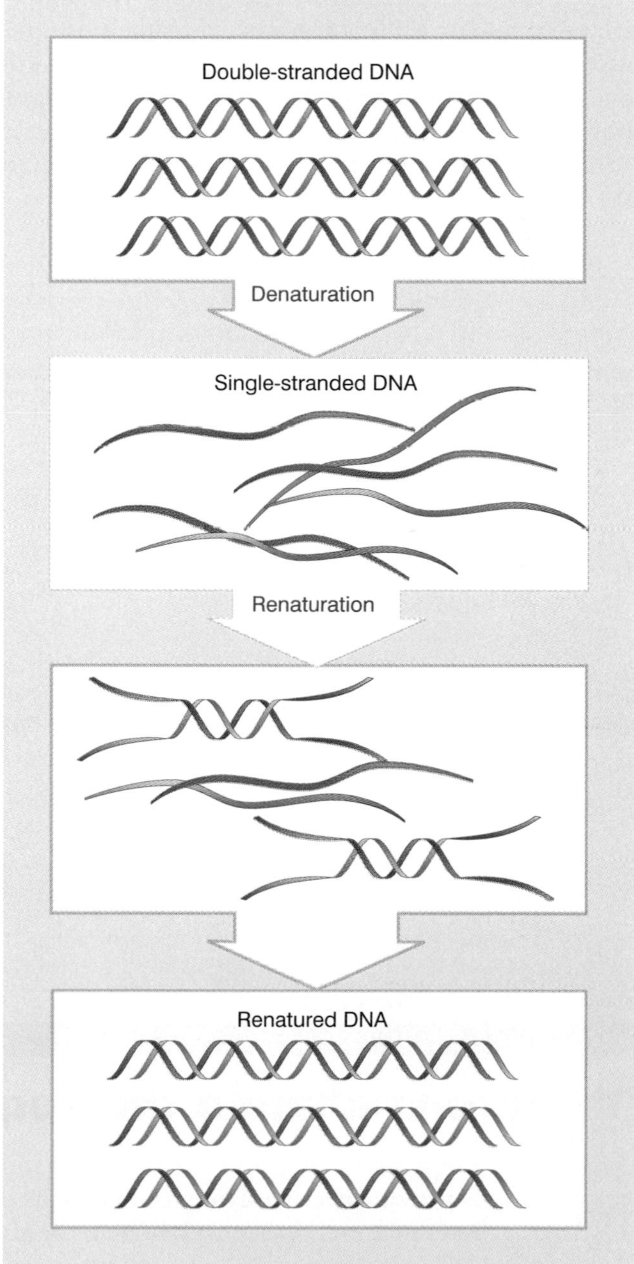

The principle of the hybridization reaction is to expose two single-stranded nucleic acid preparations to each other and then to measure the amount of double-stranded material that forms. **Figure 1.14** illustrates a procedure in which a DNA preparation is denatured and the single strands are adsorbed to a filter. Then a second denatured DNA (or RNA) preparation is added. The filter is treated so that the second preparation can adsorb to it only if it is able to base pair with the DNA that was originally adsorbed. Usually the second preparation is radioactively labeled, so that the reaction can be measured as the amount of radioactive label retained by the filter.

The extent of hybridization between two single-stranded nucleic acids is determined by their complementarity. Two sequences need not be *perfectly* complementary to hybridize. If they are closely related but not identical, an imperfect duplex is formed in which base pairing is interrupted at positions where the two single strands do not correspond.

Figure 1.14 Filter hybridization establishes whether a solution of denatured DNA (or RNA) contains sequences complementary to the strands immobilized on the filter.

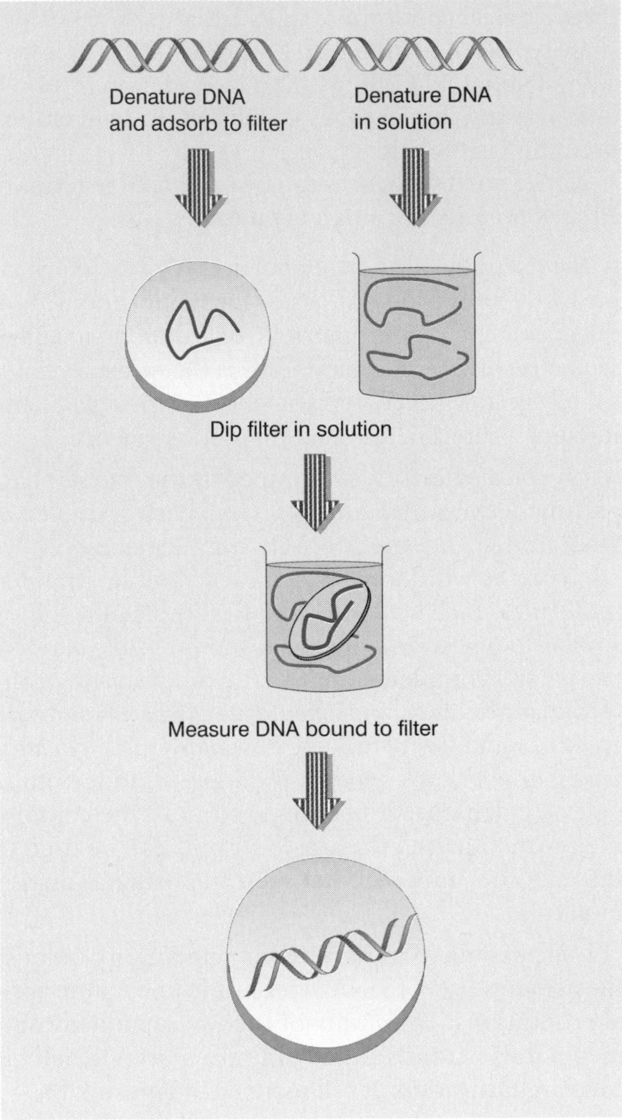
Denature DNA and adsorb to filter
Denature DNA in solution
Dip filter in solution
Measure DNA bound to filter

Mutations change the sequence of DNA

Mutations provide decisive evidence that DNA is the genetic material. When a change in the sequence of DNA causes an alteration in the sequence of a protein, we may conclude that the DNA codes for that protein. Furthermore, a change in the phenotype of the organism may allow us to identify the function of the protein. The existence of many mutations in a gene may allow many variant forms of a protein to be compared, and a detailed analysis can be used to identify regions of the protein responsible for individual enzymatic or other functions.

All organisms suffer a certain number of mutations as the result of normal cellular operations or random interactions with the environment. Such mutations are called **spontaneous;** the rate at which they occur is characteristic for any particular organism and sometimes is called the **background level.** Mutations are rare events, and of course those that damage a gene are selected against during evolution. It is therefore difficult to obtain large numbers of spontaneous mutants to study from natural populations.

The occurrence of mutations can be increased by treatment with certain compounds. These are called **mutagens,** and the changes they cause are referred to as **induced mutations.** Most mutagens act directly by virtue of an ability either to modify a particular base of DNA or to become incorporated into the nucleic acid. The effectiveness of a mutagen is judged by the degree to which it increases the rate of mutation above background. By using mutagens, it becomes possible to induce many changes in any gene.

Any base pair of DNA can be mutated. A **point mutation** changes only a single base pair, and can be caused by either of two types of event:

- Chemical modification of DNA directly changes one base into a different base.
- A malfunction during the replication of DNA causes the wrong base to be inserted into a polynucleotide chain during DNA synthesis.

Point mutations can be divided into two types, depending on the nature of the change when one base is substituted for another:

- The most common class is the **transition,** comprising the substitution of one pyrimidine by the other, or of one purine by the other. This replaces a G•C pair with an A•T pair or vice versa.
- The less common class is the **transversion,** in which a purine is replaced by a pyrimidine or vice versa, so that an A•T pair becomes a T•A or C•G pair.

The effects of nitrous acid provide a classic example of a transition caused by the chemical conversion of one base into another. **Figure 1.15** shows that nitrous acid performs an oxidative deamination that converts cytosine into uracil. In the replication cycle following the transition, the U pairs with an A, instead of with the G with which the original C would have paired. So the C•G pair is replaced by a T•A pair when the A pairs with the T in the next replication cycle. Nitrous acid also deaminates adenine, causing the reverse transition from A•T to G•C.

Transitions are also caused by **base mispairing,** when unusual partners pair in defiance of the usual restriction to Watson-Crick pairs. Base mispairing usually occurs as an aberration resulting from the introduction of an abnormal base.

Some mutagens are analogs of the usual bases that have ambiguous pairing properties; their mutagenic action results from their incorporation into DNA in place of one of the regular bases. **Figure 1.16** shows the example of bromouracil (BrdU), which is an analog of thymine that contains a bromine atom in place of the methyl group of thymine. BrdU is incorporated into DNA in place of thymine. But it has ambiguous pairing properties, because the presence of the bromine atom allows a shift to occur in which the base changes

Figure 1.15 Mutations can be induced by chemical modification of a base.

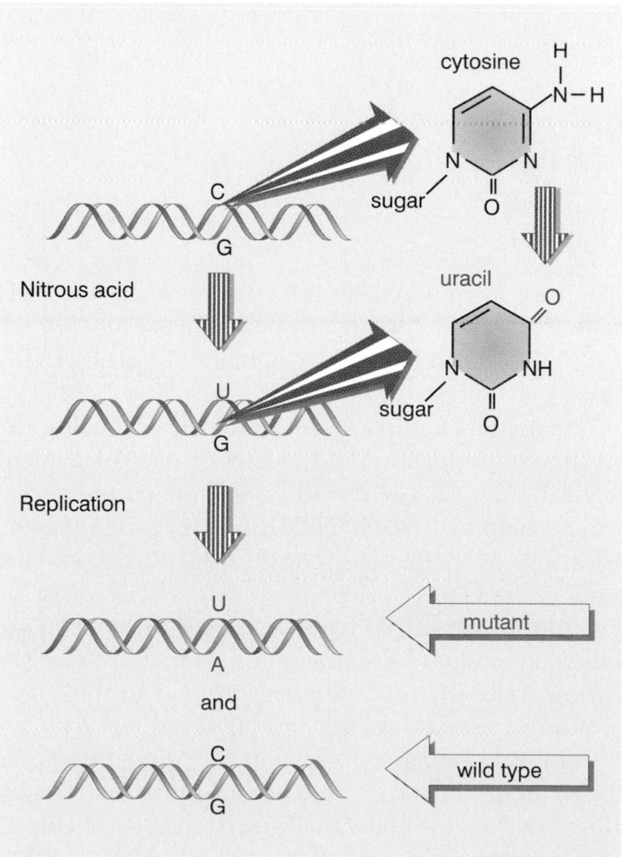

Figure 1.16 Mutations can be induced by the incorporation of base analogs into DNA.

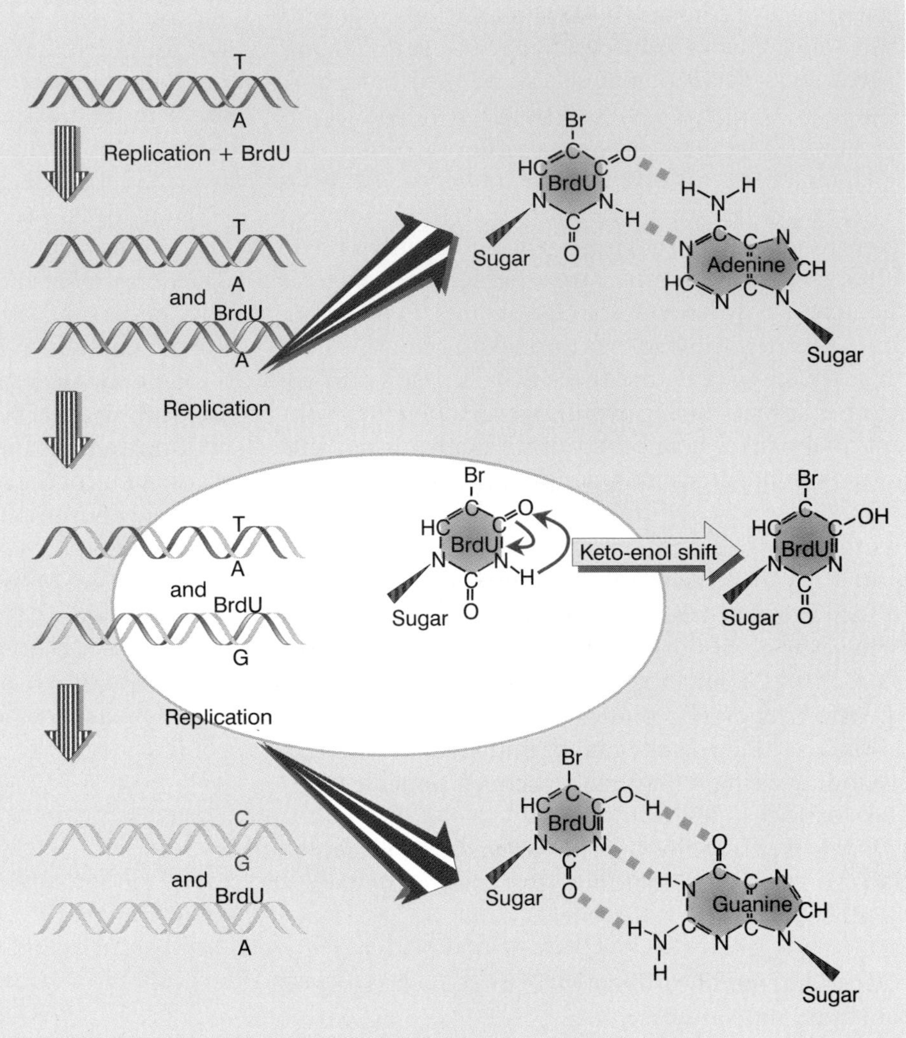

structure from a keto (=O) form to an enol (–OH) form. The enol form can base pair with guanine, which leads to substitution of the original A•T pair by a G•C pair.

The mistaken pairing can occur either during the original incorporation of the base or in a subsequent replication cycle. The transition is induced with a certain probability in each replication cycle, so the incorporation of BrdU has continuing effects on the sequence of DNA.

Mutations induced by base substitution often are **leaky:** the mutant has some residual function. This situation arises when the sequence change in the corresponding protein does not entirely abolish its activity.

Point mutations were thought for a long time to be the principal means of change in individual genes. However, we now know that **insertions** of stretches of additional material are quite frequent. The source of the inserted material lies with **transposable elements,** sequences of DNA with the ability to move from one site to another. (We discuss these elements in detail in Chapters 15–16.) An insertion usually abolishes the activity of a gene. Where such insertions have occurred, **deletions** of part or all of the inserted material, and sometimes of the adjacent regions, may subsequently occur.

A significant difference between point mutations and the insertions/deletions is that the frequency of point mutation can be increased by mutagens, whereas the occurrence of changes caused by transposable elements is not affected. However, insertions and deletions can also occur by other mechanisms—for example, involving mistakes made during replication or recombination—although probably these are less common. And a class of mutagens called the acridines introduce (very small) insertions and deletions.

The isolation of **revertants** is an important characteristic that distinguishes point mutations and insertions from deletions:

- A point mutation can revert by restoring the original sequence or by gaining a compensatory mutation elsewhere in the gene.
- An **insertion** of additional material can revert by deletion of the inserted material.
- A **deletion** of part of a gene cannot revert.

Mutations can also occur in other genes to circumvent the effects of mutation in the original gene. This effect is called **suppression,** or, more formally, intercistronic suppression. A locus in which a mutation suppresses the effect of a mutation in another locus is called a **suppressor.**

Mutations are concentrated at hotspots

So far we have dealt with mutations in terms of individual changes in the sequence of DNA that influence the activity of the genetic unit in which they occur. When we consider mutations in terms of the inactivation of the gene, most genes within a species show more or less similar rates of mutation relative to their size. This suggests that the gene can be regarded as a target for mutation, and that damage to any part of it can abolish its function. As a result, susceptibility to mutation is roughly proportional to the size of the gene. But consider the sites of mutation within the sequence of DNA; are all base pairs in a gene equally susceptible or are some more likely to be mutated than others?

What happens when we isolate a large number of independent mutations in the same gene? Many mutants are obtained. Each is the result of an individual mutational event. Then the site of each mutation is determined. Most mutations will lie at different sites, but some will lie at the same position. Two independently isolated mutations at the same site may constitute exactly the same change in DNA (in which case the same mutational event has happened on more than one occasion), or they may constitute different changes (three different point mutations are possible at each base pair).

The histogram in **Figure 1.17** shows the frequency with which mutations are found at each base pair in the *lacI* gene of *E. coli.* The statistical probability that more than one mutation occurs at a particular site is given by random-hit kinetics (as seen in the Poisson distribution). So some sites will gain one, two, or three mutations, while others will not gain any. But some sites gain far more than the number of mutations expected from a random distribution; they may have 10× or even 100× more mutations than predicted by random hits. These sites are called **hotspots.** Spontaneous mutations may occur at hotspots; and different mutagens may have different hotspots.

A major cause of spontaneous mutation in *E. coli* results from the presence of an unusual base in the DNA. In addition to the four bases that are inserted into DNA when it is synthesized, **modified bases** are sometimes found. The name reflects their origin; they are produced by chemically modifying one of the four bases already present in DNA. The most common modified base is 5-methylcytosine, generated by a methylase enzyme that adds a methyl group to a small proportion of

Figure 1.17 Spontaneous mutations occur throughout the *lacI* gene of *E. coli*, but are concentrated at a hotspot.

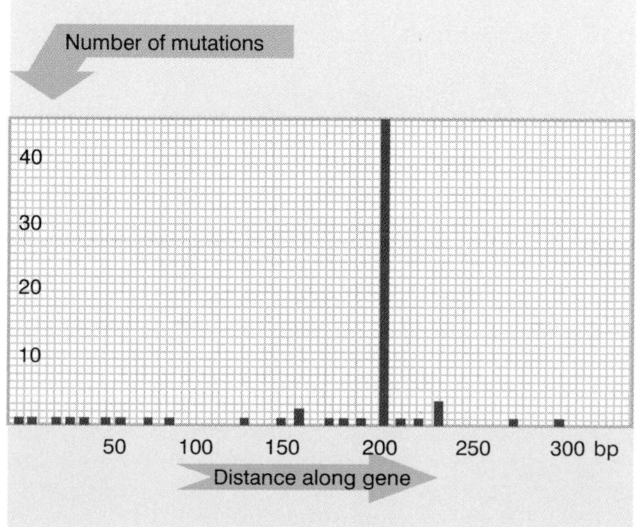

the cytosine residues (at specific sites in the DNA).

Sites containing 5-methylcytosine provide hotspots for spontaneous point mutation. In each case, the mutation takes the form of a G•C to A•T transition. The hotspots are not found in strains of *E. coli* that are unable to perform the methylation reaction.

The reason for the existence of the hotspots is that 5-methylcytosine suffers spontaneous deamination at an appreciable frequency; replacement of the amino group by a keto group converts 5-methylcytosine to thymine. **Figure 1.18** shows why deaminating the (rare) 5-methylcytosine causes a mutation, whereas deamination of the more common cytosine does not have this effect.

Figure 1.18 The deamination of 5-methylcytosine produces thymine (causing C·G to T·A transitions), while the deamination of cytosine produces uracil (which usually is removed and then replaced by cytosine).

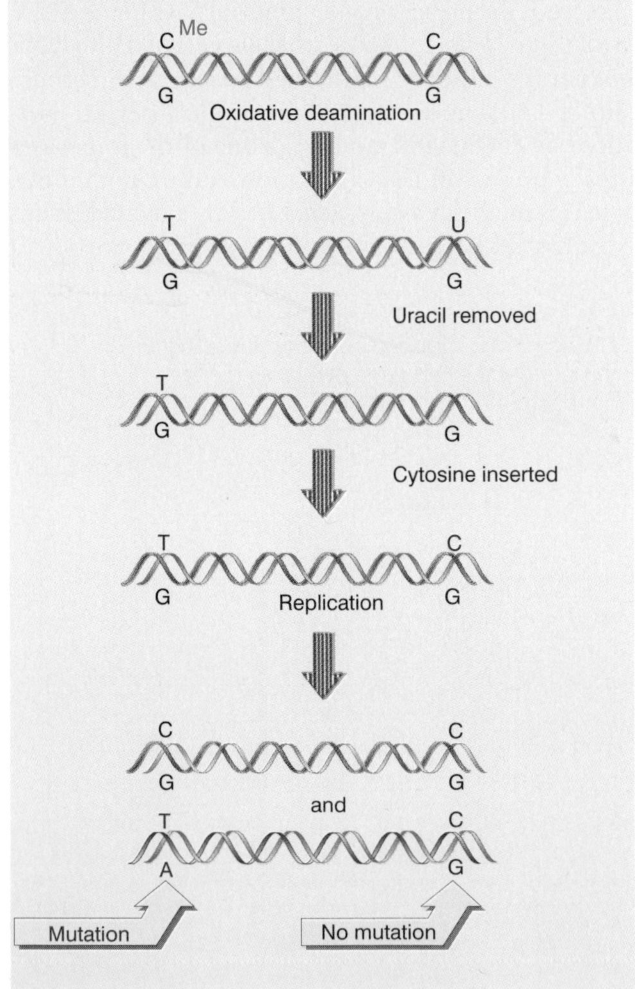

The deamination of cytosine generates uracil. However, *E. coli* contains an enzyme, uracil-DNA-glycosidase, that removes uracil residues from DNA. This action leaves an unpaired G residue, and a "repair system" then inserts a C base to partner it. The net result of these reactions is to restore the original sequence of the DNA. Presumably this system serves to protect DNA against the consequences of spontaneous deamination of cytosine (although it is not active enough to prevent the effects of nitrous acid; see Figure 1.15).

But the deamination of 5-methylcytosine leaves thymine. Because this base is a respectable constituent of DNA in its own right, the system does not recognize the change, and a mutation results. The conversion creates a mispaired G•T partnership, whose separation at the subsequent replication produces one wild-type G•C pair and one mutant A•T pair.

The operation of this system casts an interesting light on the use of T in DNA compared with U in RNA. Perhaps it relates to the need of DNA for stability of sequence; the use of T means that any deaminations of C are immediately recognized, because they generate a base (U) not usually present in the DNA.

Spontaneous mutations that inactivate gene function occur in bacteria at a rate of $\sim 10^{-5}$–10^{-6} events per locus per generation. This mutation rate corresponds to changes at individual nucleotides of 10^{-9}–10^{-10} per generation. We have no accurate measurement of the rate of mutation in eukaryotes, although usually it is thought to be somewhat similar to that of bacteria on a per-locus per-generation basis. We do not know what proportion of the spontaneous events results from point mutations.

Not all mutations in DNA lead to a detectable change in the phenotype. Mutations without apparent effect are called **silent mutations.** They fall into two types. Some involve base changes in DNA that do not cause any change in the amino acid present in the corresponding protein. Others change the amino acid, but the replacement in the protein does not affect its activity; these are called **neutral substitutions.**

Mutations that inactivate a gene are called **forward mutations.** Their effects are reversed by **back mutations,** which are of two types.

An exact reversal of the original mutation is called **true reversion.** So if an A•T pair has been replaced by a G•C pair, another mutation to restore the A•T pair will exactly regenerate the wild-type sequence.

Alternatively, another mutation may occur elsewhere in the gene, and its effects compensate for the first mutation. This is called **second-site reversion.** For example, one amino acid change in a protein may abolish gene function, but a second alteration may compensate for the first and restore protein activity.

A forward mutation results from any change that inactivates a gene, whereas a back mutation must restore function to a protein damaged by a particular forward mutation. So the demands for back mutation are much more specific than those for forward mutation. The rate of back mutation is correspondingly lower than that of forward mutation, typically by a factor of ~10.

A cistron is a single stretch of DNA

THE first systematic attempt to associate genes with enzymes showed that each stage in a metabolic pathway is catalyzed by a single enzyme and can be blocked by mutation in a different gene. This led to the **one gene : one enzyme hypothesis.** Each metabolic step is catalyzed by a particular enzyme, whose production is the responsibility of a single gene. A mutation in the gene alters the activity of the protein for which it is responsible.

Identifying which protein represents a particular gene can be a protracted task. The mutation responsible for creating Mendel's wrinkled-pea mutant was identified only in 1990 as an alteration that inactivates the gene for a starch branching enzyme!

Since a mutation is a random event with regard to the structure of the gene, the greatest probability is that it will damage or even abolish gene function. This explains the nature of recessive mutations: *they represent an absence of function, because the mutant gene has been prevented from producing its usual enzyme.* **Figure 1.19** illustrates the basis for the dominance relationship between recessive and wild-type alleles. When a heterozygote contains one wild-type and one mutant allele, the wild-type allele is able to direct production of the enzyme. The wild-type allele is therefore dominant. (This assumes that an adequate *amount* of protein is made by the single wild-type allele. When this is not true, the smaller amount made by one allele as compared to two alleles results in the intermediate phenotype of a partially dominant allele in a heterozygote.)

A modification in the hypothesis is needed to accommodate proteins that consist of more than one subunit. If the subunits are all the same, the protein is a **homomultimer,** represented by a single gene. If the subunits are different, the protein is a **heteromultimer.** Stated as a more general rule applicable to any heteromultimeric protein, the one gene : one enzyme hypothesis becomes more precisely expressed as **one gene : one polypeptide chain.**

How do we determine whether two mutations that cause a similar phenotype lie in the same gene? If they map close together, they may be alleles. However, they

Figure 1.19 Genes code for proteins; dominance is explained by the properties of mutant proteins. A recessive allele does not contribute to the phenotype because it produces no protein (or protein that is nonfunctional).

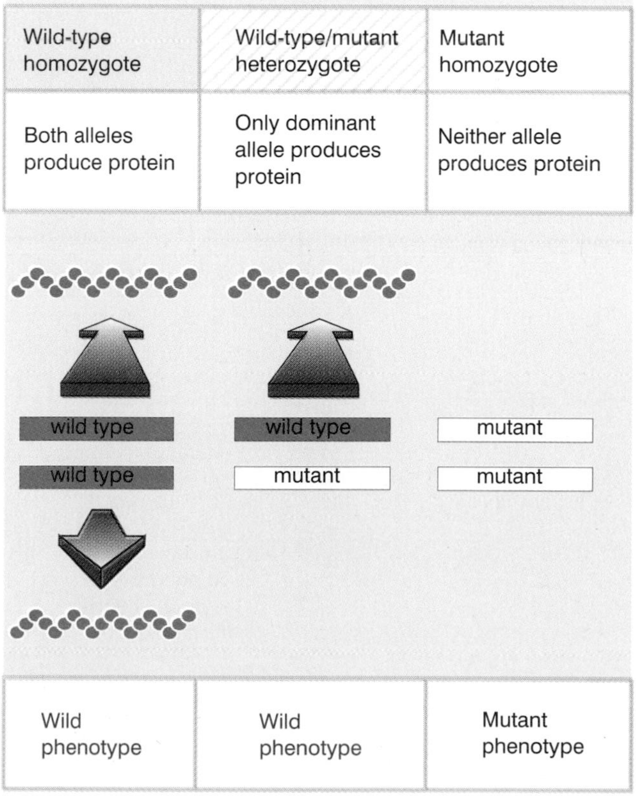

could also represent mutations in two *different* genes whose proteins are involved in the same function. The **complementation test** is used to determine whether two mutations lie in the same gene or in different genes. The test consists of making a heterozygote for the two mutations (by mating parents homozygous for each mutation).

If the mutations lie in the same gene, the parental genotypes can be represented as:

$$\frac{m_1}{m_1} \text{ and } \frac{m_2}{m_2}$$

The first parent provides an m_1 mutant allele and the second parent provides an m_2 allele, so that the heterozygote has the constitution:

$$\frac{m_1}{m_2}$$

No wild-type gene is present, so the heterozygote has mutant phenotype.

If the mutations lie in different genes, the parental genotypes can be represented as:

$$\frac{m_1 \;\; +}{m_1 \;\; +} \text{ and } \frac{+ \;\; m_2}{+ \;\; m_2}$$

Each chromosome has a wild-type copy of one gene (represented by the plus sign) and a mutant copy of the other. Then the heterozygote has the constitution:

$$\frac{m_1 \;\; +}{+ \;\; m_2}$$

in which the two parents between them have provided a wild-type copy of each gene. The heterozygote has wild phenotype; the two genes are said to **complement.**

The complementation test is shown in more detail in **Figure 1.20.** The basic test consists of the comparison shown in the top part of the figure. If two mutations lie in the same gene, we will see a difference in the phenotypes of the *trans* configuration and the *cis* configuration. The *trans* configuration is mutant, because each

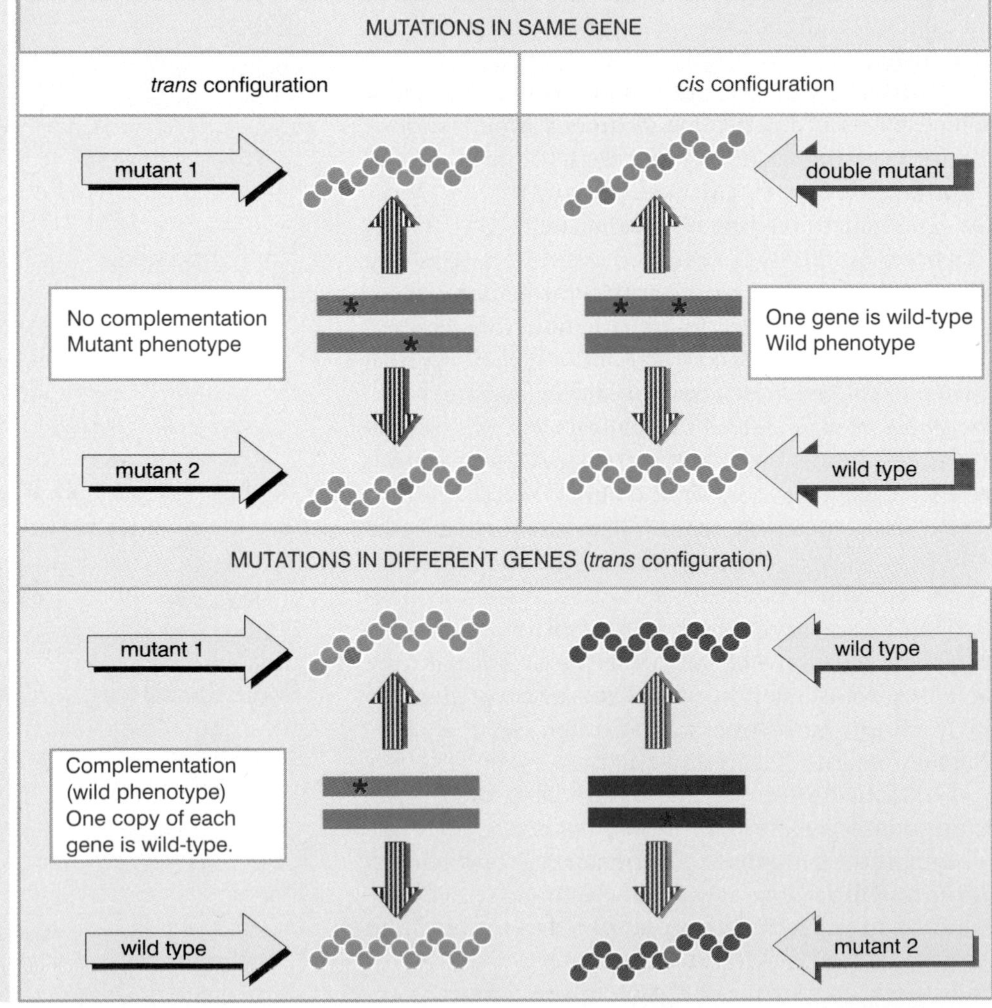

Figure 1.20 The cistron is defined by the complementation test. Genes are represented by bars; red stars identify sites of mutation.

allele has a (different) mutation. But the *cis* configuration is wild-type, because one allele has two mutations but the other allele has no mutations. However, if the two mutations lie in different genes, we always see a wild phenotype. There is always one wild-type and one mutant allele of each gene, and the configuration is irrelevant.

Failure to complement means that two mutations are part of the *same* genetic unit. Mutations that do not complement one another are said to comprise part of the same **complementation group.** Another term that is used to describe the unit defined by the complementation test is the **cistron.** This is the same as the **gene.** Basically these three terms all describe a stretch of DNA that functions as a unit to give rise to an RNA or protein product.

An exception to the rule that only different genes can complement is sometimes found when a gene represents a polypeptide that is the subunit of a homomultimeric protein. In the wild-type cell, the active protein consists of several *identical* subunits. In a cell containing two mutant alleles, however, their products can mix to form multimeric proteins that contain *both types* of subunit. Sometimes the two mutations compensate, so that the mixed-subunit protein is active, even though the proteins consisting solely of either type of mutant subunit are inactive. This effect is called **interallelic** complementation.

The nature of multiple alleles

WHEN a gene has been identified, insight into its function in principle can be gained by generating a mutant organism that entirely lacks the gene. A mutation that completely eliminates gene function, usually because the gene has been deleted, is called a **null mutation.** If a gene is essential, a null mutation is lethal.

To determine what effect a gene has upon the phenotype, it is essential to characterize a null mutant. When a mutation fails to affect the phenotype, it is always possible that this is because it is **leaky**—enough active product is made to fulfill its function, even though the activity is quantitatively reduced or qualitatively different from the wild type. But if a null mutant fails to affect a phenotype, we may safely conclude that the gene function is not necessary.

Null mutations, or other mutations that impede gene function (but do not necessarily abolish it entirely) are called **loss-of-function** mutations. A loss-of-function mutation is recessive (as in the example of Figure 1.19). Sometimes a mutation has the opposite effect and causes a protein to acquire a new function; such a change is called a **gain-of-function** mutation. A gain-of-function mutation is dominant.

If a recessive mutation is produced by every change in a gene that prevents the production of an active protein, there should be a large number of such mutations in any one gene. Many amino acid replacements may change the structure of the protein sufficiently to impede its function. Different variants of the same gene are called **multiple alleles,** and their existence makes it possible to create a heterozygote between mutant alleles.

The relationship between these multiple alleles takes various forms.

In the simplest case, a wild-type gene codes for a protein product that is functional. Mutant allele(s) code for proteins that are nonfunctional.

But there are often cases in which a series of mutant alleles have different phenotypes. For example, wild-type function of the *white* locus of *D. melanogaster* is required for development of the normal red color of the eye. The locus is named for the effect of extreme (null) mutations, which cause the fly to have a white eye in mutant homozygotes.

To describe wild-type and mutant alleles, wild genotype is indicated by a plus superscript after the name of the locus (w^+ is the wild-type allele for [red] eye color in *D. melanogaster*). Sometimes + is used by itself to describe the wild-type allele, and only the mutant alleles are indicated by the name of the locus.

An entirely defective form of the gene (or absence of phenotype) may be indicated by a minus superscript. To distinguish among a variety of mutant alleles with different effects, other superscripts may be introduced, such as w^i or w^a.

The w^+ allele is dominant over any other allele in heterozygotes. There are many different mutant alleles.

Figure 1.21 shows a (small) sample. Although some alleles have no eye color, many alleles produce some color. Each of these mutant alleles must therefore represent a different mutation of the gene, which does not eliminate its function entirely, but leaves a residual activity that produces a characteristic phenotype. These alleles are named for the color of the eye in a homozygote. (Most *w* alleles affect the quantity of pigment in the eye, and the examples in the Figure are arranged in [roughly] declining amount of color, but others, such as w^{sp}, affect the pattern in which it is deposited.)

When multiple alleles exist, an animal may be a heterozygote that carries two different mutant alleles. The phenotype of such a heterozygote depends on the nature of the residual activity of each allele. The relationship between two mutant alleles is in principle no different from that between wild-type and mutant alleles: one allele may be dominant, there may be partial dominance, or there may be codominance.

There is not necessarily a unique wild-type allele at any particular locus. Control of the human blood group system provides an example. Lack of function is represented by the null type, *O* group. But the functional alleles *A* and *B* provide activities that are codominant with one another and dominant over *O* group. The basis for this relationship is illustrated in **Figure 1.22.**

The O (or H) antigen is generated in all individuals, and consists of a particular carbohydrate group that is added to proteins. The *ABO* locus codes for a galactosyltransferase enzyme that adds a further sugar group to the O antigen. The specificity of this enzyme determines the blood group. The *A* allele produces an enzyme that uses the cofactor UDP-N-acetylgalactose, creating the A antigen. The *B* allele produces an enzyme that uses the cofactor UDP-galactose, creating the B antigen. The A and B versions of the transferase protein differ in four amino acids that presumably

Figure 1.21 The *w* locus has an extensive series of alleles, whose phenotypes extend from wild-type (red) color to complete lack of pigment.

Allele	Phenotype of Homozygote
w^+	red eye (wild type)
w^{bl}	blood
w^{ch}	cherry
w^{bf}	buff
w^h	honey
w^a	apricot
w^e	eosin
w^i	ivory
w^z	zeste (lemon-yellow)
w^{sp}	mottled, color varies
w^1	white (no color)

Figure 1.22 The ABO blood group locus codes for a galactosyltransferase whose specificity determines the blood group.

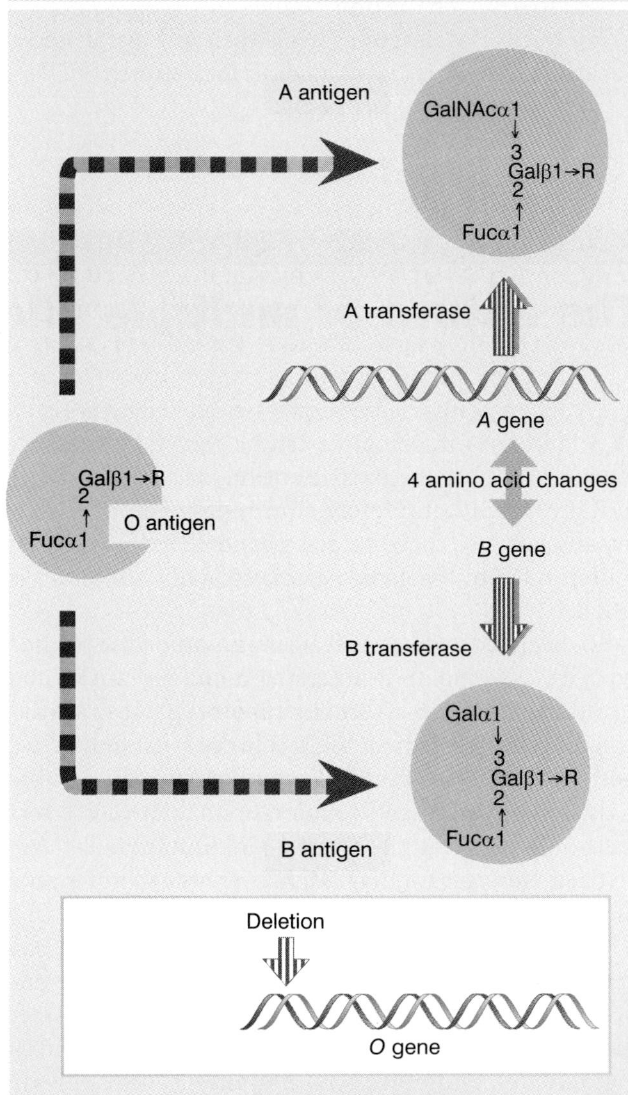

Phenotype	Genotype	Activity
O	*OO*	None
A	*AO* or *AA*	N-Ac-gal transferase
B	*BO* or *BB*	Gal transferase
AB	*AB*	N-Ac-gal- & gal-transferase

affect its recognition of the type of cofactor. The *O* allele has a mutation (a small deletion) that eliminates activity, so no modification of the O antigen occurs.

This explains why *A* and *B* alleles are dominant in the *AO* and *BO* heterozygotes: the corresponding transferase activity creates the A or B antigen. The *A* and *B* alleles are codominant in *AB* heterozygotes, because both transferase activities are expressed. The *OO* homozygote is a null that has neither activity, and therefore lacks both antigens.

Neither *A* nor *B* can be regarded as uniquely wild type, since they represent alternative activities rather than loss or gain of function. A situation such as this, in which there are multiple functional alleles in a population, is described as a **polymorphism** (see Chapter 3).

Recombination occurs by physical exchange of DNA

GENES on the same chromosome show genetic linkage because they are present on the same (very long) molecule of DNA. Linkage is revealed in the progeny of a genetic cross when the proportion of **recombinant genotypes** (where an allele of one parent is found with an allele of the other parent) is less than the number of **parental genotypes** (with the same combination of alleles as either parent). Genes on different chromosomes segregate independently, of course, as predicted by Mendel, giving 50% parental and 50% recombinant progeny. The proximity of two loci is measured by the per cent recombination between them (in formal terms a *map distance of 1 centiMorgan = 1% recombination*). When pairwise combinations of loci on the same chromosome are tested in genetic crosses, loci close to one another are linked, as defined by a map distance <50 cM. Loci that are farther apart recombine at the limit of 50%. But a **linkage map** corresponding to the chromosome can be generated by extending a series of genetic crosses in which, in effect, two loci >50 cM apart are connected because they show linkage to a locus between them.

Recombination results from a physical exchange of chromosomal material. This is visible in the form of the **crossing-over** that occurs during meiosis. Early in meiosis, at the stage when all four copies of each chromosome are organized in a **bivalent,** pairwise exchanges of material occur between the closely associated (synapsed) chromatids.

The visible result of a crossing-over event is called a **chiasma,** and is illustrated diagrammatically in **Figure 1.23.** A chiasma represents a site at which two of the chromatids in a bivalent have been broken at corresponding points. The broken ends have been rejoined crosswise, generating new chromatids. Each new chromatid consists of material derived from one chromatid on one side of the junction point, with material from the other chromatid on the opposite side. The two recombinant chromatids have reciprocal structures. The event is described as a **breakage and reunion.** Its nature explains why a single recombination event can produce only 50% recombinants: each individual recombination event involves only two of the four associated chromatids.

Figure 1.23 Chiasma formation is responsible for generating recombinants.

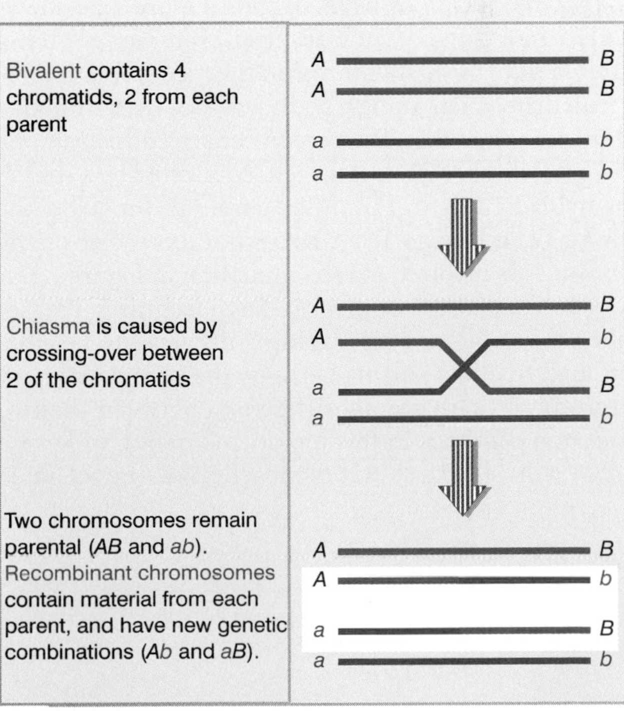

Bivalent contains 4 chromatids, 2 from each parent
A B
A B
a b
a b
Chiasma is caused by crossing-over between 2 of the chromatids
A B
A b
a B
a b
Two chromosomes remain parental (AB and ab). Recombinant chromosomes contain material from each parent, and have new genetic combinations (Ab and aB).
A B
A b
a B
a b

A crucial concept in the construction of a genetic map is that the distance between genes does not depend on the particular *alleles* that are used, but only on the genetic **loci.** *The locus defines the position occupied on the chromosome by the gene representing a particular trait. The various alternative forms of a gene*—that is, the alleles used in mapping—*all reside at the same location on its particular chromosome.* So genetic mapping is concerned with identifying the positions of genetic loci, which are fixed and lie in a linear order. In a mapping experiment, the same result is obtained irrespective of the particular combination of alleles.

The complementarity of the two strands of DNA is essential for the recombination process. Each of the chromatids shown in Figure 1.23 consists of a very long duplex of DNA. For them to be broken and reconnected without any loss of material requires a mechanism to recognize exactly corresponding positions. This is provided by complementary base pairing.

Recombination involves a process in which the single strands in the region of the crossover exchange their partners. **Figure 1.24** shows that this creates a stretch of **hybrid DNA** in which the single strand of one duplex is paired with its complement from the other duplex. The process accomplishes a result analogous to the denaturation and renaturation shown previously in Figure 1.13. The mechanism of course involves other stages (strands must be broken and resealed), and we discuss this in more detail in Chapter 14, but the crucial feature that makes precise recombination possible at all is the complementarity of DNA strands. Figure 1.24 shows only some stages of the reaction, but we see that a stretch of hybrid DNA forms in the recombination intermediate when a single strand crosses over from one duplex to the other. Each recombinant consists of one parental duplex DNA at the left, connected by a stretch of hybrid DNA to the other parental duplex at the right. Each duplex DNA corresponds to one of the chromatids involved in recombination in Figure 1.23.

Figure 1.24 Recombination involves pairing between complementary strands of the two parental duplex DNAs.

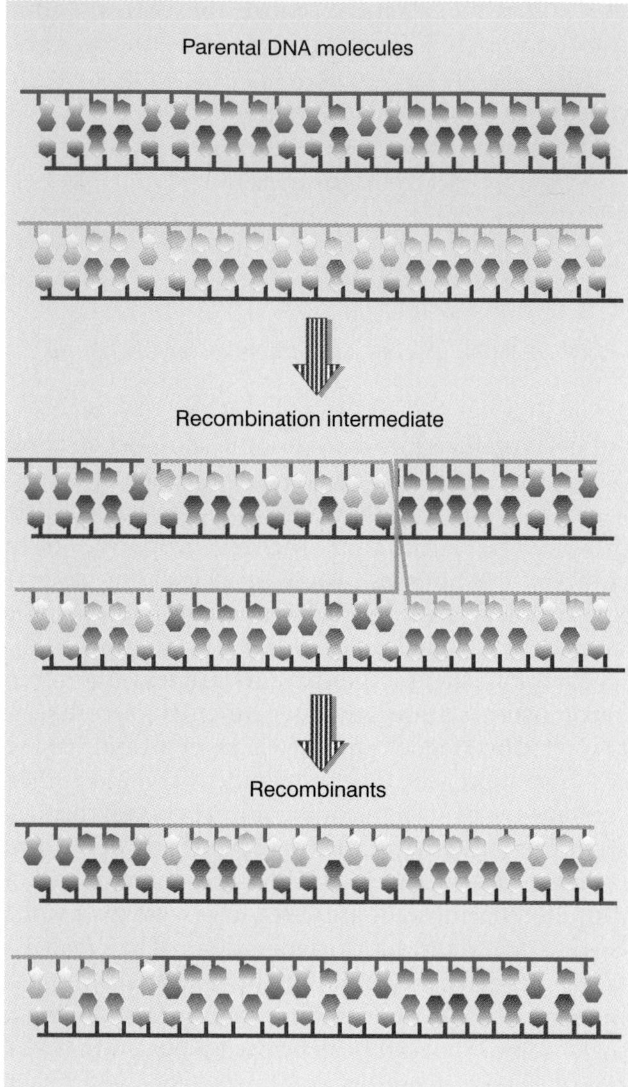

The formation of hybrid DNA requires the sequences of the two recombining duplexes to be close enough to allow pairing between the complementary strands. If there are no differences between the two parental genomes in this region, formation of hybrid DNA will be perfect. But the reaction can be tolerated even when there are small differences. In this case, the hybrid DNA has points of **mismatch,** at which a base in one strand faces a base in the other strand that is not complementary to it. The **correction** of such mismatches is another feature of genetic recombination (see Chapter 14).

The genetic code is triplet

EACH gene represents a particular polypeptide chain. The concept that each protein consists of a particular series of amino acids dates from Sanger's characterization of insulin in the 1950s. The discovery that a gene consists of DNA faces us with the issue of how a sequence of nucleotides in DNA represents a sequence of amino acids in protein.

A crucial feature of the general structure of DNA is that *it is independent of the particular sequence of its component nucleotides.* The sequence of nucleotides in DNA is important not because of its structure *per se,* but because it *codes* for the sequence of amino acids that constitutes the corresponding polypeptide. The relationship between a sequence of DNA and the sequence of the corresponding protein is called the **genetic code.**

The structure and/or enzymatic activity of each protein follows from its primary sequence of amino acids. By determining the sequence of amino acids in each protein, the gene is able to carry all the information needed to specify an active polypeptide chain. In this way, a single type of structure—the gene—is able to represent itself in innumerable polypeptide forms.

Together, the various protein products of a cell undertake the catalytic and structural activities that are responsible for establishing its phenotype. Of course, in addition to the sequences of genes that code for proteins, DNA also contains certain sequences whose function is to be recognized by regulator molecules, usually proteins. Here the function of the DNA is determined by its sequence directly, not via any intermediary code. Both types of region, genes expressed as proteins and sequences recognized as such, constitute genetic information.

The genetic code is deciphered by a complex apparatus that interprets the nucleic acid sequence. This apparatus is essential if the information carried in DNA is to have meaning. In any given region, only one of the two strands of DNA codes for protein, so we write the genetic code as a sequence of bases (rather than base pairs).

The genetic code is read in groups of three nucleotides, each group representing one amino acid. Each trinucleotide sequence is called a **codon.** A gene includes a series of codons that is read sequentially from a starting point at one end to a termination point at the other end. Written in the conventional 5′–3′ direction, the nucleotide sequence of the DNA strand that codes for protein corresponds to the amino acid sequence of the protein written in the direction from N-terminus to C-terminus.

The genetic code is read in *nonoverlapping triplets from a fixed starting point:*

- *Nonoverlapping* implies that each codon consists of three nucleotides and that successive codons are represented by successive trinucleotides.
- The use of a *fixed starting point* means that assembly of a protein must start at one end and work to the other, so that different parts of the coding sequence cannot be read independently.

The nature of the code predicts that two types of mutations will have different effects. If a particular sequence is read sequentially, such as:

UUU	AAA	GGG	CCC	(codons)
αα1	αα2	αα3	αα4	(amino acids)

then a point mutation will affect only one amino acid. For example, the substitution of an A by some other base (X) causes αα2 to be replaced by αα5:

UUU	AAX	GGG	CCC
αα1	αα5	αα3	αα4

because only the second codon has been changed.

But a mutation that inserts or deletes a single base will change the reading frame for the entire subsequent sequence. A change of this sort is called a **frameshift.** An insertion might take the form:

UUU	AAX	AGG	GCC	C
αα1	αα5	αα6	αα7	

Because the new sequence of triplets is completely different from the old one, the entire amino acid sequence of the protein is altered beyond the site of mutation. So the function of the protein is likely to be lost completely.

Frameshift mutations are induced by the **acridines,** compounds that bind to DNA and distort the structure of the double helix, causing additional bases to be incorporated or omitted during replication. Each mutagenic event sponsored by an acridine results in the addition or removal of a single base pair.

If an acridine mutant is produced by, say, addition of a nucleotide, it should revert to wild type by deletion of the nucleotide. But reversion can also be caused by deletion of a different base, at a site close to the first. Combinations of such mutations provided revealing evidence about the nature of the genetic code.

Figure 1.25 illustrates the properties of frameshift mutations. An insertion or a deletion changes the entire protein sequence following the site of mutation.

Figure 1.25 Frameshift mutations show that the genetic code is read in triplets from a fixed starting point.

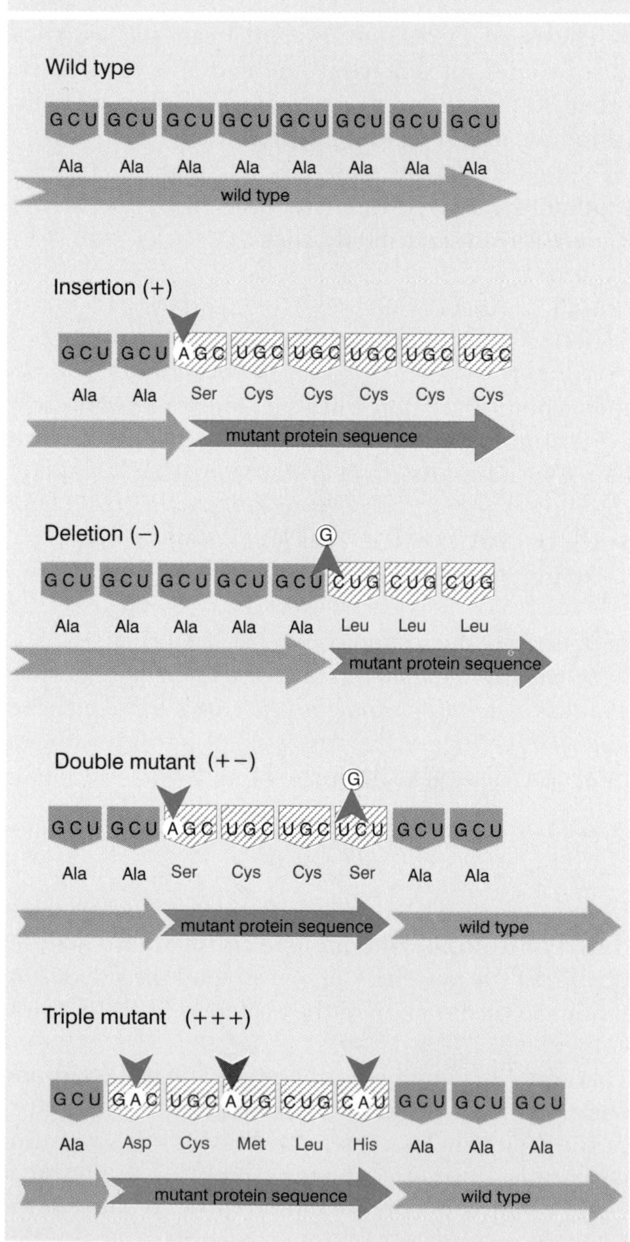

But the combination of an insertion and a deletion causes the code to be read in the incorrect frame only between the two sites of mutation; correct reading resumes after the second site.

The original analysis was performed by genetic analysis of mutations in the *rII* region of the phage T6. All acridine mutations could be classified into one of two sets, described as (+) and (–). Either type of mutation by itself causes a frameshift, the (+) type by virtue of a base addition, the (–) type by virtue of a base deletion. Double mutant combinations of the types (+ +) and (– –) continue to show mutant behavior. But combinations of the types (+ –) or (– +) suppress one another, giving rise to a description in which one mutation is described as a **suppressor** of the other. (In the context of this work, "suppressor" is used in an unusual sense, because the second mutation is in the same gene as the first.)

These results show that the genetic code must be read as a sequence that is fixed by the starting point, so additions or deletions compensate for each other, whereas double additions or double deletions remain mutant. But this does not reveal how many nucleotides make up each codon.

When triple mutants are constructed, only (+ + +) and (– – –) combinations show the wild phenotype, while other combinations remain mutant. If we take three additions or three deletions to correspond respectively to the addition or omission overall of a single amino acid, this implies that the code is read in triplets. An incorrect amino acid sequence is found between the two outside sites of mutation, and the sequence on either side remains wild type, as indicated in Figure 1.25.

If the genetic code is read in nonoverlapping triplets, there are three possible ways of translating any nucleotide sequence into protein, depending on the starting point. These are called **reading frames.** For the sequence

A C G A C G A C G A C G A C G A C G

the three possible reading frames are

ACG ACG ACG ACG ACG ACG ACG
CGA CGA CGA CGA CGA CGA CGA
GAC GAC GAC GAC GAC GAC GAC

A reading frame that consists exclusively of triplets that represent amino acids is called an **open reading frame** or **ORF.** A sequence that is translated into protein has a reading frame that starts with a special **initi-**

Figure 1.26 An open reading frame starts with AUG and continues in triplets to a termination codon. Blocked reading frames may be interrupted frequently by termination codons.

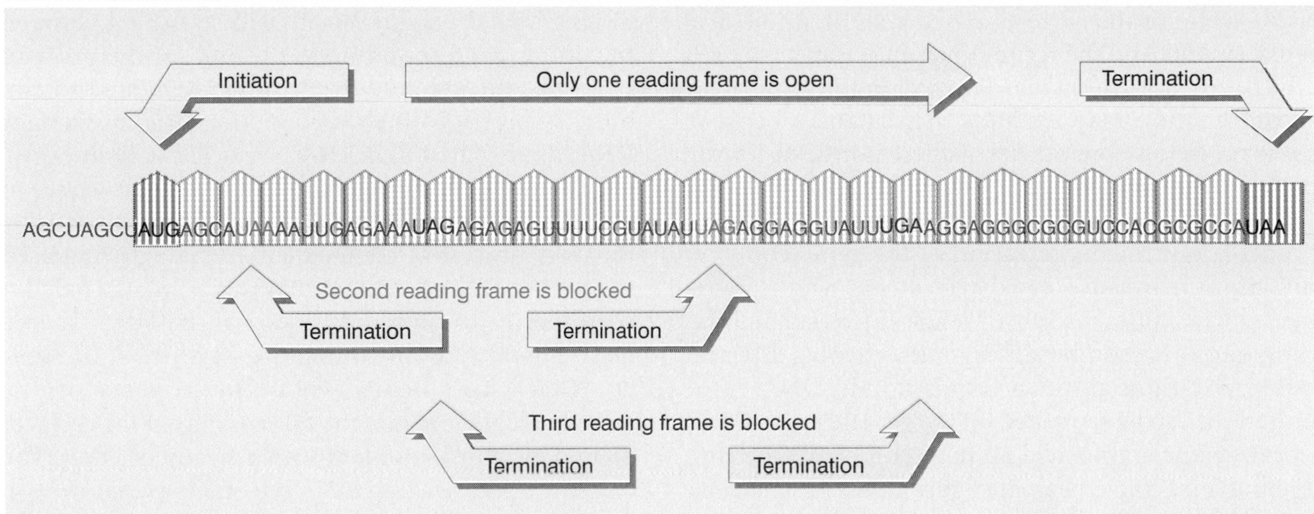

ation codon (AUG) and that extends through a series of triplets representing amino acids until it ends at one of three types of **termination codon** (see Chapter 5).

A reading frame that cannot be read into protein because termination codons occur frequently is said to be **blocked.** If a sequence is blocked in all three reading frames, it cannot have the function of coding for protein.

When the sequence of a DNA region of unknown function is obtained, each possible reading frame is analyzed to determine whether it is open or blocked. Usually no more than one of the three possible frames of reading is open in any single stretch of DNA. **Figure 1.26** shows an example of a sequence that can be read in only one reading frame, because the alternative reading frames are blocked by frequent termination codons. A long open reading frame is unlikely to exist by chance; if it were not translated into protein, there would have been no selective pressure to prevent the accumulation of termination codons. So the identification of a lengthy open reading frame is taken to be *prima facie* evidence that the sequence is translated into protein in that frame. An open reading frame (ORF) for which no protein product has been identified is sometimes called an unidentified reading frame (URF).

Bacterial genes and proteins are colinear

By comparing the nucleotide sequence of a gene with the amino acid sequence of a protein, we can determine directly whether the gene and the protein are **colinear:** whether the sequence of nucleotides in the gene corresponds exactly with the sequence of amino acids in the protein. In bacteria and their viruses, there is an exact equivalence. Each gene contains a continuous stretch of DNA whose length is directly related to the number of amino acids in the protein that it represents. A gene of 3*N* bp is required to code for a protein of *N* amino acids, according to the genetic code.

The equivalence of the bacterial gene and its product means that a physical map of DNA will exactly match an amino acid map of the protein. How well do these maps fit with the recombination map?

The colinearity of gene and protein was originally investigated in the tryptophan synthetase gene of *E. coli.* Genetic distance was measured by the percent recombination between mutations; protein distance was measured by the number of amino acids separating sites of replacement. **Figure 1.27** compares the two maps. The order of seven sites of mutation is the same as the order of the corresponding sites of amino acid replacement. And the recombination distances are relatively similar to the actual distances in the protein; the recombination map expands the distances between some mutations, but otherwise there is little distortion of the recombination map relative to the physical map.

The recombination map makes two further general points about the organization of the gene. Different mutations may cause a wild-type amino acid to be replaced with different substituents. If two such mutations cannot recombine, they must involve different point mutations at the same position in DNA. If the mutations can be separated on the genetic map, but affect the same amino acid on the upper map (the connecting lines converge in the figure), they must involve point mutations at different positions that affect the same amino acid. This happens because the unit of genetic recombination (actually 1 bp) is smaller than the unit coding for the amino acid (actually 3 bp).

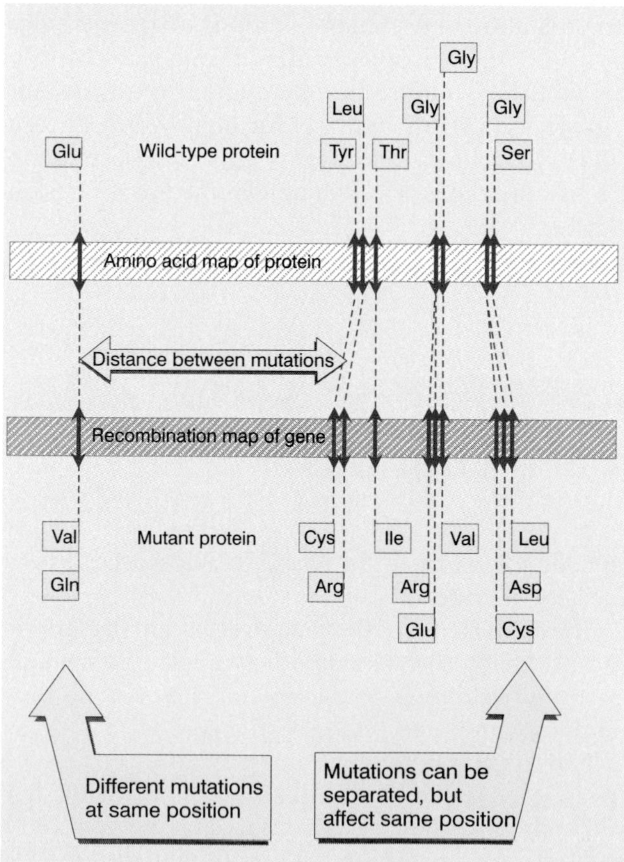

Figure 1.27 The recombination map of the tryptophan synthetase gene corresponds with the amino acid sequence of the protein.

In comparing gene and protein, we are restricted to dealing with the sequence of DNA stretching between the points corresponding to the ends of the protein. However, a gene is not directly translated into protein, but is expressed via the production of a **messenger RNA** (abbreviated to **mRNA**), a nucleic acid intermediate actually used to synthesize a protein (as we see in detail in Chapter 5). Messenger RNA is synthesized by the same process of complementary base pairing used to replicate DNA, with the important difference that it corresponds to only one strand of the DNA double helix. **Figure 1.28** shows that the sequence of messenger RNA is complementary to the sequence of one strand of DNA and is identical (apart from the replacement of T with U) with the other strand of DNA. The convention for writing DNA sequences is that the top strand runs 5′–3′, with the sequence that is the same as RNA.

A messenger RNA includes a sequence of nucleotides that corresponds with the sequence of amino acids in the protein. This part of the nucleic acid is called the **coding region.** But the messenger RNA includes additional sequences on either end; these sequences do not directly represent protein. The regions are called the **5′ nontranslated leader** and the **3′ nontranslated trailer.**

The gene is considered to include the entire sequence represented in messenger RNA. Sometimes mutations impeding gene function are found in the ad-

Figure 1.28 RNA is synthesized by using one strand of DNA as a template for complementary base pairing.

DNA consists of two base-paired strands

5' ATGCCGTTAGACCGTTAGCGGACCTGAC top strand
3' TACGGCAATCTGGCAATCGCCTGGACTG bottom strand

RNA synthesis

5' AUGCCGUUAGACCGUUAGCGGACCUGAC 3'

RNA has same sequence as DNA top strand;
is complementary to DNA bottom strand

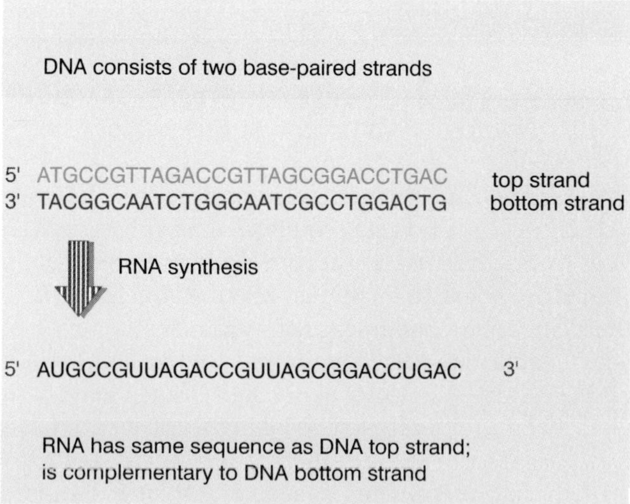

ditional, noncoding regions, confirming the view that these comprise a legitimate part of the genetic unit.

Figure 1.29 illustrates this situation, in which the gene is considered to comprise a continuous stretch of DNA, needed to produce a particular protein. It includes the sequence coding for that protein, but also includes sequences on either side of the coding region.

The process by which a gene gives rise to a protein is called gene expression. The basic stages are outlined in **Figure 1.30.** The first stage is **transcription,** when an RNA copy of one strand of the DNA is produced. For the simplest genes (and always in the case of bacteria) this RNA is in fact the mRNA. For genes of eukaryotes, the immediate transcript of the gene is a **pre-mRNA** that must be **processed** to generate the mature mRNA.

The most important stage in processing is **RNA splicing.** Many genes in eukaryotes (and a majority in higher eukaryotes) contain internal regions that do not code for protein. The process of splicing removes these regions from the pre-mRNA to generate an RNA that has a continuous open reading frame (see Figure 2.10). Other processing events that occur at this stage involve the modification of the 5′ and 3′ ends of the pre-mRNA (see Figure 5.15).

Transcription and processing of RNA occur in the nucleus. The next stage of gene expression is the **translation** of the mRNA into protein. This occurs in the cytoplasm. So it is necessary for the mRNA to be **transported** through the nuclear membrane into the cytoplasm.

Translation is accomplished by a complex apparatus that includes both protein and RNA components. The actual "machine" that undertakes the process is the **ribosome,** a large complex that includes some large RNAs (**ribosomal RNAs**, abbreviated to **rRNAs**) and many small proteins. The process of recognizing which amino acid corresponds to a particular nucleotide triplet requires an intermediate **transfer RNA** (abbreviated to **tRNA**); there is at least one tRNA species for every amino acid. Many ancillary proteins are involved.

The important point to note at this stage is that the process of gene expression involves RNA not only as the essential substrate, but also in providing components of the apparatus. The rRNA and tRNA components are coded by genes and are generated by the process of transcription (just like mRNA, except that there is no subsequent stage of translation).

Figure 1.29 The gene may be longer than the sequence coding for protein.

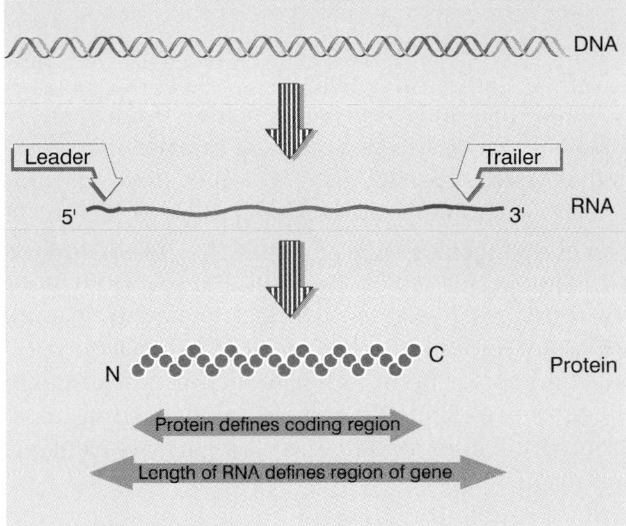

Figure 1.30 Gene expression is a multistage process.

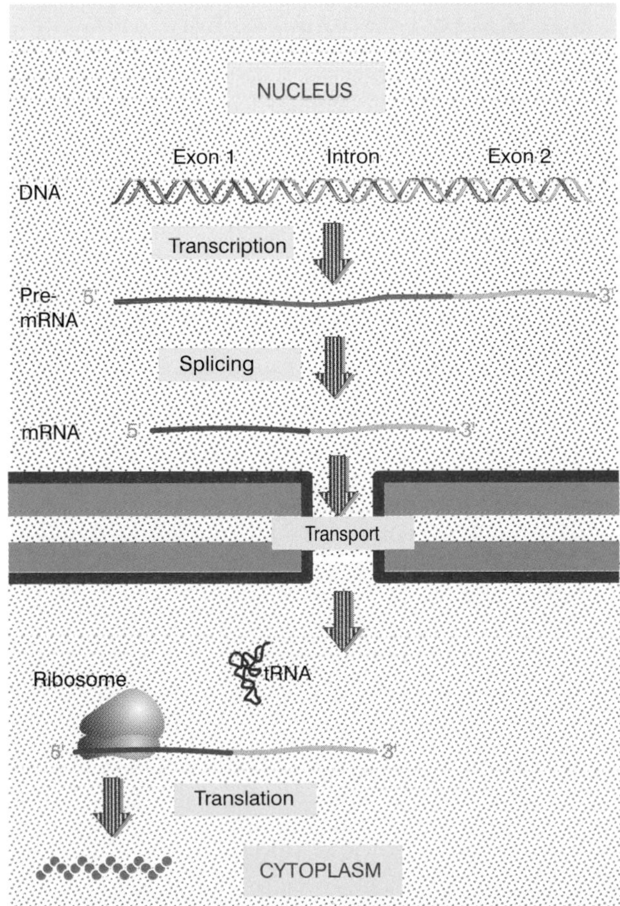

cis-acting sites and *trans*-acting molecules

THE *cis/trans* complementation test defines a gene as a unit that must be present on one chromosome. In a diploid that carries two different mutations at a particular gene locus, the distribution of the mutations determines whether a functional allele is present. If both mutations are in *cis* on one allele, the other allele has no mutations, and is wild type. But if the mutations are in *trans,* so that each allele has one mutation, both alleles are mutant (see Figure 1.20). We may now extend the concept of the difference between *cis* and *trans* effects from defining the coding region of a gene to describing the interaction between regulatory elements and a gene.

Suppose that the ability of a gene to be expressed is controlled by a protein that binds to the DNA close to the coding region. In the example depicted in **Figure 1.31,** messenger RNA can be synthesized only when the protein is bound to the DNA. Now suppose that a mutation occurs in the DNA sequence to which this protein binds, so that the protein can no longer recognize the DNA. As a result, the DNA can no longer be expressed.

So a gene can be inactivated either by a mutation in a control site or by a mutation in a coding region. The mutations cannot be distinguished genetically, because

Figure 1.31 Control sites in DNA provide binding sites for proteins; coding regions are expressed via the synthesis of RNA.

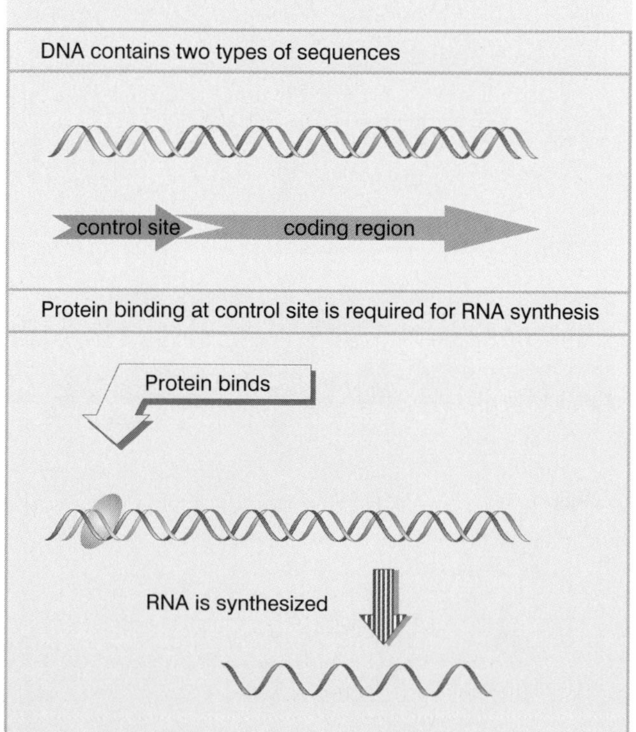
DNA contains two types of sequences
control site
coding region
Protein binding at control site is required for RNA synthesis
Protein binds
RNA is synthesized

Figure 1.32 A *cis*-acting site controls the adjacent DNA but does not influence the other allele.

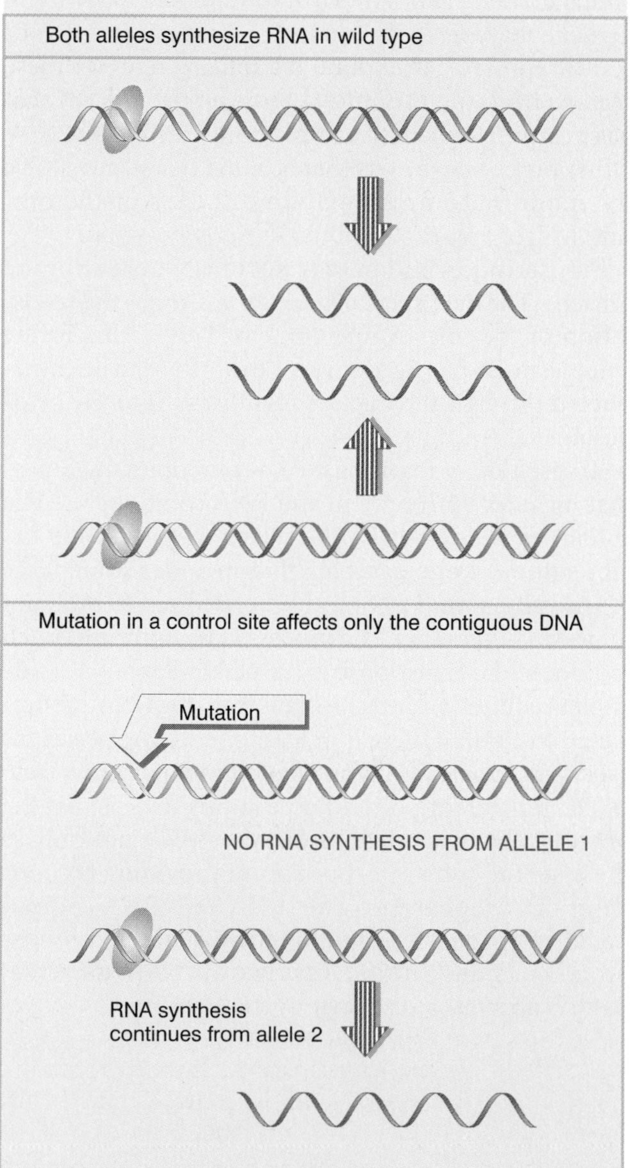
Both alleles synthesize RNA in wild type
Mutation in a control site affects only the contiguous DNA
Mutation
NO RNA SYNTHESIS FROM ALLELE 1
RNA synthesis
continues from allele 2

both have the property of acting only on the DNA sequence of the single allele in which they occur. They will therefore have identical properties in the *cis/trans* test, and a mutation in a control region is therefore defined as comprising part of the gene in the same way as a mutation in the coding region.

Figure 1.32 shows that a deficiency in the control site *affects only the coding region to which it is connected; it does not affect the ability of the other allele to be expressed.* A mutation that acts solely by affecting the properties of the contiguous sequence of DNA is called ***cis*-acting.**

We may contrast the behavior of the *cis*-acting mutation shown in Figure 1.32 with the result of a mutation in the gene coding for the regulator protein. **Figure 1.33** shows that the absence of regulator protein would prevent *both* alleles from being expressed. A mutation of this sort is said to be ***trans*-acting.**

Reversing the argument, if a mutation is *trans*-acting, we know that its effects must be exerted through some diffusible product (typically a protein) that acts on multiple targets within a cell. But if a mutation is *cis*-acting, it must function by affecting directly the properties of the contiguous DNA, which means that it is *not expressed in the form of RNA or protein.*

Figure 1.33 A *trans*-acting mutation in a protein affects both alleles of a gene that it controls.

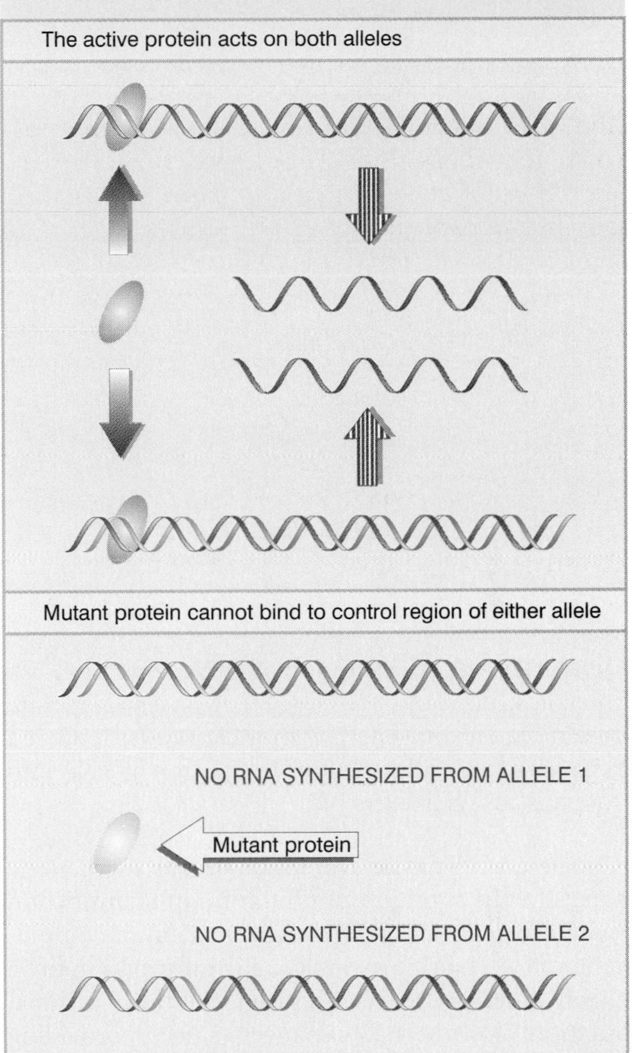
The active protein acts on both alleles
Mutant protein cannot bind to control region of either allele
NO RNA SYNTHESIZED FROM ALLELE 1
Mutant protein
NO RNA SYNTHESIZED FROM ALLELE 2

Genetic information can be provided by DNA or RNA

The **central dogma** defines the paradigm of molecular biology. Genes are perpetuated as sequences of nucleic acid, but function by being expressed in the form of proteins. Replication is responsible for the inheritance of genetic information. Transcription and translation are responsible for its conversion from one form to another.

Figure 1.34 illustrates the roles of replication, transcription, and translation, viewed from the perspective of the central dogma:

- *The perpetuation of nucleic acid may involve either DNA or RNA as the genetic material.* Cells use only DNA. Some viruses use RNA, and replication of viral RNA occurs in the infected cell.
- *The expression of cellular genetic information usually*

Figure 1.34 The central dogma states that information in nucleic acid can be perpetuated or transferred, but the transfer of information into protein is irreversible.

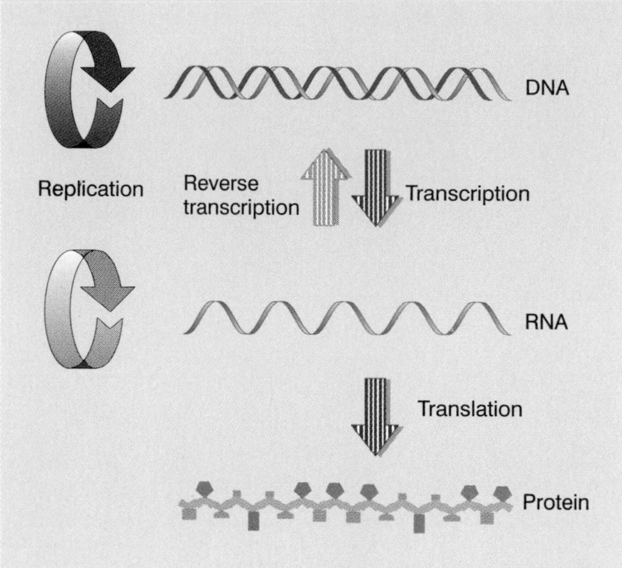

Figure 1.35 Double-stranded and single-stranded nucleic acids both replicate by synthesis of complementary strands governed by the rules of base pairing.

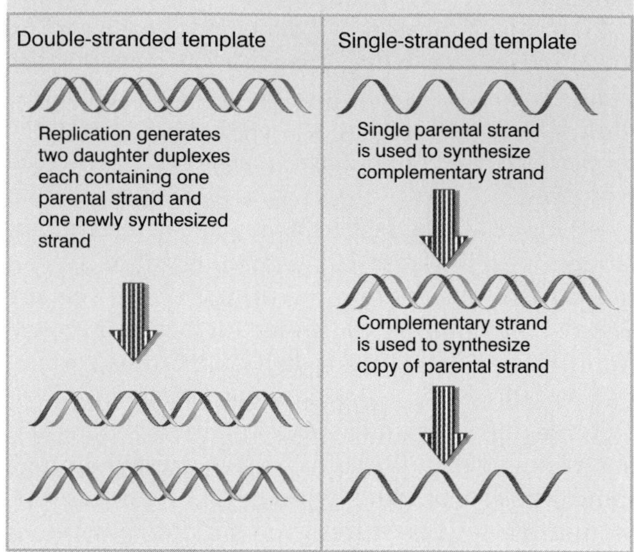

is unidirectional. Transcription of DNA generates RNA molecules that can be used further *only* to generate protein sequences; generally they cannot be retrieved for use as genetic information. Translation of RNA into protein is always irreversible.

Nucleic acid fulfills the mandate for the genetic material with a unique combination of stability and flexibility. Replication via the mechanism of complementary base pairing allows genetic information to be reliably inherited. Yet the occurrence of the occasional mutation allows heritable changes to occur, providing the substrate for evolution. The genetic code provides for expression of genetic information in the form of proteins.

These mechanisms are equally effective for the cellular genetic information of prokaryotes or eukaryotes, and for the information carried by viruses. The genomes of all living organisms consist of duplex DNA. Viruses have genomes that consist of DNA or RNA; and there are examples of each type that are double-stranded (ds) or single-stranded (ss). Details of the mechanism used to replicate the nucleic acid vary among the viral systems, but the principle of replication via synthesis of complementary strands remains the same, as illustrated in **Figure 1.35.**

Cellular genomes reproduce DNA by the mechanism of semiconservative replication. Double-stranded virus genomes, whether DNA or RNA, also replicate by using the individual strands of the duplex as templates to synthesize partner strands.

Viruses with single-stranded genomes use the single strand as template to synthesize a complementary strand; and this complementary strand in turn is used to synthesize its complement, which is, of course, identical with the original starting strand. Replication may involve the formation of stable double-stranded intermediates or use double-stranded nucleic acid only as a transient stage.

The restriction to unidirectional transfer from DNA to RNA is not absolute. It is overcome by the **retroviruses,** whose genomes consist of single-stranded RNA molecules. During the infective cycle, the RNA is converted by the process of **reverse transcription** into a single-stranded DNA, which in turn is converted into a double-stranded DNA. This duplex DNA becomes part of the genome of the cell, and is inherited like any other gene. *So reverse transcription allows a sequence of RNA to be retrieved and used as genetic information.*

The existence of RNA replication and reverse transcription establishes the general principle that *information in the form of either type of nucleic acid sequence can be converted into the other type.* In the usual course of events, however, the cell relies on the processes of DNA replication, transcription, and translation. But on rare occasions (possibly mediated by an RNA virus), information from a cellular RNA is converted into DNA and inserted into the genome. Although reverse transcription plays no role in the regular operations of the cell, it becomes a mechanism of potential importance when we consider the evolution of the genome.

The same principles are followed to perpetuate genetic information, from the massive genomes of plants or amphibians to the tiny genomes of mycoplasma and the yet smaller genetic information of DNA or RNA viruses. **Figure 1.36** summarizes some examples that illustrate the range of genome types and sizes.

The DNA genomes of all living cells consist of very long molecules; all but the bacteria and mycoplasma have more than one such molecule. Some of the larger viruses (such as influenza) have segmented genomes, consisting of more than one nucleic acid molecule. Other viral genomes consist of a single nucleic acid molecule. For the smaller viruses (SV40, ϕX, MS2), the number of functions includes overlapping genes.

Throughout the range of organisms, with genomes varying in total content over a 100,000 fold range, a common principle prevails. *The DNA codes for all the proteins that the cell(s) of the organism must synthesize; and the proteins in turn (directly or indirectly) provide the functions needed for survival.* A similar principle describes the function of the genetic information of viruses, be it DNA or RNA. *The nucleic acid codes for the protein(s) needed to package the genome and also for any functions additional to those provided by the host cell that are needed to reproduce the virus during its infective cycle.* (The smallest virus, the satellite tobacco necrosis virus [STNV], cannot replicate independently, but requires the simultaneous presence of a "helper" virus [tobacco necrosis virus, TNV], which is itself a normally infectious virus.)

Figure 1.36 The amount of nucleic acid in the genome varies over an enormous range.

Genome	Gene Number	Base Pairs
Organisms		
Plants	<50,000	$<10^{11}$
Mammals	100,000	$\sim 3 \times 10^8$
Worms	14,000	$\sim 10^8$
Flies	12,000	1.6×10^8
Fungi	6,000	1.3×10^7
Bacteria	2–4,000	$<10^7$
Mycoplasma	500	$<10^6$
dsDNA Viruses		
Vaccinia	<300	187,000
Papova (SV40)	~6	5,226
Phage T4	~200	165,000
ssDNA Viruses		
Parvovirus	5	5,000
Phage ϕX174	11	5,387
dsRNA Viruses		
Reovirus	22	23,000
ssRNA Viruses		
Coronavirus	7	20,000
Influenza	12	13,500
TMV	4	6,400
Phage MS2	4	3,569
STNV	1	1,300
Viroids		
PSTV RNA	0	359
Scrapie		
Prion	?	?

Viroids are infectious agents that cause diseases in higher plants. They are very small circular molecules of RNA. Unlike viruses, where the infectious agent consists of a **virion,** a genome encapsulated in a protein coat, *the viroid RNA is itself the infectious agent.* The viroid consists solely of the RNA, which is extensively but imperfectly base paired, forming a characteristic rod like the example shown in **Figure 1.37.** Mutations that interfere with the structure of the rod reduce infectivity.

A viroid RNA consists of a single molecular species that is replicated autonomously in infected cells. Its sequence is faithfully perpetuated in its descendants. Viroids fall into several groups. A given viroid is identified with a group by its similarity of sequence with other members of the group. For example, four viroids related to PSTV (potato spindle tuber viroid) have 70–83% similarity of sequence with it. Different isolates of a particular viroid strain vary from one another, and the change may affect the phenotype of infected cells. For example, the *mild* and *severe* strains of PSTV differ by three nucleotide substitutions.

Viroids resemble viruses in having heritable nucleic acid genomes. They fulfill the criteria for genetic information. Yet viroids differ from viruses in both structure and function. They are sometimes called **subviral pathogens.** Viroid RNA does not appear to be translated into protein. So it cannot itself code for the functions needed for its survival. This situation poses two questions. How does viroid RNA replicate? And how does it affect the phenotype of the infected plant cell?

Replication must be carried out by enzymes of the host cell, subverted from their normal function. The heritability of the viroid sequence indicates that viroid RNA provides the template.

Viroids are presumably pathogenic because they interfere with normal cellular processes. They might do

Figure 1.37 PSTV RNA is a circular molecule that forms an extensive double-stranded structure, interrupted by many interior loops. The severe and mild forms differ at three sites.

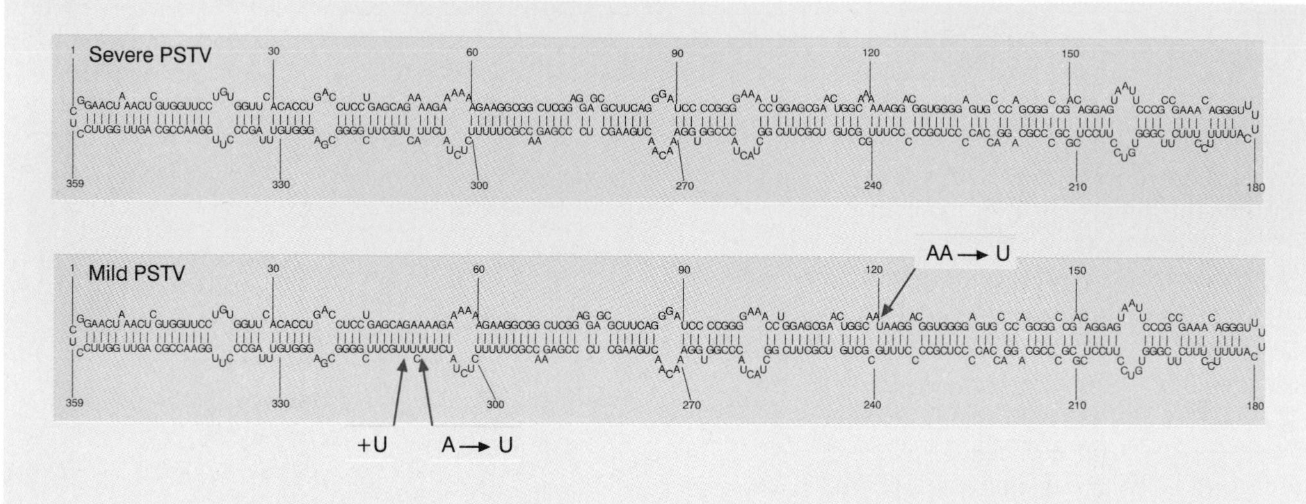

this in a relatively random way, for example, by sequestering an essential enzyme for their own replication or by interfering with the production of necessary cellular RNAs. Alternatively, they might represent abnormal regulatory molecules, with particular effects upon the expression of individual genes.

An even more unusual agent is **scrapie,** the cause of a degenerative neurological disease of sheep and goats. The disease is related to the human diseases of kuru and Creutzfeldt-Jakob syndrome, which affect brain function.

The infectious agent of scrapie does not contain nucleic acid. This extraordinary agent is called a **prion** (proteinaceous infectious agent). It is a 28 kD hydrophobic glycoprotein, **PrP.** PrP is coded by a cellular gene (conserved among the mammals) that is expressed in normal brain. The protein exists in two forms. The product found in normal brain is called PrP^{c}. It is entirely degraded by proteases. The protein found in infected brains is called PrP^{sc}. It is extremely resistant to degradation by proteases. PrP^{c} is converted to Prp^{sc} by a modification or conformational change that confers protease resistance, and which has yet to be fully defined.

As the infectious agent of scrapie, PrP^{sc} must in some way modify the synthesis of its normal cellular counterpart so that it becomes infectious instead of harmless. Mice that lack a *PrP* gene cannot be infected to develop scrapie, which demonstrates that PrP is essential for development of the disease.

Summary

Two classic experiments proved that DNA is the genetic material. DNA isolated from one strain of pneumococcus bacteria can confer properties of that strain upon another strain. And DNA is the only component that is inherited by progeny phages from the parental phages. More recently, DNA has been used to transfect new properties into eukaryotic cells.

DNA is a double helix consisting of antiparallel strands in which the nucleotide units are linked by 5′–3′ phosphodiester bonds. The backbone provides the exterior; purine and pyrimidine bases are stacked in the interior in pairs in which A is complementary to T while G is complementary to C. The strands separate and use complementary base pairing to assemble daughter strands in semiconservative replication. Complementary base pairing is also used to transcribe an RNA representing one strand of a DNA duplex.

A stretch of DNA may code for protein. The genetic code describes the relationship between the sequence of DNA and the sequence of the protein. Only one of the two strands of DNA codes for protein. A coding sequence of DNA consists of a series of codons, read from a fixed starting point. A codon consists of three nucleotides that represent a single amino acid.

A chromosome consists of an uninterrupted length of duplex DNA that contains many genes. Each gene (or cistron) is transcribed into an RNA product, which in turn is translated into a polypeptide sequence if the gene codes for protein. An RNA or protein product of a gene is said to be *trans*-acting. A gene is defined as a unit on a single stretch of DNA by the complementation test. A site on DNA that regulates the activity of an adjacent gene is said to be *cis*-acting.

A gene may have multiple alleles. Recessive alleles are caused by a loss-of-function. A null allele has total loss-of-function. Dominant alleles are caused by gain-of-function.

A mutation consists of a change in the sequence of A·T and G·C base pairs in DNA. A mutation in a coding sequence may change the sequence of amino acids in the corresponding protein. A frameshift mutation alters the subsequent reading frame by inserting or deleting a base; this causes an entirely new series of amino acids to be coded after the site of mutation. A point mutation changes only the amino acid represented by the codon in which the mutation occurs. Point mutations may be reverted by back mutation of the original mutation. Insertions may revert by loss of the inserted material, but deletions cannot revert. Mutations may also be suppressed indirectly when a mutation in a different gene counters the original defect.

The natural incidence of mutations is increased by mutagens. Mutations may be concentrated at hotspots. A type of hotspot responsible for some point mutations is caused by deamination of the modified base 5-methylcytosine.

Forward mutations occur at a rate of $\sim 10^{-6}$ per locus per generation; back mutations are rarer. Not all mutations have an effect on the phenotype.

Although all genetic information in cells is carried by DNA, viruses have genomes of double-stranded or single-stranded DNA or RNA. Viroids are subviral pathogens that consist solely of small circular molecules of RNA, with no protective packaging. The RNA does not code for protein and its mode of perpetuation and of pathogenesis is unknown. Scrapie appears to consist of a proteinaceous infectious agent.

Further reading

Reviews

Cairns, J., Stent, G., and Watson, J. (1966). *Phage and the Origins of Molecular Biology.* (Cold Spring Harbor Laboratory, New York).

Olby, R. (1974). *The Path to the Double Helix.* (MacMillan, London).

Drake, J. W. and Balz, R. H. (1976). The biochemistry of mutagenesis. *Ann. Rev. Biochem.* **45,** 11–37.

Roth, J. R. (1974). Frameshift mutations. *Ann. Rev. Genet.* **8,** 319–346.

Judson, H. (1978). *The Eighth Day of Creation.* (Knopf, New York).

Discoveries

Griffith, F. (1928). The significance of *pneuomococcal* types. *J. Hyg.* **27,** 113–159.

Avery, O. T., MacLeod, C. M., and McCarty, M. (1944). Studies on the chemical nature of the substance inducing transformation of *pneumococcal* types. *J. Exp. Med.* **98,** 451–460.

Hershey, A. D. and Chase, M. (1952). Independent functions of viral protein and nucleic acid in growth of bacteriophage. *J. Gen. Physiol.* **36,** 39–56.

Watson, J. D. and Crick, F. H. C. (1953). A structure for DNA. *Nature* **171,** 737–738.

Watson, J. D. and Crick, F. H. C. (1953). Genetic implications of the structure of DNA. *Nature* **171,** 964–967.

Wilkins, M. F. H., Stokes, A. R., and Wilson, H. R. (1953). Molecular structure of DNA. *Nature* **171,** 738–740.

Meselson, M. and Stahl, F. W. (1958). The replication of DNA in *E. coli. Proc. Nat. Acad. Sci. USA* **44,** 671–682.

Benzer, S. and Champe, S. P. (1961). Ambivalent *rII* mutants of phage T4. *Proc. Nat. Acad. Sci. USA* **47,** 403–416.

Crick, F. H. C., Barnett, L., Brenner, S., and Watts-Tobin, R. J. (1961). General nature of the genetic code for proteins. *Nature* **192,** 1227–1232.

Yanofsky, C. *et al.* (1964). On the colinearity of gene structure and protein structure. *Proc. Nat. Acad. Sci. USA* **51,** 266–272.

Yanofsky, C. *et al.* (1967). The complete amino acid sequence of the tryptophan synthetase A protein (α-subunit) and its co-linear relationship with the genetic map of the A gene. *Proc. Nat. Acad. Sci. USA* **57,** 296–298.

Coulondre, C. *et al.* (1978). Molecular basis of base substitution hotspots in *E. coli. Nature* **274,** 775–780.

McKinley, M. P., Bolton, D. C., and Prusiner, S. B. (1983). A protease-resistant protein is a structural component of the scrapie prion. *Cell* **35,** 57–62.

Oesch, B. *et al.* (1985). A cellular gene encodes scrapie PrP 27–30 protein. *Cell* **40,** 735–746.

Basler, K., Oesch, B., Scott, M., Westaway, D., Walchli, M., Groth, D. F., McKinley, M. P., Prusiner, S. B., and Weissmann, C. (1986). Scrapie and cellular PrP isoforms are encoded by the same chromosomal gene. *Cell* **46,** 417–428.

Bueler, H. *et al.* (1993). Mice devoid of PrP are resistant to scrapie. *Cell* **73,** 1339–1347.

Chapter 2

From genes to genomes

WE can think about mapping genes and genomes at several levels of resolution:

■ A genetic (or linkage) map identifies the distance between mutations in terms of recombination frequencies. It is limited by its reliance on the occurrence of mutations that affect the phenotype. Because recombination frequencies can be distorted relative to the physical distance between sites, it does not accurately represent the genetic material.

■ A linkage map can also be constructed by measuring recombination between sites in genomic DNA. These sites have sequence variations that generate differences in the susceptibility to cleavage by certain (restriction) enzymes. Because such variations are common, such a map can be prepared for any organism irrespective of the occurrence of mutants. It has the same disadvantage as any linkage map that the relative distances are based on recombination.

■ A restriction map is constructed by cleaving DNA into fragments with restriction enzymes and measuring the distances between the sites of cleavage. This represents distances in terms of the length of DNA, so it provides a physical map of the genetic material. A restriction map does not intrinsically identify sites of genetic interest. For it to be related to the genetic map, mutations have to be characterized in terms of their effects upon the restriction sites. Large changes in the genome can be recognized because they affect the sizes or numbers of restriction fragments. Point mutations are more difficult to detect.

■ The ultimate map is to determine the sequence of the DNA. From the sequence, we can identify genes and the distances between them. By analyzing the protein-coding potential of a sequence of the DNA, we can deduce whether it represents a protein. The basic assumption here is that natural selection prevents the accumulation of damaging mutations in sequences that code for proteins. Reversing the argument, we may assume that an intact coding sequence is likely to be actually used to generate a protein.

By comparing the sequence of a wild-type DNA with that of a mutant allele, we can determine the nature of a mutation and its exact site of occurrence. This defines the relationship between the genetic map (based entirely on sites of mutation) and the physical map (based on or even comprising the sequence of DNA).

Similar techniques are used to identify and sequence genes and to map the genome, although there is of course a difference of scale. In each case, the principle is to obtain a series of overlapping fragments of DNA, which can be connected into a continuous map. The crucial feature is that each segment is related to the next segment on the map by characterizing the overlap between them, so that we can be sure no segments are missing. This principle is applied both at the level of ordering restriction fragments into a map, and in connecting the sequences of the fragments.

Because the use of restriction mapping is central to the molecular analysis of both the genome and the individual gene, we review the principles of the approach briefly before we turn to the structure of the gene itself. In the next section we discuss restriction mapping as such; and the following section discusses its application to construct linkage maps. This puts us in a position to discuss the molecular organization of individual genes, relationships among groups of genes, and the identification of genes in which mutations cause human diseases. In Chapter 3, we consider the overall constitution of the genome and its total number of genes.

Genes can be mapped by restriction cleavage

ONCE a segment of DNA has been isolated, the first step to obtaining its sequence is to map the nucleic acid at the molecular level. A physical map of any DNA molecule can be obtained by breaking it at defined points whose distance apart can be accurately determined. Specific breaks are made possible by the ability of **restriction enzymes** to recognize rather short sequences of double-stranded DNA as targets for cleavage.

Each restriction enzyme has a particular target in duplex DNA, usually a specific sequence of four to six base pairs. The enzyme cuts the DNA at every point at which its target sequence occurs. Different restriction enzymes have different target sequences, and a large range of these activities (obtained from a wide variety of bacteria) is now available.

A **restriction map** represents a linear sequence of the sites at which particular restriction enzymes find their targets. Distance along such maps is measured directly in base pairs (abbreviated **bp**) for short distances; longer distances are given in **kb**, corresponding to kilobase (10^3) pairs in DNA or to kilobases in RNA. At the level of the chromosome, a map is described in megabase pairs (1 **Mb** = 10^6 bp).

When a DNA molecule is cut with a suitable restriction enzyme, it is cleaved into distinct fragments. These fragments can be separated on the basis of their size by gel electrophoresis. The cleaved DNA is placed on top of a gel made of agarose or polyacrylamide. When an electric current is passed through the gel, each fragment moves down it at a rate that is inversely related to the log of its molecular weight. This movement produces a series of bands. Each band corresponds to a fragment of particular size, decreasing down the gel.

Figure 2.1 shows an example of this technique. A

Figure 2.1 DNA can be cleaved by restriction enzymes into fragments that can be separated by gel electrophoresis.

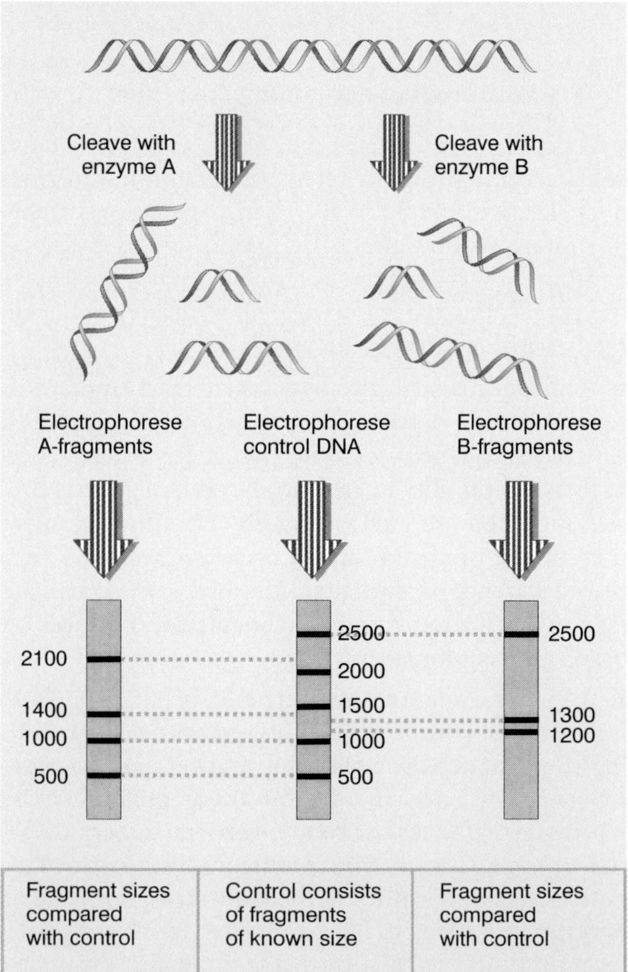

DNA molecule of length 5000 bp is incubated separately with two restriction enzymes, A and B. After cleavage the DNA is electrophoresed. The sizes of the individual fragments generated by enzyme A (left) or enzyme B (right) are determined by comparison with the positions of fragments of known size, such as the control shown in the center. This demonstrates that enzyme A has cut the substrate DNA into four fragments (lengths 2100, 1400, 1000, and 500 bp), while enzyme B has generated three fragments (lengths 2500, 1300, and 1200 bp). Can we proceed further from these data to generate a map that places the sites of breakage at defined positions on the DNA?

The patterns of cutting by the two enzymes can be related by several means. **Figure 2.2** illustrates the principle of analysis by **double digestion.** In this technique, the DNA is cleaved simultaneously with two enzymes as well as with either one by itself. The most decisive way to use this technique is to extract each fragment produced in the individual digests with either enzyme A or enzyme B and then to cleave it with the other enzyme. The products of cleavage are analyzed again by electrophoresis.

We can use these data to construct a map of the original 5000 bp molecule of DNA, as illustrated by the stages of **Figure 2.3.**

Each gel in Figure 2.2 is labeled according to the fragment that was isolated from the gel in Figure 2.1. A-2100 identifies the fragment of 2100 bp produced by degrading the original DNA molecule with enzyme A. When this fragment is retrieved and subjected to enzyme B, it is cut into fragments of 1900 and 200 bp. So one of the cuts made by enzyme B lies 200 bp from the nearest site cut by enzyme A on one side, and is 1900 bp from the site cut by enzyme A on the other side. This situation is described by the top map in Figure 2.3.

A related pattern of cuts is seen when we examine the susceptibility of fragment B-2500 to enzyme A. It is cut into fragments of 1900 and 600 bp. So the 1900 bp fragment is generated by double cuts, with an A site at one end and a B site at the other end. It can be released from either of the single-cut fragments (A-2100 or B-2500) that contain it. These single-cut fragments must therefore **overlap** in the region of the 1900 bp of the common fragment that can be generated from them. This is described in the second map of Figure 2.3, which extends our map to the right to add a cleavage site for enzyme B.

The key to restriction mapping is the use of overlapping fragments. Because of the overlap of A-2100 and B-2500 in the central region of 1900 bp, we can relate

Figure 2.2 Double digests define the cleavage positions of one enzyme with regard to the other.

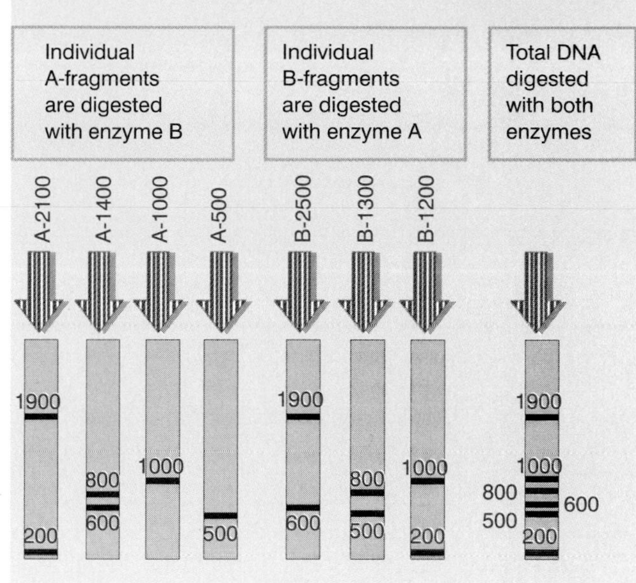

the A site 200 bp to the left of the 1900 bp region with the B site 600 bp to the right. In the same way, we can now extend the map farther on either side. The 200 bp fragment at the left is also produced by cutting B-1200 with enzyme A, so the next B site must lie 1000 bp to the left. The 600 bp fragment at the right is also produced by cutting A-1400 with enzyme B, so the next A site must lie 800 bp to the right. This gives the third map in Figure 2.3.

We can now complete the map by identifying the source of the two fragments at each end. At the left end, the 1000 bp fragment arises from B-1200 or in the form of A-1000, which is not cut by enzyme B. So A-1000 lies at the end of the map. Proceeding from the left end of the complete 5000 bp region, it is 1000 bp to the first A site and 1200 bp to the first B site. (This is why a B cut is not shown at the left end of the map above, although formally we treated the end as a B-cutting site in the analysis.)

At the right end of the map, the 800 bp double-cut fragment is generated by cutting B-1300 with enzyme A, so we must add a fragment of 500 bp to the right. This is the terminal fragment, as seen by its presence as A-500 in the single-cut A digest. So our completed map takes the form of the bottom map in Figure 2.3.

The actual construction of a restriction map usually requires recourse to several enzymes, so it becomes necessary to resolve quite a complex pattern of the overlapping fragments generated by the various enzymes. Several other techniques are used in

Figure 2.3 A restriction map can be constructed by relating the A-fragments and B-fragments through the overlaps seen with double digest fragments.

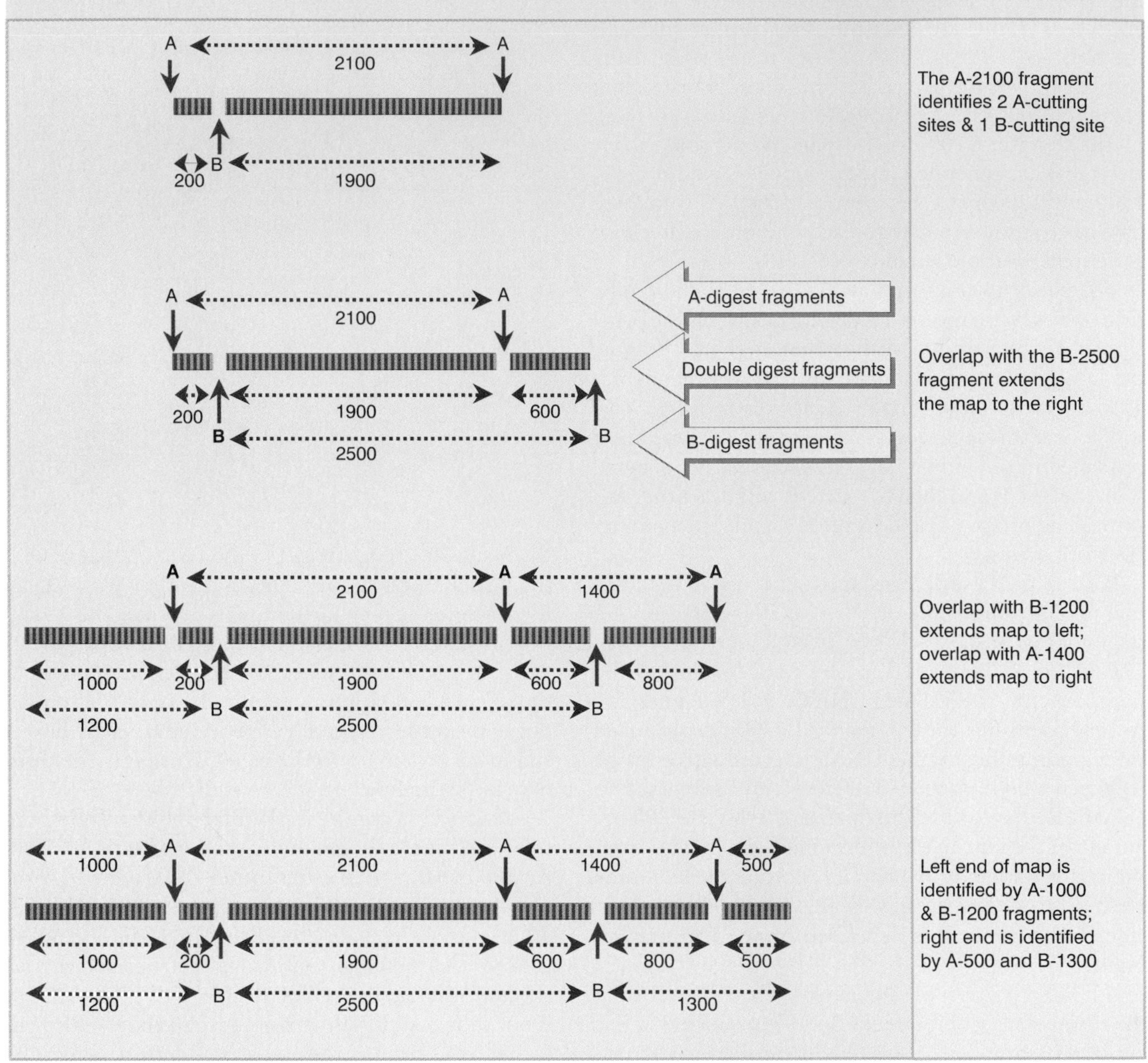

A
2100
A
200
B
1900
The A-2100 fragment identifies 2 A-cutting sites & 1 B-cutting site
A-digest fragments
Double digest fragments
B-digest fragments
600
2500
Overlap with the B-2500 fragment extends the map to the right
1400
1000
800
1200
Overlap with B-1200 extends map to left; overlap with A-1400 extends map to right
500
1300
Left end of map is identified by A-1000 & B-1200 fragments; right end is identified by A-500 and B-1300

conjunction with comparison of fragments, including **end-labeling,** in which the ends of the DNA molecule are labeled with a radioactive phosphate (certain enzymes can add phosphate moieties specifically to 5′ or to 3′ ends). **Figure 2.4** shows that this allows the fragments containing the ends to be identified directly by their radioactive label. So in the fragment A preparation, A-1000 and A-500 would be placed immediately at opposite ends of the map; similarly, fragments B-1200 and B-1300 would be identified as ends.

A restriction map of the entire 5000 bp region that was constructed in Figures 2.1–2.3 is recapitulated in its more usual form in **Figure 2.5.** The map shows the positions at which particular restriction enzymes cut DNA; the distances between the sites of cutting are measured in base pairs. *So the DNA is divided into a series of regions of defined lengths that lie between sites recognized by the restriction enzymes.* An important feature is that a restriction map can be obtained for any sequence of DNA, *irrespective of whether mutations have been identified in it,* or, indeed, whether we have any knowledge of its function.

Figure 2.4 When restriction fragments are identified by their possession of a labeled end, each fragment directly shows the distance of a cutting site from the end. Successive fragments increase in length by the distance between adjacent restriction sites.

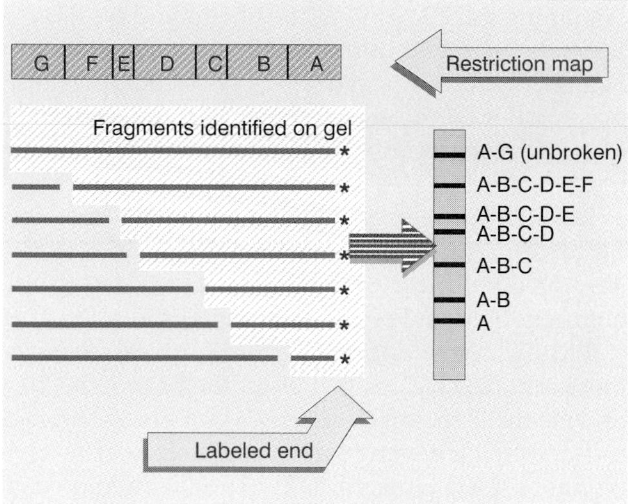

Figure 2.5 A restriction map is a linear sequence of sites separated by defined distances on DNA. The map identifies the sites cleaved by enzymes A and B, as defined by the individual fragments produced by the single and double digests.

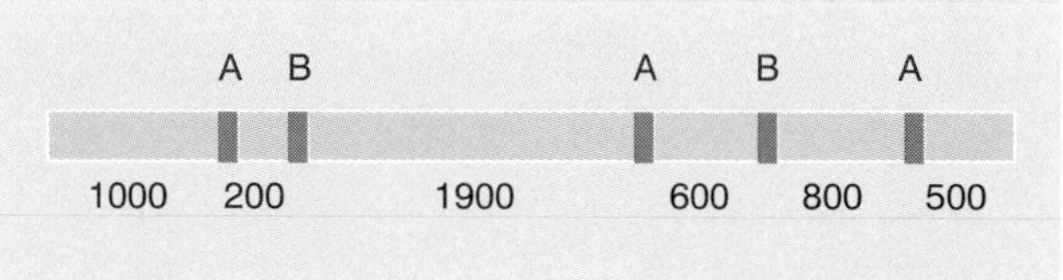

How variable are individual genomes?

THE original Mendelian view of the genome classified alleles as either wild-type or mutant. Subsequently we recognized the existence of multiple alleles, each with a different effect on the phenotype. (In some cases it may not even be appropriate to define any one allele as "wild-type.")

The coexistence of multiple alleles at a locus is called **genetic polymorphism.** Any site at which multiple alleles exist as stable components of the population is by definition polymorphic. An allele is usually defined as polymorphic if it is present at a frequency of >1% in the population.

What is the basis for the polymorphism among the mutant alleles? They possess different mutations that alter the protein function, thus producing changes in phenotype. If we compare the restriction maps or the DNA sequences of these alleles, they too will be polymorphic in the sense that each map or sequence will be different from the others.

Although not evident from the phenotype, the wild type may itself be polymorphic. Multiple versions of the wild-type allele may be distinguished by differences in sequence that do not affect their function, and which therefore do not produce phenotypic variants. A population may have extensive polymorphism at the level of genotype. Many different sequence variants may exist at a given locus; some of them are evident because they affect the phenotype, but others are hidden because they have no visible effect.

So there may be a continuum of changes at a locus, including those that change DNA sequence but do not change protein sequence, those that change protein sequence without changing function, those that create proteins with different activities, and those that create mutant proteins that are nonfunctional.

Some polymorphisms in the genome can be detected by comparing the restriction maps of different individuals. The criterion is a change in the pattern of fragments produced by cleavage with a restriction enzyme. **Figure 2.6** shows that when a target site is present in the genome of one individual and absent from another, the extra cleavage in the first genome will generate two fragments corresponding to the single fragment in the second genome.

Figure 2.6 A point mutation that affects a restriction site is detected by a difference in restriction fragments.

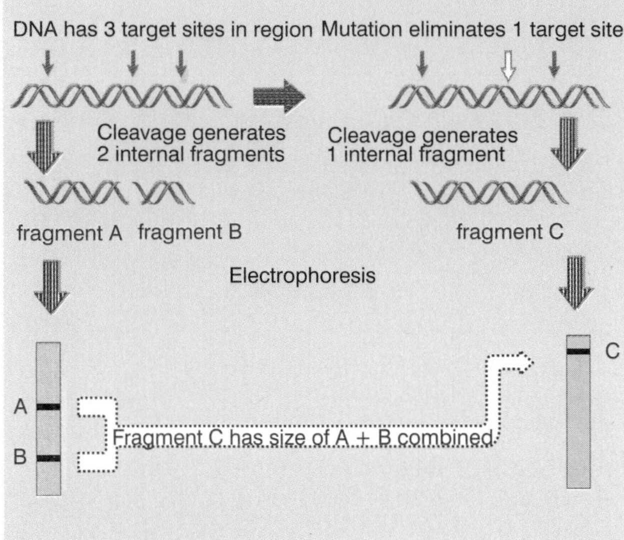

Because the restriction map is independent of gene function, a polymorphism at this level can be detected *irrespective of whether the sequence change affects the phenotype.* Probably very few of the restriction site polymorphisms in a genome actually affect the phenotype. Most involve sequence changes that have no effect on the production of proteins (for example, because they lie between genes).

A difference in restriction maps between two individuals is called a **restriction fragment length polymorphism** (RFLP). It can be used as a genetic marker in exactly the same way as any other marker. Instead of examining some feature of the phenotype, we directly assess the genotype, as revealed by the restriction map. **Figure 2.7** shows a pedigree of a restriction polymorphism followed through three generations. It displays Mendelian segregation at the level of DNA marker fragments.

Recombination frequency can be measured between a restriction marker and a visible phenotypic marker as illustrated in **Figure 2.8.** So a genetic map can include both genotypic and phenotypic markers.

Because restriction markers are not restricted to those genome changes that affect the phenotype, they provide the basis for an extremely powerful technique for identifying genetic loci at the molecular level. A typical problem concerns a mutation with known effects on the phenotype, where the relevant genetic locus can be placed on a genetic map, but for which we have no knowledge about the corresponding gene or protein. Many damaging or fatal human diseases fall into this category. For example cystic fibrosis shows Mendelian inheritance, but the molecular nature of the mutant function was unknown until it

Figure 2.7 Restriction site polymorphisms are inherited according to Mendelian rules. Four alleles for a restriction marker are found in all possible pairwise combinations, and segregate independently at each generation. Photograph kindly provided by Ray White.

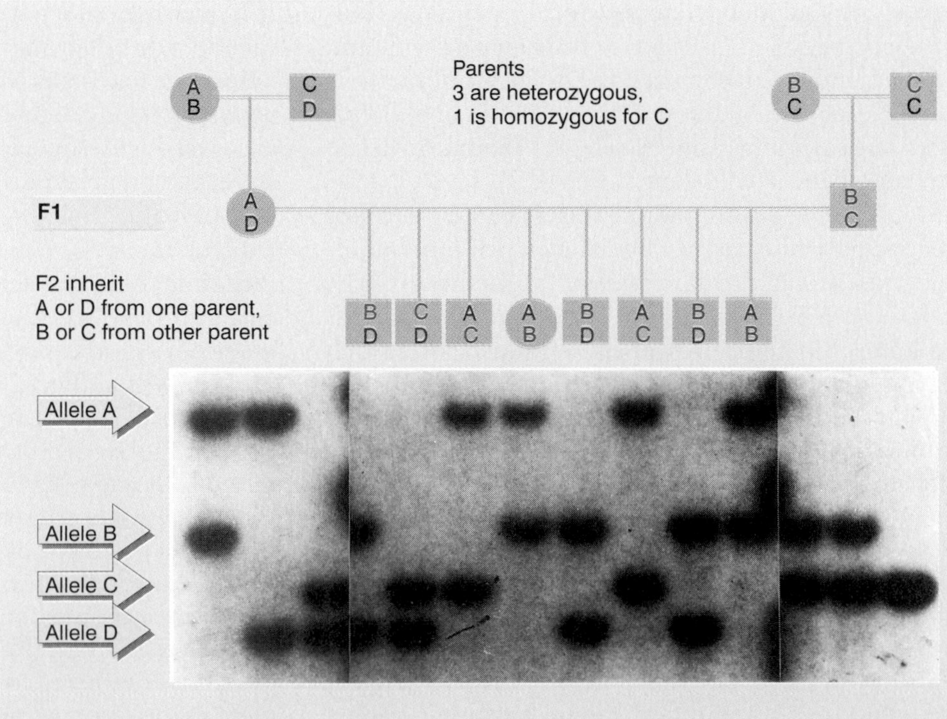

Figure 2.8 A restriction polymorphism can be used as a genetic marker to measure recombination distance from a phenotypic marker (such as eye color). The figure simplifies the situation by showing only the DNA bands corresponding to a haploid genome and omitting the bands corresponding to the allele of the other genome in a diploid.

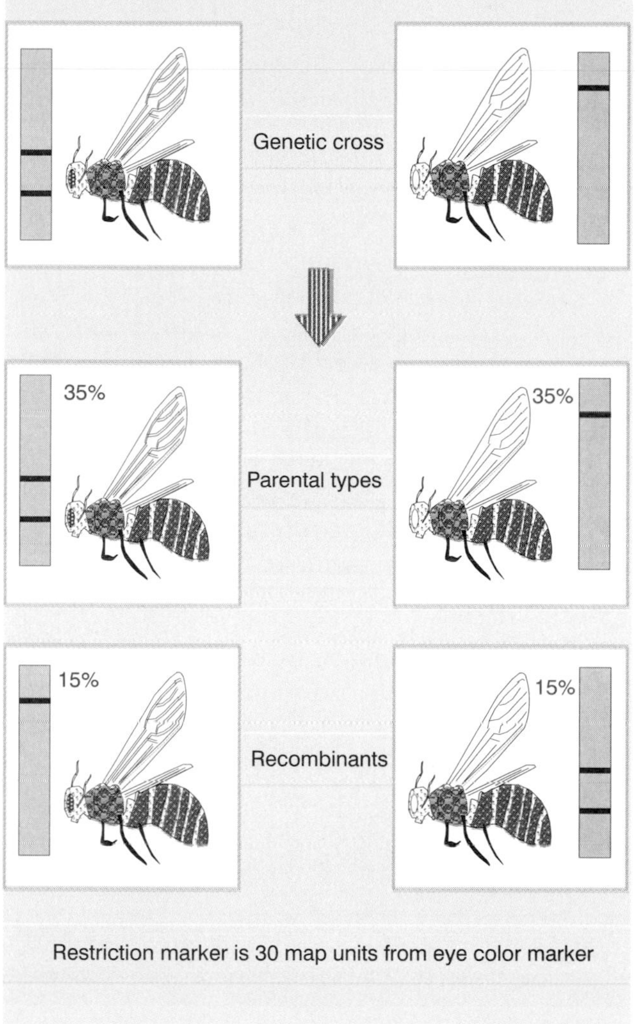

could be identified as a result of characterizing the gene.

If restriction polymorphisms occur at random in the genome, some should occur near any particular target gene. We can identify such restriction markers by virtue of their tight linkage to the mutant phenotype. If we compare the restriction map of DNA from patients suffering from a disease with the DNA of normal people, we may find that a particular restriction site is always present (or always absent) from the patients.

A hypothetical example is shown in **Figure 2.9.** This situation corresponds to finding 100% linkage between the restriction marker and the phenotype. It would imply that the restriction marker lies so close to the mutant gene that it is never separated from it by recombination.

The identification of such a marker has two important consequences:

- It may offer a diagnostic procedure for detecting the disease. Some of the human diseases that are genetically well characterized but ill defined in molecular terms cannot be easily diagnosed. If a restriction marker is reliably linked to the phenotype, then its presence can be used to diagnose the disease, either at a prenatal stage or subsequently.
- It may lead to isolation of the gene. The restriction marker must lie relatively near the gene on the genetic map if the two loci rarely or never recombine. Although "relatively near" in genetic terms can be a substantial distance in terms of base pairs of DNA, nonetheless it provides a starting point from which we can proceed along the DNA to the gene itself.

RFLPs occur frequently enough in the human genome to be useful for genetic mapping. If allelic sequences are compared between any two individual chromosomes, differences in individual base pairs occur at a frequency of >1 per 1000 bp. Those base changes that affect restriction sites can be detected as RFLPs.

Once an RFLP has been assigned to a linkage group, it can be placed in position on the genetic map, and map distances to its flanking markers determined. An effort to map RFLPs in man and mouse has led to the construction of linkage maps for both genomes. The human map contains >5000 markers separated by an average distance of 1.6 cM (1–2 Mb); the mouse map has >7000 markers with an average spacing of ~0.2 cM (~200 kb). Any unknown site can be tested for linkage to these sites and by this means rapidly placed on to the map.

The frequency of polymorphism means that every individual has a unique constellation of restriction sites. The particular combination of sites found in a specific region is called a **haplotype,** a genotype in miniature. Haplotype was originally introduced as a concept to describe the genetic constitution of the major histocompatibility locus, a region specifying proteins of importance in the immune system (see Chapter 24). The concept now has been extended to describe the particular combination of alleles or restriction sites (or any other genetic marker) present in some defined area of the genome.

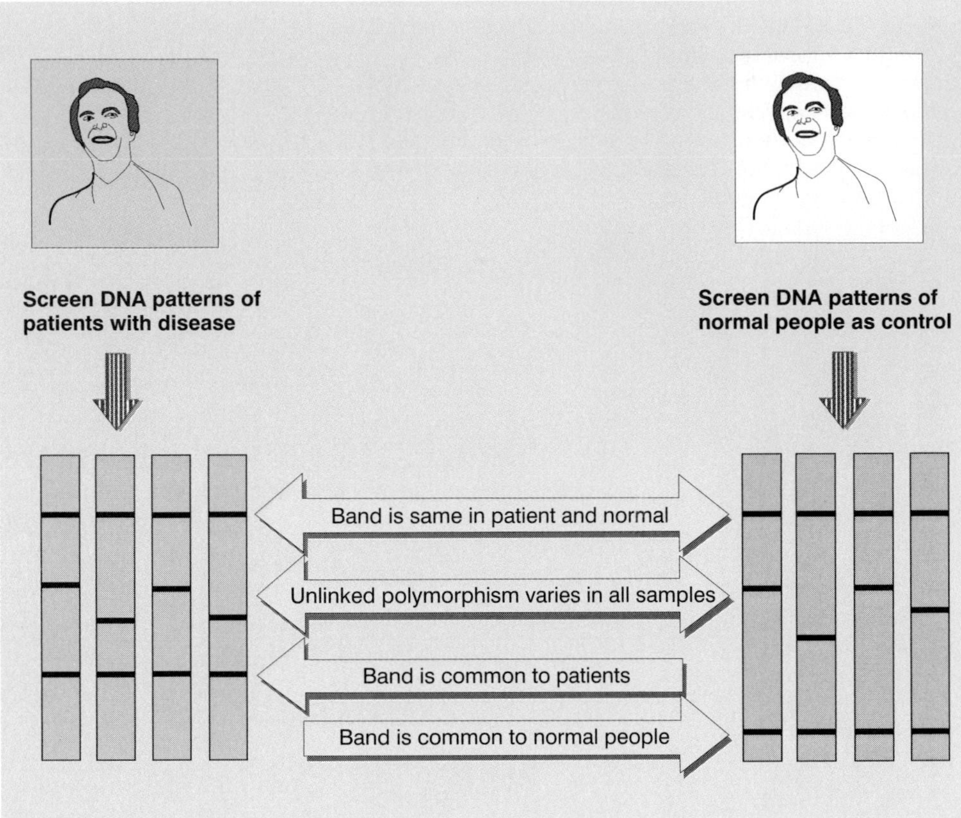
Screen DNA patterns of patients with disease
Screen DNA patterns of normal people as control
Band is same in patient and normal
Unlinked polymorphism varies in all samples
Band is common to patients
Band is common to normal people

Figure 2.9 If a restriction marker is associated with a phenotypic characteristic, the restriction site must be located near the gene responsible for the phenotype. The mutation changing the band that is common in normal people into the band that is common in patients is very closely linked to the disease gene.

The existence of RFLPs provides the basis for a technique to establish unequivocal parent-progeny relationships. In cases where parentage is in doubt, a comparison of the RFLP map in a suitable chromosome region between potential parents and child allows absolute assignment of the relationship. The use of DNA restriction analysis to identify individuals has been called **DNA fingerprinting.** We discuss in more detail in Chapter 4 the use of particularly variable "minisatellite" sequences for mapping in the human genome.

Eukaryotic genes are often interrupted

UNTIL eukaryotic genes were characterized by molecular mapping, we assumed that they would have the same organization as prokaryotic genes. We therefore expected the gene to consist of a length of DNA that is colinear with the protein. But a comparison between the structure of DNA and the corresponding mRNA shows a discrepancy in many cases. The mRNA always includes a nucleotide sequence that corresponds exactly with the protein product according to the rules of the genetic code. *But the gene includes additional sequences that lie within the coding region, interrupting the sequence that represents the protein.*

The sequences of DNA comprising an interrupted gene are divided into the two categories depicted in **Figure 2.10:**

- The **exons** are the sequences represented in the mature RNA. By definition, a gene starts and ends with exons, corresponding to the 5′ and 3′ ends of the RNA.
- The **introns** are the intervening sequences that are

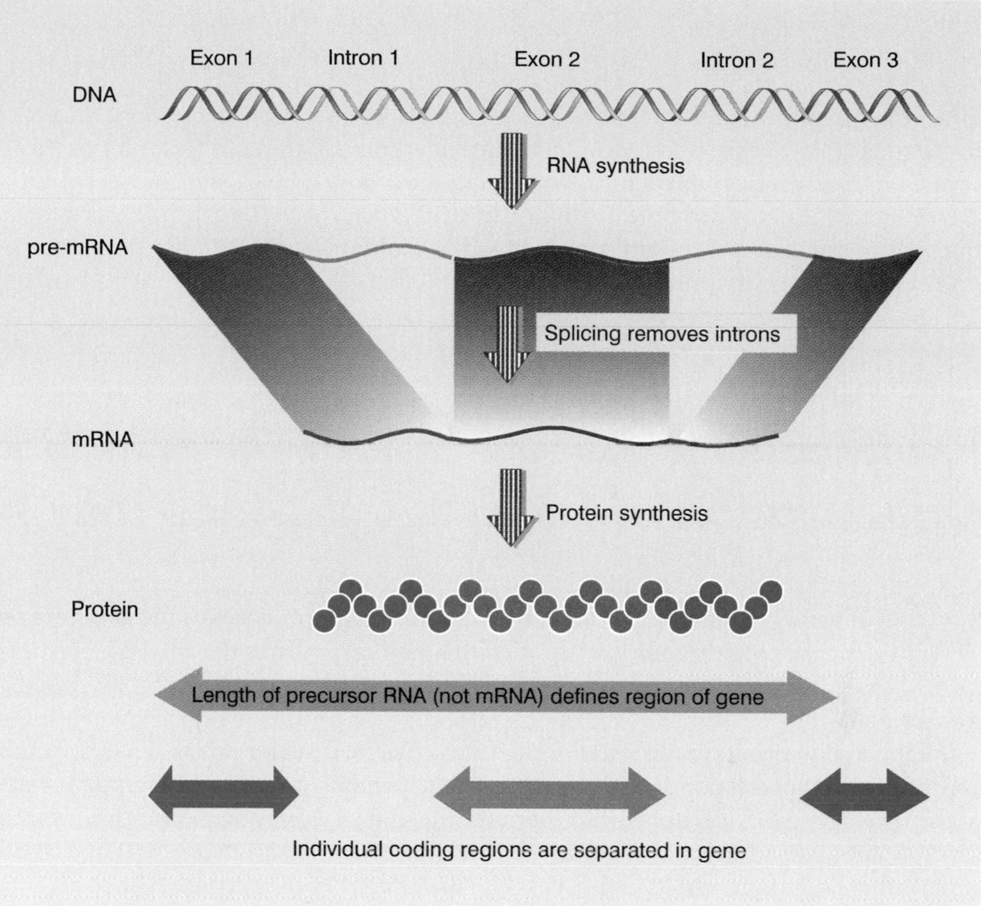

Figure 2.10 Interrupted genes are expressed via a precursor RNA. Introns are removed when the exons are spliced together. The mRNA has only the sequences of the exons.

removed when the primary transcript is processed to give the mature RNA.

The expression of interrupted genes requires an additional step that does not occur for uninterrupted genes. The DNA gives rise to an RNA copy (a **transcript**) that exactly represents the genome sequence. But this RNA is only a precursor; it cannot be used for producing protein. First the introns must be removed from the RNA to give a messenger RNA that consists only of the series of exons. This process is called **RNA splicing.** It involves a precise deletion of an intron from the primary transcript; the ends of the RNA on either side are joined to form a covalently intact molecule. We discuss the mechanisms and regulation of splicing in Chapter 22.

The **structural gene** comprises the region in the genome between points corresponding to the 5′ and 3′ terminal bases of mature mRNA. We know that transcription starts at the 5′ end of the mRNA, but probably it extends beyond the 3′end, which is generated by cleavage of the RNA (see Chapter 22). The definition of the gene can be expanded to include regulatory regions on both sides of the gene that are required for initiating and (sometimes) terminating gene expression.

How does this change our view of the gene? Following splicing, the exons are always joined together in the same order in which they lie in DNA. So the colinearity of gene and protein is maintained between the individual exons and the corresponding parts of the protein chain. The *order* of mutations in the gene remains the same as the order of amino acid replacements in the protein. But the *distances* in the gene do not correspond at all with the distances in the protein. The length of the gene is defined by the length of the initial (precursor) RNA instead of by the length of the messenger RNA.

All the exons are represented on the same molecule of RNA, and their splicing together occurs only as an *intra*molecular reaction. There is usually no joining of exons carried by *different* RNA molecules, so the mechanism excludes any splicing together of sequences representing different alleles. So mutations located in different exons of a gene cannot complement one another; thus they continue to be defined as members of the same complementation group.

What are the effects of mutations in the introns? Since the introns are not part of the messenger RNA, mutations in them cannot directly affect protein structure. However, they can prevent the production of the messenger RNA—for example, by inhibiting the splicing together of exons. A mutation of this sort acts only on the allele that carries it. So it fails to complement any other mutation in that allele, and constitutes part of the same complementation group as the exons.

Eukaryotic genes are not necessarily interrupted. Some correspond directly with the protein product in the same manner as prokaryotic genes. In yeast, most genes are uninterrupted. In higher eukaryotes, most genes are interrupted; and the introns are usually much longer than exons, creating genes that are very much larger than their coding regions.

Organization of interrupted genes may be conserved

When a gene is uninterrupted, the restriction map of its DNA corresponds exactly with the map of its mRNA (obtained by characterizing a cDNA reverse transcript).

When a gene possesses an intron, the map at each end of the gene corresponds with the map at each end of the message sequence. But within the gene, the maps diverge, because additional regions are found in the gene, but are not represented in the message. Each such region corresponds to an intron. The example of **Figure 2.11** compares the restriction maps of a β-globin gene and mRNA. There are two introns. Each intron contains a series of restriction sites that are absent from the cDNA. The pattern of restriction sites in the exons is the same in both the cDNA and the gene.

Ultimately a comparison of the nucleotide sequences of the genomic and cDNA clones precisely defines the introns. Resolution at the sequence level is necessary before we can be sure that all the segments of the gene have been identified. Short introns or exons can be missed in restriction maps if they do not happen to contain an appropriate restriction site. (An intron may pass unnoticed if it lies within a long exon, and an exon that is <50 bp long may fail to hybridize with the cDNA probe, and can therefore pass unnoticed within the introns that flank it.) But a sequence comparison is unambiguous. As indicated in **Figure 2.12,** an intron that lies within a coding region usually interrupts the integrity of the reading frame, but an intact reading frame is found in the cDNA sequence.

No particular rhyme or reason yet has been discerned in the extremely varied structures of eukaryotic

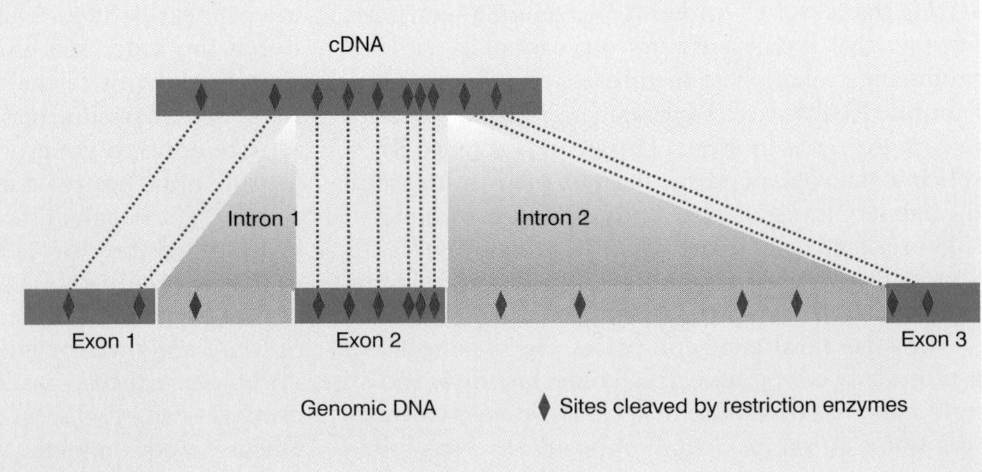

Figure 2.11
Comparison of the restriction maps of cDNA and genomic DNA for mouse β-globin shows that the gene has two additional regions not present in the cDNA. The other regions can be aligned exactly between cDNA and gene.

Figure 2.12 An intron is a sequence present in the gene but absent from the mRNA (here shown in terms of the cDNA sequence). The reading frame is indicated by the alternating open and shaded blocks; note that all three possible reading frames are blocked by termination codons in the intron.

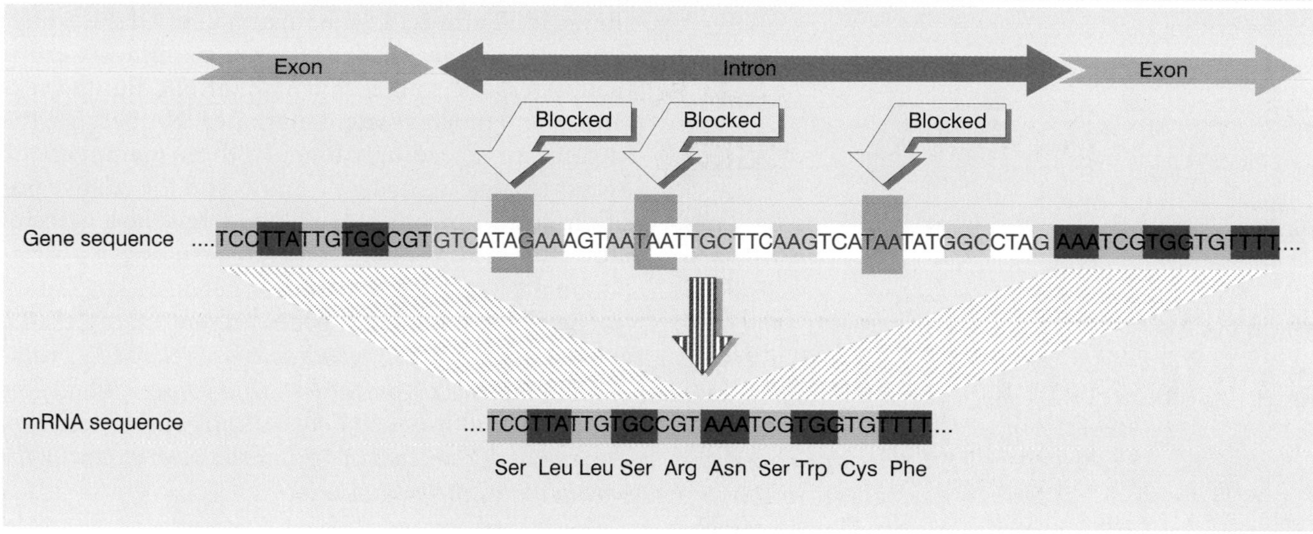

Figure 2.13 All functional globin genes have an interrupted structure with three exons. The lengths indicated in the figure apply to the mammalian β-globin genes.

	Exon 1	Intron 1	Exon 2	Intron 2	Exon 3
Length (bp)	142–145	116–130	222	573–904	216–255
Represents	5' nontranslated + coding 1–30		Amino acids 31–104		Coding 105–end + 3' nontranslated

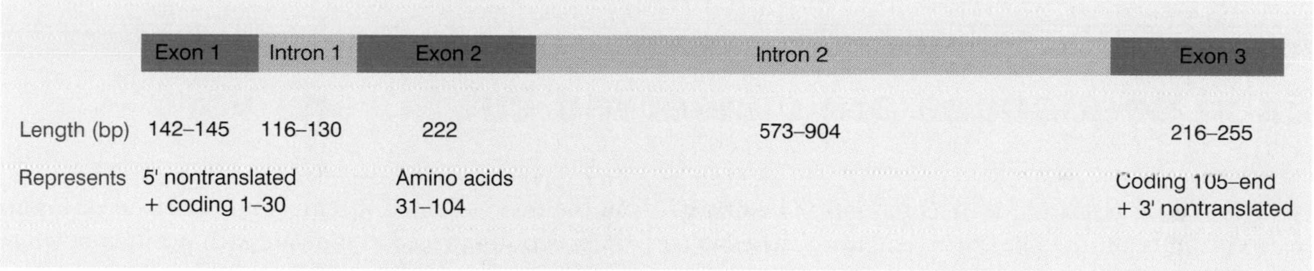

genes. Some genes are uninterrupted, so that the genomic sequence is colinear with that of the mRNA. Most higher eukaryotic genes are interrupted, but the introns vary enormously in both number and size. Introns of nuclear genes generally have termination codons in all reading frames, and have *no coding function.*

All classes of genes may be interrupted: nuclear genes coding for proteins, nucleolar genes coding for rRNA, and genes coding for tRNA. Interruptions are also found in mitochondrial genes in lower eukaryotes, and in chloroplast genes. Interrupted genes do not appear to be excluded from any class of eukaryotes, and have been found in bacteria and bacteriophages, although they are extremely rare in prokaryotic genomes.

Some interrupted genes possess only one or a few introns. The globin genes provide an extensively studied example (see Chapter 4). The two general types of globin gene, α and β, share a common type of structure. The consistency of the organization of mammalian globin genes is evident from the structure of the "generic" globin gene summarized in **Figure 2.13.**

Interruptions occur at homologous positions (relative to the coding sequence) in all known active globin genes, including those of mammals, birds, and frogs. The first intron is always fairly short, and the second is usually longer, but the actual lengths can vary. Most of the variation in overall lengths between different globin genes results from the variation in the second intron. In the mouse, the second intron in the α-globin gene is only 150 bp long, so the overall length of the gene is 850 bp, compared with the 1382 bp of the major β-globin gene. So the variation in length of the genes is

Figure 2.14 Mammalian genes for DHFR have the same relative organization of rather short exons and very long introns, but vary extensively in the lengths of corresponding introns.

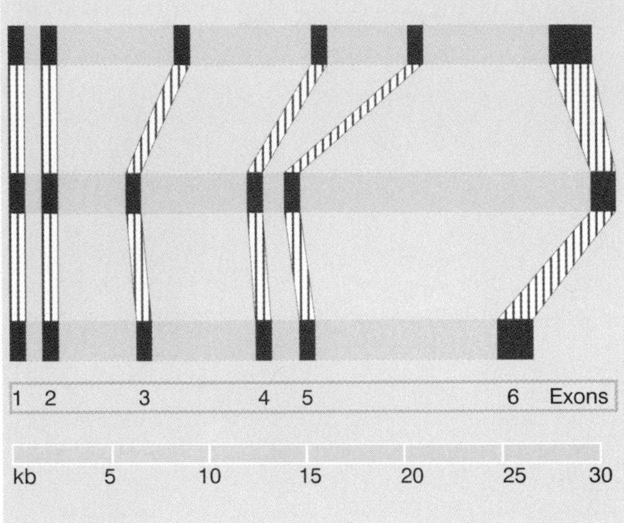

much greater than the range of lengths of the mRNAs (α-globin mRNA = 585 bases, β-globin mRNA = 620 bases).

The example of DHFR, a somewhat larger gene, is shown in **Figure 2.14.** The mammalian DHFR (dihydrofolate reductase) gene is organized into six exons that correspond to the 2000 base mRNA. But they extend over a much greater length of DNA because the introns are exceedingly long. In three mammals the exons remain essentially the same, and the relative positions of the introns are unaltered, but the lengths of individual introns vary extensively, resulting in a variation in the length of the gene from 25 to 31 kb.

The globin and DHFR genes present examples of a general phenomenon: *genes that are related by evolution have related organizations, with conservation of the positions of (at least some) of the introns. Variations in the lengths of the genes are primarily determined by the lengths of the introns.*

Exon sequences are conserved but introns vary

Is a structural gene unique in its genome? The answer can be ambiguous. The entire length of the gene is unique as such, but its exons often are related to those of other genes. As a general rule, when two genes are related, the relationship between their exons is closer than the relationship between the introns. In an extreme case, the exons of two genes may code for the same protein sequence, but the introns may be different. This implies that the two genes originated by a duplication of some common ancestral gene. Then differences accumulated between the copies, but they were restricted in the exons by the need to code for protein functions.

As we see later when we consider the evolution of the gene, exons can be considered as basic building blocks that are assembled in various combinations. A gene may have some exons that are related to exons of another gene, but the other exons may be unrelated. Usually the introns are not related at all in such cases. Such genes may arise by duplication and translocation of individual exons.

The relationship between two genes can be plotted in the form of the dot matrix comparison of **Figure 2.15.** A dot is placed to indicate each position at which the same sequence is found in each gene. The dots form a line at an angle of 45° if two sequences are identical. The line is broken by regions that lack similarity, and it is displaced laterally or vertically by deletions or insertions in one sequence relative to the other.

When the two β-globin genes of the mouse are compared, such a line extends through the three exons and through the small intron. The line peters out in the flanking regions and in the large intron. This is a typical pattern, in which coding sequences are well related, the relationship can extend beyond the boundaries of the exons, but it is lost in longer introns and the regions on either side of the gene.

The overall degree of divergence between two exons is related to the differences between the proteins. It is caused mostly by base substitutions. In the translated regions, the exons are under the constraint of needing to code for amino acid sequences, so they are limited in their potential to change sequence. Many of the changes do not affect codon meanings, because they

Figure 2.15 The sequences of the mouse α^{maj} and α^{min} globin genes are closely related in coding regions, but differ in the flanking regions and large intron. Data kindly provided by Philip Leder.

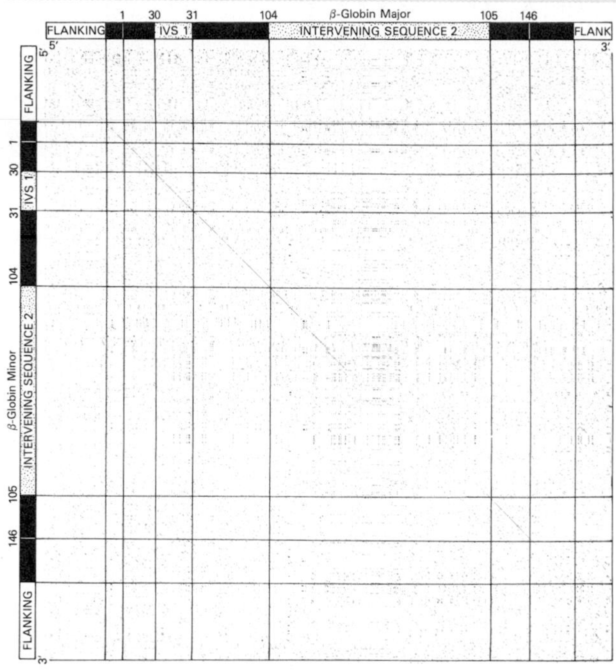

change one codon into another that represents the same amino acid. Changes occur more freely in nontranslated regions (corresponding to the 5′ leader and 3′ trailer of the mRNA).

In corresponding introns, the pattern of divergence involves both changes in size (due to deletions and insertions) and base substitutions. Introns evolve much more rapidly than exons. When a gene is compared in different species, sometimes the exons are homologous, while the introns have diverged so much that corresponding sequences cannot be recognized.

Mutations occur at the same rate in both exons and introns, but are removed more effectively from the exons by adverse selection. However, in the absence of the constraints imposed by a coding function, an intron is able quite freely to accumulate point substitutions and other changes. These changes imply that the intron does not have a sequence-specific function. Whether its presence is at all necessary for gene function is not clear.

Genes can be isolated by the conservation of exons

Some major approaches to identifying genes are based on the contrast between the conservation of exons and the variation of introns. In a region containing a gene whose function has been conserved among a range of species, the sequence representing the protein should have two distinctive properties: it must of course have an open reading frame; and it is likely to have a related sequence in other species. These features can be used to isolate genes.

Suppose we know by genetic data that a particular genetic trait is located in a given chromosomal region. If we lack knowledge about the nature of the gene product, how are we to identify the gene in a region that may be (for example) >1 Mb?

We start with a clone that lies in the general vicinity of this region and then we "walk" through the region by identifying overlapping clones from a library. As shown **in Figure 2.16,** a subfragment from one end of the first clone is used to isolate clones that extend farther along the chromosome. These clones in turn are used to isolate the next set. In each cycle, a new clone is selected because its restriction map coincides at one end with the end of the previous clone, but at the other end has new material. It is possible to walk for hundreds of kb, typically at a rate of >100 kb per month. Chromosome walking allows large contiguous regions of the chromosome to be represented in a library of clones.

Of course, it becomes much easier to identify a particular gene once the sequence of a chromosome has been obtained. This can be done either by sequencing a contiguous series of clones that were obtained by walking, or by relating the clones by other means (such as direct comparisons of sequences). With a sequence in

Figure 2.16 Chromosome walking is accomplished by successive hybridizations between overlapping genomic clones.

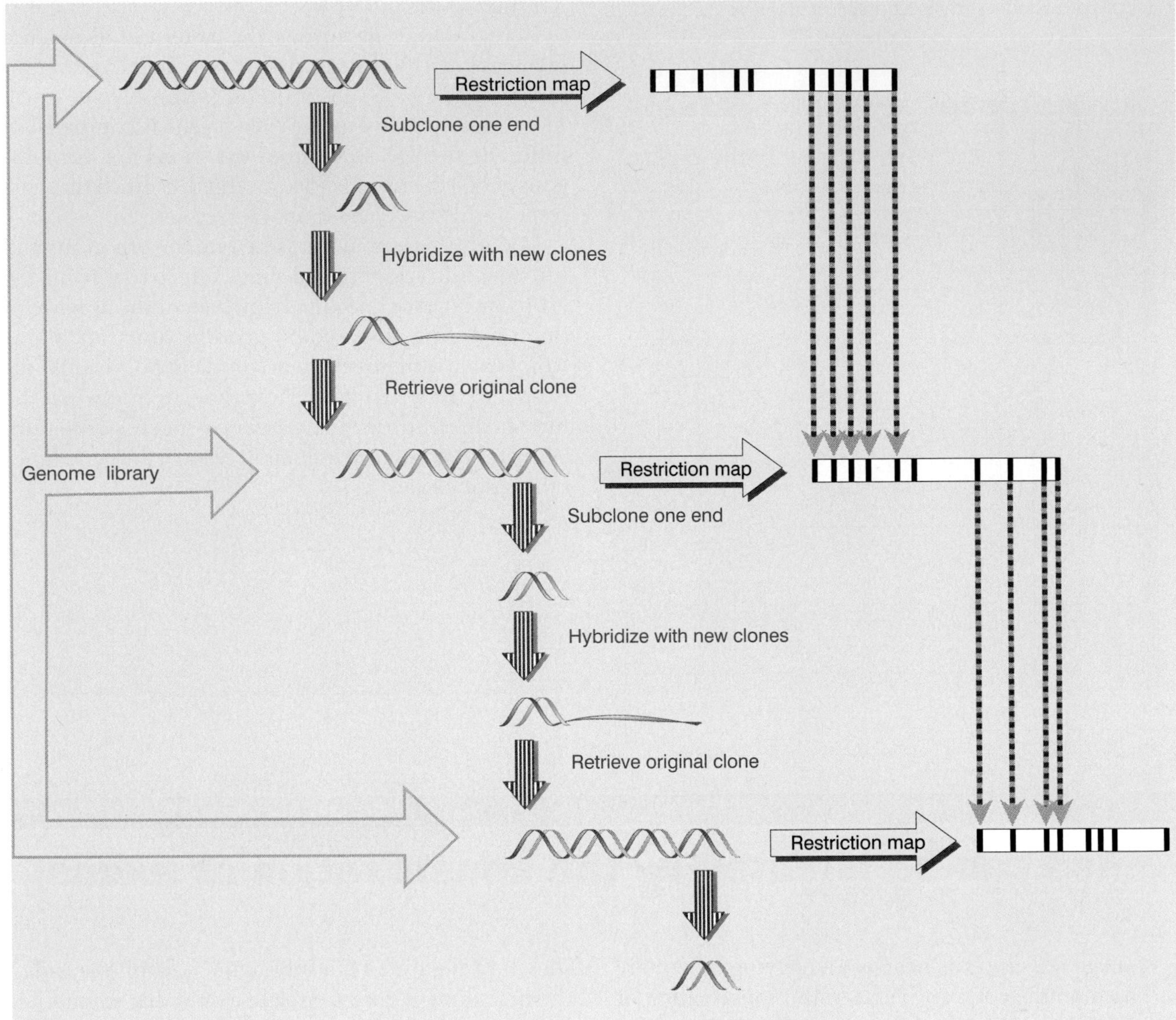

hand, a gene can be identified by comparison with either the RNA or protein product, or by the identification of a mutation in the sequence.

In the meantime, until full sequence information is available, a heroic approach that has proved successful with some genes of medical importance is to screen relatively short fragments from the region for the two properties expected of a conserved gene. First we seek to identify fragments that cross-hybridize with the genomes of other species. Then we examine these fragments for open reading frames.

The first criterion is applied by performing a **zoo blot.** We use short fragments from our chromosome walk as (radioactive) probes to test for related DNA from a variety of species by Southern blotting. If we find hybridizing fragments in several species related to that of the probe—the probe is usually human—the probe becomes a candidate for an exon of the gene.

The candidates are sequenced and, if they contain open reading frames, are used to isolate surrounding genomic regions. If these appear to be part of an exon, we may then use them to identify the entire gene, to isolate the corresponding cDNA or mRNA, and ultimately to identify the protein.

This approach is valuable for genes whose existence is implied by genetics, but whose nature is unknown.

One example is the gene *zfy* located on the human Y chromosome. **Figure 2.17** shows a zoo blot using a probe from this region. It hybridizes specifically with sex chromosomes of mammals and also with other species. It contains an open reading frame, which identifies a conserved gene.

This approach is especially important when the target gene is spread out because it has many large introns. This proved to be the case with Duchenne muscular dystrophy (DMD), a degenerative disorder of muscle, which is X-linked and affects 1 in 3500 of human male births. The steps in identifying the gene are summarized in **Figure 2.18.**

Linkage analysis localized the DMD locus to chromosomal band Xp21. Patients with the disease often have chromosomal rearrangements involving this band. By comparing the ability of X-linked DNA probes to hybridize with DNA from patients and with normal DNA, cloned fragments were obtained that correspond to the region that was rearranged or deleted in patients' DNA.

A chromosomal walk was used to construct a restriction map of the region on either side of the probe, covering a region of >100 kb. Analysis of the DNA from a series of patients identified large deletions in this region, extending in either direction. The most telling deletion is one contained entirely within the region,

Figure 2.17 A zoo blot with a probe from the human Y chromosomal gene *zfy* identifies cross-hybridizing fragments on the sex chromosomes of other mammals and birds. There is one reacting fragment on the Y chromosome and another on the X chromosome. Data kindly provided by David Page.

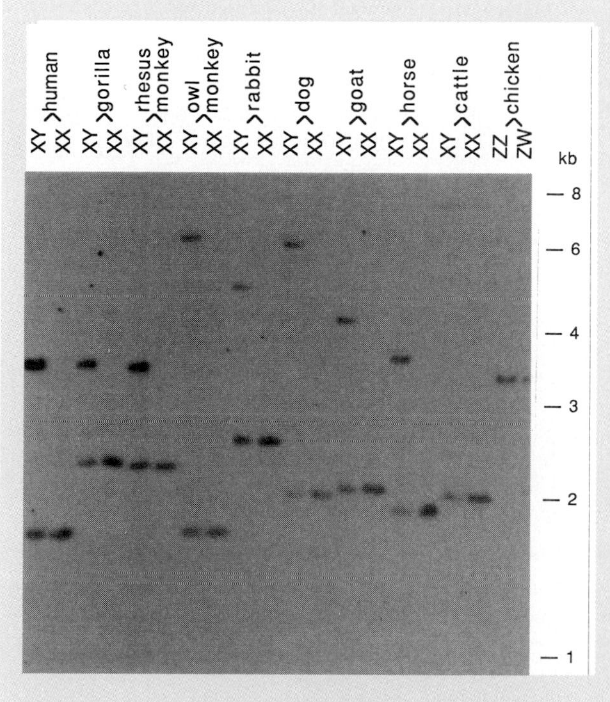

Figure 2.18 The gene involved in Duchenne muscular dystrophy has been tracked down by chromosome mapping and walking to a region in which deletions can be identified with the occurrence of the disease.

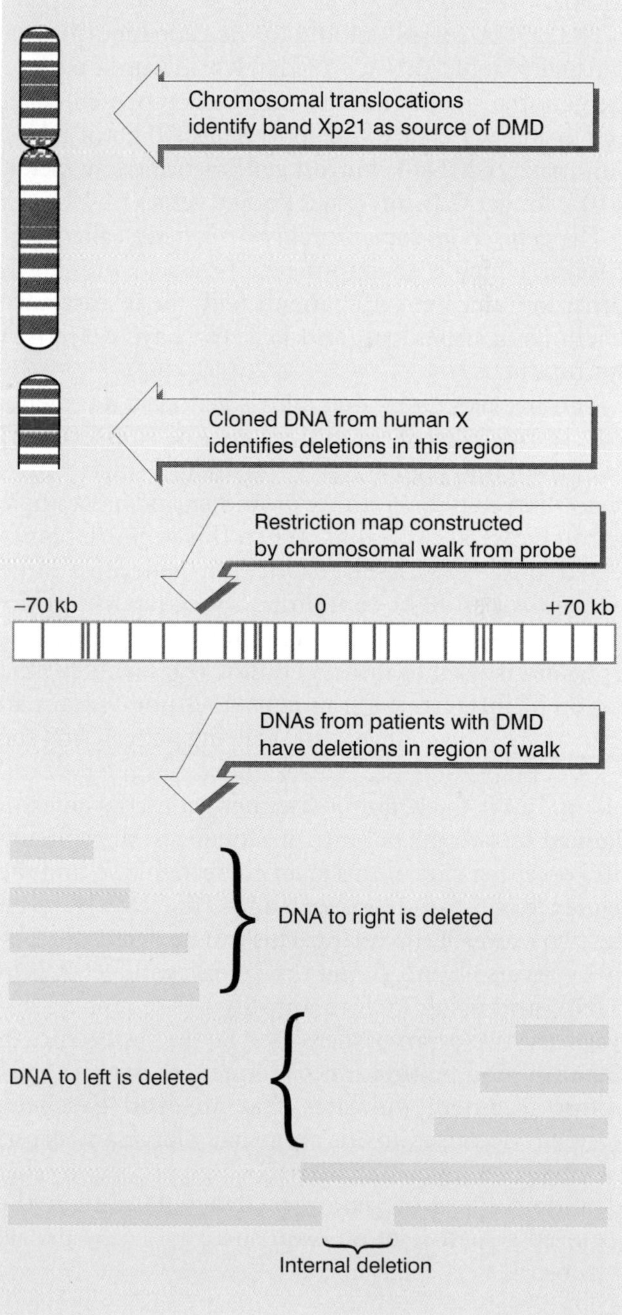

since this delineates a segment that must be important in gene function and indicates that the gene, or at least part of it, lies in this region.

Having now come into the region of the gene, we need to identify its exons and introns. A zoo blot identified fragments that cross-hybridize with the mouse X chromosome and with other mammalian DNAs. As summarized in **Figure 2.19**, these were scrutinized for open reading frames and the sequences typical of exon-intron junctions. Fragments that met these criteria were used as probes to identify homologous sequences in a cDNA library prepared from muscle mRNA.

The cDNA corresponding to the gene identifies an unusually large mRNA, ~14 kb. Hybridization back to the genome shows that the mRNA is represented in >60 exons, which are spread over ~2000 kb of DNA. This makes DMD the longest gene identified; in fact, it is 10× longer than any other known gene.

The gene codes for a protein of ~500 kD, called dystrophin, which is a component of muscle, present in rather low amounts. All patients with the disease have deletions at this locus, and lack (or have defective) dystrophin.

Another technique that allows genomic fragments to be scanned rapidly for the presence of exons is called **exon trapping. Figure 2.20** shows that it starts with a vector that contains a strong promoter, and has a single intron between two exons. When this vector is transfected into cells, its transcription generates large amounts of an RNA containing the sequences of the two exons. A restriction cloning site lies within the intron, and is used to insert genomic fragments from a region of interest. If a fragment does not contain an exon, there is no change in the splicing pattern, and the RNA contains only the same sequences as the parental vector. But if the genomic fragment contains an exon flanked by two partial intron sequences, the splicing sites on either side of this exon are recognized, and the sequence of the exon is inserted into the RNA between the two exons of the vector. This can be detected readily by reverse transcribing the cytoplasmic RNA into cDNA, and using PCR to amplify the sequences between the two exons of the vector. So the appearance in the amplified population of sequences from the genomic fragment indicates that an exon has been trapped. Because introns are usually large and exons are small in animal cells, there is a high probability that a random piece of genomic DNA will contain the required structure of an exon surrounded by partial introns.

Figure 2.19 The Duchene muscular dystrophy gene has been characterized by zoo blotting, cDNA hybridization, genomic hybridization, and identification of the protein.

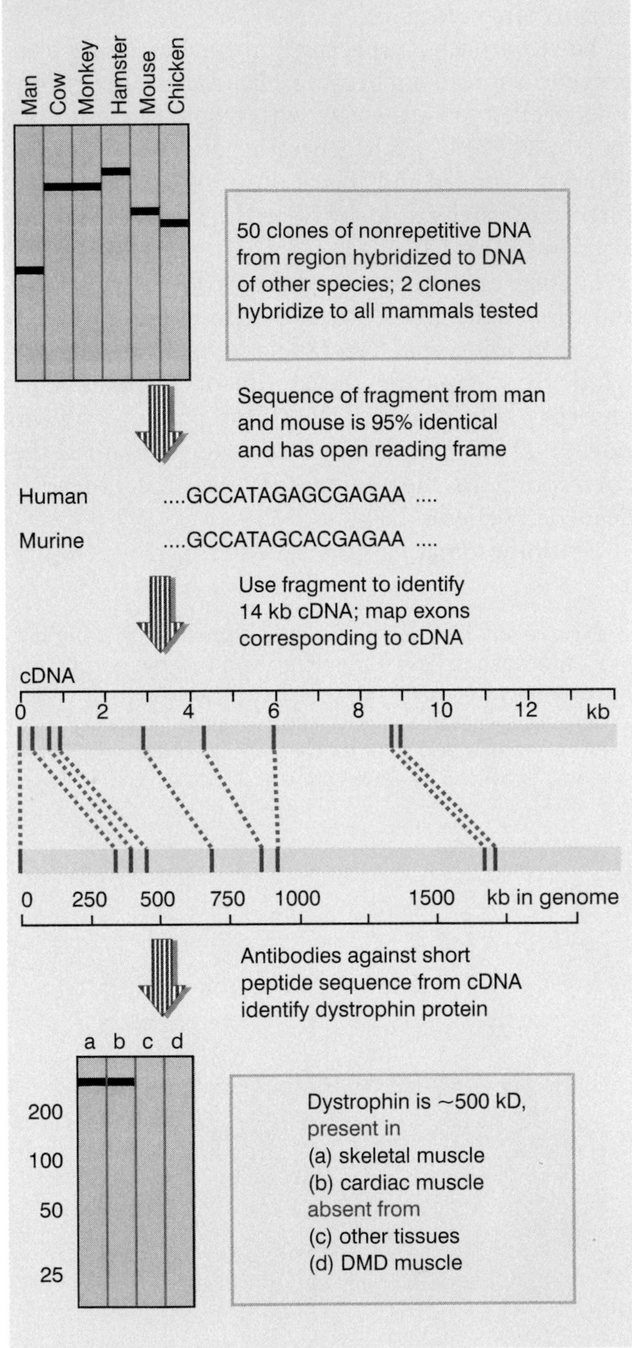
Man
Cow
Monkey
Hamster
Mouse
Chicken
50 clones of nonrepetitive DNA from region hybridized to DNA of other species; 2 clones hybridize to all mammals tested
Sequence of fragment from man and mouse is 95% identical and has open reading frame
Human ....GCCATAGAGCGAGAA ....
Murine ....GCCATAGCACGAGAA ....
Use fragment to identify 14 kb cDNA; map exons corresponding to cDNA
cDNA
0 2 4 6 8 10 12 kb
0 250 500 750 1000 1500 kb in genome
Antibodies against short peptide sequence from cDNA identify dystrophin protein
a b c d
200
100
50
25
Dystrophin is ~500 kD, present in
(a) skeletal muscle
(b) cardiac muscle
absent from
(c) other tissues
(d) DMD muscle

Figure 2.20 A special splicing vector is used for exon trapping. If an exon is present in the genomic fragment, its sequence will be recovered in the cytoplasmic RNA, but if the genomic fragment consists solely of an intron, it will be spliced out and lost.

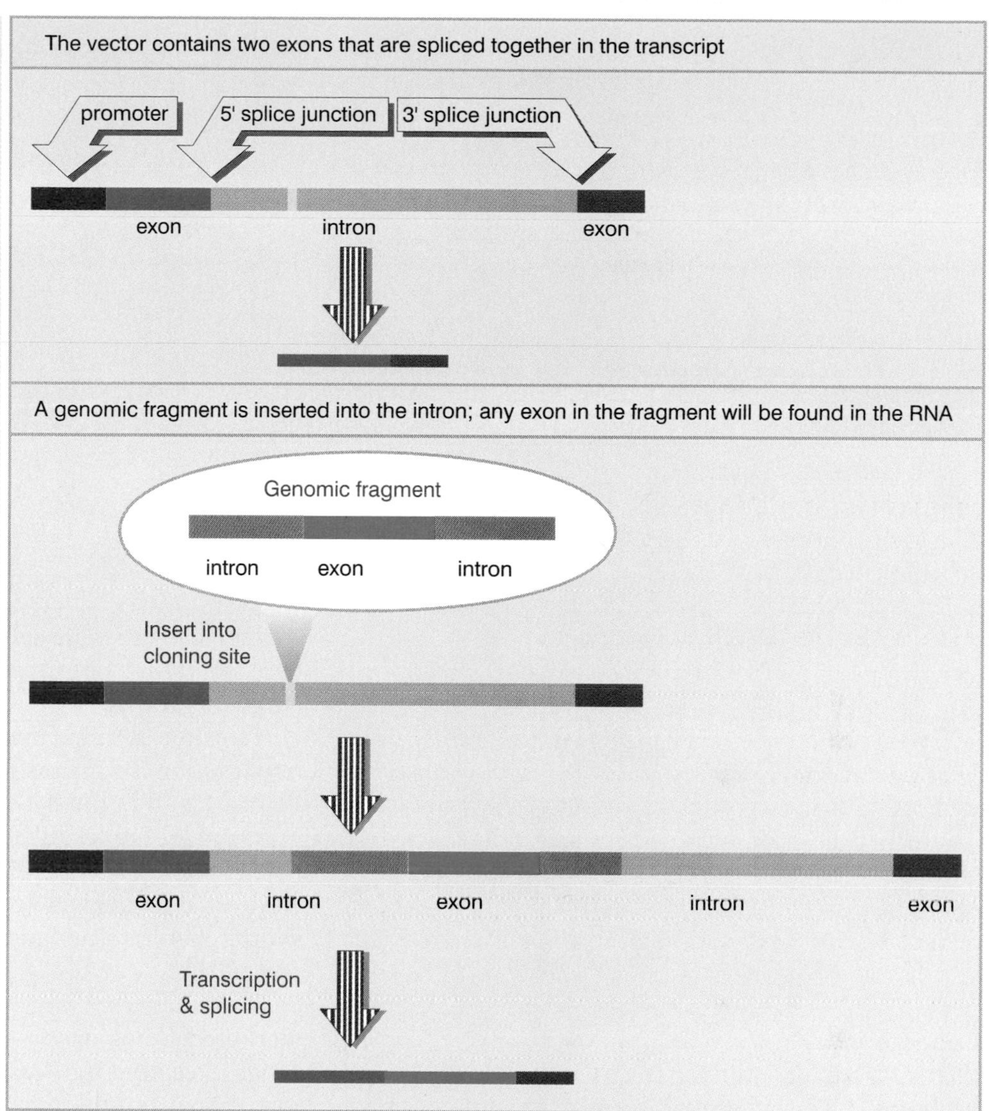
The vector contains two exons that are spliced together in the transcript
promoter
5' splice junction
3' splice junction
exon
intron
exon
A genomic fragment is inserted into the intron; any exon in the fragment will be found in the RNA
Genomic fragment
intron
exon
intron
Insert into cloning site
exon
intron
exon
intron
exon
Transcription & splicing

Genes show a wide distribution of sizes

The existence of interrupted genes makes it evident that the gene can be much larger than the unit that codes for protein. As genome size increases, the tendency is for introns to become rather large, while exons remain quite small.

Figure 2.21 shows that the exons coding for stretches of protein tend to be fairly small relative to the size of the gene. Most code for less than 100 amino acids (often less than 50 in vertebrates), and the general distribution fits well with the idea that genes have evolved by the slow addition of units that code for small, individual domains of proteins (see later). There is no very significant difference in the sizes of exons in different types of organism, except perhaps for an apparent absence of larger exons in the vertebrates. (The peak of exons coding for >300 amino acids in fungi and *Drosophila* mostly represents the presence of uninterrupted genes, that is, genes that consist of one exon.) There are some much larger exons coding for untranslated 5′ and 3′ regions (not included in the figure).

Figure 2.21 Exons coding for proteins are usually short.

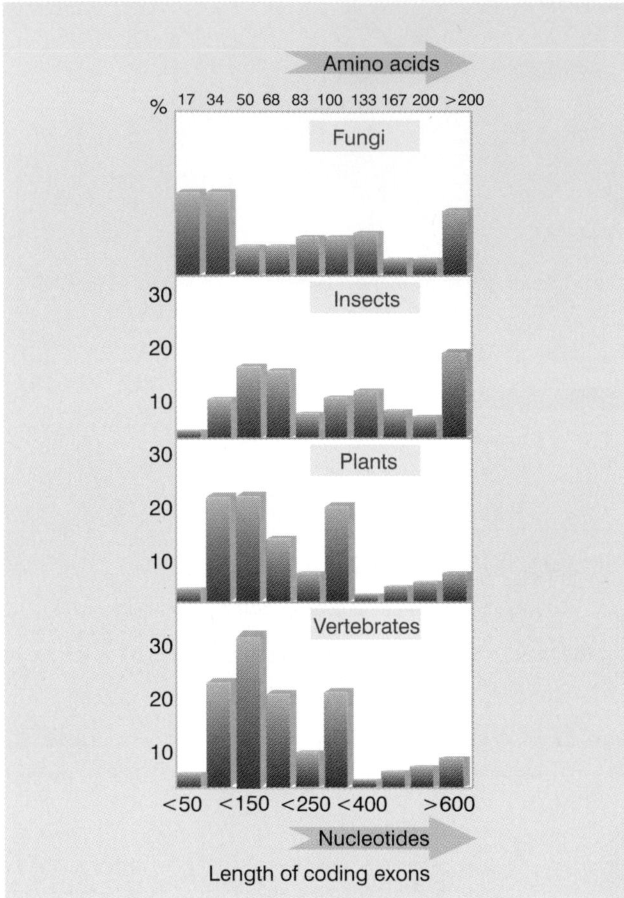

Figure 2.22 Introns in vertebrate genes range from very short to very long.

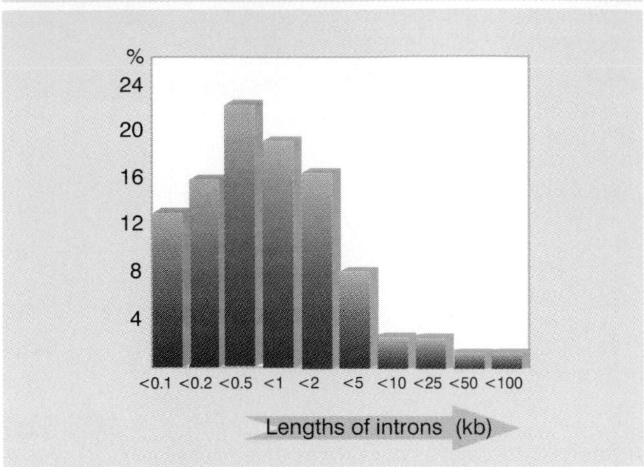

Figure 2.22 shows that introns are longer than exons. Their size distribution extends from approximately the same size as the exons (<200 bp) to lengths measured in 10s of kbs, and extending up to 50–60 kb in extreme cases.

Figure 2.23 shows the overall organization of genes in yeasts, insects, and mammals. In *S. cerevisiae,* the great majority of genes (>96%) are not interrupted, and those that have exons usually remain reasonably compact. There are virtually no *S. cerevisiae* genes with more than four exons.

In insects and mammals, the situation is reversed. Only a few genes have uninterrupted coding sequences (6% in mammals). Insect genes tend to have a fairly small number of exons, typically fewer than 10. Mammalian genes are split into more pieces, and some have several 10s of exons; ~50% of mammalian genes have >10 introns.

If we now examine the consequences of this type of organization for the overall size of the gene, we see in **Figure 2.24** that there is a striking difference between yeast and the higher eukaryotes. The average yeast gene is 1.4 kb long, and very few are longer than 5 kb. By contrast, relatively few genes in flies or mammals are shorter than 2 kb, and many have lengths between 5 kb and 100 kb.

The switch from largely uninterrupted to largely interrupted genes occurs in the lower eukaryotes. In fungi (excepting the yeasts), the majority of genes are interrupted, but they have a relatively small number of exons (<6) and are fairly short (<5 kb). The switch to long genes occurs within the higher eukaryotes, and genes become significantly larger in the insects. Perhaps genes become large at the same point where the relationship between genome complexity and organism complexity is lost (see Figure 3.1).

Very long genes are the result of very long introns, not the result of coding for longer products. There is no correlation between gene size and mRNA size in higher eukaryotes; nor is there a good correlation between gene size and the number of exons. The size of a gene therefore depends primarily on the lengths of its individual introns. In mammals, insects, and birds, the "average" gene is approximately 5× the length of its mRNA.

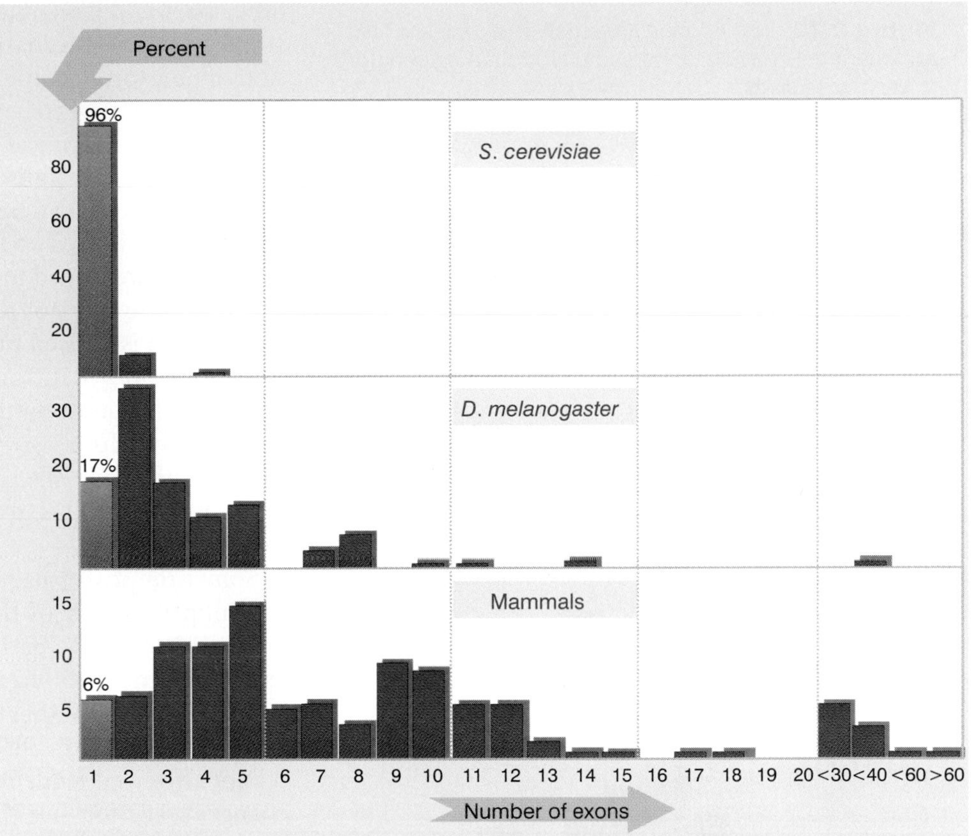

Figure 2.23 Most genes are uninterrupted in yeast, but most genes are interrupted in flies and mammals. (Uninterrupted genes have only 1 exon, and are totalled in the leftmost column.)

Some DNA sequences code for more than one protein

MOST genes consist of a sequence of DNA that is devoted solely to the purpose of coding for one protein (although the gene may include noncoding regions at either end and introns within the coding region). However, there are some cases in which a single sequence of DNA codes for more than one protein.

Overlapping genes occur in the relatively simple situation in which one gene is part of the other. The first half (or second half) of a gene is used independently to specify a protein that represents the first (or second) half of the protein specified by the full gene. This relationship is illustrated in **Figure 2.25.** The end result is much the same as though a partial cleavage took place in the protein product to generate part-length as well as full-length forms.

Two genes overlap in a more subtle manner when the same sequence of DNA is shared between two *nonhomologous* proteins. This situation arises when the same sequence of DNA is translated in more than one reading frame. In cellular genes, a DNA sequence usually is read in only one of the three potential reading frames, but in some viral and mitochondrial genes, there is an overlap between two adjacent genes that are read in different reading frames. This situation is illustrated in **Figure 2.26.** The distance of overlap is usually relatively short, so that most of the sequence representing the protein retains a unique coding function.

In some genes, *alternative* patterns of gene expression create switches in the pathway for connecting the exons. A single gene may generate a variety of mRNA products that differ in their content of exons. The difference may be that certain exons are optional—they may be included or spliced out. Or there may be exons that are treated as mutually exclusive—one or the other is included, but not both. The alternative forms produce proteins in which one part is common while the

Figure 2.24 Yeast genes are small, but genes in flies and mammals have a dispersed distribution extending to very large sizes.

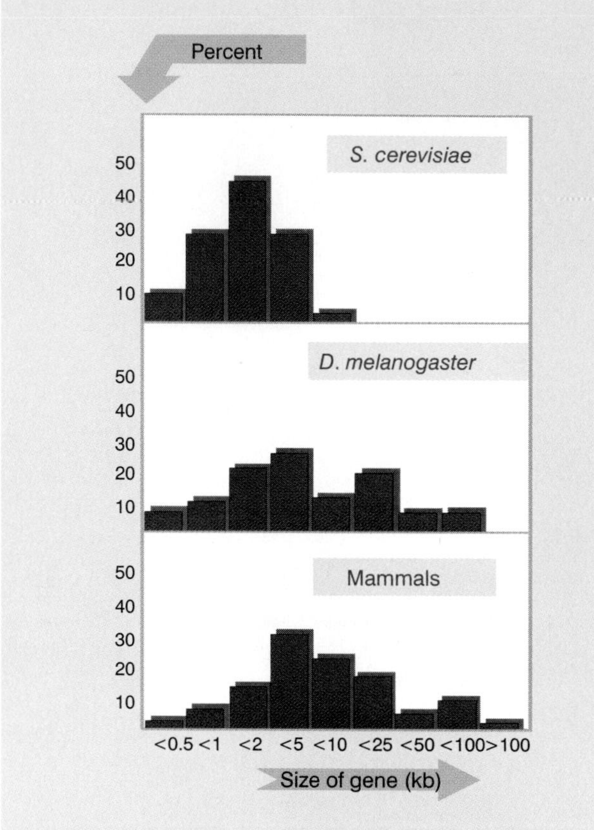

other part is different. **Figure 2.27** illustrates an example in which alternative splicing leads to the inclusion of an exon in some mRNAs, while it is left out of others. (Other types of combinations that are produced by alternative splicing are discussed in Chapter 22.)

A single type of transcript is made from the gene in Figure 2.27, but it can be spliced in either of two ways. In the first pathway, two introns are spliced out, and the three exons are joined together. In the second pathway, the second exon is not recognized. As a result, a single large intron is spliced out. This intron consists of intron 1 + exon 2 + intron 2. In effect, exon 2 has been treated in this pathway as part of the single intron. The pathways produce two proteins that are the same at their ends, but one of which has an additional sequence in the middle. So the region of DNA codes for more than one protein.

Sometimes two pathways operate simultaneously, a certain proportion of the RNA being spliced in each way; sometimes the pathways are alternatives that are expressed under different conditions, one in one cell type and one in another cell type.

In some cases, the alternative means of expression do not affect the sequence of the protein; for example, changes that affect the 5′ nontranslated leader or the 3′ nontranslated trailer may have regulatory consequences, but the same protein is made. In other cases, one exon is substituted for another, as indicated in **Figure 2.28.**

Figure 2.25 Two proteins can be generated from a single gene by starting (or terminating) expression at different points.

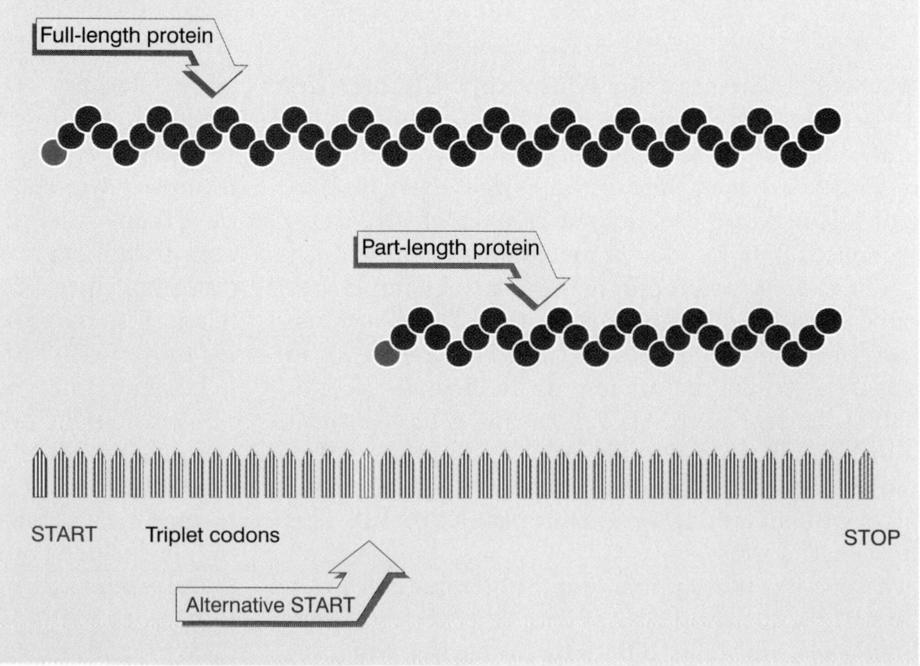

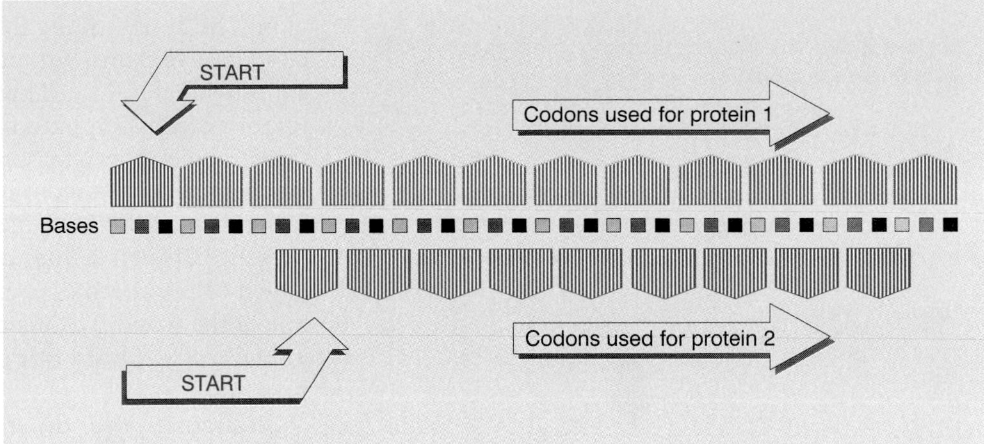

Figure 2.26 Two genes may share the same sequence by reading the DNA in different frames.

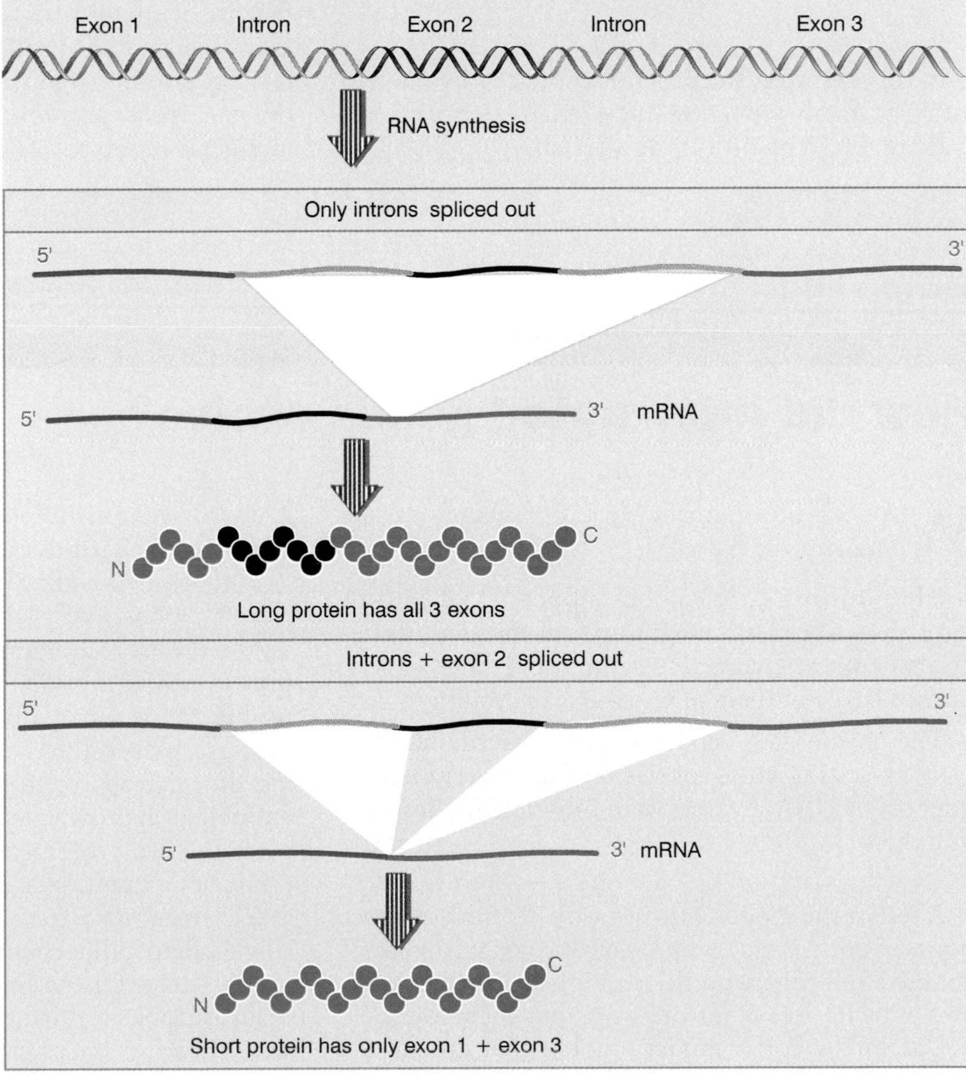

Figure 2.27 Alternative splicing uses the same pre-mRNA to generate mRNAs that have different combinations of exons.

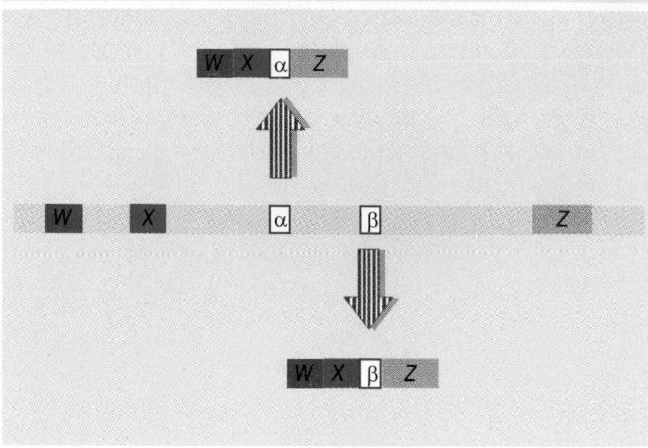

Figure 2.28 Alternative splicing generates the α and β variants of troponin T.

In this example, the proteins produced by the two mRNAs contain sequences that overlap extensively, but that are different within the alternatively spliced region. The 3′ half of the troponin T gene of rat muscle contains five exons, but only four are used to construct an individual mRNA. Three exons, *WXZ*, are the same in both expression patterns. However, in one pattern the α exon is spliced between *X* and *Z;* in the other pattern, the β exon is used. The α and β forms of troponin T therefore differ in the sequence of the amino acids present between sequences W and Z, depending on which of the alternative exons, α or β, is used. Either one of the α and β exons can be used to form an individual mRNA, but both cannot be used in the same mRNA.

So alternative (or differential) splicing can generate proteins with overlapping sequences from a single stretch of DNA. It is curious that the higher eukaryotic genome is extremely spacious in having large genes that are often quite dispersed, but at the same time it may make multiple products from an individual locus. It is not possible to say how many genes have alternative modes of expression, but on an anecdotal basis, the number seems to be a few percent.

How did interrupted genes evolve?

WHAT was the original form of genes that today are interrupted?

- The "introns early" model supposes that introns have always been an integral part of the gene. Genes originated as interrupted structures, and those without introns have lost them in the course of evolution.
- The "introns late" model supposes that the ancestral protein-coding units consisted of uninterrupted sequences of DNA. Introns were subsequently inserted into them.

A test of the models is to ask whether the difference between eukaryotic and prokaryotic genes can be accounted for by the acquisition of introns in the eukaryotes or by the loss of introns from the prokaryotes.

The introns early model suggests that the mosaic structure of genes is a remnant of an ancient approach to the reconstruction of genes to make novel proteins. Suppose that an early cell had a number of separate protein-coding sequences. One aspect of its evolution is likely to have been the reorganization and juxtaposition of different polypeptide units to build up new proteins.

If the protein-coding unit must be a continuous series of codons, every such reconstruction would require a precise recombination of DNA to place the two protein-coding units in register, end to end in the same reading frame. Furthermore, if this combination is not successful, the cell has been damaged, because it has lost the original protein-coding units.

But if an approximate recombination of DNA could place the two protein-coding units within the same transcription unit, splicing patterns could be tried out at the level of RNA to combine the two proteins into a single polypeptide chain. And if these combinations are not successful, the original protein-coding units remain available for further trials. Such an approach essentially allows the cell to try out controlled deletions in RNA without suffering the damaging instability that could occur from applying this procedure to DNA.

If current proteins evolved by combining ancestral proteins that were originally separate, the accretion of units is likely to have occurred sequentially over some period of time, with one exon added at a time. Can the different functions from which these genes were pieced together be seen in their present structure? In other words, can we equate particular functions of current proteins with individual exons?

In some cases, there is a clear relationship between the structures of the gene and protein. The example *par excellence* is provided by the immunoglobulin proteins, which are coded by genes in which every exon corresponds exactly with a known functional domain of the protein. **Figure 2.29** compares the structure of an immunoglobulin with its gene.

An immunoglobulin is a tetramer of two light chains and two heavy chains, which aggregate to generate a protein with several distinct domains. Light chains and heavy chains differ in structure, and there are several types of heavy chain. Each type of chain is expressed from a gene that has a series of exons corresponding with the structural domains of the protein.

In many instances, some of the exons of a gene can be identified with particular functions. In secretory proteins, the first exon, coding for the N-terminal region of the polypeptide, often specifies the signal sequence involved in membrane secretion. An example is insulin.

Sometimes the evolution of a gene involves the duplication of exons, creating an internally repetitious sequence in the protein. In chicken collagen, a 54 bp exon appears to have been multiplied many times, generating a series of exons that are either 54 bp or multiples of 54 bp in length.

Sequences held in common between genes that are related only in part may represent exons that have migrated or been recruited between genes. **Figure 2.30** summarizes the relationship between the receptor for human LDL (plasma low density lipoprotein) and other proteins.

In the center of the LDL receptor gene is a series of exons related to the exons of the gene for the precursor for EGF (epidermal growth factor). In the N-terminal part of the protein, a series of exons codes for a sequence related to the blood protein complement factor C9. So the LDL receptor gene was created by assembling *modules* for its various functions. These modules are also used in other proteins.

Figure 2.29 Immunoglobulin light chains and heavy chains are coded by genes whose structures (in their expressed forms) correspond with the distinct domains in the protein. Each protein domain corresponds to an exon; introns are numbered 1–5.

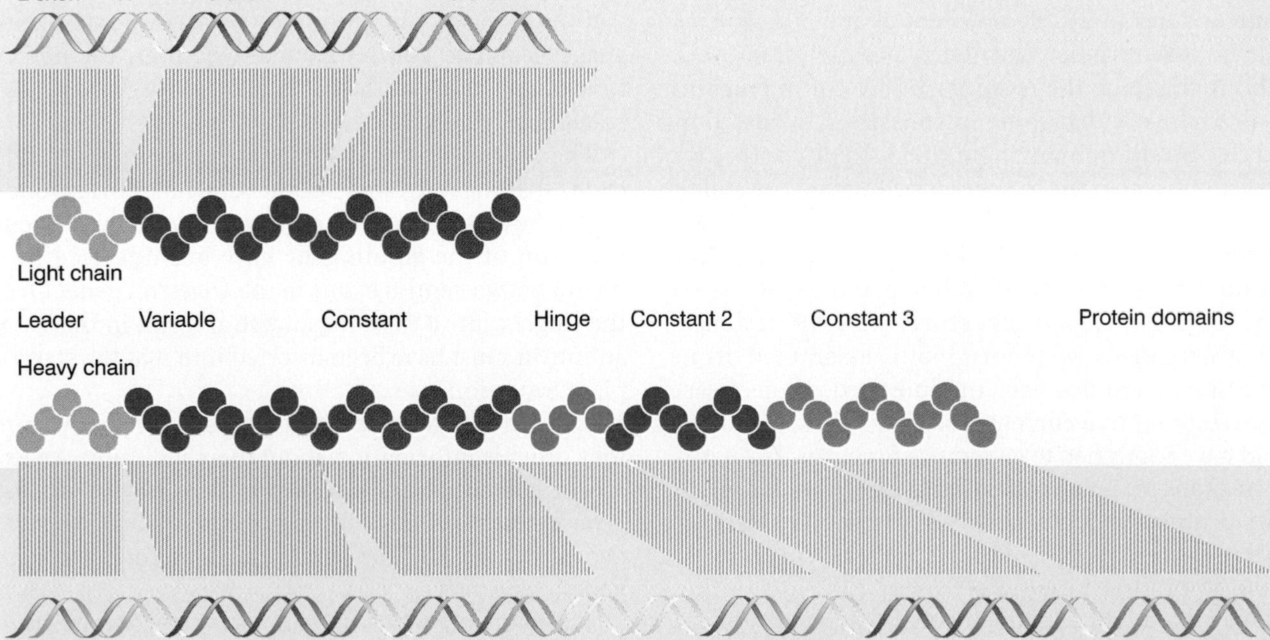

Figure 2.30 The LDL receptor gene consists of 18 exons, some of which are related to EGF precursor and some to the C9 blood complement gene. Triangles mark the positions of introns. Only some of the introns in the region related to EGF precursor are identical in position to those in the EGF gene.

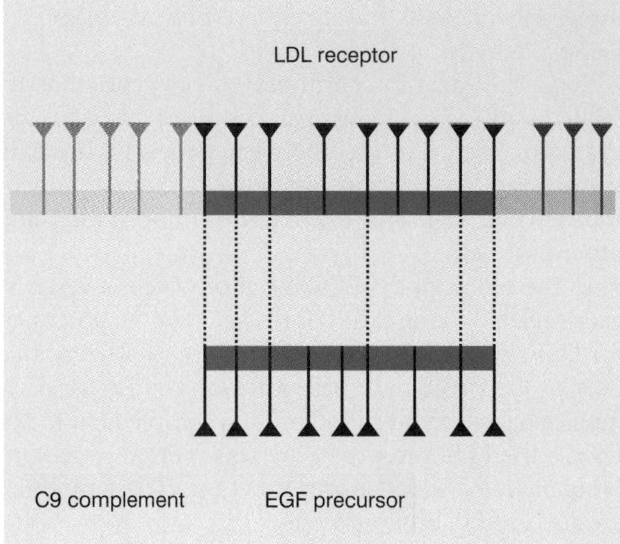

The relationship between exons and protein domains is somewhat erratic in known genes. In some cases there is a clear 1:1 relationship; in others no pattern is to be discerned. One possibility is that removal of introns has fused the adjacent exons. This means that the intron must have been precisely removed, without changing the integrity of the coding region. An alternative is that some introns arose by insertion into a coherent domain; here the difficulty is that we must suppose that the intron carried with it the ability to be spliced out.

Exons tend to be fairly small (see Figure 2.20), around the size of the smallest polypeptide that can assume a stable folded structure, ~20–40 residues. Perhaps proteins were originally assembled from rather small modules. Each module need not necessarily correspond to a current function; several modules could have combined to generate a function. The number of exons in a gene tends to increase with the length of its protein, which is consistent with the view that proteins acquire multiple functions by successively adding appropriate modules.

This idea might explain another feature of protein structure: it seems that the sites represented at exon-intron boundaries often are located at the surface of a protein. As modules are added to a protein, the connections, at least of the most recently added modules, could tend to lie at the surface.

A fascinating case of evolutionary conservation is presented by the globins, all of whose genes have three exons (see Figure 2.13). The two introns are located at constant positions relative to the coding sequence. The central exon appears to represent the heme-binding domain of the globin chain. The α- and β-globin genes have similar structures.

Another perspective on this structure is provided by the existence of two other types of protein that are related to globin. Myoglobin is a monomeric oxygen-binding protein of animals, whose amino acid sequence suggests a common (though ancient) origin with the globin subunits. Leghemoglobins are oxygen-binding proteins present in the legume class of plants; like myoglobin, they are monomeric. They too share a common origin with the other heme-binding proteins. Together, the globins, myoglobin, and leghemoglobin constitute the globin "superfamily," a set of gene families all descended from some (distant) common ancestor.

Myoglobin is represented by a single gene in the human genome, whose structure is essentially the same as that of the globin genes. The three-exon structure therefore predates the evolution of separate myoglobin and globin functions.

Leghemoglobin genes contain three introns, the first and last of which occur at points in the coding sequence that are homologous to the locations of the two introns in the globin genes. This remarkable similarity suggests an exceedingly ancient origin for the heme-binding proteins in the form of a split gene, as illustrated in **Figure 2.31**.

The central intron of leghemoglobin separates two exons that together code for the sequence corresponding to the single central exon in globin. Could the central exon of the globin gene have been derived by a fusion of two central exons in the ancestral gene? Or is the single central exon the ancestral form; in this case, an intron must have been inserted into it at the start of plant evolution?

Cases in which homologous genes differ in structure may provide information about their evolution. An example is insulin. Mammals and birds have only one gene for insulin, except for the rodents, which have two genes. **Figure 2.32** illustrates the structures of these genes.

The principle we use in comparing the organization of related genes in different species is that *a common feature identifies a structure that predated the evolutionary separation of the two species.* In chicken, the single

Figure 2.31 The exon structure of globin genes corresponds with protein function, but leghemoglobin has an extra intron in the central domain.

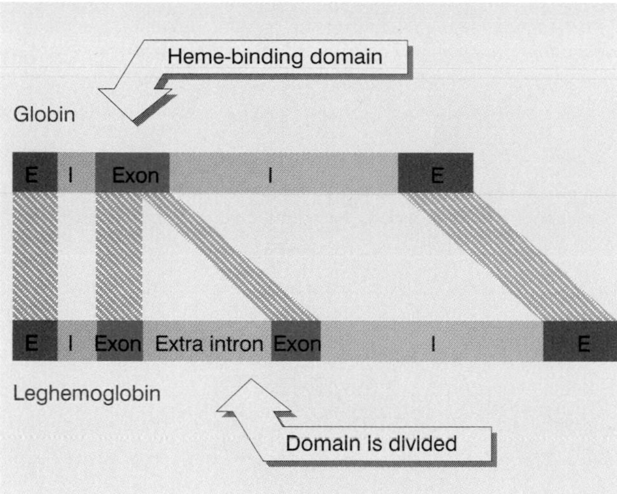

Figure 2.32 The rat insulin gene with one intron evolved by losing an intron from an ancestor with two interruptions.

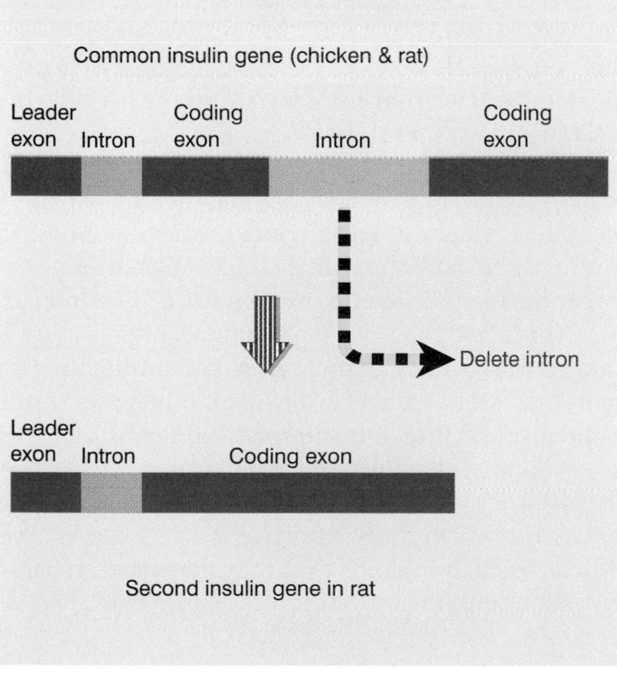

insulin gene has two introns; one of the two rat genes has the same structure. The common structure implies that the ancestral insulin gene had two introns. However, the second rat gene has only one intron. It must have evolved by a gene duplication in rodents that was followed by the precise removal of one intron from one of the copies.

The organization of some genes shows extensive discrepancies between species. In these cases, there must have been extensive removal or insertion of introns during evolution.

A well characterized case is represented by the actin genes. The typical actin gene has a nontranslated leader of <100 bases, a coding region of ~1200 bases, and a trailer of ~200 bases. Most actin genes are interrupted; the positions of the introns can be aligned with regard to the coding sequence (except for a single intron sometimes found in the leader).

Figure 2.33 shows that almost every actin gene is different in its pattern of interruptions. Taking all the genes together, introns occur at 12 different sites. However, no individual gene has more than six introns; some genes have only one intron, and one is uninterrupted altogether. How did this situation arise? If we suppose that the primordial actin gene was interrupted, and all current actin genes are related to it by loss of introns, different introns have been lost in each evolutionary branch. Probably some introns have been lost entirely, so the primordial gene could well have had 20 or more. The alternative is to suppose that a process of intron insertion continued independently in the different lines of evolution. The relationships between the intron locations found in different species may be used ultimately to construct a tree for the evolution of the gene.

The equation of at least some exons with protein domains, and the appearance of related exons in different proteins, leaves no doubt that the duplication and juxtaposition of exons has played an important role in evolution. It is possible that the number of ancestral exons, from which all proteins have been derived by duplication, variation, and recombination, could be relatively small (a few thousands or tens of thousands). By taking exons as the building blocks of evolution, this view implicitly accepts the introns early model for the origin of genes coding for proteins.

The highly interrupted structure of eukaryotic genes suggests a picture of the eukaryotic genome as a sea of introns (mostly but not exclusively unique in sequence), in which islands of exons (sometimes very short) are strung out in individual archipelagoes that constitute genes.

Alternative forms of genes for rRNA and tRNA are sometimes found, with and without introns. In the case of the tRNAs, where all the molecules conform to the same general structure, it seems unlikely that evolution brought together the two regions of the gene. After all, the different regions are involved in the base

Figure 2.33 Actin genes vary widely in their organization. The sites of introns are indicated in purple; the number identifies the codon interrupted by the intron.

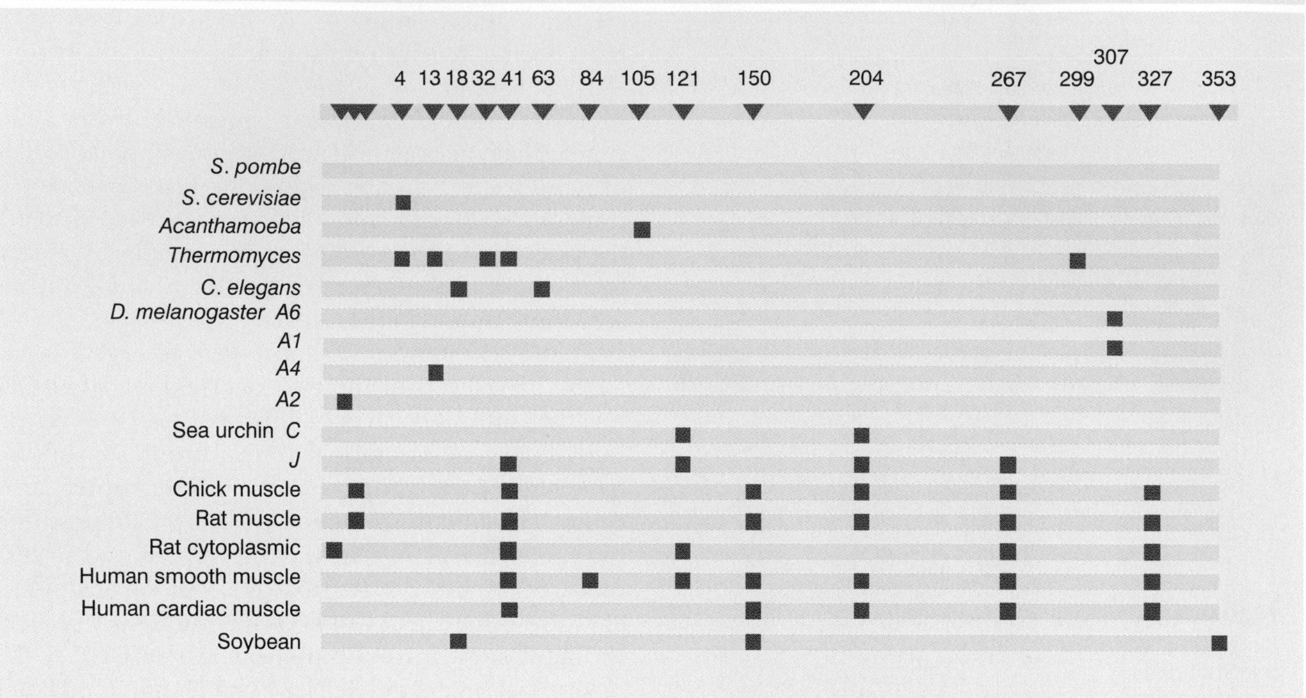

pairing that gives significance to the structure. So here it must be that the introns were inserted into continuous genes.

Organelle genomes provide some striking connections between the prokaryotic and eukaryotic worlds. Because of many general similarities between mitochondria or chloroplasts and bacteria, it seems likely that the organelles originated by an **endosymbiosis** in which an early bacterial prototype was inserted into eukaryotic cytoplasm. Yet in contrast to the resemblances with bacteria—for example, as seen in protein or RNA synthesis—some organelle genes possess introns, and therefore resemble eukaryotic nuclear genes.

Introns are found in several chloroplast genes, including some that have homologies with genes of *E. coli*. This suggests that the endosymbiotic event occurred before introns were lost from the prokaryotic line. If a suitable gene can be found, it may therefore be possible to trace gene lineage back to the period when endosymbiosis occurred.

The mitochondrial genome presents a particularly striking case. The genes of yeast and mammalian mitochondria code for virtually identical mitochondrial proteins, in spite of a considerable difference in gene organization. Vertebrate mitochondrial genomes are very small, with an extremely compact organization of continuous genes, whereas yeast mitochondrial genomes are larger and have some complex interrupted genes. Which is the ancestral form? The yeast mitochondrial introns often have the property of mobility—they are self-contained sequences that can splice out of the RNA and insert DNA copies elsewhere—which suggests that they may have arisen by insertions into the genome (see Chapter 16).

The scope of the paradigm

THE concept of the gene has evolved significantly in the past few years. The question of what's in a name is especially appropriate for the gene. We can no longer say that a gene is a sequence of DNA that continuously and uniquely codes for a particular protein. In situations in which a stretch of DNA is responsible for production of one particular protein, current usage regards the entire sequence of DNA, from the first point represented in the messenger RNA to the last point corresponding to its end, as comprising the "gene," exons, introns, and all.

When the sequences representing proteins overlap or have alternative forms of expression, we may reverse the usual description of the gene. Instead of saying "one gene-one polypeptide," we may describe the relationship as "one polypeptide-one gene." So we regard the sequence actually responsible for production of the polypeptide (including introns as well as exons) as constituting the gene, while recognizing that from the perspective of another protein, part of this same sequence also belongs to *its* gene. This allows the use of descriptions such as "overlapping" or "alternative" genes.

We can now see how far we have come from the original one gene : one enzyme hypothesis. Up to that time, the driving question was the nature of the gene. Once it was discovered that genes represent proteins, the paradigm became fixed in the form of the concept that every genetic unit functions through the synthesis of a particular protein.

This view remains the central paradigm of molecular biology: a sequence of DNA functions either by directly coding for a particular protein or by being necessary for the use of an adjacent segment that actually codes for the protein. How far does this paradigm take us beyond explaining the basic relationship between genes and proteins?

The development of multicellular organisms rests on the use of different genes to generate the different cell phenotypes of each tissue. The expression of genes is determined by a regulatory network that probably takes the form of a cascade. Expression of the first set of genes at the start of embryonic development leads to expression of the genes involved in the next stage of development, which in turn leads to a further stage, and so on until all the tissues of the adult are functioning. The molecular nature of this regulatory network is largely unknown, but we assume that it consists of genes that code for products (probably protein, perhaps sometimes RNA) that act on other genes.

While such a series of interactions is almost certainly the means by which the developmental program is executed, we can ask whether it is entirely sufficient. One specific question concerns the nature and role of **positional information.** We know that all parts of a fertilized egg are not equal; one of the features responsible for development of different tissue parts from different regions of the egg is location of information (presumably specific macromolecules) within the cell.

We do not know how these particular regions are formed. But we may speculate that the existence of positional information in the egg leads to the differential expression of genes in the cells subsequently formed in these regions, which leads to the development of the adult organism, which leads to the development of an egg with the appropriate positional information...

This possibility prompts us to ask whether some information needed for development of the organism is contained in a form that we cannot directly attribute to a sequence of DNA (although the expression of particular sequences may be needed to perpetuate the positional information). Put in a more general way, we might ask: if we could read out the entire sequence of DNA comprising the genome of some organism and interpret it in terms of proteins and regulatory regions, could we then construct an organism (or even a single living cell) by controlled expression of the proper genes?

Summary

Genes and genomes can be mapped by the use of overlapping restriction fragments. Ultimately this can be extended into a sequence. Restriction sites can be used as genetic markers. The existence of polymorphisms (RFLPs) allows linkage maps to be constructed using restriction fragments.

All types of eukaryotic genomes contain interrupted genes. The proportion of interrupted genes is low in yeasts and increases in the lower eukaryotes; few genes are uninterrupted in higher eukaryotes.

Introns are found in all classes of eukaryotic genes. The structure of the interrupted gene is the same in all tissues, exons are joined together in RNA in the same order as their organization in DNA, and the introns usually have no coding function. Introns are removed from RNA by splicing. Some genes are expressed by alternative splicing patterns, in which a particular sequence is removed as an intron in some situations, but retained as an exon in others.

Positions of introns are conserved when the organization of homologous genes is compared between species. Intron sequences vary, and may even be unrelated, although exon sequences remain well related. The conservation of exons can be used to isolate related genes in different species.

The size of a gene is determined primarily by the lengths of its introns. Introns become larger early in the higher eukaryotes, when gene sizes therefore increase significantly. The range of gene sizes in mammals is generally 1–100 kb, but it is possible to have even larger genes; the longest known case is dystrophin at 2000 kb.

Some genes share only some of their exons with other genes, suggesting that they have been assembled by addition of exons representing individual modules of the protein. Such modules may have been incorporated into a variety of different proteins. The idea that genes have been assembled by accretion of exons implies that introns were present in genes of primitive organisms. Some of the relationships between homologous genes can be explained by loss of introns from the primordial genes, with different introns being lost in different lines of descent.

Further reading

Reviews

Nathans, D. and Smith, H. O. (1975). Restriction endonucleases in the analysis and restructuring of DNA molecules. *Ann. Rev. Biochem.* **46**, 273–293.

Wilson, A. C. *et al.* (1977). Biochemical evolution. *Ann. Rev. Biochem.* **46**, 573–639.

Breathnach, R. and Chambon, P. (1981). Organization and expression of eukaryotic split genes coding for proteins. *Ann. Rev. Biochem.* **50**, 349–383.

Blake, C. C. (1985). Exons and the evolution of proteins. *Int. Rev. Cytol.* **95**, 149–185.

Wu, R. (1978). DNA sequence analysis. *Ann. Rev. Biochem.* **47**, 607–734.

White, R. *et al.* (1985). Construction of linkage maps with DNA markers for human chromosomes. *Nature* **313**, 101–105.

Diener, T. O. (1986). Viroid processing: a model involving the central conserved region and hairpin. *Proc. Nat. Acad. Sci. USA* **83**, 58–62.

Gusella, J. F. (1986). DNA polymorphism and human disease. *Ann. Rev. Biochem.* **55**, 831–854.

Interrupted genes

Wenskink, P. *et al.* (1974). A system for mapping DNA sequences in the chromosomes of *D. melanogaster. Cell* **3**, 315–325.

Berget, S. M., Moore, C., and Sharp, P. (1977). Spliced segments at the 5′ terminus of adenovirus 2 late mRNA. *Proc. Nat. Acad. Sci. USA* **74**, 3171–3175.

Chow, L. T., Gelinas, R. E., Broker, T. R., and Roberts, R. J. (1977). An amazing sequence arrangement at the 5′ ends of adenovirus 2 mRNA. *Cell* **12**, 1–8.

Glover, D. M. and Hogness, D. S. (1977). A novel arrangement of the 18S and 28S sequences in a repeating unit of *D. melanogaster* rDNA. *Cell* **10,** 167–176.

Jeffreys, A. J. and Flavell, R. A. (1977). The rabbit β-globin gene contains a large insert in the coding sequence. *Cell* **12,** 1097–1108.

RFLP maps

Danna, K. J., Sack, G. H., and Nathans, D. (1973). Studies of SV40 DNA. VII. A cleavage map of the SV40 genome. *J. Mol. Biol.* **78,** 363–376.

Donis-Keller, J. *et al.* (1987). A genetic linkage map of the human genome. *Cell* **51,** 319–337.

Dietrich, W. F. *et al.* (1996). A comprehensive genetic map of the mouse genome. *Nature* **380,** 149–152.

Dib, C. *et al.* (1996). A comprehensive genetic map of the human genome based on 5,264 microsatellites. *Nature* **380,** 152–154.

Chapter 3

How many genes are there?

CERTAIN features are unique to eukaryotic genomes compared with prokaryotic genomes. The integrity of individual genes can be interrupted, there may be multiple and (sometimes) identical copies of particular sequences, and there may be large blocks of DNA that do not code for protein. Because of the division between nucleus and cytoplasm, the arrangements for gene expression in eukaryotes must necessarily be different from those in prokaryotes.

But there is no simple definition of the "eukaryotic genome." The essence of being eukaryotic is that the major part of the genome is sequestered in the nucleus. The total amount of nuclear DNA varies enormously, the number of chromosomes into which it is segregated is extremely variable, there can be major differences in the types of sequences, and the relatively minor part of the genome contained in organelles also shows wide variations in size.

The major complication in analyzing eukaryotic genomes, especially those of higher eukaryotes, is that coding regions represent only a small proportion of the total DNA. Because genes may be interrupted, large parts of a gene may not be concerned with coding for protein. And there may also be significant lengths of DNA between genes. So it is not possible to deduce from the overall size of the genome anything about the number of genes.

We may identify the coding potential of a genome directly, by identifying regions that have open reading frames. Large scale mapping of this nature is complicated by the fact that interrupted genes may consist of many separated open reading frames. Since we do not necessarily have information about the functions of the protein products, or indeed proof that they are expressed at all, this approach is restricted to defining the potential of the genome. (However, a strong presumption exists that any conserved open reading frame is likely to be expressed; see Chapter 2.)

Another approach is to define the number of genes directly in terms of their expression in RNA or protein. This gives an assurance that we are dealing with *bona fide* genes that are expressed under known circumstances. It allows us to ask how many genes are expressed in a particular tissue or cell type, what variation exists in the relative levels of expression, and how many of the genes expressed in one particular cell are unique to that cell or are also expressed elsewhere.

Concerning the types of genes, we may ask whether a particular gene is essential: what happens to a null mutant? If a null mutation is lethal, or the organism has a visible defect, we may conclude that the gene is essential or at least conveys a selective advantage. But some genes can be deleted without apparent effect on the phenotype. Are these genes really dispensable, or does a selective disadvantage result from the absence of the gene, perhaps in other circumstances, or over longer periods of time?

Why are genomes so large?

THE total amount of DNA in the (haploid) genome is a characteristic of each living species known as its **C-value.** There is enormous variation in the range of C-values, from $<10^6$ bp for a mycoplasma to $>10^{11}$ bp for some plants and amphibians.

Figure 3.1 summarizes the range of C-values found in different evolutionary phyla. There is an increase in the minimum genome size found in each group as the complexity increases. But as absolute amounts of DNA increase in the higher eukaryotes, we see some wide variations in the genome sizes within some phyla.

Plotting the *minimum* amount of DNA required for a member of each group suggests in **Figure 3.2** that an increase in genome size is required to make more complex prokaryotes and lower eukaryotes.

Mycoplasma are the smallest prokaryotes, and have genomes only ~3× the size of a large bacteriophage (T4 is 1.7×10^5 bp). Bacteria start at about 2×10^6 bp. Unicellular eukaryotes (whose lifestyles may resemble the prokaryotic) get by with genomes that are also small, although larger than those of the bacteria. Being eukaryotic *per se* does not imply a vast increase in genome size; a yeast may have a genome size of $\sim 1.3 \times 10^7$ bp, only about twice the size of the largest bacterial genomes.

A further twofold increase in genome size is adequate to support the slime mold *D. discoideum,* able to live in either unicellular or multicellular modes. Another increase in complexity is necessary to produce the first fully multicellular organisms; the nematode worm *C. elegans* has a DNA content of 8×10^7 bp.

We can also see the steady increase in genome size with complexity in the listing in **Figure 3.3** of some of the most commonly analyzed organisms. It is necessary to increase the genome size in order to make insects, birds or amphibians, and mammals. As we climb

Figure 3.1 DNA content of the haploid genome is related to the morphological complexity of lower eukaryotes, but varies extensively among the higher eukaryotes. The range of DNA values within a phylum is indicated by the shaded area.

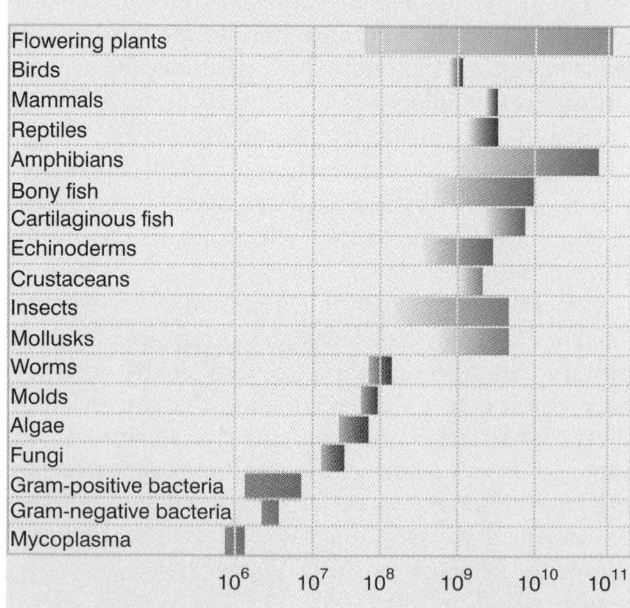

Figure 3.2 The minimum genome size found in each phylum increases from prokaryotes to mammals.

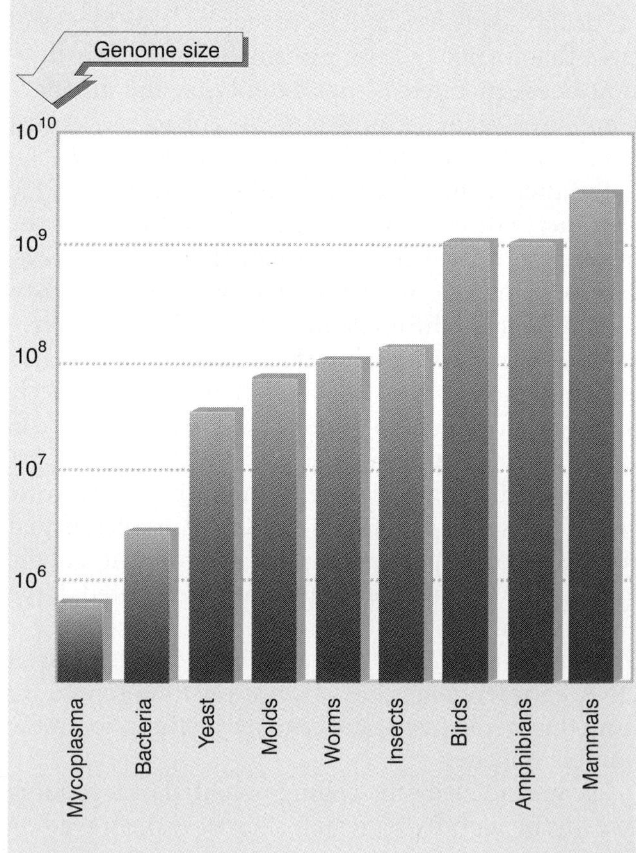

Figure 3.3 The genome sizes of some common experimental animals.

Phylum	Species	Genome (bp)
Algae	*Pyrenomas salina*	6.6×10^5
Mycoplasma	*M. pneumoniae*	1.0×10^6
Bacterium	*E. coli*	4.2×10^6
Yeast	*S. cerevisiae*	1.3×10^7
Slime mold	*D. discoideum*	5.4×10^7
Nematode	*C. elegans*	8.0×10^7
Insect	*D. melanogaster*	1.4×10^8
Bird	*G. domesticus*	1.2×10^9
Amphibian	*X. laevis*	3.1×10^9
Mammal	*H. sapiens*	3.3×10^9

the evolutionary tree, however, the relationship between complexity of the organism and content of DNA becomes obscure.

The **C-value paradox** took its name from the discrepancy between genome size and genetic complexity. It concerns both the absolute and relative amounts of DNA in eukaryotic genomes:

- Why is there an excess of DNA compared with the amount that could be expected to code for proteins? We can now account for much of the excess. We know that genes are much larger than the sequences needed to code for proteins, because exons can comprise only a small part of the total length of a gene. With this taken into account, the total coding potential is much reduced, and this aspect of the paradox has been resolved.

- Why is genome size not related to genetic complexity? For example, the toad *Xenopus* and man have genomes of essentially the same size, but we assume that man is more complex in terms of genetic development. And what allows large variations in C-value between certain species whose apparent complexities are similar (see Figure 3.1). This problem is specific to certain phyla, most notably insects, amphibians, and plants. In amphibians, the smallest genomes are $<10^9$ bp, while the largest are $\sim 10^{11}$ bp. There is unlikely to be a large difference in the number of genes needed to specify these amphibians. This implies that there must be vastly increased amounts of noncoding DNA in the larger genomes. We do not really understand why natural selection allows this DNA to accumulate. (In other phyla, this issue does not occur, and the spread of genome sizes is narrow. Birds, reptiles, and mammals all show little variation within the group, with an $\sim 2\times$ range of genome sizes.)

Eukaryotic genomes have several sequence components

Do larger genomes contain a greater number of different genes or instead contain more copies of the same genes that are present in smaller genomes? If the diversity of genes increases with genome size, we should expect the number of unique DNA sequences in the genome to increase. This will not happen if there are simply more copies of the same genes.

These questions in due course will be directly answered by genome sequences, but at present we have direct information for only a few individual (and relatively small) genomes. In the meantime, the general nature of the eukaryotic genome can be assessed by the kinetics of reassociation of denatured DNA. Reassociation between complementary sequences of DNA occurs by base pairing. This reverses the process of denaturation by which they were separated (see Figure 1.13). The kinetics of the reassociation reaction reflect the variety of sequences that are present; so the reaction can be used to quantitate genes and their RNA products.

Figure 3.4 describes the reaction. Renaturation of DNA depends on random collision of the complementary strands, and follows second-order kinetics. The reaction for any particular DNA can be characterized by conditions required for half-completion. This is the product of $C_0 \times t_{1/2}$ and is called the **Cot$_{1/2}$**. It is inversely proportional to the rate constant. Since the $Cot_{1/2}$ is the product of the concentration and time

Figure 3.4 A DNA reassociation reaction is described by the $C_0t_{1/2}$

Rate of reaction

The reaction follows the second order equation

$$\frac{dC}{dt} = -kC^2$$

C is the concentration of DNA that is single-stranded at time t

k is a reassociation rate constant.

Progress of reaction

Integrate the rate equation between the limits:
initial concentration of DNA = C_0 at time t = 0;
concentration remaining single stranded = C after time t

$$\frac{C}{C_0} = \frac{1}{1 + k.C_0t}$$

Critical parameter is $C_0t_{1/2}$

When the reaction is half complete at time t = ½

$$\frac{C}{C_0} = \frac{1}{2} = \frac{1}{1 + k.C_0t_{1/2}}$$

Therefore $C_0t_{1/2} = \dfrac{1}{k}$

required to proceed halfway, a greater $\mathrm{Cot}_{1/2}$ implies a slower reaction.

The reassociation of DNA usually is followed in the form of a **Cot curve,** which plots the fraction of DNA that has reassociated ($1 - C/C_0$) against the log of the Cot. **Figure 3.5** gives Cot curves for several simple genomes. The form of each curve is similar, with renaturation occurring over an ~100-fold range of Cot values between the points of 10% reaction and 90% reaction. But the $\mathrm{Cot}_{1/2}$ for each curve is different.

The genomes in Figure 3.5 represent a series of DNAs. Each is unique in sequence, and they become progressively longer. *The $\mathrm{Cot}_{1/2}$ is directly related to the amount of DNA in the genome.* This reflects a situation in which, as the genome becomes more complex, there are fewer copies of any particular sequence within a given mass of DNA. For example, if the C_0 of DNA is 12 pg, it will contain 3000 copies of each sequence in a bacterial genome whose size is 0.004 pg, but will contain only 4 copies of each sequence present in a eukaryotic genome of size 3 pg. So the same *absolute*

Figure 3.5 Rate of reassociation is inversely proportional to the length of the reassociating DNA.

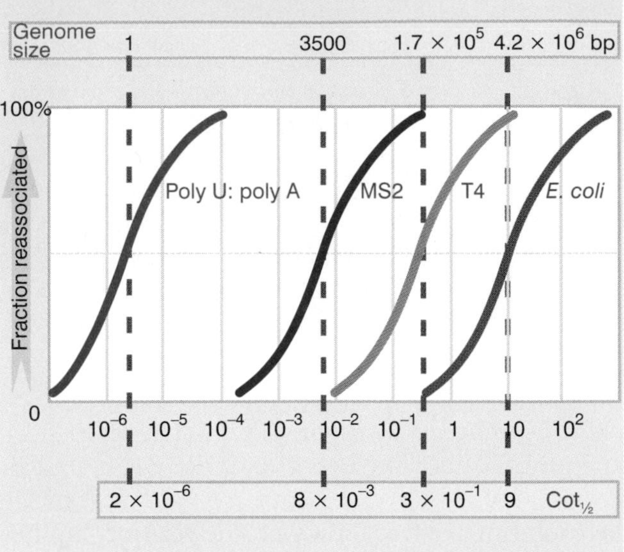

concentration of DNA measured in moles of nucleotides per liter (the C_0) will provide a concentration of each eukaryotic sequence that is 3000/4 = 750× less than that of each bacterial sequence.

Since the rate of reassociation depends on the concentration of complementary sequences, for the eukaryotic sequences to be present at the same *relative* concentration as the bacterial sequences, it is necessary to have 750× more DNA (or to incubate the same amount of DNA for 750 times longer). So the $\mathrm{Cot}_{1/2}$ of the eukaryotic reaction is 750× the $\mathrm{Cot}_{1/2}$ of the bacterial reaction.

The $\mathrm{Cot}_{1/2}$ of a reaction therefore indicates the *total length of different sequences* that are present. This is described as the **complexity,** usually given in base pairs. The $\mathrm{Cot}_{1/2}$ for the renaturation of the DNA of any genome (or part of a genome) is proportional to its complexity. The complexity of any DNA can be determined by comparing its $\mathrm{Cot}_{1/2}$ with that of a standard DNA of known complexity. Usually *E. coli* DNA is used as a standard. Assuming that the *E. coli* genome of 4.2 × 10^6 bp consists of unique sequences:

$$\frac{\mathrm{Cot}_{1/2}\ (\text{DNA of any genome})}{\mathrm{Cot}_{1/2}\ (\textit{E. coli}\ \text{DNA})} = \frac{\text{Complexity of any genome}}{4.2 \times 10^6\ \text{bp}}$$

When the DNA of a eukaryotic genome is characterized by reassociation kinetics, usually the reaction occurs over a range of Cot values spanning up to eight orders of magnitude. This is much broader than the 100-fold range expected from the examples of Figure

3.5. The reason is that each of these curves follows the equation that describes the kinetics of reassociation for a single component. *A eukaryotic genome actually includes several such components, each reassociating with its own characteristic kinetics. The Cot curve reveals a crucial difference between bacterial and eukaryotic genomes: bacterial genomes essentially consist of a single kinetic component, but eukaryotic genomes are much more complex.*

Figure 3.6 shows the reassociation of a (hypothetical) eukaryotic genome, starting at a Cot of 10^{-4} and terminating at a Cot of 10^4. The reaction falls into three distinct phases, outlined by the shaded boxes. A plateau separates the first two phases, but the second and third overlap. Each of these phases represents a different kinetic component of the genome:

- The **fast component** is the first fraction to reassociate. In this case, it represents 25% of the total DNA, renaturing between Cot values of 10^{-4} and $\sim 2 \times 10^{-2}$, with a $Cot_{1/2}$ value of 0.0013.
- The next fraction is called the **intermediate component.** This represents 30% of the DNA. It renatures between Cot values of ~0.2 and 100, with a $Cot_{1/2}$ value of 1.9.
- The **slow component** is the last fraction to renature. This is 45% of the total DNA; it extends over a Cot range from ~100 to ~10,000, with a $Cot_{1/2}$ of 630.

To calculate the complexities of these fractions, each must be treated as an independent kinetic component whose reassociation is compared with a standard DNA. The slow component represents 45% of the total DNA, so its concentration in the reassociation reaction is 0.45 of the measured C_0 (which refers to the total amount of DNA present). So the $Cot_{1/2}$ applying to the slow fraction alone is $0.45 \times 630 = 283$.

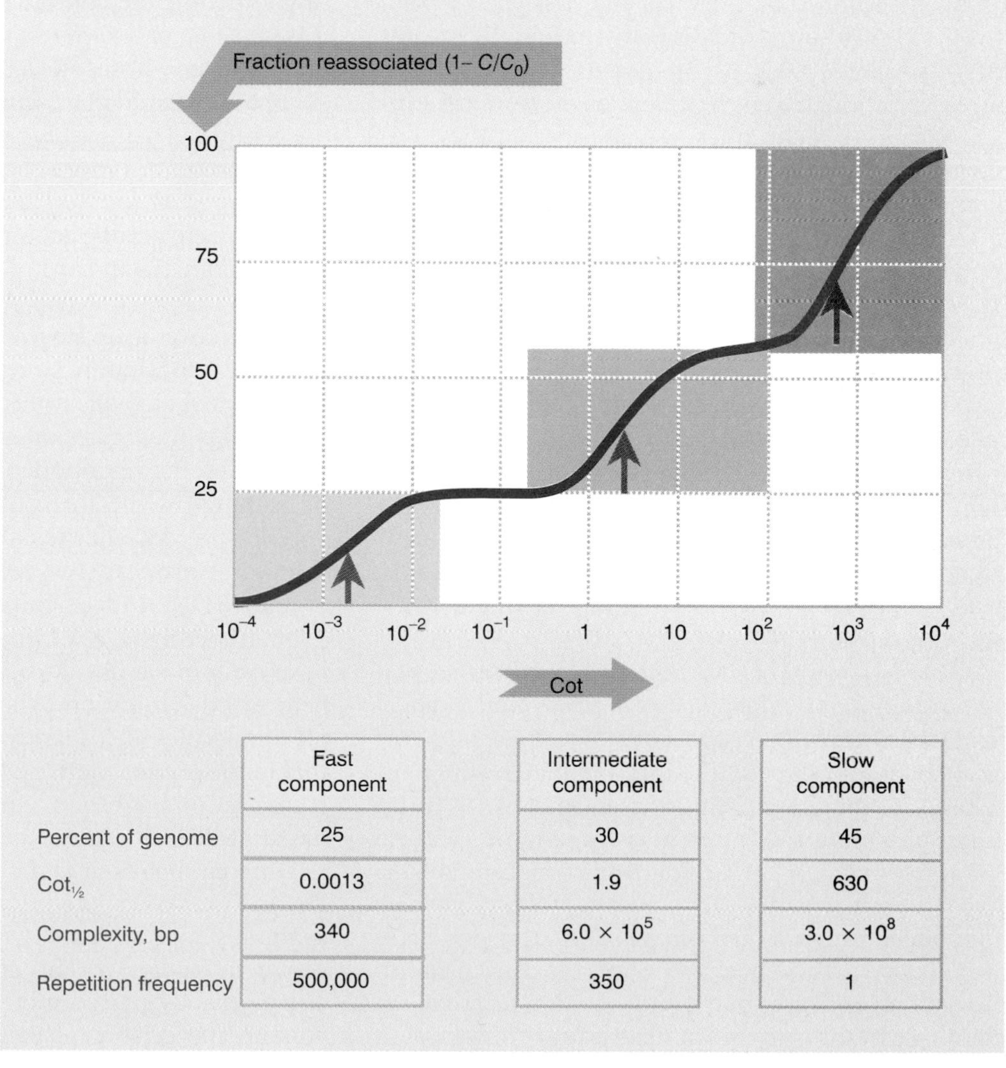

	Fast component	Intermediate component	Slow component
Percent of genome	25	30	45
$Cot_{1/2}$	0.0013	1.9	630
Complexity, bp	340	6.0×10^5	3.0×10^8
Repetition frequency	500,000	350	1

Figure 3.6 The reassociation kinetics of eukaryotic DNA show three types of component (indicated by the shaded areas). The arrows identify the $Cot_{1/2}$ values for each component.

Suppose that under these conditions, *E. coli* DNA reassociates with a $Cot_{1/2}$ of 4.0. This corresponds to a complexity for the slow fraction of 3.0×10^8 bp (= $4.2 \times 10^6 \times 283/4$). Treating the other components in a similar way shows that the intermediate component has a complexity of 6×10^5 bp, and the fast component has a complexity of only 340 bp. This provides a quantitative basis for our statement that, the faster a component reassociates, the lower is its complexity.

Reversing the argument, suppose we took three DNA preparations, each containing a unique sequence of the appropriate length (340 bp, 6×10^5 bp, and 3×10^8 bp, respectively) and mixed them in the proportions of mass 25:30:45. Each would renature as though it were a single component. Together the mixture would display the same kinetics as those determined for the whole genome of Figure 3.6.

The complexity of the slow component corresponds with its physical size. Suppose that the genome reassociating in Figure 3.6 has a haploid DNA content of 7.0×10^8 bp, determined by chemical analysis. Then 45% of it is 3.15×10^8 bp, which is only marginally greater than the value of 3.0×10^8 bp measured by the kinetics of reassociation. Given the errors of measurement inherent in both techniques, we can say that the complexity of the slow component is the same whether measured chemically or kinetically.

The slow component comprises sequences that are unique in the genome: on denaturation, *each single-stranded sequence is able to renature only with the corresponding complementary sequence.* This part of the genome is the sole component of prokaryotic DNA and is usually a major component in eukaryotes. It is called **nonrepetitive DNA.**

What is the nature of the components that renature more rapidly than the nonrepetitive (slow) DNA? In the example of Figure 3.5, the intermediate component occupies 30% of the genome. Its chemical complexity is $0.3 \times 7 \times 10^8 = 2.1 \times 10^8$ bp. But its kinetic complexity is only 6×10^5 bp.

The unique length of DNA that corresponds to the $Cot_{1/2}$ for reassociation is much shorter than the total length of the DNA chemically occupied by this component in the genome. In other words, the intermediate component behaves as though consisting of a sequence of 6×10^5 bp that is present in 350 copies in every genome (because $350 \times 6 \times 10^5 = 2.1 \times 10^8$). Following denaturation, *the single strands generated from any one of these copies are able to renature with their complements from any one of the 350 copies.* This effectively raises the concentration of reacting sequences in the reassociation reaction, explaining why the component renatures at a lower $Cot_{1/2}$.

Sequences that are present in more than one copy in each genome are called **repetitive DNA.** The number of copies present per genome is called the **repetition frequency** (f).

Repetitive DNA is often classed into two general types, corresponding approximately to the intermediate and fast components of Figure 3.6:

- **Moderately repetitive DNA** occupies the intermediate fraction, usually reassociating in a range between a Cot of 10^{-2} and that of nonrepetitive DNA.
- **Highly repetitive DNA** occupies the fast fraction, reassociating before a Cot of 10^{-2} is reached.

The behavior of a repetitive DNA component represents only an average that is useful for describing its sequences. The relevant parameters do not necessarily represent the properties of any particular sequence.

The moderately repetitive component of Figure 3.6 includes a total length of 6×10^5 bp of DNA, repeated ~350× per genome. But this does not correspond to a single, identifiable, continuous length of DNA. *Instead, it is made up of a variety of individual sequences, each much shorter,* whose total length together comes to 6×10^5 bp. These individual sequences are dispersed about the genome. Their average repetition is 350, but some will be present in more copies than this and some in fewer.

When a eukaryotic genome is analyzed by reassociation kinetics, the individual sequence components are rarely so well separated as shown in Figure 3.6. In fact, they often overlap extensively, so that in reality there is probably a continuum of repetitive components, reassociating over a range from >10× to >20,000× that of the nonrepetitive component.

The different components of eukaryotic DNA can be isolated in the form of the DNA that becomes double-stranded after renaturation to a particular Cot value. The properties of renatured nonrepetitive and repetitive DNA differ significantly.

Nonrepetitive DNA forms duplex material that behaves very much like the original preparation of DNA before its denaturation. When denatured again, the duplex molecules melt sharply at a T_m only slightly below that of the original native DNA. This shows that strand reassociation has been accurate: each unique sequence has annealed with its exact complement.

Different behavior is shown by renatured repetitive DNA. The reassociated double strands tend to melt gradually over rather a wide temperature range, as shown in **Figure 3.7**. This means that they do not consist of exactly paired molecules. Instead, they must contain appreciable mispairing. The more mispairing

Figure 3.7 The denaturation of reassociated nonrepetitive DNA takes place over a narrow temperature range close to that of native DNA, but reassociated repetitive DNA melts over a wide temperature range.

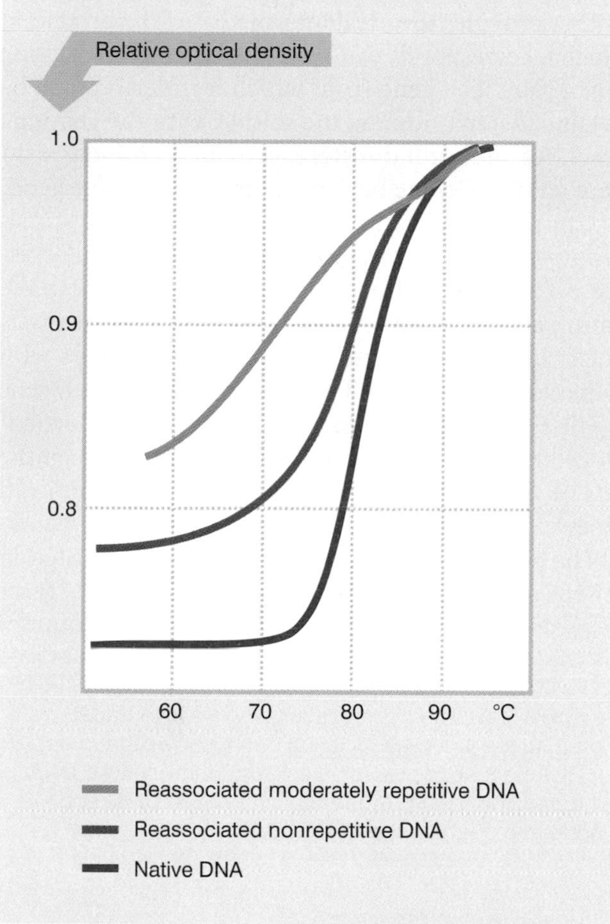

in a particular molecule, the fewer hydrogen bonds need be broken to melt it, and thus the lower the T_m.

The breadth of the melting curve shows that renatured repetitive DNA contains a spectrum of sequences, ranging from those that have been formed by reassociation between sequences that are only partially complementary, to those formed by reassociation between sequences that are very nearly or even exactly complementary. How can this happen?

Repetitive DNA components consist of families of sequences that are not exactly the same, but are related. The members of each family consist of a set of nucleotide sequences that are sufficiently similar to renature with one another. The differences between the individual members are the result of base substitutions, insertions, and deletions, all creating points within the related sequences at which the complementary strands cannot base pair. The proportion of these changes establishes the relationship between any two sequences. When two closely related members of the family renature, they form a duplex with a high T_{m}. When two more distantly related members associate, they form a duplex with a lower T_m. Overall, we see the broad range represented in the figure.

The ability of related but not identical complementary sequences to recognize each other can be controlled by the **stringency** of the conditions imposed for reassociation. A higher stringency is imposed by (for example) an increase in temperature, which requires a greater degree of complementarity to allow base pairing. So by performing the hybridization reaction at high temperatures, reassociation is restricted to rather closely related members of a family; at lower temperatures, more distantly related members may anneal. The measured size of a repetitive family is arbitrary, since it is determined by the hybridization conditions.

Moderately repetitive DNA is dispersed throughout the genome, usually in the form of relatively short individual sequences. It is responsible for the high degree of secondary structure formation in pre-mRNA, when (inverted) repeats in the introns pair to form duplex regions. Highly repetitive DNA often forms discrete clusters (see Chapter 4). Neither class represents protein.

Most structural genes lie in nonrepetitive DNA

The proportion of the genome occupied by nonrepetitive versus repetitive DNA varies widely. **Figure 3.8** summarizes the genome organization of some representative organisms. Prokaryotes, of course, contain only nonrepetitive DNA. For lower eukaryotes, most of the DNA is nonrepetitive; <20% falls into one or more moderately repetitive components. In animal cells, up to half of the DNA often is occupied by moderately and highly repetitive components. In plants and amphibians, the nonrepetitive DNA may be reduced to a minority of the genome, with the moderately and highly repetitive components accounting for up to 80%.

The length of the nonrepetitive DNA component tends to increase with overall genome size, as we proceed up to a total genome size of ~3×10^9 (characteristic of mammals). Further increase in genome size, however, generally reflects an increase in the amount and proportion of the repetitive components, so that it is rare for an organism to have a nonrepetitive DNA component of $>2 \times 10^9$. The nonrepetitive DNA content of genomes therefore accords better with our sense of the relative complexity of the organism. *E. coli* has 4.2×10^6 bp, *C. elegans* increases an order of magnitude to 6.6×10^7 bp, *D. melanogaster* increases further to ~10^8 bp, and mammals increase another order of magnitude to ~2×10^9 bp. This suggests that larger genomes have accumulated increased amounts of repetitive DNA.

Figure 3.8 The proportions of different sequence components vary in eukaryotic genomes. The absolute content of nonrepetitive DNA increases with genome size, but reaches a plateau at ~2×10^9 bp.

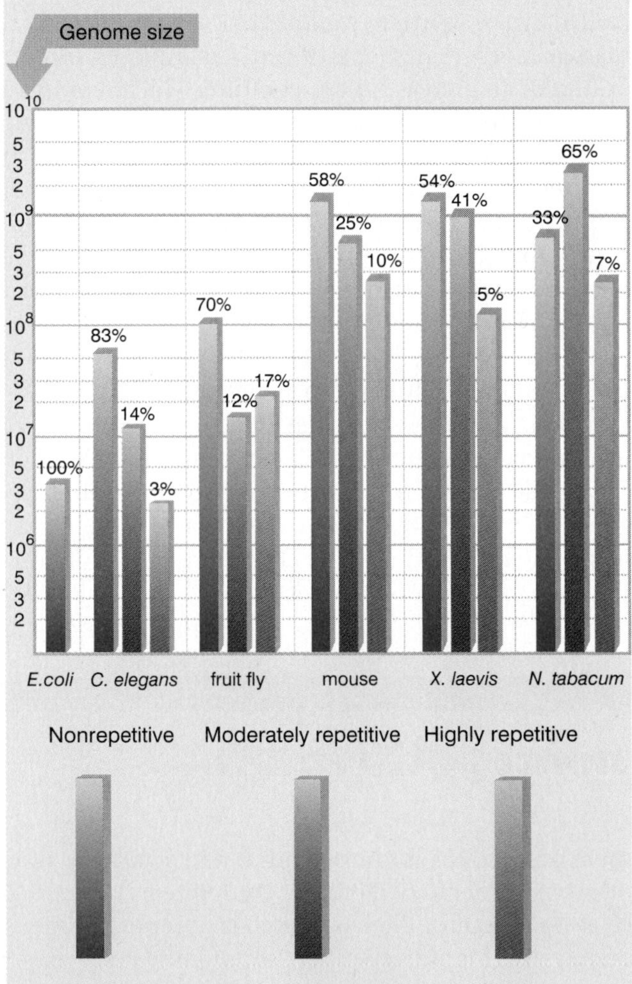

Where are structural genes to be found? For the purpose of identifying and characterizing structural genes, mRNA provides the ideal intermediate. The protein to which it corresponds can be determined by translating the mRNA. The gene from which it is derived can be obtained by hybridizing the mRNA with the genomic DNA. An individual mRNA provides a handle, as it were, for proceeding back from its protein to the gene.

A population of mRNAs isolated from polysomes defines the entire set of genes expressed in a cell or tissue. So the constitution of the mRNA reveals both the nature and number of structural genes. By means of nucleic acid hybridization, we can come to grips with some central questions. How many copies of each gene are there? How many genes are expressed in a particular cell type? How much overlap is there between the sets of genes whose expression defines different cell types?

The genome sequence components represented in mRNA can be determined by using the RNA as a **tracer** in a reassociation experiment. A very small amount of

Figure 3.9 The hybridization of an mRNA tracer preparation in a reassociation curve shows that most mRNA sequences are derived from nonrepetitive DNA, the remainder from moderately repetitive DNA, and none from highly repetitive DNA.

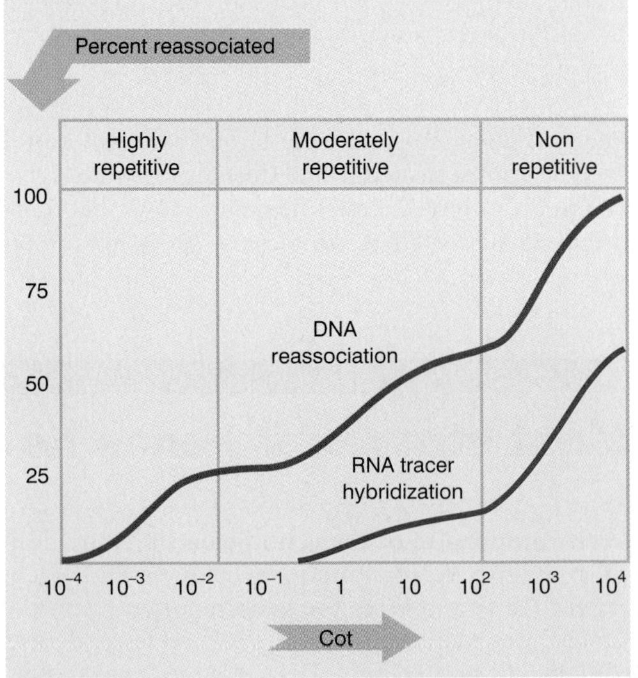

radioactively labeled RNA (or cDNA) is included together with a much larger amount of cellular DNA. The tracer RNA (or cDNA) participates in the reaction as though it were just another member of the sequence component from which it was transcribed. The Cot values at which the labeled RNA hybridizes identify the repetition frequencies of the corresponding genomic sequences.

Figure 3.9 shows a typical result for a population of mRNAs. A small proportion of the RNA, generally 10% or less, hybridizes with a $Cot_{1/2}$ corresponding to moderately repetitive sequences. The major component hybridizes with a $Cot_{1/2}$ identical with or very close to that of nonrepetitive DNA. This implies that most of the mRNA is derived from the nonrepetitive DNA component. Because of the difficulties in detecting very low degrees of repetition (especially when the repeated copies are related rather than identical), these genes are not necessarily unique, but they should be present in fewer than three or four copies per genome.

In looking at the DNA that hybridizes with mRNA, we are basically examining the exons in the genome. The conclusion therefore is that most exons are present at low repetition frequency—depending on the stringency of hybridization, they may be unique or present in a small number of copies.

Total gene number is known for several organisms

Large-scale efforts have now led to the sequencing of several genomes. A range is summarized in **Figure 3.10**. The data are based on complete sequences, except for fruit fly and man.

The sequences of the genomes of bacteria and archaea show that virtually all of the DNA (typically 85–88%) codes for RNA or protein. The bacterium with the smallest known genome, *M. genitalium*, has ~470 genes of ~1040 bp each. This identifies the minimum number of functions required to construct a cell. All classes of genes are reduced in number compared with bacteria with larger genomes, but the most significant reduction is in loci coding for enzymes concerned with metabolic functions and with regulation of gene expression. This makes *M. genitalium* more dependent on the provision of small molecules by its host.

The bacterium *Rickettsia prowazekii*, which is an obligate intracellular parasite, has a complexity about twice that of the *Mycoplasma*. A "typical" Gram-negative bacterium, *H. influenzae*, has 1,743 genes each of ~900 bp. About 60% of the genes can be identified on the basis of homology with known genes in other species, and these genes fall approximately equally into classes whose products are concerned with metabolism, cell structure or transport of components, and gene expression and its regulation. *E. coli* has a larger genome, with 4,288 genes, average size ~950 bp, and an average separation between genes of 118 bp.

The archaea have properties that are intermediate between the prokaryotes and eukaryotes. *M. jannaschii* is a methane-producing species that lives under high pressure and temperature. Its total gene number is similar to that of *H. influenzae*, but fewer of them can be identified on the basis of comparison with genes

Figure 3.10 Genome sizes, gene numbers and lethal loci.

Species	Genome (Mb)	Genes	Lethal loci
Mycoplasma genitalium	0.58	470	
Rickettsia prowazekii	1.11	834	
Haemophilus influenzae	1.83	1,743	
Methanococcus jannaschi	1.66	1,738	
B. subtilis	4.2	4,100	
E. coli	4.6	4,288	1,800
S. cerevisiae	13.5	6,034	3,600
D. melanogaster	165	12,000	3,100
C. elegans	97	19,099	
H. sapiens	3,300	100,000	

known in other organisms. Its apparatus for gene expression resembles eukaryotes more than prokaryotes, but its apparatus for cell division better resembles prokaryotes.

The most extensive data for a lower eukaryote are available from the sequence of the genome of *S. cerevisiae*. The density of genes is high: the average open reading frame is ~1.4 kb, and the average separation between genes is ~600 bp, so that ~70% of the genome is occupied by the total of ~6000 genes. About half of the genes identified by sequence were either known previously or are related to known genes. The remainder are new, which gives some indication of the number of new types of genes that may be discovered.

The identification of large genes on the basis of sequence is quite accurate. However, there are also ~600 potential small genes—with ORFs coding for <100 amino acids—which cannot be identified solely by sequence because of the high occurrence of false positives. Analysis of gene expression suggests that ~300 of these ORFs are likely to be genuine genes.

The genome of *C. elegans* DNA varies between regions rich in genes and regions in which genes are more sparsely organized. The total sequence contains ~19,000 genes. Only ~42% of the genes have putative counterparts outside the Nematoda. There may be a slightly smaller number of genes in *D. melanogaster*, although the fly genome is larger, and this comparison emphasizes the lack of a clear relationship between gene number and complexity of the organism.

Protein size increases from prokaryotes to eukaryotes. The bacteria *M. jannaschi* and *E. coli* have average protein lengths of 287 and 317 amino acids, respectively; whereas *S. cerevisiae* and *C. elegans* have average lengths of 484 and 442 amino acids, respectively. Large proteins (>500 amino acids) are rare in bacteria, but comprise a significant part (~1/3) in eukaryotes. The increase in length is due to the addition of extra domains, a typical domain constituting 100–300 amino acids. But the increase in protein size is responsible for only a very small part of the increase in genome size.

Another insight into gene number is obtained by counting the number of expressed genes. If we rely upon the estimates of the number of different mRNA species that can be counted in a cell, we would conclude that the average vertebrate cell expresses ~10,000–20,000 genes. The existence of significant overlaps between the messenger populations in different cell types would suggest that the total expressed gene number for the organism should be within a few fold of this, in the range (say) of 50,000–100,000.

Eukaryotic genes are transcribed individually, each gene producing a monocistronic messenger. There is only one general exception to this rule; in the genome of *C. elegans*, ~25% of the genes are organized into polycistronic units (which is associated with the use of *trans*-splicing to allow expression of the downstream genes in these units; see Chapter 22).

When we characterize genes corresponding to individual functions, in higher eukaryotes we often find additional copies representing unsuspected variants of the gene. The extension from individual genes to families of related genes certainly increases the number of genes in the genome. Present evidence is largely anecdotal, but it would not be surprising if on average there were three to four genes for every function that is apparently unique. Our total number of genes is therefore likely to include a somewhat smaller number of different types of functions.

Besides needing to know the density of genes to estimate the total gene number, we must also ask: is it important in itself? Are there structural constraints that make it necessary for genes to have a certain spacing, and does this contribute to the large size of eukaryotic genomes?

How many genes are essential?

NATURAL selection is the force that ensures that useful genes are retained in the genome. Mutations occur at random, and their most common effect in an open reading frame will be to damage the protein product. An organism with a damaging mutation will be at a disadvantage in evolution, and ultimately the mutation will be eliminated by the failure of organisms carrying it to compete. The frequency of a disadvantageous allele in the population is balanced between the generation of new mutations and the elimination of old mutations. Reversing this argument, whenever we see an intact open reading frame in the genome, we assume that its product plays a useful role in the organism. Natural selection must have prevented mutations from accumulating in the gene. The ultimate fate of a gene that ceases to be useful is to accumulate mutations until it is no longer recognizable.

The maintenance of a gene implies that it confers a selective advantage on the organism. But in the course of evolution, even a small relative advantage may be the subject of natural selection. However, we should like to know how many genes are actually *essential.* This means that their absence is lethal to the organism. In the case of diploid organisms, it means of course that the homozygous null mutation is lethal.

One approach to the issue of gene number is to determine the number of essential genes by mutational analysis. If we saturate some specified region of the chromosome with mutations that are lethal, the mutations should map into a number of complementation groups that corresponds to the number of lethal loci in that region. By extrapolating to the genome as a whole, we may calculate the total essential gene number.

The most extensive analyses of essential gene number have been made in *Drosophila* through attempts to correlate visible aspects of chromosome structure with the number of functional genetic units. The notion that this might be possible arose originally from the presence of bands in the polytene chromosomes of *D. melanogaster.* (These chromosomes are found at certain developmental stages and represent an unusually extended physical form, in which a series of bands [more formally called chromomeres] can be seen. We discuss their properties in Chapter 18.) From the early concept that the bands might represent a linear order of genes, we have come to the attempt to correlate the organization of genes with the organization of bands. There are ~5000 bands in the *D. melanogaster* haploid set; they vary in size over an order of magnitude, but on average there is ~20 kb of DNA per band.

The basic approach is to saturate a chromosomal region with mutations. Usually the mutations are simply collected as lethals, without analyzing the cause of the lethality. Any mutation that is lethal is taken to identify a locus that is essential for the organism. Sometimes mutations cause visible deleterious effects short of lethality, in which case we also count them as identifying an essential locus. When the mutations are placed into complementation groups, the number can be compared with the number of bands in the region, or individual complementation groups may even be assigned to individual bands. The purpose of these experiments has been to determine whether there is a consistent relationship between bands and genes; for example, does every band contain a single gene?

Totaling the analyses that have been carried out over the past 30 years, the number of lethal complementation groups is ~70% of the number of bands. It is an open question whether there is any functional significance to this relationship. But irrespective of the cause, the equivalence gives us a reasonable estimate for the lethal gene number of ~3600. If we assume that the organization of the *Drosophila* and mammalian genomes is in principle similar, then by comparing the average sizes of their genes and genomes, we would predict >75,000 lethal genes for man. By any measure, the number of lethal loci in *Drosophila* is significantly less than the total number of genes, and presumably the same is true of man.

An experiment to determine what proportion of genes are essential has produced an analogous result in the yeast *S. cerevisiae.* The genome is relatively small, and a large fraction (>50%) is transcribed, compared with higher eukaryotes. When insertions were introduced at random into the genome, only 12% were lethal, and another 14% impeded growth. The majority (70%) of the insertions had no effect. Analyses such as this suggest that only a minority of genes have lethal effects or directly impair growth. Similarly, fewer than half of the genes of *E. coli* appear to be essential (see Figure 3.3).

How do we explain the survival of genes whose deletion appears to have no effect? One possibility is that there is **redundancy**, that such genes are present in multiple copies. This is certainly true in some cases, in which multiple (related) genes must be knocked out in order to produce an effect. It is clear that there are cases in which a genome has more than one gene capable of providing a protein to fulfill a certain function, and all of them must be deleted to produce a lethal effect.

The idea that some genes are not essential (or at least cannot be shown to have serious effects upon the phenotype) raises some important questions. Does the genome contain genuinely dispensable genes, or do these genes actually have effects upon the phenotype that are significant at least during the long march of evolution? The theory of natural selection would suggest that the loss of individual genes in such circumstances produces a small disadvantage, which although not evident to us, is sufficient for the gene to be retained during the course of evolution.

Key questions that remain to be answered systematically are: What proportion of the total number of genes is essential, in how many do mutations produce at least detectable effects, and are there genes that are genuinely dispensable? Subsidiary questions about the genome as a whole are: What are the functions (if any) of DNA that does not reside in genes? What effect does a large change in total size have on the operation of the genome, as in the case of the related amphibians?

How many genes are expressed?

Reassociation analysis can be used to measure the complexity of an RNA population. One method is to hybridize nonrepetitive DNA with an excess of RNA; the proportion of the DNA that is bound at saturation identifies the complexity of the RNA population. Another method is to follow the kinetics of hybridization between a excess of an RNA population and a DNA copy prepared from it. This is exactly analogous to reassociation analysis of genomic DNA. The reaction is described in terms of the $Rot_{1/2}$ (where R_0 is the starting concentration of RNA).

Saturation analysis typically identifies ~1% of the DNA as providing a template for mRNA. From this we can calculate the number of genes so long as we know the average length of an mRNA. For a lower eukaryote such as yeast, the total number of expressed genes is ~4000. For somatic tissues of higher eukaryotes, the number usually is between 10,000 and 15,000. The value is similar for plants and for vertebrates. (The only consistent exception to this type of value is presented by mammalian brain, where much larger numbers of genes appear to be expressed, although the exact quantitation is not certain.)

Kinetic analysis typically identifies three components in a eukaryotic cell. Just as with a DNA reassociation curve, a single component hybridizes over about two decades of Rot values, and a reaction extending over a greater range must be resolved by computer curve-fitting into individual components. Again this represents what is really a continuous spectrum of sequences.

An example of an excess mRNA × cDNA reaction that generates three components is given in **Figure 3.11**:

- The first component has the same characteristics as a control reaction of ovalbumin mRNA with its DNA copy. This suggests that the first component is in fact just ovalbumin mRNA (which indeed occupies about half of the messenger mass in oviduct tissue).
- The next component provides 15% of the reaction, with a total complexity of 15 kb. This corresponds to seven to eight mRNA species of average length 2000 bases.
- The last component provides 35% of the reaction, which corresponds to a complexity of 26 Mb. This cor-

Figure 3.11 Hybridization between excess mRNA and cDNA identifies several components in chick oviduct cells, each characterized by the $Rot_{1/2}$ of reaction.

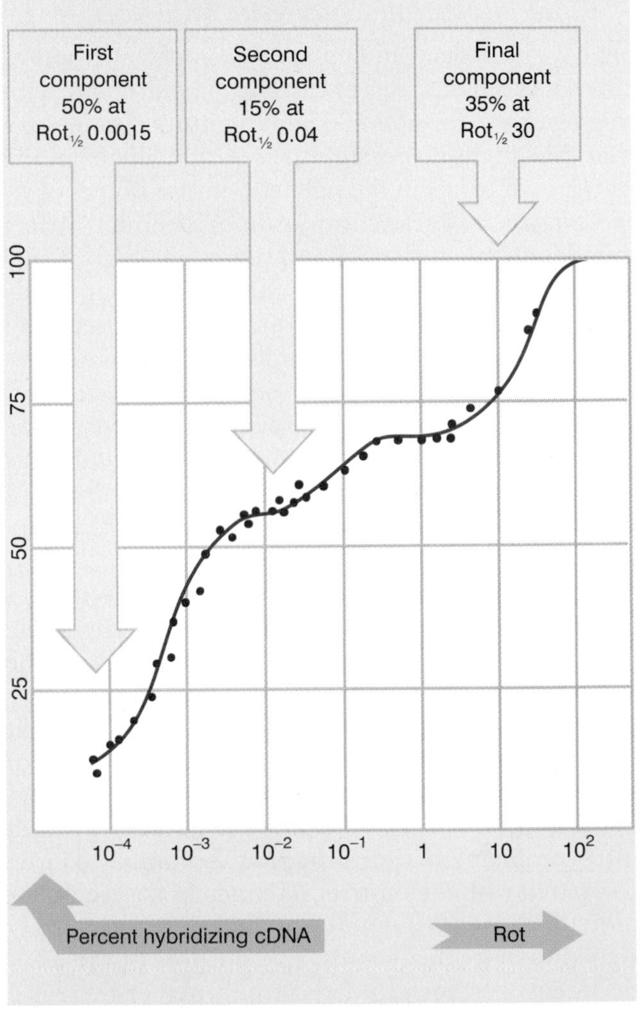

responds to ~13,000 mRNA species of average length 2000 bases.

From this analysis, we can see that about half of the mass of mRNA in the cell represents a single mRNA, ~15% of the mass is provided by a mere seven to eight mRNAs, and ~35% of the mass is divided into the large number of 13,000 mRNA species. It is therefore obvious that the mRNAs comprising each component must be present in very different amounts.

The average number of molecules of each mRNA per cell is called its **abundance.** It can be calculated quite simply if the total mass of RNA in the cell is known. In the example shown in Figure 3.11, the total mRNA can be accounted for as 100,000 copies of the first component (ovalbumin mRNA), 4000 copies of each of the 7–8 mRNAs in the second component, but only ~5 copies of each of the 13,000 mRNAs that constitute the last component.

We can divide the mRNA population into two general classes, according to their abundance:

- The oviduct is an extreme case, with so much of the mRNA represented in only one species, but most cells do contain a small number of RNAs present in many copies each. This **abundant** component typically consists of <100 different mRNAs present in 1000–10,000 copies per cell. It often corresponds to a major part of the mass, approaching 50% of the total mRNA.
- About half of the mass of the mRNA consists of a large number of sequences, of the order of 10,000, each represented by only a small number of copies in the mRNA—say, <10. This is the **scarce** or **complex** mRNA class. (It is this class that drives a saturation reaction.)

Many somatic tissues of higher eukaryotes have an expressed gene number in the range of 10,000 to 20,000. How much overlap is there between the genes expressed in different tissues? For example, the expressed gene number of chick liver is ~11,000–17,000, compared with the value for oviduct of ~13,000–15,000. How many of these two sets of genes are identical? How many are specific for each tissue?

We see immediately that there are likely to be substantial differences among the genes expressed in the abundant class. Ovalbumin, for example, is synthesized only in the oviduct, not at all in the liver. This means that 50% of the mass of mRNA in the oviduct is specific to that tissue.

But the abundant mRNAs represent only a small proportion of the number of expressed genes. In terms of the total number of genes of the organism, and of the number of changes in transcription that must be made between different cell types, we need to know the

Figure 3.12 HDA analysis allows change in expression of each gene to be measured. Each square represents one gene (top left is first gene on chromosome I, bottom right is last gene on chromosome XVI). Change in expression relative to wild type is indicated by red (reduction), white (no change), or blue (increase). Photograph kindly provided by Rick Young.

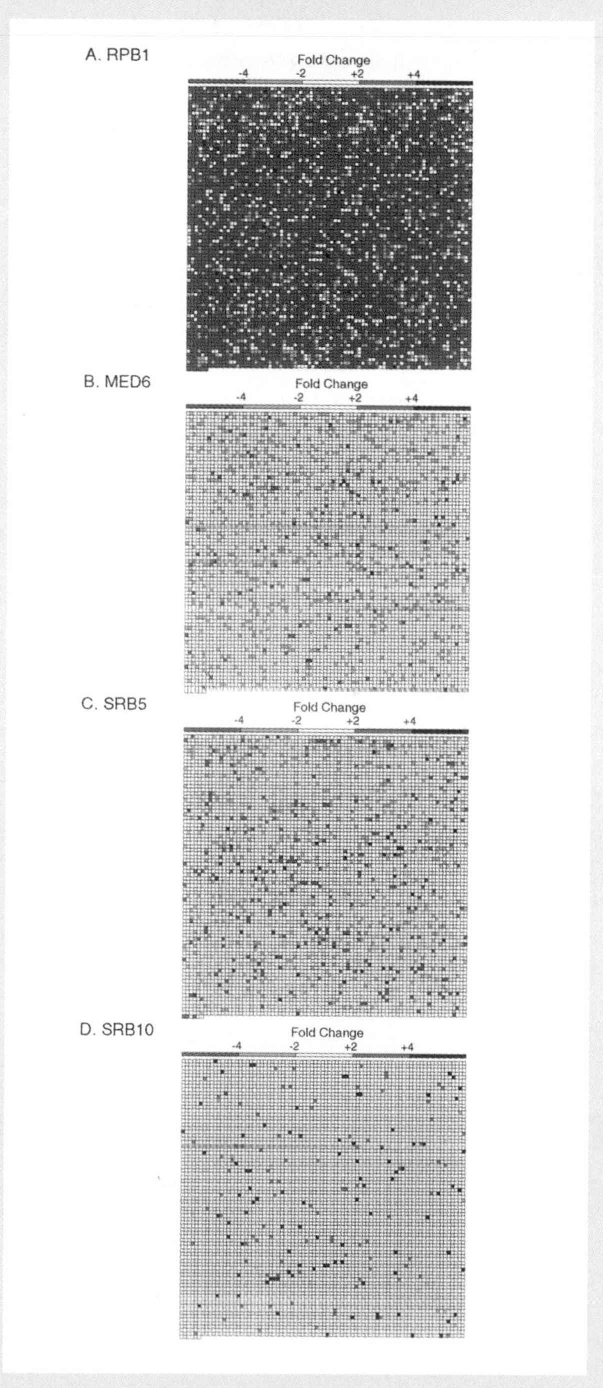

extent of overlap between the genes represented in the scarce mRNA classes of different cell phenotypes.

Comparisons between different tissues show that, for example, ~75% of the sequences expressed in liver and oviduct are the same. In other words, ~12,000 genes are expressed in both liver and oviduct, ~5000 additional genes are expressed only in liver, and ~3000 additional genes are expressed only in oviduct.

The scarce mRNAs overlap extensively. Between mouse liver and kidney, ~90% of the scarce mRNAs are identical, leaving a difference between the tissues of only 1000–2000 in terms of the number of expressed genes. The general result obtained in several comparisons of this sort is that only ~10% of the mRNA sequences of a cell are unique to it. The majority of sequences are common to many, perhaps even all, cell types.

This suggests that the common set of expressed gene functions, numbering perhaps ~10,000 in a mammal, comprise functions that are needed in all cell types. Sometimes this type of function is referred to as a **housekeeping** or **constitutive** activity. It contrasts with the activities represented by specialized functions (such as ovalbumin or globin) needed only for particular cell phenotypes. These are sometimes called **luxury** genes.

Recent technology allows more systematic and accurate estimates of the number of expressed genes. One approach (SAGE, serial analysis of gene expression) allows a unique sequence tag to be used to identify each mRNA. The technology then allows the abundance of each tag to be measured. This approach identifies 4,665 expressed genes in *S. cerevisiae* growing under normal conditions, with abundances varying from 0.3 to >200 transcripts/cell. This means that ~75% of the total gene number (~6000) is expressed under these conditions.

The most powerful new technology uses chips that contain high-density oligonucleotide arrays (HDAs). Their construction is made possibly by knowledge of the sequence of the entire genome. In the case of *S. cerevisiae*, each of 6181 ORFs is represented on the HDA by 20 25-mer oligonucleotides that perfectly match the sequence of the message and 20 mismatch oligonucleotides that differ at one base position. The expression level of any gene is calculated by subtracting the average signal of a mismatch from its perfect match partner. The entire yeast genome can be represented on four chips. This technology is sensitive enough to detect transcripts of 5460 genes (~90% of the genome), and shows that 80% of genes are expressed at low levels, with abundances of 0.1–2 transcripts/cell. An abundance of <1 transcript/cell means that not all cells have a copy of the transcript at any given moment.

The technology allows not only measurement of levels of gene expression, but also detection of differences in expression in mutant yeast strains, under different conditions of growth, and so on. The results of comparing two states are expressed in the form of a grid, in which each square represents a particular gene, and the relative change in expression is indicated by color. The upper part of **Figure 3.12** shows the effect of a mutation in RNA polymerase II, the enzyme that produces mRNA, which, as might be expected, causes the expression of most genes to be heavily reduced. By contrast, the lower part shows that a mutation in an ancillary component of the transcription apparatus (*SRB10*) has much more restricted effects, causing increases in expression of some genes.

The extension of this technology to animal cells will allow the general descriptions based on RNA hybridization analysis to be replaced by exact descriptions of the genes that are expressed, and the abundances of their products, in any given cell type.

Organelles have DNA

THE first evidence for the presence of genes outside the nucleus was provided by **nonMendelian inheritance** in plants (observed in the early years of this century, just after the rediscovery of Mendelian inheritance). NonMendelian inheritance is sometimes associated with the phenomenon of **somatic segregation.** They have a similar cause:

- NonMendelian inheritance is defined by the failure of the progeny of a mating to display Mendelian segregation for parental characters. It reflects a lack of association between the segregating character and the meiotic spindle.

- Somatic segregation describes a phenomenon in which parental characters segregate in somatic cells, and therefore display heterogeneity in the organism. This is a notable feature of plant development. It reflects a lack of association between the segregating character and the mitotic spindle.

NonMendelian inheritance and somatic segregation are therefore taken to indicate the presence of genes that reside outside the nucleus and do not utilize segregation on the meiotic and mitotic spindles to distribute replicas to gametes or to daughter cells, respectively.

The extreme form of nonMendelian inheritance is uniparental inheritance, when the genotype of only one parent is inherited and that of the other parent is permanently lost. In less extreme examples, the progeny of one parental genotype exceed those of the other genotype. Usually it is the mother whose genotype is preferentially (or solely) inherited. This effect is sometimes described as **maternal inheritance.** The important point is that the genotype contributed by the parent of one particular sex predominates, as seen in abnormal segregation ratios when a cross is made between mutant and wild type. This contrasts with the behavior of Mendelian genetics when reciprocal crosses show the contributions of both parents to be equally inherited.

The bias in parental genotypes is established at, or soon after, the formation of a zygote. There are various possible causes. The contribution of maternal or paternal information to the organelles of the zygote may be unequal; in the most extreme case, only one parent contributes. In other cases, the contributions are equal, but the information provided by one parent does not survive. Combinations of both effects are possible. Whatever the cause, the unequal representation of the information from the two parents contrasts with nuclear genetic information, which derives equally from each parent.

NonMendelian inheritance results from the presence in mitochondria and chloroplasts of DNA genomes that are inherited independently of nuclear genes. In effect, the organelle genome comprises a length of DNA that has been physically sequestered in a defined part of the cell, and is subject to its own form of expression and regulation. An organelle genome can code for some or all of the RNAs, but only codes for only some of the proteins needed to perpetuate the organelle. The other proteins are coded in the nucleus, expressed via the cytoplasmic protein synthetic apparatus, and imported into the organelle.

Genes not residing within the nucleus are generally described as **extranuclear;** they are transcribed and translated in the *same* organelle compartment (mitochondrion or chloroplast) in which they reside. By contrast, *nuclear* genes are expressed by means of *cytoplasmic* protein synthesis. (The term **cytoplasmic inheritance** is sometimes used to describe the behavior of genes in organelles. However, we shall not use this description, since it is important to be able to distinguish between events in the general cytosol and those in specific organelles.)

One type of uniparental inheritance is seen in higher animals. Maternal inheritance can be predicted by supposing that the mitochondria are contributed entirely by the ovum and not at all by the sperm. So the mitochondrial genes are derived exclusively from the mother; and in males they are discarded each generation.

Conditions in the organelle are different from those in the nucleus, and organelle DNA therefore evolves at its own distinct rate. If inheritance is uniparental, there can be no recombination between parental genomes; and usually recombination does not occur in those cases where organelle genomes are inherited from both parents. Since organelle DNA has a different replication system from that of the nucleus, the error rate during replication may be different. Mitochondrial DNA accumulates mutations more rapidly than nuclear DNA in mammals, but in plants the accumulation in the mitochondrion is slower than in the nucleus (the chloroplast is intermediate).

One consequence of maternal inheritance is that the sequence of mitochondrial DNA is more sensitive than nuclear DNA to reductions in the size of the breeding population. Comparisons of mitochondrial DNA sequences in a range of human populations allow an evolutionary tree to be constructed. The divergence among human mitochondrial DNAs spans 0.57%. A tree can be constructed in which the mitochondrial variants diverged from a common (African) ancestor. The rate at which mammalian mitochondrial DNA accumulates mutations is 2–4% per million years, >10× faster than the rate for globin. Such a rate would generate the observed divergence over an evolutionary period of 140,000–280,000 years. This implies that the human race is descended from a single female, who lived in Africa ~200,000 years ago.

Organelle genomes are circular DNAs that code for organelle proteins

Most organelle genomes take the form of a single circular molecule of DNA of unique sequence (denoted **mtDNA** in the mitochondrion and **ctDNA** in the chloroplast). There are a few exceptions where mitochondrial DNA is a linear molecule, generally in lower eukaryotes.

Usually there are several copies of the genome in the individual organelle. Since there are multiple organelles per cell, there are many organelle genomes per cell. Although the organelle genome itself is unique, it constitutes a repetitive sequence relative to any nonrepetitive nuclear sequence.

Chloroplast genomes are relatively large, usually ~140 kb in higher plants, and <200 kb in lower eukaryotes. This is comparable to the size of a large bacteriophage, for example, T4 at ~165 kb. There are multiple copies of the genome per organelle, typically 20–40 in a higher plant, and multiple copies of the organelle per cell, again 20–40.

Mitochondrial genomes vary in total size by more than an order of magnitude. Animal cells have small mitochondrial genomes, ~16.5 kb in mammals. There are several hundred mitochondria per cell. Each mitochondrion has multiple copies of the DNA. The total amount of mitochondrial DNA relative to nuclear DNA is small, <1%.

In yeast, the mitochondrial genome is much larger. In *S. cerevisiae,* the exact size varies among different strains, but is ~80 kb. There are ~22 mitochondria per cell, which corresponds to ~4 genomes per organelle. In growing cells, the proportion of mitochondrial DNA can be as high as 18%. (Some yeasts have much larger mitochondrial genomes.)

Plants show an extremely wide range of variation in mitochondrial DNA size, with a minimum of ~100 kb. The size of the genome makes it difficult to isolate intact, but restriction mapping in several plants suggests that the mitochondrial genome is usually a single sequence, organized as a circle. Within this circle, there are short homologous sequences. Recombination between these elements generates smaller, subgenomic circular molecules that coexist with the complete, "master" genome, explaining the apparent complexity of plant mitochondrial DNAs.

With mitochondrial genomes sequenced from many organisms, we can now see some general patterns in the representation of functions in mitochondrial DNA.

The major rRNAs are always coded by the mitochondrial genome. In some cases the mitochondrial genome codes a full complement of tRNAs; but in others (including animals and algae) the majority of tRNAs must be imported. Ribosomal proteins are not coded within the mammalian mitochondria; there is one exceptional r-protein coded in yeast, but many r-proteins are coded in higher plant mitochondria.

A major part of the protein-coding capacity of the mitochondrion is devoted to proteins of the respiration complexes. In plant mitochondria there are also genes coding for proteins concerned with cytochrome *c* assembly.

The total number of protein-coding genes is rather small, but does not correlate with the size of the genome: mammalian mitochondria use their 16.5 kb genomes to code for 13 proteins, whereas yeast mitochondria use their 60–80 kb genomes to code for as few as 8 proteins. Plants, with much larger mitochondrial genomes, code for more proteins. Introns are found in most mitochondrial genomes, although not in the very small mammalian genomes.

Are mitochondrial and chloroplast genomes unique or are some sequences shared between organelles or by an organelle and the nucleus? Plant mitochondria often have sequences that appear to have been acquired from chloroplast DNA. In some organisms, mitochondrial sequences have homologous regions in the nucleus. Exchanges of DNA between organelles or with the nucleus undoubtedly are rare, but occur in the course of evolution. The mechanism is unknown.

How did a situation evolve in which an organelle contains genetic information for some of its functions, while others are coded in the nucleus? Suppose that these organelles originated in **endosymbiotic** events, in which primitive cells captured bacteria that provided the functions that evolved into mitochondria and chloroplasts. At this point, the proto-organelle must have contained all of the genes needed to specify its functions. At some later time, some or most of the organelle genes must have been transferred to the nucleus. Perhaps this occurred at a period when compartments were less rigidly defined.

Sequence homologies suggest that both mitochondria and chloroplasts have evolved separately from lineages that are common to eubacteria, with mitochondria sharing an origin with α-purple bacteria, and chloroplasts sharing an origin with cyanobacteria. The closest known relative of mitochondria among the bacteria is *Rickettsia* (the causative agent of typhus), which is an obligate intracellular parasite that is probably descended from free-living bacteria. This reinforces the idea that mitochondria originated in an endosymbiotic event (involving a common ancestor with *Rickettsia*) and have since lost many of their original genes.

Mitochondrial DNA codes for few proteins

ANIMAL mitochondrial DNA is extremely compact. There are extensive differences in the detailed gene organization found in different animal phyla, but the general principle is maintained of a small genome coding for a restricted number of functions. In mammalian mitochondrial genomes, the organization is extremely compact. There are no introns, some genes actually overlap, and almost every single base pair can be assigned to a gene. With the exception of the D loop, a region concerned with the initiation of DNA replication, no more than 87 of the 16,569 bp of the human mitochondrial genome can be regarded as lying in intercistronic regions.

The complete nucleotide sequences of the human and murine mitochondrial genomes show extensive homology in organization. The map of the human genome is summarized in **Figure 3.13.** There are 13 protein-coding regions. All of the proteins are components of the apparatus concerned with respiration. These include cytochrome *b*, three subunits of cytochrome oxidase, one of the subunits of ATPase, and seven subunits (or associated proteins) of NADH dehydrogenase.

The fivefold discrepancy in size between the *S. cerevisiae* (84 kb) and mammalian (16 kb) mitochondrial genomes alone alerts us to the fact that there must be a great difference in their genetic organization in spite of their common function. The number of endogenously synthesized products concerned with mitochondrial enzymatic functions appears to be similar. Does the additional genetic material in yeast mitochondria represent other proteins, perhaps concerned with regulation, or is it unexpressed?

The map shown in **Figure 3.14** accounts for the major RNA and protein products of the yeast mitochondrion. The most notable feature is the dispersion of loci on the map.

The two most prominent loci are the interrupted genes *box* (coding for cytochrome *b*) and *oxi3* (coding for subunit 1 of cytochrome oxidase). Together these two genes are almost as long as the entire mitochondrial genome in mammals! Many of the long introns in these genes have open reading frames in register with the preceding exon. (We discuss this situation in Chapter 23.) This adds several proteins, all synthesized in low amounts, to the complement of the yeast mitochondrion.

The remaining genes appear to be uninterrupted. They correspond to the other two subunits of cytochrome oxidase coded by the mitochondrion, to the subunit(s) of the ATPase, and (in the case of *var1*) to a mitochondrial ribosomal protein. The total number of yeast mitochondrial genes is unlikely to exceed ~25.

Figure 3.13 Human mitochondrial DNA has 22 tRNA genes, 2 rRNA genes, and 13 protein-coding regions. 14 of the 15 protein-coding or rRNA-coding regions are transcribed in the same direction. 14 of the tRNA genes are expressed in the clockwise direction and 8 are read counter clockwise.

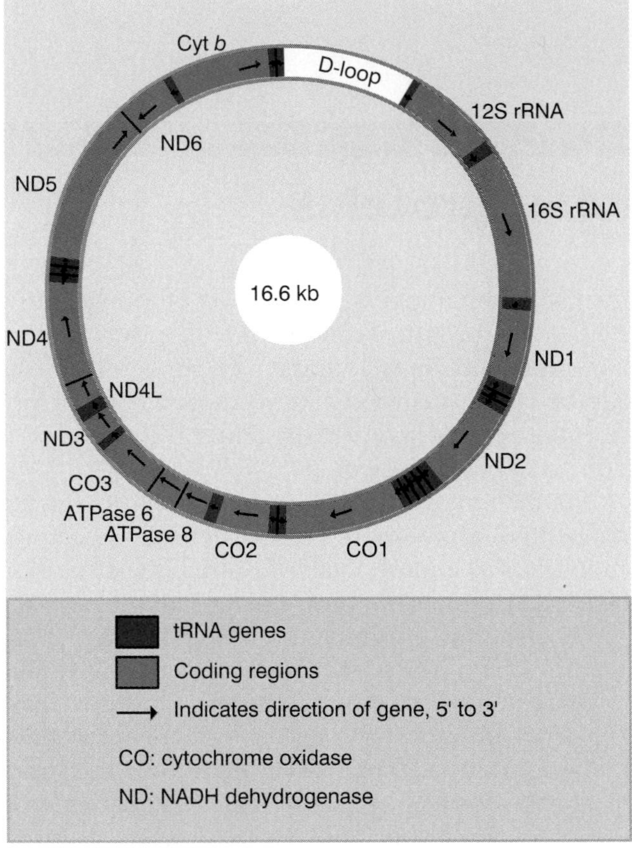

Figure 3.14 The mitochondrial genome of *S. cerevisiae* contains both interrupted and uninterrupted protein-coding genes, rRNA genes, and tRNA genes (positions not indicated). Arrows indicate direction of transcription.

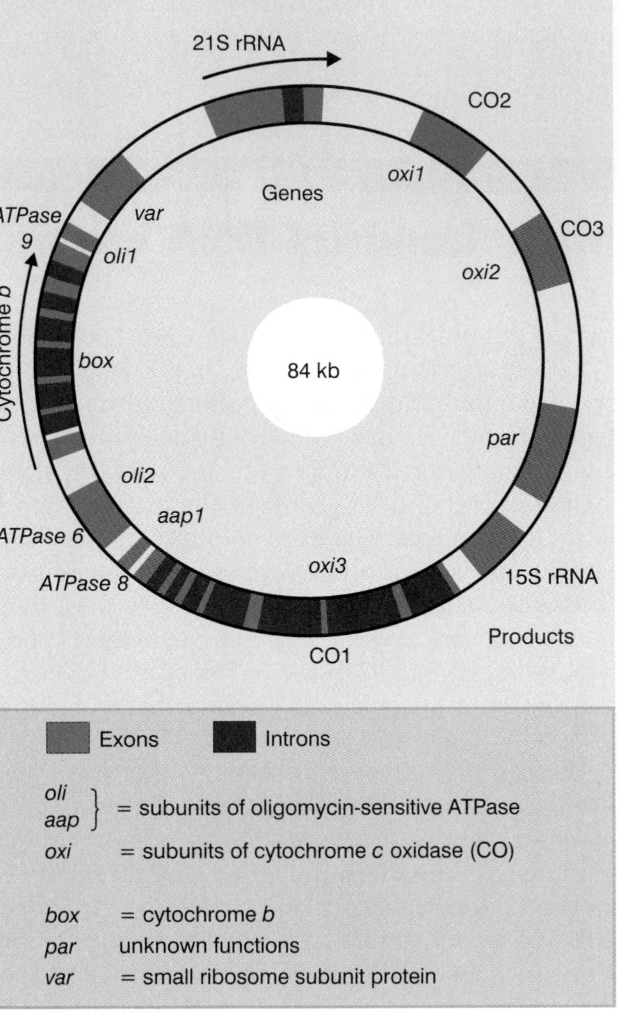

The chloroplast genome codes for ~100 proteins and RNAs

WHAT genes are carried by chloroplasts? **Figure 3.15** summarizes a situation generally similar to that of mitochondria, except that more genes are involved. The chloroplast genome codes for all the rRNA and tRNA species needed for protein synthesis. The ribosome includes two small rRNAs in addition to the major species. The tRNA set resembles that of mitochondria in including fewer species than would suffice in the cytoplasm. The chloroplast genome codes for ~50 proteins, including RNA polymerase and some ribosomal proteins. Again the rule is that organelle genes are transcribed and translated by the apparatus of the organelle.

Figure 3.15 The chloroplast genome codes for 4 rRNAs, 30 tRNAs, and ~50 proteins.

Genes	Types
RNA-coding	
16S rRNA	1
23S rRNA	1
4.5S rRNA	1
5S rRNA	1
tRNA	30
Gene Expression	
r-proteins	19
RNA polymerase	3
Others	2
Thylakoid Membranes	
Photosystem I	2
Photosystem II	7
Cytochrome *b*/*f*	3
H^+-ATPase	6
Others	
NADH dehydrogenase	6
Ferredoxin	3
Ribulose BP Cblase	1
Unidentified	29
Total	110

The complete sequence of chloroplast DNA has been determined for a liverwort (a moss) and for tobacco. In spite of a considerable difference in overall length, between 121 kb and 155 kb, the gene organization is similar, and the overall number of genes almost identical. Most of their products are components of the thylakoid membranes or concerned with redox reactions, as can be seen from Figure 3.15.

Introns in chloroplasts fall into two general classes. Those in tRNA genes are usually (although not inevitably) located in the anticodon loop, like the introns found in yeast nuclear tRNA genes (which we discuss in Chapter 22). Those in protein-coding genes resemble the introns of mitochondrial genes (see Chapter 23). This places the endosymbiotic event at a time in evolution before the separation of prokaryotes with uninterrupted genes.

The role of the chloroplast is to undertake photosynthesis. Many of its genes code for proteins of complexes located in the thylakoid membranes. The constitution of these complexes shows a different balance from that of mitochondrial complexes. Although some complexes are like mitochondrial complexes in having some subunits coded by the organelle genome and some by the nuclear genome, other chloroplast complexes are coded entirely by one genome.

The identified genes of the chloroplast show the focus of its activities. There are 45 genes coding for RNA, 27 coding for proteins concerned with gene expression, 18 coding for proteins of the thylakoid membrane, and another 10 representing functions concerned with electron transfer. The products of some 30 open reading frames remain to be identified.

Summary

The sequences comprising a eukaryotic genome can be classified in three groups: nonrepetitive sequences are unique; moderately repetitive sequences are dispersed and repeated a small number of times in the form of related but not identical copies; and highly repetitive sequences are short and usually repeated as a tandem array. The proportions of the types of sequence are characteristic for each genome, although larger genomes tend to have a smaller proportion of nonrepetitive DNA. The complexity of any class describes the length of unique sequences in it; the repetition frequency describes the number of times each sequence is repeated. The C-value paradox describes the discrepancy between coding potential and DNA content in eukaryotic genomes.

Most structural genes are located in nonrepetitive DNA. The complexity of nonrepetitive DNA is a better reflection of the complexity of the organism than the total genome complexity; nonrepetitive DNA reaches a maximum complexity of $\sim 2 \times 10^9$ bp.

The total number of genes is <1000 for *Mycoplasma* and intracellular parasites, 2000–4000 for bacteria, >6000 for yeast, >12,000 for insects, and >100,000 for mammals.

Genes are expressed at widely varying levels. There may be 10^5 copies of mRNA for an abundant gene whose protein is the principal product of the cell, 10^3 copies of each mRNA for <10 moderately abundant messages, and <10 copies of each mRNA for >10,000 scarcely expressed genes. Overlaps between the mRNA populations of cells of different phenotypes are extensive; the majority of mRNAs are present in most cells.

It is likely that not all genes are essential (defining lethal genes by the existence of devastating effects when they are mutated). The numbers of nonessential genes and essential genes could be comparable. In yeast, only 60% of genes appear to be essential; in *D. melanogaster*, there appear to be <5000 essential genes. We do not understand how nonessential genes are maintained; they may provide selective advantages that are not evident.

NonMendelian inheritance is explained by the presence of DNA in organelles in the cytoplasm. Mitochondria and chloroplasts both represent membrane-bounded systems in which some proteins are synthesized within the organelle, while others are imported. The organelle genome is usually a circular DNA that codes for all of the RNAs and for some of the proteins that are required.

Mitochondrial genomes vary greatly in size from the 16 kb minimalist mammalian genome to the 570 kb genome of higher plants. It is assumed that the larger genomes code for additional functions. Chloroplast genomes range from 120–200 kb. Those that have been sequenced have a similar organization and coding functions. In both mitochondria and chloroplasts, many of the major proteins contain some subunits synthesized in the organelle and some subunits imported from the cytosol.

Mammalian mtDNAs are transcribed into a single transcript from the major coding strand, and individual products are generated by RNA processing. Rearrangements occur in mitochondrial DNA rather frequently in yeast; and recombination between mitochondrial or between chloroplast genomes has been found. There are some tantalizing homologies between mitochondrial and chloroplast genomes.

Further reading

Reviews

Britten, R. J. and Kohne, D. E. (1968). Repeated sequences in DNA. *Science* **161**, 529–540.

Britten, R. J. and Davidson, E. H. (1971). Repetitive and nonrepetitive DNA sequences and a speculation on the origins of evolutionary novelty. *Quart. Rev. Biol.* **46**, 111–133.

Davidson, E. H. and Britten, R. J. (1973). Organization, transcription, and regulation in the animal genome. *Quart. Rev. Biol.* **48**, 565–613.

Lewin, B. (1975). Units of transcription and translation: sequence components of hnRNA and mRNA. *Cell* **4**, 77–93.

Clayton, D. A. (1984). Transcription of the mammalian mitochondrial genome. *Ann. Rev. Biochem.* **53**, 573–594.

Attardi, G. (1985). Animal mitochondrial DNA: an extreme example of economy. *Int. Rev. Cytol.* **93**, 93–146.

Palmer, J. D. (1985). Comparative organization of chloroplast genomes. *Ann. Rev. Genet.* **19**, 324–354.

Gray, M. W. (1989). Origin and evolution of mitochondrial DNA. *Ann. Rev. Cell Biol.* **5**, 25–50.

Discoveries

The number of expressed genes

Hastie, N. B. and Bishop, J. O. (1976). The expression of three abundance classes of mRNA in mouse tissues. *Cell* **9**, 761–774.

Velculescu, V. E. *et al.* (1997). Characterization of the yeast transcriptosome. *Cell* **88**, 243–251.

Holstege, F. C. P. *et al.* (1998). Dissecting the regulatory circuitry of a eukaryotic genome. *Cell* **95**, 717–728.

Mikos, G. L. G. and Rubin, G. M. (1996). The role of the genome project in determining gene function: insights from model organisms. *Cell* **86**, 521–529.

Essential genes

Judd, N. H., Shen, M. W., and Kaufman, T. C. (1972). The anatomy and function of a segment of the X chromosome of *D. melanogaster*. *Genetics* **71**, 139–156.

Goebl, M. G. and Petes, T. D. (1986). Most of the yeast genomic sequences are not essential for cell growth and division. *Cell* **46**, 983–992.

Sequencing

Oliver, S. G. *et al.* (1992). The complete DNA sequence of yeast chromosome III. *Nature* **357,** 38–46.

Dujon, B. *et al.* (1994). Complete DNA sequence of yeast chromosome XI. *Nature* **369,** 371–378.

Johnston, M. *et al.* (1994). Complete nucleotide sequence of *Saccharomyces cerevisiae* chromosome VIII. *Science* **265,** 2077–2082.

Wilson, R. *et al.* (1994). 2.2 Mb of contiguous nucleotide sequence from chromosome III of *C. elegans. Nature* **368,** 32–38.

Blattner, F. *et al.* (1997). The complete genome sequence of *E. coli K12. Science* **277,** 1453–1462.

C. elegans sequencing consortium (1998). Genome sequence of the nematode *C. elegans*: a platform for investigating biology. *Science* **282,** 2012–2022.

Andersson, S. G. E. *et al.* (1998). The genome sequence of *Rickettsia prowazekii* and the origin of mitochondria. *Nature* **396,** 133–140.

Velculescu, V. E. *et al.* (1997). Characterization of the yeast transcriptosome. *Cell* **88,** 243–251.

Holstege, F. C. P. *et al.* (1999). Dissecting the regulatory circuitry of a eukaryotic genome. Cell, 95, 717–728.

Organelle genomes

Anderson, S. *et al.* (1981). Sequence and organization of the human mitochondrial genome. *Nature* **290,** 457–465.

Ohyama, K. *et al.* (1986). Chloroplast gene organization deduced from complete sequence of liverwort *M. polymorpha* chloroplast DNA. *Nature* **322,** 572–574.

Shinozaki, K. *et al.* (1986). The complete nucleotide sequence of the tobacco chloroplast genome: its gene organization and expression. *EMBO J.* **5,** 2043–2049.

Cann, R. L., Stoneking, M., and Wilson, A. C. (1987). Mitochondrial DNA and human evolution. *Nature* **324,** 31–36.

Chapter 4

Clusters and repeats

DUPLICATION of DNA is a major force in evolution. Tandem duplication (when the duplicates remain together) may arise through errors in replication or recombination. Separation of the duplicates can occur by a chromosomal translocation. A duplicate at a new location may also be produced directly by a transposition event that is associated with the duplication of a region of DNA from the vicinity of the transposon. Duplications may apply either to intact genes or to collections of exons or even individual exons. When an intact gene is involved, the act of duplication generates two copies of a gene whose activities are indistinguishable, but then usually the copies diverge as each accumulates different mutations.

A set of genes descended by duplication and variation from some ancestral gene is called a **gene family.** Its members may be clustered together or dispersed on different chromosomes (or a combination of both). The members of a structural gene family usually have related or even identical functions, although they may be expressed at different times or in different cell types. So different globin proteins are expressed in embryonic and adult red blood cells, while different actins are utilized in muscle and nonmuscle cells.

Some gene families consist of identical members. Clustering is a prerequisite for maintaining identity between genes, although clustered genes are not necessarily identical. **Gene clusters** range from extremes where a duplication has generated two adjacent related genes to cases where hundreds of identical genes lie in a tandem array. Extensive tandem repetition of a gene may occur when the product is needed in unusually large amounts. Examples are the genes for rRNA or histone proteins. This creates a special situation with regard to the maintenance of identity and the effects of selective pressure. Gene clusters offer us an opportunity to examine the forces involved in evolution of the genome over larger regions than single genes.

Recombination is a key event in evolution of the genome. A population evolves by the classical recombination illustrated in Figures 1.23 and 1.24, in which an exact crossing-over occurs. The recombinant chromosomes have the same organization as the parental chromosome. They contain precisely the same loci in the same order. However, they contain different combinations of alleles, providing the raw material for natural selection.

How does the genome change its *content* of genes, as opposed to combination of alleles? One important mechanism is provided by **unequal crossing-over,** when a recombination event occurs between two sites that are *not* homologous. The feature that makes such events possible is the existence of repeated sequences. This allows one copy of a repeat in one chromosome to misalign for recombination with a different copy of the repeat in the homologous chromosome, instead of with the corresponding copy. When recombination occurs, this creates a deletion in one recombinant chromosome and a corresponding insertion in the other. This mechanism is responsible for the evolution of clusters of related sequences. We can trace its operation in expanding or contracting the size of an array in both gene clusters and regions of highly repeated DNA.

The highly repetitive fraction of the genome consists of multiple tandem copies of very short repeating units. These often have unusual properties. One is that they may be identified as a separate peak on a density gradient analysis of DNA, which gave rise to the name **satellite DNA.** They are often associated with inert regions of the chromosomes, and in particular with centromeres (which contain the points of attachment for segregation on a mitotic or meiotic spindle). Because of their repetitive organization, they show some of the same behavior with regard to evolution as the tandem gene clusters. In addition to the satellite sequences, there are shorter stretches of DNA that show similar behavior, called **minisatellites.** They are useful in showing a high degree of divergence between individual genomes that can be used for mapping purposes.

All of these events that change the constitution of the genome are rare, but they are significant over the course of evolution.

Gene clusters are formed by duplication and divergence

EXONS behave like modules for building genes that are tried out in the course of evolution in various combinations. At one extreme, an individual exon from one gene may be copied and used in another gene. At the other extreme, an entire gene, including both exons and introns, may be duplicated. In such a case, mutations can accumulate in one copy without attracting the adverse attention of natural selection. This copy may then evolve to a new function, perhaps to be expressed in a different time or place from the first copy, perhaps to acquire different activities.

The most common type of duplication event generates a second copy of the gene close to the first copy. In some cases, the copies remain associated, and further duplication may generate a cluster of related genes. The best characterized example of a gene cluster is presented by the globin genes, which constitute an ancient gene family, concerned with a function that is central to the animal kingdom: the transport of oxygen through the bloodstream.

The major constituent of the red blood cell is the globin tetramer, associated with its heme (iron-binding) group in the form of hemoglobin. Functional globin genes in all species have the same general structure, divided into three exons as shown previously in Figure 2.13. We conclude that all globin genes are derived from a single ancestral gene; so by tracing the development of individual globin genes within and between species, we may learn about the mechanisms involved in the evolution of gene families.

In adult cells, the globin tetramer consists of two identical α chains and two identical β chains. Embryonic blood cells contain hemoglobin tetramers that are different from the adult form. Each tetramer contains two identical α-like chains and two identical β-like chains, each of which is related to the adult polypeptide and is later replaced by it. This is an example of developmental control, in which different genes are successively switched on and off to provide alternative products that fulfill the same function at different times.

The details of the relationship between embryonic and adult hemoglobins vary with the organism. The human pathway has three stages: embryonic, fetal, and adult. The distinction between embryonic and adult is common to mammals, but the number of preadult stages varies. In man, zeta and alpha are the two α-like chains. Epsilon, gamma, delta, and beta are the β-like chains. The chains are expressed at different stages of development.

ζ is the first α-like chain to be expressed, but is soon replaced by α. In the β-pathway, ε and γ are expressed first, with δ and β replacing them later. In adults, the $\alpha_2\beta_2$ form provides 97% of the hemoglobin, $\alpha_2\delta_2$ is ~2%, and ~1% is provided by persistence of the fetal form $\alpha_2\gamma_2$.

The division of globin chains into α-like and β-like reflects the organization of the genes. Each type of globin is coded by genes organized into a single cluster. The structures of the two clusters in the higher primate genome are illustrated in **Figure 4.1.**

Stretching over 50 kb, the β cluster contains five

Figure 4.1 Each of the α-like and β-like globin gene families is organized into a single cluster that includes functional genes and pseudogenes (ψ). The organization of the clusters in higher primates is conserved. All of the active genes are transcribed from left to right.

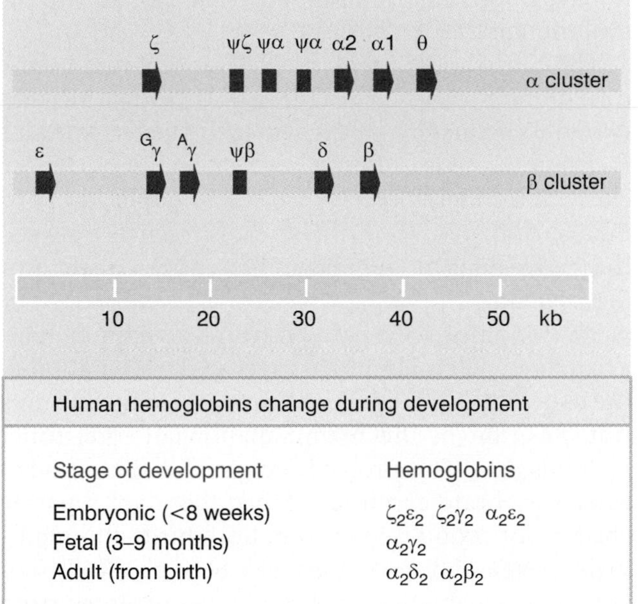

Human hemoglobins change during development

Stage of development	Hemoglobins
Embryonic (<8 weeks)	$\zeta_2\varepsilon_2$ $\zeta_2\gamma_2$ $\alpha_2\varepsilon_2$
Fetal (3–9 months)	$\alpha_2\gamma_2$
Adult (from birth)	$\alpha_2\delta_2$ $\alpha_2\beta_2$

Figure 4.2 Clusters of β-globin genes and pseudogenes are found in vertebrates. Seven mouse genes include two early embryonic, one late embryonic, two adult genes, and two pseudogenes. Rabbit and chick each have four genes.

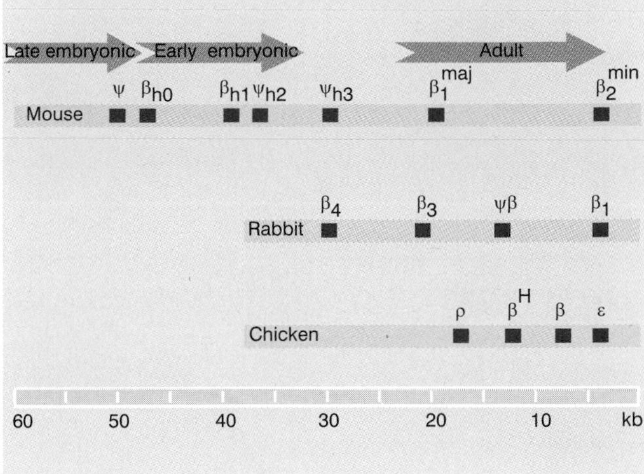

functional genes (ε, two γ, δ, and β) and one nonfunctional gene (ψβ). The two γ genes differ in their coding sequence in only one amino acid; the G variant has glycine at position 136, where the A variant has alanine.

The more compact α cluster extends over 28 kb and includes one active ζ gene, one ζ nonfunctional gene, two α genes, two α nonfunctional genes, and the θ gene of unknown function. The two α genes code for the same protein. Two (or more) identical genes present on the same chromosome are described as **nonallelic** copies.

Functional genes are defined by their expression in RNA, and ultimately by the proteins for which they code. Nonfunctional genes are defined as such by their inability to code for proteins; the reasons for inactivity vary, and the deficiencies may be in transcription or translation (or both). They are called **pseudogenes** and given the symbol ψ.

A similar general organization is found in other vertebrate globin gene clusters, but details of the types, numbers, and order of genes all vary. A location that has a pseudogene in one species may have an active gene in another; for example, ψβ of the higher primates lies at a position equivalent to an active embryonic gene in goat. Some examples of β-globin clusters are illustrated in **Figure 4.2.**

The characterization of these gene clusters makes an important general point. *There may be more members of a gene family, both functional and nonfunctional, than we would suspect on the basis of protein analysis.* The extra functional genes may represent duplicates that code for identical polypeptides; or they may be related to known proteins, although different from them (and presumably expressed only briefly or in low amounts).

With regard to the question of how much DNA is needed to code for a particular function, we see that coding for the β-like globins requires a range of 20–50 kb in different mammals. This is much greater than we would expect just from scrutinizing the known β-globin proteins or even considering the individual genes. The region of the β cluster appears to code only for the β-globins, but until more genes or gene clusters coding for particular proteins have been identified, we shall not be able to tell how often this type of arrangement occurs.

From the organization of globin genes in a variety of species, we should be able to trace the evolution of present globin gene clusters from a single ancestral globin gene. Our present view of the evolutionary descent is pictured in **Figure 4.3.**

The leghemoglobin gene of plants, which is related to the globin genes, may represent the ancestral form. The furthest back that we can trace a globin gene in modern form is provided by the sequence of the single chain of mammalian myoglobin, which diverged from the globin line of descent ~800 million years ago. The myoglobin gene has the same organization as globin

Figure 4.3 All globin genes have evolved by a series of duplications, transpositions, and mutations from a single ancestral gene.

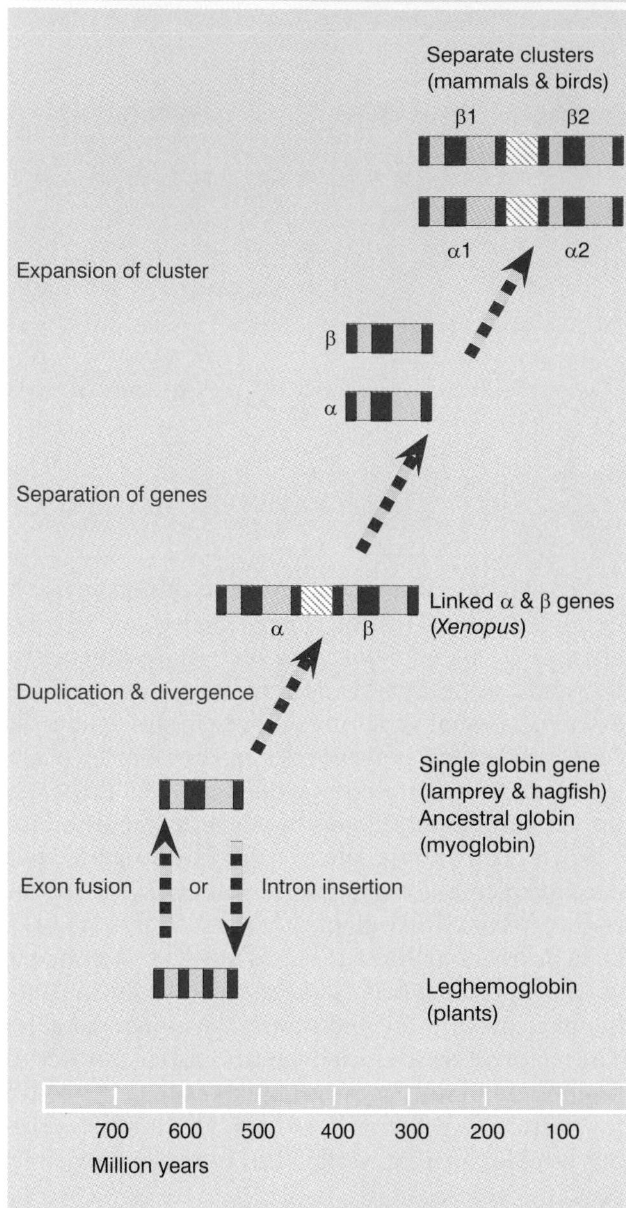

genes, so we may take the three-exon structure to represent their common ancestor.

Some "primitive fish" have only a single type of globin chain, so they must have diverged from the line of evolution before the ancestral globin gene was duplicated to give rise to the α and β variants. This appears to have occurred ~500 million years ago, during the evolution of the bony fish.

The next stage of evolution is represented by the state of the globin genes in the frog *X. laevis,* which has two globin clusters. However, each cluster contains *both* α and β genes, of both larval and adult types. The cluster must therefore have evolved by duplication of a linked α–β pair, followed by divergence between the individual copies. Later the entire cluster was duplicated.

The amphibians separated from the mammalian/avian line ~350 million years ago, so the separation of the α- and β-globin genes must have resulted from a transposition in the mammalian/avian forerunner after this time. This probably occurred in the period of early vertebrate evolution. Since there are separate clusters for α and β globins in both birds and mammals, the α and β genes must have been physically separated before the mammals and birds diverged from their common ancestor, an event that occurred probably ~270 million years ago.

Changes have occurred within the separate α and β clusters in more recent times, as we see from the description of the divergence of the individual genes in the next section.

Sequence divergence is the basis for the evolutionary clock

MOST changes in protein sequences occur by small mutations that accumulate slowly with time. Point mutations and small insertions and deletions occur by chance, probably with more or less equal probability in all regions of the genome, except for hotspots at which mutations occur much more frequently. Most mutations that change the amino acid sequence are deleterious and will be eliminated by natural selection.

Few mutations will be advantageous, but those that are may spread through the population, eventually replacing the former sequence. When a new variant replaces the previous version of the gene, it is said to have become **fixed** in the population.

A contentious issue is what proportion of mutational changes in an amino acid sequence are **neutral,** that is, without any effect on the function of the protein, and able therefore to accrue as the result of **random drift and fixation.**

The rate at which mutational changes accumulate is a characteristic of each protein, presumably depending at least in part on its flexibility with regard to change. Within a species, a protein evolves by mutational substitution, followed by elimination or fixation within the single breeding pool. Remember that when we scrutinize the gene pool of a species, we see only the variants that have survived. When multiple variants are present, they may be stable (because neither has any selective advantage) or one may in fact be transient because it is in process of being displaced.

When a species separates into two new species, each now constitutes an independent pool for evolution. By comparing the corresponding proteins in two species, we see the differences that have accumulated between them *since the time when their ancestors ceased to interbreed.* Some proteins are highly conserved, showing little or no change from species to species. This indicates that almost any change is deleterious and therefore is selected against.

The difference between two proteins is expressed as their **divergence,** the percent of positions at which the amino acids are different. The divergence between proteins can be different from that between the corresponding nucleic acid sequences. The source of this difference is the representation of each amino acid in a three-base codon, in which often the third base has no effect on the meaning.

We may divide the nucleotide sequence of a coding region into potential **replacement sites** and **silent sites:**

- At replacement sites, a mutation alters the amino acid that is coded. The effect of the mutation (deleterious, neutral, or advantageous) depends on the result of the amino acid replacement.

- At silent sites, mutation only substitutes one synonym codon for another, so there is no change in the protein. Usually the replacement sites account for 75% of a coding sequence and the silent sites provide 25%.

In addition to the coding sequence, a gene contains nontranslated regions. Here again, mutations are potentially neutral, apart from their effects on either secondary structure or (usually rather short) regulatory signals.

Although silent mutations are neutral with regard to the protein, they could affect gene expression via the sequence change in RNA. For example, a change in secondary structure might influence transcription, processing, or translation. Another possibility is that a change in synonym codons calls for a different tRNA to respond, influencing the efficiency of translation.

The mutations in replacement sites should correspond with the amino acid divergence (determined by the percent of changes in the protein sequence). A nucleic acid divergence of 0.45% at replacement sites corresponds to an amino acid divergence of 1% (assuming that the average number of replacement sites per codon is 2.25). Actually, the measured divergence underestimates the differences that have occurred during evolution, because of the occurrence of multiple events at one codon. Usually a correction is made for this.

To take the example of the human β- and δ-globin chains, there are 10 differences in 146 residues, a divergence of 6.9%. The DNA sequence has 31 changes in 441 residues. However, these changes are distributed very differently in the replacement and silent sites. There are 11 changes in the 330 replacement sites, but 20 changes in only 111 silent sites. This gives (corrected) rates of divergence of 3.7% in the replacement sites and 32% in the silent sites, almost an order of magnitude in difference.

The striking difference in the divergence of replacement and silent sites demonstrates the existence of much greater constraints on nucleotide positions that influence protein constitution relative to those that do not. So probably very few of the amino acid changes are neutral.

Suppose we take the rate of mutation at silent sites to indicate the underlying rate of mutational fixation (this assumes that there is no selection at all at the silent sites). Then over the period since the β and δ genes diverged, there should have been changes at 32% of the 330 replacement sites, a total of 105. All but 11 of them have been eliminated, which means that ~90% of the mutations did not survive.

The divergence between any pair of globin sequences is (more or less) proportional to the time since they separated. This provides an **evolutionary clock** that measures the accumulation of mutations at an apparently even rate during the evolution of a given protein.

The rate of divergence can be measured as the percent difference per million years, or as its reciprocal, the unit evolutionary period (UEP), the time in millions of years that it takes for 1% divergence to develop. Once the clock has been established by pairwise comparisons between species (remembering the practical difficulties in establishing the actual time of speciation), it can be applied to related genes *within* a species. From their divergence, we can calculate how much time has passed since the duplication that generated them.

By comparing the sequences of homologous genes in different species, the rate of divergence at both replacement and silent sites can be determined, as plotted in **Figure 4.4.**

Figure 4.4 Divergence of DNA sequences depends on evolutionary separation. Each point on the graph represents a pairwise comparison.

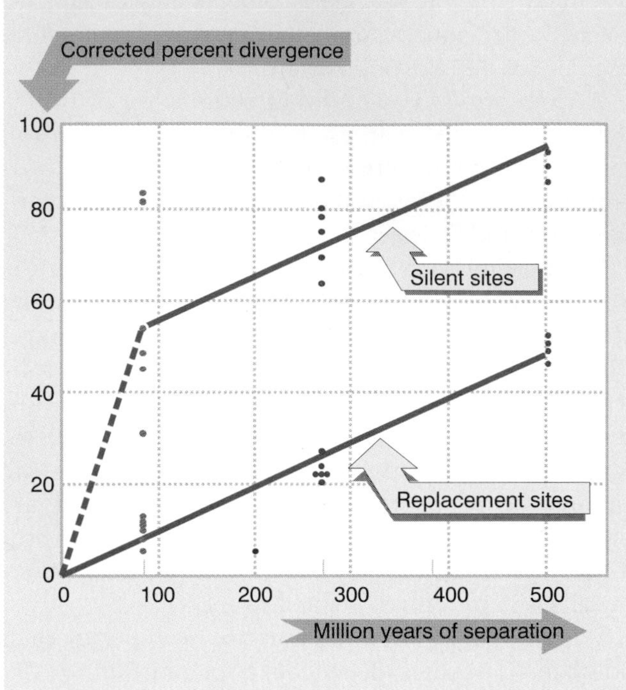

In pairwise comparisons, there is an average divergence of 10% in the replacement sites of either the α- or β-globin genes of mammals that have been separated since the mammalian radiation occurred ~85 million years ago. This corresponds to a replacement divergence rate of 0.12% per million years.

The rate is steady when the comparison is extended to genes that diverged in the more distant past. For example, the average replacement divergence between corresponding mammalian and chicken globin genes is 23%. Relative to a separation ~270 million years ago, this gives a rate of 0.09% per million years.

Going further back, we can compare the α- with the β-globin genes within a species. They have been diverging since the individual gene types separated ~500 million years ago (see Figure 4.3). They have an average replacement divergence of ~50%, which gives a rate of 0.1% per million years.

The summary of these data in Figure 4.4 shows that replacement divergence in the globin genes has an average rate of ~0.096% per million years (or a UEP of 10.4). Considering the uncertainties in estimating the times at which the species diverged, the results lend good support to the idea that there is a linear clock.

The data on silent site divergence are much less clear. In every case, it is evident that the silent site divergence is much greater than the replacement site divergence, by a factor that varies from 2 to 10. But the spread of silent site divergences in pairwise comparisons is too great to show whether a clock is applicable (so we must base temporal comparisons on the replacement sites).

From Figure 4.4, it is clear that the rate at silent sites is not linear with regard to time. *If we assume that there must be zero divergence at zero years of separation,* we see that the rate of silent site divergence is much greater for the first ~100 million years of separation. One interpretation is that a fraction of roughly half of the silent sites is rapidly (within 100 million years) saturated by mutations; this fraction behaves as neutral sites. The other fraction accumulates mutations more slowly, at a rate approximately the same as that of the replacement sites; this fraction identifies sites that are silent with regard to the protein, but that come under selective pressure for some other reason.

Now we can reverse the calculation of divergence rates to estimate the times since genes within a species have been apart. The difference between the human β and δ genes is 3.7% for replacement sites. At a UEP of 10.4, these genes must have diverged $10.4 \times 3.7 \approx 40$ million years ago—about the time of the separation of the lines leading to New World monkeys, Old World monkeys, great apes, and man. All of these higher primates have both β and δ genes, which suggests that the gene divergence commenced just before this point in evolution.

Proceeding further back, the divergence between the replacement sites of γ and ε genes is 10%, which corresponds to a time of separation ~100 million years ago. The separation between embryonic and fetal globin genes therefore may have just preceded or accompanied the mammalian radiation.

An evolutionary tree for the human globin genes is constructed in **Figure 4.5.** Features that evolved before the mammalian radiation—such as the separation of β/δ from γ—should be found in all mammals. Features that evolved afterward—such as the separation of β- and δ-globin genes—should be found in individual lines of mammals.

In each species, there have been comparatively recent changes in the structures of the clusters, since we see differences in gene number (one adult β-globin gene in man, two in mouse) or in type (we are not yet sure whether there are separate embryonic and fetal β-like globins in rabbit and mouse).

When sufficient data have been collected on the sequences of a particular gene, the arguments can be reversed, and comparisons between genes in different species can be used to assess taxonomic relationships.

Figure 4.5 Replacement site divergences between pairs of β-globin genes allow the history of the human cluster to be reconstructed. This tree accounts for the separation of classes of globin genes. Duplications of individual genes are of unknown origin. The time of the α–ζ divergence is not known.

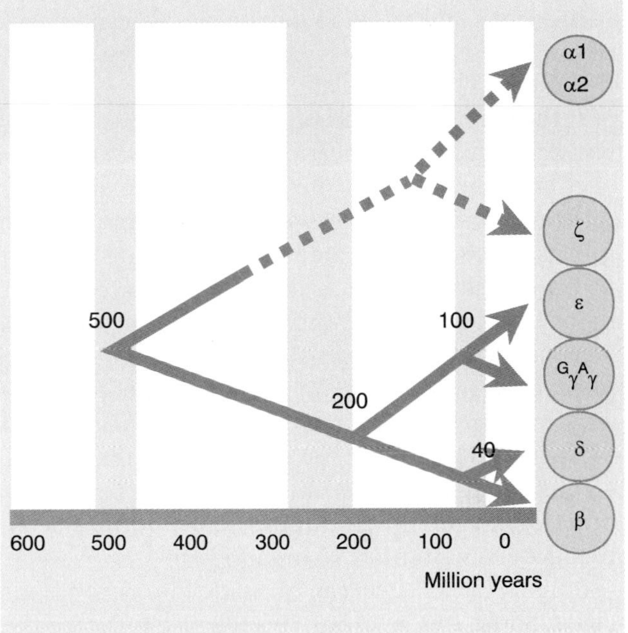

Pseudogenes are dead ends of evolution

PSEUDOGENES (ψ) are defined by their possession of sequences that are related to those of the functional genes, but that cannot be translated into a functional protein.

Some pseudogenes have the same general structure as functional genes, with sequences corresponding to exons and introns in the usual locations. They are rendered inactive by mutations that prevent any or all of the stages of gene expression. The changes can take the form of abolishing the signals for initiating transcription, preventing splicing at the exon-intron junctions, or prematurely terminating translation.

Usually a pseudogene has several deleterious mutations. Presumably once it ceased to be active, there was no impediment to the accumulation of further mutations. Pseudogenes that represent inactive versions of currently active genes have been found in many systems, including globin, immunoglobulins, and histocompatibility antigens, where they are located in the vicinity of the gene cluster, often interspersed with the active genes.

A typical example is the rabbit pseudogene, ψβ2, which has the usual organization of exons and introns, and is related most closely to the functional globin gene β1. But the deletion of a base pair at codon 20 of ψβ2 has caused a frameshift that would lead to termination shortly after. Several point mutations have changed later codons representing amino acids that are highly conserved in the β globins. Neither of the two introns any longer possesses recognizable boundaries with the exons, so probably the introns could not be spliced out even if the gene were transcribed. However, there are no transcripts corresponding to the gene, possibly because there have been changes in the 5′ flanking region.

Since this list of defects includes mutations potentially preventing each stage of gene expression, we have no means of telling which event originally inactivated this gene. However, from the divergence between the pseudogene and the functional gene, we can estimate when the pseudogene originated and when its mutations started to accumulate.

If the pseudogene had become inactive as soon as it was generated by duplication from β1, we should expect both replacement site and silent site divergence rates to be the same. (They will be different only if the gene is translated to create selective pressure on the replacement sites.) But actually there are fewer replacement site substitutions than silent site substitutions. This suggests that at first (while the gene was expressed) there was selection against replacement site substitution. From the relative extents of substitution in the two types of site, we can calculate that ψβ2 diverged from β1 ~55 million years ago, remained a functional gene for 22 million years, but has been a pseudogene for the last 33 million years.

Similar calculations can be made for other pseudogenes. Some appear to have been active for some time before becoming pseudogenes, but others appear to have been inactive from the very time of their original generation. The general point made by the structures of these pseudogenes is that each has evolved independently during the development of the globin gene cluster in each species. This reinforces the conclusion that the creation of new genes, followed by their acceptance as functional duplicates, variation to become new functional genes, or inactivation as pseudogenes, is a continuing process in the gene cluster.

The mouse ψα3 globin gene has an interesting property: it precisely lacks both introns. Its sequence can be aligned (allowing for accumulated mutations) with the α-globin mRNA. The apparent time of inactivation coincides with the original duplication, which suggests that the original inactivating event was associated with the loss of introns.

Inactive genomic sequences that resemble the RNA transcript are called **processed pseudogenes.** They originate by insertion at some random site of a product derived from the RNA, following a retrotransposition event, as discussed in Chapter 16. Their characteristic features are summarized in Figure 16.18.

How common are pseudogenes? Most gene families have members that are pseudogenes. Usually the pseudogenes represent a small minority of the total gene number. In an exceptional case, however, there is one active gene coding for a mouse ribosomal protein; and it has ~15 processed pseudogene relatives. This type of effect must be taken into account when we try to calculate the number of genes from hybridization data.

If pseudogenes are evolutionary dead ends, simply an unwanted accompaniment to the rearrangement of functional genes, why are they still present in the genome? Do they fulfill any function or are they entirely without purpose, in which case there should be no selective pressure for their retention?

We should remember that we see those genes that have survived in present populations. In past times, any number of other pseudogenes may have been eliminated. This elimination could occur by deletion of the sequence as a sudden event or by the accretion of mutations to the point where the pseudogene can no longer be recognized as a member of its original sequence family (probably the ultimate fate of any pseudogene that is not suddenly eliminated).

Even relics of evolution can be duplicated. In the β-globin genes of the goat, there are two adult species, β^A and β^C. Each of these has a pseudogene a few kb upstream of it (called $\psi\beta^Z$ and $\psi\beta^X$, respectively). The two pseudogenes are better related to each other than to the adult β-globin genes; in particular, they share several inactivating mutations. Also, the two adult β-globin genes are better related to each other than to the pseudogenes. This implies that an original ψβ-β structure was itself duplicated, giving two functional β genes (which diverged further into the β^A and β^C genes) and two nonfunctional genes (which diverged into the current pseudogenes).

The mechanisms responsible for gene duplication, deletion, and rearrangement act on all sequences that are recognized as members of the cluster, whether or not they are functional. It is left to selection to discriminate among the products.

Unequal crossing-over rearranges gene clusters

THERE are frequent opportunities for rearrangement in a cluster of related or identical genes. We can see the results by comparing the mammalian β clusters included in Figure 4.2. Although the clusters serve the same function, and all have the same general organization, each is different in size, there is variation in the total number and types of β-globin genes, and the numbers and structures of pseudogenes are different. All of these changes must have occurred since the mammalian radiation, ~85 million years ago (the last point in evolution common to all the mammals).

The comparison makes the general point that gene duplication, rearrangement, and variation is as important a factor in evolution as the slow accumulation of point mutations in individual genes. What types of mechanisms are responsible for gene reorganization?

A gene cluster can expand or contract by **unequal crossing-over,** when recombination occurs between nonallelic genes, as illustrated in **Figure 4.6.** Usually, recombination involves corresponding sequences of DNA held in exact alignment between the two homologous chromosomes. However, when there are two copies of a gene on each chromosome, an occasional misalignment allows pairing between them. (This requires some of the adjacent regions to go unpaired.)

When a recombination event occurs between the mispaired gene copies, it generates **nonreciprocal recombinant chromosomes,** one of which has a duplication of the gene and the other a deletion. The first recombinant therefore has an increase in the number of gene copies from two to three, while the second has a decrease from two to one.

In this example, we have treated the noncorresponding gene copies 1 and 2 as though they were entirely homologous. However, unequal crossing-over also can occur when the adjacent genes are well related (although the probability is less than when they are identical).

An obstacle to unequal crossing-over is presented by the interrupted structure of the genes. In a case such as the globins, the corresponding exons of adjacent gene copies are likely to be well enough related to support pairing; but the sequences of the introns have diverged appreciably. The restriction of pairing to the exons considerably reduces the continuous length of DNA that can be involved. This lowers the chance of unequal crossing-over. So divergence between introns could enhance the stability of gene clusters by hindering the occurrence of unequal crossing-over.

Thalassemias result from mutations that reduce or prevent synthesis of either α or β globin. The occurrence of unequal crossing-over in the human globin gene clusters is revealed by the nature of certain thalassemias.

Many of the most severe thalassemias result from deletions of part of a cluster. In at least some cases, the ends of the deletion lie in regions that are homologous, which is exactly what would be expected if it had been generated by unequal crossing-over.

Figure 4.7 summarizes the deletions that cause the α-thalassemias. α-thal-1 deletions are long, varying in the location of the left end, with the positions of the right ends located beyond the known genes. They eliminate both the α genes. The α-thal-2 deletions are short and eliminate only one of the two α genes. The L form removes 4.2 kb of DNA, including the α2 gene. It probably results from unequal crossing-over, because the ends of the deletion lie in homologous regions, just to the right of the ψα and α2 genes, respectively. The R form results from the removal of exactly 3.7 kb of DNA, the precise distance between the α1 and α2 genes. It appears to have been generated by unequal crossing-over between the α1 and α2 genes themselves. This is precisely the situation depicted in Figure 4.6.

Depending on the diploid combination of thalassemic chromosomes, an affected individual may have any number of α chains from zero to three. There are few differences from the wild type (four α genes) in individuals with three or two α genes. With only one α gene, the excess β chains form the unusual tetramer β_4, which causes **HbH disease.** The complete absence of α genes results in **hydrops fetalis,** which is fatal at or before birth.

The same unequal crossing-over that generated the thalassemic chromosome should also have generated a chromosome with three α genes. Individuals with such chromosomes have been identified in several populations. In some populations, the frequency of the triple α locus is about the same as that of the single α locus; in others, the triple α genes are much *less* common than single α genes. This suggests that (unknown) selective factors operate in different populations to adjust the gene levels.

Variations in the number of α genes are found rela-

Figure 4.6 Gene number can be changed by unequal crossing-over. If gene 1 of one chromosome pairs with gene 2 of the other chromosome, the other gene copies are excluded from pairing, as indicated by the extruded loops. Recombination between the mispaired genes produces one chromosome with a single (recombinant) copy of the gene and one chromosome with three copies of the gene (one from each parent and one recombinant).

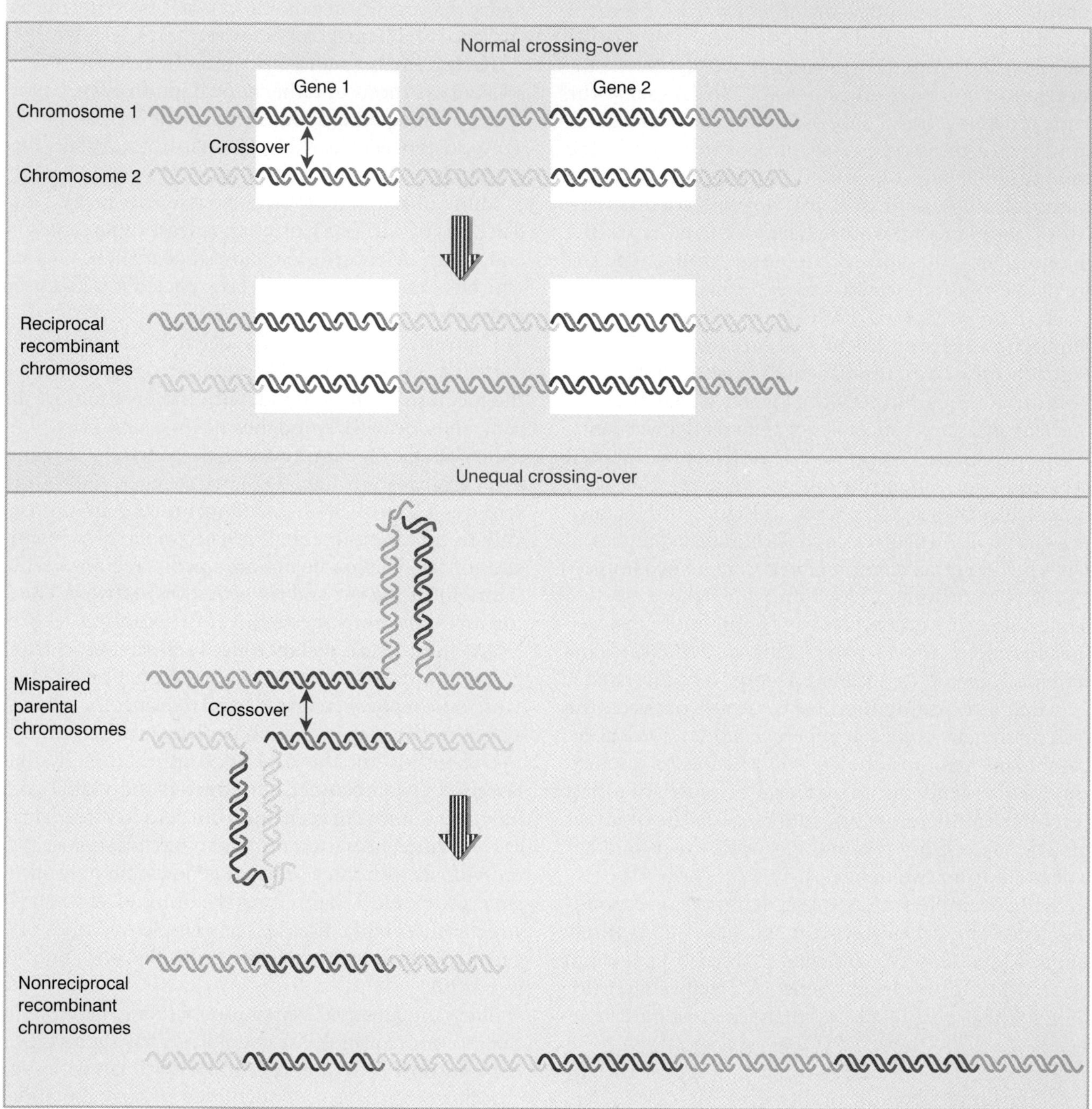

tively frequently, which argues that unequal crossing-over in the cluster must be fairly common. It occurs more often in the α cluster than in the β cluster, possibly because the introns in α genes are much shorter, and therefore present less impediment to mispairing between nonhomologous genes.

The deletions that cause β-thalassemias are summarized in **Figure 4.8.** In some (rare) cases, only the β gene is affected. These have a deletion of 600 bp, extending from the second intron through the 3′ flanking regions. In the other cases, more than one gene of the cluster is affected. Many of the deletions are very long, extending

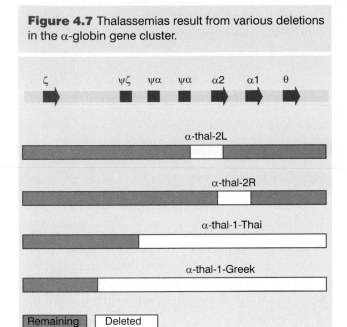

Figure 4.7 Thalassemias result from various deletions in the α-globin gene cluster.

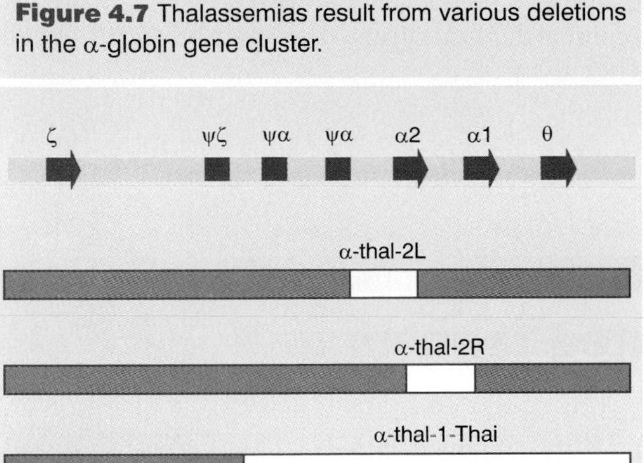

Figure 4.8 Deletions in the β-globin gene cluster cause several types of thalassemia.

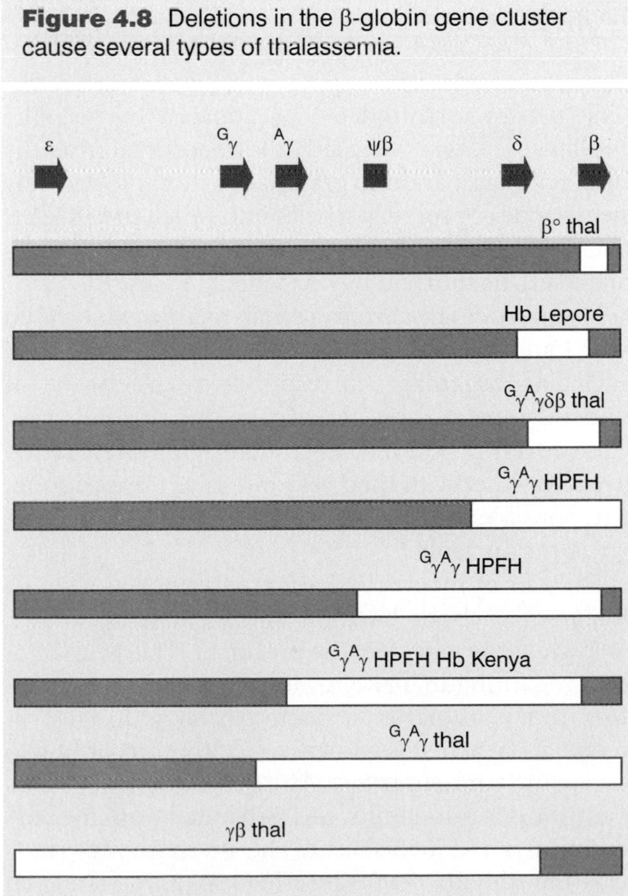

from the 5′ end indicated on the map for >50 kb toward the right.

The **Hb Lepore** type provided the classic evidence that deletion can result from unequal crossing-over between linked genes. The β and δ genes differ only ~7% in sequence. Unequal recombination deletes the material between the genes, thus fusing them together (see Figure 4.6). The fused gene produces a single β-like chain that consists of the N-terminal sequence of δ joined to the C-terminal sequence of β.

Several types of Hb Lepore are now known, the difference between them lying in the point of transition from δ to β sequences. So when the δ and β genes pair for unequal crossing-over, the exact point of recombination determines the position at which the switch from δ to β sequence occurs in the amino acid chain.

The reciprocal of this event has been found in the form of **Hb anti-Lepore,** which is produced by a gene that has the N-terminal part of β and the C-terminal part of δ. The fusion gene lies between normal δ and β genes.

Evidence that unequal crossing-over can occur between more distantly related genes is provided by the identification of **Hb Kenya,** another fused hemoglobin. This contains the N-terminal sequence of the $^{A}\gamma$ gene and the C-terminal sequence of the β gene. The fusion must have resulted from unequal crossing-over between $^{A}\gamma$ and β, which differ ~20% in sequence.

From the differences between the globin gene clusters of various mammals, we see that duplication followed (sometimes) by variation has been an important feature in the evolution of each cluster. The human thalassemic deletions demonstrate that unequal crossing-over continues to occur in both globin gene clusters. Each such event generates a duplication as well as the deletion, and we must account for the fate of both recombinant loci in the population. Deletions can also occur (in principle) by recombination between homologous sequences lying on the *same* chromosome. This does not generate a corresponding duplication.

It is difficult to estimate the natural frequency of these events, because selective forces rapidly adjust the levels of the variant clusters in the population. Generally a contraction in gene number is likely to be deleterious and selected against. However, in some populations, there may be a balancing advantage that maintains the deleted form at a low frequency.

The structures of the present human clusters show several duplications that attest to the importance of such mechanisms. The *functional* sequences include two α genes coding the same protein, fairly well related β and δ genes, and two almost identical γ genes. These comparatively recent independent duplications have survived in the population, not to mention the more distant duplications that originally generated the various types of globin genes. Other duplications may have given rise to pseudogenes or have been lost. We expect continual duplication and deletion to be a feature of all gene clusters.

Genes for rRNA form a repeated tandem unit

In the cases we have discussed so far, there are differences between the individual members of a gene cluster that allow selective pressure to act independently upon each gene. A contrast is provided by two cases of large gene clusters that contain many identical copies of the same gene or genes. Most organisms contain multiple copies of the genes for the histone proteins that are a major component of the chromosomes; and there are almost always multiple copies of the genes that code for the ribosomal RNAs. These situations pose some interesting evolutionary questions.

Ribosomal RNA is the predominant product of transcription, constituting some 80–90% of the total mass of cellular RNA in both eukaryotes and prokaryotes. The number of major rRNA genes varies from seven in *E. coli,* 100–200 in lower eukaryotes, to several hundred in higher eukaryotes. The genes for the large and small rRNA (found respectively in the large and small subunits of the ribosome) usually form a tandem pair. (The sole exception is the yeast mitochondrion.)

The lack of any detectable variation in the sequences of the rRNA molecules implies that all the copies of each gene must be identical, or at least must have differences below the level of detection in rRNA (~1%). A point of major interest is what mechanism(s) are used to prevent variations from accruing in the individual sequences.

In bacteria, the multiple rRNA gene pairs are dispersed. In most eukaryotic nuclei, the rRNA genes are contained in a tandem cluster or clusters. Sometimes these regions are called **rDNA.** (In some cases, the proportion of rDNA in the total DNA, together with its atypical base composition, is great enough to allow its isolation as a separate fraction directly from sheared genomic DNA.) An important diagnostic feature of a tandem cluster is that it generates a circular restriction map, as shown in **Figure 4.9.**

Suppose that each repeat unit has three restriction sites. In the example shown in the figure, fragments A and B are contained entirely within a repeat unit, and fragment C contains the end of one repeat and the beginning of the next. When we map these fragments by conventional means, we find that A is next to B, which is next to C, which is next to A, generating the circular map. If the cluster is large, the internal fragments (A, B, C) will be present in much greater quantities than the terminal fragments (X, Y) which connect the cluster to adjacent DNA. In a cluster of 100 repeats, X and Y would be present at 1% of the level of A, B, C. This can make it difficult to obtain the ends of a gene cluster for mapping purposes.

The region of the nucleus where rRNA synthesis occurs has a characteristic appearance, with a core of fibrillar nature surrounded by a granular cortex. The fibrillar core is where the rRNA is transcribed from the DNA template; and the granular cortex is formed by the ribonucleoprotein particles into which the rRNA is assembled. The whole area is called the **nucleolus.** Its characteristic morphology is evident in **Figure 4.10.**

The particular chromosomal regions associated with a nucleolus are called **nucleolar organizers.** Each nucleolar organizer corresponds to a cluster of tandemly repeated rRNA genes on one chromosome. The concentration of the tandemly repeated rRNA genes, together with their very intensive transcription, is responsible for creating the characteristic morphology of the nucleoli.

The pair of major rRNAs is transcribed as a single precursor in both bacteria and eukaryotic nuclei. Following transcription, the precursor is cleaved to release the individual rRNA molecules. The transcription unit is shortest in bacteria and is longest in mammals (where it is known as 45S RNA, according to its rate of sedimentation). An rDNA cluster contains many transcription units, each separated from the next by a **nontranscribed spacer.** The alternation of transcription unit and nontranscribed spacer can be seen directly in electron micrographs. The example shown

Figure 4.9 A tandem gene cluster has an alternation of transcription unit and nontranscribed spacer and generates a circular restriction map.

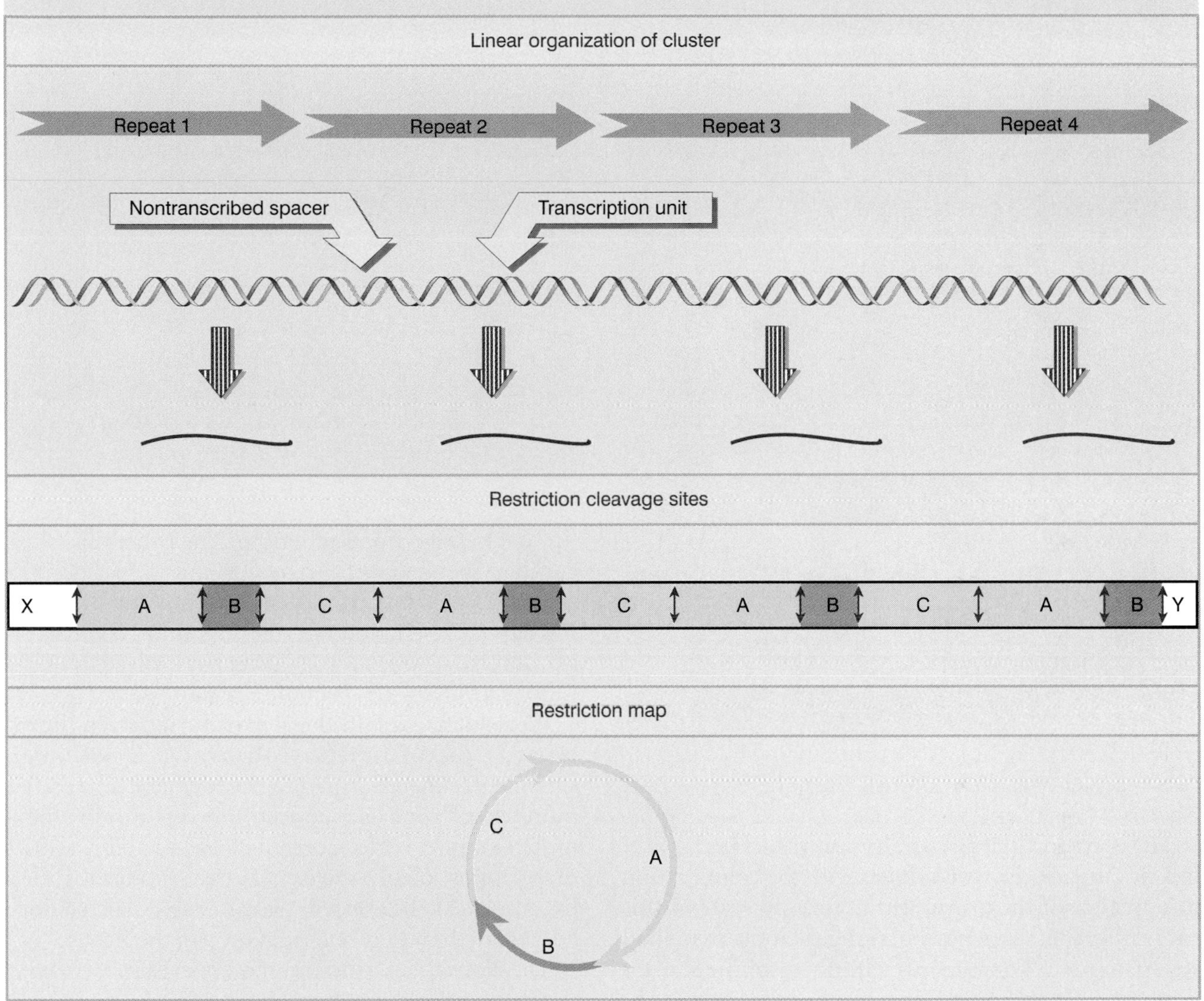

in **Figure 4.11** is taken from the newt *N. viridescens*, in which each transcription unit is intensively expressed, so that many RNA polymerases are simultaneously engaged in transcription on one repeating unit. The polymerases are so closely packed that the RNA transcripts form a characteristic matrix displaying increasing length along the transcription unit.

The nontranscribed spacer varies widely in length between and (sometimes) within species. In yeast there is a short nontranscribed spacer, relatively constant in length. In *D. melanogaster*, there is almost a twofold variation in the length of the nontranscribed spacer between different copies of the repeating unit. A similar situation is seen in *X. laevis*. In each of these cases, all of the repeating units are present as a single tandem cluster on one particular chromosome. (In the example of *D. melanogaster*, this happens to be the sex chromosome. The cluster on the X chromosome is larger than that on the Y chromosome, so female flies have more copies of the rRNA genes than male flies.)

In mammals the repeating unit is very much larger, comprising the transcription unit of ~13 kb and a nontranscribed spacer of ~30 kb. Usually, the genes lie in several dispersed clusters—in the case of man and mouse residing on five and six chromosomes, respectively. One interesting (but unanswered) question is how the corrective mechanisms that presumably function within a single cluster to ensure constancy of

Figure 4.10 The nucleolar core identifies rDNA under transcription, and the surrounding granular cortex consists of assembling ribosomal subunits. This thin section shows the nucleolus of the newt *Notopthalmus viridescens.* Photograph kindly provided by Oscar Miller.

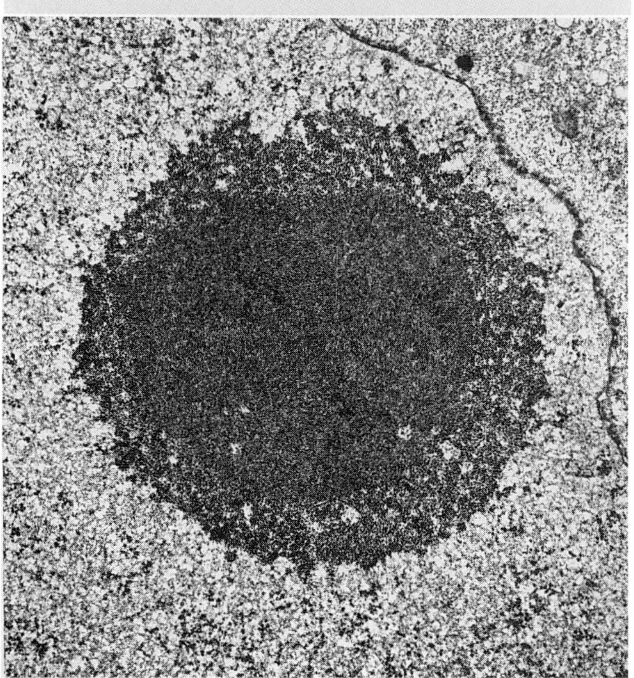

Figure 4.11 Transcription of rDNA clusters generates a series of matrices, each corresponding to one transcription unit and separated from the next by the nontranscribed spacer. Photograph kindly provided by Oscar Miller.

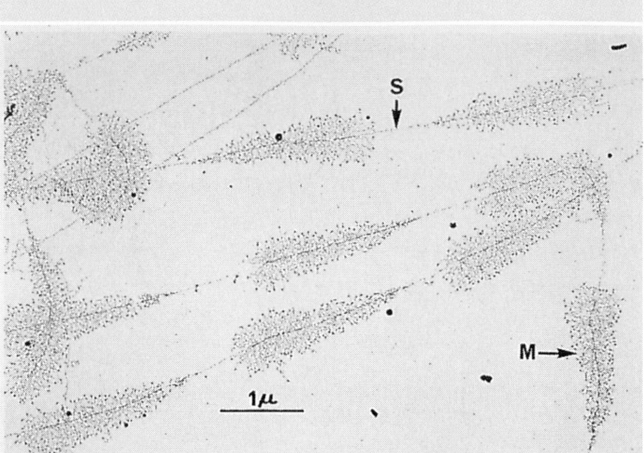

rRNA sequence are able to work when there are several clusters.

The variation in length of the nontranscribed spacer in a single gene cluster contrasts with the conservation of sequence of the transcription unit. In spite of this variation, the sequences of longer nontranscribed spacers remain homologous with those of the shorter nontranscribed spacers. This implies that each nontranscribed spacer is *internally repetitious,* so that the variation in length results from changes in the number of repeats of some subunit.

The general nature of the nontranscribed spacer is illustrated by the example of *X. laevis.* **Figure 4.12** illustrates the situation. Regions that are fixed in length alternate with regions that vary. Each of the three repetitious regions comprises a variable number of repeats of a rather short sequence. One type of repetitious region has repeats of a 97 bp sequence; the other, which occurs in two locations, has a repeating unit found in two forms, 60 bp and 81 bp long. The variation in the number of repeating units in the repetitious regions accounts for the overall variation in spacer length.

One of the fixed regions (at the start of the unit) is unique in sequence and length. The others are short constant sequences called **Bam islands.** (This description takes its name from their isolation via the use of the BamHI restriction enzyme.) From this type of organization, we see that the cluster has evolved by duplications involving the promoter region.

We need to explain the lack of variation in the expressed copies of the repeated genes. One model would suppose that there is a quantitative demand for a certain number of "good" sequences. But this would enable mutated sequences to accumulate up to a point at which their proportion of the cluster is great enough for selective pressure to be exerted. We can exclude such models because of the lack of such variation in the cluster.

The lack of variation implies the existence of selective pressure in some form that is sensitive to individual variations. One model would suppose that the entire cluster is regenerated periodically from one or from a very few members. As a practical matter, any mechanism would need to involve regeneration every generation. We can exclude such models because a regenerated cluster would not show variation in the nontranscribed regions of the individual repeats.

We are left with a dilemma. Variation in the nontranscribed regions suggests that there is frequent unequal crossing over. This will change the size of the cluster, but will not otherwise change the properties of the individual repeats. So how are mutations prevented from accumulating? We see in the next section that continuous contraction and expansion of a cluster may provide a mechanism for homogenizing its copies.

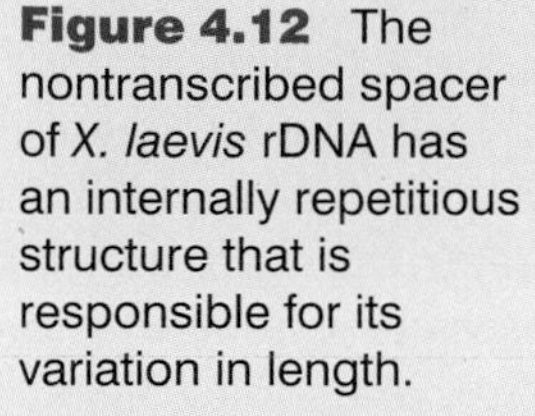

Figure 4.12 The nontranscribed spacer of *X. laevis* rDNA has an internally repetitious structure that is responsible for its variation in length.

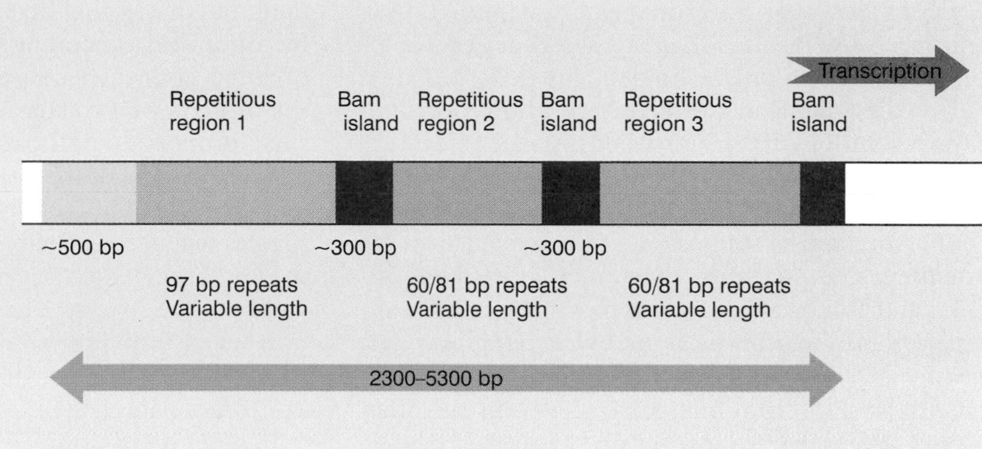

Crossover fixation could maintain identical repeats

THE same problem is encountered whenever a gene has been duplicated. How can selection be imposed to prevent the accumulation of deleterious mutations?

The duplication of a gene is likely to result in an immediate relaxation of the evolutionary pressure on its sequence. Now that there are two identical copies, a change in the sequence of either one will not deprive the organism of a functional protein, since the original amino acid sequence continues to be coded by the other copy. Then the selective pressure on the two genes is diffused, until one of them mutates sufficiently away from its original function to refocus all the selective pressure on the other.

Immediately following a gene duplication, changes might accumulate more rapidly in one of the copies, leading eventually to a new function (or to its disuse in the form of a pseudogene). If a new function develops, the gene then evolves at the same, slower rate characteristic of the original function. Probably this is the sort of mechanism responsible for the separation of functions between embryonic and adult globin genes.

Yet there are instances where duplicated genes retain the same function, coding for identical or nearly identical proteins. Identical proteins are coded by the two human α-globin genes, and there is only a single amino acid difference between the two γ-globin proteins. How is selective pressure exerted to maintain their sequence identity?

The most obvious possibility is that the two genes do not actually have identical functions, but differ in some (undetected) property, such as time or place of expression. Another possibility is that the need for two copies is quantitative, because neither by itself produces a sufficient amount of protein.

In more extreme cases of repetition, however, it is impossible to avoid the conclusion that no single copy of the gene is essential. When there are many copies of a gene, the immediate effects of mutation in any one copy must be very slight. The consequences of an individual mutation are diluted by the large number of copies of the gene that retain the wild-type sequence. Many mutant copies could accumulate before a lethal effect is generated.

Lethality becomes quantitative, a conclusion reinforced by the observation that half of the units of the rDNA cluster of *X. laevis* or *D. melanogaster* can be deleted without ill effect. So how are these units prevented from gradually accumulating deleterious mutations? And what chance is there for the rare favorable mutation to display its advantages in the cluster?

The basic principle of models to explain the maintenance of identity among repeated copies is to suppose that nonallelic genes are not independently inherited, but must be continually regenerated from *one* of the copies of a preceding generation. In the simplest case of two identical genes, when a mutation occurs in one copy, either it is by chance eliminated (because the sequence of the other copy takes over), or it is spread to both duplicates (because the mutant copy becomes the dominant version). Spreading exposes a mutation to selection. The result is that the two genes evolve together as though only a single locus existed. This is called **coincidental** or **concerted evolution** (occasionally **coevolution**). It can be applied to a pair of identical genes or (with further assumptions) to a cluster containing many genes.

One mechanism supposes that the sequences of the nonallelic genes are directly compared with one another and homogenized by enzymes that recognize any differences. This can be done by exchanging single strands between them, to form genes one of whose strands derives from one copy, one from the other copy. Any differences show as improperly paired bases, which attract attention from enzymes able to excise and replace a base, so that only A•T and G•C pairs survive. This type of event is called **gene conversion** and is associated with genetic recombination as described in Chapter 14.

We should be able to ascertain the scope of such events by comparing the sequences of duplicate genes. If they are subject to concerted evolution, we should not see the accumulation of silent site substitutions between them (because the homogenization process applies to these as well as to the replacement sites). We know that the extent of the maintenance mechanism need not extend beyond the gene itself, since there are cases of duplicate genes whose flanking sequences are entirely different. Indeed, we may see abrupt boundaries that mark the ends of the sequences that were homogenized.

We must remember that the existence of such mechanisms can invalidate the determination of the history of such genes via their divergence, *because the divergence reflects only the time since the last homogenization/regeneration event, not the original duplication.*

The **crossover fixation** model supposes that an entire cluster is subject to continual rearrangement by the mechanism of unequal crossing-over. Such events can explain the concerted evolution of multiple genes if unequal crossing-over causes all the copies to be regenerated physically from one copy.

Following the sort of event depicted in Figure 23.4, for example, the chromosome carrying a triple locus could suffer deletion of one of the genes. Of the two remaining genes, $1\frac{1}{2}$ represent the sequence of one of the original copies; only $\frac{1}{2}$ of the sequence of the other original copy has survived. Any mutation in the first region now exists in both genes and is subject to selective pressure.

Tandem clustering provides frequent opportunities for "mispairing" of genes whose sequences are the same, but that lie in different positions in their clusters. By continually expanding and contracting the number of units via unequal crossing-over, it is possible for all the units in one cluster to be derived from rather a small proportion of those in an ancestral cluster. The variable lengths of the spacers are consistent with the idea that unequal crossing-over events take place in spacers that are internally mispaired. This can explain the homogeneity of the genes compared with the variability of the spacers. The genes are exposed to selection when individual repeating units are amplified within the cluster; but the spacers are irrelevant and can accumulate changes.

In a region of nonrepetitive DNA, recombination occurs between precisely matching points on the two homologous chromosomes, generating reciprocal recombinants. The basis for this precision is the ability of two duplex DNA sequences to align exactly. We know that unequal recombination can occur when there are multiple copies of genes whose exons are related, even though their flanking and intervening sequences may differ. This happens because of the mispairing between corresponding exons in *nonallelic* genes.

Imagine how much more frequently misalignment must occur in a tandem cluster of identical or nearly identical repeats. Except at the very ends of the cluster, the close relationship between successive repeats makes it impossible even to define the exactly corresponding repeats! This has two consequences: there is continual adjustment of the size of the cluster; and there is homogenization of the repeating unit.

Consider a sequence consisting of a repeating unit "ab" with ends "x" and "y". If we represent one chromosome in black and the other in color, the exact alignment between "allelic" sequences would be

```
xababababababababababababababababy
xababababababababababababababababy
```

But probably *any* sequence **ab** in one chromosome could pair with *any* sequence **ab** in the other chromosome. In a misalignment such as

```
xababababababababababababababababy
          xababababababababababababababababy
```

the region of pairing is no less stable than in the perfectly aligned pair, although it is shorter. We do not know very much about how pairing is initiated prior to recombination, but it is very likely to start between short corresponding regions and then spread. If it starts within satellite DNA, it is more likely than not to involve repeating units that do not have exactly corresponding locations in their clusters.

Now suppose that a recombination event occurs within the unevenly paired region. The will have different numbers of repeating units. In one case, the cluster has become longer; in the other, it has become shorter,

```
xababababababababababababababababy
          ×
          xababababababababababababababababy
          ↓
xababababababababababababababababababababy
          +
          xabababababababababababy
```

where "×" indicates the site of the crossover.

If this type of event is common, clusters of tandem repeats will undergo continual expansion and contraction. This can cause a particular repeating unit to spread through the cluster, as illustrated in **Figure 4.13.** Suppose that the cluster consists initially of a sequence *abcde,* where each letter represents a repeating unit. The different repeating units are closely enough related to one another to mispair for recombination. Then by a series of unequal recombination events, the size of the repetitive region increases or decreases, and also one unit spreads to replace all the others.

The crossover fixation model predicts that *any sequence of DNA that is not under selective pressure will be taken over by a series of identical tandem repeats generated in this way.* The critical assumption is that the process of crossover fixation is fairly rapid relative to mutation, so that new mutations either are eliminated (their repeats are lost) or come to take over the entire cluster. In the case of the rDNA cluster, of course, a further factor is imposed by selection for an effective transcribed sequence.

Figure 4.13 Unequal recombination allows one particular repeating unit to occupy the entire cluster. The numbers indicate the length of the repeating unit at each stage.

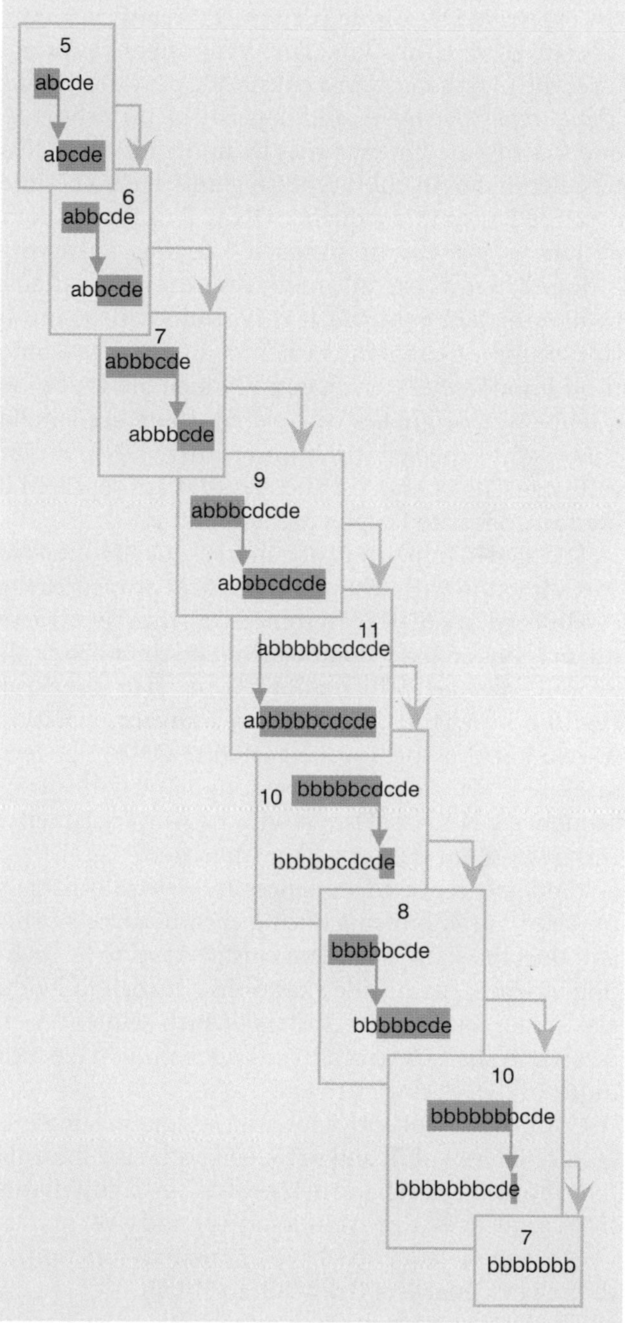

Satellite DNAs often lie in heterochromatin

REPETITIVE DNA is defined by its (relatively) rapid rate of renaturation. The component that renatures the most rapidly in a eukaryotic genome is called *highly repetitive* DNA, and consists of very short sequences repeated many times in tandem in large clusters. Because of its short repeating unit, it is sometimes described as **simple sequence DNA.** This type of component is present in almost all higher eukaryotic genomes, but its overall amount is extremely variable. In mammalian genomes it is typically <10%, but in (for example) *Drosophila virilis*, it amounts to ~50%. In addition to the large clusters in which this type of sequence was originally discovered, there are smaller clusters interspersed with nonrepetitive DNA. It typically consists of short sequences that are repeated in identical or related copies in the genome.

The tandem repetition of a short sequence often creates a fraction with distinctive physical properties that can be used to isolate it. In some cases, the repetitive sequence has a base composition distinct from the genome average, which allows it to form a separate fraction by virtue of its distinct buoyant density. A fraction of this sort is called **satellite DNA.** The term satellite DNA is essentially synonymous with simple sequence DNA. Consistent with its simple sequence, this DNA is not transcribed or translated.

Tandemly repeated sequences are especially liable to undergo misalignments during chromosome pairing, and thus the sizes of tandem clusters tend to be highly polymorphic, with wide variations between individuals. In fact, the smaller clusters of such sequences can be used to characterize individual genomes in the technique of "DNA fingerprinting".

Comparisons of corresponding regions of simple sequence DNA within and between species are informative about the mechanisms involved in manipulating DNA sequences over evolutionary periods. We may ask whether these sequences have a structural function, although evidence is still difficult to obtain.

The buoyant density of a duplex DNA depends on its G•C content according to the empirical formula

$$\rho = 1.660 + 0.00098\ (\%\text{G•C})\ \text{g-cm}^{-3}$$

Buoyant density usually is determined by centrifuging DNA through a **density gradient** of CsCl. The DNA forms a band at the position corresponding to its own density. Fractions of DNA differing in G•C content by >5% can usually be separated on a density gradient.

When eukaryotic DNA is centrifuged on a density gradient, two types of material may be distinguished:

- Most of the genome forms a continuum of fragments that appear as a rather broad peak centered on the buoyant density corresponding to the average G•C content of the genome. This is called the main band.
- Sometimes an additional, smaller peak (or peaks) is seen at a different value. This material is the satellite DNA.

Satellites are present in many eukaryotic genomes. They may be either heavier or lighter than the main band; but it is uncommon for them to represent >5% of the total DNA. A clear example is provided by mouse DNA, shown in **Figure 4.14.** The graph is a quantitative scan of the bands formed when mouse DNA is centrifuged through a CsCl density gradient. The main band contains 92% of the genome and is centered on a buoyant density of 1.701 g-cm^{-3} (corresponding to its

Figure 4.14 Mouse DNA is separated into a main band and a satellite by centrifugation through a density gradient of CsCl.

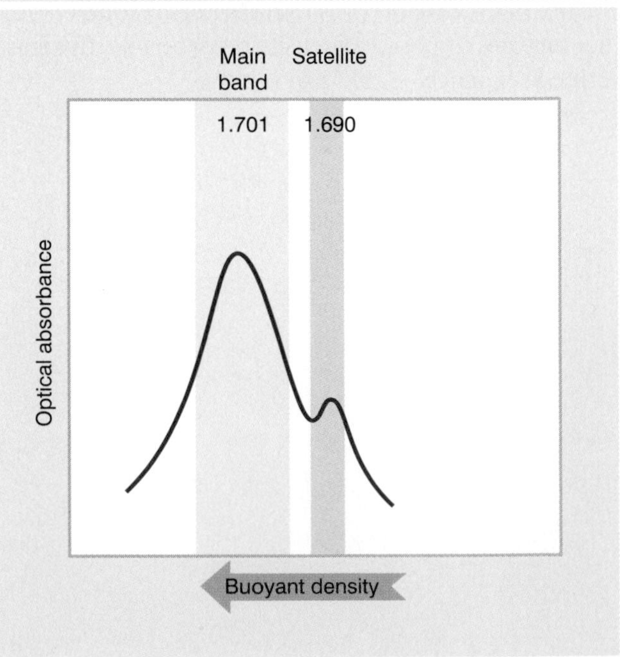

average G•C of 42%, typical for a mammal). The smaller peak represents 8% of the genome and has a distinct buoyant density of 1.690 g-cm^{-3}. It contains the mouse satellite DNA, whose G•C content (30%) is much lower than any other part of the genome.

The behavior of satellite DNA on density gradients is often anomalous. When the actual base composition of a satellite is determined, it is different from the prediction based on its buoyant density. The reason is that ρ is a function not just of base composition, but of the constitution in terms of nearest neighbor pairs. For simple sequences, these are likely to deviate from the random pairwise relationships needed to obey the equation for buoyant density. Also, satellite DNA may be methylated, which changes its density.

Often, most of the highly repetitive DNA of a genome can be isolated in the form of satellites. When a highly repetitive DNA component does not separate as a satellite, on isolation its properties often prove to be similar to those of satellite DNA. That is to say that it consists of multiple tandem repeats with anomalous centrifugation. Material isolated in this manner is sometimes referred to as a **cryptic satellite.** Together the cryptic and apparent satellites usually account for all the large, tandemly repeated blocks of highly repetitive DNA. When a genome has more than one type of highly repetitive DNA, each exists in its own satellite block (although sometimes different blocks are adjacent).

Where in the genome are the blocks of highly repetitive DNA located? An extension of nucleic acid hybridization techniques allows the location of satellite sequences to be determined directly in the chromosome complement. In the technique of *in situ* or **cytological hybridization,** the chromosomal DNA is denatured by treating cells that have been squashed on a cover slip. Then a solution containing a radioactively labeled DNA or RNA probe is added. The probe hybridizes with its complements in the denatured genome. The location of the sites of hybridization can be determined by autoradiography (see Figure 18.16).

Satellite DNAs are found in regions of **heterochromatin.** Heterochromatin is the term used to describe regions of chromosomes that are permanently tightly coiled up and inert, in contrast with the **euchromatin** that represents most of the genome (see Chapter 18). Heterochromatin is commonly found at centromeres (the regions where the kinetochores are formed at mitosis and meiosis for controlling chromosome movement). The centromeric location of satellite DNA suggests that it has some structural function in the chromosome. This function could be connected with the process of chromosome segregation.

An example of the localization of satellite DNA for the mouse chromosomal complement is shown in **Figure 4.15.** In this case, one end of each chromosome is labeled, because this is where the centromeres are located in *M. musculus* chromosomes.

Figure 4.15 Cytological hybridization shows that mouse satellite DNA is located at the centromeres. Photograph kindly provided by Mary Lou Pardue and Joe Gall.

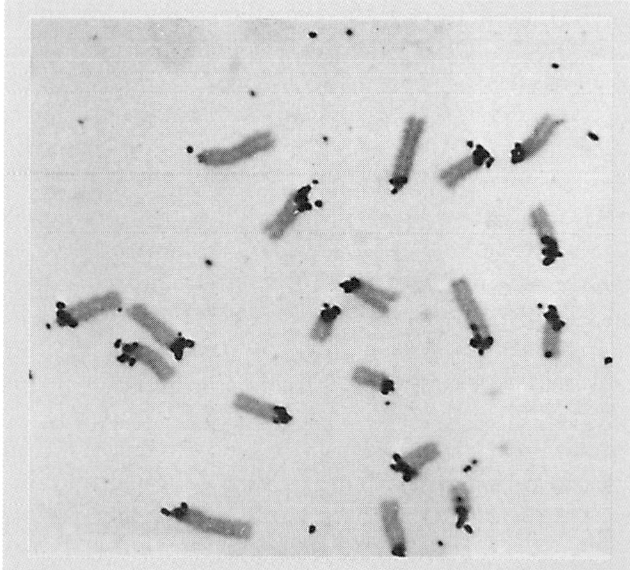

Arthropod satellites have very short identical repeats

IN the arthropods, as typified by insects and crabs, each satellite DNA appears to be rather homogeneous. Usually, a single very short repeating unit accounts for >90% of the satellite. This makes it relatively straightforward to determine the sequence.

Drosophila virilis has three major satellites and also a cryptic satellite, together representing >40% of the genome. The sequences of the satellites are summarized in **Figure 4.16.** The three major satellites have closely related sequences. A single base substitution is sufficient to generate either satellite II or III from the sequence of satellite I.

The satellite I sequence is present in other species of *Drosophila* related to *virilis,* and so may have preceded speciation. The sequences of satellites II and III seem to be specific to *D. virilis,* and so may have evolved from satellite I after speciation.

The main feature of these satellites is their very short repeating unit: only 7 bp. Similar satellites are found in other species. *D. melanogaster* has a variety of satellites, several of which have very short repeating units (5, 7, 10, or 12 bp). Comparable satellites are found in the crabs.

The close sequence relationship found among the *D. virilis* satellites is not necessarily a feature of other genomes, where the satellites may have unrelated sequences. *Each satellite has arisen by a lateral amplification of a very short sequence.* This sequence may represent a variant of a previously existing satellite (as in *D. virilis*), or could have some other origin.

Satellites are continually generated and lost from genomes. This makes it difficult to ascertain evolutionary relationships, since a current satellite could have evolved from some previous satellite that has since been lost. The important feature of these satellites is that *they represent very long stretches of DNA of very low sequence complexity, within which constancy of sequence can be maintained.*

One feature of many of these satellites is a pronounced asymmetry in the orientation of base pairs on the two strands. In the example of the *D. virilis* satellites shown in Figure 4.15, in each of the major satellites one of the strands is much richer in T and G bases. This increases its buoyant density, so that upon denaturation this **heavy strand** (H) can be separated from the complementary **light strand** (L). This can be useful in sequencing the satellite.

Figure 4.16 Satellite DNAs of *D. virilis* are related. More than 95% of each satellite consists of a tandem repetition of the predominant sequence.

Satellite	Predominant sequence	Total length	Genome proportion
I	A C A A A C T T G T T T G A	1.1×10^7	25%
II	A T A A A C T T A T T T G A	3.6×10^6	8%
III	A C A A A T T T G T T T A A	3.6×10^6	8%
Cryptic	A A T A T A G T T A T A T C		

Mammalian satellites consist of hierarchical repeats

IN the mammals, as typified by various rodents, the sequences comprising each satellite show appreciable divergence between tandem repeats. Common short sequences can be recognized by their preponderance among the oligonucleotide fragments released by chemical or enzymatic treatment. However, the predominant short sequence usually accounts for only a small minority of the copies. The other short sequences are related to the predominant sequence by a variety of substitutions, deletions, and insertions.

But a series of these variants of the short unit can constitute a longer repeating unit that is itself repeated in tandem with some variation. So mammalian satellite DNAs are constructed from a *hierarchy* of repeating units. These longer repeating units constitute the sequences that renature in reassociation analysis. They can also be recognized by digestion with restriction enzymes.

When any satellite DNA is digested with an enzyme that has a recognition site in its repeating unit, one fragment will be obtained for every repeating unit in which the site occurs. In fact, when the DNA of a eukaryotic genome is digested with a restriction enzyme, most of it gives a general smear, due to the random distribution of cleavage sites. But satellite DNA generates sharp bands, because a large number of fragments of identical or almost identical size are created by cleavage at restriction sites that lie a regular distance apart.

Determining the sequence of satellite DNA can be difficult. Using the discrete bands generated by restriction cleavage, we can attempt to obtain a sequence directly. However, if there is appreciable divergence between individual repeating units, different nucleotides will be present at the same position in different repeats, so the sequencing gels will be obscure. If the divergence is not too great—say, within ~2%—it may be possible to determine an average repeating sequence.

Individual segments of the satellite can be inserted into plasmids for cloning. A difficulty is that the satel-

Figure 4.17 The repeating unit of mouse satellite DNA contains two half-repeats, which are aligned to show the identities (in color).

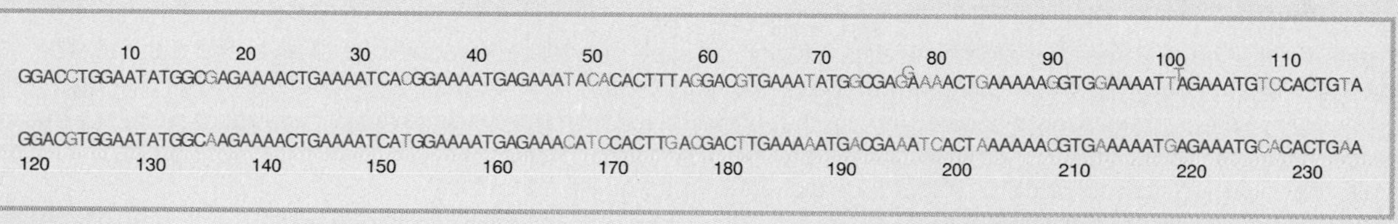

Figure 4.18 The alignment of quarter-repeats identifies homologies between the first and second half of each half-repeat. Positions that are the same in all 4 quarter-repeats are shown in color; identities that extend only through 3 quarter-repeats are indicated by grey letters in the pink area.

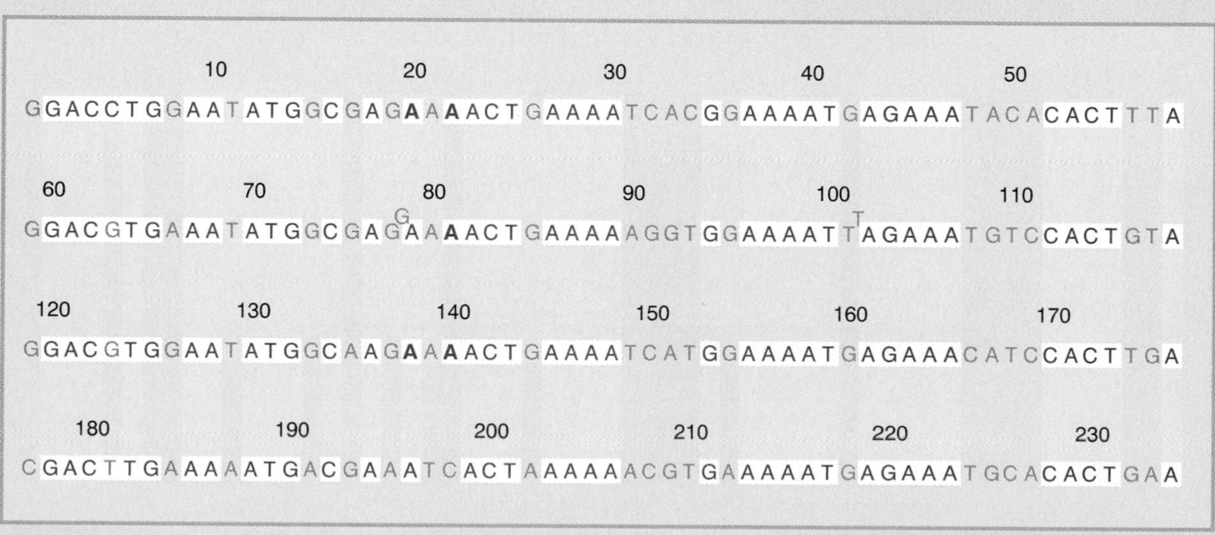

lite sequences tend to be excised from the chimeric plasmid by recombination in the bacterial host. However, when the cloning succeeds, it is possible to determine the sequence of the cloned segment unambiguously. While this gives the actual sequence of a repeating unit or units, we should need to have many individual such sequences to reconstruct the type of divergence typical of the satellite as a whole.

By either sequencing approach, the information we can gain is limited to the distance that can be analyzed on one set of sequence gels. The repetition of divergent tandem copies makes it impossible to reconstruct longer sequences by obtaining overlaps between individual restriction fragments.

The satellite DNA of the mouse *M. musculus* is cleaved by the enzyme EcoRII into a series of bands, including a predominant monomeric fragment of 234 bp. This sequence must be repeated with few variations throughout the 60–70% of the satellite that is cleaved into the monomeric band. We may analyze this sequence in terms of its successively smaller constituent repeating units.

Figure 4.17 depicts the sequence in terms of two half-repeats. By writing the 234 bp sequence so that the first 117 bp are aligned with the second 117 bp, we see that the two halves are quite well related. They differ at 22 positions, corresponding to 19% divergence. This means that the current 234 bp repeating unit must have been generated at some time in the past by duplicating a 117 bp repeating unit, after which differences accumulated between the duplicates.

Within the 117 bp unit, we can recognize two further subunits. Each of these is a quarter-repeat relative to the whole satellite. The four quarter-repeats are aligned in **Figure 4.18.** The upper two lines represent the first half-repeat of Figure 4.17; the lower two lines represent the second half-repeat. We see that the divergence between the four quarter-repeats has increased to 23 out of 58 positions, or 40%. The first three quarter-repeats are somewhat better related, and a large proportion of

Figure 4.19 The alignment of eighth-repeats shows that each quarter-repeat consists of an α and a β half. The consensus sequence gives the most common base at each position. The "ancestral" sequence shows a sequence very closely related to the consensus sequence, that could have been the predecessor to the α and β units. (The satellite sequence is continuous, so that for the purpose of deducing the consensus sequence, we can treat it as a circular permutation, as indicated by joining the last GAA triplet to the first 6 bp.)

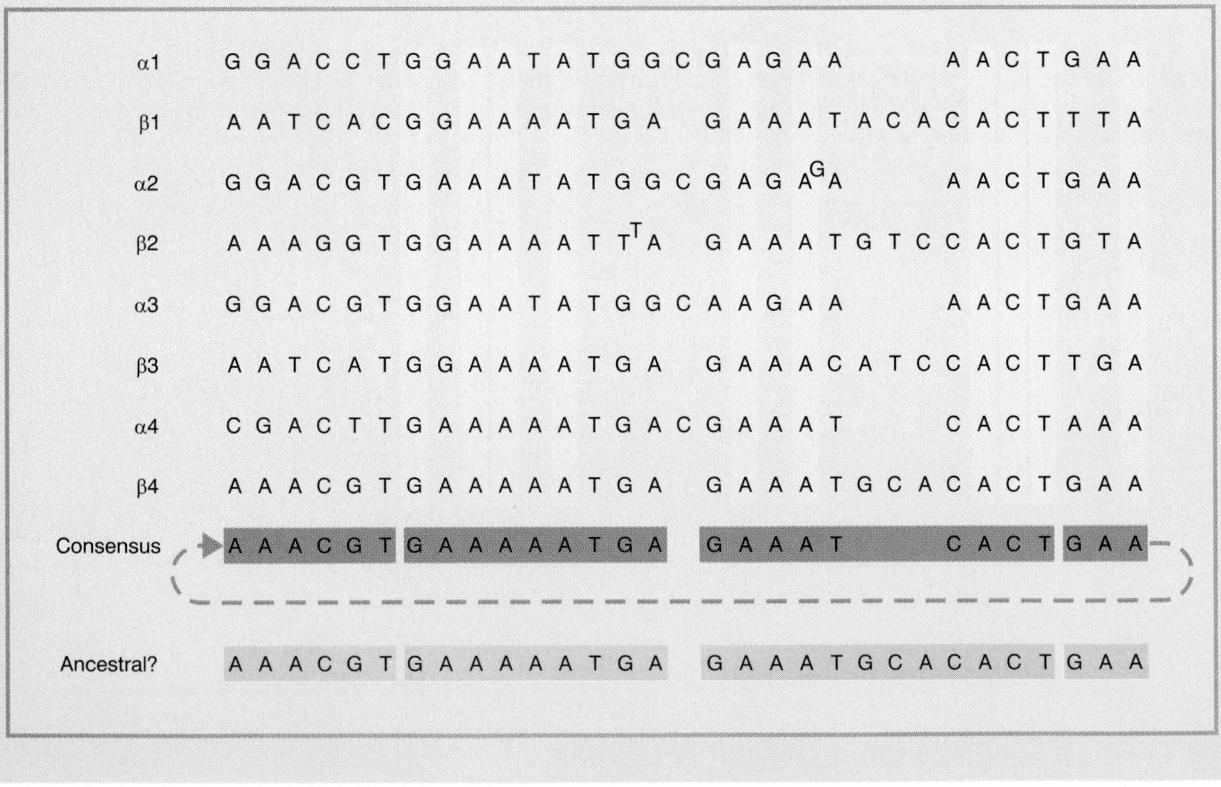

α1	G G A C C T G G A A T A T G G C G A G A A A A C T G A A
β1	A A T C A C G G A A A A T G A G A A A T A C A C A C T T T A
α2	G G A C G T G A A A T A T G G C G A G A (G) A A A C T G A A
β2	A A A G G T G G A A A A T T (T) A G A A A T G T C C A C T G T A
α3	G G A C G T G G A A T A T G G C A A G A A A A C T G A A
β3	A A T C A T G G A A A A T G A G A A A C A T C C A C T T G A
α4	C G A C T T G A A A A A T G A C G A A A T C A C T A A A
β4	A A A C G T G A A A A A T G A G A A A T G C A C A C T G A A
Consensus	A A A C G T G A A A A A T G A G A A A T C A C T G A A
Ancestral?	A A A C G T G A A A A A T G A G A A A T G C A C A C T G A A

the divergence is due to changes in the fourth quarter-repeat.

Looking within the quarter-repeats, we find that each consists of two related subunits (one-eighth-repeats), shown as the α and β sequences in **Figure 4.19.** The α sequences all have an insertion of a C, and the β sequences all have an insertion of a trinucleotide, relative to a common consensus sequence. This suggests that the quarter-repeat originated by the duplication of a sequence like the consensus sequence, after which changes occurred to generate the components we now see as α and β. Further changes then took place between tandemly repeated αβ sequences to generate the individual quarter- and half-repeats that exist today. Among the one-eighth-repeats, the present divergence is 19/31 = 61%.

The consensus sequence is analyzed directly in **Figure 4.20,** which demonstrates that the current satellite sequence can be treated as derivatives of a 9 bp sequence. We can recognize three variants of this sequence in the satellite, as indicated at the bottom of Figure 4.20. If in one of the repeats we take the next most frequent base at two positions instead of the most frequent, we obtain three well-related 9 bp sequences.

GAAAAACGT
GAAAAATGA
GAAAAAACT

The origin of the satellite could well lie in an amplification of one of these three nonamers. The overall consensus sequence of the present satellite is $GAAAAA^{AG}_{TC}T$, which is effectively an amalgam of the three 9 bp repeats.

The average sequence of the monomeric fragment of the mouse satellite DNA explains its properties. The longest repeating unit of 234 bp is identified by the restriction cleavage. The unit of reassociation between single strands of denatured satellite DNA is probably the 117 bp half-repeat, because the 234 bp fragments can anneal both in register and in half-register (in the latter case, the first half-repeat of one strand renatures with the second half-repeat of the other).

So far, we have treated the present satellite as though it consisted of identical copies of the 234 bp repeating unit. Although this unit accounts for the majority of the satellite, variants of it also are present. Some of them are scattered at random throughout the satellite; others are clustered.

The existence of variants is implied by our description of the starting material for the sequence analysis as the "monomeric" fragment. When the satellite is digested by an enzyme that has one cleavage site in the

Figure 4.20 The existence of an overall consensus sequence is shown by writing the satellite sequence in terms of a 9 bp repeat.

			G	G	A	C	C	T	
G	G	A	A	T	A	T	G	G	C
G	A	G	A	A	A	A	C	T	
G	A	A	A	A	T	C	A	C	
G	G	A	A	A	A	T	G	A	
G	A	A	A	T	C	A	C	T	
T	T	A	G	G	A	C	G	T	
G	A	A	A	T	A	T	G	G	C
G	A	G	A^G	A	A	A	C	T	
G	A	A	A	A	A	G	G	T	
G	G	A	A	A	A	T^T	T	A	
G	A	A	A	T*	C	A	C	T	
G	T	A	G	G	A	C	G	T	
G	G	A	A	T	A	T	G	G	C
A	A	G	A	A	A	A	C	T	
G	A	A	A	A	T	C	A	T	
G	G	A	A	A	A	T	G	A	
G	A	A	A	C*	C	A	C	T	
T	G	A	C	G	A	C	T	T	
G	A	A	A	A	A	T	G	A	C
G	A	A	A	T	C	A	C	T	
A	A	A	A	A	A	C	G	T	
G	A	A	A	A	A	T	G	A	
G	A	A	A	T*	C	A	C	T	
G	A	A							
G_{20}	A_{16}	A_{21}	A_{20}	A_{12}	A_{17}	T_8	G_{11}	A_5	
				T_7	C_5	A_8	C_9	T_{15}	
							C_7		

* indicates inserted triplet in β sequence
C in position 10 is extra base in α sequence

234 bp sequence, it also generates dimers, trimers, and tetramers relative to the 234 bp length. They arise when a repeating unit has lost the enzyme cleavage site as the result of mutation.

The monomeric 234 bp unit is generated when two adjacent repeats each have the recognition site. A dimer occurs when one unit has lost the site, a trimer is generated when two adjacent units have lost the site, and so on. With some restriction enzymes, most of the satellite is cleaved into a member of this repeating series, as shown in the example of **Figure 4.21.** The declining number of dimers, trimers, etc. shows that there is a random distribution of the repeats in which the enzyme's recognition site has been eliminated by mutation.

Other restriction enzymes show a different type of behavior with the satellite DNA. They continue to generate the same series of bands. But they cleave only a small proportion of the DNA, say 5–10%. This implies that a certain region of the satellite contains a concentration of the repeating units with this particular restriction site. Presumably the series of repeats in this domain are all derived from an ancestral variant that possessed this recognition site (although in the usual way, some members since have lost it by mutation).

A satellite DNA suffers unequal recombination. This has additional consequences when there is internal repetition in the repeating unit. Let us return to our cluster consisting of "ab" repeats. Suppose that the "a" and "b" components of the repeating unit are themselves sufficiently well related to pair. Then the two clusters can align in **half-register,** with the "a" sequence of one aligned with the "b" sequence of the other. How frequently this occurs will depend on the closeness of the relationship between the two halves of the repeating unit. In mouse satellite DNA, reassociation between the denatured satellite DNA strands *in vitro* commonly occurs in the half-register.

When a recombination event occurs, it changes the length of the repeating units that are involved in the reaction.

```
xababababababababababababy
                  ×
 xababababababababababababy
                  ↓
xabababababababababaababababy
                  +
 xabababababababababbabababy
```

In the upper recombinant cluster, an "ab" unit has been replaced by an "aab" unit. In the lower cluster, the "ab" unit has been replaced by a "b" unit.

Figure 4.21 Digestion of mouse satellite DNA with the restriction enzyme EcoRII identifies a series of repeating units (1, 2, 3) that are multimers of 234 bp and also a minor series (½, 1½, 2½) that includes half-repeats (see text later). The band at the far left is a fraction resistant to digestion.

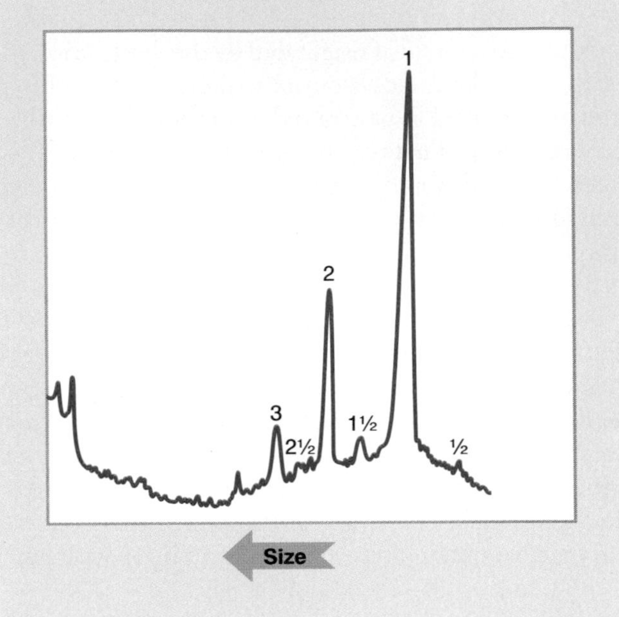

This type of event explains a feature of the restriction digest of mouse satellite DNA. Figure 4.21 shows a fainter series of bands at lengths of $\frac{1}{2}$, $1\frac{1}{2}$, $2\frac{1}{2}$, and $3\frac{1}{2}$ repeating units, in addition to the stronger integral length repeats. Suppose that in the preceding example, "ab" represents the 234 bp repeat of mouse satellite DNA, generated by cleavage at a site in the "b" segment. The "a" and "b" segments correspond to the 117 bp half-repeats.

Then in the upper recombinant cluster, the "aab" unit generates a fragment of $1\frac{1}{2}$ times the usual repeating length. And in the lower recombinant cluster, the "b" unit generates a fragment of half of the usual length. (The multiple fragments in the half-repeat series are generated in the same way as longer fragments in the integral series, when some repeating units have lost the restriction site by mutation.)

Turning the argument the other way around, the identification of the half-repeat series on the gel shows that the 234 bp repeating unit consists of two half-repeats well enough related to pair sometimes for recombination. Also visible in Figure 4.21 are some rather faint bands corresponding to $\frac{1}{4}$ and $\frac{3}{4}$-spacings. These will be generated in the same way as the $\frac{1}{2}$-spacings, when recombination occurs between clusters aligned in a quarter-register. The decreased relationship between quarter-repeats compared with half-repeats explains the reduction in frequency of the $\frac{1}{4}$ and $\frac{3}{4}$-bands compared with the $\frac{1}{2}$-bands.

Minisatellites are useful for genetic mapping

SEQUENCES that resemble satellites in consisting of tandem repeats of a short unit, but that overall are much shorter, consisting of (for example) from 5–50 repeats, are common in mammalian genomes. They were discovered by chance as fragments whose size is extremely variable in genomic libraries of human DNA. The variability is seen when a population contains fragments of many different sizes that represent the same genomic region; when individuals are examined, it turns out that there is extensive polymorphism, and that many different alleles can be found.

These sequences are called **minisatellite** or **VNTR** (variable number tandem repeat) regions. The cause of the variation is that individual alleles have different numbers of the repeating unit. For example, one such minisatellite has a repeat length of 64 bp, and is found in the population with the following distribution:

7% 18 repeats
11% 16 repeats
43% 14 repeats
36% 13 repeats
4% 10 repeats

The cause of this variation is genetic recombination between misaligned repeat units, in the same way that we have discussed already for satellite DNA. The rate of genetic exchange at minisatellite sequences is high, $\sim 10^{-4}$ per kb of DNA. (The frequency of exchanges per actual locus is assumed to be proportional to the length of the minisatellite.) This rate is ~10× greater than the rate of homologous recombination at meiosis, that is, in any random DNA sequence. So minisatellites may be hotspots for meiotic recombination.

Sometimes the presence of a minisatellite is correlated with a high rate of exchange in the vicinity, but in some cases the recombination event occurs between sister chromatids. In the latter case it changes the length of the minisatellite, but has no effect on flanking markers, because these are identical on both recombining molecules of DNA.

The high variability in minisatellites makes them especially useful for genomic mapping, because there is a high probability that individuals will vary in their alleles at such a locus. An example of mapping by minisatellites is illustrated in **Figure 4.22.** This shows an extreme case in which two individuals are both heterozygous at a minisatellite locus, and in fact all four alleles are different. All progeny gain one allele from each parent in the usual way, and it is possible unambiguously to determine the source of every allele in the progeny. In the terminology of human genetics, the meioses described in this figure are highly informative, because of the variation between alleles.

One family of minisatellites in the human genome share a common "core" sequence. The core is a G•C-rich sequence of 10–15 bp, showing an asymmetry of purine/pyrimidine distribution on the two strands. Each individual minisatellite has a variant of the core sequence, but ~1000 minisatellites can be detected on Southern blot by a probe consisting of the core sequence.

Consider the situation shown in Figure 4.22, but multiplied 1000×. The effect of the variation at individual loci is to create a unique pattern for every individual. This makes it possible to assign heredity unambiguously between parents and progeny, by showing that 50% of the bands in any individual are derived from a particular parent. This is the basis of the technique known as **DNA fingerprinting.**

Figure 4.22 Alleles may differ in the number of repeats at a minisatellite locus, so that cleavage on either side generates restriction fragments that differ in length. By using a minisatellite with alleles that differ between parents, the pattern of inheritance can be followed.

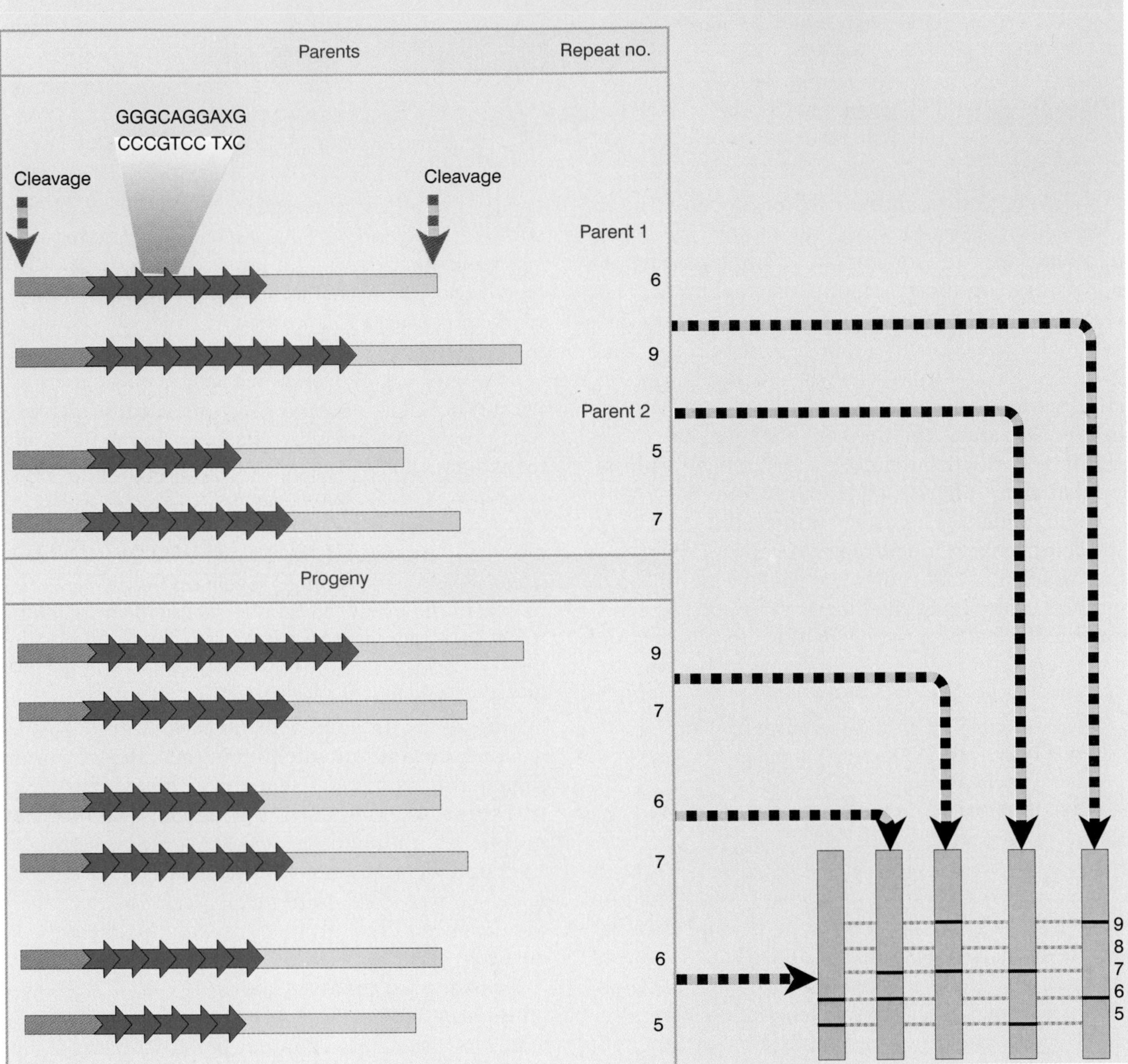

Summary

Almost all genes belong to families, defined by the possession of related sequences in the exons of individual members. Families evolve by the duplication of a gene (or genes), followed by divergence between the copies. Some copies suffer inactivating mutations and become pseudogenes that no longer have any function. Pseudogenes also may be generated as DNA copies of the mRNA sequences.

An evolving set of genes may remain together in a cluster or may be dispersed to new locations by chromosomal rearrangement. The organization of existing clusters can sometimes be used to infer the series of events that has occurred. These events act with regard to sequence rather than function, and therefore include pseudogenes as well as active genes.

Mutations accumulate more rapidly in silent sites than in replacement sites (which affect the amino acid sequence). The rate of divergence at replacement sites can be used to establish a clock, calibrated in percent divergence per million years. The clock can then be used to calculate the time of divergence between any two members of the family.

A tandem cluster consists of many copies of a repeating unit that includes the transcribed sequence(s) and a nontranscribed spacer(s). rRNA gene clusters code only for a single rRNA precursor. Maintenance of active genes in clusters depends on mechanisms such as gene conversion or unequal crossing-over that cause mutations to spread through the cluster, so that they become exposed to evolutionary pressure.

Satellite DNA consists of very short sequences repeated many times in tandem. Its distinct centrifugation properties reflect its biased base composition. Satellite DNA is concentrated in centromeric heterochromatin, but its function (if any) is unknown. The individual repeating units of arthropod satellites are identical. Those of mammalian satellites are related, and can be organized into a hierarchy reflecting the evolution of the satellite by the amplification and divergence of randomly chosen sequences.

Unequal crossing-over appears to have been a major determinant of satellite DNA organization. Crossover fixation explains the ability of variants to spread through a cluster. Minisatellites have properties similar to satellites, but are much smaller; they are useful for human genomic mapping.

Further reading

Reviews

Maniatis, T. *et al.* (1980). The molecular genetics of human hemoglobins. *Ann. Rev. Genet.* **14**, 145–178.

Smith, G. P. (1976). Evolution of repeated DNA sequences by unequal crossover. *Science* **191**, 525–535.

Maeda, N. and Smithies, O. (1986). The evolution of multigene families: human haptoglobin genes. *Ann. Rev. Genet.* **20**, 81–108.

Discoveries

Southern, E. M. (1975). Long range periodicities in mouse satellite DNA. *J. Mol. Biol.* **94**, 51–70.

Jeffreys, A. J., Wilson, V., and Thein, S. L. (1985). Hypervariable 'minisatellite' regions in human DNA. *Nature* **314**, 67–73.

Part 2

Proteins

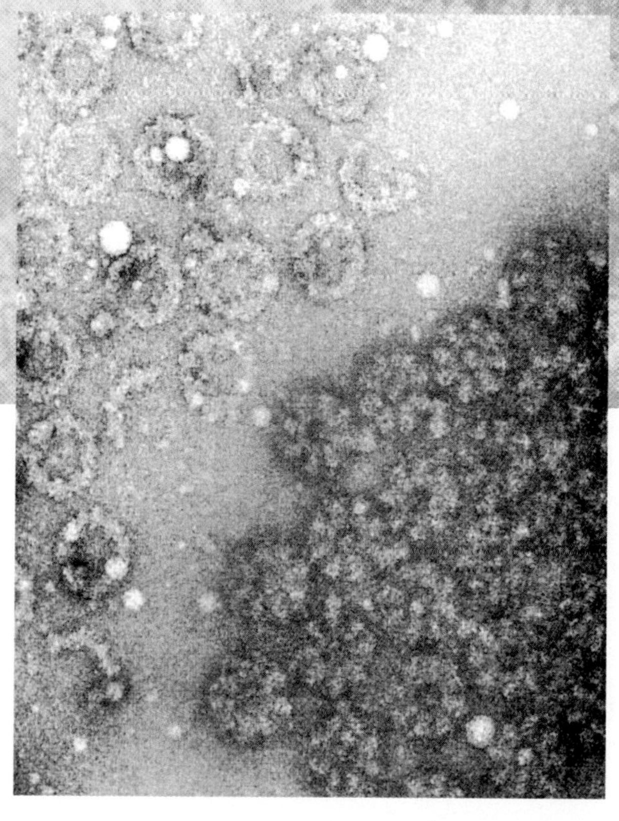

Chapter 5

Messenger RNA

GENE expression occurs by a two-stage process.

■ **Transcription** generates a single-stranded RNA identical in sequence with one of the strands of the duplex DNA. Several different types of RNA are generated by transcription. The three principal classes involved in the synthesis of proteins are:

messenger RNA (mRNA)

transfer RNA (tRNA)

ribosomal RNA (rRNA)

■ **Translation** converts the nucleotide sequence of RNA into the sequence of amino acids comprising a protein. An mRNA is translated into a protein sequence; tRNA and rRNA provide other components of the apparatus for protein synthesis. The entire length of an mRNA is not translated, but each mRNA contains at least one **coding region** that is related to a protein sequence by the genetic code: each nucleotide triplet (codon) of the coding region represents one amino acid.

Figure 5.1 *Overview*: transcription generates an RNA which is complementary to the DNA template strand and has the same sequence as the DNA coding strand. Translation reads each triplet of bases into one amino acid. Three turns of the DNA double helix contain 30 bp, which code for 10 amino acids.

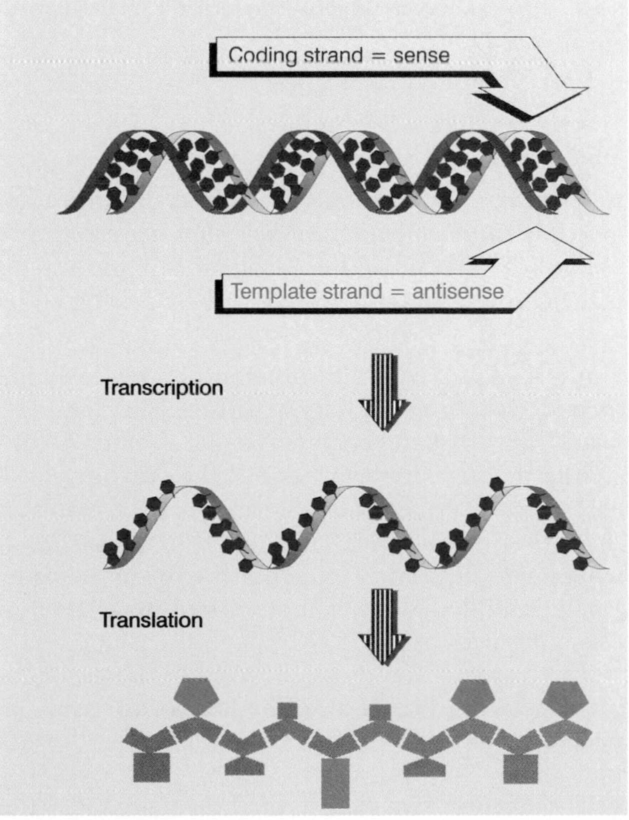

Only one strand of a DNA duplex is transcribed into a messenger RNA. We distinguish the two strands of DNA as depicted in **Figure 5.1:**

■ The strand of DNA that directs synthesis of the mRNA via complementary base pairing is called the **template strand** or **antisense strand.** (We see later that "antisense" is used as a general term to describe a sequence of DNA or RNA that is complementary to mRNA.)

- The other DNA strand bears the *same* sequence as the mRNA (except that it possesses T instead of U), and is called the **coding strand** or **sense strand.**

In this chapter we discuss mRNA and its use as a template for protein synthesis. In Chapter 6 we discuss the process of protein synthesis itself. In Chapter 7 we discuss the way the genetic code is used to interpret the meaning of a sequence of mRNA. And in Chapter 8 we turn to the question of how a protein finds its proper location in the cell when or after it is synthesized.

Transfer RNA is the adapter

MESSENGER RNA can be distinguished from the apparatus responsible for its translation by the use of cell-free systems to synthesize proteins *in vitro.* A protein-synthesizing system from one cell type can translate the mRNA from another, demonstrating that both the code and the translation apparatus are universal.

Each nucleotide triplet in the mRNA represents an amino acid. The incongruity of structure between trinucleotide and amino acid immediately raises the question of how each codon is matched to its particular amino acid. The "adapter" is **transfer RNA (tRNA)**, a small molecule whose polynucleotide chain is only 74–95 bases long.

A tRNA has two crucial properties:

- It represents a single amino acid, to which it is *covalently linked.*
- It contains a trinucleotide sequence, the **anticodon,** which is *complementary to the codon representing its amino acid.* The anticodon enables the tRNA to recognize the codon via complementary base pairing.

All tRNAs have certain common features, including their secondary and tertiary structures. They have "unusual" bases that are generated by modification of the 4 standard bases after synthesis of the polynucleotide chain. The tRNA secondary structure can be written in the form of a **cloverleaf,** illustrated in **Figure 5.2,** in which complementary base pairing forms **stems** for single-stranded **loops.** The stem-loop structures are called the **arms** of tRNA.

The construction of the cloverleaf is illustrated in more detail in **Figure 5.3.** The four major arms are named for their structure or function:

- The **acceptor arm** consists of a base-paired stem that ends in an unpaired sequence whose free 2′- or 3′-OH group can be linked to an amino acid.
- The **TψC arm** is named for the presence of this triplet sequence. (ψ stands for pseudouridine, a modified base.)
- The anticodon arm always contains the anticodon triplet in the center of the loop.
- The **D arm** is named for its content of the base dihydrouridine (another of the modified bases in tRNA).

The numbering system for tRNA illustrates the constancy of the structure. Positions are numbered from 5′ to 3′ according to the most common tRNA structure, which has 76 residues. The overall range of tRNA lengths is 74–95 bases. The variation in length is caused by differences in the structure of two of the arms.

The length of the D loop varies by up to four residues. The extra nucleotides relative to the most common structure are denoted 17:1 (lying between 17 and 18) and 20:1 and 20:2 (lying between 20 and 21). However, in the smallest D loops, residue 17 as well as these three is absent.

The most variable feature of tRNA is the so-called **extra arm,** which lies between the TψC and anticodon arms. Depending on the nature of the extra arm, tRNAs can be divided into two classes. **Class 1 tRNAs** have a small extra arm, consisting of only 3–5 bases. They represent ~75% of all tRNAs. **Class 2 tRNAs** have a large extra arm—it may even be the longest in the tRNA—with 13–21 bases, and ~5 base pairs in the stem. The additional bases are numbered from 47:1 through 47:18. The functional significance of the extra arm is unknown.

The base pairing that maintains the secondary structure is virtually invariant, corresponding to the scheme

Figure 5.2 A tRNA has the dual properties of an adaptor that recognizes both the amino acid and codon. The 3' adenosine is covalently linked to an amino acid. The anticodon base pairs with the codon on mRNA.

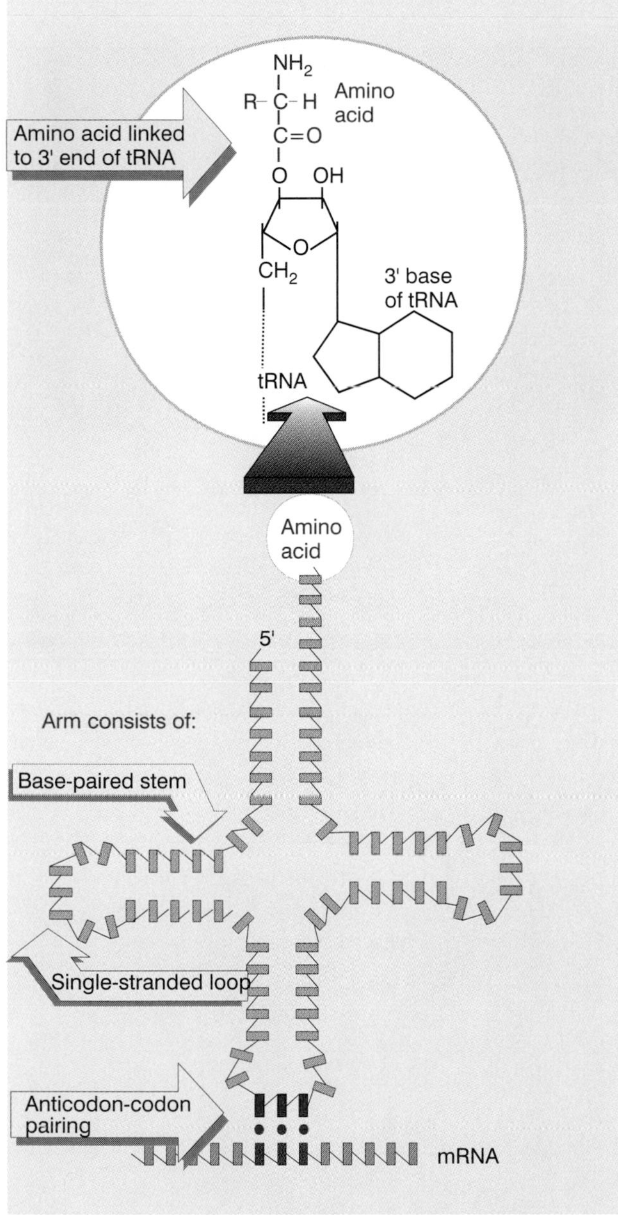

Figure 5.3 The tRNA cloverleaf has invariant and semi-invariant bases, and a conserved set of base pairing interactions.

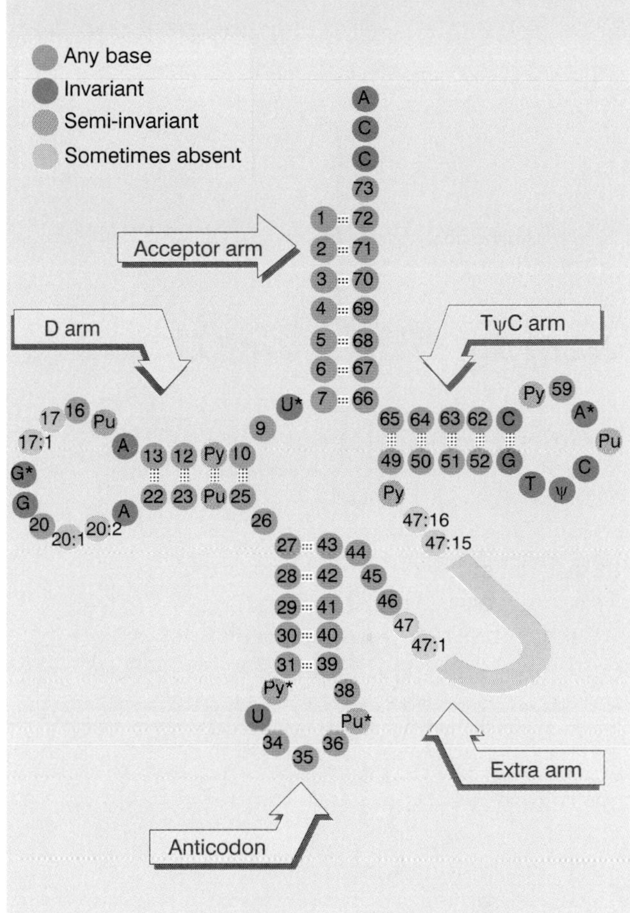

in Figure 5.3. Within a given tRNA, most of the base pairings will be conventional partnerships of A•U and G•C, but occasional G•U, G•ψ, and A•ψ pairs are found. The additional types of base pairs are less stable than the regular pairs, but still allow a double-helical structure to form in RNA.

When the sequences of tRNAs are compared, the bases found at some positions are **invariant** (or **conserved**); almost always a particular base is found at the position. These are indicated in the sequence of Figure 5.3. Actually, as more tRNAs are sequenced, positions that seemed entirely invariant do display occasional exceptions. So for practical purposes, the description of any position as invariant means that the specified base is present in >90–95% of tRNAs. Sometimes the exceptions are individual; sometimes they fall into groups representing some peculiarity of a particular cell.

Some positions are described as **semiinvariant** (or **semiconserved**) because they seem to be restricted to one type of base (purine versus pyrimidine), but either base of that type may be present.

When a tRNA is **charged** with the amino acid corresponding to its anticodon, it is called **aminoacyl-tRNA.** The amino acid is linked by an ester bond from its carboxyl group to the 2′ or 3′ hydroxyl group of the ribose of the 3′ terminal base of the tRNA (which is always adenine).

There is at least one tRNA (but usually more) for

Figure 5.4 Transfer RNA folds into a compact L-shaped tertiary structure with the amino acid at one end and the anticodon at the other end.

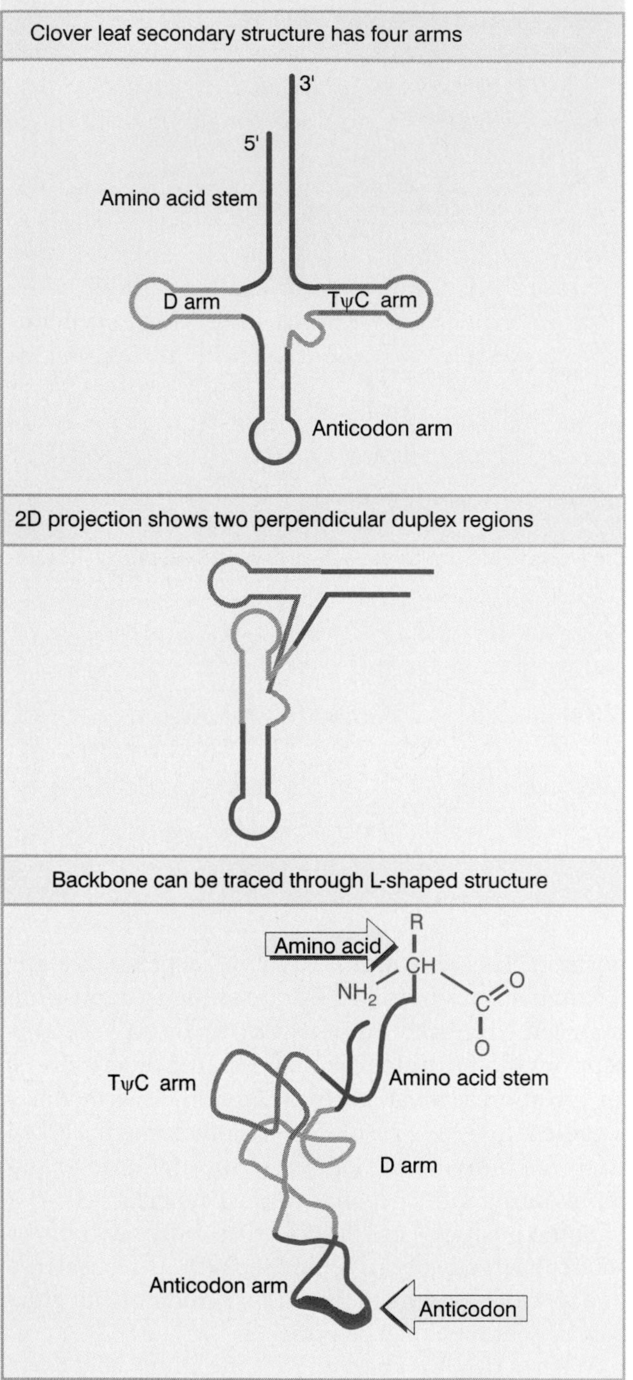

Figure 5.5 A space-filling model shows that $tRNA^{Phe}$ tertiary structure is compact. The two views of tRNA are rotated by 90°. The top view corresponds with the bottom panel in Figure 5.4. Photograph kindly provided by S. H. Kim.

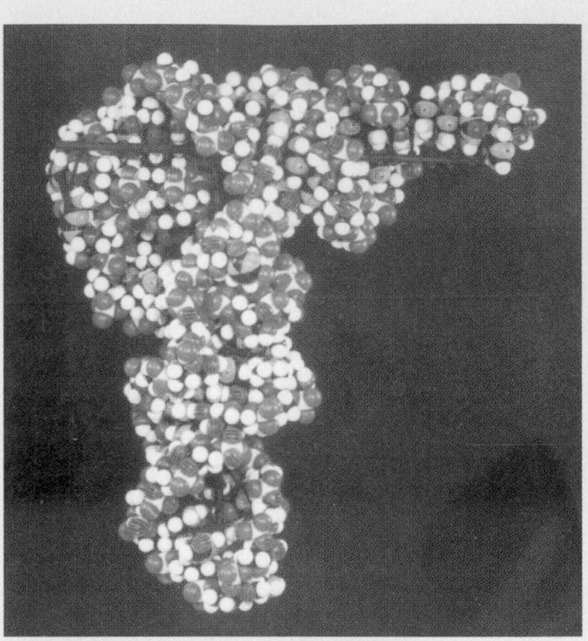

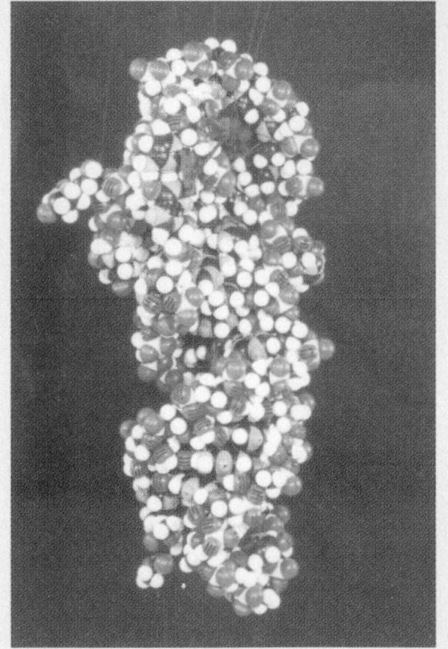

each amino acid. A tRNA is named by using the three-letter abbreviation for the amino acid as a superscript. If there is more than one tRNA for the same amino acid, subscript numerals are used to distinguish them. So two tRNAs for tyrosine would be described as $tRNA_1^{Tyr}$ and $tRNA_2^{Tyr}$. A tRNA carrying an amino acid—that is, an aminoacyl-tRNA—is indicated by a prefix that identifies the amino acid. Ala-tRNA describes $tRNA^{Ala}$ carrying its amino acid.

The secondary structure of each tRNA folds into a compact L-shaped tertiary structure in which the 3′ end that binds amino acid is distant from the anticodon that binds mRNA. All tRNAs have the same general tertiary structure, although they are distinguished by individual variations.

The base-paired double-helical stems of the secondary structure are maintained in the tertiary structure, but their arrangement in three dimensions essentially creates two double helices at right angles to each other, as illustrated in **Figure 5.4.** The acceptor stem and the TψC stem form one continuous double helix with a single gap; the D stem and anticodon stem form another continuous double helix, also with a gap. The region between the double helices, where the turn in the L-shape is made, contains the TψC loop and the D loop. So the amino acid resides at the extremity of one arm of the L-shape, and the anticodon loop forms the other end.

The tertiary structure is created by hydrogen bonding, mostly involving bases that are unpaired in the secondary structure. The bonds of the cloverleaf are described as **secondary H-bonds;** the additional bonds of the tertiary structure are called **tertiary H-bonds.** Many of the invariant and semiinvariant bases are involved in the tertiary H-bonds, which explains their conservation. Not every one of these interactions is universal: rather, they probably identify the *general* pattern for establishing tRNA structure.

A molecular model of the structure of yeast $tRNA^{Phe}$ is shown in **Figure 5.5.** Differences in the structure are found in other tRNAs, thus accommodating the dilemma that all tRNAs must have a similar shape; yet it must be possible to recognize differences between them. For example, in $tRNA^{Asp}$, the angle between the two axes is slightly greater, so the molecule has a slightly more open conformation.

The structure suggests a general conclusion about the function of tRNA. *Its sites for exercising particular functions are maximally separated.* The amino acid is as far distant from the anticodon as possible, which is consistent with their roles in protein synthesis.

The process of charging a tRNA is catalyzed by a specific enzyme, **aminoacyl-tRNA synthetase.** There are (at least) 20 aminoacyl-tRNA synthetases. Each recognizes a single amino acid and all the tRNAs onto which it can legitimately be placed.

Does the anticodon sequence alone allow aminoacyl-tRNA to recognize the correct codon? A classic experiment to test this question is illustrated in **Figure 5.6.** Reductive desulfuration converts the amino acid of cysteinyl-tRNA into alanine, generating alanyl-$tRNA^{Cys}$. The tRNA has an anticodon that responds to the codon UGU. Modification of the amino acid does not influence the specificity of the anticodon-codon interaction, so the alanine residue is incorporated into protein in place of cysteine. *Once a tRNA has been charged, the amino acid plays no further role in its specificity, which is determined exclusively by the anticodon.*

Figure 5.6 The meaning of tRNA is determined by its anticodon and not by its amino acid.

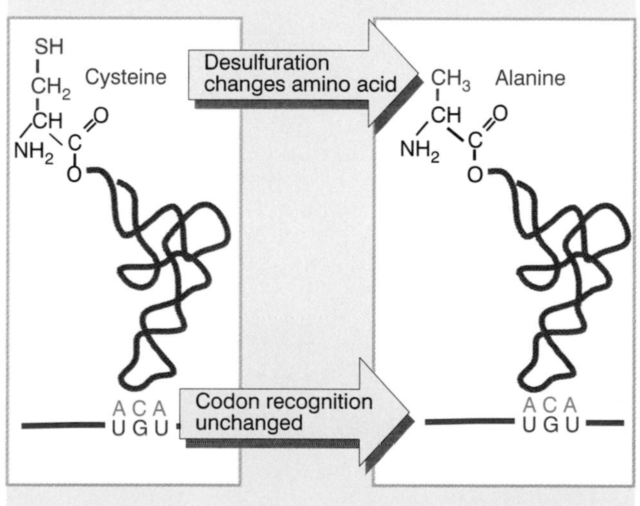

Messenger RNA is translated by ribosomes

TRANSLATION of an mRNA into a polypeptide chain is catalyzed by the **ribosome.** Ribosomes are traditionally described in terms of their (approximate) rate of sedimentation (measured in Svedbergs, in which a higher S value indicates a greater rate of sedimentation and a larger mass). Bacterial ribosomes generally sediment at ~70S. The ribosomes of the cytoplasm of higher eukaryotic cells are larger, usually sedimenting at ~80S.

The ribosome is a compact **ribonucleoprotein particle** consisting of two **subunits.** The relationship between a ribosome and its subunits is depicted in **Figure 5.7.** The two subunits dissociate *in vitro* when the concentration of Mg^{2+} ions is reduced. In each case, the **large subunit** is about twice the mass of the **small subunit.** Bacterial (70S) ribosomes have subunits that sediment at 50S and 30S. The subunits of eukaryotic cytoplasmic (80S) ribosomes sediment at 60S and 40S. The two subunits work together as part of the complete ribosome, but each undertakes distinct reactions in protein synthesis.

Figure 5.7 A ribosome consists of two subunits.

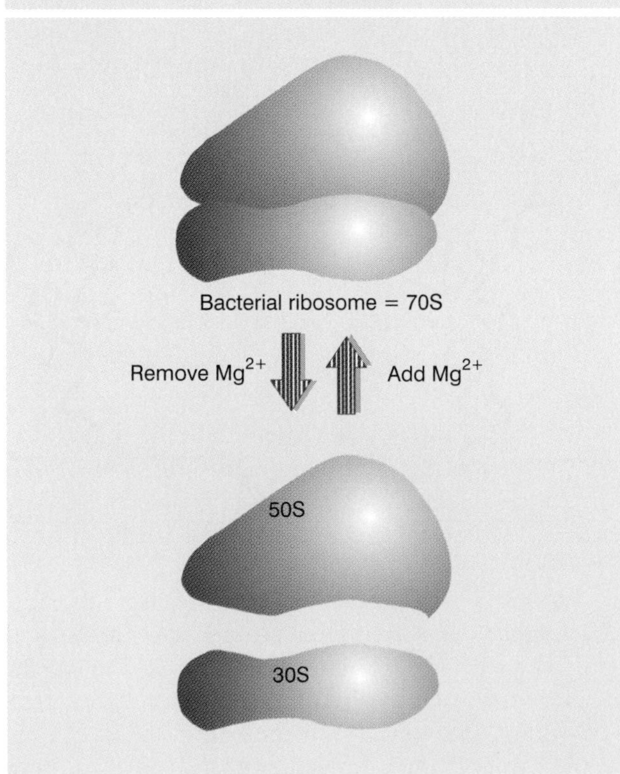

All the ribosomes of a given cell compartment are identical. *They undertake the synthesis of different proteins by associating with the different mRNAs that provide the actual coding sequences.*

The ribosome provides the environment that controls the recognition between a codon of mRNA and the anticodon of tRNA. Reading the genetic code as a series of adjacent triplets, protein synthesis proceeds from the start of a coding region to the end. *A protein is assembled by the sequential addition of amino acids in the direction from the N-terminus to the C-terminus as a ribosome moves along the mRNA.*

A ribosome begins translation at the 5′ end of a coding region; it translates each triplet codon into an amino acid as it proceeds towards the 3′ end. At each codon, the appropriate aminoacyl-tRNA associates with the ribosome, donating its amino acid to the polypeptide chain. At any given moment, the ribosome can accommodate the two aminoacyl-tRNAs corresponding to successive codons, making it possible for a peptide bond to form between the two corresponding amino acids. At each step, the growing polypeptide chain becomes longer by one amino acid.

When active ribosomes are isolated in the form of the fraction associated with newly synthesized proteins, they are found in the form of a unit consisting of an mRNA associated with several ribosomes. This is the **polyribosome** or **polysome.** The 30S subunit of each ribosome is associated with the mRNA, and the 50S subunit carries the newly synthesized protein. The tRNA spans both subunits.

Each ribosome in the polysome independently synthesizes a single polypeptide during its traverse of the messenger sequence. So the mRNA has a series of ribosomes that carry increasing lengths of the protein product, moving from the 5′ to the 3′ end, as illustrated in **Figure 5.8.** A polypeptide chain in the process of synthesis is sometimes called a **nascent protein.**

Roughly the last 30–35 amino acids added to a growing polypeptide chain are protected from the environment by the structure of the ribosome. Probably all of the preceding part of the polypeptide protrudes and is free to start folding into its proper conformation. So proteins can display parts of the mature conformation even before synthesis has been completed.

A classic characterization of polysomes is shown in the electron micrograph of **Figure 5.9.** Globin protein

Figure 5.8 A polyribosome consists of an mRNA being translated simultaneously by several ribosomes moving in the direction from 5' to 3'. Each ribosome has two tRNA molecules: one carrying the nascent protein, the second carrying the next amino acid to be added.

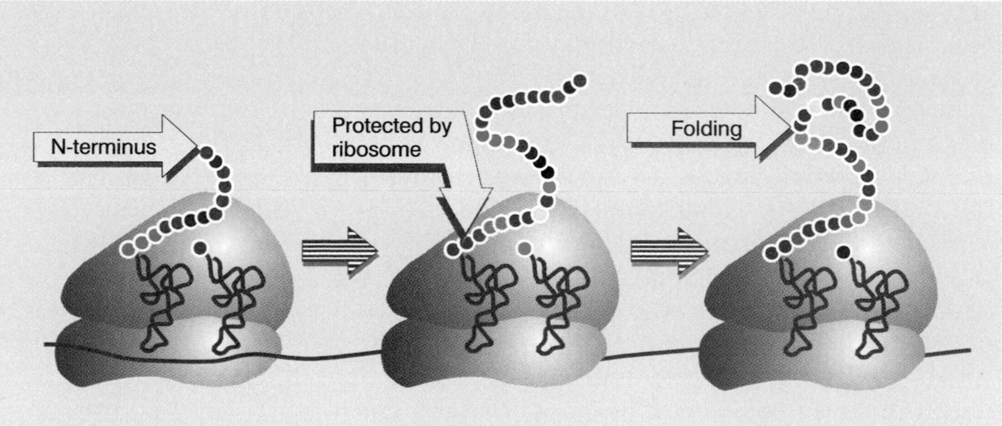
N-terminus
Protected by ribosome
Folding

Figure 5.9 Protein synthesis occurs on polysomes. Photograph kindly provided by Alex Rich.

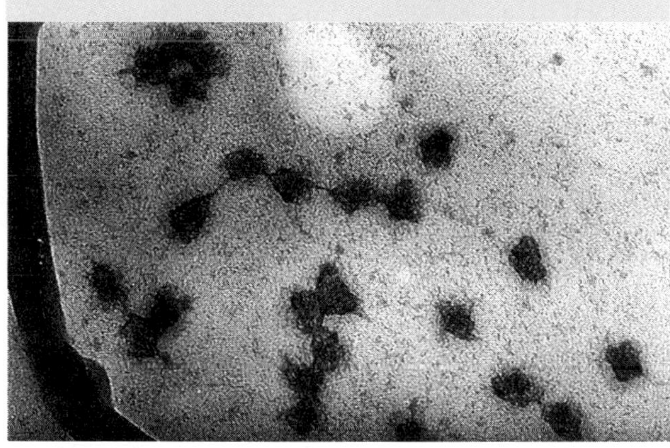

Figure 5.10 Messenger RNA is translated by ribosomes that cycle through a pool.

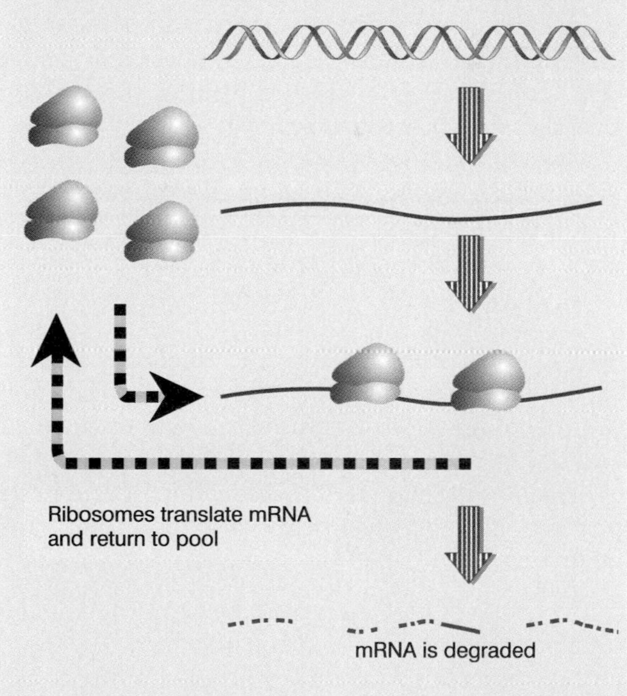
Ribosomes translate mRNA and return to pool
mRNA is degraded

is synthesized by a set of five ribosomes attached to each mRNA (pentasomes). The ribosomes appear as squashed spherical objects of ~7 nm (70 Å) in diameter, connected by a thread of mRNA. The ribosomes are located at various positions along the messenger. Those at one end have just started protein synthesis; those at the other end are about to complete production of a polypeptide chain.

The size of the polysome depends on several variables. In bacteria, it is very large, with tens of ribosomes simultaneously engaged in translation. The size is partly due to the length of the mRNA (which usually codes for several proteins) and partly due to the high efficiency with which the ribosomes attach to the mRNA.

Polysomes in the cytoplasm of a eukaryotic cell are likely to be smaller than those in bacteria; again, their size is a function both of the length of the mRNA (usually representing only a single protein in eukaryotes) and of the characteristic frequency with which ribosomes attach. An average eukaryotic mRNA probably has ~8 ribosomes attached at any one time.

The cycle of the ribosome is illustrated in **Figure 5.10.** Ribosomes are drawn from a pool (actually the pool consists of ribosomal subunits), used to translate an mRNA, and then return to the pool for further cycles. The number of ribosomes on each mRNA molecule synthesizing a particular protein is not precisely determined, in either bacteria or eukaryotes, but is a matter of statistical fluctuation, determined by the variables of mRNA size and efficiency.

An overview of the attention devoted to protein synthesis in the intact bacterium is given in **Figure 5.11**. The 20,000 or so ribosomes account for a quarter of the cell mass. There are >3000 copies of each tRNA, and, altogether, the tRNA molecules outnumber the ribosomes by almost tenfold; most of them are present as aminoacyl-tRNAs, that is, ready to be used at once in protein synthesis. Because of their instability, it is difficult to calculate the number of mRNA molecules, but a reasonable guess would be ~1500, in varying states of synthesis and decomposition. There are ~600 different types of mRNA in a bacterium. This suggests that there are usually only 2–3 copies of each mRNA per bacterium. On average, each probably codes for ~3 proteins. If there are 1850 different soluble proteins, and an average of 2.0×10^6 protein molecules per bacterium, there must on average be >1000 copies of each protein in a bacterium.

Figure 5.11 Considering *E. coli* in terms of its macromolecular components.

Component	Dry Cell Mass (%)	Molecules /Cell	Different types	Copies of each type
Wall	10	1	1	1
Membrane	10	2	2	1
DNA	1.5	1	1	1
mRNA	1	1,500	600	2–3
tRNA	3	200,000	60	>3000
rRNA	16	38,000	2	19,000
Ribosomal protein	9	10^6	52	19,000
Soluble protein	46	2.0×10^6	1850	>1000
Small molecules	3	7.5×10^6	800	

The life cycle of messenger RNA

MESSENGER RNA has the same function in all cells, but there are important differences in the details of the synthesis and structure of prokaryotic and eukaryotic mRNA.

A major difference in the production of mRNA depends on the locations where transcription and translation occur:

- In bacteria, mRNA is transcribed and translated in the single cellular compartment; and the two processes are so closely linked that they occur simultaneously. Since ribosomes attach to bacterial mRNA even before its transcription has been completed, the polysome is likely still to be attached to DNA. Bacterial mRNA usually is unstable, and is therefore translated into proteins for only a few minutes.
- In a eukaryotic cell, synthesis and maturation of mRNA occur exclusively in the nucleus. Only after these events are completed is the mRNA exported to the cytoplasm, where it is translated by ribosomes. Eukaryotic mRNA is relatively stable and continues to be translated for several hours.

Figure 5.12 shows that transcription and translation are intimately related in bacteria. Transcription begins when the enzyme RNA polymerase binds to DNA and then moves along making a copy of one strand. As soon as transcription begins, ribosomes attach to the 5′ end of the mRNA and start translation, even before the rest of the message has been synthesized. A bunch of ribosomes moves along the mRNA while it is being synthesized. The 3′ end of the mRNA is generated when transcription terminates. Ribosomes continue to translate the mRNA while it survives, but it is degraded in the overall 5′ → 3′ direction quite rapidly. The mRNA is synthesized, translated by the ribosomes and degraded, all in rapid succession. An individual molecule of mRNA survives for only a matter of minutes or even less.

Bacterial transcription and translation take place at similar rates. At 37°C, transcription of mRNA occurs at ~40 nucleotides/second. This is very close to the rate of protein synthesis, roughly 15 amino acids/second. It therefore takes ~2 minutes to transcribe and translate an mRNA of 5000 bp, corresponding to 180 kD of protein. When expression of a new gene is initiated, its mRNA typically will appear in the cell within ~2.5

Figure 5.12 *Overview*: mRNA is transcribed, translated, and degraded simultaneously in bacteria.

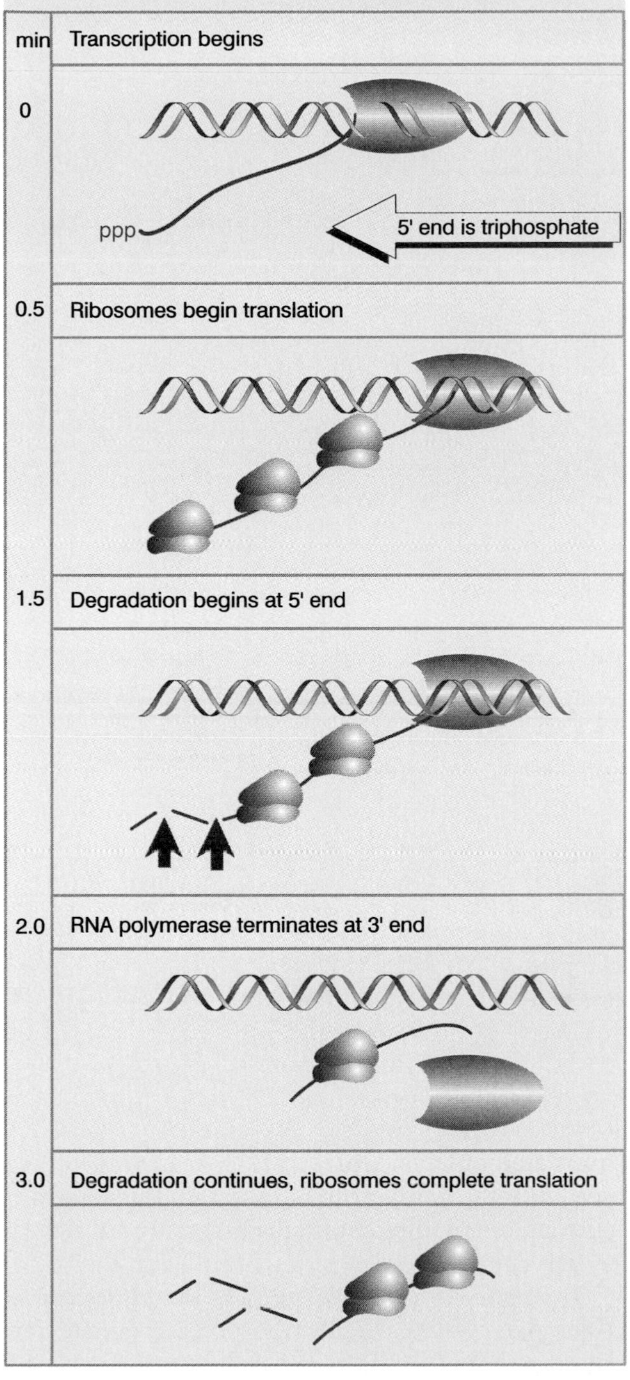
min
Transcription begins
0
ppp
5' end is triphosphate
0.5
Ribosomes begin translation
1.5
Degradation begins at 5' end
2.0
RNA polymerase terminates at 3' end
3.0
Degradation continues, ribosomes complete translation

minutes. The corresponding protein will appear within perhaps another 0.5 minute.

Bacterial translation is very efficient, and most mRNAs are translated by a large number of tightly packed ribosomes. In one example (*trp* mRNA), about 15 initiations of transcription occur every minute, and each of the 15 mRNAs is probably translated by ~30 ribosomes in the interval between its transcription and degradation.

The instability of most bacterial mRNAs is striking. Degradation of mRNA closely follows its translation. Probably it begins within 1 minute of the start of transcription. The 5′ end of the mRNA starts to decay before the 3′ end has been synthesized or translated. Degradation seems to follow the last ribosome of the convoy along the mRNA. But degradation proceeds more slowly, probably at about half the speed of transcription or translation.

The stability of mRNA has a major influence on the amount of protein that is produced. It is usually expressed in terms of the half-life. The mRNA representing any particular gene has a characteristic half-life, but the average is ~2 minutes in bacteria.

This series of events is only possible, of course, because transcription, translation, and degradation all occur in the same direction. The dynamics of gene expression have been caught *in flagrante delicto* in the electron micrograph of **Figure 5.13.** In these (unknown) transcription units, several mRNAs are under synthesis simultaneously; and each carries many ribosomes engaged in translation. (This corresponds to the stage shown in the second panel in Figure 5.12.) An RNA whose synthesis has not yet been completed is often called a **nascent RNA.**

Bacterial mRNAs vary greatly in the number of proteins for which they code. Some mRNAs represent only a single gene: they are **monocistronic.** Others (the majority) carry sequences coding for several proteins: they are **polycistronic.** In these cases, a single mRNA is transcribed from a group of adjacent genes. (Such a cluster of genes constitutes an *operon* that is controlled as a single genetic unit; see Chapter 10.)

All mRNAs contain two types of region. The **coding region** consists of a series of codons representing the amino acid sequence of the protein, starting (usually) with AUG and ending with a termination codon. But the mRNA is always longer than the coding region. In a monocistronic mRNA, extra regions are present at both ends. An additional sequence at the 5′ end, preceding the start of the coding region, is described as a **leader.** An additional sequence following the termination signal, forming the 3′ end, is called a **trailer.** Although part of the transcription unit, these sequences are not used to code for protein.

The structure of a polycistronic mRNA is illustrated in **Figure 5.14.** The **intercistronic regions** that lie between the various coding regions vary greatly in size. They may be as long as 30 nucleotides in bacterial

Figure 5.13 Transcription units can be visualized in bacteria. Photograph kindly provided by Oscar Miller.

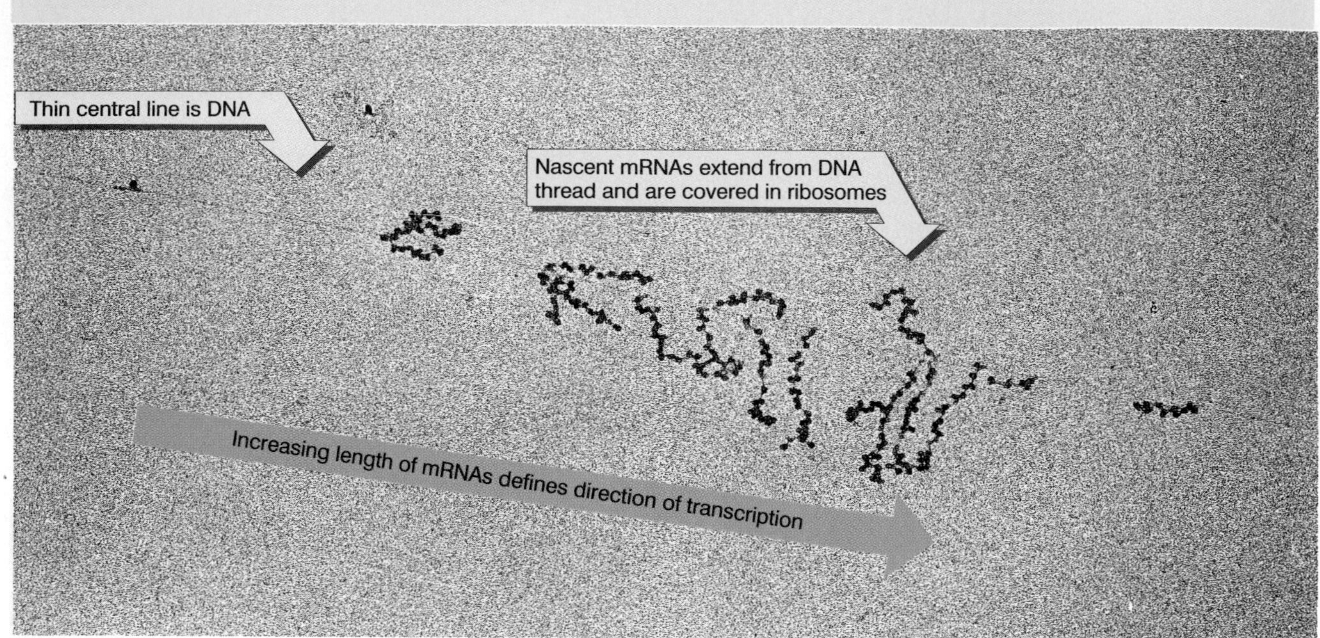
Thin central line is DNA
Nascent mRNAs extend from DNA thread and are covered in ribosomes
Increasing length of mRNAs defines direction of transcription

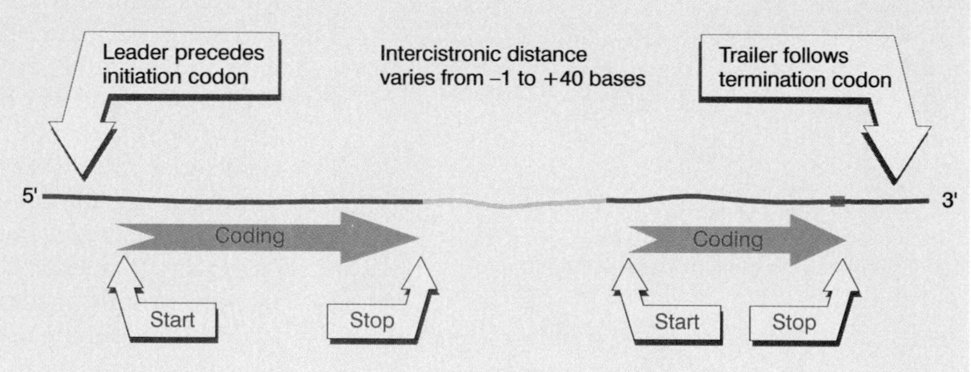
Leader precedes initiation codon
Intercistronic distance varies from –1 to +40 bases
Trailer follows termination codon
5'
3'
Coding
Coding
Start
Stop
Start
Stop

Figure 5.14 Bacterial mRNA includes non-translated as well as translated regions. Each coding region has its own initiation and termination signals. A typical mRNA may have several coding regions.

mRNAs (and even longer in phage RNAs), but they can also be very short, with as few as 1 or 2 nucleotides separating the termination codon for one protein from the initiation codon for the next. In an extreme case, two genes actually overlap, so that the last base of one coding region is also the first base of the next coding region.

The number of ribosomes engaged in translating a particular cistron depends on the efficiency of its initiation site. The initiation site for the first cistron becomes available as soon as the 5′ end of the mRNA is synthesized. How are subsequent cistrons translated? Are the several coding regions in a polycistronic mRNA translated independently or is their expression connected? Is the mechanism of initiation the same for all cistrons, or is it different for the first cistron and the internal cistrons?

Translation of a bacterial mRNA proceeds sequentially through its cistrons. Usually ribosomes terminate translation at the end of the first cistron (and dissociate into subunits), and a new ribosome assembles independently at the start of the next coding region. (We discuss the processes of initiation and termination in the next chapter; see Figure 6.15).

Translation of eukaryotic mRNA

The production of eukaryotic mRNA involves additional stages. Transcription occurs in the usual way, initiating a transcript with a 5′ triphosphate end. However, the 3′ end is generated by cleaving the transcript, rather than by terminating transcription at a fixed site. Those RNAs that are derived from interrupted genes require splicing to remove the introns, generating a smaller mRNA that contains an intact coding sequence. **Figure 5.15** shows that both ends of the transcript are modified by additions of further nucleotides (involving additional enzyme systems). The 5′ end of the RNA is modified by addition of a "cap" virtually as soon as it appears. This replaces the triphosphate of the initial transcript with a nucleotide in reverse (3′–5′) orientation, thus "sealing" the end. The 3′ end is modified by addition of a series of adenylic acid nucleotides (polyadenylic acid or poly(A)) immediately after its cleavage. Only after the completion of all modification and processing events can the mRNA be exported from the nucleus to the cytoplasm. The average delay in leaving for the cytoplasm is ~20 minutes. Once the mRNA has entered the cytoplasm, it is recognized by ribosomes and translated.

Figure 5.16 shows that the life cycle of eukaryotic mRNA is more protracted than that of bacterial mRNA. Transcription in animal cells occurs at about the same speed as in bacteria, ~40 nucleotides per second. Many eukaryotic genes are large; a gene of 10,000

Figure 5.15 Eukaryotic mRNA is modified by addition of a cap to the 5′ end and poly(A) to the 3′ end.

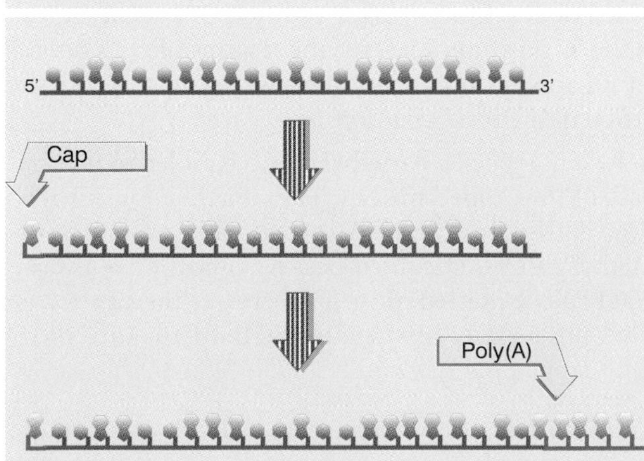

Figure 5.16 *Overview*: expression of mRNA in animal cells requires transcription, modification, processing, nucleocytoplasmic transport, and translation.

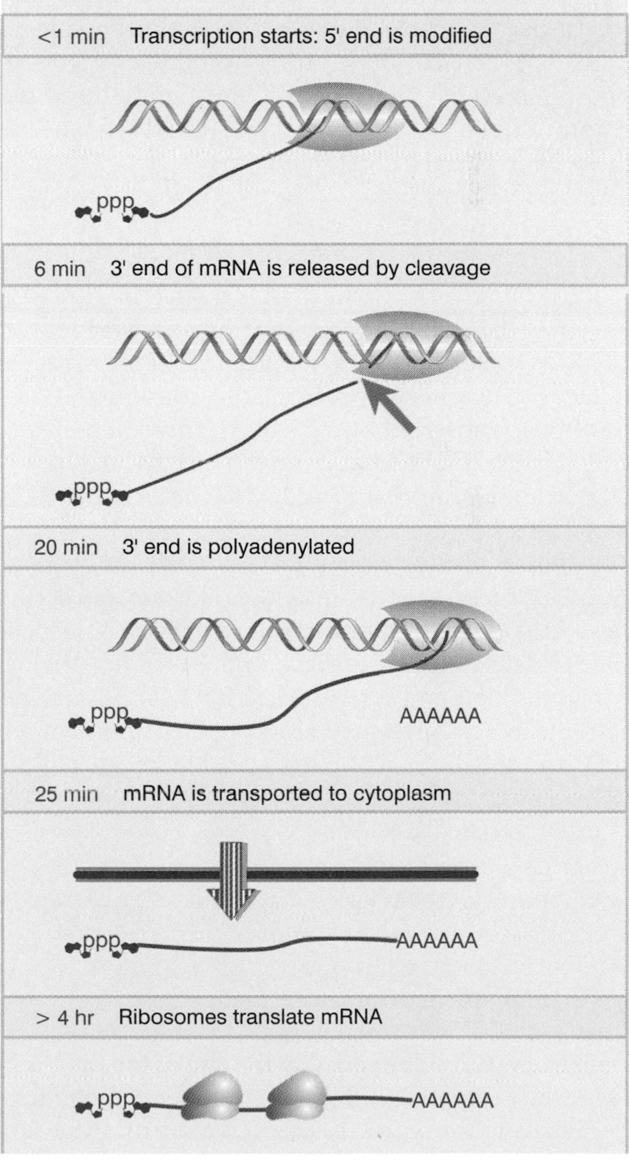

bp takes ~5 minutes to transcribe. Transcription of mRNA is not terminated by the release of enzyme from the DNA; instead, the enzyme continues past the end of the gene, and the 3′ end of the RNA is generated by a cleavage event.

Eukaryotic mRNA constitutes only a small proportion of the total cellular RNA (~3% of the mass). Half-lives are relatively short in yeast, ranging from 1–60 minutes. There is a substantial increase in stability in higher eukaryotes; animal cell mRNA is relatively stable, with half-lives ranging from 4–24 hours.

Eukaryotic polysomes are reasonably stable. The modifications at both ends of the mRNA contribute to the stability.

The 5′ end of eukaryotic mRNA is capped

TRANSCRIPTION starts with a nucleoside triphosphate (usually a purine, A or G). The first nucleotide retains its 5′ triphosphate group and makes the usual phosphodiester bond from its 3′ position to the 5′ position of the next nucleotide. The initial sequence of the transcript can be represented as:

$$5'\ ppp^{A}_{G}pNpNpNp...$$

But when the mature mRNA is treated *in vitro* with enzymes that should degrade it into individual nucleotides, the 5′ end does not give rise to the expected nucleoside triphosphate. Instead it contains *two* nucleotides, connected by a *5′–5′ triphosphate linkage* and also bearing methyl groups. The terminal base is always a guanine that is added to the original RNA molecule *after transcription.*

Addition of the 5′ terminal G is catalyzed by a nuclear enzyme, guanylyl transferase. The reaction occurs so soon after transcription has started that it is not possible to detect more than trace amounts of the original 5′ triphosphate end in the nuclear RNA. The overall reaction can be represented as a condensation between GTP and the original 5′ triphosphate terminus of the RNA. Thus:

$$\begin{array}{l} \quad 5' \quad\ 5' \\ \text{Gppp} + \textit{ppp}\text{ApNpNp...} \\ \qquad\ \downarrow \\ \quad 5'-5' \\ \text{Gpp}\textit{p}\text{ApNpNp...} + \text{pp} + \textit{p} \end{array}$$

The new G residue added to the end of the RNA is in the reverse orientation from all the other nucleotides.

This structure is called a **cap.** It is a substrate for several methylation events. **Figure 5.17** shows the full structure of a cap after all possible methyl groups have been added. Types of caps are distinguished by how many of these methylations have occurred:

- The first methylation occurs in all eukaryotes, and consists of the addition of a methyl group to the 7 position of the terminal guanine. A cap that possesses this single methyl group is known as a **cap 0.** This is as far as the reaction proceeds in unicellular eukaryotes. The enzyme responsible for this modification is called guanine-7-methyltransferase.
- The next step is to add another methyl group, to the 2′-O position of the penultimate base (which was actually the original first base of the transcript before any modifications were made). This reaction is catalyzed by another enzyme (2′-O-methyl-transferase). A cap with the two methyl groups is called **cap 1.** This is the predominant type of cap in all eukaryotes except unicellular organisms.
- In a small minority of cases in higher eukaryotes, another methyl group is added to the second base. This happens only when the position is occupied by adenine; the reaction involves addition of a methyl group at the N^6 position. The enzyme responsible acts only on an adenosine substrate that already has the methyl group in the 2′-O position.
- In some species, a methyl group is added to the third base of the capped mRNA. The substrate for this reaction is the cap 1 mRNA that already possesses two methyl groups. The third-base modification is always a 2′-O ribose methylation. This creates the **cap 2** type. This cap usually represents less than 10–15% of the total capped population.

In a population of eukaryotic mRNAs, every molecule is capped. The proportions of the different types of cap are characteristic for a particular organism. We do not know whether the structure of a particular

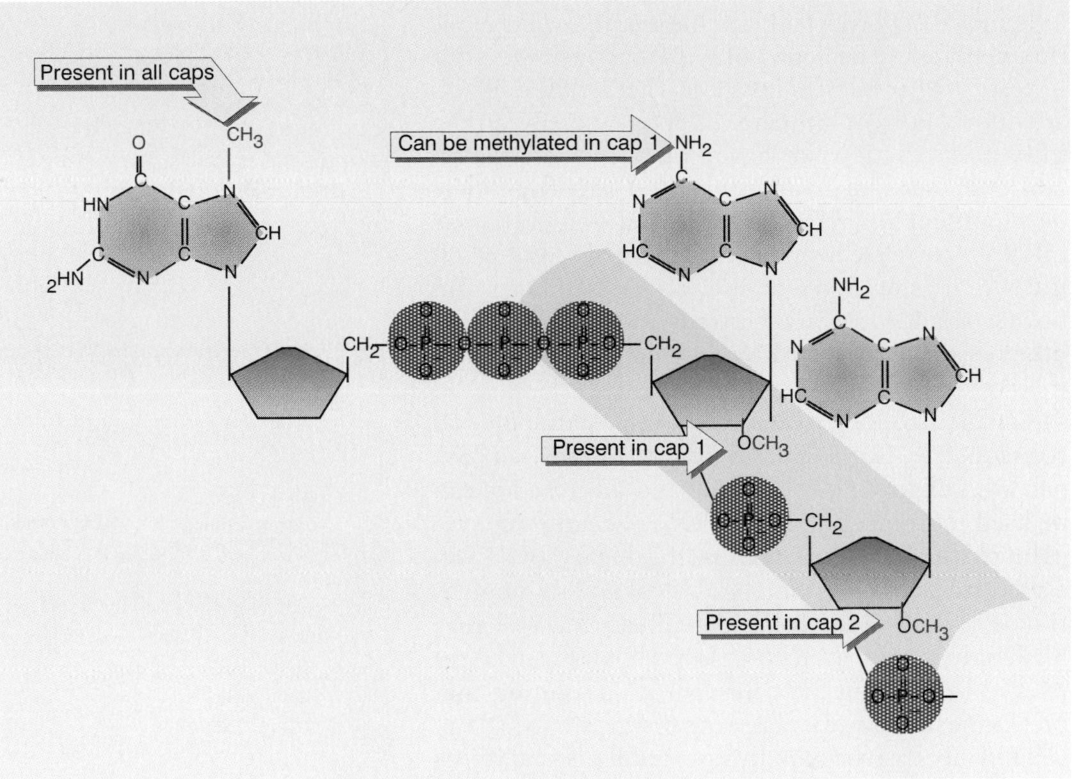

Figure 5.17 The cap blocks the 5′ end of mRNA and may be methylated at several positions.

mRNA is invariant or can have more than one type of cap.

In addition to the methylation involved in capping, a low frequency of internal methylation occurs in the mRNA only of higher eukaryotes. This is accomplished by the generation of N^6 methyladenine residues at a frequency of about one modification per 1000 bases. There are 1–2 methyladenines in a typical higher eukaryotic mRNA, although their presence is not obligatory, since some mRNAs do not have any.

The 3′ terminus is polyadenylated

The 3′ terminal stretch of A residues is often described as the **poly(A) tail;** and mRNA with this feature is denoted **poly(A)$^+$**.

The poly(A) sequence is not coded in the DNA, but is added to the RNA in the nucleus after transcription. The addition of poly(A) is catalyzed by the enzyme **poly(A) polymerase,** which adds ~200 A residues to the free 3′-OH end of the mRNA. The poly(A) tract of both nuclear RNA and mRNA is associated with a protein, the **poly(A)-binding protein (PABP).** Related forms of this protein are found in many eukaryotes. One PABP monomer of ~70 kDa is bound every 10–20 bases of the poly(A) tail. So a common feature in many or most eukaryotes is that the 3′ end of the mRNA consists of a stretch of poly(A) bound to a large mass of protein. Addition of poly(A) occurs as part of a reaction in which the 3′ end of the mRNA is generated and modified by a complex of enzymes (see Chapter 19).

What is the role of poly(A)? In several (but not all) situations, it confers stability upon mRNA. Removal of the poly(A) tail precedes degradation of certain mRNAs; stability of mRNA is likely to be connected with poly(A), although it is not clear whether the relationship is universal. The ability of the poly(A) to protect mRNA against degradation requires binding of the PABP.

Removal of poly(A) inhibits the initiation of translation *in vitro*, and depletion of PABP has the same effect in yeast *in vivo*, but it is not clear whether these effects are due to a direct influence of poly(A)-PABP on the initiation reaction or have some indirect cause. However, there are many examples in early embryonic development where polyadenylation of a particular mRNA is correlated with its translation. In some cases, mRNAs are stored in a nonpolyadenylated form, and poly(A) is added when their translation is required; in other cases, poly(A)$^+$ mRNAs are de-adenylated, and their translation is reduced.

The presence of poly(A) has an important practical consequence. The poly(A) region of mRNA can base pair with oligo(U) or oligo(dT); and this reaction can be used to isolate poly(A)$^+$ mRNA. The most convenient technique is to immobilize the oligo(U or dT) on a solid support material. Then when an RNA population is applied to the column, as illustrated in **Figure 5.18,** only the poly(A)$^+$ RNA is retained. It can be retrieved by treating the column with a solution that breaks the bonding to release the RNA.

The only drawback to this procedure is that it isolates *all* the RNA that contains poly(A). If RNA of the whole cell is used, for example, both nuclear and cytoplasmic poly(A)$^+$ RNA will be retained. If preparations of polysomes are used (a common procedure), most of the isolated poly(A)$^+$ RNA will be active mRNA. However, in addition to mRNA in polysomes, there are also ribonucleoprotein particles in the cytosol that contain poly(A)$^+$ mRNA, but which are not translated. This RNA may be "stored" for use at some other time. Isolation of total poly(A)$^+$ mRNA therefore does not correspond exactly with the active mRNA population.

The "cloning" approach for purifying mRNA uses a procedure in which the mRNA is copied to make a complementary DNA strand (known as **cDNA**). Then the cDNA can be used as a template to synthesize a DNA strand that is identical with the original mRNA sequence. The product of these reactions is a double-stranded DNA corresponding to the sequence of the mRNA. This DNA can be reproduced in large amounts.

Figure 5.18 Poly(A)$^+$ RNA can be separated from other RNAs by fractionation on Sepharose-oligo(dT).

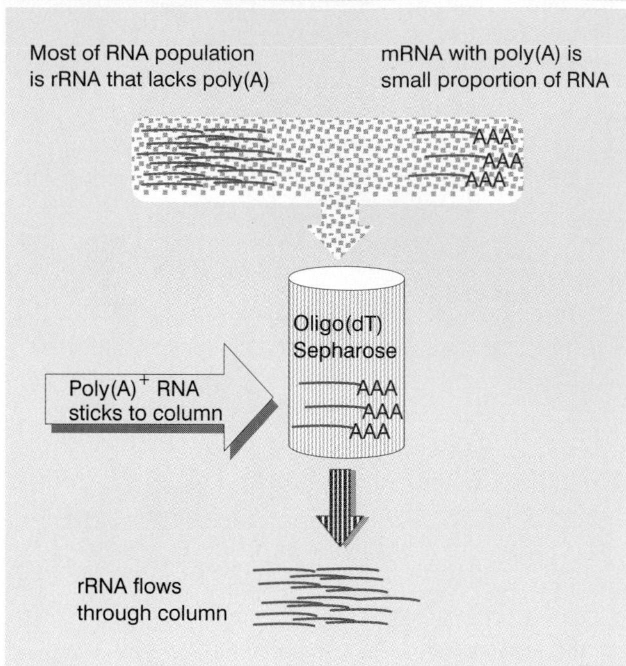
Most of RNA population is rRNA that lacks poly(A)
mRNA with poly(A) is small proportion of RNA
AAA
AAA
AAA
Oligo(dT) Sepharose
Poly(A)+ RNA sticks to column
AAA
AAA
AAA
rRNA flows through column

The availability of a cloned DNA makes it easy to isolate the corresponding mRNA by hybridization techniques. Even mRNAs that are present in only very few copies per cell can be isolated by this approach. Indeed, only mRNAs that are present in relatively large amounts can be isolated directly without using a cloning step.

Almost all cellular mRNAs possess poly(A). A significant exception is provided by the mRNAs that code for the histone proteins (a major structural component of chromosomal material). These mRNAs comprise most or all of the **poly(A)$^-$** fraction. The significance of the absence of poly(A) from histone mRNAs is not clear, and there is no particular aspect of their function for which this appears to be necessary.

Degradation pathways for mRNA

Bacterial mRNA is constantly degraded by a combination of endonucleases and exonucleases. Endonucleases cleave an RNA at an internal site. Exonucleases are involved in trimming reactions in which the extra residues are whittled away, base by base from the end. Bacterial exonucleases that act on single-stranded RNA proceed along the nucleic acid chain from the 3′ end.

For bacterial mRNA, susceptibility to degradation is conferred by sequences that are targets for endonucleolytic attack. Several 3′ ends may be generated by endonucleolytic cleavage within the mRNA. A general model for the degradation of bacterial mRNA is shown in **Figure 5.19.** The *overall* direction of degradation (as measured by loss of ability to synthesize proteins) is 5′ →3′. This probably results from a succession of endonucleolytic cleavages following the last ribosome. Degradation of the released fragments of mRNA into nucleotides proceeds by exonucleolytic attack from the free 3′-OH end toward the 5′ terminus (that is, in the opposite direction to transcription). Endonucleolytic attack releases fragments that may have different susceptibilities to exonucleases. A region of secondary structure within the mRNA may provide an obstacle to the exonuclease, thus protecting the regions on its 5′ side. The stability of each mRNA is therefore determined by the susceptibility of its particular sequence to both endo- and exonucleolytic cleavages..

There are ~12 ribonucleases in *E. coli.* Mutants in the endoribonucleases (except ribonuclease I, which is without effect) accumulate unprocessed precursors to rRNA and tRNA, but are viable. Mutants in the exonucleases often have apparently unaltered phenotypes, which suggests that one enzyme can substitute for the absence of another. Mutants lacking multiple enzymes sometimes are inviable.

Bacterial mutants that have a defective ribonuclease E have increased stability (2–3 fold) of mRNA as well as of the small 5′ ribosomal RNA. RNAase E is the enzyme that is responsible for processing 5′ rRNA from the primary transcript by a specific endonucleolytic processing event. It is also involved in the general decay of mRNA, and may be the enzyme that makes the first cleavage for many mRNAs.

Polyadenylation may play a role in initiating degradation of some mRNAs in bacteria. Poly(A) polymerase is associated with ribosomes in *E. coli,* and short (10–40 nucleotide) stretches of poly(A) are added to at least some mRNAs. Triple mutations that remove poly(A) polymerase, ribonuclease E, and polynucleotide phosphorylase (a 3′–5′ endonuclease) have a strong effect on stability. (Mutations in individual genes or pairs of genes have only a weak effect.) Poly(A) polymerase may create a poly(A) tail that acts

Figure 5.19 Degradation of bacterial mRNA is a two stage process. Endonucleolytic cleavages proceed 5'→3' behind the ribosomes. The released fragments are degraded by exonucleases that move 3'→5'.

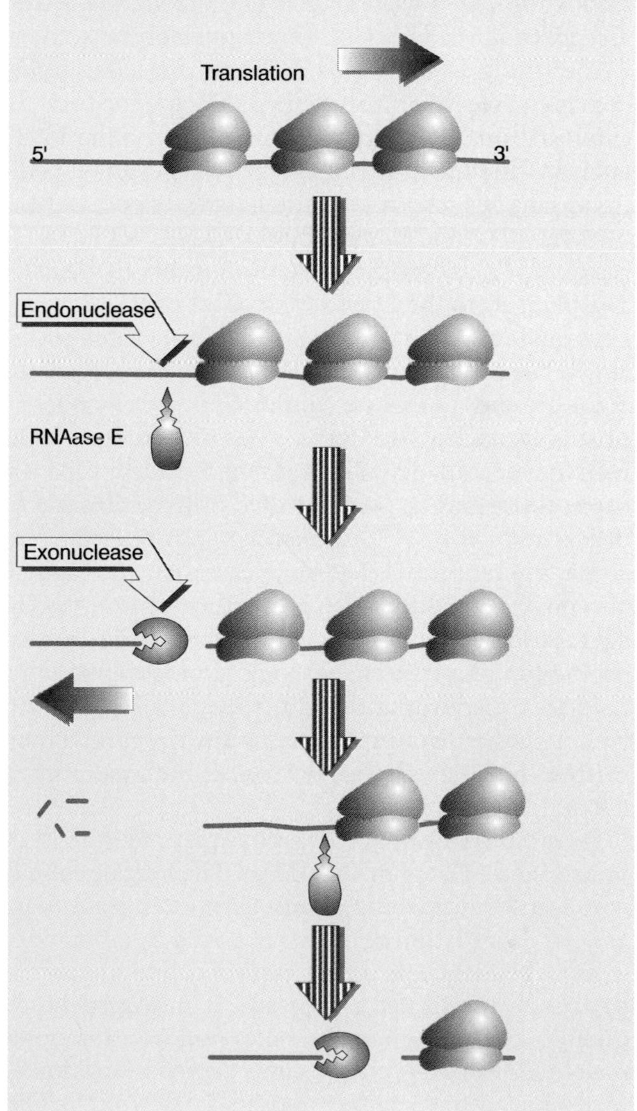

as a binding site for the nucleases. The role of poly(A) in bacteria would therefore be different from that in eukaryotic cells.

The process of degradation may be catalyzed by a multienzyme complex that includes ribonuclease E, PNPase, and a helicase. RNAase E plays dual roles. Its N-terminal domain provides an endonuclease activity. The C-terminal domain provides a scaffold that holds together the other components. The helicase unwinds the substrate RNA to make it available to PNPase. In effect, RNAase E makes the initial cut and then passes the fragments to the other components of the complex for processing.

The pathway for degradation of mRNA is different in yeast. **Figure 5.20** shows that the modifications at each end of the mRNA provide an obstacle to degradation. They must be removed in a specific order. First the mRNA is deadenylated, leaving it with a short stretch of poly(A). The mechanism is not known, but is usually pictured in terms of an exonucleolytic activity. Removal of the poly(A) is necessary for a **decapping enzyme** to cleave off the methylated cap. The basis for this relationship is that the presence of the PABP (poly(A)-binding protein) on the poly(A) prevents the decapping enzyme from binding to the 5′ end. PABP is released when the length of poly(A) falls below 10–15 residues. The decapping reaction occurs by cleavage 1–2 bases from the 5′ end. The effect of PABP on decapping implies a relationship between events occurring at the two ends of the mRNA. It is possible that the 5′ and 3′ ends in fact are connected in the ribonucleoprotein particle. Once the cap has been removed, the mRNA is degraded rapidly from the 5′ end, by the 5′–3′ exonuclease XRN1. Deadenylated mRNAs also can be degraded by the 3′–5′ exonuclease activity of the **exosome,** a complex of >5 exonucleases. The exosome is also involved in processing precursors for rRNAs. The aggregation of the individual exonucleases into the exosome complex may enable 3′–5′ exonucleolytic activities to be coordinately controlled. Yeast mutants lacking either exonucleolytic pathway degrade their mRNAs more slowly, but the loss of both pathways is lethal.

Another pathway for degradation is identified by nonsense-mediated mRNA decay. The introduction of a nonsense mutation in a yeast gene often leads to increased degradation of the mRNA. As may be expected from dependence on a termination codon, the degradation occurs in the cytoplasm. It may represent a quality control system for removing nonfunctional mRNAs. We do not yet know how degradation is linked mechanically to the failure to translate the mRNA. One possibility is that translation suppresses the use of destabilizing elements in the mRNA. Genes that are required for the process have been identified in *S. cerevisiae* (*upf* loci) and *C. elegans* (*smg* loci) by identifying suppressors of nonsense-mediated degradation, but their molecular activities have not yet been characterized.

Figure 5.20 Degradation of yeast mRNA requires deadenylation, decapping, and exonucleolysis.

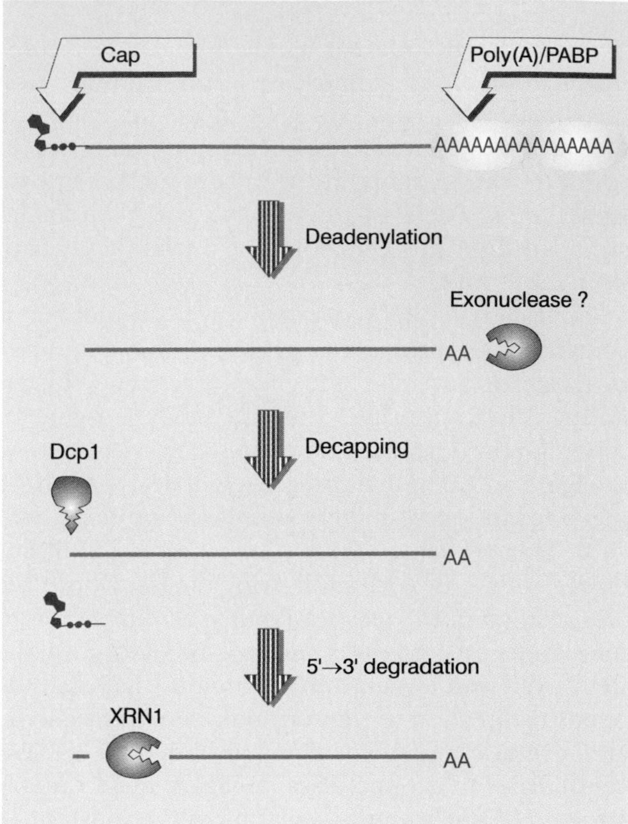

The stability of eukaryotic mRNA is often controlled by the presence (or absence) of specific destabilizing elements. The introduction of any one of these new elements into a new mRNA may cause it to be degraded. The removal of an element from an mRNA usually does not stabilize it, suggesting that an individual mRNA is likely to have more than one destabilizing element. Two types of destabilizing element have been best characterized.

A common feature in some unstable mRNAs is the presence of an AU-rich sequence of ~50 bases (called the ARE) that is found in the 3′ trailer region. The consensus sequence in the ARE is the pentanucleotide AUUUA, repeated several times. **Figure 5.21** shows that the ARE triggers destabilization by a two-stage process: first the mRNA is deadenylated; then it decays. The

Figure 5.21 An ARE in a 3' nontranslated region initiates degradation of mRNA.

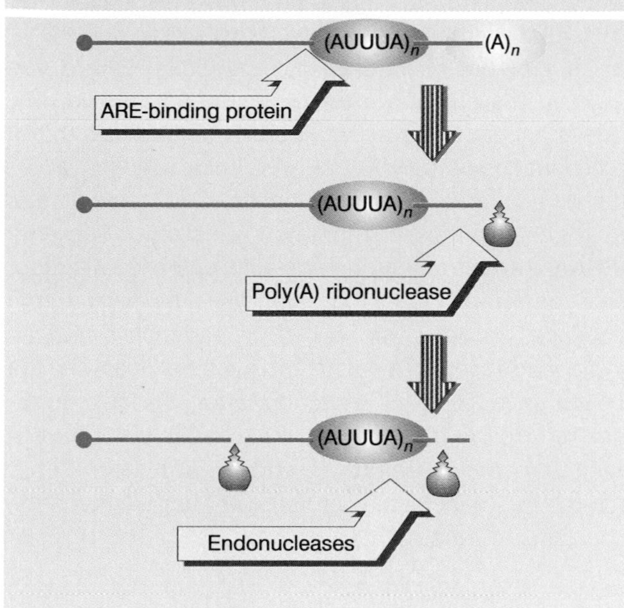

Figure 5.22 An IRE in a 3' nontranslated region controls degradation of mRNA.

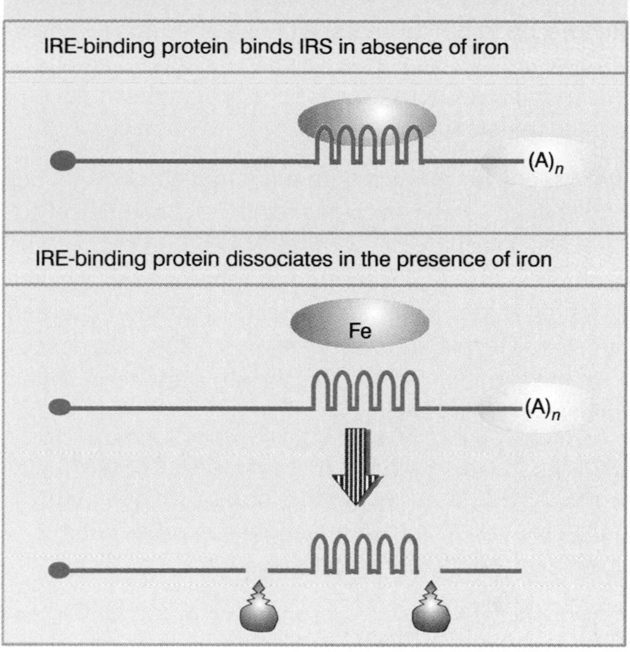

deadenylation is probably needed because it causes loss of the poly(A)-binding protein, whose presence stabilizes the 3′region.

Transferrin mRNA contains a sequence called the IRE, which controls the response of the mRNA to changes in iron concentration. The IRE is located in the 3′ nontranslated region, and contains stem-loop structures that bind a protein whose affinity for the mRNA is controlled by iron. **Figure 5.22** shows that binding of the protein to the IRE stabilizes the mRNA by inhibiting the function of (unidentified) destabilizing sequences in the vicinity. This is a general model for the stabilization of mRNA, that is, stability is conferred by inhibiting the function of destabilizing sequences.

Summary

Genetic information carried by DNA is expressed in two stages: transcription of DNA into mRNA; and translation of the mRNA into protein. Messenger RNA is transcribed from one strand of DNA and is complementary to this (noncoding) strand and identical with the other (coding) strand. The sequence of mRNA, in triplet codons 5′→3′, is related to the amino acid sequence of protein, N- to C-terminal.

The adapter that interprets the meaning of a codon is transfer RNA, which has a compact L-shaped tertiary structure; one end of the tRNA has an anticodon that is complementary to the codon, and the other end can be covalently linked to the specific amino acid that corresponds to the target codon. A tRNA carrying an amino acid is called an aminoacyl-tRNA.

The ribosome provides the apparatus that allows aminoacyl-tRNAs to bind to their codons on mRNA. The small subunit of the ribosome is bound to mRNA; the large subunit carries the nascent polypeptide. A ribosome moves along mRNA from an initiation site in the 5′ region to a termination site in the 3′ region, and the appropriate aminoacyl-tRNAs respond to their codons, unloading their amino acids, so that the growing polypeptide chain extends by one residue for each codon traversed.

The translational apparatus is not specific for tissue or organism; an mRNA from one source can be translated by the ribosomes and tRNAs from another source. The number of times any mRNA is translated is a function of the affinity of its initiation site(s) for ribosomes and its stability. There are some cases in which translation of groups of mRNAs or individual mRNAs is specifically prevented: this is called translational control.

A typical mRNA contains both a nontranslated 5′ leader and 3′ trailer as well as coding region(s). Bacterial mRNA is usually polycistronic, with nontranslated regions between the cistrons. Each cistron is represented by a coding region that starts with a specific initiation site and ends with a termination site. Ribosome subunits associate at the initiation site and dissociate at the termination site of each coding region.

A growing *E. coli* bacterium has ~20,000 ribosomes and ~200,000 tRNAs, mostly in the form of aminoacyl-tRNA. There are ~1500 mRNA molecules, representing 2–3 copies each of 600 different messengers.

Many ribosomes may translate a single mRNA simultaneously, generating a polyribosome (or polysome). Bacterial polysomes are large, typically with tens of ribosomes bound to a single mRNA. Eukaryotic polysomes are smaller, typically with fewer than 10 ribosomes; each mRNA carries only a single coding sequence.

Eukaryotic mRNA must be processed in the nucleus before it is transported to the cytoplasm for translation. A methylated cap is added to the 5′ end. It consists of a nucleotide added to the original end by a 5′–5′ bond, after which methyl groups are added. Most eukaryotic mRNA has an ~200 base sequence of poly(A) added to its 3′ terminus in the nucleus after transcription, but poly(A)$^-$ mRNAs appear to be translated and degraded with the same kinetics as poly(A)$^+$ mRNAs. Eukaryotic mRNA exists as a ribonucleoprotein particle; in some cases mRNPs that fail to be translated are stored. Bacterial mRNA has an extremely short half-life, only a few minutes. The 5′ end starts translation even while the downstream sequences are being transcribed. Degradation is initiated by endonucleases that cut at discrete sites, following the ribosomes in the 5′–3′ direction, after which exonucleases reduce the fragments to nucleotides by degrading them from the released 3′ end toward the 5′ end. Individual sequences may promote or retard degradation in bacterial mRNAs.

Yeast mRNA is degraded by multiple pathways. Poly(A)-dependent degradation involves an ordered series of events, in which poly(A) is removed from the 3′ end, causing loss of poly(A)-binding protein, the methylated cap is removed from the 5′ end, and then the mRNA is degraded from the 5′ end.

Eukaryotic mRNAs are usually stable for several hours. They may have multiple sequences that initiate degradation; examples are known in which the process is regulated.

Further reading

Reviews

Bannerjee, A. K. (1980). 5′-terminal cap structure in eukaryotic mRNAs. *Microbiol. Rev.* **44,** 175–205.

Jackson, R. J. and Standart, N. (1990). Do the poly(A) tail and 3′ untranslated region control mRNA translation? *Cell* **62,** 15–24.

Sachs, A. (1993). Messenger RNA degradation in eukaryotes. *Cell* **74,** 413–421.

Ross, J. (1995). mRNA stability in mammalian cells. *Microbiol. Rev.* **59,** 423–450.

Caponigro, G. and Parker, R. (1996). Mechanisms and control of mRNA turnover in S. cerevisiae. *Microbiol. Rev.* **60,** 233–249.

Jacobson, A. and Peltz, S. W. (1996). Interrelationships of the pathways of mRNA decay and translation in eukaryotic cells. *Ann. Rev. Biochem.* 65, 693–739.

Historical

Hoagland, M. B. *et al.* (1958). A soluble RNA intermediate in protein synthesis. *J. Biol. Chem.* **231,** 241–257.

Nirenberg, M. W. and Matthaei, H. J. (1961). The dependence of cell-free protein synthesis in E. coli upon naturally occurring or synthetic polyribonucleotides. *Proc. Nat. Acad. Sci. USA* **47,** 1588–1602.

Dintzis, H. M. (1961). Assembly of the peptide chain of hemoglobin. *Proc. Nat. Acad. Sci. USA* **47,** 247–261.

Brenner, S., Jacob, F., and Meselson, M. (1961). An unstable intermediate carrying information from genes to ribosomes for protein synthesis. *Nature* **190,** 576–581.

Chapeville, F. *et al.* (1962). On the role of soluble RNA in coding for amino acids. *Proc. Nat. Acad. Sci. USA* **48,** 1086–1092.

Slayter, H. S. *et al.* (1963). The visualization of polyribosome structure. *J. Mol. Biol.* **7,** 652–657.

Nirenberg, M. W. and Leder, P. (1964). The effect of trinucleotides upon the binding of sRNA to ribosomes. *Science* **145**, 1399–1407.

Darnell, J. *et al.* (1971). Poly(A) sequences: role in conversion of nuclear RNA into mRNA. *Science* **174**, 507–510.

Messenger stability

O'Hara, E. B. *et al.* (1995). Polyadenylation helps regulate mRNA decay in E. coli. *Proc. Nat. Acad. Sci. USA* **92**, 1807–1811.

Mitchell, P. *et al.* (1997). The exosome: a conserved eukaryotic RNA processing complex containing multiple 3′–5′ exoribonuclease activities. *Cell* **91**, 457–466.

Chapter 6

Protein synthesis

An mRNA contains a series of codons that interact with the anticodons of aminoacyl-tRNAs so that a corresponding series of amino acids is incorporated into a polypeptide chain. The ribosome provides the environment for controlling the interaction between mRNA and aminoacyl-tRNA. The ribosome behaves like a small migrating factory that travels along the template engaging in rapid cycles of peptide bond synthesis. Aminoacyl-tRNAs shoot in and out of the particle at a fearsome rate, depositing amino acids; and elongation factors cyclically associate and dissociate. Together with its accessory factors, the ribosome provides the full range of activities required for all the steps of protein synthesis.

The ribosome possesses several active centers, each of which is constructed from a particular group of proteins associated with a region of ribosomal RNA. The active centers require the direct participation of rRNA in a structural or even catalytic role. Some catalytic functions require individual proteins, but none of the activities can be reproduced by isolated proteins or groups of proteins; they function only in the context of the ribosome.

Two types of information are important in analyzing the ribosome. Mutations implicate particular ribosomal proteins or bases in rRNA in participating in particular reactions. Structural analysis, including direct modification of components of the ribosome and comparisons to identify conserved features in rRNA, identifies the physical locations of components involved in particular functions.

The basic form of the ribosome is conserved, but there are appreciable variations in the overall size and proportions of RNA and protein in the ribosomes of bacteria, eukaryotic cytoplasm, and organelles. **Figure 6.1** compares the components of bacterial and mammalian ribosomes. Both are ribonucleoprotein particles that contain more RNA than protein. The ribosomal proteins are known as **r-proteins.**

All ribosomes in a cell are identical. A ribosome always consists of two subunits, each of which contains a major rRNA and a number of small proteins. The large subunit may also contain smaller RNA(s). In *E. coli*, the small (30S) subunit consists of the 16S rRNA and 21 proteins. The large (50S) subunit contains 23S rRNA, the small 5S RNA, and 31 proteins. With the exception of one protein present at four copies per ribosome, there is one copy of each protein.

The ribosomes of higher eukaryotic cytoplasm are larger than those of bacteria. The total content of both RNA and protein is greater; the major RNA molecules are longer (called 18S and 28S rRNAs), and there are more proteins. Probably most or all of the proteins are present in stoichiometric amounts. RNA is still the predominant component by mass.

Organelle ribosomes are distinct from the ribosomes of the cytosol, and take varied forms. In some cases, they are almost the size of bacterial ribosomes and have 70% RNA; in other cases, they are only 60S and have <30% RNA.

Electron micrographs of subunits and complete

Figure 6.1 Ribosomes are large ribonucleoprotein particles that contain more RNA than protein and dissociate into large and small subunits.

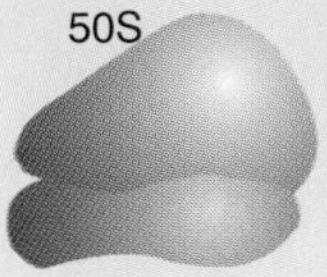

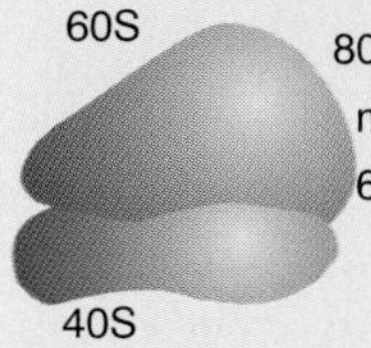

Ribosomes		rRNAs	Proteins
Bacterial 70S mass: 2.5×10^6 D 66% RNA	50S	23S = 2904 bases 5S = 120 bases	31
	30S	16S = 1542 bases	21
Mammalian 80S mass: 4.2×10^6 D 60% RNA	60S	28S = 4718 bases 5.8S = 160 bases 5S = 120 bases	49
	40S	18S = 1874 bases	33

Figure 6.2 Electron microscopic images of bacterial ribosomes and subunits reveal their shapes. Photographs kindly provided by James Lake.

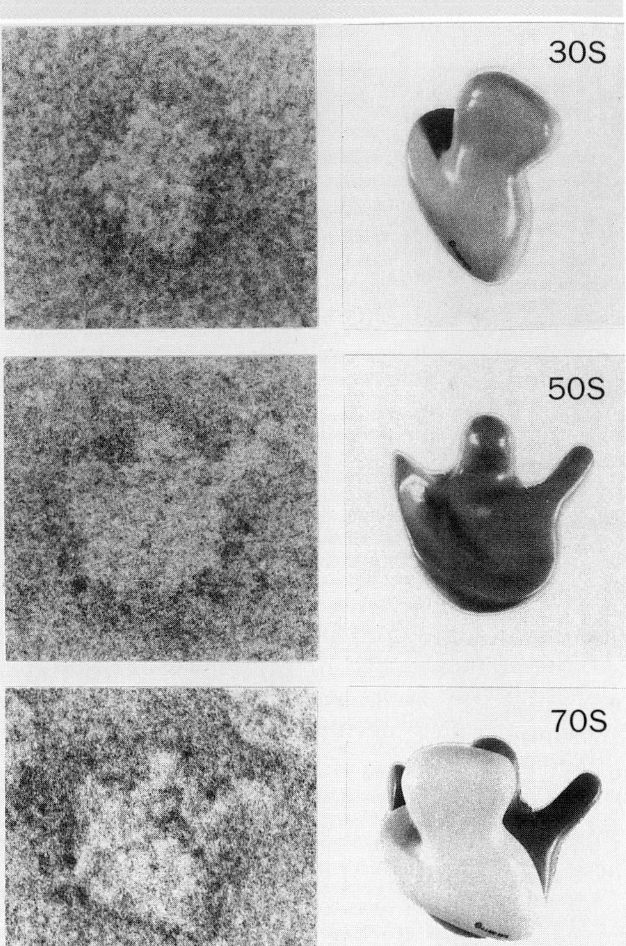

bacterial ribosomes are shown in **Figure 6.2**, together with models in the corresponding orientation. The complete 70S ribosome has an asymmetric construction. The partition between the head and body of the small subunit is aligned with the notch of the large subunit, so that the platform of the small subunit fits into the large subunit. There is a cavity between the subunits which contains some of the important sites.

The RNAs constitute the major part of the mass of the bacterial ribosome. Their presence is pervasive, and probably most or all of the ribosomal proteins actually contact rRNA. So the major rRNAs form what is sometimes thought of as the backbone of each subunit, a continuous thread whose presence dominates the structure, and which determines the positions of the ribosomal proteins.

The stages of protein synthesis

EACH ribosome subunit has specific roles in protein synthesis. Messenger RNA is associated with the small subunit; ~30 bases of the mRNA are bound at any time. One popular idea is that the mRNA fits between or close to the junction of the subunits. Two tRNA molecules are active in protein synthesis at any moment; so polypeptide elongation involves reactions taking place at just two of the (roughly) 10 codons covered by the ribosome. As evident from **Figure 6.3**, the two tRNAs are quite large relative to the ribosome; they become inserted into internal sites that stretch across the subunits. A third tRNA may remain present on the ribosome after it has been used in protein synthesis, before being recycled.

Each tRNA lies in a distinct site on the ribosome. **Figure 6.4** shows that the two sites have different features:

- An incoming aminoacyl-tRNA binds to the **A site** (or **acceptor site**). Prior to the entry of aminoacyl-tRNA, the site exposes the codon representing the next amino acid due to be added to the chain.
- The codon representing the most recent amino acid to have been added to the nascent polypeptide chain lies in the **P site** (or **donor site**). This site is occupied by **peptidyl-tRNA,** a tRNA carrying the nascent polypeptide chain.

The end of the tRNA that carries an amino acid is located on the large subunit, while the anticodon at the other end interacts with the mRNA bound by the small subunit. So the P and A sites each extend across both ribosomal subunits.

For a ribosome to synthesize a peptide bond, it must be in the state shown in *step 1* in the figure, when peptidyl-tRNA is in the P site and aminoacyl-tRNA is in the A site. Then peptide bond formation occurs when the polypeptide carried by the peptidyl-tRNA is transferred to the amino acid carried by the aminoacyl-tRNA. This reaction is catalyzed by constituents of the large subunit of the ribosome.

Transfer of the polypeptide generates the ribosome

Figure 6.3 Size comparisons show that the ribosome is large enough to bind two tRNAs (as well as 35 bases of mRNA).

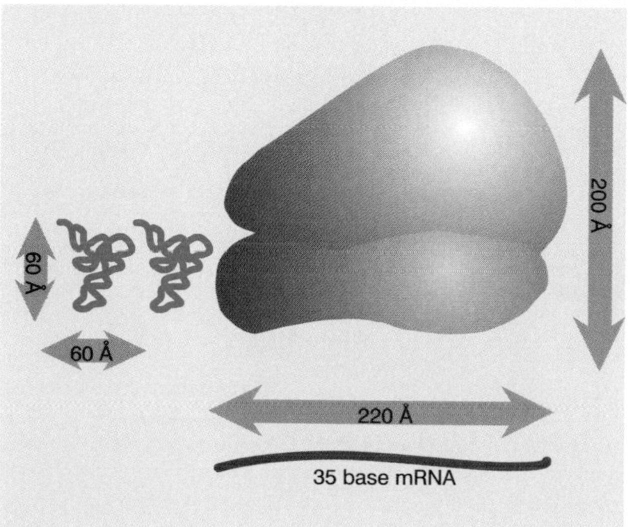
60 Å
60 Å
200 Å
220 Å
35 base mRNA

Figure 6.4 The ribosome has two sites for binding charged tRNA.

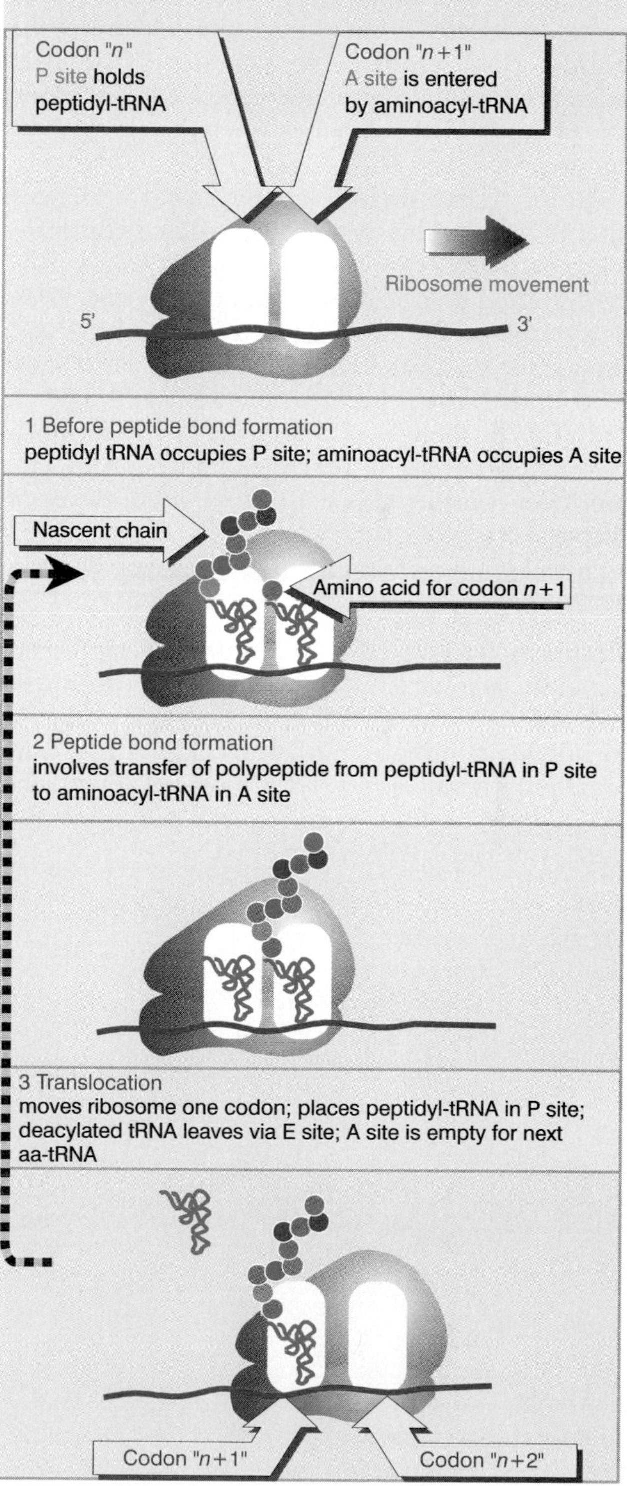
Codon "n"
P site holds peptidyl-tRNA
Codon "n+1"
A site is entered by aminoacyl-tRNA
Ribosome movement
5'
3'
1 Before peptide bond formation
peptidyl tRNA occupies P site; aminoacyl-tRNA occupies A site
Nascent chain
Amino acid for codon n+1
2 Peptide bond formation
involves transfer of polypeptide from peptidyl-tRNA in P site to aminoacyl-tRNA in A site
3 Translocation
moves ribosome one codon; places peptidyl-tRNA in P site; deacylated tRNA leaves via E site; A site is empty for next aa-tRNA
Codon "n+1"
Codon "n+2"

shown in *step 2*, in which the **deacylated tRNA,** lacking any amino acid, lies in the P site, while a new peptidyl-tRNA has been created in the A site. This peptidyl-tRNA is one amino acid residue longer than the peptidyl-tRNA that had been in the P site in *step 1*.

Then the ribosome moves one triplet along the messenger. This stage is called translocation, and we discuss it in more detail shortly. The movement transfers the deacylated tRNA out of the P site, and moves the peptidyl-tRNA into the P site (see *step 3*). The next codon to be translated now lies in the A site, ready for a new aminoacyl-tRNA to enter, when the cycle will be repeated.

The deacylated tRNA leaves the ribosome via another tRNA-binding site, the **E site**. This site is transiently occupied by the tRNA *en route* between leaving the P site and being released from the ribosome into the cytosol. So the flow of tRNA is into the A site, through the P site, and out through the E site. **Figure 6.5** compares the movement of tRNA and mRNA, which may be thought of as a sort of ratchet in which the reaction is driven by the codon-anticodon interaction. (We consider this concept in more detail in Chapter 7.)

Protein synthesis falls into the three stages shown in **Figure 6.6:**

- **Initiation** involves the reactions that precede formation of the peptide bond between the first two amino acids of the protein. It requires the ribosome to bind to the mRNA, forming an initiation complex that contains the first aminoacyl-tRNA. This is a relatively slow step in protein synthesis, and usually determines the rate at which an mRNA is translated.

Figure 6.5 tRNA and mRNA move through the ribosome in the same direction.

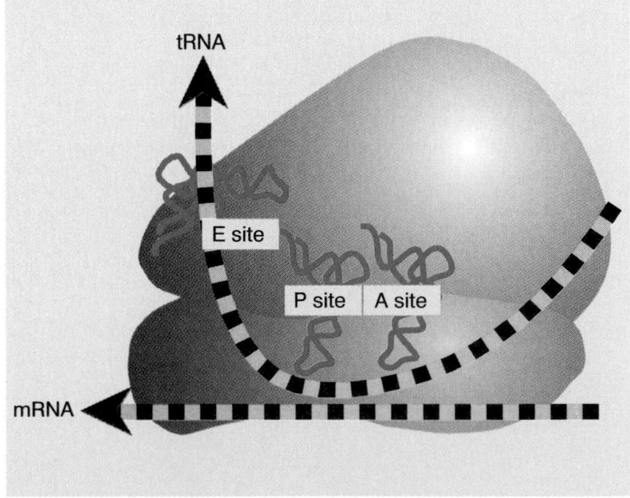

- **Elongation** includes all the reactions from synthesis of the first peptide bond to addition of the last amino acid. Amino acids are added to the chain one at a time; the addition of an amino acid is the most rapid step in protein synthesis.
- **Termination** encompasses the steps that are needed to release the completed polypeptide chain; at the same time, the ribosome dissociates from the mRNA.

Different sets of accessory factors assist the ribosome at each stage. Energy is provided at various stages by the hydrolysis of GTP.

Figure 6.6 Protein synthesis falls into three stages.

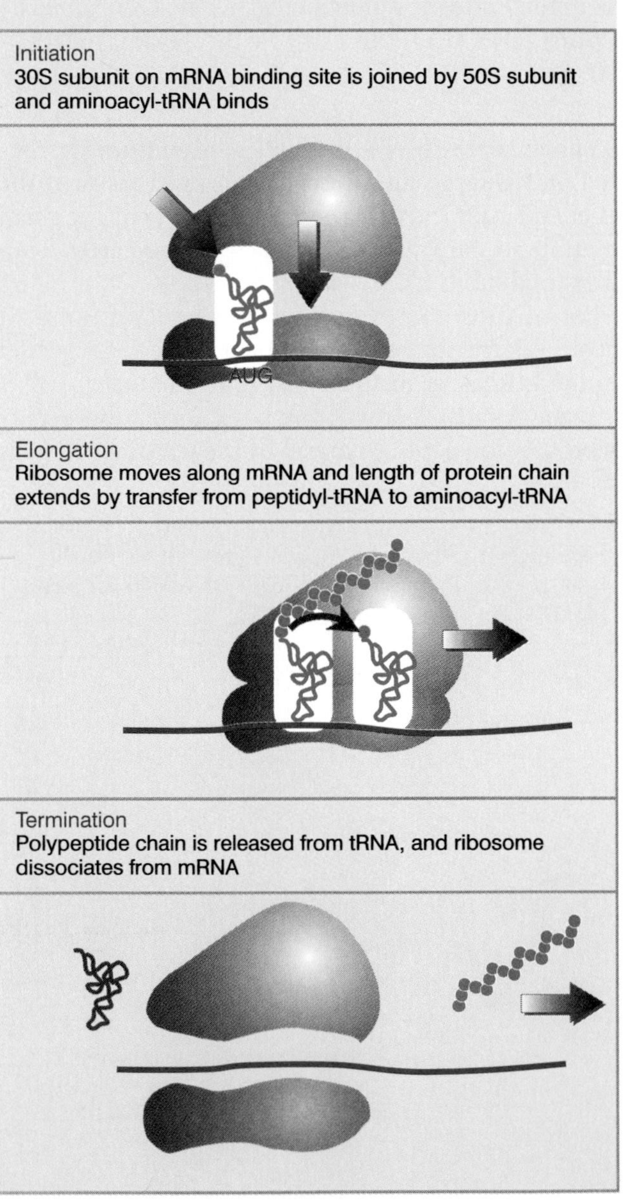

Initiation
30S subunit on mRNA binding site is joined by 50S subunit and aminoacyl-tRNA binds
AUG
Elongation
Ribosome moves along mRNA and length of protein chain extends by transfer from peptidyl-tRNA to aminoacyl-tRNA
Termination
Polypeptide chain is released from tRNA, and ribosome dissociates from mRNA

Initiation in bacteria needs 30S subunits and accessory factors

Bacterial ribosomes engaged in elongating a polypeptide chain exist as 70S particles. At termination, they are released from the mRNA to enter a pool of free ribosomes. In growing bacteria, the majority of ribosomes are synthesizing proteins; the free pool is likely to contain ~20% of the ribosomes.

Ribosomes in the free pool can dissociate into separate subunits; so free 70S ribosomes are in dynamic equilibrium with 30S and 50S subunits. *Initiation of protein synthesis is not a function of intact ribosomes, but is undertaken by the separate subunits, which reassociate during the initiation reaction.* **Figure 6.7** summarizes the ribosomal subunit cycle during protein synthesis in bacteria.

Initiation occurs at a special sequence on mRNA called the **ribosome-binding site.** This is a short sequence of bases that precedes the coding region (see Figure 6.14 later). The ribosome-binding site is a sequence at which the small and large subunits associate on mRNA to form an intact ribosome. The reaction occurs in two steps:

- Recognition of mRNA occurs when a *small subunit* binds to form an **initiation complex** at the ribosome-binding site.
- Then a *large subunit* joins the complex to generate a complete ribosome.

Figure 6.7 Initiation requires free ribosome subunits. When ribosomes are released at termination, they dissociate to generate free subunits. Initiation factors are present only on dissociated 30S subunits. When subunits reassociate to give a functional ribosome at initiation, they release the factors.

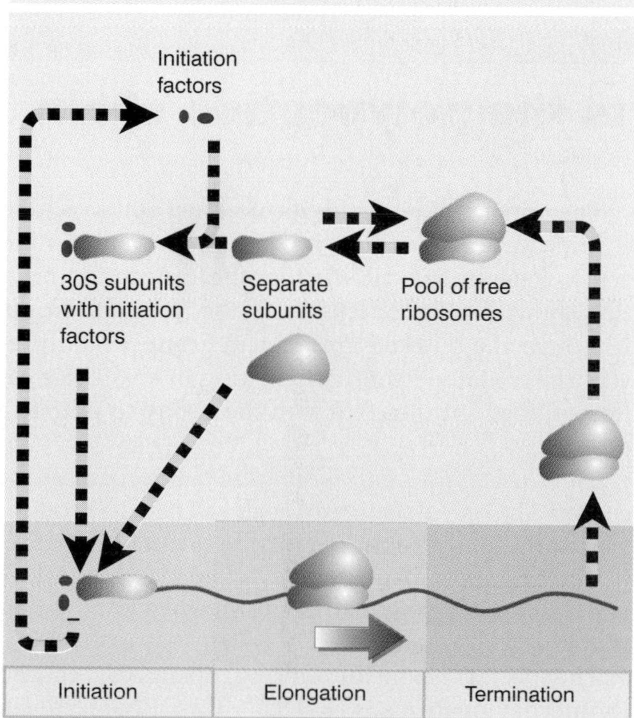

Although the 30S subunit is involved in initiation, it is not by itself competent to undertake the reactions of binding mRNA and tRNA. It requires additional proteins called **initiation factors (IF).** These factors are found only on 30S subunits, and they are released when the 30S subunits associate with 50S subunits to generate 70S ribosomes. This behavior distinguishes initiation factors from the structural proteins of the ribosome. *The initiation factors are concerned solely with formation of the initiation complex, they are absent from 70S ribosomes, and they play no part in the stages of elongation.*

Bacteria use three initiation factors, numbered **IF-1, IF-2,** and **IF-3.** They are needed for both mRNA and tRNA to enter the initiation complex:

- IF-3 is needed for 30S subunits to bind specifically to initiation sites in mRNA.
- IF-2 binds a special initiator tRNA and controls its entry into the ribosome.
- IF-1 binds to 30S subunits only as a part of the complete initiation complex, and could be involved in stabilizing it, rather than in recognizing any specific component.

The role of IF-3 is illustrated in **Figure 6.8.** The factor has dual functions: it is needed first to stabilize (free) 30S subunits; and then it enables them to bind to mRNA. IF-3 essentially controls the freedom of 30S subunits, which lasts from their dissociation from the pool of ribosomes to their reassociation with a 50S subunit at initiation.

Figure 6.8 Initiation requires 30S subunits that carry IF-3.

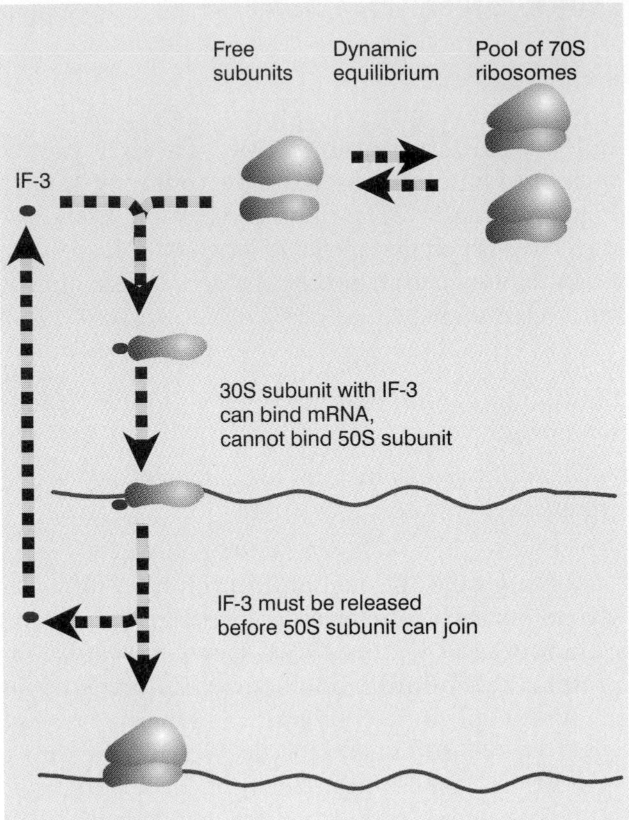
Free subunits
Dynamic equilibrium
Pool of 70S ribosomes
IF-3
30S subunit with IF-3 can bind mRNA, cannot bind 50S subunit
IF-3 must be released before 50S subunit can join

The first function of IF-3 controls the equilibrium between ribosomal states. IF-3 binds to free 30S subunits that are released from the pool of 70S ribosomes. The presence of IF-3 prevents the subunit from reassociating with a 50S subunit. In this capacity, it is an *antiassociation factor* that causes a 30S subunit to remain in the pool of free subunits. The reaction between IF-3 and the 30S subunit is stoichiometric: one molecule of IF-3 binds per subunit. There is a relatively small amount of IF-3, so its availability determines the number of free 30S subunits.

The second function of IF-3 controls the ability of 30S subunits to bind to mRNA. Small subunits must have IF-3 in order to form initiation complexes with mRNA.

IF-3 must be released from the 30S•mRNA complex in order to enable the 50S subunit to join. On its release, IF-3 immediately recycles by finding another 30S subunit.

A special initiator tRNA starts the polypeptide chain

SYNTHESIS of all proteins starts with the same amino acid: methionine. The signal for initiating a polypeptide chain is a special **initiation codon** that marks the start of the reading frame. Usually the initiation codon is the triplet AUG, but in bacteria, GUG or UUG are also used.

The AUG codon represents methionine, and two types of tRNA can carry this amino acid. One is used for initiation, the other for recognizing AUG codons during elongation.

In bacteria and in eukaryotic organelles, the initiator tRNA carries a methionine residue that has been formylated on its amino group, forming a molecule of **N-formyl-methionyl-tRNA.** The tRNA is known as **$tRNA_f^{Met}$**. The name of the aminoacyl-tRNA is usually abbreviated to fMet-$tRNA_f$.

The initiator tRNA gains its modified amino acid in a two-stage reaction. First, it is charged with the amino acid to generate Met-$tRNA_f$; then the formylation reaction shown in **Figure 6.9** blocks the free NH_2 group. Although the blocked amino acid group would prevent the initiator from participating in chain elongation, it does not interfere with the ability to initiate a protein.

This tRNA is used only for initiation. It recognizes the codons AUG or GUG (occasionally UUG). The codons are not recognized equally well: the extent of initiation declines by about half when AUG is replaced by GUG, and declines by about half yet again when UUG is employed.

The species responsible for recognizing AUG codons in internal locations is **$tRNA_m^{Met}$**. *This tRNA responds*

Figure 6.9 The initiator N-formyl-methionyl-tRNA (fMet-tRNA$_f$) is generated by formylation of methionyl-tRNA, using formyl-tetrahydrofolate as cofactor.

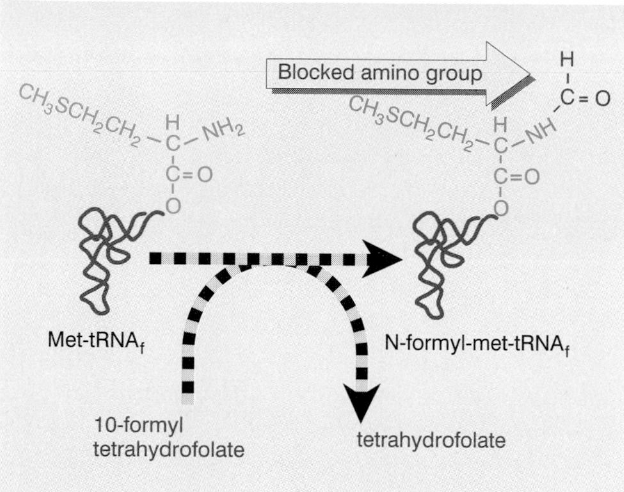

Figure 6.10 Only fMet-tRNA$_f$ can be used for initiation by 30S subunits; only other aminoacyl-tRNAs (aa-tRNA) can be used for elongation by 70S ribosomes.

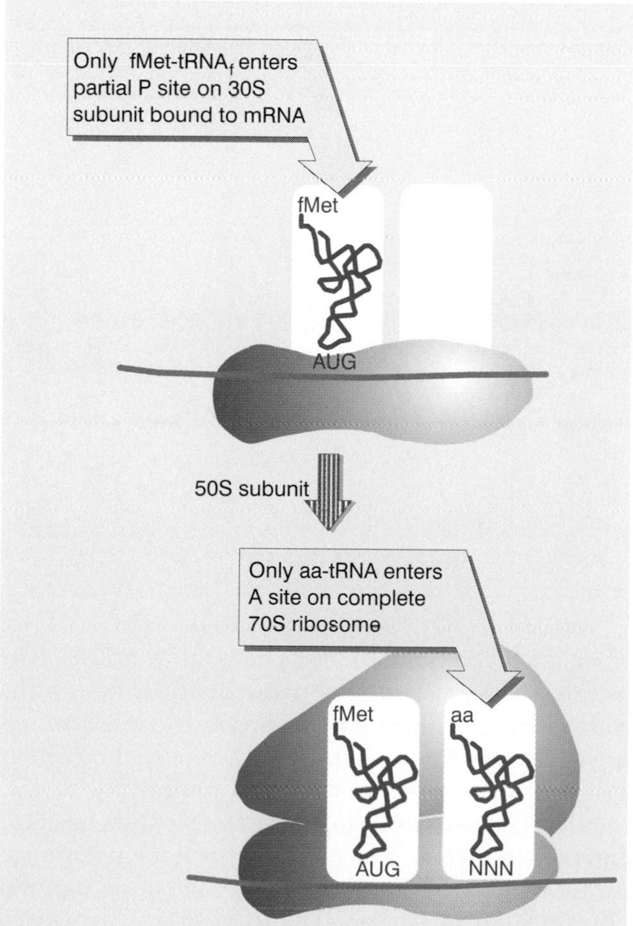

only to internal AUG codons. Its methionine cannot be formylated.

So there are two differences between the initiating and elongating Met-tRNAs: the tRNA moieties themselves are different; and the amino acids differ in the state of the amino group.

The meaning of the AUG and GUG codons depends on their context. When the AUG codon is used for initiation, it is read as formyl-methionine; when used within the coding region, it represents methionine. The meaning of the GUG codon is even more dependent on its location. When present as the *first* codon, it is read via the initiation reaction as formyl-methionine. Yet when present *within* a gene, it is read by Val-tRNA, one of the regular members of the tRNA set, to provide valine as required by the genetic code.

How is the context of AUG and GUG codons interpreted? **Figure 6.10** illustrates the decisive role of the ribosome.

In an initiation complex, the small subunit alone is bound to mRNA. The initiation codon lies within the part of the P site carried by the subunit. *The only aminoacyl-tRNA that can become part of the initiation complex is the initiator, which has the unique property of*

Figure 6.11 fMet-tRNA$_f$ has unique features that distinguish it as the initiator tRNA.

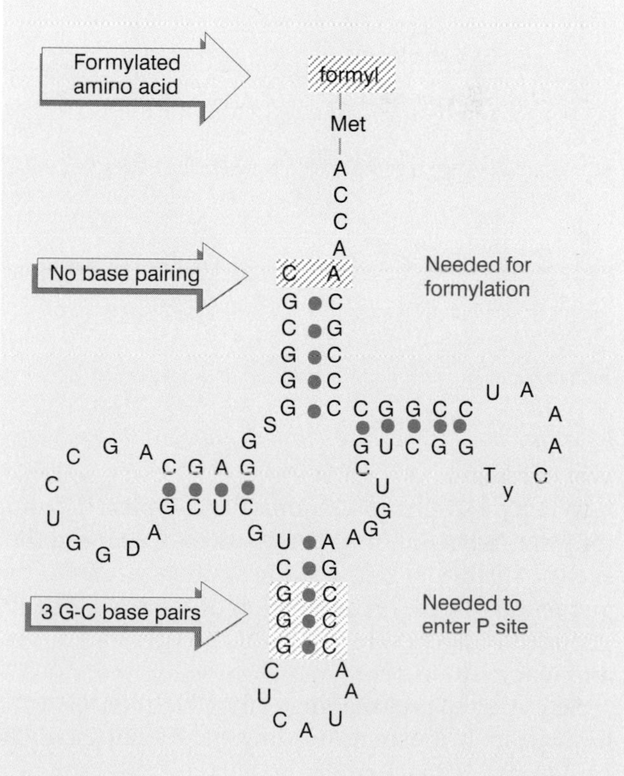

Figure 6.12 IF-2 is needed to bind fmet-tRNA$_f$ to the 30S-mRNA complex. After 50S binding, all IF factors are released and GTP is cleaved.

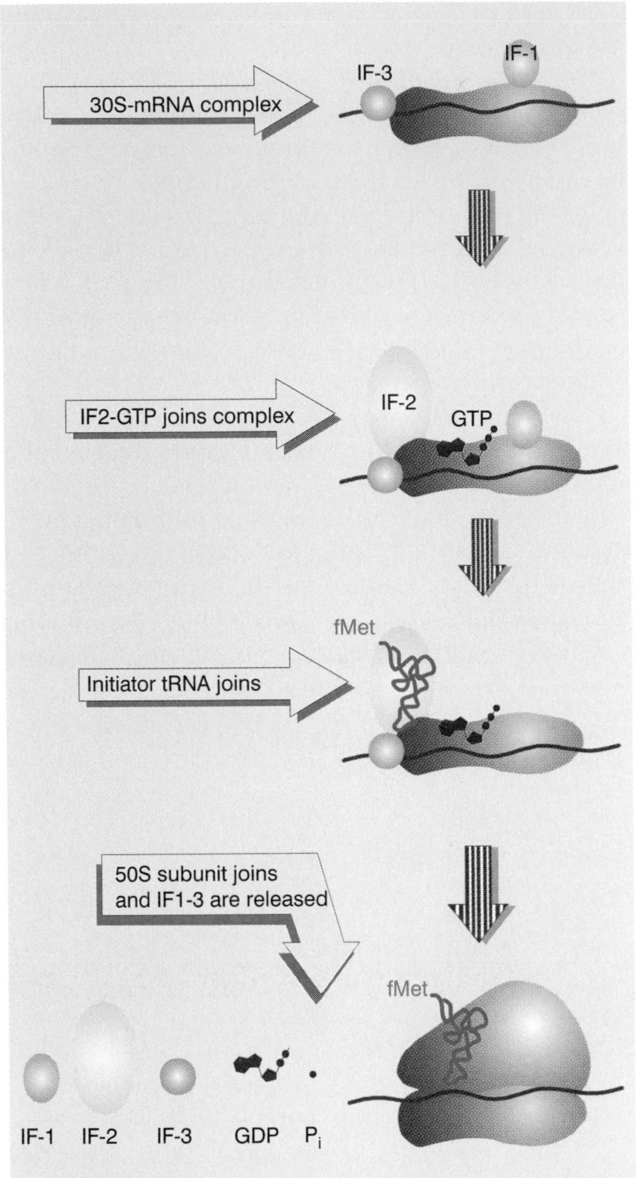

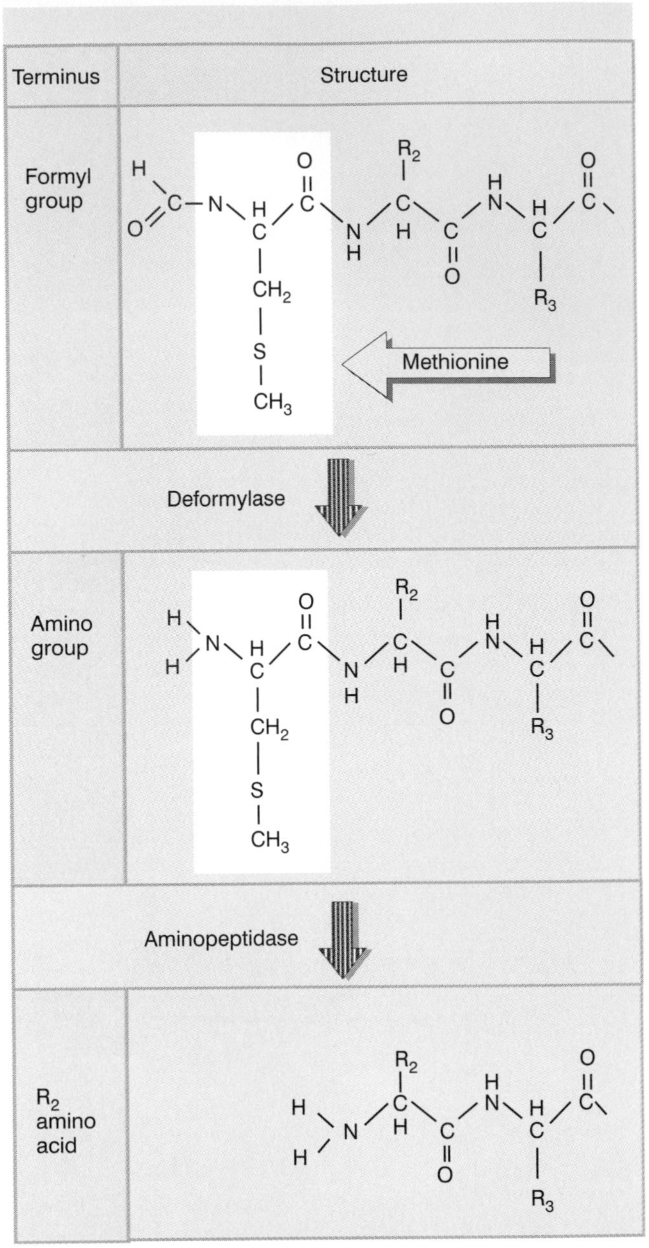

Figure 6.13 Newly synthesized proteins in bacteria start with formyl-methionine, but the formyl group, and sometimes the methionine, is removed during protein synthesis.

Terminus	Structure
Formyl group	Methionine
	Deformylase
Amino group	
	Aminopeptidase
R_2 amino acid	

being able to enter directly into the partial P site to recognize its codon.

When the large subunit joins the complex, the initiator fMet-tRNA$_f$ lies in the now-intact P site, and the A site is available for entry of the aminoacyl-tRNA complementary to the second codon of the gene. The first peptide bond forms between the initiator and the next aminoacyl-tRNA.

So initiation prevails when an AUG (or GUG) codon lies within a ribosome-binding site, because only the initiator tRNA can enter the partial P site generated when the 30S subunit binds *de novo* to the mRNA. The internal reading prevails subsequently, when the codons are encountered by a ribosome that is *continuing* to translate an mRNA, because only the regular aminoacyl-tRNAs can enter the (complete) A site.

What features distinguish the fMet-tRNA$_f$ initiator and the Met-tRNA$_m$ elongator? Some characteristic features of the tRNA sequence are important, as summarized in **Figure 6.11**. Some of these features are needed to *prevent* the initiator from being used in elongation, others are *necessary* for it to function in initiation:

- Formylation is not strictly necessary, because non-formylated Met-$tRNA_f$ can function as an initiator, but it improves the efficiency with which the Met-$tRNA_f$ is bound by IF-2.
- The bases that face one another at the last position of the stem to which the amino acid is connected are paired in all tRNAs except $tRNA_f^{Met}$. Mutations that create a base pair in this position of $tRNA_f^{Met}$ allow it to function in elongation. The absence of this pair is therefore important in preventing $tRNA_f^{Met}$ from being used in elongation. It is also needed for the formylation reaction.
- A series of 3 G•C pairs in the stem that precedes the loop containing the anticodon is unique to $tRNA_f^{Met}$. These base pairs are required to allow the fMet-$tRNA_f$ to be inserted directly into the P site.

The ability of aminoacyl-tRNAs to enter the ribosome is controlled by accessory factors. **Figure 6.12** shows how the fMet-$tRNA_f$ initiator enters the P site. The 30S subunit carries all the initiation factors, including IF-2, which is probably associated with the P site. IF-2 specifically forms a complex with fMet-$tRNA_f$. This complex places the tRNA in the partial P site.

By forming a complex specifically with fMet-$tRNA_f$, IF-2 ensures that only the initiator tRNA, and none of the regular aminoacyl-tRNAs, participates in the initiation reaction.

IF-2 remains part of the 30S subunit at this stage; it has a further role to play. This factor has a **ribosome-dependent GTPase activity**: it sponsors the hydrolysis of GTP in the presence of ribosomes, releasing the energy stored in the high-energy bond.

The GTP is hydrolyzed when the 50S subunit joins to generate a complete ribosome. The GTP cleavage could be involved in changing the conformation of the ribosome, so that the joined subunits are converted into an active 70S ribosome.

Because the same amino acid is used to start all protein chains, it is sometimes necessary to remove unwanted residues from the N-terminus, as illustrated in **Figure 6.13.** The removal reaction(s) occur rather rapidly, probably when the nascent polypeptide chain has reached a length of 15 amino acids.

In bacteria and mitochondria, the formyl residue is removed by a specific deformylase enzyme to generate a normal NH_2 terminus. If methionine is to be the N-terminal amino acid of the protein, this is the only necessary step. In about half the proteins, the methionine at the terminus is removed by an aminopeptidase, creating a new terminus of R_2 (originally the second amino acid incorporated into the chain). When both steps are necessary, they occur sequentially.

Initiation involves base pairing between mRNA and rRNA

An mRNA contains many AUG triplets: how is the initiation codon recognized as providing the starting point for translation? The sites on mRNA where protein synthesis is initiated can be identified by binding the ribosome to mRNA under conditions that block elongation. Then the ribosome remains at the initiation site. When ribonuclease is added to the blocked initiation complex, all the regions of mRNA outside the ribosome are degraded, but those actually bound to it are protected, as illustrated in **Figure 6.14.** The protected fragments can be recovered and characterized.

The initiation sequences protected by bacterial ribosomes are ~30 bases long. The ribosome-binding sites of different bacterial mRNAs display two common features:

- The AUG (or less often, GUG or UUG) initiation codon is always included within the protected sequence.
- Within 10 bases upstream of the AUG is a sequence that corresponds to part or all of the hexamer:

5′ ...A G G A G G... 3′

This polypurine stretch is known as the **Shine-Dalgarno** sequence. It is complementary to a highly

Figure 6.14 Ribosome-binding sites on mRNA can be recovered from initiation complexes.

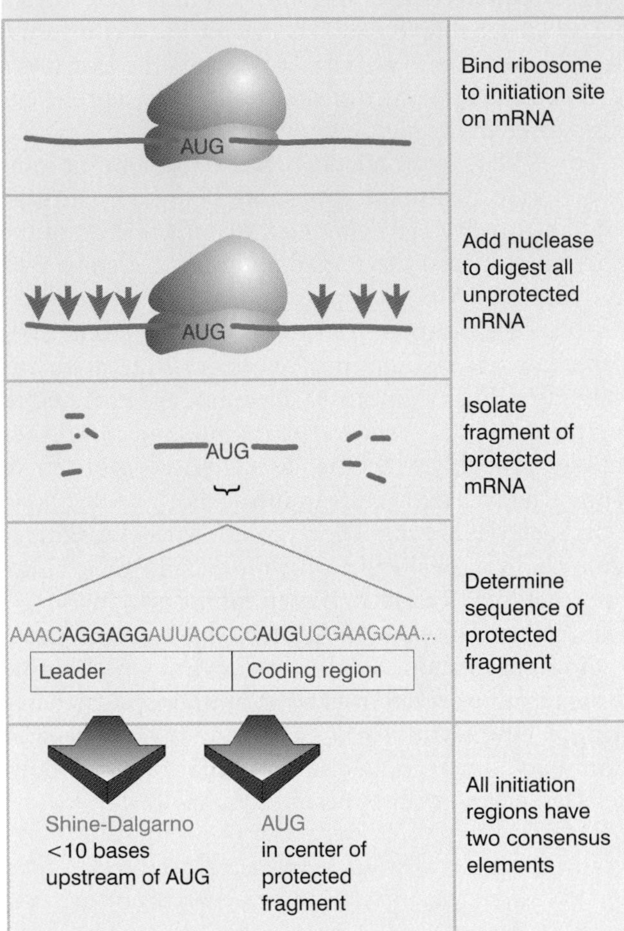
AUG
Bind ribosome to initiation site on mRNA
AUG
Add nuclease to digest all unprotected mRNA
AUG
Isolate fragment of protected mRNA
AAACAGGAGGAUUACCCCAUGUCGAAGCAA...
Leader
Coding region
Determine sequence of protected fragment
Shine-Dalgarno
<10 bases upstream of AUG
AUG
in center of protected fragment
All initiation regions have two consensus elements

conserved sequence close to the 3′ end of 16S rRNA. Written in reverse direction, the rRNA sequence is the hexamer:

3′ ... U C C U C C ... 5′

Does the Shine-Dalgarno sequence pair with its complement in rRNA during mRNA-ribosome binding? Mutations of both partners in this reaction demonstrate its importance in initiation. Point mutations in the Shine-Dalgarno sequence can prevent an mRNA from being translated. And the introduction of mutations into the complementary sequence in rRNA is deleterious to the cell and changes the pattern of protein synthesis. As further confirmation of the base-pairing reaction, a mutation in the Shine-Dalgarno sequence of an mRNA can be suppressed by a mutation in the rRNA that restores base pairing.

The sequence at the 3′ end of rRNA is conserved between prokaryotes and eukaryotes except that in all eukaryotes there is a deletion of the five-base sequence CCUCC that is the principal complement to the Shine-Dalgarno sequence. There does not appear to be base pairing between eukaryotic mRNA and 18S rRNA. This is a significant difference in the mechanism of initiation.

In bacteria, a 30S subunit binds directly to a ribosome-binding site. As a result, the initiation complex forms at a sequence surrounding the AUG initiation codon. When the mRNA is polycistronic, each coding region starts with a ribosome-binding site.

The nature of bacterial gene expression means that translation of a bacterial mRNA proceeds sequentially through its cistrons. At the time when ribosomes attach to the first coding region, the subsequent coding

Figure 6.15 Initiation occurs independently at each cistron in a polycistronic mRNA. When the intercistronic region is longer than the span of the ribosome, dissociation at the termination site is followed by independent reinitiation at the next cistron.

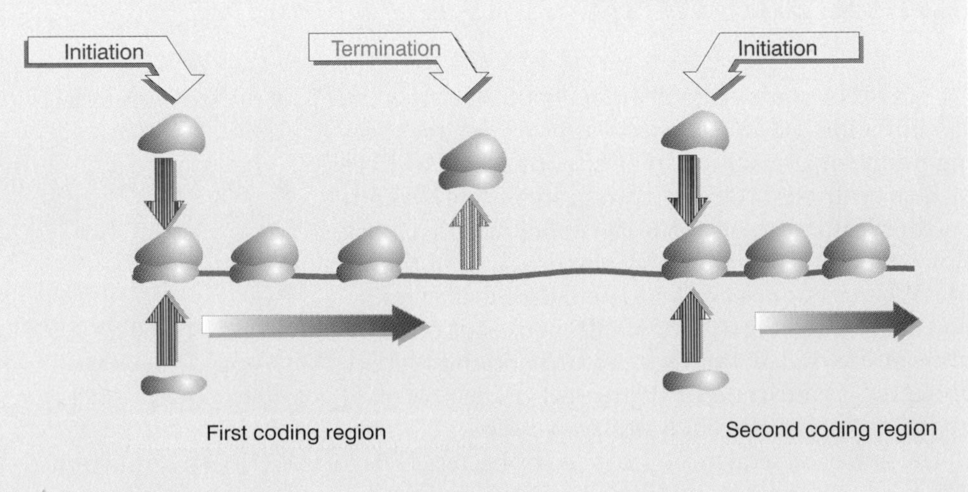
Initiation
Termination
Initiation
First coding region
Second coding region

regions have not yet even been transcribed. By the time the second ribosome site is available, translation is well under way through the first cistron.

What happens between the coding regions depends on the individual mRNA. Probably in most cases the ribosomes bind independently at the beginning of each cistron. The most common series of events is illustrated in **Figure 6.15.** When synthesis of the first protein terminates, the ribosomes leave the mRNA and dissociate into subunits. Then a new ribosome must assemble at the next coding region, and set out to translate the next cistron.

In some bacterial mRNAs, translation between adjacent cistrons is directly linked, because ribosomes gain access to the initiation codon of the second cistron *as they complete translation of the first cistron.* This effect requires the space between the two coding regions to be small. It may depend on the high local density of ribosomes; or the juxtaposition of termination and initiation sites could allow some of the usual intercistronic events to be bypassed. A ribosome physically spans ~30 bases of mRNA, so that it may simultaneously contact a termination codon and the next initiation site if they are separated by only a few bases.

Small subunits scan for initiation sites on eukaryotic mRNA

INITIATION of protein synthesis in eukaryotic cytoplasm resembles the process in bacteria, but the order of events is different, and the number of accessory factors is greater. Some of the differences in initiation are related to a difference in the way that bacterial 30S and eukaryotic 40S subunits find their binding sites for initiating protein synthesis on mRNA. In eukaryotes, small subunits first recognize the 5′ end of the mRNA, and then move to the initiation site, where they are joined by large subunits.

Virtually all eukaryotic mRNAs are monocistronic, but each mRNA usually is substantially longer than necessary just to code for its protein. The average mRNA in eukaryotic cytoplasm is 1000–2000 bases long, has a methylated cap at the 5′ terminus, and carries 100–200 bases of poly(A) at the 3′ terminus.

The nontranslated 5′ leader is relatively short, usually (but not always) less than 100 bases. The length of the coding region is determined by the size of the protein. The nontranslated 3′ trailer is often rather long, sometimes ~1000 bases. By virtue of its location, the leader cannot be ignored during initiation, but the function for the trailer is less obvious.

The first feature to be recognized during translation of a eukaryotic mRNA is the methylated cap that marks the 5′ end. Messengers whose caps have been removed are not translated efficiently *in vitro.* Binding of 40S subunits to mRNA requires several initiation factors, including proteins that recognize the structure of the cap.

Modification at the 5′ end occurs to almost all cellular or viral mRNAs, and is essential for their translation in eukaryotic cytoplasm (although it is not needed in organelles). The sole exception to this rule is provided by a few viral mRNAs (such as poliovirus) that are not capped; *only* these exceptional viral mRNAs can be translated *in vitro* without caps, so they must have some alternative feature that renders capping unnecessary.

Some viruses take advantage of this difference. Poliovirus infection inhibits the translation of host mRNAs. This is accomplished by interfering with the cap-binding proteins that are needed for initiation of cellular mRNAs, but that are superfluous for the noncapped poliovirus mRNA.

We have dealt with the process of initiation as though the ribosome-binding site is always freely available. However, its availability may be impeded by secondary structure. The recognition of mRNA requires several additional factors; an important part of their function is to remove any secondary structure in the mRNA (see Figure 6.18).

Sometimes the AUG initiation codon lies within 40 bases of the 5′ terminus of the mRNA, so that both the cap and AUG lie within the span of ribosome binding. But in many mRNAs the cap and AUG are farther apart, in extreme cases ~1000 bases distant. Yet the presence of the cap still is necessary for a stable complex to be formed at the initiation codon. How can the ribosome rely on two sites so far apart?

Figure 6.16 illustrates the "scanning" model, which supposes that the 40S subunit initially recognizes the 5′ cap and then "migrates" along the mRNA. Scanning

Figure 6.16 Eukaryotic ribosomes migrate from the 5' end of mRNA to the ribosome binding site, which includes an AUG initiation codon.

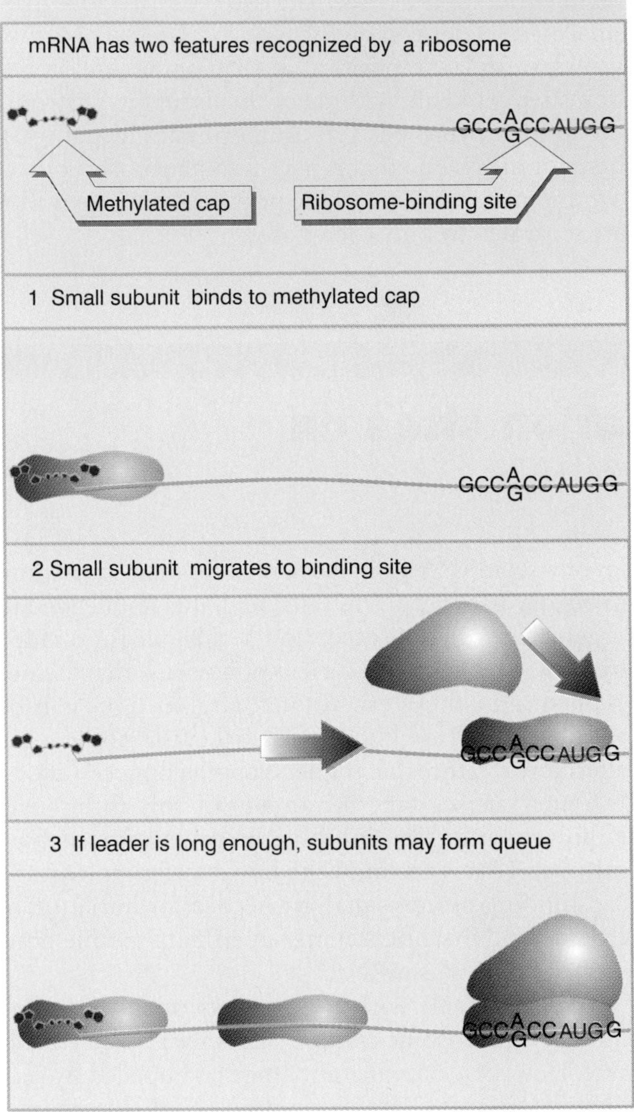

from the 5′ end is a linear process. When 40S subunits scan the leader region, they melt secondary structure hairpins with stabilities < –30 kcal, but hairpins of greater stability impede or prevent migration.

Migration stops when the 40S subunit encounters the AUG initiation codon. Usually, although not always, the first AUG triplet sequence will be the initiation codon. The AUG triplet by itself is not sufficient to halt migration; it is recognized efficiently as an initiation codon only when it is in the right context. The optimal context consists of the sequence GCC^A_GCC**AUG**G. The purine (A or G) three bases before the AUG codon, and the G immediately following it, are the most important, and influence efficiency of translation by 10×; the other bases have much smaller effects. When the leader sequence is long, further 40S subunits can recognize the 5′ end before the first has left the initiation site, creating a queue of subunits proceeding along the leader to the initiation site.

The vast majority of eukaryotic initiation events involve scanning from the 5′ cap, but there is an alternative means of initiation, used especially by certain viral RNAs, in which a 40S subunit associates directly with an internal site called an IRES. One type of IRES includes the AUG initiation codon; the other is located as much as 100 nucleotides upstream of it, requiring a 40S subunit to migrate, again probably by a scanning mechanism. Probably the same initiation factors that are used at 5′ ends are required to recognize an IRES (although obviously the RNA is recognized without involvement of a cap structure). (Use of the IRES is especially important in picornavirus infection, where the virus inhibits host protein synthesis by destroying cap structures.)

Binding is stabilized at the initiation site. When the 40S subunit is joined by a 60S subunit, the intact ribosome is located at the site identified by the protection assay. A 40S subunit protects a region of up to 60 bases; when the 60S subunits join the complex, the protected region contracts to about the same length of 30–40 bases seen in prokaryotes.

Initiation in eukaryotic cytoplasm almost always uses AUG as the initiator. The initiator tRNA is a distinct species, but its methionine does *not* become formylated. It is called **$tRNA_i^{Met}$**. So the difference between the initiating and elongating Met-tRNAs lies solely in the tRNA moiety, with Met-$tRNA_i$ used for initiation and Met-$tRNA_m$ used for elongation.

At least two features are unique to the initiator $tRNA_i^{Met}$ in yeast; it has an unusual tertiary structure; and it is modified by phosphorylation of the 2′ ribose position on base 64 (if this modification is prevented, the initiator can be used in elongation). So the principle of a distinction between initiator and elongator Met-tRNAs is maintained in eukaryotes, but its structural basis is different from that in bacteria.

Aside from this difference, the process of initiation in eukaryotes appears to be generally analogous to that in *E. coli*. There are more initiation factors—at least nine already have been found in reticulocytes (immature red blood cells), in which the most work has been done. The factors are named similarly to those in bacteria, sometimes by analogy with the bacterial factors, and are given the prefix "e" to indicate their eukaryotic origin.

Eukaryotic initiation proceeds through the formation of a **ternary complex** containing Met-tRNA$_i$, eIF2, and GTP. The complex is formed in two stages. GTP binds to eIF2; and this increases the factor's affinity for Met-tRNA$_i$, which then is bound. **Figure 6.17** shows that the ternary complex associates directly with free 40S subunits. The reaction is independent of the presence of mRNA. In fact, *the Met-tRNA$_i$ initiator must be present in order for the 40S subunit to bind to mRNA.*

Figure 6.17 In eukaryotic initiation, eIF-2 forms a ternary complex with Met-tRNA$_f$. The ternary complex binds to free 40S subunits, which attach to the 5' end of mRNA. Later in the reaction, GTP is hydrolyzed when eIF-2 is released in the form of eIF2-GDP. eIF-2B regenerates the active form.

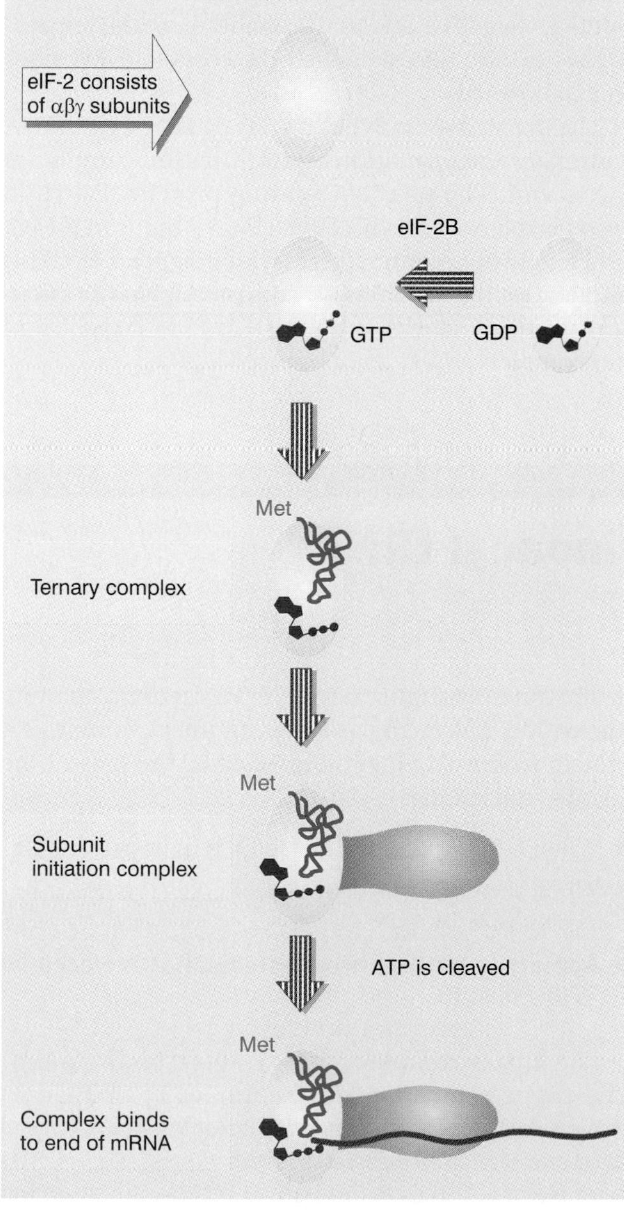

The roles of the initiation factors involved in binding the subunit initiation complex to mRNA are expanded in **Figure 6.18.**

Figure 6.18 Several eukaryotic initiation factors are required to unwind mRNA, bind the subunit initiation complex, and support joining with the large subunit.

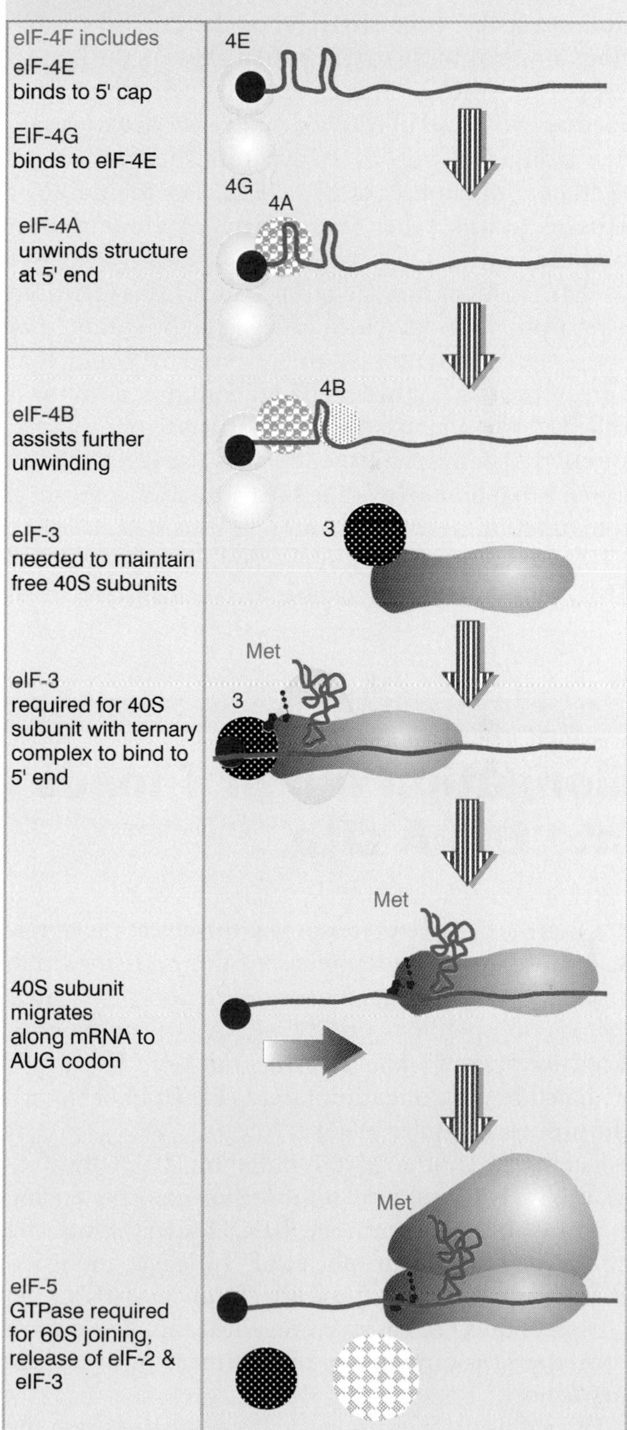

When the small subunit has bound mRNA, it migrates to (usually) the first AUG codon. Little is known about this process; we assume it requires expenditure of energy in the form of ATP. When the small subunit reaches the initiation site, it stops, and can be joined by a large subunit.

The eIF4F factor is a multimer that binds at the 5′ end. It includes the cap-binding subunit eIF4E, the RNA-dependent ATPase eIF4A, and the "scaffolding" subunit eIF4G. After eIF4E binds the cap, eIF4A unwinds any secondary structure that exists in the first 15 bases of the mRNA. Energy for the unwinding is provided by hydrolysis of ATP. Unwinding of structure farther along the mRNA is accomplished by eIF4A together with another factor, eIF4B. The main role of eIF4G is to link other components of the initiation complex.

eIF4E is a focus for regulation. Its activity is increased by phosphorylation, which is triggered by stimuli that increase protein synthesis, and reversed by stimuli that repress protein sythesis. eIF4F contains a subunit (called Mnk1) which is a kinase that acts to phosphorylate eIF4E. The availability of eIF4E is also controlled by proteins that bind to it (called 4EBP1,2,3), to prevent it from functioning in initiation.

How does the ribosome recognize the mRNA? eIF4G binds to eIF3, which is associated with the small ribosomal subunit. This provides the means by which the 40S ribosomal subunit binds to eIF4F, and thus is recruited to the complex. In effect, eIF4F functions to get eIF4G in place so that it can attract the small ribosomal subunit. Other factors (including eIF4A and eIF4B) are also involved.

Factors eIF3 and eIF6 are required to maintain subunits in their dissociated state. eIF3 is a very large factor, with 8–10 subunits. eIF3 binds to the small subunit, but eIF6 (which has most of the antiassociation activity) binds to the large ribosomal subunit. It is released when the large subunit joins the initiation complex.

Junction of the 60S subunits with the initiation complex cannot occur until eIF2 and eIF3 have been released from the initiation complex. This is mediated by eIF5, which is a GTPase. Probably all of the remaining factors are released when the complete 80S ribosome is formed.

The presence of poly(A) on the 3′ tail of an mRNA stimulates the formation of an initiation complex at the 5′ end. The poly(A)-binding protein (Pab1p in yeast) is required for this effect. Pab1p binds to eIF4G, which in turn is bound to eIF4E (see Figure 6.18). This implies that the mRNA must (transiently) have a circular organization, with both the 5′ and 3′ ends held in this complex.

Elongation factor T loads aminoacyl-tRNA into the A site

ONCE the complete ribosome is formed at the initiation codon, the stage is set for a cycle in which aminoacyl-tRNA enters the A site of a ribosome whose P site is occupied by peptidyl-tRNA. *Any aminoacyl-tRNA except the initiator can enter the A site.* Its entry is mediated by an **elongation factor** (**EF-Tu** in bacteria). The process is similar in eukaryotes.

Just like its counterpart in initiation (IF-2), this factor, EF-Tu, is associated with the ribosome only during its sponsorship of aminoacyl-tRNA entry. Once the aminoacyl-tRNA is in place, EF-Tu leaves the ribosome, to work again with another aminoacyl-tRNA. So it displays the cyclic association with, and dissociation from, the ribosome that is the hallmark of the accessory factors.

The pathway for aminoacyl-tRNA entry to the A site is illustrated in **Figure 6.19.** EF-Tu carries a guanine nucleotide. The factor provides another example of a protein whose activity is controlled by the state of the guanine nucleotide:

- When GTP is present, the factor is in its active state.
- When the GTP is hydrolyzed to GDP, the factor becomes inactive.
- Activity is restored when the GDP is replaced by GTP.

The **binary complex** of EF-Tu•GTP binds aminoacyl-tRNA to form a ternary complex of aminoacyl-tRNA•EF-Tu•GTP. *The ternary complex binds only to the A site of ribosomes whose P site is already occupied by peptidyl-tRNA.* This is the critical reaction in ensuring

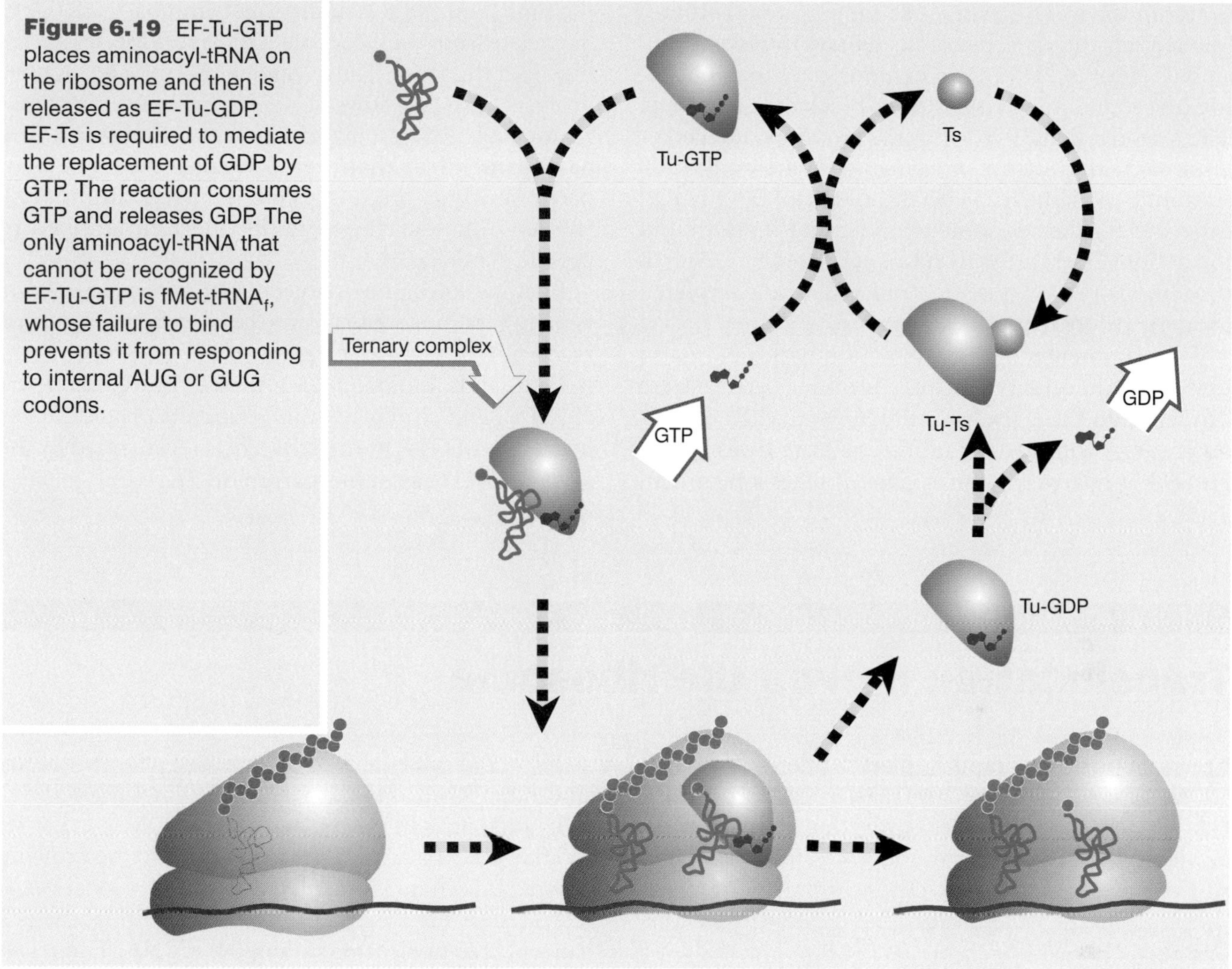

Figure 6.19 EF-Tu-GTP places aminoacyl-tRNA on the ribosome and then is released as EF-Tu-GDP. EF-Ts is required to mediate the replacement of GDP by GTP. The reaction consumes GTP and releases GDP. The only aminoacyl-tRNA that cannot be recognized by EF-Tu-GTP is fMet-tRNA$_f$, whose failure to bind prevents it from responding to internal AUG or GUG codons.

that the aminoacyl-tRNA and peptidyl-tRNA are correctly positioned for peptide bond formation.

Aminoacyl-tRNA is loaded into the A site in two stages. First the anticodon end binds to the A site of the 30S subunit. Then codon-anticodon recognition triggers a change in the conformation of EF-Tu. The GTP is cleaved. The CCA end of the tRNA now moves into the A site on the 50S subunit. The binary complex EF-Tu•GDP is released. This form of EF-Tu is inactive and does not bind aminoacyl-tRNA effectively.

Another factor, EF-Ts, mediates the regeneration of the used form, EF-Tu•GDP, into the active form, EF-Tu•GTP. First, EF-Ts displaces the GDP from EF-Tu, forming the combined factor EF-Tu•EF-Ts. Then the EF-Ts is in turn displaced by GTP, reforming EF-Tu•GTP. The active binary complex binds aminoacyl-tRNA; and the released EF-Ts can recycle.

There are ~70,000 molecules of EF-Tu per bacterium (~5% of the total bacterial protein), which approaches the number of aminoacyl-tRNA molecules. This implies that most aminoacyl-tRNAs are likely to be present in ternary complexes. There are only ~10,000 molecules of EF-Ts per cell (about the same as the number of ribosomes).

The role of GTP in the ternary complex has been studied by substituting an analog that cannot be hydrolyzed. The compound **GMP-PCP** has a methylene bridge in place of the oxygen that links the β and γ phosphates in GTP. In the presence of GMP-PCP, a ternary complex can be formed that binds aminoacyl-tRNA to the ribosome. But the peptide bond cannot be formed. Therefore, the *presence* of GTP is needed for aminoacyl-tRNA to be bound at the A site, but the *hydrolysis* is not required until later.

Kirromycin is an antibiotic that inhibits the function of EF-Tu. When EF-Tu is bound by kirromycin, it remains able to bind aminoacyl-tRNA to the A site. But the EF-Tu•GDP complex cannot be released from the ribosome. Its continued presence prevents formation of the peptide bond between the peptidyl-tRNA and the aminoacyl-tRNA. As a result, the ribosome becomes "stalled" on mRNA, bringing protein synthesis to a halt.

This effect of kirromycin demonstrates that inhibiting one step in protein synthesis blocks the next step. The reason is that EF-Tu prevents the aminoacyl end of aminoacyl-tRNA from entering the A site on the 50S subunit (see Figure 6.23). So the release of EF-Tu•GDP is needed for the ribosome to undertake peptide bond formation. The same principle is seen at other stages of protein synthesis: one reaction must be completed properly before the next can occur.

The parameters of the interaction with EF-Tu also play a role in quality control. Aminoacyl-tRNAs are brought into the A site without knowing whether their anticodons will fit the codon. The hydrolysis of EF-Tu•GTP is relatively slow: because it takes longer than the time required for an incorrect aminoacyl-tRNA to dissociate from the A site, most incorrect species are removed at this stage. The release of EF-Tu•GDP after hydrolysis also is slow, so any surviving incorrect aminoacyl-tRNAs may dissociate at this stage. The basic principle is that the reactions involving EF-Tu occur slowly enough to allow incorrect aminoacyl-tRNAs to dissociate before they become trapped in protein synthesis.

In eukaryotes, the factor **eEF-1α** is responsible for bringing aminoacyl-tRNA to the ribosome, again in a reaction that involves cleavage of a high-energy bond in GTP. It is homologous to its prokaryotic counterpart (EF-Tu), and similarly is an abundant protein. After hydrolysis of GTP, the active form is regenerated by the factor eEF-1βγ, a counterpart to EF-Ts.

Translocation moves the ribosome

The ribosome remains in place while the polypeptide chain is elongated by transferring the polypeptide attached to the tRNA in the P site to the aminoacyl-tRNA present in the A site. The reaction is shown in **Figure 6.20.** The activity responsible for synthesis of the peptide bond is called **peptidyl transferase.**

The nature of the transfer reaction is revealed by the ability of the antibiotic **puromycin** to inhibit protein synthesis. Puromycin resembles an amino acid attached to the terminal adenosine of tRNA. **Figure 6.21** shows that puromycin has an N instead of the O that joins an amino acid to tRNA. The antibiotic is treated by the ribosome as though it were an incoming aminoacyl-tRNA. Then the polypeptide attached to peptidyl-tRNA is transferred to the NH_2 group of the puromycin.

Because the puromycin moiety is not anchored to the A site of the ribosome, the polypeptidyl-puromycin adduct is released from the ribosome in the form of polypeptidyl-puromycin. This premature termination of protein synthesis is responsible for the lethal action of the antibiotic.

Peptidyl transferase is a function of the large (50S or 60S) ribosomal subunit. The transferase is part of a ribosomal site at which the ends of the peptidyl-tRNA and aminoacyl-tRNA are brought close together. Both rRNA and 50S subunit proteins are necessary for this activity. The catalytic activity may be a property of the ribosomal RNA of the 50S subunit, rather than of any individual protein (see below).

The cycle of addition of amino acids to the growing polypeptide chain is completed by the **translocation** illustrated in **Figure 6.22,** in which the ribosome advances three nucleotides along the mRNA. The result of translocation is to expel the uncharged tRNA from the P site, so that the new peptidyl-tRNA can enter. The ribosome then has an empty A site ready for entry of the aminoacyl-tRNA corresponding to the next codon.

In bacteria the discharged tRNA leaves the ribosome via another site, the E site. In eukaryotes it is expelled directly into the cytosol. The process of relocating the discharged tRNA and peptidyl-tRNA takes place in two stages. First the aminoacyl ends of the tRNAs (located in the 50S subunit) move into the new sites (while the anticodon ends remain bound to their anticodons in the 30S subunit). At this stage, the tRNAs are effectively bound in hybrid sites, consisting of the 50S E/ 30S P and the 50S P/ 30S A sites. Then movement is extended to the 30S subunits, so that the anticodon-codon pairing region finds itself in the right site. The general model for two-part movement is called the **hybrid states model.**

The figure shows two possible ways to create the hybrid states. It could involve *movement of tRNA relative to the ribosome*, so that the aminoacyl end of tRNA

Figure 6.20 Peptide bond formation takes place by reaction between the polypeptide of peptidyl-tRNA in the P site and the amino acid of aminoacyl-tRNA in the A site.

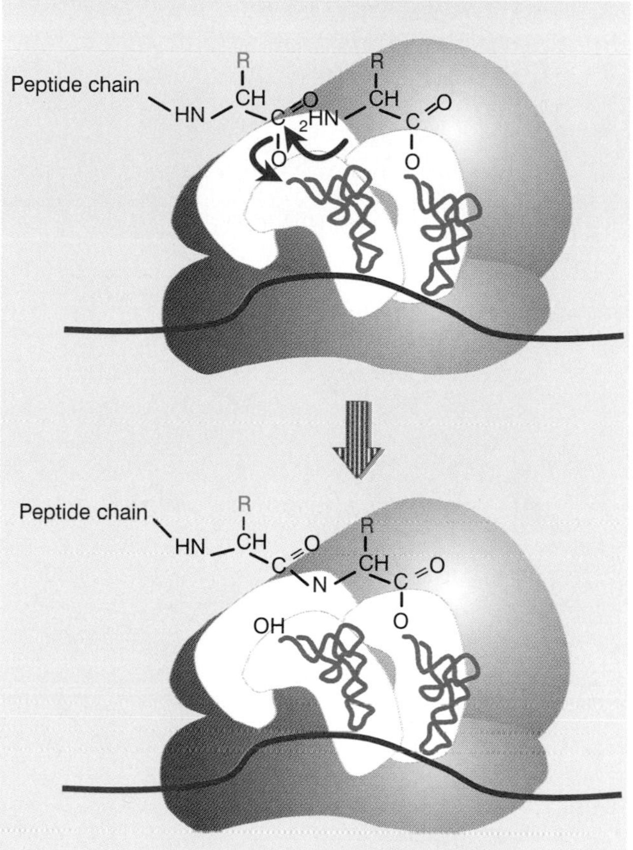

Figure 6.21 Puromycin mimics aminoacyl-tRNA because it resembles an aromatic amino acid linked to a sugar-base moiety.

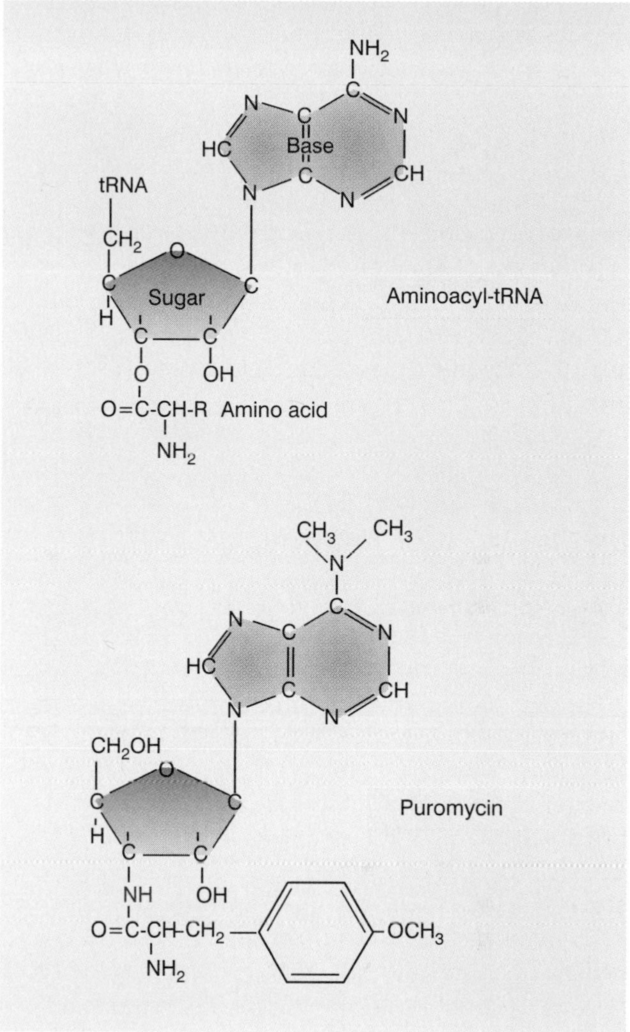

moves within the 50S subunit; the anticodon end moves later when translocation occurs. Alternatively, *the entire 50S subunit might move relative to the 30S subunit*, so that translocation in effect involves two stages, the normal structure of the ribosome being restored by the second stage.

The ribosome faces an interesting dilemma at translocation. It needs to break many of its contacts with tRNA in order to allow movement. At least some of these contacts are needed to stabilize the inherently weak codon-anticodon interaction. But at the same time it must maintain pairing between tRNA and the anticodon (breaking the pairing of the deacylated tRNA only at the right moment). One possibility is that the ribosome switches between alternative, discrete conformations. The switch could consist of changes in rRNA base pairing. The accuracy of translation is influenced by certain mutations that influence alternative base-pairing arrangements. The most likely interpretation is that the effect is mediated by the tightness of binding to tRNA of the alternative conformations.

Translocation requires GTP and another elongation factor, **EF-G**. This factor is a major constituent of the cell; it is present at a level of ~1 copy per ribosome (20,000 molecules per cell).

Ribosomes cannot bind EF-Tu and EF-G simultaneously, so protein synthesis follows the cycle illustrated in **Figure 6.23** in which the factors are alternately bound to, and released from, the ribosome. So EF-Tu•GDP must be released before EF-G can bind; and then EF-G must be released before aminoacyl-tRNA•EF-Tu•GTP can bind.

Does the ability of each elongation factor to exclude the other rely on an allosteric effect on the overall con-

Figure 6.22 Models for translocation involve two stages. First, at peptide bond formation the aminoacyl end of the tRNA in the A site becomes located in the P site. Second, the anticodon end of the tRNA becomes located in the P site.

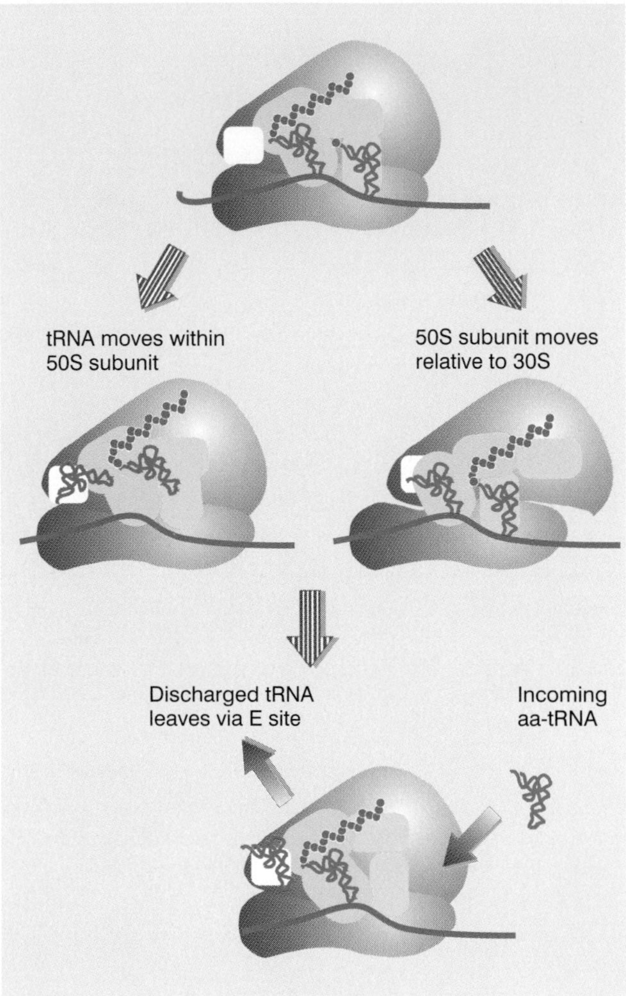

Figure 6.23 Binding of factors EF-Tu and EF-G alternates as ribosomes accept new aminoacyl-tRNA, form peptide bonds, and translocate.

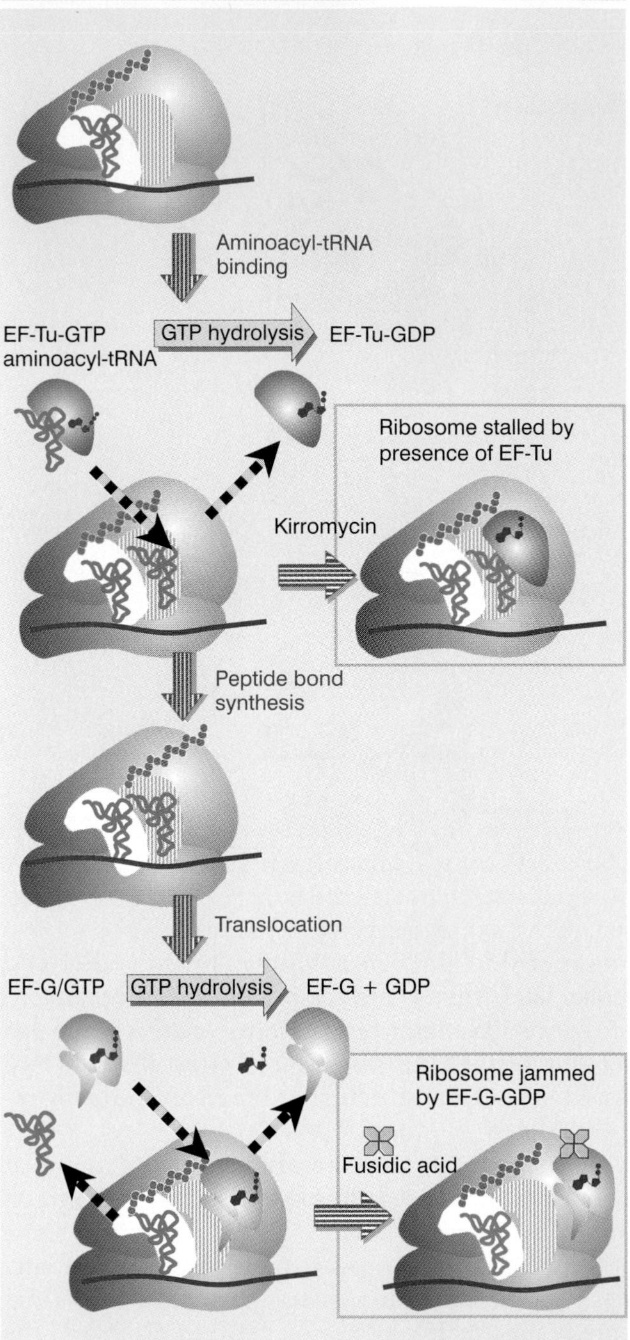

formation of the ribosome or on direct competition for overlapping binding sites? **Figure 6.24** shows an extraordinary similarity between the structures of the ternary complex of aminoacyl-tRNA•EF-Tu•GDP and EF-G. The aminoacyl-tRNA is bound by EF-Tu around its amino acceptor stem. It is striking that the structure of the protein EF-G mimics the overall structure of the protein complexed with RNA in the ternary complex. This creates the immediate assumption that they compete for the same binding site (presumably in the vicinity of the A site). The need for each factor to be released before the other can bind ensures that the events of protein synthesis proceed in an orderly manner.

Both elongation factors are typical monomeric GTP-binding proteins that are active when bound to GTP but inactive when bound to GDP. The triphosphate form is required for binding to the ribosome, which ensures that each factor obtains access to the ribosome only in the company of the GTP that it needs to fulfill its function.

EF-G binds to the ribosome to sponsor translocation; and then is released following ribosome move

Figure 6.24 The structure of the ternary complex of aminoacyl-tRNA·EF-Tu·GTP (left) resembles the structure of EF-G (right). Structurally conserved domains of EF-Tu and EF-G are in red and green; the tRNA and the domain resembling it in EF-G are in purple. Photograph kindly provided by Poul Nissen.

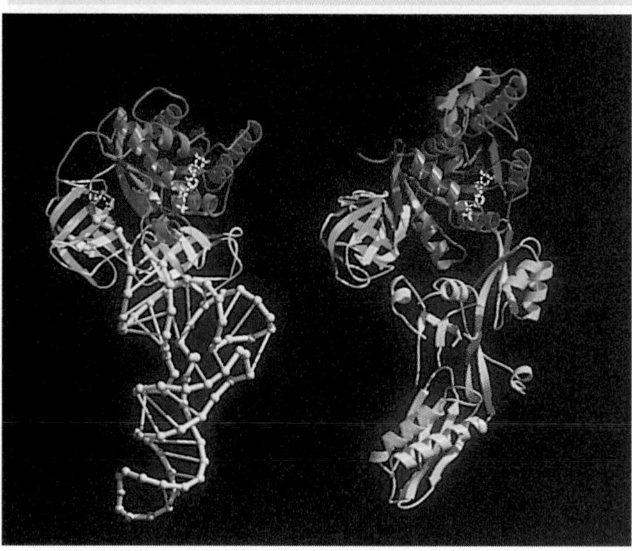

ment. It is an important part of the mechanism for translocation. The hydrolysis of GTP occurs before translocation and accelerates the ribosome movement. The most likely mechanism is that GTP hydrolysis causes a change in the structure of EF-G, which in turn forces a change in the ribosome structure. A deletion of domain I (located on the 30S subunit) allows a partial translocation of the 50S subunit to occur to create hybrid sites, suggesting that different domains of EF-G may interact with each ribosomal subunit.

EF-G can still bind to the ribosome when GMP-PCP is substituted for GTP; thus the presence of a guanine nucleotide is needed for binding, but its hydrolysis is not absolutely essential for translocation (although translocation is much slower in the absence of GTP hydrolysis). The hydrolysis of GTP is needed to release EF-G.

The need for EF-G release was discovered by the effects of the steroid antibiotic fusidic acid, which "jams" the ribosome in its post-translocation state (see Figure 6.23). In the presence of fusidic acid, one round of translocation occurs: EF-G binds to the ribosome, GTP is hydrolyzed, and the ribosome moves three nucleotides. But fusidic acid stabilizes the ribosome•EF-G•GDP complex, so that EF-G and GDP remain on the ribosome instead of being released. Because the ribosome then cannot bind aminoacyl-tRNA, no further amino acids can be added to the chain.

The eukaryotic counterpart to EF-G is the protein eEF-2, which functions in a similar manner, as a translocase dependent on GTP hydrolysis. Its action also is inhibited by fusidic acid. A stable complex of eEF-2 with GTP can be isolated; and the complex can bind to ribosomes with consequent hydrolysis of its GTP.

A unique reaction of eEF-2 is its susceptibility to diphtheria toxin. The toxin uses NAD (nicotinamide adenine dinucleotide) as a cofactor to transfer an ADPR moiety (adenosine diphosphate ribosyl) on to the eEF-2. The ADPR-eEF-2 conjugate is inactive in protein synthesis. The substrate for the attachment is an unusual amino acid, produced by modifying a histidine; it is common to the eEF-2 of many species.

The ADP-ribosylation is responsible for the lethal effects of diphtheria toxin. The reaction is extraordinarily effective: a single molecule of toxin can modify sufficient eEF-2 molecules to kill a cell.

Three codons terminate protein synthesis

Only 61 triplets are assigned to amino acids. The other three triplets are **termination codons** (or stop codons) that end protein synthesis. They have casual names from the history of their discovery. The UAG triplet is called the **amber** codon; UAA is the **ochre** codon; and UGA is sometimes called the **opal** codon.

The nature of these triplets was originally shown by a genetic test that distinguished two types of point mutation:

- A point mutation that changes a codon to represent a different amino acid is called a **missense** mutation. One amino acid replaces the other in the protein; the effect on protein function depends on the site of mutation and the nature of the amino acid replacement.

- When a point mutation creates one of the three termination codons, it causes **premature termination** of protein synthesis at the mutant codon. This is likely to abolish protein function, since only the first part of the protein is made in the mutant cell. A change of this sort is called a **nonsense** mutation.

(Sometimes the term *nonsense codon* is used to describe the termination triplets. "Nonsense" is really a misnomer, since the codons do have meaning, albeit a disruptive one in a mutant gene.)

In every gene that has been sequenced, one of the termination codons lies immediately after the codon representing the C-terminal amino acid of the wild-type sequence. Nonsense mutations show that any one of the three codons is sufficient to terminate protein synthesis within a gene. The UAG, UAA, and UGA triplet sequences are therefore necessary and sufficient to end protein synthesis, whether occurring naturally at the end of a gene or created by mutation within a coding sequence.

In bacterial genes, UAA is the most commonly used termination codon. UGA is used more heavily than UAG, although there appear to be more errors reading UGA. (An error in reading a termination codon, when an aminoacyl-tRNA responds improperly to it, results in the continuation of protein synthesis until another termination codon is encountered.)

None of the termination codons is represented by a tRNA. They function in an entirely different manner from other codons, and are recognized directly by protein factors. (Since the reaction does not depend on codon-anticodon recognition, there seems to be no particular reason why it should require a triplet sequence. Presumably this reflects the evolution of the genetic code.)

In *E. coli* two related proteins catalyze termination. They are called **release factors** (**RF**), and are specific for different sequences. **RF-1** recognizes UAA and UAG; **RF-2** recognizes UGA and UAA. The factors act at the ribosomal A site and require polypeptidyl-tRNA in the P site. The release factors are present at much lower levels than initiation or elongation factors; there are ~600 molecules of each per cell, equivalent to 1 RF per 10 ribosomes. Probably at one time there was only a single release factor, recognizing all termination codons, and later it evolved into two factors with specificities for particular codons.

RF1 and RF2 recognize the termination codons and activate the ribosome to hydrolyze the peptidyl tRNA. Cleavage of polypeptide from tRNA takes place by a reaction analogous to the usual peptidyl transfer, except that the acceptor is H_2O instead of aminoacyl-tRNA. Then RF1 or RF2 is released from the ribosome by RF-3, which is a GTP-binding protein related to EF-G. RF3 resembles the GTP-binding domains of EF-Tu and EF-G, and RF1/2 resemble the C-terminal of EF-G, which mimics tRNA. This suggests that the action of RF3 on RF1/2 utilizes the same site that is used by the elongation factors. **Figure 6.25** illustrates the basic idea that all three factors have the same general shape and bind to the ribosome successively at the same site (basically the A site or a region extensively overlapping with it).

Mutations in the RF genes reduce the efficiency of termination, as seen by an increased ability to continue protein synthesis past the termination codon. Overexpression of RF-1 or RF-2 increases the efficiency of termination at the codons on which it acts. This suggests that codon recognition by RF1 or RF2 competes with aminoacyl-tRNAs that erroneously recognize the termination codons.

The termination reaction involves release of the completed polypeptide from the last tRNA, expulsion of the tRNA from the ribosome, and dissociation of the ribosome from mRNA. The factor RRF, ribosome

Figure 6.25 Molecular mimicry enables the elongation factor Tu-tRNA complex, the translocation factor EF-G, and the release factors RF1/2-RF3 to bind to the same ribosomal site.

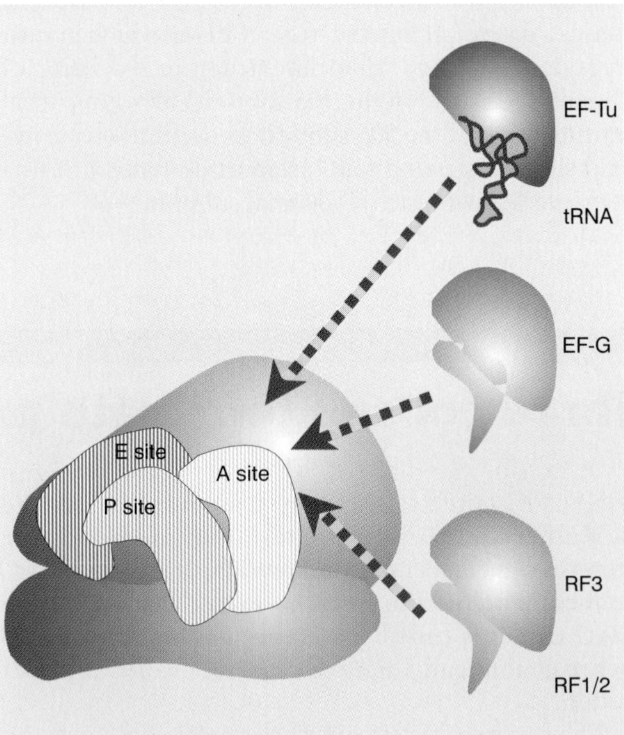

recycling factor, acts together with EF-G to cause dissociation of the 50S and 30S subunits. IF-3 is also required, which brings the wheel full circle to its original discovery, when it was proposed to be a dissociation factor! RRF acts on the 50S subunit, and IF-3 acts to remove deacylated tRNA from the 30S subunit. Once the subunits have separated, IF-3 remains necessary, of course, to prevent their reassociation.

Ribosomes have several active centers

Both major rRNAs have considerable secondary structure. Although the sequence of an RNA can be used to predict the formation of base-paired regions, in molecules as large as the rRNAs there are many alternative conformations. It is not possible to predict which would be chosen even if the molecule were simply free in solution.

The most penetrating approach to analyzing secondary structure is to compare the sequences of corresponding rRNAs in related organisms. Those regions that are important in the secondary structure retain the ability to interact by base pairing; so if a base pair is required, it can form at the same relative position in each rRNA. From such comparisons, models have been constructed for both 16S and 23S rRNA.

16S rRNA forms four general domains, in which just under half of the sequence is base paired (see Figure 6.28). The individual double-helical regions tend to be short (<8 bp). Often the duplex regions are not perfect, but contain bulges of unpaired bases. Comparable models have been drawn for mitochondrial rRNAs (which are shorter and have fewer domains) and for eukaryotic cytosolic rRNAs (which are longer and have more domains). The increase in length in eukaryotic rRNAs is due largely to the acquisition of sequences representing additional domains.

Is the structure of rRNA in the subunit invariant? Some differences in the reactivity of 16S rRNA are found when 30S subunits are compared with 70S ribosomes; also there are some differences between free ribosomes and those engaged in protein synthesis. Changes in the reactivity of the rRNA occur when mRNA is bound, when the subunits associate, or when tRNA is bound. Some changes reflect a direct interaction of the rRNA with mRNA or tRNA, while others are caused indirectly by other changes in ribosome structure. The main point is that ribosome conformation is flexible during protein synthesis.

A feature of the primary structure of rRNA is the presence of methylated residues, sometimes in regions that are well conserved. There are ~10 methyl groups in 16S rRNA (located mostly toward the 3′ end of the molecule) and ~20 in 23S rRNA. In mammalian cells, the 18S and 28S rRNAs carry 43 and 74 methyl groups, respectively, so ~2% of the nucleotides are methylated (about three times the proportion methylated in bacteria).

The large ribosomal subunit also contains a molecule of a 120 base **5S RNA** (in all ribosomes except those of mitochondria). The sequence of 5S RNA is less well conserved than those of the major rRNAs. All 5S RNA molecules display a highly base-paired structure, although the exact organization could not be settled by direct analysis of the RNA (more than 20 models were proposed). This emphasizes the difficulty of distinguishing between alternative pairing possibilities even in quite small molecules. As with the large rRNAs, a model was resolved by comparisons between the structures that can be formed in 5S RNAs of different species.

In eukaryotic cytosolic ribosomes, another small RNA is present in the large subunit. This is the **5.8S RNA.** Its sequence corresponds to the 5′ end of the prokaryotic 23S rRNA.

Some ribosomal proteins bind strongly to isolated rRNA. Some do not bind to free rRNA, but can bind after other proteins have bound. This suggests that the conformation of the rRNA is important in determining whether binding sites exist for some proteins. As each protein binds, it induces conformational changes in the rRNA that make it possible for other proteins to bind. In *E. coli*, virtually all the 30S ribosomal proteins interact (albeit to varying degrees) with 16S rRNA. The binding sites on the proteins show a wide variety of structural features, suggesting that protein-RNA recognition mechanisms may be diverse.

Progress in mapping the locations of the ribosomal proteins and rRNA is summarized in **Figure 6.26.** We have a good impression of the shapes of only a few individual proteins, so the proteins are represented by

Figure 6.26 A diagrammatic representation of the 30S subunit shows that each domain of rRNA occupies a discrete location, and the positions of all ribosomal proteins have been mapped. Note that this is a two-dimensional projection of a three- dimensional reconstruction and is not to scale.

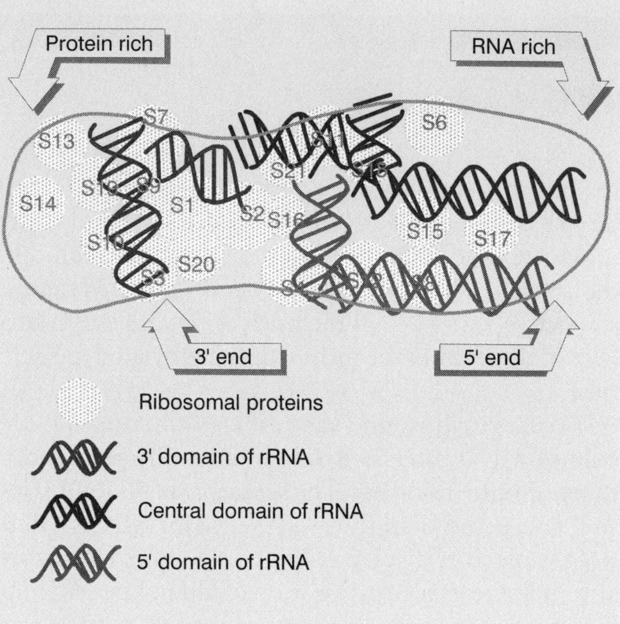

spheres in the figure. The domains of 16S rRNA identified by the secondary structure occupy relatively discrete regions of the small subunit. One interesting feature is that the centers of mass of the protein and RNA components are displaced by ~25 Å. This generates a concentration of rRNA at one end of the subunit; the subunit is relatively free of protein at the right end in Figure 6.26.

We can distinguish different locations for several ribosomal activities. The 30S subunits bind mRNA and the initiator-tRNA•initiation factor complex; then they bind 50S subunits. The 70S ribosomes possess the functionally distinct P and A sites at which tRNA is bound; the parts of each site carried on the individual subunits can be distinguished. The peptidyl transferase center is carried on the 50S subunit. The EF-G binding site, and hence responsibility for translocation, is carried on the 50S subunit.

The binding sites are large, each occupying a relatively substantial part of the ribosomal structure. They are not small, discrete regions like the active centers of enzymes. An expanded view of the ribosomal sites is drawn in **Figure 6.27**. They comprise about two-thirds of the ribosomal structure. The nascent protein debouches through the other part of the ribosome, which comprises the region in which ribosomes may be attached to membranes (see Chapter 8). A polypeptide

Figure 6.27 The ribosome has several active centers. It may be associated with a membrane. mRNA takes a turn as it passes through the A and P sites, which are angled with regard to each other. The E site lies beyond the P site. The peptidyl transferase site (not shown) stretches across the tops of the A and P sites. Part of the site bound by EF-Tu/G lies at the base of the A and P sites (not shown).

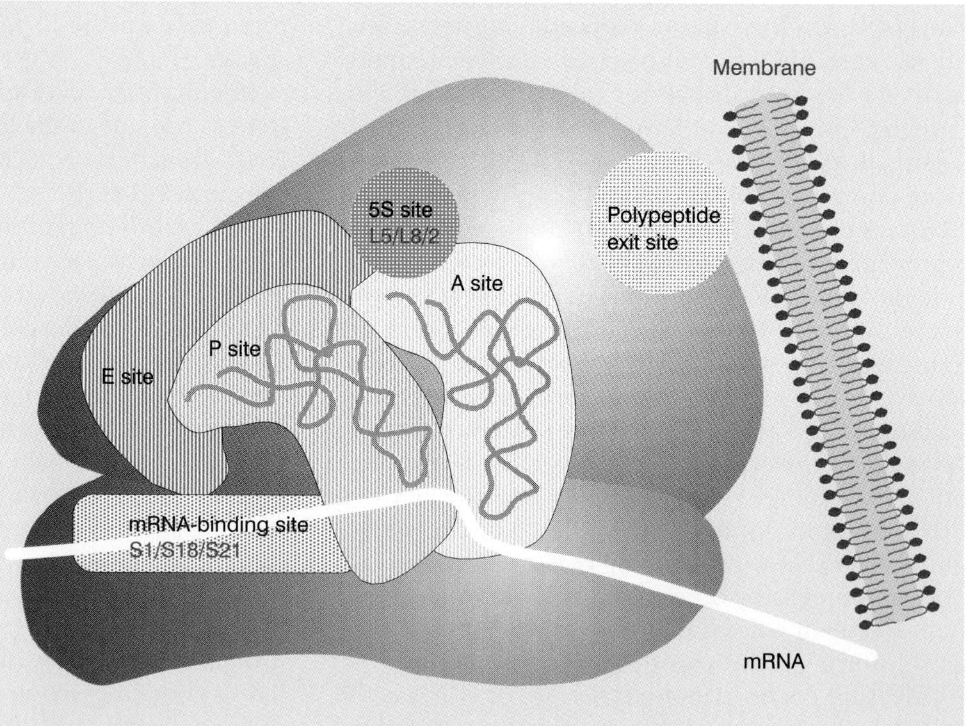

chain emerges from the ribosome through an exit channel, which leads from the peptidyl transferase site to the surface of the 50S subunit.

A tRNA enters the A site, is transferred by translocation into the P site, and then leaves the (bacterial) ribosome by the E site. The A and P sites must extend across both ribosome subunits, since tRNA is paired with mRNA in the 30S subunit, but peptide transfer takes place in the 50S subunit. The A and P sites are adjacent, enabling translocation to move the tRNA from one site into the other. The problem of how two bulky tRNAs fit into the ribosome is solved by a turn in the path for mRNA, which results in the tRNAs being angled with regard to each other. This leaves a subsidiary problem: the aminoacyl ends of the tRNAs are not close, and some conformational shift must occur to allow peptidyl transfer. The E site is located near the P site (probably representing a position *en route* to the surface of the 50S subunit).

How are these sites related to the actual topology of the ribosome? Which ribosomal components are involved in each particular function?

Initial binding of 30S subunits to mRNA requires protein S1, which has a strong affinity for single-stranded nucleic acid. It is responsible for maintaining the single-stranded state in mRNA that is bound to the 30S subunit. This action is necessary to prevent the mRNA from taking up a base-paired conformation that would be unsuitable for translation.

Analyzing the location of S1 in the ribosome is complicated by its extremely elongated structure. Crosslinking studies show that it is closely related to S18 and S21, which are among the proteins that react in affinity-labeling experiments when initiator tRNA is bound to an AUG codon. The three proteins constitute a domain that is involved in the initial binding of mRNA and in binding initiator tRNA. This would locate the mRNA-binding site in the vicinity of the cleft of the small subunit (see Figure 6.2). The 3′ end of rRNA, which pairs with the mRNA initiation site, is located in this region.

The initiation factors bind in the same region of the ribosome. IF-3 can be crosslinked to the 3′ end of the rRNA, as well as to several ribosomal proteins, including those probably involved in binding mRNA. The role of IF-3 could be to stabilize mRNA•30S subunit binding; then it would be displaced when the 50S subunit joins.

Not very much is known about the mechanism of subunit joining. It is affected by a mutation in a loop of 16S rRNA (at position 791) that is located at the subunit interface. Other nucleotides in 16S rRNA have been shown to be involved by modification/interference experiments.

The 16S rRNA is part of the A site. The rRNA has two conserved sequences in the 3′ minor domain, called the 1400 and 1500 regions, that may interact with tRNA. Crosslinking studies show that peptidyl-tRNA can be linked to a point in this sequence. When aminoacyl-tRNA binds to the ribosome, some bases in this region and elsewhere are shielded from attack by chemical probes.

The incorporation of 5S RNA into 50S subunits that are assembled *in vitro* depends on the ability of three proteins, L5, L8, and L25, to form a stoichiometric complex with it. The complex can bind to 23S rRNA, although none of the isolated components can do so. It lies in the vicinity of the P and A sites.

The peptidyl transferase site has been localized on the central protuberance by the binding of puromycin. A group of several proteins and the 23S rRNA are involved in creating the site.

The growing polypeptide chain appears to be extruded from the 50S subunit ~15 Å away from the peptidyl transferase site. It probably extends through the ribosome as an unfolded polypeptide chain until it leaves the exit domain, when it is free to start folding.

One case in which individual ribosomal proteins can be equated with a particular structural feature is that of L7/L12. L7 is a modification of L12, and has an acetyl group on the N-terminus. The L7/L12 aggregate forms the stalk of the large subunit.

Although translocation involves movement of the mRNA through the 30S subunit, the translocation factor EF-G binds to the 50S subunit. The binding site for EF-G is close to S12, one of the proteins of the mRNA-binding site on the 30S subunit. This places EF-G at the interface between subunits, in the vicinity of the L7/L12 dimers.

When the stalk is removed, the ribosomes do not support GTP hydrolysis that depends on accessory factors. Another structure that is involved with GTP hydrolysis is identified by the complex of protein L11 with a 58 base stretch of 23S rRNA. This is the binding site for some antibiotics that affect GTPase activity. Neither of these ribosomal structures actually possesses GTPase activity, but they may be necessary for it. It is still unclear whether the ribosome itself has GTPase activity or whether it just assists individual factors that act as GTPases.

An indication that the EF-G and EF-Tu sites are connected is provided by changes in 23S rRNA. Binding of EF-G to the ribosome protects bases at two locations in 23S rRNA. Bases at one of these locations, the 2660 loop, are also protected by binding of EF-Tu. This is a counterpart to the location in eukaryotic 28S rRNA that is the target for lectins that inhibit elongation (ricin depurinates rRNA and α-sarcin cleaves it). The structure of this loop is therefore important for elongation.

The functional relationships between the various ribosomal sites have yet to be defined. The ribosome probably has a highly interactive structure, in which a change at one point greatly affects the activity of another site elsewhere.

The role of ribosomal RNA in protein synthesis

The ribosome was originally viewed as a collection of proteins with various catalytic activities, held together by protein-protein interactions and by binding to rRNA. But the discovery of RNA molecules with catalytic activities (see Chapter 22) immediately suggests that rRNA might play a more active role in ribosome function. There is now evidence that rRNA interacts with mRNA or tRNA at each stage of translation, and that the proteins are necessary to maintain the rRNA in a structure in which it can perform the catalytic functions. Several interactions involve specific regions of rRNA:

- The 3′ terminus of the rRNA interacts directly with mRNA at initiation.
- Specific regions of 16S rRNA interact directly with the anticodon regions of tRNAs in both the A site and the P site; 23S rRNA interacts with the CCA terminus of peptidyl-tRNA.
- Subunit interaction involves both 16S and 23S rRNAs, although it is not yet clear whether association is mediated directly by RNA-RNA interactions or by RNA-protein interactions.

Much information about the individual steps of bacterial protein synthesis has been obtained by using antibiotics that inhibit the process at particular stages. The target for the antibiotic can be identified by the component in which resistant mutations occur. Some antibiotics act on individual ribosomal proteins, but several act on rRNA, which suggests that the rRNA is involved with many or even all of the functions of the ribosome.

The functions of rRNA have been investigated by two types of approach. Structural studies show that particular regions of rRNA are located in important sites of the ribosome, and that chemical modifications of these bases impede particular ribosomal functions. And mutations identify bases in rRNA that are required for particular ribosomal functions. **Figure 6.28** summarizes sites in 16S rRNA that have been identified by these means.

An indication of the importance of the 3′ end of 16S rRNA is given by its susceptibility to the lethal agent colicin E3. Produced by some bacteria, the colicin cleaves ~50 nucleotides from the 3′ end of the 16S rRNA of *E. coli*. The cleavage entirely abolishes initiation of protein synthesis. Several important functions require the region that is cleaved: binding the factor IF-3; recognition of mRNA; and binding of tRNA.

The 3′ end of the 16S rRNA is directly involved in the initiation reaction by pairing with the Shine-Dalgarno sequence in the ribosome-binding site of mRNA (see Figure 6.14). Another direct role for the 3′ end of 16S rRNA in protein synthesis is shown by the properties of kasugamycin-resistant mutants, which lack certain modifications in 16S rRNA. Kasugamycin blocks initiation of protein synthesis. Resistant mutants of the type *ksgA* lack a methylase enzyme that introduces four methyl groups into two adjacent adenines at a site near the 3′ terminus of the 16S rRNA. The methylation generates the highly conserved sequence G–m^6_2A–m^6_2A, found in both prokaryotic and eukaryotic small rRNA. The methylated sequence is involved in the joining of the 30S and 50S subunits, which in turn is connected also with the retention of initiator tRNA in the complete ribosome. Kasugamycin causes fMet-$tRNA_f$ to be released from the sensitive (methylated) ribosomes, but the resistant ribosomes are able to retain the initiator.

Mutations in rRNA also can influence the specificity of protein synthesis. One example is that a mutation in the 3′ major domain of 16S rRNA suppresses UGA codons. It is not clear whether this effect is specific for UGA (perhaps caused by pairing between rRNA and the UGA codon), or whether it represents a general in-

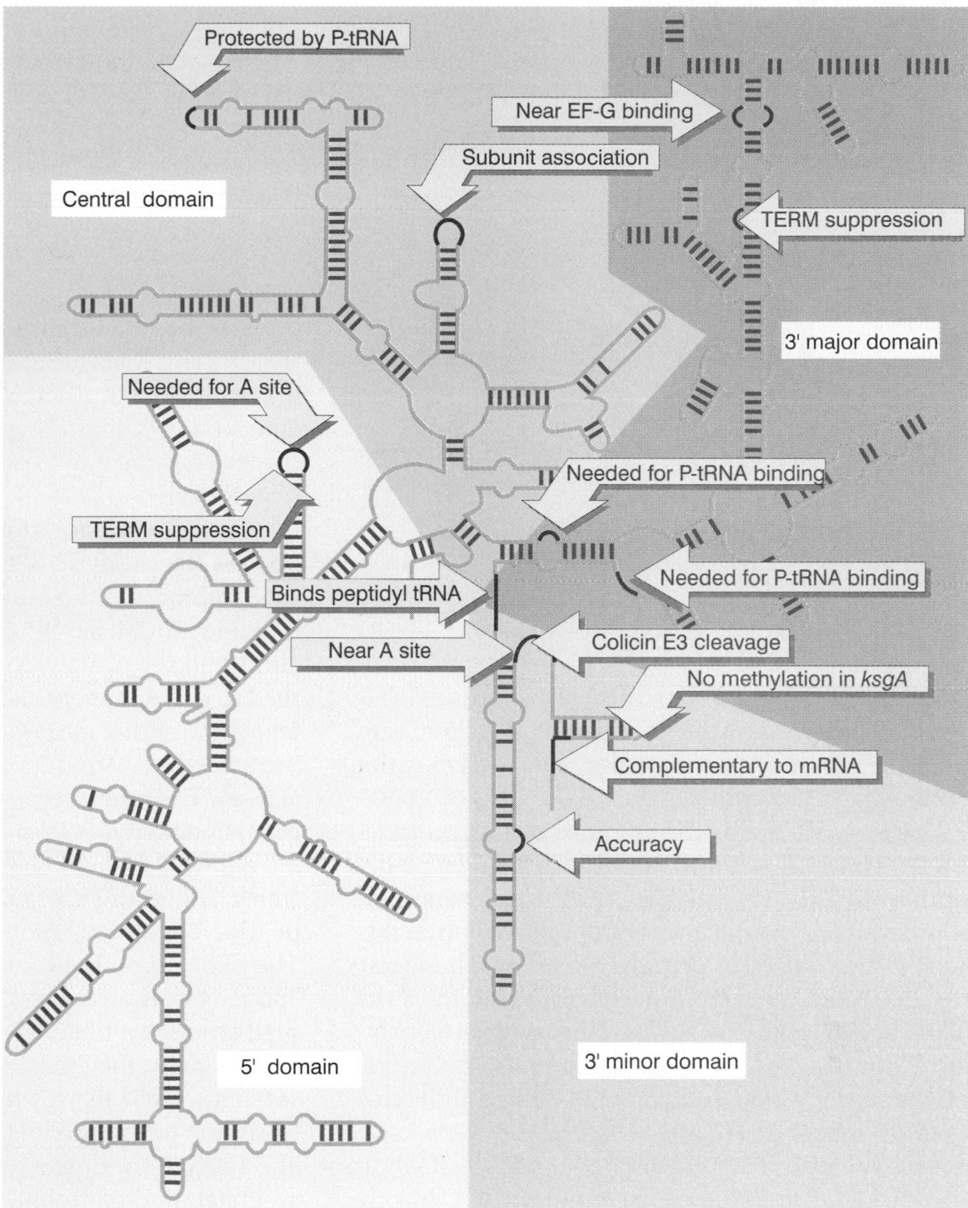

Figure 6.28 Some sites in 16S rRNA are protected from chemical probes when 50S subunits join 30S subunits or when aminoacyl-tRNA binds to the A site. Others are the sites of mutations that affect protein synthesis. TERM suppression sites may affect termination at some or several termination codons. The large colored blocks indicate the four domains of the rRNA.

terference with termination at all three termination codons. In another case, a recognition reaction between rRNA and mRNA is involved in a frameshift in which one base is skipped in reading the messenger. These results raise the possibility that rRNA is continually exposed to the sequence of mRNA as the ribosome moves along the message.

Changes in the structure of 16S rRNA occur when ribosomes are engaged in protein synthesis, as seen by protection of particular bases against chemical attack. The individual sites fall into a few groups, concentrated in the 3′ minor and central domains. Although the locations are dispersed in the linear sequence of 16S rRNA, it seems likely that base positions involved in the same function are actually close together in the tertiary structure.

Some of the changes in 16S rRNA are triggered by joining with 50S subunits, binding of mRNA, or binding of tRNA. They indicate that these events are associated with changes in ribosome conformation that affect the exposure of rRNA. They do not necessarily indicate direct participation of rRNA in these functions. One change that occurs during protein synthesis is shown in **Figure 6.29;** it involves a local movement to change the nature of a short duplex sequence.

The 16S rRNA is involved in both A site and P site function, and significant changes in its structure occur when these sites are occupied. Two distinct regions are

Figure 6.29 A change in conformation of 16S rRNA may occur during protein synthesis.

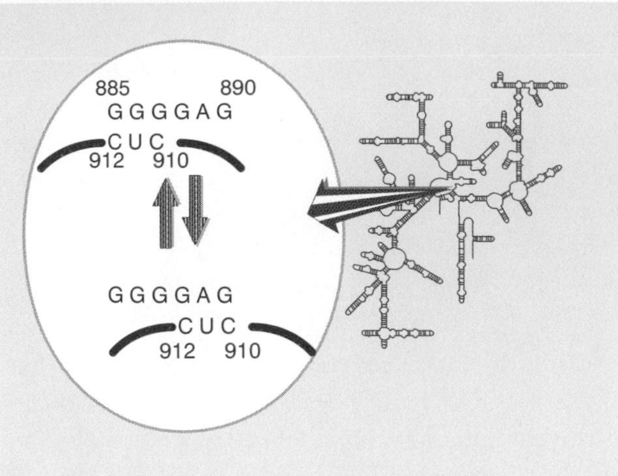

protected by tRNA bound in the A site. One is in the 3′ terminal region; the other is called the 530 loop (and also is the site of a mutation that prevents termination at the UAA, UAG, and UGA codons).

A variety of bases in different positions are protected by tRNA in the P site; probably the bases lie near one another in the tertiary structure. One prominent stretch of 16S rRNA (known as the 1400 region) can be directly cross-linked to peptidyl-tRNA, which suggests that this region is a structural component of the P site. All of the effects that tRNA binding has on 16S rRNA can be produced by the isolated oligonucleotide of the anticodon stem-loop, so that tRNA-30S subunit binding must involve this region.

The sites involved in the functions of 23S rRNA are less well identified than those of 16S rRNA, but the same general pattern is observed: bases at certain positions affect specific functions. Positions in 23S rRNA can be identified that are affected by the conformation of the A site or P site. Each of these sites therefore interacts with the 23S rRNA as well as with the 16S rRNA. In particular, oligonucleotides derived from the 3′ CCA terminus of tRNA protect a set of bases in 23S rRNA which essentially are the same as those protected by peptidyl-tRNA. This suggests that the major interaction of 23S rRNA with peptidyl-tRNA in the P site involves the 3′ end of the tRNA. In fact, the tRNA base pairs with the rRNA, as shown by the fact that a mutation in a G residue in the rRNA prevents interaction with tRNA, but interaction is restored by a compensating mutation in a C of the amino acceptor end of the tRNA.

Another site that binds tRNA is the E site, which is localized almost exclusively on the 50S subunit. Bases affected by its conformation can be identified in 23S rRNA.

What is the nature of the site on the 50S subunit that provides peptidyl transferase function? It seems likely that the rRNA at the least provides an important part of the peptidyl transferase site, because a particular region of the 23S rRNA is the site of mutations that confer resistance to antibiotics that inhibit peptidyl transferase.

A long search for ribosomal proteins that might possess the catalytic activity has been unsuccessful. More recent results suggest that the ribosomal RNA of the large subunit has the catalytic activity. Extraction of almost all the protein content of 50S subunits leaves the 23S rRNA associated largely with fragments of proteins, amounting to <5% of the mass of the ribosomal proteins. This preparation retains peptidyl transferase activity. Treatments that damage the RNA abolish the catalytic activity.

Following from these results, 23S rRNA prepared by transcription *in vitro* can catalyze the formation of a peptide bond between Ac-Phe-tRNA and Phe-tRNA. The yield of Ac-Phe-Phe is very low, suggesting that the 23S rRNA requires proteins in order to function at a high efficiency. But since the rRNA has the basic catalytic activity, the role of the proteins must be indirect, serving to fold the rRNA properly or to present the substrates to it. The reaction also works, although less effectively, if the domains of 23S rRNA are synthesized separately and then combined. In fact, some activity is shown by domain V alone, which has the catalytic center. Activity is abolished by mutations in position 2252 of domain V that lies in the P site.

Indeed, it may well be the case that certain proteins of the 50S subunit are required to enable the rRNA to form the proper structure *in vivo*. But the idea that rRNA is the catalytic component is consistent with the results discussed in Chapter 22 that identify catalytic properties in RNA that are involved with several RNA processing reactions. It fits with the notion that the ribosome evolved from functions originally possessed by RNA.

Summary

Ribosomes are ribonucleoprotein particles in which a majority of the mass is provided by rRNA. The shapes of all ribosomes are generally similar, but only those of bacteria (70S) have been characterized in detail. The small (30S) subunit has a squashed shape, with a "body" containing about two-thirds of the mass divided from the "head" by a cleft. The large (50S) subunit is more spherical, with a prominent "stalk" on the right and a "central protuberance". Locations of all proteins are known approximately in the small subunit.

Each subunit contains a single major rRNA, 16S and 23S in prokaryotes, 18S and 28S in eukaryotic cytosol. There are also minor rRNAs, most notably 5S rRNA in the large subunit. Both major rRNAs have extensive base pairing, mostly in the form of short, imperfectly paired duplex stems with single-stranded loops. Conserved features in the rRNA can be identified by comparing sequences and the secondary structures that can be drawn for rRNA of a variety of organisms. The 16S rRNA has four distinct domains; the three major domains have been mapped into regions of the small subunit. Eukaryotic 18S rRNA has additional domains. One end of the 30S subunit may consist largely or entirely of rRNA.

Each subunit has several active centers, concentrated in the translational domain of the ribosome where proteins are synthesized. Proteins leave the ribosome through the exit domain, which can associate with a membrane. The major active sites are the P and A sites, the E site, the EF-Tu and EF-G binding sites, peptidyl transferase, and mRNA-binding site. Ribosomal proteins required for the function of some of these sites have been identified, but the sites have yet to be mapped in terms of three-dimensional ribosome structure. Ribosome conformation may change at stages during protein synthesis; differences in the accessibility of particular regions of the major rRNAs have been detected.

The major rRNAs contain regions that are localized at some of these sites, most notably the mRNA-binding site and P site on the 30S subunit. The 3′ terminal region of the rRNA seems to be of particular importance. Functional involvement of the rRNA in ribosomal sites is best established for the mRNA-binding site, where mutations in 16S rRNA affect the initiation reaction. Ribosomal RNA is also the target for some antibiotics or other agents that inhibit protein synthesis. 23S rRNA appears to possess the essential catalytic activity of peptidyl transferase.

A codon in mRNA is recognized by an aminoacyl-tRNA, which has an anticodon complementary to the codon and carries the amino acid corresponding to the codon. A special initiator tRNA (fMet-tRNA$_f$ in prokaryotes or Met-tRNA$_i$ in eukaryotes) recognizes the AUG codon, which is used to start all coding sequences. In prokaryotes, GUG is also used. Only the termination (nonsense) codons UAA, UAG, and UGA are not recognized by aminoacyl-tRNAs.

Ribosomes are released from protein synthesis to enter a pool of free ribosomes that are in equilibrium with separate small and large subunits. Small subunits bind to mRNA and then are joined by large subunits to generate an intact ribosome that undertakes protein synthesis. Recognition of a prokaryotic initiation site involves binding of a sequence at the 3′ end of rRNA to the Shine-Dalgarno motif which precedes the AUG (or GUG) codon in the mRNA. Recognition of a eukaryotic mRNA involves binding to the 5′ cap; the small subunit then migrates to the initiation site by scanning for AUG codons. When it recognizes an appropriate AUG codon (usually but not always the first it encounters) it is joined by a large subunit.

A ribosome can carry two aminoacyl-tRNAs simultaneously: its P site is occupied by a polypeptidyl-tRNA, which carries the polypeptide chain synthesized so far, while the A site is used for entry by an aminoacyl-tRNA carrying the next amino acid to be added to the chain. The polypeptide chain in the P site is transferred to the aminoacyl-tRNA in the A site and then the ribosome translocates one codon along the mRNA. Translocation and several other stages of ribosome function require hydrolysis of GTP.

Protein synthesis is an expensive process. ATP is used to provide energy at several stages, including the charging of tRNA with its amino acid, and the unwinding of mRNA. It has been estimated that up to 90% of all the ATP molecules synthesized in a rapidly growing bacterium are consumed in assembling amino acids into protein!

Additional factors are required at each stage of protein synthesis. They are defined by their cyclic association with, and dissociation from, the ribosome. IF factors are involved in prokaryotic initiation. IF-3 is needed for 30S subunits to bind to mRNA and also is responsible for maintaining the 30S subunit in a free form. IF-2 is needed for fMet-tRNA$_f$ to bind to the 30S subunit and is responsible for excluding other aminoacyl-tRNAs from the initiation reaction. GTP is hydrolyzed after the initiator tRNA has been bound to the initiation complex. The initiation factors must be released in order to allow a large subunit to join the initiation complex.

Prokaryotic EF factors are involved in elongation. EF-Tu binds aminoacyl-tRNA to the 70S ribosome. GTP is hydrolyzed when EF-Tu is released, and EF-Ts is required to regenerate the active form of EF-Tu. EF-G is required for translocation. Binding of the EF-Tu and EF-G factors to ribosomes is mutually exclusive, which ensures that each step must be completed before the next can be started. RF factors are required for termination. Protein synthesis in eukaryotes is generally similar to the process in prokaryotes, but involves a more complex set of accessory factors.

Further reading

Reviews

Kozak, M. (1978). How do eukaryotic ribosomes select initiation regions in mRNA? *Cell* **15,** 1109–1123.

Maitra, U. *et al.* (1982). Initiation factors in protein biosynthesis. *Ann. Rev. Biochem.* **51,** 869–900.

Kozak, M. (1983). Comparison of initiation of protein synthesis in prokaryotes, eukaryotes, and organelles. *Microbiol. Rev.* **47,** 1–45.

Wittman, H. G. (1983). Architecture of prokaryotic ribosomes. *Ann. Rev. Biochem.* **52,** 35–65.

Noller, H. F. (1984). Structure of ribosomal RNA. *Ann. Rev. Biochem.* **53,** 119–162.

Lake, J. A. (1985). Evolving ribosome structure: domains in archaebacteria, eubacteria, eocytes, and eukaryotes. *Ann. Rev. Biochem.* **54,** 507–530.

Noller, H. F. and Nomura, M. (1987). Ribosomes. In *E. coli and S. typhimurium.* (Ed. F. C. Neidhardt, American Society for Microbiology, Washington DC, 104–105).

Hill, W. E. *et al.* (1990). *The Ribosome.* (American Society for Microbiology, Washington DC).

Hershey, J. W. B. (1991). Translational control in mammalian cells. *Ann. Rev. Biochem.* **60,** 717–755.

Noller, H. F. (1991). Ribosomal RNA and translation. *Ann. Rev. Biochem.* **60,** 191–227.

Kurland, C. G. (1992). Translational accuracy and the fitness of bacteria. *Ann. Rev. Genet.* **26,** 29–50.

Merrick, W. C. (1992). Mechanism and regulation of eukaryotic protein synthesis. *Microbiol. Rev.* **56,** 291–315.

Sachs, A., Sarnow, P., and Hentze, M. W. (1997). Starting at the beginning, middle, and end: translation initiation in eukaryotes. *Cell* **89,** 831–838.

Wilson, K. S. and Noller, H. F. (1998). Molecular movement inside the translational engine. *Cell* **92,** 337–349.

Discoveries

Ribosomes

Noller, H. F., Hoffarth, V., and Zimniak, L. (1992). Unusual resistance of peptidyl transferase to protein extraction procedures. *Science* **256,** 1416–1419.

Samaha, R. R., Green, R., and Noller, H. F. (1995). A base pair between tRNA and 23S rRNA in the peptidyl transferase center of the ribosome. *Nature* **377,** 309–314.

Lodmell, J. S. and Dahlberg, A. E. (1997). A conformational switch in *E. coli* 16S rRNA during decoding of mRNA. *Science* **277,** 1262–1267.

Protein synthesis

Moazed, D. and Noller, H. F. (1989). Intermediate states in the movement of tRNA in the ribosome. *Nature* **342,** 142–148.

Nissen, P. *et al.* (1995). Crystal structure of the ternary complex of Phe-tRNAPhe, EF-Tu, and a GTP analog. *Science* **270,** 1464–1472.

Chapter 7

Using the genetic code

THE sequence of a coding strand of DNA, read in the direction from 5′ to 3′, consists of nucleotide triplets (codons) corresponding to the amino acid sequence of a protein read from N-terminus to C-terminus. Sequencing of DNA and proteins makes it possible to compare corresponding nucleotide and amino acid sequences directly. There are 64 codons (each of 4 possible nucleotides can occupy each of the three positions of the codon, so that there are $4^3 = 64$ possible trinucleotide sequences). Each of these codons has a specific meaning in protein synthesis: 61 codons represent amino acids; 3 codons cause the termination of protein synthesis.

The meaning of a codon that represents an amino acid is determined by the tRNA that corresponds to it; the meaning of the termination codons is determined directly by protein factors.

The breaking of the genetic code originally showed that genetic information is stored in the form of nucleotide triplets, but did not reveal *how* each codon specifies its corresponding amino acid. Before the advent of sequencing, codon assignments were deduced on the basis of two types of *in vitro* studies. A system involving the translation of synthetic polynucleotides was introduced in 1961, when Nirenberg showed that polyuridylic acid [poly(U)] directs the assembly of phenylalanine into polyphenylalanine. This result means that UUU must be a codon for phenylalanine. A second system was later introduced in which a trinucleotide was used to mimic a codon, by causing the corresponding aminoacyl-tRNA to bind to a ribosome. By identifying the amino acid component of the aminoacyl-tRNA, the meaning of the codon can be found. The two techniques together assigned meaning to all of the codons that represent amino acids.

The code is summarized in **Figure 7.1**. Because there are more codons (61) than there are amino acids (20), almost all amino acids are represented by more than one codon. The only exceptions are methionine and tryptophan. Codons that have the same meaning are called **synonyms.** Because the genetic code is actually *read* on the mRNA, usually it is described in terms of the four bases present in RNA: U, C, A, and G.

Codons representing the same or related amino acids tend to be similar in sequence. Often the base in the third position of a codon is not significant, because the four codons differing only in the third base represent the same amino acid. Sometimes a distinction is made only between a purine versus a pyrimidine in this position. The reduced specificity at the last position is known as **third-base degeneracy.**

The interpretation of a codon requires base pairing with the anticodon of the corresponding aminoacyl-tRNA. The reaction occurs within the ribosome: complementary trinucleotides in isolation would usually be too short to pair in a stable manner, but the interaction is stabilized by the environment of the ribosomal A site. Also, base pairing between codon and anticodon is not solely a matter of A•U and G•C base pairing. The ribosome controls the environment in such a way that

Figure 7.1 All the triplet codons have meaning: 61 represent amino acids, and 3 cause termination (STOP).

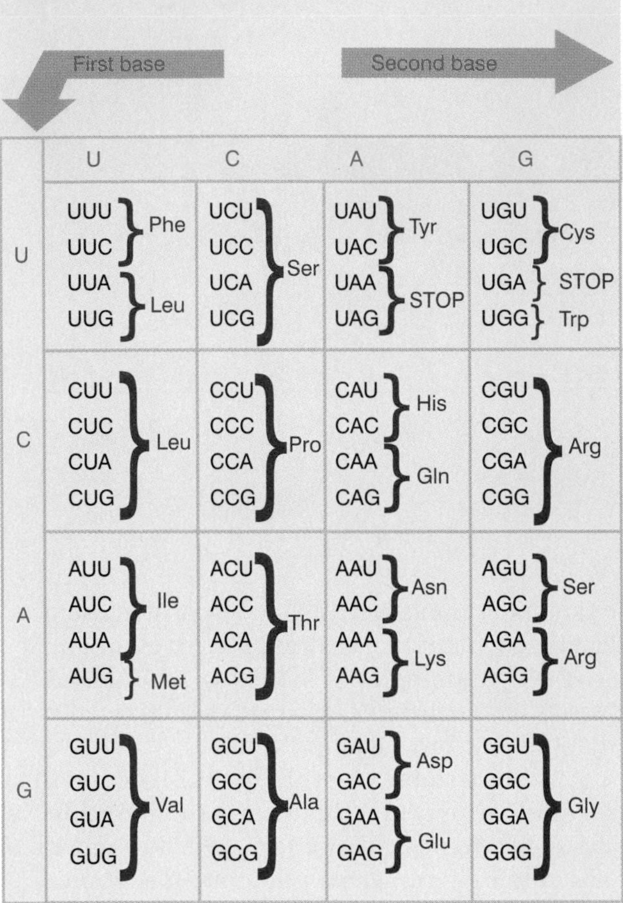

Figure 7.2 The number of codons for each amino acid does not correlate closely with its frequency of use in proteins.

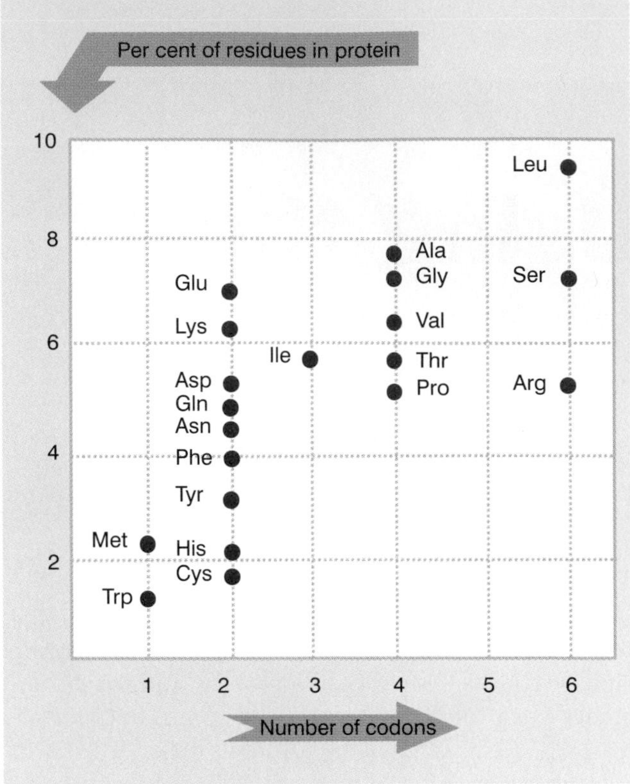

conventional pairing occurs at the first two positions of the codon, but additional reactions are permitted at the third base. As a result, a single aminoacyl-tRNA may recognize more than one codon, corresponding with the pattern of degeneracy. Furthermore, pairing interactions may also be influenced by the introduction of special bases into tRNA, especially by modification in or close to the anticodon.

The tendency for similar amino acids to be represented by related codons minimizes the effects of mutations. It increases the probability that a single random base change will result in no amino acid substitution or in one involving an amino acid of similar character. For example, a mutation of CUC to CUG has no effect, since both codons represent leucine; and a mutation of CUU to AUU results in replacement of leucine with isoleucine, a closely related amino acid.

Figure 7.2 plots the number of codons representing each amino acid against the frequency with which the amino acid is used in proteins (in *E. coli*). There is only a slight tendency for amino acids that are more common to be represented by more codons, and therefore it does not seem that the genetic code has been optimized with regard to the utilization of amino acids.

The three codons (UAA, UAG, and UGA) that do not represent amino acids are used specifically to terminate protein synthesis. One of these **stop codons** marks the end of every gene.

Is the genetic code the same in all living organisms?

Comparisons of DNA sequences with the corresponding protein sequences reveal that the *identical set of codon assignments is used in bacteria and in eukaryotic cytoplasm.* As a result, mRNA from one species usually can be translated correctly *in vitro* or *in vivo* by the protein synthetic apparatus of another species. So the codons used in the mRNA of one species have the same meaning for the ribosomes and tRNAs of other species.

The universality of the code argues that it must have been established very early in evolution. Perhaps the code started in a primitive form in which a small number of codons were used to represent comparatively few amino acids, possibly even with one codon corresponding to any member of a group of amino acids. More precise codon meanings and additional amino acids could have been introduced later. One possibility is that at first only two of the three bases in each codon were used; discrimination at the third position could have evolved later. (Originally there might have been a stereochemical relationship between amino acids and the codons representing them. Then a more complex system evolved.)

Evolution of the code could have become "frozen" at a point at which the system had become so complex that any changes in codon meaning would disrupt existing proteins by substituting unacceptable amino acids. Its universality implies that this must have happened at such an early stage that all living organisms are descended from a single pool of primitive cells in which this occurred.

Exceptions to the universal genetic code are rare. Changes in meaning in the principal genome of a species usually concern the termination codons. For example, in a mycoplasma, UGA codes for tryptophan; and in certain species of the ciliates *Tetrahymena* and *Paramecium*, UAA and UAG code for glutamine. Systematic alterations of the code have occurred only in mitochondrial DNA.

Codon-anticodon recognition involves wobbling

THE function of tRNA in protein synthesis is fulfilled when it recognizes the codon in the ribosomal A site. The interaction between anticodon and codon takes place by base pairing, but under rules that extend pairing beyond the usual G•C and A•U partnerships.

We can deduce the rules governing the interaction from the sequences of the anticodons that correspond to particular codons. The ability of any tRNA to respond to a given codon can be measured directly by the trinucleotide binding assay or by its use in an *in vitro* protein synthetic system.

The genetic code itself yields some important clues about the process of codon recognition. The pattern of third-base degeneracy is drawn in **Figure 7.3**, which shows that in almost all cases either the third base is irrelevant or a distinction is made only between purines and pyrimidines.

There are eight **codon families** in which all four codons sharing the same first two bases have the same meaning, so that the third base has no role at all in specifying the amino acid. There are seven **codon pairs** in which the meaning is the same whichever pyrimidine is present at the third position; and there are five codon pairs in which either purine may be present without changing the amino acid that is coded.

There are only three cases in which a unique meaning is conferred by the presence of a particular base at the third position: AUG (for methionine), UGG (for tryptophan), and UGA (termination). So C and U never have a unique meaning in the third position, and A never signifies a unique amino acid.

Because the anticodon is complementary to the codon, it is the *first* base in the anticodon sequence written conventionally in the direction from 5′ to 3′ that pairs with the *third* base in the codon sequence written by the same convention. So the combination:

Codon	5′ A C G 3′
Anticodon	3′ U G C 5′

is usually written as codon ACG/anticodon CGU, where the anticodon sequence must be read backwards for complementarity with the codon.

To avoid confusion, we shall retain the usual convention in which all sequences are written 5′→3′, but indicate anticodon sequences with a backward arrow as a reminder of the relationship with the codon. So the codon/anticodon pair shown above will be written as ACG and $\overleftarrow{\text{CGU}}$, respectively.

Does each triplet codon demand its own tRNA with a complementary anticodon? Or can a single tRNA respond to both members of a codon pair and to all (or at least some) of the four members of a codon family?

Often one tRNA can recognize more than one codon. *This means that the base in the first position of the anticodon must be able to partner alternative bases in the corresponding third position of the codon.* Base pairing at this position cannot be limited to the usual G•C and A•U partnerships.

The rules governing the recognition patterns are

Figure 7.3 Third bases have the least influence on codon meanings. Boxes indicate groups of codons within which third-base degeneracy ensures that the meaning is the same.

UUU	UCU	UAU	UGU
UUC	UCC	UAC	UGC
UUA	UCA	UAA	UGA
UUG	UCG	UAG	UGG
CUU	CCU	CAU	CGU
CUC	CCC	CAC	CGC
CUA	CCA	CAA	CGA
CUG	CCG	CAG	CGG
AUU	ACU	AAU	AGU
AUC	ACC	AAC	AGC
AUA	ACA	AAA	AGA
AUG	ACG	AAG	AGG
GUU	GCU	GAU	GGU
GUC	GCC	GAC	GGC
GUA	GCA	GAA	GGA
GUG	GCG	GAG	GGG

Third base relationship	Third bases with same meaning	Number of codons
third base irrelevant	U, C, A, G	32
purines differ from pyrimidines	U or C	14
	A or G	10
unique definitions	U, C, A	3
	G only	2

Figure 7.4 Codon-anticodon pairing involves wobbling at the third position.

Base in First Position of Anticodon	Base(s) Recognized in Third Position of Codon
U	A or G
C	G only
A	U only
G	C or U

summarized in the **wobble hypothesis,** which states that the pairing between codon and anticodon at the first two codon positions always follows the usual rules, but that exceptional "wobbles" occur at the third position. Wobbling occurs because the conformation of the tRNA anticodon loop permits flexibility at the first base of the anticodon.

The rules for recognition of the third base of the codon admit pairing between G and U in addition to the usual pairs (see Figure 7.6). This single change creates a pattern of base pairing in which A can no longer have a unique meaning in the codon (because the U that recognizes it must also recognize G). Similarly, C also no longer has a unique meaning (because the G that recognizes it also must recognize U). **Figure 7.4** summarizes the pattern of recognition.

It is therefore possible to recognize unique codons only when the third bases are G or U; this option is not used often, since UGG and AUG are the only examples of the first type, and there is none of the second type.

(G•U pairs are common in RNA duplex structures. But the formation of stable contacts between codon and anticodon, when only 3 base pairs can be formed, is more constrained, and thus G•U pairs can contribute only in the last position of the codon.)

tRNA contains modified bases that influence its pairing properties

TRANSFER RNA is unique among nucleic acids in its content of "unusual" bases. An unusual base is any purine or pyrimidine ring except the usual A, G, C, and U from which all RNAs are synthesized. All other bases are produced by **modification** of one of the four bases after it has been incorporated into the polyribonucleotide chain.

All classes of RNA display some degree of modification, but in all cases except tRNA this is confined to rather simple events, such as the addition of methyl groups. In tRNA, there is a vast range of modifications, ranging from simple methylation to wholesale restructuring of the purine ring. Modifications occur in all parts of the tRNA molecule. There are >50 different types of modified bases in tRNA.

Figure 7.5 shows some of the more common

Figure 7.5 Some of the modified nucleosides found in tRNA (modifications are indicated by red shading).

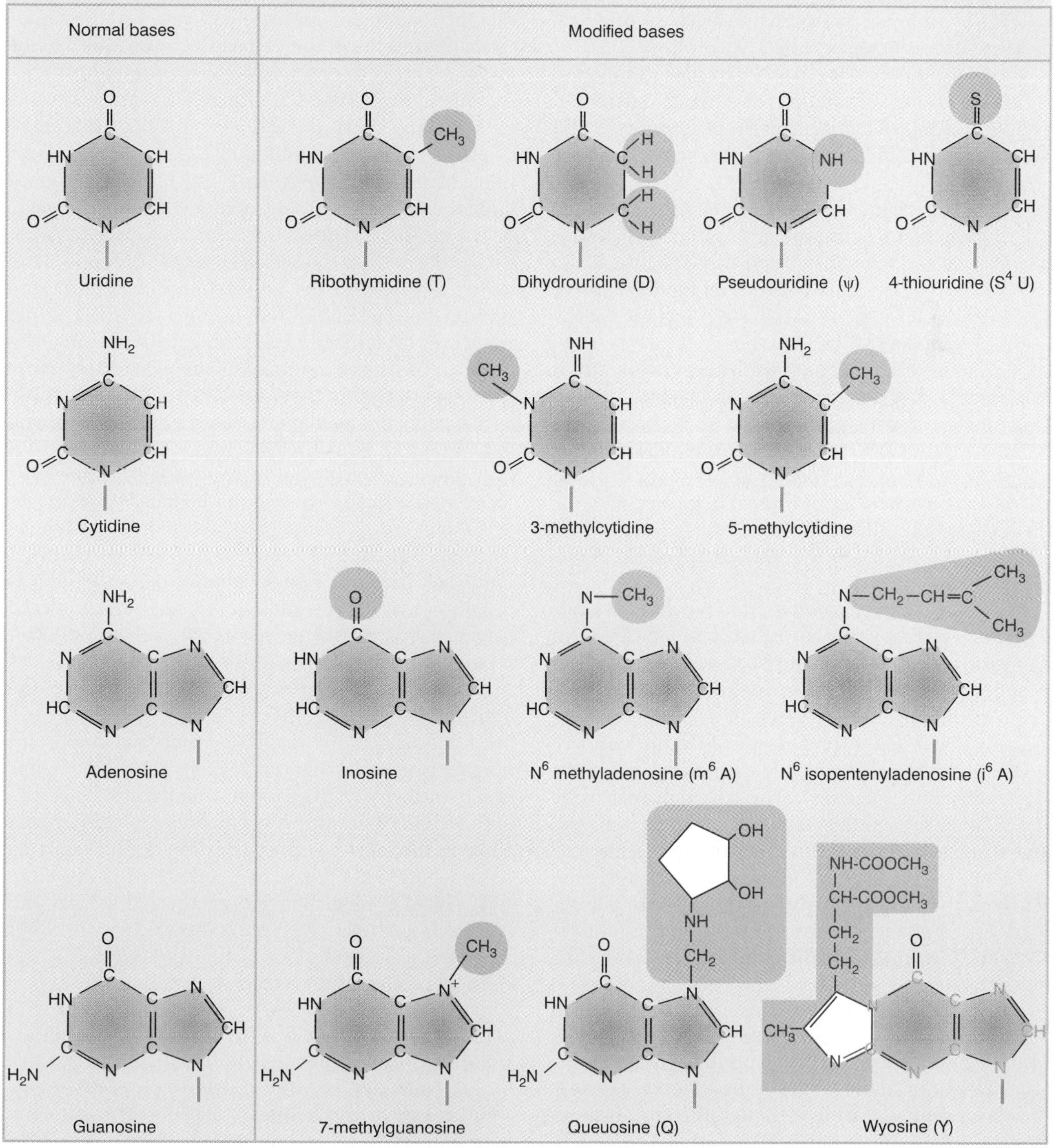

Normal bases	Modified bases
Uridine	Ribothymidine (T); Dihydrouridine (D); Pseudouridine (ψ); 4-thiouridine (S^4 U)
Cytidine	3-methylcytidine; 5-methylcytidine
Adenosine	Inosine; N^6 methyladenosine (m^6 A); N^6 isopentenyladenosine (i^6 A)
Guanosine	7-methylguanosine; Queuosine (Q); Wyosine (Y)

modified bases. Modifications of pyrimidines (C and U) are less complex than those of purines (A and G). In addition to the modifications of the bases themselves, methylation at the 2′-O position of the ribose ring also occurs.

The most common modifications of uridine are straightforward. Methylation at position 5 creates ribothymidine (T). This base is also commonly found in DNA; but here it is attached to ribose, not deoxyribose. In RNA, thymine constitutes an unusual base, originating by modification of U.

Dihydrouridine (D) is generated by the saturation of a double bond, changing the ring structure. Pseudouridine (ψ) interchanges the positions of N and C atoms. And 4-thiouridine has sulfur substituted for oxygen.

The nucleoside inosine is found normally in the cell as an intermediate in the purine biosynthetic pathway. However, it is not incorporated directly into RNA, where instead its existence depends on modification of A to create I. Other modifications of A include the addition of complex groups.

Two complex series of nucleotides depend on modification of G. The Q bases, such as queuosine, have an additional pentenyl ring added via an NH linkage to the methyl group of 7-methylguanosine. The pentenyl ring may carry various further groups. The Y bases, such as wyosine, have an additional ring fused with the purine ring itself; the extra ring carries a long carbon chain, again to which further groups are added in different cases.

The modification reaction usually involves the alteration of, or addition to, existing bases in the tRNA. An exception is the synthesis of Q bases, where a special enzyme exchanges free queuosine with a guanosine residue in the tRNA. The reaction involves breaking and remaking bonds on either side of the nucleoside.

The modified nucleosides are synthesized by specific tRNA-modifying enzymes. The original nucleoside present at each position can be determined either by comparing the sequence of tRNA with that of its gene or (less efficiently) by isolating precursor molecules that lack some or all of the modifications. The sequences of precursors show that different modifications are introduced at different stages during the maturation of tRNA.

Some modifications are constant features of all tRNA molecules—for example, the D residues that give rise to the name of the D arm, and the ψ found in the TψC sequence. On the 3′ side of the anticodon there is always a modified purine, although the modification varies widely.

Other modifications are specific for particular tRNAs or groups of tRNAs. For example, wyosine bases are characteristic of $tRNA^{Phe}$ in bacteria, yeast, and mammals. There are also some species-specific patterns.

The features recognized by the tRNA-modifying enzymes are unknown. When a particular modification is found at more than one position in a tRNA, the same enzyme does not necessarily make all the changes; for example, a different enzyme may be needed to synthesize the pseudouridine at each location. We do not know what controls the specificity of the modifying enzymes, but it is clear that there are many enzymes, with varying specificities. Some enzymes undertake single reactions with individual tRNAs; others have a range of substrate molecules. Some modifications require the successive actions of more than one enzyme.

The most direct effect of modification is seen in the anticodon, where change of sequence influences the ability to pair with the codon, thus determining the meaning of the tRNA. Modifications elsewhere in the vicinity of the anticodon also influence its pairing.

When bases in the anticodon are modified, further pairing patterns become possible in addition to those predicted by the regular and wobble pairing involving A, C, U, and G. **Figure 7.6** illustrates some of the variations in codon-anticodon pairing that are allowed by the combination of wobbling and modification.

Actually, some of the predicted regular combinations do not occur, because some bases are *always* modified. U at the first position of the anticodon is usually converted to a modified form that has altered pairing properties. There seems to be an absolute ban on the employment of A; usually it is converted to I.

Inosine (I) is often present at the first position of the anticodon, where it is able to pair with any one of three bases, U, C, and A. This ability is especially important in the isoleucine codons, where AUA codes for isoleucine, while AUG codes for methionine. Because with the usual bases it is not possible to recognize A alone in the third position, any tRNA with U starting its anticodon would have to recognize AUG as well as AUA. So AUA must be read together with AUU and AUC, a problem that is solved by the existence of tRNA with I in the anticodon.

Some modifications create preferential readings of some codons with respect to others. Anticodons with uridine-5-oxyacetic acid and 5-methoxyuridine in the first position recognize A and G efficiently as third bases of the codon, but recognize U less efficiently. Another case in which multiple pairings can occur, but with some preferred to others, is provided by the series

Figure 7.6 Wobble in base pairing allows some bases at the first position of the anticodon to recognize more than one base in the third position of the codon. Pairing between standard bases is extended from the G-C and A-U pairs to the G-U wobble pair. Base modifications may restrict or extend the pattern. Modification to 2-thiouridine restricts pairing to A alone because only one H-bond could form with G. Modification to inosine allows pairing with U, C, and A.

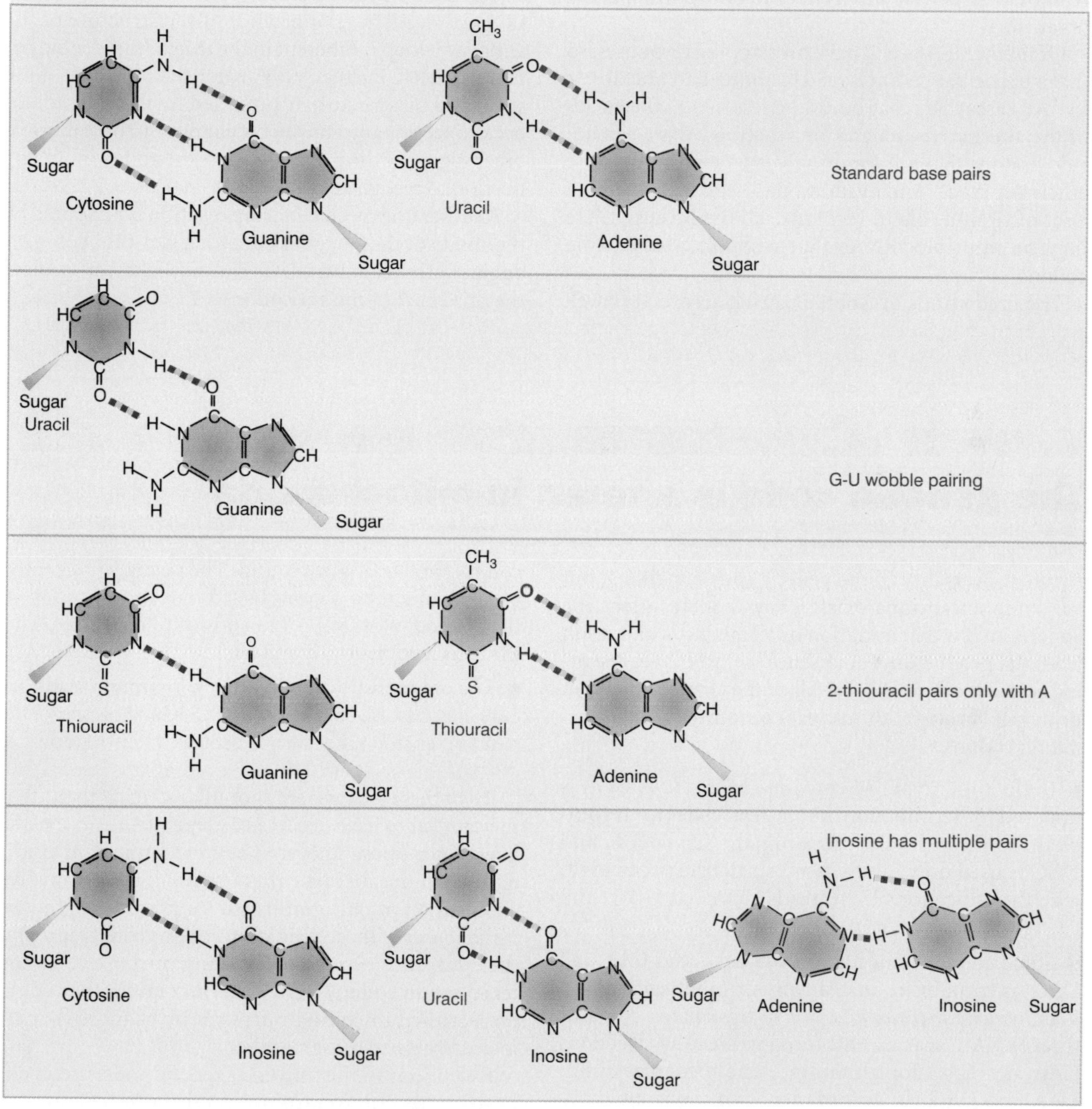

of queuosine and its derivatives. These modified G bases continue to recognize both C and U, but pair with U more readily.

A restriction not allowed by the usual rules can be achieved by the employment of 2-thiouridine in the anticodon. This modification allows the base to continue to pair with A, but prevents it from indulging in wobble pairing with G.

These and other pairing relationships make the general point that *there are multiple ways to construct a set of tRNAs able to recognize all the 61 codons representing amino acids.* No particular pattern predominates in any given organism, although the absence of a certain pathway for modification can prevent the use of some recognition patterns. So a particular codon family is read by tRNAs with different anticodons in different organisms.

Often the tRNAs will have overlapping responses, so that a particular codon is read by more than one tRNA. In such cases there may be differences in the efficiencies of the alternative recognition reactions. (As a general rule, codons that are commonly used tend to be more efficiently read.) And in addition to the construction of a set of tRNAs able to recognize all the codons, there may be multiple tRNAs that respond to the same codons.

The predictions of wobble pairing accord very well with the observed abilities of almost all tRNAs. But there are exceptions in which the codons recognized by a tRNA differ from those predicted by the wobble rules. Such effects probably result from the influence of neighboring bases and/or the conformation of the anticodon loop in the overall tertiary structure of the tRNA. Indeed, the importance of the structure of the anticodon loop is inherent in the idea of the wobble hypothesis itself. Further support for the influence of the surrounding structure is provided by the isolation of occasional mutants in which a change in a base in some other region of the molecule alters the ability of the anticodon to recognize codons (see later).

Another unexpected pairing reaction is presented by the ability of the bacterial initiator, $fMet\text{-}tRNA_f$, to recognize both AUG and GUG. This misbehavior involves the third base of the anticodon.

The genetic code is altered in mitochondria

THE universality of the genetic code is striking, but some exceptions exist. They tend to affect the codons involved in initiation or termination and result from the production (or absence) of tRNAs representing certain codons. Almost all of the changes found in principal (bacterial or nuclear) genomes affect termination codons:

- In the prokaryote *Mycoplasma capricolum*, UGA is not used for termination, but instead codes for tryptophan. In fact, it is the predominant Trp codon, and UGG is used only rarely. Two Trp-tRNA species exist, with the anticodons $\overleftarrow{UCA}$ (reads UGA and UGG) and $\overleftarrow{CCA}$ (reads only UGG).
- Some ciliates (unicellular protozoa) read UAA and UAG as glutamine instead of termination signals. *Tetrahymena thermophila*, one of the ciliates, contains three $tRNA^{Glu}$ species. One recognizes the usual codons CAA and CAG for glutamine, one recognizes both UAA and UAG (in accordance with the wobble hypothesis), and the last recognizes only UAG. We assume that a further change is that the release factor eRF has a restricted specificity, compared with that of other eukaryotes.
- In another ciliate (*Euplotes octacarinatus*), UGA codes for cysteine. Only UAA is used as a termination codon, and UAG is not found. The change in meaning of UGA might be accomplished by a modification in the anticodon of $tRNA^{Cys}$ to allow it to read UGA with the usual codons UGU and UGC.
- The only substitution in coding for amino acids occurs in a yeast (*Candida*), where CUG means serine instead of leucine (and UAG is used as a sense codon).

All of these changes are sporadic, which is to say that they appear to have occurred independently in specific lines of evolution. They may be concentrated on termination codons, because these changes do not involve substitution of one amino acid for another. So the divergent uses of the termination codons could represent their "capture" for normal coding purposes. If some termination codons were used only rarely, they could be recruited to coding purposes by changes that allowed tRNAs to recognize them.

Exceptions to the universal genetic code also occur in the mitochondria from several species. **Figure 7.7** constructs a phylogeny for the changes. It suggests that there was a universal code that was changed at various points in mitochondrial evolution. The earliest change was the employment of UGA to code for tryptophan, which is common to all (non-plant) mitochondria.

Some of these changes make the code simpler, by re-

Figure 7.7 Changes in the genetic code in mitochondria can be traced in phylogeny. The minimum number of independent changes is generated by supposing that the AUA=Met and the AAA=Asn changes each occurred independently twice, and that the early AUA=Met change was reversed in echinoderms.

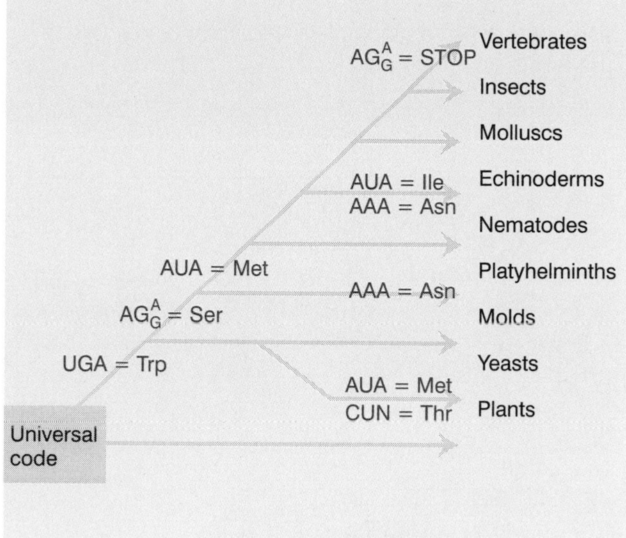

placing two codons that had different meanings with a pair that has a single meaning. Pairs treated like this include UGG and UGA (both Trp instead of one Trp and one termination) and AUG and AUA (both Met instead of one Met and the other Ile).

Why have changes been able to evolve in the mitochondrial code? Because the mitochondrion synthesizes only a small number of proteins (~10), the problem of disruption by changes in meaning is much less severe. Probably the codons that are altered were not used extensively in locations where amino acid substitutions would have been deleterious. The variety of changes found in mitochondria of different species suggests that they have evolved separately, and not by common descent from an ancestral mitochondrial code.

According to the wobble hypothesis, a minimum of 31 tRNAs (excluding the initiator) are required to recognize all 61 codons (at least 2 tRNAs are required for each codon family and one tRNA is needed per codon pair or single codon). But an unusual situation exists in (at least) mammalian mitochondria, in which there are only 22 different tRNAs. How does this limited set of tRNAs accommodate all the codons?

The critical feature lies in a simplification of codon-anticodon pairing, in which one tRNA recognizes all four members of a codon family. This reduces to 23 the minimum number of tRNAs required to respond to all usual codons. The use of AG^A_G for termination reduces the requirement by one further tRNA, to 22.

In all eight codon families, the sequence of the tRNA contains an unmodified U at the first position of the anticodon. The remaining codons are grouped into pairs in which all the codons ending in pyrimidines are read by G in the anticodon, and all the codons ending in purines are read by a modified U in the anticodon, as predicted by the wobble hypothesis. The complication of the single UGG codon is avoided by the change in the code to read UGA with UGG as tryptophan; and in mammals, AUA ceases to represent isoleucine and instead is read with AUG as methionine. This allows all the nonfamily codons to be read as 14 pairs.

The 22 identified tRNA genes therefore code for 14 tRNAs representing pairs, and 8 tRNAs representing families. This leaves the two usual termination codons UAG and UAA unrecognized by tRNA, together with the codon pair AG^A_G. Similar rules are followed in the mitochondria of fungi.

As well as these general changes in the code, specific changes in reading occur in individual genes. The specificity of such changes implies that the reading of the particular codon must be influenced by the surrounding bases (also see later).

A striking example is the incorporation of the modified amino acid seleno-cysteine at certain UGA codons within the genes that code for selenoproteins in both prokaryotes and eukaryotes. Usually these proteins catalyze oxidation-reduction reactions, and contain a single seleno-cysteine residue, which forms part of the active site. The most is known about the use of the UGA codons in three *E. coli* genes coding for formate dehydrogenase isozymes. The internal UGA codon is read by a seleno-Cys-tRNA. This unusual reaction is determined by the local secondary structure of mRNA, in particular by the presence of a hairpin loop downstream of the UGA.

Mutations in 4 *sel* genes create a deficiency in selenoprotein synthesis. *selC* codes for tRNA (with the anticodon $\overleftarrow{\text{ACU}}$) that is charged with serine. *selA* and *selD* are required to modify the serine to seleno-cysteine. *selB* codes for a guanine nucleotide-binding protein that acts as a specific translation factor for entry of seleno-Cys-tRNA into the A site; it thus provides (for this single tRNA) a replacement for factor EF-Tu. The sequence of SelB is related to both EF-Tu and IF-2.

tRNAs are charged with amino acids by individual synthetases

It is necessary for tRNAs to have certain characteristics in common, yet be distinguished by others. The crucial feature that confers this capacity is the ability of tRNA to fold into a specific tertiary structure. Changes in the details of this structure, such as the angle of the two arms of the "L" or the protrusion of individual bases, may distinguish the individual tRNAs.

All tRNAs can fit in the P and A sites of the ribosome, where at one end they are associated with mRNA via codon-anticodon pairing while at the other end the polypeptide is being transferred. Similarly, all tRNAs (except the initiator) share the ability to be recognized by the translation factors (EF-Tu or eEF-1) for binding to the ribosome. The initiator tRNA is recognized instead by IF-2 or eIF2. So the tRNA set must possess common features for interaction with elongation factors, but the initiator tRNA can be distinguished.

Amino acids enter the protein synthesis pathway through the aminoacyl-tRNA synthetases, which provide the interface for connection with nucleic acid. All synthetases function by the two-step mechanism depicted in **Figure 7.8:**

- First, the amino acid reacts with ATP to form aminoacyl~adenylate, releasing pyrophosphate. Energy for the reaction is provided by cleaving the high-energy bond of the ATP.
- Then the activated amino acid is transferred to the tRNA, releasing AMP.

Figure 7.8 An aminoacyl-tRNA synthetase charges tRNA with an amino acid.

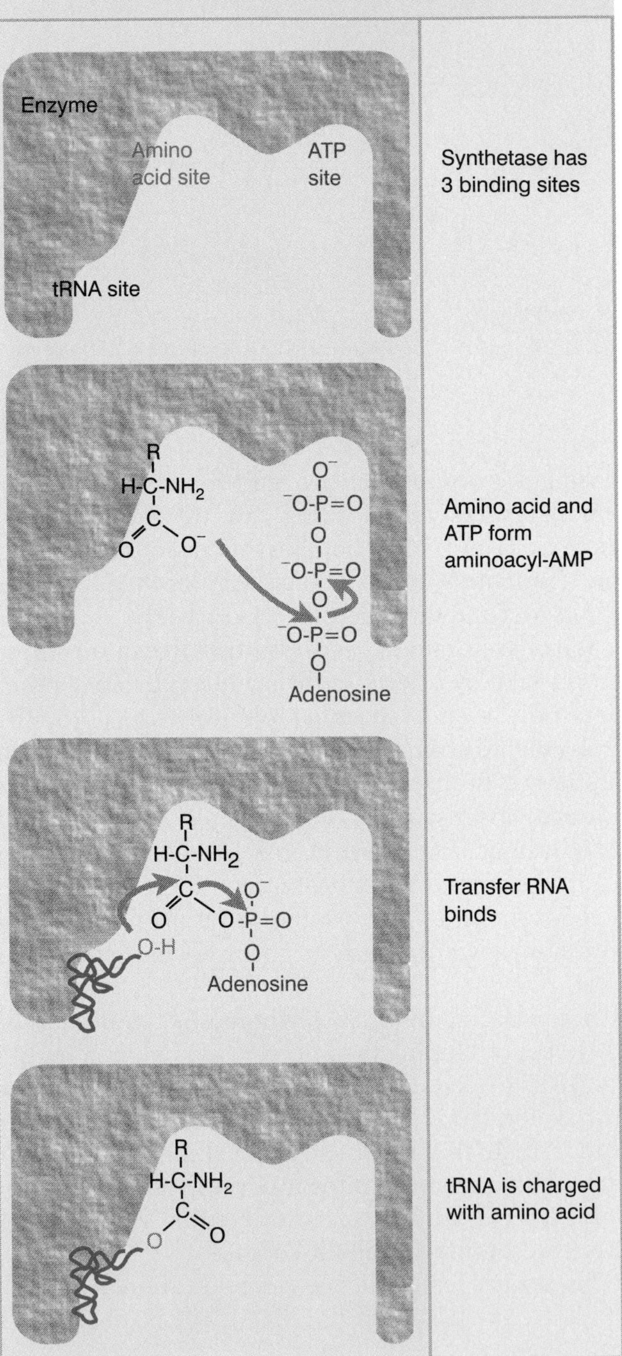

The synthetases sort the tRNAs and amino acids into corresponding sets, each synthetase recognizing a single amino acid and all the tRNAs that should be charged with it. Usually, each amino acid is represented by more than one tRNA. (There may be many such tRNAs. In addition to the several tRNAs that are needed to respond to synonym codons, sometimes there are multiple species of tRNA reacting with the same codon.) Multiple tRNAs representing the same amino acid are called **isoaccepting tRNAs;** because they are all recognized by the same synthetase, they are also described as its **cognate tRNAs.**

In spite of their common function, synthetases are a rather diverse group of proteins. The individual subunits vary from 40–110 kD, and the enzymes may be monomeric, dimeric, or tetrameric. Homologies between them are rare. Of course, the active site that recognizes tRNA comprises a rather small part of the molecule; it is interesting to compare the active sites of different synthetases.

Synthetases have been divided into two general groups, each containing 10 enzymes, on the basis of the structure of the domain that contains the active site. A general type of organization that applies to both groups is represented in **Figure 7.9.** The catalytic domain includes the binding sites for ATP and amino acid. It can be recognized as a large region that is interrupted by an insertion of the domain that binds the acceptor helix of the tRNA. This places the terminus of the tRNA in proximity to the catalytic site. A separate domain binds the anticodon region of tRNA. Those synthetases that are multimeric also possess an oligomerization domain.

Class I synthetases have an N-terminal catalytic domain that is identified by the presence of two short, partly conserved sequences of amino acids, sometimes called "signature sequences". The catalytic domain takes the form of a motif called a nucleotide-binding fold (which is also found in other classes of enzymes that bind nucleotides). The nucleotide fold consists of alternating parallel β-strands and α-helices; the signature sequence forms part of the ATP-binding site. The insertion that contacts the acceptor helix of tRNA differs widely between different class I enzymes. The C-terminal domains of the class I synthetases, which include the tRNA anticodon-binding domain and any oligomerization domain, also are quite different from one another.

Class II enzymes share three rather general similarities of sequence in their catalytic domains. The active site contains a large antiparallel β-sheet surrounded by α-helices. Again, the acceptor helix-binding domain that interrupts the catalytic domain has a structure that depends on the individual enzyme. The anticodon-binding domain tends to be N-terminal. The location of any oligomerization domain is widely variable.

The lack of any apparent relationship between the two groups of synthetases is a puzzle. Perhaps they evolved independently of one another. This makes it seem possible even that an early form of life could have existed with proteins that were made up of just the 10 amino acids coded by one type or the other.

A general model for synthetase•tRNA binding suggests that the protein binds the tRNA along the "side" of the L-shaped molecule. The same general principle applies for all synthetase•tRNA binding: the tRNA is bound principally at its two extremities, and most of the tRNA sequence is not involved in recognition by a synthetase. However, the detailed nature of the interaction is different between class I and class II enzymes, as can be seen from the models of **Figure 7.10,** which are based on crystal structures. The two types of enzyme approach the tRNA from opposite sides, with the result that the tRNA-protein models look almost like mirror images of one another.

A class I enzyme (Gln-tRNA synthetase) approaches the D-loop side of the tRNA. It recognizes the minor groove of the acceptor stem at one end of the binding site, and interacts with the anticodon loop at the other end. **Figure 7.11** is a diagrammatic representation of the crystal structure of the $tRNA^{Gln}$•synthetase complex. A revealing feature of the structure is that contacts with the enzyme change the structure of the tRNA at two important points. These can be seen by comparing the dotted and solid lines in the anticodon loop and acceptor stem:

- Bases U35 and U36 in the anticodon loop are pulled farther out of the tRNA into the protein.
- The end of the acceptor stem is seriously distorted, with the result that base pairing between U1 and A72 is disrupted. The single-stranded end of the stem pokes into a deep pocket in the synthetase protein, which also contains the binding site for ATP.

This structure explains why changes in U35, G73, or the U1-A72 base pair affect the recognition of the tRNA by its synthetase. At all of these positions, hydrogen bonding occurs between the protein and tRNA.

Figure 7.9 An aminoacyl-tRNA synthetase contains three or four regions with different functions. (Only multimeric synthetases possess an oligomerization domain.)

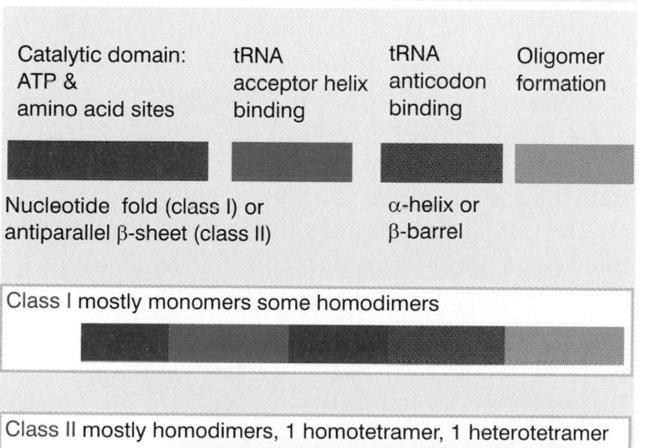

Figure 7.10 Crystal structures show that class I and class II aminoacyl-tRNA synthetases bind the opposite faces of their tRNA substrates. The tRNA is shown in red, and the protein in blue. Photographs kindly provided by Dino Moras.

Class I (Glu-tRNA synthetase)

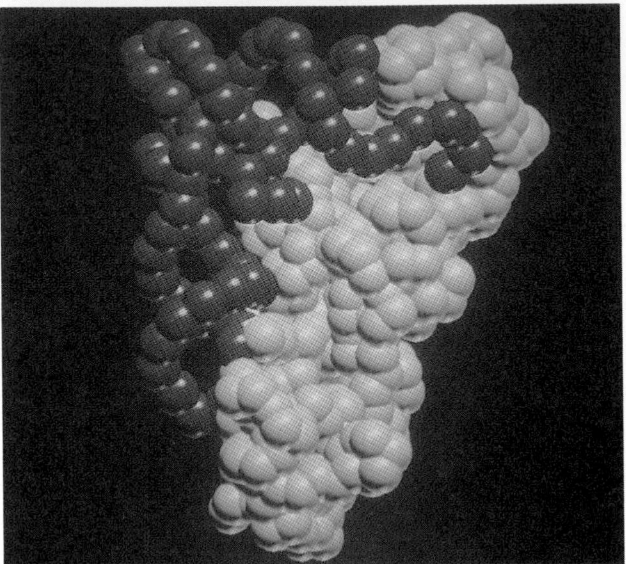

Class II (Asp-tRNA synthetase)

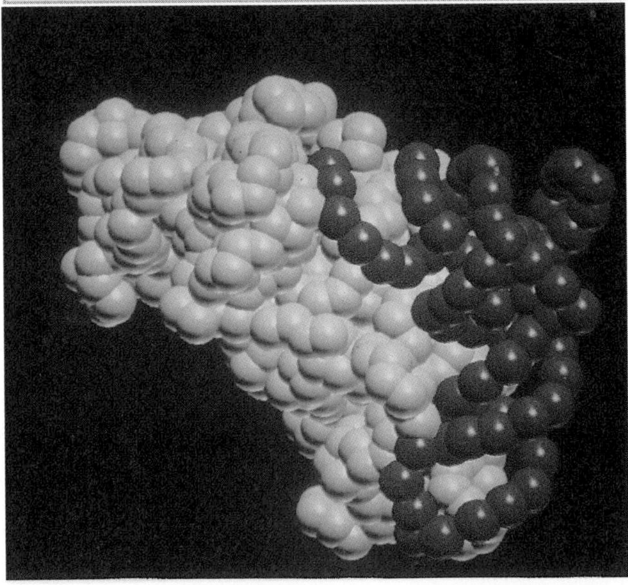

Figure 7.11 A class I tRNA synthetase contacts tRNA at the minor groove of the acceptor stem and at the anticodon.

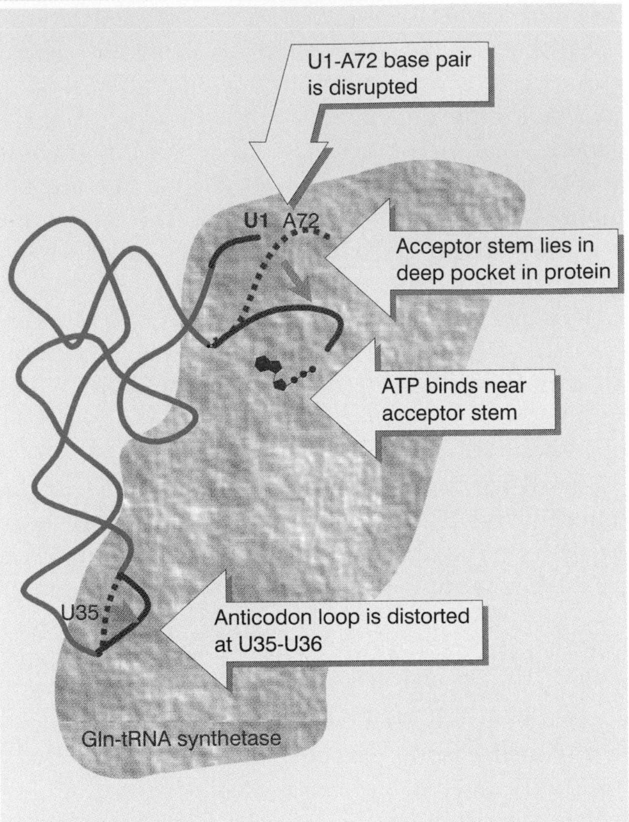

A class II enzyme (Asp-tRNA synthetase) approaches the tRNA from the other side, and recognizes the variable loop, and the major groove of the acceptor stem, as drawn in **Figure 7.12.** The acceptor stem remains in its regular helical conformation. ATP is probably bound near to the terminal adenine. At the other end of the binding site, there is a tight contact with the anticodon loop, which has a change in conformation that allows the anticodon to be in close contact with the protein.

Many attempts to deduce similarities in sequence between cognate tRNAs, or to induce chemical alterations that affect their charging, have shown that the basis for recognition is different for different tRNAs, and does not necessarily lie in some feature of primary or secondary structure alone. We know from the crystal structure that the acceptor stem and the anticodon stem make tight contacts with the synthetase, and mutations that alter recognition of a tRNA are found in these two regions. (The anticodon itself is not necessarily recognized as such; for example, the "suppressor" mutations discussed later in this chapter change a base in the anticodon, and therefore the codons to which a tRNA responds, without altering its charging with amino acids.)

A group of isoaccepting tRNAs must be charged only by the single aminoacyl-tRNA synthetase specific for their

Figure 7.12 A class II aminoacyl-tRNA synthetase contacts tRNA at the major groove of the acceptor helix and at the anticodon loop.

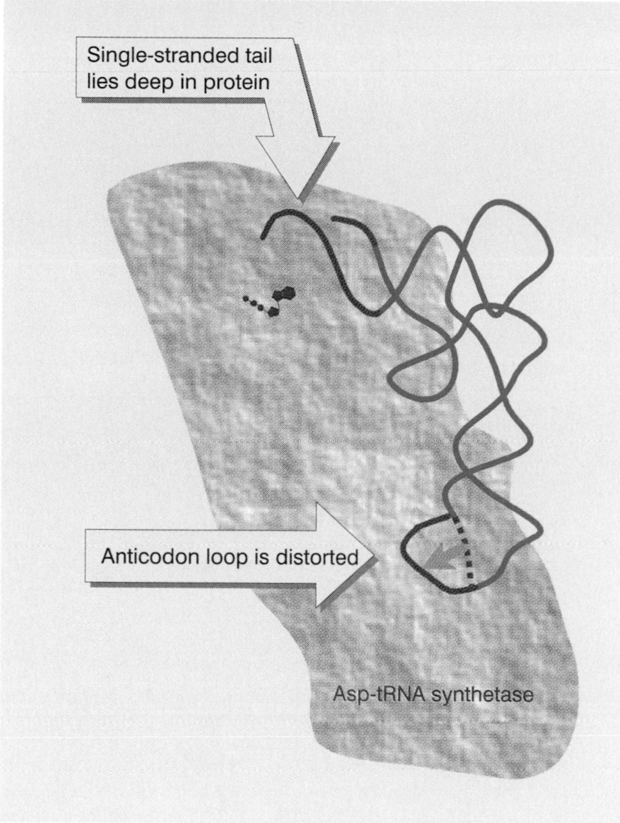
Single-stranded tail lies deep in protein
Anticodon loop is distorted
Asp-tRNA synthetase

amino acid. So isoaccepting tRNAs must share some common feature(s) enabling the enzyme to distinguish them from the other tRNAs. The entire complement of tRNAs is divided into 20 isoaccepting groups; each group is able to identify itself to its particular synthetase.

tRNAs are identified by their synthetases by contacts that recognize a small number of bases, typically from 1–5. Three types of feature commonly are used:

- Usually (but not always), at least one base of the anticodon is recognized. Sometimes all the positions of the anticodon are important.
- Often one of the last three base pairs in the acceptor stem is recognized. An extreme case is represented by alanine tRNA, which is identified by a single unique base pair in the acceptor stem.
- The so-called discriminator base, which lies between the acceptor stem and the CCA terminus, is always invariant among isoacceptor tRNAs.

No single one of these features constitutes a unique means of distinguishing 20 sets of tRNAs, or provides sufficient specificity, so it appears that recognition of tRNAs is idiosyncratic, each following its own rules.

Several synthetases can specifically charge a "minihelix" consisting only of the acceptor and TψC arms (equivalent to one arm of the L-shaped molecule) with the correct amino acid. The efficiency of aminoacylation of these substrates is much higher for the class II enzymes. For these tRNAs, specificity depends exclusively upon the acceptor stem. However, it is clear that there are significant variations between tRNAs, and in some cases the anticodon region is important. Mutations in the anticodon can affect recognition by the class II Phe-tRNA synthetase. Multiple features may be involved; minihelices from the $tRNA^{Val}$ and $tRNA^{Met}$ (where we know that the anticodon is important *in vivo*) can react specifically with their class I synthetases.

So recognition depends on an interaction between a few points of contact in the tRNA, concentrated at the extremities, and a few amino acids constituting the active site in the protein. The relative importance of the roles played by the acceptor stem and anticodon is different for each tRNA•synthetase interaction.

Accuracy depends on proofreading

The nature of discriminatory events is a general issue raised by several steps in gene expression. How do synthetases recognize just the corresponding tRNAs and amino acids? How does a ribosome recognize only the tRNA corresponding to the codon in the A site? How do the enzymes that synthesize DNA or RNA recognize only the base complementary to the template? Each case poses a similar problem: *how to distinguish one particular member from the entire set, all of which share the same general features.*

Probably any member initially can contact the active center by a random-hit process, but then the wrong members are rejected and only the appropriate one is accepted. The appropriate member is always in a minority (1 of 20 amino acids, 1 of ~40 tRNAs, 1 of 4 bases), so the criteria for discrimination must be strict. We can imagine two general ways in which the decision whether to reject or accept might be taken:

- The cycle of admittance, scrutiny, rejection/acceptance could represent a single binding step that *precedes all other stages* of whatever reaction is involved. This is tantamount to saying that the affinity of the binding site is sufficient to control the entry of substrate. In the case of synthetases, this would mean that only the cognate tRNAs could form a stable attachment at the site.
- Alternatively, *the reaction proceeds through some of its stages*, after which a decision is reached on whether the correct species is present. If it is not present, the reaction is reversed, or a bypass route is taken, and the wrong member is expelled. This sort of postbinding scrutiny is generally described as **proofreading.** In the example of synthetases, it would require that the charging reaction proceeds through certain stages even if the wrong tRNA or amino acid is present.

Synthetases use proofreading mechanisms to control the recognition of both tRNA and amino acid.

Transfer RNA binds to synthetase by the two-stage reaction depicted in **Figure 7.13.** Cognate tRNAs have a greater intrinsic affinity for the binding site, so they are bound more rapidly and dissociate more slowly. Following binding, the enzyme scrutinizes the tRNA that has been bound. If the correct tRNA is present, binding is stabilized by a conformational change in the enzyme. This allows aminoacylation to occur rapidly. If the wrong tRNA is present, the conformational change does not occur. As a result, the reaction proceeds much more slowly; this increases the chance that the tRNA will dissociate from the enzyme before it is charged. This type of control is called **kinetic proofreading.**

Specificity for amino acids varies among the synthetases. Some are highly specific for initially binding a single amino acid, but others can also activate amino acids closely related to the proper substrate. Although the analog amino acid can sometimes be converted to the adenylate form, in none of these cases is an incorrectly activated amino acid actually used to form a stable aminoacyl-tRNA.

The presence of the cognate tRNA usually is needed to trigger proofreading, even if the reaction occurs at the stage before formation of aminoacyl-adenylate.

Figure 7.13 Recognition of the correct tRNA by synthetase is controlled at two steps. First, the enzyme has a greater affinity for its cognate tRNA. Second, the aminoacylation of the incorrect tRNA is very slow.

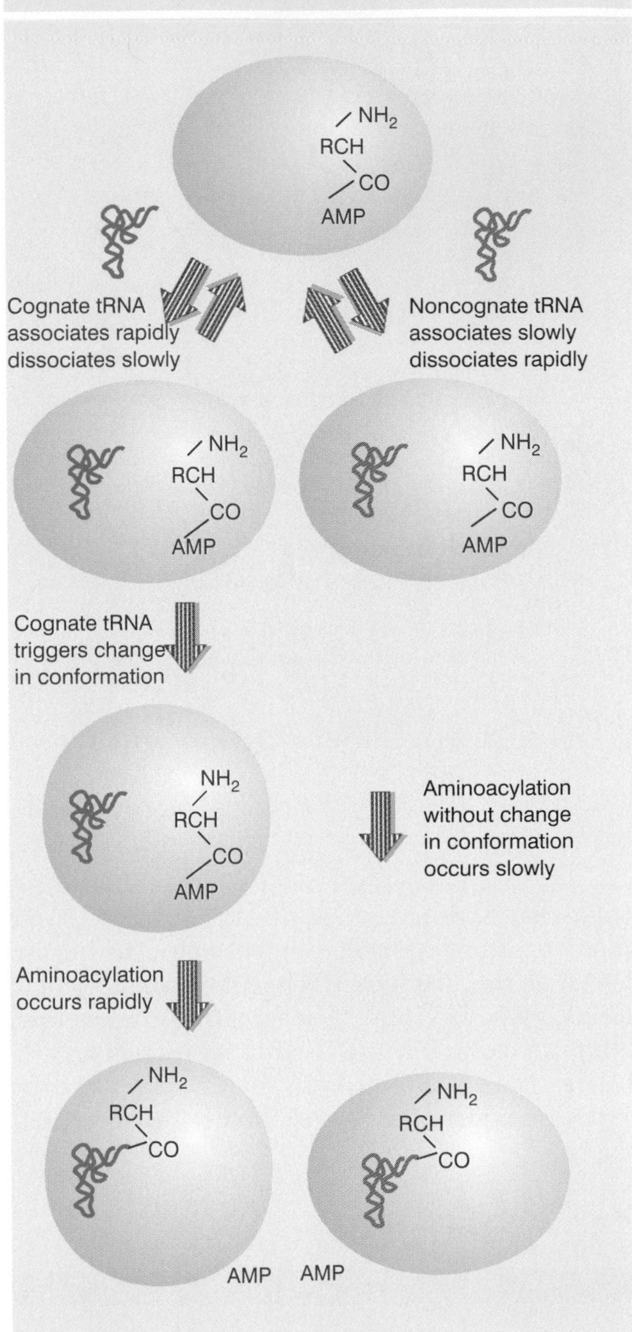

(An exception is provided by Met-tRNA synthetase, which can reject noncognate aminoacyl-adenylate complexes even in the absence of tRNA.)

There are two stages at which proofreading of an incorrect aminoacyl-adenylate may occur during formation of aminoacyl-tRNA. **Figure 7.14** shows that both use **chemical proofreading,** in which the catalytic reaction is reversed. The extent to which one pathway or the other predominates varies with the individual synthetase:

- The noncognate aminoacyl-adenylate may be hydrolyzed when the cognate tRNA binds. This mechanism is used predominantly by several synthetases, including those for methionine, isoleucine, and valine. (Usually, the reaction cannot be seen *in vivo*, but it can be followed for Met-tRNA synthetase when the incorrectly activated amino acid is homocysteine [which lacks the methyl group of methionine]. Proofreading releases the amino acid in an altered form, as homocysteine thiolactone. In fact, homocysteine thiolactone is produced in *E. coli* as a byproduct of the charging reaction of Met-tRNA synthetase. This shows that continuous proofreading is part of the process of charging a tRNA with its amino acid.)
- Some synthetases use chemical proofreading at a later stage. The wrong amino acid is actually transferred to tRNA, is then recognized as incorrect by its structure in the tRNA binding site, and so is hydrolyzed and released. The process requires a continual cycle of linkage and hydrolysis until the correct amino acid is transferred to the tRNA.

Figure 7.14 When a synthetase binds the incorrect amino acid, proofreading requires binding of the cognate tRNA. It may take place either by a conformation change that causes hydrolysis of the incorrect aminoacyl-adenylate, or by transfer of the amino acid to RNA, followed by hydrolysis.

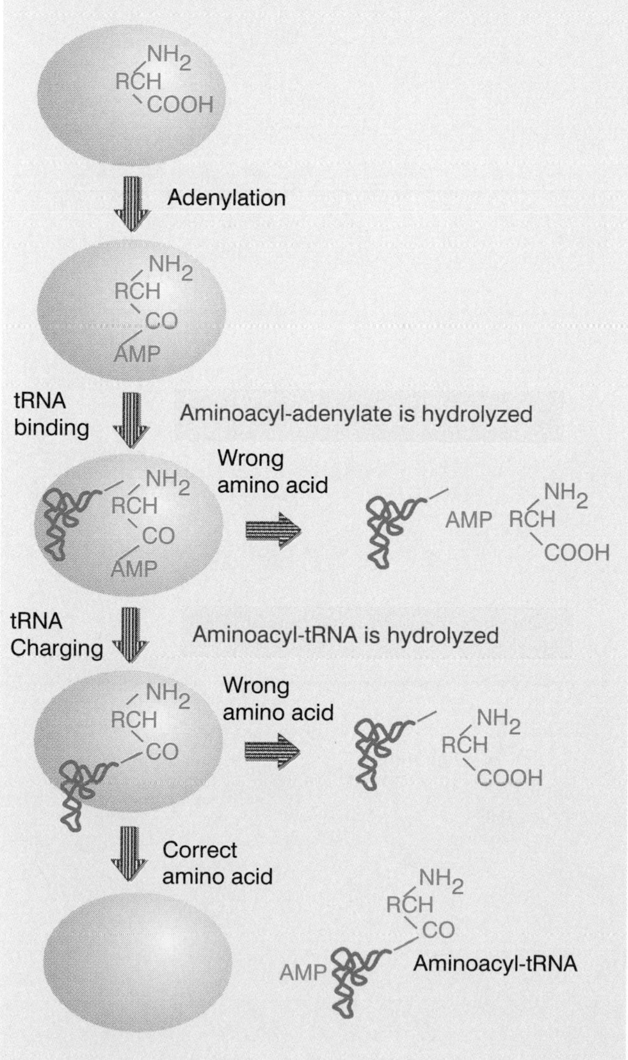

A classic example in which discrimination between amino acids depends on the presence of tRNA is provided by the Ile-tRNA synthetase of *E. coli*. The enzyme can charge valine with AMP, but hydrolyzes the valyl-adenylate when $tRNA^{Ile}$ is added. The overall error rate depends on the specificities of the individual steps, as summarized in **Figure 7.15.** The overall error rate of 1.5×10^{-5} is less than the measured rate at which valine is substituted for isoleucine (in rabbit globin), which is $2–5 \times 10^{-4}$. So mischarging probably provides only a small fraction of the errors that actually occur in protein synthesis.

In the case of Ile-tRNA synthetase, accuracy is ensured by multiple mechanisms. The crystal structure of the enzyme shows that the activation site is too small to allow leucine (a close analog of isoleucine) to enter. Adjacent to the activation site is a second site, the hydrolytic site, which is too small to allow isoleucine itself to enter. However, valine can enter this site, and as a result an incorrect $Val\text{-}tRNA^{Ile}$ is hydrolyzed. The activation and hydrolytic sites form a closed cavity within the enzyme. The combination of mechanisms provides a molecular sieve, in which size of the amino acid is important.

Figure 7.15 The accuracy of charging $tRNA^{Ile}$ by its synthetase depends on error control at two stages.

Step	Frequency of Error
Activation of valine to $Val\text{-}AMP^{Ile}$	1/225
Release of Val-tRNA	1/270
Overall rate of error	1/225 × 1/270 = 1/60,000

Suppressor tRNAs have mutated anticodons that read new codons

Isolation of mutant tRNAs has been one of the most potent tools for analyzing the ability of a tRNA to respond to its codon(s) in mRNA, and for determining the effects that different parts of the tRNA molecule have on codon-anticodon recognition.

Mutant tRNAs are isolated by virtue of their ability to overcome the effects of mutations in genes coding for proteins. We have already described the terminology in which a mutation that is able to overcome the effects of another is called a **suppressor** (see Chapter 1).

In tRNA suppressor systems, the primary mutation changes a codon in an mRNA so that the protein product is no longer functional. The secondary, suppressor mutation changes the anticodon of a tRNA, so that it recognizes the mutant codon instead of (or as well as) its original target codon. The amino acid that is now inserted restores protein function. The suppressors are named as **nonsense** or **missense,** depending on the nature of the original mutation.

In a wild-type cell, a nonsense mutation is recognized only by a release factor, terminating protein synthesis. The suppressor mutation creates an aminoacyl-tRNA that can recognize the termination codon; by inserting an amino acid, it allows protein synthesis to continue beyond the site of nonsense mutation.

This new capacity of the translation system allows a full-length protein to be synthesized, as illustrated in **Figure 7.16.** If the amino acid inserted by suppression is different from the amino acid that was originally present at this site in the wild-type protein, the activity of the protein may be altered.

Nonsense suppressors fall into three classes, one for each type of termination codon. **Figure 7.17** describes the properties of some of the best-characterized suppressors.

The easiest to characterize have been amber suppressors. In *E. coli*, at least 6 tRNAs have been mutated to recognize UAG codons. All of the amber suppressor tRNAs have the anticodon $\overleftarrow{\text{CUA}}$, in each case derived from wild type by a single base change. The site of mutation can be any one of the three bases of the anticodon, as seen from *supD*, *supE*, and *supF*. Each suppressor tRNA recognizes *only* the UAG codon, instead of its former codon(s). The amino acids inserted

Figure 7.16 Nonsense mutations can be suppressed by a tRNA with a mutant anticodon, which inserts an amino acid at the mutant codon, producing a full-length protein in which the original Leu residue has been replaced by Tyr.

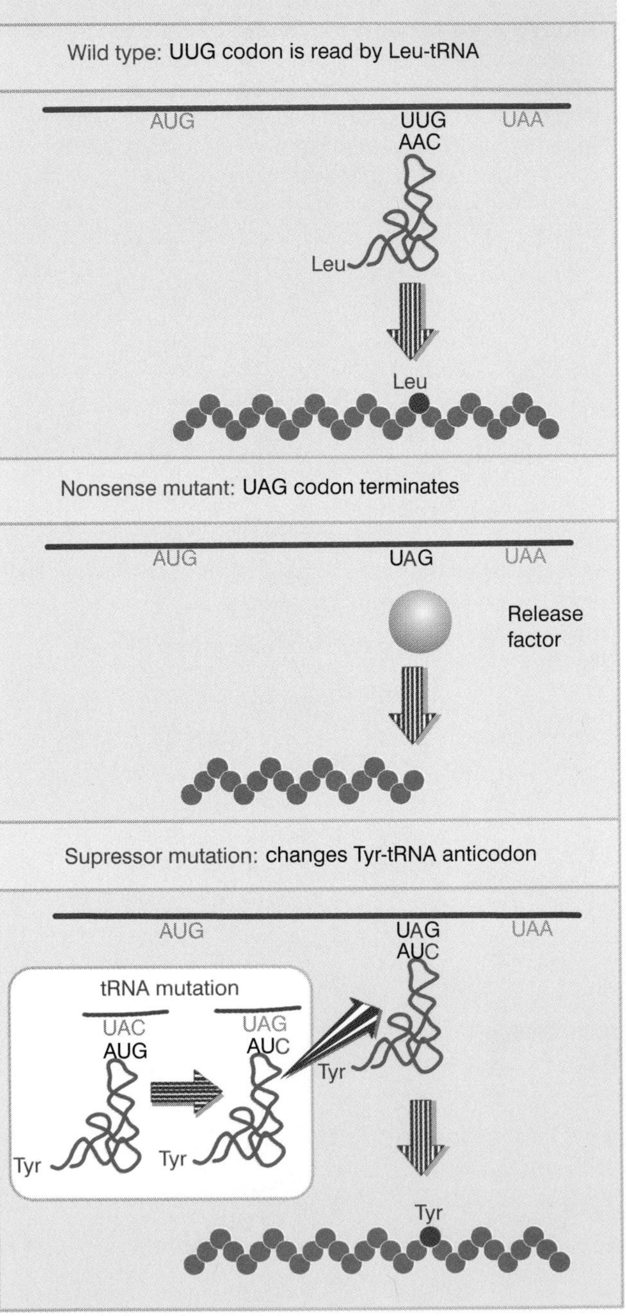

are serine, glutamine, or tyrosine, the same as those carried by the corresponding wild-type tRNAs.

Ochre suppressors also arise by mutations in the anticodon. The best known are *supC* and *supG*, which insert tyrosine or lysine in response to *both* ochre (UAA) and amber (UAG) codons. This conforms with the prediction of the wobble hypothesis that UAA cannot be recognized alone.

A UGA suppressor has an unexpected property. It is derived from tRNA^{Trp}, but its only mutation is the substitution of A in place of G *at position 24*. This change replaces a G•U pair in the D stem with an A•U pair, increasing the stability of the helix. The sequence of the anticodon remains the same as the wild type, $\overleftarrow{\text{CCA}}$. So the mutation in the D stem must in some way alter the conformation of the anticodon loop, allowing $\overleftarrow{\text{CCA}}$ to pair with UGA in an unusual wobble pairing of C with A. The suppressor tRNA continues to recognize its usual codon, UGG.

A related response is seen with a eukaryotic tRNA. Bovine liver contains a tRNA^{Ser} with the anticodon $^{\text{m}}\overleftarrow{\text{CCA}}$. The wobble rules predict that this tRNA should respond to the tryptophan codon UGG; but in fact it responds to the termination codon UGA. So it is possible that UGA is suppressed naturally in this situation.

The general importance of these observations lies in the demonstration that *codon-anticodon recognition of either wild-type or mutant tRNA cannot be predicted entirely from the relevant triplet sequences, but is influenced by other features of the molecule.*

Missense mutations change a codon representing one amino acid into a codon representing another amino acid, one that cannot function in the protein in place of the original residue. (Formally, any substitution of amino acids constitutes a missense mutation, but in practice it is detected only if it changes the activity of the protein.) The mutation can be suppressed by the insertion either of the original amino acid or of some other amino acid that is acceptable to the protein.

Figure 7.18 demonstrates that missense suppression

Figure 7.17 Nonsense suppressor tRNAs are generated by mutations in the anticodon.

Locus	tRNA	Wild Type		Suppressor	
		⟵ Codon/Anti		⟵ Anti/Codon	
supD (su1)	Ser	UCG	CGA	CUA	UAG
supE (su2)	Gln	CAG	CUG	CUA	UAG
supF (su3)	Tyr	$\text{UA}^{\text{C}}_{\text{U}}$	GUA	CUA	UAG
supC (su4)	Tyr	$\text{UA}^{\text{C}}_{\text{U}}$	GUA	UUA	$\text{UA}^{\text{A}}_{\text{G}}$
supG (su5)	Lys	$\text{AA}^{\text{A}}_{\text{G}}$	UUU	UUA	$\text{UA}^{\text{A}}_{\text{G}}$
supU (su7)	Trp	UGG	CCA	UCA	$\text{UG}^{\text{A}}_{\text{G}}$

Figure 7.18 Missense suppression occurs when the anticodon of tRNA is mutated so that it responds to the 'wrong' codon. The suppression is only partial because both the wild-type tRNA and the suppressor tRNA can respond to AGA.

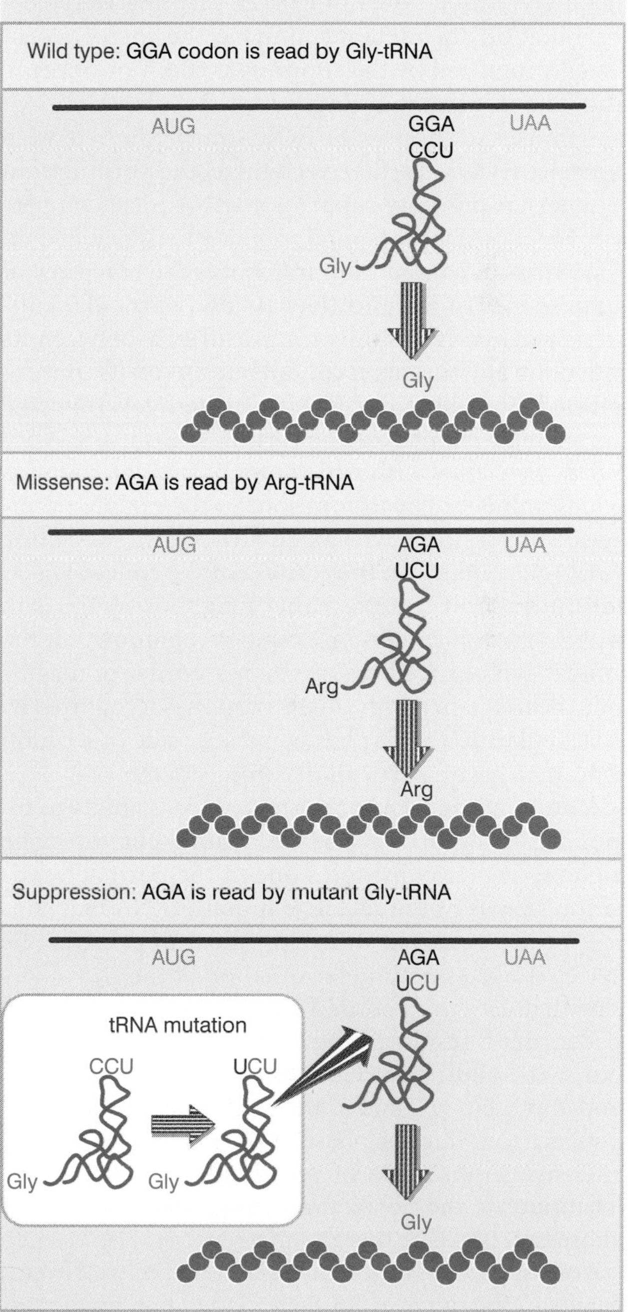

can be accomplished in the same way as nonsense suppression, by mutating the anticodon of a tRNA carrying an acceptable amino acid so that it responds to the mutant codon. So missense suppression involves a change in the meaning of the codon from one amino acid to another.

There is an interesting difference between the usual recognition of a codon by its proper aminoacyl-tRNA and the situation in which mutation allows a suppressor tRNA to recognize a new codon. In the wild-type cell, *only one meaning can be attributed to a given codon*, which represents either a particular amino acid or a signal for termination. But in a cell carrying a suppressor mutation, the mutant codon has the *alternatives* of being recognized by the suppressor tRNA or of being read with its usual meaning.

A nonsense suppressor tRNA must compete with the release factors that recognize the termination codon(s). A missense suppressor tRNA must compete with the tRNAs that respond properly to its new codon. The extent of competition influences the efficiency of suppression; so the effectiveness of a particular suppressor depends not only on the affinity between its anticodon and the target codon, but also on its concentration in the cell, and on the parameters governing the competing termination or insertion reactions.

The efficiency with which any particular codon is read is influenced by its location. So the extent of nonsense suppression by a given tRNA can vary quite widely, depending on the **context** of the codon. We do not understand the effect that neighboring bases in mRNA have on codon-anticodon recognition, but the context can change by more than an order of magnitude the frequency with which a codon is recognized by a particular tRNA. The base on the 3′ side of a codon appears to have a particularly strong effect.

A nonsense suppressor is isolated by its ability to respond to a mutant nonsense codon. But the same triplet sequence constitutes one of the normal termination signals of the cell! The mutant tRNA that suppresses the nonsense mutation must in principle be able to suppress natural termination at the end of any gene that uses this codon. **Figure 7.19** shows that this **readthrough** results in the synthesis of a longer protein, with additional C-terminal material. The extended protein will end at the next termination triplet sequence found in the phase of the reading frame. Any extensive suppression of termination is likely to be deleterious to the cell by producing extended proteins whose functions are thereby altered.

Amber suppressors tend to be relatively efficient, usually in the range of 10–50%, depending on the system. This efficiency is possible because amber codons are used relatively infrequently to terminate protein synthesis in *E. coli.*

Figure 7.19 Nonsense suppressors also read through natural termination codons, synthesizing proteins that are longer than wild-type.

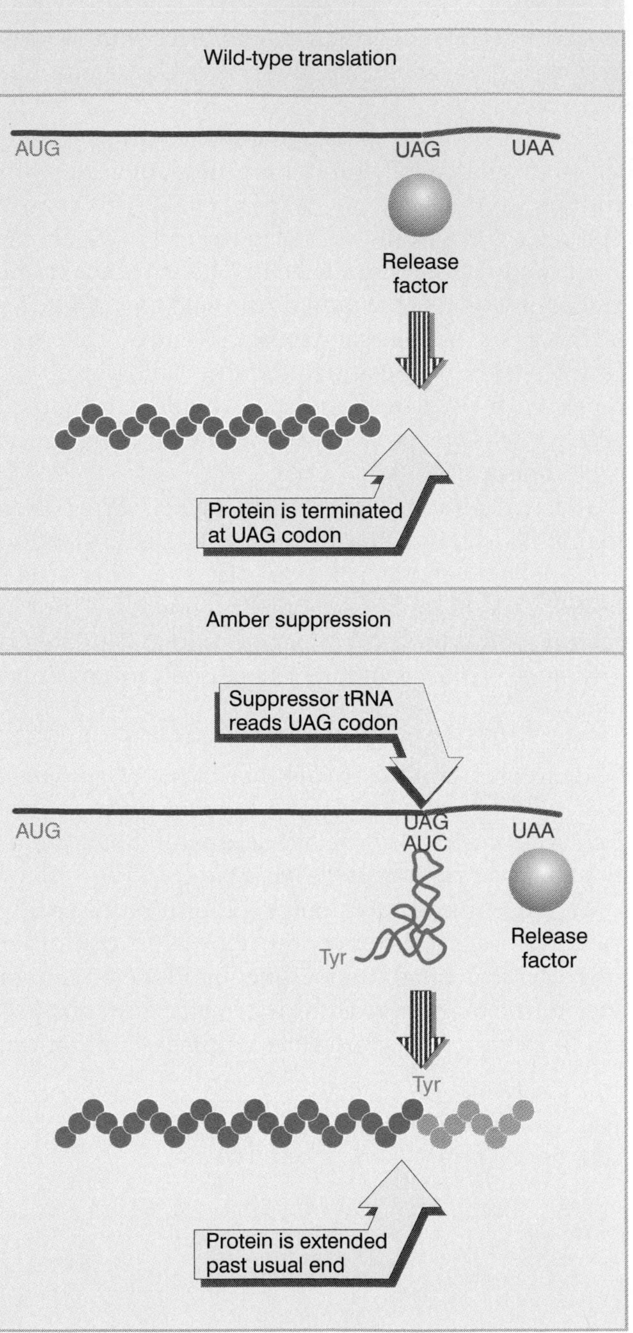

Ochre suppressors are difficult to isolate. They are always much less efficient, usually with activities below 10%. All ochre suppressors grow rather poorly, which indicates that suppression of both UAA and UAG is damaging to *E. coli*, probably because the ochre codon is used most frequently as a natural termination signal.

UGA is the least efficient of the termination codons in its natural function; it is misread by Trp-tRNA as frequently as 1–3% in wild-type situations. In spite of this deficiency, however, it is used more commonly than the amber triplet to terminate bacterial genes.

One gene's missense suppressor is likely to be another gene's mutator. A suppressor corrects a mutation by substituting one amino acid for another at the mutant site. But in other locations, the same substitution will replace the wild-type amino acid with a new amino acid. The change may inhibit normal protein function.

This poses a dilemma for the cell: it must suppress what is a mutant codon at one location, while failing to change too extensively its normal meaning at other locations. The absence of any strong missense suppressors is therefore explained by the damaging effects that would be caused by a general and efficient substitution of amino acids.

A mutation that creates a suppressor tRNA can have two consequences. First, it allows the tRNA to recognize a new codon. Second, sometimes it *prevents* the tRNA from recognizing the codons to which it previously responded. It is significant that all the high-efficiency amber suppressors are derived by mutation of one copy of a redundant tRNA set. In these cases, the cell has several tRNAs able to respond to the codon originally recognized by the wild-type tRNA. So the mutation does not abolish recognition of the old codons, which continue to be served adequately by the tRNAs of the set. (In the unusual situation in which there is only a single tRNA that responds to a particular codon, any mutation that prevents the response is lethal.)

The accuracy of translation

The lack of detectable variation when the sequence of a protein is analyzed demonstrates that protein synthesis must be extremely accurate. Very few mistakes are apparent in the form of substitutions of one amino acid for another. There are two stages in protein synthesis at which errors might be made:

- Charging a tRNA only with its correct amino acid clearly is critical. This is a function of the aminoacyl-tRNA synthetase enzyme. Probably the error rate varies with the particular enzyme, but current estimates are that mistakes occur in less than 1 in 10^5 aminoacylations.
- The specificity of codon-anticodon recognition is crucial, but puzzling. Although binding constants vary with the individual codon-anticodon reaction, the specificity is always much too low to provide an error rate of $<10^{-5}$. When free in solution, tRNAs bind to their trinucleotide codon sequences only relatively weakly. Related, but erroneous triplets (with two correct bases out of three) are recognized 10^{-1}–10^{-2} times as efficiently as the correct triplets.

Codon-anticodon base pairing therefore seems to be a weak point in the accuracy of translation. The ribosome has an important role in controlling the specificity of this interaction, functioning directly or indirectly as a "proofreader", to distinguish correct and incorrect codon-anticodon pairs, and thus amplifying the rather modest intrinsic difference. And in addition to the role of the ribosome itself, the factors that place initiator- and aminoacyl-tRNAs in the ribosome also may influence the pairing reaction.

So there must be some mechanism for stabilizing the correct aminoacyl-tRNA, allowing its amino acid to be accepted as a substrate for receipt of the polypeptide chain; contacts with an incorrect aminoacyl-tRNA must be rapidly broken, so that the complex leaves without reacting. Suppose that there is no specificity in the initial collision between the aminoacyl-tRNA•EF-Tu•GTP complex and the ribosome. If any complex, irrespective of its tRNA, can enter the A site, the number of incorrect entries must far exceed the number of correct entries. How does a ribosome assess the codon-anticodon reaction in the A site to determine whether a proper fit has been achieved?

The ability of the ribosome to influence the accuracy of translation was first shown by the effects of mutations that confer resistance to streptomycin. One effect of streptomycin is to increase the level of misreading of the pyrimidines U and C (usually one is mistaken for the other, occasionally for A). The site at which streptomycin acts is influenced by the S12 protein; the sequence of this protein is altered in resistant mutants. Ribosomes with an S12 protein derived from resistant bacteria show a reduction in the level of misreading compared with wild-type ribosomes. This compensates for the effect of streptomycin on misreading. Streptomycin probably binds the rRNA, and S12 affects this reaction by influencing the accessibility of the rRNA.

Mutations at two other loci, coding for proteins S4 and S5, influence misreading, since revertants showing the usual level of misreading can be isolated at either locus. So the accuracy of translation is controlled by the interactions of these three proteins. The level of misreading and the nature of the response to streptomycin depend on the versions of these proteins that are present. (Some combinations even make the ribosome **dependent** on the presence of streptomycin for correct translation.) We can interpret the role of these proteins in two ways:

- A direct mechanism for controlling accuracy would be for the stereochemistry of the A site to determine the latitude of codon-anticodon recognition. The geometry of the ribosome could be designed so that codon-anticodon binding is scrutinized in such a way that the criteria for accepting aminoacyl-tRNA could be made more or less precise by the structures of proteins S12, S4, and S5 (or by their effects on the structure of rRNA).
- The effect on accuracy is an indirect result of the speed of ribosome movement. Ribosome velocity determines availability for tRNA recognition, and thus the efficiency of the process. This model explains the effect of streptomycin by adjusting the kinetics of chain elongation. The relevant parameter is the speed of ribosome action relative to the time required to make and break contacts. If the velocity of peptide bond formation is increased, incorrect aminoacyl-tRNAs are more likely to be trapped by bond formation before the aminoacyl-tRNA escapes. Slowing the rate of protein synthesis gives more time to correct errors.

Mismatched aminoacyl-tRNA dissociates more rapidly than correctly matched aminoacyl-tRNA, probably by a factor of ~5×. Increasing time spent in the A site before peptide bond formation occurs therefore increases the probability that the correct aminoacyl-tRNA will be utilized. This is a kinetic proofreading mechanism. The overall error rate in protein synthesis is $\sim 5 \times 10^{-4}$ per codon, and the majority of errors probably occur by recognition of mistaken aminoacyl-tRNAs in the A site.

The cost of protein synthesis in terms of high-energy bonds may be increased by proofreading processes. An important question in calculating the cost of protein synthesis is the stage at which the decision is taken on whether to accept a tRNA. If a decision occurs immediately to release an aminoacyl-tRNA•EF-Tu•GTP complex, there is little extra cost for rejecting the large number of incorrect tRNAs that are likely (statistically) to enter the A site before the correct tRNA is recognized. But if the GTP is hydrolyzed when the complex binds, an additional high-energy bond will be cleaved for every incorrectly associating tRNA. This would increase the cost of protein synthesis well above the three high-energy bonds that are used in adding every (correct) amino acid to the chain. There is some evidence that the use of GTP *in vivo* is greater than had been expected, possibly involving an extra 3 GTP cleavages per amino acid.

The specificity of decoding has been assumed to reside with the ribosome itself, but some recent results suggest that translation factors influence the process at both the P site and the A site.

A striking case concerns initiation. Mutation of the AUG initiation codon to UUG in the yeast gene *HIS4* prevents initiation. Extragenic suppressor mutations can be found that allow protein synthesis to be initiated at the mutant UUG codon. Two of these suppressors prove to be in genes coding for the α and β subunits of eIF2, the factor that binds Met-$tRNA_i$ to the P site. The mutation in eIFβ2 resides in a part of the protein that is almost certainly involved in binding nucleic acid. It seems likely that its target is either the initiation sequence of mRNA as such or the base-paired association between the mRNA codon and $tRNA_i^{Met}$ anticodon. This suggests that eIF2 participates in the discrimination of initiation codons as well as bringing the initiator tRNA to the P site.

An indication that EF-Tu is involved in maintaining the reading frame is provided by mutants of the factor that suppress frameshifting (see next section). This probably means that EF-Tu does not merely bring aminoacyl-tRNA to the A site, but also is involved in positioning the incoming aminoacyl-tRNA relative to the peptidyl-tRNA in the P site.

tRNA may influence the reading frame

THE reading frame of a messenger usually is invariant. Translation starts at an AUG codon and continues in triplets to a termination codon. Reading takes no notice of sense: insertion or deletion of a base causes a frameshift mutation, in which the reading frame is changed beyond the site of mutation. Ribosomes and tRNAs continue ineluctably in triplets, synthesizing an entirely different series of amino acids.

There are some exceptions to the usual pattern of translation that enable a reading frame with an interruption of some sort—such as a nonsense codon or frameshift—to be translated into a full-length protein. **Recoding** events are responsible for making exceptions to the usual rules, and can involve several types of events:

- Suppression involves recognition of a codon by a tRNA that usually would respond to a different codon (as in the examples of nonsense suppression discussed earlier).
- Redefinition of the meaning of a codon occurs when an aminoacyl-tRNA is modified (as in the example of the substitution of selenocysteine discussed earlier).
- Frameshifting typically involves changing the reading frame by +1 or –1 at a specific site.

Frameshift mutations are suppressed by restoring the original reading frame. We have already discussed how this can be achieved by compensating base deletions and insertions within a gene (see Chapter 1). However, *extragenic* frameshift suppressors also can be found in the form of tRNAs with aberrant properties.

The simplest type of external frameshift suppressor corrects the reading frame when a mutation has been caused by inserting an additional base within a stretch of identical residues. For example, a G may be inserted in a run of several contiguous G bases. The frameshift suppressor is a tRNA^{Gly} that has an extra base inserted in its anticodon loop, converting the anticodon from the usual triplet sequence $\overleftarrow{\text{CCC}}$ to the quadruplet sequence $\overleftarrow{\text{CCCC}}$. The suppressor tRNA recognizes a 4-base "codon."

Some frameshift suppressors can recognize more than one 4-base "codon". For example, a bacterial tRNA^{Lys} suppressor can respond to either AAAA or AAAU, instead of the usual codon AAA. Another suppressor can read any 4-base "codon" with ACC in the first three positions; the next base is irrelevant. In these cases, the alternative bases that are acceptable in the fourth position of the longer "codon" are not related by the usual wobble rules. The suppressor tRNA probably recognizes a 3-base codon, but for some other reason—most likely steric hindrance—the adjacent base is blocked. This forces one base to be skipped before the next tRNA can find a codon.

Situations in which frameshifting is a normal event are presented by phages and viruses. Such events may affect the continuation or termination of protein synthesis, and result from the intrinsic properties of the mRNA.

In phage MS2, a frameshift causes the ribosome to recognize a termination codon at an early position in its new reading frame. The terminating ribosome then can recognize the initiation codon of the lysis gene, which lies just a few bases farther along. When the ribosome does not terminate, it reads right over the lysis gene initiation codon. *So the frameshift-dependent termination event is a prerequisite for initiation of lysis gene expression.*

In retroviruses, translation of the first gene is terminated by a nonsense codon in phase with the reading frame. The second gene lies in a different reading frame, and (in some viruses) is translated by a frameshift that changes into the second reading frame and therefore bypasses the termination codon (see Chapter 16). **Figure 7.20** illustrates the similar situation of the yeast Ty element, in which the termination codon of *tya* must be bypassed by a frameshift in order to read the subsequent *tyb* gene.

Such situations demonstrate the important point that *the rare (but predictable) occurrence of "misreading" events can be relied on as a necessary step in natural translation.*

There are two common features in this type of frameshifting:

- A "slippery" sequence allows an aminoacyl-tRNA to pair with its codon and then to move +1 (rare) or –1 base (more common) to pair with a triplet sequence that can also pair with its anticodon.
- The ribosome is delayed at the frameshifting site to allow time for the aminoacyl-tRNA to rearrange its pairing. The cause of the delay can be an adjacent codon that requires a scarce aminoacyl-tRNA, a

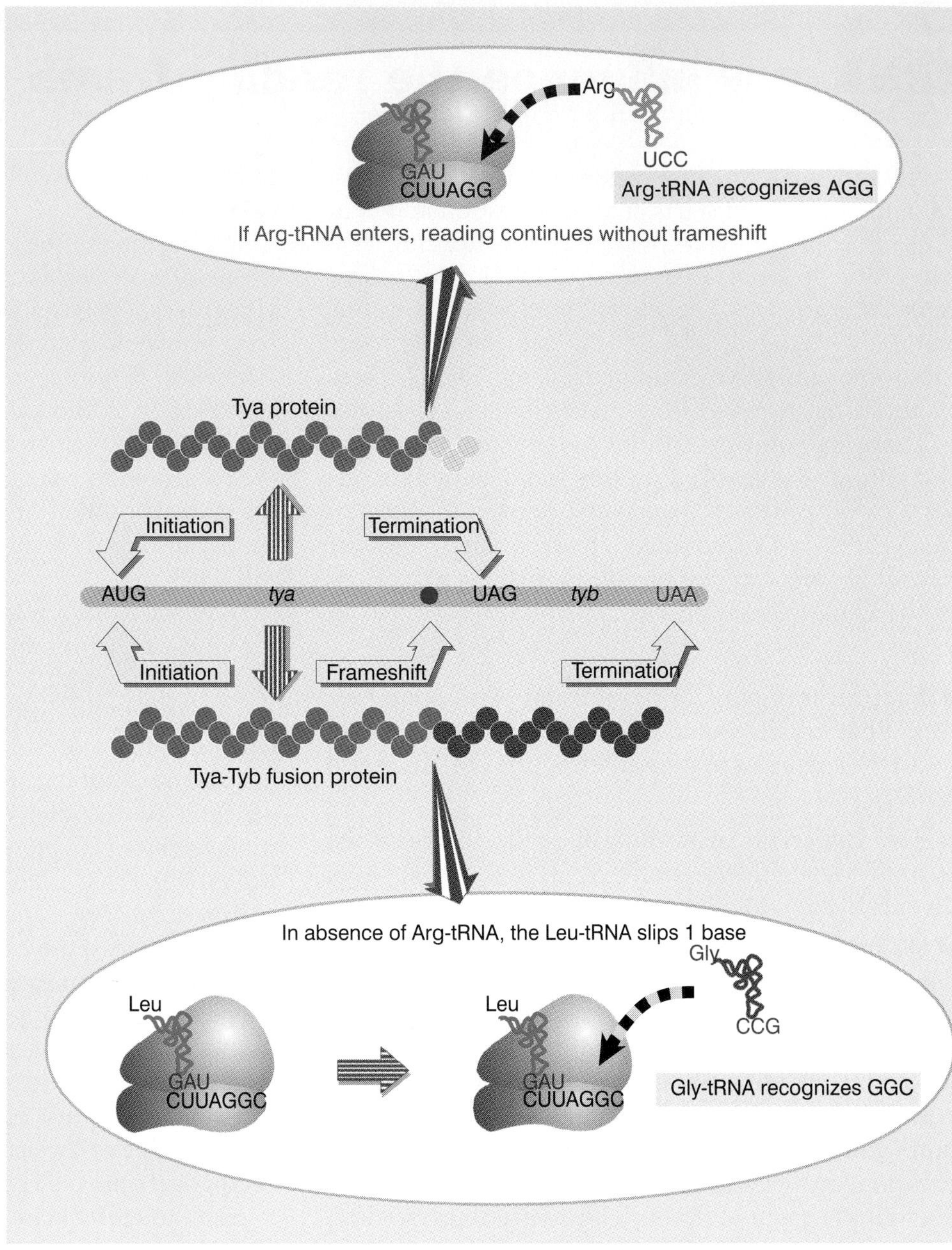

Figure 7.20 A +1 frameshift is required for expression of the *tyb* gene of the yeast Ty element. The shift occurs at a 7-base sequence in which two Leu codon(s) are followed by a scarce Arg codon.

termination codon that is recognized slowly by its release factor, or a structural impediment in mRNA (for example, a "pseudoknot", a particular conformation of RNA) that impedes the ribosome.

Figure 7.20 shows the behavior of a typical slippery sequence. The 7-nucleotide sequence CUUAGGC is usually recognized by Leu-tRNA at CUU, followed by Arg-tRNA at AGC. However, the Arg-tRNA is scarce, and when its scarcity results in a delay, the Leu-tRNA slips from the CUU codon to the overlapping UUA triplet. This causes a frameshift, because the next triplet in phase with the new pairing (GGC) is read by Gly-tRNA. Slippage usually occurs in the P site (when the Leu-tRNA actually has become peptidyl-tRNA, carrying the nascent chain).

Slippery events can involve movement in either direction; a –1 frameshift is caused when the tRNA moves backwards, and a +1 frameshift is caused when it moves forwards. In either case, the result is to expose an out-of-phase triplet in the A site for the next aminoacyl-tRNA.

Frameshifting also occurs in some cases in which the codon in the second phase cannot be reached by slipping from the codon in the original phase. One explanation is that a tRNA initially enters and binds to a codon that is out of phase. An alternative is that a tRNA with a normal anticodon occasionally recognizes a 4-base nucleotide sequence instead of a triplet. This mechanism shows the same need to delay the ribosome as frameshifting at slippery sequences.

Frameshifting at a stop codon causes readthrough of the protein. The base on the 3′ side of the stop codon influences the relative frequencies of termination and frameshifting, and thus affects the efficiency of the termination signal. This helps to explain the significance of context on termination.

Summary

The sequence of mRNA read in triplets 5′→3′ is related by the genetic code to the amino acid sequence of protein read from N- to C-terminus. Of the 64 triplets, 61 code for amino acids and 3 provide termination signals. Synonym codons that represent the same amino acids are related, often by a change in the third base of the codon. This third-base degeneracy, coupled with a pattern in which related amino acids tend to be coded by related codons, minimizes the effects of mutations. The genetic code is universal, and must have been established very early in evolution. Some changes have occurred during mitochondrial evolution; changes in nuclear genomes are rare.

Multiple tRNAs may respond to a particular codon. The set of tRNAs responding to the various codons for each amino acid is distinctive for each organism. Codon-anticodon recognition involves wobbling at the first position of the anticodon (third position of the codon), which allows some tRNAs to recognize multiple codons. All tRNAs have modified bases, introduced by enzymes that recognize target bases in the tRNA structure. Codon-anticodon pairing is influenced by modifications of the anticodon itself and also by the context of adjacent bases, especially on the 3′ side of the anticodon. Taking advantage of codon-anticodon wobble allows vertebrate mitochondria to use only 22 tRNAs to recognize all codons, compared with the usual minimum of 31 tRNAs; this is assisted by the changes in the mitochondrial code.

Each amino acid is recognized by a particular aminoacyl-tRNA synthetase, which also recognizes all of the tRNAs coding for that amino acid. Aminoacyl-tRNA synthetases have a proofreading function that scrutinizes the aminoacyl-tRNA products and hydrolyzes incorrectly joined aminoacyl-tRNAs.

Aminoacyl-tRNA synthetases vary widely, but fall into two general groups according to the structure of the catalytic domain. Synthetases of each group bind the tRNA from the side, making contacts principally with the extremities of the acceptor stem and the anticodon stem-loop; the two types of synthetases bind tRNA from opposite sides. The relative importance attached to the acceptor stem and the anticodon region for specific recognition varies with the individual tRNA.

Mutations may allow a tRNA to read different codons; the most common form of such mutations occurs in the anticodon itself. Alteration of its specificity may allow a tRNA to suppress a mutation in a gene coding for protein. A tRNA that recognizes a termination codon provides a nonsense suppressor; one that changes the amino acid responding to a codon is a missense suppressor. Suppressors of UAG and UGA codons are more efficient than those of UAA codons, which is explained by the fact that UAA is the most commonly used natural termination codon. But the efficiency of all suppressors depends on the context of the individual target codon.

Frameshifts of the +1 type may be caused by aberrant tRNAs that read "codons" of 4 bases. Frameshifts of either +1 or –1 may be caused by slippery sequences in mRNA that allow a peptidyl-tRNA to slip from its codon to an overlapping sequence that can also pair with its anticodon. This frameshifting also requires another sequence that causes the ribosome to delay. Frameshifts determined by the mRNA sequence may be required for expression of natural genes.

Further reading

Historical

Holley, R. W. *et al.* (1965). Structure of an RNA. *Science* **147,** 1462–1465.

Reviews

Lewin, B. (1974). *Gene Expression,* 1, *Bacterial Genomes.* (Wiley, New York).

Altman, S. (1978). *Transfer RNA.* (MIT Press, Cambridge).

Schimmel, P., Soll, D., and Abelson, J. (1979). *Transfer RNA: Structure, Properties and Recognition.* (Cold Spring Harbor Lab., New York).

Soll, D., Abelson, J., and Schimmel, P. (1979). *Transfer RNA: Biological Aspects.* (Cold Spring Harbor Lab., New York).

Murgola, E. J. (1985). tRNA, suppression, and the code. *Ann. Rev. Genet.* **19,** 57–80.

Bjork, G. R. *et al.* (1987). Transfer RNA modification. *Ann. Rev. Biochem.* **56,** 263–287.

Fox, T. D. (1987). Natural variation in the genetic code. *Ann. Rev. Genet.* **21,** 67–91.

Schimmel, P. (1987). Aminoacyl-tRNA synthetases: general scheme of structure-function relationships on the polypeptides and recognition of tRNAs. *Ann. Rev. Biochem.* **56,** 125–158.

Schimmel, P. (1988). Parameters for the molecular recognition of tRNAs. *Biochemistry* **28,** 2747–2759.

Eggertsson, G. and Soll, D. (1988). Transfer RNA-mediated suppression of termination codons in *E. coli. Microbiol. Rev.* **52,** 354–374.

Normanly, J. and Abelson, J. (1989). Transfer RNA identity. *Ann. Rev. Biochem.* **58,** 1029–1049.

Atkins, J. F. *et al.* (1991). Towards a genetic dissection of the basis of triplet decoding, and its natural subversion: programmed reading frameshifts and hops. *Ann. Rev. Genet.* **25,** 201–228.

Bock, A. *et al.* (1991). Selenoprotein synthesis: an expansion of the genetic code. *Trends Biochem. Sci.* **16,** 463–467.

Jakubowski, H. and Goldman, E. (1992). Editing of errors in selection of amino acids for protein synthesis. *Microbiol. Rev.* **56,** 412–429.

Osawa, S. *et al.* (1992). Recent evidence for evolution of the genetic code. *Microbiol. Rev.* **56,** 229–264.

Soll, D. and RajBhandary, U. L. (1995). *tRNA. Structure, Biosynthesis, and Function.* American Society for Microbiology, Washington.

Farabaugh, P. J. (1995). Programmed translational frameshifting. *Microbiol. Rev.* **60,** 103–134.

Gesteland, R. F. and Atkins, J. F. (1996). Recoding: dynamic reprogramming of translation. *Ann. Rev. Biochem.* **65,** 741–768.

Discoveries

tRNA•synthetase structure

Rould, M. A. *et al.* (1989). Structure of *E. coli* glutaminyl-tRNA synthetase complexed with $tRNA^{Gln}$ and ATP at 2.8Å resolution. *Science* **246,** 1135–1142.

Ruff, M. *et al.* (1991). Class II aminoacyl tRNA synthetases: crystal structure of yeast aspartyl-tRNA synthetase complexes with $tRNA^{Asp}$. *Science* **252,** 1682–1689.

Codon-anticodon recognition

Crick, F. H. C. (1966). Codon-anticodon pairing: the wobble hypothesis. *J. Mol. Biol.* **19,** 548–555.

Proofreading

Hopfield, J. J. (1974). Kinetic proofreading: a new mechanism for reducing errors in biosynthetic processes requiring high specificity. *Proc. Natl Acad. Sci. USA* **71,** 4135–4139.

Jakubowski, H. (1990). Proofreading *in vivo:* editing of homocysteine by methionyl-tRNA synthetase in *E. coli. Proc. Natl Acad. Sci USA* **87,** 4504–4508.

Nureki, O. *et al.* (1998). Enzyme structure with two catalytic sites for double sieve selection of substrate. *Science* **280,** 578–581.

Chapter 8

Protein localization

PROTEINS can be divided into two general classes with regard to localization: those that are not associated with membranes; and those that are associated with membranes. Each class can be subdivided further, depending on whether the protein associates with a particular structure in the cytosol or type of membrane. **Figure 8.1** maps the cell in terms of the possible ultimate destinations for a newly synthesized protein and the systems that transport it:

- Cytosolic (or "soluble") proteins are not localized in any particular organelle. They are synthesized in the cytosol, and remain there, where they function as individual catalytic centers, acting on metabolites that are in solution in the cytosol.
- Macromolecular structures may be located at particular sites in the cytoplasm; for example, centrioles are associated with the regions that become the poles of the mitotic spindle.
- Nuclear proteins must be transported from their site of synthesis in the cytosol through the nuclear envelope.
- Cytoplasmic organelles contain proteins synthesized in the cytosol and transported specifically to (and through) the organelle membrane, for example, to the mitochondrion or (in plant cells) to the chloroplast. (Some mitochondrial and chloroplast proteins are synthesized within the organelle.)
- The cytoplasm contains a series of membranous bodies, including endoplasmic reticulum, Golgi apparatus, endosomes, and lysosomes. This is sometimes referred to as the "reticuloendothelial system". Proteins that reside within these compartments are inserted into ER membranes, and then are directed to their particular locations by the transport system of the Golgi apparatus.
- Proteins that are secreted from the cell are transported to the plasma membrane and then must pass through it to the exterior. They start their synthesis in the same way as proteins associated with the reticuloendothelial system, but pass entirely through the system instead of halting at some particular point within it.

Proteins that are not associated with membranes are released into the cytosol when their synthesis is completed by a ribosome. Some proteins remain free in the cytosol in quasi-soluble form; others associate with macromolecular cytosolic structures, such as filaments, microtubules, centrioles, etc. This class also includes nuclear proteins (which pass into the nucleus through large aqueous pores). The ribosomes on which these proteins are synthesized are sometimes called "free ribosomes", because they fractionate separately from membranes. The "default" for a protein released from "free" ribosomes is to remain in the cytosol; to be targeted to a specific location, it requires an appropriate signal, typically a sequence motif that causes it to be assembled into a macromolecular structure or recognized by a transport system.

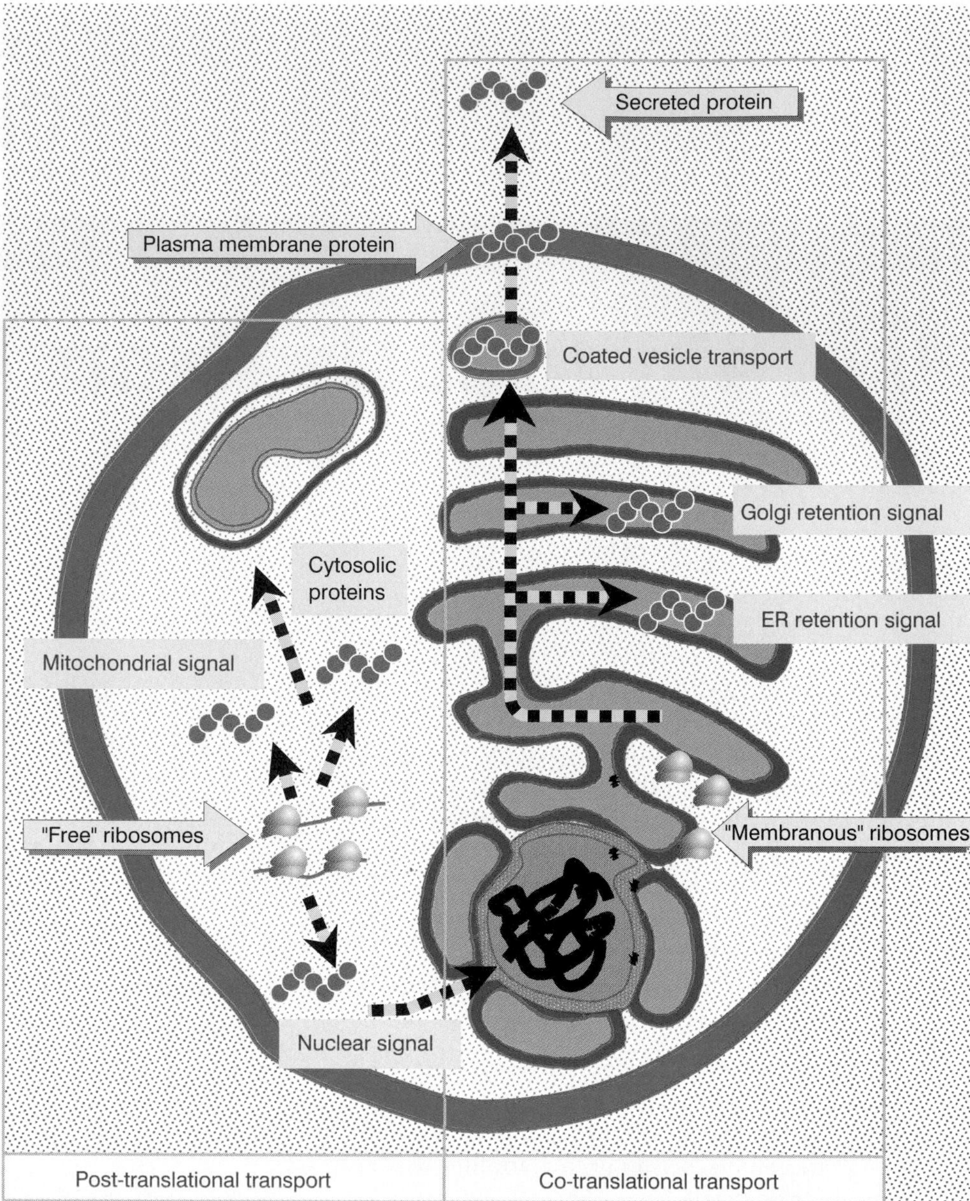

Figure 8.1 *Overview*: proteins that are localized post-translationally are released into the cytosol after synthesis on 'free' ribosomes. Some have signals for targeting to organelles such as the nucleus or mitochondria. Proteins that are localized co-translationally associate with the ER membrane during synthesis, so their ribosomes are 'membrane-bound.' The proteins pass into the endoplasmic reticulum, along to the Golgi, and then through the plasma membrane, unless they possess signals that cause retention at one of the steps on the pathway. They may also be directed to other organelles, such as endosomes or lysosomes.

Figure 8.2 summarizes some signals used by proteins released from cytosolic ribosomes. Import into the nucleus results from the presence of a variety of rather short sequences within proteins. These "nuclear localization signals" enable the proteins to pass through nuclear pores (see later). One type of signal that determines transport to the peroxisome is a very short C-terminal sequence.

The process of inserting into or passing through a membrane is called **protein translocation.** Proteins that associate with membranes follow one of two routes.

Mitochondrial and chloroplast proteins are synthesized on "free" ribosomes; after their release into the cytosol they associate with the organelle membranes by means of N-terminal sequences of ~25 amino acids in length that are recognized by receptors on the organelle envelope. Because this process takes place after synthesis of the protein has been completed, it is called **post-translational translocation.**

Proteins that reside within the reticuloendothelial system enter the endoplasmic reticulum while they are being synthesized. This process is called **co-translational translocation.** Because the ribosomes are associated with the ER membranes during synthesis of these proteins, and are therefore found in membrane fractions of the cell, they are sometimes described as "membrane-bound".

A common feature is found in proteins that use N-terminal sequences to be transported co-translationally to the ER or post-translationally to mitochondria or chloroplasts. *The N-terminal sequence is cleaved from the protein during protein translocation.* The N-terminal sequence comprises a **leader** that is not part of the mature protein. The protein carrying this leader is called a **preprotein,** and is a transient precursor to the mature protein.

Proteins that associate with membranes via N-terminal leaders use a *hierarchy* of signals to find their final destination. In the case of the reticuloendothelial system, the ultimate location of a protein depends on how it is directed as it transits the endoplasmic reticulum and Golgi apparatus. The leader sequence itself introduces the protein to the membrane; the intrinsic consequence of the interaction is for the protein to pass through the membrane into the compartment on the other side. For a protein to reside within the membrane, a further signal is required to stop passage through the membrane. Other types of signals are required for a protein to be sorted to a particular destination, that is, to remain within the membrane or lumen of some particular compartment. The general process of finding its ultimate destination by transport through successive membrane systems is called **protein sorting** or **protein trafficking,** and is discussed in Chapter 25.

Some of the destinations and signals are summarized in **Figure 8.3.** They show that some sort of permit is required to reside in any part of the membrane system. The permit most often takes the form of a short amino acid sequence, but there are also other types of information. The "default pathway" takes a protein through the ER, into the Golgi, and on to the plasma membrane. Proteins that reside in the ER possess a C-terminal tetrapeptide (KDEL, which actually provides a signal for them to return to the ER from the Golgi). The signal that diverts a protein to the lysosome is a covalent modification: the addition of a particular sugar residue. We discuss direction to these locations in Chapter 25.

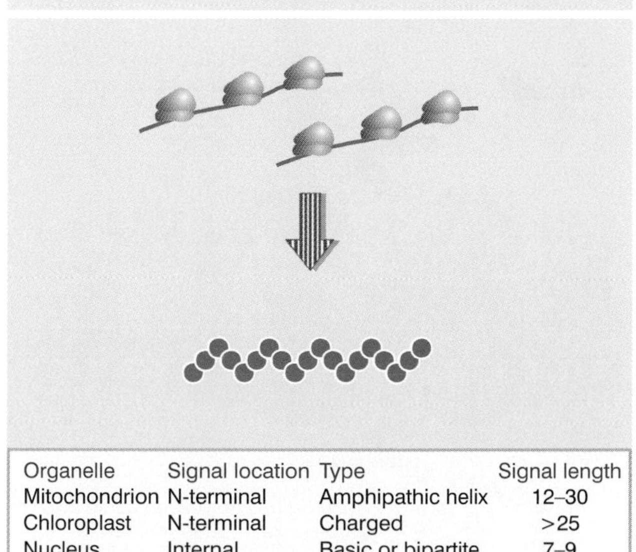

Figure 8.2 Proteins synthesized on free ribosomes in the cytosol are directed after their release to specific destinations by short signal motifs.

Organelle	Signal location	Type	Signal length
Mitochondrion	N-terminal	Amphipathic helix	12–30
Chloroplast	N-terminal	Charged	>25
Nucleus	Internal	Basic or bipartite	7–9
Peroxisome	C-terminal	SKL	3

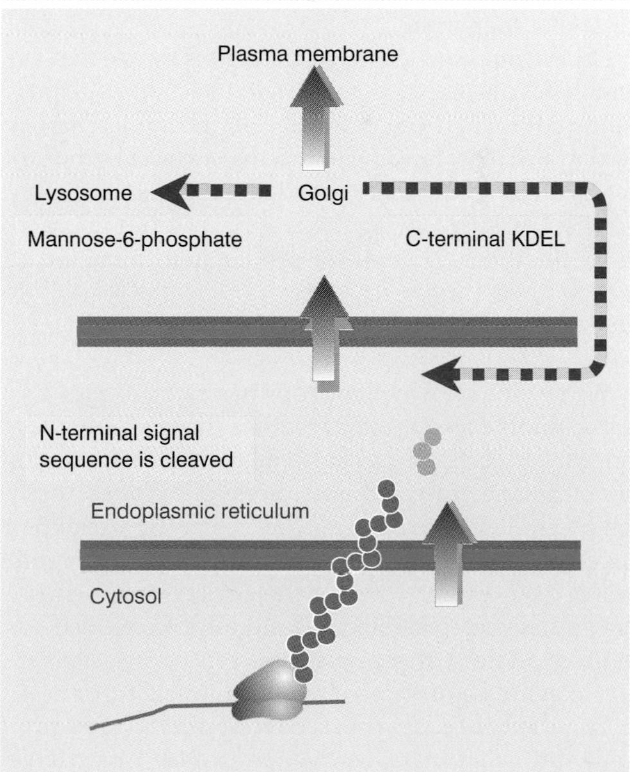

Figure 8.3 Membrane-bound ribosomes have proteins with N-terminal sequences that enter the ER during synthesis. The proteins may flow through to the plasma membrane or may be diverted to other destinations by specific signals.

Chaperones may be required for protein folding

PROTEIN folding takes place by interactions between reactive surfaces. Typically these surfaces consist of exposed hydrophobic side chains. Their interactions form a hydrophobic core. The intrinsic reactivity of these surfaces means that incorrect interactions may occur unless the process is controlled.

Some proteins are able to acquire their mature conformation spontaneously. A test for this ability is to denature the protein and determine whether it can then renature into the active form. This capacity is called **self-assembly.** A protein that can self-assemble can fold or refold into the active state from other conformations, including the condition in which it is initially synthesized. This implies that the internal interactions are intrinsically directed toward the right conformation.

When this does not happen, and alternative sets of interactions can occur, a protein may become trapped in a stable conformation that is *not* the intended final form. Proteins in this category cannot self-assemble. Their acquisition of proper structure requires the assistance of a **chaperone.**

Chaperones are proteins that mediate correct assembly by causing a target protein to acquire one possible conformation instead of others. This is accomplished by binding to reactive surfaces in the target protein that are exposed during the assembly process, and preventing those surfaces from interacting with other regions of the protein to form an incorrect conformation. *Chaperones function by preventing formation of incorrect structures rather than by promoting formation of correct structures.*

We do not know what proportion of proteins can self-assemble as opposed to those that require assistance from a chaperone. (It is not axiomatic that a protein capable of self-assembly *in vitro* actually self-assembles *in vivo*, because there may be rate differences in the two conditions, and chaperones still could be involved *in vivo*. However, there is a distinction to be drawn between proteins that can in principle self-assemble and those that in principle must have a chaperone to assist acquisition of the correct structure.)

The ability of chaperones to recognize incorrect protein conformations allows them to play two related roles concerned with protein structure:

- When a protein is initially synthesized, that is to say, as it exits the environment of the ribosome to enter the cytosol, it appears in an unfolded form. Spontaneous folding then occurs as the emerging sequence interacts with regions of the protein that were synthesized previously. Chaperones influence the folding process by controlling the accessibility of the reactive surfaces. This process is involved in initial acquisition of the correct conformation.
- When a protein is denatured, new regions are exposed and become able to interact. These interactions are similar to those that occur when a protein (transiently) misfolds as it is initially synthesized. They are recognized by chaperones as comprising incorrect folds. This process is involved in recognizing a protein that has been denatured, and either assisting renaturation or leading to its removal by degradation.

Figure 8.4 A protein is constrained to a narrow passage as it crosses a membrane.

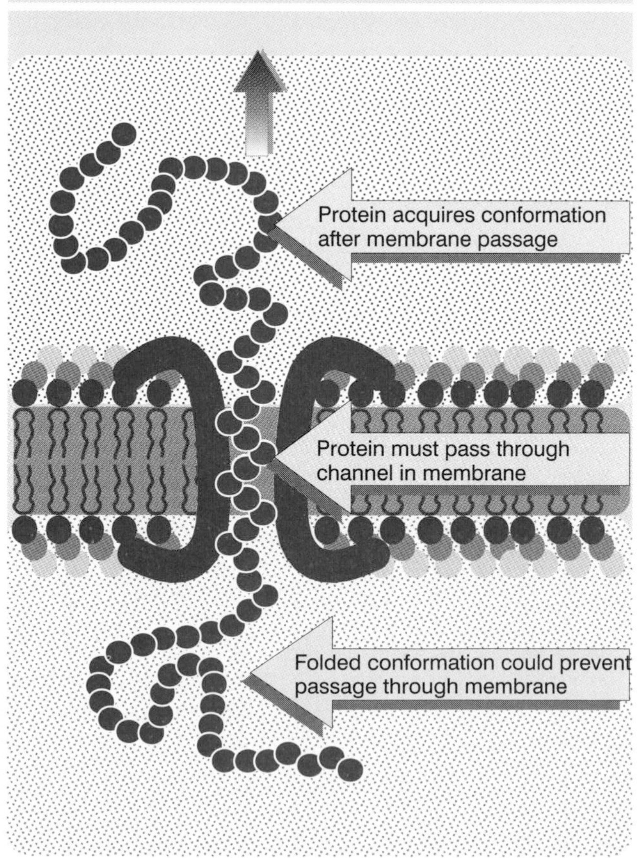
Protein acquires conformation after membrane passage
Protein must pass through channel in membrane
Folded conformation could prevent passage through membrane

Figure 8.5 Chaperone families have eukaryotic and bacterial counterparts (named in parentheses).

System	Function
Hsp70	
Hsp70 (DnaK)	ATPase
Hsp40 (DnaJ)	Stimulates ATPase
GrpE (GrpE)	Nucleotide exchange factor
Chaperonin	
Hsp60 (GroEL)	Forms two heptameric rings
Hsp10 (GroES)	Forms cap

Chaperones may also be required to assist the formation of oligomeric structures and for the transport of proteins through membranes. A persistent theme in membrane passage is that control (or delay) of protein folding is an important feature. **Figure 8.4** shows that it may be necessary to maintain a protein in an unfolded state before it enters the membrane because of the geometry of passage: the mature protein could simply be too large to fit into the available channel. Chaperones may prevent a protein from acquiring a conformation that would prevent passage through the membrane; in this capacity, their role is basically to maintain the protein in an unfolded, flexible state. Once the protein has passed through the membrane, it may require another chaperone to assist with folding to its mature conformation in much the same way that a cytosolic protein requires assistance from a chaperone as it emerges from the ribosome. The state of the protein as it emerges from a membrane is probably similar to that as it emerges from the ribosome—basically extended in a more or less linear condition.

Two major groups of chaperones have been well characterized, as summarized in **Figure 8.5:**

- The Hsp70 system consists of Hsp70, Hsp40, and GrpE. It functions on newly synthesized proteins, proteins being transported through membranes, and proteins denatured by stress. The name of the system reflects the original identification of Hsp70 as a protein induced by heat shock. The Hsp70 and Hsp40 proteins bind individually to its substrates.
- The chaperonin system consists of a large oligomeric assembly. This assembly forms a structure into which unfolded proteins are inserted.

(The connection with the heat shock response stems from the fact that increase in temperature causes production of heat shock proteins whose function is to minimize the damage caused to proteins by heat denaturation; many of the heat shock proteins are chaperones.)

The Hsp70 family is ubiquitous. It is found in bacteria, eukaryotic cytosol, in the endoplasmic reticulum, and in chloroplasts and mitochondria. Hsp70 functions in conjunction with two further components, best characterized in bacteria as DnaJ and GrpE. **Figure 8.6** shows that Hsp40 (DnaJ) binds first to a nascent protein as it emerges from the ribosome. Hsp40 contains a region called the J domain (named for DnaJ), which interacts with Hsp70. Hsp70 (DnaK) binds to both Hsp40 and to the unfolded protein. In effect, two interacting chaperones bind to the protein. The J

Figure 8.6 DnaJ assists the binding of DnaK (Hsp70), which assists the folding of nascent proteins. ATP hydrolysis drives conformational change. GrpE displaces the ADP; this causes the chaperones to be released. Multiple cycles of association and dissociation may occur during the folding of a substrate protein.

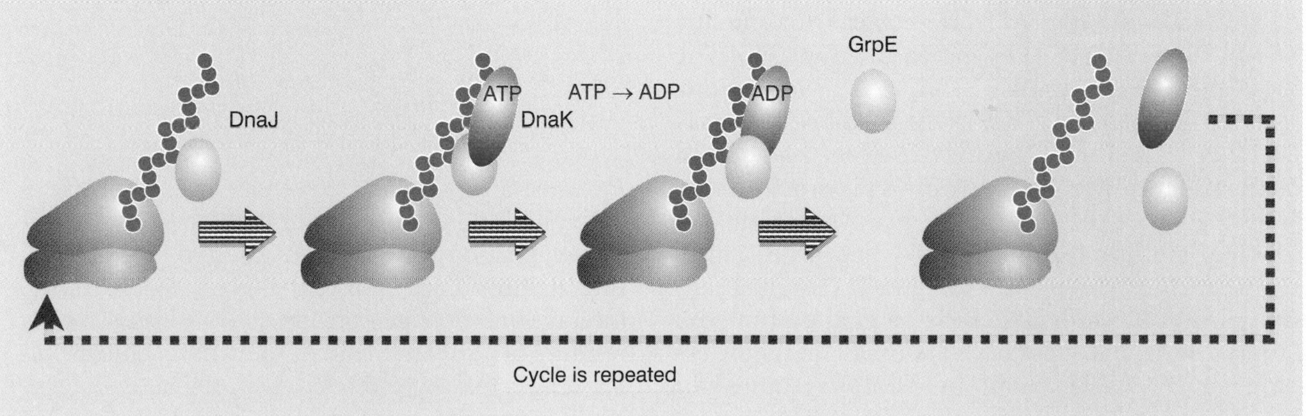

domain accounts for the specificity of the pairwise interaction, and drives a particular Hsp40 to select the appropriate partner from the Hsp70 family.

The interaction of Hsp70 (DnaK) with Hsp40 (DnaJ) stimulates the ATPase activity of Hsp70. The ADP-bound form of the complex remains associated with the protein substrate until GrpE displaces the ADP. This causes dissociation of the complex, possibly by a series of events in which Hsp40 (DnaJ) dissociates, ATP binds to Hsp70 (DnaK), and then Hsp70 dissociates.

Protein folding is accomplished by multiple cycles of association and dissociation. As the protein chain lengthens, Hsp70 (DnaK) may dissociate from one binding site and then reassociate with another, thus releasing parts of the substrate protein to fold correctly in an ordered manner. Finally, the intact protein is released from the ribosome, folded into its mature conformation.

Different members of the Hsp70 class function on appropriate types of target proteins. Cytosolic proteins (the eponymous Hsp70 and a related protein called Hsc70) act on nascent proteins on ribosomes. Variants in the ER (called BiP or Grp78 in higher eukaryotes, called Kar2 in *S. cerevisiae*), or in mitochondria or chloroplasts, function in a rather similar manner on proteins as they emerge into the interior of the organelle on passing through the membrane.

The Hsp60 class of chaperones forms a large apparatus that consists of two types of subunit. **Figure 8.7** illustrates the structure schematically. Hsp60 itself (known as GroEL in *E. coli*) forms a structure consisting of 14 subunits that are arranged in two heptameric rings stacked on top of each other in inverted orientation. This means that the top and bottom surfaces of the double ring are the same. The central hole runs all the way through the double ring, so the complete structure resembles a hollow cylinder.

This structure associates with a heptamer formed of subunits of Hsp10 (GroES in *E. coli*). A single GroES heptamer forms a dome that associates with one surface of the double ring, as shown in **Figure 8.8.** The dome sits over the central cavity, thus capping one opening of the cylinder. The dome is hollow and in effect extends the cavity into the closed surface. We can distinguish the two rings of GroEL as the proximal ring (bound to GroES) or the distal ring (not bound to GroES). The entire GroEL/GroES structure has a mass of ~10^6 daltons, comparable to a small ribosomal subunit. GroEL is sometimes called a chaperonin, and GroES a co-chaperonin, because GroEL plays the essential role in guiding folding but GroES is required for its activity.

Figure 8.7 GroEL forms an oligomer of two rings, each comprising a hollow cylinder made of 7 subunits.

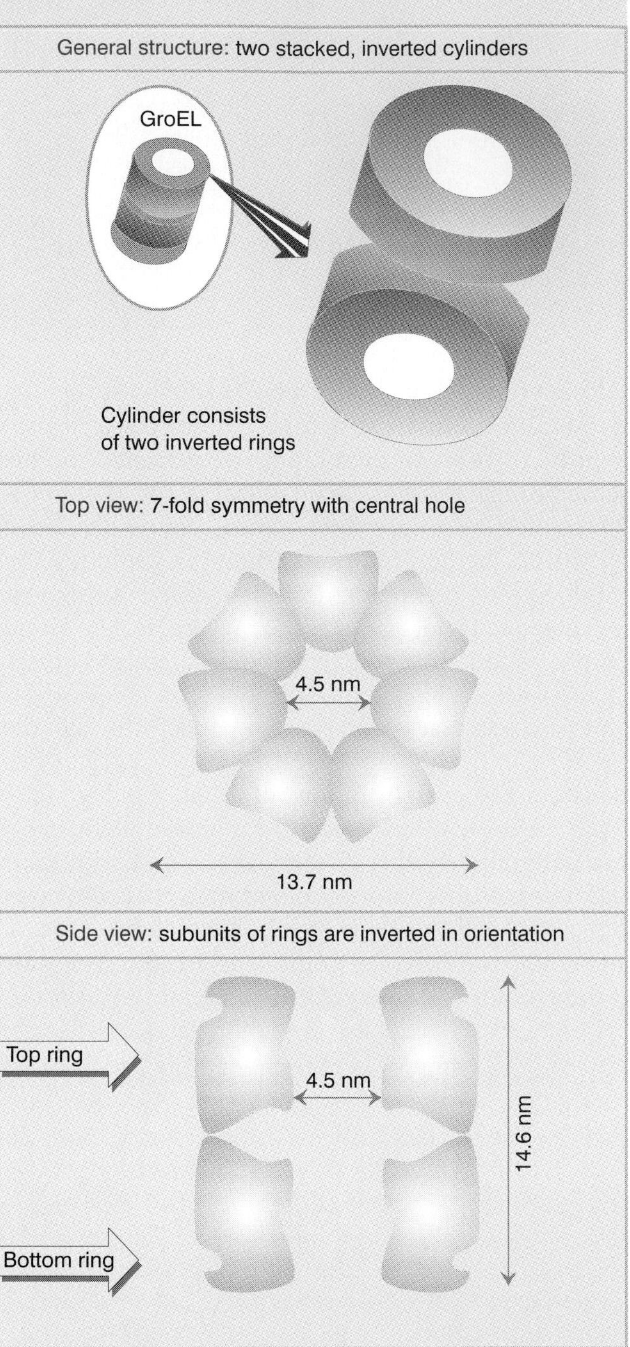

GroEL binds to many unfolded proteins, probably by recognizing a condensed "molten globule" state. Interaction with the substrate is based on hydrophobic interactions between surfaces of the substrate and residues of GroEL that are exposed in its central cavity. Substrates may be provided by proteins that have become denatured; or they may be transferred to GroEL

Figure 8.8 Two rings of GroEL associate back to back to form a hollow cylinder. GroES forms a dome that covers the central cavity on one side. Protein substrates bind to the cavity in the distal ring.

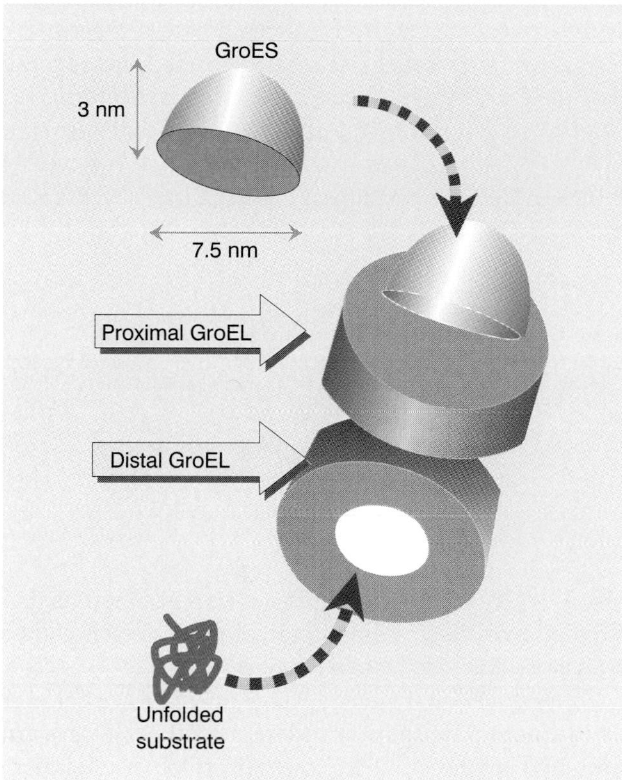

Figure 8.9 Protein folding occurs in the proximal GroEL ring and requires ATP. Release of substrate and GroES requires ATP hydrolysis in the distal ring.

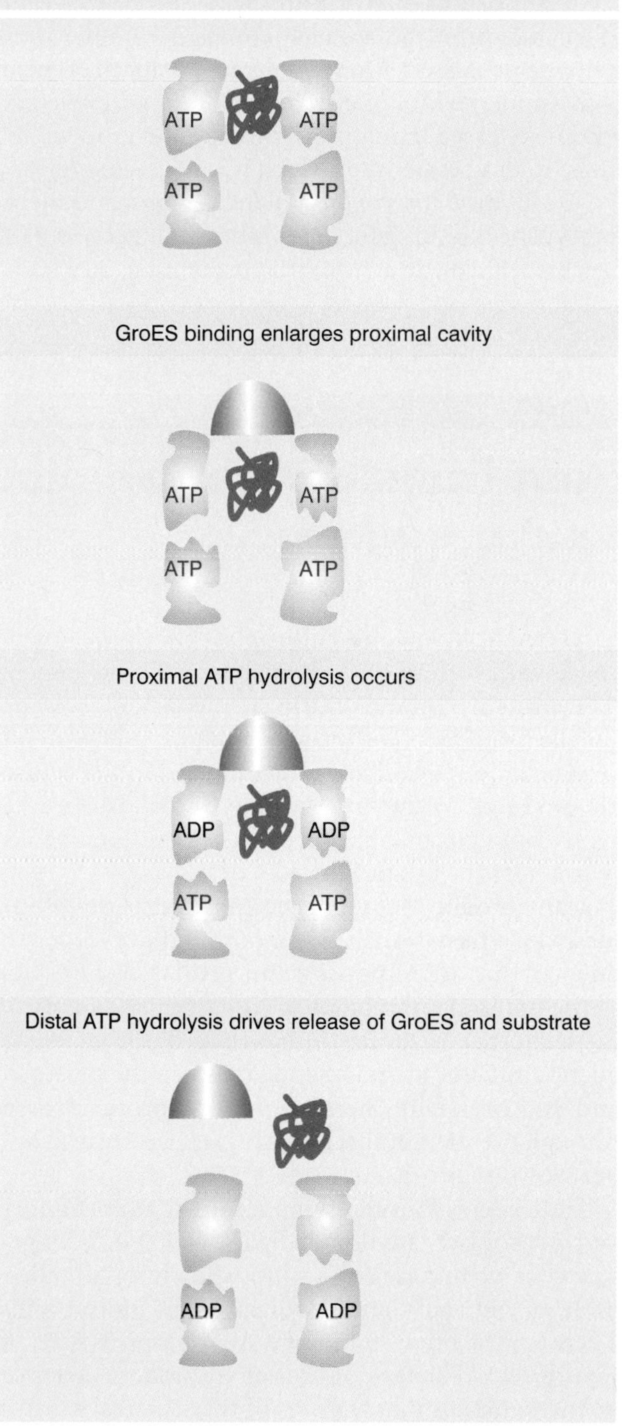

by other chaperones—for example, Hsp70 may assist a nascent protein in folding, but then pass it on to GroEL for the process to be completed when it is released from the ribosome.

The complete order of events involved in substrate binding and folding is not clear, but the key reactions are probably those illustrated in **Figure 8.9.** Folding occurs when substrate and GroES are bound to the same ring. (It is possible that substrate binds initially to a distal ring of a GroEL/GroES complex, and that GroES dissociates and rebinds to the other ring.) When GroES binds to GroEL, it induces a conformational change in the proximal GroEL ring, increasing the volume of the central cavity. This also changes the environment for the substrate. The hydrophobic residues that had previously bound substrate are involved in binding to GroES. The result is that the substrate now finds itself in a hydrophilic environment that forces a change in its conformation.

ATP plays an important role in GroEL function. Each subunit of GroEL has a molecule of ATP. The presence of ATP on the subunits of the proximal ring is required for folding to occur. Hydrolysis is not required at this stage. However, hydrolysis is required for the release of substrate, but the ATP in the proximal rings plays only an indirect role. Hydrolysis of the proximal ATP reduces the affinity of GroEL for

GroES. Hydrolysis of ATP in the distal ring then is required actually to release the GroES and eject the substrate.

An important question in the action of this (and other macromolecular) chaperones is whether their action is processive. Does a substrate enter the central cavity, undergo multiple cycles of folding within it, and become released in mature form? Or does it undergo a single folding cycle, after which it is released; typically it will still have improperly folded regions, and therefore will be bound again for another folding cycle. This process will continue until the protein has reached a mature conformation that does not offer a substrate to the chaperone.

These models have been tested by using a mutant GroEL that can bind unfolded proteins but cannot release them. When this "trap GroEL" is added to wild-type GroEL that is actively engaged with a substrate, it blocks the appearance of mature protein. This suggests that the substrate has been released before folding was completed. The simplest explanation is that substrate protein is released after each folding cycle. One cycle of folding, ATP hydrolysis, and release takes about 15 sec *in vitro*.

Post-translational membrane insertion depends on leader sequences

MITOCHONDRIA and chloroplasts both are able to synthesize all of their nucleic acids and some of their proteins. Mitochondria synthesize only ~10 organelle proteins; chloroplasts synthesize ~50 proteins. Organelle proteins that are synthesized in the cytosol are produced by the same pool of free ribosomes that synthesize cytosolic proteins. They must then be imported into the organelle.

Many proteins that enter mitochondria or chloroplasts by a post-translational process have leader sequences that are responsible for primary recognition of the outer membrane of the organelle. As shown in the simplified diagram of **Figure 8.10**, the leader sequence initiates the interaction between the precursor and the organelle membrane. The protein passes through the membrane, and the leader is cleaved by a protease on the organelle side.

The leaders of proteins imported into mitochondria and chloroplasts are usually hydrophilic, consisting of stretches of uncharged amino acids interrupted by basic amino acids, and they lack acidic amino acids. There is little other homology. An example is given in **Figure 8.11**. The lack of homology among leader sequences implies that features of the secondary or tertiary structure, or the general nature of the region, must be involved in recognition.

The leader sequence contains all the information needed to localize an organelle protein. The ability of a leader sequence can be tested by constructing an artificial protein in which a leader from an organelle protein is joined to a cytosolic protein. The experiment is performed by constructing a hybrid gene, which is then translated into the hybrid protein.

Several leader sequences have been shown by such experiments to function independently to target any attached sequence to the mitochondrion or chloroplast. For example, if the leader sequence given in Figure 8.11 is attached to the cytosolic protein DHFR (dihydrofolate reductase), the DHFR becomes localized in the mitochondrion.

The leader sequence and the transported protein represent domains that fold independently. Irrespective of the sequence to which it is attached, the leader must be able to fold into an appropriate structure to be recognized by receptors on the organelle envelope. The attached polypeptide sequence plays no part in recognition of the envelope.

Hydrolysis of ATP is required both outside and inside for translocation across the membrane. It may be involved with pushing the protein from outside and pulling from inside. In the cases of mitochondrial import and bacterial export, there is also a requirement for an electrochemical potential across the inner membrane to transfer the amino-terminal part of the leader.

Some mitochondrial proteins have internal sequences that are responsible for targeting to the organelle or for controlling localization within it, but less is known about these sequences.

What restrictions are there on transporting a hy-

Figure 8.10 Leader sequences allow proteins to recognize mitochondrial or chloroplast surfaces by a post-translational process.

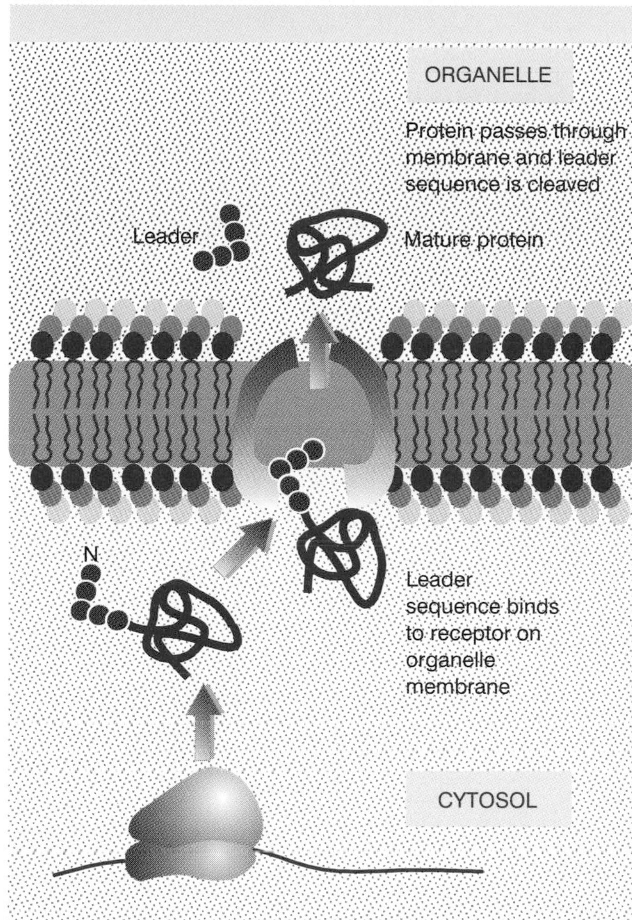

drophilic protein through the hydrophobic membrane? An insight into this question is given by the observation that methotrexate, a ligand for the enzyme DHFR, blocks transport into mitochondria of DHFR fused to a mitochondrial leader. The tight binding of methotrexate prevents the enzyme from unfolding when it is translocated through the membrane. So although the sequence of the transported protein is irrelevant for targeting purposes, in order to follow its leader through the membrane, it requires the flexibility to assume an unfolded conformation.

There are different receptors for transport through each mitochondrial membrane. They are called **TOM** and **TIM,** referring to the outer and inner membranes, respectively. The TOM complex consists of ~9 proteins, many of which are integral membrane proteins. A general model for the complex is shown in **Figure 8.12.** The TOM aggregate has a size of >500 kD, with a diameter of ~138 Å, and forms an ion-conducting channel. A complex contains 2–3 individual rings of diameter 75 Å, each with a pore of diameter 20 Å.

Tom40 is deeply imbedded in the membrane and provides the channel for translocation. It contacts preproteins as they pass through the outer membrane. It binds to three smaller proteins, Tom,5,6,7, which may be components of the channel or assembly factors. There are two subcomplexes that provide surface receptors. Tom20,22 form a subcomplex with exposed domains in the cytosol. Most proteins that are imported into mitochondria are recognized by the Tom20,22 subcomplex, which is the primary receptor and recognizes N-terminal sequences. Tom37,70,71 provides a receptor for a smaller number of proteins that have internal targeting sequences.

When a protein is translocated through the TOM complex, it passes from a state in which it is exposed to the cytosol into a state in which it is exposed to the intermembrane space. However, it is not usually released, but instead is transferred directly to the TIM complex. It is possible to trap intermediates in which the leader is cleaved by the matrix protease, while a major part of the precursor remains exposed on the cytosolic surface of the envelope. This suggests that a protein spans the two membranes during passage. The TOM and TIM complexes do not appear to interact directly (or at least

Figure 8.11 The leader sequence of yeast cytochrome *c* oxidase subunit IV consists of 25 neutral and basic amino acids. The first 12 amino acids are sufficient to transport any attached polypeptide into the mitochondrial matrix.

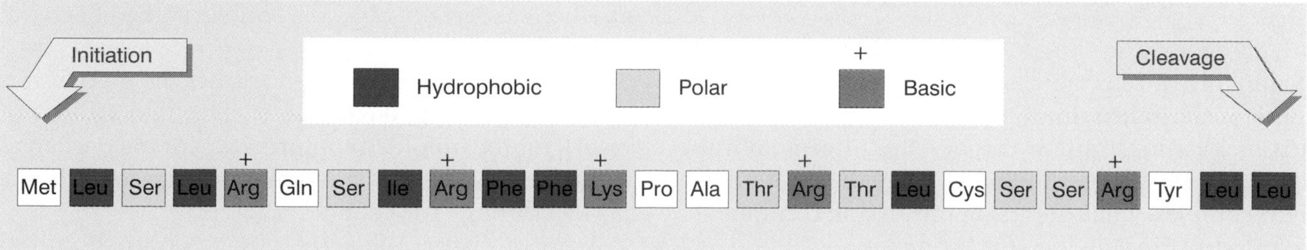

Figure 8.12 TOM proteins form receptor complex(es) that are needed for translocation across the mitochondrial outer membrane.

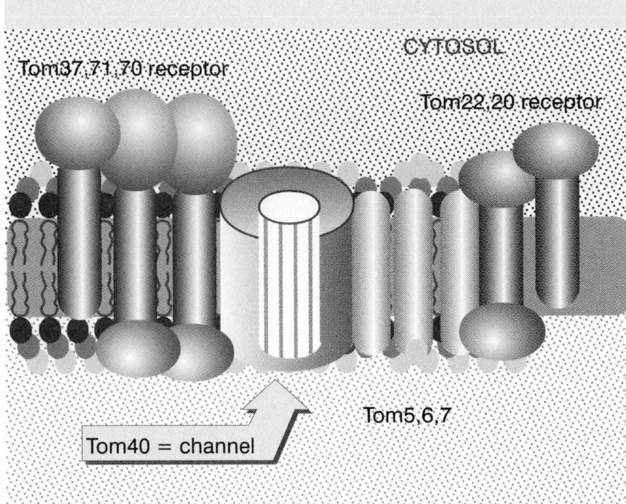

Figure 8.13 Tim proteins form the complex for translocation across the mitochondrial inner membrane.

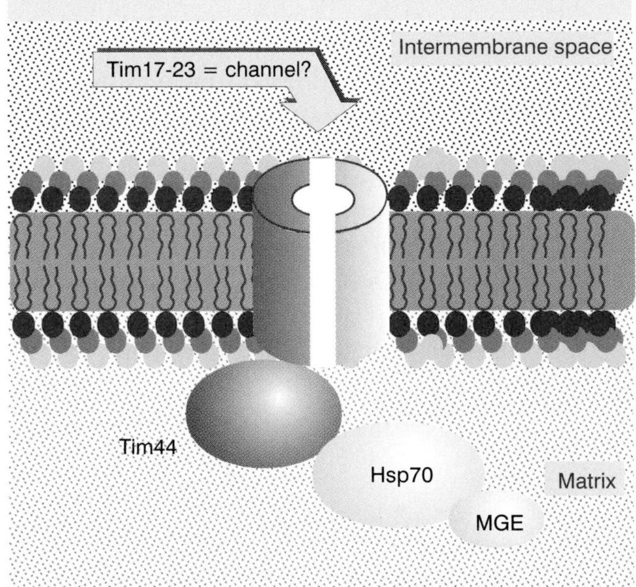

Figure 8.14 A translocating protein may be transferred directly from TOM to Tim22–50.

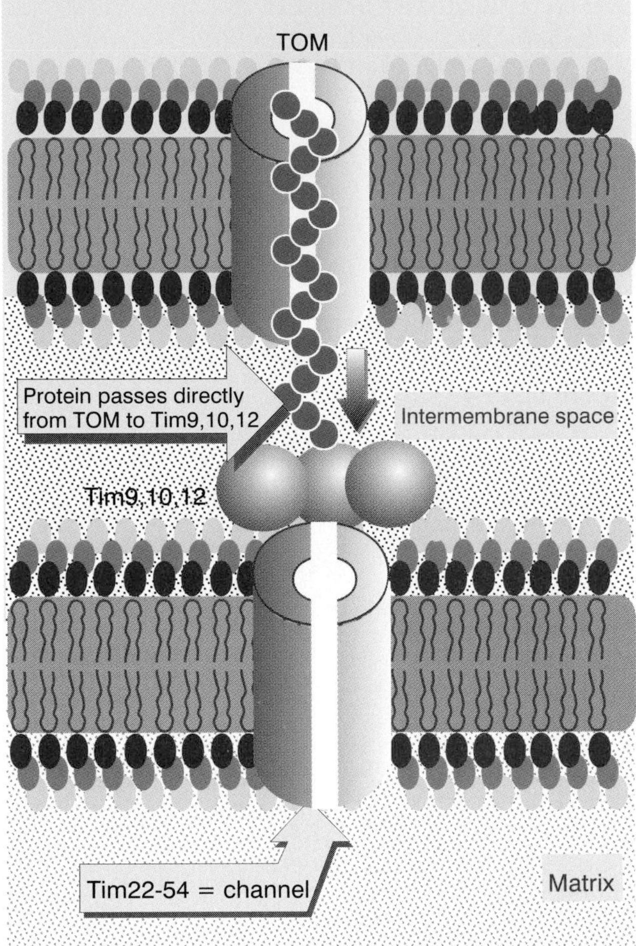

do not form a detectable stable complex), and they may therefore be linked simply by a protein in transit. When a translocating protein reaches the intermembrane space, the exposed residues may immediately bind to a TIM complex, while the rest of the protein continues to translocate through the TOM complex.

There are two TIM complexes in the inner membrane.

The Tim17-23 complex translocates proteins to the lumen. Substrates are recognized by their possession of a positively charged N-terminal signal. Tim17-23 are transmembrane proteins that comprise the channel. **Figure 8.13** shows that they are associated with Tim44 on the matrix side of the membrane. Tim 44 in turn binds the chaperone Hsp70. This is also associated with another chaperone, Mge, the counterpart to bacterial GrpE.

When the imported protein reaches the matrix, it is bound by the Hsp70 chaperone. The high affinity of Hsp70 for the unfolded conformation of the protein as it emerges from the inner membrane helps to "pull" the protein through the channel.

A major chaperone activity in the mitochondrial

matrix is provided by Hsp60 (which forms the same sort of structure as its counterpart, GroEL). Association with Hsp60 is necessary for joining of the subunits of imported proteins that form oligomeric complexes. An imported protein may be "passed on" from Hsp70 to Hsp60 in the process of acquiring its proper conformation.

The Tim22-54 complex translocates other proteins that reside in the inner membrane. The Tim22-54 complex is associated with a complex of Tim9,1,0,12, which is located in the intermembrane space. **Figure 8.14** shows that a translocating protein may pass directly from the TOM channel to the Tim9,10,12 complex, and then into the Tim 22-54 channel.

A mitochondrial protein folds under different conditions before and after its passage through the membrane. Ionic conditions and the chaperones that are present are different in the cytosol and in the mitochondrial matrix. It is possible that different conformations are achieved, that is, that a mitochondrial protein can attain its mature conformation *only* in the mitochondrion.

A hierarchy of sequences determines location within organelles

THE mitochondrion is surrounded by an envelope consisting of two membranes. Proteins imported into mitochondria may be located in the outer membrane, the intermembrane space, the inner membrane, or the matrix. A protein that is a component of one of the membranes may be oriented so that it faces one side or the other.

What is responsible for directing a mitochondrial protein to the appropriate compartment? The "default" pathway for a protein imported into a mitochondrion is to move through both membranes into the matrix. This property is conferred by the N-terminal part of the leader sequence. A protein that is localized within the intermembrane space or in the inner membrane itself requires an additional signal, which specifies its destination within the organelle. A multipart leader contains signals that function in a hierarchical manner, as summarized in **Figure 8.15**. The first part of the leader targets the protein to the organelle, and the second part is required if its destination is elsewhere than the matrix. The two parts of the leader are removed by successive cleavages.

Cytochrome *c1* is an example. It is bound to the inner membrane and faces the intermembrane space. Its leader sequence consists of 61 amino acids, and can be divided into regions with different functions. The sequence of the first 32 amino acids alone, or even the N-terminal half of this region, can transport DHFR all the way into the matrix. *So the first part of the leader sequence (32 N-terminal amino acids) comprises a matrix-targeting signal.* But the intact leader transports an attached sequence—such as murine DHFR—into the intermembrane space.

What prevents the protein from proceeding past the intermembrane space when it has an intact leader? *The region following the matrix-targeting signal (comprising 19 amino acids of the leader) provides another signal that localizes the protein at the inner membrane or within the intermembrane space.* For working purposes, we call this the *membrane-targeting signal.*

Cleavage of the matrix-targeting signal is the sole processing event required for proteins that reside in the matrix. This signal must also be cleaved from proteins that reside in the intermembrane space; but following this cleavage, the membrane-targeting signal (which is now the N-terminal sequence of the protein) directs the protein to its destination in the outer membrane, intermembrane space, or inner membrane. Then it in turn is cleaved.

The N-terminal matrix-targeting signal functions in the same manner for all mitochondrial proteins. Its recognition by a receptor on the outer membrane leads to transport through the two membranes. And the same protease is involved in cleaving the matrix-targeting signal, irrespective of the final destination of the protein. This protease is a water-soluble, Mg^{2+}-dependent enzyme that is located in the matrix. *So the N-terminal sequence must reach the matrix, even if the*

Figure 8.15 Mitochondria have receptors for protein transport in the outer and inner membranes. Recognition at the outer membrane may lead to transport through both receptors into the matrix, where the leader is cleaved. If it has a membrane-targeting signal, it may be reexported.

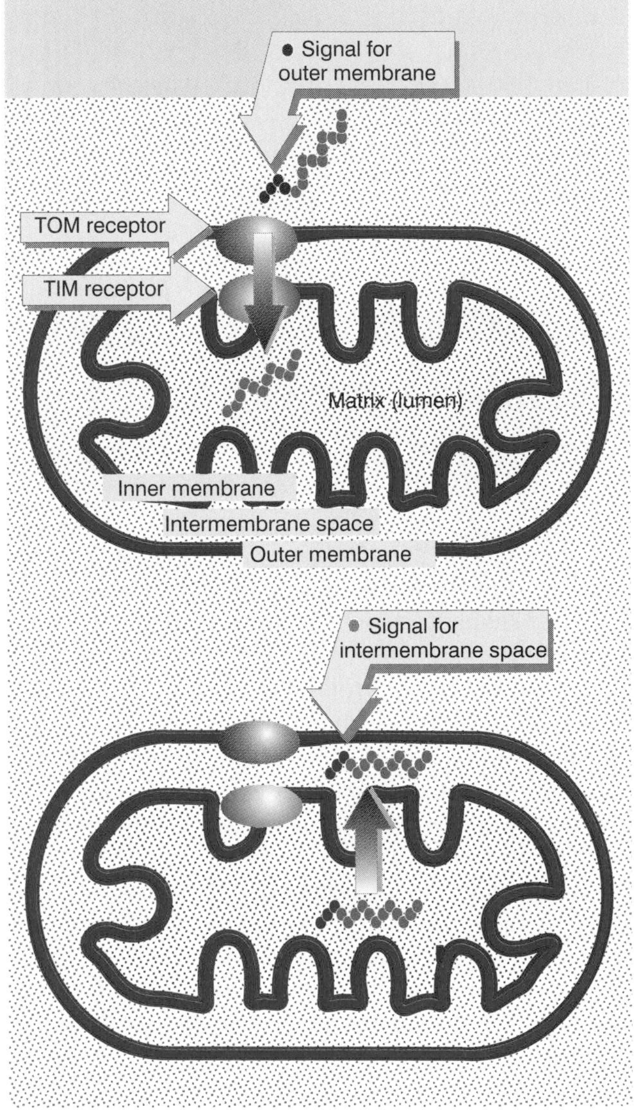

protein ultimately will reside in the intermembrane space.

Residence in the matrix occurs in the absence of any other signal. If there is a membrane-targeting signal, however, it is activated by cleavage of the matrix-targeting signal. Then the remaining part of the leader (which is now N-terminal) causes the protein to take up its final destination.

The nature of the membrane-targeting signal is controversial. One model holds that the entire protein enters the matrix, after which the membrane-targeting signal causes it to be re-exported into or through the inner membrane. An alternative model proposes that the membrane-targeting sequence simply prevents the rest of the protein from following the leader through the inner membrane into the matrix. Whichever model applies, another protease (located within the intermembrane space) completes the removal of leader sequences.

The two parts of a leader that contains both types of signal have different compositions. As indicated in **Figure 8.16,** the 35 N-terminal amino acids resemble other organelle leader sequences in the high content of uncharged amino acids, punctuated by basic amino acids. The next 19 amino acids, however, comprise an uninterrupted stretch of uncharged amino acids, long enough to span a lipid bilayer. This sequence resembles the sequences that are involved in protein translocation into membranes of the endoplasmic reticulum (see later).

Passage through chloroplast membranes is achieved in a similar manner. **Figure 8.17** illustrates the variety of locations for chloroplast proteins. They pass the outer and inner membranes of the envelope into the stroma, a process involving the same types of passage as into the mitochondrial matrix. But some proteins are transported yet further, across the stacks of the thylakoid membrane into the lumen. Proteins destined for the thylakoid membrane or lumen must cross the stroma *en route.*

Chloroplast targeting signals resemble mitochondrial targeting signals. The leader consists of ~50 amino acids, and the N-terminal half is needed to recognize the chloroplast envelope. A cleavage between positions 20–25 occurs during or following passage across the envelope, and proteins destined for the thylakoid membrane or lumen have a new N-terminal leader that guides recognition of the thylakoid membrane. There are several (at least four) different systems in the chloroplast that catalyze import of proteins into the thylakoid membrane.

The general principle governing protein transport into mitochondria and chloroplasts therefore is that *the N-terminal part of the leader targets a protein to the organelle matrix, and an additional sequence (within the leader) is needed to localize the protein at the outer membrane, intermembrane space, or inner membrane.*

Another example of targeting by means of specific amino acid sequences utilizes C-terminal regions. Peroxisomes are small bodies enclosed by a single membrane. They contain enzymes concerned with oxygen utilization. They convert oxygen to hydrogen peroxide by removing hydrogen atoms from sub-

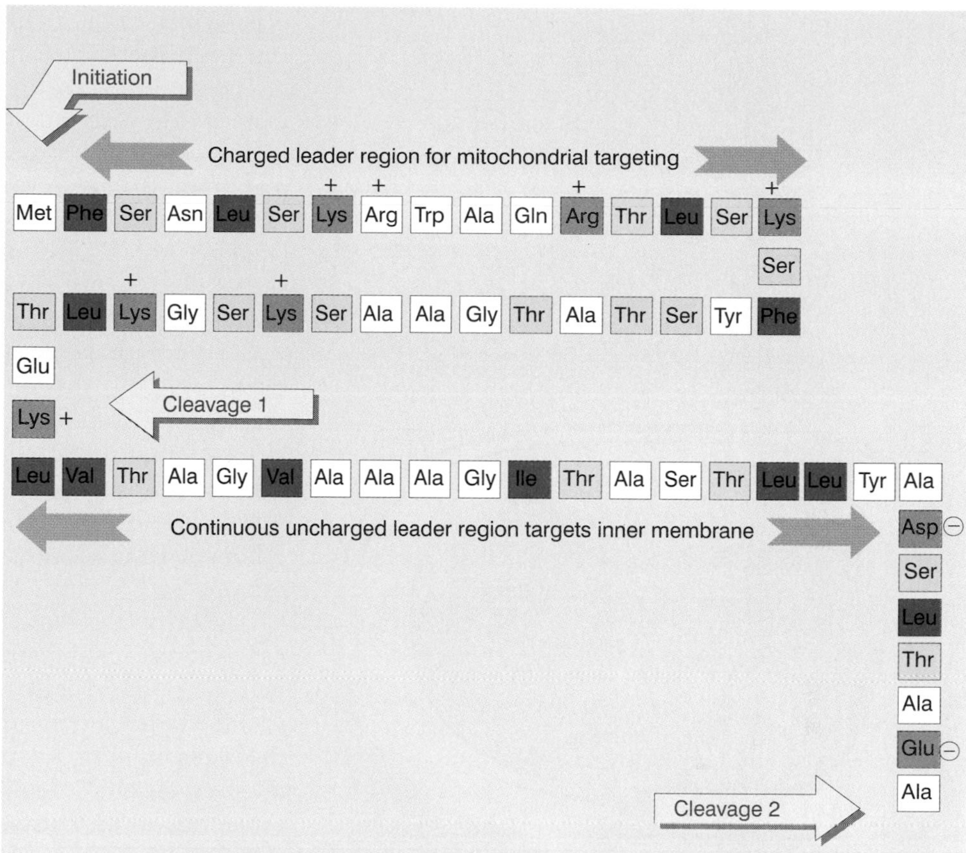

Figure 8.16 The leader of yeast cytochrome *c*1 contains an N-terminal region that targets the protein to the mitochondrion, followed by a region that targets the (cleaved) protein to the inner membrane. The leader is removed by two cleavage events, the first at an unidentified site between the two regions, the second at the start of the mature protein.

strates. Catalase then uses the hydrogen peroxide to oxidize a variety of other substrates. All of the enzymes in the peroxisome are imported from the cytosol. Like transport into the nucleus, transport into peroxisomes occurs post-translationally by means of a short sequence. Several peroxisomal enzymes have the C-terminal sequence SKL (Ser-Lys-Leu); and the addition of this tripeptide to the C-terminus of cytosolic proteins is sufficient to ensure their import into the organelle. This very short sequence therefore constitutes one means of entry; there are others.

The translocation apparatus interacts with signal sequences

PROTEINS that reside within the endoplasmic reticulum, Golgi apparatus, or plasma membrane, or that are secreted from the cell, have a common starting point for their association with membranes. The ribosomes synthesizing these proteins become associated with the endoplasmic reticulum so that the nascent protein can be co-translationally transferred to the membrane. Regions in which ribosomes are associated with the ER are sometimes called the "rough ER", in contrast with the "smooth ER" regions that lack associated polysomes and which have a tubular rather than sheet-like appearance. **Figure 8.18** shows ribosomes in the act of transferring nascent proteins to ER membranes.

The proteins synthesized at the rough endoplasmic reticulum pass from the ribosome directly to the

Figure 8.17 A protein approaches the chloroplast from the cytosol with a ~50 residue leader. The N-terminal half of the leader sponsors passage into the envelope or through it into the stroma. Cleavage occurs during envelope transit. Then the remaining part of the leader functions to direct proteins that must cross the thylakoid membrane.

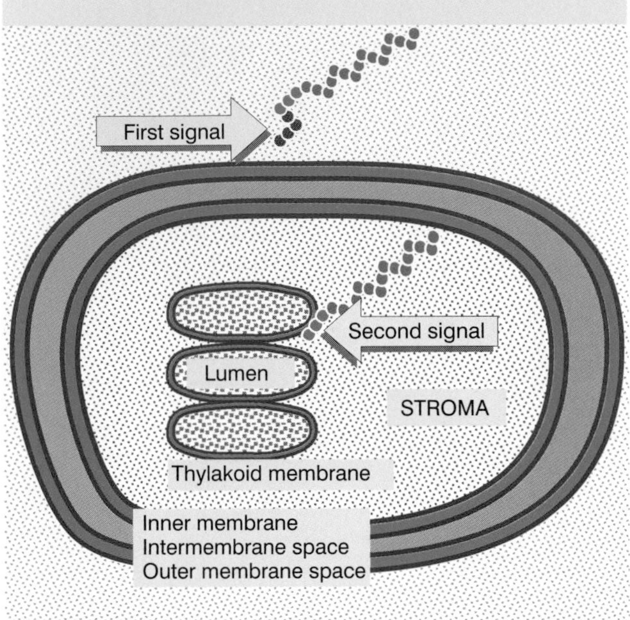

Figure 8.18 The endoplasmic reticulum consists of a highly folded sheet of membranes that extends from the nucleus. The small objects attached to the outer surface of the membranes are ribosomes. Photograph kindly provided by Lelio Orci.

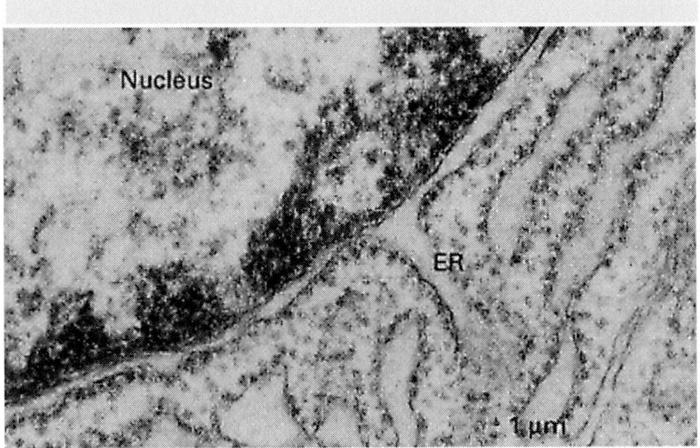

membrane. From the endoplasmic reticulum membranes, the proteins are transferred to the Golgi apparatus, and then are directed to their ultimate destination, such as the lysosome or secretory vesicle or plasma membrane. (This protein trafficking is the subject of Chapter 25.)

The model for co-translational insertion originated from work with mammalian microsomal systems (which contain ribosomes and endoplasmic reticulum membranes). These systems can package *nascent* proteins into membranes; but they do not work when isolated preproteins are added post-translationally. This shows that *the protein must associate with the membrane while it is being synthesized.*

Co-translational insertion is directed by a **signal sequence.** Usually this is a cleavable leader sequence of the 15–30 N-terminal amino acids. At or close to the N-terminus are several polar residues, and within the leader is a hydrophobic core consisting exclusively or very largely of hydrophobic amino acids. There is no other conservation of sequence. **Figure 8.19** gives an example.

Like the leader sequence of proteins targeted to organelles, the signal sequence is both necessary and sufficient to sponsor transfer of any attached polypeptide into the target membrane. A signal sequence added to the N-terminus of a globin protein, for example, causes it to be secreted through cellular membranes instead of remaining in the cytosol. However, the action of the signal sequence in co-translational transfer is different from that of the leader in post-translational transfer.

The signal sequence provides the means by which ribosomes translating the mRNA can attach to the membrane. Responsibility for the initial membrane attachment rests solely with the signal sequence; the ribosome attaches by virtue of its synthesis of the secreted protein. *So there is no intrinsic difference between ribosomes in the free fraction and ribosomes in the membrane-bound fraction.* Protein translocation can be divided into two general stages: first ribosomes carrying nascent polypeptides associate with the membranes; and then the nascent chain is transferred to the channel and translocates through it.

Salt-washed membranes cannot sponsor ribosomal attachment; but this activity can be recovered by adding back the salt wash. The active component of the salt wash is called the **signal recognition particle (SRP).** It is an 11S ribonucleoprotein complex, containing 6 proteins (total mass 240 kD) and a small (305 base, 100 kD) 7S RNA (see Figure 8.21). The 7S RNA provides the structural backbone of the particle; the individual proteins do not assemble in its absence.

The SRP has two important abilities:

- It can bind to the signal sequence of nascent secretory proteins.

Figure 8.19 The signal sequence of bovine growth hormone consists of the N-terminal 29 amino acids and has a central highly hydrophobic region, preceded or flanked by regions containing polar amino acids.

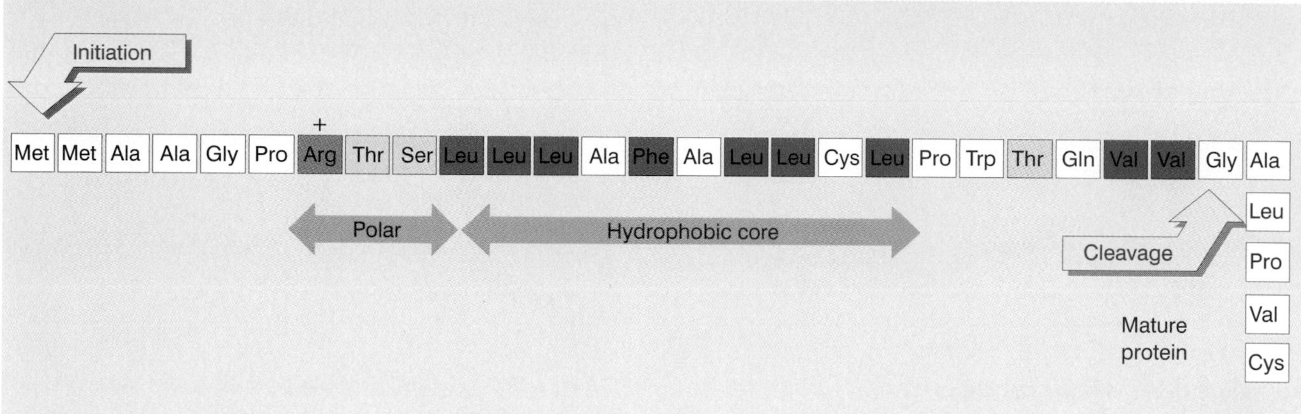

■ And it can bind to a receptor protein located in the membrane.

The SRP and SRP receptor function catalytically to transfer a ribosome carrying a nascent protein to the membrane. The first step is the recognition of the signal sequence by the SRP. Then the SRP binds to the SRP receptor and the ribosome binds to the membrane. The stages of translation of membrane proteins are summarized in **Figure 8.20.**

The role of the SRP receptor in protein translocation is transient. When the SRP binds to the signal sequence, it arrests translation. This usually happens when ~70 amino acids have been incorporated into the polypeptide chain (at this point the 25-residue leader has become exposed, with the next ~40 amino acids still buried in the ribosome).

Then when the SRP binds to the SRP receptor, the SRP releases the signal sequence. The ribosome becomes bound by another component of the membrane. At this point, translation can resume. When the ribosome has been passed on to the membrane, the role of SRP and SRP receptor has been played. They now recycle, and are free to sponsor the association of another nascent polypeptide with the membrane.

One aspect of this process may be involved with controlling the conformation of the protein. If the nascent protein were released into the cytoplasm, it could take up a conformation determined by the aqueous environment; in this conformation, it might be unable to traverse the membrane. The ability of the SRP to inhibit translation while the ribosome is being handed over to the membrane is therefore important in preventing the protein from being released into the aqueous environment.

The 7S RNA of the SRP particle is divided into two parts. The 100 bases at the 5′ end and 45 bases at the 3′ end are closely related to the sequence of Alu RNA, a common mammalian small RNA. They therefore define the **Alu domain.** The remaining part of the RNA comprises the **S domain.**

Different parts of the SRP structure depicted in **Figure 8.21** have separate functions in protein targeting. SRP54 can be crosslinked to the signal sequence of a nascent protein; it is directly responsible for recognition of the substrate protein. SRP54 is a GTPase, and the cleavage of GTP is probably needed to insert the signal sequence into the channel. The SRP68-SRP72 dimer binds to the central region of the RNA; it is needed for recognizing the SRP receptor. The SRP9-SRP14 dimer binds at the other end of the molecule; it is responsible for elongation arrest.

The SRP receptor is a dimer containing subunits SRα (72 kD) and SRβ (30 kD). The amino-terminal end of the large subunit is anchored in the endoplasmic reticulum. The bulk of the protein protrudes into the cytosol. A large part of the sequence of the cytoplasmic region of the protein resembles a nucleic-acid-binding protein, with many positive residues. This suggests the possibility that the SRP receptor recognizes the 7S RNA in the SRP. Both subunits of the SRP receptor have GTP-binding motifs; GTP hydrolysis is required to release the SRP from its receptor.

Two approaches have identified the protein components of the ER membrane that are required for protein translocation. They form a channel through the membrane called the **translocon.** Resident ER membrane proteins that are crosslinked to translocating proteins are potential subunits of the channel. And *sec* mutants in yeast (named because they fail to secrete proteins) include a class that cause precursors of secreted or

Figure 8.20 Ribosomes synthesizing secretory proteins are attached to the membrane via the signal sequence on the nascent polypeptide.

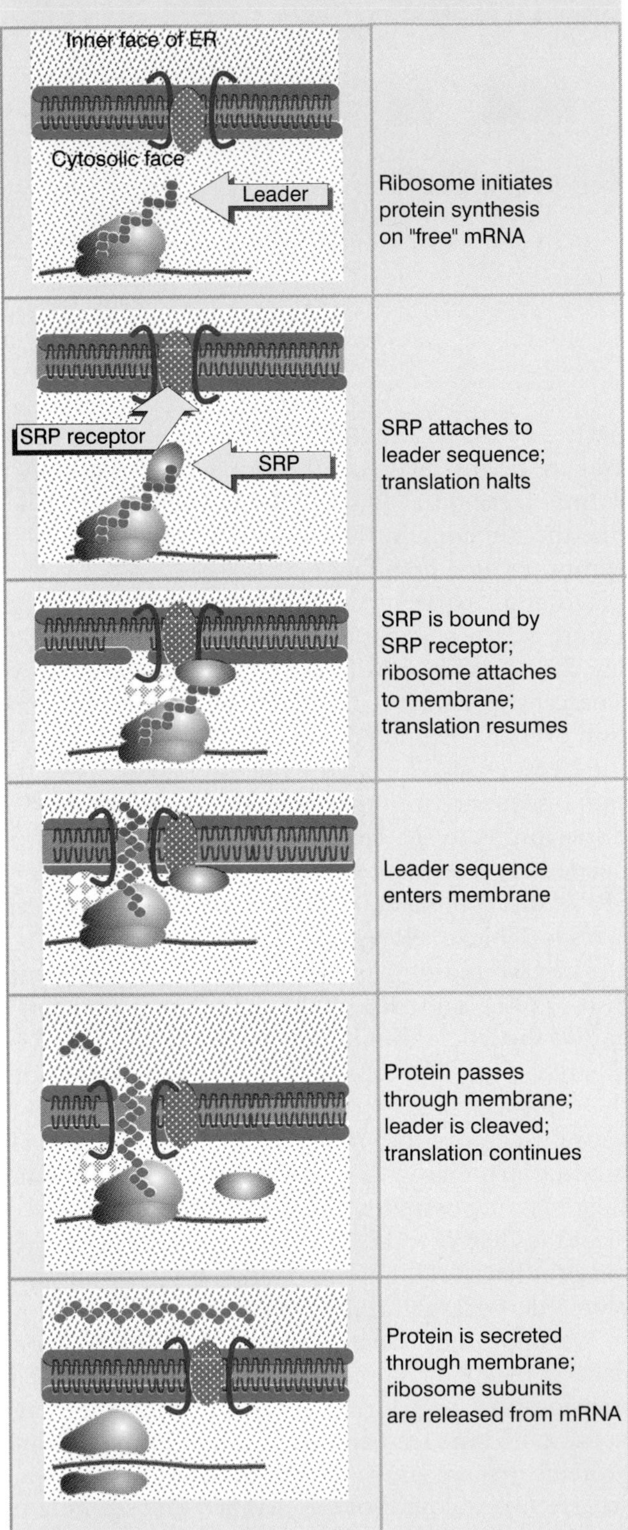
Inner face of ER
Cytosolic face
Leader
Ribosome initiates protein synthesis on "free" mRNA
SRP receptor
SRP
SRP attaches to leader sequence; translation halts
SRP is bound by SRP receptor; ribosome attaches to membrane; translation resumes
Leader sequence enters membrane
Protein passes through membrane; leader is cleaved; translation continues
Protein is secreted through membrane; ribosome subunits are released from mRNA

Figure 8.21 The two domains of the 7S RNA of the SRP are defined by its relationship to the Alu sequence. Five of the six proteins bind directly to the 7S RNA. Each function of the SRP is associated with a particular protein(s).

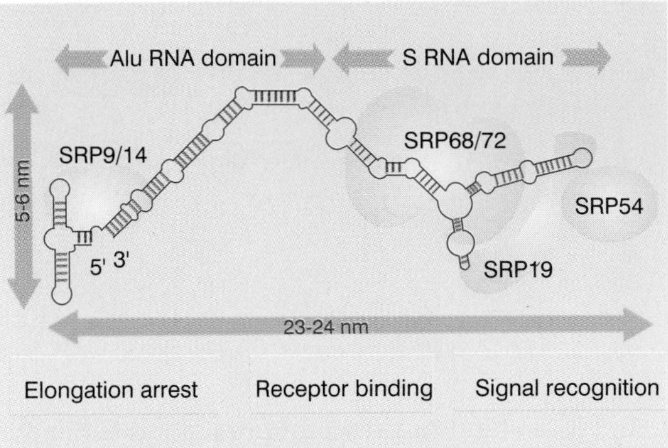
Alu RNA domain
S RNA domain
SRP9/14
SRP68/72
SRP54
SRP19
5-6 nm
5' 3'
23-24 nm
Elongation arrest
Receptor binding
Signal recognition

membrane proteins to accumulate in the cytosol. These approaches together identified the Sec61 complex, which consists of three transmembrane proteins: Sec61α,β, γ. When this complex is incorporated into artificial membranes together with the SRP receptor, it can support translocation of some nascent proteins. Other nascent proteins require the presence of an additional component, TRAM, which is a major protein that becomes crosslinked to a translocating nascent chain. TRAM stimulates the translocation of all proteins.

Sec61 is the major component of the translocon. In detergent (which provides a hydrophobic milieu that mimics the effect of a surrounding membrane), Sec61 forms cylindrical oligomers with a diameter of ~85 Å and a central pore of ~20 Å. Each oligomer consists of 3–4 heterotrimers.

The components of the translocon and their functions are summarized in **Figure 8.22.** The simplicity of this system makes several important points. We visualize Sec61 as forming the channel and also as interacting with the ribosome. When the signal sequence enters the translocon, the ribosome attaches to Sec61, forming a seal so that the pore is not exposed to the cytosol. Cleavage of the signal peptide does not occur in this system, and therefore cannot be necessary for translocation *per se*. In this system, components on the lumenal side of the membrane are not needed for translocation. Of course, the efficiency of the *in vitro* system is relatively low. Additional components could be required *in vivo* to achieve efficient transfer or to

Figure 8.22 The translocon consists of SRP, SRP receptor, Sec61, TRAM, and signal peptidase.

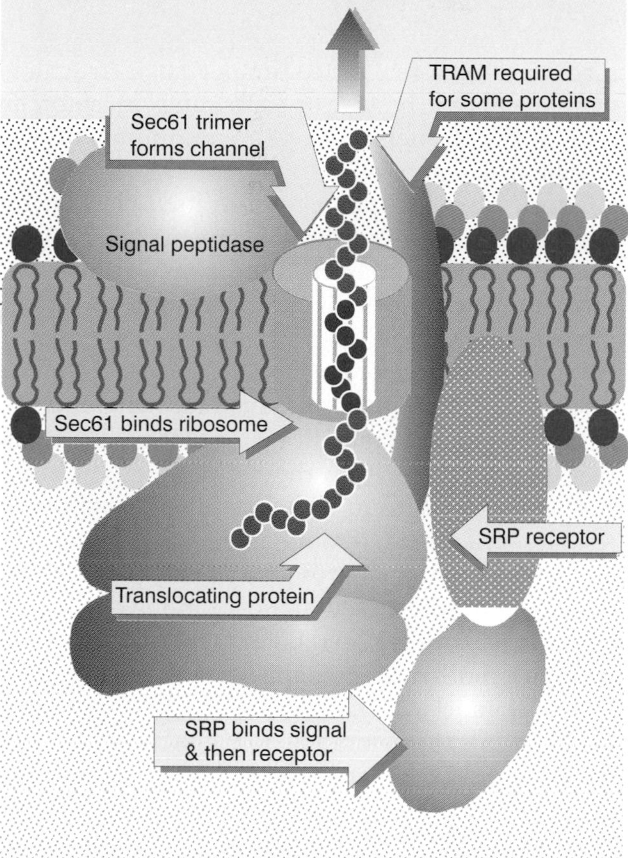
TRAM required
for some proteins
Sec61 trimer
forms channel
Signal peptidase
Sec61 binds ribosome
SRP receptor
Translocating protein
SRP binds signal
& then receptor

prevent other cellular proteins from interfering with the process.

A more complex apparatus is required in certain cases in which a protein is inserted into a membrane post-translationally. The same Sec61 complex forms the channel, but four other Sec proteins are also required, and in addition the chaperone BiP (a member of the Hsp70 class) and a supply of ATP are required on the lumenal side of the membrane. The role of BiP could be to take hold of the translocating protein and to prevent a reversal of translocation.

The signal peptide is cleaved by a complex of 5 proteins called the signal peptidase. The complex is several times more abundant than the SRP and SRP receptor. Its amount is equivalent roughly to the amount of bound ribosomes, suggesting that it functions in a structural capacity. It is located on the lumenal face of the ER membrane, which implies that the entire signal sequence must cross the membrane before the cleavage event occurs.

Several important activities occur within the endoplasmic reticulum. Proteins move through the ER *en route* to a variety of destinations (see Chapter 25). They are glycosylated and folded into their final conformations. The ER provides a "quality control" system in which misfolded proteins are identified and degraded. However, the degradation itself does not necessarily occur in the ER, but may require the protein to be exported back to the cytosol.

The first indication for cytosolic involvement in the degradation of ER proteins was provided by evidence for the involvement of the proteasome, a large protein aggregate with several proteolytic activities. Inhibitors of the proteasome prevent the degradation of aberrant ER proteins. Proteins are marked for cleavage by the proteasome when they are modified by the addition of ubiquitin, a small polypeptide chain. Ubiquitination and proteasomal activities are discussed later in this chapter; the important point to note now is that both are found in the cytosol (with a minor proportion in the nucleus).

Transport from the ER back into the cytosol occurs by a reversal of the usual process of import. The Sec61 translocon is used. The conditions are different; for example, the translocon is not associated with a ribosome. We do not know how the channel is opened to allow insertion of the protein on the ER side. Special components are presumably involved. In one particular case, human cytomegalovirus (CMV) codes for cytosolic proteins that destroy newly synthesized MHC class I (cellular major histocompatibility complex) proteins. This requires a viral protein product (US2), which is a membrane protein that functions in the ER. It interacts with the MHC proteins and probably conveys them into the translocon for reverse translocation.

The system involved in the degradation of aberrant ER proteins can be identified by mutations (in yeast) that lead to accumulation of aberrant proteins. Usually a protein that misfolds (produced by a mutated gene) is degraded instead of being transported through the ER. Yeast mutants that cannot degrade the substrate fall into two classes: some identify components of the proteolytic apparatus, such as the enzymes involved in ubiquitination; other identify components of the transport apparatus, including Sec61, BiP, and Sec63. There is also a protein in the ER membrane that functions on the cytosolic side to localize the ubiquitination enzymes at the translocon. In fact, retrograde transport into the cytosol cannot occur in the absence of this protein, which suggests that there is a mechanical link between retrograde transport and degradation.

How do proteins enter and leave membranes?

THE association of a protein with a membrane takes several forms. Proteins that are secreted from the cell must pass through a membrane, so we have to account for their ability to initiate entry into the lipid bilayer and then to pass across it. Proteins that reside in membranes may start the process in the same way, but then transfer from soluble phase to the hydrophobic environment.

The operational definition of an **integral membrane protein** is that it requires disruption of the lipid bilayer in order to be released from the membrane. A common feature in such proteins is the presence of at least one **transmembrane region,** consisting of an α-helical stretch of 21–26 hydrophobic amino acids. A sequence that fits the criteria for membrane insertion can be identified by a hydropathy plot, which measures the cumulative hydrophobicity of a stretch of amino acids. A protein that has domains exposed on both sides of the membrane is called a **transmembrane protein.**

Various forms of organization for membrane proteins are summarized in **Figure 8.23.** An important feature is the number of transmembrane regions. When there is a single transmembrane region, its position determines how much of the protein is exposed on either side of the membrane. A protein may have extensive domains exposed on both sides of the membrane or may have a site of insertion close to one end, so that little or no material is exposed on one side. The length of the N-terminal or C-terminal tail that protrudes from the membrane near the site of insertion varies from insignificant to quite bulky.

Proteins with a single transmembrane domain fall into two classes. Group I proteins in which the N-terminus faces the extracellular space are more common than group II proteins in which the orientation has been reversed so that the N-terminus faces the cytoplasm. Orientation is determined during the insertion of the protein into the endoplasmic reticulum.

When there are multiple membrane-spanning domains, an odd number means that both termini of the protein are on opposite sides of the membrane, whereas an even number implies that the termini are on the same face. The extent of the domains exposed on one or both sides is determined by the locations of the transmembrane domains. Domains at either terminus may be exposed, and internal sequences between the domains "loop out" into the extracellular space or cytoplasm. One common type of structure is the 7-membrane passage or "serpentine" receptor; another is the 12-membrane passage component of an ion channel.

Does a transmembrane domain itself play any role in protein function besides allowing the protein to insert into the lipid bilayer? In the simple group I or II proteins, it has little or no additional function; often it can be replaced by any other transmembrane domain. However, transmembrane domains play an important role in the function of proteins that make multiple passes through the membrane or that have subunits that oligomerize within the membrane. The transmembrane domains in such cases often contain polar residues, which are not found in the single membrane-spanning domains of group I and group II proteins. Polar regions in the membrane-spanning domains do not interact with the lipid bilayer, but instead interact with one another. This enables them to form a polar pore or channel within the lipid bilayer. Interaction between such transmembrane domains can create a hydrophilic passage through the hydrophobic interior of the membrane. This can allow highly charged ions or molecules to pass through the membrane, and is important for the function of ion channels and transport of ligands. Another case in which conformation of the transmembrane domains is important is provided by certain receptors that bind lipophilic ligands. In such cases, the transmembrane domains (rather than the extracellular domains) bind the ligand within the plane of the membrane.

We have a reasonable understanding of the processes by which secreted proteins pass through membranes and of how this relates to the insertion of the single-membrane- spanning group I and group II proteins. We cannot yet explain the details of insertion of proteins with multiple membrane-spanning domains.

Two models for the entry of a protein into a membrane are contrasted in **Figure 8.24.** The hydrophobic nature of the signal sequence was originally taken to suggest that it might insert directly into the lipid bilayer, as shown in the upper part of the figure. This idea was supported by the observation that short hydrophobic peptides can associate directly with membranes. According to this model, the signal sequence would break the barrier of the membrane, and transfer

Figure 8.23 The topography of membrane proteins depends on the number and arrangement of transmembrane regions.

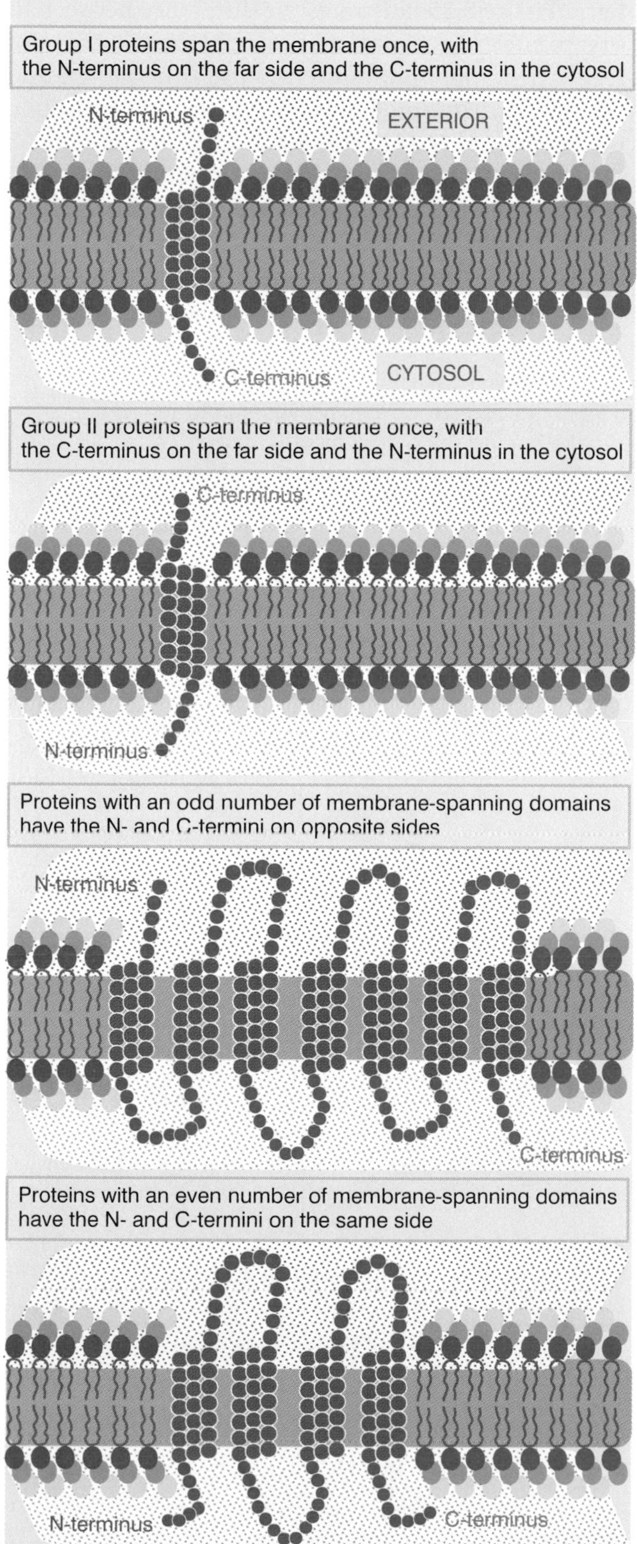
Group I proteins span the membrane once, with the N-terminus on the far side and the C-terminus in the cytosol
N-terminus
EXTERIOR
C-terminus
CYTOSOL
Group II proteins span the membrane once, with the C-terminus on the far side and the N-terminus in the cytosol
C-terminus
N-terminus
Proteins with an odd number of membrane-spanning domains have the N- and C-termini on opposite sides
N-terminus
C-terminus
Proteins with an even number of membrane-spanning domains have the N- and C-termini on the same side
N-terminus
C-terminus

Figure 8.24 Does a signal sequence interact directly with the hydrophobic environment of the lipid bilayer or does it directly enter an aqueous tunnel created by resident ER membrane proteins?

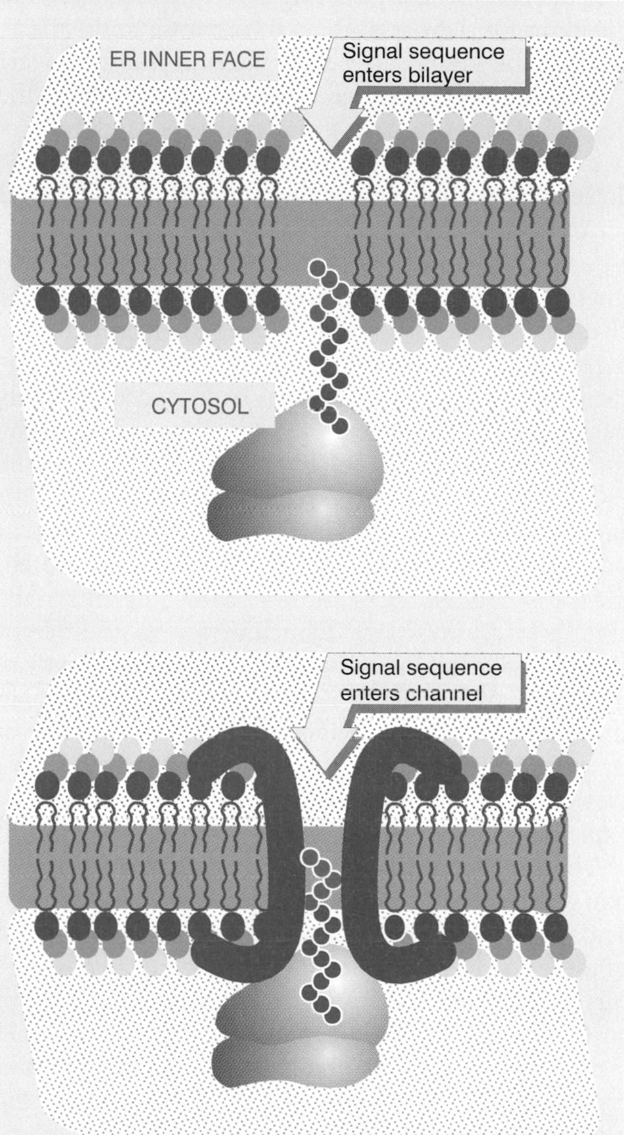
ER INNER FACE
Signal sequence enters bilayer
CYTOSOL
Signal sequence enters channel

would continue until the entire polypeptide had been translocated.

This model leaves unanswered the problem of how the rest of the protein, which is largely hydrophilic, would be able to pass through the hydrophobic membranous environment. A protein in the process of translocation across the ER membrane can be extracted by denaturants that are effective in an aqueous environment. The same denaturants do not extract proteins that are resident components of the membrane. This suggests the model for translocation illustrated in the lower part of the figure, in which proteins of the ER membrane form an aqueous channel through the bilayer. A translocating protein moves through this channel, interacting with the resident proteins rather than with the lipid bilayer.

Is the channel a preexisting structure (as implied in the figure) or is it assembled in response to the insertion of the signal sequence into the lipid bilayer? Channels can be detected by their ability to allow the passage of ions (measured as a localized change in electrical conductance). Ion-conducting channels can be detected in the ER membrane, and their state depends on protein translocation. A channel opens when a nascent polypeptide is transferred from a ribosome to the ER membrane. The translocating protein fills the channel completely, so ions cannot pass through during translocation. But if the protein is released by treatment with puromycin, then the channel becomes freely permeable. If the ribosomes are removed from the membrane, the channel closes, suggesting that the open state requires the presence of the ribosome. This suggests that the channel is a permanent structure that provides an aqueous environment in which the protein may pass through the membrane. Measurements of the abilities of fluorescence quenching agents of different sizes to enter the channel suggest that it is large, with an internal diameter of 40–60 Å. This is much larger than the diameter of an extended α-helical stretch of protein. It is also larger than the pore seen in direct views of the channel; this discrepancy remains to be explained.

The aqueous environment of an amino acid in a protein can be measured by incorporating variant amino acids that have photoreactive residues. The fluorescence of these residues indicates whether they are in an aqueous or hydrophobic environment. Experiments with such probes show that when the signal sequence is first synthesized in the ribosome, it is in an aqueous state, but is not accessible to ions in the cytosol. It remains in the aqueous state throughout its interaction with a membrane. This suggests that the translocating protein travels directly from an enclosed tunnel in the ribosome into an aqueous channel in the membrane.

In fact, access to the pore is controlled (or "gated") on both sides of the membrane. Before attachment of the ribosome, the pore is closed on the lumenal side. When the ribosome attaches, it seals the pore on the cytosolic side. When the nascent protein reaches a length of ~70 amino acids, that is, probably when it extends fully across the channel, the pore opens on the lumenal side. So at all times, the pore is closed on one side or the other, maintaining the ionic integrities of the separate compartments.

This model explains how a secreted protein passes through a membrane without any conflict between the hydrophilic protein and hydrophobic lipid bilayer. However, it poses a problem for transmembrane proteins. **Figure 8.25** illustrates the difference between the organization of a translocating protein, which is protected from the lipid bilayer by the aqueous channel, and a transmembrane protein, which has a hydrophobic segment directly in contact with the membrane. We do not know how a protein is transferred from its passage through the proteinaceous channel into the lipid bilayer itself. One possibility is that there is some mechanism for transferring hydrophobic transmembrane domains directly from the channel into the

Figure 8.25 How does a transmembrane protein make the transition from moving through a proteinaceous channel to interacting directly with the lipid bilayer?

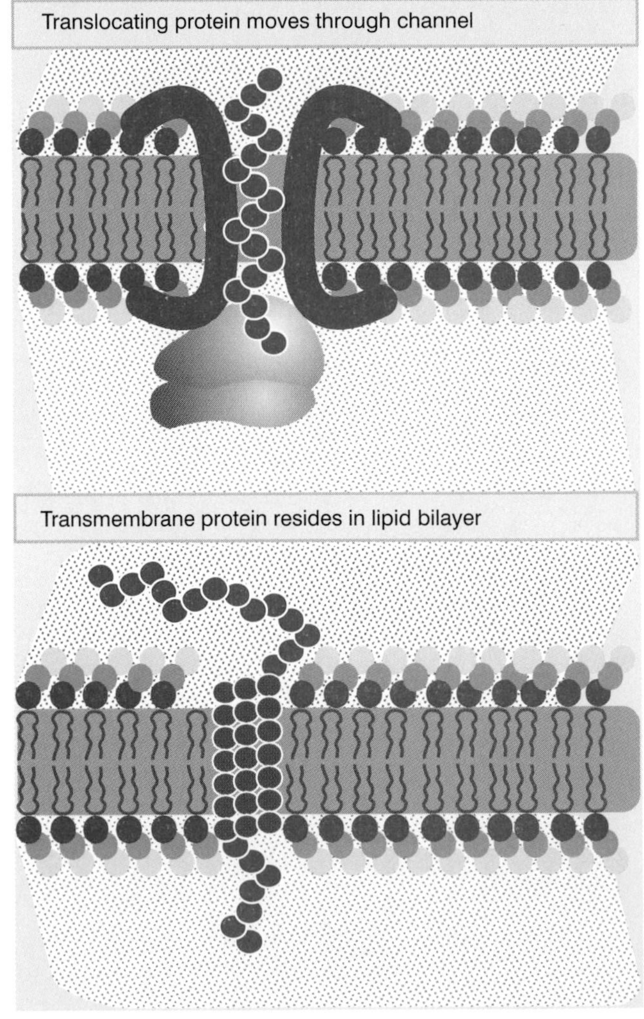
Translocating protein moves through channel
Transmembrane protein resides in lipid bilayer

membrane, by means of some rearrangement that exposes the hydrophobic region to the surrounding lipids. An alternative is that these domains cause the channel to disaggregate, exposing the hydrophobic amino acids to the lipid bilayer.

It has always been a common assumption that, whatever the exact mechanism for transferring the transmembrane segment into the membrane, it is triggered by the presence of the transmembrane sequence in the pore. However, changes in the pore occur in response to the presence of a transmembrane sequence in the ribosome. When a secreted protein passes through the pore, the channel remains sealed on the cytosolic side but opens on the lumenal side after synthesis of the first 70 residues. However, as soon as a transmembrane sequence has been fully synthesized, that is, while it is still entirely within the ribosome, the pore closes on the lumenal side. How this change relates to the transfer of the transmembrane sequence into the membrane is not clear. And we do not know whether transmembrane region(s) are transferred into the membrane during translation or only after completion of the entire protein sequence. Transfer may take place in stages, as seen by changes in the contacts made by a transmembrane region with the pore.

Proteins may also be associated with a membrane by means of covalent linkage to a fatty acid that is incorporated into the lipid bilayer. **Figure 8.26** depicts four forms of such association. In each case, a fatty acid or lipid is attached to an amino acid near to or at one terminus of the protein, with the result that the entire polypeptide chain resides on one side of the membrane, but is attached to it.

Prenylation is used to attach proteins to both the plasma membrane and internal membranes. Two types of prenyl groups have been identified: farnesyl is a 15-carbon isoprenoid (shown in the figure), and geranylgeranyl is a 20-carbon chain. They are added to cysteine residues by a thioester linkage; the cysteine is always located at the fourth position from the C terminus, as part of the sequence CAAX, where A represents aliphatic amino acids and X is methionine or serine for farnesylation, and leucine for geranylgeranylation. The usefulness of the prenyl groups for assisting

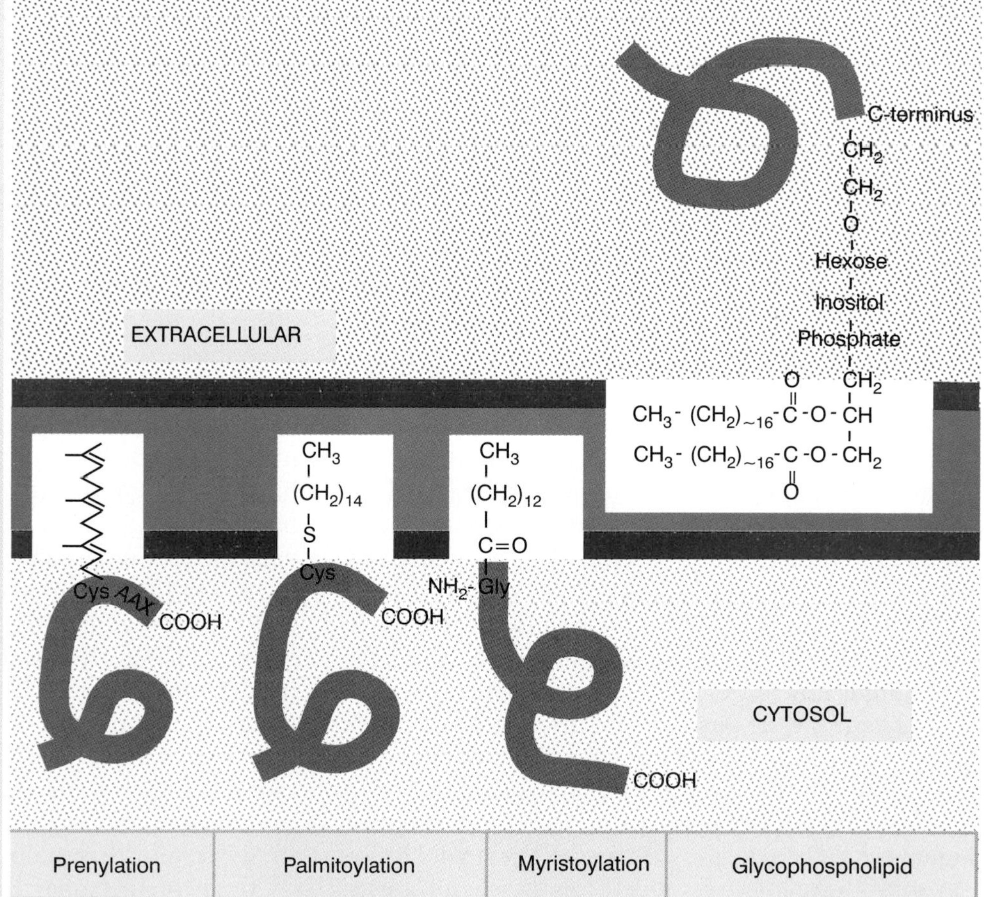

Figure 8.26 Proteins may be associated with one face of a membrane by acyl linkages to fatty acids.

attachment to the membrane is obvious, but we do not yet know how specificity is conferred with regard to the choice of membrane.

Two fatty acids are used to anchor proteins on the cytoplasmic side of the plasma membrane. Palmitic acid, a 16-carbon-chain saturated fatty acid, is linked through a sulfide bond to a cysteine residue located close to the terminus (usually the C-terminus, but sometimes the N-terminus). Myristic acid, a 14-carbon-chain saturated fatty acid, is linked to the amino group of N-terminal glycine. Myristoylated proteins are often, but not always, associated with a membrane.

The more complex structure of a glycosyl-phosphatidyl-inositol (GPI) anchor is linked to the carboxyl group of the C-terminal amino acid of protein exposed on the extracellular side of the membrane. Addition of the GPI anchor actually involves cleavage of the original polypeptide chain near the C-terminus, generating a new C-terminus that is linked to the anchor. Enzymes exist that can cleave the GPI anchor from the protein, releasing the protein into the extracellular medium.

Anchor signals are needed for membrane residence

How are proteins that are secreted *through* membranes distinguished from those that reside *within* them? The pathway by which proteins of either type I or type II are inserted into the membrane follows the same initial route as that of secretory proteins, relying on a signal sequence that functions co-translationally. But proteins that are to remain within the membrane possess a second, **stop-transfer** signal. This takes the form of a cluster of hydrophobic amino acids adjacent to some ionic residues. The cluster serves as an **anchor** that latches on to the membrane and stops the protein from passing right through.

A surprising property of anchor sequences is that they can function as signal sequences when engineered into a different location. When placed into a protein lacking other signals, such a sequence may sponsor membrane translocation. One possible explanation for these results is that the signal sequence and anchor sequence interact with some common component of the apparatus for translocation. Binding of the signal sequence initiates translocation, but the appearance of the anchor sequence displaces the signal sequence and halts transfer.

Membrane insertion starts by the insertion of a signal sequence in the form of a hairpin loop, in which the N-terminus remains on the cytoplasmic side. Two features determine the position and orientation of a protein in the membrane: whether the signal sequence is cleaved; and the location of the anchor sequence.

The insertion of type I proteins is illustrated in **Figure 8.27.** The signal sequence is N-terminal. The location of the anchor signal determines when transfer of the protein is halted. When the anchor sequence takes root in the membrane, domains on the N-terminal side will be located in the lumen, while domains on the C-terminal side are located facing the cytosol.

A common location for a stop-transfer sequence of this type is at the C-terminus. As shown in the figure, transfer is halted only as the last sequences of the protein enter the membrane. This type of arrangement is responsible for the location in the membrane of many proteins, including cell surface proteins. Most of the protein sequence is exposed on the lumenal side of the membrane, with a small or negligible tail facing the cytosol.

Type II proteins do not have a cleavable leader sequence at the N-terminus. Instead the signal sequence is combined with an anchor sequence. We imagine that the general pathway for the integration of type I proteins into the membrane involves the steps illustrated in **Figure 8.28.** The signal sequence enters the membrane, but the joint signal-anchor sequence does not pass through. Instead it stays in the membrane (perhaps interacting directly with the lipid bilayer), while the rest of the growing polypeptide continues to loop into the endoplasmic reticulum.

The signal-anchor sequence is usually internal, and its location determines which parts of the protein remain in the cytosol and which are extracellular. Essentially all the N-terminal sequences that precede the signal-anchor are exposed to the cytosol. Usually this cytosolic tail is short, ~6–30 amino acids. In effect, the N-terminus remains constrained while the rest of the protein passes through the membrane. This

Figure 8.27 Proteins that reside in membranes enter by the same route as secreted proteins, but transfer is halted when an anchor sequence passes into the membrane. If the anchor is at the C-terminus,the bulk of the protein passes through the membrane and is exposed on the far surface.

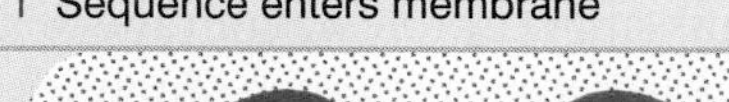
1 Sequence enters membrane

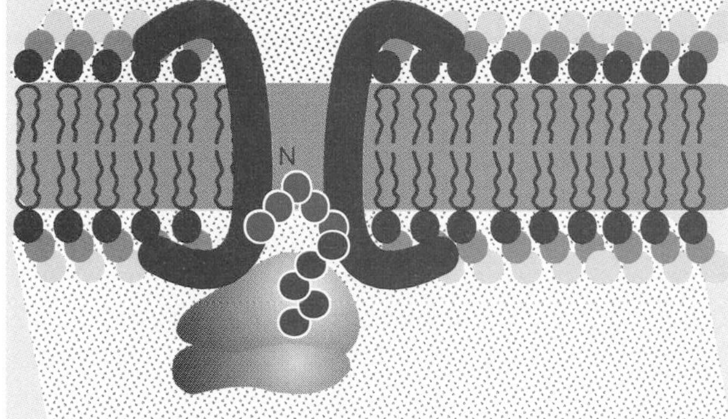
N

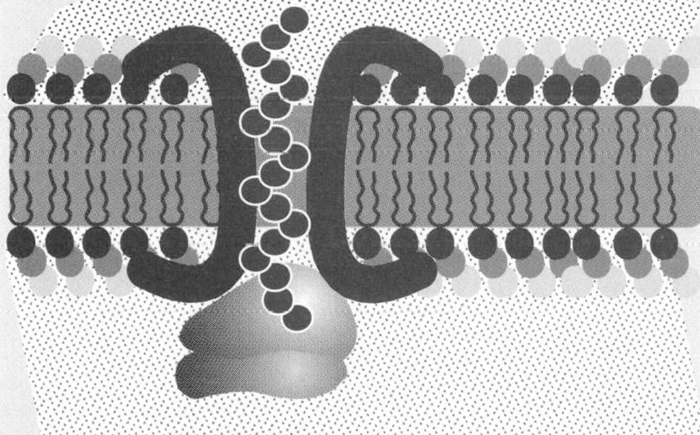
2 Signal sequence is cleaved & translocation continues

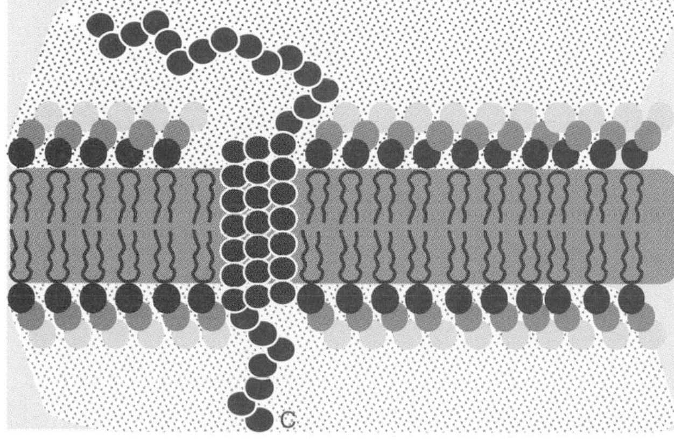
3 Anchor sequence halts transfer
C

Figure 8.28 A combined signal-anchor sequence causes a protein to reverse its orientation, so that the N-terminus remains on the inner face and the C-terminus is exposed on the outer face of the membrane.

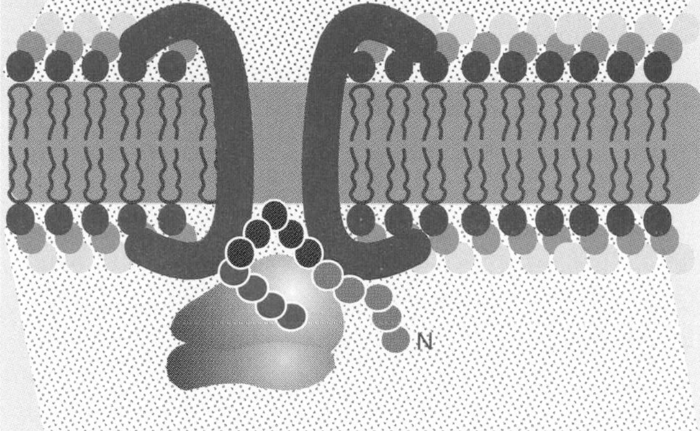
1 Signal-anchor enters membrane
N

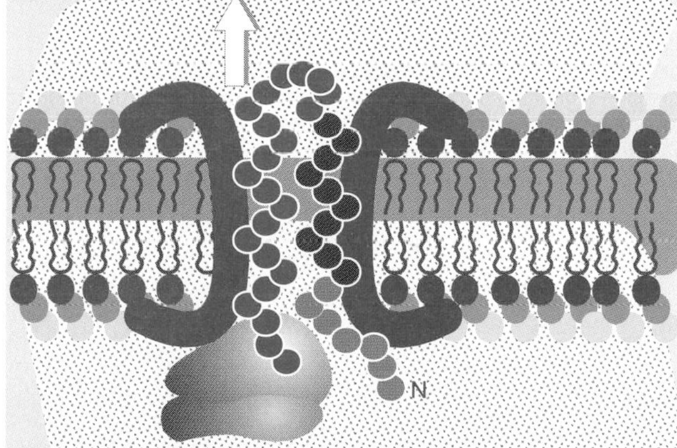
2 Signal-anchor remains & translocation continues
N

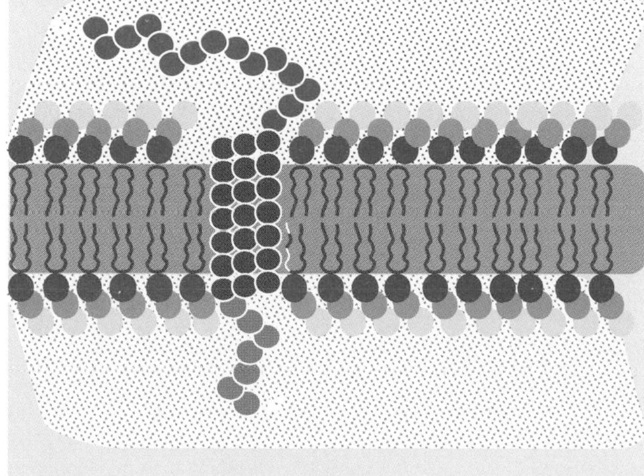
3 Translocation is completed

reverses the orientation of the protein with regard to the membrane.

The combined signal-anchor sequences of type II proteins resemble cleavable signal sequences. **Figure 8.29** gives an example. Like cleavable leader sequences, the amino acid composition is more important than the actual sequence. The regions at the extremities of the signal-anchor carry positive charges; the central region is uncharged and resembles a hydrophobic core of a cleavable leader. Mutations to introduce charged amino acids in the core region prevent membrane insertion; mutations on either side prevent the anchor from working, so the protein is secreted or located in an incorrect compartment.

The distribution of charges around the anchor sequence has an important effect on the orientation of the protein. More positive charges are usually found on the cytoplasmic side (N-terminal side in type II proteins). If the positive charges are removed by mutation, the orientation of the protein can be reversed. The effect of charges on orientation is summarized by the "positive inside" rule, which says that the side of the anchor with the most positive charges will be located in the cytoplasm. The positive charges in effect provide a hook that latches on to the cytoplasmic side of the membrane, controlling the direction in which the hydrophobic region is inserted, and thus determining the orientation of the protein.

The process of insertion into a membrane has been characterized for both type I proteins and type II proteins, in which there is a single transmembrane domain. How is a protein with multiple membrane-spanning regions inserted into a membrane? Much less is known about this process, but we assume that it relies on sequences that provide signal and/or anchor capabilities. One model is to suppose that there is an alternating series of signal and anchor sequences. Translocation is initiated at the first signal sequence and continues until stopped by the first anchor. Then it is reinitiated by a subsequent signal sequence, until stopped by the next anchor. Individual transmembrane regions accumulate in contact with the translocation pore as they are synthesized. We do not know how (and when) they move laterally into contact with the lipid bilayer.

Figure 8.29 The signal-anchor of influenza neuraminidase is located close to the N-terminus and has a hydrophobic core.

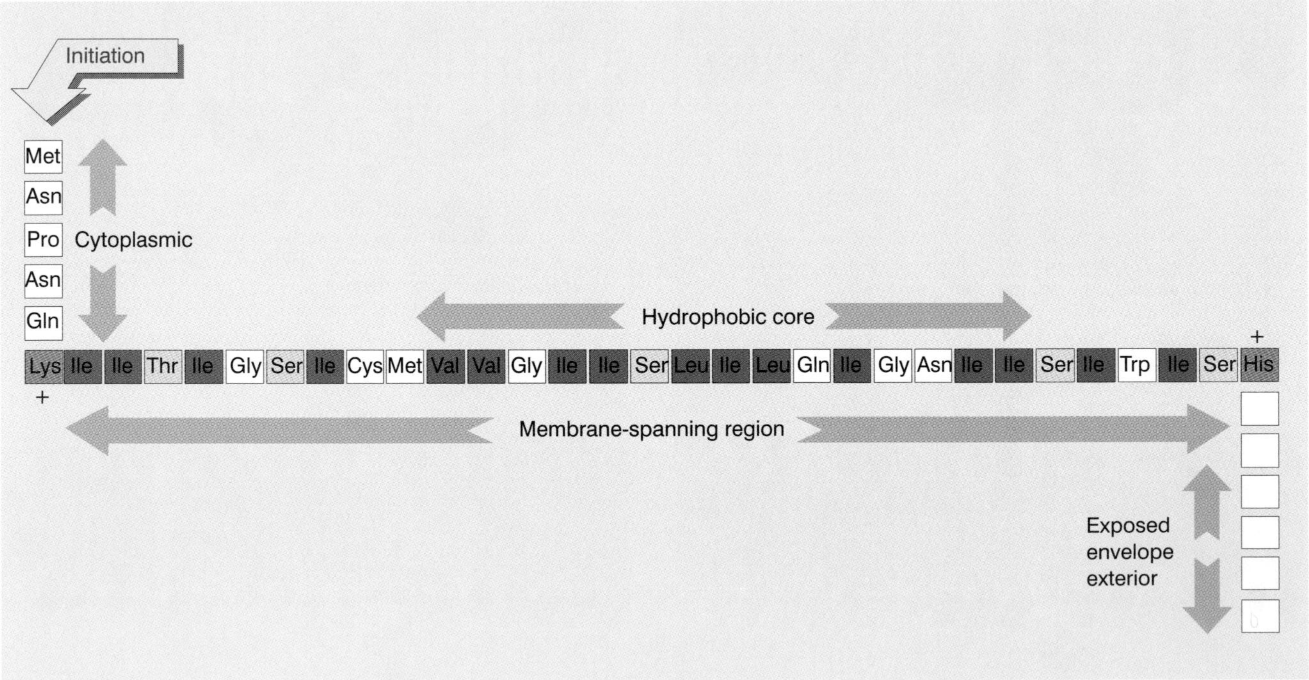

Bacteria use both co-translational and post-translational translocation

THE secretion of proteins from bacteria relies on mechanisms very similar to those characterized for eukaryotic cells. Transport from the bacterial cytoplasm passes through the inner membrane into the periplasmic space and then (sometimes) through the outer membrane into the environment. Co-translational transfer is common in *E. coli*, but is not universal. Some proteins are secreted both co-translationally and post-translationally. The relative kinetics of translation versus secretion through the membrane could determine the balance.

Exported bacterial proteins have N-terminal leader sequences, with a hydrophilic N-terminus and an adjacent hydrophobic core. Mutations in N-terminal leaders prevent secretion; they are suppressed by mutations in other genes, which are thus defined as components of the protein export apparatus. Several genes given the general description *sec* are implicated in coding for components of the secretory apparatus by the occurrence of mutations that block secretion of many or all exported proteins.

Bacterial protein translocation passes through the stages summarized in **Figure 8.30.** A chaperone binds to the nascent protein to control its folding; the protein associates with the secretion apparatus; it is translocated through the membrane; and a peptidase cleaves the leader from proteins with a cleavable N-terminus.

Several chaperones can increase the efficiency of bacterial protein export by preventing premature folding; they include trigger factor (characterized as a chaperone that assists export), GroEL (see earlier), and SecB (identified as the product of one of the *sec* mutants). Although SecB is the least abundant of these proteins, it has the major role in promoting export. It

Figure 8.30 Protein translocation in bacteria requires SecB to function as a chaperone and bind the nascent protein; then SecB transfers the protein to SecA, which is a peripheral membrane protein associated with the integral membrane proteins SecE/Y. Translocation requires hydrolysis of ATP and a protonmotive force. Leader peptidase is an integral membrane protein that cleaves the leader sequence.

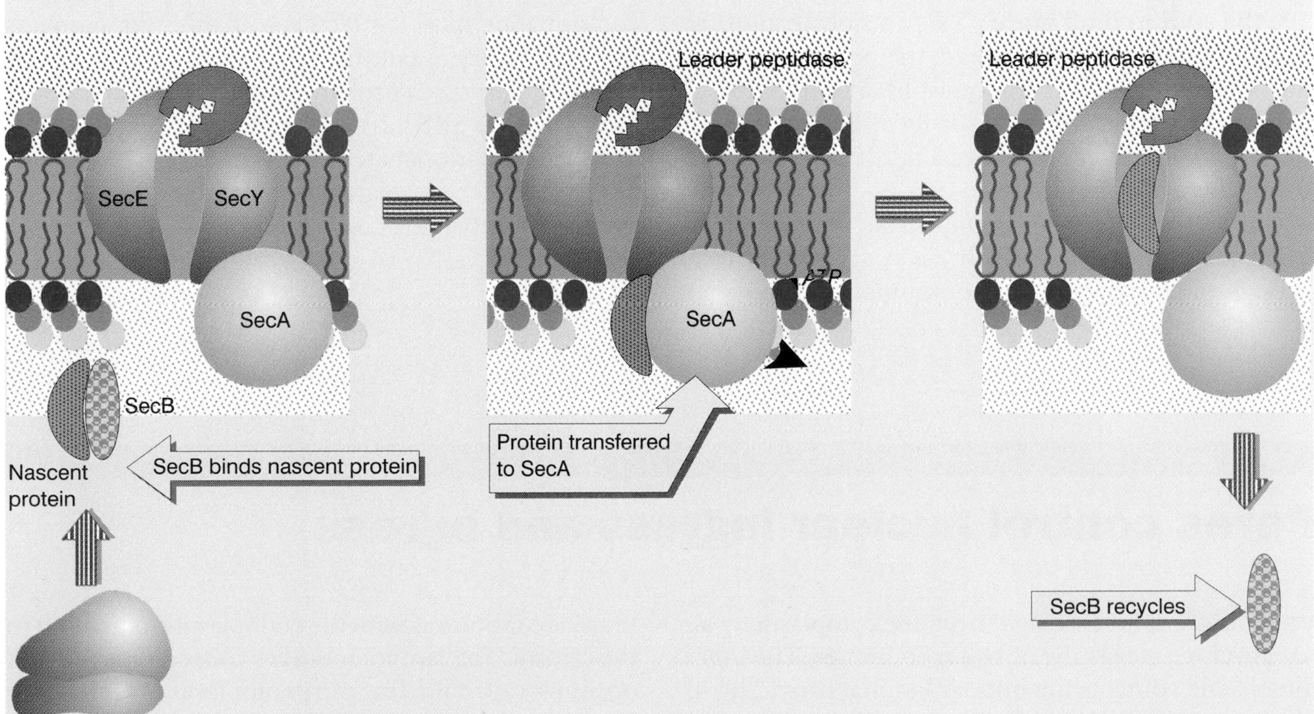

has two functions. First, it behaves as a chaperone and binds to a nascent protein to retard folding. It cannot reverse the change in structure of a folded protein, so it does not function as an unfolding factor. Its role is therefore to inhibit improper folding of the newly synthesized protein. Second, it has an affinity for the protein SecA. This allows it to target a precursor protein to the membrane.

SecA is a large peripheral membrane protein that has various means of associating with the membrane. As a peripheral membrane protein, it associates with the membrane by virtue of its affinity for acidic lipids and for the protein SecY. SecY and SecE are integral membrane proteins (SecY has 10 transmembrane segments and SecE has 3 transmembrane segments), which are part of a multisubunit complex that provides the translocase function. However, in the presence of other proteins (SecD and SecF), SecA can be found as a membrane-spanning protein.

SecA recognizes both SecB and the precursor protein that it chaperones; probably features of the mature protein sequence as well as its leader are required for recognition. SecA has an ATPase activity that depends upon binding to lipids, SecY, and a precursor protein. The ATPase functions in a cyclical manner during translocation. After SecA binds a precursor protein, it binds ATP, and ~20 amino acids are translocated through the membrane. Hydrolysis of ATP is required to release the precursor from SecA. Then the cycle may be repeated. Precursor protein is bound again to provide the spur to bind more ATP, translocate another segment of protein, and release the precursor. SecA may alternate between the peripheral and integral membrane forms during translocation; with each cycle, a 30 kDa domain of SecA may insert into the membrane and then retract.

Another process can also undertake translocation. When a precursor is released by SecA, it can be driven through the membrane by a protonmotive force (that is, an electrical potential across the membrane). This process cannot initiate transfer through the membrane, but can continue the process initiated by a cycle of SecA ATPase action. So after or between cycles of the SecA-ATP driven reaction, the protonmotive force can drive translocation of the precursor.

Many secreted bacterial proteins have cleavable N-terminal leader sequences. The leader is cleaved by an enzyme called **leader peptidase** that recognizes precursor forms of several exported proteins. The leader peptidase is an integral membrane protein, located in the inner membrane.

A notable feature of eukaryotic systems is the use of the ribonucleoprotein SRP to provide the initial recognition of a signal sequence. There is a counterpart to SRP in bacteria, although it appears to contain fewer components. *E. coli* contains a 4.5S RNA that associates with ribosomes and is homologous to the 7S RNA of the SRP. It associates with two proteins: Ffh is homologous to SRP54, and FtsY is homologous to the α subunit of the SRP receptor. In fact, FtsY replaces the functions of both the α and β SRP subunits; its N-terminal domain substitutes for SRPβ (which is an integral membrane protein) in membrane targeting, and the C-terminal domain interacts with the target protein. The complex appears to play a role in the secretion of certain, but not all, secreted proteins. The role of this complex may be to associate with the leader sequences of nascent proteins as they emerge from ribosomes, perhaps to keep the nascent protein in an appropriate conformation until it interacts with other components of the secretory apparatus. This could be the original connection between protein synthesis and secretion; in eukaryotes the SRP has acquired the additional roles of causing translational arrest and targeting to the membrane. Since SecY is related to mammalian/yeast Sec61, there are comparable components at each stage of the secretory process.

Pores control nuclear ingress and egress

THE nucleus is segregated from the cytoplasm by an envelope consisting of two membranes. The inner membrane contacts the nuclear lamina, providing in effect a surface layer for the nucleus. The outer membrane is continuous with the endoplasmic reticulum in the cytosol. The two membranes come into contact at openings called **nuclear pore complexes.** At the center of each complex is a pore that provides a water-soluble channel between nucleus and cytoplasm. This means that the nucleus and cytosol have the same ionic milieu. There are ~3000 pore complexes on the nuclear envelope of an animal cell.

Transport between nucleus and cytoplasm proceeds in both directions. Since all proteins are synthesized in the cytosol, any proteins required in the nucleus must be transported there. Since all RNA is synthesized in the nucleus, the entire cytoplasmic complement of RNA (mRNA, rRNA, tRNA, and other small RNAs) must be derived by export from the nucleus. The nuclear pores are used for both import and export of material. **Figure 8.31** summarizes the frequency with which the pores are used for some of the more prominent substrates.

We can form an impression of the magnitude of import by considering the histones, the major protein components of chromatin. In a dividing cell, enough histones must be imported into the nucleus during the period of DNA synthesis to associate with a diploid complement of chromosomes. Since histones form about half the protein mass of chromatin, we may conclude that overall about 200 chromosomal protein molecules must be imported through each pore per minute.

Uncertainties about the processing and stability of mRNA make it more difficult to calculate the number of mRNA molecules exported, but to account for the ~250,000 molecules of mRNA per cell probably requires ~1 event per pore per minute. The major RNA synthetic activity of the nucleus is of course the production of rRNA, which is exported in the form of assembled ribosomal subunits. Just to double the number of ribosomes during one cell cycle would require the export of ~5 ribosomal subunits (60S and 40S) through each pore per minute.

For ribosomal proteins to assemble with the rRNA, they must first be imported into the nucleus. So ribosomal proteins must shuttle into the nucleus as free proteins and out again as assembled ribosomal subunits. Given ~80 proteins per ribosome, their import must be comparable in magnitude to that of the chromosomal proteins.

How does a nuclear pore accommodate the transit of material of varied sizes and characteristics in either direction? Nuclear pore complexes have a uniform appearance when examined by microscopy. The pores can be released from the nuclear envelope by detergent, and **Figure 8.32** shows that they appear as annular structures, consisting of rosettes made of 8 spokes. **Figure 8.33** shows a model for the pore based on three-dimensional reconstruction of electron microscopic images. It consists of an upper ring and a lower ring, connected by a lattice of 8 structures.

The basis for the 8-fold symmetry is explained in terms of individual components in the schematic view from above shown in **Figure 8.34.** This includes the central structure of Figure 8.33, and extends it with surrounding radial arms and an internal transporter. The outside of the pore complex as such consists of a

Figure 8.31 Nuclear pores are used for import and export.

Direction	Substrate	Passages /pore/min
Import	Histones	100
	Nonhistone proteins	100
	Ribosomal proteins	150
Export	Ribosomal subunits	~5
	mRNA	<1

Figure 8.32 Nuclear pores appear as annular structures by electron microscopy. The bar is 0.5 μm. Photograph kindly provided by Ronald Milligan.

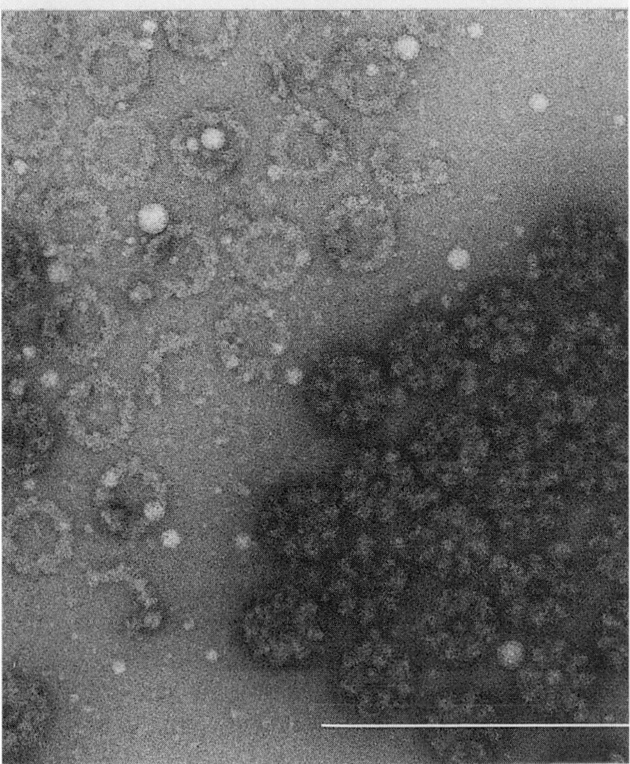

Figure 8.33 A model for the nuclear pore shows 8-fold symmetry. Two rings form the upper and lower surfaces (shown in yellow); they are connected by the spokes (shown in green on the inside and blue on the outside). Photograph kindly provided by Ronald Milligan.

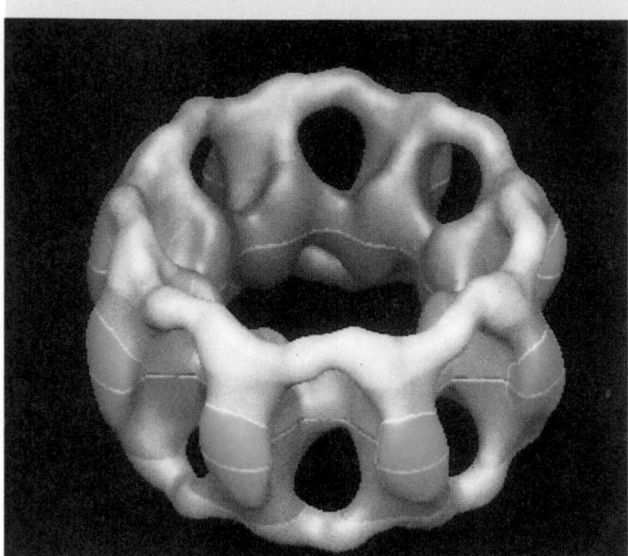

Figure 8.34 The outsides of the nuclear coaxial (cytoplasmic and nucleoplasmic) rings are connected to radial arms. The interior is connected to spokes that project towards the transporter that contains the central pore.

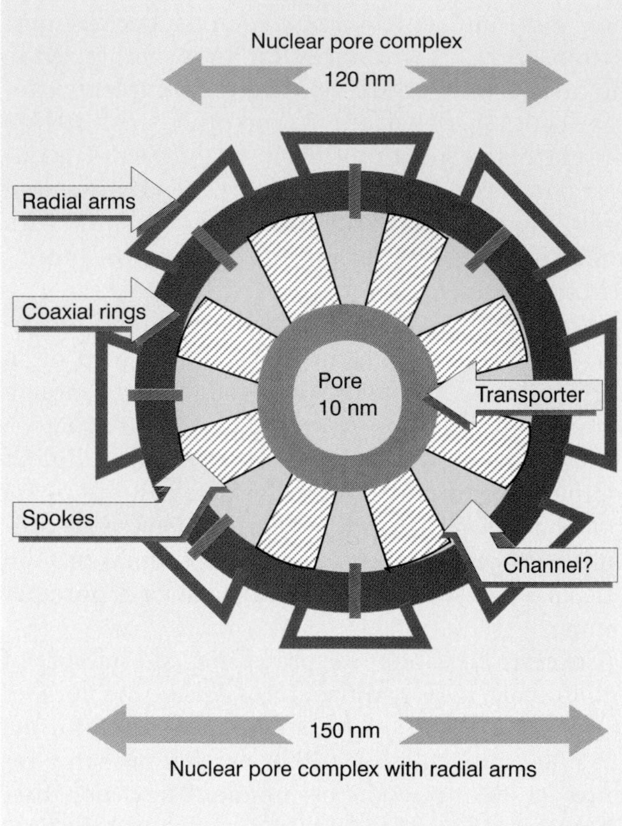

ring of diameter ~120 nm. The ring itself consists of 8 subunits. The 8 radial arms outside the ring may be responsible for anchoring the pore complex in the nuclear envelope; they penetrate the membrane. The 8 interior spokes project from the ring, closing the opening to a diameter of ~48 nm. Within this region is the transporter, which contains a pore that approximates a cylinder <10 nm in diameter.

The pore provides a passage across the outer and inner membranes of the nuclear envelope. As illustrated in **Figure 8.35,** the side view has two-fold symmetry about a horizontal axis in the plane of the nuclear envelope. There are matching annuli at the outer and inner membranes, comprising the surfaces that project into the cytosol and into the nucleus, and each is connected to the spokes, which form a central ring. (Only 2 of the 8 spokes are seen in this side view.) The spokes are symmetrical about the horizontal axis. The central pore projects for the distance across the envelope. Sometimes material can be seen within the pore, but it has been difficult to equate such sightings with the transport of particular material.

The size of the nuclear pore complex corresponds to a total mass $>100 \times 10^6$ daltons (compare this with the 80S ribosome at 4×10^6 daltons). We can identify the smallest repeating component by using the 8-fold symmetry as seen in cross-section (see Figure 8.34) and the 2-fold symmetry seen from the side (see Figure 8.35). This divides the scaffold into 16 identical units, that is, each $\sim 6 \times 10^6$ daltons. Each of these units must contain a large number of individual proteins. The central pore constitutes only a small part of the overall complex.

The ability of compounds to diffuse freely through the pores is limited by their size, and it is convenient to consider the material in three size classes:

- Molecules of <5000 daltons that are injected into the cytoplasm appear virtually instantaneously in the nucleus: we may conclude that the nuclear envelope is freely permeable to ions, nucleotides, and other small molecules.
- Proteins of 5–50 kD diffuse at a rate that is inversely related to their size, presumably determined by random contacts with, and passage through, the pore. It takes a few hours for the levels of an injected protein to equilibrate between cytoplasm and nucleus. We may conclude that small proteins can enter the nucleus by

Figure 8.35 The nuclear pore complex spans the nuclear envelope by means of a triple ring structure. The side view shows two-fold symmetry from either horizontal or perpendicular axes.

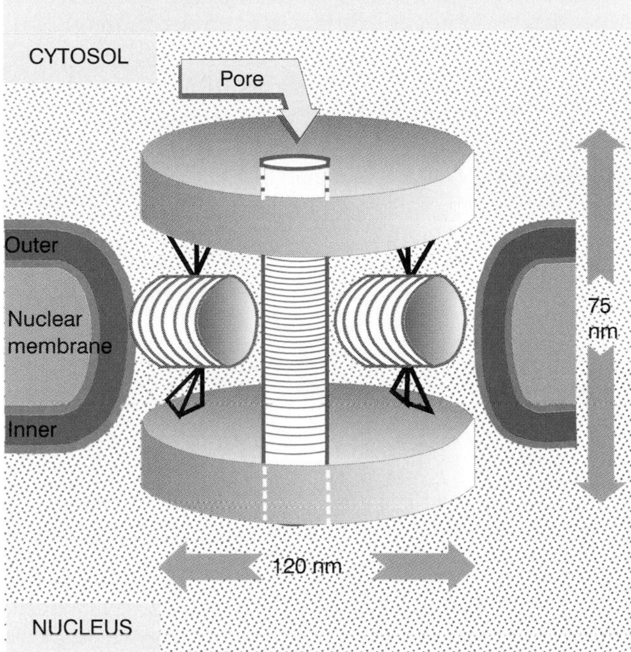

passive diffusion (but they may also be actively transported). The nuclear envelope in effect provides a mesh or molecular sieve that permits passage of material <50 kD.

■ Proteins >50 kD in size do not enter the nucleus by passive diffusion; a mechanism of active transport must be required for their passage.

The lattice-like structure of the nuclear pore suggests that different features could be responsible for active transport of large proteins versus passive transport of smaller proteins.

Eight channels are created by the open regions between the spokes, that is, around the periphery between the inner and outer annuli. These have a rigid structure. They are oval in cross-section, with a diameter of ~10 nm. We might speculate that this could provide a mesh to allow passive diffusion for small proteins. A globular protein of 50 kD in mass would have a diameter of ~5 nm if it were spherical. Presumably objects would need to be somewhat smaller than the mesh size in order to be able to pass through by diffusion.

If these channels allow free passage, material below the size limit will equilibrate in concentration between the nucleus and the cytoplasm (although, as the material approaches the size of the mesh, equilibration is slow). Proteins that enter the nucleus in this way would be retained, and therefore accumulate in the nucleus, if they participate in some nuclear function. For example, if a protein becomes incorporated into a large structure, such as a chromosome, this removes it from the equilibrating pool, and thus pulls more protein into the nucleus.

The central pore is used to transport larger material. A protein requires a specific signal to pass through the central pore. Smaller proteins may be transported in this way (as well as through the peripheral channels), and larger proteins *must* use an active transport mechanism that overcomes the apparent size restriction of the pores. But how can a protein or ribonucleoprotein with a diameter exceeding that of the pore pass through it?

Transport through the pore has been characterized by using colloidal gold particles coated with a nuclear protein. When these particles are injected into the cytoplasm, they cluster at the nuclear pores, and then accumulate in the nucleus. This suggests that the pore structure can widen to accommodate objects of the size of the coated gold particles (~20 nm). Similar experiments have shown that gold particles coated with polynucleotides can be exported from the nucleus via pores.

The rigidity of the gold particle excludes the possibility that transport through the pore requires the protein to change into a conformation with a diameter physically smaller than the pore. We conclude that the nuclear pore has a "gating" mechanism that allows the interior to expand as material passes through. Pores engaged in transporting material appear to be opened to a diameter of ~20 nm, possibly by a mechanism akin to the iris of a camera lens. It is possible that two irises, one connected with the cytoplasmic ring and one connected with the nucleoplasmic ring, open in turn as material proceeds through the pore. Very large substrates, such as exported ribonucleoprotein particles, may have to change their conformation to conform with the limit of 20 nm.

The nuclear pore complex provides a structural framework that supports the proteins actually responsible for binding and transporting material into (or out of) the nucleus; it does not include all of the active components that are involved in binding and translocation. A major question about nuclear pores is whether they are all identical, or whether their uniform appearance disguises functional differences. We should like to know whether the same pores undertake transport into and out of the nucleus or whether different classes of pore exist. We have yet to determine to what extent the functions of the pore are intrinsic as opposed to being determined by the accessory factors with which it associates.

There are multiple pathways for transport in each direction between nucleus and cytoplasm. Transport for all of the substrates for any particular pathway can be inhibited by saturating that pathway with one of its substrates. **Figure 8.36** summarizes the independent pathways. At least two different types of pathway exist for import of proteins; and each class of RNA is exported by a different system. The basic principle for each import and export system is illustrated in **Figure 8.37;** *a carrier protein takes the substrate through the pore.* (The carrier protein must then be returned across the membrane to function in another cycle.) Each carrier protein recognizes a particular type of sequence in its substrate, thus defining the specificity of the system.

The most common motif responsible for import into the nucleus is called the **nuclear localization signal (NLS).** Its presence in a cytosolic protein is necessary and sufficient to sponsor import into the nucleus. Mutation of the signal can prevent the protein from entering the nucleus.

The summary of nuclear localization signals in **Figure 8.38** shows that there is no apparent conservation of sequence of NLS signals; perhaps the shape of the region or its basicity are the important features. Many NLS sequences take the form of a short, rather basic stretch of amino. Often there is a proline residue to break α-helix formation upstream of the basic residues. Hydrophobic residues are rare. Some NLS signals are bipartite and require two separate short clusters, as found in nucleoplasmin. Competition studies suggest that NLS sequences are interchangeable, suggesting that they are all recognized by the same import system.

The first handle on the process of import was pro-

Figure 8.36 There are multiple pathways for nuclear export and import.

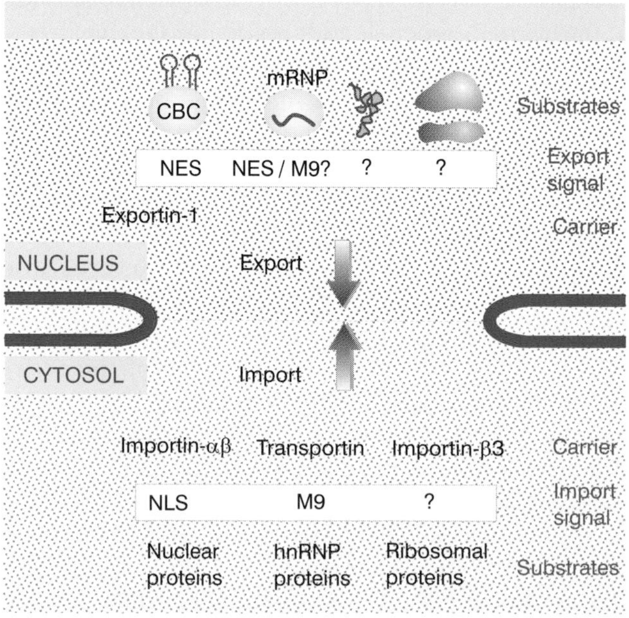

Figure 8.37 A carrier protein binds to a substrate, moves with it through the nuclear pore, is released on the other side, and must be returned for reuse.

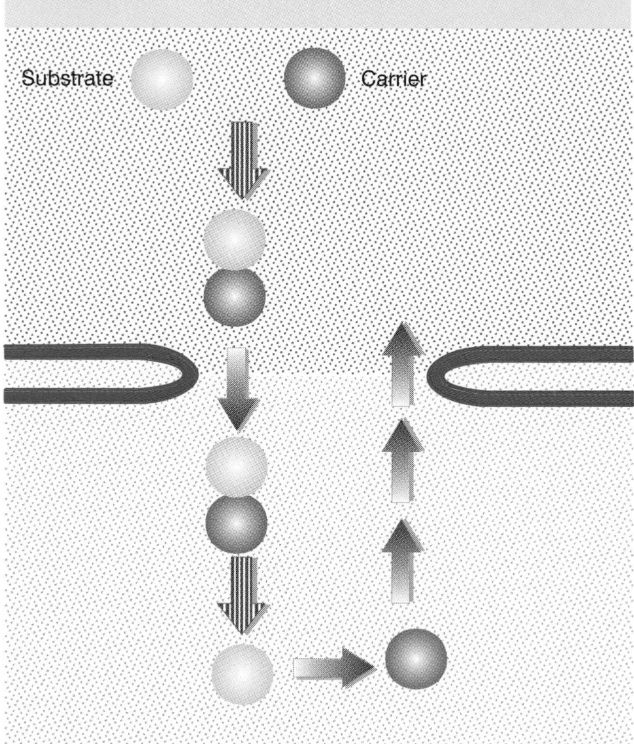

Figure 8.38 Nuclear localization signals have basic residues.

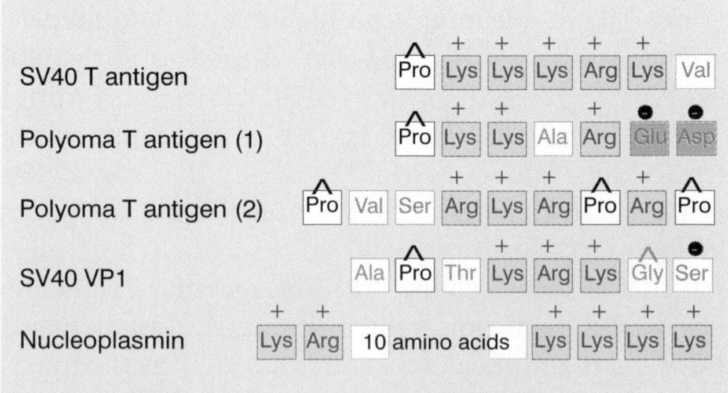

vided by systems for transport that depend upon the presence of an NLS in the substrate protein. The complexity of the system makes it impossible to develop a true *in vitro* assay for nuclear pore import, but an effective system has been developed by using permeabilized cells. When cells are treated with digitonin, the plasma membrane becomes permeable, but the nuclear envelope remains intact. Labeled proteins can be imported into the nucleus in a process that is dependent upon the provision of cytosolic components. **Figure 8.39** shows how this system has been used to characterize the transport process.

Transport can be divided into two stages: **docking,** which consists of binding to the pore; and **translocation,** which consists of movement through it. In the absence of ATP, proteins containing a nuclear import signal can bind at the pore, but they remain at the cytoplasmic face. A cytosolic fraction is needed for binding. When ATP is provided, proteins can be translocated through the pore. A different cytosolic fraction is needed to support translocation. The need for cytosolic fractions at both stages reinforces the view of the pore as a structure that provides the framework for transport, but that does not provide all of the necessary facilities for handling the substrates.

Figure 8.40 summarizes the functions of the components involved in nuclear import. The docking reaction depends on a carrier protein which binds to the substrate and to the nuclear pore. (Types of carrier proteins are summarized in Figure 8.36.) One type of carrier protein is a dimer containing the subunits importin-α and importin-β. The α subunit binds proteins that have an NLS sequence. The β subunit binds to the nuclear pore. Other import systems use carrier proteins that function as single subunits, such as transportin or importin-β3. In these cases, a single protein binds to the substrate and to the nuclear pore. These carrier proteins are related to importin-β, but they do not need an adapter like importin-α to bind the substrate, that is, they are capable of binding the substrate directly.

Translocation through the pore is inhibited by wheat germ agglutinin, a lectin (glycoprotein). The component proteins of the pore that bind to lectins are called **nucleoporins.** They are localized at or near the region of the central pore, and appear to be located on both sides of the nuclear envelope. When nucleoporins are removed, pore complexes remain normal in appearance, but can no longer function to transport large material. Material smaller than the pore size continues to be able to move through by diffusion. When the nucleoporins are added back, they restore full activity to the deficient pores. This suggests that the nucleoporins are needed for active transport of material larger than the resting diameter. Some nucleoporins have some simple peptide repeating motifs (GKFG, FG, FXFG), and it is probably these motifs that bind the importin-β carrier proteins.

The cytosolic fraction that supports translocation has two active components. One is a small GTPase called Ran; the other may be involved in targeting Ran to the nuclear pore. Ran is a typical monomeric G-protein that can exist in either the GTP-bound or GDP-bound state. The GTPase activity generates Ran-GDP, after which an exchange factor is needed to replace GDP with GTP to regenerate Ran-GTP.

The directionality of nuclear import is controlled by the state of Ran. **Figure 8.41** shows that conditions in the nucleus and cytosol differ so that typically there is Ran-GTP in the nucleus, but there is Ran-GDP in the cytosol. Export complexes are stable in the presence of Ran-GTP, whereas import complexes are stable in the presence of Ran-GDP. So export complexes are driven to form in the nucleus and dissociate in the cytosol, whereas the reverse is true of import complexes. The reaction has been best characterized for the complex of importin-αβ with an NLS-containing protein. The triple complex is stable in the presence of Ran-GDP,

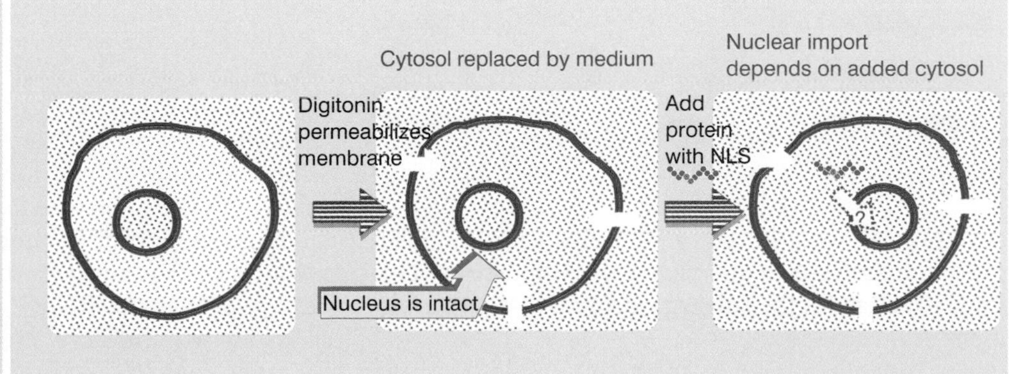
Cytosol replaced by medium
Nuclear import depends on added cytosol
Digitonin permeabilizes membrane
Add protein with NLS
Nucleus is intact

Figure 8.39 The assay for nuclear pore function uses permeabilized cells.

Figure 8.40 Nuclear import takes place in two stages. Both docking and translocation depend on cytosolic components. Translocation requires nucleoporins.

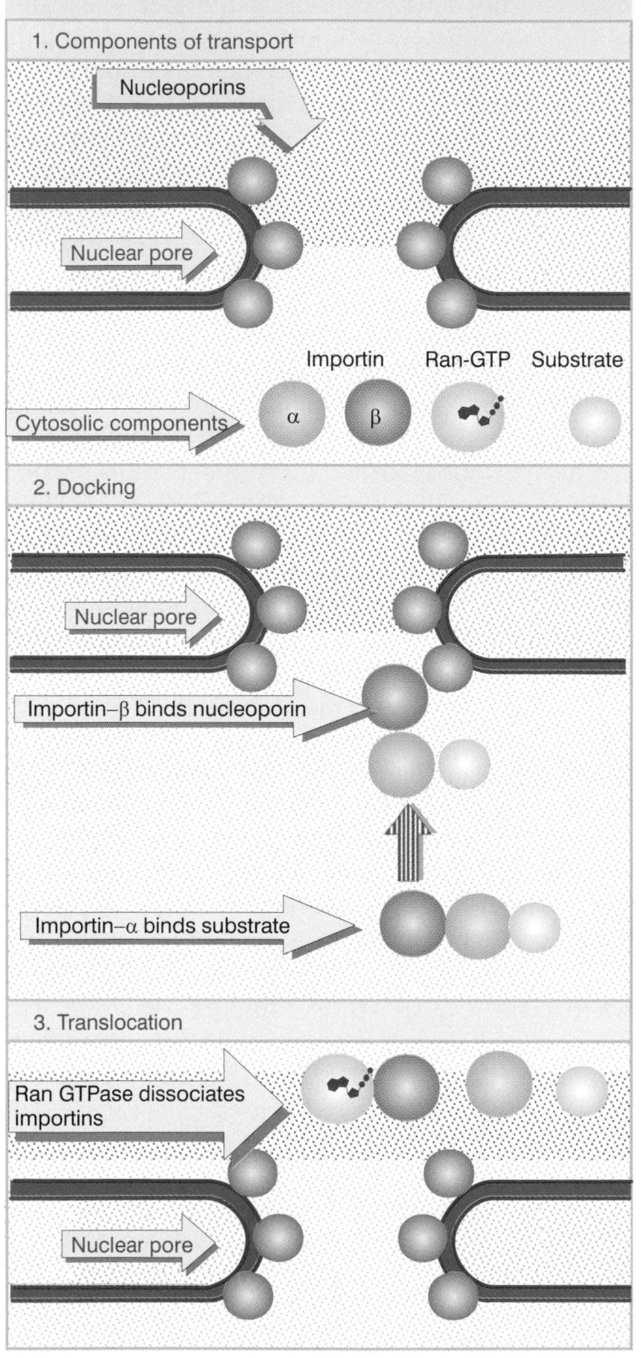

Figure 8.41 The state of the guanine nucleotide bound to Ran controls directionality of nuclear import and export.

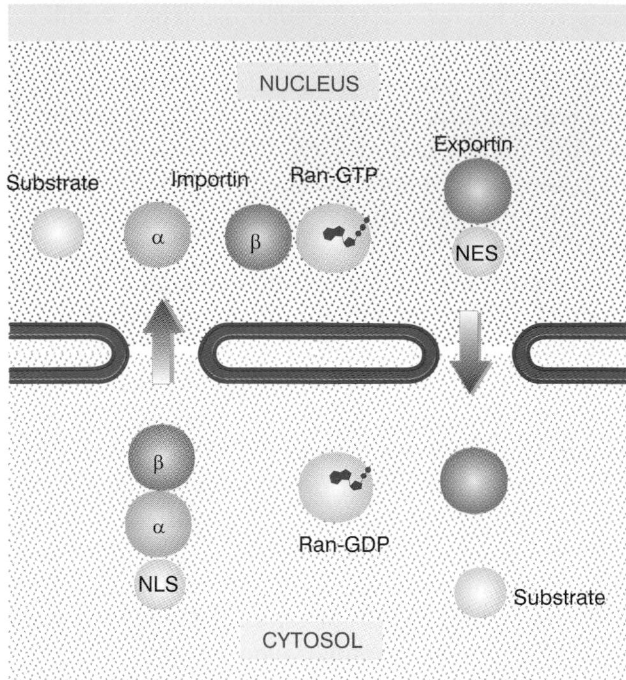

and thus can form in the cytosol. However, Ran-GTP causes importin-α to dissociate from importin-β. This leads to release of the substrate protein in the environment of the nucleus.

How does the importin-substrate complex cross the nuclear pore? Nucleoporins are distributed throughout the central pore. We do not know whether an importin-substrate complex moves across the pore continuously, or whether there is a series of docking and release events in which importin-substrate would bind to a nucleoporin, dissociate, reassociate, bind to a nucleoporin farther along the pore, and repeat the cycle. Position in the pore could be defined by the Ran-GTP/Ran-GDP gradient, but translocation as such does not appear to require GTP hydrolysis, and the source of energy for this movement is not known.

Export of mRNA has requirements similar to those of import. Mutations in a yeast nucleoporin block export of RNA from the nucleus without affecting import of proteins. This suggests the possibility that the apparatus is similar for both import and export, but could have components that confer directionality or specificity for particular substrates. There is some evidence for diversity in the export apparatus; using an assay for export of microinjected RNAs from the nucleus of the *Xenopus* oocyte, tRNAs, other small RNAs, and mRNAs appear to saturate transport only of their own class. This suggests that there are at least three groups of exported RNAs.

Export systems have similar components to import systems. **Figure 8.42** illustrates some examples of ex-

Figure 8.42 The common feature in proteins that are exported from the nucleus to the cytosol is the presence of an NES.

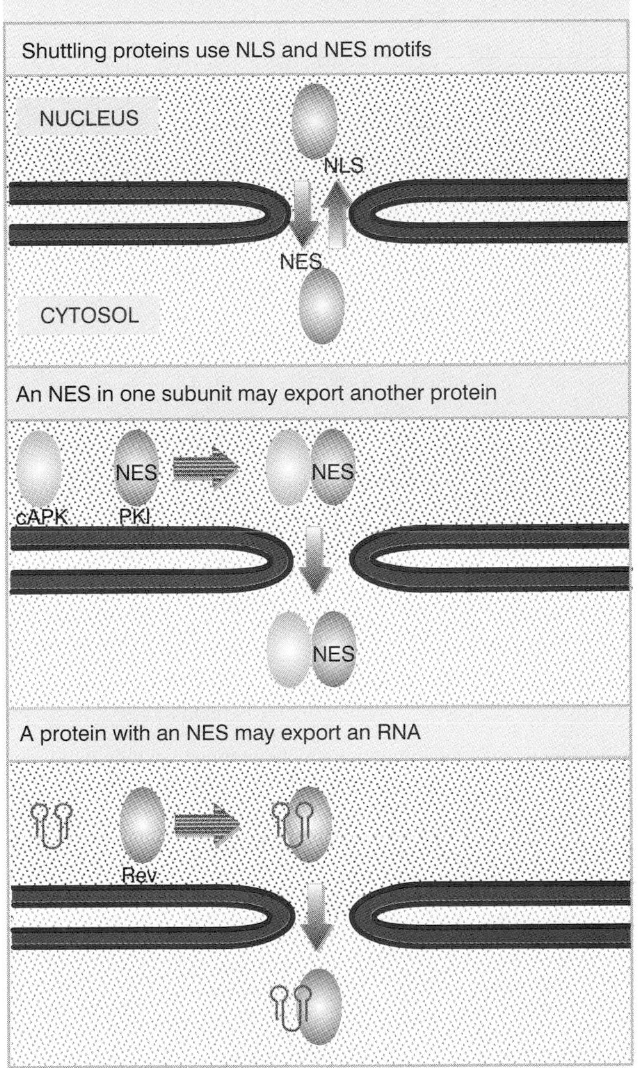

Shuttling proteins use NLS and NES motifs
NUCLEUS
NLS
NES
CYTOSOL
An NES in one subunit may export another protein
NES
NES
cAPK
PKI
NES
A protein with an NES may export an RNA
Rev

port systems. The major substrates for export from the nucleus are ribonucleoproteins—ribosomes, mRNPs, snRNPs, and tRNA-protein complexes. In the first three cases, one of the protein components of the complex may be responsible for export (for example, for snRNAs it is the cap-binding complex (CBC)). tRNA is bound directly by a specific export protein.

Some proteins are exported from the nucleus in particular circumstances; for example, the catalytic subunit of the enzyme cAPK (cyclic AMP-dependent kinase) functions in the nucleus. However, when an inhibitory subunit (PKI) is activated, it binds to the catalytic subunit, and the dimer is then exported from the nucleus. Some proteins are required to bind to RNAs and to export them from the nucleus; an example is the Rev protein of HIV, which causes export of certain RNAs from the nucleus.

Many exported proteins have a common type of signal that is necessary and sufficient for the protein to move from the nucleus to the cytosol. It is called an **NES** (nuclear export signal), and typically consists of an ~10 amino acid sequence. The only common feature in the NES sequences in different proteins is a pattern of conserved leucines. The NES binds to exportin-1, which is related in sequence to importin-β. Exportin-1 is needed for the export of U snRNAs, some proteins, and possibly (some) mRNAs. It binds the NES motif; it binds nucleoporins; and it binds Ran-GTP. The complex forms in the presence of Ran-GTP; hydrolysis of the GTP to generate Ran-GDP is accompanied by dissociation of the complex. So Ran controls directionality of export in the reverse sense from its control of import: because Ran-GTP is high in the nucleus, the complex forms there; because it becomes Ran-GDP in the cytosol, the complex dissociates there.

A protein may have both an NLS and an NES, the former used for its import into the nucleus, and the latter for its export. They may function constitutively, or their use can be regulated, for example, by association with other proteins that obscure or expose the relevant sequences.

As pointed out in Figure 8.37, in order to function more than once, an importin must return to the cytosol after taking its substrate into the nucleus. In effect, when an importin is released in the nucleus, it must become a substrate for an exportin! Such a reaction has been characterized for importin-α, which is bound in the presence of Ran-GTP by a protein called CAS. CAS behaves in a similar manner to importin-β, except that it moves in the opposite direction. Like exportins, it dissociates from its substrate (importin-α) when Ran-GTP is hydrolyzed (in the cytosol).

Some proteins "shuttle" between the cytoplasm and nucleus; they remain only briefly in either compartment before cycling back to the other. This behavior is characteristic of some proteins that are bound to poly(A)$^+$ RNA in both the nucleus and the cytoplasm. The motif responsible for transport in one such protein (M9) has a single amino acid stretch that functions as both an import and export signal, and is therefore responsible for movement in both directions.

Protein degradation by proteasomes

WHAT determines the stability of proteins? A cell contains many proteases, with varying specificities. We may divide them into three general groups:

- Some proteases are involved in specific processing events to generate mature proteins that are smaller than the precursors. Such proteases are involved in a variety of activities, including cleaving the signal sequence from a secreted protein, and cleaving cytosolic enzymes into their mature forms.
- Lysosomes are membrane-bounded organelles that degrade proteins imported into the cell; we discuss this process in the context of protein transport through membranes in Chapter 25.
- The **proteasome** is a large complex that degrades cytosolic proteins. It is involved in both general degradation (the complete conversion of a protein into small fragments) and in certain specific processing events. The major substrates for complete degradation are proteins that have been misfolded—this is basically a quality control system—and certain proteins whose degradation is a regulatory event, for example, to allow progress through the cell cycle.

Degradation of a protein by a proteasome falls into two stages: first the protein is *targeted;* and then it is *proteolysed.* Targeting is illustrated in **Figure 8.43.** A small polypeptide called **ubiquitin** is connected by a covalent link to the substrate protein that is to be degraded. There are three components of the ubiquitination system. The ubiquitin-activating enzyme, E1, utilizes the cleavage of ATP to link itself via a high-energy thiolester bond from a Cys residue to the C-terminal Gly residue of ubiquitin. The ubiquitin is then transferred to the ubiquitin-conjugating enzyme, E2, which in turn transfers the ubiquitin to form an isopeptide bond to the ε NH_2 group of a Lys in the substrate protein. The substrate protein has usually been previously bound to the ubiquitin protein ligase, E3. Ubiquitin is released from a degraded substrate by an isopeptidase.

Responsibility for choosing substrate proteins to be ubiquitinated lies with both E2 and E3. In the simple scheme shown in Figure 8.43, E3 selects the substrate. A cell may contain several E3 proteins that use different criteria for selecting substrates. There are also multiple

Figure 8.43 The ubiquitin cycle involves three activities. E1 is linked to ubiquitin. E3 binds to the substrate protein. E2 transfers ubiquitin from E1 to the substrate. Further cycles generate polyubiquitin.

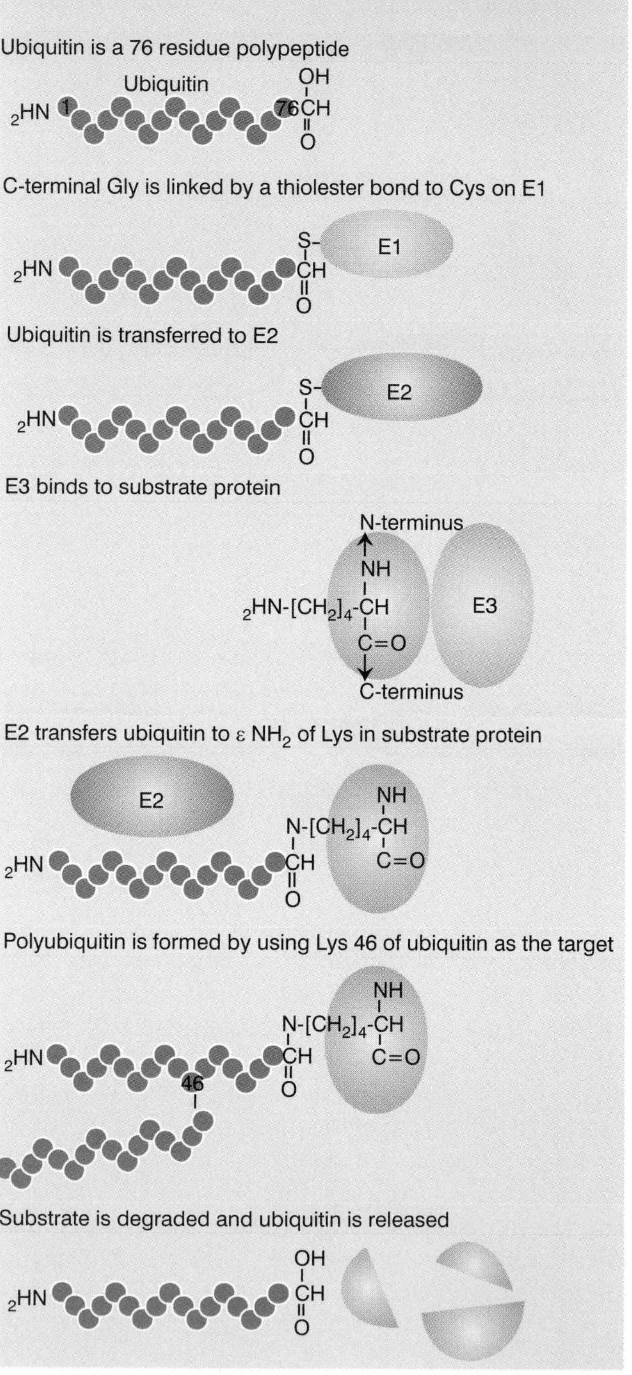

Ubiquitin is a 76 residue polypeptide
Ubiquitin
OH
2HN 1
76CH
O
C-terminal Gly is linked by a thiolester bond to Cys on E1
S-
E1
2HN
CH
O
Ubiquitin is transferred to E2
S-
E2
2HN
CH
O
E3 binds to substrate protein
N-terminus
NH
2HN-[CH2]4-CH
E3
C=O
C-terminus
E2 transfers ubiquitin to ε NH2 of Lys in substrate protein
E2
NH
N-[CH2]4-CH
2HN
CH
C=O
O
Polyubiquitin is formed by using Lys 46 of ubiquitin as the target
NH
N-[CH2]4-CH
2HN
CH
C=O
46
O
Substrate is degraded and ubiquitin is released
OH
CH
2HN
O

Figure 8.44 An archaeal 20S proteasome is a hollow cylinder consisting of rings of α and β subunits. Photographs kindly provided by Robert Huber.

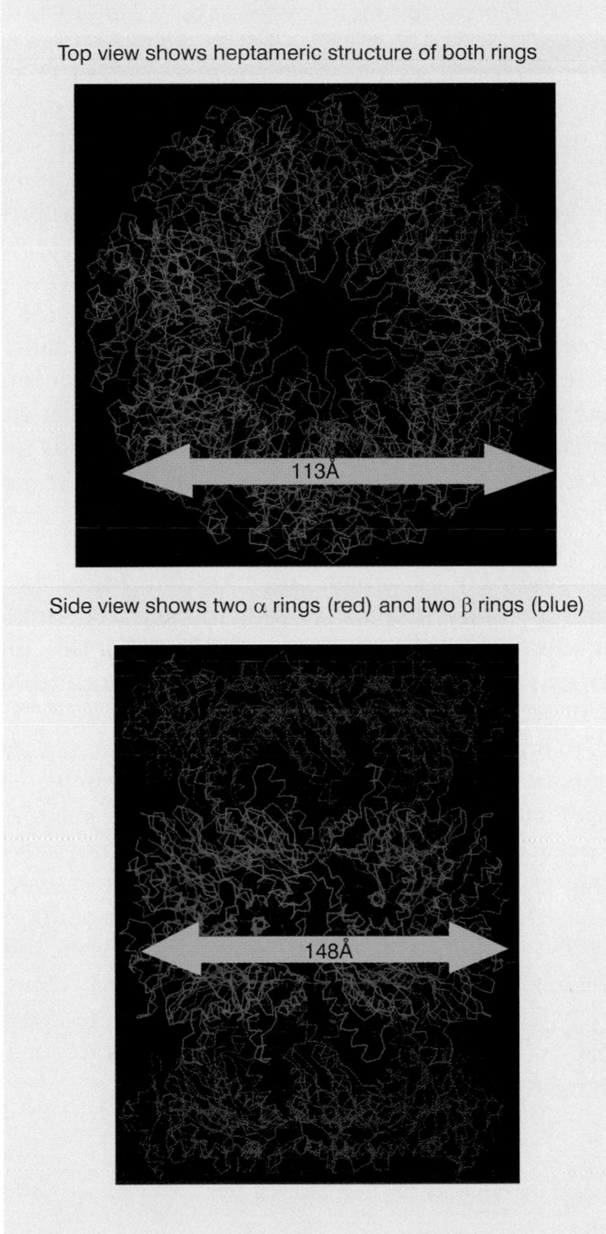
Top view shows heptameric structure of both rings
113Å
Side view shows two α rings (red) and two β rings (blue)
148Å

varieties of E2, and they also may play a role in targeting substrate proteins, sometimes independently of E3.

The addition of a single ubiquitin residue to a substrate protein is not sufficient to cause its degradation. Further ubiquitin residues are added to form a polyubiquitin chain, in which each additional ubiquitin is added to the Lys at position 46 of the preceding ubiquitin. The formation of polyubiquitin is a signal for the proteasome to degrade the protein. The control of polyubiquitination is not well understood, and there are some cases in which a single ubiquitin is added to a substrate protein, which remains stable because polyubiquitination does not occur.

The proteasome was originally discovered as a large complex that degrades proteins conjugated to ubiquitin. It exists in two forms. A 20S complex of ~700 kD contains the protease activity. Additional proteins convert the complex to a 26S form of ~2000 kD; they are regulatory subunits that confer specificity—for example, for binding to ubiquitin conjugates. ATP cleavage is required for the conversion from 20S to 26S, and is also required later in the reaction for cleaving peptide bonds, releasing the products, etc. The 20S complex takes the form of a hollow cylinder, and the additional components of the 26S complex are attached to the ends of the cylinder, making a dumbbell. Basically, the active sites are contained in the interior of a barrel, and access is obtained through relatively narrow channels, typically allowing access only to unfolded proteins. This protects normal, mature proteins from adventitious degradation.

This general type of structure is common to ATP-dependent proteases. For example, the ClpAP protease in *E. coli*, which is not related by sequence to the proteasome, has a structure in which the ClpP protease forms two rings of 7 subunits each, with the proteolytic activities contained in a central cavity. ClpA provides the ATPase activity and translocates substrates into the cavity, where they are degraded. It is hexameric (which implies an interesting symmetry mismatch in the ClpAP complex). Degradation is processive; once a substrate has been admitted to the central cavity, the reaction proceeds to its end.

The simplest proteasome is found in the archaea. It consists of two types of subunits, organized in the form α_7-β_7-β_7-α_7, where each heptamer forms a ring. **Figure 8.44** shows the backbone derived from the crystal structure of the 20S assembly. The α subunits form the two outer rings (on top and on the bottom), and the β subunits form the two inner rings. The β subunits have the protease activities, and the active sites are located at the N-terminal ends that project into the interior. The opening of ~20 Å restricts the entrance for substrates. (A yet simpler structure is found in *E. coli*, where a protein related to the β subunit, HslV, forms a structure of two six-member rings with a proteolytic core.)

The eukaryotic 20S proteasome is more complex, consisting of 7 different α subunits and 7 different β subunits. **Figure 8.45** shows that it has the same general structure of α-β-β-α rings. The rings in each half of the structure are organized in the opposite rotational

Figure 8.45 The eukaryotic 20S proteasome consists of two dimeric rings organized in counter-rotation.

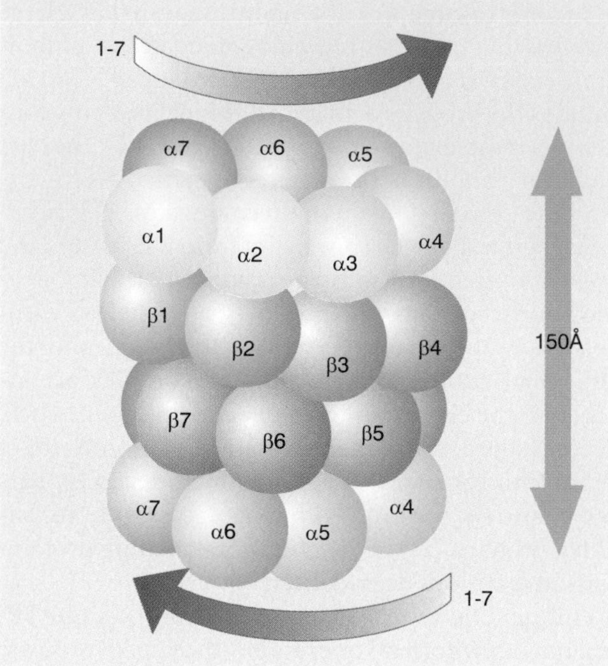

sense. A significant structural difference with the archaeal proteasome is that the central hole is occluded, so that there is no obvious entrance from the ends of the cylinder. This probably means that the structure is rearranged at some point to allow entrance from the either end.

The eukaryotic 26S proteasome is formed when the 19S caps associate with the 20S core, binding to one or both ends, to form an elongated structure of ~45 nm in length. The 19S caps are found only in eukaryotic (not archaeal or bacterial) proteasomes. The caps recognize ubiquitinated proteins, and pass them to the 20S core for proteolysis. The 19S caps contain ~18 subunits, several of which are ATPases; presumably the hydrolysis of ATP provides energy for handling the substrate proteins.

The hydrolytic mechanism of the proteasome is different from that of other proteases. The active site of a catalytic β-subunit is an N-terminal threonine; the hydroxyl group of the threonine attacks the peptide bond of the substrate. The proteasome contains several protease activities, with different specificities, for example, for cleaving after basic, acidic, or hydrophobic amino acids, allowing it to attack a variety of types of targets. Proteolytic activities with different substrate specificities may be provided by different β subunits. More than one β subunit may be needed for a particular enzymatic activity. The peptide products typically are octa- and nona-peptides; this size could reflect the distance of ~28 Å between the active sites of neighboring subunits.

Inhibitors of the proteasome block the degradation of most cellular proteins, showing that it is responsible for bulk degradation. It is also responsible for cleaving antigens in cells of the immune system to generate the small peptides that are presented on the surface of the cell to provoke the immune response (see Chapter 24). Other reactions in which target proteins are completely degraded include the removal of cell cycle regulators; in particular, cyclins are degraded during mitosis (see Chapter 27) and replication control proteins are degraded during the phase of DNA synthesis. In addition to these reactions, the proteasome may undertake specific processing events, for example, cleaving a precursor to a transcription factor to generate the active protein. The means by which these activities are regulated remain to be discovered.

Summary

Synthesis of all proteins starts on ribosomes that are "free" in the cytosol. In the absence of any particular signal, a protein is released into the cytosol when its synthesis is completed. Proteins that are imported by post-translational transfer into mitochondria or chloroplasts possess N-terminal leader sequences that target them to the organelle envelope; then they are transported through the double membrane into the matrix. Translocation requires ATP and a potential across the inner membrane. The N-terminal leader is cleaved by a protease within the organelle. Proteins that reside within the membranes or intermembrane space possess a signal (which becomes N-terminal when the first part of the leader is removed) that either causes export from the matrix to the appropriate location or which halts transfer before all of the protein has entered the matrix. Control of folding, by Hsp70 and Hsp60 in the mitochondrial matrix, is an important feature of the process. Requirements for export from bacteria also show strong dependence on control of protein conformation.

The N-terminal region of a secreted protein provides a signal sequence that causes the nascent protein and its ribosome to become attached to the membrane of the endoplasmic reticulum. The protein is translocated through the membrane by co-translational transfer. The process starts when the signal sequence is recognized by the SRP (a ribonucleoprotein particle), which interrupts translation. The SRP binds to a receptor in the ER membrane, and transfers the signal sequence to the Sec61/TRAM receptor in the membrane. Synthesis resumes, and the protein is translocated through the membrane while it is being synthesized, although there is no energetic connection between the processes. The channel through the membrane provides a hydrophilic environment, and is largely made of Sec61.

A secreted protein passes completely through the membrane into the ER lumen. For type I integral membrane proteins, the N-terminal signal sequence is cleaved, and transfer through the membrane is halted later by an anchor sequence. The protein becomes oriented in the membrane with its N-terminus on the far side and its C-terminus in the cytosol. Type II proteins do not have a cleavable N-terminal signal, but instead have a combined signal-anchor sequence, which enters the membrane and becomes embedded in it, causing the C-terminus to be located on the far side, while the N-terminus remains in the cytosol. The orientation of the signal-anchor is determined by the "positive inside" rule that the side of the anchor with more positive charges will be located in the cytoplasm. Proteins may have multiple membrane-spanning regions, with loops between them protruding on either side of the membrane. The mechanism of insertion of multiple segments is unknown.

Bacteria have components for membrane translocation that are related to those of eukaryotes, but translocation often occurs by a post-translational mechanism. SecY/E provide the translocase, and SecA associates with the channel and is involved in inserting and propelling the substrate protein. SecB is a chaperone that brings the protein to the channel.

Nuclear pore complexes are massive structures embedded in the nuclear membrane, and are responsible for all transport of protein into the nucleus and RNA out of the nucleus. Whether nuclear pore complexes are heterogeneous remains to be established. Each nuclear pore complex contains a central pore, which forms a channel of diameter <10 nm. Additional channels are present round the periphery. The central channel can be opened to a diameter of ~20 nm to allow passage of larger material, some of which may need to undergo conformational changes to fit.

Proteins that are actively transported into the nucleus require specific NLS sequences, which are short, but do not seem to share common features except for their basicity. Nuclear entry is a two-stage process, involving docking followed by an ATP-dependent translocation. The docking reaction is undertaken by the importin complex, which has subunits that bind to the substrate protein and to a nucleoporin protein in the pore, respectively. The direction of translocation is controlled by Ran. The presence of Ran-GDP in the cytosol destabilizes export complexes. The presence of Ran-GTP in the nucleus destabilizes import complexes. This ensures release of substrate on the appropriate side of the nuclear envelope.

Proteins that are exported from the nucleus have specific NES sequences, which share a pattern of leucine residues; they may bind to nucleoporins. Some nucleoporins are required specifically for RNA export from the nucleus.

The major system responsible for bulk degradation of proteins, but also for certain specific processing events, is the proteasome, a large complex that contains several protease activities. It acts upon substrate proteins that have been conjugated to ubiquitin through an isopeptide bond, and upon which a polyubiquitin chain has formed.

Further reading

Reviews

Oliver, D. (1985). Protein secretion in *E. coli*. *Ann. Rev. Microbiol.* **39,** 615–648.

Dingwall, C. and Laskey, R. A. (1986). Protein import into the cell nucleus. *Ann. Rev. Cell Biol.* **2,** 367–390.

Lee, C. and Beckwith, J. (1986). Cotranslational and posttranslational protein translocation in prokaryotic systems. *Ann. Rev. Cell Biol.* **2,** 315–336.

Walter, P. and Lingappa, V. (1986). Mechanism of protein translocation across the endoplasmic reticulum membrane. *Ann. Rev. Cell Biol.* **2,** 499–516.

Wickner, W. T. and Lodish, H. (1986). Multiple mechanisms of protein insertion into and across membranes. *Science* **230,** 400–407.

Ellis, R. J. and van der Vies, S. M. (1991). Molecular chaperones. *Ann. Rev. Biochem.* **60,** 321–347.

Baker, K. P. and Schatz, G. (1991). Mitochondrial proteins essential for viability mediate protein import into yeast mitochondria. *Nature* **349,** 205–208.

Forbes, D. J. (1992). Structure and function of the nuclear pore complex. *Ann. Rev. Cell Biol.* **8,** 495–527.

Jentsch, S. (1992). The ubiquitin-conjugation system. *Ann. Rev. Genet.* **26,** 179–207.

Georgopoulos, C. and Welch, W. J. (1993). Role of the major heat shock proteins as molecular chaperones. *Ann. Rev. Cell Biol.* **9,** 601–634.

Ciechanover, A. (1994). The ubiquitin-proteasome proteolytic pathway. *Cell* **79,** 13–21.

Walter, P. and Johnson, A. E. (1994). Signal sequence recognition and protein targeting to the endoplasmic reticulum membrane. *Ann. Rev. Cell Biol.* **10,** 87–119.

Davis, L. I. (1995). The nuclear pore complex. *Ann. Rev. Biochem.* **64,** 865–896.

Gorlich, D. and Mattaj, I. W. (1996). Nucleocytoplasmic transport. *Science* **271,** 1513–1518.

Schatz, G. and Dobberstein, B. (1996). Common principles of protein translocation across membranes. *Science* **271,** 1519–1526.

Rapoport, T. A., Jungnickel, B., and Kutay, U. (1996). Protein transport across the eukaryotic endoplasmic reticulum and bacterial inner membranes. *Ann. Rev. Biochem.* **65,** 271–303.

Coux, O., Tanaka, K., and Goldberg, A. L. (1996). Structure and functions of the 20S and 26S proteasomes. *Ann. Rev. Biochem.* **65,** 801–847.

Gorlich, D. and Mattaj, I. W. (1996). Nucleocytoplasmic transport. *Science* **271,** 1513–1518.

Hartl, F. U. (1966). Molecular chaperones in cellular protein folding. *Nature* **381,** 571–580.

Cline, K. and Henry, R. (1996). *Ann. Rev. Cell Dev. Biol.* **12,** 1–26.

Neupert, W. (1997). Protein import into mitochondria. *Ann. Rev. Biochem.* **66,** 863–917.

Bukau, B. and Horwich, A. L. (1998). The Hsp70 and Hsp60 chaperone machines. *Cell* **92,** 351–366.

Baumeister, W. *et al.* (1998). The proteasome: paradigm of a self-compartmentalizing protease. *Cell* **92,** 367–380.

Discoveries

Mitochondrial transport

van Loon, A. P. G. M. *et al.* (1986). The presequences of two imported mitochondrial proteins contain information for intracellular and intramitochondrial sorting. *Cell* **44,** 801–812.

Hartl, F.-U. *et al.* (1988). Successive translocation into and out of the mitochondrial matrix: targeting of proteins to the intermembrane space by a bipartite signal peptide. *Cell* **51,** 1027–-1037.

Chaperones and protein folding

Eilers, M. and Schatz, G. (1986). Binding of a specific ligand inhibits import of a purified precursor protein into mitochondria. *Nature* **322,** 228–232.

Collier, D. N. *et al.* (1988). The antifolding activity of SecB promotes the export of the *E. coli* maltose-binding protein. *Cell* **53,** 273–283.

Crooke, E. *et al.* (1988). ProOmpA is stabilized for membrane translocation by either purified *E. coli* trigger factor or canine signal recognition particle. *Cell* **54,** 1003–1011.

Ostermann, J. *et al.* (1989). Protein folding in mitochondria requires complex formation with Hsp60 and ATP hydrolysis. *Nature* **341,** 124–130.

Braig, K. *et al.* (1994). The crystal structure of the bacterial chaperonin GroEL at 2.8 Å. *Nature* **371,** 578–586.

Chen, S. *et al.* (1994). Location of a folding protein and shape changes in GroEL-GroES complexes imaged by cryo-electron microscopy. *Nature* **371,** 261–264.

Weissman, J. A. *et al.* (1994). GroEL-mediated protein folding proceeds by multiple rounds of binding and release of nonnative forms. *Cell* **78,** 683–702.

Weissman, J. A. *et al.* (1995). Mechanism of GroEL action: productive release of polypeptide from a sequestered position under GroES. *Cell* **83,** 577–587.

Hunt, J. F. *et al.* (1996). The crystal structure of the GroES co-chaperonin at 2.8 Å resolution. *Nature* **379**, 37–45.

Mayhew, M. *et al.* (1996). Protein folding in the central cavity of the GroEL-GroES chaperonin complex. *Nature* **379**, 420–426.

Xu, Z., Horwich, A. L., and Sigler, P. B. (1997). The crystal structure of the asymmetric GroEL-GroES-(ADP)$_7$ chaperonin complex. *Nature* **388**, 741–750.

Rye, H. S. *et al.* (1997). Distinct actions of cis and trans ATP within the double ring of the chaperonin GroEL. *Nature* **388**, 792–798.

Protein translocation

Blobel, G. and Dobberstein, B. (1975). Transfer of proteins across the membrane. I. Presence of proteolytically processed and unprocessed nascent immunoglobulin light chains on membrane-bound ribosomes of murine myeloma. *J. Cell Biol.* **67**, 835–851.

Walter, P. and Blobel, G. (1981). Translocation of proteins across the ER. III. SRP causes signal sequence and site specific arrest of chain elongation that is released by microsomal membranes. *J. Cell Biol.* **91**, 557–561.

Walter, P. and Blobel, G. (1982). Signal recognition particle contains a 7S RNA essential for protein translocation across the ER. *Nature* **299**, 691–698.

Siegel, V. and Walter, P. (1988). Each of the activities of SRP is contained within a distinct domain: analysis of biochemical mutants of SRP. *Cell* **52**, 39–49.

Brundage, L. *et al.* (1990). The purified *E. coli* integral membrane protein SecY/E is sufficient for reconstitution of SecA-dependent precursor protein translocation. *Cell* **62**, 649–657.

Simon, S. M. and Blobel, G. (1991). A protein-conducting channel in the endoplasmic reticulum. *Cell* **65**, 371–380.

Gorlich, D. and Rapoport, T. A. (1993). Protein translocation into proteoliposomes reconstituted from purified components of the endoplasmic reticulum membrane. *Cell* **75**, 615–630.

Crowley, K. S. (1994). Secretory proteins move through the ER membrane via an aqueous, gated pore. *Cell* **78**, 461–471.

Liao, S. *et al.* (1997). Both lumenal and cytosolic gating of the aqueous ER translocon pore are regulated from inside the ribosome during membrane protein integration. *Cell* **90**, 31–41.

Hanein, D. *et al.* (1997). Oligomeric rings of the Sec61p complex induced by ligands required for protein translocation. *Cell* **87**, 721–732.

Nuclear transport

Feldherr, C. M., Kallenbach, E., and Schultz, N. (1984). Movement of a karyophilic protein through the nuclear pores of oocytes. *J. Cell Biol.* **99**, 2216–2222.

Newmeyer, D. D. and Forbes, D. J. (1988). An NEM-sensitive cytosolic factor necessary for nuclear protein import: requirement in a signal-mediated binding to the nuclear pore. *Cell* **52**, 641–653.

Richardson, W. D. *et al.* (1988). Nuclear protein migration involves two steps: rapid binding at the nuclear envelope followed by slower translocation through nuclear pores. *Cell* **52**, 655–664.

Akey, C. W. and Goldfarb, D. S. (1989). Protein import through the nuclear pore complex is a multistep process. *J. Cell Biol.* **109**, 971–982.

Robbins, J. *et al.* (1991). Two interdependent basic domains in nucleoplasmin nuclear targeting sequence: identification of a class of bipartite nuclear targeting sequences. *Cell* **64**, 615–623.

Hinshaw, J. E., Carragher, B. O., and Milligan, R. A. (1992). Architecture and design of the nuclear pore complex. *Cell* **69**, 1133–1141.

Moore, M. S. and Blobel, G. (1992). The two steps of nuclear import, targeting to the nuclear envelope and translocation through the nuclear pore, require different cytosolic factors. *Cell* **69**, 939–950.

Moroianu, J., Blobel, G., and Radu, A. (1995). Previously identified protein of uncertain function is importin α and together with importin β docks import substrate at nuclear pore complexes. *Proc. Nat. Acad. Sci. USA* **92**, 2008–2011.

Radu, A., Moore, M. S., and Blobel, G. (1995). The peptide repeat domain of nucleoporin Nup98 functions as a docking site in transport across the nuclear pore complex. *Cell* **81**, 215–222.

Michael, W. M., Choi, M. and Dreyfuss, G. (1996). A nuclear export signal in hnRNP A1: a signal-mediated, temperature-dependent nuclear protein export pathway. *Cell* **83**, 415–422.

Stade, K. *et al.* (1997). Exportin 1 (Crm1p) is an essential nuclear export factor. *Cell* **90**, 1041–1050.

Fornerod, M. *et al.* (1997). CRM1 is an export receptor for leucine-rich nuclear export signals. *Cell* **90**, 1051–1060.

Kutay, U. *et al.* (1997). Export of importin α from the nucleus is mediated by a specific nuclear transport factor. *Cell* **90**, 1061–1071.

Protein degradation

Ciechanover, A. *et al.* (1980). ATP-dependent conjugation of reticulocyte proteins with the polypeptide required for protein degradation. *Proc. Nat. Acad. Sci. USA* **77**, 1365–1368.

Chau, V. *et al.* (1989). A multiubiquitin chain is confined to specific lysine in a targeted short–lived protein. *Science* **243**, 1576–1583.

Eytan, E. *et al.* (1989). ATP-dependent incorporation of 20S protease into the 26S complex that degrades proteins conjugated to ubiquitin. *Proc. Nat. Acad. Sci. USA* **86**, 7751–7755.

Lowe, J. *et al.* (1995). Crystal structure of the 20S proteasome from the archaeon *T. acidophilum* at 3.4 Å resolution. *Science* **268**, 533–539.

Wiertz, E. J. H. J. *et al.* (1996). Sec61-mediated transfer of a membrane protein from the endoplasmic reticulum to the proteasome for destruction. *Nature* **384**, 432–438.

Groll, M. *et al.* (1997). Structure of 20S proteasome from yeast at 2.4 Å resolution. *Nature* **386**, 463–471.

Part 3

mRNA

Chapter 9

Transcription

TRANSCRIPTION involves synthesis of an RNA chain representing one strand of a DNA duplex. By "representing" we mean that the RNA is *identical in sequence* with one strand of the DNA, which is called the **coding strand.** It is *complementary* to the other strand, which provides the **template** for its synthesis. This relationship between double-stranded DNA and its single-stranded RNA transcript is recapitulated in **Figure 9.1.**

RNA synthesis is catalyzed by the enzyme **RNA polymerase.** Transcription starts when RNA polymerase binds to a special region, the **promoter,** at the start of the gene. The promoter surrounds the first base pair that is transcribed into RNA, the **startpoint.** From this point, RNA polymerase moves along the template, synthesizing RNA, until it reaches a **terminator** sequence. This action defines a **transcription unit** that extends from the promoter to the terminator. The critical feature of the transcription unit, depicted in **Figure 9.2,** is that it constitutes a stretch of DNA *expressed via the production of a single RNA molecule.* A transcription unit may include more than one gene.

Sequences prior to the startpoint are described as **upstream** of it; those after the startpoint (within the transcribed sequence) are **downstream** of it. Sequences

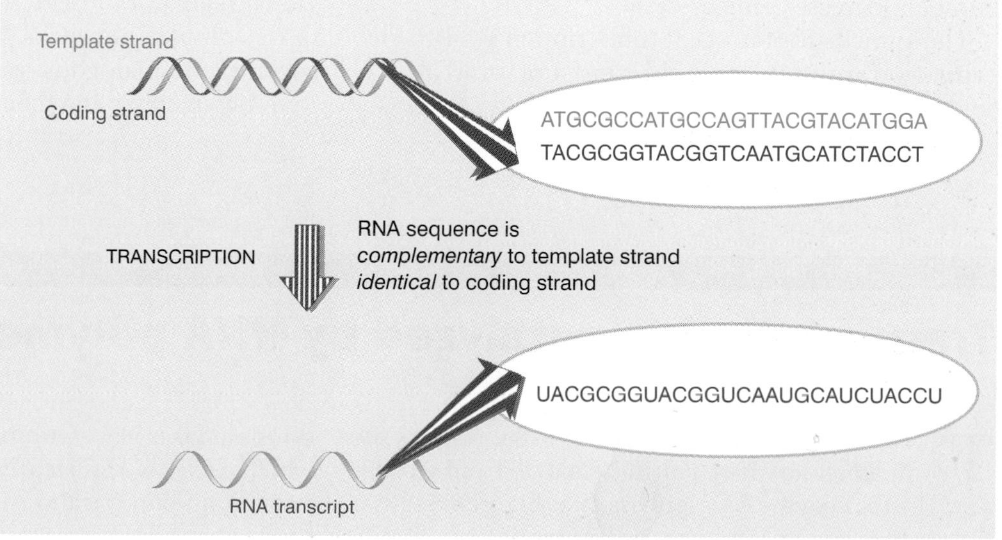

Figure 9.1 *Overview*: the function of RNA polymerase is to copy one strand of duplex DNA into RNA.

Figure 9.2 *Overview*: a transcription unit is a sequence of DNA transcribed into a single RNA, starting at the promoter and ending at the terminator.

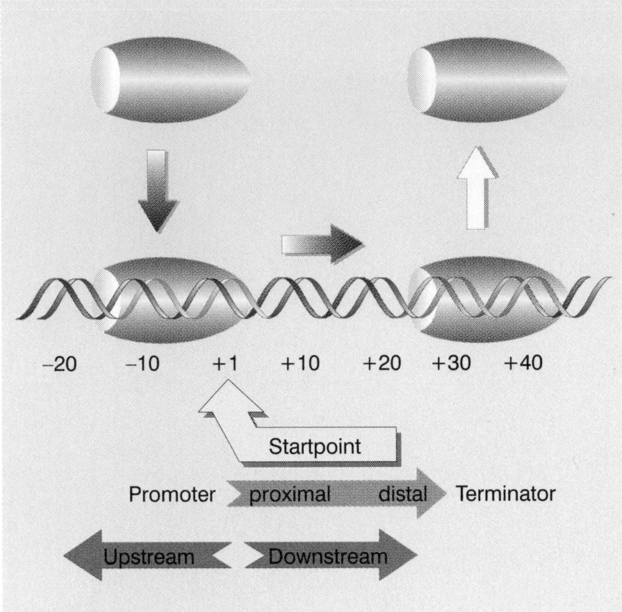

are conventionally written so that transcription proceeds from left (upstream) to right (downstream). This corresponds to writing the mRNA in the usual 5′ →3′ direction.

Often the DNA sequence is written just to show the coding strand, whose sequence is the same as the RNA. Base positions are numbered in both directions away from the startpoint, which is assigned the value +1; numbers are increased going downstream. The base before the startpoint is numbered –1, and the negative numbers increase going upstream.

The immediate product of transcription is called the **primary transcript.** It would consist of an RNA extending from the promoter to the terminator, possessing its original 5′ and 3′ ends. However, the primary transcript is almost always unstable. In prokaryotes, it is rapidly degraded (mRNA) or cleaved to give mature products (rRNA and tRNA). In eukaryotes, it is modified at the ends (mRNA) and/or cleaved to give mature products (all RNA).

Transcription is the first stage in gene expression, and the principal step at which it is controlled. Regulatory proteins determine whether a particular gene is available to be transcribed by RNA polymerase. The initial (and sometimes the only) step in regulation is the decision on whether or not to transcribe a gene. In considering the various stages of transcription, we should therefore keep in mind the opportunities they offer for regulating gene activity.

Within this context, there are two basic questions in gene expression:

- How does RNA polymerase find promoters on DNA? This is a particular example of a more general question: how do proteins distinguish their specific binding sites in DNA from other sequences?
- How do regulatory proteins interact with RNA polymerase (and with one another) to activate or to repress specific steps in the initiation, elongation, or termination of transcription?

In this chapter, we analyze the interactions of bacterial RNA polymerase with DNA, from its initial contact with a gene, through the act of transcription, culminating in its release when the transcript has been completed. Chapter 10 discusses the various means by which regulatory proteins can assist or prevent bacterial RNA polymerase from recognizing a particular gene for transcription. In Chapter 11 we consider how individual regulatory interactions can be connected into more complex networks. In Chapters 20 and 21, we consider the analogous reactions between eukaryotic RNA polymerases and their templates.

Transcription is catalyzed by RNA polymerase

TRANSCRIPTION takes place by the usual process of complementary base pairing, catalyzed and scrutinized by the enzyme RNA polymerase. **Figure 9.3** illustrates the general nature of transcription. RNA synthesis takes place within a "transcription bubble", in which DNA is transiently separated into its single strands, and one strand is used as a template for synthesis of the RNA strand. As RNA polymerase moves

Figure 9.3 Transcription takes place in a 'bubble', in which RNA is synthesized by base pairing with one strand of DNA in the transiently unwound region. As the bubble progresses, the DNA duplex reforms behind it, displacing the RNA in the form of a single polynucleotide chain.

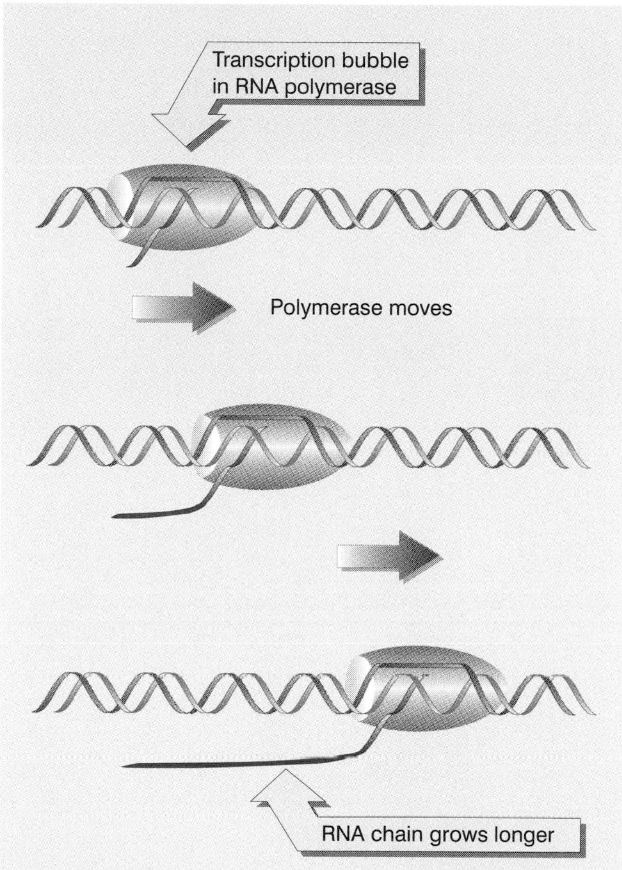

Figure 9.4 During transcription, the bubble is maintained within bacterial RNA polymerase, which unwinds and rewinds DNA, maintains the conditions of the partner and template DNA strands, and synthesizes RNA.

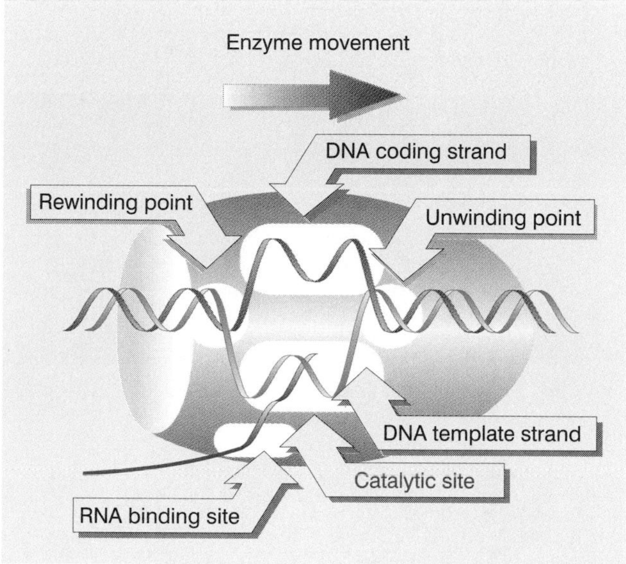

along the DNA, the bubble moves with it, and the RNA chain grows longer.

The structure of the bubble within RNA polymerase is shown in the expanded view of **Figure 9.4.** As RNA polymerase moves along the DNA template, it unwinds the duplex at the front of the bubble (the unwinding point), and rewinds the DNA at the back (the rewinding point). The length of the transcription bubble varies with the phase of the elongation reaction from 12–20 bp (see later), but the length of the RNA-DNA hybrid region within it is shorter.

The length of the RNA-DNA hybrid in the transcription bubble has not been measured directly, and remains controversial. It is clear that there is a significant change in topology extending over ~1 turn of DNA, but it is not clear how much of this region is actually base paired with RNA at any given moment. Certainly the RNA-DNA hybrid is short and transient.

As the enzyme moves on, the DNA duplex reforms, and the RNA is displaced as a free polynucleotide chain. About the last 25 ribonucleotides added to a growing chain are complexed with DNA and/or enzyme at any moment.

The classic view of transcription was that the hybrid region extends for ~12 bp, as suggested by indirect evidence about the structure of the RNA region within RNA polymerase. Attempts to make more direct measurements typically use a situation in which an elongating RNA polymerase is halted at a particular position by withholding the nucleotide that is required to add the next base to the RNA chain. Then the reactivity of bases in RNA or DNA at particular positions is tested. The results vary in suggesting a length for the RNA-DNA hybrid region between 3 and 9 base pairs.

Bacterial RNA polymerase has overall dimensions of ~90 × 95 × 160 Å. Yeast RNA polymerase is larger but less elongated. Structural analysis shows that they share a common type of structure, in which there is a "channel" or groove on the surface ~25 Å wide that could be the path for DNA. This is illustrated in **Figure 9.5** for the example of yeast RNA polymerase. The length of the groove could hold 16 bp in the bacterial enzyme, and ~25 bp in the yeast enzyme, but this represents only part of the total length of DNA bound during transcription. Aside from this general description, we

Figure 9.5 Yeast RNA polymerase has grooves that could be binding sites for nucleic acids. The pink beads show a possible path for DNA that is ~25 Å wide and 5–10 Å deep. The green beads show a narrower channel, 12–15 Å wide and ~20 Å deep, that could hold RNA. Photograph kindly provided by Roger Kornberg.

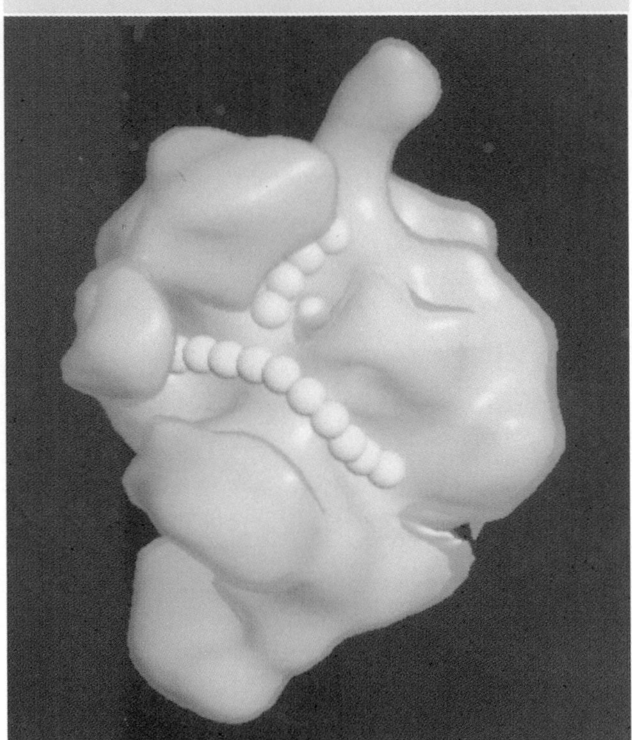

Figure 9.6 Phosphodiester bond formation involves a hydrophilic attack by the 3'-OH group of the last nucleotide of the chain on the 5' triphosphate of the incoming nucleotide, with release of pyrophosphate.

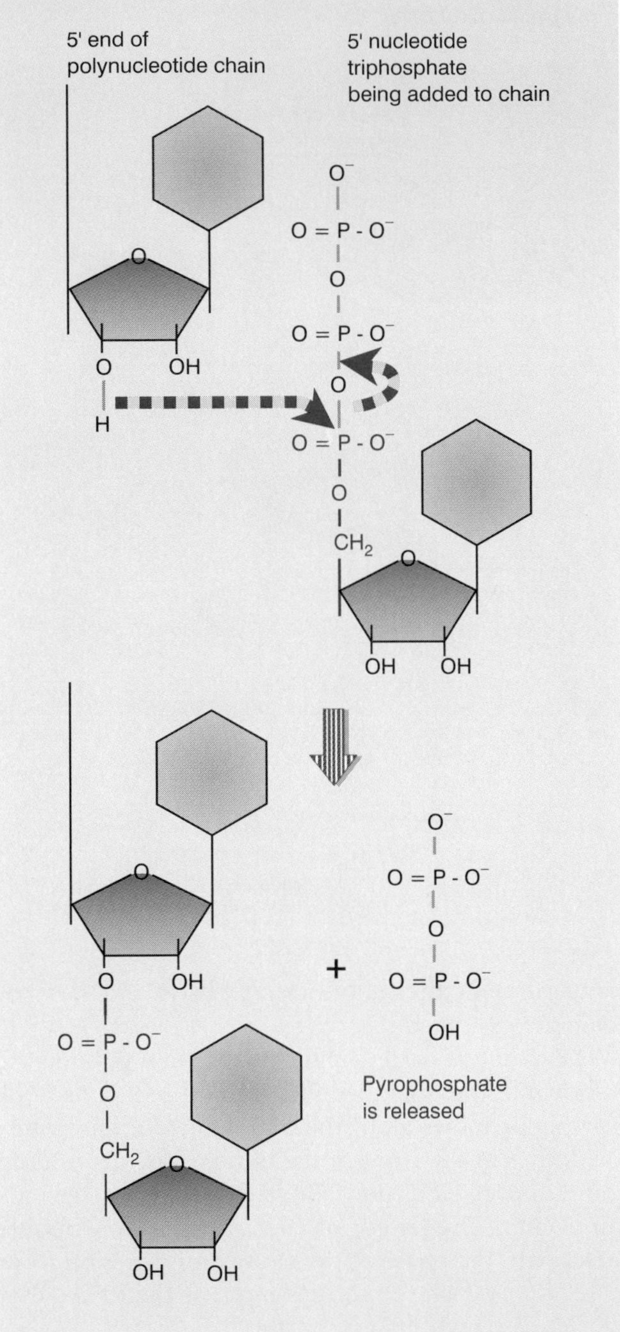

cannot yet relate the features of the enzyme to the generalized structure shown in Figure 9.4.

All nucleic acids are synthesized from nucleoside 5′ triphosphate precursors. **Figure 9.6** shows the condensation reaction between the 5′- triphosphate group of the incoming nucleotide and the 3′-OH group of the last nucleotide of the chain. The incoming nucleotide loses its terminal two phosphate groups (γ and β); its α group is used in the phosphodiester bond linking it to the chain. This reaction occurs in the catalytic site (see Figure 9.4).

The RNA chain is synthesized from the 5′ end toward the 3′ end. The overall reaction rate is ~40 nucleotides/second at 37°C (for the bacterial RNA polymerase); this is about the same as the rate of translation (15 amino acids/sec), but much slower than the rate of DNA replication (800 bp/sec).

RNA polymerase controls the entry of incoming nucleotides. Probably the enzyme allows phosphodiester bond formation to proceed only when a complementary nucleotide matches the template base. The nucleotide is expelled if it does not fit; then another can enter. The mechanism for discrimination is not known, but probably does not rely upon base pairing directly, because some base analogs that cannot pair are incorporated well.

The traditional view of elongation has been that it is a monotonic process, in which the enzyme moves forward 1 bp along DNA for every nucleotide added to the RNA chain. Changes in this pattern occur in certain circumstances, in particular when RNA polymerase pauses. One type of pattern is for the "front end" of the enzyme to remain stationary while the "back end" continues to move, thus compressing the footprint on DNA. After movement of several base pairs, the "front end" is released, restoring a footprint of full length. This gave rise to the "inchworm" model of transcription, in which the enzyme proceeds discontinuously, alternatively compressing and releasing the footprint on DNA. However, it may be the case that these events describe an aberrant situation rather than normal transcription.

One means of producing a pause is to arrest elongation (typically by omission of precursor nucleotides during an *in vitro* reaction). When the missing nucleotide is restored, the enzyme can overcome the block by cleaving the 3′ end of the RNA, to create a new 3′ terminus for chain elongation. The cleavage involves accessory factors in addition to the enzyme itself. In the case of *E. coli* RNA polymerase, the proteins GreA and GreB release the RNA polymerase from elongation arrest. Animal cell RNA polymerase requires the accessory factor (TFIIS), which enables the polymerase to cleave a few ribonucleotides from the 3′ terminus of the RNA product.

The catalytic site of RNA polymerase undertakes the actual cleavage in each case. There have been differences of opinion concerning the change in the enzyme that occurs at this time. One view is that there is an internal reorganization of structure, in which the catalytic center moves relative to the rest of the enzyme. The alternative model shown in **Figure 9.7** suggests that the enzyme as a whole "backtracks" on the DNA. The 3′ terminus of the RNA is exposed in single-stranded form, and the RNA-DNA hybrid region reverses its position. Cleavage restores a normal elongation complex. This model is supported by more recent measurements showing a constant distance between the catalytic center and the "front end".

We see therefore that RNA polymerase has the facility to unwind and rewind DNA, to hold the separated strands of DNA and the RNA product, to catalyze the addition of ribonucleotides to the growing RNA chain, and to adjust to difficulties in progressing by cleaving the RNA product and restarting RNA synthesis (with the assistance of some accessory factors).

Figure 9.7 A stalled RNA polymerase can be released by cleaving the 3' end of the transcript.

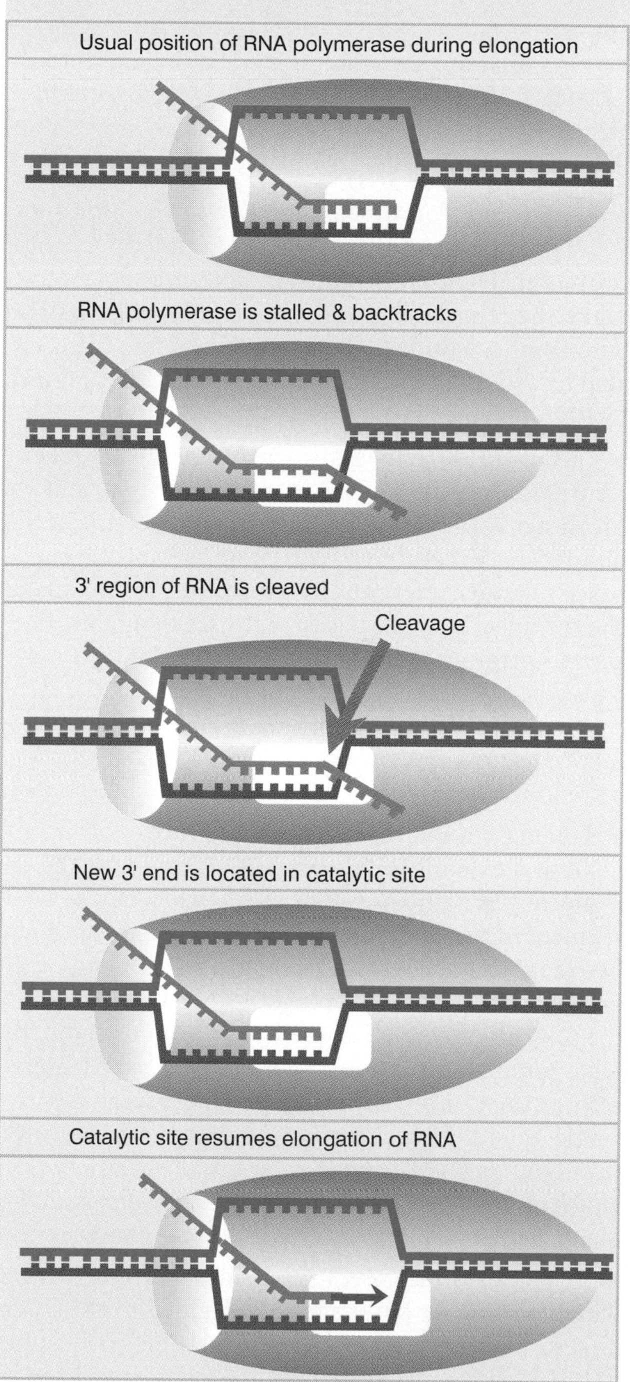
Usual position of RNA polymerase during elongation
RNA polymerase is stalled & backtracks
3' region of RNA is cleaved
Cleavage
New 3' end is located in catalytic site
Catalytic site resumes elongation of RNA

RNA polymerase consists of multiple subunits

THE transcription reaction can be divided into the stages illustrated in **Figure 9.8**, in which a bubble is created, RNA synthesis begins, the bubble moves along the DNA, and finally is terminated:

- **Template recognition** begins with the binding of RNA polymerase to the double-stranded DNA at a promoter. Then the strands of DNA are separated to make the template strand available for base pairing with ribonucleotides. The transcription bubble is created by a local unwinding that begins at the site bound by RNA polymerase.
- **Initiation** describes the synthesis of the first nucleotide bonds in RNA. The enzyme remains at the promoter while it synthesizes the first ~9 nucleotide bonds. The initiation phase is protracted by the occurrence of abortive events, in which the enzyme makes short transcripts (<9 bases), releases them, and then starts synthesis of RNA again. The initiation phase ends when the enzyme succeeds in extending the chain and escapes from the promoter. *The sequence of DNA needed for RNA polymerase to bind to the template and accomplish the initiation reaction defines the promoter.*
- During **elongation** the enzyme moves along the DNA and extends the growing RNA chain. As the enzyme moves, it unwinds the DNA helix to expose a new segment of the template in single-stranded condition. Nucleotides are covalently added to the 3′ end of the growing RNA chain, forming an RNA-DNA hybrid in the unwound region. Behind the unwound region, the DNA template strand pairs with its original partner to reform the double helix. The RNA emerges as a free single strand. *Elongation involves the movement of the transcription bubble by a disruption of DNA structure, in which the template strand of the transiently unwound region is paired with the nascent RNA at the growing point.*
- **Termination** involves recognition of the point at which no further bases should be added to the chain. To terminate transcription, the formation of phosphodiester bonds must cease, and the transcription complex must come apart. When the last base is added to the RNA chain, the transcription bubble collapses as the RNA-DNA hybrid is disrupted, the DNA reforms in duplex state, and the enzyme and RNA are both released. *The sequence of DNA required for these reactions is called the* ***terminator***.

Figure 9.8 Transcription has four stages, which involve different types of interaction between RNA polymerase and DNA. The enzyme binds to the promoter and melts DNA, remains stationary during initiation, moves along the template during elongation, and dissociates at termination.

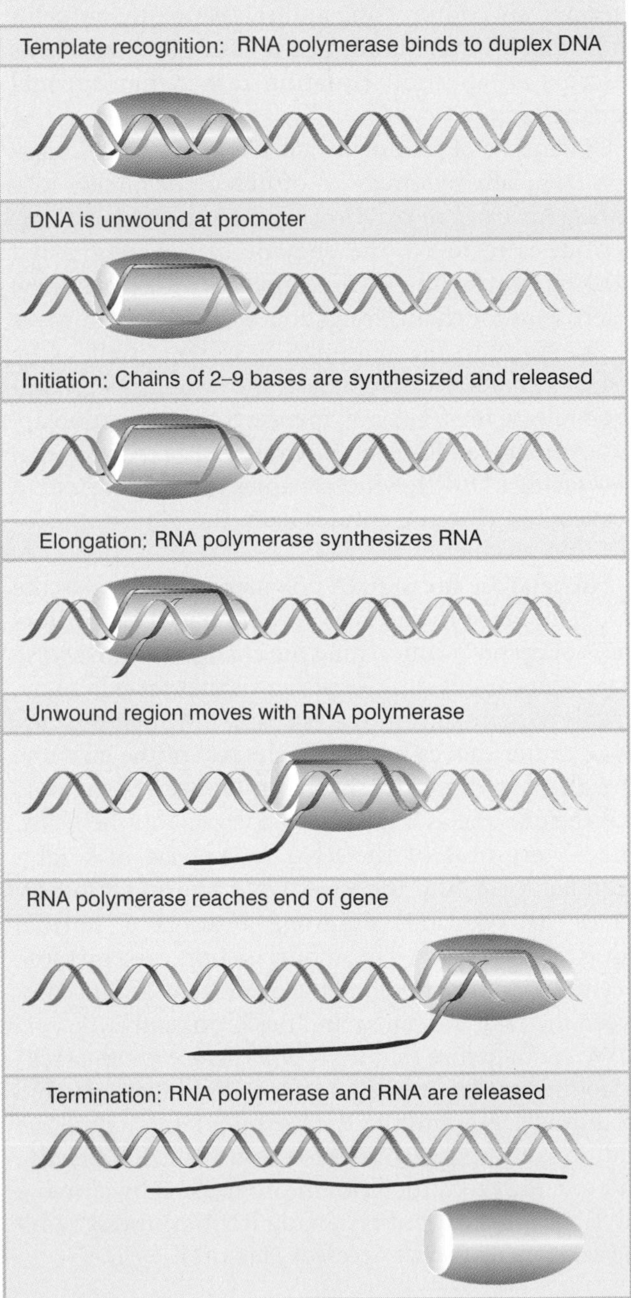
Template recognition: RNA polymerase binds to duplex DNA
DNA is unwound at promoter
Initiation: Chains of 2–9 bases are synthesized and released
Elongation: RNA polymerase synthesizes RNA
Unwound region moves with RNA polymerase
RNA polymerase reaches end of gene
Termination: RNA polymerase and RNA are released

Originally defined simply by its ability to incorporate nucleotides into RNA under the direction of a DNA template, the enzyme RNA polymerase now is seen as part of a more complex apparatus involved in transcription. *The ability to catalyze RNA synthesis defines the minimum component that can be described as RNA polymerase.* It supervises the base pairing of the substrate ribonucleotides with DNA and catalyzes the formation of phosphodiester bonds between them.

But ancillary activities are needed to initiate and to terminate the synthesis of RNA, when the enzyme must associate with, or dissociate from, a specific site on DNA. The analogy with the division of labors between the ribosome and the protein synthesis factors is obvious. Sometimes it is difficult to decide whether a particular protein that is involved in transcription at one of these stages should be considered as part of the "RNA polymerase" or as an ancillary factor.

All of the subunits of the basic polymerase that participate in elongation are necessary for initiation and termination. But transcription units differ in their dependence on additional polypeptides at the initiation and termination stages. Some of these additional polypeptides are needed at all genes, but others may be needed specifically for initiation or termination at particular genes. An additional polypeptide needed to recognize all promoters (or terminators) is likely to be classified as part of the enzyme. A polypeptide needed only for initiation (or termination) at particular genes is likely to be classified as an ancillary control factor.

The best-characterized RNA polymerases are those of eubacteria, for which *E. coli* is a typical case. *A single type of RNA polymerase appears to be responsible for almost all synthesis of mRNA, and all rRNA and tRNA, in a eubacterium.* About 7000 RNA polymerase molecules are present in an *E. coli* cell. Many of them are engaged in transcription; probably 2000–5000 enzymes are synthesizing RNA at any one time, the number depending on the growth conditions.

The **complete enzyme** or **holoenzyme** in *E. coli* has a molecular weight of ~465 kD. Its subunit composition is summarized in **Figure 9.9.**

The β and β′ subunits together make up the catalytic center; their sequences are related to those of the largest subunits of eukaryotic RNA polymerases (see Chapter 20), suggesting that there are common features to the actions of all RNA polymerases. The β subunit can be crosslinked to the template DNA, the product RNA, and the substrate ribonucleotides; mutations in *rpoB* affect all stages of transcription. Mutations in *rpoC* show that β′ also is involved at all stages.

Figure 9.9 Eubacterial RNA polymerases have four types of subunit; α, β, and β′ have rather constant sizes in different bacterial species, but σ varies more widely.

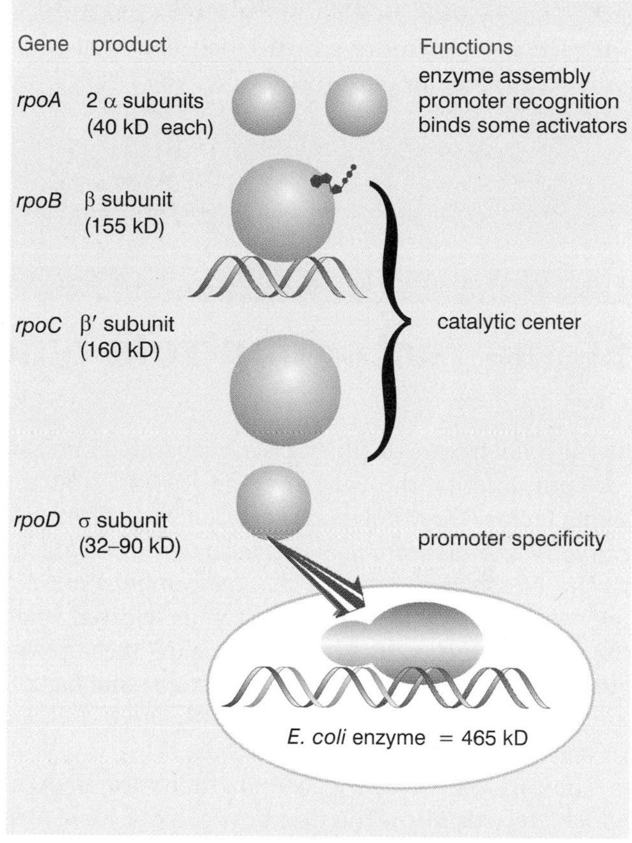

The α subunit is required for assembly of the core enzyme. When phage T4 infects *E. coli*, the α subunit is modified by ADP-ribosylation of an arginine. The modification is associated with a reduced affinity for the promoters formerly recognized by the holoenzyme, suggesting that the α subunit plays a role in promoter recognition. The α subunit also plays a role in the interaction of RNA polymerase with some regulatory factors.

The σ subunit is concerned specifically with promoter recognition, and we have more information about its functions than on any other subunit (see below).

The existence of much smaller RNA polymerases, comprising single polypeptide chains coded by certain phages, demonstrates that the apparatus required for RNA synthesis can be much smaller than that of the host enzyme. These enzymes give some idea of the "minimum" apparatus necessary for transcription. They recognize just a few promoters on the phage DNA; and they have no ability to change the set of promoters to which they respond. So they are limited to the intrinsic ability to recognize certain specific DNA-binding sequences and to synthesize RNA.

The RNA polymerases coded by the related phages T3 and T7 are single polypeptide chains of <100 kD each. They synthesize RNA at rates of ~200 nucleotides/sec at 37°C, more rapidly than bacterial RNA polymerase. The initiation reaction shows very little variation.

By contrast, the enzyme of the host bacterium can transcribe any one of many (>1000) transcription units. The host enzyme therefore requires the ability to interact with a variety of host and phage functions that modify its intrinsic transcriptional activities. The complexity of the enzyme therefore at least in part reflects its need to interact with a multiplicity of other factors, rather than any demand inherent in its catalytic activity.

Sigma factor controls binding to DNA

The holoenzyme ($\alpha_2\beta\beta'\sigma$) can be separated into two components, the **core enzyme** ($\alpha_2\beta\beta'$) and the **sigma factor** (the σ polypeptide). *Only the holoenzyme can initiate transcription. Sigma factor ensures that bacterial RNA polymerase binds in a stable manner to DNA only at promoters.* The sigma "factor" is released when the RNA chain reaches 8–9 bases, leaving the core enzyme to undertake elongation. *Core enzyme has the ability to synthesize RNA on a DNA template, but cannot initiate transcription at the proper sites.*

The core enzyme has a general affinity for DNA, in which electrostatic attraction between the basic protein and the acidic nucleic acid plays a major role. Any (random) sequence of DNA that is bound by core polymerase in this general binding reaction is described as a **loose binding site.** No change occurs in the DNA, which remains duplex. The complex at such a site is stable, with a half-life for dissociation of the enzyme from DNA of ~60 minutes. *Core enzyme does not distinguish between promoters and other sequences of DNA.*

Sigma factor introduces a major change in the affinity of RNA polymerase for DNA. *The holoenzyme has a drastically reduced ability to recognize loose binding sites*—that is, to bind to any general sequence of DNA. The association constant for the reaction is reduced by a factor of $\sim 10^4$, and the half-life of the complex is <1 second. So sigma factor destabilizes the general binding ability very considerably.

But sigma factor also *confers the ability to recognize specific binding sites.* The holoenzyme binds to promoters very tightly, with an association constant increased from that of core enzyme by (on average) 1000 times and with a half-life of several hours.

The specificity of holoenzyme for promoters compared with other sequences is $\sim 10^7$, but the association constant can be quoted only as an average, because there is wide variation in the rate at which the holoenzyme binds to different promoter sequences. This is an important factor in determining the efficiency of an individual promoter in initiating transcription. The binding constants extend from $\sim 10^{12}$ to $\sim 10^6$, reflecting promoter strengths that support initiation frequencies of ~1/sec (rRNA genes) to ~1/30 min (the *lacI* promoter).

We are now in a position to describe the stages of transcription in terms of the interactions between different forms of RNA polymerase and the DNA template. The initiation reaction can be described by the parameters that are summarized in **Figure 9.10:**

- The holoenzyme•promoter reaction starts by forming a closed binary complex. "Closed" means that the DNA remains duplex. Because the formation of the closed binary complex is reversible, it is usually described by an equilibrium constant (K_B). There is a wide range in values of the equilibrium constant for forming the closed complex.
- The closed complex is converted into an **open complex** by "melting" of a short region of DNA within the sequence bound by the enzyme. The series of events leading to formation of an open complex is called **tight binding.** For strong promoters, conversion into an open binary complex is irreversible, so this reaction is described by a rate constant (k_2). This reaction is fast.
- The next step is to incorporate the first two nucleotides; then a phosphodiester bond forms between them. This generates a **ternary complex** that contains RNA as well as DNA and enzyme. Formation of the ternary complex is described by the rate constant k_i;

Figure 9.10 RNA polymerase passes through several steps prior to elongation. A closed binary complex is converted to an open form and then into a ternary complex.

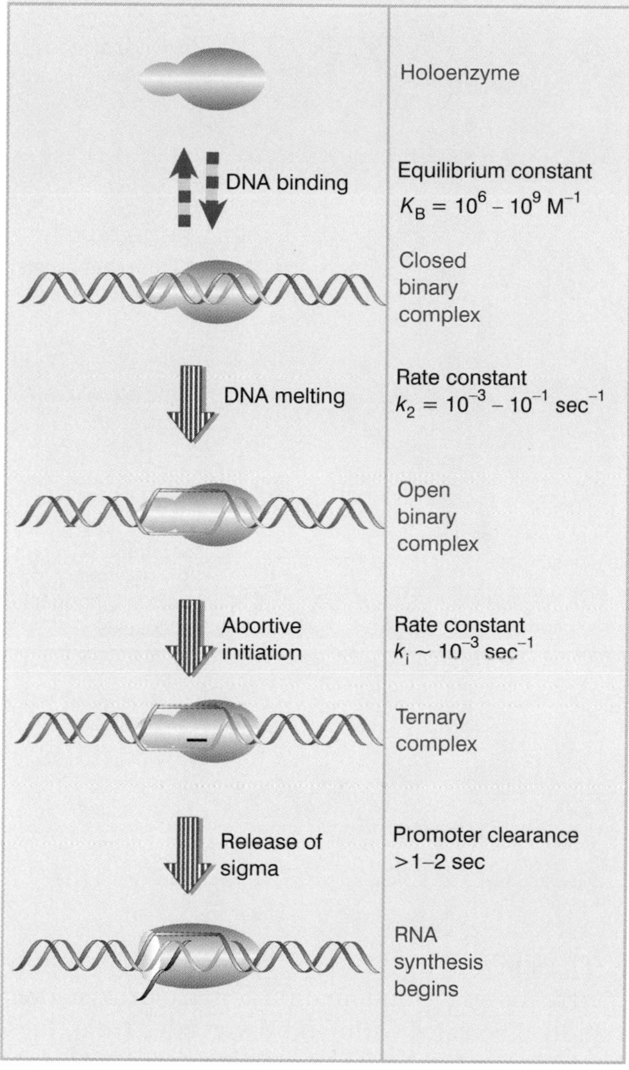

this is even faster than the rate constant k_2. Further nucleotides can be added without any enzyme movement to generate an RNA chain of up to 9 bases. After each base is added, there is a certain probability that the enzyme will release the chain. This comprises an **abortive initiation,** after which the enzyme begins again with the first base. A cycle of abortive initiations usually occurs to generate a series of short (2–9 base) oligonucleotides.

■ When initiation succeeds, the enzyme releases sigma and makes the transition to the elongation ternary complex of core polymerase•DNA•nascent RNA. The critical parameter here is *how long it takes for the polymerase to leave the promoter so that another polymerase can initiate.* This parameter is the promoter clearance time; its minimum value of 1–2 sec establishes the maximum frequency of initiation as <1 event per second. The enzyme then moves along the template, and the RNA chain extends beyond 10 bases.

When RNA polymerase binds to DNA, the elongated dimension of the protein extends along the DNA, but some interesting changes in shape occur during transcription. Transitions in shape and size identify three forms of the complex, as illustrated in **Figure 9.11:**

■ When RNA polymerase holoenzyme initially binds to DNA, it covers some 75–80 bp, extending from –55 to +20. (The long dimension of RNA polymerase (160 Å) could cover ~50 bp of DNA in extended form, which implies that binding of a longer stretch of DNA must involve some bending of the nucleic acid.)

■ The release of sigma results in a change in the shape of the RNA polymerase at the transition from initiation to elongation. The loss of sigma is associated with the loss of contacts in the –55 to –35 region, leaving only ~60 bp of DNA covered by the enzyme. This corresponds with the concept that the more upstream part of the promoter is involved in initial recognition by RNA polymerase, but is not required for the later stages of initiation (see later).

■ When the RNA chain extends to 15–20 bases, the enzyme makes a further transition, to form the complex that undertakes elongation; now it covers 30–40 bp (depending on the stage in the elongation cycle). It is not known whether the change at the start of elongation involves a significant alteration in the shape of the enzyme or whether the DNA is released into a less compact form.

How is RNA polymerase distributed in the cell? A (somewhat speculative) picture of the enzyme's situation is depicted in **Figure 9.12:**

■ Excess core enzyme exists largely as closed loose complexes, because the enzyme enters into them rapidly and leaves them slowly. There is very little, if any, free core enzyme.

■ There is enough sigma factor for about one-third of the polymerases to exist as holoenzymes, and they are distributed between loose complexes at nonspecific sites and binary complexes (mostly closed) at promoters.

■ About half of the RNA polymerases consist of core enzymes engaged in transcription.

Figure 9.11 RNA polymerase initially contacts the region from –55 to +20. When sigma dissociates, the core enzyme contracts to –30; when the enzyme moves a few base pairs, it becomes more compactly organized into the general elongation complex.

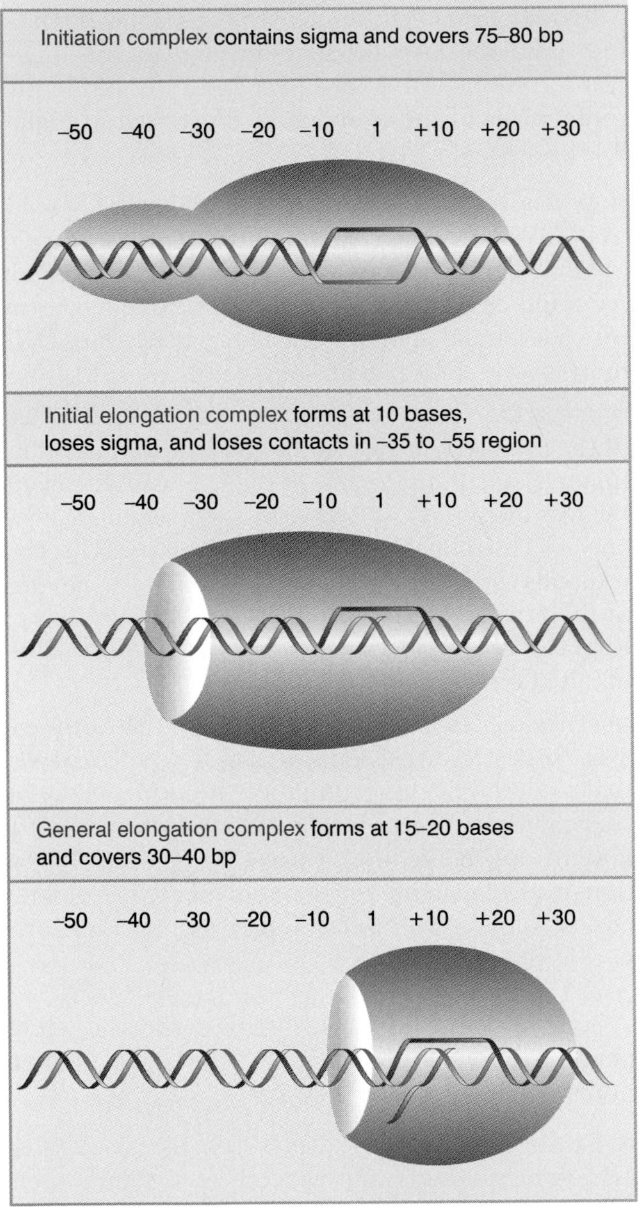
Initiation complex contains sigma and covers 75–80 bp
–50 –40 –30 –20 –10 1 +10 +20 +30
Initial elongation complex forms at 10 bases, loses sigma, and loses contacts in –35 to –55 region
–50 –40 –30 –20 –10 1 +10 +20 +30
General elongation complex forms at 15–20 bases and covers 30–40 bp
–50 –40 –30 –20 –10 1 +10 +20 +30

Figure 9.12 Core enzyme and holoenzyme are distributed on DNA, and very little RNA polymerase is free.

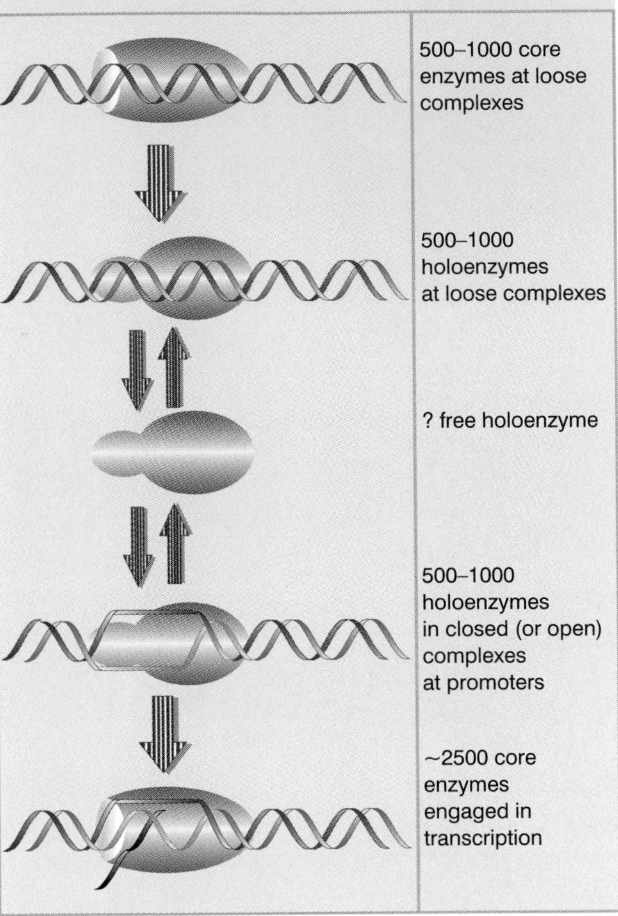
500–1000 core enzymes at loose complexes
500–1000 holoenzymes at loose complexes
? free holoenzyme
500–1000 holoenzymes in closed (or open) complexes at promoters
~2500 core enzymes engaged in transcription

■ How much holoenzyme is free? We do not know, but we suspect that the amount is very small.

RNA polymerase must find promoters within the context of the genome. Suppose that a promoter is a stretch of ~60 bp; how is it distinguished from the 4×10^6 bp that comprise the *E. coli* genome? **Figure 9.13** illustrates the principles of three models.

The simplest model is to suppose that RNA polymerase moves by random diffusion. Holoenzyme very rapidly associates with, and dissociates from, loose binding sites. So it could continue to make and break a series of closed complexes until (by chance) it encounters a promoter. Then its recognition of the specific sequence would allow tight binding to occur by formation of an open complex.

For RNA polymerase to move from one binding site on DNA to another, it must dissociate from the first site, find the second site, and then associate with it. Movement from one site to another is limited by the speed of diffusion through the medium. Diffusion sets an upper limit for the rate constant for associating with a 60 bp target of $<10^8\ M^{-1}\ sec^{-1}$. But the actual forward rate constant for some promoters *in vitro* appears to be $\sim 10^8\ M^{-1}\ sec^{-1}$, at or above the diffusion limit. *If this value applies* in vivo, *the time required for random cycles*

Figure 9.13 How does RNA polymerase find target promoters so rapidly on DNA?

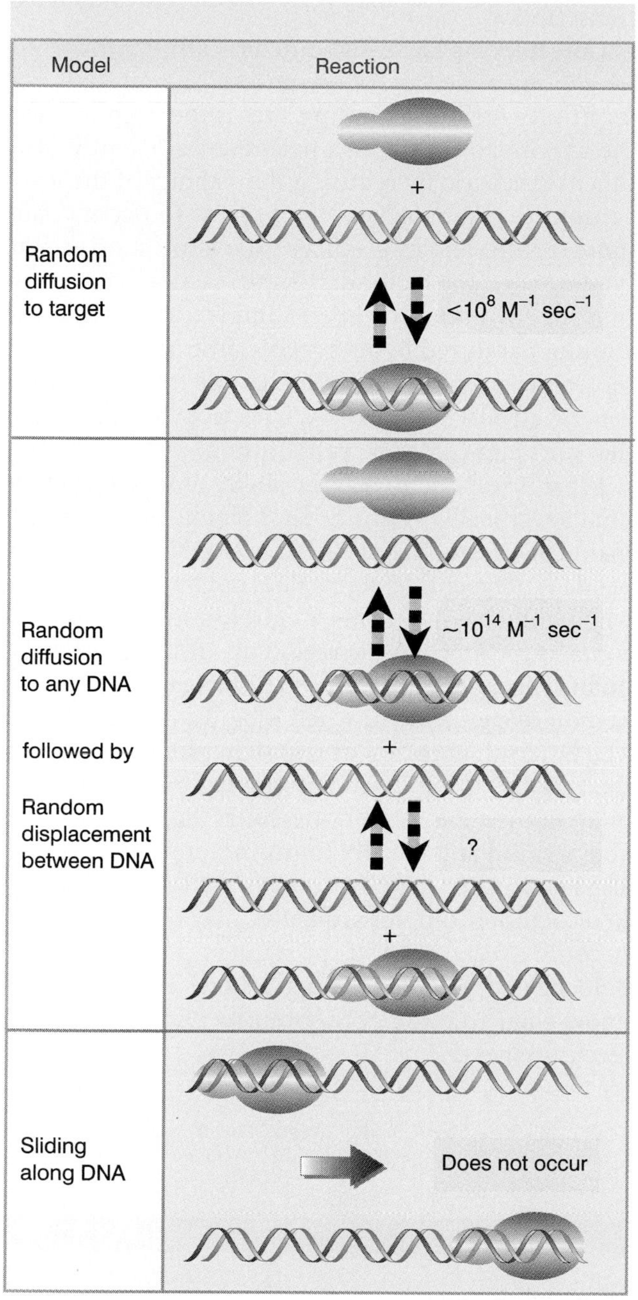

Model	Reaction
Random diffusion to target	+ $<10^8\ M^{-1}\ sec^{-1}$
Random diffusion to any DNA followed by Random displacement between DNA	$\sim10^{14}\ M^{-1}\ sec^{-1}$ + ? +
Sliding along DNA	Does not occur

Figure 9.14 Sigma factor and core enzyme recycle at different points in transcription.

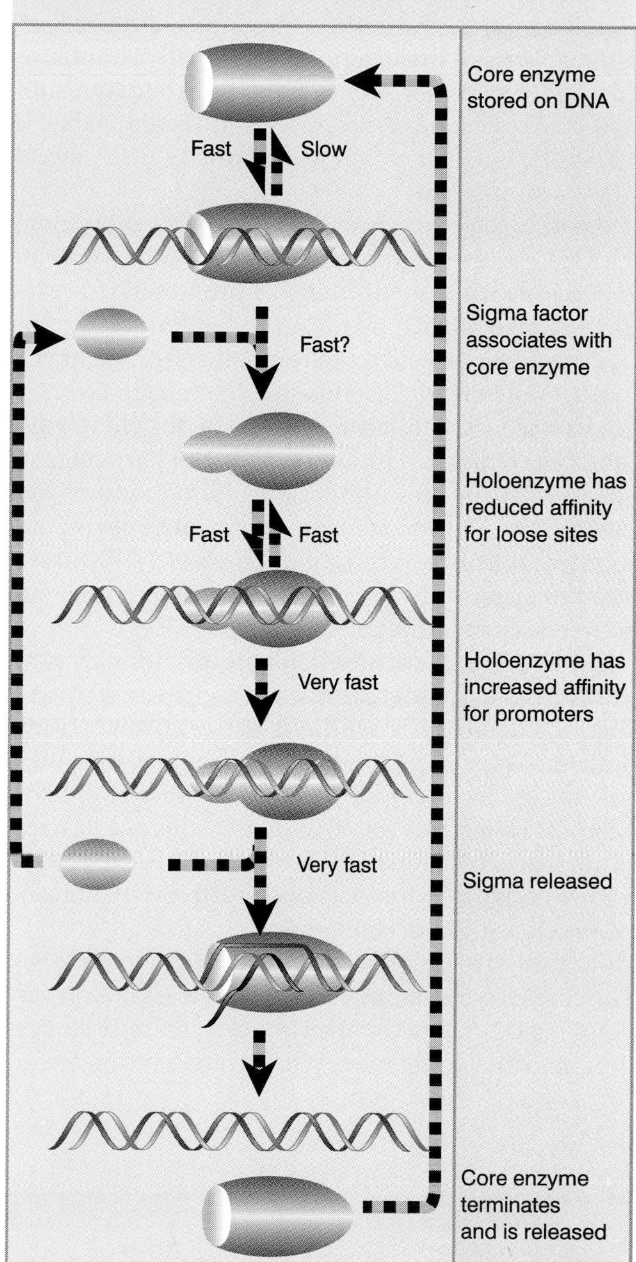
Core enzyme stored on DNA
Fast
Slow
Fast?
Sigma factor associates with core enzyme
Fast
Fast
Holoenzyme has reduced affinity for loose sites
Very fast
Holoenzyme has increased affinity for promoters
Very fast
Sigma released
Core enzyme terminates and is released

of successive association and dissociation at loose binding sites is too great to account for the way RNA polymerase finds its promoter.

RNA polymerase must therefore use some other means to seek its binding sites. The process could be speeded up if the initial target for RNA polymerase is the whole genome, not just a specific promoter sequence. By increasing the target size, the rate constant for diffusion to DNA is correspondingly increased, and is no longer limiting.

If this idea is correct, a free RNA polymerase binds DNA and then remains in contact with it. How does the enzyme move from a random (loose) binding site on DNA to a promoter? The most likely model is to suppose that the bound sequence is directly displaced by another sequence. Having taken hold of DNA, the enzyme exchanges this sequence with another sequence very rapidly, and continues to exchange sequences until a promoter is found. Then the enzyme forms a stable, open complex, after which initiation occurs. The search process becomes much faster because association and dissociation are virtually simultaneous, and time is not spent commuting between sites. Direct displacement can give a "directed walk", in which the enzyme moves preferentially from a weak site to a stronger site.

Another idea supposes that the enzyme slides along the DNA by a one-dimensional random walk, being halted only when it encounters a promoter. However, there is no evidence that RNA polymerase (or other DNA-binding proteins) can function in this manner.

RNA polymerase encounters a dilemma in reconciling its needs for initiation with those for elongation. Initiation requires tight binding *only* to particular sequences (promoters), while elongation requires close association with *all* sequences that the enzyme encounters during transcription. **Figure 9.14** illustrates how the dilemma is solved by the reversible association between sigma factor and core enzyme.

Sigma factor is involved only in initiation. It is released from the core enzyme when abortive initiation is concluded and RNA synthesis has been successfully initiated. Free sigma factor becomes immediately available for use by another core enzyme. We do not know whether sigma is released as a consequence of overcoming abortive initiation, or whether instead it is the release of sigma factor that ends abortive initiation and allows elongation to commence.

Release of sigma leaves the core enzyme on DNA. The core enzyme in the ternary complex is bound very tightly to DNA. It is essentially "locked in" until elongation has been completed. When transcription terminates, the core enzyme is released. It is then "stored" by binding to a loose site on DNA. It must find another sigma factor in order to undertake a further cycle of transcription.

Core enzyme has a high intrinsic affinity for DNA, which is increased by the presence of nascent RNA. But its affinity for loose binding sites is too high to allow the enzyme to distinguish promoters efficiently from other sequences. By reducing the stability of the loose complexes, sigma allows the process to occur much more rapidly; and by stabilizing the association at tight binding sites, the factor drives the reaction irreversibly into the formation of open complexes. To avoid becoming paralyzed by its specific affinity for the promoter, the enzyme releases sigma, and thus reverts to a general affinity for all DNA, irrespective of sequence, that suits it to continue transcription.

What is responsible for the ability of holoenzyme to bind specifically to promoters? Sigma factor has domains that recognize the promoter DNA. As an independent polypeptide, sigma does not bind to DNA, but when holoenzyme forms a tight binding complex, σ contacts the DNA in the region upstream of the startpoint. This difference is due to a change in the conformation of sigma factor when it binds to core enzyme. The N-terminal region of free sigma factor suppresses the activity of the DNA-binding region; when sigma binds to core, this inhibition is released, and it becomes able to bind specifically to promoter sequences (see also Figure 9.20 later). The inability of free sigma factor to recognize promoter sequences may be important: if σ could freely bind to promoters, it might block holoenzyme from initiating transcription. We do not know what role the core subunits play in promoter recognition.

Promoter recognition depends on consensus sequences

As a sequence of DNA whose function is to be *recognized by proteins*, a promoter differs from sequences whose role is to be transcribed or translated. The information for promoter function is provided directly by the DNA sequence: its structure is the signal. This is a classic example of a *cis*-acting site, as defined previously in Figures 1.31–32. By contrast, expressed regions gain their meaning only after the information is transferred into the form of some other nucleic acid or protein.

A key question in examining the interaction between an RNA polymerase and its promoter is how the protein recognizes a specific promoter sequence. Does the enzyme have an active site that distinguishes the chemical structure of a particular sequence of bases in the DNA double helix? How specific are its requirements?

One way to design a promoter would be for a particular sequence of DNA to be recognized by RNA polymerase. Every promoter would consist of, or at least include, this sequence. In the bacterial genome, the minimum length that could provide an adequate signal is 12 bp. (Any shorter sequence is likely to occur—just by chance—a sufficient number of additional times to provide false signals. The minimum length required for unique recognition increases with the size of genome.) The 12 bp sequence need not be contiguous. If a specific number of base pairs separates two constant shorter sequences, their combined length could be less than 12 bp, since the *distance* of separation itself provides a part of the signal (even if the intermediate *sequence* is itself irrelevant).

Attempts to identify the features in DNA that are necessary for RNA polymerase binding started by comparing the sequences of different promoters. Any essential nucleotide sequence should be present in all the promoters. Such a sequence is said to be **conserved.** However, a conserved sequence need not necessarily be conserved at every single position; some variation is permitted. How do we analyze a sequence of DNA to determine whether it is sufficiently conserved to constitute a recognizable signal?

Putative DNA recognition sites can be defined in terms of an idealized sequence that represents the base most often present at each position. A **consensus sequence** is defined by aligning all known examples so as to maximize their homology. For a sequence to be accepted as a consensus, each particular base must be reasonably predominant at its position, and most of the actual examples must be related to the consensus by rather few substitutions, say, no more than 1–2.

The striking feature in the sequence of promoters in *E. coli* is the *lack of any extensive conservation of sequence* over the 60 bp associated with RNA polymerase. The sequence of much of the binding site is irrelevant. But some short stretches within the promoter are conserved, and they are critical for its function. *Conservation of only very short consensus sequences is a typical feature of regulatory sites (such as promoters) in both prokaryotic and eukaryotic genomes.*

There are four conserved features in a bacterial promoter: the startpoint; the –10 sequence; the –35 sequence; and the separation between the –10 and –35 sequences:

- The startpoint is usually (>90% of the time) a purine. It is common for the startpoint to be the central base in the sequence CAT, but the conservation of this triplet is not great enough to be regarded as an obligatory signal.
- Just upstream of the startpoint, a 6 bp region is recognizable in almost all promoters. The center of the hexamer generally is close to 10 bp upstream of the startpoint; the distance varies in known promoters from position –18 to –9. Named for its location, the hexamer is often called the **–10 sequence.** Its consensus is **TATAAT,** and can be summarized in the form

$$T_{80}\, A_{95}\, T_{45}\, A_{60}\, A_{50}\, T_{96}$$

where the subscript denotes the percentage occurrence of the most frequently found base, varying from 45 to 96%. (A position at which there is no discernible preference for any base would be indicated by N.) If the frequency of occurrence indicates likely importance in binding RNA polymerase, we would expect the initial highly conserved TA and the final almost completely conserved T in the –10 sequence to be the most important bases.

- Another conserved hexamer is centered ~35 bp upstream of the startpoint. This is called the **–35 sequence.** The consensus is **TTGACA;** in more detailed form, the conservation is

$$T_{82}\, T_{84}\, G_{78}\, A_{65}\, C_{54}\, A_{45}$$

- The distance separating the –35 and –10 sites is between 16 and 18 bp in 90% of promoters; in the exceptions, it is as little as 15 or as great as 20 bp. *Although the actual sequence in the intervening region is unimportant, the distance is critical in holding the two sites at the appropriate separation for the geometry of RNA polymerase.*

The optimal promoter is a sequence consisting of the –35 hexamer, separated by 17 bp from the –10 hexamer, lying 7 bp upstream of the startpoint. The structure of a promoter, showing the permitted range of variation from this optimum, is illustrated in **Figure 9.15.**

A major source of information about promoter function is provided by mutations. Mutations in promoters affect the level of expression of the gene(s) they control, without altering the gene products themselves. Most are identified as bacterial mutants that have lost, or have very much reduced, transcription of the adjacent genes. They are known as **down mutations.** Less often, mutants are found in which there is increased

Figure 9.15 A typical promoter has three components, consisting of consensus sequences at –35 and –10, and the startpoint.

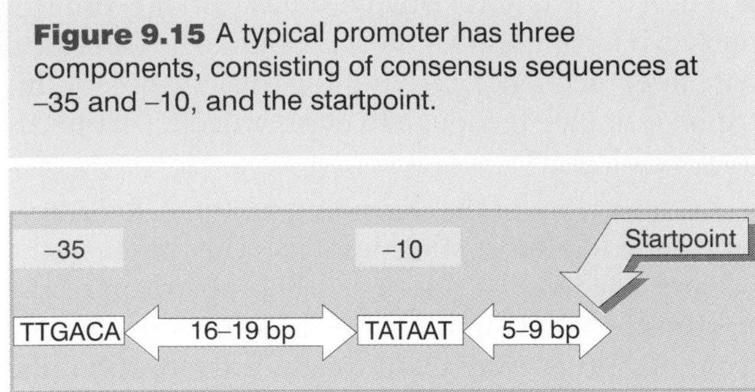

transcription from the promoter. They have **up mutations.**

It is important to remember that "up" and "down" mutations are defined relative to the *usual* efficiency with which a particular promoter functions. This varies widely. So a change that is recognized as a down mutation in one promoter might never have been isolated in another (which in its wild-type state could be even less efficient than the mutant form of the first promoter). So information gained from studies *in vivo* simply identifies the overall direction of the change caused by mutation.

Is the most effective promoter one that has the actual consensus sequences? This expectation is borne out by the simple rule that up mutations usually increase homology with one of the consensus sequences or bring the distance between them closer to 17 bp. Down mutations usually decrease the resemblance of either site with the consensus or make the distance between them further away from 17 bp. Down mutations tend to be concentrated in the most highly conserved positions, which confirms their particular importance as the main determinant of promoter efficiency. However, occasional exceptions to these rules demonstrate that promoter efficiency cannot be predicted entirely from conformity to the consensus.

To determine the absolute effects of promoter mutations, we must measure the affinity of RNA polymerase for wild-type and mutant promoters *in vitro*. There is ~100-fold variation in the rate at which RNA polymerase binds to different promoters *in vitro*, which correlates well with the frequencies of transcription when their genes are expressed *in vivo*. Taking this analysis further, we can investigate the stage at which a mutation influences the capacity of the promoter. Does it change the affinity of the promoter for binding RNA polymerase? Does it leave the enzyme able to bind but unable to initiate? Is the influence of an ancillary factor altered?

By measuring the kinetic constants for formation of a closed complex and its conversion to an open complex, as defined in Figure 9.10, we can dissect the two stages of the initiation reaction:

- Down mutations in the –35 sequence reduce the rate of closed complex formation (they reduce K_B), but do not inhibit the conversion to an open complex.
- Down mutations in the –10 sequence do not affect the initial formation of a closed complex, but they slow its conversion to the open form (they reduce k_2).

These results suggest that the function of the –35 sequence is to provide the signal for recognition by RNA polymerase, while the –10 sequence allows the complex to convert from closed to open form. We might view the –35 sequence as comprising a "recognition domain," while the –10 sequence comprises an "unwinding domain" of the promoter.

The consensus sequence of the –10 site consists exclusively of A•T base pairs, which assists the initial melting of DNA into single strands. The lower energy needed to disrupt A•T pairs compared with G•C pairs means that a stretch of A•T pairs demands the minimum amount of energy for strand separation.

The sequence immediately around the startpoint influences the initiation event. And the initial transcribed region (from +1 to +30) influences the rate at which RNA polymerase clears the promoter, and therefore has an effect upon promoter strength. So the overall strength of a promoter cannot be predicted entirely from its –35 and –10 consensus sequences.

A "typical" promoter relies upon its –35 and –10 sequences to be recognized by RNA polymerase, but one or the other of these sequences can be absent from some (exceptional) promoters. In at least some of these cases, the promoter cannot be recognized by RNA polymerase alone, and the reaction requires the intercession of ancillary proteins, which overcome the deficiency in intrinsic interaction between RNA polymerase and the promoter.

RNA polymerase binds to one face of DNA

THE ability of RNA polymerase (or indeed any protein) to recognize DNA can be characterized by **footprinting.** A sequence of DNA bound to the protein is *partially* digested with an endonuclease to attack individual phosphodiester bonds within the nucleic acid. Under appropriate conditions, any particular phosphodiester bond is broken in some, but not in all, DNA molecules. The positions that are cleaved are recognized by using DNA labeled on one strand at one end only. The principle is the same as that involved in DNA sequencing; partial cleavage of an end-labeled molecule at a susceptible site creates a fragment of unique length.

As **Figure 9.16** shows, following the nuclease treatment, the broken DNA fragments are recovered and electrophoresed on a gel that separates them according to length. Each fragment that retains a labeled end produces a radioactive band. The position of the band corresponds to the number of bases in the fragment. The shortest fragments move the fastest, so distance from the labeled end is counted up from the bottom of the gel .

In a free DNA, *every* susceptible bond position is broken in one or another molecule. But when the DNA is complexed with a protein, the region covered by the DNA-binding protein is protected in every molecule. So two reactions are run in parallel: a control of pure DNA; and an experimental mixture containing molecules of DNA bound to the protein. *When a bound protein blocks access of the nuclease to DNA, the bonds in the bound sequence fail to be broken in the experimental mixture.*

In the control, every bond is broken, generating a series of bands, one representing each base. In the figure, 31 bands can be counted. In the protected fragment, bonds cannot be broken in the region bound by the protein, so bands representing fragments of the corresponding sizes are not generated. The absence of bands 9–18 in the figure identifies a protein-binding site covering the region located 9–18 bases from the labeled end of the DNA. By comparing the control and experimental lanes with a sequencing reaction that is run in parallel it becomes possible to "read off" the corresponding sequence directly, thus identifying the nucleotide sequence of the binding site.

As described previously (see Figure 9.11), RNA polymerase initially binds the region from –50 to +20. The points at which RNA polymerase actually contacts the promoter can be identified by modifying the footprinting technique to treat RNA polymerase•promoter complexes with reagents that modify particular bases. The common feature of all the types of modification is that *they allow a breakage to be made at the corresponding bond in the polynucleotide chain.* The site of breakage can be identified by the same approach used in footprinting. We can perform the experiment in two ways:

- The direct analogy with footprinting is to treat an RNA polymerase•DNA complex with a modifying agent and to compare its susceptibility with that of free DNA. Some bands disappear, identifying sites at which the enzyme has protected the promoter against modification. Other bands increase in intensity, identifying sites at which the DNA must be held in a conformation in which it is more exposed.
- The reverse experiment can be performed by modifying the DNA *first*; then it is bound to RNA polymerase. Those DNA molecules that cannot bind RNA polymerase are recovered and treated in the usual way to generate strand breaks whose positions can be identified. This locates points at which prior modification *prevents* RNA polymerase from binding to DNA.

These changes in sensitivity reveal the geometry of the complex, as summarized in **Figure 9.17** for a typical promoter. The regions at –35 and –10 contain most of the contact points for the enzyme. Within these regions, the same sets of positions tend both to prevent binding if previously modified, and to show increased or decreased susceptibility to modification after binding. Although the points of contact do not coincide completely with sites of mutation, they occur in the same limited region.

It is noteworthy that the same *positions* in different promoters provide the contact points, even though a different base is present. This indicates that there is a common mechanism for RNA polymerase binding, although the reaction does not depend on the presence of particular bases at some of the points of contact. This model explains why some of the points of contact are not sites of mutation. Also, not every mutation lies in a point of contact; they may influence the neighborhood without actually being touched by the enzyme.

Figure 9.16 Footprinting identifies DNA-binding sites for proteins by their protection against nicking.

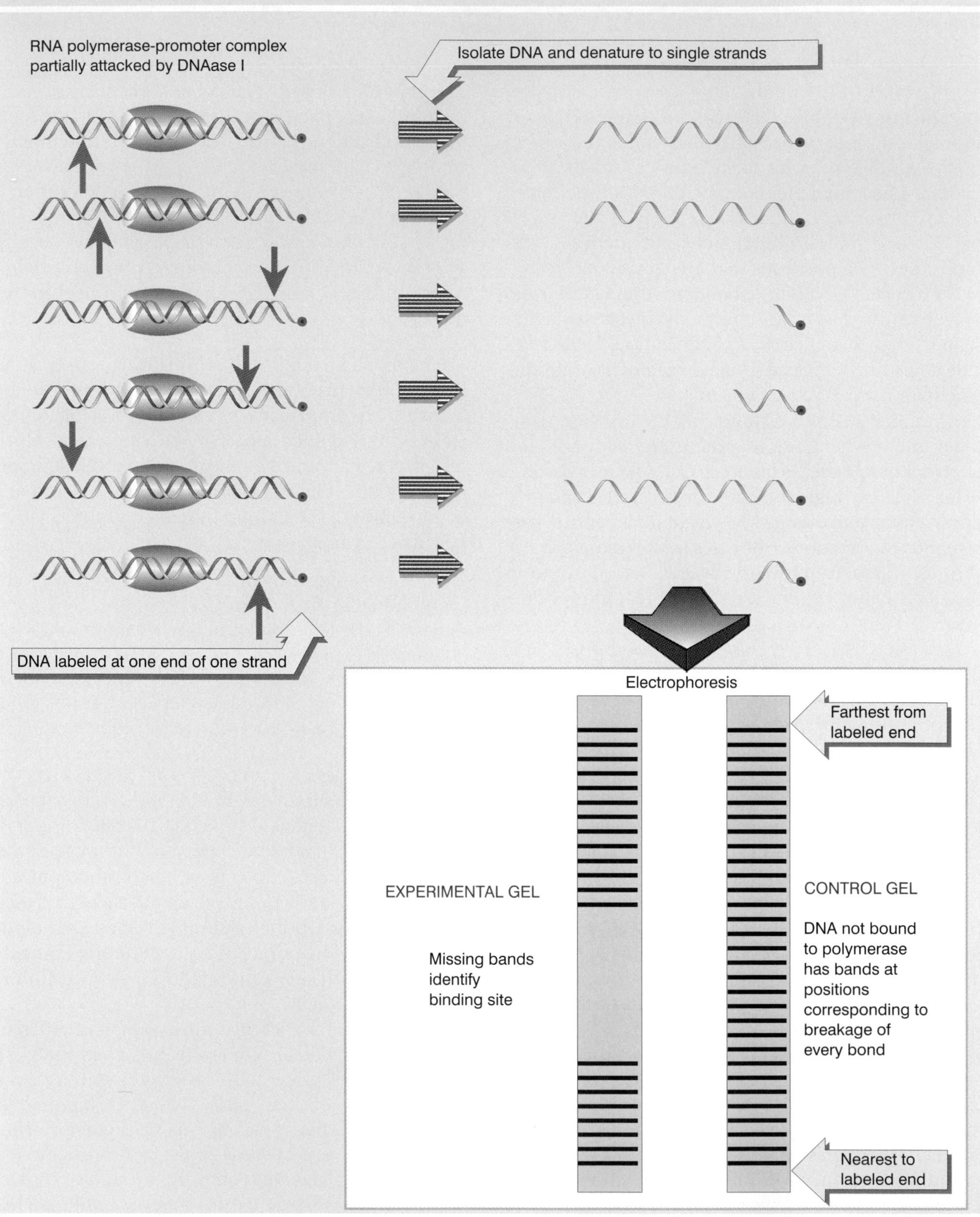
RNA polymerase-promoter complex
partially attacked by DNAase I
Isolate DNA and denature to single strands
DNA labeled at one end of one strand
Electrophoresis
Farthest from
labeled end
EXPERIMENTAL GEL
Missing bands
identify
binding site
CONTROL GEL
DNA not bound
to polymerase
has bands at
positions
corresponding to
breakage of
every bond
Nearest to
labeled end

Figure 9.17 One face of the promoter contains the contact points for RNA.

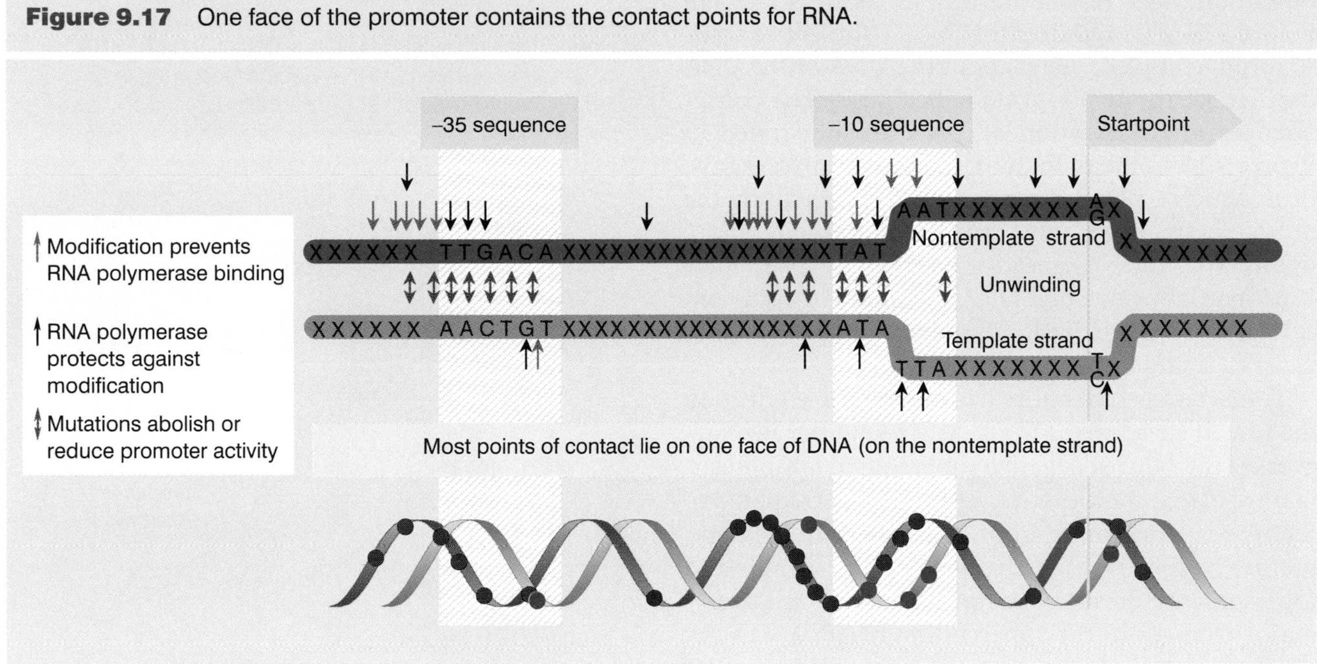
-35 sequence
-10 sequence
Startpoint
Modification prevents RNA polymerase binding
RNA polymerase protects against modification
Mutations abolish or reduce promoter activity
XXXXXX TTGACA XXXXXXXXXXXXXXXXXTAT
AATXXXXXXX
Nontemplate strand
XXXXXX
Unwinding
XXXXXX AACTGT XXXXXXXXXXXXXXXXXATA
Template strand
TTAXXXXXXX
Most points of contact lie on one face of DNA (on the nontemplate strand)

It is especially significant that the experiments with prior modification identify *only* sites in the same region that are protected by the enzyme against subsequent modification. These two experiments measure different things. Prior modification identifies all those sites that the enzyme must recognize in order to bind to DNA. Protection experiments recognize all those sites that actually make contact in the binary complex. The protected sites include all the recognition sites and also some additional positions, which suggests that the enzyme first recognizes a set of bases necessary for it to "touch down", and then extends its points of contact to additional bases.

The region of DNA that is unwound in the binary complex can be identified directly by chemical changes in its availability. When the strands of DNA are separated, the unpaired bases become susceptible to reagents that cannot reach them in the double helix. Such experiments implicate positions between –9 and +3 in the initial melting reaction. The region unwound during initiation therefore includes the right end of the –10 sequence and extends just past the startpoint.

Viewed in three dimensions, the points of contact upstream of the –10 sequence all lie on one face of DNA. This can be seen in the lower drawing in Figure 9.17, in which the contact points are marked on a double helix viewed from one side. Most lie on the coding strand. These bases are probably recognized in the initial formation of a closed binary complex. This would make it possible for RNA polymerase to approach DNA from one side and recognize that face of the DNA. As DNA unwinding commences, further sites that originally lay on the other face of DNA might be recognized and bound.

The importance of strand separation in the initiation reaction is emphasized by the effects of supercoiling. Both prokaryotic and eukaryotic RNA polymerases can initiate transcription more efficiently *in vitro* when the template is supercoiled, presumably because the supercoiled structure requires less free energy for the initial melting of DNA in the initiation complex.

The efficiency of some promoters is influenced by the degree of supercoiling. The most common relationship is for transcription to be aided by negative supercoiling. We understand in principle how this assists the initiation reaction. But why should some promoters be influenced by the extent of supercoiling while others are not? One possibility is that the dependence of a promoter on supercoiling is determined by its sequence. This would predict that some promoters have sequences that are easier to melt (and are therefore less dependent on supercoiling), while others have more difficult sequences (and have a greater need to be supercoiled). An alternative is that the location of the promoter might be important if different regions of the bacterial chromosome have different degrees of supercoiling.

Supercoiling also has a continuing involvement with transcription. As RNA polymerase transcribes DNA, unwinding and rewinding occurs, as illustrated in Figure 9.3. This requires that either the entire transcription complex rotates about the DNA or the DNA itself must rotate about its helical axis. The consequences of the rotation of DNA are illustrated in **Figure 9.18** in the **twin domain** model for transcription. As RNA polymerase pushes forward along the double helix, it generates positive supercoils (more tightly wound DNA) ahead and leaves negative supercoils (partially unwound DNA) behind. For each helical turn traversed by RNA polymerase, +1 turn is generated ahead and –1 turn behind.

Transcription therefore has a significant effect on the (local) structure of DNA. As a result, the enzymes gyrase (introduces negative supercoils) and topoisomerase I (removes negative supercoils) are required to rectify the situation in front of and behind the polymerase, respectively. If the activities of gyrase and topoisomerase are interfered with, transcription causes major changes in the supercoiling of DNA. For example, in yeast lacking an enzyme that relaxes negative supercoils, the density of negative supercoiling doubles in a transcribed region. A possible implication of these results is that transcription is responsible for generating a significant proportion of the supercoiling that occurs in the cell.

Figure 9.18 Transcription may generate more tightly wound (positively supercoiled) DNA ahead of RNA polymerase, while the DNA behind becomes less tightly wound (negatively supercoiled).

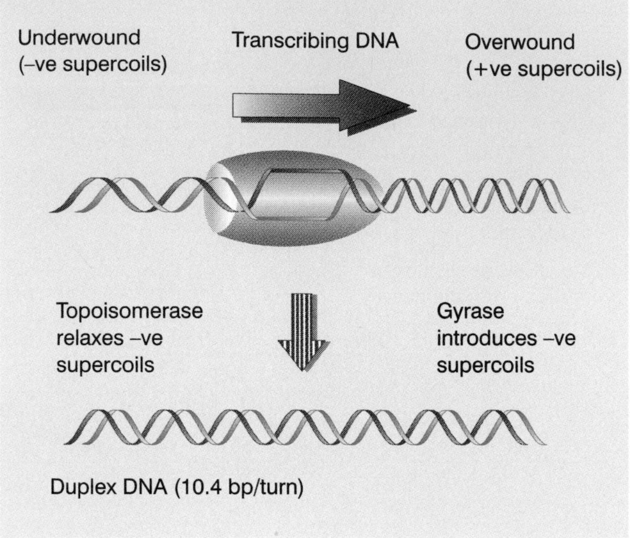

A similar situation occurs in replication, when DNA must be unwound at a moving replication fork, so that the individual single strands can be used as templates to synthesize daughter strands. Solutions for the topological constraints associated with such reactions are indicated later in Figure 14.15.

Substitution of sigma factors may control initiation

THE division of labor between a core enzyme that undertakes chain elongation and a sigma factor involved in site selection immediately raises the question of whether there is more than one type of sigma, each specific for a different class of promoter. Changes in sigma factors appear in some cases when there is a wholesale reorganization of transcription, for example, during the change in lifestyle that occurs in a sporulating bacterium (see next section). *E. coli* does not undergo such dramatic changes, but does use alternative sigma factors to respond to general environmental changes. They are listed in **Figure 9.19.** (They are named either by molecular weight of the product or for the gene.) The general factor, responsible for transcription of most genes under normal conditions, is σ^{70}. The alternative sigma factors σ^{32}, σ^{E}, and σ^{54} are activated in response to environmental changes; σ^{28} is used for expression of flagellar genes during normal growth, but its level of expression responds to changes in the environment. All the sigma factors except σ^{54} belong to the same protein family and function in the same general manner described earlier.

A common type of response to heat shock occurs in many organisms, both prokaryotic and eukaryotic. Upon an increase in temperature, synthesis of the proteins currently being made is turned off or down, and a new set of proteins is synthesized. The new proteins are the products of the **heat shock genes.** They play a role

Figure 9.19 E. coli sigma factors recognize promoters with different consensus sequences. (Numbers in the name of a factor indicate its mass.)

Gene	Factor	Use	–35 Sequence	Separation	–10 Sequence
rpoD	σ^{70}	general	TTGACA	16–18 bp	TATAAT
rpoH	σ^{32}	heat shock	CCCTTGAA	13–15 bp	CCCGATNT
rpoE	σ^{E}	heat shock	not known	not known	not known
rpoN	σ^{54}	nitrogen	CTGGNA	6 bp	TTGCA
fliA	σ^{F}	flagellar	CTAAA	15 bp	GCCGATAA

in protecting the cell against environmental stress, and are synthesized in response to other conditions as well as heat shock. Several of the heat shock proteins are chaperones. In *E. coli*, the expression of 17 heat shock proteins is triggered by changes at transcription. The gene *rpoH* is a regulator needed to switch on the heat shock response. Its product is σ^{32}, which functions as an alternative sigma factor that causes transcription of the heat shock genes.

The heat shock response is accomplished by increasing the amount of σ^{32} when the temperature increases, and decreasing its activity when the temperature change is reversed. The basic signal that induces production of σ^{32} is the accumulation of unfolded (partially denatured) proteins that results from increase in temperature. The σ^{32} protein is unstable, which is important in allowing its quantity to be increased or decreased rapidly. σ^{70} and σ^{32} can compete for the available core enzyme, so that the set of genes transcribed during heat shock depends on the balance between them.

Another group of heat-regulated genes is controlled by the factor σ^{E}. It appears to respond to more extreme temperature shifts than σ^{32}. It is induced by accumulation of unfolded proteins that are usually found in the periplasmic space or outer membrane. Less is known about this sigma factor and about the genes it controls.

Another sigma factor is used under conditions of nitrogen starvation. *E. coli* cells contain a small amount of σ^{54}, which is activated when ammonia is absent from the medium. In these conditions, genes are turned on to allow utilization of alternative nitrogen sources. Counterparts to this sigma factor have been found in a wide range of bacteria, so it represents a response mechanism that has been conserved in evolution.

Another case of evolutionary conservation of sigma factors is presented by the factor σ^{F}, which is present in small amounts and causes RNA polymerase to transcribe genes involved in chemotaxis and flagellar structure. Its counterpart in *B. subtilis* is σ^{D}, which controls flagellar and motility genes; factors with the same promoter specificity are present in many species of bacteria.

Each sigma factor causes RNA polymerase to initiate at a particular set of promoters. By analyzing the sequences of these promoters, we can show that each set is identified by unique sequence elements. Indeed, the sequence of each type of promoter ensures that it is recognized only by RNA polymerase directed by the appropriate sigma factor. We can deduce the general rules for promoter recognition from the identification of the genes responding to the sigma factors found in *E. coli* and those involved in sporulation in *B. subtilis* (which are discussed in the next section).

A significant feature of the promoters for each enzyme is that *they have the same size and location relative to the startpoint, and they show conserved sequences only around the usual centers of –35 and –10.* The consensus sequences for each set of promoters are different from one another at either or both of the –35 and –10 positions. This means that an enzyme containing a particular sigma factor can recognize only its own set of promoters, so that transcription of the different groups is mutually exclusive. So substitution of one sigma factor by another turns off transcription of the old set of genes as well as turning on transcription of a new set of genes.

The definition of a series of different consensus sequences recognized at –35 and –10 by holoenzymes containing different sigma factors carries the immediate implication that the sigma subunit must itself contact DNA in these regions. This suggests the general principle that there is a common type of relationship between sigma and core enzyme, in which the sigma factor is positioned in such a way as to make critical contacts with the promoter sequences in the vicinity of –35 and –10.

Comparisons of the sequences of several bacterial sigma factors identify regions that have been conserved. Their locations in *E. coli* σ^{70} are summarized in **Figure 9.20.**

Direct evidence that sigma contacts the promoter directly at both the –35 and –10 consensus sequences is provided by mutations in sigma that suppress mutations in the consensus sequences. When a mutation at a particular position in the promoter prevents

Figure 9.20 A map of the *E. coli* σ^{70} factor identifies conserved regions. Regions 2.1 and 2.2 contact core polymerase, 2.3 is required for melting, and 2.4 and 4.2 contact the –10 and –35 promoter elements. The N-terminal region prevents 2.4 and 4.2 from binding to DNA in the absence of core enzyme.

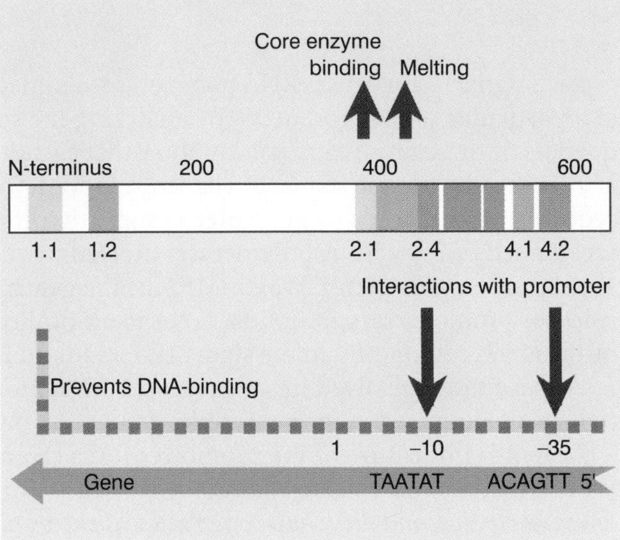

recognition by RNA polymerase, and a compensating mutation in sigma factor allows the polymerase to use the mutant promoter, we may conclude that the relevant base pair in DNA is contacted by the amino acid that has been substituted. Figure 9.20 shows that two short parts of regions 2 and 4 (named 2.4 and 4.2) are involved in contacting bases in the –10 and –35 elements, respectively. Both of these regions form short stretches of α-helix in the protein. Experiments with heteroduplexes show that σ^{70} makes contacts with bases principally on the nontemplate strand, and it continues to hold these contacts after the DNA has been unwound in this region. This suggests that sigma factor could be important in the melting reaction.

Region 2.3 resembles proteins that bind single-stranded nucleic acids, and is involved in the melting reaction. Regions 2.1 and 2.2 (which is the most highly conserved part of sigma) are involved in the interaction with core enzyme. It is assumed that all sigma factors bind the same regions of the core polymerase (ensuring that the reactions are competitive).

When the N-terminal region of σ^{70} is removed, the shortened protein becomes able to bind specifically to promoter sequences. This suggests that the N-terminal region behaves as an autoinhibition domain. It occludes the DNA-binding domains when σ^{70} is free. Association with core enzyme changes the conformation of sigma so that the inhibition is released, and the core domain becomes able to bind to DNA. When sigma binds to the core polymerase, the autoinhibition domain swings ~20 Å away from the DNA-binding domains, and the DNA-binding domains separate from one another by ~15 Å, presumably to acquire a more elongated conformation appropriate for contacting DNA. Mutations in either the –10 or –35 sequences prevent an (N-terminal-deleted) σ^{70} from binding to DNA, which suggests that σ^{70} contacts both sequences simultaneously. This implies that the sigma factor must have a rather elongated structure, extending over the ~68 Å of two turns of DNA.

The use of α-helical motifs in proteins to recognize duplex DNA sequences is common, as we see in more detail in Chapters 10 and 11. Amino acids separated by 3–4 positions lie on the same face of an α-helix and are therefore in a position to contact adjacent base pairs. **Figure 9.21** shows that amino acids lying along one face of the 2.4 region α-helix contact the bases at positions –12 to –10 of the –10 promoter sequence.

An interesting difference in behavior is found with the σ^{54} factor. This causes RNA polymerase to recognize promoters that have a distinct consensus sequence, with a conserved element at –10 and another close by at –20 (given in the "–35" column of Figure 9.19). So the geometry of the polymerase-promoter complex is different under the direction of this sigma factor. This is associated with a change in the pattern of regulation, in which sites that are rather distant from the promoter influence its activity, unlike the control of

Figure 9.21 Amino acids in the 2.4 α-helix of σ^{70} contact specific bases in the nontemplate strand of the –10 promoter sequence.

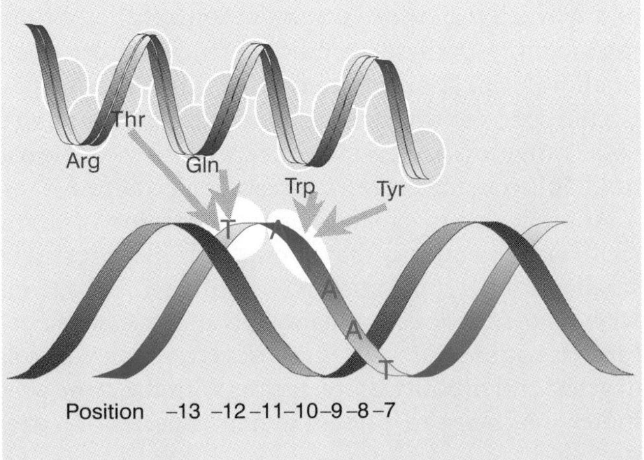

other promoters, where the regulator sites are always in close proximity to the promoter. The behavior of σ^{54} itself is different from other sigma factors, most notably in its ability to bind to DNA independently of core polymerase. In these regards, σ^{54} is more like the eukaryotic regulators we discuss in Chapter 20 than the typical prokaryotic regulators discussed in Chapters 10 and 11.

Sigma factors may be organized into cascades

SIGMA factors are used more extensively to control initiation in the bacterium *B. subtilis,* where ~10 different factors are known. Some are present in vegetative cells; others are produced only in the special circumstances of phage infection or the change from vegetative growth to sporulation.

The major RNA polymerase found in *B. subtilis* cells engaged in normal vegetative growth has the same structure as that of *E. coli,* $\alpha_2\beta\beta'\sigma$. Its sigma factor (described as σ^{43} or σ^{A}) recognizes promoters with the same consensus sequences used by the *E. coli* enzyme under direction from σ^{70}. Several variants of the RNA polymerase that contain other sigma factors are found in much smaller amounts. The variant enzymes recognize different promoters on the basis of consensus sequences at –35 and –10.

Transitions from expression of one set of genes to expression of another set are a common feature of bacteriophage infection. In all but the very simplest cases, the development of the phage involves shifts in the pattern of transcription during the infective cycle. These shifts may be accomplished by the synthesis of a phage-encoded RNA polymerase or by the efforts of phage-encoded ancillary factors that control the bacterial RNA polymerase. A well-characterized example of control via the production of new sigma factors occurs during infection of *B. subtilis* by phage SPO1.

The infective cycle of SPO1 passes through three stages of gene expression. Immediately on infection, the **early** genes of the phage are transcribed. After 4–5 minutes, the early genes cease transcription and the **middle** genes are transcribed. Then at 8–12 minutes, middle gene transcription is replaced by transcription of **late** genes.

The early genes are transcribed by the holoenzyme of the host bacterium. They are essentially indistinguishable from host genes whose promoters have the intrinsic ability to be recognized by the RNA polymerase $\alpha_2\beta\beta'\sigma^{43}$.

Expression of phage genes is required for the transitions to middle and late gene transcription. Three regulatory genes, named *28, 33,* and *34,* control the course of transcription. Their functions are summarized in **Figure 9.22.** The pattern of regulation creates a **cascade,** in which the host enzyme transcribes an early gene whose product is needed to transcribe the middle genes; and then two of the middle genes code for products that are needed to transcribe the late genes.

Mutants in the early gene *28* cannot transcribe the middle genes. The product of gene *28* (called gp28) is a protein of 26 kD that replaces the host sigma factor on the core enzyme. *This substitution is the sole event required to make the transition from early to middle gene expression.* It creates a holoenzyme that can no longer transcribe the host genes, but instead specifically transcribes the middle genes. We do not know how gp28 displaces σ^{43}, or what happens to the host sigma polypeptide.

Two of the middle genes are involved in the next transition. Mutations in either gene *33* or *34* prevent transcription of the late genes. The products of these genes together replace gp28 on the core polymerase. Again, we do not know how gp33 and gp34 exclude gp28 (or any residual host σ^{43}), *but once they have bound to the core enzyme, it is able to initiate transcription only at the promoters for late genes.*

The successive replacements of sigma factor have dual consequences. Each time the subunit is changed, the RNA polymerase becomes able to recognize a new class of genes, *and* no longer recognizes the previous class. These switches therefore constitute global changes in the activity of RNA polymerase. Probably all or virtually all of the core enzyme becomes associated with the sigma factor of the moment; and the change is irreversible.

Perhaps the most extensive example of switches in sigma factors is provided by **sporulation,** an alternative lifestyle available to some bacteria. At the end of the

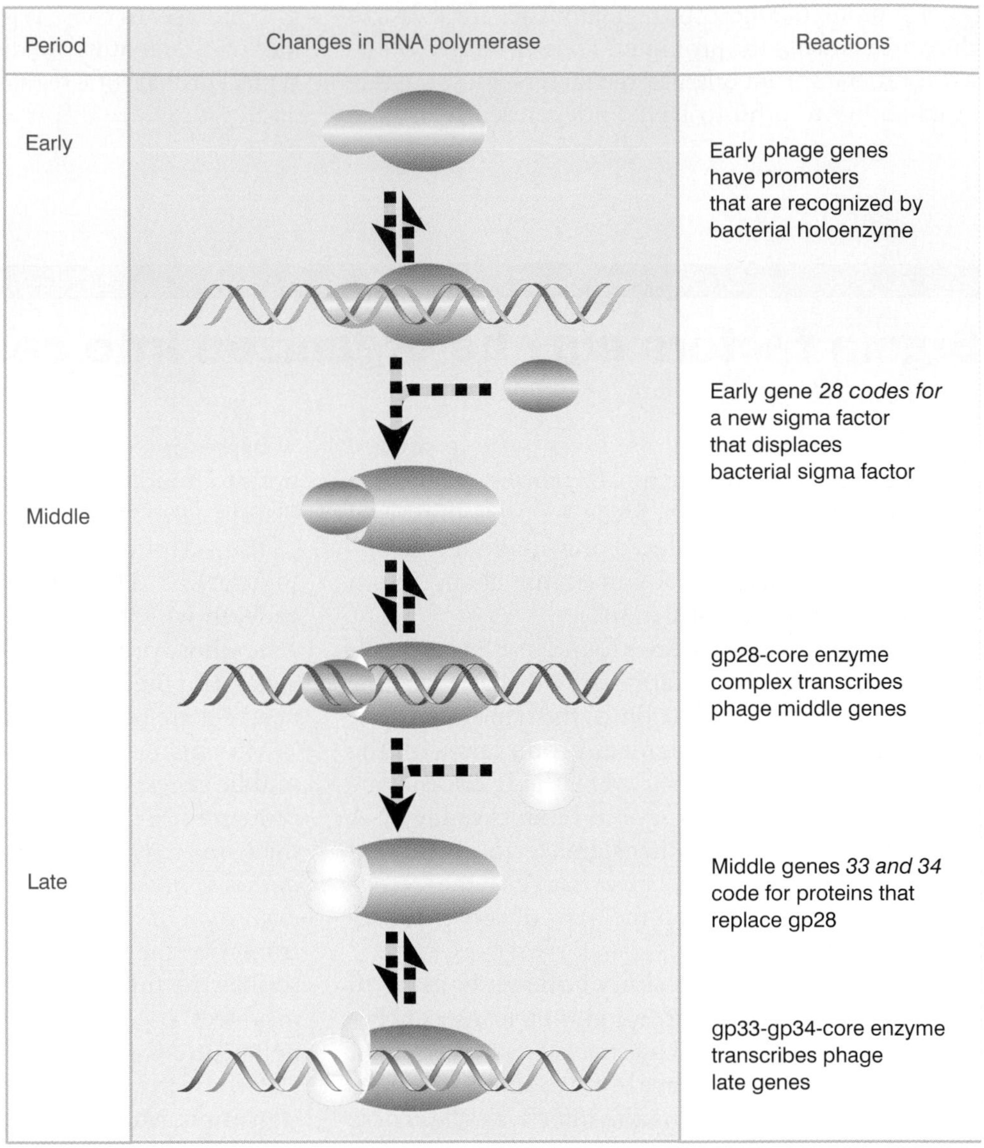

Figure 9.22 Transcription of phage SPO1 genes is controlled by two successive substitutions of the sigma factor that change the initiation specificity.

Period	Changes in RNA polymerase	Reactions
Early		Early phage genes have promoters that are recognized by bacterial holoenzyme
		Early gene *28 codes for* a new sigma factor that displaces bacterial sigma factor
Middle		gp28-core enzyme complex transcribes phage middle genes
Late		Middle genes *33 and 34* code for proteins that replace gp28
		gp33-gp34-core enzyme transcribes phage late genes

vegetative phase, logarithmic growth ceases because nutrients in the medium become depleted. This triggers sporulation, as illustrated in **Figure 9.23.** DNA is replicated, a genome is segregated at one end of the cell, and eventually it is surrounded by the tough spore coat. When the septum forms, it generates two independent compartments, the mother cell and the forespore. At the start of the process, one chromosome is attached to each pole of the cell. The growing septum traps part of one chromosome in the forespore, and then a translocase (SpoIIIE) pumps the rest of the chromosome into the forespore.

Sporulation takes ~8 hours. It can be viewed as a primitive sort of differentiation, in which a parent cell (the vegetative bacterium) gives rise to two different daughter cells with distinct fates: the mother cell is eventually lysed, while the spore that is released has an entirely different structure from the original bacterium.

Sporulation involves a drastic change in the biosynthetic activities of the bacterium, in which many genes are involved. The basic level of control lies at transcription. Some of the genes that functioned in the vegetative phase are turned off during sporulation, but most continue to be expressed. In addition, the genes specific for sporulation are expressed only during this period. At the end of sporulation, ~40% of the bacterial mRNA is sporulation-specific.

New forms of the RNA polymerase become active in sporulating cells; they contain the same core enzyme as

Figure 9.23 Sporulation involves the differentiation of a vegetative bacterium into a mother cell that is lysed and a spore that is released.

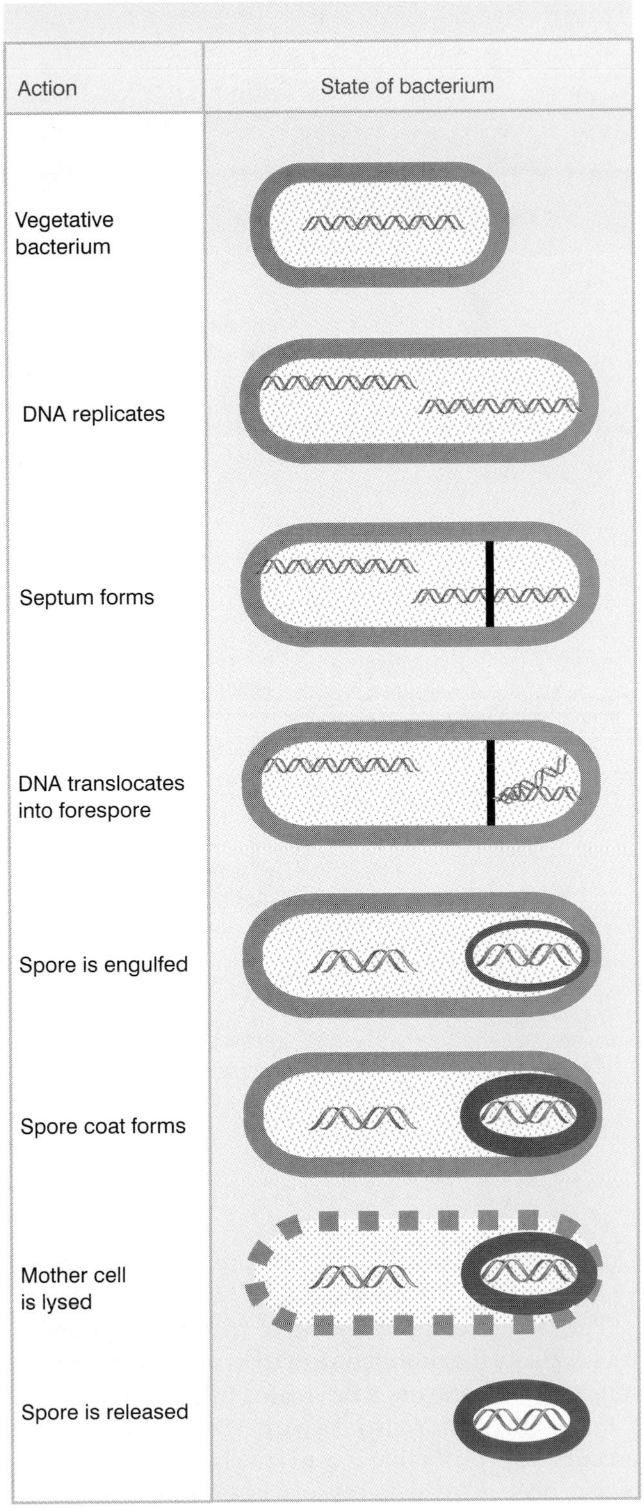

Action	State of bacterium
Vegetative bacterium	
DNA replicates	
Septum forms	
DNA translocates into forespore	
Spore is engulfed	
Spore coat forms	
Mother cell is lysed	
Spore is released	

vegetative cells, but have different proteins in place of the vegetative σ^{43}. The changes in transcriptional specificity are summarized in **Figure 9.24.** The principle is that in each compartment the existing sigma factor is successively displaced by a new factor that causes transcription of a different set of genes. Communication between the compartments occurs in order to coordinate the timing of the changes in the forespore and mother cell.

The sporulation cascade is initiated when environmental conditions trigger a "phosphorylation relay," in which a phosphate group is passed along a series of proteins until it reaches SpoOA. (Several gene products are involved in this process, whose complexity may reflect the need to avoid mistakes in triggering sporulation unnecessarily.) SpoOA is a transcriptional regulator whose activity is affected by phosphorylation. In the phosphorylated form, it activates transcription of two operons, each of which is transcribed by a different form of the host RNA polymerase. Under the direction of phosphorylated SpoOA, host enzyme utilizing the general σ^{43} transcribes the gene coding for the factor σ^F; and host enzyme under the direction of a minor factor, σ^H, transcribes the gene coding for the factor pro-σ^E. Both of these new sigma factors are produced before septum formation, but become active later.

σ^F is the first factor to become active in the forespore compartment. It is inhibited by an anti-sigma factor that binds to it; in the forespore, an anti-anti-sigma factor removes the inhibitor. This reaction is controlled by a series of phosphorylation/dephosphorylation events. The initial determinant is a phosphatase (SpoIIE) that is an integral membrane protein, and which accumulates at the pole, with the result that its phosphatase domain becomes more concentrated in the forespore. It dephosphorylates, and thereby activates, SpoIIAA, which in turn displaces the anti-sigma factor SpoIIAB from the complex of SpoIIAB-σ^F. Release of σ^F activates it.

Activation of σ^F is the start of sporulation. Under the direction of σ^F, RNA polymerase transcribes the first set of sporulation genes instead of the vegetative genes that it was previously transcribing. The replacement reaction probably affects only part of the RNA polymerase population, since σ^F is produced only in small amounts. Some vegetative enzyme remains present during sporulation. The displaced σ^{43} is not destroyed, but can be recovered from extracts of sporulating cells.

Figure 9.24 Sporulation involves successive changes in the sigma factors that control the initiation specificity of RNA polymerase. The cascades in the forespore (left) and the mother cell (right) are related by signals passed across the septum (indicated by the horizontal arrows).

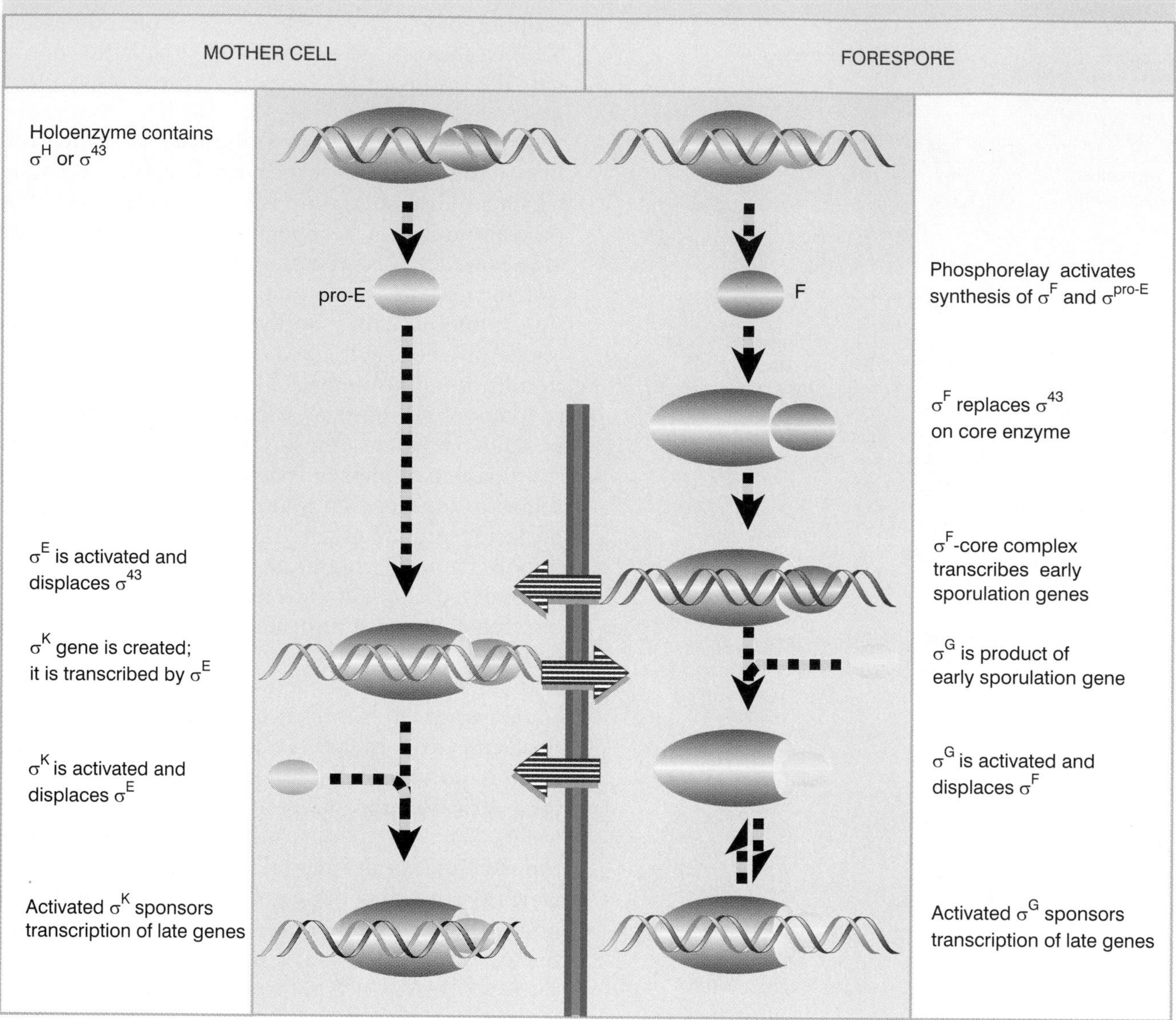

MOTHER CELL
FORESPORE
Holoenzyme contains σH or σ43
pro-E
F
Phosphorelay activates synthesis of σF and σpro-E
σF replaces σ43 on core enzyme
σE is activated and displaces σ43
σF-core complex transcribes early sporulation genes
σK gene is created; it is transcribed by σE
σG is product of early sporulation gene
σK is activated and displaces σE
σG is activated and displaces σF
Activated σK sponsors transcription of late genes
Activated σG sponsors transcription of late genes

Two regulatory events follow from the activity of σ^F, as detailed in **Figure 9.25.** In the forespore itself, another factor, σ^G, is the product of one of the early sporulation genes. σ^G is the factor that causes RNA polymerase to transcribe the late sporulation genes in the forespore. Another early sporulation gene product is responsible for communicating with the mother cell compartment. σ^F activates SpoIIR, which is secreted from the forespore. It then activates the membrane-bound protein SpoIIGA to cleave the inactive precursor pro-σ^E into the active factor σ^E in the mother cell. (Any σ^E that is produced in the forespore is degraded by forespore-specific functions.)

The cascade continues when σ^E in turn is replaced by σ^K. (Actually the production of σ^K is quite complex, because first its gene must be created by a recombination event!) This factor also is synthesized as an inactive precursor (pro-σ^K) that is activated by a protease. Once σ^K has been activated, it displaces σ^E and causes transcription of the late genes in the mother cell. The timing of these events in the two compartments is coordinated by further signals. The activity of σ^E in the

Figure 9.25 σ^F triggers synthesis of the next sigma factor in the forespore (σ^G) and turns on SpoIIR which causes SpoIIGA to cleave pro-σ^E.

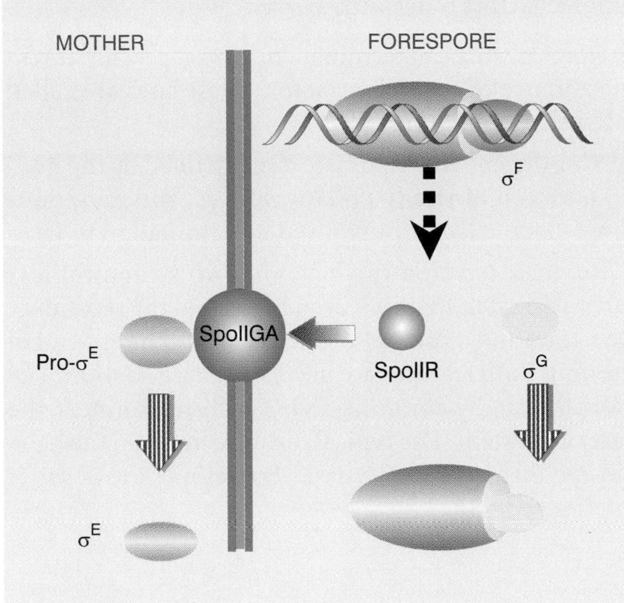

Figure 9.26 The criss-cross regulation of sporulation coordinates timing of events in the mother cell and forespore

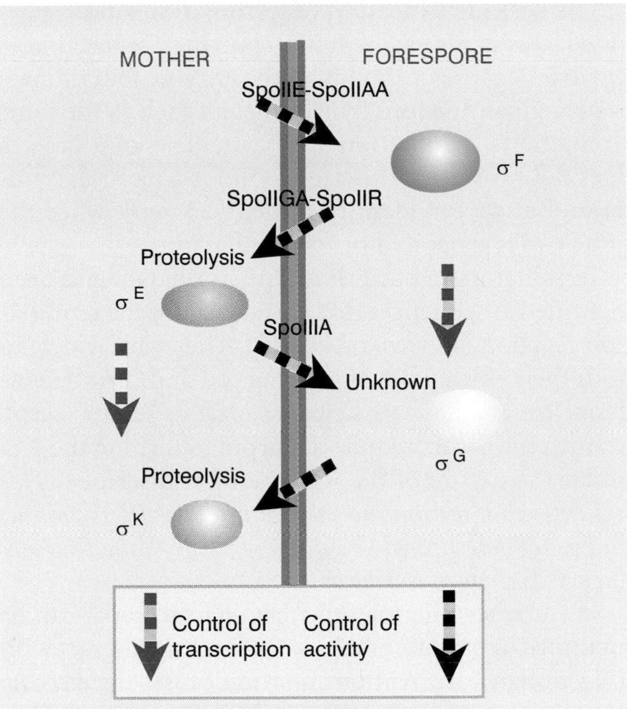

mother cell is necessary for activation of σ^G in the forespore; and in turn the activity of σ^G is required to generate a signal that is transmitted across the septum to activate σ^K.

Sporulation is thus controlled by two cascades, in which sigma factors in each compartment are successively activated, each directing the synthesis of a particular set of genes. **Figure 9.26** outlines how the two cascades are connected by the transmission of signals from one compartment to the other. As new sigma factors become active, old sigma factors are displaced, so that transitions in sigma factors turn genes off as well as on. The incorporation of each factor into RNA polymerase dictates when its set of target genes is expressed; and the amount of factor available influences the level of gene expression. More than one sigma factor may be active at any time, and the specificities of some of the sigma factors overlap. We do not know what is responsible for the ability of each sigma factor to replace its predecessor.

Bacterial RNA polymerase has two modes of termination

Once RNA polymerase has started transcription, the enzyme moves along the template, synthesizing RNA, until it meets a **terminator** (*t*) sequence. At this point, the enzyme stops adding nucleotides to the growing RNA chain, releases the completed product, and dissociates from the DNA template. (We do not know in which order the last two events occur.) Termination requires that all hydrogen bonds holding the RNA-DNA hybrid together must be broken, after which the DNA duplex reforms.

It is difficult to define the termination point of an RNA molecule that has been synthesized in the living cell. It is always possible that the 3′ end of the molecule has been generated by *cleavage* of the primary transcript, and therefore does not represent the actual site at which RNA polymerase terminated.

The best identification of termination sites is provided by systems in which RNA polymerase terminates *in vitro*. Because the ability of the enzyme to terminate is strongly influenced by parameters such as the ionic strength, its termination at a particular point *in vitro* does not prove that this same point is a natural terminator. But we can identify authentic 3′ ends when the same end is generated *in vitro* and *in vivo*.

Terminators in bacteria and their phages have been identified as sequences that are needed for the termination reaction (*in vitro* or *in vivo*). They vary widely in both their efficiencies of termination and their dependence on ancillary proteins, at least as seen *in vitro*. Many terminators require a hairpin to form in the secondary structure of the RNA being transcribed. *This indicates that termination depends on the RNA product and is not determined simply by scrutiny of the DNA sequence during transcription.*

At some terminators, the termination event can be *prevented* by specific ancillary factors that interact with RNA polymerase. **Antitermination** causes the enzyme to continue transcription past the terminator sequence, an event called **readthrough** (the same term used to describe a ribosome's suppression of termination codons).

In approaching the termination event, we must regard it not simply as a mechanism for generating the 3′ end of the RNA molecule, but as an opportunity to control gene expression. So the stages when RNA polymerase associates with DNA (initiation) or dissociates from it (termination) are both subject to specific control. There are interesting parallels between the systems employed in initiation and termination. Both require breaking of hydrogen bonds (initial melting of DNA at initiation, RNA-DNA dissociation at termination); and both require additional proteins to interact with the core enzyme. In fact, they are accomplished by alternative forms of the polymerase. However, whereas initiation relies solely upon the interaction between RNA polymerase and duplex DNA, the termination event involves recognition of signals in the transcript by RNA polymerase or by ancillary factors.

The sequences at prokaryotic terminators show no similarities beyond the point at which the last base is added to the RNA. The responsibility for termination lies with the *sequences already transcribed* by RNA polymerase. So termination relies on scrutiny of the template or product that the polymerase is currently transcribing.

Terminators have been distinguished in *E. coli* according to whether RNA polymerase requires any additional factors to terminate *in vitro:*

- Core enzyme can terminate *in vitro* at certain sites in the absence of any other factor. These sites are called **intrinsic terminators.**
- **Rho-dependent** terminators are defined by the need for addition of **rho (ρ) factor** *in vitro*; and mutations show that the factor is involved in termination *in vivo*.

Intrinsic terminators have the two structural features evident in **Figure 9.27:** a hairpin in the secondary structure; and a run of ~6 U residues at the very end of the unit. Both features are needed for termination. The hairpin usually contains a G•C-rich region near the base of the stem. The typical distance between the hairpin and the U-run is 7–9 bases. Sometimes the U-run is interrupted.

Figure 9.27 Intrinsic terminators include palindromic regions that form hairpins varying in length from 7 to 20 bp. The stem-loop structure includes a G-C-rich region and is followed by a run of U residues.

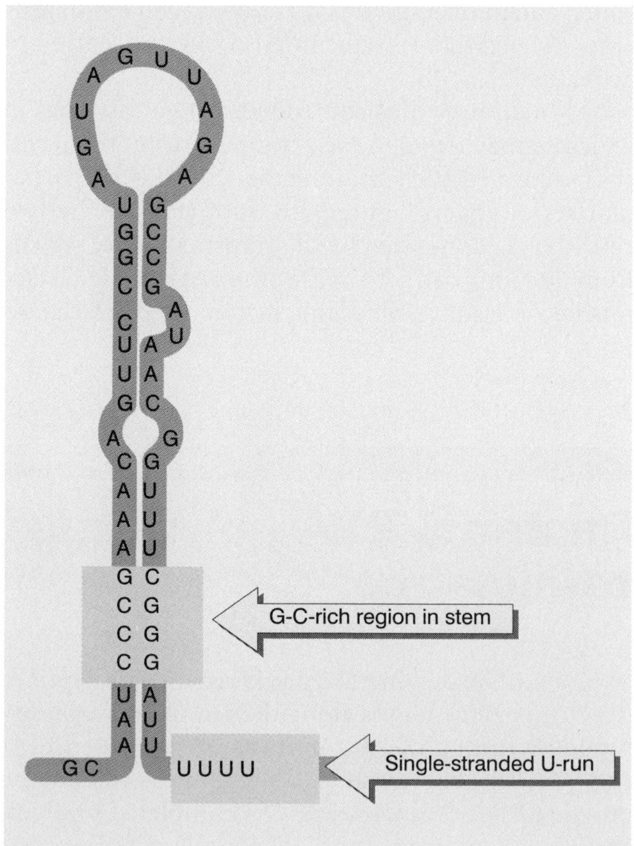

Point mutations that prevent termination occur within the stem region of the hairpin. What is the effect of a hairpin on transcription? Probably all hairpins that form in the RNA product cause the polymerase to slow (and perhaps to pause) in RNA synthesis.

Pausing creates an opportunity for termination to occur. Pausing occurs at sites that resemble terminators but have an increased separation (typically 10–11 bases) between the hairpin and the U-run. But if the pause site does not correspond to a terminator, usually the enzyme moves on again to continue transcription. The length of the pause varies, but at a typical terminator lasts ~60 seconds.

A string of U residues in the right location is necessary to allow RNA polymerase to dissociate from the template when it pauses at the hairpin. The rU•dA RNA-DNA hybrid has an unusually weak base-paired structure; it requires the least energy of any RNA-DNA hybrid to break the association between the two strands. When the polymerase pauses, the RNA-DNA hybrid unravels from the weakly bonded rU•dA terminal region. Often the actual termination event takes place at any one of several positions toward or at the end of the U-run, as though the enzyme "stutters" during termination.

The importance of the run of U bases is confirmed by making deletions that shorten this stretch; although the polymerase still pauses at the hairpin, it no longer terminates. The series of U bases corresponds to an A•T-rich region in DNA, so we see that A•T-rich regions are important in intrinsic termination as well as initiation.

Both the sequence of the hairpin and the length of the U-run influence the efficiency of termination. However, termination efficiency *in vitro* varies from 2–90%, and does not correlate in any simple way with the constitution of the hairpin or the number of U residues in intrinsic terminators. The hairpin and U-run are therefore necessary, but not sufficient, and additional parameters influence the interaction with RNA polymerase.

Less is known about the signals and ancillary factors involved in termination for eukaryotic polymerases. Each class of polymerase uses a different mechanism (see Chapter 20). However, the same features used by bacterial (core) RNA polymerase in intrinsic termination recur in eukaryotes: secondary structure and/or runs of U residues in the transcript are important.

How does rho factor work?

RHO factor is an essential protein in *E. coli*. It functions solely at the stage of termination. It is a protein of ~46 kD, probably active as a hexamer (275 kD). It functions as an ancillary factor for RNA polymerase; typically its maximum activity *in vitro* is displayed when it is present at ~10% of the concentration of the RNA polymerase.

E. coli has relatively few rho-dependent terminators; most of the known rho-dependent terminators are found in phage genomes. The sequences required for rho-dependent termination are 50–90 bases long and lie upstream of the terminator. Their common feature is that the RNA is rich in C residues and poor in G residues. An example is given in **Figure 9.28;** C is by far the most common base (41%) and G is the least common base (14%). As a general rule the efficiency of a rho-dependent terminator increases with the length of the C-rich/G-poor region.

Does rho factor act via recognizing DNA, RNA, or RNA polymerase? Rho has an ATPase activity, which is RNA-dependent and requires the presence of a polyribonucleotide, >50 or so bases long. This suggests that rho binds RNA. Probably an individual rho factor acts processively on a single RNA substrate. The "hot pursuit" model for rho action shown in **Figure 9.29** supposes that it binds to a nascent RNA chain at some point upstream of the terminator. This might require a specific sequence or type of sequence, or attachment could occur just at a free 5′ end or some other exposed sequence. Rho then translocates along the mRNA.

How does rho catch up with RNA polymerase? One possibility is that rho simply moves along the transcript faster than RNA polymerase moves along the DNA. The enzyme pauses when it reaches a terminator, and termination occurs if rho catches it there.

Does rho release the transcript by acting directly on

Figure 9.28 A rho-dependent terminator has a sequence rich in C and poor in G preceding the actual site(s) of termination.

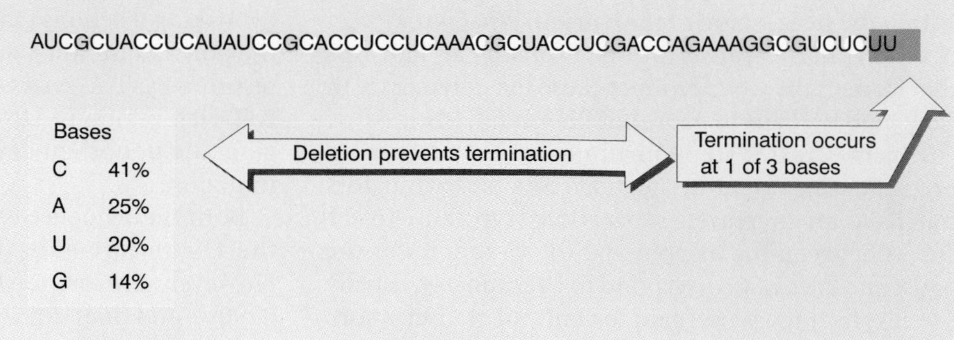

Figure 9.29 Rho factor pursues RNA polymerase along the RNA and can cause termination when it catches the enzyme pausing at a rho-dependent terminator.

RNA polymerase transcribes DNA

Rho attaches to recognition site on RNA

Rho moves along RNA, following RNA polymerase

RNA polymerase pauses at terminator and rho catches up

Rho unwinds DNA-RNA hybrid in transcription bubble

Termination: RNA polymerase, rho, and RNA are released

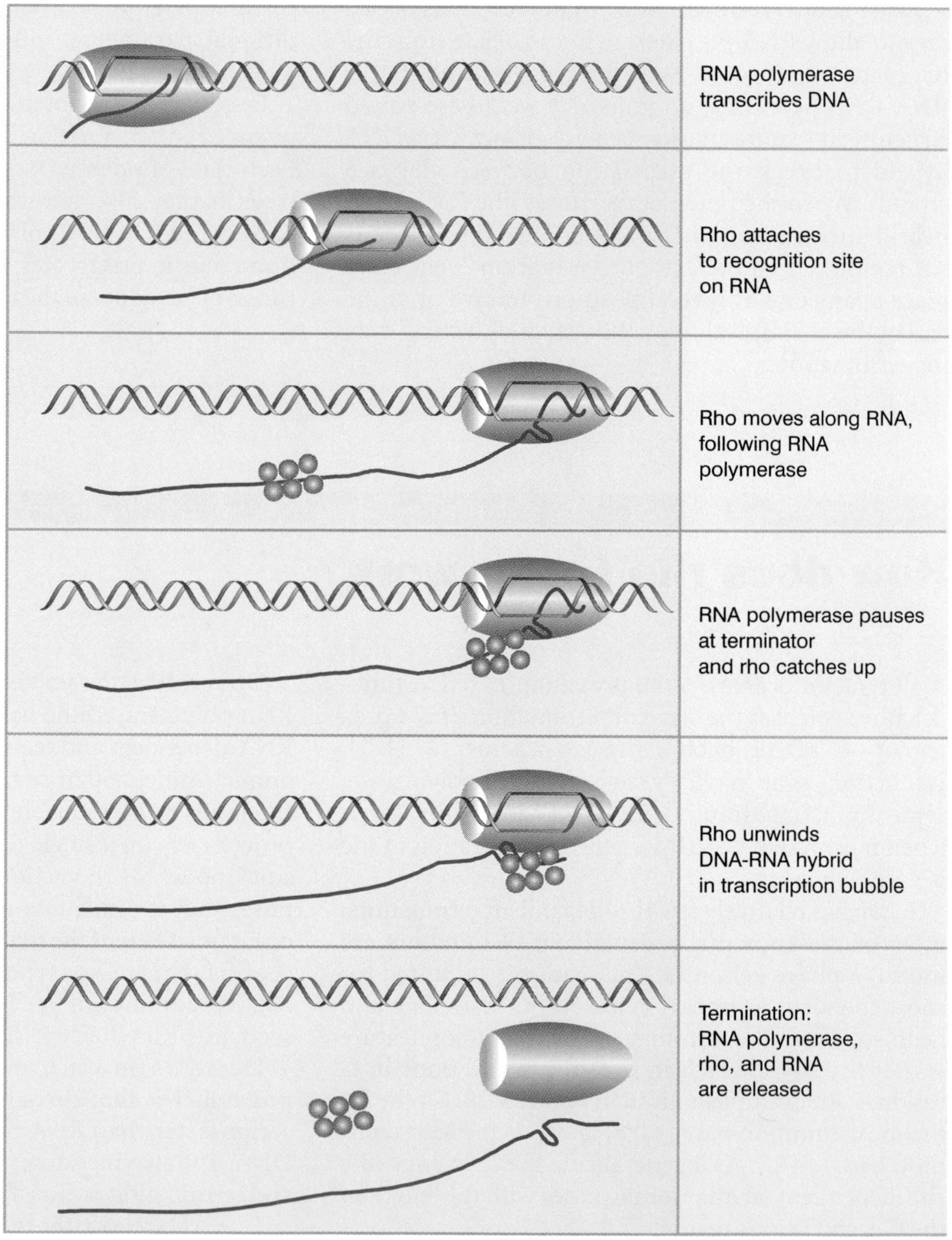

the DNA-RNA junction or indirectly by causing RNA polymerase to release RNA? In the absence of RNA polymerase, rho has a 5′–3′ helicase action that can cause an RNA-DNA hybrid to separate; hydrolysis of ATP is used to provide energy for the reaction.

These abilities suggest that rho can directly gain access to the stretch of RNA-DNA hybrid in the transcription bubble and cause it to unwind. Either as a result of this unwinding, or because of some interaction between rho and RNA polymerase, termination is completed by the release of rho and RNA polymerase from the nucleic acids.

The idea that rho moves along RNA leads to an important prediction about the relationship between transcription and translation. Rho must first have access to a binding sequence on RNA, and then must be able to move along the RNA. Either or both of these conditions may be prevented if ribosomes are translating an RNA. So the ability of rho factor to reach RNA polymerase at a terminator depends on what is happening in translation.

This model explains a puzzling phenomenon. In some cases, a nonsense mutation in one gene of a transcription unit prevents the expression of subsequent genes in the unit. This effect is called **polarity.** A common cause is the absence of the mRNA corresponding to the subsequent (distal) parts of the unit.

Suppose that there are rho-dependent terminators *within* the transcription unit, that is, before the terminator that *usually* is used. The consequences are illustrated in **Figure 9.30.** Normally these earlier terminators are not used, because the ribosomes

Figure 9.30 The action of rho factor may create a link between transcription and translation when a rho-dependent terminator lies soon after a nonsense mutation.

WILD TYPE
NONSENSE MUTANT
Ribosomes pack mRNA behind RNA polymerase
Ribosomes impede rho attachment and/or movement
Ribosomes dissociate at mutation
Rho attaches but ribosomes impede its movement
Rho obtains access to RNA polymerase
Transcription continues
Transcription terminates prematurely

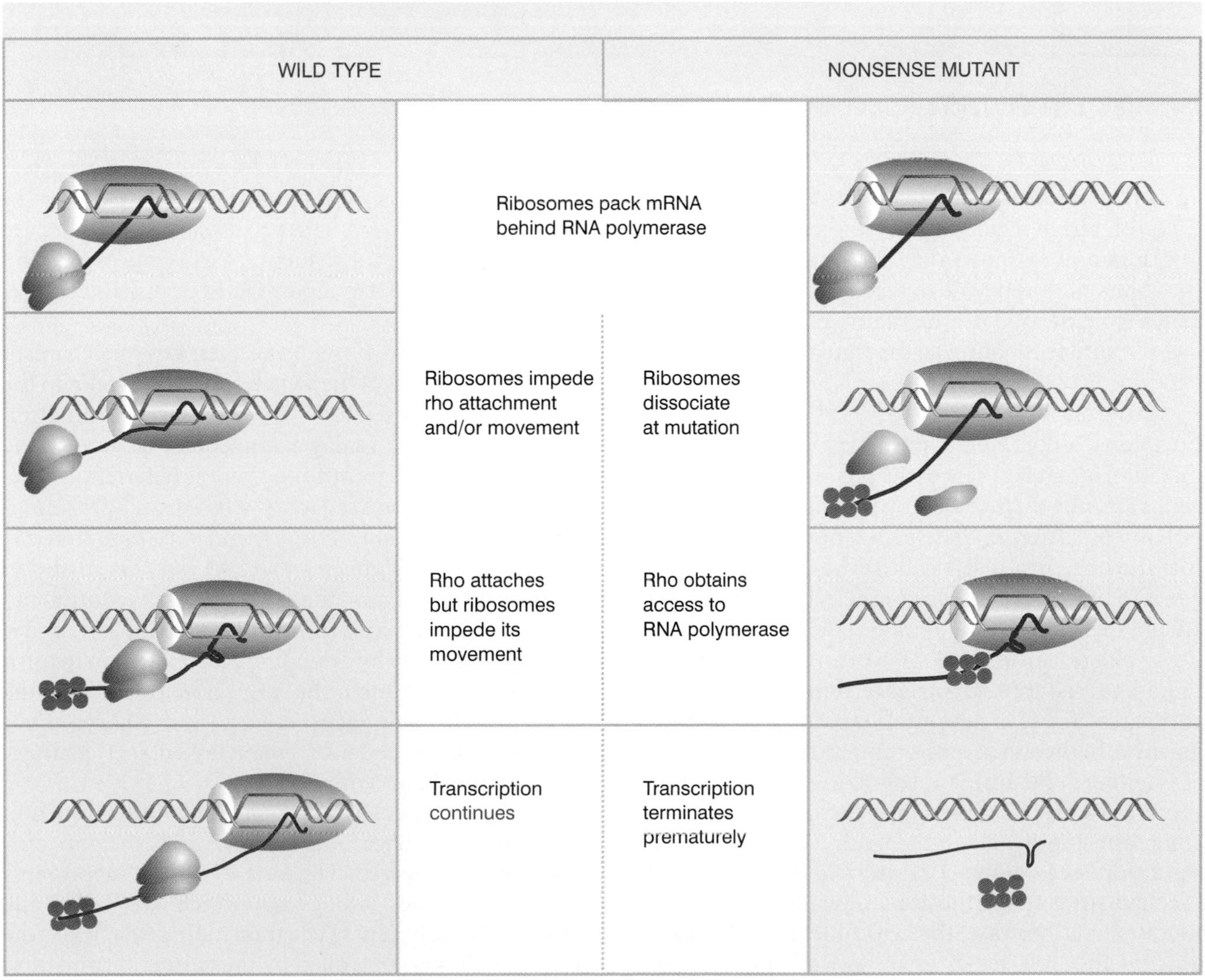

prevent rho from reaching RNA polymerase. But a nonsense mutation releases the ribosomes, so that rho is free to attach to and/or move along the mRNA, enabling it to act on RNA polymerase at the terminator. As a result, the enzyme is released, and the distal regions of the transcription unit are never expressed. (Why should there be internal terminators? Perhaps they are simply sequences that by coincidence mimic the usual rho-dependent terminator.) Some stable RNAs that have extensive secondary structure are preserved from polar effects, presumably because the structure impedes rho attachment or movement.

rho mutations show wide variations in their influence on termination. The basic nature of the effect is a failure to terminate. But the magnitude of the failure, as seen in the percentage of readthrough *in vivo*, depends on the particular target locus. Similarly, the need for rho factor *in vitro* is variable. Some (rho-dependent) terminators require relatively high concentrations of rho, while others function just as well at lower levels. This suggests that different terminators require different levels of rho factor for termination, and therefore respond differently to the residual levels of rho factor in the mutants (*rho* mutants are usually leaky).

Some *rho* mutations can be suppressed by mutations in other genes. This approach provides an excellent way to identify proteins that interact with rho. The β subunit of RNA polymerase is implicated by two types of mutation. First, mutations in the *rpoB* gene can reduce termination at a rho-dependent site. Second, mutations in *rpoB* can restore the ability to terminate transcription at rho-dependent sites in *rho* mutant bacteria.

Antitermination depends on specific sites

ANTITERMINATION is used as a control mechanism in both phage regulatory circuits and bacterial operons. **Figure 9.31** shows that antitermination controls the ability of the enzyme to read past a terminator into genes lying beyond. In the example shown in the figure, the antitermination factor regulates the expression of region 2.

Antitermination was discovered in bacteriophage infections. A common feature in the control of phage infection is that very few of the phage genes (the "early" genes) can be transcribed by the bacterial host RNA polymerase. Among these genes, however, are regulator(s) whose product(s) allow the next set of phage genes to be expressed. Two common types of action for such a regulator protein are to sponsor initiation at new (phage) promoters or to cause the host polymerase to read through phage terminators. **Figure 9.32** compares the use of new promoters as a control mechanism with the use of antitermination.

One mechanism for recognizing new phage promoters is to replace the sigma factor of the host enzyme with another factor that redirects its specificity in initiation (see Figures 9.22 and 9.24). An alternative mechanism is to synthesize a new phage RNA polymerase. In either case, the critical feature that distinguishes the new set of genes is their possession of *different promoters from those originally recognized by host RNA polymerase.* The two sets of transcripts are independent; as a consequence, early gene expression can cease after the new sigma factor or polymerase has been produced.

Antitermination provides an alternative mechanism for phages to control the switch from early genes to the next stage of expression. The use of antitermination depends on a particular arrangement of genes. The early genes lie adjacent to the genes that are to be expressed next, but are separated from them by terminator sites. *If termination is prevented at these sites, the polymerase reads through into the genes on the other side.* So in antitermination, the *same promoters* continue to be recognized by RNA polymerase. So the new genes are expressed only by extending the RNA chain to form molecules that contain the early gene sequences at the 5′ end and the new gene sequences at the 3′ end. Since the two types of sequence remain linked, early gene expression inevitably continues.

The best-characterized example of antitermination is provided by phage lambda, with which the phenomenon was discovered. The host RNA polymerase initially transcribes two genes, which are called the **immediate early** genes. The transition to the next stage of expression is controlled by preventing termination

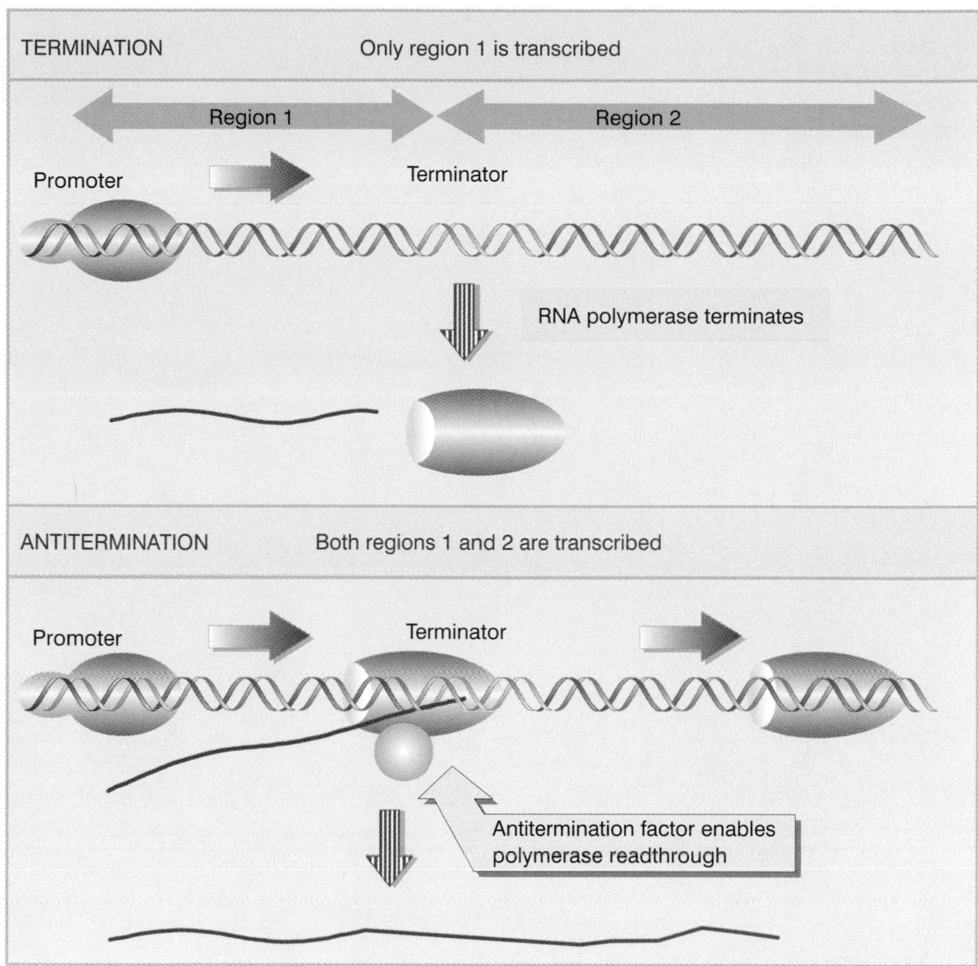
TERMINATION
Only region 1 is transcribed
Region 1
Region 2
Promoter
Terminator
RNA polymerase terminates
ANTITERMINATION
Both regions 1 and 2 are transcribed
Promoter
Terminator
Antitermination factor enables polymerase readthrough

Figure 9.31 Antitermination can be used to control transcription by determining whether RNA polymerase terminates or reads through a particular terminator into the following region.

at the ends of the immediate early genes, with the result that the **delayed early** genes are expressed. (We discuss the overall regulation of lambda development in more detail in the next chapter.)

The regulator gene that controls the switch from immediate early to delayed early expression is identified by mutations in lambda gene *N* that can transcribe *only* the immediate early genes; they proceed no further into the infective cycle. The same effect is seen when gene *28* of phage SPO1 is mutated to prevent the production of σ^{gp28}. From the genetic point of view, the mechanisms of new initiation and antitermination are similar. *Both are positive controls in which an early gene product must be made by the phage in order to express the next set of genes.*

Antitermination is the mechanism used by phage lambda to control the transition out of the early stage. It is summarized in **Figure 9.33.** There are two transcription units of immediate early genes, which are transcribed from the promoters P_L and P_R. Transcription by *E. coli* RNA polymerase itself stops at the terminators t_{L1} and t_{R1}, respectively. Both terminators depend on rho; in fact, these were the terminators with which rho was originally identified. The situation is changed by expression of the *N* gene. The product pN is an **antitermination protein** that allows RNA polymerase to read through t_{L1} and t_{R1} into the delayed early genes beyond them.

Like other phages, still another control is needed to express the late genes that code for the components of the phage particle. This switch is regulated by gene *Q*, itself one of the delayed early genes. Its product, pQ, is another antitermination protein, one that specifically allows RNA polymerase initiating at another site, the late promoter $P_{R'}$, to read through a terminator that lies between it and the late genes. So by employing antitermination proteins with different specificities, a cascade for gene expression can be constructed.

The different specificities of pN and pQ establish an important general principle: *RNA polymerase interacts with transcription units in such a way that an ancillary factor can sponsor antitermination specifically for some*

Figure 9.32 Switches in transcriptional specificity can be controlled at initiation or termination.

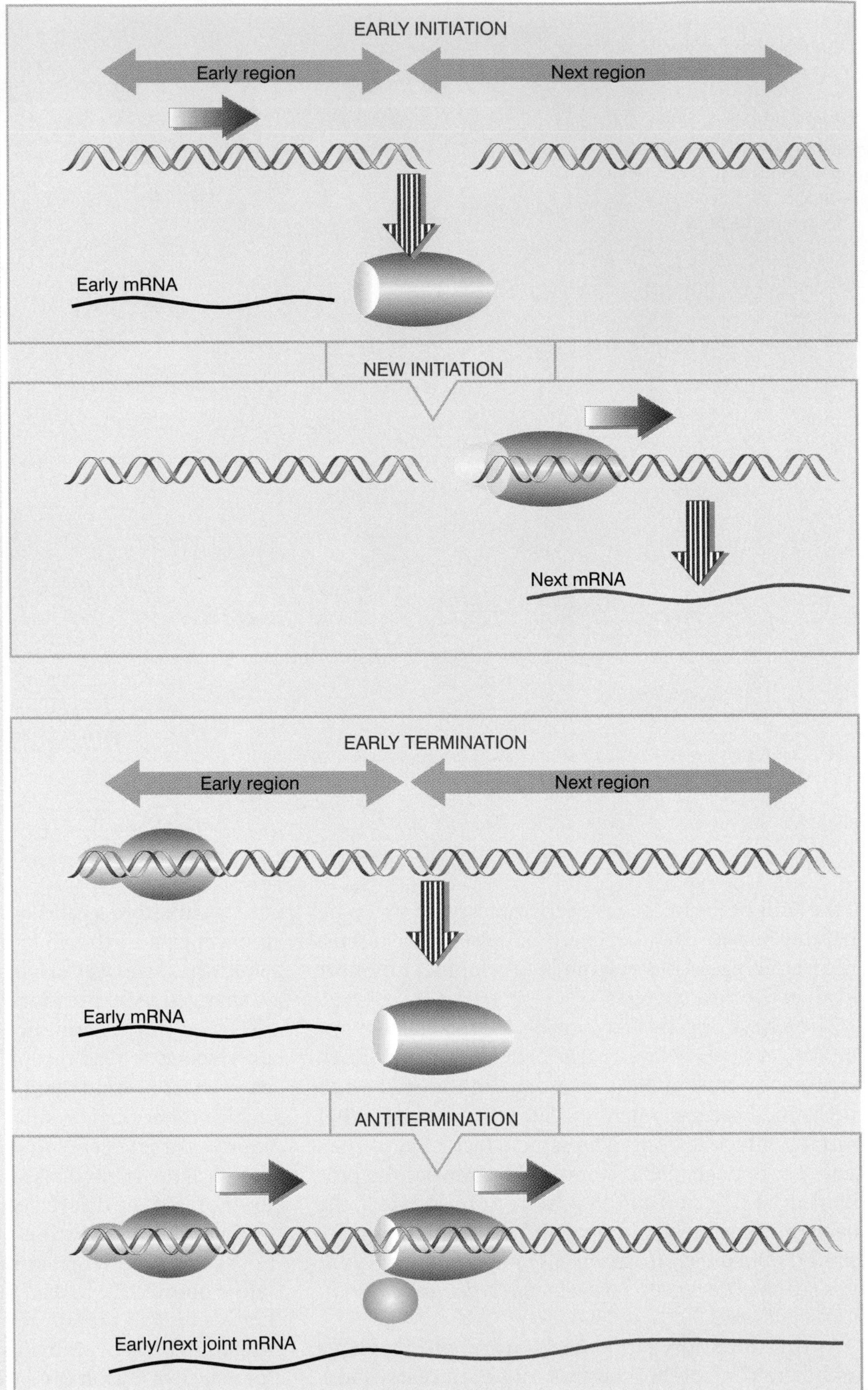

Figure 9.33 Host RNA polymerase transcribes lambda genes and terminates at *t* sites. pN allows it to read through terminators in the L and R1 units; pQ allows it to read through the R' terminator. The sites at which pN acts (*nut*) and at which pQ acts (*qut*) are located at different relative positions in the transcription units.

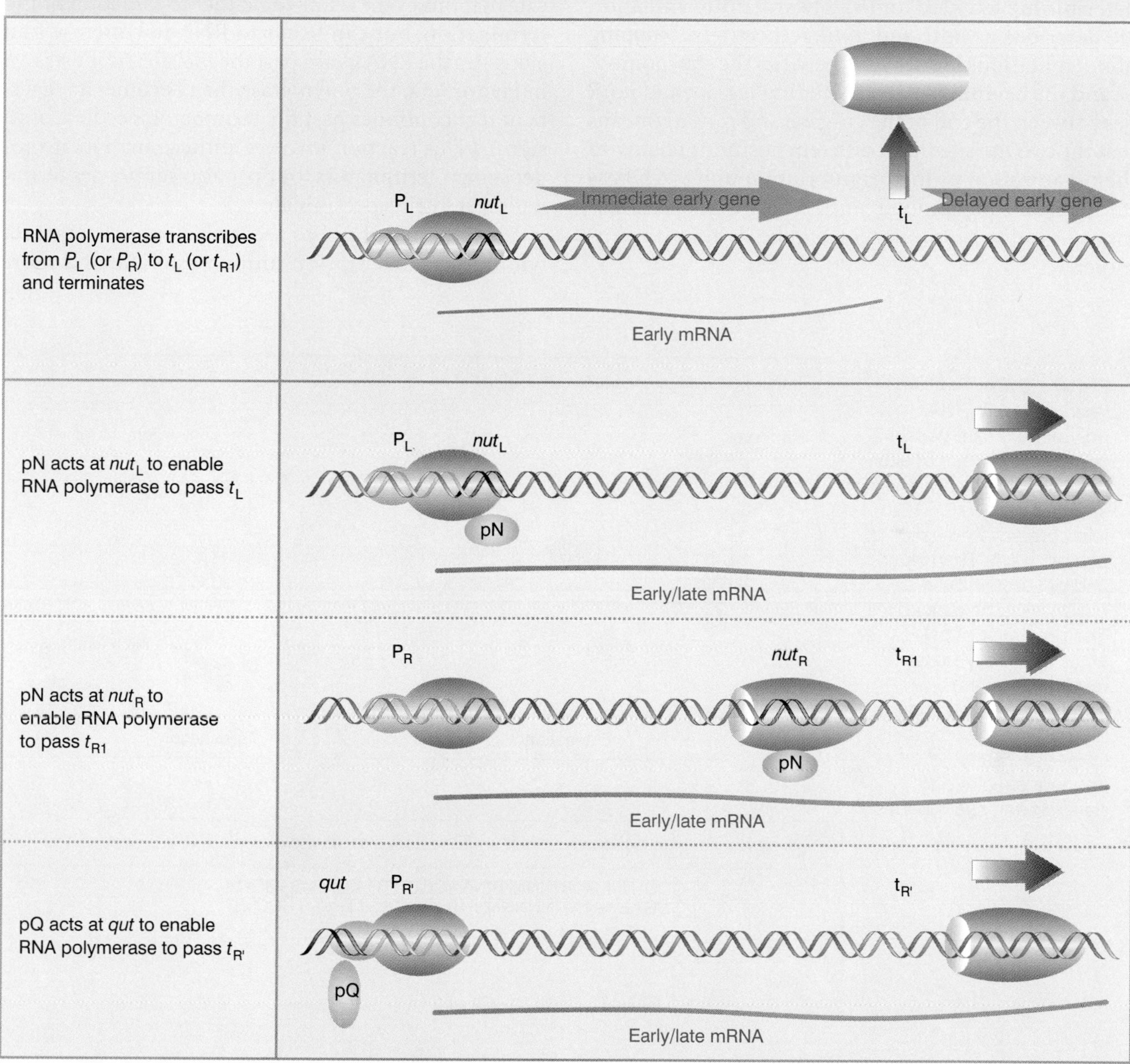

transcripts. Termination can be controlled with the same sort of precision as initiation. What sites are involved in controlling the specificity of termination?

The antitermination activity of pN is highly specific, but *the antitermination event is not determined by the terminators t_{L1} and t_{R1}; the recognition site needed for antitermination lies upstream in the transcription unit, that is, at a different place from the terminator site at which the action eventually is accomplished.* This conclusion establishes a general principle. When we know the site on DNA at which some protein exercises its effect, we cannot assume that this coincides with the

DNA sequence that it initially recognizes. They may be separate.

The recognition sites required for pN action are called *nut* (for *Nut*ilization). The sites responsible for determining leftward and rightward antitermination are described as *nutL* and *nutR*, respectively. Mapping of *nut* mutations locates *nutL* between the startpoint of P_L and the beginning of the *N* coding region; and *nutR* lies between the end of the *cro* gene and t_{L1}. This means that the two *nut* sites lie in different positions relative to the organization of their transcription units. Whereas *nutL* is near the promoter, *nutR* is near to the terminator. (*qut* is different yet again, and lies within the promoter.)

How does antitermination occur? When pN recognizes the *nut* site, it must act on RNA polymerase to ensure that the enzyme can no longer respond to the terminator. The variable locations of the *nut* sites indicate that this event is linked neither to initiation nor to termination, but can occur to RNA polymerase as it elongates the RNA chain past the *nut* site. As illustrated in **Figure 9.34,** the polymerase then becomes a juggernaut that continues past the terminator, heedless of its signal. (This reaction involves antitermination at rho-dependent terminators, but pN also suppresses termination at intrinsic terminators.)

Is the ability of pN to recognize a short sequence within the transcription unit an example of a more

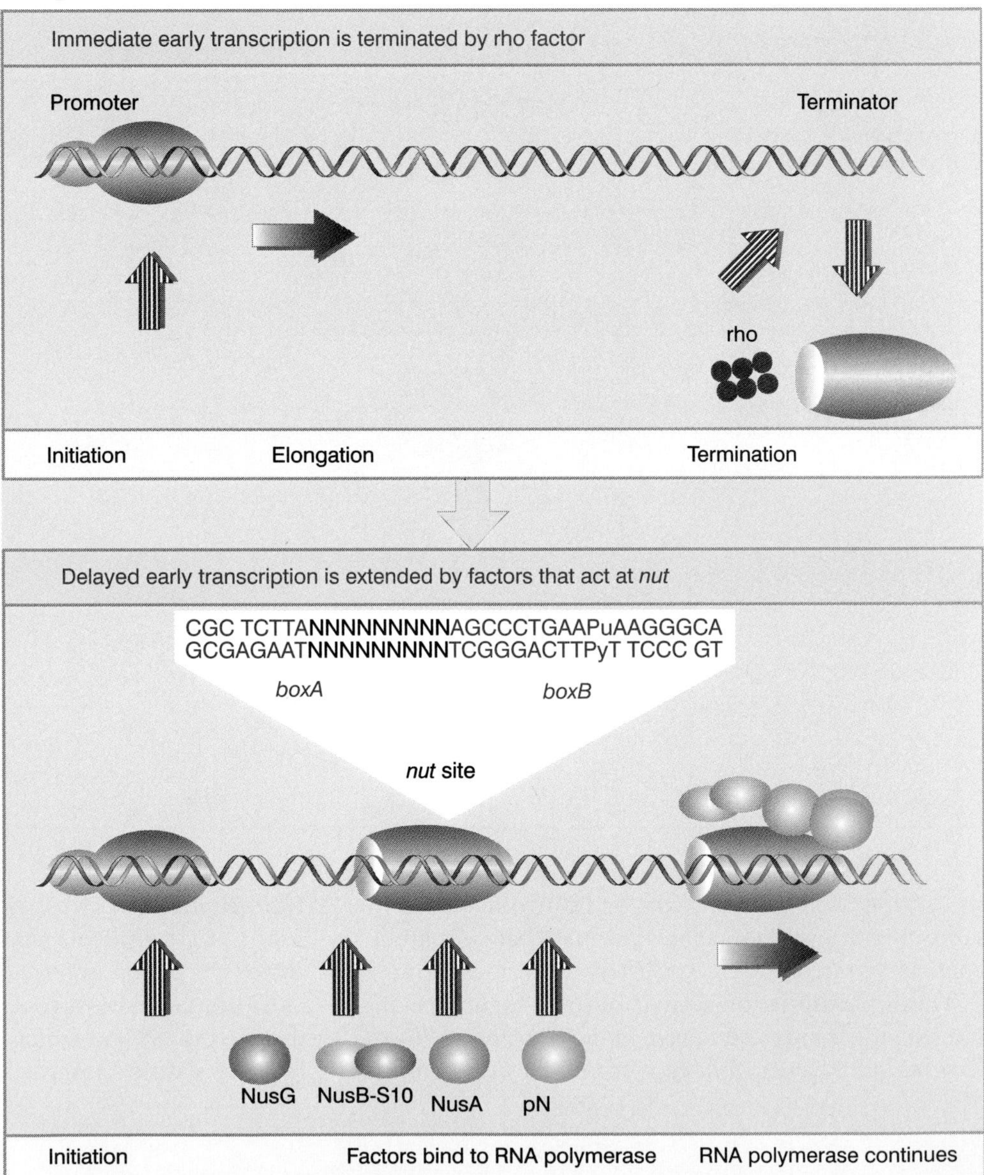
Immediate early transcription is terminated by rho factor
Promoter
Terminator
rho
Initiation
Elongation
Termination
Delayed early transcription is extended by factors that act at nut
CGC TCTTANNNNNNNNNAGCCCTGAAPuAAGGGCA
GCGAGAATNNNNNNNNNTCGGGACTTPyT TCCC GT
boxA
boxB
nut site
NusG
NusB-S10
NusA
pN
Initiation
Factors bind to RNA polymerase
RNA polymerase continues

Figure 9.34 Ancillary factors bind to RNA polymerase as it passes certain sites. The *nut* site consists of two sequences. NusB-S10 join core enzyme as it passes *boxA*. Then NusA and pN protein bind as polymerase passes *boxB*. The presence of pN allows the enzyme to read through the terminator, producing a joint mRNA that contains immediate early sequences joined to delayed early sequences.

widely used mechanism for antitermination? Other phages, related to lambda, have different *N* genes and different antitermination specificities. The region of the phage genome in which the *nut* sites lie has a different sequence in each of these phages, and each phage must therefore have characteristic *nut* sites recognized specifically by its own pN. Each of these pN products must have the same general ability to interact with the transcription apparatus in an antitermination capacity, but has a different specificity for the sequence of DNA that activates the mechanism.

pN and pQ both affect the ability of core RNA polymerase to terminate, but they interact differently with the enzyme. pN interacts with the core enzyme as it passes particular DNA sites, but pQ interacts with the holoenzyme during the initiation phase. In fact, σ^{70} is required for the interaction with pQ. This reinforces the view of RNA polymerase as an interactive structure in which conformational changes induced at one phase may affect its activity at a later phase.

More subunits for RNA polymerase

Termination and antitermination are closely connected, and involve bacterial proteins and phage proteins that interact with RNA polymerase in response to sequences within certain transcription units. A lambda *nut* site consists of two sequence elements, called *boxA* and *boxB*. Sequence elements related to *boxA* are also found in bacterial operons. *boxA* is required for binding bacterial proteins that are necessary for antitermination in both phage and bacterial operons. *boxB* is specific to the phage genome, and mutations in *boxB* abolish the ability of pN to cause antitermination.

The discovery of antitermination as a phage control mechanism led to the identification of further components of the transcription apparatus. The bacterial proteins with which pN interacts can be identified by isolating mutants of *E. coli* in which pN is ineffective. Several of these mutations lie in the *rpoB* gene. This argues that pN (like rho factor) interacts with the β subunit of the core enzyme.

Other *E. coli* mutations that prevent pN function identify the *nus* loci: *nusA*, *nusB*, *nusE*, and *nusG*. (The term "*nus*" is an acronym for *N* *u*tilization *s*ubstance.) The *nus* loci code for proteins that form part of the transcription apparatus but are not isolated with the RNA polymerase enzyme in its usual form. The *nusA*, *nusB*, and *nusG* functions are concerned solely with the termination of transcription. *nusE* codes for ribosomal protein S10; the relationship between its location in the 30S subunit and its function in termination is not clear. NusA is a general transcription factor that increases the efficiency of termination, probably by enhancing RNA polymerase's tendency to pause at terminators (and indeed at other regions of secondary structure; see below). NusB and S10 form a dimer that binds specifically to RNA containing a *boxA* sequence. NusG may be concerned with the general assembly of all the Nus factors into a complex with RNA polymerase. A distinction in the requirements for Nus functions is that NusA suffices for pN to prevent termination at intrinsic terminators, whereas all four Nus functions are required for pN action at rho-dependent terminators.

Antitermination occurs in the *rrn* (rRNA) operons of *E. coli*, and involves the same *nus* functions. The leader regions of the *rrn* operons contain *boxA* sequences; NusB-S10 dimers recognize these sequences and bind to RNA polymerase as it elongates past *boxA*. This changes the properties of RNA polymerase in such a way that it can now read through rho-dependent terminators that are present within the transcription unit.

The *boxA* sequence of lambda RNA does not bind NusB-S10, and is probably enabled to do so by the presence of NusA and pN; the *boxB* sequence could be required to stabilize the reaction. So variations in *boxA* sequences may determine which particular set of factors is required for antitermination. The consequences are the same: when RNA polymerase passes the *nut* site, it is modified by addition of appropriate factors, and fails to terminate when it subsequently encounters the terminator sites.

NusA binds to the polymerase core enzyme, but does not bind to holoenzyme. When sigma factor is added to the $\alpha_2\beta\beta'$NusA complex, it displaces the

NusA protein, thus reconstituting the $\alpha_2\beta\beta'\sigma$ holoenzyme. This suggests that RNA polymerase passes through the cycle illustrated in **Figure 9.35,** in which it exists in the alternative forms of an enzyme ready to initiate ($\alpha_2\beta\beta'\sigma$) and an enzyme ready to terminate ($\alpha_2\beta\beta'$NusA). When the holoenzyme ($\alpha_2\beta\beta'\sigma$) binds to a promoter, it releases its sigma factor, and thus generates the core enzyme ($\alpha_2\beta\beta'$) that synthesizes RNA. Then a NusA protein recognizes the core enzyme and binds to it, generating the $\alpha_2\beta\beta'$NusA complex. While the $\alpha_2\beta\beta'$NusA polymerase is bound to DNA, the Nus components cannot be displaced. But when termination occurs, the enzyme is released in a state in which the Nus factors either are released or can be displaced by sigma factor.

Core enzyme therefore alternates between associating with sigma for initiation and associating with Nus factors for termination. Sigma and the Nus factors are mutually incompatible associates of the core. There seems no reason to regard either one as any more a component than the other; we may regard them as alternative subunits. So the core enzyme represents a minimal form of RNA polymerase, competent to engage in the basic function of RNA synthesis, but lack-

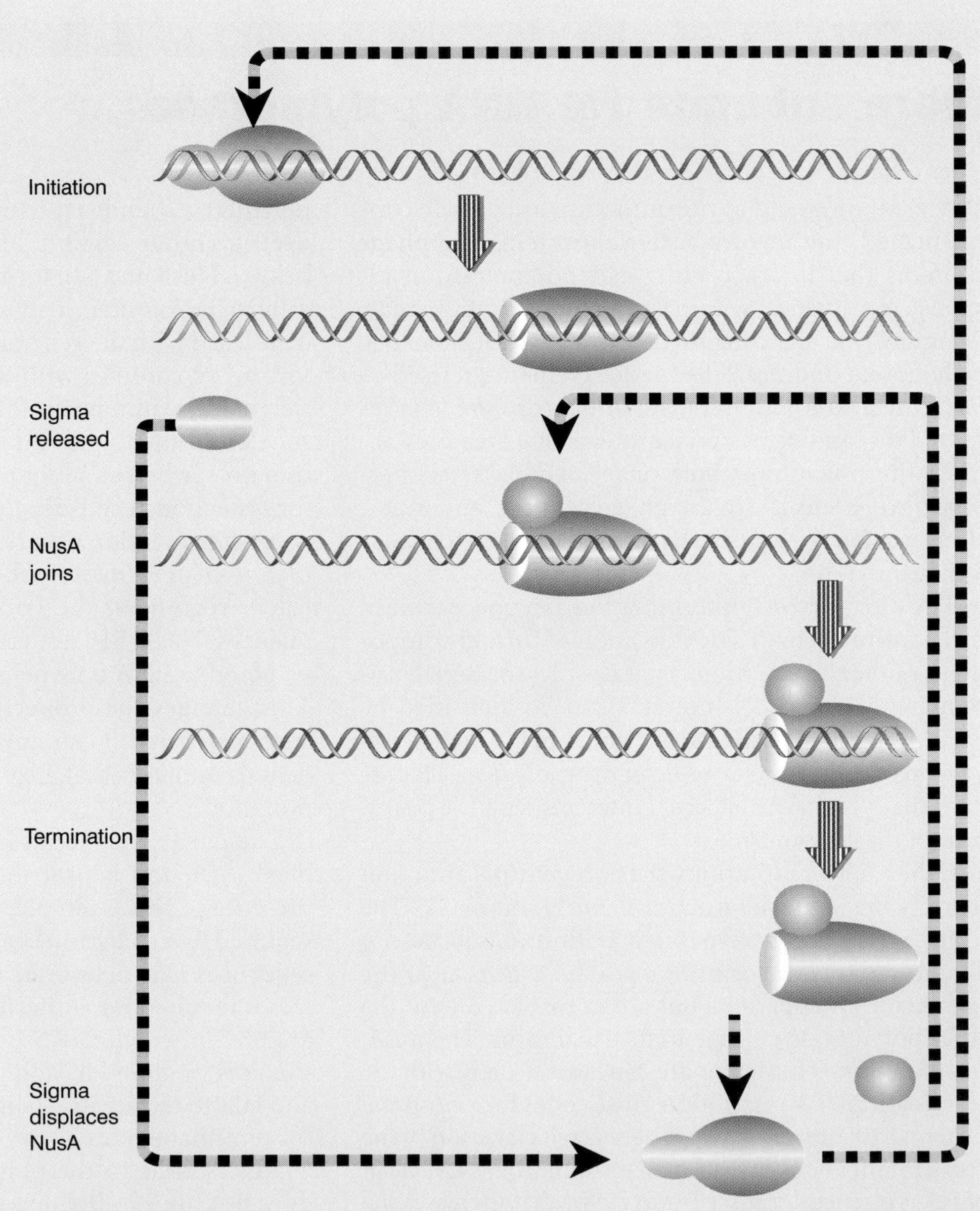

Figure 9.35 RNA polymerase may alternate between initiation-competent and termination-competent forms as sigma and Nus factors alternatively replace one another on the core enzyme.

ing subunits necessary for other functions. It is a moot point where RNA polymerase ends and the wider transcription apparatus begins.

Antitermination requires pN to bind to RNA polymerase in a manner that depends on the sequence of the transcription unit. Does pN recognize the *boxB* site in DNA or in the RNA transcript? It does not bind directly to either type of sequence, but does bind to a transcription complex when core enzyme passes the *boxB* site. Probably it recognizes the *boxB* RNA sequence, but also must make protein-protein contacts with NusA and with RNA polymerase in order to bind. After joining the transcription complex, pN remains associated with the core enzyme, in effect becoming an additional subunit whose presence changes recognition of terminators. It is possible that pN in fact continues to bind to both the *boxB* RNA sequence and to RNA polymerase, maintaining a loop in the RNA; thus the role of *boxB* RNA would partly be to tether pN in the vicinity, effectively increasing its local concentration.

What is the antiterminating action of pN? It might act to prevent RNA polymerase from pausing, thus denying rho factor the opportunity to cause termination; or it might act directly to antagonize the ability of rho to release the nascent RNA transcript. pN indeed has an antipausing activity; it is not clear yet whether it also acts on rho factor.

pQ, which prevents termination later in phage infection by acting at *qut*, has a different mode of action. The *qut* sequence lies at the start of the late transcription unit. The upstream part of *qut* lies within the promoter, while the downstream part lies at the beginning of the transcribed region. This implies that pQ action involves recognition of DNA, and implies that its mechanism of action, at least concerning the initial binding to the complex, must be different from that of pN.

The basic action of pQ is to interfere with pausing; and once pQ has acted upon RNA polymerase, the enzyme shows much reduced pausing at all sites, including rho-dependent and intrinsic terminators. So pQ does not act directly on termination *per se*, but instead allows the enzyme to hurry past the terminator, thus depriving the core polymerase and/or accessory factor of the opportunity to cause termination.

The general principle is that RNA polymerase may exist in forms that are competent to undertake particular stages of transcription, and its activities at these stages can be changed only by modifying the appropriate form. So substitutions of sigma factors may change one initiation-competent form into another; and additions of Nus factors may change the properties of termination-competent forms.

Termination seems to be inextricably connected with the mode of elongation. In its basic transcription mode, core polymerase is subject to many pauses during elongation; and pausing at a terminator site is the prerequisite for termination to occur. Under the influence of factors such as NusA, pausing becomes extended, increasing the efficiency of termination; while under the influence of pN or pQ, pausing is abbreviated, decreasing the efficiency of termination. Because recognition sites for these factors are found only in certain operons, pausing and consequently termination are altered only in those operons.

Summary

A transcription unit comprises the DNA between a promoter, where transcription initiates, and a terminator, where it ends. One strand of the DNA in this region serves as a template for synthesis of a complementary strand of RNA. The RNA-DNA hybrid region is short and transient, as the transcription "bubble" moves along DNA. The RNA polymerase holoenzyme that synthesizes bacterial RNA can be separated into two components. Core enzyme is a multimer of structure $\alpha_2\beta\beta'$ that is responsible for elongating the RNA chain. Sigma factor is a single subunit that is required at the stage of initiation for recognizing the promoter.

Core enzyme has a general affinity for DNA. The addition of sigma factor reduces the affinity of the enzyme for nonspecific binding to DNA, but increases its affinity for promoters. The rate at which RNA polymerase finds its promoters is too great to be accounted for by diffusion and random contacts with DNA; direct exchange of DNA sequences held by the enzyme may be involved.

Bacterial promoters are identified by two short conserved sequences centered at −35 and −10 relative to the startpoint. Most promoters have sequences that are closely related to the consensus sequences at these sites. The distance separating the consensus sequences is 16–18 bp. RNA polymerase initially "touches down" at the −35 sequence and then extends its contacts over the −10 region. The enzyme covers ~77 bp of DNA. The initial "closed" binary complex is converted to an "open" binary complex by melting of a sequence of ~12 bp that extends from the −10 region to the startpoint. The A•T-rich base pair composition of the −10 sequence may be important for the melting reaction.

The binary complex is converted to a ternary complex by the incorporation of ribonucleotide precursors. There are multiple cycles of abortive initiation, during which RNA polymerase synthesizes and releases RNA chains of 2–9 bases without moving from the promoter. At the end of this stage, sigma factor is released, and the core enzyme contracts to cover ~50 bp. Then core enzyme moves along DNA, synthesizing RNA. A locally unwound region of DNA moves with the enzyme. The enzyme contracts further in size to cover only 30–40 bp when the nascent chain has reached 15–20 nucleotides; then it continues to the end of the transcription unit.

The "strength" of a promoter describes the frequency at which RNA polymerase initiates transcription; it is related to the closeness with which its −35 and −10 sequences conform to the ideal consensus sequences, but is influenced also by the sequences immediately downstream of the startpoint. Negative supercoiling increases the strength of certain promoters. Transcription generates positive supercoils ahead of RNA polymerase and leaves negative supercoils behind the enzyme.

The core enzyme can be directed to recognize promoters with different consensus sequences by alternative sigma factors. In *E. coli*, such sigma factors are activated by adverse conditions, such as heat shock or nitrogen starvation. *B. subtilis* contains a single major sigma factor with the same specificity as the *E. coli* sigma factor, but also contains a variety of minor sigma factors. Another series of factors is activated when sporulation is initiated; sporulation is regulated by two cascades in which sigma factor replacements occur in the forespore and mother cell. A similar mechanism for regulating transcription is also used by phage SPO1.

The geometry of RNA polymerase-promoter recognition is similar for holoenzymes containing all sigma factors (except s54). Each sigma factor causes RNA polymerase to initiate transcription at a promoter that conforms to a particular consensus at −35 and −10. Direct contacts between sigma and DNA at these sites have been demonstrated for *E. coli* σ70. It is not clear how contacts made with sigma factor relate to contacts made by subunits of core enzyme. An unanswered question on the use of sigma factors concerns the nature of the mechanism that allows one sigma to replace another.

Bacterial RNA polymerase terminates transcription at two types of sites. Intrinsic terminators contain a G•C-rich hairpin followed by a run of U residues. They are recognized *in vitro* by core enzyme alone. Rho-dependent terminators require rho factor both *in vitro* and *in vivo*; they have a stretch of 50–90 nucleotides preceding the site of termination that is rich in C and poor in G residues. Rho factor is an essential protein that acts as an ancillary termination factor, which recognizes RNA and acts at sites where RNA polymerase has paused. The termination activity requires ATP hydrolysis.

The Nus factors increase the efficiency of rho-dependent termination, and provide the means by which antitermination factors act. The NusB-S10 dimer recognizes the *boxA* sequence in the elongating RNA; NusA joins subsequently. Factor NusA and the initiation factor sigma are mutually exclusive associates of core enzyme.

Antitermination is used by some phages to regulate progression from one stage of gene expression to the next and (less often) in bacteria. The lambda gene N codes for an antitermination protein (pN) that is necessary to allow RNA polymerase to read through the terminators located at the ends of the immediate early genes. Another antitermination protein, pQ, is required later in phage infection. pN and pQ act on RNA polymerase as it passes specific sites (*nut* and *qut*, respectively). These sites are located at different relative positions in their respective transcription units. pN recognizes RNA polymerase carrying NusA when the enzyme passes the sequence *boxB*. pN then binds to the complex and prevents termination when the polymerase reaches the rho-dependent terminator.

Further reading

Reviews

Losick, R. and Chamberlin, M. (1976). *RNA Polymerase.* (Cold Spring Harbor Laboratory, New York).

Adhya, S. and Gottesman, M. (1978). Control of transcription termination. *Ann. Rev. Biochem.* **47**, 967–996.

Reznikoff, W. S. *et al.* (1985). The regulation of transcription initiation in bacteria. *Ann. Rev. Genet.* **19**, 355–387.

Doi, R. H. and Wang, L.-F. (1986). Multiple prokaryotic RNA polymerase sigma factors. *Microbiol. Rev.* **50**, 227–243.

Losick, R. *et al.* (1986). Genetics of endospore formation in *B. subtilis. Ann. Rev. Genet.* **20**, 625–669.

Platt, T. (1986). Transcription termination and the regulation of gene expression. *Ann. Rev. Biochem.* **55**, 339–372.

Friedman, D. I., Imperiale, M. J., and Adhya, S. L. (1987). RNA 3′ end formation in the control of gene expression. *Ann. Rev. Genet.* **21**, 453–488.

Yager, T. D. and von Hippel, P. H. (1987). Transcript elongation and termination in *E. coli.* In *E. coli and S. typhimurium.* (Ed. F. C. Neidhardt, American Society for Microbiology, Washington DC, 1241–1275).

Helmann, J. D. and Chamberlin, M. (1988). Structure and function of bacterial sigma factors. *Ann. Rev. Biochem.* **57**, 839–872.

Losick, R. and Stragier, P. (1992). Crisscross regulation of cell-type specific gene expression during development in *B. subtilis. Nature* **355**, 601–604.

Errington, J. (1993). *B. subtilis* sporulation: regulation of gene expression and control of morphogenesis. *Microbiol. Rev.* **57**, 1–33.

Das, A. (1993). Control of transcription termination by RNA-binding proteins. *Ann. Rev. Biochem.* **62**, 893–930.

Haldenwang, W. G. (1995). The sigma factors of *B. subtilis. Microbiol. Rev.* **59**, 1–30.

Stragier, P. and Losick, R. (1996). Molecular Genetics of sporulation in *B. subtilis. Ann. Rev. Genet.* **30**, 297–341.

Discoveries

Sigma factors

Travers, A. A. and Burgess, R. R. (1969). Cyclic reuse of the RNA polymerase sigma factor. *Nature* **222**, 537–540.

Haldenwang, W. G. and Losick, R. (1980). A novel RNA polymerase sigma factor from B. subtilis. *Proc. Nat. Acad. Sci. USA* **77**, 7000–7004.

Haldenwang, W. G., Lang, N., and Losick, R. (1981). A sporulation-induced sigma-like regulatory protein from *B. subtilis. Cell* **23**, 615–624.

Grossman, A. D., Erickson, J. W., and Gross, C. A. (1984). The *htpR* gene product of *E. coli* is a sigma factor for heat-shock promoters. *Cell* **38**, 383–390.

RNA polymerase

Siebenlist, U., Simpson, R. B., and Gilbert, W. (1980). *E. coli* RNA polymerase interacts homologously with two different promoters. *Cell* **20**, 269–281.

Krummel, B. and Chamberlin, M. J. (1989). RNA chain initiation by *E. coli* RNA polymerase. Structural transitions of the enzyme in early ternary complexes. *Biochemistry* **28**, 7829–7842.

Rice, G. A., Kane, C. M., and Chamberlin, M. (1991). Footprinting analysis of mammalian RNA polymerase II along its transcript: an alternative view of transcription elongation. *Proc. Nat. Acad. Sci. USA* **88**, 4245–4249.

Dombrowski, A. J. *et al.* (1992). Polypeptides containing highly conserved regions of transcription initiation factor σ^{70} exhibit specificity of binding to promoter DNA. *Cell* **70**, 501–512.

Wang, D. *et al.* (1995). Discontinuous movements of DNA and RNA in RNA polymerase accompany formation of a paused transcription complex. *Cell* **81**, 341–350.

Nudler, E. *et al.* (1997). The RNA-DNA hybrid maintains the register of transcription by preventing backtracking of RNA polymerase. *Cell* **89**, 33–41.

Transcription and supercoiling

Wu, H.-Y. *et al.* (1988). Transcription generates positively and negatively supercoiled domains in the template. *Cell* **53**, 433–440.

Termination

Roberts, J. W. (1969). Termination factor for RNA synthesis. *Nature* **224**, 1168–1174.

Greenblatt, J. and Li, J. (1981). Interaction of the sigma factor and the NusA gene protein of *E. coli* with RNA polymerase in the initiation-termination cycle of transcription. *Cell* **24**, 421–428.

Chapter 10

The operon

GENE expression can be controlled at any of several stages, which we divide broadly into transcription, processing, and translation:

■ Transcription often is controlled at the stage of initiation. Transcription is not usually controlled at elongation, but may be controlled at termination to prevent transcription from proceeding past a terminator to the gene(s) beyond.

■ In eukaryotic cells, processing of the RNA product may be regulated at the stages of modification, splicing, transport, or stability. In bacteria, an mRNA is in principle available for translation as soon as it is synthesized, and these stages of control are not available.

■ Translation may be regulated, usually at the stages of initiation and termination (like transcription). Regulation of initiation is formally analogous to the regulation of transcription: the circuitry can be drawn in similar terms for regulating initiation of transcription on DNA or initiation of translation on RNA.

The basic concept for how transcription is controlled in bacteria was provided by the classic formulation of the model for control of gene expression by Jacob and Monod in 1961. They distinguished between two types of sequences in DNA: sequences that code for *trans*-**acting products;** and *cis*-**acting sequences** that function exclusively within the DNA (see Figures 1.30–32). Gene activity is regulated by the specific interactions of the *trans*-acting products (usually proteins) with the *cis*-acting sequences (usually sites in DNA). In more formal terms:

■ A gene is a sequence of DNA that codes for a diffusible product. This product may be protein (as in the case of the majority of genes) or may be RNA (as in the case of genes that code for tRNA and rRNA). *The crucial feature is that the product diffuses away from its site of synthesis to act elsewhere.* Any gene product that is free to diffuse to find its target is described as *trans*-acting.

■ The description *cis*-acting applies to any sequence of DNA that is not converted into any other form, but that functions exclusively as a DNA sequence *in situ*, affecting only the DNA to which it is physically linked. (In some cases, a *cis*-acting sequence functions in an RNA rather than in a DNA molecule.)

To help distinguish between the components of regulatory circuits and the genes that they regulate, we sometimes use the terms structural gene and regulator gene. A **structural gene** is simply any gene that codes for a protein (or RNA) product. Structural genes represent an enormous variety of protein structures and functions, including structural proteins, enzymes with catalytic activities, and regulatory proteins. A **regulator gene** describes a structural gene that codes for a protein involved in regulating the expression of other genes.

The crux of regulation is that *a regulator gene codes for a protein that controls transcription by binding to particular site(s) on DNA.* This interaction can regulate a target gene in either a positive manner (the interaction turns the gene on) or in a negative manner (the interaction turns the gene off). The sites on DNA are usually (but not exclusively) located just upstream of the target gene.

The sequences that mark the beginning and end of the transcription unit, the promoter and terminator, are examples of *cis*-acting sites. *A promoter serves to initiate transcription only of the gene or genes physically connected to it on the same stretch of DNA.* In the same way, a terminator can terminate transcription only by an RNA polymerase that has traversed the preceding gene(s). In their simplest forms, promoters and terminators are *cis*-acting elements that are recognized by the same *trans*-acting species, that is, by RNA polymerase (although other factors also participate at each site).

Additional *cis*-acting regulatory sites are often juxtaposed to, or interspersed with, the promoter. A bacterial promoter may have one or more such sites located close by, that is, in the immediate vicinity of the startpoint. A eukaryotic promoter is likely to have a greater number of sites, spread out over a longer distance.

A classic mode of control in bacteria is *negative:* a **repressor protein** prevents a gene from being expressed. **Figure 10.1** shows that the "default state" for such a gene is to be expressed via the recognition of its promoter by RNA polymerase. Close to the promoter is another *cis*-acting site called the **operator**, which is the target for the repressor protein. When the repressor binds to the operator, RNA polymerase is prevented from initiating transcription, and *gene expression is therefore turned off.*

An alternative mode of control is *positive.* This is used in bacteria (probably) with about equal frequency to negative control, and it is the most common mode of control in eukaryotes. A **transcription factor** is required to assist RNA polymerase in initiating at the promoter. **Figure 10.2** shows that the typical default state of a eukaryotic gene is inactive: RNA polymerase cannot by itself initiate transcription at the promoter. Several *trans*-acting factors have target sites in the vicinity of the promoter, and binding of some or all of these factors *enables RNA polymerase to initiate transcription.*

The unifying theme is that regulatory proteins are *trans*-acting factors that recognize *cis*-acting elements (usually) upstream of the gene. The consequences of this recognition are to activate or to repress the gene, depending on the individual type of regulatory protein. A typical feature is that the protein functions by recognizing a very short sequence in DNA, usually <10 bp in length, although the protein actually binds

Figure 10.1 *Overview*: in negative control, a *trans*-acting repressor binds to the *cis*-acting operator to turn off transcription. In prokaryotes, multiple genes are controlled coordinately.

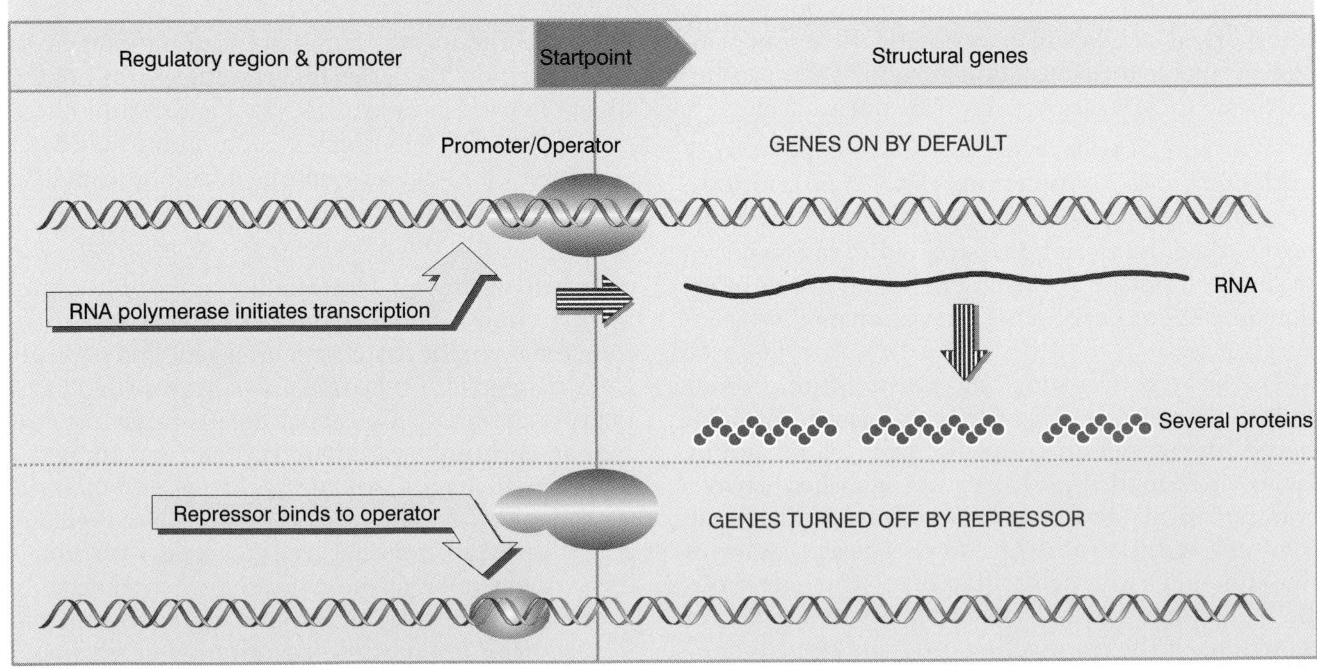

Figure 10.2 *Overview*: in positive control, *trans*-acting factors must bind to *cis*-acting sites in order for RNA polymerase to initiate transcription at the promoter. In a eukaryotic system, a structural gene is controlled individually.

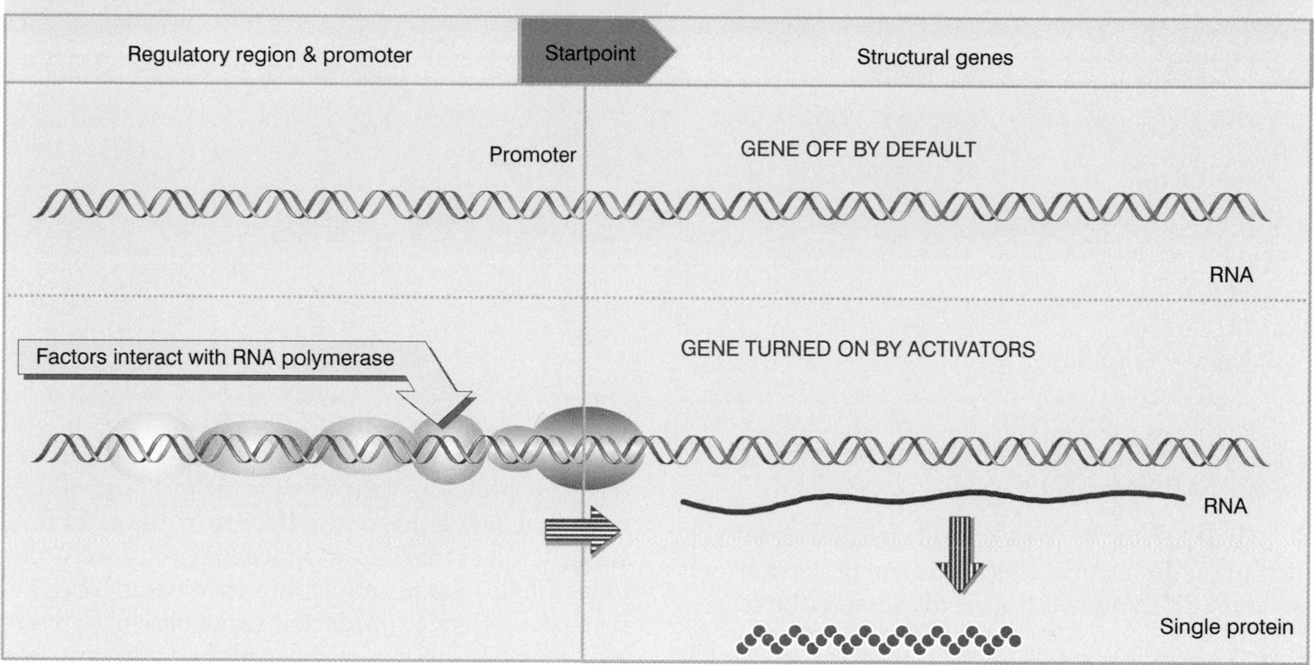

over a somewhat greater distance of DNA. The bacterial promoter is an example: although RNA polymerase covers >70 bp of DNA at initiation, the crucial sequences that it recognizes are the hexamers centered at –35 and –10.

A significant difference in gene organization between prokaryotes and eukaryotes is that structural genes in bacteria are organized in clusters, while those in eukaryotes occur individually. Clustering of structural genes allows them to be coordinately controlled by means of interactions at a single promoter: as a result of these interactions, the entire set of genes is either transcribed or not transcribed. In this chapter, we discuss this mode of control and its use by bacteria. The means employed to coordinate control of dispersed eukaryotic genes are discussed in Chapter 21.

Structural gene clusters are coordinately controlled

Bacterial structural genes are often organized into clusters that include genes coding for proteins whose functions are related. It is common for the genes for the enzymes of a metabolic pathway to be organized into such a cluster. In addition to the enzymes actually involved in the pathway, other related activities may be included in the unit of coordinate control; for example, the protein responsible for transporting the small molecule substrate into the cell.

The cluster of the three *lac* structural genes, *lacZYA*, is typical. The protein products enable cells to take up and metabolize β-galactosides, such as lactose. **Figure 10.3** summarizes the organization of the structural genes, their associated *cis*-acting regulatory elements, and the *trans*-acting regulatory gene. The roles of the three structural genes are:

- *lacZ* codes for the enzyme β-galactosidase, whose active form is a tetramer of ~500 kD. The enzyme breaks a β-galactoside into its component sugars. For

Figure 10.3 The *lac* operon occupies ~6000 bp of DNA. At the left the *lacI* gene has its own promoter and terminator. The end of the *lacI* region is adjacent to the promoter, P. The operator, O, occupies the first 26 bp of the long *lacZ* gene, followed by the *lacY* and *lacA* genes and a terminator.

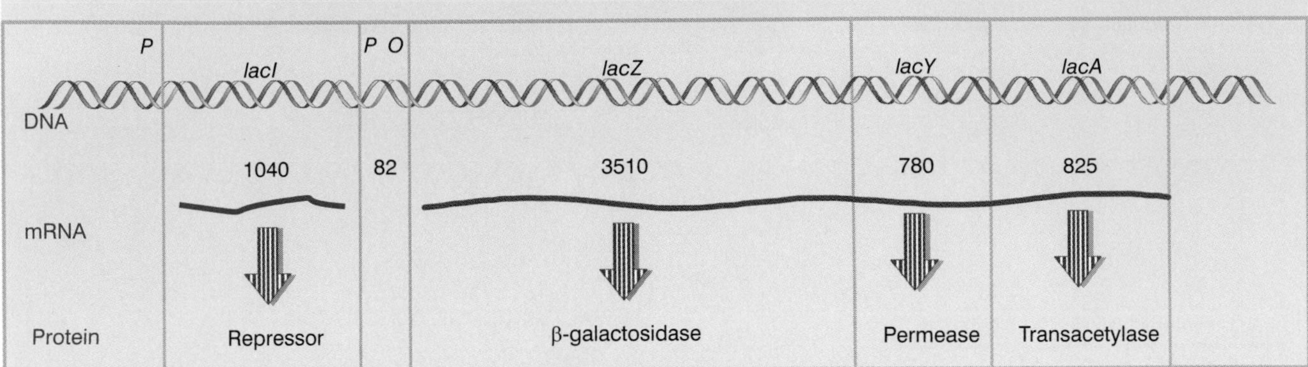

example, lactose is cleaved into glucose and galactose (which are then further metabolized).

- *lacY* codes for the β-galactoside permease, a 30 kD membrane-bound protein constituent of the transport system. This transports β-galactosides into the cell.
- *lacA* codes for β-galactoside transacetylase, an enzyme that transfers an acetyl group from acetyl-CoA to β-galactosides.

Mutations in either *lacZ* or *lacY* can create the *lac* genotype, in which cells cannot utilize lactose. (The genotypic description "*lac*" without a qualifier indicates loss-of-function.) The *lacZ* mutations abolish enzyme activity, directly preventing metabolism of lactose. The *lacY* mutants cannot take up lactose from the medium. (No defect is identifiable in *lacA* cells, which is puzzling. It is possible that the acetylation reaction gives an advantage when the bacteria grow in the presence of certain analogs of β-galactosides that cannot be metabolized, because the modification results in detoxification and excretion.)

The entire system, including structural genes and the elements that control their expression, forms a common unit of regulation; this is called an **operon.** The activity of the operon is controlled by regulator gene(s), whose protein products interact with the *cis*-acting control elements.

We can distinguish between structural genes and regulator genes by the effects of mutations. A mutation in a structural gene deprives the cell of the particular protein for which the gene codes. But a mutation in a regulator gene influences the expression of all the structural genes that it controls. The consequences of a regulatory mutation reveal the type of regulation.

Transcription of the *lacZYA* genes is controlled by a regulator protein synthesized by the *lacI* gene. It happens that *lacI* is located adjacent to the structural genes, but it comprises an independent transcription unit with its own promoter and terminator. Since *lacI* specifies a diffusible product, in principle it need not be located near the structural genes; it can function equally well if moved elsewhere, or carried on a separate DNA molecule (the classic test for a *trans*-acting regulator).

The *lac* genes are controlled by **negative regulation:** *they are transcribed unless turned off by the regulator protein.* A mutation that inactivates the regulator causes the structural genes to remain in the expressed

Figure 10.4 Repressor and RNA polymerase bind at sites that overlap around the startpoint of the *lac* operon.

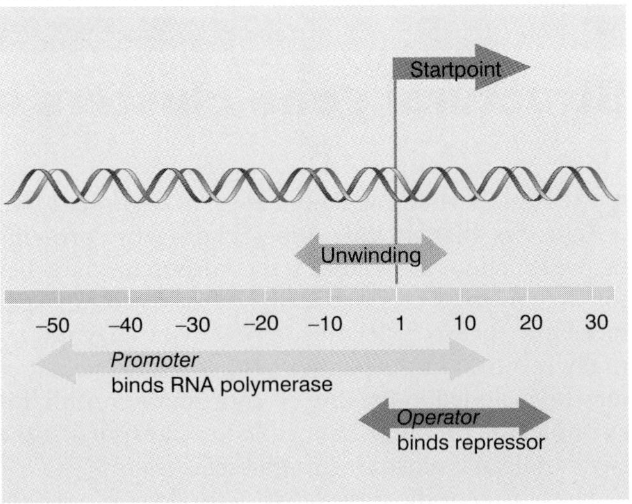

condition. The product of *lacI* is called the *lac* **repressor,** because its function is to prevent the expression of the structural genes.

The repressor is a tetramer of identical subunits of 38 kD each. There are ~10 tetramers in a wild-type cell. The regulator gene is transcribed into a monocistronic mRNA at a rate that appears to be governed simply by the affinity of its promoter for RNA polymerase.

The repressor functions by binding to an operator (formally denoted O_{lac}) at the start of the *lacZYA* cluster. The operator lies between the promoter (P_{lac}) and the structural genes (*lacZYA*). *When the repressor binds at the operator, it prevents RNA polymerase from initiating transcription at the promoter.* **Figure 10.4** expands our view of the region at the start of the *lac* structural genes. The operator extends from position –5 just upstream of the mRNA startpoint to position +21 within the transcription unit. So it overlaps the right end of the promoter. We discuss the relationship between repressor and RNA polymerase in more detail later.

Repressor is controlled by a small molecule inducer

BACTERIA need to respond swiftly to changes in their environment. Fluctuations in the supply of nutrients can occur at any time; survival depends on the ability to switch from metabolizing one substrate to another. Yet economy also is important, since a bacterium that indulges in energetically expensive ways to meet the demands of the environment is likely to be at a disadvantage. So a bacterium avoids synthesizing the enzymes of a pathway in the absence of the substrate; but is ready to produce the enzymes if the substrate should appear.

The synthesis of enzymes in response to the appearance of a specific substrate is called **induction.** This type of regulation is widespread in bacteria, and occurs also in unicellular eukaryotes (such as yeasts). The lactose system of *E. coli* provides the paradigm for this sort of control mechanism.

When cells of *E. coli* are grown in the absence of a β-galactoside, there is no need for β-galactosidase, and they contain very few molecules of the enzyme—say, <5. When a suitable substrate is added, the enzyme activity appears very rapidly in the bacteria. Within 2–3 minutes some enzyme is present, and soon there are ~5000 molecules of enzyme per bacterium. (Under suitable conditions, β-galactosidase can account for 5–10% of the total soluble protein of the bacterium.) If the substrate is removed from the medium, the synthesis of enzyme stops as rapidly as it had originally started.

Figure 10.5 summarizes the essential features of induction. Control of transcription of the *lac* genes responds very rapidly to the inducer, as shown in the upper part of the figure. In the absence of inducer, the operon is transcribed at a very low **basal level** (see

Figure 10.5 Addition of inducer results in rapid induction of *lac* mRNA, and is followed after a short lag by synthesis of the enzymes; removal of inducer is followed by rapid cessation of synthesis.

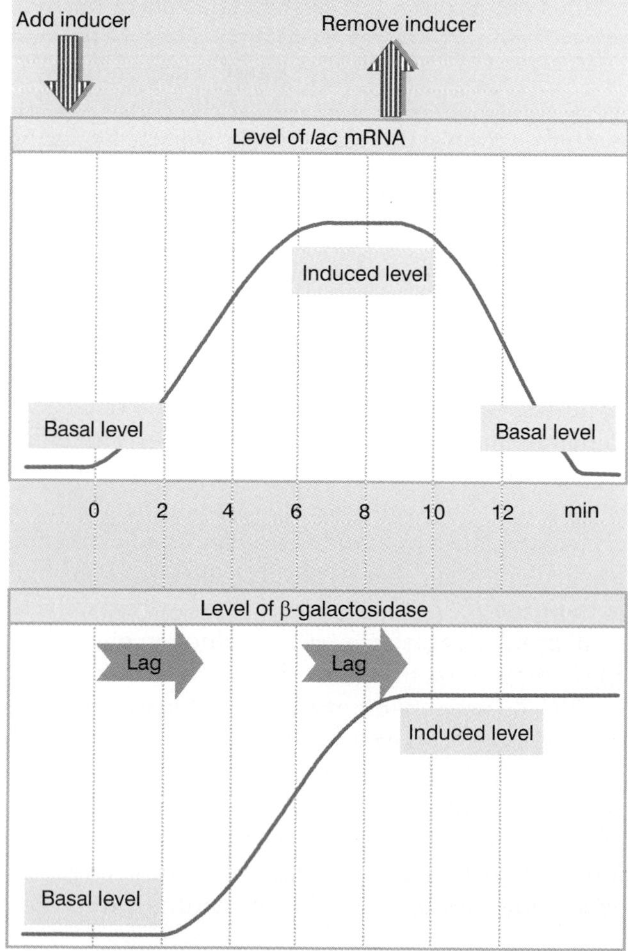

below). Transcription is stimulated as soon as inducer is added; the amount of *lac* mRNA increases rapidly to an induced level that reflects a balance between synthesis and degradation of the mRNA.

The *lac* mRNA is extremely unstable, and decays with a half-life of only ~3 minutes. This feature allows induction to be reversed rapidly. Transcription ceases as soon as the inducer is removed; and in a very short time all the *lac* mRNA has been destroyed and the cellular content has returned to the basal level.

The production of protein is followed in the lower part of the figure. Translation of the *lac* mRNA produces β-galactosidase (and the products of the other *lac* genes). There is a short lag between the appearance of *lac* mRNA and appearance of the first completed enzyme molecules (it is ~2 min after the rise of mRNA from the basal level before protein begins to increase). There is a similar lag between reaching maximal induced levels of mRNA and protein. When inducer is removed, synthesis of enzyme ceases almost immediately (as the mRNA is degraded), but the β-galactosidase in the cell is more stable than the mRNA, so the enzyme activity remains at the induced level for longer.

This type of rapid response to changes in nutrient supply not only provides the ability to metabolize new substrates, but also is used to shut off endogenous synthesis of compounds that suddenly appear in the medium. For example, *E. coli* synthesizes the amino acid tryptophan through the action of the enzyme tryptophan synthetase. But if tryptophan is provided in the medium on which the bacteria are growing, the production of the enzyme is immediately halted. This effect is called **repression.** It allows the bacterium to avoid devoting its resources to unnecessary synthetic activities.

Induction and repression represent the same phenomenon. In one case the bacterium adjusts its ability to use a given substrate for growth; in the other it adjusts its ability to synthesize a particular metabolic intermediate. The trigger for either type of adjustment is the small molecule that is the substrate for the enzyme, or the product of the enzyme activity, respectively. Small molecules that cause the production of enzymes able to metabolize them are called **inducers.** Those that prevent the production of enzymes able to synthesize them are called **corepressors.**

The ability to act as inducer or corepressor is highly specific. Only the substrate/product or a closely related molecule can serve. *But the activity of the small molecule does not depend on its interaction with the target enzyme.* Some inducers resemble the natural inducers for β-galactosidase but cannot be metabolized by the enzyme. The example *par excellence* is isopropylthiogalactoside (IPTG), one of several thiogalactosides with this property. Although it is not recognized by β-galactosidase, IPTG is a very efficient inducer of the *lac* genes.

Molecules that induce enzyme synthesis but are not metabolized are called **gratuitous inducers.** They are extremely useful because they remain in the cell in their original form. (A real inducer would be metabolized, interfering with study of the system.) The existence of gratuitous inducers reveals an important point. *The system must possess some component, distinct from the target enzyme, that recognizes the appropriate substrate; and its ability to recognize related potential substrates is different from that of the enzyme.*

The component that responds to the inducer is the repressor protein coded by *lacI.* Its role in controlling transcription of the *lacZYA* structural genes in response to the environment is summarized in **Figure 10.6.** The structural genes are transcribed into a single mRNA from a promoter just upstream of *lacZ.* The state of the repressor determines whether this promoter is turned off or on:

- In the absence of an inducer, the genes are not transcribed, because repressor protein is in an active form that is bound to the operator.
- When an inducer is added, the repressor is converted into an inactive form that leaves the operator. Then transcription starts at the promoter and proceeds through the genes to a terminator located somewhere beyond *lacA.*

The crucial features of the control circuit reside in the dual properties of the repressor: it can prevent transcription; and it can recognize the small-molecule inducer. The repressor has two binding sites, one for the operator and one for the inducer. When the inducer binds at its site, it changes the conformation of the protein in such a way as to influence the activity of the operator-binding site. The ability of one site in the protein to control the activity of another is called **allosteric control.**

Induction accomplishes a **coordinate regulation:** *all the genes are expressed (or not expressed) in unison.* The mRNA is translated sequentially from its 5′ end, which explains why induction always causes the appearance of β-galactosidase, β-galactoside permease, and β-galactoside transacetylase, in that order. Translation of a common mRNA explains why the relative amounts

Figure 10.6 Repressor maintains the *lac* operon in the inactive condition by binding to the operator; addition of inducer releases the repressor, and thereby allows RNA polymerase to initiate transcription.

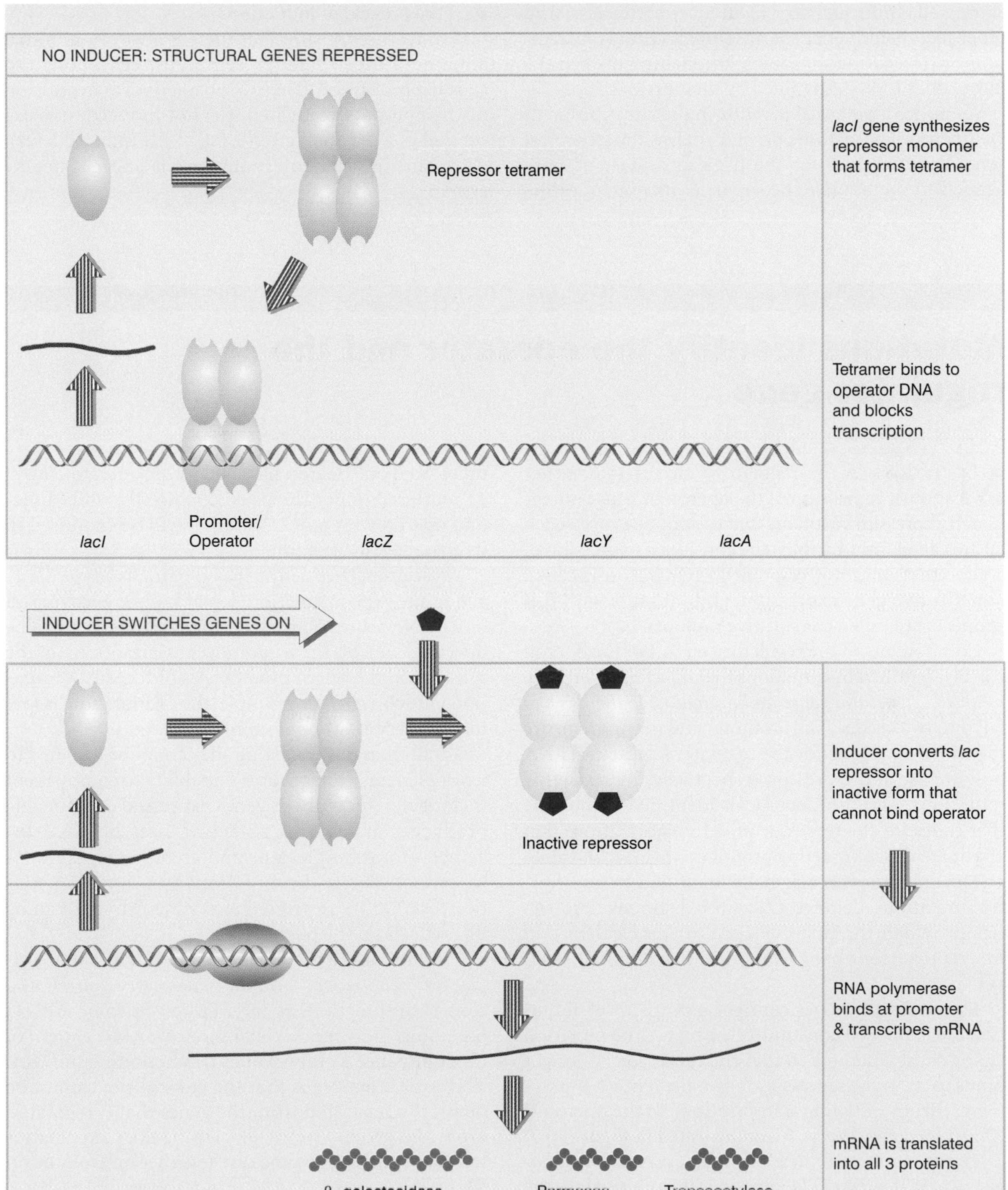
NO INDUCER: STRUCTURAL GENES REPRESSED
Repressor tetramer
lacI gene synthesizes repressor monomer that forms tetramer
Tetramer binds to operator DNA and blocks transcription
lacI
Promoter/ Operator
lacZ
lacY
lacA
INDUCER SWITCHES GENES ON
Inducer converts lac repressor into inactive form that cannot bind operator
Inactive repressor
RNA polymerase binds at promoter & transcribes mRNA
mRNA is translated into all 3 proteins
β–galactosidase
Permease
Transacetylase

of the three enzymes always remain the same under varying conditions of induction.

Induction throws a switch that causes the genes to be expressed. Inducers vary in their effectiveness, and other factors influence the absolute level of transcription or translation, but the relationship between the three genes is predetermined by their organization.

We notice a potential paradox in the constitution of the operon. The lactose operon contains the structural gene (*lacZ*) coding for the β-galactosidase activity needed to metabolize the sugar; it also includes the gene (*lacY*) that codes for the protein needed to transport the substrate into the cell. But if the operon is in a repressed state, how does the inducer enter the cell to start the process of induction?

Two features ensure that there is always a minimal amount of the protein present in the cell, enough to start the process off. There is a basal level of expression of the operon: even when it is not induced, it is expressed at a residual level (0.1% of the induced level). And some inducer enters anyway via another uptake system.

Mutations identify the operator and the regulator gene

Mutations in the regulatory circuit may either abolish expression of the operon or cause unregulated expression. Mutants that cannot be expressed at all are called **uninducible.** The continued expression of a gene that does not respond to regulation is called **constitutive** gene expression, and mutants with this property are called **constitutive mutants.**

Components of the regulatory circuit of the operon can be identified by mutations that *affect the expression of all the structural genes and map outside them.* They fall into two classes. The promoter and the operator are identified as targets for the regulatory proteins (RNA polymerase and repressor, respectively) by *cis*-acting mutations. And the locus *lacI* is identified as the gene that codes for the repressor protein by mutations that eliminate the *trans*-acting product.

The operator was originally identified by constitutive mutations, denoted O^c, whose distinctive properties provided the first evidence for *an element that functions without being represented in a diffusible product.*

The structural genes contiguous with an O^c mutation are expressed constitutively because the mutation changes the operator so that the repressor no longer binds to it. So the repressor cannot prevent RNA polymerase from initiating transcription. So the operon is transcribed constitutively, as illustrated in **Figure 10.7.**

The operator can control only the lac *genes that are adjacent to it.* If a second *lac* operon is introduced into the bacterium on an independent molecule of DNA, it has its own operator. Neither operator is influenced by the other. So if one operon has a wild-type operator, it will be repressed under the usual conditions, while a second operon with an O^c mutation will be expressed in its characteristic fashion.

These properties define the operator as a typical *cis*-acting site, whose function depends upon recognition of its DNA sequence by some *trans*-acting factor. The operator controls the adjacent genes irrespective of the presence in the cell of other alleles of the site. A mutation in such a site, for example, the O^c mutation, is formally described as *cis*-**dominant.**

A mutation in a *cis*-acting site cannot be assigned to a complementation group. (The ability to complement is characteristic only of genes expressed as diffusible products.) When two *cis*-acting sites lie close together—for example, a promoter and an operator—we cannot classify the mutations by a complementation test. We are restricted to distinguishing them by their effects on the phenotype.

Cis-dominance is a characteristic of any site that is *physically contiguous with the sequences it controls.* If a control site functions as part of a polycistronic mRNA, mutations in it will display *exactly the same pattern* of *cis*-dominance as they would if functioning in DNA. The critical feature is that the control site cannot be physically separated from the genes that it regulates. From the genetic point of view, it does not matter whether the site and genes are together on DNA or on RNA.

Constitutive transcription is also caused by mutations of the *lacI*$^-$ type, which are caused by loss of func-

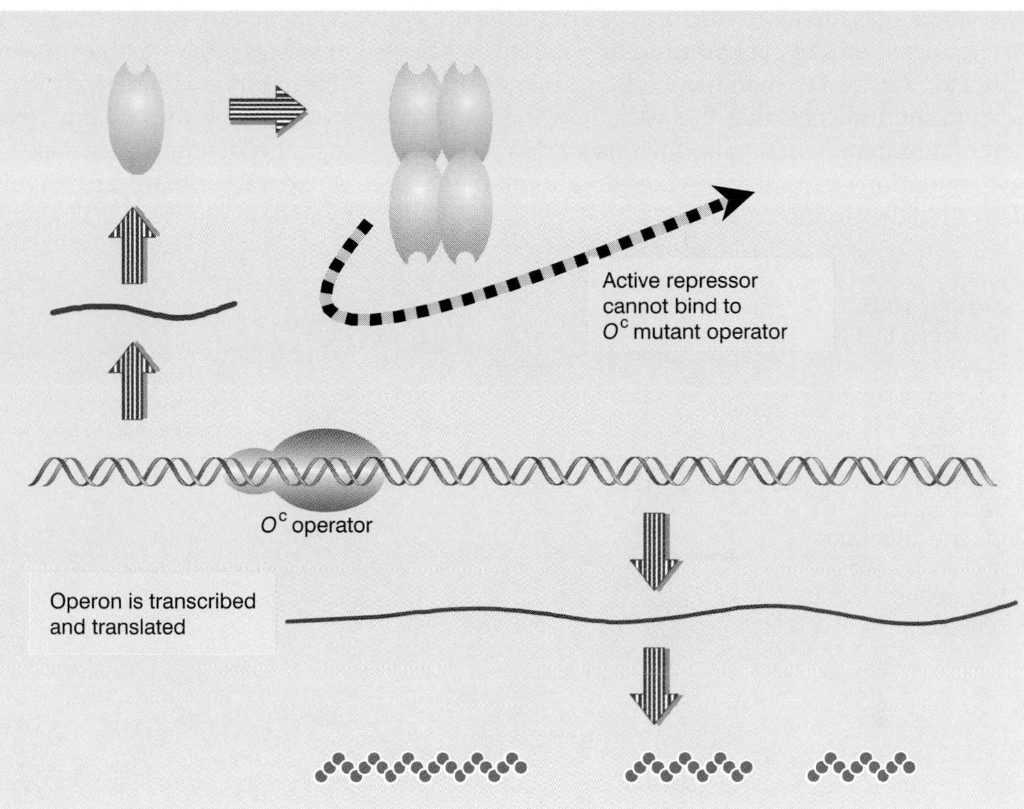
Active repressor
cannot bind to
O^c mutant operator
O^c operator
Operon is transcribed
and translated

Figure 10.7 Operator mutations are constitutive because the operator is unable to bind repressor protein; this allows RNA polymerase to have unrestrained access to the promoter. The O^c mutations are *cis*-acting, because they affect only the contiguous set of structural genes.

tion (including deletions of the gene). When the repressor is inactive or absent, transcription can initiate at the promoter. (More formally, mutations in the repressor that abolish its ability to bind to the operator will have a constitutive phenotype.) **Figure 10.8** shows that the *lacI*$^-$ mutants express the structural genes all the time, *irrespective of whether the inducer is present or absent*, because the repressor is inactive.

Mutants of the operon that are uninducible fall into the same two types of genetic classes as the constitutive mutants:

- Promoter mutations are *cis*-acting. If they prevent RNA polymerase from binding at P_{lac}, they render the operon nonfunctional because it cannot be transcribed.
- Mutations that abolish the ability of repressor to bind the inducer are described as *lacI*^{s}. They are *trans*-acting. The repressor is "locked in" to the active form that recognizes the operator and prevents transcription. The addition of inducer has no effect because its binding site is absent, and therefore it is impossible to convert the repressor to the inactive form. The mutant repressor binds to all *lac* operators in the cell to prevent their transcription, and cannot be pried off, irrespective of the properties of any wild-type repressor protein that is present, so it is genetically dominant.

The two types of mutations in *lacI* can be used to identify the individual active sites in the repressor protein. The *DNA-binding site* recognizes the sequence of the operator. It is identified by constitutive point mutations that prevent repressor from binding to DNA to block RNA polymerase. The *inducer-binding site* is identified by point mutations that cause uninducibility, because inducer cannot bind to trigger the allosteric change in the DNA-binding site.

An important feature of the repressor is that it is multimeric. Repressor subunits associate at random in the cell to form the active protein tetramer. When two different alleles of the *lacI* gene are present, the subunits made by each can associate to form a heterotetramer, whose properties differ from those of either homotetramer. This type of interaction between subunits is a characteristic feature of multimeric proteins and is described as **interallelic complementation.**

Negative complementation occurs between some repressor mutants, as seen in the combination of *lacI*$^{-d}$ with *lacI*$^+$ genes. The *lacI*$^{-d}$ mutation alone results in the production of a repressor that cannot bind the operator, and is therefore constitutive like the *lacI*$^-$ alleles. Because the *lacI*$^-$ type of mutation inactivates the repressor, it is recessive to the wild type. However, the *–d* notation indicates that this variant of the negative type is dominant when paired with a wild-type allele. Such mutations are said to be *trans*-**dominant;** they are also called **dominant negatives.**

The reason for the dominance is that the *lacI*$^{-d}$ allele produces a "bad" subunit, which is not only itself unable to bind to operator DNA, but is also able as part of a tetramer to prevent any "good" subunits from binding. This demonstrates that the repressor tetramer as a whole, rather than the individual monomer, is needed to achieve repression. The poisoning effect also can be

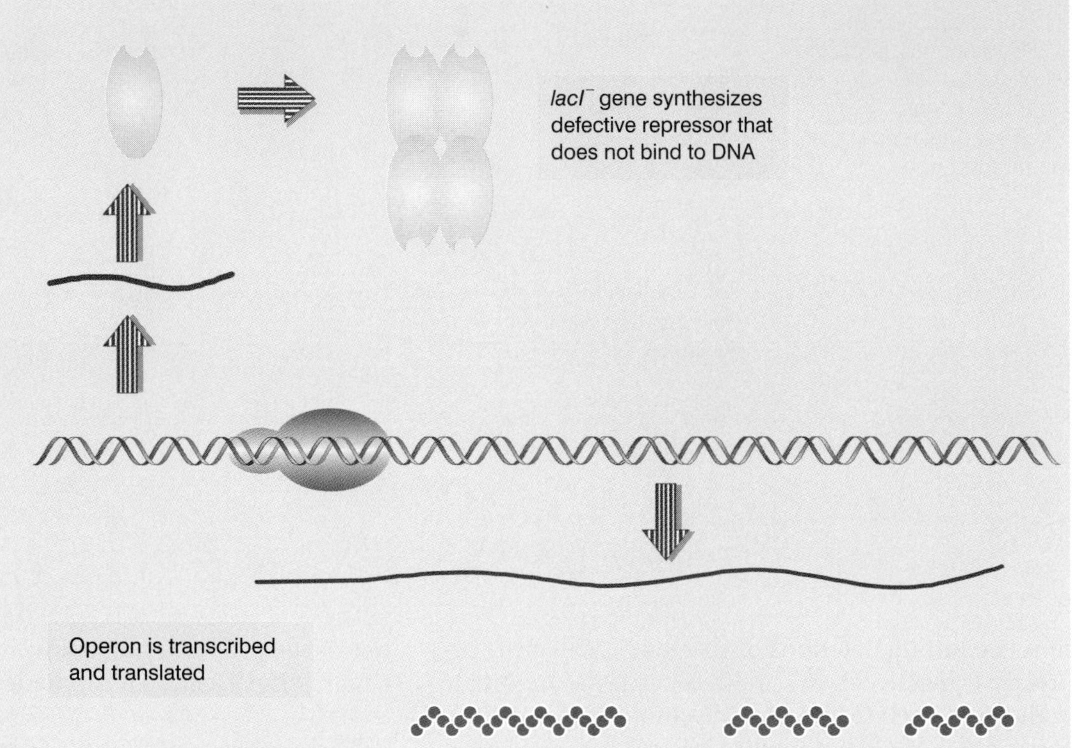
lacI⁻ gene synthesizes defective repressor that does not bind to DNA
Operon is transcribed and translated

Figure 10.8 Mutations that inactivate the *lacI* gene cause the operon to be constitutively expressed, because the mutant repressor protein cannot bind to the operator.

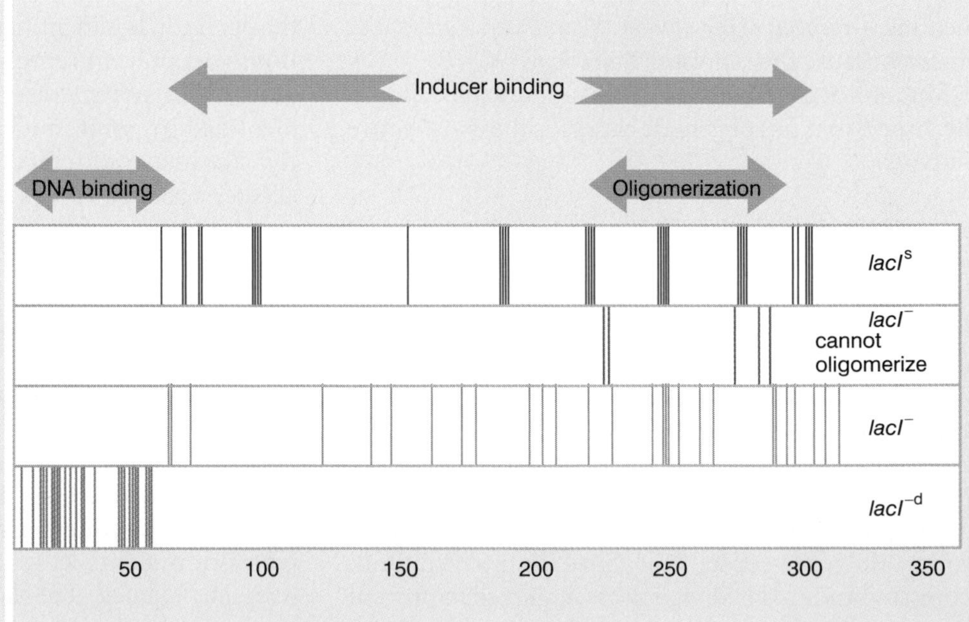
Inducer binding
DNA binding
Oligomerization
lacIˢ
lacI⁻ cannot oligomerize
lacI⁻
lacI⁻ᵈ
50
100
150
200
250
300
350

Figure 10.9 Mutations map the regions of the *lacI* gene responsible for different functions. The DNA-binding domain is identified by *lacI*$^{-d}$ mutations at the N-terminal region; *lacI*$^-$ mutations unable to form tetramers are located between residues 220–280; other *lacI*$^-$ mutations occur throughout the gene; *lacI*^{s} mutations occur in regularly spaced clusters between residues 62–300.

produced *in vitro* by mixing appropriate "good" and "bad" subunits.

The *lacI*$^{-d}$ mutations lie in the DNA-binding site of the repressor subunit. This explains their ability to prevent mixed tetramers from binding to the operator; a reduction in the number of binding sites reduces the specific affinity for the operator. The map of the *lacI* gene shown in **Figure 10.9** shows that the *lacI*$^{-d}$ mutations are clustered at the extreme left end of the gene. This identifies the immediate N-terminal region of the protein as the DNA-binding site. Mutations of the recessive *lacI*$^{-}$ type also occur elsewhere in the molecule, but could exert their effects on DNA binding indirectly.

The role of the N-terminal region in specifically binding DNA is shown also by its location as the site of occurrence of "tight binding" mutations. These increase the affinity of the repressor for the operator, sometimes so much that it cannot be released by inducer. They are rare.

Uninducible mutations of the *lacI*^{s} type render the repressor unresponsive to the inducer. This could happen either because the protein has lost its inducer-binding site, or because it has become unable to transmit its effect to the DNA-binding site.

Repressor protein binds to the operator and is released by inducer

THE repressor was isolated originally by purifying the component able to bind the gratuitous inducer IPTG. (Because the amount of repressor in the cell is so small, in order to obtain enough material it was necessary to use a promoter up mutation to increase *lacI* transcription, and to place this *lacI* locus on a DNA molecule present in many copies per cell. This results in an overall overproduction of 100–1000-fold.)

The repressor binds to double-stranded DNA containing the sequence of the wild-type *lac* operator. The repressor does not bind DNA from an *O*^{c} mutant. The addition of IPTG releases the repressor from operator DNA *in vitro*. The *in vitro* reaction between repressor protein and operator DNA therefore displays the characteristics of control inferred *in vivo*; so it can be used to establish the basis for repression.

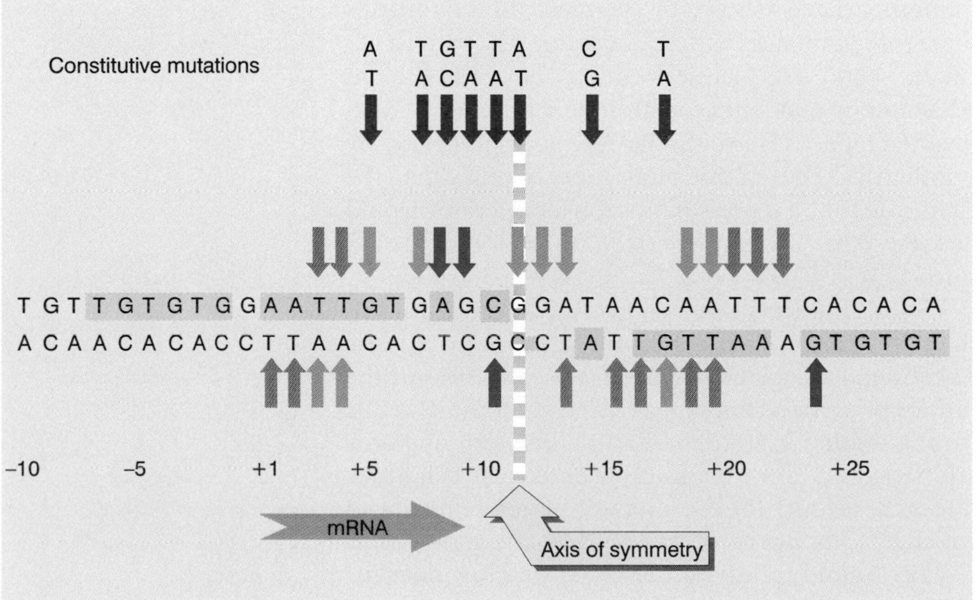

Figure 10.10 The *lac* operator has a symmetrical sequence. The sequence is numbered relative to the startpoint for transcription at +1. The regions of dyad symmetry are indicated by the shaded blocks.

Purines protected by repressor against methylation

Purines where methylation is enhanced by repressor

Thymines that can be crosslinked to repressor

How does the repressor recognize the specific sequence of operator DNA? The operator has a feature common to many recognition sites for bacterial regulator proteins: it is **palindromic**. The inverted repeats are highlighted in **Figure 10.10.** Each repeat can be regarded as a half-site of the operator.

We can use the same approaches to define the points that the repressor contacts in the operator that we used for analyzing the polymerase-promoter interaction (see Chapter 9). Deletions of material on either side define the end-points of the region; constitutive point mutations identify individual base pairs that must be crucial. Experiments in which DNA bound to repressor is compared with unbound DNA for its susceptibility to methylation or UV crosslinking identify bases that are either protected or more susceptible when associated with the protein.

The region of DNA protected from nucleases by bound repressor lies within the region of symmetry, comprising the 26 bp region from –5 to +21. The area identified by constitutive mutations is even smaller. Within a central region extending over the 13 bp from +5 to +17, there are eight sites at which single base-pair substitutions cause constitutivity. This emphasizes the same point made by the promoter mutations summarized earlier in Figure 9.17. *A small number of essential specific contacts within a larger region can be responsible for sequence-specific association of DNA with protein.*

The symmetry of the DNA sequence reflects a symmetry in the protein. Repressor is a tetramer of identical subunits, each of which must therefore have the same DNA-binding site. Each inverted repeat of the operator is contacted in the same way by a repressor monomer. This is shown by symmetry in the contacts that repressor makes with the operator (the pattern between +1 and +6 is identical with that between +21 and +16) and by matching constitutive mutations in each inverted repeat. (However, the operator is not perfectly symmetrical; the left side binds more strongly than the right side to the repressor. A stronger operator would be created by a perfect inverted duplication of the left side.)

Various inducers cause characteristic reductions in the affinity of the repressor for the operator *in vitro.* These changes correlate with the effectiveness of the inducers *in vivo.* This suggests that induction results from a reduction in the attraction between operator and repressor. So when inducer enters the cell, it reduces the affinity for the operator of any repressor to which it binds. But consider a repressor tetramer that is already bound tightly to the operator. How does inducer cause this repressor to be released?

Two models for repressor action are illustrated in **Figure 10.11:**

- The equilibrium model (left) calls for repressor bound to DNA to be in rapid equilibrium with free repressor. Inducer would bind to the free form of repressor, and thus unbalance the equilibrium by preventing reassociation with DNA.
- But the rate of dissociation of the repressor from the operator is much too slow to be compatible with this model (the half-life *in vitro* in the absence of inducer is

Figure 10.11 Does the inducer bind to free repressor to upset an equilibrium (left) or directly to repressor bound at the operator (right)?

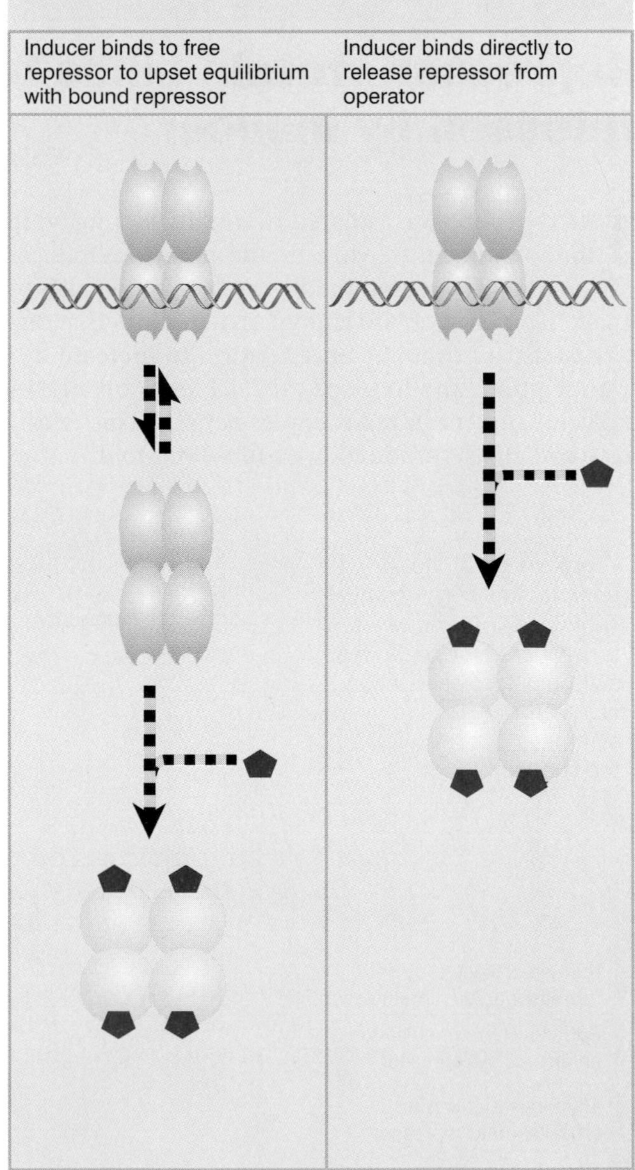
Inducer binds to free repressor to upset equilibrium with bound repressor
Inducer binds directly to release repressor from operator

>15 min). This means that instead the *inducer must bind directly to repressor protein complexed with the operator.* As indicated in the model on the right, inducer binding must produce a change in the repressor that makes it release the operator. Indeed, addition of IPTG causes an immediate destabilization of the repressor-operator complex *in vitro.*

Binding of the repressor-IPTG complex to the operator can be studied by using greater concentrations of the protein in the methylation protection/enhancement assay. The large amount compensates for the low affinity of the repressor-IPTG complex for the operator. The complex makes exactly the same pattern of contacts with DNA as the free repressor. An analogous result is obtained with mutant repressors whose affinity for operator DNA is increased; they too make the same pattern of contacts.

Overall, a range of repressor variants whose affinities for the operator span seven orders of magnitude all make the same contacts with DNA. *Changes in the affinity of the repressor for DNA must therefore occur by influencing the general conformation of the protein in binding DNA, not by making or breaking one or a few individual bonds.*

The repressor has several domains. The DNA-binding domain occupies residues 1–59. It is known as the **headpiece.** It can be cleaved from the remainder of the monomer, which is known as the **core,** by trypsin. The crystal structure illustrated in **Figure 10.12** offers a more detailed account of these regions.

Figure 10.12 The structure of a monomer of Lac repressor identifies several independent domains. Photograph kindly provided by Mitchell Lewis.

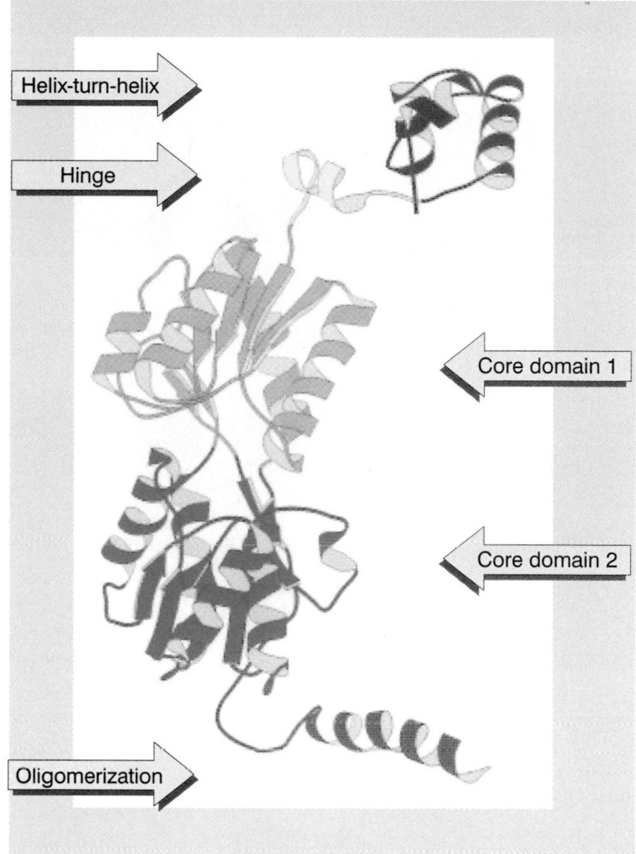

The N-terminus of the monomer consists of two α-helices separated by a turn. This is a common DNA-binding motif, known as the HTH (helix-turn-helix); the two α-helices fit into the wide groove of DNA, where they make contacts with specific bases (see Chapter 11). This region is connected by a hinge region to the main body of the protein. In the DNA-binding form of repressor, the hinge forms a small α-helix (as shown in the figure); but when the repressor is not bound to DNA, this region is disordered. The HTH and hinge together correspond to the headpiece.

The bulk of the core consists of two regions with similar structures. Each has a six-stranded parallel β-sheet sandwiched between two α-helices on either side. The inducer binds in a cleft between the two regions.

At the C-terminus, there is an α-helix that contains two leucine heptad repeats. The helices of four monomers associate to maintain the tetrameric structure.

Figure 10.13 shows the structure of the tetrameric core (using a different modeling system from Figure 10.12). It consists in effect of two dimers. The body of the dimer contains a loose interface between the N-terminal regions of the core monomers, a cleft at which inducer binds, and a hydrophobic core (top). The C-terminal regions of each monomer protrude as parallel helices. (The headpiece would join onto the N-terminal regions at the top.) Together the dimers interact to form a tetramer (center) that is held together by a C-terminal bundle of 4 helices.

Sites of mutations are shown by beads on the structure at the bottom. *lacI*^{s} mutations map in two groups: yellow shows those that affect the dimer interface, and gray shows those in the inducer-binding cleft. *lacI*$^-$ mutations that affect oligomerization map in two groups; white shows those that prevent dimer formation, and purple shows those that prevent tetramer formation from dimers.

Early work suggested a model in which the headpiece is relatively independent of the core. It can bind to operator DNA by making the same pattern of contacts with a half-site as intact repressor. However, its

Figure 10.13 The crystal structure of the core region of Lac repressor identifies the interactions between monomers in the tetramer. Each monomer is identified by a different color. Photographs kindly provided by Alan Friedman and Thomas Steitz.

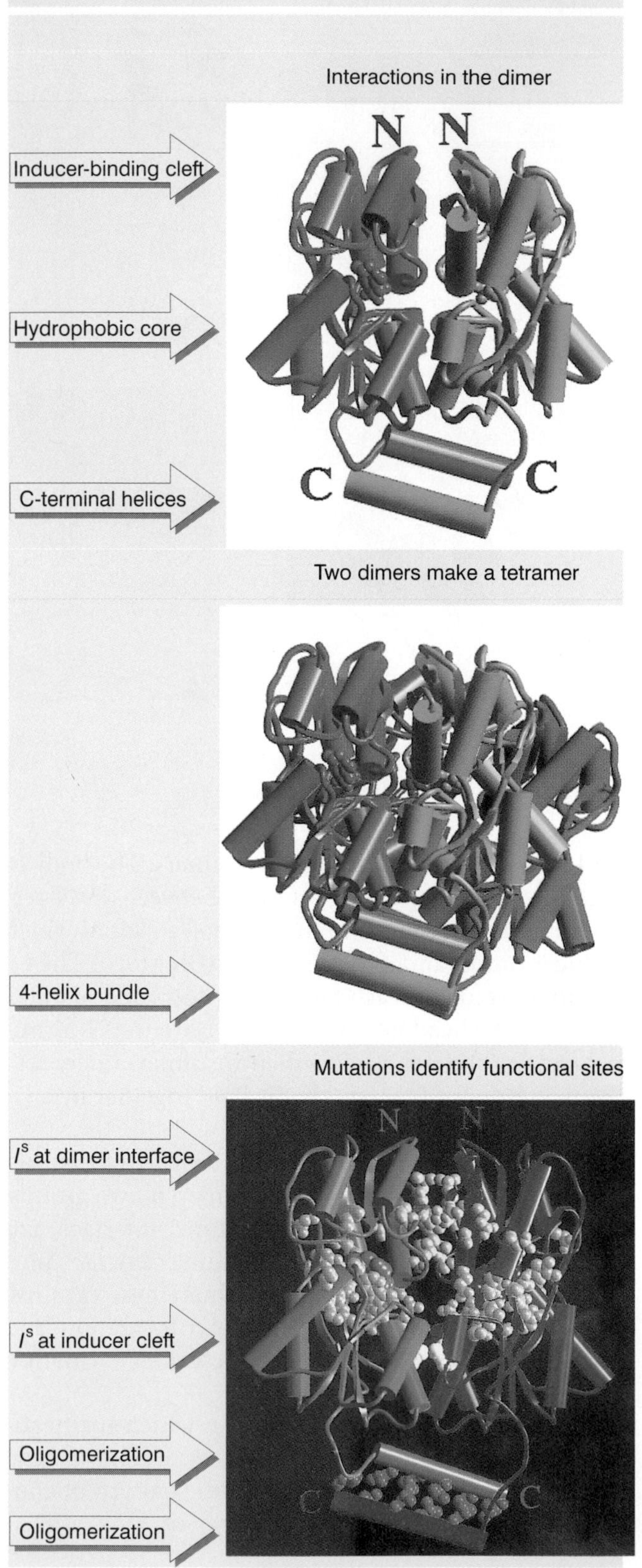

affinity for DNA is many orders of magnitude less than that of intact repressor. The reason for the difference is that the dimeric form of intact repressor allows two headpieces to contact the operator simultaneously, each binding to one half-site. This enormously increases affinity for the operator.

Binding of inducer causes an immediate conformational change in the repressor protein. Binding of two molecules of inducer to the repressor tetramer is adequate to release repression. **Figure 10.14** shows that binding of inducer changes the orientation of the headpieces relative to the core, with the result that the two headpieces in a dimer can no longer bind DNA simultaneously. This eliminates the advantage of the multimeric repressor, and reduces the affinity for the operator.

The allosteric transition that results from binding of inducer occurs in the repressor dimer. So why is a tetramer required to establish full repression? Each

Figure 10.14 Inducer changes the structure of the core so that the headpieces of a repressor dimer are no longer in an orientation that permits binding to DNA. Photographs kindly provided by Mitchell Lewis.

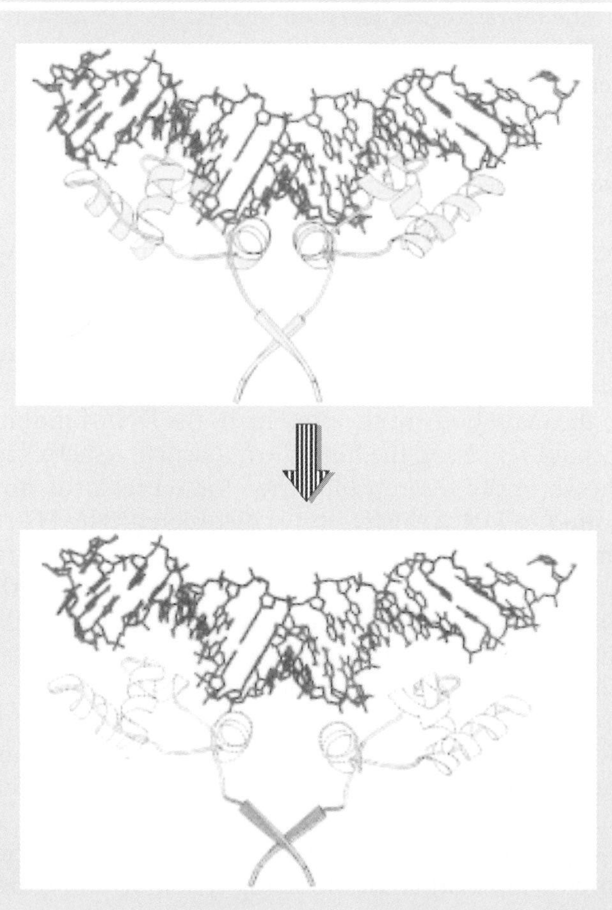

dimer can bind an operator sequence. This enables the intact repressor to bind to two operator sites simultaneously. In fact, there are two further operator sites in the initial region of the *lac* operon. The original operator, *O1*, is located just at the start of the *lacZ* gene. It has the strongest affinity for repressor. Weaker operator sequences (sometimes called pseudo-operators) are located on either side; *O2* is 410 bp downstream of the startpoint, and *O3* is 83 bp upstream of it. When Lac repressor binds simultaneously to *O1* and to one of the other repressors, it causes the DNA between them to form a loop. **Figure 10.15** shows a model for binding of tetrameric repressor to two operators.

Figure 10.15 When a repressor tetramer binds to two operators, the stretch of DNA between them is forced into a tight loop. (The blue structure in the center of the looped DNA represents CAP, another regulator protein that binds in this region). Photograph kindly provided by Mitchell Lewis.

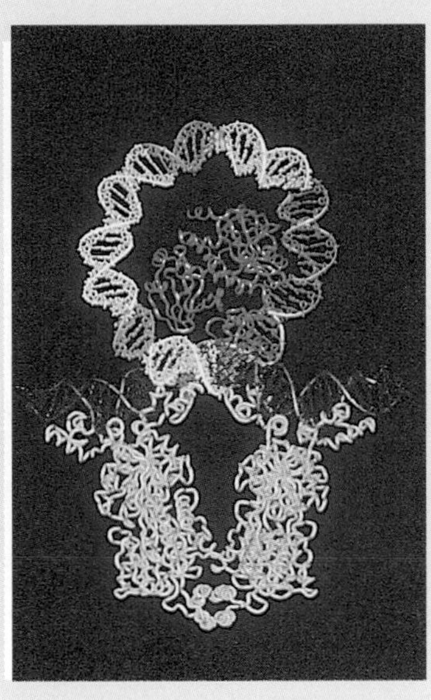

Binding at the additional operators affects the level of repression. Elimination of either the downstream operator (*O2*) or the upstream operator (*O3*) reduces the efficiency of repression by 2–4×. However, if *both O2* and *O3* are eliminated, repression is reduced 100×. *This suggests that the ability of the repressor to bind to one of the two other operators as well as to O1 is important for establishing repression.* We do not know how and why this simultaneous binding increases repression.

We know most about the direct effects of binding of repressor to the operator (*O1*). It was originally thought that repressor binding would occlude RNA polymerase from binding to the promoter. However, we now know that the two proteins may be bound to DNA simultaneously, and *the binding of repressor actually enhances the binding of RNA polymerase!* But the bound enzyme is prevented from initiating transcription.

The equilibrium constant for RNA polymerase binding alone to the *lac* promoter is $1.9 \times 10^7\ M^{-1}$. The presence of repressor increases this constant by two orders of magnitude to $2.5 \times 10^9\ M^{-1}$. In terms of the range of values for the equilibrium constant K_B given in Figure 9.11, repressor protein effectively converts the formation of closed complex by RNA polymerase at the *lac* promoter from a weak to a strong interaction.

What does this mean for induction of the operon? The higher value for K_B means that, when occupied by repressor, the promoter is 100 times more likely to be bound by an RNA polymerase. And by allowing RNA polymerase to be bound at the same time as repressor, it becomes possible for transcription to begin immediately upon induction, instead of waiting for an RNA polymerase to be captured.

The repressor in effect causes RNA polymerase to be stored at the promoter. The complex of RNA polymerase•repressor•DNA is blocked at the closed stage. When inducer is added, the repressor is released, and the closed complex is converted to an open complex that initiates transcription. The overall effect of repressor has been to speed up the induction process.

Does this model apply to other systems? The interaction between RNA polymerase, repressor, and the promoter/operator region is distinct in each system, because the operator does not always overlap with the same region of the promoter (see Figure 10.19). For example, in phage lambda, the operator lies in the upstream region of the promoter, and binding of repressor occludes the binding of RNA polymerase (see Chapter 11). So a bound repressor does not interact with RNA polymerase in the same way in all systems.

The specificity of protein-DNA interactions

PROBABLY all proteins that have a high affinity for a specific sequence also possess a low affinity for any (random) DNA sequence. A large number of low-affinity sites will compete just as well for a repressor tetramer as a small number of high-affinity sites. There is only one high-affinity site in the *E. coli* genome: the operator. The remainder of the DNA provides low-affinity binding sites. Every base pair in the genome starts a new low-affinity site. (Just moving one base pair along the genome, out of phase with the operator itself, creates a low-affinity site!) So there are 4.2×10^6 low-affinity sites.

The large number of low-affinity sites means that, even in the absence of a specific binding site, all or virtually all repressor is bound to DNA; none is free in solution.

We may describe the binding of repressor to DNA by the equilibrium:

$$K_A = \frac{[\text{Repressor-DNA}]}{[\text{Free repressor}]\,[\text{DNA}]}$$

in which [Repressor-DNA] is the concentration of repressor bound to DNA, [Free repressor] is the concentration of free repressor, and [DNA] is the concentration of nonspecific binding sites.

What proportion of total repressor protein is free? By rearranging the equation, we see that the distribution of repressor is given by:

$$\frac{[\text{Free Repressor}]}{[\text{Repressor-DNA}]} = \frac{1}{K_A \cdot [\text{DNA}]}$$

Applying the parameters for the *lac* system, we find that:

- The nonspecific equilibrium binding constant is $K_A = 2 \times 10^6\ M^{-1}$.
- The concentration of nonspecific binding sites is 4×10^6 in a bacterial volume of 10^{-15} liter, which corresponds to $[DNA] = 7 \times 10^{-3}$ M (a very high concentration).

Substituting these values gives:

$$\text{Free : Bound repressor} = 10^{-4}$$

So all but 0.01% of repressor is bound to (random) DNA. Since there are ~10 molecules of repressor per cell, this is tantamount to saying that there is no free repressor protein. This has an important implication for the interaction of repressor with the operator: it means that we are concerned with the *partitioning* of the repressor on DNA, in which the single high-affinity site of the operator *competes* with the large number of low-affinity sites.

In this competition, the absolute values of the association constants for operator and random DNA are not important; what is important is the ratio of K_{sp} (the constant for binding a specific site) to K_{nsp} (the constant for binding any random DNA sequence), that is, the specificity.

We can define the parameters that influence the ability of a regulator protein to saturate its target site by comparing the equilibrium equations for specific and nonspecific binding. As might be expected intuitively, the important parameters are:

- The size of the genome dilutes the ability of a protein to bind specific target sites.
- The specificity of the protein counters the effect of the mass of DNA.
- The amount of protein that is required increases with the total amount of DNA in the genome and decreases with the specificity.
- The amount of protein also must be in reasonable excess of the total number of specific target sites, so we expect regulators with many targets to be found in greater quantities than regulators with individual targets.

Figure 10.16 compares the equilibrium constants for *lac* repressor/operator binding with repressor/gen-

Figure 10.16 Lac repressor binds strongly and specifically to its operator, but is released by inducer. All equilibrium constants are in M^{-1}.

DNA	Repressor	Repressor + inducer
Operator	2×10^{13}	2×10^{10}
Other DNA	2×10^6	2×10^6
Specificity	10^7	10^4

eral DNA binding. From these constants, we can deduce how repressor is partitioned between the operator and the rest of DNA, and what happens to the repressor when inducer causes it to dissociate from the operator.

Repressor binds $\sim 10^7$ times better to operator DNA than to any random DNA sequence of the same length. So the operator comprises a single high-affinity site that will compete for the repressor 10^7 better than any low-affinity (random) site. How does this ensure that the repressor can maintain effective control of the operon?

Using the specificity, we can calculate the distribution between random sites and the operator, and can express this in terms of occupancy of the operator. If there are 10 molecules of *lac* repressor per cell with a specificity for the operator of 10^7, the operator will be bound by repressor 96% of the time. The role of specificity explains two features of the *lac* repressor-operator interaction:

- When inducer binds to the repressor, the affinity for the operator is reduced by $\sim 10^3$-fold. The affinity for general DNA sequences remains unaltered. So the specificity is now only 10^4, which is insufficient to capture the repressor against competition from the excess of 4.2×10^6 low-affinity sites. Only 3% of operators would be bound under these conditions.
- Mutations that reduce the affinity of the operator for the repressor by as little as 20–30× have sufficient effect to be constitutive. Within the genome, the mutant operators can be overwhelmed by the preponderance of random sites. The occupancy of the operator is reduced to ~50% if the repressor's specificity is reduced just 10×.

The consequence of these affinities is that in an uninduced cell, one tetramer of repressor usually is bound to the operator. All or almost all of the remaining tetramers are bound at random to other regions of DNA, as illustrated in **Figure 10.17.** There are likely to be very few or no repressor tetramers free within the cell.

The addition of inducer abolishes the ability of repressor to bind specifically at the operator. Those repressors bound at the operator are released, and bind to random (low-affinity) sites. So in an induced cell, the repressor tetramers are "stored" on random DNA sites. In a noninduced cell, a tetramer is bound at the operator, while the remaining repressor molecules are bound to nonspecific sites. *The effect of induction is therefore to change the distribution of repressor on DNA, rather than to generate free repressor.*

When inducer is removed, repressor recovers its ability to bind specifically to the operator, and does so very rapidly. This must involve its movement from a nonspecific "storage" site on DNA. What mechanism is used for this rapid movement? The ability to bind to the operator very rapidly is not consistent with the time that would be required for multiple cycles of dissociation and reassociation with nonspecific sites on DNA. The discrepancy excludes random-hit mechanisms for finding the operator, suggesting that the repressor can move directly from a random site on DNA to the operator. This is the same issue that we encountered previously with the ability of RNA polymerase to find its promoters (see Figure 9.13). The same solution is likely: movement could be accomplished by direct displacement from site to site (as indicated in Figure 10.17). A displacement reaction might be aided by the presence of more binding sites per tetramer (four) than are actually needed to contact DNA at any one time (two).

The parameters involved in finding a high-affinity operator in the face of competition from many low-affinity sites pose a dilemma for repressor. Under conditions of repression, there must be high specificity for the operator. But under conditions of induction, this specificity must be relieved. Suppose, for example, that there were 1000 molecules of repressor per cell. Then only 0.04% of operators would be free under conditions of repression. But upon induction only 40% of operators would become free. We therefore see an inverse correlation between the ability to achieve complete repression and the ability to relieve repression effectively. We assume that the number of repressors synthesized *in vivo* has been subject to selective forces that balance these demands.

The difference in expression of the lactose operon between its induced and repressed states *in vivo* is actually 10^3-fold. In other words, even when inducer is absent, there is a basal level of expression of ~0.1% of the induced level. This would be reduced if there were more repressor protein present, and increased if there were less. So it could be impossible to establish tight repression if there were fewer repressors than the 10 found per cell; and it might become difficult to induce the operon if there were too many.

In order to extrapolate *in vivo* from the affinity of a DNA-protein interaction *in vitro*, we need to know the effective concentration of DNA *in vivo*. The "effective concentration" differs from the mass/volume owing to several factors. The effective concentration is increased, for example, by molecular crowding, which occurs when polyvalent cations neutralize ~90% of the

Figure 10.17 Virtually all the repressor in the cell is bound to DNA.

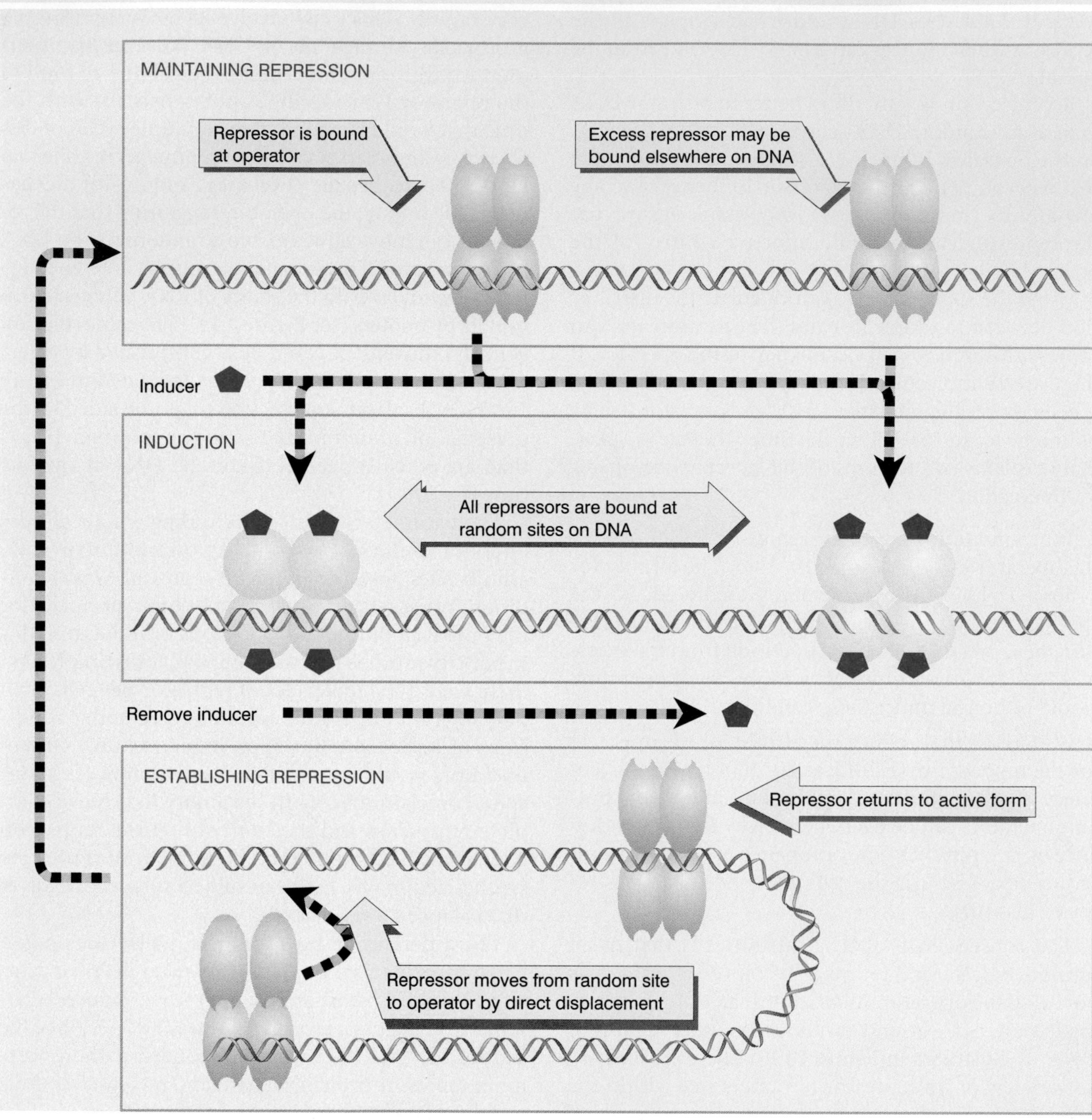
MAINTAINING REPRESSION
Repressor is bound at operator
Excess repressor may be bound elsewhere on DNA
Inducer
INDUCTION
All repressors are bound at random sites on DNA
Remove inducer
ESTABLISHING REPRESSION
Repressor returns to active form
Repressor moves from random site to operator by direct displacement

charges on DNA, and the nucleic acid collapses into condensed structures. The major force that decreases the effective concentration is the inaccessibility of DNA that results from occlusion or sequestration by DNA-binding proteins.

One way to determine the effective concentration is to compare *in vitro* and *in vivo* the rate of a reaction that depends on DNA concentration. This has been done using intermolecular recombination between two DNA molecules. To provide a control, the same reaction is followed as an intramolecular recombination, that is, the two recombining sites are presented on the same DNA molecule. We assume that concentration is the same *in vivo* and *in vitro* for the *intramolecular* reaction, and therefore any difference in the ratio of intermolecular/intramolecular recombination rates can be attributed to a change in the effective concentration *in vivo*. The results of such a comparison suggest that the effective concentration of DNA is reduced >10-fold *in vivo*.

This could affect the rates of reactions that depend on DNA concentration, including DNA recombination and protein–DNA binding. It emphasizes the problem encountered by all DNA-binding proteins in finding their targets with sufficient speed, and reinforces the conclusion that diffusion is not adequate (see Figure 9.13).

Repression can occur at multiple loci

THE *lac* repressor acts only on the operator of the *lacZYA* cluster. However, some repressors control dispersed structural genes by binding at more than one operator. An example is the *trp* repressor, which controls three unlinked sets of genes:

- An operator at the cluster of structural genes *trpEDBCA* controls coordinate synthesis of the enzymes that synthesize tryptophan from chorismic acid.
- An operator at another locus controls the *aroII* gene, which codes for one of the three enzymes that catalyze the initial reaction in the common pathway of aromatic amino acid biosynthesis.
- The *trpR* regulator gene is repressed by its own product, the *trp* repressor. So the repressor protein acts to reduce its own synthesis. This circuit is an example of **autogenous control.** Such circuits are quite common in regulatory genes, and may be either negative or positive. (We discuss examples later in this chapter and in Chapter 11.)

A related 21 bp operator sequence is present at each of the three loci at which the *trp* repressor acts. The conservation of sequence is indicated in **Figure 10.18.** Each operator contains appreciable (but not identical) dyad symmetry. The features conserved at all three operators include the important points of contact for *trp* repressor. This explains how one repressor protein acts on several loci: *each locus has a copy of a specific DNA-binding sequence recognized by the repressor* (just as each promoter shares consensus sequences with other promoters).

Figure 10.19 summarizes the variety of relationships

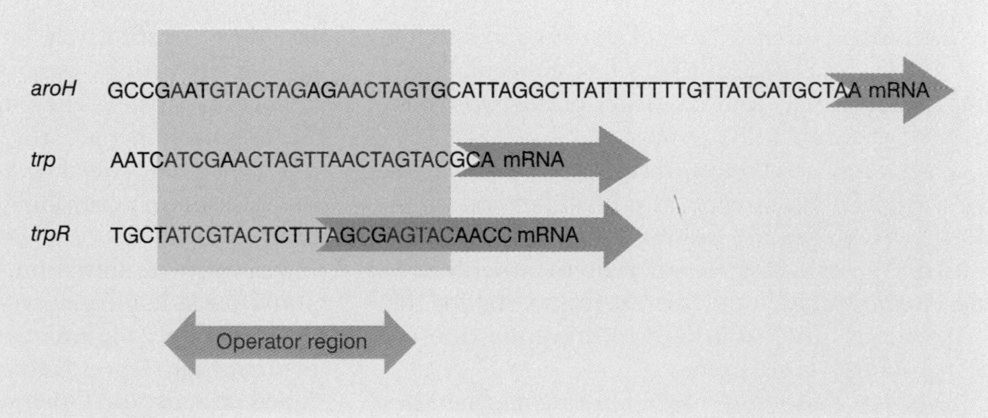

Figure 10.18 The *trp* repressor recognizes operators at three loci. Conserved bases are shown in red. The location of the mRNA varies, as indicated by the red arrows.

between operators and promoters. A notable feature of the dispersed operators recognized by TrpR is their presence at different locations within the promoter in each locus. In *trpR* the operator lies between positions –12 and +9, while in the *trp* operon it occupies positions –23 to –3, but in the *aroH* locus it lies farther upstream, between –49 and –29. In other cases, the operator lies downstream from the promoter (as in *lac*), or apparently just upstream of the promoter (as in *gal*, where the nature of the repressive effect is not quite clear). The ability of the repressors to act at operators whose positions are different in each target promoter suggests that there could be differences in the exact mode of repression, the common feature being that RNA polymerase is prevented from initiating transcription at the promoter.

Figure 10.19 Operators may lie at various positions relative to the promoter.

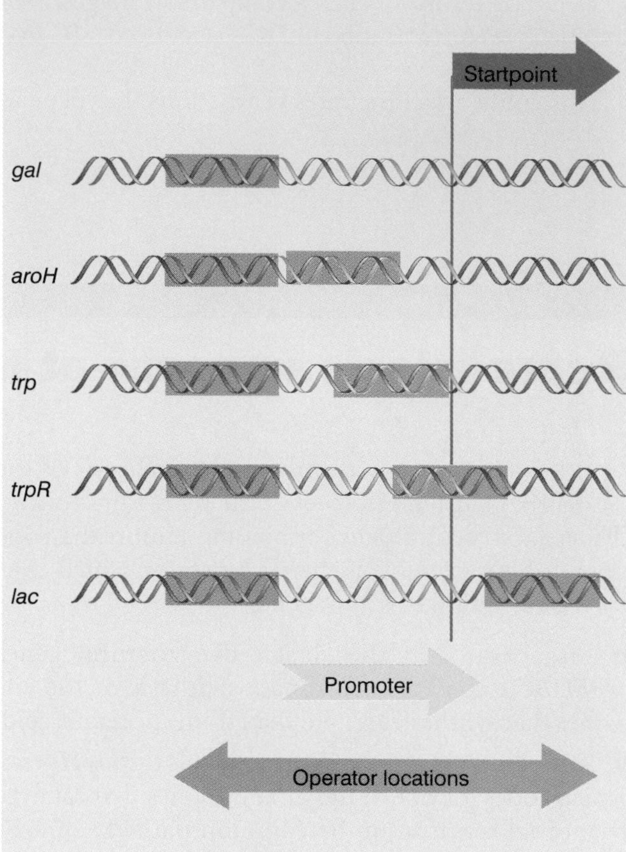

Distinguishing positive and negative control

POSITIVE and negative control systems are defined by the response of the operon when no regulator protein is present. The characteristics of the two types of control system are mirror images.

Genes under negative control are expressed unless they are switched off by a repressor protein. Any action that interferes with gene expression can provide a negative control. Typically a repressor protein either binds to DNA to prevent RNA polymerase from initiating transcription, or binds to mRNA to prevent a ribosome from initiating translation.

Negative control provides a fail-safe mechanism: if the regulator protein is inactivated, the system functions and so the cell is not deprived of these enzymes. It is easy to see how this might evolve. Originally a system functions constitutively, but then cells able to interfere specifically with its expression acquire a selective advantage by virtue of their increased efficiency.

For genes under positive control, expression is possible only when an active regulator protein is present. The mechanism for controlling an individual operon is an exact counterpart of negative control, but instead of *interfering* with initiation, the regulator protein is *essential* for it. It interacts with DNA and with RNA polymerase to *assist the initiation event.* A positive regulator protein that responds to a small molecule is usually called an **activator.** Other positive controls provide for the global substitution of sigma factors that change the selection of promoters (Chapter 9), or antitermination factors that change the recognition of terminators.

Figure 10.20 Control circuits are versatile and can be designed to allow positive or negative control of induction or repression.

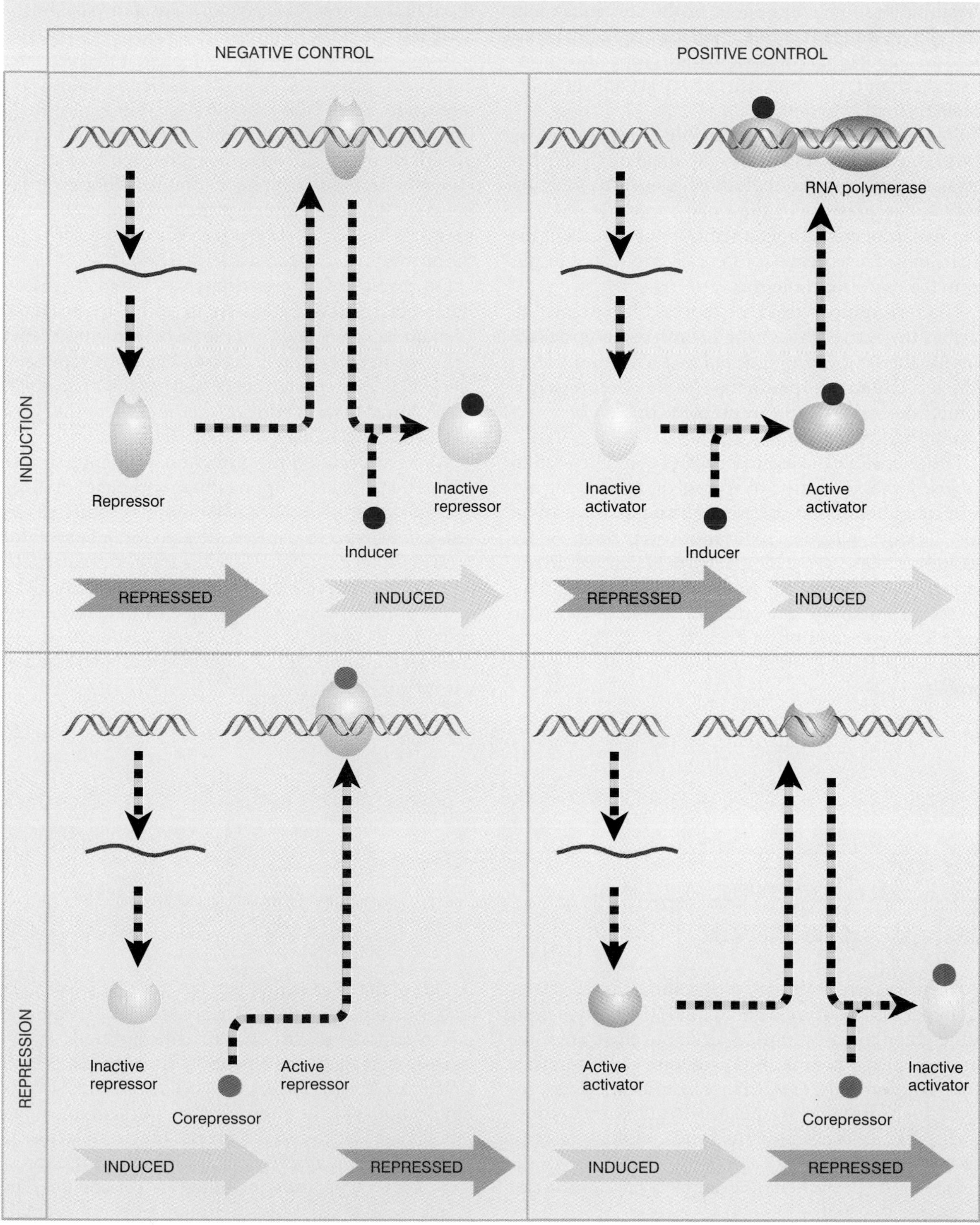

It is less obvious how positive control evolved, since the cell must have had the ability to express the regulated genes even before any control existed. Presumably some component of the control system must have changed its role. Perhaps originally it was used as a regular part of the apparatus for gene expression; then later it became restricted to act only in a particular system or systems.

Operons are defined as **inducible** or **repressible** by the nature of their response to the small molecule that regulates their expression. Inducible operons function only in the *presence* of the small-molecule inducer. Repressible operons function only in the *absence* of the small-molecule corepressor (so called to distinguish it from the repressor protein).

The terminology used for repressible systems describes the active state of the operon as **derepressed;** this has the same meaning as *induced.* The condition in which a (mutant) operon cannot be derepressed is sometimes called **super-repressed;** this is the exact counterpart of *uninducible.*

Either positive or negative control could be used to achieve either induction or repression by utilizing appropriate interactions between the regulator protein and the small-molecule inducer or corepressor. **Figure 10.20** summarizes four simple types of control circuit. Induction is achieved when an inducer inactivates a repressor protein or activates an activator protein. Repression is accomplished when a corepressor activates a repressor protein or inactivates an activator protein.

The *trp* operon is a repressible system. Tryptophan is the end product of the reactions catalyzed by a series of biosynthetic enzymes. Both the activity and the synthesis of the tryptophan enzymes are controlled by the level of tryptophan in the cell.

Tryptophan functions as a corepressor that activates a repressor protein. This is the classic mechanism for repression, one of the examples given in Figure 10.20 (lower left). In conditions when the supply of tryptophan is plentiful, the operon is repressed because the repressor protein•corepressor complex is bound at the operator. When tryptophan is in short supply, the corepressor is inactive, therefore has reduced specificity for the operator, and is stored elsewhere on DNA.

Deprivation of repressor causes ~70-fold increase in the frequency of initiation events at the *trp* promoter. Even under repressing conditions, the structural genes continue to be expressed at a low **basal** or **repressed level.** The efficiency of repression at the operator is much lower than in the *lac* operon (where the basal level is only ~1/1000 of the induced level).

We have treated both induction and repression as phenomena that rely upon allosteric changes induced in regulator proteins by small molecules. Other means also can be used to control the activities of regulator proteins. One example is OxyR, a transcriptional activator of genes induced by hydrogen peroxide. The OxyR protein is directly activated by oxidation, so it provides a sensitive measure of oxidative stress. Another common type of signal is phosphorylation of a regulator protein.

Catabolite repression involves positive regulation at the promoter

So far we have dealt with the promoter as a DNA sequence that is competent to bind RNA polymerase, which then initiates transcription. But there are some promoters at which RNA polymerase cannot initiate transcription without assistance from an ancillary protein. Such proteins are positive regulators, because their presence is necessary to switch on the transcription unit. Typically the activator overcomes a deficiency in the promoter, for example a poor consensus sequence at –35 or –10.

One of the most widely acting activators is a protein that controls the activity of a large set of operons in *E. coli* in response to carbon nutrient conditions. When glucose is available as an energy source, it is used in preference to other sugars. So when *E. coli* finds (for example) both glucose and lactose in the medium, it metabolizes the glucose and represses the use of lactose.

This choice is accomplished by preventing expression of several operons, including *lac*, *gal*, and *ara*. The effect is called **catabolite repression.** It represents a general coordinating system that exercises a preference for glucose by inhibiting the expression of the operons that code for the enzymes of alternative metabolic pathways.

Catabolite repression is set in train by the ability of glucose to reduce the level of cyclic AMP (cAMP) in the cell. Cyclic AMP is synthesized by the enzyme **adenylate cyclase.** The reaction uses ATP as substrate and introduces a 3′–5′ link via phosphodiester bonds, generating the structure drawn in **Figure 10.21.** Mutations in the gene coding for adenylate cyclase (*cya*⁻) do not respond to changes in glucose levels.

Expression of the catabolite-regulated operons shows an inverse relationship with the level of cyclic AMP, which acts by binding to the product of *cap* gene. Mutations in the *cap* gene prevent activation of the operons that normally are expressed in the absence of glucose. The protein is known as CAP (for catabolite activator protein). It is a positive control factor whose presence is necessary to initiate transcription at dependent promoters. The protein is active *only in the presence of cyclic AMP,* which behaves as the classic small-molecule inducer (see Figure 10.20; upper right).

Figure 10.22 shows that reducing the level of cyclic AMP renders the (wild-type) protein unable to bind to the control region, which in turn prevents RNA polymerase from initiating transcription. So the effect of glucose in reducing cyclic AMP levels is to deprive the relevant operons of a control factor necessary for their expression.

The CAP factor binds to DNA, and complexes of cyclic AMP•CAP•DNA can be isolated at each promoter at which it functions. The factor is a dimer of two identical subunits of 22.5 kD, which can be activated by a single molecule of cyclic AMP. A CAP monomer contains a DNA-binding region and a transcription-activating region.

A CAP dimer binds to a site of ~22 bp at a responsive promoter. The binding sites include variations of the consensus sequence given in **Figure 10.23.** Mutations preventing CAP action usually are located within the

Figure 10.21 Cyclic AMP has a single phosphate group connected to both the 3' and 5' positions of the sugar ring.

Figure 10.22 Glucose causes catabolite repression by reducing the level of cyclic AMP.

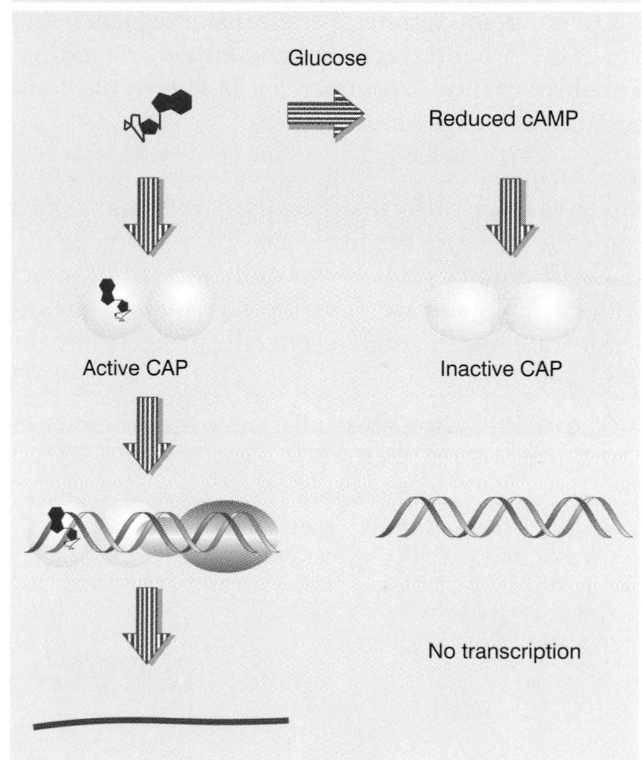

Figure 10.23 The consensus sequence for CAP contains the well-conserved pentamer TGTGA and (sometimes) an inversion of this sequence (TCANA).

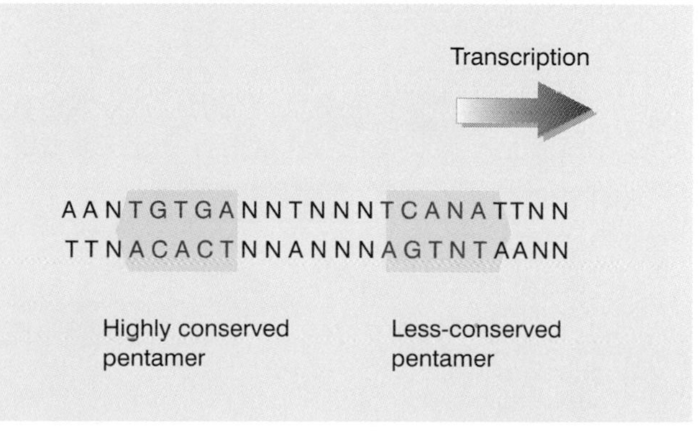

well conserved pentamer $^{\mathrm{TGTGA}}_{\mathrm{ACACT}}$, which appears to be the essential element in recognition. CAP binds most strongly to sites that contain two (inverted) versions of the pentamer, because both subunits of the dimer bind effectively to the DNA. Many binding sites lack the second pentamer, however, and in these the second subunit must bind a different sequence (if it binds to DNA). The hierarchy of binding affinities for CAP helps to explain why different genes are activated by different levels of cyclic AMP *in vivo.*

The action of CAP has the curious feature that its binding sites lie at different locations relative to the startpoint in the various operons that it regulates. And the TGTGA pentamer may lie in either orientation. The three examples summarized in **Figure 10.24** encompass the range of locations:

- Sometimes the CAP-binding site lies within the promoter, as in the *gal* locus, where the CAP-binding site is centered on –41. It is likely that only a single CAP dimer is bound, probably in quite intimate contact with RNA polymerase, since the CAP-binding site extends well into the region generally protected by the RNA polymerase.
- The CAP-binding site is adjacent to the promoter, as in the *lac* operon, in which the region of DNA protected by CAP is centered on –61. It is possible that two dimers of CAP are bound. The binding pattern is consistent with the presence of CAP largely on one face of DNA, the same face that is bound by RNA polymerase. This location would place the two proteins just about in reach of each other.
- In other operons, the CAP-binding site lies well upstream of the promoter. In the *ara* region, the binding site for a single CAP is the farthest from the startpoint, centered at –92. Here the CAP cannot be in contact with RNA polymerase, because *another* regulatory protein binds in the region between the CAP and RNA polymerase sites.

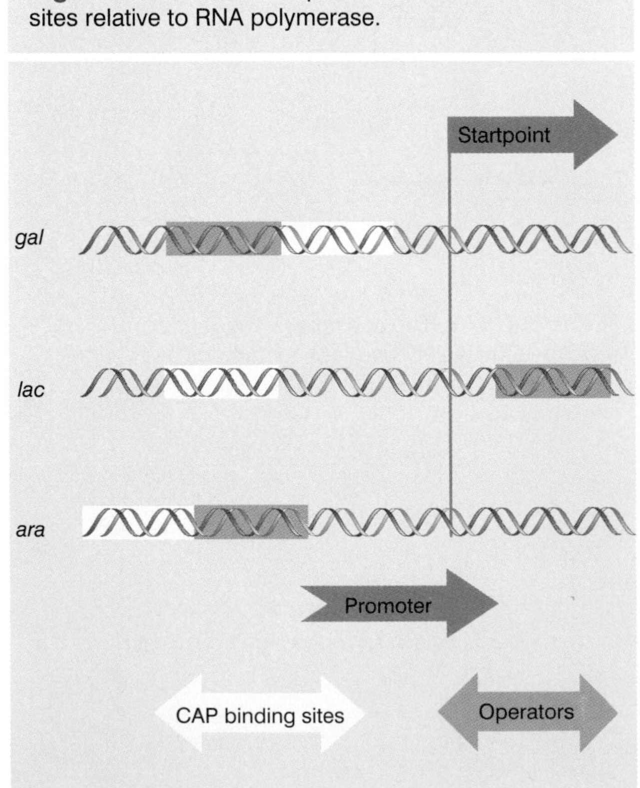

Figure 10.24 The CAP protein can bind at different sites relative to RNA polymerase.

Dependence on CAP is related to the intrinsic efficiency of the promoter. No CAP-dependent promoter has a good –35 sequence and some also lack good –10 sequences. In fact, we might argue that effective control by CAP would be difficult if the promoter had effective –35 and –10 regions that interacted independently with RNA polymerase.

There are in principle two ways in which CAP might activate transcription: it could interact directly with RNA polymerase; or it could act upon DNA to change its structure in some way that assists RNA polymerase to bind. In fact, CAP has effects upon both RNA polymerase and DNA.

Binding sites for CAP at most promoters resemble either *gal* (centered at –41 bp) or *lac* (centered at –61). When the distance is changed from either of these two standards, the ability of CAP to activate transcription is reduced. A unifying model for the ability of CAP to activate transcription in spite of its varying distance from the startpoint is suggested by the correlation that activation occurs only when this distance is an integral number of turns of the double helix. This suggests that CAP must be bound to the same face of DNA as RNA polymerase.

When the α subunit of RNA polymerase has a deletion in the C-terminal end, transcription appears normal except for the loss of ability to be activated by CAP. This suggests that CAP acts directly upon the α subunit to stimulate RNA polymerase. Experiments using CAP dimers in which only one of the subunits has a functional transcription-activating region shows that, when CAP is bound at the *lac* promoter, only the activating region of the subunit nearer the startpoint is required, presumably because it touches RNA polymerase. This offers an explanation for the lack of dependence on the orientation of the binding site: the dimeric structure of CAP ensures that one of the subunits is available to contact RNA polymerase, no matter which subunit binds to DNA and in which orientation. The activating region in CAP consists of a small exposed loop, in which point mutations block the interaction with RNA polymerase.

The effect upon RNA polymerase binding depends on the relative locations of the two proteins. When CAP binds within the promoter (as in *gal*), it increases the rate of transition from the closed to open complex. When CAP binds adjacent to the promoter (as in *lac*) its predominant effect is to increase the rate of initial binding to form a closed complex. This suggests that the exact effects of the interaction depend upon the geometry of the proteins at the individual promoter. (The interactions at the *ara* promoter may involve different interactions, involving more proteins.)

The structure of the CAP-DNA complex is interesting: *the DNA has a bend.* Proteins may distort the double helical structure of DNA when they bind, and several regulator proteins induce a bend in the axis.

Figure 10.25 illustrates a technique that can be used to measure the extent and location of a bend. A dimer of the target sequence is made, and it is cut with different restriction enzymes to generate a set of circularly permutated fragments each containing a monomeric length of DNA. The protein-binding site therefore lies at a different location in each of these fragments.

The fragments move at different speeds in an electrophoretic gel, depending on the position of the bend. (If there is no bend, all fragments move at the same rate.) The greatest impediment to motion, causing the lowest mobility, happens when the bend is in the center of the DNA fragment. The least impediment to motion, allowing the greatest mobility, happens when the bend is at one end.

The results are analyzed by plotting mobility against the site of restriction cutting. The low point on the curve identifies the situation in which the restriction enzyme has cut the sequence immediately adjacent to the site of bending.

For the interaction of CAP with the *lac* promoter, this point lies at the center of dyad symmetry. The bend

Figure 10.25 Gel electrophoresis can be used to analyze bending.

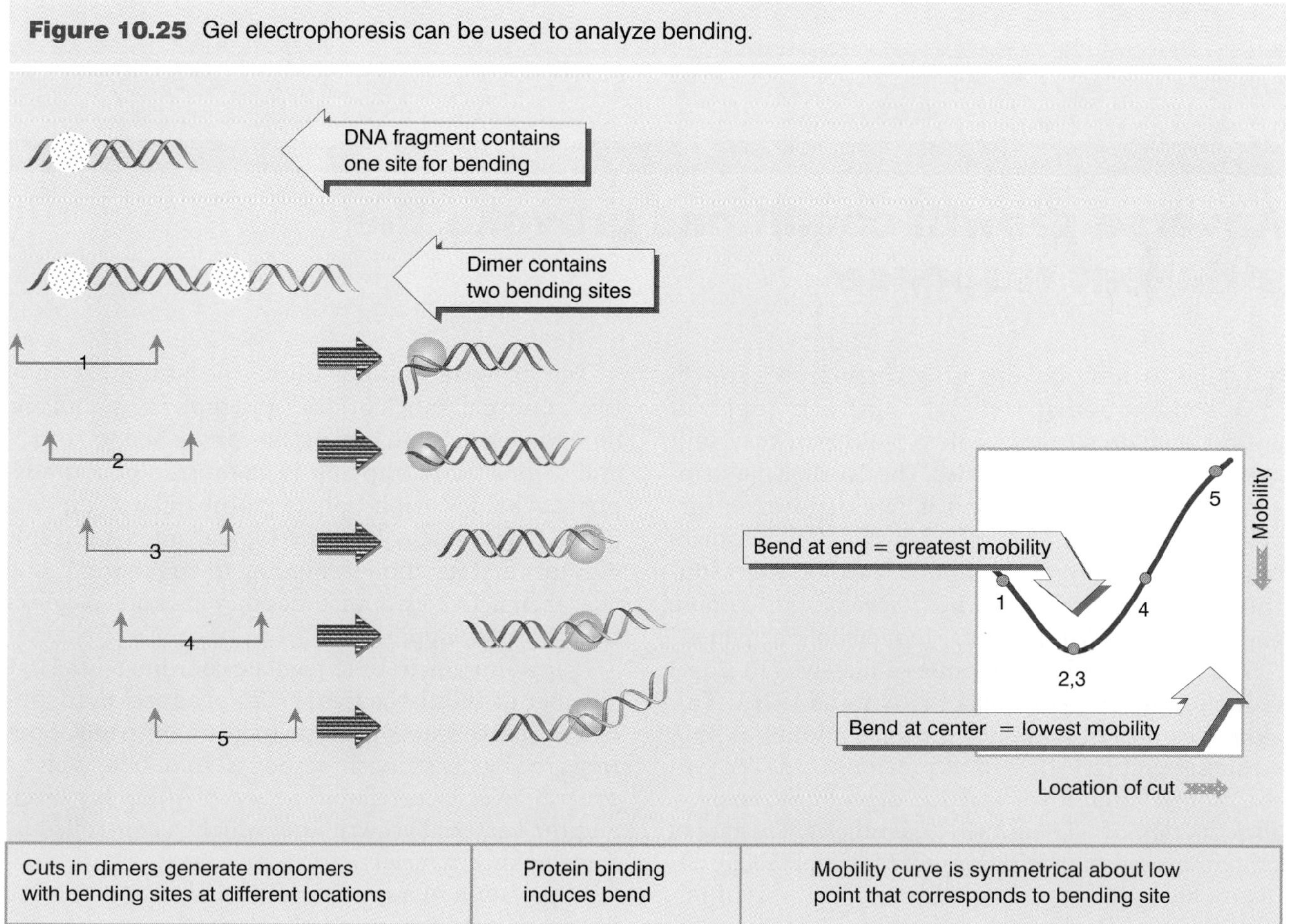
DNA fragment contains one site for bending
Dimer contains two bending sites
1
2
3
4
5
Bend at end = greatest mobility
Bend at center = lowest mobility
1
2,3
4
5
Mobility
Location of cut
Cuts in dimers generate monomers with bending sites at different locations
Protein binding induces bend
Mobility curve is symmetrical about low point that corresponds to bending site

is quite severe, >90°, as illustrated in the model of **Figure 10.26.** There is therefore a dramatic change in the organization of the DNA double helix when CAP protein binds. The mechanism of bending is to introduce a sharp kink within the TGTGA consensus sequence; the two kinks in each copy present in a palindrome cause the overall 90° bend. It is possible that the bend has some direct effect upon transcription, but it could be the case that it is needed simply to allow CAP to contact RNA polymerase at the promoter.

Whatever the exact means by which CAP activates transcription at various promoters, it accomplishes the same general purpose: to turn off alternative metabolic pathways when they become unnecessary because the cell has an adequate supply of glucose. Again, this makes the point that coordinate control, of either negative or positive type, can extend over dispersed loci by repetition of binding sites for the regulator protein.

Figure 10.26 CAP bends DNA >90° around the center of symmetry.

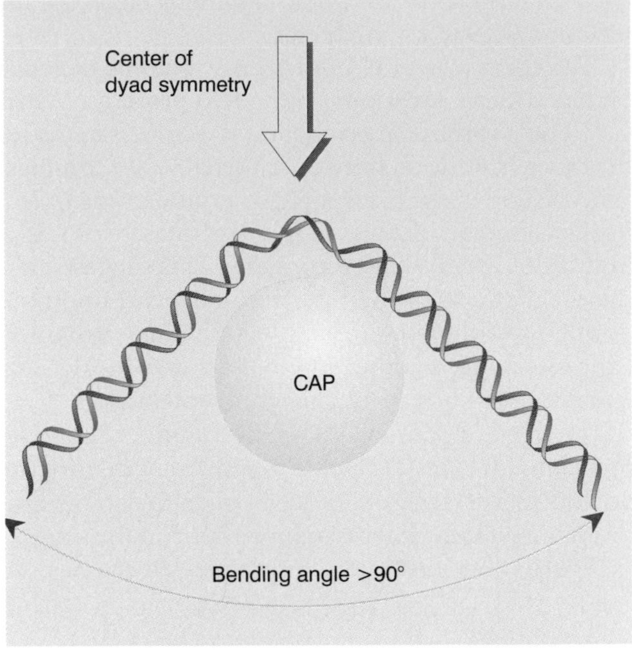

Adverse growth conditions provoke the stringent response

When bacteria find themselves in such poor growth conditions that they lack a sufficient supply of amino acids to sustain protein synthesis, they shut down a wide range of activities. This is called the **stringent response.** We can view it as a mechanism for surviving hard times: the bacterium husbands its resources by engaging in only the minimum of activities until nutrient conditions improve, when it reverses the response and again engages its full range of metabolic activities.

The stringent response causes a massive (10–20×) reduction in the synthesis of rRNA and tRNA. This alone is sufficient to reduce the total amount of RNA synthesis to only 5–10% of its previous level. The synthesis of certain mRNAs is reduced, leading to an overall reduction of ~3× in mRNA synthesis. The rate of protein degradation is increased. Many metabolic adjustments occur, as seen in reduced synthesis of nucleotides, carbohydrates, lipids, etc.

The stringent response causes the accumulation of two unusual nucleotides. **ppGpp** is guanosine tetraphosphate, with diphosphates attached to both 5′ and 3′ positions. **pppGpp** is guanosine pentaphosphate, with a 5′ triphosphate group and a 3′ diphosphate. These nucleotides are typical small-molecule effectors that function by binding to target proteins to alter their activities. Sometimes they are known collectively as (p)ppGpp.

(p)ppGpp functions to regulate coordinately a large number of cellular activities. Its production is controlled in two ways. A drastic increase in (p)ppGpp is triggered by the stringent response. And there is also a general inverse correlation between (p)ppGpp levels and the bacterial growth rate, which is controlled by some unknown means.

Deprivation of any one amino acid, or mutation to inactivate any aminoacyl-tRNA synthetase, is sufficient to initiate the stringent response. The trigger that sets the entire series of events in train is *the presence of uncharged tRNA in the A site of the ribosome.* Under normal conditions, of course, only aminoacyl-tRNA is placed in the A site by EF-Tu (see Chapter 6). But when there is no aminoacyl-tRNA available to respond to a particular codon, the uncharged tRNA becomes able to gain entry. Of course, this blocks any further progress by the ribosome; and it triggers an **idling reaction.**

The components involved in producing (p)ppGpp via the idling reaction have been identified through the existence of **relaxed (*rel*) mutants.** *rel* mutations abolish the stringent response, so that starvation for amino acids does not cause any reduction in stable RNA synthesis or alter any of the other reactions that are usually seen.

The most common site of relaxed mutation lies in the gene *relA*, which codes for a protein called the **stringent factor.** This factor is associated with the ribosomes, although the amount is rather low—say, <1 molecule for every 200 ribosomes. So perhaps only a minority of the ribosomes are able to produce the stringent response.

Ribosomes obtained from stringent bacteria can synthesize ppGpp and pppGpp *in vitro*, provided that the A site is occupied by an uncharged tRNA *specifically responding to the codon.* Ribosomes extracted from relaxed mutants cannot perform this reaction; but they are able to do so if the stringent factor is added.

Figure 10.27 shows the pathways for synthesis of (p)ppGpp. The stringent factor (RelA) is an enzyme that catalyzes the synthetic reaction in which ATP is used to donate a pyrophosphate group to the 3′ position of either GTP or GDP. The formal name for this activity is (p)ppGpp synthetase.

How is ppGpp removed when conditions return to normal? A gene called *spoT* codes for an enzyme that provides the major catalyst for ppGpp degradation. The activity of this enzyme causes ppGpp to be rapidly degraded, with a half-life of ~20 sec; so the stringent response is reversed rapidly when synthesis of (p)ppGpp ceases. *spoT* mutants have elevated levels of ppGpp, and grow more slowly as a result.

The RelA enzyme uses GTP as substrate more frequently, so that pppGpp is the predominant product. However, pppGpp is converted to ppGpp by several enzymes; among those able to perform this dephosphorylation are the translation factors EF-Tu and EF-G. The production of ppGpp via pppGpp is the most common route, *and ppGpp is the usual effector of the stringent response.*

The response of the ribosome to entry of uncharged tRNA is compared with normal protein synthesis in **Figure 10.28.** When EF-Tu places aminoacyl-tRNA in the A site, peptide bond synthesis is followed by ribosomal movement. But when uncharged tRNA is paired with the codon in the A site, the ribosome remains stationary and engages in the idling reaction.

Figure 10.27 Stringent factor catalyzes the synthesis of pppGpp and ppGpp; ribosomal proteins can dephosphorylate pppGpp to ppGpp.

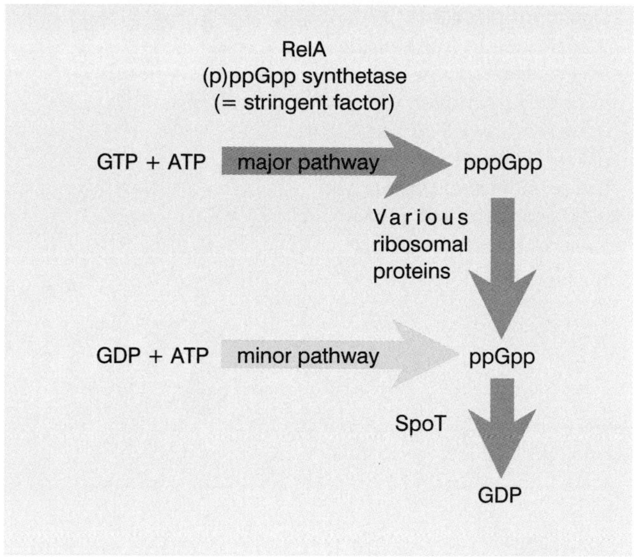

How does the state of the ribosome control the activity of RelA enzyme? An indication of the nature of the interaction is revealed by relaxed mutations in another locus, originally called *relC*, which turns out to be the same as *rplK*, which codes for the 50S subunit protein L11. This protein is located in the vicinity of the A and P sites, in a position to respond to the presence of a properly paired but uncharged tRNA in the A site. A conformational change in this protein or some other component could activate the RelA enzyme, so that the idling reaction occurs instead of polypeptide transfer from the peptidyl-tRNA.

One round of (p)ppGpp synthesis is associated with release of the uncharged tRNA from the A site, so that synthesis of (p)ppGpp is a continuing response to the level of uncharged tRNA. So under limiting conditions, a ribosome stalls when no aminoacyl-tRNA is available to respond to the codon in the A site. Entry of uncharged tRNA triggers the synthesis of a (p)ppGpp molecule, and the resulting expulsion of the uncharged tRNA allows the situation to be reassessed. Depending upon the availability of aminoacyl-tRNA, the ribosome resumes polypeptide synthesis or undertakes another idling reaction.

Figure 10.28 In normal protein synthesis, the presence of aminoacyl-tRNA in the A site is a signal for peptidyl transferase to transfer the polypeptide chain, followed by movement catalyzed by EF-G; but under stringent conditions, the presence of uncharged tRNA causes RelA protein to synthesize (p)ppGpp and to expel the tRNA.

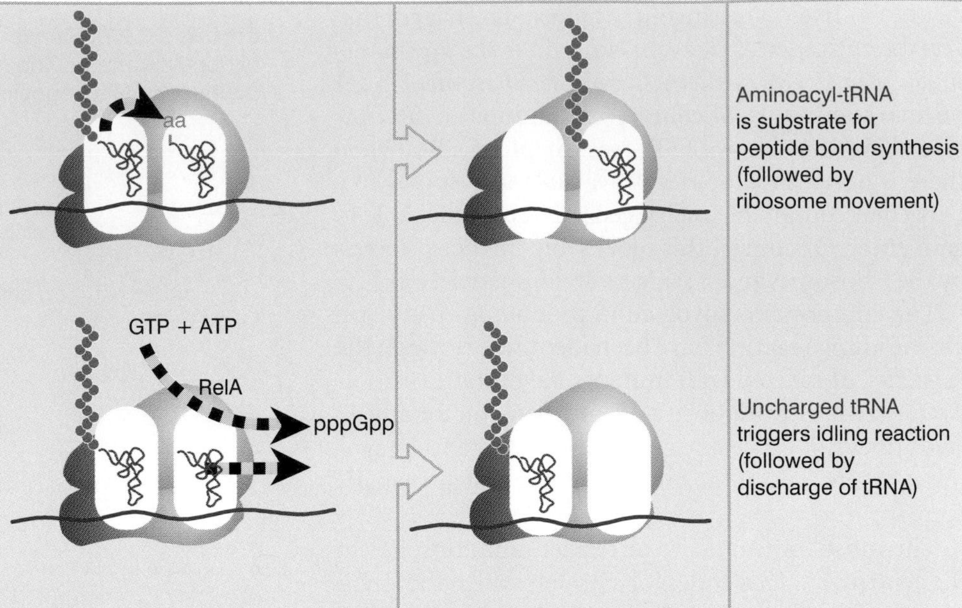
aa
Aminoacyl-tRNA is substrate for peptide bond synthesis (followed by ribosome movement)
GTP + ATP
RelA
pppGpp
Uncharged tRNA triggers idling reaction (followed by discharge of tRNA)

What does ppGpp do? It is an effector for controlling several reactions, including the inhibition of transcription. Many effects have been reported, among which two stand out:

- *Initiation of transcription is specifically inhibited at the promoters of operons coding for rRNA.* Mutations of stringently regulated promoters can abolish stringent control, which suggests that the effect requires an interaction with specific promoter sequences.
- *The elongation phase of transcription of many or most templates is reduced by ppGpp.* The cause is increased pausing by RNA polymerase. This effect is responsible for the general reduction in transcription efficiency when ppGpp is added *in vitro.*

The use of ppGpp is just one aspect of a more general regulatory network that relates production of ribosomes to the growth rate. The level of protein synthesis increases in proportion with the growth rate. This is accomplished by increasing the production of ribosomes as cells grow more rapidly. The cell therefore needs some general indicator of growth rate that can be used to control the synthesis of ribosomes. The indicator appears to be NTP levels, and the target for their action is the control of transcription of rRNA.

Figure 10.29 summarizes the systems that are used to control rRNA transcription in response to growth rate. Under conditions of starvation, ppGpp is produced, and (among its various actions) inhibits initiation at the promoters of the *rrn* loci that code for rRNA. As growth rate increases, the levels of ATP and

Figure 10.29 Nucleotide levels control initiation of rRNA transcription.

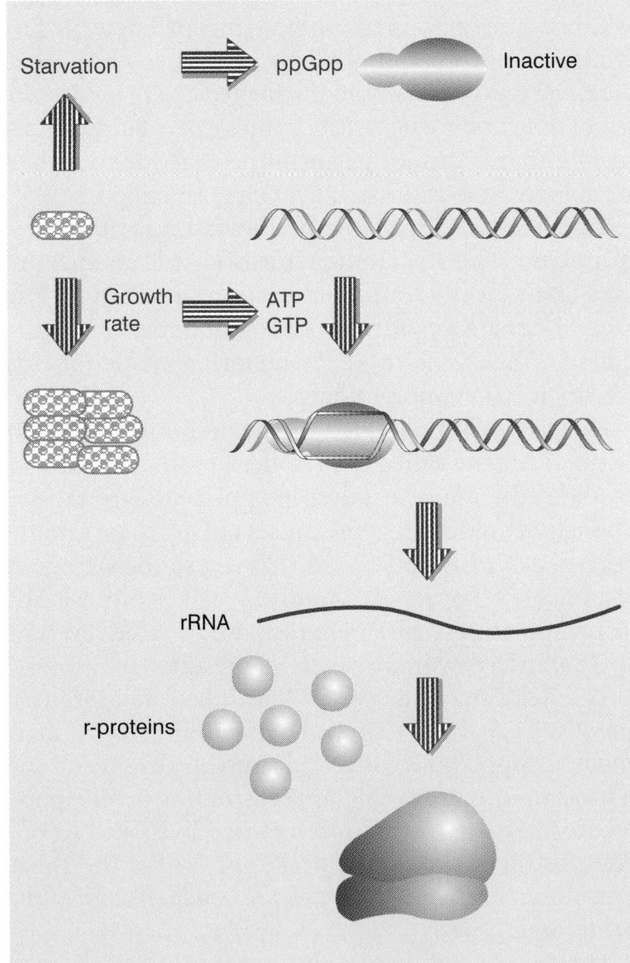
Starvation
ppGpp
Inactive
Growth rate
ATP
GTP
rRNA
r-proteins

GTP increase. These increase the rate of initiation at the *rrn* promoters.

The *rrn* promoters in *E. coli* form atypical open complexes with RNA polymerase. The open complexes are unusually unstable. The result is that the main factor governing the rate of initiation becomes the decay rate of the open complex. Increased concentration of the initiating nucleotide (which is ATP at six of the *rrn* promoters and GTP at the seventh) drives the initiation reaction forward by stabilizing the open complex.

The level of rRNA controls the production of ribosomes (by a feedback loop in which free [excess] rRNA inhibits ribosomal protein synthesis, as we see in the next section in Figure 10.34). This means that the production of ribosomes, and thus of protein synthesis, in turn responds to the levels of ATP and GTP, which reflect the nutritional condition of the cell. So concentrations of particular nucleotides control ribosome synthesis in response to normal changes in growth rate and the more extreme conditions of starvation.

Autogenous control may occur at translation

TRANSLATIONAL control is a notable feature of operons coding for components of the protein synthetic apparatus. The operon provides an arrangement for *coordinate* regulation of a group of structural genes. But, superimposed on it, further controls, such as those at the level of translation, may create *differences* in the extent to which individual genes are expressed.

A similar type of mechanism is used to achieve translational control in several systems. *Repressor function is provided by a protein that binds to a target region on mRNA to prevent ribosomes from recognizing the initiation region.* Formally this is equivalent to a repressor protein binding to DNA to prevent RNA polymerase from utilizing a promoter. **Figure 10.30** illustrates the most common form of this interaction, in which the regulator protein binds directly to a sequence that includes the AUG initiation codon, thereby preventing the ribosome from binding.

Some examples of translational repressors and their targets are summarized in **Figure 10.31.** A classic example is the coat protein of the RNA phage R17; it binds to a hairpin that encompasses the ribosome binding site in the phage mRNA. Similarly, the T4 RegA protein binds to a consensus sequence that includes the AUG initiation codon in several T4 early mRNAs; and T4 DNA polymerase binds to a sequence in its own mRNA that includes the Shine-Dalgarno element needed for ribosome binding.

Another form of translational control occurs when translation of one cistron requires changes in secondary structure that depend on translation of a preceding cistron. An effect of this nature is seen normally in the translation of the RNA phages, whose cistrons always are expressed in a set order. **Figure 10.32** shows that the phage RNA takes up a secondary structure in which only one initiation sequence is accessible; the

Figure 10.30 A regulator protein may block translation by binding to a site on mRNA that overlaps the ribosome-binding site at the initiation codon.

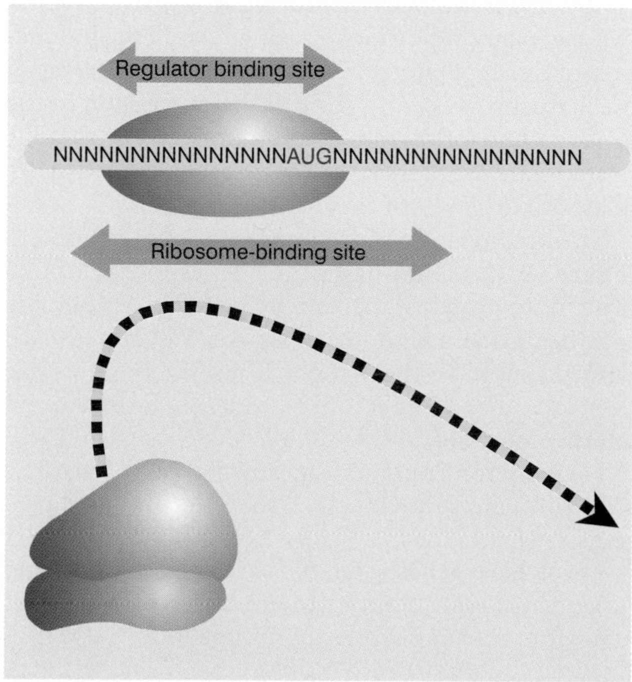

Figure 10.31 Proteins that bind to sequences within the initiation regions of mRNAs may function as translational repressors.

Repressor	Target Gene	Site of Action
R17 coat protein	R17 replicase	hairpin that includes ribosome binding site
T4 RegA	early T4 mRNAs	various sequences including initiation codon
T4 DNA polymerase	T4 DNA polymerase	Shine-Dalgarno sequence
T4 p32	gene 32	single-stranded 5' leader

second cannot be recognized by ribosomes because it is base paired with other regions of the RNA. However, translation of the first cistron disrupts the secondary structure, allowing ribosomes to bind to the initiation site of the next cistron. In this mRNA, secondary structure controls translatability.

About 70 or so proteins constitute the apparatus for bacterial gene expression. The ribosomal proteins are the major component, together with the ancillary proteins involved in protein synthesis. The subunits of RNA polymerase and its accessory factors make up the remainder. The genes coding for ribosomal proteins, protein-synthesis factors, and RNA polymerase subunits all are intermingled and organized into a small number of operons. Most of these proteins are represented only by single genes in *E. coli.*

Coordinate controls ensure that these proteins are synthesized in amounts appropriate for the growth conditions: when bacteria grow more rapidly, they devote a greater proportion of their efforts to the production of the apparatus for gene expression. An array of mechanisms is used to control the expression of the genes coding for this apparatus and to ensure that the proteins are synthesized at comparable levels that are related to the levels of the rRNAs.

The organization of six operons is summarized in **Figure 10.33.** About half of the genes for ribosomal proteins (**r-proteins**) map in four operons that lie close together. These are known as *str*, *spc*, *S10*, and α (each named simply for the first one of its functions to have been identified). The *rif* and *L11* operons lie together at another location.

Each operon contains a mélange of functions. The *str* operon has genes for small subunit ribosomal proteins as well as for EF-Tu and EF-G. The *spc* and *S10* operons have genes interspersed for both small and large ribosomal subunit proteins. The α operon has genes for proteins of both ribosomal subunits as well as for the α subunit of RNA polymerase. The *rif* locus has genes for large subunit ribosomal proteins and for the β and β′ subunits of RNA polymerase.

All except one of the ribosomal proteins are needed in equimolar amounts, which must be coordinated with the level of rRNA. The dispersion of genes whose products must be equimolar, and their intermingling with genes whose products are needed in different amounts, pose some interesting problems for coordinate regulation.

A feature common to all of the operons described in Figure 10.33 is regulation of some of the genes by one of the products. In each case, the gene coding for the regulatory product is itself one of the targets for regulation. Autogenous regulation occurs whenever a protein (or RNA) regulates its own production. In the case of the r-protein operons, the regulatory protein inhibits expression of a contiguous set of genes within the operon, so this is an example of negative autogenous regulation.

In each case, *accumulation of the protein inhibits further synthesis of itself and of some other gene products.* The effect often is exercised at the level of translation of the polycistronic mRNA. Each of the regulators is a ribosomal protein that binds directly to rRNA. *Its effect on translation is a result of its ability also to bind to its own mRNA.* The sites on mRNA at which these proteins bind either overlap the sequence where translation is initiated or lie nearby and probably influence the accessibility of the initiation site by inducing conformational changes. For example, in the S10 operon, protein L4 acts at the very start of the mRNA to inhibit translation of S10 and the subsequent genes. The inhibition may result from a simple block to ribosome access, as illustrated previously in Figure 10.31, or it may prevent a subsequent stage of translation. In two cases (including S4 in the α operon), the regulatory protein stabilizes a particular secondary structure in the mRNA that prevents the initiation reaction from continuing after the 30S subunit has bound.

Figure 10.32 Secondary structure can control initiation. Only one initiation site is available in the RNA phage, but translation of the first cistron changes the conformation of the RNA so that other initiation site(s) become available.

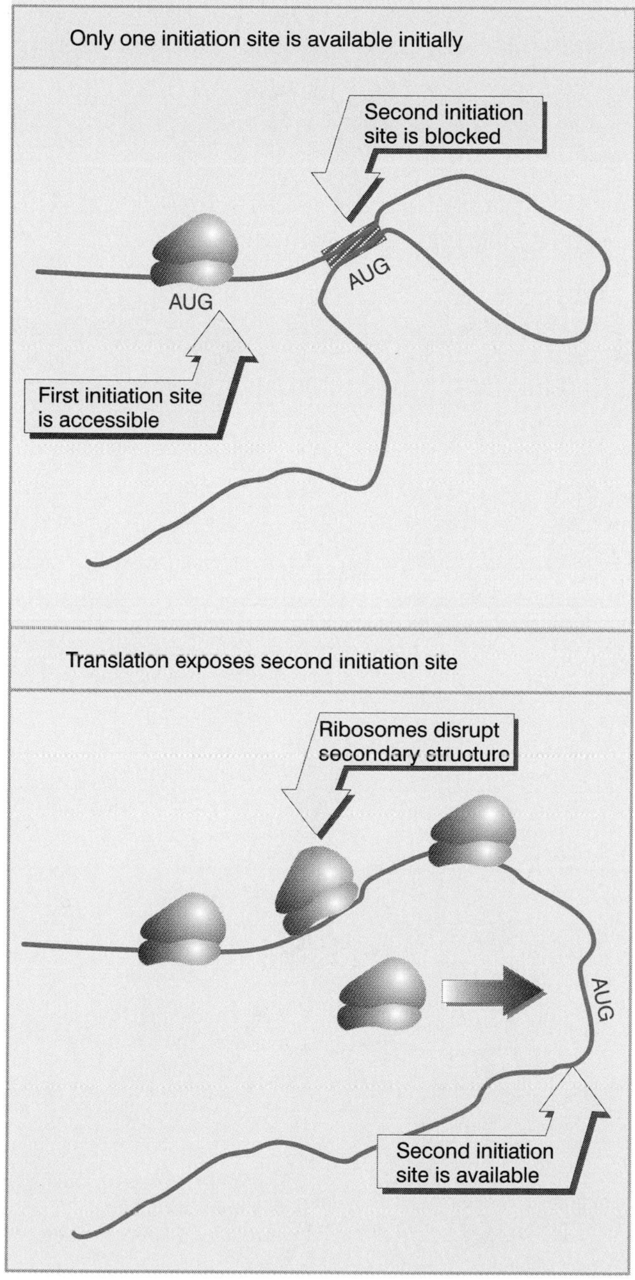
Only one initiation site is available initially
Second initiation site is blocked
AUG
AUG
First initiation site is accessible
Translation exposes second initiation site
Ribosomes disrupt secondary structure
AUG
Second initiation site is available

The use of r-proteins that bind rRNA to establish autogenous regulation immediately suggests that this provides a mechanism to link r-protein synthesis to rRNA synthesis. A generalized model is depicted in **Figure 10.34.** Suppose that the binding sites for the autogenous regulator r-proteins on rRNA are much stronger than those on the mRNAs. Then so long as any free rRNA is available, the newly synthesized r-proteins will associate with it to start ribosome assembly. There will be no free r-protein available to bind to the mRNA, so its translation will continue. But as soon as the synthesis of rRNA slows or stops, free r-proteins begin to accumulate. Then they are available to bind their mRNAs, repressing further translation. This circuit ensures that each r-protein operon responds in the same way to the level of rRNA: as soon as there is an excess of r-protein relative to rRNA, synthesis of the protein is repressed.

Autogenous regulation has been placed on a quantitative basis for gene *32* of phage T4. The protein (p32) plays a central role in genetic recombination, DNA repair, and replication, in which its function is exercised by virtue of its ability to bind to single-stranded DNA. Nonsense mutations cause the inactive protein to be overproduced. *So when the function of the protein is prevented, more of it is made.* This effect occurs at the level of translation; the gene *32* mRNA is stable, and remains so irrespective of the behavior of the protein product.

Figure 10.35 presents a model for the gene *32* control circuit. When single-stranded DNA is present in the phage-infected cell, it sequesters p32. However, in the absence of single-stranded DNA, or at least in conditions in which there is a surplus of p32, the protein prevents translation of its own mRNA. The effect is mediated directly by p32 binding to mRNA to prevent initiation of translation. This probably occurs at an A•T-rich region that surrounds the ribosome binding site.

Two features of the binding of p32 to the site on mRNA are required to make the control loop work effectively:

- The affinity of p32 for the site on gene *32* mRNA must be significantly lower than its affinity for single-stranded DNA. The equilibrium constant for binding RNA is in fact almost two orders of magnitude below that for single-stranded DNA.
- But the affinity of p32 for the mRNA must be significantly greater than the affinity for other RNA sequences. It is influenced by base composition and by secondary structure; an important aspect of the binding to gene *32* mRNA is that the regulatory region has an extended sequence lacking secondary structure.

Using the known equilibrium constants, we can plot the binding of p32 to its target sites as a function of

Figure 10.33 Genes for ribosomal proteins, protein synthesis factors, and RNA polymerase subunits are interspersed in a small number of operons that are autonomously regulated. The proteins subject to regulation are shaded in pink.

Operon	Genes and proteins	Regulator
str	*rpsL rpsG fusA tufA* S12 S7 EF-G EF-Tu	S7
spc	*rplN rplX rplE rpsN rpsH rplF rplR rpsE rplD rpmO secY-X* L14 L24 L5 S14 S8 L6 L18 S5 L30 L15 Y X	S8
S10	*rpsJ rplC rplB rplD rplW rplS rplV rpsC rpsQ rplP rpmC* S10 L3 L2 L4 L23 S19 L22 S3 S17 L16 L29	L4
α	*rpsM rpsK rpsD rpoA rplQ* S13 S11 S4 α L17	S4
L11	*rplK rplA* L11 L1	L1
rif	*rplJ rplL rpoB rpoC* L10 L7 β β'	L10

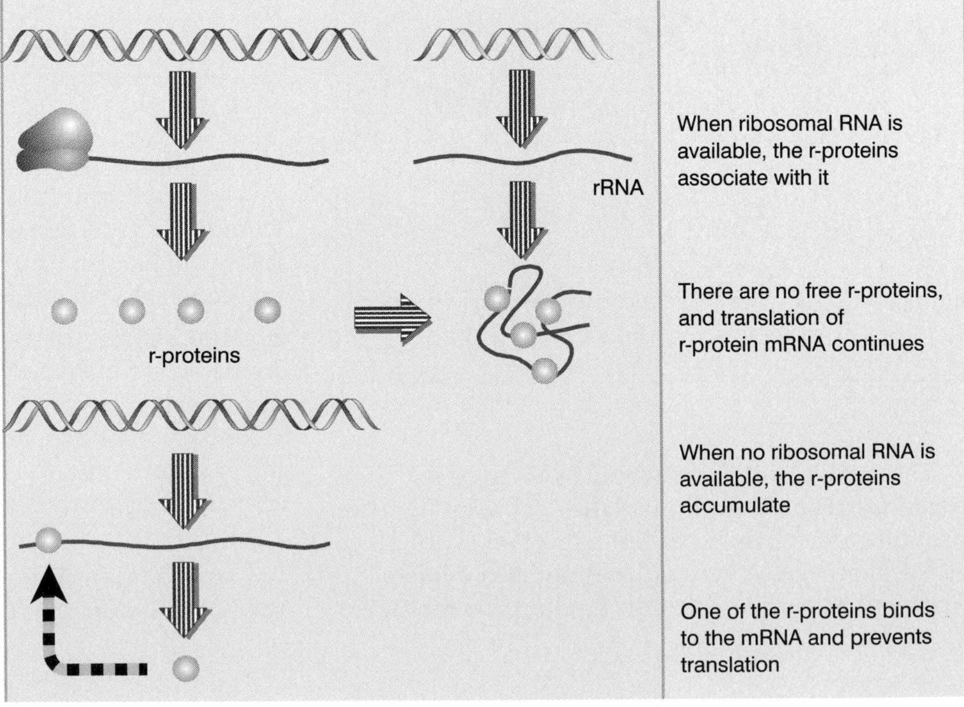

rRNA
r-proteins
When ribosomal RNA is available, the r-proteins associate with it
There are no free r-proteins, and translation of r-protein mRNA continues
When no ribosomal RNA is available, the r-proteins accumulate
One of the r-proteins binds to the mRNA and prevents translation

Figure 10.34 Translation of the r-protein operons is autogenously controlled and responds to the level of rRNA.

Figure 10.35 Excess gene *32* protein (p32) binds to its own mRNA to prevent ribosomes from initiating translation.

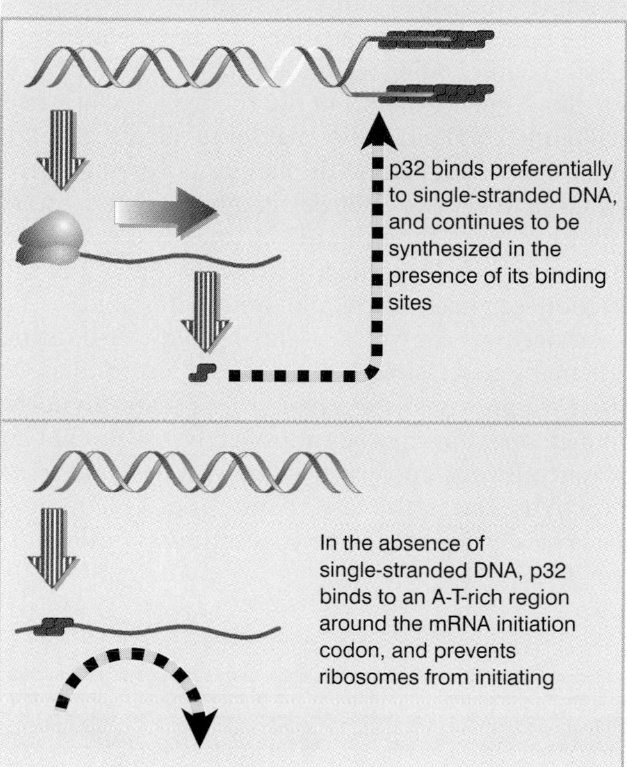

p32 binds preferentially to single-stranded DNA, and continues to be synthesized in the presence of its binding sites
In the absence of single-stranded DNA, p32 binds to an A-T-rich region around the mRNA initiation codon, and prevents ribosomes from initiating

Figure 10.36 Gene *32* protein binds to various substrates with different affinities, in the order single-stranded DNA, its own mRNA, and other mRNAs. Binding to its own mRNA prevents the level of p32 from rising $>10^{-6}$ M.

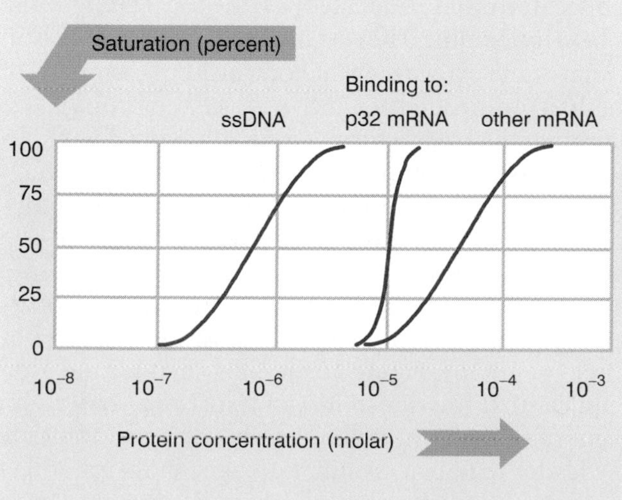

protein concentration. **Figure 10.36** shows that at concentrations below 10^{-6} M, p32 binds to single-stranded DNA. At concentrations $>10^{-6}$ M, it binds to gene *32* mRNA. At yet greater concentrations, it binds to other mRNA sequences, with a range of affinities.

These results imply that the level of p32 should be autoregulated to be $<10^{-6}$ M, which corresponds to ~2000 molecules per bacterium. This fits well with the measured level, 1000–2000 molecules/cell.

A feature of autogenous control is that each regulatory interaction is unique: a protein acts only on the mRNA responsible for its own synthesis. Phage T4 provides an example of a more general translational regulator, coded by the gene *regA*, which represses the expression of several genes that are transcribed during

Figure 10.37 Tubulin is assembled into microtubules when it is synthesized. Accumulation of excess free tubulin induces instability in the tubulin mRNA by acting at a site at the start of the reading frame in mRNA or at the corresponding position in the nascent protein.

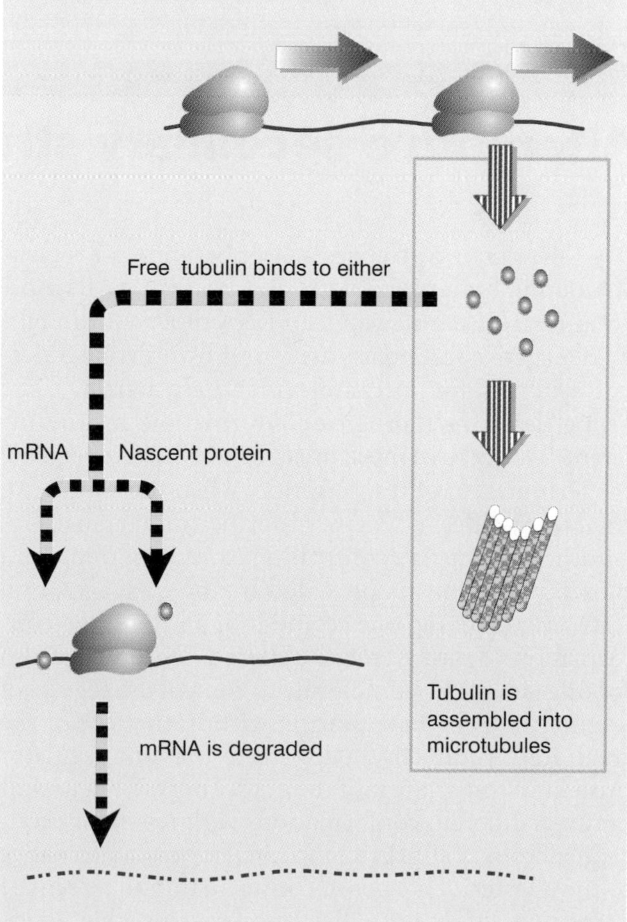

early infection. RegA protein prevents the translation of mRNAs for these genes by competing with 30S subunits for the initiation sites on the mRNA. Its action is a direct counterpart to the function of a repressor protein that binds multiple operators.

Autogenous regulation is a common type of control among proteins that are incorporated into macromolecular assemblies. The assembled particle itself may be unsuitable as a regulator, because it is too large, too numerous, or too restricted in its location. But the need for synthesis of its components may be reflected in the pool of free precursor subunits. If the assembly pathway is blocked for any reason, free subunits accumulate and shut off the unnecessary synthesis of further components.

Eukaryotic cells have a common system in which autogenous regulation of this type occurs. Tubulin is the monomer from which microtubules, which form a major filamentous system in all eukaryotic cells, are synthesized. The production of tubulin mRNA is controlled by the free tubulin pool. When this pool reaches a certain concentration, the production of further tubulin mRNA is prevented. Again, the principle is the same: tubulin sequestered into its macromolecular assembly plays no part in regulation, but the level of the free precursor pool determines whether further monomers are added to it.

The target site for regulation is a short sequence at the start of the coding region. We do not know yet what role this sequence plays, but two models are illustrated in **Figure 10.37**. Tubulin may bind directly to the mRNA; or it may bind to the nascent polypeptide representing this region. Whichever model applies, excess tubulin causes tubulin mRNA that is located on polysomes to be degraded, so the consequence of the reaction is to make the tubulin mRNA unstable.

Autogenous control is an *intrinsically* self-limiting system, by contrast with the *extrinsic* control that we discussed previously. So a repressor protein's ability to bind an operator may be controlled by the level of an extraneous small molecule, which activates or inhibits its activity. But in the case of autogenous regulation, the critical parameter is the concentration of the protein itself.

Alternative secondary structures control attenuation

RNA structure provides an opportunity for regulation in both prokaryotes and eukaryotes. Its most common role occurs when an RNA molecule can take up alternative secondary structures by utilizing different schemes for intramolecular base pairing. The properties of the alternative conformations may be different. This type of mechanism can be used to regulate the termination of transcription, when the alternative structures differ in whether they permit termination. Another means of controlling conformation (and thereby function) is provided by the cleavage of an RNA; by removing one segment of an RNA, the conformation of the rest may be altered. It is possible also for one (small) RNA molecule to control the activity of a target RNA by base pairing with it; the role of the small RNA is directly analogous to that of a regulator protein. We might regard the ability of an RNA to shift between different conformations with regulatory consequences as comprising the nucleic acid's alternative to the allosteric changes of conformation that regulate protein function. Both these mechanisms allow an interaction at one site in the molecule to affect the structure of another site.

Several operons are regulated by **attenuation,** a mechanism that controls the ability of RNA polymerase to read through an **attenuator,** which is an intrinsic terminator located at the beginning of a transcription unit. The common feature in attenuation is that some external event controls the formation of the hairpin needed for intrinsic termination. If the hairpin is allowed to form, termination prevents RNA polymerase from transcribing the structural genes. If the hairpin is prevented from forming, RNA polymerase elongates through the terminator, and the genes are expressed.

The changes in secondary structure that control attenuation are determined by the position of the ribosome on mRNA. **Figure 10.38** shows that termination requires that the ribosome can translate a leader segment of the mRNA. When the ribosome can translate the leader region, a termination hairpin forms at terminator 1. But when the ribosome is prevented from

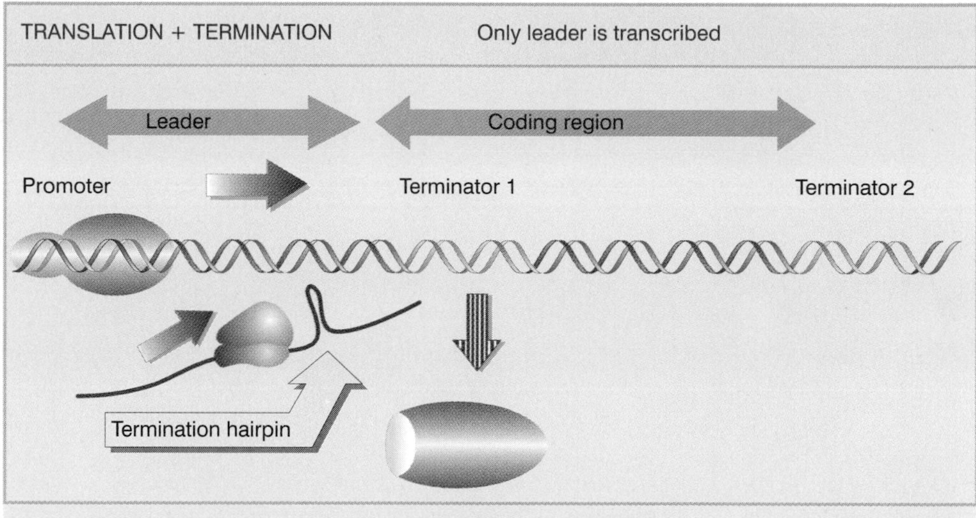

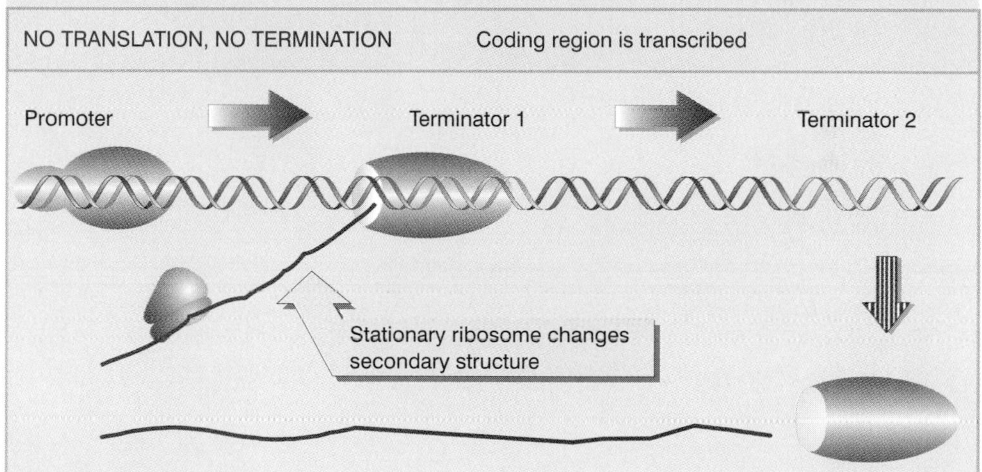

Figure 10.38 Termination can be controlled via changes in RNA secondary structure that are determined by ribosome movement.

translating the leader, the termination hairpin does not form, and RNA polymerase transcribes the coding region. *This mechanism of antitermination therefore depends upon the ability of external circumstances to influence ribosome movement in the leader region.*

The *trp* operon consists of five structural genes arranged in a contiguous series, coding for the three enzymes that convert chorismic acid to tryptophan. **Figure 10.39** shows that transcription starts at a promoter at the left end of the cluster. *trp* operon expression is controlled by two separate mechanisms. Repression of expression is exercised by a repressor protein (coded by the unlinked gene *trpR*) that binds to an operator that is adjacent to the promoter. Attenuation controls the progress of RNA polymerase into the operon by regulating whether termination occurs at a site preceding the first structural gene.

Attenuation was first revealed by the observation that deleting a sequence between the operator and the *trpE* coding region can increase the expression of the structural genes. This effect is independent of repression: both the basal and derepressed levels of transcription are increased. So this site influences events that occur *after* RNA polymerase has set out from the promoter (irrespective of the conditions prevailing at initiation).

The sequence between the promoter and the *trpE* gene has two interesting features:

- A short coding sequence could represent a **leader peptide** of 14 amino acids.
- The attenuator (intrinsic terminator) provides a barrier to transcription into the structural genes. RNA polymerase terminates there, either *in vivo* or *in vitro*, to produce a 140-base transcript.

Termination at the attenuator responds to the level of tryptophan, as illustrated in **Figure 10.40.** In the presence of adequate amounts of tryptophan, termination is efficient. But in the absence of

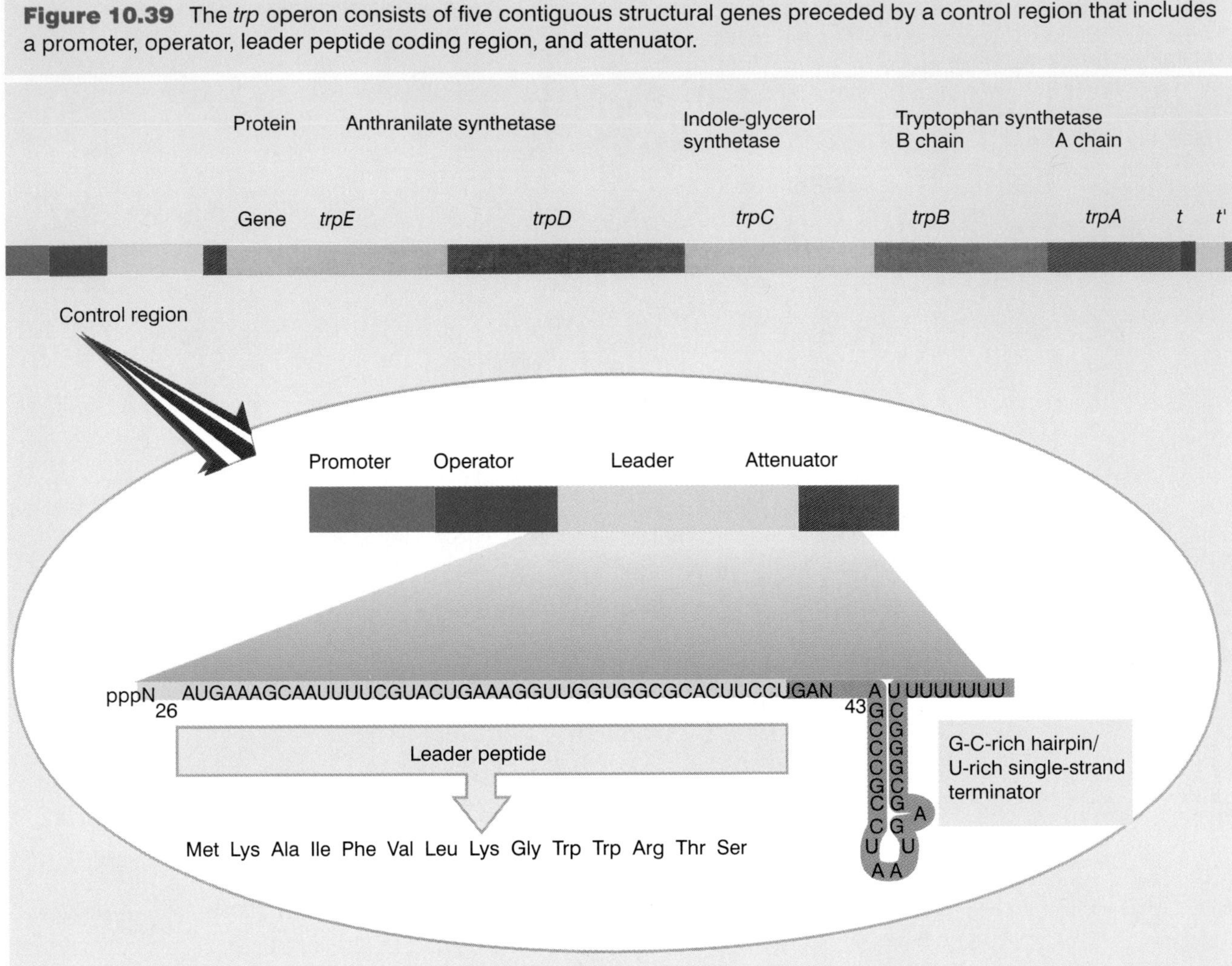

Figure 10.39 The *trp* operon consists of five contiguous structural genes preceded by a control region that includes a promoter, operator, leader peptide coding region, and attenuator.

tryptophan, RNA polymerase can continue into the structural genes.

Repression and attenuation respond in the same way to the level of tryptophan. When tryptophan is present, the operon is repressed; and most of the RNA polymerases that escape from the promoter then terminate at the attenuator. When tryptophan is removed, RNA polymerase has free access to the promoter, and also is no longer compelled to terminate prematurely.

Attenuation has ~10× effect on transcription. When tryptophan is present, termination is effective, and the attenuator allows only ~10% of the RNA polymerases to proceed. In the absence of tryptophan, attenuation allows virtually all of the polymerases to proceed. Together with the ~70× increase in initiation of transcription that results from the release of repression, this allows an ~700-fold range of regulation of the operon.

How can termination of transcription at the attenuator respond to the level of tryptophan? The sequence of the leader region suggests a mechanism. Figure 10.39 shows that it contains a ribosome binding site whose AUG codon is followed by a short coding region that contains two successive codons for tryptophan. When the cell runs out of tryptophan, ribosomes initiate translation of the leader peptide, but stop when they reach the Trp codons. The sequence of the mRNA suggests that this **ribosome stalling** influences termination at the attenuator.

The leader sequence can be written in alternative base-paired structures. The ability of the ribosome to proceed through the leader region controls transitions between these structures. The structure determines whether the mRNA can provide the features needed for termination.

Figure 10.40 An attenuator controls the progression of RNA polymerase into the *trp* genes. RNA polymerase initiates at the promoter and then proceeds to position 90, where it pauses before proceeding to the attenuator at position 140. In the absence of tryptophan, the polymerase continues into the structural genes (*trpE* starts at +163). In the presence of tryptophan there is ~90% probability of termination to release the 140-base leader RNA.

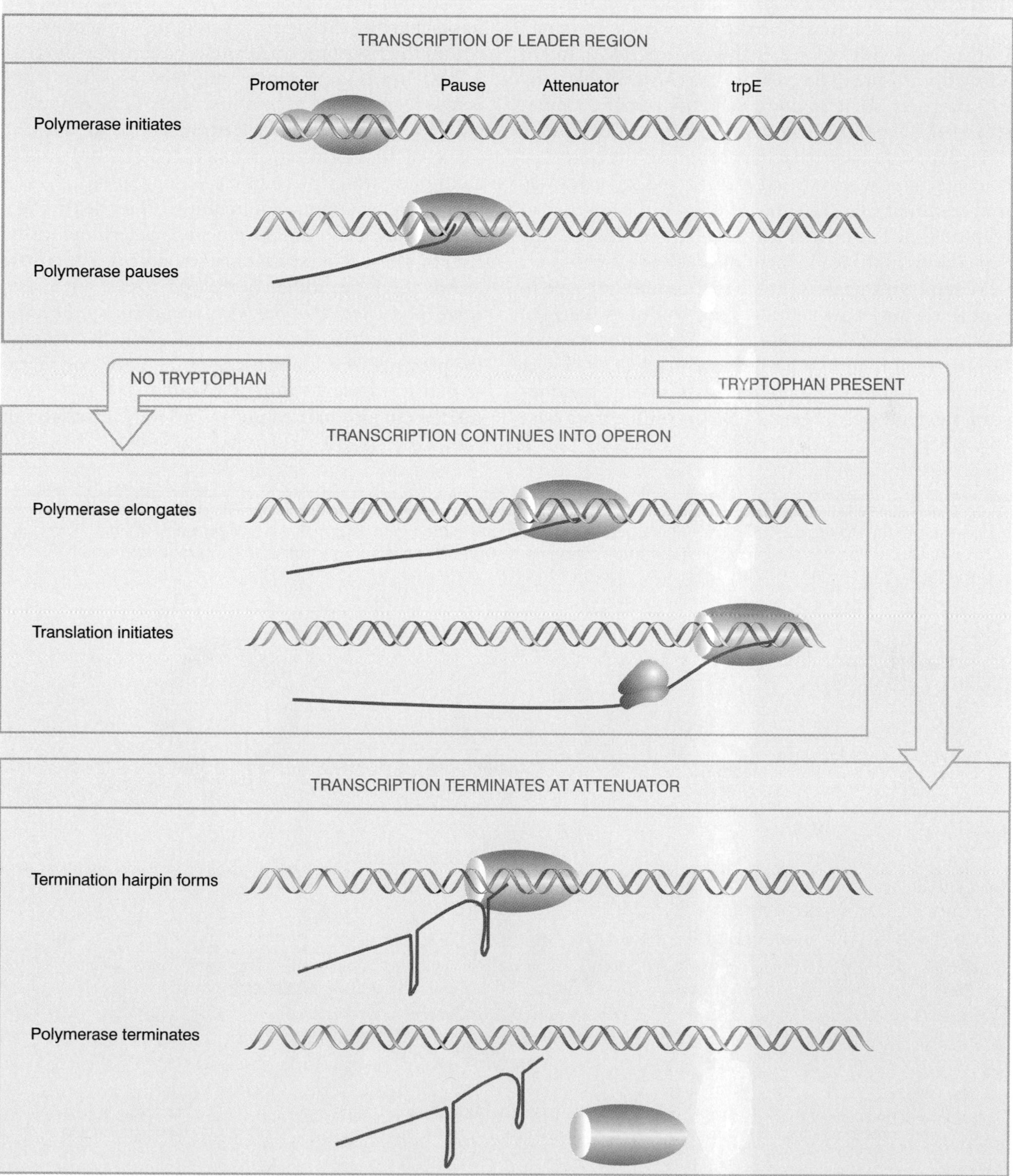

Figure 10.41 draws these structures. In the first, region 1 pairs with region 2; and region 3 pairs with region 4. The pairing of regions 3 and 4 generates the hairpin that precedes the U_8 sequence: this is the essential signal for intrinsic termination. Probably the RNA would take up this structure in lieu of any outside intervention.

A different structure is formed if region 1 is prevented from pairing with region 2. In this case, region 2 is free to pair with region 3. Then region 4 has no available pairing partner; so it is compelled to remain single-stranded. So the terminator hairpin cannot be formed.

Figure 10.42 shows that the position of the ribosome can determine which structure is formed, in such a way that termination is attenuated only in the absence of tryptophan. The crucial feature is the position of the Trp codons in the leader peptide coding sequence.

When tryptophan is present, ribosomes are able to synthesize the leader peptide. They continue along the leader section of the mRNA to the UGA codon, which lies between regions 1 and 2. As shown in the lower part of the figure, by progressing to this point, the ribosomes extend over region 2 and prevent it from base pairing. The result is that region 3 is available to base pair with region 4, generating the terminator hairpin. Under these conditions, therefore, RNA polymerase terminates at the attenuator.

When there is no tryptophan, ribosomes stall at the Trp codons, which are part of region 1, as shown in the upper part of the figure. So region 1 is sequestered within the ribosome and cannot base pair with region 2. This means that regions 2 and 3 become base paired before region 4 has been transcribed. This compels region 4 to remain in a single-stranded form. In the absence of the terminator hairpin, RNA polymerase continues transcription past the attenuator.

Control by attenuation requires a precise timing of events. For ribosome movement to determine formation of alternative secondary structures that control termination, *translation of the leader must occur at the same time when RNA polymerase approaches the terminator site.* A critical event in controlling the timing is the presence of a site that causes the RNA polymerase to pause at base 90 along the leader. The RNA polymerase remains paused until a ribosome translates the

Figure 10.41 The *trp* leader region can exist in alternative base-paired conformations. The center shows the four regions that can base pair. Region 1 is complementary to region 2, which is complementary to region 3, which is complementary to region 4. On the left is the conformation produced when region 1 pairs with region 2, and region 3 pairs with region 4. On the right is the conformation when region 2 pairs with region 3, leaving regions 1 and 4 unpaired.

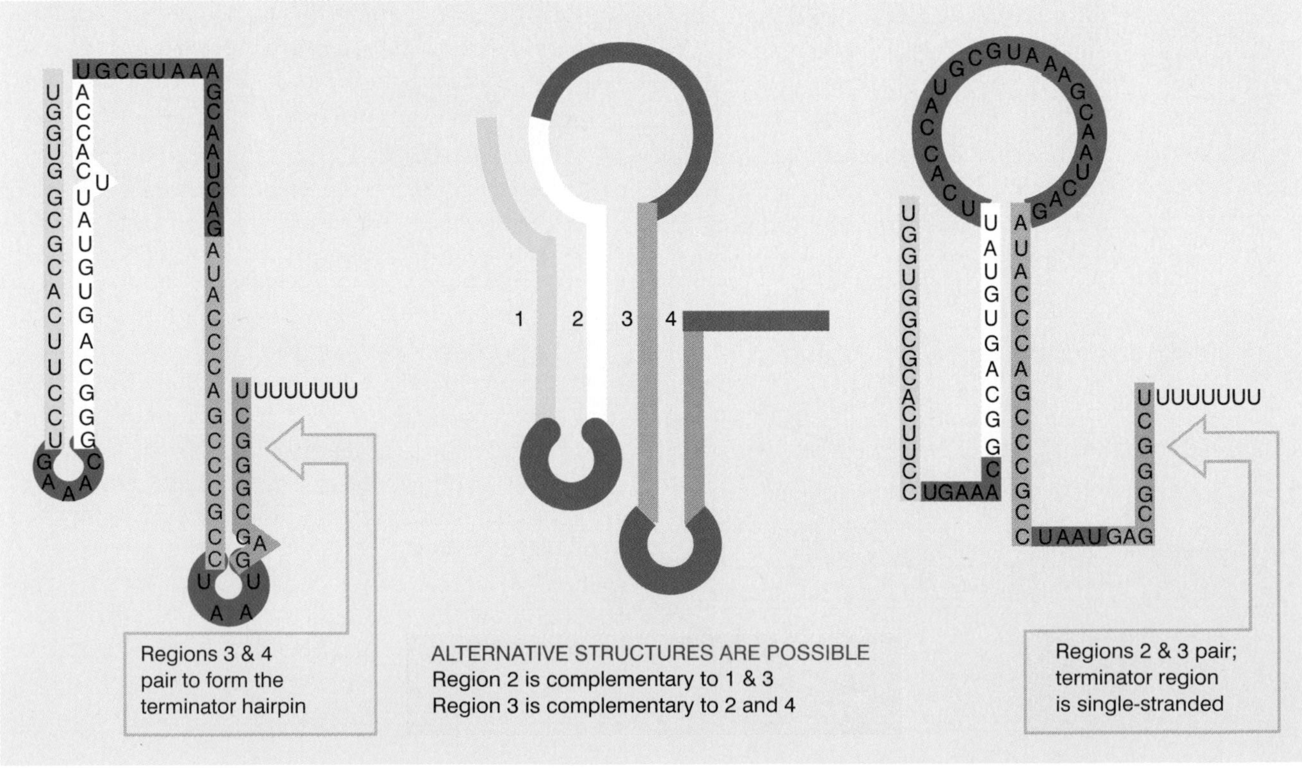
UGCGUAAA
UUUUUUUU
Regions 3 & 4 pair to form the terminator hairpin
1
2
3
4
ALTERNATIVE STRUCTURES ARE POSSIBLE
Region 2 is complementary to 1 & 3
Region 3 is complementary to 2 and 4
CUGAAA
CUAAUGAG
UUUUUUUU
Regions 2 & 3 pair; terminator region is single-stranded

leader peptide. Then the polymerase is released and moves off toward the attenuation site. By the time it arrives there, secondary structure of the attenuation region has been determined.

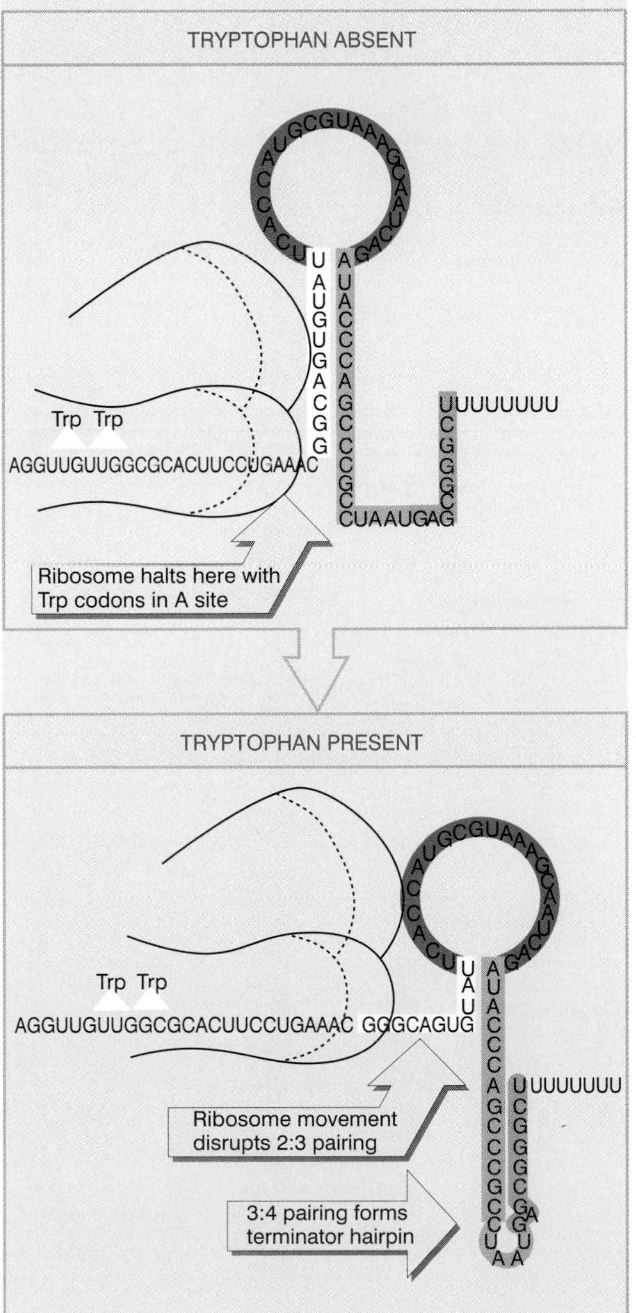

Figure 10.42 The alternatives for RNA polymerase at the attenuator depend on the location of the ribosome, which determines whether regions 3 and 4 can pair to form the terminator hairpin.

By providing a mechanism to sense the inadequacy of the supply of Trp-tRNA, attenuation responds directly to the need of the cell for tryptophan in protein synthesis.

How widespread is the use of attenuation as a control mechanism for bacterial operons? It is used in at least six operons that code for enzymes concerned with the biosynthesis of amino acids. So a feedback from the level of the amino acid available for protein synthesis (as represented by the availability of aminoacyl-tRNA) to the production of the enzymes may be common.

The use of the ribosome to control RNA secondary structure in response to the availability of an aminoacyl-tRNA establishes an inverse relationship between the presence of aminoacyl-tRNA and the transcription of the operon, equivalent to a situation in which aminoacyl-tRNA functions as a corepressor of transcription. Since the regulatory mechanism is mediated by changes in the formation of duplex regions, attenuation provides a striking example of the importance of secondary structure in the termination event, and of its use in regulation.

Each operon subject to attenuation has a leader peptide in which stalling of the ribosome at the codons representing the regulator amino acid(s) can cause the mRNA to take up a secondary structure in which a terminator hairpin cannot form. In some cases, an operon is derepressed by starvation for more than one amino acid. In these cases, codons for the various amino acids that regulate the operon are interspersed in the leader sequence in such a way that ribosome stalling at any one of them is able to prevent formation of the terminator hairpin. In some cases (such as the *his* operon), attenuation provides the only regulation of expression.

Attenuation may also be regulated by proteins that bind to RNA, either to stabilize or to destabilize formation of the hairpin required for termination. The activity of such a protein may be intrinsic or may respond to a small molecule in the same manner as a repressor protein responds to corepressor.

In the *trp* operon of *B. subtilis*, the MtrB protein binds to the leader of the transcript to promote formation of the terminator hairpin. MtrB is presumably activated by tryptophan, and prevents expression of the *trp* genes by causing attenuation. It therefore functions as a terminator protein.

The *bgl* operon is controlled by the ability of the protein BglG to bind to a sequence in the 5′nontranslated leader of the transcript. Binding of BglG sequesters part of the sequence required for formation of an intrinsic terminator hairpin. BglG therefore acts as an antiterminator.

Small RNA molecules can regulate translation

THE *trans*-acting regulators that we have discussed so far are proteins. Yet the formal circuitry of a regulatory network could equally well be constructed by using an RNA as regulator. In fact, the original model for the operon left open the question of whether the regulator might be RNA or protein.

Like a protein regulator, a small regulator RNA is an independently synthesized molecule that diffuses to a target site consisting of a specific nucleotide sequence. The target for a regulator RNA is a single-stranded nucleic acid sequence. The regulator RNA functions by complementarity with its target, at which it can form a double-stranded region.

We can imagine two general mechanisms for the action of a regulator RNA:

- Formation of a duplex region with the target nucleic acid directly prevents its ability to function, by forming or sequestering a specific site.
- Formation of a duplex region in one part of the target molecule changes the conformation of another region, thus indirectly affecting its function.

The feature common to both types of RNA-mediated regulation is that changes in secondary structure of the target control its activity.

A difference between RNA regulators and the proteins that repress operons is that the RNA does not have allosteric properties; it cannot respond to other small molecules by changing its ability to recognize its target. It can be turned on by controlling transcription of its gene or it could be turned off by an enzyme that degrades the RNA regulator product.

Antisense RNAs are used as regulators in several situations in bacterial cells. **Figure 10.43** illustrates three situations. An antisense RNA may bind to an RNA to occlude the site for ribosome binding and thus prevent initiation of protein synthesis. Antisense RNA may also directly destabilize a target mRNA by binding to it to form a duplex region that is the target for an endonuclease. And antisense RNA may affect gene expression by binding to a transcript to mimic a terminator, and thus cause premature termination of transcription. (We discuss some examples of the use of RNA-RNA interactions in specific control systems in other chapters, including control of

Figure 10.43 Antisense RNA can affect function or stability of an RNA target.

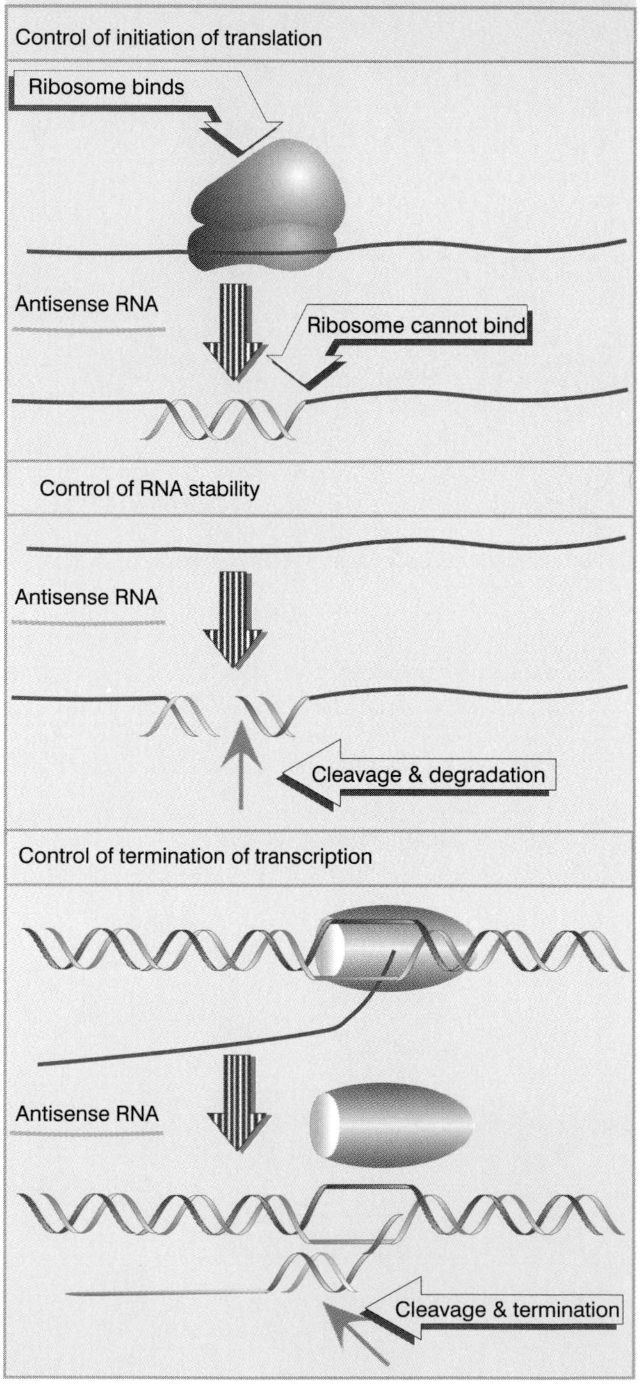

translation [see Chapter 15] and replication [see Chapter 12].)

The synthesis of a small RNA directly controls translation of the *ompF* gene of *E. coli*. The circuit is shown in **Figure 10.44.** Expression of the unlinked genes *ompC* and *ompF*, which code for two of the outer membrane proteins of *E. coli*, is controlled by the osmolarity of the medium. Increase in osmolarity activates the receptor EnvZ, which in turn activates OmpR, which activates expression of the regulator gene, *micF*.

The product of *micF* is an RNA of 174 bases. It is an example of an antisense RNA, which describes any RNA that blocks the function of an mRNA by virtue of complementarity with it.

Synthesis of *micF* RNA turns off translation of *ompF* mRNA. *MicF* RNA is complementary to a region of *ompF* mRNA that includes the ribosome binding site at which translation is initiated. The *micF* RNA could therefore function as a regulator by binding to the *ompF* mRNA and preventing its translation. It is also possible that formation of a duplex region destabilizes the *ompF* mRNA, for example, by making it sus-

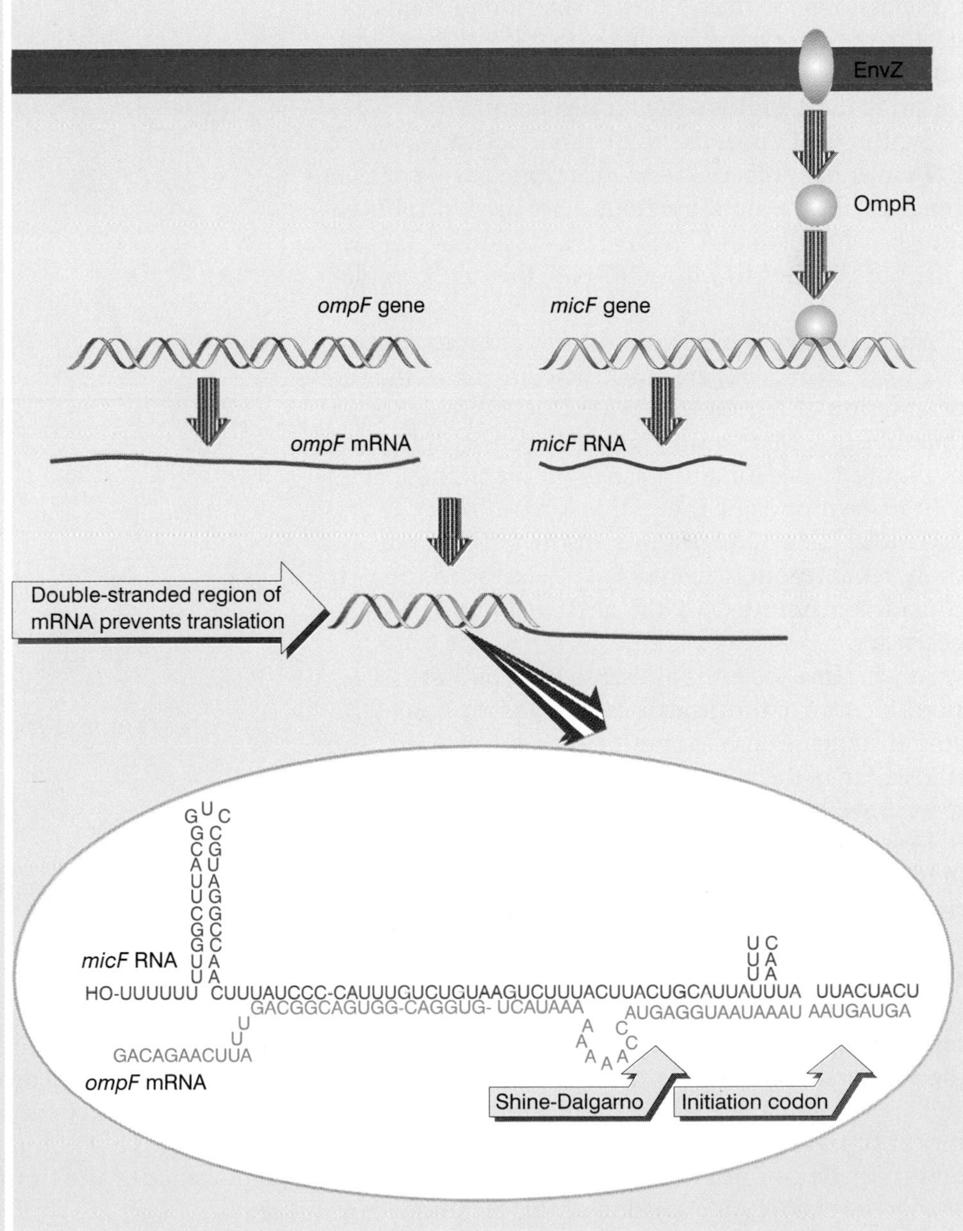

Figure 10.44 Increase in osmolarity activates EnvZ, which activates OmpR, which induces transcription of *micF* and *ompC* (not shown). *micF* RNA is complementary to the 5' region of *ompF* mRNA and prevents its translation.

ceptible to ribonucleases that act on double-stranded regions.

Figure 10.45 illustrates the behavior of a comparable system in the nematode *C. elegans*, in which a regulator gene codes for an RNA that is complementary to an mRNA. In this case, the target gene is *lin14*, which regulates larval development. Expression of *lin14* is controlled by *lin4*, which codes for two small transcripts of 22 and 61 nucleotides, which differ in their termination site. The *lin4* transcripts are complementary to a 10-base sequence that is repeated 7 times in the 3′ nontranslated region of *lin14*. We do not know how binding of the *lin4* RNA turns off the expression of *lin14*; one possibility is that duplex formation prevents export to the cytoplasm, another is that it interferes with translation. This system is especially interesting in implicating the 3′ end as a site for regulation.

Figure 10.45 *lin4* RNA regulates expression of *lin14* by binding to the 3' nontranslated region.

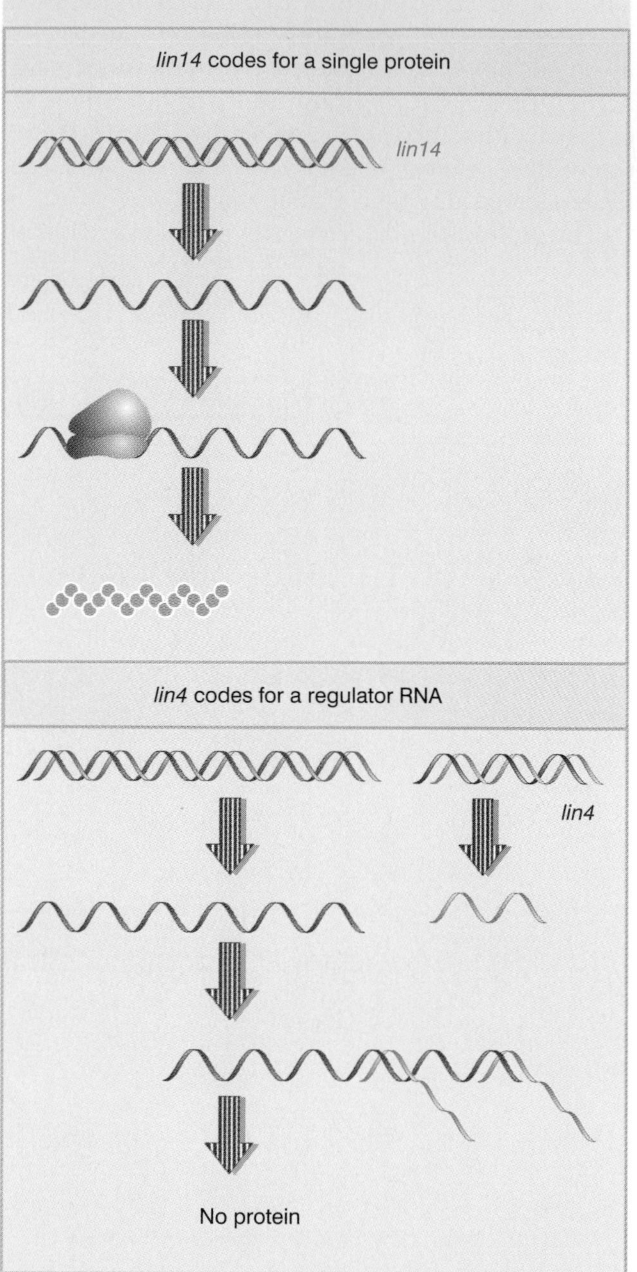
lin14 codes for a single protein
lin14
lin4 codes for a regulator RNA
lin4
No protein

Synthesis of antisense RNA can inactivate a target RNA in either prokaryotic or eukaryotic cells. Artificial genes coding for antisense RNAs have been introduced into *E. coli*, where they prevent expression of the specific target genes to whose mRNAs they are complementary.

Antisense genes have been introduced into eukaryotic cells. Such genes are constructed by reversing the orientation of a gene with regard to its promoter, so that the "antisense" strand is transcribed, as illustrated in **Figure 10.46.** An antisense thymidine kinase gene inhibits synthesis of thymidine kinase from the endogenous gene. Quantitation of the effect is not entirely reliable, but it seems that an excess (perhaps a considerable excess) of the antisense RNA may be necessary.

At what level does the antisense RNA inhibit expression? It could in principle prevent transcription of the authentic gene, processing of its RNA product, or translation of the messenger. Results with different systems show that the inhibition depends on formation of RNA•RNA duplex molecules, but this can occur either in the nucleus or in the cytoplasm. In the case of an antisense gene stably carried by a cultured cell, sense-antisense RNA duplexes form in the nucleus, preventing normal processing and/or transport of the sense RNA. In another case, injection of antisense RNA into the cytoplasm inhibits translation by forming duplex RNA in the 5′ region of the mRNA.

This technique offers a powerful approach for turning off genes at will; for example, the function of a regulatory gene can be investigated by introducing an antisense version. An extension of this technique is to place the antisense gene under control of a promoter itself subject to regulation. Then the target gene can be turned off and on by regulating the production of antisense RNA. This technique allows investigation of the importance of the timing of expression of the target gene.

Figure 10.46 Antisense RNA can be generated by reversing the orientation of a gene with respect to its promoter, and can anneal with the wild-type transcript to form duplex RNA.

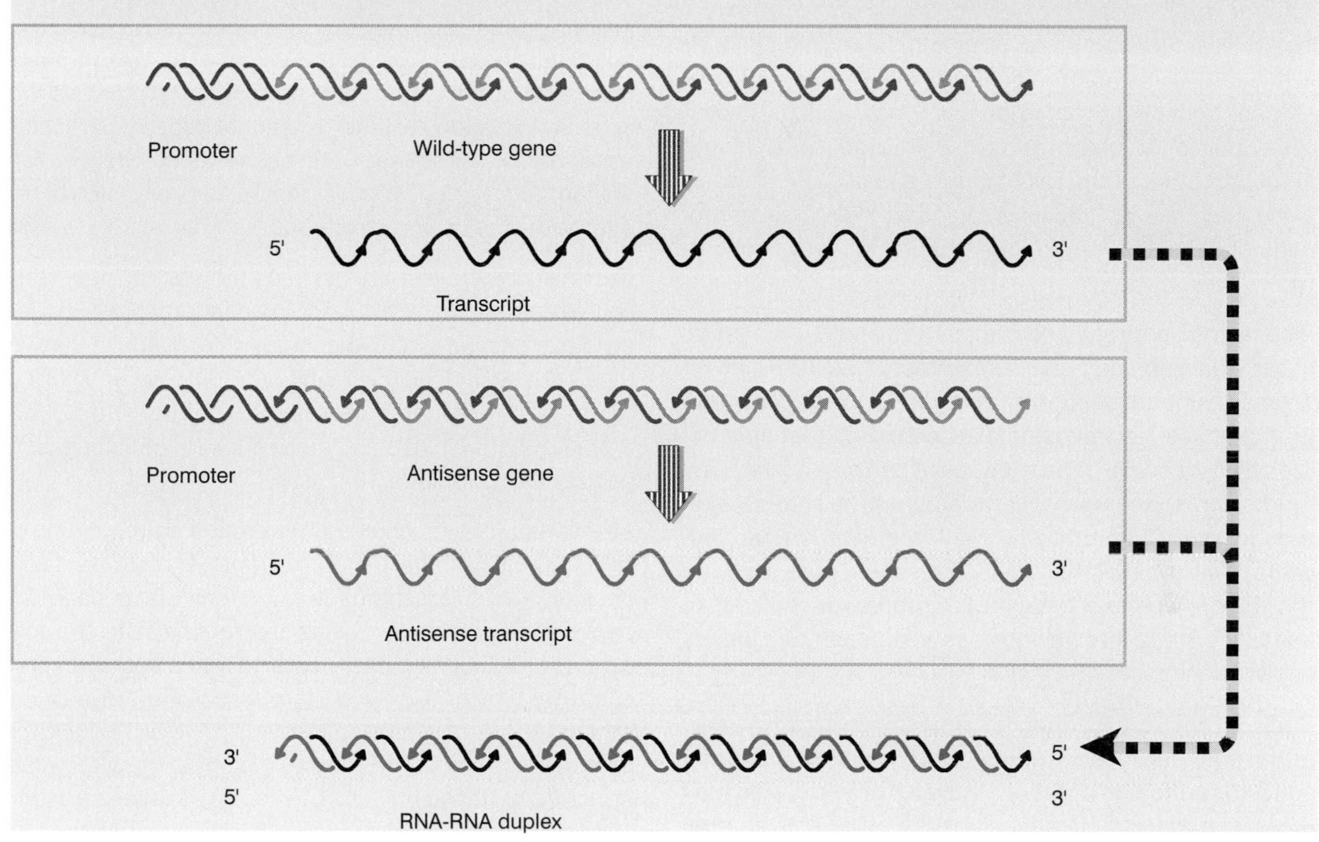

Summary

Transcription is regulated by the interaction between *trans*-acting factors and *cis*-acting sites. A *trans*-acting factor is the product of a regulator gene. It is usually protein but can be RNA. Because it diffuses in the cell, it can act on any appropriate target gene. A *cis*-acting site in DNA (or RNA) is a sequence that functions by being recognized *in situ*. It has no coding function and can regulate only those sequences that are physically contiguous with it. Bacterial genes coding for proteins whose functions are related, such as successive enzymes in a pathway, may be organized in a cluster that is transcribed into a polycistronic mRNA from a single promoter. Control of this promoter regulates expression of the entire pathway. The unit of regulation, containing structural genes and *cis*-acting elements, is called the operon.

Initiation of transcription is regulated by interactions that occur in the vicinity of the promoter. The ability of RNA polymerase to initiate at the promoter is prevented or activated by other proteins. Genes that are active unless they are turned off are said to be under negative control. Genes that are active only when specifically turned on are said to be under positive control. The type of control can be determined by the dominance relationships between wild type and mutants that are constitutive/derepressed (permanently on) or uninducible/super-repressed (permanently off).

A repressor protein prevents RNA polymerase either from binding to the promoter or from activating transcription. The repressor binds to a target sequence, the operator, that usually is located around or upstream of the startpoint. Operator sequences are short and often are palindromic. The repressor is often a homomultimer whose symmetry reflects that of its target.

The ability of the repressor protein to bind to its operator is regulated by a small molecule. An inducer prevents a repressor from binding; a corepressor activates it. Binding of the inducer or corepressor to its site produces a change in the structure of the DNA-binding site of the repressor. This allosteric reaction occurs in both free repressor proteins and directly in repressor proteins already bound to DNA.

The lactose pathway operates by induction, when an inducer β-galactoside prevents the repressor from binding its operator; transcription and translation of the *lacZ* gene then produce β-galactosidase, the enzyme that metabolizes β-galactosides. The tryptophan pathway operates by repression; the corepressor (tryptophan) activates the repressor protein so that it binds to the operator and prevents expression of the genes that code for the enzymes that biosynthesize tryptophan. A repressor can control multiple targets that have copies of an operator consensus sequence.

Some promoters cannot be recognized by RNA polymerase (or are recognized only poorly) unless a specific activator protein is present. Activator proteins also may be regulated by small molecules. The CAP activator becomes able to bind to target sequences in the presence of cyclic AMP. All promoters that respond to CAP have at least one copy of the target sequence. Binding of CAP to its target involves bending DNA. Direct contact between one subunit of CAP and RNA polymerase is required to activate transcription.

A protein with a high affinity for a particular target sequence in DNA has a lower affinity for all DNA. The ratio defines the specificity of the protein. Because there are many more nonspecific sites (any DNA sequence) than specific target sites in a genome, a DNA-binding protein such as a repressor or RNA polymerase is "stored" on DNA; probably none or very little is free. The specificity for the target sequence must be great enough to counterbalance the excess of nonspecific sites over specific sites. The balance for bacterial proteins is adjusted so that the amount of protein and its specificity allow specific recognition of the target in "on" conditions, but allow almost complete release of the target in "off" conditions.

Gene expression can be controlled at stages subsequent to transcription. Translation may be controlled by a protein that binds to a region of mRNA overlapping with the ribosome-binding site; this prevents ribosomes from initiating translation. RegA of T4 is a general regulator that functions on several target mRNAs at the level of translation. Most proteins that repress translation possess this capacity in addition to other functional roles; in particular, translation is controlled in some cases of autogenous regulation, when a gene product regulates expression of the operon containing its own gene.

The level of protein synthesis itself provides an important coordinating signal. Deficiency in aminoacyl-tRNA causes an idling reaction on the ribosome, which leads to the synthesis of the unusual nucleotide ppGpp. This is an effector that inhibits initiation of transcription at certain promoters; it also has a general effect in inhibiting elongation on all templates.

Attenuation is a mechanism that relies on regulation of termination to control transcription through bacterial operons. It is commonly used in operons that code for enzymes involved in biosynthesis of an amino acid. The polycistronic mRNA of the operon starts with a sequence that can form alternative secondary structures. One of the structures has a hairpin loop that provides an intrinsic terminator upstream of the structural genes; the alternative structure lacks the hairpin. The choice of which structure forms is controlled by the progress of translation through a short leader sequence that includes codons for the amino acid(s) that are the product of the system. In the presence of aminoacyl-tRNA bearing such amino acid(s), ribosomes translate the leader peptide, allowing a secondary structure to form that supports termination. In the absence of this aminoacyl-tRNA, the ribosome stalls, resulting in a new secondary structure in which the hairpin needed for termination cannot form. The supply of aminoacyl-tRNA therefore (inversely) controls amino acid biosynthesis. Attenuation also is controlled in some operons by regulator proteins that bind directly to the leader RNA to assist or inhibit the formation of a terminator hairpin.

An alternative to a regulator protein may be a small RNA that is complementary to a target mRNA; formation of a duplex RNA region may prevent translation by sequestering the initiation site, directly or indirectly. Regulatory RNAs that function by such means are called antisense RNAs.

Further reading

Reviews

Miller, J. and Reznikoff, W. (1978). *The Operon* (Cold Spring Harbor Laboratory, New York) including:

Barkleya, M. D. and Bourgeois, S. Repressor recognition of operator and effectors. (pp. 177–220);

Beckwith, J. *lac*: the genetic system. (pp. 11–30);

Beyreuther, K. Chemical structure and functional organization of *lac* repressor from *E. coli*. (pp. 123–154);

Miller, J. H. The *lacI* gene: its role in *lac* operon control and its use as a genetic system. (pp. 31–88);

Weber, K. and Geisler, N. *lac* repressor fragments produced *in vivo* and *in vitro*: an approach to the understanding of the interaction of repressor and DNA. (pp. 155–176);

Zabin, I. and Fowler, A. V. β-galactosidase, the lactose permease protein, and thiogalactoside transcetylase. (pp. 89–122).

Yanofsky, C. (1981). Attenuation in the control of expression of bacterial operons. *Nature* **289**, 751–758.

Gegenheimer, P. and Apirion, D. (1981). Processing of prokaryotic RNA. *Microbiol. Rev.* **45**, 502–541.

Bauer, C. E. *et al.* (1983). Attenuation in bacterial operons. In *Gene Function in Prokaryotes*. (Eds. J. Beckwith, J. E. Davies, and J. A. Gallant, Cold Spring Harbor, New York, 65–89).

Nomura, M. *et al.* (1984). Regulation of the synthesis of ribosomes and ribosomal components. *Ann. Rev. Biochem.* **53**, 75–117.

Green, J., Pines, O., and Inouye, M. (1986). The role of antisense RNA in gene regulation. *Ann. Rev. Biochem.* **55**, 569–597.

Cashel, M. and Rudd, K. E. (1987). The stringent response. In *E. coli and S. typhimurium*. (Ed. F. C. Neidhardt, American Society for Microbiology, Washington DC, 1410–1429).

King, T. C., Sirdeskmukh, R., and Schlessinger, D. (1987). Nucleolytic processing of RNA transcripts in prokaryotes. *Microbiol. Rev.* **50**, 428–451.

Landick, R. and Yanofsky, C. (1987). Transcription attenuation. In *E. coli and S. typhimurium*. (Ed. F. C. Neidhardt, American Society for Microbiology, Washington DC, 1276–1301).

Yanofsky, C. and Crawford, I. P. (1987). The tryptophan operon. In *E. coli and S. typhimurium*. (Ed. F. C. Neidhardt, American Society for Microbiology, Washington DC, 1453–1472).

Gold, L. (1988). Posttranscriptional regulatory mechanisms in *E. coli*. *Ann. Rev. Biochem.* **57**, 199–223.

Botsford, J. L. and Harman, J. G. (1992). Cyclic AMP in prokaryotes. *Microbiol. Rev.* **56**, 100–122.

Ptashne, M. (1992). *A Genetic Switch*. (Cell Press and Blackwell Scientific, Cambridge).

Kolb, A. *et al.* (1993). Transcriptional regulation by cAMP and its receptor protein. *Ann. Rev. Biochem.* **62**, 749–795.

Greenblatt, J., Nodwell, J. R., and Mason, S. W. (1993). Transcriptional antitermination. *Nature* **364**, 401–406.

Condon, C., Squires, C., and Squires, C. L. (1995). Control of rRNA transcription in *E. coli*. *Microbiol. Rev.* **59**, 623–645.

Discoveries

Operons

Jacob, F. and Monod, J. (1961). Genetic regulatory mechanisms in the synthesis of proteins. *J. Mol. Biol.* **3**, 318–356.

Repressors

Gilbert, W. and Muller-Hill, B. (1966). Isolation of the *lac* repressor. *Proc. Nat. Acad. Sci. USA* **56**, 1891–1898.

Gilbert, W. and Muller-Hill, B. (1967). The *lac* operator is DNA. *Proc. Nat. Acad. Sci. USA* **58**, 2415–2421.

Oehler, S. (1990). The three operators of the *lac* operon cooperate in repression. *EMBO J.* **9**, 973–979.

Friedman, A. M., Fischmann, T. O., and Steitz, T. A. (1995). Crystal structure of *lac* repressor core tetramer and its implications for DNA looping. *Science* **268**, 1721–1727.

Lewis, M. *et al.* (1996). Crystal structure of the lactose operon repressor and its complexes with DNA and inducer. *Science* **271**, 1247–1254.

Regulator-DNA interactions

Lin, S.-Y. and Riggs, A. D. (1975).The general affinity of *lac* repressor for *E. coli* DNA: implications for gene regulation in prokaryotes and eukaryotes. *Cell* **4**, 107–111.

Hu, M. C.-t. and Davidson, N. (1987). The inducible *lac* operator-repressor system is functional in mammalian cells. *Cell* **48**, 555–566.

Hildebrandt, E. R. *et al.* (1995). Comparison of recombination *in vitro* and in *E. coli* cells: measure of the effective concentration of DNA *in vivo*. *Cell*, **81**, 331–340.

Autogenous control

Baughman, G. and Nomura, M. (1983). Localization of the target site for translational regulation of the L11 operon and direct evidence for translational coupling in *E. coli*. *Cell* **34**, 969–988.

Cyclic AMP

Gaston, K. A. *et al.* (1990). Stringent spacing requirements for transcription activation by CRP. *Cell* **62**, 733–743.

Regulatory nucleotides

Cashel, M. and Gallant, J. (1969). Two compounds implicated in the function of the *RC* gene of *E. coli*. *Nature* **221**, 838–841.

Haseltine, W. A. and Block, R. (1973). Synthesis of guanosine tetra and pentaphosphate requires the presence of a codon specific uncharged tRNA in the acceptor site of ribosomes. *Proc. Nat. Acad. Sci. USA* **70**, 1564–1568.

Attenuation

Lee, F. and Yanofsky, C. (1977). Transcription termination at the *trp* operon attenuators of *E. coli* and *S. typhimurium*: RNA secondary structure and regulation of termination. *Proc. Nat. Acad. Sci. USA* **74**, 4365–4368.

Zurawski, G. *et al.* (1978). Translational control of transcription termination at the attenuator of the *E. coli* tryptophan operon. *Proc. Nat. Acad. Sci. USA* **75**, 5988–5991.

Antisense RNA

Izant, J. G. and Weintraub, H. (1984). Inhibition of thymidine kinase gene expression by antisense RNA: a molecular approach to genetic analysis. *Cell* **36**, 1007–1015.

Chapter 11

Phage strategies

SOME phages have only a single strategy for survival. On infecting a susceptible host, they subvert its functions to the purpose of producing a large number of progeny phage particles. As the result of this **lytic infection,** the host bacterium dies. In the typical lytic cycle, the phage DNA (or RNA) enters the host bacterium, its genes are transcribed in a set order, the phage genetic material is replicated, and the protein components of the phage particle are produced. Finally, the host bacterium is broken open (**lysed**) to release the assembled progeny particles.

Other phages have a dual existence. They are able to perpetuate themselves via the same sort of lytic cycle in what amounts to an open strategy for producing as many copies of the phage as rapidly as possible. But they also have an alternative form of existence, in which the phage genome is present in the bacterium in a latent form known as **prophage.** This form of propagation is called **lysogeny.**

In a lysogenic bacterium, the prophage is **integrated** into the bacterial genome, and is inherited in the same way as bacterial genes. By virtue of its possession of a prophage, a lysogenic bacterium has **immunity** against infection by further phage particles of the same type. Immunity is established by a single integrated prophage, so usually a bacterial genome contains only one copy of a prophage of any particular type.

Transitions occur between the lysogenic and lytic modes of existence. **Figure 11.1** shows that when a phage produced by a lytic cycle enters a new bacterial host cell, it either repeats the lytic cycle or enters the lysogenic state. The outcome depends on the conditions of infection and the genotypes of phage and bacterium.

A prophage is freed from the restrictions of lysogeny by the process called **induction,** in which it is **excised** from the bacterial genome to generate a free phage DNA that then proceeds through the lytic pathway.

The alternative forms in which these phages are propagated are determined by the regulation of transcription. Lysogeny is maintained by the interaction of a phage repressor with an operator. The lytic cycle requires a cascade of transcriptional controls. And the transition between the two lifestyles is accomplished by the establishment of repression (lytic cycle to lysogeny) or by the relief of repression (induction of lysogen to lytic phage).

Another type of existence within bacteria is represented by **plasmids.** These are autonomous units that exist in the cell as **extrachromosomal genomes.** Plasmids are self-replicating circular molecules of DNA that are maintained in the cell in a stable and characteristic number of copies; that is, the number remains constant from generation to generation.

Some plasmids also have alternative lifestyles. They can exist either in the autonomous extrachromosomal state; or they can be inserted into the bacterial chromosome, and then are carried as part of it like any other sequence. Such units are properly called **episomes** (but the terms "plasmid" and "episome" are sometimes used loosely as though interchangeable).

Figure 11.1 *Overview*: lytic development involves the reproduction of phage particles with destruction of the host bacterium, but lysogenic existence allows the phage genome to be carried as part of the bacterial genetic information.

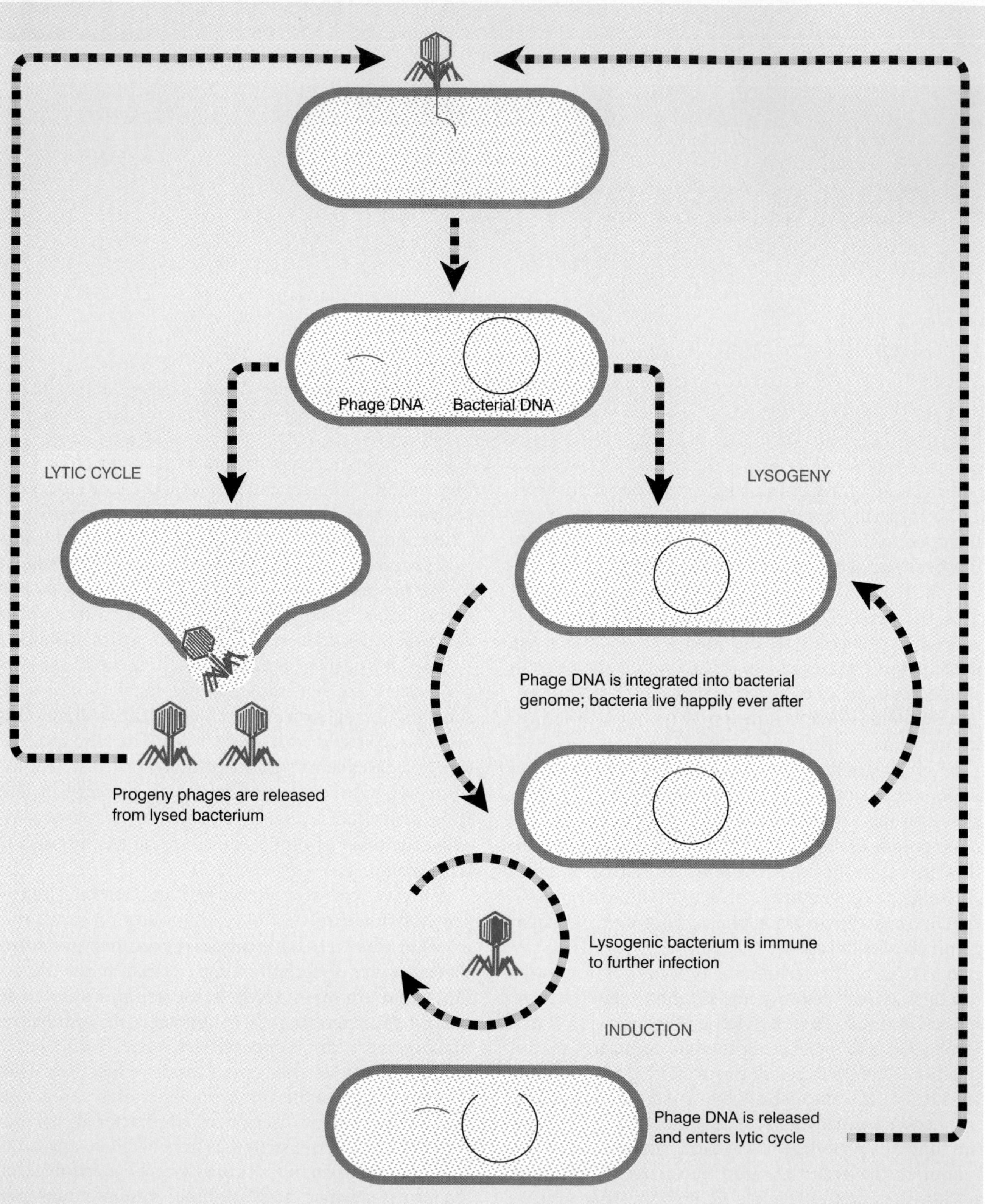
Phage DNA
Bacterial DNA
LYTIC CYCLE
LYSOGENY
Phage DNA is integrated into bacterial genome; bacteria live happily ever after
Progeny phages are released from lysed bacterium
Lysogenic bacterium is immune to further infection
INDUCTION
Phage DNA is released and enters lytic cycle

Figure 11.2 Several types of independent genetic units exist in bacteria.

Type of Unit	Genome Structure	Mode of Propagation	Consequences
Lytic phage	ds- or ss-DNA or RNA linear or circular	Infects susceptible host	Usually kills host
Lysogenic phage	ds-DNA	Linear sequence in host chromosome	Immunity to infection
Plasmid	ds-DNA circle	Replicates at defined copy number May be transmissible	Immunity to plasmids in same group
Episome	ds-DNA circle	Free circle or linear integrated	May transfer host DNA

Like lysogenic phages, plasmids and episomes maintain a selfish possession of their bacterium and often make it impossible for another element of the same type to become established. This effect also is called **immunity**, although the basis for plasmid immunity is different from lysogenic immunity. (We discuss the control of plasmid perpetuation in Chapter 12.)

Figure 11.2 summarizes the types of genetic units that can be propagated in bacteria as independent genomes. Lytic phages may have genomes of any type of nucleic acid; they transfer between cells by release of infective particles. Lysogenic phages have double-stranded DNA genomes, as do plasmids and episomes. Some plasmids and episomes transfer between cells by a conjugative process (involving direct contact between donor and recipient cells). A feature of the transfer process in both cases is that on occasion some bacterial host genes are transferred with the phage or plasmid DNA, so these events play a role in allowing exchange of genetic information between bacteria.

Lytic development is controlled by a cascade

Phage genomes of necessity are small. As with all viruses, they are restricted by the need to package the nucleic acid within the protein coat. This limitation dictates many of the viral strategies for reproduction. Typically, a virus takes over the apparatus of the host cell, which then replicates and expresses phage genes instead of the bacterial genes.

Usually the phage includes genes whose function is to ensure preferential replication of phage DNA. These genes are concerned with the initiation of replication and may even include a new DNA polymerase. Changes are introduced in the capacity of the host cell to engage in transcription. They involve replacing the RNA polymerase or modifying its capacity for initiation or termination. The result is always the same: phage mRNAs are preferentially transcribed. So far as protein synthesis is concerned, usually the phage is content to use the host apparatus, redirecting its activities principally by replacing bacterial mRNA with phage mRNA.

Lytic development is accomplished by a pathway in which the phage genes are expressed in a particular order. This ensures that the right amount of each component is present at the appropriate time. The cycle can be divided into the two general parts illustrated in **Figure 11.3:**

- **Early infection** describes the period from entry of the DNA to the start of its replication.
- **Late infection** defines the period from the start of replication to the final step of lysing the bacterial cell to release progeny phage particles.

Figure 11.3 Lytic development takes place by producing phage genomes and protein particles that are assembled into progeny phages.

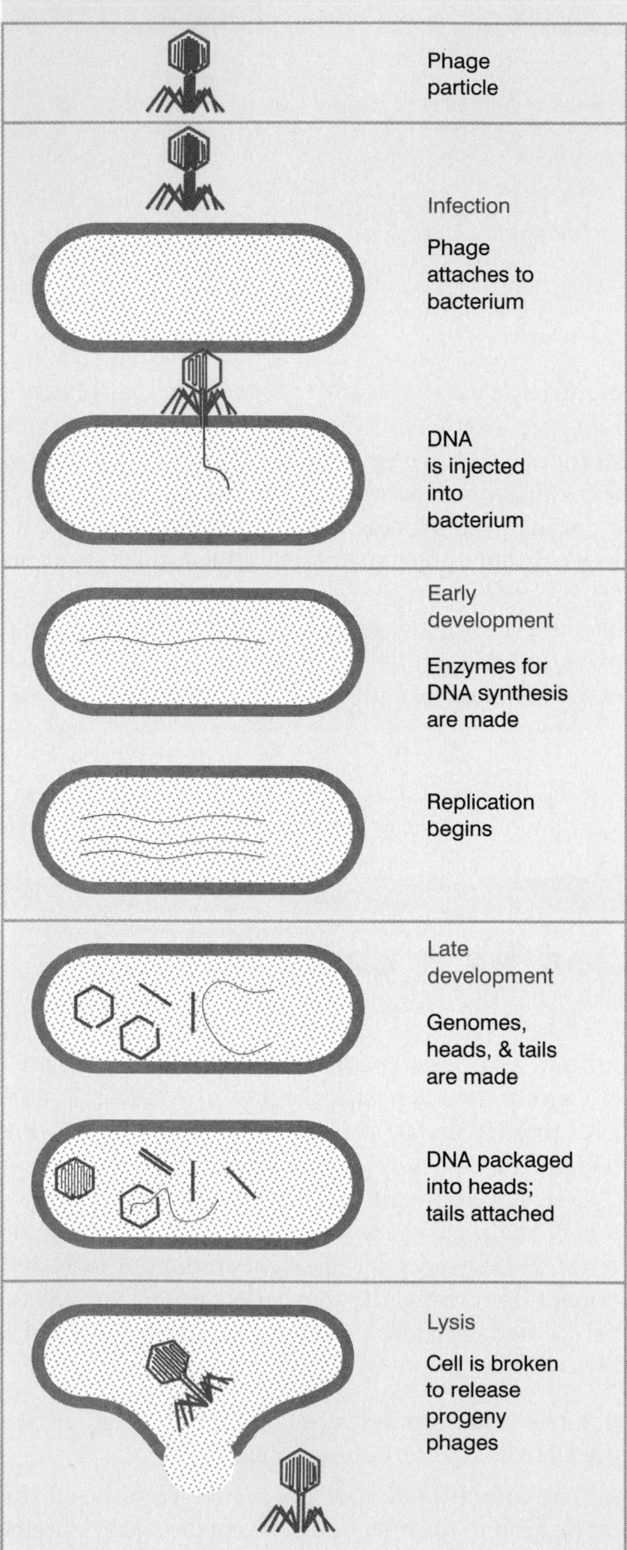

Figure 11.4 Phage lytic development proceeds by a regulatory cascade, in which a gene product at each stage is needed for expression of the genes at the next stage.

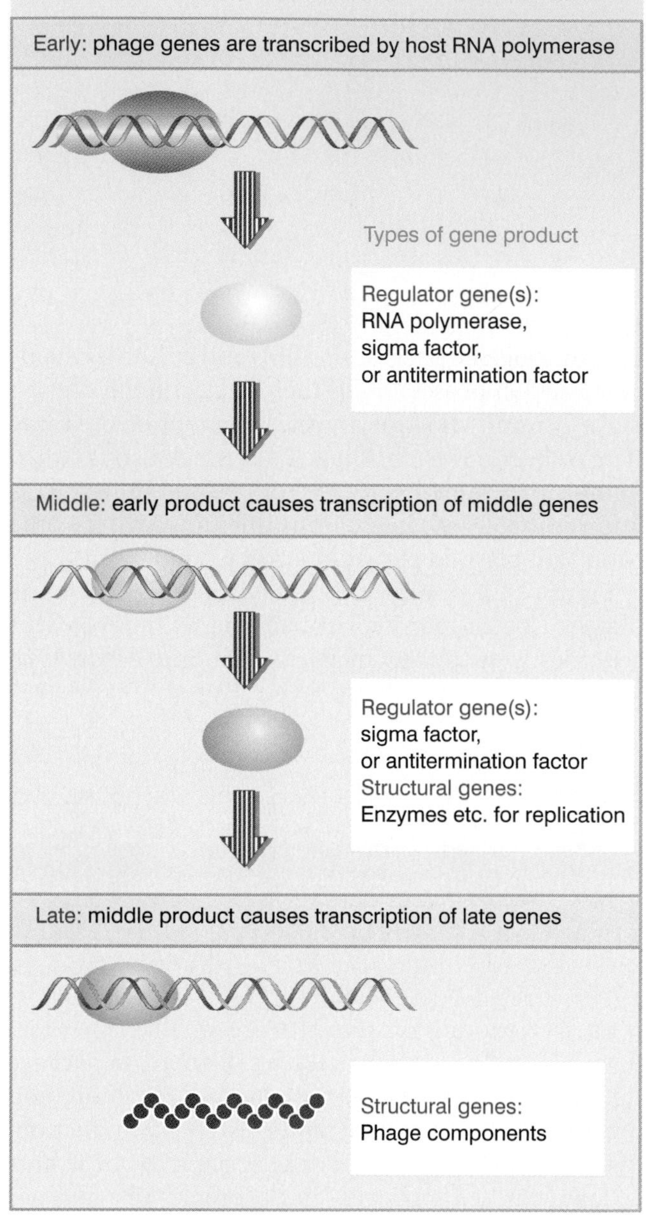

The early phase is devoted to the production of enzymes involved in the reproduction of DNA. These include the enzymes concerned with DNA synthesis, recombination, and sometimes modification. Their activities cause a **pool** of phage genomes to accumulate. In this pool, genomes are continually replicating and recombining, so that *the events of a single lytic cycle concern a population of phage genomes.*

During the late phase, the protein components of the phage particle are synthesized. Often many different proteins are needed to make up head and tail structures, so the largest part of the phage genome consists of late functions. In addition to the structural proteins, "assembly proteins" are needed to help construct the particle, although they are not themselves incorporated into it. By the time the structural components are assembling into heads and tails, replication of DNA has reached its maximum rate. The genomes then are inserted into the empty protein heads, tails are added, and the host cell is lysed to allow release of new viral particles.

The organization of the phage genetic map often reflects the sequence of lytic development. The concept of the operon is taken to somewhat of an extreme, in which the genes coding for proteins with related functions are clustered to allow their control with the maximum economy. This allows the pathway of lytic development to be controlled with a small number of regulatory switches.

The lytic cycle is under positive control, so that each group of phage genes can be expressed only when an appropriate signal is given. **Figure 11.4** shows that the regulatory genes function in a *cascade*, in which a gene expressed at one stage is necessary for synthesis of the genes that are expressed at the next stage. So at every stage of expression, one or more of the active genes is a regulator that is needed for the subsequent stage. The regulator may take the form of a new RNA polymerase, a sigma factor that redirects the specificity of the host RNA polymerase (see Chapter 9), or an antitermination factor that allows it to read a new group of genes (see Chapter 10).

The first stage of gene expression necessarily relies on the transcription apparatus of the host cell. Usually only a few genes are expressed at this stage. Their promoters are indistinguishable from those of host genes. The name of this class of genes depends on the phage. In most cases, they are known as the **early genes.** In phage lambda, they are given the evocative description of **immediate early.** Irrespective of the name, they constitute only a preliminary, representing just the initial part of the early period. Sometimes they are exclusively occupied with the transition to the next period. At all events, *one of these genes always codes for a protein that is necessary for transcription of the next class of genes.*

This second class of genes is known variously as the **delayed early** or **middle** group. Its expression typically starts as soon as the regulator protein coded by the early gene(s) is available. Depending on the nature of the control circuit, the initial set of early genes may or may not continue to be expressed at this stage (see Figure 9.32). Often the expression of host genes is reduced. Together the two sets of early genes account for all necessary phage functions except those needed to assemble the particle coat itself and to lyse the cell.

When the replication of phage DNA begins, it is time for the **late genes** to be expressed. Their transcription at this stage usually is arranged by embedding a further regulator gene within the previous (delayed early or middle) set of genes. This regulator may be another antitermination factor (as in lambda) or it may be another sigma factor (as in SPO1).

A division into three stages as shown in Figure 11.4 is common. The first stage consists of early genes transcribed by host RNA polymerase (sometimes the regulators are the only products at this stage). The second stage consists of genes transcribed under direction of the regulator produced in the first stage (most of these genes code for enzymes needed for replication of phage DNA). The final stage consists of genes for phage components, transcribed under direction of a regulator synthesized in the second stage.

The use of these successive controls, in which each set of genes contains a regulator that is necessary for expression of the next set, creates a cascade in which groups of genes are turned on (and sometimes off) at particular times. The means used to construct each phage cascade are different, but the results are similar, as the following sections show.

Functional clustering in phages T7 and T4

THE genome of phage T7 has three classes of genes, each constituting a group of adjacent loci. As **Figure 11.5** shows, the class I genes are the immediate early type, expressed by host RNA polymerase as soon as the phage DNA enters the cell. Among the products of these genes are enzymes that interfere with host gene expression and a phage RNA polymerase. The phage enzyme is responsible for expressing the class II genes (concerned principally with DNA synthesis functions) and the class III genes (concerned with assembling the mature phage particle).

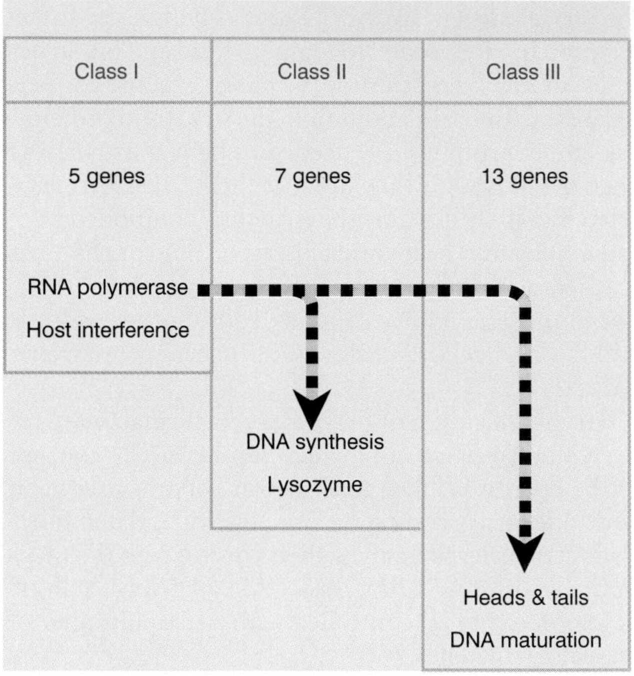

Phage T4 has one of the larger genomes (165 kb), organized with extensive functional grouping of genes. **Figure 11.6** presents the genetic map. Genes that are numbered are **essential:** a mutation in any one of these loci prevents successful completion of the lytic cycle. Genes indicated by three-letter abbreviations are **nonessential,** at least under the usual conditions of infection. We do not really understand the inclusion of many nonessential genes, but presumably they confer a selective advantage in some of T4's habitats. (In smaller phage genomes, most or all of the genes are essential.)

Figure 11.5 Phage T7 contains three classes of genes that are expressed sequentially. The genome is ~38 kb.

There are three phases of gene expression. A summary of the functions of the genes expressed at each stage is given in **Figure 11.7.** The early genes are transcribed by host RNA polymerase. The middle genes are also transcribed by host RNA polymerase, but two phage-encoded products, MotA and AsiA, are also required. The middle promoters lack a consensus –30 sequence, and instead have a binding sequence for MotA. The phage protein is an activator that compensates for the deficiency in the promoter by assisting host RNA polymerase to bind. (This is similar to a mechanism employed by phage lambda, which is illustrated later in Figure 11.25.) The early and middle genes account for virtually all of the phage functions concerned with the synthesis of DNA, modifying cell structure, and transcribing and translating phage genes.

The two essential genes in the "transcription" category fulfill a regulatory function: their products are necessary for late gene expression. Phage T4 infection depends on a mechanical link between replication and late gene expression. Only actively replicating DNA can be used as template for late gene transcription. The connection is generated by introducing a new sigma factor and also by making other modifications in the host RNA polymerase so that it is active only with a template of replicating DNA (because there is an obligatory link between RNA polymerase and the replication apparatus). This link establishes a correlation between the synthesis of phage protein components and the number of genomes available for packaging.

Figure 11.6 The map of T4 is circular. There is extensive clustering of genes coding for components of the phage and processes such as DNA replication, but there is also dispersion of genes coding for a variety of enzymatic and other functions. Essential genes are indicated by numbers; nonessential genes are identified by letters. Only some representative T4 genes are shown on the map.

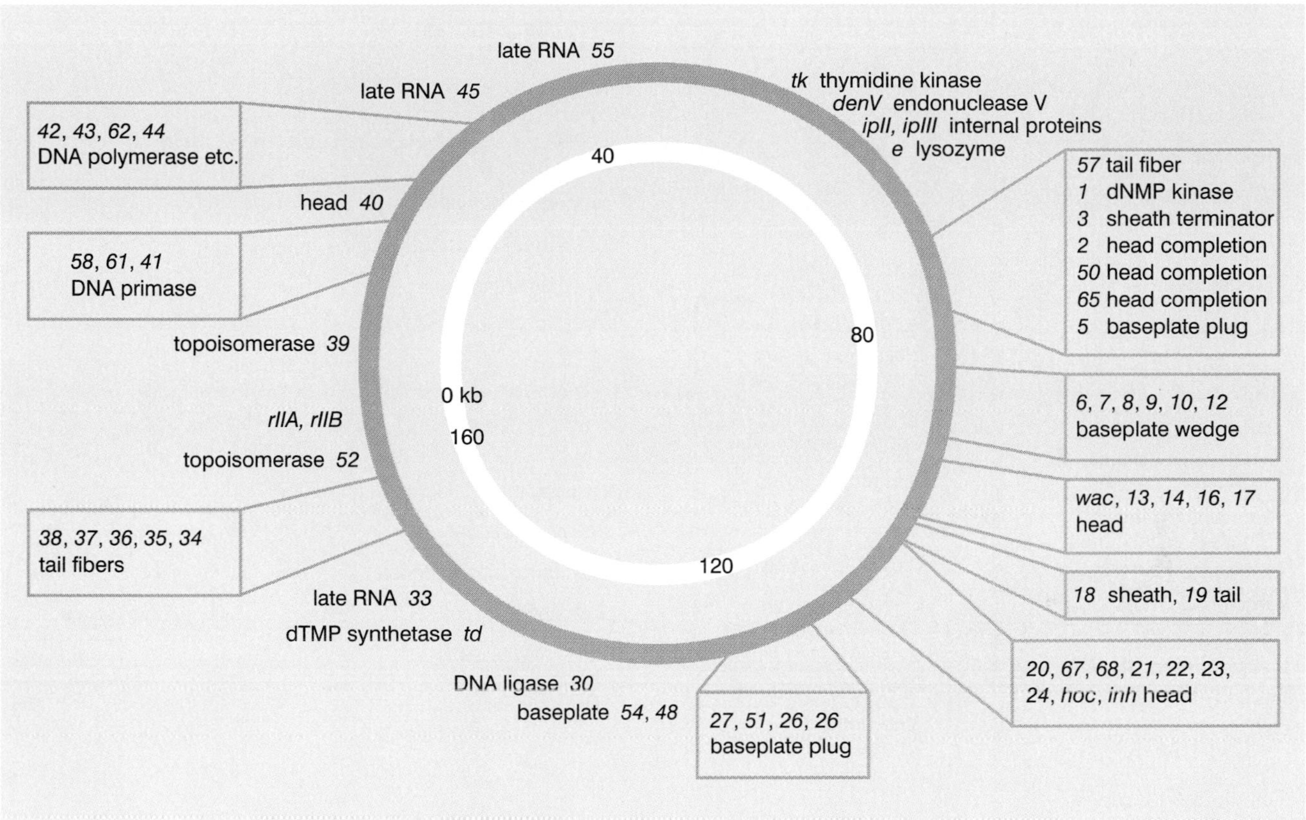

The lambda lytic cascade relies on antitermination

One of the most intricate cascade circuits is provided by phage lambda. Actually, the cascade for lytic development itself is straightforward, with two regulators controlling the successive stages of development. But the circuit for the lytic cycle is interlocked with the circuit for establishing lysogeny, as summarized in **Figure 11.8**.

When lambda DNA enters a new host cell, the lytic and lysogenic pathways start off the same way. Both require expression of the immediate early and delayed

Figure 11.7 The phage T4 lytic cascade falls into two parts: early and quasi-late functions are concerned with DNA synthesis and gene expression; late functions are concerned with particle assembly.

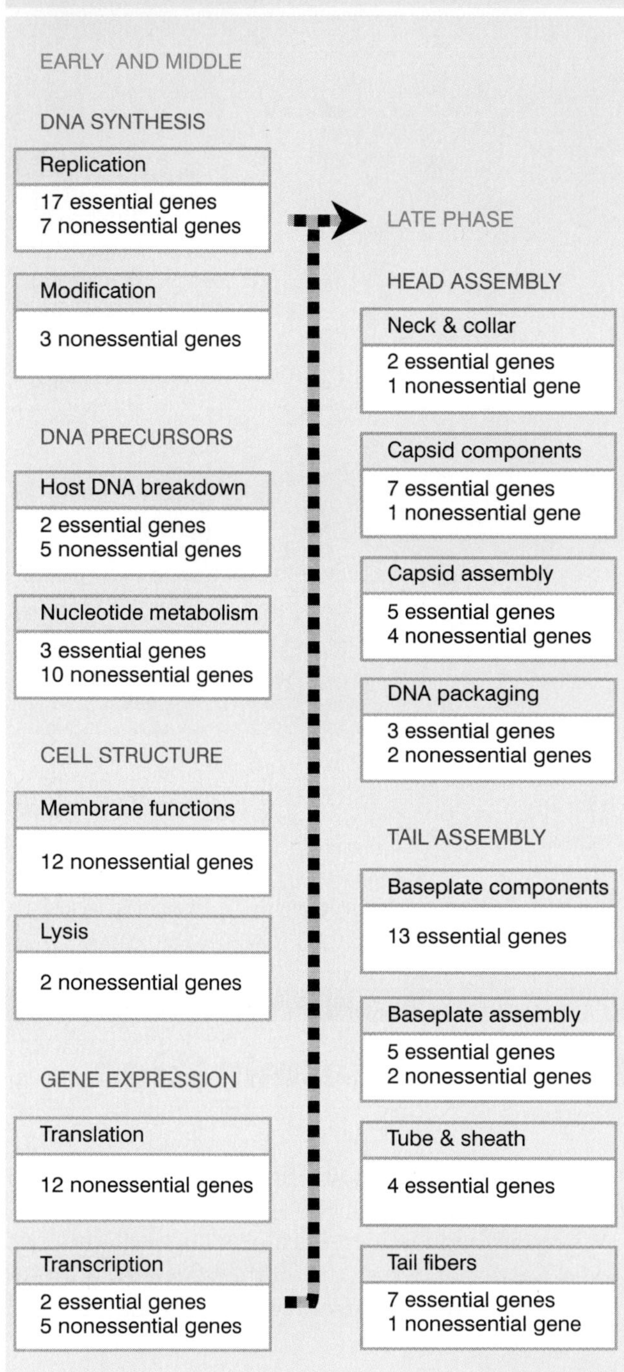

Figure 11.8 The lambda lytic cascade is interlocked with the circuitry for lysogeny.

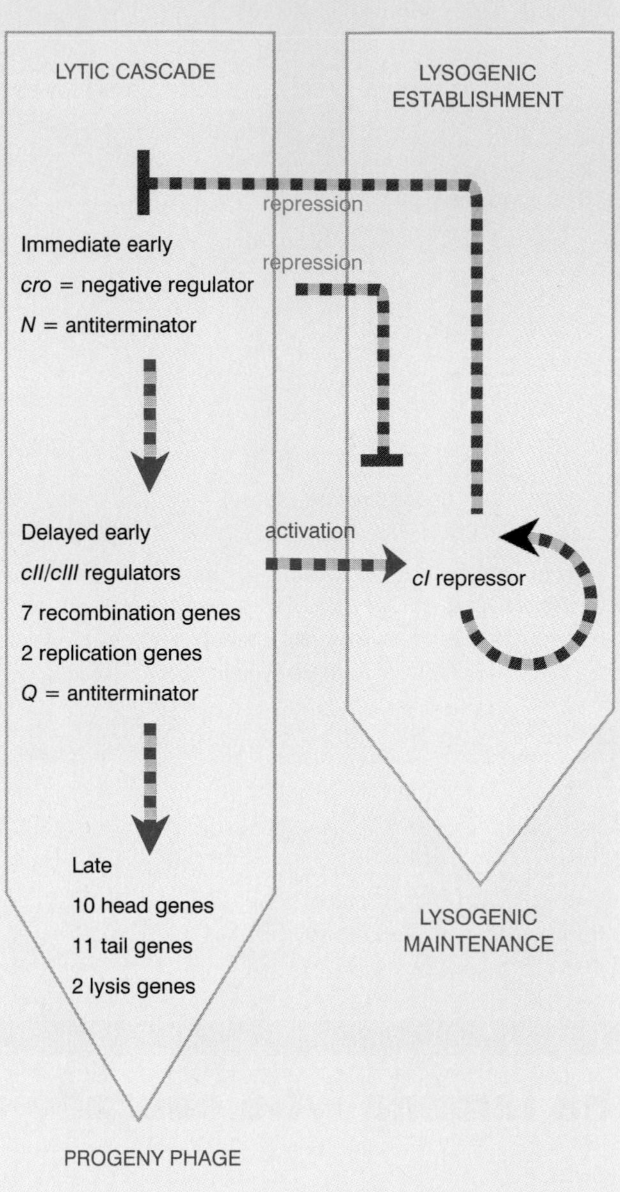

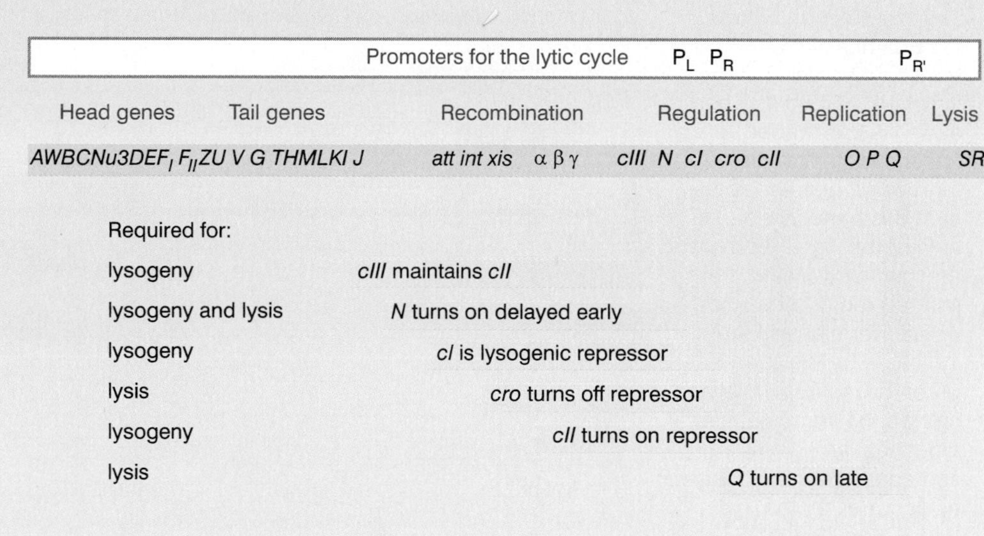

Figure 11.9 The lambda map shows clustering of related functions. The genome is 48,514 bp.

early genes. But then they diverge: lytic development follows if the late genes are expressed; lysogeny ensues if synthesis of the repressor is established.

Lambda has only two immediate early genes, transcribed independently by host RNA polymerase:

- *N* codes for an antitermination factor whose action at the *nut* sites allows transcription to proceed into the delayed early genes. We discussed the mechanisms involved in antitermination in Chapter 9.
- *cro* has dual functions: it prevents synthesis of the repressor (a necessary action if the lytic cycle is to proceed); and it turns off expression of the immediate early genes (which are not needed later in the lytic cycle).

The delayed early genes include two replication genes (needed for lytic infection), seven recombination genes (some involved in recombination during lytic infection, two necessary to integrate lambda DNA into the bacterial chromosome for lysogeny), and three regulators. The regulators have opposing functions:

- The *cII-cIII* pair of regulators is needed to start up the synthesis of repressor.
- The *Q* regulator is an antitermination factor that allows host RNA polymerase to proceed into the late genes.

So the delayed early genes serve two masters: some are needed for the phage to enter lysogeny, the others are concerned with controlling the order of the lytic cycle.

To disentangle the two pathways, first consider just the lytic cycle. **Figure 11.9** gives the map of lambda phage DNA. A group of genes concerned with regulation is surrounded by genes needed for recombination and replication. The genes coding for structural components of the phage are clustered. All of the genes necessary for the lytic cycle are expressed in polycistronic transcripts from three promoters.

Figure 11.10 shows that there are two immediate early genes, *N* and *cro*, which are transcribed by host

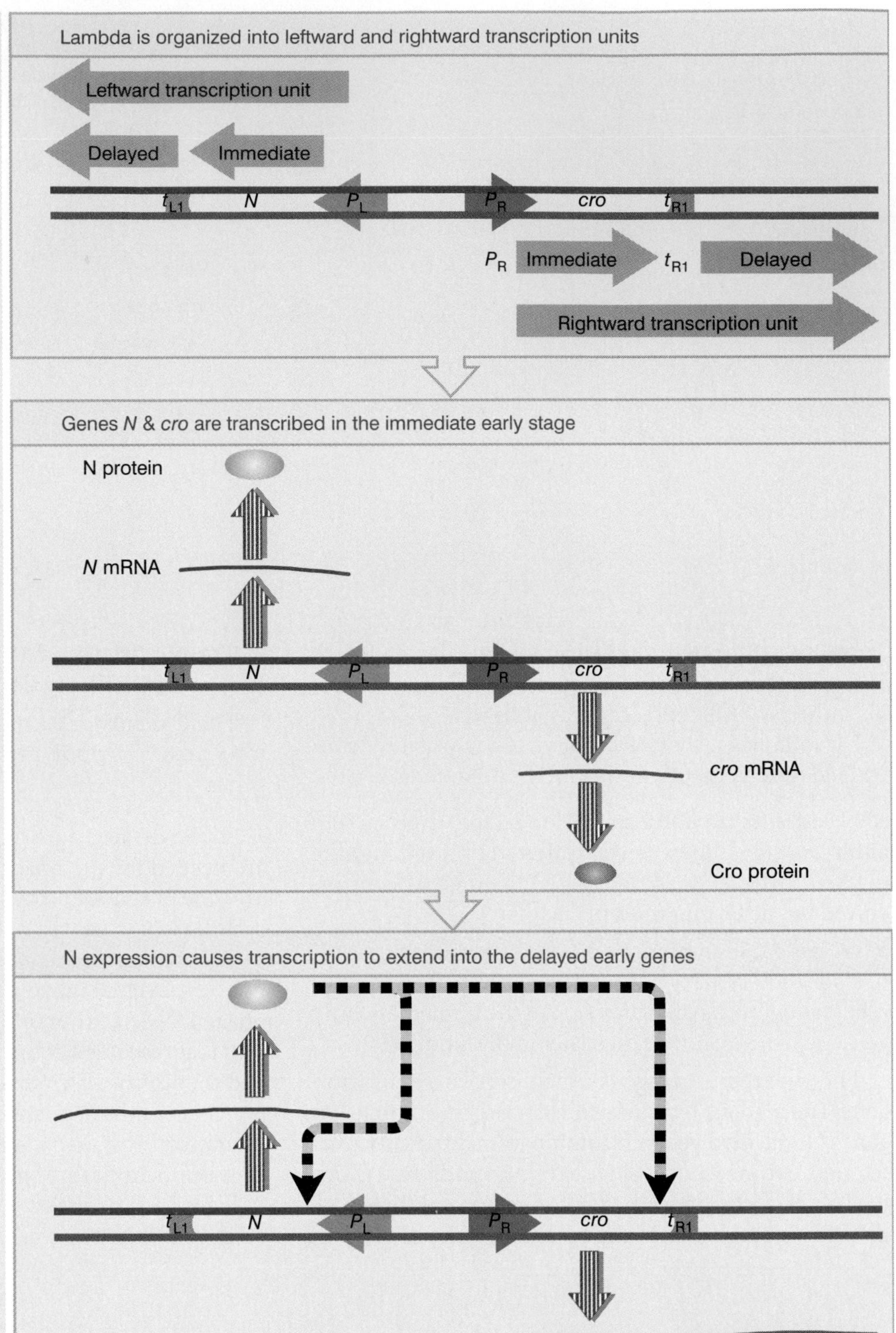

Figure 11.10 Phage lambda has two early transcription units; in the 'leftward' unit, the 'upper' strand is transcribed toward the left; in the 'rightward' unit, the 'lower' strand is transcribed toward the right. Promoters are indicated by the shaded red or blue arrowheads. Terminators are indicated by the shaded green boxes. Genes *N* and *cro* are the immediate early functions, and are separated from the delayed early genes by the terminators. Synthesis of N protein allows RNA polymerase to pass the terminators t_{L1} to the left and t_{R1} to the right.

Figure 11.11 Lambda DNA circularizes during infection, so that the late gene cluster is intact in one transcription unit.

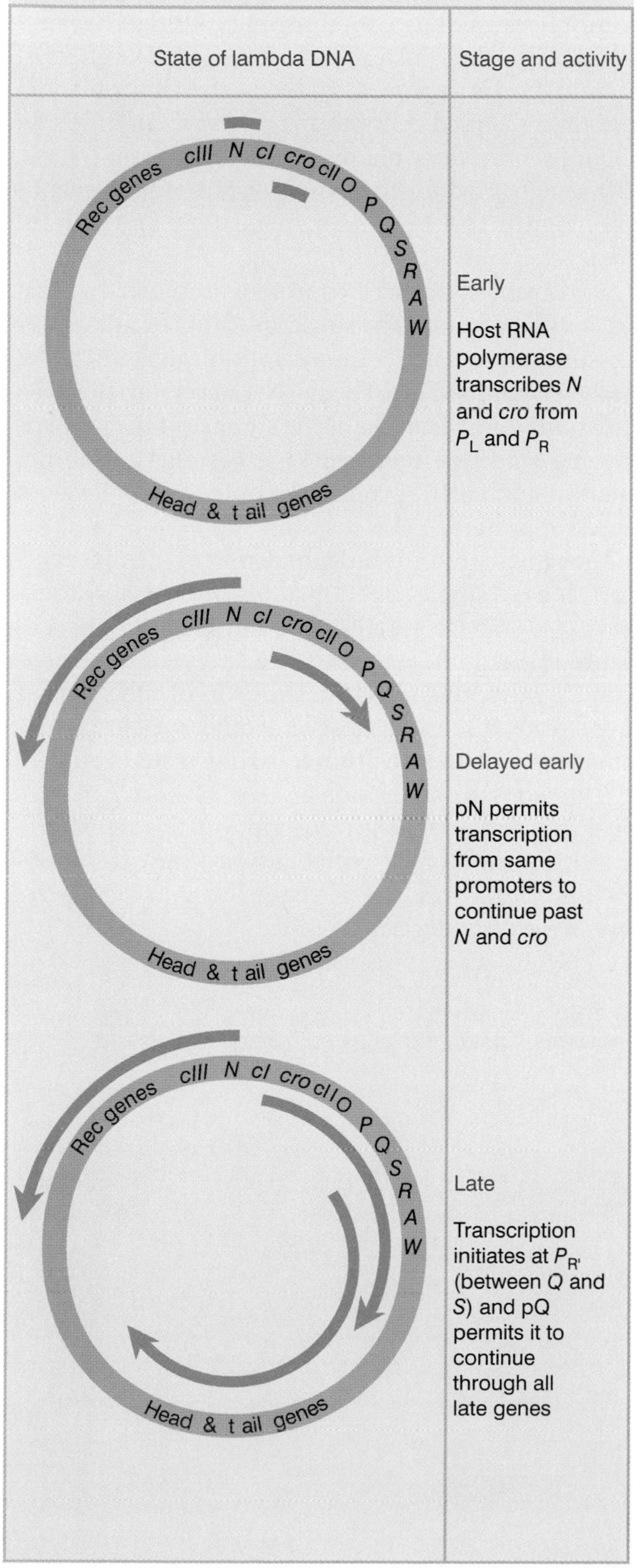

State of lambda DNA	Stage and activity
	Early Host RNA polymerase transcribes *N* and *cro* from P_L and P_R
	Delayed early pN permits transcription from same promoters to continue past *N* and *cro*
	Late Transcription initiates at $P_{R'}$ (between *Q* and *S*) and pQ permits it to continue through all late genes

RNA polymerase. *N* is transcribed toward the left, and *cro* toward the right. Each transcript is terminated at the end of the gene. pN is the regulator that allows transcription to continue into the delayed early genes. It is an antitermination factor that suppresses use of the terminators t_L and t_R. (Its mechanism is discussed in Chapter 9.) In the presence of pN, transcription continues to the left of *N* into the recombination genes, and to the right of *cro* into the replication genes.

The map in Figure 11.9 gives the organization of the lambda DNA as it exists in the phage particle. But shortly after infection, the ends of the DNA join to form a circle. **Figure 11.11** shows the true state of lambda DNA during infection. The late genes are welded into a single group, containing the lysis genes *S-R* from the right end of the linear DNA, and the head and tail genes *A-J* from the left end.

The late genes are expressed as a single transcription unit, starting from a promoter $P_{R'}$ that lies between *Q* and *S*. The late promoter is used constitutively. However, in the absence of the product of gene *Q* (which is the last gene in the rightward delayed early unit), late transcription terminates at a site t_{R3}. The transcript resulting from this termination event is 194 bases long; it is known as 6S RNA. When pQ becomes available, it suppresses termination at t_{R3} and the 6S RNA is extended, with the result that the late genes are expressed.

Late gene transcription does not seem to terminate at any specific point, but continues through all the late genes into the region beyond. A similar event happens with the leftward delayed early transcription, which continues past the recombination functions. Transcription in each direction is probably terminated before the polymerases can crash into each other.

Lysogeny is maintained by an autogenous circuit

LOOKING at the lambda lytic cascade, we see that the entire program is set in train by initiating transcription at the two promoters P_L and P_R for the immediate early genes *N* and *cro*. Because lambda uses antitermination to proceed to the next stage of (delayed early) expression, the same two promoters continue to be used right through the early period.

The expanded map of the regulatory region drawn in **Figure 11.12** shows that the promoters P_L and P_R lie on either side of the *cI* gene. Associated with each promoter is an operator (O_L, O_R) at which repressor protein binds to prevent RNA polymerase from initiating transcription. The sequence of each operator overlaps with the promoter that it controls; so often these are described as the P_L/O_L and P_R/O_R control regions.

Because of the sequential nature of the lytic cascade, the control regions provide a pressure point at which entry to the entire cycle can be controlled. *By denying RNA polymerase access to these promoters, the repressor prevents the phage genome from entering the lytic cycle.*

The repressor protein is coded by the *cI* gene. Mutants in this gene cannot maintain lysogeny, but always enter the lytic cycle. The name of the gene reflects the phenotype of the resulting infection.

When a bacterial culture is infected with a phage, the cells are lysed to generate regions that can be seen on a culture plate as small areas of clearing called **plaques.** With wild-type phages, the plaques are turbid or cloudy, because they contain some cells that have established lysogeny instead of being lysed. The effect of a *cI* mutation is to prevent lysogeny, so that the plaques contain only lysed cells. As a result, such an infection generates only **clear plaques,** and three genes (*cI, cII, cIII*) were named for their involvement in this phenotype. **Figure 11.13** compares wild-type and mutant plaques.

The *cI* gene is transcribed from a promoter P_{RM} that lies at its right end. (The subscript "RM" stands for repressor maintenance.) Transcription is terminated at the left end of the gene. The mRNA starts with the AUG initiation codon; because of the absence of the usual ribosome binding site, the mRNA is translated somewhat inefficiently, producing only a low level of repressor protein.

The repressor binds independently to the two operators. It has a single function at O_L, but has dual functions at O_R. These are illustrated in the upper part of **Figure 11.14.**

At O_L the repressor has the same sort of effect that we have already discussed for several other systems: it prevents RNA polymerase from initiating transcription at P_L. This stops the expression of gene *N*. Since P_L is used for all leftward early gene transcription, this action prevents expression of the entire leftward early transcription unit. *So the lytic cycle is stymied before it can proceed beyond the early stages.*

Figure 11.12 The lambda regulatory region contains a cluster of *trans*-acting functions and *cis*-acting elements.

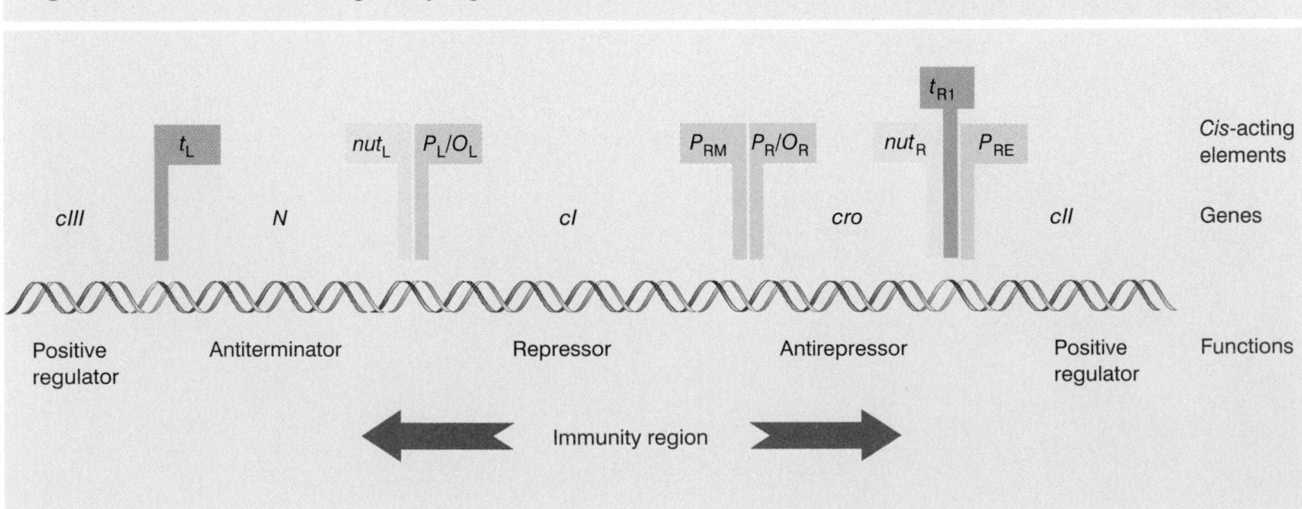

Figure 11.13 Wild-type lambda cultures form cloudy plaques (left panel); mutants that cannot lysogenize can be detected by their clear plaques (right panel). Photograph kindly provided by Dale Kaiser.

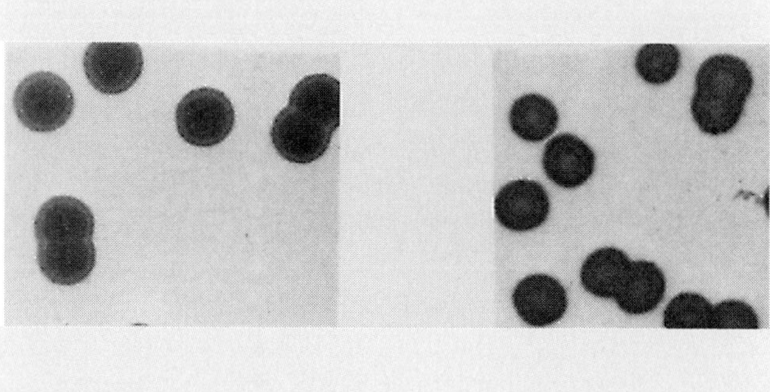

At O_R, repressor binding prevents the use of P_R. So *cro* and the other rightward early genes cannot be expressed. (We see later why it is important to prevent the expression of *cro* when lysogeny is being maintained.)

But the presence of repressor at O_R also has another effect. The promoter for repressor synthesis, P_{RM}, is adjacent to the rightward operator O_R. It turns out that *RNA polymerase can initiate efficiently at P_{RM} only when repressor is bound at O_R*. The repressor behaves as a positive regulator protein that is necessary for transcription of the *cI* gene. *Since the repressor is the product of* cI, *this interaction creates a positive autogenous circuit, in which the presence of repressor is necessary to support its own continued synthesis.*

The nature of this control circuit explains the biological features of lysogenic existence. Lysogeny is stable because the control circuit ensures that, so long as the level of repressor is adequate, there is continued expression of the *cI* gene. The result is that O_L and O_R remain occupied indefinitely. By repressing the entire lytic cascade, this action maintains the prophage in its inert form.

The presence of repressor explains the phenomenon of immunity. If a second lambda phage DNA enters a lysogenic cell, repressor protein synthesized from the resident prophage genome will immediately bind to O_L and O_R in the new genome. This prevents the second phage from entering the lytic cycle.

The operators were originally identified as the targets for repressor action by **virulent** mutations (λ*vir*). These mutations prevent the repressor from binding at O_L or O_R, with the result that the phage inevitably proceeds into the lytic pathway when it infects a new host bacterium. And λ*vir* mutants can grow on lysogens because the virulent mutations in O_L and O_R allow the incoming phage to ignore the resident repressor and thus to enter the lytic cycle. Virulent mutations in phages are the equivalent of operator-constitutive mutations in bacterial operons.

Prophage is induced to enter the lytic cycle when the lysogenic circuit is broken. This happens when the repressor is inactivated (see next section). The absence of repressor allows RNA polymerase to bind at P_L and P_R, starting the lytic cycle as shown in the lower part of Figure 11.14.

The autogenous nature of the repressor-maintenance circuit creates a sensitive response. Because the presence of repressor is necessary for its own synthesis, expression of the *cI* gene stops as soon as the existing repressor is destroyed. So no repressor is synthesized to replace the molecules that have been damaged. So the lytic cycle can start without interference from the circuit that maintains lysogeny.

The region including the left and right operators, the *cI* gene, and the *cro* gene determines the immunity of the phage. Any phage that possesses this region has the same type of immunity, because *it specifies both the repressor protein and the sites on which the repressor acts.* Accordingly, this is called the **immunity region** (as marked in Figure 11.12). Each of the four lambdoid phages ϕ80, *21*, *434*, and λ has a unique immunity region. When we say that a lysogenic phage confers immunity to any other phage of the same type, we mean more precisely that the immunity is to any other phage that has the same immunity region (irrespective of differences in other regions).

Figure 11.14 Lysogeny is maintained by an autogenous circuit (upper). If this circuit is interrupted, the lytic cycle starts (lower).

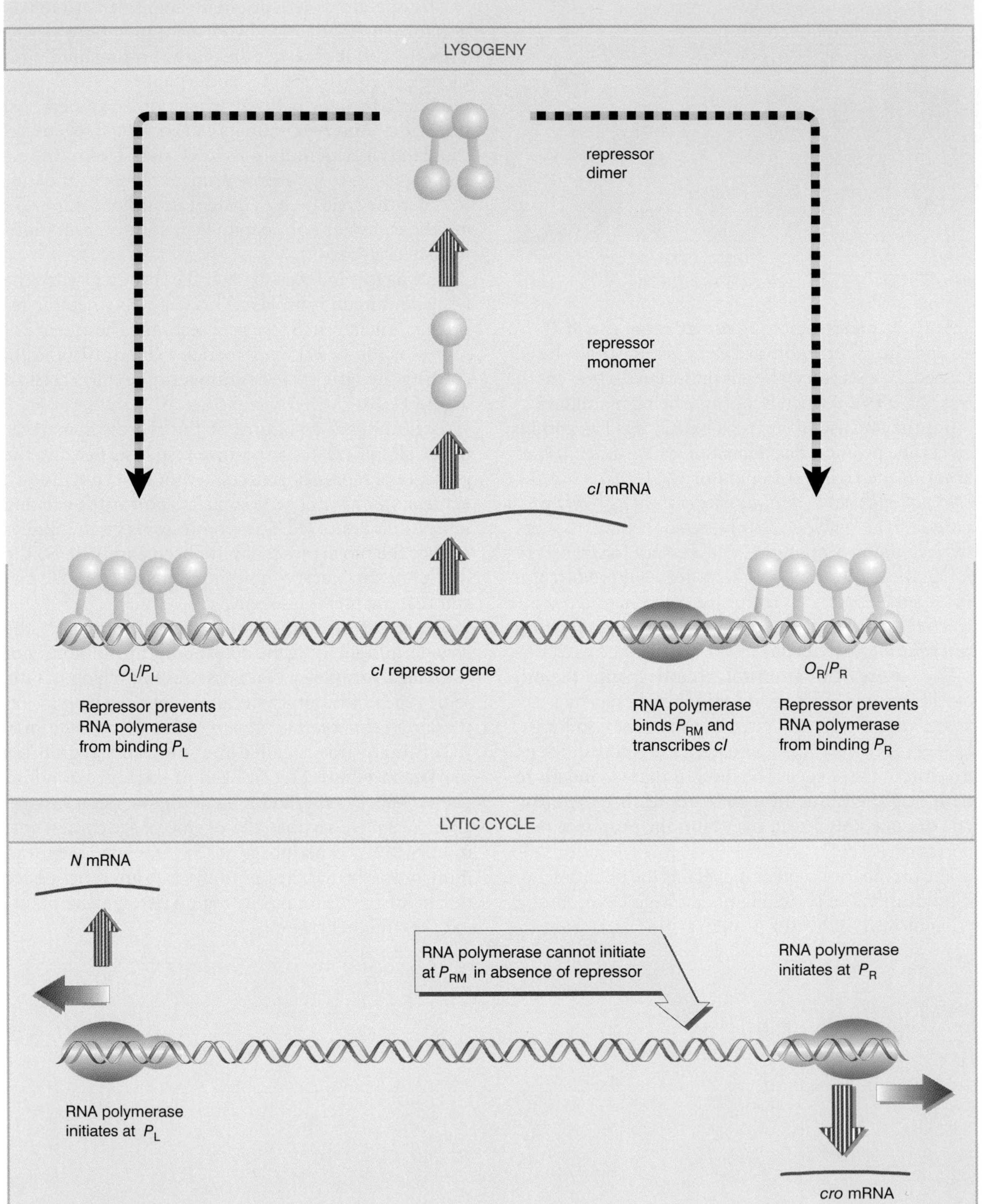
LYSOGENY
repressor dimer
repressor monomer
cI mRNA
O_L/P_L
cI repressor gene
O_R/P_R
Repressor prevents RNA polymerase from binding P_L
RNA polymerase binds P_{RM} and transcribes cI
Repressor prevents RNA polymerase from binding P_R
LYTIC CYCLE
N mRNA
RNA polymerase cannot initiate at P_{RM} in absence of repressor
RNA polymerase initiates at P_R
RNA polymerase initiates at P_L
cro mRNA

The DNA-binding form of repressor is a dimer

THE repressor subunit is a polypeptide of 27 kD with the two distinct domains summarized in **Figure 11.15.**

- The N-terminal domain, residues 1–92, provides the operator-binding site.
- The C-terminal domain, residues 132–236, is responsible for dimerization.

The two domains are joined by a connector of 40 residues. When repressor is digested by a protease, each domain is released as a separate fragment.

Each domain can exercise its function independently of the other. The C-terminal fragment can form oligomers. The N-terminal fragment can bind the operators, although with a lower affinity than the intact repressor. So the information for specifically contacting DNA is contained within the N-terminal domain, but the efficiency of the process is enhanced by the attachment of the C-terminal domain.

The dimeric structure of the repressor is crucial in maintaining lysogeny. The induction of a lysogenic prophage to enter the lytic cycle is caused by cleavage of the repressor subunit in the connector region, between residues 111 and 113. (This is a counterpart to the allosteric change in conformation that results when a small-molecule inducer inactivates the repressor of a bacterial operon, a capacity that the lysogenic repressor does not have.) Induction occurs under certain adverse conditions, such as exposure of lysogenic bacteria to UV irradiation, which leads to proteolytic inactivation of the repressor.

Figure 11.16 Repressor dimers bind to the operator. The affinity of the N-terminal domains for DNA is controlled by the dimerization of the C-terminal domains.

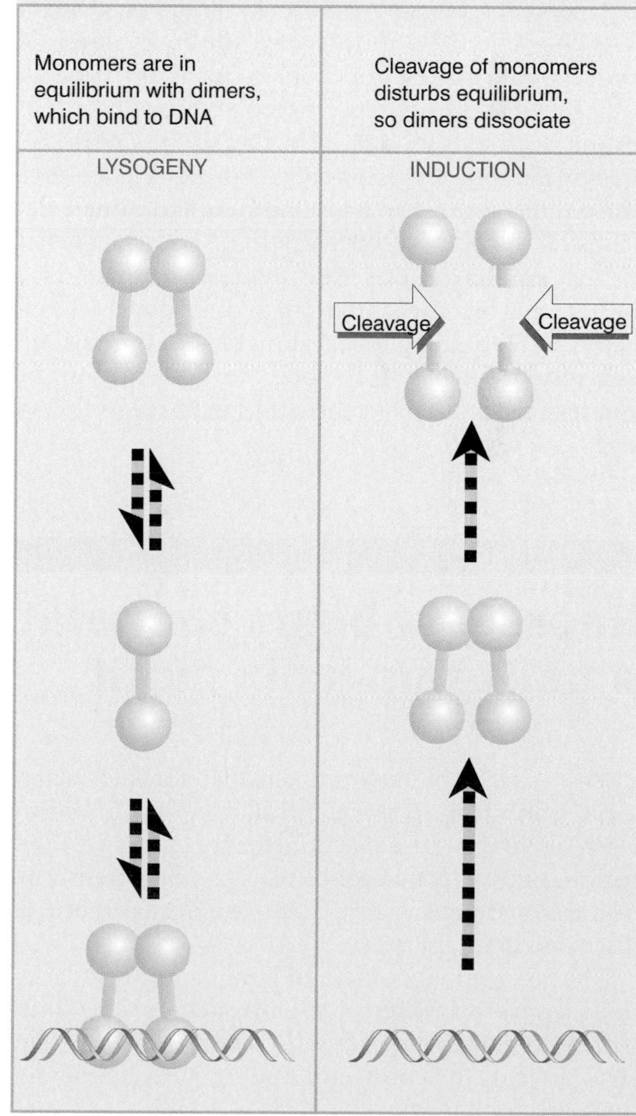

Figure 11.15 The N-terminal and C-terminal regions of repressor form separate domains. The C-terminal domains associate to form dimers; the N-terminal domains bind DNA.

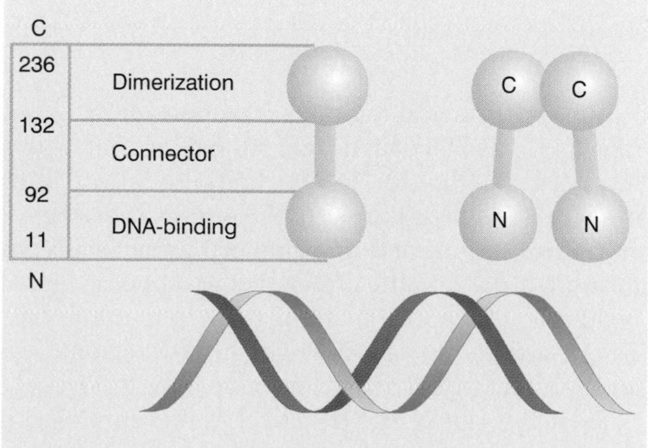

In the intact state, dimerization of the C-terminal domains ensures that when the repressor binds to DNA its two N-terminal domains each contact DNA simultaneously. But cleavage releases the C-terminal domains from the N-terminal domains. As illustrated in **Figure 11.16** this means that the N-terminal domains can no longer dimerize; this upsets the equilibrium between monomers and dimers, so that repressor dissociates from DNA, allowing lytic infection to start. (Another relevant parameter is the loss of cooperative effects between adjacent dimers: see later.)

The balance between lysogeny and the lytic cycle depends on the concentration of repressor. Intact repressor is present in a lysogenic cell at a concentration sufficient to ensure that the operators are occupied. But if the repressor is cleaved, this concentration is inadequate, because of the lower affinity of the separate N-terminal domain for the operator. Too high a concentration of repressor would make it impossible to induce the lytic cycle in this way; too low a level, of course, would make it impossible to maintain lysogeny.

The dependence of repression on repressor concentration is strongly influenced by the behavior of the repressor. Lambda repressor only binds DNA as a dimer. The equilibrium between monomers and dimers depends on the protein concentration. The consequence of the equilibrium is that repressor exists as a monomer at low concentrations and as a dimer (the effective DNA-binding form) at higher concentrations. As a result, binding to the operator depends on the square of the monomer concentration, that is, follows a second-order equilibrium, as shown in **Figure 11.17.** The usual concentration of lambda repressor in a lysogen is ~200 molecules/cell, which corresponds to ~4× 10^{-7} M, several times the concentration needed to ensure complete repression. But a drop in level of ~10× will effectively release repression and lead to induction.

Figure 11.17 Lambda repressor binds to operators with second-order kinetics.

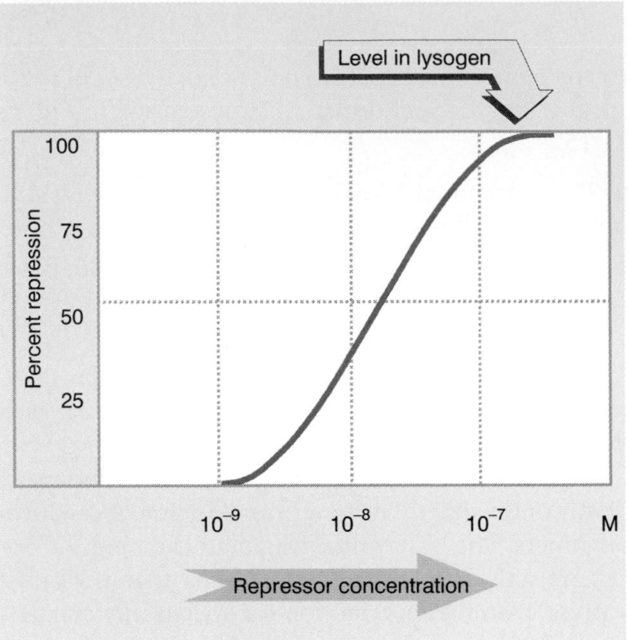

Repressor binds cooperatively at each operator using a helix-turn-helix motif

Several DNA-binding proteins that regulate bacterial transcription share a similar mode of holding DNA, in which the active domain contains two short regions of α-helix that contact DNA. (Some transcription factors in eukaryotic cells use a similar motif, as discussed in Chapter 21.)

The N-terminal domain of lambda repressor contains several stretches of α-helix, arranged as illustrated diagrammatically in **Figure 11.18.** The structures of the connector and the C-terminal domain are not known.

Two of the helical regions are responsible for binding DNA. The **helix-turn-helix model** for contact is illustrated in **Figure 11.19.** Looking at a single monomer, α-helix-3 consists of 9 amino acids, lying at an angle to the preceding region of 7 amino acids that forms α-helix-2. In the dimer, the two apposed helix-3 regions lie 34 Å apart, enabling them to fit into successive major grooves of DNA. The helix-2 regions lie at an angle that would place them across the groove.

Related forms of the α-helical motifs employed in these two regions are found in several DNA-binding proteins, including CAP, the *lac* repressor, and several phage repressors. Each helix plays a distinct role in binding the operators:

■ Contacts between helix-3 and DNA rely on hydrogen bonds between the amino acid side chains and the exposed positions of the base pairs. This helix is responsible for recognizing the specific target DNA sequence, and is therefore also known as the **recognition helix.**

■ Contacts from helix-2 to the DNA take the form of hydrogen bonds connecting with the phosphate backbone. These interactions are necessary for binding, but do not control the specificity of target recognition. In addition to these contacts, a large part of the overall energy of interaction with DNA is provided by ionic interactions with the phosphate backbone.

Figure 11.18 Lambda repressor's N-terminal domain contains five stretches of α–helix; helices 2 and 3 are involved in binding DNA.

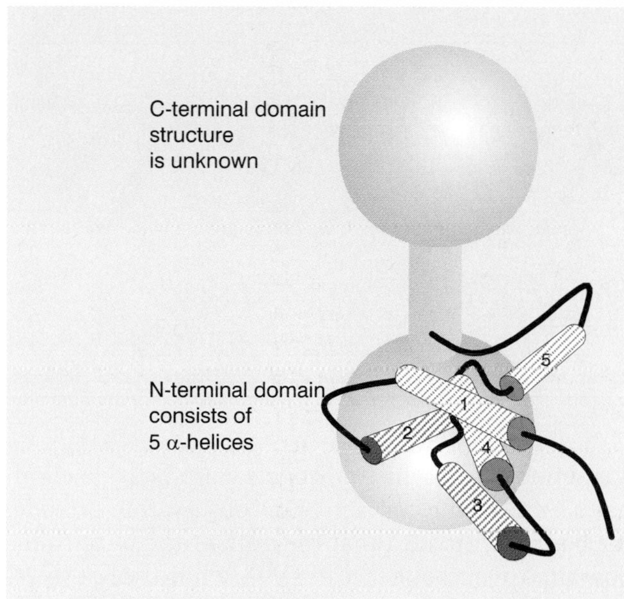

What happens if we manipulate the coding sequence to construct a new protein by substituting the recognition helix in one repressor with the corresponding sequence from a closely related repressor? The specificity of the hybrid protein is that of its new recognition helix. *The amino acid sequence of this short region determines the sequence specificities of the individual proteins, and is able to act in conjunction with the rest of the polypeptide chain.*

Figure 11.20 shows the details of the binding to DNA of two proteins that bind similar DNA sequences. Both lambda repressor and Cro protein have a similar organization of the helix-turn-helix motif, although their individual specificities for DNA are not identical:

■ Each protein uses similar interactions between hydrophobic amino acids to maintain the relationship between helix-2 and helix-3: repressor has an Ala-Val connection, while Cro has an Ala-Ile association.

■ Amino acids in helix-3 of the repressor make contacts with specific bases in the operator. Three amino acids in repressor recognize three bases in DNA; the

Figure 11.19 In the two-helix model for DNA binding, helix-3 of each monomer lies in the wide groove on the same face of DNA, and helix-2 lies across the groove.

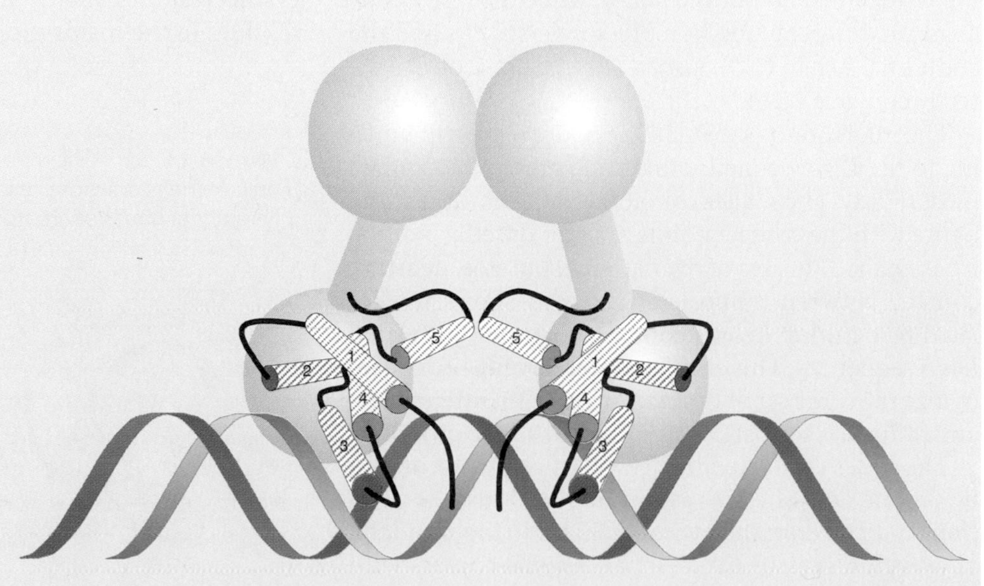

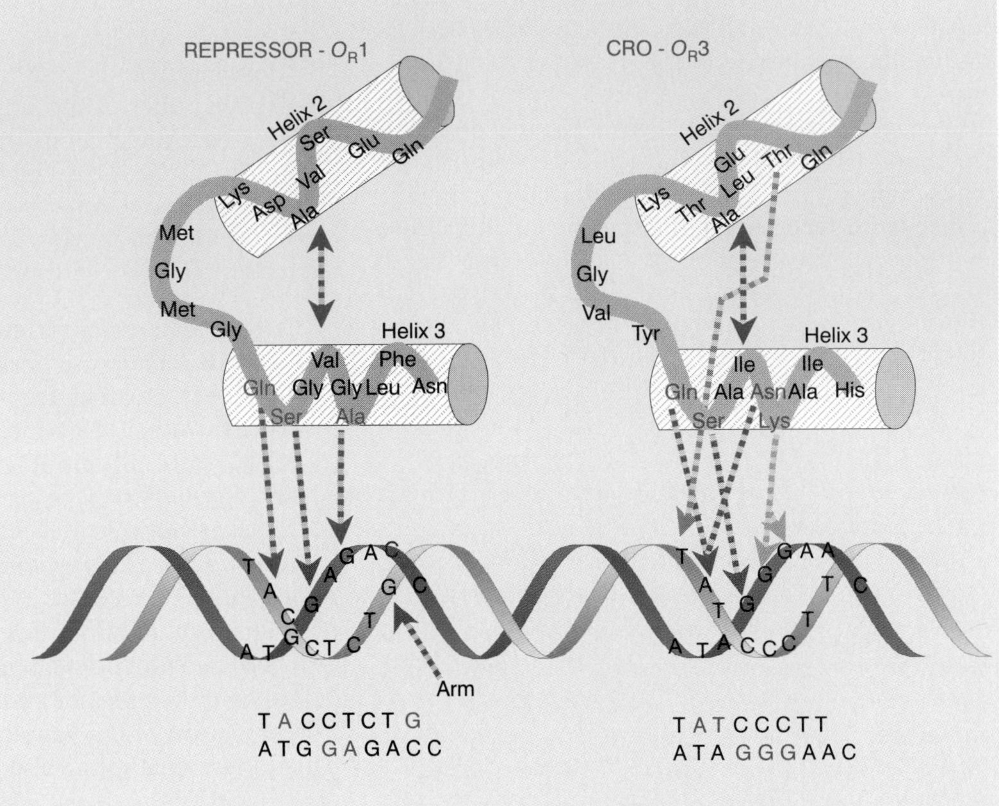

Figure 11.20 Two proteins that use the two-helix arrangement to contact DNA recognize lambda operators with affinities determined by the amino acid sequence of helix-3.

amino acids at these positions and also at additional positions in Cro recognize five (or possibly six) bases in DNA.

Two of the amino acids involved in specific recognition are identical in repressor and Cro (Gln and Ser at the N-terminal end of the helix), while the other contacts are different (Ala in repressor versus Lys and the additional Asn in Cro). Also, a Thr in helix-2 of Cro directly contacts DNA.

The interactions shown in the figure represent binding to the DNA sequence that each protein recognizes most tightly. The sequences shown at the bottom of the figure with the contact points in color differ at 3 of the 9 base pairs. The use of overlapping, but not identical, contacts between amino acids and bases shows how related recognition helices confer recognition of related DNA sequences. This enables repressor and Cro to recognize the same set of operators, but with different relative affinities for particular members of the group.

The bases contacted by helix-3 of repressor or Cro lie on one face of DNA, as can be seen from the positions indicated on the helical diagram in Figure 11.20. However, repressor makes an additional contact with the other face of DNA. Removing the last six N-terminal amino acids (which protrude from helix-1) eliminates some of the contacts. This observation provides the basis for the idea that the bulk of the N-terminal domain contacts one face of DNA, while the last six N-terminal amino acids form an "arm" extending around the back. **Figure 11.21** shows the view from the back. Lysine residues in the arm make contacts with G residues in the major groove, and also with the phos-

Figure 11.21 A view from the back shows that the bulk of the repressor contacts one face of DNA, but its N-terminal arms reach around to the other face.

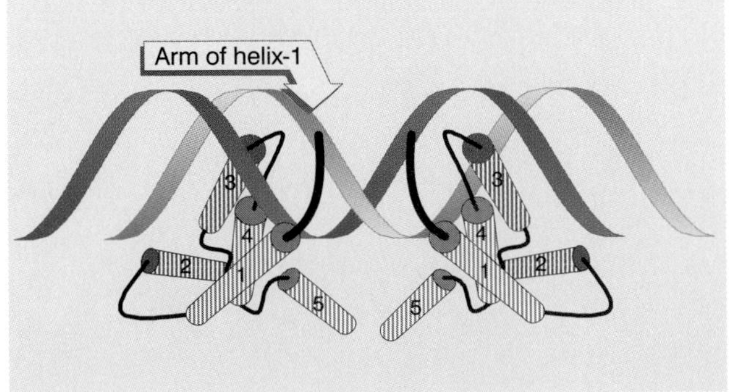

phate backbone. The interaction between the arm and DNA contributes heavily to DNA binding; the affinity of the armless repressor for DNA is reduced by ~1000-fold.

Each operator contains three repressor-binding sites. As can be seen from **Figure 11.22,** each binding site is a sequence of 17 bp displaying partial symmetry about an axis through the central base pair. No two of the six individual repressor-binding sites are identical, but they all conform with a consensus sequence. The binding sites within each operator are separated by spacers of 3–7 bp that are rich in A•T base pairs. The sites at each operator are numbered so that O_R consists of the series of binding sites O_R1–O_R2–O_R3, while O_L consists of the series O_L1–O_L2–O_L3. In each case, site 1 lies closest to the startpoint for transcription in the promoter, and sites 2 and 3 lie farther upstream.

A repressor dimer binds symmetrically at each site, so that each N-terminal domain of the dimer contacts a similar set of bases. So each individual N-terminal region contacts a half-binding site.

Bases that are not contacted directly by repressor protein may have an important effect on binding. The related phage *434* repressor binds DNA via a helix-turn-helix motif, and the crystal structure shows that helix-3 is positioned at each half-site so that it contacts the 5 outermost base pairs but not the inner 2. However, operators with A•T base pairs at the inner positions bind *434* repressor more strongly than operators with G•C base pairs. The reason is that bound *434* repressor slightly twists DNA at the center of the operator, widening the angle between the two half-sites of DNA by ~3°. This is probably needed to allow each monomer of the repressor dimer to make optimal contacts with DNA. A•T base pairs allow this twist more readily than G•C pairs, thus affecting the affinity of the operator for repressor.

Faced with the triplication of binding sites at each operator, how does repressor decide where to start binding? At each operator, site 1 has a greater affinity (roughly tenfold) than the other sites for the repressor. So the repressor always binds first to O_L1 and O_R1.

Figure 11.22 Each operator contains three repressor-binding sites, and overlaps with the promoter at which RNA polymerase binds. The orientation of O_L has been reversed from usual to facilitate comparison with O_R.

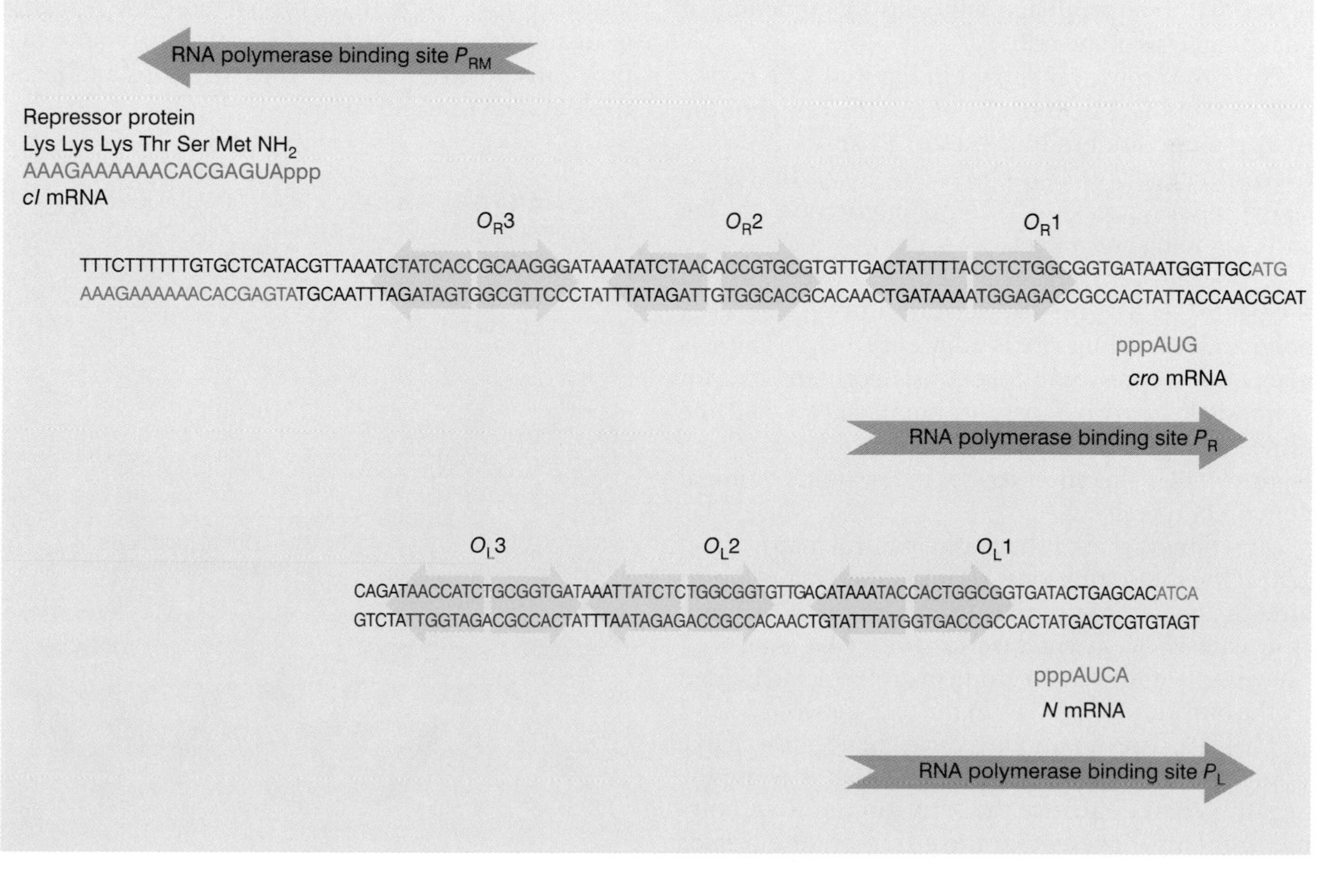

Lambda repressor binds to subsequent sites within each operator in a cooperative manner. The presence of a dimer at site 1 greatly increases the affinity with which a second dimer can bind to site 2. When both sites 1 and 2 are occupied, this interaction does *not* extend farther, to site 3. At the concentrations of repressor usually found in a lysogen, both sites 1 and 2 are filled at each operator, but site 3 is not occupied.

If site 1 is inactive (because of mutation), then repressor binds cooperatively to sites 2 and 3. That is, binding at site 2 assists another dimer to bind at site 3. This interaction occurs directly between repressor dimers and not via conformational change in DNA. Probably the connector region of the first repressor orients the C-terminal regions of the dimer in such a way that they contact the C-terminal regions of the second dimer.

A result of cooperative binding is to increase the effective affinity of repressor for the operator at physiological concentrations. This enables a lower concentration of repressor to achieve occupancy of the operator. This is an important consideration in a system in which release of repression has irreversible consequences. In an operon coding for metabolic enzymes, after all, failure of repression will merely allow unnecessary synthesis of enzymes. But failure to repress lambda prophage will lead to induction of phage and lysis of the cell.

From the sequences shown in Figure 11.22, we see that O_L1 and O_R1 lie more or less in the center of the RNA polymerase binding sites of P_L and P_R, respectively. Occupancy of O_L1–O_L2 and O_R1–O_R2 thus physically blocks access of RNA polymerase to the corresponding promoters.

A different relationship is shown between O_R and the promoter P_{RM} for transcription of *cI*. The RNA polymerase binding site is adjacent to O_R2. This explains how repressor autogenously regulates its own synthesis. When two dimers are bound at O_R1–O_R2, the dimer at O_R2 interacts with RNA polymerase (see Figure 11.14). This effect resides in the amino terminal domain of repressor.

Mutations that abolish positive control map in the *cI* gene. One interesting class of mutants remain able to bind the operator to repress transcription, but cannot stimulate RNA polymerase to transcribe from P_{RM}. They map within a small group of amino acids, located on the outside of helix-2 or in the turn between helix-2 and helix-3. The mutations reduce the negative charge of the region; conversely, mutations that increase the negative charge enhance the activation of RNA polymerase. This suggests that the group of amino acids constitutes an "acidic patch" that functions by an electrostatic interaction with a basic region on RNA polymerase.

The location of these "positive control mutations" in the repressor is indicated on **Figure 11.23.** They lie at a site on repressor that is close to a phosphate group on DNA that is also close to RNA polymerase. So the group of amino acids on repressor that is involved in positive control is in a position to contact the polymerase. The interaction between repressor and polymerase is needed for the polymerase to make the transition from a closed complex to an open complex (see Figure 11.26 later). The important principle is that *protein-protein interactions can release energy that is used to help to initiate transcription.*

What happens if a repressor dimer binds to O_R3? This site overlaps with the RNA polymerase binding site at P_{RM}. So if the repressor concentration becomes great enough to cause occupancy of O_R3, the transcription of *cI* is prevented. This leads in due course to a reduction in repressor concentration; O_R3 then becomes empty, and the autogenous loop can start up again because O_R2 remains occupied.

This mechanism could prevent the concentration of repressor from becoming too great, although it would require repressor concentration in lysogens to reach unusually high levels. In the formal sense, the repressor is an autogenous regulator of its own expression that functions positively at low concentrations and negatively at high concentrations.

Figure 11.23 Positive control mutations identify a small region at helix-2 that interacts directly with RNA polymerase.

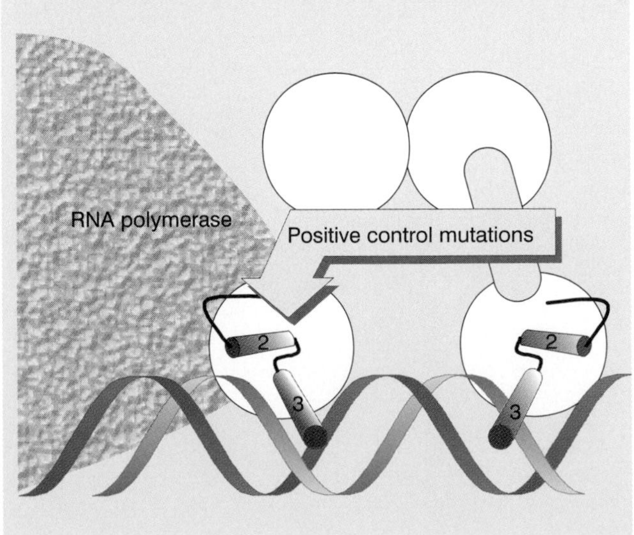

Virulent mutations occur in sites 1 and 2 of both O_L and O_R. The mutations vary in their degree of virulence, according to the extent to which they reduce the affinity of the binding site for repressor, and also depending on the relationship of the afflicted site to the promoter. Consistent with the conclusion that O_R3 and O_L3 usually are not occupied, virulent mutations are not found in these sites.

How is repressor synthesis established?

THE control circuit for maintaining lysogeny presents a paradox. *The presence of repressor protein is necessary for its own synthesis.* This explains how the lysogenic condition is perpetuated. But how is the synthesis of repressor established in the first place?

When a lambda DNA enters a new host cell, RNA polymerase cannot transcribe *cI*, because there is no repressor present to aid its binding at P_{RM}. But this same absence of repressor means that P_R and P_L are available. So the first event when lambda DNA infects a bacterium is for genes *N* and *cro* to be transcribed. Then pN allows transcription to be extended farther. This allows *cIII* (and other genes) to be transcribed on the left, while *cII* (and other genes) are transcribed on the right (see Figure 11.12).

The *cII* and *cIII* genes share with *cI* the property that mutations in them cause clear plaques. But there is a difference. The *cI* mutants can neither establish nor maintain lysogeny. The *cII* or *cIII* mutants have some difficulty in establishing lysogeny, but once established, they are able to maintain it by the *cI* autogenous circuit.

This implicates the *cII* and *cIII* genes as positive regulators whose products are needed for an alternative system for repressor synthesis. The system is needed only to *initiate* the expression of *cI* in order to circumvent the inability of the autogenous circuit to engage in *de novo* synthesis.

The CII protein acts directly on gene expression. Between the *cro* and *cII* genes is another promoter, called P_{RE}. (The subscript "RE" stands for repressor establishment.) This promoter can be recognized by RNA polymerase only in the presence of CII, whose action is illustrated in **Figure 11.24.**

The CII protein is extremely unstable *in vivo*, because it is degraded as the result of the activity of a host protein called HflA. The role of CIII is to protect CII against this degradation.

Transcription from P_{RE} promotes lysogeny in two ways. Its direct effect is that *cI* is translated into repressor protein. An indirect effect is that transcription proceeds through the *cro* gene in the "wrong" direction. So the 5′ part of the RNA corresponds to an antisense transcript of *cro*; in fact, it hybridizes to authentic *cro* mRNA, inhibiting its translation. We see in the next section that this is important because *cro* expression is needed to enter the lytic cycle.

The *cI* coding region on the P_{RE} transcript is very efficiently translated (in contrast with the weak translation of the P_{RM} transcript mentioned earlier). In fact, repressor is synthesized ~7–8 times more effectively via expression from P_{RE} than from P_{RM}. This reflects the fact that the P_{RE} transcript has an efficient ribosome-binding site, whereas the P_{RM} transcript has no ribosome-binding site and actually starts with the AUG initiation codon.

The P_{RE} promoter has a poor fit with the consensus at –10 and lacks a consensus sequence at –35. This

Figure 11.24 Repressor synthesis is established by the action of CII and RNA polymerase at P_{RE} to initiate transcription that extends from the antisense strand of *cro* through the *cI* gene.

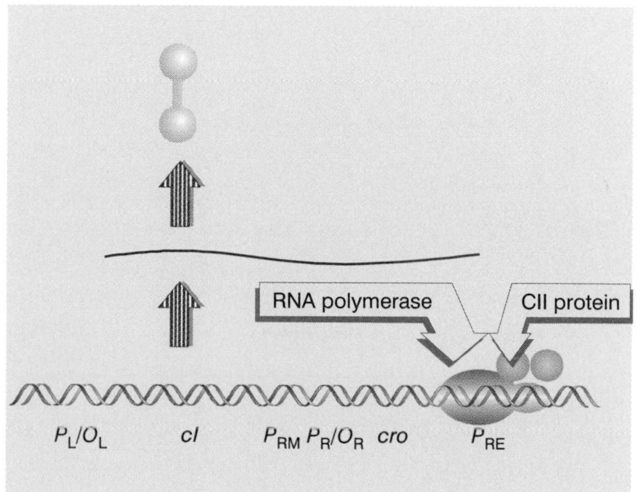

deficiency explains its dependence on *cII*. The promoter cannot be transcribed by RNA polymerase alone *in vitro*, but can be transcribed when CII is added. The regulator binds to a region extending from about –25 to –45. When RNA polymerase is added, an additional region is protected, extending from –12 to +13. As summarized in **Figure 11.25**, the two proteins bind to overlapping sites. (A similar mechanism is used by the MotA protein of phage T4, which enables *E. coli* RNA polymerase to initiate at promoters where a MotA-binding sequence is present in place of the usual –35 consensus.)

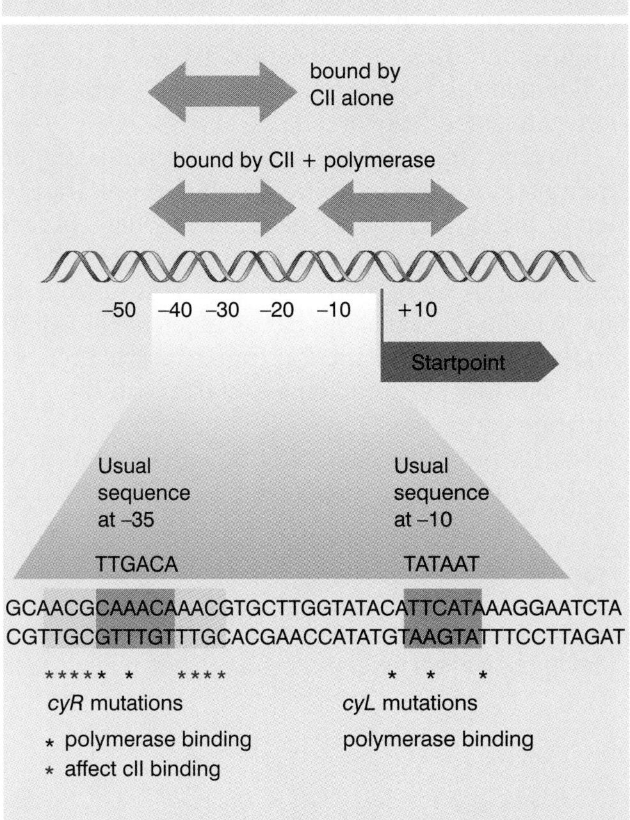

Figure 11.25 RNA polymerase binds to P_{RE} only in the presence of CII, which contacts the region around –35.

The importance of the –35 and –10 regions for promoter function, in spite of their lack of resemblance with the consensus, is indicated by the existence of *cy* mutations. These have effects similar to those of *cII* and *cIII* mutations in preventing the establishment of lysogeny; but they are *cis*-acting instead of *trans*-acting. They fall into two groups, *cyL* and *cyR*.

The *cyL* mutations are located around –10, and probably prevent RNA polymerase from recognizing the promoter.

The *cyR* mutations are located around –35, and fall into two types, affecting either RNA polymerase or CII binding. Mutations in the center of the region do not affect CII binding; presumably they prevent RNA polymerase binding. On either side of this region, mutations in short tetrameric repeats, TTGC, prevent CII from binding. Each base in the tetramer is 10 bp (one helical turn) separated from its homolog in the other tetramer, so that when CII recognizes the two tetramers, it lies on one face of the double helix.

Positive control of a promoter implies that an accessory protein has increased the efficiency with which RNA polymerase initiates transcription. **Figure 11.26** reports that either or both stages of the interaction between promoter and polymerase can be the target for regulation. Initial binding to form a closed complex or its conversion into an open complex can be enhanced.

Now we can see how lysogeny is established during an infection. **Figure 11.27** recapitulates the early stages and shows what happens as the result of expression of *cIII* and *cII*. The presence of CII allows P_{RE} to be used for transcription extending through *cI*. Repressor protein is synthesized in high amounts from this transcript. Immediately it binds to O_L and O_R.

By directly inhibiting any further transcription from P_L and P_R, repressor binding turns off the expression of all phage genes. This halts the synthesis of CII and CIII, which are unstable; they decay rapidly, with the result that P_{RE} can no longer be used. So the synthesis of

Figure 11.26 Positive regulation can influence RNA polymerase at either stage of initiating transcription.

Locus	Regulator	Polymerase Binding (equilibrium constant, K_B)	Closed-Open Conversion (rate constant, k_2)
λP_{RM}	repressor	no effect	11×
λP_{RE}	CII	100×	100×

Figure 11.27 *Summary*: a cascade is needed to establish lysogeny, but then this circuit is switched off and replaced by the autogenous repressor-maintenance circuit.

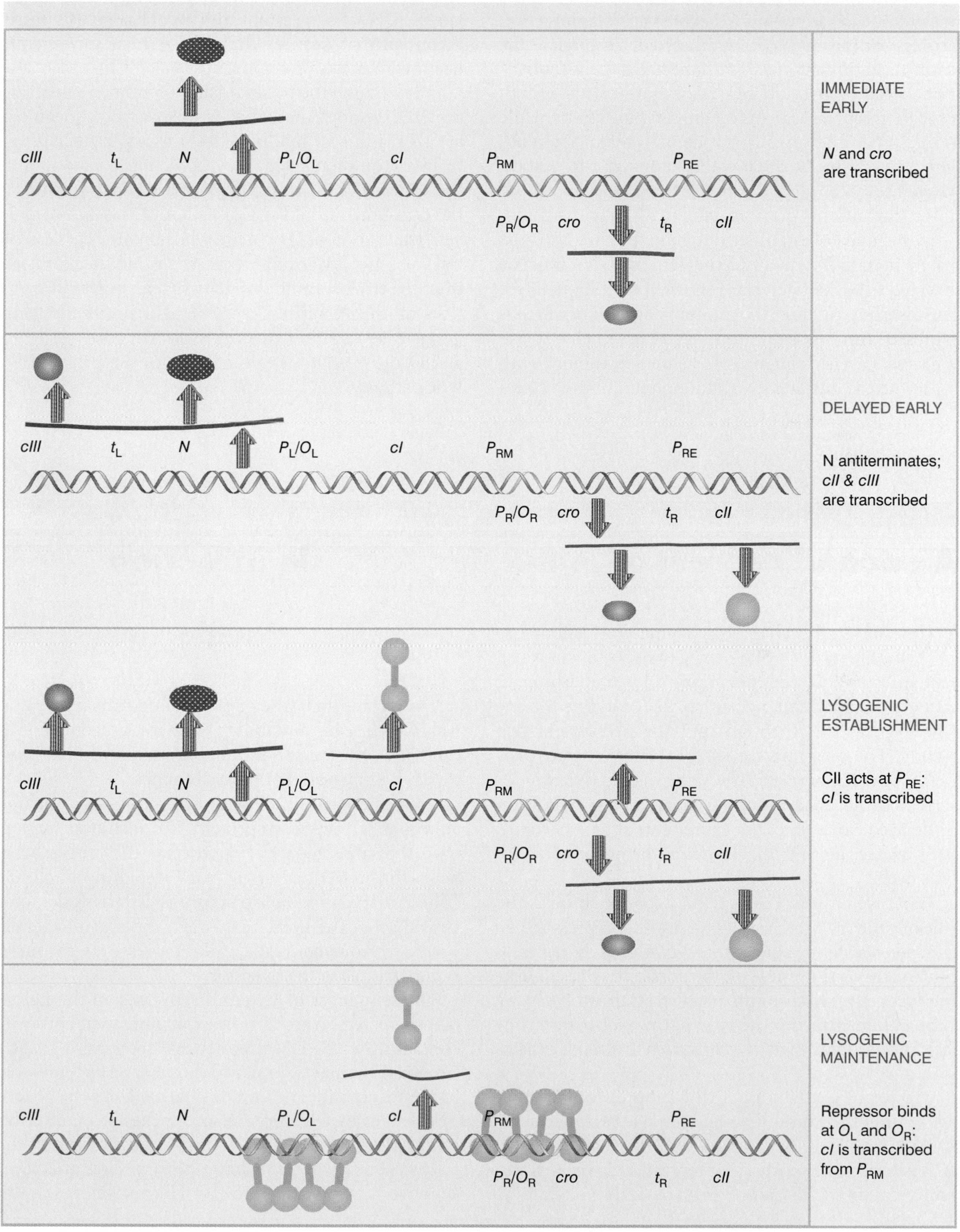

repressor via the establishment circuit is brought to a halt.

But repressor now is present at O_R. It switches on the maintenance circuit for expression from P_{RM}. Repressor continues to be synthesized, although at the lower level typical of P_{RM} function. So the establishment circuit starts off repressor synthesis at a high level; then repressor turns off all other functions, while at the same time turning on the maintenance circuit, which functions at the low level adequate to sustain lysogeny.

We shall not now deal in detail with the other functions needed to establish lysogeny, but we can just briefly remark that the infecting lambda DNA must be inserted into the bacterial genome (see Chapter 14). The insertion requires the product of gene *int*, which is expressed from its own promoter P_I, at which CII also is necessary. The sequence of P_I shows homology with P_{RE} in the CII binding site (although not in the –10 region). The functions necessary for establishing the lysogenic control circuit are therefore under the same control as the function needed physically to manipulate the DNA. So the establishment of lysogeny is under a control that ensures that all the necessary events occur with the same timing.

Emphasizing the tricky quality of lambda's intricate cascade, we now know that CII promotes lysogeny in another, indirect manner. It sponsors transcription from a promoter called $P_{anti\text{-}Q}$, which is located within the *Q* gene. This transcript is an antisense version of the *Q* region, and it hybridizes with *Q* mRNA to prevent translation of Q protein, whose synthesis is essential for lytic development. So the same mechanisms that directly promote lysogeny by causing transcription of the *cI* repressor gene also indirectly help lysogeny by inhibiting the expression of *cro* (see above) and *Q*, the regulator genes needed for the antagonistic lytic pathway.

A second repressor is needed for lytic infection

WE started this chapter by saying that lambda has the alternatives of entering lysogeny or starting a lytic infection. Lysogeny is initiated by establishing an autogenous maintenance circuit that inhibits the entire lytic cascade through applying pressure at two points. The program for establishing lysogeny proceeds through some of the same events that we described earlier in terms of the lytic cascade (expression of delayed early genes via expression of *N* is needed). We now face a problem. How does the phage enter the lytic cycle?

What we have left out of this account so far is the role of gene *cro*, which codes for another repressor. *Cro is responsible for preventing the synthesis of the repressor protein*; this action shuts off the possibility of establishing lysogeny. *cro* mutants usually establish lysogeny rather than entering the lytic pathway, because they lack the ability to switch events away from the expression of repressor.

Cro forms a small dimer (the subunit is 9 kD) that acts within the immunity region. It has two effects:

- It prevents the synthesis of repressor via the maintenance circuit; that is, it prevents transcription via P_{RM}.
- It also inhibits the expression of early genes from both P_L and P_R.

This means that, when a phage enters the lytic pathway, Cro has responsibility both for preventing the synthesis of repressor and (subsequently) for turning down the expression of the early genes.

Cro achieves its function by binding to the same operators as (*cI*) repressor protein. Cro includes a region with the same general structure as the repressor: a helix-2 is offset at an angle from recognition helix-3. (The remainder of the structure is different, demonstrating that the helix-turn-helix motif can operate within various contexts.) Like repressor, Cro binds symmetrically at the operators.

The sequences of Cro and repressor in the helix-turn-helix region are related, explaining their ability to contact the same DNA sequences (see Figure 11.20). Cro makes contacts similar to those made by repressor, but binds to only one face of DNA; it lacks the N-terminal arms by which repressor reaches around to the other side.

How can two proteins have the same sites of action, yet have such opposite effects? The answer lies in the

Figure 11.28 *Summary*: the lytic cascade requires Cro protein, which directly prevents repressor maintenance via P_{RM}, as well as turning off delayed early gene expression, indirectly preventing repressor establishment via P_{RE}.

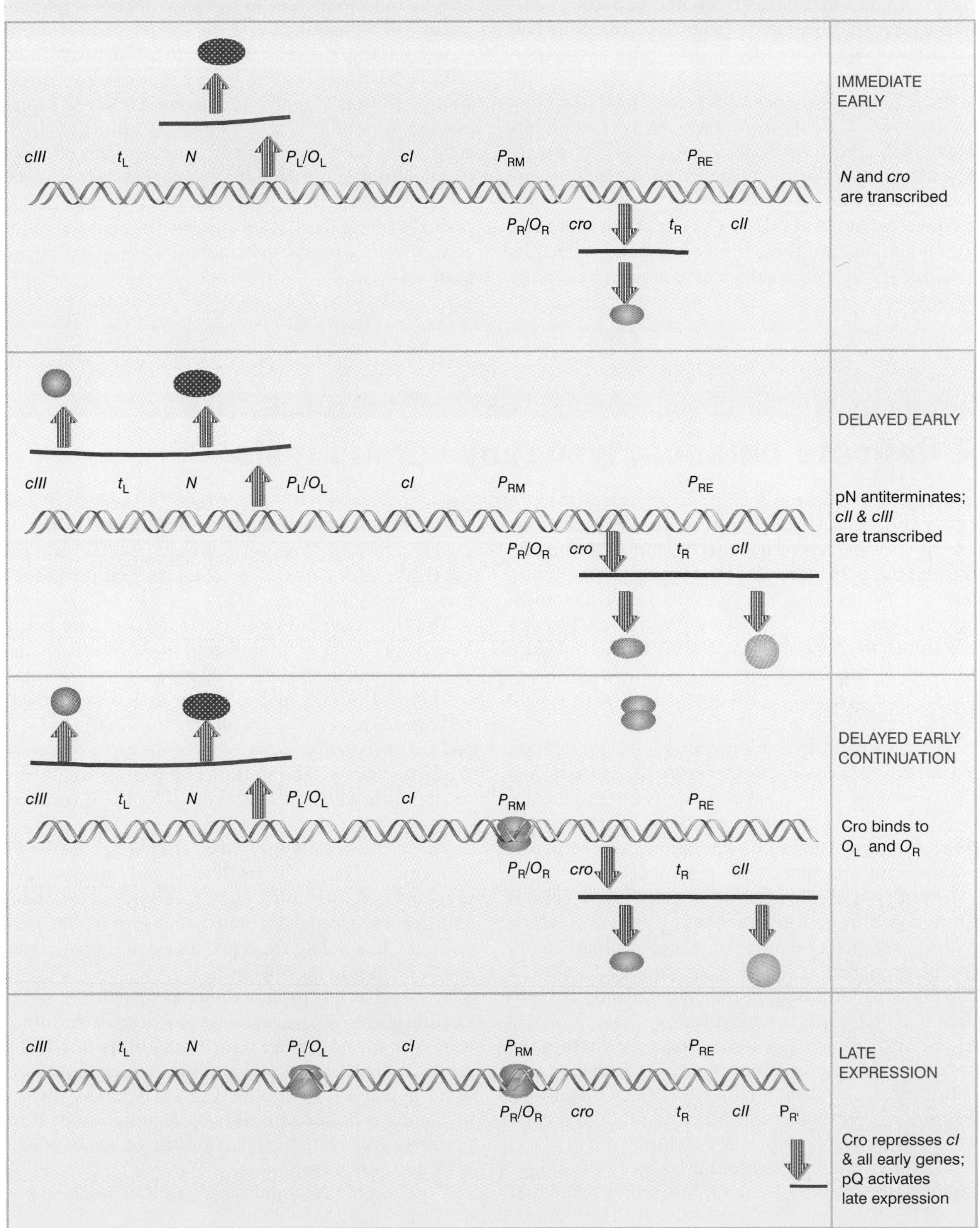
IMMEDIATE EARLY
cIII
tL
N
PL/OL
cI
PRM
PRE
PR/OR
cro
tR
cII
N and cro are transcribed
DELAYED EARLY
pN antiterminates; cII & cIII are transcribed
DELAYED EARLY CONTINUATION
Cro binds to OL and OR
LATE EXPRESSION
PR'
Cro represses cI & all early genes; pQ activates late expression

different affinities that each protein has for the individual binding sites within the operators. (Also Cro has no activating region.) Let us just consider O_R, where more is known, and where Cro exerts both its effects. The series of events is illustrated in **Figure 11.28.** (Note that the first two stages are identical to those of the lysogenic circuit shown in Figure 11.27.)

The affinity of Cro for O_R3 is greater than its affinity for O_R2 or O_R1. So it binds first to O_R3. This inhibits RNA polymerase from binding to P_{RM}. So Cro's first action is to prevent the maintenance circuit for lysogeny from coming into play.

Then Cro binds to O_R2 or O_R1. Its affinity for these sites is similar, and there is no cooperative effect. Its presence at either site is sufficient to prevent RNA polymerase from using P_R. This in turn stops the production of the early functions (including Cro itself). Because CII is unstable, any use of P_{RE} is brought to a halt. So the two actions of Cro together block *all* production of repressor.

So far as the lytic cycle is concerned, Cro turns down (although it does not completely eliminate) the expression of the early genes. Its incomplete effect is explained by its affinity for O_R1 and O_R2, which is about eight times lower than that of repressor. This effect of Cro does not occur until the early genes have become more or less superfluous, because pQ is present; by this time, the phage has started late gene expression, and is concentrating on the production of progeny phage particles.

A delicate balance: lysogeny versus lysis

THE programs for the lysogenic and lytic pathways are so intimately related that it is impossible to predict the fate of an individual phage genome when it enters a new host bacterium. Will the antagonism between repressor and Cro be resolved by establishing the autogenous maintenance circuit shown in Figure 11.27, or by turning off repressor synthesis and entering the late stage of development shown in Figure 11.28?

The same pathway is followed in both cases right up to the brink of decision. Both involve the expression of the immediate early genes and extension into the delayed early genes. The difference between them comes down to the question of whether repressor or Cro will obtain occupancy of the two operators.

The early phase during which the decision is taken is limited in duration in either case. No matter which pathway the phage follows, expression of all early genes will be prevented as P_L and P_R are repressed; and, as a consequence of the disappearance of CII and CIII, production of repressor via P_{RE} will cease.

The critical question comes down to whether the cessation of transcription from P_{RE} is followed by activation of P_{RM} and the establishment of lysogeny, or whether P_{RM} fails to become active and the pQ regulator commits the phage to lytic development.

The initial event in establishing lysogeny is the binding of repressor at O_L1 and O_R1. Binding at the first sites is rapidly succeeded by cooperative binding of further repressor dimers at O_L2 and O_R2. This shuts off the synthesis of Cro and starts up the synthesis of repressor via P_{RM}.

The initial event in entering the lytic cycle is the binding of Cro at O_R3. This stops the lysogenic-maintenance circuit from starting up at P_{RM}. Then Cro must bind to O_R1 or O_R2, and to O_L1 or O_L2, to turn down early gene expression. By halting production of CII and CIII, this action leads to the cessation of repressor synthesis via P_{RE}. The shutoff of repressor establishment occurs when the unstable CII and CIII proteins decay.

The critical influence over the switch between lysogeny and lysis is CII. If CII is active, synthesis of repressor via the establishment promoter is effective; and, as a result, repressor gains occupancy of the operators. If CII is not active, repressor establishment fails, and Cro binds to the operators.

The level of CII protein under any particular set of circumstances determines the outcome of an infection. Mutations that increase the stability of CII increase the frequency of lysogenization. Such mutations occur in *cII* itself or in other genes. The cause of CII's instability is its susceptibility to degradation by host proteases. Its level in the cell is influenced by *cIII* as well as by host functions.

The effect of the lambda protein CIII is secondary: it

helps to protect CII against degradation. Although the presence of CIII does not guarantee the survival of CII, in the absence of CIII, CII is virtually always inactivated.

Host gene products act on this pathway. Mutations in the host genes *hflA* and *hflB* increase lysogeny—*hfl* stands for *h*igh *f*requency *l*ysogenization. The mutations stabilize CII because they inactivate host protease(s) that degrade it.

The influence of the host cell on the level of CII provides a route for the bacterium to interfere with the decision-taking process. For example, host proteases that degrade CII are activated by growth on rich medium, so lambda tends to lyse cells that are growing well, but is more likely to enter lysogeny on cells that are starving (and which lack components necessary for efficient lytic growth).

Summary

Phages have a lytic life cycle in which infection of a host bacterium is followed by production of a large number of phage particles, lysis of the cell, and release of the viruses. Some phages can also exist in lysogenic form, in which the phage genome is integrated into the bacterial chromosome and is inherited in this inert, latent form like any other bacterial gene.

Lytic infection falls typically into three phases. In the first phase a small number of phage genes are transcribed by the host RNA polymerase. One or more of these genes is a regulator that controls expression of the group of genes expressed in the second phase. The pattern is repeated in the second phase, when one or more genes is a regulator needed for expression of the genes of the third phase. Genes of the first two phases code for enzymes needed to reproduce phage DNA; genes of the final phase code for structural components of the phage particle. It is common for the very early genes to be turned off during the later phases.

In phage lambda, the genes are organized into groups whose expression is controlled by individual regulatory events. The immediate early gene *N* codes for an antiterminator that allows transcription of the leftward and rightward groups of delayed early genes from the early promoters P_R and P_L. The delayed early gene *Q* has a similar antitermination function that allows transcription of all late genes from the promoter $P_{R'}$. The lytic cycle is repressed, and the lysogenic state maintained, by expression of the *cI* gene, whose product is a repressor protein that acts at the operators O_R and O_L to prevent use of the promoters P_R and P_L, respectively. A lysogenic phage genome expresses only the *cI* gene, from its promoter P_{RM}. Transcription from this promoter involves positive autogenous regulation, in which repressor bound at O_R activates RNA polymerase at P_{RM}.

Each operator consists of three binding sites for repressor. Each site is palindromic, consisting of symmetrical half-sites. Repressor functions as a dimer. Each half binding site is contacted by a repressor monomer. The N-terminal domain of repressor contains a helix-turn-helix motif that contacts DNA. Helix-3 is the recognition helix, responsible for making specific contacts with base pairs in the operator. Helix-2 is involved in positioning helix-3; it is also involved in contacting RNA polymerase at P_{RM}. The C-terminal domain is required for dimerization. Induction is caused by cleavage between the N- and C-terminal domains, which prevents the DNA-binding regions from functioning in dimeric form, thereby reducing their affinity for DNA and making it impossible to maintain lysogeny. Repressor-operator binding is cooperative, so that once one dimer has bound to the first site, a second dimer binds more readily to the adjacent site.

The helix-turn-helix motif is used by other DNA-binding proteins, including lambda Cro, which binds to the same operators, but has a different affinity for the individual operator sites, determined by the sequence of helix-3. Cro binds individually to operator sites, starting with O_{R3}, in a noncooperative manner. It is needed for progression through the lytic cycle. Its binding to O_R first prevents synthesis of repressor from P_{RM}; then it prevents continued expression of early genes, an effect also seen in its binding to O_L.

Establishment of repressor synthesis requires use of the promoter P_{RE}, which is activated by the product of the *cII* gene. The product of *cIII* is required to stabilize the *cII* product against degradation. By turning off *cII* and *cIII* expression, Cro acts to prevent lysogeny. By turning off all transcription except that of its own gene, repressor acts to prevent the lytic cycle. The choice between lysis and lysogeny depends on whether repressor or Cro gains occupancy of the operators in a particular infection. The stability of CII protein in the infected cell is a primary determinant of the outcome.

Further reading

Reviews

Friedman, D. I. and Gottesman, M. (1982). Lytic mode of lambda development. In *Lambda II.* (Eds. R. W. Hendrix, J. W. Roberts, F. W. Stahl and R. A. Weisberg, Cold Spring Harbor Laboratory, New York, 21–51).

Ptashne, M. (1992). *A Genetic Switch.* (Cell Press and Blackwell Scientific Publications, Cambridge).

Discoveries

Ptashne, M. (1967). Isolation of the λ phage repressor. *Proc. Nat. Acad. Sci. USA* **57**, 306–313.

Ptashne, M. (1967). Specific binding of the λ phage repressor to DNA. *Nature* **214**, 232–234.

Pirrotta, V., Chadwick, P., and Ptashne, M. (1970). Active form of two coliphage repressors. *Nature* **227**, 41–44.

Johnson, A. D., Meyer, B. J., and Ptashne, M. (1979). Interactions between DNA-bound repressors govern regulation by the phage λ repressor. *Proc. Nat. Acad. Sci. USA* **76**, 5061–5065.

Pabo, C. O. and Lewis, M. (1982). The operator-binding domain of λ repressor: structure and DNA recognition. *Nature* **298**, 443–447.

Sauer, R. T. *et al.* (1982). Homology among DNA-binding proteins suggests use of a conserved super-secondary structure. *Nature* **298**, 447–451.

Wharton, R. L., Brown, E. L., and Ptashne, M. (1984). Substituting an α-helix switches the sequence specific DNA interactions of a repressor. *Cell* **38**, 361–369.

Brennan, R. G. *et al.* (1990). Protein-DNA conformational changes in the crystal structure of a λ Cro-operator complex. *Proc. Nat. Acad. Sci. USA* **87**, 8165–8169.

Part 4

DNA

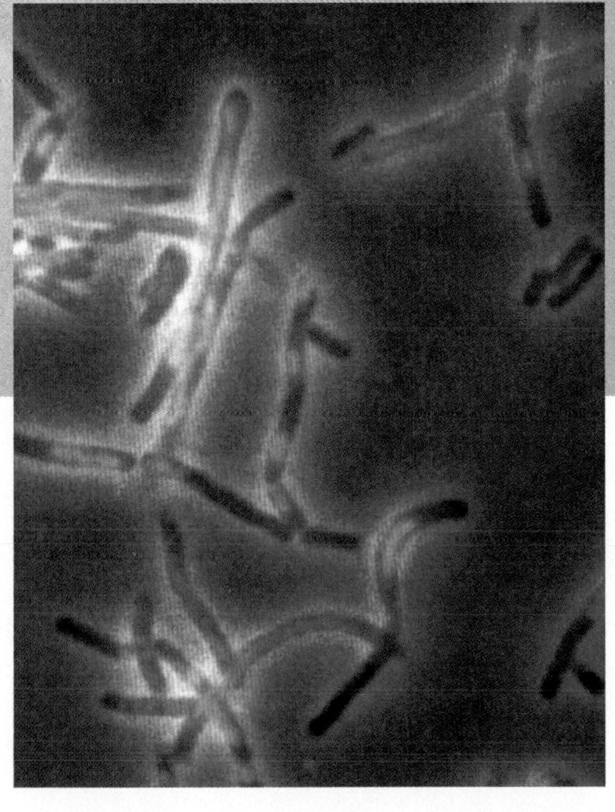

Chapter 12

The replicon

WHETHER a cell has only one chromosome (as in prokaryotes) or has many chromosomes (as in eukaryotes), the entire genome must be replicated precisely once for every cell division. How is the act of replication linked to the cell cycle?

Two general principles are used to compare the state of replication with the condition of the cell cycle:

- *Initiation of DNA replication commits the cell (prokaryotic or eukaryotic) to a further division.* From this standpoint, the number of descendants that a cell generates is determined by a series of decisions on whether or not to initiate DNA replication. Replication is controlled at the stage of initiation. *Once replication has started, it continues until the entire genome has been duplicated.*
- *If replication proceeds, the consequent division cannot be permitted to occur until the replication event has been completed.* Indeed, the completion of replication may provide a trigger for cell division. Then the duplicate genomes are segregated one to each daughter cell (via mitosis in eukaryotes). The unit of segregation is the chromosome.

In prokaryotes, the initiation of replication is a single event involving a unique site on the bacterial chromosome, and the process of division is accomplished by the development of a septum. In eukaryotic cells, initiation of replication is identified by the start of S phase, a protracted period during which DNA synthesis occurs, and which involves many individual initiation events. The act of division is accomplished by the reorganization of the cell at mitosis. In this chapter, we are concerned with the regulation of DNA replication. How is a cycle of replication initiated? What controls its progress and how is its termination signaled? In Chapter 27, we discuss the regulatory processes in eukaryotic cells that control entry into S phase and into mitosis, and also the "checkpoints" that postpone these actions until the appropriate conditions have been fulfilled.

The unit of DNA in which an individual act of replication occurs is called the **replicon**. Each replicon "fires" once and only once in each cell cycle. The replicon is defined by its possession of the control elements needed for replication. It has an **origin** at which replication is initiated. It may also have a **terminus** at which replication stops.

Any sequence attached to an origin—or, more precisely, not separated from an origin by a terminus—is replicated as part of that replicon. The origin is a *cis*-acting site, able to affect only that molecule of DNA on which it resides.

(The original formulation of the replicon [in prokaryotes] viewed it as a unit possessing both the origin *and* the gene coding for the regulator protein. Now, however, "replicon" is usually applied to eukaryotic chromosomes to describe a unit of replication that contains an origin; *trans*-acting regulator protein(s) may be coded elsewhere.)

A genome in a prokaryotic cell constitutes a single replicon; so the units of replication and segregation coincide. The largest such replicon is that of the bacterial chromosome itself. Initiation at a single origin sponsors replication of the entire genome, once for every cell division. Each haploid bacterium has a single chromosome, so this type of replication control is called **single copy**.

Bacteria may contain additional genetic information in the form of plasmids. *A plasmid is an autonomous circular DNA genome that constitutes a separate replicon* (see Figure 11.2). A plasmid replicon may replicate with the bacterial genome *pari passu* (showing single-copy control) or may be under a different type of control. When the number of copies of a plasmid is greater than that of the bacterial chromosome, it is said to replicate under **multicopy control**. Each phage or virus DNA also constitutes a replicon, able to initiate many times during an infectious cycle. Perhaps a better way to view the prokaryotic replicon, therefore, is to reverse the definition: *any DNA molecule that contains an origin can be replicated autonomously in the cell.*

A major difference in the organization of bacterial and eukaryotic genomes is seen in their replication. Each eukaryotic chromosome contains a large number of replicons. So the unit of segregation includes many units of replication. This adds another dimension to the problem of control. All the replicons on a chromosome must be fired during one cell cycle, although they are not active simultaneously, but are activated over a fairly protracted period. *Yet each of these replicons must be activated no more than once in the cell cycle.*

Some signal must distinguish replicated from nonreplicated replicons, so that replicons do not fire a second time. And because many replicons are activated independently, another signal must exist to indicate when the entire process of replicating all replicons has been completed.

We have begun to collect information about the construction of individual replicons, but we still have little information about the relationship between replicons. We do not know whether the pattern of replication is the same in every cell cycle. Are all origins always used or are some origins sometimes silent? Do origins always fire in the same order? If there are different classes of origins, what distinguishes them?

In contrast with nuclear chromosomes, which have a single-copy type of control, the DNA of mitochondria and chloroplasts may be regulated more like plasmids that exist in multiple copies per bacterium. There are multiple copies of each organelle DNA per cell, and the control of organelle DNA replication must be related to the cell cycle.

In all these systems, the key question is to define the sequences that function as origins and to determine how they are recognized by the appropriate proteins of the apparatus for replication. We start by considering the basic construction of replicons and the various forms that they take; following the consideration of the origin, we turn to the question of how replication of the genome is coordinated with bacterial division, and what is responsible for segregating the genomes to daughter bacteria.

Origins can be mapped by autoradiography and electrophoresis

A molecule of DNA engaged in replication has two types of regions. **Figure 12.1** shows that when replicating DNA is viewed by electron microscopy, the replicated region appears as an **eye** within the nonreplicated DNA. The nonreplicated region consists of the parental duplex; this opens into the replicated region where the two daughter duplexes have formed.

The point at which replication is occurring is called the **replication fork** (sometimes also known as the **growing point**). *A replication fork moves sequentially along the DNA, from its starting point at the origin.* Replication may be **unidirectional** or **bidirectional**. The type of event is determined by whether one or two replication forks set out from the origin. In unidirectional replication, one replication fork leaves the origin and proceeds along the DNA. In bidirectional replication, two replication forks are formed; they proceed away from the origin in opposite directions.

Figure 12.1 Replicated DNA is seen as a replication eye flanked by nonreplicated DNA.

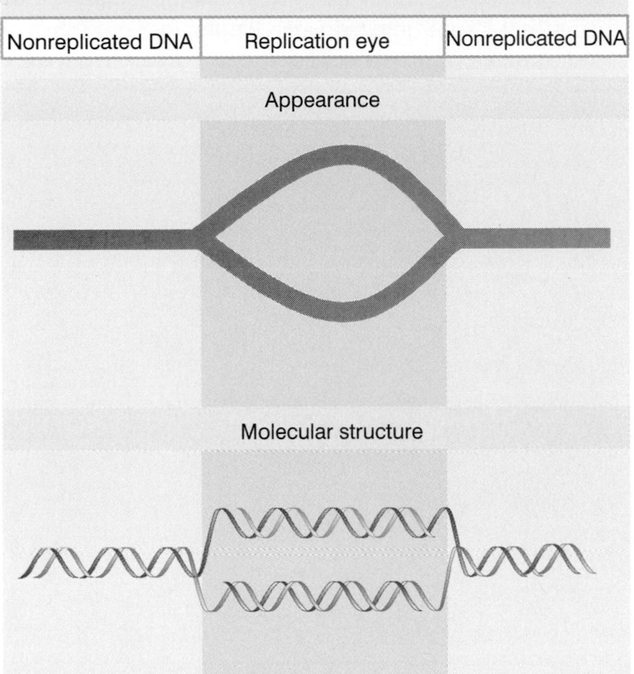

Figure 12.2 Replicons may be unidirectional or bidirectional, depending on whether one or two replication forks are formed at the origin.

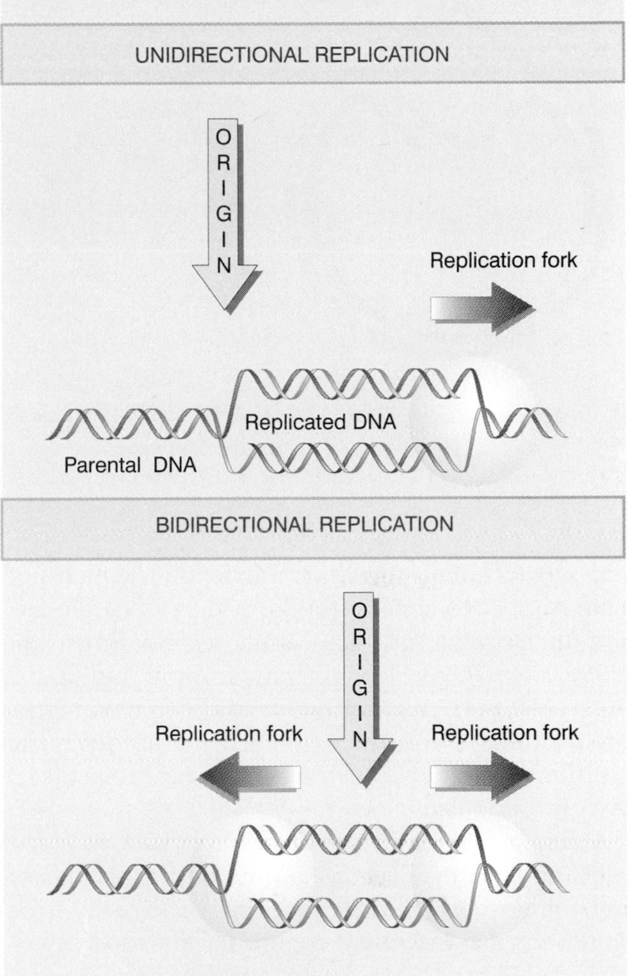

The appearance of a replication eye does not distinguish between unidirectional and bidirectional replication. As depicted in **Figure 12.2**, the eye can represent either of two structures. If generated by unidirectional replication, the eye represents one fixed origin and one moving replication fork. If generated by bidirectional replication, the eye represents a pair of replication forks. In either case, the progress of replication expands the eye until ultimately it encompasses the whole replicon.

When a replicon is circular, the presence of an eye forms the **θ-structure** drawn in **Figure 12.3**. The successive stages of replication of the circular DNA of polyoma virus are visualized by electron microscopy in **Figure 12.4**.

Whether a replicating eye has one or two replication forks can be determined in two ways. The choice of method depends on whether the DNA is a defined molecule or an unidentified region of a cellular genome.

With a defined linear molecule, we can use electron microscopy to measure the distance of each end of the eye from the end of the DNA. Then the positions of the ends of the eyes can be compared in molecules that have eyes of different sizes. If replication is unidirectional, only one of the ends will move; the other is the fixed origin. If replication is bidirectional, both will move; the origin is the point midway between them.

With undefined regions of large genomes, two successive pulses of radioactivity can be used to label the movement of the replication forks. If one pulse has a more intense label than the other, they can be distinguished by the relative intensities of labeling. These can be visualized by autoradiography. **Figure 12.5** shows that unidirectional replication causes one type of label to be followed by the other at *one* end of the eye. Bidirectional replication produces a (symmetrical) pattern at *both* ends of the eye. This is the pattern usually observed in replicons of eukaryotic chromosomes.

A more recent method for mapping origins with greater resolution takes advantage of the effects that changes in shape have upon electrophoretic migration of DNA. **Figure 12.6** illustrates the two-dimensional

Figure 12.3 A replication eye forms a theta structure in circular DNA.

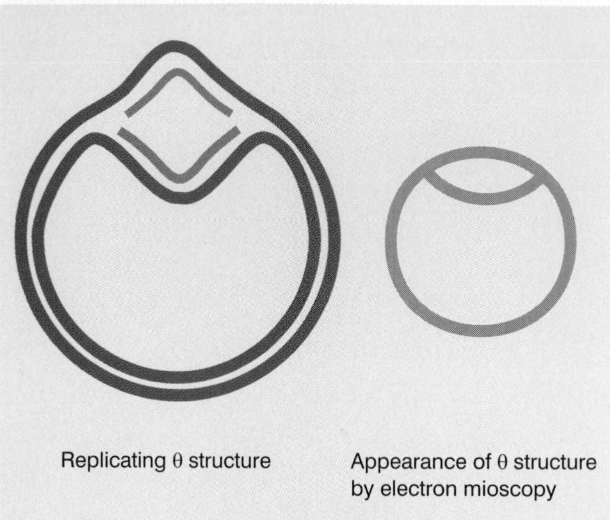

Figure 12.4 The replication eye becomes larger as the replication forks proceed along the replicon. Note that the 'eye' becomes larger than the nonreplicated segment. The two sides of the eye can be defined because they are both the same length. Photograph kindly provided by Bernard Hirt.

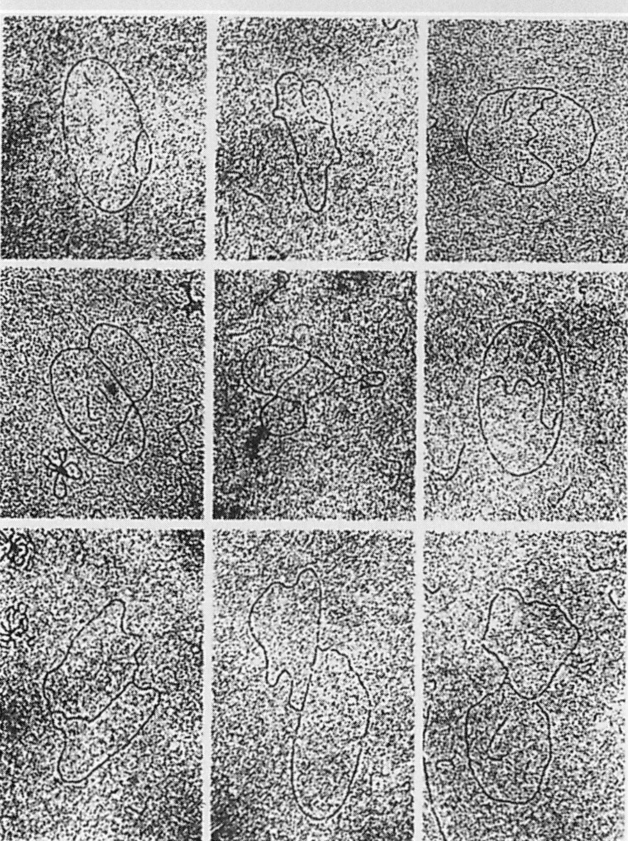

mapping technique, in which restriction fragments of replicating DNA are electrophoresed in a first dimension that separates by mass, and a second dimension where movement is determined more by shape. Different types of replicating molecules follow characteristic paths, measured by their deviation from the line that would be followed by a linear molecule of DNA that doubled in size.

A simple Y-structure, in which one fork moves along a linear fragment, follows a continuous path. An inflection point occurs when all three branches are the same length, and the structure therefore deviates most extensively from linear DNA. Analogous considerations determine the paths of double Y-structures or bubbles. An asymmetric bubble follows a discontinuous path, with a break at the point at which the bubble is converted to a Y-structure as one fork runs off the end.

Taken together, the various techniques for characterizing replicating DNA show that origins are most often used to initiate bidirectional replication. From this level of resolution, we must now proceed to the molecular level, to identify the *cis*-acting sequences that comprise the origin, and the *trans*-acting factors that recognize it.

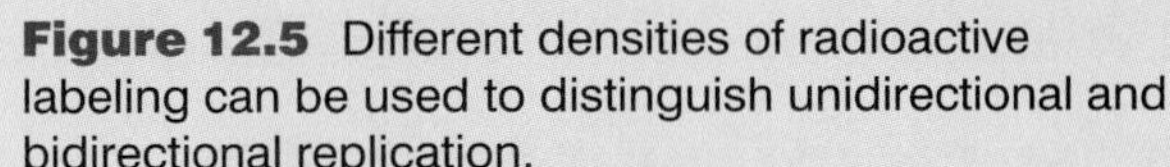

Figure 12.5 Different densities of radioactive labeling can be used to distinguish unidirectional and bidirectional replication.

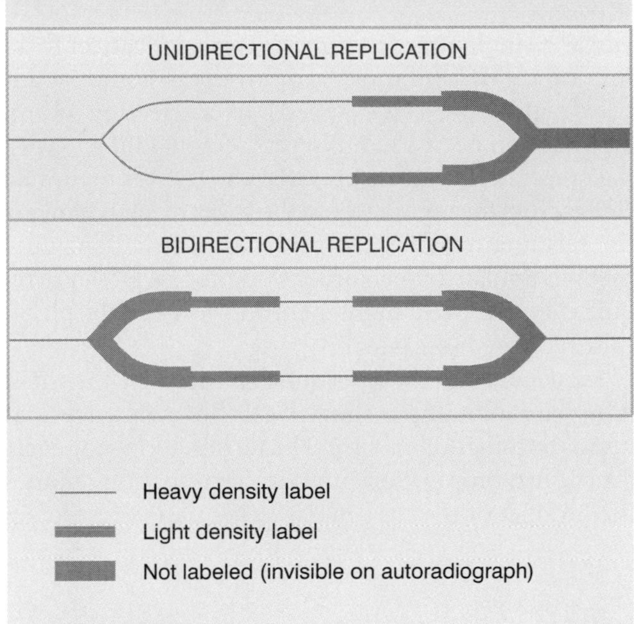

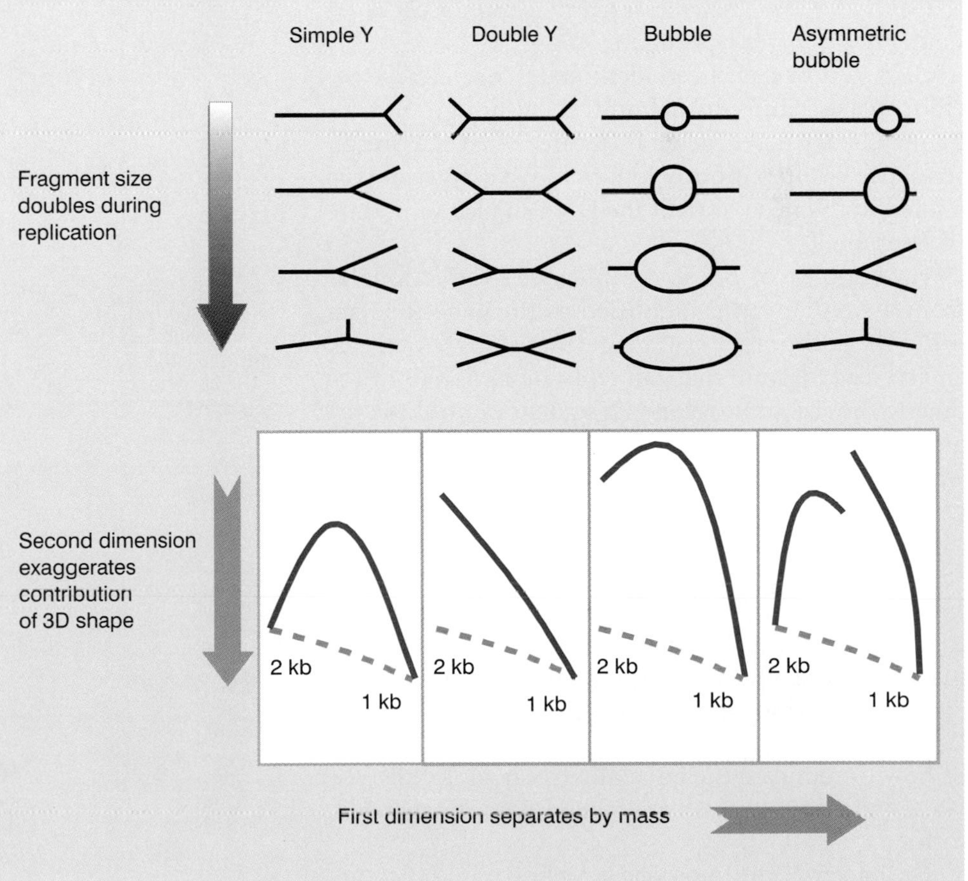

Figure 12.6 The position of the origin and the number of replicating forks determine the shape of a replicating restriction fragment, which can be followed by its electrophoretic path.

The bacterial genome is a single circular replicon

To be properly inherited, a bacterial replicon should support several functions:

- Initiating a replication cycle.
- Controlling the frequency of initiation events.
- Segregating replicated chromosomes to daughter cells.

The first two functions both are properties of the origin. Segregation could be an independent function, but in prokaryotic systems usually rests with sequences in the vicinity of the origin. Origins in eukaryotes do not function in segregation, but are concerned only with replication.

As a general principle, the DNA constituting an origin can be isolated by its ability to support replication of any DNA sequence to which it is joined. When DNA from the origin is cloned into a molecule that lacks an origin, this will create a plasmid capable of autonomous replication *only if the DNA from the origin contains all the sequences needed to identify itself as an authentic origin for replication.*

Origins now have been identified in bacteria, yeast, chloroplasts, and mitochondria, although not in higher eukaryotes. A general feature is that the overall sequence composition is A•T-rich. We assume this is related to the need to melt the DNA duplex to initiate replication.

The genome of *E. coli* is replicated bidirectionally from a single origin, identified as the genetic locus *oriC*. The addition of *oriC* to any piece of DNA creates an artificial plasmid that can replicate in *E. coli*. By reducing the size of the cloned fragment of *oriC*, the region required to initiate replication has been equated with a fragment of 245 bp. (We discuss the properties of *oriC* and its interaction with the replication apparatus in more detail in the next chapter.)

Prokaryotic replicons are usually circular—the DNA forms a closed circle with no free ends. Circular structures include the bacterial chromosome itself, all plasmids, and many bacteriophages. They are also common in chloroplast and mitochondrial DNAs. Replication of a circular molecule avoids the problem of how to replicate the ends of a linear molecule (see later), but poses the problem of how to terminate replication.

The bacterial chromosome is replicated bidirectionally as a single unit from *oriC*. Two replication forks initiate at *oriC* and move around the genome (at approximately the same speed) to a meeting point. Termination occurs in a discrete region. One interesting question is what ensures that the DNA is replicated right across the region where the forks meet. Following the termination of DNA replication itself, enzymes that manipulate higher-order structure of DNA are required for the two daughter chromosomes to be physically separated (see later).

Sequences that cause termination are called *ter* sites. A *ter* site contains a short (~23 bp) sequence that causes termination *in vitro*. The termination sequences function in only one orientation. Termination requires the product of the *tus* gene, which codes for a protein

Figure 12.7 Replication termini in *E. coli* are located beyond the point at which the replication forks actually meet.

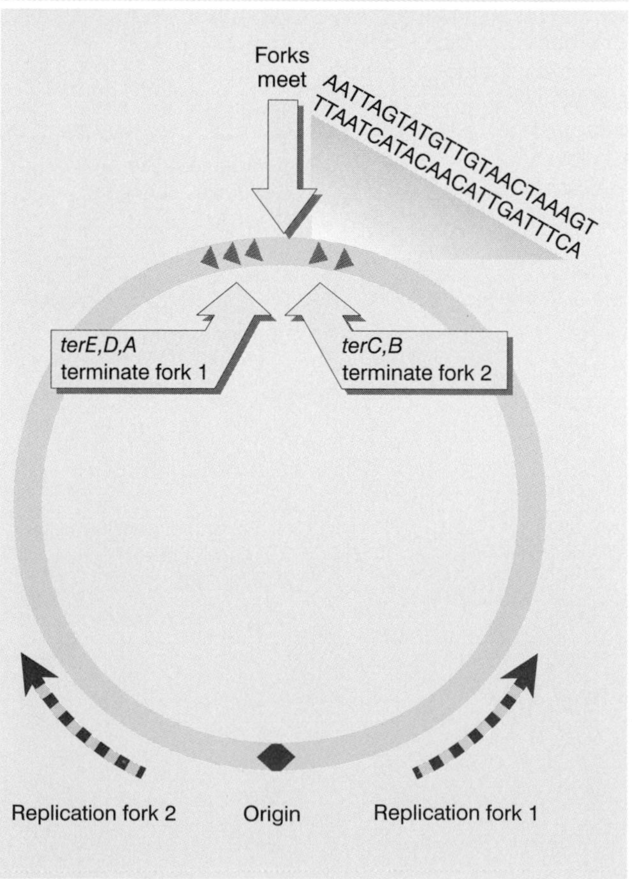

that recognizes the consensus sequence and prevents the replication fork from proceeding (see Chapter 13).

Termination in *E. coli* has the interesting features reported in **Figure 12.7**. We know that the replication forks usually meet and halt replication at a point midway round the chromosome from the origin. But two termination regions (*terE,D,A* and *terC,B*) have been identified, located ~100 kb on either side of this meeting point. Each contains multiple terminators. Each terminus is specific for one direction of fork movement, and they are arranged in such a way that each fork would have to pass the other in order to reach the terminus to which it is susceptible. This arrangement creates a "replication fork trap;" if for some reason one fork is delayed, so that the forks fail to meet at the usual central position, the more rapid fork will be trapped at the *ter* region to wait for the arrival of the slow fork.

What happens when a replication fork encounters a protein bound to DNA? We assume repressors (for example) are displaced and then reattach. A particularly interesting question is what happens when a replication fork encounters an RNA polymerase engaged in transcription. A replication fork moves >10× faster than RNA polymerase. If they are proceeding in the same direction, either the replication fork must displace the polymerase or it must slow down as it waits for the RNA polymerase to reach its terminator. It appears that a DNA polymerase moving in the same direction as an RNA polymerase can "bypass" it without disrupting transcription, but we do not understand how this happens.

A conflict arises when the replication fork meets an RNA polymerase traveling in the opposite direction, that is, toward it. Can it displace the RNA polymerase? Or do both replication and transcription come to a halt? An indication that these encounters cannot easily be resolved is provided by the organization of the *E. coli* chromosome. Almost all active transcription units are oriented so that they are expressed in the same direction as the replication fork that passes them. The exceptions all comprise small transcription units that are infrequently expressed. The difficulty of generating inversions containing highly expressed genes argues that head-on encounters between a replication fork and a series of transcribing RNA polymerases may be lethal.

Each eukaryotic chromosome contains many replicons

In eukaryotic cells, the replication of DNA is confined to part of the cell cycle. **S phase** occupies part of interphase, and usually lasts a few hours in a higher eukaryotic cell. Replication of the large amount of DNA contained in a eukaryotic chromosome is accomplished by dividing it into many individual replicons. Only some of these replicons are engaged in replication at any point in S phase. Presumably each replicon is activated at a specific time during S phase, although the evidence on this issue is not decisive.

The start of S phase is signaled by the activation of the first replicons. Over the next few hours, initiation events occur at other replicons in an ordered manner. Much of our knowledge about the properties of the individual replicons is derived from autoradiographic studies, generally using the types of protocols illustrated in Figures 12.5 and 12.6. Chromosomal replicons usually display bidirectional replication.

How large is the average replicon, and how many are there in the genome? A difficulty in characterizing the individual unit is that adjacent replicons may fuse to give large replicated eyes, as illustrated in **Figure 12.8**. The approach usually used to distinguish individual replicons from fused eyes is to rely on stretches of DNA in which several replicons can be seen to be active, presumably captured at a stage when all have initiated around the same time, but before the forks of adjacent units have met.

There is some evidence that "regional" controls might produce this sort of activation pattern, in which groups of replicons are initiated more or less coordinately, as opposed to a mechanism in which individual replicons are activated one by one in dispersed areas of the genome. Two structural features suggest the possibility of large-scale organization. Quite large regions of the chromosome can be characterized as "early

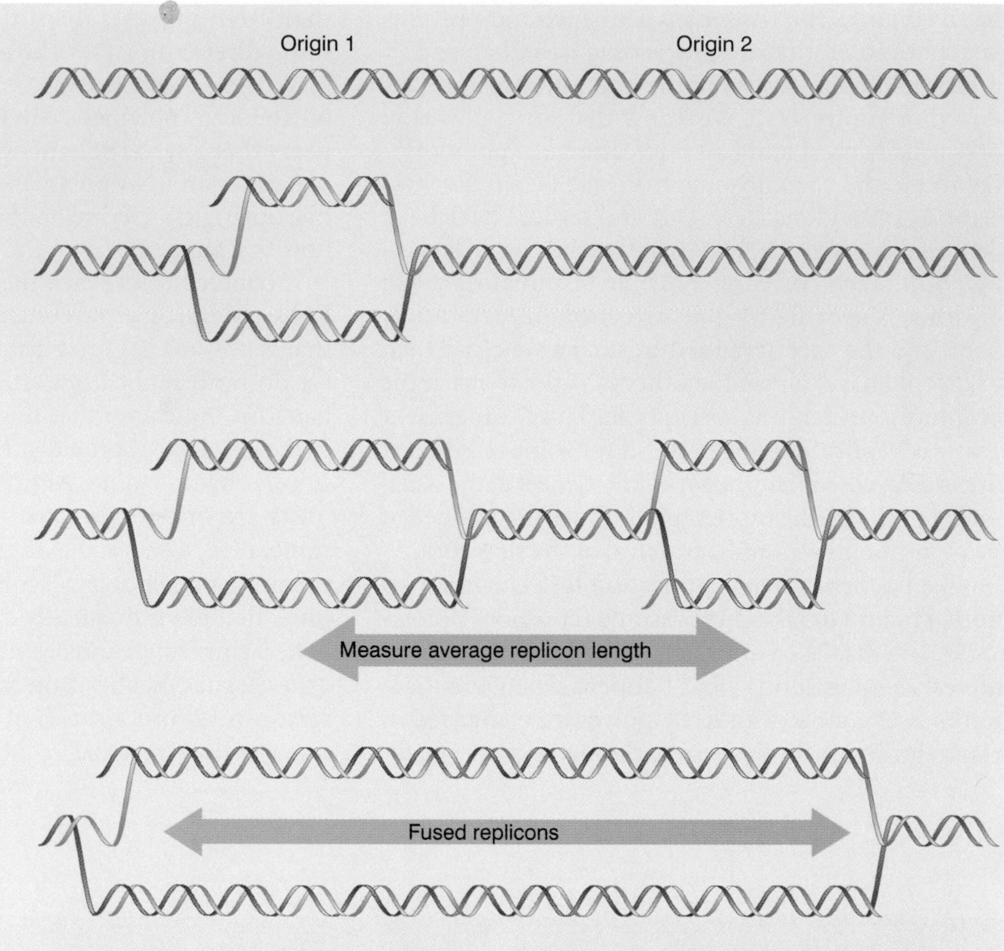

Figure 12.8 Measuring the size of the replicon requires a stretch of DNA in which adjacent replicons are active.

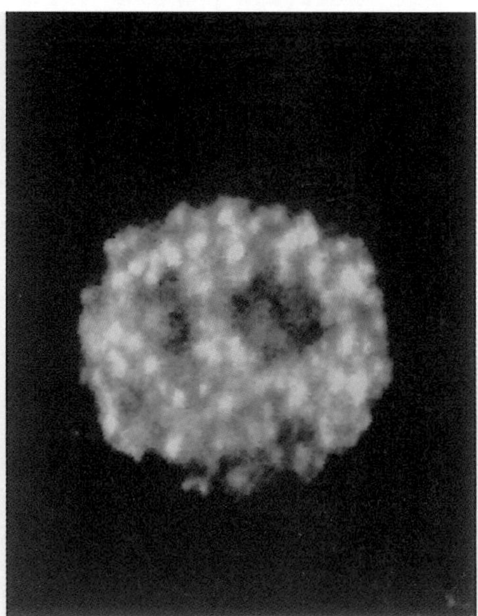

Figure 12.9 Replication forks are organized into foci in the nucleus. Cells were labeled with BrdU. The leftmost panel was stained with propidium iodide to identify bulk DNA. The right panel was stained using an antibody to BrdU to identify replicating DNA. Photographs kindly provided by A. D. Mills and Ron Laskey.

replicating" or "late replicating", implying that there is little interspersion of replicons that fire at early or late times. And visualization of replicating forks by labeling with DNA precursors identifies 100–300 "foci" instead of uniform staining; each focus shown in **Figure 12.9** probably contains >300 replication forks. The foci could represent fixed structures through which replicating DNA must move.

In groups of active replicons, the average size of the unit is measured by the distance between the origins (that is, between the midpoints of adjacent replicons). The rate at which the replication fork moves can be estimated from the maximum distance that the autoradiographic tracks travel during a given time.

Individual replicons in eukaryotic genomes are relatively small, typically ~40 kb in yeast or fly, ~100 kb in animals cells. However, they can vary >10-fold in length within a genome. The rate of replication is ~2000 bp/min, which is much slower than the 50,000 bp/min of bacterial replication fork movement.

From the speed of replication, it is evident that a mammalian genome could be replicated in ~1 hour if all replicons functioned simultaneously. But S phase actually lasts for >6 hours in a typical somatic cell, which implies that no more than 15% of the replicons are likely to be active at any given moment. (There are some exceptional cases, such as the early embryonic divisions of *Drosophila* embryos, where the duration of S phase is compressed by the simultaneous functioning of a large number of replicons.)

How are origins selected for initiation at different times during S phase? In *S. cerevisiae*, the default appears to be for origins to replicate early, but *cis*-acting sequences can cause origins linked to them to replicate at late times.

Available evidence suggests that chromosomal replicons do not have termini at which the replication forks cease movement and (presumably) dissociate from the DNA. It seems more likely that a replication fork continues from its origin until it meets a fork proceeding toward it from the adjacent replicon. We have already mentioned the potential topological problem of joining the newly synthesized DNA at the junction of the replication forks.

Isolating the origins of yeast replicons

ANY segment of DNA that has an origin should be able to replicate. So although plasmids are rare in eukaryotes, it may be possible to construct them by suitable manipulation *in vitro*. This has been accomplished in yeast, although not in higher eukaryotes.

S. cerevisiae mutants can be "transformed" to the wild phenotype by addition of DNA that carries a wild-type copy of the gene. Some yeast DNA fragments (when circularized) are able to transform defective cells very efficiently. These fragments can survive in the cell in the unintegrated (autonomous) state, that is, as self-replicating plasmids.

A high-frequency transforming fragment possesses a sequence that confers the ability to replicate efficiently in yeast. This segment is called an *ARS* (for autonomously replicating sequence). *ARS* elements are derived from authentic origins of replication; and initiation occurs at the locations of *ARS* elements in a chromosome.

Sequences with *ARS* function occur at about the same average frequency as origins of replication. Where *ARS* elements have been systematically mapped over extended chromosomal regions, it seems that only some of them are actually used to initiate replication. The others are silent, or possibly used only occasionally. If it is true that some origins have varying probabilities of being used, it follows that there can be no fixed termini between replicons. In this case, a given region of a chromosome could be replicated from different origins in different cell cycles.

An *ARS* element consists of an A•T-rich region that contains discrete sites in which mutations affect origin function. Base composition rather than sequence may be important in the rest of the region. **Figure 12.10** shows a systematic mutational analysis along the length of an origin. Origin function is abolished completely by mutations in a 14 bp "core" region, called the **A domain**, that contains an 11 bp consensus sequence consisting of A•T base pairs. This consensus sequence is the only homology between known *ARS* elements.

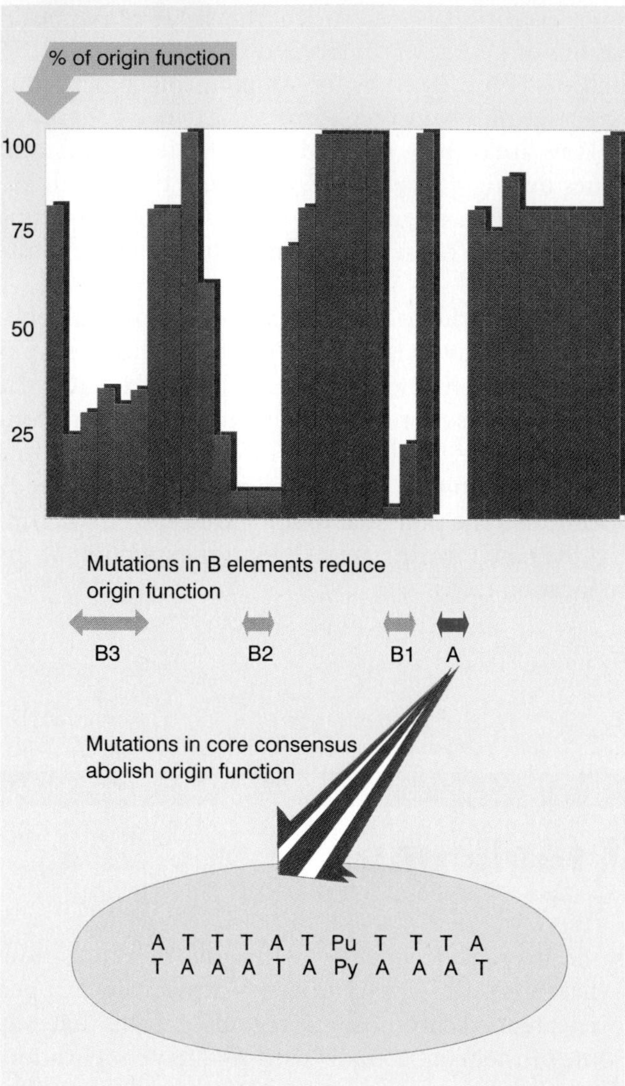

Figure 12.10 An *ARS* extends for ~50 bp and includes a consensus sequence (A) and additional elements (B1-B3).

Mutations in three adjacent elements, numbered B1–B3, reduce origin function. An origin can function effectively with any two of the B elements, so long as a functional A element is present. (Imperfect copies of the core consensus, typically conforming at 9/11 positions, are found close to, or overlapping with, each B element, but they do not appear to be necessary for origin function.)

The ORC (origin recognition complex) is a complex of six proteins with a mass of ~400 kD. ORC is associated with *ARS* elements throughout the cell cycle, so initiation may depend on changes in its condition rather than *de novo* association with an origin. We discuss this in more detail in the next chapter. Counterparts to ORC are found in higher eukaryotic cells.

ARS elements satisfy the classic definition of an origin as a *cis*-acting sequence that causes DNA replication to initiate. Are similar elements to be found in higher eukaryotes? Difficulties in finding sequences comparable to *ARS* elements that can support the existence of plasmids in higher eukaryotic cells suggest the possibility that origins may be more complex (or determined by features other than discrete *cis*-acting sequences). There are suggestions that some animal cell replicons may have complex patterns of initiation: in some cases, many small replication bubbles are found in one region, posing the question of whether there are alternative or multiple starts to replication, and whether there is a small discrete origin. It is fair to say that the nature of the higher eukaryotic origin remains to be established.

D loops maintain mitochondrial origins

The origins of replicons in both prokaryotic and eukaryotic chromosomes are static structures: they comprise sequences of DNA that are recognized in duplex form and used to initiate replication at the appropriate time. Initiation requires separating the DNA strands and commencing bidirectional DNA synthesis. A different type of arrangement is found in mitochondria.

Figure 12.11 The D loop maintains an opening in mammalian mitochondrial DNA, which has separate origins for the replication of each strand.

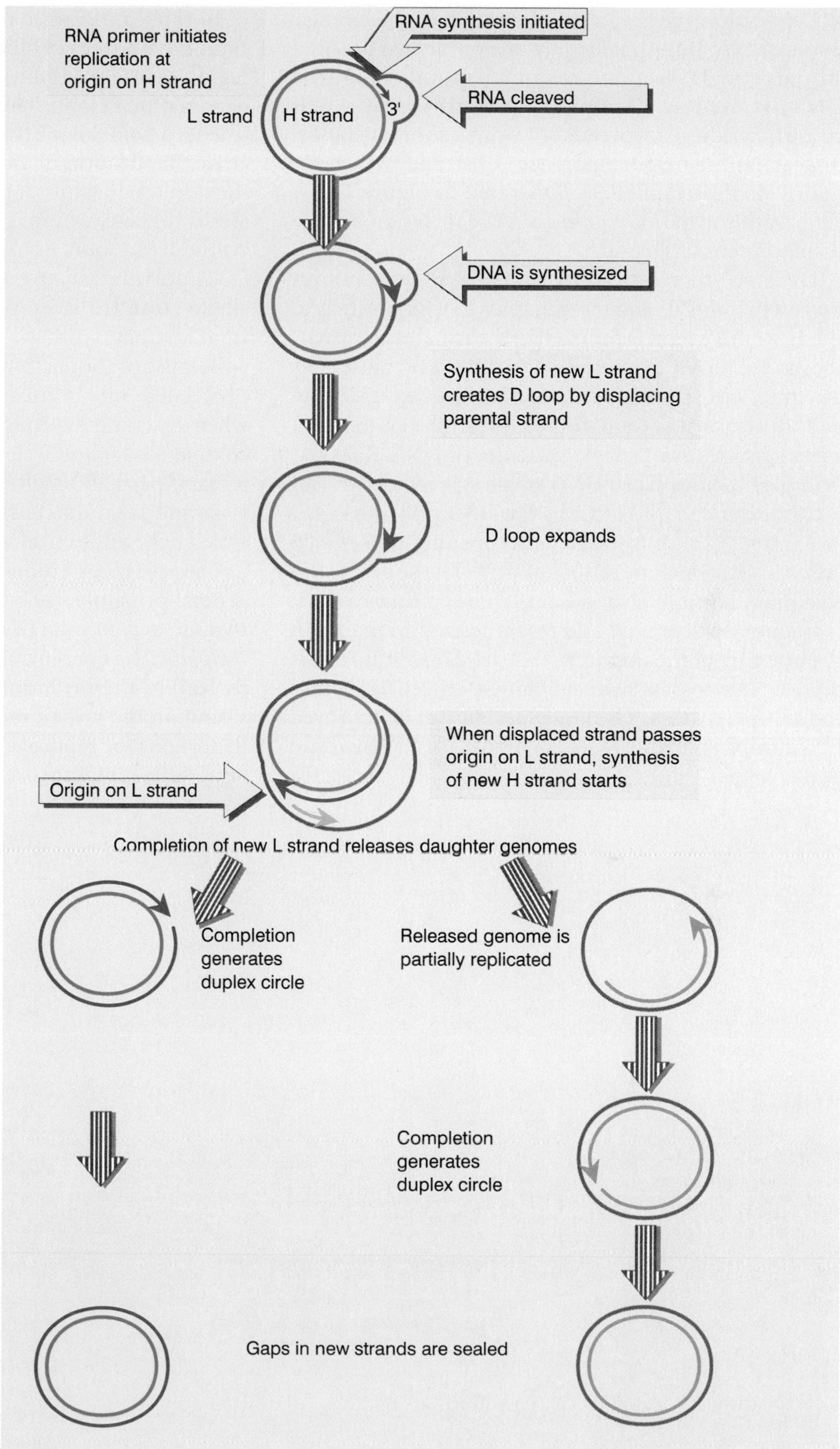
RNA primer initiates
replication at
origin on H strand
RNA synthesis initiated
RNA cleaved
3'
L strand
H strand
DNA is synthesized
Synthesis of new L strand
creates D loop by displacing
parental strand
D loop expands
When displaced strand passes
origin on L strand, synthesis
of new H strand starts
Origin on L strand
Completion of new L strand releases daughter genomes
Completion
generates
duplex circle
Released genome is
partially replicated
Completion
generates
duplex circle
Gaps in new strands are sealed

Replication starts at a specific origin in the circular duplex DNA. But initially only one of the two parental strands (the H strand in mammalian mitochondrial DNA) is used as a template for synthesis of a new strand. Synthesis proceeds for only a short distance, displacing the original partner (L) strand, which remains single-stranded, as illustrated in **Figure 12.11**. The condition of this region gives rise to its name as the **displacement** or **D loop**.

DNA polymerases cannot initiate synthesis, but require a priming 3′ end (see Chapter 13). Replication at the H strand origin is initiated in the usual way, by synthesis of an RNA primer. RNA polymerase transcribes a primer, whose 3′ ends are generated by cleavage by an endonuclease at several discrete sites. The endonuclease is specific for the triple structure of DNA-RNA hybrid plus the displaced DNA single strand. The 3′ end is then extended into DNA by the DNA polymerase.

A single D loop is found as an opening of 500–600 bases in mammalian mitochondria. The short strand that maintains the D loop is unstable and turns over; it is frequently degraded and resynthesized to maintain the opening of the duplex at this site. Some mitochondrial DNAs possess several D loops, reflecting the use of multiple origins. The same mechanism is employed in chloroplast DNA, where (in higher plants) there are two D loops.

To replicate mammalian mitochondrial DNA, the short strand in the D loop is extended. The displaced region of the original L strand becomes longer, expanding the D loop. This expansion continues until it reaches a point about two-thirds of the way around the circle. Replication of this region exposes an origin in the displaced L strand. Synthesis of an H strand initiates at this site, which is used by a special primase that synthesizes a short RNA. The RNA is then extended by DNA polymerase, proceeding around the displaced single-stranded L template in the opposite direction from L-strand synthesis.

Because of the lag in its start, H-strand synthesis has proceeded only a third of the way around the circle when L-strand synthesis finishes. This releases one completed duplex circle and one gapped circle, which remains partially single-stranded until synthesis of the H strand is completed. Finally, the new strands are sealed to become covalently intact.

The existence of rolling circles and D loops exposes a general principle. *An origin can be a sequence of DNA that serves to initiate DNA synthesis using one strand as template.* The opening of the duplex does not necessarily lead to the initiation of replication on the other strand. In the case of mitochondrial DNA replication, the origins for replicating the complementary strands lie at different locations.

Figure 12.12
Replication could run off the 3' end of a newly synthesized linear strand, but could it initiate at a 5' end?

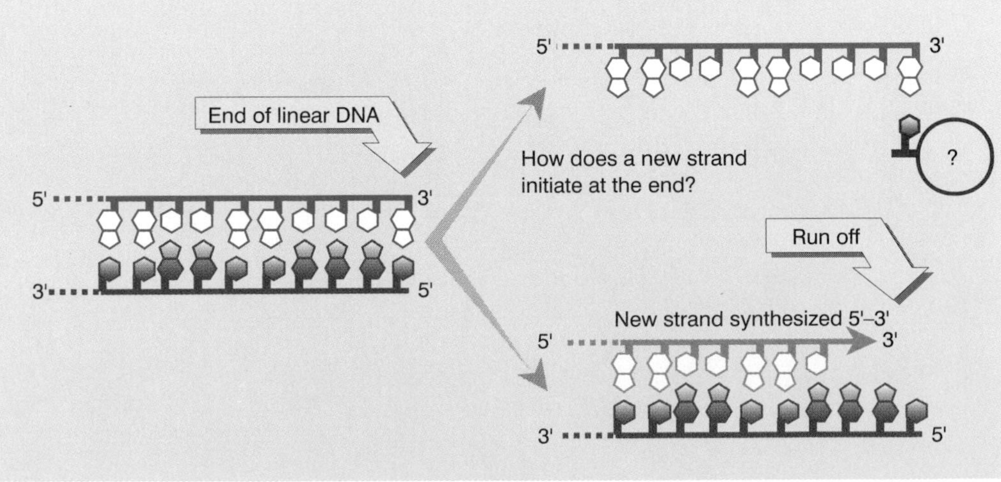

The problem of linear replicons

NONE of the replicons that we have considered so far have a linear end: either they are circular (as in the *E. coli* or mitochondrial genomes) or they are part of longer segregation units (as in eukaryotic chromosomes). But linear replicons occur, in some cases as single extrachromosomal units, and of course at the ends of eukaryotic chromosomes.

The ability of all known nucleic acid polymerases, DNA or RNA, to proceed only in the 5′→3′ direction poses a problem for synthesizing DNA at the end of a linear replicon. Consider the two parental strands depicted in **Figure 12.12**. The lower strand presents no problem: it can act as template to synthesize a daughter strand that runs right up to the end, where presumably the polymerase falls off. But to synthesize a complement at the end of the upper strand, synthesis must start right at the very last base (or else this strand would become shorter in successive cycles of replication).

We do not know whether initiation right at the end of a linear DNA is feasible. We usually think of a polymerase as binding at a site *surrounding* the position at which a base is to be incorporated. So a special mechanism must be employed for replication at the ends of linear replicons. Several types of solution may be imagined to accommodate the need to copy a terminus:

- The problem may be circumvented by converting a linear replicon into a circular or multimeric molecule. Phages such as T4 or lambda use such mechanisms (see later).
- The DNA may form an unusual structure—for example, by creating a hairpin at the terminus, so that there is no free end. Formation of a crosslink is involved in replication of the linear mitochondrial DNA of *Paramecium.*
- Instead of being precisely determined, the end may be variable. Eukaryotic chromosomes may adopt this solution, in which the number of copies of a short repeating unit at the end of the DNA changes (see Chapter 18). A mechanism to add or remove units makes it unnecessary to replicate right up to the very end.
- A protein may intervene to make initiation possible at the actual terminus. Several linear viral nucleic acids have proteins that are *covalently linked to the 5′ terminal base.* The best characterized examples are adenovirus DNA, phage ϕ29 DNA, and poliovirus RNA.

A prime example of initiation at a linear end is provided by adenovirus and ϕ29 DNAs, which actually replicate from both ends, using the mechanism of

Figure 12.13 Adenovirus DNA replication is initiated separately at the two ends of the molecule and proceeds by strand displacement.

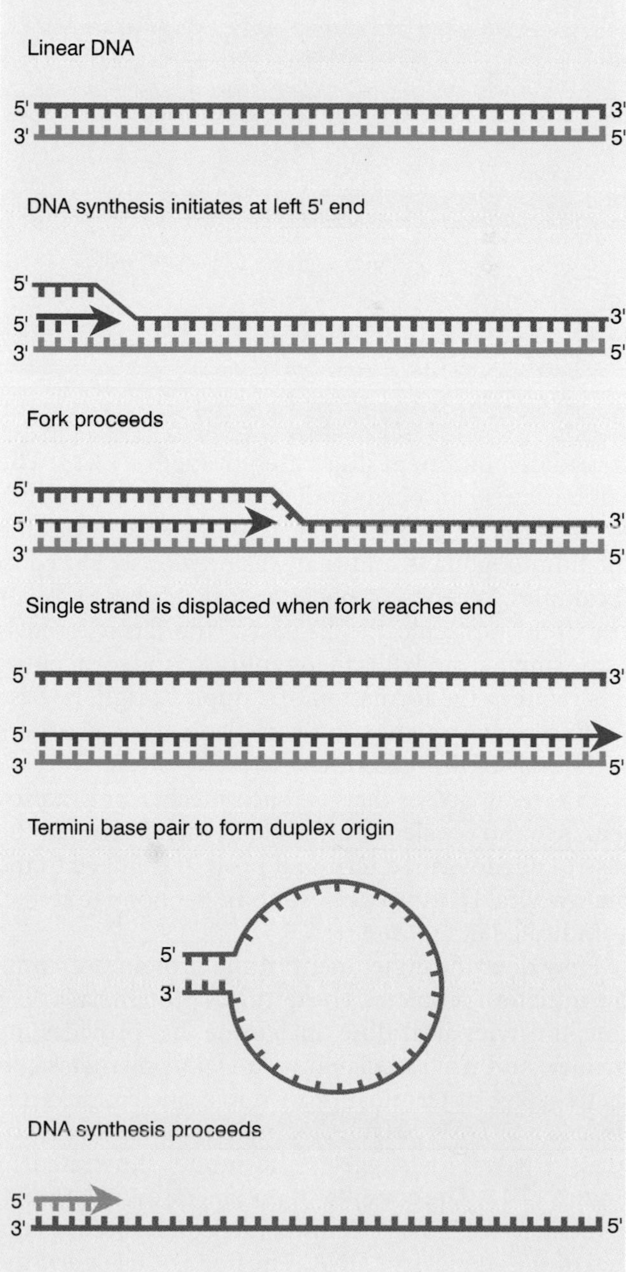
Linear DNA
5'
3'
3'
5'
DNA synthesis initiates at left 5' end
Fork proceeds
Single strand is displaced when fork reaches end
Termini base pair to form duplex origin
DNA synthesis proceeds

Figure 12.14 The 5' terminal phosphate at each end of adenovirus DNA is covalently linked to serine in the 55 kD Ad-binding protein.

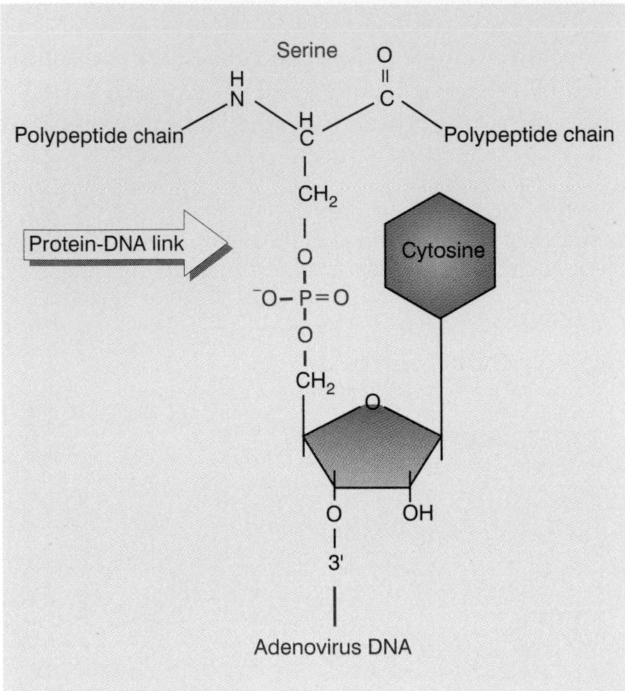

Figure 12.15 Adenovirus terminal protein binds to the 5' end of DNA and provides a C-OH end to prime synthesis of a new DNA strand.

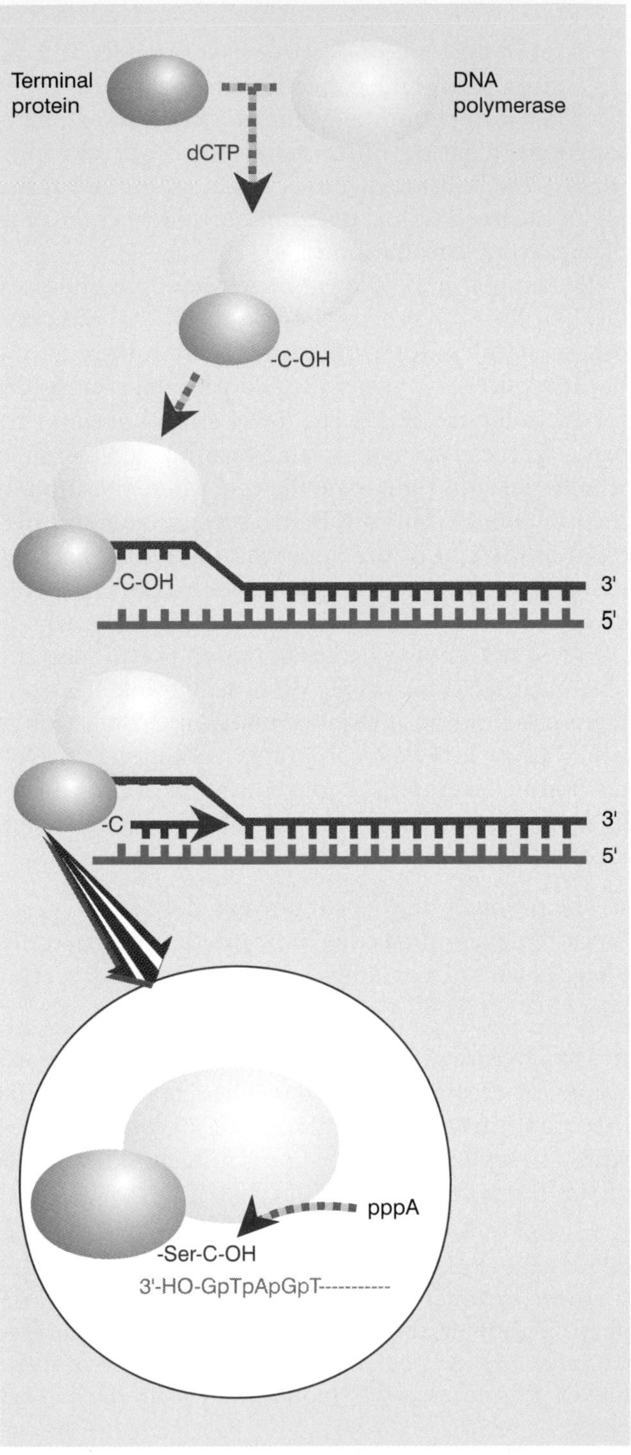

strand displacement illustrated in **Figure 12.13**. The same events can occur independently at either end. Synthesis of a new strand starts at one end, displacing the homologous strand that was previously paired in the duplex. When the replication fork reaches the other end of the molecule, the displaced strand is released as a free single strand. It is then replicated independently; this requires the formation of a duplex origin by base pairing between some short complementary sequences at the ends of the molecule.

In several viruses that use such mechanisms, a protein is found covalently attached to each 5′ end. In the case of adenovirus, a **terminal protein** is linked to the mature viral DNA via a phosphodiester bond to serine, as indicated in **Figure 12.14**.

How does the attachment of the protein overcome the initiation problem? The terminal protein has a dual role: it carries a cytidine nucleotide that provides the primer; and it is associated with DNA polymerase. In fact, linkage of terminal protein to a nucleotide is undertaken by DNA polymerase in the presence of adenovirus DNA. This suggests the model illustrated in **Figure 12.15**. The complex of polymerase and terminal protein, bearing the priming C nucleotide, binds to the end of the adenovirus DNA. The free 3′-OH end of the C nucleotide is used to prime the elongation reaction by the DNA polymerase. This generates a new strand whose 5′ end is covalently linked to the initiating C nucleotide. (The reaction actually involves displacement of protein from DNA rather than binding *de novo*. The 5′ end of adenovirus DNA is bound to the terminal protein that was used in the previous replication cycle. The old terminal protein is displaced by the new terminal protein for each new replication cycle.)

Terminal protein binds to the region located between 9 and 18 bp from the end of the DNA. The adjacent region, between positions 17 and 48, is essential for the binding of a host protein, nuclear factor I, which is also required for the initiation reaction. The initiation complex may therefore form between positions 9 and 48, a fixed distance from the actual end of the DNA.

Rolling circles produce multimers of a replicon

THE structures generated by replication depend on the relationship between the template and the replication fork. The critical features are whether the template is circular or linear, and whether the replication fork is engaged in synthesizing both strands of DNA or only one.

Replication of only one strand is used to generate copies of some circular molecules. A nick opens one strand, and then the free 3′-OH end generated by the nick is extended by the DNA polymerase. The newly synthesized strand displaces the original parental strand. The ensuing events are depicted in **Figure 12.16.**

This type of structure is called a **rolling circle,** because the growing point can be envisaged as rolling around the circular template strand. It could in principle continue to do so indefinitely. As it moves, the replication fork extends the outer strand and displaces the previous partner.

Because the newly synthesized material is covalently linked to the original material, the displaced strand has the original unit genome at its 5′ end. The original unit is followed by any number of unit genomes, synthesized by continuing revolutions of the template. Each revolution displaces the material synthesized in the previous cycle.

An example is shown in the electron micrograph of **Figure 12.17**. The rolling circle is put to several uses *in vivo*. Some pathways that are used to replicate DNA are depicted in **Figure 12.18.**

Cleavage of a unit length tail generates a copy of the original circular replicon in linear form. The linear form may be maintained as a single strand or may be converted into a duplex by synthesis of the complementary strand (which is identical in sequence to the template strand of the original rolling circle).

The rolling circle provides a means for amplifying the original (unit) replicon. This mechanism is used to generate amplified rDNA in the *Xenopus* oocyte. The genes for rRNA are organized as a large number of contiguous repeats in the genome. A single repeating unit from the genome is converted into a rolling circle. The displaced tail, containing many units, is converted into duplex DNA; later it is cleaved from the circle so that the two ends can be joined together to generate a large circle of amplified rDNA. The amplified material therefore consists of a large number of identical repeating units.

Replication by rolling circles is common among bacteriophages. Unit genomes can be cleaved from the displaced tail, generating monomers that can be packaged into phage particles or used for further replication cycles. A more detailed view of a phage replication cycle that is centered on the rolling circle is given in **Figure 12.19.** Phage ϕX174 consists of a single-stranded circular DNA, known as the plus (+) strand. A complementary strand, called the minus (−) strand, is synthesized. This action generates the duplex circle shown at the top of the figure, which is then replicated by a rolling circle mechanism.

The duplex circle is converted to a covalently closed form, which becomes supercoiled. A protein coded by the phage genome, the A protein, nicks the (+) strand of the duplex DNA at a specific site that defines the origin for replication. After nicking the origin, the A protein remains connected to the 5′ end that it generates, while the 3′ end is extended by DNA polymerase.

The structure of the DNA plays an important role in this reaction, for the DNA can be nicked *only when it is supercoiled*. The A protein is able to bind to a single-stranded decamer fragment of DNA that surrounds the site of the nick. This suggests that the supercoiling

Figure 12.16 The rolling circle generates a multimeric single-stranded tail.

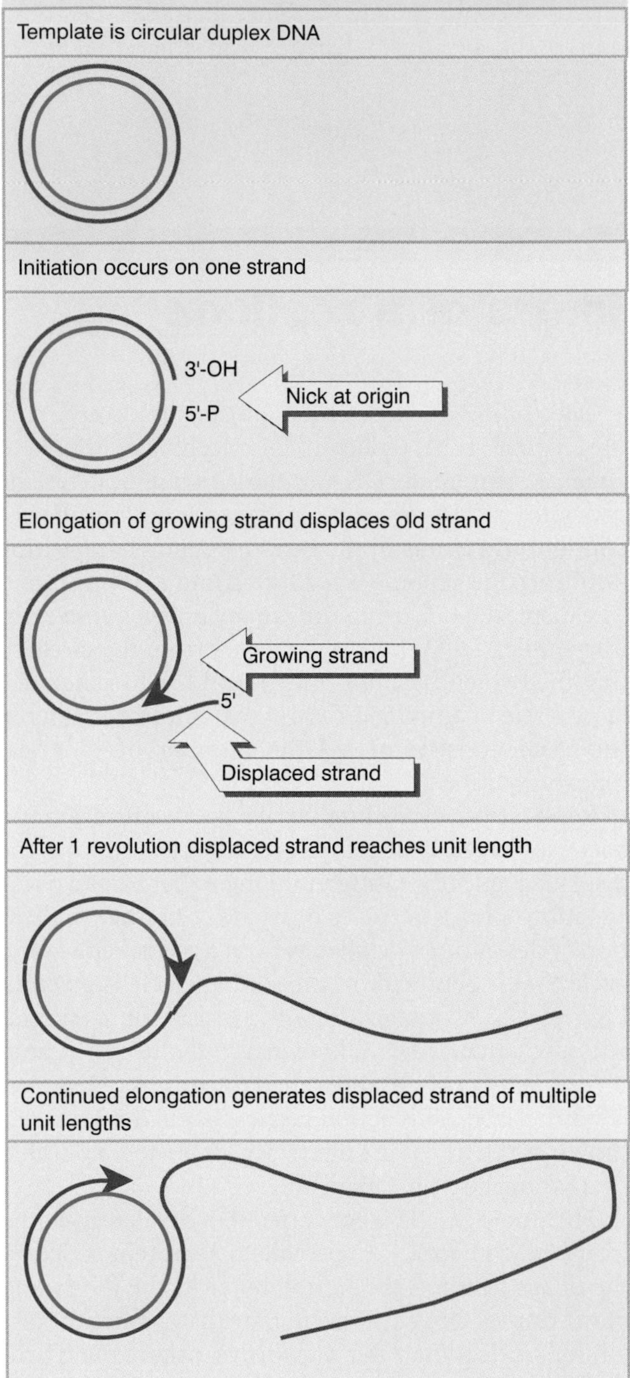
Template is circular duplex DNA
Initiation occurs on one strand
3'-OH
5'-P
Nick at origin
Elongation of growing strand displaces old strand
Growing strand
5'
Displaced strand
After 1 revolution displaced strand reaches unit length
Continued elongation generates displaced strand of multiple unit lengths

Figure 12.17 A rolling circle appears as a circular molecule with a linear tail by electron microscopy. Photograph kindly provided by David Dressler.

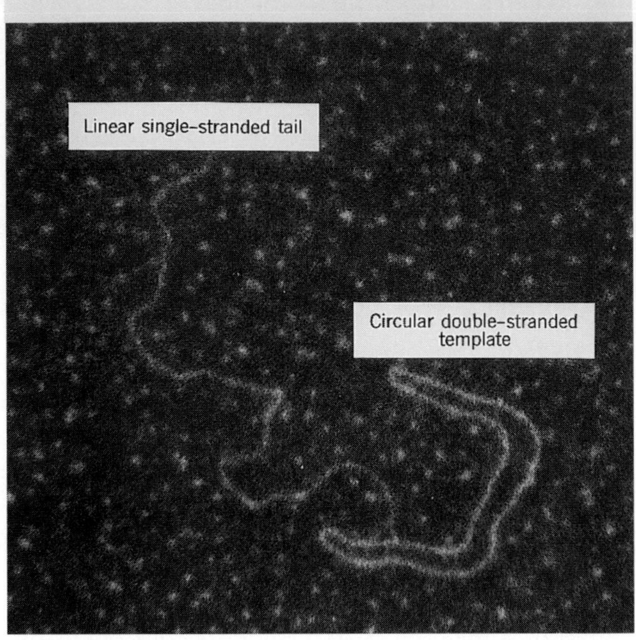
Linear single-stranded tail
Circular double-stranded template

is needed to assist the formation of a single-stranded region that provides the A protein with its binding site. (An enzymatic activity in which a protein cleaves duplex DNA and binds to a released 5′ end is sometimes called a **relaxase.**) The nick generates a 3′-OH end and a 5′-phosphate end (covalently attached to the A protein), both of which have roles to play in ϕX174 replication.

Using the rolling circle, the 3′-OH end of the nick is extended into a new chain. The chain is elongated around the circular (–) strand template, until it reaches the starting point and displaces the origin. Now the A protein functions again. It remains connected with the rolling circle as well as to the 5′ end of the displaced tail, and it is therefore in the vicinity as the growing point returns past the origin. So the same A protein is available again to recognize the origin and nick it, now attaching to the end generated by the new nick. The cycle can be repeated indefinitely.

Following this nicking event, the displaced single

Figure 12.18 Rolling circles can be used for varying purposes, depending on the fate of the displaced tail. Cleavage at unit length generates monomers, which can be converted to duplex and circular forms. Cleavage of multimers generates a series of tandemly repeated copies of the original unit. Note that the conversion to double-stranded form could occur earlier, before the tail is cleaved from the rolling circle.

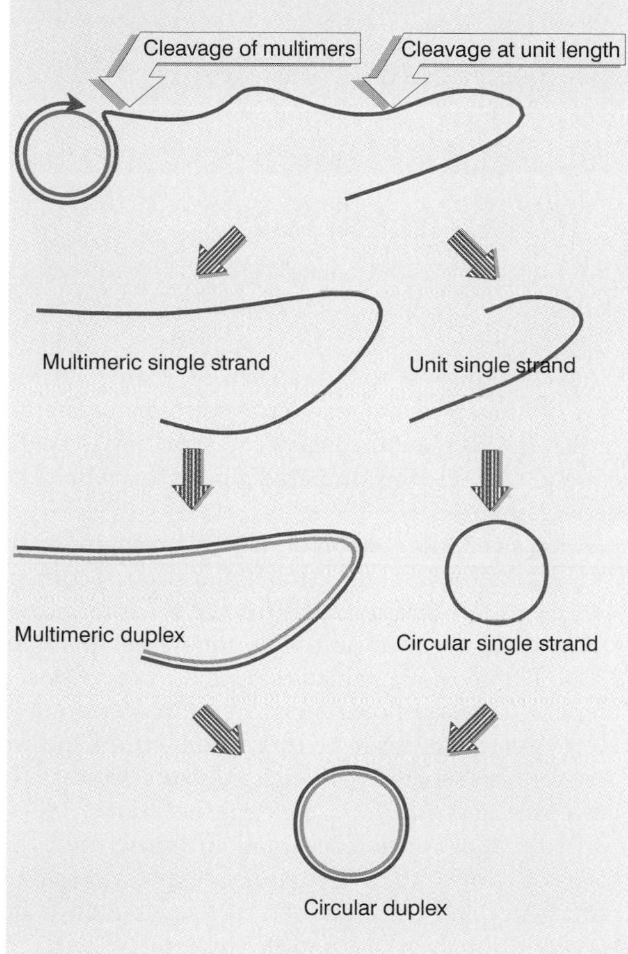

(+) strand is freed as a circle. The A protein is involved in the circularization. In fact, the joining of the 3′ and 5′ ends of the (+) strand product is accomplished by the A protein as part of the reaction by which it is released at the end of one cycle of replication, and starts another cycle.

The A protein has an unusual property that may be connected with these activities. It is *cis*-acting *in vivo.* (This behavior is not reproduced *in vitro*, as can be seen from its activity on any DNA template in a cell-free system.) *The implication is that* in vivo *the A*

Figure 12.19 ϕX174 RF DNA is a template for synthesizing single-stranded viral circles. The A protein remains attached to the same genome through indefinite revolutions, each time nicking the origin on the viral (+) strand and transferring to the new 5′ end. At the same time, the released viral strand is circularized.

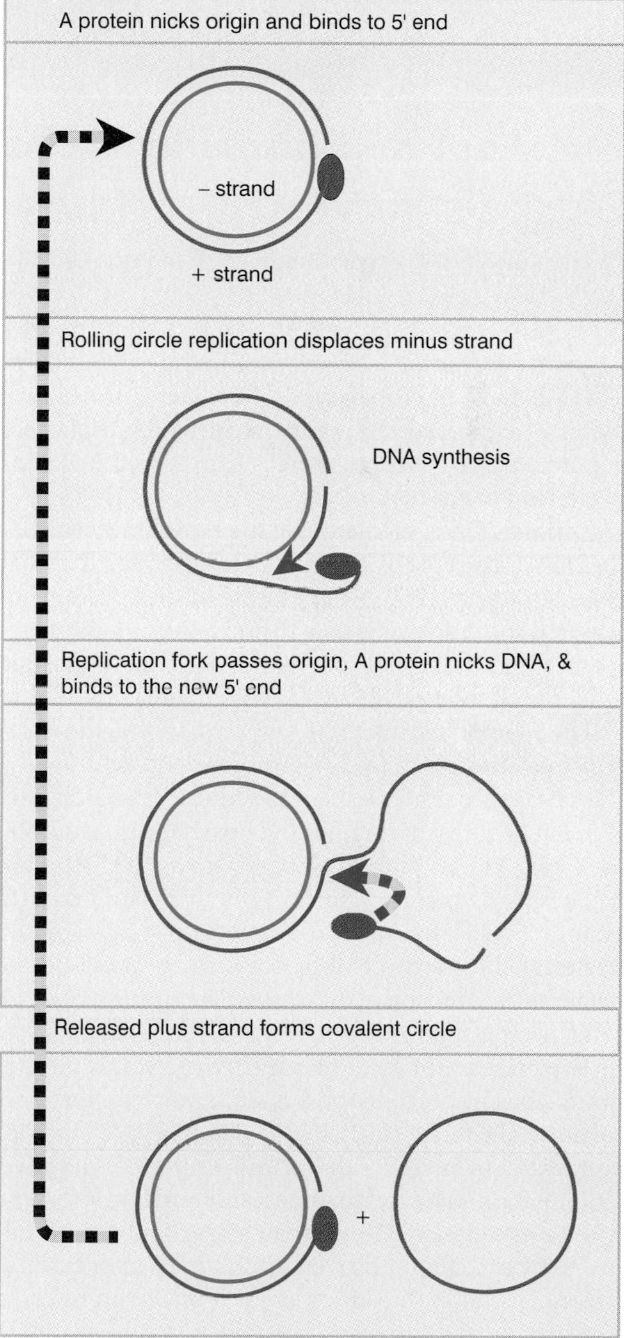

protein synthesized by a particular genome can attach only to the DNA of that genome. We do not know how this is accomplished. However, its activity *in vitro* shows how it remains associated with the same parental (–) strand template. The A protein has two active sites; this may allow it to cleave the "new" origin while still retaining the "old" origin; then it ligates the displaced strand into a circle.

The displaced (+) strand may follow either of two fates after circularization. During the replication phase of viral infection, it may be used as a template to synthesize the complementary (–) strand. The duplex circle may then be used as a rolling circle to generate more progeny. During phage morphogenesis, the displaced (+) strand is packaged into the phage virion.

Single-stranded genomes are generated for bacterial conjugation

ANOTHER example of a connection between replication and the propagation of a genetic unit is provided by bacterial **conjugation**, in which a plasmid genome or host chromosome is transferred from one bacterium to another.

Conjugation is mediated by the **F plasmid**, which is the classic example of an episome, an element that may exist as a free circular plasmid, or that may become integrated into the bacterial chromosome as a linear sequence (like a lysogenic bacteriophage). The F plasmid is a large circular DNA, ~100 kb in length.

The F factor can integrate at several sites in the *E. coli* chromosome, often by a recombination event involving certain sequences (called IS sequences; see Chapter 15) that are present on both the host chromosome and the F plasmid. In its free (plasmid) form, the F plasmid utilizes its own replication origin (*oriV*) and control system, and is maintained at a level of one copy per bacterial chromosome. When it is integrated into the bacterial chromosome, this system is suppressed, and F DNA is replicated as a part of the chromosome.

The presence of the F plasmid, whether free or integrated, has important consequences for the host bacterium. Bacteria that are F-positive are able to conjugate (or mate) with bacteria that are F-negative. Conjugation involves a contact between donor (F-positive) and recipient (F-negative) bacteria; contact is followed by transfer of the F factor. If the F factor exists as a free plasmid in the donor bacterium, it is transferred as a plasmid, and the infective process converts the F-negative recipient into an F-positive state. If the F factor is present in an integrated form in the donor, the transfer process may also cause some or all of the bacterial chromosome to be transferred. Many plasmids have conjugation systems that operate in a generally similar manner, but the F factor was the first to be discovered, and remains the paradigm for this type of genetic transfer.

A large (~33 kb) region of the F plasmid, called the **transfer region**, is required for conjugation. It contains ~40 genes that are required for the transmission of DNA; their organization is summarized in **Figure 12.20**. The genes are named as *tra* and *trb* loci. Most of them are expressed coordinately as part of a single 32 kb transcription unit (the *traY-I* unit). *traM* and *traJ* are expressed separately. *traJ* is a regulator that turns on both *traM* and *traY-I*. On the opposite strand, *finP* is a regulator that codes for a small antisense RNA that turns off *traJ*. Its activity requires expression of another gene, *finO*. Only four of the *tra* genes in the major transcription unit are concerned directly with the transfer of DNA; most are concerned with the properties of the bacterial cell surface and with maintaining contacts between mating bacteria.

F-positive bacteria possess surface appendages called **pili** that are coded by the F factor. The gene *traA* codes for the single subunit protein, **pilin**, that is polymerized into the pilus. At least 12 *tra* genes are required for the modification and assembly of pilin into the pilus. The F-pili are hair-like structures, 2–3 μm long, that protrude from the bacterial surface. A typical F-positive cell has 2–3 pili. The pilin subunits are polymerized into a hollow cylinder, ~8 nm in diameter, with a 2 nm axial hole.

Mating is initiated when the tip of the F-pilus contacts the surface of the recipient cell. **Figure 12.21**

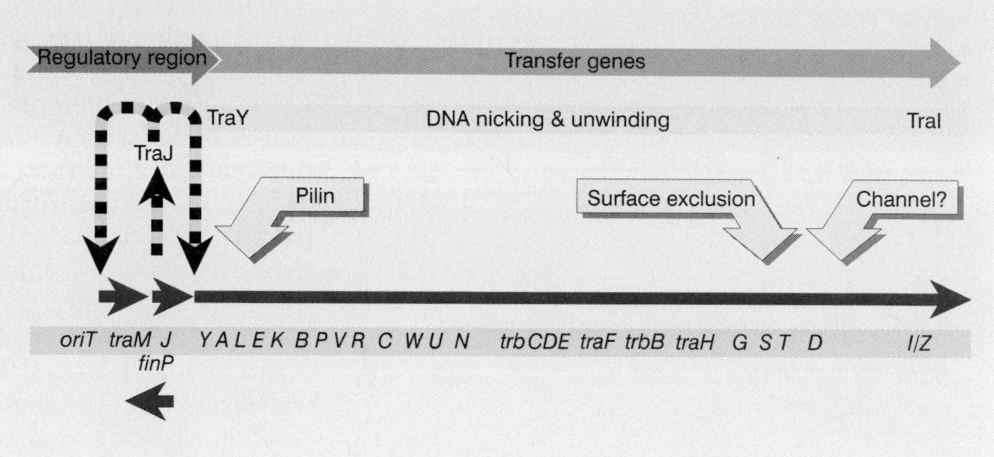

Figure 12.20 The *tra* region of the F plasmid contains the genes needed for bacterial conjugation.

shows an example of *E. coli* cells beginning to mate. A donor cell does not contact other cells carrying the F factor, because the genes *traS* and *traT* code for "surface exclusion" proteins that make the cell a poor recipient in such contacts. This effectively restricts donor cells to mating with F-negative cells. (Also note that the presence of F-pili has secondary consequences; they provide the sites to which RNA phages and some single-stranded DNA phages attach, so F-positive bacteria are susceptible to infection by these phages, whereas F-negative bacteria are resistant.)

The initial contact between donor and recipient cells is easily broken, but other *tra* genes act to stabilize the association, bringing the mating cells closer together.

Figure 12.21 Mating bacteria are initially connected when donor F pili contact the recipient bacterium. Photograph kindly provided by Ron Skurray.

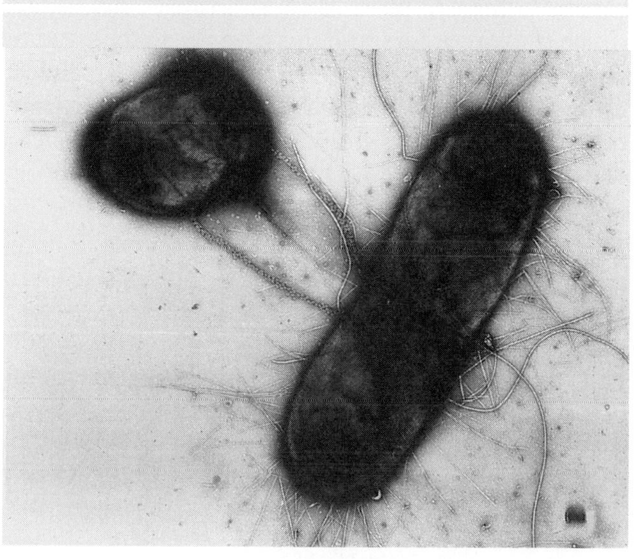

The F pili are essential for initiating pairing, but retract or disassemble as part of the process by which the mating cells are brought into close contact. There must be a channel through which DNA is transferred, but the pilus itself does not appear to provide it. TraD is an inner membrane protein in F^+ bacteria that is necessary for transport of DNA and it may provide or be part of the channel.

Transfer of the F factor is initiated at a site *oriT*, the origin of transfer, which is located at one end of the transfer region. The transfer process may be initiated when TraM recognizes that a mating pair has formed. Then TraY binds near *oriT* and causes TraI to bind. TraI is a relaxase, like ϕX174 A protein. TraI nicks *oriT* at a unique site (called *nic*), and then forms a covalent link to the 5′ end that has been generated. TraI also catalyzes the unwinding of ~200 bp of DNA (this is a helicase activity; see later). **Figure 12.22** shows that the freed 5′ end leads the way into the recipient bacterium. A complement for the transferred single strand is synthesized in the recipient bacterium, which as a result is converted to the F-positive state.

A complementary strand must be synthesized in the donor bacterium to replace the strand that has been transferred. If this happens concomitantly with the transfer process, the state of the F plasmid will resemble the rolling circle of Figure 12.16 (and will not generate the extensive single-stranded regions shown in Figure 12.22). Conjugating DNA usually appears like a rolling circle, but replication as such is not necessary to provide the driving energy, and single-strand transfer is independent of DNA synthesis. Only a single unit length of the F factor is transferred to the recipient bacterium. This implies that some (unidentified) feature terminates the process after one revolution, after which the covalent integrity of the F plasmid is restored.

Figure 12.22 Transfer of DNA occurs when the F factor is nicked at *oriT* and a single strand is led by the 5' end into the recipient. Only one unit length is transferred. Complementary strands are synthesized to the single strand remaining in the donor and to the strand transferred into the recipient.

DONOR
RECIPIENT
F-positive donors conjugate with F-negative recipients
TraY/I nick DNA at oriT
oriT
3'
5'
TraY/I multimer migrates around circle, unwinding DNA
5'
Single strand enters recipient
Complementary strands are synthesized
Donor gap is closed
Recipient circularizes

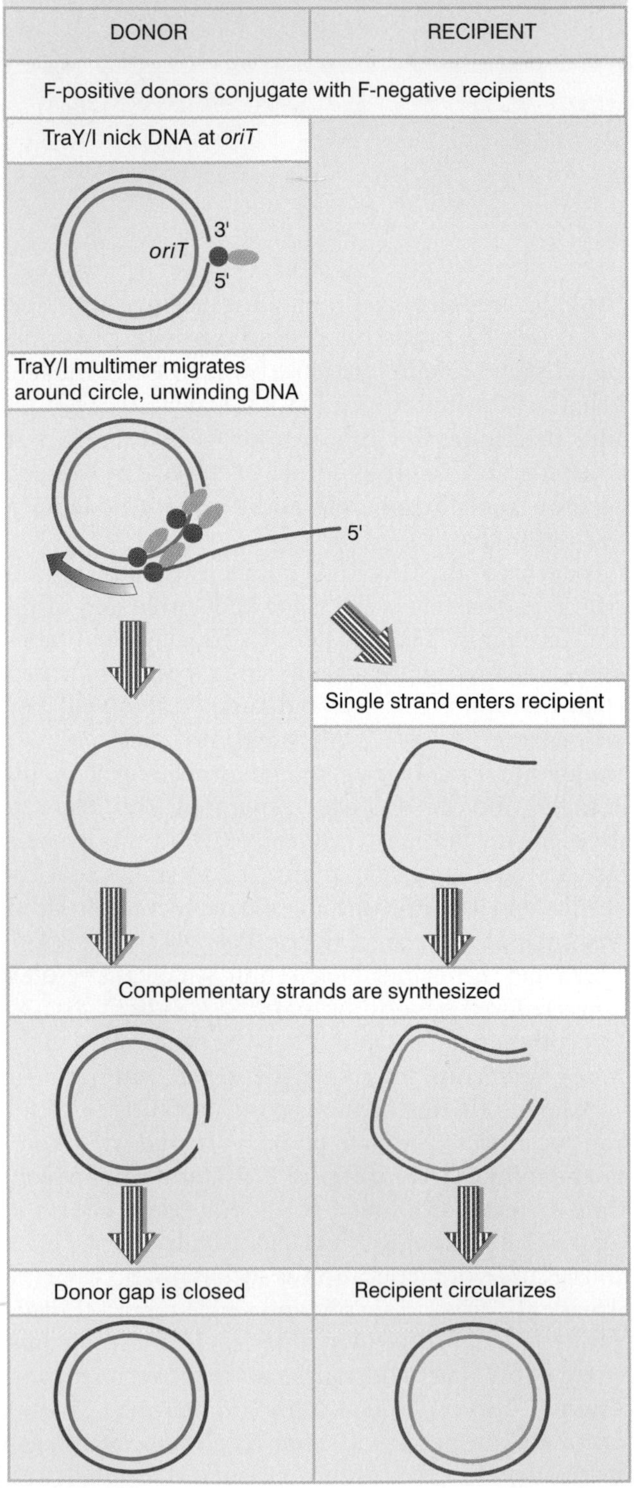

When an integrated F plasmid initiates conjugation, the orientation of transfer is directed away from the transfer region, into the bacterial chromosome. **Figure 12.23** shows that, following a short leading sequence of F DNA, bacterial DNA is transferred. The process continues until it is interrupted by the breaking of contacts between the mating bacteria. It takes ~100 minutes to transfer the entire bacterial chromosome, and under standard conditions, contact is often broken before the completion of transfer.

Donor DNA that enters a recipient bacterium is converted to double-stranded form, and may recombine with the recipient chromosome. (Note that two recombination events are required to insert the donor DNA.) So conjugation affords a means to exchange genetic material between bacteria (a contrast with their usual asexual growth). A strain of *E. coli* with an integrated F factor supports such recombination at relatively high frequencies (compared to strains that lack integrated F factors); such strains are described as **Hfr** (for high-frequency recombination). Each position of integration for the F factor gives rise to a different Hfr strain, with a characteristic pattern of transferring bacterial markers to a recipient chromosome.

Contact between conjugating bacteria is usually broken before transfer of DNA is complete. As a result, the probability that a region of the bacterial chromosome will be transferred depends upon its distance from *oriT*. Bacterial genes located close to the site of F integration (in the direction of transfer) enter recipient bacteria first, and are therefore found at greater frequencies than those located farther away that enter later. This gives rise to a gradient of transfer frequencies around the chromosome, declining from the position of F integration. Marker positions on the donor chromosome can be assayed in terms of the time at which transfer occurs, and this gave rise to the standard description of the *E. coli* chromosome as a map divided into 100 minutes. The map refers to transfer times from a particular Hfr strain; the gradient of transfer is of course different for each Hfr strain.

Figure 12.23 Transfer of chromosomal DNA occurs when an integrated F factor is nicked at *oriT*. Transfer of DNA starts with a short sequence of F DNA and continues until prevented by loss of contact between the bacteria. Following synthesis of a complementary strand, the transferred material may recombine with the bacterial chromosome in the recipient.

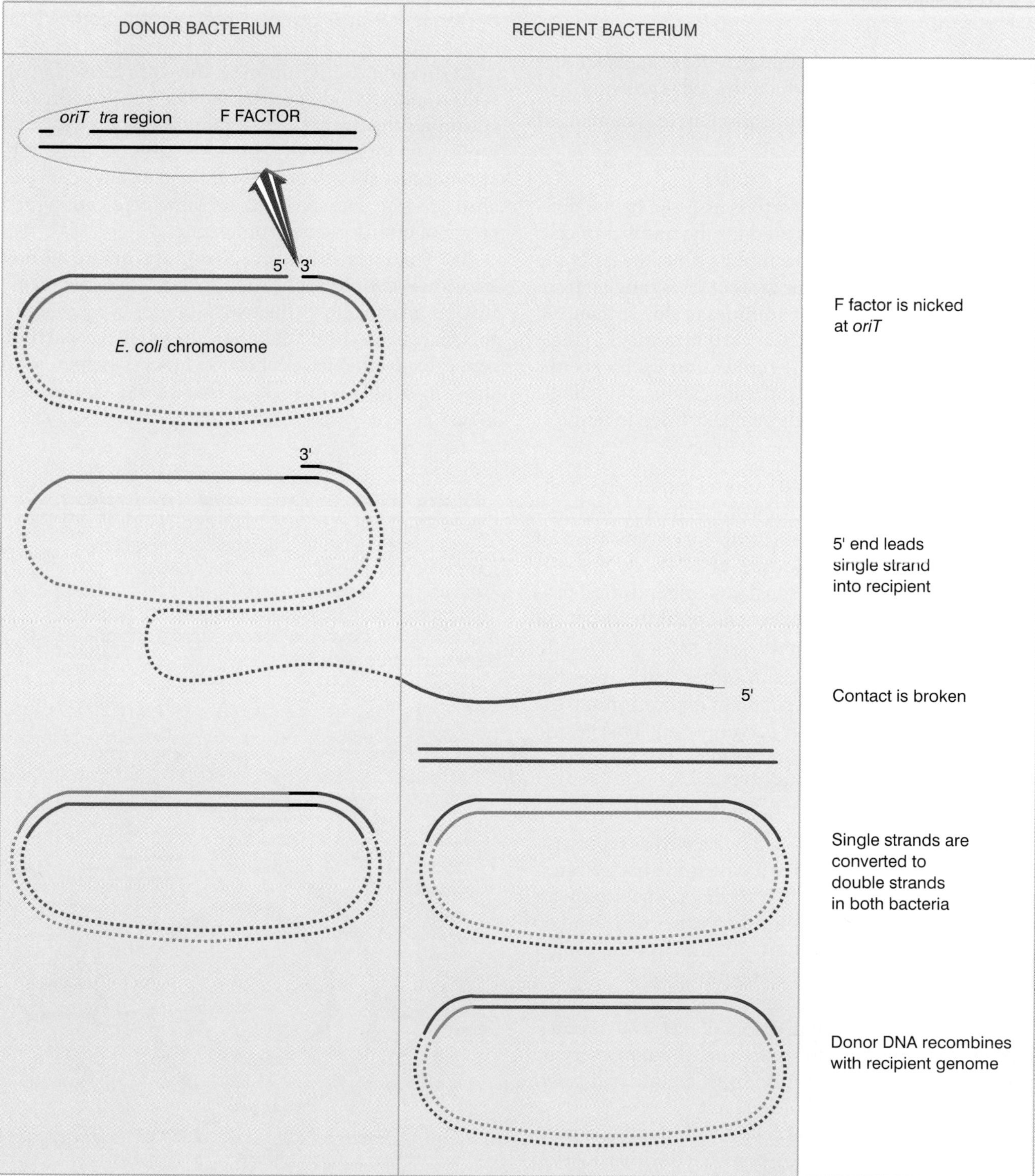
DONOR BACTERIUM
RECIPIENT BACTERIUM
oriT tra region
F FACTOR
5'
3'
E. coli chromosome
F factor is nicked at oriT
3'
5' end leads single strand into recipient
5'
Contact is broken
Single strands are converted to double strands in both bacteria
Donor DNA recombines with recipient genome

Connecting bacterial replication to the cell cycle

Bacteria have two links between replication and cell growth:

- The frequency of initiation of cycles of replication is adjusted to fit the rate at which the cell is growing.
- The completion of a replication cycle is connected with division of the cell.

The rate of bacterial growth is assessed by the **doubling time**, the period required for the number of cells to double. The shorter the doubling time, the faster the growth rate. *E. coli* cells can grow at rates ranging from doubling times as fast as 18 minutes to slower than 180 minutes. Because the bacterial chromosome is a single replicon, the frequency of replication cycles is controlled by the number of initiation events at the single origin. The replication cycle can be defined in terms of two constants:

- *C* is the fixed time of ~40 minutes required to replicate the entire bacterial chromosome. Its duration corresponds to a rate of replication fork movement of ~50,000 bp/minute. (The rate of DNA synthesis is more or less invariant at a constant temperature; it proceeds at the same speed unless and until the supply of precursors becomes limiting.)
- *D* is the fixed time of ~20 minutes that elapses between the completion of a round of replication and the cell division with which it is connected. This period may represent the time required to assemble the components needed for division.

(The constants *C* and *D* can be viewed as representing the maximum speed with which the bacterium is capable of completing these processes. They apply for all growth rates between doubling times of 18 and 60 minutes, but both constant phases become longer when the cell cycle occupies >60 minutes.)

A cycle of chromosome replication must be initiated a fixed time before a cell division, $C + D = 60$ minutes. For bacteria dividing more frequently than every 60 minutes, a cycle of replication must be initiated *before the end of the preceding division cycle.*

Consider the example of cells dividing every 35 minutes. The cycle of replication connected with a division must have been initiated 25 minutes before the preceding division. This situation is illustrated in **Figure 12.24**, which shows the chromosomal complement of a bacterial cell at 5-minute intervals throughout the cycle.

At division (35/0 minutes), the cell receives a partially replicated chromosome. The replication fork continues to advance. At 10 minutes, when this "old" replication fork has not yet reached the terminus, initiation occurs at both origins on the partially replicated chromosome. The start of these "new" replication forks creates a **multiforked chromosome**.

At 15 minutes—that is, at 20 minutes before the next division—the old replication fork reaches the terminus. Its arrival allows the two daughter chromosomes to separate; each of them has already been partially replicated by the new replication forks (which now are the *only* replication forks). These forks continue to advance.

Figure 12.24 The fixed interval of 60 minutes between initiation of replication and cell division produces multiforked chromosomes in rapidly growing cells. Note that only the replication forks moving in one direction are shown; actually the chromosome is replicated symmetrically by two sets of forks moving in opposite directions on circular chromosomes.

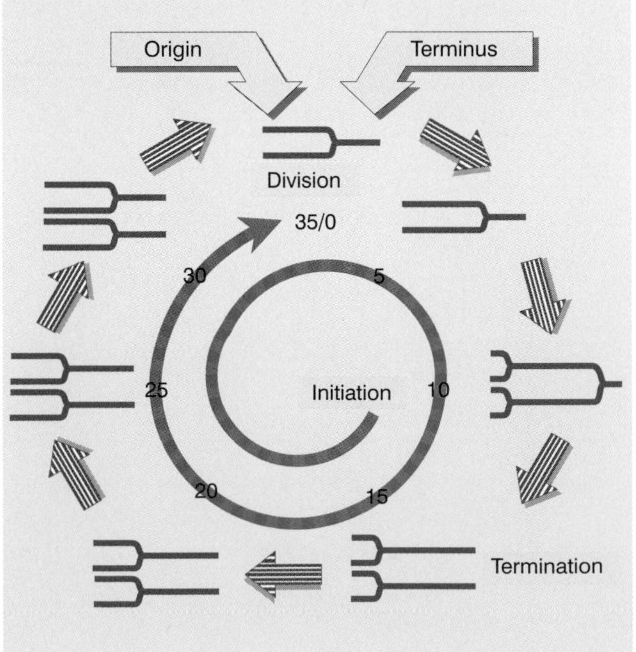

At the point of division, the two partially replicated chromosomes segregate. This re-creates the point at which we started. The single replication fork becomes "old", it terminates at 15 minutes, and 20 minutes later there is a division. We see that the initiation event occurs $1\frac{25}{35}$ cell cycles before the division event with which it is associated.

The general principle of the link between initiation and the cell cycle is that, *as cells grow more rapidly (the cycle is shorter), the initiation event occurs an increasing number of cycles before the related division.* There are correspondingly more chromosomes in the individual bacterium. This relationship can be viewed as the cell's response to its inability to reduce the periods of *C* and *D* to keep pace with the shorter cycle.

How does the cell know when to initiate the replication cycle? The initiation event occurs at a constant ratio of cell mass to the number of chromosome origins. Cells growing more rapidly are larger and possess a greater number of origins. The growth of the bacterium can be described in terms of the **unit cell**, an entity 1.7 μm long. A bacterium contains one origin per unit cell; a rapidly growing cell with two origins will be 1.7–3.4 μm long. In terms of Figure 12.24, it is at the point 10 minutes after division that the cell mass has increased sufficiently to support an initiation event at both available origins.

How is cell mass titrated?. An initiator protein could be synthesized continuously throughout the cell cycle; accumulation of a critical amount would trigger initiation. This explains why protein synthesis is needed for the initiation event. An alternative possibility is that an inhibitor protein might be synthesized at a fixed point, and diluted below an effective level by the increase in cell volume.

Cell division and chromosome segregation

CHROMOSOME segregation in bacteria is especially interesting because the DNA itself is involved in the mechanism for partition. (This contrasts with eukaryotic cells, in which segregation is achieved by the complex apparatus of mitosis.) The bacterial apparatus is quite accurate, however; anucleate cells form <0.03% of a bacterial population.

The division of a bacterium into two daughter cells is accomplished by the formation of a **septum**, a structure that forms in the center of the cell as an invagination from the surrounding envelope. The septum forms an impenetrable barrier between the two parts of the cell and provides the site at which the two daughter cells eventually separate entirely. Two related questions address the role of the septum in division: what determines the location at which it forms; and what ensures that the daughter chromosomes lie on opposite sides of it?

The formation of the septum is preceded by the organization of the **periseptal annulus**. This is observed as a zone in *E. coli* or *S. typhimurium* in which the structure of the envelope is altered so that the inner membrane is connected more closely to the cell wall and outer membrane layer. As its name suggests, the annulus extends around the cell. **Figure 12.25** summarizes its development.

The annulus is observed in a central position in a new cell. As the cell grows, two events occur. A septum forms at the midcell position defined by the annulus. And new annuli form on either side of the initial annulus. These new annuli are displaced from the center and move along the cell to positions at 1/4 and 3/4 of the cell length. These will become the midcell positions after the next division. The displacement of the periseptal annulus to the correct position may be the crucial event that ensures the division of the cell into daughters of equal size. (The mechanism of movement is unknown.) Septation begins when the cell reaches a fixed length (2*L*), and the distance between the new annuli is always *L*. We do not know how the cell measures length, but the relevant parameter appears to be linear distance as such (not area or volume).

The septum consists of the same components as the cell envelope: the periplasmic layer contains a rigid layer of peptidoglycan, made by polymerization of tri- or pentapeptide-disaccharide units in a reaction involving connections between both types of subunit (transpeptidation and transglycosylation). The rod-like shape of the bacterium is maintained by a pair of activities, PBP2 and RodA, which are responsible for extending the envelope; mutations in the gene for either protein cause the bacterium to lose its extended shape, becoming round. This demonstrates the important principle that shape and rigidity can be determined by the simple extension of a polymeric structure. The enzyme responsible for generating the peptidoglycan in the septum is PBP3 (penicillin-binding protein 3), a membrane-bound protein that has its catalytic site in the periplasm. The septum initially forms as a double layer of peptidoglycan, and the protein EnvA is required to split the covalent links between the layers, so that the daughter cells may separate.

The behavior of the periseptal annulus suggests that the mechanism for measuring position is associated with the cell envelope. It is plausible to suppose that the envelope could also be used to ensure segregation of the chromosomes. A direct link between DNA and the membrane could account for segregation. If daughter chromosomes are attached to the membrane, they could be physically separated when the septum forms. **Figure 12.26** shows that the formation of a septum could segregate the chromosomes into the different daughter cells if the origins are connected to sites that lie on either side of the periseptal annulus.

A difficulty in isolating mutants that affect cell division is that mutations in the critical functions may be lethal and/or pleiotropic. For example, if formation of the annulus occurs at a site that is essential for overall growth of the envelope, it would be difficult to distinguish mutations that specifically interfere with annulus formation from those that inhibit envelope growth generally.

Mutations that affect cell division or chromosome

Figure 12.25 Duplication and displacement of the periseptal annulus give rise to the formation of a septum that divides the cell.

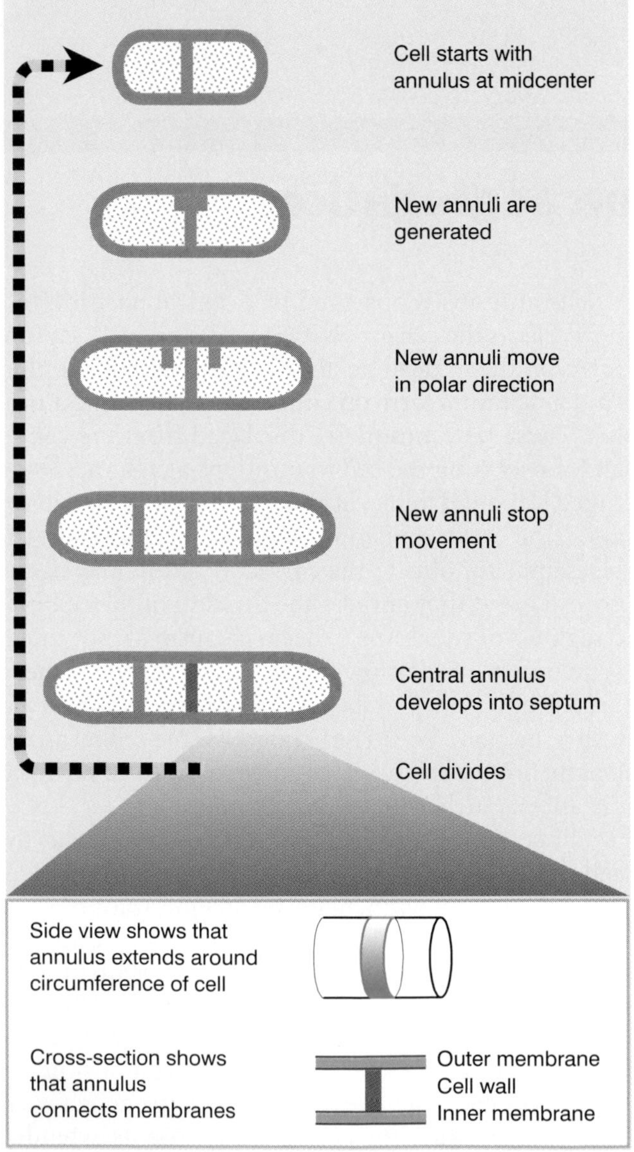
Cell starts with annulus at midcenter
New annuli are generated
New annuli move in polar direction
New annuli stop movement
Central annulus develops into septum
Cell divides
Side view shows that annulus extends around circumference of cell
Cross-section shows that annulus connects membranes
Outer membrane
Cell wall
Inner membrane

Figure 12.26 Attachment of bacterial DNA to the membrane could provide a mechanism for segregation.

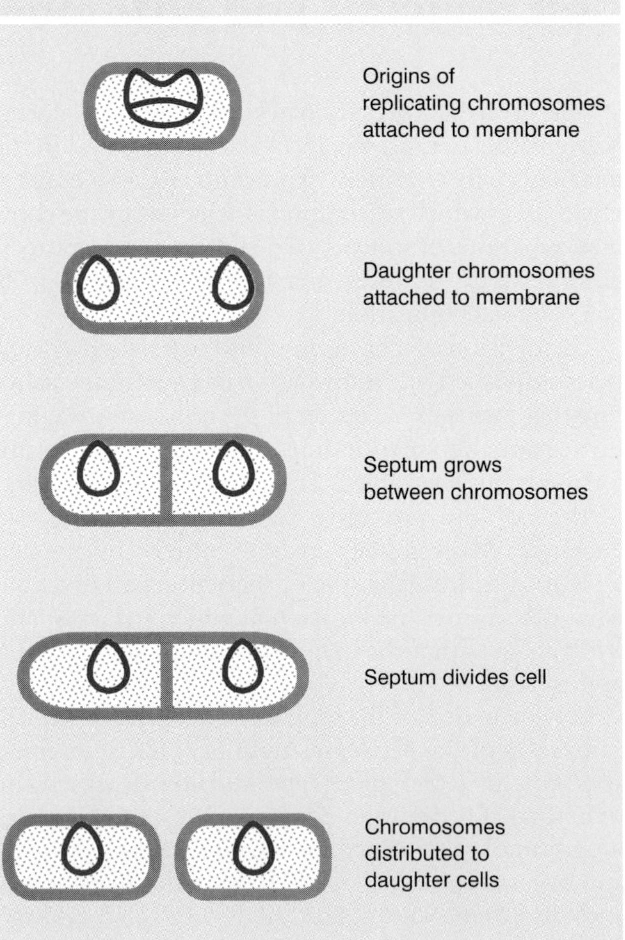
Origins of replicating chromosomes attached to membrane
Daughter chromosomes attached to membrane
Septum grows between chromosomes
Septum divides cell
Chromosomes distributed to daughter cells

segregation cause striking phenotypic changes. **Figures 12.27** and **12.28** illustrate the opposite consequences of failure in the division process and failure in segregation:

- Long **filaments** form when septum formation is inhibited, but chromosome replication is unaffected. The bacteria continue to grow, and even continue to segregate their daughter chromosomes, but septa do not form, so the cell consists of a very long filamentous structure, with the nucleoids (bacterial chromosomes) regularly distributed along the length of the cell. This phenotype is displayed by *fts* mutants (named for *t*emperature-*s*ensitive *f*ilamentation), which identify defect(s) that lie in the division process itself.

- **Minicells** form when septum formation occurs too frequently or in the wrong place, with the result that one of the new daughter cells lacks a chromosome. The minicell has a rather small size, and lacks DNA, but otherwise appears morphologically normal. **Anucleate** cells form when segregation is aberrant; like minicells, they lack a chromosome, but because septum formation is normal, their size is unaltered. This phenotype is caused by *par* (partition) mutants (named because they are defective in chromosome segregation).

The gene *ftsZ* plays a central role in division. Mutations in *ftsZ* block septum formation and generate filaments; over-expression induces minicells, by causing an increased number of septation events per unit cell mass. *ftsZ* mutants act at stages varying from the displacement of the periseptal annuli to septal morphogenesis. However, no mutations have yet been identified that affect the first stage, the generation of

Figure 12.27 Failure of cell division generates multinucleated filaments. Photograph kindly provided by Sota Hiraga.

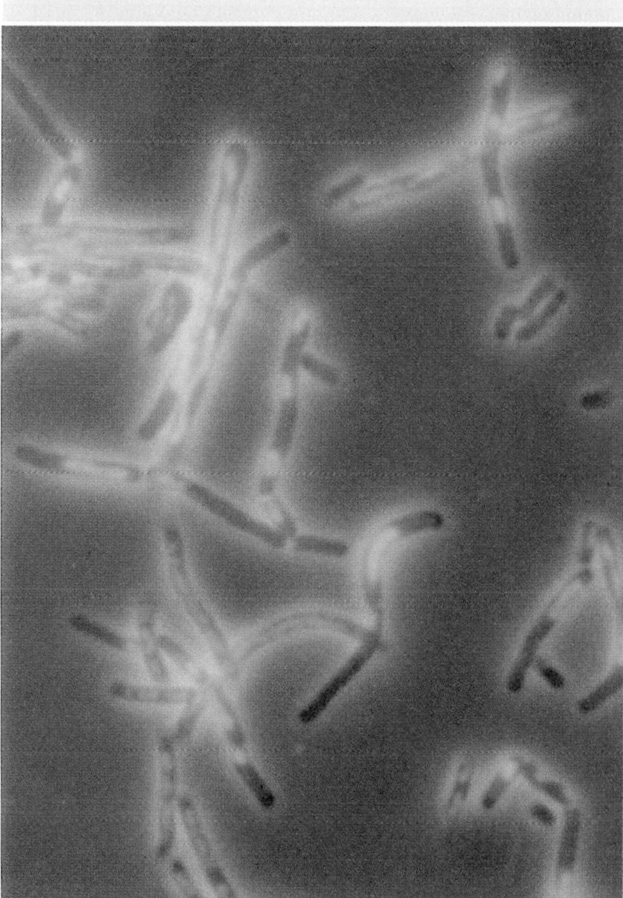

Figure 12.28 *E. coli* generate anucleate cells when chromosome segregation fails. Cells with chromosomes stain blue; daughter cells lacking chromosomes have no blue stain. This field shows cells of the *mukB* mutant; both normal and abnormal divisions can be seen. Photograph kindly provided by Sota Hiraga.

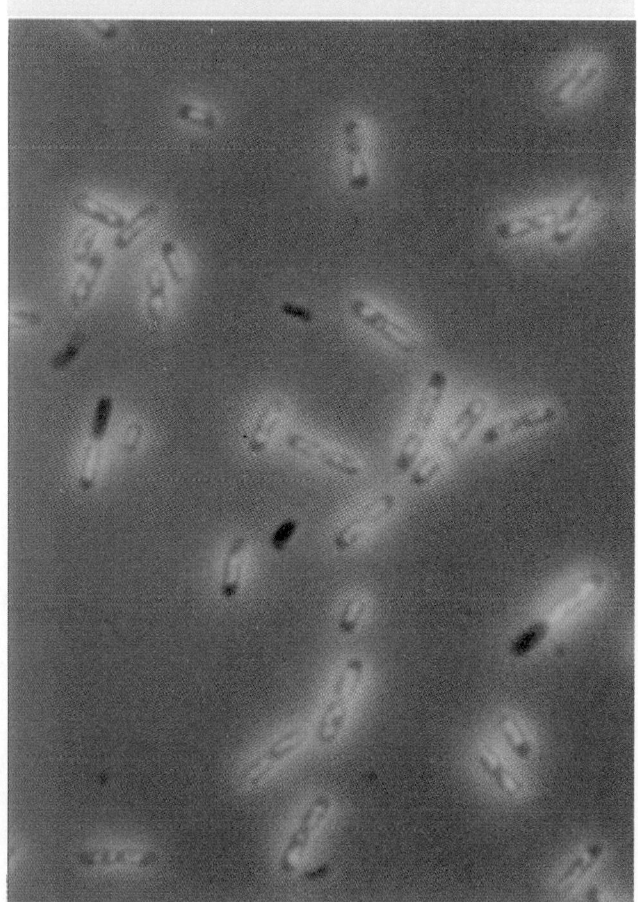

periseptal annuli. These would be the most interesting, since they would come closer to the question of how annulus formation is regulated, and whether it is connected directly with envelope growth and DNA segregation. FtsZ is therefore required for usage of pre-existing sites for septum formation, but does not itself affect their formation or localization.

FtsZ functions at an early stage of septum formation. Early in the division cycle, FtsZ is localized throughout the cytoplasm. As the cell elongates and begins to constrict in the middle, FtsZ becomes localized in a ring around the circumference, essentially in the position of the midcenter annulus in Figure 12.25. Given the dependence of division on FtsZ concentration, the formation of this ring could be the rate-limiting step in septum formation. FtsZ has GTPase activity, so one possibility is that GTP cleavage is used to support the oligomerization of FtsZ monomers into the ring structure. The structure of FtsZ resembles that of tubulin, suggesting that assembly of the ring could be akin to the formation of microtubules. FtsZ binds to ZipA, which is an integral protein that forms a ring in the inner membrane. This suggests the possibility that binding of FtsZ to ZipA could be the means by which a filamentous structure is connected to the membrane. The products of other *fts* genes are localized to the septum, and are presumed to form part of the structure.

Information about the localization of the septum has been provided by **minicell mutants**. The original minicell mutation lies in the locus *minB*; deletion of *minB* generates minicells by allowing septation to occur at the poles as well as (or instead of) at midcell. This suggests that the cell possesses the ability to initiate septum formation either at midcell or at the poles; and the role of the wild-type *minB* locus is to suppress septation at the poles. In terms of the events depicted in Figure 12.25, this implies that a new-born cell has potential septation sites associated both with the annulus at midcenter and with the poles. One pole was formed from the septum of the previous division; the other pole represents the septum from the division before that. Perhaps the poles retain remnants of the annuli from which they were derived, and these remnants can nucleate septation.

Figure 12.29 MinC/D is a division inhibitor, whose action is confined to the polar sites by MinE.

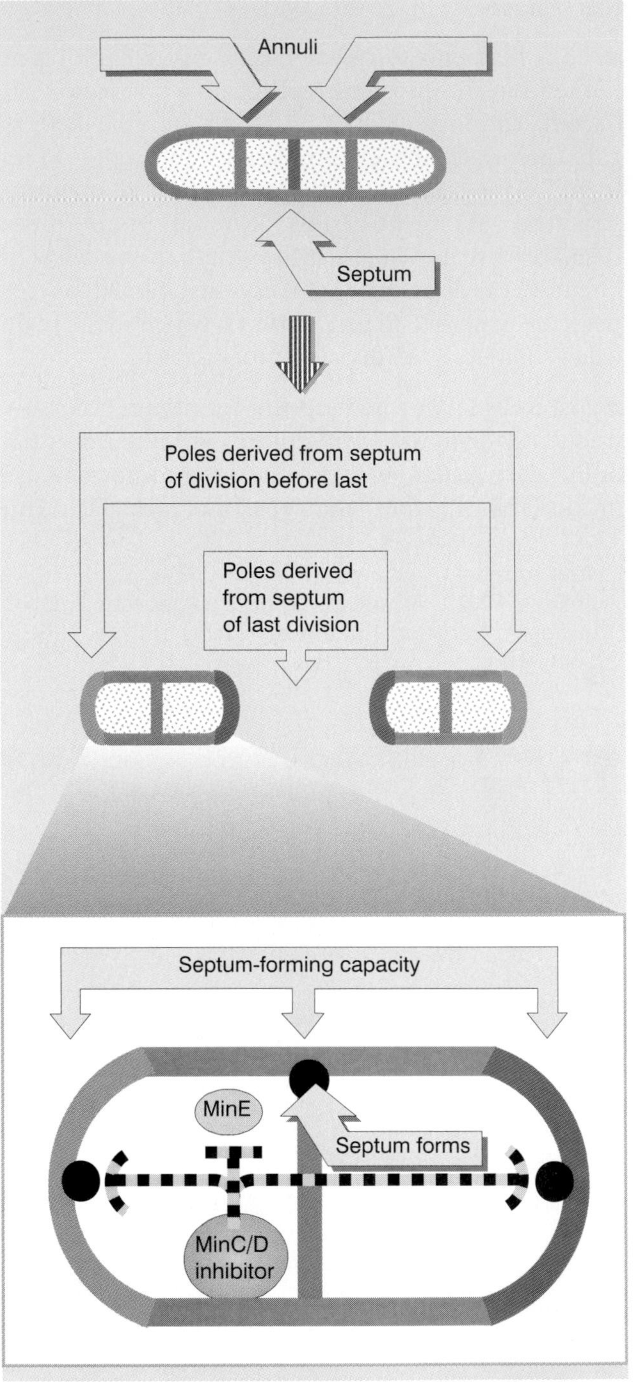

The *minB* locus consists of three genes, *minC,D,E*. Their roles are summarized in **Figure 12.29**. The products of *minC* and *minD* form (or are necessary for formation of) a division inhibitor. Expression of *minCD* in the absence of *minE*, or over-expression even in the presence of *minE*, causes a generalized inhibition of division. The resulting cells grow as long filaments without septa. Expression of *minE* at levels comparable to *minCD* confines the inhibition to the polar regions, so restoring normal growth. MinE protects the midcell sites from inhibition. Over-expression of *minE* induces minicells, because the presence of excess MinE counteracts the inhibition at the poles as well as at midcell, allowing septa to form at both locations.

The determinant of septation at the proper (midcell) site is therefore the ratio of MinCD to MinE. The wild-type level prevents polar septation, while permitting midcell septation. The effects of MinC/D and MinE are inversely related; absence of MinCD or too much MinE causes indiscriminate septation, forming minicells; too much MinCD or absence of MinE inhibits midcell as well as polar sites, resulting in filamentation.

MinE forms a ring at the septal position. Its accumulation suppresses the action of MinCD in the vicinity, thus allowing formation of the septal ring (which includes FtsZ and ZipA). Curiously, MinD is required for formation of the MinE ring.

Partitioning is the process by which the two daughter chromosomes find themselves on either side of the position at which the septum forms. Two types of event are required for proper partitioning:

- The two daughter chromosomes must be released from one another so that they can segregate following termination. This requires disentangling of DNA regions that are coiled around each other in the vicinity of the terminus. Most mutations affecting partitioning map in genes coding for topoisomerases—enzymes with the ability to pass DNA strands through one another. The mutations prevent the daughter chromosomes from segregating, with the result that the DNA is located in a single large mass at midcell. Septum formation then releases an anucleate cell and a cell containing both daughter chromosomes. This tells us that the bacterium must be able to disentangle its chromosomes topologically in order to be able to segregate them into different daughter cells.
- Mutations that affect the partition process itself are rare. We expect to find two classes. *cis*-acting mutations should occur in DNA sequences that are the targets for the partition process. *trans*-acting mutations should occur in genes that code for the protein(s) that cause segregation, which could include proteins that bind to DNA or activities that control the locations on the envelope to which DNA might be attached. (Both types of mutation have been found in the systems responsible for partitioning plasmids—see below—but only *trans*-acting functions have been found in the bacterial chromosome.)

The original form of the model for chromosome segregation shown in Figure 12.26 suggested that the envelope grows by insertion of material between the attachment sites of the two chromosomes, thus pushing them apart. But in fact the cell wall and membrane grow heterogeneously over the whole cell surface. Furthermore, the replicated chromosomes are capable of abrupt movements to their final positions at 1/4 and 3/4 cell length. If protein synthesis is inhibited before the termination of replication, the chromosomes fail to segregate and remain close to the midcell position. But when protein synthesis is allowed to resume, the chromosomes move to the quarter positions in the absence of any further envelope elongation. This suggests that an active process, requiring protein synthesis, may move the chromosomes to specific locations.

Segregation is interrupted by mutations of the *muk* class, which give rise to anucleate progeny at a much increased frequency: both daughter chromosomes remain on the same side of the septum instead of segregating. Mutations in the *muk* genes are not lethal, and may identify components of the apparatus that segregates the chromosomes. The gene *mukA* is identical with the gene for a known outer membrane protein (*tolC*), whose product could be involved with attaching the chromosome to the envelope. The gene *mukB* codes for a large (180 kD) globular protein, which has some sequence relationship to the mechanochemical enzyme dynamin, which provides a "motor" for microtubule-associated objects. This suggests the possibility that MukB is a motor that physically moves the chromosome relative to the envelope.

There have been suspicions for years that a physical link exists between bacterial DNA and the membrane, but the evidence remains indirect. Bacterial DNA can be found in membrane fractions, which tend to be enriched in genetic markers near the origin, the replication fork, and the terminus. The proteins present in these membrane fractions may be affected by mutations that interfere with the initiation of replication. The growth site could be a structure on the membrane to which the origin must be attached for initiation.

Functions involved in partition were first identified in plasmids. The components of a common system are summarized in **Figure 12.30**. Typically there are two *trans*-acting loci and a *cis*-acting element located just downstream of the two genes. ParA is an ATPase. It binds to ParB, which binds to the *parS* site on DNA. Deletions of any of the three loci prevent proper partition of the plasmid. It seems likely that the ParA-ParB oligomer binds to some cellular structure, so that *parS* effectively behaves as a centromere.

Proteins related to ParA and ParB are found in several bacteria. In *B. subtilis*, they are called Soj and SpoOJ, respectively. Mutations in these loci prevent sporulation, because of a failure to segregate one daughter chromosome into the forespore (see Figure 9.23). In sporulating cells, SpoOJ localizes at the pole

Figure 12.30 A common segregation system consists of genes *parA* and *parB* and the target site *parS*.

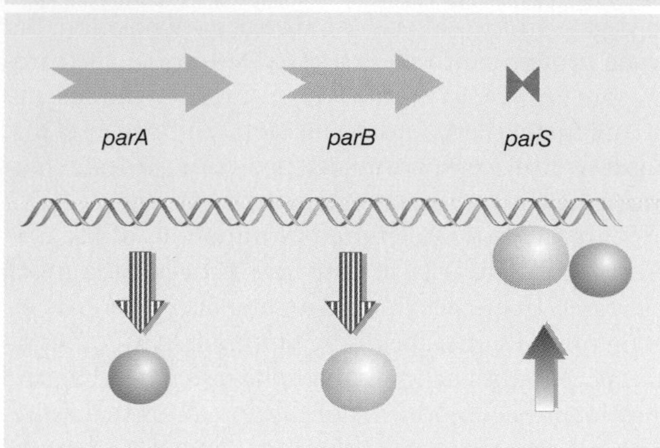

and may be responsible for localizing the origin there. SpoOJ binds to a sequence that is present in multiple copies, dispersed over ~20% of the chromosome in the vicinity of the origin. It is possible that SpoOJ binds both old and newly synthesized origins, maintaining a status equivalent to chromosome pairing, until the chromosomes are segregated to the opposite poles. In *C. crescentus*, ParA and ParB localize to the poles of the bacterium, and ParB binds sequences close to the origin, thus localizing the origin to the pole. These results suggest that a specific apparatus is responsible for localizing the origin to the pole. The next stage of the analysis will be to identify the cellular components with which this apparatus interacts.

Multiple systems ensure plasmid survival in bacterial populations

When a genome consists of a single replicon, it must also contain a system for segregating the progeny to daughter cells. Here we consider these systems and their connection with replication itself.

We have mentioned that each type of plasmid is maintained in its bacterial host at a characteristic **copy number:**

- **Single-copy** control systems resemble that of the bacterial chromosome and result in one replication per cell division. A single-copy plasmid effectively maintains parity with the bacterial chromosome.
- **Multicopy** control systems allow multiple initiation events per cell cycle, with the result that there are several copies of the plasmid per bacterium. Multicopy plasmids exist in a characteristic number (typically 10–20) per bacterial chromosome.

Copy number is primarily a consequence of the type of replication control mechanism. The system responsible for initiating replication determines how many origins can be present in the bacterium. Since each plasmid consists of a single replicon, the number of origins is the same as the number of plasmid molecules.

Single-copy plasmids have a system for replication control whose consequences are similar to that governing the bacterial chromosome. A single origin can be replicated once; then the daughter origins are segregated to the different daughter cells.

Multicopy plasmids have a replication system that allows a pool of origins to exist. If the number is great enough (in practice >10 per bacterium), an active segregation system becomes unnecessary, because even a statistical distribution of plasmids to daughter cells will result in the loss of plasmids at frequencies $<10^{-6}$.

Plasmids are maintained in bacterial populations with very low rates of loss ($<10^{-7}$ per cell division is typical, even for a single-copy plasmid). The systems that control plasmid segregation can be identified by mutations that increase the frequency of loss, but that do not act upon replication itself. Several types of mechanism are used to ensure the survival of a plasmid in a bacterial population. It is common for a plasmid to carry several systems, often of different types, all acting independently to ensure its survival. Some of these systems act indirectly, while others are concerned directly with regulating the partition event. However, in terms of evolution, all serve the same purpose: to help ensure perpetuation of the plasmid to the maximum number of progeny bacteria.

Because the multiple copies of a plasmid in a bacterium consist of the same DNA sequences, they are able to recombine. **Figure 12.31** demonstrates the consequences. A single intermolecular recombination event between two circles generates a dimeric circle; further recombination can generate higher multimeric forms. Such an event reduces the number of physically segregating units. In the extreme case of a single-copy plasmid that has just replicated, formation of a dimer by recombination means that the cell only has one unit to segregate, and the plasmid therefore must inevitably be lost from one daughter cell. To counteract this effect, plasmids often have **site-specific recombination systems** that act upon particular sequences to sponsor an *intramolecular recombination* that restores the monomeric condition. Mutations in these systems increase plasmid loss, and therefore have a phenotype that is similar to that of partition mutants. (The bacterial chromosome itself may have a similar system to deal with the consequences of recombination occurring between homologous sequences in the daughter chromosomes produced by a replication cycle.)

Addiction systems, operating on the basis that "we hang together or we hang separately", ensure that a bacterium carrying a plasmid can survive *only* so long as it retains the plasmid. There are several ways to ensure that a cell dies if it is "cured" of a plasmid, all sharing the principle illustrated in **Figure 12.32** that the plasmid produces both a poison and an antidote. The poison is a killer substance that is relatively stable, whereas the antidote consists of a substance that blocks killer action, but is relatively short lived. When the plasmid is lost, the antidote decays, and then the killer substance causes death of the cell. So bacteria that lose the plasmid inevitably die, and the population is condemned

Figure 12.31 Intermolecular recombination merges monomers into dimers, and intramolecular recombination releases individual units from oligomers.

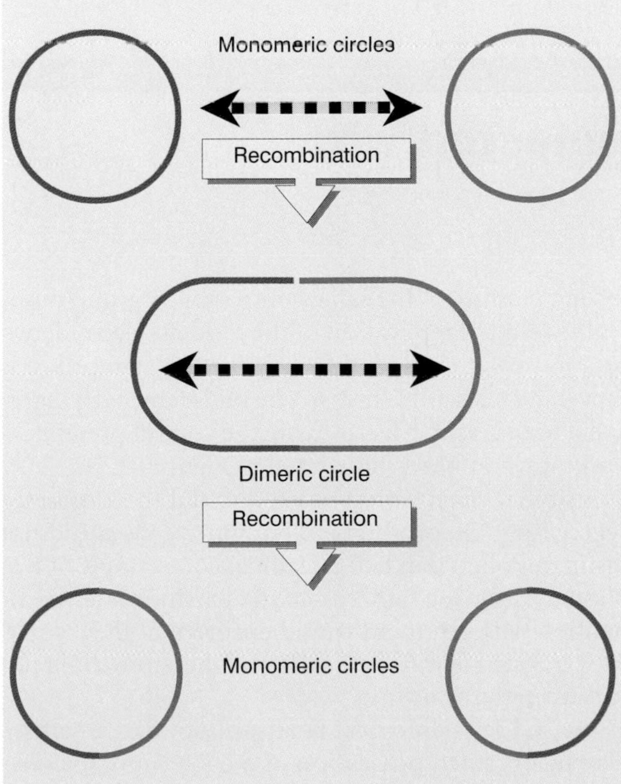

Figure 12.32 Plasmids may ensure that bacteria cannot live without them by synthesizing a long-lived killer and a short-lived antidote.

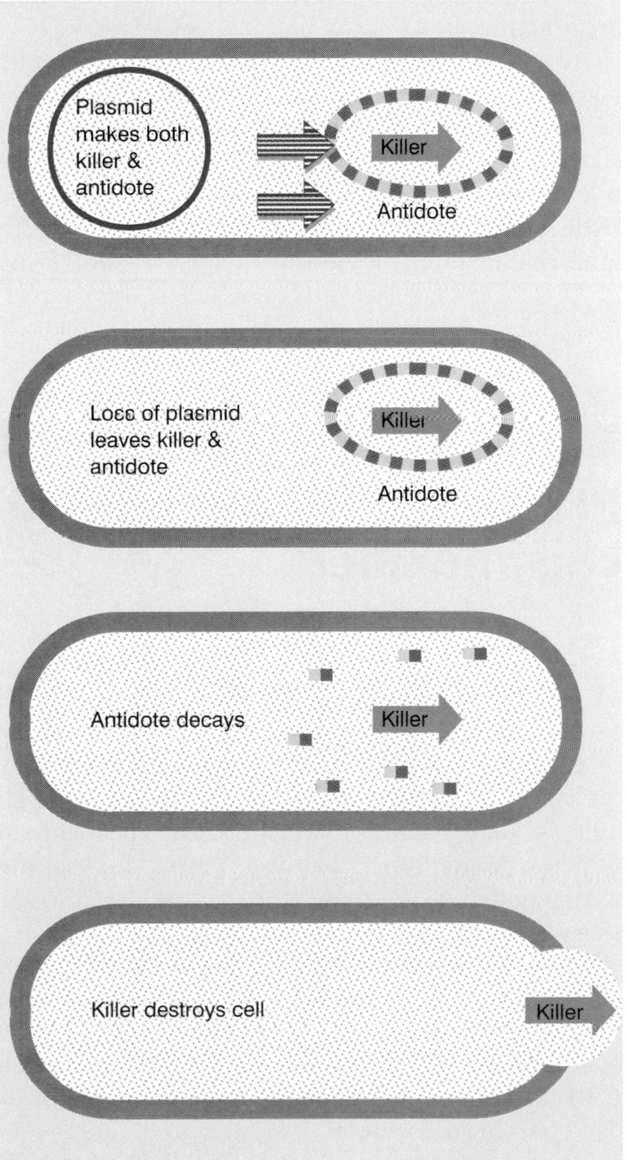

to retain the plasmid indefinitely. These systems take various forms. One specified by the F plasmid consists of killer and blocking proteins. The plasmid R1 has a killer that is the mRNA for a toxic protein, while the antidote is a small antisense RNA that prevents expression of the mRNA.

True partition systems act upon duplicate DNA molecules to ensure that they reside on either side of the septum at cell division. Probably all low copy number plasmids have such a system. Systems that have been characterized for the plasmids F, P1, and R1 have the generally similar organization depicted in Figure 12.30. There are two *trans*-acting proteins and a single *cis*-acting site. In the cases of both P1 and F, the smaller of the two proteins binds to the *cis*-acting site. In spite of their overall similarities, there are no significant sequence homologies between the corresponding genes or *cis*-acting sites.

How does a true partition system segregate replica plasmids to different daughter cells? We may imagine two general types of system:

- *Extrinsic* structures to which the plasmids must bind are limiting. The model for a single-copy plasmid is the same as illustrated previously in Figure 12.26. The cell must contain only two sites, and each copy of the plasmid binds to one. The sites could be membrane-bound or could simply constitute regions of the bacterial chromosome itself. A disadvantage of this type of model is that many different types of plasmids are found in bacterial populations, which makes it necessary to suppose that there are separate and exclusive pairs of sites for each type of plasmid.
- Segregation is an *intrinsic* process, in which the replica plasmids pass through some stage equivalent to chromosome pairing. Following association of plasmids into pairs, a partition mechanism moves one member of each pair into opposite daughter cells. The difficulty remains that the nature of this mechanism is not easy to envisage, and it is still necessary to explain how the cell can cope with partition of multiple types of plasmids.

Except for the fact that proteins required for partition are coded by the plasmid and bind to *cis*-acting sites, we know little about the mechanism of segregation. The importance to the plasmid of ensuring that all daughter cells gain replica plasmids is emphasized by the existence of multiple, independent systems in individual plasmids that ensure proper partition.

Plasmid incompatibility is connected with copy number

The phenomenon of plasmid incompatibility is related to the regulation of plasmid copy number and segregation. A **compatibility group** is defined as a set of plasmids whose members are unable to coexist in the same bacterial cell. The reason for their incompatibility is that they cannot be distinguished from one another at some stage that is essential for plasmid maintenance. DNA replication and segregation are stages at which this may apply.

The negative control model for plasmid incompatibility follows the idea that copy number control is achieved by synthesizing a repressor that measures the concentration of origins. (Formally this is the same as the titration model for regulating replication of the bacterial chromosome.)

The introduction of a new origin in the form of a second plasmid of the same compatibility group mimics the result of replication of the resident plasmid; two origins now are present. So any further replication is prevented until after the two plasmids have been segregated to different cells to create the correct prereplication copy number as illustrated in **Figure 12.33.**

A similar effect would be produced if the system for segregating the products to daughter cells could not distinguish between two plasmids. For example, if two plasmids have the same *cis*-acting partition sites, competition between them would ensure that they would be segregated to different cells, and therefore could not survive in the same line.

Viewed in teleological terms, plasmids are selfish. Having obtained possession of a bacterium, the resident plasmid will seek to prevent any other plasmid of

Figure 12.33 Two plasmids are incompatible (they belong to the same compatibility group) if their origins cannot be distinguished at the stage of initiation. The same model could apply to segregation.

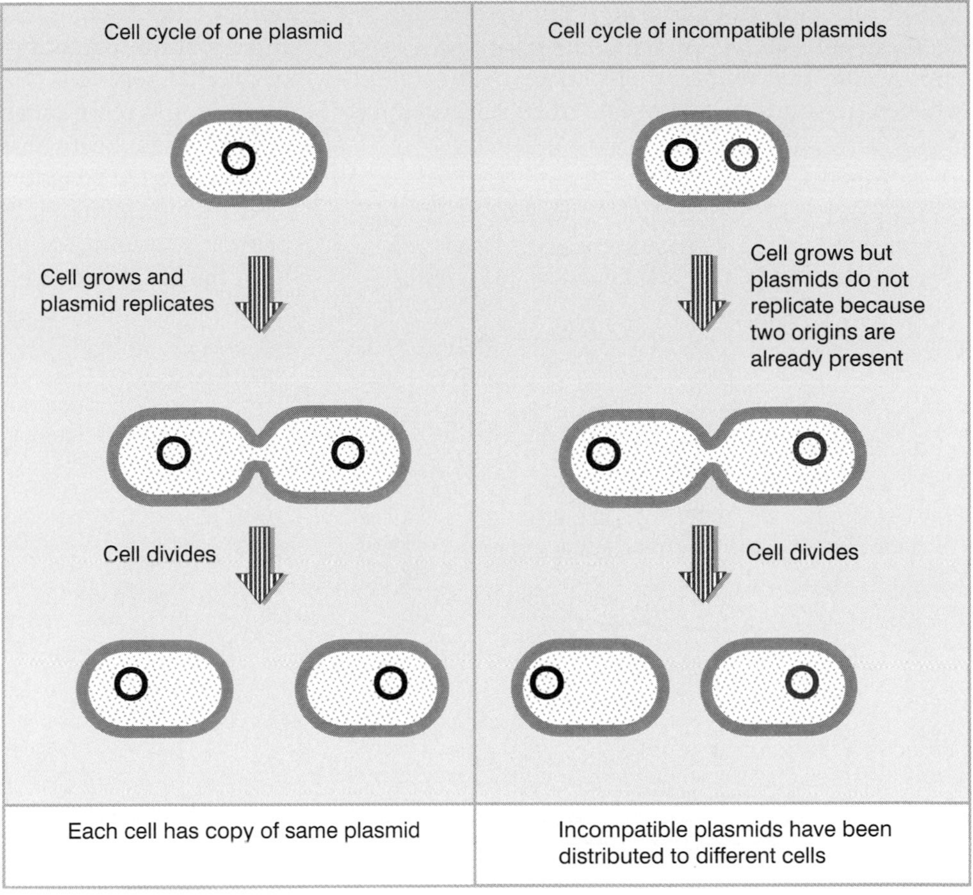
Cell cycle of one plasmid
Cell cycle of incompatible plasmids
Cell grows and plasmid replicates
Cell grows but plasmids do not replicate because two origins are already present
Cell divides
Cell divides
Each cell has copy of same plasmid
Incompatible plasmids have been distributed to different cells

the same type from establishing residence. Plasmid incompatibility is a major device used to establish these territorial rights.

The presence of a member of one compatibility group does not directly affect the survival of a plasmid belonging to a different group. Only one replicon of a given compatibility group (of a single-copy plasmid) can be maintained in the bacterium, but it does not interact with replicons of other compatibility groups (although in limiting conditions they compete for *lebensraum*).

The best characterized copy number and incompatibility system is that of the plasmid ColE1, a multicopy plasmid that is maintained at a steady level of ~20 copies per *E. coli* cell. The system for maintaining the copy number depends on the mechanism for initiating replication at the ColE1 origin, as illustrated in **Figure 12.34.**

Replication starts with the transcription of an RNA that initiates 555 bp upstream of the origin. Transcription continues through the origin. The enzyme RNAase H (whose name reflects its specificity for a substrate of RNA *h*ybridized with DNA) cleaves the transcript at the origin. This generates a 3′-OH end that is used as the "primer" at which DNA synthesis is initiated (the use of primers is discussed in more detail in Chapter 13). The primer RNA forms a persistent hybrid with the DNA. Pairing between the RNA and DNA occurs just upstream of the origin (around position –20) and also farther upstream (around position –265).

Two regulatory systems exert their effects on the RNA primer. One involves synthesis of an RNA complementary to the primer; the other involves a protein coded by a nearby locus.

The regulatory species RNA I is a molecule of ~108 bases, coded by the opposite strand from that specifying primer RNA. The relationship between the primer RNA and RNA I is illustrated in **Figure 12.35.** The RNA I molecule is initiated within the primer region and terminates close to the site where the primer RNA initiates. So RNA I is complementary to the 5′-terminal region of the primer RNA. *Base pairing between the two RNAs controls the availability of the primer RNA to initiate a cycle of replication.*

An RNA molecule such as RNA I that functions by virtue of its complementarity with another RNA coded in the same region is called a **countertranscript.** This type of mechanism, of course, is another example of the use of antisense RNA (see Chapter 10).

Mutations that reduce or eliminate incompatibility between plasmids can be obtained by selecting plasmids of the same group for their ability to coexist. Incompatibility mutations in ColE1 map in the region of overlap between RNA I and primer RNA. Because this region is represented in two different RNAs, either or both might be involved in the effect.

When RNA I is added to a system for replicating ColE1 DNA *in vitro*, it inhibits the formation of active primer RNA. But the presence of RNA I does not inhibit the initiation or elongation of primer RNA syn-

Figure 12.34 Replication of ColE1 DNA is initiated by cleaving the primer RNA to generate a 3'-OH end. The primer forms a persistent hybrid in the origin region.

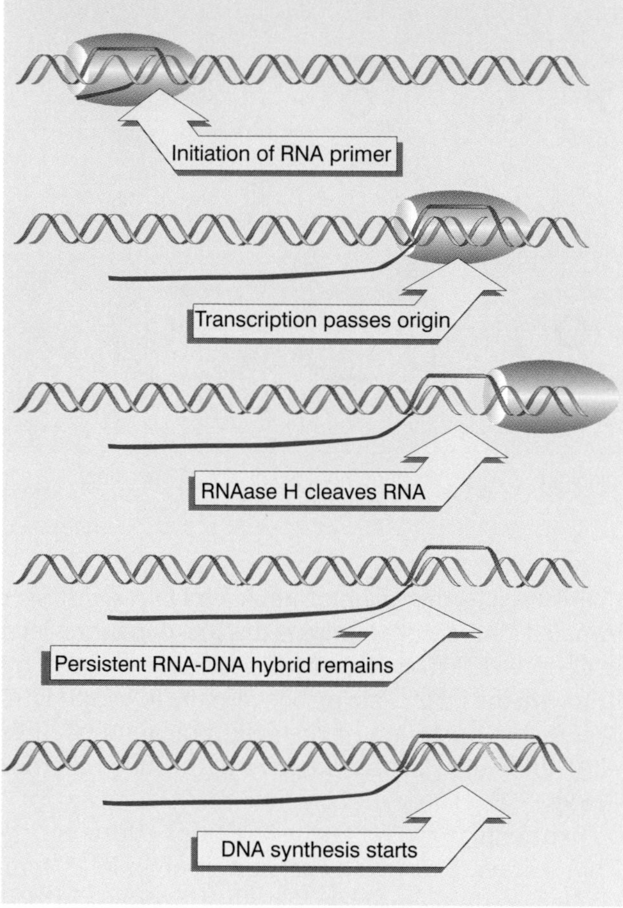

Figure 12.35 The sequence of RNA I is complementary to the 5' region of primer RNA.

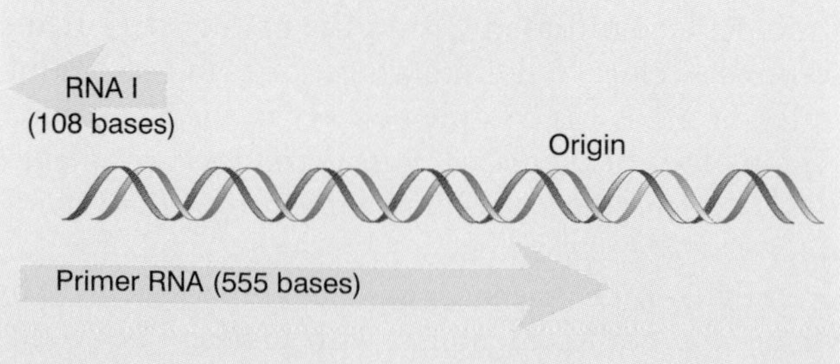

Figure 12.36 Base pairing with RNA I may change the secondary structure of the primer RNA sequence and thus prevent cleavage from generating a 3'-OH end.

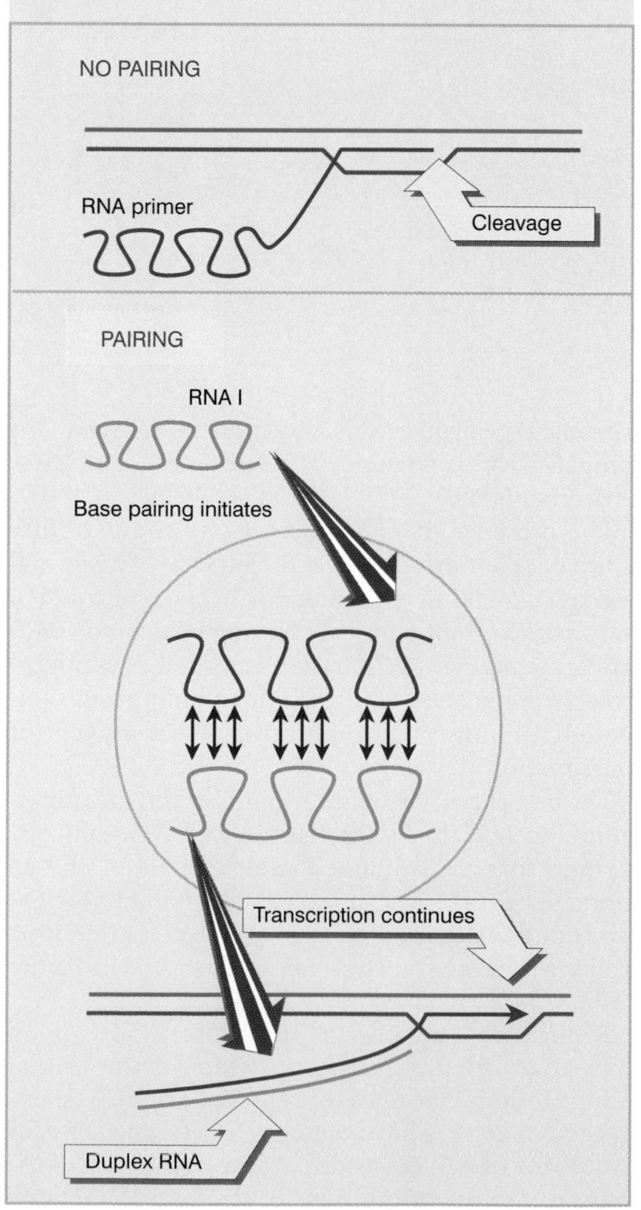

thesis. This suggests that *RNA I prevents RNAase H from generating the 3′ end of the primer RNA.* The basis for this effect lies in base pairing between RNA I and primer RNA.

Both RNA molecules have the same potential secondary structure in this region, with three duplex hairpins terminating in single-stranded loops. Mutations reducing incompatibility are located in these loops, which suggests that the initial step in base pairing between RNA I and primer RNA is contact between the unpaired loops.

How does pairing with RNA I prevent cleavage to form primer RNA? A model is illustrated in **Figure 12.36.** In the absence of RNA I, the primer RNA forms its own secondary structure (involving loops and stems). But when RNA I is present, the two molecules pair, and become completely double-stranded for the entire length of RNA I. The new secondary structure prevents the formation of the primer, probably by affecting the ability of the RNA to form the persistent hybrid.

The model is reminiscent of the mechanism involved in attenuation of transcription, in which the alternative pairings of an RNA sequence permit or prevent formation of the secondary structure needed for termination by RNA polymerase (see Chapter 10). The action of RNA I is exercised by its ability to affect distant regions of the primer precursor.

Formally, the model is equivalent to postulating a control circuit involving two RNA species. A large RNA primer precursor is a positive regulator, needed to initiate replication. The small RNA I is a negative regulator, able to inhibit the action of the positive regulator.

In its ability to act on any plasmid present in the cell, RNA I provides a repressor that prevents newly introduced DNA from functioning, analogous to the role of the lambda lysogenic repressor (see Chapter 11). Instead of a repressor protein that binds the new DNA, an RNA binds the newly synthesized precursor to the RNA primer.

Binding between RNA I and primer RNA can be influenced by the Rom protein, coded by a gene located downstream of the origin. The Rom protein enhances binding between RNA I and primer RNA transcripts of >200 bases. The result is to inhibit formation of the primer.

Figure 12.37 Mutations in the region coding for RNA I and the primer precursor need not affect their ability to pair; but they may prevent pairing with the complementary RNA coded by a different plasmid.

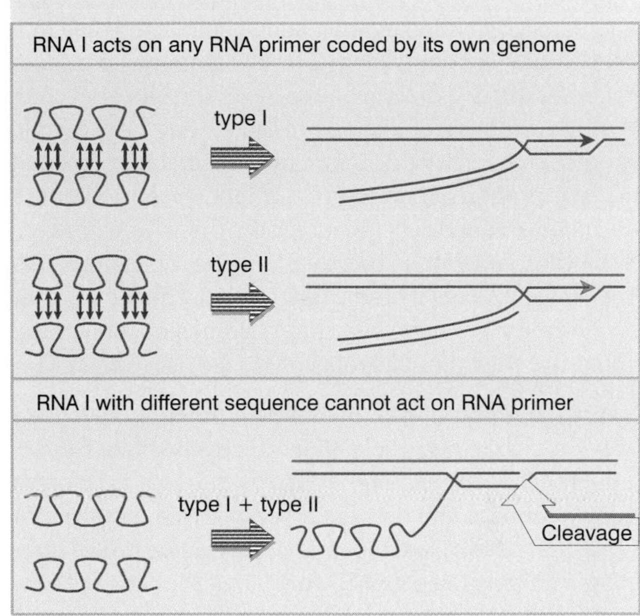

How do mutations in the RNAs affect incompatibility? **Figure 12.37** shows the situation when a cell contains two types of RNA I/primer RNA sequence. The RNA I and primer RNA made from each type of genome can interact, but RNA I from one genome does not interact with primer RNA from the other genome. This situation would arise when a mutation in the region that is common to RNA I and primer RNA occurred at a location that is involved in the base pairing between them. Each RNA I would continue to pair with the primer RNA coded by the same plasmid, but might be unable to pair with the primer RNA coded by the other plasmid. This would cause the original and the mutant plasmids to behave as members of different compatibility groups.

Summary

The entire chromosome is replicated once for every cell division cycle. Initiation of replication commits the cell to a cycle of division; completion of replication may provide a trigger for the actual division process. The bacterial chromosome consists of a single replicon, but a eukaryotic chromosome is divided into many replicons that function over the protracted period of S phase. The problem of replicating the ends of a linear replicon is solved in a variety of ways, most often by converting the replicon to a circular form. Some viruses have special proteins that recognize ends. Eukaryotic chromosomes encounter the problem at their terminal replicons.

Eukaryotic replication is (at least) an order of magnitude slower than bacterial replication. Origins sponsor bidirectional replication, and are probably used in a fixed order during S phase. The only eukaryotic origins identified at the sequence level are those of *S. cerevisiae*, which have a core consensus sequence consisting of 11 base pairs, mostly A • T.

The minimal *E. coli* origin consists of ~245 bp and initiates bidirectional replication. Any DNA molecule with this sequence can replicate in *E. coli*. Two replication forks leave the origin and move around the chromosome, apparently until they meet, although *ter* sequences that would cause the forks to terminate after meeting have been identified. Transcription units are organized so that transcription usually proceeds in the same direction as replication.

The rolling circle is an alternative form of replication for circular DNA molecules in which an origin is nicked to provide a priming end. One strand of DNA is synthesized from this end, displacing the original partner strand, which is extruded as a tail. Multiple genomes can be produced by continuing revolutions of the circle.

Rolling circles are used to replicate some phages. The A protein that nicks the ϕX174 origin has the unusual property of *cis*-action. It acts only on the DNA from which it was synthesized. It remains attached to the displaced strand until an entire strand has been synthesized, and then nicks the origin again, releasing the displaced strand and starting another cycle of replication.

Rolling circles also are involved in bacterial conjugation, when an F plasmid is transferred from a donor to a recipient cell, following the initiation of contact between the cells by means of the F-pili. A free F plasmid infects new cells by this means; an integrated F factor creates an Hfr strain that may transfer chromosomal DNA. In the case of conjugation, replication is used to synthesize complements to the single strand remaining in the donor and to the single strand transferred to the recipient, but does not provide the motive power.

A fixed time of 40 minutes is required to replicate the *E. coli* chromosome and a further 20 minutes is required before the cell can divide. When cells divide more rapidly than every 60 minutes, a replication cycle is initiated before the end of the preceding division cycle. This generates multiforked chromosomes. The initiation event depends on titration of cell mass, probably by accumulating an initiator protein. Initiation may occur at the cell membrane, since the origin is associated with the membrane for a short period after initiation.

Segregation involves additional sequences that have not yet been characterized. Most mutations that prevent proper segregation lie in ancillary functions, such as the enzymes needed for the two daughter chromosomes to disentangle. One mutation that directly affects segregation may identify a motor or a protein that connects the chromosome to the envelope. The septum that divides the cell grows at a location defined by the pre-existing periseptal annulus; a locus of three genes codes for products that regulate whether the midcell periseptal annulus or the polar sites derived from previous annuli are used for septum formation. Absence of septum formation generates multinucleated filaments; excess of septum formation generates anucleate minicells. Some bacteria have systems related to those involved in plasmid partition, in which a pair of proteins binds to a DNA sequence near the origin and causes the origin to be localized to the poles.

Plasmids have a variety of systems that ensure or assist partition, and an individual plasmid may carry systems of several types. The copy number of a plasmid describes whether it is present at the same level as the bacterial chromosome (one per unit cell) or in greater numbers. Plasmid incompatibility can be a consequence of the mechanisms involved in either replication or partition (for single-copy plasmids). Two plasmids that share the same control system for replication are incompatible because the number of replication events ensures that there is only one plasmid for each bacterial genome.

Further reading

Reviews

Clayton, D. (1982). Replication of animal mitochondrial DNA. *Cell* **28**, 693–705.

Scott, J. R. (1984). Regulation of plasmid replication. *Microbiol. Rev.* **48**, 1–23.

Ippen-Ihler, K. A. and Minkley, E. G. (1986). The conjugation system of F, the fertility factor of *E. coli*. *Ann. Rev. Genet.* **20**, 593–624.

Brewer, B. J. (1988). When polymerases collide: replication and transcriptional organization of the *E. coli* chromosome. *Cell* **53**, 679–686.

Willetts, N. and Skurray, R. (1987). Structure and function of the F factor and mechanism of conjugation. In *E. coli and S. typhimurium*. (Ed. F. C. Neidhart, American Society for Microbiology, Washington DC, 1110–1131).

Nordstrom, K. and Austin, S. J. (1989). Mechanisms that contribute to the stable segregation of plasmids. *Ann. Rev. Genet.* **23**, 37–69.

Umek, R. M., Linskens, M. H., Kowalski, D., and Huberman, J. A. (1989). New beginnings in studies of eukaryotic DNA replication origins. *Biochim. Biophys. Acta* **1007**, 1–14.

de Boer, P. A. J., Cook, W. R., and Rothfield, L. I. (1990). Bacterial cell division. *Ann. Rev. Genet.* **24**, 249–274.

Clayton, D. A. (1991). Replication and transcription of vertebrate mitochondrial DNA. *Ann. Rev. Cell Biol.* **7**, 453–478.

Fangman, W. L. and Brewer, B. J. (1991). Activation of replication origins within yeast chromosomes. *Ann. Rev. Cell Biol.* **7**, 375–402.

Hiraga, S. (1992). Chromosome and plasmid partition in *E. coli*. *Ann. Rev. Biochem.* **61**, 283–306.

DePamphlis, M. L. (1993). Eukaryotic DNA replication: anatomy of an origin. *Ann. Rev. Biochem.* **62**, 29–63.

Donachie, W. D. (1993). The cell cycle of *E. coli*. *Ann. Rev. Microbiol.* **47**, 199–230.

Coverley, D. and Laskey, R. A. (1994). Regulation of eukaryotic DNA replication. *Ann. Rev. Biochem.* **63**, 745–776.

Frost, L. S., Ippen-Ihler, K., and Skurray, R. A. (1994). Analysis of the sequence and gene products of the transfer region of the F sex factor. *Microbiol. Rev.* **58**, 162–210.

Lanka, E. and Wilkins, B. M. (1995). DNA processing reactions in bacterial conjugation. *Ann. Rev. Biochem.* **64**, 141–169.

Wake, R. G. and Errington, J. (1995). Chromosome partitioning in bacteria. *Ann. Rev. Genet.* **29**, 41–67.

Lutkenhaus, J. and Addinall, S. G. (1997). Bacterial cell division and the Z ring. *Ann. Rev. Biochem.* **66**, 93–116.

Shadel, G. S. and Clayton, D. A. (1997). Mitochondrial DNA maintenance in vertebrates. *Ann. Rev. Biochem.* **66**, 409–435.

Discoveries

Transcription and replication

Liu, B., Wong, M. L., and Alberts, B. (1994). A transcribing RNA polymerase molecule survives DNA replication without aborting its growing RNA chain. *Proc. Natl. Acad. Sci. USA* **91**, 10660–10664.

Replicons and origins

Cairns, J. (1963). The bacterial chromosome and its manner of replication as seen by autoradiography. *J. Mol. Biol.* **6**, 208–213.

Jacob, F., Brenner, S., and Cuzin, F. (1963). On the regulation of DNA replication in bacteria. *Cold Spring Harbor Symp. Quant. Biol.* **28**, 329–348.

Jacob, F., Ryter, A., and Cuzin, F. (1966). On the association between DNA and the membrane in bacteria. *Proc. R. Soc. B* **164**, 267.

Huberman, J. and Riggs, A. D. (1968). On the mechanism of DNA replication in mammalian chromosomes. *J. Mol. Biol.* **32**, 327–341.

Blumenthal, A. B., Kriegstein, H. J., and Hogness, D. S. (1973). The units of DNA replication in *D. melanogaster* chromosomes. *Cold Spring Harbor Symp. Quant. Biol.* **38**, 205–223.

Zyskind, J. W. and Smith, D. W. (1980). Nucleotide sequence of the *S. typhimurium* origin of DNA replication. *Proc. Natl. Acad. Sci. USA* **77**, 2460–2464.

Steck, T. R. and Drlica, K. (1984). Bacterial chromosome segregation: evidence for DNA gyrase involvement in decatenation. *Cell* **36**, 1081–1088.

Marahrens, Y. and Stillman, B. (1992). A yeast chromosomal origin of DNA replication defined by multiple functional elements. *Science* **255**, 817–823.

Bacterial division

Adler, H. I. *et al.* (1967). Miniature *E. coli* cells deficient in DNA. *Proc. Natl. Acad. Sci USA* **57**, 321–326.

Donachie, W. D. and Begg, K. J. (1970). Growth of the bacterial cell. *Nature* **227**, 1220–1224.

Donachie, W. D., Begg, K. J., and Vicente, M. (1976). Cell length, cell growth and cell division. *Nature* **264**, 328–333.

Hirota, Y., Ryter, A., and Jacob, F. (1968). Thermosensitive mutants of *E. coli* affected in the processes of DNA synthesis and cellular division. *Cold Spring Harbor Symp. Quant. Biol.* **33**, 677–693.

de Boer, P. A. J. *et al.* (1989). A division inhibitor and a topological specificity factor coded for by the minicell locus determine proper placement of the division septum in *E. coli. Cell* **56**, 641–649.

Mohl, D. A. and Gober, J. W. (1997). Cell cycle-dependent polar localization of chromosome partitioning proteins of *C. crescentus. Cell* **88**, 675–684.

Raskin, D. M. and de Boer, P. A. J. (1997). The MinE ring: an FtsZ-independent cell structure required for selection of the correct division site in *E. coli. Cell* **91**, 685–694.

Rolling circles

Gilbert, W. and Dressler, D. (1968). DNA replication: the rolling circle model. *Cold Spring Harbor Symp. Quant. Biol.* **33**, 473–484.

Conjugation

Ihler, G. and Rupp, W. D. (1969). Strand-specific transfer of donor DNA during conjugation in *E. coli. Proc. Natl. Acad. Sci. USA* **63**, 138–143.

Plasmid incompatibility

Tomizawa, J.-I. and Itoh, T. (1981). Plasmid ColE1 incompatibility determined by interaction of RNA with primer transcript. *Proc. Natl. Acad. Sci. USA* **78**, 6096–6100.

Masukata, H. and Tomizawa, J. (1990). A mechanism of formation of a persistent hybrid between elongating RNA and template DNA. *Cell* **62**, 331–338.

Chapter 13

DNA replication

REPLICATION of duplex DNA is a complex endeavor involving a conglomerate of enzyme activities. Different activities are involved in the stages of initiation, elongation, and termination.

- Initiation involves recognition of an origin by a complex of proteins. Before DNA synthesis begins, the parental strands must be separated and (transiently) stabilized in the single-stranded state. Then synthesis of daughter strands can be initiated at the replication fork. The act of initiating synthesis of a DNA strand is accomplished by a protein complex called the **primosome** in *E. coli*.

- Elongation is undertaken by another complex of proteins. The **replisome** *exists only as a protein complex associated with the particular structure that DNA takes at the replication fork.* It does not exist as an independent unit (for example, analogous to the ribosome). As the replisome moves along DNA, the parental strands unwind and daughter strands are synthesized.

- At the end of the replicon, joining and/or termination reactions are necessary. Following termination, the duplicate chromosomes must be separated from one another, which requires manipulation of higher-order DNA structure.

Inability to replicate DNA is fatal for a growing cell. Mutants in replication must therefore be obtained as conditional lethals, able to accomplish replication under permissive conditions (provided by the normal temperature of incubation), and displaying their defect only under nonpermissive conditions (provided by the higher temperature of 42°C). A comprehensive series of such temperature-sensitive mutants in *E. coli* identifies a set of loci called the *dna* genes. The *dna* mutants are divided into two general classes on the basis of their behavior when the temperature is raised. The major class of **quick-stop mutants** cease replication immediately on a temperature rise. They are defective in the components of the replication apparatus, typically in the enzymes needed for elongation (but also include defects in the supply of essential precursors). The smaller class of **slow-stop mutants** complete the current round of replication, but cannot start another. They are defective in the events involved in initiating a cycle of replication at the origin.

An important assay used to identify the components of the replication apparatus is called *in vitro* **complementation**. An *in vitro* system for replication is prepared from a *dna* mutant and operated under conditions in which the mutant gene product is inactive. Extracts from wild-type cells are tested for their ability to restore activity. The protein coded by the *dna* locus can be purified by identifying the active component in the extract.

Each component of the bacterial replication apparatus is now available for study *in vitro* as a biochemically pure product, and is implicated *in vivo* by mutations in its gene. Eukaryotic replication systems are highly purified, but have not yet reached the stage of identification of every single component.

DNA polymerases: the enzymes that make DNA

An enzyme that can synthesize a new DNA strand on a template strand is called a **DNA polymerase.** Both prokaryotic and eukaryotic cells contain multiple DNA polymerase activities. Only some of these enzymes actually undertake replication; sometimes they are called **DNA replicases.** The others are involved in subsidiary roles in replication and/or participate in "repair" synthesis of DNA to replace damaged sequences (see Chapter 14).

It is convenient to think of DNA-synthesizing activities in terms of the DNA polymerase enzymes; but it is a moot point whether a DNA replicase enzyme exists as a discrete entity. (We consider the same issue for RNA polymerase in Chapter 9.) Bacterial DNA replicase activity is recovered as aggregates containing various "subunits". The DNA-synthesizing activity is only one of several functions associated in the replisome.

Three DNA polymerase enzymes have been characterized in *E. coli.* DNA polymerase I (coded by *polA*) is involved in the repair of damaged DNA and, in a subsidiary role, in semiconservative replication. DNA polymerase II is also implicated in repair. DNA polymerase III, a multisubunit protein, is the replicase responsible for *de novo* synthesis of new strands of DNA.

When extracts of *E. coli* are assayed for their ability to synthesize DNA, the predominant enzyme activity is that of DNA polymerase I. Its activity is so great that it makes it impossible to detect the activities of the enzymes actually responsible for DNA replication! To develop *in vitro* systems in which replication can be followed, extracts are therefore prepared from *polA* mutant cells.

All prokaryotic and eukaryotic DNA polymerases share the same fundamental type of synthetic activity. Each can extend a DNA chain by adding nucleotides one at a time to a 3′-OH end, as illustrated diagrammatically in **Figure 13.1**. The choice of the nucleotide to add to the chain is dictated by base pairing with the template strand.

The fidelity of replication poses the same sort of problem we have encountered already in considering (for example) the accuracy of translation. It relies on the specificity of base pairing. Yet when we consider the interactions involved in base pairing, we would expect errors to occur with a frequency of $\sim 10^{-3}$ per base pair replicated. The actual rate in bacteria seems to be $\sim 10^{-8}$–10^{-10}. This corresponds to ~1 error per genome per 1000 bacterial replication cycles, or $\sim 10^{-6}$ per gene per generation.

We can divide the errors that DNA polymerase makes during replication into two classes. *Substitutions* occur when the wrong (improperly paired) nucleotide is incorporated. The error level is determined by the efficiency of **proofreading.** *Frameshifts* occur when an extra nucleotide is inserted or omitted. Fidelity with regard to frameshifts is affected by the **processivity** of the enzyme: the tendency to remain on a single template rather than to dissociate and reassociate. This is

Figure 13.1 *Overview:* DNA synthesis occurs by adding nucleotides to the 3'-OH end of the growing chain, so that the new chain is synthesized in the 5'–3' direction. The precursor for DNA synthesis is a nucleoside triphosphate, which loses the terminal two phosphate groups in the reaction.

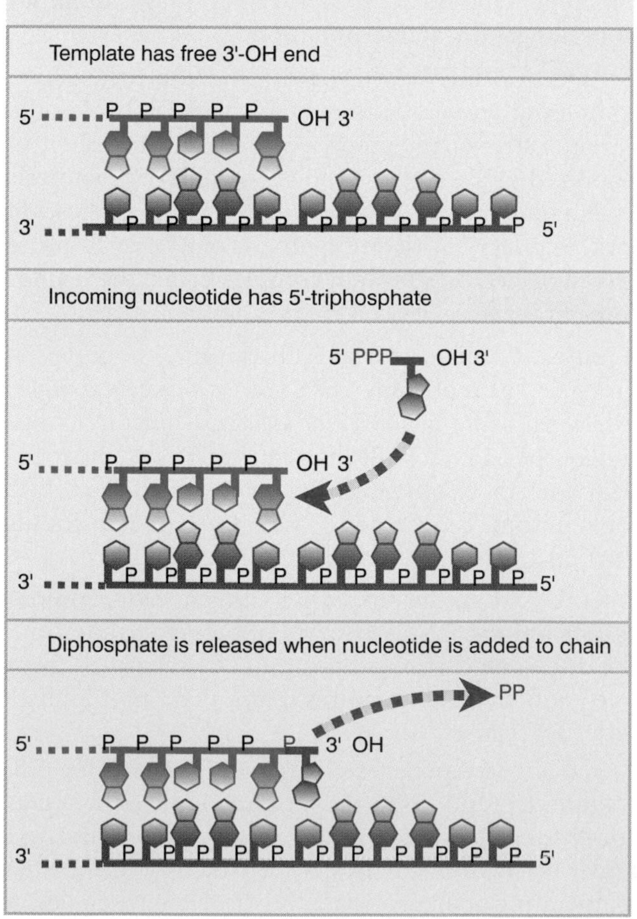

particularly important for the replication of a homopolymeric stretch, for example, a long sequence of $dT_n:dA_n$, in which "replication slippage" can change the length of the homopolymeric run. As a general rule, increased processivity reduces the likelihood of such events. In multimeric DNA polymerases, processivity is usually increased by a particular subunit that is not needed for catalytic activity *per se.*

DNA polymerase might improve the specificity of complementary base selection at either (or both) of two stages:

- It could scrutinize the incoming base for the proper complementarity with the template base; for example, by specifically recognizing matching chemical features. This would be a **presynthetic** error control.
- Or it could scrutinize the base pair *after* the new base has been added to the chain, and, in those cases in which a mistake has been made, remove the most recently added base. This would be a proofreading control.

All of the bacterial enzymes possess a 3′–5′ exonucleolytic activity that proceeds in the reverse direction from DNA synthesis. This provides the proofreading function illustrated diagrammatically in **Figure 13.2.** In the chain elongation step, a precursor nucleotide enters the position at the end of the growing chain. A bond is formed. The enzyme moves one base pair farther, ready for the next precursor nucleotide to enter. If a mistake has been made, the enzyme moves backward, excising the last base that was added, and creating a site for a replacement precursor nucleotide to enter.

Different DNA polymerases handle the relationship between the polymerizing and proofreading activities in different ways. In some cases, the activities are part of the same protein subunit, but in others they are contained in different subunits. Each DNA polymerase has a characteristic error rate that is reduced by its proofreading activity. Proofreading typically decreases the error rate in replication from $\sim 10^{-5}$ to $\sim 10^{-7}$.

The first enzyme to be characterized was DNA polymerase I, which is a single polypeptide of 103 kD. The chain can be cleaved into two regions by proteolytic treatment.

The larger cleavage product (68 kD) is called the Klenow fragment. It is used in synthetic reactions *in vitro.* It contains the polymerase and the 3′–5′ exonuclease activities. The C-terminal two-thirds of the protein contains the polymerase active site, while the N-terminal third contains the proofreading exonuclease. The active sites are ~30 Å apart in the protein, indicating that there is spatial separation between adding a base and removing one.

Figure 13.2 Bacterial DNA polymerases scrutinize the base pair at the end of the growing chain and excise the nucleotide added in the case of a misfit.

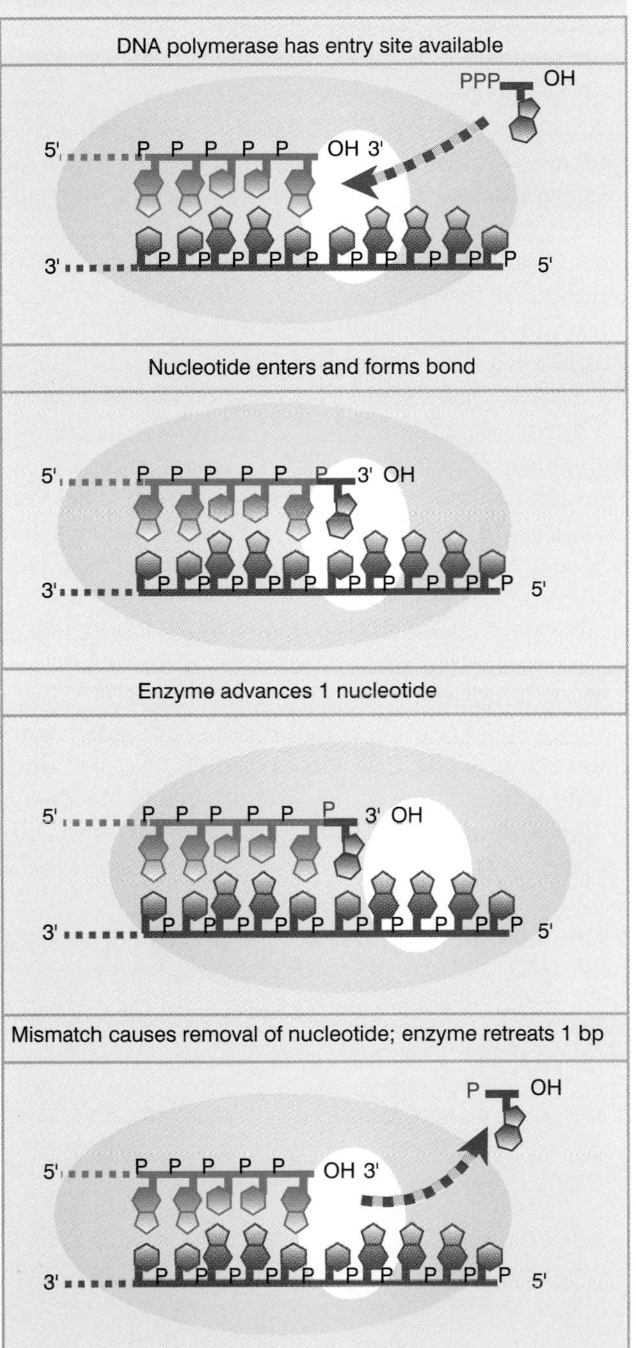

The small fragment (35 kD) possesses a 5′–3′ exonucleolytic activity, which excises small groups of nucleotides, up to ~10 bases at a time. This activity is coordinated with the synthetic/proofreading activity. It provides DNA polymerase I with a unique ability to start replication *in vitro* at a nick in DNA. (No other DNA polymerase has this ability.) At a point where a phosphodiester bond has been broken in a double-stranded DNA, the enzyme extends the 3′-OH end. As the new segment of DNA is synthesized, it *displaces the existing homologous strand in the duplex.*

This process of **nick translation** is illustrated in **Figure 13.3**. The displaced strand is degraded by the 5′–3′ exonucleolytic activity of the enzyme. The properties of the DNA are unaltered, except that a segment of one strand has been replaced with newly synthesized material, and the position of the nick has been moved along the duplex. This is of great practical use; nick translation has been a major technique for introducing radioactively labeled nucleotides into DNA *in vitro.*

The 5′–3′ synthetic/3′–5′ exonucleolytic action is probably used *in vivo* mostly for filling in short single-stranded regions in double-stranded DNA. These regions arise during replication, and also when bases that have been damaged are removed from DNA (see Chapter 14).

DNA polymerase III (the replicase) is a large enzyme complex. The replicase activity was originally discovered by a lethal mutation in the *dnaE locus*, which codes for the 130 kD α subunit that possesses the DNA synthetic activity. The 3′–5′ exonucleolytic proofreading activity is found in another subunit, ε, coded by *dnaQ.*

The ε subunit usually functions in conjunction with the α subunit, as indicated by the increase in proofreading when the two proteins are combined. The basic role of the ε subunit in controlling the fidelity of replication *in vivo* is demonstrated by the effect of mutations in *dnaQ*: the frequency with which mutations occur in the bacterial strain is increased by $>10^3$-fold.

Figure 13.3 Nick translation replaces part of a pre-existing strand of duplex DNA with newly synthesized material.

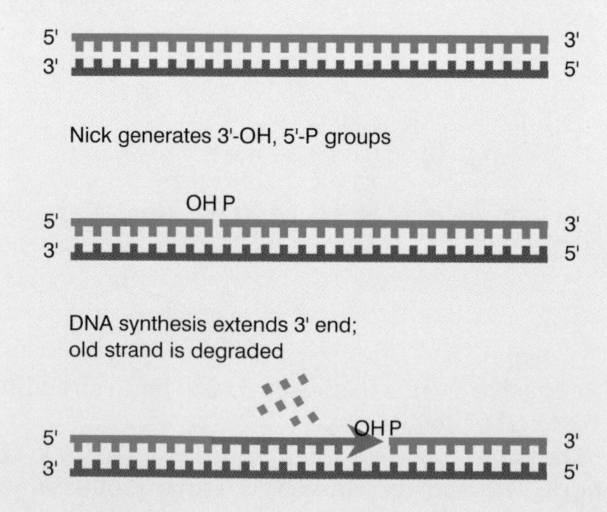

Some phages code for DNA polymerases. They include T4, T5, T7, and SPO1. The enzymes all possess 5′–3′ synthetic activities and 3′–5′ exonuclease proofreading activities. In each case, a mutation in the gene that codes for a single phage polypeptide prevents phage development. Each phage polymerase polypeptide associates with other proteins, of either phage or host origin, to make the intact enzyme.

Several classes of eukaryotic DNA polymerase have been identified. DNA polymerase δ is the nuclear replicase; DNA polymerase α is concerned with "priming" (initiating) replication. Other DNA polymerases are involved in repairing damaged nuclear DNA (β and ε) or with mitochondrial DNA replication (γ).

A common feature of all DNA polymerases is that *they cannot initiate synthesis of a chain of DNA from free nucleotides.* They require a **primer** to provide a free 3′-OH end that can be extended by addition of nucleotides. The primer can take various forms, some of which we have discussed already in Chapter 12 in the context of the replicon. Types of priming reaction are summarized in **Figure 13.4**:

- A sequence of RNA is synthesized on the template, so that the free 3′-OH end of the RNA chain is extended by the DNA polymerase. This is commonly used in replication of cellular DNA, and by some viruses. We discussed the example of ColE1 replication previously (see Figure 12.34).
- A preformed RNA pairs with the template, allowing its 3′-OH end to be used to prime DNA synthesis. This mechanism is used by retroviruses to prime reverse transcription of RNA (see Chapter 16).
- A primer terminus is generated within duplex DNA. The most common mechanism is the introduction of a nick, as used to initiate rolling circle replication (see Figure 12.16). In this case, the pre-existing strand is displaced by new synthesis. (Note the difference from nick translation shown in Figure 13.3, in which DNA polymerase I simultaneously synthesizes and degrades DNA from a nick.)
- A protein primes the reaction directly by presenting a nucleotide to the DNA polymerase. This reaction is used by certain viruses, as discussed previously in Figure 12.15.

Figure 13.4 There are several methods for providing the free 3'-OH end that DNA polymerases require to initiate DNA synthesis.

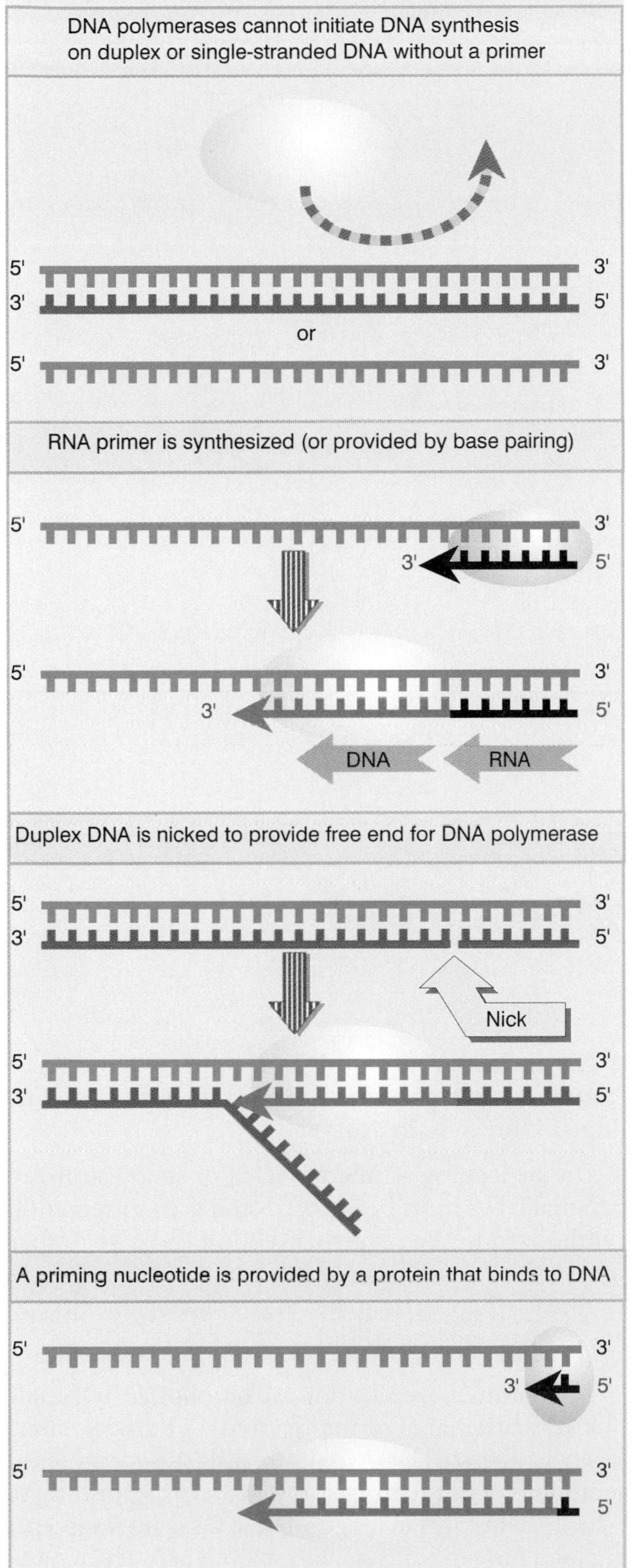

The catalytic reaction in a DNA polymerase occurs at an active site in which a nucleotide triphosphate pairs with an (unpaired) single strand of DNA. Crystal structures show that there is a common type of arrangement in DNA polymerases in the *E. coli* pol I family. The T7 replication system (which belongs to this family) consists of a catalytic subunit coded by the phage, the bacterial protein thioredoxin (which increases processivity of the catalytic subunit), and two alternative products of phage gene *4*, which provide primase and unwinding (helicase) activities.

Figure 13.5 shows the crystal structure of the T7 enzyme complexed with DNA (in the form of a primer annealed to a template strand) and an incoming nucleotide that is about to be added to the primer. The polymerase domain of the catalytic subunit has a structure that resembles a right hand. The existence of a large cleft composed of three domains is in fact common to many DNA polymerases. The DNA lies across the *palm* in a groove that is created by the *thumb* and *fingers*. The DNA is in the classic B-form duplex up to the last 2 base pairs at the 3′ end of the primer, which are in the more open A-form. A sharp turn in the DNA exposes the template base to the incoming nucleotide. The 3′ end of the primer (to which bases are added) is anchored by the fingers and palm. The DNA is held in

Figure 13.5 Crystal structure of phage T7 DNA polymerase has a "right hand" structure. DNA lies across the palm and is held by the fingers and thumb. Photograph kindly provided by Charles Richardson and Tom Ellenberger.

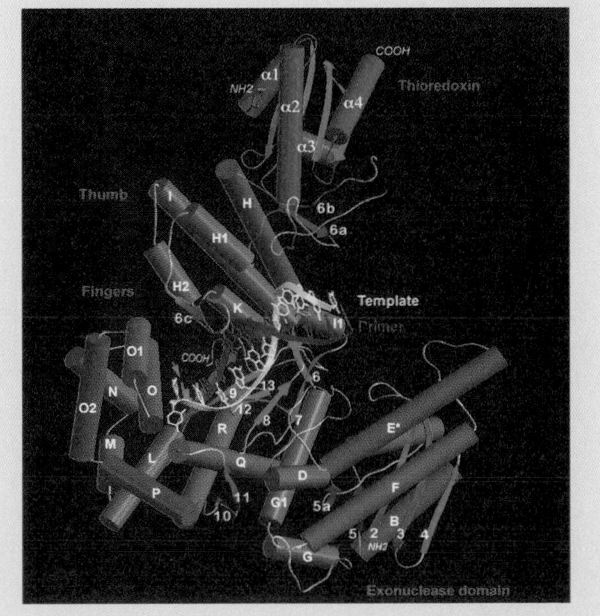

position by contacts that are made principally with the phosphodiester backbone (thus enabling the polymerase to function with DNA of any sequence). **Figure 13.6** shows the structure in diagrammatic form. The exonuclease domain responsible for proofreading is located below the palm, with a separate active site (this requires the 3′ end of the DNA chain to switch from the catalytic site to the exonuclease site when proofreading occurs).

In structures of DNA polymerases of this family complexed only with DNA (that is, lacking the incoming nucleotide), the orientation of the fingers and thumb relative to the palm is more open, with the O helix (O, O1, O2; see Figure 13.5) rotated away from the palm. This suggests that an inward rotation of the O helix by ~41° occurs to grasp the incoming nucleotide and create the active catalytic site, which is located in the palm subdomain at the bottom of the cleft. Amino acids in the active site contact the incoming base in such a way that the enzyme structure will be affected by a mismatched base; this may provide the signal that slows DNA synthesis and creates an opportunity for removal of the base by proofreading.

Figure 13.6 The catalytic domain of a DNA polymerase has a DNA-binding cleft created by three subdomains. The active site is in the palm. Proofreading is provided by a separate active site in an exonuclease domain.

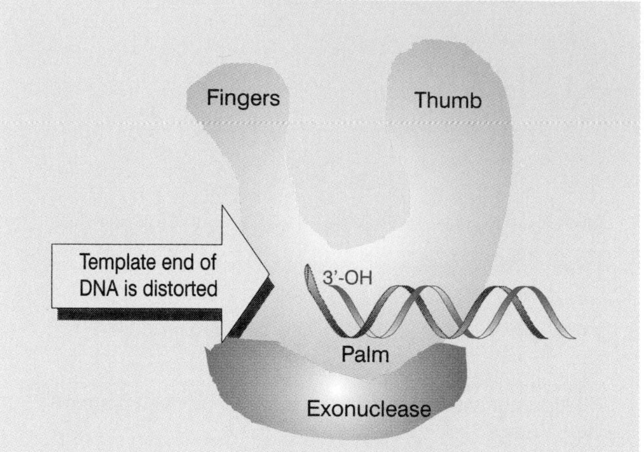

DNA synthesis is semidiscontinuous and primed by RNA

The antiparallel structure of the two strands of duplex DNA poses a problem for replication. As the replication fork advances, daughter strands must be synthesized on both of the exposed parental single strands. The fork moves in the direction from 5′→3′ on one strand, and in the direction from 3′→5′ on the other strand. Yet nucleic acids are synthesized only from a 5′ end toward a 3′ end. *The problem is solved by synthesizing the strand that grows overall from 3′ to 5′ in a series of short fragments, each actually synthesized in the "backwards" direction, that is, with the customary 5′ to 3′ polarity.*

Consider the region immediately behind the replication fork, as illustrated in **Figure 13.7**. We describe events in terms of the different properties of each of the newly synthesized strands:

- On the **leading strand** DNA synthesis can proceed continuously in the 5′ to 3′ direction as the parental duplex is unwound.
- On the **lagging strand** a stretch of single-stranded parental DNA must be exposed, and then a segment is synthesized in the reverse direction (relative to fork movement). A series of these fragments are synthesized, each 5′ to 3′; then they are joined together to create an intact lagging strand.

Discontinuous replication can be followed by the fate of a very brief label of radioactivity. The label enters newly synthesized DNA in the form of short fragments, sedimenting in the range of 7–11S, corresponding to ~1000–2000 bases in length. These **Okazaki fragments** are found in replicating DNA in both prokaryotes and

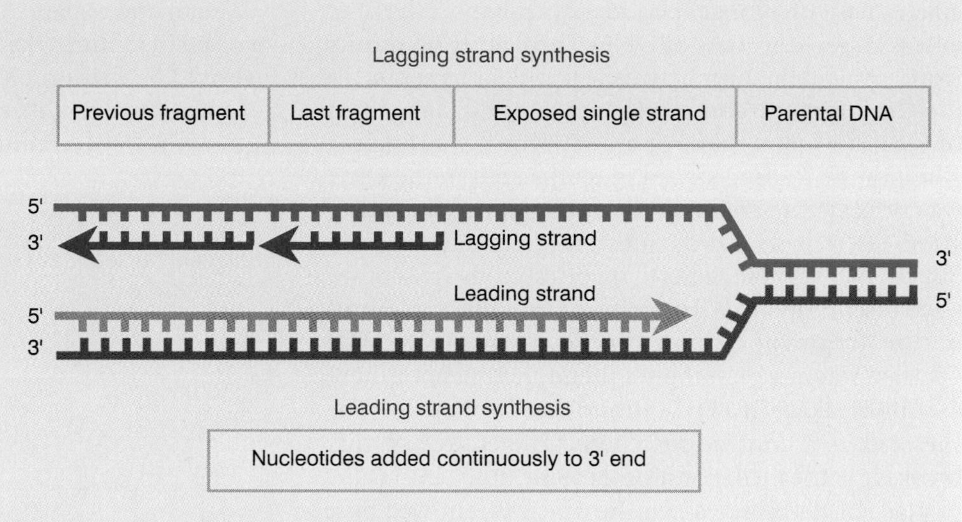

Figure 13.7 The leading strand is synthesized continuously while the lagging strand is synthesized discontinuously.

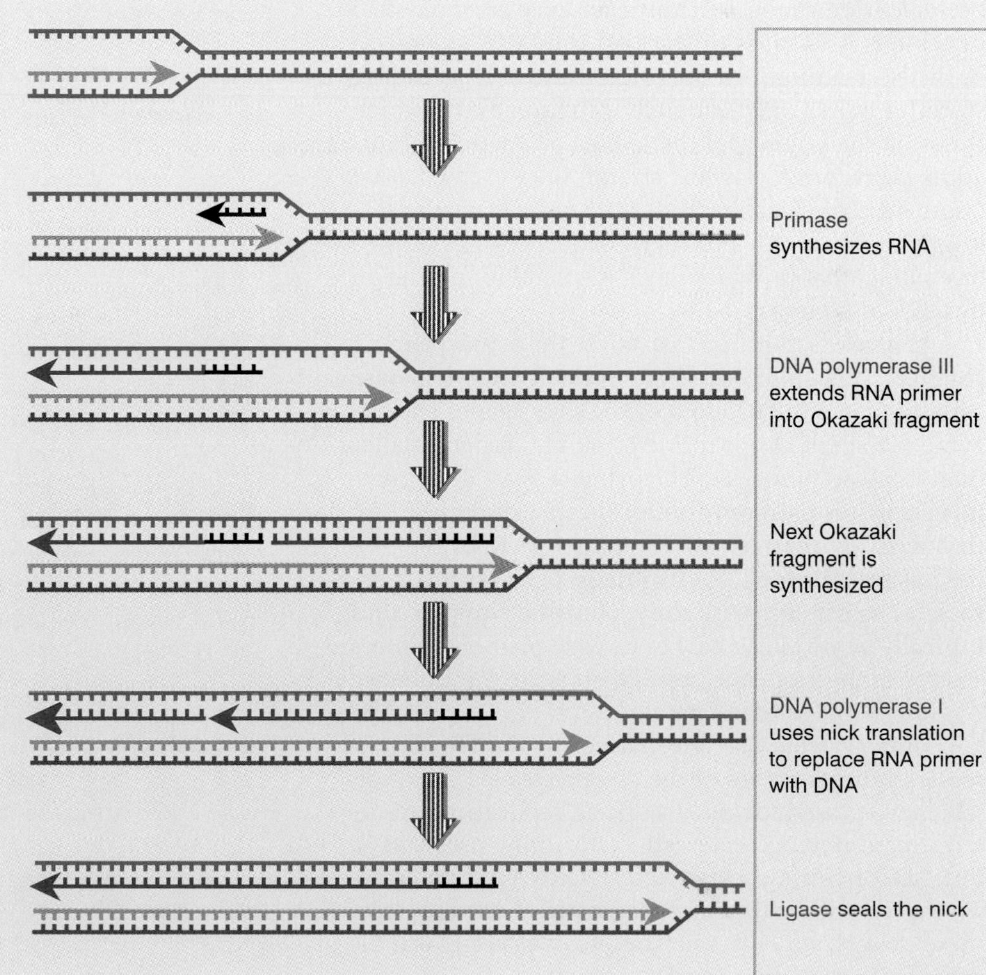

Figure 13.8 Synthesis of Okazaki fragments requires priming, extension, removal of RNA, gap filling, and nick ligation.

eukaryotes. After longer periods of incubation, the label enters larger segments of DNA. The transition results from covalent linkages between Okazaki fragments.

(The lagging strand *must* be synthesized in the form of Okazaki fragments. For a long time it was unclear whether the leading strand is synthesized in the same way or is synthesized continuously. All newly synthesized DNA is found as short fragments in *E. coli*. Superficially, this suggests that both strands are synthesized discontinuously. However, it turns out that not all of the fragment population represents *bona fide* Okazaki fragments; some are pseudofragments, generated by breakage in a DNA strand that actually was synthesized as a continuous chain. The source of this breakage is the incorporation of some uracil into DNA in place of thymine. When the uracil is removed by a repair system, the leading strand has breaks until a thymine is inserted.)

So the lagging strand is synthesized discontinuously and the leading strand is synthesized continuously. This is called **semidiscontinuous replication**.

Because DNA polymerase cannot initiate a deoxyribonucleotide chain, a priming activity is required to provide 3′-OH ends to start off the DNA chains on both the leading and lagging strands. The leading strand requires only one such initiation event, which occurs at the origin. But there must be a series of initiation events on the lagging strand, since each Okazaki fragment requires its own start *de novo*. Each Okazaki fragment starts with a primer, a sequence of RNA, ~10 bases long, that provides the 3′-OH end for extension by DNA polymerase.

A **primase** is required to catalyze the actual priming reaction. This is provided by a special RNA polymerase activity that is the product of the *dnaG* gene. The enzyme is a single polypeptide of 60 kD (much smaller than RNA polymerase). The primase is an RNA polymerase that is used only under specific circumstances, that is, to synthesize short stretches of RNA that are used as primers for DNA synthesis. DnaG primase associates transiently with the replication complex, and typically synthesizes an 11–12 base primer. Primers start with the sequence pppAG, opposite the sequence 3′-GTC-5′ in the template.

(Some systems use alternatives to the DnaG primase. In the examples of the two phages M13 and G4, which were used for early work on replication, an interesting difference emerged. G4 priming uses DnaG, but M13 priming uses bacterial RNA polymerase. There is another unusual feature in the example of these phages, which is that the site of priming is indicated by a region of secondary structure.)

We can now expand our consideration of the actions involved in joining Okazaki fragments, as illustrated in **Figure 13.8**. The complete order of events is uncertain, but must involve synthesis of RNA primer, its extension with DNA, removal of the RNA primer, its

Figure 13.9 DNA ligase seals nicks between adjacent nucleotides by employing an enzyme-AMP intermediate.

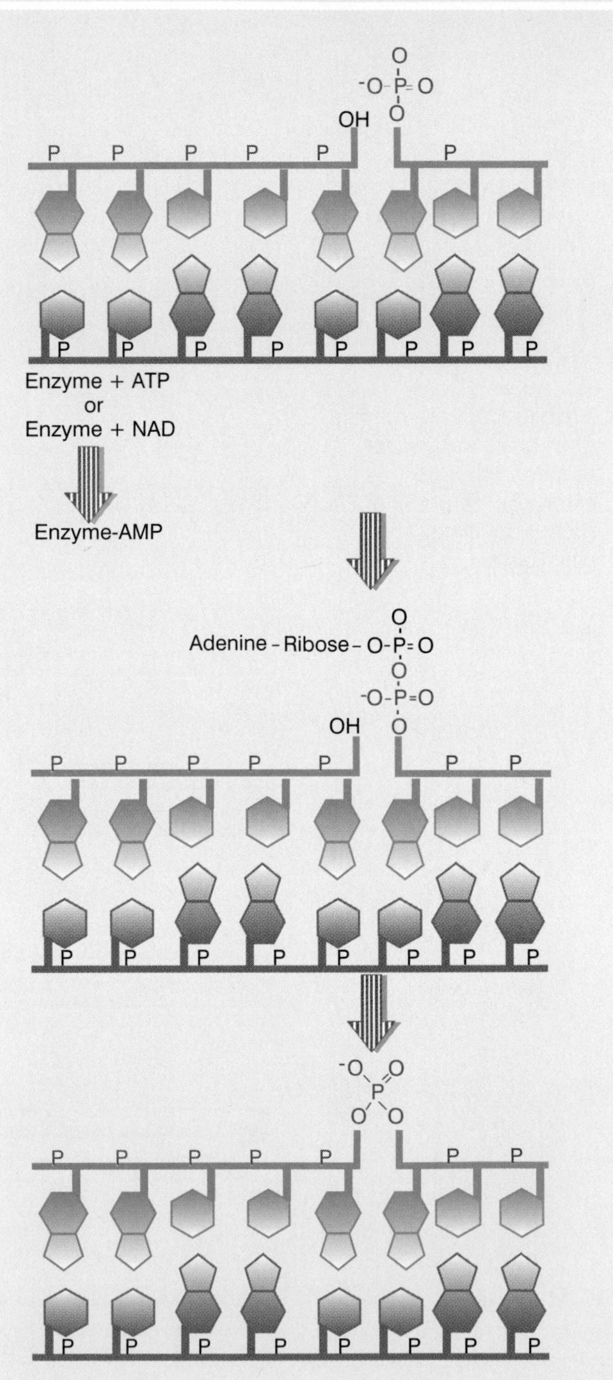

replacement by a stretch of DNA, and the covalent linking of adjacent Okazaki fragments.

The figure suggests that synthesis of an Okazaki fragment terminates just before the start of the RNA primer of the preceding fragment. When the primer is removed, there will be a gap. The gap is filled by DNA polymerase I; *polA* mutants fail to join their Okazaki fragments properly. The 5′–3′ exonuclease activity removes the RNA primer while simultaneously replacing it with a DNA sequence extended from the 3′-OH end of the next Okazaki fragment. This is equivalent to nick translation, except that the new DNA replaces a stretch of RNA rather than a segment of DNA. In mammalian systems (where the DNA polymerase does not have a 5′–3′ exonuclease activity), Okazaki fragments are removed by a two-step process. First RNAase HI (an enzyme that is specific for a DNA-RNA hybrid substrate) makes an endonucleolytic cleavage; then a 5′–3′ exonuclease called FEN1 removes the RNA.

Once the RNA has been removed and replaced, the adjacent Okazaki fragments must be linked together. The 3′-OH end of one fragment is adjacent to the 5′-phosphate end of the previous fragment. The responsibility for sealing this nick lies with the enzyme **DNA ligase.** Ligases are present in both prokaryotes and eukaryotes. Unconnected fragments persist in *lig*⁻ mutants, because they fail to join Okazaki fragments together.

The *E. coli* and T4 ligases share the property of sealing nicks that have 3′-OH and 5′-phosphate termini, as illustrated in **Figure 13.9**. Both enzymes undertake a two-step reaction, involving an enzyme-AMP complex. (The *E. coli* and T4 enzymes use different cofactors. The *E. coli* enzyme uses NAD [nicotinamide adenine dinucleotide] as a cofactor, while the T4 enzyme uses ATP.) The AMP of the enzyme complex becomes attached to the 5′-phosphate of the nick; and then a phosphodiester bond is formed with the 3′-OH terminus of the nick, releasing the enzyme and the AMP.

Coordinating synthesis of the lagging and leading strands

An Okazaki fragment is synthesized as a complement to the single-stranded region exposed by the movement of the replication fork. Two types of function are needed to convert double-stranded DNA to the single-stranded state:

- A **helicase** is an enzyme that separates the strands of DNA, usually using the hydrolysis of ATP to provide the necessary energy. A helicase is generally multimeric, often a hexamer, and typically translocates along DNA by using its multimeric structure to provide multiple DNA-binding sites. It is likely to have one conformation that binds to duplex DNA and another that binds to single-stranded DNA. Alternation between them drives the motor that melts the duplex, and requires ATP hydrolysis—typically 1–2 ATPs are hydrolyzed for each base pair that is unwound. A helicase usually initiates unwinding at a single-stranded region adjacent to a duplex, and may function with a particular polarity, preferring single-stranded DNA with a 3′ end (3′–5′ helicase) or with a 5′ end (5′–3′ helicase).
- A **single-strand binding (SSB) protein** binds to the single-stranded DNA, preventing it from re-forming the duplex state. The SSB binds as a monomer, but typically in a cooperative manner in which the binding of additional monomers to the existing complex is enhanced.

Each of these activities may be provided by an enzyme that is specific for a particular complex concerned with replication, repair, or recombination. There are ~12 helicases in *E. coli*, for example, and phages may code for their own SSB proteins rather than using the *E. coli* protein.

The conversion of ϕX174 double-stranded DNA into individual single strands illustrates the features of the process. **Figure 13.10** shows that a single strand is peeled off the circular strand, resembling the rolling circle described previously in Figure 12.16. The reaction can occur in the absence of DNA synthesis when the appropriate three proteins are provided *in vitro.*

The phage A protein nicks the viral (+) strand at the

origin of replication. In the presence of two host proteins, Rep and SSB, and ATP, the nicked DNA unwinds. The Rep protein provides a helicase that separates the strands; the SSB traps them in single-stranded form. The *E. coli* SSB is a tetramer of 74 kD that binds cooperatively to single-stranded DNA.

The significance of the cooperative mode of binding is that the binding of one protein molecule makes it much easier for another to bind. So once the binding reaction has started on a particular DNA molecule, it is

Figure 13.10 ϕX174 DNA can be separated into single strands by the combined effects of 3 functions: nicking with A protein, unwinding by Rep, and single-strand stabilization by SSB.

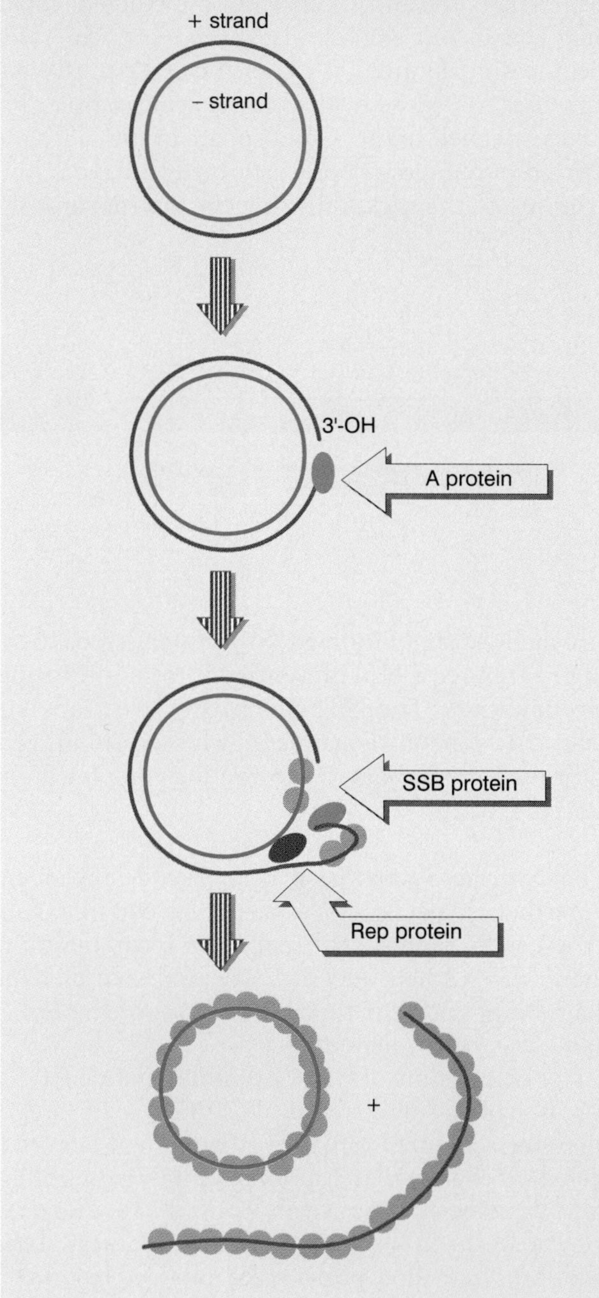

Figure 13.11 Priming requires several enzymatic activities, including helicases, single-strand binding proteins, a means of recognizing the primosome assembly sequence, and other structural proteins.

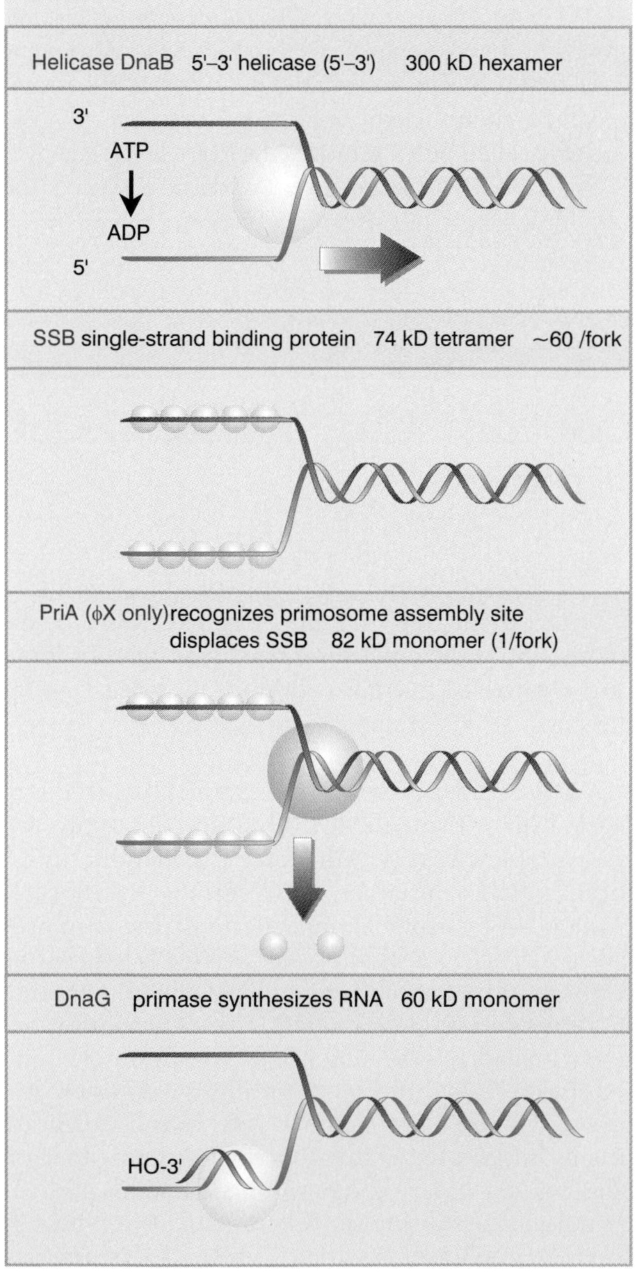

rapidly extended until *all of the single-stranded DNA is covered with the SSB protein.* Note that this protein is *not a DNA-unwinding protein*; its function is to stabilize DNA that is already in the single-stranded condition.

Under normal circumstances *in vivo*, the unwinding, coating, and replication reactions proceed in tandem. The SSB binds to DNA as the replication fork advances, keeping the two parental strands separate so that they are in the appropriate condition to act as templates. SSB is needed in stoichiometric amounts at the replication fork. It is required for more than one stage of replication; *ssb* mutants have a quick-stop phenotype, and are defective in repair and recombination as well as in replication. (Some phages use different SSB proteins, notably T4; this shows that there may be specific interactions between components of the replication apparatus and the SSB; we discuss this later.)

The formation of single-stranded DNA is prerequisite for the priming reaction. There are two types of priming in *E. coli*. The *oriC* system, named for the bacterial origin, involves similar events at the origin (where the leading strand is initiated) and at each Okazaki fragment on the lagging strand. The ϕX system, named for phage ϕX174, requires a protein complex called the primosome (see below). Sometimes replicons are referred to as being of the ϕX or *oriC* type.

The types of activities involved in the priming reaction are summarized in **Figure 13.11**. Although other replicons in *E. coli* may have alternatives for some

Figure 13.12 Leading and lagging strand polymerases move apart.

Leading and lagging enzymes start at same point on double helix
Lagging strand
5'
3'
3'
5'
5'
3'
Leading strand
Enzymes move in opposite direction and are far apart at completion of Okazaki fragment
5'
3'
3'
5'
5'
3'
Lagging enzyme must translocate to new position to start another Okazaki fragment
5'
3'
3'
5'
5'
3'

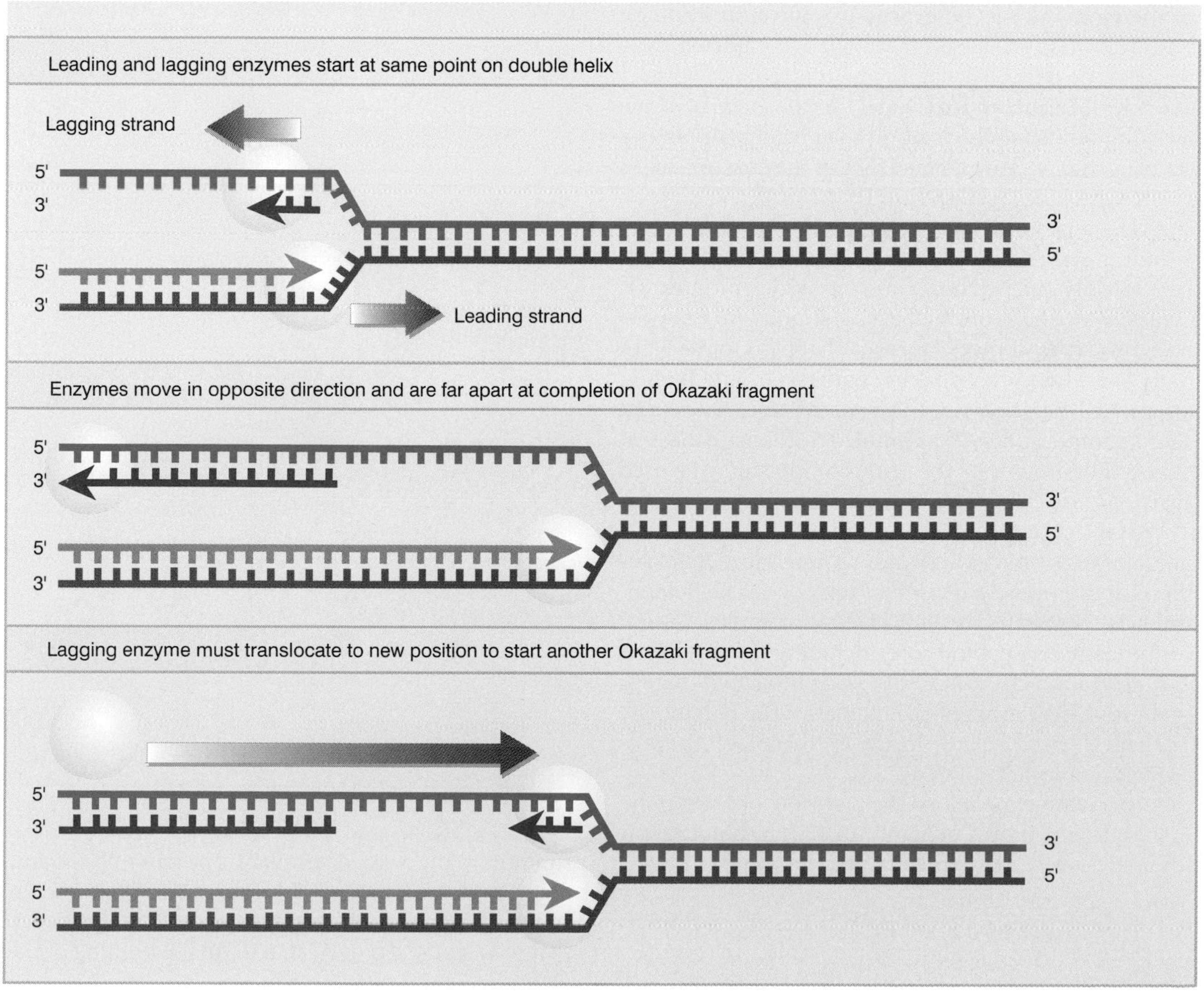

of these particular proteins, the same general types of activity are required in every case.

DnaB is the central component in both ϕX and *oriC* replicons. It provides the 5′–3′ helicase activity that unwinds DNA. Energy for the reaction is provided by cleavage of ATP. Basically DnaB is the active component of the growing point. The major difference between the two types of replicon lies in the interactions with DnaB at the priming site. The priming reaction that initiates synthesis of each Okazaki fragment occurs when DnaB activates the DnaG primase (see below).

A major problem of the semidiscontinuous mode of replication is illustrated in **Figure 13.12**; how is synthesis of the lagging strand coordinated with synthesis of the leading strand? The new DNA strands are growing in opposite directions. One enzyme unit is moving with the unwinding point and synthesizing the leading strand. The other unit is moving "backwards", relative to the DNA, along the exposed single strand. When synthesis of one Okazaki fragment is completed, synthesis of the next Okazaki fragment is required to start at a new location approximately in the vicinity of the growing point for the leading strand. This requires a translocation relative to the DNA of the enzyme that is synthesizing the lagging strand.

As the replisome moves along DNA, unwinding the parental strands, it elongates the leading strand. Periodically the primosome activity initiates an Okazaki fragment on the lagging strand. We can propose two types of model for what happens to the DNA replicase when it completes synthesis of an Okazaki fragment. It might dissociate from the template, so that a new complex must be assembled to elongate the next Okazaki fragment; or the same complex may be used again.

We can now relate the subunit structure of DNA polymerase III to the activities required for DNA synthesis and propose a model on this basis. The holoenzyme is a complex of 900 kD that contains several components: a catalytic core, including the α subunit; a dimerization component (τ) that links two cores; a processivity component (β) that keeps the polymerase on the DNA; and a clamp loader (γ) that places the processivity subunits on DNA.

A speculative model for the assembly of DNA polymerase III is depicted in **Figure 13.13**. The holoenzyme assembles on DNA in three stages:

- A β dimer plus a γ complex recognizes the primer-template to form a preinitiation complex. In this reaction, the γ complex cleaves ATP and transfers β subunits to the primed template. The pair of β subunits forms a "clamp" that binds around the DNA and ensures processivity. The γ complex is a "clamp loader" that uses hydrolysis of ATP to drive the binding of β to DNA.

Figure 13.13 DNA polymerase III holoenzyme assembles in stages, generating an enzyme complex that synthesizes the DNA of both new strands.

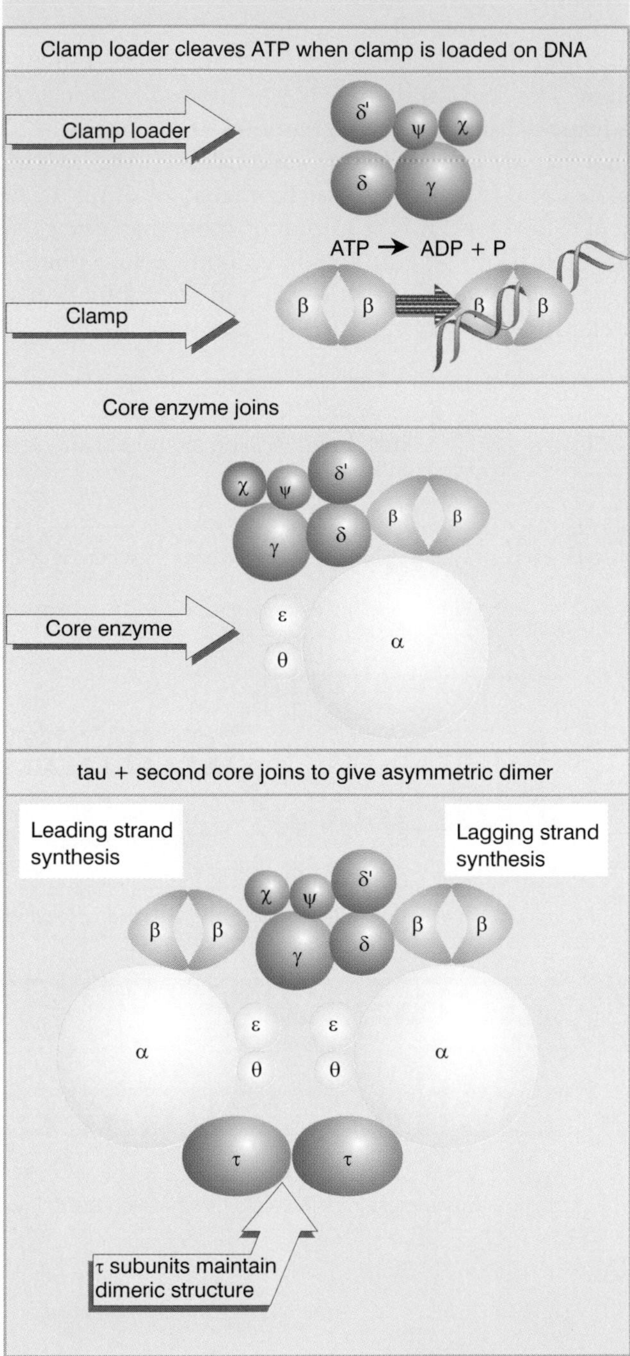

■ Binding to DNA changes the conformation of the site on β that binds to the γ complex, and as a result it now has a high affinity for the core polymerase. This enables core polymerase to bind, and this is the means by which the core polymerase is brought to DNA. The "core enzyme" contains subunits α, ε, and θ. The α subunit has a basic ability to synthesize DNA, the ε subunit has the 3′–5′ proofreading exonuclease, and θ may be required for assembly. (The processivity of the core by itself is low; but the β clamp ensures that it functions processively on the DNA.)

■ A τ dimer binds to the core polymerase, and provides a dimerization function that binds a second core polymerase (associated with another β clamp). The holoenzyme is asymmetric, because it has only 1 γ complex. This γ complex is responsible for adding a pair of β dimers to each parental strand of DNA.

Each of the core complexes of the holoenzyme synthesizes one of the new strands of DNA. Because the clamp loader is needed for unloading the β complex from DNA, the two cores have different abilities to dissociate from DNA. This corresponds to the need to synthesize a continuous leading strand (where polymerase remains associated with the template) and a discontinuous lagging strand (where polymerase repetitively dissociates and reassociates). The γ complex is associated with the core polymerase that synthesizes the lagging strand, and plays a key role in the ability to synthesize individual Okazaki fragments.

The β dimer makes the holoenzyme highly processive. β is strongly bound to DNA, but can slide along a duplex molecule. The crystal structure of β shows that it forms a ring-shaped dimer. The model in **Figure 13.14** shows the β-ring in relationship to a DNA double helix. The dimer surrounds the duplex, providing the "sliding clamp" that allows the holoenzyme to slide along DNA. The structure explains the high processivity—there is no way for the enzyme to fall off! Because the clamp is a circle of subunits surrounding DNA, its assembly or removal requires the use of an energy-dependent process by the clamp loader.

The asymmetric dimer model for DNA polymerase III suggests the structure for the replication fork that is illustrated in **Figure 13.15**. A catalytic core is associated with each template strand of DNA. The holoenzyme moves continuously along the template for the leading strand; the template for the lagging strand is "pulled through", creating a loop in the DNA. DnaB creates the unwinding point, and translocates along the DNA in the "forward" direction (moving 5′–3′ on the template for the lagging strand).

Figure 13.14 The β subunit of DNA polymerase III holoenzyme consists of a head to tail dimer (the two subunits are shown in red and orange) that forms a ring completely surrounding a DNA duplex (shown in the center). Photograph kindly provided by John Kuriyan.

Cross-section through DNA duplex surrounded by enzyme

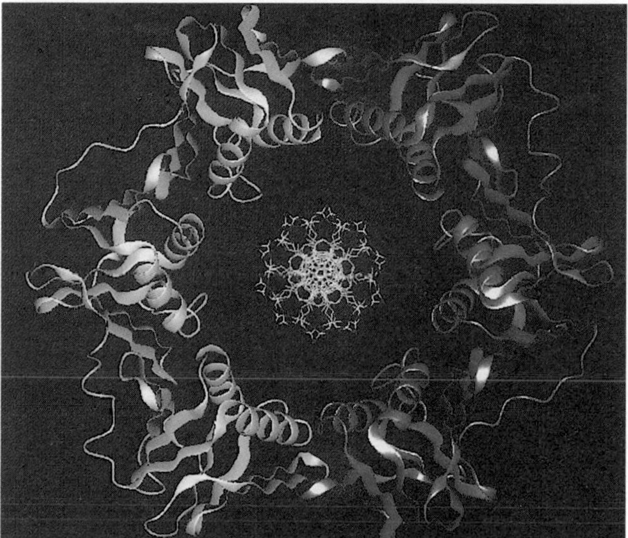

Side view of space-filling model (DNA strands in green and white)

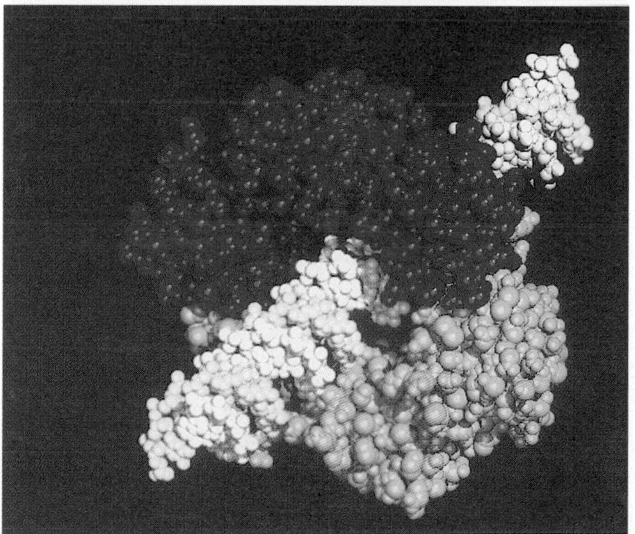

DnaB contacts the τ subunit(s) of DNA polymerase. This establishes a direct connection between the helicase-primase complex and the polymerase itself. This link has two effects. One is to increase the speed of DNA synthesis by increasing the rate of movement by DNA polymerase core by 10×. The second is to prevent the leading strand polymerase from falling off, that is, to increase its processivity.

Figure 13.15 Each catalytic core of Pol III synthesizes a daughter strand. DnaB is responsible for forward movement at the replication fork. The primosome pulls a DNA template strand through.

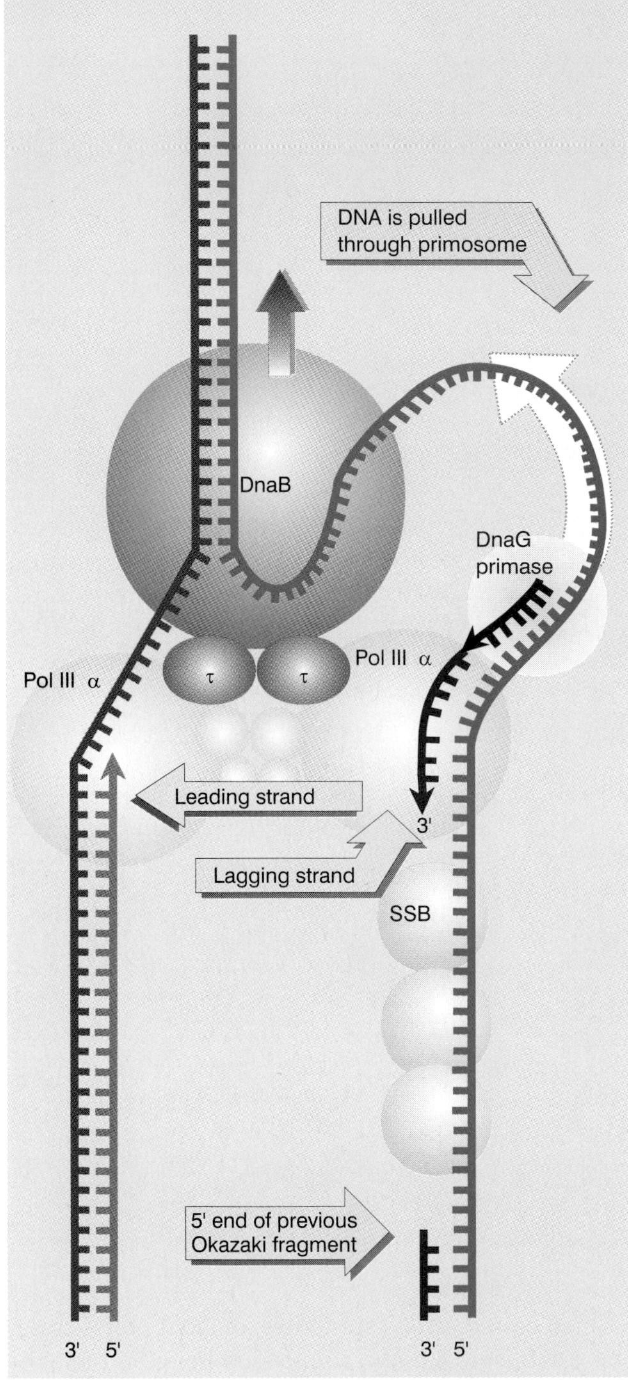

Synthesis of the leading strand creates a loop of single-stranded DNA that provides the template for lagging strand synthesis, and this loop becomes larger as the unwinding point advances. After initiation of an Okazaki fragment, the lagging strand core complex pulls the single-stranded template through the β clamp while synthesizing the new strand. The single-stranded template must extend for the length of at least one Okazaki fragment before the lagging polymerase completes one fragment and is ready to begin the next.

What happens to the loop when the Okazaki fragment is completed? When a Pol III holoenzyme meets a nick in DNA, the core complex and clamp dissociate from the β sliding clamp. The core can then reassociate with a new β subunit elsewhere. **Figure 13.16** suggests that this represents the reaction that occurs at the end of an Okazaki fragment. The core complex will dissociate when it completes synthesis of each fragment, releasing the loop. The core complex then associates with a β clamp to initiate the next Okazaki fragment. Probably a new β clamp will already be present at the next initiation site, and the β clamp that has lost its core complex will dissociate from the template (with the assistance of the γ complex) to be used again. So the lagging strand polymerase will probably transfer from one β clamp to the next in each cycle, without dissociating from the replicating complex.

What is responsible for recognizing the sites for initiating synthesis of Okazaki fragments? In *oriC* replicons, the connection between priming and the replication fork is provided by the dual properties of DnaB: it is the helicase that propels the replication fork; and it interacts with the DnaG primase at an appropriate site. Following primer synthesis, the primase is released. The length of the priming RNA is limited to 8–14 bases. Apparently DNA polymerase III is responsible for displacing the primase.

A more complex system is used for priming with the phage ϕX174 (with which in fact much of the early work was done). ϕX174 DNA is not by itself a substrate for the replication apparatus, because the naked DNA does not provide a suitable template. But once the single-stranded form has been coated with SSB, replication can proceed. A **primosome** assembles at a unique site on the single-stranded DNA, called the **primosome assembly site** (*pas*). The *pas* is the equivalent of an origin for synthesis of the complementary strand of ϕX174. The primosome consists of six proteins: PriA, PriB, PriC, DnaT, DnaB, and DnaC. The key event in localizing the primosome is the ability of PriA to displace SSB from single-stranded DNA.

Although the primosome forms initially at the *pas* on ϕX174 DNA, primers are initiated at a variety of sites. PriA translocates along the DNA, displacing SSB, to reach additional sites at which priming occurs. As in *oriC* replicons, DnaB plays a key role in unwinding and

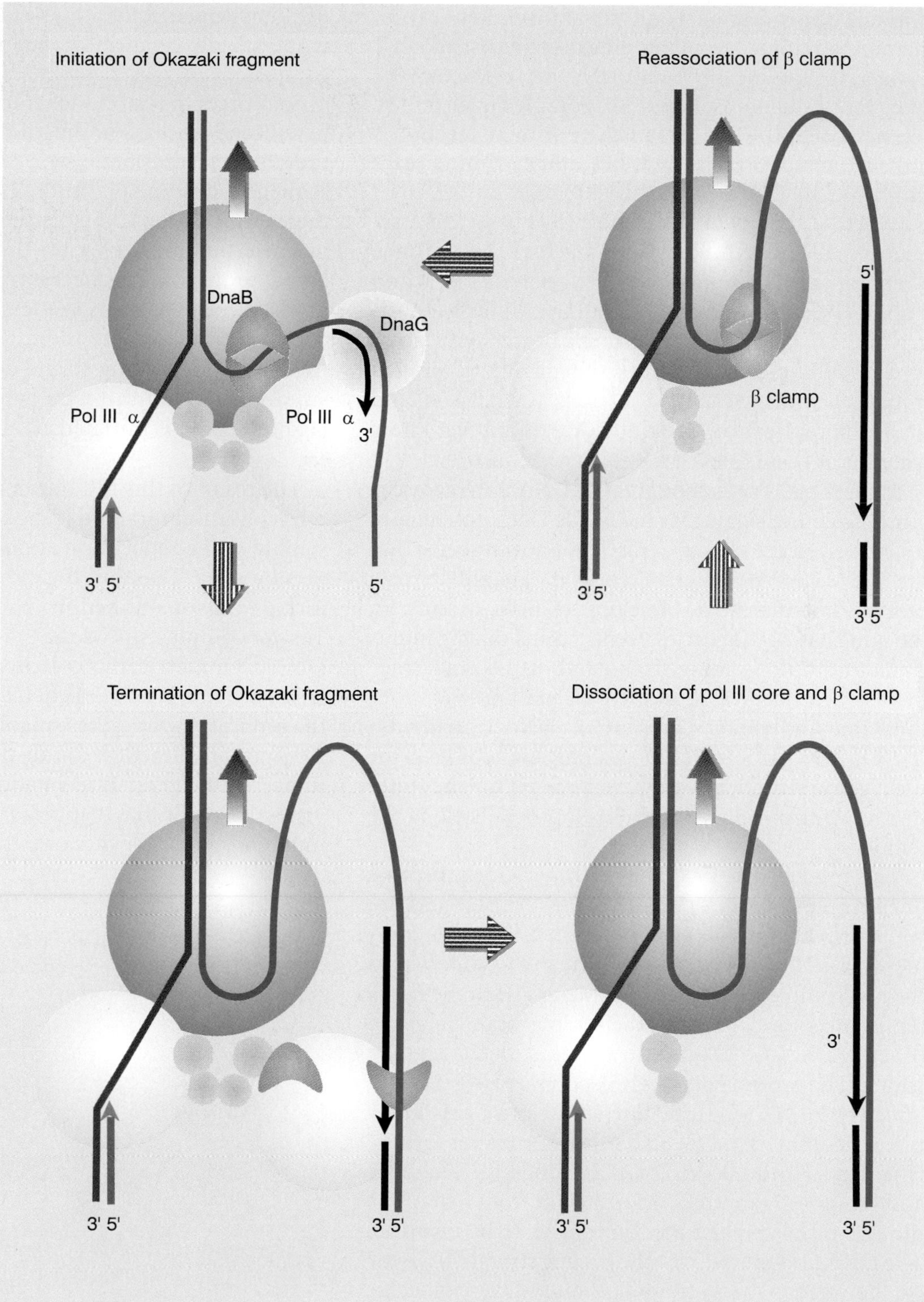

Figure 13.16 Core polymerase and the β clamp dissociate at completion of Okazaki fragment synthesis and reassociate at the beginning.

priming in φX replicons. However, different proteins are required to load it at the φX origin.

The system that synthesizes eukaryotic DNA in animal cells has two DNA polymerases. DNA polymerase α exists as a complex consisting of a 180 kD catalytic subunit, associated with other subunits, including two smaller proteins that provide a primase activity. DNA polymerase δ is less well characterized. The counterparts of these enzymes in yeast, DNA polymerases I and III, are coded by essential genes.

Initiation at the SV40 origin requires a product of the virus, the T antigen, to bind to the origin and initiate the process of strand separation. Replication factor A is a single-strand binding protein that allows T antigen to unwind the SV40 DNA more extensively. The DNA polymerase α/primer complex initiates the synthesis of both DNA strands. The priming reaction is unusual: it starts with RNA, like other priming reactions, but the RNA primer is extended by the DNA polymerase activity to give a short (3–4 base) DNA sequence, called iDNA. Replication factor C binds to the 3′ end of the iDNA and loads DNA polymerase δ and PCNA. (PCNA is called proliferating cell nuclear antigen for historical reasons; it acts as a processivity factor for DNA polymerase δ.) In this system, SV40 T antigen initiates DNA unwinding, functions as a helicase, and loads the replication apparatus; we do not yet know what fulfils these roles in the eukaryotic host cell.

At one time, it was thought that DNA polymerase α synthesized the lagging strand, while DNA polymerase δ synthesized the leading strand, but now it seems that the role of DNA polymerase α is to initiate, and the role of DNA polymerase δ is to elongate, both strands. So the same basic sequence of events occurs during initiation of both the leading strand and the Okazaki fragments of the lagging strand. We do not know whether the same individual complex of DNA polymerase α /primase that initiates the leading strand is used for the lagging strand, but we have a model for the function of a eukaryotic replisome as it proceeds along the template.

This model suggests that a replication fork contains one complex of DNA polymerase α /primase and has two complexes of DNA polymerase δ. The two complexes of DNA polymerase δ behave in the same way as the two complexes of DNA polymerase III in the *E. coli* replisome: one synthesizes the leading strand, and the other synthesizes Okazaki fragments on the lagging strand. The processivity of DNA polymerase δ is maintained by PCNA. The crystal structure of PCNA closely resembles that of the *E. coli* β subunit: a trimer forms a ring that surrounds the DNA. Although the sequence and subunit organization are different from that of the dimeric β clamp, the function is likely to be identical. The DNA polymerase δ of the lagging strand fills in the gap between Okazaki fragments after the exonuclease MF1 has removed the RNA. The enzyme DNA ligase I is specifically required to seal the nicks between the completed Okazaki fragments.

Sequences that stop movement of replication forks have been identified in the form of the *ter* elements of the *E. coli* chromosome (see Figure 12.7) or equivalent sequences in some plasmids. The common feature of these elements is a 23 bp consensus sequence that provides the binding site for the product of the *tus* gene, a 36 kD protein that is necessary for termination. Tus binds to the consensus sequence, where it provides a contra-helicase activity and stops DnaB from unwinding DNA. The leading strand continues to be synthesized right up to the *ter* element, while the nearest lagging strand is initiated 50–100 bp before reaching *ter*.

The result of this inhibition is to halt movement of the replication fork and (presumably) to cause disassembly of the replication apparatus. **Figure 13.17** reminds us that Tus stops the movement of a replication fork in only one direction. The crystal structure of a Tus-*ter* complex shows that the Tus protein binds to DNA asymmetrically; α-helices of the protein protrude around the double helix at the end that blocks the replication fork. Presumably a fork proceeding in the opposite direction can displace Tus and thus continue. A difficulty in understanding the function of the system *in vivo* is that it appears to be dispensable, since mutations in the *ter* sites or in *tus* are not lethal.

Figure 13.17 Tus binds to *ter* asymmetrically and blocks replication in only one direction.

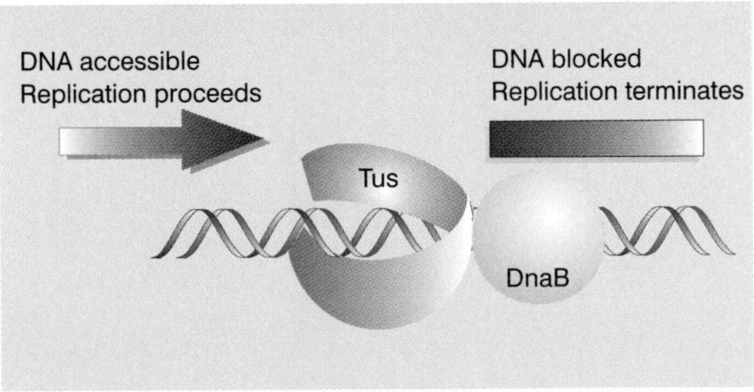

DNA accessible
Replication proceeds
DNA blocked
Replication terminates
Tus
DnaB

The replication apparatus of phage T4

When phage T4 takes over an *E. coli* cell, it provides several functions of its own that either replace or augment the host functions. Because the host DNA is degraded, the phage places little reliance on expression of host functions, except for utilizing enzymes that are present in large amounts in the bacterium. The degradation of host DNA is important in releasing nucleotides that are reused in the synthesis of phage DNA. (The phage DNA differs in base composition from cellular DNA in using hydroxymethylcytosine instead of the customary cytosine.)

The phage-coded functions concerned with DNA synthesis in the infected cell can be identified by mutations that impede the production of mature phages. Essential phage functions are identified by conditional lethal mutations, which fall into three phenotypic classes:

- Those in which there is *no DNA synthesis* at all identify genes whose products are either components of the replication apparatus or are involved in the provision of precursors (especially the hydroxymethylcytosine).
- Those in which the onset of DNA synthesis is *delayed* are concerned with the initiation of replication.
- Those in which DNA synthesis starts but then is *arrested* include regulatory functions, the DNA ligase, and some of the enzymes concerned with host DNA degradation.
- There are also nonessential genes concerned with replication; for example, including those involved in glucosylating the hydroxymethylcytosine in the DNA.

Synthesis of T4 DNA is catalyzed by a multienzyme aggregate assembled from the products of a small group of essential genes.

The gene *32* protein (gp32) is a highly cooperative single-strand binding protein, needed in stoichiometric amounts. It was the first example of its type to be characterized. The geometry of the T4 replication fork may specifically require the phage-coded protein, since the *E. coli* SSB cannot substitute. The gp32 forms a complex with the T4 DNA polymerase; this interaction could be important in constructing the replication fork.

The T4 system uses an RNA priming event that is similar to that of its host. With single-stranded T4 DNA as template, the gene *41* and *61* products act together to synthesize short primers. Their behavior is analogous to that of DnaB and DnaG in *E. coli*. The gene *41* protein is the counterpart to DnaB. It is a helicase, and also has a GTPase activity stimulated by single-stranded DNA. When the protein binds to single-stranded DNA, it can move from its initial binding site, migrating at a rate of ~400 nucleotides per second. It probably functions like DnaB, finding periodic sites at which to initiate primer synthesis. The hydrolysis of GTP is presumed to provide the energy for movement. Binding of the helicase to DNA is stimulated by the gene *59* product.

The gene *61* protein is needed in much smaller amounts than most of the T4 replication proteins. There are as few as 10 copies of gp61 per cell. (This impeded its characterization. It is required in such small amounts that originally it was missed as a necessary component, because enough was present as a contaminant of the gp32 preparation!) Gene *61* protein has the primase activity, analogous to DnaG of *E. coli*. The primase recognizes the template sequence 3′-TTG-5′ and synthesizes pentaribonucleotide primers that have the general sequence pppApCpNpNpNp. If the complete replication apparatus is present, these primers are extended into DNA chains.

The gene *43* DNA polymerase has the usual 5′–3′ synthetic activity, associated with a 3′–5′ exonuclease proofreading activity. It catalyzes DNA synthesis and removes the primers. When T4 DNA polymerase uses a single-stranded DNA as template, its rate of progress is uneven. The enzyme moves rapidly through single-stranded regions, but proceeds much more slowly through regions that have a base-paired intrastrand secondary structure. The accessory proteins assist the DNA polymerase in passing these roadblocks, and maintaining its speed.

The remaining three proteins are referred to as "polymerase accessory proteins". They increase the affinity of the DNA polymerase for the DNA, and also its processivity and speed. The gene *45* product is a dimer that probably acts as a "sliding clamp", equivalent to subunit β of *E. coli* DNA polymerase III, holding the DNA polymerase subunit more tightly on the template.

The products of genes *44* and *62* form a tight complex, which has ATPase activity. Their main role may be to link gp45 to gp43; they could be the equivalent of the γδ complex. As we have mentioned before, this type of

Figure 13.18 Similar functions are required at all replication forks.

Function	E.coli	HeLa/SV40	Phage T4
Helicase	*DnaB*	T antigen	41
Loading helicase/primase	*DnaC*	T antigen	59
Single strand maintenance	*SSB*	RPA	32
Priming	*DnaG*	Pol α/primase	61
Sliding clamp	β	PCNA	45
Clamp loading (ATPase)	*gd* complex	RFC	44/62
Catalysis	*Pol III core*	Pol δ	43
Holoenzyme dimerization	τ	?	"
RNA removal	*Pol I*	MF1	"
Ligation	*Ligase*	Ligase I	T4 ligase

intimate relationship makes it a moot point what is a component of DNA polymerase and what is an accessory factor.

We have dealt with DNA replication so far solely in terms of the progression of the replication fork. The need for other functions is shown by the DNA-delay and DNA-arrest mutants. The four genes of the DNA-delay mutants include *39*, *52*, and *60*, which code for the three subunits of T4 topoisomerase II, an activity needed for removing supercoils in the template (see Chapter 14). The essential role of this enzyme suggests that T4 DNA does not remain in a linear form, but becomes topologically constrained during some stage of replication. The topoisomerase could be needed to allow rotation of DNA ahead of the replication fork.

Comparison of the T4 apparatus with the *E. coli* apparatus suggests that DNA replication poses a set of problems that are solved in analogous ways in different systems. We may now compare the enzymatic and structural activities found at the replication fork in *E. coli*, T4, and HeLa (human) cells. **Figure 13.18** summarizes the functions and assigns them to individual proteins. We can interpret the known properties of replication complex proteins in terms of similar functions, involving the unwinding, priming, catalytic, and sealing reactions. The components of each system interact in restricted ways, as shown by the fact that phage T4 requires its own helicase, primase, clamp, etc., and the bacterial proteins cannot substitute for their phage counterparts.

Creating the replication forks at an origin

STARTING a cycle of replication of duplex DNA requires several successive activities:

- The two strands of DNA must suffer their initial separation. This is in effect a melting reaction over a short region.
- An unwinding point begins to move along the DNA; this marks the generation of the replication fork, which continues to move during elongation.
- The first nucleotides of the new chain must be synthesized into the primer. This action is required once for the leading strand, but is repeated at the start of each Okazaki fragment on the lagging strand.

Some events that are required for initiation therefore occur uniquely at the origin; others recur with the initiation of each Okazaki fragment during the elongation phase.

Plasmids carrying the *E. coli oriC* sequence have been used to develop a cell-free system for replication from this origin. Initiation of replication at *oriC in vitro* starts with formation of a complex that requires six proteins: DnaA, DnaB, DnaC, HU, Gyrase, and SSB. The reaction converts a circular supercoiled template into a new form in which there is extensive unwinding of the duplex. Of the six proteins involved in prepriming, DnaA draws our attention as the only one uniquely involved in initiation *vis-à-vis* elongation. DnaB/DnaC provides the "engine" of initiation at the origin, as it does in the ϕX primosome.

The first stage in complex formation is binding to *oriC* by DnaA protein. The reaction involves action at two types of sequences: 9 bp and 13 bp repeats. Together the 9 bp and 13 bp repeats define the limits of the 245 bp minimal origin, as indicated in **Figure 13.19**. An origin is activated by the sequence of events summarized in **Figure 13.20**, in which binding of DnaA is succeeded by association with the other proteins.

The four 9 bp consensus sequences on the right side of *oriC* provide the initial binding sites for DnaA. It binds cooperatively until 20–40 monomers have formed a central core around which *oriC* DNA is wrapped.

Then the DnaA protein acts at three 13 bp tandem repeats located in the left side of *oriC*. The DNA strands are melted at each of these sites to form an open complex. All three 13 bp repeats must be opened for the reaction to proceed to the next stage.

The sequences of the 9 bp and 13 bp repeats are distinct, and we do not know whether they are recognized by the same or different binding sites on the DnaA protein. Both types of consensus sequences are A•T-rich.

The DnaB and DnaC proteins form a complex in which six DnaC monomers bind each hexamer of DnaB. This generates a protein aggregate of 480 kD, corresponding to a sphere of radius 6 nm. The formation of a complex at *oriC* is detectable in the form of the large protein blob visualized in **Figure 13.21**.

Figure 13.19 The minimal origin is defined by the distance between the outside members of the 13-mer and 9-mer repeats.

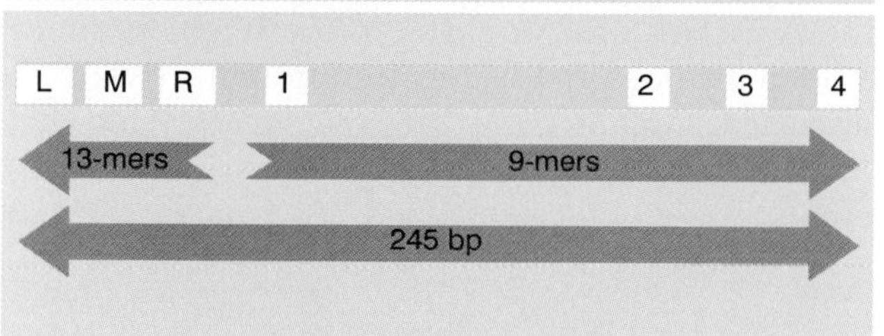
L
M
R
1
2
3
4
13-mers
9-mers
245 bp

The DnaB•DnaC complex transfers one hexamer of DnaB to form the replication fork. DnaB provides the helicase that unwinds the DNA. Probably it recognizes

Figure 13.20 Prepriming involves formation of a complex by sequential association of proteins, leading to the separation of DNA strands.

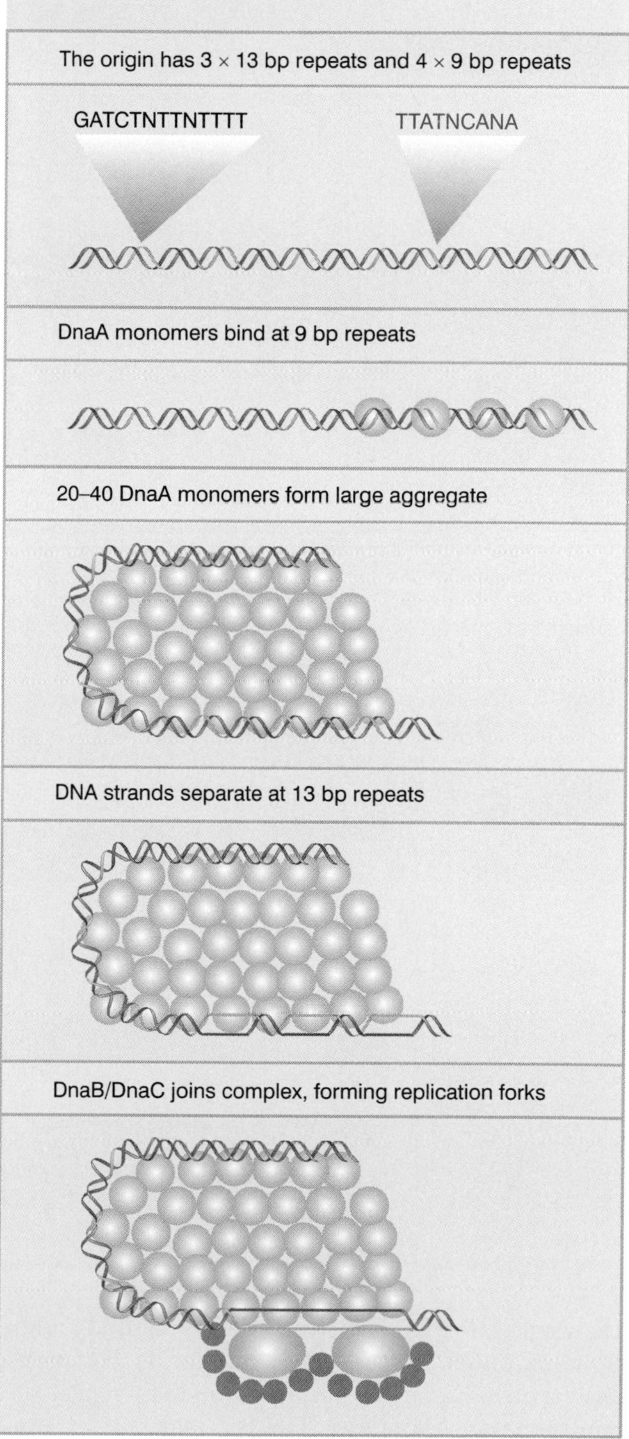
The origin has 3 × 13 bp repeats and 4 × 9 bp repeats
GATCTNTTNTTTT
TTATNCANA
DnaA monomers bind at 9 bp repeats
20–40 DnaA monomers form large aggregate
DNA strands separate at 13 bp repeats
DnaB/DnaC joins complex, forming replication forks

Figure 13.21 The complex at *oriC* can be detected by electron microscopy. Both complexes were visualized with antibodies against DnaB protein. Photographs kindly provided by Barbara Funnell.

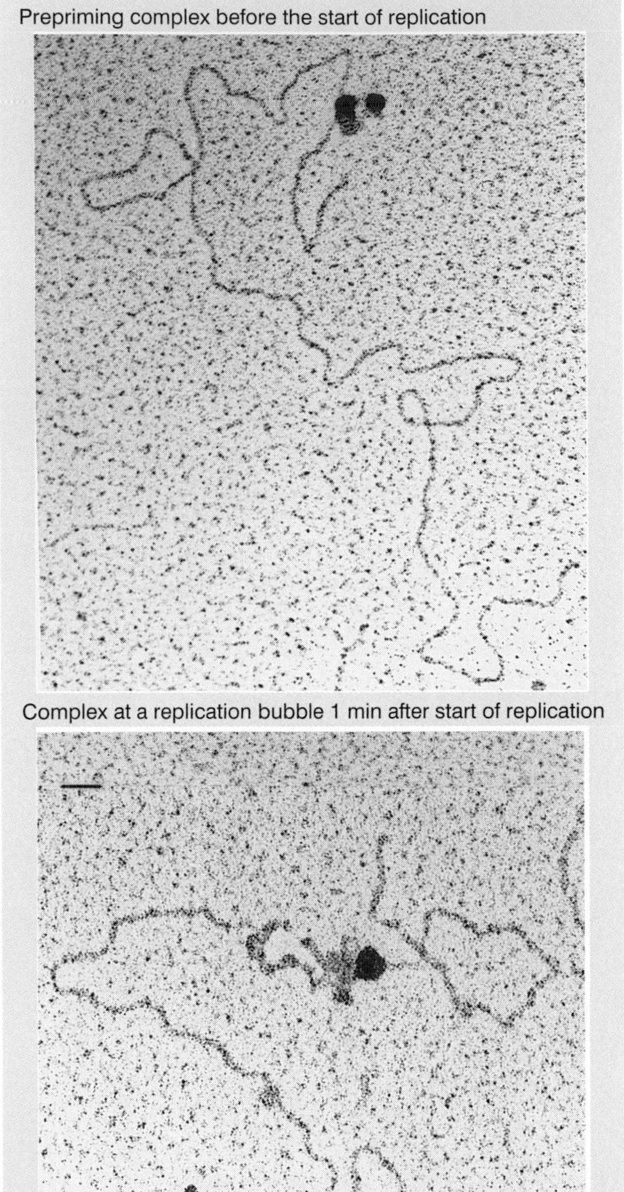

the single-stranded structure of the potential replication fork rather than the actual nucleotide sequence; at any event, it displaces DnaA from the 13 bp repeats and commences unwinding. DnaB functions in small amounts (1–2 hexamers) at the origin. DnaB also has the same ability here to activate the DnaG primase that it has for the Okazaki fragments.

Prepriming complex before the start of replication
Complex at a replication bubble 1 min after start of replication

Two further proteins are required to support the unwinding reaction. Gyrase provides a swivel that allows one strand to rotate around the other (a reaction discussed in more detail in Chapter 14); without this reaction, unwinding would generate torsional strain in the DNA. The protein SSB stabilizes the single-stranded DNA as it is formed. The length of duplex DNA that usually is unwound to initiate replication is probably <60 bp.

The protein HU is a general DNA-binding protein in *E. coli* (see Chapter 17). Its presence is not absolutely required to initiate replication *in vitro*, but it stimulates the reaction. HU has the capacity to bend DNA, and is likely to be involved in some general structural capacity.

Input of energy in the form of ATP is required for the prepriming reaction. It is required for unwinding DNA. The helicase action of DnaB depends on ATP hydrolysis; and the swivel action of gyrase requires ATP hydrolysis. ATP is also needed for the action of primase and to activate DNA polymerase III.

As the only member of the replication apparatus uniquely required at the origin, DnaA has attracted much attention. Some mutations in *dnaA* render replication asynchronous, suggesting that DnaA could be the "titrator" or "clock" that measures the number of origins relative to cell mass. Overproduction of DnaA yields conflicting results, varying from no effect to causing initiation to take place at reduced mass. However, there are no cyclic variations in DnaA concentration or expression, and current opinion has it that those cases in which *dnaA* affects cyclic control are probably due to indirect rather than direct effects.

What would be the properties of a mutation that altered the frequency of initiation? If initiation were caused by accumulation of an activator, a loss-of-function mutant should be in the slow-stop *dna* class, unable to start a new cycle. But if the regulator is an inhibitor, its loss should cause frequent cycles of initiation, perhaps leading to the accumulation of cells of reduced size. Such a mutant would not be found among the *dna* class, and would be more likely to appear among the potential segregation mutants (which we discussed in Chapter 12). However, we have yet to identify the circuit that controls the frequency of initiation events.

Common events in priming replication at the origin

Following generation of a replication fork as indicated in Figure 13.20, the priming reaction occurs to generate a leading strand. We know that synthesis of RNA is used for the priming event, but the details of the reaction are not known. Some mutations in *dnaA* can be suppressed by mutations in RNA polymerase, which suggests that DnaA could be involved in an initiation step requiring RNA synthesis *in vivo*.

RNA polymerase could be required to read into the origin from adjacent transcription units; by terminating at sites in the origin, it could provide the 3′-OH ends that prime DNA polymerase III. (An example is provided by the use of D loops at mitochondrial origins, as discussed in Chapter 12.) Alternatively, the act of transcription could be associated with a structural change that assists initiation. This latter idea is supported by observations that transcription does not have to proceed into the origin; it is effective up to 200 bp away from the origin, and can use either strand of DNA as template *in vitro*. The transcriptional event is inversely related to the requirement for supercoiling *in vitro*, which suggests that it acts by changing the local DNA structure so as to aid melting of DNA.

Another system for investigating interactions at the origin is provided by phage lambda, whose origin sponsors bidirectional replication. A map of the region is shown in **Figure 13.22**. Initiation of replication at the lambda origin requires "activation" by transcription starting from P_R. As with the events at *oriC*, this does not necessarily imply that the RNA provides a primer for the leading strand. Analogies between the systems suggest that RNA synthesis could be involved in promoting some structural change in the region.

Figure 13.22 Transcription initiating at P_R is required to activate the origin of λ DNA.

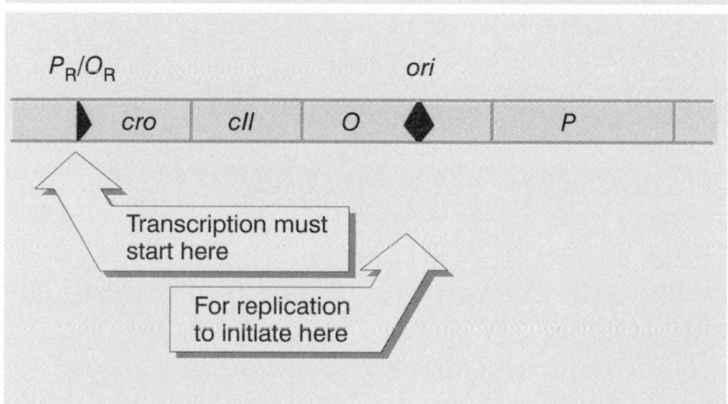

Initiation requires the products of phage genes *O* and *P*, as well as several host functions. The phage O protein binds to the lambda origin; the phage P protein interacts with the O protein and with the bacterial proteins. The origin lies within gene *O*, so the protein acts close to its site of synthesis.

Variants of the phage called λ*dv* consist of shorter genomes that carry all the information needed to replicate, but lack infective functions. λ*dv* DNA survives in the bacterium as a plasmid, and can be replicated *in vitro* by a system consisting of the phage-coded proteins O and P together with bacterial replication functions.

Lambda proteins O and P form a complex together with DnaB at the lambda origin, *ori*λ. The origin consists of two regions; as illustrated in **Figure 13.23**, a series of four binding sites for the O protein is adjacent to an A•T-rich region.

The first stage in initiation is the binding of O to generate a roughly spherical structure of diameter ~11 nm, sometimes called the O-some. The O-some contains ~100 bp or 60 kD of DNA. There are four 18 bp binding sites for O protein, which is ~34 kD. Each site is palindromic, and probably binds a symmetrical O dimer. The DNA sequences of the O-binding sites appear to be bent, and binding of O protein induces further bending.

If the DNA is supercoiled, binding of O protein causes a structural change in the origin. The A•T-rich region immediately adjacent to the O-binding sites becomes susceptible to S1 nuclease, an enzyme that specifically recognizes unpaired DNA. This suggests that a melting reaction occurs next to the complex of O proteins.

The role of the O protein is analogous to that of DnaA at *oriC*: it prepares the origin for binding of DnaB. Lambda provides its own protein, P, which substitutes for DnaC, and brings DnaB to the origin. When lambda P protein and bacterial DnaB proteins are added, the complex becomes larger and asymmetrical. It includes more DNA (a total of ~160 bp) as well as extra proteins. The λ P protein has a special role: it inhibits the helicase action of DnaB. Replication fork movement is triggered when P protein is released from the complex. Priming and DNA synthesis follow.

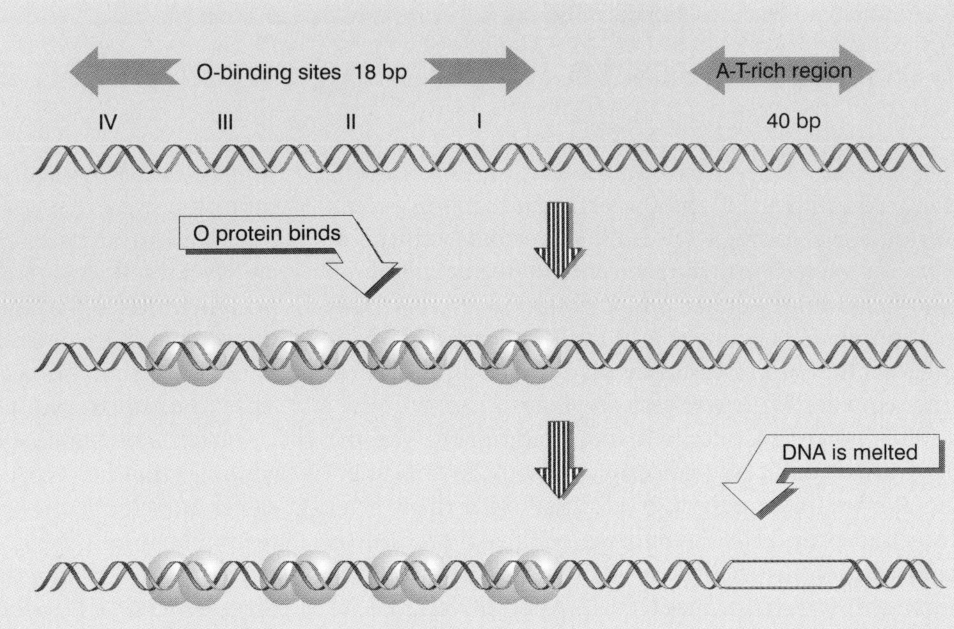

Figure 13.23 The λ origin for replication comprises two regions. Early events are catalyzed by O protein, which binds to a series of 4 sites; then DNA is melted in the adjacent A-T-rich region. Although the DNA is drawn as a straight duplex, it is actually bent at the origin.

Some proteins are essential for replication without being directly involved in DNA synthesis as such. Interesting examples are provided by the DnaK and DnaJ proteins. DnaK is a chaperone, related to a common stress protein of eukaryotes. Its ability to interact with other proteins in a conformation-dependent manner plays a role in many cellular activities, including replication. The role of DnaK/DnaJ may be to *disassemble* the pre-priming complex; by causing the release of P protein, they allow replication to begin.

The initiation reactions at *oriC* and *ori*λ are similar. The same stages are involved, and rely upon overlapping components. The first step is recognition of the origin by a protein that binds to form a complex with the DNA, DnaA for *oriC* and O protein for *ori*λ. A short region of A•T-rich DNA is melted. Then DnaB is loaded; this requires different functions at *oriC* and *ori*λ (and yet other proteins are required for this stage at other origins). When the helicase DnaB joins the complex, a replication fork is created. Finally an RNA primer is synthesized, after which replication begins.

The use of *oriC* and *ori*λ provides a general model for activation of origins. A similar series of events occurs at the origin of the virus SV40 in mammalian cells. Two hexamers of T antigen, a protein coded by the virus, bind to a series of repeated sites in DNA. In the presence of ATP, changes in DNA structure occur, culminating in a melting reaction. In the case of SV40, the melted region is rather short and is not A•T-rich, but it has an unusual composition in which one strand consists almost exclusively of pyrimidines and the other of purines. Near this site is another essential region, consisting of A•T base pairs, at which the DNA is bent; it is underwound by the binding of T antigen. An interesting difference from the prokaryotic systems is that T antigen itself possesses the helicase activity needed to extend unwinding, so that an equivalent for DnaB is not needed.

Does methylation at the origin regulate initiation?

What feature of a bacterial (or plasmid) origin ensures that it is used to initiate replication only once per cycle? Is initiation associated with some change that marks the origin so that a replicated origin can be distinguished from a nonreplicated origin?

Some sequences that are used for this purpose are included in the origin. *oriC* contains 11 copies of the sequence $\frac{\text{GATC}}{\text{CTAG}}$, which is a target for methylation at the N^6 position of adenine by the Dam methylase. The reaction is illustrated in **Figure 13.24**.

Before replication, the palindromic target site is methylated on the adenines of each strand. Replication inserts the normal (nonmodified) bases into the daughter strands, generating **hemimethylated** DNA, in which one strand is methylated and one strand is unmethylated. *So the replication event converts* dam *target sites from fully methylated to hemimethylated condition.*

What is the consequence for replication? The ability of a plasmid relying upon *oriC* to replicate in *dam⁻ E. coli* depends on its state of methylation. If the plasmid is methylated, it undergoes a single round of replication, and then the hemimethylated products accumulate, as described in **Figure 13.25**. So a hemimethylated origin cannot be used to initiate a replication cycle.

Two explanations suggest themselves: initiation may require full methylation of the Dam target sites in the origin or initiation may be inhibited by hemimethylation of these sites. The latter seems to be the case, because an origin of nonmethylated DNA can function effectively.

So hemimethylated origins cannot initiate again until the Dam methylase has converted them into fully methylated origins. The GATC sites at the origin remain hemimethylated for ~13 minutes after replication. This long period is unusual, because at typical GATC sites elsewhere in the genome, remethylation begins immediately (<1.5 minutes) following replication. One other region behaves like *oriC*; the promoter of the *dnaA* gene also shows a delay before remethylation begins.

While it is hemimethylated, the *dnaA* promoter is repressed, which causes a reduction in the level of DnaA protein. So the origin itself is inert, and production of the crucial initiator protein is repressed, during this period.

What is responsible for the delay in remethylation at *oriC* and *dnaA*? The most likely explanation is that these regions are sequestered in a form in which they are inaccessible to the *dam* methylase.

The circuit responsible for controlling reuse of origins may have been identified by mutations in the gene *seqA*, which reduce the delay in remethylation at both *oriC* and *dnaA*. The *seqA* mutants initiate DNA replication too soon, thereby accumulating an excessive number of origins. This suggests that *seqA* is part of a negative regulatory circuit that prevents origins from being remethylated. SeqA binds to hemimethylated DNA more strongly than to fully methylated DNA; it is possible that it initiates binding when the DNA becomes hemimethylated, and that its continued presence prevents formation of an open complex at the origin. SeqA does not have specificity for the *oriC* sequence, and it seems likely that this is conferred by DnaA protein, which would explain genetic interactions between *seqA* and *dnaA*.

Figure 13.24 Replication of methylated DNA gives hemimethylated DNA, which maintains its state at GATC sites until the Dam methylase restores the fully methylated condition.

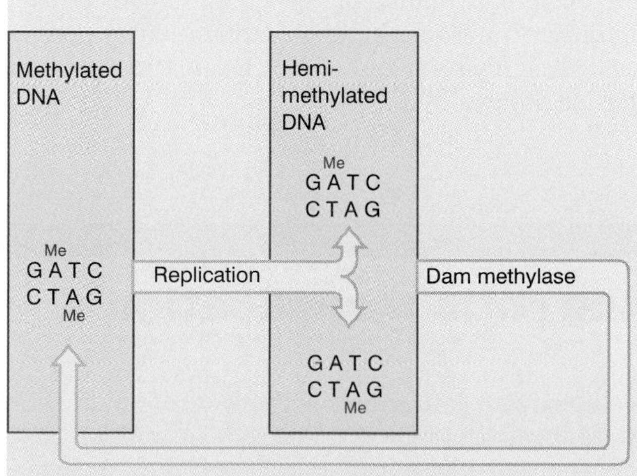

Figure 13.25 Only fully methylated origins can initiate replication; hemimethylated daughter origins cannot be used again until they have been restored to the fully methylated state.

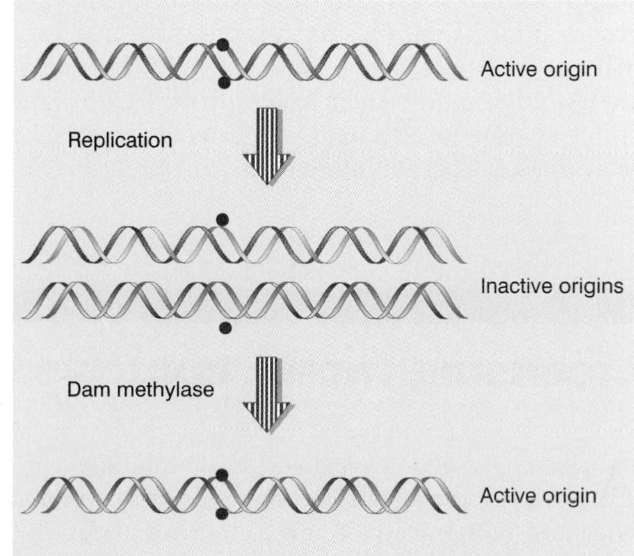

Hemimethylation of the GATC sequences in the origin is required for its association with the cell membrane *in vitro*. Hemimethylated *oriC* DNA binds to the membranes, but DNA that is fully methylated does not bind. One possibility is that membrane association is involved in controlling the activity of the origin. This function could be separate from any role that the membrane plays in segregation (see Figure 12.26). Association with the membrane could prevent reinitiation from occurring prematurely, either indirectly because the origins are sequestered or directly because some component at the membrane inhibits the reaction. The properties of the membrane fraction suggest that it includes components that regulate replication. An inhibitor is found in this fraction that competes with DnaA protein. This inhibitor can prevent initiation of replication only if it is added to an *in vitro* system before DnaA protein. This suggests the model of **Figure 13.26**, in which the inhibitor specifically recognizes hemimethylated DNA and prevents DnaA from binding. When the DNA is remethylated, the inhibitor is released, and DnaA now is free to initiate replication. If the inhibitor is associated with the membrane, then association and dissociation of DNA with the membrane may be involved in the control of replication.

Figure 13.26 A membrane-bound inhibitor binds to hemimethylated DNA at the origin, and may function by preventing the binding of DnaA. It is released when the DNA is remethylated.

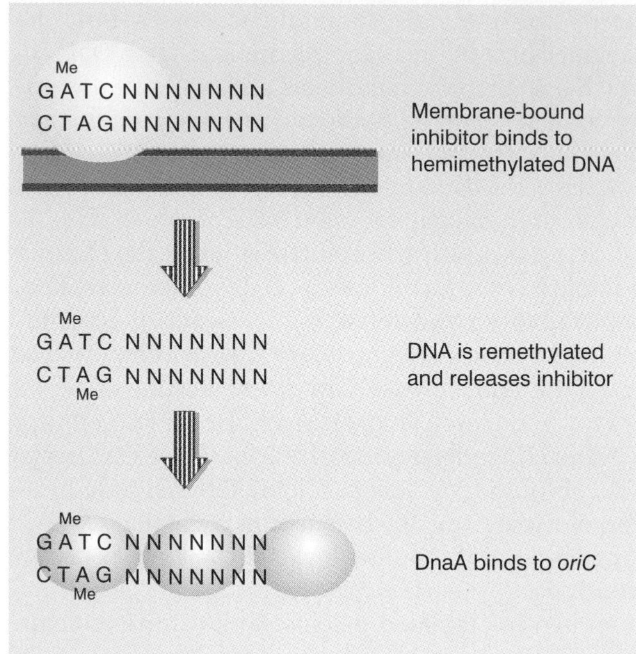

The full scope of the system used to control reinitiation is not clear, but several mechanisms may be involved: physical sequestration of the origin; delay in remethylation; inhibition of DnaA binding; repression of *dnaA* transcription. It is not immediately obvious which of these events cause the others, and whether their effects on initiation are direct or indirect.

We still have to come to grips with the central issue of which feature has the basic responsibility for *timing*. One possibility is that attachment to the membrane occurs at initiation, and that assembly of some large structure is required to release the DNA. The period of sequestration appears to increase with the length of the cell cycle, which suggests that it directly reflects the clock that controls reinitiation.

It has been extremely difficult to identify the protein component(s) that mediate membrane-attachment. A hint that this is a function of DnaA is provided by its response to phospholipids. Phospholipids promote the exchange of ATP with ADP bound to DnaA. We do not know what role this plays in controlling the activity of DnaA (which requires ATP), but the reaction implies that DnaA is likely to interact with the membrane. This would imply that more than one event is involved in associating with the membrane. Perhaps a hemimethylated origin is bound by the membrane-associated inhibitor, but when the origin becomes fully methylated, the inhibitor is displaced by DnaA associated with the membrane.

Licensing factor controls eukaryotic rereplication

A EUKARYOTIC genome is divided into multiple replicons, and the origin in each replicon is activated once and only once in a single division cycle. This could be achieved by providing some rate-limiting component that functions only once at an origin or by the presence of a repressor that prevents re-replication at origins that have been used. The critical questions about the nature of this regulatory system are how the system determines whether any particular origin has been replicated, and what protein components are involved.

Insights into the nature of the protein components have been provided by using a system in which a substrate DNA undergoes only one cycle of replication. *Xenopus* eggs have all the components needed to replicate DNA—in the first few hours after fertilization they undertake 11 division cycles without new gene expression—and they can replicate the DNA in a nucleus that is injected into the egg. **Figure 13.27** summarizes the features of this system.

The injected material can take the form of a sperm or interphase nucleus. Its DNA is replicated only once (followed by use of a density label, just like the original experiment that characterized semiconservative replication, shown previously in Figure 1.10). If protein synthesis is blocked in the egg, the membrane around the injected material remains intact, and the DNA cannot replicate again. However, in the presence of protein synthesis, the nuclear membrane breaks down just as it would for a normal cell division, and in this case subsequent replication cycles can occur. The same result can be achieved by using agents that permeabilize the nuclear membrane. This suggests that the nucleus contains a protein(s) needed for replication that is used up in some way by a replication cycle; although more of the protein is present in the egg cytoplasm, it cannot enter the nucleus, but is allowed to do so if the nuclear membrane breaks down. The system can in principle be taken further by developing an *in vitro* extract that supports nuclear replication, thus allowing the components of the extract to be isolated, and the relevant factors identified.

Figure 13.28 explains the control of reinitiation by proposing that this protein is a **licensing factor**. It is present in the nucleus prior to replication. One round of replication either inactivates or destroys the factor, and another round cannot occur until further factor is provided. Factor in the cytoplasm can gain access to the nuclear material only at the subsequent mitosis when the nuclear envelope breaks down. This regulatory system achieves two purposes. By removing a necessary component after replication, it prevents more than one cycle of replication from occurring. And it provides a feedback loop that makes the initiation of replication dependent on passing through cell division.

Components of the licensing factor are provided by the *S. cerevisiae* proteins MCM2,3,5, which are required for replication and enter the nucleus only during mitosis. Homologs are found in animal cells, where MCM3 is bound to chromosomal material before replication, but is released after replication. The animal cell MCM2,3,5 complex remains in the nucleus throughout the cell cycle, suggesting that it may be one component of the licensing factor; another component, able to enter only at mitosis, may be necessary for MCM2,3,5 to associate with chromosomal material.

An insight into the system that controls availability of licensing factor is provided by certain mutants in yeast. Mutations in the licensing factor itself could prevent initiation of replication; this is the behavior of

Figure 13.27 A nucleus injected into a *Xenopus* egg can replicate only once unless the nuclear membrane is permeabilized to allow subsequent replication cycles.

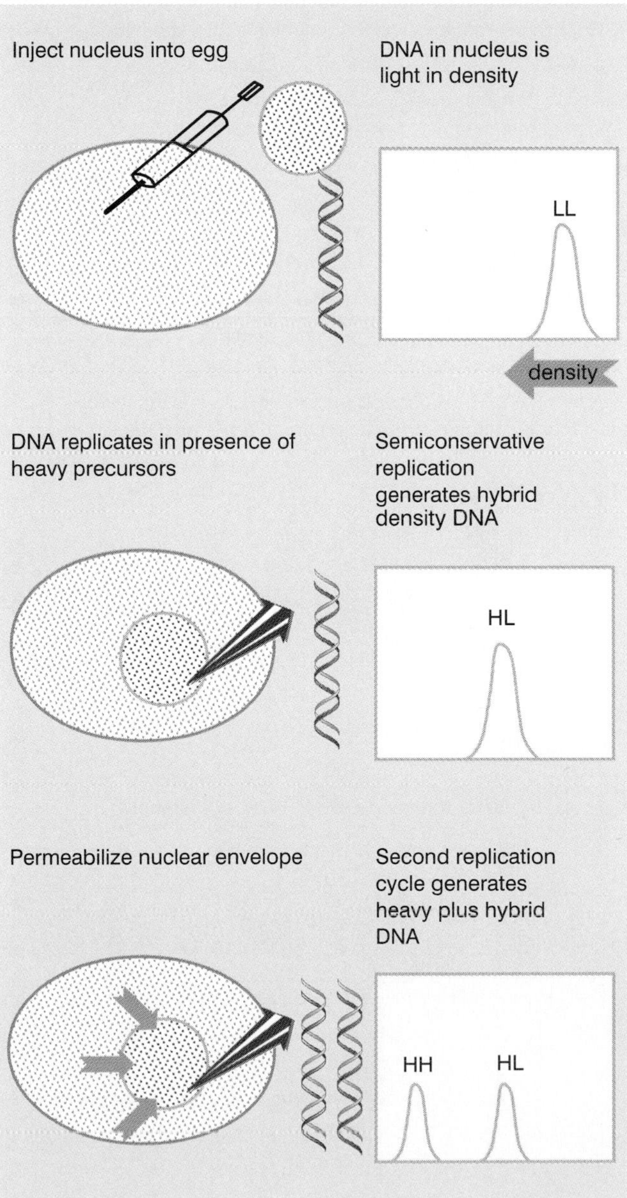
Inject nucleus into egg
DNA in nucleus is light in density
LL
density
DNA replicates in presence of heavy precursors
Semiconservative replication generates hybrid density DNA
HL
Permeabilize nuclear envelope
Second replication cycle generates heavy plus hybrid DNA
HH
HL

MCM2,3,5. Mutations in the system that inactivates licensing factor after the start of replication should allow the accumulation of excess quantities of DNA, because the continued presence of licensing factor allows rereplication to occur. Such mutations are found in genes that code for components of the ubiquitination system that is responsible for degrading certain proteins. This suggests that licensing factor may be destroyed after the start of a replication cycle.

Figure 13.28 Licensing factor in the nucleus is inactivated after replication. A new supply of licensing factor can enter only when the nuclear membrane breaks down at mitosis.

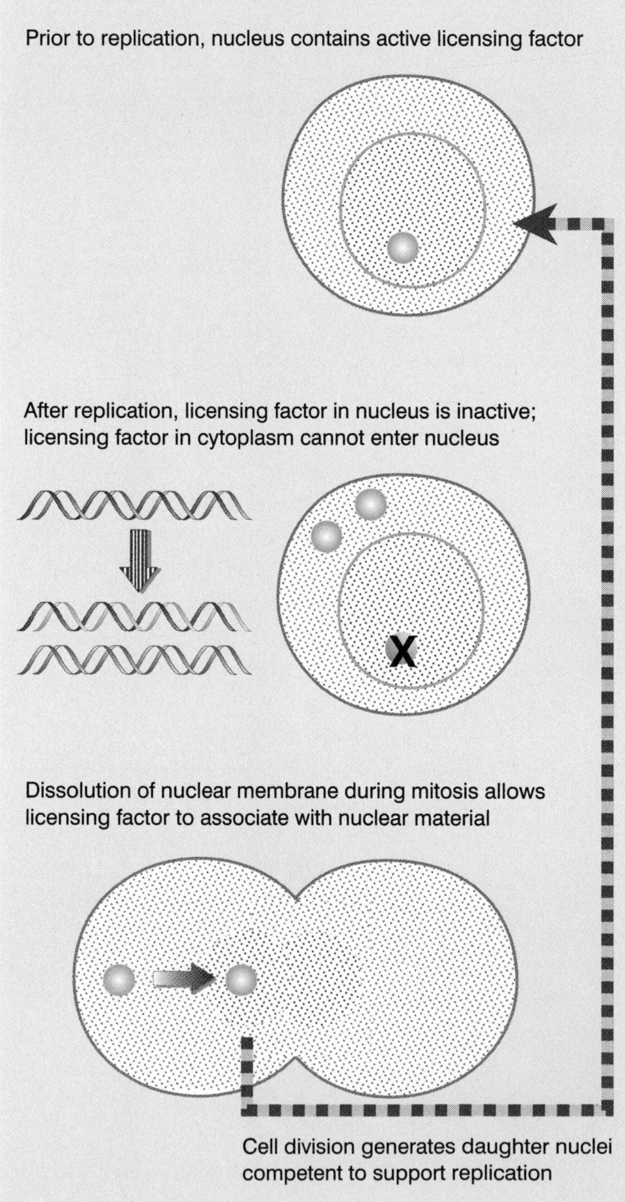
Prior to replication, nucleus contains active licensing factor
After replication, licensing factor in nucleus is inactive; licensing factor in cytoplasm cannot enter nucleus
X
Dissolution of nuclear membrane during mitosis allows licensing factor to associate with nuclear material
Cell division generates daughter nuclei competent to support replication

The state of the origin during the replication cycle has been followed in *S. cerevisiae*. The origin (*ARS*) consists of the A consensus sequence and three B elements (see Figure 12.10). The ORC complex of six proteins (all of which are coded by essential genes) binds to the A and adjacent B1 element. The transcription factor ABF1 binds to the B3 element; this has a facilitative or enhancing role on initiation, but it is the events that occur at the A and B1 elements that actually cause initiation.

The striking feature is that ORC remains bound at

Figure 13.29 Proteins at the origin control susceptibility to initiation.

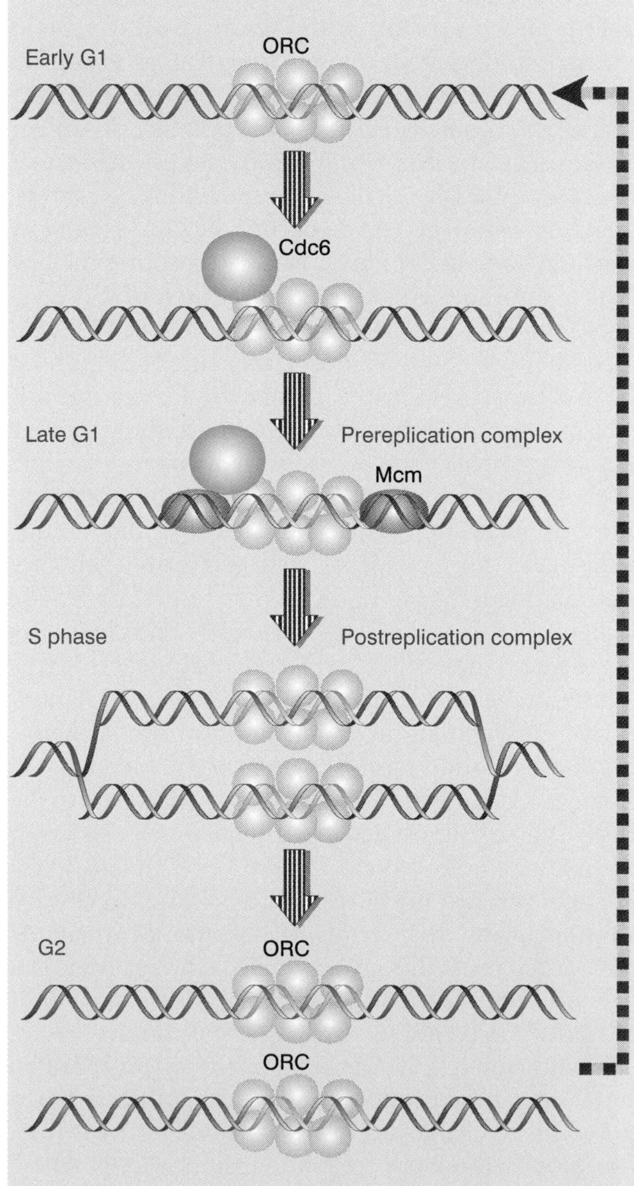
Early G1
ORC
Cdc6
Late G1
Prereplication complex
Mcm
S phase
Postreplication complex
G2
ORC
ORC

the origin through the entire cell cycle. However, changes occur in the pattern of protection of DNA as a result of the binding of other proteins to the ORC-origin complex. **Figure 13.29** summarizes the cycle of events at the origin.

At the end of the cell cycle, ORC is bound to A/B1, and generates a pattern of protection *in vivo* that is similar to that found when it binds to free DNA *in vitro*. Basically the region across A-B1 is protected against DNAase, but there is a hypersensitive site in the center of B1.

During G1, this pattern changes, most strikingly by the loss of the hypersensitive site. This is due to the binding of Cdc6 protein to the ORC. Cdc6 is a highly unstable protein (half-life <5 minutes). It is synthesized from late G1 through G1, and typically binds to the ORC between the exit from mitosis and late G1. Its rapid degradation means that no protein is available later in the cycle.

The presence of Cdc6 in turn allows Mcm proteins to bind to the complex. Their presence is necessary for initiation to occur at the origin. The origin therefore enters S phase in the condition of a **prereplication complex**, containing ORC, Cdc6, and Mcm proteins. When initiation occurs, Cdc6 and Mcm are displaced, returning the origin to the state of the **postreplication complex**, which contains only ORC. Because Cdc6 is rapidly degraded during S phase, it is not available to support reloading of Mcm proteins, and so the origin cannot be used for a second cycle of initiation during the S phase.

If Cdc6 is made available to bind to the origin during G2 (by ectopic expression), Mcm proteins do not bind until the following G1, suggesting that there is a secondary mechanism to ensure that they associate with origins only at the right time. This could be another part of licensing control. At least in *S. cerevisiae*, this control does not seem to be exercised at the level of nuclear entry, but this could be a difference between yeasts and animal cells. We discuss how the cell cycle control system regulates initiation (and reinitiation) of replication in Chapter 27.

Summary

DNA synthesis occurs by semidiscontinuous replication, in which the leading strand of DNA growing 5′–3′ is extended continuously, but the lagging strand that grows overall in the opposite 3′–5′ direction is made as short Okazaki fragments, each synthesized 5′–3′. The leading strand and each Okazaki fragment of the lagging strand initiate with an RNA primer that is extended by DNA polymerase. Bacteria and eukaryotes each possess more than one DNA polymerase activity. DNA polymerase III synthesizes both lagging and leading strands in *E. coli*. Many proteins are required for DNA polymerase III action and several constitute part of the replisome within which it functions.

The replisome contains an asymmetric dimer of DNA polymerase III; each new DNA strand is synthesized by a different core complex containing a catalytic (α) subunit. Processivity of the core complex is maintained by the β clamp, which forms a ring round DNA. The looping model for the replication fork proposes that, as one half of the dimer advances to synthesize the leading strand, the other half of the dimer pulls DNA through as a single loop that provides the template for the lagging strand. The transition from completion of one Okazaki fragment to the start of the next requires the lagging strand catalytic subunit to dissociate from DNA and then to reattach to a β clamp at the priming site for the next Okazaki fragment.

DnaB provides the helicase activity at a replication fork; this depends on ATP cleavage. DnaB may function by itself in *oriC* replicons to provide primosome activity by interacting periodically with DnaG, which provides the primase that synthesizes RNA.

Phage T4 codes for a sizeable replication apparatus, consisting of seven proteins: DNA polymerase, helicase, single-strand binding protein, priming activities, and accessory proteins. Similar functions are required in other replication systems, including a HeLa cell system that replicates SV40 DNA. Different enzymes, DNA polymerase α and DNA polymerase δ, initiate and elongate the new strands of DNA.

The common mode of origin activation involves an initial limited melting of the double helix, followed by more general unwinding to create single strands. Several proteins act sequentially at the *E. coli* origin. DnaA binds to a series of 9 bp repeats and 13 bp repeats, forming an aggregate of 20–40 monomers with DNA in which the 13 bp repeats are melted. The helicase activity of DnaB, together with DnaC, unwinds DNA further. Similar events occur at the λ origin, where phage proteins O and P are the counterparts of bacterial proteins DnaA and DnaC, respectively. In SV40 replication, several of these activities are combined in the functions of T antigen.

The ϕX priming event also requires DnaB, DnaC, and DnaT. PriA is the component that defines the primosome assembly site (*pas*) for ϕX replicons; it displaces SSB from DNA in an action that involves cleavage of ATP. PriB and PriC are additional components of the primosome.

Several sites that are methylated by the Dam methylase are present in the *E. coli* origin, including those of the 13-mer binding sites for DnaA. The origin remains hemimethylated and is in a sequestered state for ~10 minutes following initiation of a replication cycle. During this period it is associated with the membrane, and reinitiation of replication is repressed.

After cell division, nuclei of eukaryotic cells have a licensing factor that is needed to initiate replication. Its destruction after initiation of replication prevents further replication cycles from occurring in yeast. Licensing factor cannot be imported into the nucleus from the cytoplasm, and can be replaced only when the nuclear membrane breaks down during mitosis.

Further reading

Reviews

Nossal, N. G. (1983). Prokaryotic DNA replication systems. *Ann. Rev. Biochem.* **52**, 581–615.

Campbell, J. L. (1986). Eukaryotic DNA replication. *Ann. Rev. Biochem.* **55**, 733–771.

McHenry, C. S. (1988). DNA polymerase III holoenzyme of *E. coli. Ann. Rev. Biochem.* **57**, 519–550.

Kunkel, T. A. (1988). Exonucleolytic proofreading. *Cell* **53**, 837–840.

Stillman, B. (1989). Initiation of eukaryotic DNA replication *in vitro. Ann. Rev. Cell Biol.* **5**, 197–245.

Baker, T. A. and Wickner, S. H. (1992). Genetics and enzymology of DNA replication in *E. coli. Ann. Rev. Genet.* **26**, 447–477.

Hill, T. M. (1992). Arrest of bacterial DNA replication. *Ann. Rev. Microbiol.* **46**, 603–633.

Kornberg, A. and Baker, T. A. (1992). *DNA Replication.* (Freeman and Co., New York).

Marians, K. J. (1992). Prokaryotic DNA replication. *Ann. Rev. Biochem.* **61**, 673–719.

Stillman, B. (1996). Cell cycle control of DNA replication. *Science* **274**, 1659–1664.

Baker, T. A. and Bell, S. P. (1998). Polymerases and the replicosome: machines within machines. *Cell* **92**, 295–305.

Waga, S. and Stillman, B. (1998). The DNA replication fork in eukaryotic cells. *Ann. Rev. Biochem.* **67**, 721–751.

Discoveries

DNA replication

Brutlag, D. and Kornberg, A. (1972). Enzymatic synthesis of DNA: a proofreading function for the 3′–5′ exonuclease activity in DNA polymerases. *J. Biol. Chem.* **247**, 241–248.

Scott, J. F. *et al.* (1977). A mechanism of duplex DNA replication revealed by enzymatic studies of phage ϕX174: catalytic strand separation in advance of replication. *Proc. Natl. Acad. Sci. USA* **74**, 193–197.

Rowen, L. and Kornberg, A. (1978). Primase, the dnaG protein of *E. coli*: an enzyme which starts DNA chains. *J. Biol. Chem.* **253**, 758–764.

Arai, N. *et al.* (1981). Replication of ϕX174 DNA with purified enzymes. Multiplication of the duplex form by coupling of continuous and discontinuous synthetic pathways. *J. Biol. Chem.* **256**, 5239–5246.

Fuller, R. S., Kaguni, J. M., and Kornberg, A. (1981). Enzymatic replication of the origin of the *E. coli* chromosome. *Proc. Natl. Acad. Sci. USA* **78**, 7370–7374.

Arai, N. and Kornberg, A. (1981). Unique primed state of phage ϕX174 DNA replication and mobility of the primosome in a direction opposite chain synthesis. *Proc. Natl. Acad. Sci. USA* **78**, 69–73.

Baker, T. A. and Kornberg, A. (1988). Transcriptional activation of initiation of replication from the *E. coli* chromosomal origin: an RNA-DNA hybrid near *oriC*. *Cell* **55**, 113–123.

Bramhill, D. and Kornberg, A. (1988). Duplex opening by dnaA protein at novel sequences in initiation of replication at the origin of the *E. coli* chromosome. *Cell* **52**, 743–755.

Debyser, Z. *et al.* (1994). Coordination of leading and lagging strand DNA synthesis at the replication fork of bacteriophage T7. *Cell* **77**, 157–166.

Lu, M. *et al.* (1994). SeqA: a negative modulator of replication initiation in *E. coli. Cell* **77**, 413–426.

Stukenberg, P. T. *et al.* (1994). An explanation for lagging strand replication: polymerase hopping among DNA sliding clamps. *Cell* **78**, 877–887.

Waga, S. and Stillman, B. (1994). Anatomy of a DNA replication fork revealed by reconstitution of SV40 DNA replication *in vitro. Nature* **369**, 207–212.

Naktinis, V., Turner, J., and O'Donnell, M. (1996). A molecular switch in a replication machine defined by an internal competition for protein rings. *Cell* **84**, 137–145.

Kamada, K. *et al.* (1996). Structure of a replication-terminator protein complexed with DNA. *Nature* **383**, 598–603.

Doublié, S. *et al.* (1998). Crystal structure of a phage T7 DNA replication complex at 2.2Å resolution. *Nature* **391**, 251–258.

Origin recognition and licensing

Ogden, G. B., Pratt, M. J., and Schaechter, M. (1988). The replicative origin of the *E. coli* chromosome binds to cell membranes only when hemimethylated. *Cell* **54**, 127–135.

Blow, J. J. and Laskey, R. A. (1988). A role for the nuclear envelope in controlling DNA replication within the cell cycle. *Nature* **332**, 546–548.

Diffley, J. F. X. *et al.* (1994). Two steps in the assembly of complexes at yeast replication origins *in vivo. Cell* **78**, 303–316.

Tanaka, T., Knapp, D., and Nasmyth, K. (1997). Loading of an MCM protein onto DNA replication origins is regulated by Cdc6p and CDKs. *Cell* **90**, 649–660.

Chapter 14

Recombination and repair

WITHOUT genetic recombination, the content of each individual chromosome would be irretrievably fixed in its particular alleles, changeable only by mutation. The length of the target for mutation damage would be increased from the gene to the chromosome. Deleterious mutations would accumulate, eliminating each chromosome (and thereby removing any favorable mutations that have occurred).

By shuffling the genes, recombination allows favorable and unfavorable mutations to be separated and tested as individual units in new assortments. It provides a means of escape and spreading for favorable alleles, and a means to eliminate an unfavorable allele without bringing down all the other genes with which this allele is associated.

Recombination occurs between precisely corresponding sequences, so that not a single base pair is added to or lost from the recombinant chromosomes. Three types of recombination share the feature that the process involves physical exchange of material between duplex DNAs, but differ in the circumstances:

- Recombination involving reaction between homologous sequences of DNA is called **generalized** or **homologous recombination**. In eukaryotes, it occurs at meiosis, usually both in males (during spermatogenesis) and females (during oogenesis). We recall that it happens at the "four strand" stage of meiosis, and involves only two of the four strands (see Chapter 1).
- Another type of event sponsors recombination between *specific* pairs of sequences, and has been best characterized in prokaryotes. **Site-specific recombination** is responsible for the integration of phage genomes into the bacterial chromosome. The recombination event involves specific sequences of the phage DNA and bacterial DNA, which include a short stretch of homology. The enzymes involved in this event act *only* on the particular pair of target sequences. Related reactions are responsible for inverting specific regions of the bacterial chromosome.
- A different type of event allows one DNA sequence to be inserted into another without reliance on sequence homology. **Transposition** provides a means by which certain elements move from one chromosomal location to another. The mechanisms involved in transposition depend upon breakage and reunion of DNA strands, and thus are related to the processes of recombination. Transposition is the subject of Chapters 15 and 16.
- Another type of recombination is used by RNA viruses, in which the polymerase switches from one template to another while it is synthesizing RNA. As a result, the newly synthesized molecule joins sequence information from two different parents. This type of mechanism for recombination is called **copy choice**, and is discussed briefly in Chapter 16. It is not used for recombination between duplex DNA substrates.

Homologous recombination is a reaction between two duplexes of DNA. Its critical feature is that the

Figure 14.1 *Overview*: recombination occurs during the first meiotic prophase. The stages of prophase are defined by the appearance of the chromosomes, each of which consists of two replicas (sister chromatids), although the duplicated state becomes visible only at the end. The molecular interactions of any individual crossing-over event involve two of the four duplex DNAs.

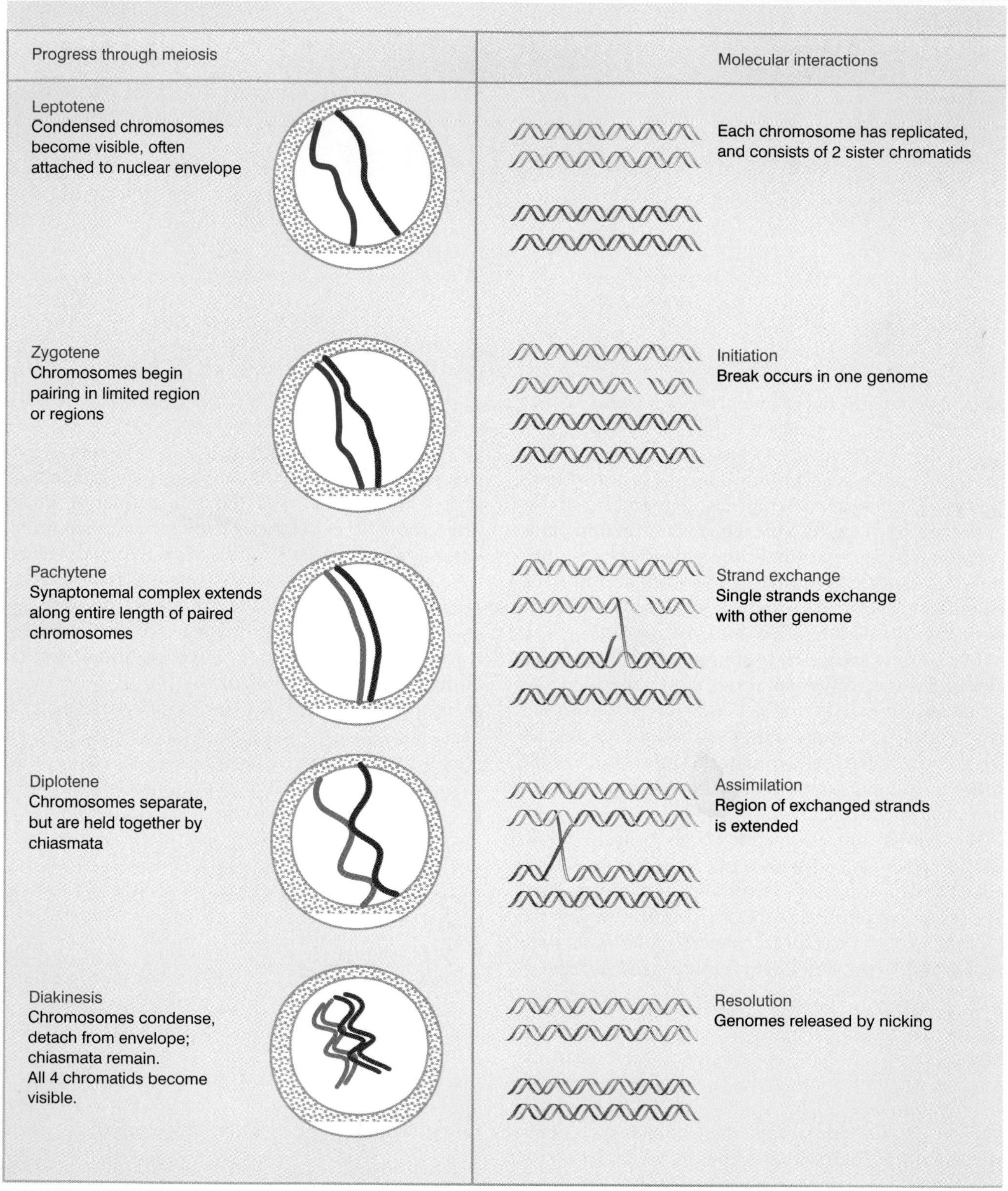

Progress through meiosis	Molecular interactions
Leptotene Condensed chromosomes become visible, often attached to nuclear envelope	Each chromosome has replicated, and consists of 2 sister chromatids
Zygotene Chromosomes begin pairing in limited region or regions	Initiation Break occurs in one genome
Pachytene Synaptonemal complex extends along entire length of paired chromosomes	Strand exchange Single strands exchange with other genome
Diplotene Chromosomes separate, but are held together by chiasmata	Assimilation Region of exchanged strands is extended
Diakinesis Chromosomes condense, detach from envelope; chiasmata remain. All 4 chromatids become visible.	Resolution Genomes released by nicking

enzymes responsible can use *any* pair of homologous sequences as substrates (although some types of sequences may be favored over others). The frequency of recombination is not constant throughout the genome, but is influenced by both global and local effects. The overall frequency may be different in oocytes and in sperm; recombination occurs twice as frequently in female as in male humans. And within the genome its frequency depends upon chromosome structure; for example, crossing-over is suppressed in the vicinity of the condensed and inactive regions of heterochromatin.

Figure 14.1 compares the visible progress of chromosomes through meiosis with the molecular interactions that are involved in exchanging material between duplexes of DNA. Meiosis starts with a protracted prophase whose five stages are summarized in the figure.

The beginning of meiosis is marked by the point at which individual chromosomes become visible. Each of these chromosomes has replicated previously, and consists of two sister chromatids, each of which contains a duplex DNA. The homologous chromosomes approach one another and begin to pair in one or more regions, forming **bivalents**. Pairing extends until the entire length of each chromosome is apposed with its homolog. The process is called **synapsis** or **chromosome pairing**. When the process is completed, the chromosomes are laterally associated in the form of a **synaptonemal complex**, which has a characteristic structure in each species, although there is wide variation in the details between species.

Recombination between chromosomes involves a physical exchange of parts, usually represented as a **breakage and reunion**, in which two nonsister chromatids (each containing a duplex of DNA) have been broken and then linked each with the other. When the chromosomes begin to separate, they can be seen to be held together at discrete sites, the **chiasmata**. The number and distribution of chiasmata parallel the features of genetic crossing-over. Traditional analysis holds that a chiasma represents the crossing-over event (see Figure 1.22). The chiasmata remain visible when the chromosomes condense and all four chromatids become evident.

What is the molecular basis for these events? Each sister chromatid contains a single DNA duplex, so each bivalent contains four duplex molecules of DNA. Recombination requires a mechanism that allows the duplex DNA of one sister chromatid to interact with the duplex DNA of a sister chromatid from the other chromosome. It must be possible for this reaction to occur between any pair of corresponding sequences in the two molecules in a highly specific manner that allows material to be exchanged with precision at the level of the individual base pair.

We know of only one mechanism for nucleic acids to recognize one another on the basis of sequence: complementarity between single strands. Figure 14.1 shows a general model for the involvement of single strands in recombination. The first step in providing single strands is to make a break in each DNA duplex. Then one or both of the strands of that duplex can be released. If (at least) one strand displaces the corresponding strand in the other duplex, the two duplex molecules will be specifically connected at corresponding sequences. If the strand exchange is extended, there can be more extensive connection between the duplex. And by exchanging both strands and later cutting them, it is possible to connect the parental duplex molecules by means of a crossover that corresponds to the demands of a breakage and reunion.

We cannot at this juncture relate these molecular events rigorously with the changes that are observed at the level of the chromosomes. There is no detailed information about the molecular events involved in recombination in higher eukaryotic cells (in which meiosis has been most closely observed). However, recently the isolation of mutants in yeast has made it possible to correlate some of the molecular steps with approximate stages of meiosis. Detailed information about the recombination process is available in bacteria, in which molecular activities are known that cause genetic exchange between duplex molecules. However, the bacterial reaction involves interaction between restricted regions of the genome, rather than an entire pairing of genomes. The synapsis of eukaryotic chromosomes remains the most difficult stage to explain at the molecular level.

Breakage and reunion involves heteroduplex DNA

THE act of connecting two duplex molecules of DNA is at the heart of the recombination process. Our molecular analysis of recombination therefore starts by expanding the view in Figure 14.1 of the use of base pairing between complementary single strands in recombination. It is useful to imagine the recombination reaction in terms of single-strand exchanges (although we shall see that this is not necessarily how it is actually initiated), because the properties of the molecules created in this way are central to understanding the processes involved in recombination.

Figure 14.2 illustrates a process that starts with breakage at the corresponding points of the homologous strands of two paired DNA duplexes. The breakage allows movement of the free ends created by the nicks. Each strand leaves its partner and crosses over to pair with its complement in the other duplex.

The reciprocal exchange creates a connection between the two DNA duplexes. The connected pair of duplexes is called a **joint molecule.** The point at which an individual strand of DNA crosses from one duplex to the other is called the **recombinant joint.**

At the site of recombination, each duplex has a region consisting of one strand from each of the parental DNA molecules. This region is called **hybrid DNA** or **heteroduplex DNA.**

An important feature of a recombinant joint is its ability to move along the duplex. Such mobility is called **branch migration. Figure 14.3** illustrates the migration of a single strand in a duplex. The branching point can migrate in either direction as one strand is displaced by the other.

Branch migration is important for both theoretical and practical reasons. As a matter of principle, it confers a dynamic property on recombining structures. As a practical feature, its existence means that the point of branching cannot be established by examining a molecule *in vitro* (because the branch may have migrated since the molecule was isolated).

Branch migration could allow the point of crossover in the recombination intermediate to move in either direction. The rate of branch migration is uncertain, but as seen *in vitro* is probably inadequate to support the formation of extensive regions of heteroduplex DNA in natural conditions. Any extensive branch migration *in vivo* must therefore be catalyzed by a recombination enzyme.

When recombination involves duplex DNA molecules, topological manipulation may be required; either the DNA duplex must be free to rotate, or equivalent relief from topological restraint must be provided (see later). If we imagine that the joint molecule of Figure 14.2 rotates one duplex relative to the other, we can visualize it in one plane as a **Holliday structure** (named for its proposer). This is illustrated in **Figure 14.4.**

The joint molecule formed by strand exchange must be **resolved** into two separate duplex molecules. Resolution requires a further pair of nicks. The outcome of the reaction depends on which pair of strands is nicked, as can be seen from Figures 14.2 and 14.4.

If the nicks are made in the pair of strands that were *not* originally nicked (the pair that did not initiate the strand exchange), all four of the original strands have been nicked. This releases **splice recombinant** DNA molecules. The duplex of one DNA parent is covalently linked to the duplex of the other DNA parent, via a stretch of heteroduplex DNA. There has been a conventional recombination event between markers located on either side of the heteroduplex region.

If the *same* two strands involved in the original nicking are nicked again, the other two strands remain intact. The nicking releases the original parental duplexes, which remain intact except that each has a residuum of the event in the form of a length of heteroduplex DNA. These are called **patch recombinants.**

These alternative resolutions of the joint molecule establish the principle that *a strand exchange between duplex DNAs always leaves behind a region of heteroduplex DNA, but the exchange may or may not be accompanied by recombination of the flanking regions.*

What is the minimum length of the region required to establish the connection between the recombining duplexes? Experiments in which short homologous sequences carried by plasmids or phages are introduced into bacteria suggest that the rate of recombination is substantially reduced if the homologous region is <75 bp. This distance is appreciably longer than the ~10 bp required for association between complementary single-stranded regions, which suggests that recombination imposes demands beyond mere annealing of complements.

Figure 14.2 Recombination between two paired duplex DNAs could involve reciprocal single-strand exchange, branch migration, and nicking.

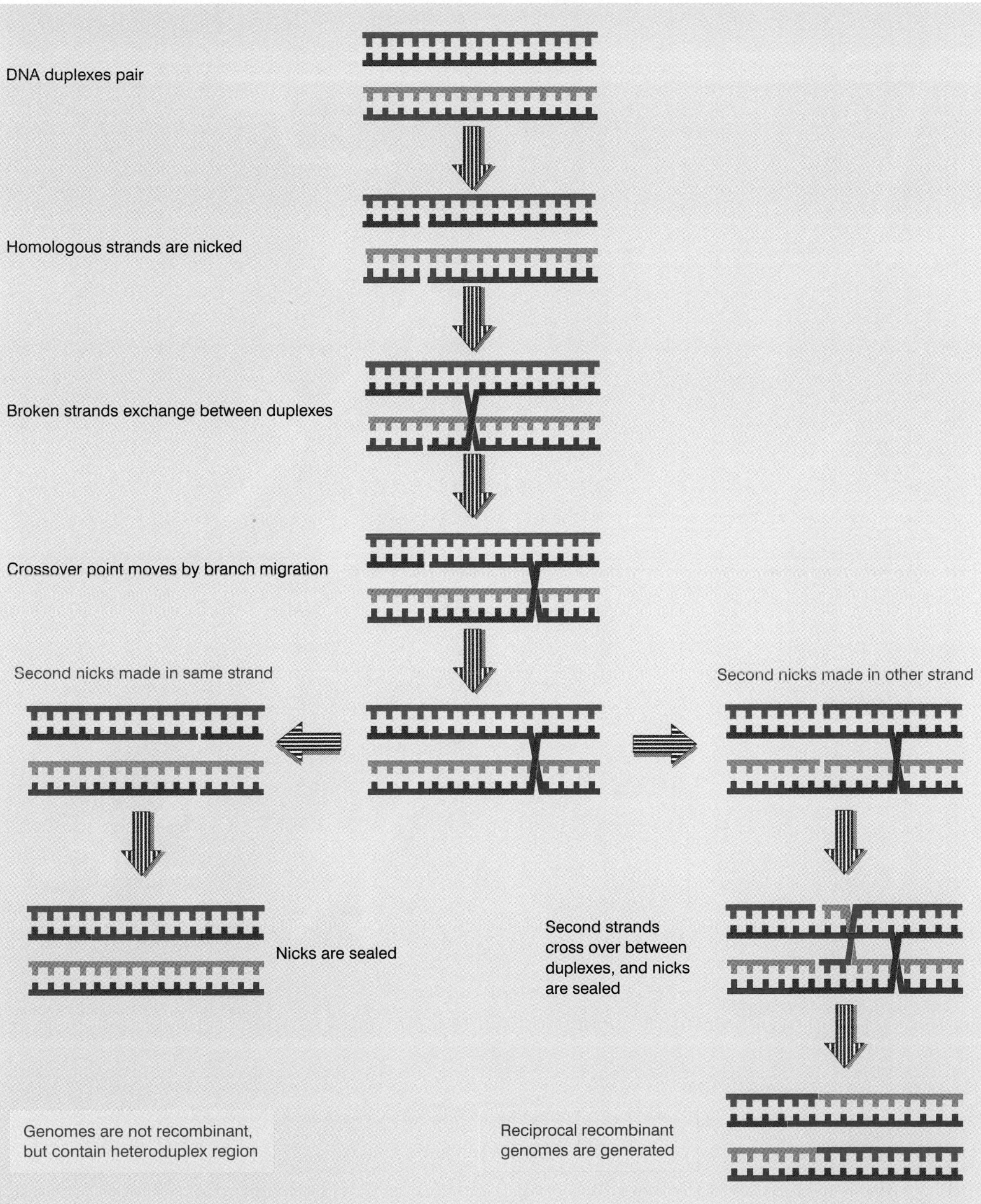
DNA duplexes pair
Homologous strands are nicked
Broken strands exchange between duplexes
Crossover point moves by branch migration
Second nicks made in same strand
Second nicks made in other strand
Nicks are sealed
Second strands cross over between duplexes, and nicks are sealed
Genomes are not recombinant, but contain heteroduplex region
Reciprocal recombinant genomes are generated

Figure 14.3 Branch migration can occur in either direction when an unpaired single strand displaces a paired strand.

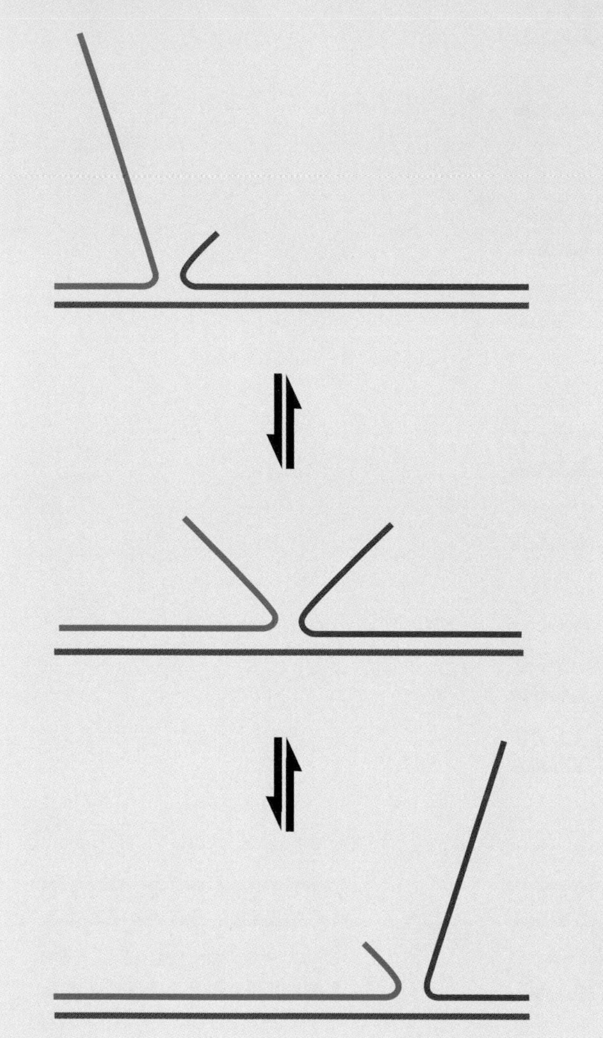

Figure 14.4 Resolution of a Holliday junction can generate parental or recombinant duplexes, depending on which strands are nicked. Both types of product have a region of heteroduplex DNA.

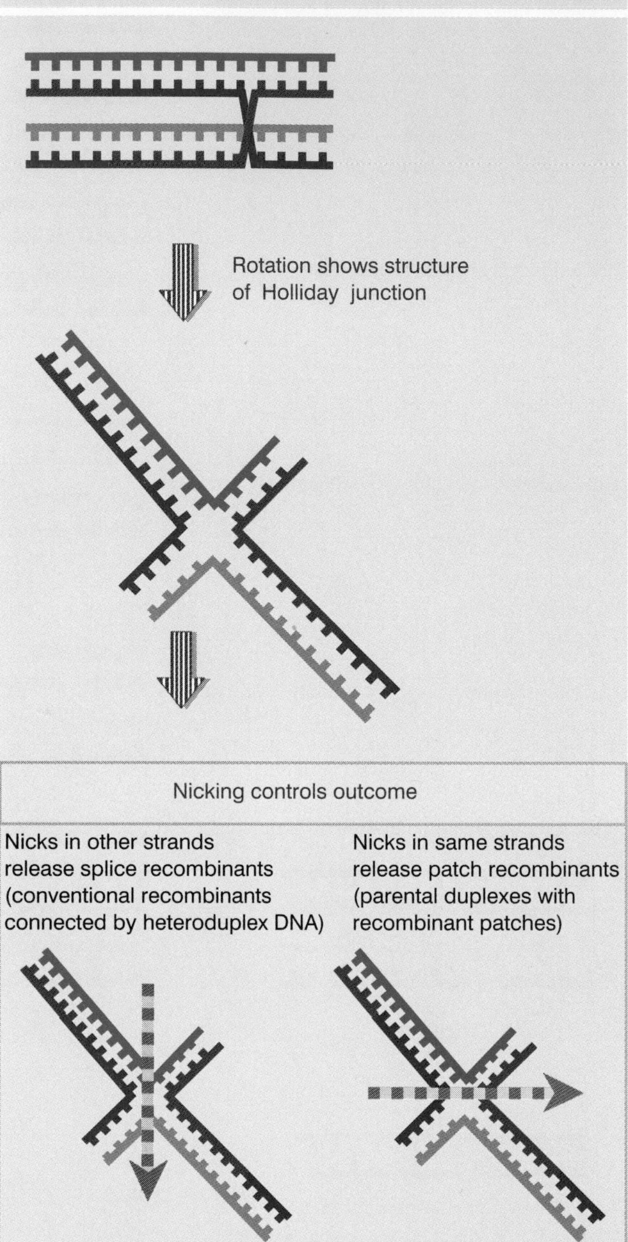
Rotation shows structure of Holliday junction
Nicking controls outcome
Nicks in other strands release splice recombinants (conventional recombinants connected by heteroduplex DNA)
Nicks in same strands release patch recombinants (parental duplexes with recombinant patches)

Double-strand breaks initiate recombination

THE general model of Figure 14.1 shows that a break must be made in one duplex in order to generate a point from which single strands can unwind to participate in genetic exchange. Both strands of a duplex must be broken to accomplish a genetic exchange. Figure 14.2 shows a model in which individual breaks in single strands occur successively. However, *genetic exchange is actually initiated by a double-strand break.* The model is illustrated in **Figure 14.5**.

Recombination is initiated by an endonuclease that cleaves one of the partner DNA duplexes, the "recipient". The cut is enlarged to a gap by exonuclease action. The exonuclease(s) nibble away one strand on either side of the break, generating 3′ single-stranded termini. One of the free 3′ ends then invades a homologous region in the other, "donor" duplex. The formation of heteroduplex DNA generates a D loop, in which one strand of the donor duplex is displaced. The D loop is extended by repair synthesis, using the free 3′ end as a primer.

Eventually the D loop becomes large enough to correspond to the entire length of the gap on the recipient chromatid. When the extruded single strand reaches the far side of the gap, the complementary single-stranded sequences anneal. Now there is heteroduplex DNA on either side of the gap, and the gap itself is represented by the single-stranded D loop.

The duplex integrity of the gapped region can be restored by repair synthesis using the 3′ end on the left side of the gap as a primer. Overall, the gap has been repaired by two individual rounds of single-strand DNA synthesis.

Branch migration converts this structure into a molecule with two recombinant joints. The joints must be resolved by cutting.

If both joints are resolved in the same way, the original noncrossover molecules will be released, each with a region of altered genetic information that is a footprint of the exchange event. If the two joints are resolved in opposite ways a genetic crossover is produced.

The structure of the two-jointed molecule before it is resolved illustrates a critical difference between the double-strand break model and models that invoke only single-strand exchanges:

- Following the double-strand break, heteroduplex DNA has been formed at each end of the region involved in the exchange. Between the two heteroduplex segments is the region corresponding to the gap, which now has the sequence of the donor DNA in both molecules (Figure 14.5). So the arrangement of heteroduplex sequences is asymmetric, and part of one molecule has been converted to the sequence of the other (which is why the initiating chromatid is called the recipient).
- Following reciprocal single-strand exchange, each DNA duplex has heteroduplex material covering the

Figure 14.5 Recombination is initiated by a double-strand break, followed by formation of single-stranded 3' ends, one of which migrates to a homologous duplex.

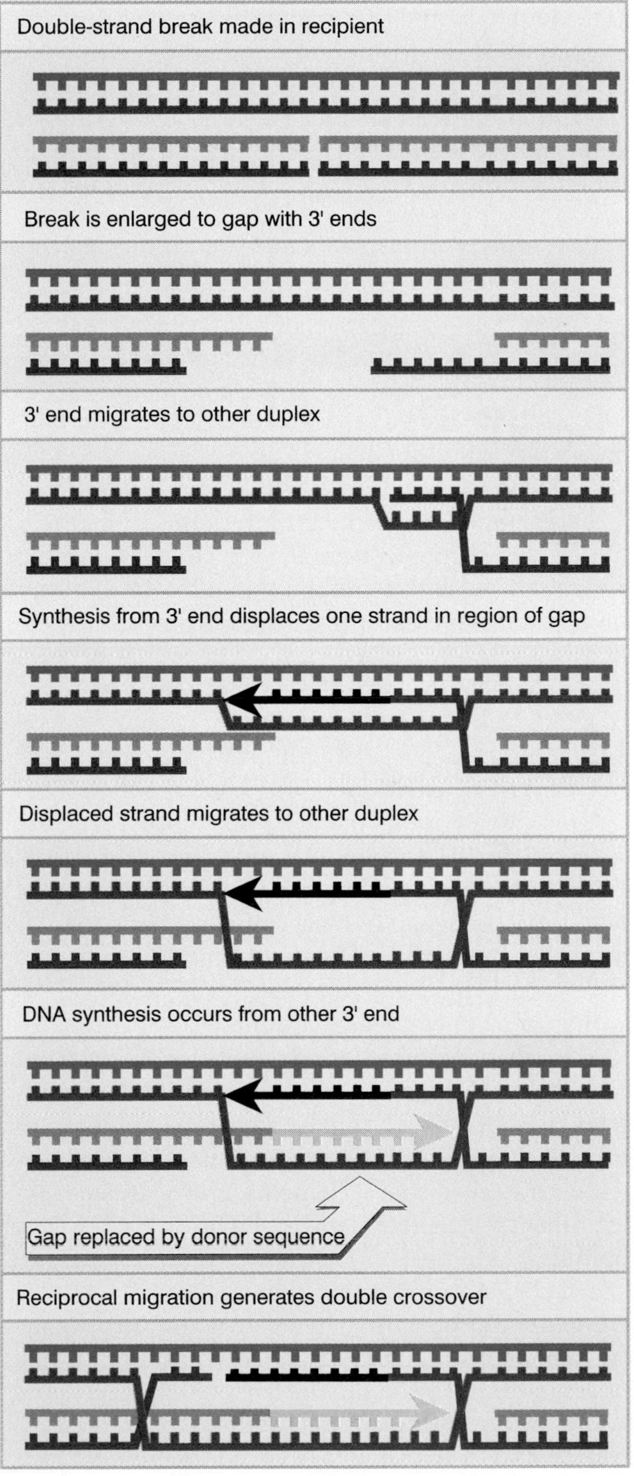
Double-strand break made in recipient
Break is enlarged to gap with 3' ends
3' end migrates to other duplex
Synthesis from 3' end displaces one strand in region of gap
Displaced strand migrates to other duplex
DNA synthesis occurs from other 3' end
Gap replaced by donor sequence
Reciprocal migration generates double crossover

region from the initial site of exchange to the migrating branch (Figure 14.2). In variants of the single-strand exchange model in which some DNA is degraded and resynthesized, the initiating chromatid is the donor of genetic information.

The double-strand break model does not reduce the importance of the formation of heteroduplex DNA, which remains the only plausible means by which two duplex molecules can interact. However, by shifting the responsibility for initiating recombination from single-strand to double-strand breaks, it influences our perspective about the ability of the cell to manipulate DNA.

The involvement of double-strand breaks seems surprising at first sight. Once a break has been made right across a DNA molecule, there is no going back. Compare the events of Figures 14.2 and 14.5. In the single-strand exchange model, at no point has any information been lost. But in the double-strand break model, the initial cleavage is immediately followed by loss of information. Any error in retrieving the information could be fatal. On the other hand, the very ability to retrieve lost information by resynthesizing it from another duplex provides a major safety net for the cell.

Double-strand breaks initiate synapsis

A basic paradox in recombination is that the parental chromosomes never seem to be in close enough contact for recombination of DNA to occur. The chromosomes enter meiosis in the form of replicated (sister chromatid) pairs, visible as a mass of chromatin. They pair to form the synaptonemal complex, and it has been assumed for many years that this represents some stage involved with recombination, possibly a necessary preliminary to exchange of DNA. A more recent view is that the synaptonemal complex is a consequence rather than a cause of recombination. However, we do not understand in either case how the structure of the synaptonemal complex relates to molecular contacts between DNA molecules.

Synapsis begins when each chromosome (sister chromatid pair) condenses around a structure called the **axial element**, which is apparently proteinaceous. Then the axial elements of corresponding chromosomes become aligned, and the synaptonemal complex forms as a tripartite structure, in which the axial elements, now called **lateral elements**, are separated from each other by a **central element**. **Figure 14.6** shows an example.

Each chromosome at this stage appears as a mass of chromatin bounded by a lateral element. The two lateral elements are separated from each other by a fine but dense central element. The triplet of parallel dense strands lies in a single plane that curves and twists along its axis. The distance between the homologous chromosomes is considerable in molecular terms, more than 200 nm (the diameter of DNA is 2 nm). So a major problem in understanding the role of the complex is that, although it aligns homologous chromosomes, it is far from bringing homologous DNA molecules into contact.

The only visible link between the two sides of the synaptonemal complex is provided by spherical or cylindrical structures observed in fungi and insects. They lie across the complex and are called **nodes** or **recombination nodules**; they occur with the same frequency and distribution as the chiasmata. Their name reflects the hope that they may prove to be the sites of recombination.

The correlation between recombination and synaptonemal complex formation is well established, and recent work has shown that all mutations that abolish chromosome pairing in *Drosophila* or in yeast also prevent recombination. There are few systems in which it is possible to compare molecular and cytological events at recombination, but recently there has been progress in analyzing meiosis in *S. cerevisiae.* The relative timing of events is summarized in **Figure 14.7**.

There is good evidence in yeast that double strands initiate recombination in both homologous and site-specific recombination. Double-strand breaks were initially implicated in the change of mating type, which involves the replacement of one sequence by another (discussed in detail in Chapter 17). Double-strand breaks also occur early in meiosis at sites that provide hotspots for recombination. Their locations are not sequence-specific. They tend to occur in promoter regions and in general to coincide with more ac-

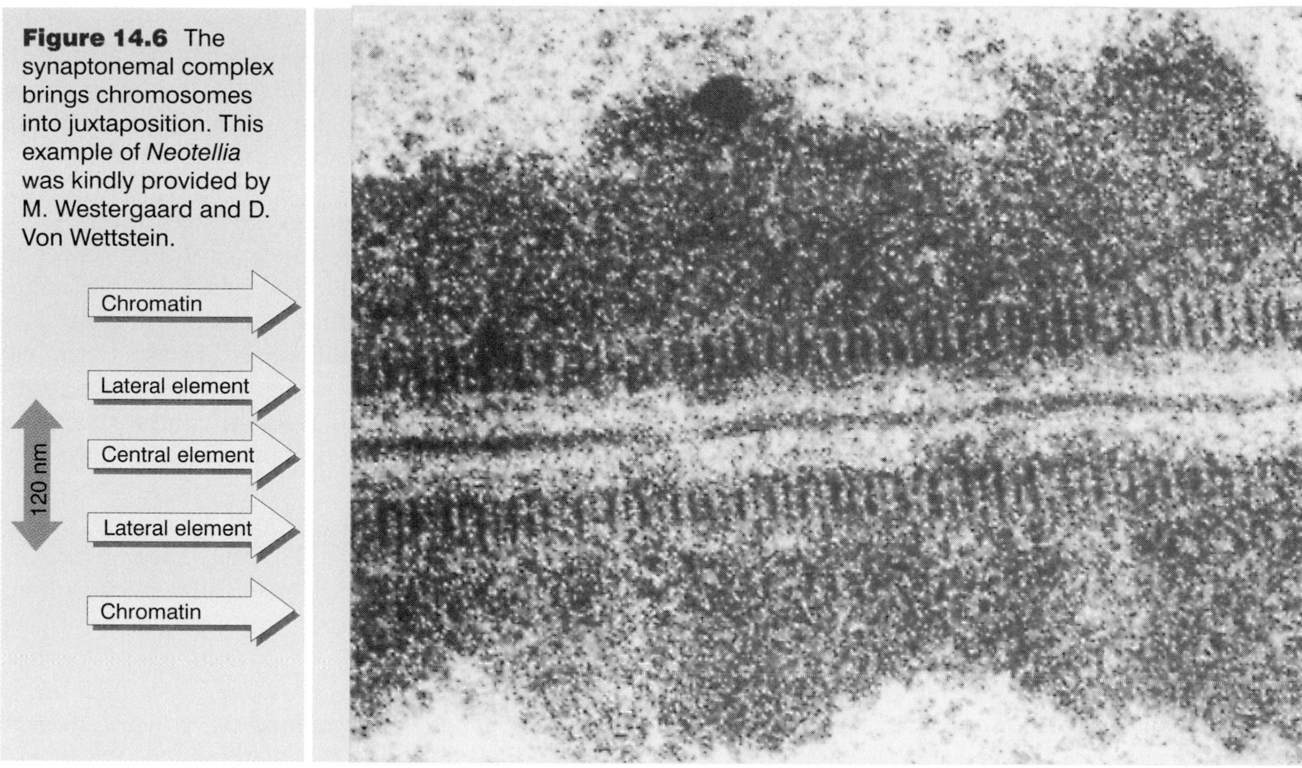

Figure 14.6 The synaptonemal complex brings chromosomes into juxtaposition. This example of *Neotellia* was kindly provided by M. Westergaard and D. Von Wettstein.

Figure 14.7 Double-strand breaks appear when axial elements form, and disappear during the extension of synaptonemal complexes. Joint molecules appear and persist until DNA recombinants are detected at the end of pachytene.

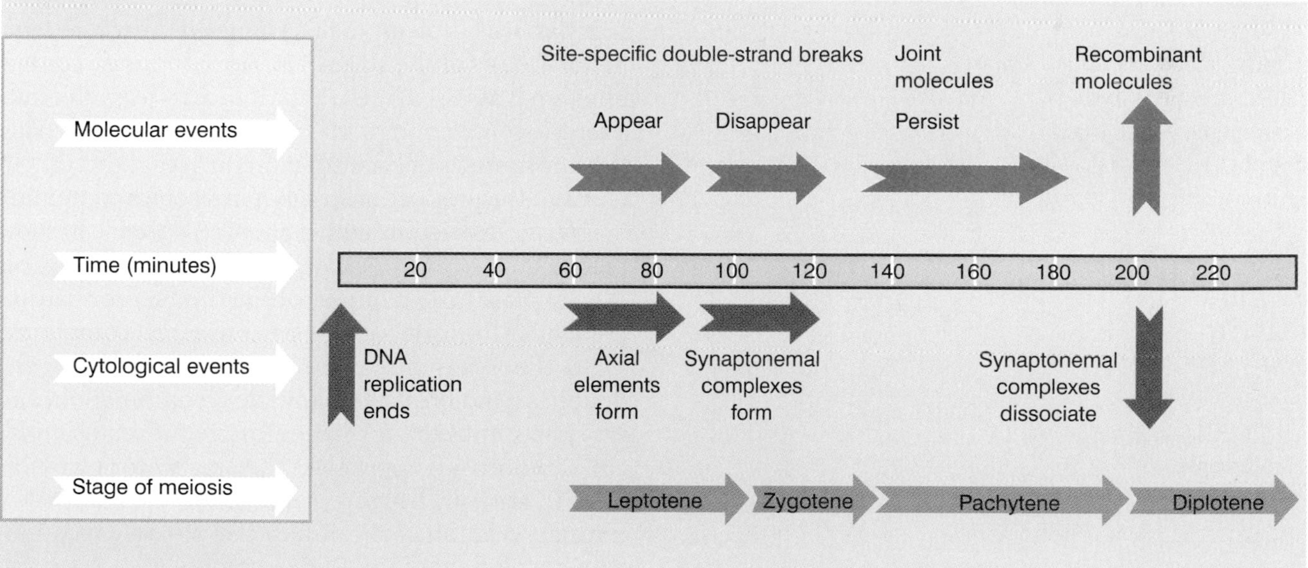

cessible regions of chromatin. The frequency of recombination declines in a gradient on one or both sides of the hotspot. The hotspot identifies the site at which recombination is initiated; and the gradient reflects the probability that the recombination events will spread from it.

We may now interpret this idea in molecular terms. The flush ends created by the double-strand break are rapidly converted on both sides into long 3′ single-stranded ends, as shown in the model of Figure 14.5. A yeast mutation (*rad50*) that blocks the conversion of the flush end into the single-stranded protrusion is defective in recombination. This suggests that double-strand breaks are necessary for recombination. The gradient is determined by the declining probability that a single-stranded region will be generated as distance increases from the site of the double-strand break.

In *rad50* mutants, the 5′ ends of the double-strand breaks are connected to the protein Spo11, which is homologous to the catalytic subunits of one family of type II topoisomerases. This suggests that the topoisomerase may be the enzyme that generates the double-strand breaks. The model for this reaction shown in **Figure 14.8** suggests that Spo11 interacts reversibly with DNA; the break is converted into a permanent structure by an interaction with another protein that dissociates the Spo11 complex. Then removal of Spo11 is followed by nuclease action. At least nine other proteins are required for the formation of double-strand breaks, and they presumably form a pathway in which target sites are identified, breaks are made, and the ends are converted into recombinogenic structures.

Figure 14.8 Spo11 is covalently joined to the 5′ ends of double-strand breaks.

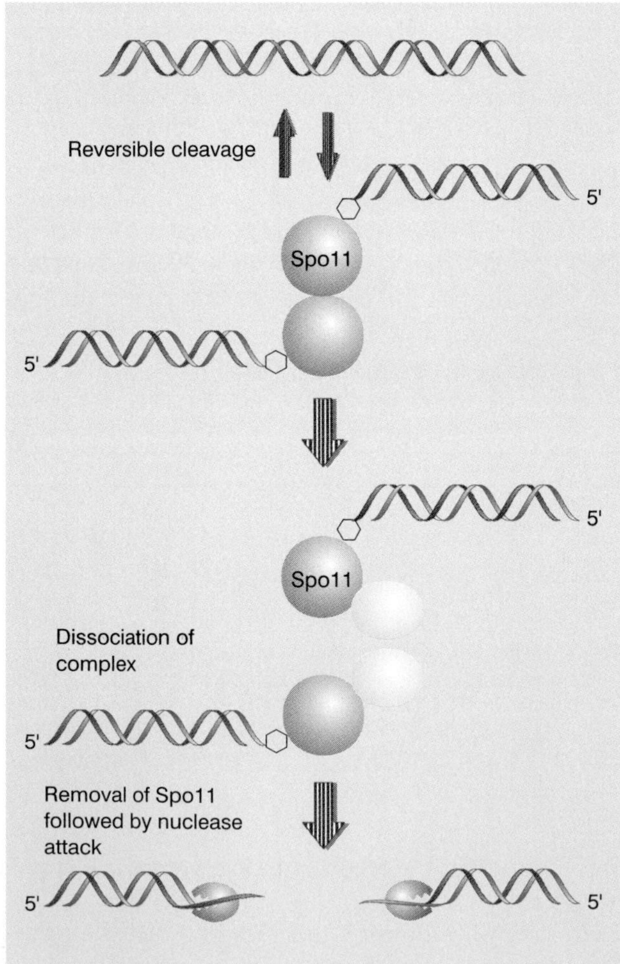

Double-strand breaks appear and then disappear over a 60 minute period. The first joint molecules, which are putative recombination intermediates, appear soon after the double-strand breaks disappear. The sequence of events suggests that double-strand breaks, individual pairing reactions, and formation of recombinant structures occur in succession at the same chromosomal site.

Double-strand breaks appear during the period when axial elements form. They disappear during the conversion of the paired chromosomes into synaptonemal complexes. This relative timing of events suggests that *formation of the synaptonemal complex results from the initiation of recombination via the introduction of double-strand breaks and their conversion into later intermediates of recombination.* This idea is supported by the observation that the *rad50* mutant cannot convert axial elements into synaptonemal complexes. This refutes the traditional view of meiosis that the synaptonemal complex represents the need for chromosome pairing to precede the molecular events of recombination.

It has been difficult to determine whether recombination occurs at the stage of synapsis, because recombination is assessed by the appearance of recombinants after the completion of meiosis. However, by assessing the appearance of recombinants in yeast directly in terms of the production of DNA molecules containing diagnostic restriction sites, it has been possible to show that recombinants appear at the end of pachytene. This clearly places the completion of the recombination event after the formation of synaptonemal complexes.

So the synaptonemal complex forms after the double-strand breaks that initiate recombination, and it persists until the formation of recombinant molecules. It does not appear to be necessary for recombination as such, because some mutants that lack a normal synaptonemal complex can generate recombinants. Mutations that abolish recombination, however, also fail to develop a synaptonemal complex. This suggests that the synaptonemal complex forms as a *consequence* of recombination, following chromosome pairing, and is required for later stages of meiosis. A yeast mutation (*zip1*) in synaptonemal complex formation has the consequence of abolishing crossover interference (the ability of one recombination event to inhibit the occurrence of another nearby). This suggests the possibility that synaptonemal complex formation may be initiated at the site of a recombination event, and that the spread of the complex along the chromosome itself inhibits the occurrence of further recombination events. In this case, recombination events would not all be initiated simultaneously, and the synaptonemal complex could form during a period while recombination is initiated, and before the process is completed. The proposal that the synaptonemal complex inhibits recombination is an ironic contrast with the earliest ideas that it might be responsible for recombination!

Mutations in proteins that are needed for axial elements to form (these are the cohesins discussed in Chapter 27) do not prevent the formation of double-strand breaks, but block formation of recombinants. The formation of the axial elements may be necessary for recombination.

We can distinguish the processes of pairing and synaptonemal complex formation by the effects of two mutations, each of which blocks one of the processes without affecting the other.

The *zip2* mutation allows chromosomes to pair, but they do not form synaptonemal complexes. So recognition between homologs is independent of recombination or synaptonemal complex formation.

The specificity of association between homologous chromosomes is controlled by the gene *hop2* in *S. cerevisiae*. In *hop2* mutants, normal amounts of synaptonemal complex form at meiosis, but the individual complexes contain nonhomologous chromosomes. This suggests that the formation of synaptonemal complexes as such is independent of homology (and therefore cannot be based on any extensive comparison of DNA sequences). The usual role of Hop2 is to prevent nonhomologous chromosomes from interacting.

Double-strand breaks form in the mispaired chromosomes in the synaptonemal complexes of *hop2* mutants, but they are not repaired. This suggests that, if formation of the synaptonemal complex requires double-strand breaks, it does not require any extensive reaction of these breaks with homologous DNA.

It is not clear what usually happens during pachytene, before DNA recombinants are observed. It may be that this period is occupied by the subsequent steps of recombination, involving the extension of strand exchange, DNA synthesis, and resolution.

At the next stage of meiosis (diplotene), the chromosomes shed the synaptonemal complex; then the chiasmata become visible as points at which the chromosomes are connected. This has been presumed to indicate the occurrence of a genetic exchange, but the molecular nature of a chiasma is unknown. It is possible that it represents the residuum of a completed exchange, or that it represents a connection between homologous chromosomes where a genetic exchange has not yet been resolved. Later in meiosis, the chiasmata move toward the ends of the chromosomes. This flexibility suggests that they represent some remnant of the recombination event, rather than providing the actual intermediate.

Recombination events occur at discrete points on meiotic chromosomes, but we cannot as yet correlate their occurrence with the discrete structures that have been observed, that is, recombination nodules and chiasmata. However, insights into the molecular basis for the formation of discontinuous structures are provided by the identification of proteins involved in yeast recombination that can be localized to discrete sites. These include MSH4 (which is related to bacterial proteins involved in mismatch-repair; see below), and Dmc1 and Rad51 (which are homologs of the *E. coli* RecA protein). The exact roles of these proteins in recombination remain to be established.

Recombination events are subject to a general control. Only a minority of interactions actually mature as crossovers, but these are distributed in such a way that typically each pair of homologs acquires only one to two crossovers, yet the probability of zero crossovers for a homolog pair is very low (<0.1%). This process is probably the result of a single **crossover control**, because the nonrandomness of crossovers is generally disrupted in certain mutants. Furthermore, the occurrence of recombination is necessary for progress through meiosis, and a "checkpoint" system (see Chapter 27) exists to block meiosis if recombination has not occurred. (The block is lifted when recombination has been successfully completed; this system provides a safeguard to ensure that cells do not try to segregate their chromosomes until recombination has occurred.)

Bacterial recombination involves single-strand assimilation

To analyze the nature of the events involved in exchange of sequences between DNA molecules, we must turn to bacterial systems. Here the recognition reaction is part and parcel of the recombination mechanism and involves restricted regions of DNA molecules rather than intact chromosomes. But the general order of molecular events is similar: a single strand from a broken molecule interacts with a partner duplex; the region of pairing is extended; and an endonuclease resolves the partner duplexes. Enzymes involved in each stage are known, although they probably represent only some of the components required for recombination.

Bacterial enzymes implicated in recombination have been identified by the occurrence of *rec*⁻ mutations in their genes. The phenotype of Rec⁻ mutants is the inability to undertake generalized recombination. Some 10–20 loci have been identified.

Bacteria do not usually exchange large amounts of duplex DNA, and there may be various routes to initiate recombination in prokaryotes. In some cases, DNA may be available with free single-stranded 3′ ends: DNA may be provided in single-stranded form (as in conjugation, discussed in Chapter 12); single-stranded gaps may be generated by irradiation damage; or single-stranded tails may be generated by phage genomes undergoing replication by a rolling circle. However, in circumstances involving two duplex molecules (as in recombination at meiosis in eukaryotes), single-stranded regions and 3′ ends must be generated.

One mechanism for generating suitable ends has been discovered as a result of the existence of certain hotspots that stimulate recombination. They were discovered in phage lambda in the form of mutants, called *chi*, that have single base-pair changes creating sites that stimulate recombination. These sites lead us to the role of other proteins involved in recombination.

These sites share a constant nonsymmetrical sequence of 8 bp:

```
5′   GCTGGTGG   3′
3′   CGACCACC   5′
```

The *chi* sequence occurs naturally in *E. coli* DNA about once every 5–10 kb. Its absence from wild-type lambda DNA, and also from other genetic elements, shows that it is not essential for recombination.

A *chi* sequence stimulates recombination in its general vicinity, say within a distance of up to 10 kb from the site. A *chi* site can be activated by a double-strand break made several kb away *on one particular side* (to the right of the sequence as written above). This dependence on orientation suggests that the recombination apparatus must associate with DNA at a broken end, and then can move along the duplex in only one direction.

chi sites are targets for the action of an enzyme coded by the genes *recBCD*. This complex exercises several activities. It is a potent nuclease that degrades DNA, originally identified as the activity exonuclease V. It has a helicase activity that can unwind duplex DNA in the presence of SSB; and it has an ATPase activity. Its role in recombination may be to provide a single-stranded region with a free 3′ end.

Figure 14.9 shows how these reactions are coordinated on a substrate DNA that has a *chi* site. When RecBCD binds DNA on the right site of *chi*, it moves along unwinding the DNA. It degrades the released single strand with the 3′ end. When it reaches the *chi* site, it pauses and cleaves one (the top) strand of the DNA at a position between four and six bases on the right. The top strand of the chi site is recognized in single-stranded form. Recognition of the *chi* site causes the RecD subunit to dissociate or become inactivated, as a result of which the enzyme loses its nuclease activity. However, it continues to function as a helicase.

RecBCD-mediated unwinding and cleavage can be used to generate ends that initiate the formation of heteroduplex joints. The enzyme RecA can take the single strand with the 3′ end that is released when RecBCD cuts at *chi*, and can use it to react with a homologous duplex sequence, thus creating a joint molecule.

RecA has two quite different types of activity: it can stimulate protease activity in the SOS response (see later); and can promote base pairing between a single strand of DNA and its complement in a duplex molecule.

RecA requires single-stranded DNA and ATP for ability to stimulate protease activity. The same substrates are required for its ability to manipulate DNA molecules. It is not yet clear exactly how the enzymatic

Figure 14.9 RecBCD nuclease approaches a *chi* sequence from one side, degrading DNA as it proceeds; at the *chi* site, it makes an endonucleolytic cut, loses RecD, and retains only the helicase activity.

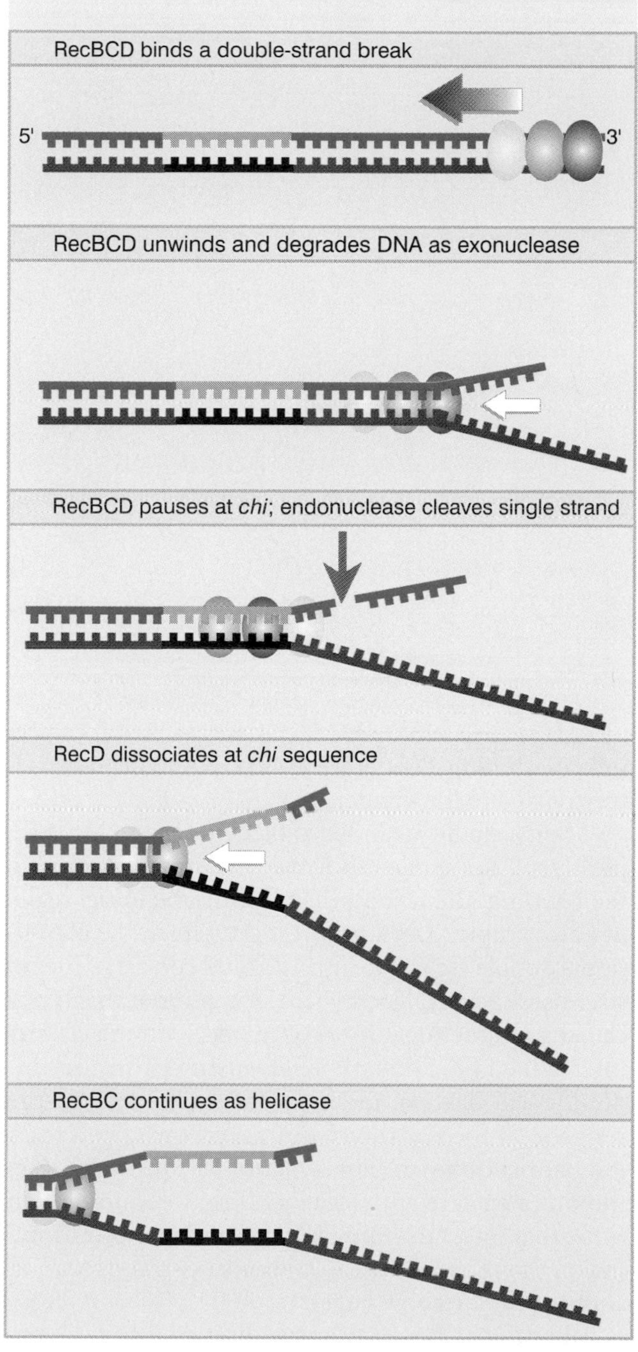
RecBCD binds a double-strand break
5'
3'
RecBCD unwinds and degrades DNA as exonuclease
RecBCD pauses at *chi*; endonuclease cleaves single strand
RecD dissociates at *chi* sequence
RecBC continues as helicase

activities of RecA are related to recombination *in vivo*, but they involve several reactions that provide useful paradigms for recombination mechanisms.

The DNA handling activity of RecA enables a single strand to displace its homolog in a duplex in a reaction that is called **single-strand uptake** or **single-strand assimilation**. The displacement reaction can occur between DNA molecules in several configurations and has three general conditions:

- One of the DNA molecules must have a single-stranded region.
- One of the molecules must have a free 3′ end.
- The single-stranded region and the 3′ end must be located within a region that is complementary between the molecules.

The reaction is illustrated in **Figure 14.10**. When a linear single strand invades a duplex, it displaces the original partner to its complement. The reaction can be followed most easily by making either the donor or recipient a circular molecule. The reaction proceeds 5′–3′ along the strand whose partner is being displaced and replaced, that is, the reaction involves an exchange in which (at least) one of the exchanging strands has a free 3′ end.

Single-strand assimilation is potentially related to the initiation of recombination. All models call for an intermediate in which one or both single strands cross over from one duplex to the other (see Figures 14.2 and 14.5). RecA could catalyze this stage of the reaction.

A mechanism for the activity of RecA in stimulating branch migration is suggested by its ability to aggregate into long filaments with single-stranded or duplex DNA. There are six RecA monomers per turn of the filament, which has a helical structure with a deep groove that contains the DNA. The stoichiometry of binding is three nucleotides (or base pairs) per RecA monomer. The DNA is held in a form that is extended 1.5 times relative to duplex B DNA, making a turn every 18.6 nucleotides (or base pairs). When duplex DNA is bound, it contacts RecA via its minor groove, leaving the major groove accessible for possible reaction with a second DNA molecule.

The interaction between two DNA molecules occurs within these filaments. When a single strand is assimilated into a duplex, the first step is for RecA to bind the single strand into a filament. Then the duplex is incorporated, probably forming some sort of triple-stranded structure. In this system, synapsis precedes physical exchange of material, because the pairing reaction can take place even in the absence of free ends, when strand exchange is impossible.

A free 3′ end is required for strand exchange. The reaction occurs within the filament, and RecA remains bound to the strand that was originally single, so that at the end of the reaction RecA is bound to the duplex molecule. Large amounts of ATP are hydrolyzed

Figure 14.10 RecA promotes the assimilation of invading single strands into duplex DNA so long as one of the reacting strands has a free end.

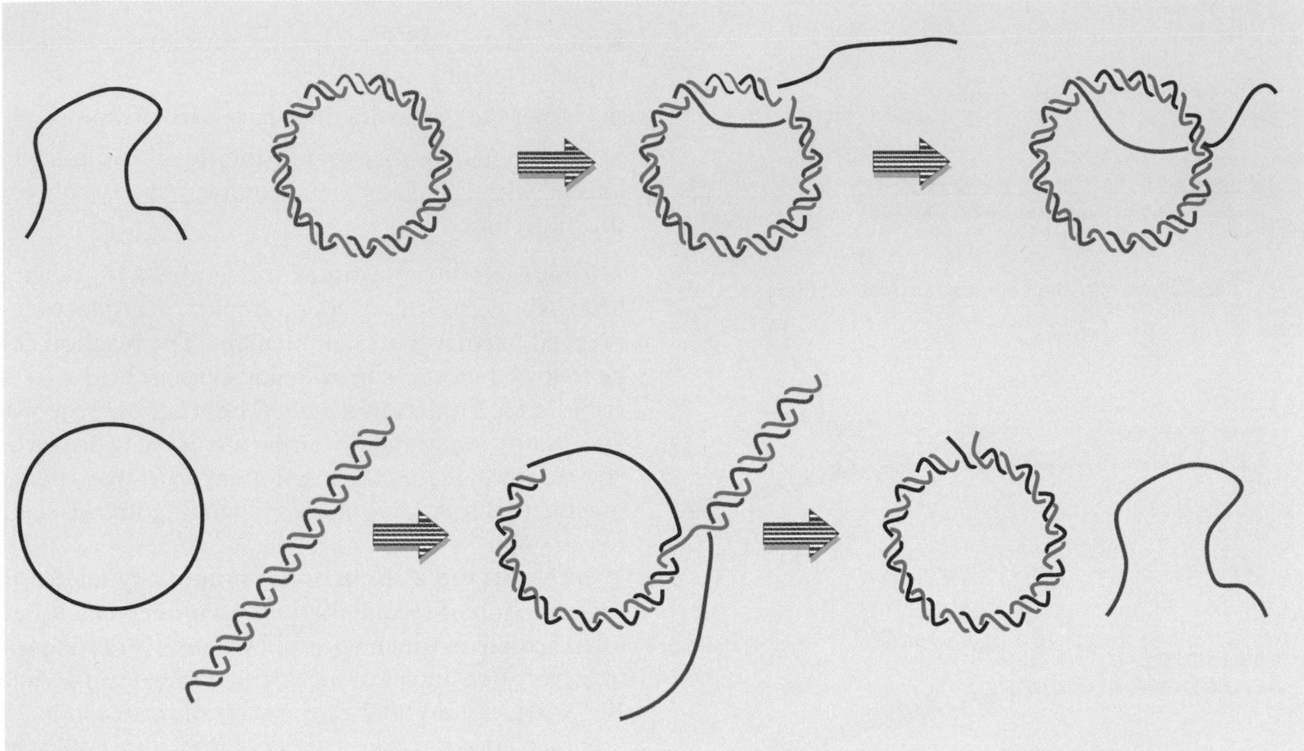

during the reaction. The ATP may act through an allosteric effect on RecA conformation. When bound to ATP, the DNA-binding site of RecA has a high affinity for DNA; this is needed to bind DNA and for the pairing reaction. Hydrolysis of ATP converts the binding site to low affinity, which is needed to release the heteroduplex DNA.

We can divide the reaction that RecA catalyzes between single-stranded and duplex DNA into three phases:

- A slow presynaptic phase in which RecA polymerizes on single-stranded DNA.
- A fast pairing reaction between the single-stranded DNA and its complement in the duplex to produce a heteroduplex joint.
- A slow displacement of one strand from the duplex to produce a long region of heteroduplex DNA.

The presence of SSB (single-strand binding protein) stimulates the reaction, by ensuring that the substrate lacks secondary structure. It is not clear yet how SSB and RecA can both act on the same stretch of DNA. Like SSB, RecA is required in stoichiometric amounts, which suggests that its action in strand assimilation involves binding cooperatively to DNA to form a structure related to the filament.

When a single-stranded molecule reacts with a duplex DNA, the duplex molecule becomes unwound in the region of the recombinant joint. The initial region of heteroduplex DNA may not even lie in the conventional double-helical form, but could consist of the two strands associated side by side. A region of this type is called a **paranemic joint** (compared with the classical intertwined **plectonemic** relationship of strands in a double helix). A paranemic joint is unstable; further progress of the reaction requires its conversion to the double-helical form. This reaction is equivalent to removing negative supercoils and may require an enzyme that solves the unwinding/rewinding problem by making transient breaks that allow the strands to rotate about each other (see later).

All of the reactions we have discussed so far represent only a part of the potential recombination event: the invasion of one duplex by a single strand. Two duplex molecules can interact with each other under the sponsorship of RecA, provided that one of them has a single-stranded region of at least 50 bases. The single-stranded region can take the form of a tail on a linear molecule or of a gap in a circular molecule.

Figure 14.11 RecA-mediated strand exchange between partially duplex and entirely duplex DNA generates a joint molecule with the same structure as a recombination intermediate.

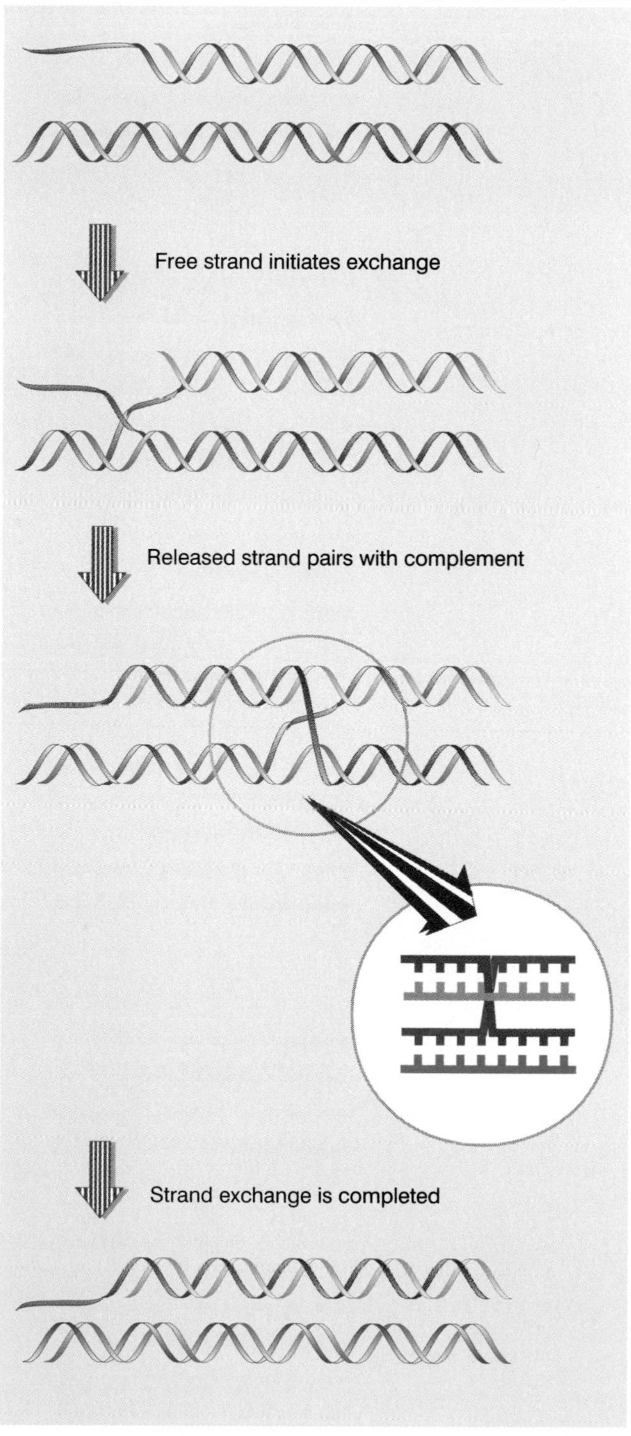

Free strand initiates exchange
Released strand pairs with complement
Strand exchange is completed

The reaction between a partially duplex molecule and an entirely duplex molecule leads to the exchange of strands. An example is illustrated in **Figure 14.11.**

Figure 14.12 RuvAB is an asymmetric complex that promotes branch migration of a Holliday junction.

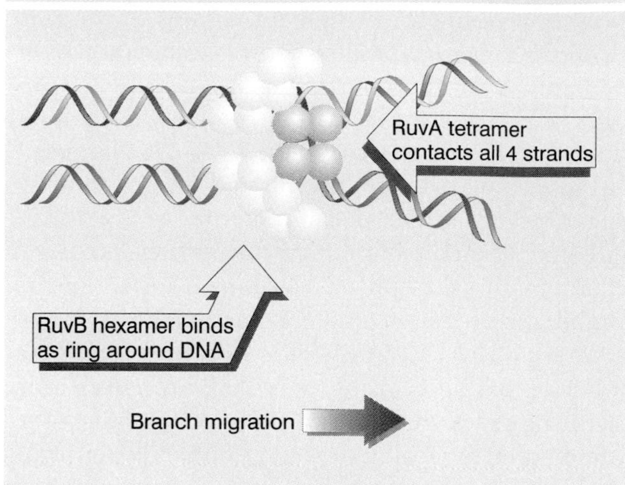

RuvA tetramer contacts all 4 strands
RuvB hexamer binds as ring around DNA
Branch migration

Assimilation starts at one end of the linear molecule, where the invading single strand displaces its homolog in the duplex in the customary way. But when the reaction reaches the region that is duplex in both molecules, the invading strand unpairs from its partner, which then pairs with the other displaced strand.

At this stage, the molecule has a structure indistinguishable from the recombinant joint in Figure 14.4. The reaction sponsored *in vitro* by RecA can generate Holliday junctions, which suggests that the enzyme can mediate reciprocal strand transfer. We know less about the geometry of four-strand intermediates bound by RecA, but presumably two duplex molecules can lie side by side in a way consistent with the requirements of the exchange reaction.

A group of three genes in *E. coli* codes for functions involved later in recombination. The products of *ruvA* and *ruvB* increase the formation of heteroduplex structures. RuvA recognizes the structure of the Holliday junction. RuvA binds to all four strands of DNA at the crossover point and forms two tetramers that sandwich the DNA. RuvB is an ATPase that functions as a hexamer; it has helicase activity that provides the motor for branch migration. Hexameric rings of RuvB bind around each duplex of DNA upstream of the crossover point. A diagram of the function of the complex is shown in **Figure 14.12.**

The RuvAB complex can cause the branch to migrate as fast as 10–20 bp/second. A similar activity is provided by the RecG helicase. RuvAB displaces RecA from DNA during its action. The RuvAB and RecG activities both can act on Holliday junctions and allow recombination to be completed, but if both are mutant, *E. coli* is completely defective in recombination activity.

The third gene, *ruvC*, codes for an endonuclease that specifically recognizes Holliday junctions. It can cleave such junctions *in vitro* to resolve recombination intermediates. A common tetranucleotide sequence provides a hotspot for RuvC to resolve the Holliday junction. The tetranucleotide (ATTG) is asymmetric, and thus may direct resolution with regard to which pair of strands is nicked. This determines whether the outcome is patch recombinant formation (no overall recombination) or splice recombinant formation (recombination between flanking markers).

We may now account for the stages of recombination in *E. coli* in terms of individual proteins. **Figure 14.13** shows the events that are involved in using recombination to repair a gap in one duplex by retrieving material from the other duplex. The major caveat in applying these conclusions to recombination in eukaryotes is that bacterial recombination generally involves interaction between a fragment of DNA and a whole chromosome. It occurs as a repair reaction that is stimulated by damage to DNA, and this is not entirely equivalent to recombination between genomes at meiosis. Nonetheless, the same molecular activities are involved in manipulating DNA.

Homologs of RecA are ubiquitous among prokaryotes, and related proteins have been found in eukaryotes. Two genes in *S. cerevisiae*, *DMC1* and *rad51*, code for proteins that are related to RecA. Mutations in these genes cause a similar phenotype; they accumulate double-strand breaks and fail to form normal synaptonemal complexes. This reinforces the idea that exchange of strands between DNA duplexes is involved in formation of the synaptonemal complex, and raises the possibility that chromosome synapsis is related to the bacterial strand assimilation reaction. However, eukaryotic homologs of RecA do not form filaments, so the mechanics of the reaction are likely to be different in eukaryotes.

Figure 14.13 Bacterial enzymes can catalyze all stages of recombination in the repair pathway following the production of suitable substrate DNA molecules.

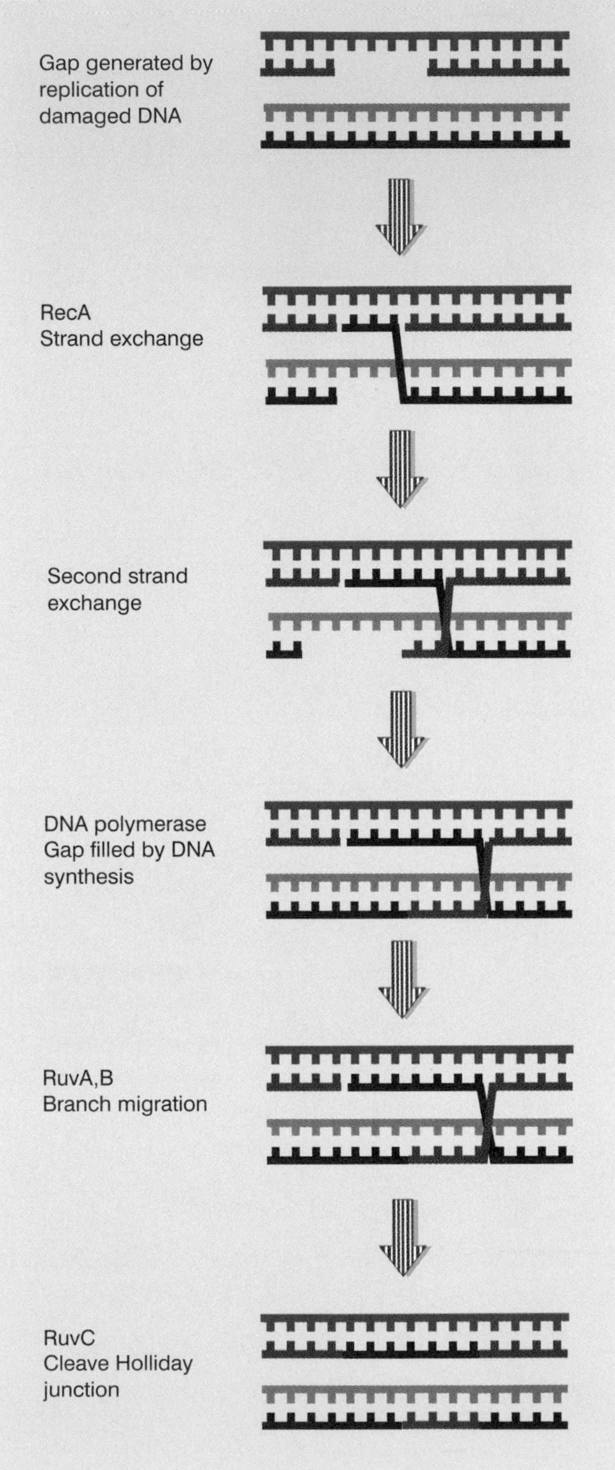

Gene conversion accounts for interallelic recombination

THE involvement of heteroduplex DNA explains the characteristics of recombination between alleles; indeed, allelic recombination provided the impetus for the development of the heteroduplex model. When recombination between alleles was discovered, the natural assumption was that it takes place by the same mechanism of reciprocal recombination that applies to more distant loci. That is to say that an individual breakage and reunion event occurs within the locus to generate a reciprocal pair of recombinant chromosomes. However, in the close quarters of a single gene, the formation of heteroduplex DNA itself is usually responsible for the recombination event.

Individual recombination events can be studied in the *Ascomycetes* fungi, because the products of a single meiosis are held together in a large cell, the ascus. Even better, the four haploid nuclei produced by meiosis are arranged in a linear order. Actually, a mitosis occurs after the production of these four nuclei, giving a linear series of eight haploid nuclei. **Figure 14.14** shows that each of these nuclei effectively represents the genetic character of one of the eight strands of the four chromosomes produced by the meiosis.

Meiosis in a heterozygote should generate four copies of each allele. This is seen in the majority of spores. But there are some spores with abnormal ratios. They are explained by the formation and correction of heteroduplex DNA in the region in which the alleles differ. The figure illustrates a recombination event in which a length of hybrid DNA occurs on one of the four meiotic chromosomes, a possible outcome of recombination initiated by a double-strand break.

Suppose that two alleles differ by a single point mutation. When a strand exchange occurs to generate heteroduplex DNA, the two strands of the heteroduplex will be mispaired at the site of mutation. So each strand of DNA carries different genetic information. If no change is made in the sequence, the strands separate at the ensuing replication, each giving rise to a duplex that perpetuates its information. This event is called

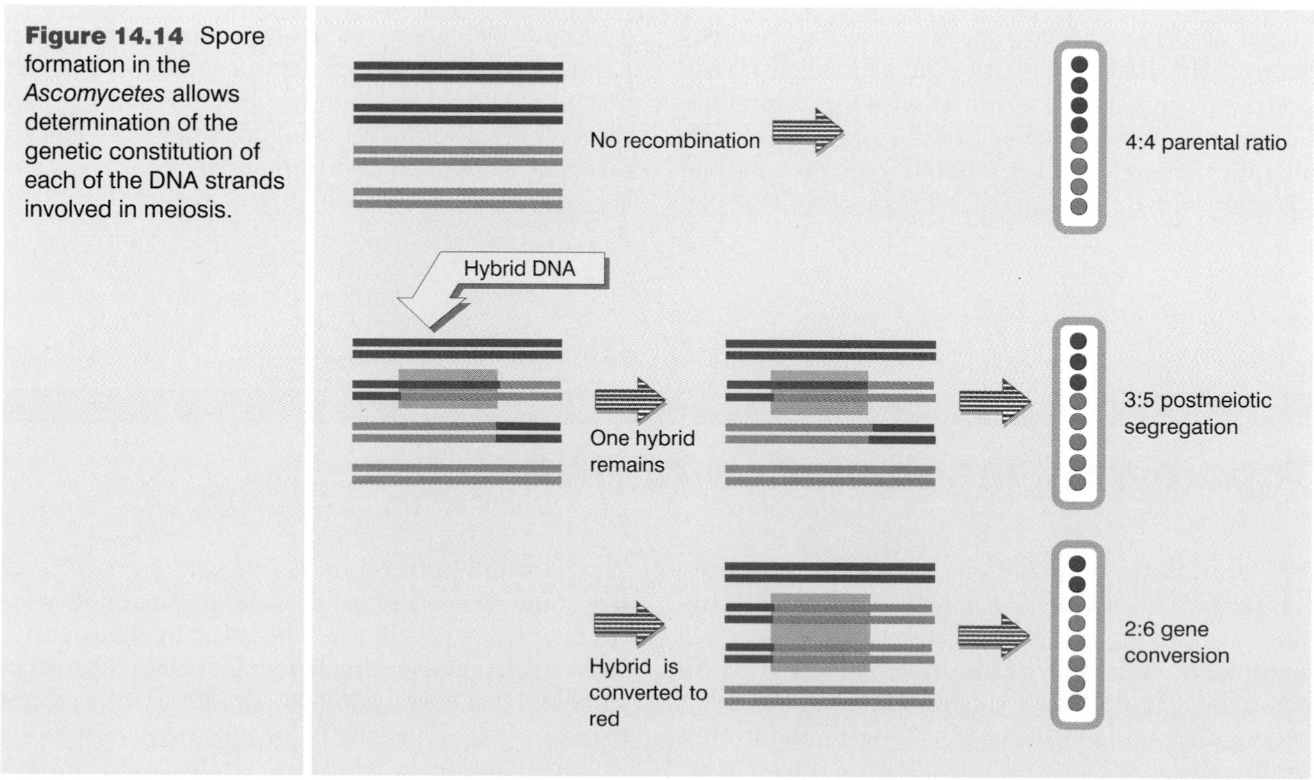

Figure 14.14 Spore formation in the *Ascomycetes* allows determination of the genetic constitution of each of the DNA strands involved in meiosis.

postmeiotic segregation, because it reflects the separation of DNA strands after meiosis. Its importance is that *it demonstrates directly the existence of heteroduplex DNA in recombining alleles.*

Another effect is seen when examining recombination between alleles: *the proportions of the alleles differ from the initial 4:4 ratio.* This effect is called **gene conversion.** It describes a *nonreciprocal transfer of information from one chromatid to another.*

Gene conversion results from exchange of strands between DNA molecules, and the change in sequence may have either of two causes at the molecular level:

- As indicated by the double-strand break model in Figure 14.5, one DNA duplex may act as a donor of genetic information that directly replaces the corresponding sequences in the recipient duplex by a process of strand exchange and gap filling.
- As part of the exchange process, heteroduplex DNA is generated when a single strand from one duplex pairs with its complement in the other duplex. Repair systems recognize mispaired bases in heteroduplex DNA, and may then excise and replace one of the strands to restore complementarity. Such an event changes the strand of DNA representing one allele into the sequence of the other allele.

Gene conversion does not depend on crossing-over, but is correlated with it. A large proportion of the aberrant asci show genetic recombination between two markers on either side of a site of interallelic gene conversion. This is exactly what would be predicted if the aberrant ratios result from initiation of the recombination process as shown in Figure 14.2 or 14.5, but with an approximately equal probability of resolving the structure with or without recombination (as indicated in Figure 14.4). The implication is that fungal chromosomes initiate crossing-over about twice as often as would be expected from the frequency of recombination between distant genes.

Various biases are seen when recombination is examined at the molecular level. Either direction of gene conversion may be equally likely, or allele-specific effects may create a preference for one direction. Gradients of recombination may fall away from hotspots. We now know that hotspots represent sites at which double-strand breaks are initiated, and the gradient is correlated with the extent to which the gap at the hotspot is enlarged and converted to long single-stranded ends (as discussed previously).

Some information about the extent of gene conversion is provided by the sequences of members of gene clusters. Usually, the products of a recombination event will separate and become unavailable for analysis at the level of DNA sequence. However, when a chromosome carries two (non-allelic) genes that are related, they may recombine by an "unequal crossing-over" event, as discussed in Chapter 4. All we need to note for now is that a heteroduplex may be formed between the two nonallelic genes. Gene conversion effectively converts one of the nonallelic genes to the sequence of the other.

The presence of more than one gene copy on the same chromosome provides a footprint to trace these events. For example, if heteroduplex formation and gene conversion occurred over part of one gene, this part may have a sequence identical with or very closely related to the other gene, while the remaining part shows more divergence. Available sequences suggest that gene conversion events may extend for considerable distances, up to a few thousand bases.

Topological manipulation of DNA

TOPOLOGICAL manipulation of DNA is a central aspect of all its functional activities—recombination, replication, and (perhaps) transcription—as well as of the organization of higher-order structure. All synthetic activities involving double-stranded DNA require the strands to separate. However, the strands do not simply lie side by side; they are intertwined. Their separation therefore requires the strands to rotate about each other in space. Some possibilities for the unwinding reaction are illustrated in **Figure 14.15**.

We might envisage the structure of DNA in terms of a free end that would allow the strands to rotate about the axis of the double helix for unwinding. Given the length of the double helix, however, this would involve

Figure 14.15 Separation of the strands of a DNA double helix could be achieved by several means.

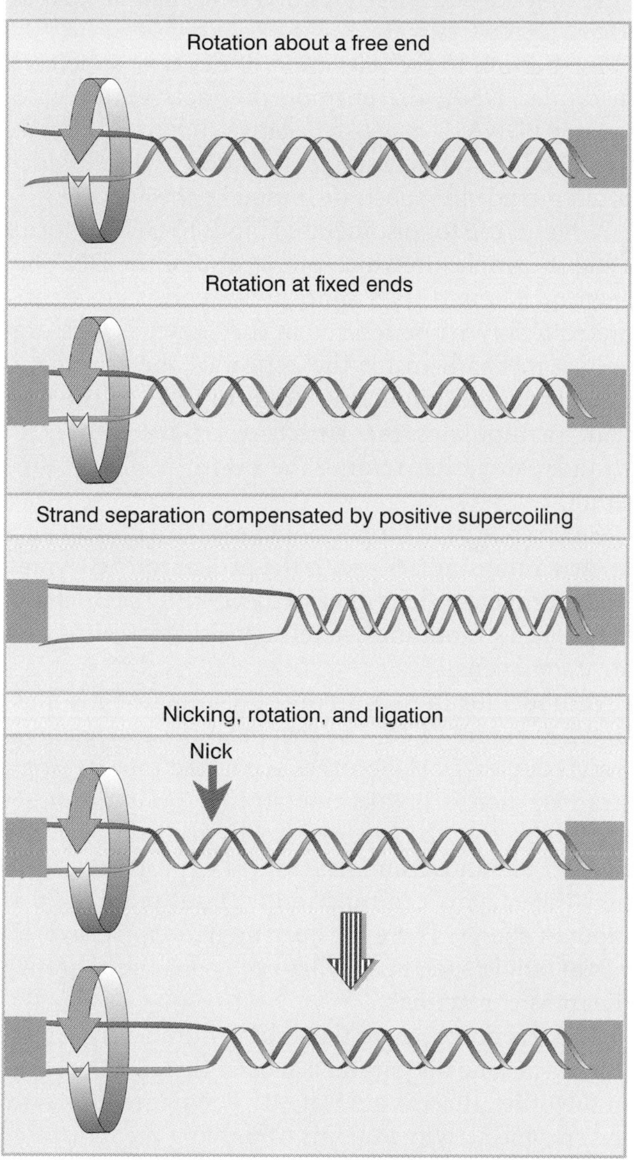

the separating strands in a considerable amount of flailing about, which seems unlikely in the confines of the cell.

A similar result is achieved by placing an apparatus to control the rotation at the free end. However, the effect must be transmitted over a considerable distance, again involving the rotation of an unreasonable length of material.

DNA actually behaves as a closed structure lacking free ends, which excludes these models as a matter of principle and brings home the severity of the topological problem. Consider the effects of separating the two strands in a molecule whose ends are not free to rotate. When two intertwined strands are pulled apart from one end, the result is to *increase their winding about each other farther along the molecule.* So movement of a replication fork would generate increasing positive supercoiling ahead of it, rapidly generating insuperable resistance to further movement. (Similar consequences ensue during transcription, as described in the twin-domain supercoiling model summarized in Figure 9.18.)

The problem can be overcome by introducing a transient nick in one strand. An internal free end allows the nicked strand to rotate about the intact strand, after which the nick can be sealed. Each repetition of the nicking and sealing reaction releases one superhelical turn.

A closed molecule of DNA can be characterized by its linking number, the number of times one strand crosses over the other in space. Closed DNA molecules of identical sequence may have different linking numbers, reflecting different degrees of supercoiling. Molecules of DNA that are the same except for their linking numbers are called **topological isomers.**

The linking number is made up of two components: the writhing number (W) and the twisting number (T).

The **twisting number**, T, is a property of the double helical structure itself, representing the rotation of one strand about the other. It represents the *total number of turns of the duplex.* It is determined by the number of base pairs per turn. For a relaxed closed circular DNA lying flat in a plane, the twist is the total number of base pairs divided by the number of base pairs per turn.

The **writhing number**, W, represents the *turning of the axis of the duplex in space.* It corresponds to the intuitive concept of supercoiling, but does not have exactly the same quantitative definition or measurement. For a relaxed molecule, $W = 0$, and the linking number equals the twist.

We are often concerned with the *change* in linking number, ΔL, given by the equation

$$\Delta L = \Delta W + \Delta T.$$

The equation states that any change in the total number of revolutions of one DNA strand about the other can be expressed as the sum of the changes of the coiling of the duplex axis in space (ΔW) and changes in the screwing of the double helix itself (ΔT). In a free DNA molecule, W and T are freely adjustable, and any ΔL (change in linking number) is likely to be expressed by a change in W, that is, by a change in supercoiling.

A decrease in linking number, that is, a change of $-\Delta L$, corresponds to the introduction of some combination of negative supercoiling and/or underwinding. An increase in linking number, measured as a change of $+\Delta L$, corresponds to a decrease in negative supercoiling/underwinding.

We can describe the change in state of any DNA by the specific linking difference, $\sigma = \Delta L/L_0$, where L_0 is the linking number when the DNA is relaxed. If all of the change in linking number is due to change in W (that is, $\Delta T = 0$), the specific linking difference equals the supercoiling density. In effect, σ as defined in terms of $\Delta L/L_0$ can be assumed to correspond to superhelix density so long as the structure of the double helix itself remains constant.

The critical feature about the use of the linking number is that *this parameter is an invariant property of any individual closed DNA molecule.* The linking number cannot be changed by any deformation short of one that involves the breaking and rejoining of strands. A circular molecule with a particular linking number can express it in terms of different combinations of T and W, but cannot change their sum so long as the strands are unbroken. (In fact, the partition of L between T and W prevents the assignment of fixed values for the latter parameters for a DNA molecule in solution.)

The linking number is related to the actual enzymatic events by which changes are made in the topology of DNA. The linking number of a particular closed molecule can be changed only by breaking a strand or strands, using the free end to rotate one strand about the other, and rejoining the broken ends. When an enzyme performs such an action, it must change the linking number by an integer; this value can be determined as a characteristic of the reaction. Then we can consider the effects of this change in terms of ΔW and ΔT.

DNA topoisomerases catalyze conversions of this type. Some topoisomerases can relax (remove) only negative supercoils from DNA; others can relax both negative and positive supercoils. Some can introduce negative supercoils.

Topoisomerases are divided into two classes, according to the nature of the mechanisms they employ. **Type I** enzymes act by making a transient break in one strand of DNA. **Type II** enzymes act by introducing a transient double-strand break. As well as those enzymes that function as general topoisomerases with DNA irrespective of sequence, enzymes involved in site-specific recombination reactions fit the definition of topoisomerases (see later).

The best characterized type I topoisomerase is the product of the *topA* gene of *E. coli*, which relaxes highly negatively supercoiled DNA. The enzyme does not act on positively supercoiled DNA. Mutations in it cause an increase in the level of supercoiling in the nucleoid (and may affect transcription, as described in Chapter 9).

In addition to the relaxation of negative supercoils in duplex DNA, the enzyme interacts with single-stranded DNA. It may like negative supercoils because they tend to stabilize single-stranded regions, which could provide the substrate bound by the enzyme.

When *E. coli* topoisomerase I binds to DNA, it forms a stable complex in which one strand of the DNA has been nicked and its 5′-phosphate end is covalently linked to a tyrosine residue in the enzyme. This suggests a mechanism for the action of the enzyme; it transfers a phosphodiester bond in DNA to the protein, manipulates the structure of the two DNA strands, and then rejoins the bond in the original strand.

Eukaryotic type I topoisomerases have no sequence or structural similarity with the prokaryotic enzymes. They form a covalent intermediate with the 3′ end of the broken strand, and can relax positive as well as negative supercoils.

A model for the action of topoisomerase I is illustrated in **Figure 14.16**. The enzyme binds to a region in which duplex DNA becomes separated into its single strands; then it breaks one strand, pulls the other strand through the gap, and finally seals the gap. The transfer of bonds from nucleic acid to protein explains how the enzyme can function without requiring any input of energy. There has been no irreversible hydrolysis of bonds; their energy has been conserved through the transfer reactions.

The reaction changes the linking number in steps of one. Each time one strand is passed through the break in the other, there is a ΔL of +1. The figure illustrates the enzyme activity in terms of moving the individual strands. In a free supercoiled molecule, the interchangeability of W and T should let the change in linking number be taken up by a change of $\Delta W = +1$, that is, by one less turn of negative supercoiling.

The reaction is equivalent to the rotation illustrated in bottom part of Figure 14.15, with the restriction that the enzyme limits the reaction to a single-strand passage per event. (By contrast, the introduction of a nick in a supercoiled molecule allows free strand rotation to relieve all the tension by multiple rotations.)

The type I topoisomerase also can pass one segment of a single-stranded DNA through another. This **single-strand passage** reaction can introduce **knots** in DNA and can **catenate** two circular molecules so that

Figure 14.16 Bacterial type I topoisomerases recognize partially unwound segments of DNA and pass one strand through a break made in the other.

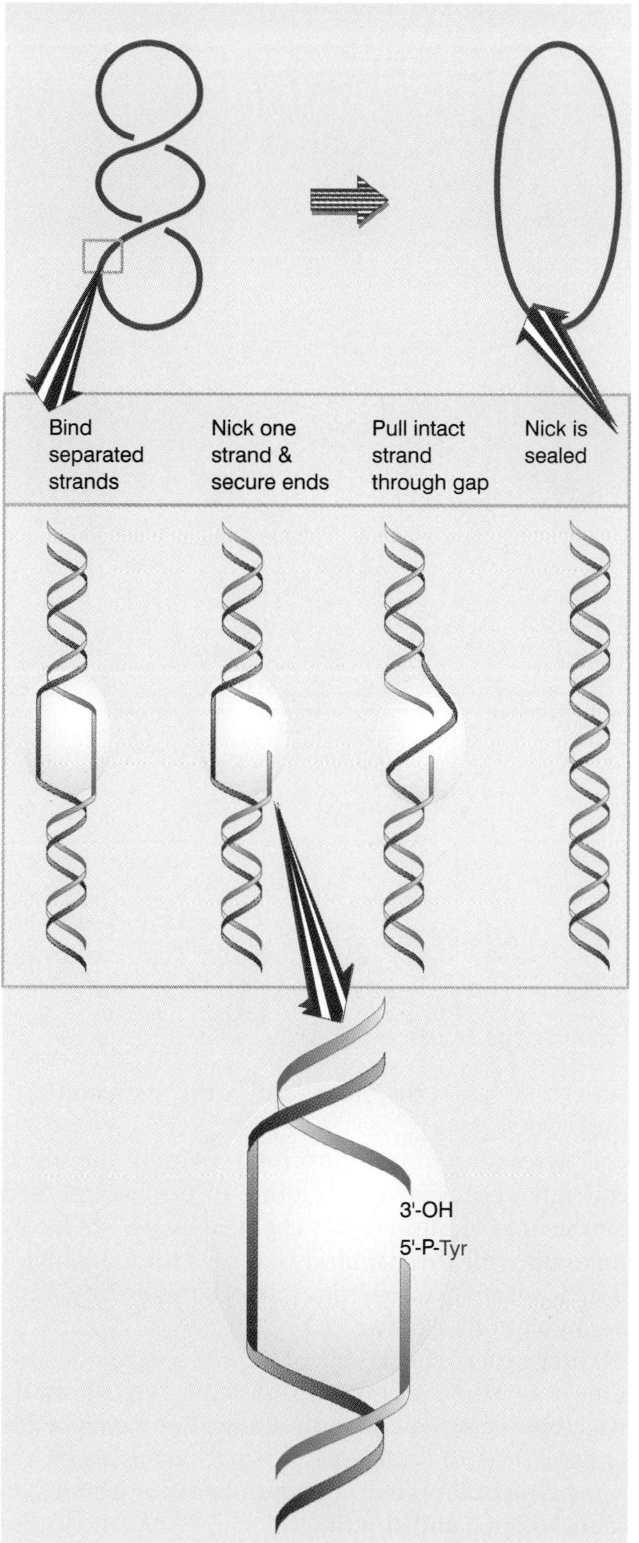
Bind
separated
strands
Nick one
strand &
secure ends
Pull intact
strand
through gap
Nick is
sealed
3'-OH
5'-P-Tyr

they are connected like links on a chain. We do not understand the uses (if any) to which these reactions are put *in vivo*.

Type II topoisomerases generally relax both negative and positive supercoils. The reaction requires ATP; probably one ATP is hydrolyzed for each catalytic event. As illustrated in **Figure 14.17**, the reaction is mediated by making a double-stranded break in one DNA duplex, and passing another duplex region through it.

A formal consequence of two-strand transfer is that the linking number is always changed in multiples of two. The topoisomerase II activity can be used also to introduce or resolve catenated duplex circles and knotted molecules.

The reaction probably represents a nonspecific recognition of duplex DNA in which the enzyme binds any two double-stranded segments that cross each other. The hydrolysis of ATP may be used to drive the enzyme through conformational changes that provide the force needed to push one DNA duplex through the break made in the other. Because of the topology of supercoiled DNA, the relationship of the crossing segments allows supercoils to be removed from either positively or negatively supercoiled circles.

Bacterial DNA gyrase is a topoisomerase of type II that is able to *introduce* negative supercoils into a relaxed closed circular molecule. DNA gyrase binds to a circular DNA duplex and supercoils it processively and catalytically: it continues to introduce supercoils into the same DNA molecule. One molecule of DNA gyrase can introduce ~100 supercoils per minute.

The supercoiled form of DNA has a higher free energy than the relaxed form, and the energy needed to accomplish the conversion is supplied by the hydrolysis of ATP. In the absence of ATP, the gyrase can *relax* negative but not positive supercoils, although the rate is more than 10 times slower than the rate of introducing supercoils.

The *E. coli* DNA gyrase is a tetramer consisting of two types of subunit, each of which is a target for antibiotics (the most often used being nalidixic acid which acts on GyrA, and novobiocin which acts on GyrB). The drugs inhibit replication, which suggests that DNA gyrase is necessary for DNA synthesis to proceed. Mutations that confer resistance to the antibiotics identify the loci that code for the subunits.

Gyrase binds its DNA substrate around the outside of the protein tetramer. Gyrase protects ~140 bp of DNA from digestion by micrococcal nuclease. The **sign inversion** model for gyrase action is illustrated in **Figure 14.18.** The enzyme binds the DNA in a crossover configuration that is equivalent to a positive

Figure 14.17 Type II topoisomerases can pass a duplex DNA through a double-strand break in another duplex.

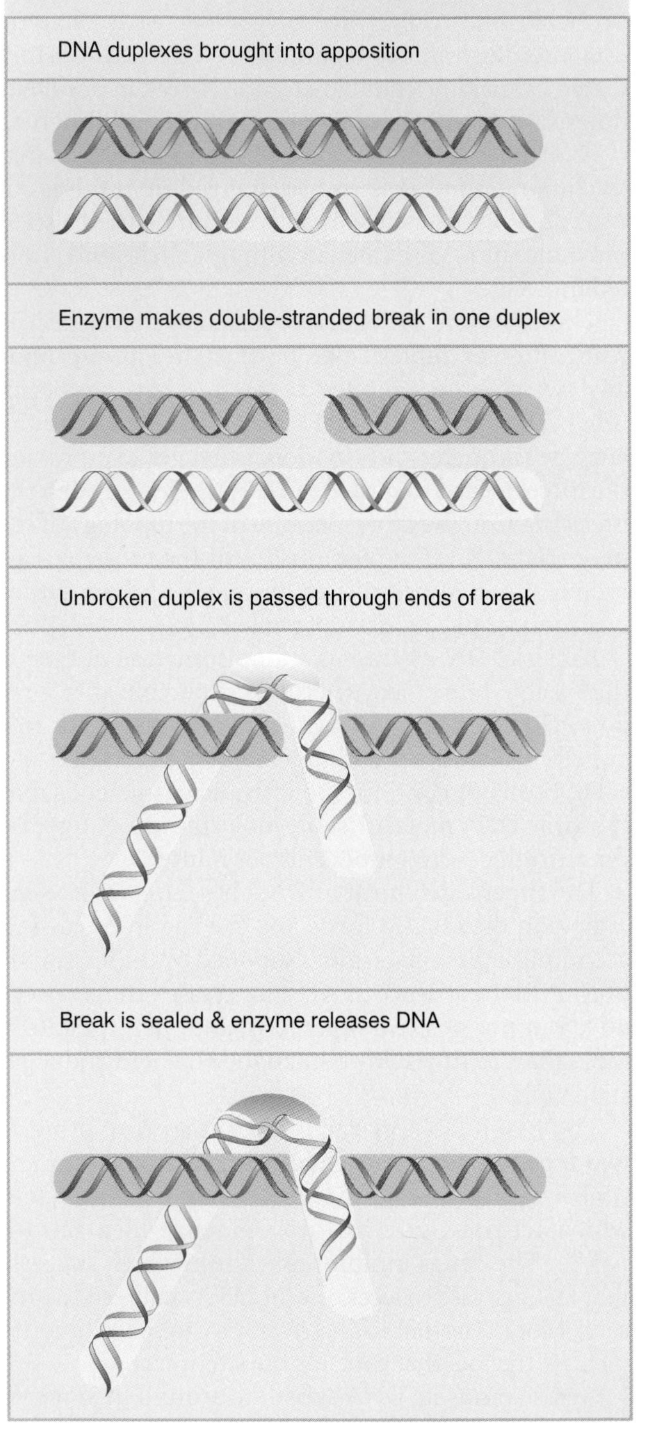
DNA duplexes brought into apposition
Enzyme makes double-stranded break in one duplex
Unbroken duplex is passed through ends of break
Break is sealed & enzyme releases DNA

Figure 14.18 DNA gyrase may introduce negative supercoils in duplex DNA by inverting a positive supercoil.

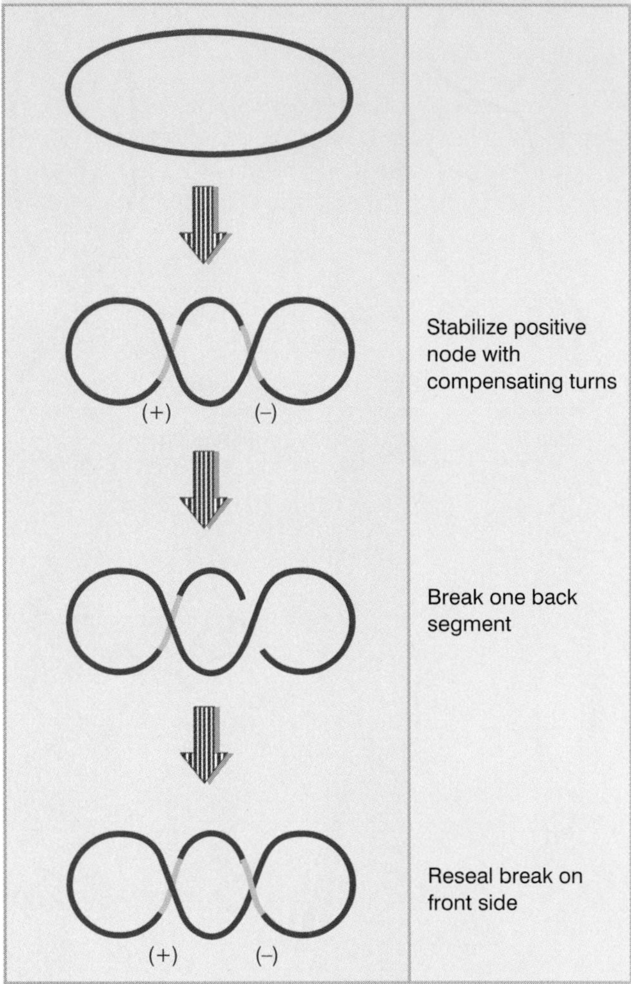
Stabilize positive node with compensating turns
(+)
(–)
Break one back segment
Reseal break on front side
(+)
(–)

supercoil. This induces a compensating negative supercoil in the unbound DNA. Then the enzyme breaks the double strand at the crossover of the positive supercoil, passes the other duplex through, and seals the break.

The reaction directly inverts the sign of the supercoil: it has been converted from a +1 turn to a –1 turn. So the linking number has changed by $\Delta L = -2$, conforming with the demand that all events involving double-strand passage must change the linking number by a multiple of two.

Gyrase then releases one of the crossing segments of the (now negative) bound supercoil; this allows the negative turns to redistribute along DNA (as change in either T or W or both), and the cycle begins again. The same type of topological manipulation is responsible for catenation and knotting.

On releasing the inverted supercoil, the conformation of gyrase changes. For the enzyme to undertake another cycle of supercoiling, its original conformation must be restored. This process is called **enzyme turnover**. It is thought to be driven by the hydrolysis of ATP, since the replacement of ATP by an analog that cannot be hydrolyzed allows gyrase to introduce only one inversion (–2 supercoils) per substrate. So it does not need ATP for the supercoiling reaction, but does need it to undertake a second cycle. Novobiocin interferes with the ATP-dependent reactions of gyrase, by preventing ATP from binding to the B subunit.

The (ATP-independent) relaxation reaction is inhibited by nalidixic acid. This implicates the A subunit in the breakage and reunion reaction. Treating gyrase with nalidixic acid allows DNA to be recovered in the form of fragments generated by a staggered cleavage across the duplex. The termini all possess a free 3′-OH group and a four-base 5′ single-strand extension covalently linked to the A subunit. The covalent linkage retains the energy of the phosphate bond; this can be used to drive the sealing reaction, explaining why gyrase can undertake relaxation without ATP. The sites of cleavage are fairly specific, occurring about once every 100 bp.

Specialized recombination involves breakage and reunion at specific sites

THE conversion of lambda DNA between its different life forms involves two types of event. The pattern of gene expression is regulated as described in Chapter 11. And the physical condition of the DNA is different in the lysogenic and lytic states:

- In the lytic lifestyle, lambda DNA exists as an independent, circular molecule in the infected bacterium.
- In the lysogenic state, the phage DNA is an integral part of the bacterial chromosome (called prophage).

Transition between these states involves site-specific recombination:

- To enter the lysogenic condition, free lambda DNA must be **integrated** into the host DNA.
- To be released from lysogeny into the lytic cycle, prophage DNA must be **excised** from the chromosome.

Integration and excision occur by recombination at specific loci on the bacterial and phage DNAs called **attachment (att) sites.** The attachment site on the bacterial chromosome is called att^{λ} in bacterial genetics. The locus is defined by mutations that prevent integration of lambda; it is occupied by prophage λ in lysogenic strains. When the att^{λ} site is deleted from the *E. coli* chromosome, an infecting lambda phage can establish lysogeny by integrating elsewhere, although the efficiency of the reaction is less than 0.1% of the frequency of integration at att^{λ}. This inefficient integration occurs at **secondary attachment sites,** which resemble the authentic *att* sequences.

For describing the integration/excision reactions, the bacterial attachment site (att^{λ}) is called *attB*, consisting of the sequence components *BOB′*. The attachment site on the phage, *attP*, consists of the components *POP′*. **Figure 14.19** outlines the recombination reaction between these sites. The sequence *O* is common to *attB* and *attP*. It is called the **core** sequence; and the recombination event occurs within it. The flanking regions *B*, *B′* and *P*, *P′* are referred to as the **arms**; each is distinct in sequence. Because the phage DNA is circular, the recombination event inserts it into the bacterial chromosome as a linear sequence. The prophage is bounded by two new *att* sites, the products of the recombination, called *attL* and *attR*.

An important consequence of the constitution of the *att* sites is that the integration and excision reactions do not involve the same pair of reacting sequences. Integration requires recognition between *attP* and *attB*; while excision requires recognition between *attL* and *attR*. The directional character of site-specific recombination is thus controlled by the identity of the recombining sites.

Although the recombination event is reversible, different conditions prevail for each direction of the reaction. This is an important feature in the life of the phage, since it offers a means to ensure that an integration event is not immediately reversed by an excision, and *vice versa.*

The difference in the pairs of sites reacting at integration and excision is reflected by a difference in the proteins that mediate the two reactions:

- Integration (*attB* × *attP*) requires the product of the phage gene *int* and a bacterial protein called integration host factor (IHF).
- Excision (*attL* × *attR*) requires the product of phage gene *xis*, in addition to Int and IHF.

So Int and IHF are required for *both* reactions. Xis plays an important role in controlling the direction; it is required for excision, but inhibits integration.

IHF is a 20 kD protein of two different subunits, coded by the genes *himA* and *himD.* IHF is not an essential protein in *E. coli*, and is not required for homologous bacterial recombination. It is one of several proteins with the ability to wrap DNA on a surface. Mutations in the *him* genes prevent lambda site-specific recombination, and can be suppressed by mutations in λ *int*, which suggests that IHF and Int interact.

Site-specific recombination can be performed *in vitro* by Int and IHF. It involves a precise breakage and reunion in the absence of any synthesis of DNA. The roles of the *att* sites can be investigated by making deletions on either side. It turns out that *attP* is much larger than *attB*. The function of *attP* requires a stretch of 240 bp, but the function of *attB* can be exercised by the 23 bp fragment extending from −11 to +11, in which there are only 4 bp on either side of the core. The disparity in their sizes suggests that *attP* and *attB* play different roles in the recombination, with *attP* providing additional information necessary to distinguish it from *attB*.

Does the reaction proceed by a concerted mechanism in which the strands in *attP* and *attB* are cut simultaneously and exchanged? Or are the strands exchanged one pair at a time, the first exchange generating a Holliday junction, the second cycle of nicking and ligation occurring to release the structure? The alternatives are depicted in **Figure 14.20**.

Figure 14.19 Circular phage DNA is converted to an integrated prophage by a reciprocal recombination between *attP* and *attB*; the prophage is excised by reciprocal recombination between *attL* and *attR*.

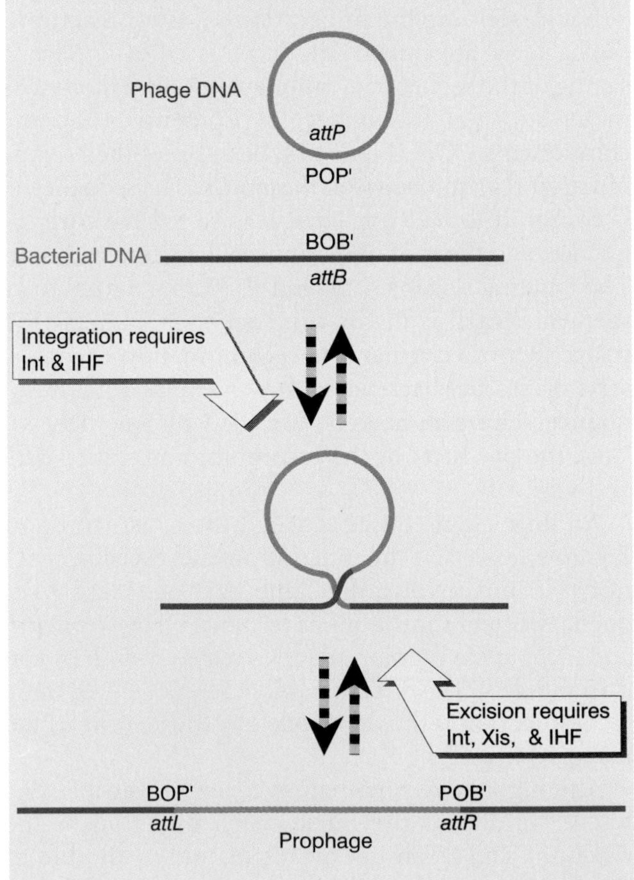

Figure 14.20 Does recombination between *attP* and *attB* proceed by sequential exchange or concerted cutting?

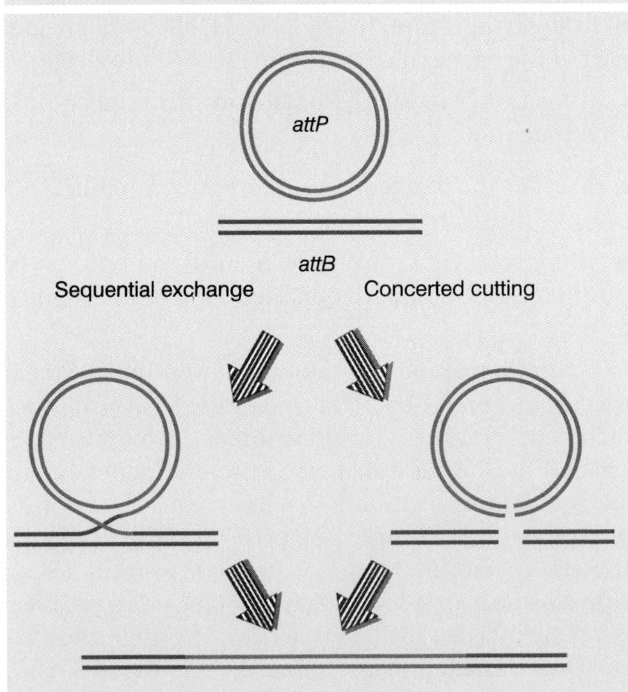

The recombination reaction has been halted at intermediate stages by the use of "suicide substrates", in which the core sequence is nicked. The presence of the nick interferes with the recombination process. This makes it possible to identify molecules in which recombination has commenced but has not been completed. The structures of these intermediates suggest that exchanges of single strands take place sequentially. Int protein can resolve Holliday junctions, and is probably responsible for the cutting and ligation reactions.

The model illustrated in **Figure 14.21** shows that if *attP* and *attB* sites each suffer the same staggered cleavage, complementary single-stranded ends could be available for crosswise hybridization. The distance between the lambda crossover points is 7 bp, and the reaction generates 3′-phosphate and 5′-OH ends. The reaction is shown for simplicity as generating overlapping single-stranded ends that anneal, but actually occurs by a process akin to the recombination event of Figure 14.2. The corresponding strands on each duplex are cut at the same position, the free 3′ ends exchange between duplexes, the branch migrates for a distance of 7 bp along the region of homology, and then the structure is resolved by cutting the other pair of corresponding strands.

The *in vitro* reaction requires supercoiling in *attP*, but not in *attB*. When the reaction is performed *in vitro* between two supercoiled DNA molecules, almost all of the supercoiling is retained by the products. So there cannot be any free intermediates in which strand rotation could occur. This is consistent with the idea that the reaction proceeds through a Holliday junction. The

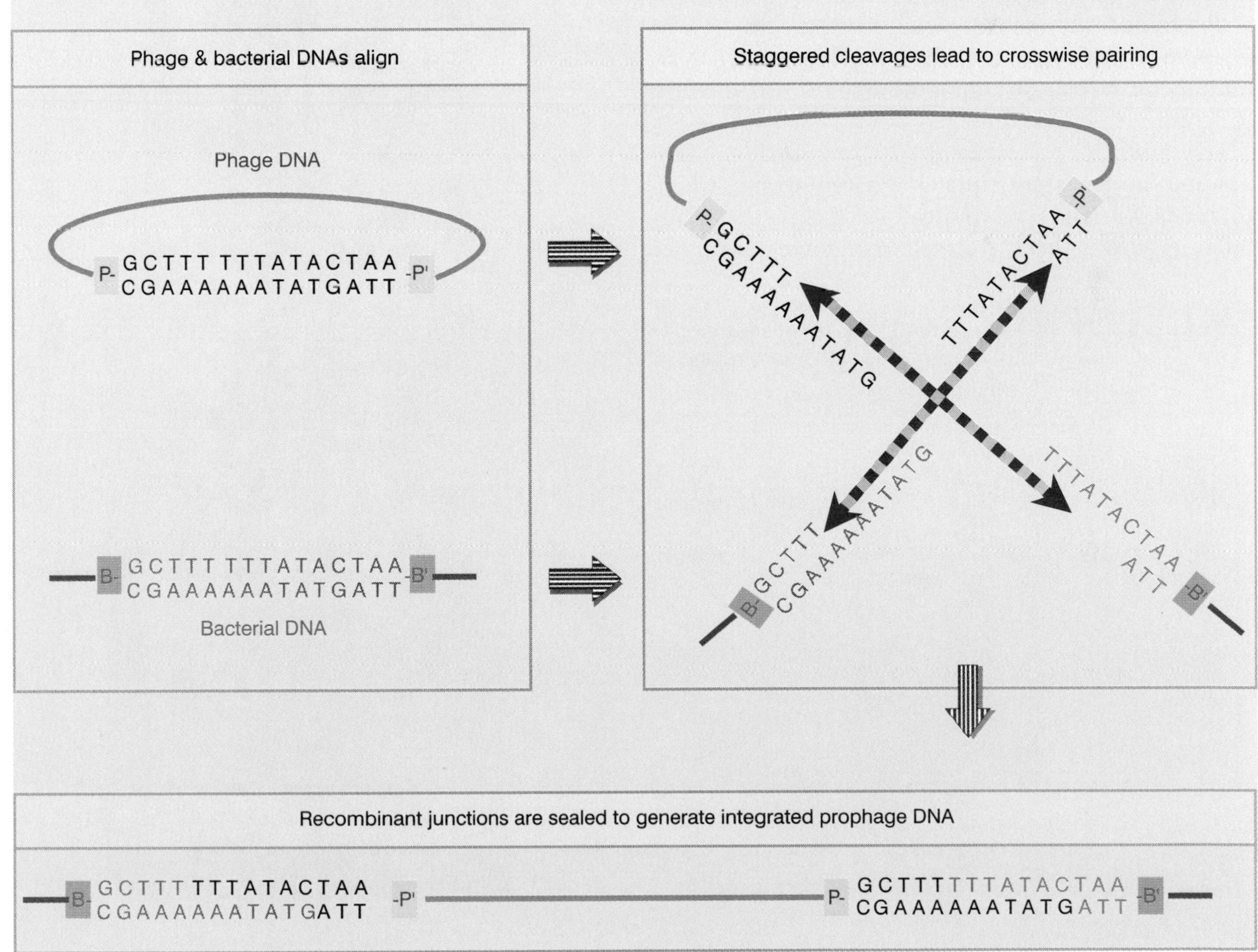

Figure 14.21 Staggered cleavages in the common core sequence of *attP* and *attB* allow crosswise reunion to generate reciprocal recombinant junctions.

breakage and reunion reaction resembles the activity of topoisomerase I, except that nicked strands from different duplexes are sealed together, instead of the ends of a broken strand from one duplex. (Int indeed has a [rather ineffectual] topoisomerase I ability to relax negatively supercoiled DNA.)

Large amounts of the Int and IHF proteins are needed for recombination *in vitro*. Int and IHF bind cooperatively to *attP*, and their affinity for the site is enhanced by supercoiling. The high stoichiometry suggests that the proteins do not function catalytically, but form a structure that supports only a single recombination event.

The proteins involved in site-specific recombination bind to specific sites in the *att* region. Int has two different modes of binding. It binds to inverted sites at the core sequence, positioning itself to make the cuts on each strand illustrated in **Figure 14.22**. These sites share a consensus sequence. It also binds to sites in the arms of *attP* that have a different consensus sequence. Different domains of Int recognize each type of sequence: an N-terminal domain recognizes the arms of *attP*, while a C-terminal domain recognizes the cores of *attP* and *attB*. The two domains probably bind DNA simultaneously, thus bringing the arms of *attP* close to the core.

IHF binds to sequences of ~20 bp in *attP*; the IHF binding sites are approximately adjacent to sites where Int binds. Xis binds to two sites located close to one another in *attP*, so that the protected region extends over 30–40 bp. Together, Int, Xis, and IHF cover virtually all of *attP*. The binding of Xis changes the organization of the DNA so that it becomes inert as a substrate for the integration reaction.

Figure 14.22 Int and IHF bind to different sites in *attP*. The Int recognition sequences in the core region include the sites of cutting.

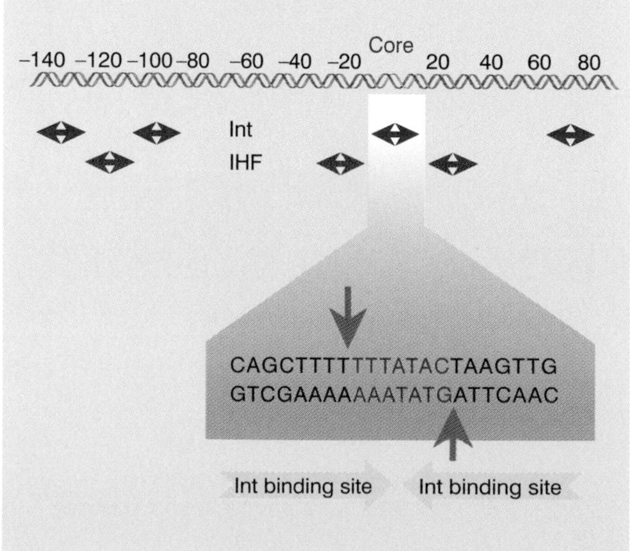

Figure 14.23 The Int binding sites in the core lie on one face of DNA. The large circles indicate positions at which methylation is influenced by Int binding; the large arrows indicate the sites of cutting. Photograph kindly provided by A. Landy.

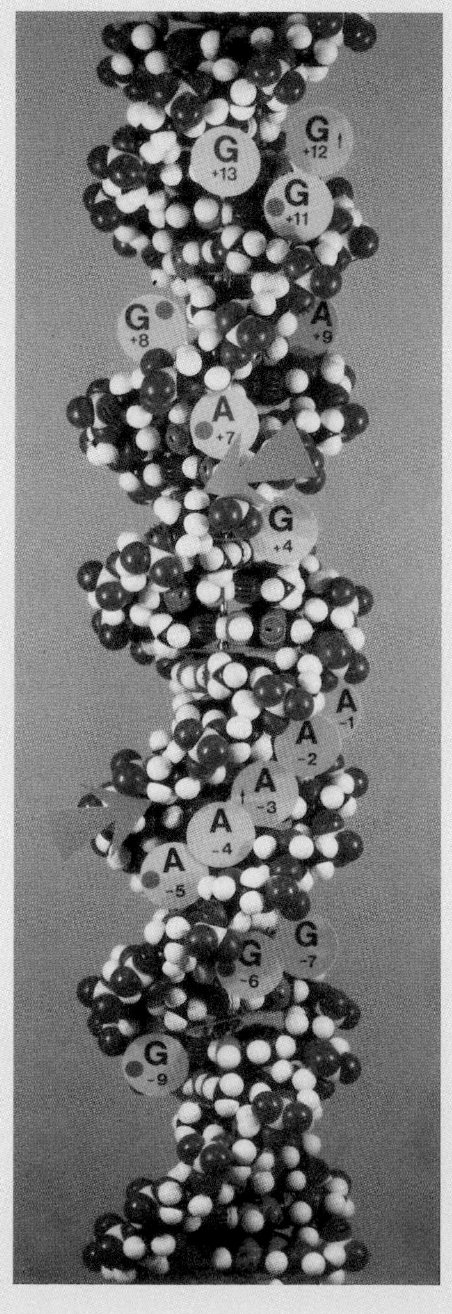

Figure 14.23 shows that when the core locations bound by Int are mapped on the double helix, virtually all of the contacts lie on one face of the DNA. The two sites of cutting are exposed in the major groove. IHF binding sites lie on the same face; and if the spacing between the Int-binding site and the flanking IHF-binding sites is altered so that it is no longer an integral number of helical turns, then integration is impeded.

When Int and IHF bind to *attP*, they generate a complex in which all the binding sites are pulled together on the surface of a protein. Supercoiling of *attP* is needed for the formation of this **intasome**.

The only binding sites in *attB* are the two Int sites in the core. But Int does not bind directly to *attB* in the form of free DNA. The intasome is the intermediate that "captures" *attB*. Probably Int molecules that are part of the intasome bind to the sites in the core of *attB*, as indicated schematically in **Figure 14.24**.

According to this model, the initial recognition between *attP* and *attB* does not depend directly on DNA homology, but instead is determined by the ability of Int proteins to recognize both *att* sequences. The two *att* sites then are brought together in an orientation predetermined by the structure of the intasome. Sequence homology becomes important at this stage, when it is required for the strand exchange reaction.

The asymmetry of the integration and excision reactions is shown by the fact that Int can form a similar complex with *attR* only if Xis is added. This complex can pair with a condensed complex that Int forms at *attL*. IHF is not needed for this reaction.

Much of the complexity of site-specific recombination may be caused by the need to regulate the reaction so that integration occurs preferentially when the virus is entering the lysogenic state, while excision is preferred when the prophage is entering the lytic cycle. By controlling the amounts of Int and Xis, the appropriate reaction will occur.

Figure 14.24 Multiple copies of Int protein may organize *attP* into an intasome, which initiates site-specific recombination by recognizing *attB* on free DNA.

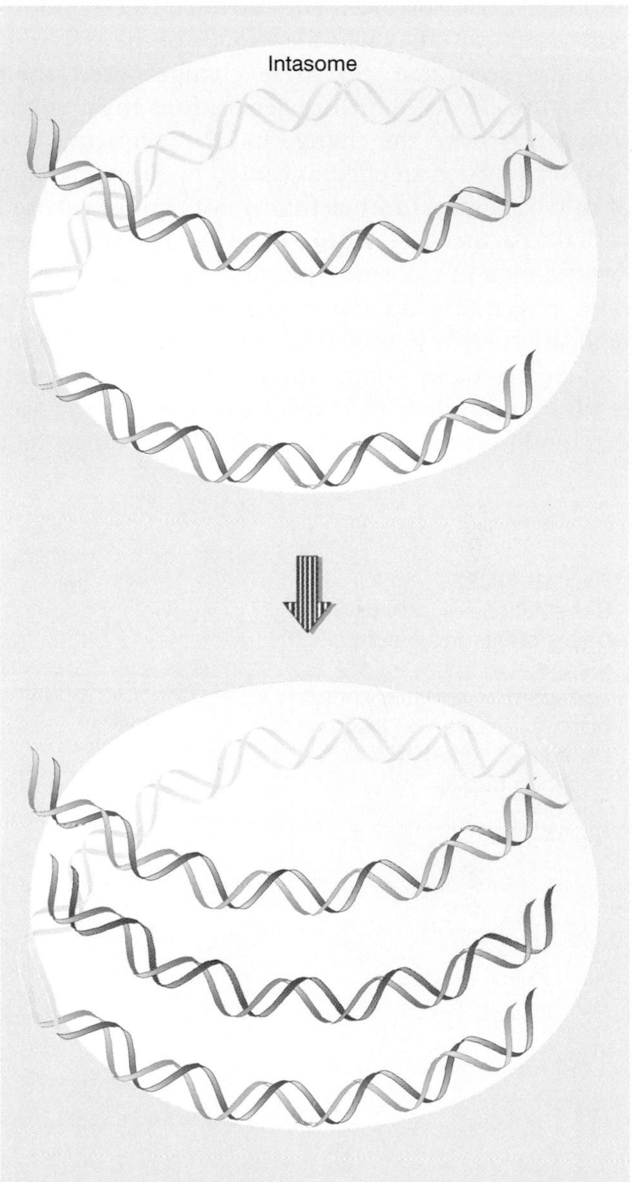

Repair systems correct damage to DNA

INJURY to DNA is minimized by systems that recognize and correct the damage. The repair systems are likely to be as complex as the replication apparatus itself, which indicates their importance for the survival of the cell. The measured rate of mutation reflects a balance between the number of damaging events occurring in DNA and the number that have been corrected (or miscorrected).

"Damage" to DNA consists of any change introducing a deviation from the usual double-helical structure. We can divide such changes into two general classes:

■ **Single base changes** affect the sequence but not the overall structure of DNA. They do not affect transcription or replication, when the strands of the DNA duplex are separated. So these changes exert their damaging effects on future generations through the consequences of the change in DNA sequence (see Chapter 1). Such an effect is caused by the conversion of one base into another that is not properly paired with the partner base. **Figure 14.25** gives two examples: deamination of cytosine (spontaneously or by chemical mutagen) creates a mismatched U•G pair; while a replication error that inserts adenine instead of cytosine creates an A•G pair. Similar consequences could result from covalent addition of a small group to a base that modifies its ability to base pair. These changes may result in very minor structural distortion (as in the case of a U•G pair) or quite significant change (as in the case of an A•G pair), but the common feature is that the mismatch persists only until the next replication.

■ **Structural distortions** may provide a physical impediment to replication or transcription. **Figure 14.26** shows some examples. Introduction of covalent links between bases on one strand of DNA or between bases on opposite strands inhibits replication and transcription. A well studied example of a structural distortion is caused by ultraviolet irradiation, which introduces covalent bonds between two adjacent thymine bases, giving the intrastrand **pyrimidine dimer** drawn in the figure. Similar consequences could result from addition of a bulky adduct to a base that distorts the structure of the double helix. A single-strand nick or the removal of a base prevents a strand from serving as a proper template for synthesis of RNA or DNA. The

Figure 14.25
Substitutions of individual bases create mismatched pairs that may be corrected by replacing one base; if uncorrected they cause a mutation in one daughter duplex.

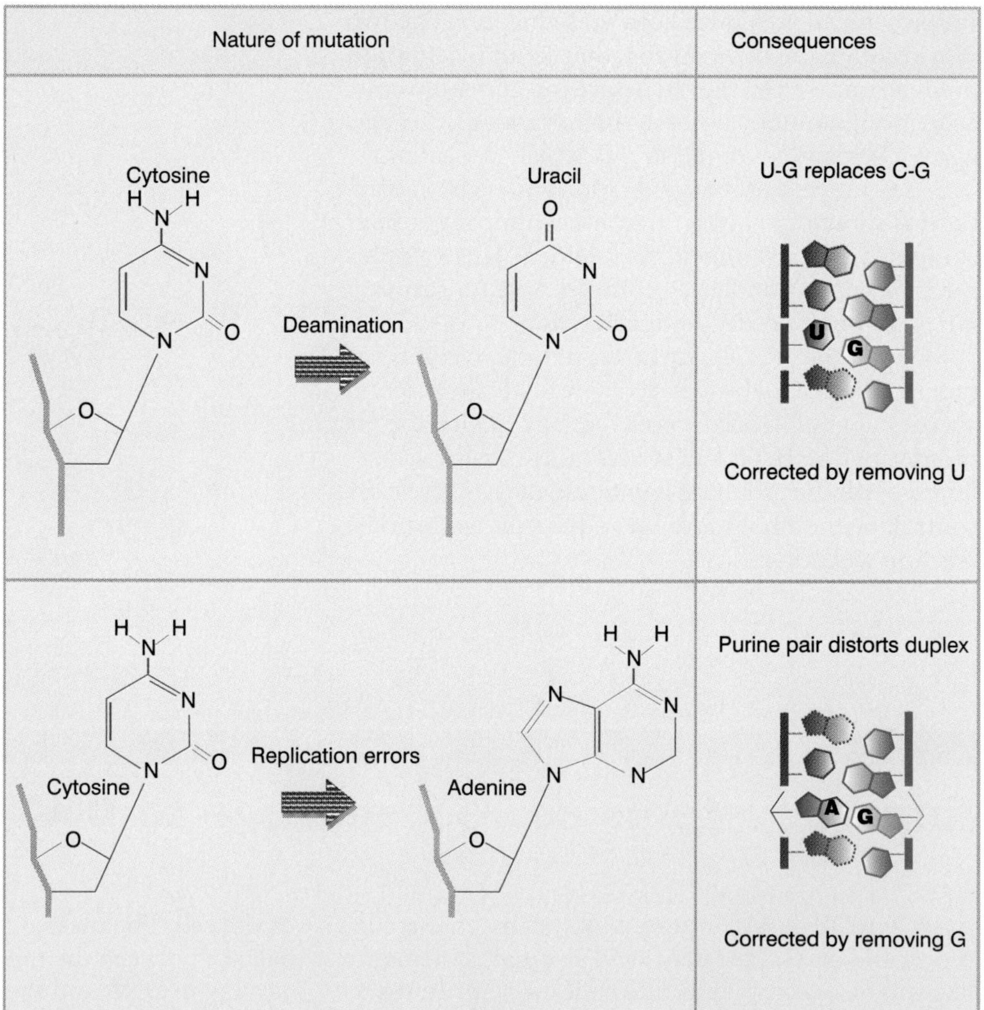

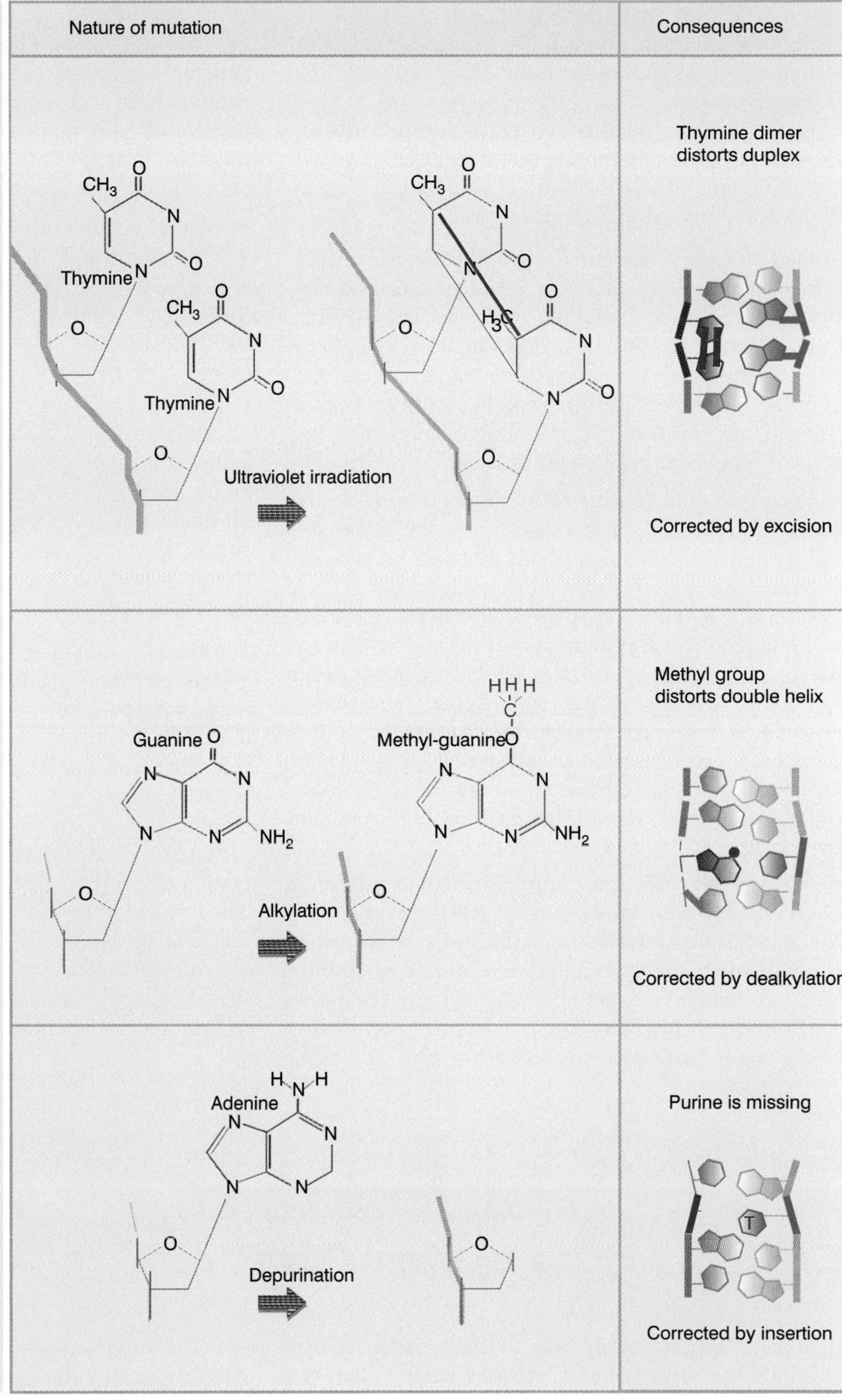

Figure 14.26 Modifications or removal of bases may cause structural defects that prevent replication or induce mutations in each replication cycle until they are removed.

common feature in all these changes is that the damaged adduct remains in the DNA, continuing to cause structural problems and/or induce mutations, until it is removed.

Repair systems can often recognize a range of distortions in DNA as signals for action, and a cell may have several systems able to deal with DNA damage. We may divide them into several general types:

- **Direct repair** is rare and involves the reversal or simple removal of the damage. **Photoreactivation** of pyrimidine dimers, in which the offending covalent bonds are reversed by a light-dependent enzyme, is the best example. This system is widespread in nature, and appears to be especially important in plants. In *E. coli* it depends on the product of a single gene (*phr*) that codes for an enzyme called photolyase.
- **Excision repair** is initiated by a recognition enzyme that sees an actual damaged base or a change in the spatial path of DNA. Recognition is followed by excision of a sequence that includes the damaged bases; then a new stretch of DNA is synthesized to replace the excised material. Such systems are common; some recognize general damage to DNA, while others act upon specific types of base damage (glycosylases remove specific altered bases; AP endonucleases remove residues from sites at which purine bases have been lost). There are often multiple excision repair systems in a single cell type, and they probably handle most of the damage that occurs.
- **Mismatch repair** is accomplished by scrutinizing DNA for apposed bases that do not pair properly. Mismatches that arise during replication are corrected by distinguishing between the "new" and "old" strands and preferentially correcting the sequence of the newly synthesized strand. Mismatches also occur when hybrid DNA is created during recombination, and their correction upsets the ratio of parental alleles (see Figure 14.14). Other systems deal with mismatches generated by base conversions, such as the result of deamination. The importance of these systems is emphasized by the fact that cancer is caused in human populations by mutation of genes related to those involved in mismatch repair in yeast.
- **Tolerance systems** cope with the difficulties that arise when normal replication is blocked at a damaged site. They provide a means for a damaged template sequence to be copied, probably with a relatively high frequency of errors. They are especially important in higher eukaryotic cells.
- **Retrieval systems** comprise another type of tolerance system. When damage remains in a daughter molecule, and replication has been forced to bypass the site, a retrieval system uses recombination to obtain another copy of the sequence from an undamaged source. These "recombination-repair" systems are well characterized in bacteria; it is not clear how important they are elsewhere.

Mutations that affect the ability of *E. coli* cells to engage in DNA repair fall into groups which correspond to several repair pathways (not necessarily all independent). The major known pathways are the *uvr* excision repair system, the *dam* replication mismatch-repair system, and the *recB* and *recF* recombination and recombination-repair pathways.

When the repair systems are eliminated, cells become *exceedingly* sensitive to ultraviolet irradiation. The introduction of UV-induced damage has been a major test for repair systems, and so in assessing their activities and relative efficiencies, we should remember that the emphasis might be different if another damaged adduct were studied.

Excision repair systems in *E. coli*

Excision repair systems vary in their specificity, but share the same general features. Each system removes mispaired or damaged bases from DNA and then synthesizes a new stretch of DNA to replace them. The main type of pathway for excision repair is illustrated in **Figure 14.27.**

In the **incision** step, the damaged structure is recognized by an endonuclease that cleaves the DNA strand on both sides of the damage.

In the **excision** step, a 5′–3′ exonuclease removes a stretch of the damaged strand.

In the **synthesis** step, the resulting single-stranded

Figure 14.27 Excision-repair removes and replaces a stretch of DNA that includes the damaged base(s).

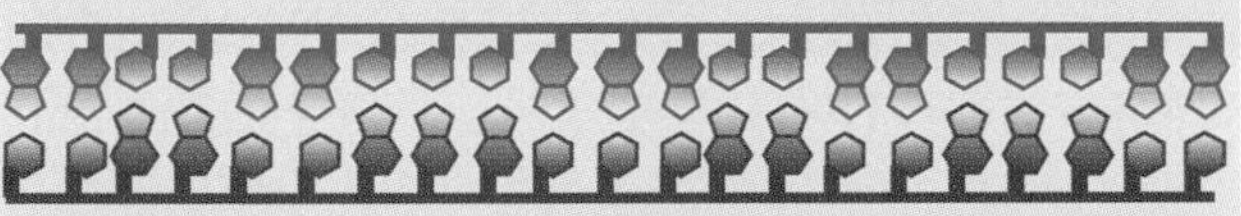

Damage
Mutant base is mismatched and/or distorts structure

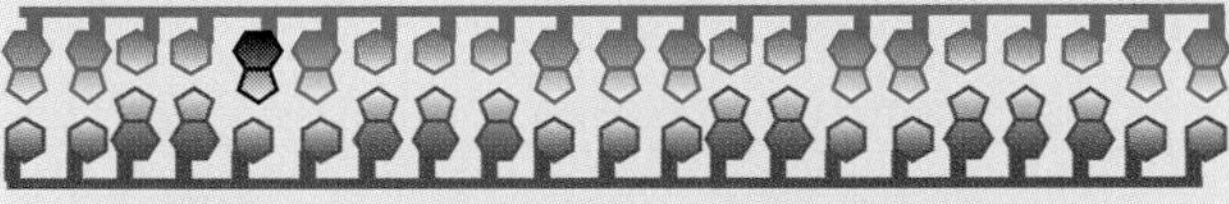

Incision
Endonuclease cleaves on both sides of damaged base

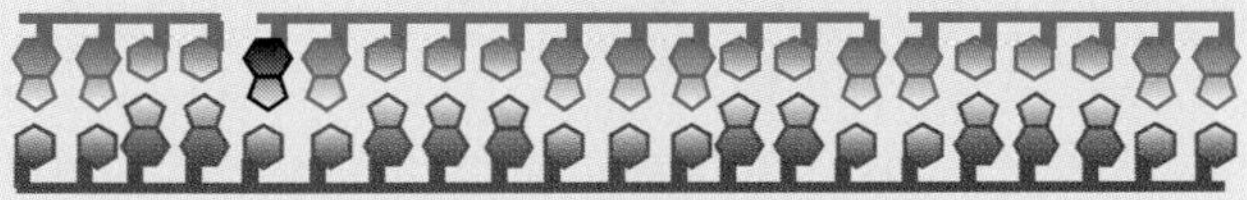

Excision
Exonuclease removes DNA between nicks

Synthesis
Polymerase synthesizes replacement DNA

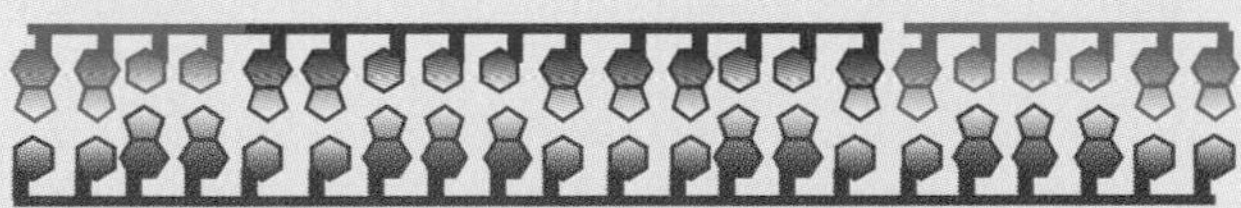

Ligase seals nick

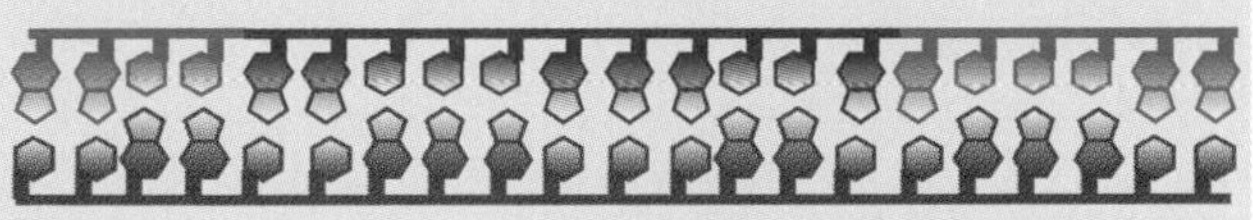

region serves as a template for a DNA polymerase to synthesize a replacement for the excised sequence. Finally, DNA ligase covalently links the 3′ end of the new material to the old material.

Different excision repair modes are identified by the heterogeneity of the lengths of the segments of repaired DNA. These pathways are described as **very short patch repair (VSP)**, **short-patch repair**, and **long-patch repair**. The VSP system deals with mismatches between specific bases (see later). The latter two excision repair systems both involve the *uvr* genes.

The *uvr* system of excision repair includes three genes, *uvrA,B,C*, that code for the components of a repair endonuclease. It functions in the stages indicated in **Figure 14.28**. First, a UvrAB combination recognizes pyrimidine dimers and other bulky lesions. Then UvrA dissociates (this requires ATP), and UvrC joins UvrB. The UvrBC combination makes an incision on each side, one seven nucleotides from the 5′ side of the damaged site, and the other three to four nucleotides away

Figure 14.28 The Uvr system operates in stages in which UvrAB recognizes damage, UvrBC nicks the DNA, and UvrD unwinds the marked region.

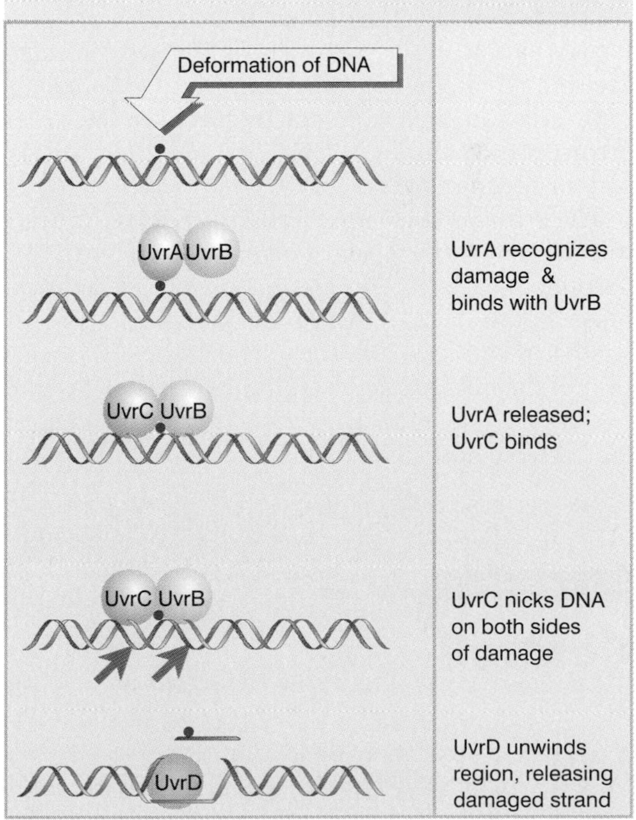

from the 3′ side. This also requires ATP. UvrD is a helicase that helps to unwind the DNA to allow release of the single strand between the two cuts. The enzyme that excises the damaged strand is probably DNA polymerase I.

The average length of excised DNA is ~12 nucleotides, which gives rise to the description of this mode as the short-patch repair. The enzyme involved in the repair synthesis probably also is DNA polymerase I (although DNA polymerases II and III can substitute for it).

For bulky lesions, short-patch repair accounts for 99% of the excision repair events. The remaining 1% involve the replacement of stretches of DNA mostly ~1500 nucleotides long, but extending up to >9000 nucleotides. This mode also requires the *uvr* genes and involves DNA polymerase I. A difference between the two modes of repair is that short-patch repair is a constitutive function of the bacterial cell, but long-patch repair must be induced by damage (see later). Long-patch repair probably acts on lesions found in regions near replication forks. We have not yet characterized the differences between these modes in terms of the involvement of different gene products.

The existence of repair systems that engage in DNA synthesis raises the question of whether their quality control is comparable with that of DNA replication. So far as we know, most systems, including *uvr*-controlled excision repair, do not differ significantly from DNA replication in the frequency of mistakes. However, **error-prone** synthesis of DNA occurs in *E. coli* under certain circumstances.

The error-prone feature was first observed when it was found that the repair of damaged λ phage DNA is accompanied by the induction of mutations if the phage is introduced into cells that had previously been irradiated with UV. This suggests that the UV irradiation of the host has activated functions that generate mutations. The mutagenic response also operates on the bacterial host DNA.

What is the actual error-prone activity? Current thinking focuses on the idea that it is caused by a component of a tolerance pathway that permits or compels replication to proceed past the site of damage. When the replicase passes any site at which it cannot insert complementary base pairs in the daughter strand, it inserts incorrect bases, which represent mutations. The error-prone activity requires DNA polymerase III, the usual replicase, which is consistent with the idea that the relevant function acts in concert with the normal replication apparatus.

Several functions are involved in this error-prone pathway. Mutations in the genes *umuD* and *umuC* abolish UV-induced mutagenesis, but do not interfere with any known enzymatic functions. The genes constitute the *umuDC* operon, whose expression is induced by DNA damage (see later). Some plasmids carry genes called *mucA* and *mucB*, which are homologs of *umuD* and *umuC*, and whose introduction into a bacterium increases resistance to UV killing and susceptibility to mutagenesis.

Genes whose products are involved in controlling the fidelity of DNA synthesis during either replication or repair may be identified by mutations that have a **mutator** phenotype. A mutator mutant has an increased frequency of spontaneous mutation. If identified originally by the mutator phenotype, a gene is described as *mut*; but often a *mut* gene is later found to be equivalent with a known replication or repair activity.

The general types of activities identified by *mut* genes fall into groups. The major group consists of components of mismatch-repair systems. They include the *dam* methylase that identifies the target for repair (see next section) and enzymes that participate directly or indirectly in the removal of particular types of damage (*mutH,U,S,L,Y*); failure to remove a damaged or mispaired base before replication allows it to induce a mutation. A smaller group, typified by *dnaQ*, is concerned with the accuracy of synthesizing new DNA.

Controlling the direction of mismatch repair

When a structural distortion is removed from DNA, the wild-type sequence is restored. In most cases, the distortion is due to the creation of a base that is not naturally found in DNA, and which is therefore recognized and removed by the repair system.

A problem arises if the target for repair is a mispaired partnership of (normal) bases created when one was mutated. The repair system has no intrinsic means of knowing which is the wild-type base and which is the mutant! All it sees are two improperly paired bases, either of which can provide the target for excision repair.

If the mutated base is excised, the wild-type sequence is restored. But if it happens to be the original (wild-type) base that is excised, the new (mutant) sequence becomes fixed. Often, however, the direction of excision repair is not random, but is biased in a way that is likely to lead to restoration of the wild-type sequence.

Some precautions are taken to direct repair in the right direction. For example, for cases such as the deamination of 5-methyl-cytosine to thymine, there is a special system to restore the proper sequence. The deamination generates a G•T pair, and the system that acts on such pairs has a bias to correct them to G•C pairs (rather than to A•T pairs).

The VSP system undertakes this reaction, and it includes the *mutL,S* system that removes T from both G•T and C•T mismatches.

Another repair activity, identified by *mutY*, replaces the A in C•A and G•A mismatches. This system functions by the direct removal of the base from DNA; *mutY* codes for an adenine glycosylase, which creates an apurinic site that is recognized by an endonuclease whose action triggers the involvement of the excision repair system.

When mismatch errors occur during replication in *E. coli*, it is possible to distinguish the original strand of DNA. Immediately after replication of methylated DNA, only the original parental strand carries the methyl groups. In the period while the newly synthesized strand awaits the introduction of methyl groups, the two strands can be distinguished.

This provides the basis for a system to correct replication errors. The *dam* gene codes for a methylase whose target is the adenine in the sequence $\frac{\text{GATC}}{\text{CTAG}}$ (see Figure 13.24). The hemimethylated state is used to distinguish replicated origins from nonreplicated origins. The same target sites are used by a replication-related repair system.

Figure 14.29 shows that DNA containing mismatched base partners is repaired preferentially by excising the strand that lacks the methylation. The excision is quite extensive; mismatches can be repaired preferentially for >1 kb within a d(GATC) site. The result is that the newly synthesized strand is corrected to the sequence of the parental strand.

E. coli dam⁻ mutants show an increased rate of spontaneous mutation. This repair system therefore helps reduce the number of mutations caused by errors in replication. It consists of several proteins, coded by the *mut* genes (which also participate in VSP repair) MutS binds to the mismatch and is joined by MutL. MutS can use two DNA-binding sites, as illustrated in **Figure 14.30**. The first specifically recognizes mismatches. The second is not specific for sequence or structure, and is

Figure 14.29 GATC sequences are targets for the Dam methylase after replication. During the period before this methylation occurs, the nonmethylated strand is the target for repair of mismatched bases.

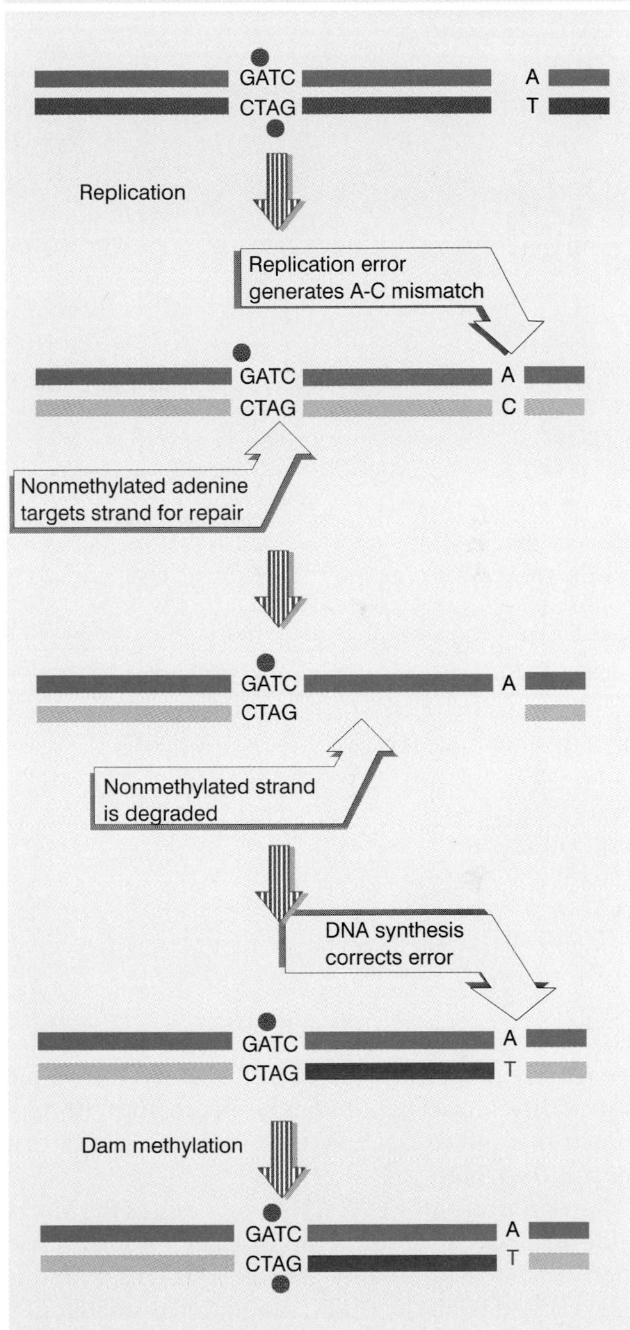

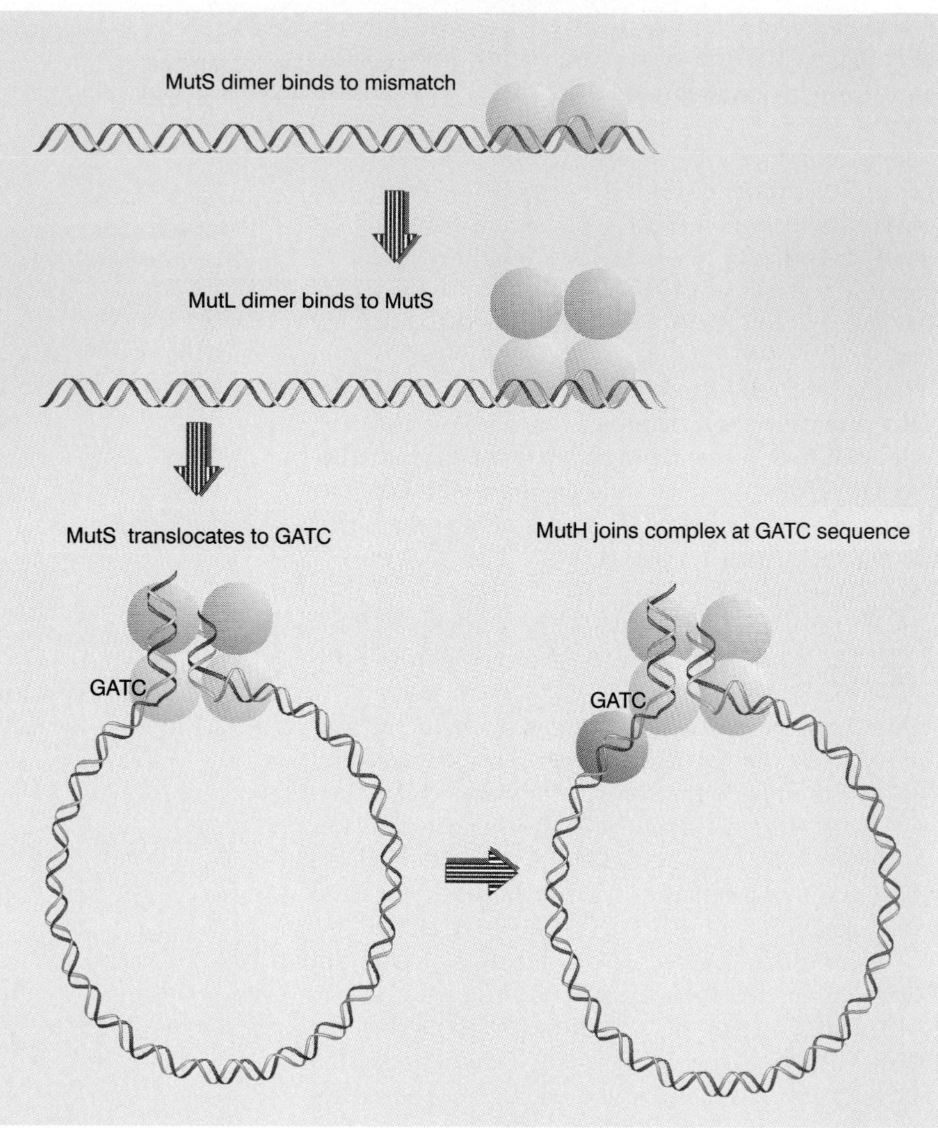
MutS dimer binds to mismatch
MutL dimer binds to MutS
MutS translocates to GATC
MutH joins complex at GATC sequence
GATC
GATC

Figure 14.30 MutS recognizes a mismatch and translocates to a GATC site. MutH cleaves the unmethylated strand at the GATC. Endonucleases degrade the strand from the GATC to the mismatch site.

used to translocate along DNA until a GATC sequence is encountered. Hydrolysis of ATP is used to drive the translocation. Because MutS is bound to both the mismatch site and to DNA as it translocates, it creates a loop in the DNA.

Recognition of the GATC sequence causes the MutH endonuclease to bind to MutSL. The endonuclease then cleaves the unmethylated strand. This strand is then excised from the GATC site to the mismatch site. The excision can occur in either the 5′–3′ direction (using RecJ or exonuclease VII) or in the 3′–5′ direction (using exonuclease I), assisted by the helicase UvrD. The new DNA strand is synthesized by DNA polymerase III.

The *msh* repair system of *S. cerevisiae* is homologous to the *E. coli mut* system. MSH2 provides a scaffold for the apparatus that recognizes mismatches. MSH3 and MSH6 provide specificity factors. The MSH2-MSH3 complex binds mismatched loops of two to four nucleotides, and the MSH2-MSH6 complex binds to single base mismatches or insertions or deletions. Other proteins are then required for the repair process itself.

Retrieval systems in *E. coli*

RETRIEVAL systems have variously been termed "post-replication repair", because they function after replication, or "recombination-repair", because the activities overlap with those involved in genetic recombination. Such systems are effective in dealing with the defects produced in daughter duplexes by replication of a template that contains damaged bases. An example is illustrated in **Figure 14.31**.

Consider a structural distortion, such as a pyrimidine dimer, on one strand of a double helix. When the DNA is replicated, the dimer prevents the damaged site from acting as a template. Replication is forced to skip past it.

DNA polymerase probably proceeds up to or close to the pyrimidine dimer. Then the polymerase ceases synthesis of the corresponding daughter strand. Replication restarts some distance farther along. A substantial gap is left in the newly synthesized strand.

The resulting daughter duplexes are different in nature. One has the parental strand containing the damaged adduct, facing a newly synthesized strand with a lengthy gap. The other duplicate has the undamaged parental strand, which has been copied into a normal complementary strand. The retrieval system takes advantage of the normal daughter.

The gap opposite the damaged site in the first duplex is filled by stealing the homologous single strand of DNA from the normal duplex. Following this **single-strand exchange**, the recipient duplex has a parental (damaged) strand facing a wild-type strand. The donor duplex has a normal parental strand facing a gap; the gap can be filled by repair synthesis in the usual way, generating a normal duplex. So the damage is confined to the original distortion (although the same recombination-repair events must be repeated after every replication cycle unless and until the damage is removed by an excision repair system).

The principal pathway for recombination-repair in *E. coli* is identified by the *rec* genes, whose activities in recombination *per se* we discussed earlier (Figures 14.9–14.11). In *E. coli* deficient in excision repair, mutation in the *recA* gene essentially abolishes all the remaining repair and recovery facilities. Attempts to replicate DNA in $uvr^{-}recA^{-}$ cells produce fragments of DNA whose size corresponds with the expected distance between thymine dimers. This result implies that the dimers provide a lethal obstacle to replication in

Figure 14.31 An *E. coli* retrieval system uses a normal strand of DNA to replace the gap left in a newly synthesized strand opposite a site of unrepaired damage.

Damage
Bases on one strand of DNA are damaged
Replication generates a copy with gap opposite damage and a normal copy
Retrieval
Gap is repaired by retrieving sequence from normal copy
Gap in normal copy is repaired

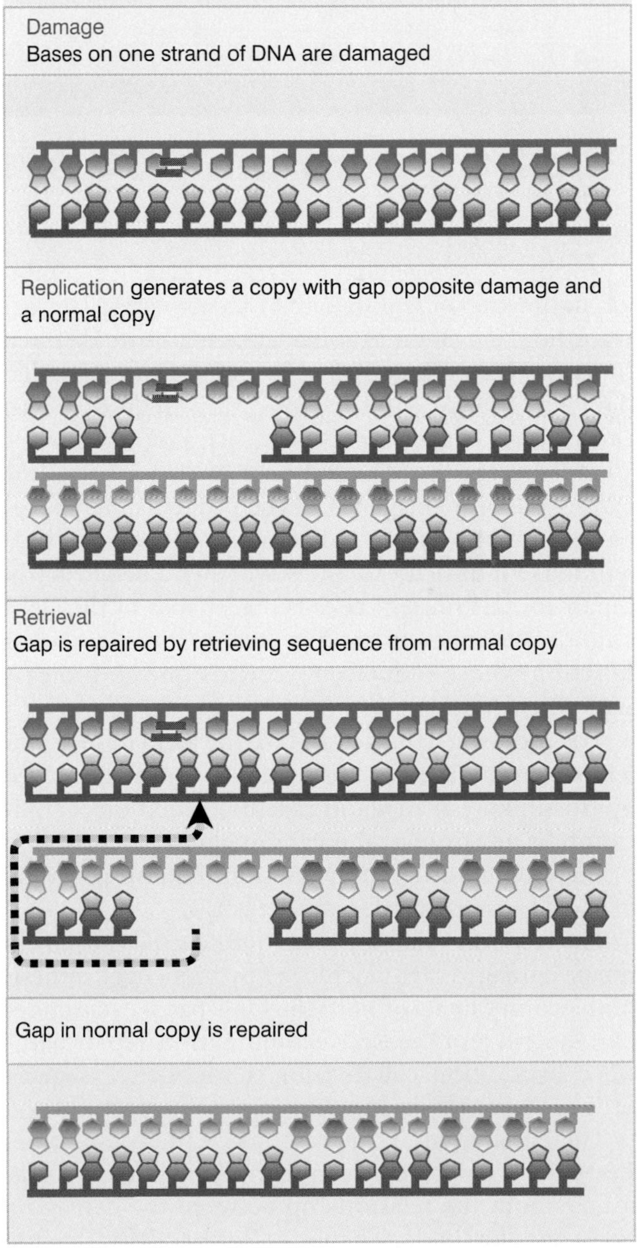

the absence of RecA function. It explains why the double mutant cannot tolerate more than one to two dimers in its genome (compared with the ability of a wild-type bacterium to handle as many as 50).

One *rec* pathway involves the *recBC* genes, and is well characterized; the other involves *recF*, and is not well defined. The ability of RecA to exchange single strands allows it to perform the retrieval step in Figure 14.31. Nuclease and polymerase activities then complete the repair action.

The designations of repair and recombination genes are based on the phenotypes of the mutants; but sometimes a mutation isolated in one set of conditions and named as a *uvr* locus turns out to have been isolated in another set of conditions as a *rec* locus. This uncertainty makes an important point. We cannot yet define how many functions belong to each pathway or how the pathways interact. The *uvr* and *rec* pathways are not entirely independent, because *uvr* mutants show reduced efficiency in recombination-repair.

We must expect to find a network of nuclease, polymerase, and other activities, constituting repair systems that are partially overlapping (or in which an enzyme usually used to provide some function can be substituted by another from a different pathway).

RecA triggers the SOS system

THE direct involvement of RecA protein in recombination-repair is only one of its activities. This extraordinary protein also has another, quite distinct, function. It can be activated by many treatments that damage DNA or inhibit replication in *E. coli.* This causes it to trigger a complex series of phenotypic changes called the **SOS response**, which involves the expression of many genes whose products include repair functions. These dual activities of the RecA protein make it difficult to know whether a deficiency in repair in *recA* mutant cells is due to loss of the DNA strand-exchange function of RecA or to some other function whose induction depends on the protease activity.

The inducing damage can take the form of ultraviolet irradiation (the most studied case) or can be caused by crosslinking or alkylating agents. Inhibition of replication by any of several means, including deprivation of thymine, addition of drugs, or mutations in several of the *dna* genes, has the same effect.

The response takes the form of increased capacity to repair damaged DNA, achieved by inducing synthesis of the components of both the long-patch excision repair system and the Rec recombination-repair pathways. In addition, cell division is inhibited. Lysogenic prophages may be induced.

The initial event in the response is the activation of RecA by the damaging treatment. We do not know very much about the relationship between the damaging event and the sudden change in RecA activity. Because a variety of damaging events can induce the SOS response, current work focuses on the idea that RecA is activated by some common intermediate in DNA metabolism.

The inducing signal could consist of a small molecule released from DNA; or it might be some structure formed in the DNA itself. *In vitro*, the activation of RecA requires the presence of single-stranded DNA and ATP. So the activating signal could be the presence of a single-stranded region at a site of damage. Whatever form the signal takes, its interaction with RecA is rapid: the SOS response occurs within a few minutes of the damaging treatment.

Activation of RecA causes proteolytic cleavage of the product of the *lexA* gene. LexA is a small (22 kD) protein that is relatively stable in untreated cells, where it functions as a repressor at many operons. The cleavage reaction is unusual; LexA has a latent protease activity that is activated by RecA. When RecA is activated, it causes LexA to undertake an autocatalytic cleavage; this inactivates the LexA repressor function, and coordinately induces all the operons to which it was bound. The pathway is illustrated in **Figure 14.32**.

The target genes for LexA repression include many repair functions. Some of these SOS genes are active only in treated cells; others are active in untreated cells, but the level of expression is increased by cleavage of LexA. In the case of *uvrB*, which is a component of the excision repair system, the gene has two promoters; one functions independently of LexA, the other is

Figure 14.32 The LexA protein represses many genes, including repair functions, *recA* and *lexA*. Activation of RecA leads to proteolytic cleavage of LexA and induces all of these genes.

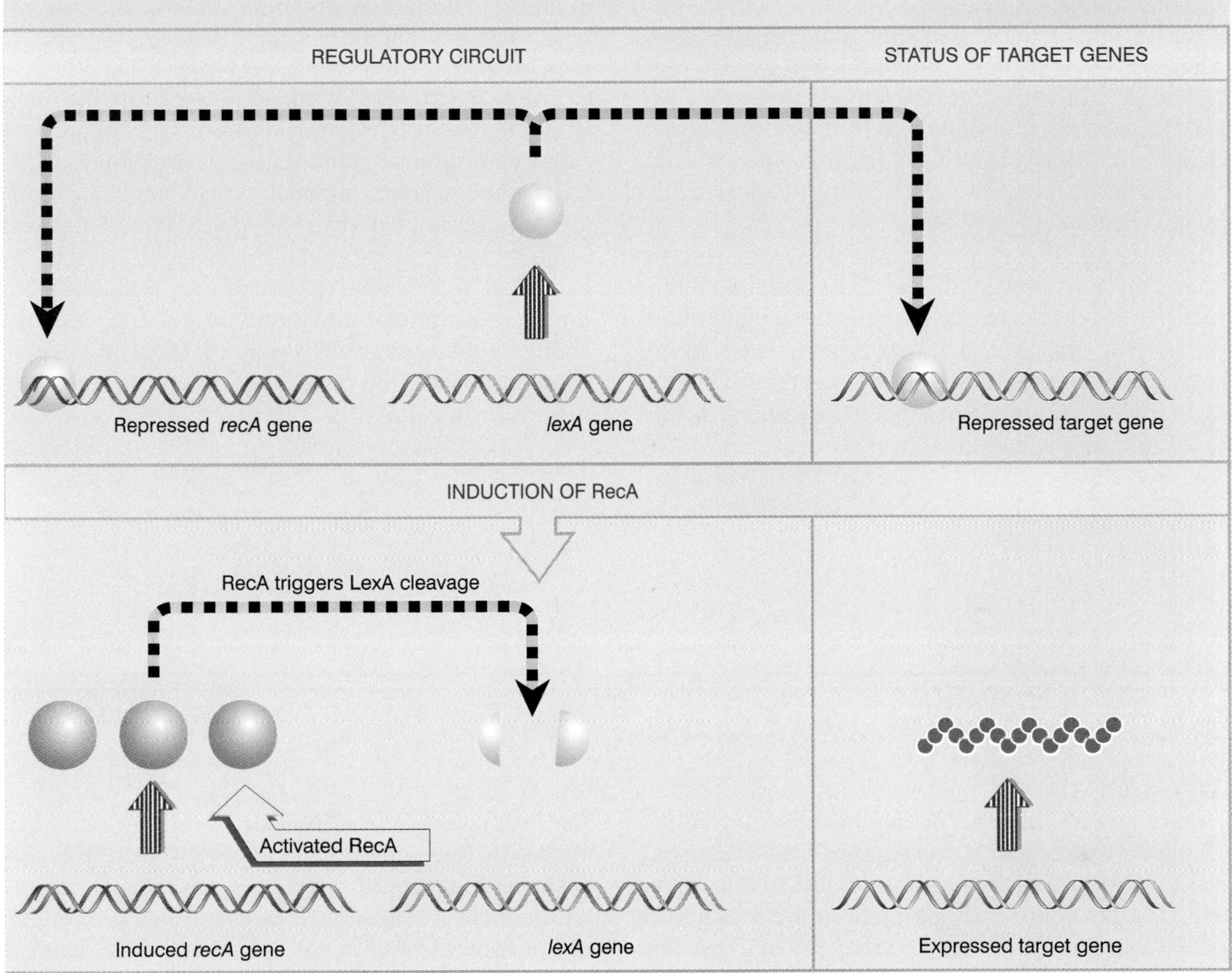

subject to its control. So after cleavage of LexA, the gene can be expressed from the second promoter as well as from the first.

LexA represses its target genes by binding to a 20 bp stretch of DNA called an **SOS box**, which includes a consensus sequence with eight absolutely conserved positions. Like other operators, the SOS boxes overlap with the respective promoters. At the *lexA* locus, the subject of autogenous repression, there are two adjacent SOS boxes.

RecA and LexA are mutual targets in the SOS circuit: RecA triggers cleavage of LexA, which represses *recA*. The SOS response therefore causes amplification of both the RecA protein and the LexA repressor. The results are not so contradictory as might at first appear.

The increase in expression of RecA protein is necessary (presumably) for its direct role in the recombination-repair pathways. On induction, the level of RecA is increased from its basal level of ~1200 molecules/cell by up to 50-fold. The high level in induced cells means there is sufficient RecA to ensure that all the LexA protein is cleaved. This should prevent LexA from re-establishing repression of the target genes.

But the main importance of this circuit for the cell lies in the ability to return rapidly to normalcy. When the inducing signal is removed, the RecA protein loses the ability to destabilize LexA. At this moment, the *lexA* gene is being expressed at a high level; in the absence of activated RecA, the LexA protein rapidly accumulates in the uncleaved form and turns off the SOS genes. This explains why the SOS response is freely reversible.

RecA also triggers cleavage of other cellular targets,

sometimes with more direct consequences. The UmuD protein is cleaved when RecA is activated; the cleavage event activates UmuD and the error-prone repair system.

Activation of RecA also causes cleavage of some other repressor proteins, including those of several prophages. Among these is the lambda repressor (with which the protease activity was discovered). This explains why lambda is induced by ultraviolet irradiation; the lysogenic repressor is cleaved, releasing the phage to enter the lytic cycle.

This reaction is not a cellular SOS response, but instead represents a recognition by the prophage that the cell is in trouble. Survival is then best assured by entering the lytic cycle to generate progeny phages. In this sense, prophage induction is piggybacking onto the cellular system by responding to the same indicator (activation of RecA).

All known proteins that are targets of RecA activation are cleaved at an Ala-Gly dipeptide sequence in the middle of the polypeptide chain. There is only limited amino acid homology on either side of the dipeptide, which suggests that the tertiary structure of the protein is an important feature in target recognition.

The two activities of RecA are relatively independent. The *recA441* mutation allows the SOS response to occur without inducing treatment, probably because RecA remains spontaneously in the activated state. Other mutations abolish the ability to be activated. Neither type of mutation affects the ability of RecA to handle DNA. The reverse type of mutation, inactivating the recombination function but leaving intact the ability to induce the SOS response, would be useful in disentangling the direct and indirect effects of RecA in the repair pathways.

Eukaryotic repair systems

BIOCHEMICAL characterization of repair systems in eukaryotic cells is less extensive. The existence of excision repair pathways can be established in cultured cells by following the actual removal of damage from DNA or by detecting the replacement of DNA segments in response to damaging treatments.

Genes involved in repair functions have been characterized genetically in yeast by virtue of their sensitivity to radiation. They are called *RAD* genes. There are three general groups of repair genes in yeast, identified by the *RAD3* group (involved in excision repair), the *RAD6* group (required for post-replication repair), and the *RAD52* group (concerned with recombination-like mechanisms).

An interesting feature of repair that has been best characterized in yeast is its connection with transcription. Transcriptionally active genes are preferentially repaired. The consequence is that the transcribed strand is preferentially repaired (removing the impediment to transcription). The cause appears to be a mechanistic connection between the repair apparatus and RNA polymerase. The RAD3 protein, which is a helicase required for the incision step, is a component of a transcription factor associated with RNA polymerase. This connection is discussed in Chapter 20.

Mammalian cells show heterogeneity in the amount of DNA resynthesized at each lesion after damage. However, the longest patches seen in mammalian cells are comparable to those of the short-patch bacterial repair system. This pathway operates on damage caused by ultraviolet irradiation or by treatments that have related effects. Another pathway introduces only 3–4 repair bases at sites of damage generated by X-irradiation or alkylation.

An indication of the existence and importance of the mammalian repair systems is given by certain human hereditary disorders. The best investigated of these is xeroderma pigmentosum (XP), a recessive disease resulting in hypersensitivity to sunlight, in particular to ultraviolet. The deficiency results in skin disorders (and sometimes more severe defects).

The disease is explicable in terms of a failing in excision repair; fibroblasts from XP patients are deficient in the excision of pyrimidine dimers and other bulky adducts. Several (~9) genetic complementation groups have been distinguished, many characterized by a deficiency at the incision step of repair. Six of the genes identifying individual XP loci have been cloned, and each has a counterpart in yeast. Six of the 11 loci in the group that includes *RAD3* are homologs of XP genes, suggesting extensive conservation of this excision repair pathway.

Double-strand breaks occur in cells in various circumstances. They initiate the process of homologous recombination and are an intermediate in the recombination of immunoglobulin genes (see Chapter 24). They also occur as the result of damage to DNA, for example, by irradiation. The major mechanism to repair these breaks is called **non-homologous end joining (NHEJ)**, and consists of ligating the blunt ends together. The same enzyme complex undertakes the process in both NHEJ and immune recombination. It consists of the proteins Ku70 and Ku80, which form a dimer that binds the DNA double-stranded end, the DNA-dependent protein kinase (DNA-PKcs), and the DNA ligase IV, which functions in conjunction with the protein XRCC4. Mutations in any of these components may render eukaryotic cells more sensitive to radiation. Some of the genes for these proteins are mutated in patients who have diseases due to deficiencies in DNA repair.

Some indirect results suggest that mammalian cells have recombination-repair systems. Again, these systems may be related to genetic recombination itself. An example is the recessive human disorder of Bloom's syndrome; an increased frequency of chromosomal aberrations, including sister chromatid exchanges, could be related to the operation of recombination systems.

Summary

Recombination involves the physical exchange of parts between corresponding DNA molecules. This results in a duplex DNA in which two regions of opposite parental origins are connected by a stretch of hybrid (heteroduplex) DNA in which one strand is derived from each parent. Correction events may occur at sites that are mismatched within the hybrid DNA. Hybrid DNA can also be formed without recombination occurring between markers on either side. Gene conversion occurs when an extensive region of hybrid DNA forms during normal recombination (or between nonallelic genes in an aberrant event) and is corrected to the sequence of only one parental strand; then one gene takes on the sequence of the other.

Recombination is initiated by a double-strand break in DNA. The break is enlarged to a gap with a single-stranded end; then the free single-stranded end forms a heteroduplex with the allelic sequence. The DNA in which the break occurs actually incorporates the sequence of the chromosome that it invades, so the initiating DNA is called the recipient. Hotspots for recombination are sites where double strand breaks are initiated. A gradient of gene conversion is determined by the likelihood that a sequence near the free end will be converted to a single strand; this decreases with distance from the break.

The recombination event is not well understood at the level of the chromosome, but the properties of yeast mutants, and the relative timing of events during meiosis, suggest that the synaptonemal complex may be the consequence of, rather than a prerequisite for, the initiation of the recombination event.

The only enzymes whose activities have been characterized in recombination are coded by the *rec* and *ruv* loci of *E. coli*. RecA has the ability to synapse homologous DNA molecules by sponsoring a reaction in which a single strand from one molecule invades a duplex of the other molecule. Heteroduplex DNA is formed by displacing one of the original strands of the duplex. The RecBCD nuclease binds to DNA on one side of a *chi* sequence, and then moves to the *chi* sequence, unwinding DNA as it progresses. A single-strand break is made at the *chi* sequence. *chi* sequences provide hotspots for recombination. RuvA and RuvB act at a heteroduplex, and RuvC cleaves Holliday junctions.

Recombination, like replication and (probably) transcription, requires topological manipulation of DNA. Topoisomerases may relax (or introduce) supercoils in DNA, and are required to disentangle DNA molecules that have become catenated by recombination or by replication.

The enzymes involved in site-specific recombination have actions related to those of topoisomerases. Phage lambda integration requires the phage Int protein and host IHF protein and involves a precise breakage and reunion in the absence of any synthesis of DNA. The reaction involves wrapping of the *attP* sequence of phage DNA into the nucleoprotein structure of the intasome, which contains several copies of Int and IHF; then the host *attB* sequence is bound, and recombination occurs. Reaction in the reverse direction requires the phage protein Xis.

Bacteria contain systems that maintain the integrity of their DNA sequences in the face of damage or errors of replication and that distinguish the DNA from sequences of a foreign source.

Repair systems can recognize mispaired, altered, or missing bases in DNA, or other structural distortions of the double helix. Excision repair systems cleave DNA near a site of damage, remove one strand, and synthesize a new sequence to replace the excised material. Three excision repair systems in *E. coli* can be distinguished by the lengths of the regions that are excised. The *dam* system is involved in correcting mismatches generated by incorporation of incorrect bases during replication, and the *uvr* genes are involved in both of the other systems for general repair. Recombination-repair systems retrieve information from a DNA duplex and use it to repair a sequence that has been damaged on both strands. The *recBC* and *recF* pathways identify two different systems. The *recA* product may be involved in repair pathways in one of its capacities, the ability to synapse molecules of DNA.

The other capacity of *recA* is the ability to induce the SOS response. RecA is activated by damaged DNA in an unknown manner. It triggers cleavage of the LexA repressor protein, thus releasing repression of many loci, and inducing synthesis of the enzymes of both excision repair and recombination-repair pathways. Genes under LexA control possess an operator SOS box. RecA also directly activates some repair activities. Cleavage of repressors of lysogenic phages may induce the phages to enter the lytic cycle.

Further reading

Reviews

Stahl, F. W. (1979). Special sites in generalized recombination. *Ann. Rev. Genet.* **13**, 7–24.

Simon, M. *et al.* (1980). Phase variation: evolution of a controlling element. *Science* **209**, 1370–1374.

Dressler, D. and Potter, H. (1982). Molecular mechanisms in genetic recombination. *Ann. Rev. Biochem.* **51**, 727–761.

Little, J. W. and Mount, D. W. (1982). The SOS regulatory system of *E. coli. Cell* **29**, 11–22.

Weisberg, R. and Landy, A. (1983). Site-specific recombination in phage lambda. In *Lambda II.* (Eds. R. Hendrix *et al.*, Cold Spring Harbor Laboratory, New York, 211–250).

Drlica, K. (1984). Biology of bacterial DNA topoisomerases. *Microbiol. Rev.* **48**, 273–289.

Walker, G. C. (1984). Mutagenesis and inducible responses to DNA damage in *E. coli. Microbiol. Rev.* **48**, 60–93.

Wang, J. (1985). DNA topoisomerases *Ann. Rev. Biochem.* **54**, 665–697.

Cox, M. M. and Lehman, I. R. (1987). Enzymes of general recombination. *Ann. Rev. Biochem.* **56**, 229–262.

Smith, G. R. (1988). Homologous recombination in prokaryotes. *Microbiol. Rev.* **52**, 1–28.

Lindahl, T. *et al.* (1988). Regulation and expression of the adaptive response to alkylating agents. *Ann. Rev. Biochem.* **57**, 133–157.

Sancar, A. and Sancar, G. B. (1988). DNA repair enzymes. *Ann. Rev. Biochem.* **57**, 29–67.

Thompson, J. F. and Landy, A. (1989). Regulation of phage lambda site specific recombination. In *Mobile DNA.* (Eds. D. E. Berg and M. M. Howe, American Society for Microbiology, Washington DC, 1–22).

Glasgow, A. C., Hughes, K. H., and Simon, M. I. (1989). Bacterial DNA inversion systems. In *Mobile DNA.* (Eds. D. E. Berg and M. M. Howe, American Society for Microbiology, Washington DC, 637–661).

West, S. C. (1992). Enzymes and molecular mechanisms of genetic recombination. *Ann. Rev. Biochem.* **61**, 603–640.

Prakash, S. *et al.* (1993). DNA repair genes and proteins of *S. cerevisiae. Ann. Rev. Genet.* **27**, 33–70.

Kowalczykowski, S. C. *et al.* (1994). Biochemistry of homologous recombination in *E. coli. Microbiol. Rev.* **58**, 401–465.

Friedberg, E. C., Walker, G. C., and Siede, W. (1995). *DNA repair and mutagenesis.* (American Society for Microbiology, Washington DC).

Wang, J. C. (1996). DNA topoisomerases *Ann. Rev. Biochem.* **65**, 635–692.

Modrich, P. and Lahue, R. (1996). Mismatch repair in replication fidelity, genetic recombination, and cancer biology. *Ann. Rev. Biochem.* **65**, 101–133.

West, S. C. (1997). Processing of recombination intermediates by the RuvABC system. *Ann. Rev. Genet.* **31**, 213–244.

Zickler, D. and Kleckner, N. (1998). The leptotene–zygotene transition of meiosis. *Ann. Rev. Genet.* **32**, 619–697.

Discoveries

Holliday, R. (1964). A mechanism for gene conversion in fungi. *Genet. Res.* 5, 282–304.

Carpenter, A. T. C. (1975). Electron microscopy of meiosis in *D. melanogaster* females. II. The recombination nodule: a recombination-associated structure at pachytene? *Proc. Natl. Acad. Sci. USA* 72, 3186–3189.

Shibata, T. R. P. *et al.* (1979). Purified *E. coli* recA protein catalyzes homologous pairing of superhelical DNA and single-stranded fragments. *Proc. Natl. Acad. Sci. USA* **76**, 1638–1642.

Szostak, J. W. *et al.* (1983). The double-strand break model for DNA recombination. *Cell* **33**, 25–35.

Sancar, A. and Rupp, W. D. (1983). A novel repair enzyme: uvrABC excision nuclease of *E. coli* cuts a DNA strand on both sides of the damaged region. *Cell* **33**, 249–260.

Alani, E., Padmore, R., and Kleckner, N. (1990). Analysis of wild-type and *rad50* mutants of yeast suggests an intimate relationship between meiotic chromosome synapsis and recombination. *Cell* **61**, 419–436.

Padmore, R., Cao, L., and Kleckner, N. (1991). Temporal comparison of recombination and synaptonemal complex formation during meiosis in *S. cerevisiae. Cell* **66**, 1239–1256.

Sun, H., Treco, D., and Szostak, J. W. (1991). Extensive 3′ overhanging, single-stranded DNA associated with the meiosis-specific double-strand breaks at the *ARG4* recombination initiation site. *Cell* **64**, 1155–1161.

Keeney, S., Giroux, C. G., and Kleckner, N. (1997). Meiosis-specific DNA double-strand breaks are catalyzed by Spo11, a member of a widely conserved protein family. *Cell* **88**, 375–384.

Roe, M. S. *et al.* (1998). Crystal structure of an octameric RuvA-Holliday junction complex. *Molecular Cell* **2**, 361–372.

Leu, J.-Y., Chua, P. R., and Roeder, G. S. (1998). The meiosis-specific Hop2 protein of *S. cerevisiae* ensures synapsis between homologous chromosomes. *Cell* **94**, 375–386.

Chapter 15

Transposons

GENOMES evolve both by acquiring new sequences and by rearranging existing sequences.

The sudden introduction of new sequences results from the ability of vectors to carry information between genomes. Extrachromosomal elements move information horizontally by mediating the transfer of (usually rather short) lengths of genetic material. In bacteria, plasmids move by conjugation (see Chapter 12), while phages spread by infection (see Chapter 11). Both plasmids and phages occasionally transfer host genes along with their own replicon. Direct transfer of DNA occurs between some bacteria by means of transformation (see Chapter 1). In eukaryotes, some viruses (notably the retroviruses discussed in Chapter 16) can transfer genetic information during an infective cycle.

Rearrangements are sponsored by processes internal to the genome. One cause is unequal recombination, which results from mispairing by the cellular systems for homologous recombination. Nonreciprocal recombination results in duplication or rearrangement of loci (see Chapter 4). Duplication of sequences within a genome provides a major source of new sequences. One copy of the sequence can retain its original function, while the other may evolve into a new function. Furthermore, significant differences between individual genomes are found at the molecular level because of polymorphic variations caused by recombination. We saw in Chapter 4 that recombination between "minisatellites" adjusts their lengths so that every individual genome is distinct.

Another major cause of variation is provided by **transposable elements** or **transposons**: these are discrete sequences in the genome that are *mobile*—they are able to transport themselves to other locations within the genome. The mark of a transposon is that it does not utilize an independent form of the element (such as phage or plasmid DNA), but moves *directly* from one site in the genome to another. Unlike most other processes involved in genome restructuring, *transposition does not rely on any relationship between the sequences at the donor and recipient sites.* Transposons are restricted to moving themselves, and sometimes additional sequences, to new sites elsewhere within the same genome; they are therefore an internal counterpart to the vectors that can transport sequences from one genome to another. They may provide the major source of mutations in the genome.

Transposons fall into two general classes. The groups of transposons reviewed in this chapter exist as sequences of DNA coding for proteins that are able directly to manipulate DNA so as to propagate themselves within the genome. The transposons reviewed in the next chapter are related to retroviruses, and the source of their mobility is the ability to make DNA copies of their RNA transcripts; the DNA copies then become integrated at new sites in the genome.

Transposons that mobilize via DNA are found in both prokaryotes and eukaryotes. Each bacterial transposon carries gene(s) that code for the enzyme activities required for its own transposition, although it may also require ancillary functions of the genome in which it resides (such as DNA polymerase or DNA gyrase). Comparable systems exist in eukaryotes, although their enzymatic functions are not so well characterized. A genome may contain both functional and nonfunctional (defective) elements; often the majority of elements in a eukaryotic genome are defective, and have lost the ability to transpose independently, although they may still be recognized as substrates for transposition by the enzymes produced by functional transposons.

Transposable elements can promote rearrangements of the genome, directly or indirectly:

- The transposition event itself may cause deletions or inversions or lead to the movement of a host sequence to a new location.
- Transposons serve as substrates for cellular recombination systems by functioning as "portable regions of homology"; two copies of a transposon at different locations (even on different chromosomes) may provide sites for reciprocal recombination. Such exchanges result in deletions, insertions, inversions, or translocations.

The intermittent activities of a transposon seem to provide a somewhat nebulous target for natural selection. This concern has prompted suggestions that (at least some) transposable elements confer neither advantage nor disadvantage on the phenotype, but could constitute "selfish DNA", concerned only with their own propagation. Indeed, in considering transposition as an event that is distinct from other cellular recombination systems, we tacitly accept the view that the transposon is an independent entity that resides in the genome.

Such a relationship of the transposon to the genome would resemble that of a parasite with its host. Presumably the propagation of an element by transposition is balanced by the harm done if a transposition event inactivates a necessary gene, or if the number of transposons becomes a burden on cellular systems. Yet we must remember that any transposition event conferring a selective advantage—for example, a genetic rearrangement—will lead to preferential survival of the genome carrying the active transposon.

Insertion sequences are simple transposition modules

TRANSPOSABLE elements were first identified in the form of spontaneous insertions in bacterial operons. Such an insertion prevents transcription and/or translation of the gene in which it is inserted. Many different types of transposable elements have now been characterized.

The simplest transposons are called **insertion sequences** (reflecting the way in which they were detected). Each type is given the prefix **IS**, followed by a number that identifies the type. (The original classes were numbered IS1–4; later classes have numbers reflecting the history of their isolation, but not corresponding to the total number of elements so far isolated!)

The IS elements are normal constituents of bacterial chromosomes and plasmids. A standard strain of *E. coli* is likely to contain several (<10) copies of any one of the more common IS elements. To describe an insertion into a particular site, a double colon is used; so λ::IS1 describes an IS1 element inserted into phage lambda.

The IS elements are autonomous units, each of which codes only for the proteins needed to sponsor its own transposition. Each IS element is different in sequence, but there are some common features in organization. The structure of a generic transposon before and after insertion at a target site is illustrated in **Figure 15.1**, which also summarizes the details of some common IS elements.

An IS element ends in short **inverted terminal repeats**; usually the two copies of the repeat are closely related rather than identical. As illustrated in the figure, the presence of the inverted terminal repeats means that the same sequence is encountered proceeding toward the element from the flanking DNA on either side of it.

When an IS element transposes, a sequence of host DNA at the site of insertion is duplicated. The nature of the duplication is revealed by comparing the sequence of the target site before and after an insertion has occurred. Figure 15.1 shows that at the site of in-

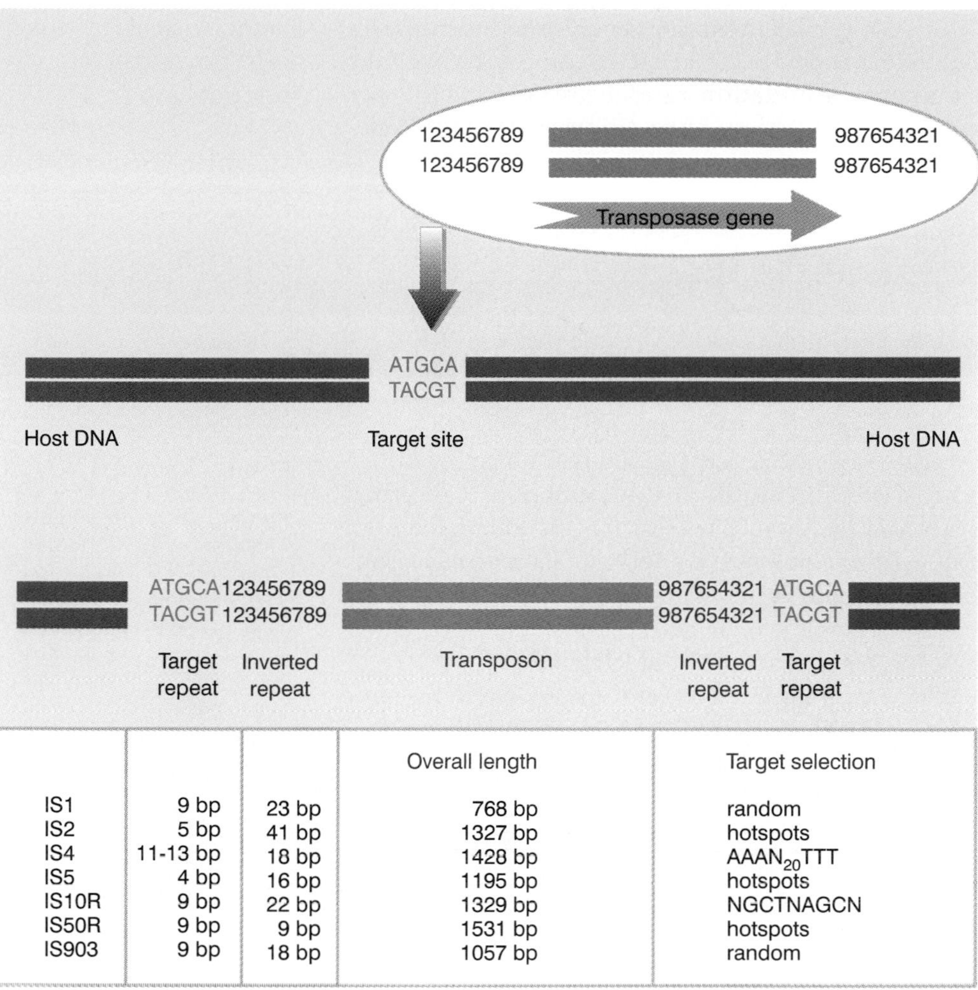

			Overall length	Target selection
IS1	9 bp	23 bp	768 bp	random
IS2	5 bp	41 bp	1327 bp	hotspots
IS4	11-13 bp	18 bp	1428 bp	$AAAN_{20}TTT$
IS5	4 bp	16 bp	1195 bp	hotspots
IS10R	9 bp	22 bp	1329 bp	NGCTNAGCN
IS50R	9 bp	9 bp	1531 bp	hotspots
IS903	9 bp	18 bp	1057 bp	random

Figure 15.1 Overview: transposons have inverted terminal repeats and generate direct repeats of flanking DNA at the target site. In this example, the target is a 5 bp sequence. The ends of the transposon consist of inverted repeats of 9 bp, where the numbers 1 through 9 indicate a sequence of base pairs.

sertion, the IS DNA is always flanked by very short **direct repeats.** (In this context, "direct" indicates that two copies of a sequence are repeated in the same orientation, not that the repeats are adjacent.) But in the original gene (prior to insertion), the target site has the sequence of only *one* of these repeats. In the figure, the target site consists of the sequence $\frac{\text{ATGCA}}{\text{TACGT}}$. *After transposition, one copy of this sequence is present on either side of the transposon.*

The *sequence* of the direct repeat varies among individual transposition events undertaken by a transposon, but the *length* is constant for any particular IS element (a reflection of the mechanism of transposition). The most common length for the direct repeats is 9 bp.

An IS element therefore displays a characteristic structure in which its ends are identified by the inverted terminal repeats, while the adjacent ends of the flanking host DNA are identified by the short direct repeats. When observed in a sequence of DNA, this type of organization is taken to be diagnostic of a transposon, and makes a *prima facie* case that the sequence originated in a transposition event.

Most IS elements insert at a variety of sites within host DNA. However, some show (varying degrees of) preference for particular hotspots.

The inverted repeats define the ends of a transposon. Recognition of the ends is common to transposition events sponsored by all types of transposon. *Cis*-acting mutations that prevent transposition are located in the ends, which are recognized by a protein(s) responsible for transposition. The protein is called a **transposase.**

All the IS elements except IS1 contain a single long coding region, starting just inside the inverted repeat at one end, and terminating just before or within the inverted repeat at the other end. This codes for the transposase. IS1 has a more complex organization, with two separate reading frames; the transposase is produced by making a frameshift during translation to allow both reading frames to be used.

The frequency of transposition varies among different elements. The overall rate of transposition is

$\sim 10^{-3}$–10^{-4} per element per generation. Insertions in individual targets occur at a level comparable with the spontaneous mutation rate, usually $\sim 10^{-5}$–10^{-7} per generation. Reversion (by precise excision of the IS element) is usually infrequent, with a range of rates of 10^{-6} to 10^{-10} per generation, $\sim 10^{3}$ times less frequent than insertion.

Composite transposons have IS modules

SOME transposons carry drug resistance (or other) markers in addition to the functions concerned with transposition. These transposons are named **Tn** followed by a number. One class of larger transposons are called **composite elements**, because a central region carrying the drug marker(s) is flanked on either side by "arms" that consist of IS elements.

The arms may be in either the same or (more commonly) inverted orientation. So a composite transposon with arms that are direct repeats has the structure

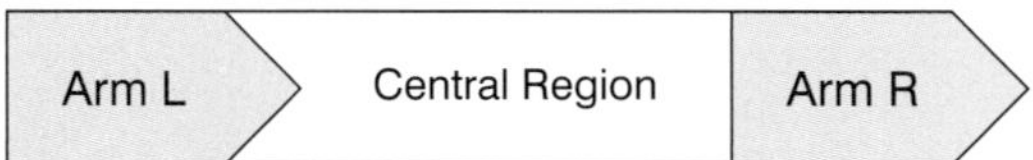

If the arms are inverted repeats, the structure is

Arm L | Central Region | Arm R

The arrows indicate the orientation of the arms, which are identified as L and R according to an (arbitrary) orientation of the genetic map of the transposon from left to right. The structure of a composite transposon is illustrated in more detail in **Figure 15.2**, which also summarizes the properties of some common composite transposons.

Since arms consist of IS modules, and each module has the usual structure ending in inverted repeats, the composite transposon also ends in the same short inverted repeats.

In some cases, the modules of a composite transposon are identical, such as Tn9 (direct repeats of IS1) or Tn903 (inverted repeats of IS903). In other cases, the modules are closely related, but not identical. So we can distinguish the L and R modules in Tn10 or in Tn5.

A functional IS module can transpose either itself or the entire transposon. When the modules of a composite transposon are identical, presumably either module can sponsor movement of the transposon, as in the

Figure 15.2 A composite transposon has a central region carrying markers (such as drug resistance) flanked by IS modules. The modules have short inverted terminal repeats. If the modules themselves are in inverted orientation (as drawn), the short inverted terminal repeats at the ends of the transposon are identical.

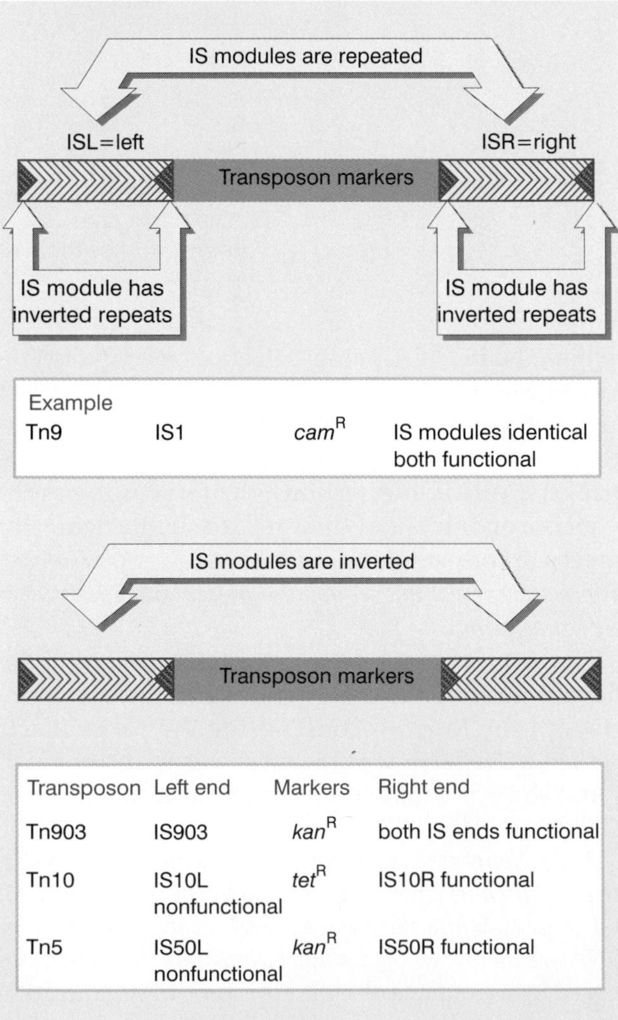

case of Tn9 or Tn903. When the modules are different, they may differ in functional ability, so transposition can depend entirely or principally on one of the modules, as in the case of Tn10 or Tn5.

We assume that composite transposons evolved when two originally independent modules associated with the central region. Such a situation could arise when an IS element transposes to a recipient site close to the donor site. The two identical modules may remain identical or diverge. The ability of a single module to transpose the entire composite element explains the lack of selective pressure for both modules to remain active.

What is responsible for transposing a composite transposon instead of just the individual module? This question is especially pressing in cases where both the modules are functional. In the example of Tn9, where the modules are IS1 elements, presumably each is active in its own right as well as on behalf of the composite transposon. Why is the transposon preserved as a whole, instead of each insertion sequence looking out for itself?

Two IS elements in fact can transpose any sequence residing between them, as well as themselves. **Figure 15.3** shows that if Tn10 resides on a circular replicon, its two modules can be considered to flank *either* the *tet*^{R} gene of the original Tn10 *or* the sequence in the other part of the circle. So a transposition event can involve either the original Tn10 transposon (marked by the movement of *tet*^{R}) or the creation of the new "inside-out" transposon with the alternative central region.

Note that both the original and "inside-out" transposons have inverted modules, but these modules evidently can function in either orientation relative to the central region. The frequency of transposition for composite transposons declines with the distance between the modules. So length dependence is a factor in determining the sizes of the common composite transposons.

A major force supporting the transposition of composite transposons is selection for the marker(s) carried in the central region. An IS10 module is free to move around on its own, and mobilizes an order of magnitude more frequently than Tn10. But Tn10 is held together by selection for *tet*^{R}; so that under selective conditions, the relative frequency of intact Tn10 transposition is much increased.

The IS elements code for transposase activities that are responsible both for creating a target site and for recognizing the ends of the transposon. *Only the ends are needed for a transposon to serve as a substrate for transposition.*

Figure 15.3 Two IS10 modules create a composite transposon that can mobilize any region of DNA that lies between them. When Tn10 is part of a small circular molecule, the IS10 repeats can transpose either side of the circle.

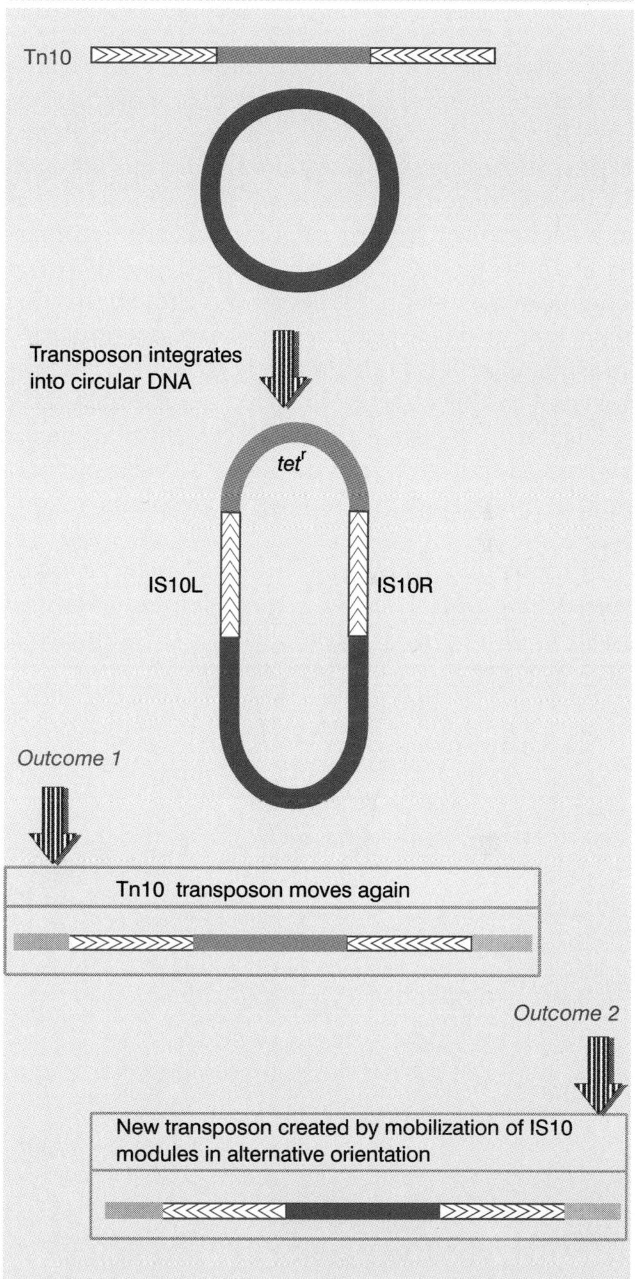

Transposition occurs by both replicative and nonreplicative mechanisms

THE insertion of a transposon into a new site is illustrated in **Figure 15.4**. It consists of making staggered breaks in the target DNA, joining the transposon to the protruding single-stranded ends, and filling in the gaps. The generation and filling of the staggered ends explain the occurrence of the direct repeats of target DNA at the site of insertion. The stagger between the cuts on the two strands determines the length of the direct repeats; so the target repeat characteristic of each transposon reflects the geometry of the enzyme involved in cutting target DNA.

The use of staggered ends is common to all means of transposition, but we can distinguish three different types of mechanism by which a transposon moves:

- In **replicative transposition**, *the element is duplicated during the reaction, so that the transposing entity is a copy of the original element.* **Figure 15.5** summarizes the results of such a transposition. The transposon is copied as part of its movement. One copy remains at the original site, while the other inserts at the new site. So transposition is accompanied by an increase in the number of copies of the transposon. Replicative transposition involves two types of enzymatic activity: a **transposase** that acts on the ends of the original transposon; and a **resolvase** that acts on the duplicated copies. A group of transposons related to TnA move only by replicative transposition (see later).
- In **nonreplicative transposition**, *the transposing element moves as a physical entity directly from one site to another, and is conserved.* This occurs by either of two types of mechanism. One utilizes the connection of donor and target DNA sequences and shares some steps with replicative transposition (see next section). The insertion sequences and composite transposons Tn10 and Tn5 use the mechanism shown in **Figure 15.6**, which involves the release of the transposon from the flanking donor DNA during transfer. This type of mechanism requires only a transposase. Both mechanisms of nonreplicative transposition cause the element to be inserted at the target site and lost from the donor site. What happens to the donor molecule after a nonreplicative transposition? Its survival requires that

Figure 15.4 The direct repeats of target DNA flanking a transposon are generated by the introduction of staggered cuts whose protruding ends are linked to the transposon.

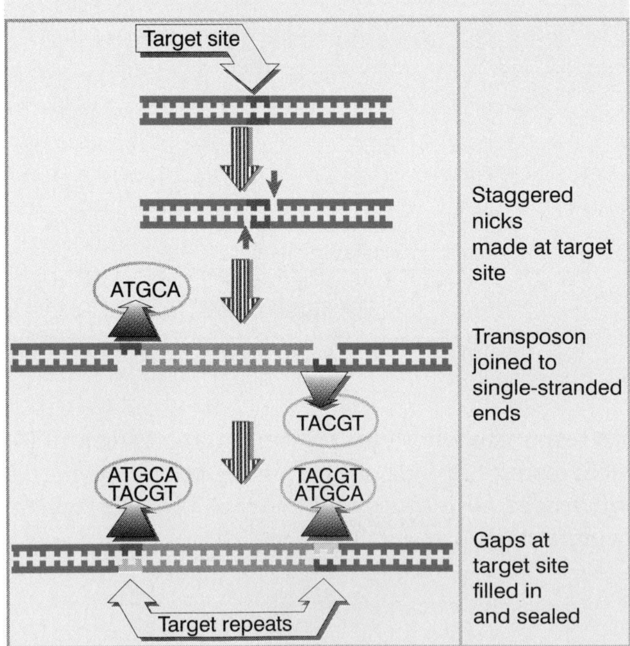
Target site
ATGCA
TACGT
ATGCA
TACGT
TACGT
ATGCA
Target repeats
Staggered nicks made at target site
Transposon joined to single-stranded ends
Gaps at target site filled in and sealed

Figure 15.5 Replicative transposition creates a copy of the transposon, which inserts at a recipient site. The donor site remains unchanged, so both donor and recipient have a copy of the transposon.

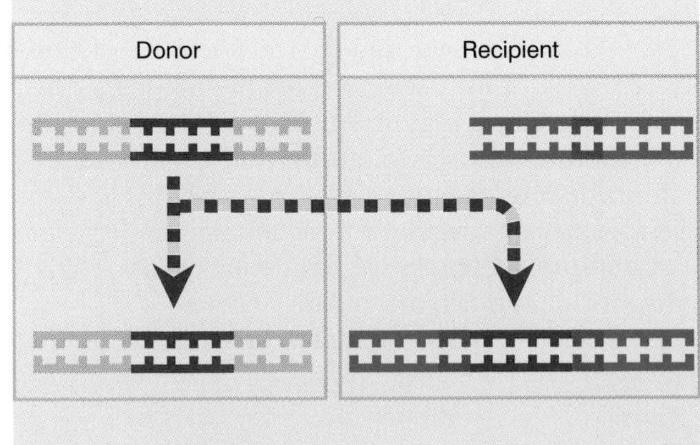
Donor
Recipient

host repair systems recognize the double-strand break and repair it.

■ **Conservative transposition** describes another sort of nonreplicative event, in which the element is excised from the donor site and inserted into a target site by a series of events in which every nucleotide bond is conserved. **Figure 15.7** summarizes the result of a conservative event. This exactly resembles the mechanism of lambda integration discussed in Chapter 11, and the transposases of such elements are related to the λ integrase family. The elements that use this mechanism are large, and can mediate transfer not only of the element itself but also of donor DNA from one bacterium to another. Although originally classified as transposons, such elements may more properly be regarded as episomes.

Figure 15.6 Nonreplicative transposition allows a transposon to move as a physical entity from a donor to a recipient site. This leaves a break at the donor site, which is lethal unless it can be repaired.

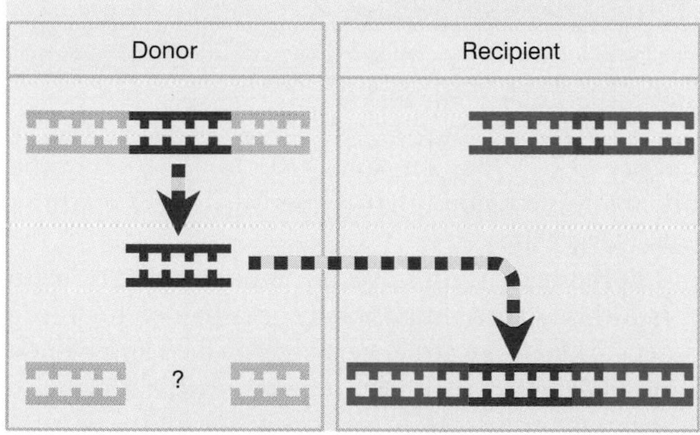

Figure 15.7 Conservative transposition involves direct movement with no loss of nucleotide bonds; compare with lambda integration and excision.

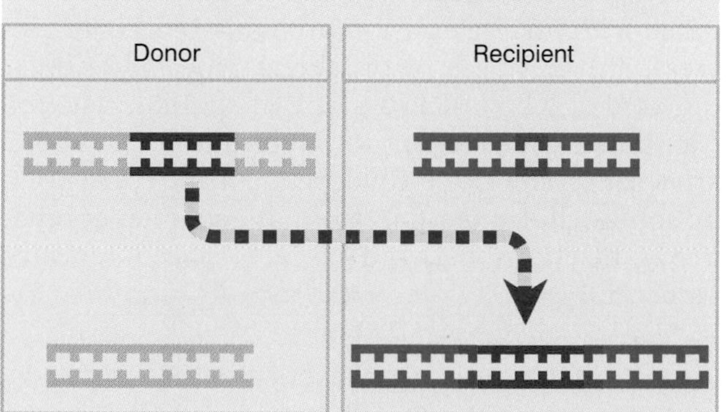

Although some transposons use only one type of pathway for transposition, others may be able to use multiple pathways. The elements IS1 and IS903 use both nonreplicative and replicative pathways, and the ability of phage Mu to turn to either type of pathway from a common intermediate has been well characterized (see later).

The same basic types of reaction are involved in all classes of transposition event. The ends of the transposon are disconnected from the donor DNA by cleavage reactions that generate 3′-OH ends. Then the exposed ends are joined to the target DNA by transfer reactions, involving transesterification in which the 3′-OH end directly attacks the target DNA. These reactions take place within a nucleoprotein complex that contains the necessary enzymes and both ends of the transposon. Transposons differ as to whether the target DNA is recognized before or after the cleavage of the transposon itself.

The choice of target site is in effect made by the transposase. In some cases, the target is chosen virtually at random. In others, there is specificity for a consensus sequence, for a structure, such as bent DNA, or for inactive regions of the chromosome.

In addition to the "simple" intermolecular transposition that results in insertion at a new site, transposons promote other types of DNA rearrangements. Some of these events are consequences of the relationship between the multiple copies of the transposon. Others represent alternative outcomes of the transposition mechanism, and they leave clues about the nature of the underlying events.

Rearrangements of host DNA may result when a transposon inserts a copy at a second site near its original location. Host systems may undertake reciprocal recombination between the two copies of the transposon; the consequences are determined by whether the repeats are the same or in inverted orientation.

Figure 15.8 illustrates the general rule that recombination between any pair of direct repeats will delete the material between them. The intervening region is excised as a circle of DNA (which is lost from the cell); the chromosome retains a single copy of the direct repeat. Note that a recombination between the directly repeated IS1 modules of the composite transposon Tn9 would replace the transposon with a single IS1 module.

Deletion of sequences adjacent to a transposon could therefore result from a two-stage process; transposition generates a direct repeat of a transposon, and recombination occurs between the repeats. However, the majority of deletions that arise in the vicinity of

Figure 15.8 Reciprocal recombination between direct repeats excises the material between them; each product of recombination has one copy of the direct repeat.

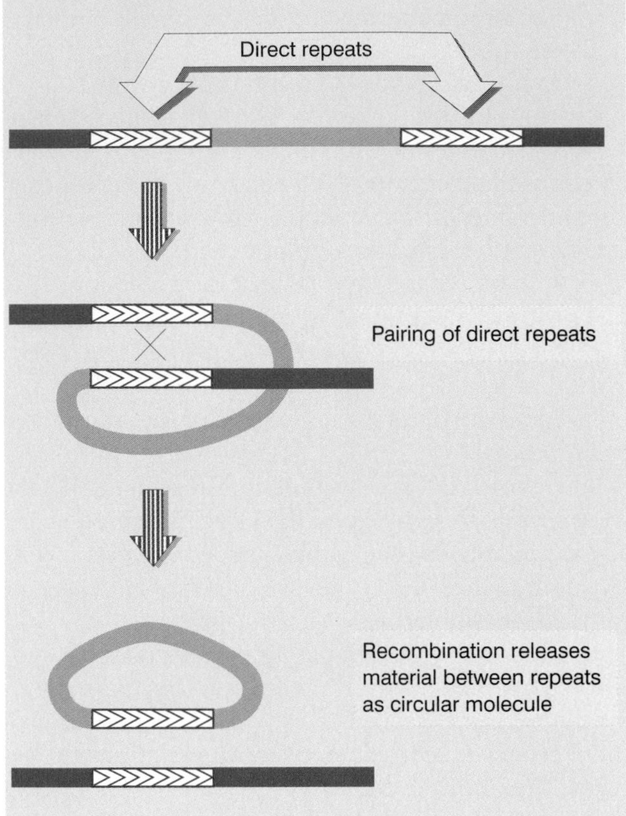
Direct repeats
Pairing of direct repeats
Recombination releases material between repeats as circular molecule

Figure 15.9 Reciprocal recombination between inverted repeats inverts the region between them.

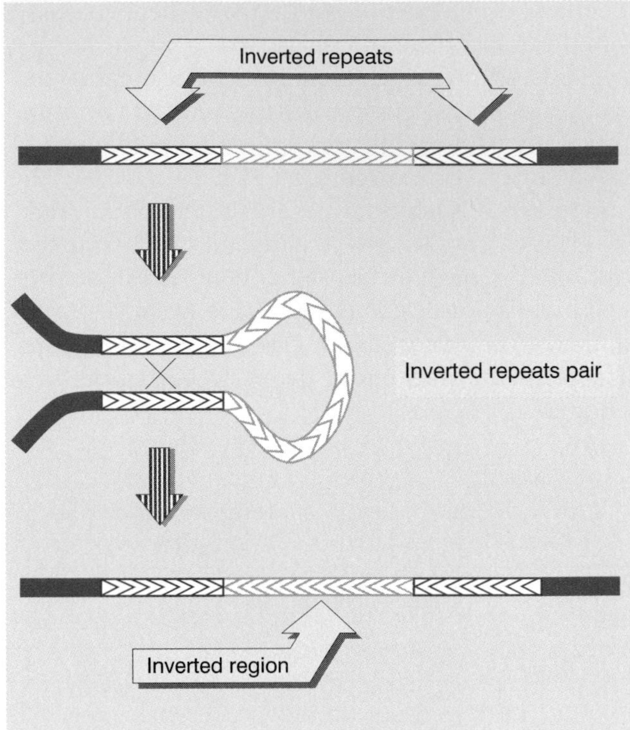
Inverted repeats
Inverted repeats pair
Inverted region

transposons probably result from a variation in the pathway followed in the transposition event itself.

Figure 15.9 depicts the consequences of a reciprocal recombination between a pair of inverted repeats. The region between the repeats becomes inverted; the repeats themselves remain available to sponsor further inversions. Note that a composite transposon whose modules are inverted is a stable component of the genome, although the direction of the central region with regard to the modules could be inverted by recombination.

Excision is not supported by transposons themselves, but may occur when bacterial enzymes recognize homologous regions in the transposons. This is important because the loss of a transposon may restore function at the site of insertion. **Precise excision** requires removal of the transposon *plus one copy of the duplicated sequence.* This is rare; it occurs at a frequency of $\sim 10^{-6}$ for Tn5 and $\sim 10^{-9}$ for Tn10. It probably involves a recombination between the 9 bp duplicated target sites.

Imprecise excision leaves a remnant of the transposon. Although the remnant may be sufficient to prevent reactivation of the target gene, it may be insufficient to cause polar effects in adjacent genes, so that a change of phenotype occurs. Imprecise excision occurs at a frequency of $\sim 10^{-6}$ for Tn10. It involves recombination between sequences of 24 bp in the IS10 modules; these sequences are inverted repeats, but since the IS10 modules themselves are inverted, they form direct repeats in Tn10.

The greater frequency of imprecise excision compared with precise excision probably reflects the increase in the length of the direct repeats (24 bp as opposed to 9 bp). Neither type of excision relies on transposon-coded functions, but the mechanism is not known. Excision is RecA-independent and could occur by some cellular mechanism that generates spontaneous deletions between closely spaced repeated sequences.

Common intermediates for transposition

Many mobile DNA elements transpose from one chromosomal location to another by a fundamentally similar mechanism. They include IS elements, prokaryotic and eukaryotic transposons, and bacteriophage Mu. Insertion of the DNA copy of retroviral RNA uses a similar mechanism (see Chapter 16). The first stages of immunoglobulin recombination also are similar (see Chapter 24).

Transposition starts with a common mechanism for joining the transposon to its target. **Figure 15.10** shows that the transposon is nicked at both ends, and the target site is nicked on both strands. The nicked ends are joined crosswise to generate a covalent connection between the transposon and the target. The two ends of the transposon are brought together in this process; for simplicity in following the cleavages, the synapsis stage is shown after cleavage, but actually occurs previously.

We know most about this process for the transposition of phage Mu, which uses the process of transposition in two ways. Upon infecting a host cell, Mu integrates into the genome by nonreplicative transposition; during the ensuing lytic cycle, the number of copies is amplified by replicative transposition. Both types of transposition involve the same type of reaction between the transposon and its target, but the subsequent reactions are different.

The initial manipulations of the phage DNA are performed by the MuA transposase. Three MuA-binding sites with a 22 bp consensus are located at each end of Mu DNA. L1, L2, and L3 are at the left end; R1, R2, and R3 are at the right end. A monomer of MuA can bind to each site. MuA also binds to an internal site in the phage genome. Binding of MuA at both the left and right ends and the internal site forms a complex. The role of the internal site is not clear, but it appears to be necessary for formation of the complex, but not for strand cleavage and subsequent steps.

Joining the Mu transposon DNA to a target site passes through the three stages illustrated in **Figure 15.11**. This involves only the two sites closest to each end of the transposon. MuA subunits bound to these sites form a tetramer. This achieves synapsis of the two ends of the transposon. The tetramer now functions in a way that ensures a coordinated reaction on both ends of Mu DNA. MuA has two sites for manipulating DNA, and their mode of action compels subunits of the transposase to act in *trans*. The *consensus-binding site*

Figure 15.10 Transposition is initiated by nicking the transposon ends and target site and joining the nicked ends into a strand transfer complex.

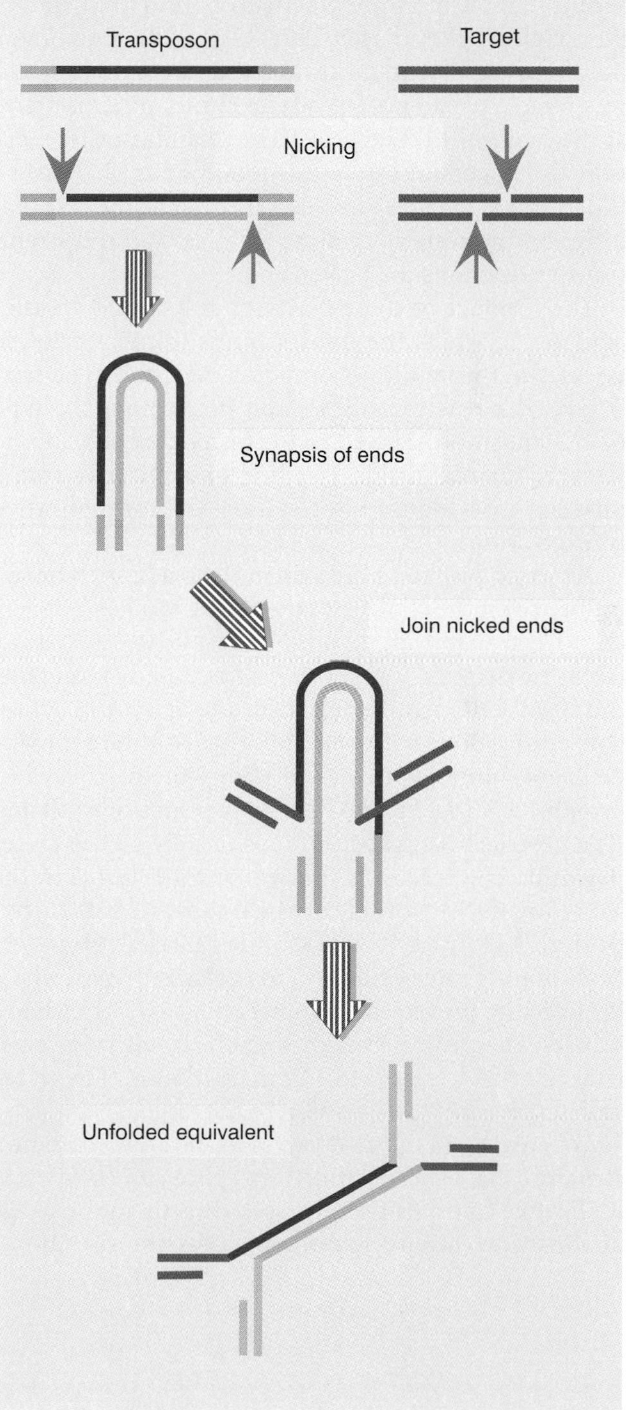

binds to the 22 bp sequences that constitute the L1, L2, R1, and R2 sites. The *active site* cleaves the Mu DNA strands at positions adjacent to the MuA-binding sites L1 and R1. But the active site cannot cleave the DNA sequence that is adjacent to the consensus sequence in the consensus-binding site. However, it can cleave the appropriate sequence on a different stretch of DNA.

The ends of the transposon are thus cleaved by MuA subunits acting in *trans*. The *trans* mode of action means that the monomers actually bound to L1 and R1 do not cleave the adjacent sites. One of the monomers bound to the left end nicks the site at the right end, and *vice versa*. (We do not know which monomer is active at this stage of the reaction.) The strand transfer reaction also occurs in *trans*; the monomer at L1 transfers the strand at R1, and *vice versa*. It could be the case that different monomers catalyze the cleavage and strand transfer reactions for a given end.

The product of these reactions is a strand transfer complex in which the transposon is connected to the target site through one strand at each end. The next step of the reaction differs and determines the type of transposition. The strand transfer complex can be a target for replication (leading to replicative transposition) or for repair (leading to nonreplicative transposition).

A second protein, MuB, assists the reaction. It has an influence on the choice of target sites. Mu has a preference for transposing to a target site >10–15 kb away from the original insertion. This is called "target immunity". It is demonstrated in an *in vitro* reaction containing donor (Mu-containing) and target (Mu-deficient) plasmids, MuA and MuB proteins, *E. coli* HU protein, and Mg^{2+} and ATP. The presence of MuB and ATP restricts transposition exclusively to the target plasmid. The reason is that when MuB binds to the MuA-Mu DNA complex, MuA causes MuB to hydrolyze ATP, after which MuB is released. However, MuB binds (nonspecifically) to the target DNA, where it stimulates the recombination activity of MuA when a transposition complex forms. In effect, the prior presence of MuA "clears" MuB from the donor, thus giving a preference for transposition to the target.

We now see in the next two sections how a common structure can be a substrate for replication (leading to replicative transposition) or used directly for breakage and reunion (leading to nonreplicative transposition).

Figure 15.11 A Mu transposon passes through three stable stages. MuA transposase forms a tetramer that synapses the ends of phage Mu. Transposase subunits act in *trans* to nick each end of the DNA; then a second *trans* action joins the nicked ends to the target DNA.

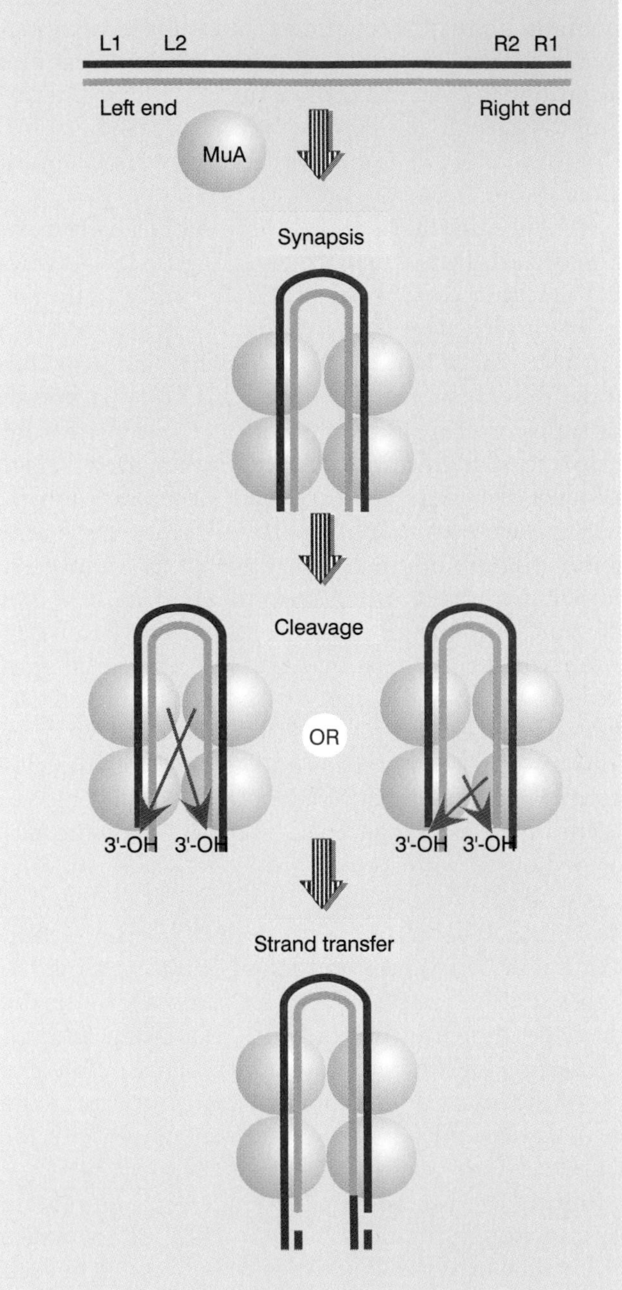

Replicative transposition proceeds through a cointegrate

THE basic structures involved in replicative transposition are illustrated in **Figure 15.12**:

- The 3′ ends of the strand transfer complex are used as primers for replication. This generates a structure called a **cointegrate**, which represents a fusion of the two original molecules. The cointegrate has two copies of the transposon, one at each junction between the original replicons, oriented as direct repeats. The crossover is formed by the transposase, as described in the previous section. Its conversion into the cointegrate requires host replication functions.
- A homologous recombination between the two copies of the transposon releases two individual replicons, each of which has a copy of the transposon. One of the replicons is the original donor replicon. The other is a target replicon that has gained a transposon flanked by short direct repeats of the host target sequence. The recombination reaction is called **resolution**; the enzyme activity responsible is called the **resolvase**.

The reactions involved in generating a cointegrate have been defined in detail for phage Mu, and are illustrated in **Figure 15.13**. The process starts with the formation of the strand transfer complex (sometimes also called a crossover complex). The donor and target strands are ligated so that each end of the transposon sequence is joined to one of the protruding single strands generated at the target site. The strand transfer complex generates a crossover-shaped structure held together at the duplex transposon. The fate of the crossover structure determines the mode of transposition.

The principle of replicative transposition is that replication through the transposon duplicates it, creating copies at both the target and donor sites. The product is a cointegrate.

The crossover structure contains a single-stranded region at each of the staggered ends. These regions are pseudoreplication forks that provide a template for DNA synthesis. (Use of the ends as primers for replication implies that the strand breakage must occur with a polarity that generates a 3′-OH terminus at this point.)

If replication continues from both the pseudoreplication forks, it will proceed through the transposon, separating its strands, and terminating at its ends. Replication is probably accomplished by host-coded functions. At this juncture, the structure has become a cointegrate, possessing direct repeats of the transposon at the junctions between the replicons (as can be seen by tracing the path around the cointegrate).

Figure 15.12 Transposition may fuse a donor and recipient replicon into a cointegrate. Resolution releases two replicons, each containing a copy of the transposon.

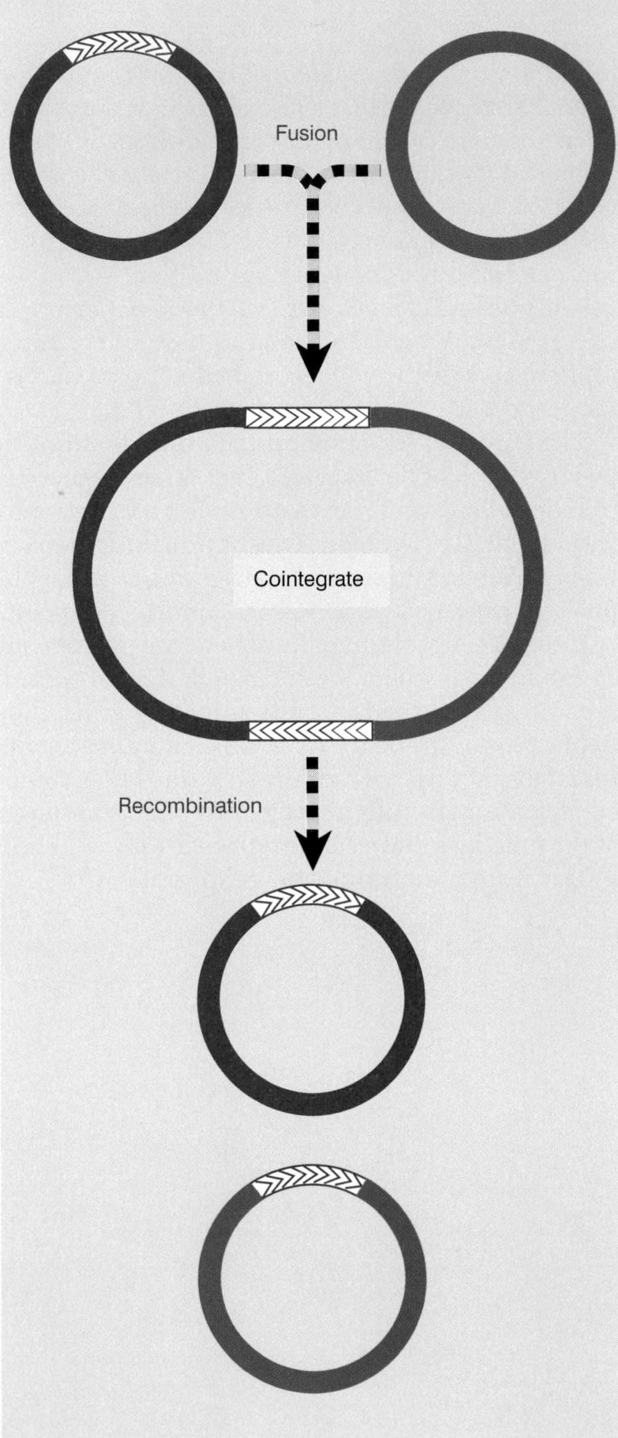
Fusion
Cointegrate
Recombination

Figure 15.13 Mu transposition generates a crossover structure, which is converted by replication into a cointegrate.

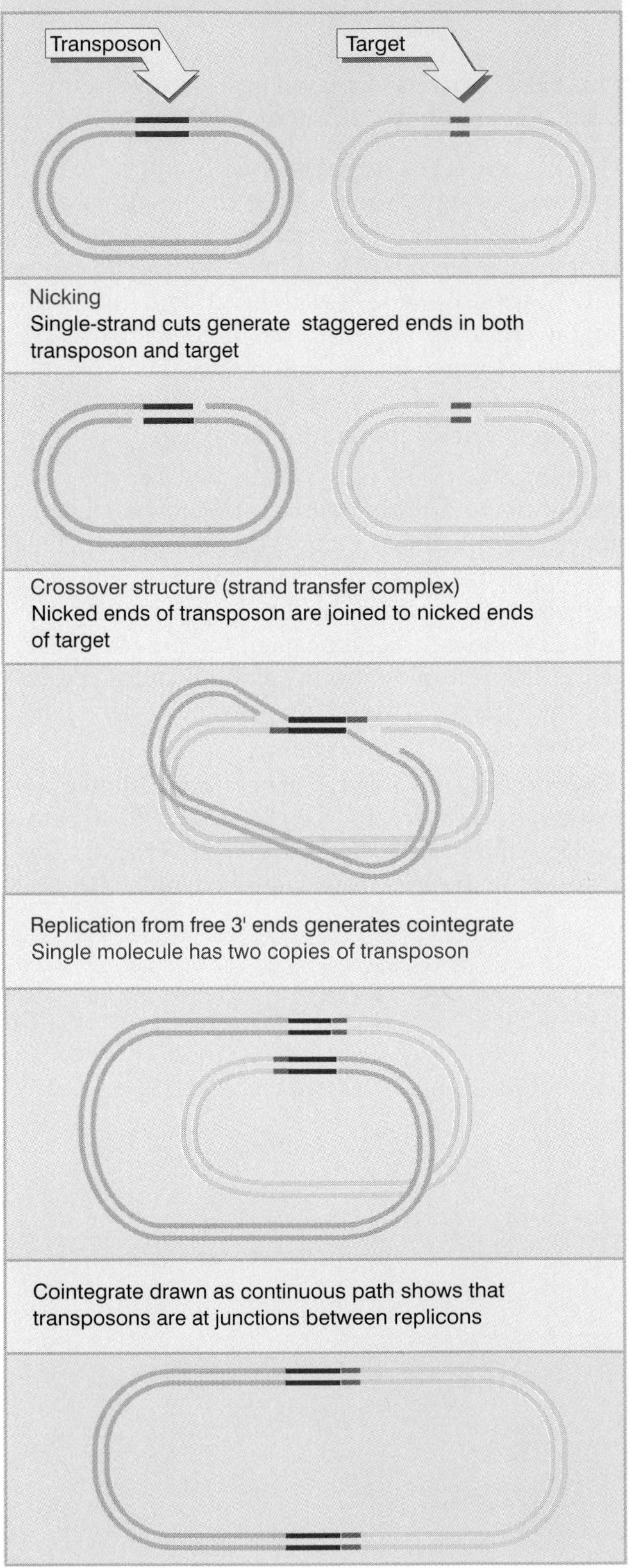
Transposon
Target
Nicking
Single-strand cuts generate staggered ends in both transposon and target
Crossover structure (strand transfer complex)
Nicked ends of transposon are joined to nicked ends of target
Replication from free 3' ends generates cointegrate
Single molecule has two copies of transposon
Cointegrate drawn as continuous path shows that transposons are at junctions between replicons

Nonreplicative transposition proceeds by breakage and reunion

The crossover structure can also be used in nonreplicative transposition. The principle of nonreplicative transposition by this mechanism is that a breakage and reunion reaction allows the target to be reconstructed; the donor remains broken. No cointegrate is formed.

Figure 15.14 shows the cleavage events that generate nonreplicative transposition of phage Mu. Once the unbroken donor strands have been nicked, the target strands on either side of the transposon can be ligated. The single-stranded regions generated by the staggered cuts must be filled in by repair synthesis. The product of this reaction is a target replicon in which the transposon has been inserted between repeats of the sequence created by the original single-strand nicks. The donor replicon has a double-strand break across the site where the transposon was originally located.

Figure 15.14 Nonreplicative transposition results when a crossover structure is released by nicking. This inserts the transposon into the target DNA, flanked by the direct repeats of the target, and the donor is left with a double-strand break.

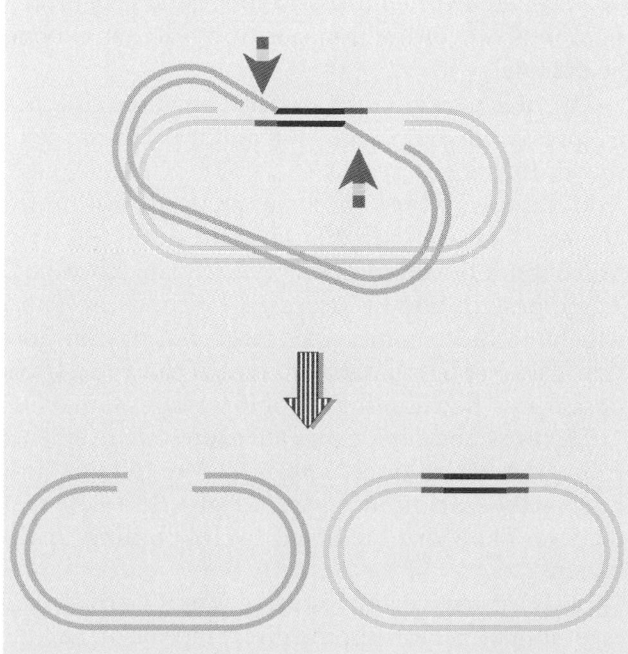

Nonreplicative transposition can also occur by an alternative pathway in which nicks are made in target DNA, but a double-strand break is made on either side of the transposon, releasing it entirely from flanking donor sequences (as envisaged in Figure 15.6). This

Figure 15.15 Both strands of Tn10 are cleaved sequentially, and then the transposon is joined to the nicked target site.

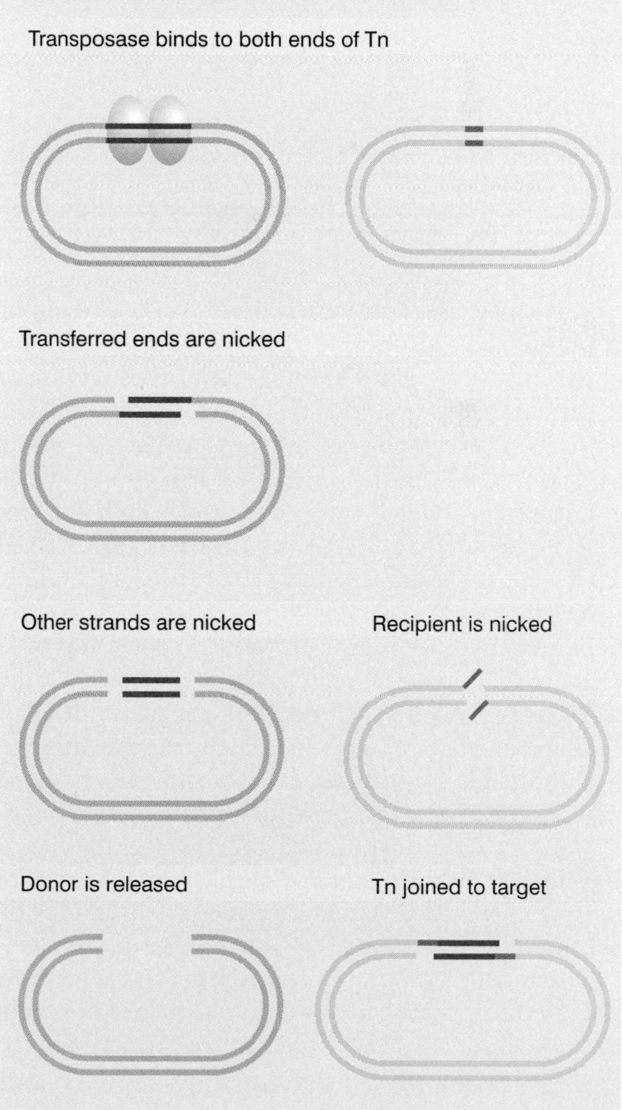
Transposase binds to both ends of Tn
Transferred ends are nicked
Other strands are nicked
Recipient is nicked
Donor is released
Tn joined to target

"cut and paste" pathway is used by Tn10, as illustrated in **Figure 15.15**.

A neat experiment to prove that Tn10 transposes nonreplicatively made use of an artificially constructed heteroduplex of Tn10 that contained single base mismatches. If transposition involves replication, the transposon at the new site will contain information from only one of the parent Tn10 strands. But if transposition takes place by physical movement of the existing transposon, the mismatches will be conserved at the new site, which proved to be the case.

The basic difference in Figure 15.15 from the model of Figure 15.14 is that both strands of Tn10 are cleaved before any connection is made to the target site. The first step in the reaction is recognition of the transposon ends by the transposase, forming a proteinaceous structure within which the reaction occurs. At each end of the transposon, the strands are cleaved in a specific order—first the transferred strand (the one to be connected to the target site) is cleaved, then the other strand (this is the same order as in the Mu transposition of Figures 15.13 and 15.14). The cleaved donor DNA is released at this point, and the transposon is joined to the nicked ends at the target site. The transposon and the target site remain constrained in the proteinaceous structure created by the transposase (and other proteins). The double-strand cleavage at each end of the transposon precludes any replicative-type transposition and forces the reaction to proceed by nonreplicative transposition, thus giving the same outcome as in Figure 15.14, but with the individual cleavage and joining steps occurring in a different order. The Tn10 transposase functions as a dimer, and each monomer has an active site that successively catalyzes the double-strand breakage of the two transposon strands and the staggered cleavage of the target site.

TnA transposition requires transposase and resolvase

REPLICATIVE transposition is the only mode of mobility of the TnA family, which consists of large (~5 kb) transposons. They are not composites relying on IS-type transposition modules, but comprise independent units carrying genes for transposition as well as for features such as drug resistance. The TnA family includes several related transposons, of which Tn3 and Tn1000 (formerly called $\gamma\delta$) are the best characterized. They have the usual terminal feature of closely related inverted repeats, generally ~38 bp in length. *Cis*-acting deletions in either repeat prevent transposition of an element. A 5 bp direct repeat is generated at the target site. They carry resistance markers such as *amp*^{r}.

The two stages of TnA-mediated transposition are accomplished by the transposase and the resolvase, whose genes, *tnpA* and *tnpR*, are identified by recessive mutations. The transposition stage involves the ends of the element, as it does in IS-type elements. Resolution requires a specific internal site, a feature unique to the TnA family.

Mutants in *tnpA* cannot transpose. The gene product is a transposase that binds to a sequence of ~25 bp located within the 38 bp of the inverted terminal repeat. A binding site for the *E. coli* protein IHF exists adjacent to the transposase binding site; and transposase and IHF bind cooperatively. The transposase recognizes the ends of the element and also makes the staggered 5 bp breaks in target DNA where the transposon is to be inserted. IHF is a DNA-binding protein that is often involved in assembling large structures in *E. coli*; its role in the transposition reaction may not be essential.

The *tnpR* gene product has dual functions. It acts as a repressor of gene expression and it provides the resolvase function.

Mutations in *tnpR* increase the transposition frequency. The reason is that TnpR represses the transcription of both *tnpA* and its own gene. So inactivation of TnpR protein allows increased synthesis of TnpA, which results in an increased frequency of transposition. This implies that the amount of the TnpA transposase must be a limiting factor in transposition.

The *tnpA* and *tnpR* genes are expressed divergently from an A•T-rich intercistronic control region, indicated in the map of Tn3 given in **Figure 15.16**. Both effects of TnpR are mediated by its binding in this region.

In its capacity as the resolvase, TnpR is involved in recombination between the direct repeats of Tn3 in a cointegrate structure. A cointegrate can in principle be

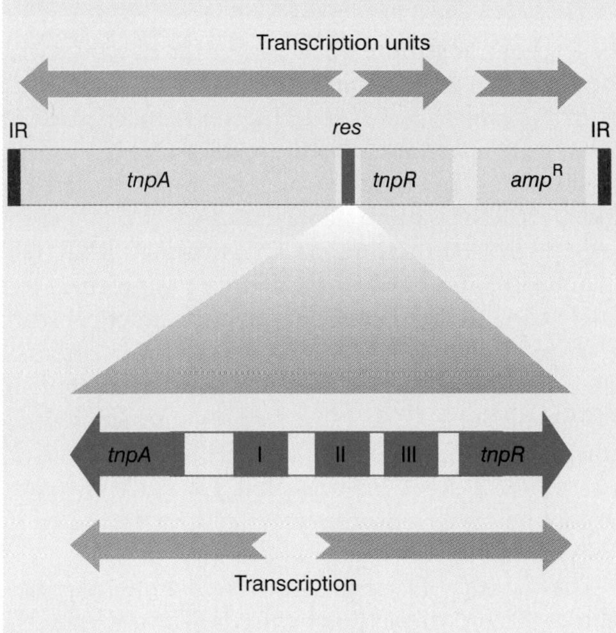

Figure 15.16 Transposons of the TnA family have inverted terminal repeats, an internal *res* site, and three known genes.

resolved by a homologous recombination between any corresponding pair of points in the two copies of the transposon. But the Tn3 resolution reaction occurs only at a specific site.

The site of resolution is called *res*. It is identified by *cis*-acting deletions that block completion of transposition, causing the accumulation of cointegrates. In the absence of *res*, the resolution reaction can be substituted by RecA-mediated general recombination, but this is much less efficient.

The sites bound by the TnpR resolvase are summarized in the lower part of Figure 15.16. Binding occurs independently at each of three sites, each 30–40 bp long. The three binding sites share a sequence homology that defines a consensus sequence with dyad symmetry.

Site I includes the region genetically defined as the *res* site; in its absence, the resolution reaction does not proceed at all. However, resolution also involves binding at sites II and III, since the reaction proceeds only poorly if either of these sites is deleted. Site I overlaps with the startpoint for *tnpA* transcription. Site II overlaps with the startpoint for *tnpR* transcription; an operator mutation maps just at the left end of the site.

Do the sites interact? One possibility is that binding at all three sites is required to hold the DNA in an appropriate topology. Binding at a single set of sites may repress *tnpA* and *tnpR* transcription without introducing any change in the DNA.

An *in vitro* resolution assay uses a cointegrate-like DNA molecule as substrate. The substrate must be supercoiled; its resolution produces two catenated circles, each containing one *res* site. The reaction requires large amounts of the TnpR resolvase; no host factors are needed. Resolution occurs in a large nucleoprotein structure. Resolvase binds to each *res* site, and then the bound sites are brought together to form a structure ~10 nm in diameter. Changes in supercoiling occur during the reaction, and DNA is bent at the *res* sites by the binding of transposase.

Resolution is a nonreplicative reaction; bonds are broken and rejoined without demand for input of energy. The products identify an intermediate stage in cointegrate resolution; they consist of resolvase covalently attached to both 5′ ends of double-stranded cuts made at the *res* site. The cleavage occurs symmetrically at a short palindromic region to generate two base extensions. Expanding the view of the crossover region located in site I, we can describe the cutting reaction as:

```
5′ T T A T A A 3′
3′ A A T A T T 5′
             ↓
5′ T T A T          +  protein—A A 3′
3′ A A—protein             T A T T 5′
```

The reaction resembles the action of lambda Int at the *att* sites (see Chapter 14). Indeed, 15 of the 20 bp of the *res* site are identical to the bases at corresponding positions in *att*. This suggests that the site-specific recombination of lambda and resolution of TnA have evolved from a common type of recombination reaction; and indeed, we shall see in Chapter 24 that recombination involving immunoglobulin genes has the same basis.

The reactions themselves are analogous in terms of manipulation of DNA, although resolution occurs only between intramolecular sites, whereas the recombination between *att* sites is intermolecular and directional (as seen by the differences in *attB* and *attP* sites). However, the mechanism of protein action is different in each case. Resolvase functions in a manner in which four subunits bind to the recombining *res* sites. Each subunit makes a single-strand cleavage. Then a reorganization of the subunits relative to one another physically moves the DNA strands, placing them in a recombined conformation. Then the nicks can be sealed.

Transposition of Tn10 has multiple controls

CONTROL of the frequency of transposition is important for the cell. A transposon must be able to maintain a certain minimum frequency of movement in order to survive; but too great a frequency could be damaging to the host cell. Every transposon appears to have mechanisms that control its frequency of transposition. A variety of mechanisms has been characterized for Tn10.

Tn10 is a composite transposon in which the element IS10R provides the active module. The organization of IS10R is summarized in **Figure 15.17.** Two promoters are found close to the outside boundary. The promoter P_{IN} is responsible for transcription of IS10R. The promoter P_{OUT} causes transcription to proceed toward the adjacent flanking DNA. Transcription usually terminates within the transposon, but occasionally continues into the host DNA; sometimes this readthrough transcription is responsible for activating adjacent bacterial genes.

The phenomenon of "multicopy inhibition" reveals that expression of the IS10R transposase gene is regulated. Transposition of a Tn10 element on the bacterial chromosome is reduced when additional copies of IS10R are introduced via a multicopy plasmid. The inhibition requires the P_{OUT} promoter, and is exercised at the level of translation. The basis for the effect lies with the overlap in the 5′ terminal regions of the transcripts from P_{IN} and P_{OUT}. *OUT* RNA is a transcript of 69 bases. It is present at >100× the level of *IN* RNA for two reasons: P_{OUT} is a much stronger promoter than P_{IN}; and *OUT* RNA is more stable than *IN* RNA.

OUT RNA functions as an antisense RNA (see Chapter 10). The level of *OUT* RNA has no effect in a single-copy situation, but has a significant effect when >5 copies are present. There are usually ~5 copies of *OUT* RNA per copy of IS10 (which corresponds to ~150 copies of *OUT* RNA in a typical multicopy situation). *OUT* RNA base pairs with *IN* RNA; and the excess of *OUT* RNA ensures that *IN* RNA is bound rapidly, before a ribosome can attach. So the paired *IN* RNA cannot be translated.

The quantity of transposase protein is often a critical feature. Tn10, whose transposase is synthesized at the low level of 0.15 molecules per cell per generation, displays several interesting mechanisms. **Figure 15.18**

Figure 15.17 Two promoters in opposite orientation lie near the outside boundary of IS10R. The strong promoter P_{OUT} sponsors transcription toward the flanking host DNA. The weaker promoter P_{IN} causes transcription of an RNA that extends the length of IS10R and is translated into the transposase.

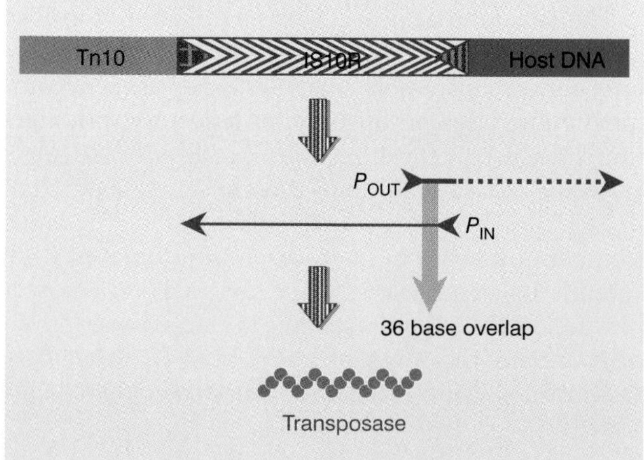

Figure 15.18 Several mechanisms restrain the frequency of Tn10 transposition, by affecting either the synthesis or function of transposase protein. Transposition of an individual transposon is restricted by methylation to occur only after replication. In multicopy situations, *cis*-preference restricts the choice of target, and OUT/IN RNA pairing inhibits synthesis of transposase.

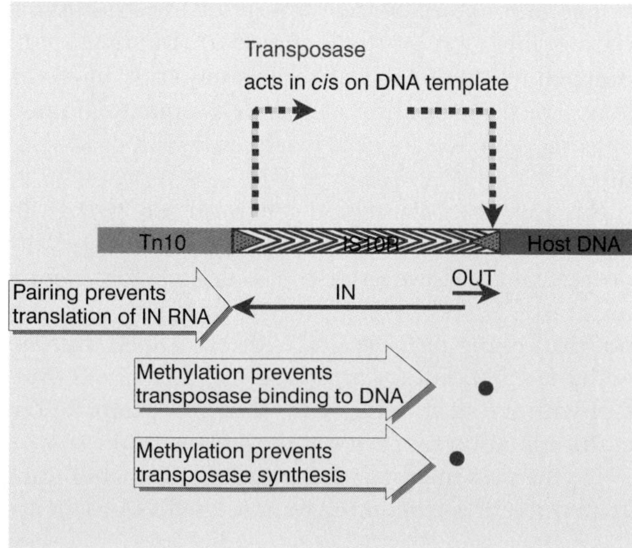

summarizes the various effects that influence transposition frequency.

A continuous reading frame on one strand of IS10R codes for the transposase. The level of the transposase limits the rate of transposition. Mutants in this gene can be complemented in *trans* by another, wild-type IS10 element, but only with some difficulty. This reflects a strong preference of the transposase for *cis*-action; the enzyme functions efficiently only with the DNA template from which it was transcribed and translated. *Cis*-preference is a common feature of transposases coded by IS elements. (Other proteins that display *cis*-preference include the A protein involved in ϕX174 replication; see Chapter 12.)

Does *cis*-preference reflect an ability of the transposase to recognize more efficiently those DNA target sequences that lie nearer to the site where the enzyme is synthesized? One possible explanation is that the transposase binds to DNA so tightly after (or even during) protein synthesis that it has a very low probability of diffusing elsewhere. Another possibility is that the enzyme may be unstable when it is not bound to DNA, so that protein molecules failing to bind quickly (and therefore nearby) never have a chance to become active.

Together the results of *cis*-preference and multicopy inhibition ensure that an increase in the number of copies of Tn10 in a bacterial genome does not cause an increased frequency of transposition that could damage the genome.

The effects of *dam* methylation provide the most important system of regulation for an individual element. They reduce the frequency of transposition and (more importantly) couple transposition to passage of the replication fork. The ability of IS10 to transpose is related to the replication cycle by the transposon's response to the state of methylation at two sites. One site is within the inverted repeat at the end of IS10R, where the transposase binds. The other site is in the promoter P_{IN}, from which the transposase gene is transcribed.

Both of these sites are methylated by the *dam* system described in Chapter 13. The Dam methylase modifies the adenine in the sequence GATC on a newly synthesized strand generated by replication. The frequency of Tn10 transposition is increased 1000-fold in *dam*$^-$ strains in which the two target sites lack methyl groups.

Passage of a replication fork over these sites generates hemimethylated sequences; this activates the transposon by a combination of transcribing the transposase gene more frequently from P_{IN} and enhancing binding of transposase to the end of IS10R. In a wild-type bacterium, the sites remain hemimethylated for a short period after replication.

Why should it be desirable for transposition to occur soon after replication? The nonreplicative mechanism of Tn10 transposition places the donor DNA at risk of being destroyed (see Figure 15.6). The cell's chances of survival may be increased if replication has just occurred to generate a second copy of the donor sequence. The mechanism is effective because only one of the two newly replicated copies gives rise to a transposition event (because it matters which strand of the transposon is unmethylated at the *dam* sites).

Since a transposon selects its target site at random, there is a reasonable probability that it may land in an active operon. Will transcription from the outside continue through the transposon and thus activate the transposase, whose overproduction may in turn lead to high (perhaps lethal) levels of transposition? Tn10 protects itself against such events by two mechanisms. Transcription across the IS10R terminus decreases its activity, presumably by inhibiting its ability to bind transposase. And the mRNA that extends from upstream of the promoter is poorly translated, because it has a secondary structure in which the initiation codon is inaccessible.

Controlling elements in maize cause breakage and rearrangements

ONE of the most visible consequences of the existence and mobility of transposons occurs during plant development, when somatic variation occurs. This is due to changes in the location or behavior of **controlling elements** (the name that transposons were given in maize before their molecular nature was discovered).

Two features of maize have helped to follow

transposition events. Controlling elements often insert near genes that have visible but nonlethal effects on the phenotype. And because maize displays clonal development, the occurrence and timing of a transposition event can be visualized as depicted diagrammatically in **Figure 15.19**.

The nature of the event does not matter: it may be an insertion, excision, or chromosome break. What is important is that it occurs in a heterozygote to alter the expression of one allele. Then the descendants of a cell that has suffered the event display a new phenotype, while the descendants of cells not affected by the event continue to display the original phenotype.

Figure 15.19 Clonal analysis identifies a group of cells descended from a single ancestor in which a transposition- mediated event altered the phenotype. Timing of the event during development is indicated by the number of cells; tissue specificity of the event may be indicated by the location of the cells.

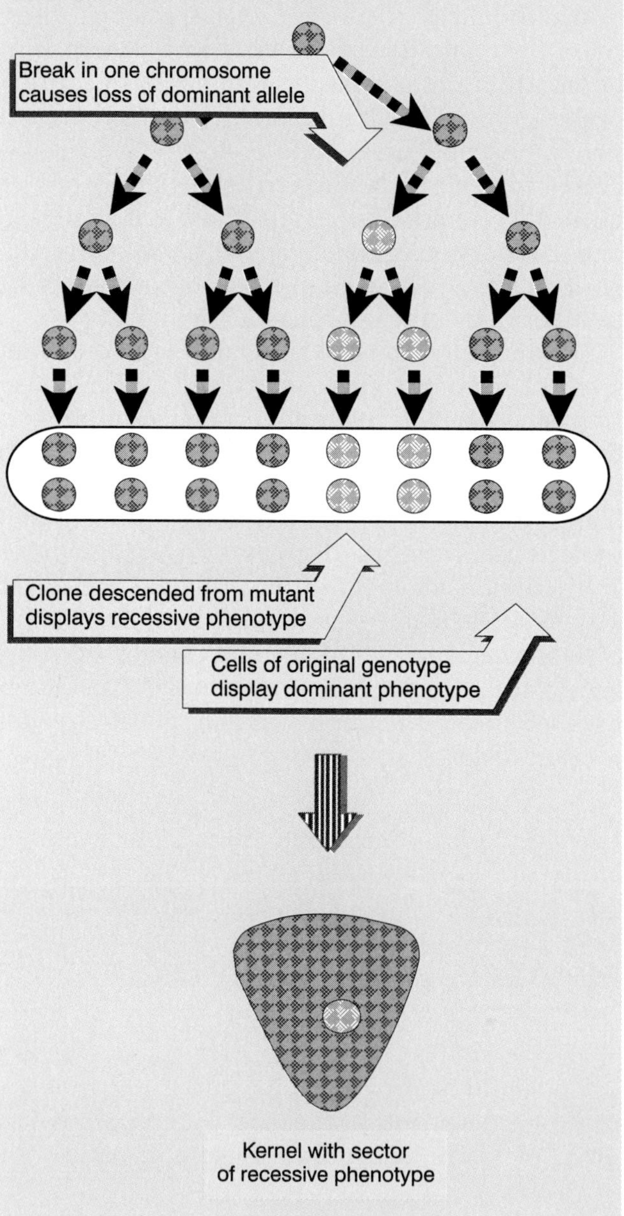

Mitotic descendants of a given cell remain in the same location and thus give rise to a **sector** of tissue. A change in phenotype during somatic development is called **variegation**; it is revealed by a sector of the new phenotype residing within the tissue of the original phenotype. The size of the sector depends on the number of divisions in the lineage giving rise to it; so the size of the area of the new phenotype is determined by the timing of the change in genotype. The earlier its occurrence in the cell lineage, the greater the number of descendants and thus the size of patch in the mature tissue. This is seen most vividly in the variation in kernel color, when patches of one color appear within another color.

Insertion of a controlling element may affect the activity of adjacent genes. Deletions, duplications, inver-

Figure 15.20 A break at a controlling element causes loss of an acentric fragment; if the fragment carries the dominant markers of a heterozygote, its loss changes the phenotype. The effects of the dominant markers, *Cl, Bz, Wx*, can be visualized by the color of the cells or by appropriate staining.

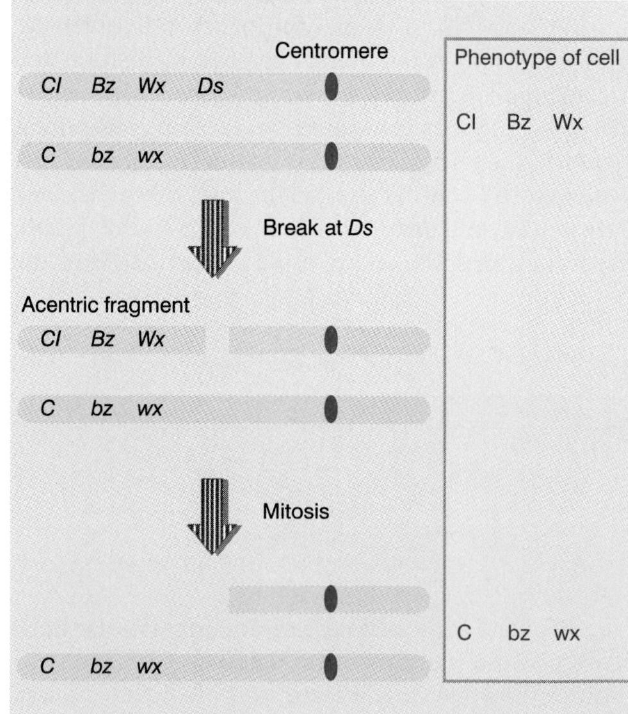

sions, and translocations all occur at the sites where controlling elements are present. Chromosome breakage is a common consequence of the presence of some elements. A unique feature of the maize system is that the activities of the controlling elements are regulated during development. The elements transpose and promote genetic rearrangements at characteristic times and frequencies during plant development.

The characteristic behavior of controlling elements in maize is typified by the *Ds* element, which was originally identified by its ability to provide a site for chromosome breakage. The consequences are illustrated in **Figure 15.20.** Consider a heterozygote in which *Ds* lies on one homolog between the centromere and a series of dominant markers. The other homolog lacks *Ds* and has recessive markers (*C, bz, wx*). Breakage at *Ds* generates an **acentric fragment** carrying the dominant markers. Because of its lack of a centromere, this fragment is lost at mitosis. So the descendant cells have only the recessive markers carried by the intact chromosome. This gives the type of situation whose results are depicted in Figure 15.19.

Figure 15.21 shows that breakage at *Ds* leads to the formation of two unusual chromosomes. These are generated by joining the broken ends of the products of replication. One is a U-shaped acentric fragment consisting of the joined sister chromatids for the region distal to *Ds* (on the left as drawn in the figure). The other is a U-shaped **dicentric chromosome** comprising the sister chromatids proximal to *Ds* (on its right in the figure). The latter structure leads to the classic **breakage-fusion-bridge** cycle illustrated in the figure.

Follow the fate of the dicentric chromosome when it attempts to segregate on the mitotic spindle. Each of its two centromeres pulls toward an opposite pole. The tension breaks the chromosome at a random site between the centromeres. In the example of the figure, breakage occurs between loci *A* and *B*, with the result that one daughter chromosome has a duplication of *A*, while the other has a deletion. If *A* is a dominant marker, the cells with the duplication will retain **A** phenotype, but cells with the deletion will display the recessive **a** phenotype.

The breakage-fusion-bridge cycle continues through further cell generations, allowing genetic changes to continue in the descendants. For example, consider the deletion chromosome that has lost *A*. In the next cycle, a break occurs between *B* and *C*, so that the descendants are divided into those with a duplication of *B* and those with a deletion. Successive losses of dominant markers are revealed by subsectors within sectors.

Figure 15.21 *Ds* provides a site to initiate the chromatid fusion-bridge-breakage cycle. The products can be followed by clonal analysis.

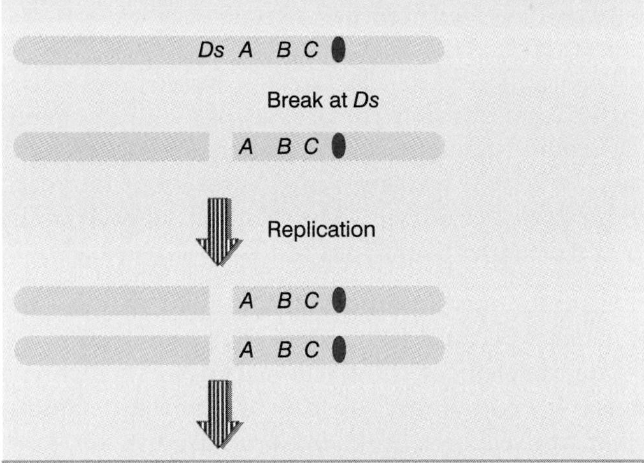

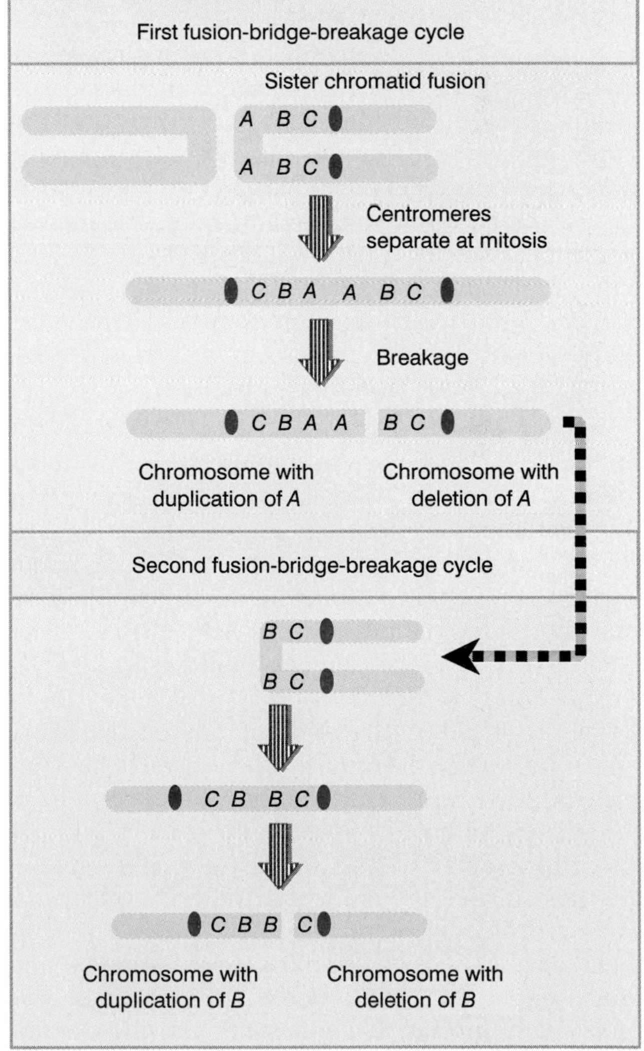

Controlling elements in maize form families of transposons

THE maize genome contains several families of controlling elements. The numbers, types, and locations of the elements are characteristic for each individual maize strain. The members of each family are divided into two classes:

- **Autonomous elements** have the ability to excise and transpose. Because of the continuing activity of an autonomous element, its insertion at any locus creates an unstable or "mutable" allele. Loss of the autonomous element itself, or of its ability to transpose, converts a mutable allele to a stable allele.
- **Nonautonomous elements** are stable; they do not transpose or suffer other spontaneous changes in condition. They become unstable only when an autonomous member of the same family is present elsewhere in the genome. When complemented in *trans* by an autonomous element, a nonautonomous element displays the usual range of activities associated with autonomous elements, including the ability to transpose to new sites. Nonautonomous elements are derived from autonomous elements by loss of *trans*-acting functions needed for transposition.

Families of controlling elements are defined by the interactions between autonomous and nonautonomous elements. A family consists of a single type of autonomous element accompanied by many varieties of nonautonomous elements. A nonautonomous element is placed in a family by its ability to be activated in *trans* by the autonomous elements. The major families of controlling elements in maize are summarized in **Figure 15.22.**

Characterized at the molecular level, the maize transposons share the usual form of organization—inverted repeats at the ends and short direct repeats in the adjacent target DNA—but otherwise vary in size and coding capacity. The best characterized families are those of the *Ac* elements and the *Spm* elements. There are typically several members (~10) of each family in a plant genome. By analyzing autonomous and nonautonomous elements of the *Ac/Ds* family, we have molecular information about many individual examples of these elements. **Figure 15.23** summarizes their structures.

Most of the length of the autonomous *Ac* element is occupied by a single gene consisting of five exons. The product is the transposase. The element itself ends in inverted repeats of 11 bp; and a target sequence of 8 bp is duplicated at the site of insertion.

Ds elements vary in both length and sequence, but are related to *Ac*. They end in the same 11 bp inverted repeats. They are shorter than *Ac*, and the length of deletion varies. At one extreme, the element *Ds9* has a deletion of only 194 bp. In a more extensive deletion, the *Ds6* element retains a length of only 2 kb, representing 1 kb from each end of *Ac*. A complex double *Ds* element has one *Ds6* sequence inserted in reverse orientation into another.

Nonautonomous elements lack internal sequences, but possess the terminal inverted repeats (and possibly other sequence features). *Nonautonomous elements are*

Figure 15.22 Each controlling element family has both autonomous and nonautonomous members. Autonomous elements are capable of transposition. Nonautonomous elements are deficient in transposition. Pairs of autonomous and nonautonomous elements can be classified in >4 families.

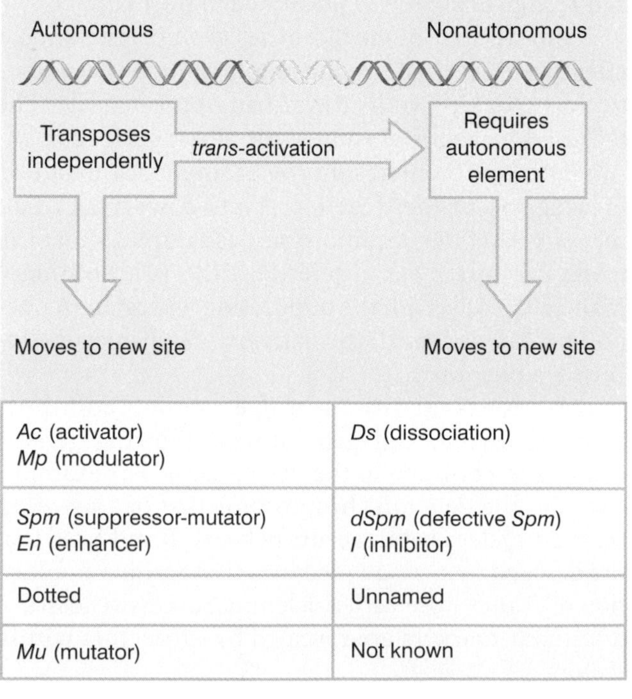

Ac (activator) *Mp* (modulator)	*Ds* (dissociation)
Spm (suppressor-mutator) *En* (enhancer)	*dSpm* (defective *Spm*) *I* (inhibitor)
Dotted	Unnamed
Mu (mutator)	Not known

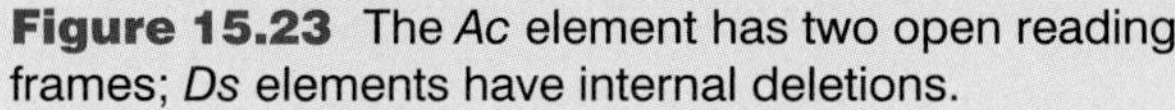

Figure 15.23 The *Ac* element has two open reading frames; *Ds* elements have internal deletions.

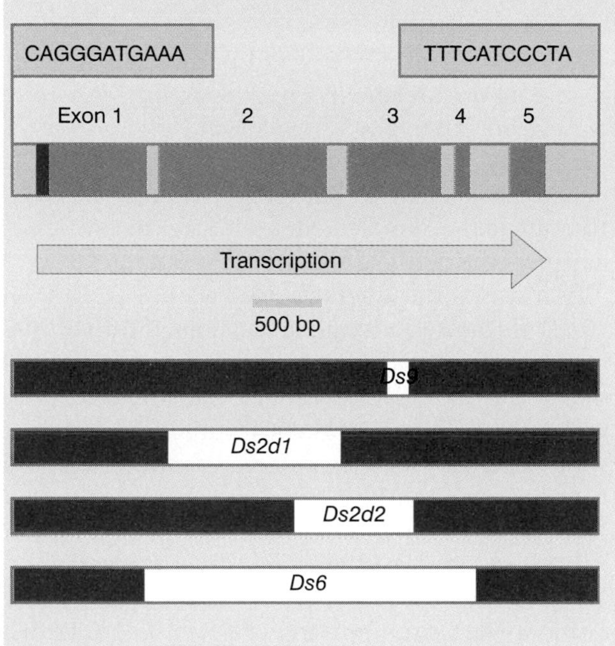

derived from autonomous elements by deletions (or other changes) that inactivate the trans*-acting transposase, but leave intact the sites (including the termini) on which the transposase acts.* Their structures range from minor (but inactivating) mutations of *Ac* to sequences that have major deletions or rearrangements.

At another extreme, the *Ds1* family members comprise short sequences whose only relationship to *Ac* lies in the possession of terminal inverted repeats. Elements of this class need not be directly derived from *Ac*, but could be derived by any event that generates the inverted repeats. Their existence suggests that the transposase recognizes only the terminal inverted repeats, or possibly the terminal repeats in conjunction with some short internal sequence.

Transposition of *Ac/Ds* occurs by a nonreplicative mechanism, and is accompanied by its disappearance from the donor location. Clonal analysis suggests that transposition of *Ac/Ds* almost always occurs after the donor element has been replicated. These features resemble transposition of the bacterial element Tn10. The recipient site is frequently on the same chromosome as the donor site, and often quite close to it.

Replication generates two copies of a potential *Ac/Ds* donor, but usually only one copy actually transposes. What happens to the donor site? The rearrangements that are found at sites from which controlling elements have been lost could be explained in terms of the consequences of a chromosome break, as illustrated previously in Figure 15.20.

Autonomous and nonautonomous elements are subject to a variety of changes in their condition. Some of these changes are genetic, others are epigenetic.

The major change is (of course) the conversion of an autonomous element into a nonautonomous element, but further changes may occur in the nonautonomous element. *Cis*-acting defects may render a nonautonomous element impervious to autonomous elements. So a nonautonomous element may become permanently stable because it can no longer be activated to transpose.

Autonomous elements are subject to "changes of phase", heritable but relatively unstable alterations in their properties. These take the form of a reversible inactivation in which the element cycles between an active and inactive condition during plant development.

Phase changes in both the *Ac* and *Mu* types of autonomous element appear to result from methylation of DNA. Comparisons of the susceptibilities of active and inactive elements to restriction enzymes suggest that the inactive form of the element is methylated in the target sequence $^{\text{CAG}}_{\text{GTC}}$. There are several target sites in each element, and we do not know which sites control the effect. We should like to know what controls the methylation and demethylation of the elements.

There may be self-regulating controls of transposition, analogous to the immunity effects displayed by bacterial transposons. An increase in the number of *Ac* elements in the genome decreases the frequency of transposition. The *Ac* element may code for a repressor of transposition; the activity could be carried by the same protein that provides transposase function.

Spm elements influence gene expression

The *Spm* and *En* autonomous elements are virtually identical; they differ at <10 positions. **Figure 15.24** summarizes the structure. The 13 bp inverted terminal repeats are essential for transposition, as indicated by the transposition-defective phenotype of deletions at the termini. Transposons related to *Spm* are found in other plants, and are defined as members of the same family by their generally similar organization. They all share nearly identical inverted terminal repeats, and generate 3 bp duplications of target DNA upon transposition. Named for the terminal similarities, they are known as the CACTA group of transposons.

A sequence of 8300 bp is transcribed from a promoter in the left end of the element. The 11 exons contained in the transcript are spliced into a 2500 base messenger. The mRNA codes for a protein of 621 amino acids. The gene is called *tnpA*, and the protein binds to a 12 bp consensus sequence present in multiple copies in the terminal regions of the element. Function of *tnpA* is required for excision, but may not be sufficient.

Figure 15.24 *Spm/En* has two genes. *tnpA* consists of 11 exons that are transcribed into a spliced 2500 base mRNA. *tnpB* may consist of a 6000 base mRNA containing ORF1 + ORF2.

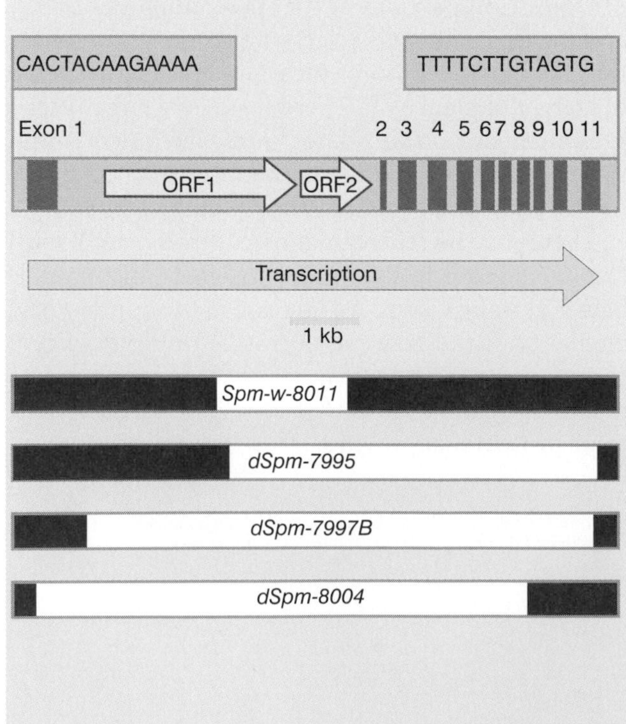

All of the nonautonomous elements of this family (denoted *dSpm* for defective *Spm*) are closely related in structure to the *Spm* element itself. They have deletions that affect the exons of *tnpA*.

Two additional open reading frames (ORF1 and ORF2) are located within the first, long intron of *tnpA*. They are contained in an alternatively spliced 6000 base RNA, which is present at 1% of the level of the *tnpA* mRNA. The (hypothetical) function containing ORFs 1 and 2 is called *tnpB*. It may provide the protein that binds to the 13 bp terminal inverted repeats to cleave the termini for transposition.

In addition to the fully active *Spm* element, there are *Spm-w* derivatives that show weaker activity in transposition. The example given in Figure 15.24 has a deletion that eliminates both ORF1 and ORF2. This suggests that the need for TnpB in transposition can be bypassed or substituted.

Spm insertions can control the expression of a gene at the site of insertion. A recipient locus may be brought under either negative or positive control. An *Spm-suppressible* locus suffers inhibition of expression. An *Spm-dependent* locus is expressed only with the aid of *Spm*. When the inserted element is a *dSpm*, suppression or dependence responds to the *trans*-acting function supplied by an autonomous *Spm*. What is the basis for these opposite effects?

A *dSpm-suppressible* allele contains an insertion of *dSpm* within an exon of the gene. This structure raises the immediate question of how a gene with a *dSpm* insertion in an exon can ever be expressed! The *dSpm* sequence can be excised from the transcript by using sequences at its termini. The splicing event may leave a change in the sequence of the mRNA, thus explaining a change in the properties of the protein for which it codes. A similar ability to be excised from a transcript has been found for some *Ds* insertions.

tnpA provides the *suppressor* function for which the *Spm* element was originally named. The presence of a defective element may reduce, but not eliminate, expression of a gene in which it resides. However, the introduction of an autonomous element, possessing a functional *tnpA* gene, may suppress expression of the target gene entirely. Suppression is caused by the ability of TnpA to bind to its target sites in the defective element, which blocks transcription from proceeding.

A *dSpm-dependent* allele contains an insertion near but not within a gene. The insertion appears to provide an enhancer that activates the promoter of the gene at the recipient locus.

Suppression and dependence at *dSpm* elements appear to rely on the same interaction between the *trans*-acting product of the *tnpA* gene of an autonomous *Spm* element and the *cis*-acting sites at the ends of the element. *So a single interaction between the protein and the ends of the element either suppresses or activates a target locus depending on whether the element is located upstream of or within the recipient gene.*

Spm elements exist in a variety of states ranging from fully active to cryptic. A cryptic element is silent and neither transposes itself nor activates *dSpm* elements. A cryptic element may be reactivated transiently or converted to the active state by interaction with a fully active *Spm* element. Inactivation is caused by methylation of sequences in the vicinity of the transcription startpoint. The nature of the events that are responsible for inactivating an element by *de novo* methylation or for activating it by demethylation (or preventing methylation) are not yet known.

The role of transposable elements in hybrid dysgenesis

CERTAIN strains of *D. melanogaster* encounter difficulties in interbreeding. When flies from two of these strains are crossed, the progeny display "dysgenic traits", a series of defects including mutations, chromosomal aberrations, distorted segregation at meiosis, and sterility. The appearance of these correlated defects is called **hybrid dysgenesis**.

Two systems responsible for hybrid dysgenesis have been identified in *D. melanogaster.* In the first, flies are divided into the types I (inducer) and R (reactive). Reduced fertility is seen in crosses of I males with R females, but not in the reverse direction. In the second system, flies are divided into the two types P (paternal contributing) and M (maternal contributing). **Figure 15.25** illustrates the asymmetry of the system; a cross between a P male and an M female causes dysgenesis, but the reverse cross does not.

Dysgenesis is principally a phenomenon of the germ cells. In crosses involving the P-M system, the F1 hybrid flies have normal somatic tissues. However, their gonads do not develop. The morphological defect in gamete development dates from the stage at which rapid cell divisions commence in the germline.

Any one of the chromosomes of a P male can induce dysgenesis in a cross with an M female. The construction of recombinant chromosomes shows that several regions within each P chromosome are able to cause dysgenesis. This suggests that a P male has a large number of **P factors**, sequences occupying many different chromosomal locations. The locations differ between individual P strains. The P factors are absent from chromosomes of M flies.

Figure 15.25 Hybrid dysgenesis is asymmetrical; it is induced by P male x M female crosses, but not by M male x P female crosses.

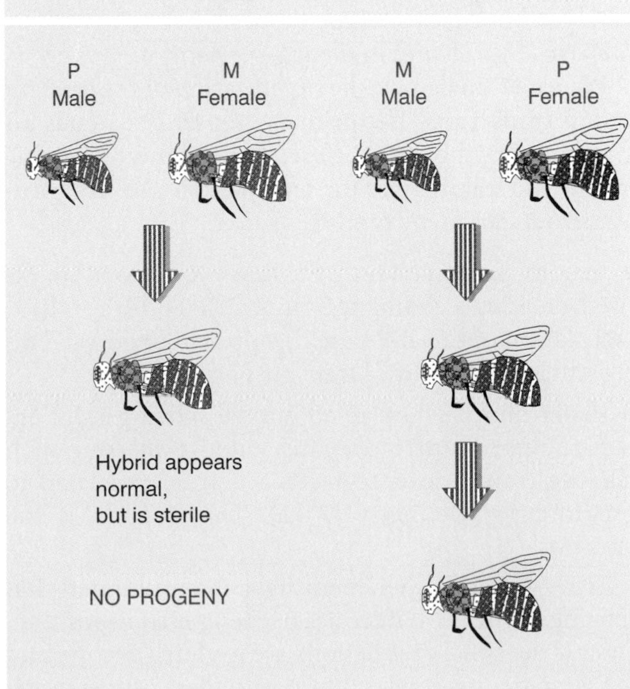

The events responsible for the induction of mutations in dysgenesis were identified by mapping the DNA of *w* mutants found among the dysgenic hybrids. All the mutations result from the insertion of DNA into the *w* locus. The inserted sequence is called the **P element**. The P insertions form a classic transposable system. Individual elements vary in length but are homologous in sequence. All P elements possess inverted terminal repeats of 31 bp, and generate direct repeats of target DNA of 8 bp upon transposition. The longest P elements are ~2.9 kb long and have four open reading frames. The shorter elements arise, apparently rather frequently, by internal deletions of a full-length P factor. At least some of the shorter P elements have lost the capacity to produce the transposase, but may be activated in *trans* by the enzyme coded by a complete P element.

A P strain carries 30–50 copies of the P factor, about a third of them full length. The factors are absent from M strains. In a P strain, the factors are carried as inert components of the genome. But they become activated to transpose when a P male is crossed with an M female.

Chromosomes from P-M hybrid dysgenic flies have P factors inserted at many new sites. The chromosome breaks typical of hybrid dysgenesis occur at hotspots that are the sites of residence of P factors. The average rate of transposition of P elements to M chromosomes is about one event per generation.

Activation of P elements is tissue-specific: it occurs only in the germline. But P elements are transcribed in both germline and somatic tissues. *Tissue-specificity is conferred by a change in the splicing pattern.*

Figure 15.26 depicts the organization of the element and its transcripts. The primary transcript extends for 2.5 kb or 3.0 kb, the difference probably reflecting merely the leakiness of the termination site. Two protein products can be produced:

- In somatic tissues, only the first two introns are excised, creating a coding region of ORF0–ORF1–ORF2. Translation of this RNA yields a protein of 66 kD. This protein is a repressor of transposon activity.
- In germline tissues, an additional splicing event occurs to remove intron 3. This connects all four open reading frames into an mRNA that is translated to generate a protein of 87 kD. This protein is the transposase.

Two types of experiment have demonstrated that splicing of the third intron is needed for transposition. First, if the splicing junctions are mutated *in vitro* and the P element is reintroduced into flies, its transposition activity is abolished. Second, if the third intron is deleted, so that ORF3 is constitutively included in the mRNA in all tissues, transposition occurs in somatic tissues as well as the germline.

So whenever ORF3 is spliced to the preceding reading frame, the P element becomes active. This is the crucial regulatory event, and usually it occurs only in the germline. What is responsible for the tissue-specific splicing? Somatic cells contain a protein that binds to sequences in exon 3 to prevent splicing of the last intron (see Chapter 21). The absence of this protein in germline cells allows splicing to generate the mRNA that codes for the transposase.

Transposition of a P element requires ~150 bp of terminal DNA. The transposase binds to 10 bp sequences that are adjacent to the 31 bp inverted repeats. Transposition occurs by a nonreplicative "cut and

Figure 15.26 The P element has four exons. The first three are spliced together in somatic expression; all four are spliced together in germline expression.

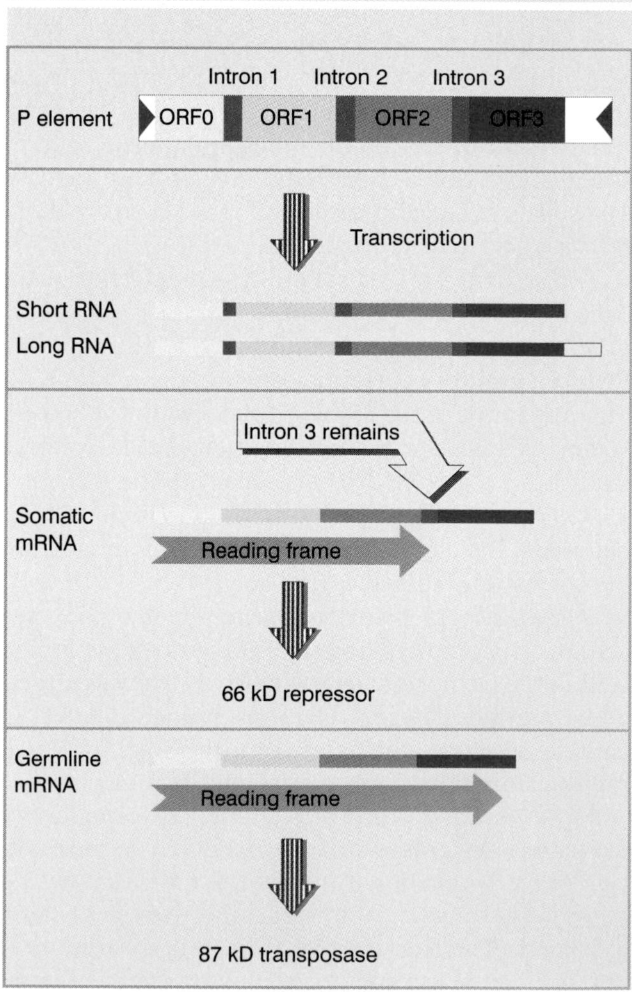

paste" mechanism resembling that of Tn10. (It contributes to hybrid dysgenesis in two ways. Insertion of the transposed element at a new site may cause mutations. And the break that is left at the donor site [as in the model of Figure 15.6] has a deleterious effect.)

It is interesting that, in a significant proportion of cases, the break in donor DNA is repaired by using the sequence of the homologous chromosome. If the homolog has a P element, the presence of a P element at the donor site may be restored (so the event resembles the result of a replicative transposition). If the homolog lacks a P element, repair may generate a sequence lacking the P element, thus apparently providing a precise excision (an unusual event in other transposable systems).

The dependence of hybrid dysgenesis on the sexual orientation of a cross shows that the cytoplasm is important as well as the P factors themselves. The contribution of the cytoplasm is described as the **cytotype**; a line of flies containing P elements has P cytotype, while a line of flies lacking P elements has M cytotype. Hybrid dysgenesis occurs only when chromosomes containing P factors find themselves in M cytotype, that is, when the male parent has P elements and the female parent does not.

Cytotype shows an inheritable cytoplasmic effect; when a cross occurs through P cytotype (the female parent has P elements), hybrid dysgenesis is suppressed for several generations of crosses with M female parents. So something in P cytotype, which can be diluted out over some generations, suppresses hybrid dysgenesis.

The effect of cytotype is explained in molecular terms by the model of **Figure 15.27**. It depends on the ability of the 66 kD protein to repress transposition.

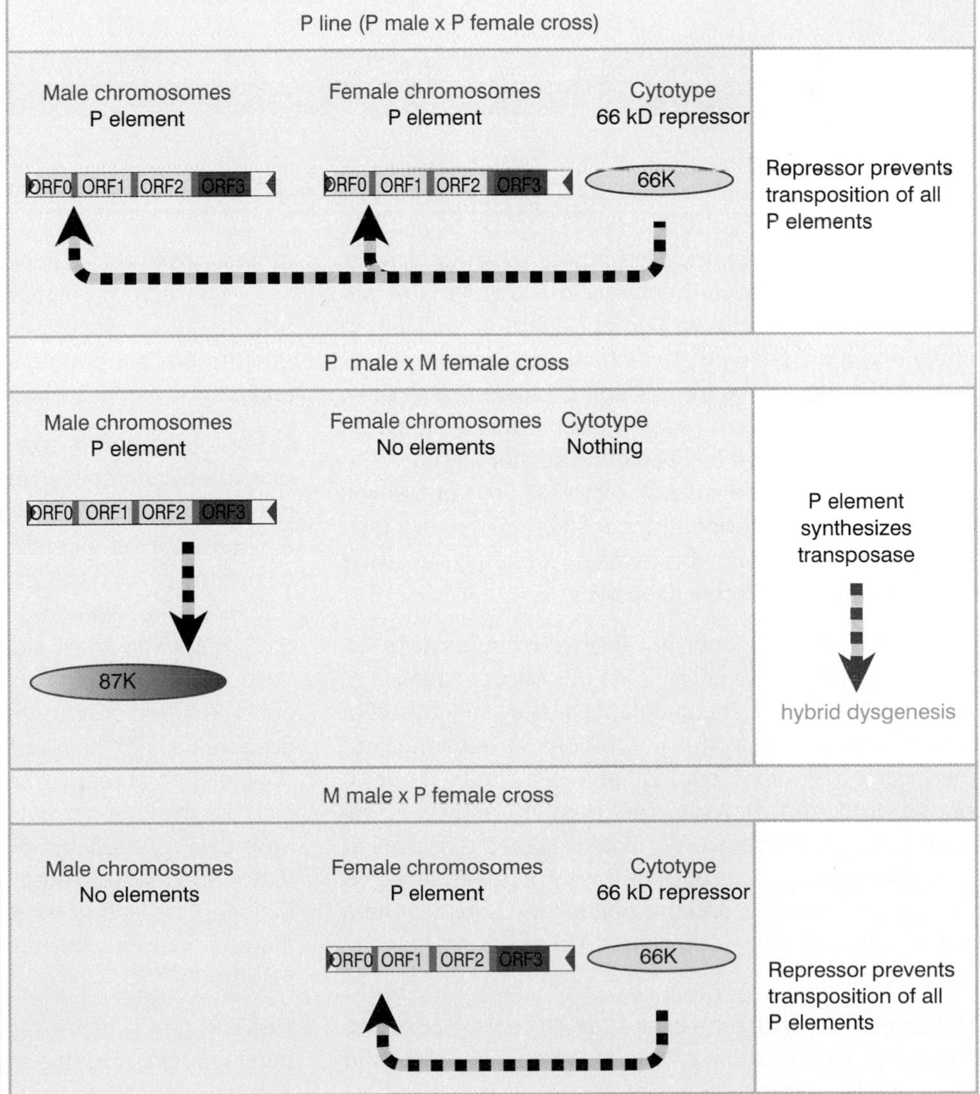

Figure 15.27 Hybrid dysgenesis is determined by the interactions between P elements in the genome and 66 kD repressor in the cytotype.

The protein is provided as a maternal factor in the egg. In a P line, there must be sufficient protein to prevent transposition from occurring, even though the P elements are present. In any cross involving a P female, its presence prevents either synthesis or activity of the transposase. But when the female parent is M type, there is no repressor in the egg, and the introduction of a P element from the male parent results in activity of transposase in the germline. The ability of P cytotype to exert an effect through more than one generation suggests that there must be enough repressor protein in the egg, and it must be stable enough, to be passed on through the adult to be present in the eggs of the next generation.

Strains of *D. melanogaster* descended from flies caught in the wild more than 30 years ago are always M. Strains descended from flies caught in the past 10 years are almost always P. Does this mean that the P element family has invaded wild populations of *D. melanogaster* in recent years? P elements are indeed highly invasive when introduced into a new population; the source of the invading element would have to be another species.

Because hybrid dysgenesis reduces interbreeding, it is a step on the path to speciation. Suppose that a dysgenic system is created by a transposable element in some geographic location. Another element may create a different system in some other location. Flies in the two areas will be dysgenic for two (or possibly more) systems. If this renders them intersterile and the populations become genetically isolated, further separation may occur. Multiple dysgenic systems therefore lead to inability to mate—and to speciation.

Summary

Prokaryotic and eukaryotic cells contain a variety of transposons that mobilize by moving or copying DNA sequences. The transposon can be identified only as an entity within the genome; its mobility does not involve an independent form. The transposon could be selfish DNA, concerned only with perpetuating itself within the resident genome; if it conveys any selective advantage upon the genome, this must be indirect. All transposons have systems to limit the extent of transposition, since unbridled transposition is presumably damaging, but the molecular mechanisms are different in each case.

The archetypal transposon has inverted repeats at its termini and generates direct repeats of a short sequence at the site of insertion. The simplest types are the bacterial insertion sequences (IS), which consist essentially of the inverted terminal repeats flanking a coding frame(s) whose product(s) provide transposition activity. Composite transposons have terminal modules that consist of IS elements; one or both of the IS modules provides transposase activity, and the sequences between them (often carrying antibiotic resistance), are treated as passengers.

The generation of target repeats flanking a transposon reflects a common feature of transposition. The target site is cleaved at points that are staggered on each DNA strand by a fixed distance (often 5 or 9 bp). The transposon is in effect inserted between protruding single-stranded ends generated by the staggered cuts. Target repeats are generated by filling in the single-stranded regions.

IS elements, composite transposons, and P elements mobilize by nonreplicative transposition, in which the element moves directly from a donor site to a recipient site. A single transposase enzyme undertakes the reaction. It occurs by a "cut and paste" mechanism in which the transposon is separated from flanking DNA. Cleavage of the transposon ends, nicking of the target site, and connection of the transposon ends to the staggered nicks, all occur in a nucleoprotein complex containing the transposase. Loss of the transposon from the donor creates a double-strand break, whose fate is not clear. In the case of Tn10, transposition becomes possible immediately after DNA replication, when sites recognized by the *dam* methylation system are transiently hemimethylated. This imposes a demand for the existence of two copies of the donor site, which may enhance the cell's chances for survival.

The TnA family of transposons mobilize by replicative transposition. After the transposon at the donor site becomes connected to the target site, replication generates a cointegrate molecule that has two copies of the transposon. A resolution reaction, involving recombination between two particular sites, then frees the two copies of the transposon, so that one remains at the donor site and one appears at the target site. Two enzymes coded by the transposon are required: transposase recognizes the ends of the transposon and connects them to the target site; and resolvase provides a site-specific recombination function.

Phage Mu undergoes replicative transposition by the same mechanism as TnA. It also can use its cointegrate intermediate to transpose by a nonreplicative mechanism. The difference between this reaction and the nonreplicative transposition of IS elements is that the cleavage events occur in a different order.

The best characterized transposons in plants are the controlling elements of maize, which fall into several families. Each family contains a single type of autonomous element, analogous to bacterial transposons in its ability to mobilize. A family also contains many different nonautonomous elements, derived by mutations (usually deletions) of the autonomous element. The nonautonomous elements lack the ability to transpose, but display transposition activity and other abilities of the autonomous element, when an autonomous element is present to provide the necessary *trans*-acting functions.

In addition to the direct consequences of insertion and excision, the maize elements may also control the activities of genes at or near the sites where they are inserted; this control may be subject to developmental regulation. Maize elements inserted into genes may be excised from the transcripts, which explains why they do not simply impede gene activity. Control of target gene expression involves a variety of molecular effects, including activation by provision of an enhancer and suppression by interference with post-transcriptional events.

Transposition of maize elements (in particular *Ac*) is nonreplicative, probably requiring only a single transposase enzyme coded by the element. Transposition occurs preferentially after replication of the element. There are probably mechanisms to limit the frequency of transposition. Advantageous rearrangements of the maize genome may have been connected with the presence of the elements.

P elements in *D. melanogaster* are responsible for hybrid dysgenesis, which could be a forerunner of speciation. A cross between a male carrying P elements and a female lacking them generates hybrids that are sterile. A P element has four open reading frames, separated by introns. Splicing of the first three ORFs generates a 66 kD repressor, and occurs in all cells. Splicing of all four ORFs to generate the 87 kD transposase occurs only in the germline, by a tissue-specific splicing event. P elements mobilize when exposed to cytoplasm lacking the repressor. They transpose by a nonreplicative "cut and paste" mechanism. The burst of transposition events inactivates the genome by random insertions. Only a complete P element can generate transposase, but defective elements can be mobilized in *trans* by the enzyme.

Further reading

Reviews

Kleckner, N. (1977). Translocatable elements in prokaryotes. *Cell* **11**, 11–23.

Calos, M. and Miller, J. H. (1980). Transposable elements. *Cell* **20**, 579–595.

Kleckner, N. (1981). Transposable elements in prokaryotes. *Ann. Rev. Genet.* **15**, 341–404.

Campbell, A. (1981). Evolutionary significance of accessory DNA elements in bacteria. *Ann. Rev. Microbiol.* **35**, 55–83.

Engels, W. R. (1983). The P family of transposable elements in *Drosophila. Ann. Rev. Genet.* **17**, 315–344.

Grindley, N. D. and Reed, R. R. (1985). Transpositional recombination in prokaryotes. *Ann. Rev. Biochem.* **54**, 863–896.

Berg, D. E. and Howe, M. (1989). *Mobile DNA* (American Society of Microbiology, Washington DC), including the following chapters:

Craig, N. L. Transposon Tn7 (pp. 211–227);
Engels, W. R. P elements in *D. melanogaster* (pp. 437–484);
Fedoroff, N. Maize transposable elements (pp. 375–412);
Galas, D. J. and Chandler, M. Bacterial insertion sequences (pp. 109–162);
Kleckner, N. Transposon Tn10 (pp. 227–268);
Sherratt, D. Tn3 and related transposable elements: site-specific recombination and transposition (pp. 163–185);
Berg, D. E. Transposon Tn5 (pp. 185–210);
Pato, M. L. Bacteriophage Mu (pp. 23–52).

Gierl, A., Saedler, H., and Peterson, P.A. (1989). Maize transposable elements. *Ann. Rev. Genet.* **23**, 71–85.

Kleckner, N. (1990). Regulation of transposition in bacteria. *Ann. Rev. Cell Biol.* **6**, 297–327.

Mizuuchi, K. (1992). Transpositional recombination: mechanistic insights from studies of Mu and other elements. *Ann. Rev. Biochem.* **61**, 1011–1051.

Reznikoff, W. S. (1993). The Tn5 transposon. *Ann. Rev. Microbiol.* **47**, 945–963.

Scott, J. R. and Churchward, G. G. (1995). Conjugative transposition. *Ann. Rev. Microbiol.* **49**, 367–397.

Craig, N. L. (1997). Target site selection in transposition. *Ann. Rev. Biochem.* **66**, 437–474.

Discoveries

Grindley, N. D. (1978). IS1 insertion generates duplication of a 9 bp sequence at its target site. *Cell* **13**, 419–426.

Johnsrud, L., Calos, M. P., and Miller, J. H. (1978). The transposon Tn9 generates a 9 bp repeated sequence during integration. *Cell* **15**, 1209–1219.

Haniford, D. B., Benjamin, H. W., and Kleckner, N. (1981). Kinetic and structural analysis of a cleaved donor intermediate and a strand transfer intermediate in Tn10 transposition. *Cell* **64**, 171–179.

Grindley, N. D. *et al.* (1982). Transposon-mediated site-specific recombination: identification of three binding sites for resolvase at the res sites of γδ and Tn3. *Cell* **30**, 19–27.

Roberts, D. *et al.* (1985). IS10 transposition is regulated by DNA adenine methylation. *Cell* **43**, 117–130.

Bender, J. and Kleckner, N. (1986). Genetic evidence that Tn10 transposes by a nonreplicative mechanism. *Cell* **45**, 801–815.

Laski, F. A., Rio, D. C., and Rubin, G. M. (1986). Tissue specificity of *Drosophila* P element transposition is regulated at the level of mRNA splicing. *Cell* **44**, 7–19.

Droge, P. *et al.* (1990). The two functional domains of gamma delta resolvase act on the same recombination site: implications for the mechanism of strand exchange. *Proc. Natl. Acad. Sci. USA* **87**, 5336–5340.

Bolland, S. and Kleckner, N. (1996). The three chemical steps of Tn10/IS10 transposition involve repeated utilization of a single active site. *Cell* **84**, 223–233.

Aldaz, H., Schuster, E., and Baker, T. A. (1996). The interwoven architecture of the Mu transposase couples DNA synthesis to catalysis. *Cell* **85**, 257–269.

Savilahti, H. and Mizuuchi, K. (1996). Mu transpositional recombination: donor DNA cleavage and strand transfer in trans by the Mu transpose. *Cell* **85**, 271–280.

Chapter 16

Retroviruses and retroposons

TRANSPOSITION that involves an obligatory intermediate of RNA is unique to eukaryotes, and is provided by the ability of **retroviruses** to insert DNA copies (proviruses) of an RNA viral genome into the chromosomes of a host cell. Some eukaryotic transposons are related to retroviral proviruses in their general organization, and they transpose through RNA intermediates. As a class, these elements are called **retroposons** (or sometimes **retrotransposons**). They range from the retroviruses themselves, able freely to infect host cells, to sequences that have transposed via RNA, but which do not themselves possess the ability to transpose. They share with all transposons the diagnostic feature of generating short direct repeats of target DNA at the site of an insertion.

Even in genomes where active transposons have not been detected, footprints of ancient transposition events are found in the form of direct target repeats flanking dispersed repetitive sequences. The features of these sequences sometimes implicate an RNA sequence as the progenitor of the genomic (DNA) sequence. We think that the RNA must have been converted into a duplex DNA copy that was inserted into the genome by a transposition-like event.

Like any other reproductive cycle, the cycle of a retrovirus or retroposon is continuous; it is arbitrary at which point we interrupt it to consider a "beginning". But our perspectives of these elements are biased by the forms in which we usually observe them, indicated in **Figure 16.1**. Retroviruses were first observed as infectious virus particles, capable of transmission between cells, and so the intracellular cycle (involving duplex DNA) is thought of as the means of reproducing the RNA virus. Retroposons were discovered as components of the genome; and the RNA forms have been mostly characterized for their functions as mRNAs. So we think of retroposons as genomic (duplex DNA) sequences that may transpose within a genome; they do not migrate between cells.

Figure 16.1 *Overview:* the reproductive cycles of retroviruses and retroposons involve alternation of reverse transcription from RNA to DNA with transcription from DNA to RNA. Only retroviruses can generate infectious particles. Retroposons are confined to an intracellular cycle.

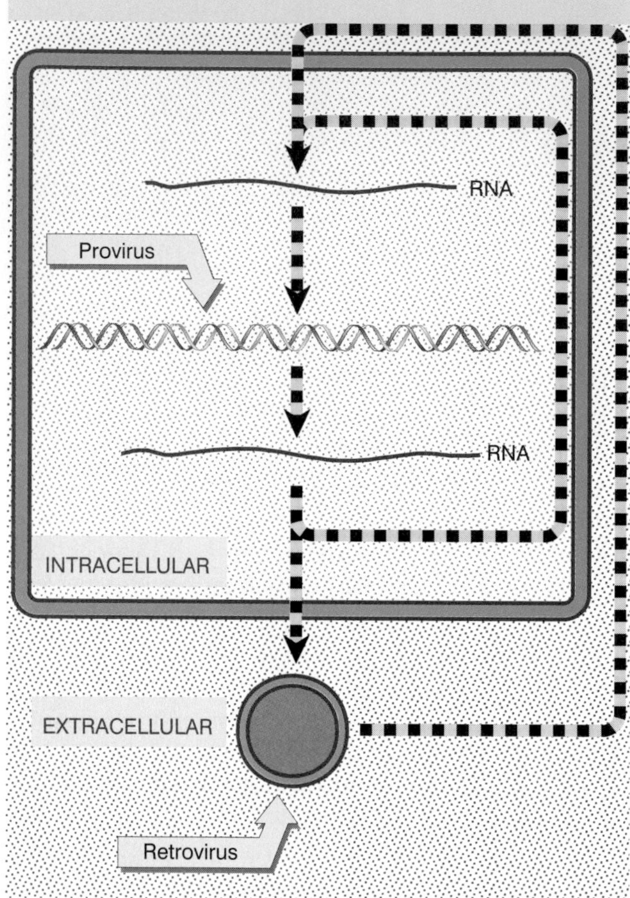

The retrovirus life cycle involves transposition-like events

RETROVIRUSES have genomes of single-stranded RNA that are replicated through a double-stranded DNA intermediate. The life cycle of the virus involves an obligatory stage in which the double-stranded DNA is inserted into the host genome by a transposition-like event that generates short direct repeats of target DNA.

The significance of this reaction extends beyond the perpetuation of the virus. Some of its consequences are that:

- A retroviral sequence that is integrated in the germline remains in the cellular genome as an **endogenous provirus**. Like a lysogenic bacteriophage, a provirus behaves as part of the genetic material of the organism.
- Cellular sequences occasionally recombine with the retroviral sequence and then are transposed with it; these sequences may be inserted into the genome as duplex sequences in new locations.

■ Cellular sequences that are transposed by a retrovirus may change the properties of a cell that becomes infected with the virus.

The particulars of the retroviral life cycle are expanded in **Figure 16.2**. The crucial steps are that the viral RNA is converted into DNA, the DNA becomes integrated into the host genome, and then the DNA provirus is transcribed into RNA.

The enzyme responsible for generating the initial DNA copy of the RNA is **reverse transcriptase**. The enzyme converts the RNA into a linear duplex of DNA in the cytoplasm of the infected cell. (The DNA also is converted into circular forms, but these do not appear to be involved in reproduction.)

Figure 16.2 The retroviral life cycle proceeds by reverse transcribing the RNA genome into duplex DNA, which is inserted into the host genome, in order to be transcribed into RNA.

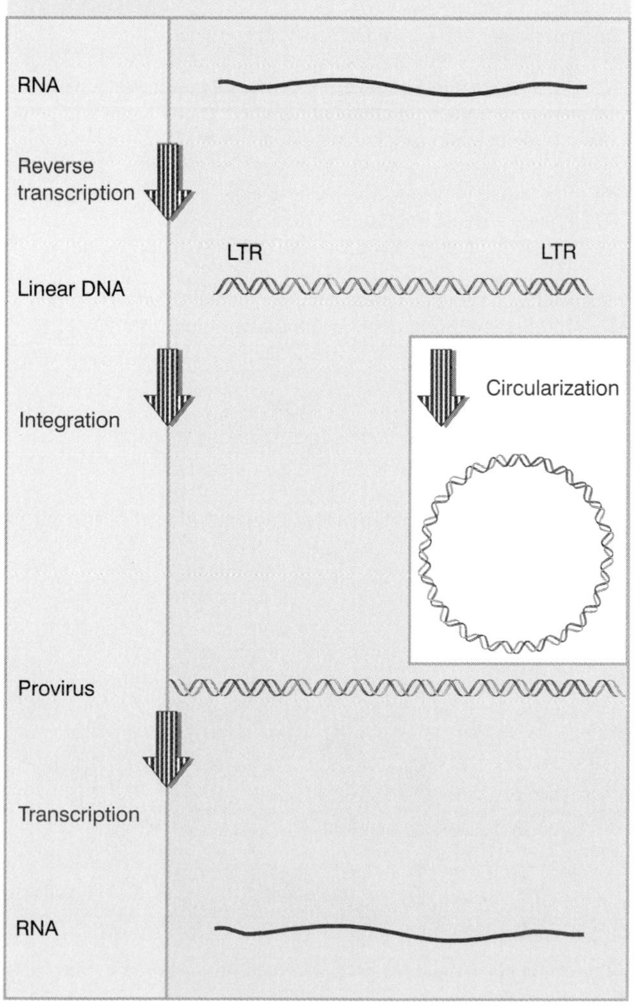

The linear DNA makes its way to the nucleus. One or more DNA copies become integrated into the host genome. A single enzyme, called **integrase**, is responsible for integration. The integrated **proviral DNA** is transcribed by the host machinery to produce viral RNAs, which serve both as mRNAs and as genomes for packaging into virions. *Integration is a normal part of the life cycle and is necessary for transcription.*

Two copies of the RNA genome are packaged into each virion, making the individual virus particle effectively diploid. When a cell is simultaneously infected by two different but related viruses, it is possible to generate heterozygous virus particles carrying one genome of each type. The diploidy may be important in allowing the virus to acquire cellular sequences. The enzymes reverse transcriptase and integrase are carried with the genome in the viral particle.

A typical retroviral sequence contains three or four "genes", the term here identifying coding regions each of which actually gives rise to multiple proteins by processing reactions. A typical retrovirus genome with three genes is organized in the sequence *gag-pol-env* as indicated in **Figure 16.3**.

Retroviral mRNA has a conventional structure; it is capped at the 5′ end and polyadenylated at the 3′ end. It is represented in two mRNAs. The full length mRNA is translated to give the Gag and Pol polyproteins. The Gag product is translated by reading from the initiation codon to the first termination codon. This termination codon must be bypassed to express Pol.

Different mechanisms are used in different viruses to proceed beyond the *gag* termination codon, depending on the relationship between the *gag* and *pol* reading frames. When *gag* and *pol* follow continuously, suppression by a glutamyl-tRNA that recognizes the termination codon allows a single protein to be generated. When *gag* and *pol* are in different reading frames, a ribosomal frameshift occurs to generate a single protein. Usually the readthrough is about 5% efficient, so Gag protein outnumbers Gag-Pol protein about 20-fold.

The Env polyprotein is expressed by another means: splicing generates a shorter **subgenomic** messenger that is translated into the Env product.

The *gag* gene gives rise to the protein components of the nucleoprotein core of the virion. The *pol* gene codes for functions concerned with nucleic acid synthesis and recombination. The *env* gene codes for components of the envelope of the particle, which also sequesters components from the cellular cytoplasmic membrane.

Both the Gag or Gag-Pol and the Env products are polyproteins that are cleaved by a protease to release the

Figure 16.3 The 'genes' of the retrovirus are expressed as polyproteins that are processed into individual products.

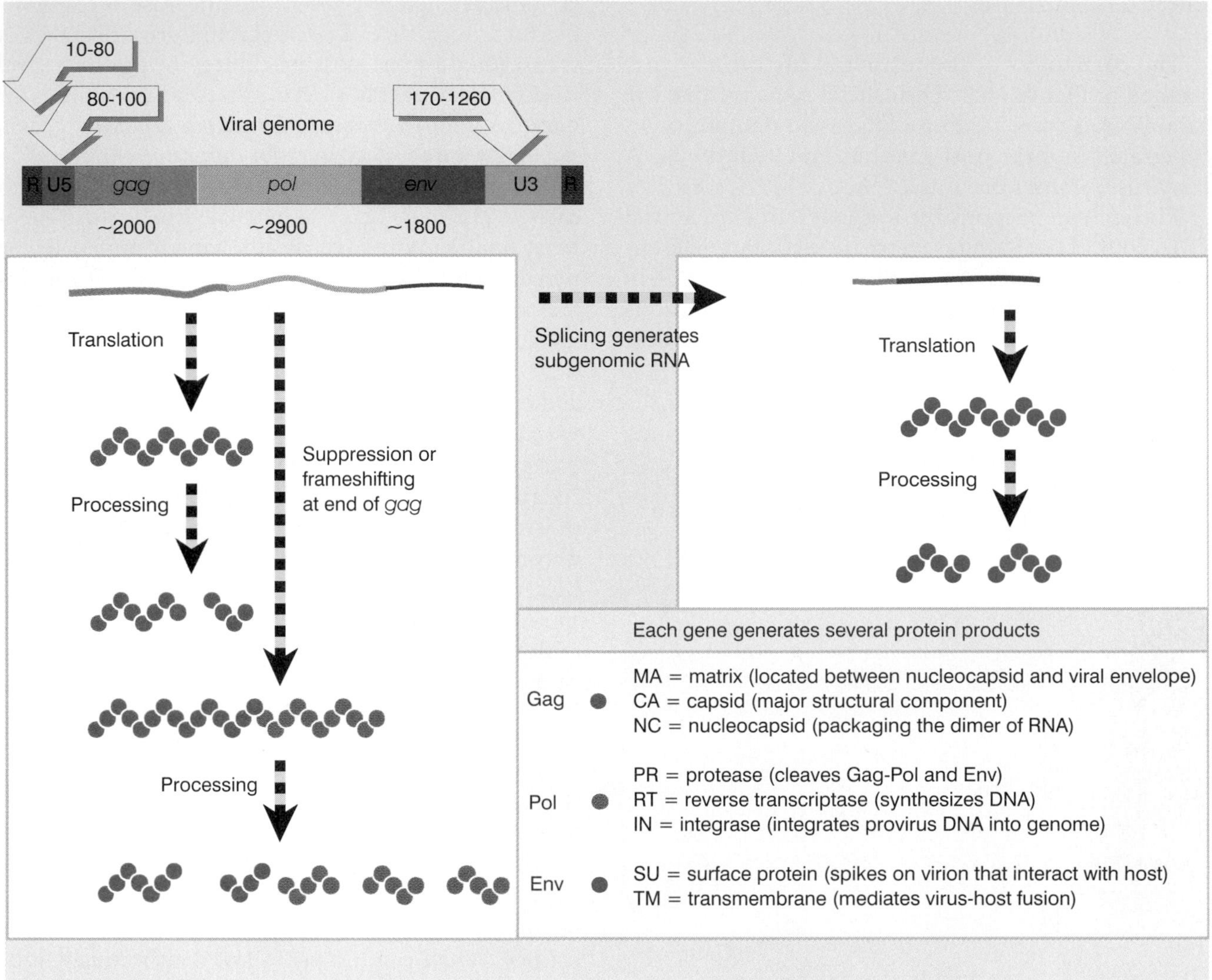

10-80

80-100

Viral genome

170-1260

R U5 *gag* *pol* *env* U3 R

~2000 ~2900 ~1800

Translation

Suppression or frameshifting at end of *gag*

Processing

Processing

Splicing generates subgenomic RNA

Translation

Processing

Each gene generates several protein products

Gag	MA = matrix (located between nucleocapsid and viral envelope) CA = capsid (major structural component) NC = nucleocapsid (packaging the dimer of RNA)
Pol	PR = protease (cleaves Gag-Pol and Env) RT = reverse transcriptase (synthesizes DNA) IN = integrase (integrates provirus DNA into genome)
Env	SU = surface protein (spikes on virion that interact with host) TM = transmembrane (mediates virus-host fusion)

Figure 16.4 Retroviruses (HIV) bud from the plasma membrane of an infected cell. Photograph kindly provided by Matthew Gonda.

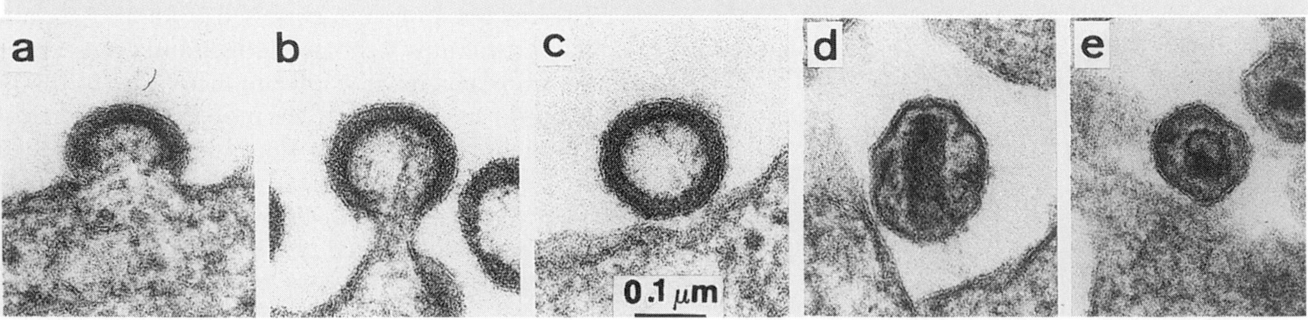

individual proteins that are found in mature virions. The protease activity is coded by the virus in various forms: it may be part of *gag* or *pol*, or sometimes takes the form of an additional independent reading frame.

The production of a retroviral particle involves packaging the RNA into a core, surrounding it with capsid proteins, and pinching off a segment of membrane from the host cell. The release of infective particles by such means is shown in **Figure 16.4**. The process is reversed during infection; a virus infects a new host cell by fusing with the plasma membrane and then releasing the contents of the virion.

Retroviruses are called **plus strand viruses**, because the viral RNA itself codes for the protein products. As its name implies, reverse transcriptase is responsible for converting the genome (plus strand RNA) into a complementary DNA strand, which is called the **minus strand DNA**. Reverse transcriptase also catalyzes subsequent stages in the production of duplex DNA. It has a DNA polymerase activity, which enables it to synthesize a duplex DNA from the single-stranded reverse transcript of the RNA. The second DNA strand in this duplex is called **plus strand DNA**. And as a necessary adjunct to this activity, the enzyme has an RNAase H activity, which can degrade the RNA part of the RNA-DNA hybrid. All retroviral reverse transcriptases share considerable similarities of amino acid sequence, and homologous sequences can be recognized in some other retroposons (see later).

The structures of the DNA forms of the virus are compared with the RNA in **Figure 16.5**. The viral RNA has direct repeats at its ends. These **R** segments vary in different strains of virus from 10–80 nucleotides. Following the R segment at the 5′ end of the virus is the **U5** region of 80–100 nucleotides, whose name indicates that it is unique to the 5′ end. Preceding the R segment at the 3′ terminus is the **U3** segment of 170–1350 nucleotides, which is unique to the 3′ end. The R segments are used during the conversion from the RNA to the DNA form to generate the more extensive direct repeats that are found in linear DNA (see Figures 16.6 and 16.7). The shortening of 2 bp at each end in the integrated form is a consequence of the mechanism of integration (see Figure 16.8).

Like other DNA polymerases, reverse transcriptase requires a primer. The native primer is tRNA. An uncharged host tRNA is present in the virion. A sequence of 18 bases at the 3′ end of the tRNA is base paired to a site 100–200 bases from the 5′ end of one of the viral RNA molecules. The tRNA may also be base paired to another site near the 5′ end of the other viral RNA, thus assisting in dimer formation between the viral RNAs.

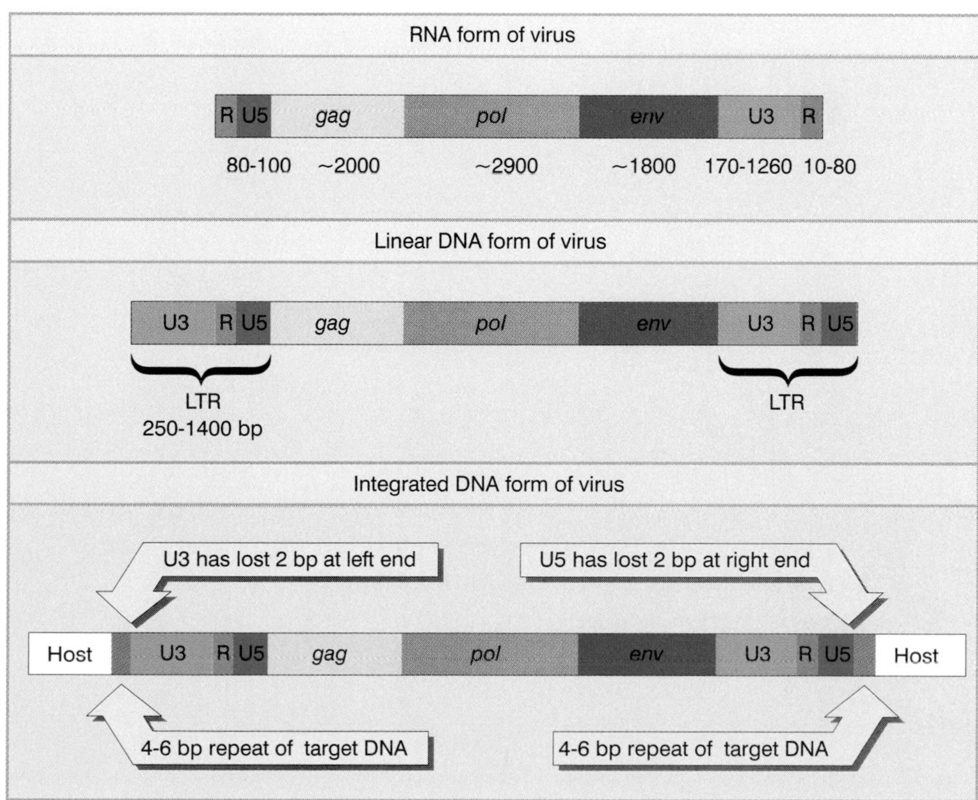

Figure 16.5 Retroviral RNA ends in direct repeats (R), the free linear DNA ends in LTRs, and the provirus ends in LTRs that are shortened by two bases each.

Figure 16.6 Minus strand DNA is generated by switching templates during reverse transcription.

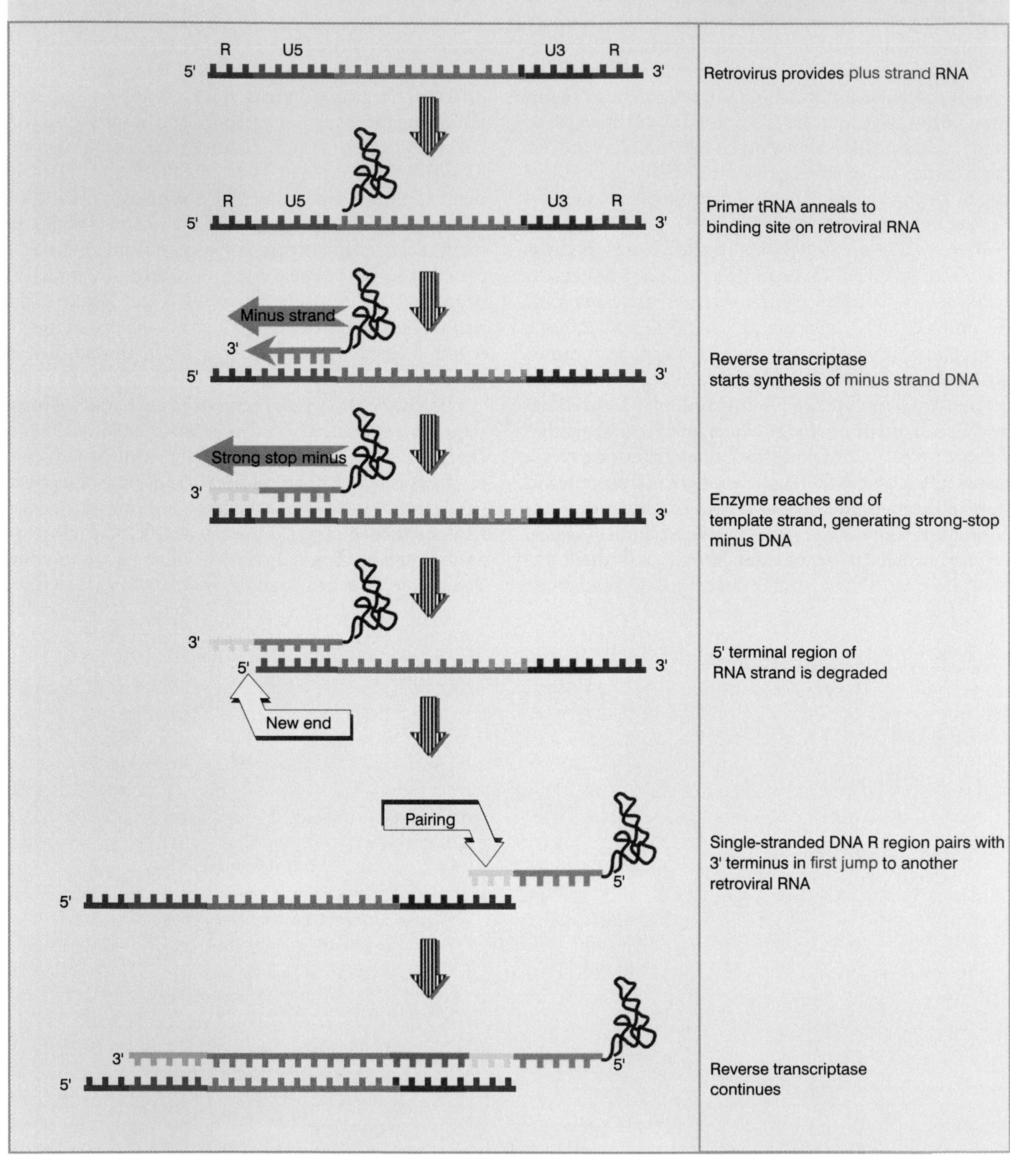

Here is a dilemma. Reverse transcriptase starts to synthesize DNA at a site *only 100–200 bases downstream from the 5′ end.* How can DNA be generated to represent the intact RNA genome? (This is an extreme variant of the general problem in replicating the ends of any linear nucleic acid; see Chapter 12.)

Synthesis *in vitro* proceeds to the end, generating a short DNA sequence called minus strong-stop DNA. This molecule is not found *in vivo* because synthesis continues by the reaction illustrated in **Figure 16.6.** *Reverse transcriptase switches templates,* carrying the nascent DNA with it to the new template. This is the first of two jumps between templates.

In this reaction, the R region at the 5′ terminus of the RNA template is degraded by the RNAase H activity of reverse transcriptase. Its removal allows the R region at a 3′ end to base pair with the newly synthesized DNA. Then reverse transcription continues through the U3 region into the body of the RNA.

The source of the R region that pairs with the minus strong-stop DNA can be either the 3′ end of the same RNA molecule (intramolecular pairing) or the 3′ end of a different RNA molecule (intermolecular pairing). The switch to a different RNA template is used in the figure because there is evidence that the sequence of the tRNA primer is not inherited in a retroposon life cycle. (If intramolecular pairing occurred, we would expect the sequence to be inherited, because it would provide the only source for the primer binding sequence in the next cycle. Intermolecular pairing allows another retroviral RNA to provide this sequence.)

The result of the switch and extension is to add a U3 segment to the 5′ end. The stretch of sequence U3-R-U5 is called the **long terminal repeat (LTR)** because a similar series of events adds a U5 segment to the 3′ end, giving it the same structure of U5-R-U3.

We now need to generate the plus strand of DNA and to generate the LTR at the other end. The reaction is shown in **Figure 16.7.** Reverse transcriptase primes synthesis of plus strand DNA from a fragment of RNA that is left after degrading the original RNA molecule. A strong-stop plus strand DNA is generated when the enzyme reaches the end of the template. This DNA is then transferred to the other end of a minus strand. Probably it is released by a displacement reaction when a second round of DNA synthesis occurs from a primer fragment farther upstream (to its left in the figure). It uses the R region to pair with the 3′ end of a minus strand DNA. This double-stranded DNA then requires completion of both strands to generate a duplex LTR at each end.

Because there are two RNA genomes in each retroviral particle, it is possible for recombination to occur during a viral life cycle. Probably two events are involved, one during minus strand synthesis and one during plus strand synthesis:

- The intermolecular pairing shown in Figure 16.6 allows a recombination to occur between sequences of the two successive RNA templates.
- Plus strand DNA may be synthesized discontinuously, in a reaction that involves several internal initiations. Strand transfer during this reaction can also occur.

The common feature of both events is that the recombination results from a change in the template during the act of DNA synthesis. This is a general example of a mechanism for recombination called **copy choice,** which for many years was regarded as a possible mechanism for general recombination. It is unlikely to be employed by cellular systems, but is a common basis for recombination during infection by RNA viruses, including those that replicate exclusively through RNA forms, such as poliovirus.

The organization of the integrated provirus resembles that of the linear DNA. The LTRs at each end of the provirus are identical. The 3′ end of U5 consists of a short inverted repeat relative to the 5′ end of U3, so the LTR itself ends in short inverted repeats. The integrated proviral DNA is like a transposon: the proviral sequence ends in inverted repeats and is flanked by short direct repeats of target DNA.

The provirus is generated by directly inserting a linear DNA into a target site. (In addition to linear DNA, there are circular forms of the viral sequences. One has two adjacent LTR sequences generated by joining the linear ends. The other has only one LTR [presumably generated by a recombination event and actually comprising the majority of circles]. Although for a long time it appeared that the intermediate is a circle [by analogy with the integration of lambda DNA], we now know that the linear form is used for integration.)

Integration of linear DNA is catalyzed by a single viral product, the integrase. Integrase acts on both the retroviral linear DNA and the target DNA. The reaction is illustrated in **Figure 16.8.**

The ends of the viral DNA are important; as is the case with transposons, mutations in the ends prevent integration. The most conserved feature is the presence of the dinucleotide sequence CA close to the end of each inverted repeat. The integrase brings the ends of the linear DNA together in a ribonucleoprotein complex, and converts the blunt ends into recessed ends by removing the bases beyond the conserved CA; usually this involves loss of two bases (see below).

Figure 16.7 Synthesis of plus strand DNA requires a second jump.

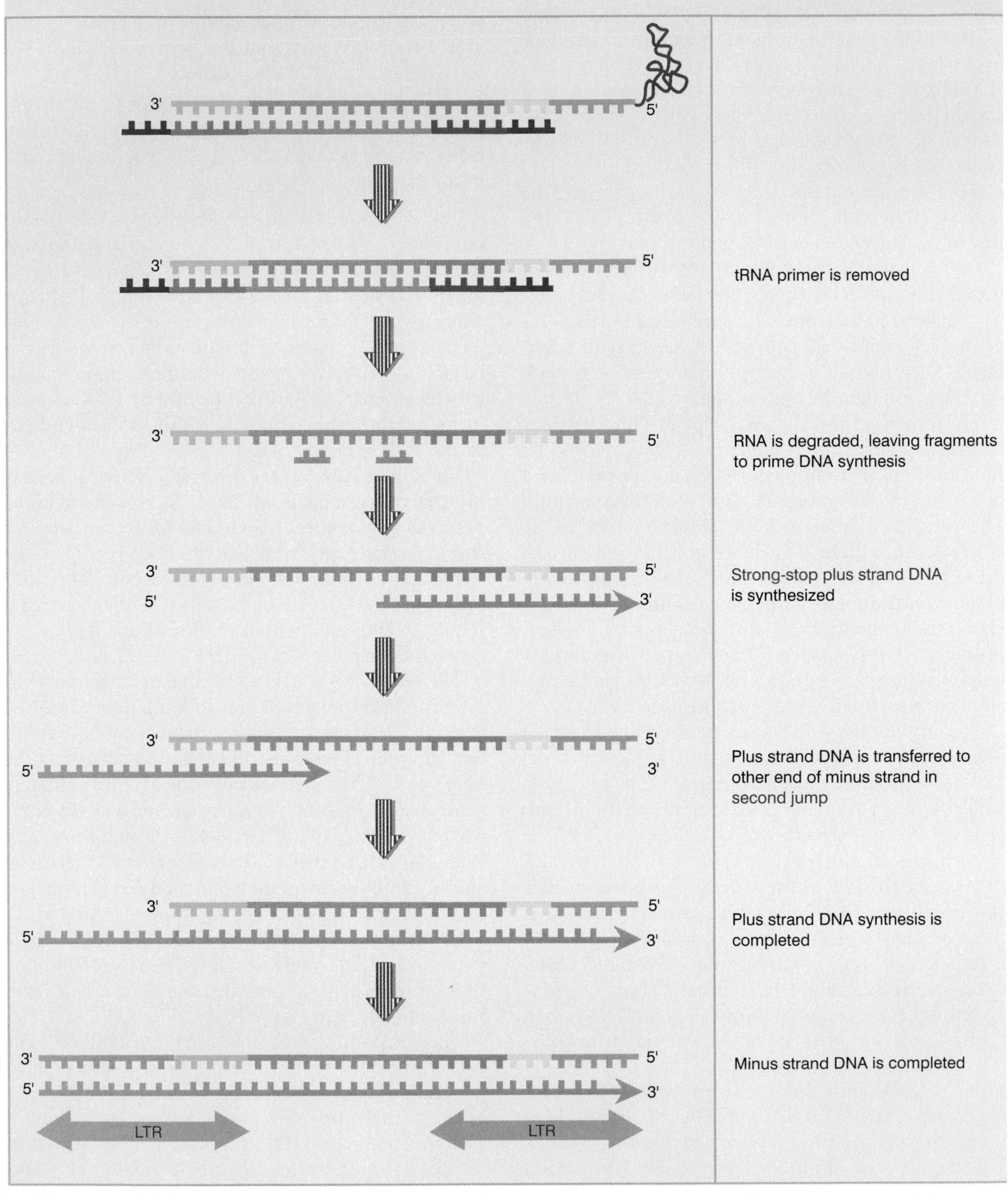
3'
5'
3'
5'
tRNA primer is removed
3'
5'
RNA is degraded, leaving fragments to prime DNA synthesis
3'
5'
5'
3'
Strong-stop plus strand DNA is synthesized
3'
5'
5'
3'
Plus strand DNA is transferred to other end of minus strand in second jump
3'
5'
5'
3'
Plus strand DNA synthesis is completed
3'
5'
5'
3'
Minus strand DNA is completed
LTR
LTR

Figure 16.8 Integrase is the only viral protein required for the integration reaction, in which each LTR loses 2 bp and is inserted between 4 bp repeats of target DNA

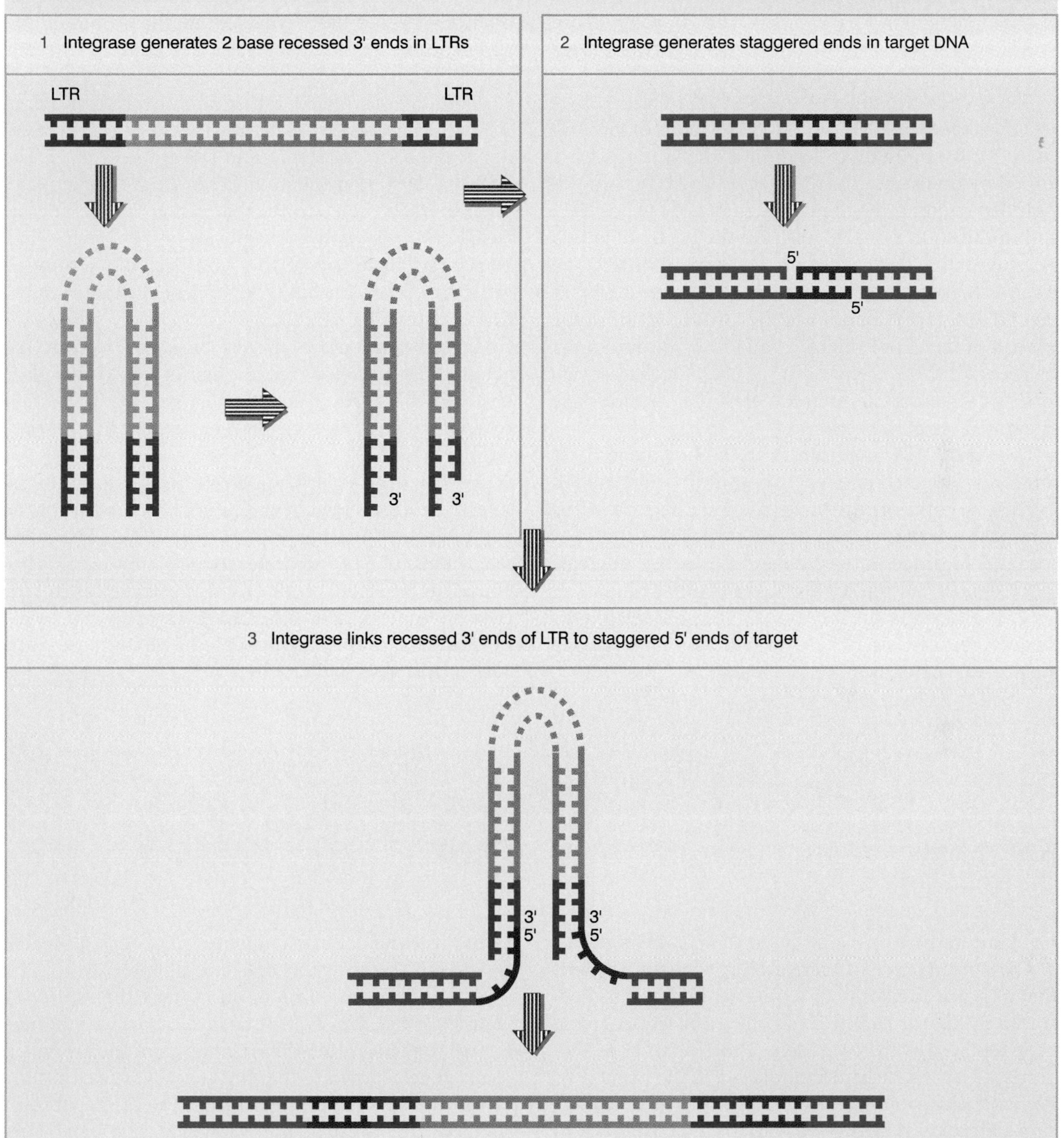
1 Integrase generates 2 base recessed 3' ends in LTRs
LTR
LTR
3'
3'
2 Integrase generates staggered ends in target DNA
5'
5'
3 Integrase links recessed 3' ends of LTR to staggered 5' ends of target
3'
5'
3'
5'

Target sites are chosen at random with respect to sequence. The integrase makes staggered cuts at a target site. In the example of Figure 16.8, the cuts are separated by 4 bp. The length of the target repeat depends on the particular virus; it may be 4, 5, or 6 bp. Presumably it is determined by the geometry of the reaction of integrase with target DNA.

The 5′ ends generated by the cleavage of target DNA are covalently joined to the 3′ recessed ends of the viral DNA. At this point, both termini of the viral DNA are joined by one strand to the target DNA. The single-stranded region is repaired by enzymes of the host cell, and in the course of this reaction the protruding two bases at each 5′ end of the viral DNA are removed. The result is that the integrated viral DNA has lost 2 bp at each LTR; this corresponds to the loss of 2 bp from the left end of the 5′ terminal U3 and loss of 2 bp from the right end of the 3′ terminal U5. There is a characteristic short direct repeat of target DNA at each end of the integrated retroviral genome.

The viral DNA integrates into the host genome at randomly selected sites. A successfully infected cell gains 1–10 copies of the provirus. (An infectious virus enters the cytoplasm, of course, but the DNA form becomes integrated into the genome in the nucleus. Retroviruses can replicate only in proliferating cells, because entry into the nucleus requires the cell to pass through mitosis, when the viral genome gains access to the nuclear material.)

The U3 region of each LTR carries a promoter. The promoter in the left LTR is responsible for initiating transcription of the provirus. Recall that the generation of proviral DNA is required to place the U3 sequence at the left LTR; so we see that the promoter is in fact generated by the conversion of the RNA into duplex DNA.

Sometimes (probably rather rarely), the promoter in the right LTR sponsors transcription of the host sequences that are adjacent to the site of integration. The LTR also carries an enhancer (a sequence that activates promoters in the vicinity) that can act on cellular as well as viral sequences. Integration of a retrovirus can be responsible for converting a host cell into a tumorigenic state when certain types of genes are activated in this way (see Chapter 28).

Can integrated proviruses be excised from the genome? Homologous recombination could take place between the LTRs of a provirus; solitary LTRs that could be relics of an excision event are present in some cellular genomes.

We have dealt so far with retroviruses in terms of the infective cycle, in which integration is necessary for the production of further copies of the RNA. However, when a viral DNA integrates in a germline cell, it becomes an inherited "endogenous provirus" of the organism. Endogenous viruses are not usually expressed, but sometimes they are activated by external events, such as infection with another virus.

Retroviruses may transduce cellular sequences

An interesting light on the viral life cycle is cast by the occurrence of **transducing viruses**, variants that have acquired cellular sequences in the form illustrated in **Figure 16.9**. Part of the viral sequence has been replaced by the *v-onc* gene. Protein synthesis generates a Gag-v-Onc protein instead of the usual Gag, Pol, and Env proteins. The resulting virus is **replication-defective;** it cannot sustain an infective cycle by itself. However, it can be perpetuated in the company of a **helper virus** that provides the missing viral functions.

Onc is an abbreviation for **oncogenesis**, the ability to **transform** cultured cells so that the usual regulation of growth is released to allow unrestricted division. Both viral and cellular *onc* genes may be responsible for creating tumorigenic cells (see Chapter 28).

A *v-onc* gene confers upon a virus the ability to transform a certain type of host cell. Loci with homologous sequences found in the host genome are called *c-onc* genes. How are the *onc* genes acquired by the retroviruses? A revealing feature is the discrepancy in the structures of *c-onc* and *v-onc* genes. The *c-onc* genes are usually interrupted by introns, but the *v-onc* genes are uninterrupted. This suggests that *the* v-onc *genes originate from spliced RNA copies of the* c-onc *genes.*

A model for the formation of transforming viruses is illustrated in **Figure 16.10**. A retrovirus has integrated near a *c-onc* gene. A deletion occurs to fuse the

Figure 16.9 Replication-defective transforming viruses have a cellular sequence substituted for part of the viral sequence. The defective virus may replicate with the assistance of a helper virus that carries the wild-type functions.

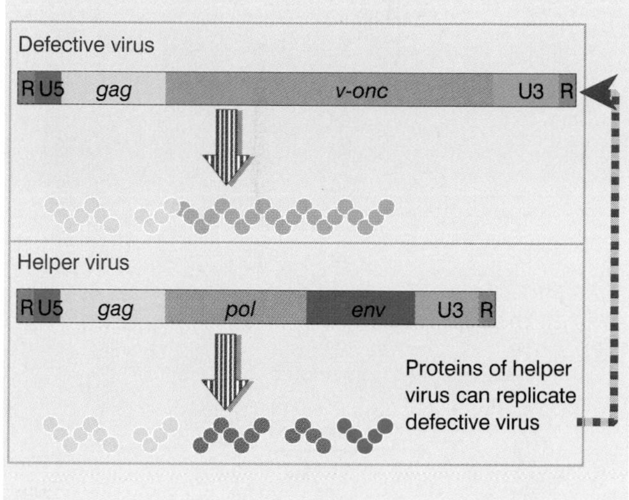

provirus to the *c-onc* gene; then transcription generates a joint RNA, containing viral sequences at one end and cellular *onc* sequences at the other end. Splicing removes the introns in both the viral and cellular parts of the RNA. The RNA has the appropriate signals for packaging into the virion; virions will be generated if the cell also contains another, intact copy of the provirus. Then some of the diploid virus particles may contain one fused RNA and one viral RNA.

A recombination between these sequences could generate the transforming genome, in which the viral repeats are present at *both* ends. (Recombination occurs at a high frequency during the retroviral infective cycle, by various means. We do not know anything about its demands for homology in the substrates, but we assume that the nonhomologous reaction between a viral genome and the cellular part of the fused RNA proceeds by the same mechanisms responsible for viral recombination.)

The common features of the entire retroviral class suggest that it may be derived from a single ancestor. Primordial IS elements could have surrounded a host gene for a nucleic acid polymerase; the resulting unit would have the form *LTR-pol-LTR*. It might evolve into an infectious virus by acquiring more sophisticated abilities to manipulate both DNA and RNA substrates, including the incorporation of genes whose products allowed packaging of the RNA. Other functions, such as transforming genes, might be incorporated later. (There is no reason to suppose that the mechanism involved in acquisition of cellular functions is unique for *onc* genes; but viruses carrying these genes may have a selective advantage because of their stimulatory effect on cell growth.)

Figure 16.10 Replication-defective viruses may be generated through integration and deletion of a viral genome to generate a fused viral-cellular transcript that is packaged with a normal RNA genome. Nonhomologous recombination is necessary to generate the replication-defective transforming genome.

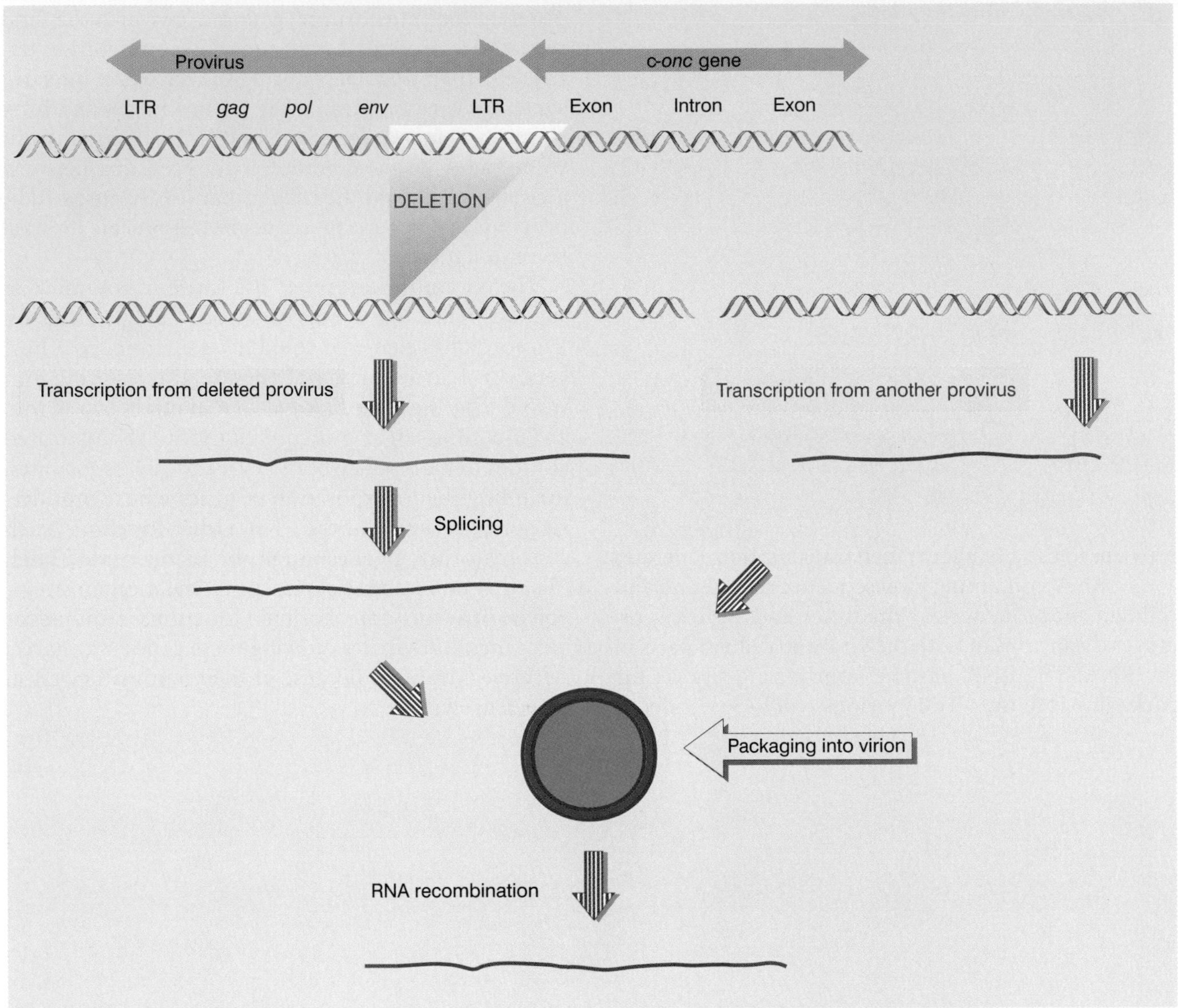

Yeast Ty elements resemble retroviruses

Ty elements comprise a family of dispersed repetitive DNA sequences that are found at different sites in different strains of yeast. Ty is an abbreviation for "transposon yeast". A transposition event creates a characteristic footprint: 5 bp of target DNA are repeated on either side of the inserted Ty element. The frequency of Ty transposition is lower than that of bacterial transposons, $\sim10^{-7}$–10^{-8}.

There is considerable divergence between individual Ty elements. Most elements fall into one of two major classes, called Ty1 and Ty917. They have the same general organization illustrated in **Figure 16.11**. Each element is 6.3 kb long; the last 330 bp at each end constitute direct repeats, called δ. Individual Ty elements of each type have many changes from the prototype of their class, including base pair substitutions, insertions, and deletions. There are ~30 copies of the Ty1 type and ~6 of the Ty917 type in a typical yeast genome. In addition, there are ~100 independent delta elements, called solo δs.

The delta sequences also show considerable heterogeneity, although the two repeats of an individual Ty element are likely to be identical or at least very closely related. The delta sequences associated with Ty elements show greater conservation of sequence than the solo delta elements, which suggests that recognition of the repeats is involved in transposition.

The Ty element is transcribed into two poly(A)$^+$ RNA species, which constitute >5% of the total mRNA of a haploid yeast cell. Both initiate within a promoter in the δ element at the left end. One terminates after 5 kb; the other terminates after 5.7 kb, within the delta sequence at the right end.

The sequence of the Ty element has two open reading frames, expressed in the same direction, but read in different phases and overlapping by 13 amino acids. The sequence of *TyA* suggests that it codes for a DNA-binding protein. The sequence of *TyB* contains regions that have homologies with reverse transcriptase, protease, and integrase sequences of retroviruses.

Figure 16.11 Ty elements terminate in short direct repeats and are transcribed into two overlapping RNAs. They have two reading frames, with sequences related to the retroviral *gag* and *pol* genes.

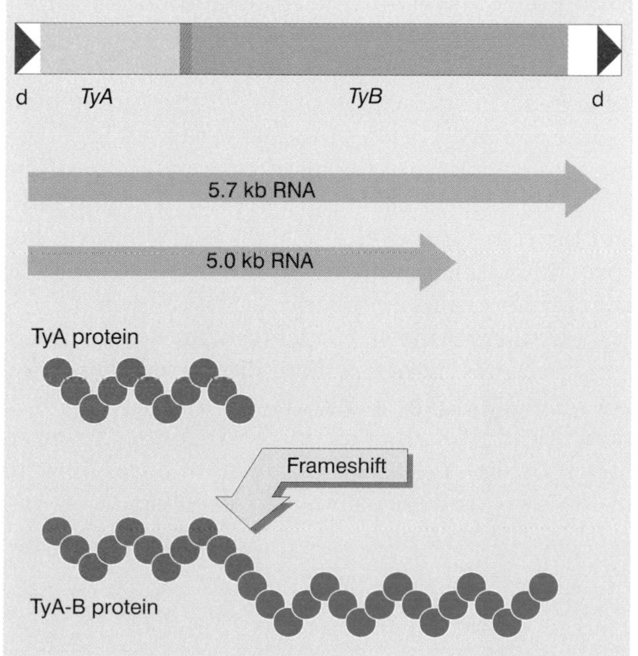

The organization and functions of *TyA* and *TyB* are analogous to the behavior of the retroviral *gag* and *pol* functions. The reading frames *TyA* and *TyB* are expressed in two forms. The TyA protein represents the *TyA* reading frame, and terminates at its end. The *TyB* reading frame, however, is expressed only as part of a joint protein, in which the *TyA* region is fused to the *TyB* region by a specific frameshift event that allows the termination codon to be bypassed (analogous to *gag-pol* translation in retroviruses).

Recombination between Ty elements seems to occur in bursts; when one event is detected, there is an increased probability of finding others. Gene conversion occurs between Ty elements at different locations, with the result that one element is "replaced" by the sequence of the other.

Ty elements can excise by homologous recombination between the directly repeated delta sequences. The large number of solo delta elements may be footprints of such events. An excision of this nature may be associated with reversion of a mutation caused by the insertion of Ty; the level of reversion may depend on the exact delta sequences left behind.

A paradox is that both delta elements have the same sequence, yet a promoter is active in the delta at one end and a terminator is active in the delta at the other end. (A similar feature is found in other transposable elements, including the retroviruses.)

Ty elements are classic retroposons, transposing through an RNA intermediate. An ingenious protocol used to detect this event is illustrated in **Figure 16.12**. An intron was inserted into an element to generate a unique Ty sequence. This sequence was placed under the control of a *GAL* promoter on a plasmid and introduced into yeast cells. Transposition results in the appearance of multiple copies of the transposon in the yeast genome; *but they all lack the intron.*

We know of only one way to remove introns: RNA splicing. This suggests that transposition occurs by the same mechanism as with retroviruses. The Ty element is transcribed into an RNA that is recognized by the splicing apparatus. The spliced RNA is recognized by a reverse transcriptase and regenerates a duplex DNA copy.

The analogy with retroviruses extends further. The original Ty element has a difference in sequence between its two delta elements. *But the transposed elements possess identical delta sequences, derived from the 5′ delta of the original element.* If we consider the delta

Figure 16.12 A unique Ty element, engineered to contain an intron, transposes to give copies that lack the intron. The copies possess identical terminal repeats, generated from one of the termini of the original Ty element.

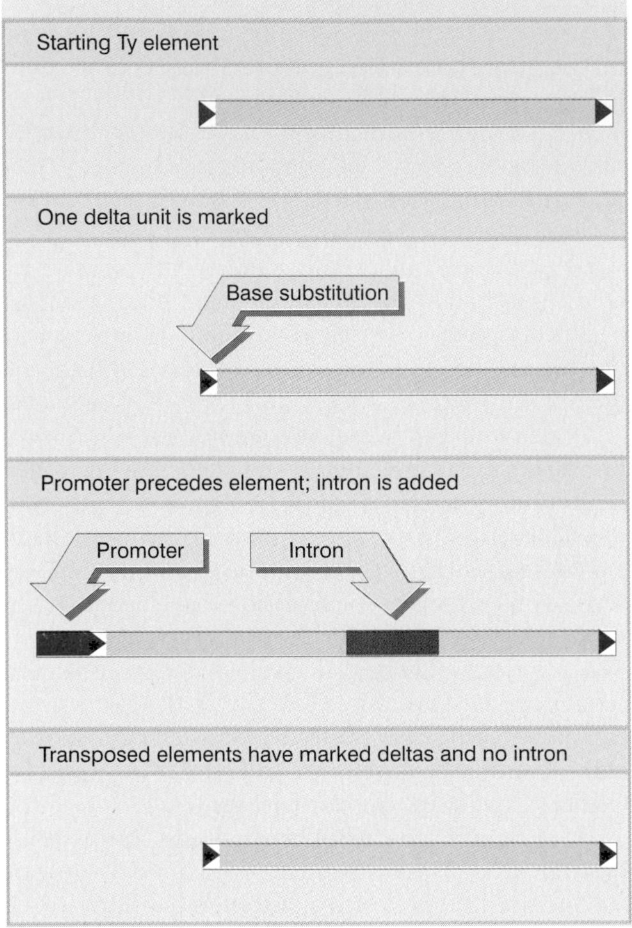
Starting Ty element
One delta unit is marked
Base substitution
Promoter precedes element; intron is added
Promoter
Intron
Transposed elements have marked deltas and no intron

Figure 16.13 Ty elements generate virus-like particles. Photograph kindly provided by Alan Kingsman.

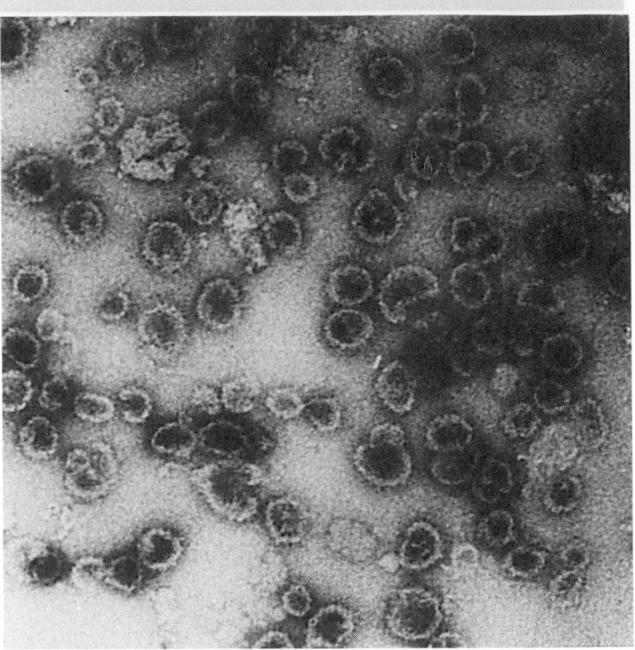

sequence to be exactly like an LTR, consisting of the regions U3-R-U5, the Ty RNA extends from R region to R region. Just as shown for retroviruses in Figures 16.3–16.6, the complete LTR is regenerated by adding a U5 to the 3′ end and a U3 to the 5′ end.

Transposition is controlled by genes within the Ty element. The *GAL* promoter used to control transcription of the marked Ty element is inducible: it is turned on by the addition of galactose. Induction of the promoter has two effects. It is necessary to activate transposition of the marked element. And its activation also increases the frequency of transposition of Ty elements on the yeast chromosome. This implies that the products of the Ty element can act in *trans* on other elements (actually on their RNAs).

Although the Ty element does not give rise to infectious particles, virus-like particles (VLPs) accumulate within the cells in which transposition has been induced. The particles can be seen in **Figure 16.13**. They contain full-length RNA, double-stranded DNA, reverse transcriptase activity, and a *TyB* product with integrase activity. The *TyA* product is cleaved like a *gag* precursor to produce the mature core proteins of the VLP. This takes the analogy between the Ty transposon and the retrovirus even further. The Ty element behaves in short like a retrovirus that has lost its *env* gene and therefore cannot properly package its genome.

Only some of the Ty elements in any yeast genome are active: most have lost the ability to transpose (and are analogous to inert endogenous proviruses). Since these "dead" elements retain the δ repeats, however, they provide targets for transposition in response to the proteins synthesized by an active element.

Many transposable elements reside in *D. melanogaster*

THE presence of transposable elements in *D. melanogaster* was first inferred from observations analogous to those that identified the first insertion sequences in *E. coli.* Unstable mutations are found that revert to wild type by deletion, or that generate deletions of the flanking material with an endpoint at the original site of mutation. They are caused by several types of transposable sequence, which are illustrated in **Figure 16.14.** They include the *copia* retroposon, the FB family of unknown type, and the P elements discussed previously in Chapter 15.

The best-characterized family of retroposons is called *copia.* Its name reflects the presence of a large number of closely related sequences that code for abundant mRNAs. The *copia* family is taken as a paradigm for several other types of elements whose sequences are unrelated, but whose structure and general behavior appear to be similar.

Figure 16.14 Three types of transposable element in *D. melanogaster* have different structures.

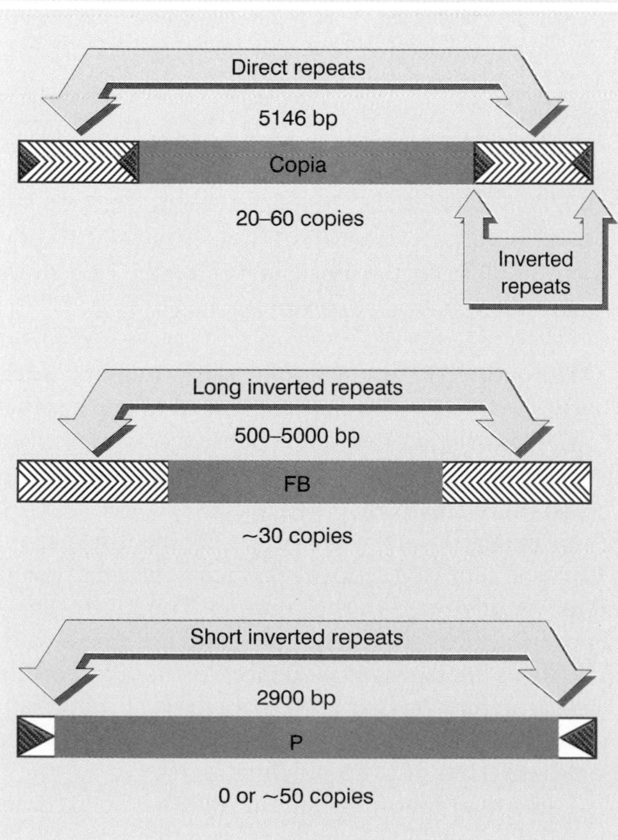

The number of copies of the *copia* element depends on the strain of fly; usually it is 20–60. The members of the family are widely dispersed. The locations of *copia* elements show a different (although overlapping) spectrum in each strain of *D. melanogaster.*

These differences have developed over evolutionary periods. Comparisons of strains that have diverged recently (over the past 40 years or so) as the result of their propagation in the laboratory reveal few changes. We cannot estimate the rate of change, but the nature of the underlying events is indicated by the result of growing cells in culture. The number of *copia* elements per genome then increases substantially, up to 2–3×. The additional elements represent insertions of *copia* sequences at new sites. Adaptation to culture in some unknown way transiently increases the rate of transposition to a range of 10^{-3} to 10^{-4} events per generation.

The *copia* element is ~5000 bp long, with identical direct terminal repeats of 276 bp. Each of the direct repeats itself ends in related inverted repeats. A direct repeat of 5 bp of target DNA is generated at the site of insertion. The divergence between individual members of the *copia* family is slight, <5%; variants often contain small deletions. All of these features are common to the other *copia*-like families, although their individual members display greater divergence.

The identity of the two direct repeats of each *copia* element implies either that they interact to permit correction events, or that both are generated from one of the direct repeats of a progenitor element during transposition. As in the similar case of Ty elements, this is suggestive of a relationship with retroviruses.

The *copia* elements in the genome are always intact; individual copies of the terminal repeats have not been detected (although we would expect them to be generated if recombination deleted the intervening material). *Copia* elements are sometimes found in the form of free circular DNA; like retroviral DNA circles, the longer form has two terminal repeats and the shorter form has only one. Particles containing *copia* RNA have been noticed.

The *copia* sequence contains a single long reading frame of 4227 bp. There are homologies between parts of the *copia* open reading frame and the *gag* and *pol* sequences of retroviruses. A notable absence from the homologies is any relationship with retroviral *env* sequences required for the envelope of the virus, which means that *copia* is unlikely to be able to generate virus-like particles.

Transcripts of *copia* are found as abundant poly(A)$^+$ mRNAs, representing both full-length and part-length transcripts. The mRNAs have a common 5′ terminus, resulting from initiation in the middle of one of the terminal repeats. Several proteins are produced, probably involving events such as splicing of RNA and cleavage of polyproteins.

Although we lack direct evidence for *copia*'s mode of transposition, there are so many resemblances with retroviral organization that the conclusion seems ineluctable that *copia* must have an origin related to the retroviruses. It is hard to say how many retroviral functions it possesses. We know, of course, that it transposes; but (as is the case with Ty elements) there is no evidence for any infectious capacity.

The members of another family of transposable elements in *D. melanogaster*, called **FB** (an abbreviation for foldback), have inverted terminal repeats of variable length. Some FB elements consist solely of juxtaposed inverted repeats; in others the inverted repeats are separated by a region of nonrepetitive DNA.

In spite of the variation in length, the inverted repeats of all members of the FB family are homologous. This feature is explained by their structure, which consists of tandem copies of a simple-sequence DNA, separated by longer stretches of more diverse sequences. Proceeding from the end into the element, the length of the simple-sequence unit increases; initially it is 10 bp, then expands to 20 bp, and finally expands again to 31 bp. The two copies of the inverted repeat in a single FB element are not identical.

The structure of the ends poses a puzzle about FB elements; we have no knowledge of what confers their ability to transpose. Sometimes two (nonidentical) FB elements apparently cooperate to transpose a large intervening segment of DNA, possibly in a manner reminiscent of composite bacterial transposons (although the length of the DNA between FB elements can be much greater, up to 200 kb). It is possible that any sequence of DNA flanked on either end by FB elements could behave as such a unit.

Retroposons fall into two classes

Two classes of retroposons are distinguished in **Figure 16.15**:

- The retroviruses are the paradigm for retroposons that have the capacity to transpose because they code for reverse transcriptase and/or integrase activities. The retroposons differ from the retroviruses themselves in not passing through an independent infectious form, but otherwise resemble them in the mechanism used for transposition. This group is called the **viral superfamily**.
- Another class of retroposons is identified by external and internal features that suggest that they originated in RNA sequences, although in these cases we can only speculate on how a DNA copy was generated. We assume that they were targets for a transposition event by an enzyme system coded elsewhere. They originated in cellular transcripts. They do not code for proteins that have transposition functions. This group is called the **nonviral superfamily**.

Mammalian genomes contain a large number of relatively short sequences that are related to one another (comprising the moderately repetitive DNA described in Chapter 4). A significant part of this component consists of retroposons. Two families account for most of this material. They were originally identified as interspersed repeated sequences; each consists of many members dispersed in the genome. The LINES comprise long interspersed sequences, and the SINES comprise short interspersed sequences. A more important distinction may be that LINES are derived from transcripts of RNA polymerase II, while SINES are derived from transcripts of RNA polymerase III.

Mammalian genomes contain 20,000–50,000 copies of a LINES called L1. The typical member is ~6500 bp

Figure 16.15 Retroposons can be divided into the viral or nonviral superfamilies.

	Viral Superfamily	Nonviral Superfamily
Common types	*Ty* (*S. cerevisiae*) *copia* (*D. melanogaster*) LINES L1 (mammals)	SINES B1/Alu (mammals) Processed pseudogenes of pol III transcripts
Termini	Long terminal repeats	No repeats
Target repeats	4–6 bp	7–21 bp
Reading frames	Reverse transcriptase and/or integrase	None (or none coding for transposon products)
Organization	May contain introns (removed in subgenomic mRNA)	No introns

long, terminating in an A-rich tract. Open reading frames may be present. One element that has been sequenced has reading frames of 1137 and 3900 bp that overlap by 14 bp. Transcripts can be found. As implied by its presence in repetitive DNA, the LINES family shows variation among individual members. However, the members of the family within a species are relatively homogeneous compared to the variation shown between species.

Figure 16.16 compares members of the viral superfamily with the retroviral paradigm. We know that an active Ty element codes for transposition function. We may infer that among the *copia* sequences in a fly genome must be some active elements coding for transposition function.

LINES elements, and some others, do not terminate in the LTRs that are typical of retroviral elements. This poses the question: how is reverse transcription primed? It does not involve the typical reaction in which a tRNA primer pairs with the LTR (see Figure 16.6). The open reading frames in these elements lack many of the retroviral functions, such as protease or integrase domains, but typically do have reverse transcriptase-like sequences, and the product may also have endonuclease activity.

Figure 16.16 Retroposons of the viral family have terminal repeats and include open reading frames.

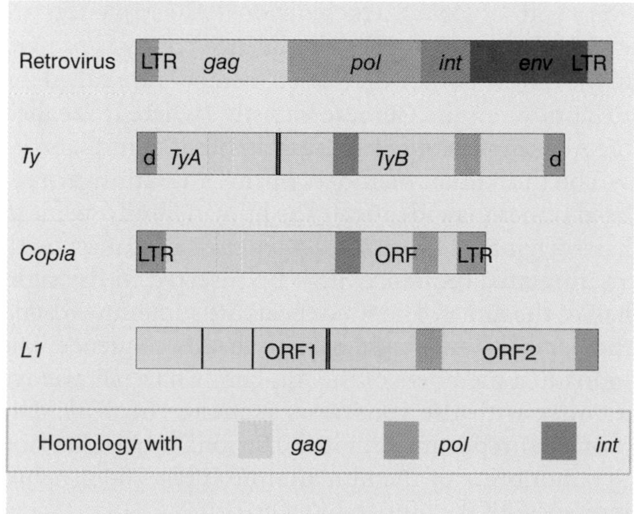

Figure 16.17 shows how these activities support transposition. A nick is made in the DNA target site by an endonuclease activity coded by the retrotransposon. The RNA product of the element associates with the protein bound at the nick. The nick provides a 3′-OH end that primes synthesis of cDNA on the RNA template. A second cleavage event is required to open the other strand of DNA, and the RNA/DNA hybrid is linked to the other end of the gap either at this stage or after it has been converted into a DNA duplex. A similar mechanism is used by some mobile introns (see Figure 23.11).

Because LINES originate from RNA polymerase II transcripts, the genomic sequences are necessarily inactive: they lack the promoter that was upstream of the original startpoint for transcription. Because they usually possess the features of the mature transcript, they are called **processed pseudogenes**.

The characteristic features of a processed pseudogene are compared in **Figure 16.18** with the features of the original gene and the mRNA. The figure shows *all* the relevant diagnostic features, only some of which are found in any individual example. Any transcript of RNA polymerase II could in principle give rise to such a pseudogene, and there are many examples, including the processed globin pseudogenes that were the first to be discovered (see Chapter 4).

The pseudogene may start at the point equivalent to the 5′ terminus of the RNA, which would be expected

Figure 16.17 Retrotransposition of non-LTR elements occurs by nicking the target to provide a primer for cDNA synthesis on an RNA template.

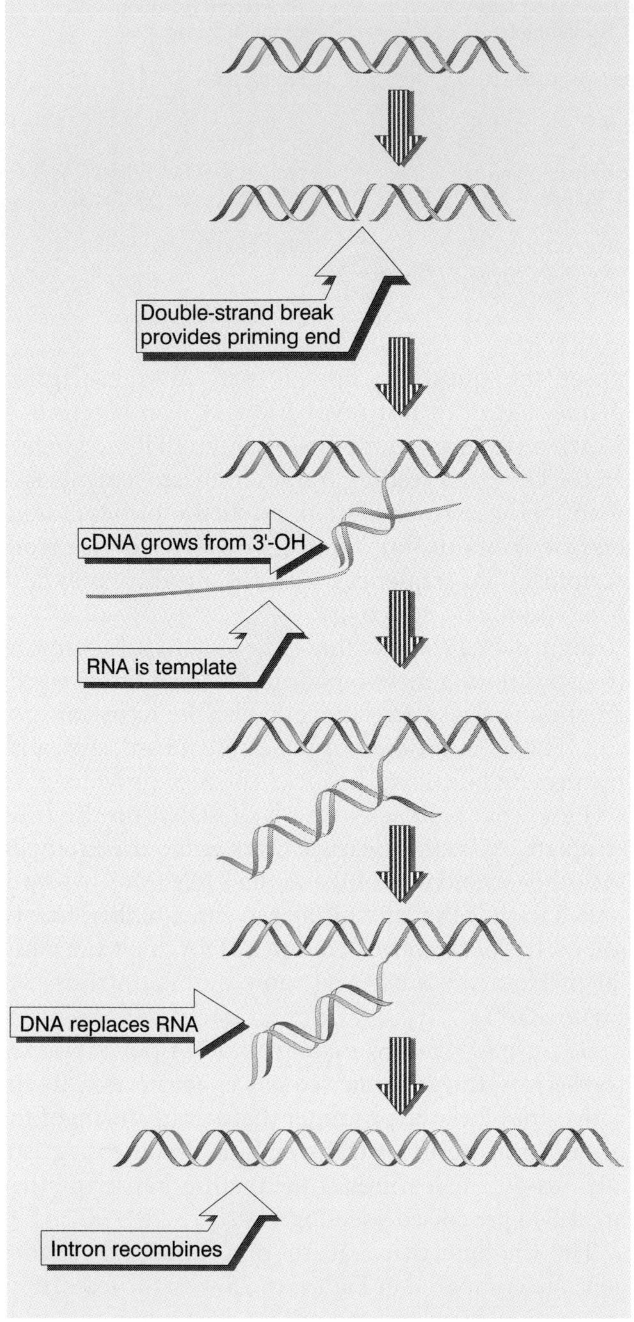
Double-strand break provides priming end
cDNA grows from 3'-OH
RNA is template
DNA replaces RNA
Intron recombines

only if the DNA had originated from the RNA. Several pseudogenes consist of precisely joined exon sequences; we know of no mechanism to recognize introns in DNA, so this feature argues for an RNA-mediated stage. The pseudogene may end in a short stretch of A•T base pairs, presumably derived from the poly(A) tail of the RNA. On either side of the pseudogene is a short direct repeat, presumed to have been generated by a transposition-like event. Processed pseudogenes reside at locations unrelated to their presumed sites of origin.

The processed pseudogenes do not carry any information that might be used to sponsor a transposition event (or to carry out the preceding reverse transcription of the RNA). Could the process have been mediated by a retrovirus? Was it accomplished by an aberrant cellular system? Perhaps the ends of the transposed sequence fortuitously resembled sequences at the ends of a transposon.

Are transposition events currently occurring in these genomes or are we seeing only the footprints of ancient systems? Note that for the transpositions to have survived, they must have occurred in the germline; presumably similar events occur in somatic cells, but do not survive beyond one generation.

The most prominent SINES comprises members of a single family. Its short length and high degree of repetition make it comparable to simple sequence DNA, except that the individual members of the family are dispersed around the genome instead of being confined to tandem clusters. Again there is significant similarity between the members within a species compared with variation between species.

In the human genome, a large part of the moderately repetitive DNA exists as sequences of ~300 bp that are interspersed with nonrepetitive DNA. At least half of the renatured duplex material is cleaved by the restriction enzyme AluI at a single site, located 170 bp along the sequence. The cleaved sequences all are members of a single family, known as the **Alu family** after the means of its identification. There are ~300,000 members in the haploid genome (equivalent to one member for every 6 kb of DNA). The individual Alu sequences are widely dispersed. A related sequence family is present in the mouse (where the 50,000 members are called the B1 family), in the Chinese hamster (where it is called the Alu-equivalent family), and in other mammals.

The individual members of the Alu family are related rather than identical. The human family seems to have originated by a 130 bp tandem duplication, with an unrelated sequence of 31 bp inserted in the right half of the dimer. The two repeats are sometimes called the "left half" and "right half" of the Alu sequence. The individual members of the Alu family have an average identity with the consensus sequence of 87%. The mouse B1 repeating unit is 130 bp long, corresponding to a monomer of the human unit. It has 70–80% homology with the human sequence.

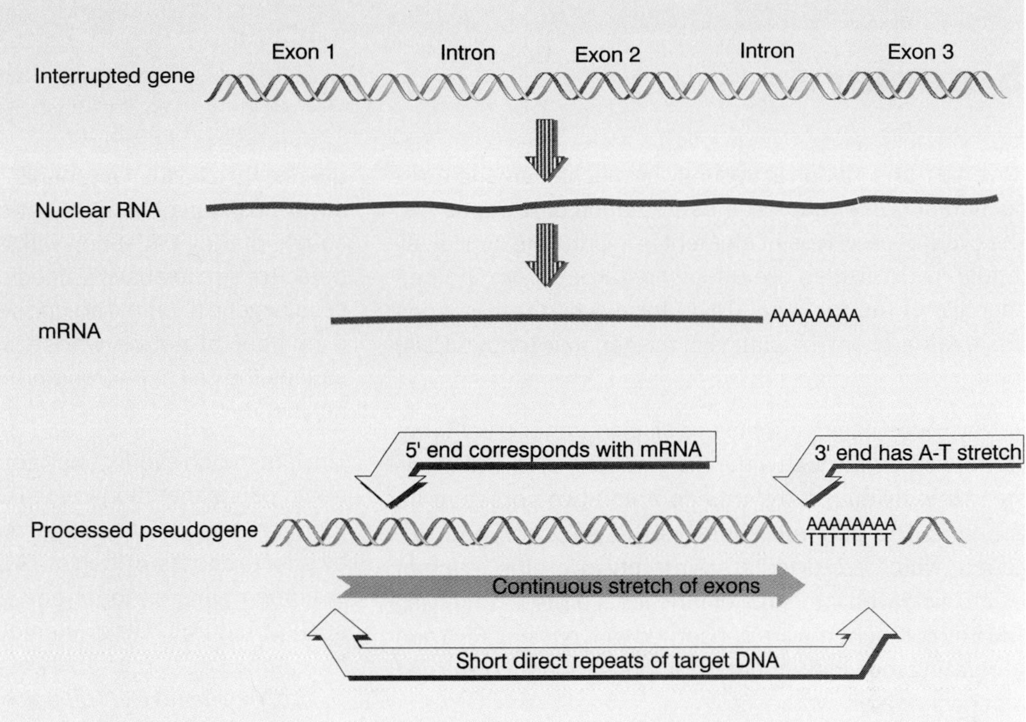

Figure 16.18 Pseudogenes could arise by reverse transcription of RNA to give duplex DNAs that become integrated into the genome.

The Alu sequence is related to 7SL RNA, a component of the signal recognition particle (see Chapter 8). The 7SL RNA corresponds to the left half of an Alu sequence with an insertion in the middle. So the 90 5′ terminal bases of 7SL RNA are homologous to the left end of Alu, the central 160 bases of 7SL RNA have no homology to Alu, and the 40 3′ terminal bases of 7SL RNA are homologous to the right end of Alu. The 7SL RNA is coded by genes that are actively transcribed by RNA polymerase III. It is possible that these genes (or genes related to them) gave rise to the inactive Alu sequences.

The members of the Alu family resemble transposons in being flanked by short direct repeats. However, they display the curious feature that the lengths of the repeats are different for individual members of the family. Because they derive from RNA polymerase III transcripts, it is possible that individual members carry internal active promoters.

A variety of properties have been found for the Alu family, and its ubiquity has prompted many suggestions for its function, but it is not yet possible to discern its true role.

At least some members of the family can be transcribed into independent RNAs. In the Chinese hamster, some (although not all) members of the Alu-equivalent family appear to be transcribed *in vivo*. Transcription units of this sort are found in the vicinity of other transcription units.

Members of the Alu family may be included within structural gene transcription units, as seen by their presence in long nuclear RNA. The presence of multiple copies of the Alu sequence in a single nuclear molecule can generate secondary structure. In fact, the presence of Alu family members in the form of inverted repeats is responsible for most of the secondary structure found in mammalian nuclear RNA.

Summary

Reverse transcription is the unifying mechanism for reproduction of retroviruses and perpetuation of retroposons. The cycle of each type of element is in principle similar, although retroviruses are usually regarded from the perspective of the free viral (RNA) form, while retroposons are regarded from the stance of the genomic (duplex DNA) form.

Retroviruses have genomes of single-stranded RNA that are replicated through a double-stranded DNA intermediate. An individual retrovirus contains two copies of its genome. The genome contains the *gag*, *pol*, and *env* genes, which are translated into polyproteins, each of which is cleaved into smaller functional proteins. The Gag and Env components are concerned with packing RNA and generating the virion; the Pol components are concerned with nucleic acid synthesis.

Reverse transcriptase is the major component of *pol*, and is responsible for synthesizing a DNA (minus strand) copy of the viral (plus strand) RNA. The DNA product is longer than the RNA template; by switching template strands, reverse transcriptase copies the 3′ sequence of the RNA to the 5′ end of the DNA, and copies the 5′ sequence of the RNA to the 3′ end of the DNA. This generates the characteristic LTRs (long terminal repeats) of the DNA. A similar switch of templates occurs when the plus strand of DNA is synthesized using the minus strand as template. Linear duplex DNA is inserted into a host genome by the integrase enzyme. Transcription of the integrated DNA from a promoter in the left LTR generates further copies of the RNA sequence.

Switches in template during nucleic acid synthesis allow recombination to occur by copy choice. During an infective cycle, a retrovirus may exchange part of its usual sequence for a cellular sequence; the resulting virus is usually replication-defective, but can be perpetuated in the course of a joint infection with a helper virus. Many of the defective viruses have gained an RNA version (*v-onc*) of a cellular gene (*c-onc*). The *onc* sequence may be any one of a number of genes whose expression in *v-onc* form causes the cell to be transformed into a tumorigenic phenotype.

The integration event generates direct target repeats (like transposons that mobilize via DNA). An inserted provirus therefore has direct terminal repeats of the LTRs, flanked by short repeats of target DNA. Mammalian and avian genomes have endogenous (inactive) proviruses with such structures. Other elements with this organization have been found in a variety of genomes, most notably in *S. cerevisiae* and *D. melanogaster*. *Ty* elements of yeast and *copia* elements of flies have coding sequences with homology to reverse transcriptase, and mobilize via an RNA form. They may generate particles resembling viruses, but do not have infectious capability. The LINES sequences of mammalian genomes are further removed from the retroviruses, but retain enough similarities to suggest a common origin.

Another class of retroposons have the hallmarks of transposition via RNA, but have no coding sequences (or at least none resembling retroviral functions). They may have originated as passengers in a retroviral-like transposition event, in which an RNA was a target for a reverse transcriptase. Processed pseudogenes arise by such events. A particularly prominent family apparently originating from a processing event is the mammalian SINES, including the human Alu family. Some snRNAs, including 7SL snRNA (a component of the SRP) are related to this family.

Further reading

Reviews

Finnegan, D. J. (1985). Transposable elements in eukaryotes. *Int. Rev. Cytol.* **93**, 281–326.

Weiner, A. M., Deininger, P. L., and Efstratiadis, A. (1986). Nonviral retroposons: genes, pseudogenes, and transposable elements generated by the reverse flow of genetic information. *Ann. Rev. Biochem.* **55**, 631–661.

Deininger, P. L. (1989). SINEs: short interspersed repeated

DNA elements in higher eukaryotes. In *Mobile DNA* (Eds. M. M. Howe and D. E. Berg., American Society for Microbiology, Washington DC, 619–636).

Hutchison, C. A. *et al.* (1989). LINEs and related retroposons: long interspersed repeated sequences in the eukaryotic genome. In *Mobile DNA* (Eds. M. M. Howe and D. E. Berg, American Society for Microbiology, Washington DC, 593–617).

Varmus, H. and Brown, P. (1989). Retroviruses. In *Mobile DNA* (Eds. M. M. Howe and D. E. Berg, American Society for Microbiology, Washington DC, 53–108).

Goff, S. P. (1992). Genetics of retroviral integration. *Ann. Rev. Genet.* **26**, 527–544.

Lai, M. M. C. (1992). RNA recombination in animal and plant viruses. *Microbiol. Rev.* **56**, 61–79.

Katz, R. A. and Skalka, A. M. (1994). The retroviral enzymes. *Ann. Rev. Biochem.* **63**, 133–173.

Discoveries

Temin, H. M. and Mizutani, S. (1970). RNA-dependent DNA polymerase in virions of Rous sarcoma virus. *Nature* **226**, 1211–1213.

Baltimore, D. (1970). RNA-dependent DNA polymerase in virions of RNA tumor viruses. *Nature* **226**, 1209–1211.

Boeke, J. D. *et al.* (1985). Ty elements transpose through an RNA intermediate. *Cell* **40**, 491–500.

Mount, S. M. and Rubin, G. M. (1985). Complete nucleotide sequence of the *Drosophila* transposable element copia: homology between copia and retroviral proteins. *Mol. Cell. Biol.* **5**, 1630–1638.

Loeb, D. D. *et al.* (1986). The sequence of a large L1Md element reveals a tandemly repeated 5′ end and several features found in retrotransposons. *Mol. Cell. Biol.* **6**, 168–182.

Craigie, R., Fujiwara, T., and Bushman, F. (1990). The IN protein of Moloney murine leukemia virus processes the viral DNA ends and accomplishes their integration *in vitro*. *Cell* **67**, 829–837.

Hu, W. S. and Temin, H. M. (1990). Retroviral recombination and reverse transcription. *Science* **250**, 1227–1233.

Luan, D. D. *et al.* (1993). Reverse transcription of R2Bm RNA is primed by a nick at the chromosomal target site: a mechanism for non-LTR retrotransposition. *Cell* 72, 595–605.

Lauermann, V. and Boeke, J. D. (1994). The primer tRNA sequence is not inherited during Ty1 retrotransposition. *Proc. Natl. Acad. Sci. USA* **91**, 9847–9851.

Chapter 17

Rearrangement of DNA

ALTHOUGH genomic DNA is usually unaltered by somatic development, there are some cases in which sequences are moved within a genome, modified, amplified, or even lost, as a natural event. In this chapter, we discuss a variety of such events in yeast, plants, and lower eukaryotes. Examples of rearrangement or loss of specific sequences are especially common in the lower eukaryotes. Usually these changes involve somatic cells; the germline remains inviolate. (However, there are organisms whose reproductive cycle involves the loss of whole chromosomes or sets of chromosomes.) We also discuss the introduction of new sequences into the genome by experimental means. Reorganization of particular sequences is rare in animals, although an extensive case is represented by the immune system. In Chapter 24, we discuss the rearrangement and expression of immunoglobulin genes.

We may distinguish two broad consequences of a gene rearrangement:

- Rearrangement may *create new genes*, needed for expression in particular circumstances, as in the case of the immunoglobulins.
- Rearrangement may be responsible for *switching expression* from one preexisting gene to another. This provides a mechanism for regulating gene expression.

Yeast mating type switching and trypanosome antigen variation share a similar type of plan in which gene expression is controlled by manipulation of DNA sequences. Phenotype is determined by the gene copy present at a particular, active locus. But the genome also contains a store of other, alternative sequences, which are silent. A silent copy can be activated only by a rearrangement of sequences in which it replaces the active gene copy. Such a substitution is equivalent to a unidirectional transposition with a specific target site.

The simplest example of this strategy is found in the yeast, *S. cerevisiae*. Haploid *S. cerevisiae* can have either of two mating types. The type is determined by the sequence present at the active mating type locus. But the genome also contains two other, silent loci, one representing each mating type. Transition between mating types is accomplished by substituting the sequence at the active locus with the sequence from the silent locus carrying the other mating type.

A range of variations is made possible by DNA rearrangement in the African trypanosomes, unicellular parasites that evade the host immune response by varying their surface antigens. The type of surface antigen is determined by the gene sequence at an active locus. This sequence can be changed, however, by substituting a sequence from any one of many silent loci. It seems fitting that the mechanism used to combat the flexibility of the immune apparatus is analogous to that used to generate immune diversity: it relies on physical rearrangements in the genome to change the sequences that are expressed.

Another means of increasing genetic capacity is employed in parasite- or symbiote-host interactions, in which exogenous DNA is introduced from a bacterium into a host cell. The mechanism resembles that of bacterial conjugation. Expression of the bacterial DNA in its new host changes the phenotype of the cell. In the example of the bacterium *Agrobacterium tumefaciens*, the result is to induce tumor formation by an infected plant cell.

Alterations in the relative proportions of components of the genome during somatic development occur to allow insect larvae to increase the number of copies of certain genes. And the occasional **amplification** of genes in cultured mammalian cells is indicated by our ability to select variant cells with an increased copy number of some genes. Initiated within the genome, the amplification event can create additional copies of a gene that survive in either intrachromosomal or extrachromosomal form.

When extraneous DNA is introduced into eukaryotic cells, it may give rise to extrachromosomal forms or may be integrated into the genome. The relationship between the extrachromosomal and genomic forms is irregular, depending on chance and to some degree unpredictable events, rather than resembling the regular interchange between free and integrated forms of bacterial plasmids.

Yet, however accomplished, the process may lead to stable change in the genome; following its injection into animal eggs, DNA may even be incorporated into the genome and inherited thereafter as a normal component, sometimes continuing to function. Injected DNA may enter the germline as well as the soma, creating a **transgenic** animal. The ability to introduce specific genes that function in an appropriate manner could become a major medical technique for curing genetic diseases.

The converse of the introduction of new genes is the ability to disrupt specific endogenous genes. Additional DNA can be introduced within a gene to prevent its expression and to generate a null allele. Breeding from an animal with a null allele can generate a homozygous "knockout", which has no active copy of the gene. This is a powerful method to investigate directly the importance and function of a gene.

Considerable manipulation of DNA sequences therefore is achieved both in authentic situations and by experimental fiat. We are only just beginning to work out the mechanisms that permit the cell to respond to selective pressure by changing its bank of sequences or that allow it to accommodate the intrusion of additional sequences.

The mating pathway is triggered by signal transduction

THE yeast *S. cerevisiae* can propagate happily in either the haploid or diploid condition. Conversion between these states takes place by mating (fusion of haploid spores to give a diploid) and by sporulation (meiosis of diploids to give haploid spores). The ability to engage in these activities is determined by the **mating type** of the strain.

The properties of the two mating types are summarized in **Figure 17.1**. We may view them as resting on the teleological proposition that there is no point in mating unless the haploids are of different genetic types; and sporulation is productive only when the diploid is heterozygous and thus can generate recombinants.

The mating type of a (haploid) cell is determined by the genetic information present at the *MAT* locus. Cells that carry the *MATa* allele at this locus are type **a**; likewise, cells that carry the *MATα* allele are type α. Cells of opposite type can mate; cells of the same type cannot.

Recognition of cells of opposite mating type is

Figure 17.1 Mating type controls several activities.

	MATa	*MATα*	*MATa/MATα*
Cell type	**a**	α	**a**/α
Mating	yes	yes	no
Sporulation	no	no	yes
Pheromone	**a** factor	α factor	none
Receptor	binds α factor	binds **a** factor	none

accomplished by the secretion of **pheromones.** α cells secrete the small polypeptide α-factor; **a** cells secrete **a**-factor. The α-factor is a peptide of 13 amino acids; the **a**-factor is a peptide of 12 amino acids that is modified by addition of a farnesyl (lipid-like) group and carboxymethylation. Each of these peptides is synthesized in the form of a precursor polypeptide that is cleaved to release the mature peptide sequence.

A cell of one mating type carries a surface receptor for the pheromone of the opposite type. When an **a** cell and an α cell encounter one another, their pheromones act on each other to arrest the cells in the G1 phase of the cell cycle, and various morphological changes occur. In a successful mating, the cell cycle arrest is followed by cell and nuclear fusion to produce an **a**/α diploid cell.

The **a**/α cell carries both the *MATa* and *MATα* alleles and has the ability to sporulate. **Figure 17.2** demonstrates how this design maintains the normal haploid/diploid life cycle. Note that *only heterozygous* diploids can sporulate; homozygous diploids (either **a**/**a** or α/α) cannot sporulate.

Much of the information about the yeast mating type pathway was deduced from the properties of mutations that eliminate the ability of **a** and/or α cells to mate. The genes identified by such mutations are called *STE* (for sterile). Mutations in the genes *STE2* and *STE3* are specific for individual mating types; but mutations in the other *STE* genes eliminate mating in *both* **a** and α cells. This situation is explained by the idea that *the events that follow the interaction of factor with receptor are identical for both types.*

Mating is a symmetrical process that is initiated by the interaction of pheromone secreted by one cell type with the receptor carried by the other cell type. The only genes that are uniquely required for the response pathway in either mating type are those coding for the receptors. Either the **a** factor-receptor or the α factor-receptor interaction switches on the same response pathway. Mutations that eliminate steps in the common pathway have the same effects in both cell types.

The initial steps in the mating-type response are summarized in **Figure 17.3**. The components are similar to those of the "classical" receptor-G protein coupled systems (see Chapter 26). The receptors are integral membrane proteins. (Ste2 is the α-receptor in the **a** cell; Ste3 is the **a**-receptor in the α cell.) When either receptor is activated, it interacts with the same G protein. The trimeric G protein consists of the subunits, α, β, and γ. The α subunit binds a guanine nucleotide. In the intact (trimeric) G protein, the α subunit carries GDP. When the pheromone receptor is activated, it causes the GDP to be displaced by GTP. As a result, the α subunit is released from the βγ dimer. This separation of subunits allows the G protein to activate the next protein in whatever pathway it is coupled to.

The most common mechanism used in such path-

Figure 17.2 *Overview*: the yeast life cycle proceeds through mating of *MAT**a*** and *MATα* haploids to give heterozygous diploids that sporulate to generate haploid spores.

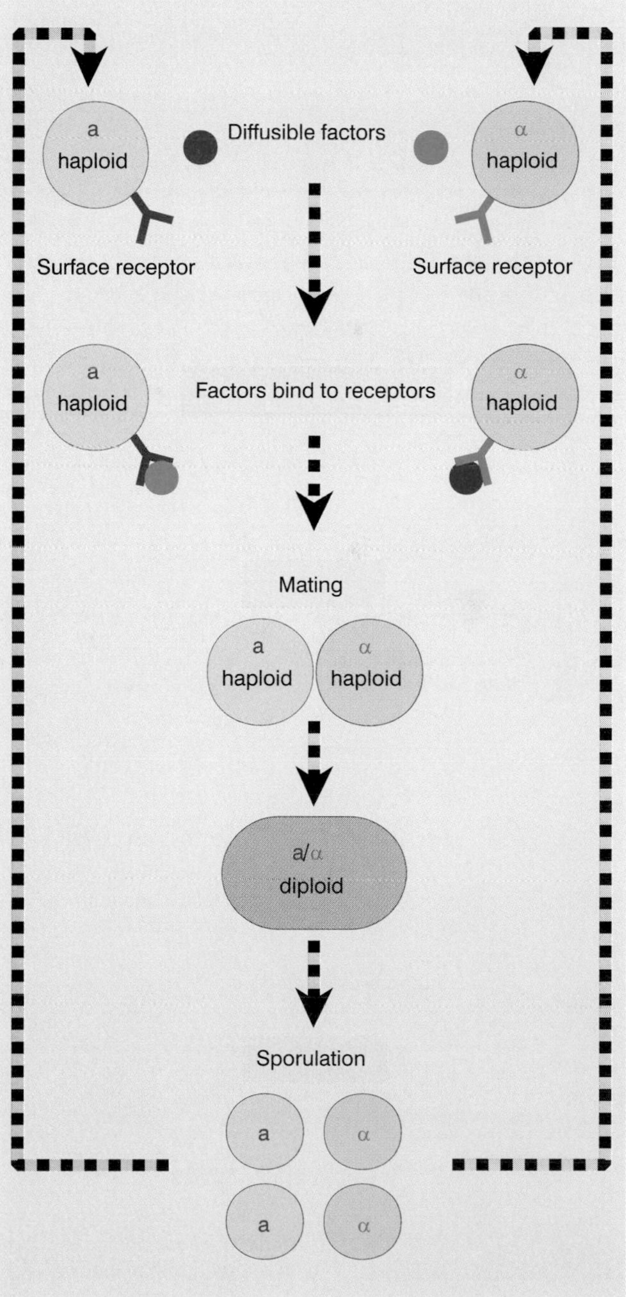

ways is for the activated α subunit to interact with the target protein. However, the situation is different in the mating type pathway, where the βγ dimer activates the next stage in the pathway. The component proteins of the G-trimer are identified by mutations in three genes, *SCG1*, *STE4*, and *STE18*, that affect the response to binding pheromone. Inactivation of *SCG1*, which codes for the G_α protein, causes constitutive expression of the pheromone response pathway (because G_α is unable to maintain $G_{\beta\gamma}$ in the inactive trimeric form). The mutation is lethal, because its effects include arrest of the cell

Figure 17.3 Either **a** or α factor/receptor interaction triggers the activation of a G protein, whose βγ subunits transduce the signal to the next stage in the pathway.

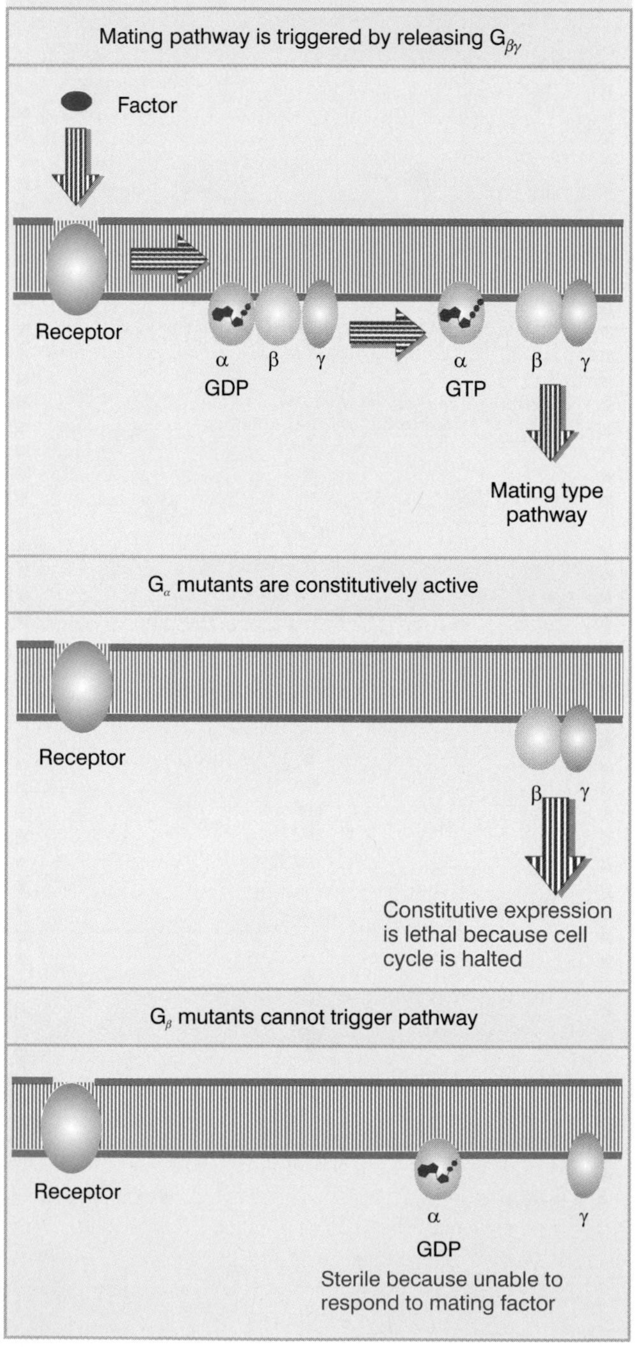

Figure 17.4 The same mating type response is triggered by interaction of either pheromone with its receptor. The signal is transmitted through a series of kinases to a transcription factor; there may be branches to some of the final functions.

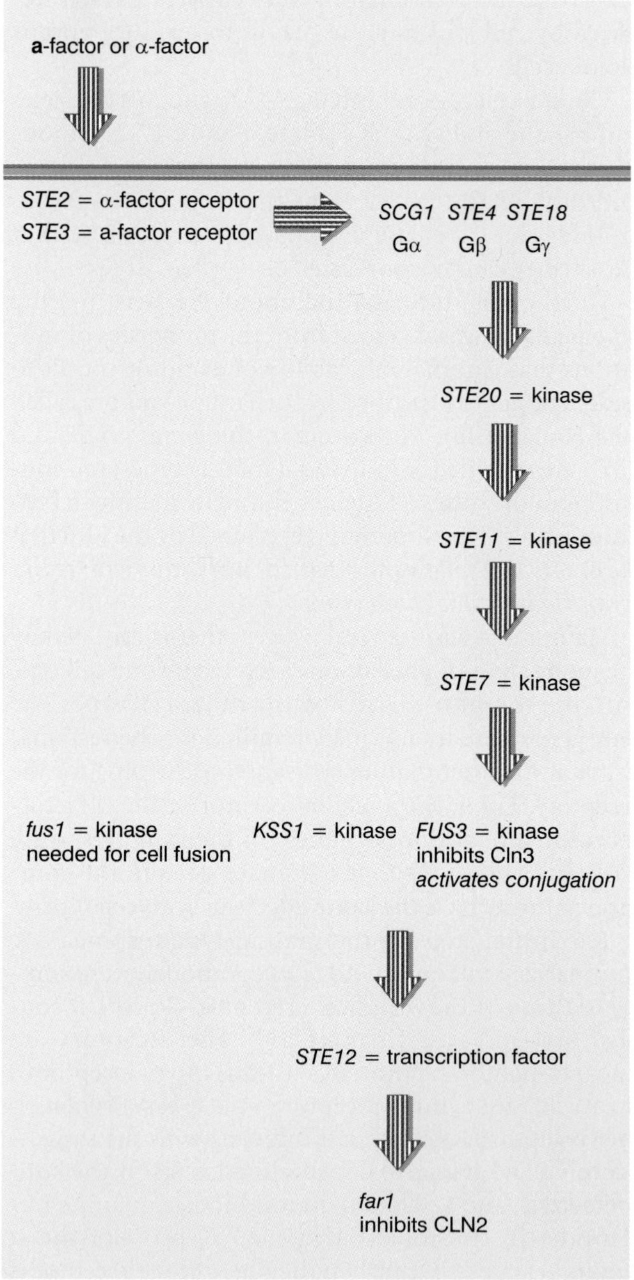

a-factor or α-factor

STE2 = α-factor receptor
STE3 = a-factor receptor

SCG1 *STE4* *STE18*
Gα Gβ Gγ

STE20 = kinase

STE11 = kinase

STE7 = kinase

fus1 = kinase
needed for cell fusion

KSS1 = kinase

FUS3 = kinase
inhibits Cln3
activates conjugation

STE12 = transcription factor

far1
inhibits CLN2

cycle. Inactivation of *STE4*, which codes for the G_β protein, or of *STE18*, which codes for the G_γ protein, creates sterility by abolishing the mating-type response (because the next step in the pathway cannot be activated).

The remaining *STE* genes identify later steps in the pathway. They form the cascade shown in **Figure 17.4**, in which the signal is passed from one to the next, ultimately activating the genes needed for mating. The genes whose products act at the beginning of the cascade code for kinases; kinases such as Ste11 and Ste7 phosphorylate the next protein in the series, thereby activating it. Eventually the transcription factor Ste12 is activated; it in turn activates genes whose products are needed for mating. (Analogous cascades are found in higher organisms, and are compared with the yeast cascade later, in Figure 26.30.)

There are several *STE* genes that have not yet been placed into the cascade, and the order of all the components is not yet certain. So genes such as *FUS3* and *KSS1*, which code for kinases, may belong within the cascade or represent branches from it. The principle, however, is clear: the signal created by interaction of pheromone with receptor is passed along a cascade that culminates by repressing functions needed for the normal cell cycle, and by activating functions needed for mating.

Some of the end targets for the cascade are direct substrates for one of the kinases; for example, Fus3 kinase acts on Cln3, which is one of three Cln proteins needed for cell cycle progression. Other targets are controlled at the level of gene expression; for example, another of the Cln proteins, Cln2, is the target for action of the protein far1, whose expression is activated by the transcription factor Ste12.

Yeast can switch silent and active loci for mating type

SOME yeast strains have the remarkable ability to **switch** their mating types. These strains carry a dominant allele *HO* and *change their mating type frequently*, as often as once every generation. Strains with the recessive allele *ho* have a stable mating type, subject to change with a frequency ~10^{-6}.

The presence of *HO* causes the genotype of a yeast population to change. Irrespective of the initial mating type, in a very few generations there are large numbers of cells of both mating types, leading to the formation of *MATa/MATα* diploids that take over the population. The production of stable diploids from a haploid population can be viewed as the *raison d'être* for switching.

The existence of switching suggests that *all cells contain the potential information needed to be either* MATa *or* MATα, *but express only one type*. Where does the information to change mating types come from? Two additional loci are needed for switching. *HMLα* is needed for switching to give a *MATα* type; *HMRa* is needed for switching to give a *MATa* type. These loci lie on the same chromosome that carries *MAT*. *HML* is far to the left, *HMR* far to the right.

The **cassette model** for mating type is illustrated in **Figure 17.5**. It proposes that *MAT* has an **active cassette** of either type α or type *a*. *HML* and *HMR* have **silent cassettes**. Usually *HML* carries an α cassette, while *HMR* carries an *a* cassette. All cassettes carry information that codes for mating type, but only the active cassette at *MAT* is expressed. Mating-type switching occurs when the active cassette is replaced by information from a silent cassette. The newly installed cassette is then expressed.

Switching is nonreciprocal; the copy at *HML* or *HMR* *replaces* the allele at *MAT*. We know this because a mutation at *MAT* is lost permanently when it is replaced by switching—it does not exchange with the copy that replaces it.

If the silent copy present at *HML* or *HMR* is mutated, switching introduces a mutant allele into the *MAT* locus. The mutant copy at *HML* or *HMR* remains there through an indefinite number of switches. Like replicative transposition, the donor element generates a new copy at the recipient site, while itself remaining inviolate.

Mating-type switching is a directed event, in which there is only one recipient (*MAT*), but two potential donors (*HML* and *HMR*). Switching usually involves replacement of *MATa* by the copy at *HMLα* or replacement of *MATα* by the copy at *HMRa*. In 80–90% of switches, the *MAT* allele is replaced by one of opposite type. This is determined by the phenotype of the cell. Cells of **a** phenotype preferentially choose *HML* as donor; cells of α phenotype preferentially choose *HMR*.

(It is possible to obtain yeast strains in which the usual orientation of the silent cassettes is reversed. When their genotypes are *HMLa* and *HMRα*, 92% of the replacements are homologous, in which an *a* cassette is replaced by another *a* cassette or an α by another α, because the choice of donor cassettes remains the same regardless of the content of the silent loci.)

Several groups of genes are involved in establishing and switching mating type. As well as the genes that directly determine mating type, they include genes needed to repress the silent cassettes, to switch mating type, or to execute the functions involved in mating.

By comparing the sequences of the two silent cassettes (*HMLα* and *HMRa*) with the sequences of the two types of active cassette (*MATa* and *MATα*), we can delineate the sequences that determine mating type. The organization of the mating type loci is summarized in **Figure 17.6**. Each cassette contains common sequences that flank a central region that differs in the *a* and α types of cassette (called Ya or Yα). On either side of this region, the flanking sequences are virtually identical, although they are shorter at *HMR*. The active cassette at *MAT* is transcribed from a promoter within the Y region.

Figure 17.5 Changes of mating type occur when silent cassettes replace active cassettes of opposite genotype; when transpositions occur between cassettes of the same type, the mating type remains unaltered.

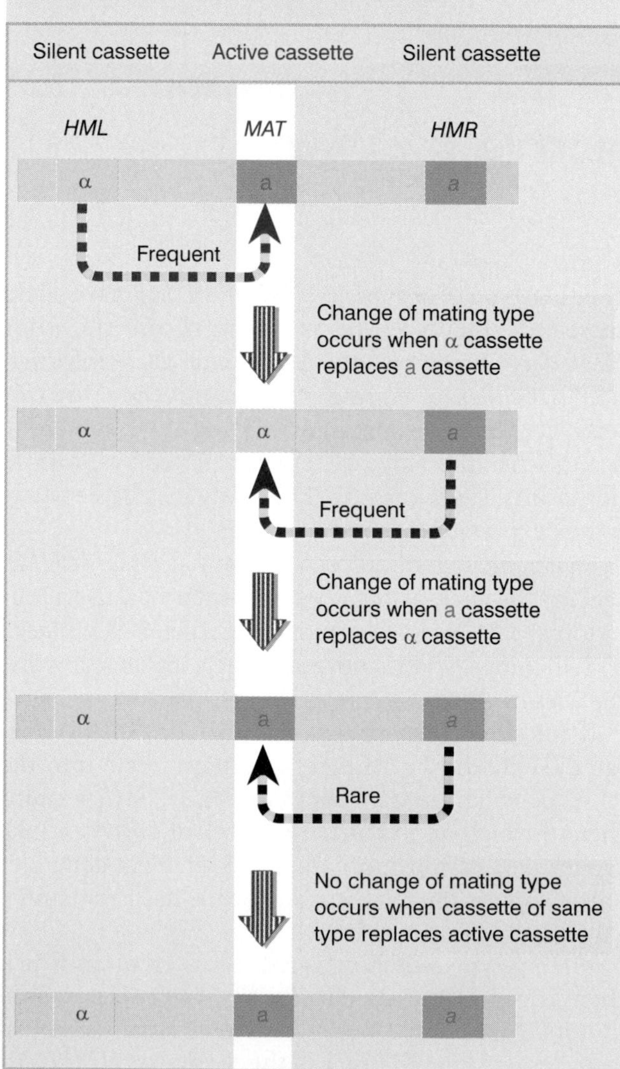

Figure 17.6 Silent cassettes have the same sequences as the corresponding active cassettes, except for the absence of the extreme flanking sequences in *HMRa*. Only the Y region changes between **a** and α types.

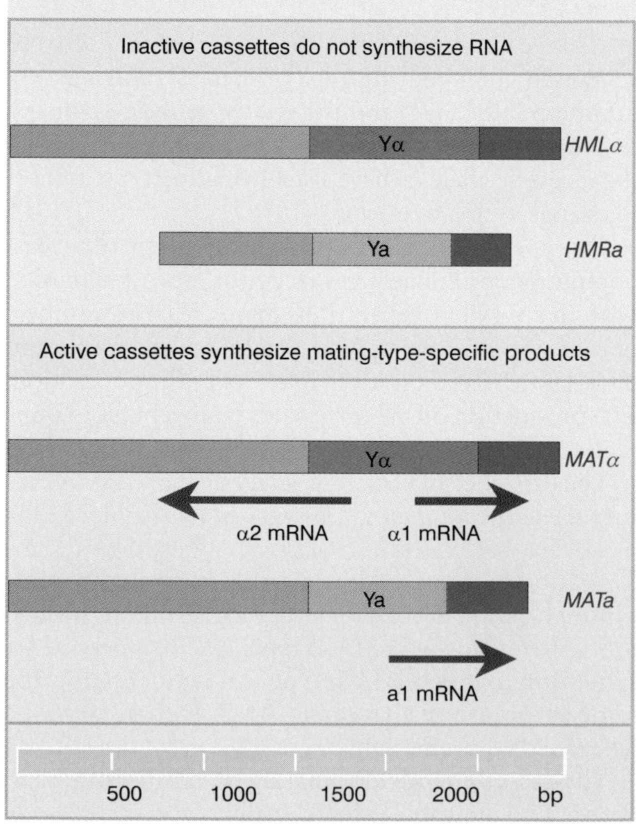

The basic function of the *MAT* locus is to control expression of pheromone and receptor genes, and other functions involved in mating. *MATα* codes for two proteins, α1 and α2. *MATa* codes for a single protein, **a1**. The **a** and α proteins directly control transcription of various target genes; they function by both positive and negative regulation. They function independently in haploids, and in conjunction in diploids. Their interactions are summarized in the table on the right of **Figure 17.7** in terms of three groups of target genes:

- **a**-specific genes are expressed constitutively in **a** cells. They are repressed in α cells. The **a**-specific genes include the **a**-factor structural gene, and *STE2*, which codes for the α-factor receptor. So the **a** phenotype is associated with readiness to recognize the pheromone produced by the opposite mating type.
- α-specific functions are induced in α cells, but are not expressed in **a** cells. They include the α-factor structural gene, and the **a**-factor receptor gene, *STE3*. Again, the expression of pheromone of one type is associated with expression of receptor for the pheromone of the opposite type.
- Haploid-specific functions include genes that are needed for transcription of pheromone and receptor genes, the *HO* gene involved in switching, and *RME*, a repressor of sporulation. They are expressed constitutively in both types of haploid, but are repressed in *a*/α diploids. As a result, the **a**-specific and α-specific functions also remain unexpressed in diploids.

We may now view the functions of the regulators and their targets from the perspective of the *MAT* functions expressed in haploid and diploid yeast cells, as outlined in the diagram on the left of Figure 17.7. The *a* and α mating types are regulated by different mechanisms:

- In *a* haploids, *a* mating functions are expressed constitutively. The functions of the products of *MATa* in the **a** cell (if any) are unknown. It may be required only to repress haploid functions in diploid cells.

Figure 17.7 In diploids the a1 and a2 proteins cooperate to repress haploid-specific functions. In *a* haploids, mating functions are constitutive. In α haploids, the α2 protein represses *a* mating functions, while α1 induces α mating functions.

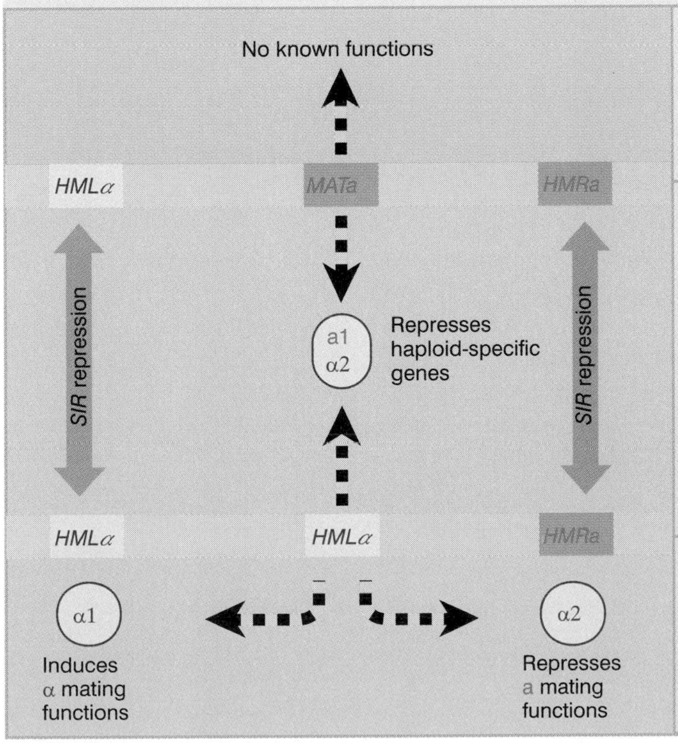

	a specific	α specific	haploid functions
a haploid	constitutive	not expressed	constitutive
diploid	not expressed	not expressed	repressed
α haploid	repressed	induced	constitutive

In α haploids, the α1 product turns on α-specific genes whose products are needed for α mating type. The α2 product represses the genes responsible for producing **a** mating type, by binding to an operator sequence located upstream of target genes.

- In diploids, the *a1* and α2 products cooperate to repress haploid-specific genes. They combine to recognize an operator sequence different from the target for α2 alone.

The abilities of the α2, **a1**, and α1 proteins to regulate transcription rely upon some interesting protein-protein interactions between themselves and with other protein(s). The pattern of gene control in **a** cells, α cells, and diploids, is summarized in **Figure 17.8.**

A protein called PRTF (which is not specific for mating type) is involved in many of these interactions. PRTF binds to a short consensus sequence called the P box. The role of PRTF in gene regulation may be quite extensive, because P boxes are found in a variety of locations. In some of these sites, the P box is required for activation of the gene; but at other loci, PRTF is needed for repression. Its effects may therefore depend on the other proteins that bind at sites adjacent to the P box.

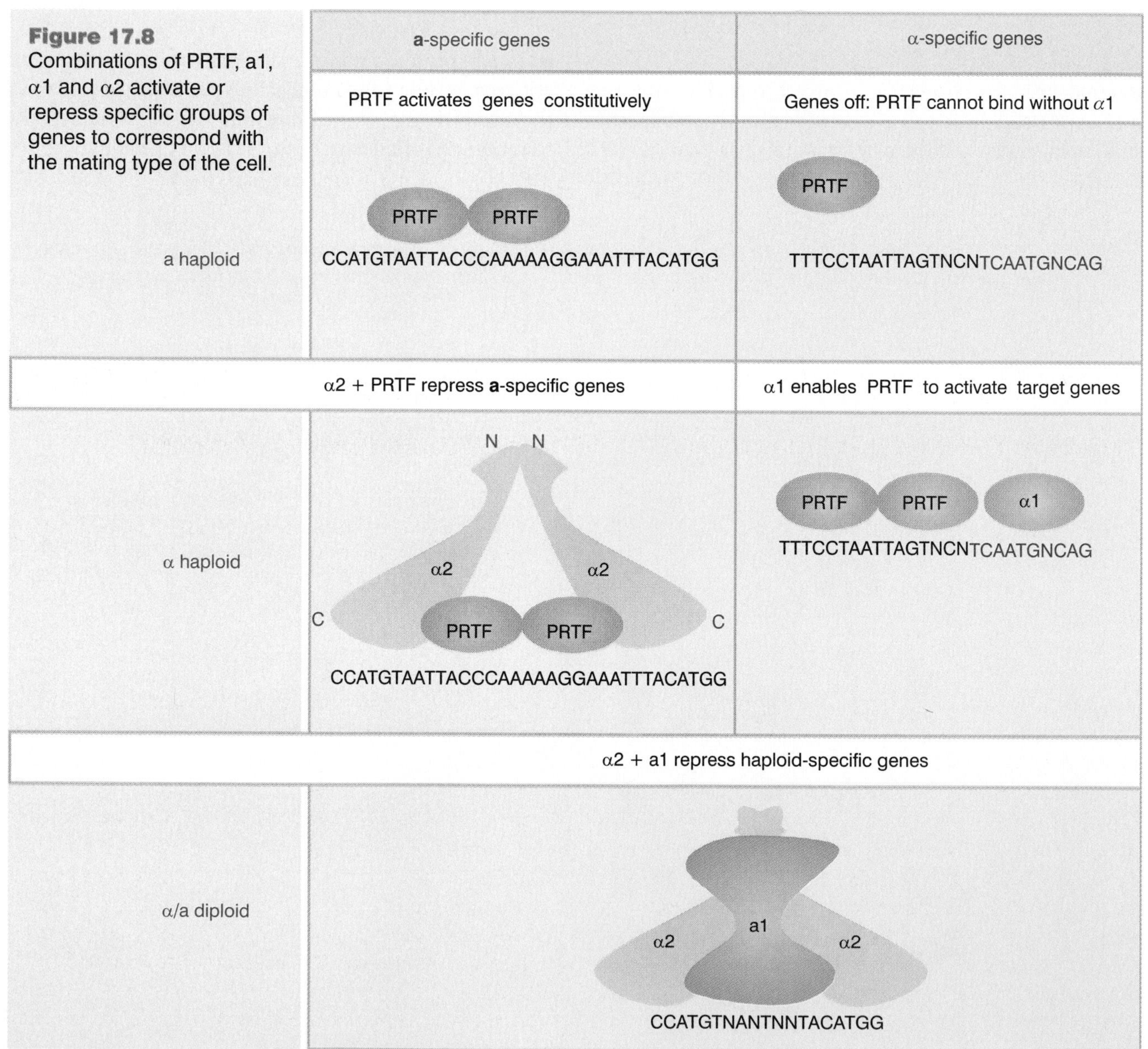

Figure 17.8
Combinations of PRTF, a1, α1 and α2 activate or repress specific groups of genes to correspond with the mating type of the cell.

Genes that are **a**-specific may be activated by PRTF alone. This is adequate to ensure their expression in an a haploid.

The **a**-specific genes are repressed in an α haploid by the combined action of the α2 protein and PRTF. The α2 protein contains two domains. The C-terminal domain binds to short palindromic elements at the ends of an operator consensus sequence of 32 bp. However, binding of this fragment to DNA does not cause repression. The N-terminal domain is needed for repression and is responsible for making contacts with PRTF. The binding site for PRTF is a P box in the center of the operator. In fact, α2 and PRTF bind to the operator cooperatively.

Expression of α-specific genes requires the α1 activator. This is another small protein, 175 amino acids long. *Cis*-acting sequences that confer α-specific transcription are 26 bp long, and can be divided into two parts. The first 16 bp form the P box, where PRTF binds; the adjacent 10 bp sequence forms the binding site for α1. The α1-factor binds *only* when PRTF is present to bind to the P box. Neither protein alone can bind to its target box, but together they can bind to DNA, presumably as a result of protein-protein interactions.

The α-specific genes are turned off by default in **a** haploids because, in the absence of α1 protein, PRTF is unable to bind to activate them.

The α2 protein can also cooperate with the **a1** protein. The combination of these proteins recognizes a different operator. The operator shares the outlying palindromic sequences with the sequence recognized by α2 alone, but is shorter because the sequence between them is different. The α2/**a1** combination represses genes with this motif in diploid cells.

The major point to be made from these results is that the phenotype of each type of cell (*a* or α haploid or *a*/α diploid) is determined by the combination of **a** and α proteins that are expressed. One aspect is the distinction between the haploid and diploid phenotypes; another is the distinction between **a** and α haploid phenotypes. The latter extends to expression of genes corresponding to the appropriate mating type and to the determination of the direction of switching of mating type (see Figure 17.5). *MATa* cells activate a recombination enhancer on the left arm of chromosome III, which increases recombination over a 40 kb region that includes *HML*. *MAT*α cells inactivate the left end of chromosome III.

Silent cassettes at *HML* and *HMR* are repressed

THE transcription map in Figure 17.6 reveals an intriguing feature. Transcription of either *MATa* or *MAT*α initiates within the Y region. Only the MAT locus is expressed; *yet the same Y region is present in the corresponding nontranscribed cassette* (*HML* or *HMR*). This implies that regulation of expression is not accomplished by direct recognition of some site overlapping with the promoter. *A site outside the cassettes must distinguish HML and HMR from MAT.*

Deletion analysis shows that sites on either side of both *HML* and *HMR* are needed to repress their expression. They are called **silencers**. The sites on the left of each cassette are called the *E* silencers, and the sites on the right side are called the *I* silencers. These control sites can function at a distance (up to 2.5 kb away from a promoter) and in either orientation. They behave like negative enhancers (enhancers are elements that activate transcription, and are discussed in Chapter 20).

Can we find the basis for the control of cassette activity by identifying genes that are responsible for keeping the cassettes silent? We would expect the products of these genes to act on the silencers. A convenient assay for mutation in such genes is provided by the fact that, when a mutation allows the usually silent cassettes at *HML* and *HMR* to be expressed, both **a** and α functions are produced, so the cells behave like *MATa*/*MAT*α diploids.

Mutations in several loci abolish silencing and lead to expression of *HML* and *HMR*. The first to be discovered were the four *SIR* loci (silent information regulators). All four wild-type *SIR* loci are needed to maintain *HML* and *HMR* in the repressed state; mutation in any one of these loci to give a *sir*⁻ allele has two effects. Both *HML* and *HMR* can be transcribed. And both the silent cassettes become targets for replacement by switching. *So the same regulatory event is involved in repressing a silent cassette and in preventing it from being a recipient for replacement by another cassette.*

Other loci required for silencing include *RAP1* (which is also required to maintain telomeric heterochromatin in its inert state) and the genes coding for histone H4. Deletions of the N-terminus of histone H4 or individual point mutations activate the silent cassettes. The effects of these mutations can be overcome either by introducing new mutations in *SIR3* or by over-expressing *SIR1*, which suggests that there is a specific interaction between histone H4 and the SIR proteins.

The general model suggested by these results is that the SIR proteins act on chromatin structure to prevent expression of the genes. Because mutations in the SIR proteins have the same effects on genes that have been inactivated by the proximity of telomeric heterochromatin, it seems likely that SIR proteins are involved generally in interacting with histones to form heterochromatic (inert) structures. We discuss this in more detail in Chapter 19.

There is an interesting connection between repression at the silencers and DNA replication. Each silencer contains an *ARS* sequence (an origin of replication). The ARS is bound by the ORC (the origin recognition complex) that is involved in initiating replication. Mutations in *ORC* genes prevent silencing, indicating that ORC protein binding at the silencer is required for silencing.

There are two separate types of connection between silencing and the replication apparatus. If a Sir1 protein is localized at the silencer (by linkage to another protein that is bound there), the binding of ORC is no longer necessary. This means that the role of ORC is solely to bring in Sir1; it is not required to initiate replication. The role of ORC could therefore be to provide an initiating center from which the silencing effect can spread. This is different from its role in replication.

However, passage through S phase is necessary for silencing to be established. This does not require initiation to occur at the *ARS* in the silencer. The effect could depend on the passage of a replication fork through the silencer, perhaps in order to allow the chromatin structure to be changed. (This would contrast with the ability to remodel chromatin at promoters without replication; see Chapter 19.)

Unidirectional transposition is initiated by the recipient *MAT* locus

A switch in mating type is accomplished by a gene conversion in which the recipient site (*MAT*) acquires the sequence of the donor type (*HML* or *HMR*). Sites needed for transposition have been identified by mutations at *MAT* that prevent switching. The unidirectional nature of the process is indicated by lack of mutations in *HML* or *HMR*.

The mutations identify a site at the right boundary of Y at *MAT* that is crucial for the switching event. The nature of the boundary is shown by analyzing the locations of these point mutations relative to the site of switching (this is done by examining the results of rare switches that occur in spite of the mutation). Some mutations lie within the region that is replaced (and thus disappear from *MAT* after a switch), while others lie just outside the replaced region (and therefore continue to impede switching). So sequences both within and outside the replaced region are needed for the switching event.

Switching is initiated by a double-strand break close to the Y-Z boundary that coincides with a site that is sensitive to attack by DNAase. (This is a common feature of chromosomal sites that are involved in initiating transcription or recombination.) It is recognized by an endonuclease coded by the *HO* locus. The *HO* endonuclease makes a staggered double-strand break just to the right of the Y boundary. Cleavage generates the single-stranded ends of four bases drawn in **Figure 17.9**. The nuclease does not attack mutant *MAT* loci that cannot switch. Deletion analysis shows that most or all of the sequence of 24 bp surrounding the Y junction is required for cleavage *in vitro*. The recognition site is relatively large for a nuclease. Probably the recognition sequence occurs only at the three mating-type cassettes.

Only the *MAT* locus and not the *HML* or *HMR* loci are targets for the endonuclease. It seems plausible that *the same mechanisms that keep the silent cassettes from*

Figure 17.9 HO endonuclease cleaves *MAT* just to the right of the Y region, generating sticky ends with a 4 base overhang.

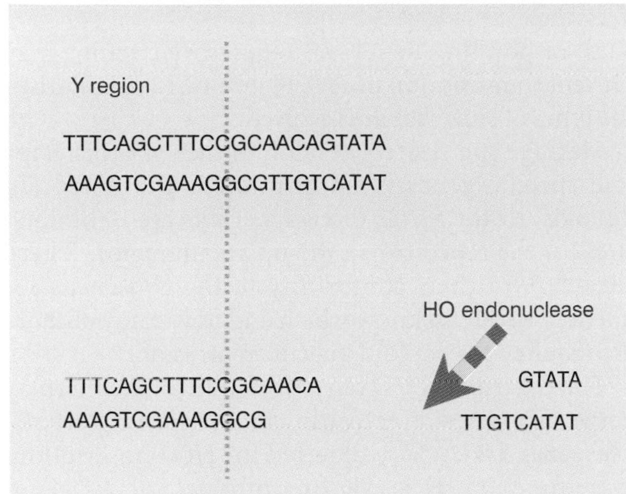

Figure 17.10 Cassette substitution is initiated by a double-strand break in the recipient (*MAT*) locus, and may involve pairing on either side of the Y region with the donor (*HMR* or *HML*) locus.

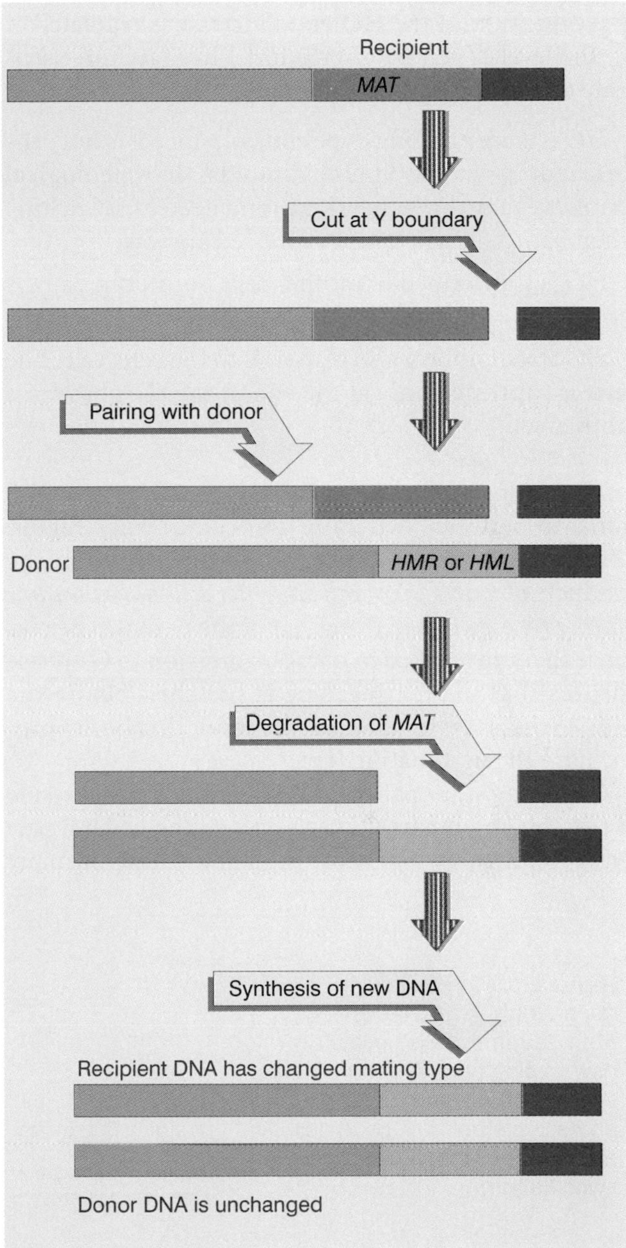

being transcribed also keep them inaccessible to the HO *endonuclease. This inaccessibility ensures that switching is unidirectional.*

The reaction triggered by the cleavage is illustrated schematically in **Figure 17.10** in terms of the general reaction between donor and recipient regions. In terms of the interactions of individual strands of DNA, it follows the scheme for recombination via a double-strand break drawn in Figure 14.5; and the stages following the initial cut require the enzymes involved in general recombination. Mutations in some of these genes prevent switching.

Suppose that the free end of *MAT* invades either the *HML* or *HMR* locus and pairs with the region of homology on the right side. The Y region of *MAT* is degraded until a region with homology on the left side is exposed. At this point, *MAT* is paired with *HML* or *HMR* at both the left side and the right side. The Y region of *HML* or *HMR* is copied to replace the region lost from *MAT* (which might extend beyond the limits of Y itself). The paired loci separate. (The order of events could be different.)

Like the double-strand break model for recombination, the process is initiated by *MAT*, the *locus that is to be replaced.* In this sense, the description of *HML* and *HMR* as donor loci refers to their ultimate role, but not to the mechanism of the process. Like replicative transposition, the donor site is unaffected, but a change in sequence occurs at the recipient; unlike transposition, the recipient locus suffers a substitution rather than addition of material.

Regulation of *HO* expression

PRODUCTION of the HO endonuclease is regulated at the level of gene transcription. There are three separate control systems:

- *HO* is under mating-type control, since it is not synthesized in *MATa/MATα* diploids. In teleological terms, we may think that there is no need for switching when *both MAT* alleles are expressed anyway.
- *HO* is transcribed in mother cells but not in daughter cells.
- *HO* transcription also responds to the cell cycle. The gene is expressed only at the end of the G1 phase of a mother cell.

The timing of nuclease production explains the relationship between switching and cell lineage. **Figure 17.11** shows that switching is detected only in the products of a division; *both daughter cells have the same mating type*, switched from that of the parent. The reason is that the restriction of *HO* expression to G1 phase ensures that the mating type is switched before the *MAT* locus is replicated, with the result that both progeny have the new mating type.

Cis-acting sites that control *HO* transcription reside in the 1500 bp upstream of the gene. The general pattern of control is that repression at any one of many sites, responding to several regulatory circuits, may prevent transcription of *HO*. **Figure 17.12** summarizes the types of sites that are involved.

Mating type control resembles that of other haploid-specific genes. Transcription is prevented (in diploids) by the **a1/α2** repressor. There are 10 binding sites for the repressor in the upstream region. These sites vary in their conformity to the consensus sequence; we do not know which and how many of them are required for haploid-specific repression.

The control of *HO* transcription involves interplay between a series of activating and repressing events. The genes *SWI1–5* are required for *HO* transcription. They function by preventing products of the genes *SIN1–6* from repressing *HO*. The *SWI* genes were discovered first, as mutants unable to switch; then the *SIN* genes were discovered for their ability to release the blocks caused by particular *SWI* mutations. SWI-SIN interactions are involved in both cell-cycle control and the restriction of expression to mother cells.

Some of the *SWI* and *SIN* genes are not specifically concerned with mating type, but are global regulators of transcription, whose functions are needed for expression of many loci. They include the activator complex SWI1,2,3 and the loci *SIN1–4* that code for chromosomal proteins. Their role in mating type ex-

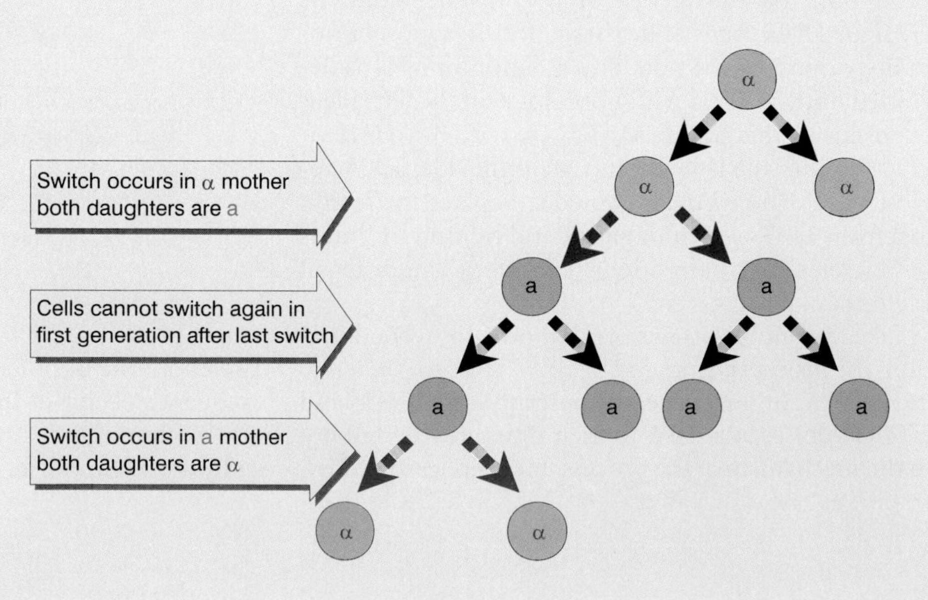
Switch occurs in α mother
both daughters are a
Cells cannot switch again in
first generation after last switch
Switch occurs in a mother
both daughters are α
α
α
α
a
a
a
a
a
a
α
α

Figure 17.11 Switching occurs only in mother cells; both daughter cells have the new mating type. A daughter cell must pass through an entire cycle before it becomes a mother cell that is able to switch again.

Figure 17.12 Three regulator systems act on transcription of the *HO* gene. Transcription occurs only when all repression is lifted.

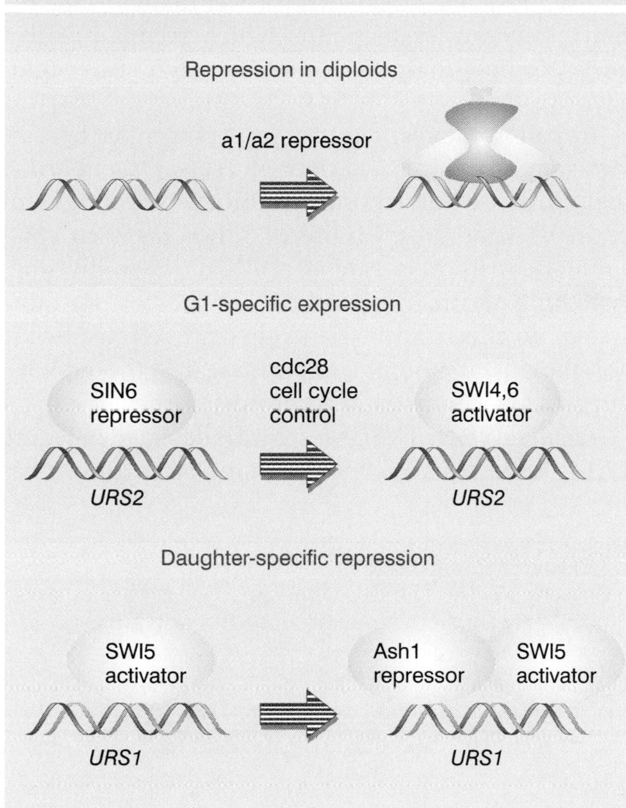

pression is incidental. The "real" regulator is therefore the SWI protein that counteracts the general repression system specifically at the *HO* locus.

Cell-cycle control is conferred by nine copies of an octanucleotide sequence called *URS2*. A copy of the consensus sequence can confer cell-cycle control on a gene to which it is attached. A gene linked to this sequence is repressed except during a transient period toward the end of G1 phase. SWI4 and SWI6 are the activators that release repression at *URS2*. Their activity depends on the function of the cell-cycle regulator *CDC28*, which executes the decision that commits the cell to divide (see Chapter 27).

The target for restricting expression to alternate generations is the activator SWI5 (which antagonizes a general repression system exercised by SIN3,4). In mutants that lack these functions, *HO* is transcribed equally well in mother and daughter cells. This system acts on *URS1* elements in the far upstream region.

SWI5 is not itself the regulator of mother-cell specificity, but is antagonized by Ash1p, a repressor that accumulates preferentially in daughter cells at the end of anaphase. Mutations in *ASH1* allow daughter cells to switch mating type. The basis for the accumulation of Ash1p is not known. The possibility that it involves transport of the protein or mRNA is suggested by the identification of mutations in an actomyosin-dependent system that block its accumulation. Its presence prevents SWI5 from activating the *HO* gene—it might work either by binding to the *URS1* promoter sequences or by binding to SWI5 protein. When the daughter cell grows to become a mother cell, the concentration of Ash1p is diluted, and it becomes possible to express the *HO* gene again.

Trypanosomes rearrange DNA to express new surface antigens

SLEEPING sickness in man (and a related disease in cows) is caused by infection with African trypanosomes. The unicellular parasite follows the life cycle illustrated in **Figure 17.13**, in which it alternates between tsetse fly and mammal. The trypanosome may be transferred either to or from the fly when it bites a mammal.

During its life cycle, the parasite undergoes several morphological and biochemical changes. The most significant biochemical change is in the **variable surface glycoprotein** (**VSG**), the major component of the surface coat. The coat covers the plasma membrane and consists of a monolayer of $(5–10) \times 10^6$ molecules of a single VSG, which is the only antigenic structure exposed on the surface. A trypanosome expresses only one VSG at any time, and its ability to change the VSG is responsible for its survival through the fly-mammal infective cycle.

Consider the cycle as starting when a fly gains a trypanosome by biting an infected mammal. The trypanosome enters the gut of the fly in the "procyclic form", and loses its VSG. After about three weeks, its progeny differentiate into the "metacyclic form", which re-acquires a VSG coat. This form is transmitted to the mammalian bloodstream during a bite by the fly. The trypanosome multiplies in the mammalian bloodstream. Its progeny continue to express the metacyclic VSG for about a week. Then a new VSG is synthesized, and further transitions occur every 1–2 weeks.

Each of the successive VSG species is immunologically distinct. As a result, the antigen presented to the mammalian immune system is constantly changing. The process of transition is called **antigenic variation.** The immune response of the organism always lags behind the change in surface antigen, so that the trypanosome evades immune surveillance, and thereby perpetuates itself indefinitely. Each transition of the VSG is accompanied by a new wave of parasitemia, with symptoms of fever, rash, etc.; the parasites eventually invade the central nervous system, after which the mammalian host becomes progressively more lethargic and eventually comatose.

Trypanosomes vary in their host range. The best investigated species is a variety of *Trypanosoma brucei* that grows well on laboratory animals (although not on man). Laboratory strains of *T. brucei* switch VSGs spontaneously at a rate of 10^{-4}–10^{-6} per division. Switching occurs independently of the host immune system. In effect, new variants then are selected by the host, because it mounts a response against the old VSG, but fails to recognize and act against the new VSG.

A general view of VSG structure is depicted in **Figure 17.14.** A nascent VSG is ~500 amino acids long; it has

Figure 17.13 *Overview:* a trypanosome passes through several morphological forms when its life cycle alternates between a tsetse fly and mammalian host.

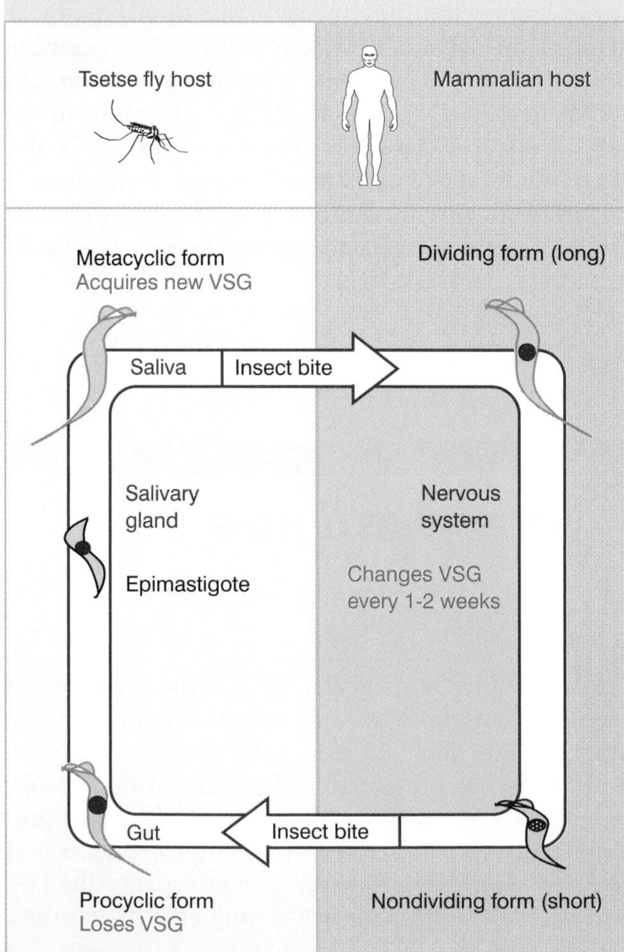

Figure 17.14 The C-terminus of VSG is cleaved and covalently linked to the membrane through a glycolipid.

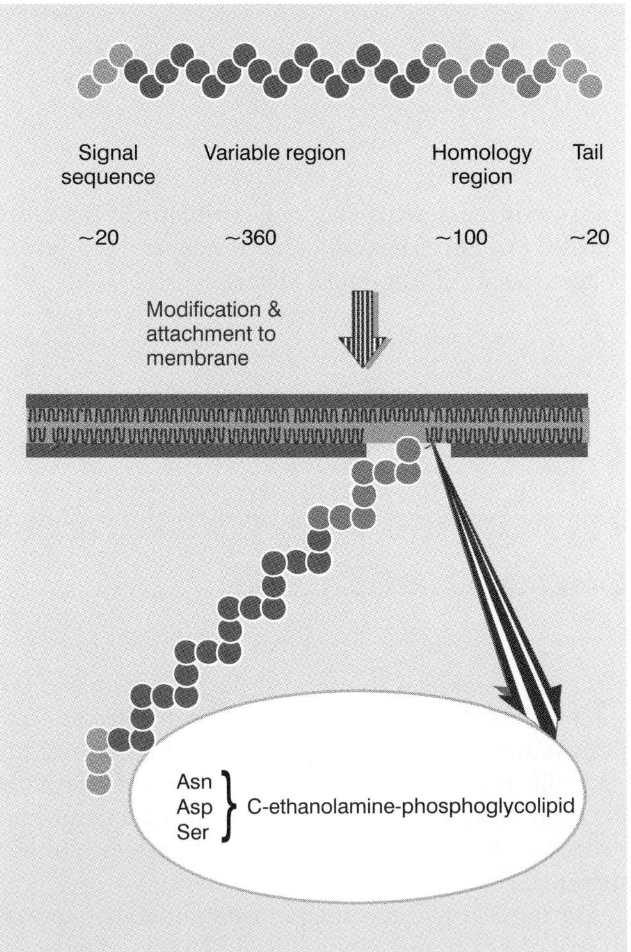

an N-terminal signal sequence, followed by a long *variable region* that provides the unique antigenic determinant, and a C-terminal *homology region* ending in a short hydrophobic tail. The nascent VSG is processed at both ends to give the mature form. The signal sequence is cleaved during secretion. The hydrophobic tail is removed before the VSG reaches the outside surface. The new C-terminus is covalently attached to the trypanosome membrane; three types of homology region are distinguished according to the C-terminal amino acid.

The VSG is attached to the membrane via a phosphoglycolipid. As a result, VSG can be released from the membrane by an enzyme that removes fatty acid. This reaction (which is used in purifying the VSG) may be important *in vivo* in allowing one VSG to be replaced by another on the surface of the trypanosome.

How many varieties of VSG can be expressed by any one trypanosome? It is not clear that any limit is encountered before death of the host. A single trypanosome can make at least 100 VSGs sufficiently different in sequence that antibodies against any one do not react against the others.

VSG variation is coded in the trypanosome genome. Every individual trypanosome carries the entire VSG repertoire of its strain. *Diversity therefore depends on changing expression from one preexisting gene to another.*

The trypanosome genome has an unusual organization, consisting of a large number of segregating units. In addition to an unknown number of chromosomes, it contains ~100 "minichromosomes", each containing ~50–150 kb of DNA. Hybridization experiments identify ~1000 VSG genes, scattered among all size classes of chromosomal material.

Each VSG is coded by a **basic-copy gene**. These genes can be divided into two classes according to their chromosomal location:

- Telomeric genes lie within 5–15 kb of a telomere. There could be >200 of these genes if every telomere has one.
- Internal genes reside within chromosomes (more formally, they lie >50 kb from a telomere).

As might be expected of a large family of genes, individual basic copies show varying degrees of relationship, presumably reflecting their origin by duplication and variation. Genes that are closely related, and which provoke the same antigenic response, are called isogenes.

How is a single VSG gene selected for expression? Only one VSG gene is transcribed in a trypanosome at a given time. The copy of the gene that is active is called the **expression-linked copy** (**ELC**). It is said to be located at an **expression site**. An expression site has a characteristic property: *it is located near a telomere.*

These features immediately suggest that the route followed to select a gene for expression depends on whether the basic copy is itself telomeric or internal. The two types of event that can create an ELC are summarized in **Figure 17.15**:

- The *expression site remains the same, but the ELC is changed.* Duplication transfers the sequence of a basic copy to replace the sequence currently occupying the expression site. Either internal or telomeric copies may be activated directly by duplication into the expression site. *The substitution of one cassette for another does not interfere with the activity of the site.*
- *The expression site is changed.* Activation *in situ* is available only to a sequence already present at a telomere. When a telomeric site is activated *in situ*, the previous expression site must cease to be active and the new site now becomes the expression site.

Internal basic copies probably can be copied into non-expressed telomeric locations as well as into expression sites. So an internal gene could be activated by a two-stage process, in which first it is transposed to a non-expressed telomere, and then this site is activated.

We can follow the fate of genes involved in activation by restriction mapping. A probe representing an expressed sequence can be derived from the mRNA. Then we can determine the status of genes corresponding to the probe. We see different results for internal and telomeric basic-copy genes:

Activation of an internal gene requires generation of new sequences. **Figure 17.16** shows that, when an internal gene is activated, a new fragment is found. The original basic-copy gene remains unaltered; the new fragment is generated by the duplication of the gene into a new context (where the sites recognized by the restriction enzyme are in the surrounding sequences and therefore generate a distinct fragment). The new fragment identifies an ELC, located close to a telomere. The ELC appears when the gene is expressed and disappears when the gene is switched off. Duplication into the ELC is the *only* pathway by which an internal basic copy can be generated.

- *Activation of a telomeric gene can occur* in situ. **Figure 17.17** shows that when a telomeric gene is activated, the gene number need not change. The structure of the gene may be essentially unaffected as detected by restriction mapping. The size of the fragment containing

Figure 17.15 VSG genes may be created by duplicative transfer from an internal or telomeric basic copy into an expression site, or by activating a telomeric copy that is already present at a potential expression site.

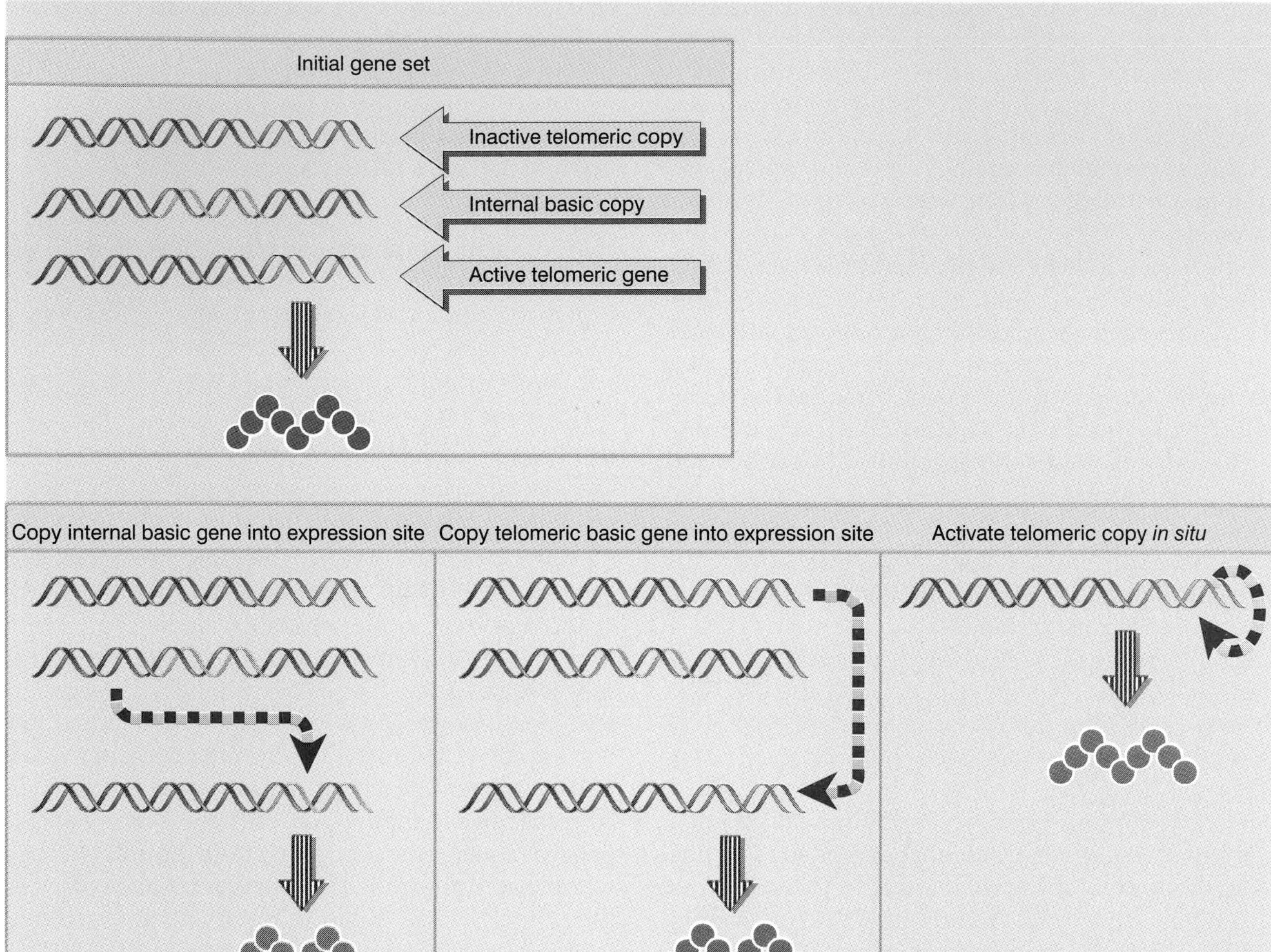

the gene may vary slightly, because the length of the telomere is constantly changing. Telomeric basic copies can also be activated by the same duplication pathway as internal copies; in this case, the basic copy remains at its telomere, while an expression-linked copy appears at another telomere (generating a new fragment as illustrated for internal basic copies in Figure 17.16).

Formation of the ELC occurs by gene conversion. Like the switch in yeast mating type, it represents the replacement of a "cassette" at the active (telomeric) locus by a stored cassette. The VSG system is more versatile in the sense that there are many potential donor cassettes (and also more than a single potential recipient site).

Almost all switches in VSG type involve replacement of the ELC by a pre-existing silent copy. Some exceptional cases have been found, however, in which the sequence of the ELC does not match any of the repertoire of silent copies in the genome. A new sequence may be created by a series of gene conversions in which short stretches of different silent copies are connected. This resembles the mechanism for generating diversity in chicken λ immunoglobulins (see Chapter 24). Although rare, such occurrences extend VSG diversity.

How many expression sites are there? So far only a few expression sites have been observed, which suggests that only a subset of telomeres can function in this capacity, but it is possible that any telomere can be used (although some may be preferred to others).

The structure of the VSG gene at the ELC is unusual, as illustrated in **Figure 17.18**. The length of DNA transferred into the ELC is 2500–3500 bp, somewhat longer

Figure 17.16 Internal basic copies can be activated only by generating a duplication of the gene at an expression-linked site.

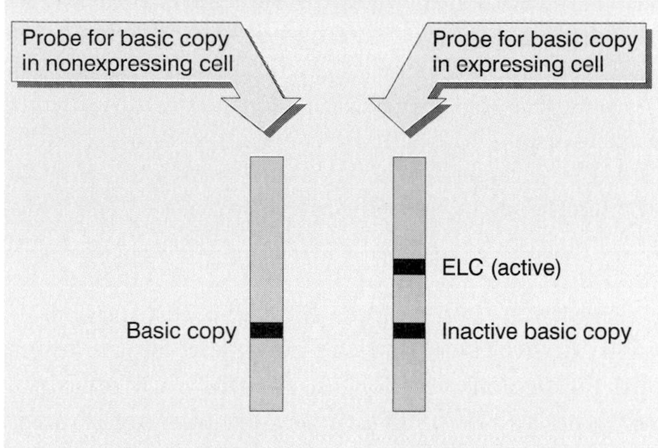

Figure 17.17 Telomeric basic copies can be activated *in situ*; the size of the restriction fragment may change (slightly) when the telomere is extended.

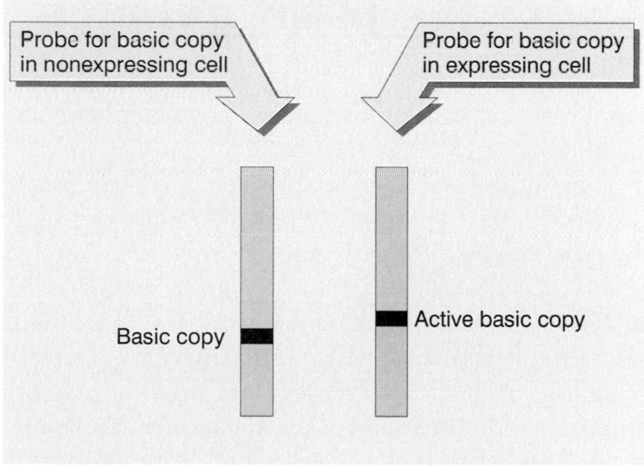

Figure 17.18 The expression-linked copy of a VSG gene contains barren regions on either side of the transposed region, which extends from ~1000 bp upstream of the VSG coding region to a site near the 3' terminus of the mRNA.

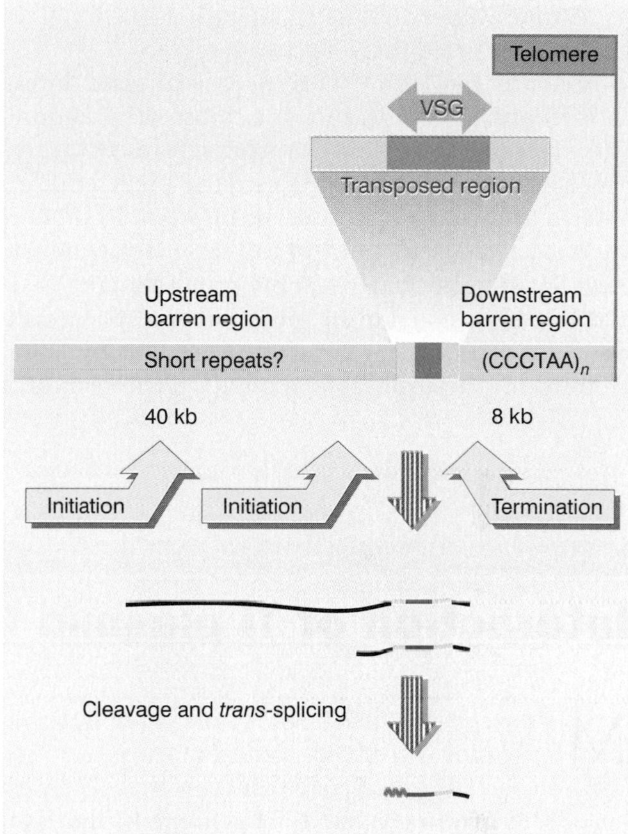

than the VSG-coding region of 1500 bp. Most of the additional length is upstream of the gene. The crossover points at which the duplicated sequence joins the ELC do not appear to be precisely determined.

Analysis of events at the 5′ end of the VSG mRNA is complicated by the fact that the mature RNA starts with a 35 base sequence coded elsewhere, and added in *trans* to the newly synthesized 5′ end (see Chapter 22). The signals for initiating and terminating transcription (and sometimes also the end of the coding region itself) are provided by the sequences flanking the transposed region. In fact, transcription may be initiated several kb upstream of the VSG gene itself. Promoters have been mapped at 4 kb and ~60 kb upstream of the VSG sequence. Use of the more distant promoter generates a transcript that contains other genes as well as the active VSG. The VSG sequence (and other gene sequences) must be released by cleavage from the transcript, after which the 35 base spliced leader is added to the 5′ end.

On either side of the transposed region are extensive regions that are not cut by restriction enzymes. These "barren regions" consist of repetitive DNA; they extend some 8 kb downstream and for up to 40 kb upstream of the ELC. Going downstream, the barren region consists largely of repeats of the sequence CCCTAA, and extends to the telomere. Proceeding upstream, it may also consist of repetitive sequences, but their nature is not yet clear. The existence of the barren regions, however, has been an impediment to characterizing ELC genes by cloning.

Activation of an expression site involves a change in the region upstream of the transcription unit. When the promoter of an ELC is substituted by another promoter, the capacity of the ELC to be inactivated and activated is not altered, and therefore cannot be the result of control by *trans*-acting factors at the promoter. The regulatory event could be an alteration in chromatin structure, initiated in the upstream region.

The order in which VSG genes are expressed during an infection is erratic, but not completely random. This may be an important feature in survival of the trypanosome. If VSG genes were used in a predetermined order, a host could knock out the infection by mounting a reaction against one of the early elements. The need for unpredictability in the production of VSGs may be responsible for the evolution of a system with many donor sequences and multiple recipients.

Antigenic variation is not a unique phenomenon of trypanosomes. The bacterium *Borrelia hermsii* causes relapsing fever in man and analogous diseases in other mammals. The name of the disease reflects its erratic course: periods of illness are spaced by periods of relief. When the fevers occur, spirochetes are found in the blood; they disappear during periods of relief, as the host responds with specific antibodies.

Like the trypanosomes, *Borrelia* survives by altering a surface protein, called the variable major protein (VMP). Changes in the VMP are associated with rearrangements in the genome. The active VMP is located near the telomere of a linear plasmid. We do not yet know the extent of the coded variants or the mechanisms used to alter their expression. It is intriguing, however, that the eukaryote *Trypanosoma* and the prokaryote *Borrelia* should both rely upon antigenic variation as a means for evading immune surveillance.

Interaction of Ti plasmid DNA with the plant genome

MOST events in which DNA is rearranged or amplified occur within a genome, but the interaction between bacteria and certain plants involves the transfer of DNA from the bacterial genome to the plant genome. **Crown gall disease** (shown in **Figure 17.19**) can be induced in most dicotyledonous plants by the soil bacterium *Agrobacterium tumefaciens.* The bacterium is a parasite that effects a genetic change in the eukaryotic host cell, with consequences for both parasite and host. It improves conditions for survival of the parasite. And it causes the plant cell to grow as a tumor.

Agrobacteria are required to induce tumor formation, but the tumor cells do not require the continued presence of bacteria. Like animal tumors, the plant cells have been transformed into a state in which new mechanisms govern growth and differentiation. *Transformation is caused by the expression within the plant cell of genetic information transferred from the bacterium.*

The *t*umor-*i*nducing principle of *Agrobacterium* resides in the **Ti plasmid**, which is perpetuated as an independent replicon within the bacterium. The plasmid carries genes involved in various bacterial and plant cell activities, including those required to generate the transformed state, and a set of genes concerned with synthesis or utilization of **opines** (novel derivatives of arginine).

Ti plasmids (and thus the *Agrobacteria* in which they reside) can be divided into four groups, according to the types of opine that are made:

- **Nopaline plasmids** carry genes for synthesizing nopaline in tumors and for utilizing it in bacteria. Nopaline tumors can differentiate into shoots with abnormal structures. They have been called **teratomas** by analogy with certain mammalian tumors that retain the ability to differentiate into early embryonic structures.
- **Octopine plasmids** are similar to nopaline plasmids, but the relevant opine is different. However, octopine tumors are usually undifferentiated and do not form teratoma shoots.
- **Agropine plasmids** carry genes for agropine metabolism; the tumors do not differentiate, develop poorly, and die early.
- **Ri plasmids** can induce hairy root disease on some plants and crown gall on others. They have agropine type genes, and may have segments derived from both nopaline and octopine plasmids.

Figure 17.19 An *Agrobacterium* carrying a Ti plasmid of the nopaline type induces a teratoma, in which differentiated structures develop. Photograph kindly provided by Jeff Schell.

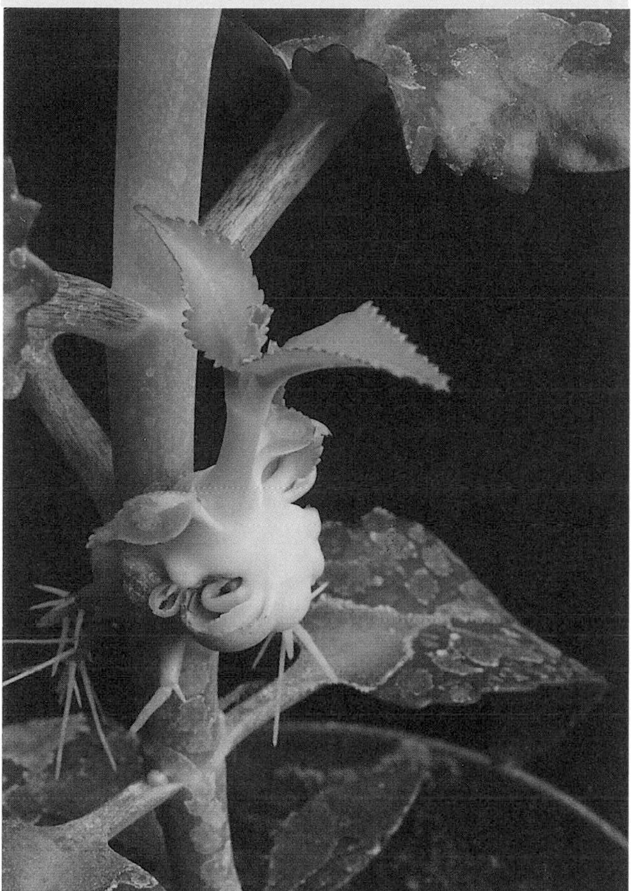

The types of genes carried by a Ti plasmid are summarized in **Figure 17.20**. Genes utilized in bacteria code for plasmid replication and incompatibility, for transfer between bacteria, sensitivity to phages, and for synthesis of other compounds, some of which are toxic to other soil bacteria. Genes used in the plant cell code for transfer of DNA into the plant, for induction of the transformed state, and for shoot and root induction.

The specificity of the opine genes depends on the type of plasmid. Genes needed for opine synthesis are linked to genes whose products catabolize the same opine; thus each strain of *Agrobacterium* causes crown gall tumor cells to synthesize opines that are useful for survival of the parasite. The opines can be used as the sole carbon and/or nitrogen source for the inducing *Agrobacterium* strain. The principle is that *the transformed plant cell synthesizes those opines that the bacterium can use.*

Figure 17.20 Ti plasmids carry genes involved in both plant and bacterial functions.

Locus	Function	Ti Plasmid
vir	DNA transfer into plant	all
shi	shoot induction	all
roi	root induction	all
nos	nopaline synthesis	nopaline
noc	nopaline catabolism	nopaline
ocs	octopine synthesis	octopine
occ	octopine catabolism	octopine
tra	bacterial transfer genes	all
Inc	incompatibility genes	all
oriV	origin for replication	all

The interaction between *Agrobacterium* and a plant cell is illustrated in **Figure 17.21**. The bacterium does not enter the plant cell, but *transfers part of the Ti plasmid to the plant nucleus.* The transferred part of the Ti genome is called **T-DNA**. It becomes integrated into the plant genome, where it expresses the functions needed to synthesize opines and to transform the plant cell.

Transformation of plant cells requires three types of function carried in the *Agrobacterium*:

- Three loci on the *Agrobacterium* chromosome, *chvA*, *chvB*, *pscA*, are required for the initial stage of binding the bacterium to the plant cell. They are responsible for synthesizing a polysaccharide on the bacterial cell surface.
- The *vir* region carried by the Ti plasmid outside the T-DNA region is required to release and initiate transfer of the T-DNA.
- The T-DNA is required to transform the plant cell.

The organization of the major two types of Ti plasmid is illustrated in **Figure 17.22**. About 30% of the ~200 kb Ti genome is common to nopaline and octopine plasmids. The common regions include genes involved in all stages of the interaction between *Agrobacterium* and a plant host, but considerable rearrangement of the sequences has occurred between the plasmids.

The T-region occupies ~23 kb. Some 9 kb is the same in the two types of plasmid. The Ti plasmids carry genes for opine synthesis (*Nos* or *Ocs*) within the T-region; corresponding genes for opine catabolism (*Noc*

Figure 17.21 T-DNA is transferred from *Agrobacterium* carrying a Ti plasmid into a plant cell, where it becomes integrated into the nuclear genome and expresses functions that transform the host cell.

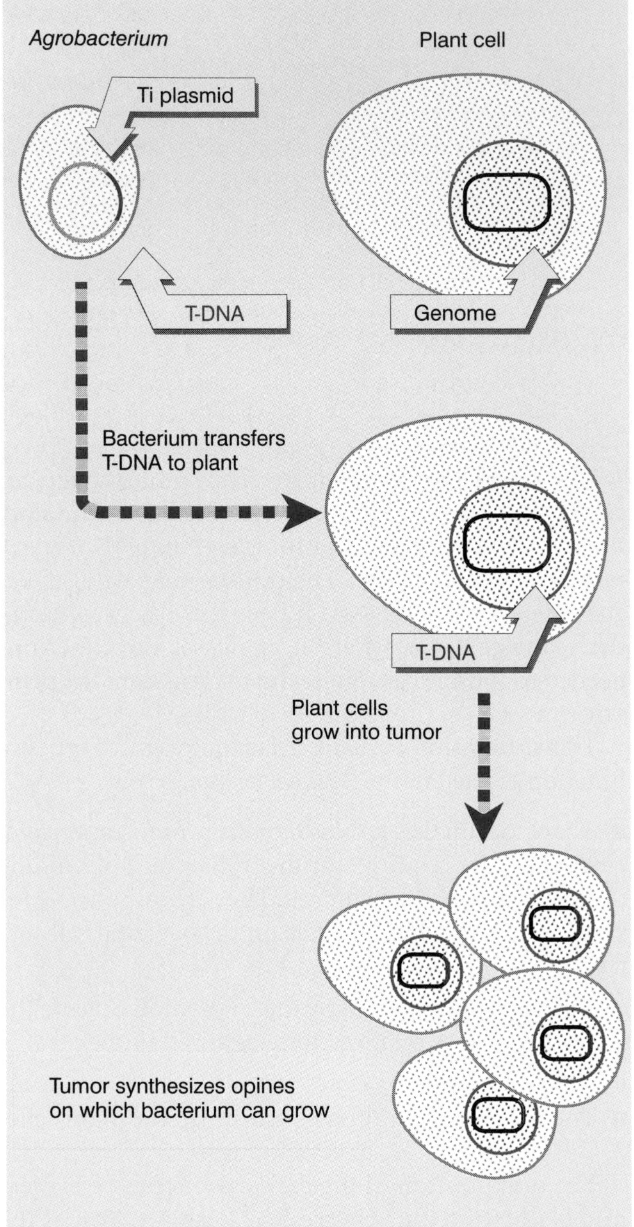

Figure 17.22 Nopaline and octopine Ti plasmids carry a variety of genes, including T-regions that have overlapping functions.

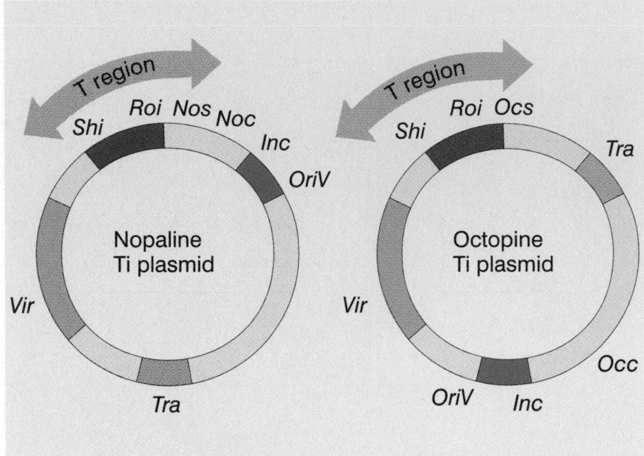

or *Occ*) reside elsewhere on the plasmid. The plasmids code for similar, but not identical, morphogenetic functions, as seen in the induction of characteristic types of tumors.

Functions affecting oncogenicity—the ability to form tumors—are not confined to the T-region. Those genes located outside the T-region must be concerned with establishing the tumorigenic state, but their products are not needed to perpetuate it. They may be concerned with transfer of T-DNA into the plant nucleus or perhaps with subsidiary functions such as the balance of plant hormones in the infected tissue. Some of the mutations are host-specific, preventing tumor formation by some plant species, but not by others.

The *virulence* genes code for the functions required for the transfer process. Six loci *virA–G* reside in a 40 kb region outside the T-DNA. Their organization is summarized in **Figure 17.23**. Each locus is transcribed as an individual unit; some contain more than one open reading frame.

We may divide the transforming process into (at least) two stages:

- *Agrobacterium* contacts a plant cell, and the *vir* genes are induced.
- *vir* gene products cause T-DNA to be transferred to the plant cell nucleus, where it is integrated into the genome.

The *vir* genes fall into two groups, corresponding to these stages. Genes *virA* and *virG* are regulators that respond to a change in the plant by inducing the other genes. So mutants in *virA* and *virG* are avirulent and cannot express the remaining *vir* genes. Genes *virB,C,D,E* code for proteins involved in the transfer of DNA. Mutants in *virB* and *virD* are avirulent in all plants, but the effects of mutations in *virC* and *virE* vary with the type of host plant.

virA and *virG* are expressed constitutively (at a rather low level). The signal to which they respond is provided by phenolic compounds generated by plants as a response to wounding. **Figure 17.24** presents an

Figure 17.23 The *vir* region of the Ti plasmid has six loci that are responsible for transferring T-DNA to an infected plant.

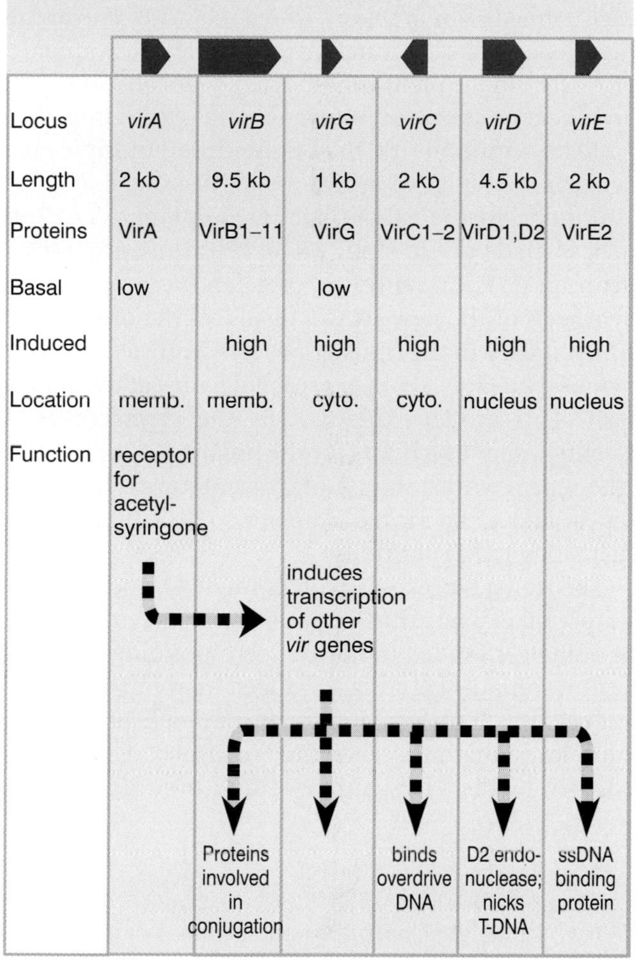

Locus	*virA*	*virB*	*virG*	*virC*	*virD*	*virE*
Length	2 kb	9.5 kb	1 kb	2 kb	4.5 kb	2 kb
Proteins	VirA	VirB1–11	VirG	VirC1–2	VirD1,D2	VirE2
Basal	low		low			
Induced		high	high	high	high	high
Location	memb.	memb.	cyto.	cyto.	nucleus	nucleus
Function	receptor for acetyl-syringone	Proteins involved in conjugation	induces transcription of other *vir* genes	binds overdrive DNA	D2 endo-nuclease; nicks T-DNA	ssDNA binding protein

Figure 17.24 Acetosyringone (4-acetyl-2,6-dimethoxyphenol) is produced by *N. tabacum* upon wounding, and induces transfer of T-DNA from *Agrobacterium*.

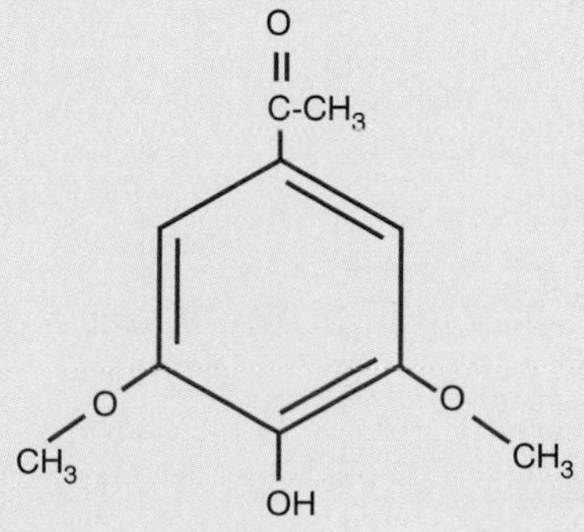

example. *N. tabacum* (tobacco) generates the molecules acetosyringone and α-hydroxyacetosyringone. Exposure to these compounds activates *virA*, which acts on *virG*, which in turn induces the expression *de novo* of *virB,C,D,E*. This reaction explains why *Agrobacterium* infection succeeds only on wounded plants.

Figure 17.25 The two-component system of VirA-VirG responds to phenolic signals by activating transcription of target genes.

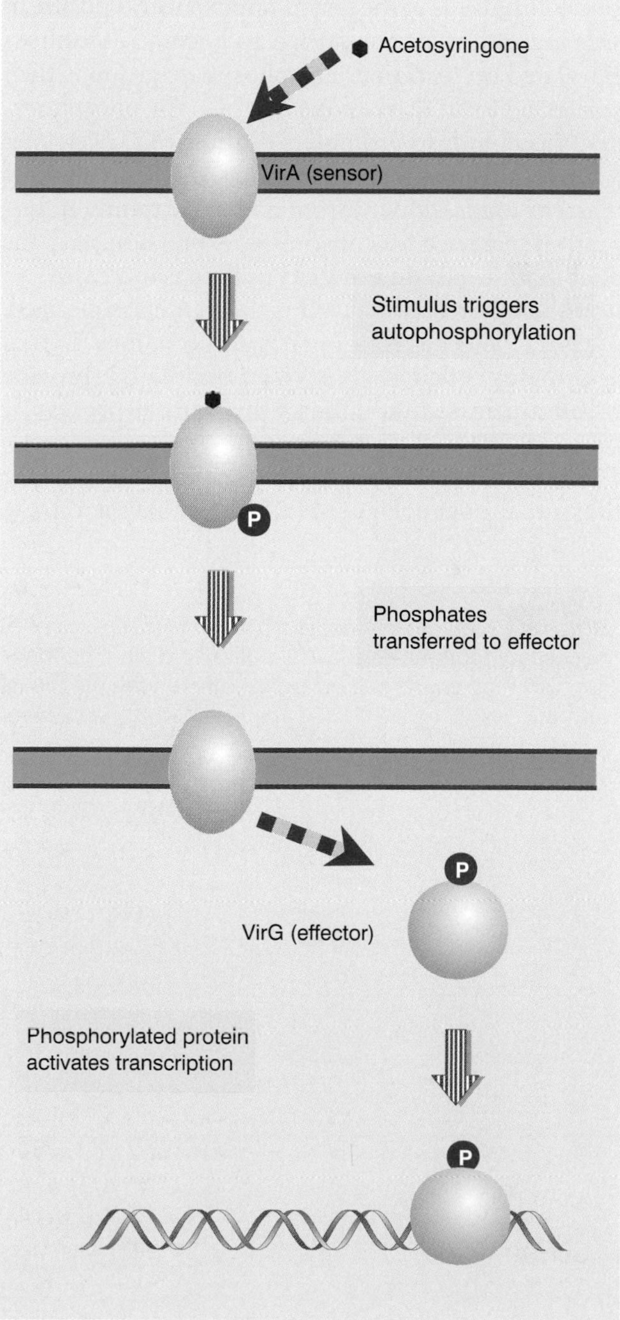
Acetosyringone
VirA (sensor)
Stimulus triggers autophosphorylation
P
Phosphates transferred to effector
P
VirG (effector)
Phosphorylated protein activates transcription
P

VirA and VirG are an example of a classic "two-component" bacterial system, in which stimulation of a sensor protein causes autophosphorylation and transfer of the phosphate to the second protein. The relationship is illustrated in **Figure 17.25**. The VirA-VirG system resembles the EnvZ-OmpR system that responds to osmolarity. The sequence of *virA* is related to that of *envZ*; and the sequences of *virG* and *ompR* are closely related, suggesting that the effector proteins function in a similar manner.

VirA forms a homodimer that is located in the inner membrane; it may respond to the presence of the phenolic compounds in the periplasmic space. Exposure to these compounds causes VirA to become autophosphorylated on histidine. The phosphate group is then transferred to an Asp residue in VirG. The phosphorylated VirG binds to promoters of the *virB,C,D,E* genes to activate transcription. When *virG* is activated, its transcription is induced from a new startpoint, different from that used for constitutive expression, with the result that the amount of VirG protein is increased.

Of the other *vir* loci, *virD* is the best characterized. The *virD* locus has four open reading frames. Two of the proteins coded at *virD*, VirD1 and VirD2, provide an endonuclease that initiates the transfer process by nicking T-DNA at a specific site.

The transfer process actually selects the T-region for entry into the plant. **Figure 17.26** shows that the T-DNA of a nopaline plasmid is demarcated from the flanking regions in the Ti plasmid by repeats of 25 bp, which differ at only two positions between the left and right ends. When T-DNA is integrated into a plant genome, it has a well-defined right junction, which retains 1–2 bp of the right repeat. The left junction is variable; the boundary of T-DNA in the plant genome may be located at the 25 bp repeat or at one of a series of sites extending over ~100 bp within the T-DNA. Sometimes multiple tandem copies of T-DNA are integrated at a single site.

A model for transfer is illustrated in **Figure 17.27**. A nick is made at the right 25 bp repeat. It provides a priming end for synthesis of a DNA single strand. Synthesis of the new strand displaces the old strand, which is used in the transfer process. Transfer is terminated when DNA synthesis reaches a nick at the left repeat. This model explains why the right repeat is essential, and it accounts for the polarity of the process (the repeat works only in one orientation). If the left repeat fails to be nicked, transfer could continue farther along the Ti plasmid.

The transfer process involves single-stranded DNA. Induction of a nopaline plasmid causes the production of a single-stranded molecule corresponding to the T-region. This T-strand is produced at about one copy per cell. It is transferred in the form of a DNA-protein complex, sometimes called the T-complex. The DNA is covered by the VirE2 single-strand binding protein,

Figure 17.26 T-DNA has almost identical repeats of 25 bp at each end in the Ti plasmid. The right repeat is necessary for transfer and integration to a plant genome. T-DNA that is integrated in a plant genome has a precise junction that retains 1–2 bp of the right repeat, but the left junction varies and may be up to 100 bp short of the left repeat.

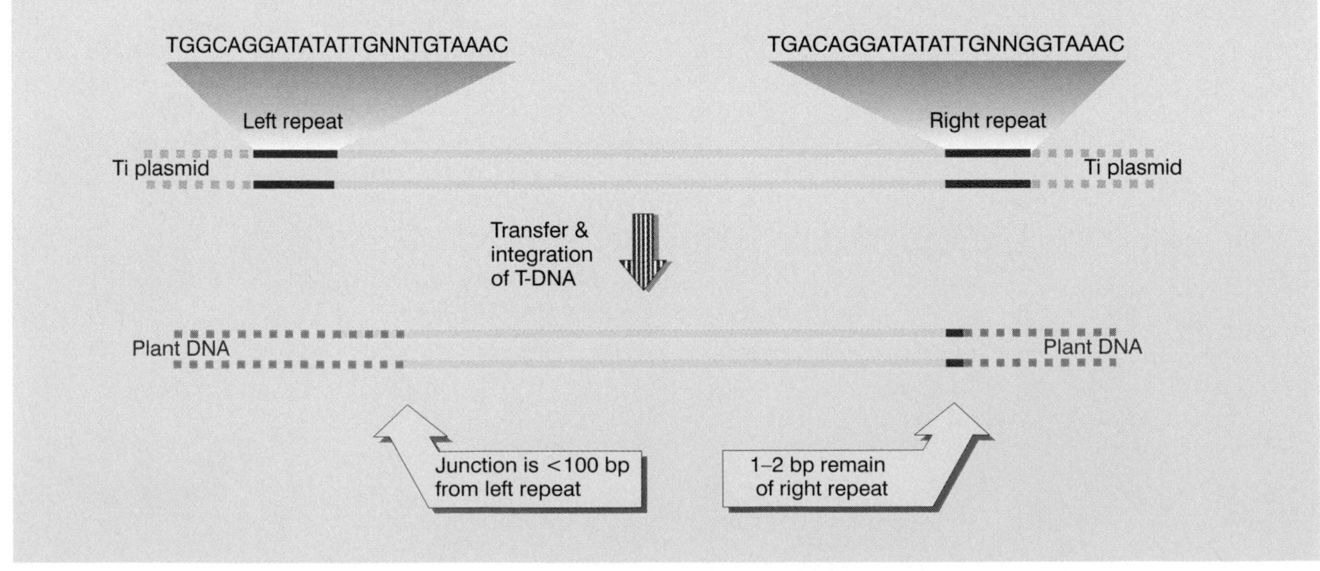

Figure 17.27 T-DNA is generated by displacement when DNA synthesis starts at a nick made at the right repeat. The reaction is terminated by a nick at the left repeat.

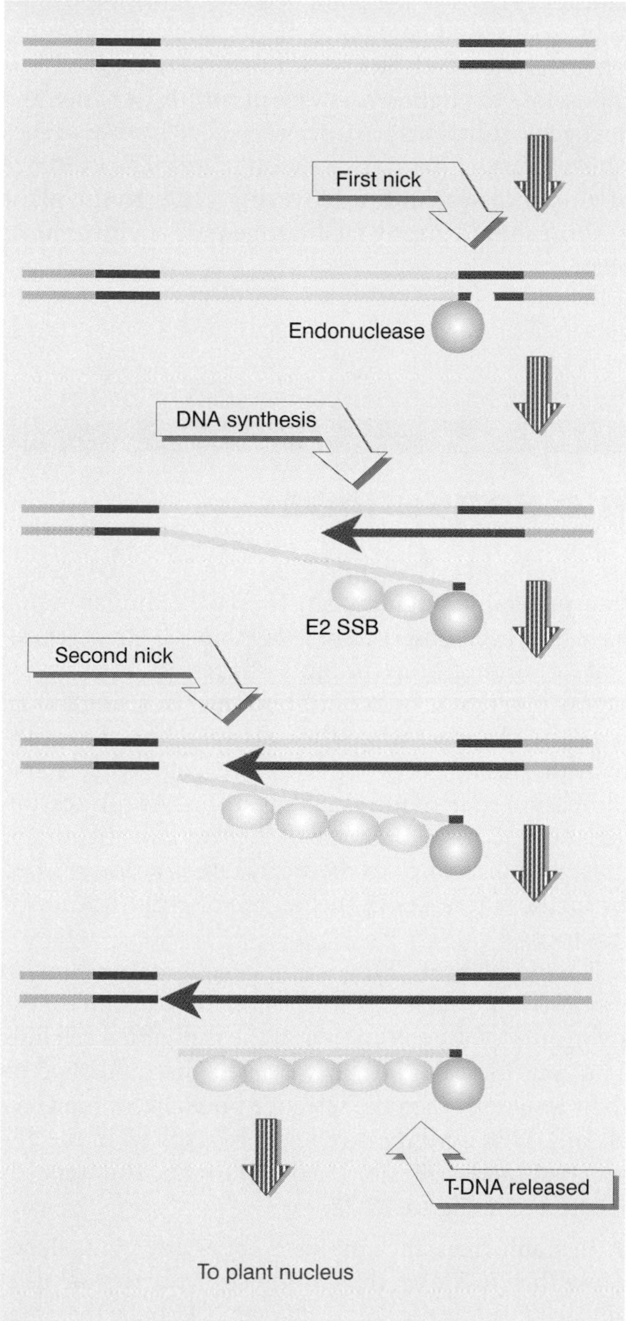

which has a nuclear localization signal and is responsible for transporting T-DNA into the plant cell nucleus. A single molecule of the D2 subunit of the endonuclease remains bound at the 5′ end. The *virB* operon codes for 11 products that are involved in the transfer reaction; they are homologous to the Tra proteins of certain bacterial plasmids that are involved in conjugation (see Chapter 12).

Outside T-DNA, but immediately adjacent to the right border, is another short sequence, called *overdrive*, which greatly stimulates the transfer process. *Overdrive* functions like an enhancer: it must lie on the same molecule of DNA, but enhances the efficiency of transfer even when located several thousand base pairs away from the border. VirC1, and possibly VirC2, may act at the *overdrive* sequence.

Octopine plasmids have a more complex pattern of integrated T-DNA than nopaline plasmids. The pattern of T-strands is also more complex, and several discrete species can be found, corresponding to elements of T-DNA. This suggests that octopine T-DNA has several sequences that provide targets for nicking and/or termination of DNA synthesis.

This model for transfer of T-DNA closely resembles the events involved in bacterial conjugation, when the *E. coli* chromosome is transferred from one cell to another in single-stranded form (see Chapter 12). A difference is that the transfer of T-DNA is (usually) limited by the boundary of the left repeat, whereas transfer of bacterial DNA is indefinite.

We do not know how the transferred DNA is integrated into the plant genome. At some stage, the newly generated single strand must be converted into duplex DNA. Circles of T-DNA that are found in infected plant cells appear to be generated by recombination between the left and right 25 bp repeats, but we do not know if they are intermediates.

Is T-DNA integrated into the plant genome as an integral unit? How many copies are integrated? What sites in plant DNA are available for integration? Are genes in T-DNA regulated exclusively by functions on the integrated segment? These questions are central to defining the process by which the Ti plasmid transforms a plant cell into a tumor.

What is the structure of the target site? Sequences flanking the integrated T-DNA tend to be rich in A•T base pairs (a feature displayed in target sites for some transposable elements). The sequence rearrangements that occur at the ends of the integrated T-DNA make it difficult to analyze the structure. We do not know whether the integration process generates new sequences in the target DNA comparable to the target repeats created in transposition.

T-DNA is expressed at its site of integration. The region contains several transcription units, each probably containing a gene expressed from an individual promoter. Their functions are concerned with the state of the plant cell, maintaining its tumorigenic properties, controlling shoot and root formation, and suppressing differentiation into other tissues. None of these genes is needed for T-DNA transfer.

The Ti plasmid presents an interesting organization of functions. Outside the T-region, it carries genes needed to initiate oncogenesis; at least some are concerned with the transfer of T-DNA, and we should like to know whether others function in the plant cell to affect its behavior at this stage. Also outside the T-region are the genes that enable the *Agrobacterium* to catabolize the opine that the transformed plant cell will produce. Within the T-region are the genes that control the transformed state of the plant, as well as the genes that cause it to synthesize the opines that will benefit the *Agrobacterium* that originally provided the T-DNA.

As a practical matter, the ability of *Agrobacterium* to transfer T-DNA to the plant genome makes it possible to introduce new genes into plants. Because the transfer/integration and oncogenic functions are separate, it is possible to engineer new Ti plasmids in which the oncogenic functions have been replaced by other genes whose effect on the plant we wish to test. The existence of a natural system for delivering genes to the plant genome should greatly facilitate genetic engineering of plants.

Selection of amplified genomic sequences

The eukaryotic genome has the capacity to accommodate additional sequences of either exogenous or endogenous origin. Endogenous sequences may be produced by **amplification** of an existing sequence. The additional sequences often take the form of a tandem array, containing many copies of a repeating unit. A gene that is contained within the repeating unit is not necessarily expressed in every copy, but expression tends to increase with the number of copies.

A tandem array of multiple copies may exist in either of two forms in a cell. If it takes the form of an extrachromosomal unit, it is inherited in an irregular manner: there is no equivalent in animal cells to the ability of a bacterial plasmid to be segregated evenly at cell division, so the entire unit is lost at a high frequency. If the array is integrated into the genome, however, it becomes a component of the genotype, and is inherited like any other genomic sequence.

Amplification of endogenous sequences is provoked by selecting cells for resistance to certain agents. The best-characterized example of amplification results from the addition of **methotrexate** (mtx) to certain cultured cell lines. This reagent blocks folate metabolism. Resistance to it is conferred by mutations that change the activity of the enzyme dihydrofolate reductase (DHFR). As an alternative to change in the enzyme itself, the amount of enzyme may be increased. The cause of this increase is an amplification of the number of *dhfr* structural genes. Amplification occurs at a frequency greater than the spontaneous point mutation rate, generally ranging from 10^{-4}–10^{-6}. Similar events have now been observed in more than 20 other genes.

A common feature in most of these systems is that highly resistant cells are not obtained in a single step, but instead appear when the cells are adapted to gradually increasing doses of the toxic reagent. So gene amplification may require several stages. Amplification generally occurs at only one of the two *dhfr* alleles; and increased resistance to methotrexate is accomplished by further increases in the degree of amplification at this locus.

The number of *dhfr* genes in a cell line resistant to methotrexate varies from 40–400, depending on the stringency of the selection and the individual cell line. The mtx^r lines fall into two classes, distinguished by their response when the selective pressure for high levels of DHFR activity is relieved by growth in the absence of methotrexate. The basis for the difference is illustrated in **Figure 17.28**.

- In **stable** lines, the amplified genes are retained, because they reside on the chromosome, at the site usually occupied by the single *dhfr* gene. Usually the other chromosome retains its normal single copy of *dhfr*.
- In **unstable** lines, the amplified genes are at least partially lost when the selective pressure is released, because the amplified genes exist as an extrachromosomal array.

Gene amplification has a visible effect on the chromosomes. In stable lines, the *dhfr* locus can be visual-

Figure 17.28 The *dhfr* gene can be amplified to give unstable copies that are extrachromosomal (double minutes) or stable (chromosomal). Extrachromosomal copies arise at early times.

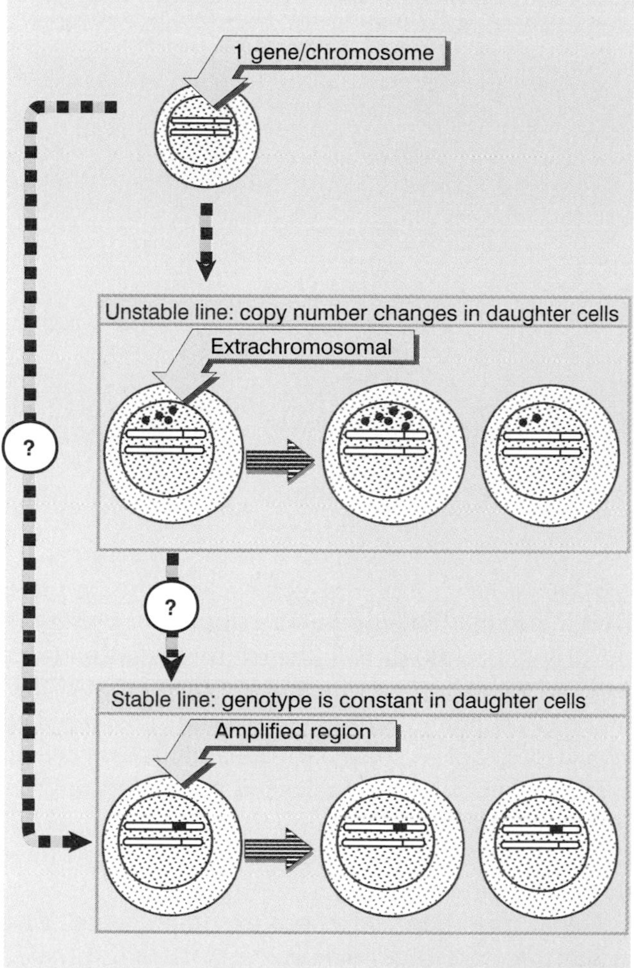

ized in the form of a **homogeneously staining region (HSR)**. An example is shown in **Figure 17.29**. The HSR takes its name from the presence of an additional region that lacks any chromosome bands after treatments such as G-banding. This change suggests that some region of the chromosome between bands has undergone an expansion.

In unstable cell lines, no change is seen in the chromosomes carrying *dhfr*. However, large numbers of elements called **double-minute chromosomes** are visible, as can be seen in **Figure 17.30**. In a typical cell line, each double minute carries two to four *dhfr* genes. The double minutes appear to be self-replicating; but they lack centromeres. As a result, they do not attach to the mitotic spindle and therefore segregate erratically, frequently being lost from the daughter cells. Notwithstanding their name, the actual status of the double minutes is regarded as extrachromosomal.

The irregular inheritance of the double minutes explains the instability of methotrexate resistance in these lines. Double minutes are lost continuously during cell divisions and, in the presence of methotrexate, cells with reduced numbers of *dhfr* genes will die. Only those cells that have retained a sufficient number of double minutes will appear in the surviving population.

The presence of the double minutes reduces the rate at which the cells proliferate. So when the selective pressure is removed, cells lacking the amplified genes have an advantage; they generate progeny more rapidly and soon take over the population. This explains why the amplified state is retained in the cell line only so long as cells are grown in the presence of methotrexate.

Because of the erratic segregation of the double minutes, increases in the copy number can occur relatively quickly as cells are selected at each division for progeny that have gained more than their fair share of the *dhfr* genes. Cells with greater numbers of copies are found in response to increased levels of methotrexate. The behavior of the double minutes explains the stepwise evolution of the *mtx*^{r} condition and the incessant fluctuation in the level of *dhfr* genes in unstable lines.

Both stable and unstable lines are found after long periods of selection for methotrexate resistance. What is the initial step in gene amplification? After short periods of selection, most or all of the resistant cells are unstable. The formation of extrachromosomal copies clearly is a more frequent event than amplification within the chromosome. At very early times in the process, amplified *dhfr* genes can be found as (small) extrachromosomal units before double minutes or any change in chromosomes can be detected. This suggests that the acquisition of resistance is most often due to generation of extrachromosomal repeats.

The amplified region is much longer than the *dhfr* gene itself. The gene has a length of ~31 kb, but the average length of the repeated unit is 500–1000 kb in the chromosomal HSR. The extent of the amplified region is different in each cell line. The amount of DNA contained in a double minute seems to lie in a range of 100–1000 kb.

How do the extrachromosomal copies arise? We know that their generation occurs without loss of the original chromosomal copy. There are two general possibilities. Additional cycles of replication could be initiated in the vicinity of the *dhfr* gene, followed by nonhomologous recombination between the copies. Or the process could be initiated by nonhomologous

Figure 17.29 Amplified copies of the *dhfr* gene produce a homogeneously staining region (HSR) in the chromosome. Photograph kindly provided by Robert Schimke.

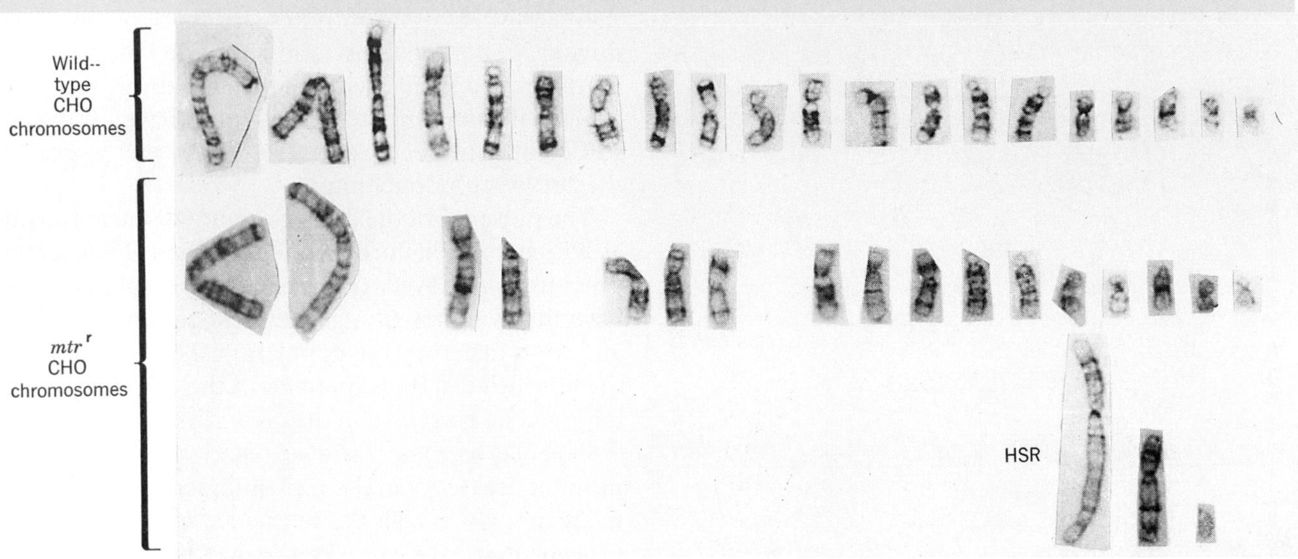

Figure 17.30 Amplified extrachromosomal *dhfr* genes take the form of double-minute chromosomes, as seen in the form of the small white dots. Photograph kindly provided by Robert Schimke.

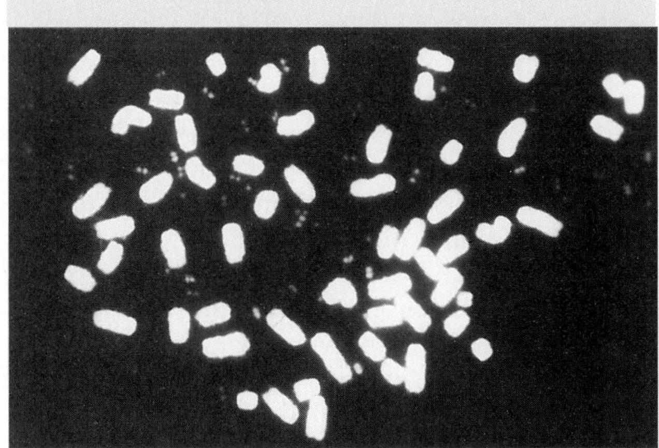

recombination between alleles. The extra copies could be released from the chromosome, possibly by some recombination-like event. Depending on the nature of this event, it could generate an extrachromosomal DNA molecule containing one or several copies. If the double minutes contain circular DNA, recombination between them in any case is likely to generate multimeric molecules.

Some information about the events involved in perpetuating the double minutes is given by an unstable cell line whose amplified genes code for a mutant DHFR enzyme. The mutant enzyme is not present in the original (diploid) cell line (so the mutation must have arisen at some point during the amplification process). Despite variations in the number of amplified genes, these cells display *only* the mutant enzyme. So the wild-type chromosomal genes cannot be continuously generating large numbers of double minutes anew, because these amplified copies would produce normal enzyme.

Once amplified extrachromosomal genes have arisen, therefore, *changes in the state of the cell are mediated through these genes and not through the original chromosomal copies.* When methotrexate is removed, the cell line loses its double minutes in the usual way. On re-exposure to the reagent, *normal* genes are amplified to give a new population of double minutes. This shows that none of the extrachromosomal copies of the mutant gene had integrated into the chromosome.

Another striking implication of these results is that the double minutes of the mutant line carried *only* mutant genes—so if there is more than one *dhfr* gene per double minute, all must be of the mutant type. This suggests that multicopy double minutes can be generated from individual extrachromosomal genes.

A major question has been whether amplified chromosomal copies arise by integration of the extrachromosomal copies or by an independent mechanism. We do not know whether intrachromosomal amplification simply proceeds less often as a *de novo* step or requires extrachromosomal amplification to occur as an intermediate step. The form taken by the amplified genes is influenced by the cell genotype; some cell lines tend to generate double minutes, while others more readily display the HSR configuration.

The type of amplification event also depends upon the particular locus that is involved. Another case of amplification is provided by resistance to an inhibitor of the enzyme transcarbamylase, which occurs by amplification of the CAD gene. (CAD is a protein which has the first three enzymatic activities of the pathway for UMP synthesis.) Amplified CAD DNA is always found within the chromosome. In this case, the amplified genes are found in the form of several dispersed amplified regions, often involving more than one chromosome.

Exogenous sequences can be introduced into cells and animals by transfection

THE procedure for introducing exogenous donor DNA into recipient cells is called **transfection**. Transfection experiments began with the addition of preparations of metaphase chromosomes to cell suspensions. The chromosomes are taken up rather inefficiently by the cells and give rise to unstable variants at a low frequency. Intact chromosomes rarely survive the procedure; the recipient cell usually gains a fragment of a donor chromosome (which is unstable because it lacks a centromere). Rare cases of stable lines may have resulted from integration of donor material into a resident chromosome.

Similar results are obtained when purified DNA is added to a recipient cell preparation. However, with purified DNA it is possible to add particular sequences instead of relying on random fragmentation of chromosomes. Transfection with DNA yields stable as well as unstable lines, with the former relatively predominant. (These experiments are directly analogous to those performed in bacterial transformation, but are described as transfection because of the historical use of transformation to describe changes that allow unrestrained growth of eukaryotic cells.)

Unstable transfectants (sometimes called **transient transfectants**) reflect the survival of the transfected DNA in extrachromosomal form; stable lines result from integration into the genome. The transfected DNA can be expressed in both cases. However, the low frequencies of transfection make it necessary to use donor markers whose presence in the recipient cells can be selected for. Most transfection experiments have used markers representing readily assayed enzymatic functions but, in principle, any marker that can be selected can be assayed. This allows the isolation of genes responsible for morphological phenomena. Most notably, transfected cells can be selected for acquisition of the transformed (tumorigenic) phenotype. This type of protocol has led to the isolation of several cellular *onc* genes (discussed in Chapter 28).

Cotransfection with more than one marker has proved informative about the events involved in transfection and extends the range of questions that we can ask with this technique. A common marker used in such experiments is the *tk* gene, coding for the enzyme thymidine kinase, which catalyzes an essential step in the provision of thymidine triphosphate as a precursor for DNA synthesis.

When tk^- cells are transfected with a DNA preparation containing both a purified tk^+ gene and the ϕX174 genome, *all the* tk^+ *transformants have both donor sequences*. This is a useful observation, because it allows unselected markers to be introduced routinely by cotransfection with a selected marker.

The arrangement of *tk* and ϕX174 sequences is different in each transfected line, but remains the same during propagation of that line. Often multiple copies of the donor sequences are present, the number varying with the individual line. Revertants lose the ϕX174 sequences together with *tk* sequences. Amplification of transfected sequences under selective pressure results in the increase of copy number of all donor sequences *pari passu*. So the two types of donor sequence become physically linked during transfection and suffer the same fate thereafter.

To perform a transfection experiment, the mass of DNA added to the recipient cells is increased by including an excess of "carrier DNA", a preparation of some other DNA (often from salmon sperm). Transfected cells prove to have sequences of the carrier DNA flanking the selected sequences on either side. Transfection therefore appears to be mediated by a large unit, consisting of a linked array of all sequences present in the donor preparation.

Since revertants for the selected marker lose all of this material, it seems likely that the *transfected cell gains only a single large unit.* The unit is formed by a concatemeric linkage of donor sequences in a reaction that is rapid relative to the other events involved in transfection. This transfecting package is ~1000 kb in length.

Because of the size of the donor unit, we cannot tell from blotting experiments whether it is physically linked to recipient chromosomal DNA (the relevant end fragments are present in too small a relative proportion). It seems plausible that the first stage is the establishment of an unstable extrachromosomal unit, followed by the acquisition of stability via integration.

In situ hybridization can be used to show that transfected cells have donor material integrated into the resident chromosomes. Any given cell line has only a single site of integration; but the site is different in each line. Probably the selection of a site for integration is a random event; sometimes it is associated with a gross chromosomal rearrangement.

The sites at which exogenous material becomes integrated usually do not appear to have any sequence relationship to the transfected DNA. The integration event involves a nonhomologous recombination between the mass of added DNA and a random site in the genome. The recombination event may be provoked by the introduction of a double-strand break into the chromosomal DNA, possibly by the action of DNA repair enzymes that are induced by the free ends of the exogenous DNA.

An exciting development of transfection techniques is their application to introduce genes into animals. An animal that gains new genetic information from the addition of foreign DNA is described as **transgenic**. The approach of directly injecting DNA can be used with mouse eggs, as shown in **Figure 17.31**. Plasmids carrying the gene of interest are injected into the germinal vesicle (nucleus) of the oocyte or into the pronucleus of the fertilized egg. The egg is implanted into a pseudopregnant mouse. After birth, the recipient mouse can be examined to see whether it has gained the foreign DNA and, if so, whether it is expressed.

The first questions we ask about any transgenic animal are how many copies it has of the foreign material, where these copies are located, and whether they are present in the germline and inherited in a Mendelian

Figure 17.31 Transfection can introduce DNA directly into the germ line of animals.

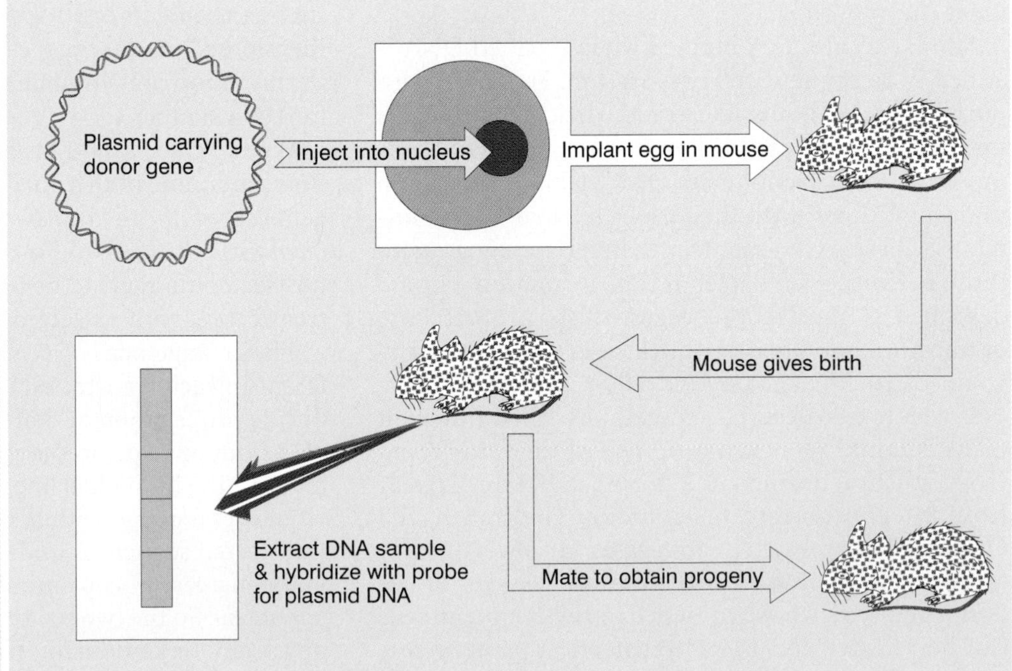
Plasmid carrying donor gene
Inject into nucleus
Implant egg in mouse
Mouse gives birth
Extract DNA sample & hybridize with probe for plasmid DNA
Mate to obtain progeny

manner. The usual result of such experiments is that a reasonable minority (say ~15%) of the injected mice carry the transfected sequence. Usually, multiple copies of the plasmid appear to have been integrated in a tandem array into a single chromosomal site. The number of copies varies from 1–150. They are inherited by the progeny of the injected mouse as expected of a Mendelian locus.

An important issue that can be addressed by experiments with transgenic animals concerns the independence of genes and the effects of the region within which they reside. If we take a gene, including the flanking sequences that contain its known regulatory elements, can it be expressed independently of its location in the genome? In other words, do the regulatory elements function independently, or is gene expression in addition controlled by other effects, for example, location in an appropriate chromosomal domain?

Are transfected genes expressed with the proper developmental specificity? The general rule now appears to be that there is a reasonable facsimile of proper control: the transfected genes are generally expressed in appropriate cells and at the usual time. There are exceptions, however, in which a transfected gene is expressed in an inappropriate tissue.

In the progeny of the injected mice, expression of the donor gene is extremely variable; it may be extinguished entirely, reduced somewhat, or even increased. Even in the original parents, the level of gene expression does not correlate with the number of tandemly integrated genes. Probably only some of the genes are active. In addition to the question of how many of the gene copies are capable of being activated, a parameter influencing regulation could be the relationship between the gene number and the regulatory proteins: a large number of promoters could dilute out any regulator proteins present in limiting amounts.

What is responsible for the variation in gene expression? One possibility that has often been discussed for transfected genes (and which applies also to integrated retroviral genomes) is that the site of integration is important. Perhaps a gene is expressed if it integrates within an active domain, but not if it integrates in another area of chromatin. Another possibility is the occurrence of epigenetic modification; for example, changes in the pattern of methylation might be responsible for changes in activity. Alternatively, the genes that happened to be active in the parents may have been deleted or amplified in the progeny.

A particularly striking example of the effects of an injected gene is provided by a strain of transgenic mice derived from eggs injected with a fusion consisting of the MT promoter linked to the rat growth hormone structural gene. Growth hormone levels in some of the transgenic mice were several hundred times greater than normal. The mice grew to nearly twice the size of normal mice, as can be seen from **Figure 17.32.**

The introduction of oncogene sequences can lead to tumor formation. Transgenic mice containing the SV40 early coding region and regulatory elements express the viral genes for large T and small t antigens only in some tissues, most often brain, thymus, and kidney. (The T/t antigens are alternatively spliced proteins coded by the early region of the virus; they have the ability to transform cultured cells to a tumorigenic phenotype; see Chapter 28.) The transgenic mice usually die before reaching 6 months, as the result of developing tumors in the brain; sometimes tumors are found also in thymus and kidney. Different oncogenes may be used to generate mice developing various cancers, thus making possible a range of model systems. For example, introduction of the *myc* gene under control of an active promoter causes the appearance of adenocarcinomas and other tumors.

Can defective genes be replaced by functional genes in the germline using transgenic techniques? One successful case is represented by a cure of the defect in the hypogonadal mouse. The *hpg* mouse has a deletion that removes the distal part of the gene coding for the

Figure 17.32 A transgenic mouse with an active rat growth hormone gene (left) is twice the size of a normal mouse (right). Photograph kindly provided by Ralph Brinster.

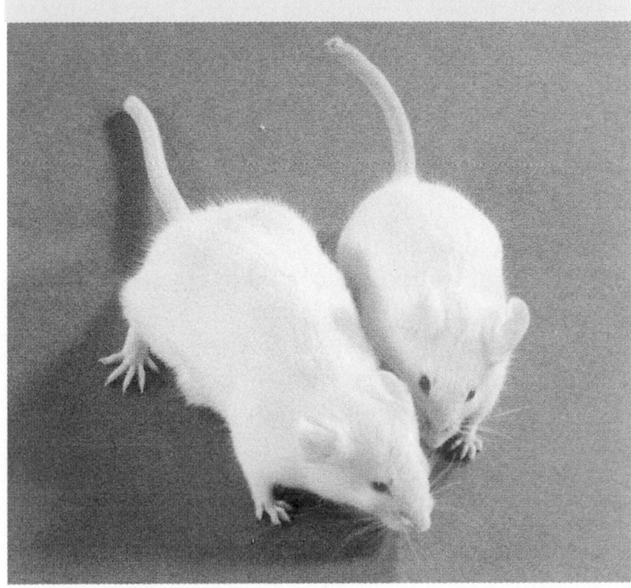

polyprotein precursor to GnRH (gonadotropin-releasing hormone) and GnRH-associated peptide (GAP). As a result, the mouse is infertile.

When an intact *hpg* gene is introduced into the mouse by transgenic techniques, it is expressed in the appropriate tissues. **Figure 17.33** summarizes experiments to introduce a transgene into *hpg/hpg* homozygous mutant mice. The resulting mice are normal. This provides a striking demonstration that expression of a transgene under normal regulatory control can be indistinguishable from the behavior of the normal allele.

Impediments to using such techniques to cure genetic defects at present are that the transgene must be introduced into the germline of the preceding generation, the ability to express a transgene is not predictable, and an adequate level of expression of a transgene may be obtained in only a small minority of the transgenic animals. Also, the large number of transgenes that may be introduced into the germline, and their erratic expression, could pose problems for the animal in cases in which over-expression of the transgene was harmful.

In the *hpg* murine experiments, for example, only 2 out of 250 mouse eggs injected with intact *hpg* genes gave rise to transgenic mice. Each transgenic animal contained >20 copies of the transgene. Only 20 of the 48 offspring of the transgenic mice retained the transgenic trait. When inherited by their offspring, however, the transgene(s) could substitute for the lack of endogenous *hpg* genes. Gene replacement via a transgene is therefore effective only under restricted conditions.

The disadvantage of direct injection of DNA is the introduction of multiple copies, their variable expression, and often difficulty in cloning the insertion site because sequence rearrangements may have been generated in the host DNA. An alternative procedure is to use a retroviral vector carrying the donor gene. A single proviral copy inserts at a chromosomal site, without inducing any rearrangement of the host DNA. It is possible also to treat cells at different stages of development, and thus to target a particular somatic tissue; however, it is difficult to infect germ cells.

A powerful technique for making transgenic mice takes advantage of embryonic stem (ES) cells, which are derived from the mouse blastocyst (an early stage of development, which precedes implantation of the egg in the uterus). **Figure 17.34** illustrates the principles of this technique.

ES cells are transfected with DNA in the usual way (most often by microinjection or electroporation). By using a donor that carries an additional sequence such as a drug resistance marker or some particular enzyme, it is possible to select ES cells that have obtained an integrated transgene carrying any particular donor trait. An alternative is to use PCR technology to assay the transfected ES cells for successful integration of the

Figure 17.33 Hypogonadism of the *hpg* mouse can be cured by introducing a transgene that has the wild-type sequence.

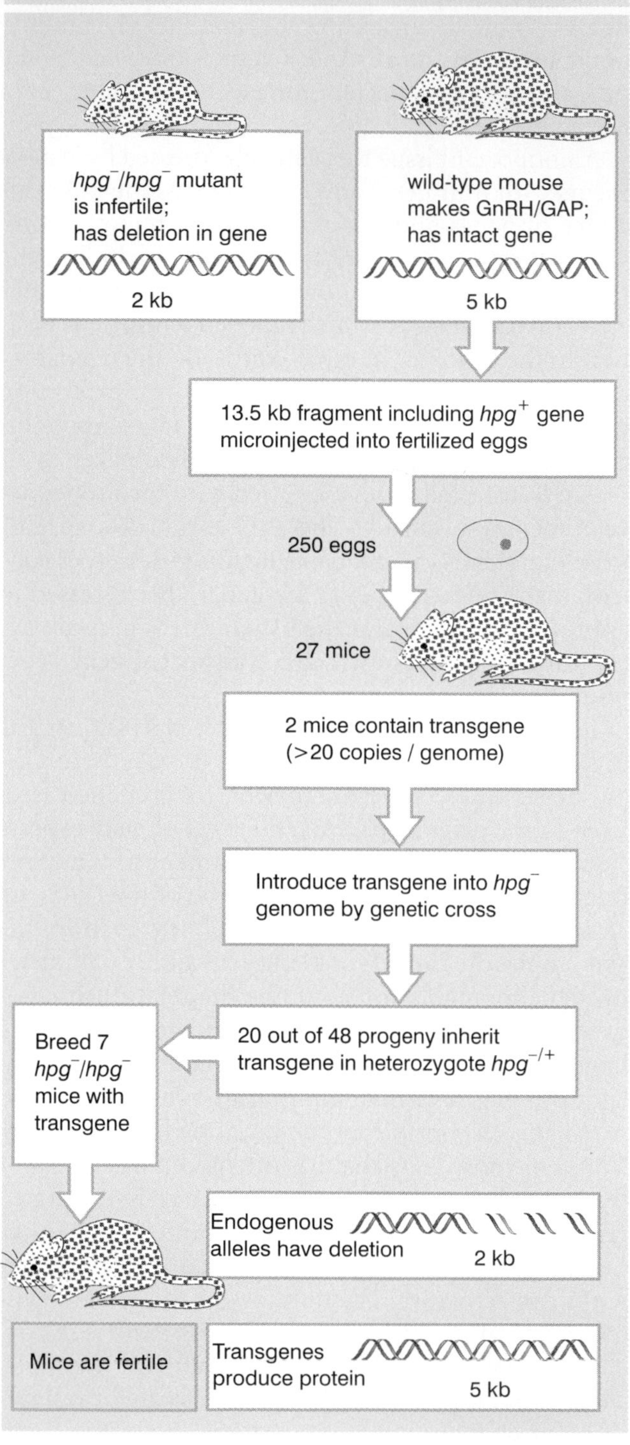

Figure 17.34 ES cells can be used to generate mouse chimeras, which breed true for the transfected DNA when the ES cell contributes to the germ line.

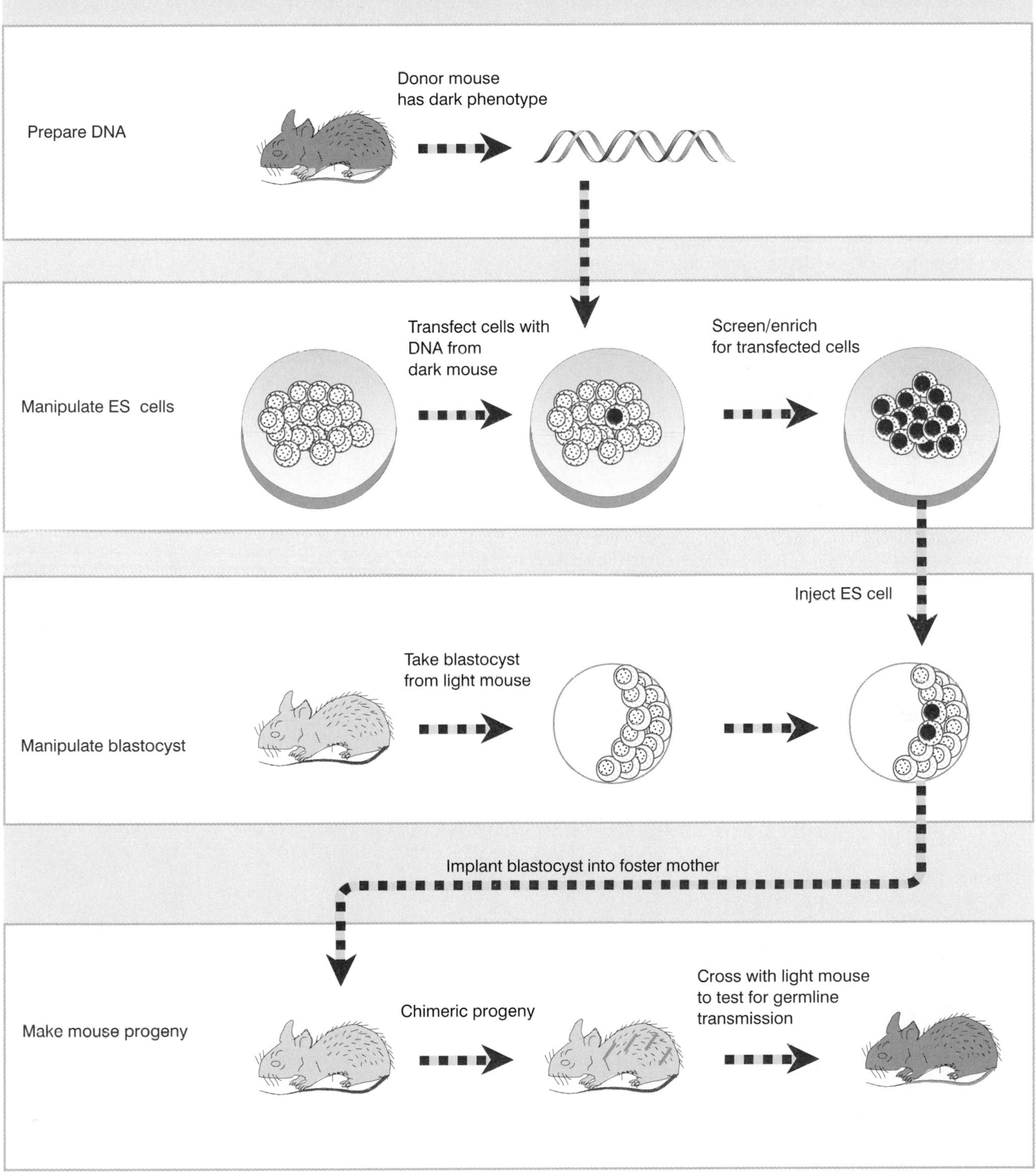

donor DNA. By such means, a population of ES cells is obtained in which there is a high proportion carrying the marker.

These ES cells are then injected into a recipient blastocyst. The ability of the ES cells to participate in normal development of the blastocyst forms the basis of the technique. The blastocyst is implanted into a foster mother, and in due course develops into a **chimeric** mouse. Some of the tissues of the chimeric mice will be derived from the cells of the recipient blastocyst; other tissues will be derived from the injected ES cells. The proportion of tissues in the adult mouse that are derived from cells in the recipient blastocyst and from injected ES cells varies widely in individual progeny; if a visible marker (such as coat color gene) is used, areas of tissue representing each type of cell can be seen.

To determine whether the ES cells contributed to the

Figure 17.35 A transgene containing *neo* within an exon and *TK* downstream can be selected by resistance to G418 and loss of TK activity.

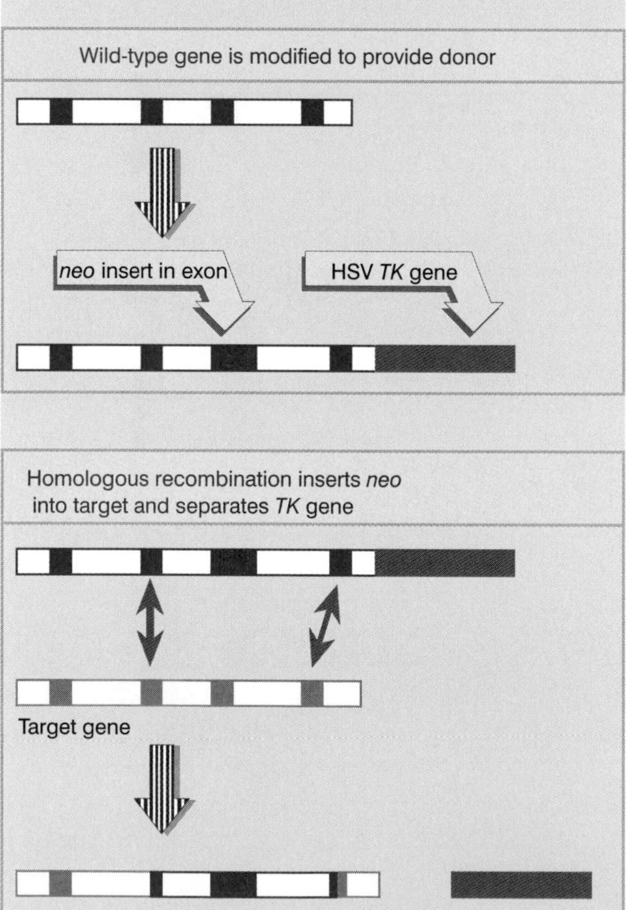
Wild-type gene is modified to provide donor
neo insert in exon
HSV *TK* gene
Homologous recombination inserts *neo* into target and separates *TK* gene
Target gene

Figure 17.36 Transgenic flies that have a single, normally expressed copy of a gene can be obtained by injecting *D. melanogaster* embryos with an active P element plus foreign DNA flanked by P element ends.

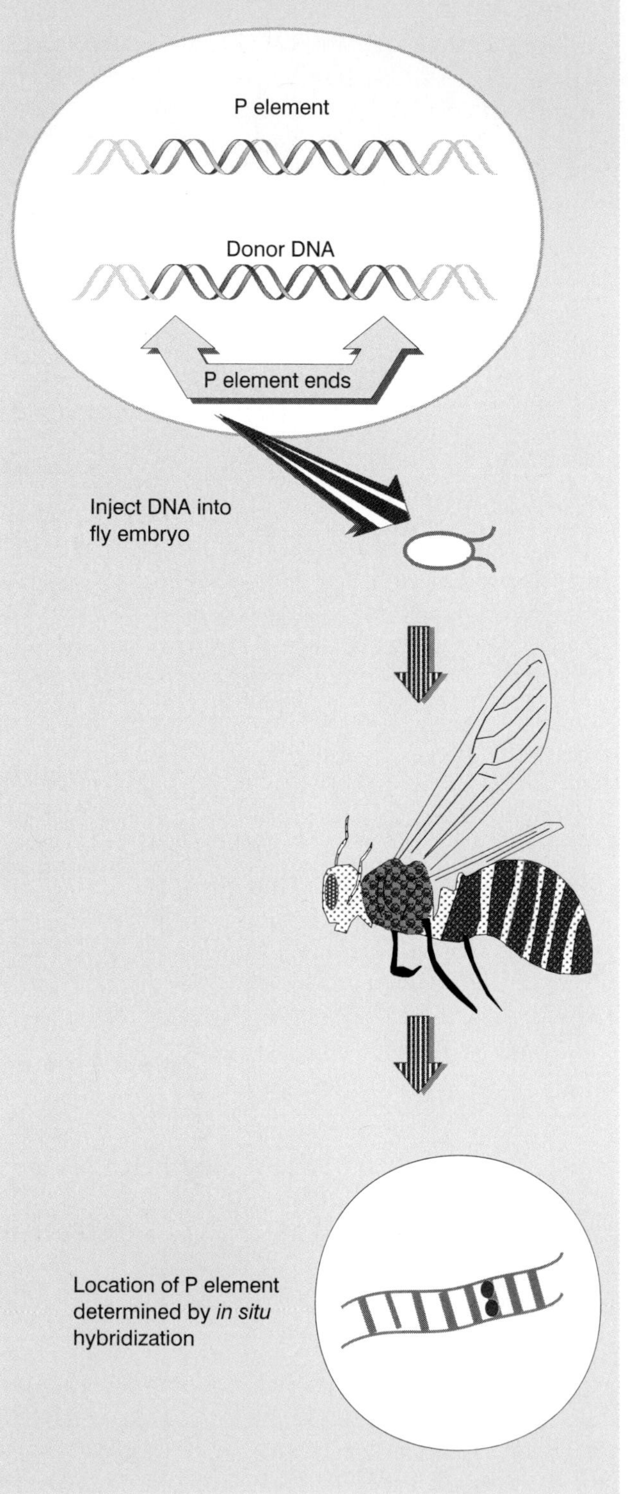
P element
Donor DNA
P element ends
Inject DNA into fly embryo
Location of P element determined by *in situ* hybridization

germline, the chimeric mouse is crossed with a mouse that lacks the donor trait. Any progeny that have the trait must be derived from germ cells that have descended from the injected ES cells. By this means, an entire mouse has been generated from an original ES cell!

A further development of these techniques makes it possible to obtain homologous recombinants. A particular use of homologous recombination is to disrupt endogenous genes, as illustrated in **Figure 17.35**. A wild-type gene is modified by interrupting an exon with a marker sequence; most often the *neo* gene that confers resistance to the drug G418 is used. Also, another marker is added on one side of the gene; for example, the *TK* gene of the herpes virus. When this DNA is introduced into an ES cell, it may be inserted into the genome by either nonhomologous or homologous recombination. A nonhomologous recombination inserts the whole unit, including the flanking *TK* sequence. But a homologous recombination requires two exchanges, as a result of which the flanking *TK* sequence is lost. Cells in which a homologous recombination has occurred can therefore be selected by the gain of *neo* resistance and absence of *TK* activity (which can be selected because *TK* causes sensitivity to the drug gancyclovir). If it is not convenient to use a selectable marker such as *TK*, cells can simply be screened by PCR assays for the absence of flanking DNA. The frequency of homologous recombination is $\sim 10^{-7}$, and probably represents <1% of all recombination events.

The presence of the *neo* gene in an exon disrupts transcription, and thereby creates a null allele. A particular target gene can therefore be "knocked out" by this means; and once a mouse with one null allele has been obtained, it can be bred to generate the homozygote. This is a powerful technique for investigating whether a particular gene is essential, and what functions in the animal are perturbed by its loss.

A sophisticated method for introducing new DNA sequences has been developed with *D. melanogaster* by taking advantage of the P element. The protocol is illustrated in **Figure 17.36**. A defective P element carrying the gene of interest is injected together with an intact P element into preblastoderm embryos. The intact P element provides a transposase that recognizes not only its own ends but also those of the defective element. As a result, either or both elements may be inserted into the genome.

Only the sequences between the ends of the P DNA are inserted; the sequences on either side are not part of the transposable element. An advantage of this technique is that only a single element is inserted in any one event, so the transgenic flies usually carry only one copy of the foreign gene, a great aid in analyzing its behavior.

Several genes that have been introduced in this way all show the same behavior. They are expressed only in the appropriate tissues and at the proper times during development, irrespective of the site of integration. So in *D. melanogaster*, all the information needed to regulate gene expression may be contained within the gene locus itself, and can be relatively impervious to external influence.

With these experiments, we see the possibility of extending from cultured cells to animals the option of examining the regulatory features. The ability to introduce DNA into the genotype allows us to make changes in it, to add new genes that have had particular modifications introduced *in vitro*, or to inactivate existing genes. So it may become possible to delineate the features responsible for tissue-specific gene expression. Ultimately it may be possible to replace defective genes in the genotype in a targeted manner.

Summary

Yeast mating type is determined by whether the MAT locus carries the a or α sequence. Expression in haploid cells of the sequence at *MAT* leads to expression of genes specific for the mating type and to repression of genes specific for the other mating type. Both activation and repression are achieved by control of transcription, and require factors that are not specific for mating type as well as the products of *MAT*. The functions that are activated in either mating type include secretion of the appropriate pheromone and expression on the cell surface of the receptor for the opposite type of pheromone. Interaction between pheromone and receptor on cells of either mating type activates a G protein on the membrane, and sets in train a common pathway that prepares cells for sporulation. Diploid cells do not express mating-type functions.

Additional, silent copies of the mating-type sequences are carried at the loci *HMLα* and *HMRa*. They are repressed by the actions of the *sir* loci. Cells that carry the HO endonuclease display a unidirectional transfer process in which the sequence at *HMLα* replaces an **a** sequence at *MAT*, or the sequence at *HMRa* replaces an α sequence at *MAT*. The endonuclease makes a double-strand break at *MAT*, and a free end invades either HMLα or HMRa. *MAT* initiates the transfer process, but is the recipient of the new sequence. The *HO* endonuclease is transcribed in mother cells but not daughter cells, and is under cell-cycle control. So switching is detected only in the products of a division, and the mating type has been switched in both daughter cells.

Trypanosomes carry >1000 sequences coding for varieties of the surface antigen. Only a single VSG is expressed in one cell, from an active site located near a telomere. The VSG may be changed by substituting a new coding sequence at the active site via a gene conversion process, or by switching the site of expression to another telomere. Switches in expression occur every 10^4–10^6 divisions.

Agrobacteria induce tumor formation in wounded plant cells. The wounded cells secrete phenolic compounds that activate *vir* genes carried by the Ti plasmid of the bacterium. The *vir* gene products cause a single strand of DNA from the T-DNA region of the plasmid to be transferred to the plant cell nucleus. Transfer is initiated at one boundary of T-DNA, but ends at variable sites. The single strand is converted into a double strand and integrated into the plant genome. Genes within the T-DNA transform the plant cell, and cause it to produce particular opines (derivatives of arginine). Genes in the Ti plasmid allow *Agrobacterium* to metabolize the opines produced by the transformed plant cell.

Endogenous sequences may become amplified in cultured cells. Exposure to methotrexate leads to the accumulation of cells that have additional copies of the *dhfr* gene. The copies may be carried as extrachromosomal arrays in the form of double-minute "chromosomes", or they may be integrated into the genome at the site of one of the *dhfr* alleles. Double-minute chromosomes are unstable, and disappear from the cell line rapidly in the absence of selective pressure. The amplified copies may originate by additional cycles of replication that are associated with recombination events.

New sequences of DNA may be introduced into a cultured cell by transfection or into an animal egg by microinjection. The foreign sequences may become integrated into the genome, often as large tandem arrays. The array appears to be inherited as a unit in a cultured cell. The sites of integration appear to be random. A transgenic animal arises when the integration event occurs into a genome that enters the germ-cell lineage. A transgene or transgenic array is inherited in Mendelian manner, but the copy number and activity of the gene(s) may change in the progeny. Often a transgene responds to tissue and temporal regulation in a manner that resembles the endogenous gene. Using conditions that promote homologous recombination, an inactive sequence can be used to replace a functional gene, thus creating a null locus. Transgenic mice can be obtained by injecting recipient blastocysts with ES cells that carry transfected DNA.

Further reading

Reviews

Mating type

Nasmyth, K. (1982). Molecular genetics of yeast mating type. *Ann. Rev. Genet.* **16**, 439–500.

Nasmyth, K. and Shore, D. (1987). Transcriptional regulation in the yeast life cycle. *Science* **237**, 1162–1170.

Kurjan, J. (1992). Pheromone response in yeast. *Ann. Rev. Biochem.* **61**, 1097–1129.

Laurenson, P. and Rine, J. (1992). Silencers, silencing, and heritable transcriptional states. *Microbiol. Rev.* **56**, 543–560.

Kurjan, J. (1993). The pheromone response pathway in *S. cerevisiae*. *Ann. Rev. Genet.* **27**, 147–179.

Haber, J. E. (1998) Mating type switching in *S. cerevisiae*. *Ann. Rev. Genet.* **32**, 561–599.

Trypanosome infection

Boothroyd, J. C. (1985). Antigenic variation in African trypanosomes. *Ann. Rev. Microbiol.* **39**, 475–502.

Donelson, J. E. and Rice-Ficht, A. C. (1985). Molecular biology of trypanosome antigenic variation. *Microbiol. Rev.* **49**, 107–125.

Borst, P. (1986). Discontinuous transcription and antigenic variation in trypanosomes. *Ann. Rev. Biochem.* **55**, 701–732.

Agrobacterium

Zambryski, P. (1988). Basic processes underlying *Agrobacterium*-mediated DNA transfer to plant cells. *Ann. Rev. Genet.* **22**, 1–30.

Zambryski, P. (1989). *Agrobacterium*-plant cell DNA transfer. In *Mobile DNA* (Eds. M. M. Berg and M. M. Howe. American Society for Microbiology, Washington DC, 309–334).

Winans, S. C. (1993). Two-way chemical signaling in *Agrobacterium*-plant interactions. *Microbiol. Rev.* **56**, 12–31.

Amplification

Pellicer, A. *et al.* (1980). Altering genotype and phenotype by DNA-mediated gene transfer. *Science* **209**, 1414–1422.

Schimke, R. T. *et al.* (1981). Chromosomal and extrachromosomal localization of amplified DHFR genes in cultured mammalian cells. *Cold Spring Harbor Symp. Quant. Biol.* **45**, 785–797.

Stark, G. R. and Wahl, G. M. (1984). Gene amplification. *Ann. Rev. Biochem.* **53**, 447–491.

Transgenic mice

Jaenisch, R. (1988). Transgenic animals. *Science* **240**, 1468–1474.

Capecchi, M. R. (1989). Altering the genome by homologous recombination. *Science* **244**, 1288–1292.

Discoveries

Mating type

Hicks, J., Strathern, J. N., and Herskowitz, I. (1977) The cassette model of mating type interconversion. In *DNA Insertion Elements* (Eds. A. Bukhari, J. Shapiro, and S. Adhya, Cold Spring Harbor Laboratory, New York, 457–462).

Strathern, J. N. *et al.* (1982). Homothallic switching of yeast mating type cassettes is initiated by a double-stranded cut in the *MAT* locus. *Cell* **31**, 183–192.

Bender, A. and Sprague, G. F., Jr. (1986). Yeast peptide pheromones, a-factor and alpha-factor, activate a common response mechanism in their target cells. *Cell* **47**, 929–937.

Bobola, N. *et al.* (1996). Asymmetric accumulation of Ash1p in postanaphase nuclei depends on a myosin and restricts yeast mating-type switching to mother cells. *Cell* **84**, 699–709.

T-DNA

Stachel, S. E., Timmerman, B., and Zambryski, P. (1986). Generation of single-stranded T-DNA molecules during the initial stages of T-DNA transfer from *A. tumefaciens* to plant cells. *Nature* **322**, 707–712.

P elements

Spradling, A. C. and Rubin, G. M. (1982). Transposition of cloned P elements into *Drosophila* germline chromosomes. *Science* **218**, 341–353.

Transgenic animals

Brinster, R. L. *et al.* (1984). Transgenic mice harboring SV40 T-antigen genes develop characteristic brain tumors. *Cell* **37**, 367–379.

Part 5

The nucleus

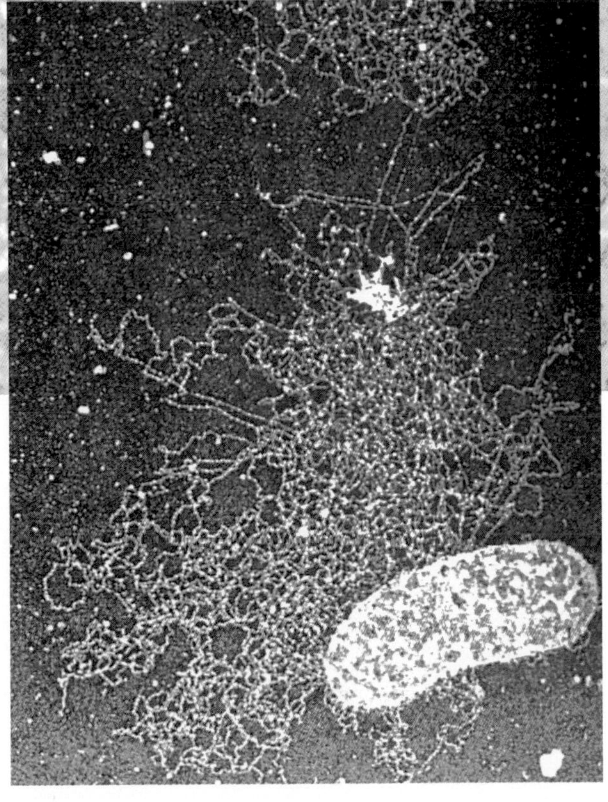

Chapter 18

Chromosomes

A general principle is evident in the organization of all cellular genetic material. It exists as a compact mass, occupying a limited volume; and its various activities, such as replication and transcription, must be accomplished within these confines. The organization of this material must accommodate transitions between inactive and active states.

The condensed state of nucleic acid results from its binding to basic proteins. The positive charges of these proteins neutralize the negative charges of the nucleic acid. The structure of the nucleoprotein complex is determined by the interactions of the proteins with the DNA (or RNA).

A common problem is presented by the packaging of DNA into phages and viruses, into bacterial cells and eukaryotic nuclei. The length of the DNA as an extended molecule would vastly exceed the dimensions of the compartment that contains it. The DNA (or in the case of some viruses, the RNA) must be compressed exceedingly tightly to fit into the space available. *So in contrast with the customary picture of DNA as an extended double helix, structural deformation of DNA to bend or fold it into a more compact form is the rule rather than exception.*

The magnitude of the discrepancy between the length of the nucleic acid and the size of its compartment is evident from the examples summarized in **Figure 18.1.** For bacteriophages and for eukaryotic viruses, whether rod-like or spherical, the nucleic acid genome, whether DNA or RNA, whether single-stranded or double-stranded, effectively fills the container.

For bacteria or for eukaryotic cell compartments, the discrepancy is hard to calculate exactly, because the DNA is contained in a compact area that occupies only part of the compartment. The genetic material is seen in the form of the **nucleoid** in bacteria and as the mass of **chromatin** in eukaryotic nuclei at interphase (between divisions).

The density of DNA in these compartments is high. In a bacterium it is ~10 mg/ml, in a eukaryotic nucleus it is ~100 mg/ml, and in the phage T4 head it is >500 mg/ml. Such a concentration in solution would be equivalent to a gel of great viscosity. We do not entirely understand the physiological implications, for example, what effect this has upon the ability of proteins to find their binding sites on DNA.

The packaging of chromatin is flexible; it changes during the eukaryotic cell cycle. At the time of division (mitosis or meiosis), the genetic material becomes even more tightly packaged, and individual **chromosomes** become recognizable.

The overall compression of the DNA can be described by the **packing ratio,** the length of the DNA divided by the length of the unit that contains it. For example, the smallest human chromosome contains ~4.6×10^7 bp of DNA (~10 times the genome size of the bacterium *E. coli*). This is equivalent to 14,000 μm (= 1.4 cm) of extended DNA. At the most condensed moment of mitosis, the chromosome is ~2 μm long. So

Figure 18.1 The length of nucleic acid is much greater than the dimensions of the surrounding compartment.

Compartment	Shape	Dimensions	Type of nucleic acid	Length
TMV	filament	0.008 × 0.3 μm	1 single-stranded RNA	2 μm = 6.4 kb
Phage ϕd	filament	0.006 × 0.85 μm	1 single-stranded DNA	2 μm = 6.0 kb
Adenovirus	icosahedron	0.07 μm diameter	1 double-stranded DNA	11 μm = 35.0 kb
Phage T4	icosahedron	0.065 × 0.10 μm	1 double-stranded DNA	55 μm = 170.0 kb
E. coli	cylinder	1.7 × 0.65 μm	1 double-stranded DNA	1.3 μm = 4.2×10^3 kb
Mitochondrion (human)	oblate spheroid	3.0 × 0.5 μm	~10 identical double-stranded DNAs	50 μm = 16.0 kb
Nucleus (human)	spheroid	6 μm diameter	46 chromosomes of double-stranded DNA	1.8 m = 6×10^6 kb

the packing ratio of DNA in the chromosome can be as great as 7000.

Packing ratios cannot be established with such certitude for the more amorphous overall structures of the bacterial nucleoid or eukaryotic chromatin. However, the usual reckoning is that mitotic chromosomes are likely to be 5–10 times more tightly packaged than interphase chromatin, which therefore has a typical packing ratio of 1000–2000.

A major unanswered question concerns the *specificity* of packaging. Is the DNA folded into a *particular* pattern, or is it different in each individual copy of the genome? How does the pattern of packaging change when a segment of DNA is replicated or transcribed?

Condensing viral genomes into their coats

FROM the perspective of packaging the *individual* sequence, there is an important difference between a cellular genome and a virus. The cellular genome is essentially indefinite in size; the number and location of individual sequences can be changed by duplication, deletion, and rearrangement. So it requires a *generalized* method for packaging its DNA, insensitive to the total content or distribution of sequences. By contrast, two restrictions define the needs of a virus. The amount of nucleic acid to be packaged is *predetermined* by the size of the genome. And it must all fit within a coat assembled from a protein or proteins coded by the viral genes.

A virus particle is deceptively simple in its superficial appearance. The nucleic acid genome is contained within a **capsid,** a symmetrical or quasi-symmetrical structure assembled from one or only a few proteins. Attached to the capsid, or incorporated into it, are other structures, assembled from distinct proteins, and necessary for infection of the host cell.

The virus particle is tightly constructed. The internal volume of the capsid is rarely much greater than the volume of the nucleic acid it must hold. The difference is usually less than twofold, and often the internal volume is barely larger than the nucleic acid.

In its most extreme form, the restriction that the capsid must be assembled from proteins coded by the virus means that the entire shell is constructed from a single type of subunit. The rules for assembly of identical subunits into closed structures restrict the capsid to one of two types. The protein subunits stack sequentially in a helical array to form a **filamentous** or rod-like shape. Or they form a pseudospherical shell, a type of structure that conforms to a polyhedron with

icosahedral symmetry. Some viral capsids are assembled from more than a single type of protein subunit, but although this extends the exact types of structures that can be formed, viral capsids still all conform to the general classes of quasi-crystalline filaments or icosahedrons.

There are two types of solution to the problem of how to construct a capsid that contains nucleic acid:

- The protein shell can be assembled around the nucleic acid, condensing the DNA or RNA by protein-nucleic acid interactions during the process of assembly.
- Or the capsid can be constructed from its component(s) in the form of an empty shell, into which the nucleic acid must be inserted, being condensed as it enters.

The capsid is assembled around the genome for single-stranded RNA viruses. The principle of assembly is that *the position of the RNA within the capsid is determined directly by its binding to the proteins of the shell.* The best characterized example is TMV (tobacco mosaic virus). Assembly starts at a duplex hairpin that lies within the RNA sequence. From this **nucleation center,** it proceeds bidirectionally along the RNA, until reaching the ends. The unit of the capsid is a two-layer disk, each layer containing 17 identical protein subunits. The disk is a circular structure, which forms a helix as it interacts with the RNA. The RNA becomes coiled in a helical array on the inside of the protein shell, as illustrated in **Figure 18.2.**

The spherical capsids of DNA viruses are assembled in a different way, as best characterized for the phages lambda and T4. In each case, an empty headshell is assembled from a small set of proteins. *Then the duplex genome is inserted into the head,* a process accompanied by a structural change in the capsid.

Figure 18.3 summarizes the assembly of lambda. It starts with a small headshell that contains a protein "core." This is converted to an empty headshell of more distinct shape. Then DNA packaging begins, the headshell expands in size though remaining the same shape, and finally the full head is sealed by addition of the tail.

Now a double-stranded DNA considered over short distances is a fairly rigid rod. Yet it must be compressed into a compact structure to fit within the capsid. We should like to know whether packaging involves a smooth coiling of the DNA into the head or requires abrupt bends.

Inserting DNA into a phage head involves two types of reaction: translocation and condensation. Both are energetically unfavorable.

Translocation is an active process in which the DNA is driven into the head by an ATP-dependent mechanism. One possibility is that the terminase enzymes that generate the ends of DNA (from longer precursor DNAs) could be involved in pushing it into the head. Another possibility is that the capsid protein(s) could pull the DNA in.

Little is known about the mechanism of condensation, except that the capsid contains "internal proteins" as well as DNA. One possibility is that they provide some sort of "scaffolding" onto which the DNA condenses. (This would be a counterpart to the use of the proteins of the shell in the plant RNA viruses.)

How specific is the packaging? It cannot depend on particular sequences, because deletions, insertions, and substitutions all fail to interfere with the assembly process. The relationship between DNA and the headshell has been investigated directly by determining which regions of the DNA can be chemically crosslinked to the proteins of the capsid. The surprising answer is that all regions of the DNA are more or less equally susceptible. This probably means that when DNA is inserted into the head, it follows a general rule for condensing, but the pattern is not determined by particular sequences.

Figure 18.2 A helical path for TMV RNA is created by the stacking of protein subunits in the virion.

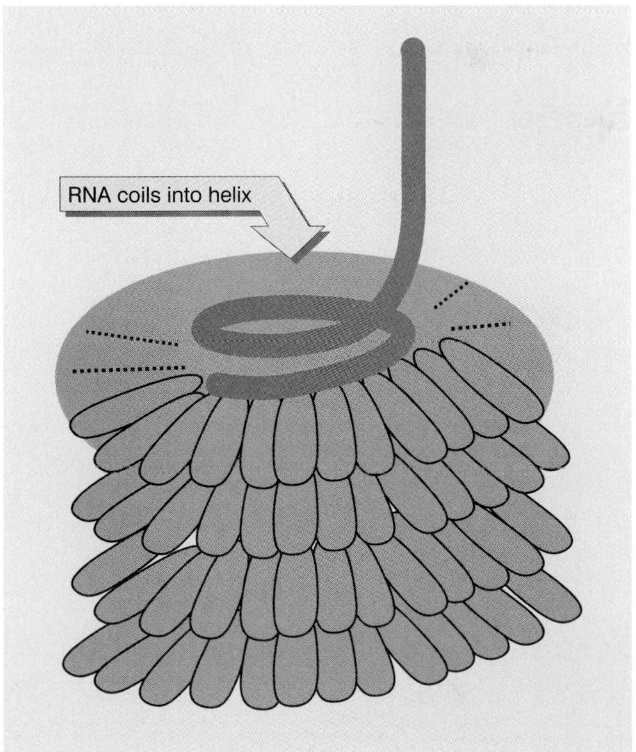

Figure 18.3 Maturation of phage lambda passes through several stages. The empty head changes shape and expands when it becomes filled with DNA. The electron micrographs show the particles at the start and end of the maturation pathway. Photographs kindly provided by A. F. Howatson.

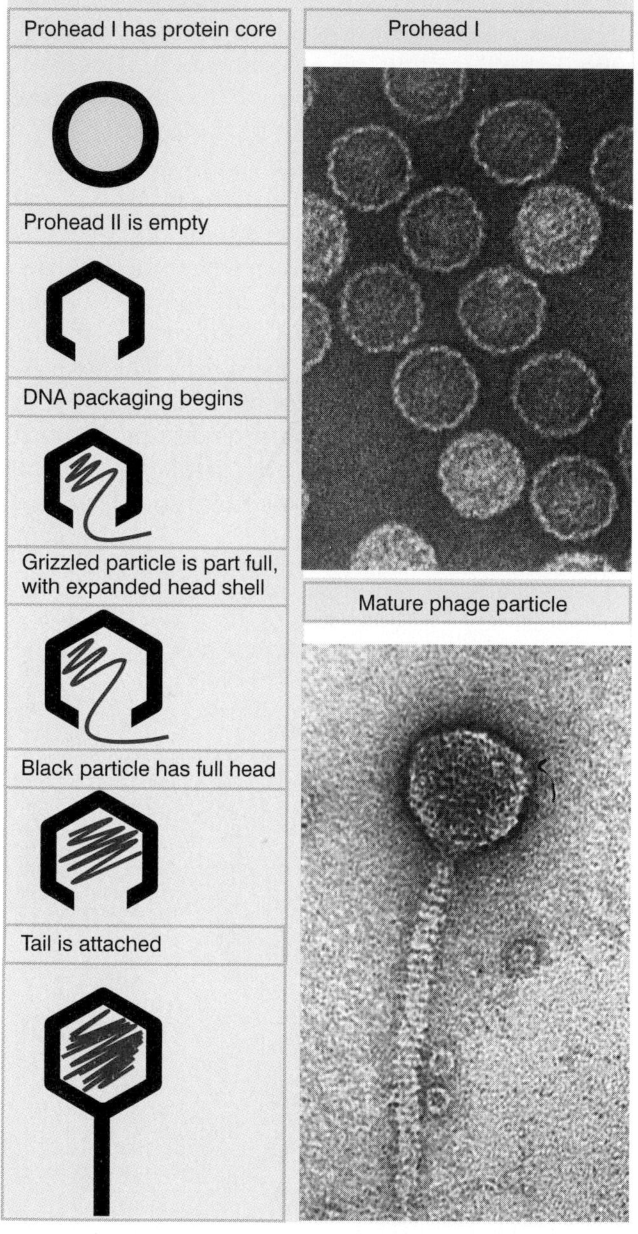
Prohead I has protein core
Prohead II is empty
DNA packaging begins
Grizzled particle is part full, with expanded head shell
Black particle has full head
Tail is attached
Prohead I
Mature phage particle

These varying mechanisms of virus assembly all accomplish the same end: packaging a single DNA or RNA molecule into the capsid. However, some viruses have genomes that consist of multiple nucleic acid molecules. Reovirus contains ten double-stranded RNA segments, all of which must be packaged into the capsid. Specific sorting sequences in the segments may be required to ensure that the assembly process selects one copy of each different molecule in order to collect a complete set of genetic information.

Some plant viruses are multipartite: their genomes consist of segments, each of which is packaged into a *different* capsid. An example is alfalfa mosaic virus, which has four different single-stranded RNAs, each packaged independently into a coat comprising the same protein subunit. A successful infection depends on the entry of one of each type into the cell.

The four components of the virus exist as particles of different sizes. This means that the same capsid protein can package each RNA into its own characteristic particle. This is a departure from the packaging of a unique length of nucleic acid into a capsid of fixed shape. The assembly pathway of viruses whose capsids have only one authentic form may be diverted by mutations that cause the formation of aberrant **monster** particles in which the head is longer than usual. These mutations show that a capsid protein(s) has an intrinsic ability to assemble into a particular type of structure, but the exact size and shape may vary. Some of the mutations occur in genes that code for **assembly proteins,** which are needed for head formation, but are not themselves part of the headshell. Such ancillary proteins limit the options of the capsid protein so that it assembles only along the desired pathway. Comparable proteins are employed in the assembly of cellular chromatin (see Chapter 19).

The bacterial genome is a nucleoid with many supercoiled loops

Although bacteria do not display structures with the distinct morphological features of eukaryotic chromosomes, their genomes nonetheless are organized into definite bodies. The genetic material can be seen as a fairly compact clump or series of clumps that occupies about a third of the volume of the cell. **Figure 18.4** displays a thin section through a bacterium in which this **nucleoid** is evident.

When *E. coli* cells are lysed, fibers are released in the form of loops attached to the broken envelope of the cell. As can be seen from **Figure 18.5**, the DNA of these loops is not found in the extended form of a free duplex, but is compacted by association with proteins.

Several DNA-binding proteins with a superficial resemblance to eukaryotic chromosomal proteins have been isolated in *E. coli*. What criteria should we apply for deciding whether a DNA-binding protein plays a structural role in the nucleoid? It should be present in sufficient quantities to bind throughout the genome. And mutations in its gene should cause some disruption of structure or of functions associated with genome survival (for example, segregation to daughter cells). None of the candidate proteins yet satisfies the genetic conditions.

Figure 18.4 A thin section shows the bacterial nucleoid as a compact mass in the center of the cell. Photograph kindly provided by Jack Griffith.

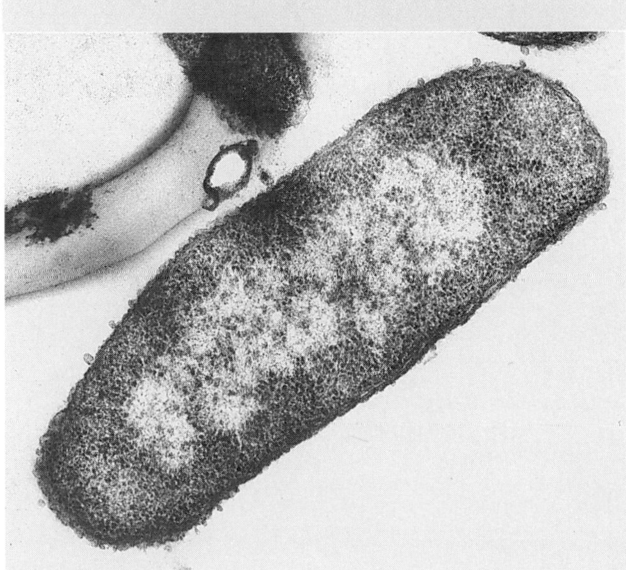

Figure 18.5 The nucleoid spills out of a lysed *E. coli* cell in the form of loops of a fiber. Photograph kindly provided by Jack Griffith.

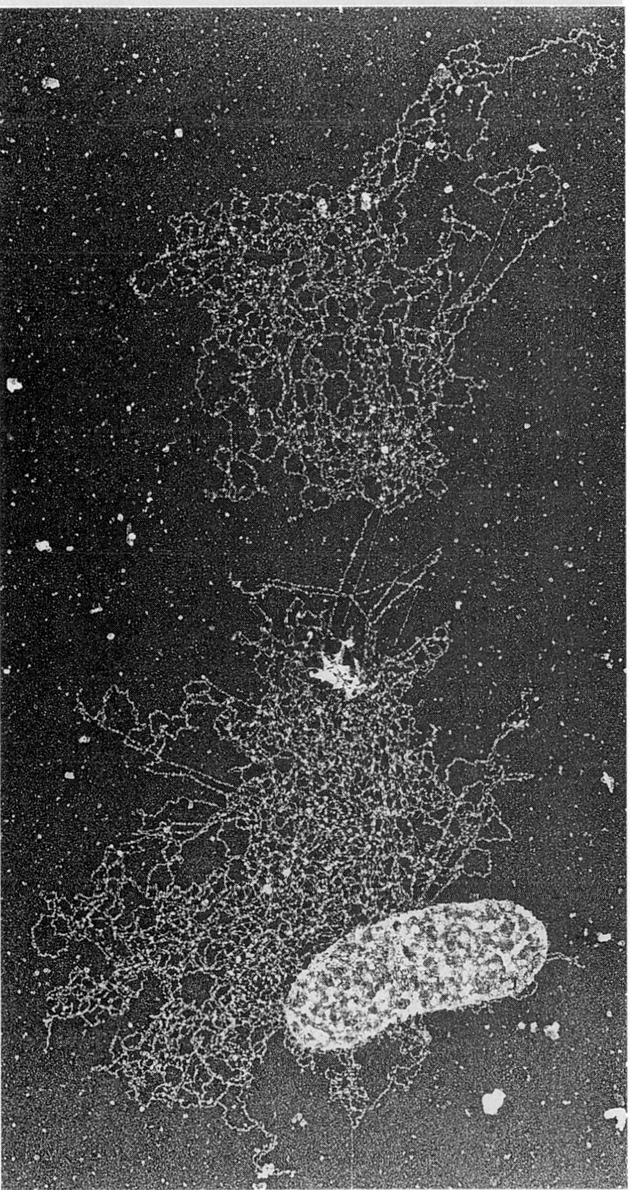

Protein HU is a dimer that condenses DNA, possibly wrapping it into a bead-like structure. It stimulates DNA replication (see Chapter 13). It is related to IHF (integration host factor), another dimer, which has a structural role in building a protein complex in some specialized recombination reactions, including the integration and excision of phage lambda (for which it is named; see Chapter 14). Null mutations in either of the genes coding for the subunits of HU (*hupA,B*) have little effect, but loss of both functions causes a cold-sensitive phenotype and some loss of superhelicity in DNA. These results raise the possibility that HU plays some general role in nucleoid condensation.

Protein H1 (also known as H-NS) binds DNA, interacting preferentially with sequences that are bent. Mutations in its gene have turned up in a variety of guises (*osmZ, bglY, pilG*), each identified as an apparent regulator of a different system. These results probably reflect the effect that H1 has on the local topology of DNA, with effects upon gene expression that depend upon the particular promoter.

We might expect that the absence of a protein required for nucleoid structure would have serious effects upon viability. Why then are the effects of deletions in the genes for proteins HU and H1 relatively restricted? One explanation is that these proteins are *redundant,* that any one can substitute for the others, so that deletions of *all* of them would be necessary to interfere seriously with nucleoid structure. Another possibility is that we have yet to identify the proteins responsible for the major features of nucleoid integrity.

The nucleoid can be isolated directly in the form of a very rapidly sedimenting complex, consisting of ~80% DNA by mass. (The analogous complexes in eukaryotes have ~50% DNA by mass; see later.) It can be unfolded by treatment with reagents that act on RNA or protein. The possible role of proteins in stabilizing its structure is evident. The role of RNA has been quite refractory to analysis.

The DNA of the compact body isolated *in vitro* behaves as a closed duplex structure, as judged by its response to ethidium bromide. This small molecule intercalates between base pairs to generate *positive* superhelical turns in "closed" circular DNA molecules, that is, molecules in which both strands have covalent integrity. (In "open" circular molecules, which contain a nick in one strand, or with linear molecules, the DNA can rotate freely in response to the intercalation, thus relieving the tension.)

In a natural closed DNA that is *negatively* supercoiled, the intercalation of ethidium bromide first removes the negative supercoils and then introduces positive supercoils. The amount of ethidium bromide needed to achieve zero supercoiling is a measure of the original density of negative supercoils.

Some nicks occur in the compact nucleoid during its isolation; they can also be generated by limited treatment with DNAase. But this does not abolish the ability of ethidium bromide to introduce positive supercoils. This capacity of the genome to retain its response to ethidium bromide in the face of nicking means that it must have many independent **domains;** *the supercoiling in each domain is not affected by events in the other domains.*

This autonomy suggests that the structure of the bacterial chromosome has the general organization depicted diagrammatically in **Figure 18.6.** Each domain consists of a loop of DNA, the ends of which are secured in some (unknown) way that does not allow rotational events to propagate from one domain to another. There are ~100 such domains per genome; each consists of ~40 kb (13 μm) of DNA, organized into some more compact fiber whose structure has yet to be characterized.

Figure 18.6 The bacterial genome consists of a large number of loops of duplex DNA (in the form of a fiber), each secured at the base to form an independent structural domain.

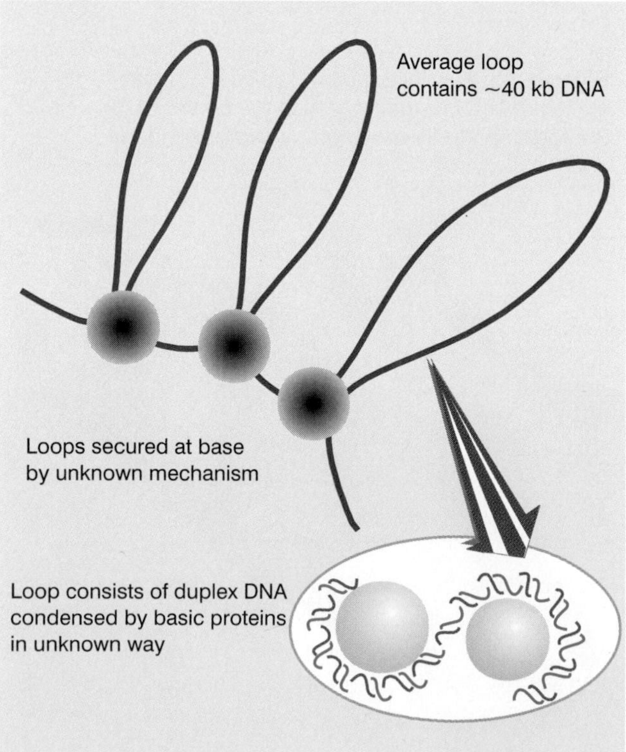
Average loop
contains ~40 kb DNA
Loops secured at base
by unknown mechanism
Loop consists of duplex DNA
condensed by basic proteins
in unknown way

The existence of separate domains could permit different degrees of supercoiling to be maintained in different regions of the genome. This could be relevant in considering the different susceptibilities of particular bacterial promoters to supercoiling (see Chapter 9).

Supercoiling in the genome can in principle take two forms:

- If a supercoiled DNA is free, its path is **unrestrained,** and negative supercoils generate a state of torsional tension that is transmitted freely along the DNA within a domain. It can be relieved by unwinding the double helix, as described in Chapter 14. The DNA is in a dynamic equilibrium between the states of tension and unwinding.
- Supercoiling can be **restrained** if proteins are bound to the DNA to hold it in a particular three-dimensional configuration. In this case, the supercoils are represented by the path the DNA follows in its fixed association with the proteins. The energy of interaction between the proteins and the supercoiled DNA stabilizes the nucleic acid, so that no tension is transmitted along the molecule.

Are the supercoils in *E. coli* DNA restrained *in vivo* or is the double helix subject to the torsional tension characteristic of free DNA? Measurements of supercoiling *in vitro* encounter the difficulty that restraining proteins may have been lost during isolation. Various approaches suggest that DNA is under torsional stress *in vivo,* although it is difficult to quantitate the level of supercoiling.

A direct approach is to use the crosslinking reagent psoralen, which binds more readily to DNA when it is under torsional tension. The reaction of psoralen with *E. coli* DNA *in vivo* corresponds to an average density of one negative superhelical turn for every 200 bp ($\sigma = -0.05$).

Another approach is to examine the ability of cells to form alternative DNA structures; for example, to generate cruciforms at palindromic sequences. From the change in linking number that is required to drive such reactions, it is possible to calculate the original supercoiling density. This approach suggests an average density of $\sigma = -0.025$, or one negative superhelical turn per 100 base pairs.

So supercoils *do* create torsional tension *in vivo.* There may be variation about an average level, and although the precise range of densities is difficult to measure, it is clear that the level is sufficient to exert significant effects on DNA structure, for example, in assisting melting in particular regions such as origins or promoters.

Many of the important features of the structure of the compact nucleoid remain to be established. What is the specificity with which domains are constructed—do the same sequences always lie at the same relative locations, or can the contents of individual domains shift? How is the integrity of the domain maintained? Biochemical analysis by itself is unable to answer these questions fully, but if it is possible to devise suitable selective techniques, the properties of structural mutants should lead to a molecular analysis of nucleoid construction.

Loops, domains, and scaffolds in eukaryotic DNA

INTERPHASE chromatin is a tangled mass occupying a large part of the nuclear volume, in contrast with the highly organized and reproducible ultrastructure of mitotic chromosomes. What controls the distribution of interphase chromatin within the nucleus?

Some indirect evidence on its nature is provided by the isolation of the genome as a single, compact body. Using the same technique just described for isolating the bacterial nucleoid, nuclei can be lysed on top of a sucrose gradient. This releases the genome in a form that can be collected by centrifugation. As isolated from *D. melanogaster,* it can be visualized as a compactly folded fiber (10 nm in diameter), consisting of DNA bound to proteins.

Supercoiling measured by the response to ethidium bromide corresponds to about one negative supercoil for every 200 bp. These supercoils can be removed by nicking with DNAase, although the DNA remains in the form of the 10 nm fiber. This suggests that the supercoiling is caused by the arrangement of the fiber in space, and represents the existing torsion.

Full relaxation of the supercoils requires one nick for

every 85 kb, identifying the average length of "closed" DNA. This region could comprise a loop or domain similar in nature to those identified in the bacterial genome. Loops can be seen directly when the majority of proteins are extracted from mitotic chromosomes. The resulting complex consists of the DNA associated with ~8% of the original protein content. As seen in **Figure 18.7**, the protein-depleted chromosomes take the form of a central **scaffold** surrounded by a halo of DNA.

The metaphase scaffold consists of a dense network of fibers. Threads of DNA emanate from the scaffold, apparently as loops of average length 10–30 μm (30–90 kb). The DNA can be digested without affecting the integrity of the scaffold, which consists of a set of specific proteins. This suggests a form of organization in which loops of DNA of ~60 kb are anchored in a central proteinaceous scaffold.

The appearance of the scaffold resembles a mitotic pair of sister chromatids. The sister scaffolds usually are tightly connected, but sometimes are separate, joined only by a few fibers. Could this be the structure responsible for maintaining the shape of the mitotic chromosomes? Could it be generated by bringing to-

Figure 18.7 Histone-depleted chromosomes consist of a protein scaffold to which loops of DNA are anchored. Photograph kindly provided by Ulrich K. Laemmli.

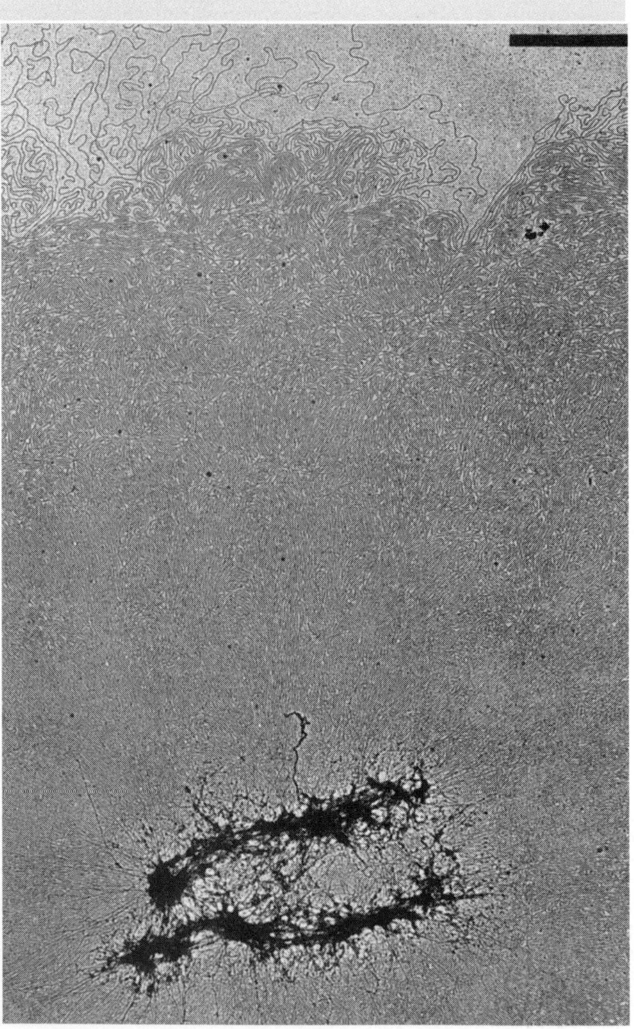

Figure 18.8 Matrix-associated regions may be identified by characterizing the DNA retained by the matrix isolated *in vivo* or by identifying the fragments that can bind to the matrix from which all DNA has been removed *in vivo*.

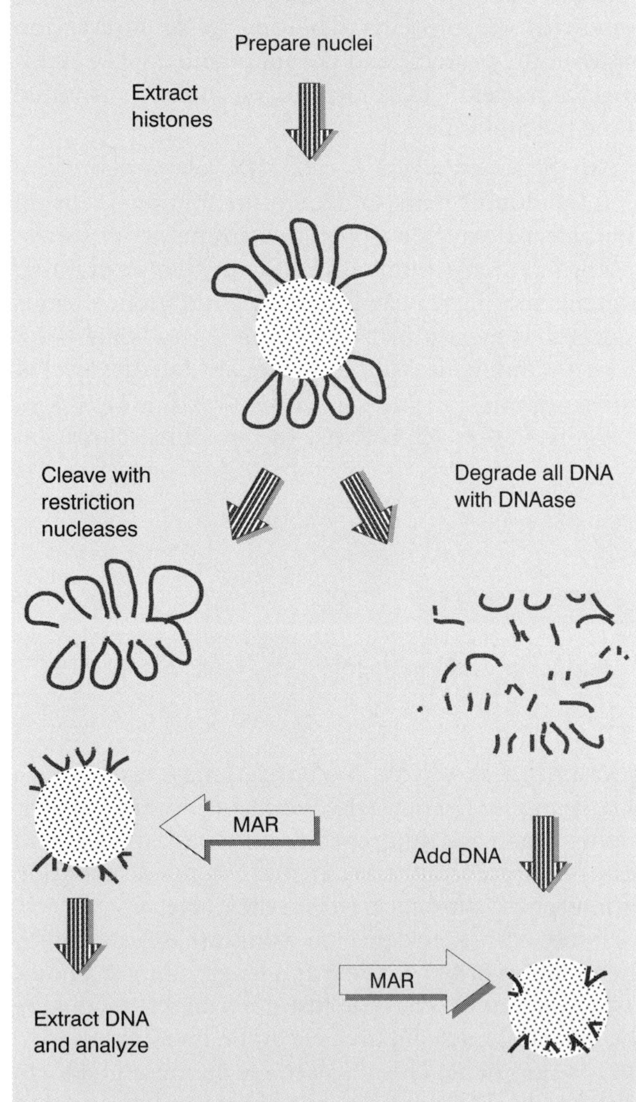

gether the protein components that usually secure the bases of loops in interphase chromatin?

Interphase cells possess a **nuclear matrix,** a filamentous structure on the interior of the nuclear membrane. Chromatin often appears to be attached to the matrix, and there have been many suggestions that this attachment is necessary for transcription or replication. When nuclei are depleted of proteins, the DNA extrudes as loops from the residual nuclear matrix.

Is DNA attached to the matrix or scaffold via specific sequences? DNA sites attached to proteinaceous structures in interphase nuclei are called **MAR** (matrix attachment regions); it is confusing that they are sometimes also called **SAR** (scaffold attachment regions), although they concern the nuclear matrix.

How might we demonstrate that particular DNA regions are genuinely associated with the matrix? *In vivo* and *in vitro* approaches are summarized in **Figure 18.8.** Both start by isolating the matrix as a crude nuclear preparation containing chromatin and nuclear proteins. Different treatments can then be used to characterize DNA in the matrix or to identify DNA able to attach to it.

To analyze the existing MAR, the chromosomal loops can be decondensed by extracting the proteins. Removal of the DNA loops by treatment with restriction nucleases leaves only the (presumptive) *in vivo* MAR sequences attached to the matrix.

The complementary approach is to remove *all* the DNA from the matrix by treatment with DNAase; then isolated fragments of DNA can be tested for their ability to bind to the matrix *in vitro.*

The same sequences should be associated with the matrix *in vivo* or *in vitro.* Once a potential MAR has been identified, the size of the minimal region needed for association *in vitro* can be determined by deletions. We can also then identify proteins that bind to the MAR sequences.

A surprising feature is the lack of conservation of sequence in MAR fragments. They are usually ~70% A•T-rich, but otherwise lack any consensus sequences. However, other interesting sequences often are in the DNA stretch containing the MAR. *Cis*-acting sites that regulate transcription are common. And a recognition site for topoisomerase II is usually present in the MAR. It is therefore possible that an MAR serves more than one function, providing a site for attachment to the matrix, but also containing other sites at which topological changes in DNA are effected.

What is the relationship between the chromosome scaffold of dividing cells and the nuclear matrix of interphase cells; are the same DNA sequences attached to both structures? In several cases, the same DNA fragments that are found with the nuclear matrix *in vivo* can be retrieved from the metaphase scaffold. And fragments that contain MAR sequences can bind to a metaphase scaffold. It therefore seems likely that DNA contains a single type of attachment site, which in interphase cells is connected to the nuclear matrix, and in mitotic cells is connected to the chromosome scaffold.

The nuclear matrix and chromosome scaffold consist of different proteins, although there are some common components. Topoisomerase II is a prominent component of the chromosome scaffold, and is a constituent of the nuclear matrix, suggesting that the control of topology is important in both cases.

The contrast between interphase chromatin and mitotic chromosomes

EACH chromosome contains a single, very long duplex of DNA. This explains why chromosome replication is semiconservative like the individual DNA molecule. (This would not necessarily be the case if a chromosome carried many independent molecules of DNA.) The single duplex of DNA is folded into a fiber that runs continuously throughout the chromosome. *So, in accounting for interphase chromatin and mitotic chromosome structure, we have to explain the packaging of a single, exceedingly long molecule of DNA into a form in which it can be transcribed and replicated, and can become cyclically more and less compressed.*

Individual eukaryotic chromosomes come into the limelight for a brief period, during the act of cell division. Only then can each be seen as a compact unit. **Figure 18.9** is an electron micrograph of a sister chromatid pair, captured at metaphase. (The sister chromatids are daughter chromosomes produced by the previous replication event, still joined together at this stage of mitosis.) Each consists of a fiber with a diameter of ~30 nm and a nubbly appearance. The DNA is 5–10× more condensed in chromosomes than in interphase chromatin.

During most of the life cycle of the eukaryotic cell, however, its genetic material occupies an area of the nucleus in which individual chromosomes cannot be distinguished. The structure of the interphase chromatin does not change visibly between divisions. No disruption is evident during the period of replication, when the amount of chromatin doubles. Chromatin is fibrillar, although the overall configuration of the fiber in space is hard to discern in detail. The fiber itself, however, is similar or identical to that of the mitotic chromosomes.

Figure 18.9 The sister chromatids of a mitotic pair each consist of a fiber (~30 nm in diameter) compactly folded into the chromosome. Photograph kindly provided by E. J. DuPraw.

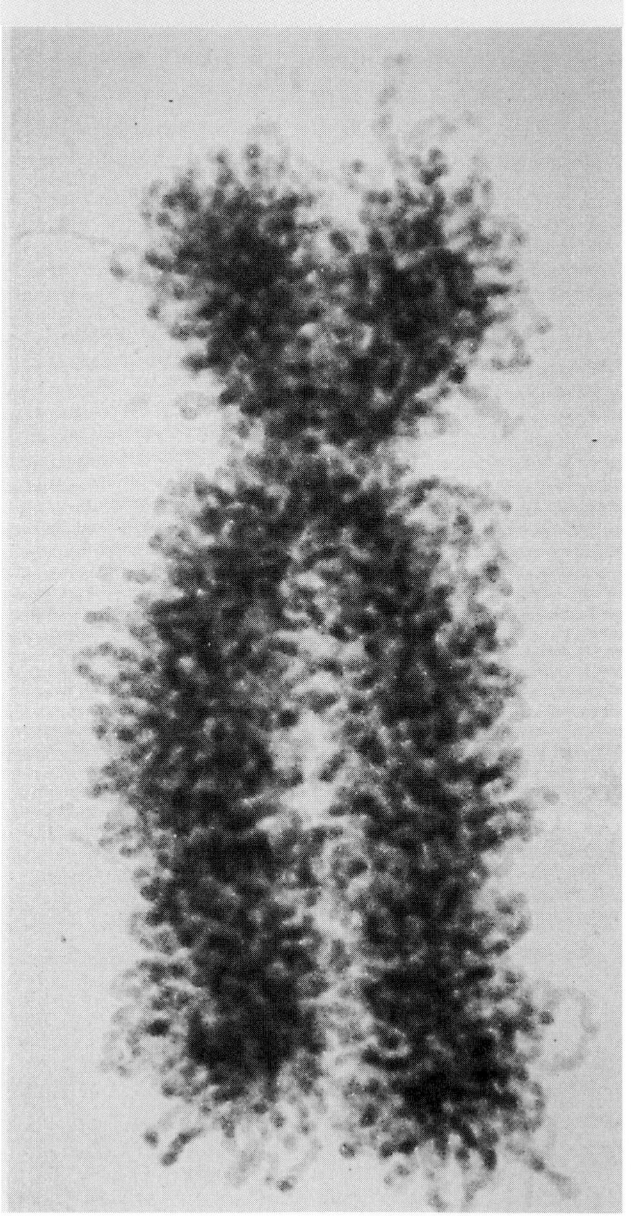

Chromatin can be divided into two types of material, which can be seen in the nuclear section of **Figure 18.10:**

- In most regions, the fibers are much less densely packed than in the mitotic chromosome. This material is called **euchromatin.** It has a relatively dispersed appearance in the nucleus, and occupies most of the nuclear region in Figure 18.10.
- Some regions of chromatin are very densely packed with fibers, displaying a condition comparable to that of the chromosome at mitosis. This material is called **heterochromatin.** It is typically found at centromeres, but occurs at other locations also. It passes through the cell cycle with relatively little change in its degree of condensation. It forms a series of discrete clumps in Figure 18.10, but often the various heterochromatic regions aggregate into a densely staining **chromocenter.**

The same fibers run continuously between euchromatin and heterochromatin, which implies that these states represent different degrees of condensation of the genetic material. In the same way, euchromatic regions exist in different states of condensation during interphase and during mitosis. So the genetic material

Figure 18.10 A thin section through a nucleus stained with Feulgen shows heterochromatin as compact regions clustered near the nucleolus and nuclear membrane. Photograph kindly provided by Edmund Puvion.

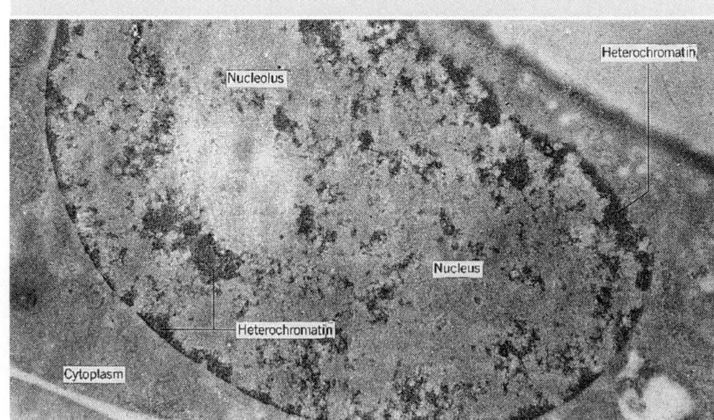

is organized in a manner that permits alternative states to be maintained side by side in chromatin, and allows cyclical changes to occur in the packaging of euchromatin between interphase and division.

The structural condition of the genetic material is correlated with its activity. Heterochromatin is not transcribed; also, it replicates late in S phase. This suggests that condensation of the genetic material is associated with (perhaps is responsible for) its inactivity. Note, however, that the reverse is not true. Active genes are contained within euchromatin; but only a small minority of the sequences in euchromatin are transcribed at any time. So location in euchromatin is *necessary* for gene expression, but is not *sufficient* for it.

Because of the diffuse state of chromatin, we cannot directly determine the specificity of its organization. But we can ask whether the structure of the chromosome is ordered. Do particular sequences always lie at particular sites, or is the folding of the fiber into the overall structure a more random event?

At the level of the chromosome, each member of the complement has a different and reproducible ultrastructure. When subjected to certain treatments and then stained with the chemical dye Giemsa, chromosomes generate a series of **G-bands.** An example of the human set is presented in **Figure 18.11.**

Until the development of this technique, chromosomes could be distinguished only by their overall size and the relative location of the centromere (see later). Now each chromosome can be identified by its characteristic banding pattern. This pattern is reproducible enough to allow translocations from one chromosome to another to be identified by comparison with the original diploid set. **Figure 18.12** shows a diagram of the bands of the human X chromosome. The bands are large structures, each ~10^7 bp of DNA, which could include many hundreds of genes.

The banding technique is of enormous practical use, but the mechanism of banding remains a mys-

Figure 18.11 G-banding generates a characteristic lateral series of bands in each member of the chromosome set. Photograph kindly provided by Lisa Shaffer.

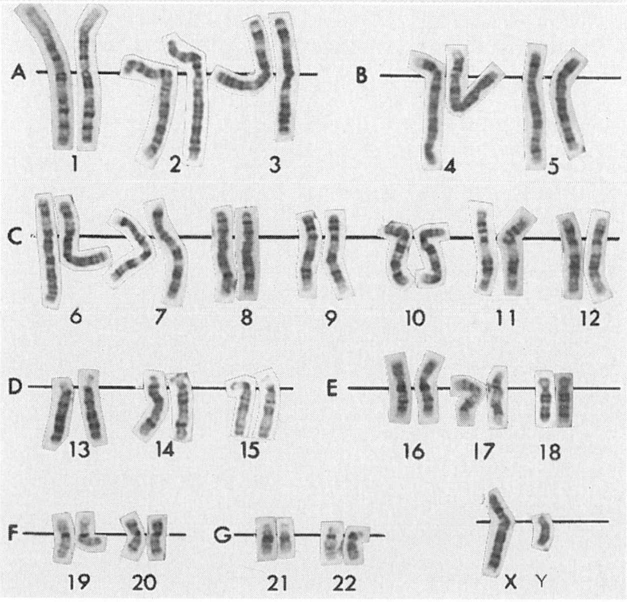

Figure 18.12 The human X chromosome can be divided into distinct regions by its banding pattern. The short arm is *p* and the long arm is *q*; each arm is divided into larger regions that are further subdivided. This map shows a low resolution structure; at higher resolution, some bands are further subdivided into smaller bands and interbands, e.g. p21 is divided into p21.1, p21.2, and p21.3.

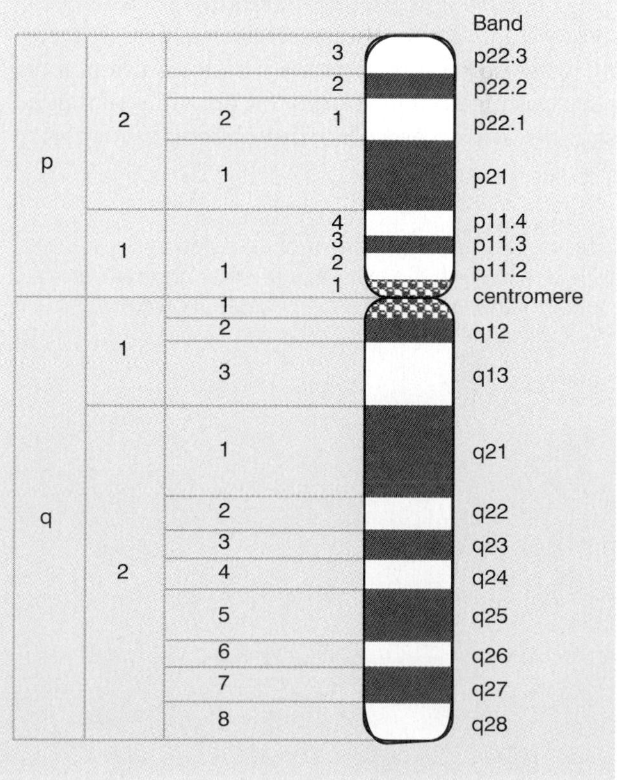

tery. All that is certain is that the dye stains untreated chromosomes more or less uniformly. So the generation of bands depends on a variety of treatments that change the response of the chromosome (presumably by extracting the component that binds the stain from the nonbanded regions). But the variety of effective treatments is so great that no common cause has yet been discerned. These results imply the existence of a definite long-range structure, but its basis is unknown.

The extended state of lampbrush chromosomes

It would be extremely useful to visualize gene expression in its natural state, to see what structural changes are associated with transcription. The compression of DNA in chromatin, coupled with the difficulty of identifying particular genes within it, makes it impossible to visualize the transcription of individual active genes.

Gene expression can be visualized directly in certain unusual situations, in which the chromosomes are found in a highly extended form that allows individual loci (or groups of loci) to be distinguished. Lateral differentiation of structure is evident in many chromosomes when they first appear for meiosis. At this stage, the chromosomes resemble a series of beads on a string. The beads are densely staining granules, properly known as **chromomeres.** However, usually there is little gene expression at meiosis, and it is not practical to use this material to identify the activities of individual genes. But an exceptional situation that allows the material to be examined is presented by **lampbrush chromosomes,** which have been best characterized in certain amphibians.

Lampbrush chromosomes are formed during an unusually extended meiosis, which can last up to several months! During this period, the chromosomes are held in a stretched-out form in which they can be visualized in the light microscope. Later during meiosis, the chromosomes revert to their usual compact size. So

Figure18.13 A lampbrush chromosome is a meiotic bivalent in which the two pairs of sister chromatids are held together at chiasmata (indicated by arrows). Photograph kindly provided by Joe Gall.

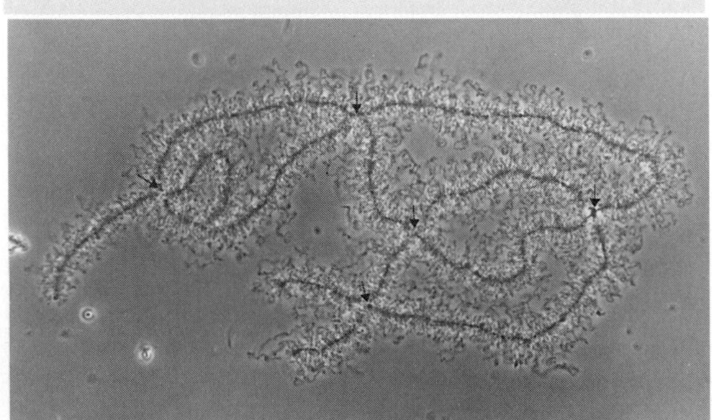

Figure 18.14 A lampbrush chromosome loop is surrounded by a matrix of ribonucleoprotein. Photograph kindly provided by Oscar Miller.

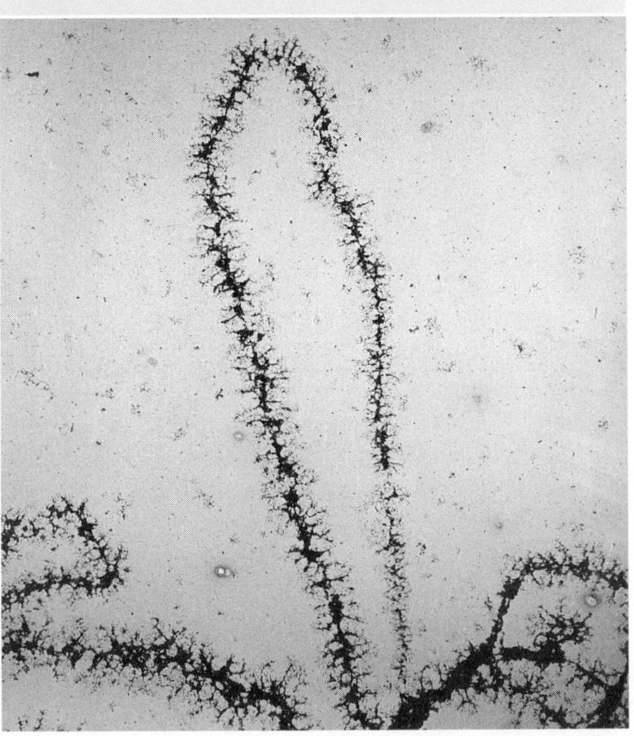

the extended state essentially proffers an unfolded version of the normal condition of the chromosome.

The lampbrush chromosomes are meiotic bivalents, each consisting of two pairs of sister chromatids. **Figure 18.13** shows an example in which the sister chromatid pairs have mostly separated so that they are held together only by chiasmata. Each sister chromatid pair forms a series of ellipsoidal chromomeres, ~1–2 μm in diameter, which are connected by a very fine thread. This thread contains the two sister duplexes of DNA, and runs continuously along the chromosome, through the chromomeres.

The lengths of the individual lampbrush chromosomes in the newt *Notophthalmus viridescens* range from 400–800 μm, compared with the range of 15–20 μm seen later in meiosis. So the lampbrush chromosomes are ~30 times less tightly packed. The total length of the entire lampbrush chromosome set is 5–6 mm, organized into about 5000 chromomeres.

The lampbrush chromosomes take their name from the lateral loops that extrude from the chromomeres at certain positions. (These resemble a lampbrush, an extinct object.) The loops extend in pairs, one from each sister chromatid. The loops are continuous with the axial thread, which suggests that they represent chromosomal material extruded from its more compact organization in the chromomere.

The loops are surrounded by a matrix of ribonucleoproteins. These contain nascent RNA chains. Often a transcription unit can be defined by the increase in the length of the RNP moving around the loop. An example is shown in **Figure 18.14.**

So the loop is an extruded segment of DNA that is being actively transcribed. In some cases, loops corresponding to particular genes have been identified. Then the structure of the transcribed gene, and the nature of the product, can be scrutinized *in situ.*

Transcription disrupts the structure of polytene chromosomes

THE interphase nuclei of some tissues of the larvae of Dipteran flies contain chromosomes that are greatly enlarged relative to their usual condition. They possess both increased diameter and greater length. **Figure 18.15** shows an example of a chromosome set from the salivary gland of *D. melanogaster.* They are called **polytene chromosomes.**

Each chromosome consists of a visible series of **bands** (more properly, but rarely, described as chromomeres). The bands range in size from the largest with a breadth of ~0.5 μm to the smallest of ~0.05 μm. (The smallest can be distinguished only under the electron microscope.) The bands contain most of the mass of DNA and stain intensely with appropriate reagents. The regions between them stain more lightly and are called **interbands.** There are ~5000 bands in the *D. melanogaster* set.

The centromeres of all four chromosomes of *D. melanogaster* aggregate to form a chromocenter that

Figure 18.15 The polytene chromosomes of *D. melanogaster* form an alternating series of bands and interbands. Photograph kindly provided by Jose Bonner.

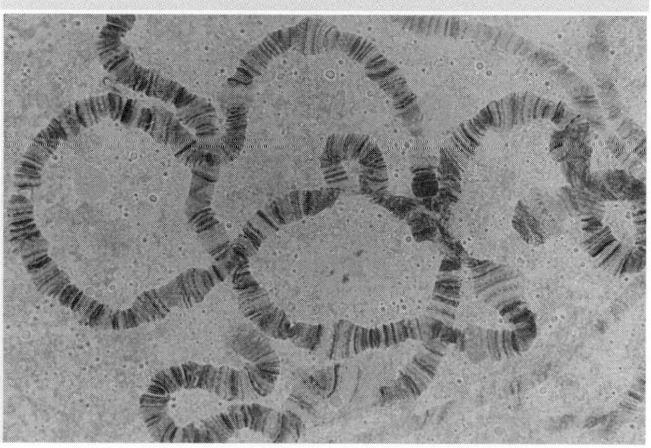

consists largely of heterochromatin (in the male it includes the entire Y chromosome). Allowing for this, ~75% of the haploid DNA set is organized into alternating bands and interbands. The length of the chromosome set is ~2000 μm. The DNA in extended form would stretch for ~40,000 μm, so the packing ratio is ~20. This demonstrates vividly the extension of the genetic material relative to the usual states of interphase chromatin or mitotic chromosomes.

What is the structure of these giant chromosomes? Each is produced by the successive replications of a synapsed diploid pair. The replicas do not separate, but remain attached to each other in their extended state. At the start of the process, each synapsed pair has a DNA content of 2C (where C represents the DNA content of the individual chromosome). Then this doubles up to 9 times, at its maximum giving a content of 1024C. The number of doublings is different in the various tissues of the *D. melanogaster* larva.

Each chromosome can be visualized as a large number of parallel fibers running longitudinally, tightly condensed in the bands, less condensed in the interbands. Probably each fiber represents a single (C) haploid chromosome. This gives rise to the name **polyteny.** The degree of polyteny is the number of haploid chromosomes contained in the giant chromosome.

The banding pattern is characteristic for each strain of *Drosophila.* The constant number and linear arrangement of the bands was first noted in the 1930s, when it was realized that they form a **cytological map** of the chromosomes. Rearrangements—such as deletions, inversions, or duplications—result in alterations of the order of bands.

The linear array of bands can be equated with the linear array of genes. So genetic rearrangements, as seen in a linkage map, can be correlated with structural rearrangements of the cytological map. Ultimately, a particular mutation can be located in a particular band. Since the total number of genes in *D. melanogaster* appears to exceed the number of bands, there are probably multiple genes in most or all bands (see Chapter 3).

The positions of particular genes on the cytological map can be determined directly by the technique of *in situ* or **cytological hybridization.** The protocol is summarized in **Figure 18.16.** A radioactive probe representing a gene (most often a labeled cDNA clone derived from the mRNA) is hybridized with the denatured DNA of the polytene chromosomes *in situ.* Autoradiography identifies the position or positions of the corresponding genes by the superimposition of grains at a particular band or bands. An example is shown in **Figure 18.17.** With this type of technique at hand, it is possible to determine directly the band within which a particular sequence lies.

Figure 18.16 Individual bands containing particular genes can be identified by *in situ* hybridization.

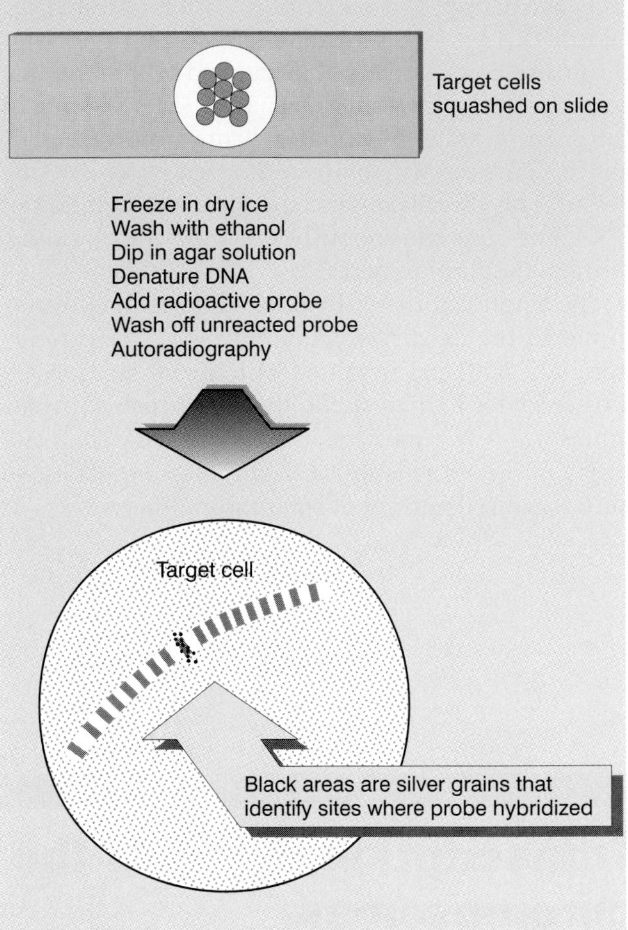
Target cells squashed on slide
Freeze in dry ice
Wash with ethanol
Dip in agar solution
Denature DNA
Add radioactive probe
Wash off unreacted probe
Autoradiography
Target cell
Black areas are silver grains that identify sites where probe hybridized

One of the intriguing features of the polytene chromosomes is that active sites can be visualized. Some of the bands pass transiently through an expanded or **puffed** state, in which chromosomal material is extruded from the axis. An example of some very large puffs (called Balbiani rings) is shown in **Figure 18.18.**

What is the nature of the puff? It consists of a region in which the chromosome fibers unwind from their usual state of packing in the band. The fibers remain continuous with those in the chromosome axis. Puffs usually emanate from single bands, although when they are very large, as typified by the Balbiani rings, the swelling may be so extensive as to obscure the underlying array of bands.

The pattern of puffs is related to gene expression. During larval development, puffs appear and regress in

Figure 18.17 A magnified view of bands 87A and 87C shows their hybridization *in situ* with labeled RNA extracted from heat-shocked cells. Photograph kindly provided by Jose Bonner.

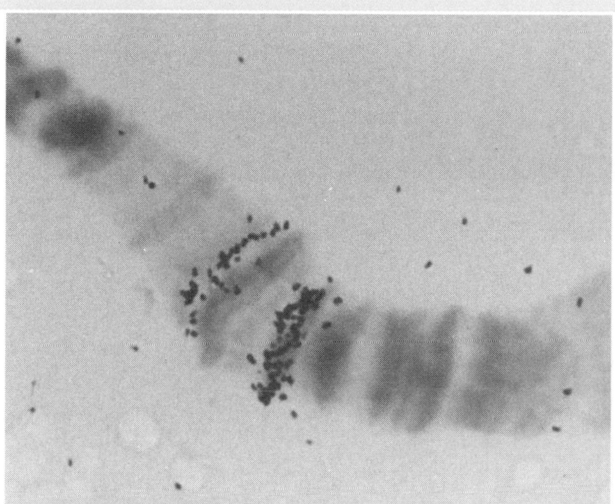

Figure 18.18 Chromosome IV of the insect *C. tentans* has three Balbiani rings in the salivary gland. Photograph kindly provided by Bertil Daneholt.

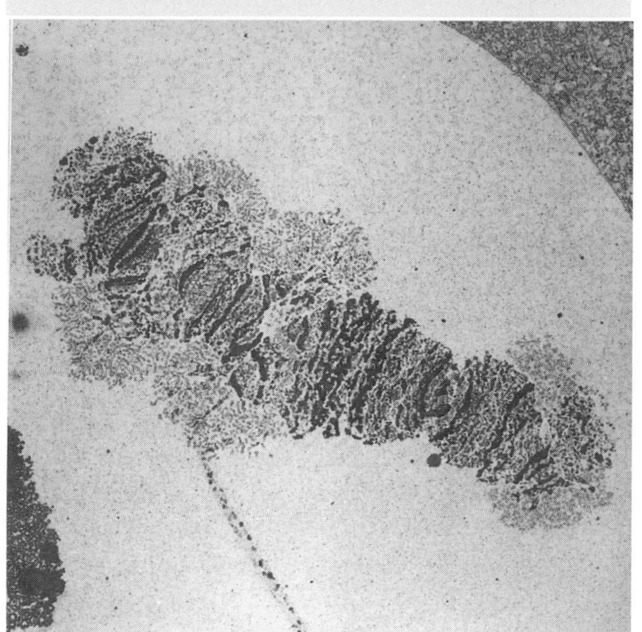

a definite, tissue-specific pattern. A characteristic pattern of puffs is found in each tissue at any given time. Puffs are induced by the hormone ecdysone that controls *Drosophila* development. Some puffs are induced directly by the hormone; others are induced indirectly by the products of earlier puffs.

The puffs are *sites where RNA is being synthesized*. The accepted view of puffing has been that expansion of the band is a consequence of the need to relax its structure in order to synthesize RNA. Puffing has therefore been viewed as a consequence of transcription. A puff can be generated by a single active gene.

The sites of puffing differ from ordinary bands in accumulating additional proteins. Characterization of these proteins at present is only rather primitive. We know that they include RNA polymerase II and other proteins associated with the act of transcription. We should like to analyze the entire set of proteins that accumulate at puffs, in particular to characterize those that are a cause rather than a consequence of the puffing. Then it should be possible to determine the nature of the molecular events responsible for the expansion of material.

The features displayed by lampbrush and polytene chromosomes suggest a general conclusion. In order to be transcribed, the genetic material is dispersed from its usual more tightly packed state. The question to keep in mind is whether this dispersion at the gross level of the chromosome mimics the events that occur at the molecular level within the mass of ordinary interphase euchromatin.

The eukaryotic chromosome as a segregation device

DURING mitosis, the sister chromatids move to opposite poles of the cell. Their movement depends on the attachment of the chromosome to microtubules, which are connected at their other end to the poles. (The microtubules comprise a cellular filamentous system, reorganized at mitosis so that they connect the chromosomes to the poles of the cell.) The sites in the two regions where microtubule ends are organized—in the vicinity of the centrioles at the poles and at the chromosomes—are called **MTOCs** (microtubule organizing centers).

The region of the chromosome that is responsible for its segregation at mitosis and meiosis is called the **centromere.** It is associated with two important features:

- It contains the site at which the sister chromatids are held together prior to the separation of the individual chromosomes. This shows as a constricted region connecting all four chromosome arms, as in the photograph of Figure 18.9, which shows the sister chromatids at the metaphase stage of mitosis.
- The term "centromere" historically has been used in both the functional and structural sense to describe the feature of the chromosome responsible for its movement. The centromere is pulled toward the pole during mitosis, and the attached chromosome is dragged along behind, as it were. The chromosome therefore provides a device for attaching a large number of genes to the apparatus for division.

The centromere is essential for segregation, as shown by the behavior of chromosomes that have been broken. A single break generates one piece that retains the centromere, and another, an **acentric fragment,** that lacks it. The acentric fragment does not become attached to the mitotic spindle; and as a result it fails to be included in either of the daughter nuclei.

(When chromosome movement relies on discrete centromeres, there can be *only* one centromere per chromosome. When translocations generate chromosomes with more than one centromere, aberrant structures form at mitosis, since the two centromeres on the *same* sister chromatid can be pulled toward different poles, breaking the chromosome. However, in some species the centromeres are "diffuse," which creates a different situation. Only discrete centromeres have been analyzed at the molecular level.)

The regions flanking the centromere often are rich in satellite DNA sequences and display a considerable amount of heterochromatin. Because the entire chromosome is condensed, centromeric heterochromatin is not immediately evident in mitotic chromosomes. However, it can be visualized by a technique called **C-banding.** In the example of **Figure 18.19,** all the centromeres show as darkly staining regions. Although it is common, heterochromatin cannot be identified around *every* known centromere, which suggests that it is unlikely to be essential for the division mechanism.

What is the feature of the centromere that is responsible for segregation? Within the centromeric region, a darkly staining fibrous object of diameter or length ~400 nm can be seen. This **kinetochore** appears to be directly attached to the microtubules. Usually it is assumed that a specific sequence of DNA in some way defines the site at which the kinetochore should be established, but so far we have made no progress toward characterizing the molecular location and organization of this structure. The kinetochore provides the MTOC on a chromosome.

Figure 18.19 C-banding generates intense staining at the centromeres of all chromosomes. Photograph kindly provided by Lisa Shaffer.

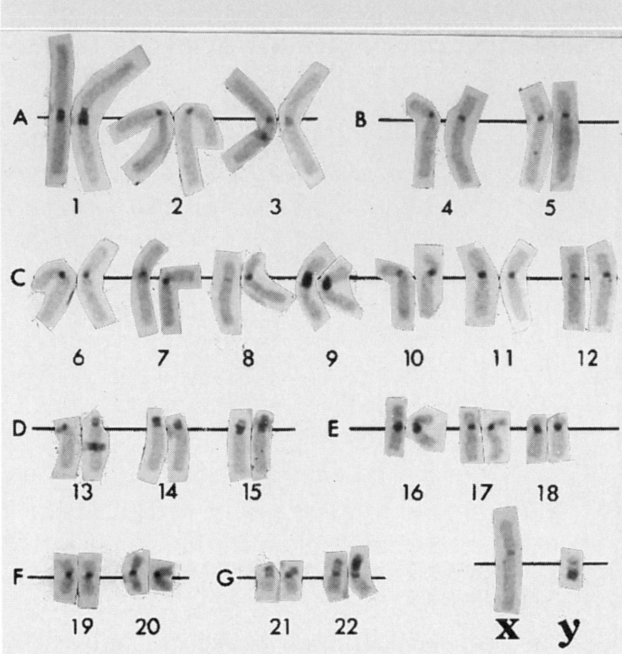

If a centromeric sequence of DNA is responsible for segregation, any molecule of DNA possessing this sequence should move properly at cell division, while any DNA lacking it will fail to segregate. This prediction has been used to isolate centromeric DNA in the yeast, *S. cerevisiae.* Yeast chromosomes do not display visible kinetochores comparable to those of higher eukaryotes, but otherwise divide at mitosis and segregate at meiosis by the same mechanisms.

Genetic engineering has produced plasmids of yeast that are replicated like chromosomal sequences (see Chapter 12). However, they are unstable at mitosis and meiosis, disappearing from a majority of the cells because they segregate erratically. Fragments of chromosomal DNA have been isolated by virtue of their ability to confer mitotic stability on these plasmids.

A *CEN* fragment is defined by its ability to confer stability upon such a plasmid. By reducing the sizes of the fragments that are incorporated into the plasmid, the minimum region necessary for mitotic centromeric function can be identified. Deletions and other changes can be made to investigate the features involved in centromeric function.

Another way to use the availability of the centromeric sequences is to modify them *in vitro* and then reintroduce them into the yeast cell, where they replace the corresponding centromere on the chromosome. This allows the sequences required for *CEN* function to be defined directly in the context of the chromosome.

A *CEN* fragment derived from one chromosome can replace the centromere of another chromosome with no apparent consequence. This result suggests that centromeres are interchangeable. *They are used simply to attach the chromosome to the spindle, and play no role in distinguishing one chromosome from another.*

The sequences required for centromeric function fall within a stretch of ~120 bp. The centromeric region is packaged into a nuclease-resistant structure, and it binds a single microtubule. We may therefore look to the *S. cerevisiae* centromeric region to identify proteins that bind centromeric DNA and proteins that connect the chromosome to the spindle.

Three types of sequence element may be distinguished in the *CEN* region, as summarized in **Figure 18.20:**

- CDE-I is a sequence of 9 bp that is conserved with minor variations at the left boundary of all centromeres.

- CDE-II is a >90% A•T-rich sequence of 80–90 bp found in all centromeres; its function could depend on its length rather than exact sequence. Its constitution is reminiscent of some short tandemly repeated (satellite) DNAs in higher eukaryotes (see Chapter 4). Its base composition may cause some characteristic distortions of the DNA double helical structure.

- CDE-III is an 11 bp sequence highly conserved at the right boundary of all centromeres. Sequences on either side of the element are less well conserved, and may also be needed for centromeric function. (CDE-III could be longer than 11 bp if it turns out that the flanking sequences are essential.)

Mutations in CDE-I or CDE-II reduce but do not inactivate centromere function, but point mutations in the central CCG of CDE-III completely inactivate the centromere.

Can we identify proteins that are necessary for the function of CEN sequences? CDE-I is bound by the homodimer CBF1; this interaction is not essential for centromere function, but in its absence the fidelity of segregation chromosome is reduced ~10×. A 240 kD complex of several proteins, called CBF3, binds to CDE-III. This interaction is essential for centromeric function. Other proteins may bind to the CBF1 and CBF3 complexes to assemble the full centromeric structure. The complex is integrated into the chromosomal structure (see Chapter 19).

Mutations in the components of the genes coding for CBF-III block chromosome movement at mitosis; and the protein complex has a microtubule-based motor activity—it is able to move itself, and objects attached to it, such as chromosomes, along microtubules. Comparable proteins with sequences related to known motor activities have been found at eukaryotic centromeres. Taken together, these observations suggest that a protein complex with motor activity may connect the centromeric region of a chromosome to microtubules, and contribute to movement on the mitotic spindle. The discovery of the CBF-III complex may be the prelude to characterizing the connection between the centromere and the apparatus for chromosome segregation.

Attempts to characterize functional centromeres from the yeast *S. pombe* have been less successful. They cannot be isolated by ability to confer stability on plasmids. However, *S. pombe* has only 3 chromosomes, and the region containing each centromere has been identified by deleting most of the sequences of each chromosome to create a stable minichromosome. This approach locates the centromeres within regions of 40–100 kb that consist largely or entirely of repetitious

Figure 18.20 Three conserved regions can be identified by the sequence homologies between yeast *CEN* elements.

TCACATGATGATATTTGATTTTATTATATTTTTAAAAAAAGTAAAAAATAAAAAGTAGTTTATTTTTAAAAAATAAAATTTAAAATATTTCACAAAATGATTTCCGAA
AGTGTACTACTATAAACTAAAATAATATAAAAATTTTTTTCATTTTTTATTTTTCATCAAATAAAAATTTTTTATTTTAAATTTTATAAAGTGTTTTACTAAAGGCTT

CDE-I *CDE-II* 80–90 bp, >90% A + T *CDE-III*

DNA. It is not clear how much of each of these rather long regions is required for chromosome segregation at mitosis and meiosis.

The significance of the difference between the short centromeric regions in *S. cerevisiae* and the long regions in *S. pombe* is not clear. The common feature is that the DNA consists of noncoding sequences that are repetitive.

Attempts to localize centromeric functions in *Drosophila* chromosomes suggest that they are dispersed in a large region, consisting of 200–600 kb. The large size of this type of centromere suggests that it is likely to contain several separate specialized functions, including sequences required for kinetochore assembly, sister chromatid pairing, etc.

The primary motif comprising the heterochromatin of primate centromeres is the α satellite DNA, which consists of tandem arrays of a 170 bp repeating unit. There is significant variation between individual repeats, although those at any centromere tend to be better related to one another than to members of the family in other locations. It is clear that the sequences required for centromeric function reside within the blocks of α satellite DNA, but it is not clear whether the α satellite sequences themselves provide this function, or whether other sequences are embedded within the α satellite arrays.

Telomeres seal the ends of chromosomes

ANOTHER essential feature in all chromosomes is the **telomere,** which "seals" the end. We know that the telomere must be a special structure, because chromosome ends generated by breakage are "sticky" and tend to react with other chromosomes, whereas natural ends are stable.

We can apply two criteria in identifying a telomeric sequence:

- It must lie at the end of a chromosome (or, at least, at the end of an authentic linear DNA molecule).
- It must confer stability on a linear molecule.

Several telomeric sequences have been obtained from linear DNA molecules present in the genomes of lower eukaryotes. The same type of sequence is found in plants and man, so the construction of the telomere seems to follow a universal principle. Each telomere consists of a long series of short, tandemly repeated sequences. All such telomeric sequences can be written in the general form $C_n(A/T)_m$, where $n > 1$ and m is 1–4.

Within the telomeric region is a specific array of discontinuities, taking the form of single-strand breaks whose structure prevents them from being sealed by the ligase enzyme that normally acts upon nicks in one DNA strand. The very terminal bases are blocked in some way—they may be organized in a hairpin so that they are not recognized by nucleases.

The problem of finding a system that offers an assay for function again has been brought to the molecular level by using yeast. All the plasmids that survive in yeast (by virtue of possessing *ARS* and *CEN* elements) are circular DNA molecules. Linear plasmids are unstable (because they are degraded). Could an authentic telomeric DNA sequence confer stability on a linear plasmid?

Fragments from yeast DNA that prove to be located at chromosome ends can be identified by such an assay. And a region from the end of a known natural linear DNA molecule—the extrachromosomal rDNA of *Tetrahymena*—is able to render a yeast plasmid stable in linear form. The nicks in the telomeric sequence are perpetuated at the same sites in yeast, a remarkable interspecies conservation.

Some indications about how a telomere functions are given by some unusual properties of the ends of linear DNA molecules. In a trypanosome population, the ends are variable in length. When an individual cell clone is followed, the telomere grows longer by 7–10 bp (1–2 repeats) per generation. Even more revealing is the fate of ciliate telomeres introduced into yeast. After replication in yeast, *yeast telomeric repeats are added onto the ends of the* Tetrahymena *repeats.*

Addition of telomeric repeats to the end of the chromosome in every replication cycle could solve the

Figure 18.21 Telomerase positions itself by base pairing between the RNA template and the protruding single-stranded DNA primer. It adds G and T bases one at a time to the primer, as directed by the template. The cycle starts again when one repeating unit has been added.

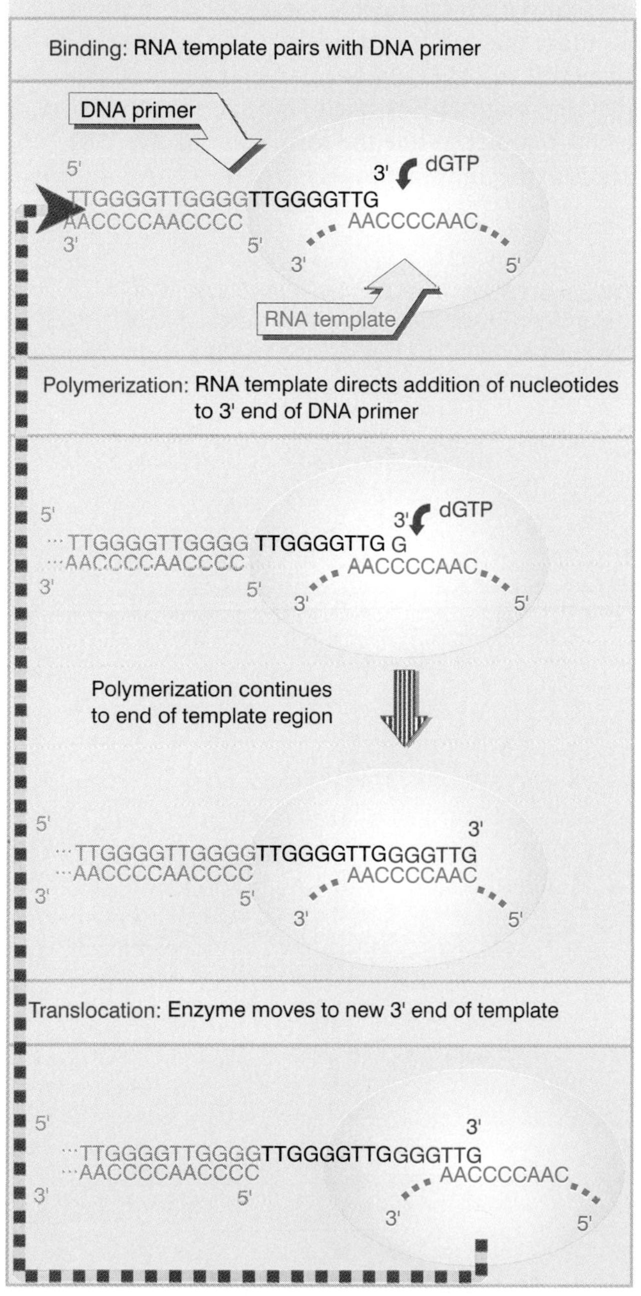

Figure 18.22 The unusual behavior of telomeric fractions may be explained by G-G interactions. In the upper model a duplex hairpin is formed by G-G pairing. In the lower model, a G quartet is formed when 1 G is contributed by each of 4 repeating units.

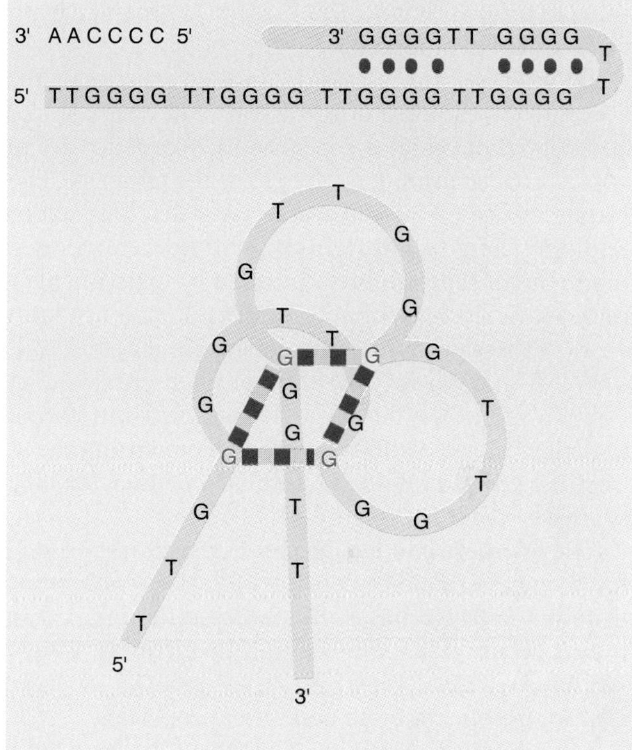

problem of replicating linear DNA molecules discussed in Chapter 12. The addition of repeats by *de novo* synthesis would counteract the loss of repeats resulting from failure to replicate up to the end of the chromosome. Extension and shortening would be in dynamic equilibrium.

If telomeres are continually being lengthened (and shortened), their exact sequence may be irrelevant. All that is required is for the end to be recognized as a suitable substrate for addition. This explains how the ciliate telomere functions in yeast.

How are the telomeric repeats synthesized? Extracts of *Tetrahymena* contain an enzyme, called **telomerase,** which uses the 3′-OH of the G+T telomeric strand as a primer for synthesis of tandem TTGGGG repeats. Only dGTP and dTTP are needed for the activity. The telomerase is a large ribonucleoprotein. It contains a short RNA component, 159 bases long in *Tetrahymena*, 192 bases long in *Euplotes*. Each RNA includes a sequence of 15–22 bases that is identical to two repeats of the C-rich repeating sequence. This RNA provides the template for synthesizing the G-rich repeating sequence. Bases are added individually, in the correct sequence, as depicted in **Figure 18.21.** The enzyme progresses discontinuously: the template RNA is positioned on the DNA primer, several nucleotides are added to the primer, and then the enzyme translocates to begin again. The telomerase is a specialized example of a reverse transcriptase, an enzyme that synthesizes a DNA sequence using an RNA template (see Chapter 16). The protein component provides the catalytic activity of reverse transcriptase, and is (presumably) confined to acting upon the RNA template provided by the nucleic acid component.

Telomerase synthesizes the individual repeats that are added to the chromosome ends, but does not itself control the number of repeats. Other proteins are involved in determining the length of the telomere. They can be identified by mutants in yeast that have altered telomere lengths. Proteins that bind telomeres in mammalian cells can be implicated by experiments to increase or decrease their amounts. We do not know yet how these proteins control telomere length.

We do not know how the complementary (C-A-rich) strand of the telomere is assembled, but we may speculate that it could be synthesized by using the 3′-OH of a terminal G-T hairpin as a primer for DNA synthesis (see top of Figure 18.22).

The telomere end has several unusual properties. It has a single-stranded extension of the G-T-rich strand, usually for 14–16 bases. But isolated telomeric fragments do not behave as though they contain single-stranded DNA; instead they show aberrant electrophoretic mobility and other properties.

A model for the structure of the end is depicted in **Figure 18.22**. It proposes the existence of a "quartet" of G residues, formed by an association of one G from each repeating unit. In the example in the figure, the second G of each of four successive T_2G_4 units forms a member of the quartet. The rest of the repeating unit is looped out. The association between the G residues requires that two of them change the orientation of the base with regard to the sugar (from the usual *anti* to the unusual *syn* configuration). Since each repeating unit has more than one G, more than one quartet could be formed if other G residues associate, in which case quartets might be stacked upon one another in a helical manner. While the formation of this structure attests to the unusual properties of the G-rich sequence

Figure 18.23 A loop forms at the end of chromosomal DNA. Photograph kindly provided by Jack Griffith.

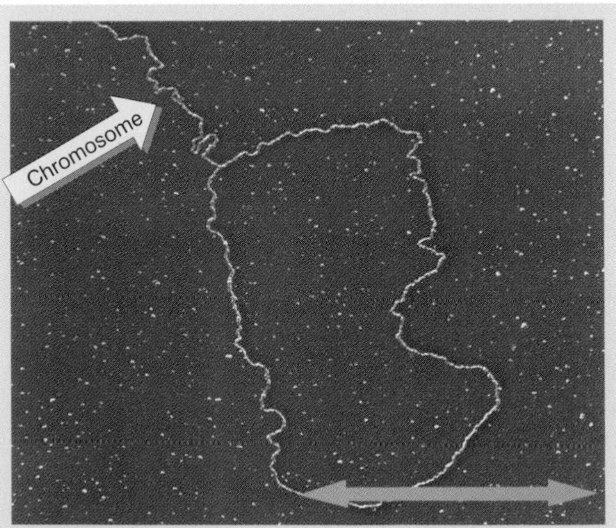

Figure 18.24 The 3' single-stranded end of the telomere $(TTAGGG)_n$ displaces the homologous repeats from duplex DNA to form a t-loop. The reaction is catalyzed by TRF2.

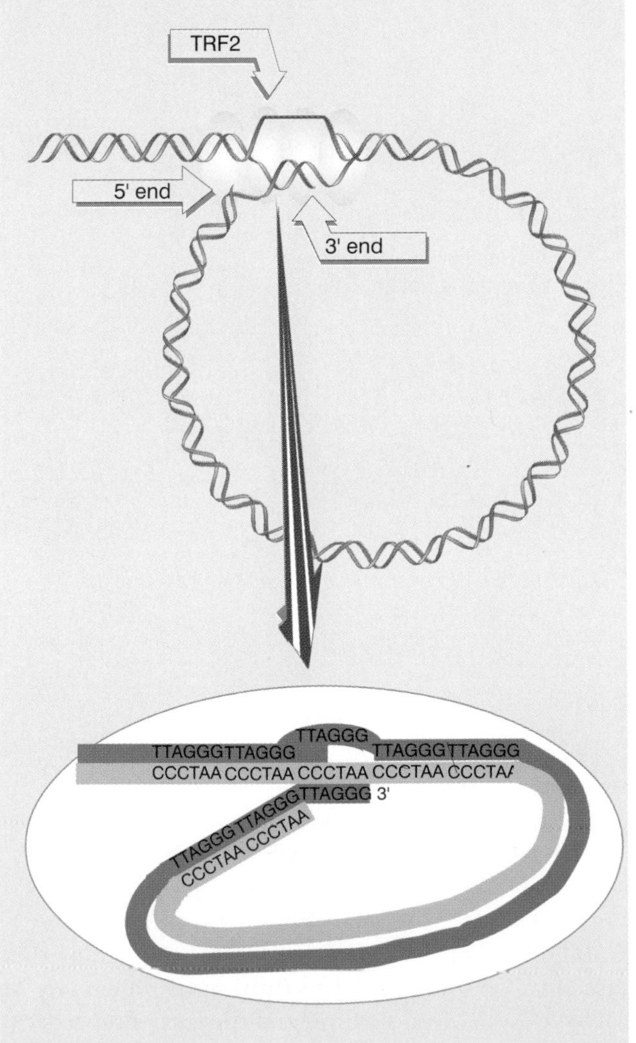

in vitro, it does not of course demonstrate whether the quartet forms *in vivo*.

What feature of the centromere is responsible for the stability of the chromosome end? **Figure 18.23** shows that a loop of DNA forms at the telomere. The absence of any free end may be the crucial feature that stabilizes the end of the chromosome. The average length of the loop in animal cells is 5–10 kb.

Figure 18.24 shows that the loop is formed when the 3′ single-stranded end of the telomere $(TTAGGG)_n$ displaces the same sequence in an upstream region of the telomere. This converts the duplex region into a structure like a D-loop, where a series of TTAGGG repeats are displaced to form a single-stranded region, and the tail of the telomere is paired with the homologous strand.

The reaction is catalyzed by the telomere-binding protein TRF2, which together with other proteins forms a complex that stabilizes the chromosome ends.

The minimum features required for existence as a chromosome are:

- Telomeres to ensure survival.
- A centromere to support segregation.
- An origin to initiate replication (see Chapter 12).

All of these elements have been put together to construct a yeast artificial chromosome (YAC). This is a useful method for perpetuating foreign sequences. It turns out that the synthetic chromosome is stable only if it is longer than 20–50 kb. We do not know the basis for this effect, but the ability to construct a synthetic chromosome offers the potential to investigate the nature of the segregation device in a controlled environment.

Summary

The genetic material of all organisms and viruses takes the form of tightly packaged nucleoprotein. Some virus genomes are inserted into preformed virions, while others assemble a protein coat around the nucleic acid. The bacterial genome forms a dense nucleoid, with about 20% protein by mass, but details of the interaction of the proteins with DNA are not known. The DNA is organized into ~100 domains that maintain independent supercoiling, with a density of unrestrained supercoils corresponding to ~1 per 100–200 bp. Interphase chromatin and metaphase chromosomes both appear to be organized into large loops. Each loop may be an independently supercoiled domain. The bases of the loops are connected to a metaphase scaffold or to the nuclear matrix by specific DNA sites.

Transcriptionally active sequences reside within the euchromatin that comprises the majority of interphase chromatin. The regions of heterochromatin are packaged ~5–10× more compactly, and are transcriptionally inert. All chromatin becomes densely packaged during cell division, when the individual chromosomes can be distinguished. The existence of a reproducible ultrastructure in chromosomes is indicated by the production of G-bands by treatment with Giemsa stain. The bands are very large regions, ~10^7 bp, that can be used to map chromosomal translocations or other large changes in structure.

Lampbrush chromosomes of amphibians and polytene chromosomes of insects have unusually extended structures, with packing ratios <100. Polytene chromosomes of *D. melanogaster* are divided into ~5000 bands, varying in size by an order of magnitude, with an average of ~25 kb. Transcriptionally active regions can be visualized in even more unfolded ("puffed") structures, in which material is extruded from the axis of the chromosome. This may resemble the changes that occur on a smaller scale when a sequence in euchromatin is transcribed.

The centromeric region contains the kinetochore, which is responsible for attaching a chromosome to the mitotic spindle. The centromere often is surrounded by heterochromatin. Centromeric sequences have been identified only in yeast, where they consist of short conserved elements and a long A•T-rich region. Proteins that bind to these sequences have been identified.

Telomeres make the ends of chromosomes stable. Almost all known telomeres consist of multiple repeats in which one strand has the general sequence $C_n(A/T)_m$, where $n > 1$ and $m = 1–4$. The other strand, $G_n(T/A)_m$, has a single protruding end that provides a template for addition of individual bases in defined order. The enzyme telomere transferase is a ribonucleoprotein, whose RNA component provides the template for synthesizing the G-rich strand. This overcomes the problem of the inability to replicate at the very end of a duplex. The telomere stabilizes the chromosome end because the overhanging single strand $G_n(T/A)_m$ displaces its homologue in earlier repeating units in the telomere to form a loop, so there are no free ends.

Further reading

Reviews

Blackburn, E. H. and Szostak, J. W. (1984). The molecular structure of centromeres and telomeres. *Ann. Rev. Biochem.* **53**, 163–194.

Clarke, L. and Carbon, J. (1985). The structure and function of yeast centromeres. *Ann. Rev. Genet.* **19**, 29–56.

Drlica, K. and Rouviere-Yaniv, J. (1987). Histone-like proteins of bacteria. *Microbiol. Rev.* **51**, 301–319.

Brock, T. D. (1988). The bacterial nucleus: a history. *Microbiol. Rev.* **52**, 397–411.

Black, L. W. (1989). DNA packaging in dsDNA bacteriophages. *Ann. Rev. Microbiol.* **43**, 267–292.

Zakian, V. A. (1989). Structure and function of telomeres. *Ann. Rev. Genet.* **23**, 579–604.

Schulman, I. and Bloom, K. S. (1991). Centromeres: an integrated protein/DNA complex required for chromosome movement. *Ann. Rev. Cell Biol.* **7**, 311–336.

Blackburn, E. H. (1991). Structure and function of telomeres. *Nature* **350**, 569–573.

Blackburn, E. H. (1992). Telomerases. *Ann. Rev. Biochem.* **61**, 113–129.

Hyman, A. A. and Sorger, P. K. (1995). Structure and function of kinetochores in budding yeast. *Ann. Rev. Cell Biol.* **11**, 471–495.

Zakian, V. A. (1995). Telomeres: beginning to understand the end. *Science* **270**, 1601–1607.

Zakian, V. A. (1996). Structure, function, and replication of *S. cerevisiae* telomeres. *Ann. Rev. Genet.* **30**, 141–172.

Discoveries

Viruses

Fraenkel-Conrat, H. and Williams, R. C. (1955). Reconstitution of active tobacco mosaic virus from its inactive protein and nucleic acid components. *Proc. Nat. Acad. Sci. USA* **41**, 690–698.

Caspar, D. L. D. and Klug, A. (1962). Physical principles in the construction of regular viruses. *Cold Spring Harbor Quant. Biol.* **27**, 1–24.

Centromeres

Bloom, K. S. and Carbon, J. (1982). Yeast centromere DNA is in a unique and highly ordered structure in chromosomes and small circular minichromosomes. *Cell* **29**, 305–317.

Murray, A. and Szostak, J. W. (1983). Construction of artificial chromosomes in yeast. *Nature* **305**, 189–193.

Telomeres

Greider, C. and Blackburn, E. H. (1987). The telomere terminal transferase of *Tetrahymena* is a ribonucleoprotein enzyme with two kinds of primer specificity. *Cell* **51**, 887–898.

Henderson, E., Hardin, C. H., Walk, S. K., Tinoco, I., and Blackburn, E. H. (1987). Telomeric oligonucleotides form novel intramolecular structures containing guanine-guanine base pairs. *Cell* **51**, 899–908.

Williamson, J. R., Raghuraman, K. R., and Cech, T. R. (1989). Monovalent cation-induced structure of telomeric DNA: the G-quartet model. *Cell* **59**, 871–880.

Shippen-Lentz, D. and Blackburn, E. H. (1990) Functional evidence for an RNA template in telomerase. *Science* **247**, 546–552.

J. D. Griffith *et al.* (1999). Mammalian telomeres end in a large duplex loop. *Cell* **97**, 503–514.

Chapter 19

Nucleosomes

CHROMATIN has a compact organization in which most DNA sequences are structurally inaccessible and functionally inactive. Within this mass is the minority of active sequences. What is the general structure of chromatin, and what is the difference between active and inactive sequences? The high overall packing ratio of the genetic material immediately suggests that DNA cannot be directly packaged into the final structure of chromatin. There must be *hierarchies* of organization.

The fundamental subunit of chromatin has the same type of design in all eukaryotes. The **nucleosome** contains ~200 bp of DNA, organized by an octamer of small, basic proteins into a bead-like structure. The protein components are **histones.** They form an interior core; the DNA lies on the surface of the particle. Nucleosomes are an invariant component of euchromatin and heterochromatin in the interphase nucleus, and of mitotic chromosomes. The nucleosome provides the first level of organization, giving a packing ratio of ~6. Its components and structure are well characterized.

The second level of organization is the coiling of the series of nucleosomes into a helical array to constitute the ~30 nm fiber that is found in both interphase chromatin and mitotic chromosomes (see Figure 18.9). In chromatin this brings the packing ratio of DNA to ~40. The structure of this fiber requires additional proteins, but is not well defined.

The final packing ratio is determined by the third level of organization, the packaging of the fiber itself. This gives an overall packing ratio of ~1000 in euchromatin, cyclically interchangeable with packing into mitotic chromosomes to achieve an overall ratio of ~10,000. Heterochromatin generally has a packing ratio ~10,000 in both interphase and mitosis.

We need to work through these levels of organization to characterize the events involved in cyclical packaging, replication, and transcription. We assume that association with additional proteins, or modifications of existing chromosomal proteins, are involved in changing the structure of chromatin. We do not know the individual targets for controlling cyclical packaging. Both replication and transcription require unwinding of DNA, and thus must involve an unfolding of the structure that allows the relevant enzymes to manipulate the DNA. This is likely to involve changes in all levels of organization.

When chromatin is replicated, the nucleosomes must be reproduced on both daughter duplex molecules. As well as asking how the nucleosome itself is assembled, we must inquire what happens to other proteins present in chromatin. Since replication disrupts the structure of chromatin, it both poses a problem for maintaining regions with specific structure and offers an opportunity to change the structure.

The mass of chromatin contains up to twice as much protein as DNA. Approximately half of the protein mass is accounted for by the nucleosomes. The mass of RNA is less than 10% of the mass of DNA. Much of the

RNA consists of nascent transcripts still associated with the template DNA.

As their dysphonious name suggests, the **nonhistones** include all the proteins of chromatin except the histones. They are presumed to be more variable between tissues and species—although good evidence still is lacking on the extent of the variability—and they comprise a smaller proportion of the mass than the histones. They also comprise a much larger number of proteins, so that any individual protein is present in amounts much smaller than any histone.

The functions of nonhistone proteins include control of gene expression and higher-order structure. So RNA polymerase may be considered to be a prominent nonhistone. The HMG (high-mobility group) proteins comprise a discrete and well-defined subclass of nonhistones (at least some of which are transcription factors). A major problem in working with other nonhistones is that they tend to be contaminated with other nuclear proteins, and so far it has proved difficult to obtain those nonhistone proteins responsible for higher-order structures.

The nucleosome is the subunit of all chromatin

When interphase nuclei are suspended in a solution of low ionic strength, they swell and rupture to release fibers of chromatin. **Figure 19.1** shows a lysed nucleus in which fibers are streaming out. In some regions, the fibers consist of tightly packed material, but in regions that have become stretched, they can be seen to consist of discrete particles. These are the nucleosomes. In especially extended regions, individual nucleosomes are connected by a fine thread, a free duplex of DNA. *A continuous duplex thread of DNA runs through the series of particles.*

Individual nucleosomes can be obtained by treating chromatin with the endonuclease **micrococcal nuclease.** It cuts the DNA thread at the junction between nucleosomes. First, it releases groups of particles; finally, it releases single nucleosomes. Individual nucleosomes can be seen in **Figure 19.2** as compact particles. They sediment at ~11S.

The nucleosome contains ~200 bp of DNA associated with a histone octamer that consists of two copies each of H2A, H2B, H3, and H4. These are known as the **core histones.** Their association is illustrated diagrammatically in **Figure 19.3.** This model explains the stoichiometry of the core histones in chromatin: H2A, H2B, H3, and H4 are present in equimolar amounts, with 2 molecules of each per ~200 bp of DNA.

Histones H3 and H4 are among the most conserved proteins known. This suggests that their functions are identical in all eukaryotes. The types of H2A and H2B can be recognized in all eukaryotes, but show appreciable species-specific variation in sequence.

Histone H1 comprises a set of closely related proteins that show appreciable variation between tissues and between species (and are absent from yeast). The role of H1 is different from the core histones. It is present in half the amount of a core histone and can be extracted more readily from chromatin (typically with dilute salt [0.5 M] solution). *All of the H1 can be removed without affecting the structure of the nucleosome, which suggests that its location is external to the particle.*

The shape of the nucleosome corresponds to a flat disk or cylinder, of diameter 11 nm and height 6 nm. The length of the DNA is roughly twice the ~34 nm circumference of the particle. Immediately these dimensions suggest that DNA could not be squashed within the particle, but must lie on the outside.

The DNA follows a symmetrical path around the octamer. **Figure 19.4** shows the DNA path diagrammatically as a helical coil that makes two turns around the cylindrical octamer. Note that the DNA "enters" and "leaves" the nucleosome at points close to one another. Histone H1 may be located in this region (see later).

Considering this model in terms of a cross-section through the nucleosome, in **Figure 19.5** we see that the two circumferences made by the DNA lie close to one another. The height of the cylinder is 6 nm, of which 4 nm is occupied by the two turns of DNA (each of diameter 2 nm).

The pattern of the two turns has a possible functional consequence. Since one turn around the nucleosome takes ~80 bp of DNA, two points separated by 80 bp in the free double helix may actually be close on the nucleosome surface, as illustrated in **Figure 19.6.**

Figure 19.1 Chromatin spilling out of lysed nuclei consists of a compactly organized series of particles. The bar is 100 nm. Photograph kindly provided by Pierre Chambon.

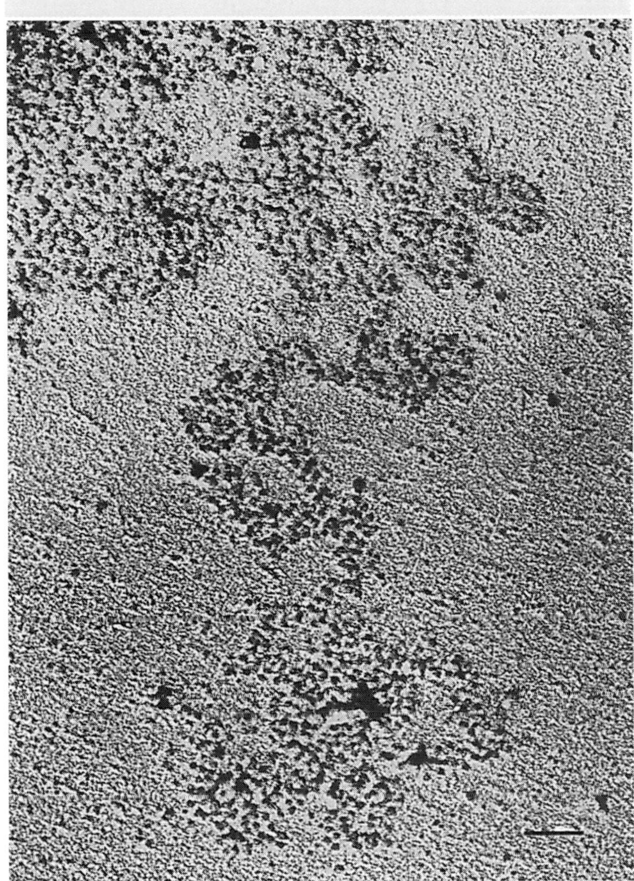

Figure 19.2 Individual nucleosomes are released by digestion of chromatin with micrococcal nuclease. The bar is 100 nm. Photograph kindly provided by Pierre Chambon.

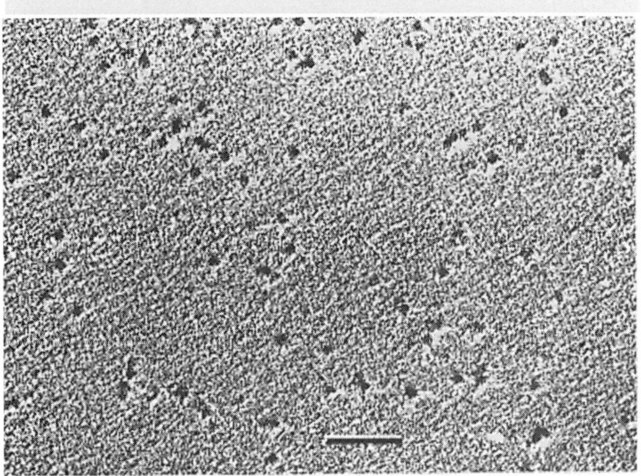

Figure 19.3 The nucleosome consists of approximately equal masses of DNA and histones (including H1). The predicted mass of the nucleosome is 262 kD.

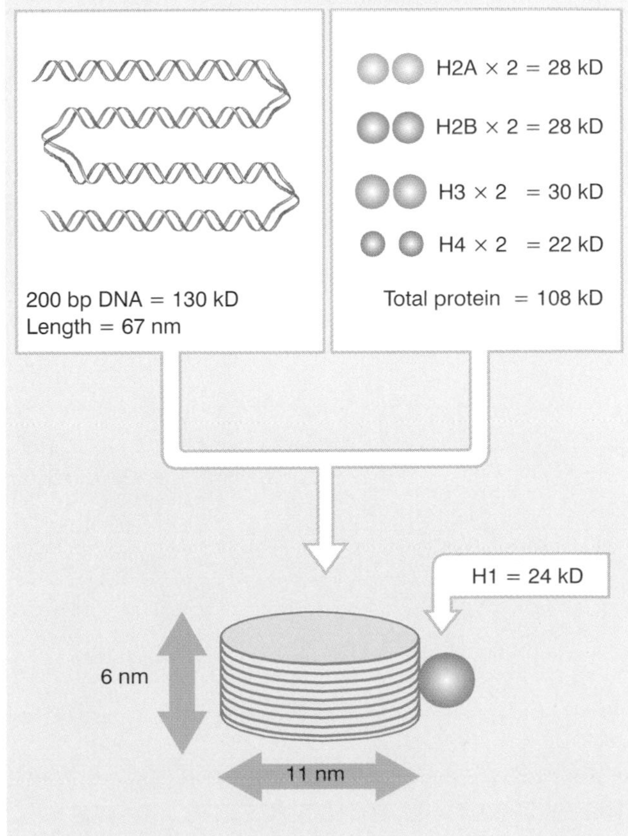

Figure 19.4 The nucleosome may be a cylinder with DNA organized into two turns around the surface.

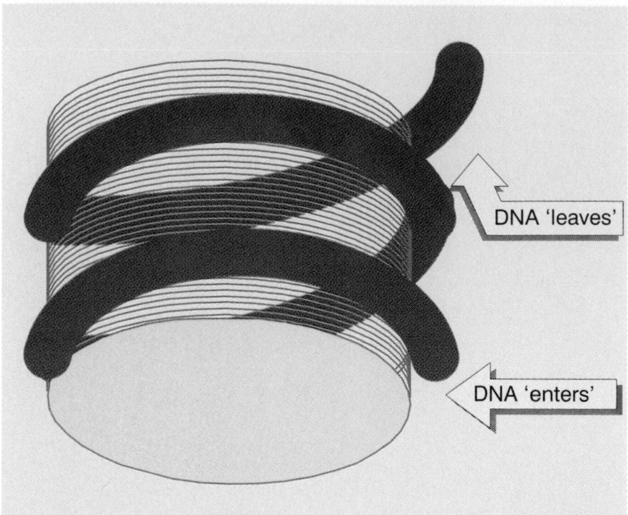
DNA 'leaves'
DNA 'enters'

Figure 19.5 The two turns of DNA on the nucleosome lie close together.

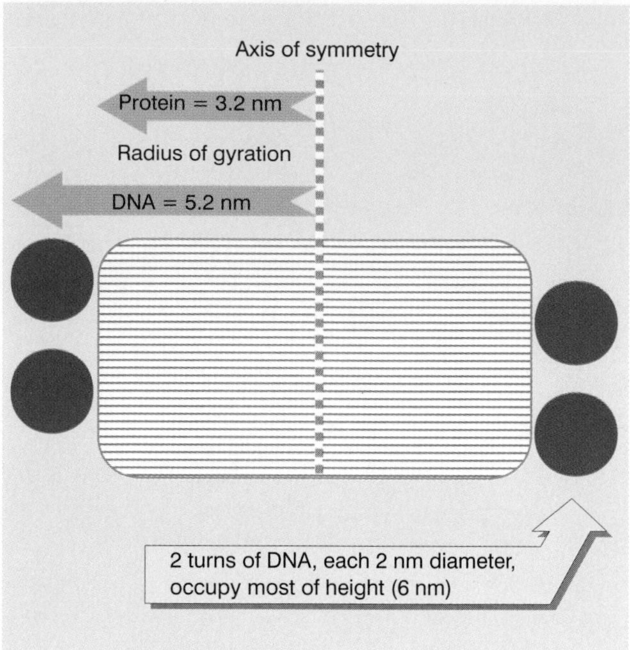
Axis of symmetry
Protein = 3.2 nm
Radius of gyration
DNA = 5.2 nm
2 turns of DNA, each 2 nm diameter, occupy most of height (6 nm)

Figure 19.6 Sequences on the DNA that lie on different turns around the nucleosome may be close together.

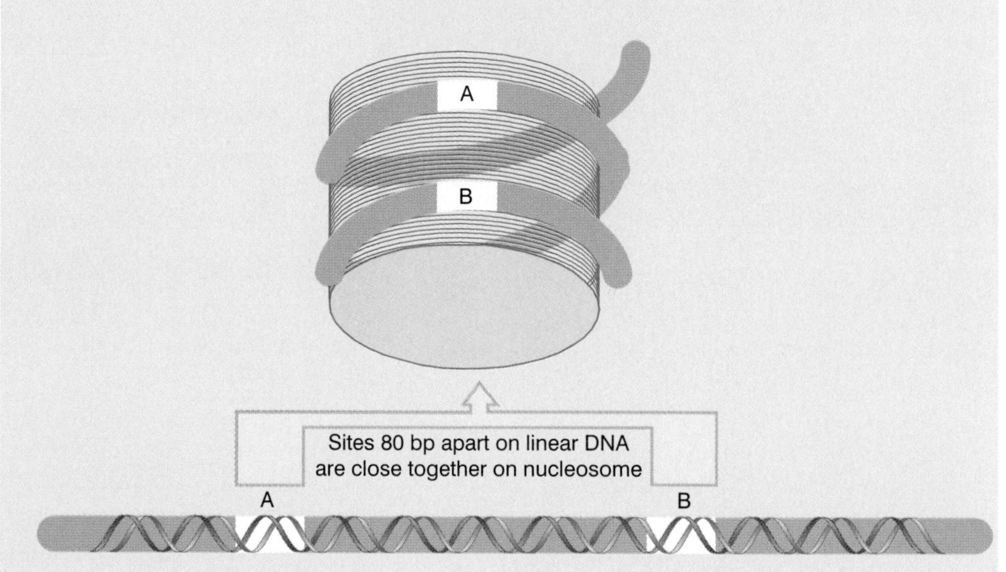
A
B
Sites 80 bp apart on linear DNA are close together on nucleosome
A
B

DNA is coiled in arrays of nucleosomes

When chromatin is digested with the enzyme micrococcal nuclease, the DNA is cleaved into integral multiples of a unit length. Fractionation by gel electrophoresis reveals the "ladder" presented in **Figure 19.7**. Such ladders extend for ~10 steps, and the unit length, determined by the increments between successive steps, is ~200 bp.

Figure 19.8 shows that the ladder is generated by groups of nucleosomes. When nucleosomes are fractionated on a sucrose gradient, they give a series of discrete peaks that correspond to monomers, dimers, trimers, etc. When the DNA is extracted from the individual fractions and electrophoresed, each fraction yields a band of DNA whose size corresponds with a step on the micrococcal nuclease ladder. The monomeric nucleosome contains DNA of the unit length, the nucleosome dimer contains DNA of twice the unit length, and so on.

So each step on the ladder represents the DNA derived from a discrete number of nucleosomes. *We therefore take the existence of the 200 bp ladder in any chromatin to indicate that the DNA is organized into nucleosomes.* The micrococcal ladder is generated when only ~2% of the DNA in the nucleus is rendered acid-soluble (degraded to small fragments) by the enzyme. *So a small proportion of the DNA is specifically attacked; it must represent especially susceptible regions.*

Figure 19.7 Micrococcal nuclease digests chromatin in nuclei into a multimeric series of DNA bands that can be separated by gel electrophoresis. Photograph kindly provided by Markus Noll.

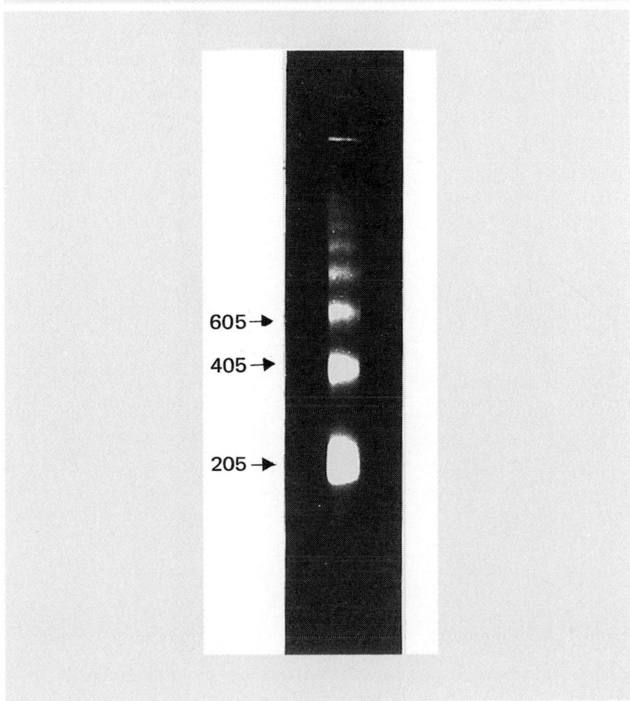

When chromatin is spilled out of nuclei, we often see a series of nucleosomes connected by a thread of free DNA (the beads on a string). However, the need for tight packaging of DNA *in vivo* suggests that probably there is usually little (if any) free DNA.

This view is confirmed by the fact that >*90% of the DNA of chromatin can be recovered in the form of the 200 bp ladder.* Almost all DNA must therefore be organized in nucleosomes. In their natural state, nucleosomes are likely to be closely packed, with DNA passing directly from one to the next. Free DNA is probably generated by the loss of some histone octamers during isolation.

The length of DNA present in the nucleosome varies somewhat from the "typical" value of 200 bp. The chromatin of any particular cell type has a characteristic average value (±5 bp). The average most often is between 180 and 200, but there are extremes as low as 154 bp (in a fungus) or as high as 260 bp (in a sea urchin sperm). The average value may be different in individual tissues of the adult organism. And there can be differences between different parts of the genome in a single cell type. Variations from the genome average include tandemly repeated sequences, such as clusters of 5S RNA genes.

A common structure underlies the varying amount of DNA that is contained in nucleosomes of different sources. The association of DNA with the histone octamer forms a **core particle** containing 146 bp of DNA, irrespective of the total length of DNA in the nucleosome. The variation in total length of DNA per nucleosome is superimposed on this basic core structure.

The core particle is defined by the effects of micrococcal nuclease on the nucleosome monomer. The initial reaction of the enzyme is to cut between nucleosomes, but if it is allowed to continue after monomers have been generated, then it proceeds to digest some of the DNA of the individual nucleosome. This occurs by a reaction in which DNA is "trimmed" from the ends of the nucleosome.

Figure 19.8 Each multimer of nucleosomes contains the appropriate number of unit lengths of DNA. Photograph kindly provided by John Finch.

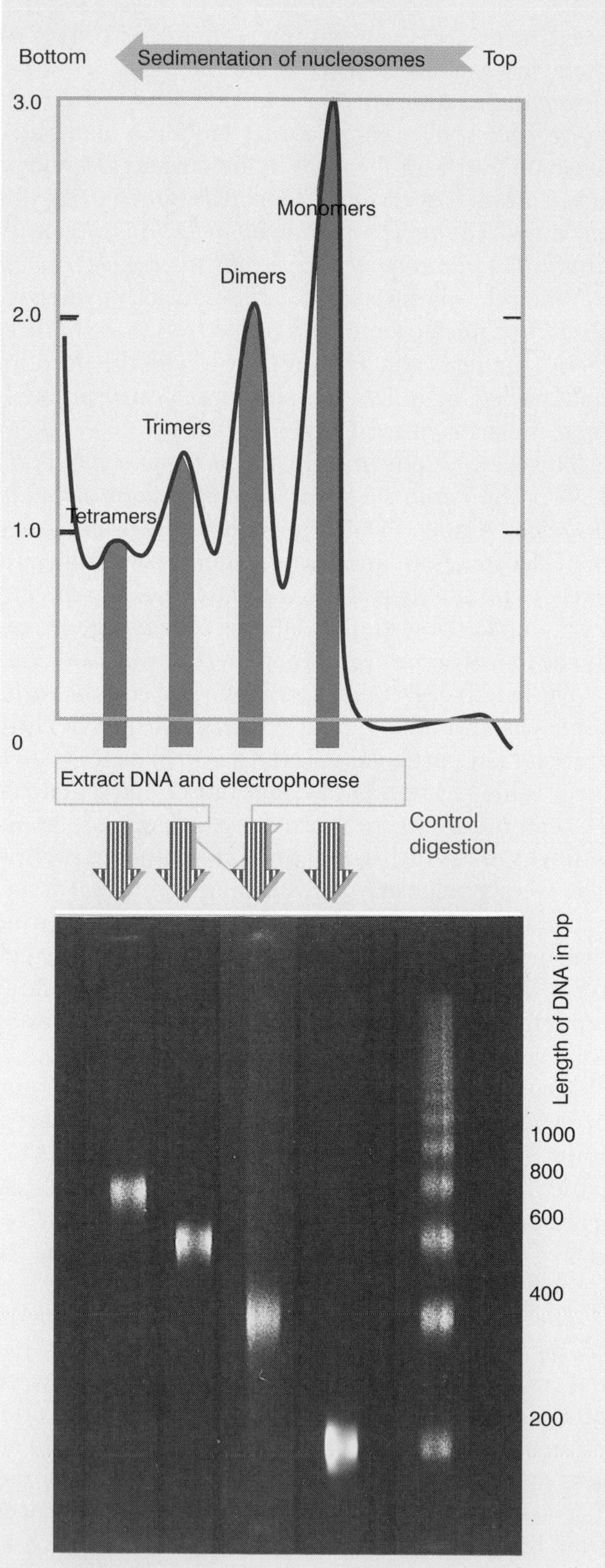

The length of the DNA is reduced in discrete steps, as shown in **Figure 19.9.** With rat liver nuclei, the nucleosome monomers initially have 205 bp of DNA. Then some monomers are found in which the length of DNA has been reduced to ~165 bp. Finally this is reduced to the length of the DNA of the core particle, 146 bp. (The core is reasonably stable, but continued digestion generates a **limit digest,** in which the longest fragments are the 146 bp DNA of the core, while the shortest are as small as 20 bp.)

This analysis suggests that the nucleosomal DNA can be divided into two regions:

- **Core DNA** has an invariant length of 146 bp, and is relatively resistant to digestion by nucleases.
- **Linker DNA** comprises the rest of the repeating unit. Its length varies from as little as 8 bp to as much as 114 bp per nucleosome.

The sharp size of the band of DNA generated by the initial cleavage with micrococcal nuclease suggests that the region immediately available to the enzyme is restricted. It represents only part of each linker. (If the

Figure 19.9 Micrococcal nuclease reduces the length of nucleosome monomers in discrete steps. Photograph kindly provided by Roger Kornberg.

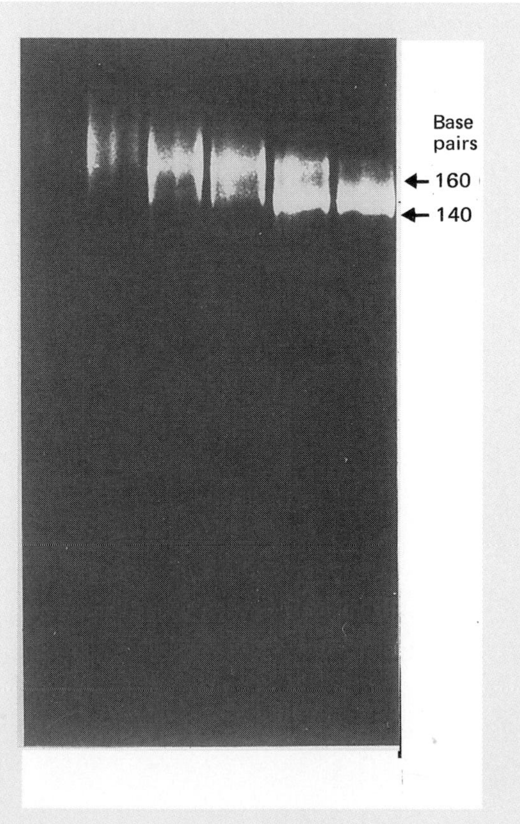

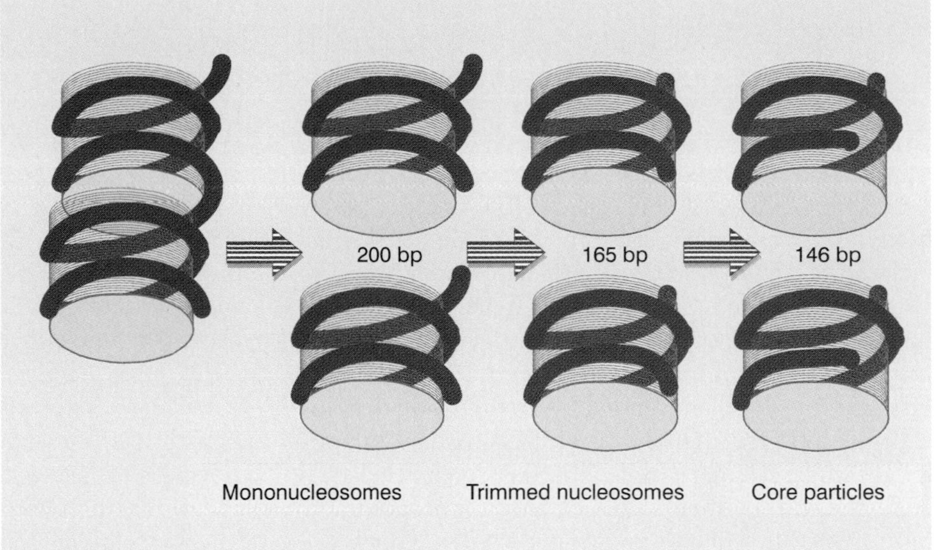

Figure 19.10 Micrococcal nuclease initially cleaves between nucleosomes. Mononucleosomes typically have ~200 bp DNA. End-trimming reduces the length of DNA first to ~165 bp, and then generates core particles with 146 bp.

entire linker DNA were susceptible, the band would range from 146 bp to >200 bp.) But once a cut has been made in the linker DNA, the rest of this region becomes susceptible, and it can be removed relatively rapidly by further enzyme action. The connection between nucleosomes is represented in **Figure 19.10.**

Core particles have properties similar to those of the nucleosomes themselves, although they are smaller. Their shape and size are similar to nucleosomes, which suggests that the essential geometry of the particle is established by the interactions between DNA and the protein octamer in the core particle. Because core particles are more readily obtained as a homogeneous population, they are often used for structural studies in preference to nucleosome preparations. (Nucleosomes tend to vary because it is difficult to obtain a preparation in which there has been no end-trimming of the DNA.)

What is the physical nature of the core and the linker regions? *These terms are operational definitions that describe the regions in terms of their relative susceptibility to nuclease treatment.* This description does not make any implication about their actual structure. In fact, the path of DNA on the histone octamer appears to be continuous. It takes 165 bp to make the two turns around the octamer. This is an invariant feature of nucleosomes. The transition from one nucleosome to the next is made within the additional length of DNA, and there could be differences in the path in this region depending on the length of DNA per nucleosome.

The existence of linker DNA depends on factors extraneous to the four core histones. Reconstitution experiments *in vitro* show that histones have an intrinsic ability to organize DNA into core particles, but do not form nucleosomes with the proper unit length. The degree of supercoiling of the DNA is an important factor. Histone H1 and/or nonhistone proteins influence the length of linker DNA associated with the histone octamer in a natural series of nucleosomes. And "assembly proteins" that are not part of the nucleosome structure are involved *in vivo* in constructing nucleosomes from histones and DNA (see later).

Where is histone H1 located? The H1 is lost during the degradation of nucleosome monomers. It can be retained on monomers that still have 165 bp of DNA; but is always lost with the final reduction to the 146 bp core particle. This suggests that H1 could be located in the region of the linker DNA immediately adjacent to the core DNA.

If H1 is located at the linker, it could "seal" the DNA in the nucleosome by binding at the point where the nucleic acid enters and leaves (see Figure 19.4). The idea that H1 lies in the region joining adjacent nucleosomes is consistent with old results that H1 is removed the most readily from chromatin, and that H1-depleted chromatin is more readily "solubilized". And it is easier to obtain a stretched-out fiber of beads on a string when the H1 has been removed.

DNA structure varies on the nucleosomal surface

THE exposure of DNA on the surface of the nucleosome explains why it is accessible to cleavage by certain nucleases. The reaction with nucleases that attack single strands has been especially informative. The enzymes DNAase I and DNAase II make single-strand nicks in DNA; they cleave a bond in one strand, but the other strand remains intact at this point. So no effect is visible in the double-stranded DNA. But upon denaturation, short fragments are released instead of full-length single strands. If the DNA has been labeled at its ends, the end fragments can be identified by autoradiography as summarized in **Figure 19.11**. (This is exactly analogous to the restriction mapping technique shown in Figure 2.4.)

When DNA is free in solution, it is nicked (relatively) at random. The DNA on nucleosomes also can be nicked by the enzymes, *but only at regular intervals.* When the points of cutting are determined by using radioactively end-labeled DNA and then DNA is denatured and electrophoresed, a ladder of the sort displayed in **Figure 19.12** is obtained.

The interval between successive steps on the ladder is 10–11 bases. The ladder extends for the full distance of core DNA. The cleavage sites are numbered as S1 through S13 (where S1 is ~10 bases from the labeled 5′ end, S2 is ~20 bases from it, and so on). Their positions relative to the DNA superhelix are illustrated in **Figure 19.13.**

Not all sites are cut with equal frequency: some are cut rather effectively, others are cut scarcely at all. The enzymes DNAase I and DNAase II generate the same ladder, although with some differences in the intensities of the bands. This shows that the pattern of cutting represents a unique series of targets in DNA, determined by its organization, with only some slight preference for particular sites imposed by the individual enzyme. The same cutting pattern is obtained by cleaving with a hydroxyl radical, which argues that the pattern reflects the structure of the DNA itself, rather than any sequence preference.

The sensitivity of nucleosomal DNA to nucleases is analogous to a footprinting experiment. So we can as-

Figure 19.11 Nicks in double-stranded DNA are revealed by fragments when the DNA is denatured to give single strands. If the DNA is labeled at (say) 5' ends, only the 5' fragments are visible by autoradiography. The size of the fragment identifies the distance of the nick from the labeled end.

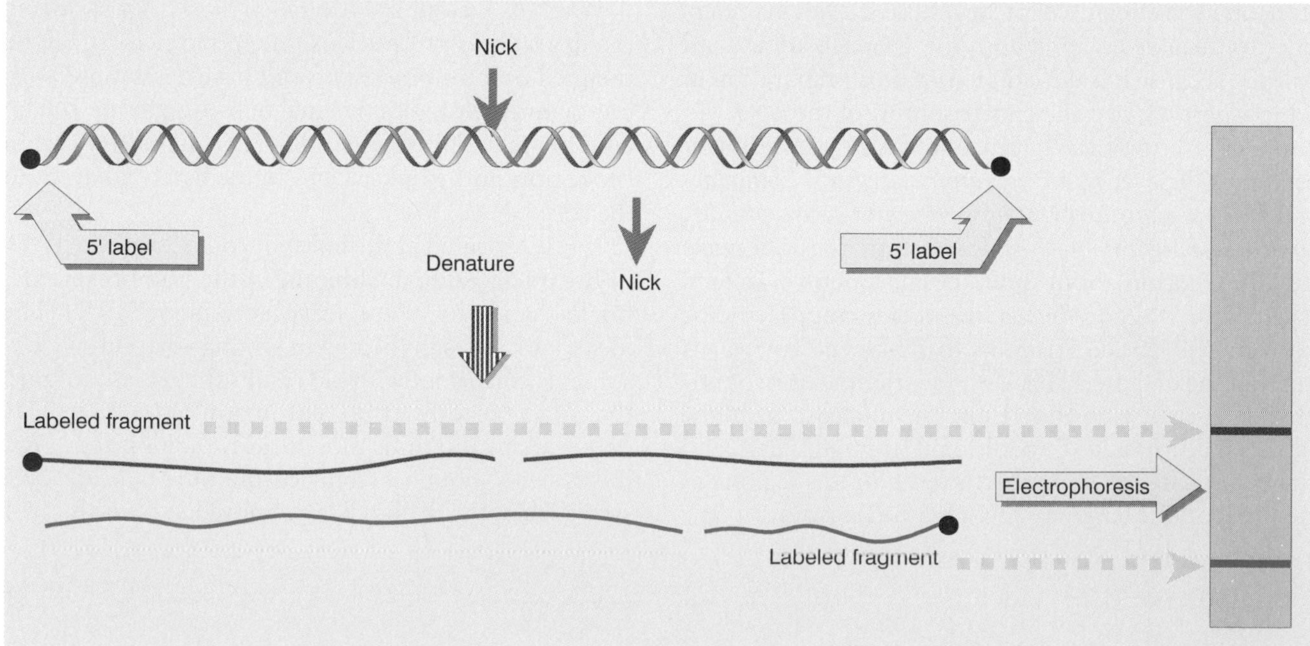

sign the lack of reaction at particular target sites to the structure of the nucleosome, in which certain positions on DNA are rendered inaccessible.

Since there are two strands of DNA in the core particle, in an end-labeling experiment both 5′ (or both 3′) ends are labeled, one on each strand. So the cutting pattern includes fragments derived from both strands. This is implied in Figure 19.11, where each labeled fragment is derived from a different strand. The corollary is that, in an experiment, each labeled band in fact represents two fragments, generated by cutting the *same* distance from *either* of the labeled ends.

How then should we interpret discrete preferences at particular sites? One view is that the path of DNA on the particle is symmetrical (about a horizontal axis through the nucleosome drawn in Figure 19.4). So if (for example) no 80-base fragment is generated by DNAase I, this must mean that the position at 80 bases from the 5′ end of *either* strand is not susceptible to the enzyme. The second numbering scheme used in Figure 19.13 reflects this view, and identifies S7 = site 0 as the center of symmetry.

When DNA is immobilized on a flat surface, sites are cut with a regular separation. **Figure 19.14** suggests that this reflects the recurrence of the exposed site with the helical periodicity of B-form DNA. The **cutting periodicity** (the spacing between cleavage points) coincides with, indeed, is a reflection of, the **structural periodicity** (the number of base pairs per turn of the double helix). So the distance between the sites corresponds to the number of base pairs per turn.

Figure 19.12 Sites for nicking lie at regular intervals along core DNA, as seen in a DNAase I digest of nuclei. Photograph kindly provided by Leonard Lutter.

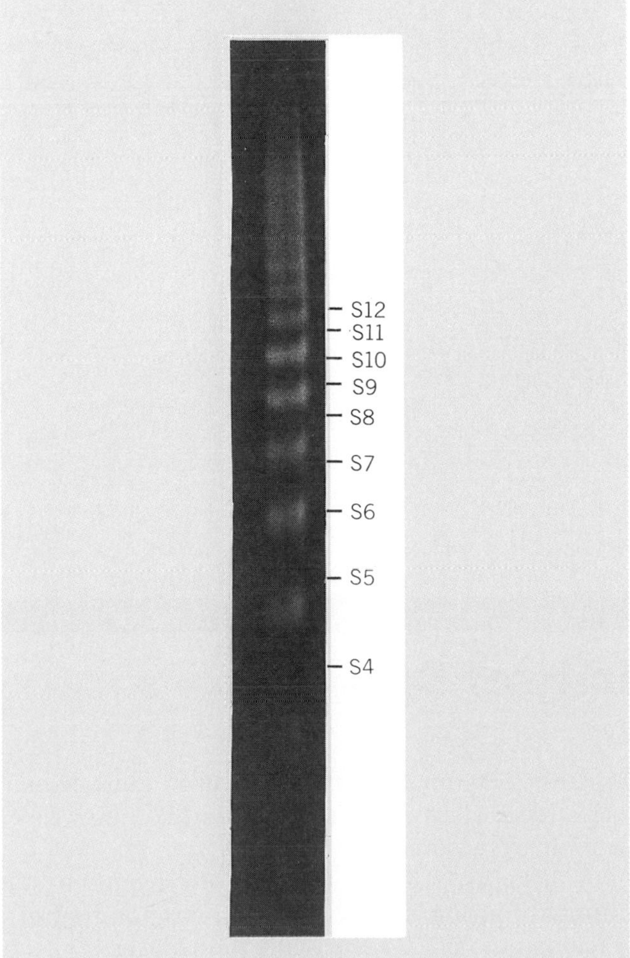

Figure 19.13 Two numbering schemes divide core particle DNA into 10 bp segments. Sites may be numbered S1 to S13 from one end; or taking S7 to identify coordinate 0 of the dyad symmetry, they may be numbered −7 to +7.

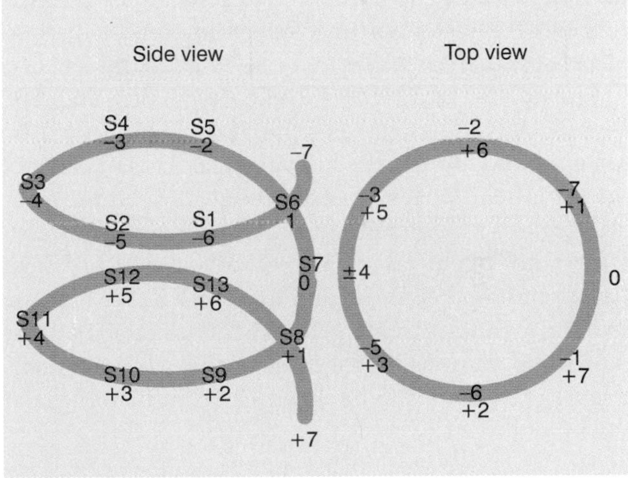

Figure 19.14 The most exposed positions on DNA recur with a periodicity that reflects the structure of the double helix. (For clarity, sites are shown for only one strand.)

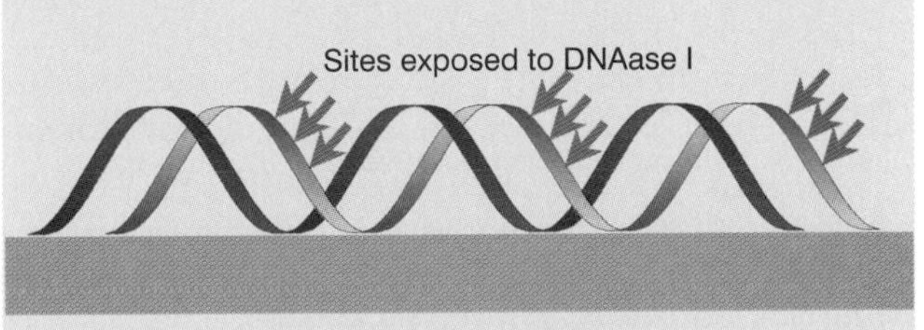

Measurements of this type suggest that the average value for double-helical B-type DNA is 10.5 bp/turn.

What is the nature of the target sites on the nucleosome? **Figure 19.15** shows that each site has 3–4 positions at which cutting occurs; that is, the cutting site is defined ±2 bp. So a cutting site represents a short stretch of bonds on both strands, exposed to nuclease action over 3–4 base pairs. The relative intensities indicate that some sites are preferred to others.

From this pattern, we can calculate the "average" point that is cut. At the ends of the DNA, pairs of sites from S1 to S4 or from S10 to S13 lie apart a distance of 10.0 bases each. In the center of the particle, the separation from sites S4 to S10 is 10.7 bases. (Because this analysis deals with *average* positions, sites need not lie an integral number of bases apart.)

The variation in cutting periodicity along the core DNA (10.0 at the ends, 10.7 in the middle) means that there is variation in the structural periodicity of core DNA. The DNA has more bp/turn than its solution value in the middle, but has fewer bp/turn at the ends. The average periodicity over the nucleosome is less than the 10.5 bp/turn of DNA in solution; it is in the range of 10.2–10.4 bp/turn, depending on the method of measurement.

The crystal structure of the core particle suggests that DNA is organized as a flat superhelix, with 1.65 turns wound around the histone octamer. The pitch of the superhelix varies, with a discontinuity in the middle. Regions of high curvature are arranged symmetrically, and occur at positions ±1 and ±4. These correspond to S6 and S8, and to S3 and S11, which are the sites least sensitive to DNAase I. The high curvature is probably responsible for these changes, but their precise nature remains to be determined at the molecular level.

Figure 19.15 High resolution analysis shows that each site for DNAase I consists of several adjacent susceptible phosphodiester bonds as seen in this example of sites S4 and S5 analyzed in end-labeled core particles. Photograph kindly provided by Leonard Lutter.

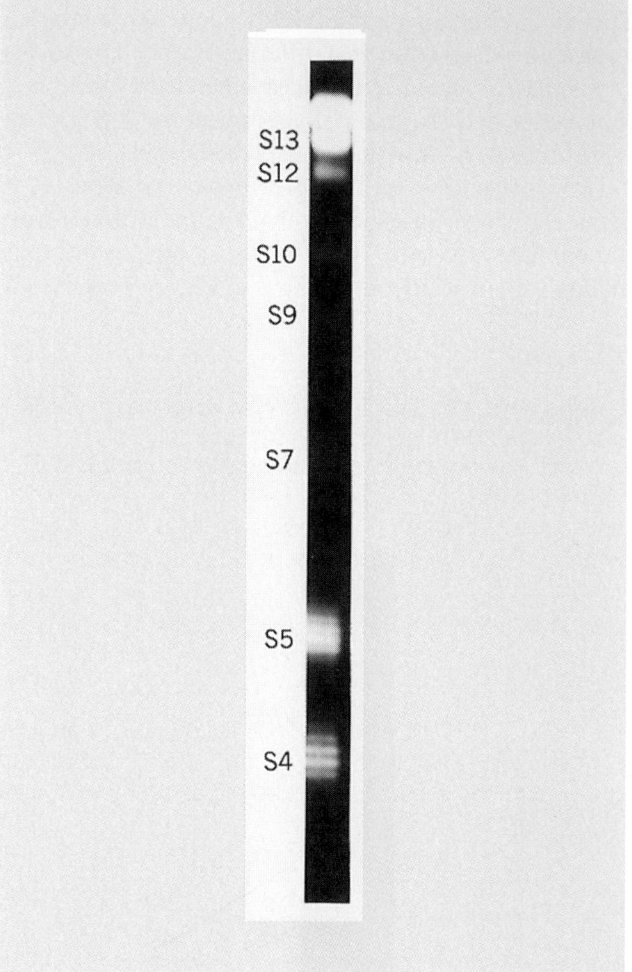

Supercoiling and the periodicity of DNA

Some insights into the structure of nucleosomal DNA emerge when we compare predictions for supercoiling in the path that DNA follows with actual measurements of supercoiling of nucleosomal DNA. Much work on the structure of sets of nucleosomes has been carried out with the virus SV40. The DNA of SV40 is a circular molecule of 5200 bp, with a contour length ~1500 nm. In both the virion and infected nucleus, it is packaged into a series of nucleosomes, called a **minichromosome.**

As usually isolated, the contour length of the minichromosome is ~210 nm, corresponding to a packing ratio of ~7 (essentially the same as the ~6 of the nucleosome itself). Changes in the salt concentration can convert it to a flexible string of beads with a much lower overall packing ratio. This emphasizes the point that nucleosome strings can take more than one form *in vitro,* depending on the conditions.

The degree of supercoiling on the individual nucleosomes of the minichromosome can be measured as illustrated in **Figure 19.16.** First, the free supercoils of the minichromosome itself are relaxed, so that the nucleosomes form a circular string with a superhelical density of 0. Then the histone octamers are extracted. This releases the DNA to follow a free path. Every supercoil that was present but restrained in the minichromosome will appear in the deproteinized DNA as –1 turn. So now the total number of supercoils in the SV40 DNA is measured.

The observed value is close to the number of nucleosomes. The reverse result is seen when nucleosomes are assembled *in vitro* on to a supercoiled SV40 DNA: the formation of each nucleosome removes ~1 negative supercoil.

So the DNA follows a path on the nucleosomal surface that generates ~1 negative supercoiled turn when the restraining protein is removed. But the path that DNA follows on the nucleosome corresponds to –1.65 superhelical turns (see Figure 19.4). This discrepancy is sometimes called the **linking number paradox.**

The discrepancy is at least partly explained by the difference between the 10.2 average bp/turn of nucleosomal DNA and the 10.5 bp/turn of free DNA. In a nucleosome of 200 bp, there are $200/10.2 \approx 19.6$ turns. When DNA is released from the nucleosome, it now has $200/10.5 \approx 19.0$ turns. The path of the more tightly wound DNA on the nucleosome absorbs –0.6 turns, and this at least in part explains the discrepancy between the physical path of –1.65 and the measurement of –1.0 superhelical turns. In effect, some of the torsional strain in nucleosomal DNA goes into increasing the number of bp/turn; only the rest is left to be measured as a supercoil. Given the difficulties of making these measurements, it is possible that the change in periodicity of nucleosomal DNA accounts entirely for the linking number paradox.

Figure 19.16 The supercoils of the SV40 minichromosome can be relaxed to generate a circular structure, whose loss of histones then generates supercoils in the free DNA.

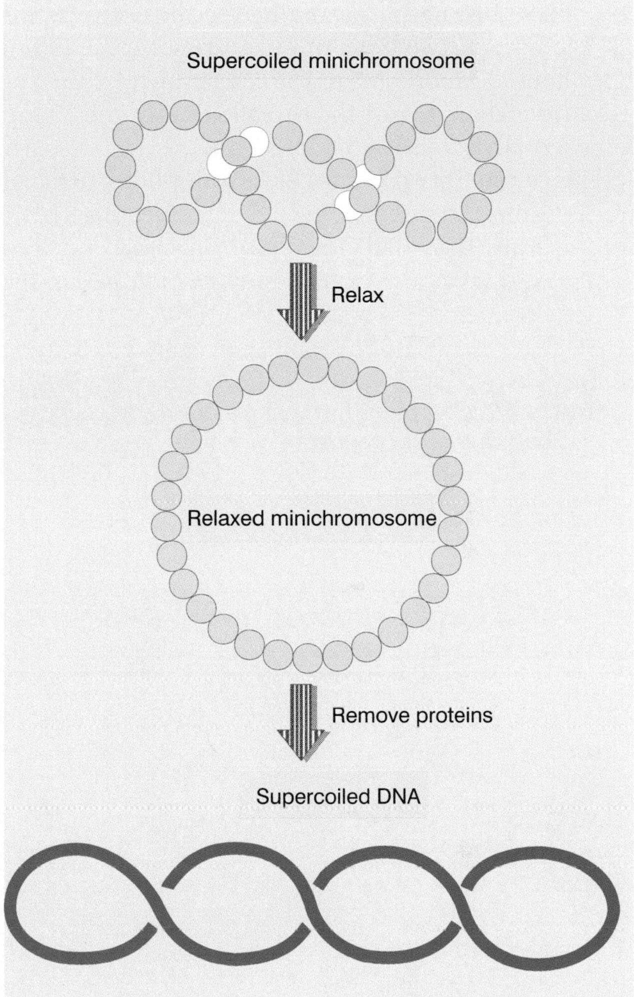

The path of nucleosomes in the chromatin fiber

When chromatin is examined in the electron microscope, two types of fibers are seen: the 10 nm fiber and 30 nm fiber. They are described by the approximate diameter of the thread (that of the 30 nm fiber actually varies from ~25–30 nm).

The **10 nm fiber** is essentially a continuous string of nucleosomes. Sometimes, indeed, it runs continuously into a more stretched-out region in which nucleosomes are seen as a string of beads, as indicated in the example of **Figure 19.17.** The 10 nm fibril structure is obtained under conditions of low ionic strength and does not require the presence of histone H1. This means that it is a function strictly of the nucleosomes themselves. It may be visualized essentially as a continuous series of nucleosomes, as in **Figure 19.18.** It is not clear whether such a structure exists *in vivo* or is simply a consequence of unfolding during extraction *in vitro.*

When chromatin is visualized in conditions of greater ionic strength the **30 nm fiber** is obtained. An example is given in **Figure 19.19.** The fiber can be seen to have an underlying coiled structure. It has ~6 nucleosomes for every turn, which corresponds to a packing ratio of 40 (that is, each μm along the axis of the fiber contains 40 μm of DNA). The presence of H1 is required. This fiber is the basic constituent of both interphase chromatin and mitotic chromosomes.

The most likely arrangement for packing nucleosomes into the fiber is a solenoid, illustrated in **Figure**

Figure 19.17 The 10 nm fiber in partially unwound state can be seen to consist of a string of nucleosomes. Photograph kindly provided by Barbara Hamkalo.

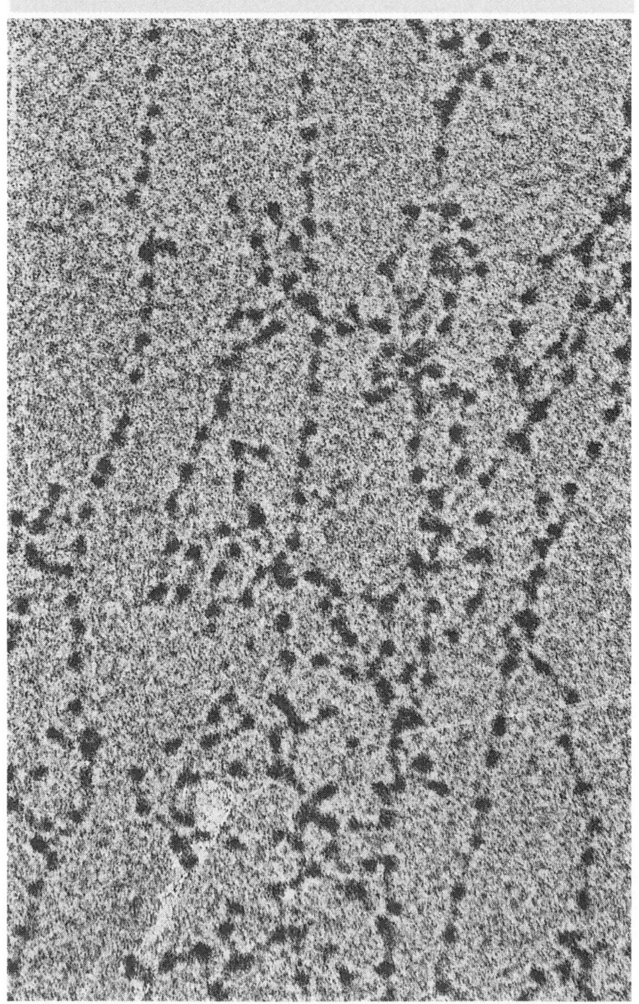

Figure 19.18 The 10 nm fiber is a continuous string of nucleosomes.

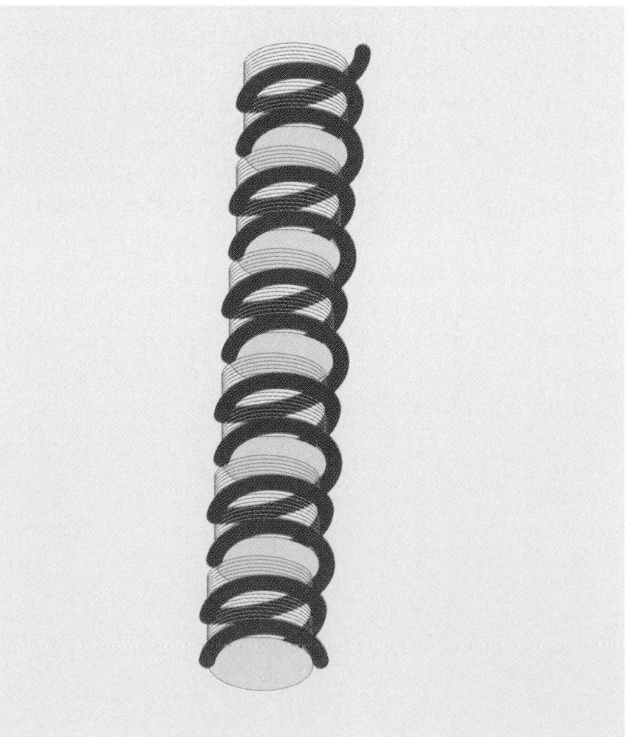

Figure 19.19 The 30 nm fiber has a coiled structure (shown at the same magnification as the 10 nm fiber of Figure 19.17). Photograph kindly provided by Barbara Hamkalo.

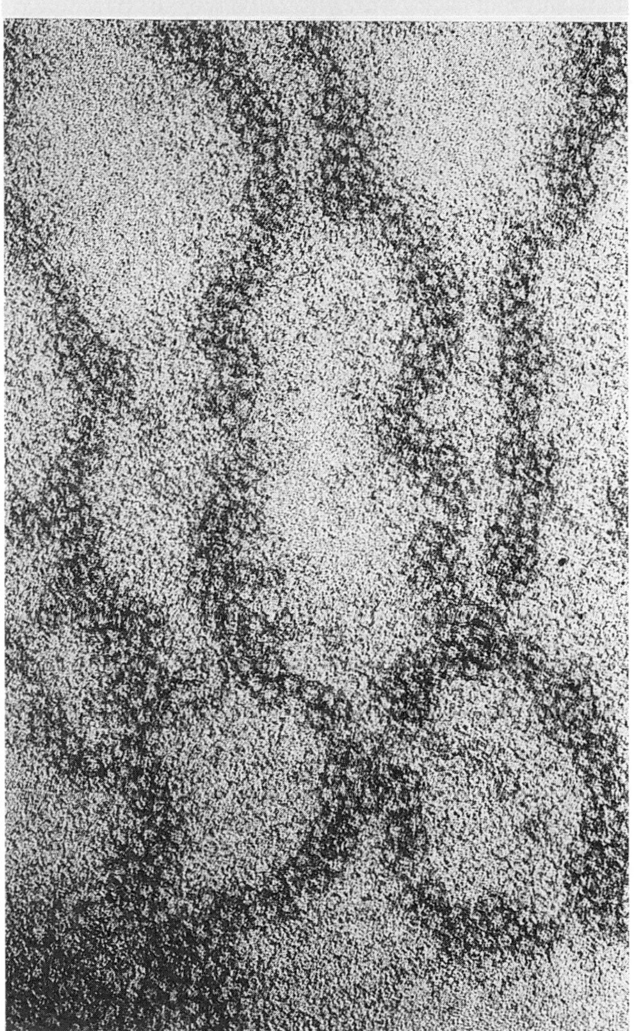

19.20. The nucleosomes turn in a helical array, with an angle of ~60° between the faces of adjacent nucleosomes.

The 30 nm and 10 nm fibers can be reversibly converted by changing the ionic strength. This suggests that the linear array of nucleosomes in the 10 nm fiber is coiled into the 30 nm structure at higher ionic strength and in the presence of H1.

Figure 19.20 The 30 nm fiber may have a helical coil of 6 nucleosomes per turn, organized radially.

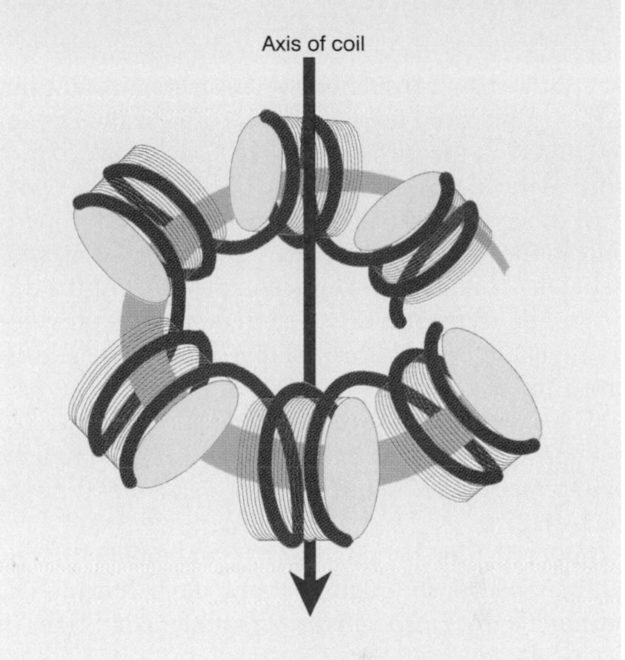

Although the presence of H1 is necessary for the formation of the 30 nm fiber, information about its location is conflicting. Its relative ease of extraction from chromatin seems to argue that it is present on the outside of the superhelical fiber axis. But diffraction data, and the fact that it is harder to find in 30 nm fibers than in 10 nm fibers that retain it, would argue for an interior location.

How do we get from the 30 nm fiber to the specific structures displayed in mitotic chromosomes? And is there any further specificity in the arrangement of interphase chromatin; do particular regions of 30 nm fibers bear a fixed relationship to one another or is their arrangement random?

Organization of the histone octamer

So far we have considered the construction of the nucleosome from the perspective of how the DNA is organized on the surface. From the perspective of protein, we need to know how the histones interact with each other and with DNA. Do histones react properly only in the presence of DNA, or do they possess an independent ability to form octamers? Most of the evidence about histone-histone interactions is provided by their abilities to form stable complexes, and by crosslinking experiments with the nucleosome.

The core histones form two types of complexes. H3 and H4 form a tetramer ($H3_2 \cdot H4_2$). Various complexes are formed by H2A and H2B, in particular a dimer (H2A•H2B).

Intact histone octamers can be obtained either by extraction from chromatin or (with more difficulty) by letting histones associate *in vitro* under conditions of high-salt and high-protein concentration. The octamer can dissociate to generate a hexamer of histones that has lost an H2A•H2B dimer. Then the other H2A•H2B dimer is lost separately, leaving the $H3_2 \cdot H4_2$ tetramer. This argues for a form of organization in which the nucleosome has a central "kernel" consisting of the $H3_2 \cdot H4_2$ tetramer. The tetramer can organize DNA *in vitro* into particles that display some of the properties of the core particle.

Crosslinking studies extend these relationships to show which pairs of histones lie near each other in the nucleosome. (A difficulty with such data is that usually only a small proportion of the proteins becomes crosslinked, so it is necessary to be cautious in deciding whether the results typify the major interactions.) From these data, a model has been constructed for the organization of the nucleosome. It is shown in diagrammatic form in **Figure 19.21.**

Structural studies show that the overall shape of the isolated histone octamer is similar to that of the core particle. This suggests that the histone-histone interactions establish the general structure. The positions of the individual histones have been assigned to regions of the octameric structure on the basis of their interaction behavior and response to crosslinking.

The crystal structure (at 3.1 Å resolution) suggests the model for the histone octamer shown in **Figure 19.22.** Tracing the paths of the individual polypeptide backbones in the crystal structure suggests that the histones are not organized as individual globular proteins, but that each is interdigitated with its partner, H3 with H4, and H2A with H2B. So the model distinguishes the $H3_2 \cdot H4_2$ tetramer (white) from the H2A•H2B dimers (blue), but does not show individual histones.

The top view represents the same perspective that was illustrated schematically in Figure 19.21. The $H3_2 \cdot H4_2$ tetramer accounts for the diameter of the octamer. It forms the shape of a horseshoe. The H2A•H2B pairs fit in as two dimers, but only one can be seen in this view. The side view represents the same perspective that was illustrated in Figure 19.4. Here the responsibilities of the $H3_2 \cdot H4_2$ tetramer and of the separate H2A•H2B dimers can be distinguished. The protein forms a sort of spool, with a superhelical path that could correspond to the binding site for DNA, which would be wound in almost two full turns in a nucleosome. The model displays two fold symmetry about an axis that would run perpendicular through the side view.

Figure 19.21 In a symmetrical model for the nucleosome, the $H3_2$-$H4_2$ tetramer provides a kernel for the shape. One H2A-H2B dimer can be seen in the top view; the other is underneath.

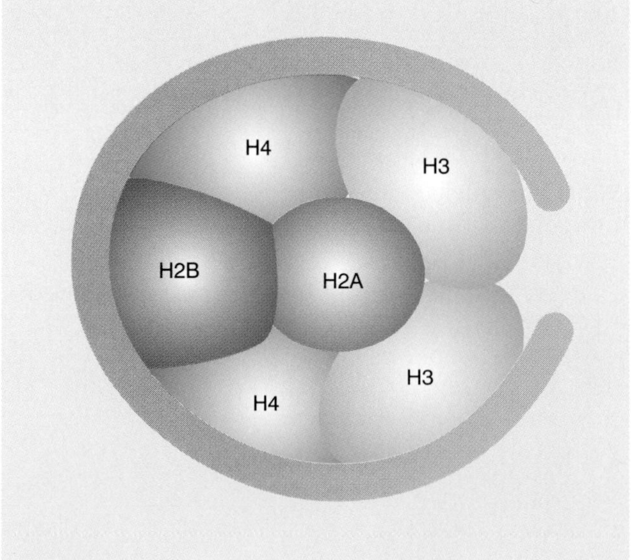

A more detailed view of the positions of the histones, based on a crystal structure at 2.8Å, is summarized in **Figure 19.23.** The upper view shows the position of one histone of each type relative to one turn around the nucleosome (numbered from 0 to +7). All four core histones show a similar type of structure in which three α-helices are connected by two loops: this is called the **histone fold.** These regions interact to

Figure 19.22 The crystal structure of the histone core octamer is represented in a space-filling model with the $H3_2$-$H4_2$ tetramer shown in white and the H2A-H2B dimers shown in blue. Only one of the H2A-H2B dimers is visible in the top view, because the other is hidden underneath. The potential path of the DNA is shown in the top view as a narrow tube (one quarter the diameter of DNA), and in the side view by the parallel lines in a 20 Å wide bundle. Photographs kindly provided by Evangelos Moudrianakis.

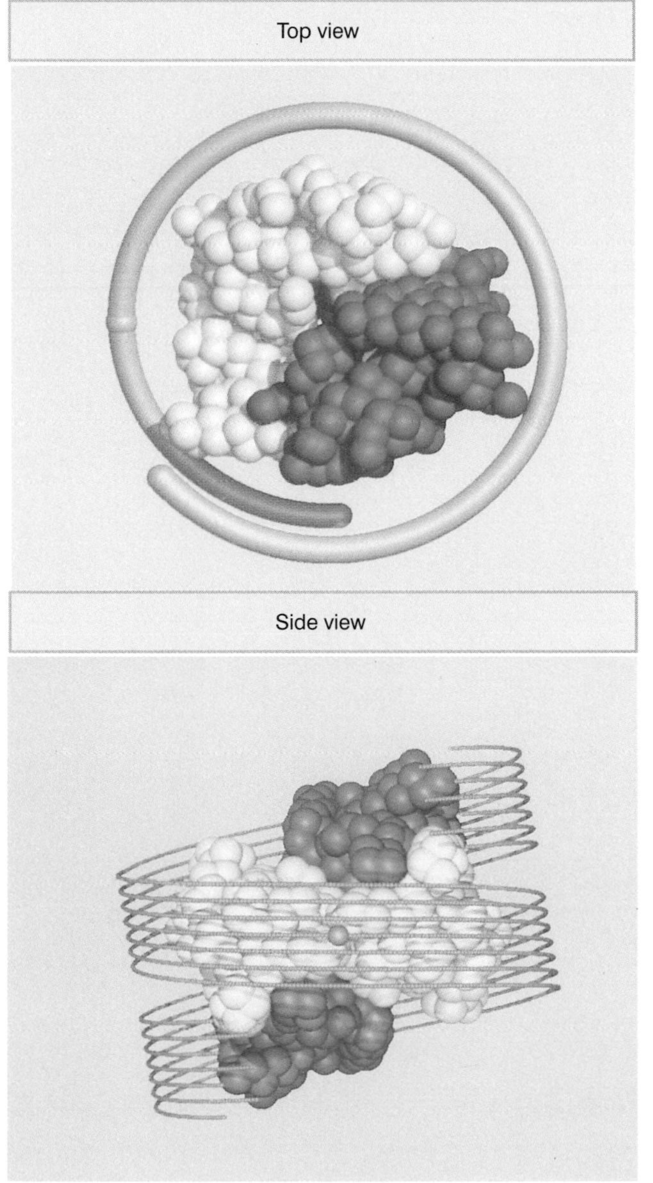

Figure 19.23 Histone positions in a top view show H3-H4 and H2A-H2B pairs in a 'half nucleosome;' the symmetrical organization can be seen in the superimposition of both halves.

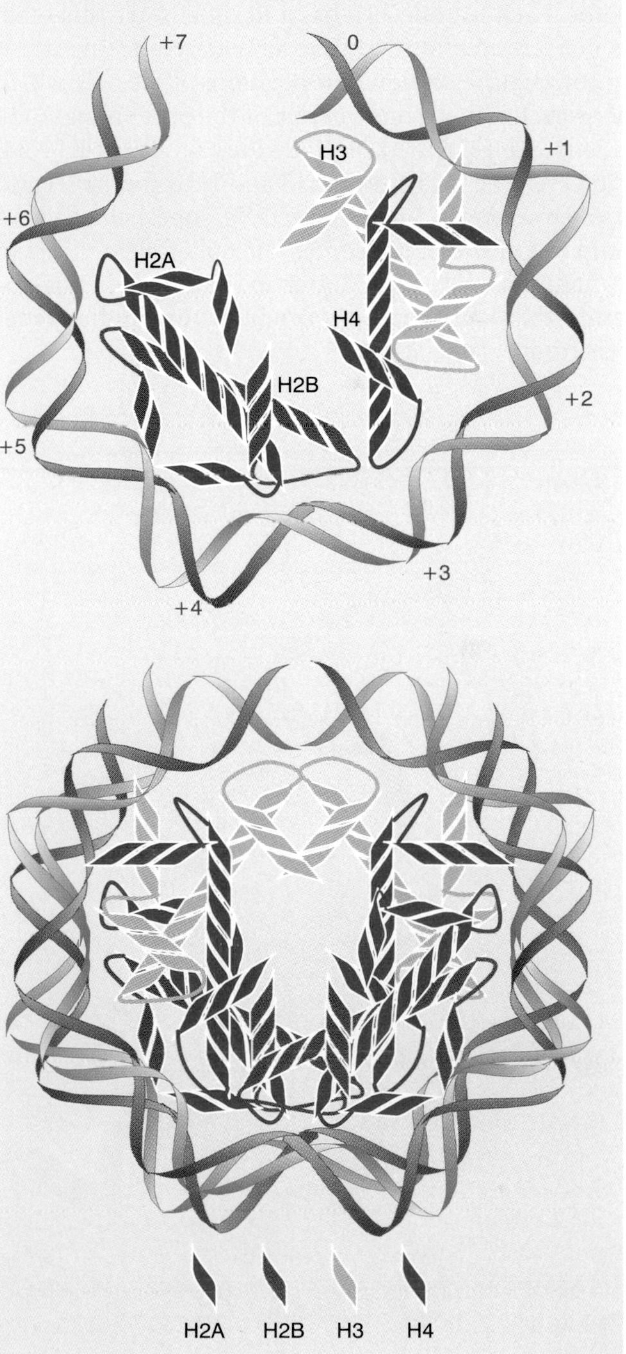

form crescent-shaped heterodimers; each heterodimer binds 2.5 turns of the DNA double helix (H2A•H2B binds at +3.5–6; H3•H4 binds at +0.5–3 for the circumference that is illustrated). Binding is mostly to the phosphodiester backbones (consistent with the need to package any DNA irrespective of sequence). The $H3_2$•$H4_2$ tetramer is formed by interactions between the two H3 subunits, as can be seen in the lower part of the figure.

Each of the core histones has a globular body that contributes to the central protein mass of the nucleosome. Each histone also has a flexible N-terminal tail, which has sites for modification that may be important in chromatin function. The positions of the tails, which account for about one quarter of the protein mass, are not so well defined, as indicated in **Figure 19.24.** However, the tails of both H3 and H2B can be seen to pass between the turns of the DNA superhelix and extend out of the nucleosome. The tail of H4 appears to contact an H2A•H2B dimer in an adjacent nucleosome; this could be an important feature in the overall structure.

Figure 19.24 The globular bodies of the histones are localized in the histone octamer of the core particle, but the locations of the N-terminal tails, which carry the sites for modification, are not known, and could be more flexible.

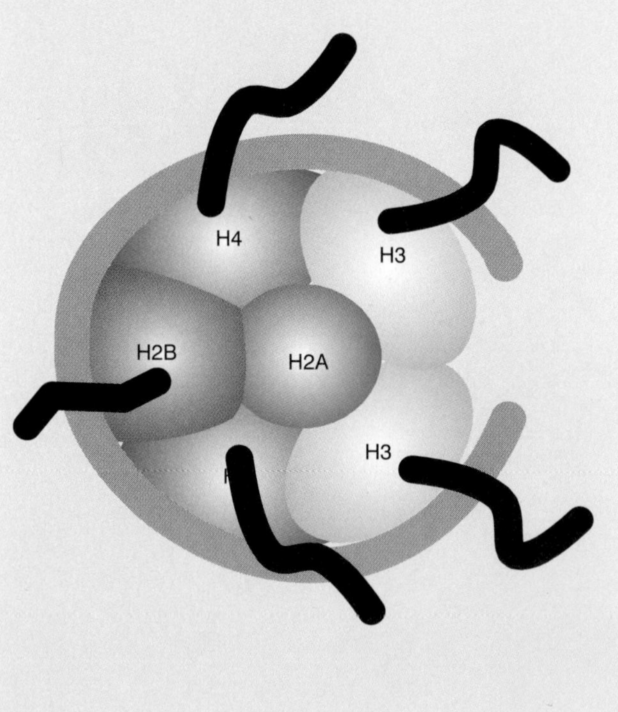

All of the histones are modified by covalently linking extra moieties to the free groups of certain amino acids. Acetylation and methylation occur on the free (ε) amino group of lysine. As seen in **Figure 19.25,** this removes the positive charge that resides on the NH_3^+ form of the group. Methylation also occurs on arginine and histidine. Phosphorylation occurs on the hydroxyl group of serine and also on histidine. This introduces a negative charge in the form of the phosphate group.

These modifications are transient. Because they change the charge of the protein molecule, they are potentially able to change the functional properties of the octamers. Modification of histones is associated with structural changes that occur in chromatin at replica-

Figure 19.25 Acetylation of lysine or phosphorylation of serine reduces the overall positive charge of a protein.

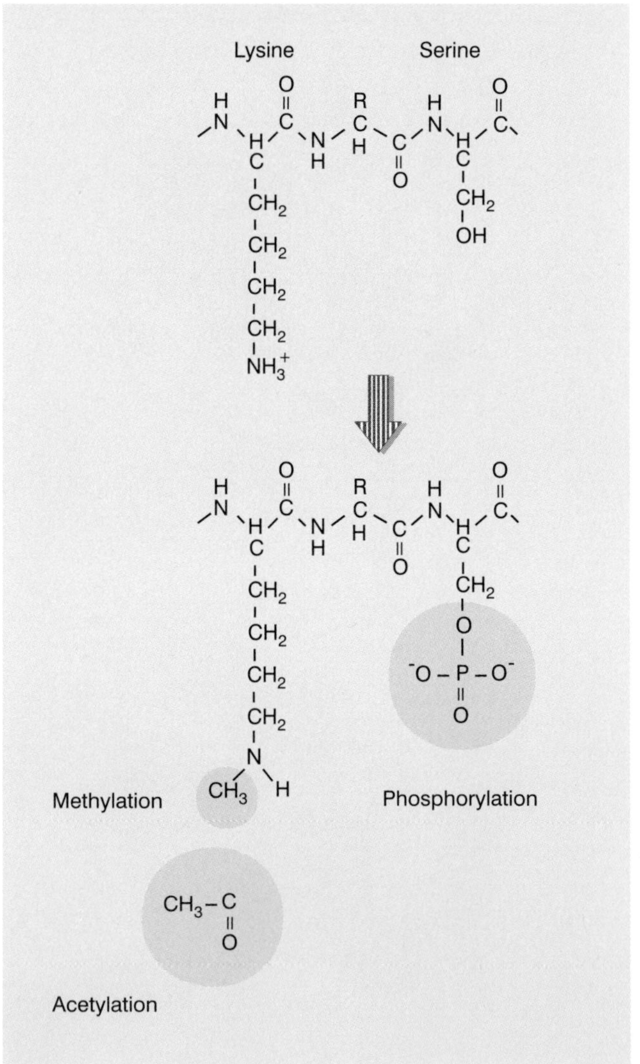

tion and transcription. Phosphorylations on specific positions and on different histones may be required for particular processes, for example, the Ser^{10} position of H3 is phosphorylated when chromosomes condense at mitosis.

In synchronized cells in culture, both the pre-existing and newly synthesized core histones appear to be acetylated and methylated during S phase (when DNA is replicated and the histones also are synthesized). During the cell cycle, the modifying groups are later removed.

The coincidence of modification and replication suggests that acetylation (and methylation) could be connected with nucleosome assembly. One speculation has been that the reduction of positive charges on histones might lower their affinity for DNA, allowing the reaction to be better controlled. The idea has lost some ground in view of the observation that nucleosomes can be reconstituted, at least *in vitro,* with unmodified histones.

A cycle of phosphorylation and dephosphorylation occurs with H1, but its timing is different from the modification cycle of the other histones. With cultured mammalian cells, one or two phosphate groups are introduced at S phase. But the major phosphorylation event is the later addition of more groups at mitosis, to bring the total number up to as many as six. All the phosphate groups are removed at the end of the process of division. The phosphorylation of H1 is catalyzed by the M-phase kinase that provides an essential trigger for mitosis (see Chapter 27). In fact, this enzyme is now often assayed in terms of its H1 kinase activity. Not much is known about phosphatase(s) that remove the groups later.

The timing of the major H1 phosphorylation has prompted speculation that it is involved in mitotic condensation. However, in *Tetrahymena* (a protozoan) it is possible to delete all the genes for H1 without significantly affecting the overall properties of chromatin. There is a relatively small effect on the ability of chromatin to condense at mitosis. Some genes are activated and others are repressed by this change, suggesting that there are alterations in local structure. Mutations that eliminate sites of phosphorylation in H1 have no effect, but mutations that mimic the effects of phosphorylation produce a phenotype that resembles the deletion. This suggests that the effect of phosphorylating H1 is to eliminate its effects on local chromatin structure.

Reproduction of chromatin requires assembly of nucleosomes

THE description of chromatin as a thread of duplex DNA coiled around a series of histone octamers to form nucleosomes is the crucial first step toward visualizing the state of the genetic material in the nucleus. This somewhat static view accounts for the structure of the individual subunit and (to some degree) for its relationship with the adjacent subunit. However, the organization of nucleosomes must be *flexible* enough to satisfy the various structural and functional demands made on chromatin.

Cyclical changes in packing affect the entire mass of euchromatin. During cell division, euchromatin must become more tightly packaged in mitotic chromosomes. The transition is likely to be controlled by changes in proteins that are widely distributed throughout chromatin.

Replication and transcription are local events that require some dispersion of structure. Replication occurs as a series of individual events in local regions (replicons), generating duplicate double-stranded DNA regions each associated with a set of histone octamers. The events involved in reproducing the nucleosome particle have yet to be defined. We should like to know what happens to the nucleosome during replication, and how new nucleosomes are assembled.

It seems inevitable that the separation of parental DNA strands must disrupt the structure of the 30 nm fiber. We should like to know the extent of this disruption. Is it confined to the immediate vicinity of the point where DNA is being synthesized, or does it extend farther? Are there discernible structural differences between regions that have replicated and those that have yet to do so?

The transience of the replication event is a major difficulty in analyzing the structure of a particular region while it is being replicated. The structure of the replication fork is distinctive. It is more resistant to micrococcal nuclease and is digested into bands that differ in size from nucleosomal DNA. This suggests that a large protein complex is engaged in replicating the DNA, but the nucleosomes reform more or less immediately behind as it moves along.

Reproduction of chromatin does not involve any protracted period during which the DNA is free of histones. Once DNA has been replicated, nucleosomes are quickly generated on both the duplicates. This point is illustrated by the electron micrograph of **Figure 19.26**, which shows a recently replicated stretch of DNA, already packaged into nucleosomes on both daughter duplex segments.

How histones associate with DNA to generate nucleosomes has been a vexed and confusing question. Do the histones *preform* a protein octamer around which the DNA is subsequently wrapped? Or does an $H3_2$•$H4_2$ kernel bind DNA, after which H2A•H2B dimers are added?

Self-assembly *in vitro* is a slow process, limited by the tendency of the assembling particles to precipitate. It is difficult to know which conditions mimic the physiological. Both pathways can be used *in vitro* to assemble nucleosomes, as illustrated in **Figure 19.27.**

Accessory proteins are involved in assisting histones to associate with DNA. Candidates for this role can be identified by using extracts that assemble histones and exogenous DNA into nucleosomes. Accessory proteins may act as "molecular chaperones" that bind to the histones in order to release either individual histones or complexes (H3•H4 or H2A•H2B) to the DNA in a controlled manner. This could be necessary because the histones, as basic proteins, have a general high affinity for DNA. *Such interactions allow histones to form nucleosomes without becoming trapped in other kinetic intermediates (that is, other complexes resulting from indiscreet binding of histones to DNA).*

Attempts to produce nucleosomes *in vitro* began by

Figure 19.26 Replicated DNA is immediately incorporated into nucleosomes. Photograph kindly provided by S. MacKnight.

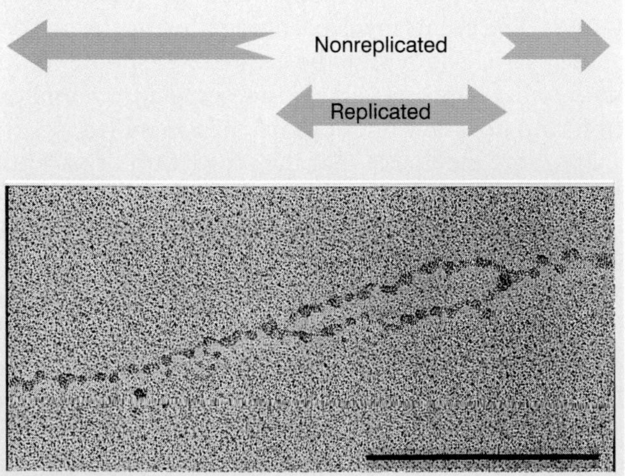

Figure 19.27 *In vitro*, DNA can either interact directly with an intact (crosslinked) histone octamer or can assemble with the $H3_2$-$H4_2$ tetramer, after which two H2A-H2B dimers are added.

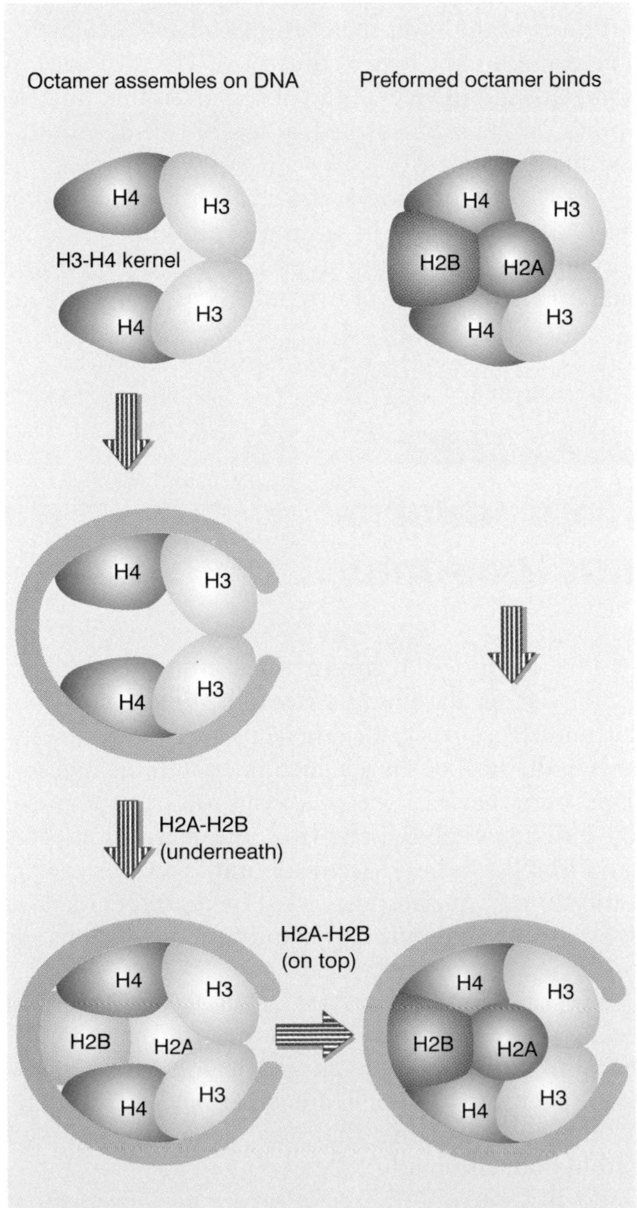

Figure 19.28 If histone octamers were conserved, old and new octamers would band at different densities when replication of 'heavy' octamers occurs in 'light' amino acids (left); but actually the octamers band diffusely between heavy and light densities, suggesting disassembly and reassembly (right).

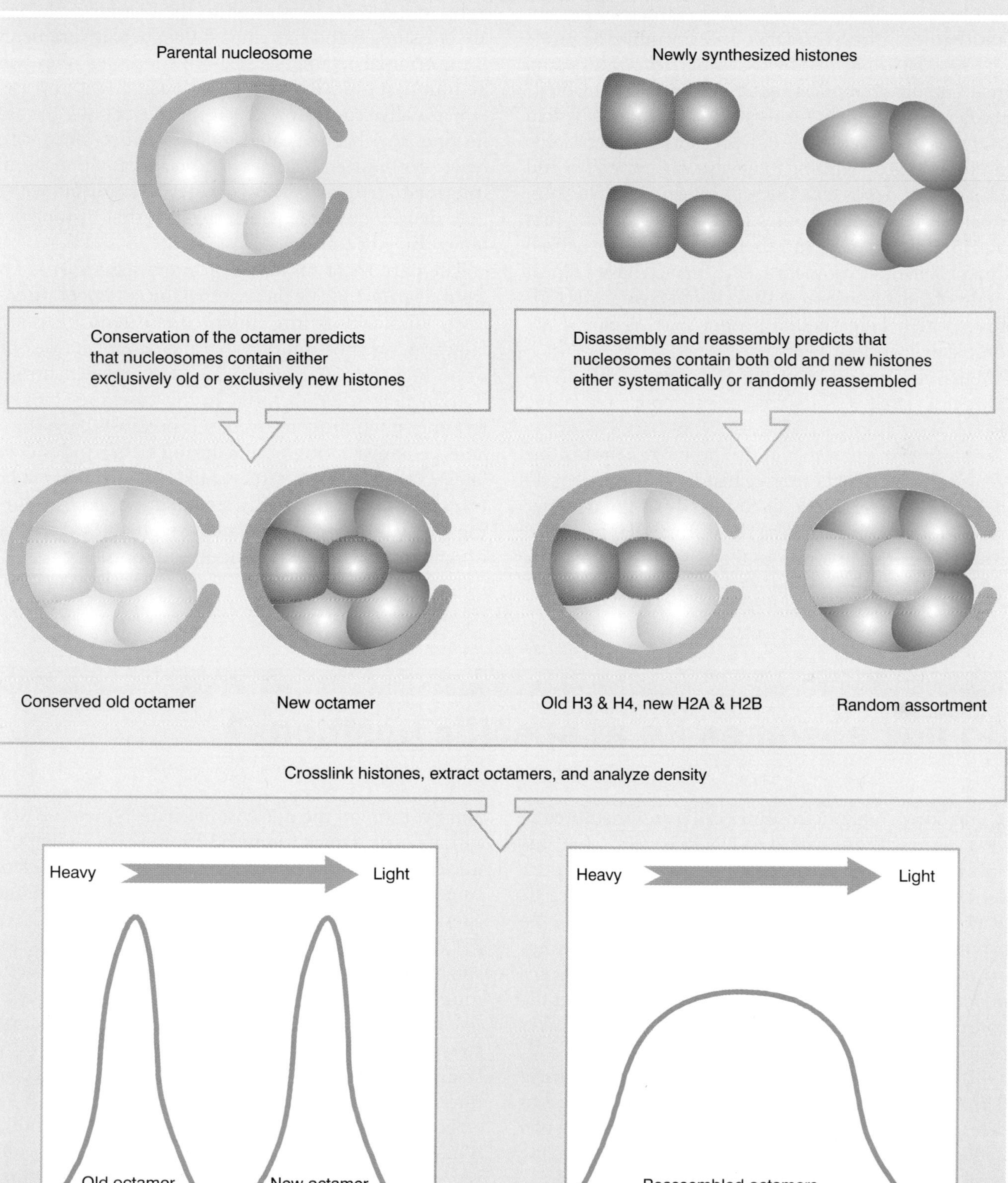

considering a process of assembly between free DNA and histones. But nucleosomes form *in vivo* only when DNA is replicated. A system that mimics this requirement has been developed by using extracts of human cells that replicate SV40 DNA and assemble the products into chromatin. The assembly reaction occurs preferentially on replicating DNA. It requires an ancillary factor, CAF-1, that consists of >5 subunits, with a total mass of 238 kD. CAF-1 acts stoichiometrically, and functions by binding to newly synthesized H3 and H4, which it deposits at the replication fork. This suggests that new nucleosomes form by assembling first the H3•H4 tetramer, and then adding the H2A•H2B dimers. The nucleosomes that are formed have a repeat length of 200 bp, although they do not have any H1 histone, which suggests that proper spacing can be accomplished without H1.

When chromatin is reproduced, a stretch of DNA *already associated with nucleosomes* is replicated, giving rise to two daughter duplexes. What happens to the pre-existing nucleosomes at this point? Are the histone octamers dissociated into free histones for reuse, or do they remain assembled? The integrity of the octamer can be tested by crosslinking the histones. **Figure 19.28** summarizes the possible outcomes from an experiment in which cells are grown in the presence of heavy amino acids to identify the histones before replication. Then replication is allowed to occur in the presence of light amino acids. At this point the histone octamers are crosslinked and centrifuged on a density gradient. If the original octamers have been conserved, they will be found at a position of high density, and new octamers will occupy a low density position; but if the old histones have been released and then reassembled with newly synthesized histones, the octamers will have an intermediate density. Little material is found at the high density position, which suggests that histone octamers are not conserved.

The pattern of disassembly and reassembly is far from clear. It could be the case that the octamers are entirely dissociated into their constituent histones. However, one possibility is that the octamers lose one or both H2A•H2B dimers, which are replaced at random by new or old material. In this case, the $H3_2 \cdot H4_2$ tetramer might be conserved. It is possible that a similar type of disruption occurs during transcription (see later). The $H3_2 \cdot H4_2$ tetramer could have an ability to be transiently associated with a single strand of DNA during replication; it may in fact have an increased chance of remaining with the leading strand for reuse.

Do nucleosomes lie at specific positions?

We know that nucleosomes can be reconstituted *in vitro* without regard to DNA sequence, but this does not mean that their formation *in vivo* is independent of sequence. Does a particular DNA sequence always lie in a certain position *in vivo* with regard to the topography of the nucleosome? Or are nucleosomes arranged randomly on DNA, so that a particular sequence may occur at any location, for example, in the core region in one copy of the genome and in the linker region in another?

To investigate this question, it is necessary to use a defined sequence of DNA; more precisely, we need to determine the position relative to the nucleosome of a defined point in the DNA. **Figure 19.29** illustrates the principle of a procedure used to achieve this.

Suppose that the DNA sequence is organized into nucleosomes in only one particular configuration, so that each site on the DNA always is located at a particular position on the nucleosome. This type of organization is called **nucleosome positioning** (or sometimes nucleosome phasing). In a series of positioned nucleosomes, the linker regions of DNA comprise unique sites.

Consider the consequences for just a single nucleosome. Cleavage with micrococcal nuclease generates a monomeric fragment that constitutes a *specific sequence.* If the DNA is isolated and cleaved with a restriction enzyme that has only one target site in this fragment, it should be cut at a unique point. This produces two fragments, each of unique size.

The products of the micrococcal/restriction double digest are separated by gel electrophoresis. A probe representing the sequence on one side of the restriction site is used to identify the corresponding fragment in the double digest. This technique is called **indirect end labeling.**

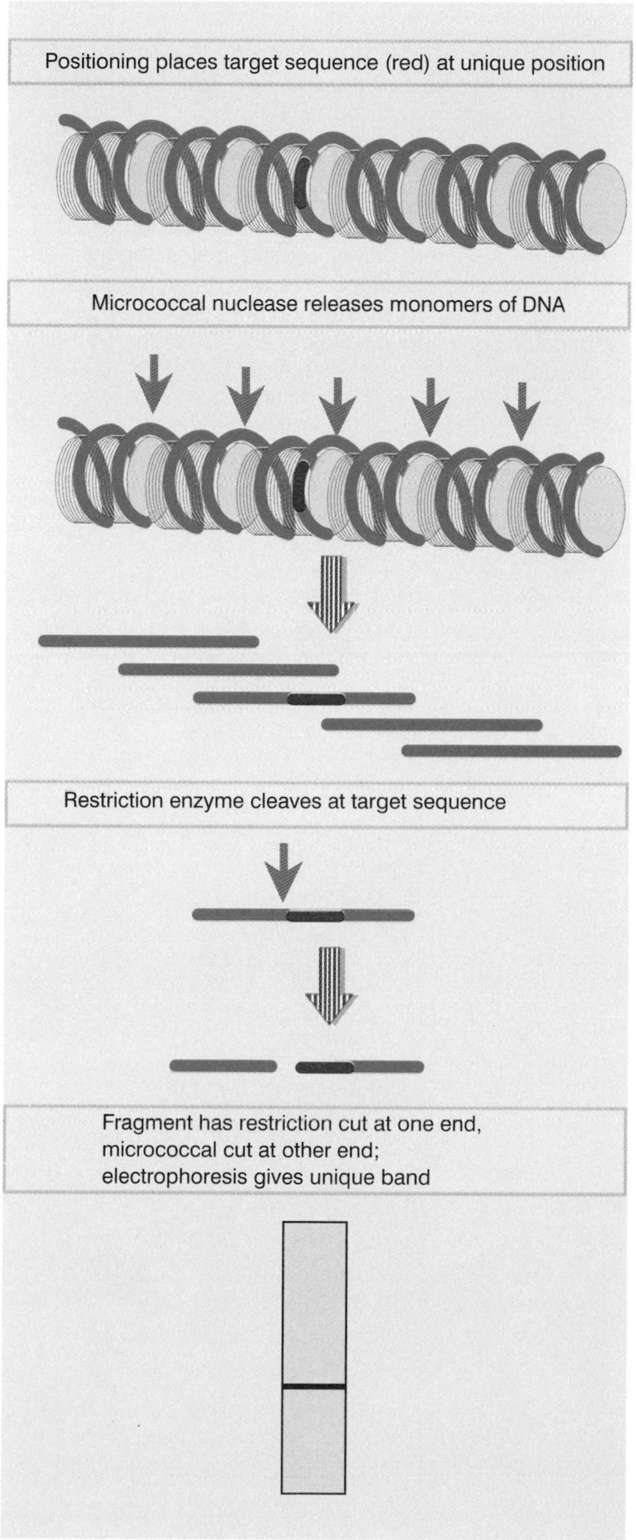

Reversing the argument, the identification of a single sharp band demonstrates that the position of the restriction site is uniquely defined with respect to the end of the nucleosomal DNA (as defined by the micrococcal nuclease cut). So the nucleosome has a unique sequence of DNA.

What happens if the nucleosomes do *not* lie at a single position? Now the linkers consist of *different* DNA sequences in each copy of the genome. So the restriction site lies at a different position each time; in fact, it lies at all possible locations relative to the ends of the monomeric nucleosomal DNA. **Figure 19.30** shows

Figure 19.29 Nucleosome positioning places restriction sites at unique positions relative to the linker sites cleaved by micrococcal nuclease.

Positioning places target sequence (red) at unique position
Micrococcal nuclease releases monomers of DNA
Restriction enzyme cleaves at target sequence
Fragment has restriction cut at one end, micrococcal cut at other end; electrophoresis gives unique band

Figure 19.30 In the absence of nucleosome positioning, a restriction site lies at all possible locations in different copies of the genome. Fragments of all possible sizes are produced when a restriction enzyme cuts at a target site (red) and micrococcal nuclease cuts at the junctions between nucleosomes (green).

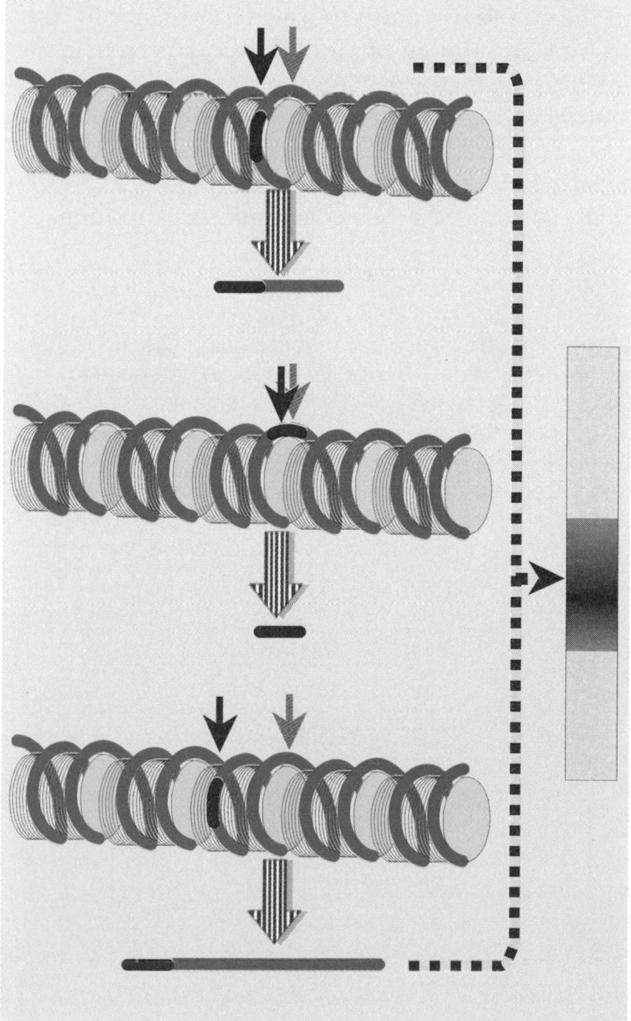

that the double cleavage then generates a broad smear, ranging from the smallest detectable fragment (~20 bases) to the length of the monomeric DNA.

In discussing these experiments, we have treated micrococcal nuclease as an enzyme that cleaves DNA at the exposed linker regions without any sort of sequence specificity. However, the enzyme actually does have some sequence specificity (biased toward selection of A•T-rich sequences). So we cannot assume that the existence of a specific band in the indirect end-labeling technique represents the distance from a restriction cut to the linker region. It could instead represent the distance from the restriction cut to a preferred micrococcal nuclease cleavage site!

This possibility is controlled by treating the naked DNA in exactly the same way as the chromatin. If there are preferred sites for micrococcal nuclease in the particular region, specific bands are found. Then this pattern of bands can be compared with the pattern generated from chromatin.

A *difference* between the control DNA band pattern and the chromatin pattern provides evidence for nucleosome positioning. Some of the bands present in the control DNA digest may disappear from the nucleosome digest, indicating that preferentially cleaved positions are unavailable. New bands may appear in the nucleosome digest when new sites are rendered preferentially accessible by the nucleosomal organization.

Figure 19.31 Translational positioning describes the linear position of DNA relative to the histone octamer. Displacement of the DNA by 10 bp changes the sequences that are in the more exposed linker regions, but does not alter which face of DNA is protected by the histone surface and which is exposed to the exterior. DNA is really coiled around the nucleosomes, and is shown in linear form only for convenience.

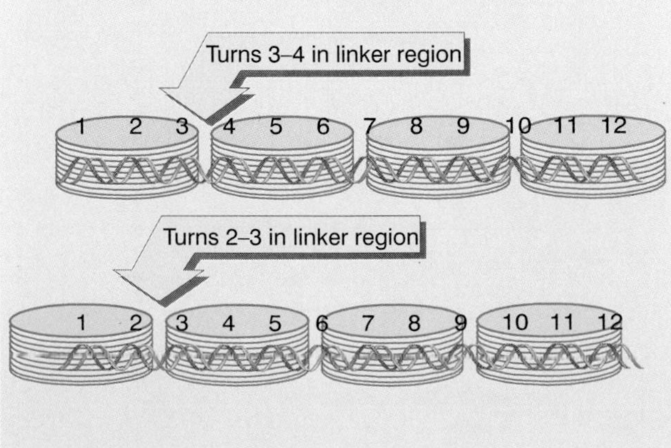

Nucleosome positioning might be accomplished in either of two ways:

- It is intrinsic: *every nucleosome is deposited specifically at a particular DNA sequence.* This modifies our view of the nucleosome as a subunit able to form between any sequence of DNA and a histone octamer.
- It is extrinsic: *the first nucleosome in a region is preferentially assembled at a particular site.* A preferential

Figure 19.32 Rotational positioning describes the exposure of DNA on the surface of the nucleosome. Any movement that differs from the helical repeat (~10.2 bp/turn) displaces DNA with reference to the histone surface. Nucleotides on the inside are more protected against nucleases than nucleotides on the outside.

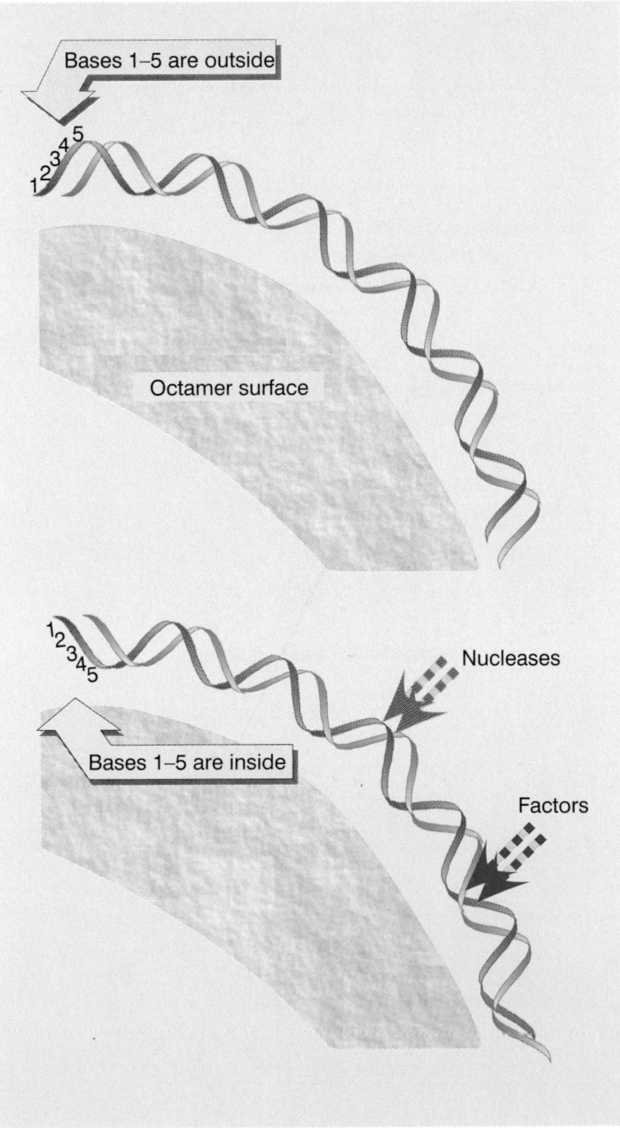

starting point for nucleosome positioning results from the presence of a region from which nucleosomes are excluded. The excluded region provides a *boundary* that restricts the positions available to the adjacent nucleosome. Then a series of nucleosomes may be assembled sequentially, with a defined repeat length.

It is now clear that the deposition of histone octamers on DNA is not random with regard to sequence. The pattern is intrinsic in some cases, in which it is determined by structural features in DNA. It is extrinsic in other cases, in which it results from the interactions of other proteins with the DNA and/or histones.

Certain structural features of DNA affect placement of histone octamers. DNA has intrinsic tendencies to bend in one direction rather than another; thus A•T-rich stretches locate so that the minor groove faces in towards the octamer, whereas G•C-rich stretches place so that the minor groove points out. Long runs of dA•dT (>8 bp) avoid positioning in the central superhelical turn of the core. It is not yet possible to sum all of the relevant structural effects and thus entirely to predict the location of a particular DNA sequence with regard to the nucleosome. Sequences that cause DNA to take up more extreme structures may have effects such as the exclusion of nucleosomes, and thus could cause boundary effects.

Positioning of nucleosomes near boundaries is common. If there is some variability in the construction of nucleosomes—for example, if the length of the linker can vary by, say, 10 bp—the specificity of location would decline proceeding away from the first, defined nucleosome at the boundary. In this case, we might expect the positioning to be maintained rigorously only relatively near the boundary.

The location of DNA on nucleosomes can be described in two ways. **Figure 19.31** shows that **translational positioning** describes the position of DNA with regard to the boundaries of the nucleosome. In particular, it determines which sequences are found in the linker regions. Shifting the DNA by 10 bp brings the next turn into a linker region. So translational positioning determines which regions are more accessible (at least as judged by sensitivity to micrococcal nuclease).

Because DNA lies on the outside of the histone octamer, one face of any particular sequence is obscured by the histones, but the other face is accessible. Depending upon its positioning with regard to the nucleosome, a site in DNA that must be recognized by a regulator protein could be inaccessible or available. The exact position of the histone octamer with respect to DNA sequence may therefore be important. **Figure 19.32** shows the effect of **rotational positioning** of the double helix with regard to the octamer surface. If the DNA is moved by a partial number of turns (imagine the DNA as rotating relative to the protein surface), there is a change in the exposure of sequence to the outside.

Both translational and rotational positioning can be important in controlling access to DNA. The best characterized cases of positioning involve the specific placement of nucleosomes at promoters. Translational positioning and/or the exclusion of nucleosomes from a particular sequence may be necessary to allow a transcription complex to form. Some regulatory factors can bind to DNA only if a nucleosome is excluded to make the DNA freely accessible, and this creates a boundary for translational positioning. In other cases, regulatory factors can bind to DNA on the surface of the nucleosome, but rotational positioning is important to ensure that the face of DNA with the appropriate contact points is exposed. We discuss the connection between nucleosomal organization and transcription in Chapter 21.

Are transcribed genes organized in nucleosomes?

ATTEMPTS to visualize genes during transcription have produced conflicting results. The next two figures show each extreme.

Heavily transcribed chromatin can be seen to be rather extended (too extended to be covered in nucleosomes). In the intensively transcribed genes coding for rRNA, shown in **Figure 19.33**, the extreme packing of RNA polymerases makes it hard to see the DNA. We cannot directly measure the lengths of the rRNA transcripts because the RNA is compacted by proteins, but we know (from the sequence of the rRNA) how long the transcript must be. The length of the transcribed

Figure 19.33 The extended axis of an rDNA transcription unit alternates with the only slightly less extended non-transcribed spacer. Photograph kindly provided by Charles Laird.

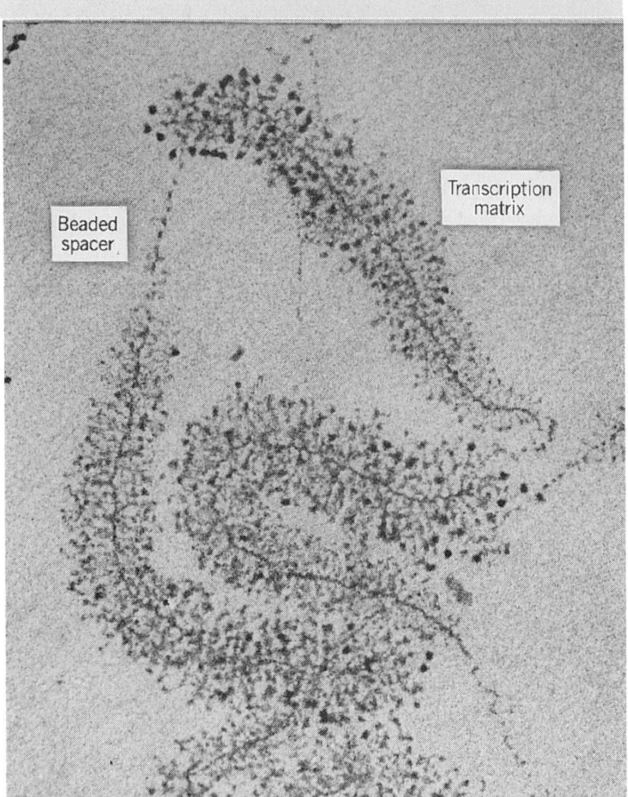

Figure 19.34 An SV40 minichromosome can be transcribed. Photograph kindly provided by Pierre Chambon.

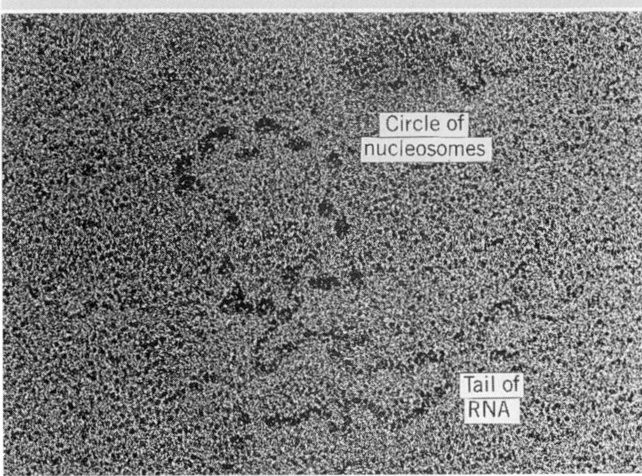

Figure 19.35 RNA polymerase is comparable in size to the nucleosome and might encounter difficulties in following the DNA around the histone octamer.

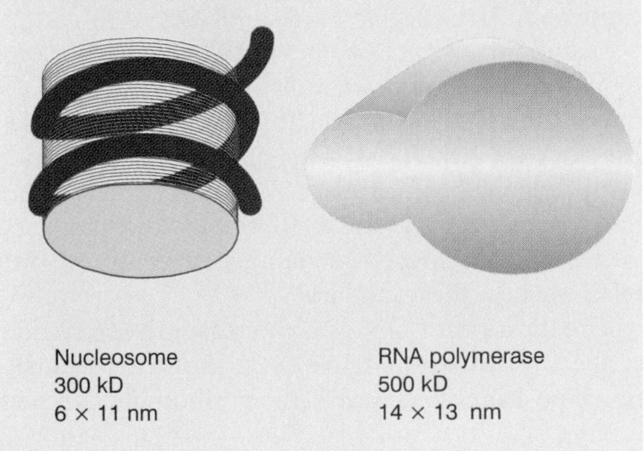

DNA segment, measured by the length of the axis of the "Christmas tree," is ~85% of the length of the rRNA. This means that the DNA is almost completely extended.

On the other hand, transcription complexes of SV40 minichromosomes can be extracted from infected cells. They contain the usual complement of histones and display a beaded structure. Chains of RNA can be seen to extend from the minichromosome, as in the example of **Figure 19.34.** This argues that transcription can proceed while the SV40 DNA is organized into nucleosomes. Of course, the SV40 minichromosome is transcribed less intensively than the rRNA genes.

Transcription involves the unwinding of DNA, and may require the fiber to unfold in restricted regions of chromatin. A simple-minded view suggests that some "elbow-room" must be needed for the process. The features of polytene and lampbrush chromosomes described in Chapter 18 offer hints that a more expansive structural organization is associated with gene expression.

In thinking about transcription, we must bear in mind the relative sizes of RNA polymerase and the nucleosome. The eukaryotic enzymes are large multisubunit proteins, typically >500 kD. Compare this with the ~260 kD of the nucleosome. **Figure 19.35** illustrates the approach of RNA polymerase to nucleosomal DNA. Even without detailed knowledge of the interaction, it is evident that it involves the approach of two comparable bodies.

Consider the two turns that DNA makes around the nucleosome. Would RNA polymerase have sufficient access to DNA if the nucleic acid were confined to this path? During transcription, as RNA polymerase moves along the template, it binds tightly to a region of ~50 bp, including a locally unwound segment of ~12 bp. The need to unwind DNA makes it seem unlikely that the segment engaged by RNA polymerase could remain on the surface of the histone octamer.

It therefore seems inevitable that transcription must involve a structural change. So the first question to ask about the structure of active genes is whether DNA being transcribed remains organized in nucleosomes. If the histone octamers are displaced, do they remain attached in some way to the transcribed DNA?

One experimental approach is to digest chromatin with micrococcal nuclease, and then to use a probe to some specific gene or genes to determine whether the corresponding fragments are present in the usual 200 bp ladder at the expected concentration. The conclusions that we can draw from these experiments are limited but important. *Genes that are being transcribed contain nucleosomes at the same frequency as nontranscribed sequences.* So genes do not necessarily enter an alternative form of organization in order to be transcribed.

But since the average transcribed gene probably only has a single RNA polymerase at any given moment, this does not reveal what is happening at sites actually engaged by the enzyme. Perhaps they retain their nucleosomes; more likely the nucleosomes are temporarily displaced as RNA polymerase passes through, but reform immediately afterward.

Figure 19.36 A protocol to test the effect of transcription on nucleosomes shows that the histone octamer is displaced from DNA and rebinds at a new position.

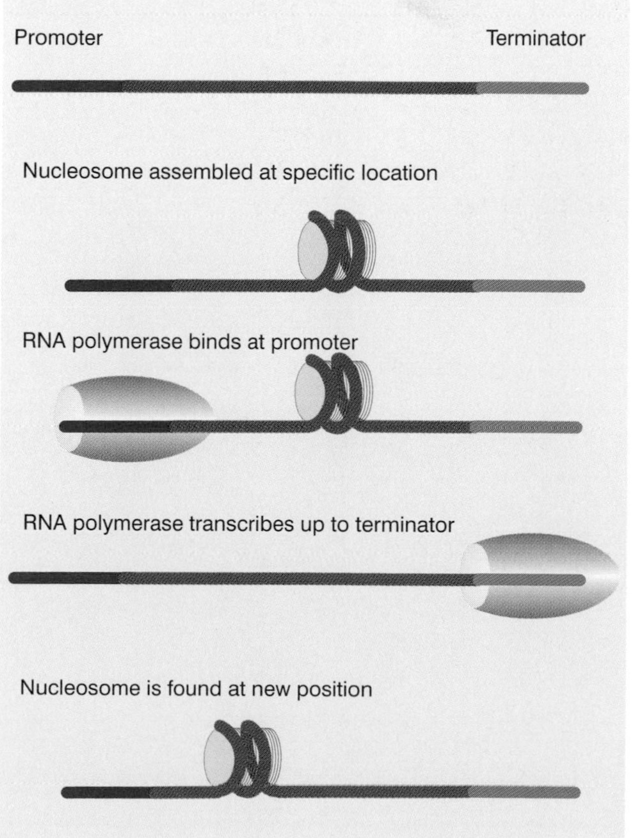
Promoter
Terminator
Nucleosome assembled at specific location
RNA polymerase binds at promoter
RNA polymerase transcribes up to terminator
Nucleosome is found at new position

Experiments to test whether an RNA polymerase can transcribe directly through a nucleosome suggest that the histone octamer is displaced by the act of transcription. **Figure 19.36** shows what happens when an RNA polymerase transcribes a short piece of DNA containing a single octamer core *in vitro*. The core remains associated with the DNA, but is found in a different location. The core is most likely to rebind to the same DNA molecule from which it was displaced.

Figure 19.37 shows a model for polymerase progression. DNA is displaced as the polymerase enters the nucleosome, but the polymerase reaches a point at which the DNA loops back and reattaches, forming a closed region. As polymerase advances further, unwinding the DNA, it creates positive supercoils in this loop; the effect could be dramatic, because each base pair through which the polymerase advances makes a significant addition to the supercoiling in a closed loop of ~80 bp. In fact, the polymerase progresses easily for the first 30 bp into the nucleosome. Then it proceeds more slowly, as though encountering increasing difficulty in progressing. Pauses occur every 10 bp, suggesting that the structure of the loop imposes a constraint related to rotation around each turn of DNA. When the polymerase reaches the midpoint of the nucleosome (the next bases to be added are essentially at the axis of dyad symmetry), pausing ceases, and the polymerase advances rapidly. This suggests that the midpoint of the nucleosome marks the point at which the octamer is displaced (possibly because positive supercoiling has reached some critical level that expels the octamer from DNA). This releases tension ahead of the polymerase and allows it to proceed. The octamer then binds to the DNA behind the polymerase and no longer presents an obstacle to progress. Probably the octamer changes position without ever completely losing contact with the DNA.

Is the octamer released as an intact unit? Crosslinking the proteins of the octamer does not create an obstacle to transcription. Transcription can continue even when crosslinking is extensive enough to ensure that the central regions of the core histones have been linked. This implies that transcription does not require dissociation of the octamer into its component histones, nor is it likely to require any major unfolding of the central structure. However, addition of histone H1 to this system causes a rapid decline in transcription. This suggests two conclusions: the histone octamer (whether remaining present or displaced) functions as an intact unit; and it may be necessary to remove H1 from active chromatin or modify its interactions in some way.

The organization of nucleosomes may be changed by transcription. **Figure 19.38** shows what happens to the yeast *URA3* gene when it is transcribed under control of an inducible promoter. The rate of transcription is high (overcoming the criticism that the target genes being characterized *en masse* may not actually be suffering an act of transcription). Positioning is examined by using micrococcal nuclease to examine cleavage sites relative to a restriction site at the 5′ end of the gene. Initially the gene displays a pattern of nucleosomes that are organized from the promoter for a significant distance across the gene; positioning is lost in the 3′ regions. When the gene is expressed, a general smear replaces the positioned pattern of nucleosomes. So, nucleosomes are present at the same density but are no longer organized in phase. This suggests that transcription destroys the nucleosomal positioning. When

Figure 19.37 RNA polymerase displaces DNA from the histone octamer as it advances. The DNA loops back and attaches (to polymerase or to the octamer) to form a closed loop. As the polymerase proceeds, it generates positive supercoiling ahead. This displaces the octamer, which keeps contact with DNA and/or polymerase, and is inserted behind the RNA polymerase.

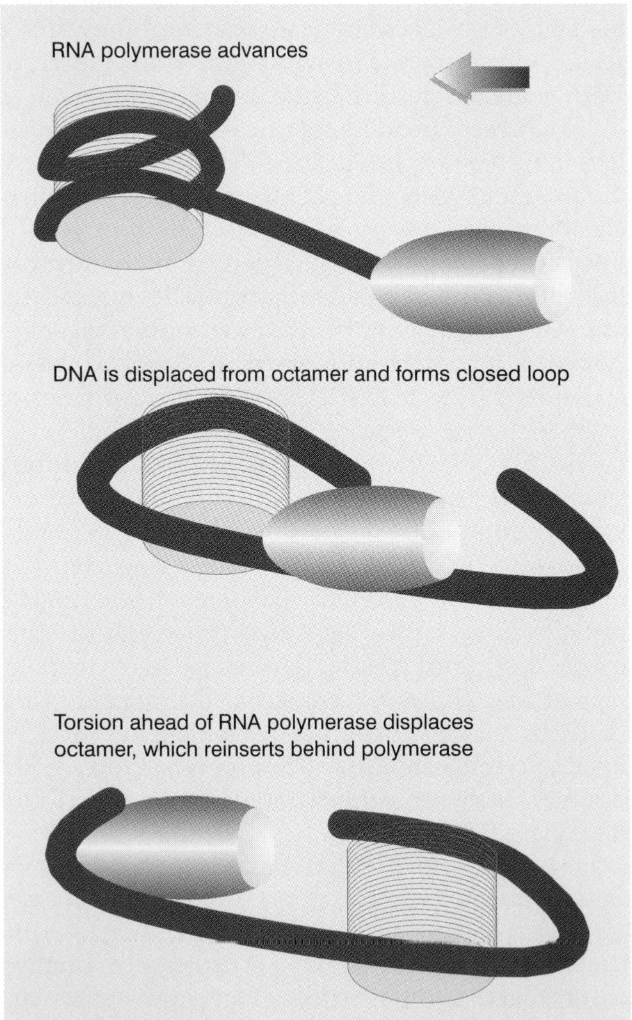
RNA polymerase advances
DNA is displaced from octamer and forms closed loop
Torsion ahead of RNA polymerase displaces octamer, which reinserts behind polymerase

Figure 19.38 The *URA3* gene has translationally positioned nucleosomes before transcription. When transcription is induced, nucleosome positions are randomized. When transcription is repressed, the nucleosomes resume their particular positions. Photograph kindly provided by Fritz Thoma.

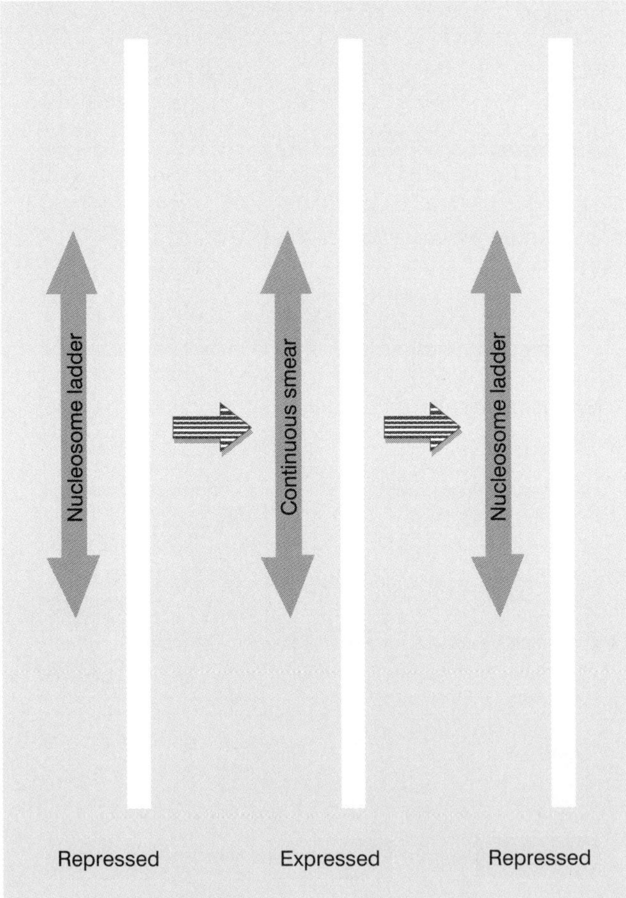
Nucleosome ladder
Continuous smear
Nucleosome ladder
Repressed
Expressed
Repressed

repression is reestablished, positioning appears within 10 min (although it is not complete). This result makes the interesting point that the positions of the nucleosomes can be adjusted without replication.

The unifying model is to suppose that RNA polymerase displaces histone octamers as it progresses. If the DNA behind the polymerase is available, the octamer reattaches there (possibly or probably never having ever totally lost contact with the DNA. It remains a puzzle how an octamer could retain contact with DNA, without unfolding or losing components, as an object of even larger size than itself proceeds along the DNA. Perhaps the octamer is "passed back" by making contacts with RNA polymerase). If the DNA is not available, for example, because another polymerase continues immediately behind the first, then the octamer may be permanently displaced, and the DNA may persist in an extended form.

DNAase hypersensitive sites change chromatin structure

Chromatin consists of DNA wound around a series of nucleosomes, more loosely packaged in regions that contain active genes. In addition to the general changes that occur in active or potentially active regions, structural changes occur at specific sites associated with initiation of transcription or with certain structural features in DNA. These changes were first detected by the effects of digestion with very low concentrations of the enzyme DNAase I.

When chromatin is digested with DNAase I, the first effect is the introduction of breaks in the duplex at specific, **hypersensitive sites.** Since susceptibility to DNAase I reflects the availability of DNA in chromatin, we take these sites to represent chromatin regions in which the DNA is particularly exposed because it is not organized in the usual nucleosomal structure. A typical hypersensitive site is 100× more sensitive to enzyme attack than bulk chromatin. These sites are also hypersensitive to other nucleases and to chemical agents. They represent stretches of DNA devoid of nucleosomes.

Hypersensitive sites are created by the (tissue-specific) structure of chromatin. Their locations can be determined by the technique of indirect end labeling that we introduced earlier in the context of nucleosome positioning. This application of the technique is recapitulated in **Figure 19.39.** In this case, cleavage at the hypersensitive site by DNAase I is used to generate one end of the fragment, and its distance is measured from the other end that is generated by cleavage with a restriction enzyme.

Many of the hypersensitive sites are related to gene expression. Every active gene has a site, or sometimes more than one site, in the region of the promoter. *Most hypersensitive sites are found only in chromatin of cells in which the associated gene is being expressed;* they do not occur when the gene is inactive. The 5′ hypersensitive site(s) appear before transcription begins; and the DNA sequences contained within the hypersensitive sites are required for gene expression, as seen by mutational analysis.

A particularly well-characterized nuclease-sensitive region lies on the SV40 minichromosome. A short segment near the origin of replication, just upstream of the promoter for the late transcription unit, is cleaved preferentially by DNAase I, micrococcal nuclease, and other nucleases (including restriction enzymes).

The state of the SV40 minichromosome can be visualized by electron microscopy. In up to 20% of the samples, a "gap" is visible in the nucleosomal organization, as evident in **Figure 19.40.** The gap is a region of ~120 nm in length (about 350 bp), surrounded on either side by nucleosomes. The visible gap corresponds with the nuclease-sensitive region. This shows directly that increased sensitivity to nucleases is associated with the exclusion of nucleosomes.

A hypersensitive site is not necessarily uniformly sensitive to nucleases. **Figure 19.41** shows the maps of two hypersensitive sites.

Within the SV40 gap of ~300 bp, there are two hypersensitive DNAase I sites and a "protected" region. The protected region presumably reflects the association of (nonhistone) protein(s) with the DNA. The gap is associated with the DNA sequence elements that are necessary for promoter function.

Figure 19.39 Indirect end-labeling identifies the distance of a DNAase hypersensitive site from a restriction cleavage site. The existence of a particular cutting site for DNAase I generates a discrete fragment, whose size indicates the distance of the DNAase I hypersensitive site from the restriction site.

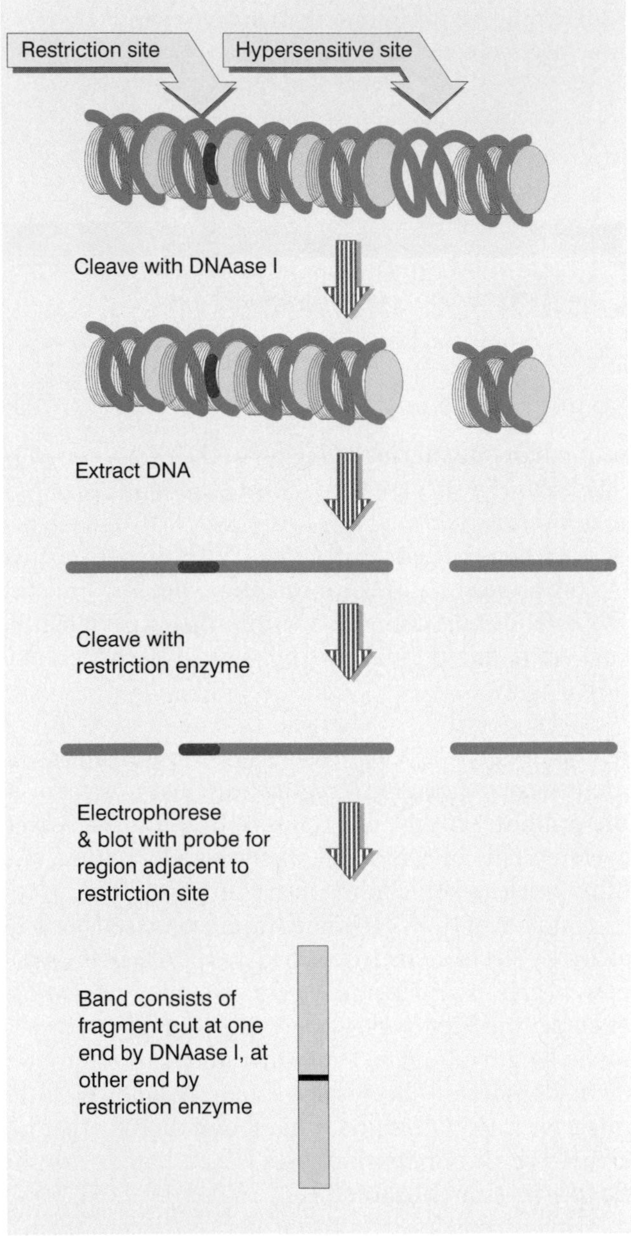
Restriction site
Hypersensitive site
Cleave with DNAase I
Extract DNA
Cleave with restriction enzyme
Electrophorese & blot with probe for region adjacent to restriction site
Band consists of fragment cut at one end by DNAase I, at other end by restriction enzyme

Figure 19.40 The SV40 minichromosome has a nucleosome gap. Photograph kindly provided by Moshe Yaniv.

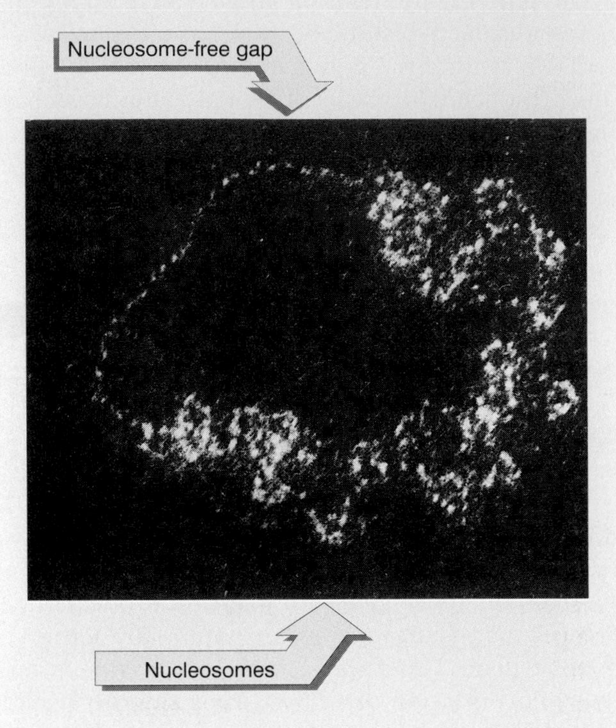
Nucleosome-free gap
Nucleosomes

The hypersensitive site at the β-globin promoter is preferentially digested by several enzymes, including DNAase I, DNAase II, and micrococcal nuclease. The enzymes have preferred cleavage sites that lie at slightly different points in the same general region. So a region extending from about –70 to –270 is preferentially accessible to nucleases when the gene is transcribable. What is the structure of the hypersensitive site? Its preferential accessibility to nucleases indicates that it is not protected by histone octamers, but this does not necessarily imply that it is free of protein. A region of free DNA might be vulnerable to damage; and in any case, why should it exclude nucleosomes? We assume that the hypersensitive site results from the binding of specific regulatory proteins that exclude nucleosomes. Indeed, the binding of such proteins is probably the basis for the existence of the protected region within the hypersensitive site.

Hypersensitive sites associated with transcription may be generated by transcription factors. When a plasmid containing the hypersensitive region of the adult chick β-globin gene is recombined with histones in the presence of an extract from red blood cell nuclei, the relevant region becomes hypersensitive. The extract cannot confer hypersensitivity if it is added *after* the histones, which suggests that it must recognize DNA directly and in some way change the organization of the region prior to or during the deposition of nucleosomes. In Chapter 21, we discuss similar events at

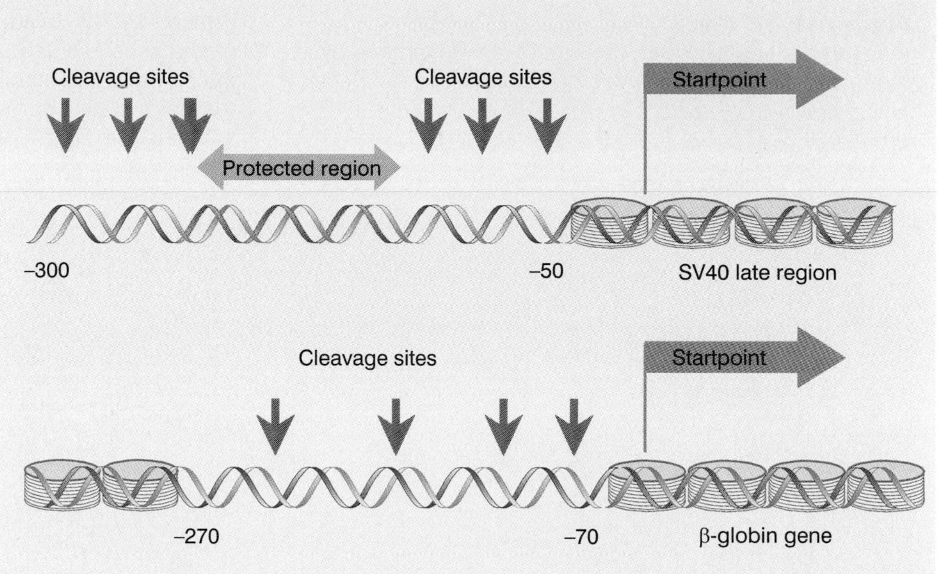

Figure 19.41 The SV40 gap includes hypersensitive sites, sensitive regions, and a protected region of DNA. The hypersensitive site of a chicken β-globin gene comprises a region that is susceptible to several nucleases.

other loci, where transcription factors may have the ability to prevent histones from associating with DNA or (in some cases) to displace them.

The proteins that generate hypersensitive sites are likely to be regulatory factors of various types, since hypersensitive sites are found associated with promoters, other elements that regulate transcription, origins of replication, centromeres, and sites with other structural significance. In some cases, they are associated with more extensive organization of chromatin structure. A hypersensitive site may provide a boundary for a series of positioned nucleosomes.

The stability of hypersensitive sites is revealed by the properties of chick fibroblasts transformed with temperature-sensitive tumor viruses. These experiments take advantage of an unusual property: although fibroblasts do not belong to the erythroid lineage, transformation of the cells at the normal temperature leads to activation of the globin genes. The activated genes have hypersensitive sites. If transformation is performed at the higher (nonpermissive) temperature, the globin genes are not activated; and hypersensitive sites do not appear. When the globin genes have been activated by transformation at low temperature, they can be inactivated by raising the temperature. But the hypersensitive sites are retained through at least the next 20 cell doublings.

This result demonstrates that acquisition of a hypersensitive site is only one of the features necessary to initiate transcription; and it implies that the events involved in establishing a hypersensitive site are distinct from those concerned with perpetuating it. Once the site has been established, it is perpetuated through replication in the absence of the circumstances needed for induction. Could some specific intervention be needed to abolish a hypersensitive site?

Domains define regions that contain active genes

A region of the genome that contains an active gene may have an altered structure. The change in structure precedes, and is different from, the disruption of nucleosome structure that may be caused by the actual passage of RNA polymerase.

One indication of the change in structure of transcribed chromatin is provided by its increased susceptibility to degradation by DNAase I. DNAase I sensitivity defines a chromosomal **domain,** a region of altered structure including at least one active transcription

Figure 19.42 Sensitivity to DNAase I can be measured by determining the rate of disappearance of the material hybridizing with a particular probe.

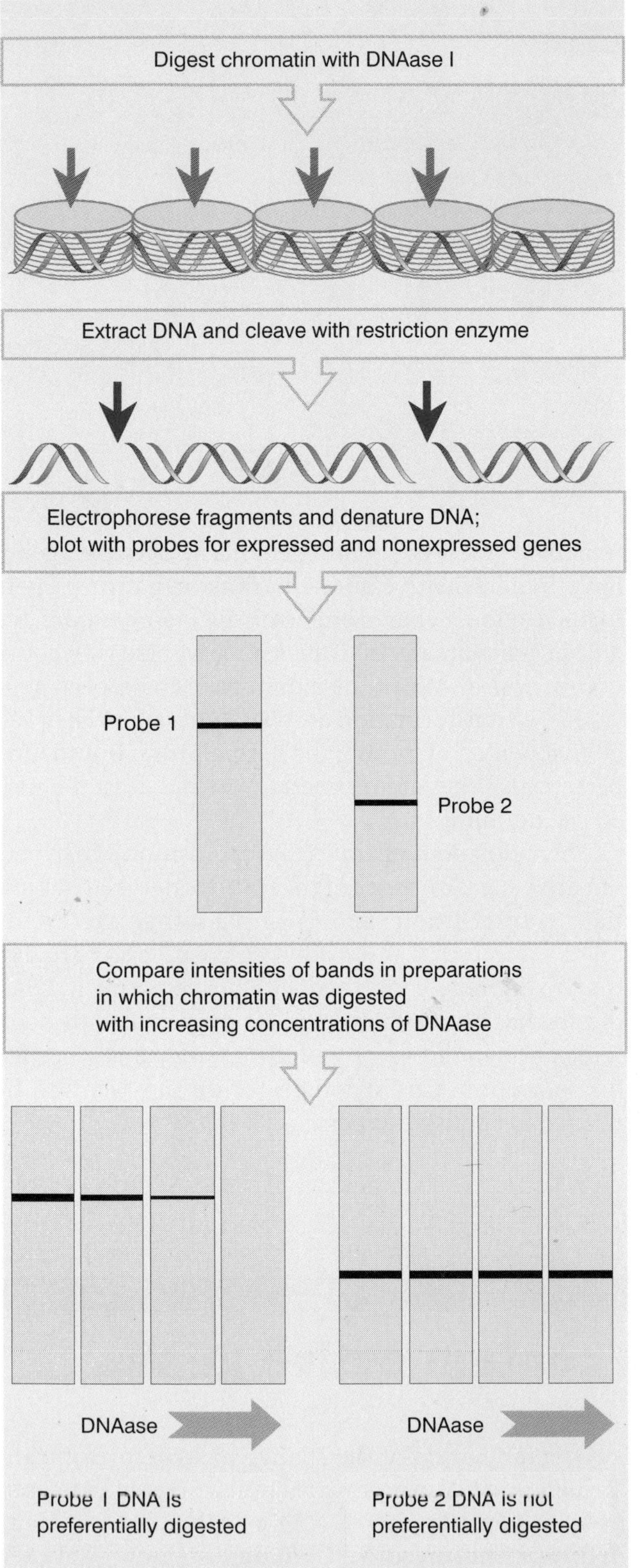
Digest chromatin with DNAase I
Extract DNA and cleave with restriction enzyme
Electrophorese fragments and denature DNA;
blot with probes for expressed and nonexpressed genes
Probe 1
Probe 2
Compare intensities of bands in preparations
in which chromatin was digested
with increasing concentrations of DNAase
DNAase
DNAase
Probe 1 DNA is
preferentially digested
Probe 2 DNA is not
preferentially digested

Figure 19.43 In adult erythroid cells, the adult β-globin gene is highly sensitive to DNAase I digestion, the embryonic β-globin gene is partially sensitive (probably due to spreading effects), but ovalbumin is not sensitive. Data kindly provided by Harold Weintraub.

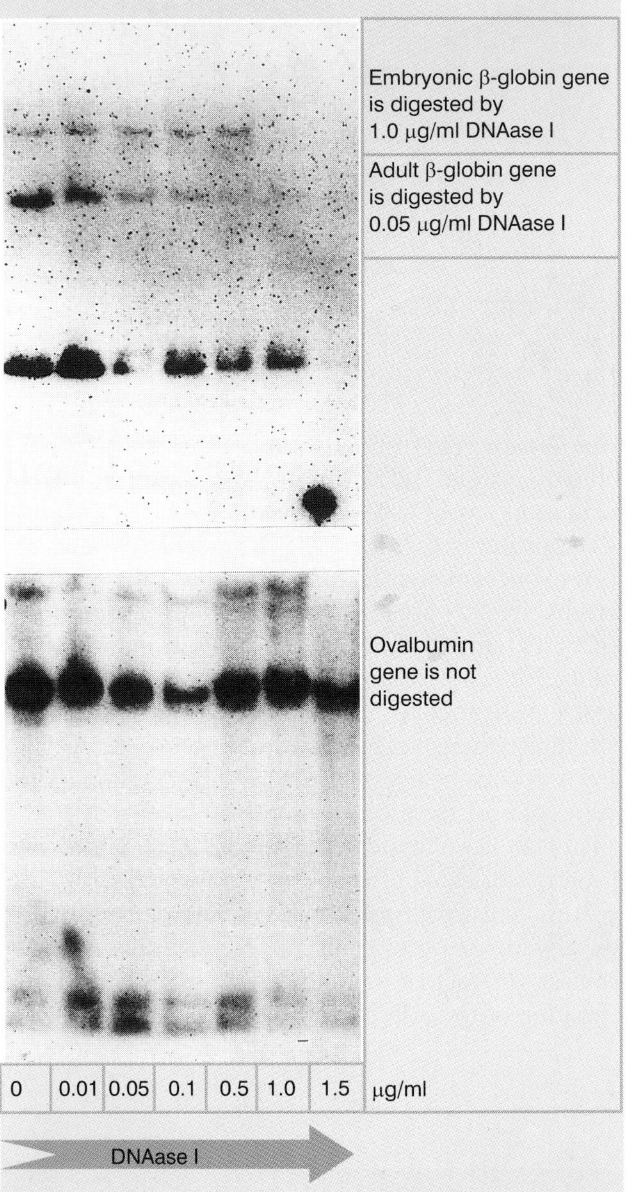
Embryonic β-globin gene
is digested by
1.0 μg/ml DNAase I
Adult β-globin gene
is digested by
0.05 μg/ml DNAase I
Ovalbumin
gene is not
digested
0
0.01
0.05
0.1
0.5
1.0
1.5
μg/ml
DNAase I

unit, and sometimes extending farther. (Note that use of the term "domain" does not imply any necessary connection with the structural domains identified by the loops of chromatin or chromosomes.)

When chromatin is digested with DNAase I, it is eventually degraded into acid-soluble material (very small fragments of DNA). The progress of the overall reaction can be followed in terms of the proportion of DNA that is rendered acid soluble. *When only 10% of the total DNA has become acid soluble, more than 50% of the DNA of an active gene has been lost.* This suggests that active genes are preferentially degraded.

The fate of individual genes can be followed by quantitating the amount of DNA that survives to react with a specific probe. The protocol is outlined in **Figure 19.42.** The principle is that the loss of a particular band indicates that the corresponding region of DNA has been degraded by the enzyme.

Figure 19.43 shows what happens to β-globin genes and an ovalbumin gene in chromatin extracted from chicken red blood cells (in which globin genes are expressed and the ovalbumin gene is inactive). The restriction fragments representing the β-globin genes are rapidly lost, while those representing the ovalbumin gene show little degradation. (The ovalbumin gene in fact is digested at the same rate as the bulk of DNA.)

So the bulk of chromatin is relatively resistant to DNAase I and contains nonexpressed genes (as well as other sequences). *A gene becomes relatively susceptible to the enzyme specifically in the tissue(s) in which it is expressed.*

Is preferential susceptibility a characteristic only of rather actively expressed genes, such as globin, or of all active genes? Experiments using probes representing the entire cellular mRNA population suggest that all active genes, whether coding for abundant or for rare mRNAs, are preferentially susceptible to DNAase I. (However, there are variations in the degree of susceptibility.) Since the rarely expressed genes are likely to have very few RNA polymerase molecules actually engaged in transcription at any moment, this implies that the sensitivity to DNAase I does not result from the act of transcription, but is a feature of *genes that are able to be transcribed.*

What is the extent of the preferentially sensitive region? This can be determined by using a series of probes representing the flanking regions as well as the transcription unit itself. The sensitive region always extends over the entire transcribed region; an additional region of several kb on either side may show an intermediate level of sensitivity (probably as the result of spreading effects).

The critical concept implicit in the description of the domain is that a region of high sensitivity to DNAase I extends over a considerable distance. Often we think of regulation as residing in events that occur at a discrete site in DNA—for example, in the ability to initiate transcription at the promoter. Even if this is true, such regulation must determine, or must be accompanied by, a more wide-ranging change in structure. This is a difference between eukaryotes and prokaryotes.

Heterochromatin depends on interactions with histones

The overall properties of chromatin are influenced by interactions with proteins of the **SMC** (structural maintenance of chromosome) family. They are ATPases that fall into two functional groups. **Condensins** are involved with the control of overall structure. **Cohesins** are concerned with connections between sister chromatids that must be released at mitosis (see Chapter 27).

Condensins form complexes that have a core of the heterodimer SMC2•SMC4 associated with other (non-SMC) proteins. (Cohesins have a similar organization based on the heterodimeric core of SMC1•SMC3.) The condensin complex is named for its ability to cause chromatin to condense *in vitro*. It has an ability to introduce positive supercoils into DNA in an action that uses hydrolysis of ATP and depends on the presence of topoisomerase I. This ability is controlled by the phosphorylation of the non-SMC subunits, which occurs at mitosis. We do not know yet how this connects with other modifications of chromatin, for example, the phosphorylation of histones. The activation of the condensin complex specifically at mitosis makes it questionable whether it is also involved in the formation of interphase heterochromatin.

An interphase nucleus contains both euchromatin and heterochromatin. The condensation state of heterochromatin is close to that of mitotic chromosomes. Heterochromatin is inert. It remains condensed in interphase, is transcriptionally repressed, replicates late in S phase, and may be localized to the nuclear periphery. Centromeric heterochromatin typically consists of satellite DNAs. However, the formation of heterochromatin is not rigorously defined by sequence. When a gene is transferred, either by a chromosomal translocation or by transfection and integration, into a position adjacent to heterochromatin, it may become inactive as the result of its new location.

Such inactivation is the result of an **epigenetic** effect (see also later). It may differ between individual cells in an animal, and results in the phenomenon of **position effect variegation,** in which genetically identical cells have different phenotypes. This has been well characterized in *Drosophila.* **Figure 19.44** shows an example of position effect variegation in the fly eye, in which some regions lack color while others are red, because the *white* gene is inactivated by adjacent heterochromatin in some cells, while it remains active in other cells.

Figure 19.44 Position effect variegation in eye color results when the *white* gene is integrated near heterochromatin. Cells in which *white* is inactive give patches of white eye, while cells in which *white* is active give red patches. The severity of the effect is determined by the closeness of the integrated gene to heterochromatin. Photograph kindly provided by Steve Henikoff.

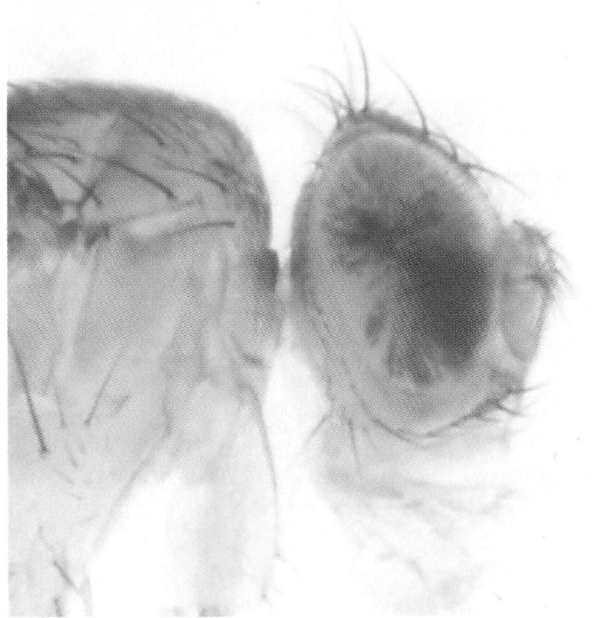

The explanation for this effect is shown in **Figure 19.45.** Inactivation spreads from heterochromatin into the adjacent region for a variable distance. In some cells it goes far enough to inactivate a nearby gene, but in others it does not. This happens at a certain point in embryonic development, and after that point the state of the gene is inherited by all the progeny cells. Cells descended from an ancestor in which the gene was inactivated form patches corresponding to the phenotype of loss-of-function (in the case of *white,* absence of color).

The closer a gene lies to heterochromatin, the higher the probability that it will be inactivated. This suggests that the formation of heterochromatin may be a two-stage process: a *nucleation* event occurs at a specific sequence; and then the inactive structure *propagates* along the chromatin fiber. The distance for which the inactive structure extends is not precisely determined (for example, by boundary elements) and may be stochastic, being influenced by parameters such as the quantities of limiting protein components.

Genes that are closer to heterochromatin are more likely to be inactivated, and will therefore be inactive in a greater proportion of cells. On this model, the boundaries of a heterochromatic region might be terminated by exhausting the supply of one of the proteins that is required.

The effect of **telomeric silencing** in yeast is analogous to position effect variegation in *Drosophila;* genes translocated to a telomeric location show the same sort of variable loss of activity. This results from a spreading effect that propagates from the telomeres.

A second form of silencing occurs in yeast. Yeast mating type is determined by the activity of a single active locus (*MAT*), but the genome contains two other copies of the mating type sequences (*HML* and *HMR*), which are maintained in an inactive form. We discuss the contribution of these loci to yeast mating type in Chapter 17. The silent loci *HML* and *HMR* share many properties with heterochromatin, and could be regarded as constituting regions of heterochromatin in miniature.

The existence of a common basis for both types of silencing in yeast is suggested by their common re-

Figure 19.45 Extension of heterochromatin inactivates genes. The probability that a gene will be inactivated depends on its distance from the heterochromatin region.

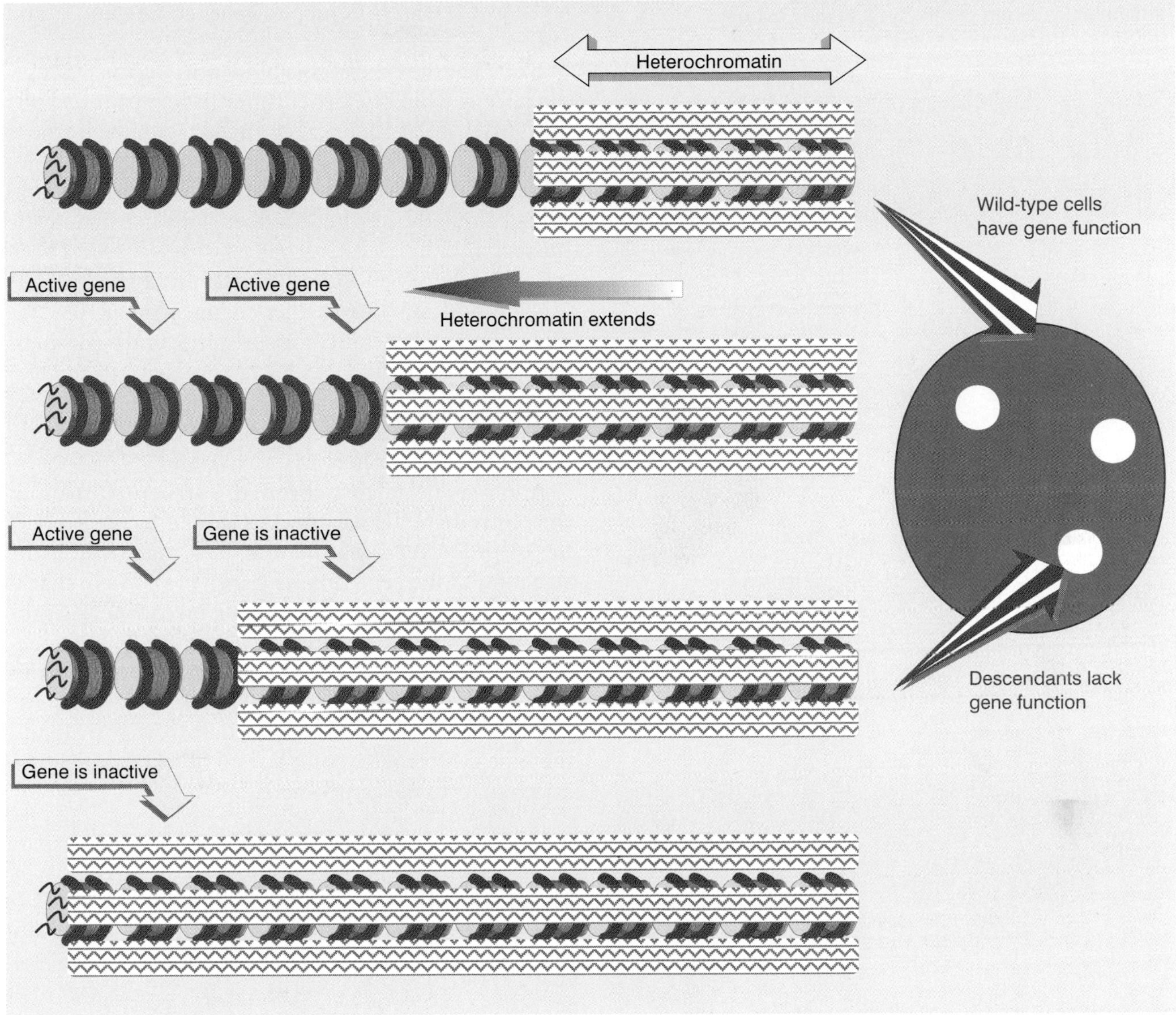

liance on the same set of genetic loci. Mutations in any one of a number of genes cause *HML* and *HMR* to become activated, and also relieve the inactivation of genes that have been integrated near telomeric heterochromatin. The products of these loci therefore function to maintain the inactive state of both types of heterochromatin.

Figure 19.46 proposes a model for actions of these proteins. Only one of them is a sequence-specific DNA-binding protein. This is RAP1, which binds to the $C_{1-3}A$ repeats at the telomeres, and also binds to the *cis*-acting silencer elements that are needed for repression of *HML* and *HMR*. The proteins SIR3 and SIR4 interact with RAP1 and also with one another (they may function as a heteromultimer). SIR3/SIR4 interact with the N-terminal tails of the histones H3 and H4.

Figure 19.46 Formation of heterochromatin is initiated when RAP1 binds to DNA. SIR3/4 bind to RAP1 and also to histones H3/H4. The complex polymerizes along chromatin and may connect telomeres to the nuclear matrix.

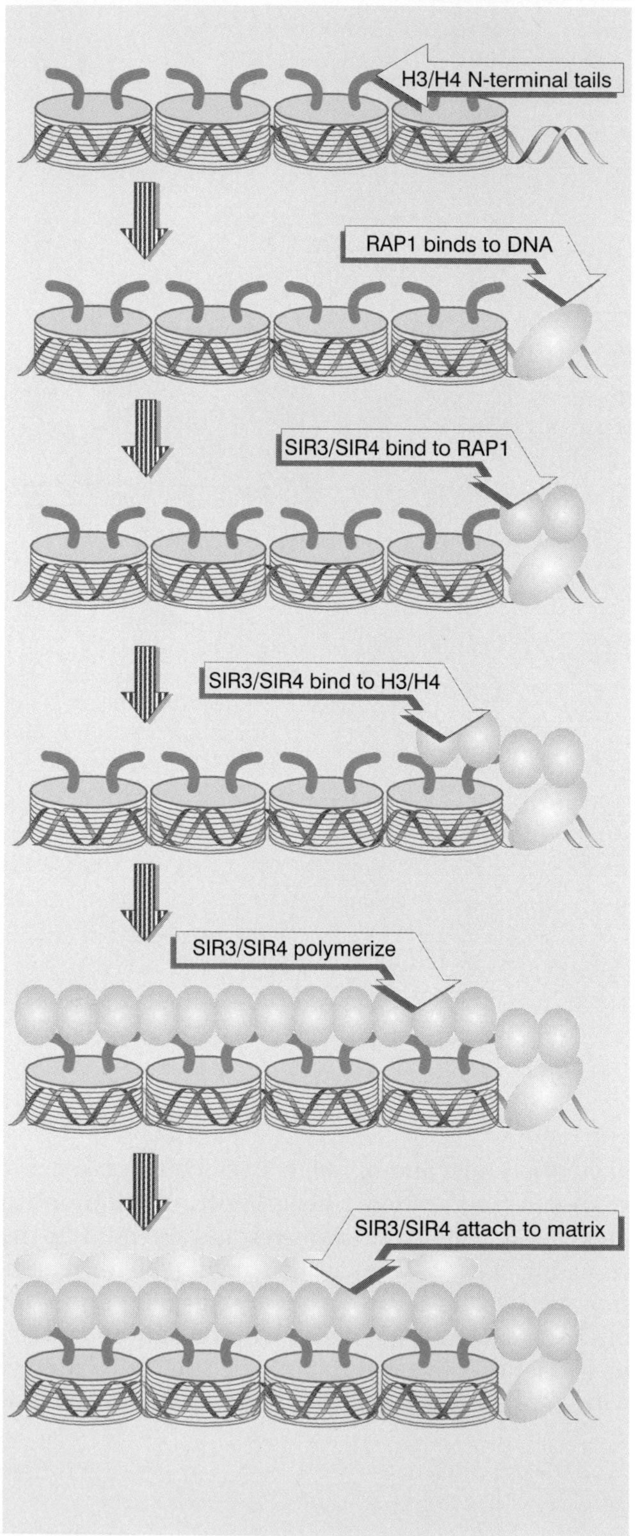
H3/H4 N-terminal tails
RAP1 binds to DNA
SIR3/SIR4 bind to RAP1
SIR3/SIR4 bind to H3/H4
SIR3/SIR4 polymerize
SIR3/SIR4 attach to matrix

(In fact, the first evidence that histones might be involved directly in formation of heterochromatin was provided by the discovery that mutations abolishing silencing at *HML/HMR* map to genes coding for H3 and H4.)

RAP1 has the crucial role of identifying the DNA sequences at which heterochromatin forms. It recruits SIR3/SIR4, and they interact directly with the histones H3/H4. Once SIR3/SIR4 have bound to histones H3/H4, the complex may polymerize further, and spread along the chromatin fiber. This may inactivate the region, either because coating with SIR3/SIR4 itself has an inhibitory effect, or because binding to histones H3/H4 induces some further change in structure. We do not know what limits the spreading of the complex. The C-terminus of SIR3 has a similarity to nuclear lamin proteins (constituents of the nuclear matrix) and may be responsible for tethering heterochromatin to the nuclear periphery.

Another specialized chromatin structure forms at the centromere. Its nature is suggested by the properties of an *S. cerevisiae* mutation, *cse4*, that disrupts the structure of the centromere. Cse4p is a protein that is related to histone H3. A mammalian centromeric protein, CENP-A, has a related sequence. Genetic interactions between *cse4* and CDE-II, and between *cse4* and a mutation in the H4 histone gene, suggest that a histone octamer may form around a core of Cse4p•H4, and then the centromeric complexes CBF1 and CBF3 may attach to form the centromere.

Global changes in X chromosomes

Sex presents an interesting problem for gene regulation, because of the variation in the number of X chromosomes. If X-linked genes were expressed equally well in each sex, females would have twice as much of each product as males. The importance of avoiding this situation is shown by the existence of **dosage compensation,** which equalizes the level of expression of X-linked genes in the two sexes. Mechanisms used in different species are summarized in **Figure 19.47:**

- In mammals, one of the two female X chromosomes is inactivated completely. The result is that females have only one active X chromosome, which is the same situation found in males. The active X chromosome of females and the single X chromosome of males are expressed at the same level.
- In *Drosophila,* the expression of the single male X chromosome is doubled relative to the expression of each female X chromosome.
- In *C. elegans,* the expression of each female X chromosome is halved relative to the expression of the single male X chromosome.

The common feature in all these mechanisms of dosage compensation is that *the entire chromosome is the target for regulation.* A global change occurs that quantitatively affects all of the promoters on the chromosome. We know most about the inactivation of the X chromosome in mammalian females, where the entire chromosome becomes heterochromatin.

Figure 19.47 Different means of dosage compensation are used to equalize X chromosome expression in male and female.

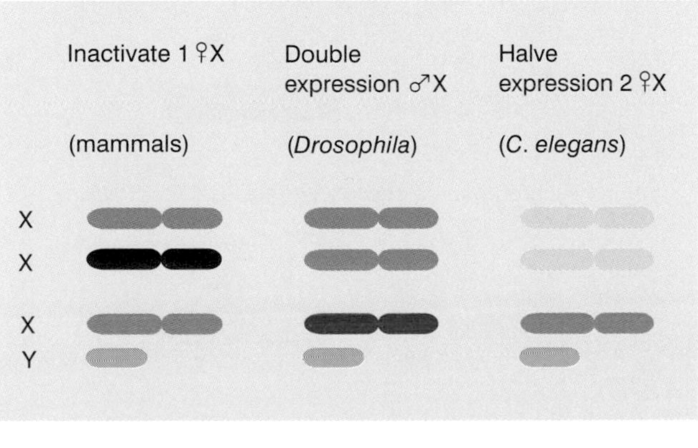

The twin properties of heterochromatin are its condensed state and associated inactivity. It can be divided into two types:

- **Constitutive heterochromatin** contains specific sequences that have no coding function. Typically these include satellite DNAs, and are often found at the centromeres. These regions are invariably heterochromatic because of their intrinsic nature.
- **Facultative heterochromatin** takes the form of entire chromosomes that are inactive in one cell lineage, although they can be expressed in other lineages. The example *par excellence* is the mammalian X chromosome. The inactive X chromosome is perpetuated in a heterochromatic state, while the active X chromosome is part of the euchromatin. So *identical DNA sequences are involved in both states.* Once the inactive state has been established, it is inherited by descendant cells. This is an example of epigenetic inheritance, because it does not depend on the DNA sequence.

Our basic view of the situation of the female mammalian X chromosomes was formed by the **single X hypothesis** in 1961. Female mice that are heterozygous for X-linked coat color mutations have a variegated phenotype in which some areas of the coat are wild-type, but others are mutant. **Figure 19.48** shows that this can be explained *if one of the two X chromosomes is inactivated at random in each cell of a small precursor population.* Cells in which the X chromosome carrying the wild-type gene is inactivated give rise to progeny that express only the mutant allele on the active chromosome. Cells derived from a precursor where the other chromosome was inactivated have an active wild-type gene. In the case of coat color, cells descended from a particular precursor stay together and thus form a patch of the same color, creating the pattern of visible variegation. In other cases, individual cells in a population will express one or the other of X-linked alleles; for example, in heterozygotes for the X-linked locus G6PD, any particular red blood cell will express only one of the two allelic forms. (Random inactivation of one X chromosome occurs in eutherian mammals. In marsupials, the choice is directed: it is always the X chromosome inherited from the father that is inactivated.)

Inactivation of the X chromosome in females is

governed by the **n–1 rule:** however many X chromosomes are present, all but one will be inactivated. In normal females there are of course 2 X chromosomes, but in rare cases where nondisjunction has generated a 3X or greater genotype, only one X chromosome remains active. This suggests a general model in which a specific event is limited to one X chromosome and protects it from an inactivation mechanism that applies to all the others.

A single locus on the X chromosome is sufficient for inactivation. When a translocation occurs between the X chromosome and an autosome, this locus is present on only one of the reciprocal products, and only that product can be inactivated. By comparing different translocations, it is possible to map this locus, which is called the *Xic* (X-inactivation center). A cloned region of 450 kb contains all the properties of the *Xic*. When this sequence is inserted as a transgene on to an autosome, the autosome becomes subject to inactivation (in a cell culture system).

Xic is a *cis*-acting locus that contains the information necessary to count X chromosomes and inactivate all copies but one. Inactivation spreads from *Xic* along the entire X chromosome. When *Xic* is present on an X chromosome–autosome translocation, inactivation spreads into the autosomal regions (although the effect is not always complete).

Xic contains a gene, called *Xist*, that is expressed only on the *inactive* X chromosome. The behavior of this gene is effectively the opposite from all other loci on the chromosome, which are turned off. Deletion of *Xist* prevents an X chromosome from being inactivated. However, it does not interfere with the counting mechanism (because other X chromosomes can be inactivated). So we can distinguish two features of *Xic*: an unidentified element(s) required for counting; and the *Xist* gene required for inactivation.

Figure 19.49 illustrates the role of *Xist* RNA in X-inactivation. *Xist* codes for an RNA that lacks open reading frames. The *Xist* RNA "coats" the X chromo-

Figure 19.48 X-linked variegation is caused by the random inactivation of one X chromosome in each precursor cell.

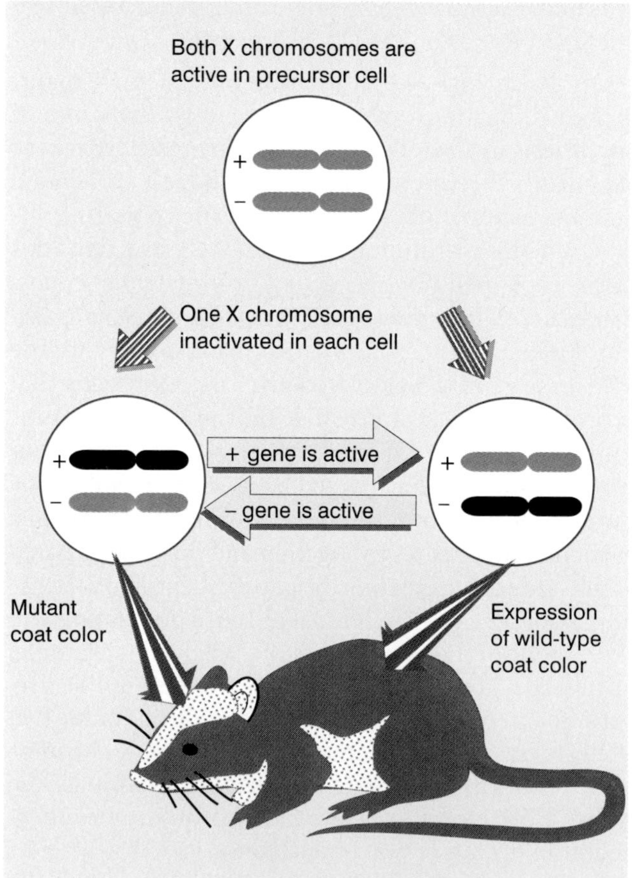

Figure 19.49 X-inactivation involves stabilization of *XIST* RNA, which coats the inactive chromosome.

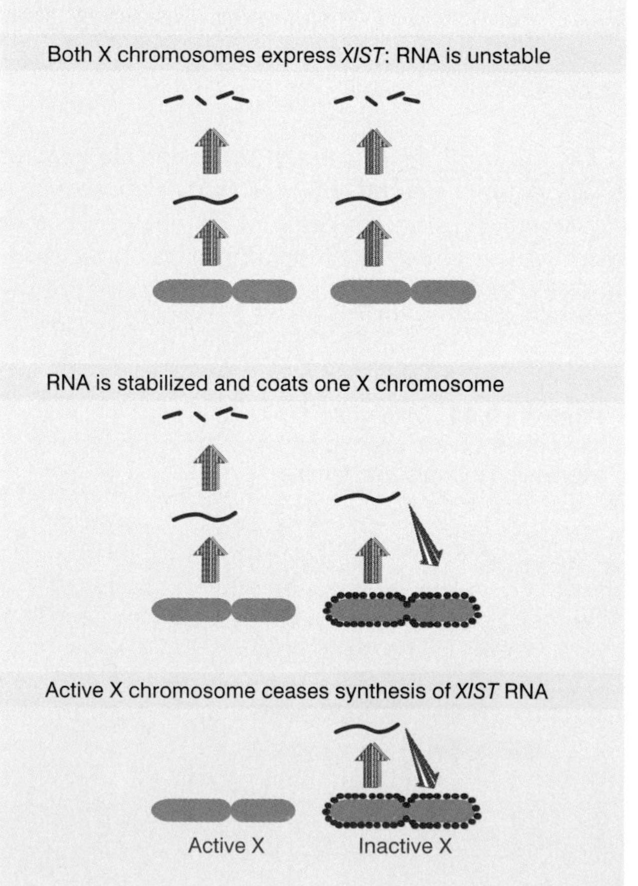

some from which it is synthesized, suggesting that it has a structural role. Prior to X-inactivation, it is synthesized by both female X chromosomes. Following inactivation, the RNA is found only on the inactive X chromosome. The transcription rate remains the same before and after inactivation, so the transition depends on post-transcriptional events.

Prior to X-inactivation, *Xist* RNA decays with a half life of ~2 hrs. X-inactivation is mediated by stabilizing the *Xist* RNA on the inactive X chromosome. The *Xist* RNA shows a punctate distribution along the X chromosome, suggesting that association with proteins to form particulate structures may be the means of stabilization. We do not know yet what other factors may be involved in this reaction and how the *Xist* RNA is limited to spreading in *cis* along the chromosome. The characteristic features of the inactive X chromosome, which include a lack of acetylation of histone H4, and methylation of CpG sequences (see below), presumably occur later as part of the mechanism of inactivation.

The n–1 rule suggests that stabilization of *Xist* RNA is the "default," and that some blocking mechanism prevents stabilization at one X chromosome (which will be the active X). This means that, although *Xic* is necessary and sufficient for a chromosome to be *inactivated*, the products of other loci may be necessary for the establishment of an *active* X chromosome.

Silencing of *Xist* expression is necessary for the active X. Deletion of the gene for DNA methyltransferase prevents silencing of *Xist*, probably because methylation at the *Xist* promoter is necessary for cessation of transcription (the effects of methylation on transcription are discussed in Chapter 21).

Global changes occur in other types of dosage compensation. In *Drosophila*, a complex of proteins is found in males, where it localizes on the X chromosome. In *C. elegans*, a protein complex associates with both X chromosomes in XX embryos, but the protein components remain diffusely distributed in the nuclei of XO embryos. The protein complex contains an SMC core, and is similar to the condensin complexes that are associated with mitotic chromosomes in other species. This suggests that it has a structural role in causing the chromosome to take up a more condensed, inactive state. Multiple sites on the X chromosome may be needed for the complex to be fully distributed along it.

Changes affecting all the genes on a chromosome, either negatively (mammals and *C. elegans*) or positively (*Drosophila*) are therefore a common feature of dosage compensation. However, the components of the dosage compensation apparatus may vary (RNA in mammals, proteins in *Drosophila* and *C. elegans*), and the means by which it is localized to the chromosome are different, as of course is its mechanism of action.

Methylation is responsible for imprinting

METHYLATION of DNA occurs at specific sites. In bacteria, it is associated with identifying the particular bacterial strain, and also with distinguishing replicated and nonreplicated DNA (see Chapter 12). In eukaryotes, its principal known function is connected with the control of transcription. We discuss the role of methylation in controlling transcription in Chapter 21, assuming for the present that methylation is associated with gene inactivation.

From 2–7% of the cytosines of animal cell DNA are methylated (the value varies with the species). Most of the methyl groups are found in CG "doublets," and, in fact, the majority of the CG sequences are methylated. Usually the C residues on both strands of this short palindromic sequence are methylated, giving the structure

$$\begin{array}{l} 5'\ {}^{m}\text{CpG}\ 3' \\ 3'\ \ \text{GpC}^{m}\ 5' \end{array}$$

Such a site is described as **fully methylated.** But consider the consequences of replicating this site. **Figure 19.50** shows that each daughter duplex has one methylated strand and one unmethylated strand. Such a site is called **hemimethylated.**

The perpetuation of the methylated site now depends on what happens to hemimethylated DNA. If

Figure 19.50 The state of methylated sites could be perpetuated by an enzyme that recognizes only hemimethylated sites as substrates.

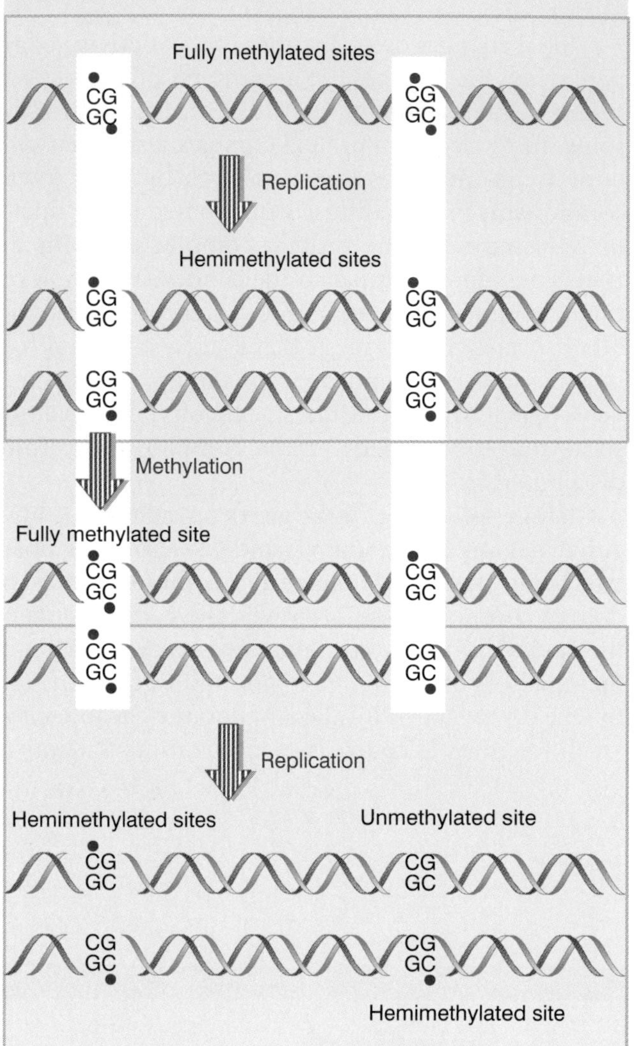

Figure 19.51 The state of methylation is controlled by three enzymes.

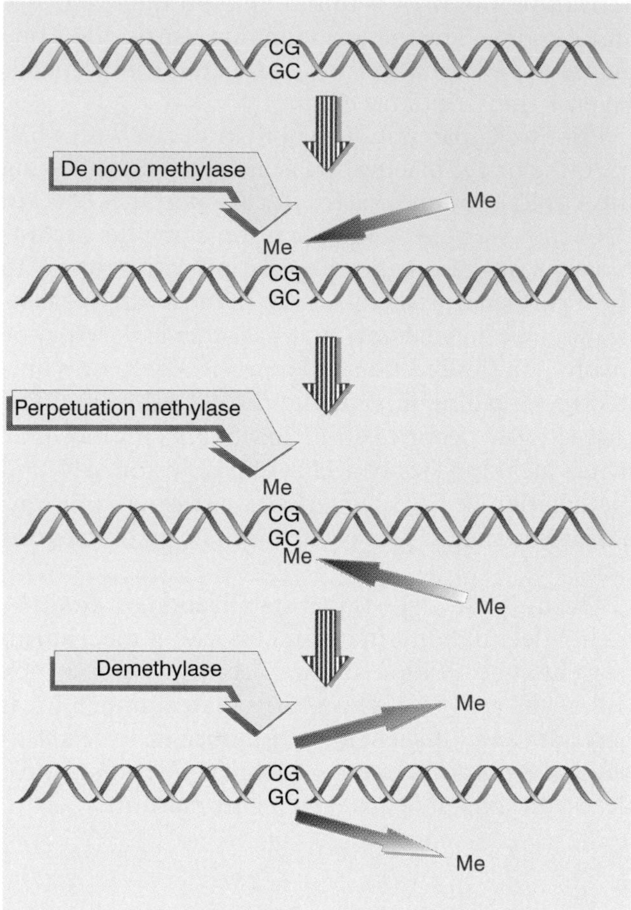

methylation of the unmethylated strand occurs, the site is restored to the fully methylated condition. However, if replication occurs next, the hemimethylated condition will be perpetuated on one daughter duplex, but the site will become unmethylated on the other daughter duplex. **Figure 19.51** shows that the state of methylation of DNA is controlled by **methylases,** which add methyl groups to the 5 position of cytosine, and **demethylases,** which remove the methyl groups.

There are two types of DNA methylase, whose actions are distinguished by the state of the methylated DNA. To modify DNA at a new position requires the action of the ***de novo* methylase,** which presumably recognizes DNA by virtue of a specific sequence. This enzyme has not yet been characterized, but we infer its existence from the existence of *de novo* methylation. (We do not know whether it adds methyl groups to both DNA strands or only to one strand.)

The only known methylase activity in mammalian cells acts as a **maintenance methylase:** it acts constitutively on hemimethylated sites to convert them to fully methylated sites. The enzyme is ubiquitous. Its existence means that any methylated site is perpetuated after replication. The methylase is essential: mouse embryos in which its gene has been disrupted do not survive past early embryogenesis.

So the situation shown on the left of Figure 19.50 usually prevails *in vivo;* methyl groups are rarely lost by a failure to convert hemimethylated sites to fully methylated sites. So if a *de novo* methylation occurs on one allele but not on the other, this difference will be perpetuated through ensuing cell divisions, maintaining a difference between the alleles that does not depend on their sequences.

The model for constitutive perpetuation of methylated sites is supported by the observation that, when methylated DNA is introduced into a cell, it continues to be methylated through an indefinite number of replication cycles, with a fidelity ~95% per site. If nonmethylated DNA is introduced, it is not methylated *de novo*. This implies that the enzyme recognizes *only* the hemimethylated sites.

The state of methylation may change at the promoters of genes that are subject to tissue-specific control. The promoters are methylated when the gene is inactive, but unmethylated when it is active.

Both methylation and demethylation occur during embryogenesis. A typical pattern of methylation is established in each sex during gametogenesis. In males, the pattern develops in two stages. Spermatocytes display the methylation pattern that is characteristic of mature sperm, so no further changes occur during spermatogenesis. But further changes are made in this pattern after fertilization. In females, the maternal pattern is imposed during oogenesis, when oocytes mature through meiosis after birth. As may be expected from the inactivity of genes in gametes, the typical state is to be methylated. However, there can be differences between the two sexes, with the result that the paternal and maternal alleles have different patterns of methylation.

A change in methylation pattern occurs during embryogenesis. All allelic differences are lost when primordial germ cells develop in the embryo; irrespective of sex, the previous patterns of methylation are erased, and a typical gene is then unmethylated. The methylation pattern of germ cells is therefore established by a two stage process: first the previous pattern is erased by a genome-wide demethylation; then the pattern specific for each sex is imposed. A major question is how the specificity of methylation is determined in the male and female gametes.

Systematic changes occur in early embryogenesis. Some sites will continue to be methylated, but others will be specifically unmethylated in cells in which a gene is expressed. From this pattern of changes, we may infer that individual sequence-specific demethylation events occur during somatic development of the organism as particular genes are activated.

The specific pattern of methyl groups in germ cells is responsible for the phenomenon of **imprinting,** which describes a difference in behavior between the alleles inherited from each parent. The expression of certain genes in mouse embryos depends upon the sex of the parent from which they were inherited. For example, the allele coding for IGF-II (insulin-like growth factor II) that is inherited from the father is expressed, but the allele that is inherited from the mother is not expressed. The IGF-II gene of oocytes is methylated, but the IGF-II gene of sperm is not methylated, so that the two alleles behave differently in the zygote. This is the most common pattern, but the dependence on sex is reversed for some genes. In fact, the opposite pattern (expression of maternal copy) is shown for IGF-IIR, the receptor for IGF-II.

Imprinted genes are sometimes clustered; of the 17 known imprinted genes in mouse, more than half are contained in two particular regions, each containing both maternally and paternally expressed genes. This suggests the possibility that imprinting mechanisms may function over long distances. Some insights into this possibility come from deletions in the human population that cause the Prader-Willi and Angelman diseases. Most cases are caused by the same 4 Mb deletion, but the syndromes are different, depending on which parent contributed the deletion. The reason is that the deleted region contains at least one gene that is paternally imprinted and at least one that is maternally imprinted. There are some rare cases, however, with much smaller deletions. Prader-Willi syndrome can be caused by a 20 kb deletion that silences genes that are distant on either side of it. The basic effect of the deletion is to prevent a father from resetting the paternal mode to a chromosome inherited from his mother. The result is that these genes remain in maternal mode, so that the paternal as well as maternal alleles are silent in the offspring. The inverse effect is found in some small deletions that cause Angelman's syndrome. The implication is that this region comprises some sort of "imprint center" that acts at a distance to switch one parental type to the other.

Imprinting is a classic example of epigenetic inheritance. Although the paternal and maternal alleles have identical sequences, they display different properties, depending on which parent provided them. These properties are inherited through meiosis and the subsequent somatic mitoses.

This sex-specific mode of inheritance requires that the pattern of methylation is established specifically during each gametogenesis. The fate of the two Igf2 alleles in a mouse is illustrated in **Figure 19.52.** In the early embryo, the paternal allele is nonmethylated and expressed, and the maternal allele is methylated and silent. What happens when this mouse itself forms gametes? If it is a male, the allele contributed to the sperm

Figure 19.52 The parental alleles of Igf2 are differentially methylated in the early embryo, but the patterns of methylation are reset when gametes are formed by the adult.

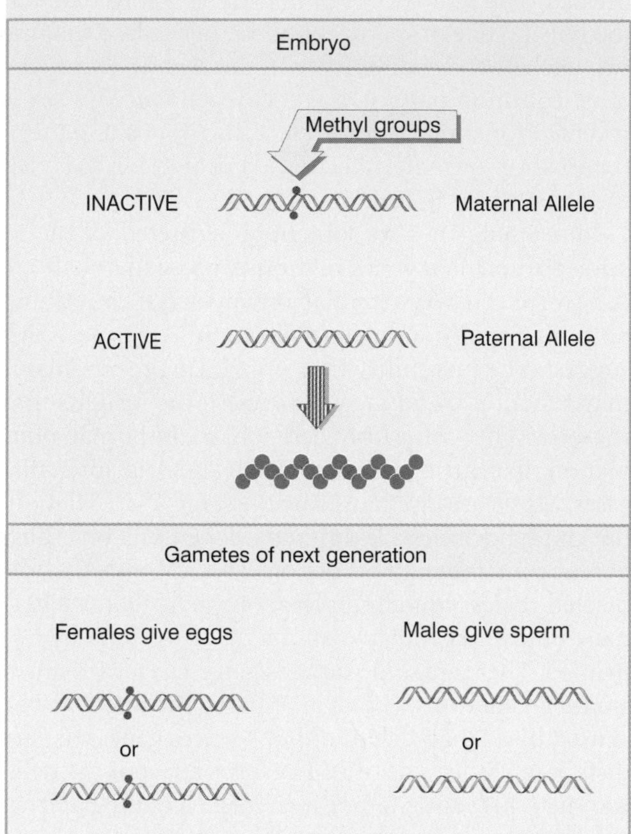

must be nonmethylated, irrespective of whether it was originally methylated or not. So when the maternal allele finds itself in a sperm, it must be demethylated. If the mouse is a female, the allele contributed to the egg must be methylated; so if it was originally the paternal allele, methyl groups must be added.

The behavior of *Igf2*, in which methylation creates an inactive imprinted state, is the most common. This reflects a direct effect of methylation on promoter activity. However, in some cases, methylation marks the active state of an imprinted gene. In these cases, the effect of methylation is indirect; for example, methylation could inactivate a silencing element.

By using an assay to detect sequences that undergo *de novo* methylation very early in embryonic development, a potential imprinting control site has been identified in *Igf2R*. It consists of an ~100 bp sequence in a CpG island in the intron of the gene. The site contains a *cis*-acting *DNS* (*de novo* methylation signal) element at one end that is required for *de novo* methylation to occur. It contains a sequence at the other end in which mutations allow the paternal allele to be methylated *de novo*. This suggests that this *ADS* (allele discriminating signal) is bound by a protein that specifically protects the paternal allele against *de novo* methylation. These two elements are necessary, but not sufficient, for methylation of sites in the vicinity that are differentially methylated.

Modes of epigenetic inheritance

EPIGENETIC **inheritance** describes the ability of different states, which may have different phenotypic consequences, to be inherited without any change in the sequence of DNA. How can this occur?

We can divide epigenetic mechanisms into two general classes:

- DNA may be modified by the covalent attachment of a moiety that is then perpetuated. Two alleles with the same sequence may have different states of methylation that confer different properties.
- Or a self perpetuating protein state may be established. This might involve assembly of a protein complex, modification of specific protein(s), or establishment of an alternative protein conformation.

Methylation, as we have just seen, establishes epigenetic inheritance so long as the maintenance methylase acts constitutively to restore the methylated state after each cycle of replication, as shown in Figure 19.50. A state of methylation can be perpetuated through an indefinite series of somatic mitoses. In fact, the "default" situation is for methylation to be maintained unless a demethylase specifically removes the methyl groups. Methylation can also be perpetuated through meiosis: for example, in the fungus *Ascobolus* there are epigenetic effects that can be transmitted through both mitosis and meiosis by maintaining the state of methylation. In mammalian cells, epigenetic effects are created by resetting the state of methylation differently in male and female meioses.

Situations in which epigenetic effects appear to be maintained by means of protein states are less well understood in molecular terms. Position effect variegation shows that constitutive heterochromatin may extend for a variable distance, and the structure is then perpetuated through somatic divisions. Since there is no methylation of DNA in either *Saccharomyces* or *Drosophila*, the inheritance of epigenetic states of position effect variegation or telomeric silencing is likely to be due to the perpetuation of protein structures.

Figure 19.53 considers two extreme possibilities for the fate of a protein complex at replication.

- A complex could perpetuate itself if it splits symmetrically, so that half complexes associate with each daughter duplex. If the half complexes have the capacity to nucleate formation of full complexes, the original state will be restored. This is basically analogous to the maintenance of methylation. The problem with this model is that there is no evident reason why protein complexes should behave in this way.
- A complex could be maintained as a unit and segregate to one of the two daughter duplexes. The problem with this model is that it requires a new complex to be assembled *de novo* on the other daughter duplex, and it is not evident why this should happen.

Figure 19.53 What happens to protein complexes on chromatin during replication?

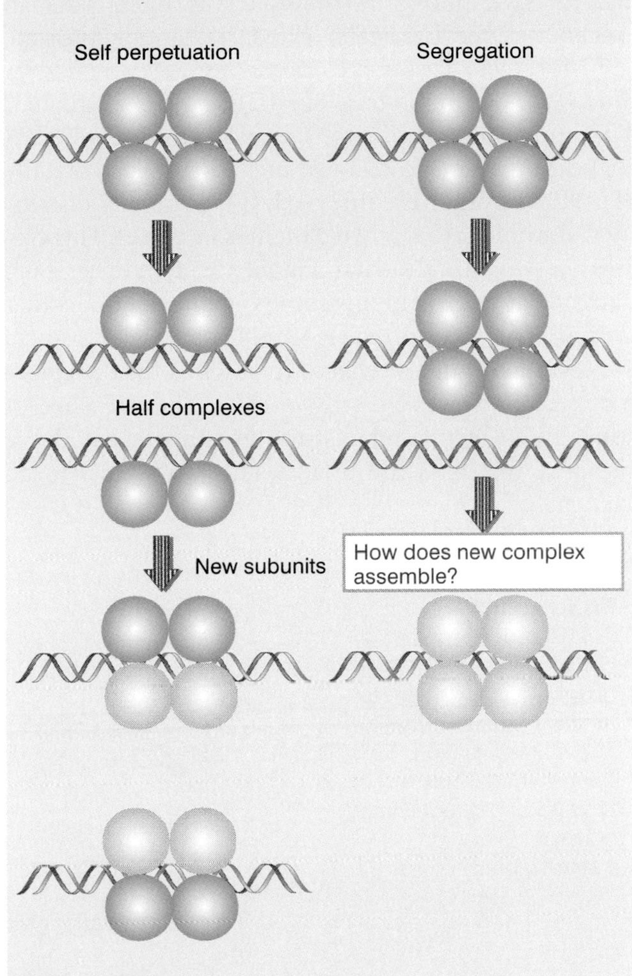

Consider now the need to perpetuate a heterochromatic structure consisting of protein complexes. Suppose that a protein is distributed more or less continuously along a stretch of heterochromatin, as implied in Figure 19.45. If individual subunits are distributed at random to each daughter duplex at replication, the two daughters will continue to be marked by the protein, although its density will be reduced to half of the level before replication. If the protein has a self-assembling property that causes new subunits to associate with it, the original situation may be restored. *Basically, the existence of epigenetic effects forces us to the view that a protein responsible for such a situation must have some sort of self-templating or self-assembling capacity.*

In some cases, it may be the state of protein modification, rather than the presence of the protein *per se*, that is responsible for an epigenetic effect. There is a general correlation between the activity of chromatin and the state of acetylation of the histones, in particular the acetylation of histones H3 and H4, which occurs on their N-terminal tails. Activation of transcription is associated with acetylation in the vicinity of the promoter; and repression of transcription is associated with deacetylation (see Chapter 21). The most dramatic correlation is that the inactive X chromosome in mammalian female cells is underacetylated on histone H4.

The inactivity of constitutive heterochromatin may require that the histones are not acetylated. If a histone acetyltransferase is tethered to a region of telomeric heterochromatin in yeast, silenced genes become active. When yeast is exposed to trichostatin to inhibit deacetylation, centromeric heterochromatin becomes acetylated, and silenced genes in centromeric regions may become active. *The effect may persist even after trichostatin has been removed.* In fact, it may be perpetuated through mitosis and meiosis. This suggests that an epigenetic effect has been created by changing the state of histone acetylation.

How might the state of acetylation be perpetuated? The situation regarding the fate of histone octamers at replication is not clear. However, it seems likely that the $H3_2 \cdot H4_2$ tetramer is conserved. It is probably distributed at random to the two daughter duplexes. This creates the situation shown in **Figure 19.54,** in which each daughter duplex contains some histone octamers that are fully acetylated on the H3 and H4 tails, while others are completely unacetylated. To account for the epigenetic effect, we could suppose that the presence of some fully acetylated histone octamers provides a signal that causes the unacetylated octamers to be acetylated.

(The actual situation is probably more complicated than shown in the figure, because transient acetylations occur during replication. If they are simply reversed following deposition of histones into nucleosomes, they may be irrelevant. An alternative possibility is that the usual deacetylation is prevented, instead of, or as well as, inducing acetylation.)

One of the clearest cases of the dependence of epigenetic inheritance on the condition of a protein is provided by the behavior of **prions**—proteinaceous infectious agents. They have been characterized in two circumstances: by genetic effects in yeast; and as the causative agents of neurological diseases in mammals, including man.

Figure 19.55 summarizes the effects of the Sup35 protein in yeast. In wild-type cells, which are characterized as [*psi*$^-$], the gene *sup35* codes for a translation termination factor. Cells of the mutant [*PSI*$^+$] type have abnormal termination of protein synthesis. (This was

Figure 19.54 Acetylated cores are conserved and distributed at random to the daughter chromatin fibers at replication. Each daughter fiber has a mixture of old (acetylated) cores and new (unacetylated) cores.

Histone tails are acetylated in parental chromatin
Acetylated cores are distributed randomly at replication
What is responsible for restoring the acetylated state?

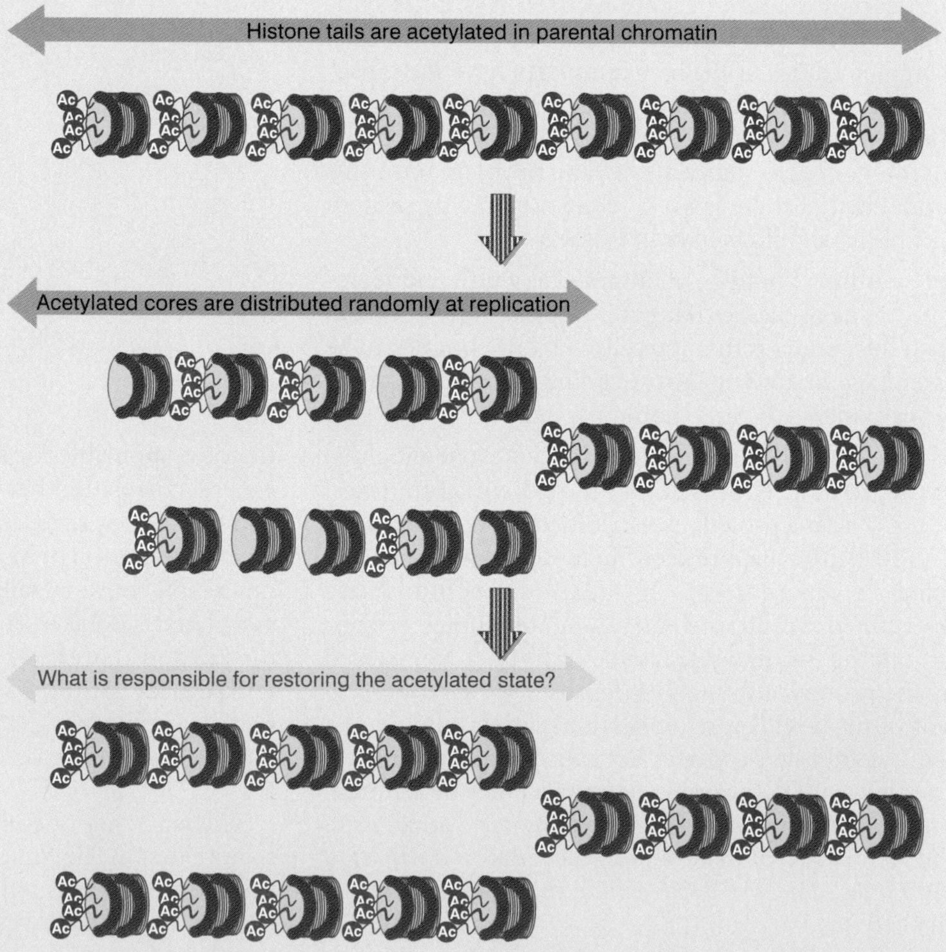

originally detected by the lethal effects of the enhanced efficiency of suppressors of ochre codons in [PSI^+] strains.)

[PSI^+] strains have unusual genetic properties. When a [psi^-] strain is crossed with a [PSI^+] strain, *all of the progeny are* [PSI^+]. This is a pattern of inheritance that would be expected of an extrachromosomal agent, but the [PSI^+] trait cannot be mapped to any such nucleic acid. The [PSI^+] trait is metastable, which means that, although it is inherited by most progeny, it is lost at a higher rate than is consistent with mutation. Similar behavior is shown also by the locus *URE2*, which codes for a protein required for nitrogen-mediated repression of certain catabolic enzymes. When a yeast strain is converted into an alternative state, called [*URE3*], the Ure2 protein is no longer functional.

The [PSI^+] state is determined by the conformation of the Sup35 protein. In a wild-type [psi^-] cell, the protein displays its normal function. But in a [PSI^+] cell, the protein is present in an alternative conformation in which its normal function has been lost. To explain the unilateral dominance of [PSI^+] over [psi^-] in genetic crosses, we must suppose that *the presence of protein in the [PSI^+] state causes all the protein in the cell to enter this state.* This requires an interaction between the [PSI^+] protein and newly synthesized protein, probably reflecting the generation of an oligomeric state in which the [PSI^+] protein has a nucleating role, as illustrated in **Figure 19.56.**

Figure 19.55 The state of the Sup35 protein determines whether termination of translation occurs.

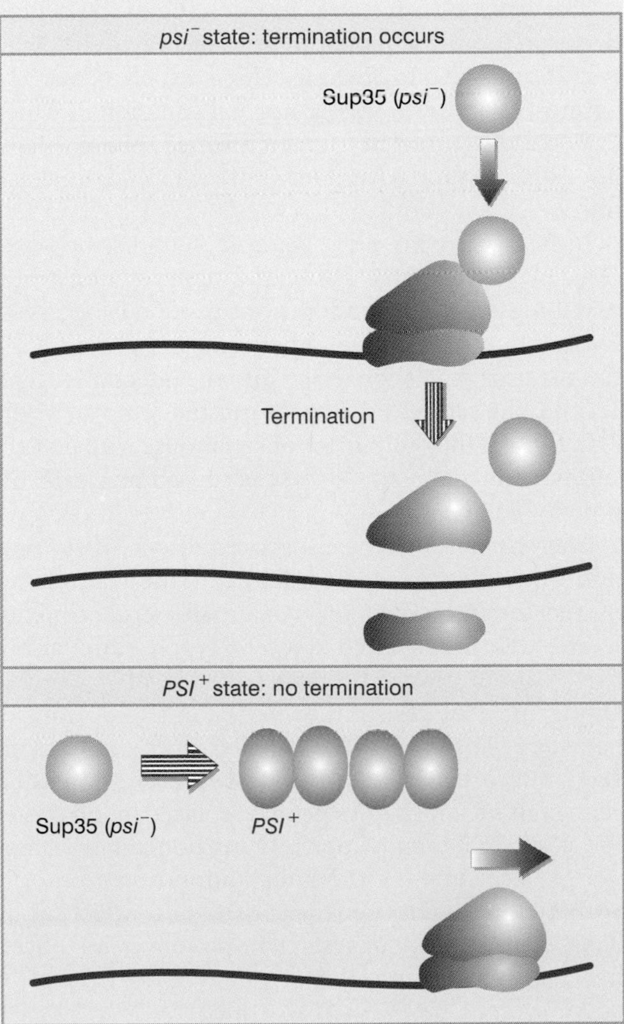

Figure 19.56 Newly synthesized Sup35 protein is converted into the [PSI^+] state by the presence of pre-existing [PSI^+] protein.

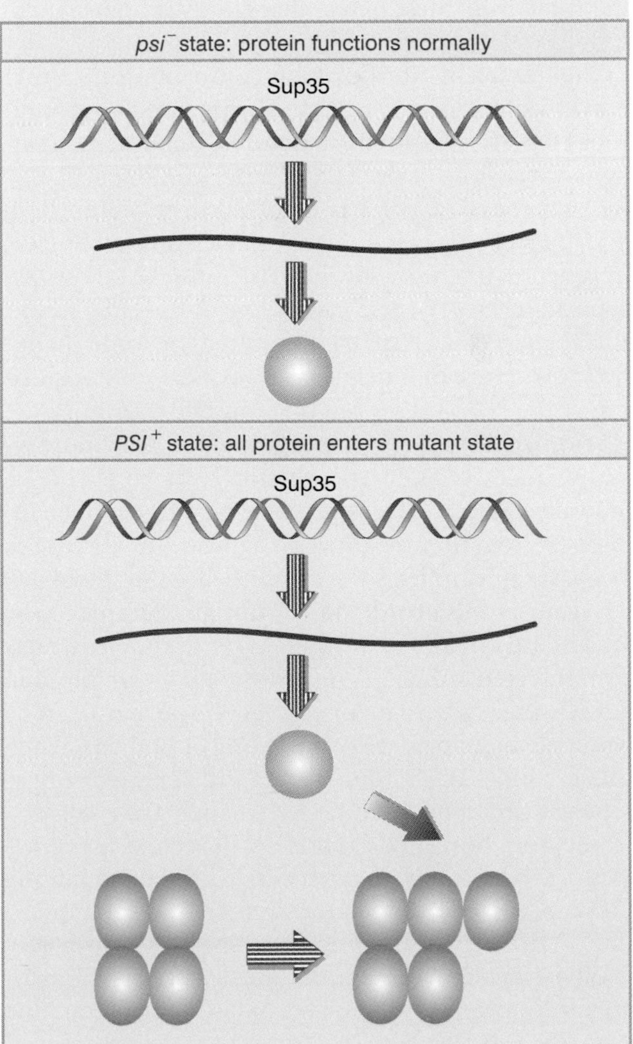

Inheritance of the [*psi*$^-$] and [*PSI*$^+$] conditions is epigenetic, *because the sequence of the gene is the same in both states.* A switch in condition occurs at a low frequency as the result of a spontaneous acquisition or loss of the alternative state.

A common feature in both the Sup35 and Ure2 proteins is that each consists of two domains that function independently. The C-terminal domain is sufficient for the activity of the protein. The N-terminal domain is sufficient for formation of the structures that confer the inactive condition. So yeast in which the N-terminal domain of Sup35 has been deleted cannot acquire the [*PSI*$^+$] state; and the presence of a [*PSI*$^+$] N-terminal domain is sufficient to maintain Sup35 protein in the [*PSI*$^+$] condition.

Loss of function in the [*PSI*$^+$] state is due to the sequestration of the protein in an oligomeric complex. Sup35 protein in [*PSI*$^+$] cells is clustered in discrete foci, whereas the protein in [*psi*$^-$] cells is diffused in the cytosol. Sup35 protein from [*PSI*$^+$] cells forms amyloid fibers *in vitro*—these have a characteristic high content of β sheet structures.

The involvement of protein conformation (rather than modification) is suggested by the effects of conditions that affect protein structure. Denaturing treatments cause loss of the [*PSI*$^+$] state. And in particular, the chaperone Hsp104 is involved in inheritance of [*PSI*$^+$]. Its effects are paradoxical. Deletion of *HSP104* prevents maintenance of the [*PSI*$^+$] state. And overexpression of Hsp104 also causes loss of the [*PSI*$^+$] state. This suggests that Hsp104 is required for some change in the structure of Sup35 that is necessary for acquisition of the [*PSI*$^+$] state, but that must be transitory.

Prion diseases occur in sheep and man, and, more recently, in cows. The basic phenotype is an ataxia—a neurodegenerative disorder that is manifested by an inability to remain upright. The name of the disease in sheep, **scrapie,** reflects the consequence that the sheep rub against walls in order to stay upright. Scrapie can be perpetuated by inoculating sheep with tissue extracts from infected animals. The disease **kuru** was found in New Guinea, where it appeared to be perpetuated by cannibalism, in particular the eating of brains. Related diseases in Western populations with a pattern of genetic transmission include Gerstmann-Straussler syndrome; and the related Creutzfeldt-Jakob disease (CJD) occurs sporadically. Most recently, disease resembling CJD appears to have been transmitted by consumption of meat from cows suffering from "mad cow" disease.

When tissue from scrapie-infected sheep is inoculated into mice, the disease occurs in a period ranging from 75–150 days. Purification of the active component in this preparation identified a protease-resistant protein, which can form amyloid-like structures. The protein is coded by a gene that is normally expressed in brain. The form of the protein in normal brain, called PrP^C, is sensitive to proteases. Its conversion to the resistant form, called PrP^{Sc}, is associated with occurrence of the disease. The infectious preparation has no detectable nucleic acid, is more sensitive to UV irradiation at wave lengths that damage protein than nucleic acid, but has a low infectivity (1 infectious unit / 10^5 PrP^{Sc} proteins). This corresponds to an epigenetic inheritance in which there is no change in genetic information, because normal and diseased cells have the same *PrP* gene sequence, but the PrP^{Sc} form of the protein is the infectious agent, whereas PrP^C is harmless.

The basis for the difference between the PrP^{Sc} and PrP^C forms is not known. Both proteins are glycosylated and linked to the membrane by a GPI-linkage. No changes in these modifications have been found. The PrP^{Sc} form has a high content of β sheets, which is absent from the PrP^C form.

The assay for infectivity in mice allows the dependence on protein sequence to be tested. **Figure 19.57** illustrates the results of some critical experiments. The leftmost column shows the normal situation in which PrP^{Sc} protein extracted from an infected mouse will induce disease (and ultimately kill) when it is injected into a recipient mouse.

If the *PrP* gene is "knocked out", a mouse becomes resistant to infection. This experiment demonstrates two things. First, the endogenous protein is necessary for an infection, presumably because it provides the raw material that is converted into the infectious agent. Second, the cause of disease is not the removal of the PrP^C form of the protein, because a mouse with no PrP^C survives normally: the disease is caused by a gain-of-function in PrP^{Sc}.

The existence of species barriers allows hybrid proteins to be constructed to delineate the features required for infectivity. The original preparations of scrapie were perpetuated in several types of animal, but these cannot always be transferred readily. For example, mice are resistant to infection from prions of hamsters. This means that hamster-PrP^{Sc} cannot convert mouse-PrP^C to PrP^{Sc}. However, the situation changes if the mouse *PrP* gene is replaced by a hamster *PrP* gene. (This can be done by introducing the hamster *PrP* gene into the *PrP* knockout mouse.) A mouse with a hamster *PrP* gene is sensitive to infection by hamster PrP^{Sc}. This suggests that the conversion of cellular PrP^C protein into the Sc state requires that the Sc and C proteins have matched sequences.

Figure 19.57 A PrP^{Sc} protein can only infect an animal that has the same type of endogenous PrP^{C} protein.

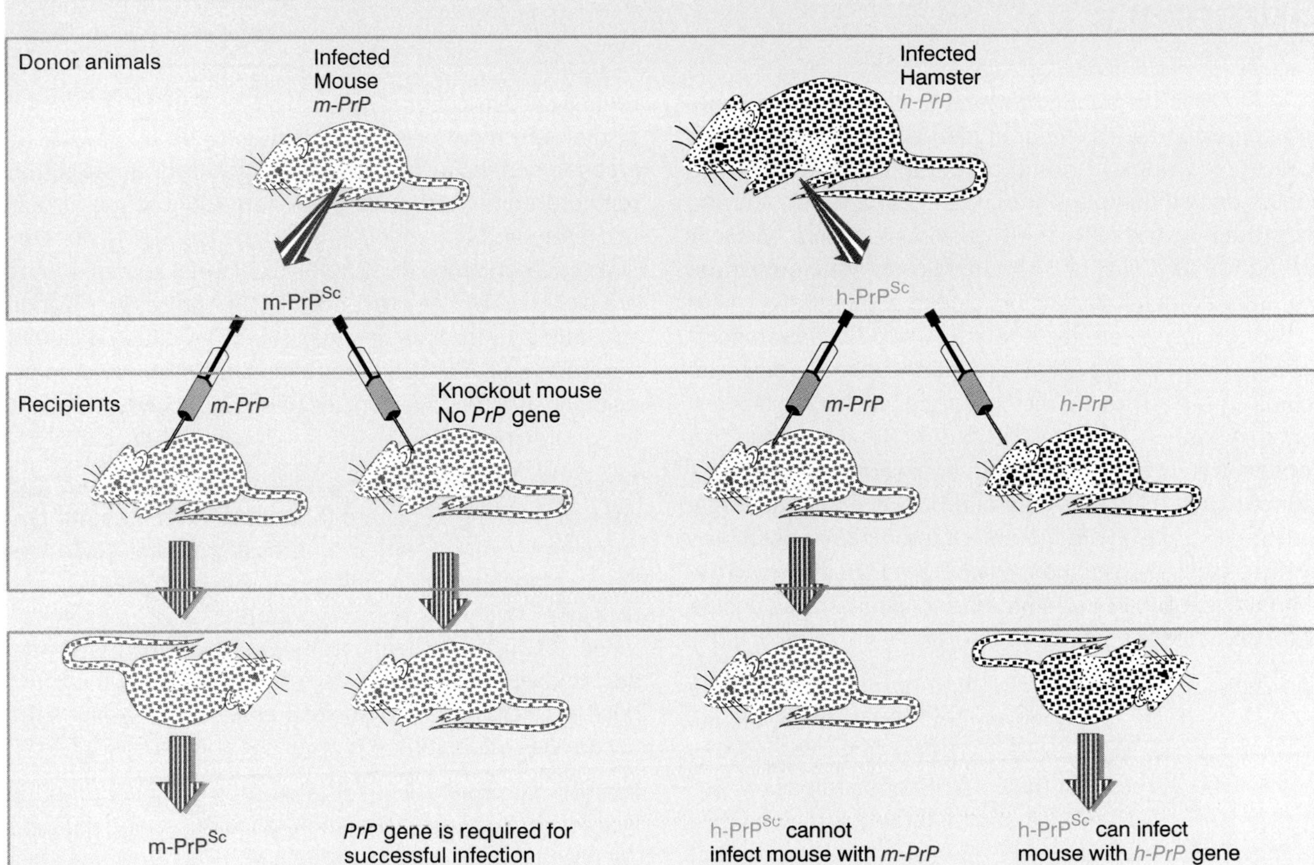

The mouse and hamster PrP proteins differ in 16 amino acids out of 254. Chimeric genes can be tested for the ability to confer susceptibility to infection. The central part of the hamster protein (codons 94–188), which differs at 5 positions from the corresponding mouse sequence, is necessary for sensitivity.

There are different "strains" of PrP^{Sc}, which are distinguished by characteristic incubation periods upon inoculation into mice. This implies that the protein is not restricted solely to alternative states of PrP^{C} and PrP^{Sc}, but that there may be multiple Sc states. These differences must depend on some self-propagating property of the protein other than its sequence. If conformation is the feature that distinguishes PrP^{Sc} from PrP^{C}, then there must be multiple conformations, each of which has a self-templating property when it converts PrP^{C}.

The probability of conversion from PrP^{C} to PrP^{Sc} is affected by the sequence of PrP. Gerstmann-Straussler syndrome in man is caused by a single amino acid change in PrP. This is inherited as a dominant trait. If the same change is made in the mouse PrP gene, mice develop the disease. This suggests that the mutant protein has an increased probability of spontaneous conversion into the Sc state.

The prion offers an extreme case of epigenetic inheritance, in which the infectious agent is a protein that can adopt multiple conformations, each of which has a self-templating property. This property is likely to involve the state of aggregation of the protein.

Summary

All eukaryotic chromatin consists of nucleosomes. A nucleosome contains a characteristic length of DNA, usually ~200 bp, wrapped around an octamer containing two copies each of histones H2A, H2B, H3, and H4. A single H1 protein is associated with each nucleosome. Virtually all genomic DNA is organized into nucleosomes. Treatment with micrococcal nuclease shows that the DNA packaged into each nucleosome can be divided operationally into two regions. The linker region is digested rapidly by the nuclease; the core region of 146 bp is resistant to digestion. Histones H3 and H4 are the most highly conserved and an H3 • H4 tetramer accounts for the diameter of the particle. The H2A and H2B histones are organized as two H2A • H2B dimers. Octamers are assembled by the successive addition of two H2A • H2B dimers to the $H3_2 \cdot H4_2$ kernel, but it is not known whether this occurs before or after association with DNA.

The path of DNA around the histone octamer creates –1.65 supercoils. The DNA "enters" and "leaves" the nucleosome in the same vicinity, and could be "sealed" by histone H1. Removal of the core histones releases –1.0 supercoils. The difference can be largely explained by a change in the helical pitch of DNA, from an average of 10.2 bp/turn in nucleosomal form to 10.5 bp/turn when free in solution. There is variation in the structure of DNA from a periodicity of 10.0 bp/turn at the nucleosome ends to 10.7 bp/turn in the center. There are kinks in the path of DNA on the nucleosome.

Nucleosomes are organized into a fiber of 30 nm diameter which has 6 nucleosomes per turn and a packing ratio of 40. Removal of H1 allows this fiber to unfold into a 10 nm fiber that consists of a linear string of nucleosomes. The 30 nm fiber probably consists of the 10 nm fiber wound into a solenoid. The 30 nm fiber is the basic constituent of both euchromatin and heterochromatin; nonhistone proteins are responsible for further organization of the fiber into chromatin or chromosome ultrastructure.

RNA polymerase displaces histone octamers during transcription. The octamers reattach to DNA after the polymerase has passed, unless transcription is very intensive (such as in rDNA) when they may be displaced completely.

Two types of changes in sensitivity to nucleases are associated with gene activity. Chromatin capable of being transcribed has a generally increased sensitivity to DNAase I, reflecting a change in structure over an extensive region that can be defined as a domain containing active or potentially active genes. Hypersensitive sites in DNA occur at discrete locations, and are identified by greatly increased sensitivity to DNAase I. They occur at sites that regulate transcription, at origins for replication, at centromeres, and other locations. A hypersensitive site consists of a sequence of ~200 bp from which nucleosomes are excluded by the presence of other proteins. A hypersensitive site forms a boundary that may cause adjacent nucleosomes to be restricted in position. Nucleosome positioning may be important in controlling access of regulatory proteins to DNA.

Formation of heterochromatin may be initiated at certain sites and then propagated for a distance that is not precisely determined. When a heterochromatic state has been established, it is inherited through subsequent cell divisions. This gives rise to a pattern of epigenetic inheritance, in which two identical sequences of DNA may be associated with different protein structures, and therefore have different abilities to be expressed. This explains the occurrence of position effect variegation in *Drosophila*.

Inactive chromatin at yeast telomeres and silent mating type loci appears to have a common cause, and involves the interaction of certain proteins with the N-terminal tails of histones H3 and H4. Formation of the inactive complex may be initiated by binding of one protein to a specific sequence of DNA; the other components may then polymerize in a cooperative manner along the chromosome.

Inactivation of one X chromosome in female (eutherian) mammals occurs at random. The *Xic* locus is necessary and sufficient to count the number of X chromosomes. The n–1 rule ensures that all but one X chromosome are inactivated. *Xic* contains the gene *Xist*, which codes for an RNA that is expressed only on the inactive X chromosome. Stabilization of *Xist* RNA is the mechanism by which the inactive X chromosome is distinguished.

Methylation of DNA is inherited epigenetically. Replication of DNA creates hemimethylated products, and a maintenance methylase restores the fully methylated state. Some methylation events depend on parental origin. Sperm and eggs contain specific and different patterns of methylation, with the result that paternal and maternal alleles are differently expressed in the embryo. This is responsible for imprinting, in which the nonmethylated allele inherited from one parent is essential because it is the only active allele; the allele inherited from the other parent is silent. Patterns of methylation are reset during gamete formation in every generation.

Prions are proteinaceous infectious agents that are responsible for the disease of scrapie in sheep and for related diseases in man. The infectious agent is a variant of a normal cellular protein. The PrP^{Sc} form has an altered conformation that is self-templating: the normal PrP^{C} form does not usually take up this conformation, but does so in the presence of PrP^{Sc}. A similar effect is responsible for inheritance of the *PSI* element in yeast.

Further reading

Reviews

Kornberg, R. D. (1977). Structure of chromatin. *Ann. Rev. Biochem.* **46**, 931–954.

McGhee, J. D. and Felsenfeld, G. (1980). Nucleosome structure. *Ann. Rev. Biochem.* **49**, 1115–1156.

Razin, A. and Riggs, A. D. (1980). DNA methylation and gene function. *Science* **210**, 604–610.

Prusiner, S. (1982). Novel proteinaceous infectious particles cause scrapie. *Science* **216**, 136–144.

Wang, J. (1982). The path of DNA in the nucleosome. *Cell* **29**, 724–726.

Felsenfeld, G. and McGhee, J. D. (1986). Structure of the 30 nm chromatin fiber. *Cell* **44**, 375–377.

Travers, A. A. and Klug, A. (1987). The bending of DNA in nucleosomes and its wider implications. *Phil. Trans. Roy. Soc. B* **317**, 537–561.

Gross, D. S. and Garrard, W. T. (1988). Nuclease hypersensitive sites in chromatin. *Ann. Rev. Biochem.* **57**, 159–197.

Kornberg, R. D. and Lorch, Y. (1992). Chromatin structure and transcription. *Ann. Rev. Cell Biol.* **8**, 563–587.

Thompson, J. S., Hecht, A., and Grunstein, M. (1993). Histones and the regulation of heterochromatin in yeast. *Cold Spring Harbor Symp. Quant. Biol.* **58**, 247–256.

Loo, S. and Rine, J. (1995). Silencing and heritable domains of gene expression. *Ann. Rev. Cell Biol.* **11**, 519–548.

Wickner, R. B. (1996). Prions and RNA viruses of *S. cerevisiae. Ann. Rev. Genet.* **30**, 109–139.

Lindquist, S. (1997). Mad cows meet psi-chotic yeast: the expansion of the prion hypothesis. *Cell* **89**, 495–498.

Horwich, A. L. and Weissman, J. S. (1997). Deadly conformations—protein misfolding in prion disease. *Cell* **89**, 499–510.

Prusiner, S. B. and Scott, M. R. (1997). Genetics of prions. *Ann. Rev. Genet.* **31**, 139–175.

Bartolomei, M. S. and Tilghman, S. (1997). Genomic imprinting in mammals. *Ann. Rev. Genet.* **31**, 493–525.

Workman, J. L. and Kingston, R. E. (1998). Alteration of nucleosome structure as a mechanism of transcriptional regulation. *Ann. Rev. Biochem.* **67**, 545–580.

Discoveries

Models for the nucleosome

Kornberg, R. D. (1974). Chromatin structure: a repeating unit of histones and DNA. *Science* **184**, 868–871.

Finch, J. T. *et al.* (1977). Structure of nucleosome core particles of chromatin. *Nature* **269**, 29–36.

Richmond, T. J., Finch, J. T., Rushton, B., Rhodes, D., and Klug, A. (1984). Structure of the nucleosome core particle at 7 Å resolution. *Nature* **311**, 532–537.

Arents, G., Burlingame, R. W., Wang, B.-C., Love, W. E., and Moudrianakis, E. N. (1991). The nucleosomal core histone octamer at 3.1 Å resolution: a tripartite protein assembly and a left-handed superhelix. *Proc. Nat. Acad. Sci. USA* **88**, 10148–10152.

Luger, K. *et al.* (1997). Crystal structure of the nucleosome core particle at 2.8Å resolution. *Nature* **389**, 251–260.

Nucleosomes and transcription

Weintraub, H. and Groudine, M. (1976). Chromosomal subunits in active genes have an altered conformation. *Science* **193**, 848–856.

Schmid, V. M., Fascher, K.-D., and Horz, W. (1992). Nucleosome disruption at the yeast PHO5 promoter upon PHO5 induction occurs in the absence of DNA replication. *Cell* **71**, 853–864.

Cavalli, G. and Thoma, F. (1993). Chromatin transitions during activation and repression of galactose-regulated genes in yeast. *EMBO J.* **12**, 4603–4613.

O'Neill, T. E., Smith, J. G., and Bradbury, E. M. (1993). Histone octamer dissociation is not required for transcript elongation through arrays of nucleosome cores by phage T7 RNA polymerase *in vitro*. *Proc. Nat. Acad. Sci. USA*, **90**, 6203–6207.

Studitsky, V. M., Clark, D. J., and Felsenfeld, G. (1994). A histone octamer can step around a transcribing polymerase without leaving the template. *Cell* **76**, 371–382.

Hypersensitive sites

Varshavsky, A. J., Sundin, O., and Bohn, M. J. (1978). SV40 viral minichromosome: preferential exposure of the origin of replication as probed by restriction endonucleases. *Nuc. Acids Res.* **5**, 3469–3479.

Scott, W. A. and Wigmore, D. J. (1978). Sites in SV40 chromatin which are preferentially cleaved by endonucleases. *Cell* **15**, 1511–1518.

Wu, C., Bingham, P. M., Livak, K. J., Holmgren, R., and Elgin, S. C. R. (1979). The chromatin structure of specific genes: evidence for higher order domains of defined DNA sequence. *Cell* **16**, 797–806.

Stalder, J. *et al.* (1980). Tissue-specific DNA cleavage in the globin chromatin domain introduced by DNAase I. *Cell* **20**, 451–460.

Groudine, M. and Weintraub. H. (1982). Propagation of globin DNAase I-hypersensitive sites in absence of factors required for induction: a possible mechanism for determination. *Cell* **30**, 131–139.

van Assendelft, G. B., Hanscombe, O., Grosveld, F., and Greaves, D. R. (1989). The β-globin dominant control region activates homologous and heterologous promoters in a tissue-specific manner. *Cell* **56**, 969–977.

Histone modification

Turner, B. M., Birley, A. J., and Lavender, J. (1992). Histone H4 isoforms acetylated at specific lysine residues define individual chromosomes and chromatin domains in *Drosophila* polytene nuclei. *Cell* **69**, 375–384.

Jeppesen, P. and Turner, B. M. (1993). The inactive X chromosome in female mammals is distinguished by a lack of histone H4 acetylation, a cytogenetic marker for gene expression. *Cell* **74**, 281–289.

Shen, X. *et al.* (1995). Linker histones are not essential and affect chromatin condensation *in vivo*. *Cell* **82**, 47–56.

Heterochromatin

Shore, D. and Nasmyth, K. (1987). Purification and cloning of a DNA-binding protein from yeast that binds to both silencer and activator elements. *Cell* **51**, 721–732.

Kayne, P. S., Kim, U. J., Han. M., Mullen, R. J., Yoshizaki, F., and Grunstein, M. (1988). Extremely conserved histone H4 N terminus is dispensable for growth but essential for repressing the silent mating loci in yeast. *Cell* **55**, 27–39.

Palladino, F., Laroche, T., Gilson, E., Axelrod, A., Pillus, L., and Gasser, S. M. (1993). SIR3 and SIR4 proteins are required for the positioning and integrity of yeast telomeres. *Cell* **75**, 543–555.

Moretti, P., Freeman, K., Coodly, L., and Shore, D. (1994). Evidence that a complex of SIR proteins interacts with the silencer and telomere-binding protein RAP1. *Genes Dev.* **8**, 2257–2269.

Hecht, A., Laroche, T., Strahl-Bolsinger, S., Gasser, S. M., and Grunstein, M. (1994). Histone H3 and H4 N-termini interact with the silent information regulators SIR3 and SIR4: a molecular model for the formation of heterochromatin in yeast. *Cell* **80**, 583–592.

Meluh, P. B. *et al.* (1998). *Cse4p* is a component of the core centromere of *S. cerevisiae*. *Cell* **94**, 607–613.

X chromosome inactivation

Lyon, M. F. (1961). Gene action in the X chromosome of the mouse. *Nature* **190**, 372–373.

Penny, G. D. *et al.* (1996). Requirement for *Xist* in X chromosome inactivation. *Nature* **379**, 131–137.

Lee, J. T. *et al.* (1996). A 450 kb transgene displays properties of the mammalian X-inactivation center. *Cell* **86**, 83–94.

Panning, B., Dausman, J., and Jaenisch, R. (1997). X chromosome inactivation is mediated by *Xist* RNA stabilization. *Cell* **90**, 907–916.

Methylation and imprinting

Chaillet, J. R., Vogt, T. F., Beier, D. R., and Leder, P. (1991). Parental-specific methylation of an imprinted transgene is established during gametogenesis and progressively changes during embryogenesis. *Cell* **66**, 77–83.

Colot, V., Maloisel, L., and Rossignol, J.-L. (1996). Interchromosomal transfer of epigenetic states in *Ascobolus*: transfer of DNA methylation is mechanistically related to homologous recombination. *Cell* **86**, 855–864.

Prions

McKinley, M. P., Bolton, D. C., and Prusiner, S. B. (1983). A protease-resistant protein is a structural component of the scrapie prion. *Cell* **35**, 57–62.

Oesch, B. *et al.* (1985). A cellular gene encodes scrapie PrP 27–30 protein. *Cell* **40**, 735–746.

Basler, K., Oesch, B., Scott, M., Westaway, D., Walchli, M., Groth, D. F., McKinley, M. P., Prusiner, S. B., and Weissmann, C. (1986). Scrapie and cellular PrP isoforms are encoded by the same chromosomal gene. *Cell* **46**, 417–428.

Hsiao, K. *et al.* (1989). Linkage of a prion protein missense variant to Gerstmann-Straussler syndrome. *Nature* **338**, 342–345.

Bueler, H. *et al.* (1993). Mice devoid of PrP are resistant to scrapie. *Cell* **73**, 1339–1347.

Scott, M. *et al.* (1993). Propagation of prions with artificial properties in transgenic mice expressing chimeric PrP genes. *Cell* **73**, 979–988.

Wickner, R. B. (1994). [URE3] as an altered URE2 protein: evidence for a prion analog in *S. cerevisiae. Science* **264**, 566–569.

Chernoff, Y. O. *et al.* (1995). Role of the chaperone protein Hsp104 in propagation of the yeast prion-like factor [PSI$^+$]. *Science* **268**, 880–884.

Masison, D. C. and Wickner, R. B. (1995). Prion-inducing domain of yeast Ure2p and protease resistance of Ure2p in prion-containing cells. *Science* **270**, 95–95.

Glover, J. R. *et al.* (1997). Self-seeded fibers formed by Sup35, the protein determinant of [*PSI*$^+$], a heritable prion-like factor of *S. cerevisiae. Cell* **89**, 911–819.

Chapter 20

Initiation of transcription

TRANSCRIPTION in eukaryotic cells is divided into three classes. Each class is transcribed by a different RNA polymerase:

- RNA polymerase I transcribes rRNA
- RNA polymerase II transcribes mRNA
- RNA polymerase III transcribes tRNA and other small RNAs.

Accessory factors are needed for initiation, but are not required subsequently. The balance of responsibilities *vis-à-vis* the accessory factors is similar for all eukaryotic RNA polymerases. The *factors*, rather than the enzymes themselves, are principally responsible for recognizing the promoter. This contrasts with the *modus operandi* of bacterial RNA polymerase, in which a basic enzyme recognizes the promoters, assisted in certain cases by accessory factors.

The promoters for RNA polymerases I and II are (mostly) upstream of the startpoint, but some promoters for RNA polymerase III lie downstream of the startpoint. Each promoter contains characteristic sets of short conserved sequences that are recognized by the appropriate class of factors. RNA polymerases I and III each recognize a relatively restricted set of promoters, and rely upon a small number of accessory factors.

Promoters utilized by RNA polymerase II show more variation in sequence, and are modular in design. Short sequence elements that are recognized by transcription factors lie upstream of the startpoint. These *cis*-acting sites usually are spread out over a region of >200 bp. Some of these elements and the factors that recognize them are common: they are found in a variety of promoters and are used constitutively. Others are specific: they identify particular classes of genes and their use is regulated. The elements occur in different combinations in individual promoters.

The number of factors that can act in conjunction with RNA polymerase II is large. We may divide them into three general groups. We consider the first two groups in this chapter, and the third group in the next chapter:

- The **general factors** are required for the mechanics of initiating RNA synthesis at all promoters. They join with RNA polymerase to form a complex surrounding the startpoint, and they determine the site of initiation. The general factors together with RNA polymerase constitute the **basal transcription apparatus.**
- The upstream factors are DNA-binding proteins that recognize specific short consensus elements located upstream of the startpoint. The activity of these factors is not regulated; they are ubiquitous, and act upon any promoter that contains the appropriate binding site on DNA. They increase the efficiency of initiation, and are required for a promoter to function at an adequate level. The precise set of such factors that is required for full expression is characteristic of any particular promoter.
- The inducible factors function in the same general way as the upstream factors, but have a regulatory role.

They are synthesized or activated at specific times or in specific tissues, and they are therefore responsible for the control of transcription patterns in time and space. The sequences that they bind are called response elements.

A promoter that contains only elements recognized by general and upstream factors should be transcribed in any cell type. Such promoters may be responsible for expression of cellular genes that are constitutively expressed (sometimes called housekeeping genes). No element/factor combination is an essential component of the promoter, which suggests that initiation by RNA polymerase II may be sponsored in many different ways. The common feature is that the upstream or inducible transcription factors bind to sequence elements located upstream of the startpoint. Binding of the factors to DNA is associated with the construction of a complex in which protein-protein interactions are important. The upstream and inducible factors function by interacting with the basal transcription apparatus, typically with certain general factors.

Sequence components of the promoter are defined operationally by the demand that they must be located in the general vicinity of the startpoint and are required for initiation. The **enhancer** is another type of site involved in initiation. It is identified by sequences that stimulate initiation, but that are located a considerable distance from the startpoint. Enhancer elements are often targets for tissue-specific or temporal regulation. **Figure 20.1** illustrates the general properties of promoters and enhancers.

The components of an enhancer resemble those of the promoter; they consist of a variety of modular elements. However, the elements are organized in a closely packed array. The elements in an enhancer function like those in the promoter, but the enhancer does not need to be near the startpoint. However, proteins bound at enhancer elements interact with proteins bound at promoter elements. The distinction between promoters and enhancers is operational, rather than implying a fundamental difference in mechanism. This view is fortified by the fact that some types of element are found in both promoters and enhancers.

Any protein that is needed for the initiation of transcription, but which is not itself part of RNA polymerase, is defined as a transcription factor. Many transcription factors act by recognizing *cis*-acting sites that are classified as comprising parts of promoters or enhancers. However, binding to DNA is not the only

Figure 20.1 *Overview:* a typical gene transcribed by RNA polymerase II has a promoter that extends upstream from the site where transcription is initiated. The promoter contains several short (<10 bp) sequence elements that bind transcription factors, dispersed over >200 bp. An enhancer containing a more closely packed array of elements that also bind transcription factors may be located several kb distant. (DNA may be coiled or otherwise rearranged so that transcription factors at the promoter and at the enhancer interact to form a large protein complex.)

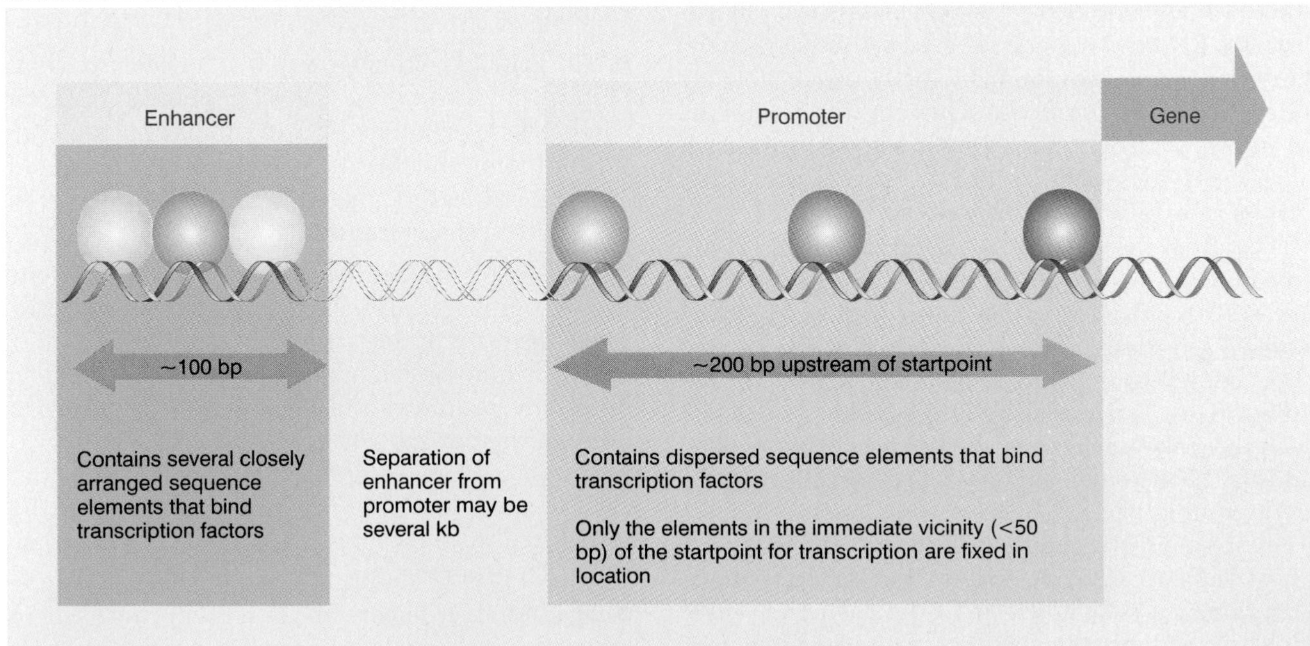

means of action for a transcription factor. A factor may recognize another factor, or may recognize RNA polymerase, or may be incorporated into an initiation complex only in the presence of several other proteins. The ultimate test for membership of the transcription apparatus is functional: a protein must be needed for transcription to occur at a specific promoter or set of promoters.

A significant difference in the transcription of eukaryotic and prokaryotic mRNAs is that initiation at a eukaryotic promoter involves a large number of factors that bind to a variety of *cis*-acting elements. The promoter is defined as the region containing all these binding sites, that is, which can support transcription at the normal efficiency and with the proper control. So the major feature defining the promoter for a eukaryotic mRNA is the location of binding sites for transcription factors. RNA polymerase itself binds around the startpoint, but does not directly contact the extended upstream region of the promoter. By contrast, the bacterial promoters discussed in Chapter 9 are largely defined in terms of the binding site for RNA polymerase in the immediate vicinity of the startpoint. Other sequences nearby regulate the promoter, but are generally considered distinct from it.

The common mode of regulation of eukaryotic transcription is positive: a transcription factor is provided under tissue-specific control to activate a promoter or set of promoters that contain a common target sequence. Regulation by specific repression of a target promoter is less common.

A eukaryotic transcription unit generally contains a single gene, and termination occurs beyond the end of the coding region. We should like to define the mechanism of termination, but it lacks the regulatory importance that applies in prokaryotic systems. RNA polymerases I and III terminate at discrete sequences in defined reactions, but the mode of termination by RNA polymerase II is not clear. However, the significant event in generating the 3′ end of an mRNA is not the termination event itself, but instead results from a cleavage reaction in the primary transcript (see Chapter 22).

Eukaryotic RNA polymerases consist of many subunits

The three eukaryotic RNA polymerases have different locations in the nucleus, corresponding with their responsibilities.

The most prominent activity is the enzyme RNA polymerase I, which resides in the nucleolus and is responsible for transcribing the genes coding for rRNA. It accounts for most cellular RNA synthesis.

The other major enzyme is RNA polymerase II, located in the nucleoplasm (the part of the nucleus excluding the nucleolus). It represents most of the remaining cellular activity and is responsible for synthesizing heterogeneous nuclear RNA (hnRNA), the precursor for mRNA.

RNA polymerase III is a minor enzyme activity. This nucleoplasmic enzyme synthesizes tRNAs and other small RNAs.

A major distinction between the eukaryotic enzymes is drawn from their response to the bicyclic octapeptide **α-amanitin.** In cells from origins as divergent as animals, plants, and insects, the activity of RNA polymerase II is rapidly inhibited by low concentrations of α-amanitin; RNA polymerase I is not inhibited. The response of RNA polymerase III to α-amanitin has not been so well conserved; in animal cells it is inhibited by high levels, but in yeast and insects it is not inhibited.

All eukaryotic RNA polymerases are large proteins, appearing as aggregates of 500 kD or more. They typically have 8–14 subunits. The purified enzyme can undertake template-dependent transcription of RNA, but is not able to initiate selectively at promoters. The general constitution of a eukaryotic RNA polymerase II enzyme as typified in *S. cerevisiae* is illustrated in **Figure 20.2.** The three largest subunits have homology to subunits of bacterial RNA polymerase; the two largest probably carry the catalytic site. Three of the remaining subunits are common to all the RNA polymerases, that is, they are also components of RNA polymerases I and III.

The largest subunit in RNA polymerase II has a car-

Figure 20.2 Eukaryotic RNA polymerase II has >10 subunits.

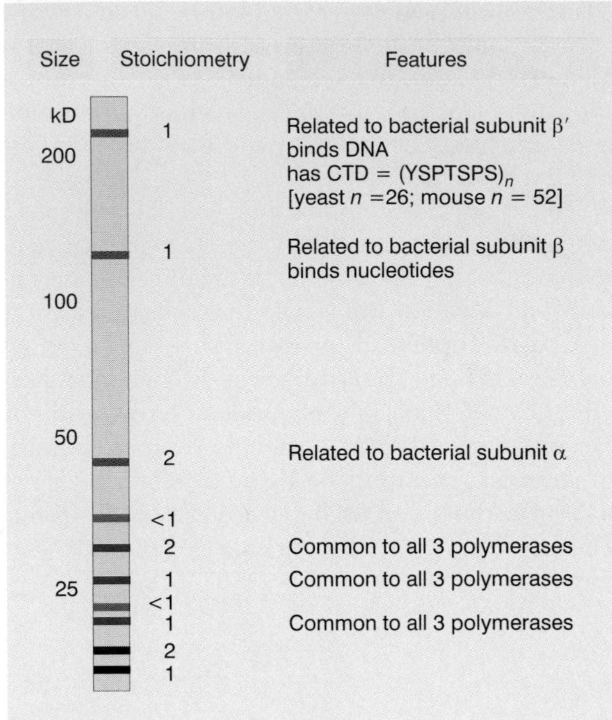

boxy-terminal domain (**CTD**), which consists of multiple repeats of a consensus sequence of 7 amino acids. The sequence is unique to RNA polymerase II. There are ~26 repeats in yeast and ~50 in mammals. The number of repeats is important, because deletions that remove (typically) more than half of the repeats are lethal (in yeast). The CTD can be highly phosphorylated on serine or threonine residues; this is involved in the initiation reaction (see later).

The RNA polymerase activities of mitochondria and chloroplasts are smaller, and resemble bacterial RNA polymerase rather than any of the nuclear enzymes. Of course, the organelle genomes are much smaller, the resident polymerase needs to transcribe relatively few genes, and the control of transcription is likely to be very much simpler (if existing at all). So these enzymes are analogous to the phage enzymes that have a single fixed purpose and do not need the ability to respond to a more complex environment.

Promoter elements are defined by mutations and footprinting

PROMOTERS have been defined in terms of their abilities to initiate transcription in suitable test systems. Having identified sequences that are needed for promoter function, we may then characterize the proteins that bind them. Several types of system have been used:

- The **oocyte system** follows the principles established for translation, and relies on injection of a suitable DNA template into the nucleus of the *X. laevis* oocyte. The RNA transcript can be recovered and analyzed. The main limitation of this system is that it is restricted to the conditions that prevail in the oocyte. It allows characterization of DNA sequences, but not of the factors that normally bind them.

- **Transfection systems** allow exogenous DNA to be introduced into a cultured cell and expressed. (The procedure is discussed in Chapter 17.) The system is genuinely *in vivo* in the sense that transcription is accomplished by the same apparatus responsible for expressing the cell's own genome. However, it differs from the natural situation because the template consists of a gene that would not usually be transcribed in the host cell. The usefulness of the system may be extended by using a variety of host cells. **Cotransfection** with two (or more) DNAs allows assay of the interactions between two factors.

- **Transgenic systems** involve the addition of a gene to the germline of an animal. Expression of the **transgene** can be followed in any or all of the tissues of the animal. Some common limitations apply to transgenic systems and to transfection: the additional gene often is present in multiple copies, and is integrated at a different location from the endogenous gene.

- The *in vitro* **system** takes the classic approach of purifying all the components and manipulating conditions until faithful initiation is seen. "Faithful" initiation is defined as production of an RNA starting at the site corresponding to the 5′ end of mRNA (or rRNA or tRNA precursors).

We start with a particular fragment of DNA that can initiate transcription in one of these systems. Then the boundaries of the sequence constituting the promoter can be determined by reducing the length of the fragment from either end, until at some point it ceases to be active, as illustrated in **Figure 20.3.** The boundary upstream can be identified by progressively removing material from this end until promoter function is lost. To test the boundary downstream, it is necessary to reconnect the shortened promoter to the sequence to be transcribed (since otherwise there is no product to assay).

Several precautions are required to avoid extraneous effects. To ensure that the promoter is always in the same context, the same long upstream sequence is always placed next to it. Because termination does not occur properly in the *in vitro* systems, the template is cut at some distance from the promoter (usually ~500 bp downstream), to ensure that all polymerases "run off" at the same point, generating an identifiable transcript.

Once the boundaries of the promoter have been defined, the importance of particular bases within it can be determined by introducing point mutations or other rearrangements in the sequence. As with bacterial RNA polymerase, these can be characterized as *up* or *down* mutations. Some of these rearrangements affect only the *rate* of initiation; others influence the *site* at which initiation occurs, as seen in a change of the startpoint. To be sure that we are dealing with comparable products, in each case it is necessary to characterize the 5′ end of the RNA.

We can apply several criteria in identifying the sequence components of a promoter (or any other site in DNA):

- Mutations in the site prevent function *in vitro* or *in vivo*. (Many techniques now exist for introducing point mutations at particular base pairs, and in principle every position in a promoter can be mutated, and the mutant sequence tested *in vitro* or *in vivo*.)

Figure 20.3 Promoter boundaries can be determined by making deletions that progressively remove more material from one side. When one deletion fails to prevent RNA synthesis but the next stops transcription, the boundary of the promoter must lie between them.

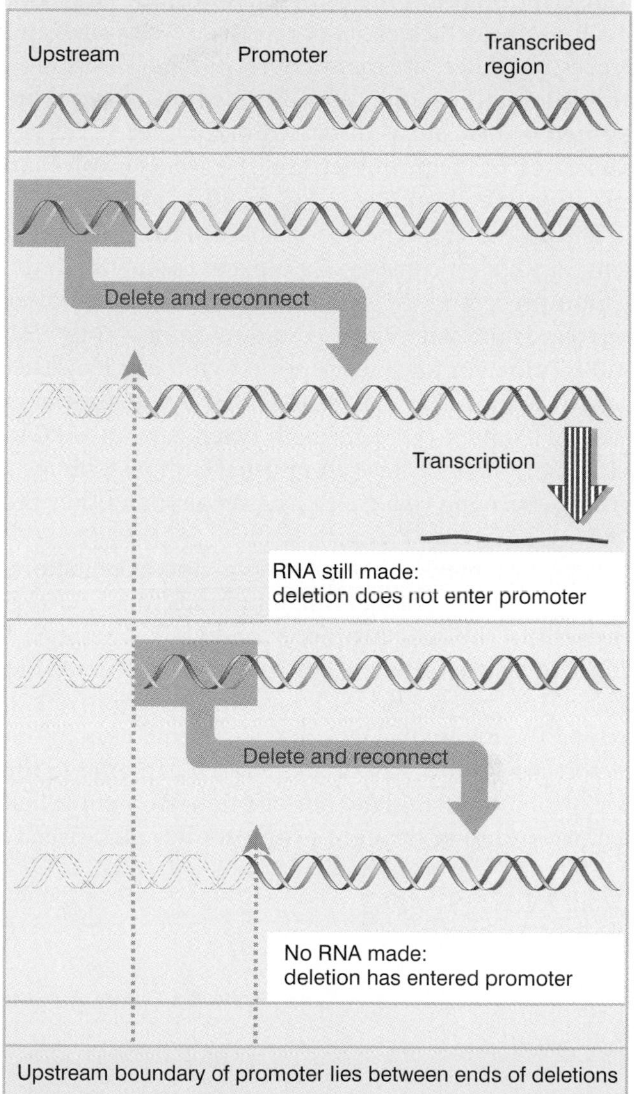

- Proteins that act by binding to a site may be footprinted on it. There should be a correlation between the ability of mutations to prevent promoter function and to prevent binding of the factor.

- When a site recognized by a particular factor is present at multiple promoters, it should be possible to derive a consensus sequence that is bound by the factor. A new promoter should become responsive to this factor when an appropriate copy of the element is introduced.

RNA polymerase I has a bipartite promoter

RNA polymerase I transcribes only the genes for ribosomal RNA, from a single type of promoter. The transcript includes the sequences of both large and small rRNAs, which are later released by cleavages and processing. There are many copies of the transcription unit, alternating with nontranscribed spacers, and organized in a cluster as discussed in Chapter 4. The organization of the promoter, and the events involved in initiation, are illustrated in **Figure 20.4.**

The promoter has been best characterized in human cells, in which it consists of a bipartite sequence in the region preceding the startpoint. The **core promoter** surrounds the startpoint, extending from –45 to +20, and is sufficient for transcription to initiate. However, its efficiency is very much increased by the upstream control element (UCE), which extends from –180 to –107. Both regions have an unusual composition for a promoter, being rich in G•C base pairs; and they are ~85% identical.

RNA polymerase I requires two ancillary factors. UBF1 is a single polypeptide that binds to a G•C-rich element in the core promoter and UCE. Factor SL1 does not by itself have specificity for the promoter, but once UBF1 has bound, SL1 can bind cooperatively to extend the region of DNA that is covered. Once both factors are bound, RNA polymerase I can bind to the core promoter to initiate transcription. We assume that factors bound at the core promoter interact directly with RNA polymerase I, but we do not know how binding of the same factors at the UCE stimulates initiation in the core region. (We see later that action at a distance involving enhancers is a prominent feature of initiation at promoters for RNA polymerase II, in which it has been investigated in more detail.)

SL1 consists of 4 proteins. One of them, called TBP, is a factor that is required also for initiation by RNA polymerases II and III. We discuss its role in initiation by those polymerases shortly. TBP does not bind directly to G•C-rich DNA, so DNA-binding is probably the responsibility of the other components of SL1. It is likely that TBP interacts with RNA polymerase, possibly with a common subunit or a feature that has been conserved among polymerases.

The behavior of SL1 resembles a bacterial sigma factor. As an isolated protein complex, it does not bind specifically to the promoter, but in conjunction with other components, specific promoter regions are bound. It may have primary responsibility for ensuring that the RNA polymerase is properly localized at the startpoint. We see shortly that a comparable function is provided for RNA polymerases II and III by a factor that consists of TBP associated with other proteins. So a common feature in initiation by all three polymerases is a reliance on a "positioning" factor that consists of TBP associated with proteins that are specific for each type of promoter.

Figure 20.4 Transcription units for RNA polymerase I have a core promoter separated by ~70 bp from the upstream control element. UBF1 binds to both regions, after which SL1 can bind. RNA polymerase I then binds to the core promoter. The nature of the interaction between the factors bound at the upstream control element and those at the core promoter is not known.

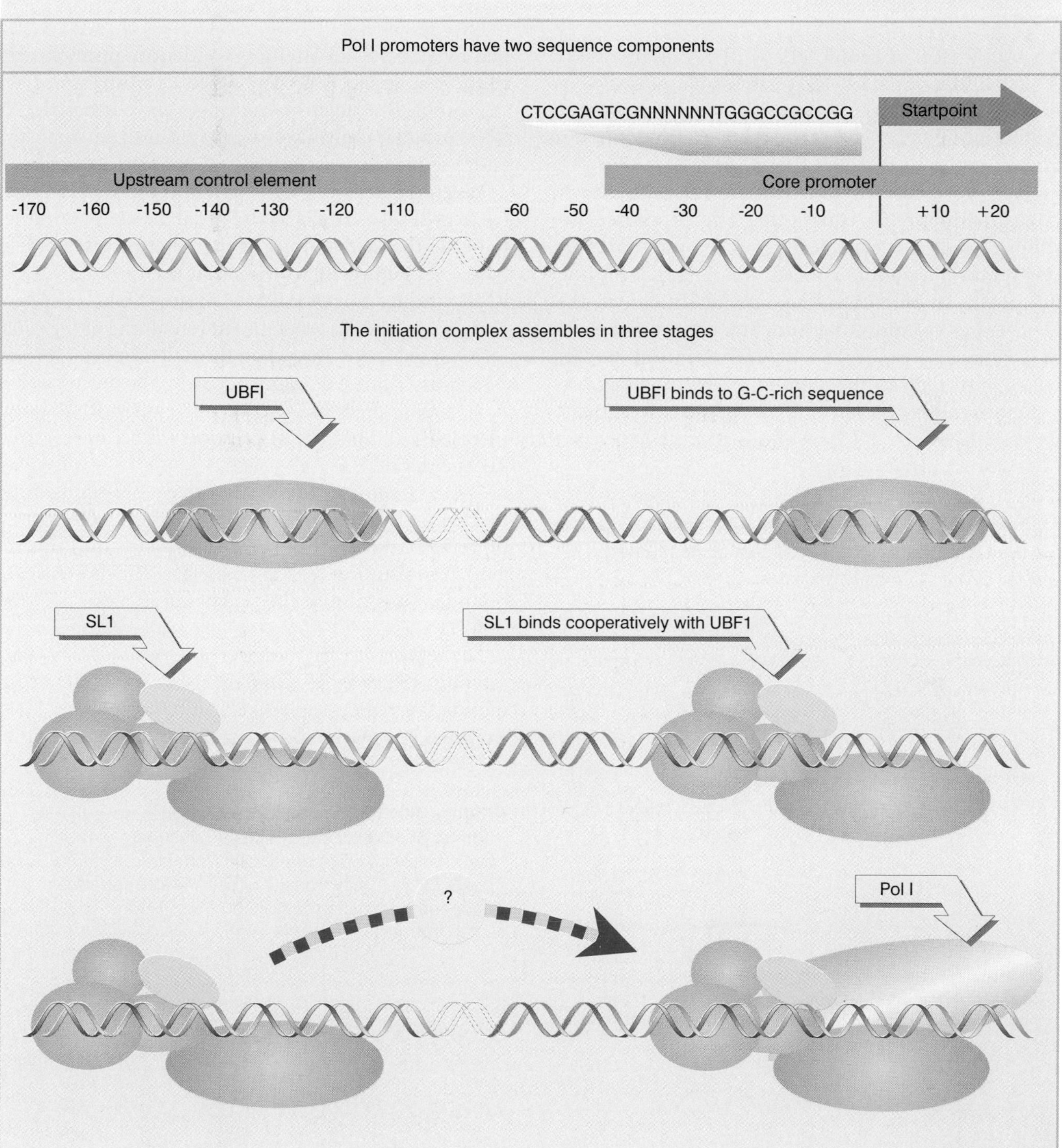

RNA polymerase III uses both downstream and upstream promoters

RECOGNITION of promoters by RNA polymerase III illustrates strikingly the relative roles of transcription factors and the polymerase enzyme. The promoters fall into two general classes that are recognized in different ways by different groups of factors. The promoters for 5S and tRNA genes are *internal*; they lie downstream of the startpoint. The promoters for snRNA (small nuclear RNA) genes lie upstream of the startpoint in the more conventional manner of other promoters. In both cases, the individual elements that are necessary for promoter function consist exclusively of sequences recognized by transcription factors, which in turn direct the binding of RNA polymerase.

Before the promoter of 5S RNA genes was identified in *X. laevis*, all attempts to identify promoter sequences assumed that they would lie upstream of the startpoint. But deletion analysis showed that the 5S RNA product continues to be synthesized when the entire sequence upstream of the gene is removed!

When the deletions continue into the gene, a product very similar in size to the usual 5S RNA continues to be synthesized so long as the deletion ends before base +55. **Figure 20.5** shows that the first part of the RNA product corresponds to plasmid DNA; the second part represents the segment remaining of the usual 5S RNA sequence. But when the deletion extends past +55, transcription does not occur. So the promoter lies *downstream of position +55*, but causes RNA polymerase III to initiate transcription a more or less fixed distance away.

When deletions extend into the gene from its distal end, transcription is unaffected so long as the first 80 bp remain intact. Once the deletion cuts into this region, transcription ceases. This places the downstream boundary position of the promoter at about position +80.

So the promoter for 5S RNA transcription lies between positions +55 and +80 within the gene. A fragment containing this region can sponsor initiation of any DNA in which it is placed, from a startpoint ~55 bp farther

Figure 20.5 Deletion analysis shows that the promoter for 5S RNA genes is internal; initiation occurs a fixed distance (~55 bp) upstream of the promoter.

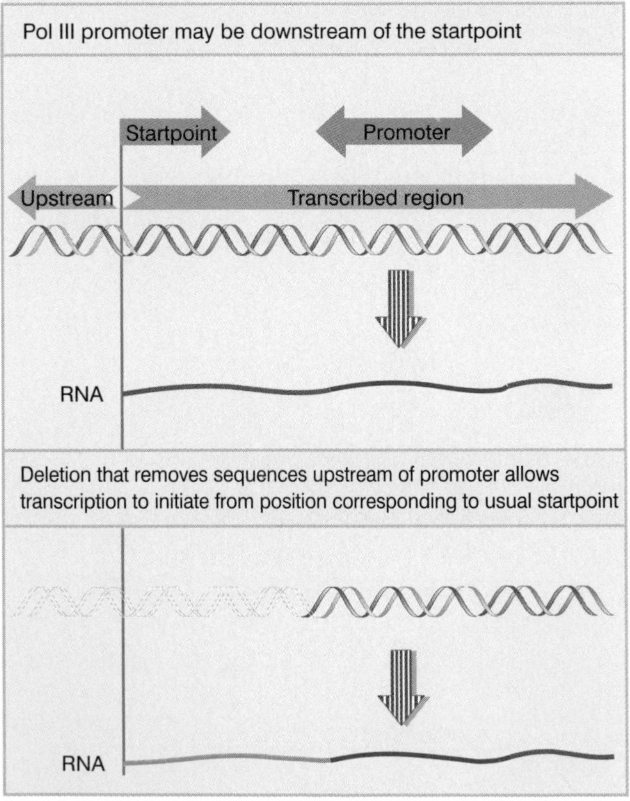

Figure 20.6 Promoters for RNA polymerase III may consist of bipartite sequences downstream of the startpoint, with boxA separated from either boxC or boxB. Or they may consist of separated sequences upstream of the startpoint (Oct, PSE, TATA).

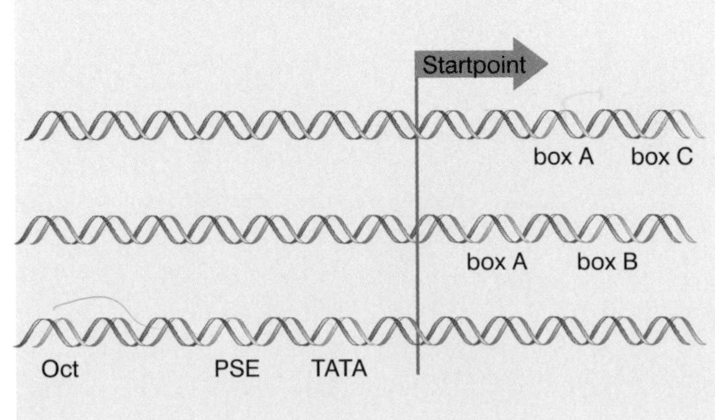

upstream. (The wild-type startpoint is unique; in deletions that lack it, transcription initiates at the purine base nearest to the position 55 bp upstream of the promoter.)

The structures of three types of promoters for RNA polymerase III are summarized in **Figure 20.6.** There are two types of internal promoter. Each contains a bipartite structure, in which two short sequence elements are separated by a variable sequence. Type 1 consists of a boxA sequence separated from a boxC sequence, and type 2 consists of a boxA sequence separated from a boxB sequence. The distance between boxA and boxB in a type 2 promoter can vary quite extensively, but the boxes usually cannot be brought too close together without abolishing function. We discuss the organization of the upstream type of promoter later.

Figure 20.7 summarizes the stages of reaction at

Figure 20.7 Initiation via the internal pol III promoters involves the assembly factors TFIIIA and TFIIIC, the initiation factor TFIIIB, and RNA polymerase III.

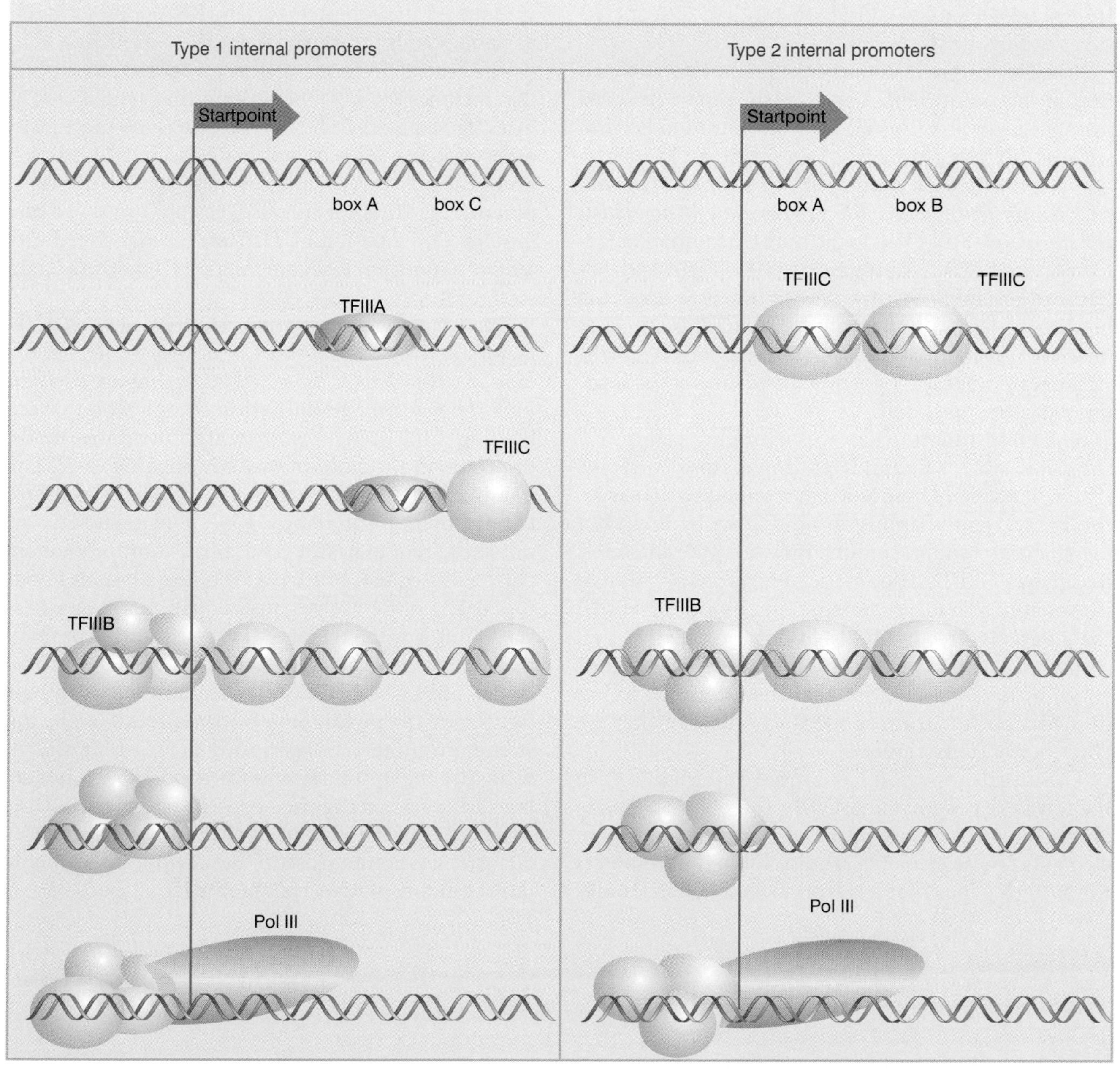

internal promoters. Three accessory factors are involved. TFIIIA is a member of an interesting class of zinc finger proteins that we discuss later. TFIIIB consists of TBP and two other proteins. TFIIIC is a large protein complex (>500 kD), comparable in size to RNA polymerase itself, and containing at least 5 subunits.

We do not fully understand all the interactions that occur at the pol III promoters, but the principle is clear. At type 2 promoters (on the right side of the figure), TFIIIC recognizes boxB, but binds to a more extensive region including both boxes A and B. At type 1 promoters (on the left side of the figure), TFIIIA binds to a sequence that includes boxC, and this is required to enable TFIIIC to bind. In both cases, the binding of TFIIIC in turn enables TFIIIB to bind to a sequence surrounding the startpoint.

A crucial feature in defining the roles of the factors is that, at this point, TFIIIA and TFIIIC can be removed from the promoter (by high salt concentration *in vitro*) without affecting the initiation reaction. *TFIIIB remains bound in the vicinity of the startpoint and its presence is sufficient to allow RNA polymerase III to bind at the startpoint.* So TFIIIB is the only true initiation factor required by RNA polymerase III. TFIIIA and TFIIIC are **assembly factors,** whose role is to assist the binding of TFIIIB at the right location. This sequence of events explains how the promoter boxes downstream can cause RNA polymerase to bind at the startpoint, farther upstream.

So TFIIIB functions as a "positioning factor," responsible for localizing RNA polymerase correctly. Like SL1 at the pol I promoter, it resembles a sigma factor, in lacking the ability to bind DNA by itself, but being able to bind in conjunction with other proteins. Recall that TFIIIB includes the same protein, TBP, that is present in SL1; this could be the subunit of TFIIIB that interacts directly with RNA polymerase III.

Although the ability to transcribe these genes is conferred by the internal promoter, changes in the region immediately upstream of the startpoint can alter the efficiency of transcription.

The upstream region has a more important role in the third class of polymerase III promoters. In the example shown in Figure 20.6, there are three upstream elements. These elements are also found in promoters for snRNA genes that are transcribed by RNA polymerase II. (Genes for some snRNAs are transcribed by RNA polymerase II, while others are transcribed by RNA polymerase III.) The upstream elements function in a similar manner in promoters for both polymerases II and III. The TATA element appears to confer specificity for the type of polymerase.

Initiation at an upstream promoter for RNA polymerase III can occur on a short region that immediately precedes the startpoint and contains only the TATA element. However, efficiency of transcription is much increased by the presence of the PSE and OCT elements. The factors that bind at these elements interact cooperatively. (The PSE element may be essential at promoters used by RNA polymerase II, whereas it is stimulatory in promoters used by RNA polymerase III; its name stands for proximal sequence element.)

The crucial TATA element is recognized by a factor that includes the TBP, the subunit that actually recognizes the sequence in DNA. The TBP is associated with other proteins, some of which are specific for promoters of the pol III type, and this may explain why RNA polymerase III is specifically recruited to these promoters. The function of TBP and its associated proteins is to position RNA polymerase III correctly at the startpoint.

There is an underlying unity in the functions of the factors at RNA polymerase III promoters. *The factors bind at the promoter before RNA polymerase itself can bind.* They form a **preinitiation complex** that directs binding of the RNA polymerase. The alternative modes of promoter recognition by RNA polymerase III emphasize the importance of the factors for establishing the startpoint for initiation. RNA polymerase III does not itself seem to have a great intrinsic affinity for any particular sequence of DNA. It binds adjacent to factors that are themselves bound just upstream of the startpoint. For the type 1 and type 2 internal promoters, the assembly factors ensure that TFIIIB (which includes TBP) is bound just upstream of the startpoint, to provide the positioning information. For the upstream promoters, transcription factors that directly recognize the upstream sites form a complex (including TBP) that is recognized by RNA polymerase III. So irrespective of the location of the promoter sequences, factor(s) are bound close to the startpoint in order to direct binding of RNA polymerase III.

The basal apparatus consists of RNA polymerase II and general factors

RNA polymerase II cannot initiate transcription itself, but is absolutely dependent on auxiliary transcription factors. The enzyme together with these factors constitutes the basal transcription apparatus that is needed to transcribe any promoter. Our starting point for considering promoter organization is therefore to define a "generic" promoter, the shortest sequence at which RNA polymerase II can initiate transcription, and to characterize the enzyme subunits and transcription factors that are needed to recognize it.

A generic promoter can in principle be expressed in any cell: it does not depend on sequences whose use is under tissue-specific control. The accessory proteins that are required for polymerase II to initiate at such a promoter define the general transcription factors involved in the mechanics of binding to DNA and initiating transcription. The general factors are described as **TFIIX,** where "X" is a letter that identifies the individual factor. A generic promoter functions at only a low efficiency; additional upstream factors are required for a proper level of function. The upstream and inducible factors are not described systematically, but have casual names reflecting their histories of identification.

We may expect any sequence components involved in the binding of RNA polymerase and general transcription factors to be conserved at most or all promoters. As with bacterial promoters, when promoters for RNA polymerase II are compared, homologies in the regions near the startpoint are restricted to rather short sequences. These elements correspond with the sequences implicated in promoter function by mutation.

At the startpoint, there is no extensive homology of sequence, but there is a tendency for the first base of mRNA to be A, flanked on either side by pyrimidines. (This description is also valid for the CAT start sequence of bacterial promoters.) This region is called the **initiator** (**Inr**), and may be described in the general form Py_2CAPy_5. The Inr is contained between positions −3 and +5. A promoter consisting only of the Inr has the simplest possible form recognizable by RNA polymerase II.

Most promoters have a sequence called the **TATA box,** usually located ~25 bp upstream of the startpoint. It constitutes the only upstream promoter element that has a relatively fixed location with respect to the startpoint. It is found in all eukaryotes. The 8 bp consensus sequence consists entirely of A•T base pairs (at two positions the orientation is variable), and in only a minority of actual cases is a G•C pair present. The TATA box tends to be surrounded by G•C-rich sequences, which could be a factor in its function. It is almost identical with the −10 sequence found in bacterial promoters; in fact, it could pass for one except for the difference in its location at −25 instead of −10.

Single base substitutions in the TATA box act as strong down mutations. Some mutations reverse the orientation of an A•T pair, so base composition alone is not sufficient for its function. So the TATA box comprises an element whose behavior is analogous to our concept of the bacterial promoter: a short, well-defined sequence just upstream of the startpoint, which is necessary for transcription. The minority of promoters that do not contain a TATA element are called **TATA-less promoters.**

The first step in complex formation at a promoter containing a TATA box is binding of the factor **TFIID** to a region that extends upstream from the TATA sequence. TFIID contains two types of component. Recognition of the TATA box is conferred by the **TATA-binding protein** (**TBP**), a small protein of ~30 kD. The other subunits are called **TAFs** (for TBP-associated factors). Some TAFs are stoichiometric with TBP; others are present in lesser amounts. TFIIDs containing different TAFs could recognize different promoters. Some (substoichiometric) TAFs are tissue-specific. The total mass of TFIID typically is ~800 kD, containing TBP and 11 TAFs, varying in mass from 30–250 kD. The TAFs in TFIID are named in the form $TAF_{II}00$, where "00" gives the molecular mass of the subunit. The TAF_{II}s are not confined exclusively to TFIID; certain TAF_{II}s are found also in protein complexes that act to modify the structure of chromatin prior to transcription (see Chapter 21).

TFIID is ubiquitous, but not unique. Another complex contains the factor TRF (related to TBP) which associates with its own set of additional factors (nTAFs), and which may function specifically in neuronal cells (in *Drosophila*) as an alternative to TFIID, possibly at specific promoters.

The idea that TBP is associated with different TAFs serves as a general description for its employment at each class of promoters. Genetic evidence from yeast, and biochemical analysis of animal transcription extracts, suggest that TBP is required for all initiation by RNA polymerase III. So TFIIIB and SL1 may both be viewed as consisting of TBP associated with a particular group of proteins that substitute for the TAFs that are found in TFIID. TBP is the key component of the "positioning factor," incorporated at each type of promoter by a different mechanism. In the case of promoters for RNA polymerase II, the key feature in positioning is the fixed distance of the TATA box from the startpoint.

The means by which the factor recognizes the promoter is different in each case. **Figure 20.8** summarizes its utilization. TFIIIB binds adjacent to TFIIIC, while SL1 binds in conjunction with UBF1. TFIID is solely responsible for recognizing a promoter for RNA polymerase II. At a promoter that has a TATA element, TBP binds specifically to DNA, but at other promoters it may be incorporated by association with other proteins that bind to DNA. Whatever its means of entry into the initiation complex, it has the common purpose of interaction with the RNA polymerase.

Any individual molecule of TBP itself is not necessarily available for all promoters, but may in effect be sequestered by its associated proteins to be used continuously by a specific class of promoter. TBP must have the capacity to interact appropriately with the variety of factors and/or polymerases that are employed at each type of promoter.

TBP has the unusual property of binding to DNA in the minor groove. (Virtually all known DNA-binding proteins bind in the wide groove.) The crystal structure of TBP suggests a detailed model for its binding to DNA. **Figure 20.9** shows that it surrounds one face of DNA, forming a "saddle" around the double helix. In effect, the inner surface of TBP binds to DNA, and the larger outer surface is available to extend contacts to other proteins. The DNA-binding site consists of sequences that are conserved between species, while the variable N-terminal tail is exposed to interact with other proteins.

Binding of TBP may be inconsistent with the presence of nucleosomes. Because nucleosomes form preferentially by placing A•T-rich sequences with the minor grooves facing inward, they could prevent binding of TBP. This may explain why the presence of nucleosomes prevents initiation of transcription.

TBP binds DNA in an unusual way. It not only sits in the minor groove, but also bends the DNA by ~80°, as illustrated in **Figure 20.10.** The TATA box bends towards the major groove, widening the minor groove. The distortion is restricted to the 8 bp of the TATA box; at each end of the sequence, the minor groove has its usual width of ~5 Å, but at the center of the sequence

Figure 20.8 RNA polymerases are positioned at all promoters by a factor that contains TBP.

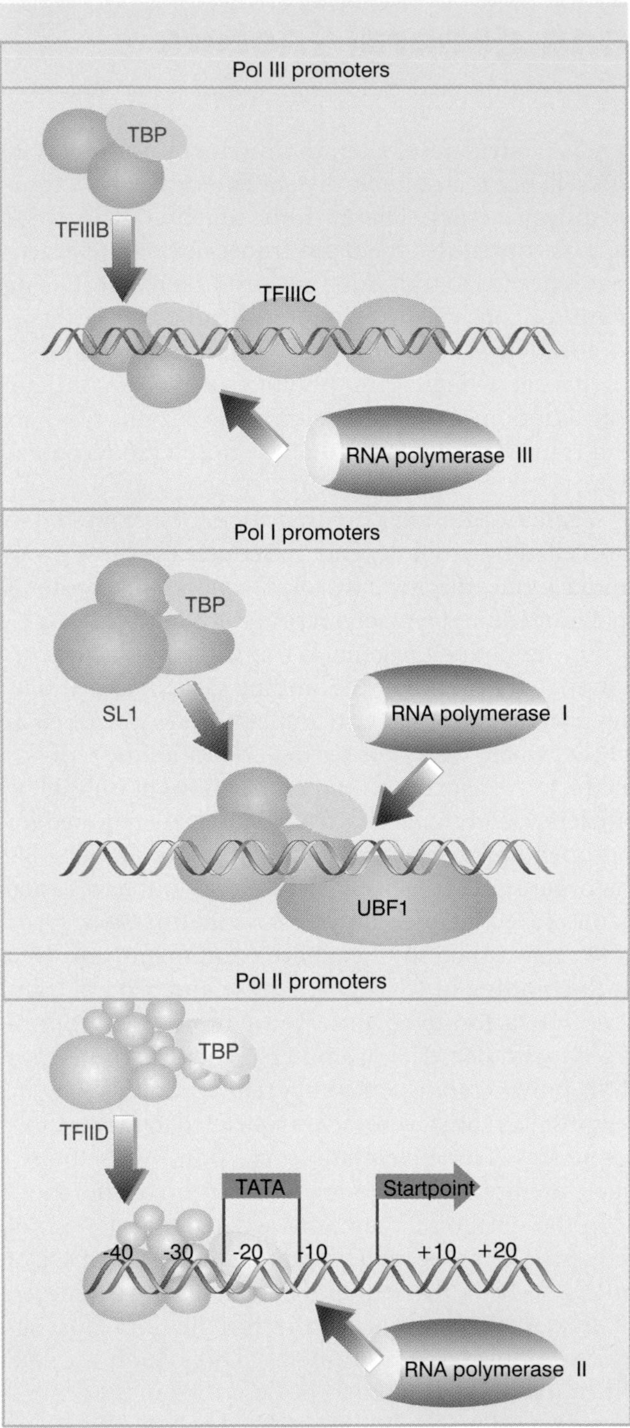

Figure 20.9 A view in cross-section shows that TBP surrounds DNA from the side of the narrow groove. TBP consists of two related (40% identical) conserved domains, which are shown in light and dark blue. The N-terminal region varies extensively and is shown in green. The two strands of the DNA double helix are in light and dark grey. Photograph kindly provided by Stephen Burley.

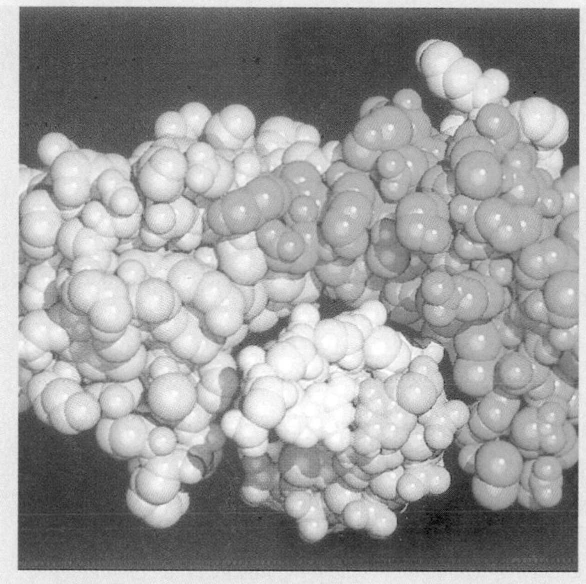

Figure 20.10 The cocrystal structure of TBP with DNA from -40 to the startpoint shows a bend at the TATA box that widens the narrow groove where TBP binds. Photograph kindly provided by Stephen Burley.

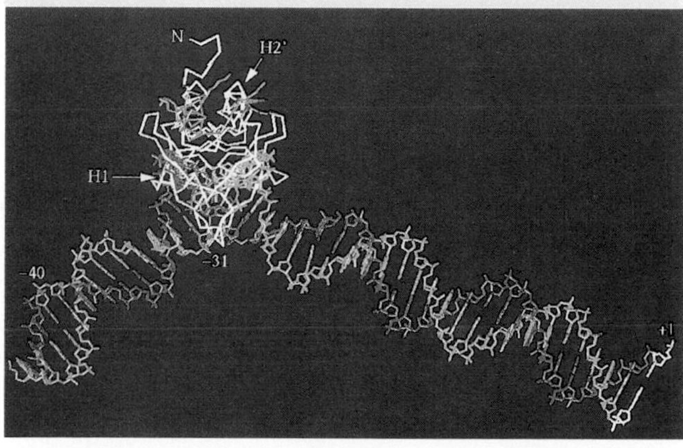

the minor groove is >9 Å. This is a deformation of the structure, but does not actually separate the strands of DNA, because base pairing is maintained.

This structure has several functional implications. By changing the spatial organization of DNA on either side of the TATA box, it allows the transcription factors and RNA polymerase to form a closer association than would be possible on linear DNA. The bending at the TATA box corresponds to unwinding of about 1/3 of a turn of DNA, and is compensated by a positive writhe. We do not know yet how this relates to the initiation of strand separation.

The presence of TBP in the minor groove, combined with other proteins binding in the major groove, creates a high density of protein-DNA contacts in this region. Binding of purified TBP to DNA *in vitro* protects ~1 turn of the double helix at the TATA box, typically extending from –37 to –25; but binding of the TFIID complex in the initiation reaction regularly protects the region from –45 to –10, and also extends farther upstream beyond the startpoint. TBP is the only general transcription factor that makes sequence-specific contacts with DNA.

Within TFIID as a free protein complex, the factor $TAF_{II}230$ binds to TBP, where it occupies the concave DNA-binding surface. In fact, the structure of the binding site, which lies in the N-terminal domain of $TAF_{II}230$, mimics the surface of the minor groove in DNA. This molecular mimicry allows $TAF_{II}230$ to control the ability of TBP to bind to DNA; the N-terminal domain of $TAF_{II}230$ must be displaced from the DNA-binding surface of TBP in order for TFIID to bind to DNA.

Some TAFs resemble histones; in particular $TAF_{II}42$ and $TAF_{II}62$ appear to be (distant) homologs of histones H3 and H4, and they form a heterodimer using the same motif (the histone fold) that histones use for the interaction. Together with other TAFs, they may form the basis for a structure resembling a histone octamer which is involved in the nonspecific interactions of TFIID with DNA. Histone folds are also used in pairwise interactions between other TAF_{II}s.

Initiation requires the transcription factors to act in a defined order to build a complex that is joined by RNA polymerase. The series of events can be followed by the increasing size of the protein complex associated with DNA. Footprinting of the DNA regions protected by each complex suggests the model summarized in **Figure 20.11.** As each TFII factor joins the complex, an increasing length of DNA is covered. RNA polymerase is incorporated at a late stage.

Commitment to a promoter is initiated when TFIID binds the TATA box. When TFIIA joins the complex, TFIID becomes able to protect a region extending farther upstream. TFIIA may activate TBP by relieving the repression that is caused by the $TAF_{II}230$.

Addition of TFIIB gives some partial protection of

Figure 20.11 An initiation complex assembles at promoters for RNA polymerase II by an ordered sequence of association with transcription factors.

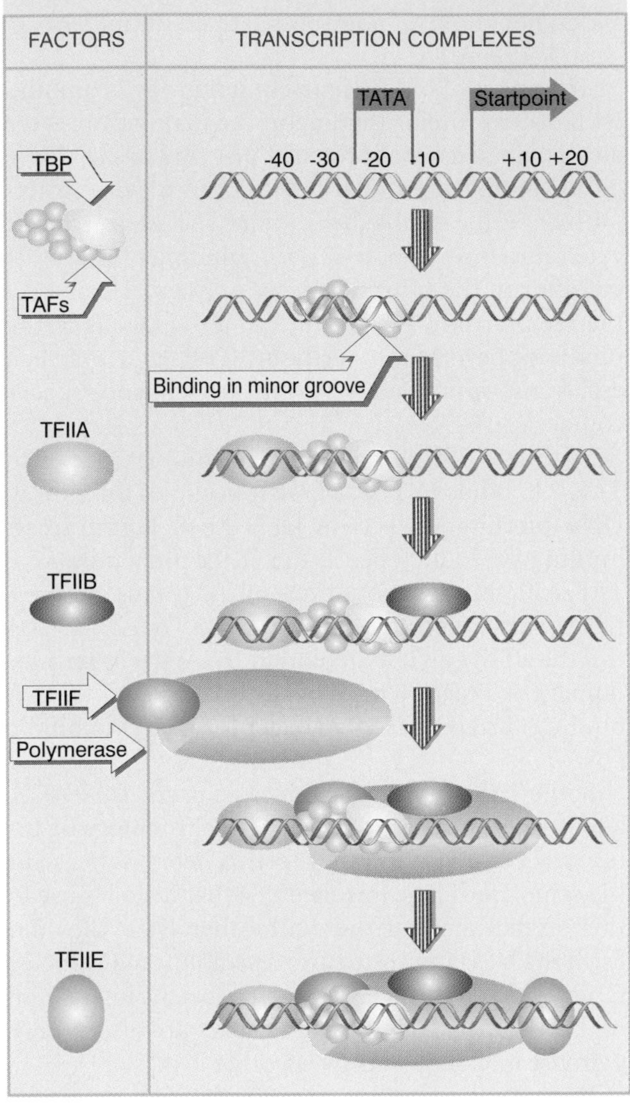

Figure 20.12 Two views of the ternary complex of TFIIB-TBP-DNA show that TFIIB binds along the bent face of DNA. The two strands of DNA are green and yellow, TBP is blue, and TFIIB is red and purple. Photograph kindly provided by Stephen Burley.

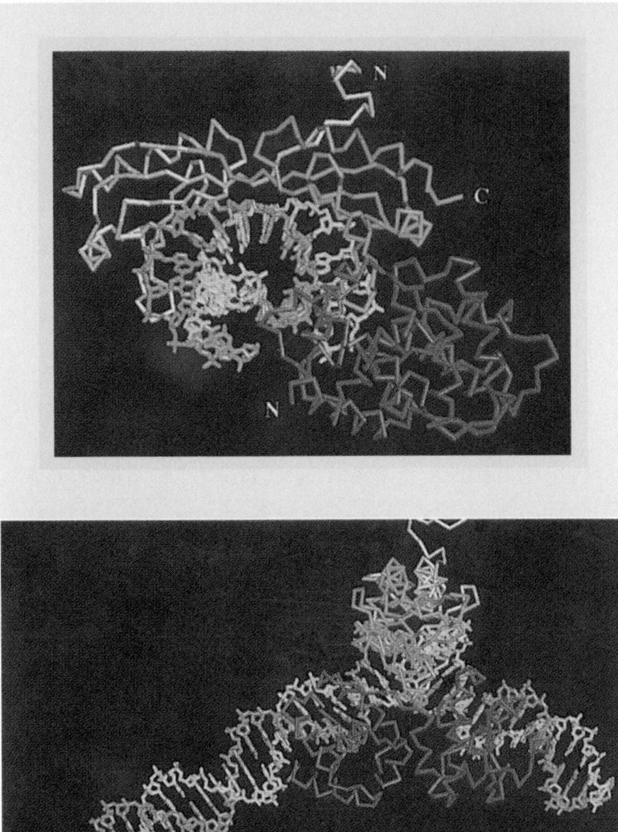

the region of the template strand in the vicinity of the startpoint, from –10 to +10. This suggests that TFIIB is bound downstream of the TATA box, perhaps loosely associated with DNA and asymmetrically oriented with regard to the two DNA strands. The crystal structure shown in **Figure 20.12** confirms this model. TFIIB binds adjacent to TBP, extending contacts along one face of DNA. It may provide the surface that is in turn recognized by RNA polymerase. (In archaea, the homologue of TFIIB actually makes sequence-specific contacts with the promoter.)

The factor TFIIF consists of two subunits. The larger subunit (RAP74) has an ATP-dependent DNA helicase activity that could be involved in melting the DNA at initiation. The smaller subunit (RAP38) has some homology to the regions of bacterial sigma factor that contact the core polymerase; it binds tightly to RNA polymerase II. TFIIF may bring RNA polymerase II to the assembling transcription complex and provide the means by which it binds. The complex of TBP and TAFs may interact with the CTD tail of RNA polymerase, and interaction with TFIIB may also be important when TFIIF/polymerase joins the complex.

Polymerase binding extends the sites that are protected downstream to +15 on the template strand and +20 on the nontemplate strand. The enzyme extends the full length of the complex, since additional protection is seen at the upstream boundary.

The initiation reaction, as defined by formation of the first phosphodiester bond, can occur at this stage. Some further general factors, TFIIE and TFIIH, are required for promoter clearance—to allow RNA polymerase to commence movement away from the promoter.

Binding of TFIIE causes the boundary of the region protected downstream to be extended by another turn of the double helix, to +30. Two further factors, TFIIH and TFIIJ, join the complex after TFIIE. They do not change the pattern of binding to DNA. TFIIH has several activities, including an ATPase, a helicase, and a kinase activity that can phosphorylate the CTD tail of RNA polymerase II; it is also involved in repair of damage to DNA (see next section).

Most of the TFII factors are released before RNA polymerase II leaves the promoter. **Figure 20.13** proposes a model in which phosphorylation of the tail is needed to release RNA polymerase II from the transcription factors so that it can make the transition to the elongating form. TFIIH is an exceptional factor that may play a role also in elongation.

The CTD may coordinate processing of RNA with transcription. The capping enzyme (guanylyl transferase), which adds the G residue to the 5′ end of newly synthesized mRNA, binds to the phosphorylated CTD: this may be important in enabling it to modify the 5′ end as soon as it is synthesized. Some splicing factors bind to the CTD and so do some components of the cleavage/polyadenylation apparatus, suggesting that it may be a general focus for connecting other processes with transcription.

The general process of initiation is similar to that catalyzed by bacterial RNA polymerase. Binding of RNA polymerase generates a closed complex, which is converted at a later stage to an open complex in which the DNA strands have been separated. In the bacterial reaction, formation of the open complex completes the necessary structural change to DNA; a difference in the eukaryotic reaction is that further unwinding of the template is needed after this stage.

On a linear template, ATP hydrolysis, TFIIE, and the helicase activity of TFIIH are required for polymerase movement. This requirement is bypassed with a supercoiled template. This suggests that TFIIE and TFIIH are involved in an extension of the unwound region of DNA to allow polymerase movement to begin.

What happens at TATA-less promoters? The same general transcription factors, including TFIID, are needed. The Inr provides the positioning element; TFIID binds to it via an ability of one or more of the TAFs to recognize the Inr directly. The function of TBP at these promoters is more like that at promoters for RNA polymerase I and at internal promoters for RNA polymerase III.

Figure 20.13 Phosphorylation of the CTD by the kinase activity of TFIIH may be needed to release RNA polymerase to start transcription.

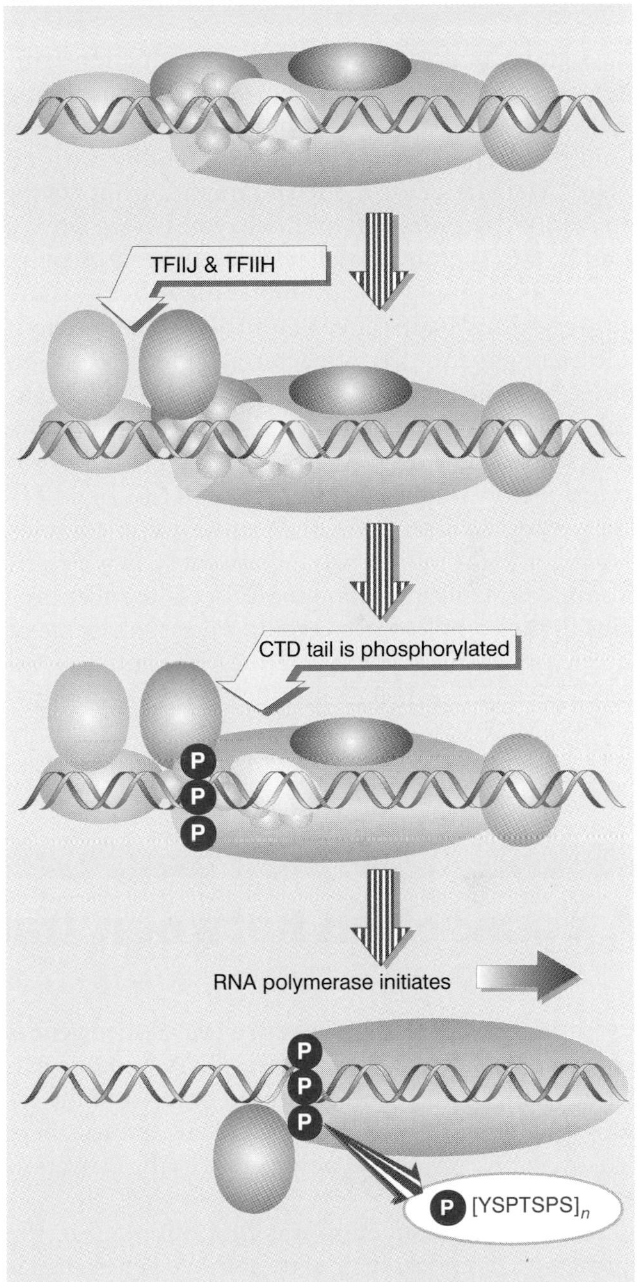

Many of the general factors consist of multiple subunits, so the total number of polypeptides involved in the basal apparatus is rather large. There are probably

~20 polypeptides with a total mass of ~500 kD. Remember that RNA polymerase II itself has ~10 subunits with a mass of ~500 kD, so we see that initiation involves the assembly of an extremely large complex.

Assembly of the RNA polymerase II initiation complex provides an interesting contrast with prokaryotic transcription. Bacterial RNA polymerase is essentially a coherent aggregate with intrinsic ability to bind DNA; the sigma factor, needed for initiation but not for elongation, becomes part of the enzyme before DNA is bound, although it is later released. But RNA polymerase II can bind to the promoter only after separate transcription factors have bound. The factors play a role analogous to that of bacterial sigma factor—to allow the basic polymerase to recognize DNA specifically at promoter sequences—but have evolved more independence. Indeed, the factors are primarily responsible for the specificity of promoter recognition. The process of assembling the transcription complex reminds us of ribosome subunit assembly, in which ribosomal proteins must bind to rRNA (or to other proteins in the complex) in a certain order. Only some of the factors participate in protein-DNA contacts (and only TBP makes sequence-specific contacts); thus protein-protein interactions are important in the assembly of the complex.

The sequences in the vicinity of the startpoint comprise a "core" promoter at which the basal transcription apparatus is assembled. When a TATA box is present, it determines the location of the startpoint. Its deletion causes the site of initiation to become erratic, although any overall reduction in transcription is relatively small. Indeed, some TATA-less promoters lack unique startpoints; initiation occurs instead at any one of a cluster of startpoints. The TATA box aligns the RNA polymerase (via the interaction with TFIID and other factors) so that it initiates at the proper site. This explains why its location is fixed with respect to the startpoint. Binding of TBP to TATA is the predominant feature in recognition of the promoter, but two large TAFS (TAF_{250} and TAF_{150}) also contact DNA in the vicinity of the startpoint and influence the efficiency of the reaction.

Although assembly can take place just at the core promoter *in vitro*, this reaction is not sufficient for transcription *in vivo*, where interactions with other factors that recognize the more upstream elements are required. These factors interact with the basal apparatus at various stages during its assembly.

A connection between transcription and repair

The template strand of DNA in a transcribed gene is preferentially repaired when DNA is damaged. What identifies the template strand of DNA to the repair apparatus? In both bacteria and eukaryotes, there is a direct link from RNA polymerase to the activation of repair.

The basic phenomenon was first observed in the form of preferential repair of transcribed genes; then it was discovered that it is only the template strand of DNA that is the target—the nontemplate strand is repaired at the same rate as bulk DNA. In bacteria, the repair activity is provided by the *uvr* excision-repair system (see Chapter 14). Preferential repair is abolished by mutations in the gene *mfd*, whose product provides the link from RNA polymerase to the Uvr enzymes.

Figure 20.14 shows a model for the link between transcription and repair. When RNA polymerase encounters DNA damage in the template strand, it stalls because it cannot use the damaged sequences as a template to direct complementary base pairing. This explains the specificity of the effect for the template strand (damage in the nontemplate strand does not impede progress of the RNA polymerase).

The Mfd protein has two roles. First, it displaces the ternary complex of RNA polymerase from DNA. Second, it causes the UvrABC enzyme to bind to the damaged DNA. This leads to repair of DNA by the excision-repair mechanism. After the DNA has been repaired, the next RNA polymerase to traverse the gene is able to produce a normal transcript.

A similar mechanism, although relying on different components, is used in eukaryotes. The template strand of a transcribed gene is preferentially repaired following UV-induced damage. The general transcription factor TFIIH may be involved. TFIIH is found in alternative forms, which consist of a core associated with other subunits.

Figure 20.15 shows that the basic factor involved in transcription consists of a core (of 5 subunits) associated with other subunits that have a kinase activity; this complex also includes a repair subunit. The kinase catalytic subunit that phosphorylates the CTD of RNA polymerase belongs to a group of kinases that are involved in cell cycle control (see Chapter 27). It is possible that this connection influences transcription in response to the stage of the cell cycle.

Another complex consists of the core associated with a large group of proteins all of which are coded by repair genes. These include a subunit that recognizes damaged DNA and an endonuclease. This provides the coupling function that enables a template strand to be preferentially repaired when RNA polymerase becomes stalled at damaged DNA. Homologous proteins are found in the complexes in yeast (where they are often identified by *rad* mutations that are defective in repair) and in man (where they are identified by mutations that cause diseases resulting from deficiencies in repairing damaged DNA).

Figure 20.14 Mfd recognizes a stalled RNA polymerase and directs DNA repair to the damaged template strand.

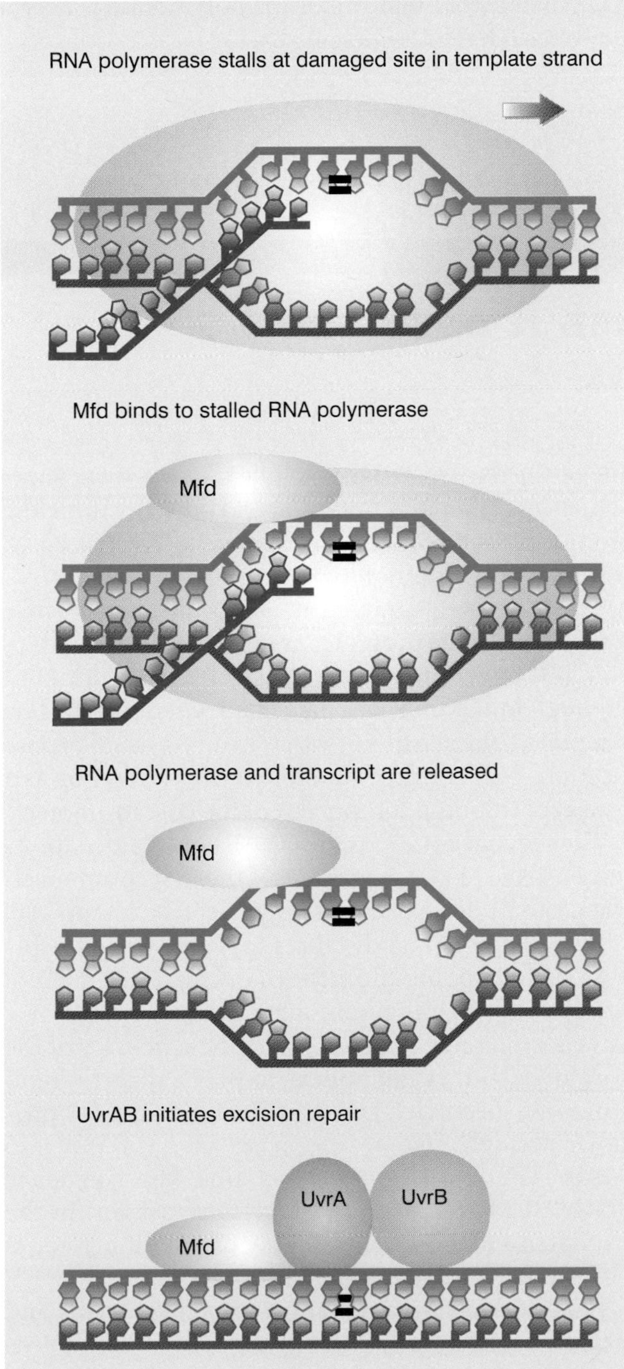
RNA polymerase stalls at damaged site in template strand
Mfd binds to stalled RNA polymerase
Mfd
RNA polymerase and transcript are released
Mfd
UvrAB initiates excision repair
UvrA
UvrB
Mfd

Figure 20.15 The TFIIH core may associate with a kinase at initiation and associate with a repair complex when damaged DNA is encountered.

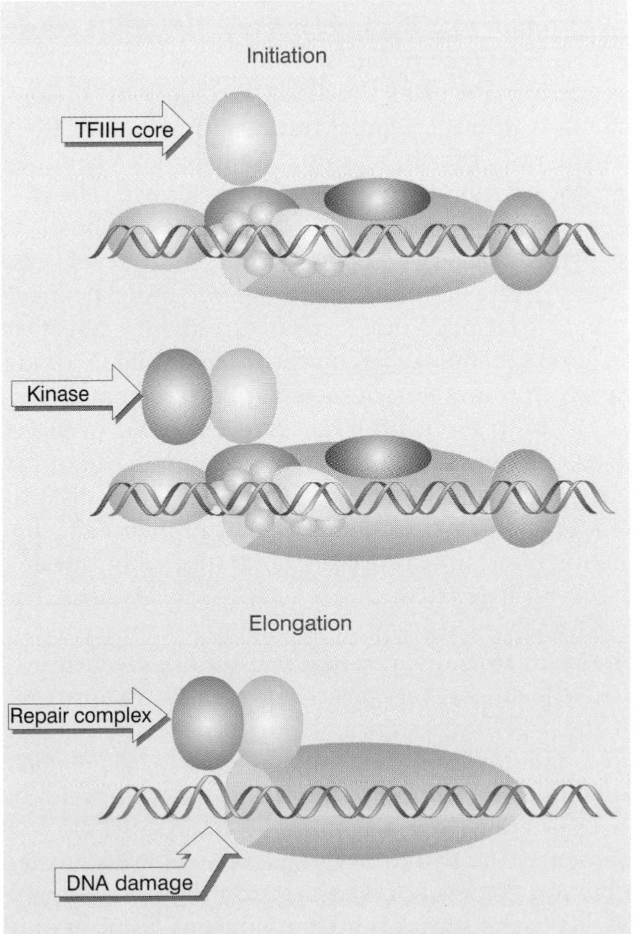
Initiation
TFIIH core
Kinase
Elongation
Repair complex
DNA damage

The kinase complex and the repair complex can associate and dissociate reversibly from the core TFIIH. This suggests a model in which the first form of TFIIH is required for initiation, but may be replaced by the other form (perhaps in response to encountering DNA damage). TFIIH dissociates from RNA polymerase at an early stage of elongation (after transcription of ~50 bp); its reassociation at a site of damaged DNA may require additional coupling components, such as Rad26, which is required for the preferential repair of transcribed as opposed to nontranscribed strands.

The repair function may require modification or degradation of RNA polymerase. The large subunit of RNA polymerase is degraded when the enzyme stalls at sites of UV damage. Degradation follows ubiquitination and is undertaken by the proteasome (see Chapter 8). This process is deficient in cells from patients with Cockayne's syndrome (a repair disorder). Cockayne's syndrome is caused by mutations in either of two genes (*CSA* and *CSB*), both of whose products appear to be part of or bound to TFIIH. We do not yet understand the connection between the transcription/repair apparatus as such and the degradation of RNA polymerase. It is possible that removal of the polymerase is necessary once it has become stalled.

Promoters for RNA polymerase II have short sequence elements

A promoter for RNA polymerase II consists of two types of region. The startpoint itself is identified by the Inr and/or by the TATA box close by. In conjunction with the general transcription factors, RNA polymerase II forms an initiation complex surrounding the startpoint, as we have just described. The efficiency and specificity with which a promoter is recognized, however, depend upon short sequences, farther upstream, which are recognized by upstream or inducible factors. These sequences and the factors that recognize them may be common, and found in a wide variety of promoters, or they may be specific, and particular for transcription in a restricted time or place. Usually these sequences are ~100 bp upstream of the startpoint, but sometimes they are more distant. Binding of factors at these sites may influence the formation of the initiation complex at (probably) any one of several stages.

An analysis of a typical promoter is summarized in **Figure 20.16.** Individual base substitutions were introduced at almost every position in the 100 bp upstream of the β-globin startpoint. The striking result is that *most mutations do not affect the ability of the promoter to initiate transcription.* Down mutations occur in three locations, corresponding to three short discrete elements. The two upstream elements have a greater effect on the level of transcription than the element closest to the startpoint. Up mutations occur in only one of the elements. We conclude that the three short sequences centered at –30, –75, and –90 constitute the promoter. Each of them corresponds to the consensus sequence for a common type of promoter element.

The TATA box (centered at –30) is the least effective component of the promoter as measured by the reduction in transcription that is caused by mutations. But although initiation is not prevented when a TATA box is mutated, the startpoint varies from its usual precise location. This confirms the role of the TATA box as a crucial positioning component of the core promoter.

The sequence at –75 is the **CAAT box.** Named for its consensus sequence, it was one of the first common elements to be described. It is often located close to –80, but it can function at distances that vary considerably from the startpoint. It functions in either orientation. Susceptibility to mutations suggests that the CAAT box plays a strong role in determining the efficiency of the promoter. It does not appear to play a direct role in promoter specificity, but its inclusion increases promoter strength.

The **GC box** at –90 contains the sequence GGGCGG. Often multiple copies are present in the promoter, and they occur in either orientation. It too is a relatively common promoter component.

Promoters are organized on a principle of "mix and match." A variety of elements can contribute to promoter function, but none is essential for all promoters.

Figure 20.16 Saturation mutagenesis of the upstream region of the β-globin promoter identifies three short regions (centered at -30, -75, and -90) that are needed to initiate transcription. These correspond to the TATA, CAAT, and GC boxes.

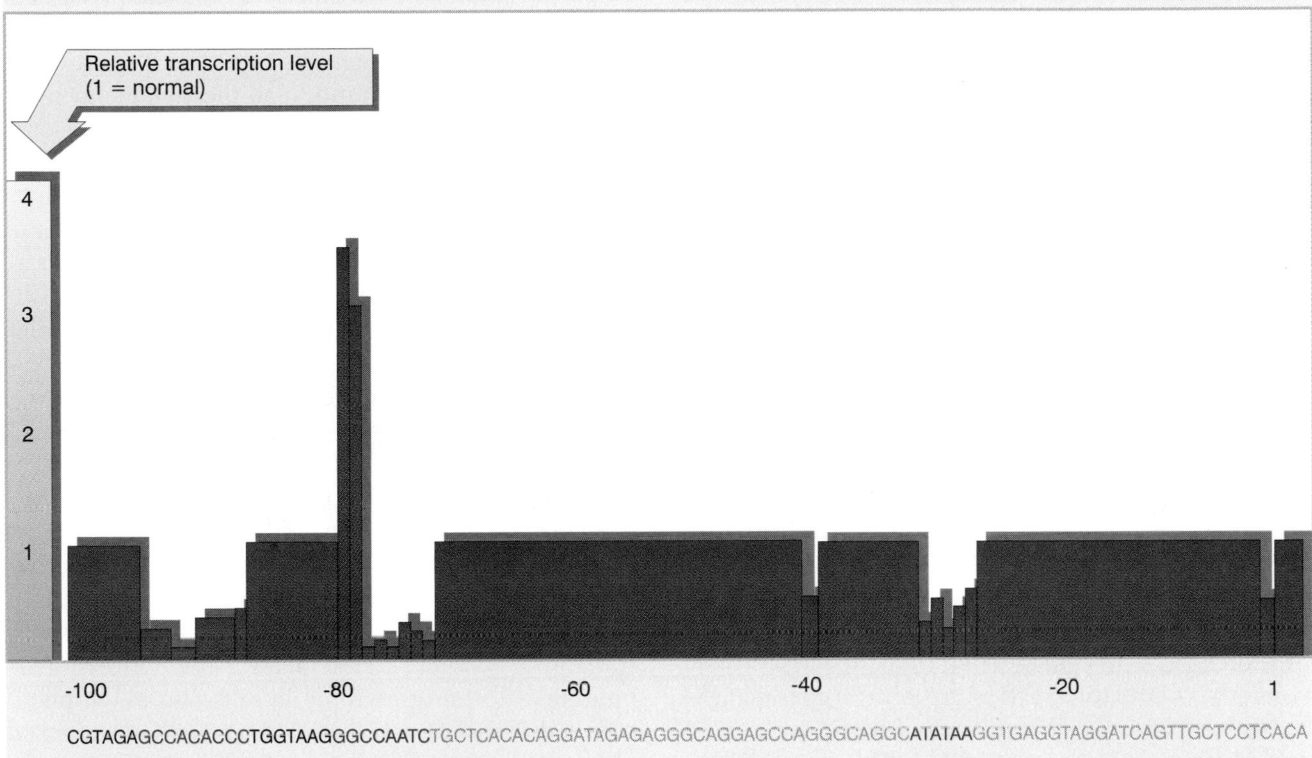

Some examples are summarized in **Figure 20.17**. Four types of element are found altogether in these promoters: TATA, GC boxes, CAAT boxes, and the octamer (an 8 bp element). The elements found in any individual promoter differ in number, location, and orientation. No element is common to all of the promoters. One of the puzzles of promoter organization is that the promoter conveys directional information (transcription proceeds only in the downstream direction), but the GC and CAAT boxes seem to be able to function in either orientation (although their sequences are asymmetrical).

Factors that are more or less ubiquitous are assumed to be available to any promoter that has a copy of the element that they recognize. This common availability distinguishes the upstream factors from the inducible factors that we discuss later. Elements in the upstream category include the CAAT box, GC box, and the octamer. All promoters probably require one or more of these elements in order to function efficiently.

The GC box is recognized by the factor SP1. This interaction illustrates the demands that can be placed on a single factor. The closest GC box usually is 40–70 bp

Figure 20.17 Promoters contain different combinations of TATA boxes, CAAT boxes, GC boxes, and other elements.

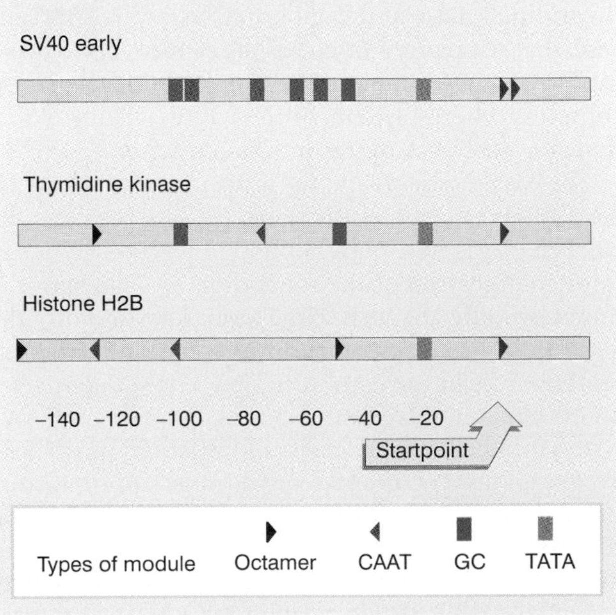

upstream of the startpoint, but the context of the GC boxes is different in every promoter. So in the thymidine kinase promoter, GC boxes are adjacent to a CAAT box and a TATA box, but in the SV40 promoter, a tandemly repeated series of GC boxes is upstream of a TATA box. The subunit of SP1 is a monomer of 105 kD, which contacts one strand of DNA over a ~20 bp binding site that includes at least one 6 bp GC box. In the SV40 promoter, the multiple boxes between –70 and –110 all are bound, so that the whole region is protected by SP1. In the thymidine kinase promoter, however, SP1 presumably interacts with a factor bound at the CAAT box on one side, and with TFIID bound at the TATA box on the other side.

The sequences to which the factors bind as characterized by footprinting are typically longer than the consensus sequences identified by comparing promoters. They usually cover ~20 bp of DNA, whereas the consensus sequences are <10 bp. Given the sizes of the factors, and the length of DNA each covers, we expect that the various proteins will together cover the entire region upstream of the startpoint in which the elements reside.

The diversity of elements from which a functional promoter may be constructed, and the variation in their locations relative to the startpoint, argues that the factors have an ability to interact with one another by protein-protein interactions in multiple ways. There appear to be no constraints on the potential relationships between the elements. The modular nature of the promoter is illustrated by experiments in which equivalent regions of different promoters have been exchanged. Hybrid promoters, for example, between thymidine kinase and β-globin, work well. This suggests that the main purpose of the elements is to bring the factors they bind into the vicinity of the initiation complex, where protein-protein interactions determine the efficiency of the initiation reaction.

The basal elements and the more upstream elements have different types of functions. We have seen that the basal elements (the TATA box and Inr) primarily determine the location of the startpoint, but can sponsor initiation only at a rather low level. They identify the *location* at which the general transcription factors assemble to form the basal complex. The sequence elements farther upstream, such as the GC or CAAT boxes, influence the *frequency* of initiation, most likely by acting directly on the general transcription factors to enhance the efficiency of assembly into an initiation complex (see later).

How can initiation be influenced by sites spread over a length of DNA that is greater than RNA polymerase could directly contact? Initiation involves a hierarchy of interactions, in which factors bound at upstream elements interact with general factors, which in turn interact directly with RNA polymerase (see later). This helps to explain the flexibility with which elements may be arranged, and the distance over which they can be dispersed, since it relieves us of the obligation to suppose that factors bound to all these elements must interact directly with RNA polymerase.

The most common use of promoter elements is for a particular consensus sequence to be recognized by a corresponding transcription factor (or by a member of a family of factors). However, some elements can be recognized by more than one factor. For example, the CAAT box can interact with factors of the CTF family, the factors CP1 and CP2, and the factors C/EBP and ACF. CAAT boxes in different promoters are recognized by different factors. The exact details of recognition are not so important as the fact that a variety of factors recognize CAAT boxes.

Another example of an element that is recognized by more than one factor is presented by the octamer sequence. A ubiquitous transcription factor, Oct-1, binds to the octamer to activate the histone H2B (and presumably also other) genes. Oct-1 is the only octamer-binding factor in nonlymphoid cells. But in lymphoid cells, a different factor, Oct-2, binds to the octamer to activate the immunoglobulin κ light gene. So Oct-2 is a tissue-specific activator, while Oct-1 is ubiquitous.

The use of the same octamer in the ubiquitously expressed H2B gene and the lymphoid-specific immunoglobulin genes poses a paradox. Why does the ubiquitous Oct-1 fail to activate the immunoglobulin genes in nonlymphoid tissues? The *context* must be important: Oct-2 rather than Oct-1 may be needed to interact with other proteins that bind at the promoter. These results mean that we cannot predict whether a gene will be activated by a particular factor simply on the basis of the presence of particular elements in its promoter.

There are also cases in which a particular protein can recognize more than one type of sequence. The best characterized example is the protein C/EBP, which binds to the CAAT box, but which also binds to another quite different sequence element.

A pertinent factor in considering transcription *in vitro* is that the template exists as an accessible DNA molecule. *In vivo* it is organized into nucleosomes, which suggests that its recognition by RNA polymerase is subject to different constraints. This may influence the geometry of the interactions of transcription factors with DNA, with one another, and with RNA polymerase. To investigate the formation of an active transcription complex in natural circumstances, we need really to use a template consisting of DNA assembled into chromatin rather than free DNA.

Repression of transcription in eukaryotes is generally accomplished at the level of influencing chromatin structure; regulator proteins that function like *trans*-acting bacterial repressors to block transcription are relatively rare, but some examples are known. One case is the global repressor Dr1/DRAP1, a heterodimer that binds to TBP to prevent it from interacting with other components of the basal apparatus. The importance of this interaction is suggested by the lethality of null mutations in the genes that code for the repressor in yeast.

In a more specific case, the CAAT sequence is a target for regulation. Two copies of this element are found in the promoter of a gene for histone H2B (see Figure 20.17) that is expressed only during spermatogenesis in a sea urchin. CAAT-binding factors can be extracted from testis tissue and also from embryonic tissues, but only the former can bind to the CAAT box. In the embryonic tissues, another protein, called the CAAT-displacement protein (CDP), binds to the CAAT boxes, *preventing the transcription factor from recognizing them.*

Figure 20.18 illustrates the consequences for gene expression. In testis, the promoter is bound by transcription factors at the TATA box, CAAT boxes, and octamer sequences. In embryonic tissue, the exclusion of the CAAT-binding factor from the promoter prevents a transcription complex from being assembled. The analogy with the effect of a bacterial repressor in preventing RNA polymerase from initiating at the promoter is obvious. These results also make the point that the function of a protein in binding to a known promoter element cannot be assumed: it may be an activator, a repressor, or even irrelevant to gene transcription.

Figure 20.18 A transcription complex involves recognition of several elements in the sea urchin H2B promoter in testis. Binding of the CAAT displacement factor in embryo prevents the CAAT-binding factor from binding, so an active complex cannot form.

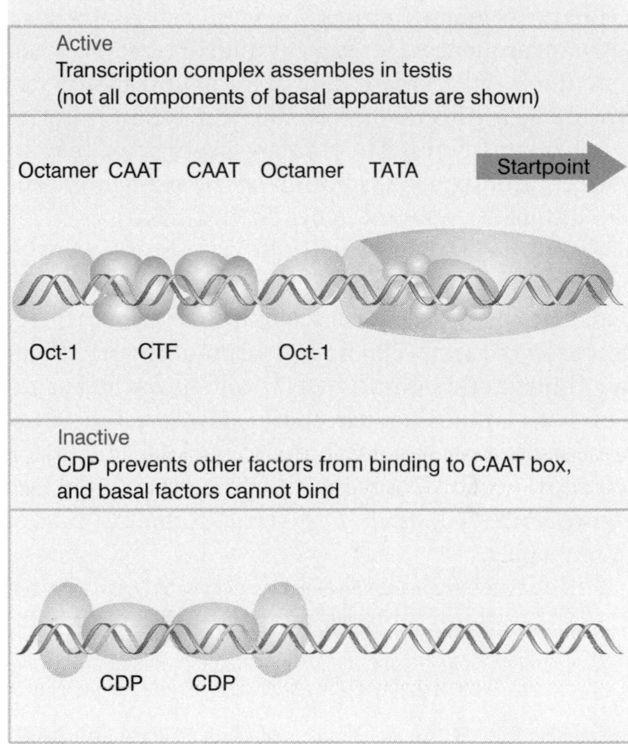

Enhancers contain bidirectional elements that assist initiation

We have considered the promoter so far as an isolated region responsible for binding RNA polymerase. But eukaryotic promoters do not necessarily function alone. In at least some cases, the activity of a promoter is enormously increased by the presence of an **enhancer,** which consists of another group of elements, but located at a variable distance from those regarded as comprising part of the promoter itself.

The concept that the enhancer is distinct from the promoter reflects two characteristics. The position of the enhancer relative to the promoter need not be fixed, but can vary substantially. And it can function in either orientation. Manipulations of DNA show that an enhancer can stimulate any promoter placed in its vicinity.

For operational purposes, it is sometimes useful to

define the promoter *as a sequence or sequences of DNA that must be in a (relatively) fixed location with regard to the startpoint.* By this definition, the TATA box and other upstream elements are included, but the enhancer is excluded. This is, however, a working definition rather than a rigid classification.

Elements analogous to enhancers, called upstream activator sequences (**UAS**), are found in yeast. They can function in either orientation, at variable distances upstream of the promoter, but cannot function when located downstream. They have a regulatory role: in several cases the UAS is bound by the regulatory protein(s) that activates the genes downstream.

An enhancer in the virus SV40 was one of the first to be characterized. It is located in a region of the genome that contains two identical sequences of 72 bp each, repeated in tandem ~200 bp upstream of the startpoint of a transcription unit. These **72 bp repeats** lie in a region with an unusual chromatin structure, where the presence of a site hypersensitive to nuclease identifies a region in which DNA is more exposed than usual (see Figures 19.39–40). Each 72 bp repeat contains a copy of the enhancer.

A difference between the enhancer and a typical promoter is presented by the density of regulatory elements. **Figure 20.19** summarizes the susceptibility of the SV40 enhancer to damage by mutation; we see that a much greater proportion of its sites directly influences its function than is the case with the promoter analyzed in the same way in Figure 20.16. There is a corresponding increase in the density of protein-binding sites. Many of these sites are common elements in promoters; for example, AP1 and the octamer.

Enhancers often show redundancy in function. The SV40 enhancer can be separated into two halves; they function poorly as enhancers by themselves, but constitute an effective enhancer together or even when they are separated by introducing sequences between them. Mutations that inactivate the left element can be compensated by duplicating other regions in the enhancer. Although these regions are different in sequence, they appear to play similar roles, since an active enhancer can be created by the combination of a sufficient *number* of wild-type domains, irrespective of their types. Such redundancy is common in enhancers; the result is that multiple mutations are required, to eliminate more than one element, before an enhancer is inactivated. In the SV40 enhancer, no individual mutation decreases activity by as much as 10×.

Cellular enhancers have similar properties. An

Figure 20.19 An enhancer contains several structural motifs. The histogram plots the effect of all mutations that reduce enhancer function to <75% of wild type. Binding sites for proteins are indicated below the histogram.

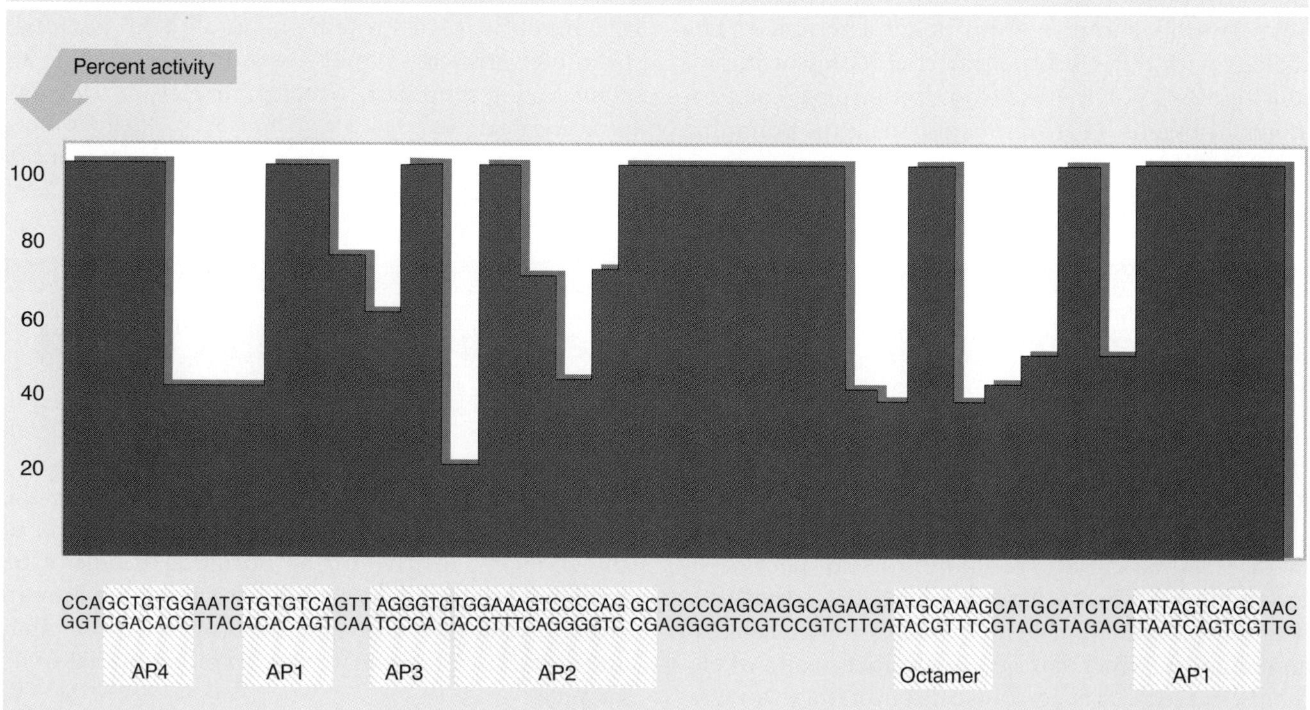

enhancer works upon the promoter that is nearest to it, but the enhancer may be either upstream or downstream of the promoter. Responsibility for tissue-specific transcription may lie with either a promoter or enhancer. A promoter may be specifically regulated, and a nearby enhancer used to increase the efficiency of initiation; or a promoter may lack specific regulation, but become active only when a nearby enhancer is specifically activated. An example is provided by immunoglobulin genes, which carry enhancers *within* the transcription unit. The immunoglobulin enhancers appear to be active only in the B lymphocytes in which the immunoglobulin genes are expressed. Such enhancers provide part of the regulatory network by which gene expression is controlled.

Reconstruction experiments in which the enhancer sequence is removed from the DNA and then is inserted elsewhere show that normal transcription can be sustained so long as it is present *anywhere* on the DNA molecule. If a β-globin gene is placed on a DNA molecule that contains an enhancer, its transcription is increased *in vivo* more than 200-fold, even when the enhancer is several kb upstream or downstream of the startpoint, in either orientation. We have yet to discover at what distance the enhancer fails to work.

How can an enhancer stimulate initiation at a promoter that can be located at apparently any distance away on either side of it? When enhancers were first discovered, several possibilities were considered for their action as elements distinctly different from promoters:

- An enhancer could change the overall structure of the template—for example, by influencing the density of supercoiling.
- It could be responsible for locating the template at a particular place within the cell—for example, attaching it to the nuclear matrix.
- An enhancer could provide an "entry site," a point at which RNA polymerase (or some other essential protein) associates with chromatin.

Now we take the view that enhancer function involves the same sort of interaction with the basal apparatus as the interactions sponsored by upstream promoter elements. Enhancers are modular, like promoters. Some elements are found in both enhancers and promoters. Some individual elements found in promoters share with enhancers the ability to function at variable distance and in either orientation. So the distinction between enhancers and promoters is blurred: enhancers might be viewed as containing promoter elements that are grouped closely together, with the ability to function at increased distances from the startpoint.

If the enhancer represents an extreme case of the ability to mix and match promoter elements, it might be considered a part of the promoter in a more distant location. The essential role of the enhancer may be to increase the concentration of transcription factors in the vicinity of the promoter (vicinity in this sense being a relative term). Two types of experiment illustrated in **Figure 20.20** suggest that this is the case.

A fragment of DNA that contains an enhancer at one end and a promoter at the other is not effectively transcribed, but the enhancer can stimulate transcription from the promoter when they are connected by a protein bridge. Since structural effects, such as changes in supercoiling, could not be transmitted across such a bridge, this suggests that the critical feature is bringing the enhancer and promoter into close proximity.

A bacterial enhancer provides a binding site for the regulator NtrC, which acts upon RNA polymerase using promoters recognized by σ^{54}. When the enhancer is placed upon a circle of DNA that is catenated (interlocked) with a circle that contains the promoter, initiation is almost as effective as when the enhancer and promoter are on the same circular molecule. But there is no initiation when the enhancer and promoter are on separated circles. Again this suggests that the critical feature is localization of the protein bound at the enhancer, to increase its chance of contacting a protein bound at the promoter.

If proteins bound at an enhancer several kb distant from a promoter interact directly with proteins bound in the vicinity of the startpoint, the organization of DNA must be flexible enough to allow the enhancer and promoter to be closely located. This requires the intervening DNA to be extruded as a large "loop." Such loops have been directly observed in the case of the bacterial enhancer.

The generality of enhancement is not yet clear. We do not know what proportion of cellular promoters require an enhancer to achieve their usual level of expression. Nor do we know how often an enhancer provides a target for regulation. Some enhancers are activated only in the tissues in which their genes function, but others could be active in all cells.

A difference between enhancers and promoters may be that an enhancer shows greater cooperativity between the binding of factors. A complex that assembles at the enhancer that responds to IFN (interferon) γ assembles cooperatively to form a functional structure called the **enhanceosome.** Binding of the nonhistone

Figure 20.20 An enhancer may function by bringing proteins into the vicinity of the promoter. An enhancer does not act on a promoter at the opposite end of a long linear DNA, but becomes effective when the DNA is joined into a circle by a protein bridge. An enhancer and promoter on separate circular DNAs do not interact, but can interact when the two molecules are catenated.

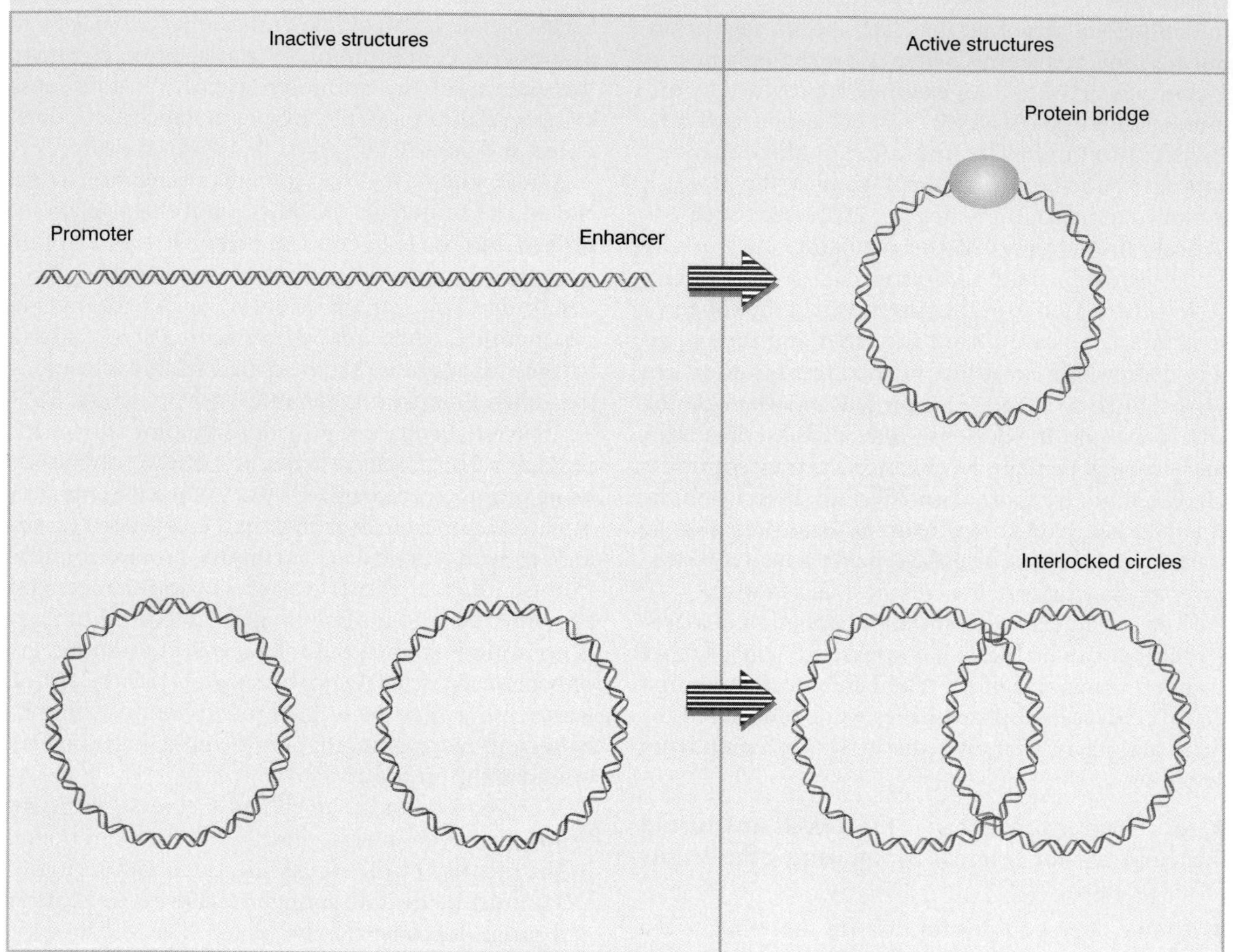

protein HMGI(Y) bends the DNA into a structure that then binds several transcription factors (NF-κB, IRF, ATF-Jun). In contrast with the "mix and match" construction of promoters, all of these components are required to create an active structure at the enhancer. These components do not themselves directly bind to RNA polymerase, but they create a surface that binds a **coactivating complex.** The coactivator binds RNA polymerase II and recruits it to the pre-initiation complex of basal transcription factors that is assembling at the promoter. We discuss the function of coactivators in more detail later.

Independent domains bind DNA and activate transcription

TRANSCRIPTION factors and other regulatory proteins require two types of ability:

- They recognize specific target sequences located in enhancers, promoters, or other regulatory elements that affect a particular target gene.
- Having bound to DNA, a transcription factor, or a positive regulatory protein, exercises its function by binding to other components of the transcription apparatus.

Can we characterize domains in the transcription factors that are responsible for these activities? Often a factor has separate domains that bind DNA and activate transcription. Each domain behaves as a separate module that functions independently when it is linked to a domain of the other type. The geometry of the overall transcription complex must allow the activating domain to contact the basal apparatus irrespective of the exact location and orientation of the DNA-binding domain.

Upstream promoter elements may be an appreciable distance from the startpoint, and in many cases may be oriented in either direction. Enhancers may be even farther away and always show orientation independence. This organization has implications for both the DNA and proteins. The DNA may be looped or condensed in some way to allow the formation of the transcription complex. And the domains of the transcription factor may be connected in a flexible way, as illustrated diagrammatically in **Figure 20.21**.

The modular nature of a transcriptional activator has been most thoroughly characterized for the example of the yeast activator GAL4, which controls the transcription of genes whose products are responsible for metabolizing galactose. Regulation is exercised at UAS_G (an upstream activating sequence). The GAL4 protein has three functions: it binds to a DNA sequence in the UAS_G; it activates transcription; and it binds another regulator protein, called GAL80. (GAL80 prevents GAL4 from activating transcription: GAL80 is released when galactose is present, thus allowing GAL4 to activate its target genes.) **Figure 20.22** shows that each of these functions can be localized in a particular region of the GAL4 protein. In fact, GAL4 has two activating domains, each able to activate transcription. Another domain confers the ability to dimerize.

Binding to DNA naturally is prerequisite for activating transcription. But does activation depend on the *particular* DNA-binding domain?

Figure 20.23 illustrates an experiment to answer this question. The bacterial repressor LexA has an

Figure 20.21 DNA-binding and activating functions in a transcription factor may comprise independent domains of the protein.

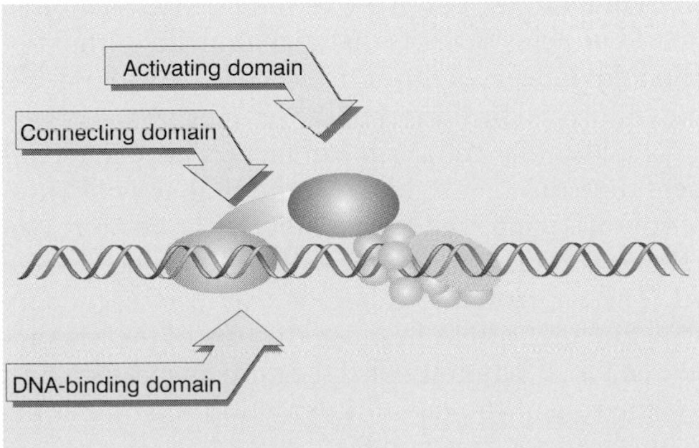

Figure 20.22 The GAL4 protein has independent regions that bind DNA, activate transcription (2 regions), dimerize, and bind the regulator GAL80.

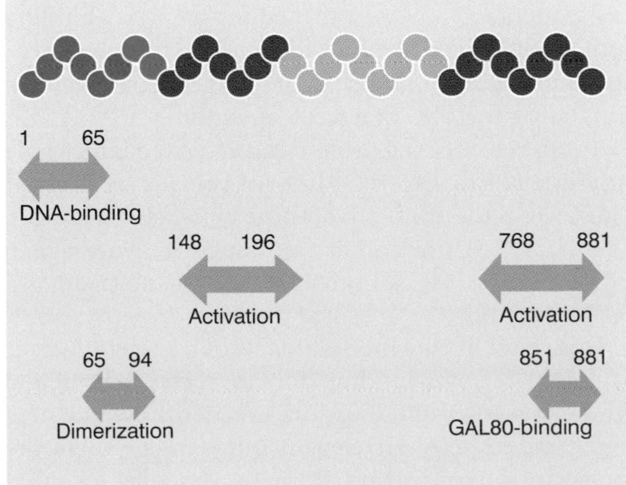

Figure 20.23 The ability of GAL4 to activate transcription is independent of its specificity for binding DNA. When the GAL4 DNA-binding domain is replaced by the LexA DNA-binding domain, the hybrid protein can activate transcription when a LexA operator is placed near a promoter.

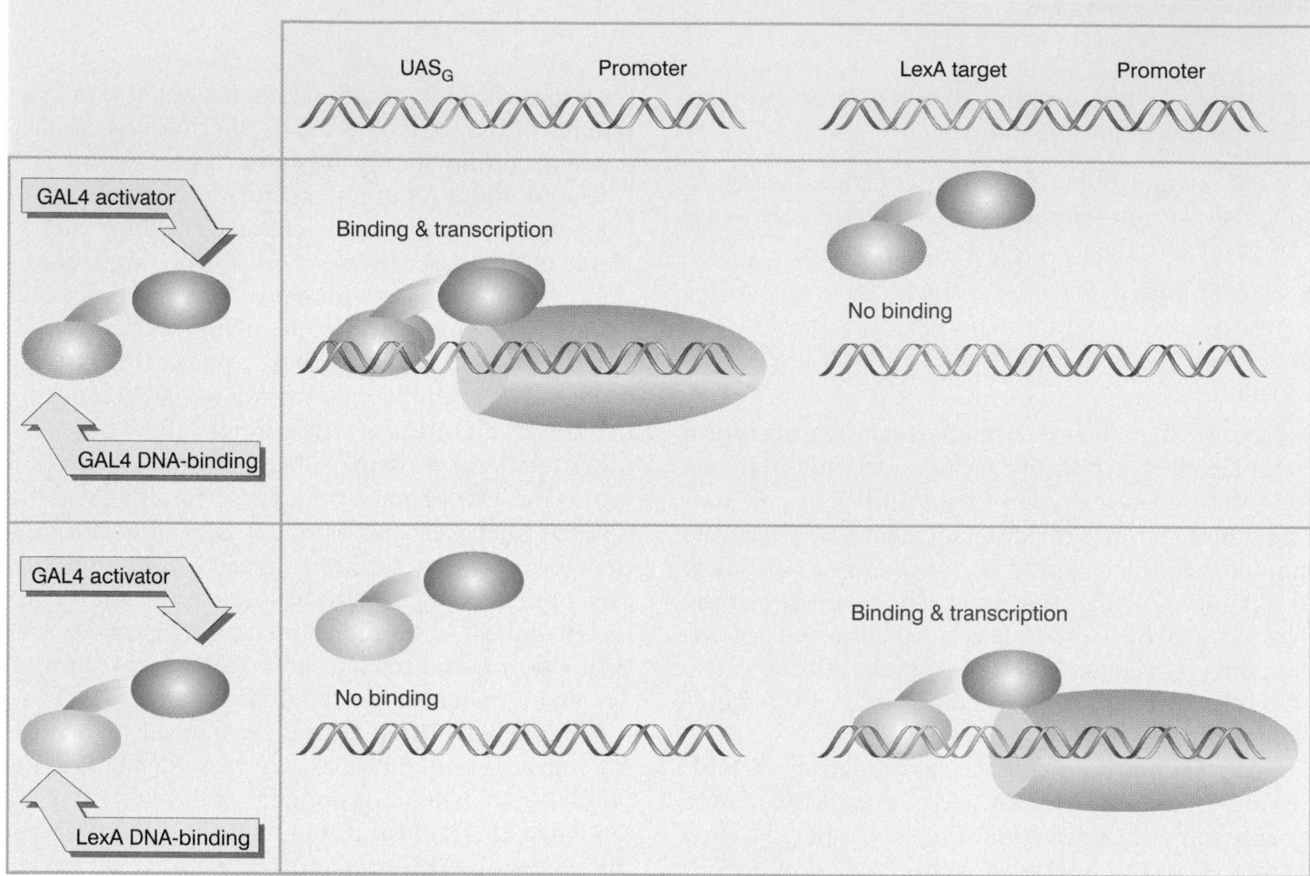

N-terminal DNA-binding domain that recognizes a specific operator; binding of LexA at this operator represses the adjacent promoter. In a "swap" experiment, this sequence can be substituted for the DNA-binding domain of GAL4. The hybrid gene can then be introduced into yeast together with a target gene that contains either the *UAS* or a LexA operator.

An authentic GAL4 protein can activate a target gene only if it has a *UAS*. The LexA repressor by itself of course lacks the ability to activate either sort of target. The LexA-GAL4 hybrid can no longer activate a gene with a *UAS*, but it can now activate a gene that has a LexA operator!

This result fits the modular view of transcription activators. The DNA-binding domain serves to bring the protein into the right location. Precisely how or where it is bound to DNA is irrelevant, but, once it is there, the transcription-activating domain can play its role. According to this view, it does not matter whether the transcription-activating domain is brought to the vicinity of the promoter by recognition of a *UAS* via the DNA-binding domain of GAL4 or by recognition of a LexA operator via the LexA specificity module. The ability of the two types of module to function in hybrid proteins suggests that each domain of the protein folds independently into an active structure that is not influenced by the rest of the protein.

The idea that transcription factors have independent domains that bind DNA and that activate transcription is reinforced by the ability of the tat protein of HIV to stimulate initiation without binding DNA at all. The tat protein binds to a region of secondary structure in the RNA product; the part of the RNA required for tat action is called the *tar* sequence. A model for the role of the tat-*tar* interaction in stimulating transcription is shown in **Figure 20.24.**

Figure 20.24 The activating domain of the tat protein of HIV can stimulate initiation if it is tethered in the vicinity by binding to the RNA product of a previous round of transcription. Activation is independent of the means of tethering, as shown by the substitution of a DNA-binding domain for the RNA-binding domain.

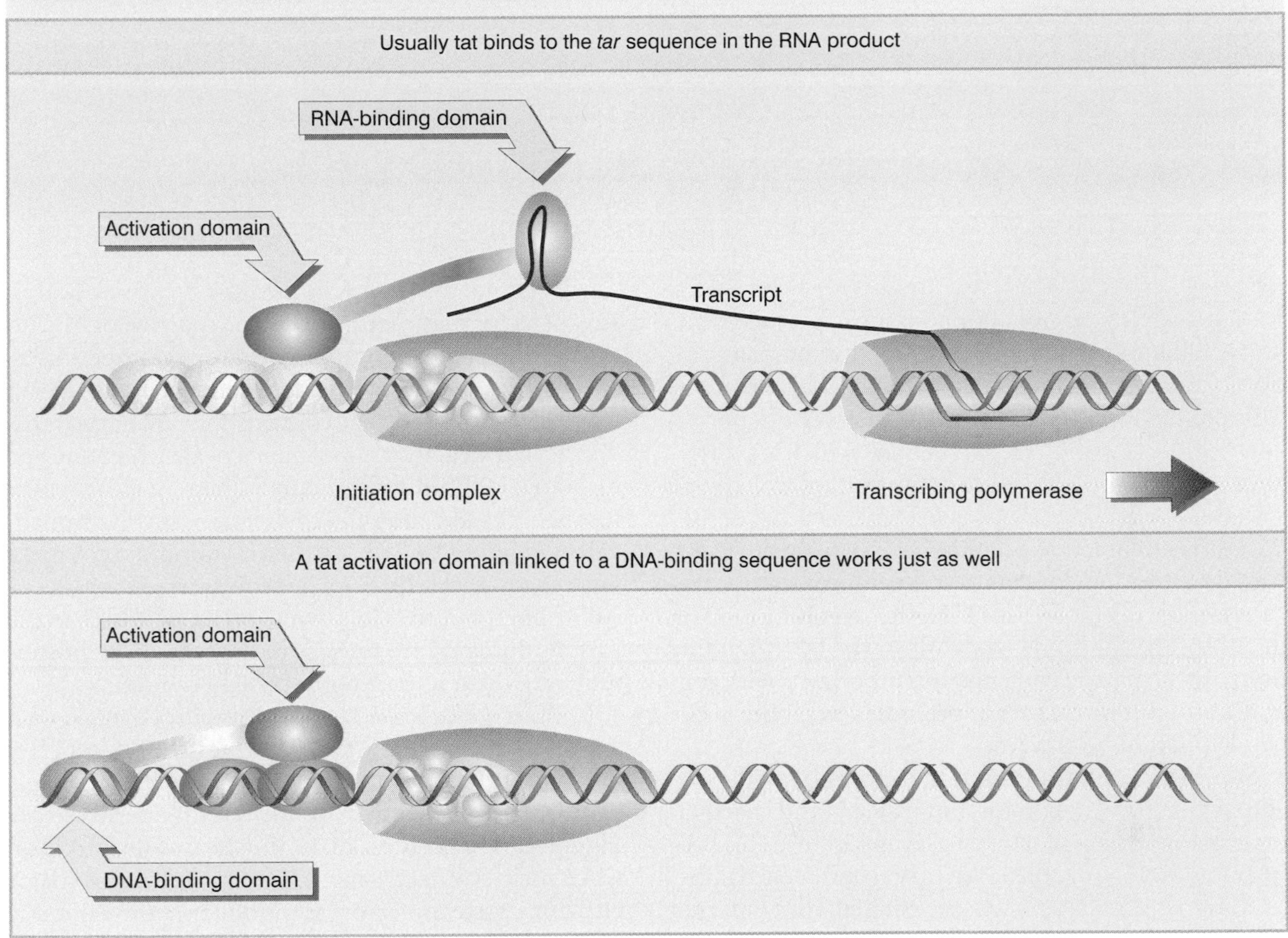

The *tar* sequence is located just downstream of the startpoint, so that when tat binds to *tar*, it is brought into the vicinity of the initiation complex. This is sufficient to ensure that its activation domain is in close enough proximity to the initiation complex. The activation domain interacts with one or more of the transcription factors bound at the complex in the same way as an upstream transcription factor. (Of course, the first transcript must be made in the absence of tat in order to provide the binding site.)

An extreme demonstration of the independence of the localizing and activating domains is indicated by some constructs in which tat was engineered so that the activating domain was connected to a DNA-binding domain instead of to the usual *tar*-binding sequence. When an appropriate target site is placed into the promoter, the tat activating-domain could activate transcription. This suggests that we should think of the DNA-binding (or in this case the RNA-binding) domain as providing a "tethering" function, *whose main purpose is to ensure that the activating domain is in the vicinity of the initiation complex.*

The notion of tethering is a more specific example of the general idea that initiation requires a high concentration of transcription factors in the vicinity of the promoter. This may be achieved when factors bind to enhancers in the general vicinity, when factors bind to upstream promoter components, or in an extreme case by tethering to the RNA product. The common requirement of all these situations is flexibility in the exact three dimensional arrangement of DNA and proteins. The principle of independent domains is common in transcriptional activators.

We might view the function of the DNA-binding domain as *bringing the activating domain into the vicinity of the startpoint.* This explains why the exact locations of DNA-binding sites can vary within the promoter.

Interaction of upstream factors with the basal apparatus

TRANSCRIPTION factors bound at upstream elements or enhancers influence the initiation of transcription by contacting other factors in the basal apparatus. This contact may be direct or indirect. It does not involve contacts with RNA polymerase, which is a difference with the mechanism of activation in prokaryotic systems.

A direct interaction occurs when an upstream factor uses its activating domain to contact the basal apparatus. As indicated in Figure 20.21, the activating domain in effect functions independently of the DNA-binding domain. By making protein-protein contacts with general transcription factors, it promotes assembly of the basal apparatus (see below).

Some upstream transcription factors do not directly contact the basal apparatus, but instead bind **coactivators** that in turn contact the basal apparatus. This is illustrated in **Figure 20.25.** We may regard coactivators as transcription factors whose specificity is conferred by the ability to bind to DNA-binding transcription factors instead of directly to DNA. A particular upstream factor may require a specific coactivator. But although the protein components are organized differently, the mechanism is the same. An upstream transcription factor that contacts the basal apparatus directly has an activation domain covalently connected to the DNA-binding domain. When a factor works through a coactivator, the connections involve noncovalent binding between protein subunits (compare Figures 20.21 and 20.24). The same interactions are responsible for activation, irrespective of whether the various domains are present in the same protein subunit or divided into multiple protein subunits.

Figure 20.25 An upstream transcription factor may bind a coactivator that contacts the basal apparatus.

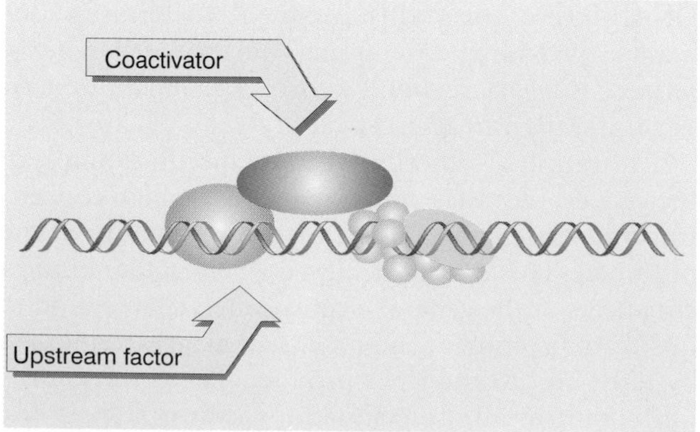

Contact with the basal apparatus may be made with any one of several basal factors, typically TFIID, TFIIB, or TFIIA. All of these factors participate in a relatively early stage of assembly of the basal apparatus (see Figure 20.11). **Figure 20.26** illustrates the situation when such a contact is made. It suggests that the major effect of upstream factors is to influence the assembly of the basal apparatus, presumably by increasing some rate-limiting step (as opposed to acting after assembly of the basal apparatus to influence its activity).

TFIID may be the most common target for upstream transcription factors, which may contact any one of several TAFs. In fact, we may regard a major role of the TAFs as providing the connection from the basal apparatus to upstream transcription factors. Different TAFs in TFIID may provide surfaces that interact with different upstream factors. Some upstream factors interact only with individual TAFs; others interact with multiple TAFs. We assume that the interaction either assists binding of TFIID to the TATA box or assists the binding of other factors around the TFIID-TATA box complex. In either case, the interaction stabilizes the basal transcription complex; this speeds the process of initiation, and thereby increases use of the promoter.

Repressors also may work at this stage. A repressor may block an upstream transcription factor from in-

Figure 20.26 Upstream activators may work at different stages of initiation, by contacting the TAFs of TFIID or contacting TFIIB.

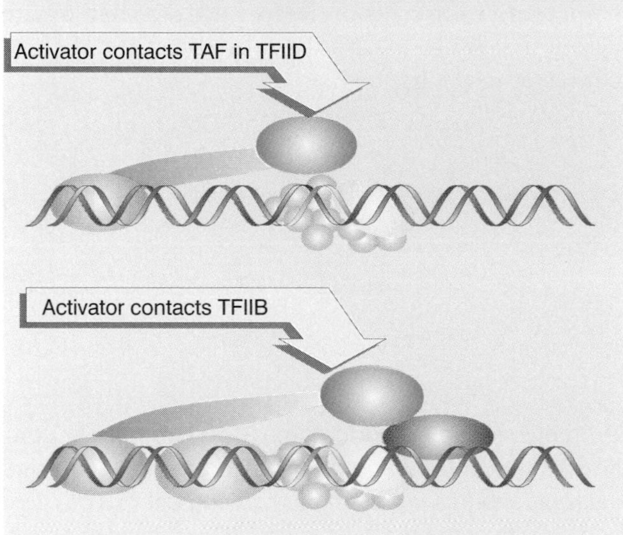
Activator contacts TAF in TFIID
Activator contacts TFIIB

teracting with the basal apparatus. The repressor DR1 prevents TFIIB from joining the TFIID complex; it could block the function of TFIID by replacing a TAF. Repressors that work in this way have an active role in inhibiting basal apparatus function (compared with a bacterial repressor that directly blocks RNA polymerase binding or movement).

The activating domains of the yeast transcription factors GAL4 and GCN4 have multiple negative charges, giving rise to their description as "acidic activators." Another particularly effective activator of this type is carried by the VP16 protein of the Herpes Simplex Virus. (VP16 does not itself have a DNA-binding domain, but interacts with the transcription apparatus via an intermediary protein.) Experiments to characterize acidic activator functions have often made use of the VP16 activating region linked to a DNA-binding motif.

It is not clear whether the negative charges are necessary for the function of "acidic" activators. On the one hand, protein sequences that have activating capacity when linked to the DNA-binding domain of GAL4 have in common only their possession of negative charges. This suggested that the activating region is an "acid blob" that interacts in a relatively nonspecific manner with some other protein involved in transcription. And mutations that remove acidic amino acids of GCN4 reduce activation ability, while mutations that remove basic amino acids generate more effective activators. However, systematic mutation to remove the charged groups in GAL4 has no effect upon activating capacity. Perhaps lack of positive charges is important.

Acidic activators function by enhancing the ability of TFIIB to join the basal initiation complex. Experiments *in vitro* show that binding of TFIIB to an initiation complex at an adenovirus promoter is stimulated by the presence of GAL4 or VP16 acid activators; and the VP16 activator can bind directly to TFIIB. Assembly of TFIIB into the complex at this promoter is therefore a rate-limiting step that is stimulated by the presence of an acidic activator.

The resilience of an RNA polymerase II promoter to the rearrangement of elements, and its indifference even to the particular elements present, suggests that the events by which it is activated are relatively general in nature. Any upstream factor whose activating region is brought within range of the basal initiation complex may be able to stimulate its formation. Some striking illustrations of such versatility have been accomplished by constructing promoters consisting of new combinations of elements. For example, when a yeast UAS_G element is inserted near the promoter of a higher eukaryotic gene, this gene can be activated by GAL4 in a mammalian cultured cell. Whatever means GAL4 uses to activate the promoter seems therefore to have been conserved between yeast and higher eukaryotes. The GAL4 protein must recognize some feature of the mammalian transcription apparatus that resembles its normal contacts in yeast.

How does an activator stimulate transcription? We can imagine two general types of model:

- The recruitment model argues that its sole effect is to increase the binding of RNA polymerase to the promoter.
- An alternative model is to suppose that it induces some change in the transcriptional complex, for example, in the conformation of the enzyme, which increases its efficiency.

A test of these models in one case in yeast showed that recruitment can account for activation. When the concentration of RNA polymerase was increased sufficiently, the activator failed to produce any increase in transcription, suggesting that its sole effect is to increase the effective concentration of RNA polymerase at the promoter.

Several forms of RNA polymerase have been found in which the enzyme is associated with various transcription factors. One such "holoenzyme complex" includes a subset of general transcription factors and a large number of SRB proteins (so named because many of their genes were originally identified as suppressors of mutations in *R*NA polymerase *B*.) The complex of SRB proteins interacts with the CTD domain of RNA polymerase. It is sometimes called the **mediator.** The name was suggested by its ability to mediate the effects on RNA polymerase of activators such as VP16. In fact, the mediator may be associated with proteins involved in activation and also with proteins involved in repression, suggesting that it may influence RNA polymerase either negatively or positively depending on the circumstances. And in addition to these complexes, which consist of factors acting directly on transcription, there are complexes in which RNA polymerase and transcription factors are associated with factors that act to manipulate the structure of chromatin (see next Chapter).

Summary

Of the three eukaryotic RNA polymerases, RNA polymerase I transcribes rDNA and accounts for the majority of activity, RNA polymerase II transcribes structural genes for mRNA and has the greatest diversity of products, and RNA polymerase III transcribes small RNAs. The enzymes have similar structures, with two large subunits and many smaller subunits; there are some common subunits among the enzymes.

None of the three RNA polymerases recognize their promoters directly. A unifying principle is that transcription factors have primary responsibility for recognizing the characteristic sequence elements of any particular promoter, and they serve in turn to bind the RNA polymerase and to position it correctly at the startpoint. At each type of promoter, the initiation complex is assembled by a series of reactions in which individual factors join (or leave) the complex. The factor TBP is required for initiation by all three RNA polymerases. In each case it provides one subunit of a transcription factor that binds in the vicinity of the startpoint.

The TATA box (if there is one) near the startpoint, and the initiator region immediately at the startpoint, are responsible for selection of the exact startpoint at promoters for RNA polymerase II. TBP binds directly to the TATA box when there is one; in TATA-less promoters it is located near the startpoint by other means. After binding of TFIID, the general transcription factors for RNA polymerase II assemble the basal transcription apparatus at the promoter, and are mostly released when RNA polymerase begins elongation.

RNA polymerase is found as part of much larger complexes that contain factors that interact with activators and repressors. A common point of contact for RNA polymerase with these proteins is its CTD, which is phosphorylated during the initiation reaction. TFIID and SRB proteins both may interact with the CTD. It may also provide a point of contact for proteins that modify the RNA transcript, including the 5′ capping enzyme.

Promoters for RNA polymerase II contain a variety of short *cis*-acting elements, each of which is recognized by a *trans*-acting factor. The *cis*-acting elements are located upstream of the TATA box and may be present in either orientation and at a variety of distances with regard to the startpoint. The upstream elements are recognized by transcription factors that interact with the basal transcription complex to determine the efficiency with which the promoter is used. Some upstream factors interact directly with components of the basal apparatus; others interact via coactivators. The targets in the basal apparatus are the TAFs of TFIID, or TFIIB or TFIIA. The interaction stimulates assembly of the basal apparatus.

Promoters may be stimulated by enhancers, sequences that can act at great distances and in either orientation on either side of a gene. Enhancers also consist of sets of elements, although they are more compactly organized. Some elements are found in both promoters and enhancers. Enhancers probably function by assembling a protein complex that interacts with the proteins bound at the promoter, requiring that DNA between is "looped out."

Further reading

Reviews

McKnight, S. and Tjian, R. (1986). Transcriptional selectivity of viral genes in mammalian cells. *Cell* **46,** 795–805.

Bird, A. P. (1986). A fraction of the mouse genome that is derived from islands of nonmethylated, CpG-rich DNA. *Nature* **321,** 209–213.

Maniatis, T., Goodbourn, S., and Fischer, J. A. (1987). Regulation of inducible and tissue-specific gene expression. *Science* **236,** 1237–1245.

Muller, M. M., Gerster, T., and Schaffner, W. (1988). Enhancer sequences and the regulation of gene transcription. *Eur. J. Biochem.* **176,** 485–495.

Geiduschek, E. P. and Tocchini-Valentini, G. P. (1988). Transcription by RNA polymerase III. *Ann. Rev. Biochem.* **57,** 873–914.

Mitchell. P. and Tjian, R. (1989). Transcriptional regulation in mammalian cells by sequence-specific DNA-binding proteins. *Science* **245,** 371–378.

Young, R. A. (1991). RNA polymerase II. *Ann. Rev. Biochem.* **60,** 689–715.

Zawel, L. and Reinberg, D. (1993). Initiation of transcription by RNA polymerase II: a multi-step process. *Prog. Nucleic Acids Res. Mol. Biol.* **44,** 67–108.

Selby, C. P. and Sancar, A. (1994). Mechanisms of transcription-repair coupling and mutation frequency decline. *Mut. Res.* **58,** 317–329.

Burley, S. K. and Roeder, R. G. (1997). Biochemistry and structural biology of TFIID. *Ann. Rev. Biochem.* **65,** 769–799.

Nikolov, D. B. and Burley, S. K. (1997). RNA polymerase II transcription initiation: a structural view. *Proc. Nat. Acad. Sci. USA* **94,** 15–22.

Discoveries

RNA polymerase III

Sakonju, S., Bogenhagen, D. F., and Brown, D. D. (1980). A control region in the center of the 5S RNA gene directs specific initiation of transcription: I. the 5′ border of the region. *Cell* **19,** 13–25.

Bogenhagen, D. F., Sakonju, S., and Brown, D. D. (1980). A control region in the center of the 5S RNA gene directs specific initiation of transcription: II. the 3′ border of the region. *Cell* **19,** 27–35.

Kassavatis, G. A., Braun, B. R., Nguyen, L. H., and Geiduschek, E. P. (1990). *S. cerevisiae* TFIIIB is the transcription initiation factor proper of RNA polymerase III, while TFIIIA and TFIIIC are assembly factors. *Cell* **60,** 235–245.

RNA polymerase II and general factors

Buratowski, S., Hahn, S., Guarente, L., and Sharp, P. A. (1989). Five intermediate complexes in transcription initiation by RNA polymerase II. *Cell* **56,** 549–561.

Pugh, B. F. and Tjian, R. (1990). Mechanism of transcriptional activation by Sp1: evidence for coactivators. *Cell* **61,** 1187–1197.

Dynlacht, B. D., Hoey, T., and Tjian, R. (1991). Isolation of coactivators associated with the TATA-binding protein that mediate transcriptional activation. *Cell* **66,** 563–576.

Nikolov, D. B. *et al.* (1992). Crystal structure of TFIID TATA-box binding protein. *Nature* **360,** 40–46.

Kim, Y. *et al.* (1993). Crystal structure of a yeast TBP/TATA box complex. *Nature* **365,** 512–520.

Kim, J. L., Nikolov, D. B., and Burley, S. K. (1993). Cocrystal structure of TBP recognizing the minor groove of a TATA element. *Nature* **365,** 520–527.

Goodrich, J. A. and Tjian, R. (1994). Transcription factors IIE and IIH and ATP hydrolysis direct promoter clearance by RNA polymerase II. *Cell* **77,** 145–156.

Martinez, E. *et al.* (1994). TATA-binding protein-associated factor(s) in TFIID function through the initiator to direct basal transcription from a TATA-less class II promoter. *EMBO J.* **13,** 3115–3126.

Nikolov, D. B. *et al.* (1995). Crystal structure of a TFIIB-TBP-TATA-element ternary complex. *Nature* **377,** 119–128.

Verrijzer, C. P. *et al.* (1995). Binding of TAFs to core elements directs promoter selectivity by RNA polymerase II. *Cell* **81,** 1115–1125.

Liu, D. *et al.* (1998). Solution structure of a TBP-TAFII230 complex: protein mimicry of the minor groove surface of the TATA box unwound by TBP. *Cell* **94,** 573–583.

Upstream transcription factors

Kadonaga, J. *et al.* (1987). Isolation of cDNA encoding transcription factor Sp1 and functional analysis of the DNA binding domain. *Cell* **51,** 1079–1090.

Horikoshi, M. *et al.* (1988). Transcription factor ATD interacts with a TATA factor to facilitate establishment of a preinitiation complex. *Cell* **54,** 1033–1042.

Ma, J. and Ptashne, M. (1987). A new class of yeast transcriptional activators. *Cell* **51,** 113–119.

Chen, J.-L. *et al.* (1994). Assembly of recombinant TFIID reveals differential coactivator requirements for distinct transcriptional activators. *Cell* **79,** 93–105.

Hörlein, A. J. *et al.* (1995). Ligand-independent repression by the thyroid hormone receptor mediated by a nuclear receptor corepressor. *Nature* **377**, 397–404.

Transcription repair coupling

Selby, C. P. and Sancar, A. (1993). Molecular mechanism of transcription-repair coupling. *Science* **260,** 53–58.

Schaeffer, L. *et al.* (1993). DNA repair helicase: a component of BTF2 (TFIIH) basic transcription factor. *Science* **260,** 58–63.

Svejstrup, J. Q. *et al.* (1995). Different forms of TFIIH for transcription and DNA repair: holo-TFIIH and a nucleotide excision repairosome. *Cell* **80,** 21–28.

Bregman, D. *et al.* (1996). UV-induced ubiquitination of RNA polymerase II: a novel modification deficient in Cockayne syndrome cells. *Proc. Nat. Acad. Sci. USA* **93,** 11586–11590.

Enhancers

Banerji, J., Rusconi, S., and Schaffner, W. (1981). Expression of β-globin gene is enhanced by remote SV40 DNA sequences. *Cell* **27**, 299–308.

Zenke, M. *et al.* (1986). Multiple sequence motifs are involved in SV40 enhancer function. *EMBO J.* **5**, 387–397.

Mueller-Storm, H. P., Sogo, J. M., and Schaffner, W. (1989). An enhancer stimulates transcription in *trans* when attached to the promoter via a protein bridge. *Cell* **58,** 767–777.

Chapter 21

Regulation of transcription

THE phenotypic differences that distinguish the various kinds of cells in a higher eukaryote are largely due to differences in the expression of genes that code for proteins, that is, those transcribed by RNA polymerase II. In principle, the expression of these genes might be regulated at any one of several stages. The concept of the "level of control" implies that gene expression is not necessarily an automatic process once it has begun. It could be regulated in a gene-specific way at any one of several sequential steps. We can distinguish (at least) five potential control points, forming the series:

Activation of gene structure
↓
Initiation of transcription
↓
Processing the transcript
↓
Transport to cytoplasm
↓
Translation of mRNA

The existence of the first step is implied by the discovery that genes may exist in either of two structural conditions. Genes are found in an "active" state only in the cells in which they are expressed (see Chapter 19). The change of structure is distinct from the act of transcription, and indicates that the gene is "transcribable." This suggests that acquisition of the "active" structure must be the first step in gene expression.

Transcription of a gene in the active state is controlled at the stage of initiation, that is, by the interaction of RNA polymerase with its promoter. For most genes, this is a major control point; probably it is the most common level of regulation.

There is at present no evidence for control at subsequent stages of transcription in eukaryotic cells, for example, via antitermination mechanisms.

The primary transcript is modified by capping at the 5′ end, and usually also by polyadenylation at the 3′ end. Introns must be excised from the transcripts of interrupted genes. The mature RNA must be exported from the nucleus to the cytoplasm. Regulation of gene expression by selection of sequences at the level of nuclear RNA might involve any or all of these stages, but the one for which we have most evidence concerns changes in splicing; some genes are expressed by means of alternative splicing patterns whose regulation controls the type of protein product (see Chapter 22).

Finally, the translation of an mRNA in the cytoplasm can be specifically controlled. There is little evidence for the employment of this mechanism in adult somatic cells, but it occurs in some embryonic situations. The mechanism is presumed to involve the blocking of initiation of translation of certain mRNAs by specific protein factors.

But having acknowledged that control of gene expression can occur at multiple stages, and that production of RNA cannot inevitably be equated with production of protein, it is clear that the overwhelming majority of regulatory events occur at the initiation of transcription. Regulation of tissue-specific gene transcription lies at the heart of eukaryotic differentiation; indeed, we see examples in Chapter 29 in which proteins that regulate embryonic development prove to be transcription factors. A regulatory transcription factor serves to provide common control of a large number of target genes, and we seek to answer two questions about this mode of regulation: how does the transcription factor identify its group of target genes; and how is the activity of the transcription factor itself regulated in response to intrinsic or extrinsic signals?

Response elements identify genes under common regulation

The principle that emerges from characterizing groups of genes under common control is that *they share a promoter (or enhancer) element that is recognized by a regulatory transcription factor.* An element that causes a gene to respond to such a factor is called a **response element;** examples are the **HSE** (heat shock response element), **GRE** (glucocorticoid response element), **SRE** (serum response element).

Response elements have the same general characteristics as upstream elements of promoters or enhancers. They contain short consensus sequences, and copies of the response elements found in different genes are closely related, but not necessarily identical. The region bound by the factor extends for a short distance on either side of the consensus sequence. In promoters, the elements are not present at fixed distances from the startpoint, but are usually <200 bp upstream of it. The presence of a single element usually is sufficient to confer the regulatory response, but sometimes there are multiple copies.

Response elements may be located in promoters or in enhancers. Some types of elements are typically found in one rather than the other: usually an HSE is found in a promoter, while a GRE is found in an enhancer. We assume that all response elements function by the same general principle. *A gene is regulated by a sequence at the promoter or enhancer that is recognized by a specific protein. The protein functions as a transcription factor needed for RNA polymerase to initiate. Active protein is available only under conditions when the gene is to be expressed.*

An example of a situation in which many genes are controlled by a single factor is provided by the heat shock response. This is common to a wide range of prokaryotes and eukaryotes and involves multiple controls of gene expression: an increase in temperature turns off transcription of some genes, turns on transcription of the **heat shock genes,** and causes changes in the translation of mRNAs. The control of the heat shock genes illustrates the differences between prokaryotic and eukaryotic modes of control. In bacteria, a new sigma factor is synthesized that directs RNA polymerase holoenzyme to recognize an alternative –10 sequence common to the promoters of heat shock genes (see Chapter 9). In eukaryotes, the heat shock genes also possess a common consensus sequence (HSE), but it is located at various positions relative to the startpoint, and is recognized by an independent transcription factor, HSTF. The activation of this factor therefore provides a means to initiate transcription at the specific group of ~20 genes that contains the appropriate target sequence at its promoter.

All the heat shock genes of *D. melanogaster* contain multiple copies of the HSE. The HSTF binds cooperatively to adjacent response elements. Both the HSE and HSTF have been conserved in evolution, and it is striking that a heat shock gene from *D. melanogaster* can be activated in species as distant as mammals or sea urchins. The HSTF proteins of fruit fly and yeast appear similar, and show the same footprint pattern on DNA containing HSE sequences. Yeast HSTF becomes phosphorylated when cells are heat-shocked; this modification is responsible for activating the protein.

The metallothionein (MT) gene provides an example of how a single gene may be regulated by many different circuits. The metallothionein protein protects the cell against excess concentrations of heavy metals, by binding the metal and removing it from the cell. The gene is expressed at a basal level, but is induced to greater levels of expression by heavy metal ions (such as cadmium) or by glucocorticoids. The control region combines several different kinds of regulatory element, and suggests the principle that *when a promoter is regulated in more than one way, each regulatory event depends on binding of its own protein to a particular sequence.*

The organization of the promoter for a MT gene is summarized in **Figure 21.1.** A major feature of this map is the high density of elements that can activate transcription. The two "constitutive" promoter elements are the TATA box and GC box, located at their usual positions fairly close to the startpoint. Also needed for the basal level of constitutive expression are the two basal level elements (BLE), which fit the formal description of enhancers. Although located near the startpoint, they can be moved elsewhere without loss of effect. They contain sequences related to those found in other enhancers, and are bound by proteins that bind the SV40 enhancer.

The TRE is a consensus sequence that is present in several enhancers, including one BLE of metallothionein and the 72 bp repeats of the virus SV40. The TRE has a binding site for factor AP1; this interaction is part of the mechanism for constitutive expression, in line with our previous description of AP1 as an upstream factor. However, AP1 binding also has a second function. The TRE confers a response to phorbol esters such as TPA (an agent that promotes tumors), and this response is mediated by the interaction of AP1 with the TRE. This binding reaction is one (not necessarily the sole) means by which phorbol esters trigger a series of transcriptional changes.

The inductive response to metals is conferred by the multiple MRE sequences, which function as promoter elements. The presence of one MRE confers the ability to respond to heavy metal; a greater level of induction is achieved by the inclusion of multiple elements.

The response to steroid hormones is governed by a GRE, located 250 bp upstream of the startpoint, which behaves as an enhancer. Deletion of this region does not affect the basal level of expression or the level induced by metal ions. But it is absolutely needed for the response to steroids.

The regulation of metallothionein illustrates the general principle that *any one of several different elements, located in either an enhancer or promoter, can independently activate the gene.* The absence of an element needed for one mode of activation does not affect activation in other modes. The variety of elements, their independence of action, and the apparently unlimited flexibility of their relative arrangements, suggest that a factor binding to any one element is able independently to increase the efficiency of initiation by the basal transcription apparatus, probably by virtue of protein-protein interactions that stabilize or otherwise assist formation of the initiation complex.

Figure 21.1 The regulatory region of a human metallothionein gene contains regulator elements in both its promoter and enhancer. The promoter has elements for metal induction; an enhancer has an element for response to glucocorticoid. Promoter elements are shown above the map, and proteins that bind them are indicated below.

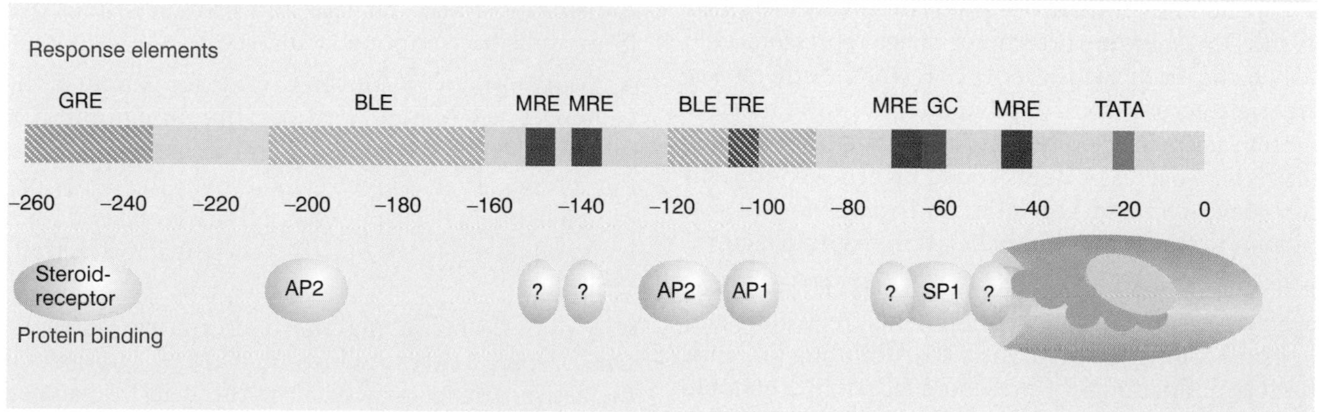

There are many types of DNA-binding domains

COMPARISONS between the sequences of many transcription factors suggest that common types of **motifs** can be found that are responsible for binding to DNA. The motifs are usually quite short and comprise only a small part of the protein structure. Motifs have also been identified that are responsible for activating transcription via interactions between proteins of the transcription apparatus.

We have detailed information about several groups of proteins that regulate transcription by using particular motifs to bind DNA:

- The **steroid receptors** are defined as a group by a functional relationship: each receptor is activated by binding a particular steroid. The glucocorticoid receptor is the most fully analyzed. Together with other receptors, such as the thyroid hormone receptor or the retinoic acid receptor, the steroid receptors are members of a superfamily of transcription factors with the same general *modus operandi.*
- The **zinc finger** motif comprises a DNA-binding domain. It was originally recognized in factor TFIIIA, which is required for RNA polymerase III to transcribe 5S rRNA genes. It has since been identified in several other transcription factors (and presumed transcription factors). A distinct form of the motif is found also in the steroid receptors.
- The **helix-turn-helix** motif was originally identified as the DNA-binding domain of phage repressors. One α-helix lies in the wide groove of DNA; the other lies at an angle across DNA. A related form of the motif is present in the **homeodomain,** a sequence first characterized in several proteins coded by genes concerned with developmental regulation in *Drosophila.* It is also present in genes for mammalian transcription factors.
- The amphipathic **helix-loop-helix (HLH)** motif has been identified in some developmental regulators and in genes coding for eukaryotic DNA-binding proteins. Each amphipathic helix presents a face of hydrophobic residues on one side and charged residues on the other side. The length of the connecting loop varies from 12–28 amino acids. The motif enables proteins to dimerize, and a basic region near this motif contacts DNA.
- **Leucine zippers** consist of a stretch of amino acids with a leucine residue in every seventh position. A leucine zipper in one polypeptide interacts with a zipper in another polypeptide to form a dimer. Adjacent to each zipper is a stretch of positively charged residues that is involved in binding to DNA.

The activity of an inducible transcription factor may be regulated in any one of several ways, as illustrated schematically in **Figure 21.2:**

- A factor is tissue-specific because it is synthesized only in a particular type of cell. This is typical of factors that regulate development, such as homeodomain proteins.
- The activity of a factor may be directly controlled by modification. HSTF is converted to the active form by phosphorylation. AP1 (a heterodimer between the subunits Jun and Fos) is converted to the active form by phosphorylating the Jun subunit.
- A factor is activated or inactivated by binding a ligand. The steroid receptors are prime examples. Ligand binding may influence the localization of the protein (causing transport from cytoplasm to nucleus), as well as determining its ability to bind to DNA.
- One transcription factor is produced as a protein bound to the nuclear envelope and endoplasmic reticulum. The absence of sterols (such as cholesterol) causes the cytosolic domain to be cleaved; it then translocates to the nucleus and provides the active form of the transcription factor.
- Availability of a factor may vary; for example, the factor NF-κB (which activates immunoglobulin κ genes in B lymphocytes) is present in many cell types. But it is sequestered in the cytoplasm by the inhibitory protein I-κB. In B lymphocytes, NF-κB is released from I-κB and moves to the nucleus, where it activates transcription.
- A dimeric factor may have alternative partners. One partner may cause it to be inactive; synthesis of the active partner may displace the inactive partner. We see later that such situations may be amplified into networks in which various alternative partners pair with one another, especially among the HLH proteins.

(We note *en passant* that mutations of the transcription factors in some of these classes give rise to factors

Figure 21.2 The activity of a regulatory transcription factor may be controlled by synthesis of protein, covalent modification of protein, ligand binding, or binding of inhibitors that sequester the protein or affect its ability to bind to DNA.

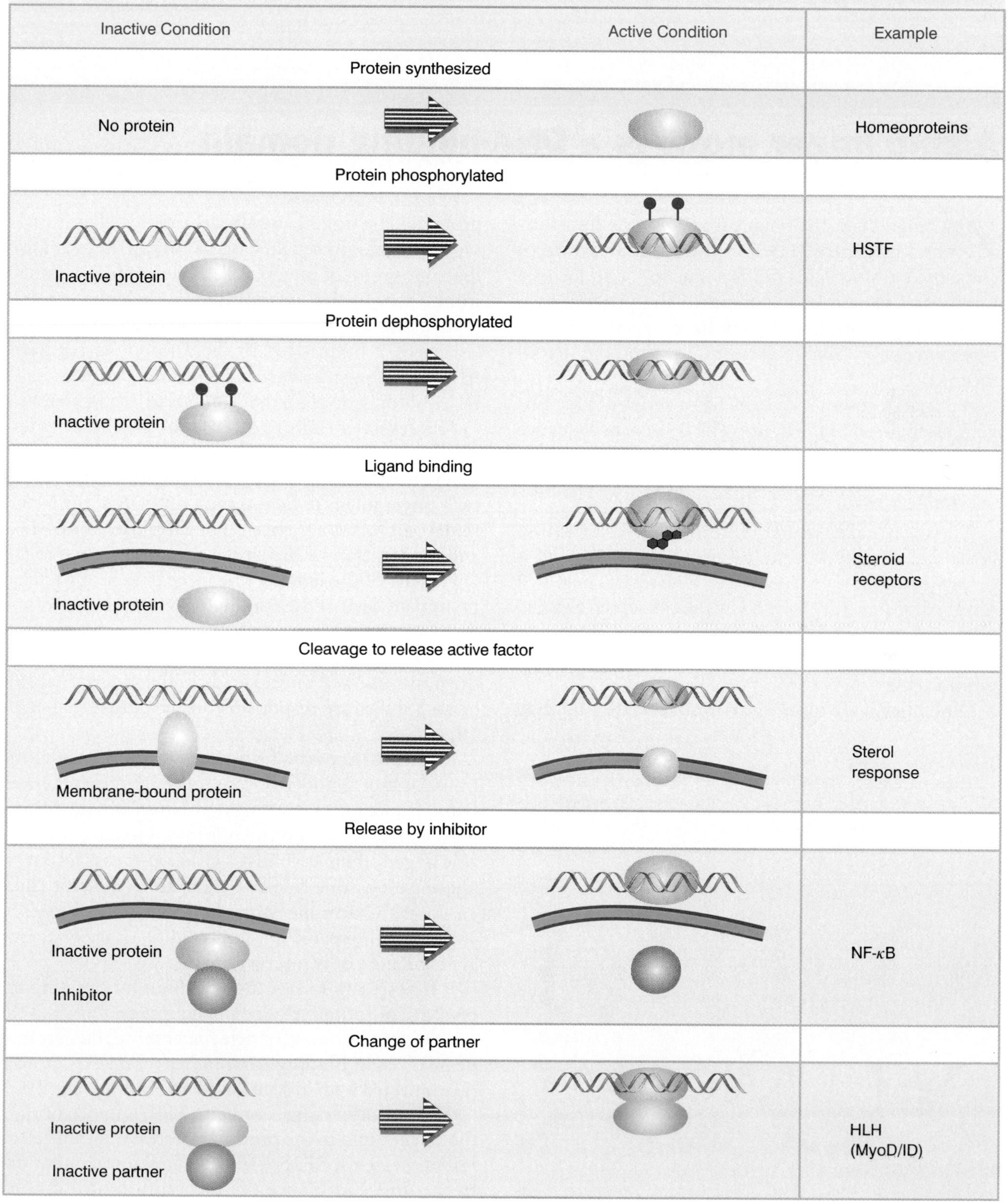

Inactive Condition	Active Condition	Example
Protein synthesized		
No protein		Homeoproteins
Protein phosphorylated		
Inactive protein		HSTF
Protein dephosphorylated		
Inactive protein		
Ligand binding		
Inactive protein		Steroid receptors
Cleavage to release active factor		
Membrane-bound protein		Sterol response
Release by inhibitor		
Inactive protein Inhibitor		NF-κB
Change of partner		
Inactive protein Inactive partner		HLH (MyoD/ID)

that inappropriately activate, or prevent activation, of transcription; their roles in generating tumors are discussed in Chapter 28, and Figure 28.19 should be compared with Figure 21.2.)

We now discuss in more detail the DNA-binding and activation reactions that are sponsored by some of these classes of proteins. In many cases, binding to DNA is undertaken by a short region of α-helix that makes contacts with either the bases or phosphate backbone in the major groove.

A zinc finger motif is a DNA-binding domain

ZINC fingers take their name from the structure illustrated in **Figure 21.3**, in which a small group of conserved amino acids binds a zinc ion, and forms a relatively independent domain in the protein. Two types of DNA-binding proteins have structures of this type: the classic "zinc finger" proteins; and the steroid receptors.

A "finger protein" typically has a series of zinc fingers, as depicted in the figure. The consensus sequence of a single finger is:

$$\text{Cys-X}_{2\text{–}4}\text{-Cys-X}_3\text{-Phe-X}_5\text{-Leu-X}_2\text{-His-X}_3\text{-His}$$

The motif takes its name from the loop of amino acids that protrudes from the zinc-binding site and is described as the Cys_2/His_2 finger. The zinc is held in a tetrahedral structure formed by the conserved Cys and His residues. The finger itself comprises ~23 amino acids, and the linker between fingers is usually 7–8 amino acids.

Zinc fingers are a common motif in DNA-binding proteins. The fingers usually are organized as a single series of tandem repeats; occasionally there is more than one group of fingers. The stretch of fingers ranges from 9 repeats that occupy almost the entire protein (as in TFIIIA) to providing just one small domain consisting of 2 fingers (as in the *Drosophila* regulator ADR1). The general transcription factor Sp1 has a DNA-binding domain that consists of 3 zinc fingers.

Figure 21.3 Transcription factor SP1 has a series of three zinc fingers, each with a characteristic pattern of cysteine and histidine residues that constitute the zinc-binding site.

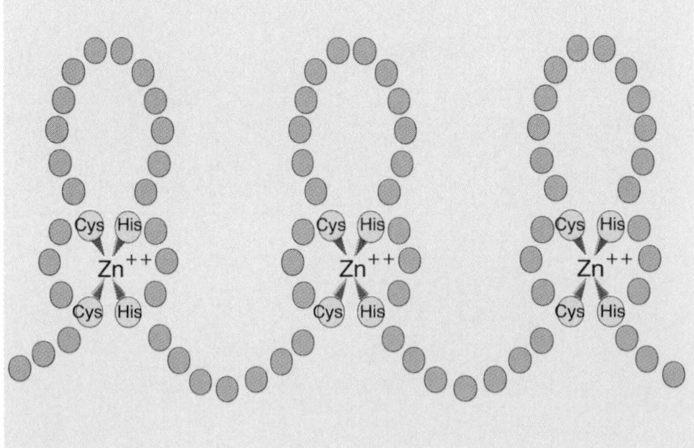

The crystal structure of DNA bound by a protein with three fingers suggests the structure illustrated schematically in **Figure 21.4**. The C-terminal part of each finger forms α-helices that bind DNA; the N-terminal part forms a β-sheet. (For simplicity, the β-sheet and the location of the zinc ion are not shown in the lower part of the figure.) The three α-helical stretches fit into one turn of the major groove; each α-helix (and thus each finger) makes two sequence-specific contacts with DNA (indicated by the arrows). We expect that the nonconserved amino acids in the C-terminal side of each finger are responsible for recognizing specific target sites.

Knowing that zinc fingers are found in authentic transcription factors that assist both RNA polymerases II and III, we may view finger proteins from the reverse perspective. When a protein is found to have multiple zinc fingers, there is at least a *prima facie* case for investigating a possible role as a transcription factor. Such an identification has suggested that several loci involved in embryonic development of *D. melanogaster* are regulators of transcription.

It is necessary to be cautious about interpreting the presence of (putative) zinc fingers, especially when the protein contains only a single finger motif. Fingers may be involved in binding RNA rather than DNA or even unconnected with any nucleic acid binding activity.

The prototype finger protein, TFIIIA, binds both to the 5S gene and to the product, 5S rRNA. A translation initiation factor, eIF2β, has a zinc finger; and mutations in the finger influence the recognition of initiation codons. Retroviral capsid proteins have a motif related to the finger that may be involved in binding the viral RNA.

Steroid receptors (and some other proteins) have another type of finger. The structure is based on a sequence with the zinc-binding consensus:

$$\text{Cys-X}_2\text{-Cys-X}_{13}\text{-Cys-X}_2\text{-Cys}$$

These are called Cys_2/Cys_2 fingers. They are distinct from Cys_2/His_2 fingers. Proteins with Cys_2/Cys_2 fingers often have nonrepetitive fingers, in contrast with the tandem repetition of the Cys_2/His_2 type. Binding sites in DNA (where known) are short and palindromic. Mutational analyses show that the regions of the regulator proteins that bind DNA include the finger motifs.

The glucocorticoid and estrogen receptors each have two fingers, each with a zinc atom at the center of a tetrahedron of cysteines. The two fingers form α-helices that fold together to form a large globular domain. The aromatic sides of the α-helices form a hydrophobic center together with a β-sheet that connects the two helices. One side of the N-terminal helix makes contacts in the major groove of DNA. Two glucocorticoid receptors dimerize upon binding to DNA, and each engages a successive turn of the major groove. This fits with the palindromic nature of the response element (see later).

Each finger controls one important property of the receptor. **Figure 21.5** identifies the relevant amino

Figure 21.4 Zinc fingers may form α-helices that insert into the major groove, associated with β-sheets on the other side.

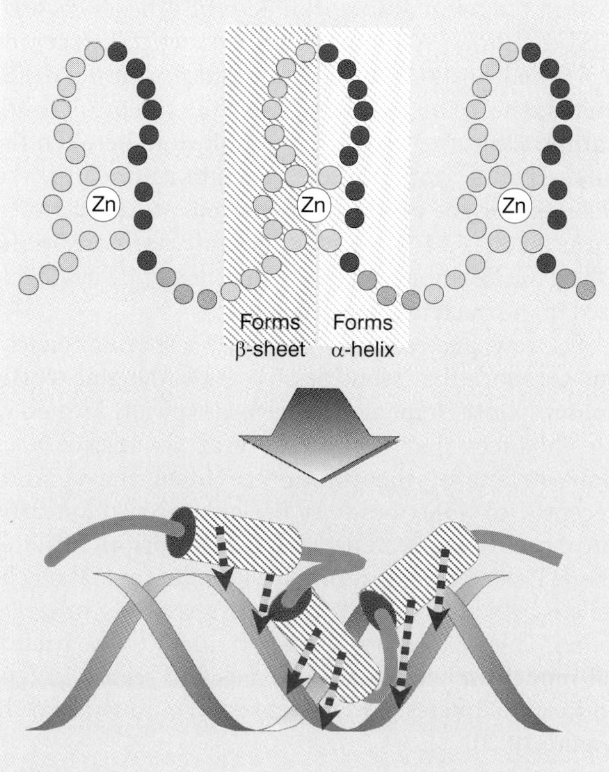

Figure 21.5 The first finger of a steroid receptor controls specificity of DNA-binding (positions shown in red); the second finger controls specificity of dimerization (positions shown in blue). The expanded view of the first finger shows that discrimination between GRE and ERE target sequences rests on two amino acids at the base.

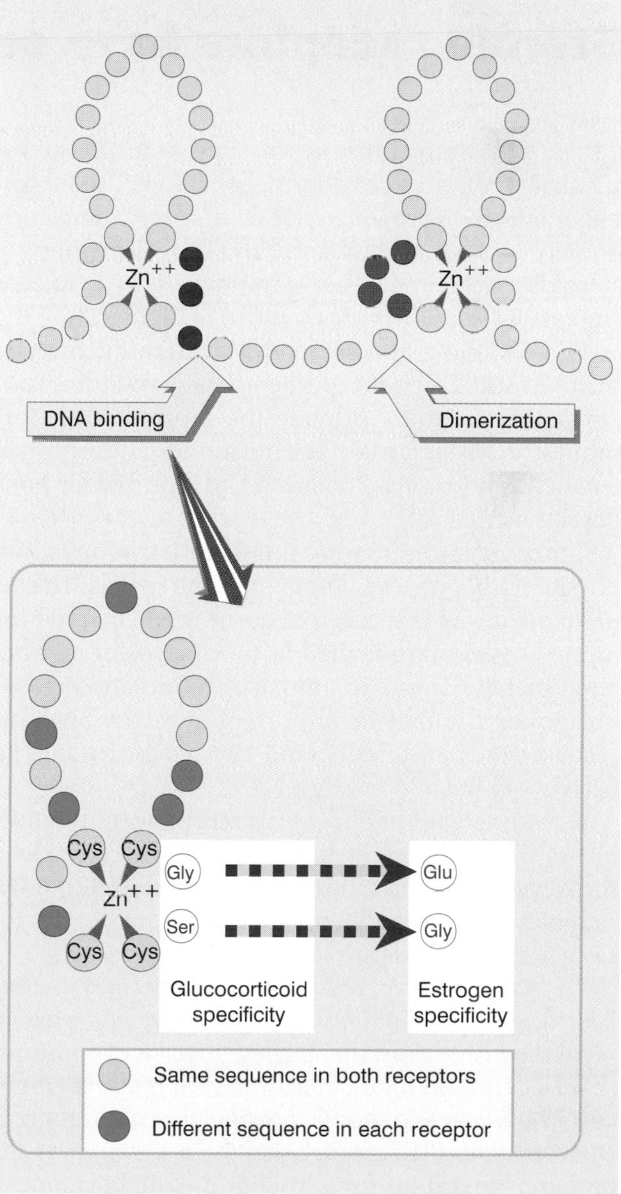

acids. Those on the right side of the first finger control binding to DNA; those on the left side of the second finger control the ability to form a dimer (which we discuss in the next section).

Direct evidence that the first finger binds DNA was obtained by a "specificity swap" experiment. The finger of the estrogen receptor was deleted and replaced by the sequence of the glucocorticoid receptor. The new protein recognized the GRE sequence (the usual target of the glucocorticoid receptor) instead of the ERE (the usual target of the estrogen receptor). This region therefore establishes the specificity with which DNA is recognized.

The differences between the sequences of the glucocorticoid receptor and estrogen receptor fingers lie mostly at the base of the finger. The substitution at two positions shown in Figure 21.5 allows the glucocorticoid receptor to bind at an ERE instead of a GRE.

Steroid receptors have several independent domains

STEROID hormones are synthesized in response to a variety of neuroendocrine activities, and exert major effects on growth, tissue development, and body homeostasis in the animal world. The major groups of steroids and some other compounds with related (molecular) activities are classified in **Figure 21.6.**

The adrenal gland secretes >30 steroids, the two major groups being the glucocorticoids and mineralocorticoids. Steroids provide the reproductive hormones (androgen male sex hormones and estrogen female sex hormones). Vitamin D is required for bone development.

Other hormones, with unrelated structures and physiological purposes, function at the molecular level in a similar way to the steroid hormones. Thyroid hormones, based on iodinated forms of tyrosine, control basal metabolic rate in animals. Steroid and thyroid hormones also may be important in metamorphosis (ecdysteroids in insects, and thyroid hormones in frogs).

Retinoic acid (vitamin A) is a morphogen responsible for development of the anterior-posterior axis in the developing chick limb bud. Its metabolite, 9-*cis* retinoic acid, is found in tissues that are major sites for storage and metabolism of vitamin A.

We may account for these various actions in regulating body development and function in terms of pathways for regulating gene expression. These diverse compounds share a common mode of action: *each is a small molecule that binds to a specific receptor that activates gene transcription.* ("Receptor" may be a misnomer: the protein is a receptor for steroid or thyroid hormone in the same sense that *lac* repressor is a receptor for a β-galactoside: it is not a receptor in the sense of comprising a membrane-bound protein that is exposed to the cell surface.)

We know most about the interaction of glucocorticoids with their receptor, whose action is illustrated in **Figure 21.7.** A steroid hormone can pass through the cell membrane to enter the cell by simple diffusion. Within the cell, a glucocorticoid binds the glucocorticoid receptor. (Work on the glucocorticoid receptor has relied on the synthetic steroid hormone, dexamethasone.) The localization of free receptors is not entirely clear; they may be in equilibrium between the nucleus and cytoplasm. But when hormone binds to the receptor, the protein is converted into an activated form that has a 10× increased affinity for nonspecific DNA; the hormone-receptor complex is always localized in the nucleus.

The activated receptor recognizes a specific consensus sequence that identifies the GRE, the glucocorticoid response element. The GRE is typically located in an enhancer that may be several kb upstream or downstream of the promoter. When the steroid-receptor complex binds to the enhancer, the nearby promoter is activated, and transcription initiates there. Enhancer activation provides the general mechanism by which steroids regulate a wide set of target genes. This action corresponds formally to the bacterial model for induction by a positive regulator (coinducer activates inducer protein), illustrated in Figure 10.20.

Receptors for the diverse groups of steroid

Figure 21.6 Several types of hydrophobic small molecules activate transcription factors. Corticoids and steroid sex hormones are synthesized from cholesterol, vitamin D is a steroid, thyroid hormones are synthesized from tyrosine, and retinoic acid is synthesized from isoprene (in fish liver).

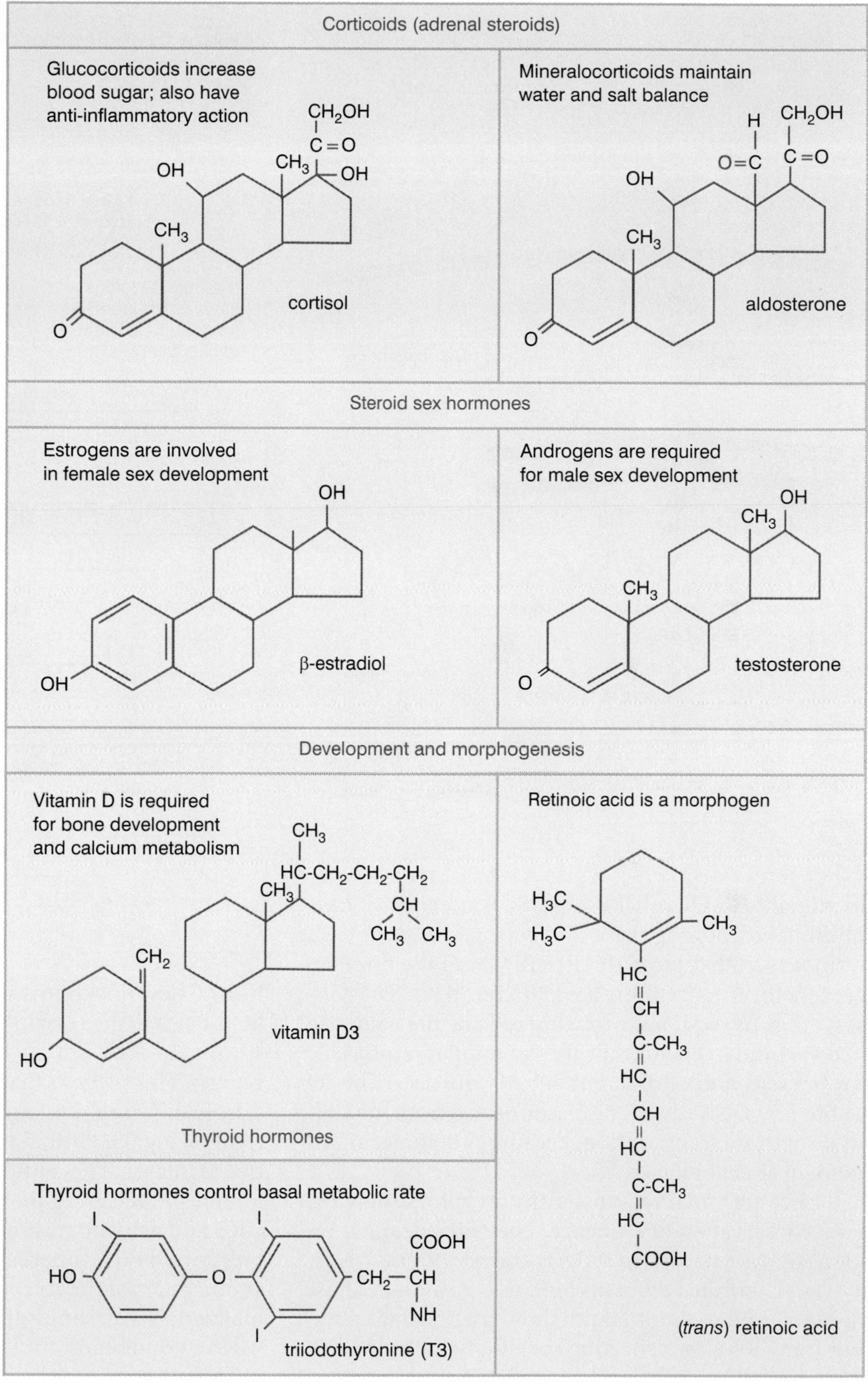

hormones, thyroid hormones, and retinoic acid represent a new "superfamily" of gene regulators, the ligand-responsive transcription factors. All the receptors have independent domains for DNA-binding and hormone binding, in the same relative locations. Their general organization is summarized in **Figure 21.8.**

The central DNA-binding domains are well related for the various steroid receptors, and remain

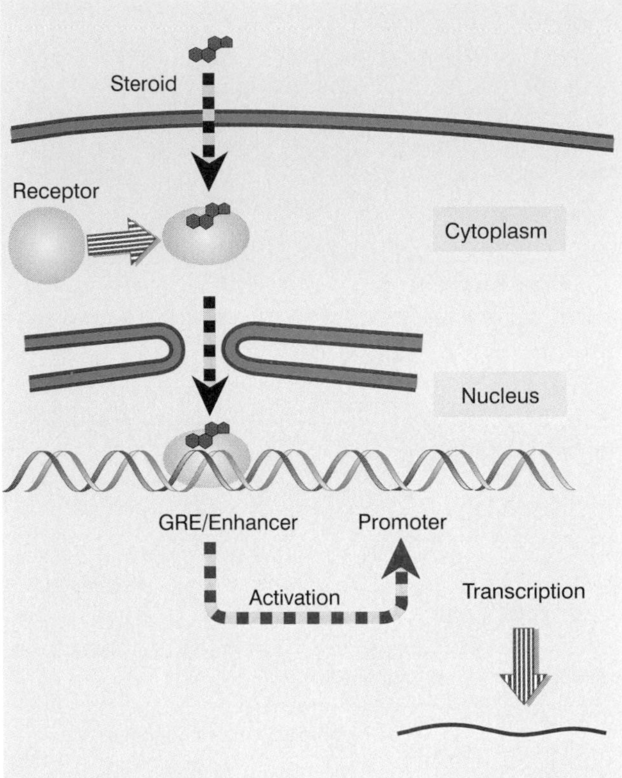

Figure 21.7 Glucocorticoids regulate gene transcription by causing their receptor to bind to an enhancer whose action is needed for promoter function.

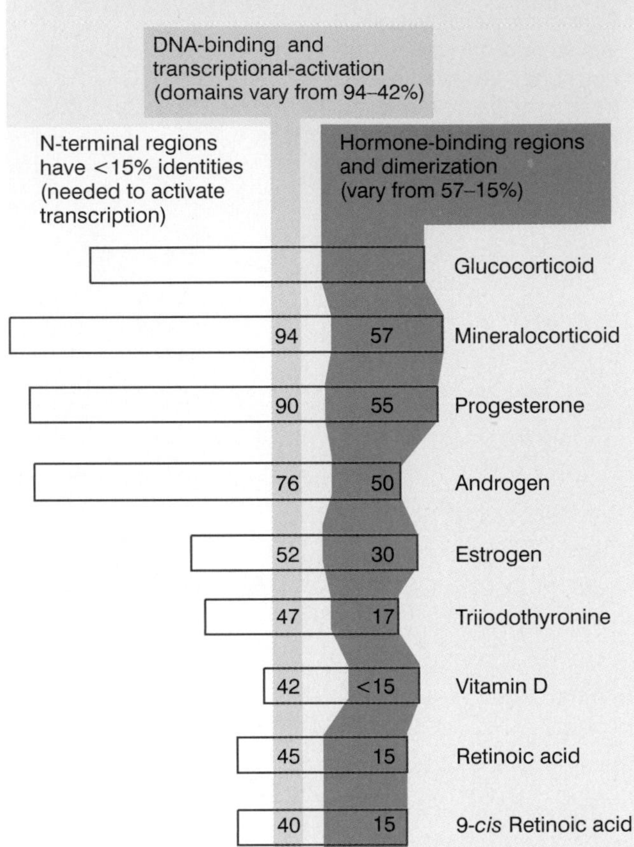

Figure 21.8 Receptors for many steroid and thyroid hormones have a similar organization, with an individual N-terminal region, conserved DNA-binding region, and a C-terminal hormone-binding region.

identifiable in the other receptors. Some ligands have multiple receptors that are closely related, such as the 3 retinoic acid receptors (RARα, β, γ) and the three receptors for 9-*cis*-retinoic acid (RXRα, β, γ). The conservation of sequence probably reflects the common need to bind to DNA, while the variation is responsible for the selection of different target sequences. The act of binding DNA cannot be disconnected from the ability to activate transcription, because mutations in this domain affect both activities.

The N-terminal regions of the receptors show the least conservation of sequence. They include other regions that are needed to activate transcription.

The C-terminal domains bind the hormones. Those in the steroid receptor family show relationships ranging from 30–57%, reflecting specificity for individual hormones. Their relationships with the other receptors are minimal, reflecting specificity for a variety of compounds—thyroid hormones, vitamin D, retinoic acid, etc.

The C-terminal region regulates the activity of the receptor in a way that varies for the individual receptor. If the C-terminal domain of the glucocorticoid receptor is deleted, the remaining N-terminal protein is constitutively active: it no longer requires steroids for activity. This suggests that, in the absence of steroid, the steroid-binding domain prevents the receptor from recognizing the GRE; it functions as an internal negative regulator. The addition of steroid inactivates the inhibition, releasing the receptor's ability to bind the GRE and activate transcription. The basis for the repression could be internal, relying on interactions with another part of the receptor. Or it could result from an interaction with some other protein, which is displaced when steroid binds.

The interaction between the domains is different in the estrogen receptor. If the hormone-binding domain is deleted, the protein is unable to activate transcription, although it continues to bind to the ERE. This region is therefore required to activate rather than to repress activity.

Each receptor recognizes response elements that are related to a consensus. These response elements have a characteristic feature: each consensus consists of two short repeats (or **half sites**). This immediately suggests that the receptor binds as a multimer, so that each half of the consensus is contacted by one subunit (reminiscent of the λ operator-repressor interaction described in Chapter 11).

The response elements for the various receptors may be either palindromes or direct repeats in which the half sites are separated by 0–4 base pairs whose sequence is irrelevant. Only two types of half site are used by the various receptors. Their orientation and spacing determine which receptor recognizes the response element. This behavior allows response elements that have restricted consensus sequences to be recognized specifically by a variety of receptors. The rules that govern recognition are not absolute, but may be modified by context, and there are also cases in which palindromic response elements are recognized permissively by more than one receptor.

The receptors fall into two groups:

- Glucocorticoid (GR), mineralocorticoid (MR), androgen (AR), and progesterone (PR) receptors all form homodimers. They recognize response elements whose half sites have the consensus sequence TGTTCT. The half sites are arranged as palindromes, and the spacing between the sites determines the type of element. The estrogen (ER) receptor functions in the same way, but has the half site sequence TGACCT.
- The thyroid (T3R), vitamin D (VDR), retinoic acid (RAR), and 9-*cis*-retinoic acid (RXR) receptors form heterodimers, which recognize half elements with the sequence TGACCT. The half sites are arranged as direct repeats, and recognition is influenced by their separation as follows:

  1 bp – RXR
  3 bp – VDR
  4 bp – T3R
  5 bp – RAR.

This pattern of recognition in the second group applies for dimeric receptors in which one subunit is the receptor on the list, and the other is RXR (so the first on the list is actually a homodimer). The requirement for heterodimer formation was discovered because retinoic acid and thyroid hormone receptors (RARs and T3Rs) bind much more efficiently to their target sites in the presence of RXR. The heterodimeric receptors respond to the ligand for the first subunit, and apparently do not require RXR to bind its ligand. These receptors can also form homodimers, which recognize palindromic sequences.

Now we are in a position to understand the basis for specificity of recognition. Recall that Figure 21.5 shows how recognition of the sequence of the half site is conferred by the amino acid sequence in the first finger. Specificity for dimer formation is carried by amino acids in the second finger. The formation of dimers in turn determines the distance between the subunits that sit in successive turns of the major groove, and thus controls the response to the spacing of half sites.

There is of course a difference in structure between the homodimers (such as GR or ER) and the heterodimers (VDR, T3R, or RAR with RXR). The homodimers form "head to head" symmetrical structures that recognize palindromic sequences in DNA. The heterodimers form "head to tail" structures that are not symmetrical, and in which the dimerization interface controls the spacing with which the repeated sites are

Figure 21.9 TR and RAR bind the SMRT corepressor in the absence of ligand. The promoter is not expressed. When SMRT is displaced by binding of ligand, the receptor binds a coactivator complex. This leads to activation of transcription by the basal apparatus.

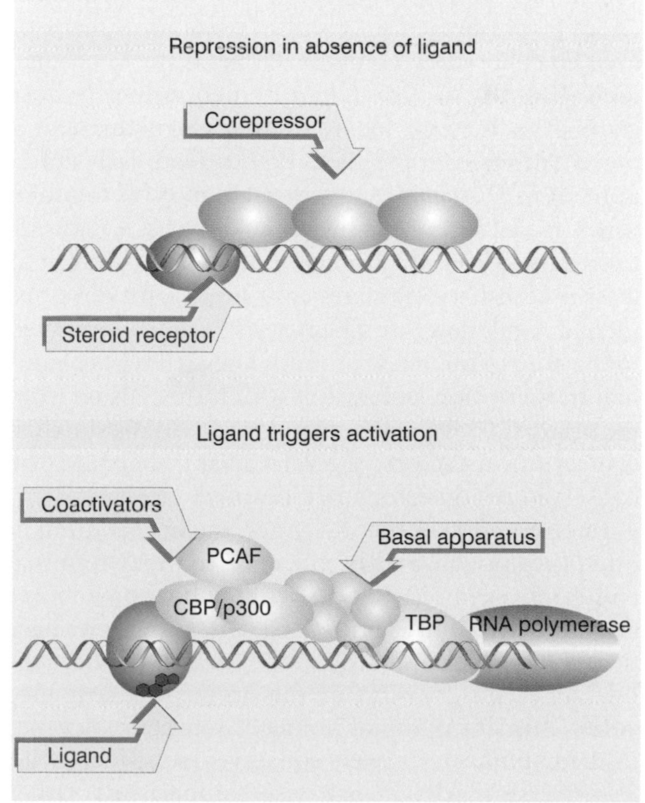

recognized on DNA. Nevertheless, the same principles of organization apply: each receptor subunit has two DNA-binding modules, the first of which is responsible for specificity in contacts with the DNA sequence, the second of which is responsible for dimerization. (The exact positions of the residues responsible for dimerization differ in individual pairwise combinations.) Other residues in both modules contact the phosphate backbone of DNA.

How do the steroid receptors activate transcription? They do not act directly on the basal apparatus, but function via a coactivating complex. The coactivator includes various activities. including the common component CBP/p300, one of whose functions is to modify the structure of chromatin by acetylating histones (see Figure 21.19 later).

All receptors in the superfamily are ligand-dependent activators of transcription. However, some are also able to repress transcription. The TR and RAR receptors, in the form of heterodimers with RXR, bind to certain loci in the *absence* of ligand and repress transcription by means of their ability to interact with a corepressor protein. The corepressor functions by the reverse of the mechanism used by coactivators: it inhibits the function of the basal transcription apparatus, one of its actions being the deacetylation of histones (see Figure 21.20 later). We do not know the relative importance of the repressor activity *vis-à-vis* the ligand-dependent activation in the physiological response to hormone.

The effect of ligand binding on the receptor is to convert it from a repressing complex to an activating complex, as shown in **Figure 21.9.** In the absence of ligand, the receptor is bound to a corepressor complex. The component of the corepressor that binds to the receptor is SMRT. Binding of ligand causes a conformational change that displaces SMRT. This allows the coactivator to bind.

Homeodomains bind related targets in DNA

THE homeobox is a sequence that codes for a domain of 60 amino acids present in proteins of many or even all eukaryotes. Its name derives from its original identification in *Drosophila* homeotic loci (whose genes determine the identity of body structures). It is present in many of the genes that regulate early development in *Drosophila,* and a related motif is found in genes in a wide range of higher eukaryotes. It is attractive to think that the homeodomain identifies (or at least is common in) genes concerned with developmental regulation (see Chapter 29). Sequences related to the homeodomain are found in several types of animal transcription factors, but with the extension from the original *Drosophila* homeodomains to mammalian transcription factors, the relationship between the conserved regions drops significantly.

In *Drosophila* homeotic genes, the homeodomain often (but not always) occurs close to the C-terminal end. Some examples of genes containing homeoboxes are summarized in **Figure 21.10.** Often the genes have little conservation of sequence except in the homeobox. The conservation of the homeobox sequence varies. A major group of homeobox-containing genes in *Drosophila* has a well conserved sequence, with 80–90% similarity in pairwise comparisons. Other genes have less well related homeoboxes. The homeodomain is sometimes combined with other motifs in animal transcription factors. One example is presented by the Oct (octamer-binding) proteins, which represent a group in which a conserved stretch of 75 amino acids called the Pou region is located close to a region resembling the homeodomain. The corresponding sequences in homeoboxes of the pou group of proteins are the least well related to the original group, and thus comprise the farthest extension of the family.

The homeodomain is responsible for binding to DNA, and experiments to swap homeodomains between proteins suggest that the specificity of DNA recognition lies within the homeodomain, but (like the situation with phage repressors) no simple code relating protein and DNA sequences can be deduced. The C-terminal region of the homeodomain shows homology with the helix-turn-helix motif of prokaryotic repressors. We recall from Chapter 11 that the λ repressor has a "recognition helix" (α-helix-3) that makes contacts in the major groove of DNA, while the other helix (α-helix-2) lies at an angle across the DNA. The homeodomain can be organized into three potential helical regions; the sequences of three examples are compared in **Figure 21.11.** The best conserved part of

Figure 21.10 The homeodomain may be the sole DNA-binding motif in a transcriptional regulator or may be combined with other motifs. It represents a discrete (60 residue) part of the protein.

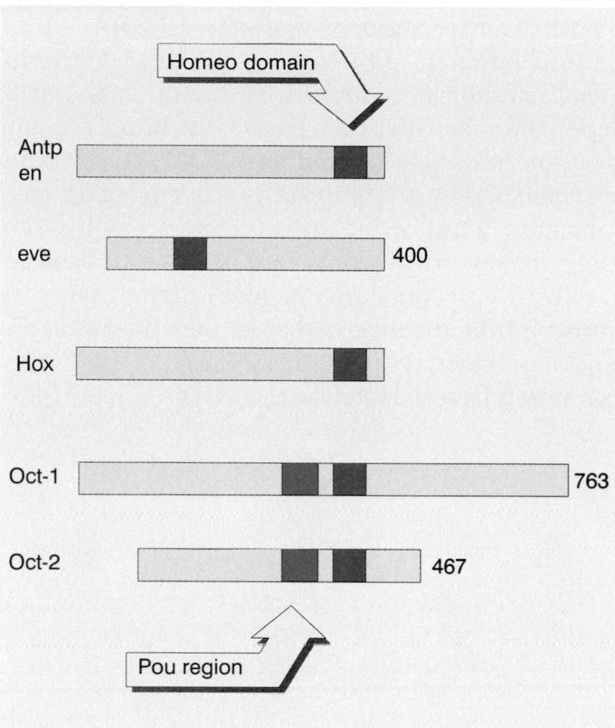

the sequence lies in the third helix. The difference between these structures and the prokaryotic repressor structures lies in the length of the helix that recognizes DNA, helix 3, which is 17 amino acids long in the homeodomain, compared to 9 residues long in the λ repressor.

The crystal structure of the homeodomain of the product of the *D. melanogaster engrailed* gene is represented schematically in **Figure 21.12.** Helix 3 binds in the wide groove of DNA and makes the majority of the contacts between protein and nucleic acid. Many of the contacts that orient the helix in the major groove are made with the phosphate backbone, so they are not specific for DNA sequence. They lie largely on one face of the double helix, and flank the bases with which specific contacts are made. The remaining contacts are made by the N-terminal arm of the homeodomain, the sequence that just precedes the first helix. It projects into the minor groove. So the N-terminal and C-terminal regions of the homeodomain are primarily responsible for contacting DNA.

A striking demonstration of the generality of this model derives from a comparison of the crystal structure of the homeodomain of engrailed with that of the $\alpha 2$ mating protein of yeast. The DNA-binding domain of this protein resembles a homeodomain, and can form three similar helices: its structure in the DNA

Figure 21.11 The homeodomain of the *Antennapedia* gene represents the major group of genes containing homeoboxes in *Drosophila*; *engrailed* (*en*) represents another type of homeotic gene; and the mammalian factor Oct-2 represents a distantly related group of transcription factors. The homeodomain is conventionally numbered from 1 to 60. It starts with the N-terminal arm, and the three helical regions occupy residues 10–22, 28–38, and 42–58.

	1	N-terminal arm							10				Helix 1						20		
En	Glu	Lys	Arg	Pro	Arg	Thr	Ala	Phe	Ser	Ser	Glu	Gln	Leu	Ala	Arg	Leu	Lys	Arg	Glu	Phe	Asn	Glu
Antp	Arg	Lys	Arg	Gly	Arg	Gln	Thr	Tyr	Thr	Arg	Tyr	Gln	Thr	Leu	Glu	Leu	Glu	Lys	Glu	Phe	His	Phe
Oct-2	Arg	Arg	Lys	Lys	Arg	Thr	Ser	Ile	Glu	Thr	Asn	Val	Arg	Phe	Ala	Leu	Glu	Lys	Ser	Phe	Leu	Ala

							30				Helix 2					40
En	Asn	Arg	Tyr	Leu	Thr	Glu	Arg	Arg	Arg	Glu	Glu	Leu	Ser	Ser	Glu	Leu	Gly	Leu
Antp	Asn	Arg	Tyr	Leu	Thr	Arg	Arg	Arg	Arg	Ile	Glu	Ile	Ala	His	Ala	Leu	Cys	Leu
Oct-2	Asn	Glu	Lys	Pro	Thr	Ser	Glu	Glu	Ile	Leu	Leu	Ile	Ala	Glu	Gln	Leu	His	Met

	41									50				Helix 3						60
En	Asn	Glu	Ala	Gln	Ile	Lys	Ile	Trp	Phe	Gln	Asn	Lys	Arg	Ala	Lys	Ile	Lys	Lys	Ser	Asn
Antp	Thr	Glu	Arg	Gln	Ile	Lys	Ile	Trp	Phe	Gln	Asn	Arg	Arg	Met	Lys	Trp	Lys	Lys	Glu	Asn
Oct-2	Glu	Lys	Glu	Val	Ile	Arg	Val	Trp	Phe	Cys	Asn	Arg	Arg	Gln	Lys	Glu	Lys	Arg	Ile	Asn

Figure 21.12 Helix 3 of the homeodomain binds in the major groove of DNA, with helices 1 and 2 lying outside the double helix. Helix 3 contacts both the phosphate backbone and specific bases. The N-terminal arm lies in the minor groove, and makes additional contacts.

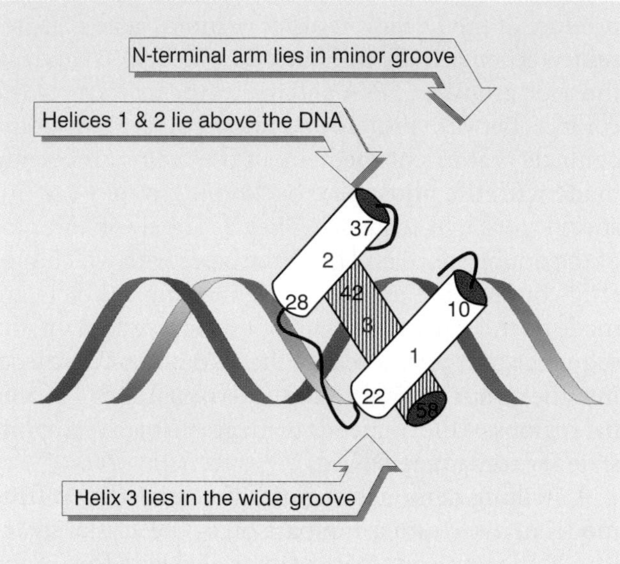
N-terminal arm lies in minor groove
Helices 1 & 2 lie above the DNA
37
2
28
42
3
10
1
22
58
Helix 3 lies in the wide groove

groove can be superimposed almost exactly on that of the engrailed homeodomain. These similarities suggest that all homeodomains bind to DNA in the same manner. This means that a relatively small number of residues in helix 3 and in the N-terminal arm are responsible for specificity of contacts with DNA.

Homeodomain proteins can be either transcriptional activators or repressors. The nature of the factor depends on the other domain(s)—the homeodomain is responsible solely for binding to DNA. The activator or repressor domains both act by influencing the basal apparatus. Activator domains may interact with coactivators that in turn bind to components of the basal apparatus. Repressor domains also interact with the transcription apparatus (that is, they do not act by blocking access to DNA as such). The repressor Eve, for example, interacts directly with TFIID.

Helix-loop-helix proteins interact by combinatorial association

Two common features in DNA-binding proteins are the presence of helical regions that bind DNA, and the ability of the protein to dimerize. Both features are represented in the group of **helix-loop-helix** proteins that share a common type of sequence motif: a stretch of 40–50 amino acids contains two amphipathic α-helices separated by a linker region (the loop) of varying length. The proteins in this group form both homodimers and heterodimers by means of interactions between the hydrophobic residues on the corresponding faces of the two helices. The helical regions are 15–16 amino acids long, and each contains several conserved residues. Two examples are compared in **Figure 21.13.** The ability to form dimers resides with these amphipathic helices, and is common to all HLH proteins. The loop is probably important only for allowing the freedom for the two helical regions to interact independently of one another.

Most HLH proteins contain a region adjacent to the HLH motif itself that is highly basic, and which is needed for binding to DNA. There are ~6 conserved residues in a stretch of 15 amino acids (see Figure 21.13). Members of the group with such a region are called **bHLH proteins.** A dimer in which both subunits have the basic region can bind to DNA. The HLH domains probably correctly orient the two basic regions contributed by the individual subunits. We do not yet know much about the means by which HLH proteins activate transcription.

The bHLH proteins fall into two general groups. Class A consists of proteins that are ubiquitously expressed, including mammalian E12/E47. Class B consists of proteins that are expressed in a tissue-specific manner, including mammalian MyoD, myogenin, and Myf-5 (a group of transcription factors that are involved in myogenesis [muscle formation]). A common

Figure 21.13 All HLH proteins have regions corresponding to helix 1 and helix 2, separated by a loop of 10–24 residues. Basic HLH proteins have a region with conserved positive charges immediately adjacent to helix 1.

MyoD	Ala Asp Arg Arg Lys Ala Ala Thr Met Arg Gln Arg Arg Arg	Basic region
Id	Arg Leu Pro Ala Leu Leu Asp Gln Glu Glu Val Asn Val Leu	6 conserved residues are absent from Id
MyoD	Leu Ser Lys Val Asn Gln Ala Phe Gln Thr Leu Lys Arg Cys Thr	Helix 1
Id	Leu Tyr Asp Met Asn Gly Cys Tyr Ser Arg Leu Lys Gln Leu Val	
		Conserved residues are found in both MyoD and Id
MyoD	Lys Val Gln Ile Leu Arg Asn Ala Ile Arg Tyr Ile Gln Gly Leu Glu	Helix 2
Id	Lys Val Gln Ile Leu Glu His Val Ile Asp Tyr Ile Arg Asp Leu Glu	

modus operandi for a tissue-specific bHLH protein may be to form a heterodimer with a ubiquitous partner. There is also a group of gene products that specify development of the nervous system in *D. melanogaster* (where **Ac-S** is the tissue-specific component and **da** is the ubiquitous component). The Myc proteins (which are the cellular counterparts of oncogene products and are involved in growth regulation) form a separate class of bHLH proteins, whose partners and targets are different.

Dimers formed from bHLH proteins differ in their abilities to bind to DNA. For example, E47 homodimers, E12-E47 heterodimers, and MyoD-E47 heterodimers all form efficiently and bind strongly to DNA; E12 homodimerizes well but binds DNA poorly, while MyoD homodimerizes only poorly. So both dimer formation and DNA binding may represent important regulatory points. At this juncture, it is possible to define groups of HLH proteins whose members form various pairwise combinations, but not to predict from the sequences the strengths of dimer formation or DNA binding. All of the dimers in this group that bind DNA recognize the same consensus sequence, but we do not know yet whether different homodimers and heterodimers have preferences for slightly different target sites that are related to their functions.

Differences in DNA-binding result from properties of the region in or close to the HLH motif; for example, E12 differs from E47 in possessing an inhibitory region just by the basic region, which prevents DNA binding by homodimers. Some HLH proteins lack the basic region and/or contain proline residues that appear to disrupt its function. The example of the protein Id is shown in Figure 21.13. Proteins of this type have the same capacity to dimerize as bHLH proteins, but a dimer that contains one subunit of this type can no longer bind to DNA specifically. This is a forceful demonstration of the importance of doubling the DNA-binding motif in DNA-binding proteins.

The importance of the distinction between the nonbasic HLH and bHLH proteins is suggested by the properties of two groups of HLH proteins: the **da-Ac-S/emc** group and the MyoD/Id type. A model for their functions in forming a regulatory network is illustrated in **Figure 21.14.**

In *D. melanogaster,* the gene *emc* (*extra-macrochaetae*) is required to establish the normal spatial pattern of adult sensory organs. It functions by *suppressing* the functions of several genes, including *da* (*daughterless*) and the *achaete-scute* complex (*Ac-S*), which otherwise cause additional cells to take up this fate. *Ac-S* and *da* are genes of the bHLH type. The suppressor *emc* codes for an HLH protein that lacks the basic region. We suppose that, in the absence of *emc* function, the **da** and **Ac-S** proteins form dimers that activate transcription of appropriate target genes, but the production of **emc** protein causes the formation of heterodimers that cannot bind to DNA. So production of **emc** protein in the appropriate cells is necessary to suppress the function of **Ac-S/da.**

The formation of muscle cells is triggered by a change in the transcriptional program that requires several bHLH proteins, including MyoD. MyoD is produced specifically in myogenic cells; and, indeed, overexpression of MyoD in certain other cells can induce them to commence a myogenic program. The trigger for muscle differentiation is probably a heterodimer consisting of MyoD-E12 or MyoD-E47, rather than a MyoD homodimer. Before myogenesis begins, a member of the nonbasic HLH type, the Id protein, may bind

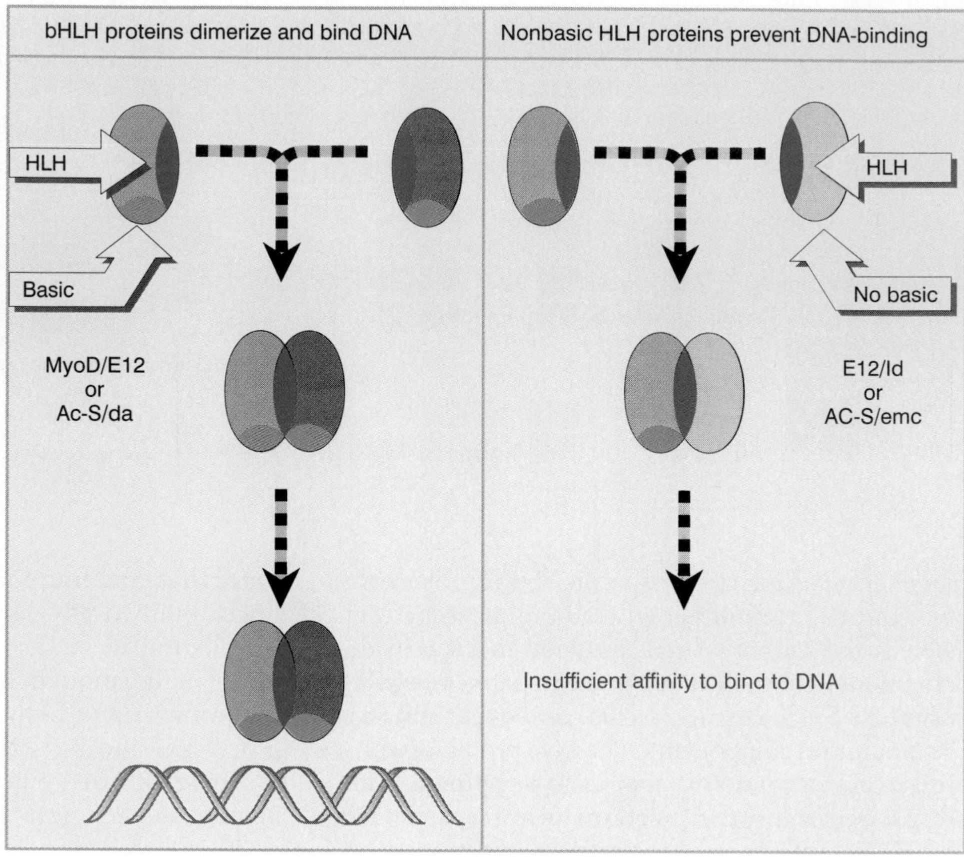

Figure 21.14 An HLH dimer in which both subunits are of the bHLH type can bind DNA, but a dimer in which one subunit lacks the basic region cannot bind DNA.

to MyoD and/or E12 and E47 to form heterodimers that cannot bind to DNA. It binds to E12/E47 better than to MyoD, and so might function by sequestering the ubiquitous bHLH partner. Over-expression of Id can prevent myogenesis. So the removal of Id could be the trigger that releases MyoD to initiate myogenesis.

The behavior of the HLH proteins therefore illustrates two general principles of transcriptional regulation. A small number of proteins form combinatorial associations; it is possible that particular combinations have different functions with regard to DNA binding and transcriptional regulation, but we do not yet entirely understand the relative functions of the tissue-specific and ubiquitous bHLH proteins. And members of the same general class of proteins function as suppressors by participating in the combinatorial associations. Differentiation may depend either on the presence or on the removal of particular suppressor proteins of the nonbasic HLH class.

Leucine zippers are involved in dimer formation

INTERACTIONS between proteins are a common theme in building a transcription complex, and a motif found in several transcription factors (and other proteins) is involved in both homo- and heteromeric interactions. The **leucine zipper** is a stretch of amino acids rich in leucine residues that provide a dimerization motif. Dimer formation itself has emerged as a common principle in the action of proteins that recognize specific DNA sequences and, in the case of the leucine zipper, its relationship to DNA binding is especially clear, because we can see how dimerization juxtaposes the DNA-binding regions of each subunit. The reaction is depicted diagrammatically in **Figure 21.15.**

An amphipathic α-helix has a structure in which the hydrophobic groups (including leucine) face one side, while charged groups face the other side. A leucine zipper forms an amphipathic helix in which the leucines of the zipper on one protein could protrude from the α-helix and interdigitate with the leucines of the zipper of another protein in parallel to form a coiled coil. The two right-handed helices wind around each other, with 3.5 residues per turn, so the pattern repeats integrally every 7 residues.

How is this structure related to DNA binding? The region adjacent to the leucine repeats is highly basic in each of the zipper proteins, and could comprise a DNA-binding site. The two leucine zippers in effect form a Y-shaped structure, in which the zippers comprise the stem, and the two basic regions bifurcate symmetrically to form the arms that bind to DNA. This is known as the **bZIP** structural motif. It explains why the target sequences for such proteins are inverted repeats with no separation.

Zippers may be used to sponsor formation of homodimers or heterodimers. They are lengthy motifs. Leucine occupies every seventh residue in the potential zipper. There are 4 repeats in the protein C/EBP (a factor that binds as a dimer to both the CAAT box and the SV40 core enhancer), and 5 repeats in the factors Jun and Fos (which form the heterodimeric transcription factor, AP1).

AP1 was originally identified by its binding to a DNA sequence in the SV40 enhancer (see Figure 20.19). The active preparation of AP1 includes several polypeptides. A major component is Jun, the product of the gene *c-jun,* which was identified by its relationship with the oncogene *v-jun* carried by an avian sarcoma virus (see Chapter 28). The mouse genome contains a family of related genes, *c-jun* (the original isolate) and *junB* and *junD* (identified by sequence homology with *jun*). There are considerable sequence similarities in the three Jun proteins; they have leucine zippers that can interact to form homodimers or heterodimers.

Figure 21.15 The basic regions of the bZIP motif are held together by the dimerization at the adjacent zipper region when the hydrophobic faces of two leucine zippers interact in parallel orientation.

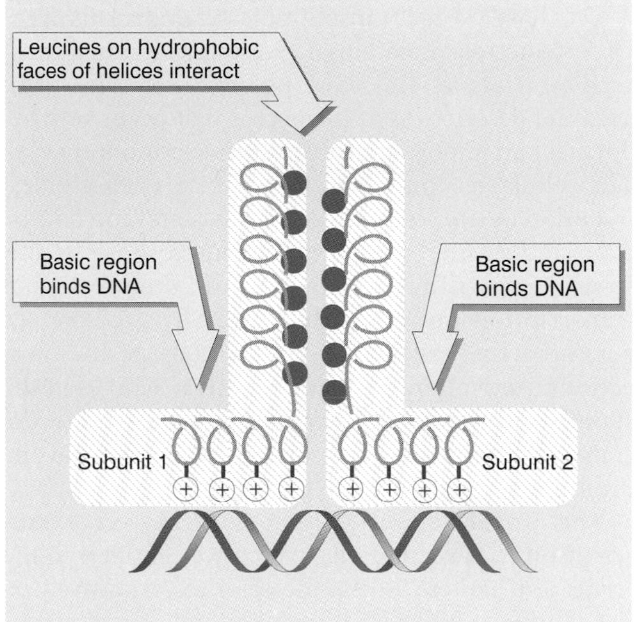

The other major component of AP1 is the product of another gene with an oncogenic counterpart. The *c-fos* gene is the cellular homologue to the oncogene *v-fos* carried by a murine sarcoma virus. Expression of *c-fos* activates genes whose promoters or enhancers possess an AP1 target site. The *c-fos* product is a nuclear phosphoprotein that is one of a group of proteins. The others are described as Fos-related antigens (FRA); they constitute a family of Fos-like proteins.

Fos also has a leucine zipper. Fos cannot form homodimers, but can form a heterodimer with Jun. A leucine zipper in each protein is required for the reaction. The ability to form dimers is a crucial part of the interaction of these factors with DNA. Fos cannot by itself bind to DNA, possibly because of its failure to form a dimer. But the Jun-Fos heterodimer can bind to DNA with same target specificity as the Jun-Jun dimer; and this heterodimer binds to the AP1 site with an affinity ~10× that of the Jun homodimer.

Chromatin remodeling is an active process

We have treated transcription so far in terms of interactions involving DNA and individual transcription factors and RNA polymerases. This is an accurate description of the events that occur *in vitro,* but lacks an important feature of transcription *in vivo.* The cellular genome is packaged into nucleosomes; initiation of transcription generally is prevented if the promoter region is organized into nucleosomes. In this sense, histones function as generalized repressors of transcription (a rather old idea), although we see below that they are also involved in more specific interactions. Activation of a gene requires changes in the state of chromatin: the essential issue is how the transcription factors gain access to the DNA of the promoter.

This provides a contrast with the regulation of transcription in bacterial cells, which we can view in the terms of Chapters 9–11 as determined exclusively by the actions of individual repressor and activator proteins. Whether a bacterial gene is transcribed can be predicted from the sum of the concentrations of the various factors that either activate or repress the individual gene. Changes in these concentrations *at any time* will change the state of expression accordingly.

The stability of chromatin implies that this type of model cannot apply in a eukaryotic cell. Two extreme types of model for changing chromatin structure have been considered historically. The **pre-emptive model** of **Figure 21.16** proposes that *competition* exists between histones and transcription factors to obtain possession of DNA, and whichever takes possession cannot then be displaced. The displacement of histone octamers at replication would provide an opportunity for transcription factors to bind to DNA. The transcription factors would then have to be maintained at their sites through subsequent replication cycles to prevent histone octamers from reassembling.

Early *in vitro* experiments produced results consistent with this model. The transcription factor TFIIIA, required for RNA polymerase III to transcribe 5S rRNA genes, cannot activate the target genes *in vitro* if they are complexed with histones. However, the factor can form the necessary complex with free DNA; then the addition of histones does not prevent the gene from remaining active. Once the factor has bound, it remains at the site, allowing a succession of RNA polymerase molecules to initiate transcription. Whether the factor or histones get to the control site first may be the critical factor.

A similar situation is seen with the TFIID complex at promoters for RNA polymerase II. A plasmid containing an adenovirus promoter can be transcribed *in vitro* by RNA polymerase II in a reaction that requires TFIID and other transcription factors. The template can be assembled into nucleosomes by the addition of histones. If the histones are added *before* the TFIID, transcription cannot be initiated. But if the TFIID is added first, the template still can be transcribed in its chromatin form. So TFIID can recognize free DNA, but either cannot recognize or cannot function on nucleosomal DNA. Only the TFIID must be added before the histones; the other transcription factors and RNA polymerase can be added later. This suggests that binding of TFIID to the promoter creates a structure to

Figure 21.16 The pre-emptive model for transcription of chromatin proposes that if nucleosomes form at a promoter, transcription factors (and RNA polymerase) cannot bind. If transcription factors (and RNA polymerase) bind to the promoter to establish a stable complex for initiation, histones are excluded.

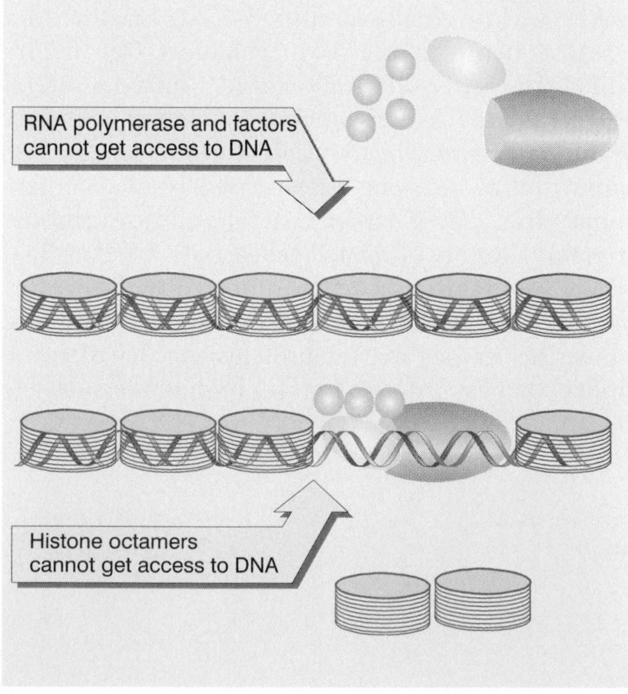

which the other components of the transcription apparatus can bind.

An important feature of this model is that the continued presence of the factor on DNA may be needed to exclude formation of nucleosomes. If a factor is not available to bind DNA at the time of replication, nucleosomes may form and preclude transcription. In these examples of pre-emptive mechanisms for binding to DNA, the factors do not bind effectively to DNA that has been incorporated into nucleosomes. In interpreting results obtained with *in vitro* systems, however, we should note that often the conditions use disproportionate quantities of components, which might upset or create new equilibria. The major importance of these results, therefore, is not that they demonstrate the mechanism used *in vivo,* but that they establish the principle that transcription factors, once present, may exclude nucleosomes from the promoter.

Molecular mechanisms for changes in chromatin structure are suggested by the more recent discovery of mechanisms for displacing histones that depend on the input of energy. The large number of protein-protein and protein-DNA contacts that need to be disrupted to release histones from chromatin makes it seem unlikely that free histones could be in spontaneous equilibrium with octamers. There is no free ride: the energy must come from somewhere to disrupt these contacts. **Figure 21.17** illustrates the operation of a **dynamic model** by a factor that hydrolyzes ATP.

Figure 21.17 The dynamic model for transcription of chromatin relies upon factors that can use energy provided by hydrolysis of ATP to displace nucleosomes from specific DNA sequences.

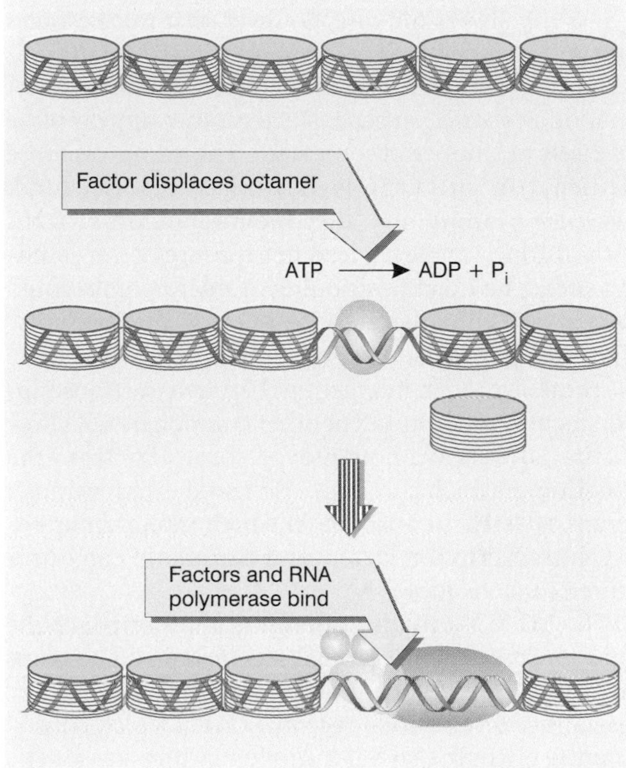
Factor displaces octamer
ATP → ADP + P_i
Factors and RNA polymerase bind

The general process of inducing changes in chromatin structure is called **chromatin remodeling.** Changes at a single nucleosome (typically a change in rotational positioning) may be detected by loss of the DNAaseI 10 base ladder. Displacement of a nucleosome(s) is usually detected by a change in the micrococcal nuclease ladder that results from the absence of protection against cleavage. So changes in chromatin structure may extend from altering the positions of nucleosomes to removing them altogether.

Some transcription factors can disrupt nucleosomes when they bind to DNA and/or to establish a boundary that causes nucleosomes to be positioned around the binding site. The process of has been reconstructed *in vitro* for the hsp70 *Drosophila* promoter. Binding of the transcription factor GAGA to four $(CT)_n$-rich sites on the promoter disrupts the nucleosomes, creates a hypersensitive region, and causes the adjacent nucleosomes to be rearranged so that they occupy preferential instead of random positions. Disruption is an energy-dependent process that requires hydrolysis of ATP. This suggests that GAGA disrupts nucleosome(s) at its binding site, and the creation of a boundary then induces a realignment in which the adjacent nucleosomes become positioned. GAGA can exert its effects even when added after assembly of nucleosomes, suggesting that it can change existing nucleosome structure. Its effect is lost in the absence of ATP, suggesting an energy-dependent dynamic mechanism.

The *PHO* system was one of the first in which it was shown that a change in nucleosome organization is involved in gene activation. At the *PHO5* promoter, the bHLH regulator PHO4 responds to phosphate starvation by inducing the disruption of four precisely positioned nucleosomes. This event is independent of transcription (it occurs in a $TATA^-$ mutant) and independent of replication. There are two binding sites for PHO4 at the promoter, one located between nucleosomes, which can be bound by the isolated DNA-binding domain of PHO4, and the other within a nucleosome, which cannot be recognized. Disruption of the nucleosome to allow DNA binding at the second site is necessary for gene activation. This action requires the presence of the transcription-activating domain. The activator sequence of VP16 can substitute for the PHO4 activator sequence in nucleosome disruption. This suggests that disruption occurs by protein-protein interactions that involve the same region that makes protein-protein contacts to activate transcription.

It is not always the case, however, that nucleosomes must be excluded in order to permit initiation of transcription. Some transcription factors can bind to DNA on a nucleosomal surface. Nucleosomes appear to be precisely positioned at some steroid hormone response elements in such a way that receptors can bind. Receptor binding may alter the interaction of DNA with histones, and even lead to exposure of new binding sites. The exact positioning of nucleosomes could be required either because the nucleosome "presents" DNA in a particular rotational phase or because there are protein-protein interactions between the transcription factors and histones or other components of chromatin. So we have now moved some way from the situation in which chromatin is viewed exclusively as a repressive structure to one in which interactions between transcription factors and chromatin can be required for activation.

The MMTV promoter presents an example of the need for specific nucleosomal organization. It contains an array of 6 partly palindromic sites, each bound by one dimer of hormone receptor (HR), which constitute the HRE. It also has a single binding site for the factor NF1, and two adjacent sites for the factor OTF. HR and NF1 cannot bind simultaneously to their sites in free DNA. **Figure 21.18** shows how the nucleosomal structure controls binding of the factors.

Figure 21.18 Hormone receptor and NF1 cannot bind simultaneously to the MMTV promoter in the form of linear DNA, but can bind when the DNA is presented on a nucleosomal surface.

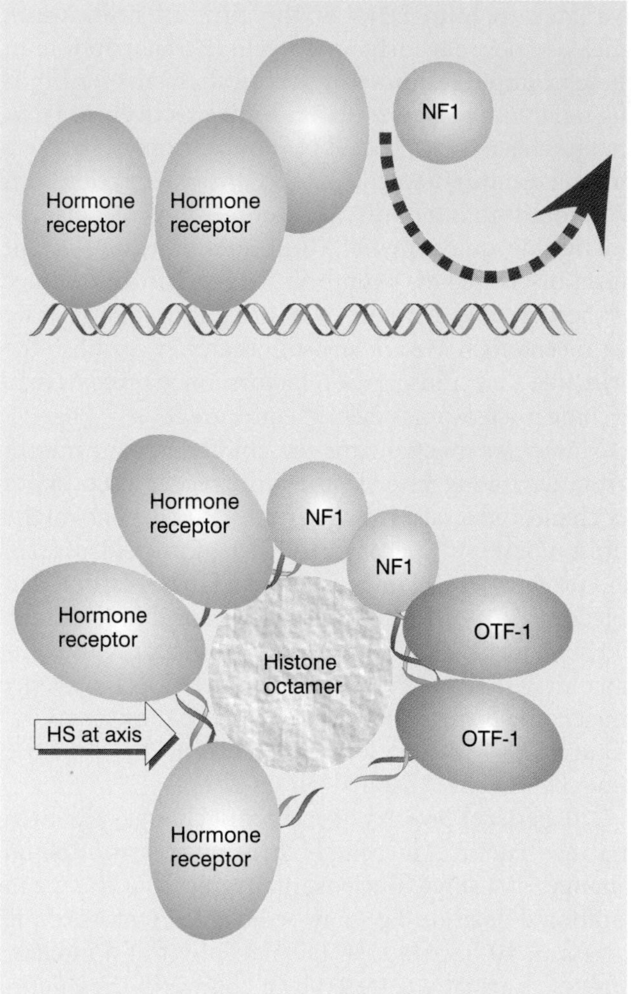

The HR protects its binding sites at the promoter when hormone is added, but does not affect the micrococcal nuclease-sensitive sites that mark either side of the nucleosome. This suggests that HR is binding to the DNA on the nucleosomal surface. However, the rotational positioning of DNA on the nucleosome prior to hormone addition allows access to only two of the four sites. Binding to the other two sites requires a change in rotational positioning on the nucleosome. This can be detected by the appearance of a sensitive site at the axis of dyad symmetry (which is in the center of the binding sites that constitute the HRE). NF1 can be footprinted on the nucleosome after hormone induction, so these structural changes may be necessary to allow NF1 to bind, perhaps because they expose DNA and abolish the steric hindrance by which HR blocks NF1 binding to free DNA.

Functions concerned with activating many promoters were first identified by the characterization of a group of genes, *SWI1,2,3*, which are needed for expression of the HO nuclease involved in mating-type switching in yeast (see Chapter 17). Mutations in these loci are pleiotropic, and the range of defects is similar to those shown by mutants that have lost the CTD tail of RNA polymerase II. These mutations also show genetic interactions with mutations in genes that code for components of chromatin, in particular *SIN1*, whose product resembles a nonhistone protein, and *SIN2*, which codes for histone H3. The demonstration that these gene functions are required for expression of a variety of individual loci suggests that they might code for components of a complex that is involved in interacting with chromatin structure.

The **SWI/SNF complex** contains many proteins coded by genes originally identified by *SWI* or *SNF* mutations. It comprises ~10 proteins with a combined molecular weight ~2 × 10^6. It has an ATPase activity that is conferred by the SWI2 subunit. It is presumed to act catalytically because there are only ~150 complexes per cell. The basic role of the SWI/SNF complex is chromatin remodeling—causing nucleosomes to change structure and/or position so as to allow transcriptional activators to have access to their target sites.

The ability of the GAL4 activator to regulate *GAL1* is influenced by SWI/SNF. The SWI/SNF complex stimulates binding of GAL4 to its target site on nucleosomal DNA, ~30× *in vitro*. The stimulation requires ATP hydrolysis and is abolished by a mutation in *SWI2* that inactivates its ATPase activity. The SWI/SNF interaction alters nucleosomal sensitivity to DNAase I in an ATP-dependent reaction. SWI/SNF cannot stimulate GAL4 binding if the histone octamers are crosslinked, which suggests that disruption of nucleosomes may be involved. The structural effects may therefore influence both protein-DNA and protein-protein interactions in the nucleosome, but have not yet been characterized in detail.

One puzzle about the action of the SWI/SNF complex is its sheer size. It dwarfs RNA polymerase and the nucleosome, making it difficult to understand how all of these components could interact with DNA retained on the nucleosomal surface. However, a transcription complex with full activity, called RNA polymerase II holoenzyme, can be found that contains the RNA polymerase itself, all the TFII factors except TBP and TFIIA, and the SWI/SNF complex, which is associated with the CTD tail of the polymerase. In fact, virtually all of the SWI/SNF complex may be present in holoenzyme preparations. This suggests that the remodeling of chromatin and recognition of promoters is undertaken in a coordinated manner by a single complex.

There are also other chromatin remodeling complexes with similar activities. RSC is a 15-subunit complex of yeast. Some of its subunits are homologous to subunits of the SWI/SNF complex. RSC is ~10× more abundant than SWI/SNF, but is not found associated with RNA polymerase. NURF is another large complex that is required for a transcription factor to generate a hypersensitive site in a nucleosome array. All these complexes use ATP hydrolysis to provide the energy for chromatin remodeling.

Histone acetylation and deacetylation control chromatin activity

It has long been known that the state of histone acetylation is correlated with the state of gene expression. All the core histones are acetylated. Histone acetylation appears to be increased in a domain containing active genes, and acetylated chromatin is more sensitive to DNAase I and (possibly) to micrococcal nuclease. Histone acetylation also occurs during S phase, when it may be associated with the incorporation of histones into nucleosomes. The most striking change in histone modification found to date is the underacetylation of histone H4 on the inactive X chromosome in female mammals. This suggests that absence of acetyl groups may be a prerequisite for a more condensed, inactive structure.

Enzymes that can acetylate histones are called **histone acetyltransferases** or **HATs**; the acetyl groups are removed by **histone deacetylases** or **HDACs**. There are two groups of HAT enzymes: group A describes those that are involved with transcription; group B describes those involved with nucleosome assembly. Two inhibitors have been useful in analyzing acetylation. Trichostatin and butyric acid inhibit histone deacetylases, and cause acetylated nucleosomes to accumulate. The use of these inhibitors has supported the general view that acetylation is associated with gene expression; in fact, the ability of butyric acid to cause changes in chromatin resembling those found upon gene activation was one of the first indications of the connection between acetylation and gene activity.

The breakthrough in analyzing the role of histone acetylation was provided by the characterization of the acetylating and deacetylating enzymes, and their association with other proteins that are involved in specific events of activation and repression. A basic change in our view of histone acetylation was caused by the discovery that HATs are not necessarily dedicated enzymes associated with chromatin: rather it turns out that known activators of transcription have HAT activity.

The connection was established when the catalytic subunit of a group A HAT was identified as a homologue of the yeast regulator protein GCN5. Then it was shown that GCN5 itself has HAT activity on histones H3 and H4. GCN5 is part of an adaptor complex that is necessary for the interaction between certain enhancers and their target promoters. Its HAT activity is required for activation of the target gene.

One of the first general activators to be characterized as an HAT was p300/CBP. (Actually, p300 and CBP are different proteins, but they are so closely related that they are often referred to as a single type of activity.) p300/CBP is a coactivator that links an upstream transcription factor to the basal apparatus (see Figure 20.25). p300/CBP interacts with various transcription factors, including hormone receptors, AP-1 (c-Jun and c-Fos), and MyoD. The interaction is inhibited by the viral regulator proteins adenovirus E1A and SV40 T antigen, which bind to p300/CBP to prevent the interaction with transcription factors; this explains how these viral proteins inhibit cellular transcription. (This inhibition is important for the ability of the viral proteins to contribute to the tumorigenic state; see Chapter 28).

p300/CBP acetylates the N-terminal tails of H4 in nucleosomes. Another coactivator, called PCAF, preferentially acetylates H3 in nucleosomes. p300/CBP and PCAF form a complex that functions in transcriptional activation. In some cases yet another HAT is involved: the coactivator ACTR, which functions with hormone receptors, is itself an HAT that acts on H3 and H4, and also recruits both p300/CBP and PCAF to form a coactivating complex. One explanation for the presence of multiple HAT activities in a coactivating complex is that each HAT has a different specificity, and that multiple different acetylation events are required for activation. This enables us to redraw our picture for the action of coactivators as shown in **Figure 21.19,** where RNA polymerase is bound at a hypersensitive site and coactivators are acetylating histones on the nucleosomes in the vicinity.

The acetylases are found in large protein complexes. In yeast, GCN5 is part of the 1.8 MDa SAGA complex, which contains several proteins that are implicated in transcription by mutations in their genes that block

Figure 21.19
Coactivators may have HAT activities that acetylate the tails of nucleosomal histones.

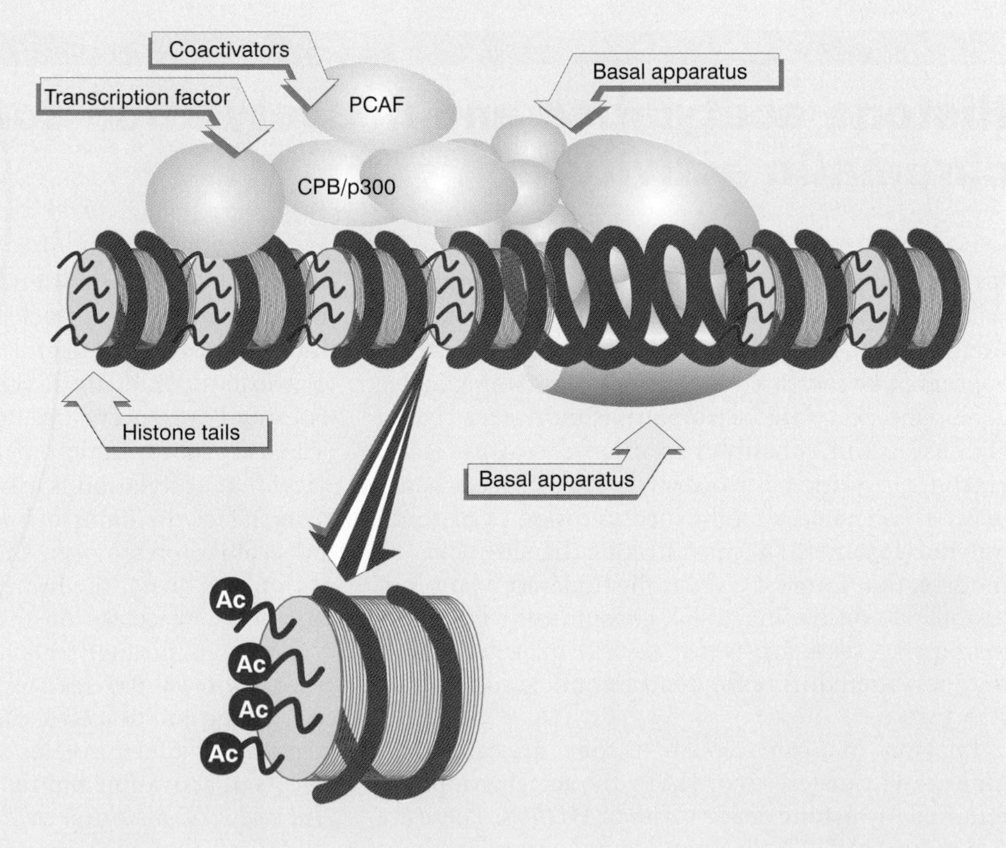

initiation. Among these proteins are several TAF_{II}s. The mammalian coactivator PCAF has a comparable structure. In both cases, the TAF_{II}s include histone-like species that probably form the basis for a structure resembling the histone octamer.

Deacetylation is associated with repression of gene activity. In yeast, mutations in *SIN3* and *Rpd3* behave as though these loci repress a variety of genes. They form a complex with the DNA-binding protein Ume6, and this complex represses transcription at promoters that have the *URS1* element that is bound by Ume6, as illustrated in **Figure 21.20.** Rpd3 has histone deacetylase activity; we do not know whether the function of Sin3 is just to bring Rpd3 to the promoter or whether it has an additional role in repression.

A similar system for repression is found in mammalian cells. The bHLH family of transcription regulators includes activators that function as heterodimers, including MyoD (see earlier). It also includes repressors, in particular the heterodimer Mad:Max, where Mad can be any one of a group of closely related proteins. The Mad:Max heterodimer (which binds to specific DNA sites) interacts with a homologue of Sin3 (called mSin3 in mouse and hSin3 in man). mSin3 is part of a repressive complex that includes histone binding proteins and the histone deacetylases HDAC1 and HDAC2. Deacetylase activity is required for repression. The modular nature of this system is emphasized by other means of employment: a corepressor (SMRT), which enables retinoid hormone receptors to repress certain target genes, functions by binding mSin3, which in turns brings the HDAC activities to the site. Another means of bringing HDAC activities to the site may be a connection with MeCP2, a protein that binds to methylated cytosines (see below).

When a coactivator acetylates DNA to assist the initiation of transcription, only the local region—~1 kb upstream and downstream of the promoter—is affected. We do not know how this influences transcription.

Acetylation occurs at both replication (when it is transient) and at transcription (when it is maintained while the gene is active). Is it playing the same role in each case? One possibility is that the important effect is on nucleosome structure. Acetylation may be necessary to "loosen" the nucleosome core. At replication, acetylation of histones could be necessary to allow them to be incorporated into new cores more easily. At

Figure 21.20 A repressor complex contains three components: a DNA binding subunit, a corepressor, and a histone deacetylase.

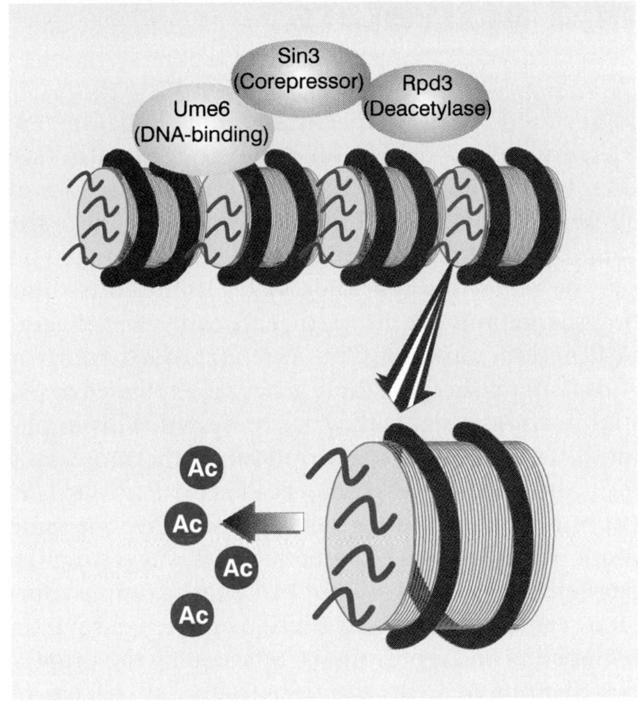

transcription, a similar effect could be necessary to allow a related change in structure, possibly even to allow the histone core to be displaced from DNA. Alternatively, acetylation could generate binding sites for other proteins that are required for transcription. In either case, deacetylation would reverse the effect.

Is the effect of acetylation quantitative or qualitative? One possibility is that a certain number of acetyl groups are required to have an effect, and the exact positions at which they occur are largely irrelevant. An alternative is that individual acetylation events have specific effects. We might interpret the existence of complexes containing multiple HAT activities in either way—if individual enzymes have different specificities, we may need multiple activities either to acetylate a sufficient number of different positions or because the individual events are necessary for different effects upon transcription. At replication, it appears, at least with respect to histone H4, that acetylation at any two of three available positions is adequate, favoring a quantitative model in this case.

Polycomb and trithorax are antagonistic repressors and activators

Chromatin can be specifically repressed. One example is the formation of heterochromatin (see Chapter 18). Another is provided by the genetics of homeotic genes in *Drosophila,* which have led to the identification of a protein complex that may maintain certain genes in a repressed state. *Pc* mutants show transformations of cell type that are equivalent to gain-of-function mutations in *Antennapedia* (*Antp*) or *Ultrabithorax,* because these genes are expressed in tissues in which usually they are repressed. This implicates *Pc* in regulating transcription. Furthermore, *Pc* is the prototype for a class of loci called the *Pc* group (*Pc-G*); mutations in these genes generally have the same result of derepressing homeotic genes, suggesting the possibility that the group of proteins has some common regulatory role. A connection between chromatin remodeling and repression is indicated by the properties of *brahma*, a fly counterpart to *SWI2*. Loss of *brahma* function suppresses mutations in *Polycomb.*

Consistent with the pleiotropy of *Pc* mutations, Pc is a nuclear protein that can be visualized at ~80 sites on polytene chromosomes. These sites include the *Antp* gene. Another member of the *Pc-G, polyhomeotic,* is visualized at a set of polytene chromosome bands that are identical with those bound by Pc. The two proteins coimmunoprecipitate in a complex of ~2.5×10^6 D that contains 10–15 polypeptides. The relationship between these proteins and the products of the ~30 *Pc-G* genes remains to be established; one possibility is that many of these gene products form a general repressive complex that is modified by some of the others for specific loci.

The Pc-G proteins are not conventional repressors. They are not responsible for determining the initial pattern of expression of the genes on which they act. In the absence of Pc-G proteins, these genes are initially repressed as usual, but later in development the repression is lost without Pc-G group functions. This suggests that the Pc-G proteins in some way recognize the state of repression when it is established, and they then act to perpetuate it through cell division of the daughter cells. **Figure 21.21** shows a model in which Pc-G proteins bind in conjunction with a repressor, but the Pc-G proteins remain bound after the repressor is no longer available. This is necessary to maintain repression, so that if Pc-G proteins are absent, the gene becomes activated.

A region of DNA that is sufficient to enable the response to the *Pc-G* genes is called a PRE (*Polycomb* response element). It can be defined operationally by the property that it confers repression of enhancers in its vicinity throughout development. The assay for a PRE is to insert it close to a reporter gene that is controlled by an enhancer that is repressed in early development, and then to determine whether the reporter becomes expressed subsequently in the descendants. An effective PRE will prevent such re-expression.

The PRE is a complex structure, of the order of 10 kb. No individual member of the Pc-G proteins has yet been shown to bind to specific sequences in the PRE, so the basis for the assembly of the complex is still unknown. When a locus is repressed by Pc-G proteins, however, the proteins appear to be present over a much larger length of DNA than the PRE itself. Polycomb is found locally over a few kilobases of DNA surrounding a PRE.

This suggests that the PRE may provide a nucleation center, from which a structural state depending on Pc-G proteins may propagate. This model is supported by the observation of effects related to position effect variegation (see Figure 19.45), that is, a gene near to a locus whose repression is maintained by Pc-G may become heritably inactivated in some cells but not others. In one typical situation, crosslinking experiments *in vivo* showed that Pc protein is found over large regions of the *bithorax* complex that are inactive, but the protein is excluded from regions that contain active genes. The idea that this could be due to cooperative interactions within a multimeric complex is supported by the existence of mutations in *Pc* that change its nuclear distribution and abolish the ability of other *Pc-G* members to localize in the nucleus. The role of Pc-G proteins in maintaining, as opposed to establishing, repression must mean that the formation of the complex at the PRE also depends on the local state of gene expression.

A connection between the Pc-G complex and more general structural changes in chromatin is suggested by a homology between a 37 amino acid region near the N-terminus of Pc and a nonhistone protein, HP1, that

Figure 21.21 Pc-G proteins do not initiate repression, but are responsible for maintaining it.

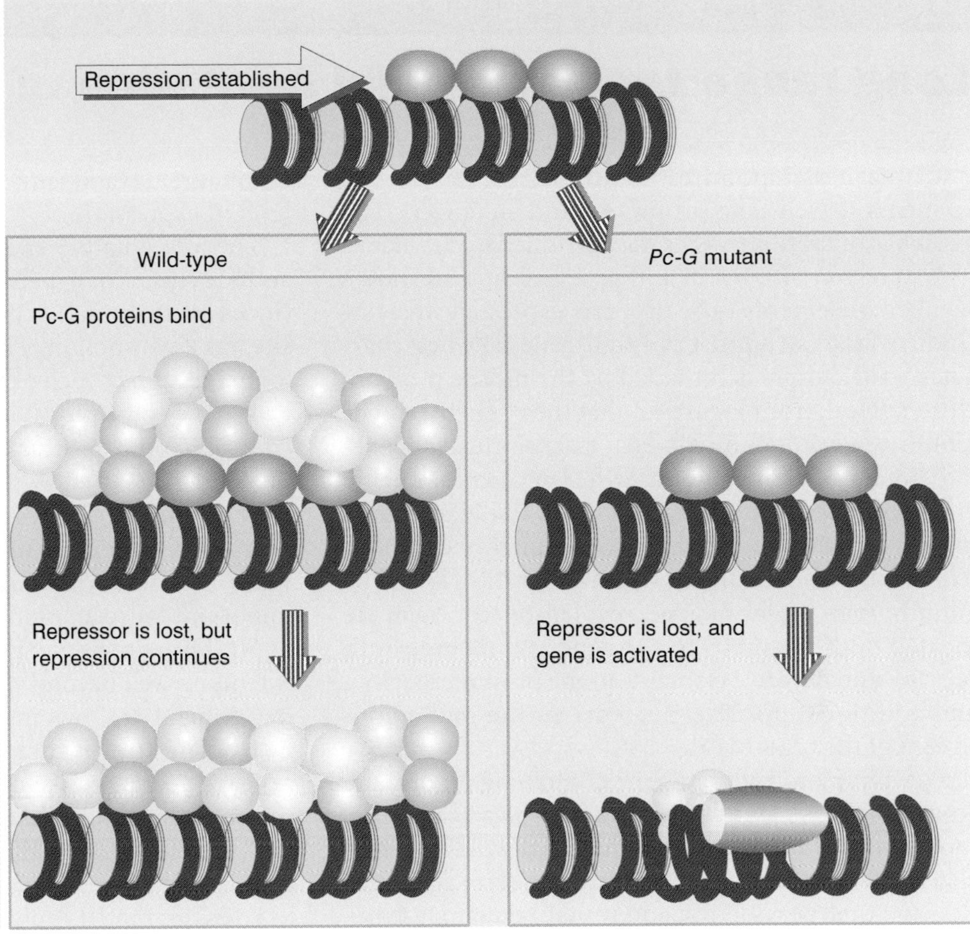

Repression established
Wild-type
Pc-G mutant
Pc-G proteins bind
Repressor is lost, but repression continues
Repressor is lost, and gene is activated

is associated with heterochromatin. The common motif is called the chromodomain. HP1 is coded by the gene *Su(var)205*, a suppressor of position-effect variegation. Since variegation is probably caused by the spreading of inactivity from constitutive heterochromatin, it is possible that the chromodomain is used by Pc and HP1 to interact with common components that are involved in inducing the formation of heterochromatic or inactive structures. This model implies that similar mechanisms are used to repress individual loci or to create heterochromatin.

The trithorax group (trxG) of proteins have the opposite effect to the Pc-G proteins: they act to maintain genes in an active state. There may be some similarities in the actions of the two groups: mutations in some loci prevent both Pc-G and trx from functioning, suggesting that they could rely on common components. A factor coded by the *trithorax-like* gene, called GAGA because it binds to GA-rich consensus sequences, has binding sites in the PRE. In fact, the sites where Pc binds to DNA coincide with the sites where GAGA factor binds.

What does this mean? GAGA is probably needed for activating factors, including trxG members, to bind to DNA. The genetics suggest that it is needed for activation by trxG. Is it also needed for PcG proteins to bind and exercise repression? This is not yet clear, but such a model would demand that something other than GAGA determines whether PcG or trxG complexes subsequently assemble at the site.

Long range regulation and insulation of domains

Some genes respond to additional regulatory elements as well as to promoters and enhancers. The existence of these elements was identified by the inability of a region of DNA including a gene and all known regulatory elements to be properly expressed when introduced into an animal as a transgene. The best characterized example is provided by the mouse β-globin cluster. Recall from Figure 4.1 that the α-globin and β-globin genes in mammals each exist as clusters of related genes, expressed at different times during embryonic and adult development. These genes are provided with a large number of regulatory elements, which have been analyzed in detail. In the case of the adult human β-globin gene, regulatory sequences are located both 5′ and 3′ to the gene and include both positive and negative elements in the promoter region, and additional positive elements within and downstream of the gene.

But a human β-globin gene containing all of these control regions is never expressed in a transgenic mouse within an order of magnitude of wild-type levels. Some further regulatory sequence is required. Regions that provide the additional regulatory function are identified by DNAase I hypersensitive sites that are found at the ends of the cluster. The map of **Figure 21.22** shows that the 20 kb upstream of the ε-gene contains a group of 5 sites; and there is a single site 30 kb downstream of the β-gene. Transfecting various constructs into mouse erythroleukemia cells shows that sequences between the individual sites in the 5′ region can be removed without much effect, but that removal of any of the sites reduces the overall level of expression.

Figure 21.22 A globin domain is marked by hypersensitive sites at either end. The group of sites at the 5' side constitutes the LCR and is essential for the function of all genes in the cluster.

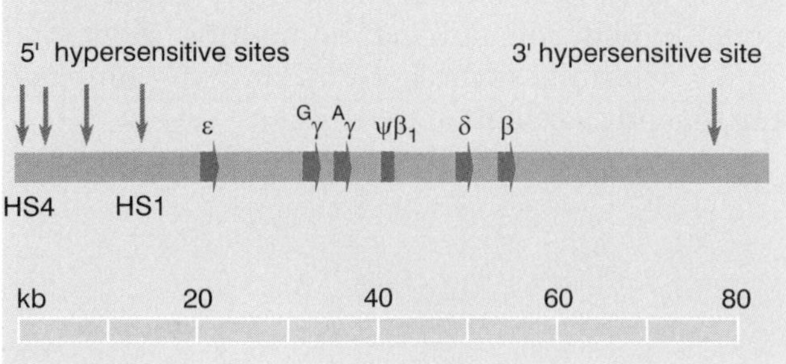

The 5′ regulatory sites are the primary regulators, and the cluster of hypersensitive sites is called the **LCR** (locus control region). We do not know whether the 3′ site has any function. The LCR is absolutely required for expression of each of the globin genes in the cluster. Each gene is then further regulated by its own specific controls. Some of these controls are autonomous: expression of the ε- and γ-genes appears intrinsic to those loci in conjunction with the LCR. Other controls appear to rely upon position in the cluster, which provides a suggestion that *gene order* in a cluster is important for regulation.

The entire region containing the globin genes, and extending well beyond them, constitutes a chromosomal **domain.** It shows increased sensitivity to digestion by DNAase I (see Figure 19.42). Deletion of the 5′ LCR restores normal resistance to DNAase over the whole region. It is possible that the LCR is required to "open up" the whole domain and make it available for transcription.

Does this model apply to other gene clusters? The α-globin locus has a similar organization of genes that are expressed at different times, with a group of hypersensitive sites at one end of the cluster, and increased sensitivity to DNAase I throughout the region. Only a small number of other cases are known in which an LCR controls a group of genes.

Does a domain have a discrete boundary that marks its end; or do effects emanating from discrete points within it decline gradually at the ends of the region?

Elements that prevent the passage of activating or inactivating effects are called **insulators,** and have been identified in two circumstances. When an insulator is placed between an enhancer and a promoter, it prevents the enhancer from activating the promoter. This may explain how the action of an enhancer is limited to a particular promoter. When an insulator is placed between an active gene and heterochromatin, it protects the gene against the inactivating effect that spreads from the heterochromatin. Those insulators that have been characterized so far possess both these properties, suggesting that they affect the general organization of chromatin.

The leftmost hypersensitive site of the chick β-globin LCR (HS4) is an insulator that marks the 5′ end of the functional domain. HS4 lies within a 250 bp region that has the properties of an (unmethylated) CpG island (although it is not associated with a promoter). We do not know yet what identifies the other end of the domain.

The first information about the delineation of domains was provided by the analysis of *D. melanogaster* genes summarized in **Figure 21.23.** Two genes for the protein hsp70 lie within an 18 kb region that constitutes band 87A7. The boundaries of the band are marked by sequences, denoted *scs* and *scs'* (specialized chromatin structures). They consist of 350 bp in *scs* or 200 bp in *scs'* of a region highly resistant to degradation by DNAase I, flanked on either side by hypersensitive sites, spaced at about 100 bp. The cleavage pattern at these sites is altered when the genes are turned on by heat shock.

Figure 21.23 Specialized chromatin structures that include hypersensitive sites mark the ends of a domain in the *D. melanogaster* genome and insulate genes between them from the effects of surrounding sequences.

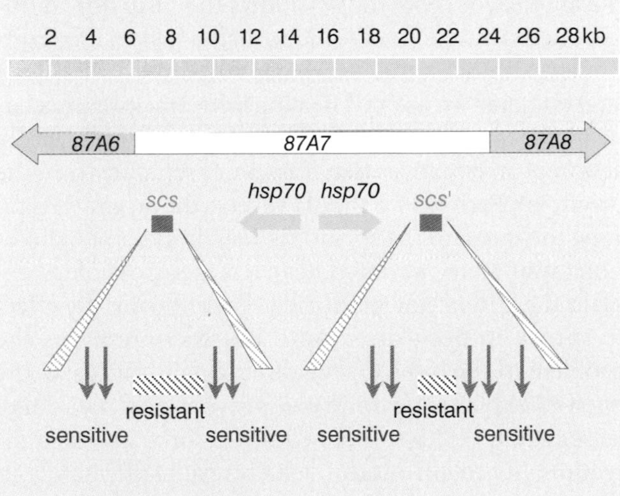

The *scs* elements insulate the hsp70 genes from the effects of surrounding regions. If we take *scs* units and place them on either side of a *white* gene, the gene can function anywhere it is placed in the genome, even in sites where it would normally be repressed by context, for example, in heterochromatic regions.

The *scs* units do not seem to play either positive or negative roles in controlling gene expression, but just restrict effects from passing from one region to the next. If adjacent regions have repressive effects, however, the *scs* elements might be needed to block the spread of such effects, and therefore could be essential for gene expression. In this case, deletion of the *scs* elements could eliminate the function of the band.

The generality of this type of element is indicated by the properties of a protein, BEAF-32, that binds to *scs'*. It shows discrete localization within the nucleus, but the most remarkable data derive from its localization on polytene chromosomes. **Figure 21.24** shows that an anti-BEAF-32 antibody stains ~50% of the interbands of the polytene chromosomes. This suggests indeed that the band is a functional unit, and that elements in the interbands prevent activating or inactivating effects from propagating from one band to the next.

Insulators may have directional properties. Insertions of the transposon *gypsy* into the *yellow* (*y*) locus of *D. melanogaster* cause loss of gene function in some tissues, but not in others. The reason is that the *y* locus is regulated by four enhancer elements, as shown in **Figure 21.25.** When *gypsy* is inserted just upstream of the promoter, it blocks activation in the tissues wing blade and body cuticle (from the upstream enhancers), but does not block expression in bristles and tarsal claws (from the downstream enhancers). The sequence

Figure 21.24 A protein that binds to the insulator *scs'* is localized at interbands in *Drosophila* polytene chromosomes. Red staining identifies the DNA (the bands) on both the upper and lower samples; green staining identifies BEAF-32 (often at interbands) on the upper sample. Yellow shows coincidence of the two labels. Some of the more prominent stained interbands are marked by white lines. Photograph kindly provided by Uli Laemmli.

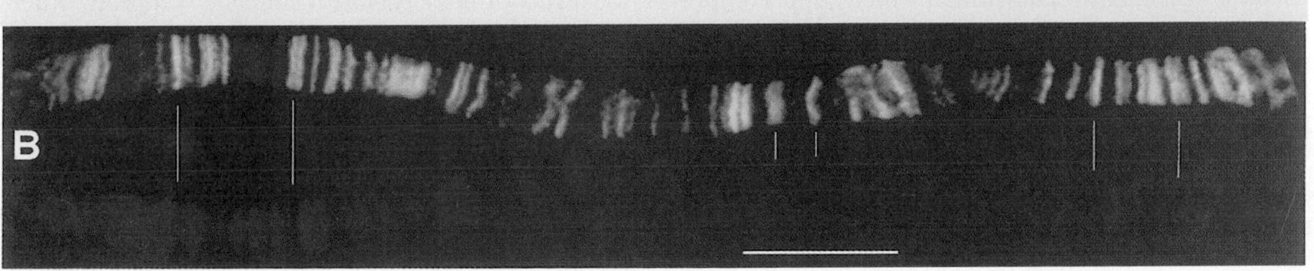

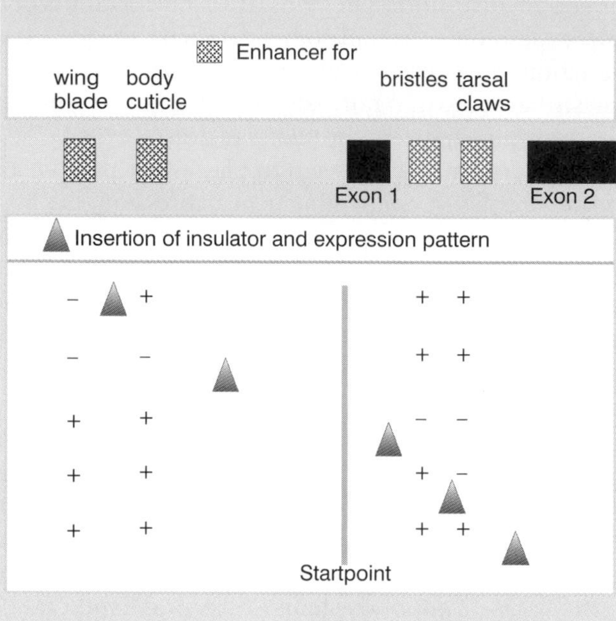

Figure 21.25 The insulator of the *gypsy* transposon blocks the action of an enhancer when it is placed between the enhancer and the promoter.

responsible for this effect is an insulator that lies at one end of the transposon. A systematic analysis in which the insulator was moved along the *y* locus showed that it blocks expression of any enhancer that it separates from the promoter.

Since some of the enhancers are upstream of the promoter and others are downstream, the effect cannot depend on position with regard to the promoter, nor can it require transcription to occur through the insulator. Directionality does not depend on the orientation of the binding site itself, since all of the insertions can be reversed in orientation, but retain the same insulating properties. This is difficult to explain in terms of looping models for enhancer-promoter interaction, which essentially predict the irrelevance of the intervening DNA. The directionality is difficult to explain in terms of a competitive model, in which the insulator has binding sites that compete with the enhancer for connection to the promoter. The obvious model to invoke is a tracking mechanism, in which some component must move unidirectionally from the enhancer to the promoter, but this is inconsistent with previous characterizations of the independence of enhancers from such effects.

Proteins that act upon the insulator have been identified or inferred through the existence of two other loci that affect insulator function in a *trans*-acting manner. Mutations in *su(Hw)* abolish insulation: *y* is expressed in all tissues in spite of the presence of the insulator. This suggests that *su(Hw)* codes for a protein that recognizes the insulator and is necessary for its action. Su(Hw) is a nuclear protein; mapping to polytene chromosomes shows that it is bound at a large number of sites. The insulator contains 12 binding sites for su(Hw); it is not known how many are required for its action.

The second locus is *mod(mdg4)*, in which mutations have the opposite effect. This is observed by the loss of directionality. These mutations increase the effectiveness of the insulator by extending its effects so that it blocks utilization of enhancers on both sides. *su(Hw)* is epistatic to *mod(mdg4)*, which implies that mod(mdg4) acts through su(Hw). The basic role of the wild-type protein from the *mod(mdg4)* locus is therefore to impose directionality on the ability of su(Hw) to insulate promoters from the boundary.

Binding of su(Hw) to DNA, followed by binding of mod(mdg4) to su(Hw), therefore creates a unidirectional block to activation of a promoter. The fact that the absence of mod(mdg4) allows the insulator to impede activity in both directions makes it even more difficult to understand the basis for directionality. It suggests that we are not dealing with tracking mechanisms, because in this situation the insulator blocks the action of an enhancer that it does not separate from the promoter. Perhaps we should reverse the argument and view the insulator as spreading inactivity in one direction (away from the promoter); it is then possible to explain the absence of mod(mdg4) as allowing the effect to spread in both directions. But there remains the problem of the basis for the directionality and also the need to explain the ability to prevent position effect variegation. Perhaps there could be some intrinsic directionality to chromatin, which results ultimately in the incorporation of su(Hw), mod(mdg4), or some other component in one orientation, presumably by virtue of an interaction with some component of chromatin that is itself preferentially oriented. Any such directionality would need to reverse at the promoter. But there is no good answer at present to the questions of what provides the basis for the directionality and how the insulator interferes with enhancer-promoter interactions.

Sometimes elements with different *cis*-acting properties are combined to generate regions with complex regulatory effects. The *Fab-7* region is defined by deletions in the *bithorax* locus of *Drosophila*. This locus contains a series of *cis*-acting regulatory elements that control the activities of three transcription units (see Figure 29.30). The relevant part of the locus is drawn in

Figure 21.26. The regulatory elements *iab-6* and *iab-7* control expression of the adjacent gene *Abd-B* in successive regions of the embryo (segments A6 and A7), which is necessary for the differences between them to develop. A deletion of *Fab-7* causes A6 to develop like A7, instead of in the usual way. This is a dominant effect, which suggests that the cause is that *iab-7* has taken over control from *iab-6*. We can interpret this in molecular terms by supposing that *Fab-7* provides a boundary that prevents *iab-7* from acting when *iab-6* is usually active.

Like other boundary elements, *Fab-7* contains a distinctive chromatin structure that is marked by a series of hypersensitive sites. The region can be divided into two types of element by smaller deletions and by testing fragments for their ability to provide a boundary. A sequence of ~3.3 kb behaves as an insulator when it is placed in other constructs. A sequence of ~0.8 kb behaves as a repressor that acts on *iab-7*. The presence of these two elements explains the complicated genetic behavior of *Fab-7* (which we have not described in detail).

Figure 21.26 *Fab-7* is a boundary element that is necessary for the independence of regulatory elements *iab-6* and *iab-7*.

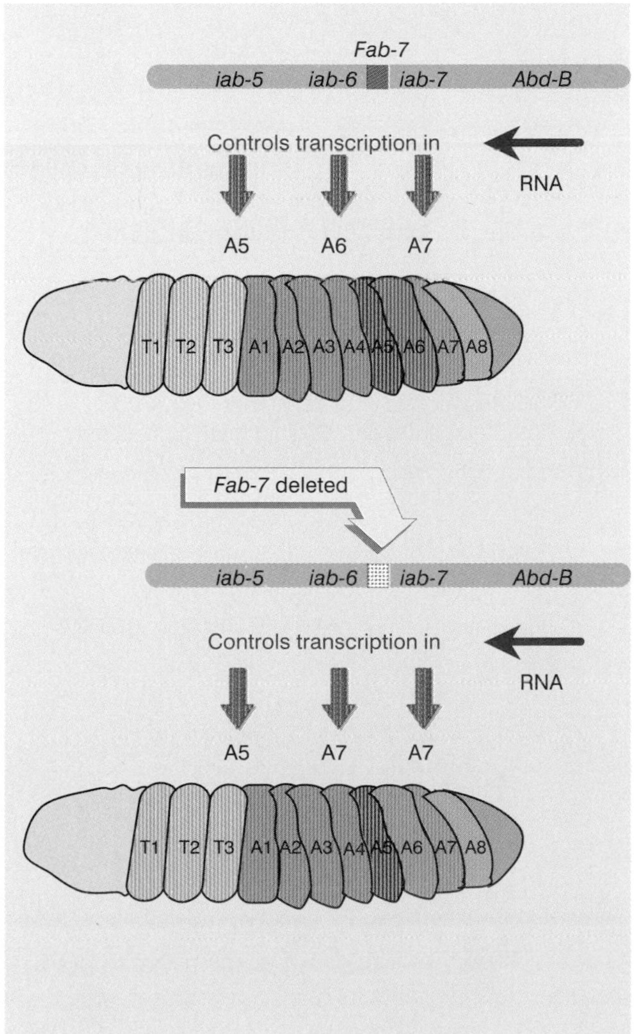

An insight into the action of the boundary element is provided by the effects of substituting other insulators for *Fab-7*. The effect of *Fab-7* is simply to prevent interaction between *iab-6* and *iab-7;* it does not affect the ability of *iab-6* to act on *Abd-B*. When an *Su(Hw)* binding site is introduced, a stronger effect is seen: *iab-6* is no longer able to control *Abd-B*. As a result, *iab-5* takes over. And when an *scs* element is used, the effect extends to blocking the action of *iab-5*. This suggests a scheme in which stronger elements can block the actions of regulatory sequences that lie farther away.

This conclusion introduces a difficulty for explaining the action of boundary elements. They cannot be functioning in this instance simply by preventing the transmission of effects past the boundary. This argues against models based on simple tracking or inhibiting the linear propagation of structural effects. It suggests that there may be some sort of competitive effect, in which the strength of the element determines how far its effect can stretch.

This relates to another property of insulators. All known insulators have two effects. They can prevent an enhancer from activating a promoter, typically when they are placed between the two elements, although their effects are not necessarily restricted to such locations. And they can prevent heterochromatin from spreading to inactivate nearby genes. But there is no reason to expect these two situations to have the same basis. We do not have a definitive view of the action of enhancers on promoters, but generally it appears to involve some sort of direct interaction, not a tracking along chromatin or propagation of structure (see Chapter 20). The propagation of heterochromatin, by contrast, does seem to involve a linear spread from an activating center. The ability of insulators to act in both types of situation could mean that they may have some sort of supraorganizational capacity that we do not understand. It would be interesting to see whether these two capacities could be separated, for example, by mutations that affect one but not the other, or whether they really reside in the same molecular function.

If we now put together the various types of structures that have been found in different systems, we can think about the possible nature of a chromosomal domain. **Figure 21.27** summarizes the structures that

Figure 21.27 Domains may possess three types of sites: insulators to prevent effects from spreading between domains; MARs to attach the domain to the nuclear matrix; and LCRs that are required for initiation of transcription.

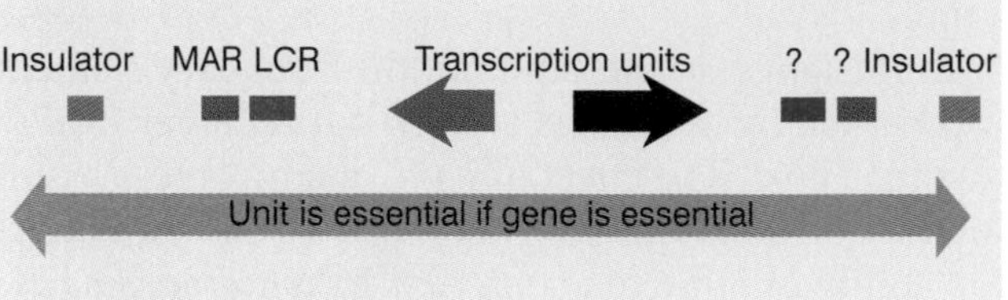

might be involved in defining a domain. A domain (perhaps this might be a band in the case of *Drosophila*) might contain several transcription units. The **LCR** itself is an essential structure whose function at a distance is essential for the ability of any and all genes in a domain to be expressed. Several types of *cis*-acting structures could be required for function. As defined originally, the property of the LCR rests with an **enhancer-like** hypersensitive site that is needed for the full activity of promoter(s) within the domain. An **insulator** may stop activating or repressing effects spreading beyond the functional unit. A **matrix attachment site** (**MAR**) may be responsible for physical attachment to a site on the nuclear periphery. These elements might be found at one end or at both ends of a domain.

Perhaps this type of organization could also help to explain the large size of the genome. A certain amount of space could be required for such a structure to operate, for example, to allow chromatin to become decondensed and to become accessible. Although the exact sequences of much of the unit might be irrelevant, there might be selection for the overall amount of DNA within it, or at least selection might prevent the various transcription units from becoming too closely spaced.

Gene expression is associated with demethylation

METHYLATION of DNA is one of the parameters that controls transcription. Methylation in the vicinity of the promoter is associated with the absence of transcription. This is one of several regulatory events that influence the activity of a promoter; like the other regulatory events, typically this will apply to both (allelic) copies of the gene. However, methylation also occurs as an epigenetic event that can distinguish alleles whose sequences are identical. This can result in differences in the expression of the paternal and maternal alleles. In this chapter we are concerned with the means by which methylation influences transcription.

The distribution of methyl groups can be examined by taking advantage of restriction enzymes that cleave target sites containing the CG doublet. Two types of restriction activity are compared in **Figure 21.28.** These **isoschizomers** are enzymes that cleave the same target sequence in DNA, but have a different response to its state of methylation.

The enzyme HpaII cleaves the sequence CCGG (writing the sequence of only one strand of DNA). But if the second C is methylated, the enzyme can no

Figure 21.28 The restriction enzyme MspI cleaves all CCGG sequences whether or not they are methylated at the second C, but HpaII cleaves only nonmethylated CCGG tetramers.

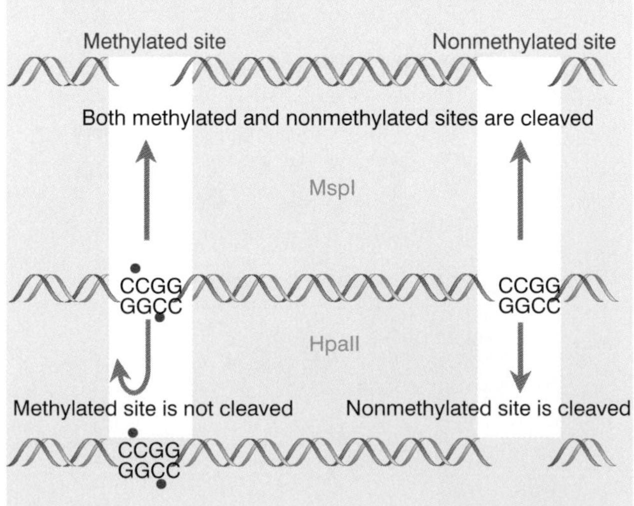

longer recognize the site. However, the enzyme MspI cleaves the same target site *irrespective* of the state of methylation at this C. So MspI can be used to identify all the CCGG sequences; and HpaII can be used to determine whether or not they are methylated.

With a substrate of nonmethylated DNA, the two enzymes would generate the same restriction bands. But in methylated DNA, the modified positions are not cleaved by HpaII. For every such position, one larger HpaII fragment replaces two MspI fragments. An example is given in **Figure 21.29.**

Many genes show a pattern in which the state of methylation is constant at most sites, but varies at others. Some of the sites are methylated in all tissues examined; some sites are unmethylated in all tissues. *A minority of sites are methylated in tissues in which the gene is not expressed, but are not methylated in tissues in which the gene is active.* So an active gene may be described as *undermethylated.*

Experiments with the drug 5-azacytidine produce indirect evidence that demethylation can result in gene expression. The drug is incorporated into DNA in place of cytidine, and cannot be methylated, because the 5′ position is blocked. This leads to the appearance of demethylated sites in DNA as the consequence of replication (following the scheme on the right of Figure 13.24).

The phenotypic effects of 5-azacytidine include the induction of changes in the state of cellular differentiation; for example, muscle cells are induced to develop from nonmuscle cell precursors. The drug also activates genes on a silent X chromosome, which raises the possibility that the state of methylation could be connected with chromosomal inactivity.

As well as examining the state of methylation of resident genes, we can compare the results of introducing methylated or nonmethylated DNA into new host cells. Such experiments show a clear correlation: *the methylated gene is inactive, but the nonmethylated gene is active.*

What is the extent of the undermethylated region? In the chicken α-globin gene cluster in adult erythroid cells, the undermethylation is confined to sites that extend from ~500 bp upstream of the first of the two adult α genes to ~500 bp downstream of the second. Sites of undermethylation are present in the entire region, including the spacer between the genes. The region of undermethylation coincides with the region of maximum sensitivity to DNAase I. This argues that undermethylation is a feature of a domain that contains a transcribed gene or genes.

Methylation at the 5′ end of a gene may be directly related to expression. Many genes are not methylated at the 5′ end when they are expressed, although they remain methylated at the 3′ end. As with other changes in chromatin, it seems likely that the absence of methyl groups is associated with the *ability to be transcribed* rather than with the act of transcription itself.

Our problem in interpreting the general association between undermethylation and gene activation is that only a minority (sometimes a small minority) of the methylated sites are involved. It is likely that the state of methylation is critical at specific sites or in a restricted region; for example, demethylation at the promoter may be necessary to make it available for the initiation of transcription. It is also possible that a reduction in the level of methylation (or even the complete removal of methyl groups from some stretch of DNA) is part of some structural change needed to permit transcription to proceed.

In the γ-globin gene, for example, the presence of methyl groups in the region around the startpoint, between −200 and +90, suppresses transcription. Removal of the 3 methyl groups located upstream of the startpoint or of the 3 methyl groups located downstream does not relieve the suppression. But removal of all methyl groups allows the promoter to function. Transcription may therefore require a methyl-free region at the promoter. There are exceptions to the general relationship we have described.

Some genes can be expressed even when they are extensively methylated. Any connection between methylation and expression thus is not universal in an organism, but the general rule is that methylation prevents gene expression and demethylation is required for expression.

The presence of **CpG-rich islands** in the 5′ regions

Figure 21.29 The results of MspI and HpaII cleavage are compared by gel electrophoresis of the fragments.

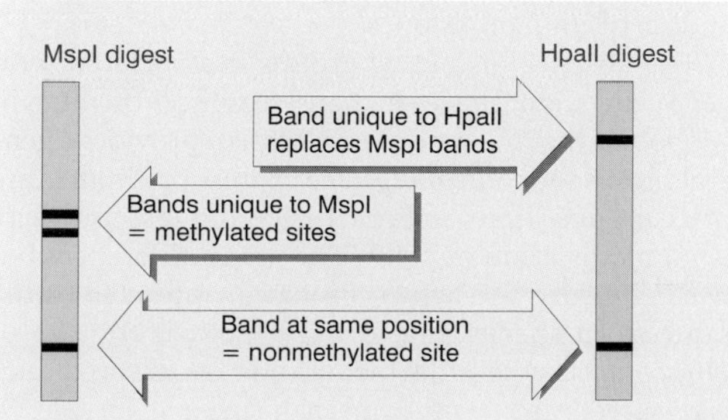
MspI digest
HpaII digest
Band unique to HpaII replaces MspI bands
Bands unique to MspI = methylated sites
Band at same position = nonmethylated site

of some genes is connected with the effect of methylation on gene expression. These islands are detected by the presence of an increased density of the dinucleotide sequence, CpG.

The CpG doublet occurs in vertebrate DNA at only ~20% of the frequency that would be expected from the proportion of G•C base pairs. In certain regions, however, the density of CpG doublets reaches the predicted value; in fact, it is increased by 10× relative to the rest of the genome.

These CpG-rich islands have an average G•C content of ~60%, compared with the 40% average in bulk DNA. They take the form of stretches of DNA typically 1–2 kb long. There are ~45,000 such islands in the human genome. As diagnosed by restriction analysis, the islands are unmethylated. The nucleosomes at the islands have a reduced content of histone H1 (which probably means that their packaging is relaxed), the other histones are extensively acetylated (a feature that tends to be associated with gene expression), and there are hypersensitive sites (as would be expected of active promoters).

In several cases, CpG-rich islands begin just upstream of a promoter and extend downstream into the transcribed region before petering out. **Figure 21.30** compares the density of CpG doublets in a "general" region of the genome with a CpG island identified from the DNA sequence. The CpG island surrounds the 5′ region of the APRT gene, which is constitutively expressed.

All of the "housekeeping" genes that are constitutively expressed have CpG islands; this accounts for about half of the islands altogether. The other half of the islands occur at the promoters of tissue-regulated genes; only a minority (<40%) of these genes have islands. In these cases, the islands are unmethylated irrespective of the state of expression of the gene. The presence of unmethylated CpG-rich islands may be necessary, but therefore is not sufficient, for transcription. So the presence of unmethylated CpG islands may be taken as an indication that a gene is potentially active, rather than inevitably transcribed. Many islands that are nonmethylated in the animal become methylated in cell lines in tissue culture, and this could be connected with the inability of these lines to express all of the functions typical of the tissue from which they were derived.

Methylation of a CpG island at a promoter usually prevents expression of the gene. Repression is caused by either of two proteins that bind to methylated CpG sequences. The protein MeCP1 requires the presence of several methyl groups to bind to DNA, while MeCP2 can bind to a single methylated CpG base pair. This explains why a methylation-free zone is required for initiation of transcription. Binding of either these proteins prevents transcription *in vitro* by a nuclear extract.

Figure 21.30 The typical density of CpG doublets in mammalian DNA is ~1/100 bp, as seen for a γ-globin gene. In a CpG-rich island, the density is increased to >10 doublets/100 bp. The island in the APRT gene starts ~100 bp upstream of the promoter and extends ~400 bp into the gene. Each vertical line represents a CpG doublet.

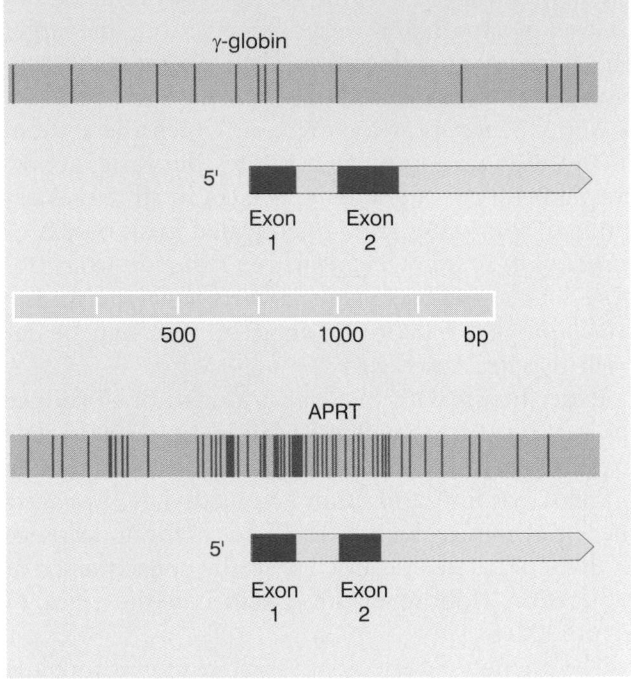

MeCP2, which directly represses transcription by interacting with complexes at the promoter, is bound also to the Sin3 repressor complex, which contains histone deacetylase activities (see Figure 21.20). This observation provides a direct connection between two types of repressive modifications: methylation of DNA and acetylation of histones.

The absence of methyl groups is associated with gene expression. However, there are some difficulties in supposing that the state of methylation provides a general means for controlling gene expression. In the case of *D. melanogaster* (and other Dipteran insects), there is no methylation of DNA. The other differences between inactive and active chromatin appear to be the same as in species that display methylation. So in *Drosophila*, any role that methylation has in vertebrates is replaced by some other mechanism.

We have described three changes that occur in active genes:

- A hypersensitive site(s) is established near the promoter.
- The nucleosomes of a domain including the transcribed region become more sensitive to DNAase I.
- The DNA of the same region is undermethylated.

All of these changes are necessary for transcription.

Summary

Some regulatory promoter elements are present in many genes and are recognized by ubiquitous factors; others are present in a few genes and are recognized by tissue-specific factors. Elements that uniquely identify particular groups of genes that are regulated in response to certain transcription factors are called response elements (REs). Binding of factors to specific sequences is followed by protein-protein interactions with other components of the general transcription apparatus. Transcription factors often have a modular construction, in which there are independent domains responsible for binding to DNA and for activating transcription. The main function of the DNA-binding domain may be to tether the activating domain in the vicinity of the initiation complex.

Several groups of transcription factors have been identified by sequence homologies. The homeodomain is a 60 residue sequence found in genes that regulate development in insects and worms and in mammalian transcription factors. It is related to the prokaryotic helix-turn-helix motif and provides the motif by which the factors bind to DNA.

Another motif involved in DNA-binding is the zinc finger, which is found in proteins that bind DNA or RNA (or sometimes both). A finger has cysteine residues that bind zinc. One type of finger is found in multiple repeats in some transcription factors; another is found in single or double repeats in others.

Steroid receptors were the first members identified of a group of transcription factors in which the protein is activated by binding a small hydrophobic hormone. The activated factor becomes localized in the nucleus, and binds to its specific response element, where it activates transcription. The DNA-binding domain has zinc fingers.

The leucine zipper contains a stretch of amino acids rich in leucine that are involved in dimerization of transcription factors. An adjacent basic region is responsible for binding to DNA.

HLH (helix-loop-helix) proteins have amphipathic helices that are responsible for dimerization, adjacent to basic regions that bind to DNA.

Many transcription factors function as dimers, and it is common for there to be multiple members of a family that form homodimers and heterodimers. This creates the potential for complex combinations to govern gene expression. In some cases, a family includes inhibitory members, whose participation in dimer formation prevents the partner from activating transcription.

The existence of a preinitiation complex signals that the gene is in an "active" state, ready to be transcribed. The complex is stable, and may remain in existence through many cycles of replication. The ability to form a preinitiation complex could be a general regulatory mechanism. By binding to a promoter to make it possible for RNA polymerase in turn to bind, the factor in effect switches the gene on.

The variety of situations in which hypersensitive sites occur suggests that their existence reflects a general principle. *Sites at which the double helix initiates an activity are kept free of nucleosomes.* A transcription factor, or some other nonhistone protein concerned with the particular function of the site, modifies the properties of a short region of DNA so that nucleosomes are excluded. The structures formed in each situation need not necessarily be similar (except that each, by definition, creates a site hypersensitive to DNAase I).

Genes whose control regions are organized in nucleosomes usually are not expressed. In the absence of specific regulatory proteins, promoters and other regulatory regions are organized by histone octamers into a state in which they cannot be activated. This may explain the need for nucleosomes to be precisely positioned in the vicinity of a promoter, so that essential regulatory sites are appropriately exposed. Some transcription factors have the capacity to recognize DNA on the nucleosomal surface, and a particular positioning of DNA may be required for initiation of transcription.

Acetylation of histones occurs at both replication and transcription and could be necessary to form a less compact chromatin structure. Some coactivators, which connect transcription factors to the basal apparatus, have histone acetylase activity. Conversely, repressors may be associated with deacetylases.

Active chromatin and inactive chromatin are not in equilibrium. Sudden, disruptive events are needed to convert one to the other. Chromatin remodeling complexes have the ability to displace histone octamers by a mechanism that involves hydrolysis of ATP. A typical form of this chromatin remodeling is to displace one or more histone octamers from specific sequences of DNA, creating a boundary that results in the precise or preferential positioning of adjacent nucleosomes. Chromatin remodeling may also involve changes in the positions of nucleosomes.

A group of hypersensitive sites upstream of the cluster of β-globin genes forms a locus control region (LCR) that is required for expression of all of the genes in the domain. The ends of domains may be marked by sequences that block the transmission of regulatory effects. Domains may also possess sites for attachment to the nuclear matrix.

CpG islands contain concentrations of CpG doublets and often surround the promoters of constitutively expressed genes, although they are also found at the promoters of regulated genes. The island including a promoter must be unmethylated for that promoter to be able to initiate transcription. A specific protein binds to the methylated CpG doublets and prevents initiation of transcription.

The formation of heterochromatin occurs by proteins that bind to specific chromosomal regions (such as telomeres) and that interact with histones. The formation of an inactive structure may propagate along the chromatin thread from an initiation center. Similar events occur in silencing of the inactive yeast mating type loci. Repressive structures that are required to maintain the inactive states of particular genes are formed by the Pc-G protein complex in *Drosophila.* They share with heterochromatin the property of propagating from an initiation center.

Further reading

Reviews

Brown, D. D. (1984). The role of stable complexes that repress and activate eukaryotic genes. *Cell* **37**, 359–365.

Weintraub, H. (1985). Assembly and propagation of repressed and derepressed chromosomal states. *Cell* **42**, 705–711.

Yamamoto, K. R. (1985). Steroid receptor regulated transcription of specific genes and gene networks. *Ann. Rev. Genet.* **19**, 209–252.

Guarente, L. (1987). Regulatory proteins in yeast. *Ann. Rev. Genet.* **21**, 425–452.

Evans, R. M. (1988). The steroid and thyroid hormone receptor superfamily. *Science* **240**, 889–895.

Ptashne, M. (1988). How eukaryotic transcriptional activators work. *Nature* **335**, 683–689.

Evans, R. M. and Arriza, J. L. (1989). A molecular framework for the actions of glucocorticoid hormones in the nervous system. *Neuron* **2**, 1105–1112.

Grunstein, M. (1990). Histone function in transcription. *Ann. Rev. Cell Biol.* **6**, 643–678.

Harrison, S. C. (1991). A structural taxonomy of DNA-binding proteins. *Nature* **353**, 715–719.

Weintraub, H. *et al.* (1991). The *MyoD* gene family: nodal point during specification of the muscle cell lineage. *Science* **251**, 760–766.

Felsenfeld, G. (1992). Chromatin as an essential part of the transcriptional mechanism. *Nature* **355**, 219–224.

Pabo, C. T. and Sauer, R. T. (1992). Transcription factors: structural families and principles of DNA recognition. *Ann. Rev. Biochem.* **61**, 1053–1095.

Gehring, W. J. *et al.* (1994). Homeodomain-DNA recognition. *Cell* **78**, 211–223.

Tsai, M.-J. and O'Malley, B. W. (1994). Molecular mechanisms of action of steroid/thyroid receptor superfamily members. *Ann. Rev. Biochem.* **63**, 451–486.

Discoveries

Combinatorial associations of HLH proteins

Davis, R. L. *et al.* (1987). Expression of a single transfected cDNA converts fibroblasts to myoblasts. *Cell* **51,** 987–1000.

Davis, R. L. *et al.* (1990). The MyoD DNA binding domain contains a recognition code for muscle-specific gene activation. *Cell* **60,** 733–746.

Benezra, R. *et al.* (1990). The protein Id: a negative regulator of helix-loop-helix DNA-binding proteins. *Cell* **61,** 49–59.

Lassar, A. B. *et al.* (1991). Functional activity of myogenic HLH proteins requires hetero-oligomerization with E12/E47-like proteins *in vivo. Cell* **66,** 305–315.

Transcription factor structures

Miller, J. *et al.* (1985). Repetitive zinc binding domains in the protein transcription factor IIIA from *Xenopus* oocytes. *EMBO J.* **4,** 1609–1614.

Landschulz, W. H., Johnson, P. F., and McKnight, S. L. (1988). The leucine zipper: a hypothetical structure common to a new class of DNA binding proteins. *Science* **240,** 1759–1764.

Han, K., Levine, M. S., and Manley, J. L. (1989). Synergistic activation and repression of transcription by Drosophila homeobox proteins. *Cell* **56,** 573–583.

Murre, C., McCaw, P. S., and Baltimore, D. (1989). A new DNA binding and dimerization motif in immunoglobulin enhancer binding, daughterless, MyoD, and myc proteins. *Cell* **56,** 777–783.

Vinson, C. R., Sigler, P. B., and McKnight, S. L. (1989). Scissors-grip model for DNA recognition by a family of leucine zipper proteins. *Science* **246,** 911–916.

Pavletich, N. P. and Pabo, C. O. (1991). Zinc finger-DNA recognition: crystal structure of a Zif268-DNA complex at 2.1 Å. *Science* **252,** 809–817.

Wolberger, C. *et al.* (1991). Crystal structure of a MATα2 homeodomain-operator complex suggests a general model for homeodomain-DNA interactions. *Cell* **67,** 517–528.

Rastinejad, F., Perlmann, T., Evans, R. M., and Sigler, P. B. (1995). Structural determinants of nuclear receptor assembly on DNA direct repeats. *Nature* **375,** 203–211.

Activation of transcription

Bogenhagen, D. F., Wormington, W. M., and Brown, D. D. (1982). Stable transcription complexes of *Xenopus* 5S RNA genes: a means to maintain the differentiated state. *Cell* **28,** 413–421.

Workman, J. L. and Roeder, R. G. (1987). Binding of transcription factor TFIID to the major late promoter during *in vitro* nucleosome assembly potentiates subsequent initiation by RNA polymerase II. *Cell* **51,** 613–622.

Cairns, B. R., Kim, Y.-J., Sayre, M. H., Laurent, B. C., and Kornberg, R. (1994). A multisubunit complex containing the *SWI/ADR6, SWI2/1, SWI3, SNF5,* and *SNF6* gene products isolated from yeast. *Proc. Nat. Acad. Sci. USA* **91,** 1950–1954.

Franke, A., DeCamillis, M., Zink, D., Cheng, N., Brock, H. W., and Paro, R. (1992). *Polycomb* and *polyhomeotic* are constituents of a multimeric protein complex in chromatin of *D. melanogaster. EMBO J.* **11,** 2941–2950.

Peterson, C. L. and Herskowitz, I. (1992). Characterization of the yeast *SWI1, SWI2,* and *SWI3* genes, which encode a global activator of transcription. *Cell* **68,** 573–583.

Tamkun, J. W., Deuring, R., Scott, M. P., Kissinger, M., Pattatucci, A. M., Kaufman, T. C., and Kennison, J. A. (1992). *brahma*: a regulator of *Drosophila* homeotic genes structurally related to the yeast transcriptional activator SNF2/SWI2. *Cell* **68,** 561–572.

Kamakaka, R. T., Bulger, M., and Kadonaga, J. T. (1993). Potentiation of RNA polymerase II transcription by GAL4-VP16 during but not after DNA replication and chromatin assembly. *Genes Dev.* **7,** 1779–1795.

McPherson, C. E., Shim, E.-Y., Friedman, D. S., and Zaret, K. S. (1993). An active tissue-specific enhancer and bound transcription factors existing in a precisely positioned nucleosomal array. *Cell* **75,** 387–398.

Coté, J., Quinn, J., Workman, J. L., and Peterson, C. L. (1994). Stimulation of GAL4 derivative binding to nucleosomal DNA by the yeast SWI/SNF complex. *Science* **265,** 53–60.

Kwon, H., Imbaizano, A. N., Khavari, P. A., Kingston, R. E., and Green, M. R. (1994). Nucleosome disruption and enhancement of activator binding of human SWI/SNF complex. *Nature* **370,** 477–481.

Truss, M., Barstch, J., Schelbert, A., Hache, R. J. G., and Beato, M. (1994). Hormone induces binding of receptors and transcription factors to a rearranged nucleosome on the MMTV promoter *in vivo. EMBO J.* **14,** 1737–1751.

Tsukiyama, T., Becker, P. B., and Wu, C. (1994). ATP-dependent nucleosome disruption at a heat shock promoter mediated by binding of GAGA transcription factor. *Nature* **367,** 525–532.

Chen, H. *et al.* (1997). Nuclear receptor coactivator ACTR is a novel histone acetyltransferase and forms a multimeric activation complex with P/CAF and CP/p300. *Cell* **90,** 569–580.

Modification of chromatin

Brownell, J. E. *et al.* (1996). *Tetrahymena* histone acetyltransferase A: a homologue to yeast Gcn5p linking histone acetylation to gene activation. *Cell* **84,** 843–851.

Schnitzler, G., Sif, S., and Kingston, R. E. (1998). Human SWI/SNF interconverts a nucleosome between its base state and a stable remodeled state. *Cell* **94,** 17–27.

Ogryzko, V. V. *et al.* (1998). Histone-like TAFs within the PCAF histone acetylase complex. *Cell* **94,** 35–44.

Grant, P. A. *et al.* (1998). A subset of TAF_{II}s are integral components of the SAGA complex required for nucleosome acetylation and transcriptional stimulation. *Cell* **94,** 45–53.

Repression of chromatin

Zink, B. and Paro, R. (1989). *In vivo* binding patterns of a *trans*-regulator of the homeotic genes in *D. melanogaster. Nature* **337,** 468–471.

Eissenberg, J. C., James, T. C., Fister-Hartnett, D. M., Hartnett, T., Ngan, V., and Elgin, S. C. R. (1990). Mutation in a heterochromatin-specific chromosomal protein is associated with suppression of position-effect variegation in *D. melanogaster. Proc. Nat. Acad. Sci. USA* **87,** 9923–9927.

Orlando, V. and Paro, R. (1993). Mapping Polycomb-repressed domains in the *bithorax* complex using *in vivo* formaldehyde cross-linked chromatin. *Cell* **75,** 1187–1198.

Chan, C.-S., Rastelli, L., and Pirrotta, V. (1994). A *Polycomb* response element in the *Ubx* gene that determines an epigenetically inherited state of repression. *EMBO J.* **13,** 2553–2564.

Horlein, A. J. *et al.* (1995). Ligand-independent repression by the thyroid hormone receptor mediated by a nuclear receptor corepressor. *Nature* **377,** 397–404.

Strutt, H., Cavalli, G., and Paro, R. (1997). Colocalization of Polycomb protein and GAGA factor on regulatory elements responsible for the maintenance of homeotic gene expression. *EMBO J.* **16,** 3621–3632.

Insulators

Kellum, R. and Schedl, P. (1991). A position-effect assay for boundaries of higher order chromosomal domains. *Cell* **64,** 941–950.

Geyer, P. K. and Corces, V. G. (1992). DNA position-specific repression of transcription by a *Drosophila* zinc finger protein. *Genes Dev.* **6,** 1865–1873.

Chung, J. H., Whiteley, M., and Felsenfeld, G. (1993). A 5′ element of the chicken β-globin domain serves as an insulator in human erythroid cells and protects against position effect in *Drosophila. Cell* **74,** 505–514.

Roseman, R. R., Pirrotta, V., and Geyer, P. K. (1993). The su(Hw) protein insulates expression of the *D. melanogaster white* gene from chromosomal position-effects. *EMBO J.* **12,** 435–442.

Hagstrom, K., Muller, M., and Schedl, P. (1996). *Fab-7* functions as a chromatin domain boundary to ensure proper segment specification by the *Drosophila bithorax* complex. *Genes Dev.* **10,** 3202–3215.

Mihaly, J. *et al.* (1997). In situ dissection of the *Fab-7* region of the bithorax complex into a chromatin domain boundary and a *Polycomb*-response element. *Development* **124,** 1809–1820.

Harrison, D. A., Gdula, D. A., Cyne, R. S., and Corces, V. G. (1993). A leucine zipper domain of the suppressor of hairy-wing protein mediates its repressive effect on enhancer function. *Genes Dev.* **7,** 1966–1978.

Methylation

Bird, A. *et al.* (1985). A fraction of the mouse genome that is derived from islands of nonmethylated, Cp-G-rich DNA. *Cell* **40,** 91–99.

Boyes, J. and Bird, A. (1991). DNA methylation inhibits transcription indirectly via a methyl-CpG binding protein. *Cell* **64,** 1123–1134.

Chapter 22

Nuclear splicing

INTERRUPTED genes are found in all classes of organisms. They represent a minor proportion of the genes of the very lowest eukaryotes, but the vast majority of genes in higher eukaryotic genomes. Genes vary widely according to the numbers and lengths of introns, but a typical mammalian gene has 7–8 exons spread out over ~16 kb. The exons are relatively short (~100–200 bp), and the introns are relatively long (>1 kb) (see Chapter 2).

The discrepancy between the interrupted organization of the gene and the uninterrupted organization of its mRNA requires processing of the primary transcription product. The primary transcript has the same organization as the gene, and is sometimes called the **pre-mRNA.** Removal of the introns from pre-mRNA leaves a typical messenger of ~2.2 kb. The process by which the introns are removed is called **RNA splicing.**

One of the first clues about the nature of the discrepancy in size between nuclear genes and their products in higher eukaryotes was provided by the properties of nuclear RNA. Its average size is much larger than mRNA, it is very unstable, and it has a much greater sequence complexity. Taking its name from its broad size distribution, it was called **heterogeneous nuclear RNA (hnRNA).** It includes pre-mRNA, but could also include other transcripts (that is, those which are not ultimately processed to mRNA).

The physical form of hnRNA is a ribonucleoprotein particle (**hnRNP**), in which the hnRNA is bound by proteins. As characterized *in vitro,* an hnRNP particle takes the form of beads connected by a fiber. The structure is summarized in **Figure 22.1.** The most abundant proteins in the particle are the core proteins, but other proteins are present at lower stoichiometry, making a total of ~20 proteins. The proteins typically are present at ~10^8 copies per nucleus, compared with ~10^6 molecules of hnRNA. The exact structure of the hnRNP and

Figure 22.1 hnRNA exists as a ribonucleo-protein particle organized as a series of beads.

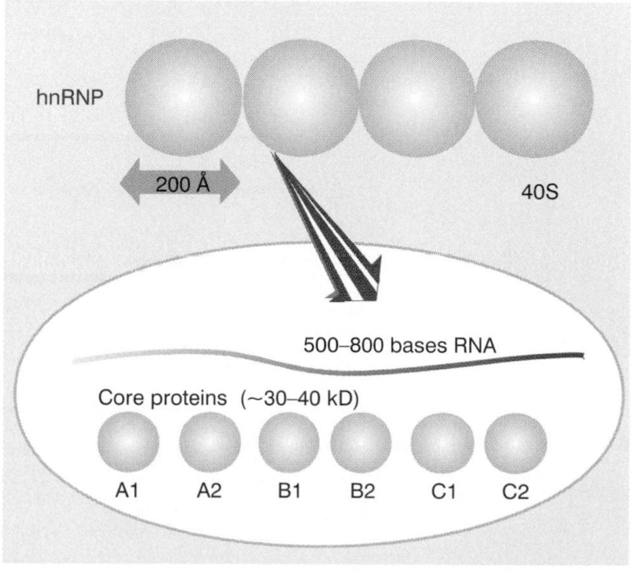

Figure 22.2 *Overview*: RNA is modified in the nucleus by additions to the 5' and 3' ends and by splicing to remove the introns. The splicing event requires breakage of the exon-intron junctions and joining of the ends of the exons; the expanded illustration shows the principle schematically, but not the actual order of events. Mature mRNA is transported through nuclear pores to the cytoplasm, where it is translated.

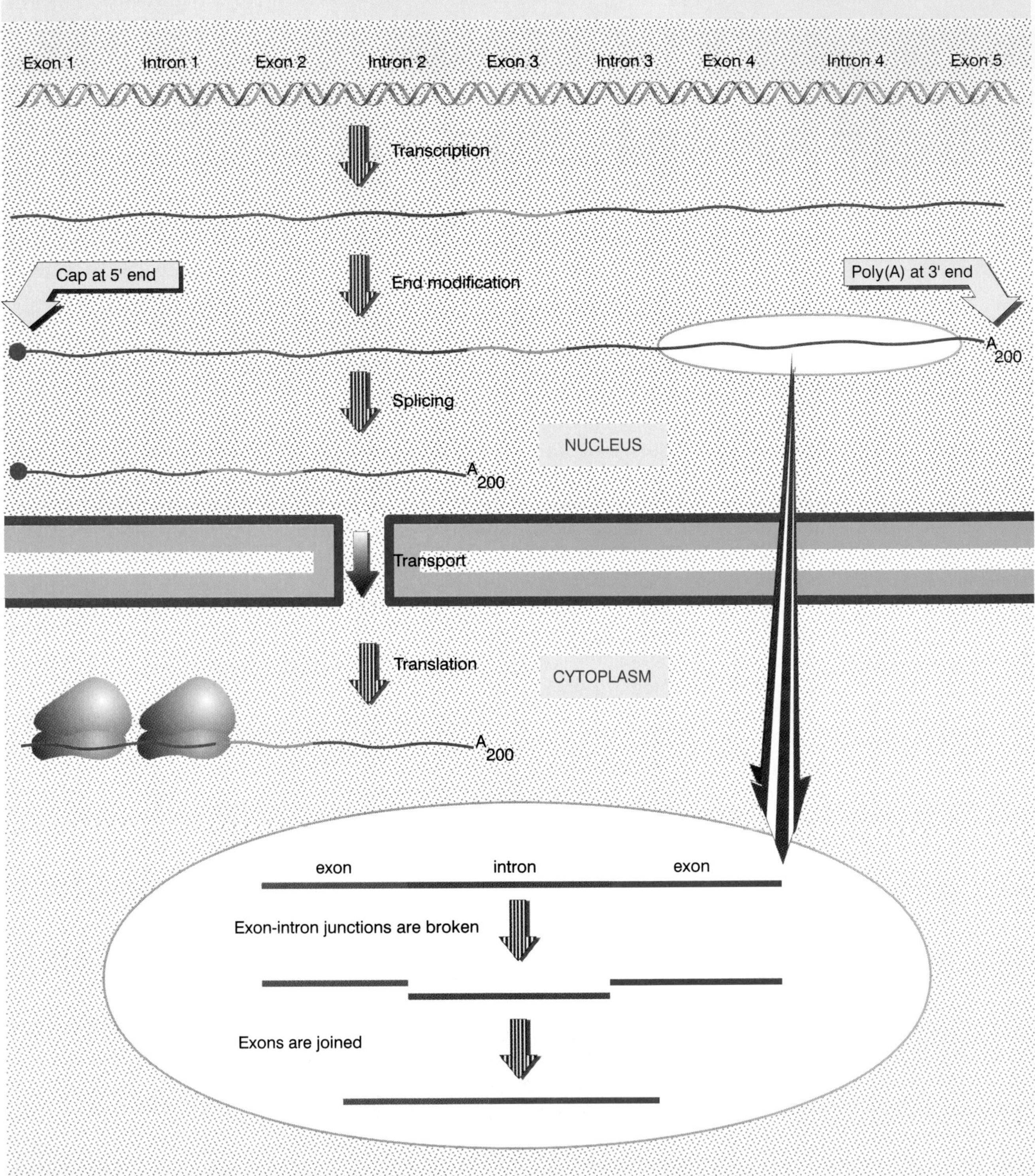

the functional implications of RNA's packaging in this manner remain to be determined.

Splicing occurs in the nucleus, together with the other modifications that are made to newly synthesized RNAs. The process of expressing an interrupted gene is reviewed in **Figure 22.2.** The transcript is capped at the 5′ end (as we saw in Chapter 5), has the introns removed (as we see in this chapter), and is polyadenylated at the 3′ end (this chapter). The RNA is then transported through nuclear pores to the cytoplasm, where it is available to be translated.

With regard to the various processing reactions that occur in the nucleus, we should like to know at what point splicing occurs *vis-à-vis* the other modifications of RNA. Does splicing occur at a particular location in the nucleus; and is it connected with other events, for example nucleocytoplasmic transport? Does the lack of splicing make an important difference in the expression of uninterrupted genes?

With regard to the splicing reaction itself, one of the main questions is how its specificity is controlled. What ensures that the ends of each intron are recognized in pairs so that the correct sequence is removed from the RNA? Are introns excised from a precursor in a particular order? Is the maturation of RNA used to *regulate* gene expression by discriminating among the available precursors or by changing the pattern of splicing?

We can identify several types of splicing systems:

- Introns are removed from the nuclear RNAs of higher eukaryotes by a system that recognizes only short consensus sequences conserved at exon-intron boundaries and within the intron. This reaction requires a large splicing apparatus, which takes the form of an array of proteins and ribonucleoproteins that functions as a large particulate complex (the spliceosome).

- Excision of certain introns is an autonomous property of the RNA itself. Two groups of introns with this capacity are found in diverse locations. Each forms a characteristic type of secondary/tertiary structure. The sequences of one group are related to those of nuclear introns. The ability of RNA to show enzymatic activities is seen also in the self-cleavage of viroid RNAs and the catalytic activity of RNAase P. We discuss these reactions in Chapter 23.
- The removal of introns from yeast nuclear tRNA precursors involves enzymatic activities whose dealings with the substrate resemble those of the tRNA processing enzymes, in which a critical feature is the conformation of the tRNA precursor.

Two fundamental types of mechanism are employed in splicing. All introns except those of nuclear pre-tRNAs are excised by transesterification reactions, although the catalytic and other components that are involved are distinct in each case. The introns of nuclear pre-tRNA are removed by cleavage and ligation.

Many of the introns with autonomous capacity to splice are mobile, that is, they have the ability to insert copies at new locations. This implies that they are likely to have originated by insertion into pre-existing genes (see Chapter 23). It remains speculative whether introns of higher eukaryotic nuclear genes originated by insertion or were part of the original construction of the gene. It seems inevitable that introns of nuclear pre-tRNA genes must have originated by insertion into pre-existing genes.

Nuclear splice junctions are interchangeable but are read in pairs

To focus on the molecular events involved in nuclear intron splicing, we must consider the nature of the **splice sites,** the two exon-intron boundaries that include the sites of breakage and reunion.

By comparing the nucleotide sequence of mRNA with that of the structural gene, the junctions between exons and introns can be assigned. There is no extensive homology or complementarity between the two ends of an intron. However, the junctions have well-conserved, though rather short, consensus sequences.

It is possible to assign a specific end to every intron by relying on the conservation of exon-intron junctions. They can all be aligned to conform to the

Figure 22.3 The ends of nuclear introns are defined by the GT-AG rule.

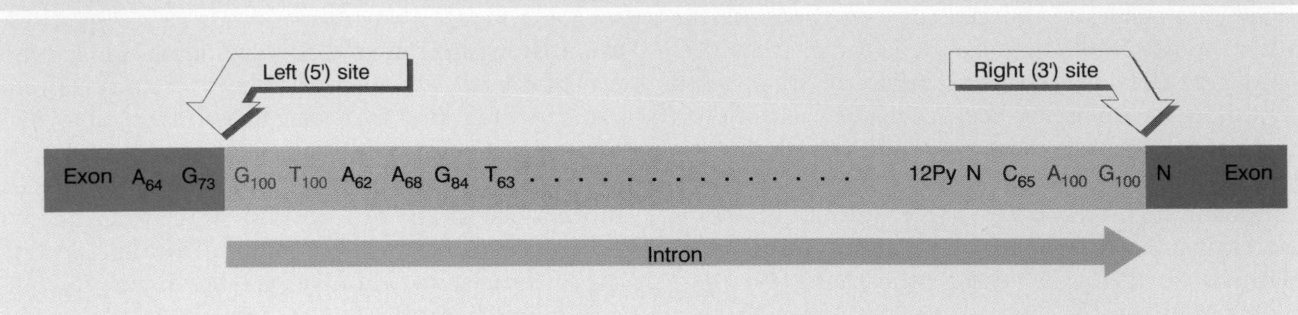

Figure 22.4 Splicing junctions are recognized only in the correct pairwise combinations.

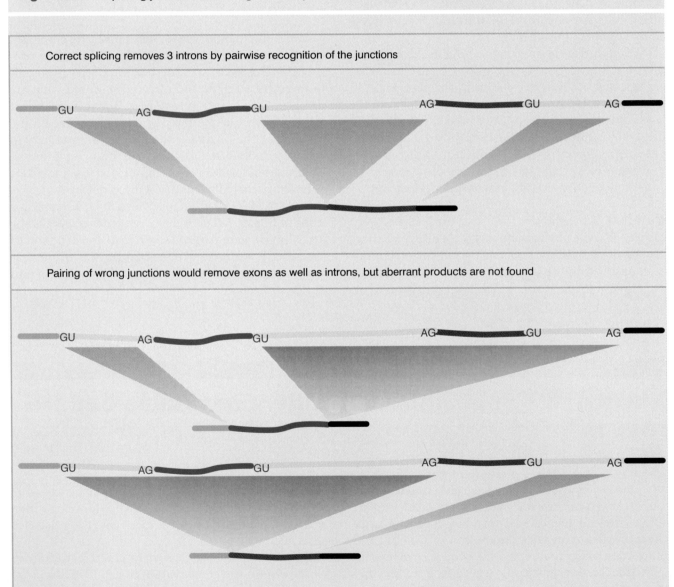

Correct splicing removes 3 introns by pairwise recognition of the junctions

GU AG GU AG GU AG

Pairing of wrong junctions would remove exons as well as introns, but aberrant products are not found

GU AG GU AG GU AG

GU AG GU AG GU AG

consensus sequence given in **Figure 22.3.** (In this as in other cases, we write just the sequence of the DNA strand that is identical with the RNA product.)

The subscripts indicate the percent occurrence of the specified base (or type of base) at each consensus position. High conservation is found only *immediately within the intron* at the presumed junctions. This identifies the sequence of a generic intron as:

GT... ...AG

Because the intron defined in this way starts with the dinucleotide GT and ends with the dinucleotide AG, the junctions are often described as conforming to the **GT-AG rule.** (The actual sequences in the RNA are of course GU-AG.)

Note that the two sites have different sequences and so they define the ends of the intron *directionally.* They are named proceeding from left to right along the intron, that is, as the **left (or 5′)** and **right (or 3′)** splice sites. Sometimes they are called the **donor** and **acceptor** sites. The consensus sequences are implicated as the sites recognized in splicing by point mutations that prevent splicing *in vivo* and *in vitro.*

A typical mammalian mRNA has many introns. The basic problem of nuclear splicing results from the simplicity of the splice sites, and is illustrated in **Figure 22.4:** what ensures that the correct pairs of sites are spliced together? The corresponding GU-AG pairs must be connected across great distances (some introns are >10 kb long), avoiding the connection of incorrect pairs. We can imagine two types of principle that might be responsible for pairing the appropriate 5′ and 3′ sites:

- It could be an *intrinsic property* of the RNA to connect the sites at the ends of a particular intron. This would require matching of specific sequences or structures.
- Or all 5′ sites may be functionally equivalent and all 3′ sites may be similarly indistinguishable, but splicing could follow *rules* that ensure a 5′ site is always connected to the 3′ site that comes next in the RNA.

Neither the splice sites nor the surrounding regions have any sequence complementarity, which excludes models for pairing between intron ends. Experiments to construct hybrid RNA precursors show that any 5′ splice site can in principle be connected to any 3′ splice site. For example, when the first exon of the early SV40 transcription unit is linked to the third exon of mouse β globin, the hybrid intron can be excised to generate a perfect connection between the SV40 exon and the β-globin exon. Such experiments make two general points:

- *Splice sites are generic:* they do not have specificity for individual RNA precursors, and individual precursors do not convey specific information (such as secondary structure) that is needed for splicing.
- *And the apparatus for splicing is not tissue specific;* an RNA can usually be properly spliced by any cell, whether or not it is usually synthesized in that cell. (We

Figure 22.5 Northern blotting of nuclear RNA with an ovomucoid probe identifies discrete precursors to mRNA. The contents of the more prominent bands are indicated. Photograph kindly provided by Bert O'Malley.

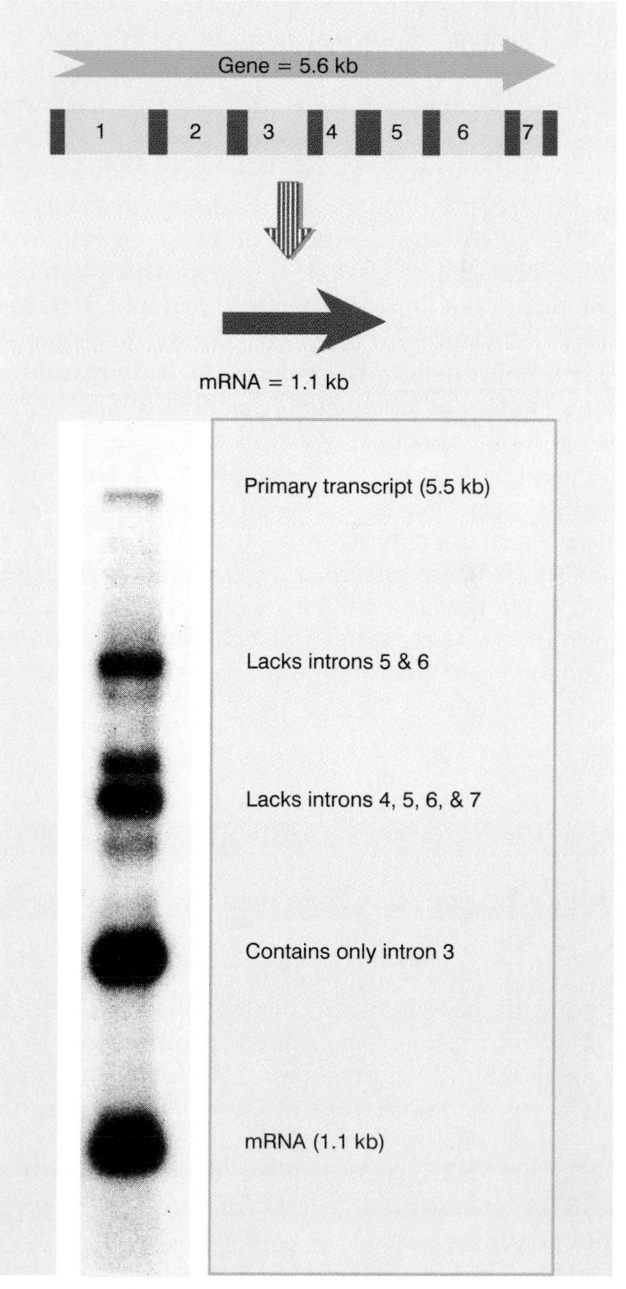

discuss exceptions in which there are tissue-specific alternative splicing patterns later.)

Here is a paradox. Probably all 5′ splice sites look similar to the splicing apparatus, and all 3′ splice sites look similar to it. *In principle any 5′ splice site may be able to react with any 3′ splice site.* But in the usual circumstances splicing occurs only between the 5′ and 3′ sites of the *same* intron. *What rules ensure that recognition of splice sites is restricted so that only the 5′ and 3′ sites of the same intron are spliced?*

Are introns removed in a specific *order* from a particular RNA? Using RNA blotting, we can identify nuclear RNAs that represent intermediates from which some introns have been removed. **Figure 22.5** shows a blot of the precursors to ovomucoid mRNA. There is a discrete series of bands, which suggests that splicing occurs via definite pathways. (If the seven introns were removed in an entirely random order, there would be more than 300 precursors with different combinations of introns, and we should not see discrete bands.)

There does not seem to be an *obligatory* pathway, since intermediates can be found in which different combinations of introns have been removed. However, there is evidence for a *preferred* pathway or pathways. When only one intron has been lost, it is virtually always 5 or 6. But either can be lost first. When two introns have been lost, 5 and 6 are again the most frequent, but there are other combinations. Intron 3 is never or very rarely lost at one of the first three splicing steps. From this pattern, we see that there is a preferred pathway in which introns are removed in the order 5/6, 7/4, 2, 3/1. But clearly there are other pathways, as, for example, there are some molecules in which 4 or 7 is lost last. A caveat in interpreting these results is that we do not have proof that all these intermediates actually lead to mature mRNA.

The general conclusion suggested by this analysis is that the conformation of the RNA influences the accessibility of the splice sites. As particular introns are removed, the conformation changes, and new pairs of splice sites become available. But the ability of the precursor to remove its introns in more than one order suggests that alternative conformations are available at each stage. Of course, the longer the molecule, the more structural options become available; and when we consider larger genes, it becomes difficult to see how specific secondary structures could control the reaction. One important conclusion of this analysis is that the reaction does not proceed sequentially along the precursor.

A simple model to control recognition of splice sites would be for the splicing apparatus to act in a processive manner. Having recognized a 5′ site, the apparatus might be compelled to scan the RNA in the appropriate direction until it meets the next 3′ site. This would restrict splicing to adjacent sites. But this model is excluded by experiments that show that splicing can occur in *trans* as an intermolecular reaction under special circumstances (see later) or in RNA molecules in which part of the nucleotide chain is replaced by a polyethanolglycol linker. This means that there cannot be a requirement for strict scanning along the RNA from the 5′ splice site to the 3′ splice site. Another problem with the scanning model is that it cannot explain the existence of alternative splicing patterns, such as those seen in SV40, where a common 5′ site is spliced to more than one 3′ site. The basis for proper recognition of correct splice site pairs remains unknown.

Nuclear splicing proceeds through a lariat

THE mechanism of splicing has been characterized *in vitro,* using systems in which introns can be removed from RNA precursors. Nuclear extracts can splice purified RNA precursors, which shows that the action of splicing is not linked to the process of transcription. Splicing is also independent of modification of RNA, and can occur to RNAs that are neither capped nor polyadenylated.

The stages of splicing *in vitro* are illustrated in the pathway of **Figure 22.6.** We discuss the reaction in terms of the individual RNA species that can be identified, but we should remember that *in vivo* the species containing exons are not released as free molecules, but remain held together by the splicing apparatus.

In the first stage, a cut is made at the 5′ splice site, separating the left exon and the right intron-exon molecule. The left exon takes the form of a linear molecule. The right intron-exon molecule forms a **lariat,** in which the 5′ terminus generated at the end of the intron becomes linked by a 5′–2′ bond to a base within the intron. The target base is an A in a sequence that is called the **branch site.**

In the second stage, cutting at the 3′ splice site releases the free intron in lariat form, while the right exon is ligated (spliced) to the left exon. The cleavage and ligation reactions are shown separately in the figure for illustrative purposes, but actually occur as one coordinated transfer.

The lariat is then "debranched" to give a linear excised intron, which is rapidly degraded.

Figure 22.6 Splicing occurs in two stages, in which the 5' exon is separated and then is joined to the 3' exon.

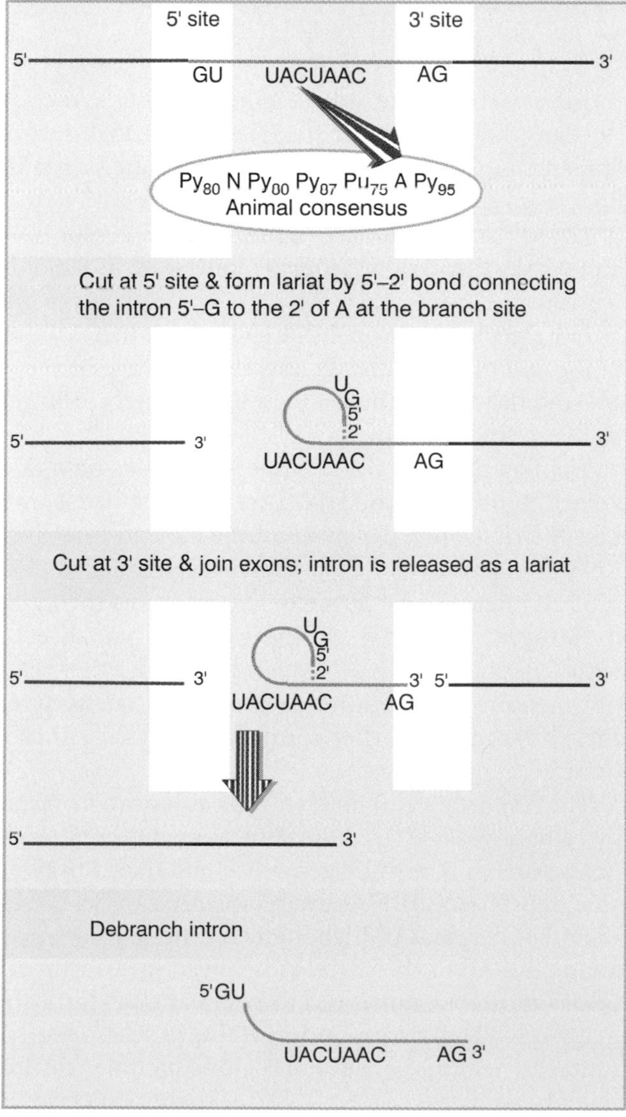

The sequences needed for splicing are the short consensus sequences at the 5′ and 3′ splice sites and at the branch site. Together with the knowledge that most of the sequence of an intron can be deleted without impeding splicing, this indicates that there is no demand for specific conformation in the intron (or exon).

The branch site provides the means by which the 3′ splice site is identified. The branch site in yeast is highly conserved, and has the consensus sequence UACUAAC. The branch site in higher eukaryotes is not well conserved, but has a preference for purines or pyrimidines at each position and retains the target A nucleotide (see Figure 22.6).

The branch site lies 18–40 nucleotides upstream of the 3′ splice site. Mutations or deletions of the branch site in yeast prevent splicing. In higher eukaryotes, the relaxed constraints in its sequence result in the ability to use related sequences in the vicinity when the authentic branch is deleted. Proximity to the 3′ splice site appears to be important, since the cryptic site is always close to the authentic site. When a cryptic branch sequence is used in this manner, splicing otherwise appears to be normal; and the exons give the same

Figure 22.7 Nuclear splicing occurs by two transesterification reactions in which a free OH end attacks a phosphodiester bond.

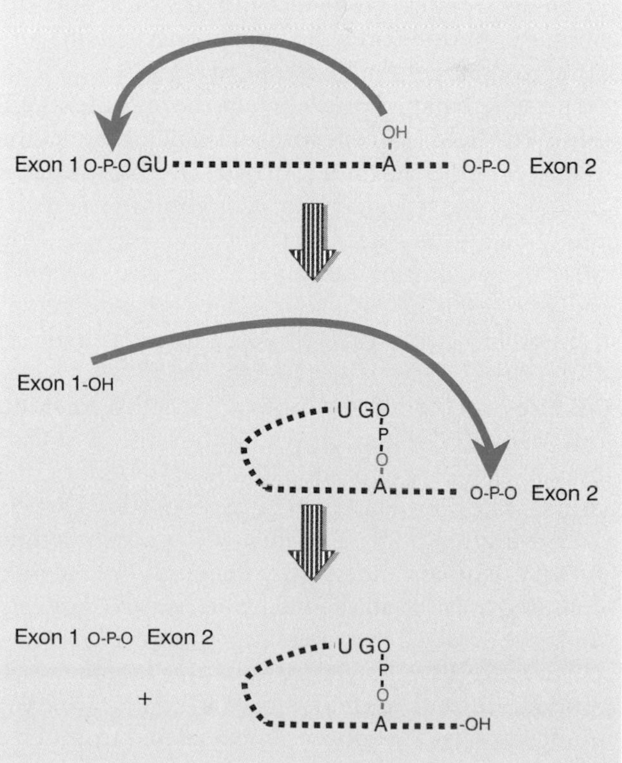

products as wild type. *The role of the branch site therefore is to identify the nearest 3′ splice site as the target for connection to the 5′ splice site.*

The bond that forms the lariat goes from the 5′ position of the invariant G that was at the 5′ end of the intron to the 2′ position of the invariant A in the branch site. This corresponds to the third A residue in the yeast UACUAAC box.

The chemical reactions proceed by **transesterification:** a bond is in effect *transferred* from one location to another. **Figure 22.7** shows that the first step is a nucleophilic attack by the 2′-OH of the invariant A of the UACUAAC sequence on the 5′ splice site. In the second step, the free 3′-OH of the exon that was released by the first reaction now attacks the bond at the 3′ splice site. Note that the number of phosphodiester bonds is conserved. There were originally two 5′–3′ bonds at the exon-intron splice sites; one has been replaced by the 5′–3′ bond between the exons, and the other has been replaced by the 5′–2′ bond that forms the lariat.

The spliceosome contains snRNAs

The 5′ and 3′ splice sites and the branch sequence are recognized by components of the splicing apparatus that assemble to form a large complex. This complex brings together the consensus sequences before any reaction occurs, explaining why a deficiency in any one of the sites may prevent the reaction from initiating. The complex assembles sequentially, and several intermediates can be recognized by fractionating complexes of different sizes. Splicing occurs only after all the components have assembled.

The splicing apparatus contains both proteins and RNAs. The RNAs take the form of small molecules that exist as ribonucleoprotein particles. Both the nucleus and cytoplasm of eukaryotic cells contain many discrete small RNA species. They range in size from 100–300 bases in higher eukaryotes, and extend in length to ~1000 bases in yeast. They vary considerably in abundance, from 10^5–10^6 molecules per cell to concentrations too low to be detected directly.

Those restricted to the nucleus are called **small nuclear RNAs (snRNAs)**; those found in the cytoplasm are called **small cytoplasmic RNAs (scRNAs)**. In their natural state, they exist as ribonucleoprotein particles (snRNP and scRNP). Colloquially, they are sometimes known as **snurps** and **scyrps.** There is also a class of small RNAs found in the nucleolus, called SnoRNAs, which are involved in processing ribosomal RNA.

The snRNPs involved in splicing, together with some additional proteins, form a large particulate complex, called the **spliceosome.** Isolated from the *in vitro* splicing systems, it comprises a 50–60S ribonucleoprotein particle. The spliceosome may be formed in stages as the snRNPs join, proceeding through several "presplicing complexes." The spliceosome is a large body, equivalent in size to a ribosomal subunit.

The spliceosome forms on the intact precursor RNA and passes through an intermediate state in which it contains the individual 5′ exon linear molecule and the right lariat-intron-exon. Little spliced product is found in the complex, which suggests that cleavage of the 3′ site and ligation of the exons is immediately followed by release of the product.

We may think of the snRNP particles as being involved in building the structure of the spliceosome. Like the ribosome, the spliceosome depends on RNA-RNA interactions as well as protein-RNA and protein-protein interactions. Some of the reactions involving the snRNPs require their RNAs to base pair directly with sequences in the RNA being spliced; other reactions require recognition between snRNPs or between their proteins and other components of the spliceosome.

The importance of snRNA molecules can be tested directly in yeast by making mutations in their genes. Mutations in 5 snRNA genes are lethal and prevent splicing. These snRNAs are relatively scarce in yeast; their counterparts in higher eukaryotic cells are more abundant. All of the snRNAs involved in splicing can be recognized in conserved forms in animal, bird, and insect cells. The corresponding RNAs in yeast are often rather larger, but conserved regions include features that are similar to the snRNAs of higher eukaryotes.

There are also many other snRNAs, and judging by the lack of effects of mutations in yeast, several of the more abundant snRNAs are dispensable.

The snRNPs involved in splicing are U1, U2, U5, U4, and U6. They are named according to the snRNAs that are present. Each snRNP contains a single snRNA and several (<20) proteins. The U4 and U6 snRNPs are usually found as a single (U4/U6) particle. The snRNPs together contain ~40 individual proteins. A common structural core for each snRNP consists of a group of 8 proteins, all of which are recognized by an autoimmune antiserum called **anti-Sm;** conserved sequences in the proteins form the target for the serum. The other proteins in each snRNP are unique to it. The Sm proteins bind to the conserved sequence PuAU$_{3-6}$Gpu, which is present in all snRNAs except U6. The U6 snRNP contains instead a set of Sm-like (Lsm) proteins. The Sm proteins must be involved in the autoimmune reaction, although their relationship to the phenotype of the autoimmune disease is not clear.

Some of the proteins in the snRNPs may be involved directly in splicing; others may be required in structural roles or just for assembly or interactions between the snRNP particles. Most of the proteins that are involved in splicing are components of the snRNPs; some additional proteins are included in the spliceosome and are generally described as "splicing factors."

Splicing can be broadly divided into two stages: first the consensus sequences are recognized and the complex assembles; then the cleavage and ligation reactions change the structure of the substrate RNA.

Recognition of the consensus sequences involves both RNAs and proteins. Certain snRNAs have sequences that are complementary to the consensus sequences or to one another, and base pairing between snRNA and pre-mRNA, or between snRNAs, plays an important role in splicing.

The human U1 snRNP contains 8 proteins as well as the RNA. The probable secondary structure of the U1 snRNA is drawn in **Figure 22.8.** It contains several domains. The Sm-binding site is required for interaction with the common snRNP proteins. Domains identified by the individual stem-loop structures provide binding sites for proteins that are unique to U1 snRNP.

U1 snRNA base pairs directly with the 5′ splice site. The 5′-terminal 11 nucleotides are single-stranded and include a stretch complementary to the consensus sequence at the 5′ site of the intron. The extent of complementarity between U1 snRNA and actual sites is usually 4–6 bp (because actual sites rarely conform perfectly to the consensus).

Mutational analysis can test directly whether it is necessary for U1 snRNA to pair with the 5′ splice site. The results of such an experiment are illustrated in **Figure 22.9.** The wild-type sequence of the splice site of the 12S adenovirus pre-mRNA pairs at 5 out of 6 positions with U1 snRNA. A mutant in the 12S RNA that cannot be spliced has two sequence changes; the GG residues at positions 5–6 in the intron are changed to AU. A particularly informative aspect of this mutant is that this sequence change does not alter the *overall* extent of pairing, because complementarity is lost at one position and gained at the other. This means that the total number of base pairs formed between U1 RNA and the 5′ splice site is not the sole determinant of efficiency, and pairing at some positions may be more important than others.

When a mutation is introduced into U1 snRNA that restores pairing at position 5, normal splicing is regained. Other cases in which corresponding mutations are made in U1 snRNA to see whether they can suppress the mutation in the splice site suggest the general rule: *complementarity between U1 snRNA and the 5′*

Figure 22.8 U1 snRNA has a base paired structure that creates several domains. The 5' end remains single-stranded and can base pair with the 5' splicing site.

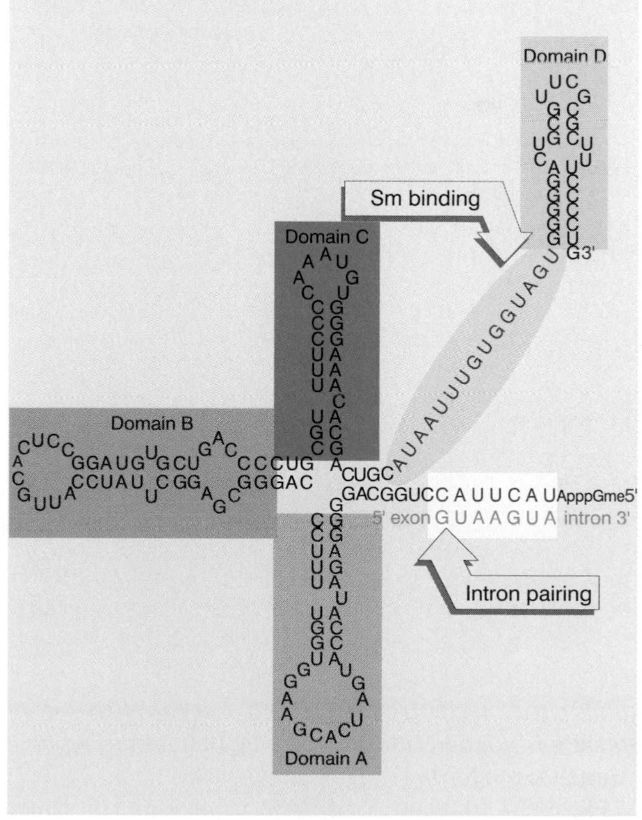

Figure 22.9 Mutations that abolish function of the 5' splicing site can be suppressed by compensating mutations in U1 snRNA that restore base pairing.

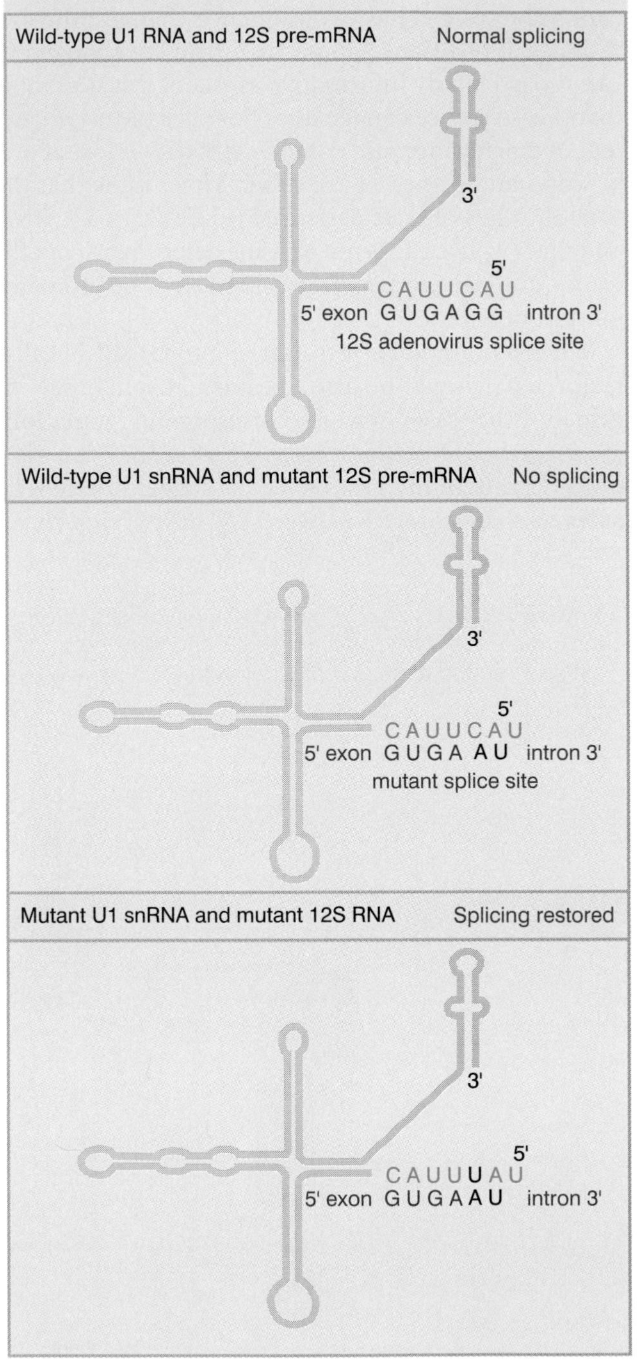

splice site is necessary for splicing, but the efficiency of splicing is not determined solely by the number of base pairs that can form.

Figure 22.10 relates the stages of splicing to the functions of the snRNPs. The first stage in splicing is formation of the E (early presplicing) complex, which contains U1 snRNP, the splicing factor U2AF, and some other proteins. The E complex is sometimes called the commitment complex, because its formation identifies a pre-mRNA as a substrate for formation of the splicing complex.

Binding of U1 snRNP by base pairing with the 5′ splice site is the first step. The recruitment of U1 snRNP involves an interaction between one of its proteins (U1-70k) and the protein ASF/SF2 (a general splicing factor in the SR class: see below). The U2AF splicing factor binds to a pyrimidine tract downstream of the branch site. The name of U2AF reflects its original isolation as the U2 auxiliary factor. Both U1 snRNP and U2AF are needed for U2 snRNP to bind to the branch site. The U2 snRNA includes sequences complementary to the branch site. A sequence near the 5′ end of the snRNA base pairs with the branch sequence in the intron. In yeast this typically involves formation of a 7 bp duplex with the entire UACUAAC box (see Figure 22.12 below). Several proteins of the U2 snRNP are bound to the substrate RNA just upstream of the branch site. The addition of U2 snRNP to the E complex generates the A presplicing complex. The binding of U2 snRNP requires ATP hydrolysis, and commits a pre-mRNA to the splicing pathway.

There may be more than one way to form the E complex. **Figure 22.11** illustrates some possibilities. The most direct reaction is for both splice sites to be recognized across the intron (shown on the left). The presence of U1 snRNP at the 5′ splice site is necessary for U2AF to bind near the branch site, making it possible that the 5′ and 3′ ends of the intron are brought together in this complex. The E complex is converted to the A complex when U2 snRNP binds at the branch site. *The basic feature of this route for splicing is that the two splice sites are recognized without requiring any sequences outside of the intron.*

The complex contains members of a family called "SR proteins," which comprise an important group of splicing factors and regulators. About 6 members are well characterized, but there are many others. They take their name from the presence of an Arg-Ser-rich region that is variable in length. They bind to RNA with low sequence specificity, but are an essential component of the spliceosome, possibly forming a framework on the RNA substrate. They connect U2AF to U1, as shown in the figure. In an extreme case, the SR proteins may enable U2AF/U2 snRNP to bind *in vitro* in the absence of U1, raising the possibility that there could be a U1-independent pathway for splicing.

Some 3′ splice sites are "weak" and do not bind U2AF and U2 snRNP effectively. Additional sequences are needed to bind the SR proteins, which assist U2AF in binding to the branch site. Such sequences are called "splicing enhancers," and they are most commonly found in the exon downstream of the 3′ splice site.

An alternative route to form the spliceosome may be followed when the introns are long and the splice sites

Figure 22.10 The splicing reaction proceeds through discrete stages in which spliceosome formation involves the interaction of components that recognize the consensus sequences.

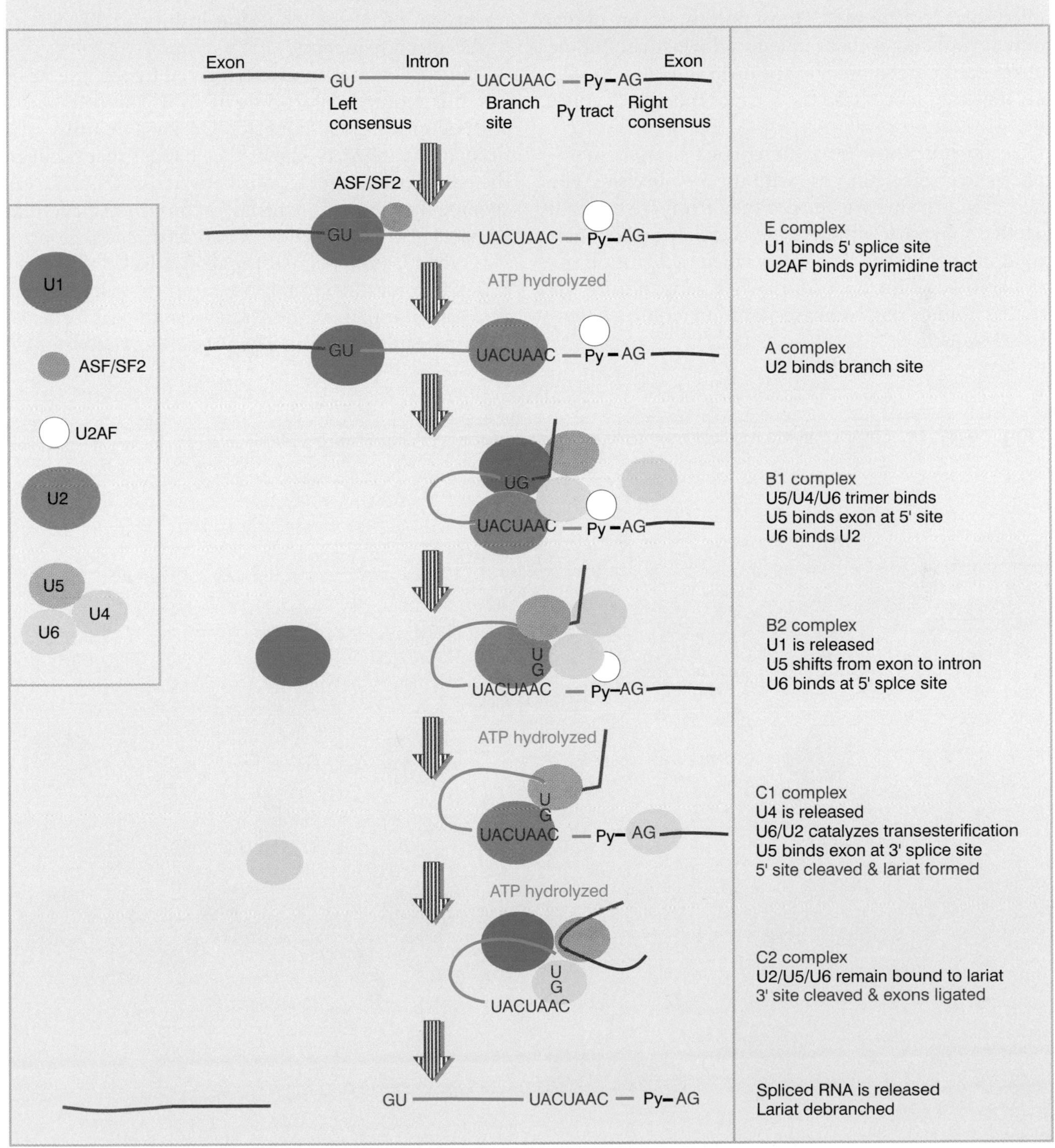

are weak. As shown on the right of the figure, the 5′ splice site is recognized by U1 snRNA in the usual way. However, the 3′ splice site is recognized as part of a complex that forms across the *next exon*, in which the next 5′ splice site is also bound by U1 snRNA. This U1 snRNA is connected by SR proteins to the U2AF at the branch site. When U2 snRNP joins to generate the A complex, there is a rearrangement, in which the correct (leftmost) 5′ splice site displaces the downstream 5′ splice site in the complex. The important feature of this route for splicing is that sequences downstream of the intron itself are required. In the form shown in the figure, these sequences take the form of the next 5′ splice site.

We do not know how the correct branch site/3′ splice site is selected to use with the 5′ splice site. The stage at which the two ends of the intron are brought together may vary with the type of splicing reaction. It could happen in the E complex when recognition occurs across the intron, but may not happen until the later B complex stage when recognition requires downstream sequences.

The B1 complex is formed when a trimer containing the U5 and U4/U6 snRNPs binds (see Figure 22.10). This complex is regarded as a spliceosome, since it contains the components needed for the splicing reaction. It is converted to the B2 complex when U1 is released. The dissociation of U1 is necessary to allow other components to come into juxtaposition with the 5′ splice site, most notably U6 snRNA. At this point U5 snRNA changes its position; initially it is close to exon sequences at the 5′ splice site, but it shifts to the vicinity of the intron sequences.

The catalytic reaction is triggered by the release of U4; this requires hydrolysis of ATP. The role of U4 snRNA may be to sequester U6 snRNA until it is needed. **Figure 22.12** shows the changes that occur in the base pairing interactions between snRNAs during splicing. In the U6/U4 snRNP, a continuous length of 26 bases of U6 is paired with two separated regions of U4. When U4 dissociates, the region in U6 that is released becomes free to take up another structure. The first part of it pairs with U2; the second part forms an intramolecular hairpin. The interaction between U4

Figure 22.11 There may be multiple routes for initial recognition of 5' and 3' splice sites.

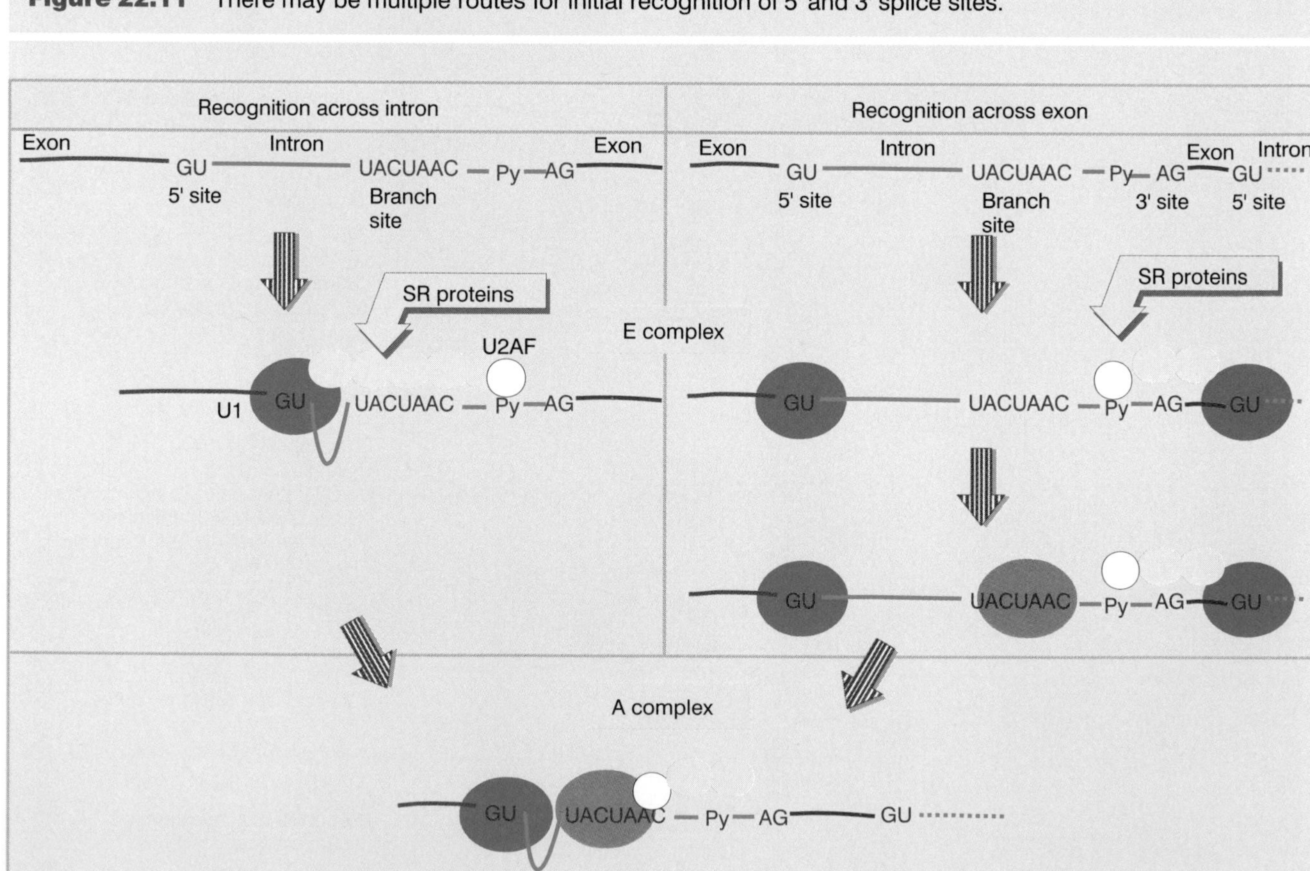

Figure 22.12 U6-U4 pairing is incompatible with U6–U2 pairing. When U6 joins the spliceosome it is paired with U4. Release of U4 allows a conformational change in U6; one part of the released sequence forms a hairpin (dark grey), and the other part (black) pairs with U2. Because an adjacent region of U2 is already paired with the branch site, this brings U6 into juxtaposition with the branch. Note that the substrate RNA is reversed from the usual orientation and is shown 3'–5'.

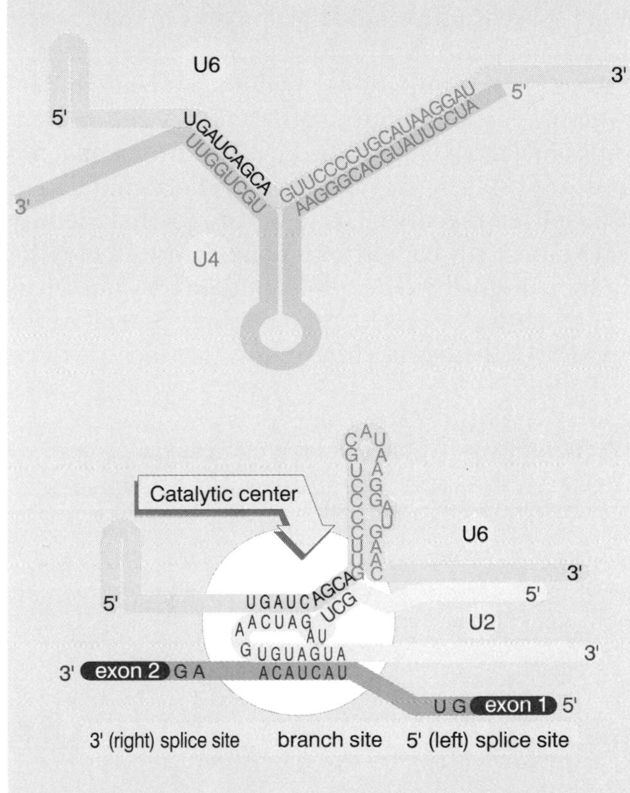

and U6 is mutually incompatible with the interaction between U2 and U6, and thus the release of U4 controls the ability of the spliceosome to proceed.

Although for clarity the figure shows the RNA substrate in extended form, the 5′ splice site is actually close to the U6 sequence immediately on the 5′ side of the stretch bound to U2. This sequence in U6 snRNA contacts the conserved GU at the 5′ splice site, and in some cases the pairing may extend to sequences downstream of the GU (mutations that enhance such pairing improve the efficiency of splicing).

We see therefore that several pairing reactions between snRNAs and the substrate RNA occur in the course of splicing; they are summarized in **Figure 22.13.** The snRNPs have sequences that pair with the substrate and with one another. They also have single-stranded regions in loops that are in close proximity to

Figure 22.13 Splicing utilizes a series of base pairing reactions between snRNAs and splice sites.

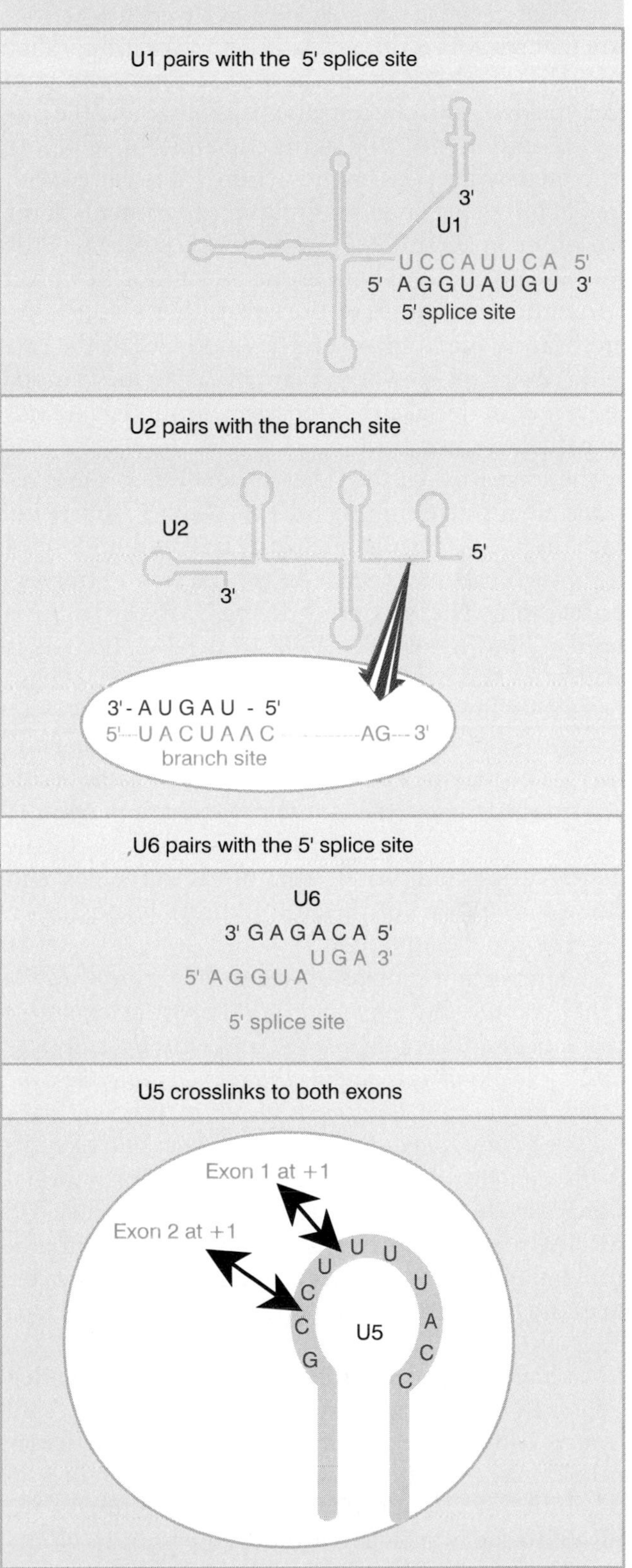

sequences in the substrate, and which play an important role, as judged by the ability of mutations in the loops to block splicing.

The base pairing between U2 and U6 creates a structure that resembles the active center of the group II introns that are described in the next section (see Figure 22.17 below). This suggests the possibility that the catalytic component could comprise an RNA structure generated by the U2-U6 interaction. U6 is paired with the 5′ splice site, and crosslinking experiments show that a loop in U5 snRNA is immediately adjacent to the first base positions in both exons. But although we can define the proximities of the substrate (5′ splice site and branch site) and snurps (U2 and U6) at the catalytic center (as shown in Figure 22.12), the components that undertake the transesterifications have not been directly identified.

The formation of the lariat at the branch site is responsible for determining the use of the 3′ splice site, since the 3′ consensus sequence nearest to the 3′ side of the branch becomes the target for the second transesterification. The second splicing reaction follows rapidly. Binding of U5 snRNP to the 3′ splice site is needed for this reaction, but we do not know whether the snRNP functions by using its RNA component to recognize the consensus sequence by base pairing. There is no region of complementarity available in single-stranded form. Perhaps the protein components of the snRNP are involved. This idea is supported by the identification of a protein that binds to the site, and which may be part of the snRNP particle, possibly a loosely bound component.

The important conclusion suggested by these results is that *the snRNA components of the splicing apparatus interact both among themselves and with the substrate RNA by means of base pairing interactions, and these interactions allow for changes in structure that may bring reacting groups into apposition and may even create catalytic centers.* Furthermore, the conformational changes in the snRNAs are reversible; for example, U6 snRNA is not used up in a splicing reaction, and at completion must be released from U2, so that it can reform the duplex structure with U4 to undertake another cycle of splicing.

We have described individual reactions in which each snRNP participates, but as might be expected from a complex series of reactions, any particular snRNP may play more than one role in splicing. So the ability of U1 snRNP to promote binding of U2 snRNP to the branch site is independent of its ability to bind to the 5′ splice site. Similarly, different regions of U2 snRNA can be defined that are needed to bind to the branch site and to interact with other splicing components.

Figure 22.14 shows an electron micrograph of mammalian spliceosomes, which contain U1, U2, U4, U5, and U6 snRNPs, and additional proteins. The spliceosomes appear to be particles ~25 nm × 50 nm, possibly consisting of discrete domains connected by rigid structures. Individual snRNP particles have diameters >8 nm, so that four types of snRNP could between them account for most of the mass and volume of the spliceosome. Could the domains in the particle be individual snRNPs?

An extensive mutational analysis has been undertaken in yeast to identify both the RNA and protein components of the spliceosome. Mutations in genes needed for splicing are identified by the accumulation of unspliced precursors. A series of loci that identify genes potentially coding for proteins involved in splicing were originally called *RNA,* but are now known as *PRP* mutants (for pre-RNA processing). Several of the products of these genes have motifs that identify them

Figure 22.14 Spliceosomes are ellipsoidal particles with several discrete regions. The bar is 50 nm. Photograph kindly provided by Tom Maniatis.

as RNA-binding proteins, and some appear to be related to a family of ATP-dependent RNA helicases. We suppose that, in addition to RNA-RNA interactions, protein-RNA interactions are important in creating or releasing structures in the pre-mRNA or snRNA components of the spliceosomes.

Some of the PRP proteins are components of snRNP particles, but others function as independent factors. One interesting example is PRP16, a helicase that hydrolyzes ATP, and associates transiently with the spliceosome to participate in the second catalytic step. Another example is PRP22, another ATP-dependent helicase, which is required to release the mature mRNA from the spliceosome. The conservation of bonds during the splicing reaction means that input of energy is not required to drive bond formation *per se*, which implies that the ATP hydrolysis is required for other purposes. The use of ATP by PRP16 and PRP22 may be examples of a more general phenomenon: the use of ATP hydrolysis to drive conformational changes that are needed to proceed through splicing.

In addition to the predominant class of GT-AG introns, there is a minor class of introns marked by the ends AT-AC (comprising ~1/10,000 introns). The first of these introns to be discovered required an alternative splicing apparatus, consisting of U11 and U12 (related to U1 and U2, respectively), a U5 variant, and the $U4_{atac}$ and $U6_{atac}$ snRNAs. The splicing reaction is essentially similar to that at GT-AG introns, and the snRNAs play analogous roles. Whether there are differences in the protein components of this apparatus is not known. It now turns out that the dependence on the type of spliceosome is also influenced by sequences in the intron, so that there are some AT-AC introns spliced by U2-type spliceosomes, and some GT-AG introns spliced by U12-type spliceosomes. A strong consensus sequence defines the U12-dependent type of intron: $^{G}_{A}$TATCCTTT...PyA$^{G}_{C}$. In fact, most U12-dependent introns have the GT...AG termini. The two types of introns coexist in a variety of genomes, and in some cases are found in the same gene, where in fact U12-dependent introns tend to be flanked by U2-dependent introns. What is known about the phylogeny of these introns suggests that AT-AC U12-dependent introns may once have been more common, but tend to be converted to GT-AG termini, and to U2-dependence, in the course of evolution.

The involvement of snRNPs in splicing is only one example of their involvement in RNA processing reactions. Various snRNPs are involved in polyadenylation and in the generation of authentic 3′ ends of *Xenopus* histone mRNA (see later). Other snRNPs are required for the processing of nuclear RNA to mature rRNAs. Especially in view of the demonstration that group I introns are self-splicing, and that the RNA of ribonuclease P has catalytic activity (as discussed in Chapter 23), it is plausible to think that RNA-RNA reactions are important in most or all RNA processing events.

Group II introns autosplice via lariat formation

INTRONS in protein-coding genes (in fact, in all genes except nuclear tRNA-coding genes) can be divided into three general classes. Nuclear pre-mRNA introns have in common only the possession of the GT...AG dinucleotides at the 5′ and 3′ ends. Group I and group II introns are found in organelles and in bacteria. (Group I introns are found also in the nucleus in lower eukaryotes.) Group I and group II introns are classified according to their internal organization. Each can be folded into a typical type of secondary structure.

Figure 22.15 shows that three classes of introns are excised by two successive transesterifications (shown previously for nuclear introns in Figure 22.6). In the first reaction, the 5′ exon-intron junction is attacked by a free hydroxyl group (provided by an internal 2′-OH position in nuclear and group II introns, and by a free guanine nucleotide in group I introns). In the second reaction, the free 3′-OH at the end of the released exon in turn attacks the 3′ intron-exon junction.

Group I introns are more common than group II introns. There is little relationship between the two classes, but they share the striking property that the RNA can perform the splicing reaction *in vitro* by itself, without requiring enzymatic activities provided by proteins; however, proteins are almost certainly required *in vivo* to assist with folding. We discuss the

Figure 22.15 Three classes of splicing reactions proceed by two transesterifications. First, a free OH group attacks the exon 1–intron junction. Second, the OH created at the end of exon 1 attacks the intron–exon 2 junction.

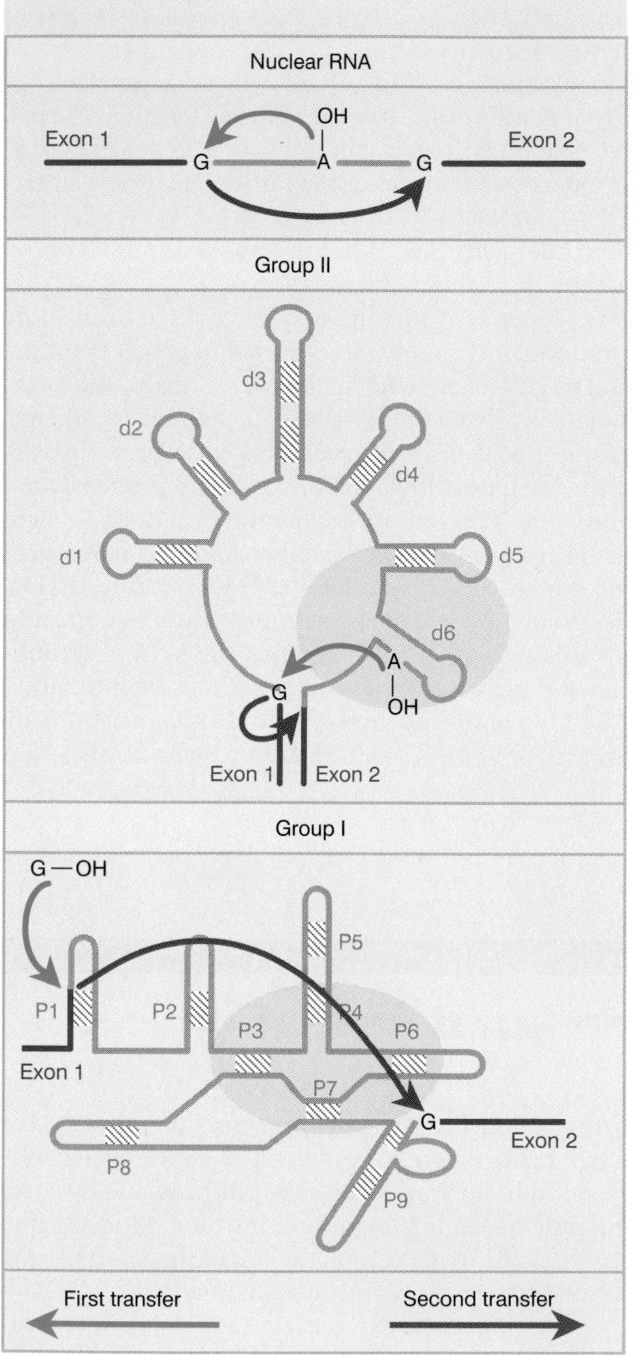

catalytic reaction of group I introns in Chapter 23, but now we examine the parallels between group II introns and nuclear splicing.

Group II mitochondrial introns have splice sites that resemble nuclear splice sites. They are excised by the same mechanism as nuclear pre-mRNAs, via a lariat that is held together by a 5′–2′ bond. An example of a lariat produced by splicing a group II intron is shown in **Figure 22.16.** When an isolated group II RNA is incubated *in vitro* in the absence of additional components, it is able to perform the splicing reaction. This means that the two transesterification reactions shown in Figure 22.15 can be performed by the group II intron RNA sequence itself. Because the number of phosphodiester bonds is conserved in the reaction, an external supply of energy is not required; this could have been an important feature in the evolution of splicing.

A group II intron forms into a secondary structure that involves closely juxtaposing two of its domains.

Figure 22.16 Mitochondrial group II introns are released by splicing in the form of stable lariats. Photograph kindly provided by Leslie Grivell and Annika Arnberg.

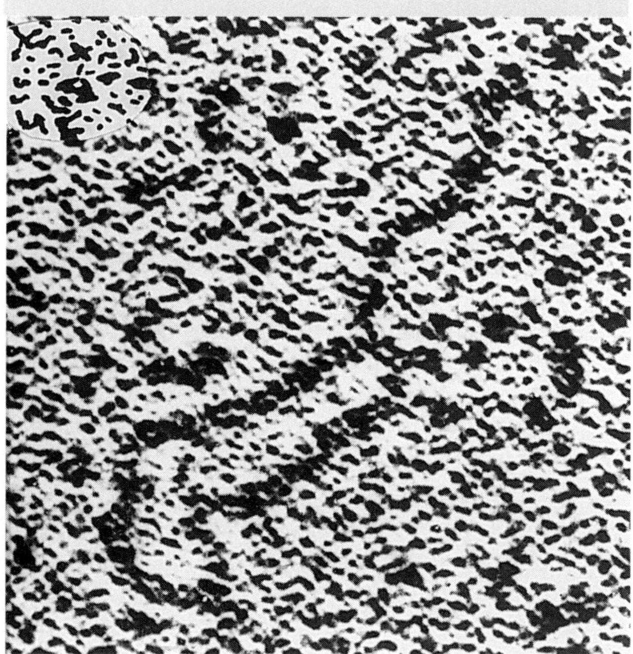

Domain 5 is separated by 2 bases from domain 6, which contains an A residue that donates the 2′-OH group for the first transesterification. This constitutes a catalytic domain in the RNA. **Figure 22.17** compares this secondary structure with the structure formed by the combination of U6 with U2 and of U2 with the branch site. The similarity is the basis for suggesting that U6 may have a catalytic role.

The features of group II splicing suggest that splicing evolved from an autocatalytic reaction undertaken by an individual RNA molecule, in which it accomplished a controlled deletion of an internal sequence. Probably such a reaction requires the RNA to fold into a specific conformation, or series of conformations, and would occur exclusively in *cis* conformation.

The ability of group II introns to remove themselves by an autocatalytic splicing event stands in great contrast to the requirement of nuclear introns for a complex splicing apparatus. We may regard the snRNAs of the spliceosome as compensating for the lack of sequence information in the intron, and providing the information required to form particular structures in RNA. The functions of the snRNAs may have evolved from the original autocatalytic system. These snRNAs act in *trans* upon the substrate pre-mRNA; we might imagine that the ability of U1 to pair with the 5′ splice site, or of U2 to pair with the branch sequence, replaced a similar reaction that required the relevant sequence to be carried by the intron. So the snRNAs may undergo reactions with the pre-mRNA substrate and with one another that have substituted for the series of conformational changes that occur in RNAs that splice by group II mechanisms. In effect, these changes have relieved the substrate pre-mRNA of the obligation to carry the sequences needed to sponsor the reaction. Of course, as the splicing apparatus has become more complex (and as the number of potential substrates has increased) some of the reactions that used to be catalyzed by RNA have been taken over by proteins.

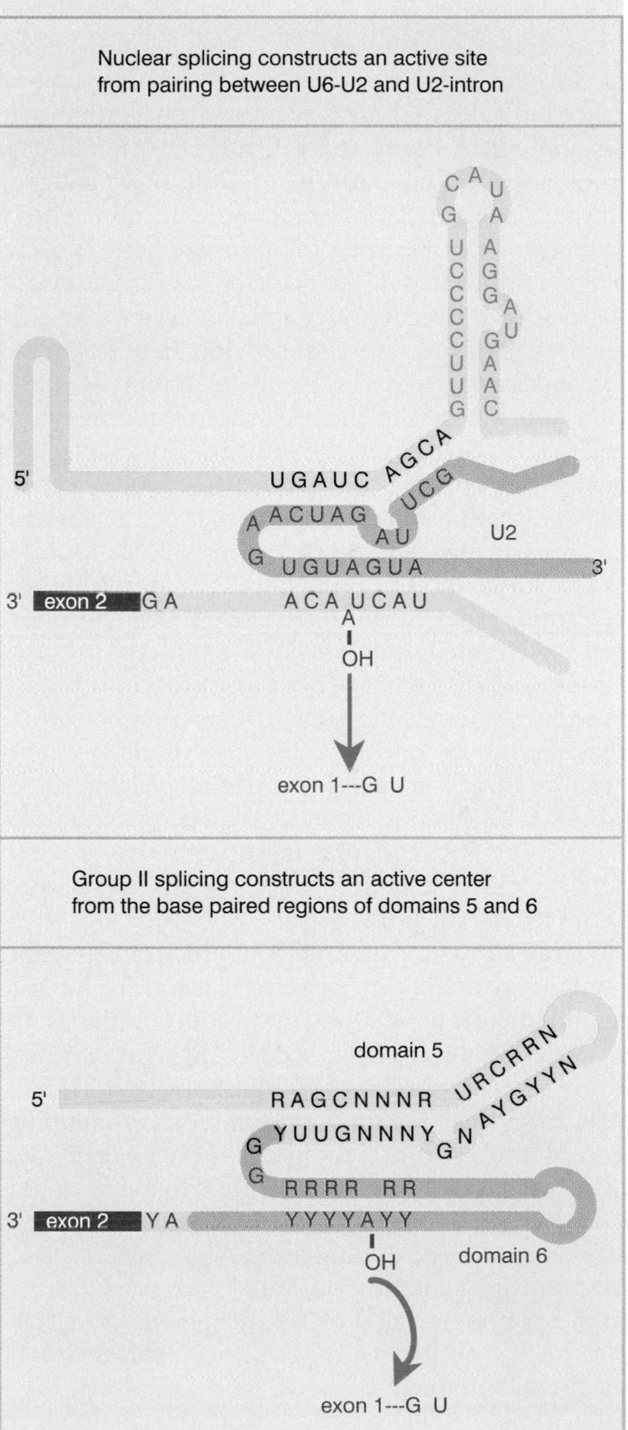

Figure 22.17 Nuclear splicing and group II splicing involve the formation of similar secondary structures. The sequences are more specific in nuclear splicing; group II splicing uses positions that may be occupied by either purine (R) or pyrimidine (Y).

Alternative splicing involves differential use of splice junctions

THE majority of interrupted genes are transcribed into an RNA that gives rise to a single type of spliced mRNA: in these cases, there is no variation in assignment of exons and introns. But the RNAs of some genes follow patterns of **alternative splicing,** when a single gene gives rise to more than one mRNA sequence. In some cases, the ultimate pattern of expression is dictated by the primary transcript, because the use of different startpoints or termination sequences alters the pattern of splicing. In other cases, a single primary transcript is spliced in more than one way, and internal exons are substituted, added, or deleted. In some cases, the multiple products all are made in the same cell, but in others the process is regulated so that particular splicing patterns occur only under particular conditions.

One of the most pressing questions in splicing is to determine what controls the use of such alternative pathways. Proteins that intervene to bias the use of alternative splice sites have been identified in two ways. In some mammalian systems, it has been possible to characterize alternative splicing *in vitro,* and to identify proteins that are required for the process. In *D. melanogaster,* aberrations in alternative splicing may be caused either by mutations in the genes that are alternatively spliced or in the genes whose products are necessary for the reaction.

Figure 22.18 shows examples in which one splice site remains constant, but the other varies. The large T/small t antigens of SV40 and the products of the adenovirus E1A region are generated by connecting a varying 5′ site to a constant 3′ site. In the case of the T/t antigens, the 5′ site used for T antigen removes a termination codon that is present in the t antigen mRNA, so that T antigen is larger than t antigen. In the case of the E1A transcripts, one of the 5′ sites connects to the last exon in a different reading frame, again making a significant change in the C-terminal part of the protein. In these examples, all the relevant splicing events take place in every cell in which the gene is expressed, so all the protein products are made.

Figure 22.18 Alternative forms of splicing may generate a variety of protein products from an individual gene. Changing the splice sites may introduce termination codons (shown by asterisks) or change reading frames.

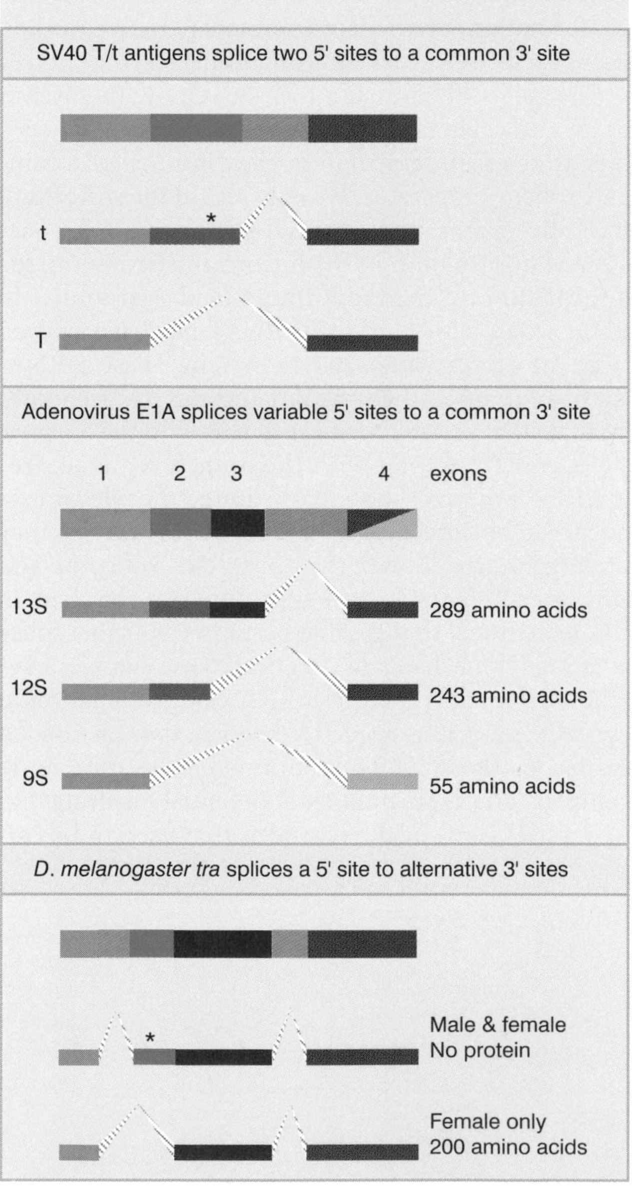
SV40 T/t antigens splice two 5' sites to a common 3' site
t
*
T
Adenovirus E1A splices variable 5' sites to a common 3' site
1
2
3
4
exons
13S
289 amino acids
12S
243 amino acids
9S
55 amino acids
D. melanogaster tra splices a 5' site to alternative 3' sites
*
Male & female
No protein
Female only
200 amino acids

There are differences in the ratios of T/t antigens in different cell types. Extracts from cells that produce relatively more small t antigen during an infection also produce more RNA with its characteristic splicing pattern in an *in vitro* system. A protein extracted from these cells can cause preferential production of small t RNA in extracts from other cell types. This protein, which was called ASF (alternative splicing factor), turns out to be the same as the splicing factor SF2, which is required for early steps in spliceosome assembly and for the first cleavage-ligation reaction (see Figure 22.10). ASF/SF2 is an RNA-binding protein in the SR family. When a pre-mRNA has more than one 5′ splice site preceding a single 3′ splice site, increased concentrations of ASF/SF2 promote use of the 5′ site nearest to the 3′ site at the expense of the other site. This effect of ASF/SF2 can be counteracted by another splicing factor, SF5. The exact molecular roles of these factors are not yet known, but we see in general terms that alternative splicing involving different 5′ sites may be influenced by proteins involved in spliceosome assembly.

The pathway of sex determination in *D. melanogaster* involves interactions between a series of genes in which alternative splicing events distinguish male and female. The pathway takes the form illustrated in **Figure 22.19**, in which the ratio of X chromosomes to autosomes determines the expression of *sxl*, and changes in expression are passed sequentially through the other genes to *dsx*, the last in the pathway.

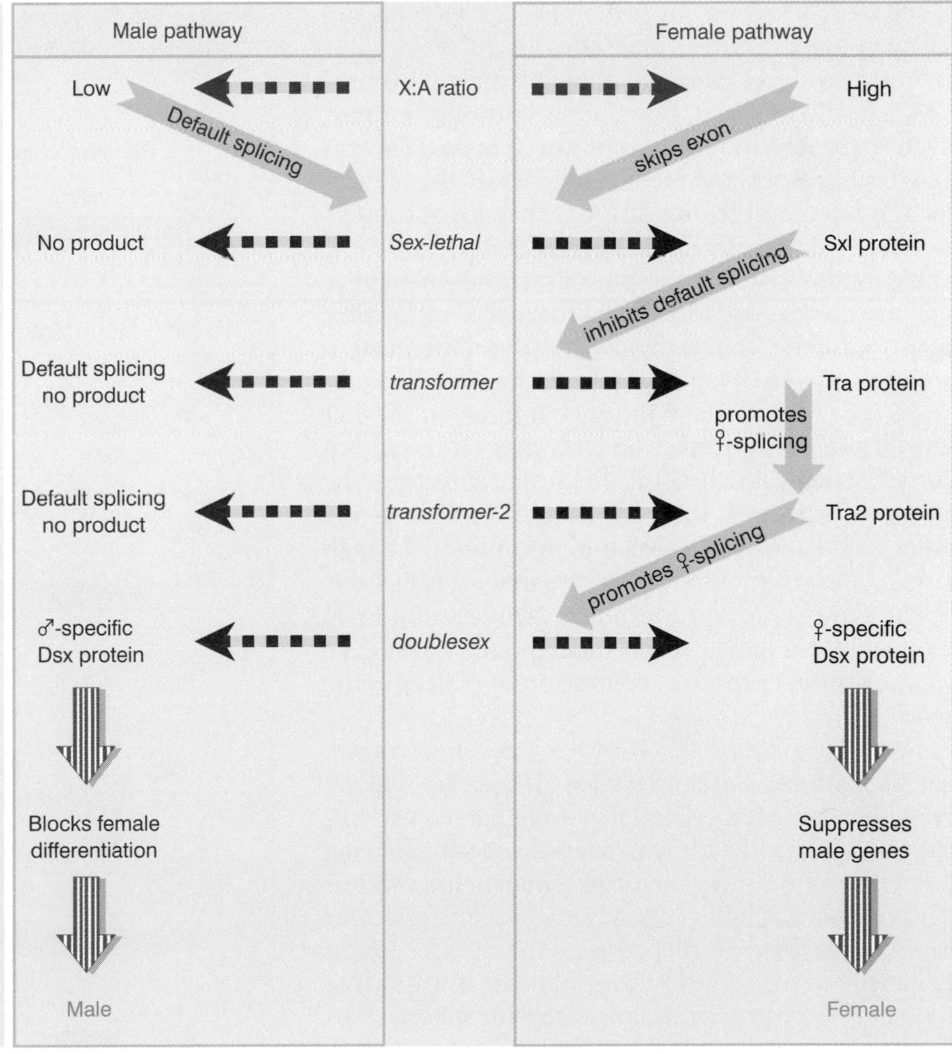

Figure 22.19 Sex determination in *D. melanogaster* involves a pathway in which different splicing events occur in females. Blocks at any stage of the pathway result in male development.

The pathway starts with sex-specific splicing of *sxl*, following the same pattern that is shown below in Figure 22.20 for *dsx*. Exon 3 of the *sxl* gene contains a termination codon that prevents synthesis of functional protein. This exon is included in the mRNA produced in males, but is skipped in females. As a result, only females produce Sxl protein. The protein has a concentration of basic amino acids that resembles other RNA-binding proteins.

The presence of Sxl protein changes the splicing of the *transformer (tra)* gene. Figure 22.18 shows that this involves splicing a constant 5′ site to alternative 3′ sites. One splicing pattern occurs in both males and females, and results in an RNA that has an early termination codon. The presence of Sxl protein inhibits usage of the normal 3′ splice site; when this site is skipped, the next 3′ site is used, leaving out a whole exon. This generates a female-specific mRNA that codes for a protein.

So *tra* produces a protein only in females; this protein is a splicing regulator. Similarly *tra2* is productively expressed in females but not in males. The Tra and Tra2 proteins are SR splicing factors that act directly upon the target transcripts. Tra and Tra2 cooperate (in females) to affect the splicing of *dsx*.

Figure 22.20 shows examples of cases in which splice sites are used to add or to substitute exons or introns, again with the consequence that different protein products are generated. In the *doublesex (dsx)* gene, females splice the 5′ site of intron 3 to the 3′ site of that intron; as a result translation terminates at the end of exon 4. Males splice the 5′ site of intron 3 directly to the 3′ site of intron 4, thus omitting exon 4 from the mRNA, and allowing translation to continue through exon 6. The result of the alternative splicing is that different proteins are produced in each sex: the male product blocks female sexual differentiation, while the female product represses expression of male-specific genes.

Sex determination therefore has a pleasing symmetry: the pathway starts with a female-specific splicing event that causes omission of an exon that has a termination codon, and ends with a female-specific splicing event that causes inclusion of an exon that has a termination codon. The events have different molecular bases. At the first control point, Sxl inhibits the default splicing pattern. At the last control point, Tra and Tra2 cooperate to promote the female-specific splice.

Alternative splicing of *dsx* RNA is controlled by competition between 3′ splice sites. *dsx* RNA has an element downstream of the leftmost 3′ splice site that is bound by Tra2; Tra and SR proteins associate with Tra2 at the site, which becomes an enhancer that assists binding of U2AF at the adjacent branch site. This commits the formation of the spliceosome to use this 3′ site in females rather than the alternative 3′ site. The proteins recognize the enhancer cooperatively, possibly relying on formation of some secondary structure as well as sequence *per se*.

The Tra and Tra2 proteins are not needed for normal splicing, because in their absence flies develop as normal males. As specific regulators, they need not necessarily participate in the mechanics of the splicing reaction; in this respect they differ from SF2, which is a factor required for general splicing, but can also influence choice of alternative splice sites.

Competition between 5′ splice sites also may be con-

Figure 22.20 Alternative splicing events that involve both sites may cause exons to be added or substituted.

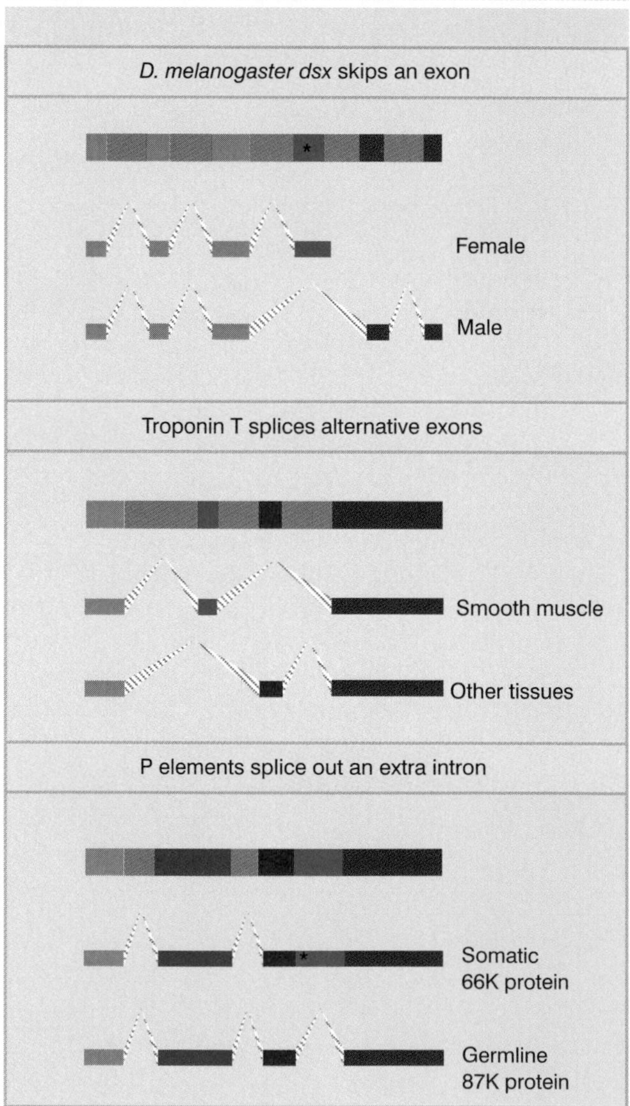
D. melanogaster dsx skips an exon
Female
Male
Troponin T splices alternative exons
Smooth muscle
Other tissues
P elements splice out an extra intron
Somatic
66K protein
Germline
87K protein

trolled by specific regulators. In one case, binding of SR proteins near the site favors its use, presumably by increasing the affinity for U1. This may identify a type of enhancer that functions on the 5′ site.

Alternative splicing also may be influenced by repression of one site. Exons 2 and 3 of the mouse troponin T gene are mutually exclusive; exon 2 is used in smooth muscle, but exon 3 is used in other tissues. Smooth muscle contains proteins that bind to repeated elements located on either side of exon 3, and which prevent use of the 3′ and 5′ sites that are needed to include it.

P elements of *D. melanogaster* show a tissue-specific splicing pattern. In somatic cells, there are two splicing events, but in germline an additional splicing event removes another intron. Because a termination codon lies in the germline-specific intron, a longer protein (with different properties) is produced in germline. We discuss the consequences for control of transposition in Chapter 15, and note for now that the tissue specificity results from differences in the splicing apparatus.

The default splicing pathway when the RNA is subjected to a heterologous (human) splicing extract is the germline pattern, in which intron 3 is excised. But extracts of somatic cells of *D. melanogaster* contain a protein that inhibits excision of this intron. The protein binds to sequences in exon 3; if these sequences are deleted, the intron is excised. The function of the protein is therefore probably to repress association of the spliceosome with the 5′ site of intron 3.

cis-splicing and *trans*-splicing reactions

In both mechanistic and evolutionary terms, splicing has been viewed as an *intramolecular* reaction, amounting essentially to a controlled deletion of the intron sequences at the level of RNA. In genetic terms, splicing occurs only in *cis*. This means that *only sequences on the same molecule of RNA can be spliced together.* The upper part of **Figure 22.21** shows the normal situation. The introns can be removed from each RNA molecule, allowing the exons of that RNA molecule to be spliced together, but there is no *intermolecular* splicing of exons between different RNA molecules. We cannot say that *trans* splicing never occurs between pre-mRNA transcripts of the same gene, but we know that it must be exceedingly rare, because if it were prevalent the exons of a gene would be able to complement one another genetically instead of belonging to a single complementation group.

Some manipulations can generate *trans*-splicing. In the example illustrated in the lower part of Figure 22.21, complementary sequences were introduced into the introns of two RNAs. Base pairing between the complements should create an H-shaped molecule. This molecule could be spliced in *cis*, to connect exons that are covalently connected by an intron, or it could be spliced in *trans*, to connect exons of the juxtaposed RNA molecules. Both reactions occur *in vitro*.

Another situation in which *trans*-splicing is possible *in vitro* occurs when substrate RNAs are provided in the form of one containing a 5′ splice site and the other containing a 3′ splice site together with appropriate downstream sequences (which may be either the next 5′ splice site or a splicing enhancer). In effect, this mimics the splicing reaction that is used for long introns and short exons (see the right side of Figure 22.11), and shows that *in vitro* it is not necessary for the left and right splice sites to be on the same RNA molecule.

These results show that there is no *mechanistic* impediment to *trans*-splicing. They exclude models for splicing that require processive movement of a spliceosome along the RNA. It must be possible for a spliceosome to recognize the 5′ and 3′ splice sites of different RNAs when they are in close proximity.

Although *trans*-splicing is rare, it occurs *in vivo* in some special situations. One is revealed by the presence of a common 35 base leader sequence at the end of numerous mRNAs in the trypanosome. But the leader sequence is not coded upstream of the individual transcription units. Instead it is transcribed into an independent RNA, carrying additional sequences at its 3′ end, from a repetitive unit located elsewhere in the genome. **Figure 22.22** shows that this RNA carries the 35 base leader sequence followed by a 5′ splice site sequence. The sequences coding for the mRNAs carry a 3′ splice site just preceding the sequence found in the mature mRNA.

Figure 22.21 Splicing usually occurs only in *cis* between exons carried on the same physical RNA molecule, but *trans* splicing can occur when special constructs are made that support base pairing between introns.

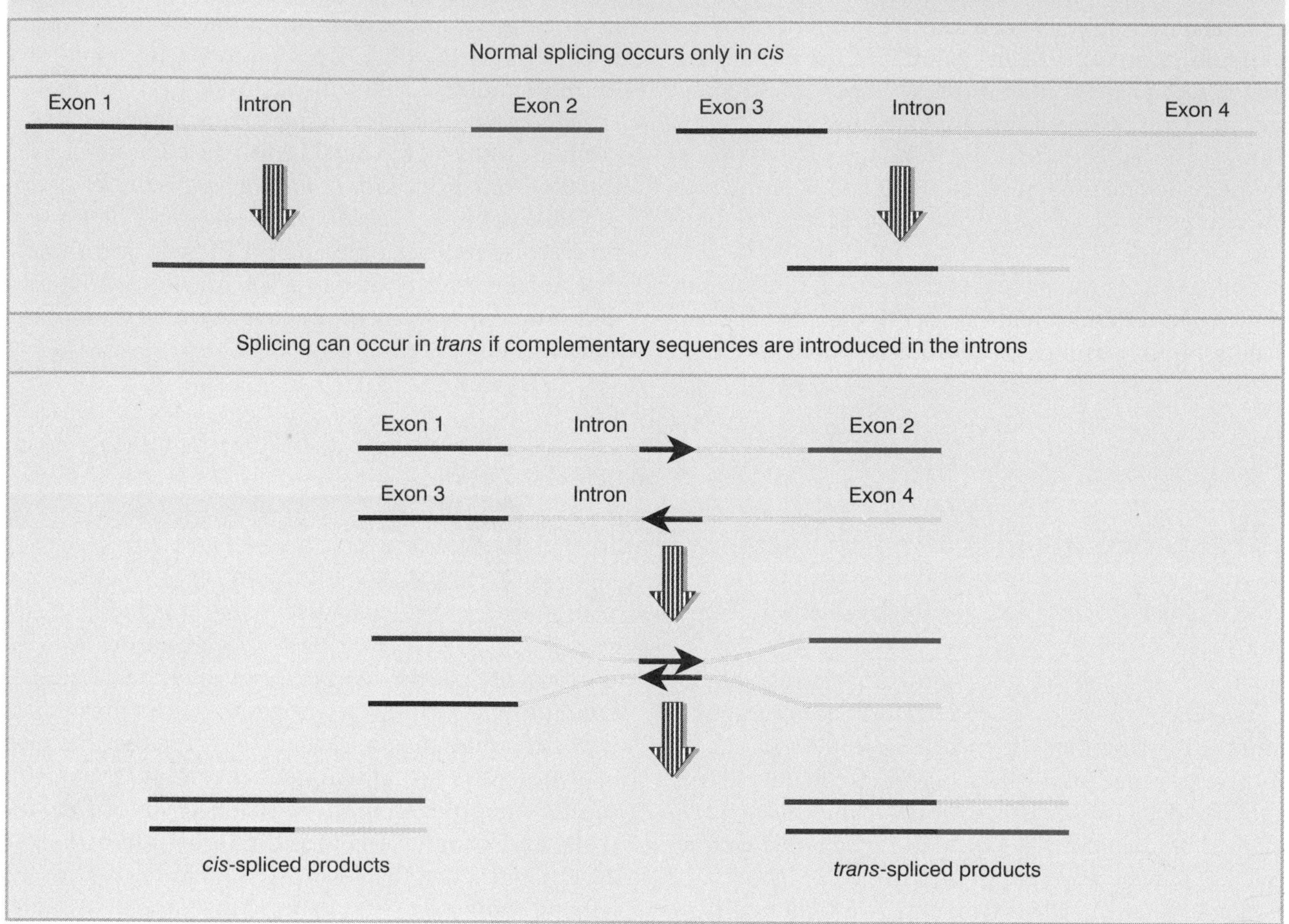

If the leader and the mRNA are connected by a *trans*-splicing reaction, the 3′ region of the leader RNA and the 5′ region of the mRNA will in effect comprise the 5′ and 3′ halves of an intron. If splicing occurs by the usual nuclear mechanism, a 5′–2′ link should form by a reaction between the GU of the 5′ intron and a branch sequence near the AG of the 3′ intron. Because the two parts of the intron are not covalently linked, this would generate a Y-shaped molecule instead of a lariat. Molecules of this type have been detected, supporting the model.

A similar situation is presented by the expression of actin genes in *C. elegans*. Three actin mRNAs (and some other RNAs) share the same 22 base leader sequence at the 5′ terminus. The leader sequence is not coded in the actin gene, but is transcribed independently as part of a 100 base RNA coded by a gene elsewhere. *Trans*-splicing also occurs in chloroplasts.

The RNA that donates the 5′ exon for *trans*-splicing is called the **SL RNA** (spliced leader RNA). The SL RNAs found in several species of trypanosomes and also in the nematode (*C. elegans*) have some common features. They fold into a common secondary structure that has three stem-loops and a single-stranded region that resembles the Sm-binding site. The SL RNAs therefore exist as snRNPs that count as members of the

Figure 22.22 The SL RNA provides an exon that is connected to the first exon of an mRNA by *trans*-splicing. The reaction involves the same interactions as nuclear *cis*-splicing, but generates a Y-shaped RNA instead of a lariat, because the intron is in two separate parts.

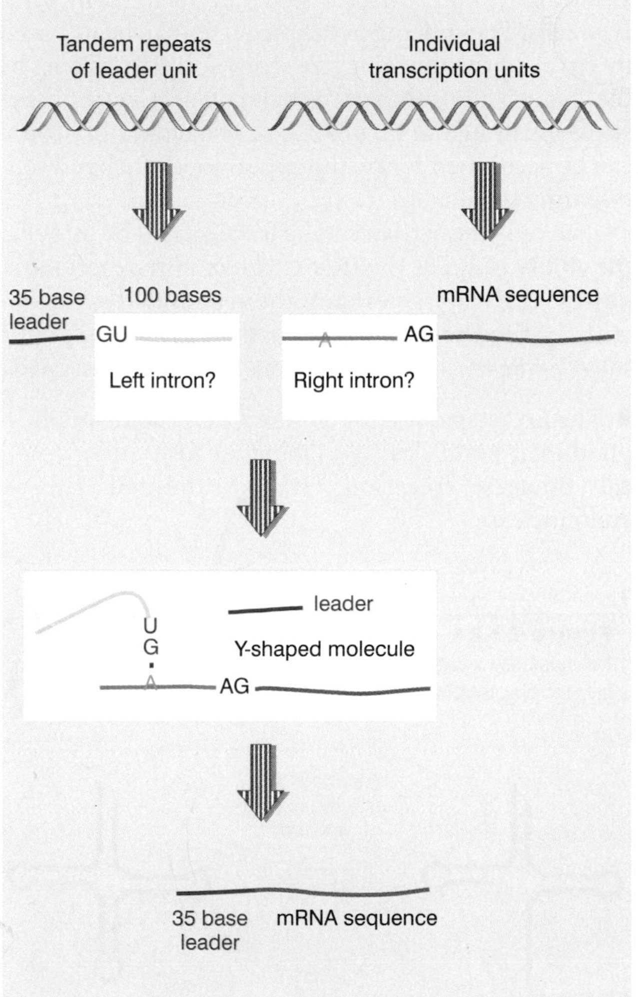

Sm snRNP class. Trypanosomes possess the U2, U4, and U6 snRNAs, but do not have U1 or U5 snRNAs. The absence of U1 snRNA can be explained by the properties of the SL RNA, which can carry out the functions that U1 snRNA usually performs at the 5′ splice site; thus SL RNA in effect consists of an snRNA sequence possessing U1 function, linked to the exon-intron site that it recognizes.

There are two types of SL RNA in *C. elegans.* SL1 RNA (the first to be discovered) is used for splicing to coding sequences that are preceded only by a 5′ non-translated region (the most common situation). SL2 RNA is used in cases in which a pre-mRNA contains two coding sequences; it is spliced to the second sequence, thus releasing it from the first, and allowing it to be used as an mRNA.

The *trans*-splicing reaction of the SL RNA may represent a step towards the evolution of the full nuclear splicing apparatus. The SL RNA provides in *cis* the ability to recognize the 5′ splice site, and this probably depends upon the specific conformation of the RNA. The remaining functions required for splicing are provided by independent snRNPs. The SL RNA can function without participation of proteins like those in U1 snRNP, which suggests that the recognition of the 5′ splice site depends directly on RNA.

Yeast tRNA splicing involves cutting and rejoining

The splicing reactions that we have discussed so far depend upon short consensus sequences and occur by transesterification reactions in which breaking and making of bonds is coordinated. The splicing of tRNA genes is achieved by a different mechanism that relies upon separate cleavage and ligation reactions.

About 40 of the ~400 nuclear tRNA genes in yeast are interrupted. Each has a single intron, located just one nucleotide beyond the 3′ side of the anticodon. The introns vary in length from 14 to 46 bp. Those in related tRNA genes are related in sequence, but the introns in tRNA genes representing different amino acids are unrelated. *There is no consensus sequence that could be recognized by the splicing enzymes.* The same general rules apply to the nuclear tRNA genes of plants, amphibians, and mammals.

All the introns include a sequence that is complementary to the anticodon of the tRNA. This creates an alternative conformation for the anticodon arm in which the anticodon is base paired to form an extension of the usual arm. An example is drawn in **Figure 22.23.** Only the anticodon arm is affected—the rest of the molecule retains its usual structure.

The exact sequence and size of the intron is not important. Most mutations in the intron do not prevent splicing. *Splicing of tRNA depends principally on recognition of a common secondary structure in tRNA rather than a common sequence of the intron.* Regions in various parts of the molecule are important, including the stretch between the acceptor arm and D arm, in the TψC arm, and especially the anticodon arm. This is reminiscent of the structural demands placed on tRNA for protein synthesis (see Chapter 6).

The intron is not entirely irrelevant, however. Pairing between a base in the intron loop and an unpaired base in the stem is required for splicing. Mutations at other positions that influence this pairing (for example, to generate alternative patterns for pairing) influence splicing. The rules that govern availability of tRNA precursors for splicing resemble the rules that govern recognition by aminoacyl-tRNA synthetases (as discussed in Chapter 6).

In a temperature-sensitive mutant of yeast that fails to remove the introns, the interrupted precursors accumulate in the nucleus. The precursors can be used as substrates for a cell-free system extracted from wild-type cells. The splicing of the precursor can be followed by virtue of the resulting size reduction. This is seen by the change in position of the band on gel electrophoresis, as illustrated in **Figure 22.24.** The reduction in size can be accounted for by the appearance of a band representing the intron.

The cell-free extract can be fractionated by assaying the ability to splice the tRNA. The *in vitro* reaction requires ATP. Characterizing the reactions that occur with and without ATP shows that the *two separate stages of the reaction are catalyzed by different enzymes.*

- The first step does not require ATP. It involves phosphodiester bond cleavage, taking the form of an atypical nuclease reaction. It is catalyzed by an endonuclease.

Figure 22.23 The intron in yeast $tRNA^{Phe}$ base pairs with the anticodon to change the structure of the anticodon arm. Pairing between an excluded base in the stem and the intron loop in the precursor may be required for splicing.

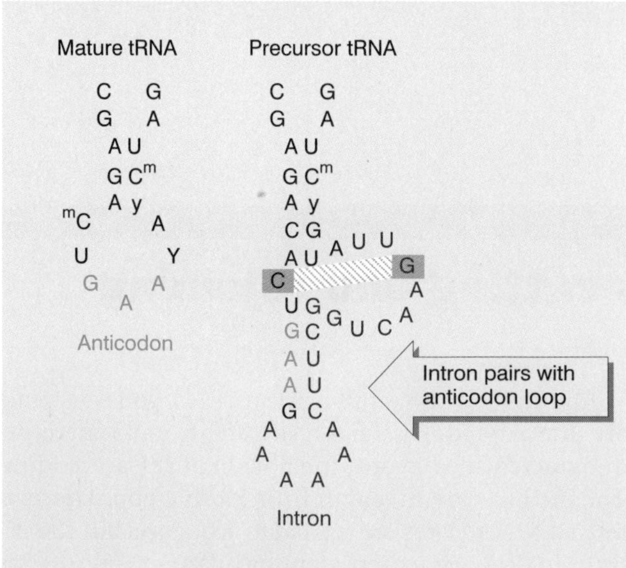

Figure 22.24 Splicing of yeast tRNA *in vitro* can be followed by assaying the RNA precursor and products by gel electrophoresis.

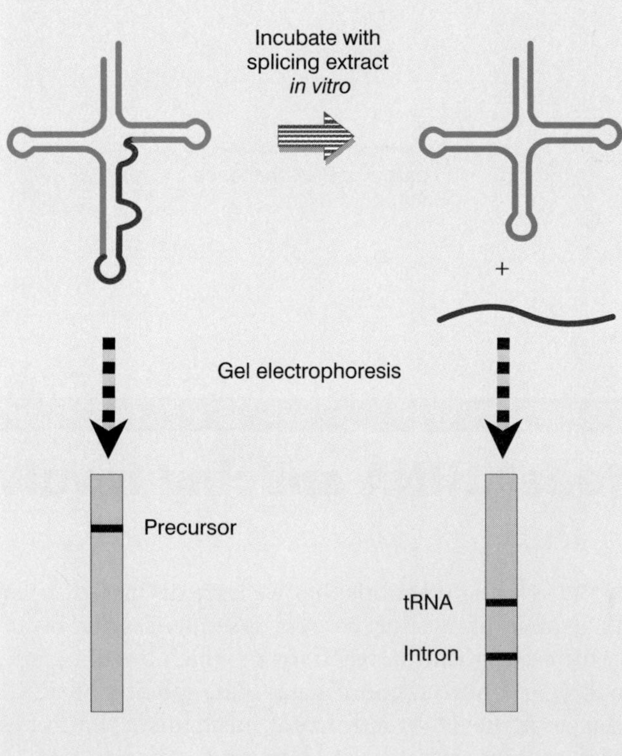

- The second step requires ATP and involves bond formation; it is a ligation reaction, and the responsible enzyme activity is described as an **RNA ligase.**

The endonuclease is responsible for the specificity of intron recognition. It cleaves the precursor at both ends of the intron. The yeast endonuclease is a multimeric protein. Its activities are illustrated in **Figure 22.25.** The subunits Sen34 and Sen2 cleave the 3′ and 5′ splice sites, respectively. Subunit Sen54 may determine the sites of cleavage by "measuring" distance from the mature domain. Although it was originally thought that the sequence of the intron was irrelevant, it participates in base pairing that is required for cleavage. The AI base pair that forms between the first base in the anticodon loop and the base preceding the 3′ splice site is required for 3′ splice site cleavage.

An interesting insight into the evolution of tRNA splicing is provided by the endonucleases of archaea. These are homodimers or homotetramers, in which each subunit has an active site (although only two of

Figure 22.25 The 3' and 5' cleavages in *S. cerevisiae* pre-tRNA are catalyzed by different subunits of the endonuclease. Another subunit may determine location of the cleavage sites by measuring distance from the mature structure. The AI base pair is also important.

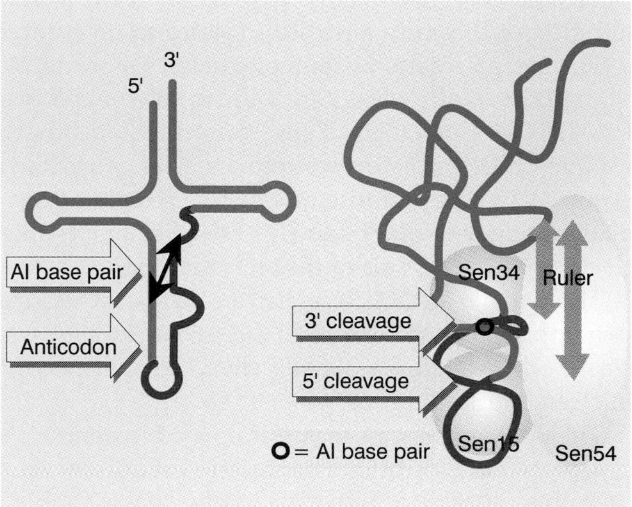

Figure 22.26 Splicing of tRNA requires separate nuclease and ligase activities. The exon-intron boundaries are cleaved by the nuclease to generate 2', 3'-cyclic phosphate and 5'-OH termini. The cyclic phosphate is opened to generate 3'-OH and 2'-phosphate groups. The 5'-OH is phosphorylated. After releasing the intron, the tRNA half molecules fold into a tRNA-like structure that now has a 3'-OH, 5'-P break. This is sealed by a ligase.

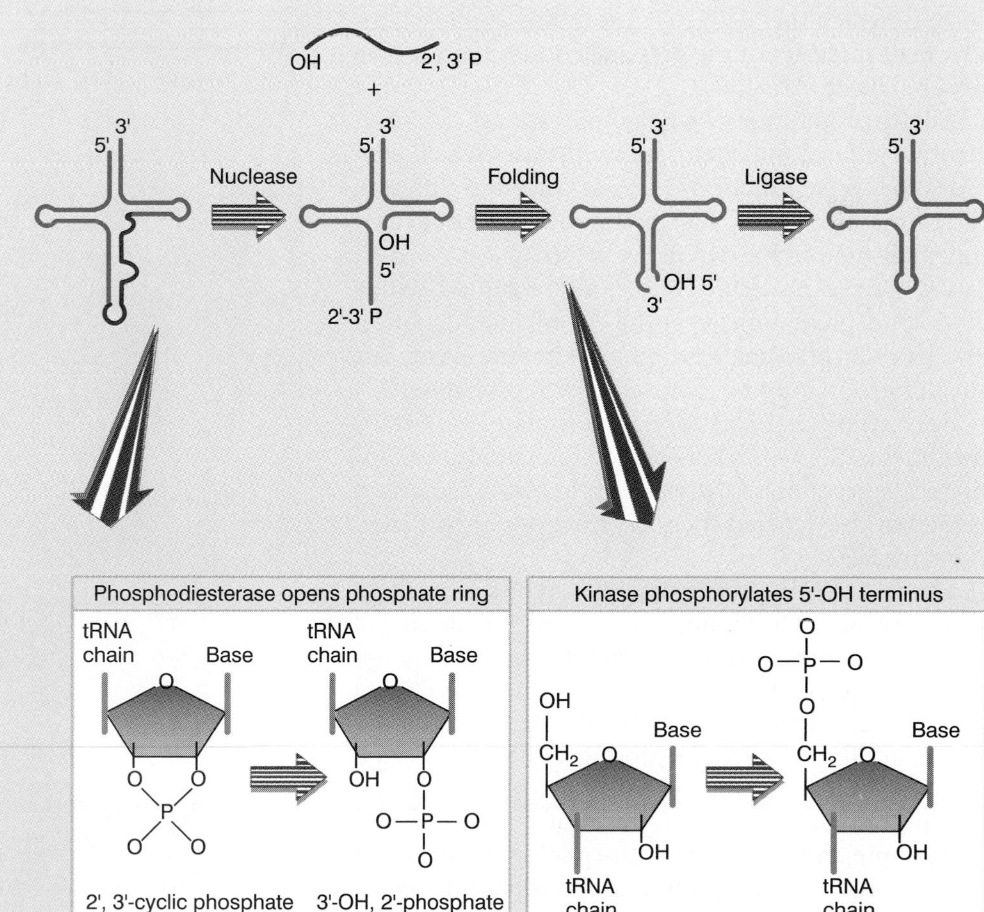

the sites function in the tetramer) that cleaves one of the splice sites. The subunit has sequences related to the sequences of the active sites in the Sen34 and Sen2 subunits of the yeast enzyme. So the origin of splicing of tRNA precedes the separation of the archaea and the eukaryotes. If it originated by insertion of the intron into tRNAs, this must have been a very ancient event.

The overall reaction is summarized in **Figure 22.26.** The products of cleavage are a linear intron and two half-tRNA molecules. These intermediates have unique ends. Each 5′ terminus ends in a hydroxyl group; each 3′ terminus ends in a 2′,3′-cyclic phosphate group. (All other known RNA splicing enzymes cleave on the other side of the phosphate bond.)

The two half-tRNAs base pair to form a tRNA-like structure. When ATP is added, the second reaction occurs. Both of the unusual ends generated by the endonuclease must be altered.

The cyclic phosphate group is opened to generate a 2′-phosphate terminus. This reaction requires cyclic phosphodiesterase activity. The product has a 2′-phosphate group and a 3′-OH group.

The 5′-OH group generated by the nuclease must be phosphorylated to give a 5′-phosphate. This generates a site in which the 3′-OH is next to the 5′-phosphate. Covalent integrity of the polynucleotide chain is then restored by ligase activity.

All three activities—phosphodiesterase, polynucleotide kinase, and adenylate synthetase (which provides the ligase function)—are arranged in different functional domains on a single protein. They act sequentially to join the two tRNA halves.

The spliced molecule is now uninterrupted, with a 5′–3′ phosphate linkage at the site of splicing, but it also has a 2′-phosphate group marking the event. The surplus group must be removed by a phosphatase.

Generation of a 2′,3′-cyclic phosphate also occurs during the tRNA-splicing reaction in plants and mammals. The reaction in plants seems to be the same as in yeast, but the detailed chemical reactions are different in mammals.

The yeast tRNA precursors also can be spliced in an extract obtained from the germinal vesicle (nucleus) of *Xenopus* oocytes. This shows that the reaction is not species-specific. *Xenopus* must have enzymes able to recognize the introns in the yeast tRNAs.

The ability to splice the products of tRNA genes is therefore well conserved, but is likely to have a different origin from the other splicing reactions (such as that of nuclear pre-mRNA). The tRNA-splicing reaction uses cleavage and synthesis of bonds and is determined by sequences that are external to the intron. Other splicing reactions use transesterification, in which bonds are transferred directly, and the sequences required for the reaction lie within the intron.

An unusual splicing system that is related to tRNA splicing mediates the response to unfolded proteins in yeast. The accumulation of unfolded proteins in the lumen of the ER triggers a response pathway that leads to increased transcription of genes coding for chaperones that assist protein folding in the ER. A signal must therefore be transmitted from the lumen of the ER to the nucleus.

The sensor that activates the pathway is the protein Ire1p. It is an integral membrane protein (Ser/Thr) kinase that has domains on each side of the ER

Figure 22.27 The unfolded protein response occurs by activating special splicing of HAC1 mRNA to produce a transcription factor that recognizes the UPR.

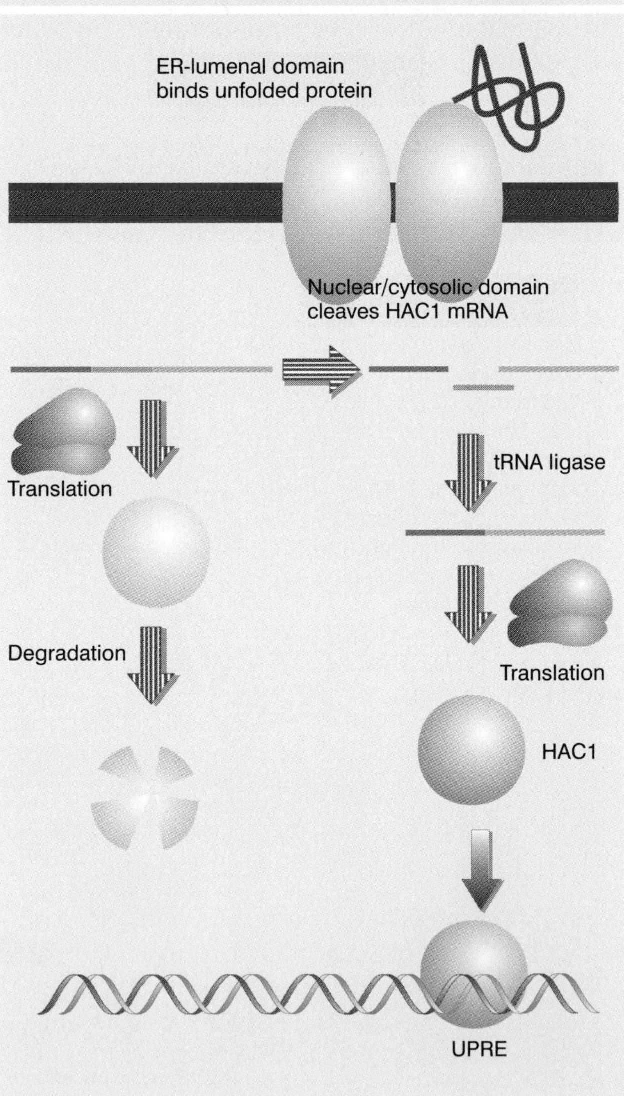

membrane. The N-terminal domain in the lumen of the ER detects the presence of unfolded proteins, presumably by binding to exposed motifs. This causes aggregation of monomers and activates the C-terminal domain on the other side of the membrane by autophosphorylation.

Genes that are activated by this pathway have a common promoter element, the UPRE. The transcription factor Hac1p binds to the UPRE, and is produced in response to accumulation of unfolded proteins. The trigger for production of Hac1p is the action of Ire1p on Hac1 mRNA.

The operation of the pathway is summarized in **Figure 22.27.** Under normal conditions, when the pathway is not activated, Hac1 mRNA is translated into a protein that is rapidly degraded. The activation of Ire1p results in the splicing of the Hac1 mRNA to change the sequence of the protein to a more stable form. This form provides the functional transcription factor that activates genes with the UPRE.

Unusual splicing components are involved in this reaction. Ire1P has an endonuclease activity that acts directly on Hac1 mRNA to cleave the two splicing junctions. The two junctions are ligated by the tRNA ligase that acts in the tRNA splicing pathway.

Where does the modification of Hac1 mRNA occur? Ire1p is probably located in the inner nuclear membrane, with the N-terminal sensor domain in the ER lumen, and the C-terminal kinase/nuclease domain in the nucleus. This would enable it to act directly on Hac1 RNA before it is exported to the cytoplasm. It also would allow easy access by the tRNA ligase. There is no apparent relationship between the Ire1p nuclease activity and the tRNA splicing endonuclease, so it is not obvious how this specialized system would have evolved.

3′ ends are generated by termination and by cleavage reactions

INFORMATION about the termination reaction for eukaryotic RNA polymerases is much less detailed than our knowledge of initiation. RNA polymerases I and III have discrete termination events (like bacterial RNA polymerase), but it is not clear whether RNA polymerase II usually terminates in this way.

For RNA polymerase I, the sole product of transcription is a large precursor that contains the sequences of the major rRNA. The precursor is subjected to extensive processing. Termination occurs at a discrete site >1000 bp downstream of the mature 3′ end, which is generated by cleavage. Termination involves recognition of an 18 base terminator sequence by an ancillary factor.

With RNA polymerase III, transcription *in vitro* generates molecules with the same 5′ and 3′ ends as those synthesized *in vivo*. The termination reaction resembles intrinsic termination by bacterial RNA polymerase. Termination usually occurs at the second U within a run of 4 U bases, but there is heterogeneity, with some molecules ending in 3 or even 4 U bases. The same heterogeneity is seen in molecules synthesized *in vivo*, so it seems to be a *bona fide* feature of the termination reaction.

Just like the prokaryotic terminators, the U run is embedded in a G•C-rich region. Although sequences of dyad symmetry are present, they are not needed for termination, since mutations that abolish the symmetry do not prevent the normal completion of RNA synthesis. Nor are any sequences beyond the U run necessary, since all distal sequences can be replaced without any effect on termination.

The U run itself is not sufficient for termination, because regions of 4 successive U residues exist within transcription units read by RNA polymerase III. (However, there are no internal U_5 runs, which fits with the greater efficiency of termination when the terminator is a U_5 rather than a U_4 sequence.) The critical feature in termination must therefore be the recognition of a U_4 sequence in a context that is rich in G•C base pairs.

How does the termination reaction occur? It cannot rely on the weakness of the rU-dA RNA-DNA hybrid region that lies at the end of the transcript, because often only the first two U residues are transcribed. Perhaps the G•C-rich region plays a role in slowing down the enzyme, but there does not seem to be a counterpart to the hairpin involved in prokaryotic termination. We remain puzzled as to how the enzyme can respond so specifically to such a short signal. And in contrast with the initiation reaction, which RNA polymerase III cannot accomplish alone, termination seems to be a function of the enzyme itself.

Similar termination sites have been found in some RNA polymerase II transcription units, but their significance is not certain. It is possible that termination by RNA polymerase II is only loosely specified. In some transcription units, termination occurs >1000 bp downstream of the site corresponding to the mature 3′ end of the mRNA (which is generated by cleavage at a specific sequence). Instead of using specific terminator sequences, the enzyme ceases RNA synthesis within multiple sites located in rather long "terminator regions." The nature of the individual termination sites is not known.

The 3′ ends of mRNAs are generated by cleavage followed by polyadenylation. Addition of poly(A) to nuclear RNA can be prevented by the analog **3′-deoxyadenosine,** also known as **cordycepin.** Although cordycepin does not stop the transcription of nuclear RNA, its addition prevents the appearance of mRNA in the cytoplasm. This shows that polyadenylation is *necessary* for the maturation of mRNA from nuclear RNA.

Generation of the 3′ end to which poly(A) is added is illustrated in **Figure 22.28.** RNA polymerase transcribes past the site corresponding to the 3′ end, and sequences in the RNA are recognized as targets for an endonucleolytic cut followed by polyadenylation.

A common feature of mRNAs in higher eukaryotes (but not in yeast) is the presence of the sequence AAUAAA in the region from 11 to 30 nucleotides upstream of the site of poly(A) addition. The sequence is highly conserved; only occasionally is even a single base different. Deletion or mutation of the AAUAAA hexamer prevents generation of the polyadenylated 3′ end. The signal is needed for both cleavage and polyadenylation.

The development of a system in which polyadenylation occurs *in vitro* opens the route to analyzing the reactions. The formation and functions of the complex that undertakes 3′ processing are illustrated in **Figure 22.29.** Generation of the proper 3′ terminal structure requires an **endonuclease** (consisting of the components CFI and CFII) to cleave the RNA, a **poly(A) polymerase** (PAP) to synthesize the poly(A) tail, and a

Figure 22.28 The sequence AAUAAA is necessary for cleavage to generate a 3' end for polyadenylation.

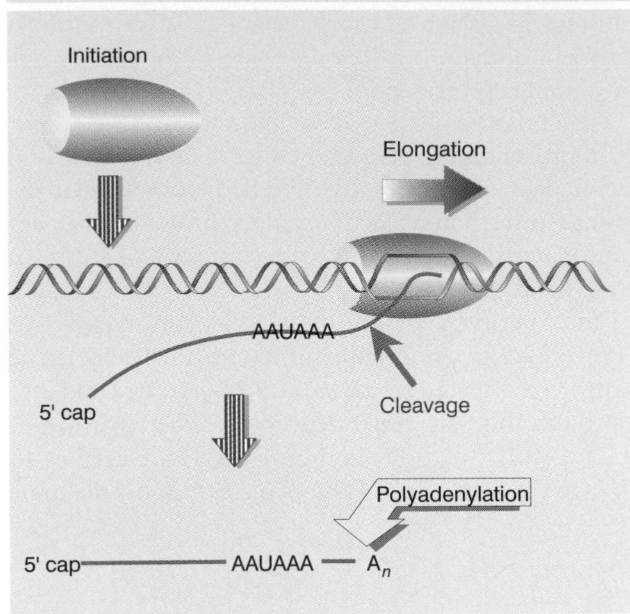

specificity component (CPSF) that recognizes the AAUAAA sequence and directs the other activities. A stimulatory factor, CstF, binds to a G-U-rich sequence that is downstream from the cleavage site itself.

The specificity factor contains 4 subunits, which together bind specifically to RNA containing the sequence AAUAAA. The individual subunits are proteins that have common RNA-binding motifs, but which by themselves bind nonspecifically to RNA. Protein-protein interactions between the subunits may be needed to generate the specific AAUAAA-binding site. CPSF binds strongly to AAUAAA only when CstF is also present to bind to the G-U-rich site.

The specificity factor is needed for both the cleavage and polyadenylation reactions. It exists in a complex with the endonuclease and poly(A) polymerase, and this complex usually undertakes cleavage followed by polyadenylation in a tightly coupled manner.

The two components CFI and CFII (cleavage factors I and II), together with the specificity factor, are necessary and sufficient for the endonucleolytic cleavage.

The poly(A) polymerase has a nonspecific catalytic activity. When it is combined with the other components, the synthetic reaction becomes specific for RNA containing the sequence AAUAAA. The polyadenylation reaction passes through two stages. First, a rather short oligo(A) sequence (~10 residues) is added to the 3′ end. This reaction is absolutely dependent on the

Figure 22.29 The 3' processing complex consists of several activities. CPSF and CstF each consist of several subunits; the other components are monomeric. The total mass is >900 kD.

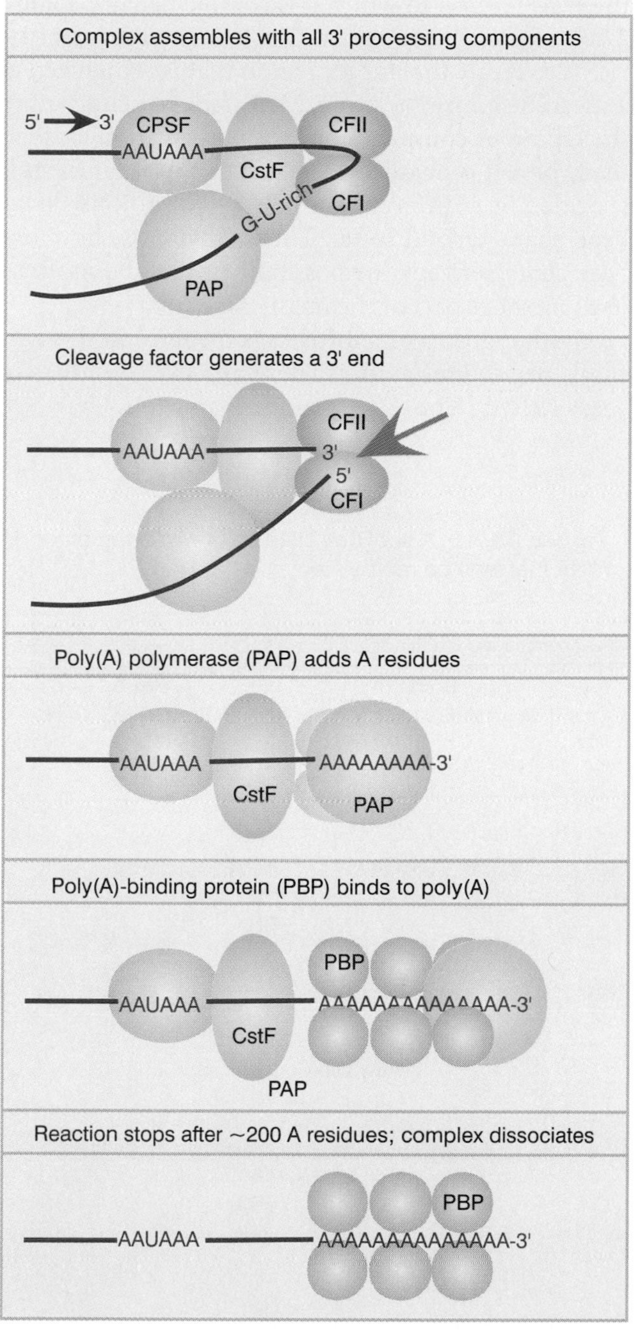

AAUAAA sequence, and poly(A) polymerase performs it under the direction of the specificity factor. In the second phase, the oligo(A) tail is extended to the full ~200 residue length. This reaction requires another stimulatory factor that recognizes the oligo(A) tail and directs poly(A) polymerase specifically to extend the 3′ end of a poly(A) sequence.

The poly(A) polymerase by itself adds A residues individually to the 3′ position. Its intrinsic mode of action is distributive; it dissociates after each nucleotide has been added. However, in the presence of CPSF and PABP (poly(A)-binding protein), it functions processively to extend an individual poly(A) chain. The PABP is a 33 kD protein that binds stoichiometrically to the poly(A) stretch. The length of poly(A) is controlled by the PABP, which in some way limits the action of poly(A) polymerase to ~200 additions of A residues. The limit may represent the accumulation of a critical mass of PABP on the poly(A) chain. PABP remains a component of mRNA; its exact role in mRNA metabolism has yet to be defined.

Some mRNAs are not polyadenylated. The formation of their 3′ ends is therefore different from the coordinated cleavage/polyadenylation reaction. The most prominent members of this mRNA class are the mRNAs coding for histones. Formation of their 3′ ends depends upon secondary structure. The RNA terminates in a stem-loop structure, and mutations that prevent formation of the duplex stem prevent formation of the end of the RNA. Secondary mutations that restore duplex structure (though not necessarily the original sequence) behave as revertants. This suggests that *formation of the secondary structure is more important than the exact sequence.*

Either or both of the DNA strands could in principle be involved in forming secondary structure. They can be distinguished by using templates consisting of heteroduplex molecules, in which the two strands of DNA are not identical. It turns out that it is important to be able to write a duplex structure for the *coding strand,* not the strand used as template. This suggests that the secondary structure exerts its effect by forming in the RNA as it is transcribed.

An *in vitro* system has been developed that generates authentic 3′ ends in *X. laevis* histone mRNAs. When the histone H3 gene is injected into the *Xenopus* oocyte, it is faithfully initiated and transcribed, but termination occurs at variable sites. However, when a nuclear extract from the sea urchin is simultaneously injected with the gene, the transcribed mRNA has the proper 3′ end.

The nuclear extract has three active components: a heat-labile factor (of unknown function); a factor that binds to a hairpin just upstream of the cleavage site; and a 56 base RNA called **U7 snRNA** whose 5′ terminus is complementary to a short sequence just downstream of the 3′ cleavage site.

The reaction between histone H3 mRNA and U7 snRNA is drawn in **Figure 22.30.** The upstream hairpin and the downstream sequence that pairs with U7 snRNA are conserved in histone H3 mRNAs of several species. The U7 snRNA has sequences towards its 5′ end that could pair with the histone mRNA consensus sequences, and has an extensive hairpin at its 3′ end.

3′ processing is inhibited by mutations in the downstream histone consensus sequence that reduce ability to pair with U7 snRNA. Compensatory mutations in U7 snRNA that restore complementarity also restore 3′ processing. This suggests that U7 snRNA functions by base pairing with the histone mRNA. Cleavage to generate a 3′ terminus occurs a fixed distance from the site recognized by U7 snRNA, which suggests that the snRNA functions at this stage.

Processing and modification of rRNA requires a class of small RNAs called snoRNAs (small nucleolar RNAs). They are associated with the protein fibrillarin, which is an abundant component of the nucleolus (the region of the nucleus within which the rRNA genes are transcribed). Some snoRNAs are required for cleavage of the precursor to rRNA; one example is U3 snoRNA, which is required for the first cleavage event in the 5′ transcribed spacer region in both yeast and *Xenopus*. Other snoRNAs are required for the modifications that are made to bases in the rRNA.

Vertebrate rRNAs contain >100 2′-O-methyl groups, which are found at conserved locations. These methylations require the C/D group of snoRNAs, so-called because they possess two short conserved sequence motifs called boxes C and D. Most of these snoRNAs contain sequences that are complementary to regions of the 18S or 28S rRNAs that are methylated. Loss of a particular snoRNA prevents methylation in the rRNA region to which it is complementary. **Figure 22.31** suggests that the snoRNA base pairs with the rRNA to create the duplex region that is recognized as a substrate for methylation. Methylation occurs within the region of complementarity, at a position close to the D box. It is possible that each methylation event is specified by a different snoRNA; ~40 snoRNAs have been characterized so far. The methylase(s) have not been characterized; one possibility is that the snoRNA itself provides part of the methylase activity.

Another group of snoRNAs is involved in the synthesis of pseudouridine. There are 43 Ψ residues in yeast rRNAs. The ACA group of snoRNAs has >20

Figure 22.30 Generation of the 3′ end of histone H3 mRNA depends on a conserved hairpin and a sequence that base pairs with U7 snRNA.

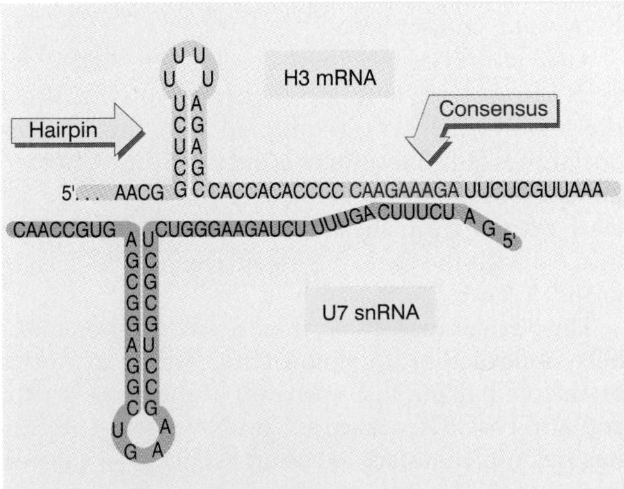

Figure 22.31 A snoRNA base pairs with a region of rRNA that is to be methylated.

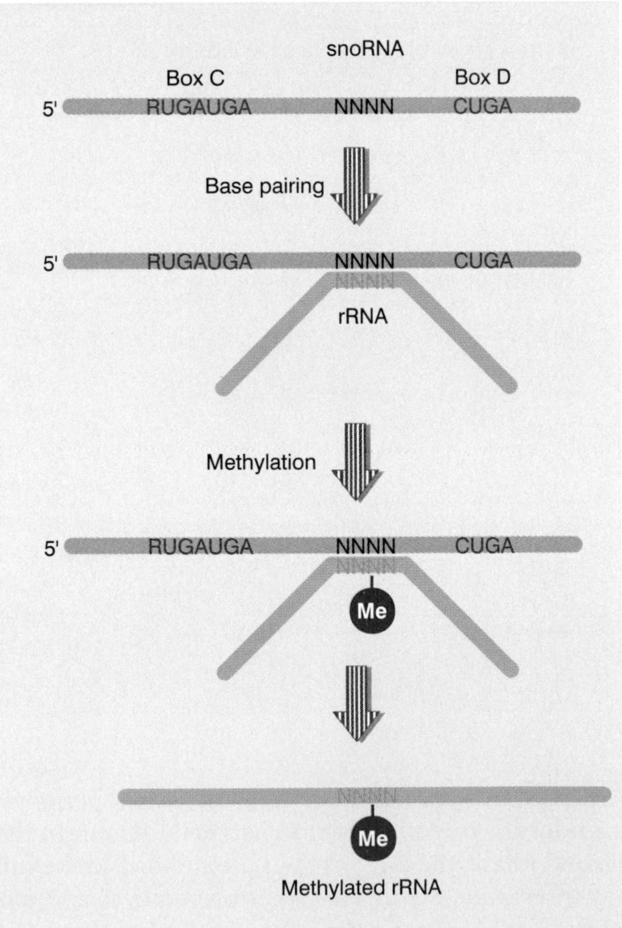

members and is named for the presence of an ACA triplet 3 nucleotides from the 3′ end. Some of the ACA snoRNAs can be implicated in individual pseudouridine processing events: when the snoRNA is deleted or mutated, the pseudouridine is not generated. Their properties suggest the model shown in **Figure 22.32**. The snoRNA folds into a secondary structure in which the ACA triplet is at one end of a hairpin that has some unpaired regions. The unpaired regions (A and B) pair with complementary sequences in the rRNA. The Ψ is synthesized at a position 15 bases away from the ACA triplet.

Pseudouridine is synthesized by conversion of uridine; the reaction involves breaking the bond that connects ribose to the N1 of uridine, and reconnecting the ribose to position C5 of the uridine. The known uridine synthases are proteins that function without an RNA cofactor. Synthases that could be involved in snoRNA-mediated pseudouridine synthesis have not been identified.

The involvement of the U7 snRNA in 3′ end generation, and the role of snoRNAs in rRNA processing and modification, is consistent with the view we develop in Chapter 23 that many—perhaps all—RNA processing events depend on RNA-RNA interactions. As with splicing reactions, the snRNA probably functions in the form of a ribonucleoprotein particle containing proteins as well as the RNA. It is common (although not the only mechanism of action) for the RNA of the particle to base pair with a short sequence in the substrate RNA.

Figure 22.32 An ACA group snoRNA base pairs with rRNA to determine the position of pseudouridine modification.

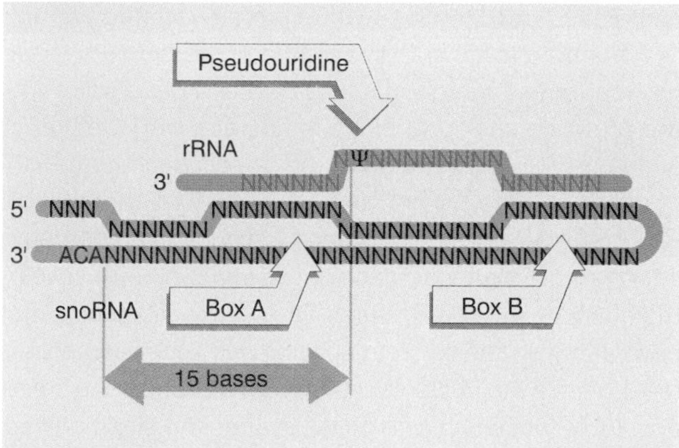

Summary

Splicing accomplishes the removal of introns and the joining of exons into the mature sequence of RNA. There are at least four types of reaction, as distinguished by their requirements *in vitro* and the intermediates that they generate. The systems include eukaryotic nuclear introns, group I and group II introns, and tRNA introns. Each reaction involves a change of organization within an individual RNA molecule, and is therefore a *cis*-acting event.

Nuclear splicing follows preferred but not obligatory pathways. Only very short consensus sequences are necessary; the rest of the intron appears irrelevant. All 5′ splice sites are probably equivalent, as are all 3′ splice sites. We do not know how 5′ and 3′ sites are linked only in the proper pairs. The required sequences are given by the GT-AG rule, which describes the ends of the intron. The UACUAAC branch site of yeast, or a less well-conserved consensus in mammalian introns, is also required. The reaction with the 5′ splice site involves formation of a lariat that joins the GU end of the intron via a 5′–2′ linkage to the A at position 6 of the branch site. Then the 3′-OH end of the exon attacks the 3′ splice site, so that the exons are ligated and the intron is released as a lariat. Both reactions are transesterifications in which bonds are conserved. Several stages of the reaction require hydrolysis of ATP, probably to drive conformational changes in the RNA and/or protein components. Lariat formation is responsible for choice of the 3′ splice site. Alternative splicing patterns are caused by protein factors that either stimulate use of a new site or block use of the default site.

Nuclear splicing requires formation of a spliceosome, a large particle that assembles the consensus sequences into a reactive conformation. The spliceosome contains the U1, U2, U4/U6, and U5 snRNPs and some additional splicing factors. The U1, U2, and U5 snRNPs each contain a single snRNA and several proteins; the U4/U6 snRNP contains 2 snRNAs and several proteins. Some proteins are common to all snRNP particles. The snRNPs recognize consensus sequences. U1 snRNA base pairs with the 5′ splice site, U2 snRNA base pairs with the branch sequence, U5 snRNP acts at the 5′ splice site. When U4 releases U6, the U6 snRNA base pairs with U2, and this may create the catalytic center for splicing. An alternative set of snRNPs provides analogous functions for splicing the subclass of ATAC introns. The snRNA molecules may have catalytic-like roles in splicing and other processing reactions. In the nucleolus, two groups of snoRNAs are responsible for pairing with rRNAs at sites that are modified; group C/D snoRNAs indicate target sites for methylation, and group ACA snoRNAs identify sites where uridine is converted to pseudouridine.

Splicing is usually intramolecular, but some cases have been found of *trans*- (intermolecular) splicing. These reactions probably occur by spliceosome formation with the appropriate site sequences on each molecule.

Group II introns share with nuclear introns the use of a lariat as intermediate, but are able to perform the reaction as a self-catalyzed property of the RNA. These introns follow the GT-AG rule, but form a characteristic secondary structure that holds the reacting splice sites in the appropriate apposition.

Yeast tRNA splicing involves separate endonuclease and ligase reactions. The endonuclease recognizes the secondary (or tertiary) structure of the precursor and cleaves both ends of the intron. The two half-tRNAs released by loss of the intron can be ligated in the presence of ATP.

The termination capacity of RNA polymerase II has not been characterized, and 3′ ends of its transcripts are generated by cleavage. The sequence AAUAAA, located 11–30 bases upstream of the cleavage site, provides the signal for both cleavage and polyadenylation. An endonuclease and the poly(A) polymerase are associated in a complex with other factors that confer specificity for the AAUAAA signal.

Further reading

Reviews

Birnstiel, M. L. *et al.* (1985). Transcription termination and 3′ processing: the end is in site. *Cell* **41**, 349–359.

Padgett, R. A. *et al.* (1986). Splicing of messenger RNA precursors. *Ann. Rev. Biochem.* **55**, 1119–1150.

Maniatis, T. and Reed, R. (1987). The role of small nuclear ribonucleoprotein particles in pre-mRNA splicing. *Nature* **325**, 673–678.

Sharp, P. A. (1987). Splicing of mRNA precursors. *Science* **235**, 766–771.

Guthrie, C. and Patterson, B. (1988). Spliceosomal snRNAs. *Ann. Rev. Genet.* **22**, 387–419.

Green, M. R. (1991). Biochemical mechanisms of constitutive and regulated pre-mRNA splicing. *Ann. Rev. Cell Biol.* **7**, 559–599.

Guthrie, C. (1991). Messenger RNA splicing in yeast: clues to why the spliceosome is a ribonucleoprotein. *Science* **253**, 157–163.

Wahle, E. and Keller, W. (1992). The biochemistry of 3′-end cleavage and polyadenylation of messenger RNA precursors. *Ann. Rev. Biochem.* **61**, 419–440.

Dreyfuss, G. *et al.* (1993). hnRNP proteins and the biogenesis of mRNA. *Ann. Rev. Biochem.* **62**, 289–321.

Nilsen, T. (1993). *trans*-splicing of nematode pre-mRNA. *Ann. Rev. Microbiol.* **47**, 413–440.

Weiner, A. (1993). mRNA splicing and autocatalytic introns: distant cousins or the products of chemical determinism. *Cell* **72**, 161–164.

Madhani, H. D. and Guthrie, C. (1994). Dynamic RNA-RNA interactions in the spliceosome. *Ann. Rev. Genet.* **28**, 1–26.

Sharp, P. A. (1994). Split genes and RNA splicing. *Cell* **77**, 805–815.

Krämer, A. (1996). The structure and function of proteins involved in mammalian pre-mRNA splicing. *Ann. Rev. Biochem.* **65**, 367–409.

Discoveries

Splice site recognition

Reed, R. and Maniatis, T. (1985). Intron sequences involved in lariat formation during pre-mRNA splicing. *Cell* **41**, 95–105.

Reed, R. and Maniatis, T. (1986). A role for exon sequences and splice-site proximity in splice-site selection. *Cell* **46,** 681–690.

Zhuang, Y. and Weiner, A.M. (1986). A compensatory base change in U1 snRNA suppresses a 5′ splice site mutation. *Cell* **46,** 827–835.

Parker, R., Siliciano, P. G., and Guthrie, C. (1987). Recognition of the TACTAAC box during mRNA splicing in yeast involves base pairing to the U2-like snRNA. *Cell* **49,** 229–239.

Patterson, B. and Guthrie, C. (1987). An essential yeast snRNA with a U5-like domain is required for splicing *in vivo*. *Cell* **49,** 613–624.

Zhuang, Y. A., Goldstein, A. M., and Weiner, A. M. (1989). UACUAAC is the preferred branch site for mammalian mRNA splicing. *Proc. Nat. Acad. Sci. USA* **86,** 2752–2756.

Burgess, S., Couto, J. R., and Guthrie, C. (1990). A putative ATP binding protein influences the fidelity of branchpoint recognition in yeast splicing. *Cell* **60,** 705–717.

Spliceosomes

Grabowski, P. J., Seiler, S. R,. and Sharp, P. A. (1985). A multicomponent complex is involved in the splicing of messenger RNA precursors. *Cell* **42,** 345–353.

Lamond, A. I. (1988). Spliceosome assembly involves the binding and release of U4 small nuclear ribonucleoprotein. *Proc. Nat. Acad. Sci. USA* **85,** 411–415.

Reed, R., Griffith, J., and Maniatis, T. (1988). Purification and visualization of native spliceosomes. *Cell* **53,** 949–961.

Newman, A. and Norman, C. (1991). Mutations in yeast U5 snRNA alter the specificity of 5′ splice site cleavage. *Cell* **65,** 115–123.

Madhani, H. D. and Guthrie, C. (1992). A novel base-pairing interaction between U2 and U6 snRNAs suggests a mechanism for the catalytic activation of the spliceosome. *Cell* **71,** 803–817.

Lesser, C. F. and Guthrie, C. (1993). Mutations in U6 snRNA that alter splice site specificity: implications for the active site. *Science* **262,** 1982–1988.

Sontheimer, E. J. and Steitz, J. A. (1993). The U5 and U6 small nuclear RNAs as active site components of the spliceosome. *Science* **262,** 1989–1996.

Tarn, W.-Y. and Steitz, J. (1996). A novel spliceosome containing U11, U12, and U5 snRNPs excises a minor class (AT-AC) intron *in vitro*. *Cell* **84,** 801–811.

Dietrich, R. C., Incorvaia, R., and Padgett, R. A. (1997). Terminal intron dinucleotide sequences do not distinguish between U2- and U12-dependent introns. *Molecular Cell* **1,** 151–160.

Burge, C. B., Padgett, R. A., and Sharp, P. A. (1998). Evolutionary fates and origins of U12-type introns. *Molecular Cell* **2,** 773–785.

trans-splicing

Murphy, W. J., Watkins, K. P., and Agabian, N. (1986). Identification of a novel Y branch structure as an intermediate in trypanosome mRNA processing: evidence for *trans*-splicing. *Cell* **47,** 517–525.

Sutton, R. and Boothroyd, J. C. (1986). Evidence for *trans*-splicing in trypanosomes. *Cell* **47,** 527–535.

Krause, M. and Hirsh, D. (1987). A *trans*-spliced leader sequence on actin mRNA in *C. elegans*. *Cell* **49,** 753–761.

Huang, X. Y. and Hirsh, D. (1989). A second *trans*-spliced RNA leader sequence in the nematode *C. elegans*. *Proc. Nat. Acad. Sci. USA* **86,** 8640–8644.

Hannon, G. J. *et al.* (1990). *trans*-splicing of nematode pre-mRNA *in vitro*. *Cell* **61,** 1247–1255.

Bruzik, J. P. and Maniatis, T. (1995). Enhancer-dependent interaction between 5′ and 3′ splice sites in *trans*. *Proc. Nat. Acad. Sci. USA* **92,** 7056–7059.

tRNA splicing

Mattoccia, E. *et al.* (1988). Site selection by the tRNA splicing endonuclease of *X. laevis*. *Cell* **55,** 731–738.

Reyes, V. M. and Abelson, J. (1988). Substrate recognition and splice site determination in yeast tRNA splicing. *Cell* **55,** 719–730.

Baldi, I. M. *et al.* (1992). Participation of the intron in the reaction catalyzed by the *Xenopus* tRNA splicing endonuclease. *Science* **255,** 1404–1408.

Other splicing systems

Sidrauski, C., Cox, J. S., and Walter, P. (1996). tRNA ligase is required for regulated mRNA splicing in the unfolded protein response. *Cell* **87,** 405–413.

Sidrauski, C. and Walter, P. (1997). The transmembrane kinase Ire1p is a site-specific endonuclease that initiates mRNA splicing in the unfolded protein response. *Cell* **90,** 1031–1039.

Nucleolar RNA processing

Kass, S. *et al.* (1990). The U3 small nucleolar ribonucleoprotein functions in the first step of preribosomal RNA processing. *Cell* **60,** 897–908.

Kiss-László, Z. *et al.* (1996). Site-specific ribose methylation of preribosomal RNA: a novel function for small nucleolar RNAs. *Cell* **85,** 1068–1077.

Ni, J., Tien, A. L., and Fournier, M. J. (1997). Small nucleolar RNAs direct site-specific synthesis of pseudouridine in rRNA. *Cell* **89,** 565–573.

3′ end generation

Galli, G. *et al.* (1983). Biochemical complementation with RNA in the *Xenopus* oocyte: a small RNA is required for the generation of 3′ histone mRNA termini. *Cell* **34,** 823–828.

Conway, L. and Wickens, M. (1985). A sequence downstream of AAUAAA is required for formation of SV40 late mRNA 3′ termini in frog oocytes. *Proc. Nat. Acad. Sci. USA* **82,** 3949–3953.

Gil, A. and Proudfoot, N. (1987). Position-dependent sequence elements downstream of AAUAAA are required for efficient rabbit β-globin mRNA 3′ end formation. *Cell* **49,** 399–406.

Takagaki, Y., Ryner, L. C., and Manley, J. L. (1988). Separation and characterization of a poly(A) polymerase and a cleavage/specificity factor required for pre-mRNA polyadenylation. *Cell* **52,** 731–742.

Chapter 23

Catalytic RNA

THE idea that only proteins have enzymatic activity was deeply rooted in biochemistry. (Yet devotées of protein function once thought that only proteins could have the versatility to be the genetic material!) A rationale for the identification of enzymes with proteins lies in the view that only proteins, with their varied three-dimensional structures and variety of side-groups, have the flexibility to create the active sites that catalyze biochemical reactions. But the characterization of systems involved in RNA processing has shown this view to be an oversimplification.

Several types of catalytic reactions are now known to reside in RNA. **Ribozyme** has become a general term used to describe an RNA with catalytic activity, and it is possible to characterize the enzymatic activity in the same way as a more conventional enzyme. Some RNA catalytic activities are directed against separate substrates, while others are intramolecular (which limits the catalytic action to a single cycle):

- The enzyme ribonuclease P is a ribonucleoprotein that contains a single RNA molecule bound to a protein. The RNA possesses the ability to catalyze cleavage in a tRNA substrate, while the protein component plays an indirect role, probably to maintain the structure of the catalytic RNA.
- Small RNAs of the virusoid class have the ability to perform a self-cleavage reaction. Although this reaction is intramolecular, the molecule can be divided into an "enzymatic" and a "substrate" part, and engineering of related sequences can create "enzymes" that act upon independent "substrates."
- Introns of the group I and group II classes possess the ability to splice themselves out of the pre-mRNA that contains them. Engineering of group I introns has generated RNA molecules that have several other catalytic activities related to the original activity.

The common theme of these reactions is that the RNA can perform an intramolecular or intermolecular reaction that involves cleavage or joining of phosphodiester bonds *in vitro*. Although the specificity of the reaction and the basic catalytic activity is provided by RNA, proteins associated with the RNA may be needed for the reaction to occur efficiently *in vivo*.

RNA splicing is not the only means by which changes can be introduced in the informational content of RNA. In the process of **RNA editing,** changes are introduced at individual bases, or bases are added at particular positions within an mRNA. The insertion of bases (most commonly uridine residues) occurs for several genes in the mitochondria of certain lower eukaryotes; like splicing, it involves the breakage and reunion of bonds between nucleotides, but also requires a template for coding the information of the new sequence.

Group I introns undertake self-splicing by transesterification

GROUP I introns are found in diverse locations. They occur in the genes coding for rRNA in the nuclei of the lower eukaryotes *Tetrahymena thermophila* (a ciliate) and *Physarum polycephalum* a (slime mold). They are common in the genes of fungal mitochondria. They are present in three genes of phage T4 and also are found in bacteria. Group I introns have an intrinsic ability to splice themselves. This is called **self-splicing** or **autosplicing.** (This property is found also in the group II introns discussed in Chapter 22.)

Self-splicing was discovered as a property of the transcripts of the rRNA genes in *T. thermophila.* The genes for the two major rRNAs follow the usual organization, in which both are expressed as part of a common transcription unit. The product is a 35S precursor RNA with the sequence of the small rRNA in the 5′ part, and the sequence of the larger (26S) rRNA toward the 3′ end.

In some strains of *T. thermophila,* the sequence coding for 26S rRNA is interrupted by a single, short intron. When the 35S precursor RNA is incubated *in vitro,* splicing occurs as an autonomous reaction. The intron is excised from the precursor and accumulates as a linear fragment of 400 bases, which is subsequently converted to a circular RNA. These events are summarized in **Figure 23.1.**

The reaction requires only a monovalent cation, a divalent cation, and a guanine nucleotide cofactor. No other base can be substituted for G; but a triphosphate is not needed; GTP, GDP, GMP, and guanosine itself can all be used, so there is no net energy requirement. The guanine nucleotide must have a 3′-OH group.

The fate of the guanine nucleotide can be followed by using a radioactive label. The radioactivity initially enters the excised linear intron fragment. The G residue becomes linked to the 5′ end of the intron by a normal phosphodiester bond.

Figure 23.2 shows that three transfer reactions occur. In the first transfer, the guanine nucleotide behaves as a cofactor that provides a free 3′-OH group that attacks the 5′ end of the intron. This reaction creates the G-intron link and generates a 3′-OH group at the end of the exon. The second transfer involves a similar chemical reaction, in which this 3′-OH then attacks the second exon. The two transfers are connected; no free exons have been observed, so their ligation might occur as part of the same reaction that releases the intron. The intron is released as a linear molecule, but the third transfer reaction converts it to a circle.

Each stage of the self-splicing reaction occurs by a transesterification, in which one phosphate ester is converted directly into another, without any intermediary hydrolysis. Bonds are exchanged directly, and energy is conserved, so the reaction does not require input of energy from hydrolysis of ATP or GTP. (There is a parallel for the transfer of bonds without net input of energy in the DNA nicking-closing enzymes discussed in Chapter 14.)

Figure 23.1 Splicing of the *Tetrahymena* 35S rRNA precursor can be followed by gel electrophoresis. The removal of the intron is revealed by the appearance of a rapidly moving small band. When the intron becomes circular, it electrophoreses more slowly, as seen by a higher band.

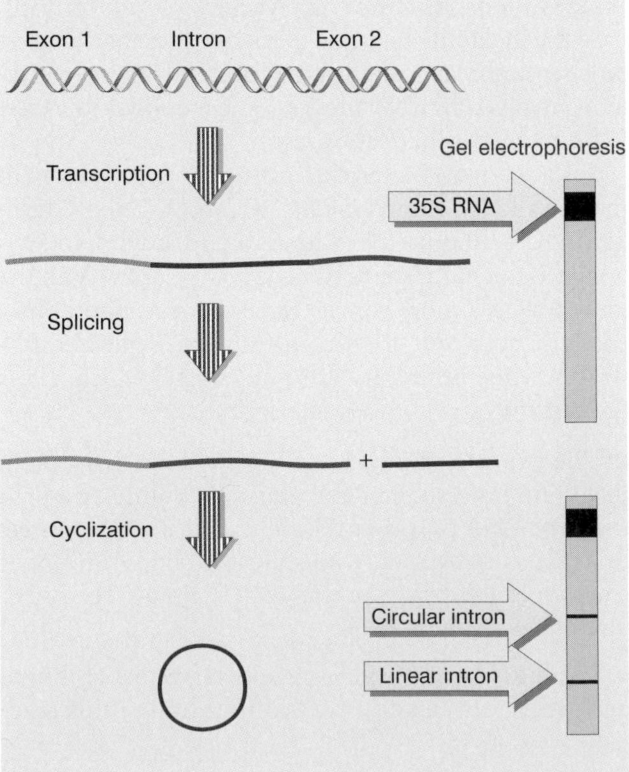

Figure 23.2 Self-splicing occurs by transesterification reactions in which bonds are exchanged directly. The bonds that have been generated at each stage are indicated by the shaded boxes.

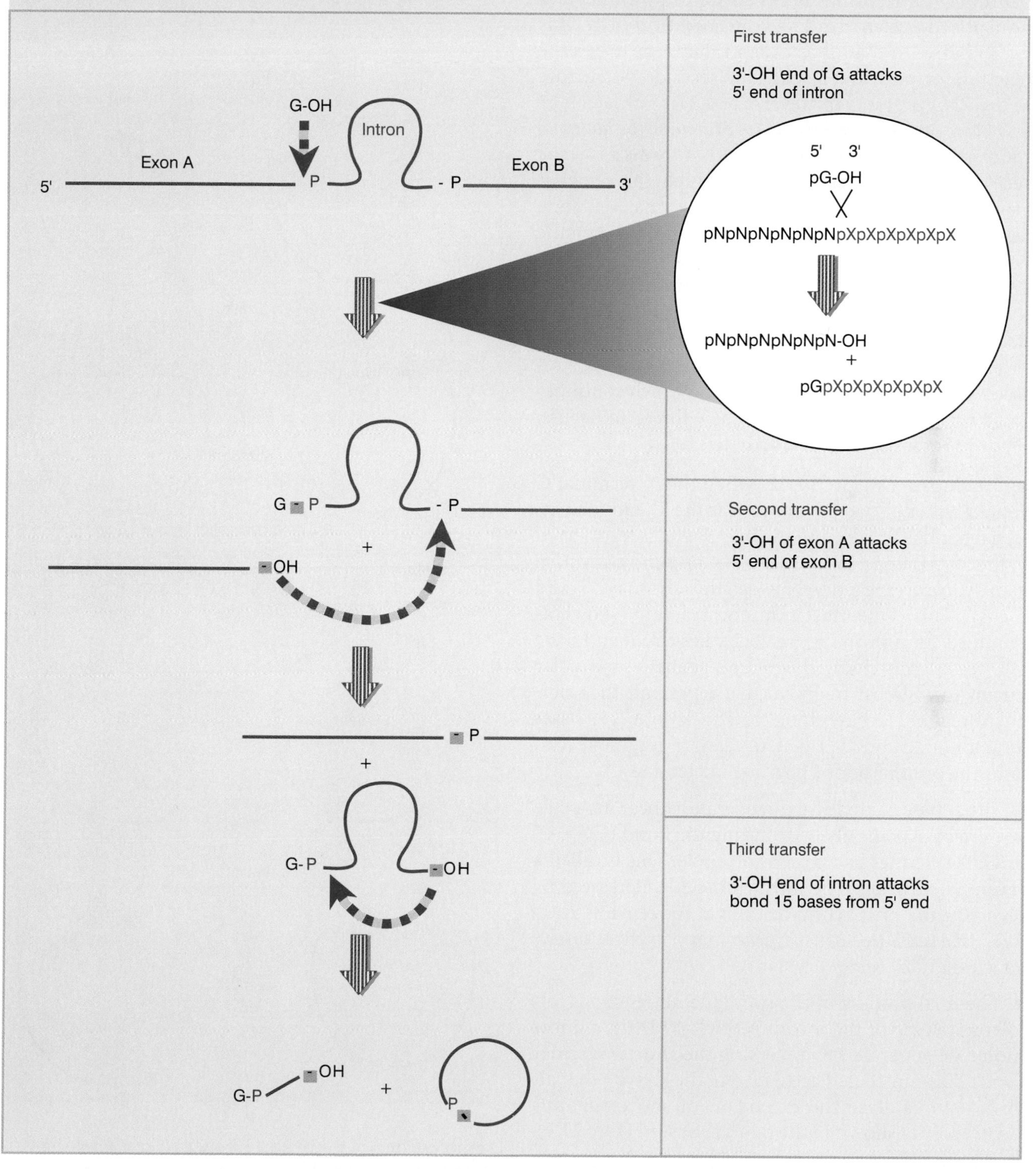

If each of the consecutive transesterification reactions involves no net change of energy, why does the splicing reaction proceed to completion instead of coming to equilibrium between spliced product and nonspliced precursor? The concentration of GTP is high relative to that of RNA, and therefore drives the reaction forward; and a change in secondary structure in the RNA prevents the reverse reaction (see later).

The in vitro *system includes no protein so the ability to splice is intrinsic to the RNA.* The RNA forms a specific secondary/tertiary structure in which the relevant groups are brought into juxtaposition so that a guanine nucleotide can be bound to a specific site and then the bond breakage and reunion reactions shown in Figure 23.2 can occur. Although a property of the RNA itself, the reaction is assisted *in vivo* by proteins, which stabilize the RNA structure.

The ability to engage in these transfer reactions resides with the sequence of the intron, which continues to be reactive after its excision as a linear molecule. **Figure 23.3** summarizes its activities:

- The intron can circularize when the 3′ terminal G attacks either of two positions near the 5′ end. The internal bond is broken and the new 5′ end is transferred to the 3′-OH end of the intron. The *primary cyclization* usually involves reaction between the terminal G^{414} and the A^{16}. This is the most common reaction (shown as the third transfer in Figure 23.2). Less frequently, the G^{414} reacts with U^{20}. Each reaction generates a circular intron and a linear fragment that represents the original 5′ region (15 bases long for attack on A^{16}, 19 bases long for attack on U^{20}). The released 5′ fragment contains the original added guanine nucleotide.
- Either type of circle can regenerate a linear molecule *in vitro* by specifically hydrolyzing the bond (G^{414}–A^{16} or G^{414}–U^{20}) that had closed the circle. This is called a *reverse cyclization.* The linear molecule generated by reversing the primary cyclization at A^{16} remains reactive, and can perform a secondary cyclization by attacking U^{20}.
- The final product of the spontaneous reactions following release of the intron is the L-19 RNA, a linear molecule generated by reversing the shorter circular form. This molecule has an enzymatic activity that allows it to catalyze the extension of short oligonucleotides (not shown in the figure, but see Figure 23.8).
- The reactivity of the released intron extends beyond merely reversing the cyclization reaction. Addition of the oligonucleotide UUU reopens the primary circle by reacting with the G^{414}–A^{16} bond. The UUU (which resembles the 3′ end of the 15-mer released by the primary cyclization) becomes the 5′ end of the linear molecule that is formed. This is an *intermolecular* reaction, and thus demonstrates the ability to connect together two different RNA molecules.

Figure 23.3 The excised intron can form circles by using either of two internal sites for reaction with the 5′ end, and can reopen the circles by reaction with water or oligonucleotides.

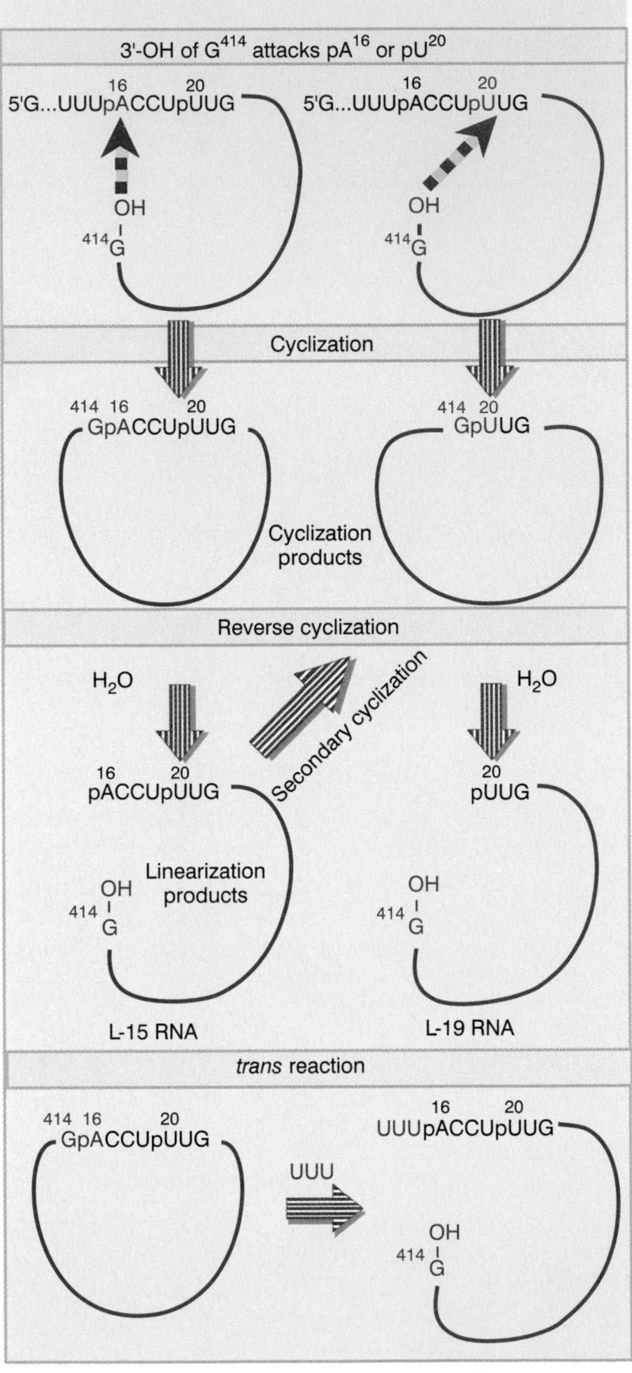

This series of reactions demonstrates vividly that the autocatalytic activity reflects a generalized ability of the RNA molecule to form an active center that can bind guanine cofactors, recognize oligonucleotides, and bring together the reacting groups in a conformation that allows bonds to be broken and rejoined. Other group I introns have not been investigated in as much detail as the *Tetrahymena* intron, but their properties are generally similar.

The autosplicing reaction is an intrinsic property of RNA *in vitro*, but to what degree are proteins involved *in vivo?* Some indications for the involvement of proteins are provided by mitochondrial systems, where splicing of group I introns requires the *trans*-acting products of other genes. One striking case is presented by the *cyt18* mutant of *N. crassa*, which is defective in splicing several mitochondrial group I introns. The product of this gene turns out to be the mitochondrial tyrosyl-tRNA synthetase! One possible implication is that the intron might take up a tRNA-like tertiary structure that is recognized by the synthetase; binding of the enzyme might be directly involved in splicing, but more probably has an indirect effect such as stabilizing the RNA conformation.

This relationship between the synthetase and splicing is consistent with the idea that splicing originated as an RNA-mediated reaction, subsequently assisted by RNA-binding proteins that originally had other functions. The *in vitro* self-splicing ability may represent the basic biochemical interaction; the RNA structure creates the active site, but is able to function efficiently *in vivo* only when assisted by a protein complex.

Group I introns form a characteristic secondary structure

All group I introns can be organized into a characteristic secondary structure, with 9 helices (P1–P9). **Figure 23.4** shows a model for the secondary structure of the *Tetrahymena* intron. Two of the base-paired regions are generated by pairing between conserved sequence elements that are common to group I introns. P4 is constructed from the sequences *P* and *Q;* P7 is formed from sequences *R* and *S*. The other base-paired regions vary in sequence in individual introns. Mutational analysis identifies an intron "core," containing P3, P4, P6, and P7, which provides the minimal region that can undertake a catalytic reaction. The lengths of group I introns vary widely, and the consensus sequences are located a considerable distance from the actual splice junctions.

Some of the pairing reactions are directly involved in bringing the splice junctions into a conformation that supports the enzymatic reaction. P1 includes the 3′ end of the left exon. The sequence within the intron that pairs with the exon is called the IGS, or internal guide sequence. (Its name reflects the fact that originally the region immediately 3′ to the IGS sequence shown in the figure was thought to pair with the 3′ splice junction, thus bringing the two junctions together. This interaction may occur, but does not seem to be essential.) A very short sequence, sometimes as short as 2 bases, between P7 and P9, base pairs with the sequence that immediately precedes the reactive G (position 414 in *Tetrahymena*) at the 3′ end of the intron.

The importance of base pairing in creating the necessary core structure in the RNA is emphasized by the properties of *cis*-acting mutations that prevent splicing of group I introns. Such mutations have been isolated for the mitochondrial introns through mutants that cannot remove an intron *in vivo*, and they have been isolated for the *Tetrahymena* intron by transferring the splicing reaction into a bacterial environment. The construct shown in **Figure 23.5** allows the splicing reaction to be followed in *E. coli*. The self-splicing intron is placed at a location that interrupts the tenth codon of the β-galactosidase coding sequence. The protein can therefore be successfully translated from an RNA only after the intron has been removed.

The synthesis of β-galactosidase in this system indicates that splicing can occur in conditions quite distant from those prevailing in *Tetrahymena* or even *in vitro*. One interpretation of this result is that self-splicing can occur in the bacterial cell. Another possibility is that there are bacterial proteins that assist the reaction.

Using this assay, we can introduce mutations into

Figure 23.4 Group I introns have a common secondary structure that is formed by 9 base paired regions. The sequences of regions P4 and P7 are conserved, and identify the individual sequence elements P, Q, R, and S. P1 is created by pairing between the end of the left exon and the IGS of the intron; a region between P7 and P9 pairs with the 3' end of the intron.

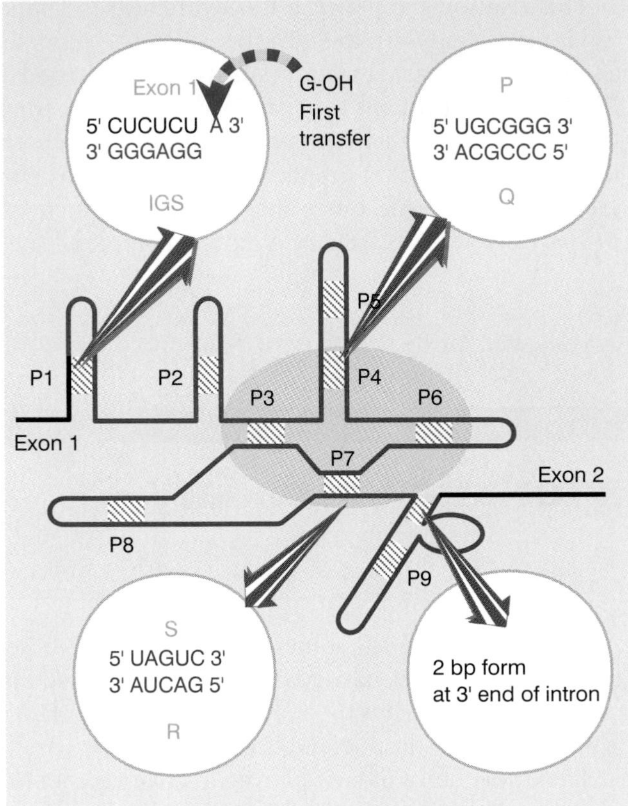

Figure 23.5 Placing the *Tetrahymena* intron within the β-galactosidase coding sequence creates an assay for self-splicing in *E. coli.* Synthesis of β-galactosidase can be tested by adding a compound that is turned blue by the enzyme. The sequence is carried by a bacteriophage, so the presence of blue plaques indicates successful splicing.

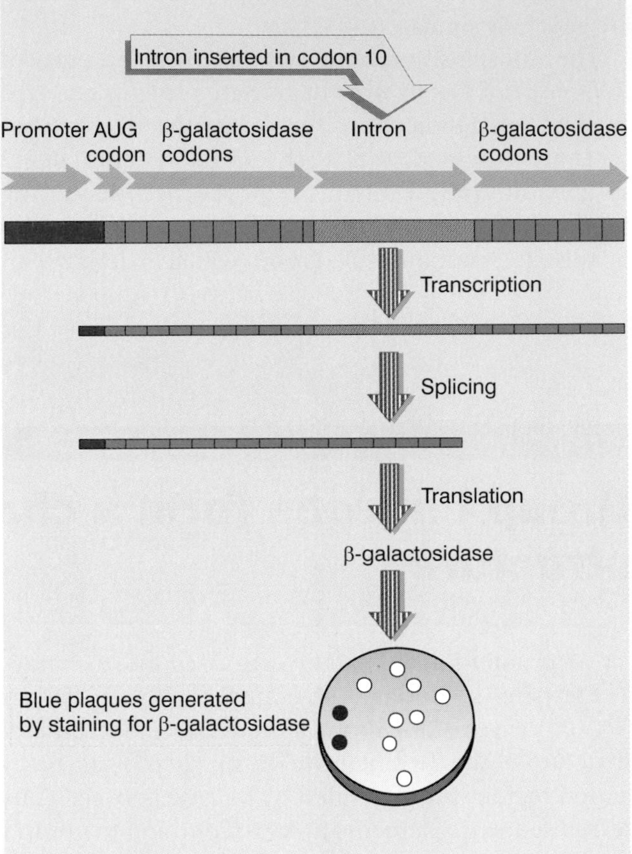

the intron to see whether they prevent the reaction. Mutations in the group I consensus sequences that disrupt their base pairing stop splicing. The mutations can be reverted by making compensating changes that restore base pairing.

Mutations in the corresponding consensus sequences in mitochondrial group I introns have similar effects. A mutation in one consensus sequence may be reverted by a mutation in the complementary consensus sequence to restore pairing; for example, mutations in the R consensus can be compensated by mutations in the S consensus.

Together these results suggest that the group I splicing reaction depends on the formation of secondary structure between pairs of consensus sequences within the intron. The principle established by this work is that *sequences distant from the splice junctions themselves are required to form the active site that makes self-splicing possible.*

Ribozymes have various catalytic activities

THE catalytic activity of group I introns was discovered by virtue of their ability to autosplice, but they are able to undertake other catalytic reactions *in vitro.* All of these reactions are based on transesterifications. We analyze these reactions in terms of their relationship to the splicing reaction itself.

The catalytic activity of a group I intron is conferred by its ability to generate a particular secondary and tertiary structure that creates active sites, equivalent to the active sites of a conventional (proteinaceous) enzyme. **Figure 23.6** illustrates the splicing reaction in terms of these sites (this is the same series of reactions shown previously in Figure 23.2).

The P1 helix represents the formation of a substrate-binding site, in which the 3′ end of the first intron base pairs with the IGS in an intermolecular reaction. A guanosine-binding site is formed by sequences in P7. This site may be occupied either by a free guanosine nucleotide or by the G residue in position 414. In the first transfer reaction, it is used by free guanosine nucleotide; but it is subsequently occupied by G^{414}. The second transfer releases the joined exons. The third transfer creates the circular intron.

Binding to the substrate involves a change of conformation; before substrate binding, the 5′ end of the IGS is close to P2 and P8, but after binding, when it forms the P1 helix, it is close to conserved bases that lie between P4 and P5. The reaction is visualized by contacts that are detected in the secondary structure in **Figure 23.7.** In the tertiary structure, the two sites alternatively contacted by P1 are 37 Å apart, which implies a substantial movement in the position of P1.

The L-19 RNA is generated by opening the circular intron (the last stage of the intramolecular rearrangements shown in Figure 23.3). It still retains enzymatic abilities. These resemble the activities involved in the original splicing reaction, and we may consider ribozyme function in terms of the ability to bind an intramolecular sequence complementary to the IGS in the substrate-binding site, while binding either the terminal G^{414} or a free G-nucleotide in the G-binding site.

Figure 23.8 illustrates the mechanism by which it extends the oligonucleotide C_5 to generate a C_6 chain. The C_5 oligonucleotide binds in the substrate-binding site, while G^{414} occupies the G-binding site. By transesterification reactions, a C is transferred from C_5 to the 3′-terminal G, and then back to a new C_5 molecule.

Figure 23.6 Excision of the group I intron in *Tetrahymenar* RNA occurs by successive reactions between the occupants of the guanosine-binding site and substrate-binding site. The left exon is red, and the right exon is purple. The part of the intron that includes the reactive 5' end is pink.

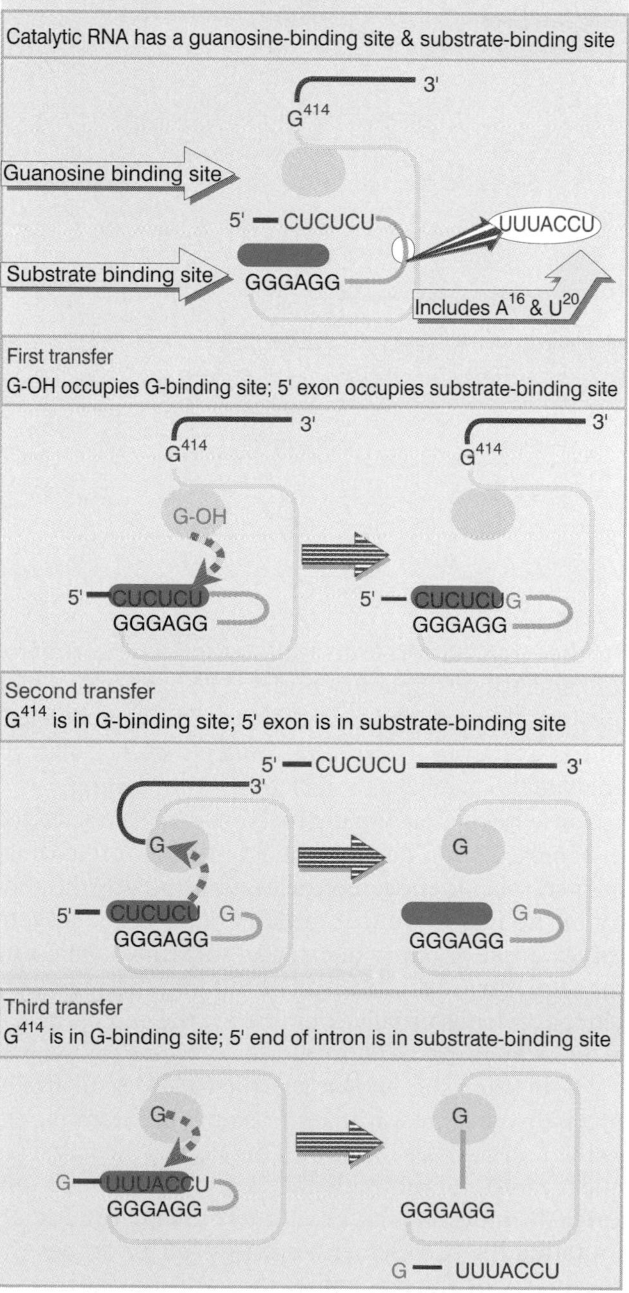

Figure 23.7 The position of the IGS in the tertiary structure changes when P1 is formed by substrate binding.

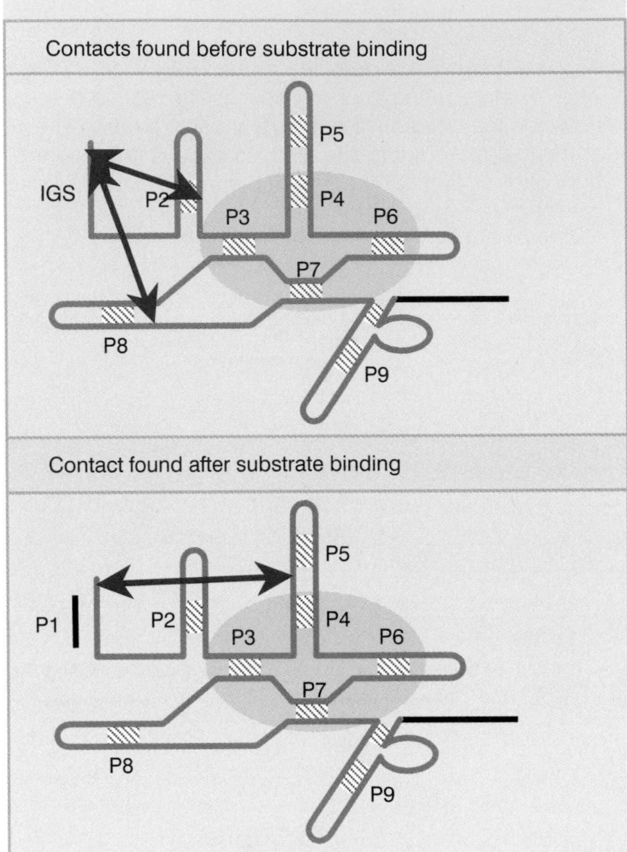

Figure 23.8 The L-19 linear RNA can bind C in the substrate-binding site; the reactive G-OH 3' end is located in the G-binding site, and catalyzes transfer reactions that convert 2 C_5 oligonucleotides into a C_4 and a C_6 oligonucleotide.

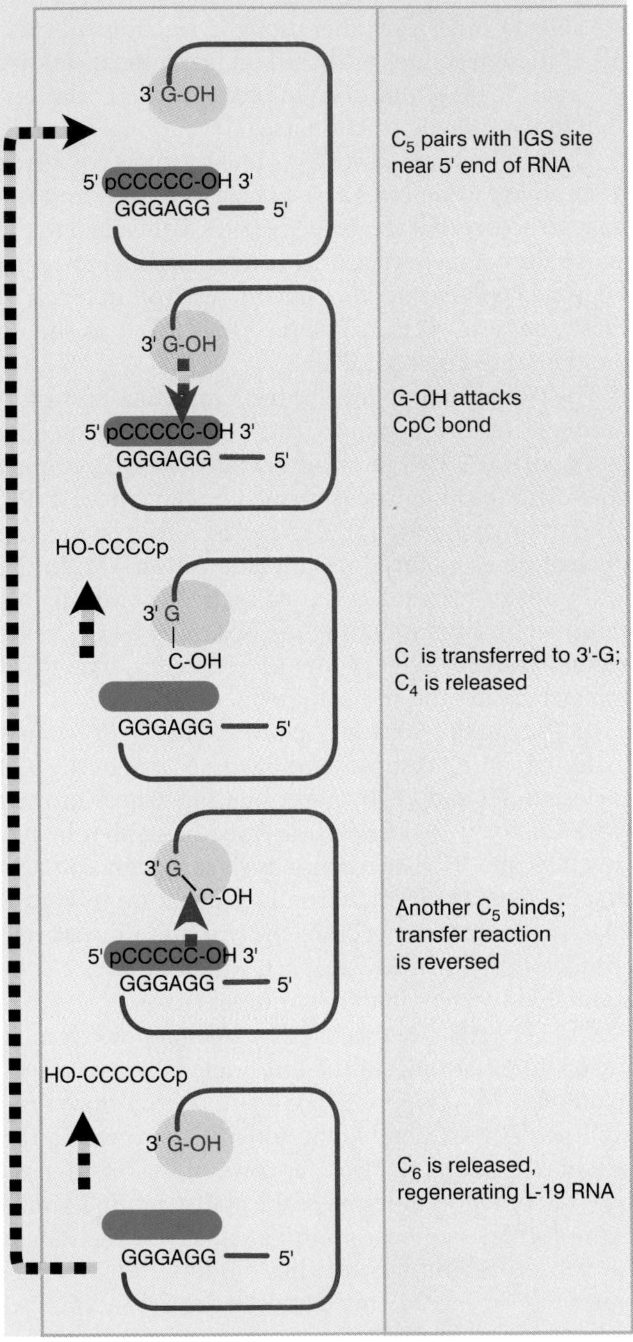

Further transfer reactions lead to the accumulation of longer cytosine oligonucleotides. The reaction is a true catalysis, because the L-19 RNA remains unchanged, and is available to catalyze multiple cycles. The ribozyme is behaving as a nucleotidyl transferase.

Some further enzymatic reactions are characterized in **Figure 23.9.** The ribozyme can function as a sequence-specific endoribonuclease by utilizing the ability of the IGS to bind complementary sequences. In this example, it binds an external substrate containing the sequence CUCU, instead of binding the analogous sequence that is usually contained at the end of the left exon. A guanine-containing nucleotide is present in the G-binding site, and attacks the CUCU sequence in precisely the same way that the exon is usually attacked in the first transfer reaction. This cleaves the target sequence into a 5′ molecule that resembles the left exon, and a 3′ molecule that bears a terminal G residue. By mutating the IGS element, it is possible to change the specificity of the ribozyme, so that it recognizes sequences complementary to the new sequence at the IGS region.

Altering the IGS, so that the specificity of the substrate-binding site is changed, and other RNA targets are enabled to enter, can be used to generate a ligase activity. An RNA terminating in a 3′-OH is bound in the substrate site, and an RNA terminating in a 5′-G residue is bound in the G-binding site. An attack by the hydroxyl on the phosphate bond connects the two RNA molecules, with the loss of the G residue.

The phosphatase reaction is not directly related to the splicing transfer reactions. An oligonucleotide sequence that is complementary to the IGS and terminates in a 3′-phosphate can be attacked by the G^{414}. The phosphate is transferred to the G^{414}, and an oligonucleotide with a free 3′-OH end is then released. The phosphate can then be transferred either to an oligonucleotide terminating in 3′-OH (effectively reversing the reaction) or indeed to water (releasing inorganic phosphate and completing an authentic phosphatase reaction).

Figure 23.9 Catalytic reactions of the ribozyme involve transesterifications between a group in the substrate-binding site and a group in the G-binding site.

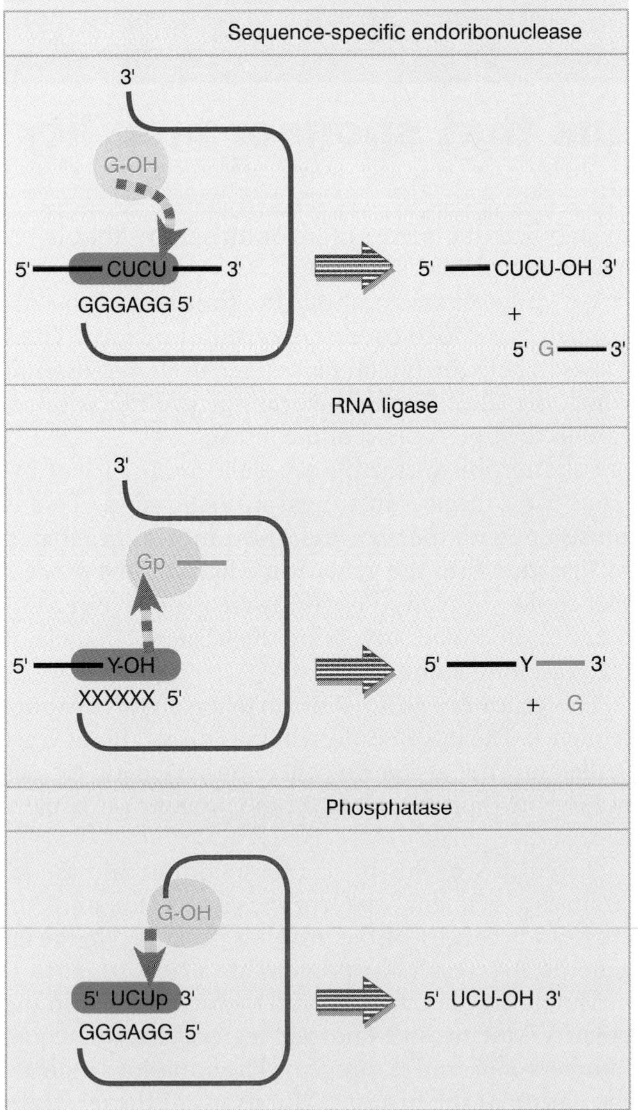

The reactions catalyzed by RNA can be characterized in the same way as classical enzymatic reactions in terms of Michaelis-Menten kinetics. **Figure 23.10** analyzes the reactions catalyzed by RNA. The K_M values for RNA-catalyzed reactions are low, and therefore imply that the RNA can bind its substrate with high specificity. The turnover numbers are low, which reflects a low catalytic rate. In effect, the RNA molecules behave in the same general manner as traditionally defined for enzymes, although they are relatively slow compared to protein catalysts (where a typical range of turnover numbers is 10^3–10^6).

How does RNA provide a catalytic center? Its ability seems reasonable if we think of an active center as a surface that exposes a series of active groups in a fixed relationship. In a protein, the active groups are provided by the side chains of the amino acids, which have appreciable variety, including positive and negative ionic groups and hydrophobic groups. In an RNA, the available moieties are more restricted, consisting primarily of the exposed groups of bases. Short regions are held in a particular structure by the secondary/tertiary conformation of the molecule, providing a surface of active groups able to maintain an environment in which bonds can be broken and made in another molecule. It seems inevitable that the interaction between the RNA catalyst and the RNA substrate will

Figure 23.10 Reactions catalyzed by RNA have the same features as those catalyzed by proteins, although the rates are slower. The K_M gives the concentration of substrate required for half-maximum velocity; this is an inverse measure of the affinity of the enzyme for substrate. The turnover number gives the number of substrate molecules transformed in unit time by a single catalytic site.

Enzyme	Substrate	K_M (mM)	Turnover (/min)
19 base virusoid	24 base RNA	0.0006	0.5
L-19 Intron	CCCCCC	0.04	1.7
RNAase P RNA	pre-tRNA	0.00003	0.4
RNAase P complete	pre-tRNA	0.00003	29
RNAase T1	GpA	0.05	5,700
β-galactosidase	lactose	4.0	12,500

rely on base pairing to create the environment. Divalent cations (typically Mg^{2+}) play an important role in structure, typically being present at the active site where they coordinate the positions of the various groups. They play a direct role in the endonucleolytic activity of virusoid ribozymes (see later).

The evolutionary implications of these discoveries are intriguing. The split personality of the genetic apparatus, in which RNA is present in all components, but proteins undertake catalytic reactions, has always been puzzling. It seems unlikely that the very first replicating systems could have contained both nucleic acid and protein.

But suppose that the first systems contained only a self-replicating nucleic acid with primitive catalytic activities, just those needed to make and break phosphodiester bonds. If we suppose that the involvement of 2′-OH bonds in current splicing reactions is derived from these primitive catalytic activities, we may argue that the original nucleic acid was RNA, since DNA lacks the 2′-OH group and therefore could not undertake such reactions. Proteins could have been added for their ability to stabilize the RNA structure. Then the greater versatility of proteins could have allowed them to take over catalytic reactions, leading eventually to the complex and sophisticated apparatus of modern gene expression.

Some introns code for proteins that sponsor mobility

CERTAIN introns of both the group I and group II classes contain open reading frames that are translated into proteins. Expression of the proteins allows the intron (either in its original DNA form or as a DNA copy of the RNA) to be **mobile:** it is able to insert itself into a new genomic site. Introns of both groups I and II are extremely widespread, being found in both prokaryotes and eukaryotes.

Three types of protein functions have been identified in open reading frames of these introns. They are concerned with either DNA or RNA metabolism:

- *endonucleases,* which cleave at target sites in DNA to allow the insertion of intron sequences;
- *reverse transcriptases,* which are involved in generating DNA copies of the intronic RNA;
- *maturases,* which are needed for splicing out the particular intron from the pre-mRNA.

Group I introns that have open reading frames code for endonucleases; sometimes a maturase activity is also associated with the protein. Group II introns that have open reading frames show both endonuclease and reverse transcriptase-like sequences; again, a maturase activity may be associated with the protein. In neither case is there any information about how the maturase function is related to the other enzymatic activities, but the most likely function for the maturase is to stabilize the intron in a conformation that is required for splicing.

We know the most about the function of the endonucleases coded by certain group I introns. Their role is to help the intron perpetuate itself in crosses in which the alleles for the relevant gene differ with regard to their possession of the intron.

Polymorphisms for the presence or absence of introns are common in fungal mitochondria. This is consistent with the view that these introns originated by insertion into the gene. Some light on the process that could be involved is cast by an analysis of recombination in crosses involving the large rRNA gene of the yeast mitochondrion.

This gene has a group I intron that contains a coding sequence. The intron is present in some strains of yeast (called ω^+) but absent in others (ω^-). Genetic crosses between ω^+ and ω^- are *polar:* the progeny are usually ω^+.

If we think of the ω^+ strain as a donor and the ω^- strain as a recipient, we form the view that in $\omega^+ \times \omega^-$ crosses a new copy of the intron is generated in the ω^- genome. As a result, the progeny are all ω^+.

Mutations can occur in either parent to abolish the polarity. Mutants show normal segregation, with equal numbers of ω^+ and ω^- progeny. The mutations indicate the nature of the process. Mutations in the ω^- strain occur close to the site where the intron would be inserted. Mutations in the ω^+ strain lie in the reading frame of the intron and prevent production of the protein. This suggests the model of **Figure 23.11,** in which the protein coded by the intron in an ω^+ strain recognizes the site where the intron should be inserted in an ω^- strain and causes it to be preferentially inherited.

What is the action of the protein? The product of the ω intron is an endonuclease *that recognizes the ω^- gene as a target for a double-strand break.* The endonuclease recognizes an 18 bp target sequence that contains the site where the intron is inserted. The target sequence is cleaved on each strand of DNA, 2 bases to the 3′ side of the insertion site. So the cleavage sites are 4 bp apart, and generate overhanging single strands.

This type of cleavage is related to the cleavage characteristic of transposons when they migrate to new sites (see Chapter 15). The double-strand break probably initiates a gene conversion process in which the sequence of the ω^+ gene is copied to replace the sequence of the ω^- gene. The reaction involves transposition by a duplicative mechanism, and occurs solely at the level of DNA. Note that the insertion of the intron interrupts the sequence recognized by the endonuclease, thus ensuring stability.

Other group I introns that contain open reading frames also are mobile. The general mechanism of intron perpetuation appears to be the same: the intron codes for an endonuclease that cleaves a specific target site where the intron will be inserted. There are differences in the details of insertion; for example, the endonuclease coded by the phage T4 *td* intron cleaves a target site that is 24 bp upstream of the site at which the intron is itself inserted.

In spite of the common mechanism for intron mobility, there is no homology between the sequences of the target sites or the intron coding regions. We assume that the introns have a common evolutionary origin, but evidently they have diverged greatly. The target sites are among the longest and therefore the most specific known for any endonucleases. The specificity ensures that the intron perpetuates itself only by insertion into a single target site and not elsewhere in the genome. This is called **intron homing.**

Introns carrying sequences that code for endonucleases are found in a variety of bacteria and lower

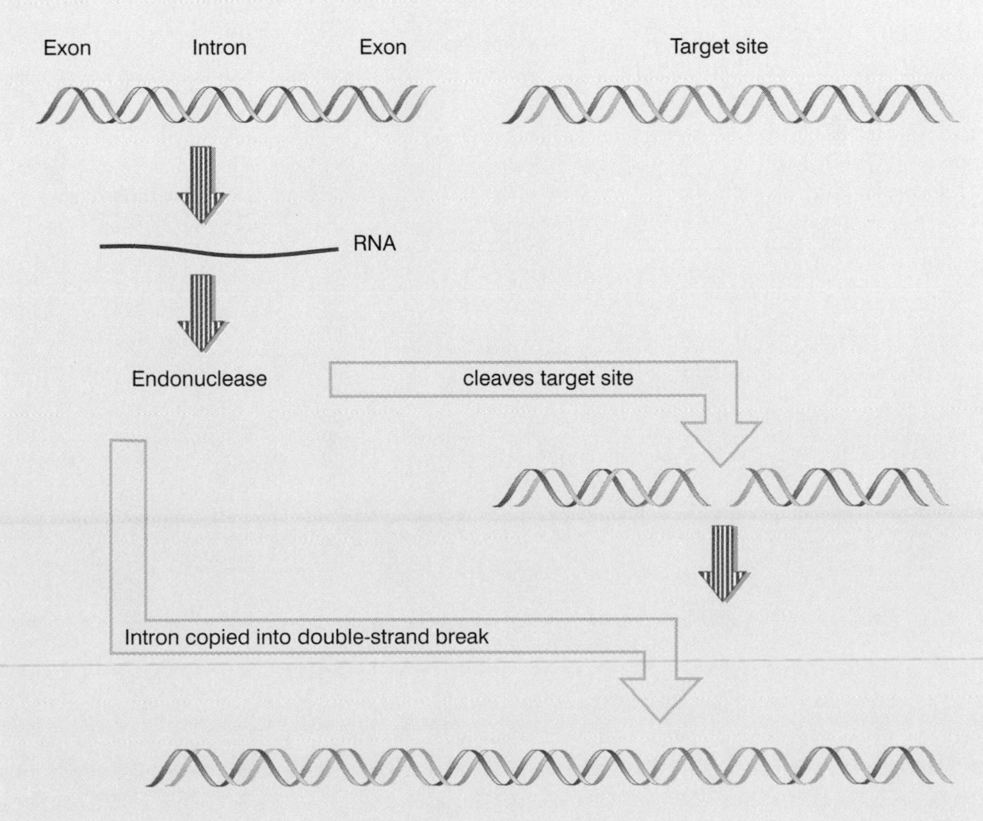
Exon
Intron
Exon
Target site
RNA
Endonuclease
cleaves target site
Intron copied into double-strand break

Figure 23.11 An intron codes for an endonuclease that makes a double-strand break in DNA. The sequence of the intron is duplicated and then inserted at the break.

eukaryotes. These results strengthen the view that introns carrying coding sequences originated as independent elements that coded for a function involved in the ability to be spliced out of RNA or to migrate between DNA molecules. Consistent with this idea, the pattern of codon usage is somewhat different in the intron coding regions from that found in the exons.

Most of the open reading frames contained in group II introns have regions that are related to reverse transcriptases (in addition to the endonuclease regions). Introns of this type are found in organelles of lower eukaryotes and also in some bacteria. The reverse transcriptase activity is specific for the intron, and is involved in homing. The reverse transcriptase generates a DNA copy of the intron from the pre-mRNA, and thus allows the intron to become mobile by a mechanism resembling that of retroviruses (discussed in Chapter 16). The type of retrotransposition involved in this case resembles that of a group of retroposons that lack LTRs, and which generate the 3′-OH needed

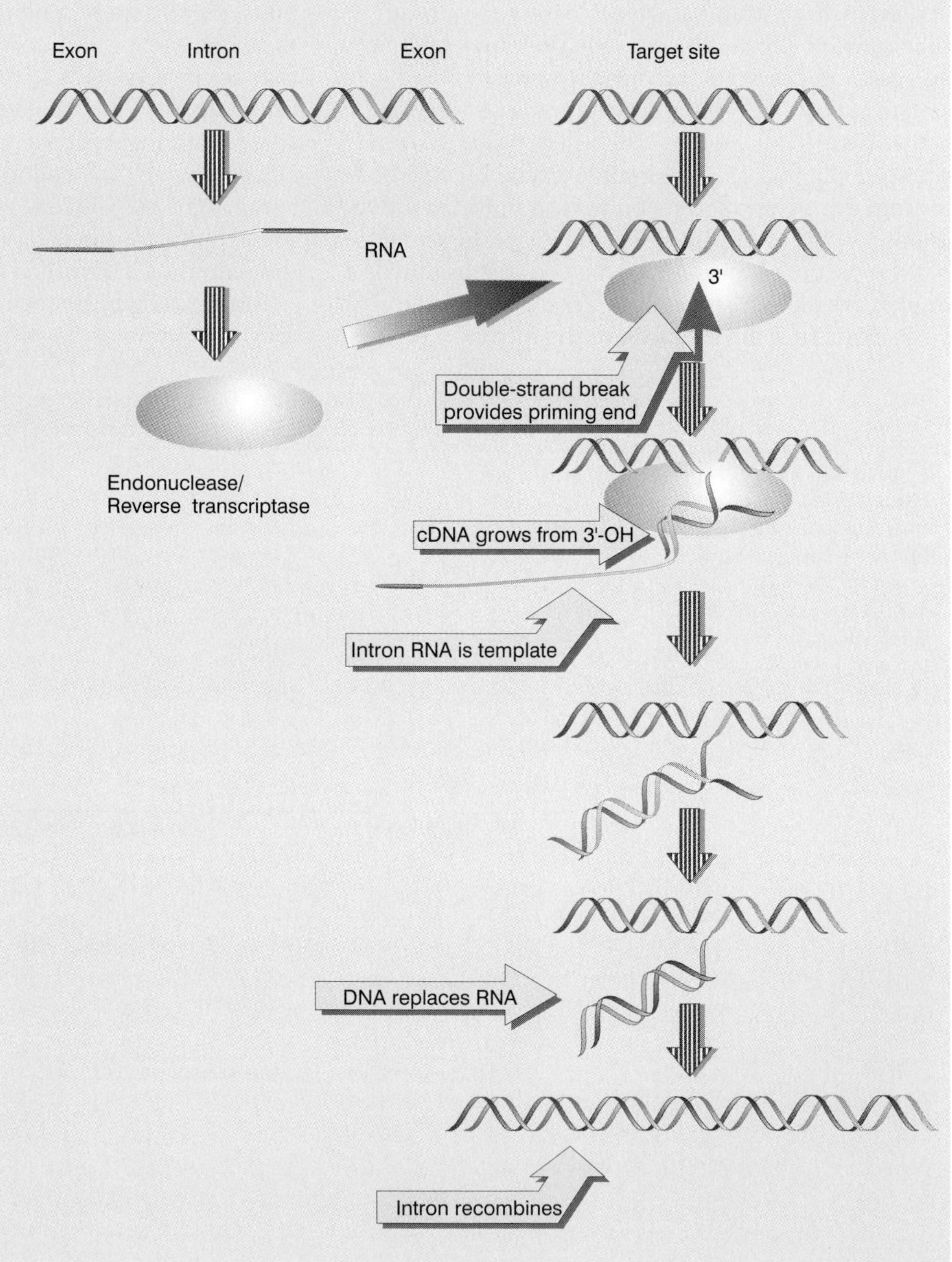
Exon
Intron
Exon
Target site
RNA
3'
Double-strand break provides priming end
Endonuclease/ Reverse transcriptase
cDNA grows from 3'-OH
Intron RNA is template
DNA replaces RNA
Intron recombines

Figure 23.12 Reverse transcriptase coded by an intron allows a copy of the RNA to be inserted at a target site generated by a double-strand break.

for priming by making a nick in the target (see Figure 16.17).

Figure 23.12 illustrates the reaction for the example of a group II intron. As before, the endonuclease makes a double-strand break at the target site. A 3′ end is generated at the site of the break, and provides a primer for the reverse transcriptase. The intron RNA provides the template for the synthesis of cDNA. Because the RNA includes exon sequences on either side of the intron, the cDNA product is longer than the region of the intron itself, so that it can span the double-strand break, allowing the cDNA to repair the break. The result is the insertion of the intron.

An *in vitro* system for mobility can be generated by incubating a ribonucleoprotein preparation with a substrate DNA. The ribonucleoprotein includes the RNA containing a group II intron and its protein product. It contains an endonuclease activity that makes a staggered double-strand break at the appropriate target site. Both the RNA and protein components of the ribonucleoprotein are required for cleavage, possibly both in catalytic capacities.

Mobile introns are likely to have been inserted into pre-existing genes. They could have evolved into nuclear pre-mRNA introns. Introns with self-splicing capacity could have played an important role in the evolution of splicing; for example, they could have migrated into the nucleus and been the progenitors of snRNAs.

RNA can have ribonuclease activities

ONE of the first demonstrations of the capabilities of RNA was provided by the dissection of ribonuclease P, an *E. coli* tRNA-processing endonuclease. Ribonuclease P can be dissociated into its two components, the 375 base RNA and the 20 kD polypeptide. Under the conditions initially used to characterize the enzyme activity *in vitro,* both components were necessary to cleave the tRNA substrate.

But a change in ionic conditions, an increase in the concentration of Mg^{2+}, renders the protein component superfluous. *The RNA alone can catalyze the reaction!* Analyzing the results as though the RNA were an enzyme, each "enzyme" catalyzes the cleavage of multiple substrates. Although the catalytic activity resides in the RNA, the protein component greatly increases the speed of the reaction, as seen in the increase in turnover number (see Figure 23.10).

Because mutations in either the gene for the RNA or the gene for protein can inactivate RNAase P *in vivo,* we know that both components are necessary for natural enzyme activity. Originally it had been assumed that the protein provided the catalytic activity, while the RNA filled some subsidiary role; for example, assisting in the binding of substrate (it has some short sequences complementary to exposed regions of tRNA). But these roles are reversed!

The catalytic activity of RNAase P requires RNA to function as a catalyst with an external substrate. Another example of the ability of RNA to function as an endonuclease is provided by some small plant RNAs (~350 bases) that undertake a self-cleavage reaction. As with the case of the *Tetrahymena* group I intron, however, it is possible to engineer constructs that can function on external substrates.

These small plant RNAs fall into two general groups: viroids and virusoids. The **viroids** are infectious RNA molecules that function independently, without encapsidation by any protein coat. The **virusoids** are similar in organization, but are encapsidated by plant viruses, being packaged together with a viral genome. The virusoids cannot replicate independently, but require assistance from the virus. The virusoids are sometimes called **satellite RNAs.**

Viroids and virusoids both replicate via rolling circles (see Figure 12.16). The strand of RNA that is packaged into the virus is called the plus strand. The complementary strand, generated during replication of the RNA, is called the minus strand. Multimers of both plus and minus strands are found. Both types of monomer are generated by cleaving the tail of a rolling circle; circular plus strand monomers are generated by ligating the ends of the linear monomer.

Both plus and minus strands of viroids and virusoids undergo self-cleavage *in vitro.* The cleavage reaction is promoted by divalent metal cations; it generates 5′-OH and 2′,3′-cyclic phosphodiester termini. Some of the RNAs cleave *in vitro* under physiological conditions. Others do so only after a cycle of heating and cooling; this suggests that the isolated RNA has an inappropriate conformation, but can generate an active conformation when it is denatured and renatured.

The viroids and virusoids that undergo self-cleavage form a "hammerhead" secondary structure at the cleavage site, as drawn in the upper part of **Figure 23.13.** The sequence of this structure is sufficient for cleavage. When the surrounding sequences are deleted, the need for a heating-cooling cycle is obviated, and the small RNA self-cleaves spontaneously. This suggests that the sequences beyond the hammerhead usually interfere with its formation.

The active site is a sequence of only 58 nucleotides. The hammerhead contains three stem-loop regions whose position and size are constant, and 13 conserved nucleotides, mostly in the regions connecting the center of the structure. The conserved bases and duplex stems generate an RNA with the intrinsic ability to cleave.

An active hammerhead can also be generated by pairing an RNA representing one side of the structure with an RNA representing the other side. The lower part of Figure 23.13 shows an example of a hammerhead generated by hybridizing a 19 base molecule with a 24 base molecule. The hybrid mimics the hammerhead structure, with the omission of loops I and III. When the 19 base RNA is added to the 24 base RNA, cleavage occurs at the appropriate position in the hammerhead.

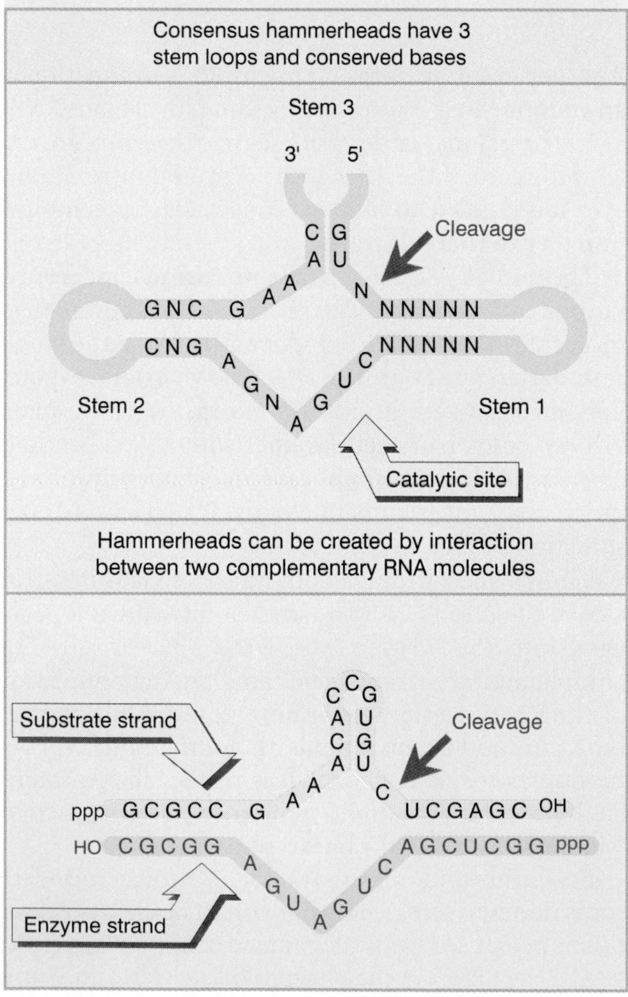

Figure 23.13 Self-cleavage sites of viroids and virusoids have a consensus sequence and form a hammerhead secondary structure by intramolecular pairing. Hammerheads can also be generated by pairing between a 'substrate' strand and an 'enzyme' strand.

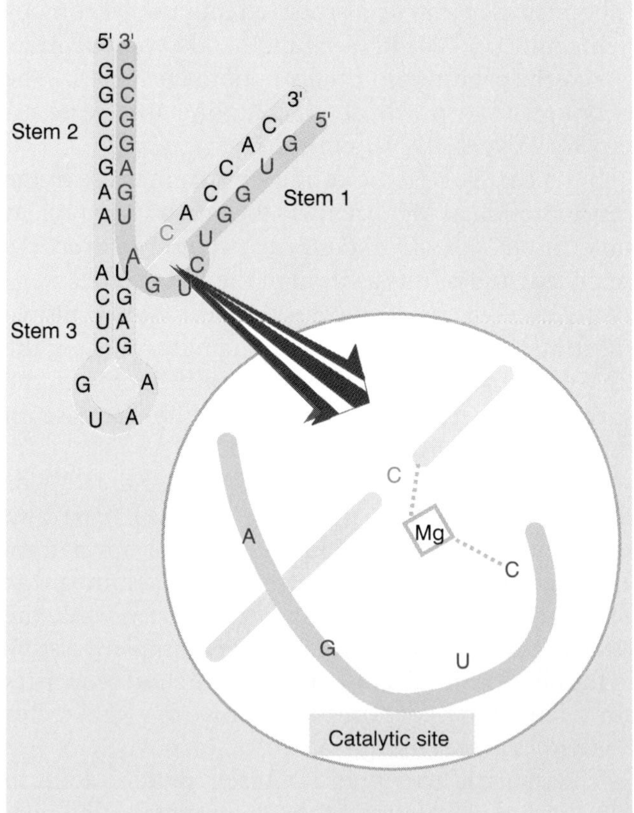

Figure 23.14 A hammerhead ribozyme forms a V-shaped tertiary structure in which stem 2 is stacked upon stem 3. The catalytic center lies between stem 2/3 and stem 1. It contains a magnesium ion that initiates the hydrolytic reaction.

We may regard the top (24 base) strand of this hybrid as comprising the "substrate," and the bottom (19 base) strand as comprising the "enzyme." When the 19 base RNA is mixed with an excess of the 24 base RNA, multiple copies of the 24 base RNA are cleaved. This suggests that there is a cycle of 19 base–24 base pairing, cleavage, dissociation of the cleaved fragments from the 19 base RNA, and pairing of the 19 base RNA with a new 24 base substrate. The 19 base RNA is therefore a ribozyme with endonuclease activity. The parameters of the reaction are similar to those of other RNA-catalyzed reactions (see Figure 23.10).

The crystal structure of a hammerhead shows that it forms a compact V-shape, in which the catalytic center lies in a turn, as indicated diagrammatically in **Figure 23.14.** An Mg^{2+} ion located in the catalytic site plays a crucial role in the reaction. It is positioned by the target cytidine and by the cytidine at the base of stem 1; it may also be connected to the adjacent uridine. It extracts a proton from the 2′-OH of the target cytidine, and then directly attacks the labile phosphodiester bond. Mutations in the hammerhead sequence that affect the transition state of the cleavage reaction occur in both the active site and other locations, suggesting that there may be a substantial rearrangement of structure prior to cleavage.

It is possible to design enzyme-substrate combinations that can form hammerhead structures, and these have been used to demonstrate that introduction of the appropriate RNA molecules into a cell can allow the enzymatic reaction to occur *in vivo.* A ribozyme designed in this way essentially provides a highly specific restriction-like activity directed against an RNA target. By placing the ribozyme under control of a regulated promoter, it can be used in the same way as (for example) antisense constructs specifically turn off expression of a target gene under defined circumstances.

RNA editing utilizes information from several sources

A prime axiom of molecular biology is that the sequence of an mRNA can only represent what is coded in the DNA. The central dogma envisaged a linear relationship in which a continuous sequence of DNA is transcribed into a sequence of mRNA that is in turn directly translated into protein. The occurrence of interrupted genes and the removal of introns by RNA splicing introduces an additional step into the process of gene expression: the coding sequences (exons) in DNA must be reconnected in RNA. But the process remains one of information transfer, in which the actual coding sequence in DNA remains inviolate.

Changes in the information coded by DNA occur in some exceptional circumstances, most notably in the generation of new sequences coding for immunoglobulins in mammals and birds. These changes occur specifically in the somatic cells (B lymphocytes) in which immunoglobulins are synthesized (see Chapter 24). New information is generated in the DNA of an individual during the process of reconstructing an immunoglobulin gene; and information coded in the DNA is changed by somatic mutation. The information in DNA continues to be faithfully transcribed into RNA.

RNA editing is a process in which *information changes at the level of mRNA.* It is revealed by situations in which the coding sequence in an RNA differs from the sequence of DNA from which it was transcribed. RNA editing occurs in two different situations, with different causes. In mammalian cells there are cases in which a substitution occurs in an individual base in mRNA, causing a change in the sequence of the protein that is coded. In trypanosome mitochondria, more widespread changes occur in transcripts of several genes, when bases are systematically added or deleted.

Figure 23.15 summarizes the sequences of the apolipoprotein-B gene and mRNA in mammalian intestine and liver. The genome contains a single (interrupted) gene whose sequence is identical in all tissues, with a coding region of 4563 codons. This gene is transcribed into an mRNA that is translated into a protein of 512 kD representing the full coding sequence in the liver.

A shorter form of the protein, ~250 kD, is synthesized in intestine. This protein consists of the N-terminal half of the full-length protein. It is translated from an mRNA whose sequence is identical with that of liver except for a change from C to U at codon 2153. This substitution changes the codon CAA for glutamine into the ochre codon UAA for termination.

What is responsible for this substitution? No alternative gene or exon is available in the genome to code for the new sequence, and no change in the pattern of splicing can be discovered. We are forced to conclude that a change has been made directly in the sequence of the transcript.

Editing of this sort is rare, but apo-lipo-B is not unique. Another example is provided by glutamate receptors in rat brain. Editing at one position changes a glutamine codon in DNA into a codon for arginine in RNA; the change affects the conductivity of the channel and therefore has an important effect on controlling ion flow through the neurotransmitter. At another position in the receptor, an arginine codon is converted to a glycine codon.

The editing event in apo-B causes C_{2153} to be changed to U; both changes in the glutamate receptor are from A to I (inosine). These events are *deaminations* in which the amino group on the nucleotide ring is removed. Such events are catalyzed by enzymes called cytidine and adenosine deaminases, respectively.

What controls the specificity of an editing reaction? Enzymes that undertake deamination as such often have broad specificity—for example, the best characterized adenosine deaminase acts on any A residue in a duplex RNA region. Editing enzymes are related to the general deaminases, but have other regions or additional subunits that control their specificity. In the case of apo-B editing, the catalytic subunit of an editing complex is related to bacterial cytidine deaminase, but has an additional RNA-binding region that helps to recognize the specific target site for editing. A special adenosine deaminase enzyme recognizes the target sites in the glutamate receptor RNA, and similar events occur in a serotonin receptor RNA.

The complex may recognize a particular region of secondary structure in a manner analogous to tRNA-modifying enzymes or could directly recognize a nucleotide sequence. The development of an *in vitro* system for the apo-B editing event suggests that a relatively small sequence (~26 bases) surrounding the editing site provides a sufficient target. **Figure 23.16** shows that in the case of the GluR-B RNA, a base-

Figure 23.15 The sequence of the apo-B gene is the same in intestine and liver, but the sequence of the mRNA is modified by a base change that creates a termination codon in intestine.

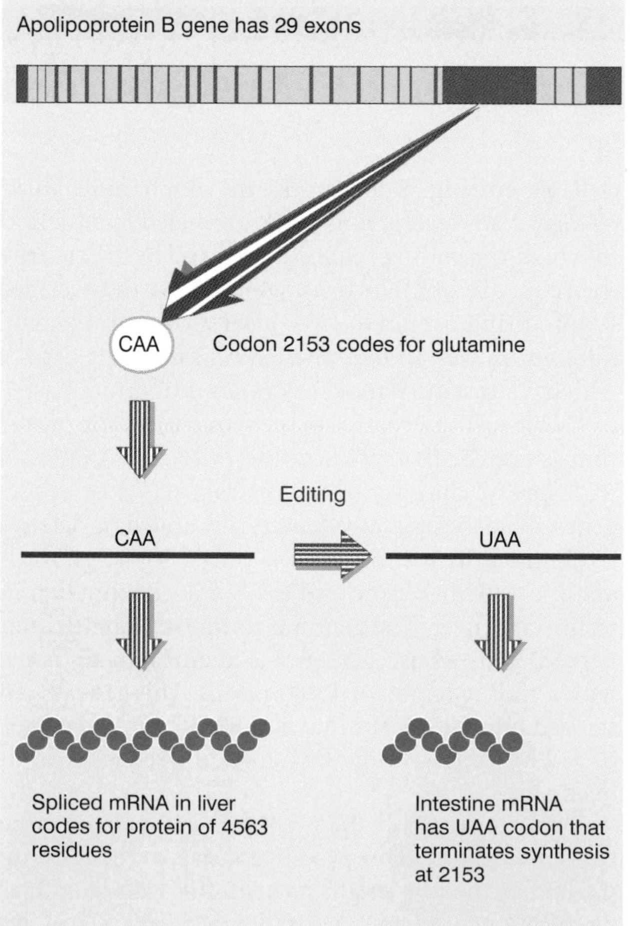

Figure 23.16 Editing of mRNA occurs when a deaminase acts on an adenine in an imperfectly paired RNA duplex region

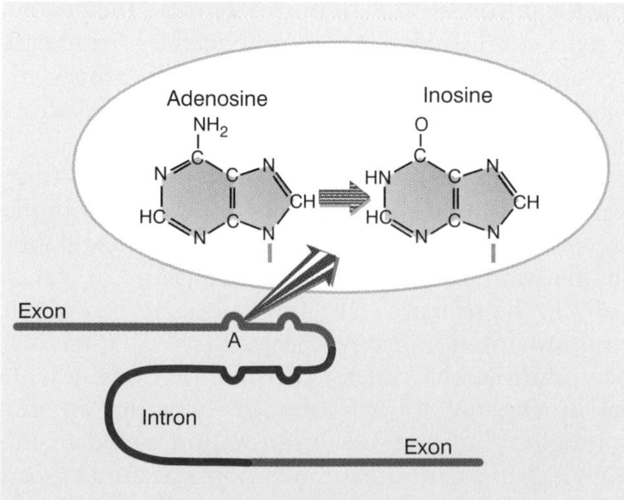

paired region that is necessary for recognition of the target site is formed between the edited region in the exon and a complementary sequence in the downstream intron. A pattern of mispairing within the duplex region is necessary for specific recognition. So different editing systems may have different types of requirement for sequence specificity in their substrates.

Dramatic changes in sequence occur in the products of several genes of trypanosome mitochondria. In the first case to be discovered, the sequence of the cytochrome oxidase subunit II protein has a −1 frameshift relative to the sequence of the *coxII* gene. The sequences of the gene and protein given in **Figure 23.17** are conserved in several trypanosome species. How does this gene function?

The *coxII* mRNA has an insert of an additional four nucleotides (all uridines) around the site of frameshift. The insertion restores the proper reading frame; it inserts an extra amino acid and changes the amino acids on either side. No second gene with this sequence can be discovered, and we are forced to conclude that the extra bases are inserted during or after transcription. A similar discrepancy between mRNA and genomic sequences is found in genes of the SV5 and measles paramyxoviruses, in these cases involving the addition of G residues in the mRNA.

Similar editing of RNA sequences occurs for other genes, and includes deletions as well as additions of uridine. The extraordinary case of the *coxIII* gene of *T. brucei* is summarized in **Figure 23.18.**

More than half of the residues in the mRNA consist of uridines that are not coded in the gene. Comparison between the genomic DNA and the mRNA shows that no stretch longer than 7 nucleotides is represented in the mRNA without alteration; and runs of uridine up to 7 bases long are inserted.

What provides the information for the specific insertion of uridines? A **guide RNA** contains a sequence that is complementary to the correctly edited mRNA. **Figure 23.19** shows a model for its action in the cytochrome *b* gene of *Leishmania.*

The sequence at the top shows the original transcript, or pre-edited RNA. Gaps show where bases will

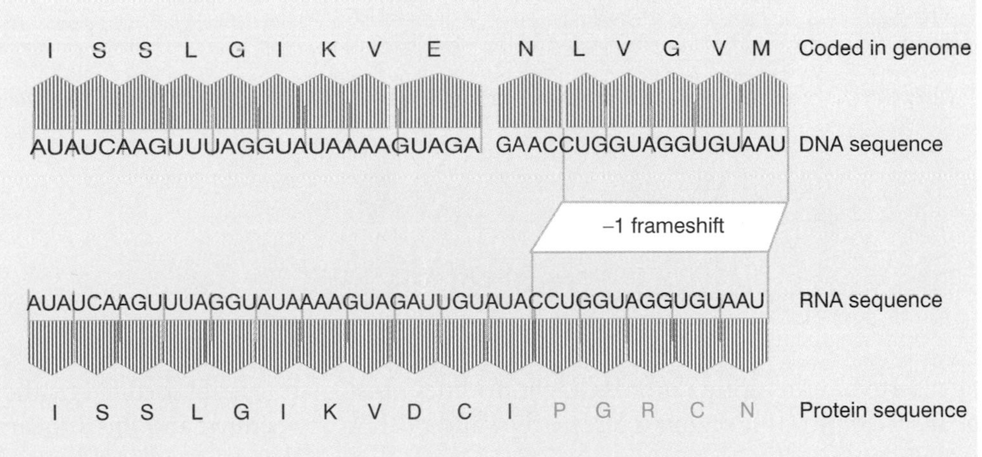

Figure 23.17 The mRNA for the trypanosome *coxII* gene has a −1 frameshift relative to the DNA; the correct reading frame is created by the insertion of 4 uridines.

Figure 23.18 Part of the mRNA sequence of *T. brucei coxIII* shows many uridines that are not coded in the DNA (shown in red) or that are removed from the RNA (shown as T).

T
UAUAUGUUUUGUUGUUUAUUAUGUGAUUAUGGUUUUGUUUUUUAUUGGUAUUUUUUAGAUUUAUUUAAUUUGUUGAUA
TT
AAUACAUUUUAUUUGUUUGUUAAUUUUUUUGUUUUGUGUUUUGGUUUAGGUUUUUUUGUUGUUGUUGUUUUGUAUUAU
TTTT TT
GAUUGAGUUUGUUGUUUGGUUUUUUGUUUUUUGUGAAACCAGUUAUGAGAGUUUGCAUUGUUAUUUAUUACAUUAAGU
TTTT
UGGUGUUUUUGGUUC

Figure 23.19 Pre-edited RNA base pairs with a guide RNA on both sides of the region to be edited. The guide RNA provides a template for the insertion of uridines. The mRNA produced by the insertions is complementary to the guide RNA.

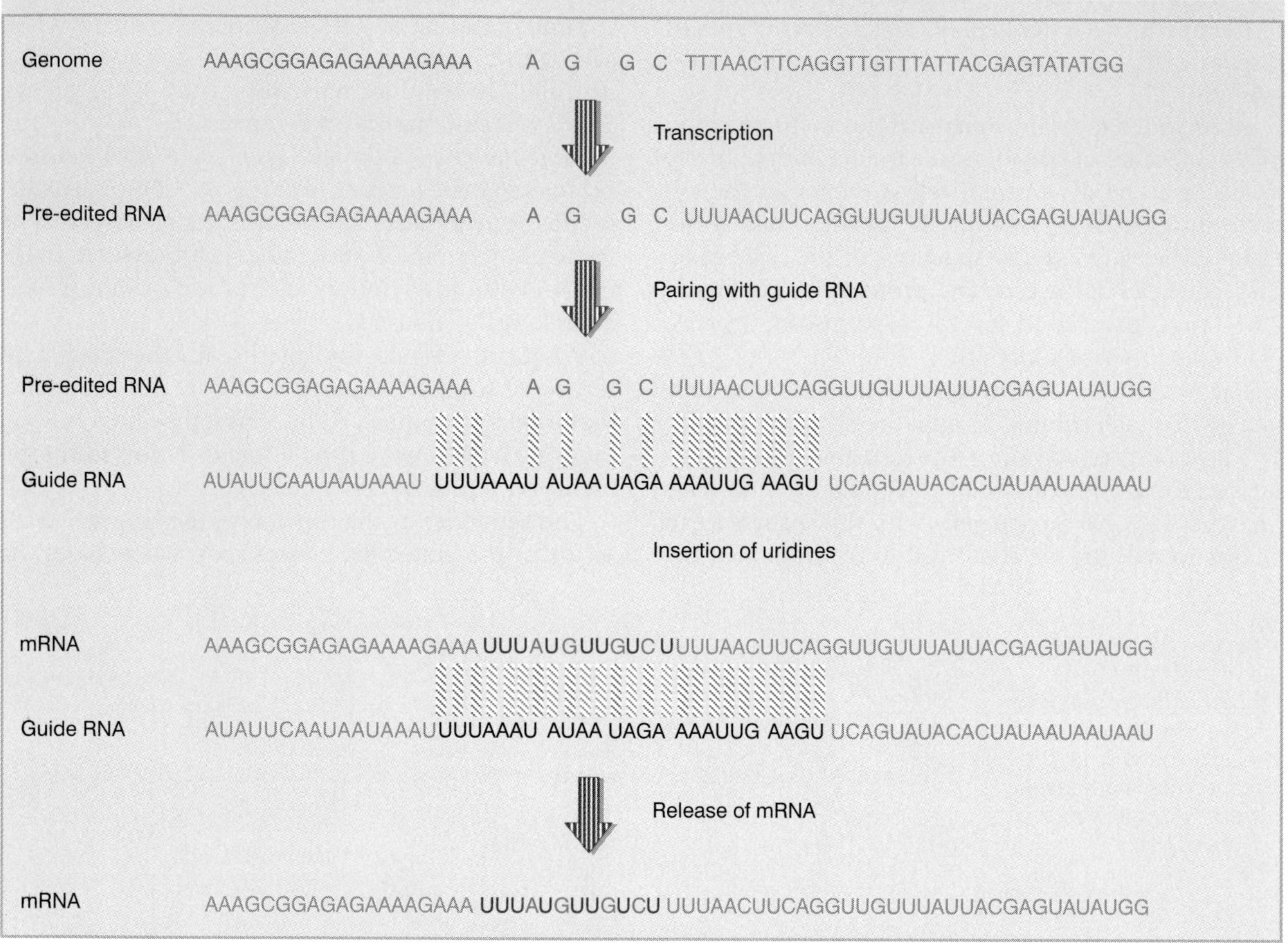

be inserted in the editing process. Eight uridines must be inserted into this region to create the valid mRNA sequence.

The guide RNA is complementary to the mRNA for a significant distance including and surrounding the edited region. Typically the complementarity is more extensive on the 3′ side of the edited region and is rather short on the 5′ side. Pairing between the guide RNA and the pre-edited RNA leaves gaps where unpaired A residues in the guide RNA do not find complements in the pre-edited RNA. The guide RNA provides a template that allows the missing U residues to be inserted at these positions. When the reaction is completed, the guide RNA separates from the mRNA, which becomes available for translation.

Specification of the final edited sequence can be quite complex; in this example, a lengthy stretch of the transcript is edited by the insertion altogether of 39 U residues, and this appears to require two guide RNAs that act at adjacent sites. In such cases, editing usually proceeds from the 3′ end of the transcript toward the 5′ end. The first guide RNA pairs at the 3′-most site, and the edited sequence then becomes a substrate for further editing by the next guide RNA.

The guide RNAs are encoded as independent transcription units. **Figure 23.20** shows a map of the relevant region of the *Leishmania* mitochondrial DNA. It includes the "gene" for cytochrome *b*, which codes for the pre-edited sequence, and two regions that specify guide RNAs. Genes for the major coding regions and for their guide RNAs are interspersed.

In principle, a mutation in either the "gene" or one of its guide RNAs could change the primary sequence of the mRNA, and thus of the protein. By genetic criteria,

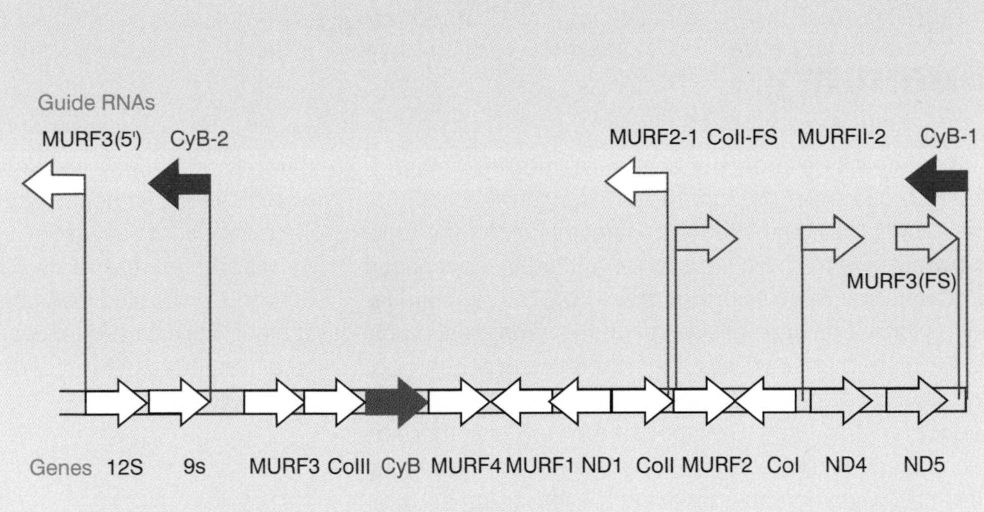

Figure 23.20 The *Leishmania* genome contains 'genes' coding for pre-edited RNAs interspersed with units that code for the guide RNAs required to generate the correct mRNA sequences. Some genes have multiple guide RNAs.

each of these units could be considered to comprise part of the "gene." Since the units are independently expressed, they should of course complement in *trans*. If mutations were available, we should therefore find that 3 complementation groups were needed to code for the primary sequence of a single protein.

The characterization of intermediates that are partially edited suggests that the reaction proceeds along the pre-edited RNA in the 3′–5′ direction. The guide RNA dictates specificity of uridine insertions by its pairing with the pre-edited RNA.

Editing of uridines is catalyzed by a 20S enzyme complex that contains an endonuclease, a terminal uridyltransferase (TUTase), and an RNA ligase, as illustrated in **Figure 23.21**. It is directed to the appropriate target sites by pairing between the substrate RNA and the guide RNA. The substrate RNA is cleaved at a site that is (presumably) identified by the absence of pairing with the guide RNA, a uridine is inserted or deleted to base pair with the guide RNA, and then the substrate RNA is ligated. UTP provides the source for the uridyl residue. It is added by the TUTase activity; it is not clear whether this activity, or a separate exonuclease, is responsible for deletion. (At one time it was thought that a stretch of U residues at the end of guide RNA might provide the source for added U residues or a sink for deleted residues, but transfer of U residues to guide RNAs appears to be an aberrant reaction that is not responsible for editing.)

The structures of partially edited molecules suggest that the U residues are added one a time, and not in groups. It is possible that the reaction proceeds through successive cycles in which U residues are added, tested for complementarity with the guide RNA, retained if acceptable and removed if not, so that the construction of the correct edited sequence occurs gradually. We do not know whether the same types of reaction are involved in editing reactions that add C residues.

Figure 23.21 Addition or deletion of U residues occurs by cleavage of the RNA, removal or addition of the U, and ligation of the ends. The reactions are catalyzed by a complex of enzymes under the direction of guide RNA.

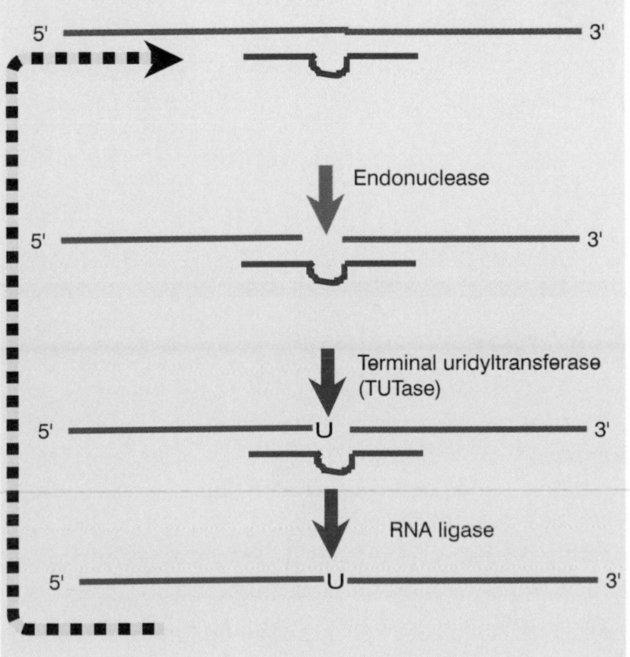

Summary

Self-splicing is a property of group I introns, found in *Tetrahymena* and *Physarum* nuclei, in fungal mitochondria, and in phage T4. The information necessary for the reaction resides in the intron sequence (although the reaction is actually assisted by proteins *in vivo*). The reaction requires formation of a specific secondary (and presumably tertiary) structure involving short consensus sequences. The RNA creates a structure in which the substrate sequence is held by the IGS region of the intron, and other conserved sequences generate a guanine nucleotide binding site. It occurs by a transesterification involving a guanosine residue as cofactor. No input of energy is required. The guanosine breaks the bond at the 5′ exon-intron junction and becomes linked to the intron; the hydroxyl at the free end of the exon then attacks the 3′ exon-intron junction. The intron cyclizes and loses the guanosine and the terminal 15 bases. A series of related reactions can be catalyzed via attacks by the terminal G-OH residue of the intron on internal phosphodiester bonds. By providing appropriate substrates, it has been possible to engineer ribozymes that perform a variety of catalytic reactions, including nucleotidyl transferase activities.

Some group I and some group II mitochondrial introns have open reading frames. The proteins coded by some group I introns are endonucleases that make double-stranded cleavages in target sites in DNA; the cleavage initiates a gene conversion process in which the sequence of the intron itself is copied into the target site. This could mean that these types of introns originated by insertion events.

Catalytic reactions are undertaken by the RNA component of the RNAase P ribonucleoprotein. Virusoid RNAs can undertake self-cleavage at a "hammerhead" structure. Hammerhead structures can form between a substrate RNA and a ribozyme RNA, allowing cleavage to be directed at highly specific sequences. These reactions support the view that RNA can form specific active sites that have catalytic activity.

RNA editing changes the sequence of an RNA after or during its transcription. The changes are required to create a meaningful coding sequence. Substitutions of individual bases occur in mammalian systems; they take the form of deaminations in which C is converted to U, or A is converted to I. A catalytic subunit related to cytidine or adenosine deaminase functions as part of a larger complex that has specificity for a particular target sequence.

Additions and deletions (most usually of uridine) occur in trypanosome mitochondria and in paramyxoviruses. Extensive editing reactions occur in trypanosomes in which as many as half of the bases in an mRNA are derived from editing. The editing reaction uses a template consisting of a guide RNA that is complementary to the mRNA sequence. The reaction is catalyzed by an enzyme complex that includes an endonuclease, terminal uridyltransferase, and RNA ligase, using free nucleotides as the source for additions or releasing cleaved nucleotides following deletion.

Further reading

Reviews

Cech, T. R. (1985). Self-splicing RNA: implications for evolution. *Int. Rev. Cytol.* **93**, 3–22.

Cech, T. R. (1987). The chemistry of self-splicing RNA and RNA enzymes. *Science* **236**, 1532–1539.

Cech, T. R. (1990). Self-splicing of group I introns. *Ann. Rev. Biochem.* **59**, 543–568.

Symons, R. H. (1992). Small catalytic RNAs. *Ann. Rev. Biochem.* **61**, 641–671.

Lambowitz, A. M. and Belfort, M. (1993). Introns as mobile genetic elements. *Ann. Rev. Biochem.* **62**, 587–622.

Michel, F. and Ferat, J.-L. (1995). Structure and activities of group II introns. *Ann. Rev. Biochem.* **64**, 435–461.

Discoveries

Catalytic introns

Cech, T. R. *et al.* (1981). *In vitro* splicing of the rRNA precursor of *Tetrahymena*: involvement of a guanosine nucleotide in the excision of the intervening sequence. *Cell* **27**, 487–496.

Kruger, K. *et al.* (1982). Self-splicing RNA: autoexcision and autocyclization of the rRNA intervening sequence of *Tetrahymena. Cell* **31**, 147–157.

Belfort, M., Pedersen-Lane, J., West, D., Ehrenman, K., Maley, G., Chu, F., and Maley, F. (1985). Processing of the intron-containing thymidylate synthase (td) gene of phage T4 is at the RNA level. *Cell* **41**, 375–382.

Been, M. D. and Cech, T. R. (1986). One binding site determines sequence specificity of *Tetrahymena* pre-rRNA self-splicing, trans-splicing, and RNA enzyme activity. *Cell* **47**, 207–216.

Burke, J. M. *et al.* (1986). Role of conserved sequence elements 9L and 2 in self-splicing of the *Tetrahymena* ribosomal RNA precursor. *Cell* **45**, 167–176.

Michel, F. and Wetshof, E. (1990). Modeling of the three-dimensional architecture of group I catalytic introns based on comparative sequence analysis. *J. Mol. Biol.* **216**, 585–610.

Intron ORFs and mobility

Carignani, G. *et al.* (1983). An RNA maturase is encoded by the first intron of the mitochondrial gene for the subunit I of cytochrome oxidase in *S. cerevisiae. Cell* **35**, 733–742.

Zimmerly, S. *et al.* (1995). Group II intron mobility occurs by target DNA-primed reverse transcription. *Cell* **82**, 545–554.

Zimmerly, S. *et al.* (1995). A group II intron is a catalytic component of a DNA endonuclease involved in intron mobility. *Cell* **83**, 529–538.

RNA ribonucleases

Guerrier-Takada, C. *et al.* (1983). The RNA moiety of ribonuclease P is the catalytic subunit of the enzyme. *Cell* **35**, 849–857.

Forster, A. C. and Symons, R. H. (1987). Self-cleavage of virusoid RNA is performed by the proposed 55-nucleotide active site. *Cell* **50**, 9–16.

Scott, W. G., Finch, J. T., and Klug, A. (1995). The crystal structure of an all-RNA hammerhead ribozyme: a proposed mechanism for RNA catalytic cleavage. *Cell* **81**, 991–1002.

Vertebrate editing

Powell, L. M., Wallis, S. C., Pease, R. J., Edwards, Y. H., Knott, T. J., and Scott, J. (1987). A novel form of tissue-specific RNA processing produces apolipoprotein-B48 in intestine. *Cell* **50**, 831–840.

Sommer, B. *et al.* (1991). RNA editing in brain controls a determinant of ion flow in glutamate-gated channels. *Cell* **67**, 11–19.

Higuchi, M. *et al.* (1993). RNA editing of AMPA receptor subunit GluR-B: a base-paired intron-exon structure determines position and efficiency. *Cell* **75**, 1361–1370.

Navaratnam, N. *et al.* (1995). Evolutionary origins of apoB mRNA editing: catalysis by a cytidine deaminase that has acquired a novel RNA-binding motif at its active site. *Cell* **81**, 187–195.

Trypanosome editing

Benne, R., Van den Burg J., Brakenhoff, J. P., Sloof, P., Van Boom, J. H., and Tromp, M. C. (1986). Major transcript of the frameshifted *coxII* gene from trypanosome mitochondria contains four nucleotides that are not encoded in the DNA. *Cell* **46**, 819–826.

Feagin, J. E., Abraham, J. M., and Stuart, K. (1988). Extensive editing of the cytochrome *c* oxidase III transcript in *Trypanosoma brucei. Cell* **53**, 413–422.

Blum, B., Bakalara, N., and Simpson, L. (1990). A model for RNA editing in kinetoplastid mitochondria: "guide" RNA molecules transcribed from maxicircle DNA provide the edited information. *Cell* **60**, 189–198.

Seiwert, S. D., Heidmann, S. and Stuart, K. (1996). Direct visualization of uridylate deletion *in vitro* suggests a mechanism for kinetoplastid editing. *Cell* **84**, 831–841.

Chapter 24

Immune diversity

It is an axiom of genetics that the genetic constitution created in the zygote by the combination of sperm and egg is inherited by all somatic cells of the organism. We look to differential control of gene expression, rather than to changes in DNA content, to explain the different phenotypes of particular somatic cells.

Yet there are exceptional situations in which the reorganization of certain DNA sequences is used to regulate gene expression or to create new genes. The immune system provides a striking and extensive case in which the content of the genome changes, when recombination creates active genes in lymphocytes. Other cases are represented by the substitution of one sequence for another to change the mating type of yeast or to generate new surface antigens by trypanosomes (see Chapter 17).

The **immune response** of vertebrates provides a protective system that distinguishes foreign proteins from the proteins of the organism itself. Foreign material (or part of the foreign material) is recognized as comprising an **antigen.** Usually the antigen is a protein (or protein-attached moiety) that has entered the bloodstream of the animal—for example, the coat protein of an infecting virus. Exposure to an antigen initiates production of an immune response that *specifically recognizes the antigen and destroys it.*

Immune reactions are the responsibility of white blood cells—the B and T lymphocytes, and macrophages. The lymphocytes are named after the tissues that produce them. In mammals, **B cells** mature in the bone marrow, while **T cells** mature in the thymus. *Each class of lymphocyte uses the rearrangement of DNA as a mechanism for producing the proteins that enable it to participate in the immune response.*

The immune system has many ways to destroy an antigenic invader, but it is useful to consider them in two general classes. Which type of response the immune system mounts when it encounters a foreign structure depends partly on the nature of the antigen. The response is defined according to whether it is executed principally by B cells or T cells.

The **humoral response** depends on B cells. It is mediated by the secretion of **antibodies,** which are **immunoglobulin** proteins. *Production of an immunoglobulin specific for a foreign molecule is the primary event responsible for recognition of an antigen.* Recognition requires the antibody to bind to a small region or structure on the antigen.

The function of antibodies is represented in **Figure 24.1.** Foreign material circulating in the bloodstream (for example, a toxin or pathogenic bacterium) has a surface that presents antigens. The antigen(s) are recognized by the antibodies, which form an antigen-antibody complex. This complex then attracts the attention of other components of the immune system.

The humoral response depends on these other components in two ways. First, B cells need signals provided by T cells to enable them to secrete antibodies. These T cells are called **helper T cells,** because they assist the B cells. Second, antigen-antibody formation is a trigger for the antigen to be destroyed. The major pathway is provided by the action of **complement,** a component whose name reflects its ability to "complement" the action of the antibody itself. Complement consists of a set of ~20 proteins that function through a cascade of proteolytic actions. If the target antigen is part of a cell, for example an infecting bacterium, the action of complement culminates in lysing the target cell. The action of complement also provides a means of attracting macrophages, which scavenge the target cells or their products. Alternatively, the antigen-antibody complex may be taken up directly by macrophages (scavenger cells) and destroyed.

The **cell-mediated response** is executed by a class of T lymphocytes called **cytotoxic (killer) T cells.** The basic function of the T cell in recognizing a target antigen is indicated in **Figure 24.2.** A cell-mediated response typically is elicited by an intracellular parasite, such as a virus that infects the body's own cells. As a result of the viral infection, foreign (viral) antigens are displayed on the surface of the cell. These antigens are recognized by the **T-cell receptor (TCR),** which is the T cells' equivalent of the antibody produced by a B cell.

A crucial feature of this recognition reaction is that *the antigen must be presented by a cellular protein that is a member of the MHC (major histocompatibility complex).* The MHC protein has a groove on its surface that binds a peptide fragment derived from the foreign

Figure 24.1 *Overview:* humoral immunity is conferred by the binding of free antibodies to antigens to form antigen-antibody complexes that are removed from the bloodstream by macrophages or that are attacked directly by the complement proteins.

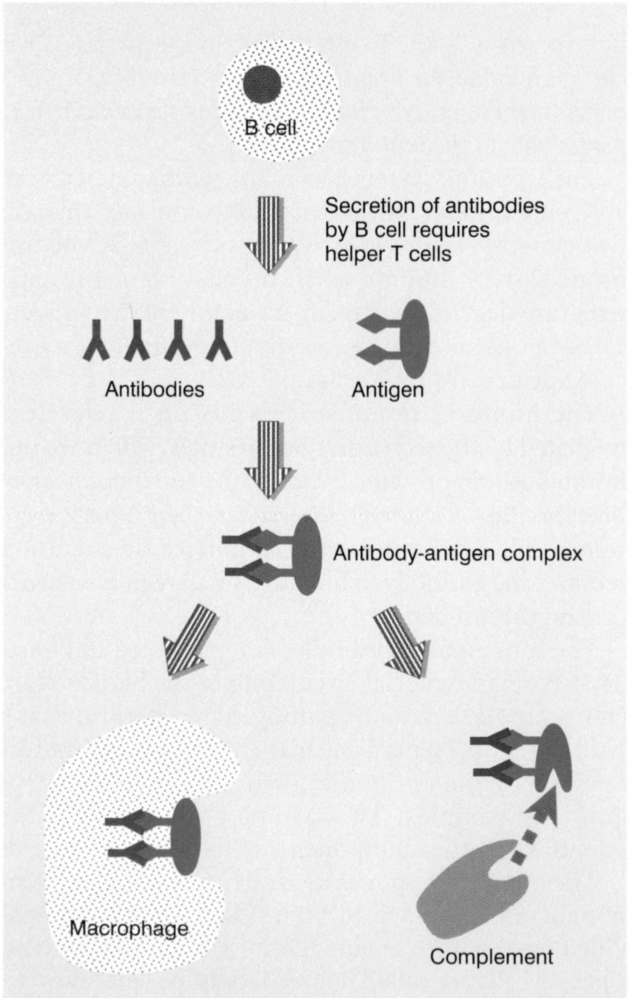

Figure 24.2 *Overview*: in cell-mediated immunity, killer T cells use the T-cell receptor to recognize a fragment of the foreign antigen which is 'presented' on the surface of the target cell by the MHC protein.

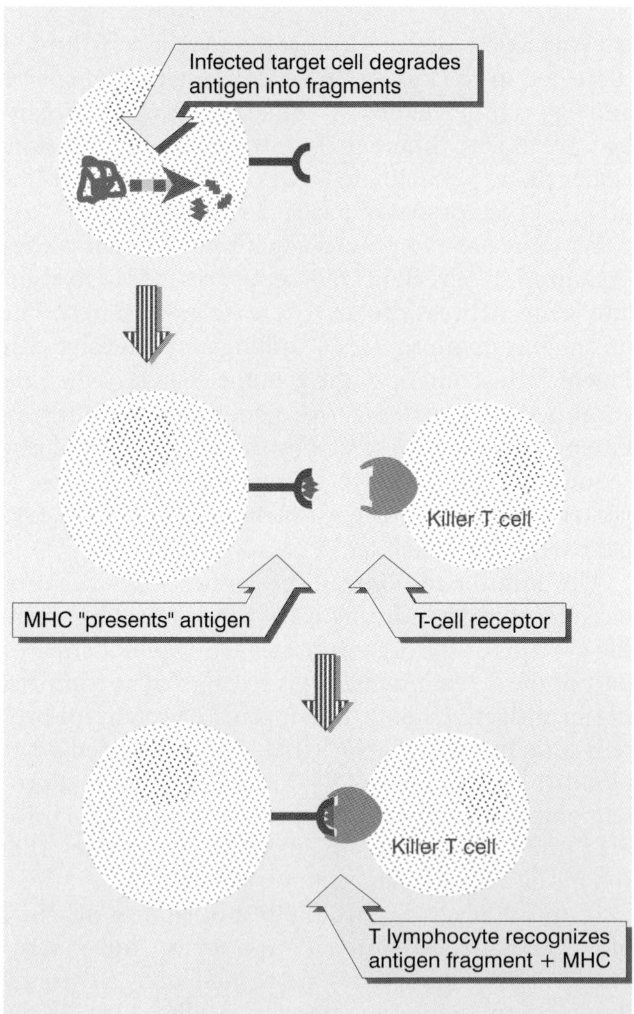

antigen. The combination of peptide fragment and MHC protein is recognized by the T-cell receptor. Every individual has a characteristic set of MHC proteins. They are important in graft reactions; a graft of tissue from one individual to another is rejected because of the difference in MHC proteins between the donor and recipient, an issue of major medical importance. The demand that the T lymphocytes recognize both foreign antigen and MHC protein ensures that the cell-mediated response acts only on host cells that have been infected with a foreign antigen. (We discuss the division of MHC proteins into the general types of class I and class II later in this chapter.)

The purpose of each type of immune response is to attack a foreign target. Target recognition is the prerogative of B-cell immunoglobulins and T-cell receptors. A crucial aspect of their function lies in the ability to distinguish "self" from "nonself." Proteins and cells of the body itself must *never* be attacked. Foreign targets must be *destroyed entirely.* The property of failing to attack "self" is called **tolerance.** Loss of this ability results in an **autoimmune disease,** in which the immune system attacks its own body, often with disastrous consequences.

What prevents the lymphocyte pool from responding to "self" proteins? Tolerance probably arises early in lymphocyte cell development when B and T cells that recognize "self" antigens are destroyed. This is called **clonal deletion.** In addition to this negative selection, there is also positive selection for T cells carrying certain sets of T-cell receptors.

A corollary of tolerance is that it can be difficult to obtain antibodies against proteins that are closely related to those of the organism itself. As a practical matter, therefore, it may be difficult to use (for example) mice or rabbits to obtain antibodies against human proteins that have been highly conserved in mammalian evolution. The tolerance of the mouse or rabbit for its own protein may extend to the human protein in such cases.

Each of the three groups of proteins required for the immune response—immunoglobulins, T-cell receptors, MHC proteins—is diverse. Examining a large number of individuals, we find many variants of each protein. Each protein is coded by a large family of genes; and in the case of antibodies and the T-cell receptors, the diversity of the population is increased by DNA rearrangements that occur in the relevant lymphocytes.

Immunoglobulins and T-cell receptors are direct counterparts, each produced by its own type of lymphocyte. The proteins are related in structure, and their genes are related in organization. The sources of variability are similar. The MHC proteins also share some common features with the antibodies, as do other lymphocyte-specific proteins. In dealing with the genetic organization of the immune system, we are therefore concerned with a series of related gene families, indeed a **superfamily** that may have evolved from some common ancestor representing a primitive immune response.

Clonal selection amplifies lymphocytes that respond to individual antigens

THE name of the immune response describes one of its central features. After an organism has been exposed to an antigen, it becomes *immune* to the effects of a new infection. Before exposure to a particular antigen, the organism lacks adequate capacity to deal with any toxic effects. This ability is acquired during the immune response. After the infection has been defeated, the organism retains the ability to respond rapidly in the event of a re-infection.

These features are accommodated by the **clonal selection theory** illustrated in **Figure 24.3.** The pool of lymphocytes contains B cells and T cells carrying a large variety of immunoglobulins or T-cell receptors. *But any individual B lymphocyte produces one immunoglobulin, which is capable of recognizing only a single antigen; similarly any individual T lymphocyte produces only one particular T-cell receptor.*

In the pool of immature lymphocytes, the unstimulated B cells and T cells are morphologically indistinguishable. But on exposure to antigen, a B cell whose antibody is able to bind the antigen, or a T cell whose receptor can recognize it, is stimulated to divide,

Figure 24.3 The pool of immature lymphocytes contains B cells and T cells making antibodies and receptors with a variety of specificities. Reaction with an antigen leads to clonal expansion of the lymphocyte with the antibody (B cell) or receptor (T cell) that can recognize the antigen.

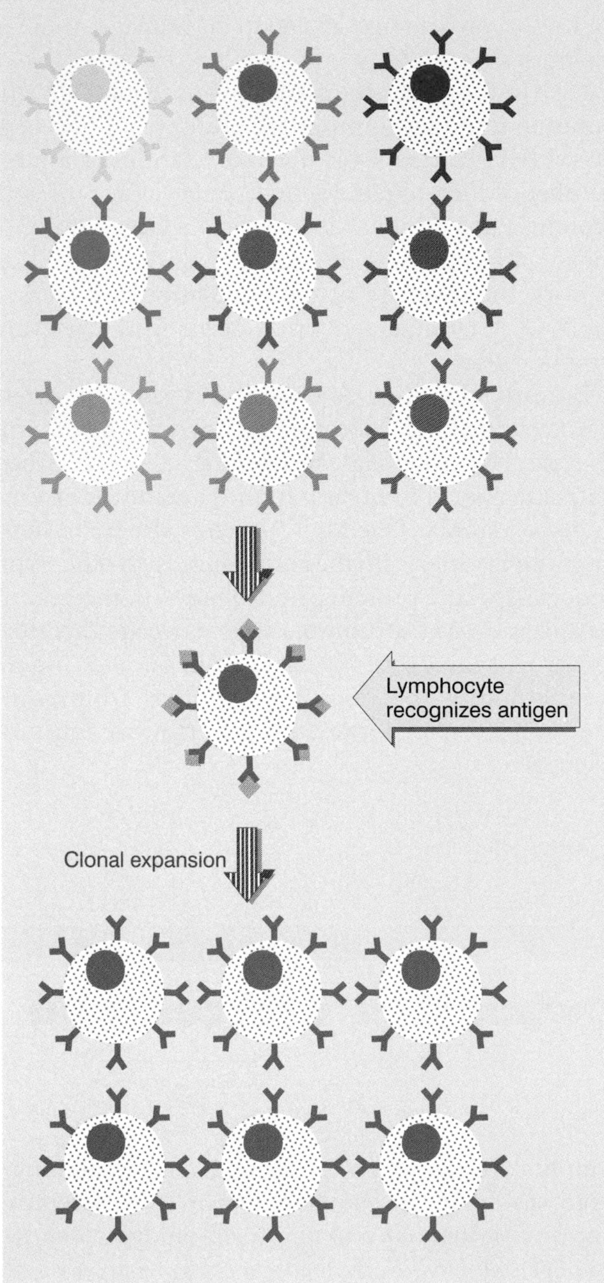
Lymphocyte recognizes antigen
Clonal expansion

probably by some feedback from the surface of the cell, where the antibody/receptor-antigen reaction occurs. The stimulated cells acquire the features characteristic of mature B or T lymphocytes, involving (for example) an increase in cell size (especially pronounced for B cells).

The initial expansion of a specific B- or T-cell population upon first exposure to an antigen is called the **primary immune response.** Large numbers of B or T lymphocytes with specificity for the offending antigen are produced. Each population represents a clone of the original responding cell. Antibody is secreted from the B cells in large quantities, and it may even come to dominate the antibody population.

After a successful primary immune response has been mounted, the organism retains B cells and T cells carrying the corresponding antibody or receptor. These **memory cells** represent an intermediate state between the immature cell and the mature cell. They have not acquired all of the features of the mature cell, but they are long-lived, and can rapidly be converted to mature cells. Their presence allows a **secondary immune response** to be mounted rapidly if the animal is exposed to the same antigen again.

The pool of immature lymphocytes in a mammal contains $\sim 10^{12}$ cells. This pool contains some lymphocytes that have unique specificities (because a corresponding antigen has never been encountered), while others are represented by up to 10^6 cells (because clonal selection has expanded the pool to respond to an antigen).

What features are recognized in an antigen? Antigens are usually macromolecular. Although small molecules may have antigenic determinants and can be recognized by antibodies, usually they are not effective in provoking an immune response (because of their small size). But they do provoke a response when conjugated with a larger carrier molecule (usually a protein). A small molecule that is used to provoke a response by such means is called a **hapten.**

Only a small part of the surface of a macromolecular antigen is actually recognized by any one antibody. The binding site consists of only 5–6 amino acids. Of course, any particular protein may have more than one such binding site, in which case it provokes antibodies with specificities for different regions. The region provoking a response is called an **antigenic determinant** or **epitope.** When an antigen contains several epitopes, some may be more effective than others in provoking the immune response; in fact, they may be so effective that they entirely dominate the response.

How do lymphocytes find target antigens and where does their maturation take place? Lymphocytes are peripatetic cells. They develop from immature stem cells that are located in the adult bone marrow. They migrate to the peripheral lymphoid tissues (spleen, lymph nodes) either directly via the bloodstream (if they are B cells) or via the thymus (where they become T cells). The lymphocytes recirculate between blood and lymph; the process of dispersion ensures that an antigen will be exposed to lymphocytes of all possible specificities. When a lymphocyte encounters an antigen that binds its antibody or receptor, clonal expansion begins the immune response.

Immunoglobulin genes are assembled from their parts in lymphocytes

A remarkable feature of the immune response is an animal's ability to produce an appropriate antibody whenever it is exposed to a new antigen. How can the organism be prepared to produce antibody proteins each designed specifically to recognize an antigen whose structure cannot be anticipated?

For practical purposes, we usually reckon that a mammal has the ability to produce 10^6–10^8 different antibodies. Each antibody is an immunoglobulin tetramer consisting of two identical **light (L) chains** and two identical **heavy (H) chains.** If any light chain can associate with any heavy chain, to produce 10^6–10^8 potential antibodies requires 10^3–10^4 different light chains and 10^3–10^4 different heavy chains.

There are 2 types of light chain and ~10 types of heavy chain. Different classes of immunoglobulins have different effector functions. *The class is determined by the heavy chain constant region, which exercises the effector function* (see Figure 24.17). The small number of heavy chains means that the regions of the molecule with fixed functions are relatively conserved.

The structure of the immunoglobulin tetramer is illustrated in **Figure 24.4.** Light chains and heavy chains share the same general type of organization in which each protein chain consists of two principal regions: the N-terminal **variable (V) region;** and the C-terminal **constant (C) region.** They were defined originally by comparing the amino acid sequences of different immunoglobulin chains. As the names suggest, the variable regions show considerable changes in sequence from one protein to the next, while the constant regions show substantial homology.

Corresponding regions of the light and heavy chains associate to generate distinct domains in the immunoglobulin protein:

- The variable (V) domain is generated by association between the variable regions of the light chain and heavy chain. *The V domain is responsible for recognizing the antigen.* An immunoglobulin has a Y-shaped structure in which the arms of the Y are identical, and each arm has a copy of the V domain. Production of V domains of different specificities creates the ability to respond to diverse antigens. The total number of variable

Figure 24.4 Heavy and light chains combine to generate an immunoglobulin with several discrete domains.

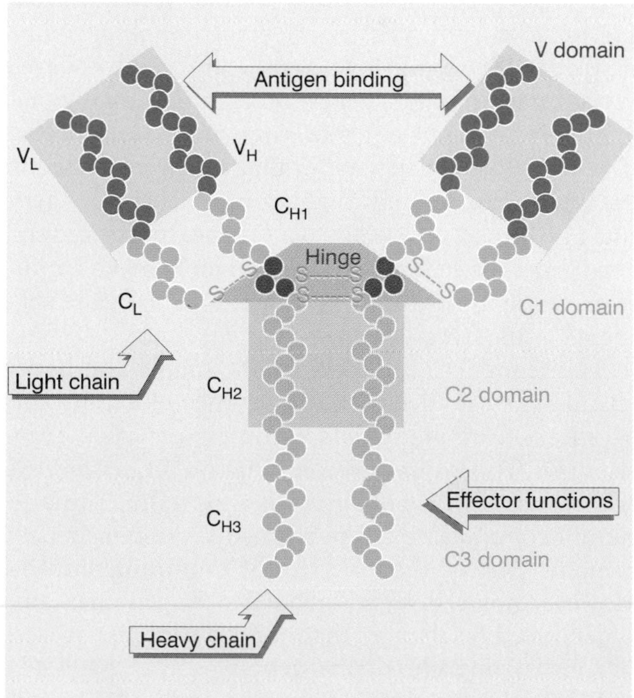

regions for either light- or heavy-chain proteins is measured in hundreds. *So the protein displays the maximum versatility in the region responsible for binding the antigen.*

■ The number of constant regions is vastly smaller than the number of variable regions—typically there are only 1–10 C regions for any particular type of chain. The association of constant regions in the immunoglobulin tetramer generates several individual C domains. The first domain results from association of the constant region of the light chain (C_L) with the C_{H1} part of the heavy-chain constant region. The two copies of this domain complete the arms of the Y-shaped molecule. Association between the C regions of the heavy chains generates the remaining C domains, which vary in number depending on the type of heavy chain.

Comparing the characteristics of the variable and constant regions, we see the central dilemma in immunoglobulin gene structure. How does the genome code for a set of proteins in which any individual polypeptide chain must have one of <10 possible C regions, but can have any one of several hundred possible V regions? It turns out that the number of coding sequences for each type of region reflects its variability. There are many genes coding for V regions, but only a few genes coding for C regions.

In this context, *"gene" means a sequence of DNA coding for a discrete part of the final immunoglobulin polypeptide* (heavy or light chain). So **V genes** code for variable regions and **C genes** code for constant regions, although *neither type of gene is expressed as an independent unit.* To construct a unit that can be expressed in the form of an authentic light or heavy chain, a V gene must be joined physically to a C gene. In this system, two "genes" code for one polypeptide. To avoid confusion, we shall now refer to these units as "gene segments" rather than "genes."

The sequences coding for light chains and heavy chains are assembled in the same way: *any one of many V gene segments may be joined to any one of a few C gene segments.* This **somatic recombination** occurs *in the B lymphocyte in which the antibody is expressed.* The large number of available V gene segments is responsible for a major part of the diversity of immunoglobulins. However, not all diversity is coded in the genome; some is generated by changes that occur during the process of constructing a functional gene.

Essentially the same description applies to the formation of functional genes coding for the protein chains of the T-cell receptor. Two types of receptor are found on T cells, one consisting of two types of chain called α and β, the other consisting of γ and δ chains. Like the genes coding for immunoglobulins, the genes coding for the individual chains in T-cell receptors consist of separate parts, including V and C regions, that are brought together in an active T cell (see later).

The crucial fact about the synthesis of immunoglobulins, therefore, is that *the arrangement of V gene segments and C gene segments is different in the cells producing the immunoglobulins (or T-cell receptors) from all other somatic cells or germ cells.*

The construction of a functional immunoglobulin or T-cell receptor gene might seem to be a Lamarckian process, representing a change in the genome that responds to a particular feature of the phenotype (the antigen). At birth, the organism does not possess the functional gene for producing a particular antibody or T-cell receptor. It possesses a large number of V gene segments and a smaller number of C gene segments. The subsequent construction of an active gene from these parts allows the antibody/receptor to be synthesized so that it is available to react with the antigen. The clonal selection theory requires that this rearrangement of DNA occurs *before the exposure to antigen,* which then results in *selection* for those cells carrying a protein able to bind the antigen. The entire process occurs in somatic cells and does not affect the germline; so the response to an antigen is not inherited by progeny of the organism.

Recombination between V and C gene segments to give functional loci occurs in a population of immature lymphocytes. A B lymphocyte usually has only one productive rearrangement of light-chain gene segments and one of heavy-chain gene segments; a T lymphocyte productively rearranges an α gene and a β gene, or one δ gene and one γ gene. The antibody or T-cell receptor produced by any one cell is determined by the particular configuration of V gene segments and C gene segments that have been joined.

There are two families of immunoglobulin light chains, κ and λ, and one family containing all the types of heavy chain (H). Each family resides on a different chromosome, and consists of its own set of both V gene segments and C gene segments. This is called the **germline** pattern, and is found in the germline and in somatic cells of all lineages other than the immune system.

But in a cell expressing an antibody, each of its chains—one light type (either κ or λ) and one heavy type—is coded by a single intact gene. The recombination event that brings a V gene segment to partner a C gene segment creates an active gene consisting of exons that correspond precisely with the functional domains of the protein. The introns are removed in the usual way by RNA splicing.

The principles by which functional genes are assembled are the same in each family, but there are differences in the details of the organization of the V and C gene segments, and correspondingly of the recombination reaction between them. In addition to the V and C gene segments, other short DNA sequences (including J segments and D segments) are included in the functional somatic loci.

A λ light chain is assembled from two parts, as illustrated in **Figure 24.5.** The V gene segment consists of the leader exon (L) separated by a single intron from the variable (V) segment. The C gene segment consists of the J segment separated by a single intron from the constant (C) exon.

The name of the **J segment** is an abbreviation for joining, since it identifies the region to which the V segment becomes connected. So the joining reaction does not directly involve V and C gene segments, but occurs via the J segment; when we discuss the joining of "V and C gene segments" for light chains, we really mean V-JC joining.

The J segment is short and codes for the last few (13) amino acids of the variable region, as defined by amino acid sequences. In the intact gene generated by recombination, the V-J segment constitutes a single exon coding for the entire variable region.

The consequences of the κ joining reaction are

Figure 24.5 The λ C gene segment is preceded by a J segment, so that V-J recombination generates a functional λ light-chain gene.

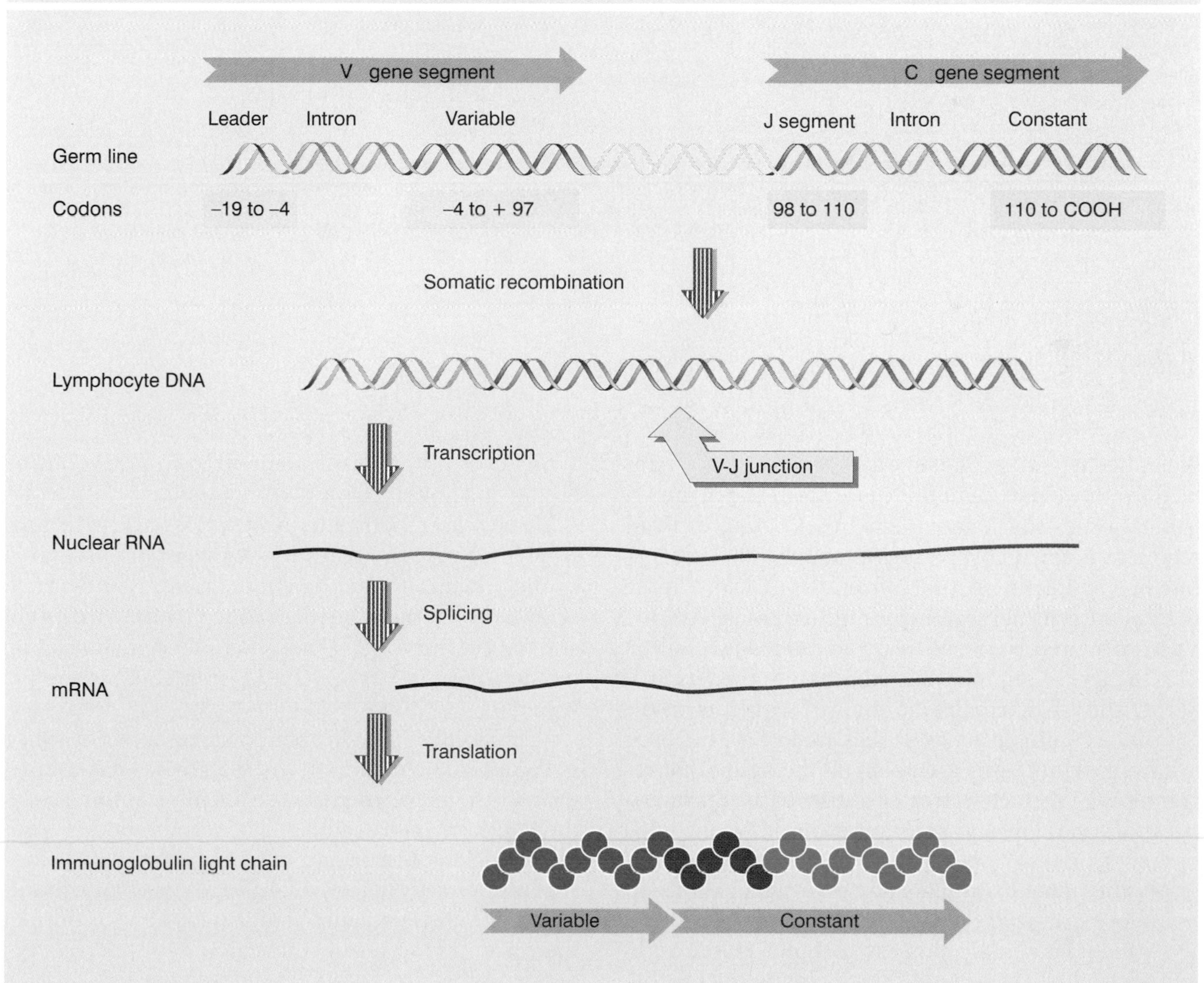

Figure 24.6 The C_κ gene segment is preceded by multiple J segments in the germ line; V-J joining may recognize any one of the J segments, which is then spliced to the C gene segment during RNA processing.

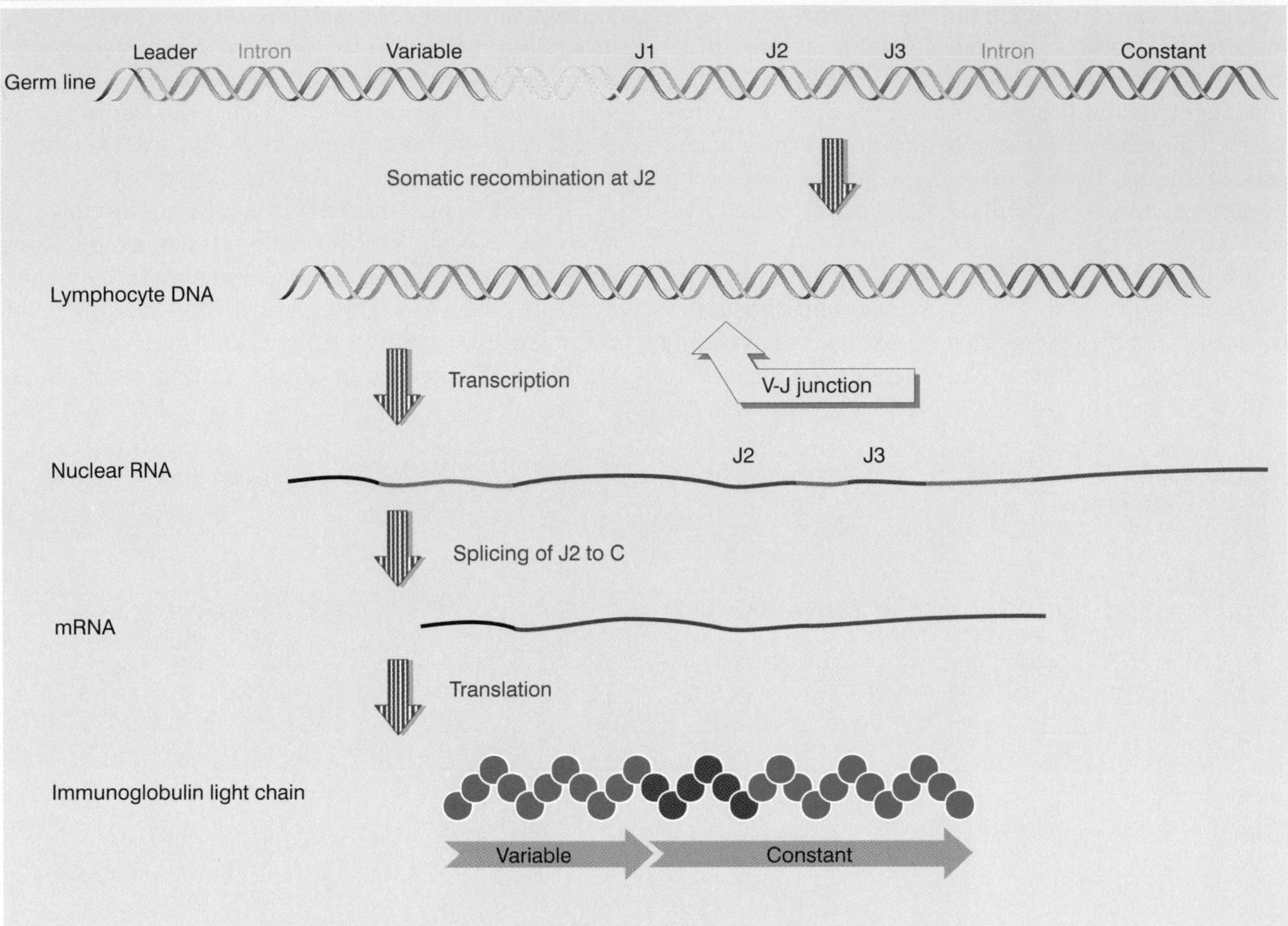

illustrated in **Figure 24.6.** A κ light chain also is assembled from two parts, but there is a difference in the organization of the C gene segment. A group of five J segments is spread over a region of 500–700 bp, separated by an intron of 2–3 kb from the C_κ exon. In the mouse, the central J segment is nonfunctional (ψJ3). A V_κ segment may be joined to any one of the J segments.

Whichever J segment is used becomes the terminal part of the intact variable exon. Any J segments on the left of the recombining J segment are lost (J1 has been lost in the figure). Any J segment on the right of the recombining J segment is treated as part of the intron between the variable and constant exons (J3 is included in the intron that is spliced out in the figure).

All functional J segments possess a signal at the left boundary that makes it possible to recombine with the V segment; and they possess a signal at the right boundary that can be used for splicing to the C exon. Whichever J segment is recognized in DNA joining uses its splicing signal in RNA processing.

Heavy chain construction involves an additional segment. The **D** (for diversity) segment was discovered by the presence in the protein of an extra 2–13 amino acids between the sequences coded by the V segment and the J segment. An array of >10 D segments lies on the chromosome between the V_H segments and the 4 J_H segments.

V-D-J joining takes place in two stages, as illustrated in **Figure 24.7.** First one of the D segments recombines with a J_H segment; then a V_H segment recombines with the DJ_H combined segment. The reconstruction leads to expression of the adjacent C_H segment (which consists of several exons). (We come later to the use of different C_H gene segments; it suffices for now to consider the reaction in terms of connection to one of several J segments that precede a C_H gene segment.)

Figure 24.7 Heavy genes are assembled by sequential joining reactions. First a D segment is joined to a J segment; then a V gene segment is joined to the D segment.

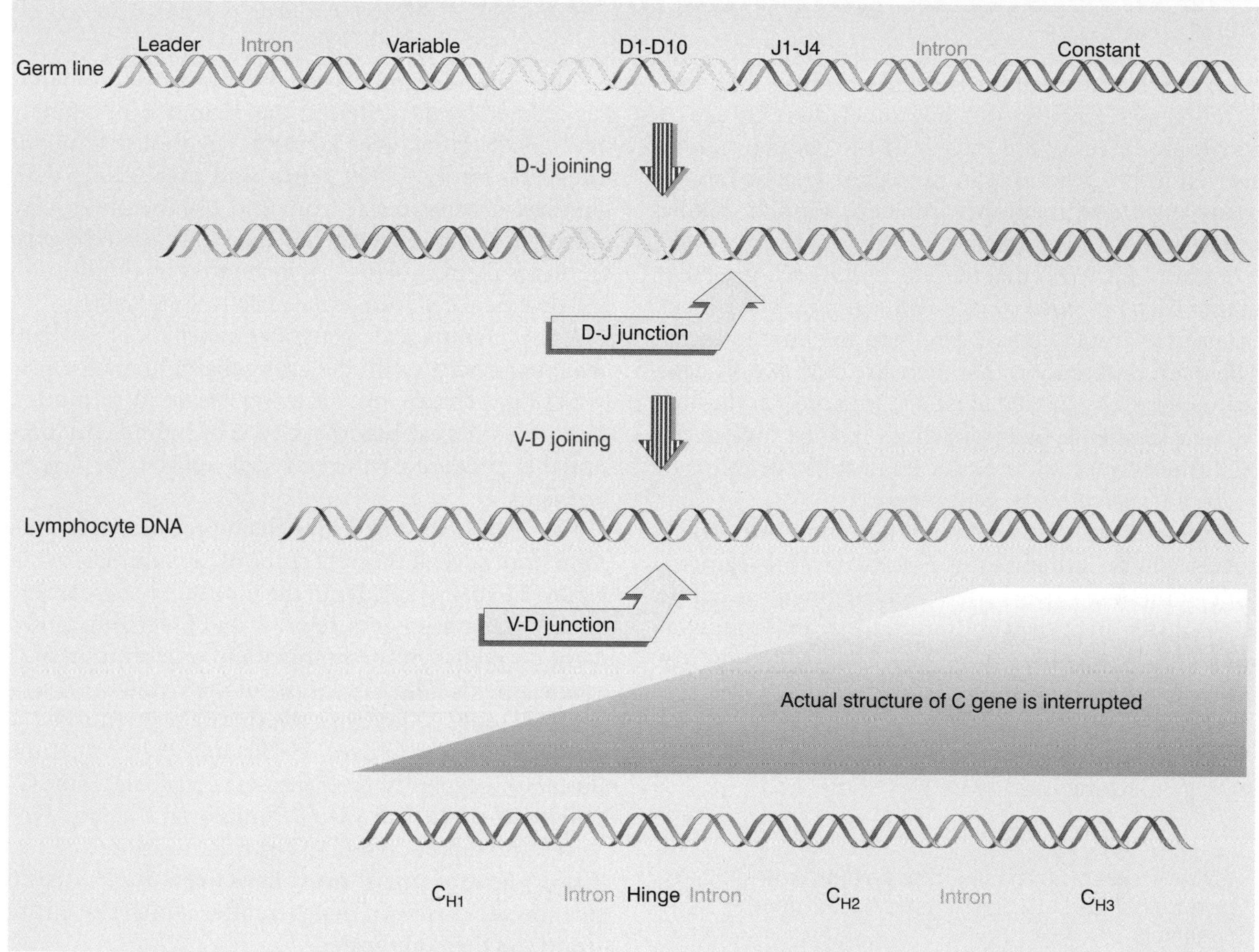

The D segments are organized in a tandem array. The mouse heavy-chain locus contains 12 D segments of variable length; the human locus has ~30 D segments (not all necessarily active). Some unknown mechanism must ensure that the *same* D segment is involved in the D-J joining and V-D joining reactions. (When we discuss joining of V and C gene segments for heavy chains, we assume the process has been completed by V-D and D-J joining reactions.)

The V gene segments of all three immunoglobulin families are similar in organization. The first exon codes for the signal sequence (involved in membrane attachment), and the second exon codes for the major part of the variable region itself (<100 codons long). The remainder of the variable region is provided by the D segment (in the H family only) and by a J segment (in all three families).

The structure of the constant region depends on the type of chain. For both κ and λ light chains, the constant region is coded by a single exon (which becomes the third exon of the reconstructed, active gene). For H chains, the constant region is coded by several exons; corresponding with the protein chain shown in Figure 24.4, separate exons code for the regions C_{H1}, hinge, C_{H2}, and C_{H3}. Each C_H exon is ~100 codons long; the hinge is shorter. The introns usually are relatively small (~300 bp).

The diversity of germline information

Now we must examine the different types of V and C gene segments to see how much diversity can be accommodated by the variety of the coding regions carried in the germline. In each light Ig gene family, many V gene segments are linked to a much smaller number of C gene segments.

Figure 24.8 shows that the λ locus has ~6 C gene segments, each preceded by its own J segment. The λ locus in mouse is much less diverse than the human locus. The main difference is that there are only two V_λ gene segments; each is linked to two J-C regions. Of the four C_λ gene segments, one is inactive. At some time in the past, the mouse suffered a catastrophic deletion of most of its germline V_λ gene segments.

Figure 24.9 shows that the κ locus has only one C gene segment, although it is preceded by 5 J segments (one of them inactive). The V_κ gene segments occupy a large cluster on the chromosome, upstream of the constant region. The human cluster has two regions. Just preceding the C_κ gene segment, a region of 600 kb contains the 5 J_κ segments and 40 V_κ gene segments. A gap of 800 kb separates this region from another group of 36 V_κ gene segments.

Figure 24.8 The λ family consists of V gene segments linked to a small number of J-C gene segments.

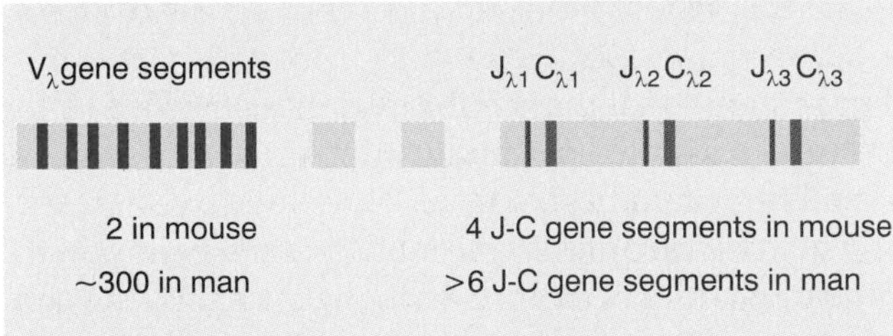

Figure 24.9 The human and mouse κ families consist of V gene segments linked to 5 J segments connected to a single C gene segment.

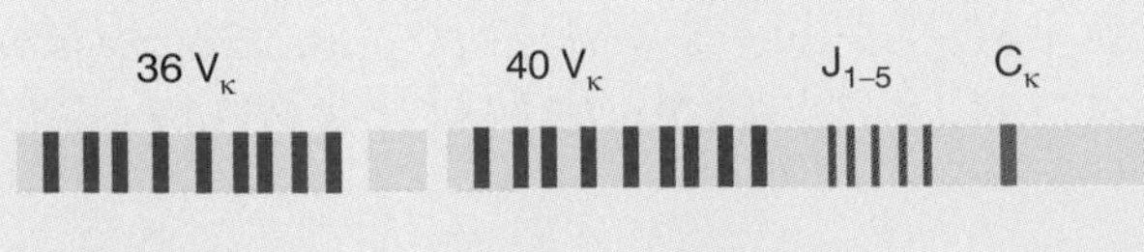

The V_κ gene segments can be subdivided into families, defined by the criterion that members of a family have >80% amino acid identity. The mouse family is unusually large, ~1000 genes, and there are ~18 V_κ families, varying in size from 2 to 100 members. Like other families of related genes, therefore, related V gene segments form subclusters, generated by duplication and divergence of individual ancestral members.

A given lymphocyte generates *either* a κ *or* a λ light chain to associate with the heavy chain. In man, ~60% of the light chains are κ and ~40% are λ. In mouse, 95% of B cells express the κ type of light chain, presumably because of the reduced number of λ gene segments.

The single locus for heavy chain production in man consists of several discrete sections, as summarized in **Figure 24.10.** It is similar in the mouse, where there are more V_H gene segments, fewer D and J segments, and a slight difference in the number and organization of C gene segments. The 3′ member of the V_H cluster is separated by only 20 kb from the first D segment. The D segments are spread over ~50 kb, and then comes the cluster of J segments. Over the next 220 kb lie all the C_H gene segments. There are 9 functional C_H gene segments and 2 pseudogenes. The organization suggests that a γ gene segment must have been duplicated to give the subcluster of γ-γ-ε-α, after which the entire group was then duplicated.

How far is the diversity of germline information responsible for V region diversity in immunoglobulin proteins? By combining any one of ~300 V gene segments with any one of 4–5 J segments, a typical light-chain locus has the potential to produce some 1200–1500 chains. There is even greater diversity in the H chain locus; by combining any one of ~300 V_H gene segments, 20 D segments, and 4 J segments, the genome potentially can produce 4000 variable regions to accompany any C_H gene segment. In mammals, this is the starting point for diversity, but additional mechanisms introduce further changes. *When closely related variants of immunoglobulins are examined, there often are more proteins than can be accounted for by the number of corresponding V gene segments.* The new members are created by somatic changes in individual genes during or after the recombination process. We consider the relevant mechanisms in later sections.

The avian immune system presents a different case.

An extreme example of reliance on diversity coded in the genome is presented by the chicken, in which a similar mechanism is used by both the single light-chain locus (of the λ type) and the H chain locus. The organization of the λ locus is drawn in **Figure 24.11**. It has only one functional V gene segment, J segment, and C gene segment. Upstream of the functional $V_{\lambda 1}$ gene segment lie 25 V_λ pseudogenes, organized in either orientation. They are classified as pseudogenes because either the coding segment is deleted at one or both ends, or proper signals for recombination are missing (or both). This assignment is confirmed by the fact that only the $V_{\lambda 1}$ gene segment recombines with the J-C_λ gene segment.

Figure 24.10 A single gene cluster in man contains all the information for heavy-chain gene assembly.

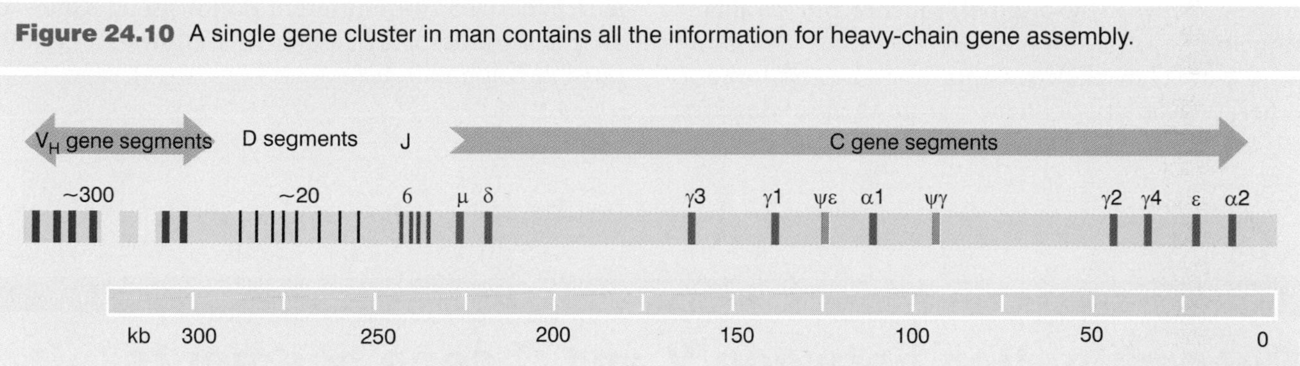

Figure 24.11 The chicken λ light locus has 25 V pseudogenes upstream of the single functional V-J-C region. But sequences derived from the pseudogenes are found in active rearranged V-J-C genes.

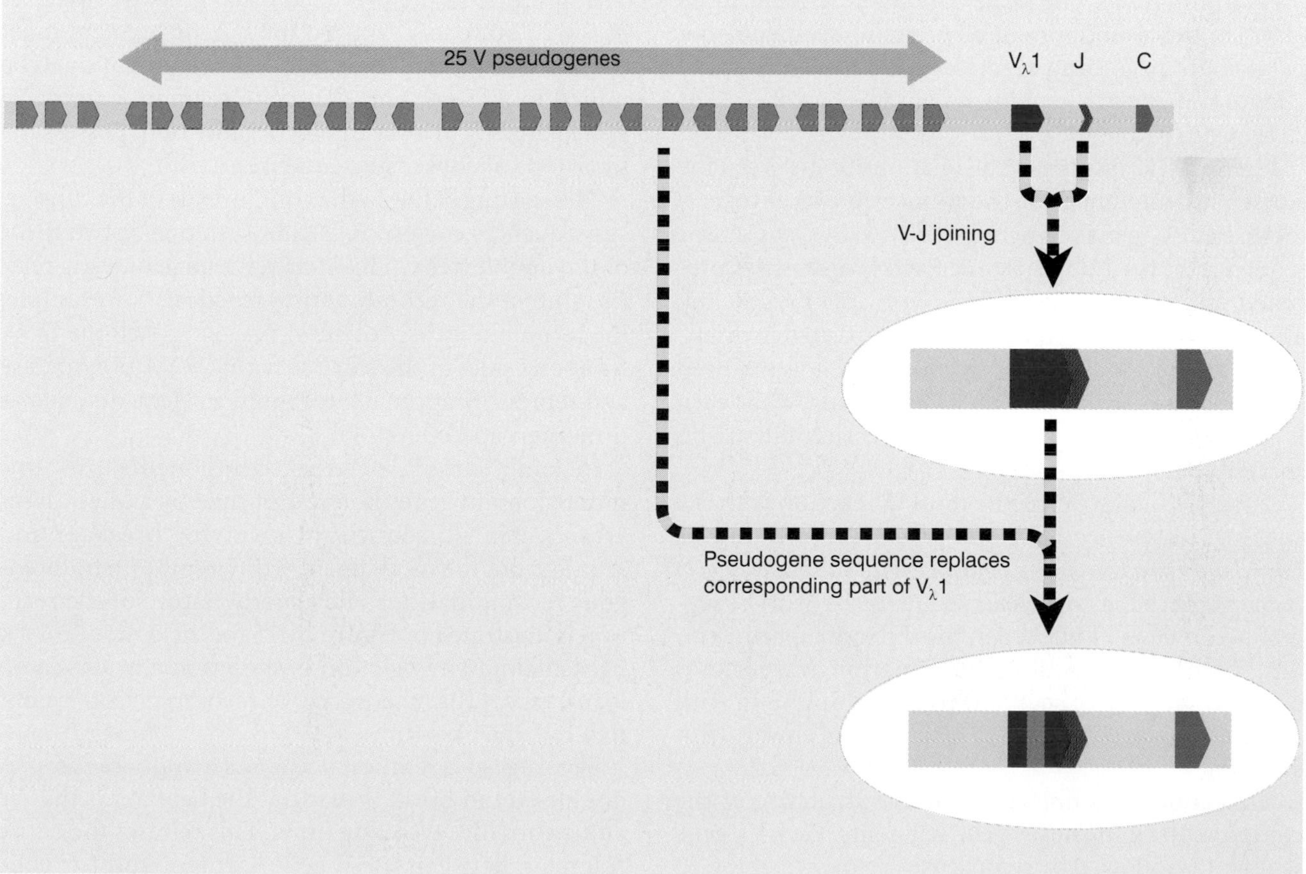

But sequences of active rearranged V_λ-J-C_λ gene segments show considerable diversity! A rearranged gene has one or more positions at which a cluster of changes has occurred in the sequence. A sequence identical to the new sequence can almost always be found in one of the pseudogenes (which themselves remain unchanged). The exceptional sequences that are not found in a pseudogene always represent changes at the junction between the original sequence and the altered sequence.

So a novel mechanism appears to be employed to generate diversity. Sequences from the pseudogenes, between 10 and 120 bp in length, are substituted into the active $V_{\lambda 1}$ region, presumably by gene conversion. The unmodified $V_{\lambda 1}$ sequence is not expressed, even at early times during the immune response. A successful conversion event probably occurs every 10–20 cell divisions to every rearranged $V_{\lambda 1}$ sequence. At the end of the immune maturation period, a rearranged $V_{\lambda 1}$ sequence has 4–6 converted segments spanning its entire length, derived from different donor pseudogenes. If all pseudogenes participate, this allows 2.5×10^8 possible combinations!

Recombination between V and C gene segments generates deletions and rearrangements

Assembly of light- and heavy-chain genes involves the same mechanism (although the number of parts is different). The same consensus sequences are found at the boundaries of all germline segments that participate in joining reactions. Each consensus sequence consists of a heptamer separated by either 12 or 23 bp from a nonamer.

Figure 24.12 illustrates the relationship between the consensus sequences at the mouse Ig loci. At the κ locus, each V_κ gene segment is followed by a consensus sequence with a 12 bp spacing. Each J_κ segment is preceded by a consensus sequence with a 23 bp spacing. The V and J consensus sequences are inverted in orientation. At the λ locus, each V_λ gene segment is followed by a consensus sequence with 23 bp spacing, while each J_λ gene segment is preceded by a consensus of the 12 bp spacer type.

The rule that governs the joining reaction is that *a consensus sequence with one type of spacing can be joined only to a consensus sequence with the other type of spacing.* Since the consensus sequences at V and J segments can lie in either order, the different spacings do not impart any directional information, but serve to prevent one V gene segment from recombining with another, or one J segment from recombining with another.

This concept is borne out by the structure of the components of the heavy gene segments. Each V_H gene segment is followed by a consensus sequence of the 23 bp spacer type. The D segments are flanked on either side by consensus sequences of the 12 bp spacer type. The J_H segments are preceded by consensus sequences of the 23 bp spacer type. So the V gene segment must be joined to a D segment; and the D segment must be joined to a J segment. A V gene segment cannot be joined directly to a J segment, because both possess the same type of consensus sequence.

The spacing between the components of the consensus sequences corresponds almost to one or two turns of the double helix. This may reflect a geometric relationship in the recombination reaction. For example, the recombination protein(s) may approach the DNA from one side, in the same way that RNA polymerase and repressors approach recognition elements such as promoters and operators.

Recombination of the components of immunoglobulin genes is accomplished by a physical rearrangement of sequences, involving breakage and reunion, but the mechanism is different from homologous recombination. The general nature of the reaction is illustrated in **Figure 24.13** for the example of a κ light chain. (The reaction is similar at a heavy-chain locus, except that there are two recombination events: first D-J, then V-DJ.)

Breakage and reunion occur as separate reactions. A double-strand break is made at the heptamers that lie at the ends of the coding units. This releases the entire fragment between the V gene segment and J-C gene

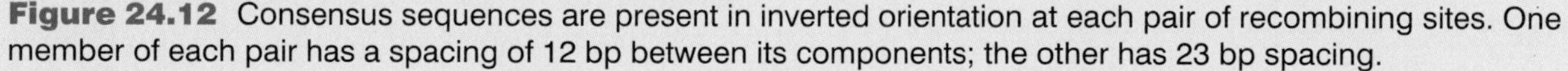

Figure 24.12 Consensus sequences are present in inverted orientation at each pair of recombining sites. One member of each pair has a spacing of 12 bp between its components; the other has 23 bp spacing.

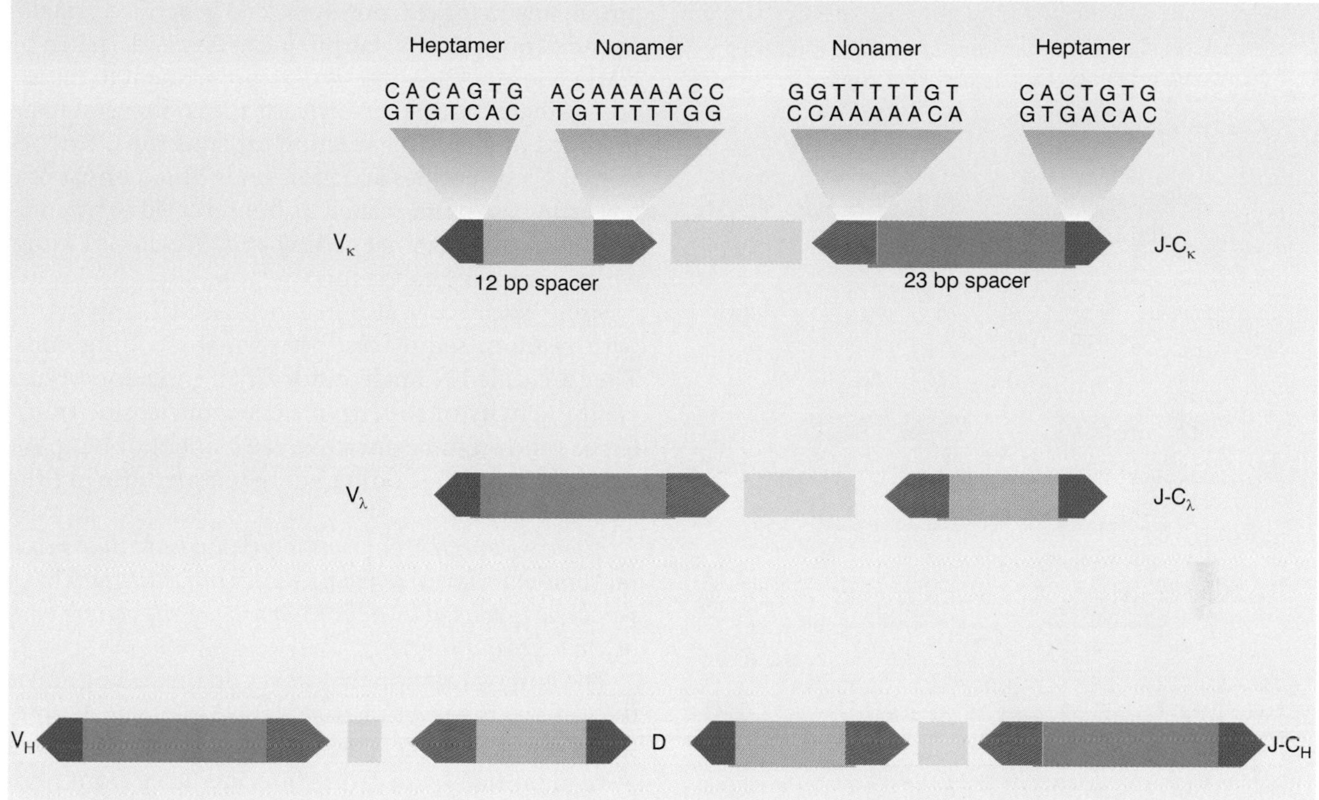

segment; the cleaved termini of this fragment are called **signal ends.** The cleaved termini of the V and J-C loci are called **coding ends.** The two coding ends are covalently linked to form a coding joint; this is the connection that links the V and J segments. If the two signal ends are also connected, the excised fragment would form a circular molecule.

The joining reaction causes changes in sequence that affect the amino acid coded at the V-J junction in light chains or at the V-D and D-J junctions in heavy chains. These changes are a consequence of the enzymatic mechanisms involved in breaking and rejoining the DNA. Base pairs are lost or inserted at the V_H-D or D-J or both junctions during the recombination process. The nature of the reaction and the sources of additional base pairs are shown in **Figure 24.14.** Deletion also occurs in V_L-J_L joining, but insertion at these joints is unusual.

The proteins RAG1 and RAG2 are necessary and sufficient to cleave DNA for V(D)J recombination.. They are coded by two genes, separated by <10 kb on the chromosome, whose transfection into fibroblasts causes a suitable substrate DNA to undergo the V(D)J joining reaction. Mice that lack either *RAG1* or *RAG2* are unable to recombine their immunoglobulins or T-cell receptors, and as a result have immature B and T lymphocytes.

RAG1 recognizes the heptamer/nonamer signals with the appropriate 12/23 spacing and recruits RAG2 to the complex; the nonamer provides the site for initial recognition, and the heptamer directs the site of cleavage. The complex then nicks one strand at each junction. The nick has 3′-OH and 5′-P ends. The free 3′-OH end then attacks the phosphate bond at the corresponding position *in the other strand of the duplex.* This creates a hairpin at the coding end, in which the 3′ end of one strand is covalently linked to the 5′ end of the other strand; it leaves a blunt double-strand break at the signal end.

This second cleavage is a transesterification reaction in which bond energies are conserved. It resembles the topoisomerase-like reactions catalyzed by the resolvase proteins of bacterial transposons (see Chapter 15). The parallel with these reactions is supported further by a

Figure 24.13 Breakage and reunion at consensus sequences generates immunoglobulin genes.

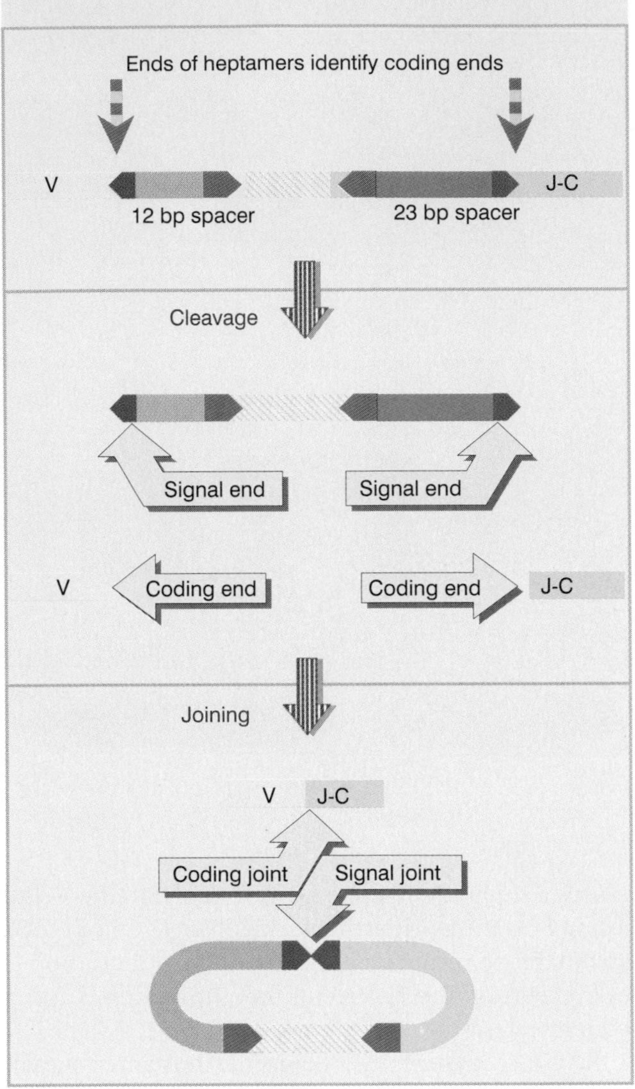

homology between RAG1 and bacterial invertase proteins (which invert specific segments of DNA by similar recombination reactions). In fact, the RAG proteins can insert a donor DNA whose free ends consist of the appropriate signal sequences (heptamer—12/23—spacer nonamer) into an unrelated target DNA in an *in vitro* transposition reaction. This suggests that somatic recombination of immune genes evolved from an ancestral transposon. It also suggests that the RAG proteins are responsible for chromosomal translocations in which Ig or TCR loci are connected to other loci (see Chapter 28).

The hairpins at the coding ends provide the substrate for the next stage of reaction. If a single-strand break is introduced into one strand close to the hairpin, an unpairing reaction at the end generates a single-stranded protrusion. Hairpin nicking is undertaken by RAG1 and RAG2. Synthesis of a complement to the exposed single strand then converts the coding end to an extended duplex. This reaction explains the introduction of **P nucleotides** at coding ends; they consist of a few extra base pairs, related to, but reversed in orientation from, the original coding end. We do not know which enzymes catalyze this reaction.

Some extra bases also may be inserted, apparently with random sequences, between the coding ends. They are called **N nucleotides.** Their insertion occurs via the activity of the enzyme deoxynucleoside transferase (known to be an active component of lymphocytes) at a free 3′ coding end generated during the joining process.

These various mechanisms together ensure that a coding joint may have a sequence that is different from what would be predicted by a direct joining of the coding ends of the V, D, and J regions.

The joining reaction that works on the coding end is the same as the reaction that is used to repair double-strand breaks in cells. The initial stages of the reaction were identified by isolating intermediates from lymphocytes of mice with the *SCID* mutation, which results in a much reduced activity in immunoglobulin and TCR recombination. *SCID* mice accumulate broken molecules that terminate in double-strand breaks at the coding ends, and are thus deficient in completing some aspect of the joining reaction. The *SCID* mutation inactivates a DNA-dependent protein kinase (DNA-PKcs). The kinase is recruited to DNA by the Ku70 and Ku80 proteins, which bind to the DNA ends. The actual ligation is undertaken by DNA ligase IV and also requires the protein XRCC4. These proteins are also involved in the reaction of nonhomologous end-joining, which occurs in DNA repair (see Chapter 14). As a result, mutations in the Ku proteins or in XRCC4 or DNA ligase IV are found among human patients who have diseases caused by deficiencies in DNA repair that result in increased sensitivity to radiation.

Changes in the sequence at the junction make it possible for a great variety of amino acids to be coded at this site. It is interesting that the amino acid at position 96 is created by the V-J joining reaction. It forms part of the antigen-binding site and also is involved in making contacts between the light and heavy chains. So the maximum diversity is generated at the site that contacts the target antigen.

Figure 24.14 Processing of coding ends introduces variability at the junction.

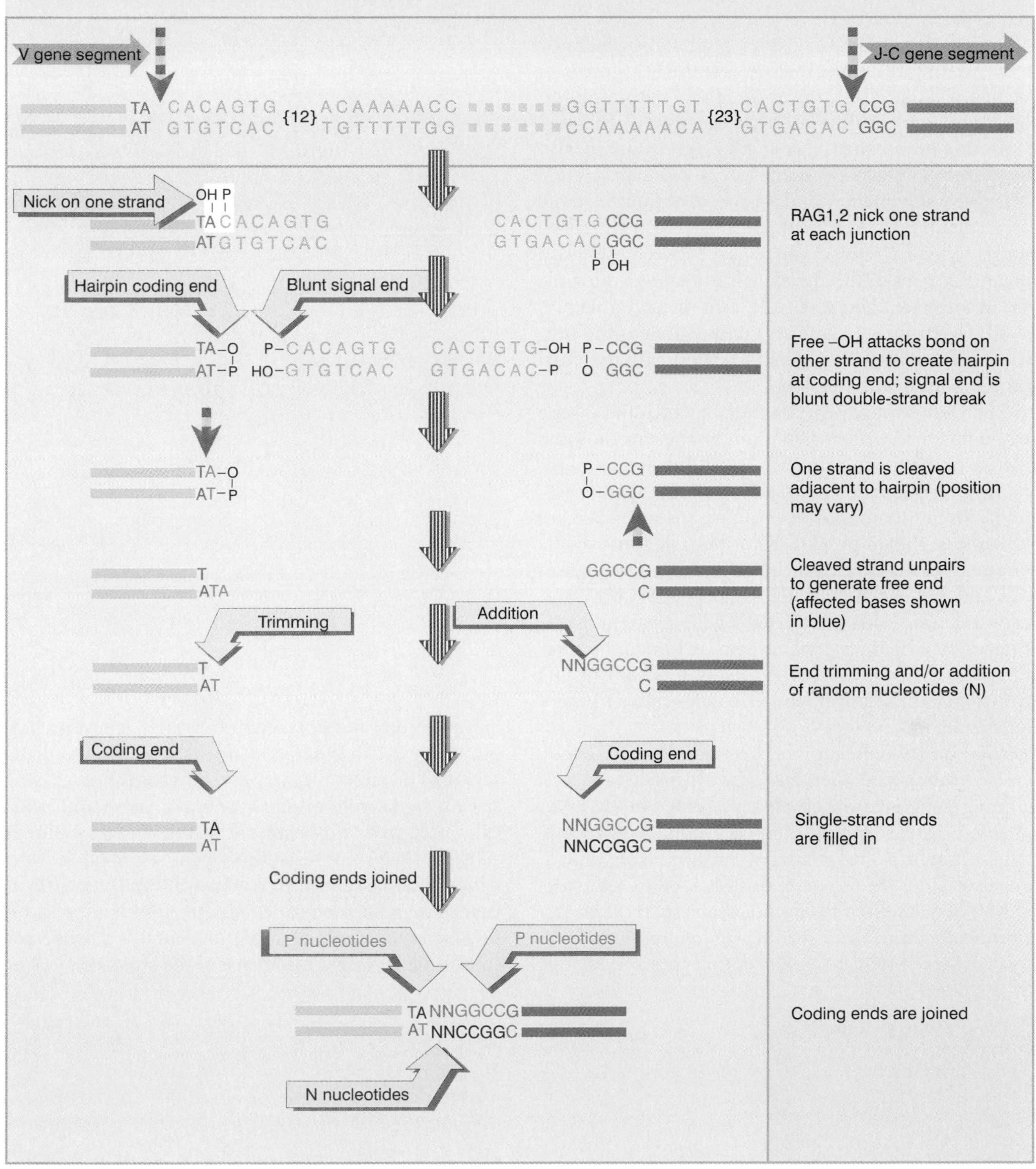

Changes in the number of base pairs at the coding joint affect the reading frame. The joining process appears to be random with regard to the reading frame, so that probably only one third of the joined sequences retain the proper frame of reading through the junctions. If the V-J region is joined so that the J segment is out of phase, translation is terminated prematurely by a nonsense codon in the incorrect frame. We may think of the formation of aberrant genes as comprising the price the cell must pay for the increased diversity that it gains by being able to adjust the sequence at the joining site.

Similar although even greater diversity is generated in the joining reactions that involve the D segment of the heavy chain. The same result is seen with regard to the reading frame; nonproductive genes are generated by joining events that place J and C out of phase with the preceding V gene segment.

We have shown the V and J-C loci as organized in the same orientation. As a result, the cleavage at each consensus sequence releases the region between them as a linear fragment. If the signal ends are joined, it is converted into a circular molecule, as indicated in Figure 24.13. Deletion to release an excised circle is the predominant mode of recombination at the immunoglobulin and TCR loci.

There are some cases, however, in which the V gene segment is inverted in orientation on the chromosome relative to the J-C loci. In such a case, breakage and reunion inverts the intervening material instead of deleting it. The outcomes of deletion versus inversion are the same as shown previously for homologous recombination between direct or inverted repeats in Figures 15.8 and 15.9. There is one further proviso, however; recombination with an inverted V gene segment makes it necessary for the signal ends to be joined, because otherwise there is a break in the locus. Inversion occurs in TCR recombination, and also sometimes in the κ light-chain locus.

What is the connection between joining of V and C gene segments and their activation? Unrearranged V gene segments are not actively represented in RNA. But when a V gene segment is joined productively to a C_κ gene segment, the resulting unit is transcribed. However, since the sequence upstream of a V gene segment is not altered by the joining reaction, *the promoter must be the same in unrearranged, nonproductively rearranged, and productively rearranged genes.*

Figure 24.15 A V gene promoter is inactive until recombination brings it into the proximity of an enhancer in the C gene segment. The enhancer is active only in B lymphocytes.

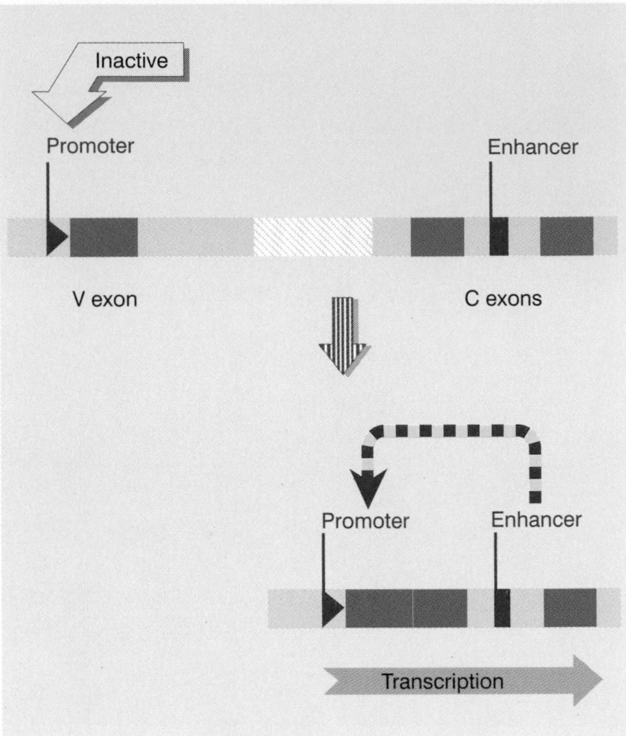

A promoter lies upstream of every V gene segment, but is inactive. It is activated by its relocation to the C region. The effect must depend on sequences downstream. What role might they play? An enhancer located within or downstream of the C gene segment activates the promoter at the V gene segment. The enhancer is tissue specific; it is active only in B cells. Its existence suggests the model illustrated in **Figure 24.15**, in which the V gene segment promoter is activated as soon as it is brought into reach of the enhancer.

Allelic exclusion is triggered by productive rearrangement

Each B cell expresses a single type of light chain and a single type of heavy chain, because only a single productive rearrangement of each type occurs in a given lymphocyte, to produce one light- and one heavy-chain gene. Because each event involves the genes of only *one* of the homologous chromosomes, *the alleles on the other chromosome are not expressed in the same cell.* This phenomenon is called **allelic exclusion.**

The occurrence of allelic exclusion complicates the analysis of somatic recombination. A probe reacting with a region that has rearranged on one homolog will also detect the allelic sequences on the other homolog. We are therefore compelled to analyze the different fates of the two chromosomes together.

The usual pattern displayed by a rearranged active gene can be interpreted in terms of a deletion of the material between the recombining V and C loci.

Two types of gene organization are seen in active cells:

- Probes to the active gene may reveal both the rearranged and germline patterns of organization. We assume then that joining has occurred on one chromosome, while the other chromosome has remained unaltered.
- Two different rearranged patterns may be found, indicating that the chromosomes have suffered independent rearrangements. In some of these instances, material between the recombining V and C gene segments is entirely absent from the cell line. This is most easily explained by the occurrence of independent deletions on each chromosome.

When two chromosomes both lack the germline pattern, usually only one of them has passed through a **productive rearrangement** to generate a functional gene. The other has suffered a **nonproductive rearrangement;** this may take several forms, but in each case the gene sequence cannot be expressed as an immunoglobulin chain. (It may be incomplete, for example because D-J joining has occurred but V-D joining has not followed, or it may be aberrant, with the process completed, but failing to generate a gene that codes for a functional protein.)

The coexistence of productive and nonproductive rearrangements suggests the existence of a feedback loop to control the recombination process. A model is outlined in **Figure 24.16.** Suppose that each cell starts with two loci in the unrearranged germline configuration Ig^0. Either of these loci may be rearranged to generate a productive gene Ig^+ or a nonproductive gene Ig^-.

If the rearrangement is productive, the synthesis of an active chain provides a trigger to prevent rearrangement of the other allele. The active cell has the configuration Ig^0/Ig^+.

If the rearrangement is nonproductive, it creates a cell with the configuration Ig^0/Ig^-. There is no impediment to rearrangement of the remaining germline allele. If this rearrangement is productive, the expressing cell has the configuration Ig^+/Ig^-. Again, the presence of an active chain suppresses the possibility of further rearrangements.

Two successive nonproductive rearrangements produce the cell Ig^-/Ig^-. In some cases an Ig^-/Ig^- cell can try yet again. Sometimes the observed patterns of DNA can only have been generated by successive rearrangements.

The crux of the model is that the cell keeps trying to recombine V and C gene segments until a productive rearrangement is achieved. Allelic exclusion is caused by the suppression of further rearrangement as soon as an active chain is produced. The use of this mechanism *in vivo* is demonstrated by the creation of transgenic mice whose germline has a rearranged immunoglobulin gene. Expression of the transgene in B cells suppresses the rearrangement of endogenous genes.

Allelic exclusion is independent for the heavy- and light-chain loci. Heavy chain genes usually rearrange first. Allelic exclusion for light chains must apply equally to both families (cells may have *either* active κ or λ light chains). It is likely that the cell rearranges its κ genes first, and tries to rearrange λ only if both κ attempts are unsuccessful.

There is an interesting paradox in this series of events. The same consensus sequences and the same V(D)J recombinase are involved in the recombination reactions at H, κ, and λ loci. Yet the three loci rearrange in a set order. What ensures that heavy rearrangement

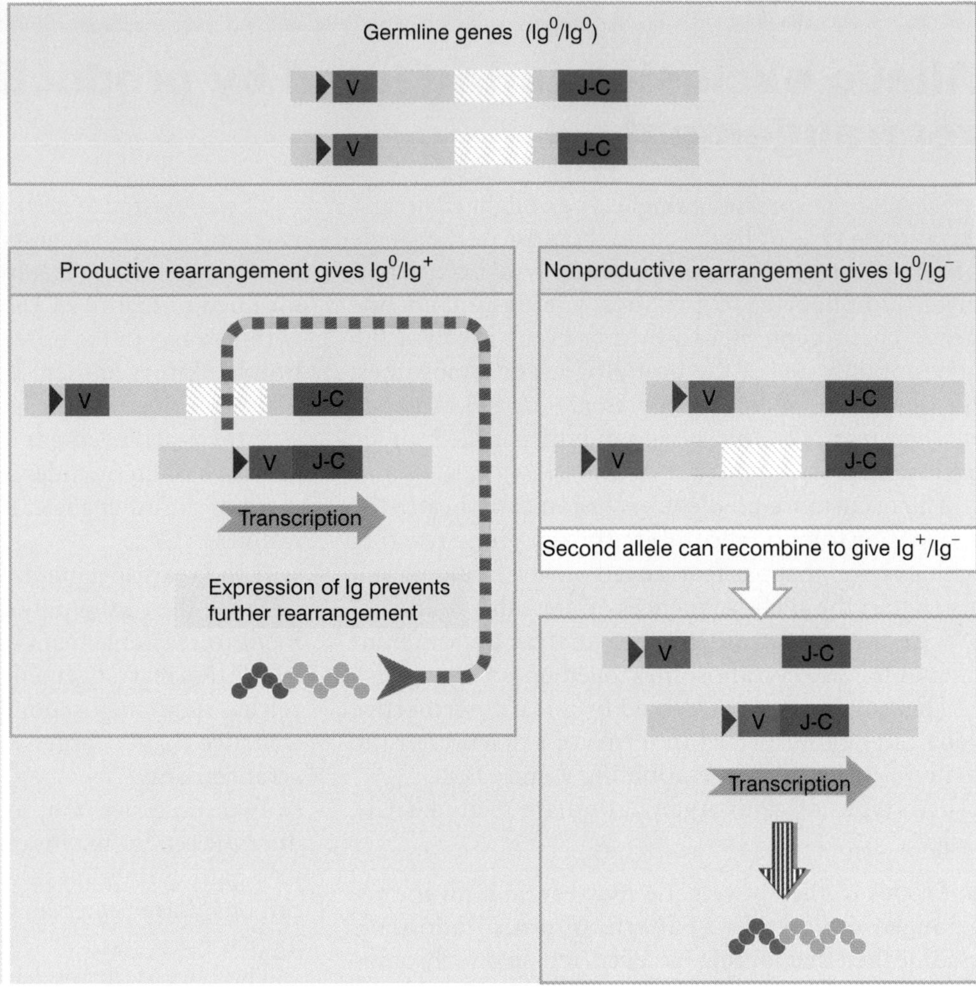

Figure 24.16 A successful rearrangement to produce an active light or heavy chain suppresses further rearrangements of the same type, and results in allelic exclusion.

precedes light rearrangement, and that κ precedes λ? The loci may become accessible to the enzyme at different times, possibly as the result of transcription. Transcription occurs even before rearrangement, although of course the products have no coding function. The transcriptional event may change the structure of chromatin, making the consensus sequences for recombination available to the enzyme.

DNA recombination causes class switching

THE **class** of immunoglobulin is defined by the type of C_H region. **Figure 24.17** summarizes the five Ig classes. IgM (the first immunoglobulin to be produced by any B cell) and IgG (the most common immunoglobulin) possess the central ability to activate complement, which leads to destruction of invading cells. IgA is found in secretions (such as saliva), and IgE is associated with the allergic response and defense against parasites.

All lymphocytes start productive life as immature cells engaged in synthesis of IgM. Cells expressing IgM have the germline arrangement of the C_H gene segment cluster shown in Figure 24.10. The V-D-J joining reaction triggers expression of the C_μ gene segment. A

Figure 24.17 Immunoglobulin type and function is determined by the heavy chain. J is a 'joining protein' in IgM; all other Ig types exist as tetramers.

Type	IgM	IgD	IgG	IgA	IgE
Heavy chain	μ	δ	γ	α	ε
Structure	$(\mu_2L_2)_5J$	δ_2L_2	γ_2L_2	$(\alpha_2L_2)_2J$	ε_2L_2
Proportion	5%	1%	80%	14%	<1%
Effector function	Activates complement	Development of tolerance (?)	Activates complement	Found in secretions	Allergic response

lymphocyte generally produces only a single class of immunoglobulin at any one time, but the class may change during the cell lineage. A change in expression is called **class switching.** It is accomplished by a substitution in the type of C_H region that is expressed. Switching can be stimulated by environmental effects; for example, the growth factor TGFβ causes switching from C_μ to C_α.

Switching involves only the C_H gene segment; the same V_H gene segment continues to be expressed. So a given V_H gene segment may be expressed successively in combination with more than one C_H gene segment. The same light chain continues to be expressed throughout the lineage of the cell. Class switching therefore allows the type of effector response (mediated by the C_H region) to change, while maintaining a constant facility to recognize antigen (mediated by the V regions).

Changes in the expression of C_H gene segments are made in two ways. The majority occur via further DNA recombination events, involving a system different from that concerned with V-D-J joining (and able to operate only later during B cell development). Another type of change occurs at the level of RNA processing, but generally this is involved with changing the C-terminal sequence of the C_H region rather than its class (see next section).

Cells expressing downstream C_H gene segments have deletions of C_μ and the other gene segments preceding the expressed C_H gene segment. Class switching is accomplished by a recombination to bring a new C_H gene segment into juxtaposition with the expressed V-D-J unit. The sequences of switched V-D-J-C_H units show that the sites of switching lie upstream of the C_H gene segments themselves. The switching sites are called **S regions. Figure 24.18** depicts two successive switches.

In the first switch, expression of C_μ is succeeded by expression of $C_{\gamma 1}$. The $C_{\gamma 1}$ gene segment is brought into the expressed position by recombination between the sites S_μ and $S_{\gamma 1}$, deleting the material between. The S_μ site lies between V-D-J and the C_μ gene segment. The $S_{\gamma 1}$ site lies upstream of the $C_{\gamma 1}$ gene segment.

The linear deletion model imposes a restriction on the heavy-gene locus: *once a class switch has been made, it becomes impossible to express any C_H gene segment that used to reside between C_μ and the new C_H gene segment.* In the example of Figure 24.18, cells expressing $C_{\gamma 1}$ should be unable to give rise to cells expressing $C_{\gamma 3}$, which has been deleted.

However, it should be possible to undertake another switch to any C_H gene segment *downstream* of the expressed gene. The figure shows a second switch to C_α expression, accomplished by recombination between $S_{\alpha 1}$ and the switch region $S_{\mu,\gamma 1}$ that was generated by the original switch.

We assume that all of the C_H gene segments have S regions upstream of the coding sequences. We do not know whether there are any restrictions on the use of S regions. Sequential switches do occur, but we do not know whether they are optional or an obligatory means to proceed to later C_H gene segments. We should like to know whether IgM can switch directly to *any* other class.

We know that switch sites are not uniquely defined, because different cells expressing the same C_H gene segment prove to have recombined at different points. Switch regions vary in length (as defined by the limits of the sites involved in recombination) from 1–10 kb. They contain groups of short homologous repeats, with repeating units that vary from 20–80 nucleotides in length. An S region typically is located ~2 kb upstream of a C_H gene segment. The switching reaction releases the excised material between the switch sites as a circular DNA molecule. Two of the proteins required for the joining phase of VDJ recombination (and also for general nonhomologous end-joining), Ku and

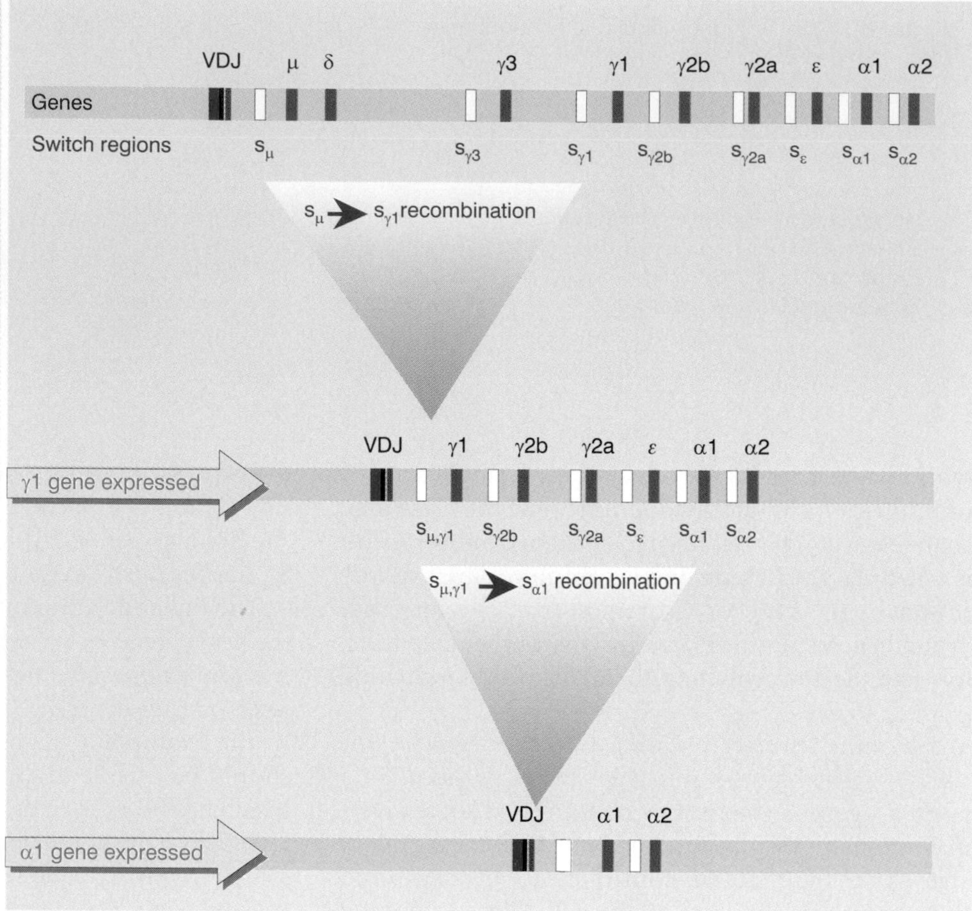

Figure 24.18 Class switching of heavy genes may occur by recombination between switch regions (S), deleting the material between the recombining S sites. Successive switches may occur.

DNA-PKcs, are required, suggesting that there may be some similarities with the reaction for nonhomologous end-joining.

A promoter lies immediately upstream of each switch region, and switching may be activated by transcription from the promoter (which itself may be responsive to activators that respond to environmental conditions, such as stimulation by cytokines, thus creating a mechanism to regulate switching). Because the S regions lie within the introns that precede the C_H coding regions, switching does not alter the translational reading frame.

Early heavy chain expression can be changed by RNA processing

The period of IgM synthesis that begins lymphocyte development falls into two parts, during which different versions of the μ constant region are synthesized:

- As a stem cell differentiates to a pre-B lymphocyte, an accompanying light chain is synthesized, and the IgM molecule ($L_2\mu_2$) appears at the surface of the cell. This form of IgM contains the μ_m version of the constant region (*m* indicates that IgM is located in the membrane). The membrane location may be related to the need to initiate cell proliferation in response to the initial recognition of an antigen.

■ When the B lymphocyte differentiates further into a plasma cell, the μ_s version of the constant region is expressed. The IgM actually is secreted as a pentamer IgM_5J, in which J is a joining polypeptide (no connection with the J region) that forms disulfide linkages with μ chains. Secretion of the protein is followed by the humoral response depicted in Figure 24.1.

The μ_m and μ_s versions of the μ heavy chain differ only at the C-terminal end. The μ_m chain ends in a hydrophobic sequence that probably secures it in the membrane. This sequence is replaced by a shorter hydrophilic sequence in μ_s; the substitution allows the μ heavy chain to pass through the membrane. The change of C-terminus is accomplished by an alternative splicing event, which is controlled by the 3′ end of the nuclear RNA, as illustrated in **Figure 24.19.**

At the membrane-bound stage, the RNA terminates after exon M2, and the constant region is produced by splicing together six exons. The first four exons code for the four domains of the constant region. The last two exons, M1 and M2, code for the 41-residue hydrophobic C-terminal region and its nontranslated trailer. The 5′ splice junction within exon 4 is connected to the 3′ splice junction at the beginning of M1.

At the secreted stage, the nuclear RNA terminates after exon 4. The 5′ splice junction within this exon that had been linked to M1 in the membrane form is ignored. This allows the exon to extend for an additional 20 codons.

A similar transition from membrane to secreted forms is found with other constant regions. The conservation of exon structures suggests that the mechanism is the same.

Figure 24.19 The 3' end controls the use of splicing junctions so that alternative forms of the heavy gene are expressed.

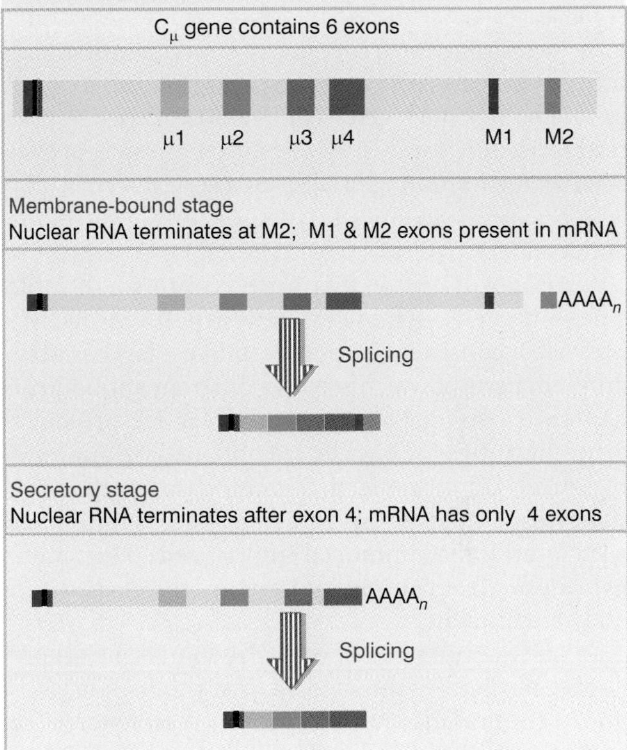

Somatic mutation generates additional diversity

Comparisons between the sequences of expressed immunoglobulin genes and the corresponding V gene segments of the germline show that new sequences appear in the expressed population. We have seen that some of this additional diversity results from sequence changes at the V-J or V-D-J junctions. However, other changes occur upstream at locations within the variable domain; they represent **somatic mutations** induced specifically in the active lymphocyte.

A probe representing an expressed V gene segment can be used to identify all the corresponding fragments in the germline. Their sequences should identify the complete repertoire available to the organism. Any expressed gene whose sequence is different must have been generated by somatic changes.

One difficulty is to ensure that every potential contributor in the germline V gene segments actually has been identified. This problem is overcome by the simplicity of the mouse λ chain system. A survey of several myelomas producing λ_1 chains showed that many have the sequence of the single germline gene segment. *But others have new sequences that must have been generated by mutation of the germline gene segment.*

To determine the frequency of somatic mutation in other cases, we need to examine a large number of cells in which the same V gene segment is expressed. A practical procedure for identifying such a group is to characterize the immunoglobulins of a series of cells, all of which express an immune response to a particular antigen.

(Epitopes used for this purpose are small molecules—haptens—whose discrete structure is likely to provoke a consistent response, unlike a large protein, different parts of which provoke different antibodies. A hapten is conjugated with a nonreactive protein to form the antigen. The cells are obtained by immunizing mice with the antigen, obtaining the reactive lymphocytes, and sometimes fusing these lymphocytes with a myeloma [immortal tumor] cell to generate a **hybridoma** that continues to express the desired antibody indefinitely.)

In one example, 10 out of 19 different cell lines producing antibodies directed against the hapten phosphorylcholine had the same V_H sequence. This sequence was the germline V gene segment T15, one of 4 related V_H genes. The other 9 expressed gene segments differed from each other and from all 4 germline members of the family. They were more closely related to the T15 germline sequence than to any of the others, and their flanking sequences were the same as those around T15. This suggested that they arose from the T15 member by somatic mutation.

Sequence changes extend through and beyond the V gene segment. They take the form of substitutions of individual nucleotide pairs. The variation is different in each case. It represents ~3–15 substitutions, corresponding to <10 amino acid substitutions in the protein. Only some of the mutations affect the amino acid sequence, since others lie in third-base coding positions as well as in nontranslated regions.

The large proportion of ineffectual mutations suggests that somatic mutation occurs more or less at random in a region including the V gene segment and extending beyond it. There is a tendency for some mutations to recur on multiple occasions. These may represent hotspots as a result of some intrinsic preference in the system.

Somatic mutation requires the enhancer that activates transcription at each Ig locus. There is a correlation between the occurrence of transcription and the induction of mutations, but the molecular basis for somatic mutation remains unknown.

Somatic mutation occurs during clonal proliferation, apparently at a rate $\sim 10^{-3}$ per bp per cell generation. Approximately half of the progeny cells gain a mutation; as a result, cells expressing mutated antibodies become a high fraction of the clone.

In many cases, a single family of V gene segments is used consistently to respond to a particular antigen. Upon exposure to an antigen, presumably the V region with highest intrinsic affinity provides a starting point. Then somatic mutation increases the repertoire. Random mutations have unpredictable effects on protein function; some inactivate the protein, others confer high specificity for a particular antigen. The proportion and effectiveness of the lymphocytes responding is increased by selection among the lymphocyte population for those cells bearing antibodies in which mutation has increased the affinity for the antigen.

B cell development and memory

We are now in a position to summarize the relationship between the generation of high-affinity antibodies and the differentiation of the B cell. **Figure 24.20** shows that B cells are derived from a self-renewing population of stem cells in the bone marrow. Maturation to give B cells depends upon Ig gene rearrangement, which requires the functions of the *SCID* and *RAG1,2* (and other) genes. If gene rearrangement is blocked, mature B cells are not produced. The antibodies carried by the B cells have specificities determined by the particular combinations of V(D)J regions, and any additional nucleotides incorporated during the joining process.

Exposure to antigen triggers two aspects of the

Figure 24.20 B cell differentiation is responsible for acquired immunity. Pre-B cells are converted to B cells by Ig gene rearrangement. Initial exposure to antigen provokes both the primary response and storage of memory cells. Subsequent exposure to antigen provokes the secondary response of the memory cells.

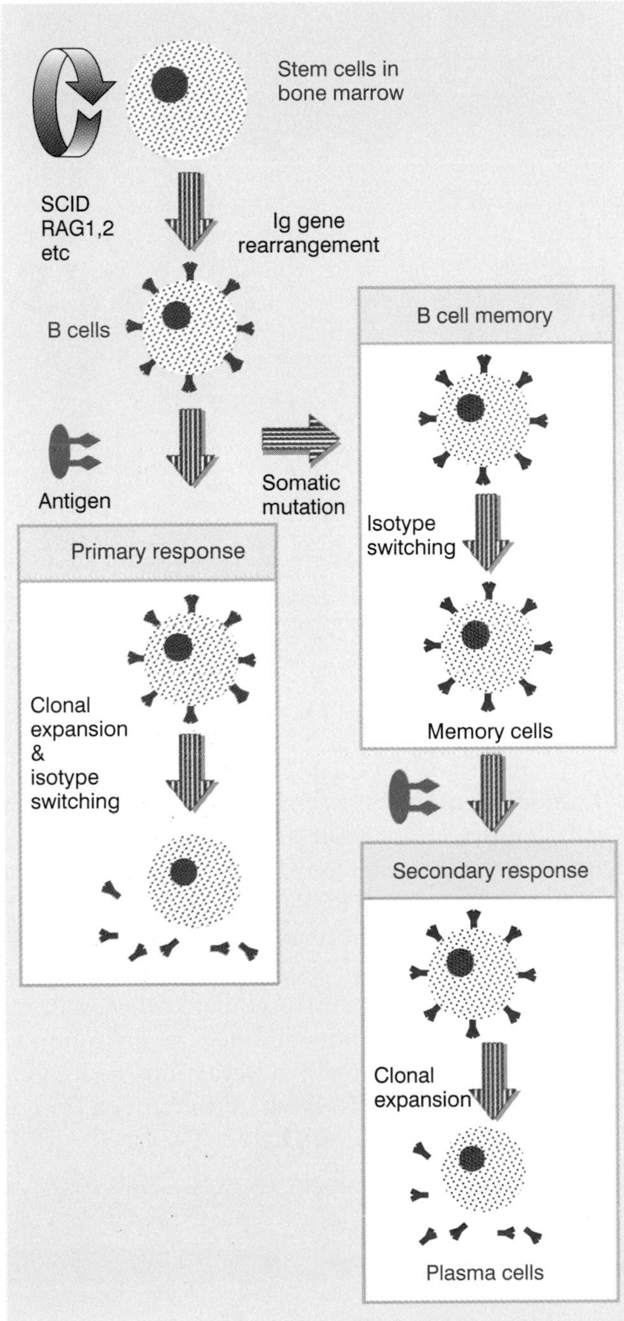

Figure 24.21 B cell development proceeds through sequential stages.

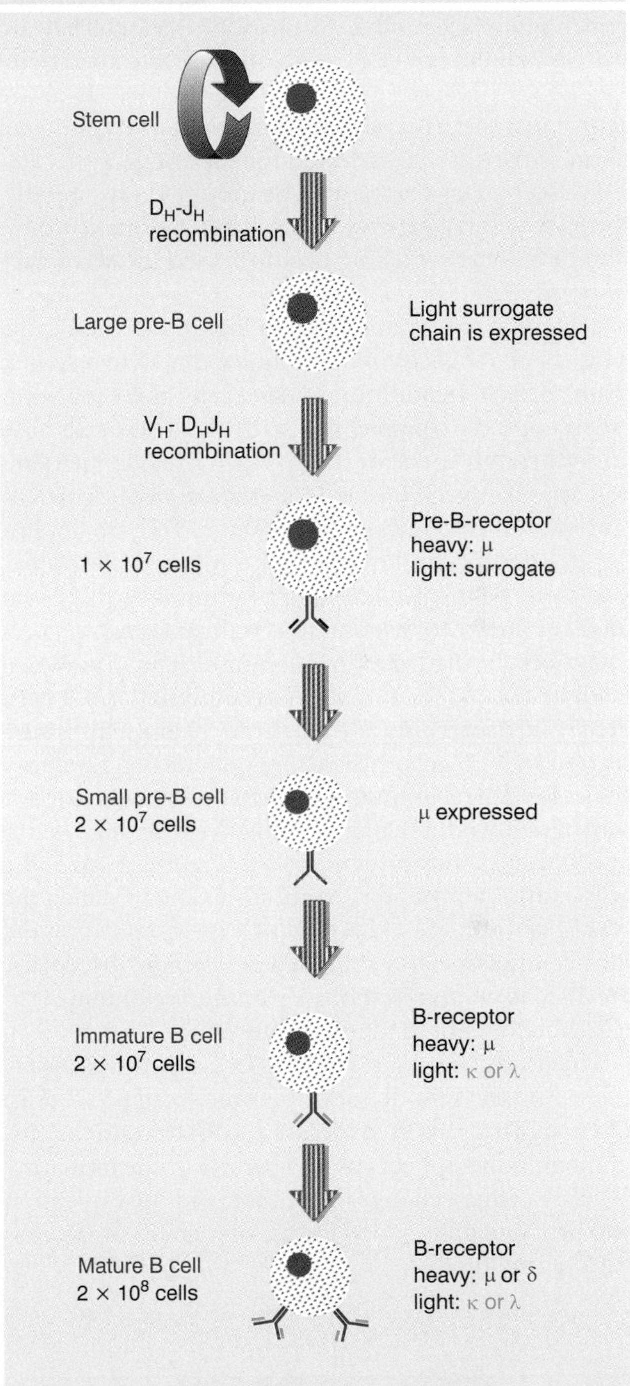

immune response. The **primary response** occurs by clonal expansion of B cells responding to the antigen. This generates a large number of plasma cells that are specific for the antigen; isotype switching occurs to generate the appropriate type of effector response. The population of cells concerned with the primary response is a dead end; these cells do not live beyond the primary response itself.

Provision for a **secondary response** is made through the phenomenon of **B-cell memory.** Somatic mutation generates B cells that have increased affinity for the antigen. These cells do not trigger an immune response at this time, although they may undergo isotype switching to select other forms of C_H region. They are stored as memory cells, with appropriate specificity and effector response type, but are inactive. They are activated if there is a new exposure to the same antigen. Because they are pre-selected for the antigen, they enable a secondary response to be mounted very rapidly, simply by clonal expansion; no further somatic mutation or isotype switching occurs during the secondary response.

The pathways summarized in Figure 24.20 show the development of acquired immunity, that is, in response to an antigen. In addition to these cells, there is a separate set of B cells, named the Ly-1 cells. These cells have gone through the process of V gene rearrangement, and apparently are selected for expression of a particular repertoire of antibody specificities. They do not undergo somatic mutation or the memory response. They may be involved in natural immunity, that is, an intrinsic ability to respond to certain antigens.

A more detailed view of B cell development is shown in **Figure 24.21.** The first step is recombination between the D and J segments of the μ heavy chain. This is succeeded by V-D recombination, generating a μ heavy chain. Several recombination events, involving a succession of nonproductive and productive rearrangements, may occur, as shown previously in Figure 24.16. These cells express a protein resembling a λ chain, called the surrogate light (SL) chain, which is expressed on the surface and associates with the μ heavy chain to form the pre-B-receptor. It resembles an immunoglobulin complex, but does not function as one.

The production of μ chain represses synthesis of SL chain, and the cells divide to become small pre-B cells. Then light chain is expressed and functional immunoglobulin appears on the surface of the immature B cells. Further cell divisions occur, and the expression of δ heavy chain is added to that of μ chain, as the cells mature into B cells.

Figure 24.22 The B cell antigen receptor consists of an immunoglobulin tetramer (H_2L_2) linked to two copies of the signal-transducing heterodimer (Igαβ).

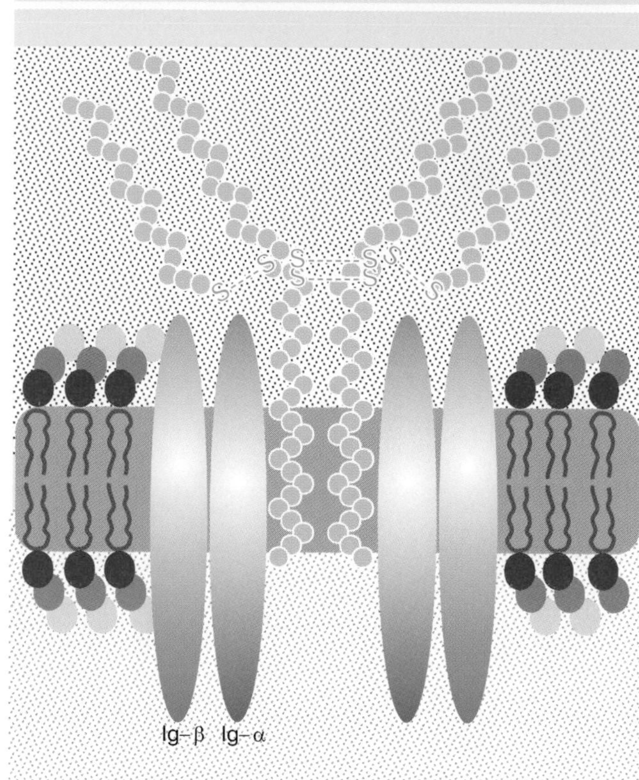

Immunoglobulins function both by secretion from B cells and by surface expression. **Figure 24.22** shows that the active complex on the cell surface is called the **B-cell receptor (BCR),** and consists of an immunoglobulin associated with transmembrane proteins called **Igα and Igβ.** They provide the signaling components that trigger intracellular pathways in response to antigen-antibody binding. The activation of the BCR is also influenced by interactions with other receptors, for example to mediate the interaction of antigen-activated B cells with helper T cells.

T-cell receptors are related to immunoglobulins

THE lymphocyte lineage presents an example of evolutionary opportunism: a similar procedure is used in both B cells and T cells to generate proteins that have a variable region able to provide significant diversity, while constant regions are more limited and account for a small range of effector functions. T cells produce either of two types of T-cell receptor. The different T-cell receptors are synthesized at different times during T-cell development, as summarized in **Figure 24.23.**

The γδ receptor is found on <5% of T lymphocytes. It is synthesized only at an early stage of T-cell development. In mice, it is the only receptor detectable at <15 days of gestation, but has virtually been lost by birth at day 20.

TCR αβ is found on >95% of lymphocytes. It is synthesized later in T-cell development than γδ. In mice, it first becomes apparent at 15–17 days after gestation. By birth it is the predominant receptor. It is synthesized by a separate lineage of cells from those involved in TCR γδ synthesis, and involves independent rearrangement events.

T cells with αβ receptors are divided into several subtypes that have a variety of functions connected with interactions between cells involved in the immune response. **Cytotoxic (killer) T cells** possess the capacity to lyse an infected target cell. **Helper T cells** assist T cell-mediated target killing or B cell-mediated antibody-antigen interaction.

Figure 24.23 The γδ receptor is synthesized early in T-cell development. TCR αβ is synthesized later and is responsible for 'classical' cell-mediated immunity, in which target antigen and host histocompatibility antigen are recognized together.

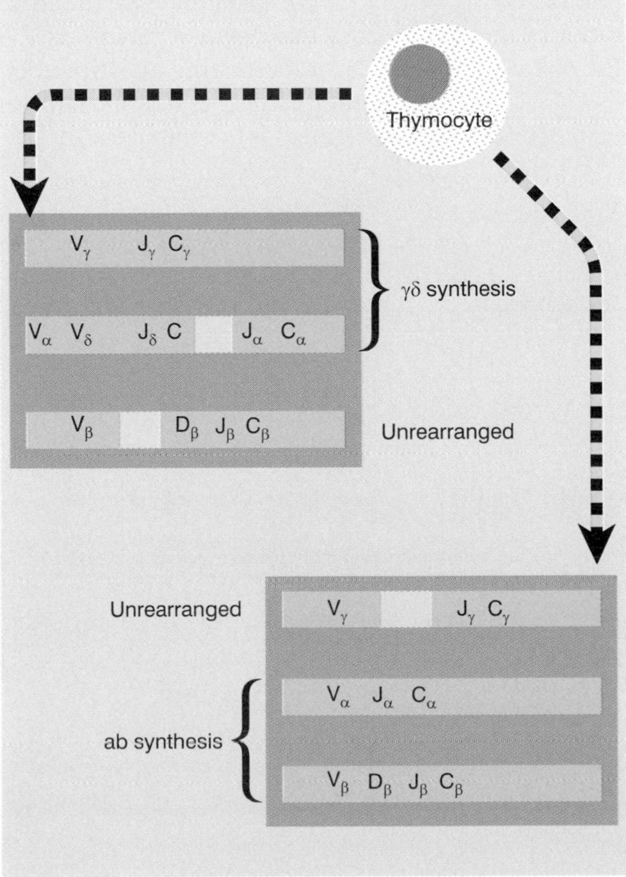

The immune response requires a T cell to recognize a host cell displaying a fragment of a foreign protein on its surface. To do so, the T cell simultaneously recognizes the foreign antigen and an MHC protein carried by the presenting cell, as illustrated previously in Figure 24.2. Both helper T cells and killer T cells work in this way, but they have different requirements for the presentation of antigen; different types of MHC protein are used in each case (see later). Helper T cells require antigen to be presented by an MHC class II protein, while killer T cells require antigen to be presented by an MHC class I protein.

The TCR αβ receptor is responsible for helper T cell function in humoral immunity and for killer T cell function in cell-mediated immunity. This places upon it the responsibility of recognizing both the foreign antigen and the host MHC protein. The MHC protein binds a short peptide derived from the foreign antigen, and the TCR then recognizes the peptide in a groove on the surface of the MHC. The MHC is said to **present** the peptide to the TCR. A given TCR has specificity for a particular MHC as well as for the foreign antigen. The basis for this dual capacity is one of the most interesting issues to be defined about the αβ TCR.

Like immunoglobulins, a TCR must recognize a foreign antigen of unpredictable structure. A common view has been that Nature might well solve the problem of antigen recognition by B cells and T cells in the same way, in which case we might expect the organization of the T-cell receptor genes to resemble the immunoglobulin genes in the use of variable and constant regions. *Each locus is organized in the same way as the immunoglobulin genes, with separate segments that are brought together by a recombination reaction specific to the lymphocyte.* The components are the same as those found in the three Ig families.

The organization of the TCR proteins resembles that of the immunoglobulins. The V sequences have the same general internal organization in both Ig and TCR proteins. The TCR C region is related to the constant Ig regions and has a single constant domain followed by transmembrane and cytoplasmic portions. Exon-intron structure is related to protein function.

The resemblance of the organization of TCR genes with the Ig genes is striking. As summarized in **Figure 24.24,** the organization of TCR α resembles that of Ig κ, with V gene segments separated from a cluster of J segments that precedes a single C gene segment. The organization of the locus is similar in both man and mouse, with some differences only in the number of V_α gene segments and J_α segments. (In addition to the α segments, this locus also contains δ segments, which we discuss shortly.)

The components of TCR β resemble those of IgH. **Figure 24.25** shows that the organization is different, with V gene segments separated from two clusters each containing a D segment, several J segments, and a C gene segment. Again the only differences between man and mouse are in the numbers of the V_β and J_β units.

Diversity is generated by the same mechanisms as in immunoglobulins. Intrinsic diversity results from the combination of a variety of V, D, J, and C segments; some additional diversity results from the introduction of new sequences at the junctions between these components (in the form of P and N nucleotides, as described previously in Figure 24.14). Some TCR β chains incorporate two D segments, generated by D-D joins (directed by an appropriate organization of the nonamer and heptamer sequences). A difference between TCR and Ig is that somatic mutation does not occur at the TCR loci.

The same mechanisms are likely to be involved in the reactions that recombine Ig genes in B cells and TCR genes in T cells. The recombining TCR segments are surrounded by nonamer and heptamer consensus sequences identical to those used by the Ig genes. This argues strongly that the same enzymes are involved. Most rearrangements probably occur by the deletion model (see Figure 24.13). How is the process controlled so that Ig loci are rearranged in B cells, while T-cell receptors are rearranged in T cells?

The organization of the γ locus resembles that of Ig λ, with V gene segments separated from a series of J-C segments. **Figure 24.26** shows that this locus has relatively little diversity, with ~8 functional V segments. The organization is different in man and mouse. Mouse has 3 functional J-C loci, but some segments are inverted in orientation. Man has multiple J segments for each C gene segment.

The δ subunit is coded by segments that lie at the TCR α locus, as illustrated previously in Figure 24.24. The segments D_δ-D_δ-J_δ-C_δ lie between the V gene segments and the J_α-C_α segments. Both of the D segments may be incorporated into the δ chain to give the structure VDDJ. The nature of the V gene segments used in the δ rearrangement is an interesting question. Very few V sequences are found in active TCR δ chains. In man, only one V gene segment is in general use for δ re-

Figure 24.24 The human TCRα locus has interspersed α and δ segments. A Vδ segment is located within the Vα cluster. The D-J-Cδ segments lie between the V gene segments and the J-Cα segments. The mouse locus is similar, but has more Vδ segments.

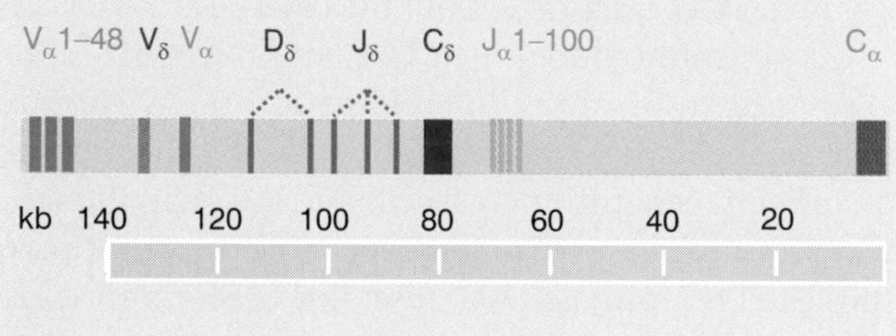

Figure 24.25 The TCRβ locus contains many V gene segments spread over ~500 kb, and lying ~280 kb upstream of the two D-J-C clusters.

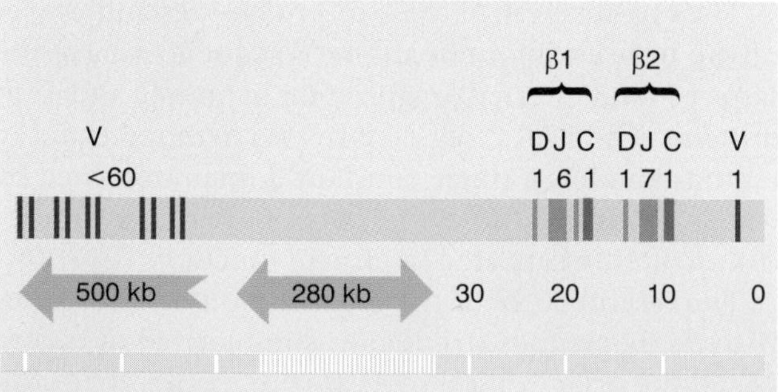

Figure 24.26 The TCRγ locus contains a small number of functional V gene segments (and also some pseudogenes; not shown), lying upstream of the J-C loci.

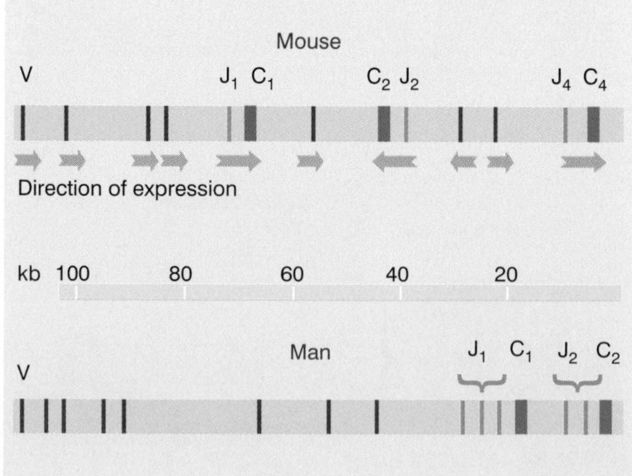

arrangement. In mouse, several V_δ segments are found; some are unique for δ rearrangement, but some are also found in α rearrangements. The basis for specificity in choosing V segments in α and δ rearrangement is not known. One possibility is that many of the V_α gene segments can be joined to the DDJ_δ segment, but that only some (therefore defined as V_δ) can give active proteins.

While for the present we have labeled the V segments that are found in δ chains as V_δ gene segments, we must reserve judgment on whether they are really unique to δ rearrangement. The interspersed arrangement of genes implies that synthesis of the TCR αβ receptor and the γδ receptor is mutually exclusive at any one allele, because the δ locus is lost entirely when the V_α-J_α rearrangement occurs.

Rearrangements at the TCR loci, like those of immunoglobulin genes, may be productive or nonproductive. The β locus shows allelic exclusion in much the same way as immunoglobulin loci; rearrangement is suppressed once a productive allele has been generated. The α locus may be different; several cases of continued rearrangement suggest the possibility that substitution of V_α sequences may continue after a productive allele has been generated.

Recombination to generate functional TCR chains is linked to the development of the T lymphocyte, as summarized in **Figure 24.27.** The first stage is rearrangement to form an active TCR β chain. This binds a nonrearranging surrogate α chain, called preTCRα. At this stage, the lymphocyte has not expressed either of the surface proteins CD4 or CD8. Formation of the preTCR heterodimer triggers proliferation and expression of these markers, so that the lymphocyte is converted from $CD4^-CD8^-$ to $CD4^+CD8^+$. The preTCR

Figure 24.27 T cell development proceeds through sequential stages.

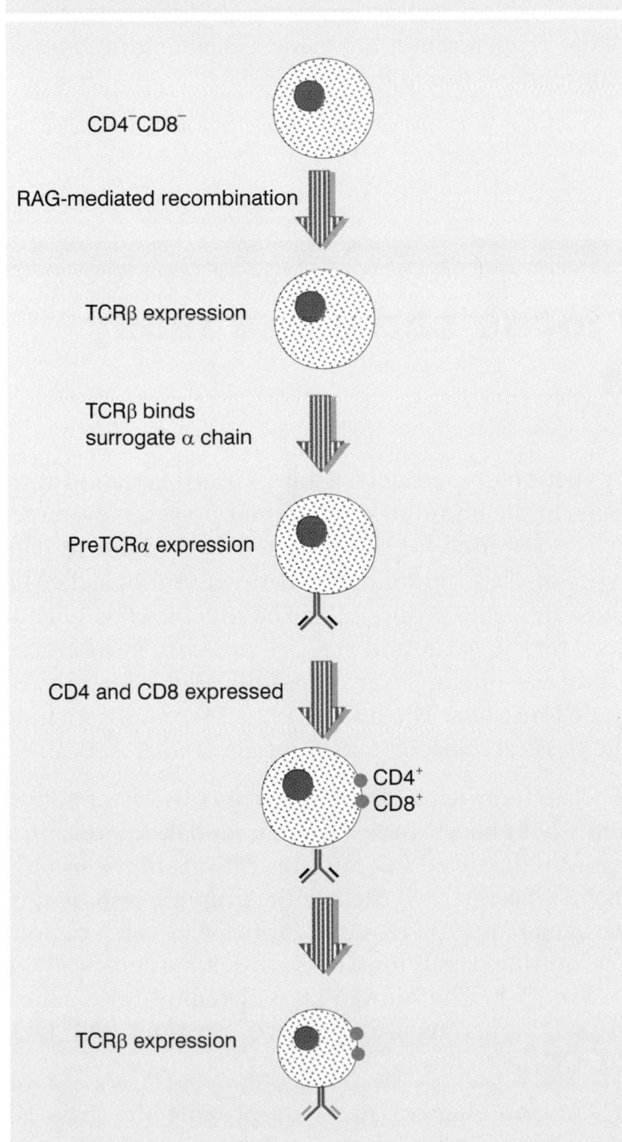

Figure 24.28 The two chains of the T-cell receptor associate with the polypeptides of the CD3 complex. The variable regions of the TCR are exposed on the cell surface. The cytoplasmic domains of the ζ chains of CD3 provide the effector function.

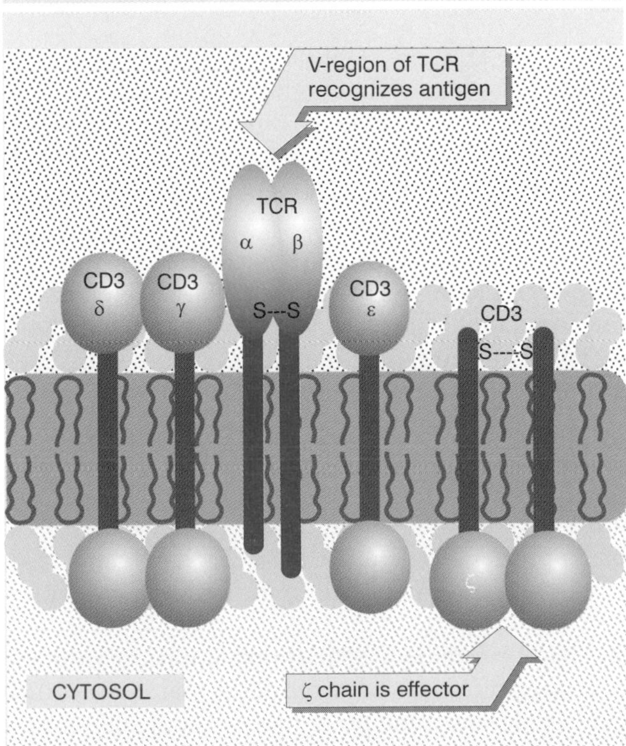

complex also signals the next stage in rearrangement: it causes cessation of β chain rearrangement and the start of α chain rearrangement. By contrast, α chain rearrangement is signaled to halt only much later, when the surface TCR αβ heterodimer becomes crosslinked on the surface during positive selection (which rescues the thymocytes from cell death, and then, if they survive the subsequent negative selection, allows them to give rise to the separate T lymphocyte classes which are $CD4^+CD8^-$ and $CD4^-CD8^+$.)

The T-cell receptor is associated with a complex of proteins called **CD3**, which is involved in transmitting a signal from the surface of the cell to the interior when its associated receptor is activated by binding antigen. Our present picture of the components of the receptor complex on a T cell is illustrated in **Figure 24.28.** The important point is that the interaction of the TCR variable regions with antigen causes the ζ subunits of the CD3 complex to activate the T-cell response. The activation of CD3 provides the means by which either αβ or γδ TCR signals that it has recognized an antigen. This is comparable to the constitution of the B cell receptor, in which immunoglobulin associates with the Igαβ signaling chains (see Figure 24.22).

A central dilemma about T-cell function remains to be resolved. Cell-mediated immunity requires two recognition processes. Recognition of the foreign antigen requires the ability to respond to novel structures. Recognition of the MHC protein is of course restricted to one of those coded by the genome, but, even so, there are many different MHC proteins. So considerable diversity is required in both recognition reactions. Although helper and killer T cells rely upon different classes of MHC proteins, they use the same pool of α and β gene segments to assemble their receptors. Even allowing for the introduction of additional variation during the TCR recombination process, it is not clear how enough different versions of the T-cell receptor are made available to accommodate all these demands.

The major histocompatibility locus codes for many genes of the immune system

THE major histocompatibility locus occupies a small segment of a single chromosome in the mouse (where it is called the ***H2* locus**) and in man (called the ***HLA* locus**). Within this segment are many genes coding for functions concerned with the immune response. At those individual gene loci whose products have been identified, many alleles have been found in the population; the locus is described as highly **polymorphic**, meaning that individual genomes are likely to be different from one another. Genes coding for certain other functions also are located in this region.

Histocompatibility antigens are classified into three types by their immunological properties. In addition, other proteins found on lymphocytes and macrophages have a related structure and are important in the function of cells of the immune system:

- **Class I proteins** are the **transplantation antigens.** They are present on every cell of the mammal. As their name suggests, these proteins are responsible for the rejection of foreign tissue, which is recognized as such by virtue of its particular array of transplantation antigens. In the immune system, their presence on target cells is required for the cell-mediated response. The types of class I proteins are defined serologically (by their antigenic properties). The murine class I genes code for the H2-K and H2-D/L proteins. Each mouse strain has one of several possible alleles for each of these functions. The human class I functions include the classical transplantation antigens, HLA-A, B, C.

- **Class II proteins** are found on the surfaces of both B and T lymphocytes as well as macrophages. These proteins are involved in communications between cells that are necessary to execute the immune response; in particular, they are required for helper T cell function. The murine class II functions are defined genetically as I-A and I-E. The human class II region (also called HLA-D) is arranged into four subregions, DR, DQ, DZ/DO, DP.

- The **complement proteins** provide the class III MHC. Their genetic locus is also known as the S region; S stands for serum, indicating that the proteins are components of the serum. Their role is to interact with antibody-antigen complexes to cause the lysis of cells in the classical pathway of the humoral response.

- The *Qa* and *Tla* loci proteins are found on murine hematopoietic cells. They are known as differentiation antigens, because each is found only on a particular subset of the blood cells, presumably related to their function. They are structurally related to the class I H2 proteins, and like them are polymorphic.

We can now relate the types of proteins to the organization of the genes that code for them.

The murine MHC locus is summarized on the map of **Figure 24.29.** The classical *H2* region occupies 0.3 map units; the adjacent region occupies another 1.0 map units. In molecular terms, this "small segment" of the chromosome is sizable; the 1.3 map units together potentially represent ~2000 kb of DNA.

The $H2^K$ genes map at the left end, and the $H2^{D/L}$ genes map at the right end. The class II and class III genes map between. Most of the polymorphism in individual genes occurs in those of the *H2* type. The adjacent region extends for another map unit; within it are genes coding for the differentiation antigens. We may regard them as extending the region of the chromosome devoted to functions concerned with the development of lymphocytes and macrophages. Variation in the number of genes between different mouse strains seems to occur largely in the *Qa* and *Tla* loci.

The class I mouse genes reside in clusters. The genes in each cluster usually are oriented in the same direction; adjacent genes tend to be more closely related, which suggests that they have originated by ancestral tandem duplications. Other genes also lie in the MHC locus. Within the D/L class I region lie the genes for the subunits of tumor necrosis factor (TNF), a protein that is involved in inflammatory diseases.

Next to the H2-D/L locus are the *Qa* and *Tla* loci. The ~10 *Qa* genes are closely related to the *H2* genes. The *Tla* region contains ~20 genes, but they are less well related to the classical *H2* sequences.

The human MHC locus is >3800 kb, about twice the length of the murine locus. As outlined in **Figure 24.30,** it contains similar functions, although not in the identical order. The major difference is that the class I HLA-A, B and C genes all are located in the same region (extending over >1000 kb), contrasted with the separation between murine H2-K and H2-D/L. The relative organization of the class I, II, and III genes is otherwise generally similar; moving from the telomere toward the centromere, there are the class I genes, followed by the class III genes and the class II genes for DR, DQ, DP.

All MHC proteins are dimers located in the plasma membrane, with a major part of the protein protruding on the extracellular side. The structures of class I and class II MHC proteins are related, although they have different components, as summarized in **Figure 24.31.**

Class II antigens consist of two chains, α and β, whose combination generates an overall structure in which there are two extracellular domains.

Figure 24.29 The histocompatibility locus of the mouse contains several loci that were originally defined genetically. Each locus contains many genes. Spaces between clusters that have not been connected are indicated by queries.

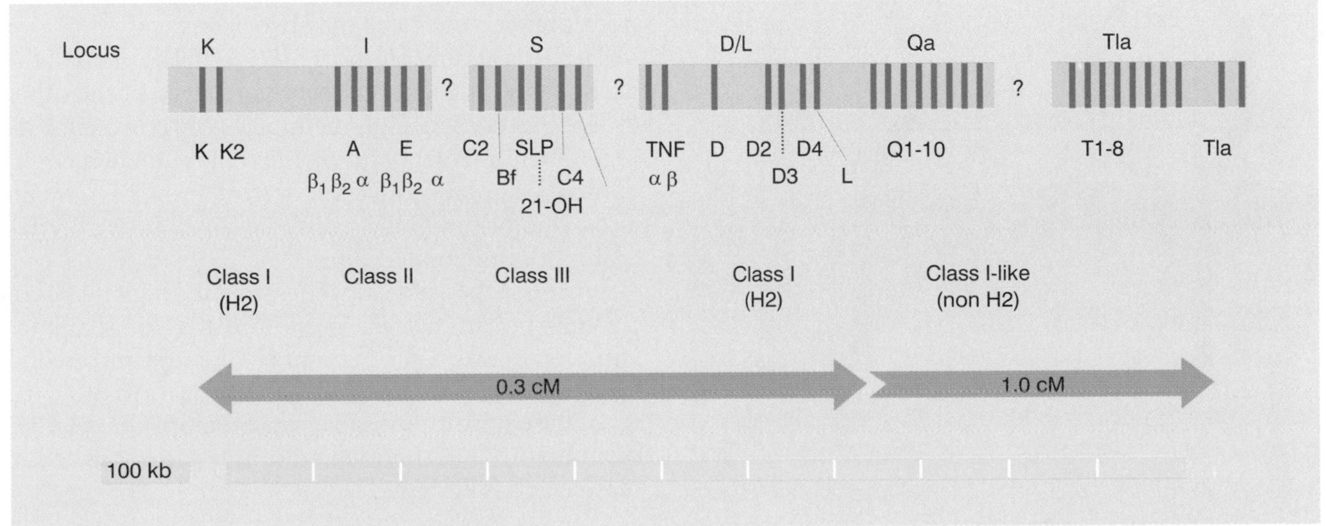

Figure 24.30 The human major histocompatibility locus codes for similar functions to the murine locus, although its detailed organization is different. Genes concerned with nonimmune functions also have been located in this region.

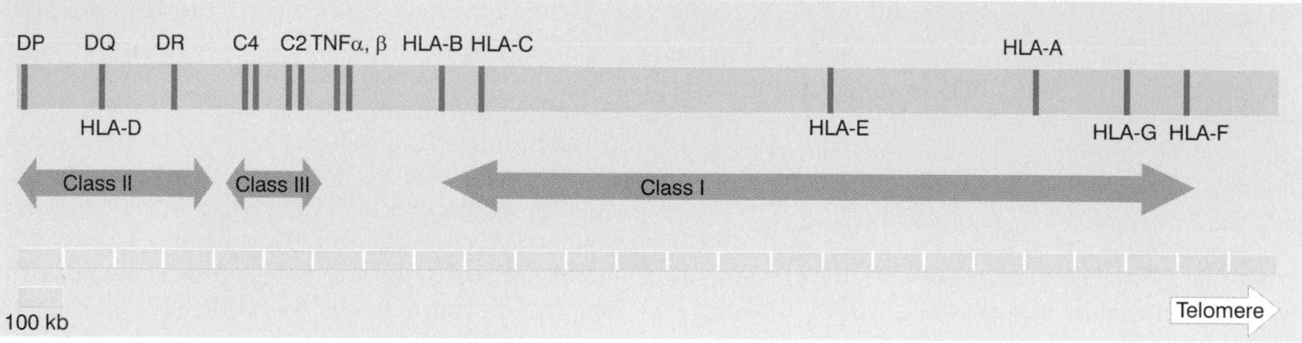

Figure 24.31 Class I and class II histocompatibility antigens have a related structure. Class I antigens consist of a single (α) polypeptide, with three external domains (α1, α2, α3), that interacts with β_2 microglobulin (β2 m). Class II antigens consist of two (α and β) polypeptides, each with two domains (α1 & α2, β1 & β2) with a similar overall structure.

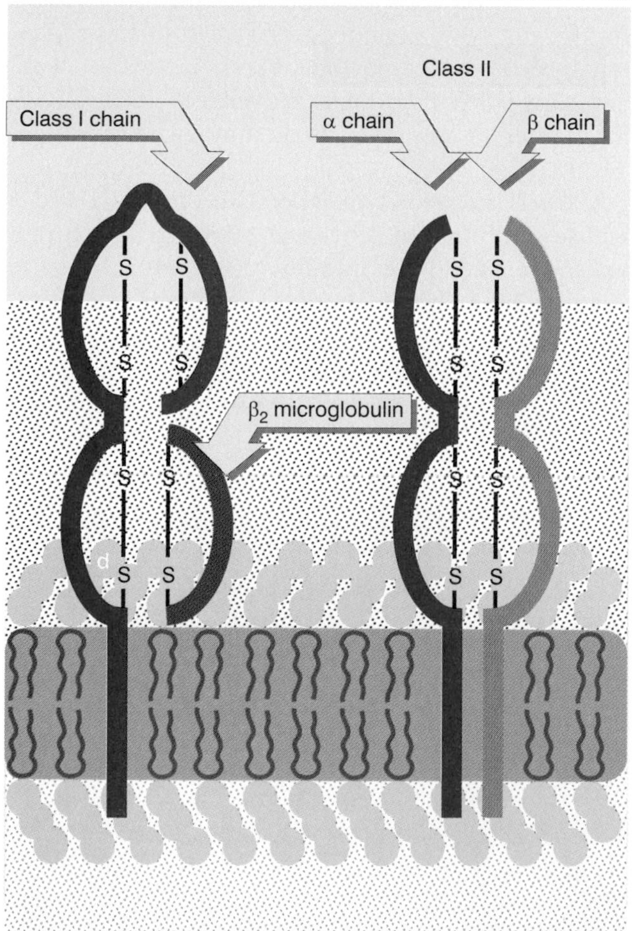

All class I MHC proteins consist of a dimer between the class I chain itself and the β2-microglobulin protein. The class I chain is a 45 kD transmembrane component that has three **external domains** (each ~90 amino acids long, one of which interacts with β2 microglobulin), a **transmembrane region** of ~40 residues, and a short **cytoplasmic domain** of ~30 residues that resides within the cell.

The β2 microglobulin is a secreted protein of 12 kD. It is needed for the class I chain to be transported to the cell surface. Mice that lack the β2 microglobulin gene have no MHC class I antigen on the cell surface.

The organization of class I genes summarized in **Figure 24.32** coincides with the protein structure. The first exon codes for a signal sequence (cleaved from the protein during membrane passage). The next three exons code for each of the external domains. The fifth exon codes for the transmembrane domain. And the last three rather small exons together code for the cytoplasmic domain. The only difference in the genes for human transplantation antigens is that their cytoplasmic domain is coded by only two exons.

The exon coding for the third external domain of the class I genes is highly conserved relative to the other exons. The conserved domain probably represents the region that interacts with β2 microglobulin, which explains the need for constancy of structure. This domain also exhibits homologies with the constant region domains of immunoglobulins.

What is responsible for generating the high degree of polymorphism in these genes? Most of the sequence variation between alleles occurs in the first and second external domains, sometimes taking the form of a cluster of base substitutions in a small region. One mechanism involved in their generation is gene conversion between class I genes.

Figure 24.32 Each class of MHC genes has a characteristic organization, in which exons represent individual protein domains.

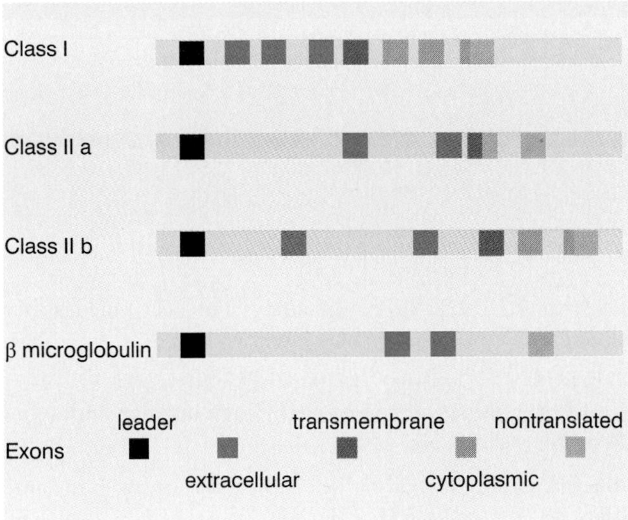

Pseudogenes are present as well as functional genes; at present we have some way to go before estimating the total number of active genes in the region.

Like the class I genes, the class II and class III genes also are interrupted, with the exons related to protein domains (see Figure 24.31). There are fewer genes in these classes, ~10 each.

The gene for β2 microglobulin is located on a separate chromosome. It has four exons, the first coding for a signal sequence, the second for the bulk of the protein (from amino acids 3 to 95), the third for the last four amino acids and some of the nontranslated trailer, and the last for the rest of the trailer.

The length of β2 microglobulin is similar to that of an immunoglobulin V gene; there are certain similarities in amino acid constitution; and there are some (limited) homologies of nucleotide sequence between β2 microglobulin and Ig constant domains or type I gene third external domains. All the groups of genes that we have discussed in this chapter may have descended from a common ancestor that coded for a primitive domain.

Summary

Immunoglobulins and T-cell receptors are proteins that play analogous functions in the roles of B cells and T cells in the immune system. An Ig or a TCR protein is generated by rearrangement of DNA in a single lymphocyte; exposure to an antigen recognized by the Ig or TCR leads to clonal expansion to generate many cells which have the same specificity as the original cell. Many different rearrangements occur early in the development of the immune system, creating a large repertoire of cells of different specificities.

Each immunoglobulin protein is a tetramer containing two identical light chains and two identical heavy chains. A TCR is a dimer containing two different chains. Each polypeptide chain is expressed from a gene created by linking one of many V segments via D and J segments to one of a few C segments. Ig L chains (either κ or λ) have the general structure V-J-C, Ig H chains have the structure V-D-J-C, TCR α and γ have components like Ig L chains, and TCR δ and β are like Ig H chains.

Each type of chain is coded by a large cluster of V genes separated from the cluster of D, J, and C segments. The numbers of each type of segment, and their organization, are different for each type of chain, but the principle and mechanism of recombination appear to be the same. The same nonamer and heptamer consensus sequences are involved in each recombination; the reaction always involves joining of a consensus with 23 bp spacing to a consensus with 12 bp spacing. Although considerable diversity is generated by joining different V, D, J segments to a C segment, additional variations are introduced in the form of changes at the junctions between segments and (in the case of immunoglobulins) by somatic mutation.

Allelic exclusion ensures that a given lymphocyte synthesizes only a single Ig or TCR. A productive rearrangement inhibits the occurrence of further rearrangements. Although the use of the V region is fixed by the first productive rearrangement, B cells switch use of C_H genes from the initial μ chain to one of the H chains coded farther downstream. This process involves a different type of recombination in which the sequences between the VDJ region and the new C_H gene are deleted. More than one switch occurs in C_H gene usage. At an earlier stage of Ig production, switches occur from synthesis of a membrane-bound version of the protein to a secreted version. These switches are accomplished by alternative splicing of the transcript.

Further reading

Reviews

Steinmetz, M. and Hood, L. (1983). Genes of the MHC complex in mouse and man. *Science* **222**, 727–732.

Tonegawa, S. (1983). Somatic generation of antibody diversity. *Nature* **302**, 575–581.

Flavell, R. A., Allen, H., Burkly, L. C., Sherman, D. H., Waneck, G. L., and Widera, G. (1986). Molecular biology of the H-2 histocompatibility complex. *Science* **233**, 437–443.

Kronenberg, M., Siu, G., Hood, L. E., and Shastri, N. (1986). The molecular genetics of the T-cell antigen receptor and T-cell antigen recognition. *Ann. Rev. Immunol.* **4**, 529–591.

Yancopoulos, G. D. and Alt, F. W. (1986). Regulation of the assembly and expression of variable-region genes. *Ann. Rev. Immunol.* **4**, 339–68.

Alt, F. W., Blackwell, T. K., and Yancopoulos, G. D. (1987). Development of the primary antibody repertoire. *Science* **238**, 1079–1087.

Marrack, P. and Kappler, J. (1987). The T-cell receptor. *Science* **238**, 1073–1079.

Storb, U. (1987). Transgenic mice with immunoglobulin genes. *Ann. Rev. Immunol.* **5**, 151–174.

Blackwell, T. K. and Alt, F. W. (1989). Mechanism and developmental program of immunoglobulin gene rearrangement in mammals. *Ann. Rev. Genet.* **23**, 605–636.

French, D. L., Laskov, R., and Scharff, M. D. (1989). The role of somatic hypermutation in the generation of antibody diversity. *Science* **244**, 1152–1157.

Kocks, C. and Rajewsky, K. (1989). Stable expression and somatic hypermutation of antibody V regions in B-cell developmental pathways. *Ann. Rev. Immunol.* **7**, 537–559.

Raulet, D. H. (1989). The structure, function, and molecular genetics of the gamma/delta T-cell receptor. *Ann. Rev. Immunol.* **7**, 175–207.

Davis, M. M. (1990). T-cell receptor gene diversity and selection. *Ann. Rev. Biochem.* **59**, 475–496.

Gellert, M. (1992). Molecular analysis of V(D)J recombination. *Ann. Rev. Genet.* **26**, 425–446.

Schatz, D. G., Oettinger, M. A., and Schlissel, M. S. (1992). V(D)J recombination: molecular biology and regulation. *Ann. Rev. Immunol.* **10**, 359–383.

Rajewsky, K. (1996). Clonal selection and learning in the antibody system. *Nature* **381**, 751–758.

Discoveries

Hozumi, N. and Tonegawa, S. (1976). Evidence for somatic rearrangement of immunoglobulin genes coding for variable and constant regions. *Proc. Nat. Acad. Sci. USA* **73**, 3628–3632.

Max, E. E., Seidman, J. G., and Leder, P. (1979). Sequences of five potential recombination sites encoded close to an immunoglobulin κ constant region gene. *Proc. Nat. Acad. Sci. USA* **76**, 3340–3344.

Hood, L., Kronenberg, M., and Hunkapiller, T. (1985). T cell antigen receptors and the immunoglobulin supergene family. *Cell* **40**, 225–229.

Lewis, S., Gifford, A., and Baltimore, D. (1985). DNA elements are asymmetrically joined during the site-specific recombination of kappa immunoglobulin genes. *Science* **228**, 677–685.

Reynaud, C. A., Anquez, V., Grimal, H., and Weill, J. C. (1987). A hyperconversion mechanism generates the chicken light chain preimmune repertoire. *Cell* **48**, 379–388.

Roth, D. B., Menetski, J. P., Nakajima, P. B., Bosma, M. J., and Gellert, M. (1992). V(D)J recombination: broken DNA molecules with covalently sealed (hairpin) coding ends in *SCID* mouse thymocytes. *Cell* **70**, 983–991.

McBlane, J. F. *et al.* (1995). Cleavage at a V(D)J recombination signal requires only RAG1 and RAG2 proteins and occurs in two steps. *Cell* **83**, 387–395.

Hiom, K., Melek, M., and Gellert, M. (1998). DNA transposition by the RAG1 and RAG2 proteins: a possible source of oncogenic translocations. *Cell* **94**, 463–470.

Part 6

Cells

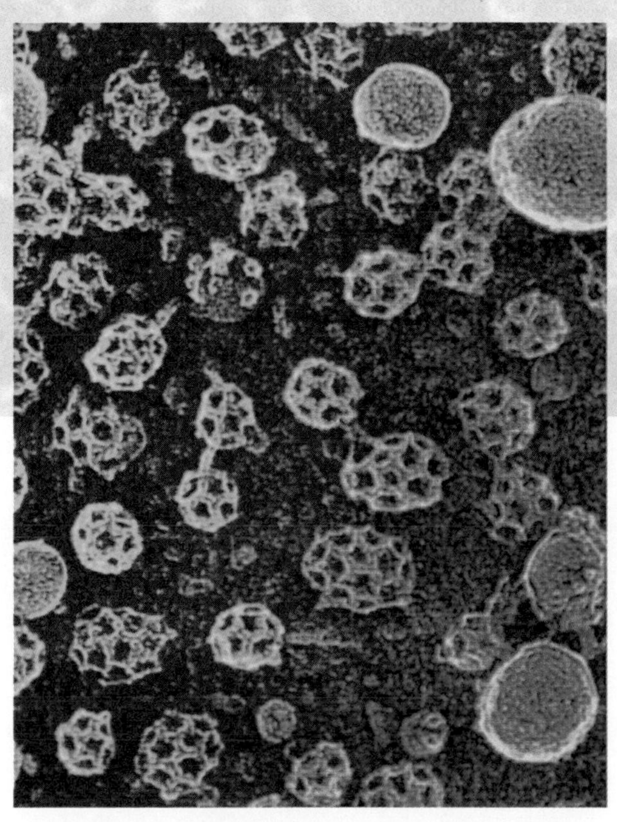

Chapter 25

Protein trafficking

A variety of molecules move out of and into the cell. At one extreme of the size range, proteins may be secreted from the cell into the extracellular fluid or may be internalized from the cell surface. At the other extreme, ions such as K^+, Na^+, and Ca^{2+} may be pumped out of or into the cell. In this chapter, we are concerned with the processes by which proteins are physically transported through membranous systems to the plasma membrane or other organelles, or from the cell surface to organelles within the cell. In the next chapter, we discuss the pathways of signal transduction by which an interaction at the surface can trigger internal pathways.

Proteins enter the pathway that leads to secretion by co-translational transfer to the membranes of the endoplasmic reticulum (ER) (see Figure 8.18). They are then transferred to the Golgi apparatus, where they are **sorted** according to their final intended destination. **Figure 25.1** summarizes the routes by which proteins are carried forward or diverted to other organelles. Their destinations are determined by specific signals, which take the form of short sequences of amino acids or covalent modifications that are made to the protein.

The transport machinery consists of small membranous vesicles. A soluble protein is carried within the lumen of a vesicle, and an integral membrane protein is carried within its membrane. **Figure 25.2** illustrates the nature of the budding and fusion events by which the vesicles move between adjacent compartments. A vesicle buds off from a donor surface and then fuses with a target surface. Its proteins are released into the lumen or into the membrane of the target compartment, depending on their nature, and must be loaded into new vesicles for transport to the next compartment. The series of events is repeated at each transition between membrane surfaces, for example, during passage from the ER to the Golgi, or between cisternae of the Golgi stacks.

Once a protein enters a membranous environment, it remains in the membrane until it reaches its final destination. A membrane protein that enters the endoplasmic reticulum is inserted into the membrane with the appropriate orientation (N-terminal lumenal, C-terminal cytosolic for group I proteins, the reverse for group II proteins). The orientation is retained as it moves through the system. The process starts in the same way irrespective of whether the protein is destined to reside in the Golgi, in lysosomes, or in the plasma membrane. In each case, it is transported in membrane vesicles along the secretory pathway to the appropriate destination, where some structural feature of the protein is recognized and it is permanently secured (or secreted from the cell).

Two important changes occur to a protein in the endoplasmic reticulum: it becomes folded into its proper conformation; and it is modified by glycosylation.

A protein is translocated into the ER in unfolded form. Folding occurs as the protein enters the lumen; probably a series of domains each folds independently as the protein passes through the membrane. Folding

Figure 25.1 *Overview*: proteins that enter the endoplasmic reticulum are transported to the Golgi and towards the plasma membrane. Specific signals cause proteins to be returned from the Golgi to the ER, to be retained in the Golgi, to be retained in the plasma membrane, or to be transported to endosomes and lysosomes. Proteins may be transported between the plasma membrane and endosomes.

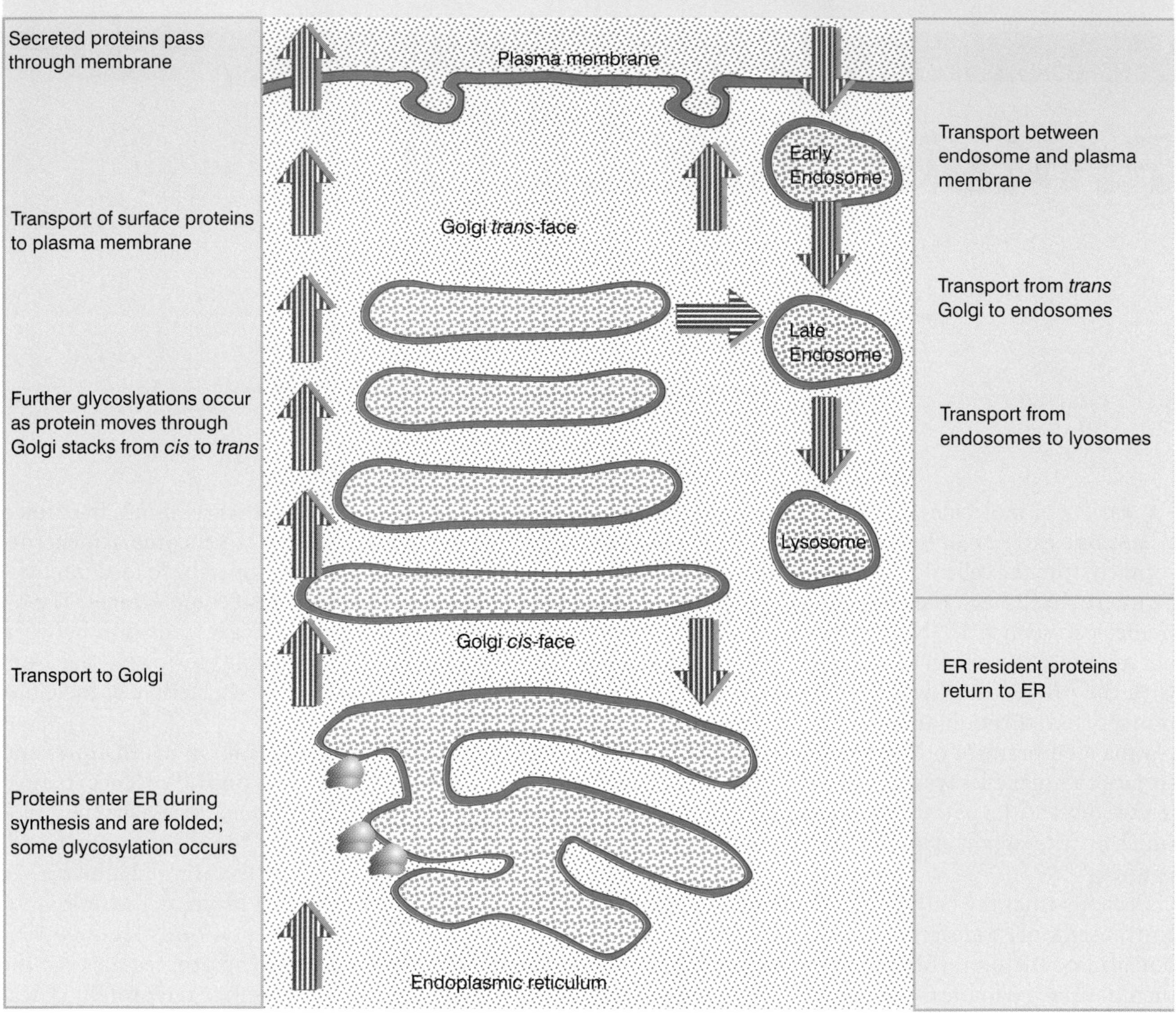

of a 50 kD protein is complete in <3–4 minutes, compared with the ~1 minute required to synthesize the chain.

Folding in the ER is associated with modification and is assisted by accessory proteins. Addition of carbohydrate may be required for correct folding; in fact, this may be an important function of the modification. Reshuffling of disulfide bonds by the enzyme protein disulfide isomerase (PDI) may be involved, and association with specific ER proteins may be necessary. The presence of chaperones in the ER may be necessary to recognize the partially folded forms of proteins as they emerge from transport through the membrane, and to assist them in acquiring their proper conformation. Some or all of these activities could be exercised by a complex of enzymes as a protein enters the ER; that is, the necessary functions all could associate with the protein at the site of translocation through the membrane. Calculations of the spontaneous rates of folding and oligomerization suggest that these accessory activities are needed to catalyze the process to enable it to occur rapidly enough in the cell.

Figure 25.2 Vesicles are released when they bud from a donor compartment and are surrounded by coat proteins (left). During fusion, the coated vesicle binds to a target compartment, is uncoated, and fuses with the target membrane, releasing its contents (right).

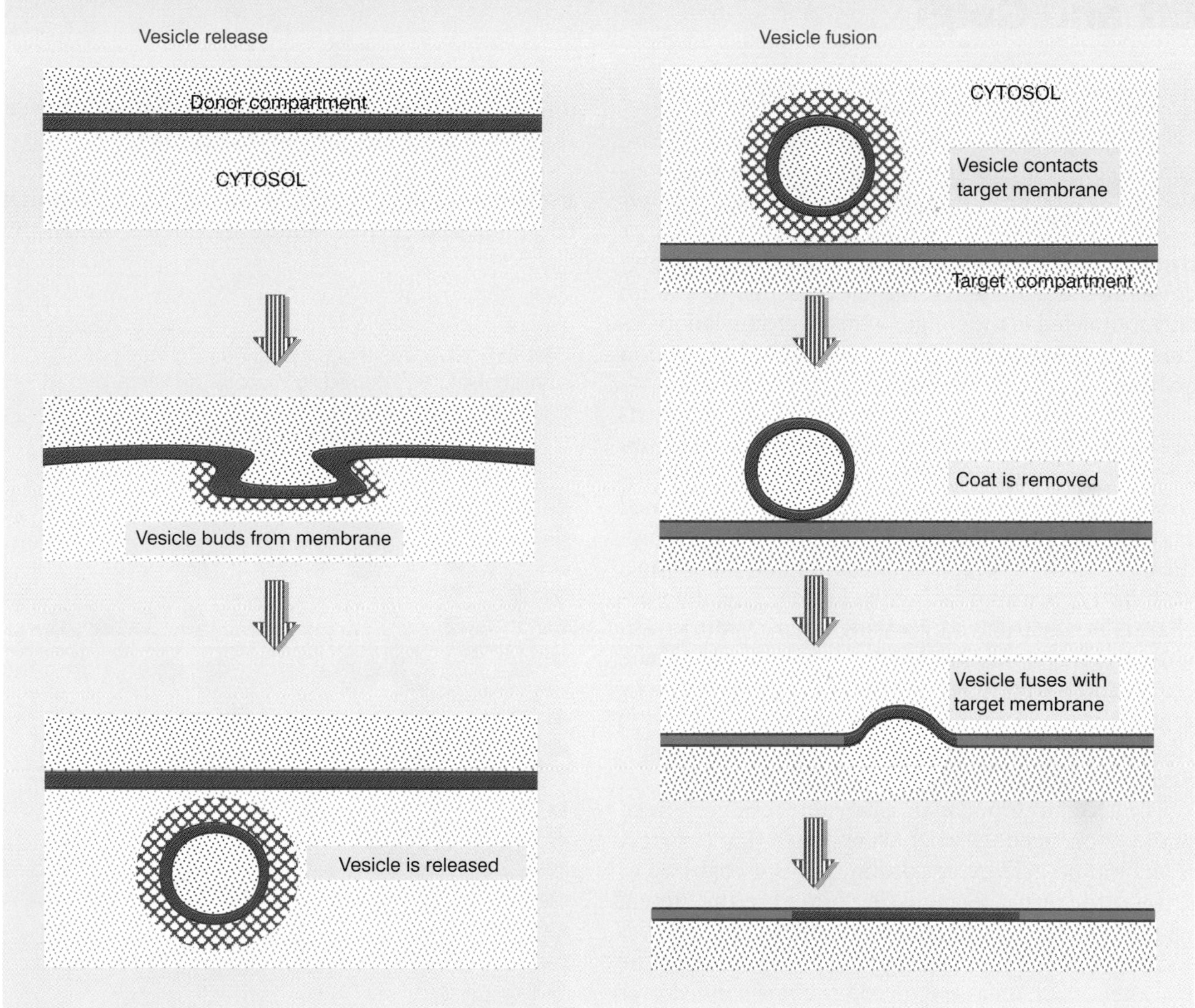
Vesicle release
Donor compartment
CYTOSOL
Vesicle buds from membrane
Vesicle is released
Vesicle fusion
CYTOSOL
Vesicle contacts target membrane
Target compartment
Coat is removed
Vesicle fuses with target membrane

Multimeric glycoproteins usually oligomerize in the ER. In fact, oligomerization may be necessary for further transport. Oligomers are rapidly transported from the ER to the Golgi, but unassembled subunits or misassembled proteins are held back. Misfolded proteins are often associated with ER-specific chaperones. In due course, they are removed by degradation. So a protein is allowed to move forward into the Golgi only if it has been properly folded previously in the ER.

One protein with a function in folding is BiP, a member of the Hsp70 family of chaperones. BiP facilitates oligomerization and/or folding of proteins in the lumen of the ER. BiP has two functions: to assist in folding newly translocated proteins; and to remove misfolded proteins. These activities could result from the same basic mode of action. Suppose that BiP recognizes certain peptide sequences that are inaccessible within the conformation of a mature, properly folded protein. These sequences are exposed and attract BiP when the protein enters the ER lumen in an essentially one-dimensional form. And if a protein is misfolded or denatured, these sequences may become exposed on its surface instead of being properly buried.

Oligosaccharides are added to proteins in the ER and Golgi

VIRTUALLY all proteins that pass through the secretory apparatus are glycosylated. Glycoproteins are generated by the addition of oligosaccharide groups either to the NH_2 group of asparagine (N-linked glycosylation) or to the OH group of serine, threonine, or hydroxylysine (O-linked glycosylation). N-linked glycosylation is initiated in the endoplasmic reticulum and completed in the Golgi; O-linked glycosylation occurs in the Golgi alone. The stages of N-glycosylation are illustrated in the next three figures.

The addition of all N-linked oligosaccharides starts in the ER by a common route, as illustrated in **Figure 25.3.** An oligosaccharide containing 2 N-acetyl glucosamine, 9 mannose, and 3 glucose residues is formed on a special lipid, **dolichol.** Dolichol is a highly hydrophobic lipid that resides within the ER membrane, with its active group facing the lumen. The oligosaccharide is constructed by adding sugar residues individually; it is linked to dolichol by a pyrophosphate group, and is transferred as a unit to a target protein by a membrane-bound glycosyl transferase enzyme whose active site is exposed in the lumen of the endoplasmic reticulum.

The acceptor group is an asparagine residue, located within the sequence Asn-X-Ser or Asn-X-Thr (where X is any amino acid except proline). It is recognized as soon as the target sequence is exposed in the lumen, when the nascent protein crosses the ER membrane.

Some trimming of the oligosaccharide occurs in the ER, after which a nascent glycoprotein is handed over to the Golgi. The oligosaccharide structures generated during transport through the ER and the Golgi fall into two classes, determined by the fate of the mannose residues. Mannose residues are added only in the ER, although they can be trimmed subsequently.

■ **High mannose oligosaccharides** are generated by trimming the sugar residues in the ER. **Figure 25.4** shows that almost immediately after addition of the oligosaccharide, the three glucose residues are removed by the enzymes glucosidases I and II. For proteins that reside in the ER, a mannosidase removes some of the mannose residues to generate the final structure of the oligosaccharide. The ER mannosidase attacks the first mannose quickly, and the next three more slowly; the total number of mannose residues that is removed varies with the individual substrate protein.

■ **Complex oligosaccharides** result from additional trimming and further additions carried out in the

Figure 25.3 An oligosaccharide is formed on dolichol and transferred by glycosyl transferase to asparagine of a target protein.

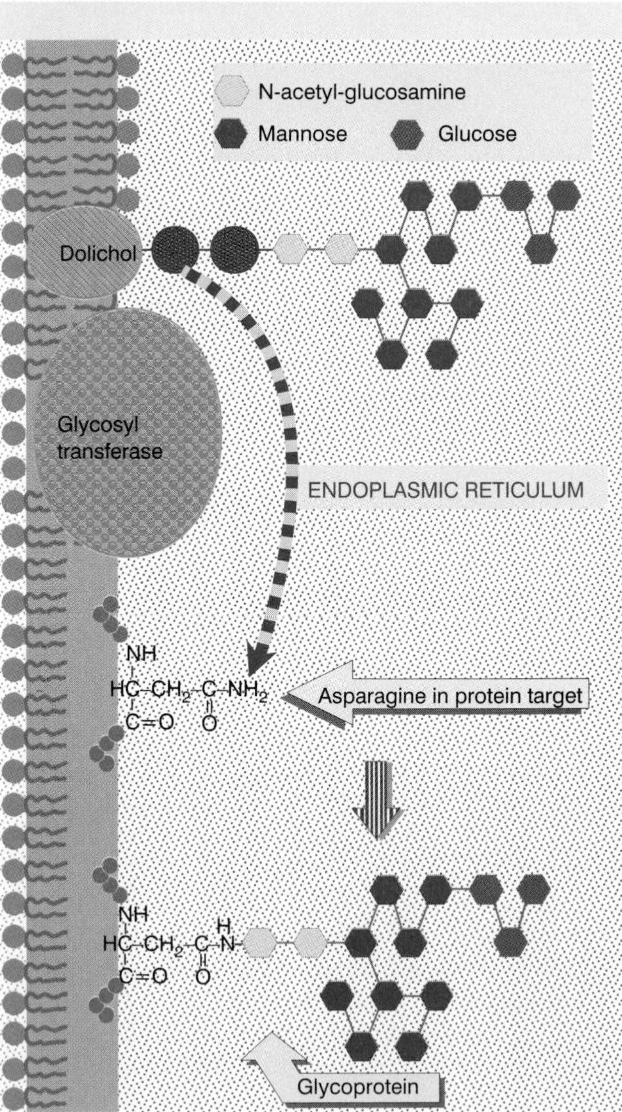

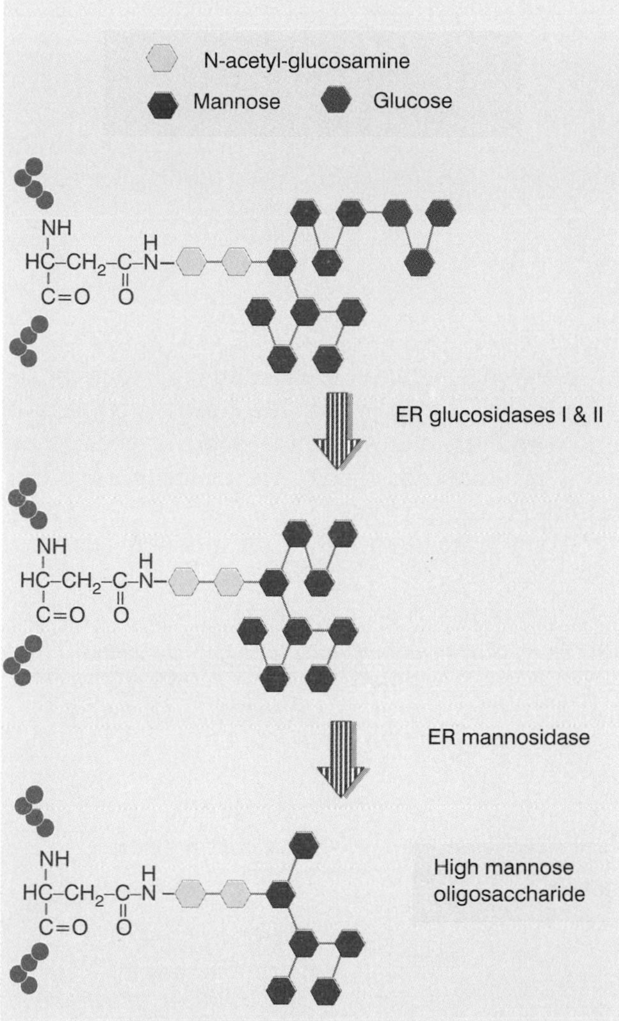

Figure 25.4 Sugars are removed in the ER in a fixed order, initially comprising 3 glucose and 1–4 mannose residues. This trimming generates a high mannose oligosaccharide.

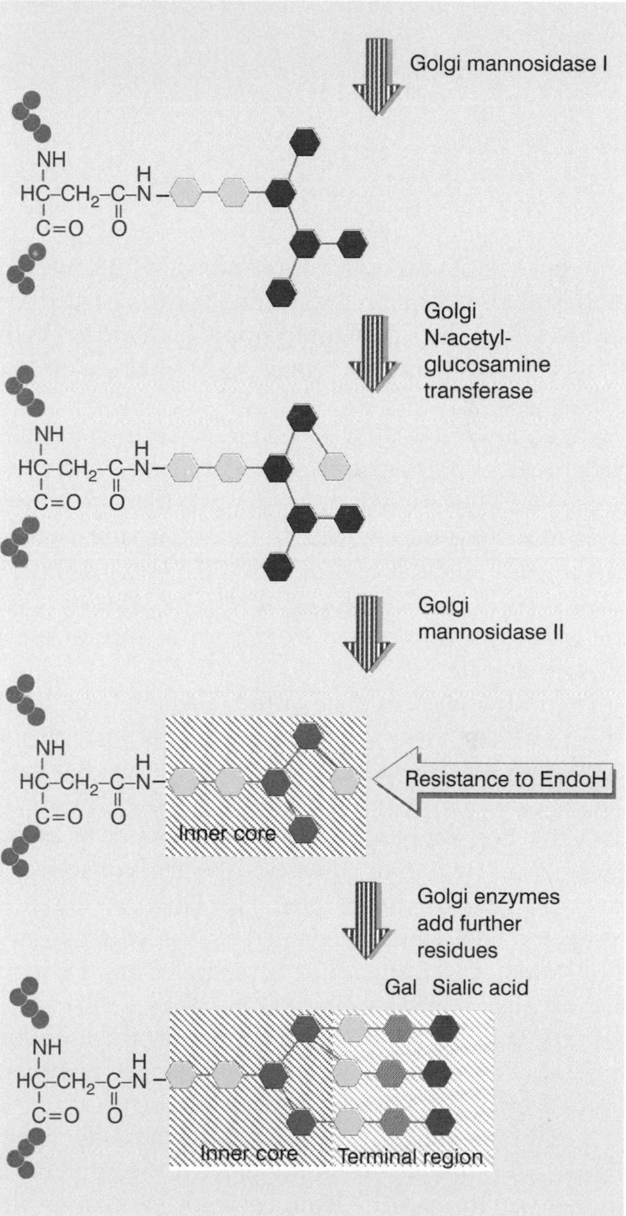

Figure 25.5 Processing for a complex oligosaccharide occurs in the Golgi and trims the original preformed unit to the inner core consisting of 2 N-acetyl-glucosamine and 3 mannose residues. Then further sugars can be added, in the order in which the transfer enzymes are encountered, to generate a terminal region containing N-acetyl-glucosamine, galactose, and sialic acid.

Golgi. Golgi modifications occur in the fixed order illustrated in **Figure 25.5**. The first step is further trimming of mannose residues by Golgi mannosidase I. Then a single sugar residue is added by the enzyme N-acetyl-glucosamine transferase. Then Golgi mannosidase II removes further mannose residues. This generates a structure called the **inner core,** consisting of the sequence NAc-Glc·NAc-Glc·Man_3. At this point, the oligosaccharide becomes resistant to degradation by the enzyme endoglycosidase H (Endo H). *Susceptibility to Endo H is therefore used as an operational test to determine when a glycoprotein has left the ER.*

Additions to the inner core generate the **terminal region.** The residues that can be added to a complex oligosaccharide include N-acetyl-glucosamine, galactose, and sialic acid (N-acetyl-neuraminic acid). The pathway for processing and glycosylation is highly

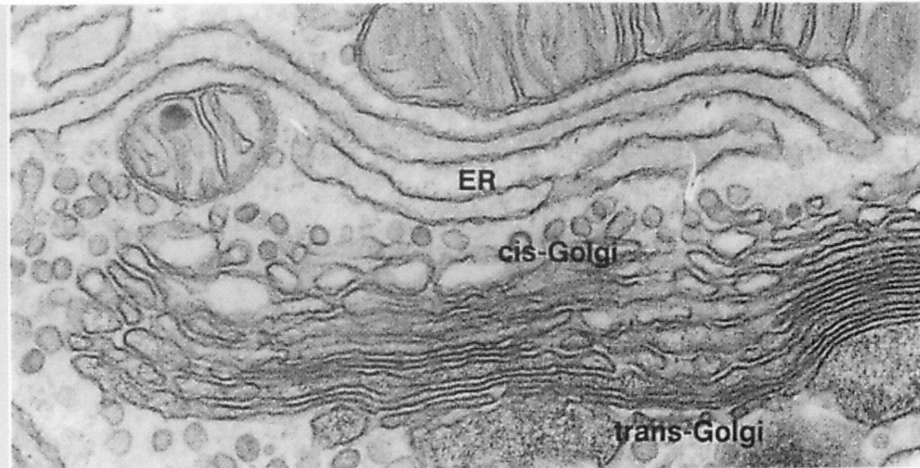

Figure 25.6 The Golgi apparatus consists of a series of individual membrane stacks. Photograph kindly provided by Alain Rambourg.

ordered, and the two types of reaction are interspersed in it. Addition of one sugar residue may be needed for removal of another, as in the example of the addition of N-acetyl-glucosamine before the final mannose residues are removed.

We do not know what determines how each protein undergoes its particular characteristic pattern of processing and glycosylation. We assume that the necessary information resides in the structure of the polypeptide chain; it cannot lie in the oligosaccharide, as all proteins subject to N-linked glycosylation start the pathway by addition of the same (preformed) oligosaccharide.

The individual cisternae of the Golgi are organized into a series of **stacks,** somewhat resembling a pile of plates. A typical stack consists of 4–8 flattened cisternae. **Figure 25.6** shows an example. A major feature of the Golgi apparatus is its *polarity.* The *cis* side faces the endoplasmic reticulum; the *trans* side in a secretory cell faces the plasma membrane. The Golgi consists of compartments, which are named *cis, medial, trans,* and *TGN (trans-Golgi network),* proceeding from the *cis* to the *trans* face. Proteins enter a Golgi stack at the *cis* face and are modified during their transport through the successive cisternae of the stack. When they reach the *trans* face, they are directed to their destination.

Membrane structure changes across the Golgi stack. The main difference is an increase in the content of cholesterol proceeding from *cis* to *trans.* As a result, fractionation of Golgi preparations generates a gradient in which the densest fractions represent the *cis* cisternae, and the lightest fractions represent the *trans* cisternae. The positions of enzymes on the gradient, and *in situ* immunochemistry with antibodies against individual enzymes, suggest that certain enzymes are differentially distributed proceeding from *cis* to *trans.* The difference between the *cis* and *trans* faces of the Golgi is clear, but it is not clear how the concept of compartments relates to individual cisternae; there may in fact be a continuous series of changes proceeding through the cisternae.

Nascent proteins encounter the modifying enzymes

Figure 25.7 A Golgi stack consists of a series of cisternae, organized with *cis* to *trans* polarity. Protein modifications occur in order as a protein moves from the *cis* face to the *trans* face.

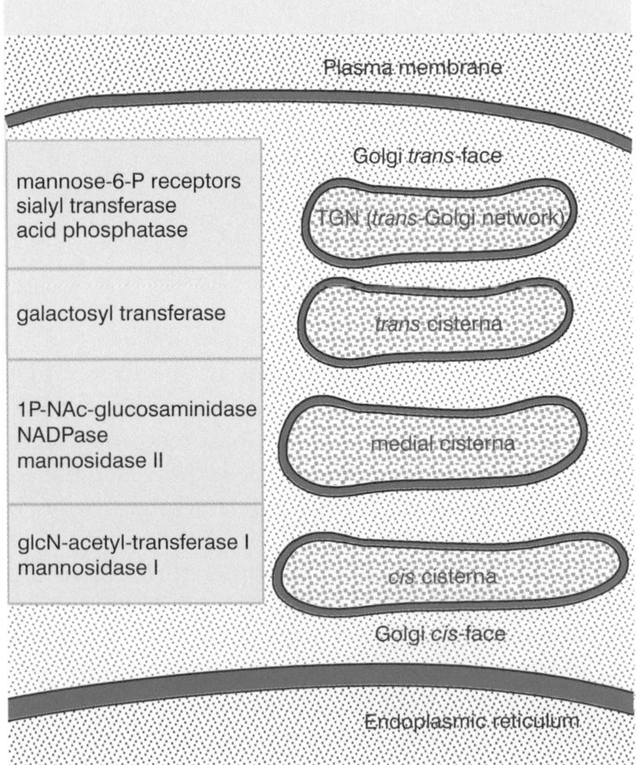

as they are transported through the Golgi stack. **Figure 25.7** illustrates the order in which the enzymes function. This may be partly determined by the fact that the modification introduced by one enzyme is needed to provide the substrate for the next, and partly by the availability of the enzymes proceeding through the cisternae.

The addition of a complex oligosaccharide can change the properties of a protein significantly. Glycoproteins often have a mass with a significant proportion of oligosaccharide. What is the significance of these extensive glycosylations? In some cases, the saccharide moieties play a structural role, for example in the behavior of surface proteins that are involved in cell adhesion. Another possible role could be in promoting folding into a particular conformation. At least one modification—the addition of mannose-6-phosphate—confers a targeting signal. Are other carbohydrate moieties used to direct protein sorting? None is yet known.

Coated vesicles transport both exported and imported proteins

SECRETED and transmembrane proteins start on the route to localization when they are translocated into the ER during synthesis. Transport from the ER, through the Golgi, to the plasma membrane occurs in vesicles. A protein is incorporated into a vesicle at one membrane surface, and is released from the vesicle at the next membrane surface. Progress through the system requires a series of such transport events. A protein

Figure 25.8 Proteins are transported in coated vesicles. Constitutive (bulk flow) transport from ER through the Golgi takes place by COP-coated vesicles. Clathrin-coated vesicles are used for both regulated exocytosis and endocytosis.

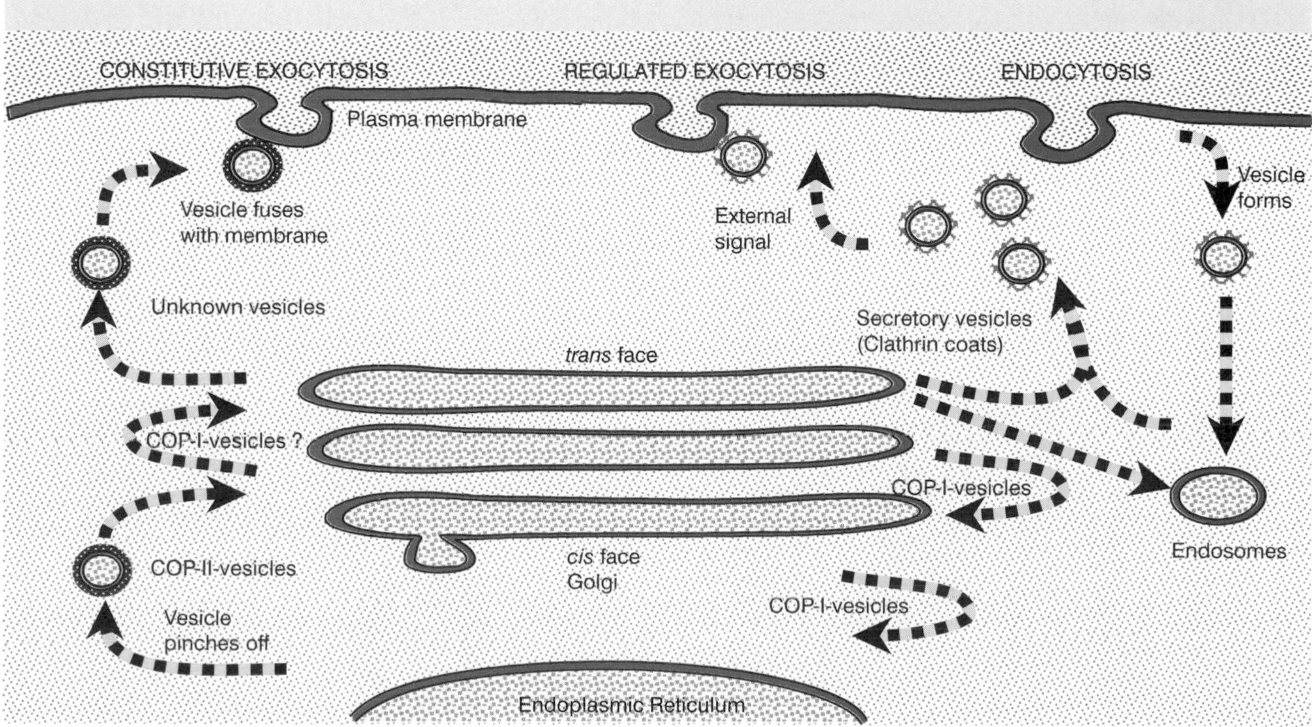

changes its state of glycosylation as it passes from *cis* to *trans* compartments through the Golgi.

Vesicles are used to transport proteins both out of the cell and into the cell. The secretion of proteins is called **exocytosis;** internalization of proteins is called **endocytosis.** The pathways for vesicle movement are pictured in **Figure 25.8.** The cycle for each type of vesicle is similar, whether they are involved in export or import of proteins: *budding from the donor membrane is succeeded by fusion with the target membrane.*

Vesicles involved in transporting proteins have a protein layer surrounding their membranes, and for this reason are called **coated vesicles.** Examples are shown in the electron micrograph of **Figure 25.9.** We discuss the types of protein coats below. The coat serves two purposes: it is involved with the processes of budding and fusion; and it enables the type of vesicle to be identified, so that it is directed to the appropriate target membrane. It is possible that the coat is also concerned with the selection of proteins to be transported.

One of the most remarkable features of protein trafficking is the conservation of the vesicular apparatus, including structural components of the vesicles, and proteins required for budding or fusion. Many of these functions have been identified through mutations of the *sec* genotype in *S. cerevisiae*, which are unable to export proteins through the ER-Golgi pathway. Many of the genes identified by *sec* mutants in yeast have direct counterparts in animal cells. In particular, the proteins

Figure 25.9 Coated vesicles are released from the *trans* face of the Golgi. The diameter of a vesicle is ~70 nm. Photograph kindly provided by Lelio Orci.

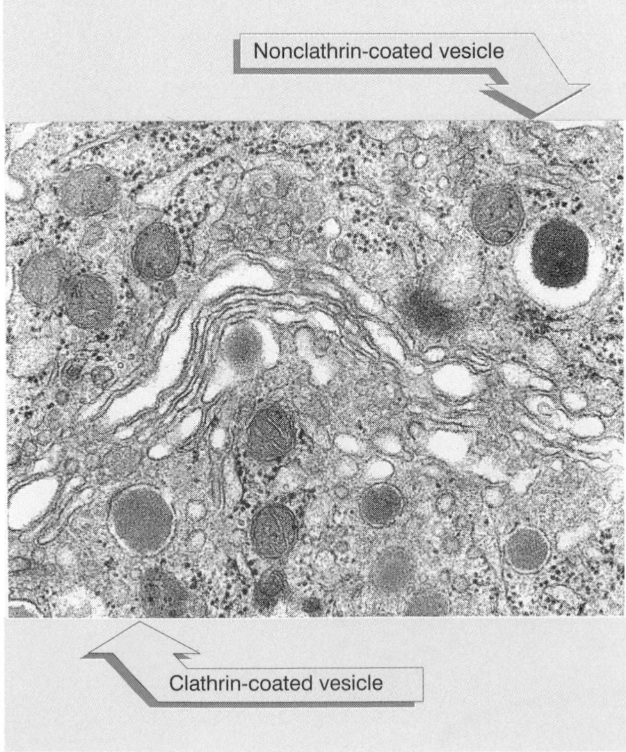
Nonclathrin-coated vesicle
Clathrin-coated vesicle

Figure 25.10 Vesicle formation results when coat proteins bind to a membrane, deform it, and ultimately surround a membrane vesicle that is pinched off.

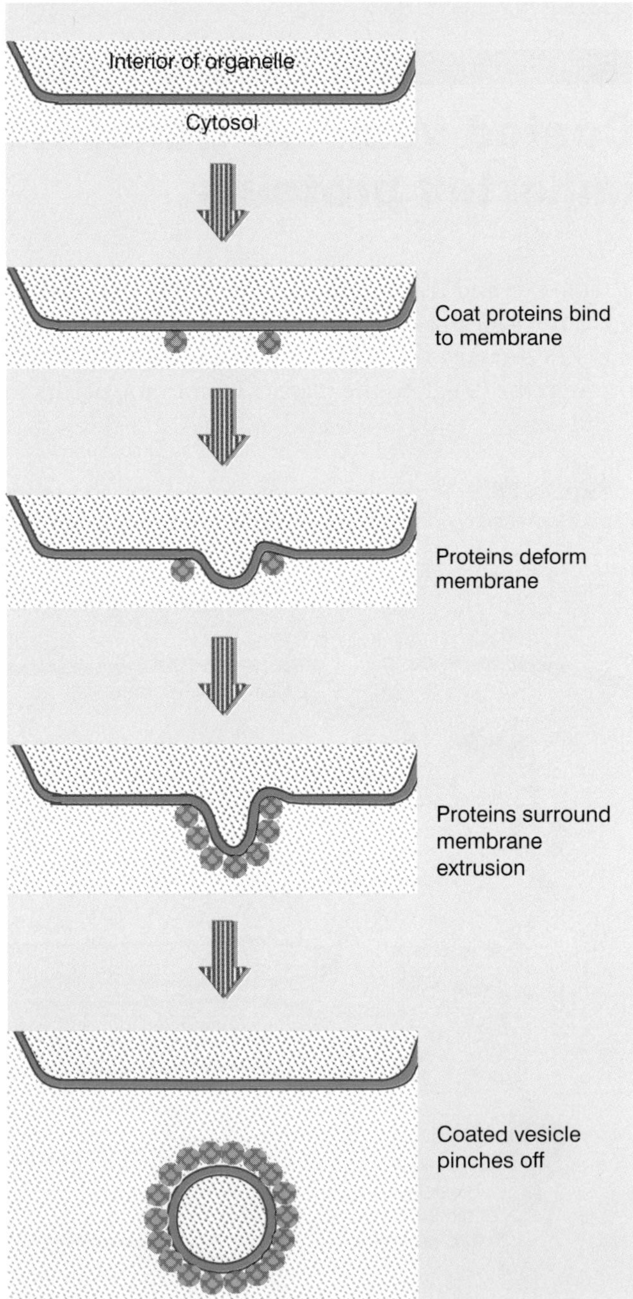
Interior of organelle
Cytosol
Coat proteins bind to membrane
Proteins deform membrane
Proteins surround membrane extrusion
Coated vesicle pinches off

involved in budding, fusion, and targeting in mammalian brain (where release of proteins from the cell provides the means of propagating nerve impulses) have homologues in the yeast secretory pathway.

The process of generating a vesicle requires a membrane bilayer to protrude a vesicle that eventually pinches off as a bud (see Figure 25.2). Such events require deformation of the membrane, as illustrated in **Figure 25.10.** Proteins concerned with this process are required specifically for budding, and become part of the coat.

In the reverse reaction, fusion is a property of membrane surfaces, and, in order to fuse with a target membrane, a vesicle must be "uncoated" by removal of the protein layer. A coated vesicle recognizes its destination by a reaction between a protein in the vesicle membrane and a receptor in the target membrane.

A vesicle therefore follows a cycle in which it gains its coat, is released from a donor membrane, moves to the next membrane, becomes uncoated, and fuses with the target membrane. When a vesicle is generated, it carries proteins that were resident in (or associated with) the stretch of membrane that was pinched off. The interior of the vesicle has the constitution of the lumen of the organelle from which it was generated. When the vesicle fuses with its target membrane, its components become part of that membrane or the lumen of the compartment. Proteins that are transported by vesicles (that is, which are not part of the structure of the vesicle itself) are called the **cargo.** Proteins must be sorted at each stage, when they either remain in the compartment or are incorporated into new vesicles and transported farther along the system.

The dynamic state of transport through the ER-Golgi system poses a dilemma. There is continuous movement of vesicles carrying proteins from the ER and through the Golgi. Movement in this direction is called **forward** or **anterograde transport.** In the course of ~20 minutes, a typical protein can pass through the system. A significant proportion of the membrane surface of the ER and Golgi is incorporated into vesicles that move to the plasma membrane. Such a flow of membrane should rapidly denude the Golgi apparatus and enormously enlarge the plasma membrane, yet both are stable in size. The net amount (and types) of lipid in each membrane must remain unperturbed in spite of vesicle movement.

The need to maintain the structure of the reticulo-endothelial system suggests that there is a pathway for returning membrane segments from the Golgi to ER, so that there is no net flow of membrane. Movement in this direction is called **retrograde transport.** We do not yet understand the balance of forward and retrograde flow. One possibility is that some vesicles engaged in retrograde movement do not carry cargo, except for returning components to earlier parts of the system. Alternatively, reverse flow might occur by structures that have a high surface-to-volume ratio, such as tubules, which could thus return large amounts of material.

Different groups of coated vesicles can be identified by the types of transport they undertake. In some cases, the vesicles can be distinguished by the biochemical components of their coats.

Newly synthesized proteins enter the ER and may be transported along the ER-Golgi system. The apparatus is present in all cells. This transport is undertaken by **transition vesicles.** Two different types of transition vesicles have been identified on the basis of their coats.

- The vesicles that were originally identified in transport between Golgi cisternae are now called **COP-I-coated vesicles.** (COP is an acronym for coat protein.) They are involved in retrograde transport; it is an open question whether they also undertake anterograde transport.
- Transition vesicles that proceed from the ER to the Golgi have a different coat, called **COP-II.** Their major role appears to be forward transport.
- Some proteins are **constitutively secreted,** moving from the *trans*-Golgi to the plasma membrane. The vesicles that undertake this process have not been characterized biochemically.

The export of some proteins is regulated. These proteins are packaged into **secretory vesicles.** These vesicles provide a storage medium, and release their contents at the plasma membrane only following receipt of a particular signal (triggered, for example, by a hormone or Ca^{2+}). This occurs in cells that are specialized to produce the appropriate proteins:

- Vesicles that form at the *trans*-face of the Golgi for use in the regulated pathway may fuse to form **secretory granules.** They may also transport their cargoes to endosomes. Secretory vesicles may form at endosomes to transport proteins to the plasma membrane. The most common route for regulated transport is probably via the endosome.
- Proteins enter the cell by packaging into **endocytic vesicles,** which are released from the plasma membrane, and transport their contents toward the interior of the cell. The cargo is released when the endocytic vesicle fuses with the membrane of a target compartment such as an endosome.

Endocytic and secretory vesicles have **clathrin** as the most prominent protein in their coats, and are therefore known as **clathrin-coated vesicles.** Their structure is known in some detail. The 180 kD chain of clathrin, together with a smaller chain of 35 kD, forms a polyhedral coat on the surface of the coated vesicle. The sub-unit of the coat consists of a **triskelion,** a three-pronged protein complex consisting of 3 light and 3 heavy chains. The triskelions form a lattice-like network on the surface of the coated vesicle, as revealed in the electron micrograph of **Figure 25.11.** Endocytic vesicles form and are coated at invaginations of the plasma membrane that are called **coated pits.** Similar structures can be observed at the *trans* face of the Golgi, where vesicles destined for endosomes and secretory vesicles originate.

The structure of clathrin-coated vesicles is shown in outline in **Figure 25.12.** The inner shell of the coat is made by proteins called **adaptors,** which bind both to clathrin and to integral membrane proteins of the vesicle. Two different types of adaptors correspond to coated vesicles with different origins. The AP2 adaptor is found at plasma membrane-coated pits and characterizes endocytic vesicles. These vesicles also contain an additional adaptor protein, AP180, which controls the size of the vesicle. The AP1 adaptor is found at Golgi-coated pits, and identifies vesicles that are targeted for endosomes. Each AP is a heterotetramer; the individual subunits are called **adaptins.** The β subunits in each adaptor bind clathrin. The α or γ subunits are involved in interactions with the membrane where the vesicles are formed; that is, they are responsible for assembly of the full AP complex at the appropriate membrane, after which a clathrin coat assembles.

Formation of a vesicle is energetically unfavorable—the membrane must be deformed and finally a small sphere is pinched off. What provides the energy? The small GDP/GTP-binding protein **dynamin** is required for formation of clathrin-coated vesicles at endocytosis. The GDP-bound form of dynamin binds to the clathrin lattice. Replacement of the GDP by GTP (which is catalyzed by an exchange factor) causes the dynamin to form a ring around the neck of the forming vesicle; this may be the basis for the closure reaction that releases the vesicle.

A significant proportion of the clathrin and the

Figure 25.11 Coated vesicles have a polyhedral lattice on the surface, created by triskelions of clathrin. Photograph kindly provided by Tom Kirchhausen.

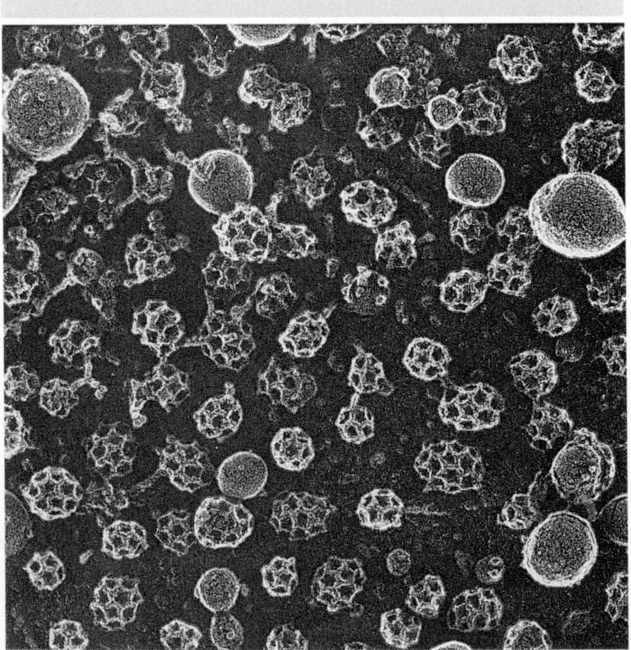

Figure 25.12 Clathrin-coated vesicles have a coat consisting of two layers: the outer layer is formed by clathrin, and the inner layer is formed by adaptors, which lie between clathrin and the integral membrane proteins.

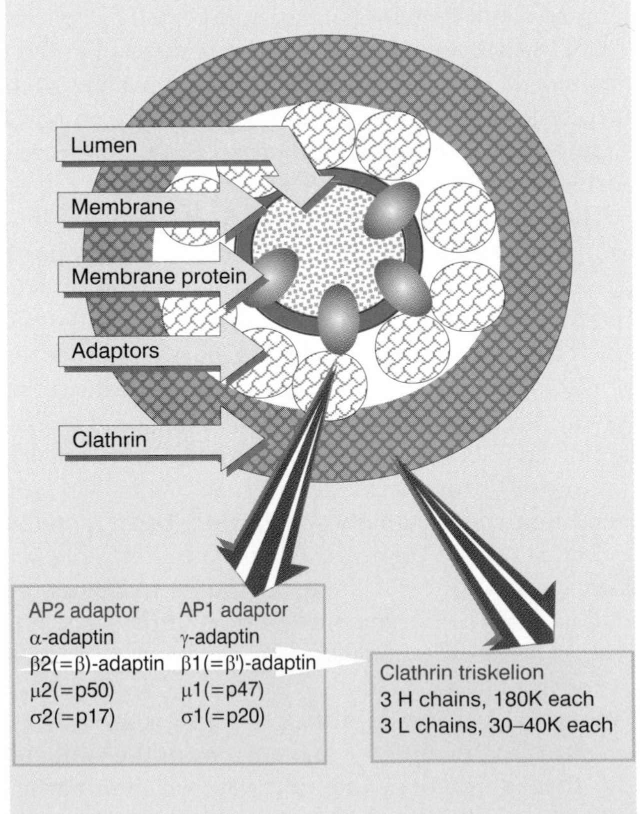

adaptors in the cell are found in a pool of free molecules, which suggests that both components are removed when vesicles become uncoated prior to fusion with their membrane targets.

Clathrin-coated vesicles are used to transport proteins to a variety of destinations. Clathrin is common to endocytic and secretory coated vesicles, for example, so other proteins in the coat must distinguish the vesicles. The adaptors bind to the cytoplasmic tails of membrane proteins that are carried by the vesicles, and thus appear to have responsibility for recognizing the appropriate cargo proteins to load into the vesicle.

How many types of cargo protein can be carried by a single endocytic vesicle? It is not yet clear how many types of vesicle exist and what variety they display on the coats and in their cargo. We do know that some endocytic vesicles carry more than one type of cargo protein. Generally they are viewed as fairly specific carriers.

The coats of the COP-I-coated transition vesicles have 7 major protein components, called COPs. They exist as a high molecular weight complex (~700kD), called **coatomer,** which is the precursor to the COP coat. The β-COP component has some homology to β- and β′-adaptins. This suggests a similar organization in which β-COP plays a similar role to the β- and β′-adaptins in connecting an outer coat protein (an analog of clathrin) to the membrane proteins in the vesicle.

COP-I-coated vesicles appear to be capable of performing both anterograde and retrograde transport. Such vesicles can be found carrying types of cargo that are transported in either direction; but any individual vesicle carries only anterograde or retrograde cargo, not both. Certain mutations in COP proteins block retrograde transport, which suggests that COP-I vesicles provide the sole (or at least major) capacity for retrograde transport. We do not know how COP-I vesicles moving in one direction are distinguished from those moving in the opposite direction: presumably there is some further component that has yet to be identified. Both directions of transport are probably supported through every level of the Golgi stack.

The coats of COP-II-coated vesicles consist of the protein complexes Sec23p/Sec24p (found as a 400 kD tetramer), Sec13p/Sec31p (which form a 700 kD complex), and Sar1p. There is no homology between the Sec protein components and the components of COP-I-coated vesicles.

Another class of vesicle has the AP3 coat, whose subunits (δ, β3, μ3, σ3) are related to those of the AP1 and AP2 adaptor complexes. This coat complex is found on some synaptic vesicles, which form at endosomes. This type of coat complex is also found on vesicles that transport cargo from the Golgi to lysosomes, and on storage vesicles.

Are coated vesicles responsible for all transport between membranous systems? There are conflicting models for the nature of forward transport from the ER, through Golgi cisternae, and then from the TGN to the plasma membrane.

The vesicular model for anterograde transport proposes that the Golgi cisternae are fixed structures that gain and lose proteins by the processes of vesicle fusion and budding. The process starts when COP-II-coated vesicles bud at the ER and transport cargo to the Golgi. It is not clear whether the vesicles that move between Golgi cisternae would also have COP-II coats. The nature of the coat(s) of vesicles that proceed from the Golgi to the plasma membrane remains unknown.

An alternative model for anterograde transport suggests that there is **cisternal maturation.** Instead of being fixed structures, cisternae move from the *cis* side of the Golgi to the *trans* side, maturing into more *trans*-like types of cisternae by changes in their protein constitution. Evidence for cisternal maturation has been provided by following the fate of a substrate protein that is too large to be incorporated into vesicles. Procollagen type I assembles into rod-like triple helices that are ~300 nm long in the lumen of the ER. These rods can be followed as they move into the *cis*-Golgi and through the Golgi to the TGN. Because they remain intact, and are too large to be incorporated into vesicles (COP-coated vesicles are 60–90 nm in diameter), this means that the membrane-bound compartment containing the rods must itself have moved from the *cis* to the *trans* side of the Golgi. This shows at least the plausibility of cisternal maturation, although it does not demonstrate whether normal cargo proteins are carried by vesicles or also move by cisternal maturation.

To take the model for cisternal maturation to its extremes, the *cis*-Golgi could be formed by fusion between COP-II-coated vesicles that bud from the ER; this process might also involve larger tubules. The *cis*-Golgi cisternae would move steadily forward until they mature into the *trans*-Golgi cisternae. At the TGN, secretory vesicles might form by fragmenting into tubular structures, without requiring any special type of coat. Of course, as cisternae mature, proteins that belong to more *cis*-like cisternae must be retrieved; this would occur by COP-I-mediated retrograde vesicular transport.

The outstanding question is the relative quantitative importance of cisternal maturation and vesicular transport for the anterograde direction.

Budding and fusion reactions

BUDDING and fusion are essentially reversible reactions. Budding occurs when coat proteins assemble on a patch of membrane, ultimately causing its release as an independent vesicle. Fusion occurs when the coat proteins are removed, exposing the membrane surface, which can then fuse with a target membrane. Whether the coat proteins assemble or disassemble is controlled by the state of a small GTP-binding protein.

Budding of COP-I, clathrin-coated, and AP3 vesicles is initiated by ARF (ADP-ribosylation factor). Sar1p is a closely related protein that serves the same role for COP-II-coated vesicles. ARF is myristoylated at the N-terminus and can insert spontaneously into lipid bilayers. ARF-GTP is the active form; ARF-GDP is inactive. This suggests that ARF's activity (and ability to recycle) is controlled by GTP hydrolysis. One possibility is that the type of guanine nucleotide controls the conformation, so that the myristoylated N-terminus is exposed only when GTP is bound. ARF/Sar1p is the key component that triggers both the budding and fusion processes in response to the condition of its guanine nucleotide.

Figure 25.13 shows that budding is initiated when ARF/Sar1p is converted to the GTP-bound form. It inserts into the membrane and enables coat proteins to bind in a stoichiometric manner. ARF/Sar1p is recruited to the appropriate membranes by interacting with a receptor that promotes nucleotide exchange (that is, generates the active GTP-bound form) of the factor. The receptor does not become a component of the vesicle. The function of ARF/Sar1p is to provide the binding sites at which the other coat proteins can assemble. Coat proteins surround the membrane as a prerequisite for budding, but another function may be needed to complete the process of "pinching off."

What controls the specificity of the cargo carried by the vesicle? It is necessary to distinguish those proteins that should be transported out of the compartment from resident proteins that should remain there. In the case of COP-II vesicles, this may be a function of the coat. The COP-II coat can cause vesicles to bud when liposomes are mixed with the coat proteins. When the liposomes contain membrane proteins that are resident in the ER and other membrane proteins that are involved in targeting vesicles to the Golgi, only the latter class enter the vesicles. This suggests that specificity may be determined by a direct interaction with the coat proteins.

Fusion is a reversal of budding. The coat of the vesicle is an impediment to fusion, and must be removed. If we suppose that uncoated vesicles would have an ability to fuse spontaneously with membranes in the cell, we can view the coat as a protective layer that preserves the vesicle until it reaches its destination. **Figure 25.14** shows that dissociation of the coat is triggered by hydrolysis of the GTP bound to ARF. This causes ARF to withdraw from the membrane of the vesicle. The coat of COP-I-coated vesicles is unstable in the absence of ARF, so the coatomers then dissociate from the vesicle. In the case of clathrin-coated vesicles, the coat is stable, and further components, including a chaperone-like protein and an ATPase, are necessary to remove it.

Figure 25.13 ARF and coatomer are sufficient for the budding of COP-I-coated vesicles.

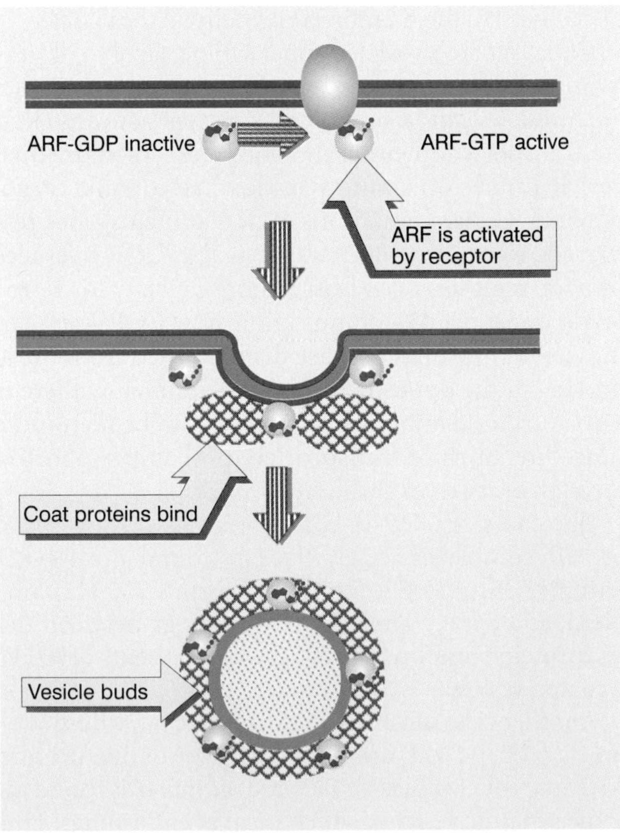
ARF-GDP inactive
ARF-GTP active
ARF is activated by receptor
Coat proteins bind
Vesicle buds

Components involved in the fusion between the donor membrane of the vesicle and the target membrane were identified via a mammalian protein called NSF, identified by its sensitivity to the sulfhydryl agent NEM (N-ethyl-maleimide). NSF is the homolog for the product of yeast gene *sec18*, which is required for vesicle fusion during movement of transition vesicles. Fusion requires a 20S complex that consists of NSF (a soluble ATPase), a SNAP (*S*oluble *N*SF-*A*ttachment *P*rotein), and SNAREs (*SNAP*-*re*ceptors) located in the membrane. The fusion particle is a basic part of the vesicular apparatus; it functions at all surfaces where vesicles fuse in the secretory and endocytic pathways. It contains the components necessary for vesicle-membrane recognition and uses ATP hydrolysis to allow them to recycle.

What controls the specificity of vesicle targeting? When a vesicle buds from a particular membrane, it has a specific target: vesicles leaving the ER have the *cis*-Golgi as their destination, vesicles leaving the *trans*-Golgi fuse with the plasma membrane, etc. The apparatus for budding and fusion is ubiquitous, so some additional component must allow a vesicle to recognize the appropriate target membrane. The **SNARE hypothesis** proposes that recognition results from the interaction of a v-SNARE membrane protein carried by the vesicle with a t-SNARE membrane protein that is present on the target membrane.

Figure 25.15 illustrates the interaction between SNAREs, for the example of a synaptic system (involving exocytosis by neurons: see below). The v-SNARE is a transmembrane protein carried by the vesicle. The

Figure 25.14 Vesicle uncoating is triggered by hydrolysis of GTP bound to ARF.

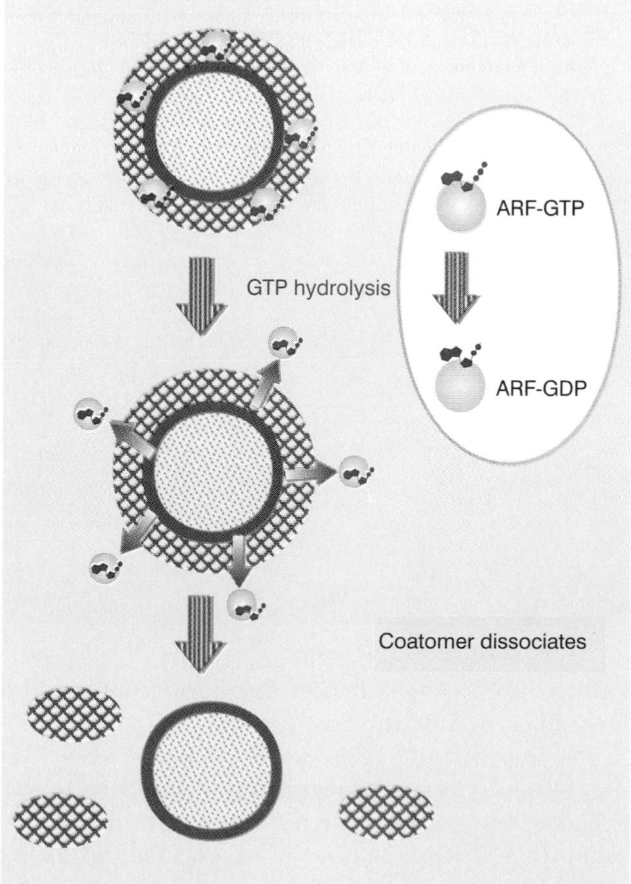

Figure 25.15 Specificity for docking is provided by SNAREs. The v-SNARE carried by the vesicle binds to the t-SNARE on the plasma membrane to form a SNAREpin. NSF and SNAP remain bound to the far end of the SNAREpin during fusion. After fusion, ATP is hydrolyzed and NSF and SNAP dissociate to release the SNAREs.

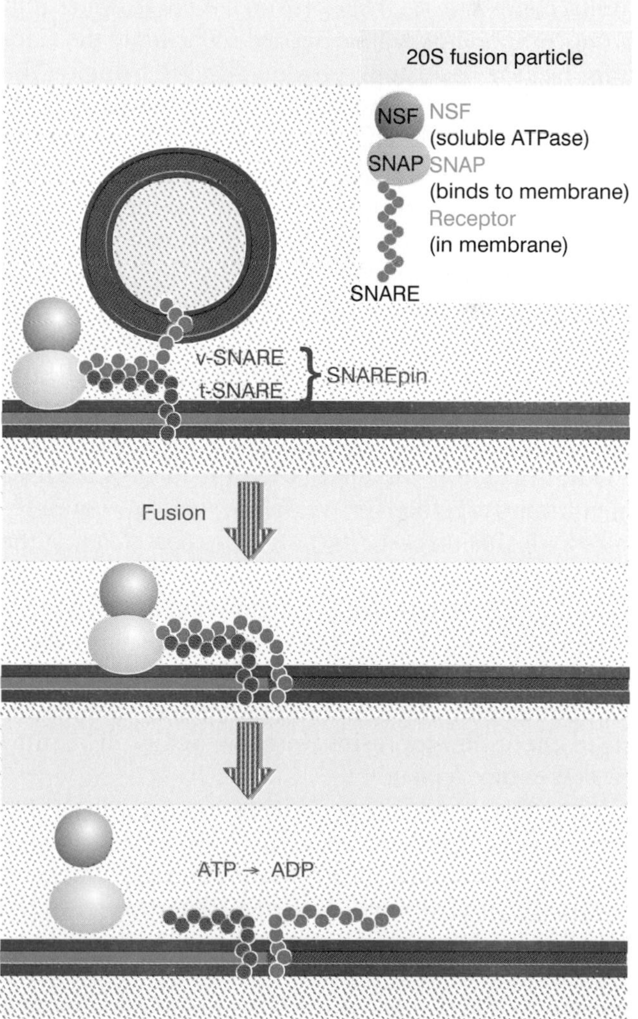

t-SNARE includes two proteins; syntaxin is a transmembrane protein, and SNAP-25 is connected to the membrane by a fatty acyl linkage. (The name of SNAP-25 has an independent origin, and it has no connection with the SNAPs of the fusion particle.) Homologs to these SNAREs are found in other systems, including other animal cell types and yeast cells.

The major part of each SNARE is exposed in the cytoplasm, and includes an extensive coiled-coil structure. Such structures commonly participate in protein-protein interactions. In fact, v-SNAREs can bind directly to the t-SNAREs *in vitro*, even without the other components of the fusion particle. The interaction between v-SNARE and t-SNARE is sufficient to sponsor membrane fusion. Liposomes containing v-SNAREs can fuse with liposomes containing t-SNAREs in the absence of any additional components. An energy source is not required, suggesting that activation energy is provided by changes in the conformation of the proteins. This *in vitro* reaction is slow, occurring over a time course of minutes. Comparison with the millisecond time course of fusion *in vivo* implies that other components will be needed to facilitate the reaction. But the basic apparatus involved in bringing the membranes together consists of the SNAREs.

A SNARE complex has a rod-like structure (~4 × 14 nm) in which the v-SNARE and t-SNARE are bound in parallel. Their membrane anchors are at the same end, implying that the rod must lie in a plane between the two membrane surfaces. This structure is called a SNAREpin. **Figure 25.16** is based on the crystal structure, which shows that the complex consists of a 4-helix bundle. **Figure 25.17** shows a model for the SNAREpin superimposed at the appropriate scale on an electron micrograph of the complex.

The formation of the SNAREpin brings the two membranes together, but we do not yet know how it relates to the fusion event itself. It is possible that first the vesicle must be brought into the proximity of the target membrane by a "tethering" reaction that involves other components. It is also possible that other steps may be necessary between SNAREpin formation and fusion of the membrane surfaces. Pairing between SNAREs is clearly necessary for fusion, but whether it can be sufficient is controversial.

The other components of the fusion particle bind to the far end of the complex. Hydrolysis of ATP and dissociation of the fusion complex is probably necessary not for fusion as such, but in order to release the SNAREs from the SNAREpin and to allow them to be used again. (The 20S fusion complex was originally envisaged to play a role prior to fusion, possibly in providing the energy for fusion, but now we believe it is more likely to function post-fusion.)

The interaction between v-SNARE and t-SNAREs may also be a target for regulation. A synaptic protein called synaptotagmin can bind to a complex of all 3 synaptic SNAREs; its binding is mutually exclusive with that of SNAP. So synaptotagmin could prevent fu-

Figure 25.16 A SNAREpin forms by a 4-helix bundle. Photograph kindly provided by Axel Brunger.

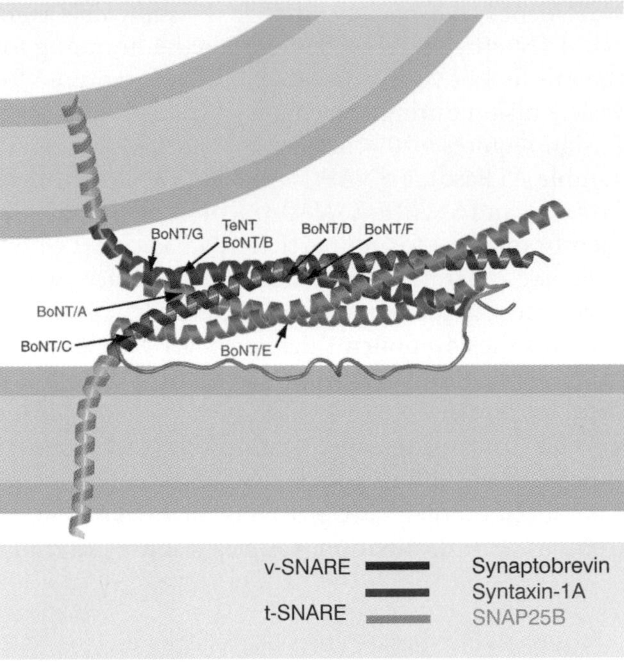

Figure 25.17 A SNAREpin complex protrudes parallel to the plane of the membrane. Photograph kindly provided by James Rothman.

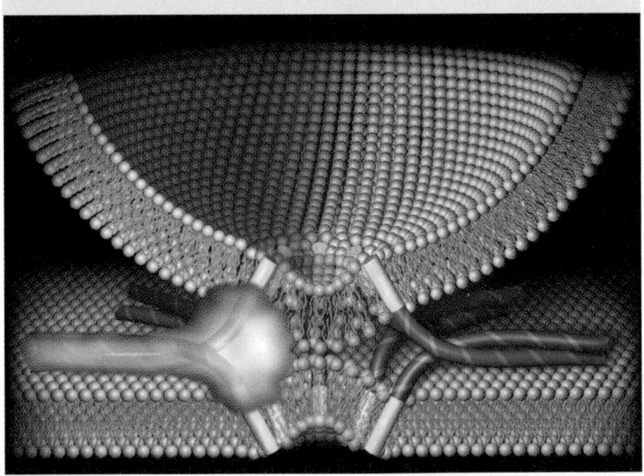

sion by blocking the formation of a fusion particle; its release could be the event that triggers exocytosis.

The synapse has been especially useful for investigating fusion because it offers the advantage of large numbers of vesicles of the same type which fuse with the plasma membrane when a specific trigger is applied.

Impulses in the nervous system are propagated by the passage of material from a donor (or presynaptic) cell to a recipient (or postsynaptic) cell. **Figure 25.18** illustrates a nerve terminal. An impulse in the donor cell triggers the exocytic pathway. Stored coated vesicles (called synaptic vesicles) move to the plasma membrane and release their contents of neurotransmitters into the extracellular fluid. The neurotransmitters in turn act upon receptors at the plasma membrane of the recipient cell.

Exocytosis would lead to the accumulation of vesicle components in the plasma membrane if there were no means to retrieve them. There are two possibilities for the recycling of these vesicles, both of which may occur.

In the "kiss and run" model illustrated in **Figure 25.19**, a vesicle does not completely fuse with the plasma membrane, but contacts it transiently. The neurotransmitter is released through some sort of pore; then the vesicle re-forms. Major questions about this pathway arc how the vesicle maintains its integrity and what sort of structure forms the pore.

In the fusion model illustrated in **Figure 25.20**, the vesicle fuses with the plasma membrane in the conventional manner, releasing its contents into the extracellular space. Recycling occurs by the formation of clathrin-coated vesicles at coated pits, that is, by the endocytic pathway. This may occur at large invaginations of the plasma membrane. The importance of endocytosis in this pathway is emphasized by the fact that inhibition of the formation of the clathrin-coated vesicles affects neurotransmitter release from synaptic vesicles. A major question about the pathway is the relationship between the endocytic and exocytic vesicles. The synaptic vesicles are not clathrin-coated. It is probable that the clathrin-coated endocytic vesicles give rise to synaptic vesicles by losing their clathrin coats, but synaptic vesicles may also form by other pathways (as in the case of AP3-coated vesicles). It is probably true that removal of the clathrin coat takes place quite soon after budding for all classes of clathrin-coated vesicles; the process of removal is not well defined.

Figure 25.18 Neurotransmitters are released from a donor (presynaptic) cell when an impulse causes exocytosis. Synaptic (coated) vesicles fuse with the plasma membrane, and release their contents into the extracellular fluid. The contents are neurotransmitters that act upon the target (postsynaptic) cell.

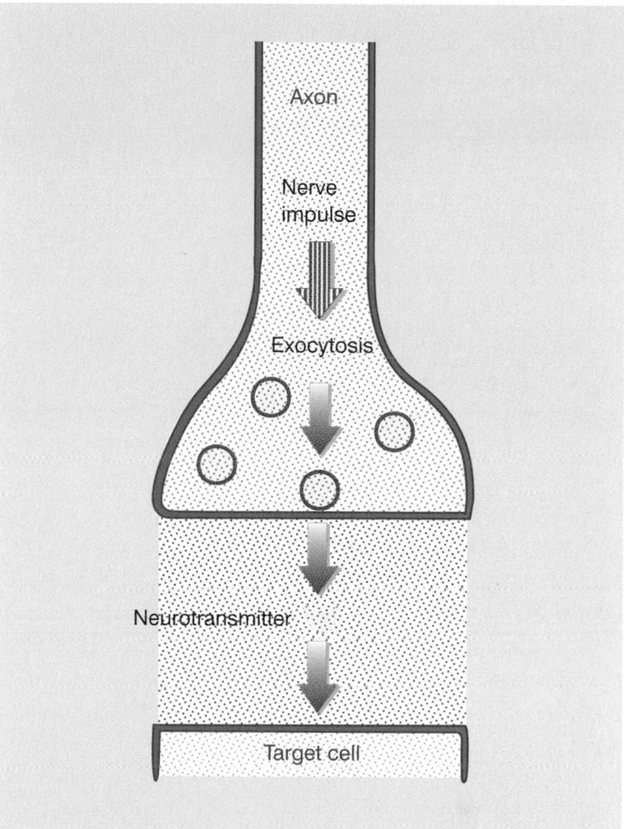

Figure 25.19 The 'kiss and run' model proposes that a synaptic vesicle touches the plasma membrane transiently, releases its contents through a pore, and then re-forms.

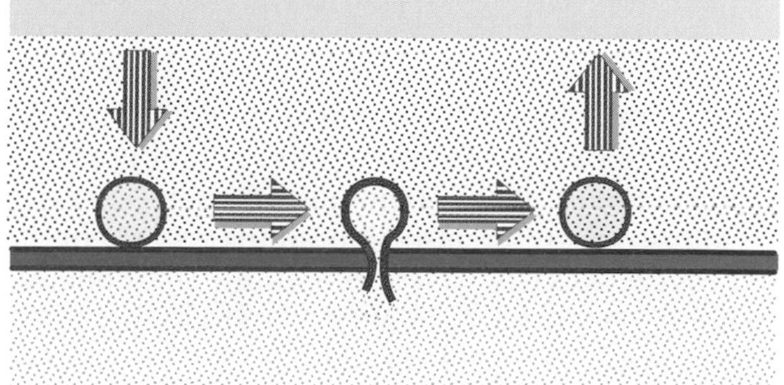

Figure 25.20 When synaptic vesicles fuse with the plasma membrane, their components are retrieved by endocytosis of clathrin-coated vesicles.

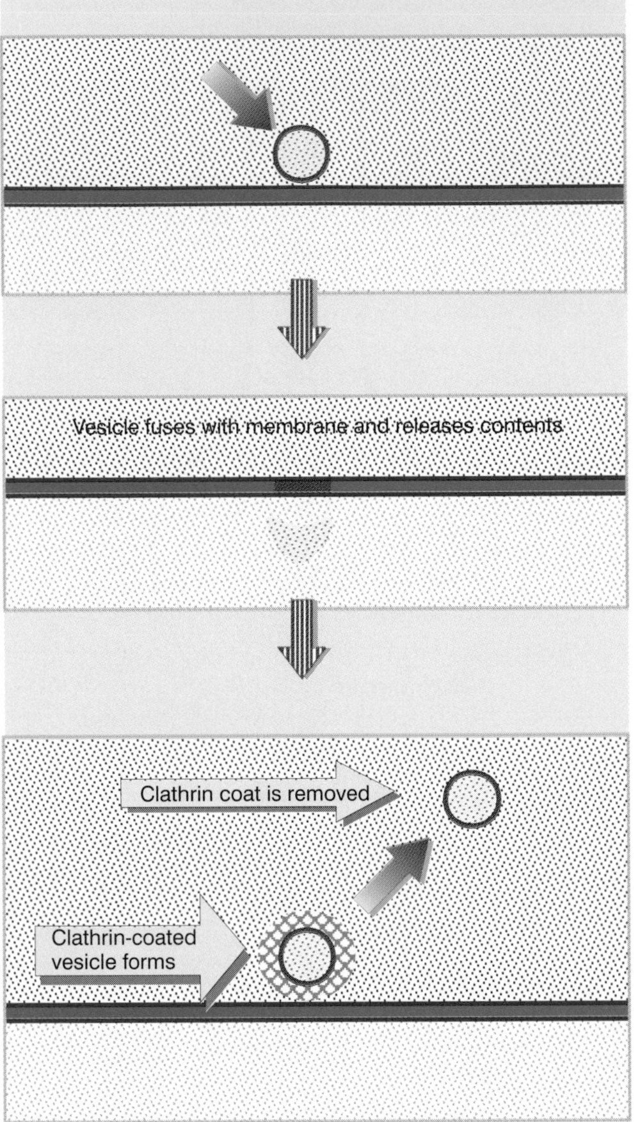

Another class of proteins that act only at particular stages of protein transport consists of **Rabs**. They are attached to membranes via the addition of prenyl or palmityl groups at the C-terminus. There are ~30 Rabs, distributed to different membranous systems in the cell. **Figure 25.21** summarizes their distribution: different Rabs are involved in ER to Golgi transport, in the constitutive and regulated pathways from the Golgi to the plasma membrane, and in stages of transport between endosomes (see also below). For example, mutations in the yeast genes *YPT1* or *SEC4* that code for two such (related) proteins block transport and cause the accumulation of vesicles in the Golgi stacks or between Golgi and plasma membrane, respectively.

Figure 25.21 Rab proteins affect particular stages of vesicular transport.

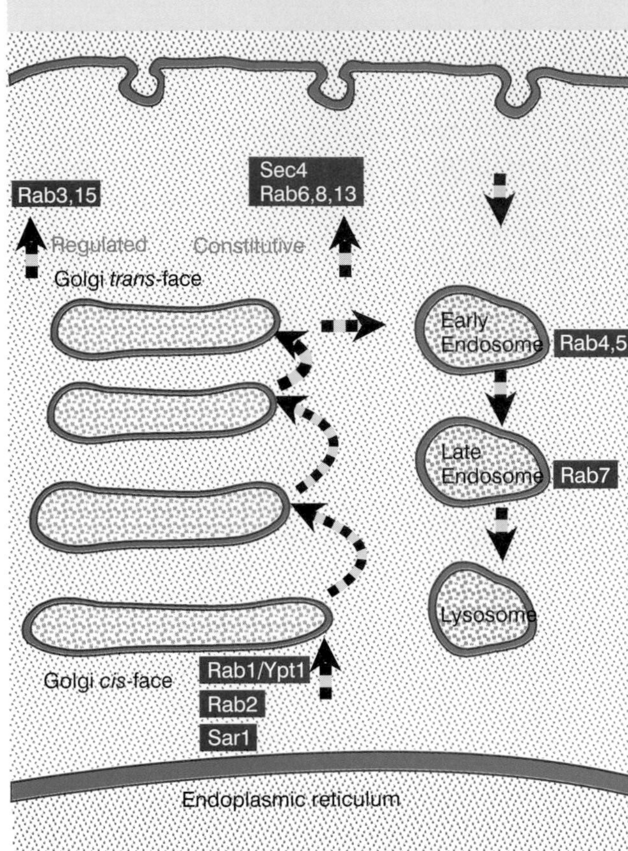

The Rabs are GTP-binding proteins that are active in the form bound to GTP; but hydrolysis of the GTP converts the protein to an inactive form. As with other monomeric GTP-binding proteins, their activities are affected by other proteins that influence the hydrolysis of GTP (see Chapter 26). There may be GAP (GTP-hydrolyzing) activities specific for certain Rabs, GEF proteins that stimulate dissociation of the guanine nucleotide, and GDI proteins that prevent dissociation of guanine nucleotide.

The involvement of at least some Rab proteins in specific transport stages suggests that the Rabs could be involved in targeting, but their role is not yet clear.

Protein localization depends on further signals

Various types of signals influence transport through the ER-Golgi system. A protein that has no special signals will presumably enter vesicles at a rate determined by its concentration in the compartment, and may move in the anterograde direction by bulk flow. However, most proteins appear to have specific signals that facilitate or retard transport.

A typical cargo protein has a **transport signal** that is responsible for its entry into budding vesicles. **Figure 25.22** shows that the transport signal in a transmembrane protein is usually a region in its cytoplasmic domain that binds to an adaptor protein of the vesicle coat. **Figure 25.23** shows that the transport signal in a soluble cargo protein (for example, a secreted protein that passes through the lumen) is a region that binds to the lumenal domain of a transmembrane cargo receptor, which in turn has a cytoplasmic domain that binds an adaptor protein. Interaction between the cargo and the coat thus directly or indirectly determines specificity of transport. Such mechanisms control anterograde transport from the ER to the cell surface and other destinations.

A protein may be prevented from leaving a compartment by a **retention signal.** Such signals are often found in transmembrane regions, perhaps because aggregation between them creates a structure that is too large to be incorporated into a budding vesicle.

We have detailed information about several types of signal: a conformation that is required for proteins to be internalized by endocytosis; an amino acid sequence that targets proteins to the ER; and a modification that targets proteins to **lysosomes** (small membranous bodies, where proteins are degraded; see later).

Internalization of receptors via coated pits requires information in their cytoplasmic tails. The sequence NPXY (Asn-Pro-X-Tyr) is located close to the C-terminus. Although this is a basic signal for internalization, other sequences in the cytoplasmic tail influence the efficiency.

Figure 25.22 A transport signal in a transmembrane cargo protein interacts with an adaptor protein.

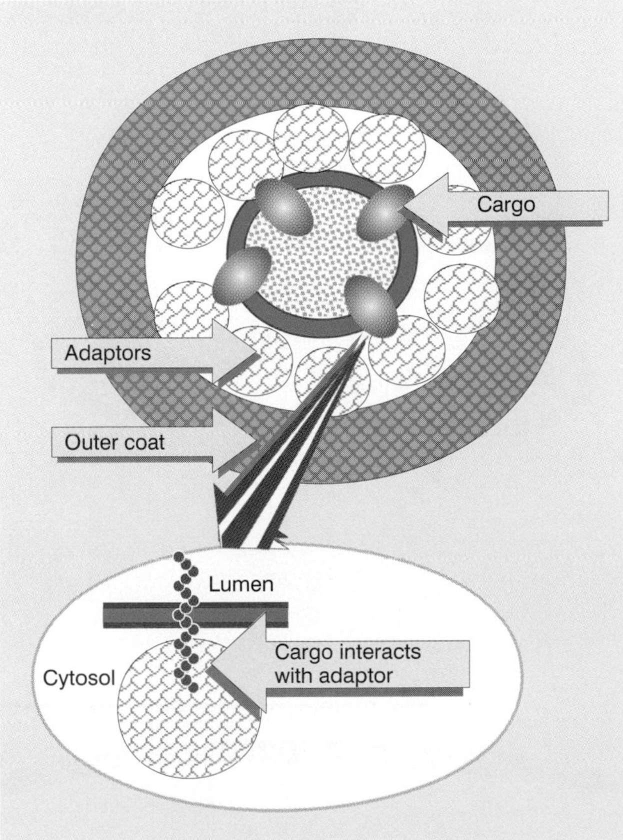

Figure 25.23 A transport signal in a luminal cargo protein interacts with a transmembrane receptor that interacts with an adaptor protein.

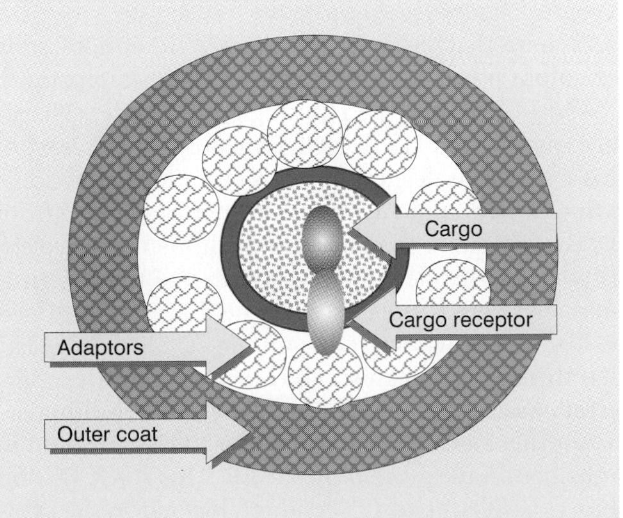

Enzymes that will be transported to lysosomes are recognized as targets for high mannose glycosylation, and are trimmed in the ER as described in Figure 25.5. Then **mannose-6-phosphate** residues are generated by a two-stage process in the Golgi. First the moiety N-acetyl-glucosamine-1-phosphate is added to the 6 position of mannose by GlcNAc-phosphotransferase; then a glucosaminidase removes the N-acetyl-glucosamine (GlcNAc).

The action of the phosphotransferase provides the critical step in marking a protein for lysosomal transport. It occurs early in ER-Golgi transfer, possibly between the ER and the *cis*-Golgi. The basis for the enzyme's specificity is its ability to recognize a structure that is common to lysosomal proteins. The structure consists of two short sequences, which are separated in the primary sequence, but form a common surface in the tertiary structure. Each of these sequences has a crucial lysine residue. The nature of this signal explains how proteins with little identity of sequence may share a common pathway for localization.

Lysosomal proteins continue to be transported along the Golgi stacks until they encounter receptors for mannose-6-phosphate. Recognition of mannose-6-phosphate targets a protein for transport in a coated vesicle to the lysosome. This final stage of sorting for the lysosome occurs in the *trans*-Golgi, where the proteins are collected by specific transport vesicles that are coated with clathrin. The vesicles transport the lysosomal proteins to the late endosome, where they join the pathway for movement to the lysosome. A single pool of mannose-6-phosphate receptors is probably used for directing proteins to the lysosome whether they are newly synthesized or endocytosed. Most of the receptors in fact are located on endosomes, where they could recognize endocytosed proteins.

Proteins that reside in the lumen of the endoplasmic reticulum possess a short sequence at the C-terminus, Lys-Asp-Glu-Leu (KDEL in single letter code). The alternative signals HDEL or DDEL are used in yeast. If this sequence is deleted, or if it is extended by the addition of other amino acids, the protein is secreted from the cell instead of remaining in the lumen. Conversely, if this tetrapeptide sequence is added to the C-terminus of lysozyme, the enzyme is held in the ER lumen instead of being secreted from the cell. This suggests that there is a mechanism to recognize the C-terminal tetrapeptide and cause it to be localized in the lumen.

Another signal is responsible for the localization of transmembrane proteins in the ER. This is KKXX, and thus consists of two Lys residues, located in the cytoplasmic tail just prior to the C-terminus.

An interesting question emerges from the behavior of proteins that have an ER-localization signal. Does this signal cause a protein to be held so that it cannot pass beyond the ER or is it the target for a more active localization process? The model shown in **Figure 25.24** suggests that *the KDEL sequence causes a protein to be returned to the ER* from an early Golgi stack. The same experiment has been performed with KKXX proteins, with similar results.

Because the modification of proteins as they pass through the Golgi is ordered, we can use the types of sugar groups that are present on any particular species as a marker for its progress on the exocytic pathway. When a KDEL sequence is added to a protein that usu-

Figure 25.24 An (artificial) protein containing both lysosome and ER-targeting signals reveals a pathway for ER-localization. The protein becomes exposed to the first but not to the second of the enzymes that generates mannose-6-phosphate in the Golgi, after which the KDEL sequence causes it to be returned to the ER.

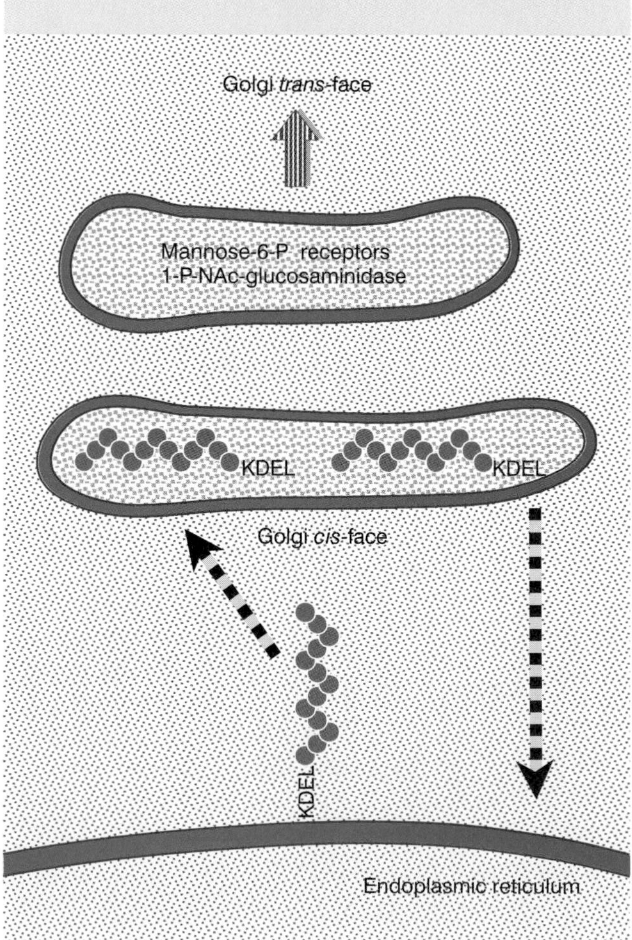

ally is targeted to the lysosome (because its oligosaccharide gains 6-mannose-P residues), it causes the protein to be held in the ER. But the protein is modified by the addition of GlcNAc-P, which happens only in the Golgi. The GlcNAc is not removed, so the protein cannot have proceeded far enough through the Golgi stacks to encounter the second of the enzymes in the mannose-6-P pathway. This suggests that KDEL is recognized by a receptor located after entering the Golgi, but before the stack containing the second enzyme.

Mutations in the *S. cerevisiae* genes *ERD1* and *ERD2* prevent retention of proteins with the HDEL signal in the ER; instead the proteins are secreted from the cell. The products of both these genes are integral membrane proteins. The *ERD1* mutation causes a general defect in the Golgi; this supports the idea that sorting of the ER proteins occurs by salvage from the Golgi. The *ERD2* mutation identifies the receptor for the HDEL sequence. One model for its role is that it cycles between the Golgi salvage compartment and the ER. This idea is supported by the localization of the corresponding receptor in mammalian cells: it is found largely in the Golgi, but over-expression of a protein with KDEL sequence causes it to concentrate in the ER. So binding of a KDEL-protein causes the receptor to move from Golgi to ER. It may have a high affinity for the HDEL sequence under the conditions prevailing in the Golgi, but a low affinity in the ER. This could enable ERD2 to seize proteins by their HDEL tails in the Golgi, and take them back to the ER, where they are then released.

The dilysine motif of KKXX proteins binds to the β'- and α-COP components of coatomer. Yeast mutants that affect β'-, α-, or γ-COP are defective in retrieval of KKXX proteins from the Golgi. This suggests that vesicles with the COP-I coat are involved in retrieving proteins from the Golgi and returning them to the ER; that is, COP-I vesicles are responsible for retrograde transport.

Overall protein transport is a unidirectional process: proteins enter the ER and are transported through the Golgi, unless stopped *en route*. As mentioned previously, COP-II-coated vesicles are thought to provide the major capacity for anterograde transport from the ER to the Golgi. COP-I-coated vesicles provide transport capacity along the Golgi stacks. However, both COP-I- and COP-II-coated vesicles can be observed to bud from the ER, so there is the possibility that they are involved at multiple stages (perhaps in transporting different types of cargoes).

Retrograde transport usually is obscured by anterograde transport, but is revealed when cells are treated with the drug brefeldin A (BFA), which specifically blocks the forward direction of transport. BFA blocks conversion of ARF from the GDP-bound to the GTP-bound form, and therefore prevents budding of coated vesicles. As a result of the block, a network of tubules forms between the cisternae of the Golgi (abolishing their usual independence) and joins them to the endoplasmic reticulum. There is resorption of most of the membranes of the *cis*-medial Golgi into the endoplasmic reticulum, which is accompanied by the redistribution of Golgi proteins into the ER, effecting a retrograde transport.

This happens because the COP-II vesicles involved in forward movement are more sensitive to the drug than the COP-I vesicles involved in retrograde movement. Retrograde transport may serve to retrieve membrane components to compensate for anterograde movement, and of course also provides for the retrieval of ER-proteins from the Golgi. There is also the possibility of other transport systems: certain toxins that are endocytosed at the plasma membrane can be found in the ER, but this retrograde transport does not appear to depend on the known systems.

The effects of brefeldin on ER-Golgi transport are universal, but in addition it inhibits other transport processes differently in different cells. In some cell types it inhibits transcytosis (transport from the basolateral surface to the apical surface in polarized cells); in other cells it inhibits transport from the *trans*-Golgi network to endosomes. A related phenomenon is revealed by isolating cells that can grow in the presence of brefeldin. This identifies mutants in which transport is resistant in particular locations (such as endosomes or Golgi) but remains sensitive in others. Brefeldin acts by binding to a common domain (the Sec7 domain) in the exchange factors (GEFs) that are responsible for regenerating ARF-GTP from ARF-GDP. BFA stabilizes the association of the GEF with the GDP-bound form of the ARF. This causes the ARF to remain in its inactive state. The differing effects of BFA on individual transport processes probably mean that there is a variety of GEFs that act on ARFs on different membrane surfaces, so that the apparatus involved in assembling coated vesicles is specific for individual types of surface. The characteristic susceptibility of each GEF explains the effect of brefeldin on budding from its particular membrane.

Receptors recycle via endocytosis

THE systems involved in importing proteins into the cell are closely related to those used for exporting secreted proteins. Ingestion of receptors starts by a common route, which leads to several pathways in which receptors have different fates. Some receptors are internalized continuously, but others remain exposed on the surface until a ligand is bound, after which they become susceptible to endocytosis. The signals that trigger internalization are different for ligand-independent and ligand-induced endocytosis.

In either case, the receptors slide laterally into coated pits, the indented regions of the plasma membrane surrounded by clathrin. It is not clear whether simple lateral diffusion can adequately explain the movement into coated pits or whether some additional force is required. Coated pits invaginate into the cytoplasm and pinch off to form clathrin-coated vesicles. These vesicles move to early endosomes, are uncoated, fuse with the target membrane, and release their contents. The process is called **receptor-mediated endocytosis.**

The immediate destination for endocytic (clathrin-coated) vesicles is the **endosome,** a rather heterogeneous structure consisting of membrane-bounded tubules and vesicles. There are at least two types of endosome, as indicated in **Figure 25.25. Early endosomes** lie just beneath the plasma membrane and are reached by endocytosed proteins within ~1 minute. **Late endosomes** are closer to the nucleus, and are reached within 5–10 minutes.

The early endosome provides the main location for sorting proteins on the endocytic pathway. Its role is a counterpart to that played by the Golgi for newly synthesized proteins. The interior of the endosome is acidic, with a pH <6. Proteins that are transported to the endosome change their structure in response to the lowering of pH; this change is important in determining their fate.

Receptors that have been endocytosed to the early endosome behave in one of two ways. They may return to the plasma membrane (by vesicular transport). Or they may be transported further to the lysosome, where they are degraded. Transport to the lysosome is the default pathway, and applies to any material that does not possess a signal specifically directing it elsewhere.

The lysosome contains the cellular supply of hydrolytic enzymes, which are responsible for degradation of macromolecules. Like endosomes, the lysosome is an acidic compartment, whose pH is maintained at the low level of 5.

The relationship between the various types of endosomes and lysosomes is not yet clear. Vesicles may be used to transport proteins along the pathway from one pre-existing structure to the next; or early endosomes may "mature" into late endosomes, which in turn "mature" into lysosomes. At all events, the pathway is

Figure 25.25 Endosomes sort proteins that have been endocytosed and provide one route to the lysosome. Proteins are transported via clathrin-coated vesicles from the plasma membrane to the early endosome, and may then either return to the plasma membrane or proceed further to late endosomes and lysosomes. Newly synthesized proteins may be directed to late endosomes (and then to lysosomes) from the Golgi stacks. The common signal in lysosomal targeting is the recognition of mannose-6-phosphate by a specific receptor.

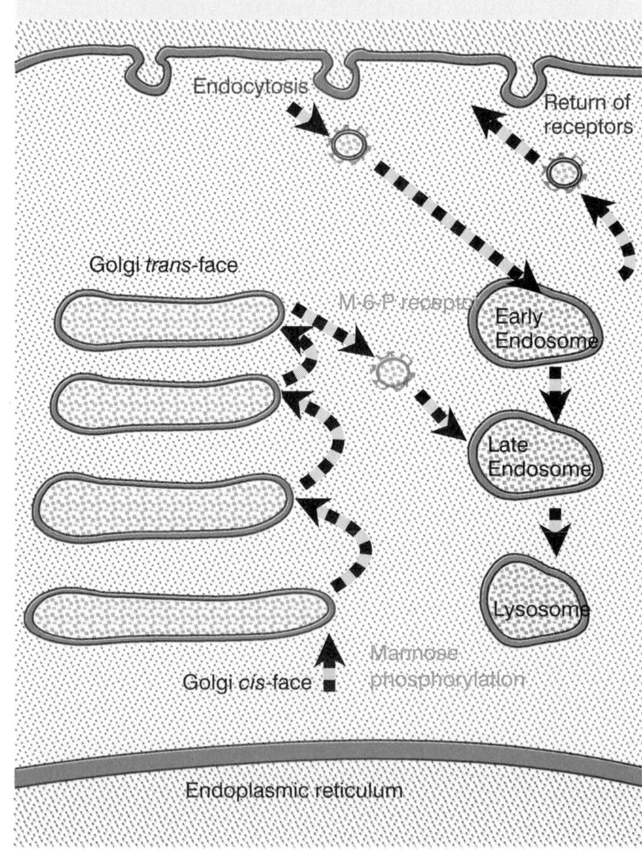

unidirectional, and a protein that has left the early endosome for the late endosome will end up in the lysosome.

There are two routes to the lysosome. Proteins endocytosed from the plasma membrane may be directed via the early endosome to the late endosome. Newly synthesized proteins may be directed from the *trans*-Golgi via the late endosome, as described above.

The fate of a receptor-ligand complex depends upon its response to the acidic environment of the endosome. Exposure to low pH changes the conformation of the external domain of the receptor, causing its ligand to be released, and/or changes the structure of the ligand. But the receptor must avoid becoming irreversibly denatured by the acid environment; the presence of multiple disulfide bridges in the external domain may play an important role in maintaining this unusual stability.

Four possible fates for a receptor-ligand complex are described in the alternative pathways of Figures 25.26–29:

■ *Receptor recycles to the surface in coated vesicles, while the ligand is degraded.* **Figure 25.26** shows that this pathway is used by receptors that transport ligands into cells at high rates. A receptor recycles every 1–20 minutes, and can undertake >100 cycles during its lifetime of ~20 hours. The classic example of this pathway is the LDL receptor, whose ligands are the plasma low density lipoproteins apolipoprotein E and apolipoprotein B (collectively known as the LDLs). Apo-B is a very large (500 kD) protein that carries cholesterol and cholesterol esters. The LDL is released from its receptor in the endosome. The receptor recycles to the surface to be used again. The LDL and its cholesterol separate in the endosome; the LDL is sent on to the lysosome, where it is degraded, and the cholesterol is released for use by the cell. This constitutes the major route for removing cholesterol from the circulation, and people with mutations in the LDL receptor accumulate large amounts of plasma cholesterol that cause the disease of familial hypercholesterolemia.

■ *Receptor and ligand both recycle.* The transferrin receptor provides the classic example of this pathway, illustrated in **Figure 25.27.** The ligand for the receptor is the iron-carrying form of transferrin. When this reaches the endosome, the acid environment causes transferrin to release the iron. The iron-free ligand, called apo-transferrin, remains bound to the transferrin receptor, and recycles to the plasma membrane. In the neutral pH of the plasma membrane, apo-transferrin dissociates from the receptor. This leaves apo-transferrin free to bind another iron, while the transferrin receptor is available to internalize another iron-carrying transferrin. Again this cycle is quite intensively used; a transferrin receptor recycles every 15–20 minutes, and has a half-life of >30 hours. It provides the cell with the means of taking up iron.

■ *Receptor and ligand both are degraded.* The EGF receptor binds its ligand as a requirement for internalization. Although EGF and its receptor appear to dissociate at low pH, they are both carried on to the lysosome, where they are degraded, as indicated in

Figure 25.26 LDL receptor transports apo-B (and apo-E) into endosomes, where receptor and ligand separate. The receptor recycles to the surface, apo-B (or apo-E) continues to the lysosome and is degraded, and cholesterol is released (not shown).

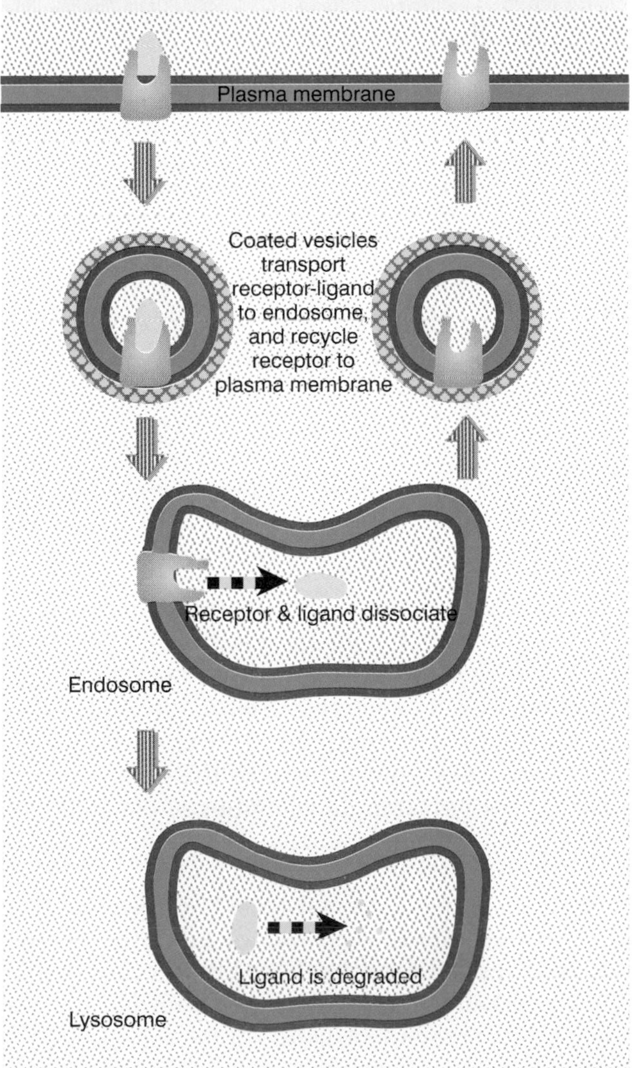

Figure 25.27 Transferrin receptor bound to transferrin carrying iron releases the iron in the endosome; the receptor now bound to apo-transferrin (lacking iron) recycles to the surface, where receptor and ligand dissociate.

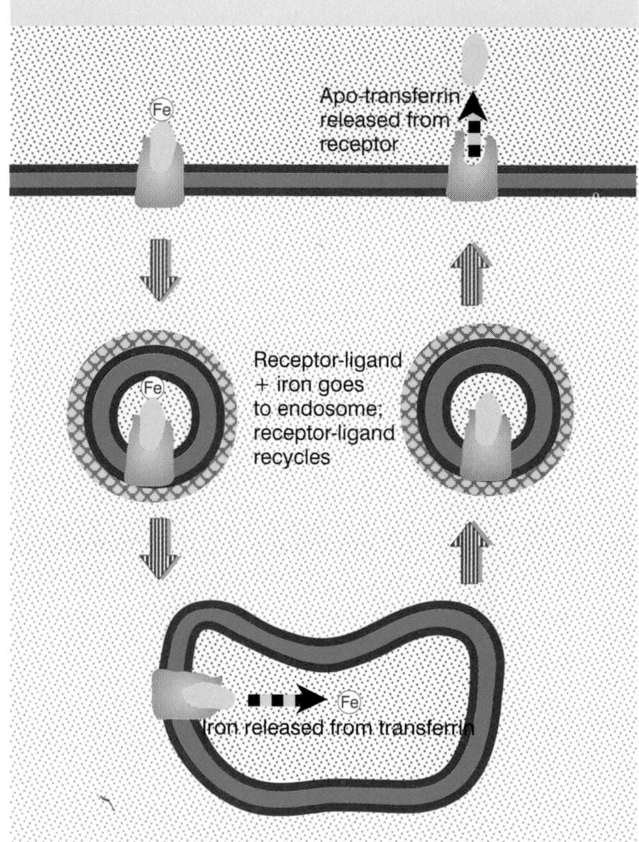

Figure 25.28 EGF receptor carries EGF to the lysosome where both the receptor and ligand are degraded.

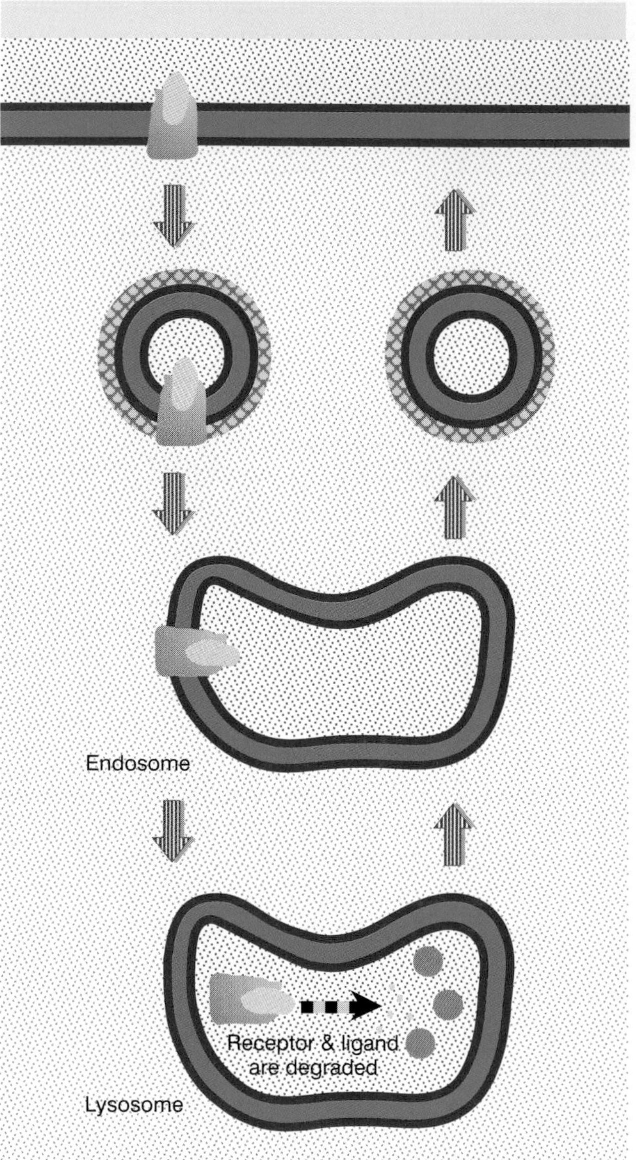

Figure 25.28. We do not know how and whether these events are related to the ability of EGF to change the phenotype of a target cell via binding to the receptor.

■ *Receptor and ligand are transported elsewhere.* The route illustrated in **Figure 25.29** is available in certain polarized cells. A receptor-ligand combination is taken up at one cell surface, transported to the endosome, and then released for transport to the far surface of the cell. This is called **transcytosis.** By this means, receptors can transport immunoglobulins across epithelial cells.

Rapid recycling in general occurs for receptors that bring ligands into the cell, not for those that trigger pathways of signal transduction. Receptors involved in signaling changes from the surface are usually degraded if they are endocytosed.

What features of protein structure are required for endocytosis? Mutations that prevent internalization can be used to identify the relevant sequences in the receptors. In fact, the characterization of an internalization defect provided evidence that entry into coated pits is needed for receptor-mediated endocytosis of LDL. In cells from human patients with such defects in the LDL receptor, the receptor gathers in small clusters over the plasma membrane, and cannot enter coated pits in the manner observed for wild-type cells.

The mutations responsible for this type of defect all

Figure 25.29 Ig receptor transports immunoglobulin across the cell from one surface to the other.

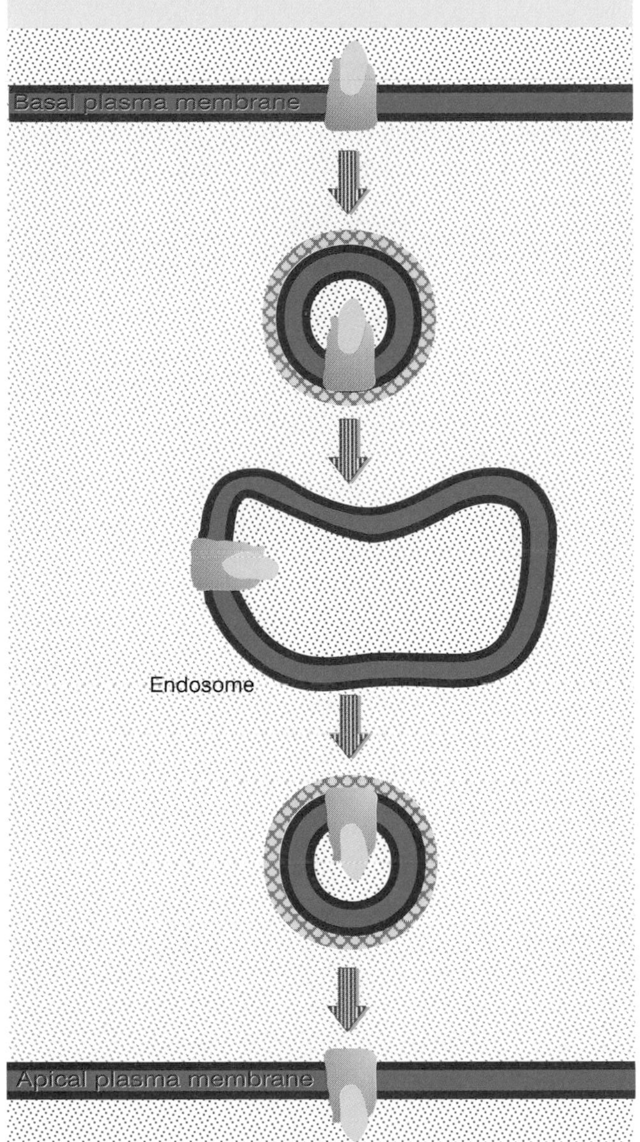

Figure 25.30 The cytoplasmic domain of an internalized receptor interacts with proteins of the inner layer of a coated pit.

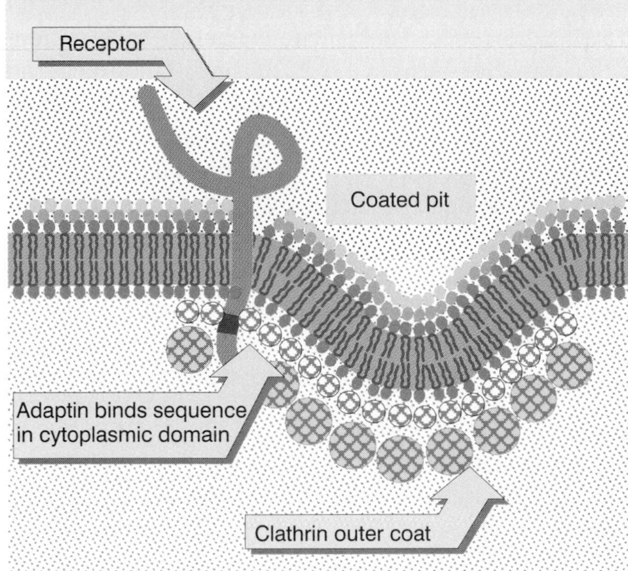

affect the *cytoplasmic domain* of the receptor, which functions independently in sponsoring endocytosis. A recombinant protein whose extracellular domain is derived from influenza virus hemagglutinin, which is not usually endocytosed, can be internalized if it is provided with a cytoplasmic domain from an endocytosed receptor.

We do not yet have a clear view of all of the signals that mediate endocytosis, but two features are common. The relevant region of the protein comprises a relatively short part of the cytoplasmic tail close to the plasma membrane. And the presence of a tyrosine residue in this region often is necessary. Removal of the tyrosine prevents internalization; conversely, substituting tyrosine in the relevant region of a protein that is not endocytosed (such as influenza virus hemagglutinin) allows it to be internalized.

Mutational analysis of the LDL receptor shows that the short amino acid sequence NPXY (Asn-Pro-X-Tyr) is required for internalization. The essential tyrosine can be replaced by other aromatic amino acids, but not by other types of amino acids. The NPXY motif is found in several group I proteins that are internalized, and is usually located close to the plasma membrane. It could be a general signal for endocytosis. It is not present in all internalized proteins. In proteins that are internalized in response to ligand binding, the internalization signal may be generated by a change in conformation as a result of the binding.

Do internalized receptors interact directly with proteins on the vesicles that transport them? Clathrin forms an outer polyhedral layer on clathrin-coated vesicles, but other proteins form an inner layer. The adaptins recognize the appropriate sequences in the cytoplasmic domains of receptors that are to be internalized (see Figure 25.12). **Figure 25.30** illustrates a model in which, as a coated pit forms, the adaptins bind to the receptor cytoplasmic domain, immobilizing the receptor in the pit. As a result, the receptor is retained by the coated vesicle when it pinches off from the plasma membrane.

Summary

Proteins that reside within the reticuloendothelial system or that are secreted from the plasma membrane enter the ER by co-translational transfer directly from the ribosome. They are transported through the Golgi in the anterograde (forward) direction. Specific signals may cause them to be retained in the ER or a Golgi stack, or directed to other organelles such as endosomes. The default pathway is to be transported to the plasma membrane. Retrograde transport is less well characterized, but proteins that reside in the ER are retrieved from the Golgi by virtue of specific signals; an example is the C-terminal KDEL.

Proteins are transported between membranous surfaces as cargoes in membrane-bound coated vesicles. The vesicles form by budding from donor membranes; they unload their cargoes by fusing with target membranes. The protein coats are added when the vesicles are formed and must be removed before they can fuse with target membranes. Anterograde transport does not result in any net flow of membrane from the ER to the Golgi and/or plasma membrane, so membrane moving with anterograde transport must be returned to the ER by a retrograde mechanism.

Modification of proteins by addition of a preformed oligosaccharide starts in the ER. High mannose oligosaccharides are trimmed. Complex oligosaccharides are generated by further modifications that are made during transport through the Golgi, determined by the order in which the protein encounters the enzymes localized in the various Golgi stacks. Proteins are sorted for different destinations in the *trans*-Golgi. The signal for sorting to lysosomes is the presence of mannose-6-phosphate.

Different types of vesicles are responsible for transport to and from different membrane systems. The vesicles are distinguished by the nature of their protein coats.

COP-I-coated vesicles are responsible for retrograde transport from the Golgi to the ER. COP-I vesicles are coated with coatomer. One of the proteins of coatomer, β-COP, is related to the β-adaptin of clathrin-coated vesicles, suggesting the possibility of a common type of structure between COP-I-coated and clathrin-coated vesicles.

COP-II vesicles undertake forward movement from the ER to Golgi. Vesicles that transport proteins along the Golgi stacks have not yet been identified. Vesicles responsible for constitutive (bulk) movement from the Golgi to the plasma membrane also have not been identified. An alternative model for anterograde transport proposes that *cis*-Golgi cisternae actually become *trans*-Golgi cisternae, so that there is a continuous process of cisternal maturation from the *cis* to the *trans* face.

In the pathway for regulated secretion of proteins, proteins are sorted into clathrin-coated vesicles at the Golgi *trans* face. Some vesicles may fuse into (larger) secretory granules. Vesicles also move to endosomes, which control trafficking to the cell surface. Secretory vesicles are stimulated to unload their cargoes at the plasma membrane by extracellular signals. Similar vesicles are used for endocytosis, the pathway by which proteins are internalized from the cell surface. The predominant protein in the outer coat of these vesicles is clathrin. The inner coat contains adaptins, which bind to clathrin; β-adaptin is found in endocytic vesicles, and β'-adaptin characterizes vesicles that move from Golgi to endosomes.

Budding and fusion of all types of vesicles is controlled by a small GTP-binding protein. This is ARF for clathrin and COP-I-coated vesicles and Sar1P for COP-II-coated vesicles. When activated by GTP, ARF/Sar1p inserts into the membrane and causes coat proteins to assemble. This leads ultimately to pinching off a vesicle (for which further proteins may also be required). When inactivated because GTP is hydrolyzed to GDP, ARF withdraws from the membrane and the coat proteins either disassemble spontaneously (COP-coated vesicles) or are caused to do so by other proteins (clathrin-coated vesicles).

Vesicles recognize appropriate target membranes because a v-SNARE on the vesicle pairs specifically with a t-SNARE on the target membrane. Pairing occurs by a coiled-coil interaction in which the SNARE complex lies parallel to the membrane surface. This is an essential part of the fusion process, but it is not clear whether other subsequent steps are also required. The other proteins associated with the SNAREs in the fusion complex are the soluble ATPase NSF and SNAP, which is responsible for binding NSF. Hydrolysis of ATP is probably necessary to release the SNAREs after pairing to allow them to recycle.

Receptors may be internalized either continuously or as the result of binding to an extracellular ligand. Receptor-mediated endocytosis initiates when the receptor moves laterally into a coated pit. The cytoplasmic domain of the receptor has a signal that is recognized by proteins that are presumed to be associated with the coated pit. An exposed tyrosine located near the transmembrane domain is a common signal; it may be part of the sequence NPXY. When a receptor has entered a pit, the clathrin coat pinches off a vesicle, which then migrates to the early endosome.

The acid environment of the endosome causes some receptors to release their ligands; the ligand are carried to lysosomes, where they are degraded, and the receptors are recycled back to the plasma membrane by means of coated vesicles. A ligand that does not dissociate may recycle with its receptor. In some cases, the receptor-ligand complex is carried to the lysosome and degraded.

Further reading

Reviews

Farquhar, M. G. (1985). Progress in unraveling pathways of Golgi traffic. *Ann. Rev. Cell Biol.* **1**, 447–488.

Goldstein, J. L., Brown, M. S., Anderson, R. G., Russell, D. W., and Schneider, W. J. (1985). Receptor-mediated endocytosis: concepts emerging from the LDL receptor system. *Ann. Rev. Cell Biol.* **1**, 1–39.

Griffiths, G. and Simons, K. (1986). The *trans* Golgi network: sorting at the exit site of the Golgi complex. *Science* **234**, 438–443.

von Figura, K. and Hasilik, A. (1986). Lysosomal enzymes and their receptors. *Ann. Rev. Biochem.* **55**, 167–193.

Pfeffer, S. R. and Rothman, J. E. (1987). Biosynthetic protein transport and sorting by the endoplasmic reticulum and Golgi. *Ann. Rev. Biochem.* **56**, 829–852.

Rose, J. K. and Doms, R. W. (1988). Regulation of protein export from the endoplasmic reticulum. *Ann. Rev. Cell Biol.* **4**, 257–288.

Hurtley, S. M. and Helenius, A. (1989). Protein oligomerization in the endoplasmic reticulum. *Ann. Rev. Cell Biol.* **5**, 277–307.

Kornfeld, S. and Mellman, I. (1989). The biogenesis of lysosomes. *Ann. Rev. Cell Biol.* **5**, 483–525.

Pelham, H. R. (1989). Control of protein exit from the endoplasmic reticulum. *Ann. Rev. Cell Biol.* **5**, 1–23.

Rothman, J. E. (1989). Polypeptide chain binding proteins: catalysts of protein folding and related processes in cells. *Cell* **59**, 591–601.

Pearse, B. M. and Robinson, M. S. (1990). Clathrin, adaptors, and sorting. *Ann. Rev. Cell Biol.* **6**, 151–171.

Rothman, J. E. and Orci, L. (1992). Molecular dissection of the secretory pathway. *Nature* **355**, 409–415.

Nuoffer, C. and Balch, W. E. (1994). GTPases: multifunctional molecular switches regulating vesicular traffic. *Ann. Rev. Biochem.* **63**, 949–990.

Rothman, J. E. (1994). Mechanisms of intracellular protein transport. *Nature* **372**, 55–68.

Rothman, J. E. (1996). Protein sorting by transport vesicles. *Science* **272**, 227–234.

Schmid, S. L. (1997). Clathrin-coated vesicle formation and protein sorting: an integrated process. *Ann. Rev. Biochem.* **66**, 511–548.

Discoveries

Vesicles

Novick, P., Field, C., and Schekman, R. (1980). Identification of 23 complementation groups required for posttranslational events in the yeast secretory pathway. *Cell* **21**, 205–215.

Lippincott-Schwartz, J., Yuan, L. C., Bonifacino, J. S., and Klausner, R. D. (1989). Rapid redistribution of Golgi proteins into the ER in cells treated with brefeldin A: evidence for membrane cycling from Golgi to ER. *Cell* **56**, 801–813.

Malhotra, V., Serafini, T., Orci, L., Shepherd, J. C., and Rothman, J. E. (1989). Purification of a novel class of coated vesicles mediating biosynthetic protein transport through the Golgi stack. *Cell* **58**, 329–336.

Barlowe, C. *et al.* (1994). COP-II: a membrane coat formed by Sec proteins that drive vesicle budding from the ER. *Cell* **77**, 895–907.

Letourneur, F. *et al.* (1994). Coatomer is essential for retrieval of dilysine-tagged proteins to the ER. *Cell* **79**, 1199–1207.

Orci, L. *et al.* (1997). Bidirectional transport by distinct populations of COP-I-coated vesicles. *Cell* **90**, 335–349.

Bonfanti, L. *et al.* (1998). Procollagen traverses the Golgi stack without leaving the lumen of cisternae: evidence for cisternal maturation. *Cell* **95**, 993–1003.

Faundez, V., Horng, J.-T., and Kelly, R. (1998). A function for the AP3 coat complex in synaptic vesicle formation from endosomes. *Cell* **93**, 423–432.

Fusion mechanisms

Clary, D. O., Griff, I. C., and Rothman, J. E. (1990). SNAPs, a family of NSF attachment proteins involved in intracellular membrane fusion in animals and yeast. *Cell* **61**, 709–721.

Wilson, D. W., Whiteheart, S. W., Wiedmann, M., Brunner, M., and Rothman, J. E. (1992). A multisubunit particle implicated in membrane fusion. *J. Cell Biol.* **117**, 531–538.

Ostermann, J. *et al.* (1993). Stepwise assembly of functionally active transport vesicles. *Cell* **75**, 1015–1025.

Sollner, T *et al.* (1993). SNAP receptors implicated in vesicle targeting and fusion. *Nature* **362**, 318–324.

Weber, T. *et al.* (1998). SNAREpins: minimal machinery for membrane fusion. *Cell* **92**, 759–772.

Protein targeting

Munro, S. and Pelham, H. R. (1987). A C-terminal signal prevents secretion of luminal ER proteins. *Cell* **48**, 899–907.

Griffiths, G., Hoflack, B., Simons, K., Mellman, I., and Kornfeld, S. (1988). The mannose 6-phosphate receptor and the biogenesis of lysosomes. *Cell* **52**, 329–341.

Collawn, J. F. *et al.* (1990). Transferrin receptor internalization sequence YXRF implicates a tight turn as the structural recognition motif for endocytosis. *Cell* **63**, 1061–1072.

Lewis, M. J., Sweet, D. J., and Pelham, H. R. B. (1990). The *ERD2* gene determines the specificity of the luminal ER protein retention system. *Cell* **61**, 1359–1363.

Semenza, J. C., Hardwick, K. G., Dean, N., and Pelham, H. R. B. (1990). *ERD2*, a yeast gene required for the receptor-mediated retrieval of luminal ER proteins from the secretory pathway. *Cell* **61**, 1349–1357.

Chapter 26

Signal transduction

THE plasma membrane separates a cell from the surrounding environment. It is permeable only to small lipid-soluble molecules, such as the steroid hormones, which can diffuse through it into the cytoplasm. It is impermeable to water-soluble material, including ions, small inorganic molecules, and polypeptides or proteins. The response to hydrophilic material depends on an interaction on the extracellular side of the cell with a protein component of the plasma membrane. The extracellular molecule typically is called the **ligand,** and the plasma membrane protein that binds it is called the **receptor.**

Two fundamental types of response to an external stimulatory molecule that cannot cross the membrane are reviewed in **Figure 26.1:**

■ *Material*—molecular or macromolecular—is physically transmitted from the outside of the membrane to the inside by transport through the lipid bilayer.

■ *A signal* is transmitted by means of a change in the properties of a membrane protein that activates its cytosolic domain.

The physical transfer of material extends from ions to small molecules such as sugars, and to macromolecules such as proteins. Three major transport routes controlled by plasma membrane proteins are reviewed in **Figure 26.2:**

■ **Channels** control the passage of ions: different channels exist for potassium, sodium, and calcium ions. By opening and closing in response to appropriate signals, the channels establish ionic levels within the cell (a feature of particular significance for cells of the neural network).

■ One means to import small molecules is for a receptor itself to transport the molecule from one side of the membrane to the other. **Transporters** are responsible for the import of small molecules (such as sugars) across the membrane. The target molecule binds to the receptor on the extracellular side, but then is released on the cytoplasmic side.

■ Ligand-binding may trigger the process of **internalization,** in which the receptor-ligand combination is brought into the cell by the process of **endocytosis.** In due course, the receptor and ligand are separated; the receptor may be returned to the surface for another cycle, or may be degraded. As described in Chapter 25, endocytosis involves the passage of membrane proteins from one surface to another via coated vesicles.

The transmission of a signal involves the interaction of an extracellular ligand with a transmembrane protein that has domains on both sides of the membrane. Binding of ligand converts the receptor from an inactive to an active form. The basic principle of this interaction is that ligand binding on the extracellular side influences the activity of the receptor domain on the cytoplasmic side. The process is called **signal transduction,** because a signal has in effect been transduced

Figure 26.1 *Overview*: information may be transmitted from the exterior to the interior of the cell by movement of a ligand or by signal transduction.

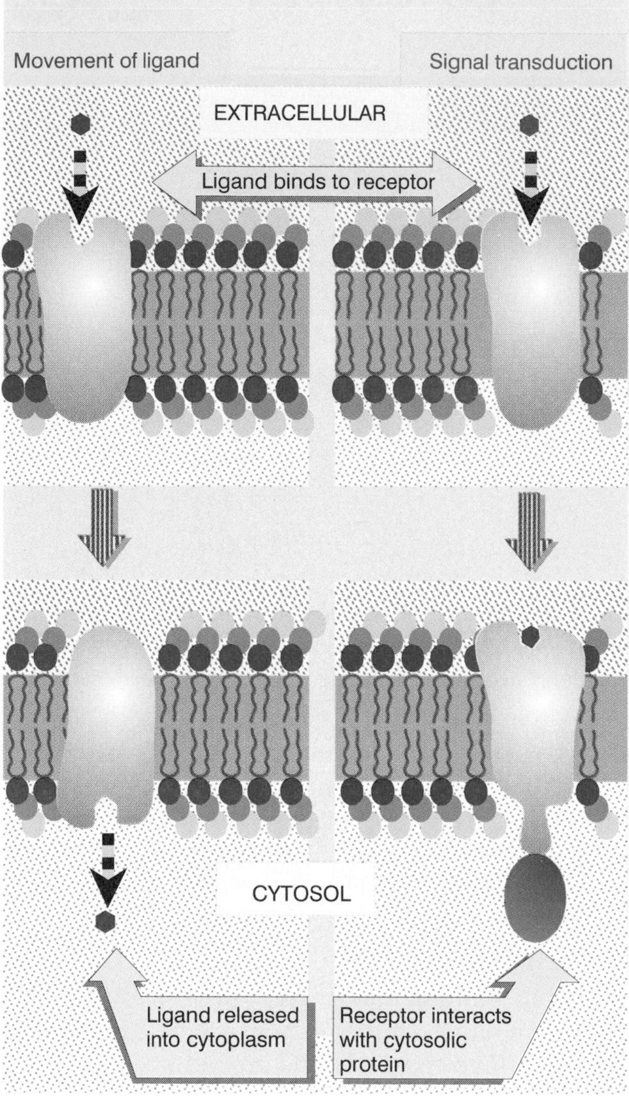

across the membrane. Signal transduction provides a means for **amplification** of the original signal.

The principle of signal transduction is that the active form of a receptor triggers a catalytic activity in the cytosol. The amplitude of the cytosolic signal is much greater than the original extracellular signal (the ligand). The cytosolic signal may take the form of directly activating a series of proteins or it may be accomplished by increasing the quantity of a small molecule inside the cell. A molecule produced in response to transduction of an extracellular signal is called a **second messenger** (by contrast with the first messenger, which was the extracellular ligand).

Two major types of signal transduction are reviewed in **Figure 26.3:**

- The receptor has a protein kinase activity in its cytosolic domain. The activity of this kinase is activated when ligand binds to the extracellular domain. The kinase phosphorylates its own cytoplasmic domain; this **autophosphorylation** enables the receptor to associate with and activate a target protein, which in turn acts upon new substrates within the cell. The most common kinase receptors are tyrosine kinases, but there are also some serine/threonine kinase receptors.

- The receptor may interact with a **G protein** that is associated with the cytosolic face of the membrane. G proteins are named for their ability to bind guanine nucleotides. The inactive form of the G protein is a trimer bound to GDP. Upon binding ligand, a receptor acts upon the G protein to cause the GDP to be replaced with GTP; as a result, the G protein dissociates into a single subunit carrying GTP and a dimer of the two other subunits. Either the monomer or the dimer then acts upon a target protein, often also associated with the membrane, which in turn reacts with a target(s) in the cytoplasm. This chain of events often stimulates the production of second messengers, the classic example being the production of cyclic AMP.

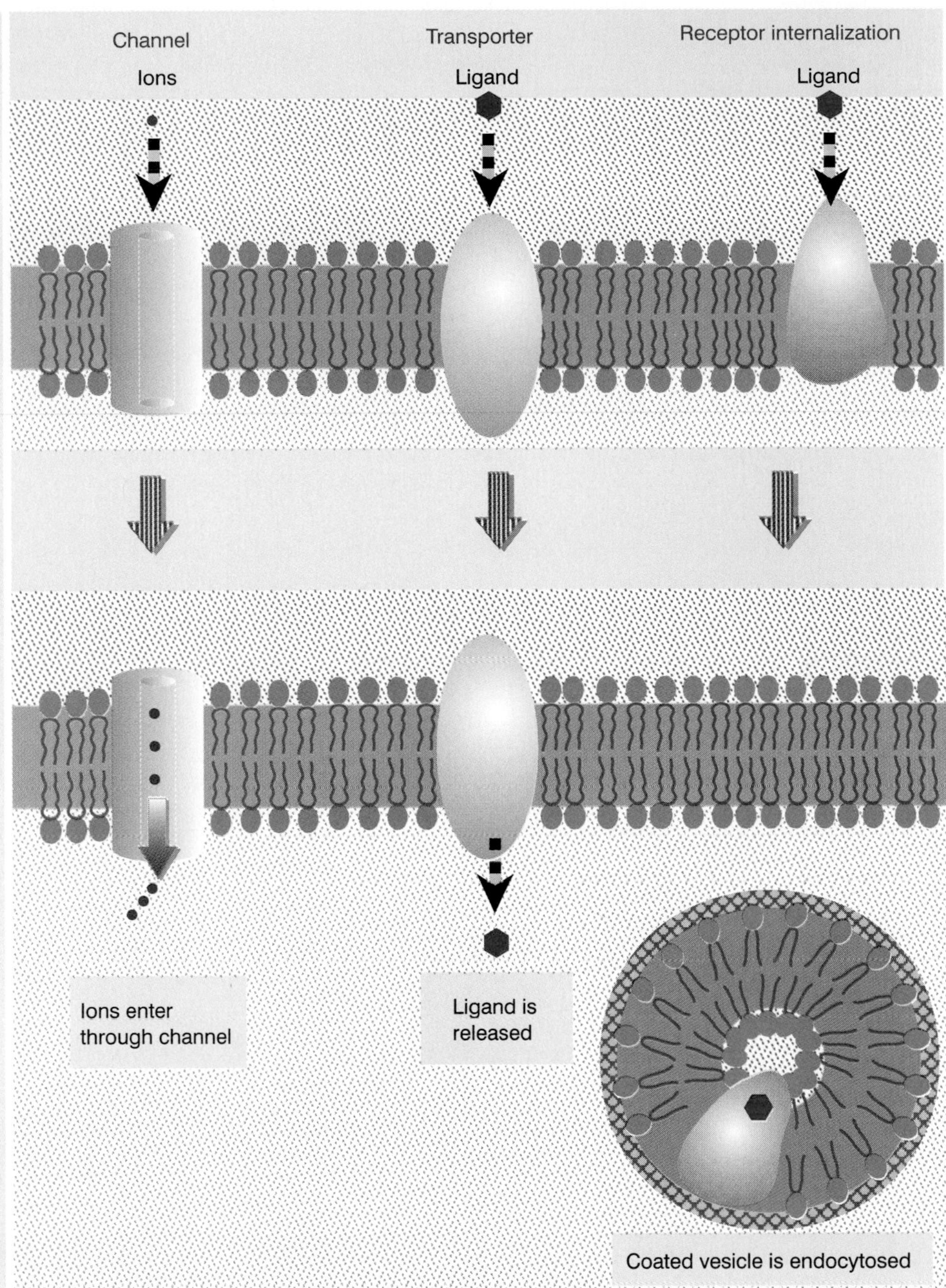
Channel
Ions
Transporter
Ligand
Receptor internalization
Ligand
Ions enter through channel
Ligand is released
Coated vesicle is endocytosed

Figure 26.2 Three means for transferring material of various sizes into the cell are provided by ion channels, receptor-mediated ligand transport, and receptor internalization.

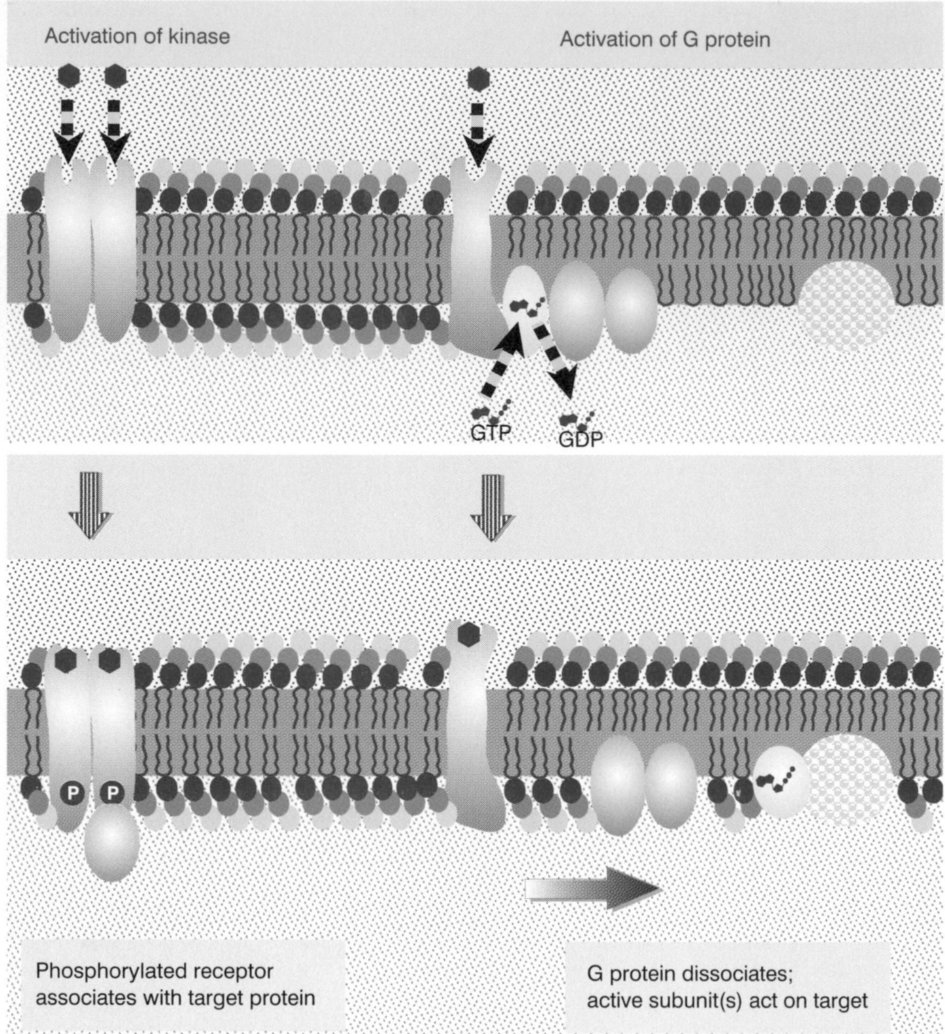
Activation of kinase
Activation of G protein
GTP
GDP
P
P
Phosphorylated receptor associates with target protein
G protein dissociates; active subunit(s) act on target

Figure 26.3 A signal may be transduced by activating the kinase activity of the cytoplasmic domain of a transmembrane receptor or by dissociating a G protein into subunits that act on target proteins on the membrane.

Carriers and channels form water-soluble paths through the membrane

THE impermeability of the plasma membrane to water-soluble compounds enables different aqueous conditions to be maintained on either side. The ionic environment of the cytosol is quite different from the extracellular ionic milieu. Within the cytoplasm, different organelles offer different ionic environments. A striking example is the maintenance of an acid pH in endosomes and lysosomes, with immediate implications for the functions of the proteins that enter them (see Chapter 25). Another example is the maintenance of a store of Ca^{2+} in the endoplasmic reticulum. (A major exception from the ability of organelles to differ from the cytosol is provided by the nucleus, which essentially is subjected to the same conditions as the cytosol. This happens because the nuclear pores form relatively large openings in the nuclear envelope, through which ions and other small molecules can diffuse freely.)

A notable feature of the cytosolic environment is that there are more free cations (positively charged) (~150 mM) than anions (~10 mM). The reason is that many cellular constituents are negatively charged—for example, nucleic acids have multiple negative charges for every phosphate group in the phosphodiester backbone. The superfluity of cations therefore establishes electrical neutrality by balancing these fixed charges.

The intracellular concentrations of Na^+ and Cl^- are low (~10 mM) while those outside the cell are high (>100 mM); and the situation for K^+ is reversed. This creates a **concentration gradient** across the membrane for each ion.

The plasma membrane is electrically charged (due to the different phospholipid compositions of the inner and outer leaflets). There is an **electrical gradient** in which the inside is negative compared to the outside. This voltage difference favors the entry of cations and opposes the entry of anions.

Together the concentration gradient and electrical gradient constitute the **electrochemical gradient,** which is characteristic for each solute. A solute whose gradient is favorable can enter the cell when a channel opens; the gradient is sufficient to drive **passive transport** of a solute such as Na^+ or Cl^- into the cell. But a solute that faces an unfavorable gradient requires **active transport** in which energy is used to pump it into the cell against the gradient.

The passage of ions (and other small solutes) through the plasma membrane is mediated by resident transmembrane proteins. A common feature of these proteins is their large size and the presence of multiple membrane-spanning regions, features which together argue that they provide a relatively static feature of the membrane. **Figure 26.4** illustrates two general means of transport across the membrane:

- A **carrier protein** binds a solute on one side of the membrane and then experiences a conformational change that transports the solute to the other side of the membrane. By binding the solute on one side and releasing it on the other, the carrier in effect directly transports the solute across the membrane. Several types of carriers are distinguished by the number of solutes that they transport, and the directions in which they transport them. Carriers that transport a single solute across the membrane are called **uniporters;** carriers that simultaneously or sequentially transport two different solutes are called **symporters;** and carriers that transport one solute in one direction while transporting a different solute in the opposite direction are called **antiporters.**

Carrier proteins may be used for passive transport or linked to an energy source to provide active transport. Energy for active transport is provided by hydrolysis of ATP, the classic example being the Na^+-K^+ pump that functions as an antiporter, pumping sodium out of the cell and potassium into it. Another source of energy is the electrochemical gradient itself; a symporter brings Na^+ into the cell together with some other

Figure 26.4 A carrier (porter) transports a solute into the cell by a conformational change that brings the solute-binding site from the exterior to the interior, while an ion channel is controlled by the opening of a gate (which might in principle be located on either side of the membrane).

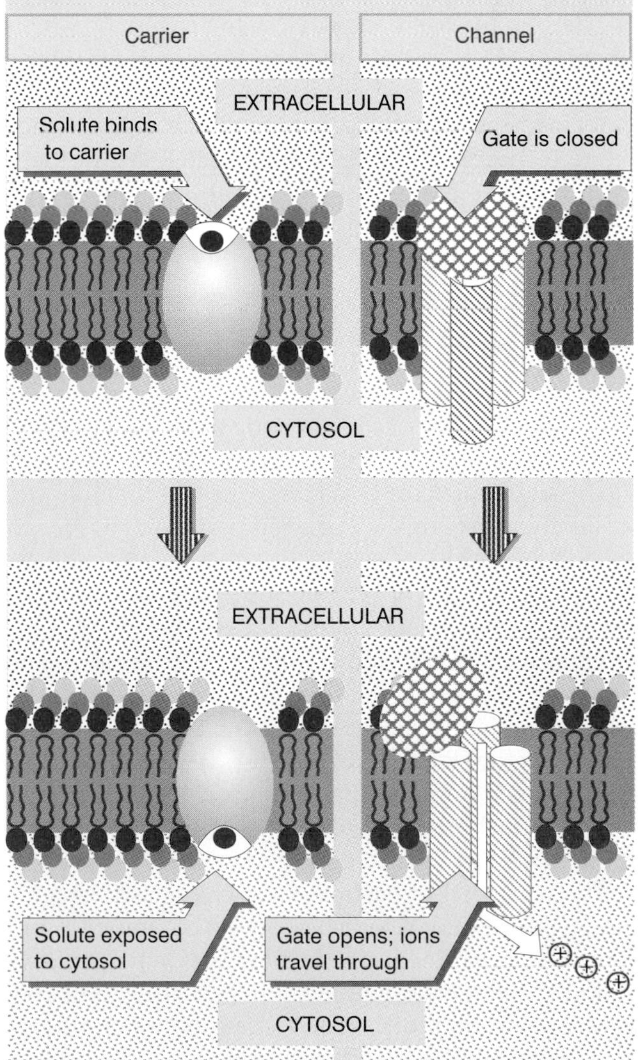

solute, using the favorable gradient of sodium to overcome the unfavorable gradient of the other solute.

■ An **ion channel** comprises a water-soluble pore in the membrane. Its activity is controlled by regulation of the opening and closing of the channel. When it is open, ions can diffuse passively, as driven by the electrochemical gradient. Ion channels allow *only* passive transport. The resting state of an ion channel is closed, and the **gates** that control channel activity usually open only briefly, in response to a specific signal. **Ligand-gated** channels are receptors that respond to binding of particular molecules, amongst which the neurotransmitters acetylcholine, glycine, GABA (γ-amino-butyric acid), and glutamate are prominent examples. **Voltage-gated** channels respond to electric changes, again a prominent feature of the neural system. **Second-messenger** gated channels provide yet another means for signal transduction, one interesting example comprising channels that respond to activation of G proteins.

The structures of both carriers and channels present a paradox. They are transmembrane proteins that have multiple membrane-spanning domains, each consisting of a stretch of amino acids of sufficient hydrophobicity to reside in the lipid bilayer. Yet within these hydrophobic regions must be a highly selective, water-filled path that permits ions to travel through the membrane.

One solution to this problem lies in the structure of the transmembrane regions. Instead of comprising unremittingly hydrophobic stretches like those of single membrane-pass proteins, they contain some polar amino acids. They are likely to be organized as illustrated in **Figure 26.5** as amphipathic helices in which the hydrophobic face associates with the lipid bilayer, while the polar faces are aligned with one another to create the channel.

The importance of the interior of the channel is indicated by the **ion selectivity.** Different channels permit the passages of different ions or groups of ions. The channels are extremely narrow, so ions must be stripped of their associated water molecules in order to pass through. The channel possesses a "filter" at the entrance to the pore that has specificity for the desired ion, presumably based upon its geometry and electrostatic charge.

The structures of particular ion channels are beginning to reveal their general features. A common feature is that the constituent proteins are large and have several membrane-spanning regions. A channel probably consists of a "ring" of 4, 5, or 6 subunits, organized in a symmetrical or quasi-symmetrical manner. The water-filled pore is found at the central axis of symmetry. The size of the pore generally increases with the number of subunits in the ring. The subunits are always related in structure, and sometimes are identical. They may consist of separate proteins or of related domains in a single large protein.

Figure 26.5 A channel may be created by amphipathic helices, which present their hydrophobic faces to the lipid bilayer, while juxtaposing their charged faces away from the bilayer. In this example, the channel is lined with positive charges, which would encourage the passage of anions.

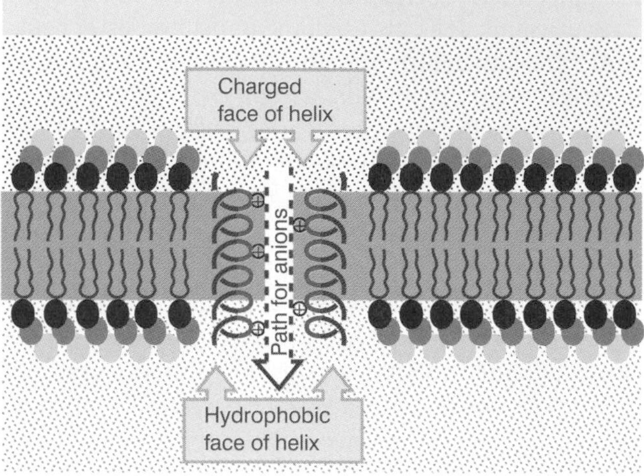

Voltage-gated sodium channels have a single type of subunit, a protein of 1820 amino acids with a repetitive structure that consists of four related domains. Each domain has several membrane-spanning regions. The four domains are probably arranged in the membrane in a pseudo-symmetrical structure. Two smaller subunits are associated with the large protein.

Potassium channels have a smaller subunit, equivalent to one of the domains of the sodium channel; four identical subunits associate to create the channel. Six transmembrane domains are identified in the protein subunit by hydrophobicity analysis; they are numbered S1–S6. The S4 domain has an unusual structure for a transmembrane region: it is highly positively charged, with arginine or lysine residues present at every third or fourth position. The S4 motif is found in voltage-gated K^+, Na^+, and Ca^{2+} channels, so it seems likely that it is involved with a common property, thought to be channel opening. Some potassium channels have only the S5–S6 membrane-spanning domains, and they appear to be basically shorter versions of the protein.

Analysis of the *shaker* potassium channel of the fly has revealed some novel features, illustrated in **Figure 26.6.** The region that forms the pore has been identi-

Figure 26.6 A potassium channel has a pore consisting of unusual transmembrane regions, with a gate whose mechanism of action resembles a ball and chain.

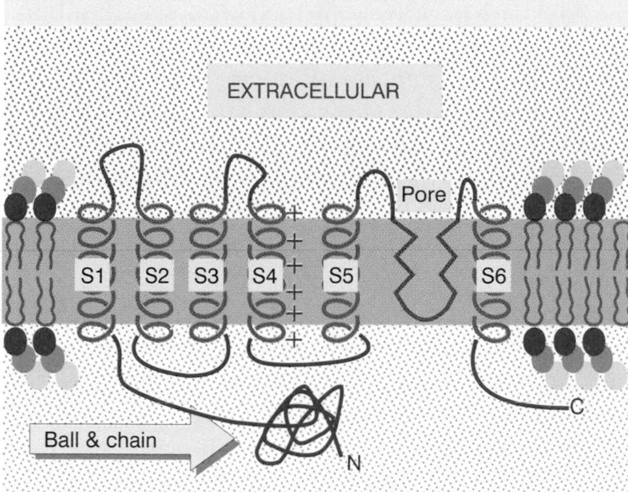

fied by mutations that alter the response to toxins that inhibit channel function. It occupies the region between transmembrane domains S5 and S6, forming two membrane-spanning stretches that are not organized in the usual hydrophobic α-helical structure. The structure could be a rather extended β-hairpin. The state of the channel (open or closed) is controlled by the N-terminal end, which resembles a ball on a chain. The ball is in effect tethered to the channel by a chain, and plugs it on the cytoplasmic side. The length of the chain controls the rate with which the ball can plug the channel after it has been opened.

A major question about potassium channels is how their selectivity is maintained. K^+ and Na^+ ions are (positively charged) spheres of 1.33 Å and 0.95 Å, respectively. K^+ ions are selected over Na^+ ions by a margin of $10^4\times$, but at the same time, up to 10^8 ions per second can move through the pore, basically close to the diffusion limit. The salient features of the pore of a potassium channel, based on the crystal structure, are summarized in **Figure 26.7**, and shown as a cutaway model in **Figure 26.8.** The pore is ~45 Å long and consists of three regions. It starts inside the cell with a long internal pore, opens out into a central cavity of ~10 Å diameter, and then passes to the extracellular space with a narrow selectivity filter. The lining of the inner pore and central cavity is hydrophobic, providing a relatively inert surface to a diffusing potassium ion. The central cavity is aqueous, and may serve to lower the electrostatic barrier to crossing the membrane (which is at its maximum in the center). The selectivity filter has negative charges and is lined with the polypeptide backbone. When a K^+ ion loses its hydrating water on entering the filter, the contacts that it made with the

Figure 26.7 The pore of a potassium channel consists of three regions.

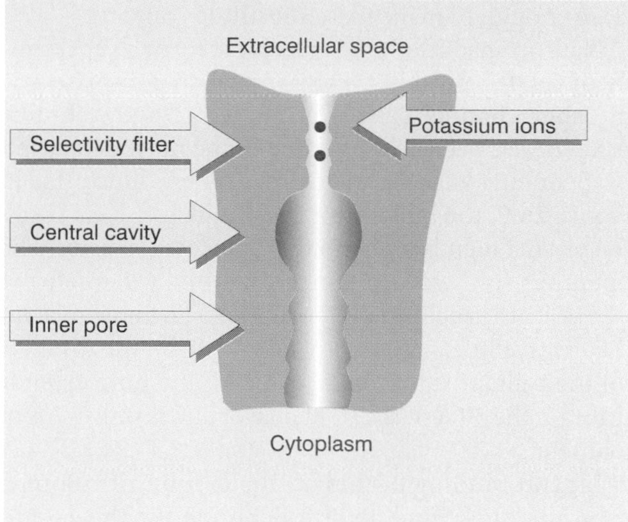

Figure 26.8 A model of the potassium channel pore shows electrostatic charge (blue = positive, white = neutral, red = negative) and hydrophobicity (= yellow). Photograph kindly provided by Rod MacKinnon.

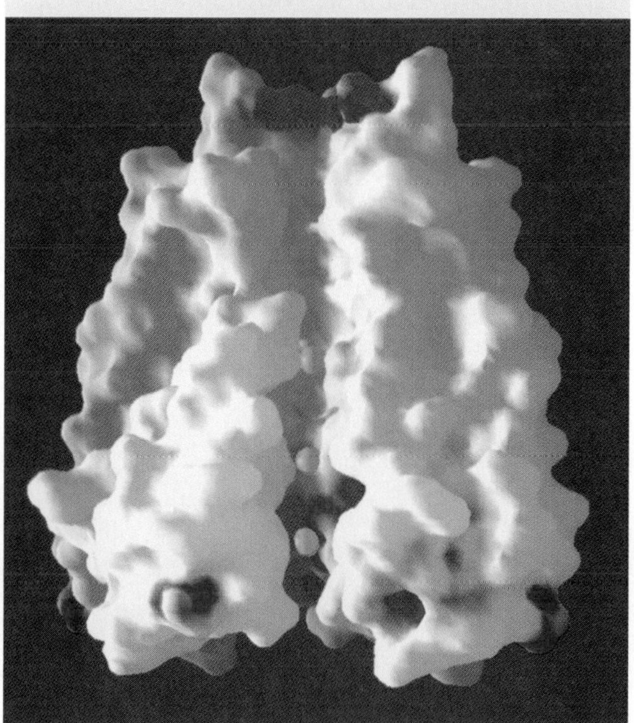

water will be replaced by contacts with the oxygens of the polypeptide carbonyl groups. The size of the pore may be set so that a smaller sodium ion would not be close enough to make these substitute contacts.

Neurotransmitter-gated receptors form a superfamily of related proteins. They appear to comprise a 5-member channel class. The nicotinic acetylcholine receptor has been characterized in the most detail, and is a pentamer with the structure $\alpha_2\beta\delta\gamma$. As illustrated in **Figure 26.9,** the bulk of the 5 subunits projects above the plasma membrane into the extracellular space. The openings to the channel narrow from a diameter of ~25 Å until reaching the pore itself. The entrance on the extracellular side is very deep, ~60 Å; the distance on the cellular side is shorter, 20 Å. The pore extends through the 30 Å of the lipid bilayer and is only ~7 Å in diameter.

Ligand binding occurs on the α subunits. Both α subunits must bind an acetylcholine for the gate to open. Where is the gate? Since the channel is really narrow only in the region within the lipid bilayer, the gate seems likely to be located well within the receptor. Structural changes that occur upon opening seem greatest just by the cytoplasmic side of the lipid bilayer, so it is possible that the gate is located at the level of the phospholipid heads on the cytoplasmic boundary. So the acetylcholine receptor, like many other receptors, must transmit information about ligand binding internally, from the extracellular acetylcholine binding site to the near-cytoplasmic gate.

How does the gate function? It might consist of an electrostatic repulsion, in which positive groups are extruded into the channel to prevent passage of cations. Or it may take the form of a physical impediment to passage, in which a conformational change brings bulky groups to block the pore.

Ion selectivity may be determined by the walls of the wide entry passage. The walls lining the entrances to the pore have negatively charged groups; each subunit carries ~10 negative charges in its extracellular region. These charge clusters could modify the ionic environment at the entrance to the channel, concentrating the desired ions and diluting ions that are selected against. The structure of the acetylcholine receptor allows passage of Na^+, K^+, or Ca^{2+} ions, but because of the prevailing gradients, its main use in practice is to allow the entry of Na^+ into the cell.

The acetylcholine receptor is an example of a superfamily of receptors gated by neurotransmitters. All appear to have the same general organization, consisting of 5 subunits whose structures are related to one another. All the subunits are about the same size (~50 kD), and each is probably organized in the membrane as a bundle of 4 helices (each helix containing a transmembrane domain). In each case, one of the four transmembrane domains (called M2) has an amphipathic structure and seems likely to be involved in lining the walls of the pore itself. The presence of serine and threonine residues, and some paired acid-basic residues, may assist ion passage. The sequences of subunits of the glycine and GABA receptors are related to the acetylcholine receptor subunits. Some changes in the sequences seem likely to reflect the ion selectivity. So the glycine and GABA receptors have positively charged groups in the entrance walls, consistent with their transport of anions such as Cl^-.

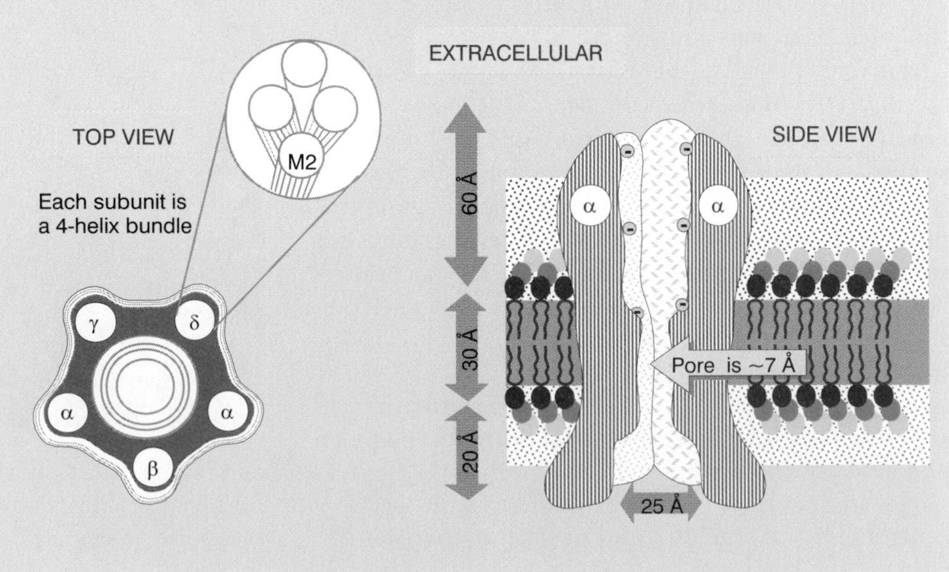

Figure 26.9 The acetylcholine receptor consists of a ring of 5 subunits, protruding into the extracellular space, and narrowing to form an ion channel through the membrane.

G proteins may activate or inhibit target proteins

G proteins transduce signals from a variety of receptors to a variety of targets. The components of the general pathway can be described as:

- The **receptor** is a resident membrane protein that is activated by an extracellular signal.
- A **G protein** is converted into active form when an interaction with the activated receptor causes its bound GDP to be replaced with GTP.
- An **effector** is the target protein that is activated (or, less often, inhibited) by the G protein; sometimes it is another membrane-associated protein.
- A **second messenger** is a small molecule that is released as the result of activation of (certain types) of effectors.

Another terminology that is sometimes used to describe the relationship of the components of the transduction pathway is to say that the receptor is *upstream* of the G protein, while the effector is *downstream*.

The effectors linked to different types of G proteins are summarized in **Figure 26.10**. The important point is that there is a large variety of G proteins, activated by a wide variety of receptors. The activation of an individual G protein may cause it to stimulate or to inhibit a particular effector; and some G proteins act upon multiple effectors (causing the activation in turn of multiple pathways). Two of the classic G proteins are G_s, which stimulates adenylate cyclase (increasing the level of cAMP), and G_t, which stimulates cGMP phosphodiesterase (decreasing the level of cGMP). The cyclic nucleotides are a major class of second messengers; another important group consists of small lipid molecules, such as inositol phosphate or DAG (diacylglycerol).

Although the receptors that couple to G proteins respond to a wide variety of ligands, they have a common type of structure and mode of binding the ligand. They are **serpentine** receptors, with 7 transmembrane regions, and function as monomers. The greatest conservation of sequence is found in hydrophobic transmembrane regions, which in fact are used to classify the serpentine receptors into individual families.

The binding sites for small hydrophobic ligands lie in the transmembrane domains, so that the ligand becomes bound in the plane of the membrane. The smallest ligands, such as biogenic amines, may be bound by a single transmembrane segment. Larger ligands, such as extended peptides, may have more extensive binding sites in which extracellular domains provide additional points of contact. Large peptide hormones may be bound mainly by the extracellular domains.

When the ligand binds to its site, it triggers a conformational change in the receptor that causes it to interact with a G protein. A well-characterized (although not typical) case is that of rhodopsin, which contains a retinal chromophore covalently linked to an amino

Figure 26.10 Classes of G proteins are distinguished by their effectors and are activated by a variety of transmembrane receptors.

G protein	Effector function	Second messenger	Example of receptor
s	Stimulates adenylyl cyclase	↑ cAMP	β-adrenergic
olf	Stimulates adenylyl cyclase	↑ cAMP	Odorant
i	Inhibits adenylate cyclase	↓ cAMP	Somatostatin
	Opens K^+ channels	↑ Membrane potential	Somatostatin
o	Closes Ca^{2+} channels	↓ Membrane potential	m2 acetylcholine
t (transducin)	Stimulates cGMP phosphodiesterase	↓ cGMP	Rhodopsin
q	Activates phospholipase Cb	↑ InsP3, DAG	m1 acetylcholine

acid in a transmembrane domain. Exposure to light converts the retinal from the 11-*cis* to the all-*trans* conformation, which triggers a conformational change in rhodopsin that causes its cytoplasmic domain to associate with the G_t protein (transducin).

G proteins are trimers whose function depends on the ability to dissociate into an α monomer and a βγ dimer. The dissociation is triggered by the activation of an associated receptor.

In the trimeric state, a G protein is inactive. **Figure 26.11** illustrates the classic model for G protein action. The basic principle is that the activity of the G protein is controlled by its guanine nucleotide. The inactive trimer has GDP bound to the α subunit. The active form consists of an α monomer bound to GTP, dissociated from a βγ dimer.

The inactive trimer is constitutively associated with the receptor. When the receptor is activated, its conformation changes, and this in turn induces a change in the conformation of the α subunit. This causes GDP to be released. Because the concentration of GTP in the cytosol is much greater than that of GDP, the vacant nucleotide binding site is filled with GTP. Binding of GTP induces the α subunit to dissociate from both the receptor and the βγ complex.

The duration of the activation of a G protein is controlled by the α subunit. All α subunits are GTPases, and each hydrolyzes its GTP at a characteristic (slow) rate; when the GTP is hydrolyzed to GDP, the α subunit reassociates with the βγ dimer.

The interaction between receptor and G protein is catalytic. After a G protein has dissociated from an activated receptor, the receptor binds another (inactive) trimer, and the cycle starts again. So one ligand-receptor complex can activate many G protein molecules in a short period, amplifying the original signal.

The most common pathway for the next stage in the pathway, as illustrated in the figure, calls for the activated α subunit to interact with the effector. In the case of G_s, the α_s subunit activates adenylate cyclase; in the case of G_t, the α_t subunit activates cGMP phosphodiesterase. In other cases, however, it is the βγ dimer that interacts with the effector protein. In some cases, *both* the α subunit and the βγ dimer interact with effectors.

Consistent with the idea that it is more often the α subunits that interact with effectors, there are more varieties of α subunits (16 known in mammals) than of β (5) or γ subunits (11). However, irrespective of whether the α or βγ subunits carry the signal, *the common feature in all of these reactions is that a G protein acts upon an effector enzyme that in turn changes the concentration of some small molecule(s) in the cell.*

Figure 26.11 When a receptor is activated by hormone binding, it causes GTP to replace GDP on a Gα subunit. The Gα subunit dissociates from the βγ dimer, and activates an effector such as adenylate cyclase.

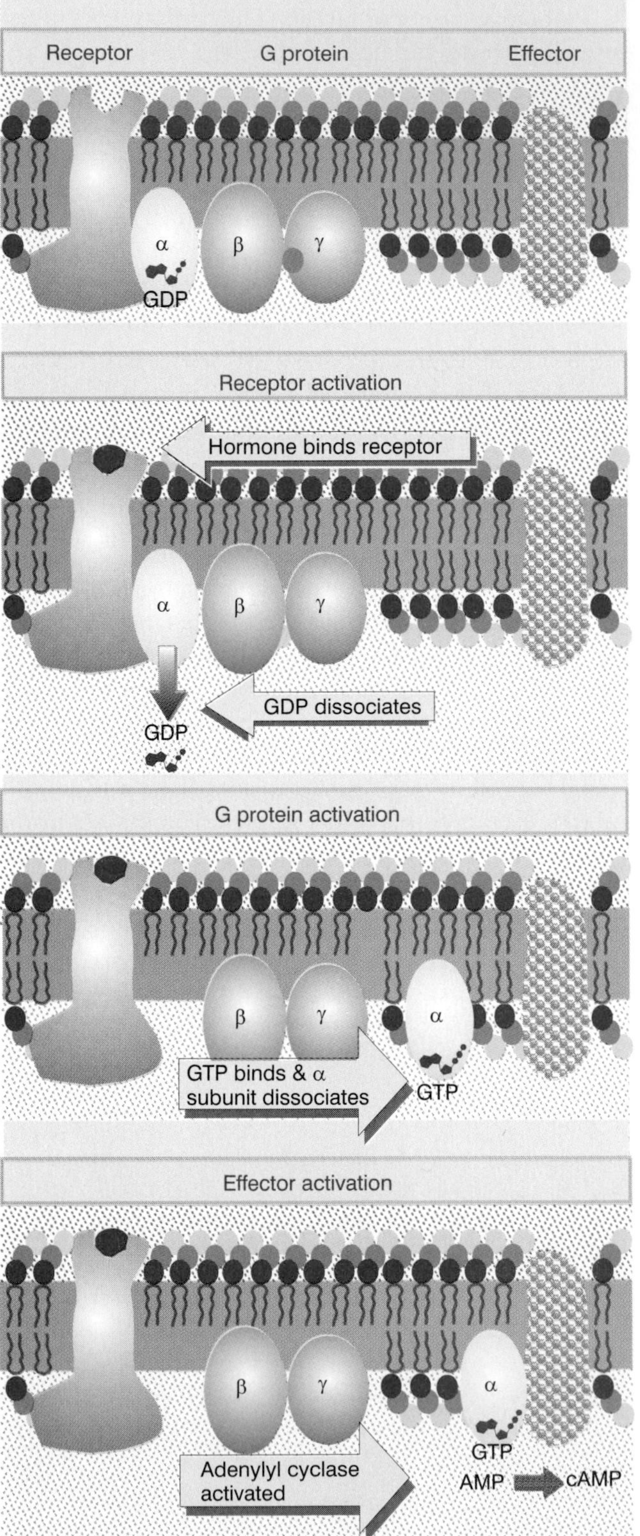

In either the intact or dissociated state, G proteins are associated with the cytoplasmic face of the plasma membrane. But the individual subunits are quite hydrophilic, and none of them appears to have a transmembrane domain. The $\beta\gamma$ dimer has an intrinsic affinity for the membrane because the γ subunit is prenylated. The α_i and α_o types of subunit are myristoylated, which explains their ability to remain associated with the membrane after release from the $\beta\gamma$ dimer. The α_s subunit is palmitoylated.

Because several receptors can activate the same G proteins, and (at least in some cases) a given G protein has more than one effector, we must ask how specificity is controlled. The most common model is to suppose that receptors, G proteins, and effectors all are free to diffuse in the plane of the membrane. In this case, the concentrations of the components of the pathway, and their relative affinities for one another, are the important parameters that regulate its activity. We might imagine that an activated α-GTP subunit scurries along the cytoplasmic face of the membrane from receptor to effector. But it is also possible that the membrane constrains the locations of the proteins, possibly in a way that restricts interactions to local areas. Such compartmentation could allow localized responses to occur.

Protein tyrosine kinases induce phosphorylation cascades

GROWTH factor receptors take their names from the nature of their ligands, which usually are small polypeptides (casually called growth factors, more properly called **cytokines**) that stimulate the growth of particular classes of cells. The factors have a variety of effects, including changes in the uptake of small molecules, initiation or stimulation of the cell cycle, and ultimately cell division. The ligands most usually are secreted from one cell to act upon the receptor of another cell. Examples of secreted cytokines are EGF (epidermal growth factor), PDGF (platelet-derived growth factor), and insulin. In some cases, ligands instead take the form of components of the extracellular matrix, or membrane proteins on the surface of another cell.

The receptors share a general characteristic structure: they are group I integral membrane proteins, spanning the membrane once, with an N-terminal protein domain on the extracellular side of the membrane, and the C-terminal domain on the cytoplasmic side. Some receptors, such as those for EGF or PDGF, consist of single polypeptide chains; others, such as the insulin receptor, are disulfide-bonded dimers (each dimer being a group I protein).

These receptors have a common mode of function. Their cytosolic domains have an enzymatic activity. They are **protein kinases.** There are many types of protein kinases, but they all have some features in common. The basic activity is the ability to add a phosphate group to an amino acid in a target protein. The phosphate is provided by hydrolyzing ATP to ADP. A protein kinase has an ATP-binding site and a catalytic center that can bind to the target amino acid.

Two groups of protein kinases are distinguished by their locations: the receptor protein kinases reside in the membrane; cytosolic protein kinases are free in the cytosol. Each group includes two major types of kinases, defined by the amino acids that they phosphorylate in the protein target:

- **Protein tyrosine kinases** are the predominant type of receptors with kinase activity, although there are also many cytosolic tyrosine kinases. More than 50 receptor tyrosine kinases are known.
- **Protein serine/threonine kinases** are the most common type of cytosolic kinases, and are responsible for the vast majority of phosphorylation events in the cell; there are some receptor kinases of the Ser/Thr type.
- A third type of kinase is found among the cytosolic enzymes: **dual specificity kinases** can phosphorylate target proteins on either tyrosine or serine/threonine.

There are phosphatases with specificity for the appropriate amino acids to match each type of kinase. Most phosphatases are cytosolic, although there are some receptor phosphatases. One way to terminate an activation event is for a phosphatase (typically a cytosolic phosphatase) to reverse the phosphorylation event caused by a receptor kinase.

The effector pathways that are activated by receptor tyrosine kinases (sometimes abbreviated to **RTKs**) fall into two groups:

- *An enzymatic activity is activated that leads to the production of a small molecule second messenger.* The second messenger may be the immediate product of an enzyme that is activated directly by the receptor, or may be produced later in the pathway. Lipids are common second messengers in these pathways. The enzymes include phospholipases (which cleave lipids from larger substrates) and kinases that phosphorylate lipid substrates. Some common pathways are summarized in **Figure 26.12.** The second messengers that are released in each pathway act in the usual way to activate or inactivate target proteins.
- *The effector pathway is a cascade that involves a series of interactions between macromolecular components.* The most common components of such pathways are protein kinases; each kinase activates the next kinase in the pathway by phosphorylating it, and the ultimate kinases in the pathway typically act on proteins such as transcription factors that may have wide-ranging effects upon the cell phenotype.

The basic principle underlying the function of all types of effector pathway is that the signal is amplified as it passes from one component of the pathway to the next. When some components have multiple targets, the pathway branches, thus creating further diversity in the response to the original stimulus.

Receptor tyrosine kinases have some common features. The extracellular domain often has characteristic repeating motifs. It contains a ligand-binding site. The catalytic domain is large (~250 amino acids), and often occupies the bulk of the cytoplasmic region. Certain conserved features are characteristic of all kinase catalytic domains. Sometimes the catalytic domain is broken into two parts by an interruption of some other sequence (which may have an important function in selecting the substrate).

When a ligand binds to the extracellular domain of a growth factor receptor, the catalytic activity of the cytoplasmic domain is activated. *Phosphorylation of tyrosine is identified as the key event by which the growth factor receptors function because mutants in the tyrosine kinase domain are biologically inactive, although they continue to be able to bind ligand.*

A key question in the concept of how a signal is transduced across a membrane is how binding of the ligand to the extracellular domain activates the catalytic domain in the cytoplasm. *The general principle is that a conformational change is induced that affects the overall organization of the receptor.* An important factor in this interaction is that membrane proteins have a restricted ability to diffuse laterally (in contrast with the continuous motion of the lipids in the bilayer). This enables their state of aggregation to be controlled by external events.

Lateral movement plays a key role in transmitting information from one side of the membrane to the

Figure 26.12 Effectors for receptor tyrosine kinases include phospholipases and kinases that act on lipids to generate second messengers.

Effector	Substrate	Products
PLC (phospolipase C) (3 families, PLCα, β, γ)	PIP2 (phosphatidylinositol 4,5-diphosphate)	DAG (diacylglycerol) + IP3 (inositol 1,4,5-triphosphate) DAG activates protein kinase C IP3 mobilizes Ca^{2+}
PLA2 (phospholipase A2)	Phospholipids (phosphatidylcholine, phosphatidylethanolamine, phosphatidylinositol)	Arachidonic acid Converted to prostaglandins & leukotrienes
PI3 kinase (phosphatidylinositol-3 kinase)	Phosphatidyl inositol	PI3 (phosphatidyl inositol-3 phosphate)
PI4 kinase (phosphatidylinositol-4 kinase)	Phosphatidyl inositol	PI4 (phosphatidyl inositol-4 phosphate) Converted to PIP2 (phosphatidyl diphosphate)

other. **Figure 26.13** shows that binding of ligand induces a conformation change in the N-terminal region of a group I receptor that causes the extracellular domains to dimerize. This causes the transmembrane domains to diffuse laterally, bringing the cytoplasmic domains into juxtaposition. The stabilization of contacts between the C-terminal cytosolic domains causes a change in conformation that activates the kinase activity. In some cases, phosphorylation also causes the receptor to interact with proteins present on the cytoplasmic surface of a coated pit, leading to endocytosis of the receptor. An extreme case of lateral diffusion is seen in certain cases of receptor internalization, when receptors of a given type aggregate into a "cap" in response to an extracellular stimulus.

Figure 26.14 shows that the dimerization can take several forms. A ligand binds to one or to both monomers to induce them to dimerize, a dimeric ligand binds to two monomers to bring them together, or a ligand binds to a dimeric receptor (one stabilized by extracellular disulfide bridges) to cause an intramolecular change of conformation. A major consequence of

Figure 26.13 The principle underlying signal transduction by a tyrosine kinase receptor is that ligand binding to the extracellular domain triggers dimerization; this causes a conformational change in the cytoplasmic domain that activates the tyrosine kinase catalytic activity.

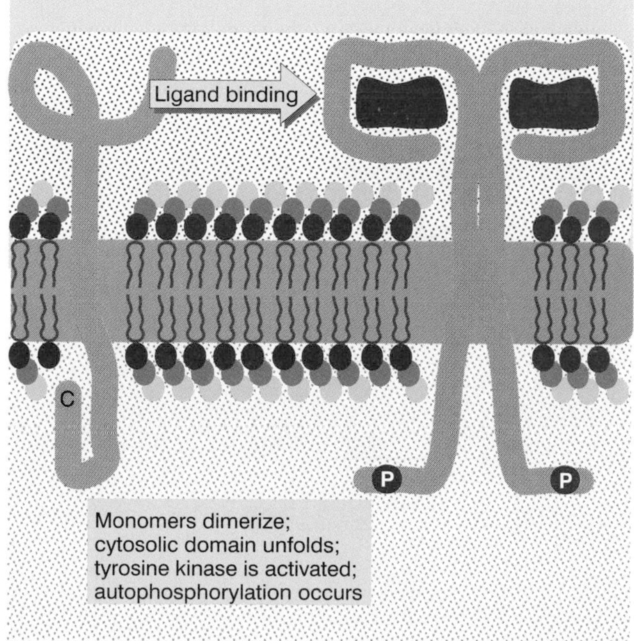
Ligand binding
C
P
P
Monomers dimerize;
cytosolic domain unfolds;
tyrosine kinase is activated;
autophosphorylation occurs

Figure 26.14 Binding of ligand to the extracellular domain can induce aggregation in several ways. The common feature is that this causes new contacts to form between the cytoplasmic domains.

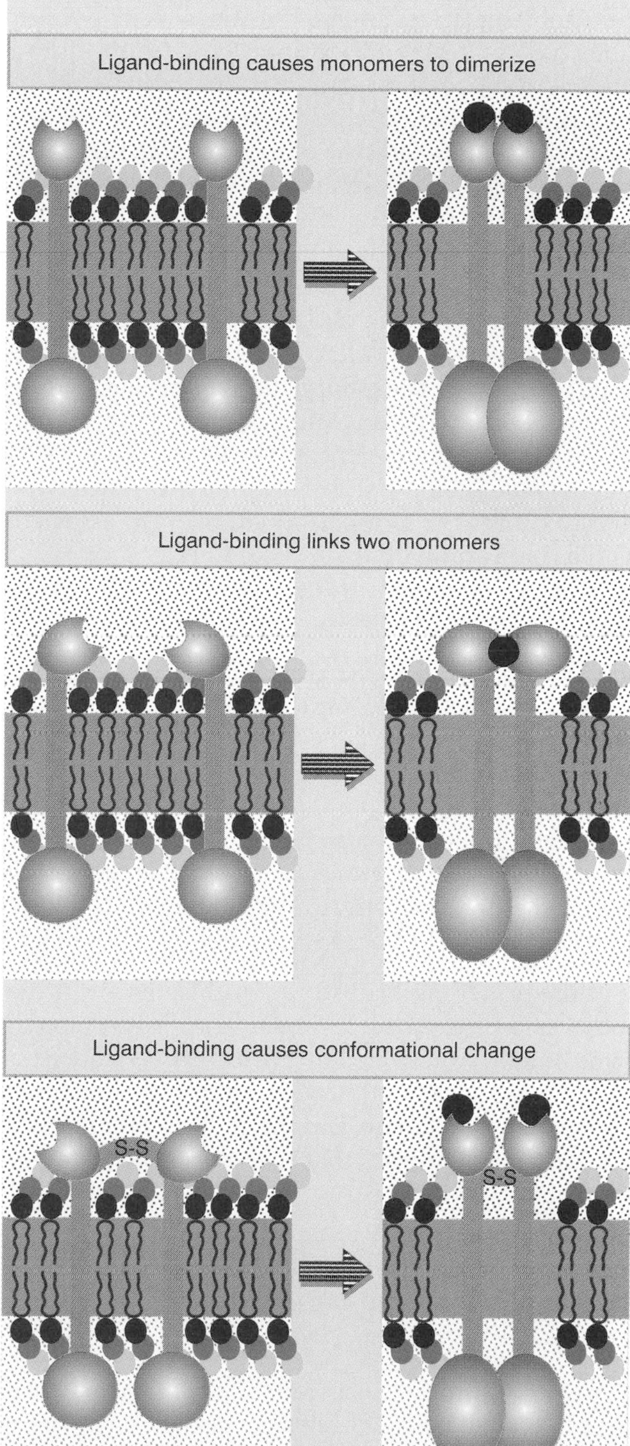
Ligand-binding causes monomers to dimerize
Ligand-binding links two monomers
Ligand-binding causes conformational change
S-S
S-S

this mechanism is to allow transmission of a conformational change from the extracellular domain to the cytoplasmic domain without requiring a change in the structure of the transmembrane region.

The key event that triggers the signaling pathway in the cytoplasm is activation of the kinase activity when dimerization causes an "activation loop" to move into a tethered conformation. This transition causes an **autophosphorylation,** in which the kinase activity of one subunit phosphorylates the other subunit in the dimer. It is necessary for *both* subunits to have kinase activity for the receptor to be activated; if one subunit is defective in kinase activity, the dimer cannot be activated.

Autophosphorylation has two consequences. First, phosphorylation within the kinase region increases the catalytic activity, and therefore comprises a positive feedback. Second, phosphorylation at tyrosine residues elsewhere in the cytoplasmic domain provides the means for passing the signal to the next component in the pathway. The existence of the phosphorylated tyrosine(s) causes the cytoplasmic domain to associate with its target protein(s).

We can distinguish three types of proteins with which the activated receptor may interact:

- The protein may be a *target* that is activated by its association with the receptor, but which is not itself phosphorylated. If the target in turn causes activation of an enzyme, which is usually the case, the pathway continues through an amplification step. Targets may be adaptor molecules which themselves have no catalytic activity (for example, Grb2; see below), or may be enzymes that are activated by binding to the receptor (for example, PI3 kinase; see Figure 26.12).

- If the protein is a substrate for the enzyme, it becomes phosphorylated. If the substrate is itself an enzyme, it may be activated by the phosphorylation (for example, c-Src or PLCγ; see Figure 26.12). Sometimes the substrate is a kinase, and the pathway is continued by a cascade of kinases that successively activate one another.

- Some substrates may be end-targets, such as cytoskeletal proteins, whose phosphorylation changes their properties and causes assembly of a new structure.

Two motifs found in a variety of cytoplasmic proteins that are involved in signal transduction are used to connect proteins to the components that are upstream and downstream of them in a signaling pathway. The domains are named **SH2** and **SH3,** for *S*rc *h*omology, because they were originally described in the c-Src cytosolic tyrosine kinase (see Chapter 28).

Their presence in various proteins is summarized in **Figure 26.15.** The cytoplasmic tyrosine kinases comprise one group of proteins that have these domains; other prominent members are phospholipase Cγ and the regulatory subunit (p85) of PI3 kinase (both targets for activation by receptor tyrosine kinases; see Figure 26.12). The extreme example of a protein with these domains is Grb2/sem5, which consists *solely* of an SH2 domain flanked by two SH3 domains (see below).

The SH2 domain is a region of ~100 amino acids that interacts with a target site in other proteins. The target site is called an **SH2-binding site. Figure 26.16** shows an example of a reaction in which SH2 domains are involved. Activation of a tyrosine kinase receptor causes autophosphorylation of a site in the cytosolic tail. Phosphorylation converts the site into an SH2-binding site. So a protein with a corresponding SH2 domain binds to the receptor only when the receptor is phosphorylated.

An SH2 domain specifically binds to a particular SH2-binding site. The specificity of each SH2 domain is different (except for a group of kinases related to Src, which seem to share the same specificity). The typical SH2-binding site is only 3–5 amino acids long, consisting of a phosphotyrosine and the amino acids on its C-terminal side. SH2 binding is a high-affinity interaction—as much as $10^3\times$ tighter compared to a typical kinase-substrate binding reaction.

Figure 26.15 Several types of proteins involved in signaling have SH2 and SH3 domains.

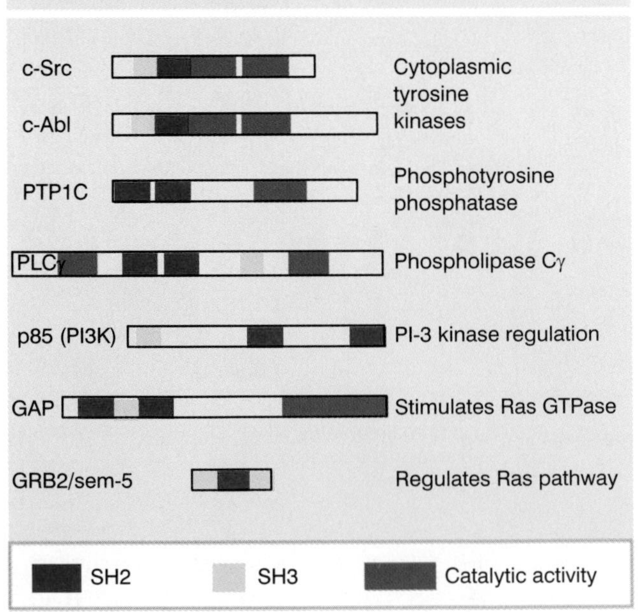

Figure 26.16 Phosphorylation of tyrosine in an SH2-binding domain creates a binding site for a protein that has an SH2 domain.

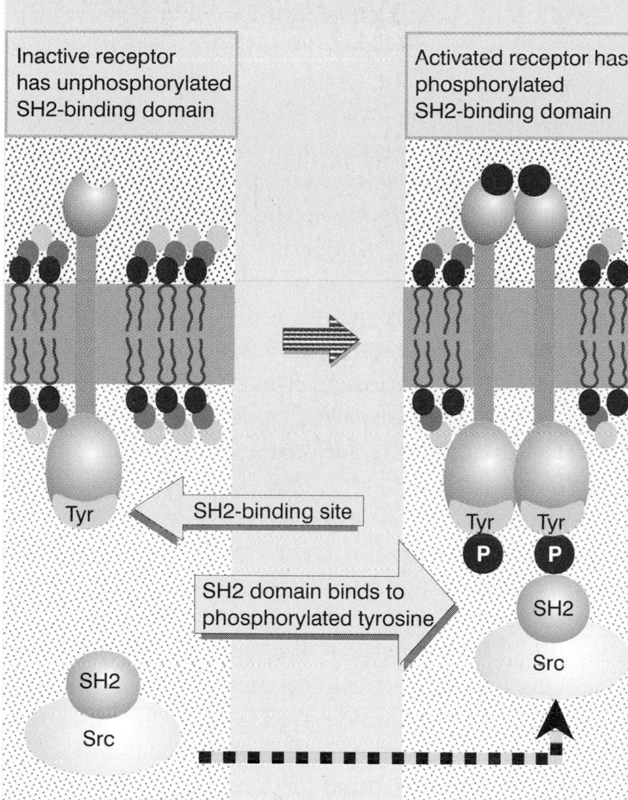

Some proteins contain multiple SH2 domains, which increases their affinity for binding to phosphoproteins or confers the ability to bind to different phosphoproteins. A receptor may contain different SH2-binding sites, enabling it to activate a variety of target proteins. **Figure 26.17** summarizes the organization of the cytoplasmic domain of the PDGF receptor, which has ~10 distinct SH2-binding sites, each created by a different phosphorylation event. Different pathways may be triggered by the proteins that bind to the various phosphorylated residues.

The SH2 domain has a globular structure in which its N-terminal and C-terminal ends are close together, so that its structure is relatively independent of the rest of the protein. The phosphotyrosine binds to a pocket in the SH2 domain, as illustrated in **Figure 26.18.**

A protein that contains an SH2 domain is activated when it binds to an SH2-binding site. The reaction can take the form of activating enzymatic activities, typically kinases, phosphatases, and phospholipases. The activation may involve the SH2-containing protein directly (when it itself has enzymatic activity) or may be

Figure 26.17 Autophosphorylation of the cytosolic domain of the PDGF receptor creates SH2-binding sites for several proteins. Some sites can bind more than one type of SH2 domain. Some SH2-containing proteins can bind to more than one site. The kinase domain consists of two separated regions (shown in blue), and is activated by the phosphorylation site in it.

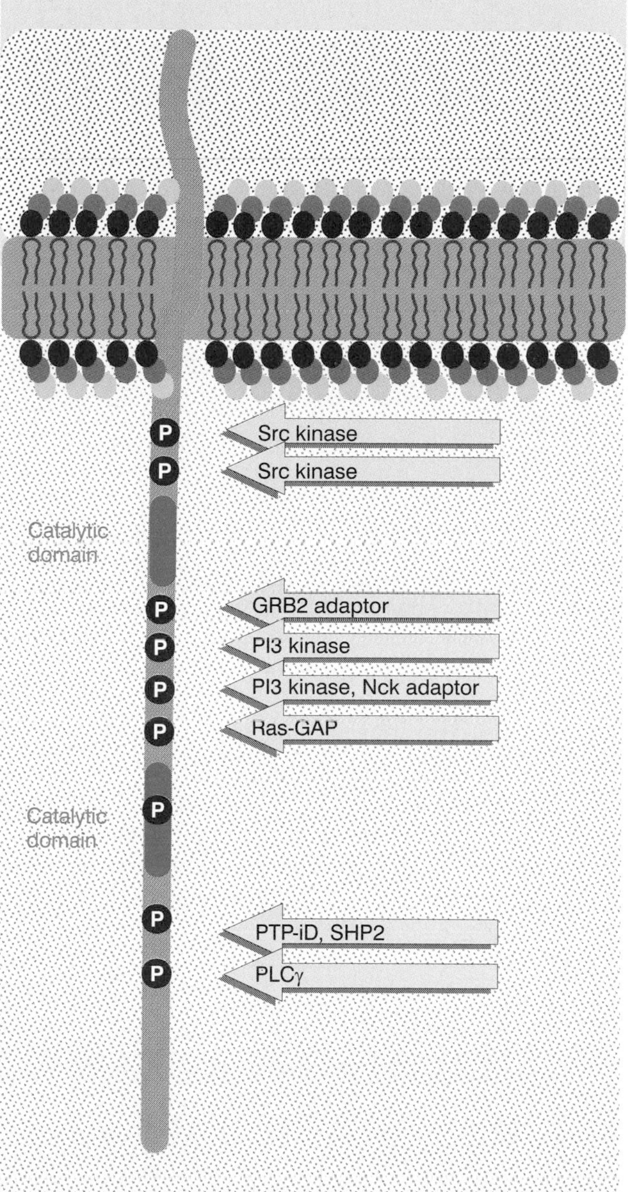

indirect. An example of a protein containing an SH2 domain that does not have a catalytic activity is provided by p85, the regulatory subunit of PI3 kinase; when p85 binds to a receptor, it is the associated PI3K catalytic subunit that is activated.

The SH3 domain provides the effector function by which some of the SH2-containing proteins bind to a

Figure 26.18 The crystal structure of an SH2 domain (purple strands) bound to a peptide containing phosphotyrosine shows that the P-Tyr (white) fits into the SH2 domain, and the 4 C-terminal amino acids in the peptide (backbone yellow, side-chains green) also make contact. Photograph kindly provided by John Kuriyan.

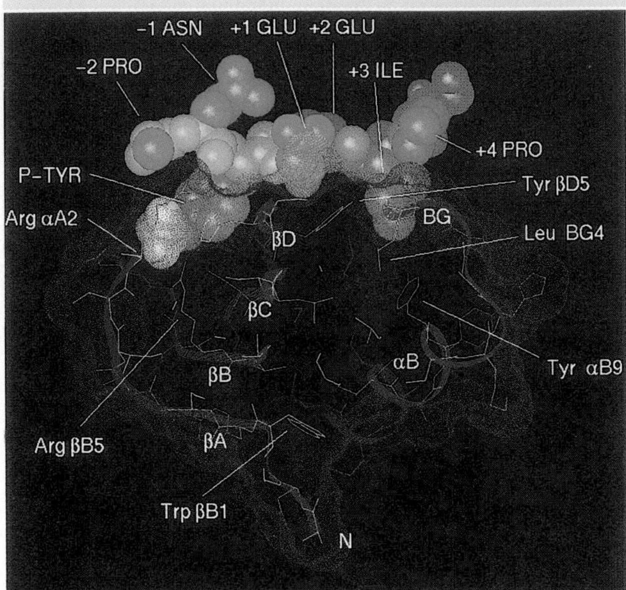

downstream component. The case of the "adaptor" Grb2 strengthens this idea; consisting only of SH2 and SH3 domains, it uses the SH2 domain to contact the component upstream in the pathway, and the SH3 domain to contact the component downstream. SH3 binds the motif PXXP in a sequence-specific manner. It is the PXXP-containing target for Grb2 that is activated when Grb2 binds to the activated receptor.

A receptor tyrosine kinase can initiate a signaling cascade at the membrane. However, in many cases, the activation of the kinase is followed by its internalization, that is, it is removed from the membrane and transported to the interior of the cell by endocytosis of a vesicle carrying a patch of plasma membrane. The relationship between kinase activity and endocytosis is unclear. Phosphorylation at particular residues may be needed for endocytosis; whether the kinase activity as such is needed may differ for various receptors. It is possible that endocytosis of receptor kinases serves principally to clear receptor (and ligand) from the surface following the response to ligand binding (thus terminating the response). However, in some cases, movement of receptors to coated pits followed by internalization could be necessary for them to act on the target proteins.

Because growth factor receptors generate signals that lead to cell division, their activation in the wrong circumstances is potentially damaging to an organism, and can lead to uncontrolled growth of cells. Many of the growth factor receptor genes are represented in the **oncogenes,** a class of mutant genes active in cancers. The mutant genes are derived by changes in cellular genes; often the mutant protein is truncated in either or both of its N-terminal or C-terminal regions. The mutant protein usually displays two properties: the tyrosine kinase has been activated; and there is no longer any response to the usual ligand. As a result, the tyrosine kinase activity of the receptor is either increased or directed against new targets. The nature of these changes in generating tumorigenic phenotypes in cells is the subject of Chapter 28.

The Ras/MAPK pathway

THE best characterized pathway that is initiated by receptor tyrosine kinases passes through the activation of a monomeric GTP-binding protein to activate a cascade of cytosolic kinases. Although there are still some gaps in the pathway to fill in, and branches that have not yet been identified, the broad outline is clear, as illustrated in **Figure 26.19.** In mammalian cells, the cascade is often initiated by activation of a tyrosine kinase receptor, such as the EGF or PDGF receptors. The receptor activates the Ras pathway by means of an "adaptor" protein. The activation of Ras leads to the activation of the Raf Ser/Thr kinase, which in turn activates the kinase MEK (formerly known as MAP kinase kinase); its name reflects the fact that it is the kinase that phosphorylates, and thereby activates, a MAP kinase. The name of the family of MAP kinases reflects

Figure 26.19 Autophosphorylation triggers the kinase activity of the cytoplasmic domain of a receptor. The target protein may be recognized by an SH2 domain. The signal may subsequently be passed along a cascade of kinases.

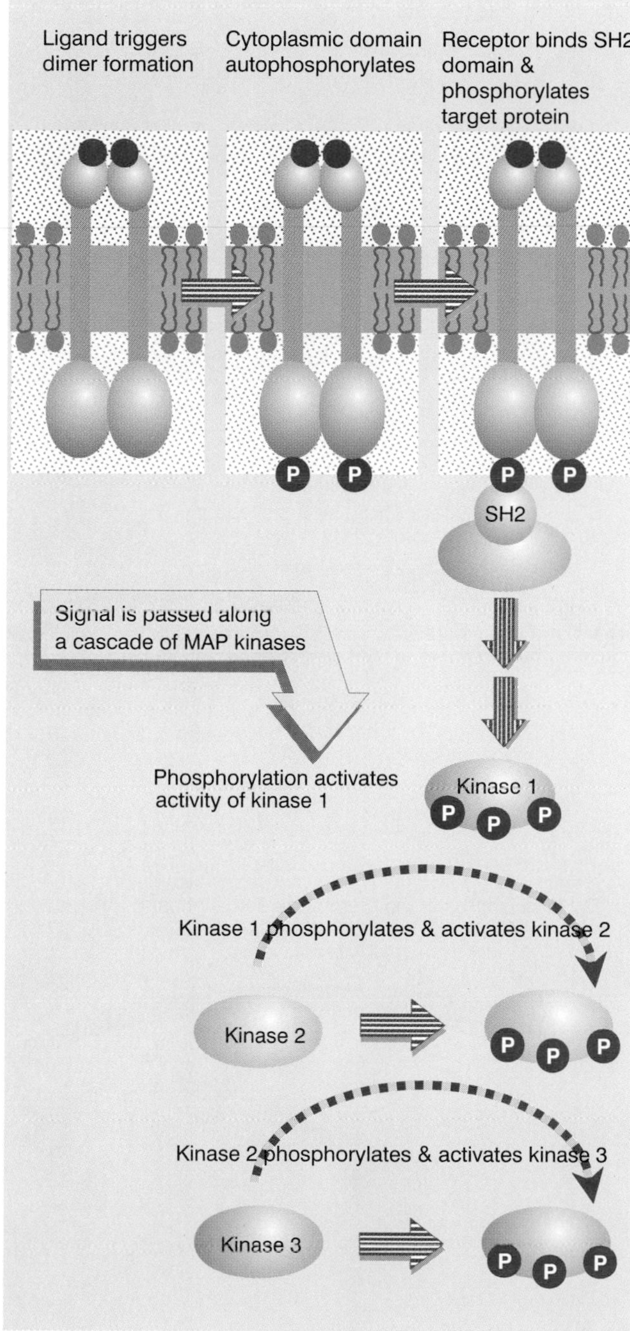

Ligand triggers dimer formation
Cytoplasmic domain autophosphorylates
Receptor binds SH2 domain & phosphorylates target protein
P
P
P
P
SH2
Signal is passed along a cascade of MAP kinases
Phosphorylation activates activity of kinase 1
Kinase 1
P
P
P
Kinase 1 phosphorylates & activates kinase 2
Kinase 2
P
P
P
Kinase 2 phosphorylates & activates kinase 3
Kinase 3
P
P
P

their identification as mitogen-activated kinases; one MAPK family has also been called ERKs, for extracellular signal-regulated. Some major effects of Ras are conveyed via this pathway, but there is also a branch at Ras, which involves the activation of other monomeric G proteins.

The cascade from MEK to the end products is sometimes known as the MAP kinase pathway. Each kinase in this part of the cascade phosphorylates its target kinase, and the phosphorylation event activates the kinase activity of the target enzyme, as illustrated in **Figure 26.20.** The cascade of phosphorylation events leads ultimately to the phosphorylation of transcription factors that trigger changes in cell phenotype varying from growth to differentiation, depending on the cell type. Other targets for the kinases include cytoskeletal proteins that may directly influence cell structure.

The relationship between components of the pathway can be tested by investigating the effects of one component upon the action of another. For example, a mutation that inactivates one component should make it impossible for the pathway to be activated by any components that act earlier. Using such tests allows components to be ordered in a pathway, and to determine whether one component is upstream or downstream of another.

The pathway has been characterized in several situations (see Figure 26.30): in terms of biochemical components responsible for growth of mammalian cultured cells, as the pathway involved in eye development in the fly *D. melanogaster,* as the pathway of vulval development in the worm *C. elegans,* and as the response to mating in the yeast *S. cerevisiae.*

The striking feature is that the pathway is activated by different means in each case (appropriate to the individual system), and it has different end effects in each system, but many of the intermediate components can be recognized as playing analogous roles. It is much as though Nature has developed a signal transduction cascade that can be employed wholesale by means of connecting the beginning to an appropriate stimulus and the end to an appropriate effector. The total pathway is sometimes known as the Ras pathway (named after one of the earlier components) or the Ras/MAPK pathway. Several of the components of this pathway in mammals are related to oncogenes, which suggests that the aberrant activation of this pathway at any one of various stages has a powerful potential to cause tumors.

Figure 26.21 shows how the events initiating the cascade occur at the plasma membrane. The activated receptor tyrosine kinase associates with the adaptor protein Grb2, which binds to the receptor but is not phosphorylated. Grb2 binds to the protein SOS, and the activation of Grb2 activates SOS, which then

Figure 26.20 A common signal transduction cascade passes from a receptor tyrosine kinase through an adaptor to activate Ras, which triggers a series of Ser/Thr phosphorylation events. Finally, activated MAP kinases enter the nucleus and phosphorylate transcription factors. Missing components are indicated by successive arrows.

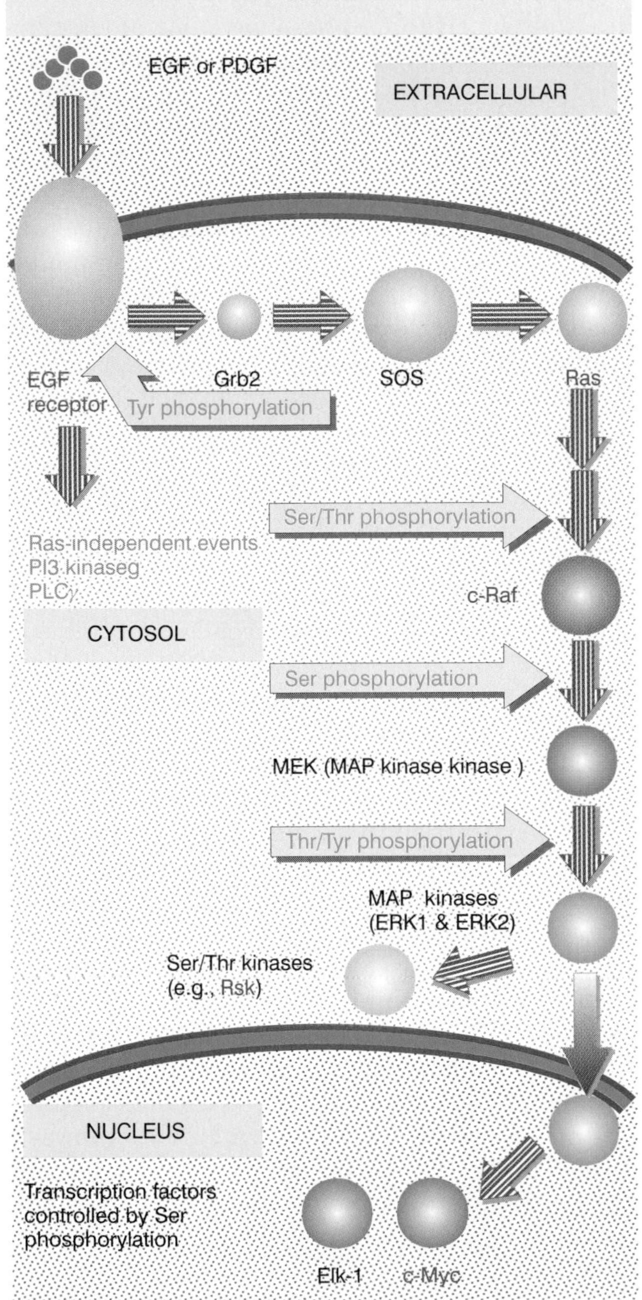

Figure 26.21 The Ras cascade is initiated by a series of activation events that occur on the cytoplasmic face of the plasma membrane.

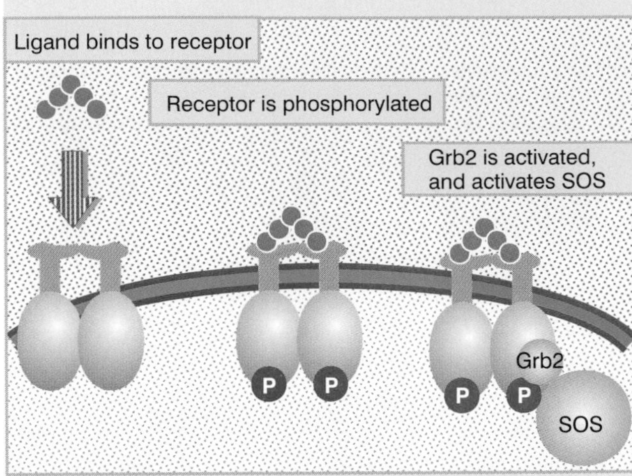

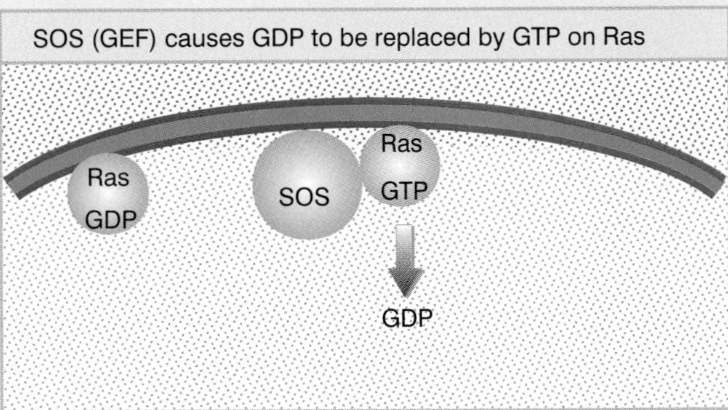

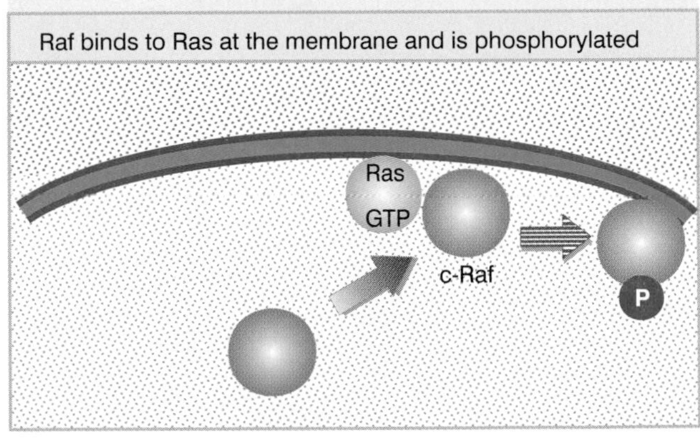

activates Ras. In fact, the sole role of Grb2 in activating SOS appears to be fulfilled by binding to it. The binding reaction brings SOS to the membrane, and thus into the vicinity of Ras, which is sufficient to activate Ras. (Grb2 is not the only adaptor that can activate Ras; an alternative pathway is provided by the adaptor SHC. Which adaptor is used depends on the cell type.)

Grb2 uses its SH3 domain to contact SOS. SH3 domains may be used for other protein–protein interactions. In particular, they may provide connections to small GTP-binding proteins (of which Ras is the paradigm). Another role that has been proposed for SH3 domains (and in particular for the SH3 domain of c-Src) is the ability to interact with proteins of the cytoskeleton, thus triggering changes in cell structure.

When a tyrosine kinase receptor is activated, its intracellular domain may be phosphorylated at more than one site, and each site may trigger a different pathway (see Figure 26.17). The most common consequence of a phosphorylation is to activate a signal transduction pathway, but in some cases it may have a negative effect, providing a feedback loop to limit the action of the pathway. These effects may be direct or indirect. **Figure 26.22** illustrates an example of a system in which two phosphorylations counteract each other. Torso is a receptor tyrosine kinase that activates the Ras pathway during *Drosophila* embryogenesis. Two sites become phosphorylated when it is activated. Phosphorylation of Y630 is required to activate the downstream pathway. Phosphorylation of Y918 provides negative regulation.

The receptor binds the regulator RasGAP to the phosphorylated site Y918. This keeps RasGAP in an activated state in which it prevents Ras from functioning (the interaction between RasGAP and Ras is discussed in more detail below). Y918 is phosphorylated constitutively (or when Torso is activated at a low level). Under these circumstances, the pathway is turned off.

High activation of Torso results in phosphorylation of Y630. This creates a binding site for the cytosolic phosphatase corkscrew (CSW). Corkscrew then dephosphorylates Y918. The result is to release RasGAP, which becomes ineffective, allowing Ras to function. So corkscrew is required for Torso to activate Ras.

Corkscrew may have a second role in the pathway, which is to recruit the adaptor that in turn binds to SOS, which activates Ras.

We see from this example that phosphorylated sites may influence the signaling pathway positively or negatively. The state of one phosphorylated site may in fact control the state of another site. An activating site may act indirectly (to inactive an inactivating site) as well as directly to recruit components of the signaling pathway.

We turn now to the events involved in activating Ras. Ras is an example of a monomeric guanine nucleotide binding protein. Other examples are found in protein trafficking (such as the Rabs) or in protein synthesis (such as EF-Tu). The general principles by which such proteins are controlled are illustrated in **Figure 26.23.**

Figure 26.22 Phosphorylation at different sites on a receptor tyrosine kinase may either activate or inactivate the signal transduction pathway.

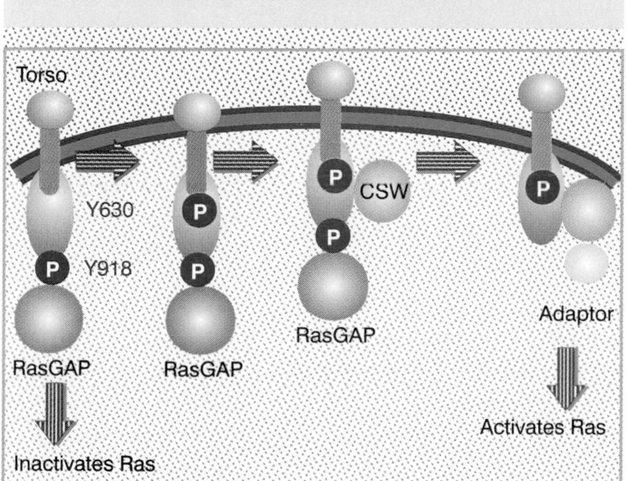

Figure 26.23 Monomeric G proteins are active when bound to GTP and inactive when bound to GDP. Their activity is controlled by other proteins; inactivating functions are shown in blue, and activating functions are shown in red.

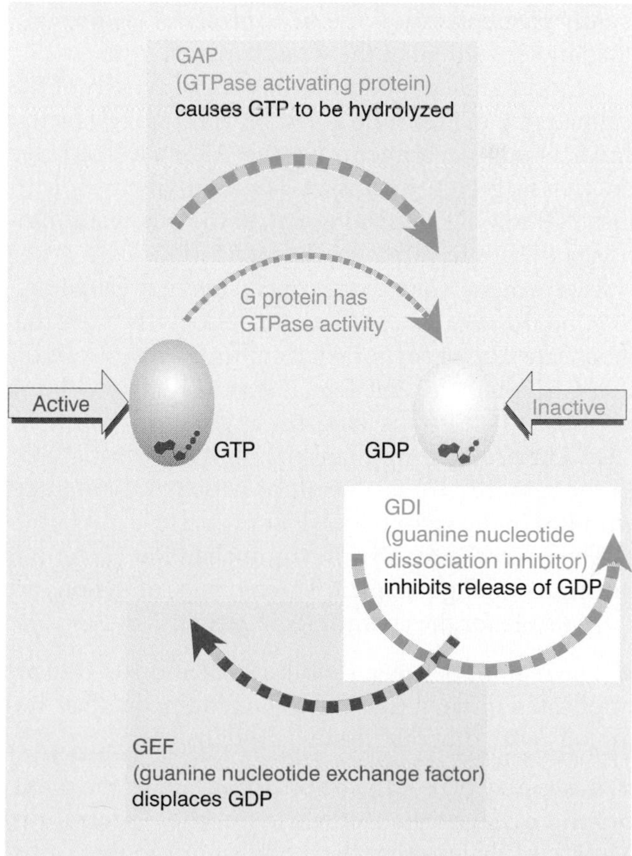

The activity of the G protein depends on whether it is bound to GTP (active state) or to GDP (inactive state). Like trimeric G proteins, a monomeric G protein possesses an intrinsic GTPase activity that converts it from the active state to the inactive state. Three types of protein either stimulate or reverse this reaction:

- The GTPase activity can be stimulated by a GAP (GTPase activating protein); thus *GAP activity inactivates the G protein.* Different GAPs have specificities for different GTP-binding proteins; they are typically named as *Protein*-GAP, where *Protein* is the substrate on which they act.
- The GDP bound to an inactive G protein can be displaced by reaction with a GEF (guanine nucleotide exchange factor). Release of the GDP creates an empty site. Because the concentration of GTP in the cytosol is greater than that of GDP, the site is then filled with GTP. *The overall result is to replace GDP with GTP, thus activating the protein.* GEFs have the same sort of specificity as GAPs, and similarly are named in the form *Protein*-GEF.
- The displacement reaction can be blocked by a GDI (guanine nucleotide dissociation inhibitor), *thus maintaining the G protein in the inactive state.*

Any particular monomeric G protein may be regulated by some or all of these mechanisms.

A more detailed model for the cycle of activity of Ras is illustrated in **Figure 26.24.** Ras carrying GTP is active and acts upon its target molecule. After Ras has acted upon its target, the intrinsic GTPase activity hydrolyzes the GTP to GDP, returning Ras to the inactive condition, and thus terminating the interaction.

Figure 26.24 The relative amounts of Ras-GTP and Ras-GDP are controlled by two proteins. Ras-GAP inactivates Ras by stimulating hydrolysis of GTP. SOS (GEF) activates Ras by stimulating replacement of GDP by GTP, and is responsible for recycling of Ras after it has been inactivated.

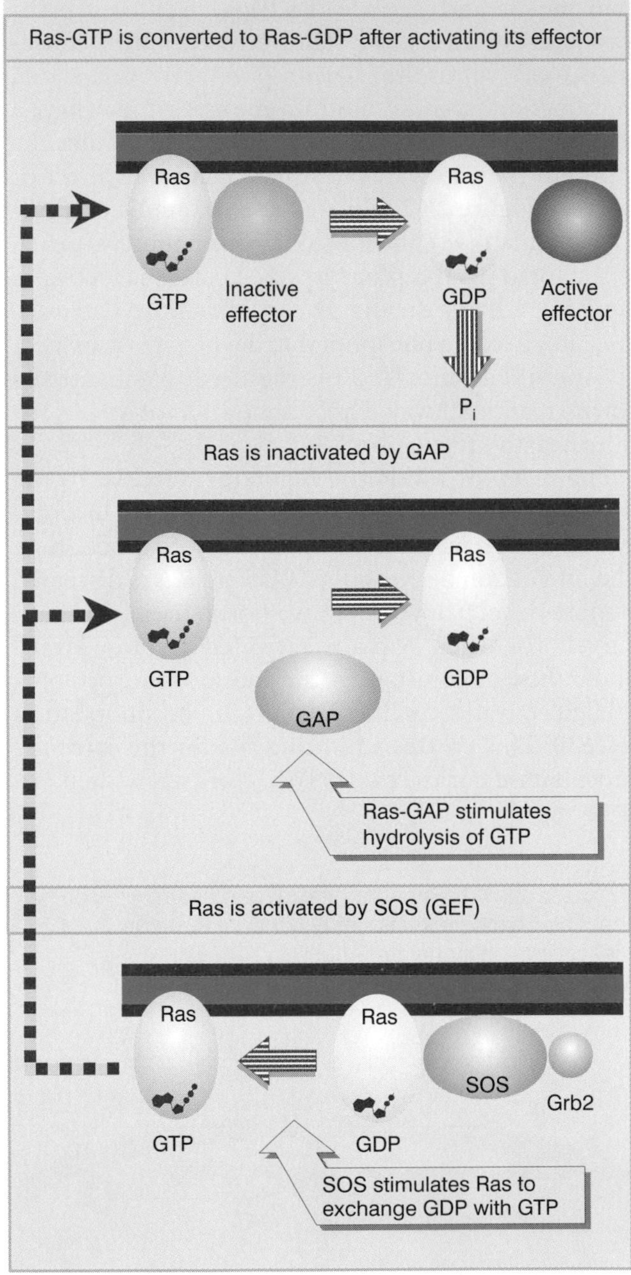

Two proteins control the conversion between the active and inactive states of Ras. Ras-GAP is the GAP that inactivates Ras in mammalian cultured cells. SOS is the Ras-GEF that activates Ras. It is activated by Grb2 in response to phosphorylation of a receptor tyrosine kinase. The probable mechanism is that Grb2 brings SOS to the membrane, as a result of which SOS interacts with Ras in the vicinity (see Figure 26.21).

The general structure of mammalian Ras proteins is illustrated in **Figure 26.25.** Three groups of regions are responsible for the characteristic activities of Ras:

- The regions between residues 5–22 and 109–120 are implicated in guanine nucleotide binding by their homology with other G-binding proteins.
- Ras is attached to the cytoplasmic face of the membrane by farnesylation close to the C-terminus. Mutations that prevent the modification abolish oncogenicity, showing that membrane location is important for Ras function. After the farnesylation, the three C-terminal amino acids are cleaved from the protein, and the carboxyl group of the (now C-terminal) Cys^{186} is methylated; also, other Cys residues in the vicinity are

Figure 26.25 Discrete domains of Ras proteins are responsible for guanine nucleotide binding, effector function, and membrane attachment.

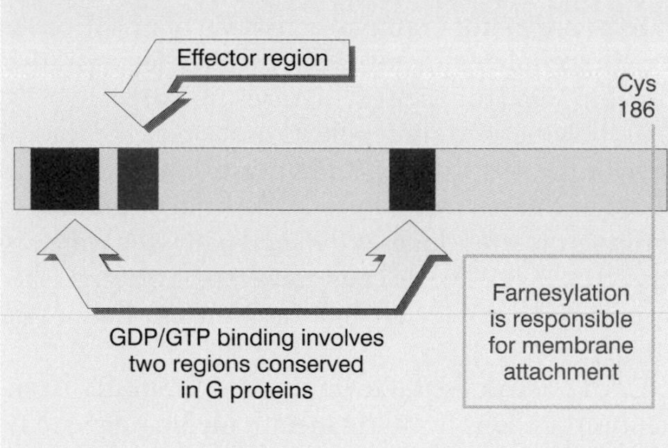

reversibly palmitoylated. These changes further increase affinity for the membrane.

- The effector domain (residues 30–40) is the region that reacts with the target molecule when Ras has been activated. This region is required for the oncogenic activity of Ras proteins that have been activated by mutation at position 12. The same region is required for the interaction with Ras-GAP.

The crystal structure of Ras protein is illustrated schematically in **Figure 26.26.** The regions close to the guanine nucleotide include the domains that are conserved in other GTP-binding proteins. The potential effector loop is located near the phosphates; it consists of hydrophilic residues, and is potentially exposed in the cytoplasm.

When GTP is hydrolyzed, there is a switch in the conformation of Ras protein. The change involves L4, which includes position 61, at which some oncogenic

Figure 26.26 The crystal structure of Ras protein has 6 β strands, 4 α helices, and 9 connecting loops. The GTP is bound by a pocket generated by loops L9, L7, L2, and L1. The effector region from 30 to 40 is relatively well exposed, but so are other regions.

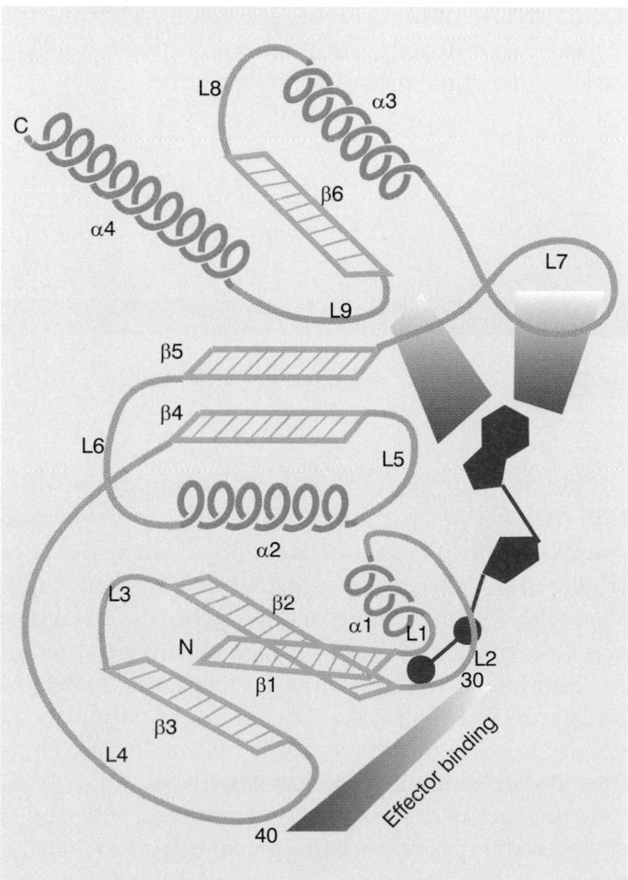

Figure 26.27 Changes in cell structure that occur during growth or transformation are mediated via monomeric G proteins.

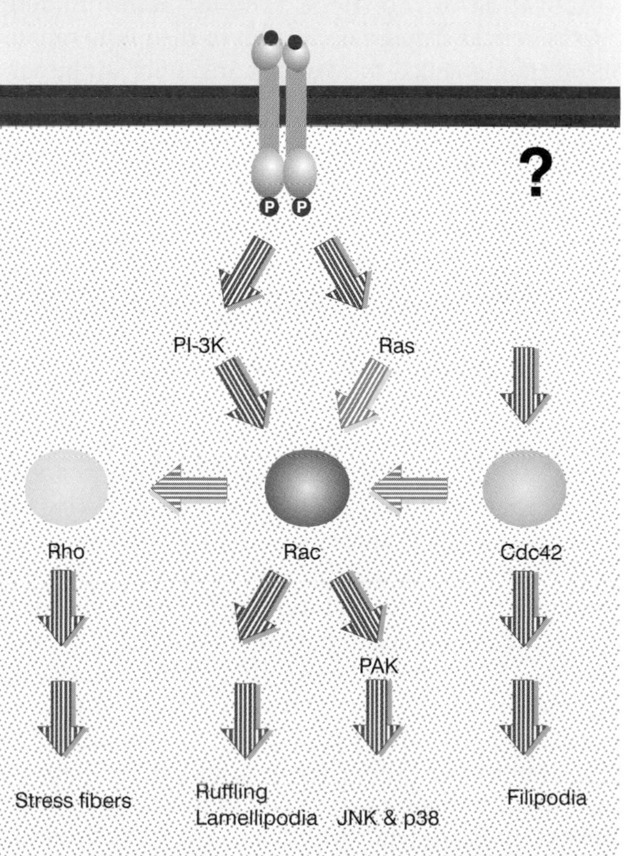

mutations occur. Mutations that activate Ras constitutively (these are oncogenic as discussed in Chapter 28) occur at position 12 in loop 1, and directly affect binding to GTP. The changes between the wild-type and oncogenic forms are restricted to these regions, and impede the ability of the mutant Ras to make the conformational switch on GTP hydrolysis. The primary basis for the oncogenic property, therefore, lies in the reduced ability to hydrolyze GTP.

When mitogenesis is triggered by activation of a growth factor, or when a cell is transformed into the tumorigenic state (see Chapter 28), there is a series of coordinated events, including changes in transcription and changes in cell structure. Activation of the Ras/MAPK pathway activates transcription factors that are responsible for one important set of changes. Other changes are triggered by the activation of a group of monomeric G proteins. Each member of this group is responsible for particular types of structural change, as summarized in **Figure 26.27.**

The complete set of relationships that activates these factors is not known, but these structural changes generally can occur independently of the activation of Ras, suggesting that there are other pathways from growth factor receptors to the other monomeric G proteins.

Activation of rho triggers the formation of actin stress fibers and their connection to the plasma membrane at sites called focal adhesions. Rho can be activated in response to addition of the lipid LPA (a component of serum), through activation of growth factors.

Rac can be activated by the activation of PI-3 kinase (a kinase that phosphorylates a small lipid messenger) in response to (for example) activation of PDGF receptor. It stimulates membrane ruffling, formation of lamellipodia (transient structures that are driven by actin polymerization/depolymerization at the leading edge of the membrane), and progression into the G1 phase of the cell cycle. By an independent pathway it activates the stress kinases JNK and p38 (which are discussed below). (The two pathways can be distinguished by mutations in Rac that fail to activate one but not the other.)

Cdc42 activates the formation of filopodia (transient protrusions from the membrane that depend on actin polymerization), although we do not yet know in detail the pathway by which it is itself activated.

There is some crosstalk between these pathways, both laterally and vertically, as shown by the (grey) arrows in Figure 26.27. Rac can be activated by Ras. And Cdc42 can activate Rac, which in turn can activate Rho. This may help the coordination of the events they control; for example, lamellipodia often form along the membrane between two filopodia (making a web-like structure). All of these events are necessary for the full response to mitogenic stimulation, implying the activation of multiple monomeric G proteins.

Activating MAP kinase pathways

SIGNAL transduction generates differential responses to stimuli that vary qualitatively (by activating different pathways) or quantitatively (by activating pathways with different intensities or for different durations). An individual stimulus may activate one or more pathways. The strength of activation of any particular pathway may influence the response, since there are cases in which more intense or long-term stimulation of a single pathway gives a different response from less intense or short-term stimulation. One of our major aims is to understand how differences in such stimuli are transduced into the typical cellular responses.

One of the important features of signal transduction pathways is that they both diverge and converge, thus allowing different but overlapping responses to be triggered in different circumstances. Divergence may start with the initiating event. Activation of a receptor tyrosine kinase may itself trigger multiple pathways: for example, activation of EGF receptor activates the Ras pathway and also Ras-independent pathways involving second messengers (see Figure 26.20). There may also be "branches" later in a pathway.

Convergence of pathways is illustrated by the ability of different types of initiating signal to lead to the activation of MAP kinases. The original paradigm and best characterized example of a pathway leading to MAP kinase involves the activation of Ras, as summarized in Figure 26.19. Returning to the early events in this pathway, the next component after Ras is the Ser/Thr (cytosolic) kinase, Raf. The relationship between Ras and Raf has been puzzling. We know that Ras and Raf are on the same pathway, because both of them are required for the phosphorylation of the proteins later in the pathway (such as MAP kinase). Ras must be upstream of Raf because it is required for the activation of Raf in response to extracellular ligands. Similarly, Raf must be downstream of Ras because the pathway triggered by Ras can be suppressed by expression of a dominant-negative (kinase deficient) mutant of Raf. Ras is localized on the cytoplasmic side of the plasma membrane, and its activation results in binding of Raf, which as a result is itself brought to the vicinity of the plasma membrane. However, the events that then activate Raf, and in particular the kinase that phosphorylates it, are not yet known. The present model is that Ras activates Raf indirectly, perhaps because some kinase associated with the membrane is constitutively active (see Figure 26.21). The importance of localization of enzymatic activities is emphasized by the abilities of components both upstream and downstream of Ras (that is, SOS and Raf) to exercise their activating functions as a consequence of being brought from the cytosol to the plasma membrane.

Raf activity leads to the activation of MEK. Raf directly phosphorylates MEK, which is activated by phosphorylation on two serine residues. MEK is an unusual enzyme with dual specificity, which can phosphorylate both threonine and tyrosine. Its target is the ERK MAP kinase.

Both types of phosphorylation are necessary to convert a MAP kinase into the active state. There are at least 3 MAP kinase families, and they provide important switching points in their pathways. They are activated in response to a wide variety of stimuli, including stimulation of cell growth, differentiation, etc., and appear to play central roles in controlling changes in cell phenotype. The MAP kinases are serine/threonine kinases. After this point in the pathway, all the activating events take the form of serine/threonine phosphorylations.

The ultimate effect of the MAP kinase pathway is a change in the pattern of transcription. So the initiating event occurs at the cell surface, but the final readout occurs in the nucleus, where transcription factors are activated (or inactivated). This type of response requires a nuclear localization step. General possibilities for this step are illustrated in **Figure 26.28.** In the classic MAP kinase pathway, it is accomplished by the movement of a MAP kinase itself to the nucleus, where it phosphorylates target transcription factors. An alternative pathway is to phosphorylate a cytoplasmic factor; this may be a transcription factor that then moves to the nucleus or a protein that regulates a transcription factor (for example, by releasing it to go to the nucleus).

The MAP kinases have several targets, including

Figure 26.28 A signal transduction cascade passes to the nucleus by translocation of a component of the pathway or of a transcription factor. The factor may translocate directly as a result of phosphorylation or may be released when an inhibitor is phosphorylated.

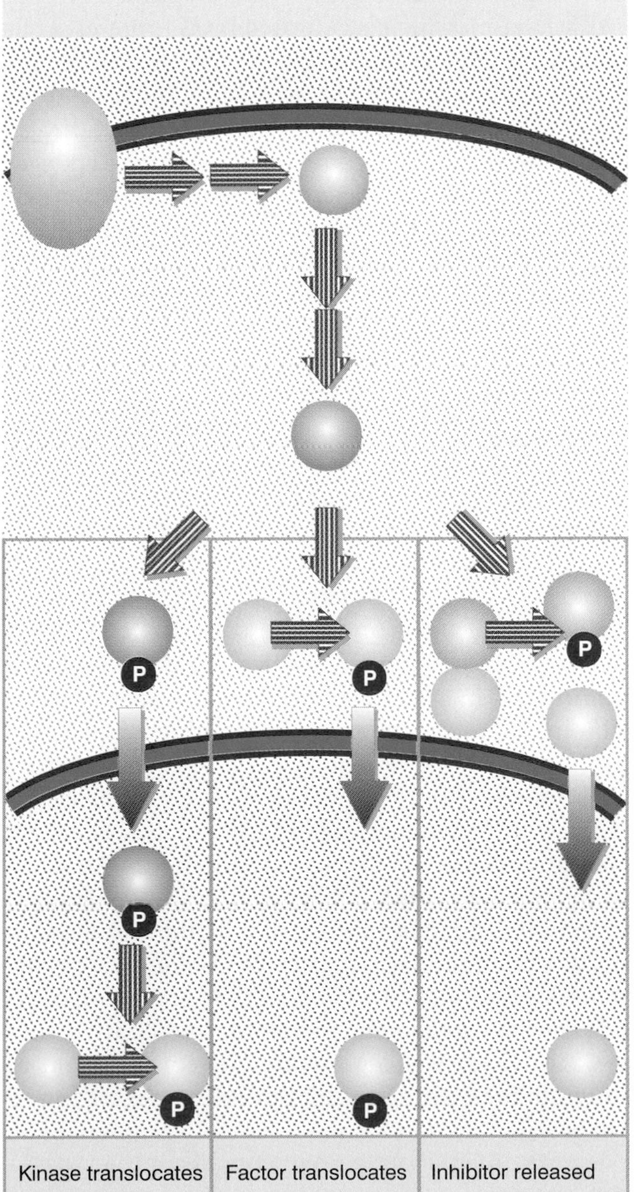

other kinases, such as Rsk, which extend the cascade along various branches. The ability of some MAP kinases to translocate into the nucleus after activation extends the range of substrates. In the classic pathway, ERK1 and ERK2 are the targets of MEK, and ERK2 translocates into the nucleus after phosphorylation. The direct end of one branch of the cascade is provided by the phosphorylation of transcription factors, including c-Myc and Elk-1 (which cooperates with SRF [serum response factor]). This enables the cascade to regulate the activity of a wide variety of genes. (The important transcription factor c-Jun is phosphorylated by another MAPkinase, called JNK; see later.)

In the MAP kinase pathway, MEK provides a convergence point. Ras activates Raf, which in turn activates MEK. Another kinase that can activate MEK is MEKK (MEK kinase), which is activated by G proteins, as illustrated in **Figure 26.29.** (We have not identified the component(s) that link the activated G protein to the MEKK.) So two principal types of stimulus at the cell surface—activation of receptor tyrosine kinases or of trimeric G proteins—both can activate the MAP kinase cascade. Formally, Raf and MEKK provide analogous functions in parallel pathways.

Figure 26.29 Pathways activated by receptor tyrosine kinases and by serpentine receptors converge upon MEK.

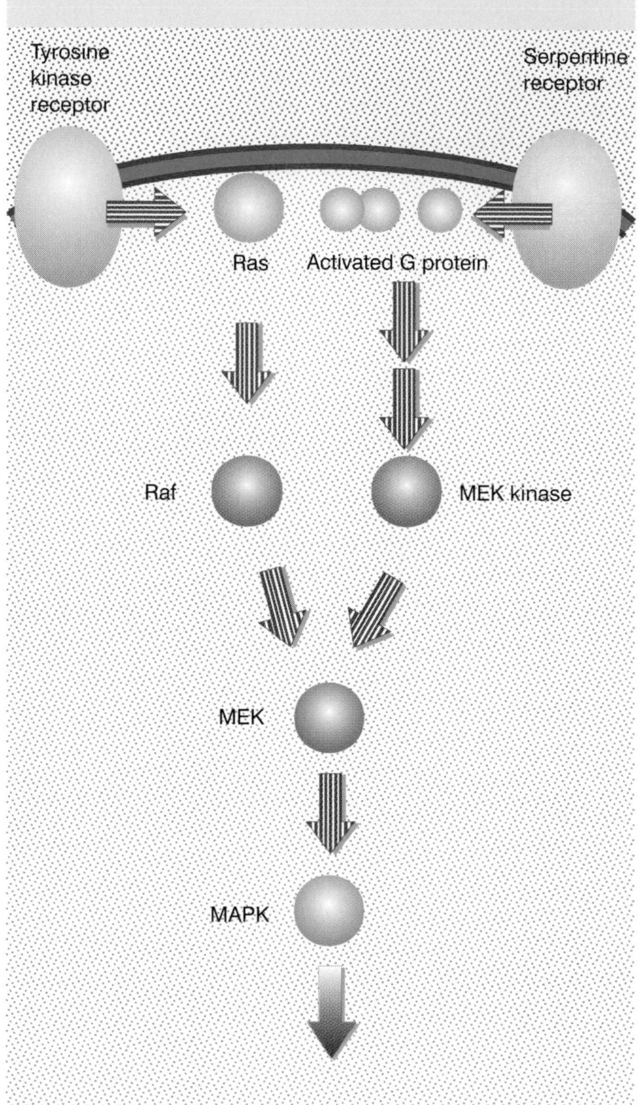

The counterparts for the components of the pathways in several organisms are summarized in **Figure 26.30.**

In mammals, fly, and worm, it starts by the activation of a receptor tyrosine kinase; in mammals the ligand is a polypeptide growth factor, in *D. melanogaster* retina it is a surface transmembrane protein on an adjacent cell (a "counter-receptor"), and in *C. elegans* vulval induction it is not known. The pathway continues through Grb2 in mammals, and through close homologs in the worm and fly. At the next stage, a homolog of SOS functions in the fly in the same way as in mammals. The pathway continues through Ras-like proteins (that is, monomeric guanine nucleotide-binding proteins) in all three higher eukaryotes. Mutations in a homolog of GAP also may influence the pathway in *D. melanogaster,* suggesting that there are alternative regulatory circuits, at least in flies. An interesting feature is that, although the Ras-dependent pathway is utilized in a variety of cells, the mutations in the SOS and GAP functions in *Drosophila* are specific for eye development; this implies that a common pathway may be regulated by components that are tissue-specific. There is a high degree of conservation of function; for example, Grb2 can substitute for Sem-5 in worms.

In yeast, the initiating event consists of the interaction of a polypeptide mating factor with a trimeric G protein, whose $\beta\gamma$ dimer (STE2,3) activates the kinase STE20, which activates the MEKK, STE11. We do not know whether there are other components in addition to STE20 between $G_{\beta\gamma}$ and STE11, but the yeast pathway at present provides the best characterization of the route from a G protein to the MAP kinase cascade. The pathway then continues through components all of which have direct counterparts in yeast and mammals. *STE7* is homologous to MEK, and *FUS3* and *KSS1* code for kinases that share with MAP kinase the requirement for activation by phosphorylation on both threonine and tyrosine. Their targets in turn directly execute the consequences of the cascade.

Figure 26.30 Homologous proteins are found in signal transduction cascades in a wide variety of organisms.

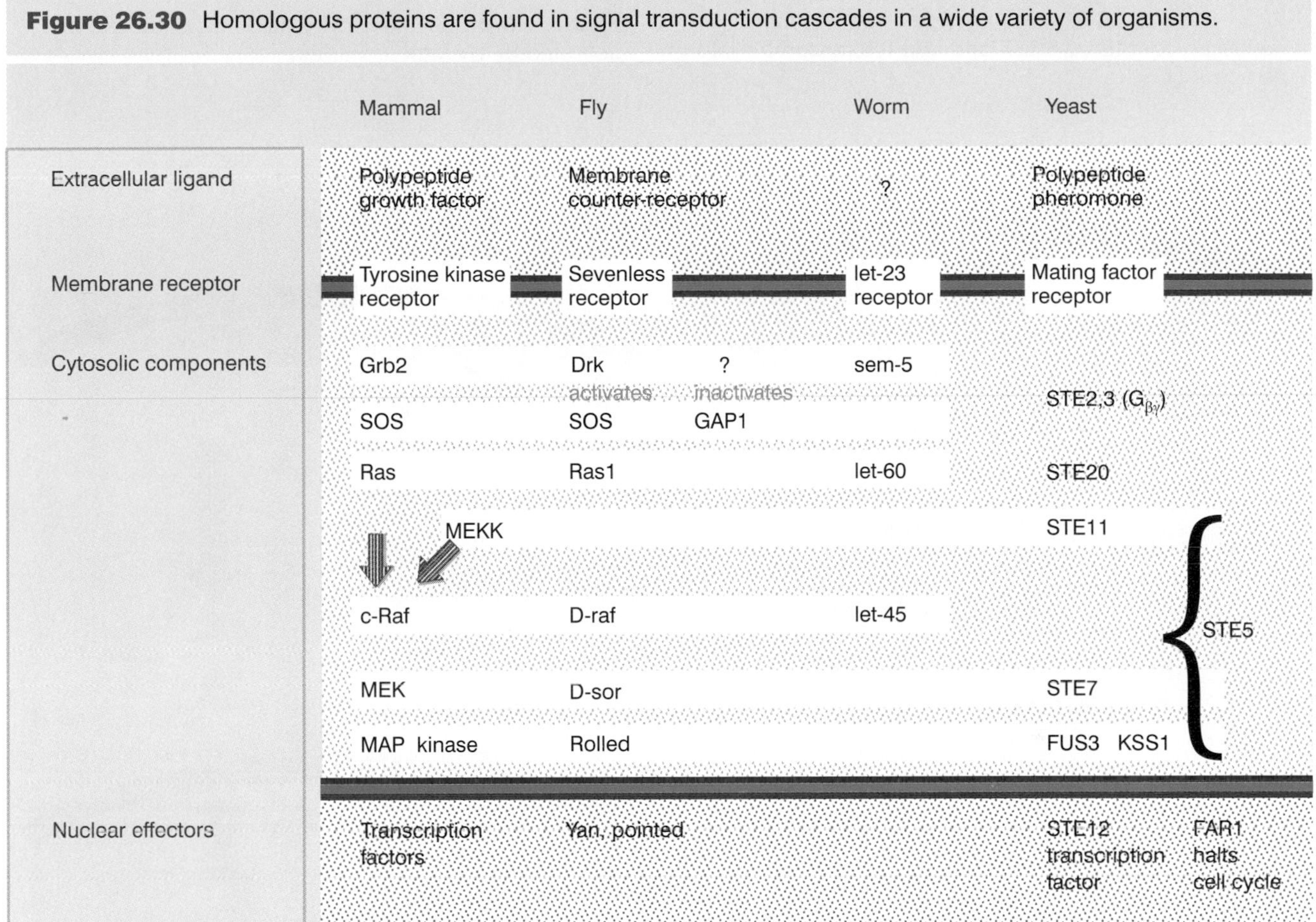

Mammal
Fly
Worm
Yeast
Extracellular ligand
Polypeptide growth factor
Membrane counter-receptor
?
Polypeptide pheromone
Membrane receptor
Tyrosine kinase receptor
Sevenless receptor
let-23 receptor
Mating factor receptor
Cytosolic components
Grb2
Drk
?
sem-5
activates
inactivates
SOS
SOS
GAP1
STE2,3 ($G_{\beta\gamma}$)
Ras
Ras1
let-60
STE20
MEKK
STE11
c-Raf
D-raf
let-45
STE5
MEK
D-sor
STE7
MAP kinase
Rolled
FUS3 KSS1
Nuclear effectors
Transcription factors
Yan, pointed
STE12 transcription factor
FAR1 halts cell cycle

The MAP kinase cascade shown in Figure 26.30 is the best characterized, but there are also other, parallel cascades with related components. In yeast, in addition to the mating response pathway, cascades containing kinases homologous to MEKK, MEK, and MAPK respond to signaling initiated by changes in osmolarity, or activation of PKC (protein kinase C) in *S. cerevisiae.*

How is specificity established in the cascade: what prevents a component of one MAP kinase cascade activating the enzyme that corresponds to the next stage in a parallel pathway? One possible explanation is that there may be a restriction in the localization of the components. *STE5* in yeast is implicated in the cascade between *STE2,3* and *FUS3, KSS1*, but does not place in a single position. **Figure 26.31** shows that STE5 binds to three of the kinases, STE11 (MEKK), STE7 (MEK), and FUS3 (MAPK), suggesting that this complex has to form before each kinase can activate the next kinase in the pathway. Each of the kinases binds to a different region on STE5, which provides a **scaffold.** If the kinases can only function in the context of the scaffold, they may be prevented from acting upon kinases in other pathways, and therefore act only upon one another.

What degree of amplification is achieved through the Ras/MAPK pathway? Typically an ~10× amplification of signal can be achieved at each stage of a kinase cascade, allowing an overall amplification of $>10^4$ through the pathway. However, the combination of the last three kinases into one complex would presumably restrict amplification at these stages. In mammalian cells, the pathway can be fully activated by very weak signals; for example, the ERK1,2 MAP kinases are fully activated when <5% of the Raf protein molecules bind to Ras.

Another example of a pathway that proceeds through a MAP kinase is provided by the activation of the transcription factor Jun in response to stress signals. **Figure 26.32** shows that activation of the kinase JNK involves both convergence and divergence. JNK is regulated by two classes of extracellular signals: UV light (typical of a stress response); and also as a consequence of activation of Ras (by an unidentified branch

Figure 26.31 STE5 may provide a scaffold that is necessary for MEKK, MEK, and MAPK to assemble into an active complex.

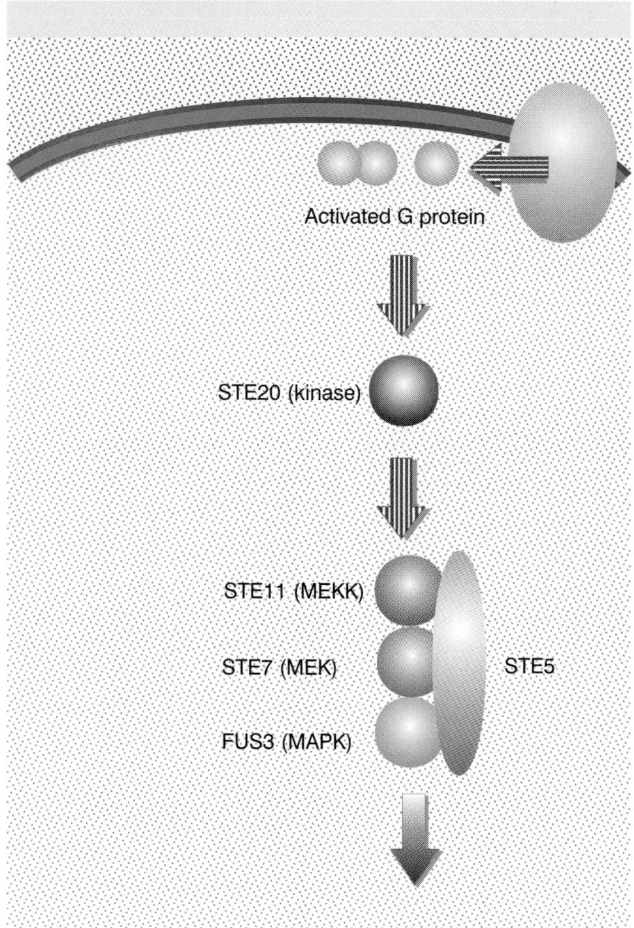

Figure 26.32 JNK is a MAP-like kinase that can be activated by UV light or via Ras.

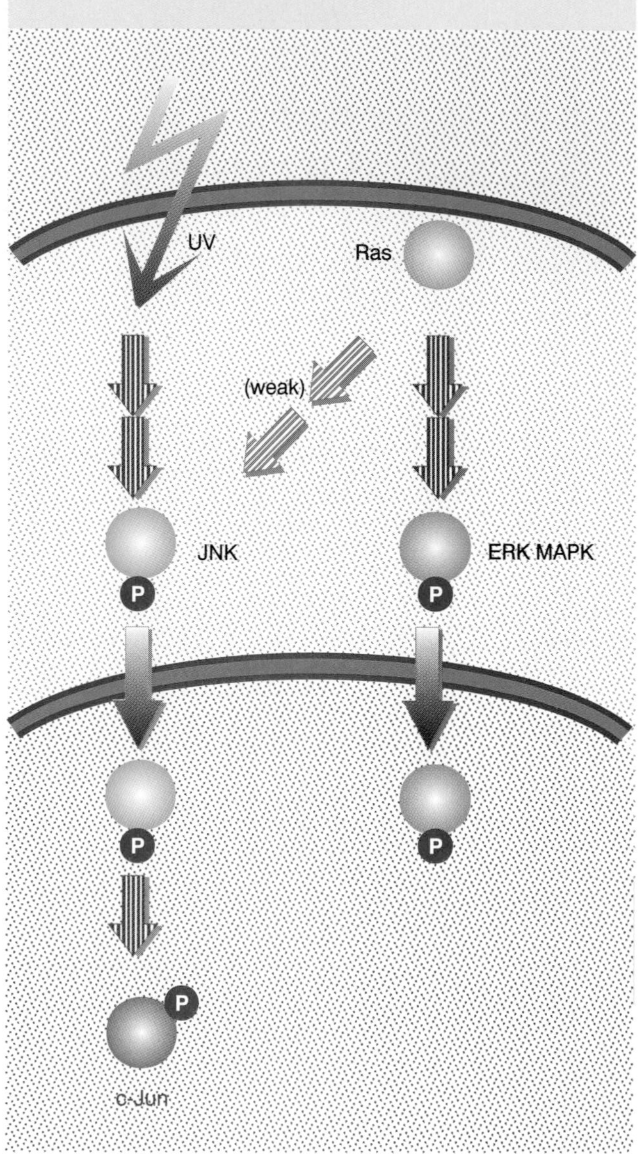

of the Ras pathway). JNK is a (distant) relative of MAP kinases such as the ERKs, showing the classic features of being activated by phosphorylation of Thr and Tyr, and phosphorylating its targets on Ser. The proteins JIP1,2 provide a scaffold that may ensure the integrity of the pathway leading to JNK activation.

The presence of multiple MAPK signaling pathways with analogous components is common. **Figure 26.33** summarizes the mammalian pathways. Each pathway functions in a linear manner, as indicated previously, but in addition there may be "crosstalk" between the pathways, when a component in one pathway can activate the subsequent component in other pathways as well as its own. Usually these "lateral" signals are weaker than those propagating down the pathway. At the very start of the pathway, there is also signaling from Ras to Rac. The strengths of these lateral signals, as well as the extent of activation of an individual pathway, may be important in determining the biological response.

A puzzling feature of the Ras/MAPK pathway is that activation of the same pathway under different circumstances can cause different outcomes. When PC12 cells are treated with the growth factor NGF, they differentiate (by becoming neuronal-like) and stop dividing. When they are treated with EGF, however, they receive a signal for continued proliferation. In both cases, the principal signal transduction event is the activation of the ERK MAP kinase pathway. The differences in outcome might be explained, of course, by other (uniden-

Figure 26.33 Three MAP kinase pathways have analogous components. Crosstalk between the pathways is shown by grey arrows.

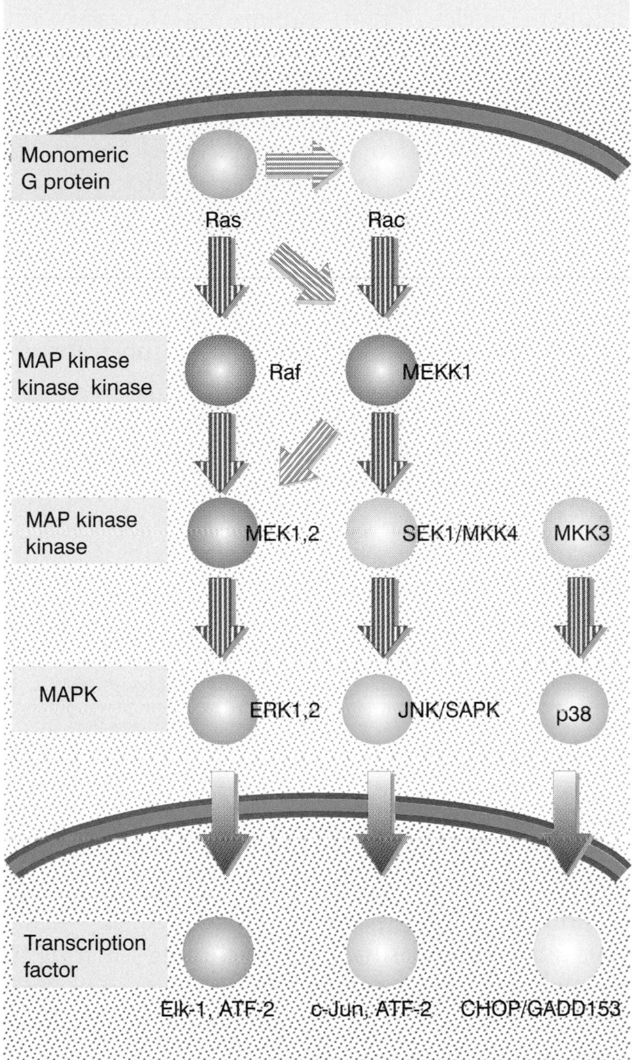

tified) pathways that are activated by the respective receptors. However, the major difference in the two situations is that NGF stimulation causes prolonged elevation of Ras-GTP, whereas EGF stimulation produces only a transient effect. (One reason for this difference is that EGF receptor is more susceptible to feedback mechanisms that reverse its activation.)

The idea that duration of the stimulus to the ERK MAPK pathway may be the critical parameter is supported by results showing that a variety of conditions that cause persistent activation of ERK MAP kinase, including constitutive activation of the EGF receptor, all cause differentiation. By contrast, all conditions in which activation is transient lead instead to proliferation. More direct proof of the role of the ERK MAPK pathway is provided by showing that mutations constitutively activating MEK cause differentiation of PC12 cells. So activation of the ERK MAPK pathway is sufficient to trigger the differentiation response. Another point is made by the fact that the same MEK mutation has different effects in a different host cell; in fibroblasts, it stimulates proliferation. This is another example of the ability of a cell to connect the same signal transduction pathway to different readouts.

How might the duration of the signal determine the type of outcome? The concentration of some active component in the pathway could increase with the duration of activation, and at some point would exceed a threshold at which it triggered a new response. One model for such an action is suggested by *Drosophila* development, in which increasing concentrations of a transcription factor activate different target genes, as the result of combinatorial associations with other factors that depend upon relative concentrations (see Chapter 29). Another possibility is suggested by the fact that prolonged activation is required before ERK2 translocates to the nucleus. The mechanism is unknown, but could mean that transient stimulation does not support the phosphorylation and activation of nuclear transcription factors, so the expression of new functions (such as those needed for differentiation) could depend upon the stimulus lasting long enough to cause translocation of ERK2.

Cyclic AMP and activation of CREB

Cyclic AMP is the classic second messenger, and its connection to transcription is by the activation of CREB (cAMP response element binding protein). **Figure 26.34** shows how the pathway proceeds through the Ser/Thr kinase, PKA.

The initial step in the pathway is activation of

Figure 26.34 When cyclic AMP binds to the R subunit of PKA, the C subunit is released; some C subunits diffuse to the nucleus, where they phosphorylate CREB.

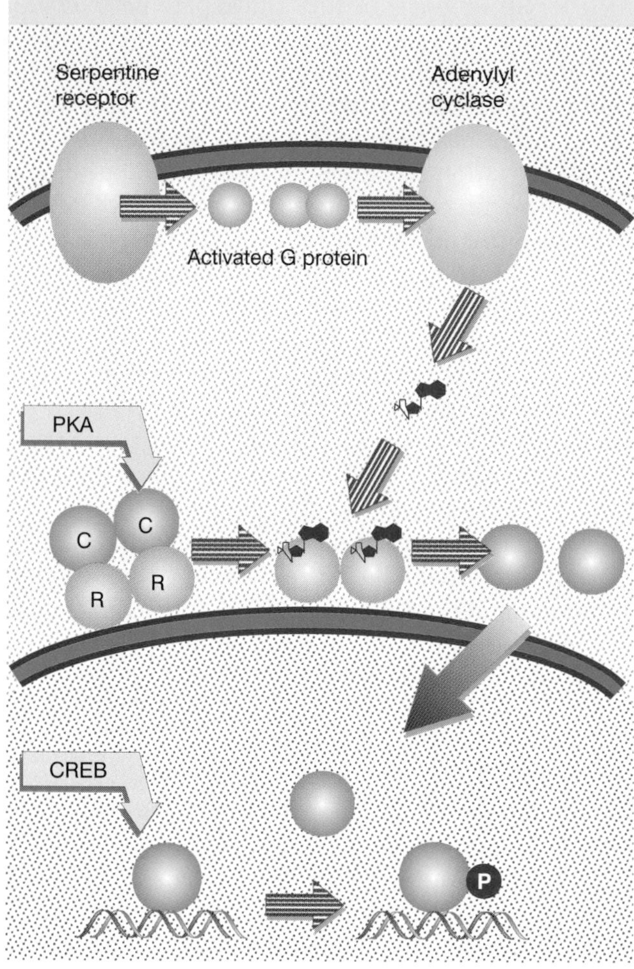

adenylate cyclase at the plasma membrane by an activated G protein (see Figure 26.10). An increase in the level of cAMP causes binding to the regulatory R subunit of PKA, which is anchored to membranes in the perinuclear region. The cAMP-R subunit releases the catalytic (C) subunit of PKA, which is free to translocate to the nucleus. Translocation occurs by passive diffusion, and involves only a proportion of the released C subunits, which have targets for phosphorylation in both the cytosol and nucleus.

The circuitry also has some feedback loops. The end-targets for PKA are also substrates for the phosphatase PPase I, which in effect reverses the action of PKA. However, PKA also has as a target a protein whose phosphorylation converts it into an inhibitor of PPase I, thus preventing the reversal of phosphorylation.

CREB is one of the major nuclear substrates for PKA. Phosphorylation at a single Ser residue greatly increases the activity of CREB bound to the response element CRE, which is found in genes whose transcription is induced by cAMP (see Chapter 21). The rate of transcription of these genes is directly proportional to the concentration of phosphorylated CREB in the nucleus. The kinetics of the response are limited by the relatively slow rate at which the free C subunit diffuses into the nucleus. Typically, the phosphorylated C subunit reaches a maximum level in the nucleus after ~30 minutes, and then is slowly dephosphorylated (over several hours). Several circuits may be involved in the dephosphorylation, including direct control of phosphatases and indirect control by the entry into the nucleus of the protein PKI, which binds to the C subunit and causes it to be re-exported to the cytoplasm. The kinetics of activating PKC in the nucleus may be important in several situations, including learning, in which a weak stimulus of cAMP has only short-term effects, whereas a strong stimulus is required for long-term effects, including changes in transcription. This parallels the different consequences of short-term and long-term stimulation of the MAPK pathway (see above).

The JAK-STAT pathway

Some signal transduction pathways have large numbers of components (permitting a high degree of amplification) and many feedback circuits (permitting sensitive control of the duration and strength of the signal). The JAK-STAT pathway is much simpler, and consists of three components that function as illustrated in **Figure 26.35.**

JAK-STAT pathways are activated by several cytokine receptors. These receptors do not possess intrinsic kinase activities. However, binding of a cytokine

Figure 26.35 Cytokine receptors associate with and activate JAK kinases. STATs bind to the complex and are phosphorylated. They dimerize and translocate to the nucleus. The complex binds to DNA and activates transcription.

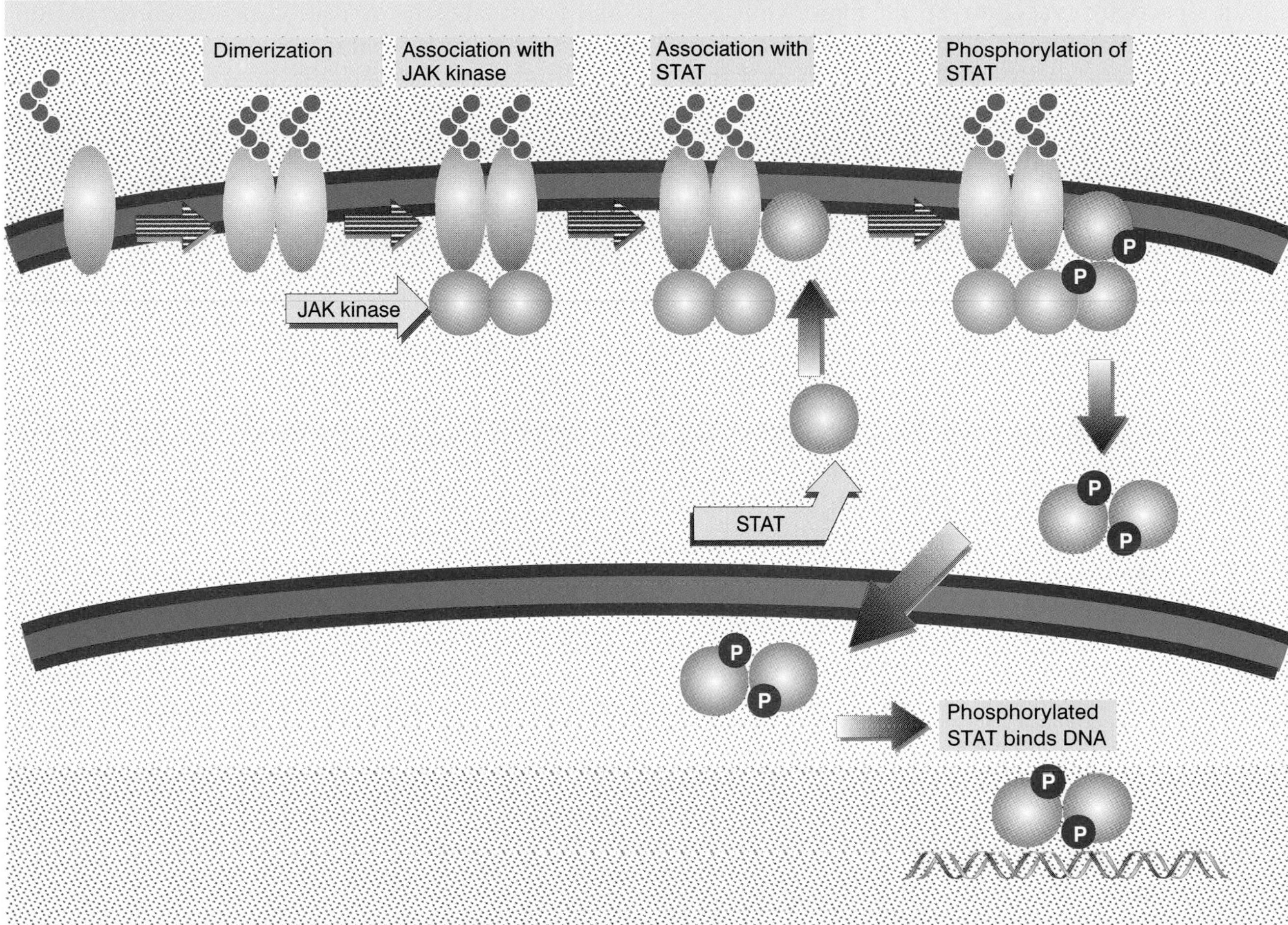

causes its receptor to dimerize, which provides the signal to associate with and activate a JAK kinase. The JAK kinases take their name (originally Janus kinases) from the characteristic presence of two kinase domains in each molecule. Several members of the family are known (JAK1,2,3, etc.); each associates with a specific set of cytokine receptors. The interaction between the activated (dimeric) cytokine receptor and JAK kinase(s) in effect produces the same result as the ligand-induced dimerization of a tyrosine kinase receptor: the difference is that the receptor and kinase activities are held in different proteins instead of in the same protein.

The JAK kinases are tyrosine kinases whose major substrates are transcription factors called STATs. There are >7 STATs; each STAT is phosphorylated by a particular set of JAK kinases. The phosphorylation occurs while the JAK is associated with the receptor at the plasma membrane. A pair of JAK kinases associates with an activated receptor, and both may be necessary for the pathway to function. An example is that stimulation by the interferon IFNγ requires both JAK1 and JAK2.

STAT phosphorylation leads to the formation of both homodimers and heterodimers. The basis for dimerization is a reciprocal interaction between an SH2 domain in one subunit and a phosphorylated Tyr in the other subunit.

The STAT dimers translocate to the nucleus, and in some cases associate with other proteins. They bind to specific recognition elements in target genes, whose transcription is activated.

Given a multiplicity of related cytokine receptors, JAK kinases, and STAT transcription factors, how is specificity achieved? The question is sharpened by the fact that many receptors can activate the same JAKs, but activate different STATs. Control of specificity lies with formation of a multipartite complex containing the receptor, JAKs, and STATs. The STATs interact directly with the receptor as well as with the JAKs, and an SH2 domain in a particular STAT recognizes a binding site in a particular receptor. So the major control of specificity lies with the STAT.

Stimulation of a JAK-STAT pathway is only transient. Its activation may be terminated by the action of a phosphatase. An example is the pathway activated by binding of erythropoietin (red blood cell hormone) to its receptor. This activates JAK2 kinase. Recruitment of another component terminates the reaction; the phosphatase SH-PTP1 binds via its SH2 domain to a phosphotyrosine site in the erythropoietin receptor. This site in the receptor is probably phosphorylated by JAK2. The phosphatase then dephosphorylates JAK2 and terminates the activation of the corresponding STATs. This creates a simple feedback circuit: erythropoietin receptor activates JAK2, JAK2 acts on a site in the receptor, and this site is recognized by the phosphatase that in turn acts on JAK2. This again emphasizes the way in which formation of a multicomponent complex may be used to ensure specificity in controlling the pathway.

TGFβ signals through Smads

ANOTHER pathway in which phosphorylation at the membrane triggers migration of a transcription factor to the nucleus is provided by TGFβ signaling. The TGFβ family contains many related polypeptide ligands. They bind to receptors that consist of two types of subunits, as illustrated in **Figure 26.36.** Both subunits have serine/threonine kinase activity. (Actually all serine/threonine receptor kinases are members of the TGFβ receptor family.)

Figure 26.36 Activation of TGFβ receptors causes phosphorylation of a Smad, which is imported into the nucleus to activate transcription.

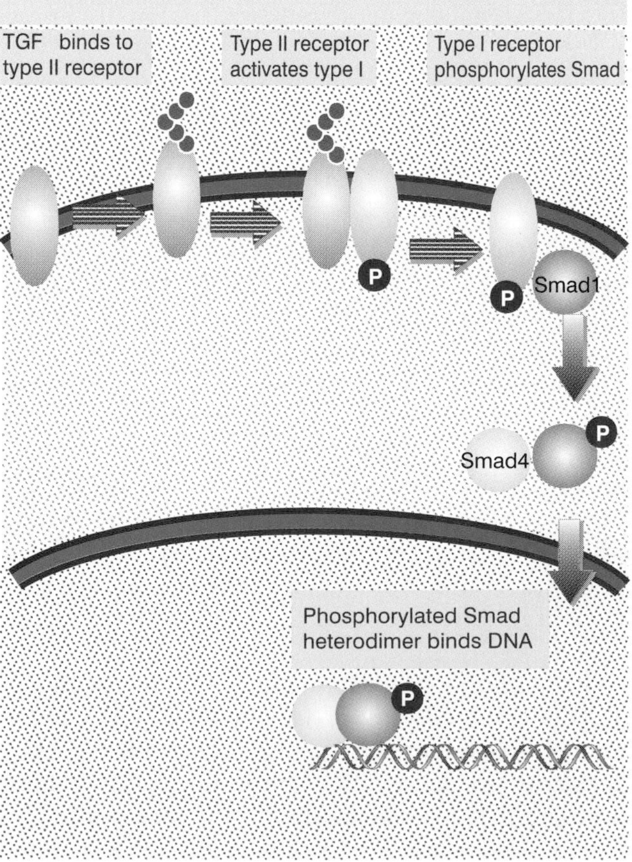
TGF binds to type II receptor
Type II receptor activates type I
Type I receptor phosphorylates Smad
P
P
Smad1
P
Smad4
Phosphorylated Smad heterodimer binds DNA
P

The ligand binds to the type II receptor, creating a receptor-ligand combination that has high affinity for the type I receptor. A tetrameric complex is formed in which the type II receptor phosphorylates the type I receptor. (A variation occurs in a subset of these receptors: when binding BMPs—bone morphogenetic proteins, which are related to TGFβ—both type I and type II subunits have low affinity for the ligand, but the combination of subunits has high affinity.)

Once the active complex has formed, the type I receptor phosphorylates a member of the cytosolic Smads family. Typically a Smad activator is phosphorylated at the motif SSXS at the C-terminus. This causes it to form a dimer with the common partner Smad4. The heterodimer is imported into the nucleus, where it binds to DNA and activates transcription.

The nine Smad proteins fall into three functional categories. The pathway-specific activators are Smad2,3 (which mediate TGFβ/activin signaling) and Smad1,5 (which activate BMP signaling). Smad4 is a universal partner which can dimerize with all of the pathway-specific Smads. Inhibitory Smads act as competitive inhibitors of the activator Smads, providing another level of complexity to the pathway. Each ligand in the TGFβ superfamily activates a particular receptor that signals through a characteristic combination of Smads proteins. Various other proteins bind to the Smads dimers and influence their capacity to act on transcription.

Signaling systems of this type are important in early embryonic development, where they are part of the pathways that lead to development of specific tissues (typically bone formation and the development of mesoderm). Also, because TGFβ is a powerful growth inhibitor, this pathway is involved in tumor suppression. The TGFβ type II receptor is usually inactivated in hereditary nonpolyposis colorectal cancers, and mutations in Smad4 occur in 50% of human pancreatic cancers.

One striking feature of the JAK-STAT and TGFβ pathways is the simplicity of their organization, compared (for example) with the Ras-MAPK pathway. The specificity of these pathways depends on variation of the components that assemble at the membrane—different combinations of JAK-STATs in the first case, different Smad proteins in the second. Once the pathway has been triggered, it functions in a direct linear manner. The component that is phosphorylated at the plasma membrane (STAT in the JAK-STAT pathway, Smad in the TGFβ pathway) itself provides the unit that translocates to the nucleus to activate transcription—perhaps the ultimate demonstration of the role of localization.

Summary

Integral proteins of the plasma membrane offer several means for communication between the extracellular milieu and the cytoplasm. They include ion channels, transporters, and receptors. All such proteins reside in the plasma membrane by means of hydrophobic domains. Lipids may cross the plasma membrane, but specific transport mechanisms are required to promote the passage of hydrophilic molecules.

Ions may be transported by carrier proteins, which may utilize passive diffusion or may be connected to energy sources to undertake active diffusion. The detailed mechanism of movement via a carrier is not clear, but is presumed to involve conformational changes in the carrier protein that directly or indirectly allow a substrate to move from one side of the membrane to the other. Ion channels can be used for passive diffusion (driven by the gradient). They may be gated (controlled) by voltage, extracellular ligands, or cytoplasmic second messengers. Channels typically have multiple subunits, each with several transmembrane domains; hydrophilic residues within the transmembrane domains face inward so as to create a hydrophilic path through the membrane.

Receptors typically are group I proteins, with a single transmembrane domain, consisting exclusively of uncharged amino acids, connecting the extracellular and cytosolic domains. Many receptors for growth factors are protein tyrosine kinases. Such receptors have a binding site for their ligand in the extracellular domain, and a kinase activity in their cytoplasmic domain. When a ligand binds to the receptor, it causes the extracellular domain to dimerize; most often the product is a homodimer, but there are some cases where heterodimers are formed. The dimerization of the extracellular domains causes the transmembrane domains to diffuse laterally within the membrane, bringing the cytoplasmic domains into contact. This results in an autophosphorylation in which each monomeric subunit phosphorylates the other.

The phosphorylation creates a binding site for the SH2 motif of a target protein. Specificity in the SH2-binding site typically is determined by the phosphotyrosine in conjunction with the 4–5 neighboring amino acids on its C-terminal side. The next active component in the pathway may be activated indirectly or directly. Some target proteins are adaptors that are activated by binding to the phosphorylated receptor, and they in turn activate other proteins. An adaptor typically uses its SH2 domain to bind the receptor and uses an SH3 domain to bind the next component in the pathway. Other target proteins are substrates for phosphorylation, and are activated by the acquisition of the phosphate group.

One group of effectors consists of enzymes that generate second messengers, most typically phospholipases and kinases that generate or phosphorylate small lipids. Another type of pathway consists of the activation of a kinase cascade, in which a series of kinases successively activate one another, leading ultimately to the phosphorylation and activation of transcription factors in the nucleus. The MAP kinase pathway is the paradigm for this type of response.

The connection from receptor tyrosine kinases to the MAP kinase pathway passes through Ras. An adaptor (Grb2 in mammalian cells) is activated by binding to the phosphorylated receptor. Grb2 binds to SOS, and SOS causes GDP to be replaced by GTP on Ras. Ras is anchored to the cytoplasmic face of the membrane. The activated Ras binds the Ser/Thr kinase Raf, thus bringing Raf to the membrane, which causes Raf to be activated, probably because it is phosphorylated by a kinase associated with the membrane. Raf phosphorylates MEK, which is a dual-specificity kinase that phosphorylates ERK MAP kinases on both tyrosine and threonine. ERK MAP kinases activate other kinases; ERK2 MAP kinase also translocates to the nucleus, where it phosphorylates transcription factors that trigger pathways required for cell growth (in mammalian cells) or differentiation (in fly retina, worm vulva, or yeast mating).

An alternative connection to the MAP kinase cascade exists from serpentine receptors. Activation of a trimeric G protein causes MEKK to be activated. One component in the pathway between $G_{\beta\gamma}$ and MEKK in *S. cerevisiae* is the kinase STE20. The MEKK (STE11), MEK (STE7), and MAPK (FUS3) form a complex with the scaffold protein STE5 that may be necessary for the kinases to function.

The cyclic AMP pathway for activating transcription proceeds by releasing the catalytic subunit of PKA in the cytosol. It diffuses to the nucleus, where it phosphorylates the transcription factor CREB. The activity of this factor is responsible for activating cAMP-inducible genes. The response is downregulated by phosphatases that dephosphorylate CREB and by an inhibitor that exports the C subunit back to the cytosol.

JAK-STAT pathways are activated by cytokine receptors. The activated receptor associates with a JAK kinase and activates it. The target for the kinase is a STAT(s); STATs associate with a receptor-JAK kinase complex, are phosphorylated by the JAK kinase, dimerize, translocate to the nucleus, and form a DNA-binding complex that activates transcription at a set of target genes. In an analogous manner, TGFβ ligands activate type II/type I receptor systems that phosphorylate Smad proteins, which then are imported into the nucleus to activate transcription.

Further reading

Reviews

Hunter, T. and Cooper, J. A. (1985). Protein-tyrosine kinases. *Ann. Rev. Biochem.* **54**, 897–930.

Hunter, T. (1987). A thousand and one protein kinases. *Cell* **50**, 823–829.

Neer, E. J. and Clapham, D. E. (1988). Roles of G protein subunits in transmembrane signaling. *Nature* **333**, 129–134.

Miller, C. (1989). Genetic manipulation of ion channels: a new approach to structure and mechanism. *Neuron* **2**, 1195–1205.

Unwin, N. (1989). The structure of ion channels in membranes of excitable cells. *Neuron* **3**, 665–676.

Ullrich, A. and Schlessinger, J. (1990). Signal transduction by receptors with tyrosine kinase activity. *Cell* **61**, 203–212.

Koch, C. A. *et al.* (1991). SH2 and SH3 domains: elements that control interactions of cytoplasmic signaling proteins. *Science* **252**, 668–674.

Boguski, M. S. and McCormick, F. (1993). Proteins regulating Ras and its relatives. *Nature* **366**, 643–654.

Clapham, D. E. and Neer, E. J. (1993). New roles of G protein βγ-dimers in transmembrane signaling. *Nature* **365**, 403–406.

Darnell, J. E., Kerr, I. M., and Stark, G. R. (1994). JAK-STAT pathways and transcriptional activation in response to IFNα and other extracellular signaling proteins. *Science* **264**, 1415–1421.

Strader, D. *et al.* (1994). Structure and function of G protein-coupled receptors. *Ann. Rev. Biochem.* **63**, 101–132.

van der Geer, P., Hunter, T., and Lindberg, R. A. (1994). Receptor protein-tyrosine kinases and their signal transduction pathways. *Ann. Rev. Cell Biol.* **10**, 251–337.

Cohen, G. B., Ren, R., and Baltimore, D. (1995). Molecular binding domains in signal transduction proteins. *Cell* **80**, 237–248.

Divecha, N. and Irvine, R. F. (1995). Phospholipid signaling. *Cell* **80**, 269–278.

Heldin, C.-H. (1995). Dimerization of cell surface receptors in signal transduction. *Cell* **80**, 213–223.

Herskowitz, I. (1995). MAP kinase pathways in yeast: for mating and more. *Cell* **80,** 187–198.

Hill, C. S. and Treisman, R. (1995). Transcriptional regulation by extracellular signals: mechanisms and specificity. *Cell* **80,** 199–212.

Hunter, T. (1995). Protein kinases and phosphatases: the Yin and Yang of protein phosphorylation and signaling. *Cell* **80,** 237–248.

Marshall, C. J. (1995). Specificity of receptor tyrosine kinase signaling: transient versus sustained extracellular signal-regulated kinase activation. *Cell* **80,** 179–186.

Neer, E. J. (1995). Heterotrimeric G proteins: organizers of transmembrane signals. *Cell* **80,** 249–257.

Schindler, C. and Darnell, J. E. (1995). Transcriptional responses to polypeptide ligands: the JAK-STAT pathway. *Ann. Rev. Biochem.* **64,** 621–651.

Massagué, J. (1996). TGFβ signaling: receptors, transducers, and Mad proteins. *Cell* **85,** 947–950.

Spring, S. R. (1997). G protein mechanisms: insights from structural analysis. *Ann. Rev. Biochem.* **66,** 639–678.

Massagué, J. (1998). TGFβ signal transduction. *Ann. Rev. Biochem.* **67,** 753–791.

Discoveries

Receptor kinases

Hafen, E. *et al.* (1987). *Sevenless,* a cell-specific homeotic gene of *Drosophila,* encodes a putative transmembrane receptor with a tyrosine kinase domain. *Science* **236,** 55–63.

Aroian, R. V. *et al.* (1990). The *let-23* gene necessary for *C. elegans* vulval induction encodes a tyrosine kinase of the EGF receptor subfamily. *Nature* **348,** 693–699.

Cunningham, B. C. *et al.* (1991). Dimerization of the extracellular domain of the human growth hormone receptor by a single hormone molecule. *Science* **254,** 821–825.

Fantl, W. J. *et al.* (1992). Distinct phosphotyrosines on a growth factor receptor bind to specific molecules that mediate different signaling pathways. *Cell* **69,** 413–423.

Ras and Raf

Simon, M. A. *et al.* (1991). Ras1 and a putative guanine nucleotide exchange factor perform crucial steps in signaling by the sevenless protein tyrosine kinase. *Cell* **67,** 701–716.

Lowenstein, E. J. *et al.* (1992). The SH2 and SH3-domain containing protein Grb2 links receptor tyrosine kinases to ras signaling. *Cell* **70,** 431–442.

Buday, L. and Downward, J. (1993). EGF regulates $p21^{ras}$ through the formation of a complex of receptor, Grb2 adaptor protein, and SOS nucleotide exchange factor. *Cell* **73,** 611–620.

Chardin, P. *et al.* (1993). Human SOS1: a guanine nucleotide exchange factor for Ras that binds to Grb2. *Science* **260,** 1338–1343.

Vojtek, A. B., Hollenberg, S. M., and Cooper, J. A. (1993). Mammalian Ras interacts directly with the serine/threonine kinase Raf. *Cell* **74,** 205–214.

Aronheim, A. *et al.* (1994). Membrane targeting of the nucleotide exchange factor SOS is sufficient for activating the Ras signaling pathway. *Cell* **78,** 949–961.

Leevers, S. J., Paterson, H. F., and Marshall, C. J. (1994). Requirement for Ras in Raf activation is overcome by targeting Raf to the plasma membrane. *Nature* **369,** 411–414.

Protein-protein interactions

Dale, T. C. *et al.* (1989). Rapid activation by interferon α of a latent DNA-binding protein present in the cytoplasm of untreated cells. *Proc. Nat. Acad. Sci. USA* **86,** 1203–1207.

Velazquez, L. *et al.* (1992). A protein tyrosine kinase in the interferon α/β signaling pathway. *Cell* **70,** 313–322.

Booker, G. W. *et al.* (1993). Solution structure and ligand-binding site of the SH3 domain of the p85α subunit of phosphatidylinositol 3-kinase. *Cell* **73,** 813–822.

Fu, X.-Y. and Zhang, J.-J. (1993). Transcription factor p91 interacts with the EGF receptor and mediates activation of the c-fos gene promoter. *Cell* **74,** 1135–1145.

Songyang, Z. *et al.* (1993). SH2 domains recognize specific phosphopeptide sequences. *Cell* **72,** 767–778.

Choi, K.-Y. *et al.* (1994). Ste5 tethers multiple protein kinases in the MAP kinase cascade required for mating in *S. cerevisiae. Cell* **78,** 499–512.

Cowley, S. *et al.* (1994). Activation of MAP kinase kinase is necessary and sufficient for PC12 differentiation and for transformation of NIH 3T3 cells. *Cell* **77,** 841–852.

Shuai, K. *et al.* (1994). Interferon activation of the transcription factor STAT91 involves dimerization through SH2-phosphotyrosyl peptide interactions. *Cell* **76,** 821–828.

Klingmuller, U. *et al.* (1995). Specific recruitment of SH-PTP1 to the erythropoietin receptor causes inactivation of JAK2 and termination of proliferative signals. *Cell* **80,** 729–738.

Macias-Silva, M. *et al.* (1996). Madr2 is a substrate of the TGFβ receptor and its phosphorylation is required for nuclear accumulation and signaling. *Cell* **87,** 1215–1224.

Kinase activation

Hagiwara, M. *et al.* (1992). Transcriptional attenuation following cAMP induction requires PPA-mediated dephosphorylation of CREB. *Cell* **70,** 105–113.

Wood, K. W. *et al.* (1992). Ras mediates nerve growth factor receptor modulation of three signal-transducing protein kinases: MAP kinase, Raf-1, and RSK. *Cell* **68,** 1041–1050.

Hagiwara, M. *et al.* (1993). Coupling of hormonal stimulation and transcription via the cAMP-responsive factor CREB is rate limited by nuclear entry of PKA. *Mol. Cell. Biol.* **13,** 4852–4859.

Lange-Carter, C. A. *et al.* (1993). A divergence in the MAP kinase regulatory network defined by MEK kinase and Raf. *Science* **260,** 315–319.

Dérijard, B. *et al.* (1994). JNK1: a protein kinase stimulated by UV light and Ha-Ras that binds and phosphorylates the c-Jun activation domain. *Cell* **76,** 1025–1037.

Monomeric G proteins

Ridley, A. J. and Hall, A. (1992). The small GTP-binding protein rho regulates the assembly of focal adhesions and actin stress fibers in response to growth factors. *Cell* **70,** 389–399.

Ridley, A. J. *et al.* (1992). The small GTP-binding protein rac regulates growth factor-induced membrane ruffling. *Cell* **70,** 401–410.

Nobes, C. D. and Hall, A. (1995). Rho, Rac, and Cdc42 GTPases regulate the assembly of multimolecular focal complexes associated with actin stress fibers, lamellipodia, and filopodia. *Cell* **81,** 53–62.

Lamarche, N. *et al.* (1996). Rac and Cdc42 induce actin polymerization and G cell cycle progression independently of p^{65}PAK and the JNK/SAPK MAP kinase cascade. *Cell* **87,** 519–529.

Channels

Doyle, D. A. *et al.* (1998). The structure of the potassium channel: molecular basis of K^+ conduction and selectivity. *Science* **280,** 69–77.

Chapter 27

Cell cycle and growth regulation

THE act of division is the culmination of a series of events that have occurred since the last time a cell divided. The period between two mitotic divisions defines the somatic **cell cycle.** The time from the end of one mitosis to the start of the next is called **interphase.** The period of actual division, corresponding to the visible mitosis, is called **M phase.**

In order to divide, a eukaryotic somatic cell must double its mass and then apportion its components equally between the two **daughter cells.** Doubling of size is a continuous process, resulting from transcription and translation of the genes that code for the proteins constituting the particular cell phenotype. By contrast, reproduction of the genome occurs only during a specific period of DNA synthesis.

Mitosis of a somatic cell generates two identical daughter cells, each bearing a diploid complement of chromosomes. Interphase is divided into periods that are defined by reference to the timing of DNA synthesis, as summarized in **Figure 27.1:**

- Cells are released from mitosis into **G1 phase,** during which RNAs and proteins are synthesized, but there is no DNA replication.
- The initiation of DNA replication marks the transition from G1 phase to the period of **S phase.** S phase is defined as lasting until all of the DNA has been replicated. During S phase, the total content of DNA in-

Figure 27.1 *Overview*: interphase is divided into the G1, S, and G2 periods. One cell cycle is separated from the next by mitosis (M). Cells may withdraw from the cycle into G0 or re-enter from it.

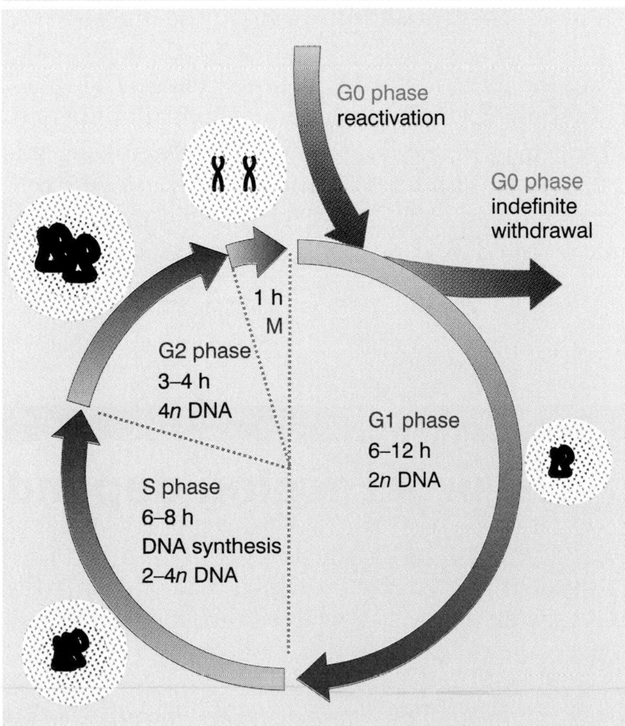

creases from the diploid value of $2n$ to the fully replicated value of $4n$.

- The period from the end of S phase until mitosis is called **G2 phase;** during this period, the cell has two complete diploid sets of chromosomes.

(S phase was so called as the synthetic period when DNA is replicated, G1 and G2 standing for the two "gaps" in the cell cycle when there is no DNA synthesis.)

The changes in cellular components are summarized in **Figure 27.2.** During interphase, there is little visible change in the appearance of the cell. The more or less continuous increase of RNA and protein contrasts with the discrete doubling of DNA. The nucleus increases in size predominantly during S phase, when proteins accumulate to match the production of DNA. Chromatin remains a compact mass in which no change of state is visible.

Mitosis segregates one diploid set of chromosomes to each daughter cell. Individual chromosomes become visible only during this period, when the nuclear envelope dissolves, and the cell is reorganized on a **spindle**. The mechanism for specific segregation of material applies only to chromosomes, and other components are apportioned essentially by the flow of cytoplasm into the two daughter cells. Virtually all synthetic activities come to a halt during mitosis.

In a cycling somatic animal cell, this sequence of events is repeated every 18–24 hours. Figure 27.1 shows that G1 phase usually occupies the bulk of this period, varying from ~6 hours in a fairly rapidly growing animal cell to ~12 hours in a more slowly growing cell. The duration of S phase is determined by the length of time required to replicate all the genome, and a period of 6–8 hours is typical. G2 phase is usually the shortest part of interphase, possibly comprising the preparations for mitosis. M phase (mitosis) is a brief interlude in the cell cycle, usually <1 hour in duration.

Figure 27.2 Synthesis of RNA and proteins occurs continuously, but DNA synthesis occurs only in the discrete period of S phase. The units of mass are arbitrary.

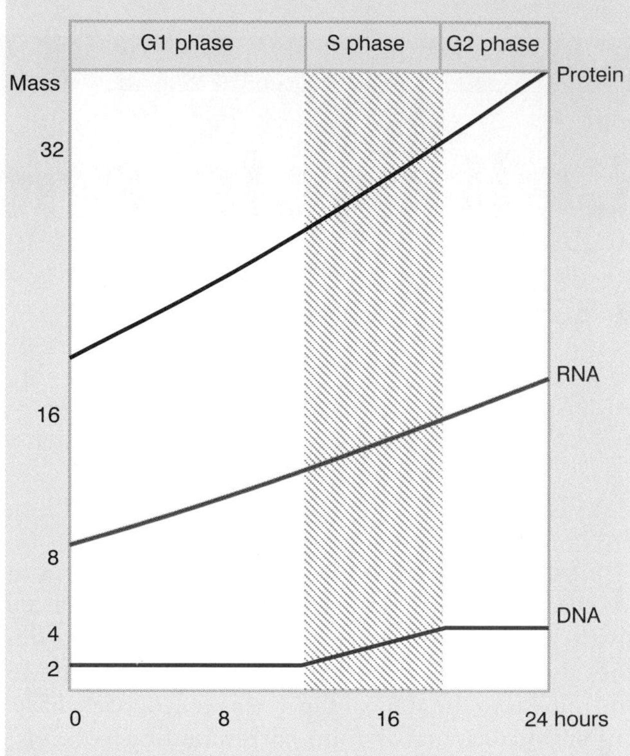

Cycle progression depends on discrete control points

THERE are two points at which a decision may be taken on whether to proceed through another cell cycle:

- *Commitment to chromosome replication occurs in G1 phase.* If conditions to pass a "commitment point" are satisfied, there is a lag period and then a cell enters S phase. The commitment point has been defined most clearly in yeast cells, where it is called **START.** The comparable feature in animal cells is called the **restriction point.**

- *Commitment to mitotic division occurs at the end of G2.* If the cell does not divide at this point, it remains in the condition of having twice the normal complement of chromosomes.

How do cells use these two control points?

For animal cells growing in culture, G1 control is the major point of decision, and G2/M control is subsidiary. Cells spend the longest part of their cycle in G1, and it is the length of G1 that is adjusted in response to growth conditions. When a cell proceeds past G1, barring accidents it will complete S phase, proceed through G2, and divide. Cultured cells do not halt in G2. Control at G1 is probably typical of most diploid cells, in culture or *in vivo.*

Some cell phenotypes do not divide at all. These cells are often considered to have withdrawn from the cell cycle into another state, resembling G1 but distinct from it because they are unable to proceed into S phase. This noncycling state is called **G0.** Certain types of cells can be stimulated to leave G0 and re-enter a cell cycle. Withdrawal from, or re-entry into, the cell cycle can occur before the restriction point in G1 (see Figure 27.1).

Some cell types do halt in G2. In the diploid world, these are usually cells likely to be called upon to divide again; for example, nuclei at some stages of insect embryogenesis divide and rest in the tetraploid state. In the haploid world, it is more common for cells to rest in G2; this affords some protection against damage to DNA, since there are two copies of the genome instead of the single copy present in G1. Some yeasts can use either G1 or G2 as the primary control point, depending on the nutritional conditions. Some (haploid) mosses usually use G2 as the control point.

The upper part of **Figure 27.3** (in blue) identifies the critical points in the cell cycle:

- For a cell with a cycle determined through G1

Figure 27.3 Checkpoints control the ability of the cell to progress through the cycle by determining whether earlier stages have been completed successfully. A horizontal red bar indicates the stage at which a checkpoint blocks the cycle.

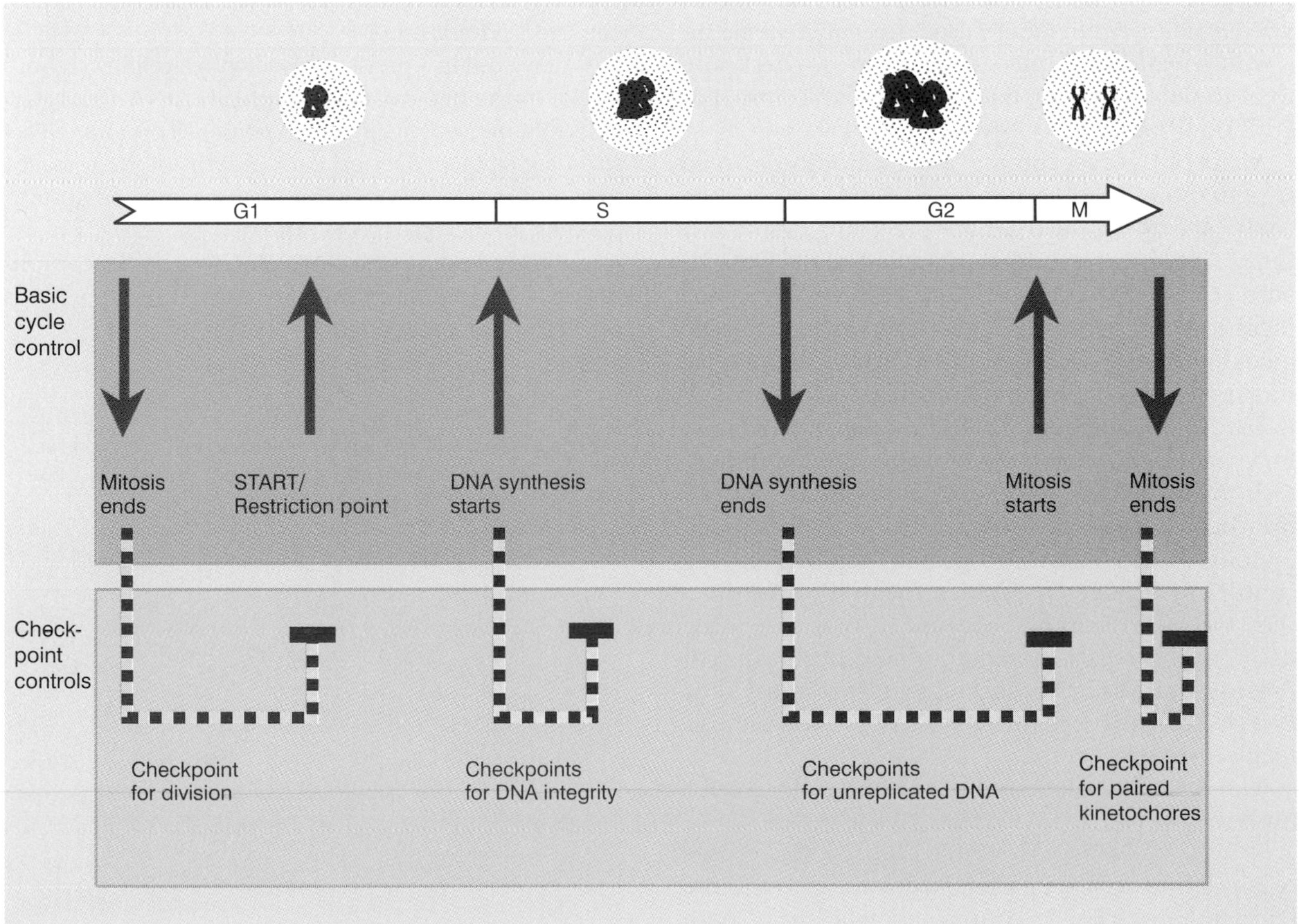

control, START marks the point at which the cell takes the basic decision: do I divide? Various parameters influence the ability of a cell to take this decision, including the response to external stimuli (such as supply of nutrients), and an assessment of whether cell mass is sufficient to support a division cycle. (Generally a cell is permitted to divide at a mass that is not absolute but is determined by a control that itself responds to growth rate.)

- The beginning of S phase is marked by the point at which the replication apparatus begins to synthesize DNA.
- The start of mitosis is identified by the moment at which the cell begins to reorganize for division.

Each of these events represents a discrete moment at which a molecular change occurs in a regulatory molecule. Once a cell has taken the decision to proceed at START, the other events will follow in order as the result of the cell cycle pathway.

However, superimposed on this pathway are **checkpoints** that assess the readiness of the cell to proceed. The lower part of Figure 27.3 (in red) shows that *each checkpoint represents a control loop that makes the initiation of one event in the cell cycle dependent on the successful completion of an earlier event.* A checkpoint works by acting directly on the factors that control progression through the cell cycle.

Many of the checkpoints operate at mitosis, to ensure that the cell does not try to divide unless it has completed all of the necessary preceding events. One important checkpoint establishes that all the DNA has been replicated; other checkpoints determine whether there is any damaged DNA and postpone mitosis until it has been repaired. This explains why a common feature in the cycle of probably all somatic eukaryotic cells is that completion of DNA replication is a prerequisite for cell division. There are also checkpoints within mitosis to ensure that one phase does not start until the previous phase has been completed. Checkpoints also operate within S phase to prevent replication from continuing if there are problems with the integrity of the DNA (for example, because of breaks or other damage). There is a checkpoint that confirms that the previous mitosis has been successfully completed before a cell can proceed through START to commit itself to another cycle.

Two types of cycle must be coordinated for a cell to divide:

- A cell must replicate every sequence of DNA once and only once. This is controlled by licensing factor (see Chapter 13). Having begun replication, it must complete it; *and it must not try to divide until replication has been completed.* This control is accomplished by the checkpoint at mitosis.
- The mass of the cell must double, so that there is sufficient material to apportion to the daughter cells. So *a cell must not try to start a replication cycle unless its mass will be sufficient to support division.* Cell mass influences ability to proceed through START and may also have a checkpoint at mitosis.

Some embryonic cycles bypass some of these controls and respond instead to a timer or an oscillator. So the control of the cell cycle can be coupled as required to time, growth rate, mass, and the completion of replication.

The existence of different regulators at different stages of the cell cycle was revealed by early experiments that fused together cells in different stages of the cycle. As illustrated in **Figure 27.4,** cell fusion is performed by mixing the cells in the presence of either a chemical or viral agent that causes their plasma membranes to fuse, generating a hybrid cell (called a **heterokaryon**) that contains two (or more) nuclei in a common cytoplasm.

When a cell in S phase is fused with a cell in G1, both nuclei in the heterokaryon replicate DNA. This suggests that *the cytoplasm of the S phase cell contains an activator of DNA replication.* The *quantity* of the activator may be important, because in fusions involving multiple cells, an increase in the ratio of S phase to G1 phase nuclei increases the rate at which G1 nuclei enter replication. The regulator identified by these fusions is called the **S phase activator.**

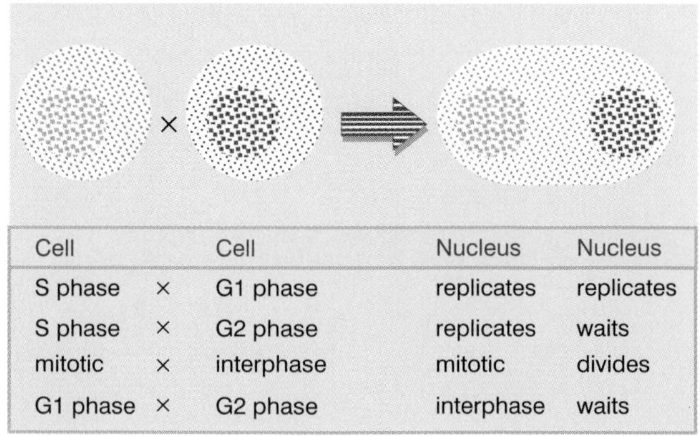

Cell		Cell	Nucleus	Nucleus
S phase	×	G1 phase	replicates	replicates
S phase	×	G2 phase	replicates	waits
mitotic	×	interphase	mitotic	divides
G1 phase	×	G2 phase	interphase	waits

Figure 27.4 Cell fusions generate heterokaryons whose nuclei behave in a manner determined by the states of the cells that were fused.

When an S phase cell is fused with a G2 cell, the S phase nucleus continues to replicate, but the G2 nucleus does not replicate. This suggests that DNA that has replicated becomes refractory to the effects of the S phase activator, a feature to ensure that each sequence of DNA replicates only once. However, a compromise in the timing of mitosis is seen in such fusions. The S phase nucleus enters M phase sooner than it would

Figure 27.5 The phases of the cell cycle are controlled by discrete events that happen during G1, at S phase, and at mitosis.

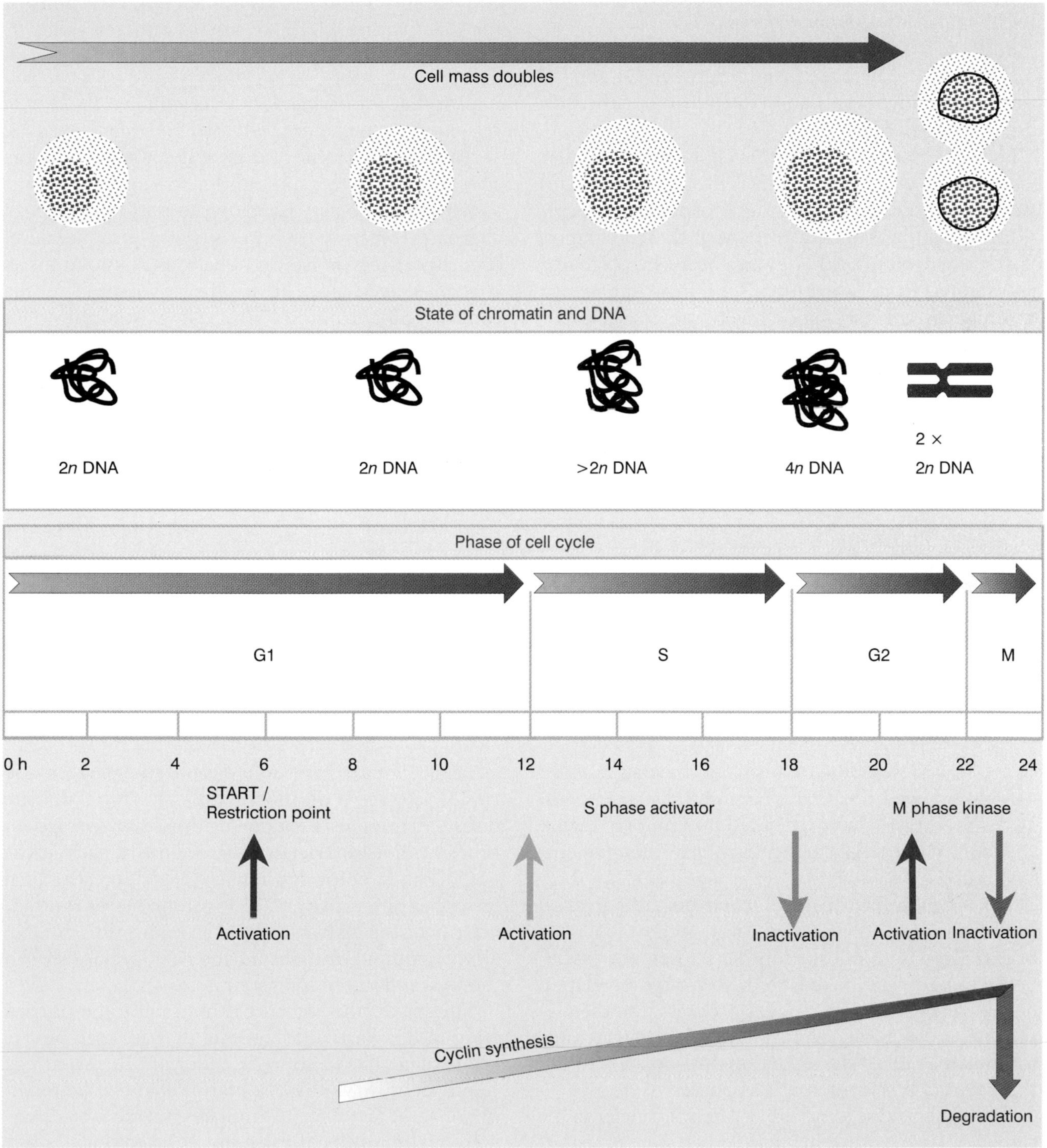

have in its previous cytoplasm, but *the G2 cell does not enter mitosis until after the S phase nucleus has completed replication.* This could mean that some regulator in the S phase cell—perhaps the S phase activator itself—inhibits the start of mitosis.

When a mitotic cell is fused with a cell at any stage of interphase, it causes the interphase nucleus to enter a pseudo-mitosis, characterized by a premature condensation of its chromosomes. This suggests that *an M phase inducer is present in dividing cells.*

Both the S phase and M phase inducers are present only transiently, because fusions between G1 and G2 cells do not induce replication or mitosis in either nucleus of the heterokaryon.

Figure 27.5 relates the cell cycle to the molecular basis for the regulatory cyclical events that control transitions between phases. Some of these events require the synthesis of new proteins or the degradation of existing proteins; other events occur by reversible activation or inactivation of pre-existing components. A minimum of three molecular activities must exist:

- During G1, an event whose molecular basis is unknown commits the cell to enter the cycle, that is, to pass START. We do not understand the nature of the lag period before S phase actually begins.
- The period of S phase is marked by the presence of the S phase activator. This is a protein kinase. It is related to the kinase that activates mitosis (which was identified first and is better characterized).
- Mitosis depends upon the activation of a pre-existing protein, the **M phase kinase,** which has two subunits. One is a kinase catalytic subunit that is activated by modification at the start of M phase. The other subunit is a **cyclin,** so named because it accumulates by continuous synthesis during interphase, but is destroyed during mitosis. Its destruction is responsible for inactivating M phase kinase and releasing the daughter cells to leave mitosis.

A striking feature of cell cycle regulation is that similar regulatory activities are employed in (probably) all eukaryotic systems. Some of these systems have cycles that superficially appear quite different from the normal somatic cycle. So very rapid divisions in which S phase alternates directly with mitosis characterize the development of the *Xenopus* egg, where entry into mitosis is controlled by M phase kinase, the very same factor that controls somatic mitosis. Yeast cells exist in a unicellular state, and certain species divide by an asymmetrical budding process; but control of entry into S phase and control of mitosis are determined by procedures and proteins related to those employed in the *Xenopus* egg. Homologous genes play related roles in organisms as distant as yeasts, insects, and mammals.

M phase kinase regulates entry into mitosis

THE early development of eggs of the toad *X. laevis* has provided a particularly powerful system to analyze the features that drive a cell into mitosis. **Figure 27.6** summarizes the early divisions of *Xenopus* embryogenesis.

In its immature form, the *Xenopus* oocyte is arrested in its first meiotic division cycle, just at the initial stage of chromosome condensation. The closest correspondence to a somatic cycle is to the G2 stage. Ovulation occurs when hormones trigger progress into the second meiotic division. When the egg is laid, it is arrested towards the end of the second meiosis in a condition that corresponds to a somatic M phase.

The egg is a large structure (~1 mm diameter in the case of *X. laevis*), which contains a vast store of material needed for early divisions. Fertilization triggers a series of very rapid division cycles. The initial division takes ~90 minutes; then during the **cleavage stage,** another 11 divisions occur synchronously, each lasting ~30 minutes. These divisions in effect represent an alternation of S phase and mitosis; the major synthetic activity of the cleavage egg is the replication of DNA, all the required proteins having been previously synthesized and stored in the oocyte.

The size of the egg allows material to be purified from it. It is particularly useful that arrested oocytes (equivalent to G2 somatic cells) and arrested eggs (equivalent to M phase somatic cells) can be readily obtained. A factor that stimulates mitosis was discovered by the results of injecting arrested oocytes with

Figure 27.6 MPF induces M phase in early *Xenopus* development

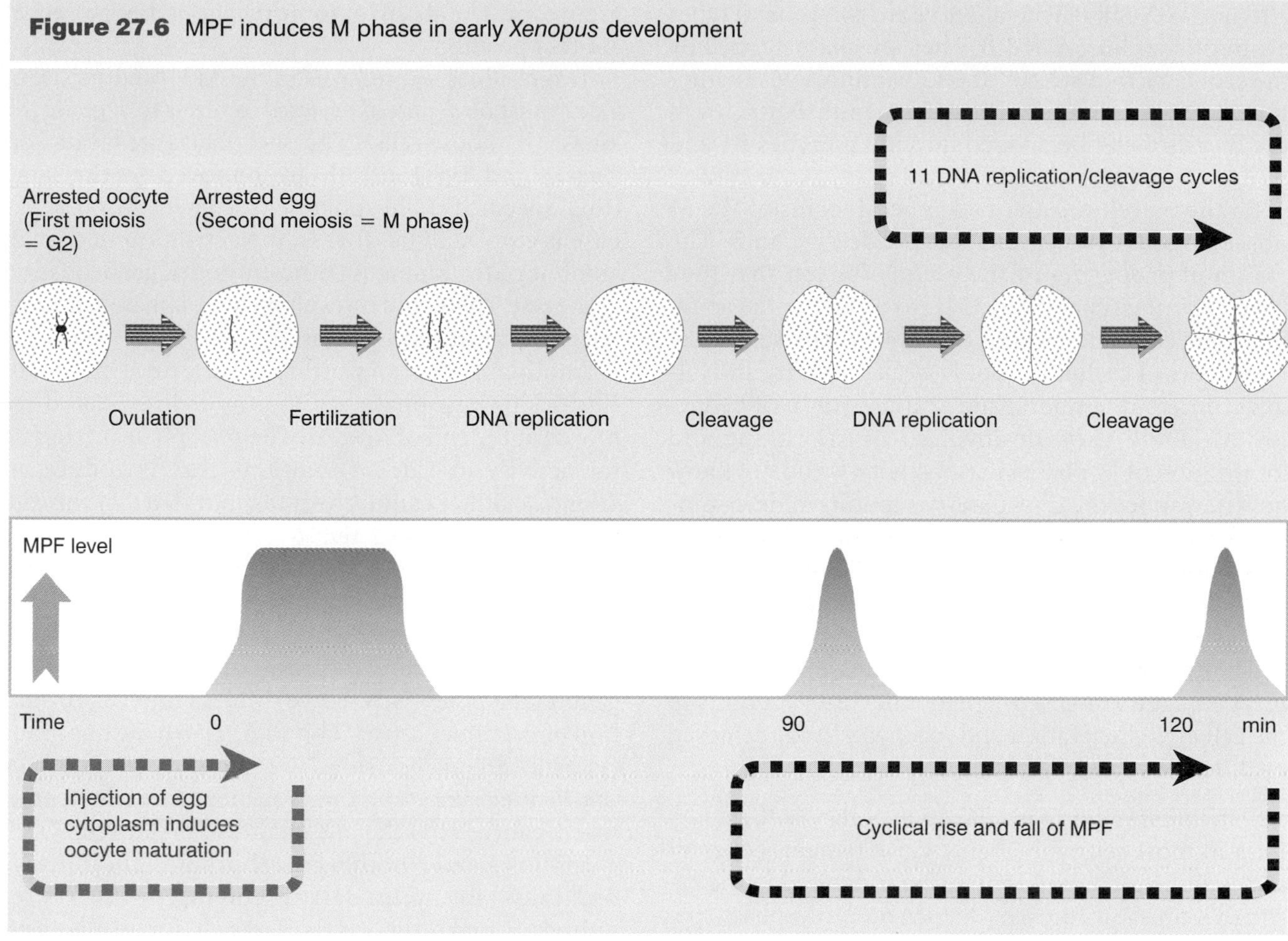

cytoplasm extracted from arrested eggs. The injection releases the immature oocytes from the G2 block, and induces them to enter M phase. The active component of the extract was therefore called **maturation promoting factor (MPF).** (Although MPF is present in arrested eggs, other factors prevent division from initiating.)

Because MPF turns out to have a general responsibility for causing somatic cells to enter M phase, MPF is now understood to stand for **M phase promoting factor**. The importance of MPF in inducing mitosis can be seen from its cyclical increase and decrease at the next stage of development, in the cleaving egg. As a synchronous wave of mitoses occurs in the cleavage egg, the level of MPF activity rises; as the mitoses are completed and S phase occurs, the MPF activity disappears.

MPF causes germinal vesicle (nuclear envelope) breakdown when injected into *Xenopus* oocytes, and induces several mitotic events in a cell-free system, including nuclear envelope disaggregation, chromosome condensation, and spindle formation. MPF is a kinase that can phosphorylate a variety of protein substrates. This immediately suggests a mode for its action: *by phosphorylating target proteins at a specific point in the cell cycle, MPF controls their ability to function.* In fact, the activity of the enzyme is most often assayed by its ability to phosphorylate target proteins (rather than by the induction of mitosis), and so the name **M phase kinase** provides a better description. It consists of two types of subunit with different functions:

- Cdc2 is the *catalytic subunit* which phosphorylates serine and threonine residues in target proteins. It is named for the homologous protein in *S. pombe.*
- Its partner is a cyclin; this is a *regulatory subunit* which is necessary for the kinase to function with appropriate substrates.

The crystal structure of a dimer of this type shows that the cyclin induces a change in the conformation of its partner that is necessary to create the active kinase site. So a catalytic subunit by itself is inactive, and can be active only when joined by a cyclin partner.

An M phase kinase typically has a single type of catalytic subunit (Cdc2), but may have any one of several alternative cyclin partners. There are two general types of mitotic cyclins, A and B. They are characterized by the sequence of a stretch of ~150 amino acids (sometimes known as the cyclin box). In mammals and frogs, the B cyclins can be divided into the subtypes B1 and B2.

So there are (at least) two general forms of the M phase kinase: Cdc2-cyclin A and Cdc2-cyclin B. The common properties of the cyclins suggest that they have the same type of function: to influence the activity of the catalytic subunit of M phase kinase. Yet the two classes of cyclins have only weak similarity and follow a different temporal and spatial pattern of behavior. Although they are involved in the timing and localization of M phase kinase activity, we do not know how they function as regulatory subunits in determining the specificity of the catalytic subunit. We should like to know in particular whether Cdc2-cyclin A and Cdc2-cyclin B recognize different proteins as substrates.

The events that activate M phase kinase at G2/M and inactivate it during M phase identify crucial points in the cell cycle. Activation and inactivation are achieved by different types of action:

- *Activation requires modification of the catalytic subunit.* In most cells, the level of Cdc2 remains constant through the cycle, and is in excess compared to the cyclins. Cyclins are necessary to turn on the M phase specific kinase activity of Cdc2. However, they cannot provide the activating event because they accumulate to a maximum level before the kinase activity appears. The intact Cdc2-cyclin dimer accumulates in an inactive form; modification of Cdc2 is the critical event that triggers the G2/M transition.
- *Inactivation is achieved by the physical destruction of the cyclin.* Cyclins were originally named for their property of accumulating continuously through the cell cycle; then they are destroyed abruptly by proteolysis during mitosis (see Figure 27.5). The timing of cyclin destruction is characteristic; typically A precedes B (by a few minutes in embryonic divisions, by rather longer in a cultured cell cycle), and this difference appears to be common to all cells.

What activates the catalytic activity of the M phase kinase? Cdc2 is itself a phosphoprotein, and its state of phosphorylation is a crucial determinant of activity. The events involved in the cycle of activation and inactivation of M phase kinase are summarized in **Figure 27.7**. For M phase kinase to be active, phosphate groups must be absent at some positions, but present at another position.

Two residues located within the ATP-binding site of Cdc2 must be *dephosphorylated* in order to activate the kinase (in mouse cells). The phosphates are located on Thr-14 and Tyr-15; both are removed by the same (dual specificity) phosphatase (see later). M phase kinase is autocatalytic, that is, the activation of a small amount of the kinase is sufficient to trigger activation of the rest. This could be explained if M phase kinase itself activates the phosphatase.

Another phosphorylation occurs on Thr-161 of Cdc2. This phosphate group is added in G2 and removed at the end of mitosis. The phosphate is *required* for activity of Cdc2; mutations that introduce an amino acid that cannot be phosphorylated at this site inactivate the kinase activity.

Association of the catalytic and cyclin subunits, and phosphorylation and dephosphorylation, occur in a specific order. Cyclin B associates specifically with the tyrosine-*dephosphorylated* form of Cdc2 *in vitro*. This generates a potentially active dimer. However, formation of the dimer causes Thr-14/Tyr-15 to be phosphorylated. So *association with cyclin induces the inactivating event*. The dimer is then maintained in its inactive form until the phosphates are removed.

Cyclins A and B both have a short motif near the N-terminus—the cyclin destruction box—which is required to make the cyclin a target for proteolysis. Cyclins are degraded by a common proteolytic system, the proteasome (a complex containing proteolytic activities that recognizes its targets when ubiquitin is added to them; see Chapter 8).

Destruction of the cyclin subunits is responsible for inactivating M phase kinase during mitosis, and indeed is necessary for cells to exit mitosis. A truncated cyclin B that lacks the N-terminal region is resistant to proteolysis. When this protein is synthesized in *Xenopus* eggs, or in a cell-free extract that undertakes some of the typical cycling reactions, it causes anaphase arrest. This is the basis for concluding that *loss of kinase activity is a prerequisite for completing mitosis*.

An important question is how the roles of cyclin A and cyclin B differ in mitosis. A hint that their functions are different is provided by differences in the timing of their synthesis and destruction, but there is as yet no direct evidence to show whether both cyclins are required for passage through mitosis in an animal cell or whether they are redundant with one another.

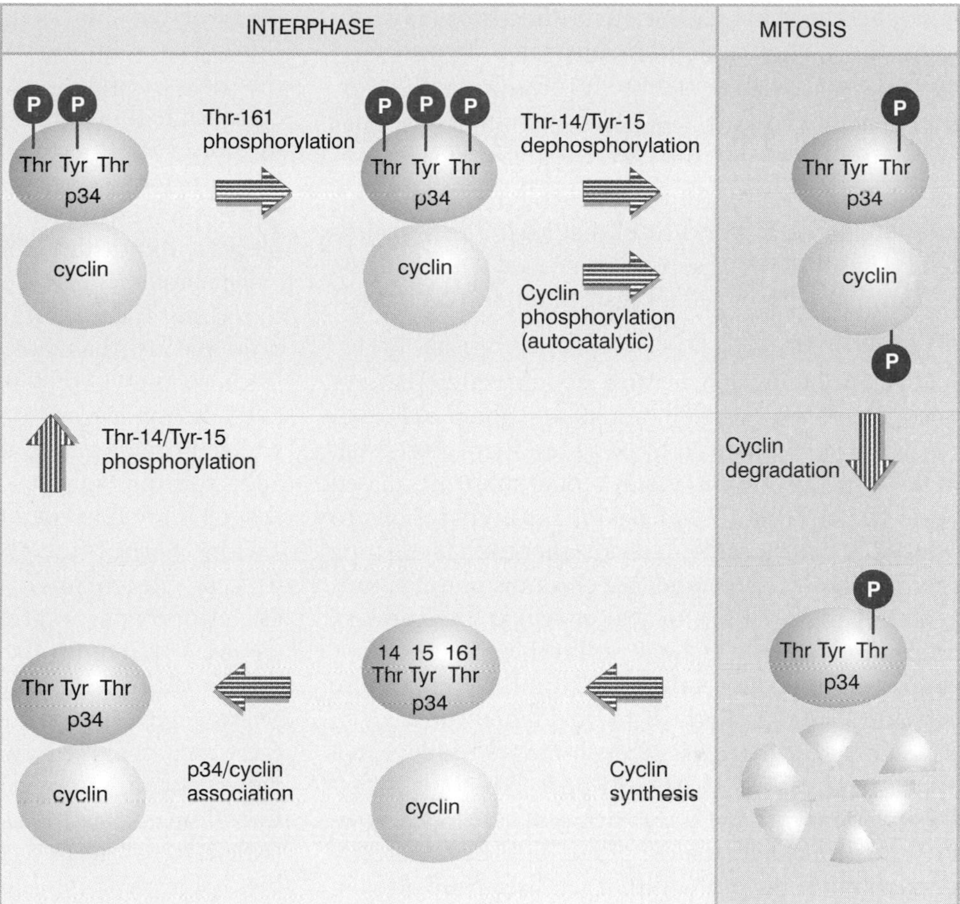

Figure 27.7 The activity of M phase kinase is regulated by phosphorylation, dephosphorylation, and protein proteolysis. The 3 phosphorylated amino acids are: Thr-14, Tyr-15, Thr-161. The first two are in the ATP-binding site. The relative timing of all the events is not certain.

Protein phosphorylation and dephosphorylation control the cell cycle

PHOSPHORYLATION (catalyzed by kinases) and dephosphorylation (catalyzed by phosphatases) are the critical events that regulate the cell cycle. *They are used both to control the activities of the regulatory circuit itself and to control the activities of the substrates that execute the decisions of the regulatory circuit.*

The cell cycle regulatory circuit consists of a series of kinases and phosphatases that respond to external signals and checkpoints by phosphorylating or dephosphorylating the next member of the pathway. The ultimate readout of the circuit is to determine the activity of M phase kinase (or the S phase kinase) by controlling its state of phosphorylation.

Activation of M phase kinase is the event that triggers onset of M phase. Inactivation is necessary to exit M phase. This suggests that the events regulated by M phase kinase are reversible: *phosphorylation of substrates is required for the reorganization of the cell into a mitotic spindle, and dephosphorylation of the same substrates is required to return to an interphase organization.*

What are the targets for M phase kinase? A major reorganization of the cell occurs at mitosis, and the ability of MPF to induce mitosis implies that the M phase kinase, directly or indirectly, triggers these activities. We discuss the reorganization of structure later, but ask now whether M phase kinase acts directly or indirectly upon the various potential substrates. Two general models could be proposed for its role:

- It may be a "master regulator" that phosphorylates target proteins that in turn act to regulate other necessary functions—a classic cascade.
- Or it may be a "workhorse" that itself directly phosphorylates the crucial substrates needed to execute the regulatory events or cell reorganization involved in the cycle.

The only common feature in substrates that are phosphorylated by M phase kinase is the presence of the duo Ser-Pro, flanked by basic residues (most often in the form Ser-Pro-X-Lys). Potential substrates, based upon the ability of M phase kinase preparations to phosphorylate targets *in vitro,* include H1 histone (perhaps required to condense chromosomes), lamins (possibly required for nuclear envelope breakdown), nucleolin (potentially involved in the arrest of ribosome synthesis), and other structural and enzymatic activities. The strength of the evidence varies as to which of these targets is phosphorylated *in vivo* in a cyclic manner and whether M phase kinase is in fact the active enzyme. From the variety of substrates, however, it seems likely that M phase kinase acts directly upon many of the proteins that are directly involved in the change in cell structure at mitosis.

What criteria should be applied to conclude that a potential substrate is an authentic target for Cdc2 in the cell cycle? The same sites should be phosphorylated by Cdc2 *in vitro* that are cyclically phosphorylated *in vivo* at time(s) when Cdc2 is known to be active. Ideally it should be possible to show that a mutation in Cdc2 kinase activity blocks the phosphorylation *in vivo,* but this is at present practical only in yeast. To conclude that the phosphorylation is a significant event in the cycle, some function of the protein must be altered by the presence of phosphate. This can be tested by making mutations at the amino acid that is phosphorylated to determine whether the absence of phosphorylation blocks a mitotic function.

The best characterized substrate for Cdc2 kinase is the H1 histone (one of the 5 histones that are the major protein constituents of chromatin; see Chapter 19). It has been known for a long time that H1 is phosphorylated during the cell cycle, with 2 phosphate groups added during S phase, and 4 further phosphate groups added during mitosis. The major H1 kinase activity of the cell is provided by M phase kinase.

What purpose the phosphorylation of H1 serves in the cell cycle remains a matter for speculation, since no effects upon chromatin structure have been directly demonstrated. It is reasonable to suppose that it might be connected with chromosome condensation at M phase. Not enough is known about the timing of modification at S phase to wonder whether it is concerned with preparations for replication (which might require uncoiling) or with the consequences of replication (when preparations for mitosis could begin). But H1 histone is an exceedingly good substrate for kinases based on the Cdc2 engine, with the result that H1 kinase activity has become the usual means by which this enzyme is assayed *in vitro.* An illustration of the appeal of this assay is its application to *S. cerevisiae,* where H1 kinase activity is routinely measured to assess the cyclic activity of M phase kinase, although this yeast in fact is unusual in containing no H1 histone!

Cdc2 is the key regulator in yeasts

To define the circuit for cell cycle control, we need to identify the genes that regulate progression through the cycle. We define a mutation in the cell cycle as one that *blocks the cell cycle at a specific stage,* a definition that excludes mutations in genes that control continuous processes of growth and metabolism. But the approach of characterizing mutants that are unable to proceed through the cycle has not been straightforward to develop. A mutation that prevents a cell from dividing will be lethal; and the reverse is also true, insofar as any lethal mutation will stop cells from dividing. It has therefore been difficult to devise procedures to isolate cell cycle mutants of animal cells, or, indeed, to demonstrate that potential mutants have specific blocks in the cell cycle.

Because of their nature, cell cycle mutants must be obtained as conditional lethals, so that although they are unable to grow under the conditions of isolation, they can be maintained by growth in other conditions. A series of such mutants has been isolated in two yeasts, in which the block to the cell cycle can be seen to affect the visible phenotype of the cell.

The mitotic cycles of fission yeast (*S. pombe*) and baker's yeast (*S. cerevisiae*) are summarized in **Figure 27.8.**

The cell cycle of *S. pombe* is divided into the usual phases. The cell grows longitudinally and then divides; progress through the cell cycle can be assessed (approximately) by the physical length of the cell (which doubles from 7–8 μm to 14–16 μm).

The cycle of *S. cerevisiae* is unusual. Cells proceed almost directly from S phase into division, so there is effectively very little or no G2 phase. (A short G2 phase is shown in this and subsequent figures for the purpose of localizing the relative timing of events concerned with the transition into M phase.) And instead of an equal division of the cell, the daughter cell grows as a **bud** off the mother cell, eventually obtaining its independence when it is released as a small separate cell. Again the cell cycle can be followed visually in terms of the growth of the bud. Mitosis itself shares some unusual features in both types of yeast; the nuclear membrane does not break down, and segregation of chromosomes therefore is compelled to occur within the nucleus.

Extensive screens for **cell division cycle** or *cdc* mutants have been performed in both yeasts. Initial isolation relies upon the criterion that cells accumulate at a particular stage of the cell cycle at an elevated temperature (36°C), but continue normally through the cycle at 23°C. The mutant phenotype allows cells to continue growing while the cycle is blocked, causing an obvious aberration; in *S. pombe* the cells become highly elongated, and in *S. cerevisiae* they fail to bud. **Figure 27.9** compares cell cycle mutants with wild-type *S. pombe.* The left panel shows a group of normal cells; the center panel shows a *cdc* mutant that is blocked in the ability to divide, and has therefore become elongated, because it has continued to grow.

Because the mutants are temperature sensitive, the time at which a mutation takes effect can be determined by temperature shift protocols in which cells are shifted up in temperature at a specific point in the cycle. If this point is prior to the point at which the mutation takes effect, the cells halt at the execution point, but if the point is past the time when the mutation acts, the cells continue their cycle.

Of the order of 80 *cdc* genes have been identified in each species, but not all of these loci are concerned with regulating the cell cycle. A significant number represents genes whose products are needed for structural purposes; for example, absence of enzymes that replicate DNA or synthesize the nucleotide precursors can block progression through S phase.

Yeast cells may exist in either haploid or diploid form. They have two forms of life cycle, as illustrated in **Figure 27.10.** Haploid cells double by mitosis. A haploid cell has a **mating type** of either **a** or α. Haploids of opposite types enter the sexual mating pathway, in which they conjugate (fuse) to form diploid cells. The

Figure 27.8 *S. pombe* lengthens and then divides during a conventional cell cycle, while *S. cerevisiae* buds during a cycle in which G2 is absent or very brief, and M occupies the greatest part.

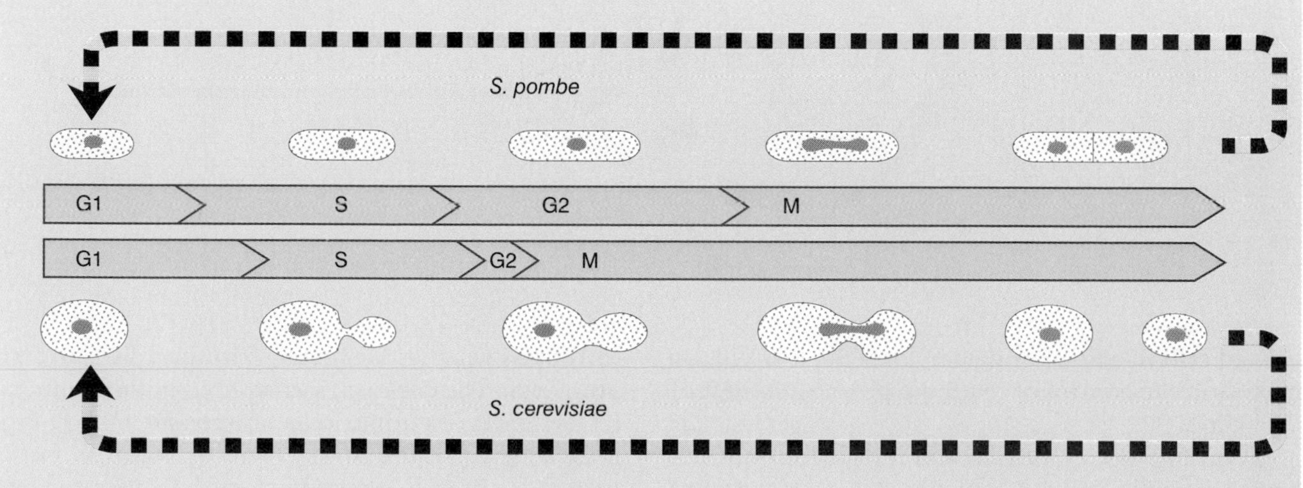

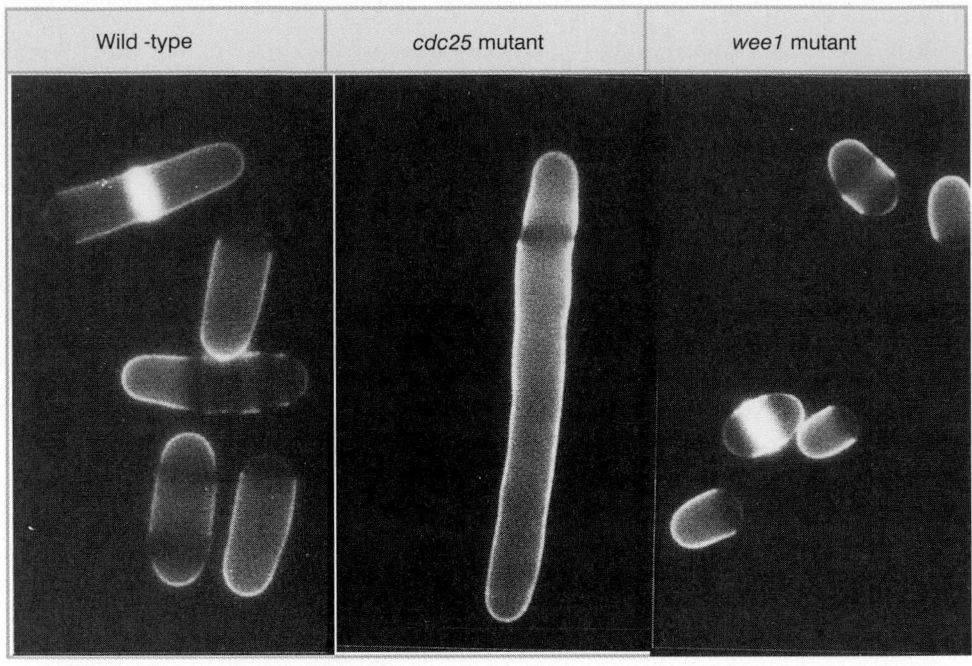

Figure 27.9 *S. pombe* cells are stained with calcofluor to identify the cell wall (surrounding the yeast cell) and the septum (which forms a central division when a cell is dividing). Wild-type cells double in length and then divide in half, but *cdc* mutants grow much longer, and *wee* mutants divide at a much smaller size. Photograph kindly provided by Paul Nurse.

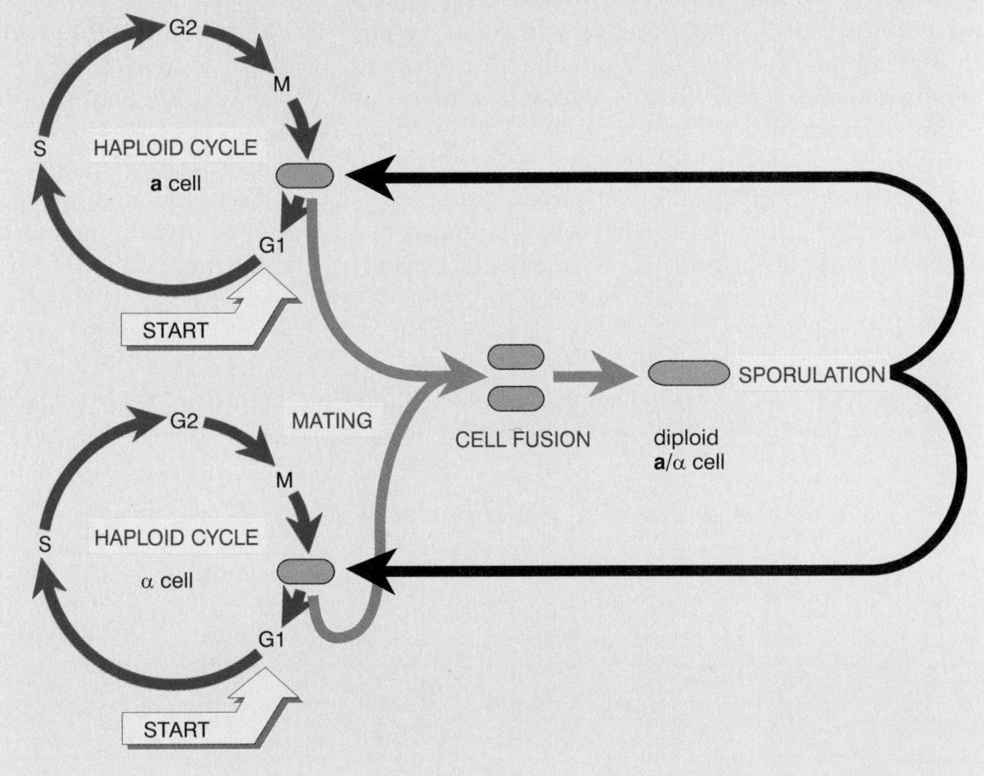

Figure 27.10 Haploid yeast cells of either **a** or α mating type may reproduce by a mitotic cycle. Cells of opposite type may mate to form an **a**/α diploid. The diploid may sporulate to generate haploid spores of both **a** and α types.

diploid cells in turn sporulate to form haploid cells by meiosis. (The control of mating type is the subject of Chapter 17.)

A crucial point in the cell cycle is defined by the behavior of haploid cells. A haploid cell decides at a point early in G1 whether to proceed through a division cycle or to mate. The decision is influenced by environmental factors; for example, cells of opposite mating type must be present for conjugation to occur. In fact, a mating factor (a polypeptide hormone) secreted by a cell of one type causes a cell of the other type to arrest its cycle (see Chapter 17). But a mating factor can divert a cell into the mating pathway *only early in G1;* later in G1 the cell becomes committed to the division cycle and cannot be stimulated to enter the mating pathway. This assay was how the commitment point in G1 (START) was originally identified.

The events and genes involved in the cycle of *S. pombe* are summarized in **Figure 27.11.** Various *cdc* mutants of *S. pombe* block the cell cycle at one of two stages: at the boundary between G2 and M phase; or in G1 at START.

cdc2 is identified as a crucial regulator by its involvement at *both* stages of cell cycle block: mutants of *cdc2* may be blocked prior to START or prior to M phase (depending on the point a cell had reached in the cycle when the mutation took effect).

The homologous gene in *S. cerevisiae* is called *CDC28.* A deficiency in the gene of either type of yeast can be corrected by the homolog from the other yeast. In both yeasts, proteins that resemble B-type cyclins associate with Cdc2/CDC28 at G2/M.

Figure 27.11 The cell cycle in *S. pombe* requires *cdc* genes to pass specific stages, but may be retarded by genes that respond to cell size (*wee1*). Cells may be diverted into the mating pathway early in G1.

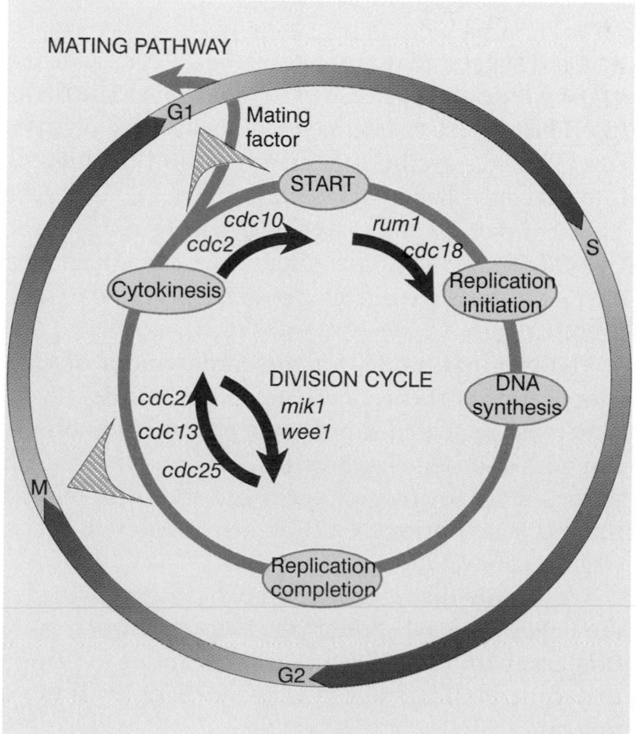

A major insight into the nature and function of M phase kinase was provided by the observation that *Xenopus Cdc2* is the homolog of *cdc2* of *S. pombe* and *CDC28* of *S. cerevisiae.* (When this discovery was made, the *Xenopus* protein was called p34, after its size, but then it was renamed Cdc2 after its name in *S. pombe.*)

The existence of a Cdc2 catalytic subunit, in organisms as diverse in evolution as yeasts, frogs, and mammals, identifies the key feature of cell cycle control. Conservation of function is indicated by the ability of the cloned human gene to complement the deficiency in *cdc2* mutants of *S. pombe.*

(Ability to compensate for the deficiency of a specific yeast mutant has been used with great effect to identify higher eukaryotic genes homologous to several cell cycle regulators of yeast. The assay is essentially to introduce the cloned gene into a yeast mutant and to identify cells that resume growth. It is quite remarkable that the control of the cell cycle has been so well conserved as to make this possible. However, it should be remembered that this is sometimes a lax assay that leads to the identification of a gene that is only loosely related to the mutant function.)

Cdc2 in *S. pombe* has different partners at each stage of the cell cycle. At mitosis, its partner is the product of *cdc13*, generating an M phase kinase that resembles the Cdc2-B cyclin dimer of animal cells. During G1 in *S. pombe*, the active form of Cdc2 has a different partner, the B-like cyclin cig2 (there is also a related cyclin, cig1).

The activity of the Cdc2 catalytic subunit in these dimers (and of the equivalent CDC28 in *S. cerevisiae*) is controlled by phosphorylation in the same way as Cdc2 in animal cells. A difference is that in yeast there is no Thr-14, so there are only two relevant sites: Tyr-15 where phosphorylation is inhibitory; and Thr-161 where phosphorylation is required.

Figure 27.12 summarizes changes in phosphorylation of the alternative dimers during the cell cycle.

The upper part of the figure summarizes the condition of the M phase kinase. At mitosis, the Cdc2 subunit of the Cdc2/Cdc13 dimer is in the active state that lacks the phosphate at Tyr-15 and has the phosphate at Thr-161. At the end of mitosis, kinase activity is lost when Cdc13 is degraded. The state of phosphorylation of Cdc2 does not change at this point. As new Cdc13 is synthesized, it associates with Cdc2, but the dimeric complex is maintained in an inactive state by another protein (see Figure 27.16). After START, however, activity of the Cdc2/Cdc13 dimer is inhibited by addition of the phosphate to Tyr-15. Removal of the inhibitory phosphate is the trigger that activates mitosis.

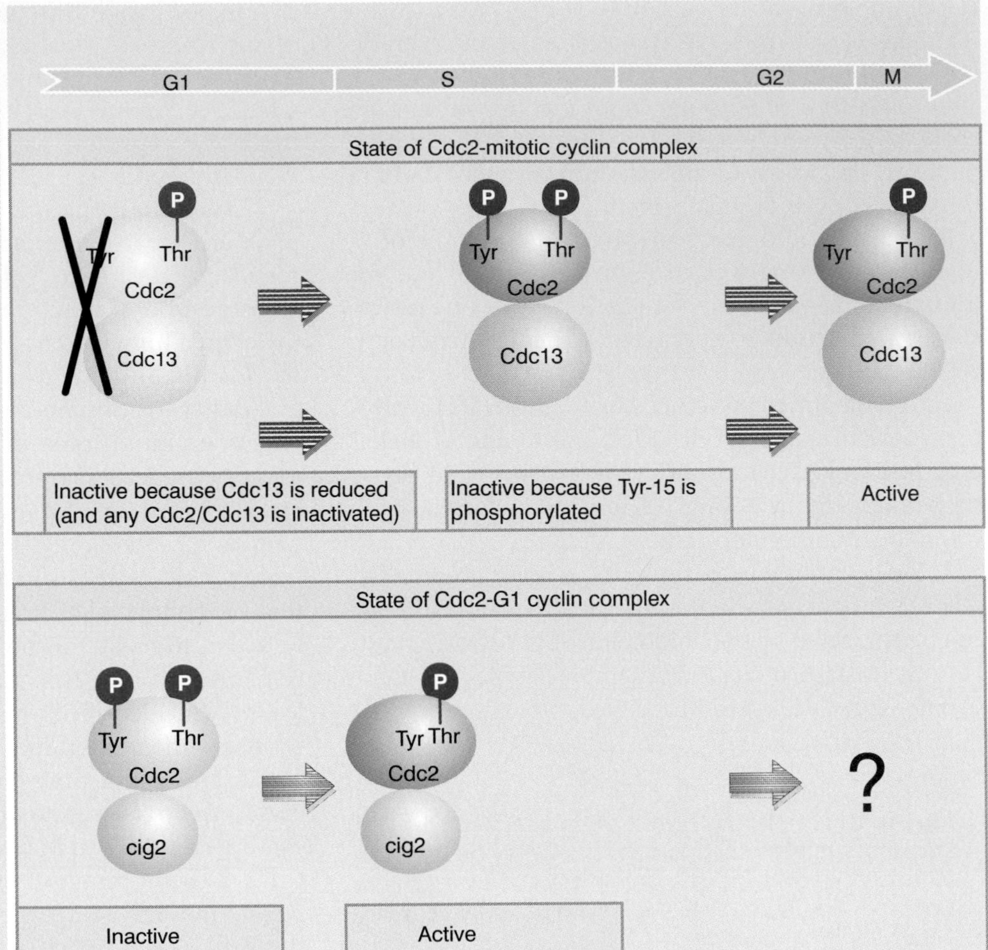

Figure 27.12 The state of the cell cycle in *S. pombe* is defined by the forms of the Cdc2-cyclin complexes.

The lower part of the figure shows that Cdc2 forms a dimer with cig2 during G1. This forms a kinase with the same general structure as the M phase kinase. This kinase is in effect a counterpart to the M phase kinase, and it controls progression through the earlier part of the cell cycle. cig2 resembles a B-type cyclin. The dimer is converted from the inactive state to the active state by dephosphorylation of the Tyr-15 residue of Cdc2 at or after START. We do not know what happens to the Cdc2-cig2 dimer at the transition from S phase into G2.

The existence of alternative dimers during the cell cycle suggests the concept that *cyclins can be classified according to their period of active partnership with Cdc2* as **G1 cyclins** or **mitotic (or G2) cyclins.** (Note that the cyclins defined in this way by their partnership with Cdc2 or a Cdc2-like protein do not necessarily have the property of cyclic degradation by which the original cyclins were defined. Cyclic degradation is a general property specifically for mitotic cyclins in animal cells.)

We may take the concept of cyclin classification further and suggest that *the phase of the cell cycle is defined by the nature of the active Cdc2-cyclin dimer.* The period of G1 and S phase is defined by the presence of an active Cdc2-G1 cyclin dimer. Progression through G1 into S is controlled by activation of the Cdc2-G1 cyclin. The period of G2 and mitosis is characterized by an active Cdc2-mitotic cyclin dimer. The transition from G2 into mitosis is controlled by activation of the Cdc2-mitotic cyclin.

The original model for the employment of Cdc2 proposed that the available form of the Cdc2-cyclin dimer was regulated simply by replacement of one cyclin with another. However, it seems now that the mitotic and G1 forms may coexist in the yeast cell, but that their activities are differently regulated at each stage of the cycle.

We can divide *cdc* genes into two classes, defined by the stage of the cell cycle at which the effects of a mutation are manifested. We know most about the circuit that controls the state of the Cdc2/Cdc13 dimer at mitosis.

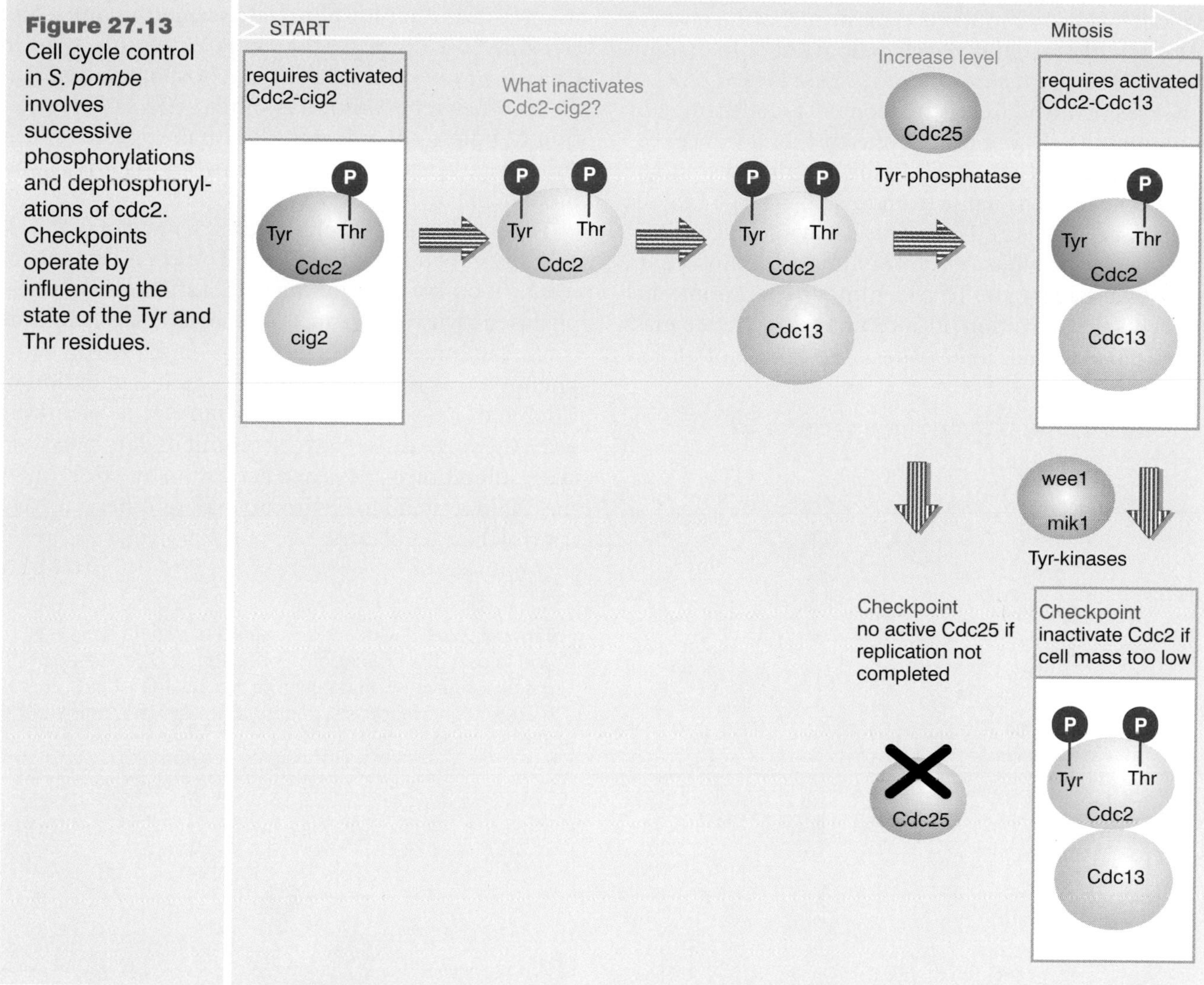

Figure 27.13 Cell cycle control in *S. pombe* involves successive phosphorylations and dephosphorylations of cdc2. Checkpoints operate by influencing the state of the Tyr and Thr residues.

The state of Cdc2 complexes and the enzymes that act upon them during the *S. pombe* cell cycle are summarized in **Figure 27.13.** We do not know which enzymes act upon the Cdc2-cig2 dimer during G1. By G2, the predominant form is the Cdc2/Cdc13 dimer whose activity is controlled by antagonism between kinases and phosphatases that respond to environmental signals or to checkpoints.

cdc25 codes for a tyrosine phosphatase that is required to dephosphorylate Cdc2 in the Cdc2/Cdc13 dimer. It is probably responsible for the key dephosphorylating event in activating the M phase kinase. It is not a very powerful phosphatase (its sequence is atypical), and the quantities of Cdc25 and Cdc2 proteins are comparable, so the reaction appears almost to be stoichiometric rather than catalytic.

The level of Cdc25 increases at mitosis. Its accumulation over a threshold level could be important. In mutants of *cdc2* that do not require *cdc25,* or in strains that over-express *cdc25,* blocking DNA replication does not impede mitosis. (Wild-type cells arrest in the cell cycle if DNA synthesis is prevented, for example, by treatment with hydroxyurea. But these mutant cells attempt to divide in spite of the deficiency in replication, with lethal consequences.) This suggests that Cdc25 executes the checkpoint that ensures S phase is completed before M phase can be activated.

Under normal conditions, the cell division cycle is related to the size of the cell. In poor growth conditions, when the cells increase in size more slowly, G1 becomes longer, because START does not occur until the cells attain a critical size. This is a protection against starting a cell division cycle and risking division before the amount of material is adequate to support two daughter cells. *cdc* mutants typically delay the onset of mitosis and lead either to cell cycle arrest or to division at increased size (as shown for *cdc25* in Figure 27.9.)

Genes involved in cell size control are identified by mutants with the opposite property: they advance cells into mitosis and therefore divide at reduced size. The *wee1* gene takes its name from this phenotype (see the right panel of Figure 27.9). This behavior suggests that *wee1* usually inhibits cells from initiating mitosis until their size is adequate. This identifies a checkpoint that prevents the activation of Cdc2 until an adequate mass has been attained. *wee1* codes for a kinase of an unusual sort: it can phosphorylate serine/threonine and tyrosine (we discuss other kinases of this type in Chapter 26). It inhibits Cdc2 by phosphorylating Tyr-15. Another gene, *mik1*, has similar effects. The deletion of both *wee1* and *mik1* is lethal, suggesting that either gene can fulfill the same function. This is a common theme in the yeast cell cycle.

The products of *wee1* and *cdc25* play antagonistic roles, as shown in Figure 27.13. The kinase activity of *wee1* acts on Tyr-15 to inhibit Cdc2 function. The phosphatase activity of *cdc25* acts on the same site to activate Cdc2. Mutants that over-express *cdc25* have the same phenotype as mutants that lack *wee1*. Regulation of Cdc2 activity is therefore important for determining *when* the yeast cell is ready to commit itself to a division cycle; inhibition by *wee1* and activation by Cdc25 allow the cell to respond to environmental or other cues that control these regulators.

Figure 27.14 DNA damage triggers the G2 checkpoint.

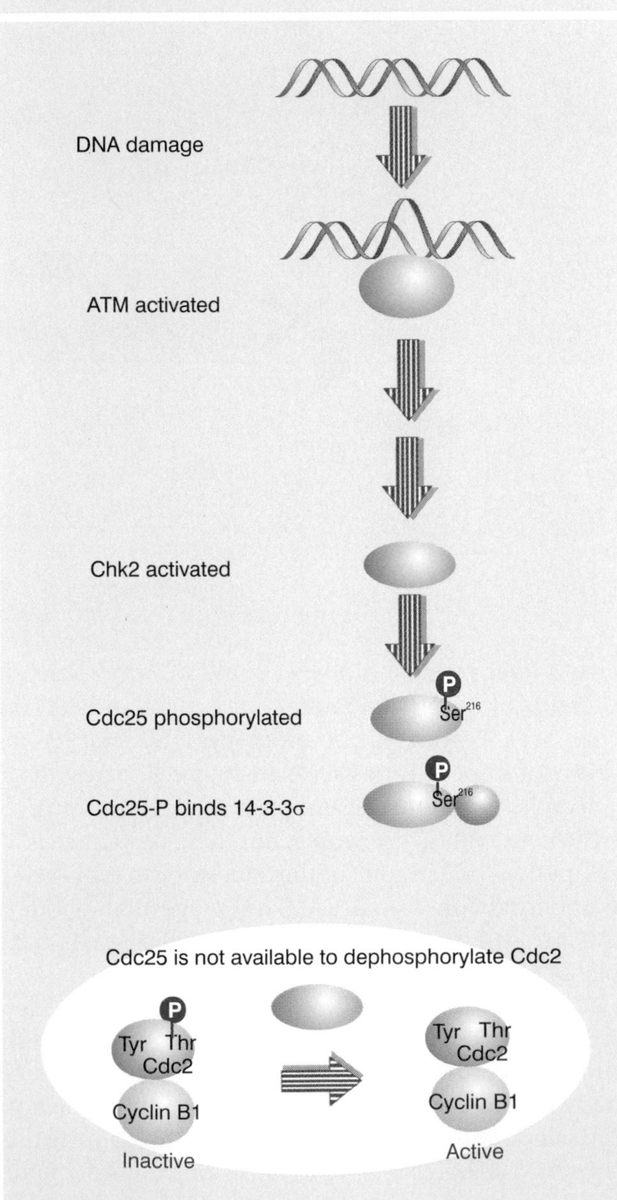

Figure 27.15 Cdc18 is required to initiate S phase and is part of a checkpoint for ordering S phase and mitosis. Active M phase kinase inactivates Cdc18, generating a reciprocal relationship between mitosis and S phase.

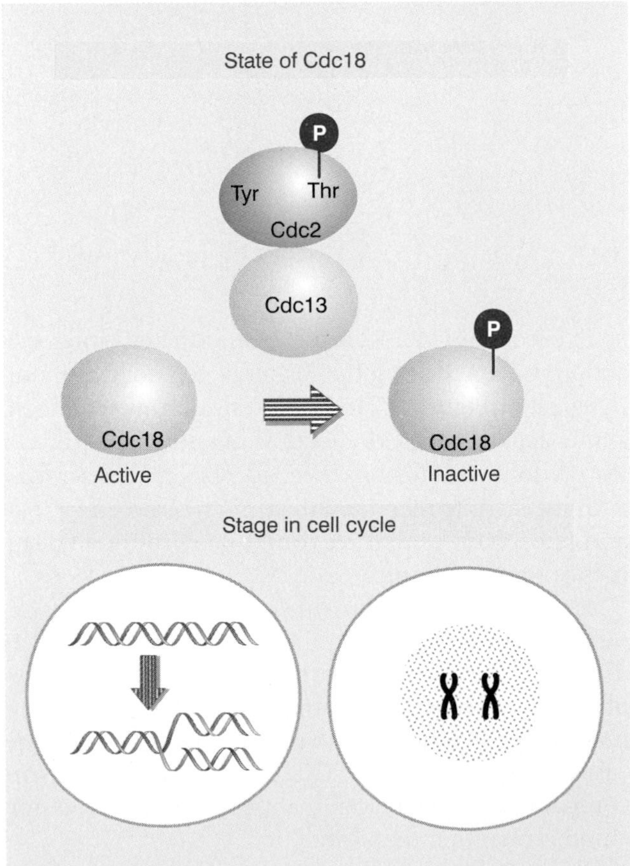

It is striking that all of the genes known to affect the G2/M boundary appear to have been widely conserved in evolution. Extending beyond the conservation of the components of the M phase kinase, *cdc25* has a counterpart in the *string* gene of *D. melanogaster*, and analogous proteins are found in amphibian and mammalian cells.

The basic principle established by this work is that the activity of the key regulator, Cdc2, is controlled by kinase and phosphatase activities that themselves respond to other signals. Cdc2 is the means by which all of these various signals are ultimately integrated into a decision on whether to proceed through the cycle.

Treatment with agents that damage DNA causes many types of cell to halt in G2 as the result of a checkpoint that prevents progress into mitosis until the damage has been repaired. **Figure 27.14** shows that the G2 checkpoint involves both kinases and phosphatases. One of the sensors of DNA damage in mammalian cells is the kinase ATM. This in turn activates (directly or indirectly) the kinase Chk2. Chk2 (and also the related kinase Chk1) phosphorylates Cdc25 on the residue Ser^{216}. This causes Cdc25 to bind to the protein 14-3-3σ, which maintains it in an inactive state. As a result, Cdc25 cannot dephosphorylate Cdc2, so M phase cannot be activated. A similar pathway applies the checkpoint in *S. pombe*, although in *S. cerevisiae* the target is different (the apparatus that releases cells from anaphase).

Activation of the G1/S form of the kinase (Cdc2/cig2 in *S. pombe*) is required to enter S phase, but inactivity

Figure 27.16 Rum1 inactivates the M phase kinase, preventing it from blocking the initiation of S phase.

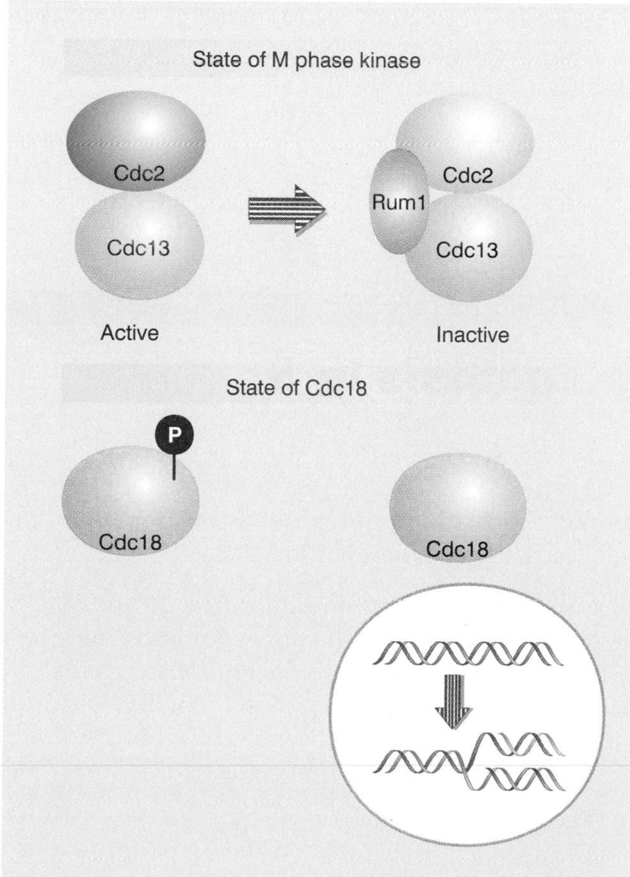

Figure 27.17 Checkpoints can arrest the cell cycle at many points in response to intrinsic or extrinsic conditions.

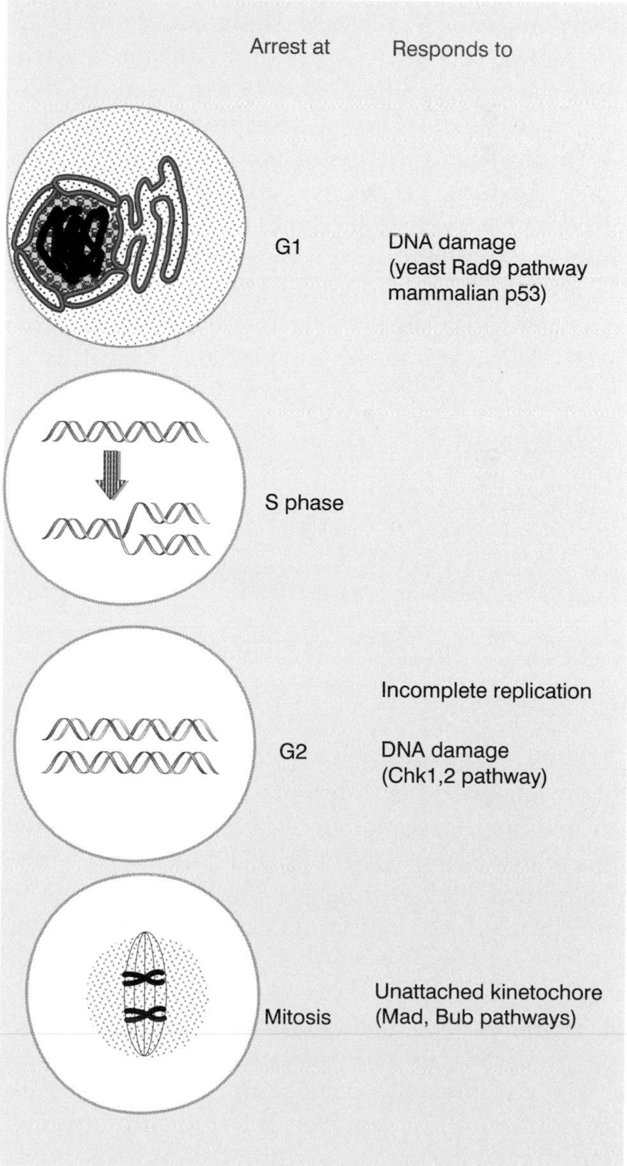

of the M phase form (Cdc2/Cdc13) is also required. Mutants in *cdc13* fail to enter mitosis (of course), but also undergo multiple cycles of DNA replication, suggesting that the M phase kinase usually inhibits S phase. This provides a checkpoint that ensures the alternation of S phase and mitosis. Activation of the M phase kinase during G2 prevents further rounds of DNA replication occurring before mitosis; and inactivation of the M phase kinase prevents another mitosis from occurring before the next S phase has occurred.

The target protein through which this circuit functions is probably Cdc18. Transcription of *cdc18* is activated as a consequence of passing START, and Cdc18 is required to enter S phase. Over-expression of *cdc18* allows multiple cycles of DNA replication to occur without mitosis. If Cdc18 is a target for Cdc2/Cdc13 M phase kinase, and is inactivated by phosphorylation, the circuit will take the form shown in **Figure 27.15.** For Cdc18 to be active, Cdc2/Cdc13 must be inactive. When the M phase kinase is activated, it causes Cdc18 to be inactive (possibly by phosphorylating it), and thereby prevents initiation of another S phase.

Activity of the Cdc2/Cdc13 M phase kinase is itself influenced by the factor Rum1, which controls entry into S phase. When *rum1* is over-expressed, multiple rounds of replication occur, and cells fail to enter mitosis. When *rum1* is deleted, cells enter mitosis prematurely. These properties suggest that Rum1 is an inhibitor of the M phase kinase. It is expressed between G1 and G2 and keeps any M phase kinase in an inactive state. (This is important during G1, before the inhibitory phosphate is added to Tyr-15; see Figure 27.12.) It also represses the level of Cdc13 protein. The overall effect is to minimize M phase kinase activity, which is necessary to allow S phase to proceed. The consequences of the production of Rum1 on the state of M phase kinase, and consequently on the state of Cdc18, are illustrated in **Figure 27.16,** which suggests a model for the overall circuit to control S phase.

A general theme emerges from these results: the circuits that control the cell cycle have interlocking feedback loops to ensure orderly progression. And giving dual roles to a single component in which its activity is necessary to promote one event but to block another creates an intrinsic alternation of events. So an active M phase kinase simultaneously promotes mitosis as a legitimate event and inhibits S phase as an illegitimate event. This creates an **intrinsic checkpoint:** one event cannot be initiated until the state of the component responsible for the prior event has been reversed. By contrast, the pathway of Figure 27.14 is an example of an **extrinsic checkpoint:** in response to specific conditions, a pathway is activated whose end result is to alter the state of regulatory components in the cell cycle pathway. **Figure 27.17** summarizes the stages at which some important checkpoints act.

CDC28 acts at both START and mitosis in *S. cerevisiae*

THE analysis of *cdc* mutants in *S. cerevisiae* shows that there is more than one type of cycle, although the cycles are connected at crucial points. (Remember that mitosis in *S. cerevisiae* is unusual morphologically, and chromosome segregation occurs within the intact nucleus; see Figure 27.8.) **Figure 27.18** shows how the three cycles of *S. cerevisiae* relate to the conventional phases of the overall cell cycle:

- The **chromosome cycle** comprises the events required to duplicate and separate the chromosomes, consisting of the initiation, continuation, and completion of S phase, and nuclear division. So a mutation such as *cdc8* stops this cycle in S phase.
- Mutations in the chromosome cycle do not stop the **cytoplasmic cycle,** which consists of bud emergence and nuclear migration into the bud (visualized at the start of M phase in Figure 27.8). This cycle can be halted before bud emergence by *cdc24,* but the mutation does not prevent chromosome replication.
- The **centrosome cycle** consists of the events associated with the duplication and then separation of the spindle pole body (SPB), which in effect substitutes for

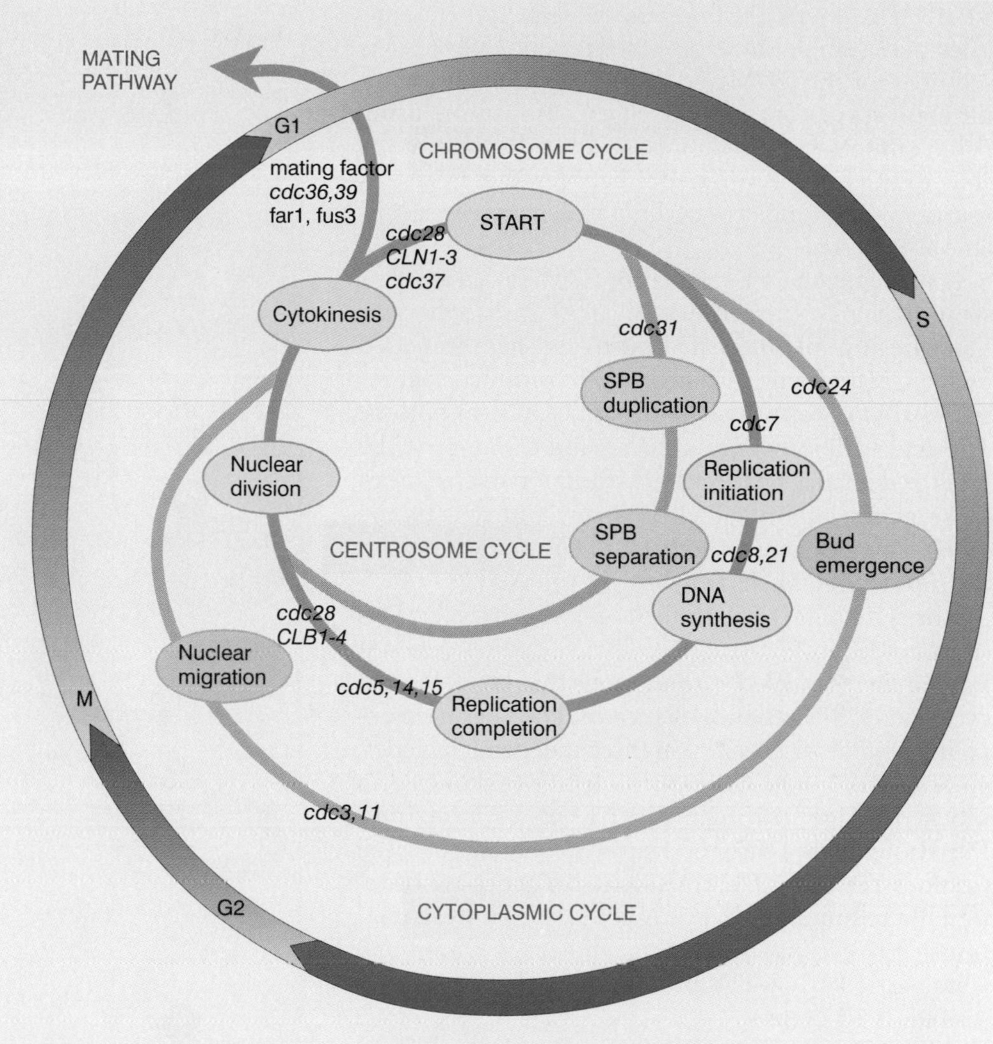

Figure 27.18 The cell cycle in *S. cerevisiae* consists of three cycles that separate after START and join before cytokinesis. Cells may be diverted into the mating pathway early in G1.

the centrosome and organizes microtubules to allow chromosome segregation within the nucleus. Blocking this cycle, for example with *cdc31,* does not prevent S phase or bud emergence.

Completion of an entire cell cycle requires all three constituent cycles to be functional, since nuclear division requires both the chromosome and centrosome cycles, and cytokinesis requires these and the cytoplasmic cycle.

The decision on whether to initiate a division cycle is made before the point START. The crucial gene in passing START is *CDC28* (the homolog of *cdc2* of *S. pombe*). Mutations in *CDC28* prevent all three cycles from proceeding. The ability to pass START is determined by environmental conditions, since stationary phase populations that are limited by nutrients are arrested at START. Morphologically, the cells of such a population are arrested at a point after cell separation and prior to SPB duplication, bud emergence, and DNA replication, which is to say that when *S. cerevisiae* exhaust their nutrients, they complete the current cycle and arrest all three cycles before START. In terms of phases of the cell cycle, this corresponds to a point early in G1.

Most mutants in *CDC28* are blocked at START, which has therefore been characterized as the major point for *CDC28* action. This contrasts with the better characterized function of M phase kinase at M phase. However, one mutant in *CDC28* passes START normally, but is inhibited in mitosis, suggesting that CDC28 is required to act at both points in the cycle. In fact, the function of *CDC28* is required at three stages: classically before START; then during S phase; and (of course) prior to M phase.

There is a difference in emphasis concerning control of the cell cycle in fission yeast and bakers' yeast, since in *S. pombe* we know most about control of mitosis, and in *S. cerevisiae* we know most about control of

START. However, in both yeasts, the same principle applies that a single catalytic subunit (Cdc2/CDC28) is required for the G2/M and the G1/S transitions. It has different regulatory partners at each transition, using B-like (mitotic) cyclins at mitosis and G1 cyclins at the start of the cycle. A difference between these yeasts is found in the number of cyclin partners that are available at each stage.

A single regulatory partner for Cdc2 is used at mitosis in *S. pombe*, the product of *cdc13*. In *S. cerevisiae*, there are multiple alternative partners. These are coded by the *CLB1–4* genes, which code for products that resemble B cyclins and associate with CDC28 at mitosis. The genes fall into pairs, *CLB1–2* and *CLB3–4*, on the basis of sequence relationships. Mutation in any one of these genes fails to block division, but loss of the *CLB1–2* pair of genes is lethal. Constitutive expression of *CLB1* prevents cells from exiting mitosis.

The state of CDC28 in the *S. cerevisiae* cell cycle is summarized in **Figure 27.19.** The G1 cyclins were not immediately revealed by mutations that block the cell cycle in G1. The absence of such mutants has been explained by the discovery that three independent genes, *CLN1, CLN2, and CLN3, all must be inactivated* to block passage through START in *S. cerevisiae.* Mutations in any one or even any two of these genes fail to block the cell cycle; thus the *CLN* genes are **functionally redundant.** The *CLN* genes show a weak relationship to cyclins (resembling neither the A nor the B class particularly well), although they are usually described as G1 cyclins.

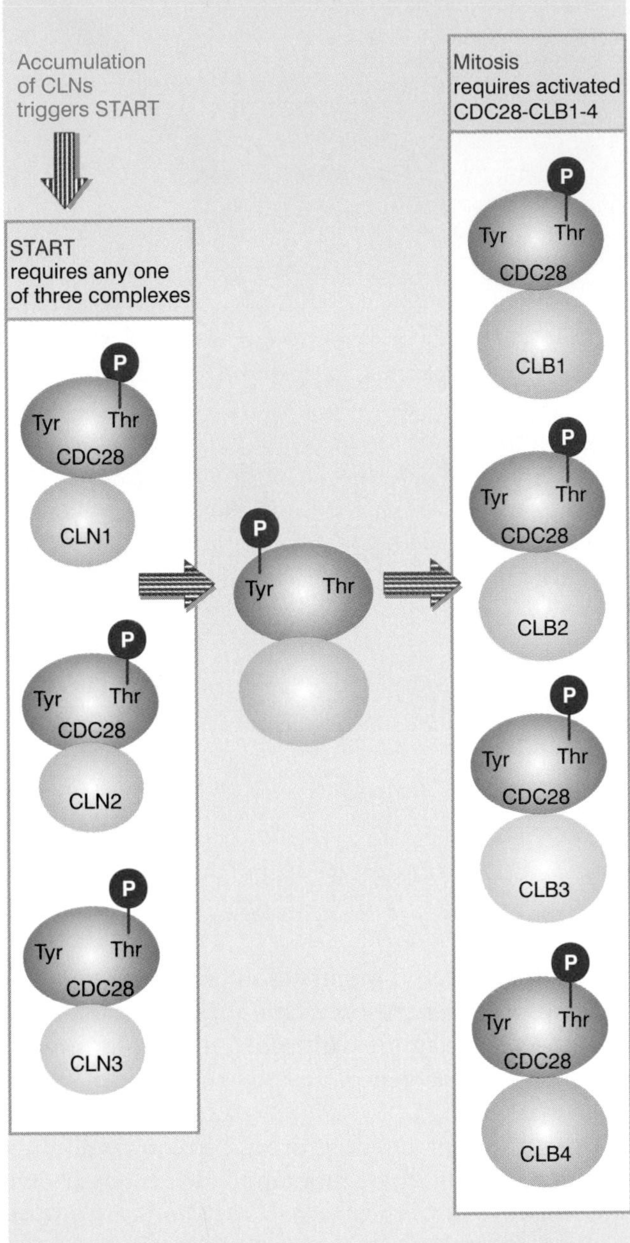

Figure 27.19 Cell cycle control in *S. cerevisiae* involves association of CDC28 with redundant cyclins at both START and G2/M.

Accumulation of the CLN proteins is the rate-limiting step for controlling the G1/S transition. Blockage of protein synthesis arrests the cycle by preventing the proteins from accumulating. The half-life of CLN2 protein is ~15 minutes; its accumulation to exceed a critical threshold level could be the event that triggers passage. Its instability presents a different type of control from that shown by the abrupt destruction of the cyclin A and B types. Dominant mutations that truncate the protein by removing the C-terminal stretch (which contains sequences that target the protein for degradation) stabilize the protein, and as a result G1 phase is basically absent, with cells proceeding directly from M phase into S phase. Similar behavior is shown by the product of *CLN3.*

The redundancy of the *CLN* genes and the *CLB* genes is a feature found at several other stages of the cell cycle. The inhibition of the G2/M transition by *wee1* and *mik1* identifies a pair of redundant kinases; and redundant phosphatases have been found at mitosis in *S. pombe.* In each case, the hallmark is that deletion of an individual gene produces a cyclic phenotype, but deletion of both or all members of the group is lethal. Members of a group may play overlapping rather than identical functions. This form of organization has the practical consequence of making it difficult to identify mutants in the corresponding function.

The animal cell cycle is controlled by many cdk-cyclin complexes

The conservation of the structure of the M phase kinase between animal cells and yeasts and the involvement of related dimers that regulate START in both *S. pombe* and *S. cerevisiae* suggest that a similar scheme controls the cycle in animal cells. The major difference is that animal cells have more variation in the subunits of the kinases. Instead of using the same catalytic subunit at both START and G2/M, animal cells use different catalytic subunits at each stage. They also have a large number of cyclins.

The components of the regulatory kinases for both G1/S and G2/M in yeasts and animal cells are summarized in **Figure 27.20**. The unifying theme is that the kinase consists of a catalytic subunit and a regulatory (cyclin) partner, but there are several types of variation.

At mitosis in animal cells, the Cdc2 catalytic subunit is provided by a single gene. The regulatory partner at mitosis usually is not unique, but is provided by a family of B-type cyclins, and also A-type cyclins in animal cells.

During G1, animal cells have multiple kinases involved in cell cycle control, and they vary in *both* the catalytic subunit and the regulatory (cyclin) subunit. This contrasts with the retention of a common catalytic subunit in yeasts.

Just as families of cyclins can be defined by ability to interact with Cdc2, so may families of catalytic subunits be defined by the ability to interact with cyclins. Catalytic subunits that associate with cyclins are called **cyclin-dependent kinases (cdks)**. Higher eukaryotes possess a large number of genes (~10) related to the true *cdc2* homolog. It is not entirely clear how many of these gene products are involved with the cell cycle and how many code for kinases with other functions. The cdk/cyclin dimers have the same general type of kinase activity as the Cdc2-cyclin dimers, and are often assayed in the same way, by H1 histone kinase activity. The involvement of the Cdc2/cdk kinase engines (and/or others related to them) at two regulatory points *in vivo* is consistent with an increase in H1 kinase activity at S phase as well as at M phase.

The pairwise associations between the catalytic and regulatory subunits are not exclusive, and a particular cyclin may associate with several potential catalytic subunits, while a catalytic subunit may associate with several potential cyclins. The trick is to determine which of these pairwise combinations form in the cell and are concerned with regulating the cell cycle.

Two of the *cdk* genes, *cdk2* and *cdk4,* code for proteins that form pairwise combinations with potential G1 cyclins. The first and best characterized is *cdk2*; it has 66% similarity to the *cdc2* homolog in the same organism.

G1 cyclins were originally identified as genes that could overcome the deficiency of *CLN* mutants in *S. cerevisiae.* New types of cyclins (D and E) were identified by this means. They are distantly related to one another and to other cyclins. Cyclin E accumulates in a periodic manner through the cycle, but is not

Figure 27.20 Similar or overlapping components are used to construct M phase kinase and a G1 counterpart.

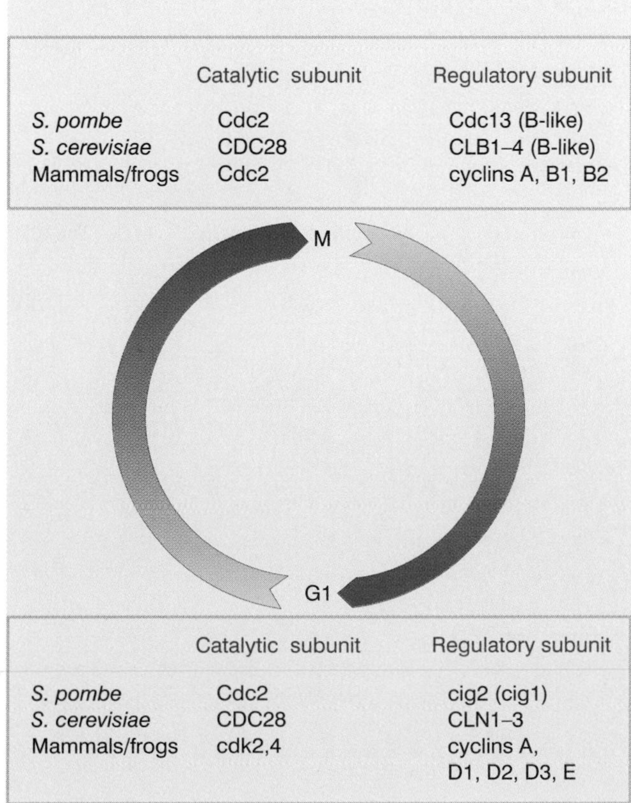

	Catalytic subunit	Regulatory subunit
S. pombe	Cdc2	Cdc13 (B-like)
S. cerevisiae	CDC28	CLB1–4 (B-like)
Mammals/frogs	Cdc2	cyclins A, B1, B2

	Catalytic subunit	Regulatory subunit
S. pombe	Cdc2	cig2 (cig1)
S. cerevisiae	CDC28	CLN1–3
Mammals/frogs	cdk2,4	cyclins A, D1, D2, D3, E

regulated by periodic destruction of protein. There are 3 D-type cyclins; they form dimers with cdk4 and cdk6.

Proteins described as "cyclins" are therefore now significantly more diverse than the A and B classes encompassed by the original definition. For working purposes, we class as cyclins various proteins that have some sequence relationship to the original class, and which can participate in formation of a kinase by pairing with a Cdc2 or cdk2 or related catalytic subunit. It is not yet certain that all of the proteins presently classed as cyclins in this way in fact have this function *in vivo*.

The timing of activity of the various forms of cdk- and Cdc2-cyclins during the animal cell cycle suggests a model in which cdk2-G1 cyclin dimers function to regulate progression through G1 and S phase, while Cdc2-cyclin A,B dimers regulate passage through mitosis. We know most about the details of controlling the G2/M transition, but the principles are likely to be similar for G1/S, since the Cdc2 and cdk catalytic subunits conserve the residues that are involved in regulation.

Figure 27.21 shows that the components and events in an animal cell mitosis are similar to those in yeast cells (compare with Figure 27.13). The lower part of the figure shows the changes in M phase kinase; the upper part shows the enzymes that catalyze these changes. A cell leaves mitosis with Cdc2 monomers and no mitotic cyclins (because cyclins A and B were

Figure 27.21 Control of mitosis in animal cells requires phosphorylations and dephosphorylations of M phase kinase by enzymes that themselves are under similar control or respond to M phase kinase.

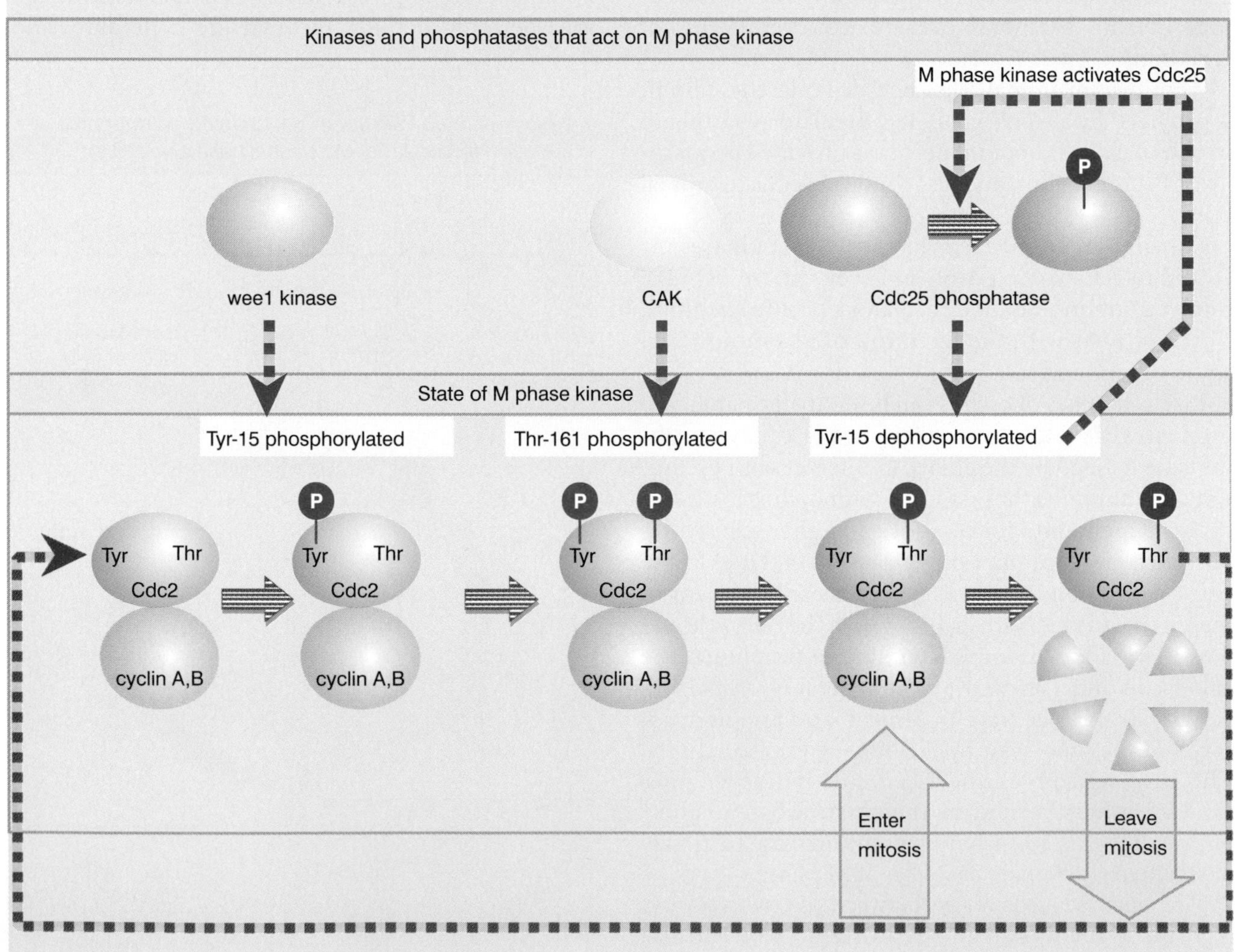

degraded during mitosis). Cyclins are then resynthesized. After a lag period, their level reaches a threshold at which they form dimers with Cdc2. But this does not activate the kinase activity; as we saw previously in Figure 27.7, the activity of the dimer is controlled by the state of certain Tyr and Thr residues:

- The phosphate that is necessary at Thr-161 is added by CAK (the Cdc2-activating kinase). CAK activity is probably constitutive.
- The wee1 kinase is a counterpart to the enzyme of *S. pombe* and phosphorylates Tyr-15 to maintain the M phase kinase in inactive form.
- The Cdc25 phosphatase is a counterpart to the yeast enzyme and removes the phosphate from Tyr-15. Cdc25 is itself activated by phosphorylation; and M phase kinase can perform this phosphorylation, creating a positive feedback loop. Removal of the phosphate from Tyr-15 is the event that triggers the start of mitosis. Cdc25 is itself regulated by several pathways, including phosphatases that inactivate it, but these pathways are not yet well defined.
- Separate kinases and phosphatases that act on Thr-14 have not been identified; in some cases, the same enzyme may act on Thr-14 and Tyr-15.

This means of control is common, but not universal, since in some cells tyrosine phosphorylation does not seem to be critical, and in these cases other means must therefore be used to control the activity of M phase kinase.

Different cdk-cyclin dimers may regulate entry into S and progression through S in animal cells. Some may be concerned with entering the cycle from G0 or exiting to it. The pairwise combinations of dimers that form during G1 and S are summarized in **Figure 27.22.** All of these dimers require phosphorylation on Thr-161 by CAK to generate the active form.

The synthesis of D cyclins is activated when growth factors stimulate cells to re-enter the cycle from G0. The D cyclins have short half-lives, and their levels decline rapidly when the growth stimulus is removed. They may be involved with triggering re-entry of quiescent cells into the cycle. Loss of D cyclins could be a trigger for a cell to leave the cell cycle for the G0 state.

The activity of D cyclins is required during the latter part of G1, but not close to the G1/S boundary. Their functions may be partly redundant, but there are some differences between the D cyclins in their susceptibilities to inhibitors of the cell cycle. The significance of the ability of each D cyclin to associate with 3 different cdk subunits is not clear.

Activity of the cdk2-cyclin E complex is necessary to enter S phase. Cyclin E is synthesized during a period that spans the G1/S transition, but we do not yet know how and when it is inactivated or at what point it becomes dispensable. Cyclin E clearly has a unique role.

Progression through S phase requires the cdk2-cyclin A complex. Cyclin A is also required to associate with Cdc2 for entry into mitosis. The dual use of cyclin A in animal cells appears to be the only case in which a cyclin is used for both G1/S and G2/M transitions.

The states of cyclin-cdk complexes may influence the licensing system that prevents reinitiation of replication (see Figure 13.28). Cyclin B-cdk complexes prevent Cdc6 from loading on to the origin; an orderly

Figure 27.22 Several cdk-cyclin complexes are active during G1 and S phase. The shaded arrows show the duration of activity.

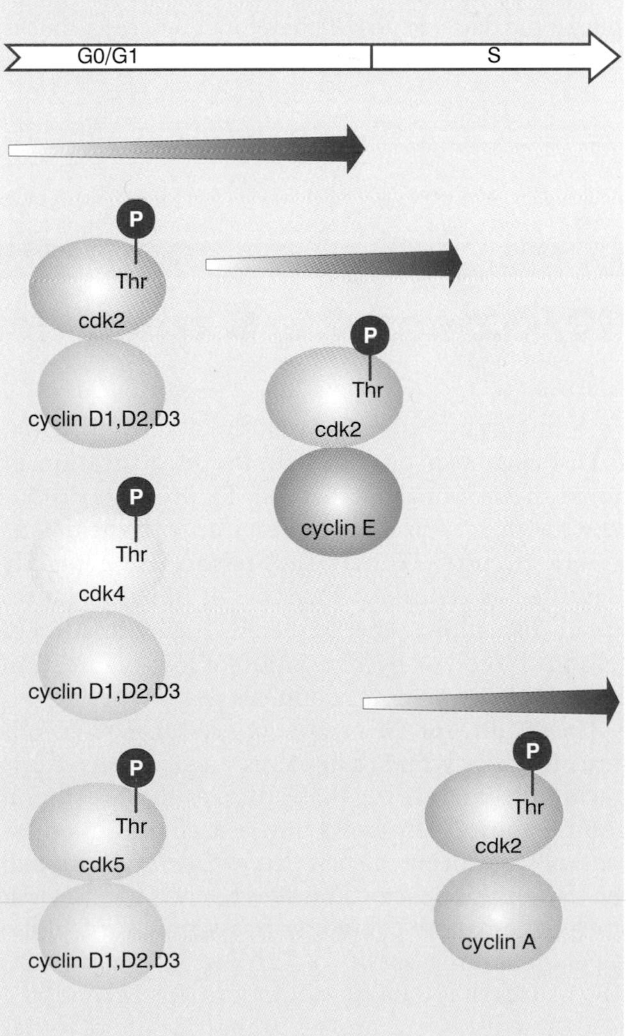

succession of events is thus determined, as the degradation of cyclin B in mitosis releases the block and allows the procedure to start for forming a pre-replication complex at the origin.

Both the beginning and end of S phase are important points in the cell cycle. Just as a cell must know when it is ready to initiate replication, so it must have some means of recognizing the successful completion of replication. This may be accomplished by examining the state of DNA.

Inhibitors of DNA replication can block the cell cycle. The effect may depend on the presence of replication complexes on DNA. We do not know how they block the activation of M phase kinase.

Another interesting connection between DNA and the cell cycle has been revealed by the properties of the *RAD9* mutant of *S. cerevisiae.* Wild-type yeast cells cannot progress from G2 into M if they have damaged DNA. This can be caused by X-irradiation or by the result of replication in a mutant such as *cdc9* (DNA ligase). Mutation of *RAD9* allows these cells to divide in spite of the damage. *RAD9* therefore exercises a checkpoint that inhibits mitosis in response to the presence of damaged DNA. The reaction may be triggered by the existence of double-strand breaks. *RAD9*-dependent arrest and recovery from arrest can occur in the presence of cycloheximide, suggesting that the *RAD9* pathway functions at the post-translational level. At least 6 other genes are involved in this pathway; its ultimate regulatory target remains to be found.

Other controls that ensure an orderly progression through the cell cycle and that can be visualized in the context of checkpoints include monitoring the completion of DNA replication and checking cell size. It may be necessary to bypass some of these checkpoints in early embryogenesis; for example, early divisions in *Drosophila* in effect are nuclear only (because there are no cellular compartments: see Chapter 29). It may be significant that the product of *string* (the counterpart of *S. pombe cdc25* which exercises a size-dependent control) is present at high levels in these early cycles. After the fourteenth cycle, division becomes dependent on *string* expression. It is at this point that cellular compartments develop, and it becomes appropriate for there to be a checkpoint for cell size. *string* may provide this checkpoint.

G0/G1 and G1/S transitions involve cdk inhibitors

An important insight into control of the cell cycle at G1 has been provided by the identification of tumor suppressor genes that code for products that interact with cdk-cyclin complexes or with the downstream circuitry. Tumor suppressors are generally identified as genes in which loss of function causes tumor formation, either as seen by transformation of cells in culture, or by association of loss-of-function mutations with tumors in animals (see Chapter 28).

The product of the tumor suppressor *RB* is a substrate for cdk-cyclin D complexes, and exerts its effects during the part of G1 that precedes the restriction point. **Figure 27.23** shows the basic circuit. In quiescent cells, or during the first part of G1, RB is bound to the transcription factor E2F. This has two effects. First, some genes whose products are essential for S phase depend upon the activity of E2F. By sequestering E2F, RB ensures that S phase cannot initiate. Second, the E2F-RB complex represses transcription of other genes. This may be the major effect in RB's ability to arrest cells in G1 phase.

The nonphosphorylated form of RB forms a complex with cdk-cyclins. The complex with cdk4,6-cyclin D1,2,3 is the most prominent, but RB is also a substrate for cdk2-cyclin E. At or close to the restriction point, RB is phosphorylated by cdk4,6-cyclin D kinases. The phosphorylation causes RB to release E2F, which then activates transcription of the genes whose functions are required for S phase, and also releases repression of genes by the E2F-RB complex. The importance of E2F is seen by the result that expression of E2F in quiescent cells enables them to synthesize DNA.

There is an especially close relationship between RB and cyclin D1. Over-expression of D1 causes cells to enter S phase early. Inhibition of expression of D1 arrests cells before S phase. The sole role of cyclin D1 could be to inactivate RB and permit entry into S phase.

Figure 27.23 A block to the cell cycle is released when RB is phosphorylated by cdk-cyclin.

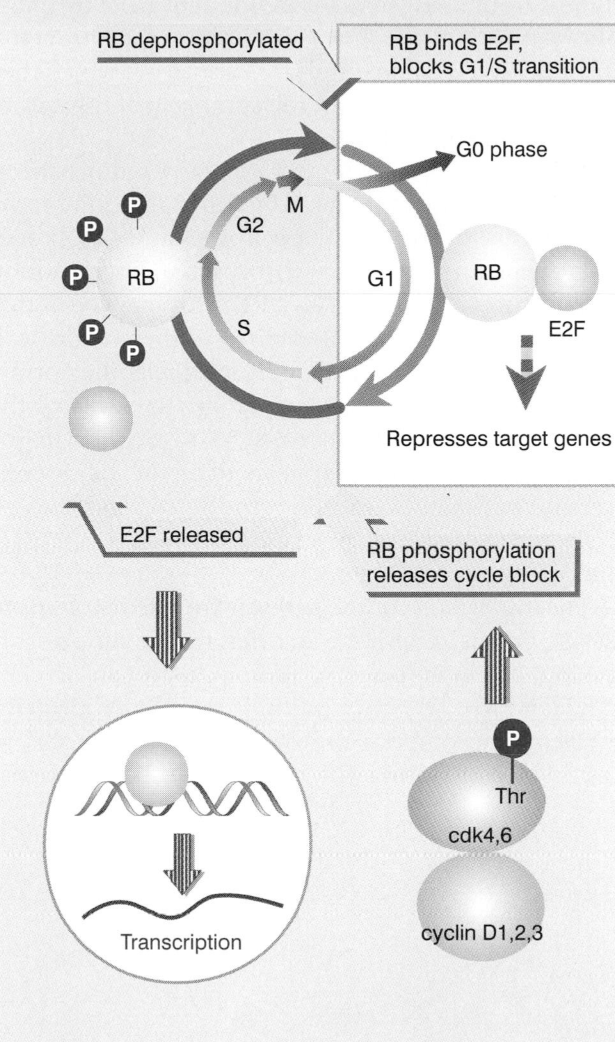

Figure 27.24 p16 binds to cdk4 and cdk6 and to cdk4,6-cyclin dimers. By inhibiting cdk-cyclin D activity, p16 prevents phosphorylation of RB and keeps E2F sequestered so that it is unable to initiate S phase.

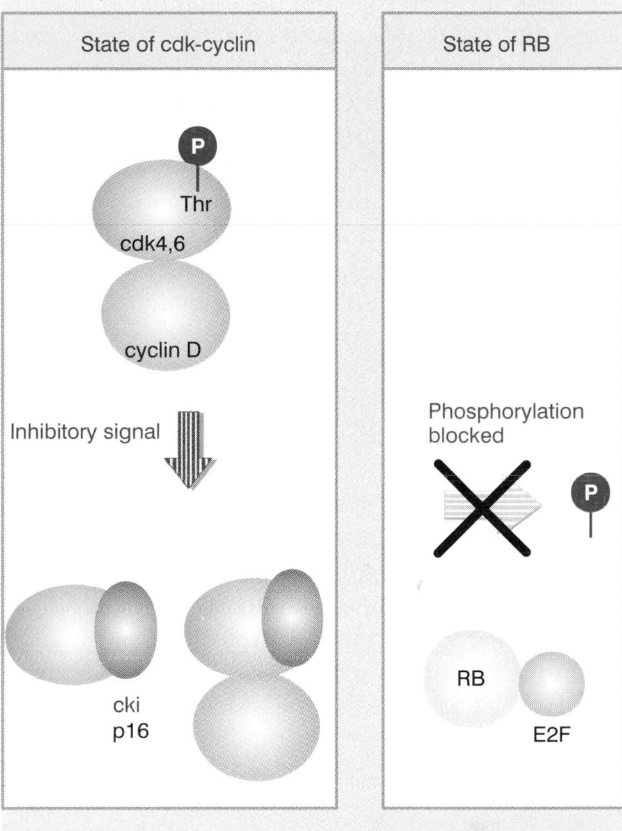

There are several related transcription factors in the E2F family, sharing the property that all recognize genes with the same consensus element. RB binds three of these factors. Two further proteins, p107 and p130, which are related to RB, behave in a similar way, and bind the other members of the E2F group. So together RB and p107 may control the activity of the E2F group of factors.

RB is a target for several pathways that inhibit growth, and may be the means by which growth inhibitory signals maintain cells in G1 (or G0). Several of these signals, including the growth inhibitory factor TGFβ, act through inhibitors of cdk-cyclin kinases. The inhibitors are called **ckis**. They are found as proteins bound to cdk-cyclin dimers in inactive complexes, for example, in quiescent cells. By maintaining the cdk-cyclin complexes in inactive form, they prevent the phosphorylation of RB, making it impossible to release cells to enter S phase.

The cki proteins fall into two classes. The INK4 family is specific for cdk4 and cdk6, and has four members: $p16^{INK4A}$, $p15^{INK4B}$, $p18^{INK4C}$, and $p19^{INK4D}$. The Kip family inhibit all G1 and S phase cdk enzymes, and have three members: $p21^{Cip1/WAF1}$, $p27^{Kip1}$, and $p57^{Kip2}$. (Each protein is identified by its size, with the casual name used as a superscript.)

INK4 proteins bind specifically to cdk4 and cdk6. This suggests a connection with the G0/G1 transition. p16 cannot inhibit proliferation of cells that lack RB, which suggests that it functions by preventing cdk-cyclin kinase activity from using RB as a substrate, as illustrated in **Figure 27.24.** By binding to the cdk subunits, INK4 proteins inhibit both cdk4-cyclin D and cdk6-cyclin D activities. As exemplified by p16 and p19, they bind next to the ATP binding site of cdk6. This both inhibits catalytic activity and triggers a conformational change that prevents cyclin from binding (the conformational change is propagated to the cyclin-binding site).

p21 is a universal cdk inhibitor, binding to all complexes of cdk2,4,6. This suggests that it is likely to block progression through all stages of G1/S. In primary cultured cells (taken directly from the animal), cdk-cyclin dimers are usually found in the form of quaternary complexes that contain two further components. One is PCNA, a component of DNA polymerase δ, which may provide a connection with DNA replication. The other is the inhibitor p21. It may seem paradoxical that an inhibitor is consistently associated with the cdk-cyclin dimer, but it turns out that at a stoichiometry of 1:1 the p21 is not inhibitory. An increase in the number of p21 subunits associated with the cdk-cyclin dimer inhibits kinase function. In transformed cells (from lines that have been successfully perpetuated in culture), cdk-cyclin complexes lack p21 and PCNA. This suggests the possibility that p21 is involved in G1/S control, and that relaxation of this control is necessary for cells to be perpetuated in culture.

p27 has a sequence that is partly related to p21, and also binds promiscuously to cdk-cyclin complexes. Over-expression of p27 blocks progression through S phase, and levels of p27 are increased when cells are sent into a quiescent state by treatment with TGFβ. p21 and p27 block the catalytic subunit of cdk-cyclin dimers from being a substrate for activation by phosphorylation by CAK. They also prevent catalytic activity of the cdk-cyclin complex. The stages at which they function are illustrated in the summary of inhibitory pathways in **Figure 27.25.**

p21 and p27 are probably partially redundant in their functions. The pathway by which they inhibit the cell cycle is not entirely clear, but we know that it does not depend on controlling RB, because they can inhibit proliferation of cells that lack RB. This may mean that their inhibition of cdk2-cyclin E dimers is critical. Since both are present in proliferating cells, the normal progress of the cell cycle may require the levels of the cdk-cyclin dimers to increase to overcome an inhibitory threshold. p27 appears to be the major connection between extracellular mitogens and the cell cycle, with an inverse correlation between p27 activity and ability to proliferate.

The importance of the pathway from cki proteins to RB is emphasized by the fact that tumor suppressors are found at every stage, including cki proteins, cyclins D1,2, and RB. The implication is that the cki proteins are needed to suppress unrestrained growth of cells. In terms of controlling the cell cycle, this pathway is

Figure 27.25 p21 and p27 inhibit assembly and activity of cdk4,6-cyclin D and cdk2-cyclin E by CAK. They also inhibit cycle progression independent of RB activity. p16 inhibits both assembly and activity of cdk4, 6-cyclin D.

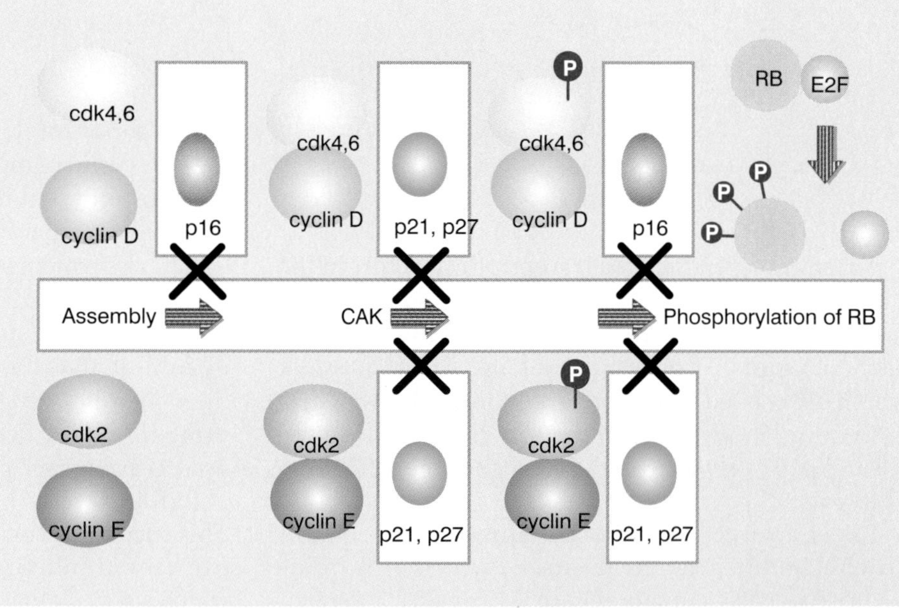

clearly central. It may be the key pathway by which cells are enabled to undertake a division cycle (by passing the restriction point in G1 if they are already cycling, or by re-entering G1 if they are quiescent in G0).

The cki proteins provide another level of control. In *S. cerevisiae*, the cki Sic1 is bound to CDC28-CLB during G1, and this maintains the kinase in an inactive state. Entry into S phase requires degradation of Sic1 to release the kinase. **Figure 27.26** shows how Sic1 is targeted for degradation by a ubiquitinating system (see Figure 8.43 for a general description). The Sic1 target is recognized by a complex called the SCF, which functions as an E3 ligase. The SCF complex includes Cdc53, Skp1, and Cdc4. Cdc4 is the targeting component, which, together with Skp1, binds to Sic1. For this reason, the complex is described as SCF^{Cdc4}. Skp1 is the connection to Cdc53, which interacts with the E2 ligase (Cdc34). The E2 ligase adds ubiquitin to Sic1, causing it to be degraded.

Cdc4 is a member of a class of proteins called F-box proteins. It uses the F-box motif to bind to Skp1. This is a general paradigm for the construction of SCF complexes. Other SCF complexes exist in which the targeting subunit is a different F-box protein, but the Cdc53 and Skp1 components remain the same. An example relevant to the cell cycle is SCF^{Grr1} in which the F-box protein Grr1 provides the targeting subunit, and causes the degradation of G1 cyclins.

There are further layers of control in this system. The substrates for the SCF must be phosphorylated to be recognized. The kinases that perform the phosphorylation are the cdk-cyclin complexes that are active at the appropriate stage of the cell cycle. The abundance of the SCF complexes is itself controlled by degrading the F-box subunits. SCF^{Cdc4} targets Cdc4, thus creating an autoregulatory limitation on its activity. The consequence of such feedbacks is to maintain a supply of the Cdc53-Skp1 cores that can be recruited as appropriate by the F-box subunits.

Figure 27.26 The SCF is an E3 ligase that targets the inhibitor Sic1.

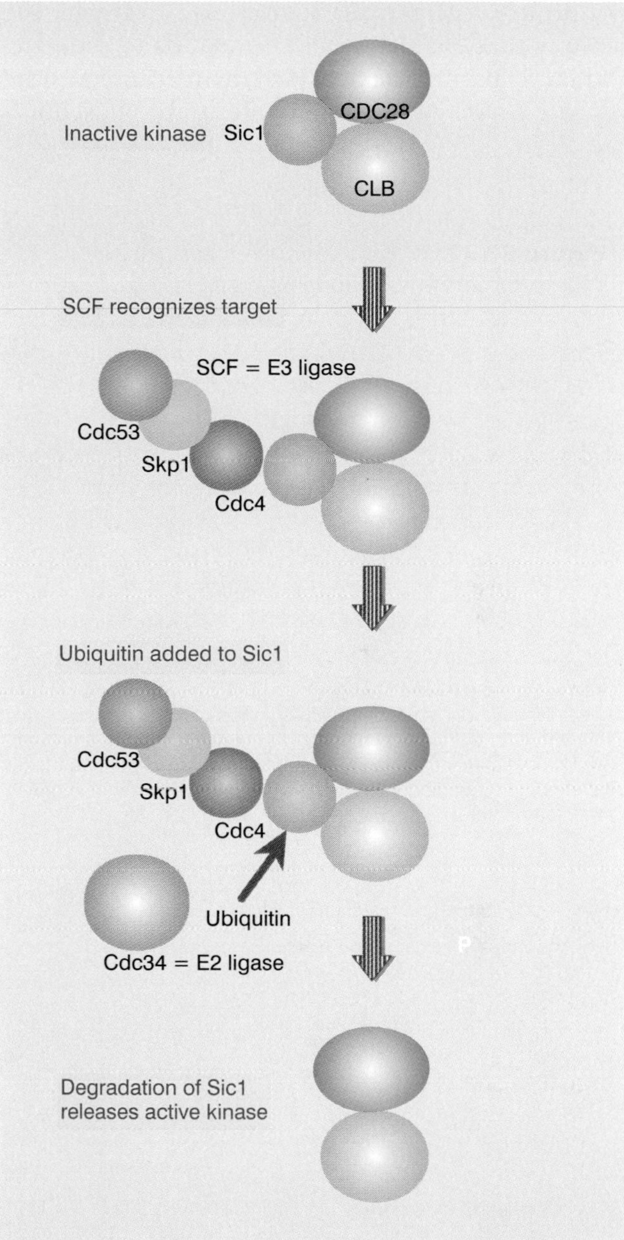

Protein degradation is important in mitosis

The timing of the events that regulate mitosis is summarized in **Figure 27.27**. Mitosis is initiated, as we have seen, by the activation of M phase kinase. Progress through mitosis requires degradation of cyclins, and also of other proteins. The separation of chromosomes at anaphase requires the activity of the proteasomal system for protein degradation, but does not directly require cyclin degradation. This suggests that the degradation system has (at least) three targets. The first event to occur is the degradation of cyclin A at metaphase. Then two targets are degraded at anaphase. The protein Pds1p must be degraded to trigger the pathway that enables sister chromatids to separate. And B cyclins must be degraded in order to inactivate M phase kinase. At the end of mitosis, the phosphorylations of substrate proteins performed by M phase kinase must be reversed.

A large complex of 8 subunits is responsible for selecting the substrates that are degraded in anaphase. It is called the **anaphase promoting complex (APC)** (sometimes also known as the **cyclosome**). Similar complexes are found in both yeasts and vertebrates. The APC becomes active specifically during mitosis. It functions as an E3 ubiquitin ligase (see Figure 8.43), which is responsible for binding to the substrate pro-

Figure 27.27 Progress through mitosis requires destruction of cyclins and other targets.

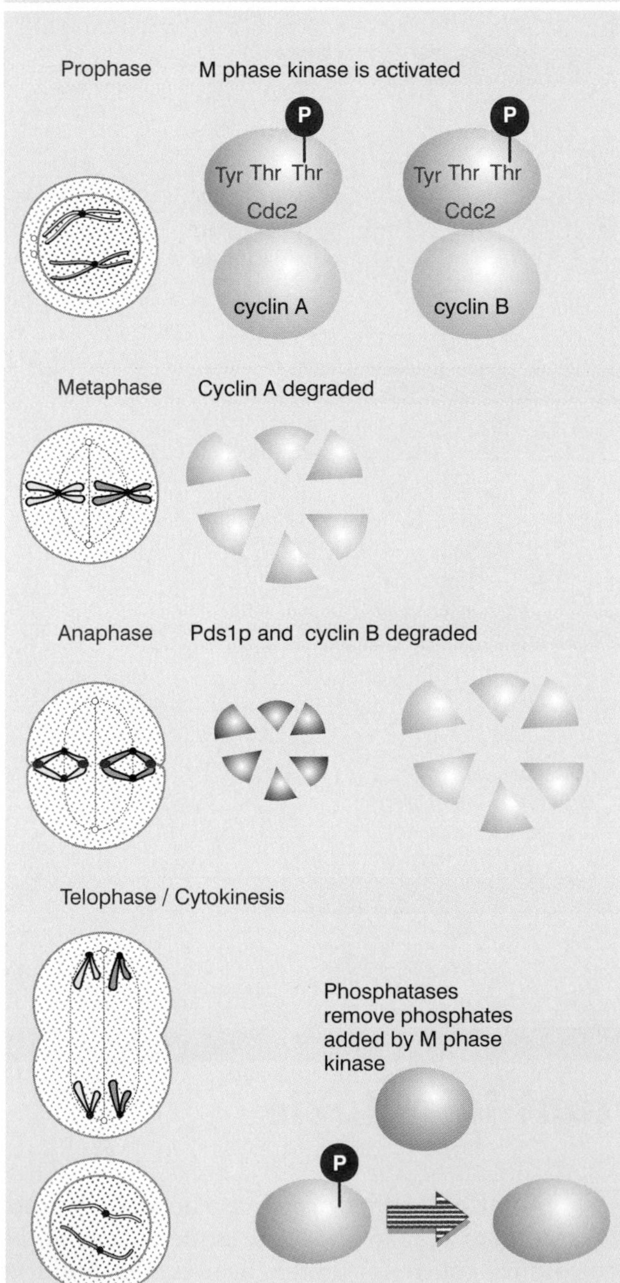

Figure 27.28 Two versions of the APC are required to pass through anaphase.

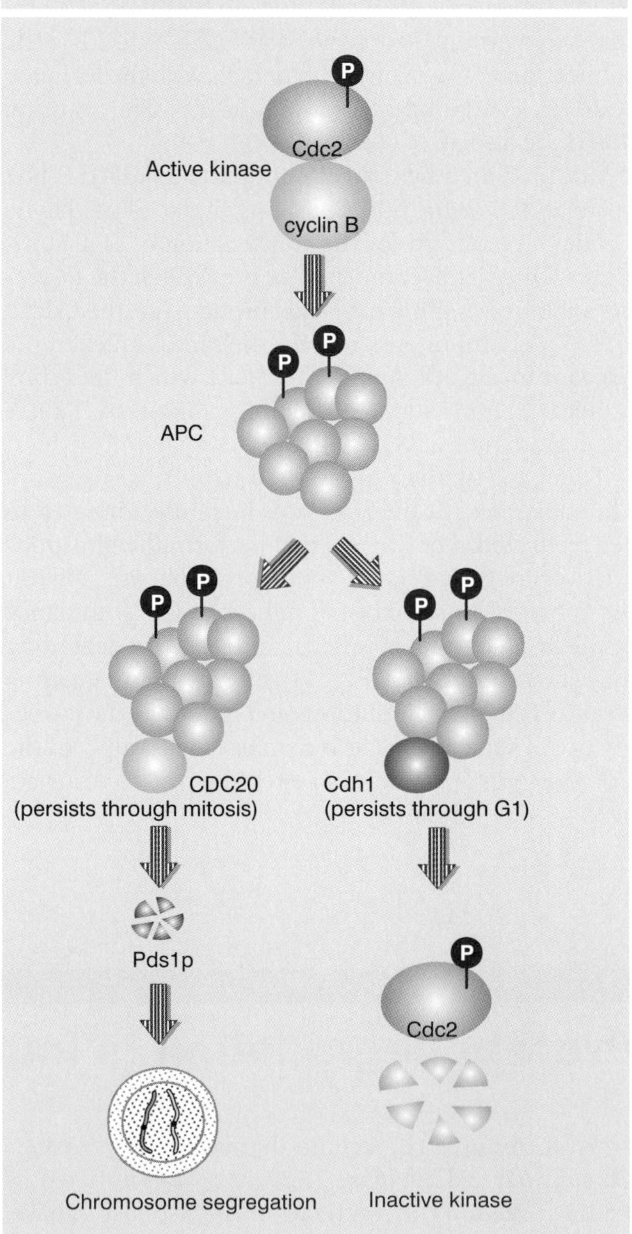

tein so that ubiquitin is transferred to it. The ubiquitinated substrate is then degraded by the proteasome.

There are two separate routes to activating the APC, and each causes it to target a particular substrate. **Figure 27.28** shows that the regulatory factors CDC20 and Cdh1 bind to the APC and activate its ubiquitination activity. CDC20 is necessary for the APC to degrade Pds1p. Cdh1 is necessary for the APC to degrade Clb2 (yeast cyclin B). The timing of activation and persistence of each type of complex is different.

The route to releasing sister chromatids so that they may segregate to opposite poles is indirect, as illustrated in **Figure 27.29**. Prior to anaphase, Pds1p binds to the protein Esp1p and keeps it in an inactive state. When the APC degrades Pds1p, Esp1p is released. It causes the protein Scc1p to be released from the sister chromatids. Scc1p is a component of the **cohesin** complex, a "glue" that holds the chromatids together. The core of the cohesin complex is a heterodimer of SMC proteins. (Other SMC proteins form the condensins, which are involved in controlling chromosome condensation; see Chapter 19). Together with Scc1p (and other proteins), the cohesin complex contains Smc1p and Smc3p, which are (putative) ATPases with coiled-coil domains. When Scc1p is released by the action of Esp1p, the cohesin complex can no longer hold the chromatids together, and they therefore become free to segregate on the spindle. Scc1p is degraded later in anaphase by the APC.

Figure 27.29 Anaphase progression requires the APC to degrade Pds1p to allow Esp1p to remove Scc1 from sister chromatids.

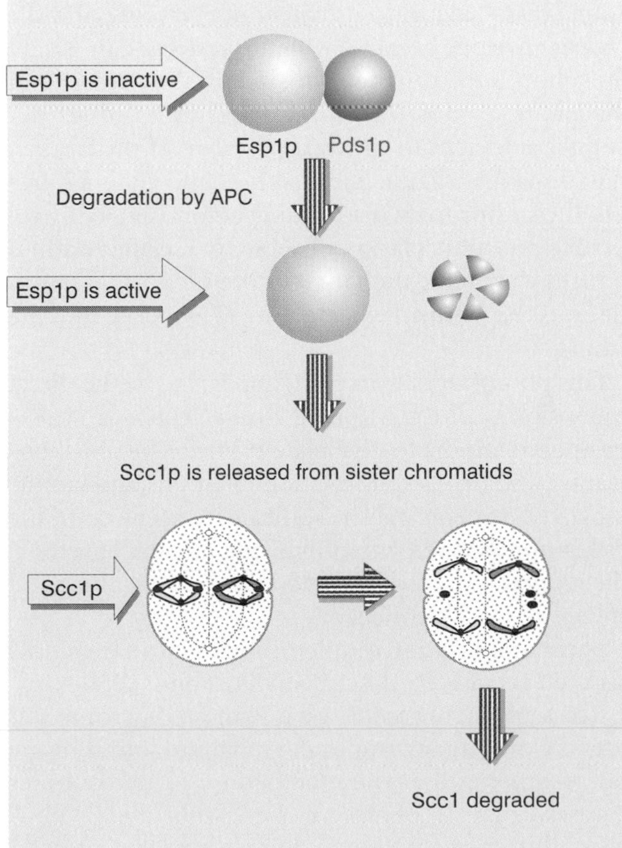

Formation of the cohesin complex (or possibly its association with the chromosomes) occurs during S phase. Mutants in the locus *ctf7* have sister chromatids that never associate. The gene acts during the period of DNA replication. It is possible that the establishment of cohesion is triggered by replication in some way.

Smc3p is involved in meiosis as well as mitosis. It is required for sister chromatid cohesion, together with a protein (Rec8p) that is related to Scc1p. This suggests that a cohesin complex, related to that of mitosis, may form at meiosis. Both of these components are required for synaptonemal complexes to form (see Chapter 14).

The cohesion system is the target for a checkpoint. Progression through mitosis requires all kinetochores to be paired with their homologs. **Figure 27.30** shows

Figure 27.30 An unattached kinetochore causes the Mad pathway to inhibit CDC20 activity.

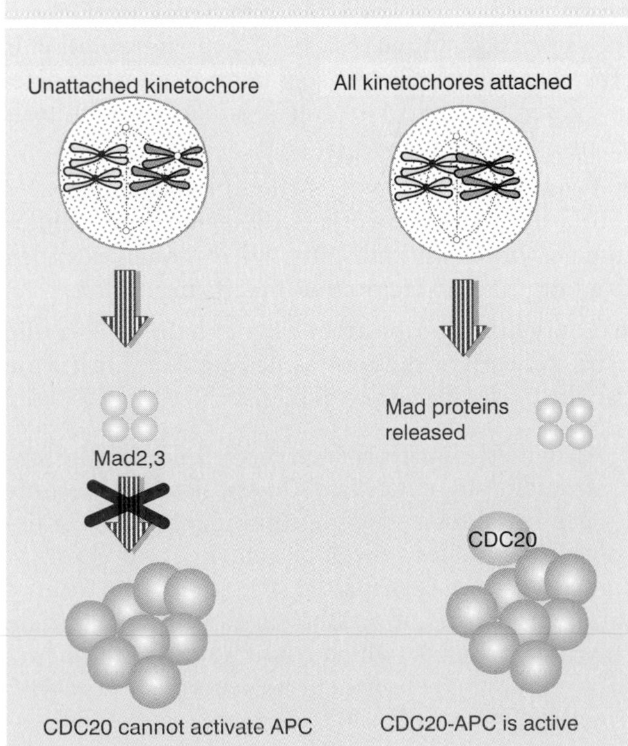

that the checkpoint consists of a surveillance system that is triggered by the presence of an unattached kinetochore. Mutations in the *Mad* and *Bub* genes allow mitosis to continue (aberrantly) in the presence of unpaired kinetochores. The Mad proteins control the system for chromatid segregation. They bind to CDC20 and prevent it from activating the APC. When the kinetochores are all attached, some (unidentified) signal causes Mad proteins to be released from CDC20, which now activates the APC, allowing anaphase to continue. A similar role is played by Bub proteins in controlling the ability of Cdh1 to activate the APC.

Reorganization of the cell at mitosis

THE culmination of the cell cycle is the act of division, when the chromosomes segregate into two diploid sets and the other components of the cell are partitioned between the two daughter cells. The change in cell structure is dramatic, as summarized in **Figure 27.31**. The division between nucleus and cytoplasm is abolished, and the cytoskeleton is entirely reorganized. The relevant events include:

- Condensation of chromatin to give recognizable chromosomes.
- Dissolution of the nuclear lamina and breakdown of the nuclear envelope. The lamina dissociates into individual lamin subunits, and the nuclear envelope, endoplasmic reticulum, and Golgi apparatus break down into small membrane vesicles. (Nuclear dissolution is typical of animal cells but does not occur, for example, in the yeasts, *S. cerevisiae* and *S. pombe*, where mitosis involves nuclear division.)
- Dissociation and reconstruction of microtubules into a spindle. Microtubules dissociate into tubulin dimers, which reassemble into microtubules extended from the mitotic microtubule organizing centers.
- Reorganization of actin filaments to replace the usual network by the contractile ring that pinches the daughter cells apart at cytokinesis.

All of these changes are reversible; following the separation of daughter cells, the actin filaments resume their normal form, the microtubular spindle is dissolved, the nuclear envelope reforms, and chromosomes take a more dispersed structure in the form of interphase chromatin. Modification of appropriate substrate proteins (which could be either the structural subunits themselves or proteins associated with them) provides a plausible means to control the passage of mitosis. The question then becomes how the mitotic changes and their reversal depend upon the activation and inactivation of M phase kinase.

The best characterized substrate for M phase kinase is histone H1. As noted previously, we do not know what role the phosphorylation of H1 plays at either G1/S or M phase transitions. It is a reasonable assumption, however, that M phase kinase acts directly on chromatin by phosphorylating H1 and other target proteins, and that this is the cause of chromosomal condensation.

Nuclear integrity is abandoned when the underlying lamina dissociates into its constituent lamins, and the nuclear envelope breaks down into vesicles. Lamins are phosphorylated during mitosis, and the presence of phosphate groups on only two serine residues per lamin is sufficient to cause dissociation of the lamina. Mutations that change these serines into alanines prevent the lamina from dissociating at mitosis. So the reversible phosphorylation of these two serine residues induces a structural change in the individual lamin subunits that controls their ability to associate into the lamina.

This phosphorylation event appears to be the direct responsibility of the M phase kinase, which can cause the nuclear lamina to dissociate *in vitro*. The phosphorylation and dephosphorylation of lamins and lamin-associated proteins may be sufficient to account for the dissolution and reformation of the nuclear envelope. The reorganization of the endoplasmic reticulum and Golgi is not well defined.

The reorganization of microtubules into the spindle has been extensively described, but cannot yet be connected to the action of M phase kinase. Microtubules extend from organizing centers that are found in the centrosomes and at the kinetochores. Microtubules themselves consist of dimers of α-tubulin and β-tubulin. Within the centrosome there is a related protein,

Figure 27.31 Major changes in the cell at mitosis involve the chromosomes, nuclear envelope, nuclear lamina, and microtubules.

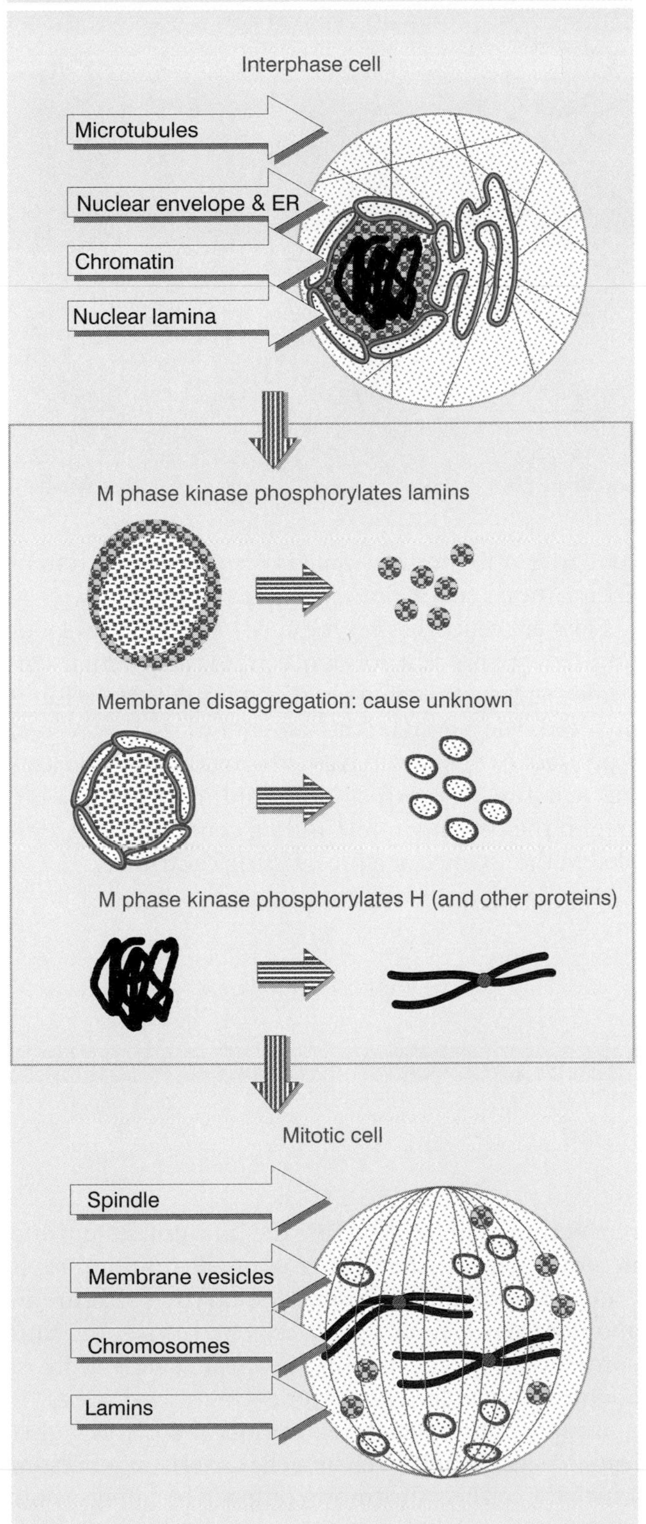

γ-tubulin, which is part of a complex that may provide the actual nucleating source. The direct interaction of this complex with the αβ tubulin dimers of microtubules remains to be characterized.

The structure of centrosomes is not well defined, but in animal cells a centrosome contains a pair of **centrioles,** surrounded by a dense amorphous region. The centriole is a small hollow cylinder whose wall consists of a series of triplet fused microtubules. A centriole is shown in cross-section in **Figure 27.32.**

The function of the centriole in mitosis is not clear. Originally it was thought that it might provide the structure to which microtubules are anchored at the pole, but the fibers seem instead to terminate in the amorphous region around the centrioles. It is possible that the centriole is concerned with orienting the spindle; it may also have a role in establishing directionality for cell movement. However, there are cell types in which centrosomes do not appear to contain centrioles.

Centrioles have their own cycle of duplication. When born at mitosis, a cell inherits two centrioles. During interphase they reproduce, so that at the start

Figure 27.32 The centriole consists of nine microtubule triplets, apparent in cross-section as the wall of a hollow cylinder. Photograph kindly provided by A. Ross.

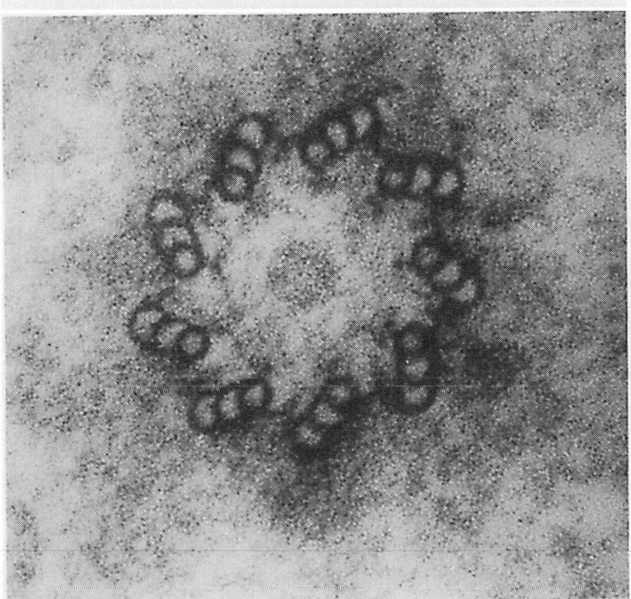

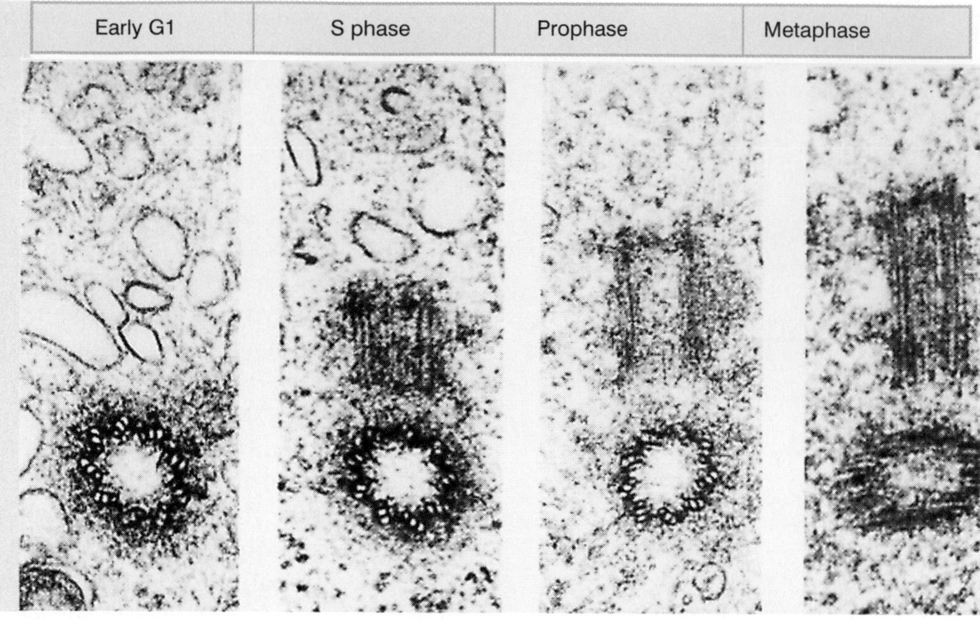

Figure 27.33 A centriole reproduces by forming a procentriole on a perpendicular axis; the procentriole is subsequently extended into a mature centriole. Photograph kindly provided by J. B. Rattner and S. G. Phillips.

of mitosis there are four centrioles, two at each pole. Probably only the parental centriole is functional.

The centriole cycle is illustrated in **Figure 27.33.** Soon after mitosis, a **procentriole** is elaborated perpendicular to the parental centriole. It has the same structure as the mature parental centriole, but is only about half its length. Later during interphase, it is extended to full length. It plays no role in the next mitosis, but becomes a parental centriole when it is distributed to one of the daughter cells. The orientation of the parental centriole at the mitotic pole is responsible for establishing the direction of the spindle.

How are centrioles reproduced? The precise elaboration of the procentriole adjacent to the parental centriole suggests that some sort of template function is involved. The parental centriole cannot itself be seen to reproduce or divide, but it could provide some nucleating structure onto which tubulin dimers assemble to extend the procentriole. Could a centriole be assembled in the absence of a pre-existing centriole?

Apoptosis

During development of a multicellular eukaryotic organism, some cells are required to die. Unwanted cells are eliminated during embryogenesis, metamorphosis, and tissue turnover. This process is called **programmed cell death** or **apoptosis.** It provides a crucial control over the total cell number. In *C. elegans*, in which somatic cell lineages have been completely defined, 131 of the 1090 somatic cells formed during adult development undergo programmed cell death—cells die predictably at a defined time and place in each animal. Similar, although less precisely defined, cell deaths occur during vertebrate development; the most prominent locations are the immune system and nervous system. The proper control of apoptosis is crucial in probably all higher eukaryotes. Failure to apoptose allows tumorigenic cells to survive, and thus contributes to cancer. Inappropriate activation of apoptosis is involved in neurodegenerative diseases.

Apoptosis involves the activation of a pathway that leads to suicide of the cell by a characteristic process in which the cell becomes more compact, blebbing occurs at the membranes, chromatin becomes condensed, and DNA is fragmented. The pathway is an active process that depends on RNA and protein synthesis by the dying cell. The typical features of a cell as it becomes heteropycnotic (condensed with a small, fragmented nucleus) are shown in **Figure 27.34,** and the course of fragmentation of DNA is shown in **Figure 27.35.** Ultimately the dead cells become fragmented into membrane-bound pieces, and may be engulfed by surrounding cells.

Apoptosis can be triggered by a variety of stimuli, including withdrawal of essential growth factors, treatment with glucocorticoids, γ-irradiation, and activation of certain receptors, as summarized in **Figure 27.36.** These all involve a molecular insult to the cell. Another means of initiating apoptosis is used in the immune system, where cytotoxic T lymphocytes attack target cells. And the ability of the tumor suppressor p53 to trigger apoptosis is an important defense against cancer (see Chapter 28). Apoptosis is impor-

Figure 27.34 Cell structure changes during apoptosis. The top panel shows a normal cell. The lower panel shows an apoptosing cell; arrows indicate condensed nuclear fragments. Photographs kindly provided by Shigekazu Nagata.

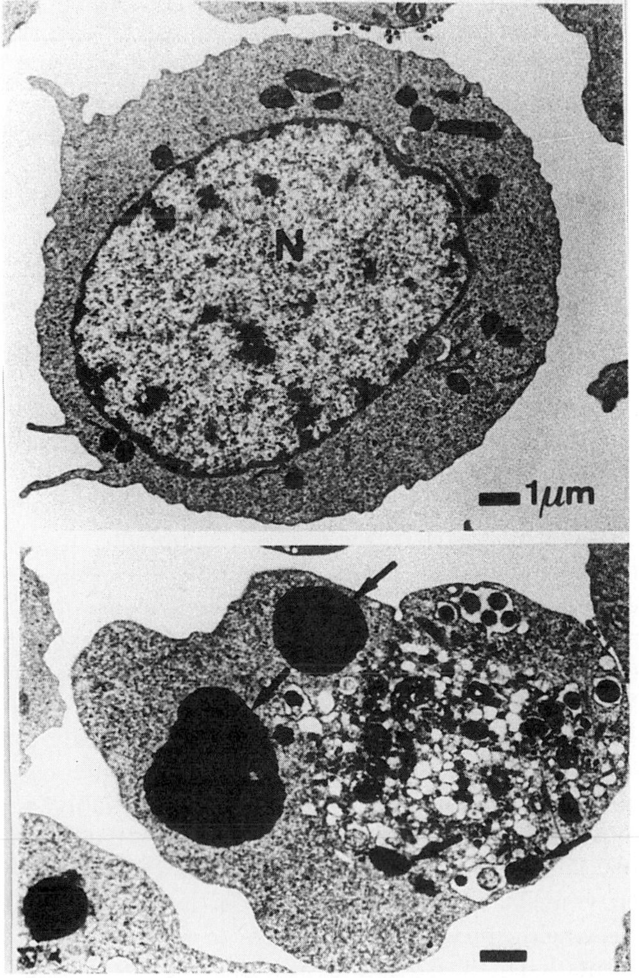

Figure 27.35 Fragmentation of DNA occurs ~2 hours after apoptosis is initiated in cells in culture. Photograph kindly provided by Shigekazu Nagata.

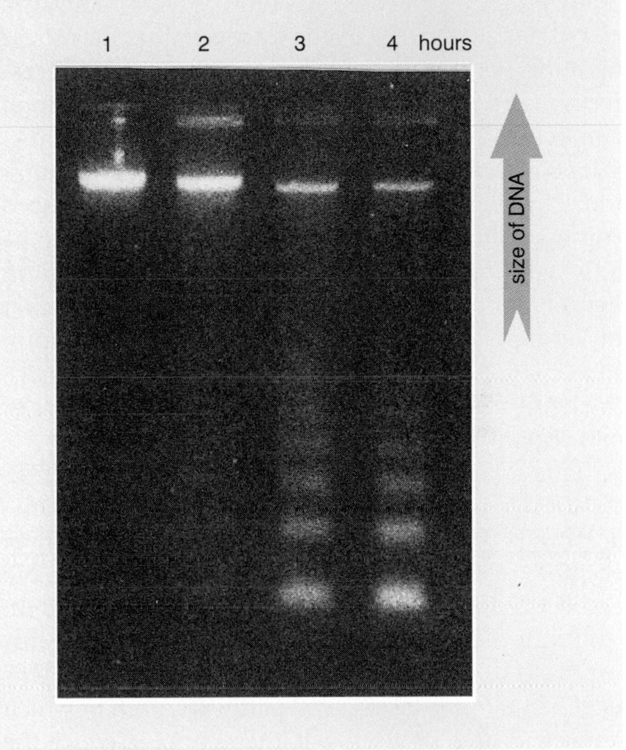

Figure 27.36 Apoptosis is triggered by a variety of pathways.

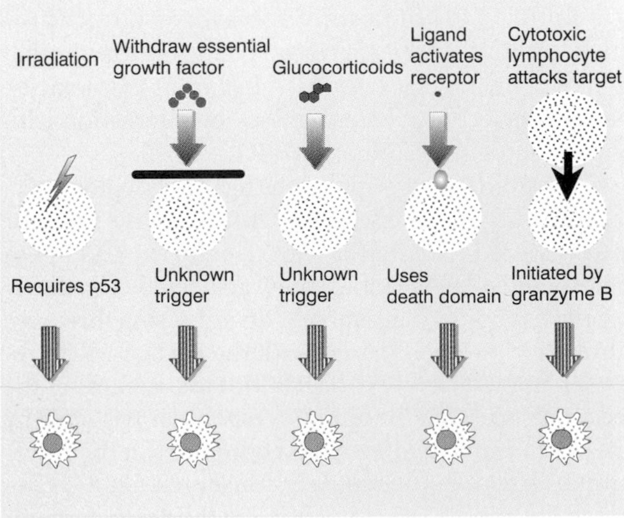

tant, therefore, not only in tissue development, but in the immune defense and in the elimination of cancerous cells.

A major pathway that triggers apoptosis in mammalian cells was discovered via the properties of the **Fas receptor.** An antibody directed against Fas protein kills cells that express it on their surface. The reason is that the antibody-Fas reaction activates Fas, which triggers a pathway for apoptosis. The validity of this pathway *in vivo* was demonstrated by the discovery that the mouse mutation *lpr* results in deficiency in Fas. *lpr* is a recessive mutation whose effect is to cause proliferation of lymphocytes, which results in a complex immune disorder affecting both B cells and T cells. Another mutation with similar effects is *gld* (generalized lymphoproliferative disease), and this turns out to lie in the gene that codes the ligand for the Fas receptor. The related properties of these two loci suggest that this apoptotic pathway is triggered by an interaction between the ligand (*gld* product) and Fas (*lpr* product). The pathway is required for limiting the numbers of mature lymphocytes.

Fas is a cell surface receptor related to the TNF (tumor necrosis factor) receptor. Its ligand is a transmembrane protein related to TNF. There is a family of receptors related to TNF that, in addition to two TNF receptors and Fas, includes several receptors found on T lymphocytes. A corresponding family of ligands comprises a series of transmembrane proteins. This suggests that there are several pathways, each of which can be triggered by a cell-cell interaction, in which the "ligand" on one cell surface interacts with the receptor on the surface of the other cell.

The Fas and TNF receptors both can activate apoptosis. In addition to its membrane-bound form, TNF also exists as cleaved, soluble protein, which functions as a soluble growth factor. In this form, largely produced by macrophages, TNF is a pleiotropic factor that signals many cellular responses, including cytotoxicity. Most of its responses are triggered by interaction with one of the TNF receptors, TNF-R1.

An assay for the capacity of the ligand-receptor interaction to trigger apoptosis is to introduce the receptor into cultured cells that do not usually express it. On treatment with the ligand, the transfected cells die by apoptosis, but the parental cells do not. Using this assay, similar results are obtained with the *gld* ligand/Fas receptor and with TNF/TNF-R1. Mutant versions of the receptor show that the apoptic response is triggered by an ~80 amino acid intracellular domain near the C-terminus. This region is loosely conserved (~28%) between Fas and TNF-R1, and is called the **death domain.**

Several proteins have been found that interact specifically with the Fas and/or TNF receptors, and which trigger apoptosis when transfected into a recipient cell. All of these proteins themselves have death domains, and it is possible that a homomeric interaction between two death domains provides the means by which the signal is passed from the receptor to the next component of the pathway.

The "classical" pathway for apoptosis is summarized in **Figure 27.37.** A ligand-receptor interaction triggers the activation of a protease, which leads to the release of cytochrome *c* from mitochondria, which in turn activates a series of proteases, whose actions culminate in the destruction of cell structures.

A complex containing several components forms at the receptor. TNF receptor binds a protein called TRADD, which in turn binds a protein called FADD. Fas receptor also binds FADD. In either case, FADD binds the protein caspase-8 (also known as FLICE), which has a death domain as well as protease catalytic activity, and which may then trigger a common pathway.

Members of the **caspase** family (cysteine aspartate proteases) are important downstream components of the pathway. The prototype (ICE = caspase-1) is the IL-1β-converting enzyme, which cleaves the pro-IL-1β precursor into its active form. Transfection of ICE into cultured cells causes apoptosis; and the process is inhibited by CrmA (a product of cowpox virus). All caspases are inhibited by CrmA, although each caspase has a characteristic sensitivity. CrmA inhibits apoptosis triggered in several different ways, which demonstrates that the caspases play an essential role in the pathway. However, it turns out that ICE is not itself the protease commonly involved in apoptosis, because inactivation of the gene for ICE does not block general apoptosis in the mouse. (But ICE may be needed specifically for apoptosis of one pathway in lymphocytes.)

Caspases have a catalytic cysteine, and cleave their targets at an aspartate. There are ~8 mammalian members of the caspase family. One that is commonly required for apoptosis is caspase-3. The mechanisms of action of caspase-3 and ICE are similar; both cleave at tetrapeptide sequences in their substrates (recognizing YVAD and DEVD, respectively).

The event that activates a caspase is usually an oligomerization. The trigger to activate the first caspase in the pathway (caspase-8) is probably an oligomerization caused by association with the receptor. The mechanism of activation is to cleave a proenzyme. Caspase-3, for example, is activated by cleaving the pro-caspase precursor to release two fragments that then associate to form an active heterodimer.

Figure 27.37 Apoptosis can be triggered by activating surface receptors. Caspase proteases are activated in the pathway. Apoptosis can be blocked by Bcl2.

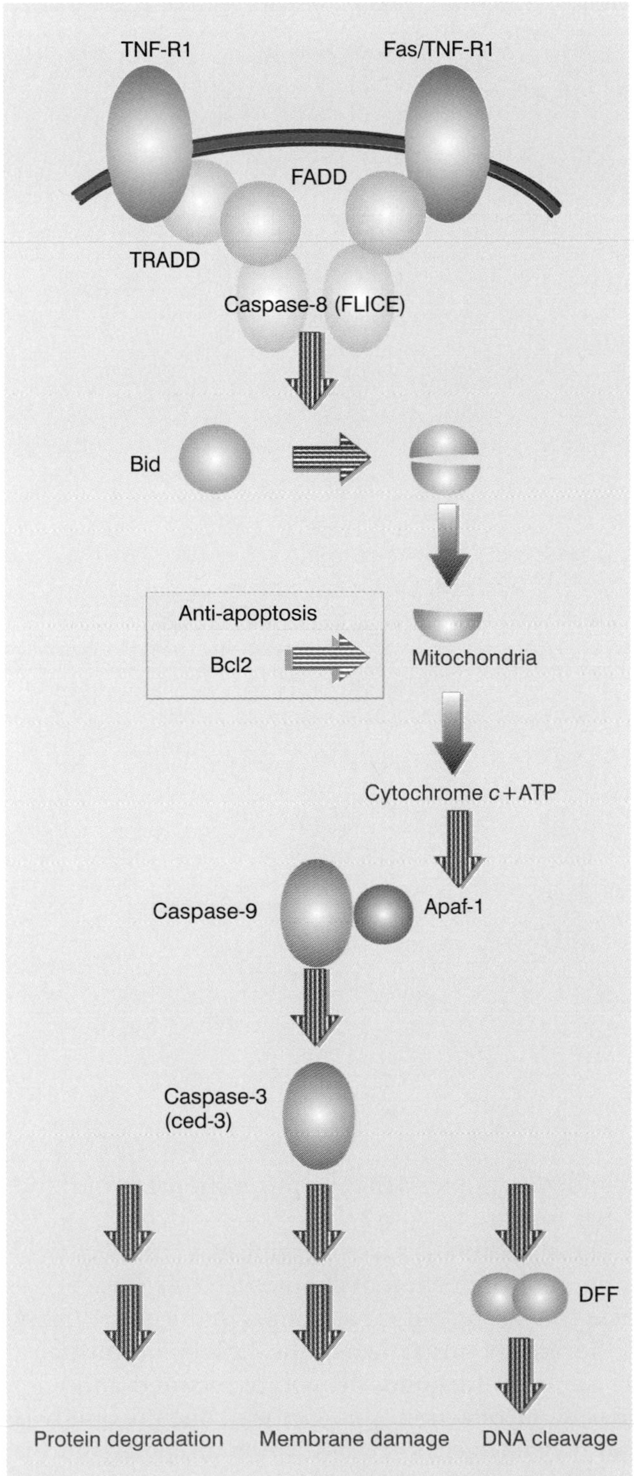

Changes in mitochondria occur during apoptosis (and also during other forms of cell death). These are typically detected by changes in permeability. In association with these changes, cytochrome *c* is released from mitochondria. Its role is to provide a cofactor that is needed to activate caspase-9.

To move the pathway from the plasma membrane to the mitochondrion, Caspase-8 cleaves a protein called Bid, releasing the C-terminal domain, which then translocates to the mitochondrial membrane. This causes cytochrome *c* to be released. Cytochrome *c* triggers the interaction of the cytosolic protein Apaf-1 with caspase-9. When Apaf-1 oligomerizes with procaspase-9, this causes the autocleavage that activates caspase-9. Caspase-9 in turn cleaves pro-caspase-3 to generate caspase-3 (which is in fact the best characterized component of the downstream pathway. It is the homolog of the *C. elegans* protein ced-3; see below).

The release of cytochrome *c* is a crucial control point in the pathway. Bid is a member of the important Bcl2 family. Some members of this family are required for apoptosis, while others counteract it. The eponymous Bcl2 inhibits apoptosis in many cells. It has a C-terminal membrane anchor, and is found on the outer mitochondrial, nuclear, and ER membranes. It prevents the release of cytochome c, which suggests that in some way it counteracts the action of Bid.

bcl2 was originally discovered as a proto-oncogene that is activated in lymphomas by translocations that result in its over-expression. (As discussed in more detail in Chapter 28, this means that Bcl2 is a member of a class of proteins that causes proliferation or tumorigenesis when inappropriately expressed.) Its role as an inhibitor of apoptosis was discovered when it was shown that its addition protects cultured lymphoid and myeloid cells from dying when the essential factor IL-3 is withdrawn.

Mammalian cells that are triggered into apoptosis by a wide variety of stimuli, including activation of the Fas/TNF-R1 pathways, can be rescued by expression of Bcl2. This suggests that these pathways converge on a single mechanism of cell killing, and that Bcl2 functions at a late, common stage of cell death. There are some systems in which Bcl2 cannot block apoptosis, so the pathway that it blocks may be common, but is not the sole one.

Caspase-3 acts at what might be called the effector stage of the pathway. We have not identified all of the targets of the protease activity that are essential for apoptosis. One known target is the enzyme PARP (poly[ADP-ribose] polymerase). Its degradation is not essential, but is a useful diagnostic for apoptosis. Caspase-3 cleaves PARP but not IL-1β; and ICE (caspase-1) cleaves IL-1β but not PARP.

The pathway that leads to DNA fragmentation has been identified. Caspase-3 cleaves one subunit of a dimer called DFF (DNA fragmentation factor). The other subunit then activates a nuclease that degrades DNA.

The pathway shown in Figure 27.37 is the prototypical pathway for activation of apoptosis via a protease cascade. However, Fas can also activate apoptosis by a pathway that involves the kinase JNK, whose most prominent substrate is the transcription factor c-Jun (see Figure 26.32). This leads by undefined means to the activation of proteases. **Figure 27.38** shows that this pathway is mediated by the protein Daxx (which does not have a death domain). Binding of FADD and Daxx to Fas is independent: each adaptor recognizes a different site on Fas. The two pathways function independently after Fas has engaged the adaptor. The TNF receptor also can activate JNK by means of distinct adaptor proteins.

In the normal course of events, activation of Fas probably activates both pathways. Over-expression experiments show that either pathway can cause apoptosis. The relative importance of the two pathways may vary with the individual cell type, in response to other signals that affect each pathway. For example, JNK is activated by several forms of stress independently of the Fas-activated pathway. This pathway is not inhibited by Bcl2, which may explain the variable ability of cells to resist apoptosis in response to Bcl2.

Bcl2 belongs to a family whose members can homodimerize and heterodimerize. Two other members are bcl-x (characterized in chicken) and Bax (characterized in man). bcl-x is produced in alternatively spliced forms that have different properties. When transfected into recipient cells, bcl-x_L mimics Bcl2, and inhibits apoptosis. But bcl-x_S counteracts the ability of Bcl2 to protect against apoptosis. Bax behaves in the same way as bcl-x_S. This suggests that the formation of Bcl2 homodimers may be needed to provide the protective form, and that Bcl2/Bax or Bcl2/bcl-x_S heterodimers may fail to protect. Whether Bax or bcl-x_S homodimers actively assist apoptosis, or are merely permissive, remains to be seen. The general conclusion suggested by these results is that combinatorial associations between members of the family may produce dimers with different effects on apoptosis, and the relative proportions of the family members that are expressed may be important. The susceptibility of a cell to undergo apoptosis may be proportional to the ratio of Bax to Bcl2.

Figure 27.38 Fas can activate apoptosis by a JNK-mediated pathway.

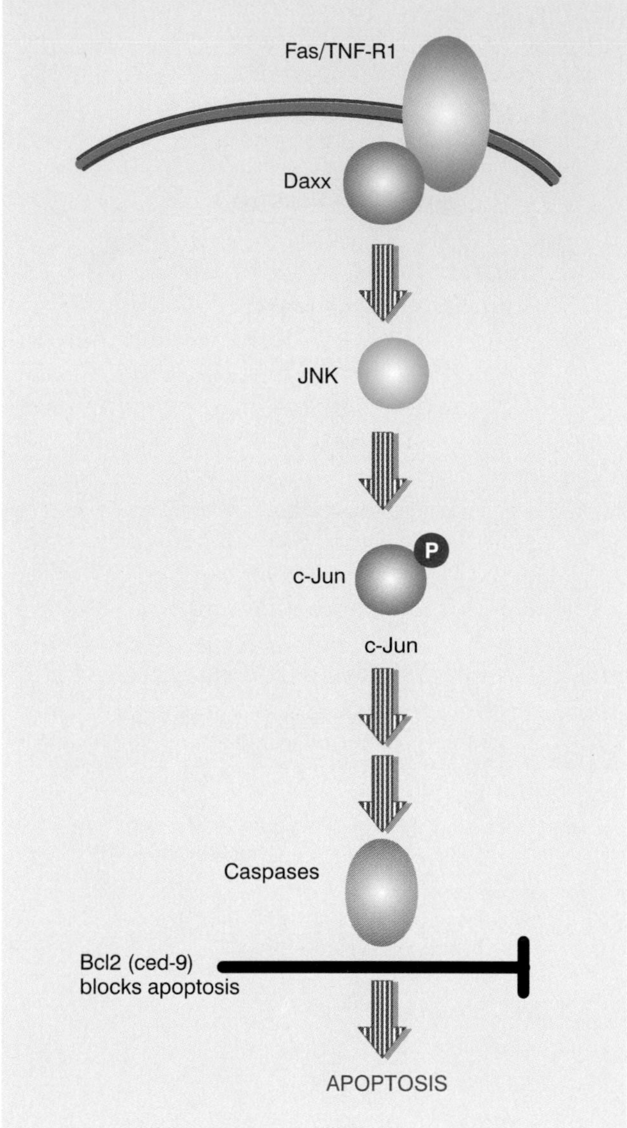

The mitochondrion is emerging as a crucial control point in the induction of apoptosis. The release of cytochrome *c* is preceded by changes in the permeability of the mitochondrial membrane. Bcl2 family members act at the mitochondrial membrane, and although their mode of action is not known, one possibility is that they form channels in the membrane. Apoptosis involves localization (or perhaps increased concentration) of Bcl2 family members at the mitochondrial membrane, including Bid (required to release cytochrome *c*) and Bax (perhaps involved in membrane permeability changes).

Another apoptotic pathway is triggered by cytotoxic T lymphocytes, which kill target cells by a process that involves the release of granules containing serine proteases and other lytic components. One such component is perforin, which can make holes in the target cell membrane, and under some conditions can kill target cells. The serine proteases in the granules are called granzymes. In the presence of perforin, granzyme B can induce many of the features of apoptosis, including fragmentation of DNA. It activates a caspase called Ich-3, which is necessary for apoptosis in this pathway.

The control of apoptosis involves components that inhibit the pathway as well as those that activate it. This first became clear from the genetic analysis of cell death in *C. elegans*, when mutants were found that either activate or inactivate cell death. Mutations in *ced-3* and *ced-4* cause the survival of cells that usually die, demonstrating that these genes are essential for cell death. *ced-3* codes for the protease activity (and was in fact the means by which caspases were first implicated in apoptosis). It is the only protease of this type in *C. elegans*. *ced-4* codes for the homolog to Apaf-1.

ced-9 inhibits apoptosis. It codes for the counterpart of Bcl2. A mutation that inactivates *ced-9* is lethal, because it causes the death of cells that should survive. This process requires *ced-3* and *ced-4*, and this was the original basis for the idea that *ced-9* blocks the apoptotic pathway(s) in which *ced-3* and *ced-4* participate. This relationship makes an important point: *ced-3* and *ced-4* are not expressed solely in cells that are destined to die, but are expressed also in other cells, where normally their action is prevented by *ced-9*. The proper control of apoptosis may therefore involve a balance between activation and inhibition of this pathway.

The properties of mice lacking Apaf-1 or caspase-9 throw some light upon the generality of apoptotic pathways. Lack of caspase-9 is lethal, because the mice have a malformed cerebrum as the result of the failure of apoptosis. Apoptosis is also reduced in thymocytes (immune precursors to lymphocytes). Apaf-1 deficient mice have less severe defects in brain development, implying that there are alternative means for activating caspase-9. Both types of deficient mice continue to show Fas-mediated apoptosis, implying that Fas has alternative means of triggering apoptosis.

The existence of mechanisms to inhibit as well as to activate apoptosis suggests that many (possibly even all) cells possess the intrinsic capacity to apoptose. If the components of the pathway are ubiquitous, the critical determinant of whether a cell lives or dies may depend on the regulatory mechanisms that determine whether the pathway is activated or repressed.

Summary

THE cell cycle consists of transitions from one regulatory state to another. The change in regulatory state is separated by a lag period from the subsequent changes in cell phenotype. The transitions take the form of activating or inactivating a kinase(s), which modifies substrates that determine the physical state of the cell. Checkpoints can retard a transition until some intrinsic or extrinsic condition has been satisfied.

The two key control points in the cell cycle are in G1 and at the end of G2. During G1, a commitment is made to enter a replication cycle; the decision is identified by the restriction point in animal cells, and by START in yeast cells. After this decision has been taken, cells are committed to beginning an S phase, although there is a lag period before DNA replication initiates. The end of G2 is marked by a decision that is executed immediately to enter mitosis.

A unifying feature in the cell cycles of yeasts and animals is the existence of an M phase kinase, consisting of two subunits: Cdc2, with serine/threonine protein kinase catalytic activity; and a mitotic cyclin of either the A or B class. Homologous subunits can be recognized in (probably) all eukaryotic cells. The genes that code for the catalytic subunit in yeasts are the eponymous *cdc2* in *S. pombe* and *CDC28* in *S. cerevisiae*. Animal cells usually contain multiple mitotic cyclins (A, B1, B2); in *S. pombe,* there is only a single cyclin at M phase, a B class coded by *cdc13,* although *S. pombe* has several CLB proteins.

The activity of the M phase kinase is controlled by the phosphorylation state of the catalytic subunit. The active form requires dephosphorylation on Tyr-15 (in yeasts) or Thr-14/Tyr-15 (in animal cells) and phosphorylation on Thr-161. The cyclins are also phosphorylated, but the significance of this modification is not known. In animal cells, the kinase is inactivated by degradation of the cyclin component, which occurs abruptly during mitosis. Cyclins of the A type are typically degraded before cyclins of the B type. Destruction of at least the B cyclins, and probably of both classes of cyclin, is required for cells to exit mitosis.

A comprehensive analysis of genes that affect the cell cycle has identified *cdc* mutants in both *S. pombe* and *S. cerevisiae*. The best characterized mutations are those that affect the components or activity of M phase kinase. Mutations *cdc25* and *wee1* in *S. pombe* have opposing effects in regulating M phase kinase in response to cell size (and other signals). wee1 is a kinase that acts on Tyr-15 and maintains Cdc2 in an inactive state; Cdc25 is a phosphatase that acts on Tyr-15 and activates Cdc2. The existence of *wee1* and *cdc25* homologs in higher eukaryotes suggests that the apparatus for cell cycle control is widely conserved in evolution.

By phosphorylating appropriate substrates, the kinase provides MPF activity, which stimulates mitosis or meiosis (as originally defined in *Xenopus* oocytes). A prominent substrate is histone H1, and H1 kinase activity is now used as a routine assay for M phase kinase. Phosphorylation of H1 could be concerned with the need to condense chromatin at mitosis. Another class of substrates comprises the lamins, whose phosphorylation causes the dissolution of the nuclear lamina. A general principle governing these (and presumably other) events is that the state of the substrates is controlled reversibly in response to phosphorylation, so that the phosphorylated form of the protein is required for mitotic organization, while the dephosphorylated form is required for interphase organization. Phosphatases are required to reverse the modifications introduced by M phase kinase.

Transition from G1 into S phase requires a kinase related to the M phase kinase. In yeasts, the catalytic subunit is identical with that of the M phase kinase, but the cyclins are different (the combinations being CDC28-cig1,2 in *S. pombe*, Cdc2-CLN1,2,3 in *S. cerevisiae*). Activity of the G1/S phase kinase and inactivity of the M phase kinase are both required to proceed through G1. Initiation of S phase in *S. pombe* requires Rum1 to inactivate Cdc2/Cdc13 in order to allow the activation of Cdc18, which may be the S phase activator.

In mammalian cells, a family of catalytic subunits is provided by the *cdk* genes, named because they code for the catalytic subunits of cyclin-dependent kinases. There are ~10 *cdk* genes in an animal genome. Aside from the classic Cdc2, the best characterized product is cdk2 (which is well related to Cdc2). In a normal cell cycle, cdk2 is partnered by cyclin E during the G1/S transition and by cyclin A during the progression of S phase. cdk2, cdk4, and cdk5 all partner the D cyclins to form kinases that are involved with the transition from G0 to G1. These cdk-cyclin complexes phosphorylate RB, causing it to release the transcription factor E2F, which then activates genes whose products are required for S phase. A group of cki (inhibitor) proteins that are activated by treatments that inhibit growth can bind to cdk-cyclin complexes, and maintain them in an inactive form.

Checkpoints control progression of the cell cycle. One checkpoint responds to the presence of unreplicated or damaged DNA by blocking mitosis. Others control progress through mitosis, for example, detecting unpaired kinetochores.

Apoptosis is achieved by an active pathway that executes a program for cell death. The components of the pathway may be present in many or all higher eukaryotic cells. Apoptosis may be triggered by various stimuli. A common pathway involves activation of caspase-8 by oligomerization at an activated surface receptor. Caspase-8 cleaves Bid, which triggers release of cytochrome *c* from mitochondria. The cytochrome *c* causes Apaf-1 to oligomerize with caspase-9. The activated caspase-9 cleaves procaspase-3, whose two subunits then form the active protease. This cleaves various targets that lead to cell death. The pathway is inhibited by Bcl2 at the stage of release of cytochrome *c*. An alternative pathway for triggering apoptosis that does not pass through Apaf-1 and caspase-9, and which is not inhibited by Bcl2, involves the activation of JNK. Different cells use these pathways to differing extents. Apoptosis was first shown to be necessary for normal development in *C. elegans*, and knockout mutations in mice show that this is also true of vertebrates. Every cell may contain the components of the apoptotic pathway and be subject to regulation of the balance between activation and repression of cell death.

Further reading

Historical

Howard, A. and Pelc, S. (1953). Synthesis of DNA in normal and irradiated cells and its relation to chromosome breakage. *Heredity Suppl.* **6**, 261–273.

Hartwell, L. *et al.* (1974). Genetic control of the cell division cycle in yeast. *Science* **183**, 46–51.

Reviews

Murray, A. W. and Szostak, J. W. (1985). Chromosome segregation in mitosis and meiosis. *Ann. Rev. Cell Biol.* **1**, 289–315.

Mitchison, T. J. (1988). Microtubule dynamics and kinetochore function in mitosis. *Ann. Rev. Cell Biol.* **4**, 527–549.

Hartwell, L. H. and Weinert, T. A. (1989). Checkpoints: controls that ensure the order of cell cycle events. *Science* **246**, 629–634.

McIntosh, J. R. and Koonce, M. P. (1989). Mitosis. *Science* **246**, 622–628.

Murray, A. W. and Kirschner, M. W. (1989). Dominoes and clocks: the union of two views of the cell cycle. *Science* **246**, 614–621.

Nurse, P. (1990). Universal control mechanism regulating onset of M phase. *Nature* **344**, 503–508.

Ellis, R. E., Yuan, J., and Horvitz, H. R. (1991). Mechanisms and functions of cell death. *Ann. Rev. Cell Biol.* **7**, 663–698.

Forsburg, S. L. and Nurse, P. (1991). Cell cycle regulation in the yeasts *Saccharomyces cerevisiae* and *Schizosaccharomyces pombe. Ann. Rev. Cell Biol.* **7**, 227–256.

Norbury, C. and Nurse, P. (1992). Animal cell cycles and their control. *Ann. Rev. Biochem.* **61**, 441–470.

Reed, S. I. (1992). The role of p34 kinases in the G1 to S phase transition. *Ann. Rev. Cell Biol.* **8**, 529–561.

Herr, C. J. (1993). Mammalian G1 cyclins. *Cell* **73**, 1059–1065.

Hunter, T. and Pines, J. (1994). Cyclins and cancer II: cyclin D and CDK inhibitors come of age. *Cell* **79**, 573–582.

King, R. W., Jackson, P. K., and Kirschner, M. W. (1994). Mitosis in transition. *Cell* **79**, 563–571.

Nurse, P. (1994). Ordering S phase and M phase in the cell cycle. *Cell* **79**, 547–550.

Sherr, C. J. (1994). G2 phase progression: cycling on cue. *Cell* **79**, 551–555.

Sherr, C. J. and Roberts, J. M. (1995). Inhibitors of mammalian G1 cyclin-dependent kinases. *Genes & Dev.* **9**, 1149–1163.

Weinberg, R. A. (1995). The retinoblastoma protein and cell cycle control. *Cell* **81**, 323–330.

Chao, D. T. and Korsmeyer, S. J. (1998). Bcl2 family: regulators of cell death. *Ann. Rev. Immunol.* **16**, 395–419.

Discoveries

M phase kinase and mitosis

Evans, T. *et al.* (1983). Cyclin: a protein specified by maternal mRNA in sea urchin eggs that is destroyed at each cleavage division. *Cell* **33**, 389–396.

Simanis, V. and Nurse, P. (1986). The cell cycle control gene *cdc2*^{C} of fission yeast encodes a protein kinase potentially regulated by phosphorylation. *Cell* **45**, 261–268.

Arion, D., Meijer, L., Brizuela, L., and Beach, D. (1988). cdc2 is a component of the M phase-specific histone H1 kinase: evidence for identity with MPF. *Cell* **55**, 371–378.

Dunphy, W. G., Brizuela, L., Beach, D., and Newport, J. (1988). The *Xenopus* cdc2 protein is a component of MPF, a cytoplasmic regulator of mitosis. *Cell* **54**, 423–431.

Gautier, J., Norbury, C., Lohka, M., Nurse, P., and Maller, J. (1988). Purified maturation-promoting factor contains the product of a *Xenopus* homologue of the fission yeast cell cycle control gene *cdc2*$^{+}$. *Cell* **54**, 433–439.

Draetta, G., Luca, F., Westendorf, J., Brizuela, L., Ruderman, J., and Beach, D. (1989). cdc2 protein kinase is complexed with both cyclin A and B: evidence for proteolytic inactivation of MPF. *Cell* **56**, 829–38.

Labbe, J. C., Picard, A., Peaucellier, G., Cavadore, J. C., Nurse, P., and Doree, M. (1989). Purification of MPF from starfish: identification as the H1 histone kinase p34^{cdc2} and a possible mechanism for its periodic activation. *Cell* **57**, 253–263.

Murray, A. W., Solomon, M. J., and Kirschner, M. W. (1989). The role of cyclin synthesis and degradation in the control of maturation promoting factor activity. *Nature* **339**, 280–286.

Riabowol, K., Draetta, G., Brizuela, L., Vandre, D., and Beach, D. (1989). The cdc2 kinase is a nuclear protein that is essential for mitosis in mammalian cells. *Cell* **57**, 393–401.

Cyclin-dependent kinases and S phase

Blow, J. J. and Nurse, P. (1990). A cdc2-like protein is involved in the initiation of DNA replication in Xenopus egg extracts. *Cell* **62**, 855–862.

Hayles, J. *et al.* (1994). Temporal order of S phase and mitosis in fission yeast is determined by the state of the p34^{cdc2} mitotic B cyclin complex. *Cell* **78**, 813–822.

Control of cdk-cyclin complexes

Gould, K. L. and Nurse, P. (1989). Tyrosine phosphorylation of the fission yeast cdc2^{+} protein kinase regulates entry into mitosis. *Nature* **342,** 39–45.

Gautier, J., Solomon, M. J., Booher, R. N., Bazan, J. F., and Kirschner, M. W. (1991). cdc25 is a specific tyrosine phosphatase that directly activates p34^{cdc2}. *Cell* **67,** 197–211.

Xiong, Y., Zhang, H., and Beach, D. (1993). Subunit rearrangement of the cyclin-dependent kinases is associated with cellular transformation. *Genes & Dev.* **7,** 1572–1583.

Fisher, R. P. and Morgan, D. O. (1994). A novel cyclin associates with MO15/cdk7 to form the cdk-activating kinase. *Cell* **78,** 713–724.

Jeffrey, P. D. *et al.* (1995). Mechanism of cdk activation revealed by the structure of a cyclin A-cdk2 complex. *Nature* **376,** 313–320.

Substrates for M phase kinase

Peter, M. *et al.* (1990). *In vitro* disassembly of the nuclear lamina and M phase-specific phosphorylation of lamins by cdc2 kinase. *Cell* **61,** 591–602.

Foisner, R. and Gerace, L. (1993). Integral membrane proteins of the nuclear envelope interact with lamins and chromosomes, and binding is modulated by mitotic phosphorylation. *Cell* **73,** 1267–1279.

Checkpoints and chromosome segregation

Weinert, T. A. and Hartwell, L. H. (1988). The *RAD9* gene controls the cell cycle response to DNA damage in *Saccharomyces cerevisiae. Science* **241,** 317–322.

Michaelis, C., Ciosk, R., and Nasmyth, K. (1997). Cohesins: chromosomal proteins that prevent premature separation of sister chromatids. *Cell* **91,** 35–45.

Mechanisms of proteolysis

Glotzer, M., Murray, A. W., and Kirschner, M. W. (1991). Cyclin is degraded by the ubiquitin pathway. *Nature* **349,** 132–138.

Holloway, S. L. *et al.* (1993). Anaphase is initiated by proteolysis rather than by the inactivation of MPF. *Cell* **73,** 1393–1402.

King, R. W. *et al.* (1995). A 20S complex containing CDC27 and CDC16 catalyzes the mitosis-specific conjugation of ubiquitin to cyclin B. *Cell* **81,** 279–288.

Skowyra, D., *et al.* (1997). F-box proteins are receptors that recruit phosphorylated substrates to the SCF ubiquitin-ligase complex. *Cell* **91,** 209–219.

Ciosk, R. *et al.* (1998). An ESP1/PDS1 complex regulates loss of sister chromatid cohesion at the metaphase to anaphase transition in yeast. *Cell* **93,** 1067–1076.

Fang, G., Yu, H., and Kirschner, M. W. (1998). Direct binding of CDC20 protein family members activates the anaphase-promoting complex in mitosis and G1. *Molecular Cell* **2,** 163–171.

Apoptosis

Ito, N. *et al.* (1991). The polypeptide encoded by the cDNA for human cell surface antigen Fas can mediate apoptosis. *Cell* **66,** 233–243.

Watanabe-Fukunaga, R. *et al.* (1992). Lymphoproliferation disorder in mice explained by defects in Fas antigen that mediates apoptosis. *Nature* **356,** 314–317.

Miura, M. *et al.* (1993). Induction of apoptosis in fibroblasts by IL-1β-converting enzyme, a mammalian homologue of the *C. elegans* death gene *ced-3. Cell* **75,** 653–660.

Suda, T. *et al.* (1993). Molecular cloning and expression of the Fas ligand, a novel member of the TNF family. *Cell* **76,** 1169–1178.

Tartaglia, L. A. *et al.* (1993). A novel domain within the 55 kD TNF receptor signals cell death. *Cell* **74,** 845–853.

Li, P. *et al.* (1997). Cytochrome *c* and ATP-dependent formation of Apaf-1/caspase-9 complex initiates an apoptotic protease cascade. *Cell* **91,** 479–489.

Liu, X. *et al.* (1997). DFF, a heterodimeric protein that functions downstream of caspase-3 to trigger DNA fragmentation during apoptosis. *Cell* **89,** 175–184.

Yang, X. *et al.* (1997). Daxx, a novel Fas-binding protein that activates JNK and apoptosis. *Cell* **89,** 1067–1076.

Chapter 28

Oncogenes and cancer

A major feature of all higher eukaryotes is the defined life span of the organism, a property that extends to the individual somatic cells, whose growth and division are highly regulated. A notable exception is provided by cancer cells, which arise as variants that have lost their usual growth control. Their ability to grow in inappropriate locations or to propagate indefinitely may be lethal for the individual organism in which they occur.

Three types of changes that occur when a cell becomes tumorigenic are summarized in **Figure 28.1**:

- **Immortalization** describes the property of indefinite growth (without any other changes in the phenotype necessarily occurring).
- **Transformation** describes the failure to observe the normal constraints of growth; for example, transformed cells become independent of factors usually needed for cell growth.
- **Metastasis** describes the stage at which the cancer cell gains the ability to invade normal tissue, so that it can move away from the tissue of origin and establish a new colony elsewhere in the body.

To characterize the aberrant events that enable cells to bypass normal control and generate tumors, we need to compare the growth characteristics of normal and transformed cells *in vitro*. Transformed cells can be grown readily, but it is much more difficult to grow their normal counterparts.

When cells are taken from a vertebrate organism and placed in culture, they grow for several divisions, but then enter a senescent stage, in which growth ceases. This is followed by a **crisis**, in which most of the cells die. The survivors that emerge are capable of dividing indefinitely, but their properties have changed in the act of emerging from crisis. The nature of crisis is poorly defined, and we do not understand the molecular changes that adapt a cell to growth in culture, but in principle this comprises the process of immortalization. (The features of crisis depend on both the species and tissue. Typically, mouse cells pass through crisis at ~12 generations. Human cells enter crisis at ~40 generations, although it is rare for human cells to emerge from it, and only some types of human cells in fact can do so.)

The limitation of the life span of most cells by crisis restricts us to two options in studying nontransformed cells, neither entirely satisfactory:

- **Primary cells** are the immediate descendants of cells taken directly from the organism. They faithfully mimic the *in vivo* phenotype, but in most cases survive for only a relatively short period, because the culture dies out at crisis.
- Cells that have passed through crisis become **established** to form a (nontumorigenic) cell line. They can be perpetuated indefinitely, but their properties have changed in passing through crisis, and may indeed continue to change during adaptation to culture. These

Figure 28.1 *Overview*: three types of properties distinguish a cancer cell from a normal cell. Sequential changes in cultured cells can be correlated with changes in tumorigenicity.

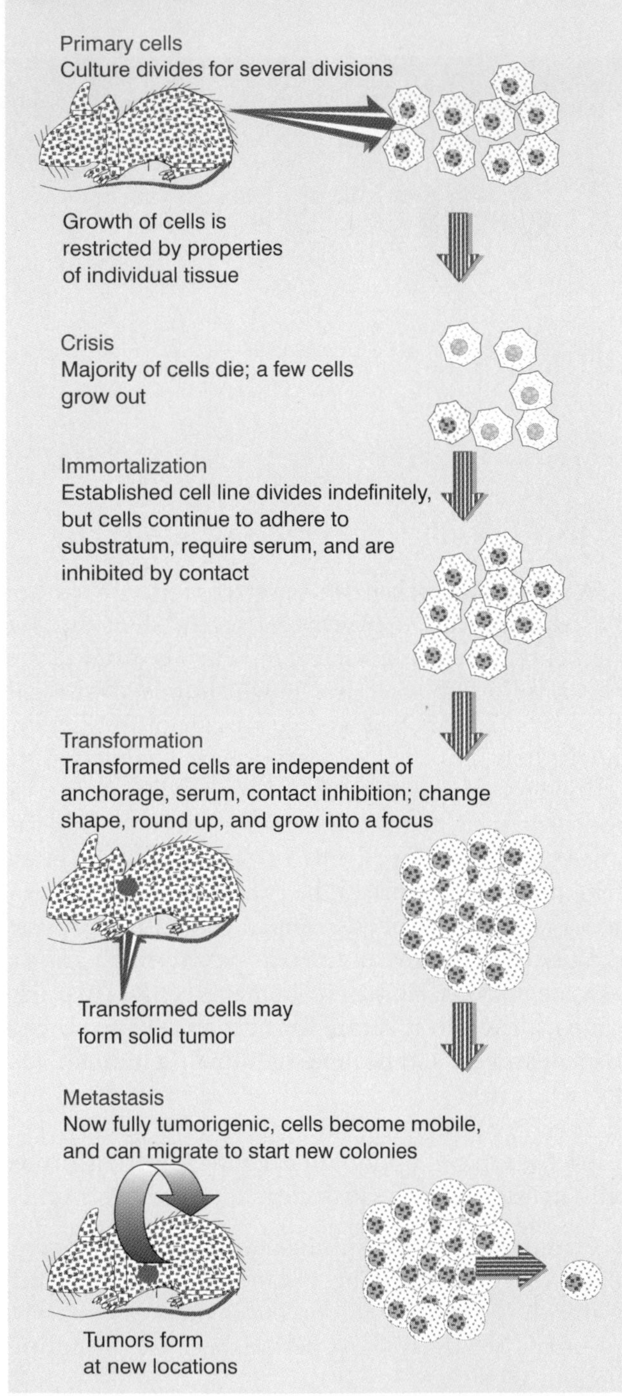

changes may partly resemble those involved in tumor formation, which reduces the usefulness of the cells.

Primary cells
Culture divides for several divisions
Growth of cells is
restricted by properties
of individual tissue
Crisis
Majority of cells die; a few cells
grow out
Immortalization
Established cell line divides indefinitely,
but cells continue to adhere to
substratum, require serum, and are
inhibited by contact
Transformation
Transformed cells are independent of
anchorage, serum, contact inhibition; change
shape, round up, and grow into a focus
Transformed cells may
form solid tumor
Metastasis
Now fully tumorigenic, cells become mobile,
and can migrate to start new colonies
Tumors form
at new locations

An established cell line by definition has become immortalized, but usually is not tumorigenic. Non-tumorigenic established cell lines display characteristic features similar to those of primary cultures, often including:

- **Anchorage dependence**—a solid or firm surface is needed for the cells to attach to.
- **Serum (or growth factor) dependence**—serum is needed to provide essential growth factors.
- **Density-dependent inhibition**—cells grow only to a limited density, because growth is inhibited, perhaps by processes involving cell-cell contacts.
- **Cytoskeletal organization**—cells are flat and extended on the surface on which they are growing, and have a typical elongated network of stress fibers (consisting of actin filaments).

The consequence of these properties is that the cells grow as a **monolayer** (that is, a layer one cell thick) on a substratum.

These properties provide parameters by which the normality of the cell may be judged. Of course, any established cell line provides only an approximation of *in vivo* control. The need for caution in analyzing the genetic basis for growth control in such lines is emphasized by the fact that almost always they suffer changes in the chromosome complement and are not true diploids. A cell whose chromosomal constitution has changed from the true diploid is said to be **aneuploid.**

Cells cultured from tumors instead of from normal tissues show changes in some or all of these properties. They are said to be **transformed.** A transformed cell grows in a much less restricted manner. It has reduced serum-dependence, does not need to attach to a solid surface (so that individual cells "round-up" instead of spreading out) and the cells pile up into a thick mass of cells (called a **focus**) instead of growing as a surface monolayer. Furthermore, the cells may form tumors when injected into appropriate test animals. **Figure 28.2** compares a "normal" fibroblast growing in culture with a "transformed" variant.

It would be naive to suppose that there is a uniform basis for cancer cell formation; many types of changes in the cellular constitution confer the ability to form a tumor. However, the joint changes of immortalization and transformation of cells in culture provide a paradigm for the formation of animal tumors. By comparing transformed cell lines with normal cells, we hope to identify the genetic basis for tumor formation and also to understand the phenotypic processes that are involved in the conversion.

Certain events convert normal cells into transformed cells, and provide models for the processes in-

Figure 28.2 Normal fibroblasts grow as a layer of flat, spread-out cells, whereas transformed fibroblasts are rounded up and grow in cell masses. The cultures on the left contain normal cells, those on the right contain transformed cells. The top views are by conventional microscopy, the bottom by scanning electron microscopy. Photographs kindly provided by Hidesaburo Hanafusa and J. Michael Bishop.

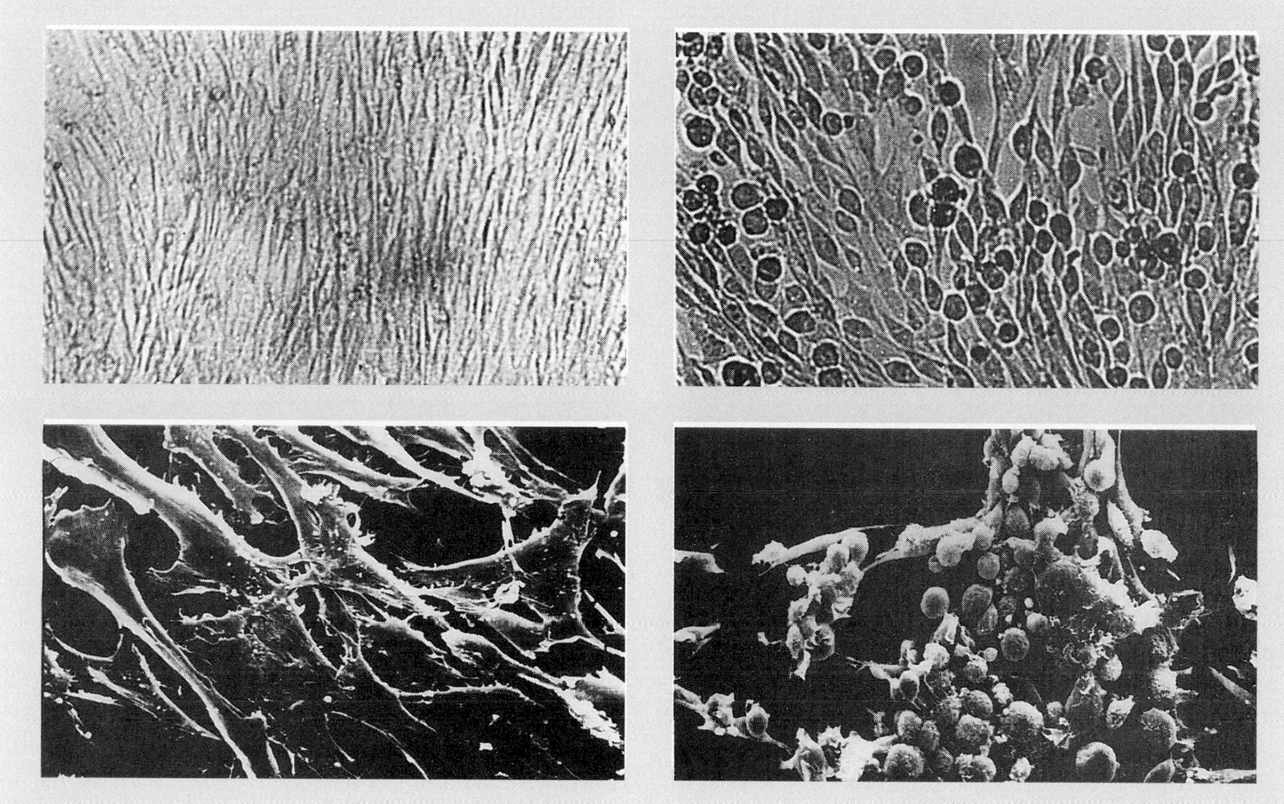

volved in tumor formation. Usually, multiple genetic changes are necessary to create a cancer; and sometimes tumors gain increased virulence as the result of a progressive series of changes. The incidence of human cancers with age suggests that typically 6–7 events are required over a span of 20–40 years to induce a cancer. In certain (rare) cases, propensity to cancer is inherited as a Mendelian trait, implying that a single genetic change is an important or necessary component (although other changes are also necessary).

A variety of agents increase the frequency with which cells (or animals) are converted to the transformed condition; they are said to be **carcinogenic.** Sometimes these **carcinogens** are divided into those that "initiate" and those that "promote" tumor formation, implying the existence of different stages in cancer development. Carcinogens may cause epigenetic changes or (more often) may act, directly or indirectly, to change the genotype of the cell.

There are two classes of genes in which mutations cause transformation:

- **Oncogenes** were initially identified as genes carried by viruses that cause transformation of their target cells. A major class of the viral oncogenes have cellular counterparts that are involved in normal cell functions. The cellular genes are called **proto-oncogenes,** and in certain cases their mutation or aberrant activation in the cell is associated with tumor formation. About 100 oncogenes have been identified. The oncogenes fall into several groups, representing different types of activities ranging from transmembrane proteins to transcription factors, and the definition of these functions may therefore lead to an understanding of the types of changes that are involved in tumor formation. The generation of an oncogene represents a gain-of-function in which a cellular proto-oncogene is inappropriately activated. This can involve a mutational change in the protein, or constitutive activation, over-expression, or failure to turn off expression at the appropriate time.

- **Tumor suppressors** are detected by deletions (or other inactivating mutations) that are tumorigenic.

The most compelling evidence for their nature is provided by certain hereditary cancers, in which patients with the disease develop tumors that have lost both alleles, and therefore lack an active gene. There is also now evidence that changes in these genes may be associated with the progression of a wide range of cancers. About 10 tumor suppressors are known at present. They represent loss-of-function in genes that usually impose some constraint on the cell cycle or cell growth; the release of the constraint is tumorigenic.

In the first part of this chapter, we consider how these two classes of genes are identified, and we ask how oncogenes are activated, and how tumor suppressors are inactivated. Then we consider the molecular basis for these events, and how oncogenes are connected into pathways that extend from signal transduction at the cell surface to activation of transcription factors in the nucleus.

Transforming viruses carry oncogenes

TRANSFORMATION may occur spontaneously, may be caused by certain chemical agents, and, most notably, may result from infection with **tumor viruses.** There are many classes of tumor viruses, including both DNA and RNA viruses, and they occur widely in the avian and animal kingdoms.

The transforming activity of a tumor virus resides in a particular gene or genes carried in the viral genome. Oncogenes were given their name by virtue of their ability to convert cells to a tumorigenic (or oncogenic) state. An oncogene initiates a series of events that is executed by cellular proteins. In effect, the virus throws a regulatory switch that changes the growth properties of its target cell. **Figure 28.3** summarizes the general properties of the major classes of transforming viruses. The oncogenes carried by the DNA viruses specify proteins that inactivate tumor suppressors, so their action in part mimics loss-of-function of the tumor suppressors. The oncogenes carried by retroviruses are derived from cellular genes and therefore may mimic the behavior of gain-of-function mutations in animal proto-oncogenes.

Polyomaviruses and adenoviruses have been isolated from a variety of mammals. Although perpetuated in the wild in a single host species, a virus may be able to grow in culture on a variety of cells from different species. The response of a cell to infection depends on its species and phenotype and falls into one of two classes, as illustrated in **Figure 28.4:**

- **Permissive cells** are productively infected. The virus proceeds through a lytic cycle that is divided into the usual early and late stages. The cycle ends with release of progeny viruses and (ultimately) cell death.
- **Nonpermissive cells** cannot be productively infected, and viral replication is abortive. Some of the in-

Figure 28.3 Transforming viruses may carry oncogenes.

Viral class	Type of virus	Genome size	Oncogenes	Origin of oncogene	Action of oncogene
Polyoma	dsDNA	5–6 kb	T antigens	Early viral gene	Inactivates tumor suppressor
HPV	dsDNA	~8 kb	E6 & E7	Early viral gene	Inactivates tumor suppressor
Adeno	dsDNA	~37 kb	E1A & E1B	Early viral gene	Inactivates tumor suppressor
Retrovirus (acute)	ssRNA	6–9 kb	Individual	Cellular	Activates oncogenic pathway

Figure 28.4 Permissive cells are productively infected by a DNA tumor virus that enters the lytic cycle, while nonpermissive cells are transformed to change their phenotype.

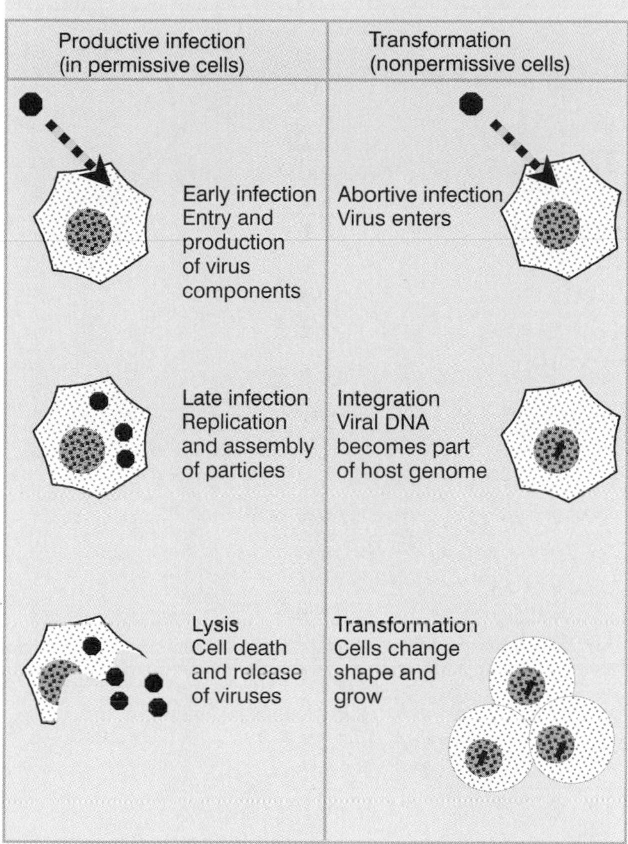

Figure 28.5 Cells transformed by polyomaviruses or adenoviruses have viral sequences that include the early region integrated into the cellular genome. Sites of integration are random.

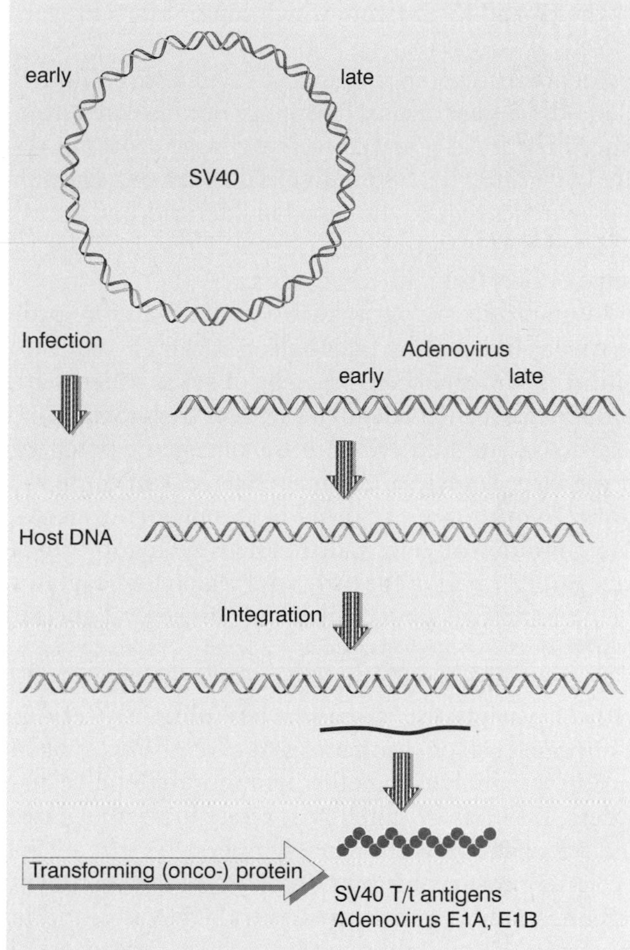

fected cells are transformed; in this case, the phenotype of the individual cell changes and the culture is perpetuated in an unrestrained manner.

A common mechanism underlies transformation by DNA tumor viruses. *Oncogenic potential resides in a single function or group of related functions that are active early in the viral lytic cycle. When transformation occurs, the relevant gene(s) are integrated into the genomes of transformed cells and expressed constitutively.* This suggests the general model for transformation by these viruses illustrated in **Figure 28.5,** in which the constitutive expression of the oncogene generates transforming protein(s) (oncoproteins).

Polyomaviruses are small. Polyomavirus itself is common in mice, the analogous virus SV40 (simian virus 40) was isolated from rhesus monkey cells, and more recently the human viruses BK and JC have been characterized. All of the polyomaviruses can cause tumors when injected into newborn rodents.

During a productive infection, the early region of each virus uses alternative splicing to synthesize overlapping proteins called T antigens. (The name reflects their isolation originally as the proteins found in tumor cells.) The various T antigens have a variety of functions in the lytic cycle. They are required for expression of the late region and for DNA replication of the virus.

Cells transformed by polyomaviruses contain integrated copies of part or all of the viral genome. The integrated sequences always include the early region. The T antigens have transforming activity, which rests upon their ability to interact with cellular proteins. This is independent of their ability to interact directly with the viral genome. SV40 requires "big T" and "little t" antigens, and polyoma requires "T" and "middle T" antigens for transformation.

Papillomaviruses are small DNA viruses that cause

epithelial tumors; there are ~75 human papillomaviruses (HPVs); most are associated with benign growths (such as warts), but some are associated with cancers, in particular cervical cancers. Two virus-associated products are expressed in cervical cancers; these are the E6 and E7 proteins, which can immortalize target cells.

Adenoviruses were originally isolated from human adenoids; similar viruses have since been isolated from other mammals. They comprise a large group of related viruses, with >80 individual members. Human adenoviruses remain the best characterized, and are associated with respiratory diseases. They can infect a range of cells from different species.

Human cells are permissive and are therefore productively infected by adenoviruses, which replicate within the infected cell. But cells of some rodents are nonpermissive. All adenoviruses can transform nonpermissive cultured cells, but the oncogenic potential of the viruses varies; the most effective can cause tumors when they are injected into newborn rodents. The genomes of cells transformed by adenoviruses have gained a part of the early viral region that contains the E1A and E1B genes, which code for several nuclear proteins.

Epstein-Barr is a human herpes virus associated with a variety of diseases, including infectious mononucleosis, nasopharyngeal carcinoma, African Burkitt lymphoma, and other lymphoproliferative disorders. EBV has a limited host range for both species and cell phenotype. Human B lymphocytes that are infected *in vitro* become immortalized, and some rodent cell lines can be transformed. Viral DNA is found in transformed cells, although it has been controversial whether it is integrated. It remains unclear exactly which viral genes are required for transformation.

Retroviruses present a different situation from the DNA tumor viruses. They can transfer genetic information both horizontally and vertically, as illustrated in **Figure 28.6.** Horizontal transfer is accomplished by the normal process of viral infection, in which increasing numbers of cells become infected in the same host. Vertical transfer results whenever a virus becomes integrated in the germline of an organism as an endogenous provirus; like a lysogenic bacteriophage, it is inherited as a Mendelian locus by the progeny (see Chapter 11).

The retroviral life cycle propagates genetic information through both RNA and DNA templates. A retroviral infection proceeds through the stages illustrated previously in Figure 16.2, in which the RNA is reverse-transcribed into single-stranded DNA, then converted into double-stranded DNA, and finally integrated into the genome, where it may be transcribed again into infectious RNA. Integration into the genome leads to vertical transmission of the provirus; expression of the provirus may generate retroviral particles that are horizontally transmitted. Integration is a normal part of the life cycle of every retrovirus, whether it is nontransforming or transforming.

Figure 28.6 Retroviruses transfer genetic information horizontally by infecting new hosts; information is inherited vertically if a virus integrates in the genome of the germline.

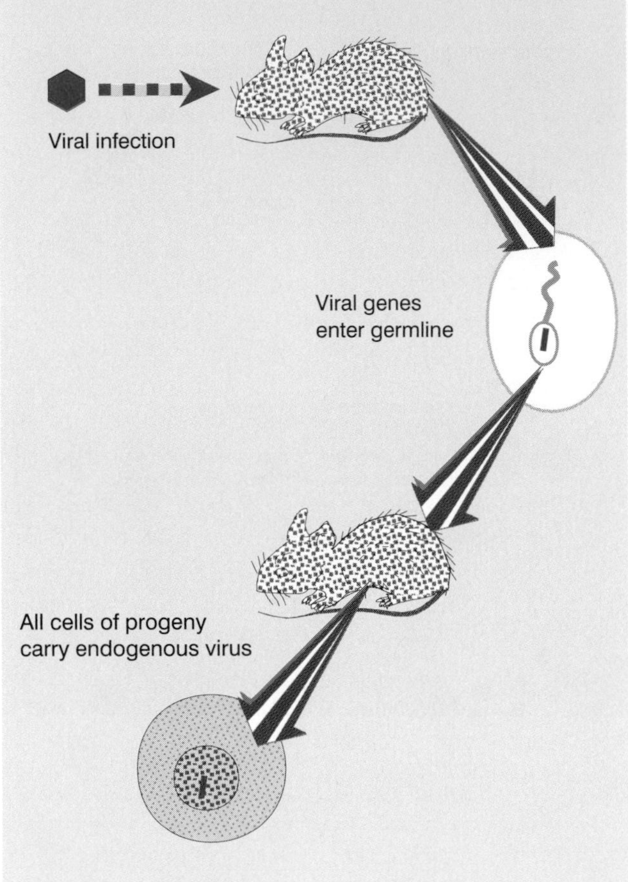

The tumor retroviruses fall into two general groups with regard to the origin of their tumorigenicity:

- **Nondefective viruses** follow the usual retroviral life cycle. They provide infectious agents that have a long latent period, and often are associated with the induction of leukemias. Two classic models are FeLV (feline leukemia virus) and MMTV (mouse mammary tumor virus). *Tumorigenicity does not rely upon an individual viral oncogene, but upon the ability of the virus to activate a cellular proto-oncogene(s).*

■ **Acute transforming viruses** *have gained new genetic information in the form of an oncogene.* This gene is not present in the ancestral (nontransforming virus); it originated as a cellular gene that was captured by the virus by means of a transduction event during an infective cycle. These viruses usually induce tumor formation *in vivo* rather rapidly, and they can transform cultured cells *in vitro.* Reflecting the fact that each acute transforming virus has specificity toward a particular type of target cell, these viruses are divided into classes according to the type of tumor that is caused in the animal: leukemia, sarcoma, carcinoma, etc.

When a retrovirus captures a cellular gene by exchanging part of its own sequence for a cellular sequence (see Chapter 16), it generates the structure summarized in **Figure 28.7.** This type of event is rare, but creates a transducing virus that has two important properties:

■ Usually it cannot replicate by itself, because viral genes needed for reproduction have been lost by the exchange with cellular sequences. So almost all of these viruses are replication-defective. But they can propagate in a simultaneous infection with a wild-type "helper" virus that provides the functions that were lost in the recombination event. (RSV is an exceptional transducing virus that retains the ability to replicate.)

■ During an infection, the transducing virus carries with it the cellular gene(s) that were obtained in the recombination event, and their expression may alter the phenotype of the infected cell. Any transducing virus whose cellular genetic information assists the growth of its target cells could have an advantage in future infective cycles. If a virus gains a gene whose product stimulates cell growth, the acquisition may enable the virus to spread by stimulating the growth of the particular cells that it infects. (This is important also because a retrovirus can replicate only in a proliferating cell; see Chapter 11.) After a virus has collected a cellular gene, the gene may gain mutations that enhance its ability to influence cell phenotype.

Figure 28.7 A transforming retrovirus carries a copy of a cellular sequence in place of some of its own gene(s).

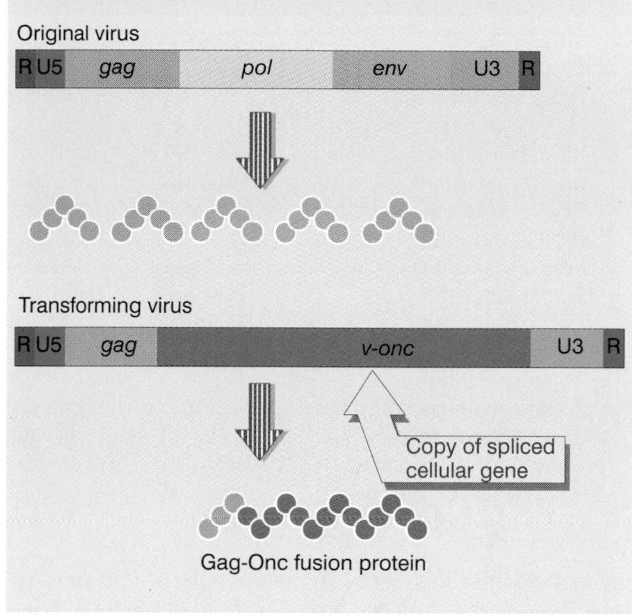

Of course, transformation is not the only mechanism by which retroviruses affect their hosts. A notable example is the HIV-1 retrovirus, which belongs to the retroviral group of lentiviruses. The virus infects and kills T lymphocytes carrying the CD4 receptor, devastating the immune system of the host, and inducing the disease of AIDS. The virus carries the usual *gag-pol-env* regions, and also has an additional series of reading frames, which overlap with one another, to which its lethal actions are attributed.

Retroviral oncogenes have cellular counterparts

ONCOGENES of some retroviruses are summarized in **Figure 28.8.** The type of tumor results from the combination of the particular oncogene with the time and place in which it is expressed. It is striking that usually the oncogenic activity resides in a single gene. AEV is one of a very few exceptions in which a retrovirus carries more than one oncogene.

The new sequences present in an acute transforming retrovirus can be delineated by comparing the sequence of the virus with that of the parental

Figure 28.8 Each transforming retrovirus carries an oncogene derived from a cellular gene. Viruses have names and abbreviations reflecting the history of their isolation and the types of tumor they cause. This list shows some representative examples of the retroviral oncogenes.

Virus	Name	Species	Tumor	Oncogene
Rous sarcoma	RSV	chicken	sarcoma	*src*
Harvey murine sarcoma	Ha-MuSV	rat	sarcoma & erythroleukemia	*H-ras*
Kirsten murine sarcoma	Ki-MuSV	rat	sarcoma & erythroleukemia	*K-ras*
Moloney murine sarcoma	Mo-MuSV	mouse	sarcoma	*mos*
FBJ murine osteosarcoma	FBJ-MuSV	mouse	chondrosarcoma	*fos*
Simian sarcoma	SSV	monkey	sarcoma	*sis*
Feline sarcoma	PI-FeSV	cat	sarcoma	*sis*
Feline sarcoma	SM-FeSV	cat	fibrosarcoma	*fms*
Feline sarcoma	ST-FeSV	cat	fibrosarcoma	*fes*
Avian sarcoma	ASV-17	chicken	fibrosarcoma	*jun*
Fujinami sarcoma	FuSV	chicken	sarcoma	*fps*
Avian myelocytomatosis	MC29	chicken	carcinoma, sarcoma, & myelocytoma	*myc*
Abelson leukemia	MuLV	mouse	B cell lymphoma	*abl*
Reticuloendotheliosis	REV-T	turkey	lymphatic leukemia	*rel*
Avian erythroblastosis	AEV	chicken	erythroleukemia & fibrosarcoma	*erbB, erbA*
Avian myeloblastosis	AMV	chicken	myeloblastic leukemia	*myb*

(nontumorigenic) virus. Invariably there is a new region that is closely related to a sequence in the cellular genome. The normal cellular sequence itself is not oncogenic—if it were, the organism could scarcely have survived—but it defines a **proto-oncogene**, a cellular gene whose capture by the retrovirus and subsequent modification created the oncogene.

The viral oncogenes and their cellular counterparts are described by using prefixes *v* for viral and *c* for cellular. So the oncogene carried by Rous sarcoma virus is called *v-src*, and the proto-oncogene related to it in cellular genomes is called *c-src*. Comparisons between *v-onc* and *c-onc* genes can be used to identify the features that confer oncogenicity.

More than 30 *c-onc* genes have been identified so far by their representation in retroviruses. Sometimes the same *c-onc* gene is represented in different transforming viruses; for example, the monkey virus SSV and the PI strain of the feline virus FeSV both carry a *v-onc* derived from *c-sis*. Some viruses carry related *v-onc* genes, such as in the Harvey and Kirsten strains of MuSV, which carry *v-ras* genes derived from two different members of the cellular *c-ras* gene family. In other cases the *v-onc* genes of related viruses represent unrelated cellular progenitors; for example, three different isolates of FeSV may have been derived from the same original (nontransforming) virus, but have transduced the *sis*, *fms*, and *fes* oncogenes. The events involved in formation of a transducing virus can be complex; some viruses include sequences derived from more than one cellular gene.

Given the rarity of the transducing event, it is significant that multiple independent isolates occur representing the same *c-onc* gene. For example, several viruses carry *v-myc* genes. They are all derived from a single *c-myc* gene, but the *v-myc* genes differ in their exact ends and in individual point mutations. The existence of such isolates probably means that we have identified most of the genes of the *c-onc* type that can be activated by viral transduction.

Direct evidence that expression of the *v-onc* sequence accomplishes transformation was first obtained with RSV. Temperature-sensitive mutations in *v-src* allow the transformed phenotype to be reverted by increase in temperature, and regained by decrease in temperature. This shows clearly that in this case the *v-src* gene is needed both to initiate and maintain the transformed state.

Two general types of theory might explain the difference in properties between *v-onc* genes and *c-onc* genes:

- A quantitative model proposes that viral genes are functionally indistinguishable from the cellular genes, but are oncogenic because they are expressed in much greater amounts or in inappropriate cell types, or because their expression cannot be switched off.
- A qualitative model supposes that the *c-onc* genes intrinsically lack oncogenic properties, but may be converted by mutation into oncogenes whose devastating effects reflect the acquisition of new properties (or loss of old properties).

How closely related are *v-onc* genes to the corresponding *c-onc* genes? In some cases, the only changes are a very small number of point mutations. The *mos, sis,* and *myc* genes offer examples in which the entire *c-onc* gene has been gained by the virus; in this case, the small number of amino acid substitutions do not seem to affect function of the protein, and in fact are not required for transforming activity. So the *v-onc* product is likely to fulfill the same enzymatic or other functions as the *c-onc* product, but with some change in its regulation; in these cases, over-expression is responsible for oncogenicity. A good example is *c-myc,* where oncogenicity may be caused by over-expression either by a *v-myc* gene carried by a transforming retrovirus or by changes in the cellular genome that cause over-expression of *c-myc.*

Two cases in which point mutations play a critical role in creating an oncogenic protein are presented by *ras* and *src.*

In the case of *ras,* changes in the regulation of Ras activity that activate the protein can be directly attributed to the individual point mutations that have occurred in the *v-onc* gene. Over-expression of *c-ras* may have weak oncogenic effects, but full oncogenicity requires sequence changes in the protein.

In some cases, a *v-onc* gene is truncated by the loss of sequences from the N-terminus or C-terminus (or both) of the *c-onc* gene, probably as a result of the sites involved in the recombination event that generated it. Loss of these regions may remove some regulatory constraint that normally limits the activity of the *c-onc* product. Such sequence changes are required for oncogenicity of *src. v-src* is oncogenic at low levels of protein, but *c-src* is not oncogenic at high protein levels (>10× normal). The viral and cellular *src* genes are coextensive, but *v-src* has replaced the C-terminal 19 amino acids of *c-src* with a different sequence of 12 amino acids. This has an important regulatory consequence, in activating the Src protein constitutively. In cases where *v-onc* genes are truncations of *c-onc* genes, point mutations may also contribute to the oncogenicity of the *v-onc* product. In the case of Src, changes in two tyrosine residues that are targets for phosphorylation have strong effects on oncogenicity (see later).

The characterization of transforming retroviruses played an important role in the definition of oncogenes. The concept that oncogenes arise by activation of proto-oncogenes is an important paradigm for animal cancers. However, most events involved in human cancers do not involve viral intermediates, and other mechanisms are responsible for generating oncogenes.

Ras proto-oncogenes can be activated by mutation

SOME oncogenes can be detected by using a direct assay for transformation in which "normal" recipient cells are transfected with DNA obtained from animal tumors. (Actually the established mouse NIH 3T3 fibroblast line usually is used as recipient.) The procedure is illustrated in **Figure 28.9.** If a specific gene contributes to the transformed state, its introduction into new cells transforms them. Another assay that can be used is to inject cells into "nude" mice (which lack the ability to reject such transplants immunologically). The ability to form tumors can then be measured directly in the animal.

When a cell is transformed in a 3T3 culture (or some other "normal" culture), its descendants pile up into a focus. The appearance of foci is used as a measure of the transforming ability of a DNA preparation. Starting with a preparation of DNA isolated from tumor cells, the efficiency of focus formation is low. However, once the transforming gene has been isolated and cloned, greater efficiencies can be obtained. In fact, the transforming "strength" of a gene can be characterized by the efficiency of focus formation by the cloned sequence.

DNA with transforming activity can be isolated only from tumorigenic cells; it is not present in normal DNA. The transforming genes isolated by this assay have two revealing properties:

- *They have closely related sequences in the DNA of normal cells.* This argues that transformation was caused by mutation of a normal cellular gene (a proto-oncogene) to generate an oncogene. The change may take the form of a point mutation or more extensive reorganization of DNA around the *c-onc* gene.
- *They may have counterparts in the oncogenes carried by known transforming viruses.* This suggests that the

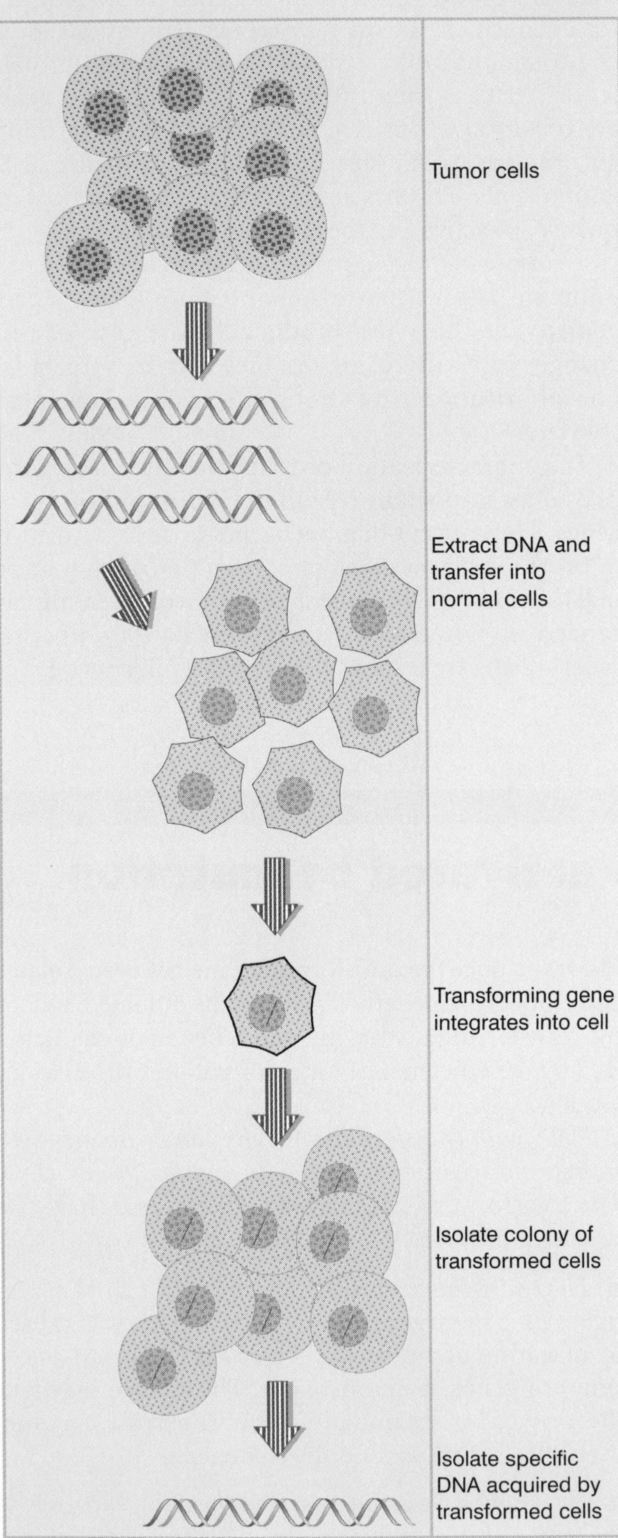
Tumor cells
Extract DNA and transfer into normal cells
Transforming gene integrates into cell
Isolate colony of transformed cells
Isolate specific DNA acquired by transformed cells

Figure 28.9 The transfection assay allows (some) oncogenes to be isolated directly by assaying DNA of tumor cells for the ability to transform normal cells into tumorigenic cells.

repertoire of proto-oncogenes is limited, and probably the same genes are targets for mutations to generate oncogenes in the cellular genome or to become viral oncogenes.

Oncogenes derived from the *c-ras* family are often detected in the transfection assay. The family consists of several active genes in both man and rat, dispersed in the genome. (There are also some pseudogenes.) The individual genes, *N-ras, H-ras,* and *K-ras,* are closely related, and code for protein products ~21 kD and known collectively as $p21^{ras}$.

The *H-ras* and *K-ras* genes have *v-ras* counterparts, carried by the Harvey and Kirsten strains of murine sarcoma virus, respectively (see Figure 28.8). Each *v-ras* gene is closely related to the corresponding *c-ras* gene, with only a few individual amino acid substitutions. The Harvey and Kirsten virus strains must have originated in independent recombination events in which a progenitor virus gained the corresponding *c-ras* sequence.

Oncogenic variants of the *c-ras* genes are found in transforming DNA preparations obtained from various primary tumors and tumor cell lines. Each of the *c-ras* proto-oncogenes can give rise to a transforming oncogene by a single base mutation. *The mutations in several independent human tumors cause substitution of a single amino acid, most commonly at position 12 or 61, in one of the Ras proteins.*

Position 12 is one of the residues that is mutated in the *v-H-ras* and *v-K-ras* genes. *So mutations occur at the same positions in* v-ras *genes in retroviruses and in mutant* c-ras *genes in multiple rat and human tumors.* This suggests that the normal c-Ras protein can be converted into a tumorigenic form by a mutation in one of a few codons in rat or man (and perhaps any mammal).

The general principle established by this work is that *substitution in the coding sequence can convert a cellular proto-oncogene into an oncogene.* Such an oncogene can be associated with the appearance of a spontaneous tumor in the organism. It may also be carried by a retrovirus, in which case a tumor is induced by viral infection.

The *ras* genes appear to be finely balanced at the edge of oncogenesis. Almost any mutation at either position 12 or 61 can convert a *c-ras* proto-oncogene into an active oncogene. All three *c-ras* genes have glycine at position 12. If it is replaced *in vitro* by any other of the 19 amino acids except proline, the mutated *c-ras* gene can transform cultured cells. The particular substitution influences the strength of the transforming ability.

Position 61 is occupied by glutamine in wild-type

c-ras genes. Its change to another amino acid usually creates a gene with transforming potential. Some substitutions are less effective than others; proline and glutamic acid are the only substitutions that have no effect.

When the expression of a normal *c-ras* gene is increased, either by placing it under control of a more active promoter or by introducing multiple copies into transfected cells, recipient cells are transformed. Some mutant *c-ras* genes that have changes in the protein sequence also have a mutation in an intron that increases the level of expression (by increasing processing of mRNA ~10×). Also, some tumor lines have amplified *ras* genes. A 20-fold increase in the level of a nontransforming Ras protein is sufficient to allow the transformation of some cells. The effect has not been fully quantitated, but it suggests the general conclusion that oncogenesis depends on over-activity of Ras protein, and is caused either by increasing the amount of protein or (more efficiently) by mutations that increase the activity of the protein.

Transfection by DNA can be used to transform only certain cell types. Limitations of the assay explain why relatively few oncogenes have been detected by transfection. This system has been most effective with *ras* genes, where there is extensive correlation between mutations that activate *c-ras* genes in transfection and the occurrence of tumors.

Ras is a monomeric guanine nucleotide-binding protein that is active when bound to GTP and inactive when bound to GDP. It has an intrinsic GTPase activity. **Figure 28.10** reviews the discussion of Chapter 26 in which we saw that the conversion between the two forms of Ras is catalyzed by other proteins. GAP proteins stimulate the ability of Ras to hydrolyze GTP, thus converting active Ras into inactive Ras. GEF proteins stimulate the replacement of GDP by GTP, thus reactivating the protein.

Constitutive activation of Ras could be caused by mutations that allow the GDP-bound form of Ras to be active or that prevent hydrolysis of GTP. What are the effects of the mutations that create oncogenic *ras* genes? Many mutations that confer transforming activity inhibit the GTPase activity. GAP cannot increase the GTPase activity of Ras proteins that have been activated by oncogenic mutations. In other words, Ras has become refractory to the interaction with GAP that turns off its activity. Inability to hydrolyze GTP causes Ras to remain in a permanently activated form; its continued action upon its target protein is responsible for its oncogenic activity.

Figure 28.10 Pathways that rely on Ras could function by controlling either GNRF or GAP. Oncogenic Ras mutants are refractory to control, because Ras remains in the active form.

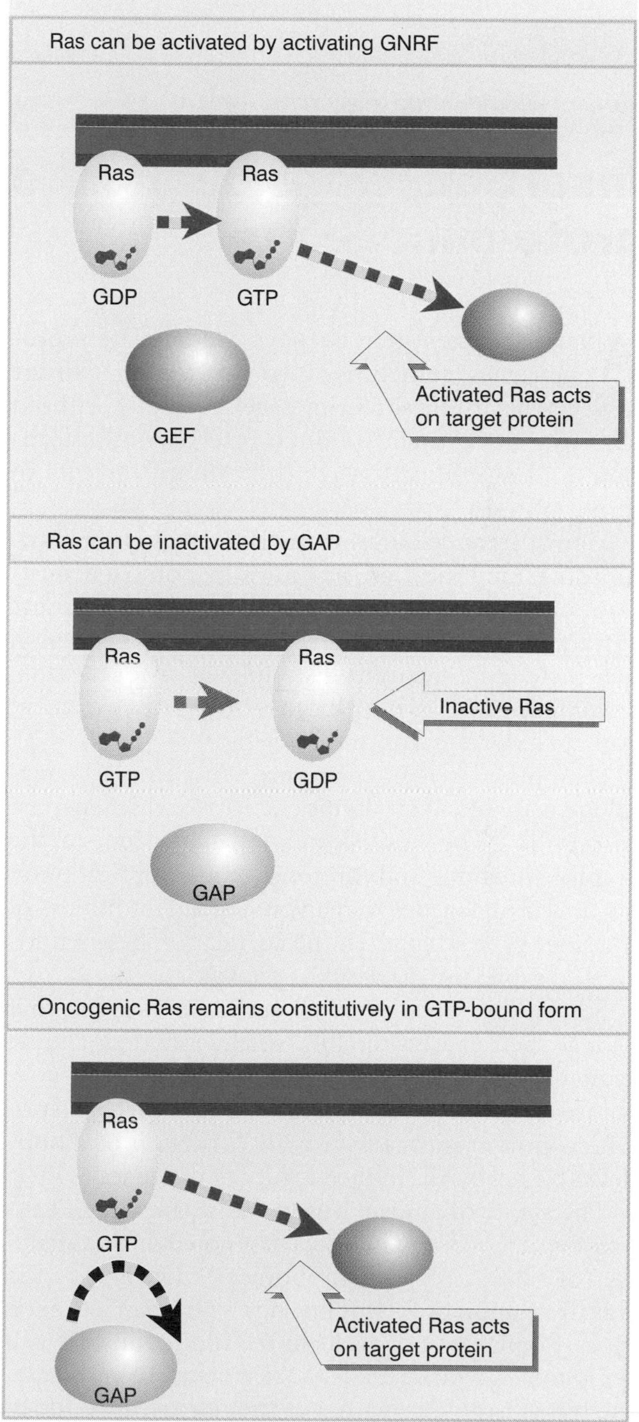
Ras can be activated by activating GNRF
Ras
Ras
GDP
GTP
GEF
Activated Ras acts on target protein
Ras can be inactivated by GAP
Ras
Ras
Inactive Ras
GTP
GDP
GAP
Oncogenic Ras remains constitutively in GTP-bound form
Ras
GTP
Activated Ras acts on target protein
GAP

This establishes an important principle: *constitutive activation of a cellular protein may be oncogenic.* In the case of Ras, its effects result from activating the ERK MAP kinase pathway and (possibly) other pathways. The level of expression is finely balanced, since overstimulation of Ras by either increase in expression or mutation of the protein has oncogenic consequences (although mutation is required for a full effect).

Insertion, translocation, or amplification may activate proto-oncogenes

A variety of genomic changes can activate proto-oncogenes, sometimes involving a change in the target gene itself, sometimes activating it without changing the protein product. Insertion, translocation, and amplification can be causative events in tumorigenesis.

Many tumor cell lines have visible regions of chromosomal amplification, as shown by homogeneously staining regions (see Figure 17.29) or double minute chromosomes (see Figure 17.30). The amplified region may include an oncogene. Examples of oncogenes that are amplified in various tumors include *c-myc, c-abl, c-myb, c-erbB, c-K-ras,* and *Mdm2.*

Established cell lines are prone to amplify genes (along with other karyotypic changes to which they are susceptible). The presence of known oncogenes in the amplified regions and the consistent amplification of particular oncogenes in many independent tumors of the same type strengthen the correlation between increased expression and tumor growth.

Some proto-oncogenes are activated by events that change their expression, but which leave their coding sequence unaltered. The best characterized is *c-myc,* whose expression is elevated by several mechanisms. One common mechanism is the insertion of a nondefective retrovirus in the vicinity of the gene.

The ability of a retrovirus to transform without expressing a *v-onc* sequence was first noted during analysis of the bursal lymphomas caused by the transformation of B lymphocytes with avian leukemia virus. Similar events occur in the induction of T-cell lymphomas by murine leukemia virus. In each case, the transforming potential of the retrovirus is due to the ability of its LTR (the long terminal repeat of the integrated form) to cause expression of cellular gene(s).

In many independent tumors, the virus has integrated into the cellular genome within or close to the *c-myc* gene. **Figure 28.11** summarizes the types of insertions. The retrovirus may be inserted at a variety of locations relative to the *c-myc* gene.

The gene consists of three exons; the first represents a long nontranslated leader, and the second two code for the c-Myc protein. The simplest insertions to explain are those that occur within the first intron. The LTR provides a promoter, and transcription reads through the two coding exons. Transcription of *c-myc* under viral control differs from its usual control: the level of expression is increased (because the LTR provides an efficient promoter); expression cannot be switched off in B or T cells in response to the usual differentiation signals; and the transcript lacks its usual nontranslated leader (which may usually limit expression). All of these changes add up to increased constitutive expression.

Activation of *c-myc* in the other two classes of insertions reflects different mechanisms. The retroviral genome may be inserted within or upstream of the first intron, but in reverse orientation, so that its promoter points in the wrong direction. The retroviral genome also may be inserted downstream of the *c-myc* gene. In these cases, the enhancer in the viral LTR may be responsible for activating transcription of c-Myc, either from its normal promoter or from a fortuitous promoter.

In all of these cases, the coding sequence of c-myc *is unchanged, so oncogenicity is attributed to the loss of normal control and increased expression of the gene.*

Other oncogenes that are activated in tumors by the insertion of a retroviral genome include *c-erbB, c-myb, c-mos, c-H-ras,* and *c-raf.* Up to 10 other cellular genes (not previously identified as oncogenes by their presence in transforming viruses) are implicated as potential oncogenes by this criterion. The best characterized among this latter class are *wnt1* and *int2*. The *wnt1* gene codes for a protein involved in early embryogenesis that is related to the *wingless* gene of *Drosophila*; *int2* codes for an FGF (fibroblast growth factor).

Translocation to a new chromosomal location is another of the mechanisms by which oncogenes are activated. A **reciprocal translocation** occurs when an illegitimate recombination occurs between two chromosomes as illustrated in **Figure 28.12.** The involvement of such events in tumorigenesis was discovered via a connection between the loci coding immunoglobulins and the occurrence of certain tumors. Specific chromosomal translocations are often associated with tumors that arise from undifferentiated B lymphocytes. The common feature is that an oncogene on one chromosome is brought by translocation into the proximity of an Ig locus on another chromosome. Similar events occur in T lymphocytes to bring oncogenes into the proximity of a TCR locus.

In both man and mouse, the nonimmune partner is

Figure 28.11 Insertions of ALV at the *c-myc* locus occur at various positions, and activate the gene in different ways.

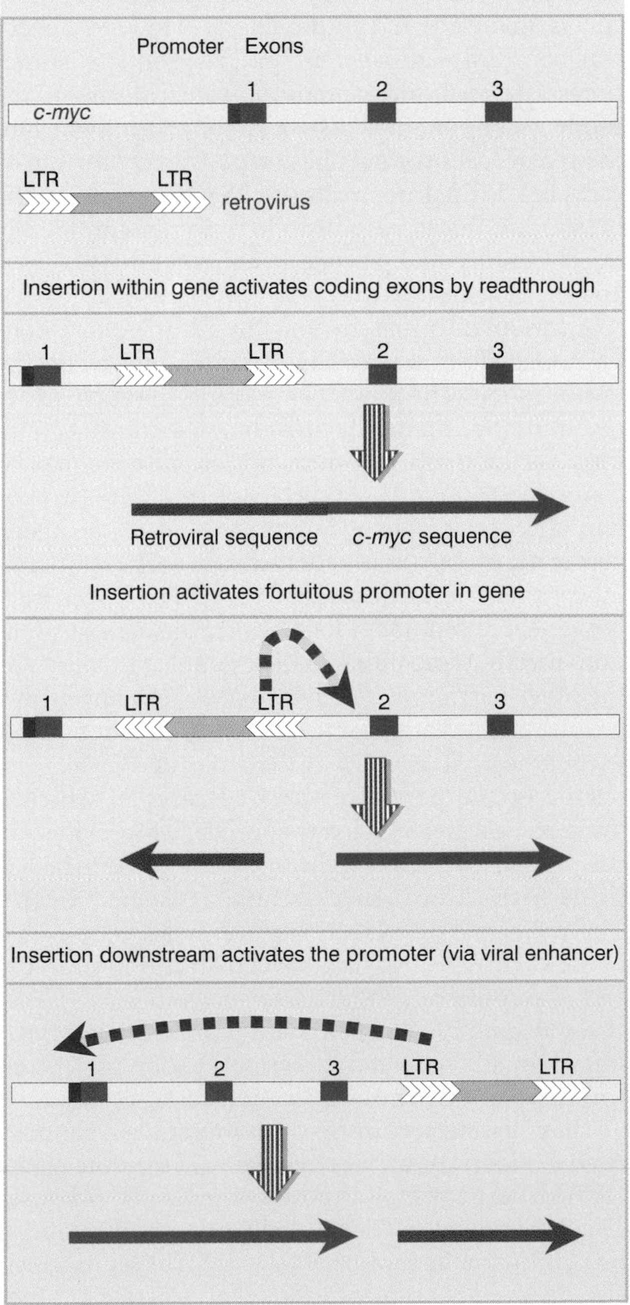

Figure 28.12 A chromosomal translocation is a reciprocal event that exchanges parts of two chromosomes. Translocations that activate the human *c-myc* proto-oncogene involve Ig loci in B cells and TCR loci in T cells.

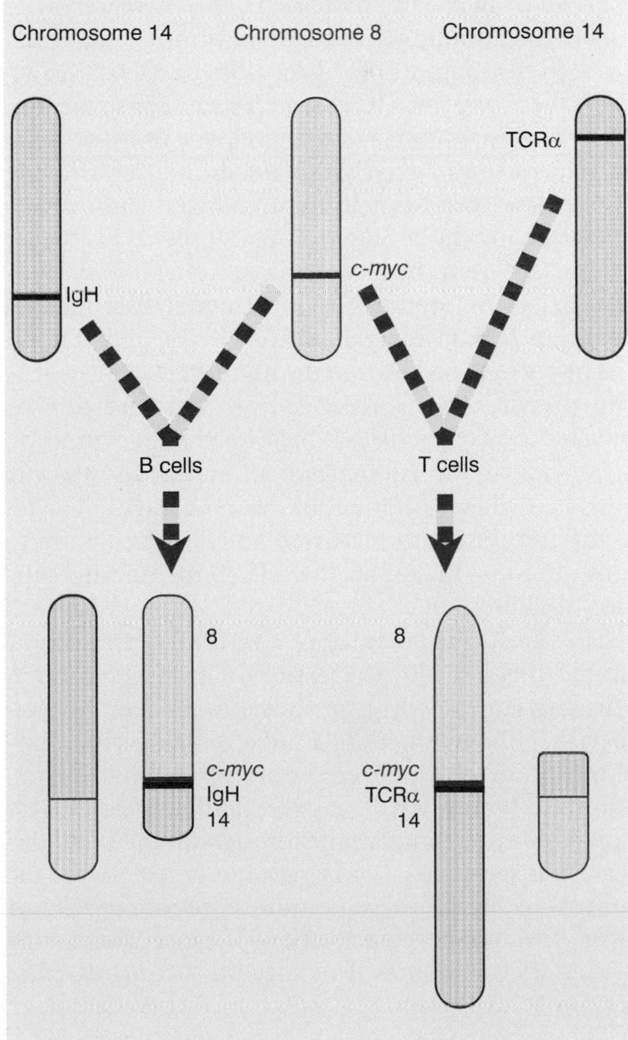

often the *c-myc* locus. In man, the translocations in B-cell tumors usually involve chromosome 8, which carries *c-myc*, and chromosome 14, which carries the IgH locus; ~10% involve chromosome 8 and either chromosome 2 (kappa locus) or chromosome 22 (lambda locus). The translocations in T-cell tumors often involve chromosome 8, and either chromosome 14 (which has the TCR α locus at the other end from the Ig locus) or chromosome 7 (which carries the TCR β locus). Analogous translocations occur in the mouse.

Translocations in B cells fall into two classes, reflecting the two types of recombination that occur in immunoglobulin genes. One type is similar to those involved in somatic recombination to generate the active genes, involving the consensus sequences used for V-D-J recombination. These can occur at all the Ig loci. In the other type, the translocation occurs at a switching site at the IgH locus, presumably reflecting the operation of the system for class switching.

When *c-myc* is translocated to the Ig locus, its level of expression is usually increased. The increase varies considerably among individual tumors, generally being in the range 2–10×. Why does translocation activate the *c-myc* gene? The event has two consequences: *c-myc* is brought into a new region, one in which an Ig or TCR gene was actively expressed; and the structure of the *c-myc* gene may itself be changed (but usually not involving the coding regions). It seems likely that several different mechanisms can activate the *c-myc* gene in its new location (just as retroviral insertions activate *c-myc* in a variety of ways).

The correlation between the tumorigenic phenotype and the activation of *c-myc* by either insertion or translocation suggests that continued high expression of c-Myc protein is oncogenic. Expression of *c-myc* must be switched off to enable immature lymphocytes to differentiate into mature B and T cells; failure to turn off *c-myc* maintains the cells in the undifferentiated (dividing) state.

The oncogenic potential of *c-myc* has been demonstrated directly by the creation of transgenic mice. Mice carrying a *c-myc* gene linked to a B lymphocyte-specific enhancer (the IgH enhancer) develop lymphomas. The tumors represent both immature and mature B lymphocytes, suggesting that over-expression of *c-myc* is tumorigenic throughout the B cell lineage. Transgenic mice carrying a *c-myc* gene under the control of the LTR from a mouse mammary tumor virus, however, develop a variety of cancers, including mammary carcinomas. This suggests that increased or continued expression of *c-myc* transforms the type of cell in which it occurs into a corresponding tumor.

c-myc exhibits three means of oncogene activation: retroviral insertion, chromosomal translocation, and gene amplification. The common thread among them is deregulated expression of the oncogene rather than a qualitative change in its coding function, although in at least some cases the transcript has lost the usual (and possibly regulatory) nontranslated leader. *c-myc* provides the paradigm for oncogenes that may be effectively activated by increased (or possibly altered) expression.

Translocations are now known in many types of tumors. Often a specific chromosomal site is commonly involved, creating the supposition that a locus at that site is involved in tumorigenesis. However, every translocation generates reciprocal products; sometimes a known oncogene is activated in one of the products, but in other cases it is not evident which of the reciprocal products has responsibility for oncogenicity. Also, it is not axiomatic that the gene(s) at the breakpoint have responsibility; for example, the translocation could provide an enhancer that activates another gene nearby.

A variety of translocations found in B and T cells have identified new oncogenes. In some cases, the translocation generates a hybrid gene, in which an active transcription unit is broken by the translocation. This has the result that the exons of one gene may be connected to another. In such cases, there are two potential causes of oncogenicity. The proto-oncogene part of the protein may be activated in some way that is independent of the other part, for example, because it is over-expressed under its new management (a situation directly comparable to the example of *c-myc*). Or the other partner in the hybrid gene may have some positive effect that generates a gain-of-function in the part of the protein coded by the proto-oncogene.

One of the best characterized cases in which a translocation creates a hybrid oncogene is provided by the *Philadelphia* (PH^1) chromosome present in patients with chronic myelogenous leukemia (CML). This reciprocal translocation is too small to be visible in the karyotype, but links a 5000 kb region from the end of chromosome 9 carrying *c-abl* to the *bcr* gene of chromosome 22. The *bcr* (breakpoint cluster region) was originally named to describe a region of ~5.8 kb within which breakpoints occur on chromosome 22.

The consequences of this translocation are summarized in **Figure 28.13.** The *bcr* region lies within a large (>90 kb) gene, which is now known as the *bcr* gene. The breakpoints in CML usually occur within one of two introns in the middle of the gene. The same gene is also involved in translocations that generate another

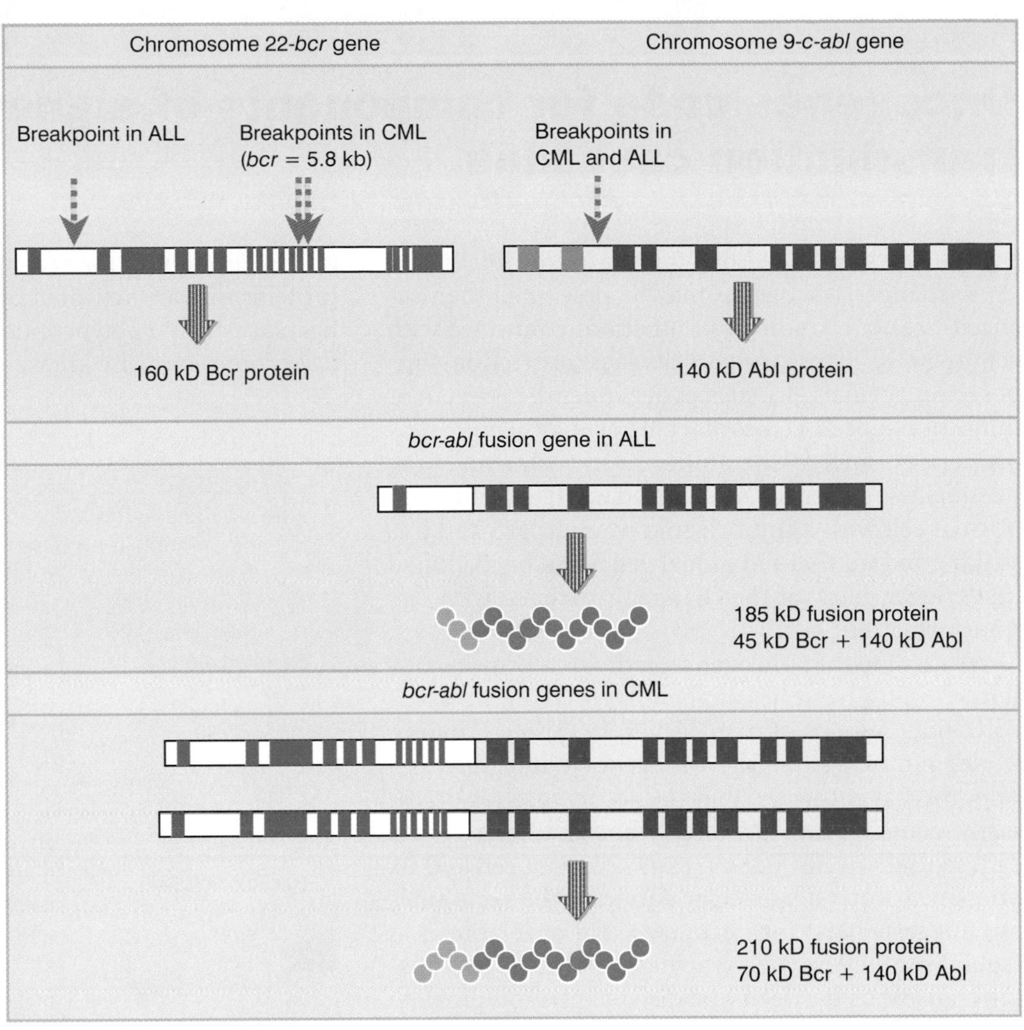

Figure 28.13 Translocations between chromosome 22 and chromosome 9 generate Philadelphia chromosomes that synthesize *bcr-abl* fusion transcripts that are responsible for two types of leukemia.

disease, ALL (acute lymphoblastic leukemia); in this case, the breakpoint in the *bcr* gene occurs in the first intron.

The *c-abl* gene is expressed by alternative splicing that uses either of the first two exons. The breakpoints in both CML and ALL occur in the intron that precedes the first common exon. Although the exact breakpoints on both chromosomes 9 and 22 vary in individual cases, the common outcome is the production of a transcript coding for a Bcr-Abl fusion protein, in which N-terminal sequences derived from *bcr* are linked to *c-abl* sequences. In ALL, the fusion protein has ~45 kD of the Bcr protein; in CML the fusion protein has ~70 kD of the Bcr protein. In each case, the fusion protein contains ~140 kD of the usual ~145 kD c-Abl protein, that is, it has lost just a few N-terminal amino acids of the c-Abl sequence.

Why is the fusion protein oncogenic? The Bcr-Abl protein appears to activate the Ras pathway for transformation. It may have multiple ways of doing so, including activation of the adaptors Grb2 and Shc (see Chapter 26). Both the Bcr and Abl regions of the joint protein may be important in transforming activity.

Changes at the N-terminus are involved in activating the oncogenic activity of *v-abl*, a transforming version of the gene carried in a retrovirus. The *c-abl* gene codes for a tyrosine kinase activity; this activity is essential for transforming potential in oncogenic variants. Deletion (or replacement) of the N-terminal region activates the kinase activity and transforming capacity. So the N-terminus provides a domain that usually regulates kinase activity; its loss may cause inappropriate activation.

Oncogenes code for components of signal transduction cascades

Whether activated by quantitative or qualitative changes, oncogenes may be presumed to influence (directly or indirectly) functions connected with cell growth. Transformed cells lack restrictions imposed on normal cells, such as dependence on serum or inhibition by cell-cell contact. They may acquire new properties, such as the ability to metastasize. Many phenotypic properties are changed when we compare a normal cell with a tumorigenic counterpart, and it is striking indeed that individual genes can be identified that trigger many of the changes associated with this transformation.

We assume that oncogenes, individually or in concert, set in train a series of phenotypic changes that involve the products of many genes. In this description, we see at once a similarity with genes that regulate developmental pathways: they do not themselves necessarily code for the products that characterize the differentiated cells, but they may direct a cell and its progeny to enter a particular pathway. The same analogy suggests itself for oncogenes and developmental regulators: they provide switches responsible for causing transitions between one discrete phenotypic state and another.

Taking this relationship further, we may ask what activities the products of proto-oncogenes play in the normal cell, and how are they changed in the transformed cell? Could some proto-oncogenes be regulators of normal development whose malfunction results in aberrations of growth that are manifested as tumors? We have stumbled across some examples of such relationships, but do not yet have any systematic understanding of the connection.

Oncoproteins are organized according to their types of function in **Figure 28.14.** The left part of the figure groups the oncogenes according to the locations of their products. The boxes on the right give details of the corresponding proto-oncogenes. The functions of many oncogenes remain unknown, and further groups will no doubt be identified:

- Growth factors are proteins secreted by one cell that act on another. The oncoprotein counterparts can only transform cells bearing the appropriate receptor. (This is called autocrine transformation.)

- The growth factor receptors are transmembrane proteins that are activated by binding an extracellular ligand (usually a polypeptide). Most often the receptor is a protein tyrosine kinase. Oncogenicity may result

Figure 28.14 Oncogenes may code for secreted proteins, transmembrane proteins, cytoplasmic proteins, or nuclear proteins.

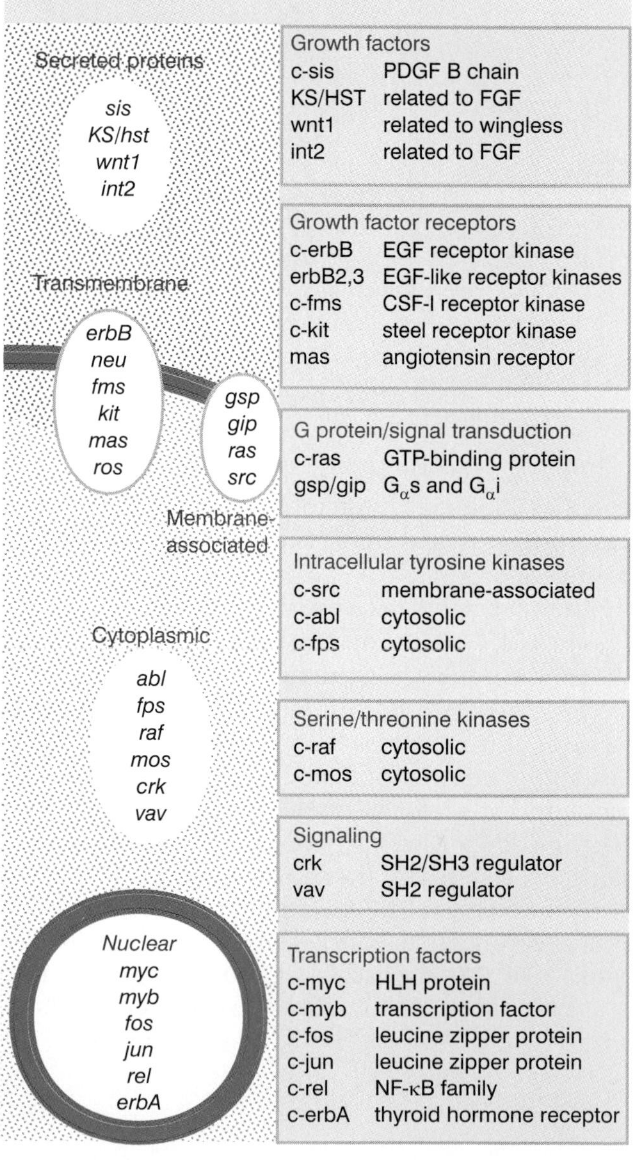

from constitutive (that is, ligand-independent) activation of the kinase activity.

■ One group of intracellular protein kinases phosphorylate tyrosine residues in target proteins. c-Src, which associates with the cytoskeleton as well as with the membrane, is the prototype of a family of kinases with similar catalytic activities (including c-Yes, c-Fgr, Lck, c-Fps, Fyn). We understand the effects of oncogenic mutations on the Src kinase activity in some detail, although we have yet to explain why the altered kinase activity is oncogenic. Other protein tyrosine kinases in the intracellular group are cytosolic; c-Abl is found in both cytosol and nucleus.

■ Signal transduction pathways are often involved in oncogenesis. The best characterized example is c-Ras, which plays a central role in transmitting the signal from receptor tyrosine kinases (see Chapter 26). Oncogenic mutations change the regulation of Ras activity. Other stages in signal transduction are identified by Gsp and Gip, which are mutant forms of the α subunits of the G_s and G_i trimeric G proteins. Crk and Vav are proteins associated with later stages of signaling.

■ A group of cytosolic enzymes are protein serine/threonine kinases, that is, they phosphorylate target proteins on serine or threonine. Little is known about the effects of oncogenic mutations beyond the fact that they probably increase or constitutively activate the kinase activities. Mos is an example which can activate ERK MAPK.

■ Nuclear proteins include transcription factors of several types. The functions of these proto-oncoproteins are rather well described (see Chapter 21). Generally we understand what effects the oncogenic mutations have on the factors, but we cannot yet relate these changes to the activation or repression of a set of target genes that defines the oncogenic state.

The common feature is that each type of protein is in a position to trigger general changes in cell phenotypes, either by initiating or responding to changes associated with cell growth, or by changing gene expression directly. Before we consider in detail the potential of each group for initiating a series of events that has an oncogenic outcome, we need to consider how many independent pathways are identified by these factors.

Recall the example of the best characterized mitogenic pathway, the MAPK pathway which consists of the following stages:

growth factor
↓
growth factor receptor (tyrosine kinase)
↓
Ras
↓
kinase cascade (serine/threonine kinases)
↓
transcription factor(s)

When a growth factor interacts with its receptor, it activates the tyrosine kinase activity. The signal is passed (via an adaptor) to Ras. At this point, the pathway switches to a series of serine/threonine kinases. The targets at the end of the pathway may be controlled directly or indirectly by phosphorylation, and include transcription factors, which are in a position to make widespread changes in the pattern of gene expression.

If a pathway functions in a linear manner, in which the signal passes directly from one component to the next, the same results should be achieved by constitutive activation of any component (so that it no longer needs to be activated by a signal from an earlier component).

A signal transduction pathway, of course, is likely to branch at several stages, so that an initial stimulus may trigger a variety of responses. The activation of components that are downstream will therefore activate a smaller number of end-functions than the activation of components at the start of the pathway. But we can analyze any individual part of the pathway by tracing it back to the beginning as though it were strictly linear.

In the example of the Ras pathway, we know that it is activated by many growth factors to generate a mitogenic response. Mutations in the early part of this pathway, including the *ras* and *raf* genes, may be oncogenic. But oncogenic mutations are not found in the following components of the cascade, the MEK and MAP kinases. This suggests that there may be a branch in the pathway at the stage of *ras* or *raf*, and that activation of this branch is also necessary for oncogenicity. Ras activates a cytoskeletal GTPase called Rac, which may identify this branch. Of course, the ERK MAPK pathway terminates in the activation of several "immediate early" genes, including *fos* and *jun*, which themselves have oncogenic counterparts, suggesting that the targets of the MAPK pathway can be sufficient for oncogenicity.

The central role of this pathway is indicated by

the number of its components that are coded by proto-oncogenes. One explanation of the discrepancies between the susceptibilities of MAP kinases and other components to oncogenic mutation may be that the *level* or *duration* of expression is important. It could be the case that mutations in MEK or MAP kinases do not activate the enzymes sufficiently to be oncogenic. Alternatively, the oncogenic mutations (which, after all, represent gain-of-function) may cause new targets to be activated in addition to the usual pathway. The general principle is clear: *aberrant activation of mitogenic pathways can contribute to oncogenicity,* but we cannot yet relate the activation of these pathways to individual responses in terms of immortalization or transformation.

Growth factor receptor kinases and cytoplasmic tyrosine kinases

THE protein tyrosine kinases constitute a major class of oncoproteins, and fall into two general groups: transmembrane receptors for growth factors; and cytoplasmic enzymes. We have more understanding about the biological functions of the receptors, because we know the general nature of the signal transduction cascades that they initiate, and we can see how their inappropriate activation may be oncogenic. The normal roles in the cell of the cytoplasmic tyrosine kinases are not so well defined, but in several cases it appears that they provide catalytic functions for receptors that themselves lack kinase activity; that is, the activation of the receptor leads to activation of the cytoplasmic tyrosine kinase. We have a great deal of information about their enzymatic activities and the molecular effects of oncogenic mutations, although it has been more difficult to identify their physiological targets.

Receptors for many growth factors have kinase activity. They tend to be large integral membrane proteins, with domains assembled in modular fashion from a variety of sources. We discussed the general nature of transmembrane receptors and the means by which they are activated to initiate signal transduction cascades in Chapter 26. The EGF receptor is the paradigm for tyrosine kinase receptors. It is a group I receptor with its N-terminus on the extracellular surface, has a single transmembrane region, and has its C-terminus inside the cell. The extracellular N-terminal region binds the ligand that activates the receptor. The intracellular C-terminal region includes a domain that has tyrosine kinase activity. Most of the receptors that are coded by cellular proto-oncogenes have a similar form of organization.

Dimerization of the extracellular domain of a receptor activates the tyrosine kinase activity of the intracellular domain. Various forms of this reaction were summarized in Figure 26.14. When the cytoplasmic domains of the monomers are brought into contact, they trigger an autophosphorylation reaction, in which each monomer phosphorylates the other.

A (generalized) relationship between a growth factor receptor and an oncogenic variant is illustrated in **Figure 28.15.** The wild-type receptor is regulated by ligand binding. In the absence of ligand, the monomers do not interact. Growth factor binding triggers an interaction, allowing the receptor to form dimers. This in turn activates the receptor, and triggers signal transduction. By contrast, the oncogenic variant spontaneously forms dimers that are constitutively active. Different types of events may be responsible for the constitutive dimerization and activation in different growth factor receptors, and multiple events may be involved in any one receptor.

The oncogene *v-erb* is a truncated version of *c-erbB,* the gene coding for the EGF receptor. The oncoprotein retains the tyrosine kinase and transmembrane domains, but lacks the N-terminal part of the protein that binds EGF, and does not have the C-terminus. The deletions at both ends may be needed for oncogenicity. The change in the extracellular N-terminal domain allows the protein to dimerize spontaneously; and the C-terminal deletion removes a cytosolic domain that inhibits transforming activity. There is also an activat-

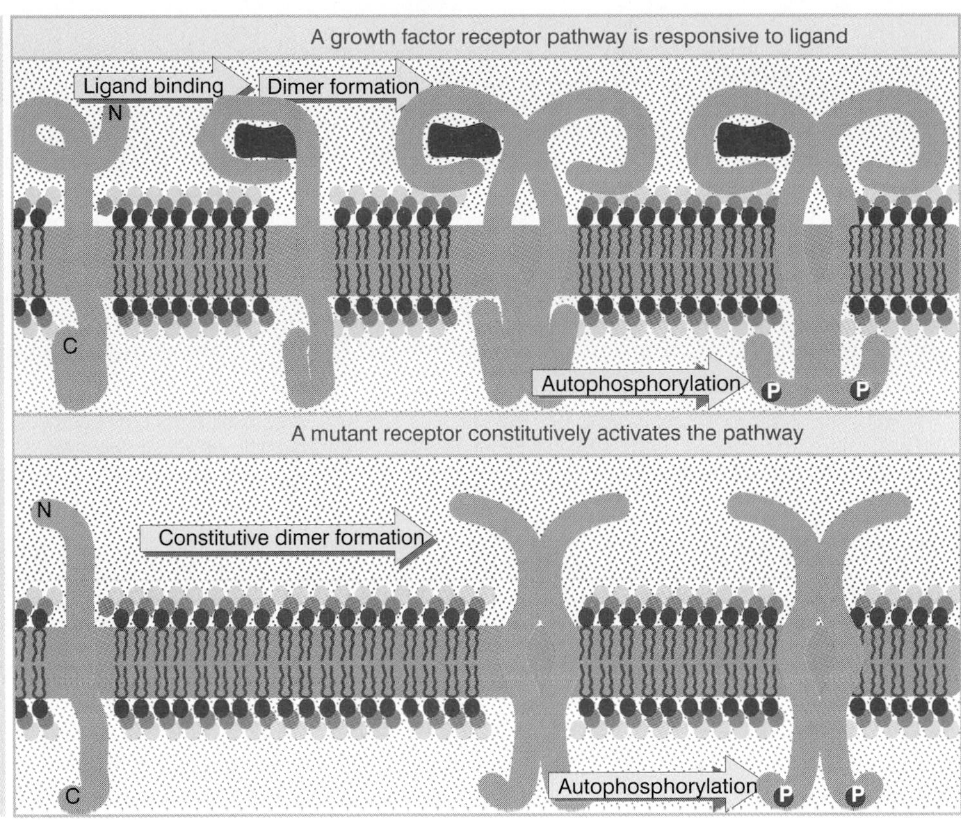

Figure 28.15 Activation of a growth factor receptor involves ligand binding, dimerization, and autophosphorylation. A truncated oncogenic receptor that lacks the ligand-binding region is constitutively active because it is not repressed by the N-terminal domain.

ing mutation in the catalytic domain. So the basis for oncogenicity is the combination of mutations that activate the receptor constitutively.

The general principle that constitutive or altered activity may be responsible for oncogenicity applies to the group of growth factor receptors summarized previously in Figure 28.14. Another example of an activation event is provided by *erbB2,* which codes for a receptor closely related to the EGF receptor. An oncogenic form has a key mutation in its transmembrane region; this increases the propensity of the receptor monomers to form dimers.

Some proto-oncogenes code for receptors or factors involved in the development of particular cell types. Mutation of such a receptor (or growth factor) may promote unrestricted growth of cells of the appropriate type. The proto-oncogene *c-fms* codes for the CSF-I receptor, which mediates the action of colony stimulating factor I, a macrophage growth factor that stimulates the growth and maturation of myeloid precursor cells. *c-fms* can be rendered oncogenic by a mutation in the extracellular domain; this increases dimerization and makes the protein constitutively active in the absence of CSF-I. Oncogenicity is enhanced by C-terminal mutations, which could be inactivating an inhibitory intracellular domain.

The cellular action and basis for oncogenicity of the cytoplasmic group of protein tyrosine kinases is more obscure. The cytoplasmic group is characterized by the viral oncogenes *src, yes, fgr, fps/fes, abl, ros.* (c-Src is actually associated with membranes.) A major stretch of the sequences of all these genes is related, corresponding to residues 80–516 of *c-src.* This includes the SH2 and SH3 domains and the catalytic domain responsible for kinase activity. Presumably the regions outside this domain control the activities of the individual members of the family. In few cases, however, do we know the cellular function of a *c-onc* member of this group.

The paradigm for a cytoplasmic tyrosine kinase in search of a role is presented by the Src proteins. Since its isolation by Rous in 1911, RSV has been perpetuated under a variety of conditions, and there are now several "strains," carrying variants of *v-src.* The common feature in the sequence of *v-src* is that the C-terminal sequence of *c-src* has been replaced. The various strains contain different point mutations within the *src* sequence.

Proteins in the Src family were the first oncoproteins of the kinase type to be characterized. Src was also the first example of a kinase whose target is a tyrosine residue in protein. The level of phosphotyrosine is increased about 10× in cells that have been transformed

Figure 28.16 An Src protein has an N-terminal domain that associates with the membrane, a modulatory domain that includes SH2 and SH3 motifs, a kinase catalytic domain, and (c-Src only) a suppressor domain.

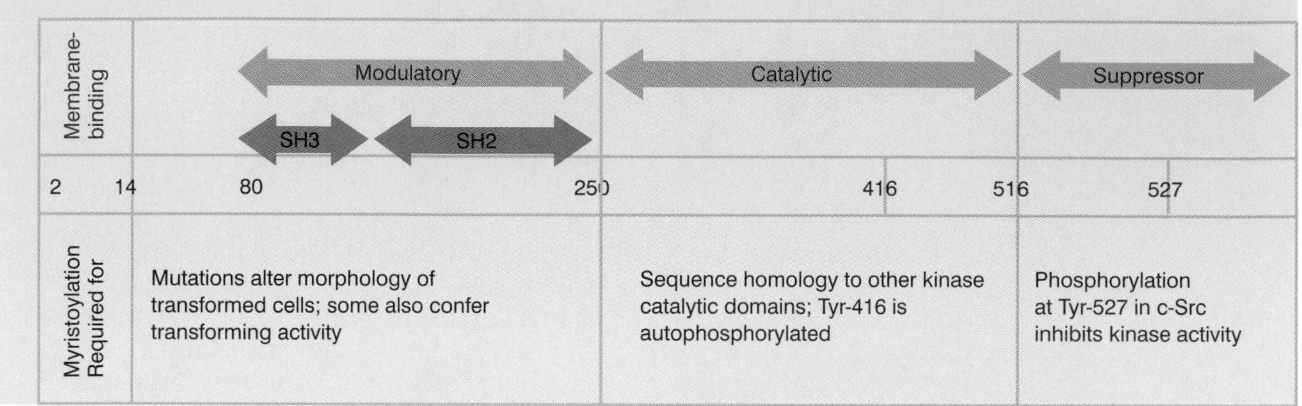

by RSV. In addition to acting on other proteins, Src is able to phosphorylate itself.

Src proteins have several interesting features. **Figure 28.16** summarizes their activities in terms of protein domains.

Both v-Src and c-Src are modified at the N-terminus. The N-terminal amino acid is cleaved, and myristic acid (a rare fatty acid of 14 carbon residues) is covalently added to the N-terminal glycine. Myristoylation enables Src proteins to attach to the cytosolic face of membranes in the cytoplasm. Most of the protein is associated with the cytoplasmic face of the endosomes, and it is enriched in regions of cell to cell contact and adhesion plaques.

Myristoylation is essential for oncogenic activity of v-Src, since N-terminal mutants that cannot be myristoylated have reduced tumorigenicity. The simplest explanation for the dependence of transformation on the membrane location of v-Src is that important substrates for Src are located in the membrane.

The major difference between v-Src and c-Src lies in their kinase activities. The activity of v-Src is ~20× greater than that of c-Src. The transforming activity of *src* mutants is correlated with the level of kinase activity, and we believe that oncogenicity results from phosphorylation of target protein(s). We do not know whether the increased activity is itself responsible for oncogenicity or whether there is also a change in the specificity with which target proteins are recognized.

Kinase activity plays two roles in Src function. First, attempts to identify a function for the phosphorylation in cell transformation have concentrated on identifying cellular substrates that may be targets for v-Src (especially those that may not be recognized by c-Src). A variety of substrates have been identified, but none has yet been equated with the cause of transformation. Second, the state of phosphorylation of Src itself may be important for its transforming activity.

Two sites in Src control its kinase activity. The c-Src protein is phosphorylated *in vivo* by the kinase Csk at tyrosine residue 527, which is part of the C-terminal sequence of 19 amino acids that is missing from v-Src. The v-Src protein is phosphorylated *in vivo* at Tyr-416, which is located in the catalytic domain. This position is not normally phosphorylated *in vivo* in c-Src.

The importance of these phosphorylations can be tested by mutating the tyrosine residues at 416 and 527 to prevent addition of phosphate groups. The mutations have opposite effects, as summarized in **Figure 28.17:**

- Mutation of Tyr-527 to the related amino acid phenylalanine activates the transforming potential of c-Src. The protein c-Src$^{\text{Phe-527}}$ becomes phosphorylated on Tyr-416, has its kinase activity increased ~10×, and it transforms target cells, although not as effectively as v-Src. *Phosphorylation of Tyr-527 therefore represses the oncogenicity of c-Src. Removal of this residue when the C-terminal region was lost in generating v-Src contributes significantly to the oncogenic activity of the transforming protein.*
- Mutation of Tyr-416 in c-Src eliminates its residual ability to transform. This mutation also greatly reduces the activity of the c-Src$^{\text{Phe-527}}$ mutant. It also reduces the transforming potential of v-Src, but less effectively. *Phosphorylation at Tyr-416 therefore activates the oncogenicity of Src proteins.*

Figure 28.17 Two tyrosine residues are targets for phosphorylation in Src proteins. Phosphorylation at Tyr-527 of c-Src suppresses autophosphorylation at Tyr-416, which is associated with transforming activity. Only Tyr-416 is present in v-Src. Transforming potential of c-Src may be activated by removing Tyr-527 or repressed by removing Tyr-416.

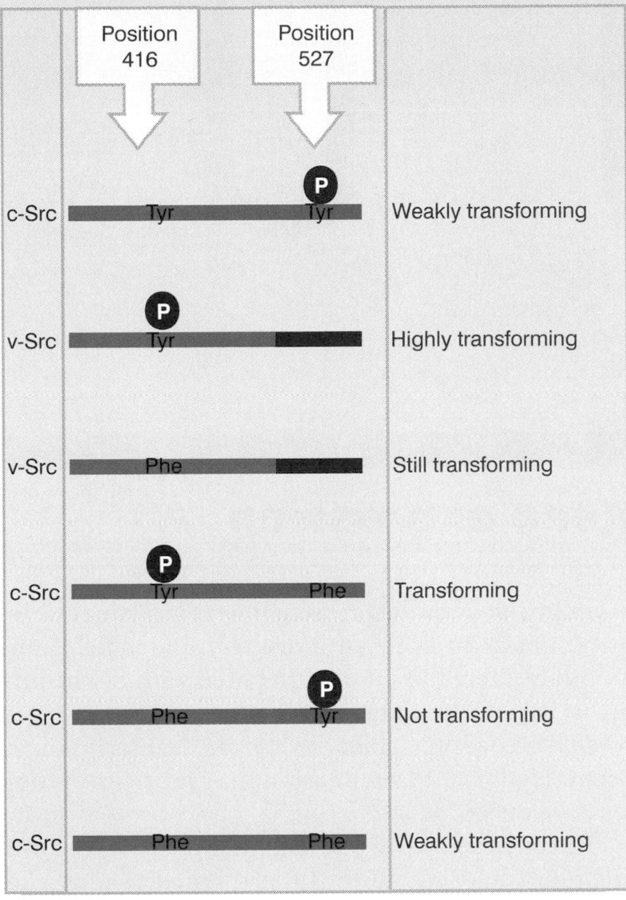

Point mutations at other positions in c-Src support a correlation in which oncogenicity is associated with decreased phosphorylation at Tyr-527 and increased phosphorylation at Tyr-416. The state of these tyrosines may therefore be a general indicator of the oncogenic potential of c-Src. The reduced phosphorylation at Tyr-527 may be responsible for the increased phosphorylation at Tyr-416, which may be the crucial event. However, v-Src is less dependent on the state of Tyr-416, and mutants at this position retain transforming activity; presumably *v-src* has accumulated other mutations that increase transforming potential.

What is the function of c-Src; and how is it related to the oncogenicity of v-Src? The c-Src and v-Src proteins are very similar: they share N-terminal modification, cellular location, and protein tyrosine kinase activity. c-Src is expressed at high levels in several types of terminally differentiated cells, which suggests that it is not involved in regulating cell proliferation. But we have so far been unable to determine the normal function of c-Src.

The modulatory region of c-Src contains two motifs that are found in a variety of other cytoplasmic proteins that are involved in signal transduction: these may connect a protein to the components that are upstream and downstream of it in a signaling pathway. The names of these two domains, SH2 and SH3, reflect their original identification as regions of *S*rc *h*omology. We discussed their functions in Chapter 26.

How is c-Src usually activated? Most mutations in the SH2 region reduce transforming activity (suggesting that the SH2 function is required to activate c-Src), and most mutations in SH3 increase transforming activity (suggesting that SH3 has a negative regulatory role). **Figure 28.18** shows a more detailed autoregulatory model for the function of the SH2 domain. The state of phosphorylation at Tyr-527 is critical. In the inactive state, Tyr-527 is phosphorylated, and this enables the C-terminal region of c-Src itself to bind to the N-terminal SH2 domain. When an appropriate receptor tyrosine kinase (such as PDGF receptor) is activated, the autophosphorylation reaction creates a phosphopeptide sequence that binds to the c-Src SH2 region, releasing the region containing Tyr-527, which may be dephosphorylated. The following events are not entirely clear; one possibility is that this leads to the phosphorylation of Tyr-416, and activation of kinase activity. In this way, c-Src kinase activity responds to the activation of the receptor kinase. In any case, the oncogenic v-Src protein, of course, lacks Tyr-527 and is constitutively active.

Alternative ways for activating c-Src may be involved in some oncogenic reactions. For example, the polyoma middle T antigen activates c-Src by binding to the C-terminal region including Tyr-527 and prevents its phosphorylation. Some mutations in the SH2 domain of c-Src can activate the kinase activity (with oncogenic consequences), presumably because they prevent it from sequestering Tyr-527. Mutations in the SH2 and SH3 domains of c-Src can influence its specificity with regard to transforming different types of target cells, which suggests that these regions provide the connections to other (cell specific) proteins in the pathway.

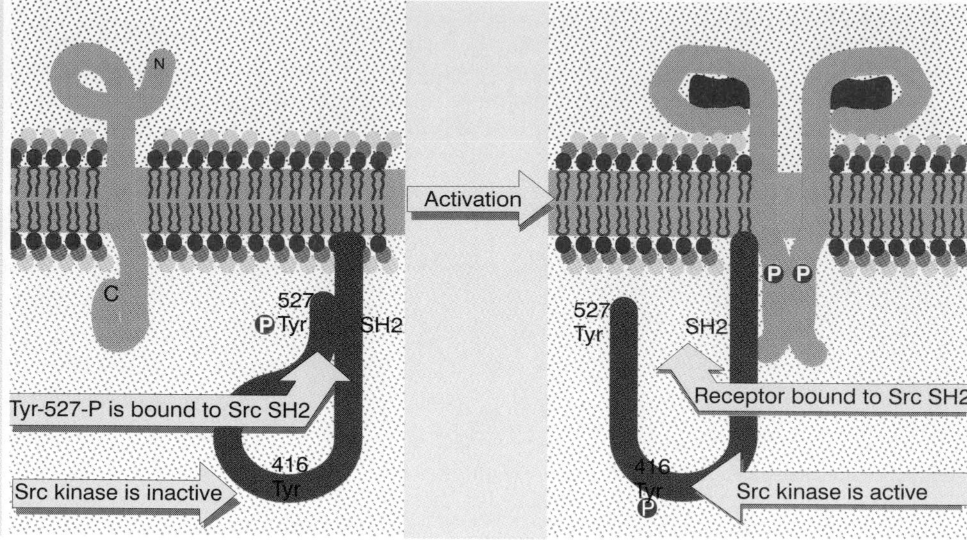

Figure 28.18 When a receptor tyrosine kinase is activated, autophosphorylation generates a binding site for the Src SH2 domain, Tyr-527 is released and dephosphorylated, Tyr-416 becomes phosphorylated, and Src kinase is activated.

Oncoproteins may regulate gene expression

It goes almost without saying that it is necessary to make changes in gene expression in order to convert a cell to the transformed phenotype. Many oncogenes act at early stages in pathways that lead ultimately to changes in gene expression. Some act directly at the level of transcription. Retroviral oncogenes include examples derived from the major classes of cellular transcription factors. Several prominent gene families coding for transcription factors are identified by *v-onc* genes: *rel, jun, fos, erbA, myc,* and *myb.* In the cases of Rel, Jun, and ErbA proteins, there are differences in transcriptional activity between the c-Onc and v-Onc proteins that may be related to transforming capacity.

The actions of *v-onc* genes may in principle be quantitative or qualitative; and those that affect transcription might either increase or decrease expression of particular genes. By virtue of increased expression or activity they could turn up transcription of genes whose products can be tolerated only in small amounts. Failure to respond to normal regulation of activity by other cellular factors also might lead to increased gene expression. A less likely possibility is the acquisition of specificity for new target promoters. Alternatively, if the oncoproteins are defective in the ability to activate transcription, they might function as dominant negative suppressors of the cellular transcription factors. The first steps towards distinguishing these possibilities lie with determining which functions are altered in v-Onc compared with c-Onc proteins: is DNA-binding altered either quantitatively or qualitatively; is the ability to activate transcription altered? **Figure 28.19** summarizes the properties of some of these oncoproteins.

The oncogene *v-rel* was identified as the transforming function of the avian (turkey) reticuloendotheliosis virus. The retrovirus is highly oncogenic in chickens, where it causes B-cell lymphomas. *v-rel* is a truncated version of *c-rel,* lacking the ~100 C-terminal amino acids, and has a small number of point mutations in the remaining sequence.

The *rel* gene defines a family with several members, which form various pairwise combinations that regulate transcription. The best characterized family member is the transcription factor NF-κB. This is a dimer of two subunits, p65 and p50, which is held in the cytoplasm by a regulator, I-κB (which masks the nuclear localization sequence). When I-κB is phosphorylated, it is degraded and therefore releases NF-κB, which enters the nucleus and activates transcription of target genes whose promoters or enhancers have the κB motif (see Chapter 21). This regulatory story is essentially recapitulated in *Drosophila* development, where *dorsal*

Figure 28.19 Oncogenes that code for transcription factors have mutations that inactivate transcription (*v-erbA* and possibly *v-rel*) or that activate transcription (*v-jun* and *v-fos*).

Mechanism of regulation of cellular transcription factor
Function of oncogenic factor
v-Rel has lost regions needed to stay in cytoplasm
NF-κB
p65-p50 dimer
I-κB
Phosphorylation of I-κB releases NF-κB
v-Rel may prevent p65/p50 from forming dimer and/or activating transcription
v-Jun and v-Fos have terminal truncations and point mutations
Jun Fos
AP1 may be regulated by phosphorylation and interaction with other factors
v-Jun and v-Fos may activate target genes without responding to usual controls
v-ErbA has lost TH-binding site
Ligand binding activates thyroid hormone receptor (as dimer with RXR)
v-ErbA cannot activate transcription, and also inhibits its dimeric partner

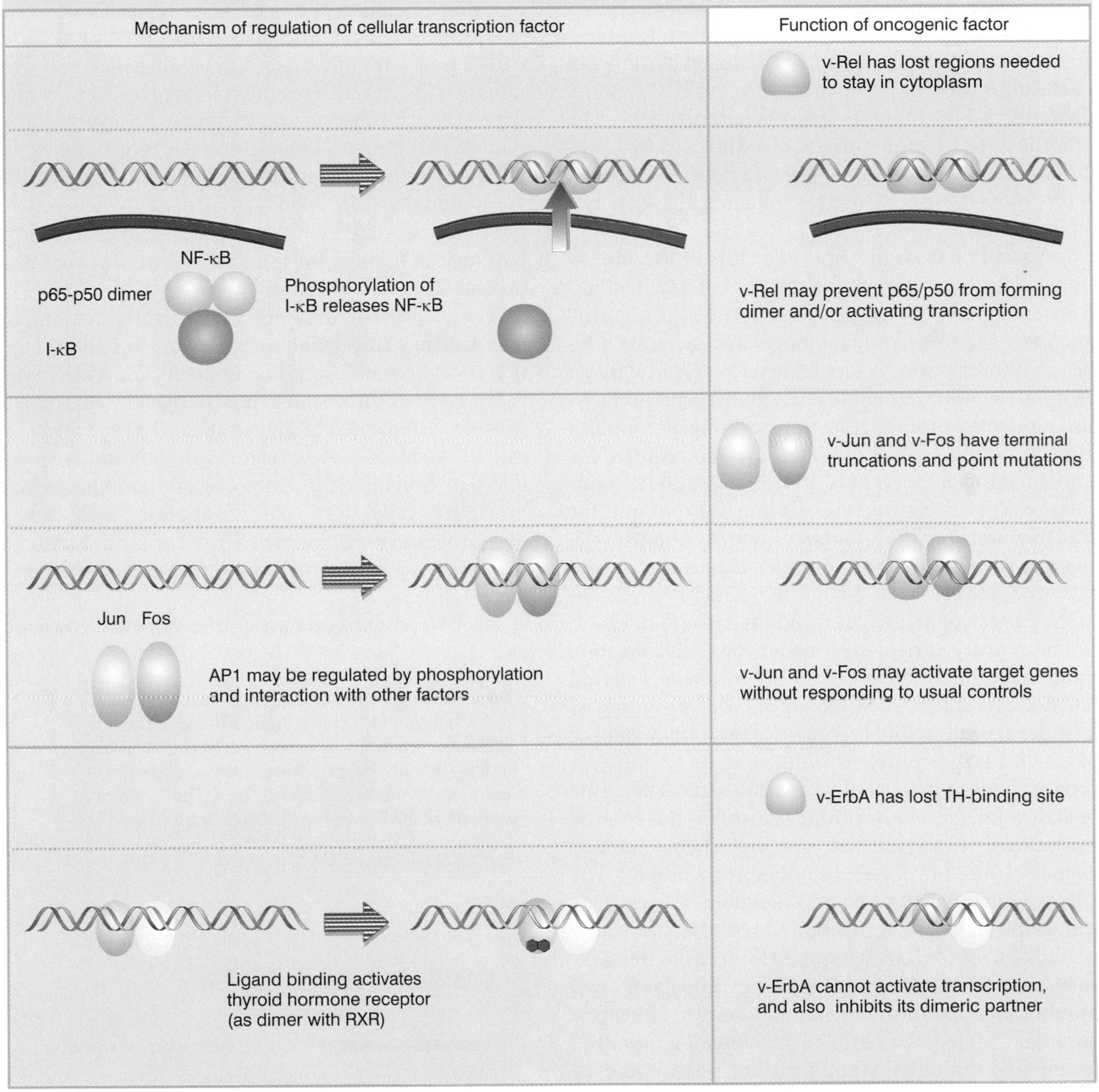

codes for an NF-κB homolog that is held in the cytoplasm by the *cactus* product, an Iκ-B homolog (see Chapter 29). There is sequence relationship between the two subunits of NF-κB, and *c-rel* has 60% similarity with p50.

NF-κB is one of the most pleiotropic transcription factors; indeed, it has been suggested that it may constitute a general second messenger. Many types of stimulus to the cell result in activation of NF-κB; and a broad range of genes is activated via the presence of κB binding sites. It remains to be seen what changes in *v-rel* make it oncogenic, and how they affect the transcriptional regulation exercised by the Rel-related proteins. One possibility is that the *v-rel* product forms dimers with cellular family members, and that these dimers are inactive, so that v-Rel prevents NF-κB or related factors from functioning. v-Rel is exclusively nuclear, because it has lost the sequences required for export to the cytoplasm.

The transcription factor AP1 is the nuclear factor required to mediate transcription induced by phorbol ester tumor promoters (such as TPA). An AP1 binding site confers TPA-inducibility upon a target gene. The canonical AP1 factor consists of a dimer of two subunits, coded by the genes *c-jun* and *c-fos,* which activates genes whose promoters or enhancers have an AP1 binding site.

Jun and Fos are transcription factors of the leucine zipper class. Each protein is a member of a family, and a series of pairwise interactions between Jun family members and Fos family members may generate a series of transcription factors related to AP1. Mutations of *v-jun* or *v-fos* that abolish the ability to bind DNA also render the product non-transforming, providing a direct proof that ability to bind to DNA is required for transforming activity. Beyond this, however, it is not clear what changes in transcription are required for transformation: they could involve quantitative changes (over-expression or under-expression of particular target genes) or qualitative changes (alteration in the pattern of genes that responds to the factor).

The cellular gene *c-erbA* codes for a thyroid hormone receptor, a member of the general class of steroid hormone receptors (see Chapter 21). Upon binding its ligand, a typical steroid receptor activates expression of particular target genes by binding to its specific response element in a promoter or enhancer. The mode of action for thyroid hormone is distinct: it is located permanently in the nucleus, and, indeed, may bind its response element whether or not ligand is present. The effect of hormone binding may therefore be to activate transcription by previously bound receptor.

Ability to bind DNA is required for transforming capacity. *v-erbA* is truncated at both ends and has a small number of substitutions relative to *c-erbA*. Hormone binding is altered; the *c-erbA* product binds triiodothyronine (T_3) with high affinity, but the *v-erbA* product has little or no affinity for the ligand in mammalian cells. This suggests that loss of the ligand-binding capacity (perhaps together with other changes) may create a protein whose function has become independent of the hormone. The consequence of losing the response to ligand is that the factor can no longer be stimulated to activate transcription.

These results place *v-erbA* as a dominant negative oncogene, one that functions by overcoming the action of its normal cellular counterpart. The implication is that genes usually activated by c-ErbA act to *suppress* transformation. In this particular case, it seems likely that these genes usually promote differentiation; blocking this action allows the cells to proliferate.

c-jun, c-fos, c-rel, and also *c-myc* are "immediate early" genes, members of a class of genes that are rapidly induced when resting cells are treated with mitogens, which suggests that they may be involved in a cascade that initiates cycling. So their targets are likely to be concerned with initiating or promoting growth. We should therefore expect an increase in their activities to be associated with oncogenesis, an expectation that may be fulfilled for *v-fos* and *v-myc,* but does not explain the behavior of *v-rel.*

The adenovirus oncogene E1A provides an example of a protein that regulates gene expression indirectly, that is, without itself binding to DNA. The E1A region is expressed as three transcripts, derived by alternative splicing, as indicated in **Figure 28.20.** The 13S and 12S mRNAs code for closely related proteins and are produced early in infection. They possess the ability to immortalize cells, and can cooperate with other oncoproteins (notably Ras) to transform primary cells (see later). No other viral function is needed for this activity.

The E1A proteins exercise a variety of effects on gene

Figure 28.20 The adenovirus E1A region is spliced to form three transcripts that code for overlapping proteins. Domain 1 is present in all proteins, domain 2 in the 289 and 243 residue proteins, and domain 3 is unique to the 289 residue protein. The C-terminal domain of the 55 residue protein is translated in a different reading frame from the common C-terminal domains of the other two proteins.

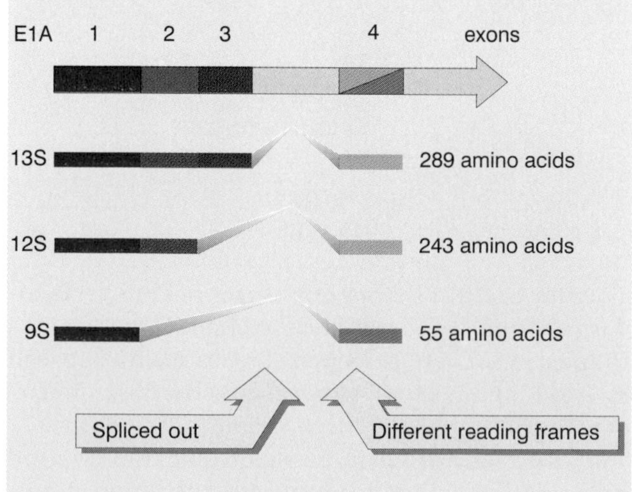

expression. They activate the transcription of some genes, but repress others. Loci that are activated include genes transcribed by RNA polymerase III as well as RNA polymerase II.

Mutation of the E1A proteins suggests that transcriptional activation requires only the short region of domain 3, found only in the 289 amino acid protein coded by 13S mRNA. Repression of transcription, induction of DNA synthesis, and morphological transformation all require domains 1 and 2, common to both the 289 and 243 amino acid proteins. This suggests that repression of target genes is required to cause transformation. E1A proteins act by binding to several cellular proteins that in turn repress or activate transcription of appropriate target genes. Among these targets are the CBP and p300 coactivators, the TBP basal transcription factor, and the cell cycle regulators RB and p27.

RB is a tumor suppressor that controls the cell cycle

THE common theme in the role of oncogenes in tumorigenesis is that increased or altered activity of the gene product is oncogenic. Whether the oncogene is introduced by a virus or results from a mutation in the genome, it is dominant over its allelic proto-oncogene(s). A mutation that activates a single allele is tumorigenic. Tumorigenesis then results from gain of a function.

Certain tumors are caused by a different mechanism: loss of both alleles at a locus is tumorigenic. Propensity to form such tumors may be inherited through the germline; it also occurs as the result of somatic change in the individual. Such cases identify tumor suppressors: genes whose products are needed for normal cell function, and whose loss of function causes tumors. The two best characterized genes of this class code for the proteins RB and p53.

Retinoblastoma is a human childhood disease, involving a tumor of the retina. It occurs both as a heritable trait and sporadically (by somatic mutation). It is often associated with deletions of band q14 of human chromosome 13. The *RB* gene has been localized to this region by molecular cloning.

Figure 28.21 summarizes the situation. Retinoblastoma arises when both copies of the *RB* gene are inactivated. In the inherited form of the disease, one parental chromosome carries an alteration in this region. A somatic event in retinal cells that causes loss of the other copy of the *RB* gene causes a tumor. In the sporadic form of the disease, the parental chromosomes are normal, and both *RB* alleles are lost by (individual) somatic events.

The cause of retinoblastoma is therefore loss of

Figure 28.21 Retinoblastoma is caused by loss of both copies of the *RB* gene in chromosome band 13q14. In the inherited form, one chromosome has a deletion in this region, and the second copy is lost by somatic mutation in the individual. In the sporadic form, both copies are lost by individual somatic events.

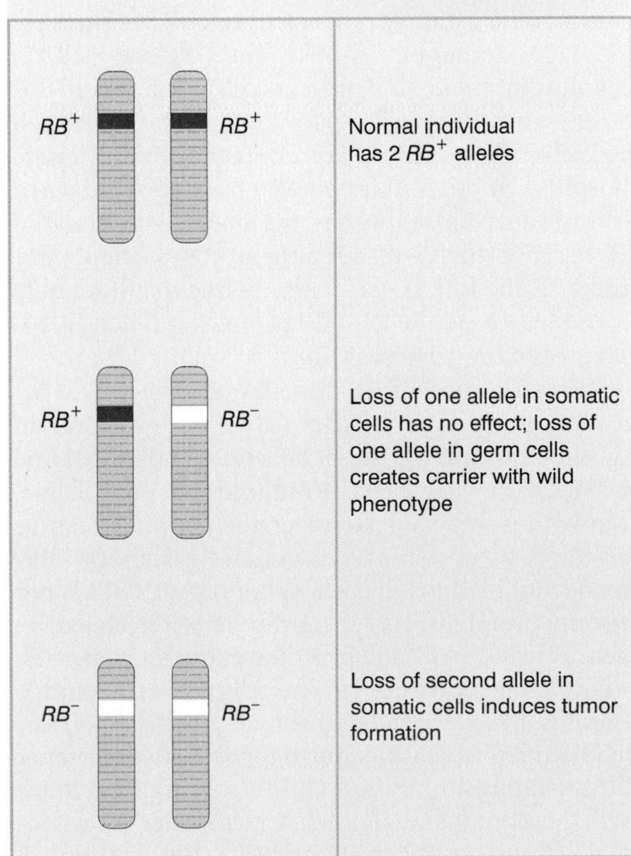
RB^+
RB^+
Normal individual has 2 RB^+ alleles
RB^+
RB^-
Loss of one allele in somatic cells has no effect; loss of one allele in germ cells creates carrier with wild phenotype
RB^-
RB^-
Loss of second allele in somatic cells induces tumor formation

protein function, usually resulting from mutations that prevent gene expression (as opposed to point mutations that affect function of the protein product). Loss of *RB* is involved also in other forms of cancer, including osteosarcomas and small cell lung cancers.

RB is a nuclear phosphoprotein that influences the cell cycle, as discussed previously in Chapter 27. In resting (G0/G1) cells, RB is not phosphorylated. RB is phosphorylated during the cell cycle by cyclin/cdk complexes, most particularly at the end of G1; it is dephosphorylated during mitosis. The nonphosphorylated form of RB specifically binds several proteins, and these interactions therefore occur only during part of the cell cycle (prior to S phase). Phosphorylation releases these proteins.

The target proteins include the E2F group of transcription factors, which activate target genes whose products are essential for S phase. Binding to RB inhibits the ability of E2F to activate transcription, which suggests that RB may repress the expression of genes dependent on E2F. In this way, RB indirectly prevents cells from entering S phase. Also, the RB-E2F complex directly represses some target genes, so its dissociation allows them to be expressed.

Certain viral tumor antigens bind specifically to the nonphosphorylated form of RB. The best characterized are SV40 T antigen and adenovirus E1A. This suggests the model shown in **Figure 28.22**. Nonphosphorylated RB prevents cell proliferation; this activity must be suppressed in order to pass through the cell cycle, which is accomplished by the cyclic phosphorylation. And it may also be suppressed when a tumor antigen sequesters the nonphosphorylated RB. Because the RB-tumor antigen complex does not bind E2F, the E2F is permanently free to allow entry into S phase (and the RB-E2F complex is not available to repress its target genes).

Over-expression of RB impedes cell growth. An indication of the importance of RB for cell proliferation is given by the properties of an osteosarcoma cell line that lacks RB; when RB is introduced into this cell line, its growth is impeded. However, the inhibition can be overcome by expression of D cyclins, which form cdk-cyclin combinations that phosphorylate RB. RB is not the only protein of its type: proteins with related sequences, called p107 and p130, have similar properties.

The connection between the cell cycle and tumorigenesis is illustrated in **Figure 28.23**. Several regulators are identified as tumor suppressors by the occurrence of inactivating mutations in tumors. In addition to RB itself, there are the small inhibitory proteins (most notably p16 and possibly p21), and D cyclin(s). Although these proteins (most notably RB) play a role in the cycle of a proliferating cell, the role that is relevant for tumorigenesis is more probably their function in the quiescent (G0) state. In quiescent cells, RB is not phosphorylated, D cyclin levels are low or absent, and p16, p21, and p27 ensure inactivity of cdk-cyclin complexes. The loss of this circuit is necessary for unrestrained growth.

Figure 28.22 A block to the cell cycle is released when RB is phosphorylated (in the normal cycle) or when it is sequestered by a tumor antigen (in a transformed cell).

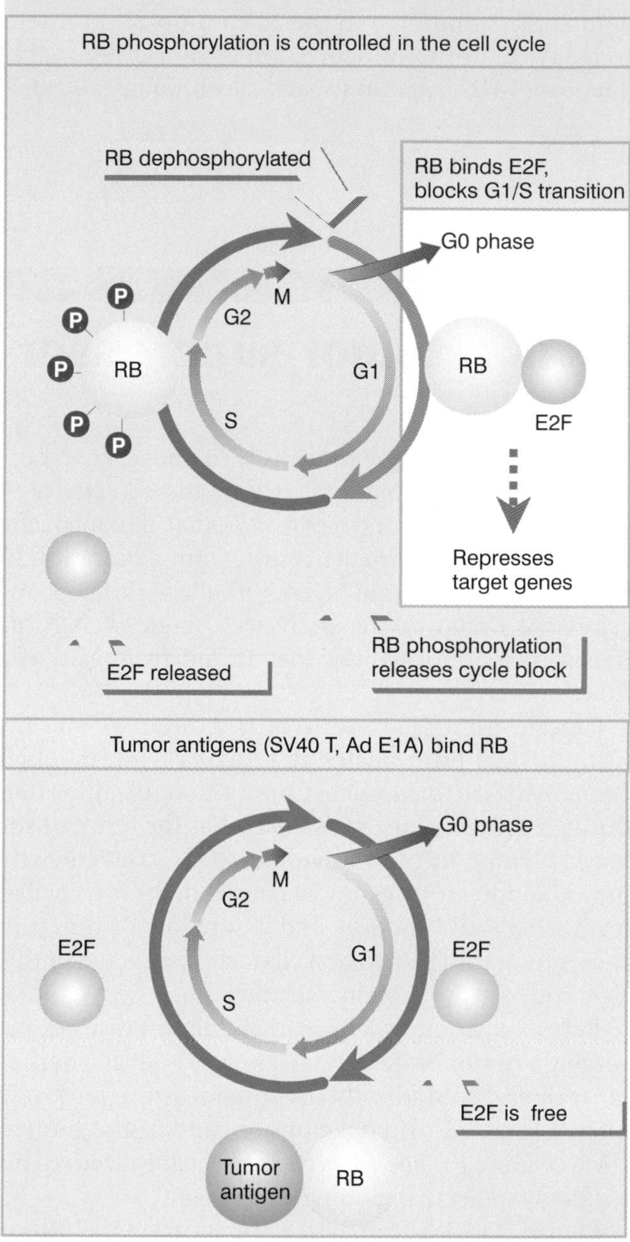

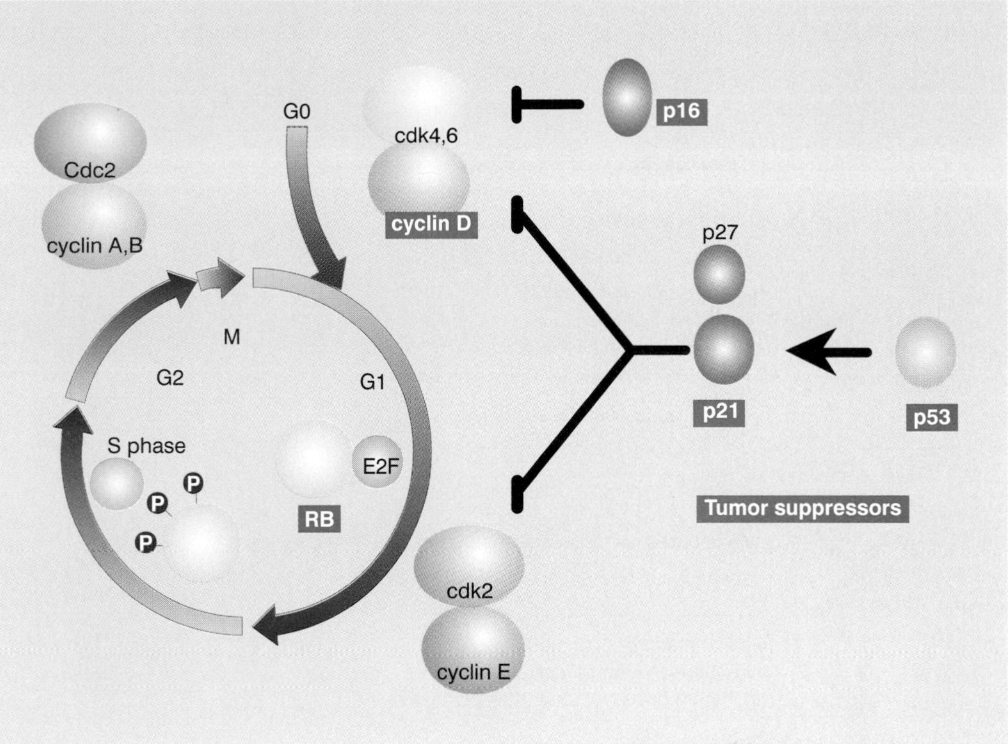

Figure 28.23
Several components concerned with G0/G1 or G1/S cycle control are found as tumor suppressors.

Tumor suppressor p53 suppresses growth or triggers apoptosis

THE most important tumor suppressor is p53 (named for its molecular size). More than half of all human cancers either have lost p53 protein or have mutations in the gene. p53 is a nuclear phosphoprotein. It was originally discovered in SV40-transformed cells, where it is associated with T antigen. A large increase in the amount of p53 protein is found in many transformed cells or lines derived from tumors. In early experiments, the introduction of cloned p53 was found to immortalize cells. These experiments caused p53 to be classified as an oncogene, with the usual trait of dominant gain-of-function.

But all the transforming forms of p53 turned out to be mutant forms of the protein! They fall into the category of dominant negative mutants, which function by overwhelming the wild-type protein and preventing it from functioning. The most common form of a dominant negative mutant is one that forms a heteromeric protein containing both mutant and wild-type subunits, in which the wild-type subunits are unable to function. p53 probably exists as a tetramer. When mutant and wild-type subunits of p53 associate, the tetramer takes up the mutant conformation.

Figure 28.24 shows that the same phenotype is produced either by the deletion of both alleles or by a missense point mutation in one allele that produces a dominant negative subunit. Both situations are found in human cancers. Mutations in p53 accumulate in many types of human cancer, probably because loss of p53 provides a growth advantage to cells; that is, wild-type p53 restrains growth. The diversity of these cancers suggests that p53 is not involved in a tissue-specific event, but in some general and rather common control of cell proliferation; and the loss of this control may be a secondary event that occurs to assist the growth of many tumors. Mutant p53 cells also have an increased

Figure 28.24 Wild-type p53 is required to restrain cell growth. Its activity may be lost by deletion of both wild-type alleles or by a dominant mutation in one allele.

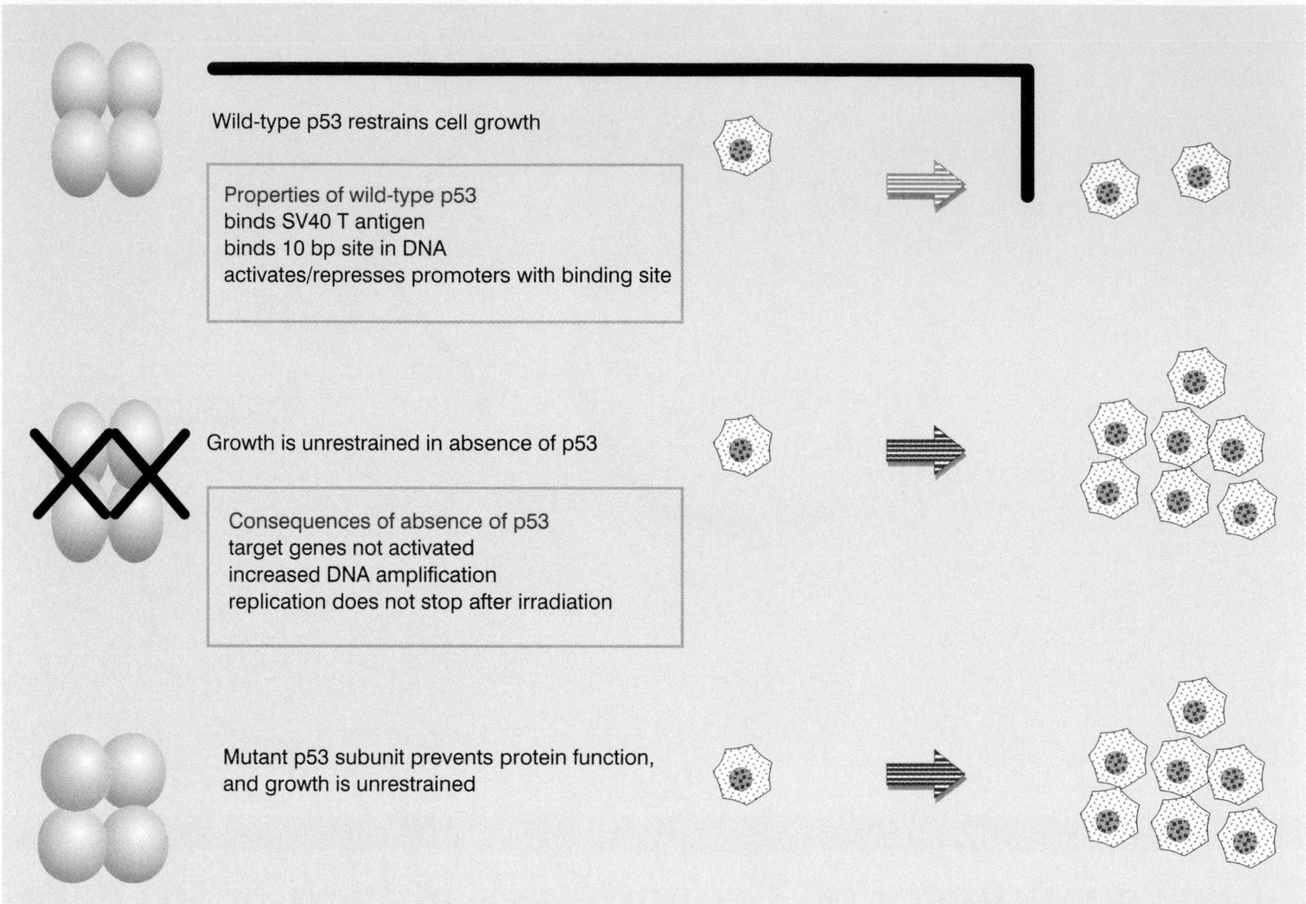

propensity to amplify DNA, which is likely to reflect p53's role in the characteristic instability of the genome that is found in cancer cells.

Wild-type p53 restrains cell growth
Properties of wild-type p53
binds SV40 T antigen
binds 10 bp site in DNA
activates/represses promoters with binding site
Growth is unrestrained in absence of p53
Consequences of absence of p53
target genes not activated
increased DNA amplification
replication does not stop after irradiation
Mutant p53 subunit prevents protein function, and growth is unrestrained

This interpretation implies that a normal cell has a capacity to grow in an unrestrained manner that usually is inhibited by p53. Mutant mice that lack p53 are viable, but develop a variety of tumors rather early. This confirms that it is loss of p53 function that contributes to the tumorigenic phenotype. p53 is defined as a tumor suppressor also by the fact that wild-type p53 can suppress or inhibit the transformation of cells in culture by various oncogenes.

Mutation in p53 is the cause of Li-Fraumeni syndrome, which is a rare form of inherited cancer. Affected individuals display cancers in a variety of tissues. They are heterozygotes that have missense mutations in one allele. These mutations behave as dominant negatives, overwhelming the function of the wild-type allele. This explains the occurrence of the disease as an autosomal dominant.

All normal cells have low levels of p53. A paradigm for p53 function is provided by systems in which it becomes activated, the most usual cause being irradiation or other treatments that damage DNA. This results in a large increase in the amount of p53. Two types of event can be triggered by the activation of p53: growth arrest and apoptosis (cell death). The outcome depends in part on which stage of the cell cycle has been reached. **Figure 28.25** shows that in cells early in G1, p53 triggers a checkpoint that blocks further progression through the cell cycle. This allows the damaged DNA to be repaired before the cell tries to enter S phase. But if a cell is committed to division, then p53 triggers a program of cell death. The typical results of this apoptosis are the collapse of the cell into a small heteropycnotic mass and the fragmentation of nuclear DNA (see Chapter 27). The stage of the cell cycle is not the only determinant of the outcome; for example, some cell types are more prone to show an apoptotic response than others.

We may rationalize the existence of these two

outcomes by supposing that damage to DNA can activate oncogenic pathways, and that the purpose of p53 is to protect the organism against the consequences. If it is possible, a checkpoint is triggered to allow the damage to be repaired, but if this is not possible, the cell is destroyed. We do not know in molecular terms how p53 triggers one pathway or the other, depending on the conditions, but we have an understanding of individual activities of p53 that may be relevant to these pathways.

Figure 28.25 Damage to DNA activates p53. The outcome depends on the stage of the cell cycle. Early in the cycle, p53 activates a checkpoint that prevents further progress until the damage has been repaired. If it is too late to exercise the checkpoint, p53 triggers apoptosis.

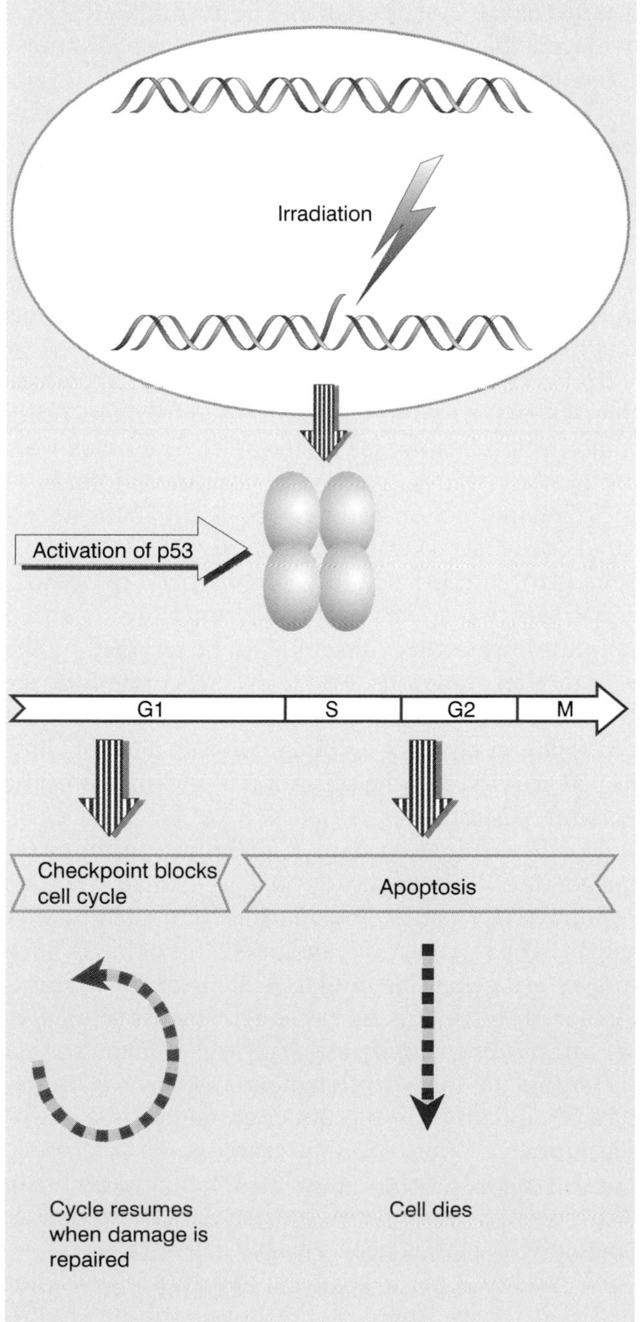

p53 has a variety of molecular activities. **Figure 28.26** summarizes the responsibilities of individual domains of the protein for these activities:

- p53 is a DNA-binding protein that recognizes an interrupted palindromic 10 bp motif. The ability to bind to its specific target sequences is conferred by the central domain.
- p53 activates transcription at promoters that contain multiple copies of this motif. The immediate N-terminal region provides the transactivator domain. p53 may repress other genes; the mechanism is unknown.

Figure 28.26 Different domains are responsible for each of the activities of p53.

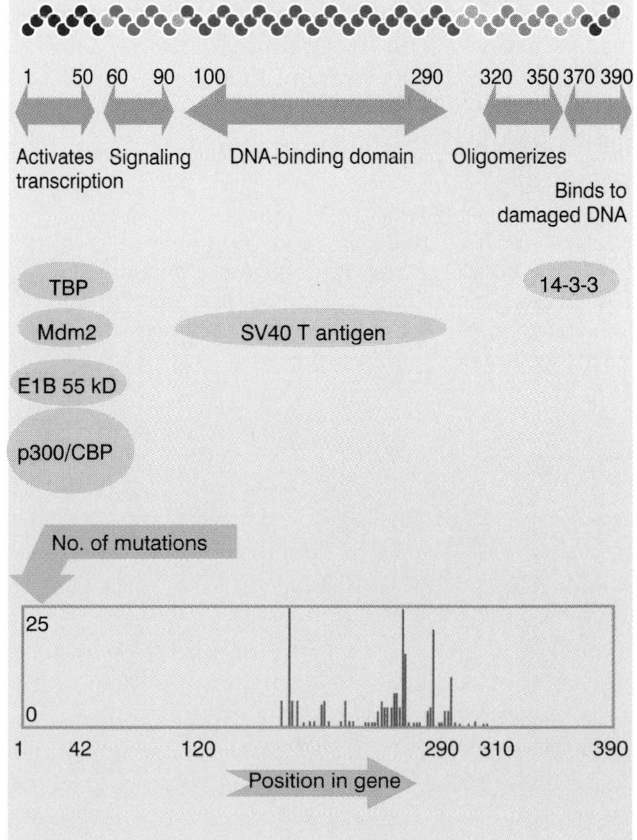

- p53 also has the ability to bind to damaged DNA. The C-terminal domain recognizes single-stranded regions in DNA.
- p53 is a tetramer (oligomerization is a prerequisite for mutants to behave in a dominant negative manner). Oligomerization requires the C-terminal region.
- A (putative) signaling domain contains copies of the sequence PXXP, which forms a binding site for SH3 domains.

Mutations in p53 have various effects on its properties, including increasing its half-life from 20 minutes to several hours, causing a change in conformation that can be detected with an antibody, changing its location from the nucleus to the cytoplasm, preventing binding to SV40 T antigen, and preventing DNA-binding. As shown in Figure 28.26, the majority of these mutations map in the central DNA-binding domain, suggesting that this is an important activity.

p53 activates various pathways through its role as a transcription factor. The pathways can be divided into the three groups summarized in **Figure 28.27.** The major pathway leading to inhibition of the cell cycle at G1 is mediated via activation of p21, which is a cki (cell cycle inhibitor) that is involved with preventing cells from proceeding through G1 (see Figures 27.25 and 28.23). Activation of GADD45 identifies the pathway that is involved with maintaining genome stability. GADD45 is a repair protein that is activated also by other pathways that respond to irradiation damage. The pathway leading to apoptosis remains to be identified.

Figure 28.27 p53 activates several independent pathways. Activation of cell cycle arrest together with inhibition of genome instability is an alternative to apoptosis.

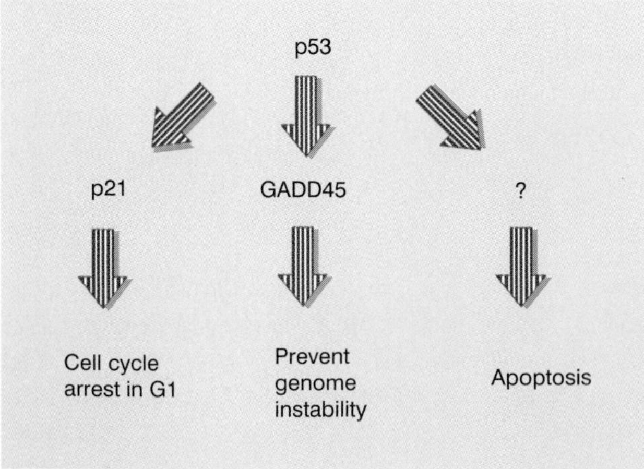

When it functions as a transcription factor, p53 uses the central domain to bind to its target sequence. The N-terminal transactivation domain interacts directly with TBP (the TATA box-binding protein). This region of p53 is also a target for several other proteins. An interaction with E1B 55 kD enables adenovirus to block p53 action, which is an essential part of its transforming capacity. Other regions of p53 can also be targets for inhibition; the SV40 T antigen binds to the specific DNA-binding region, thereby preventing the recognition of target genes

The stability of p53 is an important parameter. It usually has a short half-life. The response to DNA damage stabilizes the protein and activates p53's transactivation activity.

The cellular oncoprotein Mdm2 inhibits p53 activity. p53 induces transcription of Mdm2, so the interaction between p53 and Mdm2 forms a negative feedback loop in which the two components limit each other's activities. The circuitry that controls p53's activity is illustrated in the upper part of **Figure 28.28.** Proteins that activate p53 behave as tumor suppressors; proteins that inactivate p53 behave as oncogenes.

To function as a transcription factor, p53 requires the coactivators p300/CBP (also used by many other transcription factors: see Chapter 21). The coactivator binds to the transactivation (N-terminal) domain of p53. The interaction between p53 and p300 is necessary in order for Mdm2 to bind to p53.

Control of p53's stability is influenced by Mdm2, which has two effects on p53. It functions as an E3 ubiquitin ligase that causes p53 to be targeted by the degradation apparatus. And it also acts directly at the N-terminus to inhibit the transactivation activity of p53. The consequence of this circuit is that Mdm2 limits p53 activity; and the activation of p53 increases the amount of Mdm2.

The C-terminal domain of p53 binds without sequence-specificity to short (<40 base) single-stranded regions of DNA and to mismatches generated by very short (1–3 base) deletions and insertions of bases. Such targets are generated by DNA damage. The consequence of this interaction is to activate the sequence-specific binding activity of the central domain, so that p53 stimulates transcription of its target genes. The nature of this connection is not clear, but may be a two-stage process. When p53 binds through its C-terminal domain to a damaged site on DNA, a change occurs in its properties; it then dissociates from the damaged site and binds to a target gene, which it activates.

The ability of p53 to trigger apoptosis is less well understood. It may depend on the transactivation of a different set of target genes from those involved in activating the G1 checkpoint. The two activities can be separated by the response to adenovirus E1B 19 kD protein, which blocks the apoptotic activity of p53, but does not block its activity to activate target genes. The independence of the effects of p53 on growth arrest and apoptosis is emphasized by the fact that the E1B 55 kD protein blocks transactivation capacity but does not interfere with apoptosis.

The importance of the connection between tumorigenesis and loss of apoptosis is also shown by the properties of the *bcl2* oncogene. *bcl2* was originally identified as a target that is activated by translocations in certain tumors. It turns out to have the property of inhibiting most pathways for apoptosis (see Chapter 27). This suggests that apoptosis plays an important role in inhibiting tumorigenesis, probably because it eliminates potentially tumorigenic cells. When apoptosis is prevented because *bcl2* is activated, these cells survive instead of dying.

Figure 28.28 p53 activity is antagonized by Mdm2, which is neutralized by p19ARF.

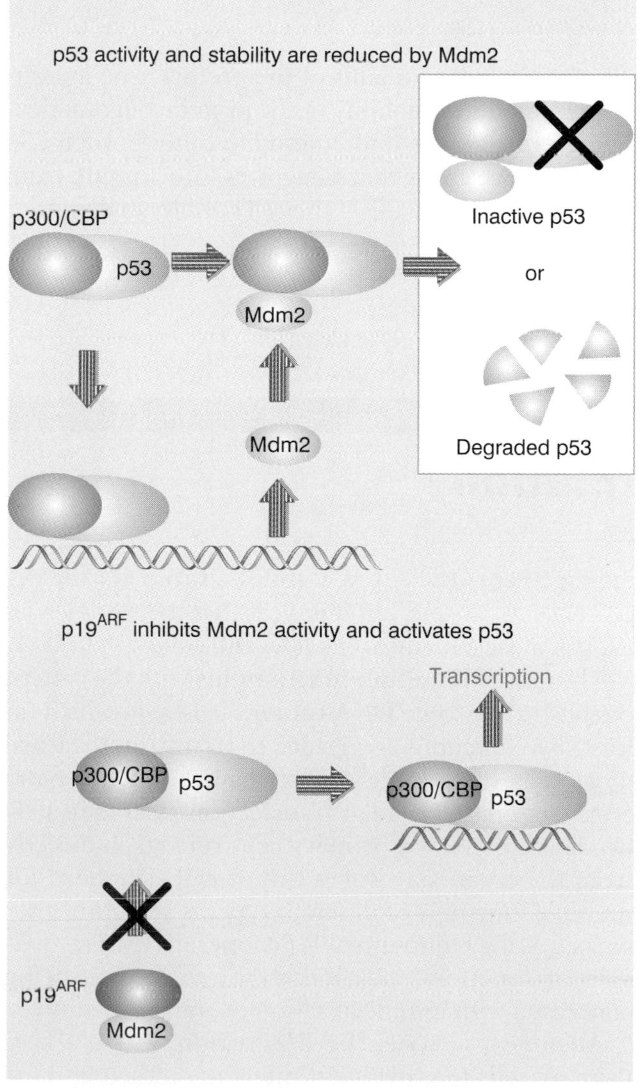

Cells with defective p53 function have a variety of phenotypes; this pleiotropy makes it difficult to determine which (if any) of these effects is directly connected to the tumor suppressor function. Most of our knowledge about p53 action comes from situations in which it has been activated. We assume that the pathways it triggers—growth arrest or apoptosis—are connected to its ability to suppress tumors. Certainly it is clear that the failure of p53 to respond to DNA damage is likely to increase susceptibility to mutational changes that are oncogenic. However, we do not know whether this is the sole role played by p53. $p53^-$ mice have normal survival, implying that p53's role is not essential for development.

The general definition of their properties shows that both RB and p53 are tumor suppressors that in some way usually control cell proliferation; their absence removes this control, and contributes to tumor formation. Both of these proteins have functions that are connected with cell cycle progression, but it remains to be proven formally that this is the activity responsible for tumor suppression, and to characterize how its absence permits unrestrained growth.

Both the RB and p53 tumor suppressors are activated by multiple pathways. One important locus that influences both these tumor suppressors is INK4A-ARF. The transcript is alternatively spliced to give two mRNAs that code for proteins with no sequence relationship. $p16^{INK4A}$ is upstream of RB. $p19^{ARF}$ is upstream of p53. Deletions of the locus are common in human cancers (almost as common as mutations in p53), and have a highly significant effect, because they eliminate both $p16^{INK4A}$ and $p19^{ARF}$ and therefore lead to loss of both the RB and p53 tumor suppressor pathways.

$p16^{INK4A}$ inhibits the cdk4/6 kinase (see Figure 27.25). So it prevents the kinase from phosphorylating RB. In the absence of this phosphorylation, progress through the cell cycle (and therefore growth) is inhibited. $p16^{INK4A}$ is often inhibited by point mutations in human tumors.

$p19^{ARF}$ antagonizes Mdm2, as shown in the lower part of Figure 28.28. This in turn leads to stabilization (and therefore increased activity) of p53. In effect, therefore, $p19^{ARF}$ functions as a tumor suppressor by inhibiting the inhibitor of the p53 tumor suppressor. The nature of the interaction between $p19^{ARF}$ and

Mdm2 is not entirely clear: $p19^{ARF}$ may promote degradation of Mdm2 or directly block its interaction with p53. At all events, $p19^{ARF}$ arrests the cell cycle in a p53-dependent manner. Loss of $p19^{ARF}$ or loss of p53 have similar effects on cell growth (and tumors usually lose one or the other but not both), suggesting that they function in the same pathway; that is, $p19^{ARF}$ in effect functions exclusively through p53. The cellular oncogene C-myc and the adenoviral oncogene E1 both act via $p19^{ARF}$ to activate p53-dependent pathways.

p53 responds to environmental signals that affect cell growth, and many of these signals act by causing specific sites on p53 to be modified. The most common form of modification is the phosphorylation of serine, but acetylation of lysine also occurs. Different pathways lead to the modification of different amino acid residues in p53, as summarized in **Figure 28.29.** There is often overlap between the various residues activated by each pathway. For example, ionizing radiation activates the kinase ATM, which phosphorylates S^{15}; and through unknown pathways, ionizing radiation also causes phosphorylation of S^{33}, dephosphorylation of S^{376}, and acetylation of L^{382}. UV radiation shares a pathway with ionizing radiation to phosphorylate S^{15} and S^{33}, but also causes phosphorylation of S^{392}. The target sites for these various pathways are located in the terminal regulatory domains of the protein. The modifications may affect stability of the protein, oligomerization, DNA-binding, and binding to other proteins. So p53 acts as a sensor that integrates information from many pathways that affect the cell's ability to divide.

Figure 28.29 Each pathway that activates p53 causes modification of a particular set of residues.

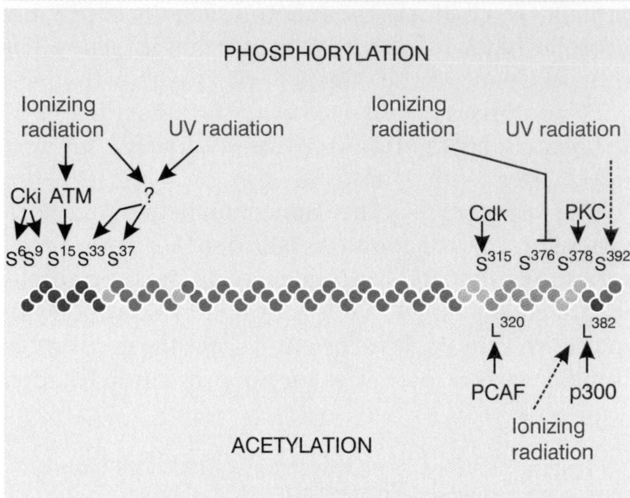

Immortalization and transformation

Most tumors arise as the result of multiple events. It is likely that some of these events involve the activation of oncogenes, while others take the form of inactivation of tumor suppressors. The requirement for multiple events reflects the fact that normal cells have multiple mechanisms to regulate their growth and differentiation, and several separate changes may be required to bypass these controls. Indeed, the existence of many genes in which single mutations were tumorigenic would no doubt be deleterious to the organism, and has been selected against. Nonetheless, oncogenes and tumor suppressors define genes in which mutations create a predisposition to tumors, that is, they represent one of the necessary events. It is an open question as to whether the oncogenes and tumor suppressor genes identified in available assays are together sufficient to account entirely for the occurrence of cancers, but it is clear that their properties explain at least many of the relevant events.

The need for multiple functions fits with the pattern established by some DNA tumor viruses, in which (at least) two functions are needed to transform the usual target cells. In the same way, expression of two or more oncogenes in the cellular transfection assay is usually needed to convert a primary cell (one taken directly from the organism) into a tumor cell. The need for multiple functions of different types is sometimes described as the requirement for **cooperativity.** The division of functions may loosely be viewed as being concerned with immortalization or transformation.

Adenovirus carries the *E1A* region, which allows primary cells to grow indefinitely in culture, and the

E1B region, which causes the morphological changes characteristic of the transformed state.

Polyoma produces three T antigens; large T elicits indefinite growth, middle T is responsible for morphological transformation, and small T is without known function. Large T and middle T together can transform primary cells.

Consistent with the classification of oncogenic functions, adenovirus E1A together with polyoma middle T can transform primary cells. This suggests that one function of each type is needed.

Several cellular oncogenes have been identified by transforming ability in the 3T3 transfection assay; 10–20% of spontaneous human tumors have DNA with detectable transforming activity in this assay. Of course, 3T3 cells have been adapted to indefinite growth in culture over many years, and have passed through some of the changes characteristic of tumor cells. The exact nature of these changes is not clear, but generally they can be classified as involving functions concerned with immortalization. Oncogenic activity in this assay therefore depends on the ability to induce further changes in an established cell line.

The principal products of 3T3 transfection assays are mutated *c-ras* genes. They do not have the ability to transform primary cells *in vitro,* and this supports the implication that their functions are concerned with the act of transforming cells that have previously been immortalized. *ras* oncogenes clearly provide one major pathway for transforming immortalized cells; we do not know how many other transforming pathways may exist that are independent of *ras.*

Whatever functions are required for immortalization can be provided (or circumvented) by other oncogenes. Although *ras* oncogenes alone cannot transform primary fibroblasts, dual transfection with *ras* and another oncogene can do so. The ability to transform primary cells in conjunction with *ras* provides a general assay for oncogenes that have an immortalization-like function. This group includes several retroviral oncogenes, *v-myc, v-jun,* and *v-fos.* It also includes adenovirus E1A and polyoma large T. Mutant p53 genes have the same effect, suggesting that loss of p53 constitutes one route to immortalization. However, in many cases the distinction between immortalizing and transforming proteins is blurred. For example, although E1A is classified as having an immortalizing function, it has (some) of the functions usually attributed to transforming proteins, and loss of p53 confers some properties that are usually considered transforming.

One way to investigate the oncogenic potential of individual oncogenes independently of the constraints that usually are involved in their expression is to create transgenic animals in which the oncogene is placed under control of a tissue-specific promoter. A general pattern is that increased proliferation often occurs in the tissue in which the oncogene is expressed. Oncogenes whose expression have this effect with a variety of tissues include SV40 T antigen, *v-ras,* and *c-myc.*

Increased proliferation (hyperplasia) is often damaging and sometimes fatal to the animal (usually because the proportion of one cell type is increased at the expense of another). However, the expression of a single oncogene does not usually cause malignant transformation (neoplasia), with the production of tumors that kill the animal. Tumors resulting from the introduction of an oncogene (for example, in transgenic mice) are probably due to the occurrence of a second event.

The need for two types of event in malignancy is indicated by the difference between transgenic mice that carry either the *v-ras* or activated *c-myc* oncogene, and mice that carry both oncogenes. Mice carrying either oncogene develop malignancies at rates of 10% for *c-myc* and 40% for *v-ras;* mice carrying both oncogenes develop 100% malignancies over the same period. These results with transgenic mice are even more striking than the comparable results on cooperation between oncogenes in cultured cells.

An interesting convergence is seen in the properties of the tumor antigens of the DNA tumor viruses: the antigens can bind to the cellular tumor suppressor products RB and p53. The two cellular proteins are recognized independently. Either different T antigens of the virus bind separately to RB and to p53, or different domains of the same antigen do so. So adenovirus E1A binds RB, while E1B binds p53; HPV E7 binds RB, while E6 binds p53. SV40 T antigen can bind both RB and p53. The consequences for function of the tumor suppressor are especially clear in the case of HPV E6, which targets p53 for degradation. In effect, HPV converts a target cell into a $p53^-$ state. It seems likely that the loss of p53 (and/or RB) is a major step in the transforming action of DNA tumor viruses, and explains some significant part of the action of the T antigens. The critical events are inhibition of p53's ability to activate transcription, and loss of RB's ability to bind substrates such as E2F. Loss of the tumor suppressors may be one major route in the immortalization pathway.

The changes required for "immortalization" remain to be characterized in terms of changes in cellular properties. It is important to remember that immortalization is required for cells to be perpetuated indefinitely in tissue culture, and we do not know how the relevant events relate to the formation of a tumor *in vivo*. It is clear that some of the same tumor suppressors and oncogenes that are associated with immortalization in culture play roles in tumorigenesis. However, it is not necessarily the case that the events involved in passing through "crisis" to become immortalized in culture have any exact parallel in the formation of a tumor.

With this caveat in mind, it is interesting that p53 provides an important function in immortalization. Established cell lines have usually lost p53 function, which suggests that the role of p53 is connected with the acquisition of ability to support prolonged growth. However, loss of the known functions of p53 is not enough by itself to explain immortalization, since, for example, a $p53^-$ mouse is viable, and therefore is able to undergo the usual pattern of cell cycle arrest and differentiation. Primary cells from a $p53^-$ mouse can pass into the established state more readily than cells that have p53 function, which suggests that loss of p53 activity facilitates or is required for immortalization.

In this context, we might view p53 as an immortalizing function, since the inability to trigger either growth arrest or apoptosis clearly must lead to continued growth. We do not know whether only one or both of these activities are required for immortalization *in vitro*. Also, it is important to realize that these activities of p53 have been described largely in terms of the response to irradiation. It is certainly consistent that cells that are unable to arrest or apoptose when damaged by DNA should develop the chromosomal abnormalities that are often associated with tumor cells. However, we should not assume that the failure in the response to irradiation is responsible for all the deficiencies seen in $p53^-$ cells. For example, growth of normal cells in the body is controlled by a variety of signals, involving diffusible factors (such as TGFβ), inhibition by cell-cell contacts, etc. Failure to respond to such signals may contribute to tumorigenesis, but we do not know how p53 is involved. We know that more than the growth arrest pathway is needed for p53's contribution to tumorigenesis, because a $p21^-$ mouse shows deficiencies in the G1 checkpoint (as would be expected) but does not develop tumors. The contrast with the increased susceptibility of a $p53^-$ mouse to tumors shows that other functions of p53 are involved besides its control of p21.

In some systems, immortalization may be connected with an inability of the cells to differentiate. Growth and differentiation are often mutually exclusive, because a cell must stop dividing in order to differentiate. An oncoprotein that blocks differentiation may allow a cell to continue proliferating (in a sense resembling the immortalization of cultured cells); continued proliferation in turn may provide an opportunity for other oncogenic mutations to occur. This may explain the occurrence among the oncoproteins of products that usually regulate differentiation.

A connection between differentiation and tumorigenesis is shown by avian erythroblastosis virus (AEV). The AEV-H strain carries only *v-erbB*, but the AEV-E54 strain carries two oncogenes, *v-erbB* and *v-erbA*. The major transforming activity of AEV is associated with *v-erbB*, a truncated form of the EGF receptor, which is equivalent to the single oncogene carried by other tumor retroviruses: it can transform erythroblasts and fibroblasts. The other gene, *v-erbA*, cannot transform target cells alone, but it increases the transforming efficiency of *v-erbB*. Expression of *v-erbA* itself has two phenotypic effects upon target cells: it prevents the spontaneous differentiation (into erythrocytes) of erythroblasts that have been transformed by *v-erbB;* and it expands the range of conditions under which transformed erythroblasts can propagate. *v-erbA* may therefore contribute to tumorigenicity by a combination of inhibiting differentiation and stimulating proliferation. In fact, *v-erbA* has a similar effect in extending the efficacy of transformation by other oncogenes that induce sarcomas, notably *v-src*, *v-fps*, and *v-ras*.

Correlations between the activation of oncogenes and the successful growth of tumors are strong in some cases, but by and large the nature of the initiating event remains open. It seems clear that oncogene activity assists tumor growth, but activation could occur (and be selected for) after the initiation event and during early growth of the tumor. We hope that the functions of *c-onc* genes will provide insights into the regulation of cell growth in normal as well as aberrant cells, so that it will become possible to define the events needed to initiate and establish tumors.

An important influence on the ability of a cell to grow is provided by telomerase, the ribonucleoprotein enzyme that is responsible for extending telomeres. The function of telomerase is to compensate for the shortening of telomeres that occurs at each replication cycle. Telomerase is turned off in many somatic cells, typically when differentiation occurs. However, its activity usually is reactivated in tumors. The limiting step in production of telomerase is the transcription of the catalytic subunit, which is repressed in differentiated somatic cells and restored in tumor cells. Similarly, telomerase activity is reactivated in cultured cells that have emerged from crisis. The critical question is whether telomerase activity is essential for tumor formation and at what stage it might be necessary.

Continued division in cells that lack telomerase activity (for example, when primary cells are placed into culture) will cause the telomeres to shorten in each generation. One possible cause of crisis in cultured cells is that telomere lengths become too short to ensure stability at the ends of the chromosomes. This could explain the frequent occurrence of chromosomal abnormalities in cultured cells. When cells enter senescence as the result of telomere shortening, p53 is activated, leading to growth arrest or apoptosis. The trigger that activates p53 is the loss of the telomere-binding protein TRF2 from the chromosome ends. The loss of p53 allows the cells to survive and divide, although of course chromosomal abnormalities result from the lack of proper telomeres.

This suggests that a critical parameter for immortalization might be the reactivation of telomerase. Telomerase activity can be restored by transfecting the gene for the catalytic subunit into target cells, and this allows them to be perpetuated in culture without passing through crisis. We might view the finite replicative capacity of primary human cells in general, or their inability to continue propagation once telomere lengths have become too short in particular, as a tumor suppression mechanism that in effect prevents cells from undertaking the indefinite replication that is needed to make a tumor.

Mice in which the gene coding for the telomerase RNA has been inactivated can survive for 6 generations. Mouse telomeres are exceptionally long, in the range from 10 to 60 kb. In the absence of telomerase, telomeres shorten at 50–100 bp per cell division. There are ~60 divisions in sperm cell production, and ~25 divisions in oocyte production, which fits with the observed rate of shortening of ~4.8 kb per male mouse generation. By the sixth generation, chromosomal abnormalities become more frequent, the ability to form tumors declines, and the mice become infertile (due to the inability to produce sperm). The effects of lack of telomerase are first seen in tissues consisting of highly proliferative cells (as might be expected). All of these observations demonstrate the importance of telomerase for continued cell division. However, cells from the telomerase-negative mice can pass through crisis and can be transformed to give tumorigenic cells, so the presence of telomerase is not essential for, or at least is not the only means of, supporting an immortal state (although reactivation of telomerase is by far the most common mechanism).

There is a curious inconsistency between the results obtained with cultured cells and the survival of telomerase-negative mice. Crisis of mouse cells occurs typically after 10–20 divisions in culture, but one would not expect the telomeres to have reached a limiting length at this point. Mice of the first telomerase-negative generation have passed a greater number of cell divisions without telomerase, and without suffering any ill effects.

Lack of telomerase is clearly associated with inability to continue growth, and reactivation of telomerase is one means by which cells can behave as immortal. It is not clear whether telomerase is the only relevant factor in driving cells into crisis and to what extent other mechanisms might be able to compensate for lack of telomerase. We do not know what pathway is responsible for controlling telomerase production *in vivo*, and how it is connected to pathways that control cell growth.

Summary

A tumor cell is distinguished from a normal cell by its immortality, morphological transformation, and (sometimes) ability to metastasize. Oncogenes are identified by genetic changes that represent gain-of-functions associated with the acquisition of these properties. An oncogene may be derived from a proto-oncogene by mutations that affect its function or level of expression. Tumor suppressors are identified by loss-of-function mutations that allow increased cell proliferation. The mutations may either eliminate function of the tumor repressor or create a dominant negative version.

DNA tumor viruses carry oncogenes without cellular counterparts. Their oncogenes may work by inhibiting the activities of cellular tumor suppressors. RNA tumor viruses carry *v-onc* genes that are derived from the mRNA transcripts of cellular (*c-onc*) genes. Some *v-onc* oncogenes represent the full length of the *c-onc* proto-oncogene, but others are truncated at one or both ends. Most are expressed as fusion proteins with a retroviral product. Src is an exception in which the retrovirus (RSV) is replication-competent, and the protein is expressed as an independent entity.

Some *v-onc* genes are qualitatively different from their *c-onc* counterparts, since the *v-onc* gene is oncogenic at low levels of protein, while the *c-onc* gene is not active even at high levels. In such cases, proto-oncogenes are activated efficiently only by changes in the protein coding sequence. Other proto-oncogenes can be activated by large (>10×) increases in the level of expression; *c-myc* is an example that can be activated quantitatively by a variety of means, including translocations with the Ig or TCR loci or insertion of retroviruses.

c-onc genes have counterpart *v-onc* genes in retroviruses, but some proto-oncogenes have been identified only by their association with cellular tumors. The transfection assay detects some activated *c-onc* sequences by their ability to transform rodent fibroblasts. *ras* genes are the predominant type identified by this assay. The creation of transgenic mice directly demonstrates the transforming potential of certain oncogenes.

Cellular oncoproteins may be derived from several types of genes. The common feature is that each type of gene product is likely to be involved in pathways that regulate growth, and the oncoprotein has lack of regulation or increased activity.

Growth factor receptors located in the plasma membrane are represented by truncated versions in *v-onc* genes. The cellular receptors often have protein tyrosine kinase activity. The oncogenic versions have constitutive activity or altered regulation. In the same way, mutation of genes for polypeptide growth factors gives rise to oncogenes, because a receptor becomes inappropriately activated.

Some oncoproteins are cytoplasmic tyrosine kinases; their targets are largely unknown. They may be activated in response to the autophosphorylation of tyrosine kinase receptors. The molecular basis for the difference between c-Src and v-Src lies in the phosphorylation states of two tyrosines. Phosphorylation of Tyr-527 in the C-terminal tail of c-Src suppresses phosphorylation of Tyr-416. The phosphorylated Tyr-527 binds to the SH2 domain of Src. However, when the SH2 domain recognizes the phosphopeptide sequence created by autophosphorylation of PDGF receptor; the PDGF receptor displaces the C-terminal region of Src, thus allowing dephosphorylation of Tyr-527, with the consequent phosphorylation of Tyr-416 and activation of the kinase activity. v-Src has lost the repressive C-terminus that includes Tyr-527, and therefore has permanently phosphorylated Tyr-416, and is constitutively active.

Ras proteins can bind GTP and are related to the α subunits of G proteins involved in signal transduction across the cell membrane. Oncogenic variants have reduced GTPase activity, and therefore are constitutively active. Activation of Ras is an obligatory step in a signal transduction cascade that is initiated by activation of a tyrosine kinase receptor such as the EGF receptor; the cascade passes to the ERK MAP kinase, which is a serine/threonine kinase, and terminates with the nuclear phosphorylation of transcription factors including Fos.

Nuclear oncoproteins may be involved directly in regulating gene expression, and include Jun and Fos, which are part of the AP1 transcription factor. v-ErbA is derived from another transcription factor, the thyroid hormone receptor, and is a dominant negative mutant that prevents the cellular factor from functioning. v-Rel is related to the common factor NF-κB, but its mode of oncogenic action is not known.

Retinoblastoma (RB) arises when both copies of the *RB* gene are deleted or inactivated. The *RB* product is a nuclear phosphoprotein whose state of phosphorylation controls entry into S phase. Nonphosphorylated RB sequesters the transcription factor E2F. The RB-E2F complex represses certain target genes. E2F is released when RB is phosphorylated by cyclin/cdk complexes; E2F can then activate genes whose products are needed for S phase. Loss of RB prevents repression by RB-E2F, and means that E2F is constitutively available. The cell cannot be restrained from proceeding through the cycle. Adenovirus E1A and papova virus T antigens bind to nonphosphorylated RB, and thus prevent it from binding to E2F.

p53 was originally classified as an oncogene because missense mutations in it are oncogenic. It is now classified as a tumor suppressor because the missense mutants in fact function by inhibiting the activity of wild-type p53. The same phenotype is produced by loss of both wild-type alleles. The level of p53 is usually low, but in response to damage to DNA, p53 activity increases, and triggers either of two pathways, depending upon the stage of the cell cycle and the cell phenotype. Early in the cycle, it provides a checkpoint that prevents further progress; this allows damaged DNA to be repaired before replication. Later in the cycle, it causes apoptosis, so that the cell with damaged DNA dies instead of perpetuating itself. Loss of p53 function is common in established cell lines and may be important in immortalization *in vitro*. Absence of p53 is common in human tumors and may contribute to the progression of a wide variety of tumors, without specificity for cell type.

p53 has a sequence-specific DNA-binding domain that recognizes a palindromic ~10 bp sequence; genes whose promoters have this sequence and which are activated by p53 include the cdk inhibitor p21 and the protein GADD45 (which is activated by several pathways for response to DNA damage). Activation of these and other genes (involving a transactivation domain that interacts directly with TBP) is probably the means by which p53 causes cell cycle arrest. The ability of p53 to activate these target genes is increased after it has bound to damaged DNA, for which it uses a different (non sequence-specific) DNA-binding domain. p53 has a less well characterized ability to repress some genes. Mutant p53 lacks these activities, and therefore allows the perpetuation of cells with damaged DNA. Loss of p53 may be associated with increased amplification of DNA sequences.

p53 is bound by viral oncogenes such as SV40 T antigen, whose oncogenic properties result, at least in part, from the ability to block p53 function. It is also bound by the cellular proto-oncogene, Mdm2, which inhibits its activity. p53 and Mdm2 are mutual antagonists.

The locus INK4A contains two tumor suppressors that together control both major tumor suppressor pathways. $p19^{ARF}$ inhibits Mdm2, so that p19 in effect turns on p53. $p16^{INK4A}$ inhibits the cdk4/6 kinase, which phosphorylates RB. Deletion of INK4A therefore blocks both tumor suppressor pathways by leading to activation of Mdm2 (inhibiting p53) and activation of cdk4/6 (inhibiting RB).

Loss of p53 may be necessary for immortalization, because both the G1 checkpoint and the trigger for apoptosis are inactivated. Telomerase is usually turned off in differentiating cells, which provides a mechanism of tumor suppression by preventing indefinite growth. Reactivation of telomerase is usually necessary to allow continued proliferation of tumor cells.

Further reading

Reviews

Bishop, J. M. (1983). Cellular oncogenes and retroviruses. *Ann. Rev. Biochem.* **52,** 301–354.

Heldin, C.-H. and Westermark, B. (1984). Growth factors: mechanism of action and relation to oncogenes. *Cell* **37,** 9–20.

Varmus, H. (1984). The molecular genetics of cellular oncogenes. *Ann. Rev. Genet.* **18,** 553–612.

Bishop, J. M. (1985). Viral oncogenes. *Cell* **42,** 23–38.

Berk, A. J. (1986). Adenovirus promoters and E1A transactivation. *Ann. Rev. Genet.* **20,** 45–79.

Barbacid, M. (1987). Ras genes. *Ann. Rev. Biochem.* **56,** 779–827.

Haluska, F. G., Tsujimoto, Y., and Croce, C. M. (1987). Oncogene activation by chromosome translocation in human malignancy. *Ann. Rev. Genet.* **21,** 321–335.

Jove, R. and Hanafusa, H. (1987). Cell transformation by the viral src oncogene. *Ann. Rev. Cell Biol.* **3,** 31–56.

Nevins, J. R. (1987). Regulation of early adenovirus gene expression. *Microbiol. Rev.* **51,** 419–430.

Showe, L. C. and Croce, C. M. (1987). The role of chromosomal translocations in B- and T-cell neoplasia. *Ann. Rev. Immunol.* **5,** 253–277.

Cory, S. and Adams, J. M. (1988). Transgenic mice and oncogenesis. *Ann. Rev. Immunol.* **6,** 25–48.

Hanahan, D. (1988). Dissecting multistep tumorigenesis in transgenic mice. *Ann. Rev. Genet.* **22,** 479–519.

Adams, J. M. and Cory, S. (1991). Transgenic models of tumor development. *Science* **254,** 1161–1167.

Cantley, L. C. *et al.* (1991). Oncogenes and signal transduction. *Cell* **64,** 281–302.

Cross, M. and Dexter, T. M. (1991). Growth factors in development, transformation, and tumorigenesis. *Cell* **64,** 271–280.

Hunter, T. (1991). Cooperation between oncogenes. *Cell* **64,** 249–270.

Levine, A. J., Momand, J., and Finlay, C. A. (1991). The p53 tumor suppressor gene. *Nature* **351,** 453–456.

Marshall, C. J. (1991). Tumor suppressor genes. *Cell* **64,** 313–326.

Lowy, D. R. (1993). Function and regulation of Ras. *Ann. Rev. Biochem.* **62,** 851–891.

Levine, A. J. (1997). p53, the cellular gatekeeper for growth and division. *Cell* **88,** 323–331.

Discoveries

Types of oncogenes

Brugge, J. and Erikson, R. L. (1977). Identification of a transformation-specific antigen induced by an avian sarcoma virus. *Nature* **269,** 346–348.

Collet, M. S. and Erikson, R. L. (1978). Protein kinase activity associated with the avian sarcoma virus *src* gene product. *Proc. Nat. Acad. Sci. USA* **75,** 2021–2024.

Hunter, T. and Sefton, B. (1980). Transforming gene product of Rous sarcoma virus phosphorylates tyrosine. *Proc. Nat. Acad. Sci. USA* **77,** 1311–1315.

Waterfield, M. D. *et al.* (1983). Platelet derived growth factors is structurally related to the putative transforming protein p28^{sis} of simian sarcoma virus. *Nature* **304,** 35–39.

Bohmann, D., Bos, T. J., Admon, A., Nishimura, T., Vogt, P. K., and Tjian, R. (1987). Human proto-oncogene c-jun encodes a DNA binding protein with structural and functional properties of transcription factor AP1. *Science* **238,** 1386–1392.

Angel, P. *et al.* (1988). Oncogene jun encodes a sequence-specific trans-activator similar to AP1. *Nature* **332,** 166–171.

Bos, T. J., Bohmann, D., Tsuchie, H., Tjian, R., and Vogt, P. K. (1988). v-jun encodes a nuclear protein with enhancer binding properties of AP1. *Cell* **52,** 705–712.

Oncogenic pathways and cooperativity

Stewart, T. A., Pattengale, P. K., and Leder, P. (1984). Spontaneous mammary adenocarcinomas in transgenic mice that carry and express MTV/*myc* fusion genes. *Cell* **38,** 627–637.

Sinn, E., Muller, W., Pattengale, P., Tepler, I., Wallace, R., and Leder, P. (1987). Coexpression of MMTV/v-Ha-ras and MMTV/*c-myc* genes in transgenic mice: synergistic action of oncogenes *in vivo.* *Cell* **49,** 465–475.

Vaux, D. L., Cory, S., and Adams, J. M. (1988). Bcl2 gene promotes hematopoietic cell survival and cooperates with c-myc to immortalize pre-B cells. *Nature* **335,** 440–442.

Howe, L. R., Leevers, S. J., Gomez, N., Nakielny, S., Cohen, P., and Marshall, C. J. (1992). Activation of the MAP kinase pathway by the protein kinase raf. *Cell* **71,** 335–342.

Tumor suppressors

Linzer, D. I. H. and Levine, A. J. (1979). Characterization of a 54kD cellular SV40 tumor antigen present in SV40-transformed cells and uninfected embryonal carcinoma cells. *Cell* **14,** 43–52.

Sarnow, P. *et al.* (1982). Adenovirus E1B-58 kD tumor antigen and SV40 large T antigen are physically associated with the same 54 kD cellular protein in transformed cells. *Cell* **28,** 387–394.

Cavanee, W. K. *et al.* (1983). Expression of recessive alleles by chromosomal mechanisms in retinoblastoma. *Nature* **305,** 779–784.

Finlay, C. A., Hinds, P. W., and Levine, A. J. (1989). The p53 proto-oncogene can act as a suppressor of transformation. *Cell* **57,** 1083–1093.

Malkin, D. *et al.* (1990). Germ line p53 mutations in a familial syndrome of breast cancer, sarcomas, and other neoplasms. *Science* **250,** 1233–1238.

Donehower, L. A. *et al.* (1992). Mice deficient for p53 are developmentally normal but susceptible for spontaneous tumors. *Nature* **356,** 215–221.

Momand, J., Zambetti, G. P., Olson, D. C., George, D., and Levine, A. J. (1992). The mdm-2 oncogene product forms a complex with the p53 protein and inhibits p53-mediated transactivation. *Cell* **69,** 1237–1245.

Seto, E. *et al.* (1992). Wild-type p53 binds to the TATA-binding protein and represses transcription. *Proc. Nat. Acad. Sci. USA* **89,** 12028–12032.

Immortalization

Blasco, M. A. *et al.* (1997). Telomere shortening and tumor formation by mouse cells lacking telomerase RNA. *Cell* **91,** 25–34.

Meyerson, M. *et al.* (1997). hEST2, the putative human telomerase catalytic subunit gene, is up-regulated in tumor cells and during immortalization. *Cell* **90,** 785–795.

Chapter 29

Gradients, cascades, and signaling pathways

DEVELOPMENT begins with a single fertilized egg, but gives rise to cells that have different developmental fates. The problem of early development is to understand how this asymmetry is introduced: how does a single initial cell give rise within a few cell divisions to progeny cells that have different properties from one another? The means by which asymmetry is generated varies with the type of organism. The egg itself may be homogeneous, with the acquisition of asymmetry depending on the process of the initial division cycles, as in the case of mammals. Or the egg may have an initial asymmetry in the distribution of its cytoplasmic components, which in turn gives rise to further differences as development proceeds, as in the case of *Drosophila.*

Early development is defined by the formation of **axes.** By whatever means are used to develop asymmetry, the early embryo develops differences along the **anterior-posterior axis** (head-tail) and along the **dorsal-ventral axis** (top-bottom.) At the stage of interpreting the axial information, a relatively restricted set of signaling pathways is employed, and essentially the same pathways are found in flies and mammals.

The paradigm for considering the molecular basis for development is to suppose that each cell type may be characterized by its pattern of gene expression, that is, by the particular gene products that it produces. The principal level for controlling gene expression is at transcription, and components of pathways regulating transcription provide an important class of developmental regulators. We may include a variety of activities within the rubric of transcriptional regulators, which could act to change the structure of a promoter region, to initiate transcription at a promoter, to regulate the activity of an enhancer, or indeed sometimes to repress the action of transcription factors. However, the regulators of transcription most often prove to be DNA-binding proteins that activate transcription at particular promoters or enhancers.

Fly development uses a cascade of transcription factors

The systematic manner in which these regulators are turned on and off to form circuits that determine body parts has been worked out in *D. melanogaster.* The basic principle is that a series of events resulting from the initial asymmetry of the egg is translated into the control of gene expression so that specific regions of the egg acquire different properties. The means by which asymmetry is translated into control of gene expression differ for each of four systems that have been characterized in the insect egg. It may involve localization of factors that control transcription or translation within the egg, or localized control of the activities of such factors. But the end result is the same: spatial and temporal regulation of gene expression.

This initial stage of development is succeeded by a stage at which the identities of parts of the embryo are determined: regions are defined whose descendants will form particular body parts. The genes that regulate this process are identified by loci in which mutations cause a body part to be absent, to be duplicated, or to develop as another body part. Such loci are prime candidates for genes whose function is to provide regulatory "switches." Most of these genes code for regulators of transcription. They act upon one another in a hierarchical manner, but they act also upon other genes whose products are actually responsible for the formation of pattern. The ultimate targets are genes that code for kinases, cytoskeletal elements, secreted proteins, and transmembrane receptors.

Viewing the process as a whole, we see that the establishment of differences in the patterns of transcription in different regions of the embryo leads to a *cascade* of control, when regulatory events are connected so that a gene turned on (or off) at one stage itself controls expression of other genes at the next stage. Formally, such a cascade resembles those described previously for bacteriophages or for bacterial sporulation (as discussed previously in Chapter 10), although it is more complex in the case of eukaryotic development. In this paradigm, the common feature of regulatory proteins is that they are transcription factors that regulate the expression of other transcription factors (as well as other target proteins). As in the case of prokaryotic regulation, the basic relationship between the regulator protein and the target gene is that the regulator recognizes a short sequence in the DNA of the promoter (or an enhancer) of a target gene. All of the targets for a particular regulator are identified by their possession of a copy of the appropriate consensus sequence.

The development of an adult organism from a fertilized egg follows a predetermined pathway, in which specific genes are turned on and off at particular times. From the perspective of mechanism, we have most information about the control of transcription. However, subsequent stages of gene expression are also targets for regulation. And, of course, the cascade of gene regulation is connected to other types of signaling, including cell-cell interactions that define boundaries between groups of cells.

The mechanics of development in terms of cellular events are different in different types of species, but we assume that the principle established with *Drosophila* will hold in all cases: that a regulatory cascade determines the appropriate pattern of gene expression in cells of the embryo and ultimately of the adult. Indeed, homologous genes in distantly related organisms play related roles in development. The same pathways are found in (for example) flies and mammals, although the consequences of their employment are rather different in terms of the structures that develop.

Genes involved in regulating development are identified by mutations that are lethal early in development or that cause the development of abnormal structures. A mutation that affects the development of a particular body part attracts our attention because a single body part is a complex structure, requiring expression of a particular set of many genes. Single mutations that influence the structure of the entire body part therefore identify potential regulator genes that switch or select between developmental pathways.

In *Drosophila,* the body part that is analyzed is the segment, the basic unit that can be seen by looking at the adult fly. Mutations fall into (at least) three groups, defined by their effect on the segmental structure:

- **Maternal genes** are expressed during oogenesis by the mother. They may act upon or within the maturing oocyte.
- **Segmentation genes** are expressed after fertilization.

Mutations in these genes alter the number or polarity of segments. Three groups of segmentation genes act sequentially to define increasingly smaller regions of the embryo.

- **Homeotic genes** control the identity of a segment, but do not affect the number, polarity, or size of segments. Mutations in these genes cause one body part to develop the phenotype of another part.

The genes in each group act successively to define the properties of increasingly more restricted parts of the embryo. The maternal genes define broad regions in the egg; differences in the distribution of maternal gene products control the expression of segmentation genes; and the homeotic genes act on segment identity at around the time that the last group of segmentation genes are defining the segments.

A gradient must be converted into discrete compartments

THE basic question of *Drosophila* development is illustrated in **Figure 29.1** in terms of three stages of development: the egg; the larva; and the adult fly.

At the start of development, **gradients** are established in the egg along two axes, **anterior-posterior** and **dorsal-ventral**. The anterior end of the egg becomes the head of the adult; the posterior end becomes the tail. The dorsal side is on top (looking down on a larva); the ventral side is underneath. The gradients consist of molecules (proteins or RNAs) that are differentially distributed in the cytoplasm. The gradient responsible for anterior-posterior development is established soon after fertilization; the dorsal-ventral gradient is established a little later. It is only a modest oversimplification to say that the anterior-posterior systems control positional information along the larva, while the dorsal-ventral system regulates tissue differentiation (that is, the specification of distinct embryonic tissues, including mesoderm, neuroectoderm, and dorsal ectoderm).

Insect development involves two quite different types of structures. The first part of development is concerned with elaborating the larva; then the larva metamorphoses into the fly. This means that the structure of the embryo (the larva) is distinct from the structure of the adult (the fly), in contrast with development of (for example) mammals, where the embryo develops the same body parts that are found in the adult. As the larva develops, it forms some body parts that are exclusively larval (they will not give rise to adult tissues; often they are polyploid), while other body parts are the progenitors that will metamorphose into adult structures (they are usually diploid). In spite of the differences between insect development and vertebrate development, the same general principles appear to govern both processes, and we discover relationships between *Drosophila* regulators and mammalian regulators.

Discrete regions in the embryo correspond to parts of the adult body. They are shown in terms of the superficial organization of the larva in the middle panel of Figure 29.1. Bands of **denticles** (small hairs) are found in a particular pattern on the surface (cuticle) of the larva. The cuticular pattern has features determined by both the anterior-posterior axis and the dorsal-ventral axis:

- Along the anterior-posterior axis, the denticles form discrete bands. Each band corresponds to a segment of the adult fly: in fact, the 11 bands of denticles correspond on a 1:1 basis with the 11 segments of the adult.
- Along the dorsal-ventral axis, the denticles that extend from the ventral surface are coarse; those that extend from the dorsal surface are much finer.

Although the cuticle represents only the surface body layer, its structure is diagnostic of the overall organization of the embryo in both axes. Much of the analysis of phenotypes of mutants in *Drosophila* development has therefore been performed in terms of the distortion of the denticle patterns along one axis or the other.

The difference in form between the gradients of the egg and the segments of the adult poses some prime questions. How are gradients established in the egg? And how is a continuous gradient converted into discrete differences that define individual cell types? How can a large number of separate compartments develop from a single gradient?

Figure 29.1 *Overview:* gradients in the egg are translated into segments on the anterior-posterior axis and into specialized structures on the dorsal-ventral axis of the larva, and then into the segmented structure of the adult fly.

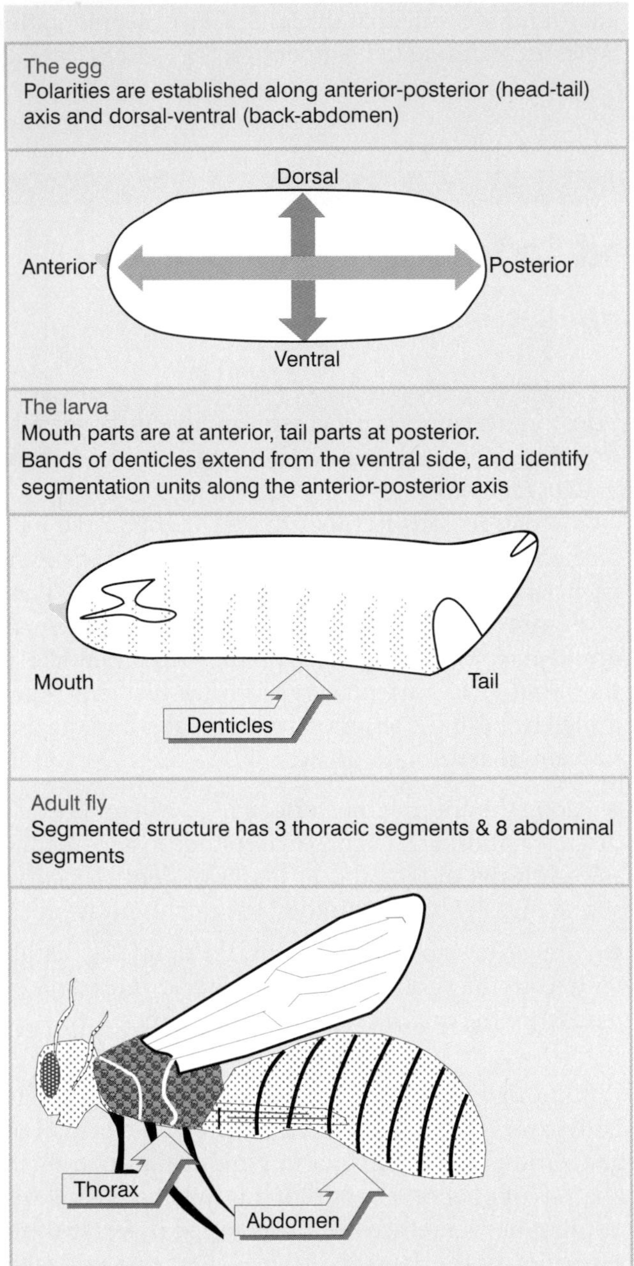

The egg
Polarities are established along anterior-posterior (head-tail) axis and dorsal-ventral (back-abdomen)
Dorsal
Anterior
Posterior
Ventral
The larva
Mouth parts are at anterior, tail parts at posterior.
Bands of denticles extend from the ventral side, and identify segmentation units along the anterior-posterior axis
Mouth
Tail
Denticles
Adult fly
Segmented structure has 3 thoracic segments & 8 abdominal segments
Thorax
Abdomen

Figure 29.2 The early development of the *Drosophila* egg occurs in a common cytoplasm until the stage of cellular blastoderm.

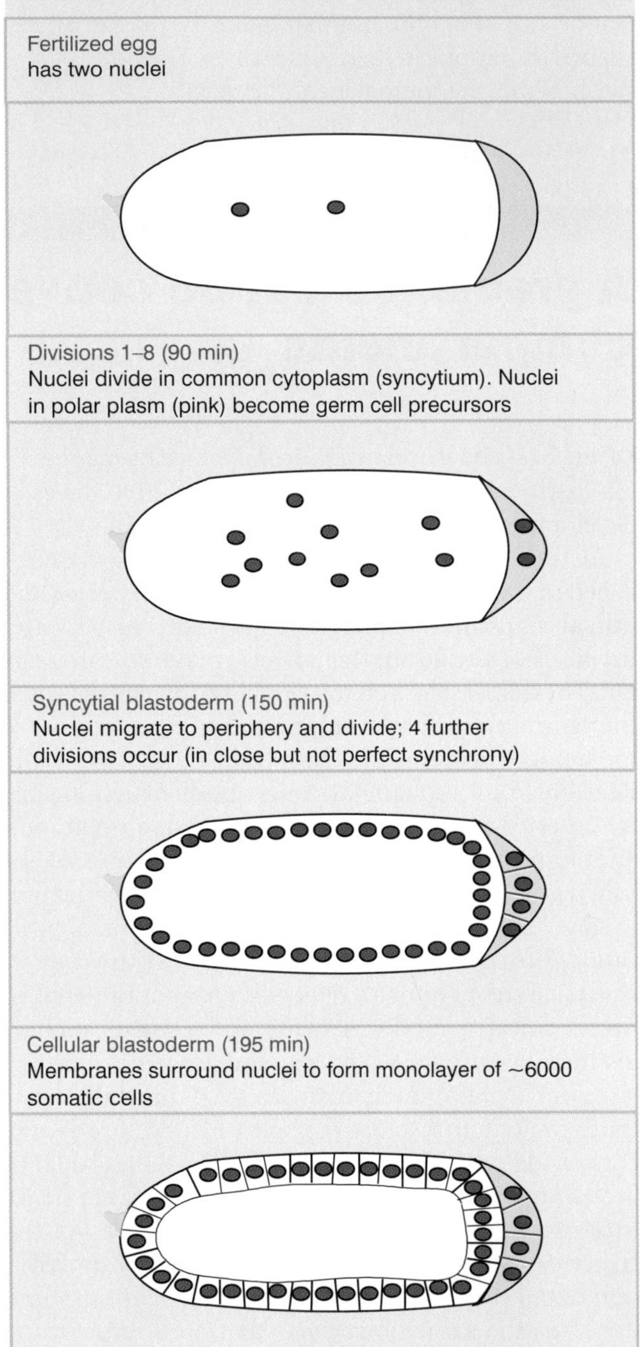

Fertilized egg
has two nuclei
Divisions 1–8 (90 min)
Nuclei divide in common cytoplasm (syncytium). Nuclei in polar plasm (pink) become germ cell precursors
Syncytial blastoderm (150 min)
Nuclei migrate to periphery and divide; 4 further divisions occur (in close but not perfect synchrony)
Cellular blastoderm (195 min)
Membranes surround nuclei to form monolayer of ~6000 somatic cells

The nature of the gradients and their ability to affect the development of a variety of cell types located throughout the embryo depend upon some idiosyncratic features of *Drosophila* development. The early stages are summarized in **Figure 29.2**.

At fertilization the egg possesses the two parental nuclei and is distinguished at the posterior end by the presence of a region called polar plasm. For the first 9 divisions, the nuclei divide in the common cytoplasm. Material can diffuse in this cytoplasm (although there are probably constraints imposed by cytoskeletal organization). At division 7, some nuclei migrate into the polar plasm, where they become precursors to germ cells. After division 9, nuclei migrate and divide to form a layer at the surface of the egg. Then they divide 4 times, after which membranes surround them to form somatic cells.

Up to the point of cellularization, the nuclei effectively reside in a common cytoplasm. At the stage of the cellular blastoderm, the first discrete compartments become evident, and at this time particular regions of the egg are **determined** to become particular types of adult structures. (Determination is progressive and gradual; over the next few cell divisions, the fates of individual regions of the egg become increasingly restricted.) At the start of this process, nuclei migrate to the surface to form the monolayer of the blastoderm, but they do not do so in any predefined manner. It is therefore the location in which the nuclei find themselves at this stage that determines what types of cells their descendants will become. *A nucleus determines its position in the embryo by reference to the anterior-posterior and dorsal-ventral gradients, and behaves accordingly.*

Maternal gene products establish gradients in early embryogenesis

An initial asymmetry is imposed on the *Drosophila* oocyte during oogenesis. **Figure 29.3** illustrates the structure of a follicle in the *Drosophila* ovary. A single progenitor undergoes four successive mitoses to generate 16 interconnected cells. The connections are known as "cytoplasmic bridges" or "ring canals." Individual cells have 2, 3, or 4 such connections. One of the two cells that has 4 connections undergoes meiosis to become the oocyte; the other 15 cells become "nurse cells." Cytoplasmic material, including protein and RNA, passes from the nurse cells to the oocyte; the accumulation of such material accounts for a considerable part of the volume of the egg. The cytoplasmic connections are made at one end of the oocyte, and this end becomes the anterior end of the egg.

Figure 29.3 A *Drosophila* follicle contains an outer surface of follicle cells that surround nurse cells that are in close contact with the oocyte. Nurse cells are connected by cytoplasmic bridges to each other and to the anterior end of the oocyte. Follicle cells are somatic; nurse cells and the oocyte are germline in origin.

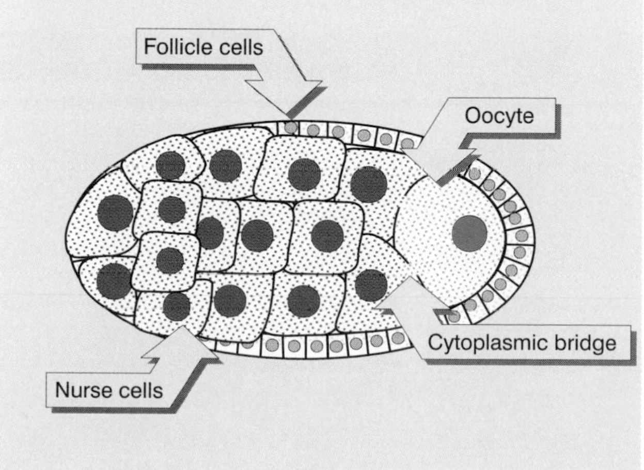

Genes that are expressed within the mother fly are important for early development. These maternal genes are identified by **female sterile** mutations. They do not affect the mother itself, but are required in order to have progeny. Females with such mutations lay eggs that fail to develop into adults; the embryos can be recognized by defects in the cuticular pattern, and they die during development.

The common feature in all maternal genes is that they are expressed prior to fertilization (although their products may either act at the time of expression or be stored for later use). The maternal genes are divided into two classes, depending on their site of expression. Genes that are expressed in somatic cells of the mother that affect egg development are called **maternal somatic genes.** For example, they may act in the follicle cells. Genes that are expressed within the germline are called **maternal germline genes.** These genes may act in either the nurse cell or the oocyte. Some genes act at both stages.

Four groups of genes concerned with the development of particular regions of the embryo can be identified by mutations in maternal genes. The genes in each group can be organized into a pathway that reflects their order of action, by conventional genetic tests (such as comparing the properties of double mutants with the individual mutants) or by biochemical assays (showing which mutants contain components that can bypass the stages that are blocked in other mutants).

The components of these pathways are summarized in **Figure 29.4,** which shows that there is a common principle to their operation. *Each pathway is initiated by localized events outside the egg; this results in the localization of a signal within the egg.* This signal takes the form of a protein with an asymmetric distribution; this is called a **morphogen.** Formally, we may define a morphogen as a protein whose local concentration (or activity) causes the surrounding region to take up a particular structure or fate. In each of these systems, the morphogen either is a transcriptional regulator or leads to the activation of a transcription factor in the localized region. Three systems are concerned with the anterior-posterior axis, and one with the dorsal-ventral axis:

- The **anterior system** is responsible for development of the head and thorax. The maternal germline products are required to localize the *bicoid* product at the anterior end of the egg. In fact, *bicoid* mRNA is transcribed in nurse cells and transported into the oocyte. Bicoid protein is the morphogen: it functions as a transcriptional regulator, and controls expression of the gene *hunchback* (and probably also other segmentation and homeotic genes).
- The **posterior system** is responsible for the segments of the abdomen. The nature of the initial asymmetrical event is not clear. A large number of products act to

Figure 29.4 Each of the four maternal systems that functions in the egg is initiated outside the egg. The pathway is carried into the egg, where each pathway has a localized product that is the morphogen. This may be a receptor or a regulator of gene expression. The final component is a transcription factor, which acts on zygotic targets that are responsible for the next stage of development.

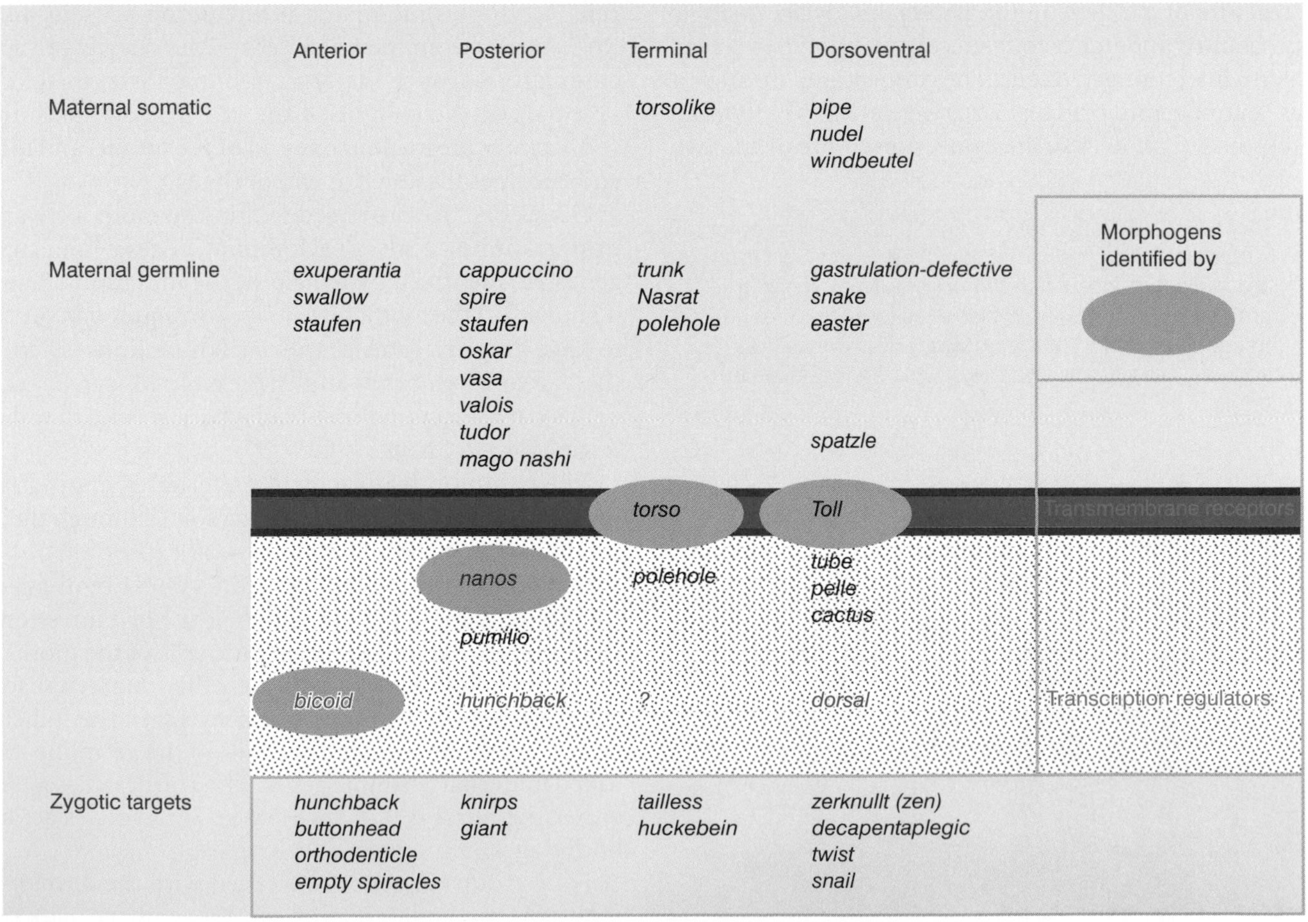

	Anterior	Posterior	Terminal	Dorsoventral	Morphogens identified by
Maternal somatic			*torsolike*	*pipe* *nudel* *windbeutel*	
Maternal germline	*exuperantia* *swallow* *staufen*	*cappuccino* *spire* *staufen* *oskar* *vasa* *valois* *tudor* *mago nashi*	*trunk* *Nasrat* *polehole*	*gastrulation-defective* *snake* *easter* *spatzle*	
			torso	*Toll*	Transmembrane receptors
		nanos *pumilio*	*polehole*	*tube* *pelle* *cactus*	
	bicoid	*hunchback*	?	*dorsal*	Transcription regulators
Zygotic targets	*hunchback* *buttonhead* *orthodenticle* *empty spiracles*	*knirps* *giant*	*tailless* *huckebein*	*zerknullt (zen)* *decapentaplegic* *twist* *snail*	

cause the localization of the product of *nanos*, which is the morphogen. This leads to localized repression of expression of *hunchback* (via control of translation of the mRNA).

- The **terminal system** is responsible for development of the specialized structures at the unsegmented ends of the egg (the acron at the head, and the telson at the tail). As indicated by the dependence on maternal somatic genes, the initial events that create asymmetry occur in the follicle cells. They lead to localized activation of the transmembrane receptor coded by *torso;* the end product of the pathway has yet to be identified.
- The fourth system is responsible for **dorsal-ventral** development. The pathway is initiated by a signal from a follicle cell on the ventral side of the egg. It is transmitted through the transmembrane receptor coded by *Toll*. This leads to a gradient of activation of the transcription factor produced by *dorsal* (by controlling its localization within the cell).

About 30 maternal genes involved in pattern formation have been identified. All of the components of the four pathways are maternal, so we see that the systems for establishing the initial pattern formation all depend on events that occur prior to fertilization. The two body axes are established independently. Mutations that affect polarity cause posterior regions to develop as anterior structures, or ventral regions to develop in dorsal form. On the anterior-posterior axis, the anterior and posterior systems provide opposing gradients, with sources at the anterior and posterior ends of the embryo, respectively, that control development of the segments of the body. Defects in either system affect the body segments. The terminal and dorsal-ventral systems operate independently of the other systems.

The existence of localized concentrations of materials needed for development is indicated by the success of the rescue protocol summarized in **Figure 29.5**. Material is removed from a wild-type embryo and injected into the embryo of a mutant that is defective in early development. If the mutant embryo develops normally, we may conclude that the mutation causes a deficiency of material that is present in the wild-type embryo. This allows us to distinguish components that are necessary for morphogenesis, or that are upstream in the pathway, from the morphogen itself—only the morphogen has the property of localized rescue.

Figure 29.5 Mutant embryos that cannot develop can be rescued by injecting cytoplasm taken from a wild-type embryo. The donor can be tested for time of appearance and location of the rescuing activity; the recipient can be tested for time at which it is susceptible to rescue and the effects of injecting material at different locations.

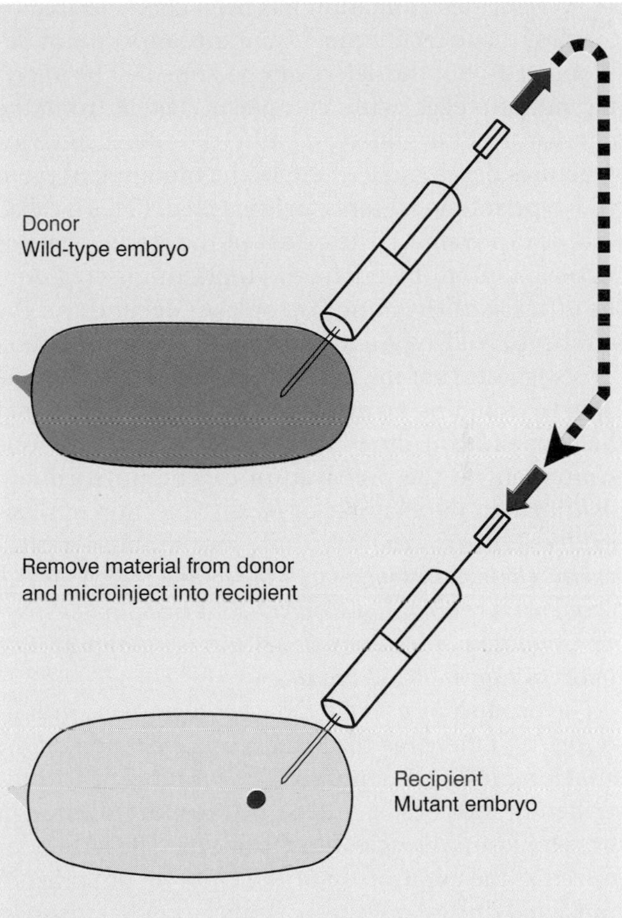

Anterior-posterior development uses localized gene regulators

ANTERIOR determination has been characterized by the rescue technique. *bicoid* mutants do not develop heads; but the defect can be remedied by injecting mutant eggs with cytoplasm taken from the anterior tip of a wild-type embryo. Indeed, anterior structures develop elsewhere in the mutant embryo if wild-type anterior cytoplasm is injected! (This is called **ectopic** expression.) The extent of the rescue depends on the amount of wild-type cytoplasm injected. And the efficacy of the donor cytoplasm depends on the number of wild-type *bicoid* genes carried by the donor. This suggests that the anterior region of a wild-type embryo contains a concentration of some product that depends on the *bicoid* gene dosage. The active component in the preparation can be purified and identified; in the example of *bicoid,* injection of purified *bicoid* mRNA can substitute for the anterior cytoplasm. *These results imply that the components on which* bicoid *acts are ubiquitous, and all that is required to trigger formation of anterior structures is an appropriate concentration of* bicoid *product.*

The product of *bicoid* establishes a gradient with its source (and therefore the highest concentration) at the anterior end of the embryo. The *bicoid* gene is transcribed in nurse cells; mRNA is transported through the cytoplasmic bridges into the oocyte. The RNA is localized at the anterior tip of the embryo, but it is not translated during oogenesis. Translation begins soon after fertilization. The protein then establishes a gradient along the embryo, as indicated in **Figure 29.6.** The gradient could be produced by diffusion of the protein product from the localized source at the anterior tip. The gradient is established by division 7, and remains stable until after the blastoderm stage. Gradients in the same direction, but more shallow, are found also for the other genes, *swa* and *exu*, needed for anterior development. The functions of these genes may be concerned with transporting *bicoid* mRNA into the oocyte, or with limiting its diffusion from the anterior end. The localization of the *bicoid* RNA to the anterior end depends upon sequences in the 3′ untranslated region.

What is the consequence of establishing the bicoid gradient? The gradient can be increased or decreased by changing the number of functional gene copies in the mother. The concentration of bicoid protein is correlated with the development of anterior structures. Weakening the gradient causes anterior segments to develop more posterior-like characteristics; strengthening the gradient causes anterior-like structures to extend farther along the embryo. So the bicoid protein behaves as a morphogen that determines anterior-posterior position in the embryo in a concentration-dependent manner.

The fate of cells in the anterior part of the embryo is determined by the concentration of bicoid protein in which they find themselves. The *bicoid* product is a sequence-specific DNA-binding protein that regulates transcription of genes to whose promoters it binds. The immediate effect of *bicoid* is exercised on other genes that in turn regulate the development of yet further genes. A major target for *bicoid* is the gene *hunch-*

Figure 29.6 Bicoid protein forms a gradient during *D. melanogaster* development that extends for ~200 mm along the egg of 500 mm.

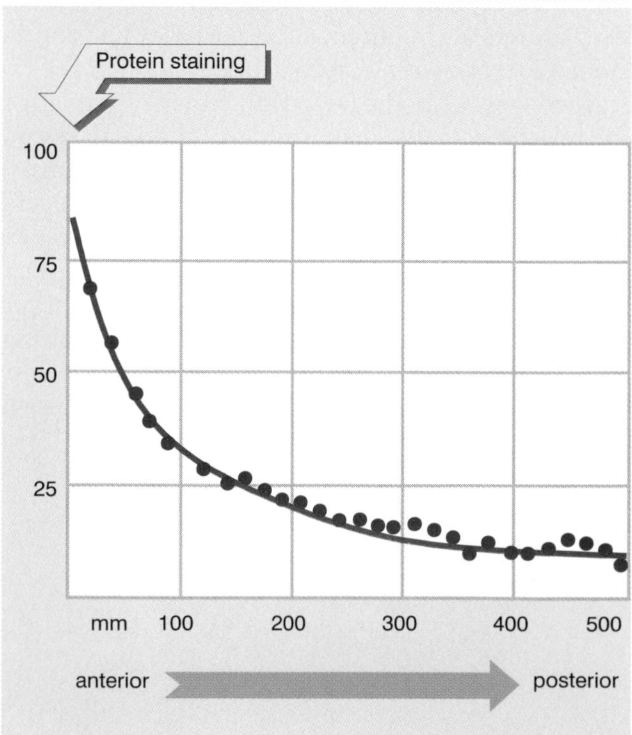

back (see later). Transcription of *hunchback* is turned on by *bicoid* in a dose-dependent manner, that is, *hunchback* is activated above a certain threshold of bicoid protein. This allows a gradient to provide a spatial on-off switch that affects gene expression. *In this way, quantitative differences in the amount of the morphogen (bicoid protein) are transformed into qualitatively different states (cell structures) during embryonic development. bicoid* plays an **instructive** role in anterior development, since it is a positive regulator that is *needed for expression* of genes that in turn determine the synthesis of anterior structures. The effect of bicoid on hunchback is to produce a band of expression that occupies the anterior part of the embryo (see Figure 29.15).

Posterior development depends on the expression of a large group of genes. Embryos produced by females who are mutant for any one of these genes develop normal head and thoracic segments, but lack the entire abdomen. Some of these genes are concerned with exporting material from the nurse cells to the egg; others are required to transport or to localize the material within the egg.

The posterior pathway functions by a series of events in which one product is responsible for localizing the next. **Figure 29.7** correlates the order of genes in the genetic pathway with the activities of their products in the embryo. The functions *spire* and *cappuccino* are needed for Staufen protein to be localized at the pole. Staufen protein in turn localizes *oskar* RNA; possibly a complex of Staufen protein and *oskar* RNA is assembled. These functions are needed to localize Vasa, which is an RNA-binding protein. Its specificity and targets are not known.

If *oskar* is over-expressed or mislocalized in the embryo, it induces germ cell formation at ectopic sites. It requires only the products of *vasa* and *tudor*. This implies that all of the activities that precede *oskar* in the pathway are needed only to localize *oskar* RNA. The ability to form both pole cells and induce abdominal structures is possessed by *oskar,* in conjunction with *vasa* and *tudor* (and of course any components that are ubiquitously expressed in the egg). One effect of *oskar* function is to localize Vasa protein at the posterior end. The functions of *valois* and *tudor* are not known, but it is possible that *valois* is off the main pathway.

Two types of pattern-determining events occur at the posterior pole, and the pathway branches at *tudor.*

Figure 29.7 The posterior pathway has two branches, responsible for abdominal development and germ cell formation.

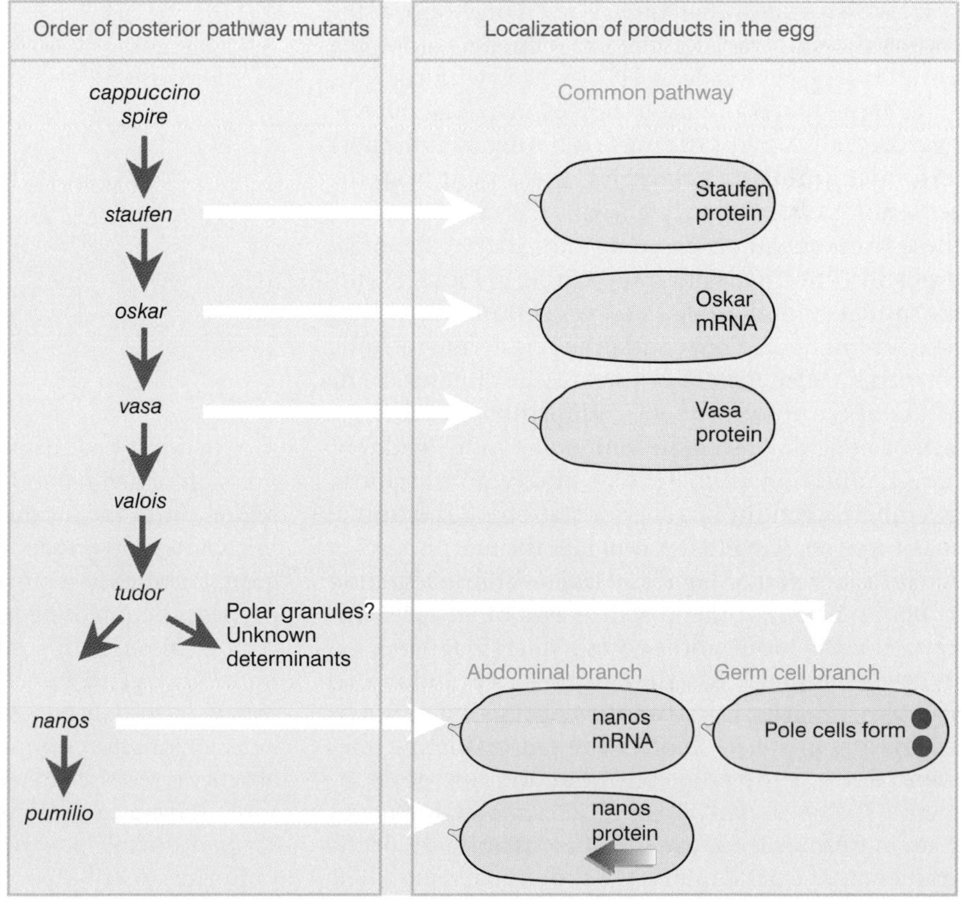

The polar plasm contains two morphogens: the posterior determinant (nanos) controls abdominal development; and an unidentified signal controls formation of the pole cells, which will give rise to the germline (see Figure 29.2). All of the posterior genes except *nanos* and *pumilio* are required for both processes, that is, they are defective in both abdominal development and pole cell formation. *nanos* and *pumilio* identify the abdominal branch. We do not know whether there are additional functions representing a separate branch for germ cell formation, or whether the pathway up to *tudor* is by itself sufficient.

The posterior system resembles the anterior system in the basic nature of the morphogenetic event: a maternal mRNA is localized at the posterior pole. This is the product of *nanos,* and provides the morphogen. There are two important differences between the systems. Localization is more complex than in the case of the anterior system, because posterior determinants that originate in the nurse cells must be transported the full length of the oocyte to the far pole. And nanos protein acts to *prevent* translation of a transcription factor (hunchback). Its role is said to be **permissive,** since it functions to repress genes whose products would interfere with posterior development.

How do we know that *nanos* is the morphogen at the end of the pathway? Rescue experiments (along the lines shown previously in Figure 29.5) with the mutants in the posterior group showed that in all but one case the cytoplasm of the nurse cell contained the posterior determinant (although it was absent from the posterior end of the oocyte itself). This indicates that these mutants all act in some subsidiary role, most probably concerned with transporting or localizing the morphogen in the egg. The exception was *nanos,* whose mutants did not contain any posterior-rescuing activity. Purified *nanos* RNA can rescue mutants in any of the other posterior genes, indicating that it is the last, or most downstream, component in the pathway. Indeed, injection of *nanos* RNA into ectopic locations in embryos can induce the formation of abdominal structures, showing that it provides the morphogen.

The upper part of **Figure 29.8** shows the localization of *nanos* mRNA at the posterior end of an early embryo. But the localization poses a dilemma: *nanos* activity is required for development of abdominal segments, that is, for structures occupying approximately the posterior half of the embryo. How does *nanos* RNA at the pole control abdominal development? The lower part of Figure 29.8 shows that nanos protein diffuses from the site of translation to form a gradient that extends along the abdominal region. The production of nanos protein also is controlled by repression of any *nanos* mRNA outside of the posterior region. Both the localization and repression of *nanos* mRNA depend on sequences in the 3′ untranslated region. Pumilio is required in some unknown way for proper function of nanos.

Figure 29.8 *nanos* products are localized at the posterior end of a *Drosophila* embryo. The upper photograph shows the tightly localized RNA in the very early embryo (at the time of the 3rd nuclear division). The lower photograph shows the spreading of *nanos* protein at the 8th nuclear division. Photographs kindly provided by Ruth Lehmann.

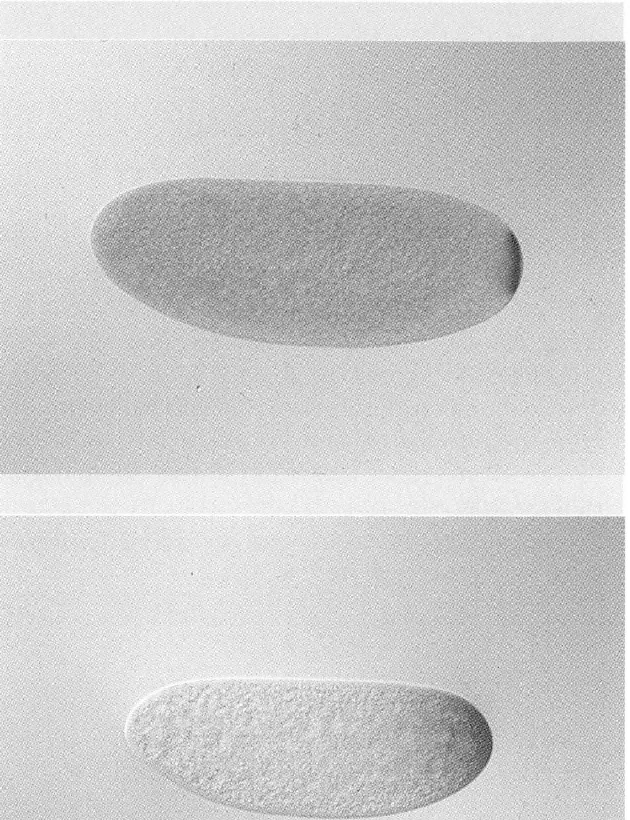

Both *bicoid* and *nanos* act on the expression of the *hunchback* gene. *hunchback* codes for a repressor of transcription: its presence is needed for formation of anterior structures (in the region of the thorax), and its absence is required for development of posterior structures. It has a complex pattern of expression. It is transcribed during oogenesis to give an mRNA that is uniformly distributed in the egg. After fertilization, the *hunchback* pattern is changed in two ways. The bicoid gradient activates synthesis of *hunchback* RNA in the anterior region. And nanos prevents translation of *hunchback* mRNA in the posterior region; a result of this inhibition is that the mRNA is degraded. The anterior and posterior systems together therefore enhance hunchback levels in the anterior half of the egg, and remove it from the posterior half. We see later that the significance of this distribution lies with the genes that hunchback regulates. It represses the genes *knirps* and (probably) *giant,* which are needed to form abdominal structures. So *the basic role of* hunchback *is to repress formation of abdominal structures by preventing the expression of* knirps *and* giant *in more anterior regions.*

Dorsoventral development uses localized receptor-ligand interactions

DORSAL-ventral development displays a complex interplay between the oocyte and follicle cells. The formation of pattern starts with the expression of genes in the oocyte that are needed for proper development of the follicle cells. And then expression of genes in the follicle cells transmits a signal to the oocyte that results in development of ventral structures. Another pathway is responsible for development of dorsal structures in the developing egg. Each of these systems functions by activating a localized ligand-receptor interaction that triggers a signal transduction pathway.

The process starts with the localization of *gurken* mRNA. This passes through two stages, the first playing a role in anterior-posterior patterning, the second in dorsal-ventral patterning. These are key events that define the spatial asymmetry of the egg chamber. (The requirement of *gurken* for both pathways is the only feature that breaches their independence.)

First *gurken* mRNA is localized on the posterior side of the oocyte. This results in a signal that causes adjacent follicle cells to become posterior. The follicle cells signal back to the oocyte in a process that results in the establishment of a polarized network of microtubules. This is necessary for the localization of the maternal transcripts of *bicoid* and *oskar* to opposite poles, as discussed earlier.

Dorsal-ventral polarity is established later when *gurken* mRNA becomes localized on the dorsal side of the oocyte. **Figure 29.9** illustrates the pathway and its consequences. The products of *cornichon* and *brainiac* are needed for proper localization of the *gurken* mRNA or for activation of the protein. Of the group of loci that act earlier, the products of *K10* and *squid* are needed to localize the RNA; and *cappuccino* and *spire* mutants have an array of defects that suggest their products have a general role in organizing the cytoskeleton of the oocyte. Accordingly, *cappuccino* and *spire* are required also for the earlier localization of *gurken* mRNA involved in anterior-posterior patterning.

gurken codes for a protein that resembles the growth factor TGFα. The next locus in the pathway is *torpedo*, which codes for the *Drosophila* EGF receptor. It is expressed in the follicle cells. So the pathway moves from oocyte to follicle cells when the ligand (Gurken), possibly in a transmembrane form that exposes the extracellular domain on the oocyte, interacts with the receptor (Torpedo) on the plasma membrane of a follicle cell.

An interesting and general principle emerges from the activation of Torpedo, which is a typical receptor tyrosine kinase. Activation of Torpedo leads to the activation of a Ras signaling pathway, which proceeds through Raf and D-mek (the equivalent of MAPKK), to activate a classic MAP kinase pathway. The ultimate readout of this pathway is not known, but its effect is to prevent activation on the dorsal side of the embryo of the ventral-determining pathway that we discuss next.

The utilization of this pathway shows that similar pathways may be employed in different circumstances to produce highly specific effects. The trigger to activate the pathway in the oocyte-follicle cell interaction is the specific localization of Gurken. The consequence is a change in the properties of follicle cells that prevents them from acquiring ventral fates. The basic components of the pathway, however, are the same as those employed in signal transduction of proliferation signals in vertebrate systems. The same pathway is employed again in the specific development of retinal cells in *Drosophila* itself, where another receptor-counter receptor interaction activates the Ras pathway, with

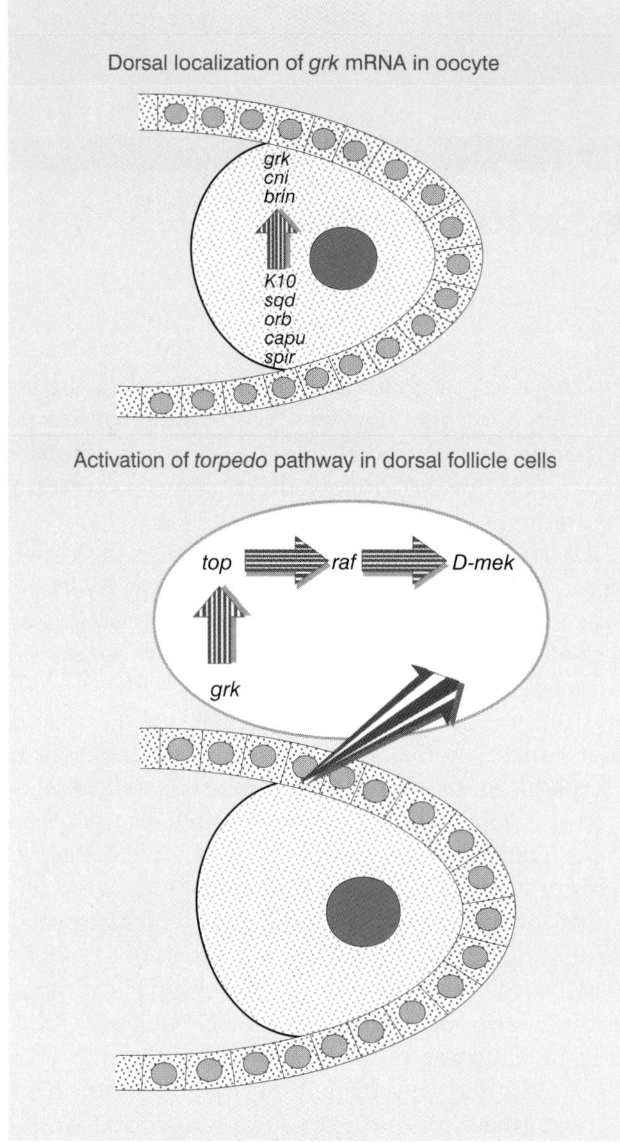

Figure 29.9 Dorsal and ventral identities are first distinguished when *grk* mRNA is localized on the dorsal side of the oocyte. Synthesis of Grk activates the receptor coded by *torpedo,* which triggers an MAPK pathway in the follicle cells.

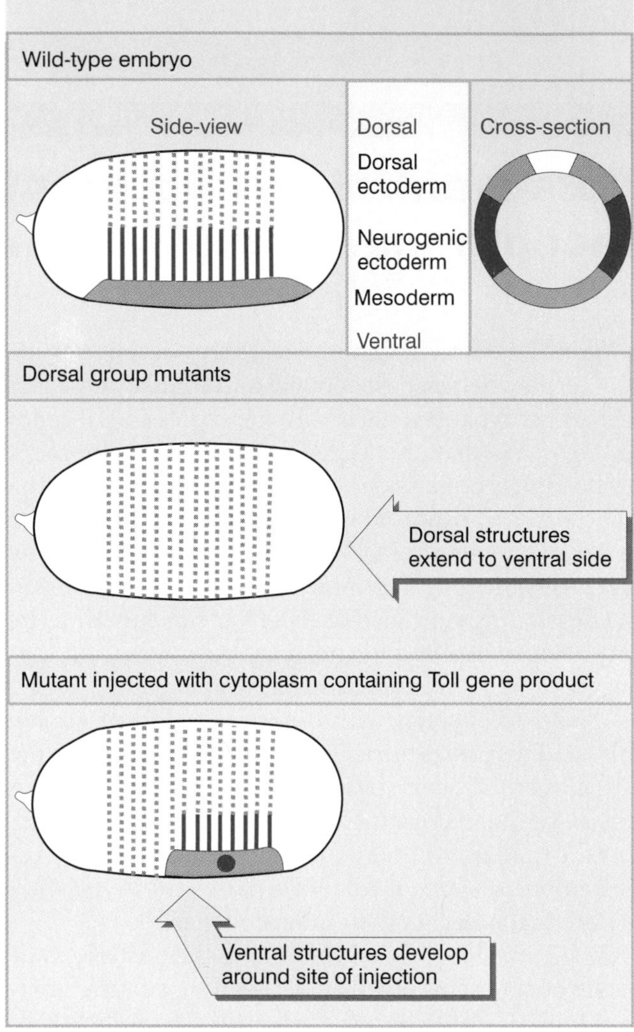

Figure 29.10 Wild-type *Drosophila* embryos have distinct dorsal and ventral structures. Mutations in genes of the *dorsal* group prevent the appearance of ventral structures, and the ventral side of the embryo is dorsalized. Ventral structures can be restored by injecting cytoplasm containing the Toll gene product.

specific, but very different, effects on cell differentiation. So essentially the same pathway can be employed to interpret an initial signal and produce a response that is predetermined by the cell phenotype.

Development of ventral structures requires a group of 11 maternal genes whose products establish the dorsal-ventral axis between the time of fertilization and cellular blastoderm (see Figure 29.4). **Figure 29.10** shows that the dorsal-ventral pattern can be viewed from the side by the phenotype of the cuticle, and can be seen in cross-section to represent the formation of different types of tissues. The *dorsal* system is necessary for the development of ventral structures including the mesoderm and neurogenic ectoderm. (The system was named for the effects of mutations [to dorsalize], rather than for the role of the gene products [to ventralize].) Mutants in any genes of the *dorsal* group lack ventral structures, and have dorsal structures on the ventral side, as indicated in the figure. But injecting wild-type cytoplasm into mutant embryos rescues the defect and allows ventral structures to develop.

The ventral-determining pathway also begins in the follicle cell and ends in the oocyte. The pathway is summarized in **Figure 29.11**. The initial steps are not well

Figure 29.11 The dorsal-ventral pathway is summarized on the right and shown in detail on the left. It involves interactions between follicle cells and the oocyte. The pathway moves into the oocyte when spatzle binds to Toll and activates the morphogen. The pathway is completed by transporting the transcription factor dorsal into the nucleus.

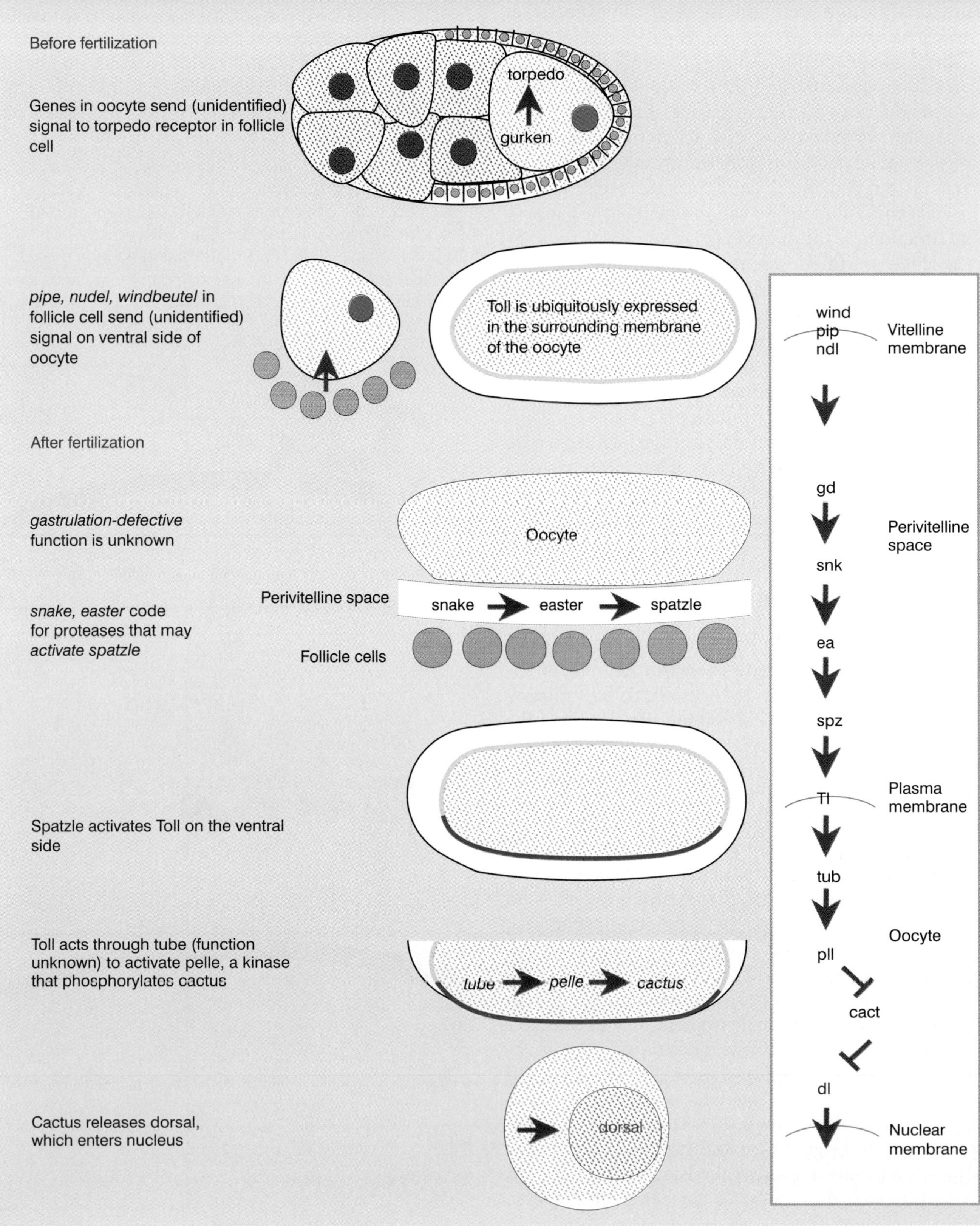

defined, and require the expression of three loci in the follicle cells on the ventral side. These loci function before fertilization, but the egg does not receive the signal until after fertilization. The nature of the signal is not known, although we know most of the events that ensue in the oocyte.

The signal leads to a series of proteolytic cleavages that occur in the perivitelline space (the outermost layer of the oocyte). The product of *snake* probably cleaves the product of *easter,* which in turn is activated to cleave the *spatzle* product. Cleavage activates spatzle, so that it can provide a ligand for a receptor coded by the *Toll* gene. Toll is the first component of the pathway that functions in the oocyte.

Rescue experiments identify *Toll* as the crucial gene that conveys the signal into the oocyte. *Toll*$^-$ mutants lack any dorsal-ventral gradient, and injection of Toll induces the formation of dorsal-ventral structures. The other genes of the dorsal group code for products that either regulate or are required for the action of *Toll,* but they do not establish the primary polarity.

There is a paradox in the distribution of Toll protein. *Toll* gene product activity is found in all parts of a donor embryo when cytoplasm is extracted and tested by injection. Yet it induces ventral structures only in the appropriate location in normal development. An initial general distribution of *Toll* gene product must therefore in some way be converted into a concentration of active product by local events.

Toll is a transmembrane protein homologous to the vertebrate interleukin-1 (IL1) receptor. Binding of ligand is sufficient to activate the ventral-determining pathway. The reaction occurs on the ventral side of the perivitelline space. The spatzle ligand either cannot diffuse far from the site of cleavage, or perhaps it binds to Toll very rapidly, with the result that Toll is activated only on the ventral side of the embryo. Loss-of-function mutations in Toll are dorsalized, because the receptor cannot be activated. There are also dominant (*Toll*^{D}) mutations, which confer ventral properties on dorsal regions; these are gain-of-function mutations, which are ventralized because the receptor is constitutively active. Genetic analysis shows that toll acts via tube and pelle. Tube is probably an adaptor protein that recruits the kinase pelle to the activated receptor. The target for the pelle kinase is not proven, but its activation leads to phosphorylation of the product of cactus, which is the final regulator of the transcription factor coded by dorsal. **Figure 29.12** shows the parallels between the toll signaling pathway in flies and the IL1 vertebrate pathway (where the biochemistry is well characterized). Activation of the receptor causes a complex to assemble that includes adaptor proteins (several in vertebrates), which bind a kinase. Activation of the vertebrate kinase (IRAK) in turn activates the kinase NIK, which phosphorylates I-κB. It is not clear whether the fly kinase (pelle) acts directly on cactus (the equivalent of Iκ-B) or through an intermediate.

At all events, dorsal and cactus form an interacting pair of proteins that are related to the transcription factor NF-κB and its regulator IκB. NF-κB consists of two

Figure 29.12 Activation of IL1 receptor triggers formation of a complex containing adaptor(s) and a kinase. The IRAK kinase activates NIK, which phosphorylates I-κB. This triggers degradation of I-κB, releasing NF-κB, which translocates to the nucleus to activate transcription.

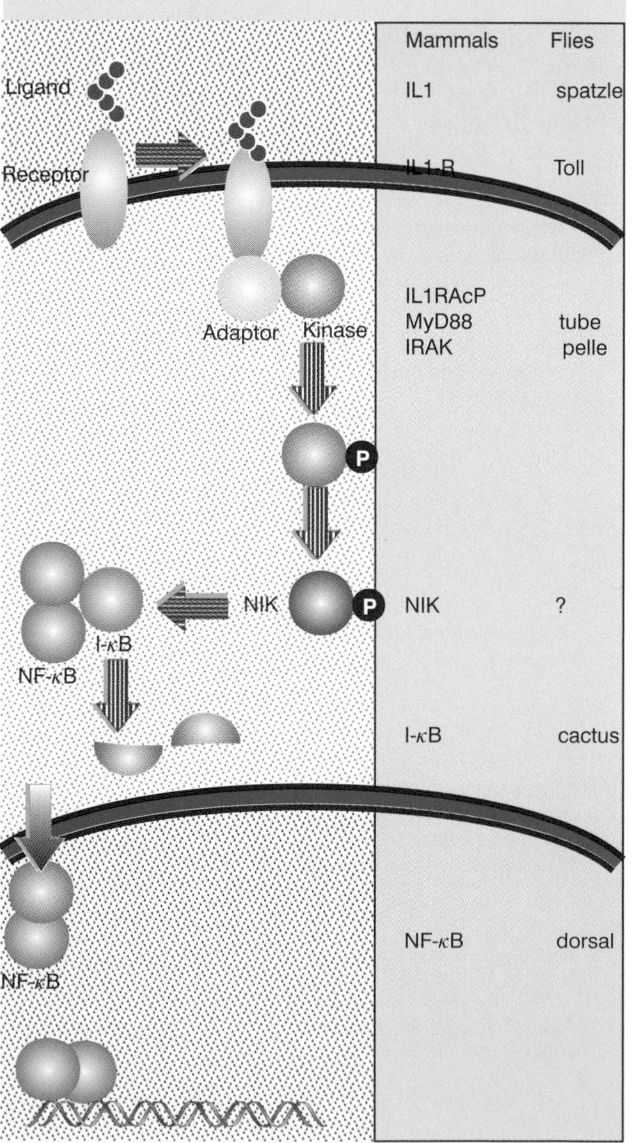

subunits (related in sequence) that are bound by IκB in the cytoplasm. When IκB is phosphorylated, it releases NF-κB, which then moves into the nucleus, where it functions as a transcription factor of genes whose promoters have the κB sequence motif. (An example of the pathway is illustrated in Figure 21.2.) It seems likely that cactus regulates dorsal in the same way that IκB regulates NF-κB. A cactus-dorsal complex is inert in the cytoplasm, but when cactus is phosphorylated, it releases dorsal protein, which enters the nucleus. The pathway is therefore conserved from receptor to effector, since activation of interleukin-1 receptor has as a principal effect the activation of NF-κB, and activation of Troll leads to activation of dorsal.

As a result of the activation of Toll, a gradient of dorsal protein in the nucleus is established, from ventral to dorsal side of the embryo. On the ventral side, dorsal protein is released to the nucleus, but on the dorsal side of the embryo it remains in the cytoplasm. A steep gradient is established at the stage of syncytial blastoderm, and becomes sharper during the transition to cellular blastoderm. The proportion of dorsal protein that is in the nucleus correlates with the ventral phenotype that will be displayed by this region. An example of a gradient visualized by staining with antibody against dorsal protein is shown in **Figure 29.13.** The total amount of dorsal protein in the embryo does not change: the gradient is established solely by a redistribution of the protein between nucleus and cytoplasm.

Dorsal both activates and represses gene expression. It activates the genes *twist* and *snail,* which are required for the development of ventral structures. And it represses the genes *dpp* and *zen,* which are required for the development of dorsal structures; as a result, these genes are expressed only in the 40% most dorsal of the embryo. We discuss the determination of dorsoventral structures by the Dpp pathway in the next section.

One of the crucial aspects of dorsal-ventral development is the relationship between the systems. This is summarized in **Figure 29.14.** The ability of one system to repress the next is responsible for restricting the localized activities to the appropriate part of the embryo. The initial interaction between gurken and torpedo leads to the repression of spatzle activity on the dorsal side of the embryo. This restricts the activation of dorsal protein to the ventral side of the embryo. Nuclear localization of dorsal protein in turn represses the expression of *dpp*, so that it forms a gradient diffusing from the dorsal side. In this way, ventral structures are formed in the nuclear gradient of dorsal protein, and dorsal structures are formed in the gradient of dpp protein.

The terminal system is initiated in a way that is similar to the dorsal-ventral system. A transmembrane receptor, coded by the *torso* gene, is produced by translation of a maternal RNA after fertilization. The receptor is localized throughout the embryo. It is activated at the poles by local production of an extracellular ligand. Torso protein has a kinase activity, which initiates a cascade that leads to local expression of the *tailless* and *huckebein* RNAs, which code for factors that regulate transcription.

Figure 29.13 Dorsal protein forms a gradient of nuclear localization from ventral to dorsal side of the embryo. On the ventral side (lower) the protein identifies bright nuclei; on the dorsal side (upper) the nuclei lack protein and show as dark holes in the bright cytoplasm. Photograph kindly provided by Michael Levine.

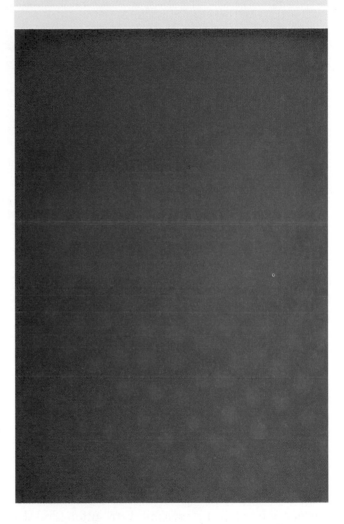

The pattern of regulators at each stage of development for each of the systems is summarized in **Figure 29.15.** Two types of mechanism are used to create the initial asymmetry. For the anterior-posterior axis, an RNA is localized at one end of the egg (bicoid for the

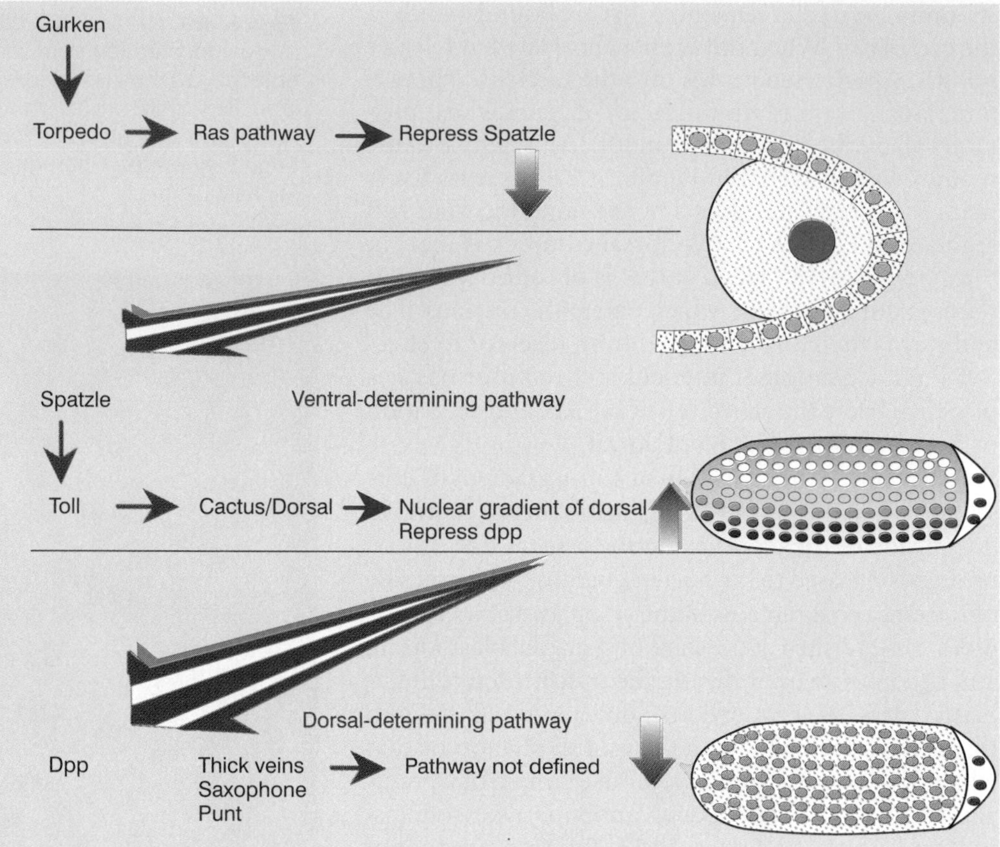

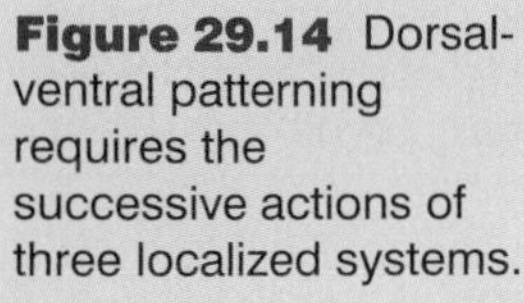
Figure 29.14 Dorsal-ventral patterning requires the successive actions of three localized systems.

anterior system, nanos for the posterior system); localization depends upon the interaction of sequences in the 3′ end of the RNA with maternal proteins. In the case of the dorsal-ventral and terminal systems, a receptor protein is specifically activated in a localized manner, as the result of the limited availability of its ligand. All of these interactions depend on RNAs and/or proteins expressed from maternal genes.

The local event leads to the production of a morphogen, which forms a gradient, either quantitatively (bicoid) or by nucleocytoplasmic distribution (dorsal), or is localized in a broad restricted region (nanos). The extent of the region in which the morphogen is active is ~50% across the egg for each of these systems. The morphogens are translated from maternal RNAs, and development is therefore still dependent on maternal genes up to this stage.

Establishing anterior-posterior and dorsal-ventral gradients is the first step in determining orientation and spatial organization of the embryo. Under the direction of maternal genes, gradients form across the common cytoplasm and influence the behavior of the nuclei located in it. The next step is the development of discrete regions that will give rise to different body parts. This requires the expression of the zygotic genome, and the loci that now become active are called *zygotic genes.* Genes involved at this stage are identified by segmentation mutants.

The products of the segmentation genes form bands that distinguish individual regions on the anterior-posterior axis. When we consider the results of the anterior and posterior systems together, we see that there are several broad regions (the two regions generated by the anterior system are adjacent to the two regions defined by the posterior system; see below). On the dorsal-ventral axis, there are three rather broad bands that define the regions in which the mesoderm, neuroectoderm, and dorsal ectoderm form (proceeding from the ventral to the dorsal side).

Figure 29.15 In each axis-determining system, localized products in the egg cause other maternal RNAs or proteins to be broadly localized at syncytial blastoderm, and zygotic RNAs are transcribed in bands at cellular blastoderm.

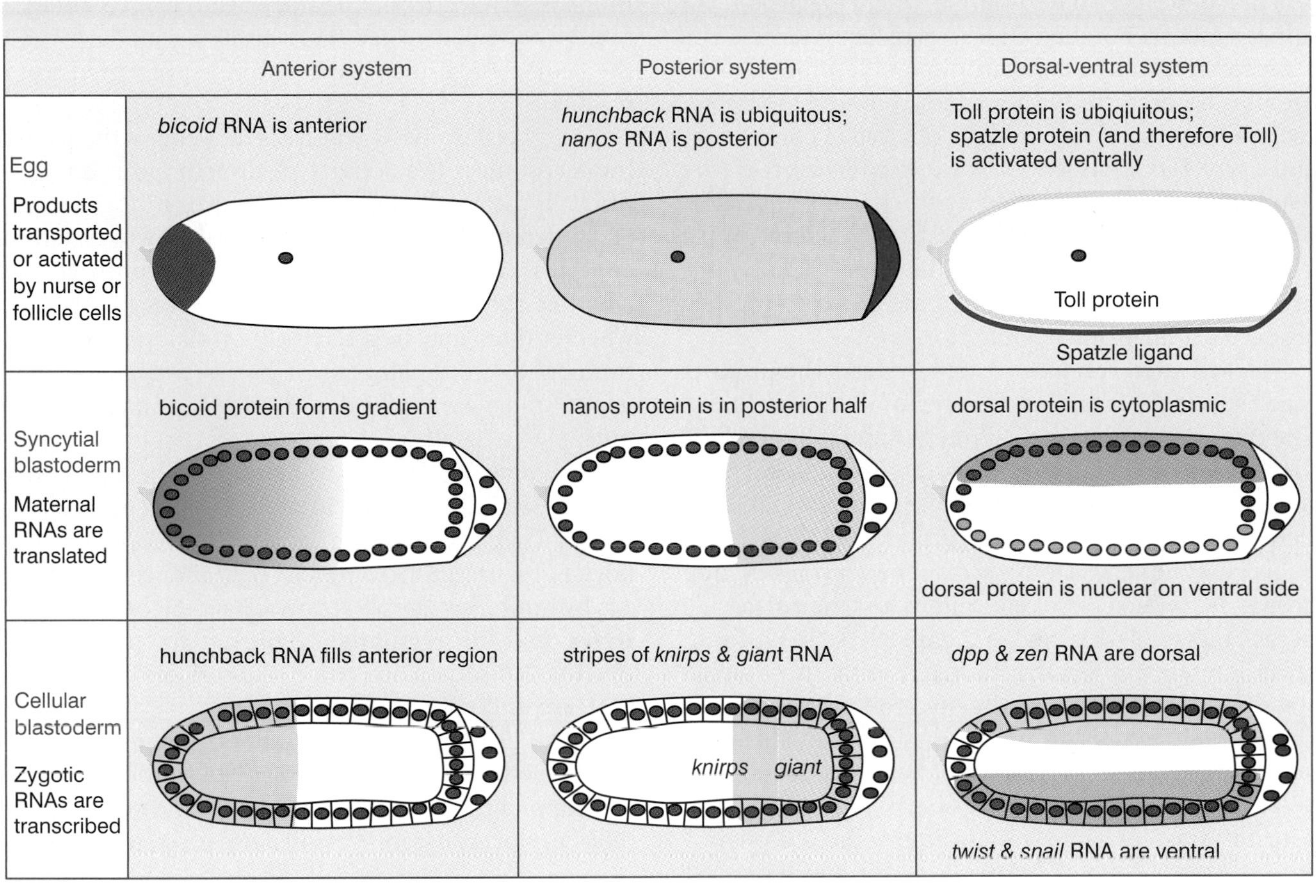

	Anterior system	Posterior system	Dorsal-ventral system
Egg Products transported or activated by nurse or follicle cells	*bicoid* RNA is anterior	*hunchback* RNA is ubiquitous; *nanos* RNA is posterior	Toll protein is ubiqiuitous; spatzle protein (and therefore Toll) is activated ventrally Toll protein Spatzle ligand
Syncytial blastoderm Maternal RNAs are translated	bicoid protein forms gradient	nanos protein is in posterior half	dorsal protein is cytoplasmic dorsal protein is nuclear on ventral side
Cellular blastoderm Zygotic RNAs are transcribed	hunchback RNA fills anterior region	stripes of *knirps* & *giant* RNA *knirps* *giant*	*dpp* & *zen* RNA are dorsal *twist* & *snail* RNA are ventral

TGFβ/BMPs are diffusible morphogens

The principle of dorsoventral development in flies, amphibians, and mammals is the same. On one side of the animal, neural structures (including the CNS) develop. This is the ventral side in flies, and the dorsal side in vertebrates. On the other side, mesenchymal structures develop. This is the dorsal side in flies, and the ventral side in vertebrates. The important point here is that the same relative development is seen from one side of the animal to the other, but its absolute direction is reversed between flies and vertebrates. This must mean that the dorsoventral axis was inverted at some point during evolution, causing the CNS to be displaced from the ventral side to the dorsal side. This idea is supported by the fact that the same signaling pathway is initiated on the dorsal side of flies and on the ventral side of vertebrate embryos.

Mesenchymal (non-neural) structures are determined by diffusible factors in the TGFβ/BMP family. These factors are small polypeptide ligands for receptors that activate the Smads transcription factors (see Figure 26.36). Formation of neural structures requires counteracting activities that also diffuse from a center; they prevent the TGFβ/BMP ligands from activating the target receptors. (The names reflect the histories of their discoveries as transforming growth factor β and bone morphogenetic proteins; but in fact the most important role of these polypeptides is as morphogens in development.)

The involvement of this pathway in development was first described in Drosophila, where the product of dpp is a member of the TGFfl growth factor family. The receptors typically are heterodimers that form transmembrane proteins with serine/threonine kinase activity. The heterodimer consists of a type I component and a type II component. In the dpp pathway, there are two type I members (coded by thick veins and saxophone) and a single type II member (coded by punt). Mutations in tkv and punt have the same phenotype as mutations in dpp, suggesting that the tkv/punt heterodimer is the principal receptor.

The activated receptor phosphorylates the product mad. This is the founding member of the Smad family. The typical pattern of activation in mammalian cells is for the regulated Smad to associate with a general partner to form a heterodimer that is imported into the nucleus, where it activates transcription.

Because the gene is repressed on the ventral side, dpp protein is secreted from cells only across the dorsal side of the embryo, as depicted in **Figure 29.16**. So Dpp is in effect the morphogen that induces synthesis of dorsal structures. Several loci influence the production of Dpp, largely by post-translational mechanisms. The net result is to increase dpp activity on the dorsal side, and to repress it on the ventral side, of the embryo. The concentration of Dpp directly affects the cell phenotype, the most dorsal phenotypes requiring the greatest concentration.

The same pathway is involved in inducing the analogous structures in frog or mouse, but it is inverted with regard to the dorsoventral axis. **Figure 29.17** shows that Bmp4 is secreted from one side of the egg. It is antagonized by a variety of factors. Neural tissues develop in the (dorsal) regions which Bmp4 is prevented from reaching.

The crucial unifying feature is that neural tissues are induced unless the activity of dpp/Bmp4 is antagonized. Typically the Dpp/Bmp diffuses from a source, and different phenotypes may be produced by different concentrations of the morphogen. It is controversial whether the morphogen diffuses extracellularly or whether there may be a relay system that propagates it from cell to cell. Analogous pathways, triggered by different Bmps, are involved in the development of many organs.

Antagonists to Dpp/Bmp may act at the level either of sequestering the ligand itself or of acting upon its receptor. They take the form of large proteinsóchordin, noggin, follistatin in Xenopus. They induce neural tissue by inhibiting Bmp4 action. The antagonists themselves may be regulated, typically by proteolysis. Inactive Bmp/chordin complexes are cleaved by the protease Xolloid to release active Bmp.

Figure 29.18 compares the pathways in fly and frog. The fly pathway is well characterized for dorsoventral development. Mutants in sog and tolloid suggest that their products antagonize and assist Dpp, respectively.

Figure 29.16 The morphogen dpp forms a gradient originating on the dorsal side of the fly embryo. This prevents the formation of neural structures and induces mesenchymal structures.

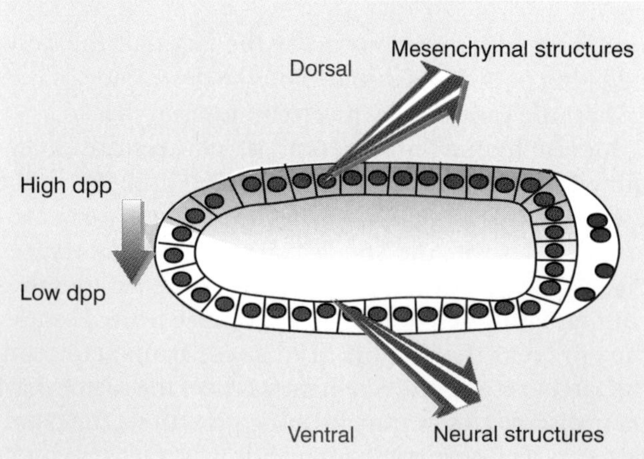

Figure 29.17 Two common pathways are used in early development of Xenopus. The Niewkoop center uses the Wnt pathway to induce the Spemann organizer. The organizer diffuses dorsalizing factors that counteract the effects of the ventralizing BMPs.

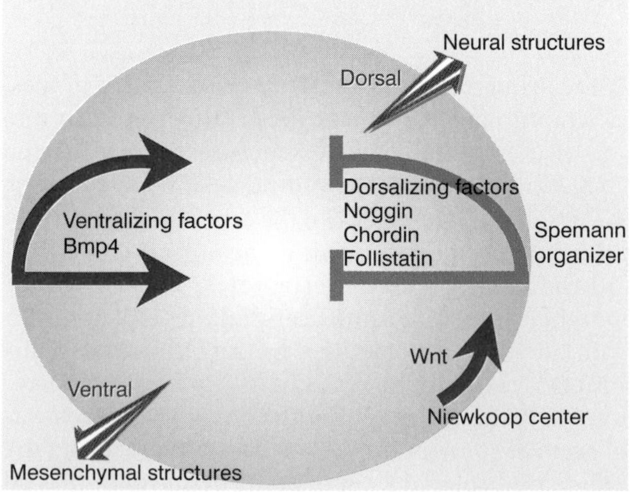

Figure 29.18 The TGFβ/Bmp signaling pathway is conserved in evolution. The ligand may be sequestered by an antagonist, which is cleaved by a protease. Ligand binds to a dimeric receptor, causing the phosphorylation of a specific Smad, which together with a Co-Smad translocates to the nucleus to activate gene expression.

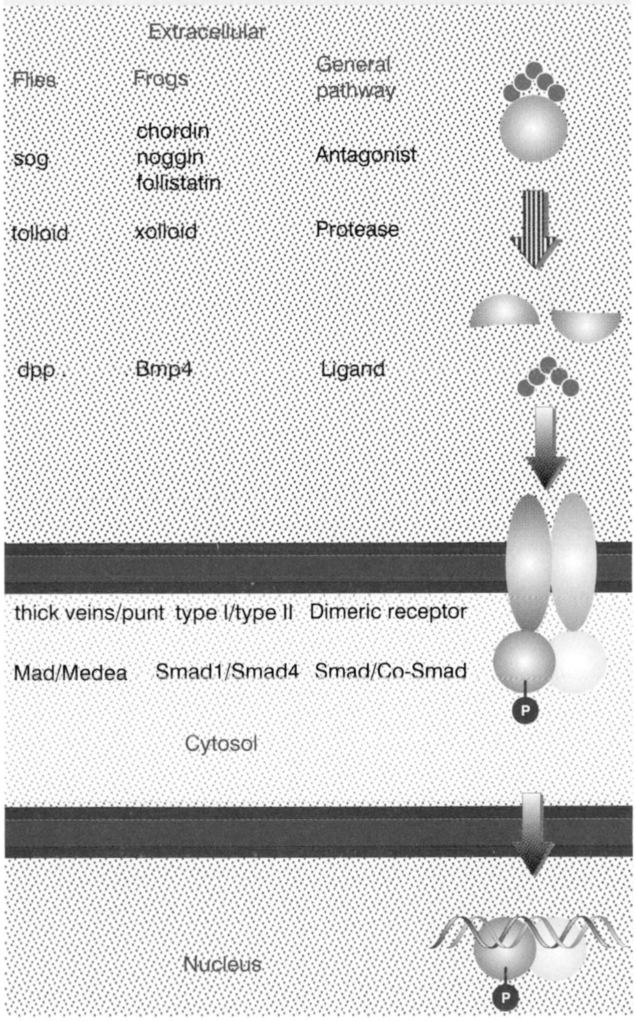

The biochemical reactions actually have been better characterized for the corresponding frog proteins (Chordin is related to Sog). Frogs may have several such pathways, with a variety of Bmp ligands that interact in an overlapping manner with a family of receptors. The frog pathway shown in the figure is for the ventralizing effects of Bmp4, but the others are similar, although their specific effects on morphogenetic determination are of course different. There can be variation in specificity at each stage of the pathway. The antagonists, ligands, and receptors may be expressed in different places and times, providing specificity with regard to local concentrations of the morphogen, but there may also be partial redundancy. Both noggin and chordin must be knocked out in mouse to produce a phenotype. Each receptor has specificity for certain Smads, so that different target genes can be activated in different tissues.

Cell fate is determined by compartments that form by the blastoderm stage

A "fate map" of the embryo can be drawn at the blastoderm stage to identify regions in terms of the adult segments that will develop from the descendants of the embryonic cells. This is possible because, by this stage, cells have begun to acquire information about the pathways they will follow and the structures they will therefore form. This information derives from the maternal regulators described above, and then it is further refined by the actions of zygotic genes. The concept that is intrinsic in the fate map is that a region identified at blastoderm consists of a "compartment" of cells that will give rise specifically to a particular adult structure.

We can consider the development of *D. melanogaster* in terms of the two types of unit depicted in **Figure 29.19:** the segment and parasegment.

- The **segment** is a visible morphological structure. The adult fly consists of a series of clearly demarcated segments, and the larva has a series of corresponding segments separated by grooves. We are concerned primarily with the three thoracic (T) and eight abdominal (A) segments, about whose development most is known. The pattern of segmental units is determined by blastoderm, when the main mass of the embryo is divided into a series of alternating anterior (A) and posterior (P) compartments. So a segment consists of an A compartment succeeded by a P compartment; segment A3, for example, consists of compartments A3A and A3P.
- Another type of classification originates earlier, when divisions can first be seen at gastrulation. The embryo can be divided into **parasegments,** each consisting of a P compartment succeeded by an A compartment. Parasegment 8, for example, consists of compartments A2P and A3A. In the 5–6-hour embryo, shallow grooves on the surface separate the adjacent parasegments. When segments form at around 9 hours, the grooves deepen and move, so that each segmental boundary represents the center of a parasegment. So the anterior part of the segment is derived from one parasegment, and the posterior part of the segment is derived from the next parasegment. In effect, the segmental units are initially evident as P-A pairs in parasegments, and then are recognized as A-P pairs in segments.

Figure 29.19 *Drosophila* development proceeds through formation of compartments that define parasegments and segments.

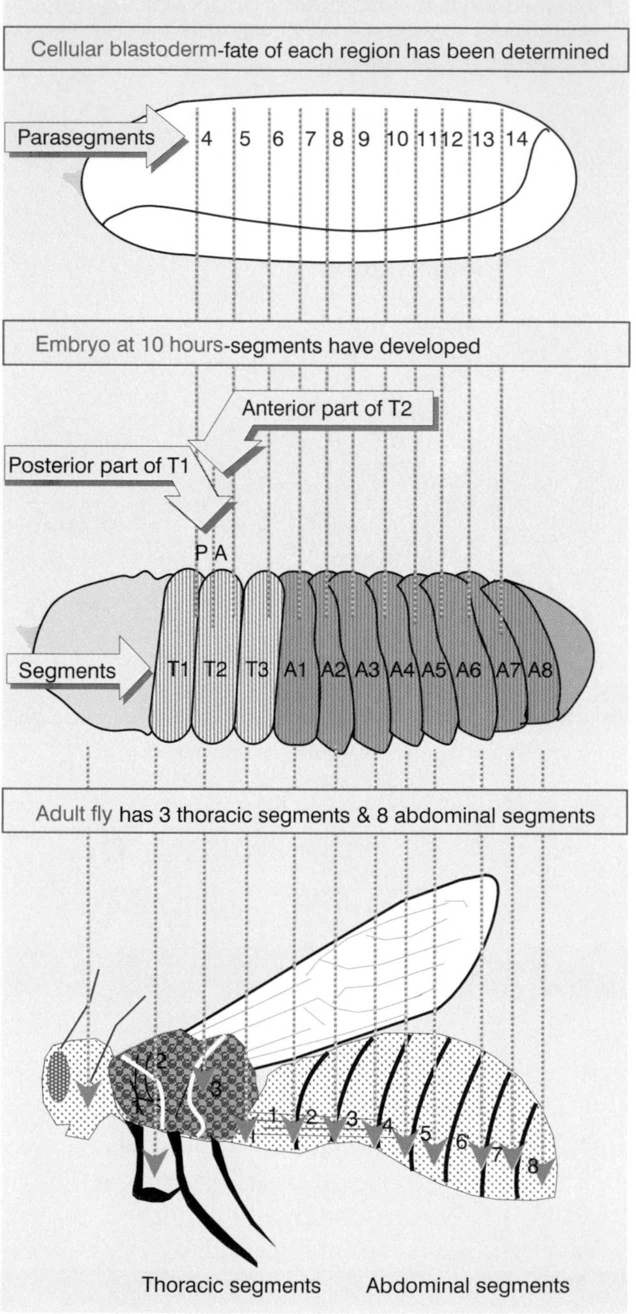

How are these compartments defined during embryogenesis? The general nature of segmentation mutants suggests that the functions of segmentation genes are to establish "rules" by which segments form; *a mutation changes a rule in such a way as to cause many or all segments to form improperly.* The drastic consequences of segment malformation make these mutants embryonic lethals—they die at various stages before metamorphosis into adults.

Probably ~30 loci are involved in segment formation. **Figure 29.20** shows that they can be classified according to the size of the unit that they affect:

- **Gap** mutants have a group of several adjacent segments deleted from the final pattern. Four gap genes are involved in formation of the major body segments, and others are concerned with the head and tail structures.
- **Pair-rule** mutants have corresponding parts of the pattern deleted in every other segment. The afflicted

Figure 29.20 Segmentation genes affect the number of segments and fall into three groups.

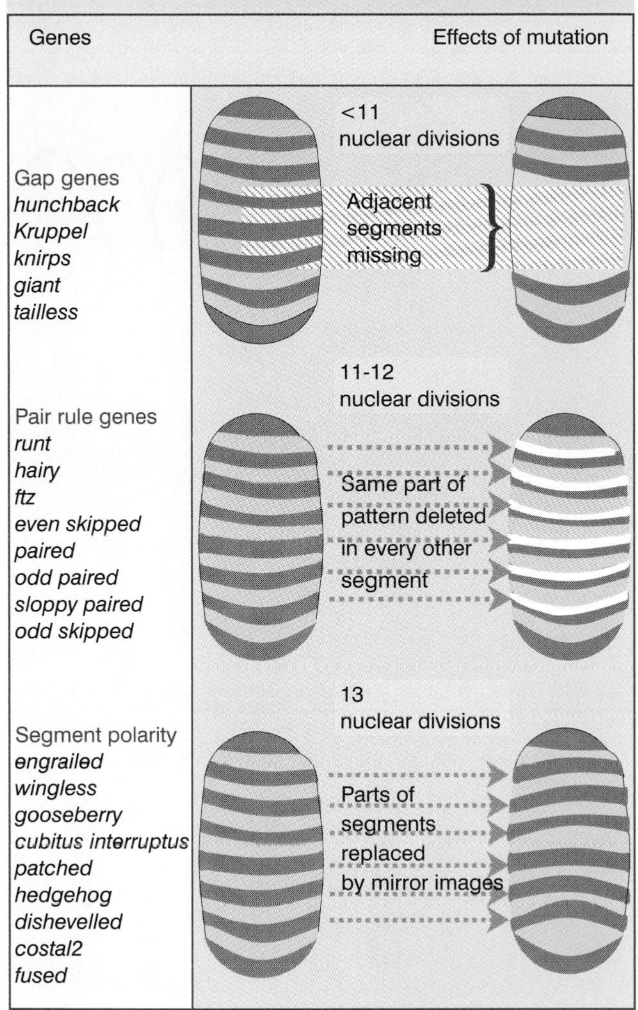

Figure 29.21 Maternal and segmentation genes act progressively on smaller regions of the embryo.

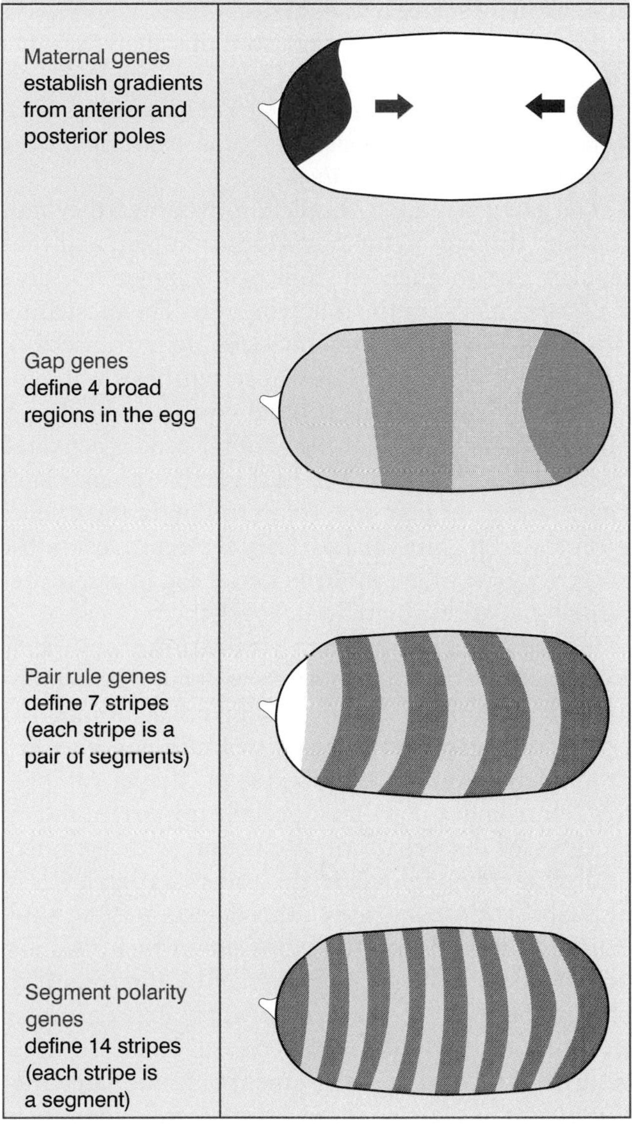

segments may be even-numbered or odd-numbered. There are 8 pair-rule genes.

- **Segment polarity** mutants most often lose part of the P compartment of each segment, and it is replaced by a mirror image duplication of the A compartment. Some mutants cause loss of A compartments or middle segments. There are ~16 segment polarity genes.

These groups of genes are expressed at successive periods during development; and they define increasingly restricted regions of the egg, as can be seen from **Figure 29.21.** The maternal genes establish gradients from the anterior and posterior ends. The maternal gradients either activate or repress the gap genes, which are amongst the earliest to be transcribed following fertilization (following the eleventh nuclear division); they divide the embryo into four broad regions. The gap genes regulate the pair-rule genes, which are transcribed slightly later; their target regions are restricted to *pairs* of segments. The pair-rule genes in turn regulate the segment polarity genes, which are expressed during the thirteenth nuclear division, and by now the target size is the *individual* segment.

Many of the maternal genes, the gap genes, and the pair-rule genes are regulators of transcription. Their effects may be either to activate or to repress transcription; in some cases, a given protein may activate some target genes and repress other target genes, depending on its level or the context. The genes in any one class regulate one another as well as regulating the genes of the next class. It is only when we reach the level of segment polarity genes that products are found that act in other ways; for example, secretion of a protein from one cell to influence its neighbor.

The principle that emerges from this analysis is that at each stage *a small number of maternal, gap, and pair-rule regulator proteins is used in combinatorial associations to specify the pattern of gene expression in a particular region of the embryo.*

The gap genes are controlled in two ways: they may respond directly to the bicoid morphogen; and they regulate one another. The four bands shown in Figure 29.21 are created in the following way. The most anterior band consists of hunchback protein, transcribed from the hunchback RNA whose synthesis was activated by bicoid. The next band consists of Kruppel protein; transcription of the *Kruppel* gene is activated by hunchback protein. The next two bands consist of knirps and giant proteins. Transcription of these genes is repressed by hunchback. They are expressed in the posterior part of the embryo because nanos has prevented the expression of hunchback there.

Figure 29.22 examines the transition from the 4-band to the 7-striped stage in more detail. The detailed interactions among the gap proteins are determined by examining the pattern of the distribution of other gap proteins in a mutant lacking one particular gap protein. Hunchback plays an especially important role. It is expressed in a broad anterior region, with a gradient of decline in the middle of the embryo. High levels of hunchback repress *Kruppel;* this determines the anterior boundary of Kruppel expression, which rises just as hunchback falls off, in parasegment 3. But some level of hunchback is needed for *Kruppel* expression, so when the level of hunchback decreases further, *Kruppel* is turned off, around parasegment 5. In the same way, expression of *giant* responds to successive changes in the level of hunchback; and *knirps* expression requires the absence of hunchback.

The control is refined further by interactions among the proteins. *The general principle is that one interaction may be required to express a protein in a particular region, and other interactions may be required to repress its expression at the boundaries.* The effects are worked out by examining pairwise interactions. For example, over-expression of *giant* causes the Kruppel band to become much narrower, suggesting that giant contributes to repressing the boundaries of Kruppel. The posterior margins of knirps and giant are determined by the operation of the terminal system. Altogether, these interactions mean that, as we proceed along the egg from anterior to posterior, any particular position can be defined by the levels of the various gap proteins.

Figure 29.22 Expression of the gap genes defines adjacent regions of the embryo. The gap genes control the pair-rule genes, each of which is expressed in 7 stripes.

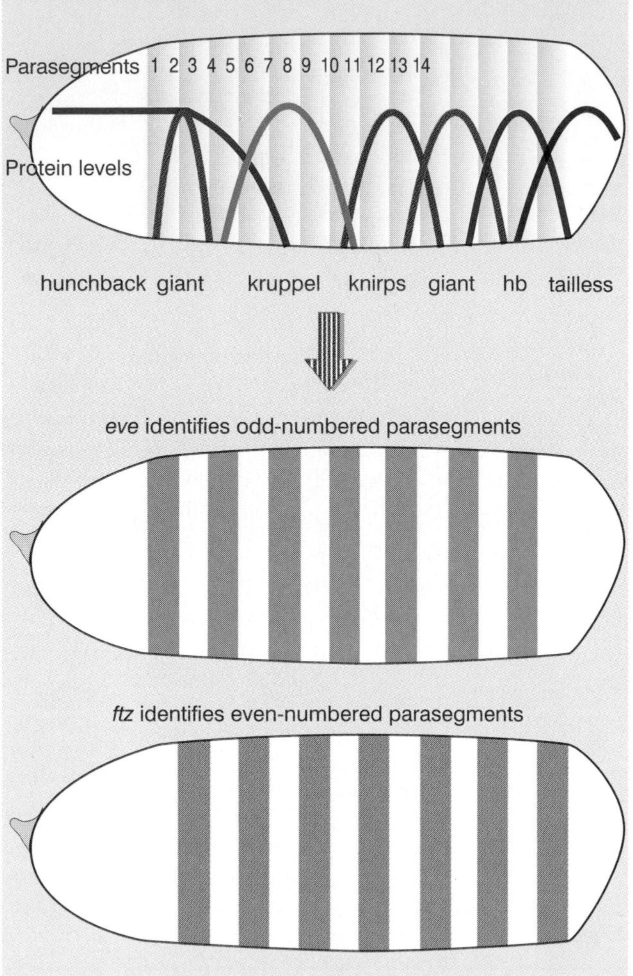

All the gap proteins are regulators of transcription, and in addition to regulating one another, they regulate the expression of the pair-rule genes that function at the next stage. Each pair-rule protein is found in a pattern of 7 "stripes" along the embryo, and Figure 29.22 shows the approximate positions that these stripes will take as the result of expression of the gap genes. (Of course, the parasegments have not developed yet, and are shown just to relate their positions to the protein distribution.) The 7 stripes of a pair-rule gene identify either all the odd-numbered parasegments (like *eve*) or all the even-numbered parasegments (like *ftz*). Two of the pair-rule genes, *hairy* and *eve,* are called primary pair-rule genes, because they are expressed first, and their pattern of expression influences the expression of the other pair-rule genes.

Recall that mutations in pair-rule genes delete half the segments. **Figure 29.23** compares the segmentation patterns of wild-type and *fushi tarazu* (*ftz*) larvae. The mutant has only half the number of segments, because every other segment is missing.

The *ftz* mRNA is present from early blastoderm to gastrula stages of development. **Figure 29.24** shows the locations of the transcripts, visualized *in situ* at blastoderm in wild type. *The gene is expressed in 7 stripes, each 3–4 cells wide, running across the embryo.* As shown previously in Figure 29.22, the stripes correspond to even-numbered parasegments (4 = T1P/T2A, 6 = T3P/A1A, 8 = A2P/A3A, etc.).

This pattern suggests a function for the *ftz* gene: *it must be expressed at blastoderm for the structures that will be descended from the even-numbered parasegments to develop.* Mutants in which *ftz* is defective lack these parasegments because the gene product is absent during the period when they must be formed. In other words, expression of *ftz* is required for survival of the cells in the regions in which it is expressed.

The expression of *ftz* is an example of the general rule that the stripes in which a pair-rule gene is expressed correspond to the regions that are missing from the embryo when the gene is mutated. *Compartments are therefore determined by the pattern of expression of segmentation genes.* The width of the stripe in which a gene is expressed corresponds to the size of the segmental unit that it affects. Different mechanisms are used to specify the expression patterns of different pair-rule genes; we have the most information about *ftz* and *eve*.

Figure 29.23 *ftz* mutants have half the number of segments present in wild-type. Photographs kindly provided by Walter Gehring.

Wild type

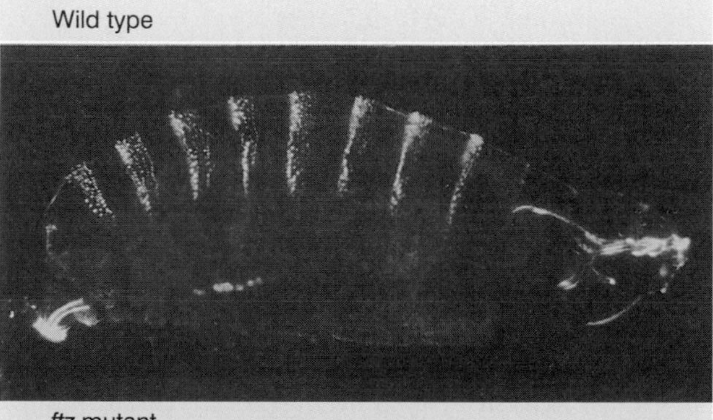

ftz mutant

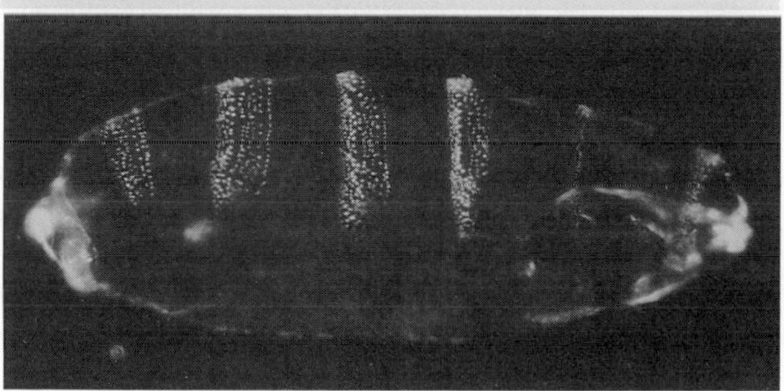

Figure 29.24 Transcripts of the *ftz* gene are localized in stripes corresponding to even numbered parasegments. The expressed regions correspond to the regions that are missing in the *ftz* mutant of Figure 29.23. Photograph kindly provided by Walter Gehring.

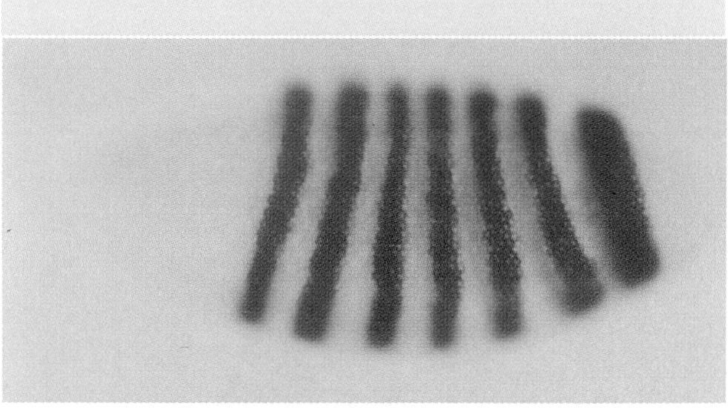

In the early embryo, *ftz* is uniformly expressed. If protein synthesis is blocked before the stripes develop, the embryo retains the initial pattern. So the development of stripes depends on the specific degradation of *ftz* RNA in the regions between the bands and at the anterior and posterior ends of the embryo. Once the stripes have developed, transcription of *ftz* ceases in the interbands and at the ends of the embryo. The specificity of transcription depends on regions upstream of the *ftz* promoter, and also on the function of several other segmentation genes. The transcription of *ftz* responds to other pair-rule genes (and perhaps gap genes) through elements that act on all stripes.

The expression pattern of *eve* is complementary to *ftz*, but has a different basis: it is controlled separately in each stripe. A detailed reconstruction using subregions of the *eve* promoter shows that the information for localization in each stripe is coded in a separate part of the promoter; the promoter can be divided into regions that correspond to the local levels of gap gene products, in particular parasegments. For example, the promoter region that is responsible for *eve* expression in parasegment 3 has binding sites for the gap proteins bicoid, hunchback, giant, and Kruppel. **Figure 29.25**

Figure 29.25 The eve stripe in parasegment 3 is activated by hunchback and bicoid. Repression by giant sets the anterior boundary; repression by Kruppel sets the posterior boundary. Mulltiple binding sites for these proteins in a 480 bp region of the promoter control expression of the gene.

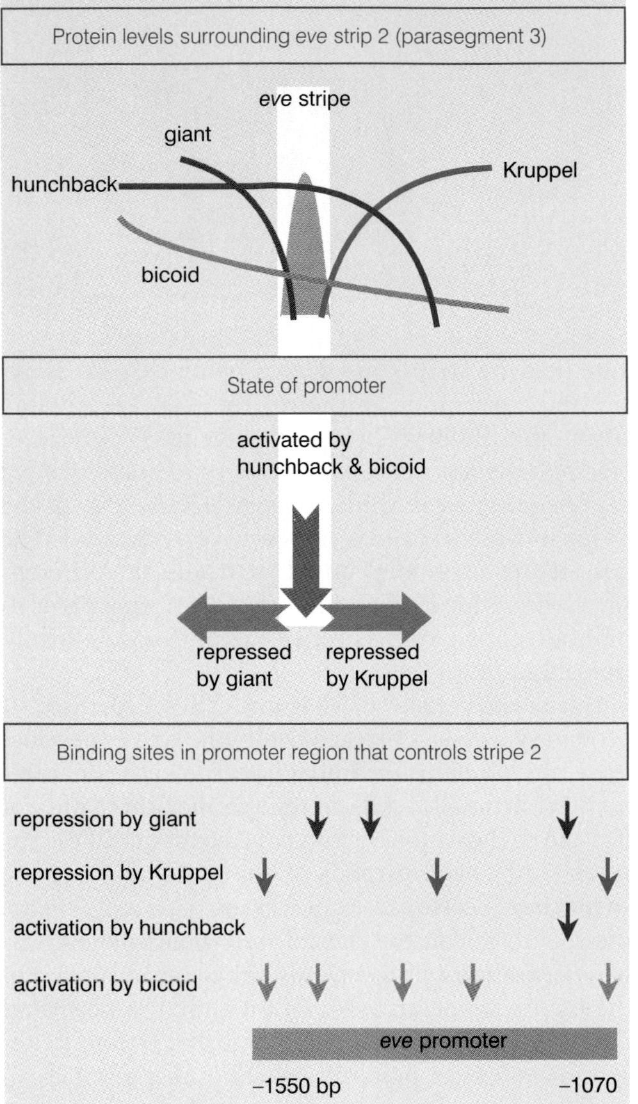

shows that this part of the promoter extends for 480 bp. It works in the following way. *eve* transcription is activated by hunchback and bicoid. The two boundaries are determined because the promoter is repressed by giant on the anterior side and by Kruppel on the posterior side (see also Figure 29.22). Other parts of the promoter respond to the protein levels in other parts of the embryo. So the different stripes of the primary pair-rule gene products are regulated by separate pathways, each of which is susceptible to activation by a particular combination of gap gene products and other regulators.

This illustrates in miniature the principle that combinations of proteins control gene expression in local areas. The general principle is that generally distributed proteins (such as bicoid or hunchback) are needed for activation, whereas the borders are formed by selective repression (by giant and Kruppel in this particular example). We should emphasize that the hierarchy of gene control is not exclusively restricted to interactions between successive stages of control (maternal → gap → pair-rule). For example, the involvement of bicoid protein in regulating *eve* transcription in parasegment 3 shows that a maternal gene may have a direct effect on a pair-rule gene.

The stripes of *eve* and *ftz* are fuzzy to begin with, and become sharper as development proceeds, corresponding to more finely defined units. **Figure 29.26** shows an example of an embryo simultaneously stained for expression of *ftz* and *eve.* Initially there is a series of alternating fuzzy stripes, but the stripes narrow from the posterior margin and sharpen on the anterior side as they intensify during development. This may depend on an autoregulatory loop, in which the expression of the gene is regulated by its own product.

The pair-rule genes control the expression of the segment polarity genes, which are expressed in 14 stripes. Each stripe identifies a segment. The compartmental pattern in which segment polarity genes are expressed is exceedingly precise. Perhaps the ultimate demonstration of precision is provided by the pattern gene *engrailed.* The function coded by *engrailed* is needed in all segments and is concerned with the distinction between the A and P compartments. *engrailed* is expressed in every P compartment, but not in A compartments. Mutants in this gene do not distinguish between anterior and posterior compartments of the segments.

Antibodies against the protein coded by *engrailed* react against the nucleus of cells expressing it. The regions in which *engrailed* is expressed form a pattern of stripes. When the stripes of engrailed protein first become apparent, *they are only one cell wide.* **Figure 29.27** shows the pattern at a stage when each segment has a stripe just 1 cell in width, with the stripe beginning to widen into several cells.

In fact, the pattern of stripes becomes established over a 30 minute period, moving along the embryo from anterior to posterior. Initially one stripe is apparent; then every other segment has a stripe; and finally the complete pattern has a stripe 3–4 cells wide corresponding to the P compartment of every segment.

Figure 29.26 Simultaneous staining for *ftz* (brown) and *eve* (grey) shows that they are first expressed as broad alternating stripes at the time of blastoderm (upper), but narrow during the next 1 hour of development (lower). Photographs kindly provided by Peter Lawrence.

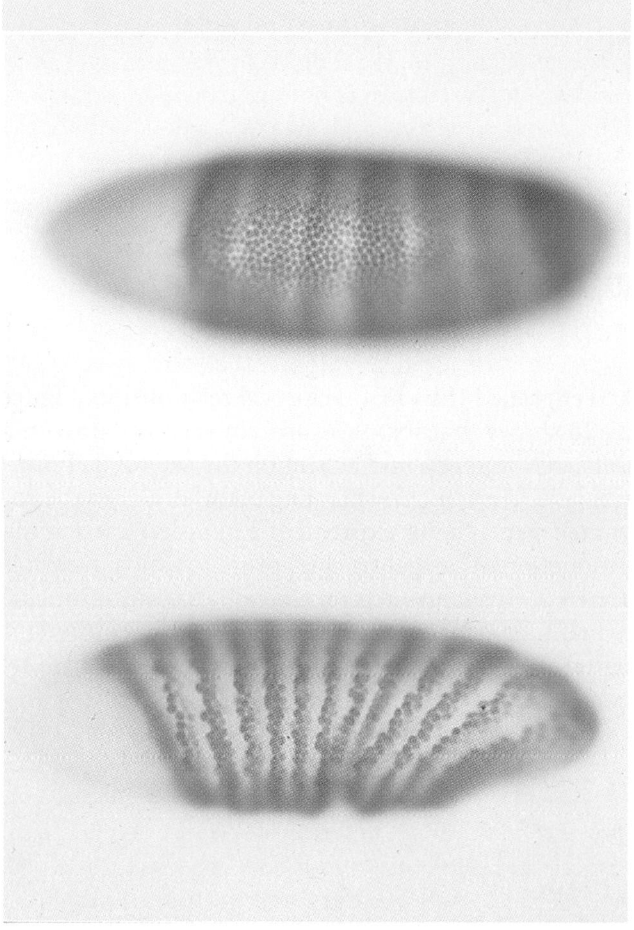

The expression of *engrailed* is of particular importance, because it defines the boundaries of the actual compartments from which adult structures will be derived. The initial 1-cell-wide stripes of engrailed protein form at the anterior boundaries of both the *ftz* and *eve* stripes, and delineate what will become the anterior boundary of every P compartment. Why is *engrailed* initially transcribed exclusively in this anterior edge, within the broader stripes of *ftz* and *eve* expression? This question is a specific example of a more general question: how can a broad stripe be subdivided into more restricted, narrower stripes? We can consider two general types of model:

■ A combinatorial model supposes that different genes are expressed in overlapping patterns of stripes. A pattern of stripes develops for each of the pair-rule genes. The different pair-rule gene stripes overlap, because they are out of phase with one another. As a result of these patterns, different cells in the cellular blastoderm express different combinations of pair-rule genes. Each compartment is defined by the particular combination of the genes that are expressed, and these combinations determine the responses of the cells at the next stage of development. In other words, the segmentation genes are controlled by the pair-rule genes in the same general manner that the pair-rule genes are controlled by the gap genes.

■ A boundary model supposes that a compartment is defined by the striped pattern of expression, but that interactions involving cell-cell communication at the boundaries cause subdivisions to arise within the compartment. In the case of *engrailed,* we would suppose that some unique event is triggered by the juxtaposition of cells possessing ftz (or eve) with cells that do not, and this is necessary to trigger *engrailed* expression.

Figure 29.27 Engrailed protein is localized in nuclei and forms stripes as precisely delineated as 1cell in width. Photograph kindly provided by Patrick O'Farrell.

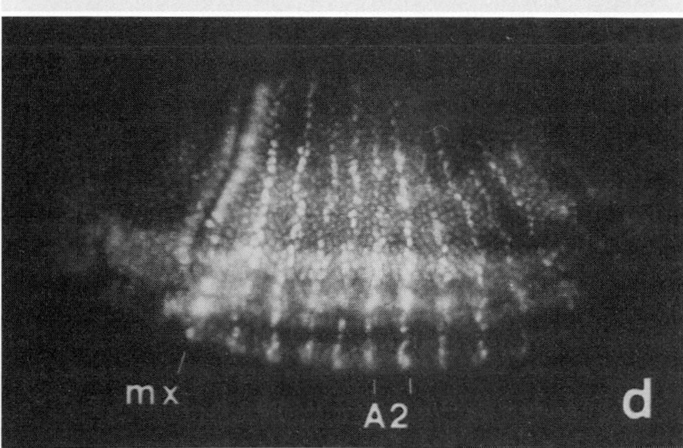

Each of the 14 segments is subdivided further by functions of the segment polarity genes. The type of subdivision imposed by a segment polarity gene is repeated in every segment. For example, *engrailed* distinguishes the A and P compartments. *engrailed* is a transcription factor, but other segment polarity genes have different types of functions. *wingless* codes for a protein that is secreted and taken up by nearby cells, so this allows cell-cell interactions to play a role. *wingless* is initially expressed in a row of cells immediately adjacent to the anterior side of the cells expressing *engrailed;* thus *wingless* comes to identify the posterior boundary of the preceding parasegment. The initial expression of *engrailed* in response to ftz and eve is shortly replaced by an autoregulatory loop in which secretion of wingless protein from the adjacent cells is needed for expression of *engrailed;* and expression of *engrailed* is needed for expression of *wingless*. This keeps the boundary sharp. We do not yet have a good understanding of how the other segment polarity genes function to delineate the segments or parts of segments. However, their products include secreted proteins, transmembrane proteins, kinases, cytoskeletal proteins, as well as transcription factors, suggesting that at this stage cell-cell communications become important.

The wingless/wnt signaling pathway

The wingless signaling pathway is one of the most interesting, and has close parallels in all animal development. Like other signaling pathways utilized in development, it is initiated by an extracellular ligand, and results in the expression of a transcription factor, although the interactions between components of the pathway are somewhat unusual.

In fly embryonic development at the stage of segmental definition, the cells that define the boundaries of the A and P compartments express wingless (Wg) and engrailed (En) in a reciprocal relationship. **Figure 29.28** shows that wingless protein is secreted from a cell at a boundary, and acts upon the cell on its posterior side. The *wingless* signaling pathway causes the *engrailed* gene to be expressed. Engrailed causes the production of hedgehog (Hg) protein, which in turn is secreted. Hedgehog acts on the cell on its anterior side to maintain *wingless* expression. Wg is also required for patterning of adult eyes, legs, and wings (hence its name).

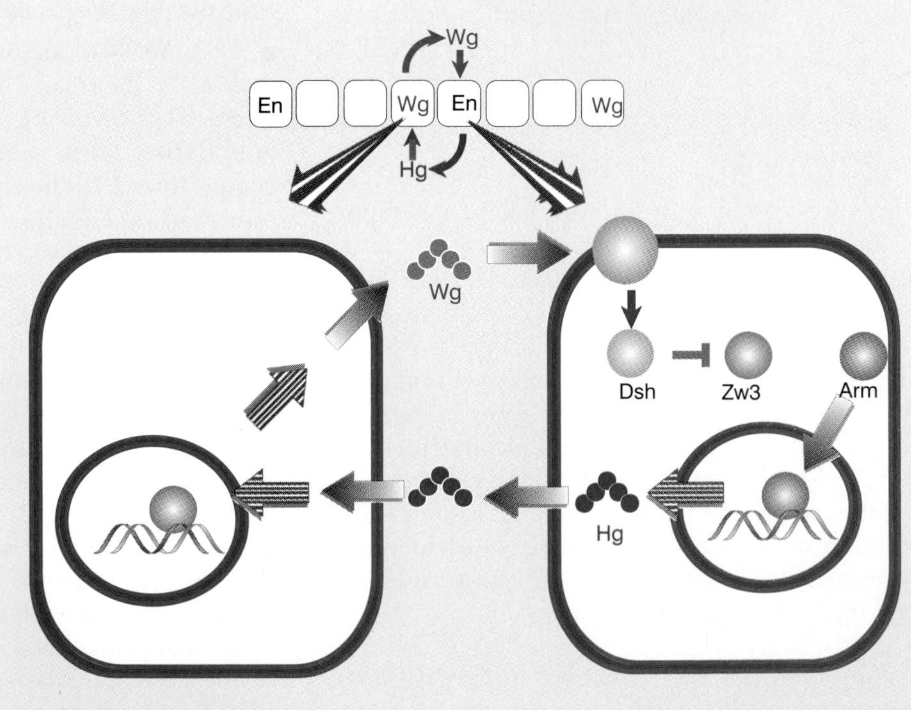

Figure 29.28
Reciprocal interactions maintain Wg and Hg signaling between adjacent cells. Wg activates a receptor, which activates a pathway leading to translocation of Arm to the nucleus. This leads to expression of Hedgehog protein, which is secreted to act on the neighboring cell, where it maintains Wg expression.

Mutants in other genes that have segment polarity defects similar to wg mutants identify other components of the pathway. They signal positively to execute the pathway. Mutants in *zw3* have an opposite phenotype; *zw3* functions to block the pathway. **Figure 29.29** shows the results of ordering the genes genetically and defining the biochemical interactions between their products.

Figure 29.29 Wg secretion is assisted by porc. Wg activates the Dfz2 receptor, which inhibits Zw3 kinase. Active Zw3 causes turnover of Arm. Inhibition of Zw3 stabilizes Arm, allowing it to translocate to the nucleus. In the nucleus, Arm partners Pan, and activates target genes (including *engrailed)*. A similar pathway is found in vertebrate cells (components named in blue).

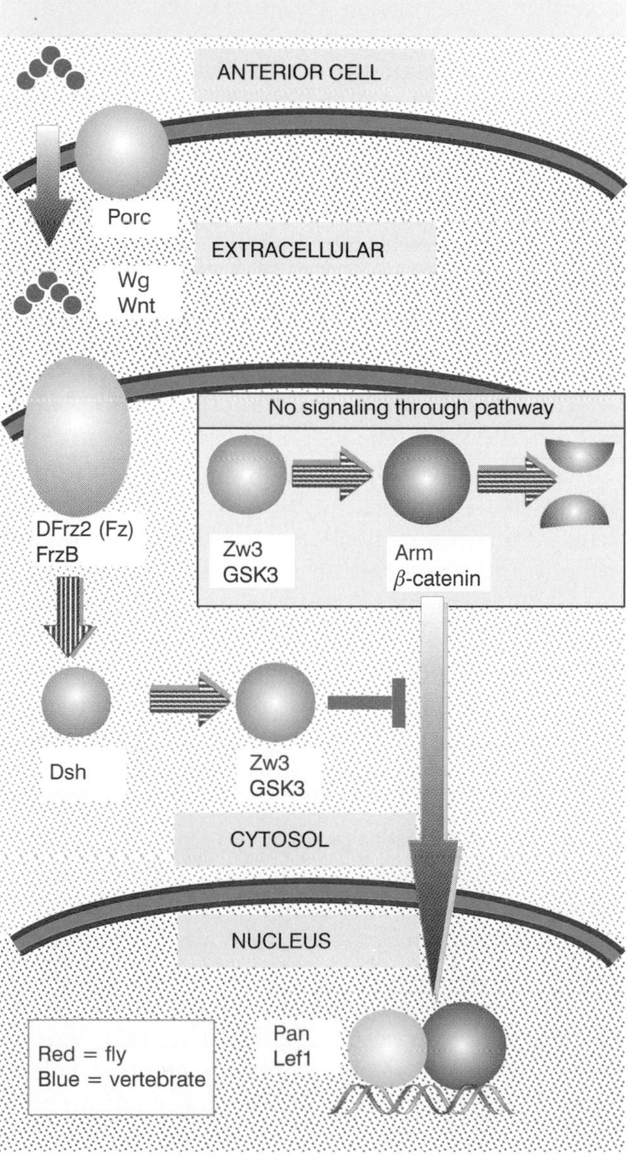

Porc (coded by *porcupine*) is a transmembrane protein that is required to assist the secretion of Wg from the anterior cell. The identification of the receptor for Wg on the posterior cell has actually been very difficult. Wg interacts with frizzled *in vitro*, but mutational analysis suggests that the related protein, DFz2 (*Drosophila frizzled-2*) may be the receptor. It is possible that these may play redundant roles and/or that some other unidentified coreceptor is necessary. The frizzled family members are 7–membrane pass proteins, with the appearance of classical receptors (although there is no evidence for an involvement of G proteins).

Whatever the details of the ligand-receptor interaction, it results in activation of Dsh (coded by *Dishevelled*). Dsh is a cytosolic phosphoprotein. Its detailed action is not known, but, directly or indirectly, it inhibits the Ser/Thr kinase Zw3. When Zw3 is active, it inhibits the action of the next component in the pathway, Arm (coded by *Armadillo*). The inhibition of Zw3 releases the action of Arm, which translocates to the nucleus. Arm binds to its partner Pan, and activates a set of target genes.

The pathway in vertebrate cells is basically the same. The ligands that trigger the pathway are members of the Wnt family (which has ~8 members). They are related to Wg. Different members of the Wnt family are expressed in different times and places in the vertebrate embryo, where the pathway is involved in patterning events. One example was shown previously in Figure 29.17, where the Niewkoop center triggers a Wnt pathway with the effect of inducing the Spemann organizer.

The transduction pathway has homologous components. Dsh signals to the kinase GSK3 (named for its historical identification as glycogen synthase kinase, but in fact a homologue of Zw3). GSK3 is constitutively active, but the result of Dsh's action is to inhibit it. This allows β-catenin (the homologue of Arm) to translocate to the nucleus, where it partners the transcription factor Tcf/LEF1 (named depending on the system). The β-catenin subunit activates transcription at promoters that are bound by the Tcf subunit. The homeobox gene siamois is a major target.

The most surprising feature of this pathway is the nature of the Arm/β-catenin protein. It has two unconnected activities. It is a component of a complex that links the cytoskeleton at adhesion complexes. β-catenin binds to cadherin. Mutations of armadillo that disrupt the cadherin-binding site show a defect in cell adhesion. A separate domain of Arm/β-catenin has a transactivation function when the protein translocates to the nucleus. Nuclear translocation appears to be a consequence of the increase in the levels of Arm/

β-catenin that result from the inhibition of Zw3/GSK3. The phosphorylation of Arm/β-catenin destabilizes it, probably by making it a target for proteasomal degradation.

This signaling pathway is also implicated in colon cancer. Mutations in *APC* (adenomatous polyposis coli) are common in colon cancer. APC binds to β-catenin, and its usual effect is to destabilize it. The mutant proteins found in colon cancer allow levels of β-catenin to increase. One possibility is that APC is the direct target for GSK3, and that its phosphorylation causes it to destabilize β-catenin. However, it is not yet clear whether APC is required for morphogenetic pathways in flies and vertebrates.

Complex loci are extremely large and involved in regulation

SEGMENT polarity genes control the pattern within each segment, including the polarity. **Homeotic genes** impose the program that determines the unique differentiation of each segment. Homeotic mutations cause cells of one compartment to develop the phenotype of a different cell compartment. Some homeotic mutants "transform" part of a segment or an entire segment into another type of segment. Most homeotic genes are expressed in a spatially restricted manner that corresponds to parasegments.

Homeotic genes interact in complicated interlocking patterns. Many homeotic genes code for transcription factors that act upon other homeotic genes as well as upon other target loci. As a result, a mutation in one homeotic gene influences the expression of other homeotic genes. The consequence is that *the final appearance of a mutant depends not only on the loss of one homeotic gene function, but also on how other homeotic genes change their spatial patterns in response to the loss.*

Homeotic genes act during embryogenesis. Their expression depends on the prior expression of the segmentation genes; we might regard the homeotic genes as integrating the pattern of signals established by the segmentation genes. Homeotic mutations cause one segment of the abdomen to develop as another, legs to develop in place of antennae, or wings to develop in place of eyes. Note that homeotic genes do not *create* patterns *de novo;* they modify cell fates that are determined by genes such as the segment polarity genes, by switching the set of genes that functions in a particular place. Indeed, the segment polarity genes are active at about the same time as the peak of expression of the homeotic genes.

The genetic properties of some homeotic mutations are unusual and led to the identification of **complex loci.** A conventional gene—even an interrupted one—is identified at the level of the genetic map by a cluster of noncomplementing mutations. In the case of a large gene, the mutations might map into individual clusters corresponding to the exons. A hallmark of complex loci is that, in addition to rather well-spaced groups of mutations, extending over a relatively large map distance, there are complex patterns of complementation, in which some pairwise combinations complement but others do not. The individual mutations may have different and complex morphological effects on the phenotype. These relationships are caused by the existence of an array of regulatory elements extending far beyond the transcription units they control, which may be extremely large. Many of the bizarre results that are obtained in complementation assays turn out to result from mutations in promoters or enhancers that affect expression in one cell type but not another. We now recognize that complex loci do not have any novel features of genetic organization, apart from the fact that they have many regulatory elements that control expression in different parts of the embryo.

Two of the complex loci are involved in regulating development of the adult insect body. The *ANT-C* and *BX-C* complex loci together provide a continuum of functions that specify the identities of all of the segmented units of the fly. Each of these complexes contains several homeotic genes. The two separate complexes may have evolved from a split in a single ancestral complex, as suggested by the evolution of the corresponding genes in other species. In the beetle

Figure 29.30 The homeotic genes of the *ANT-C* complex confer identity on the most anterior segments of the fly. The genes vary in size, and are interspersed with other genes. The *antp* gene is very large and has alternative forms of expression.

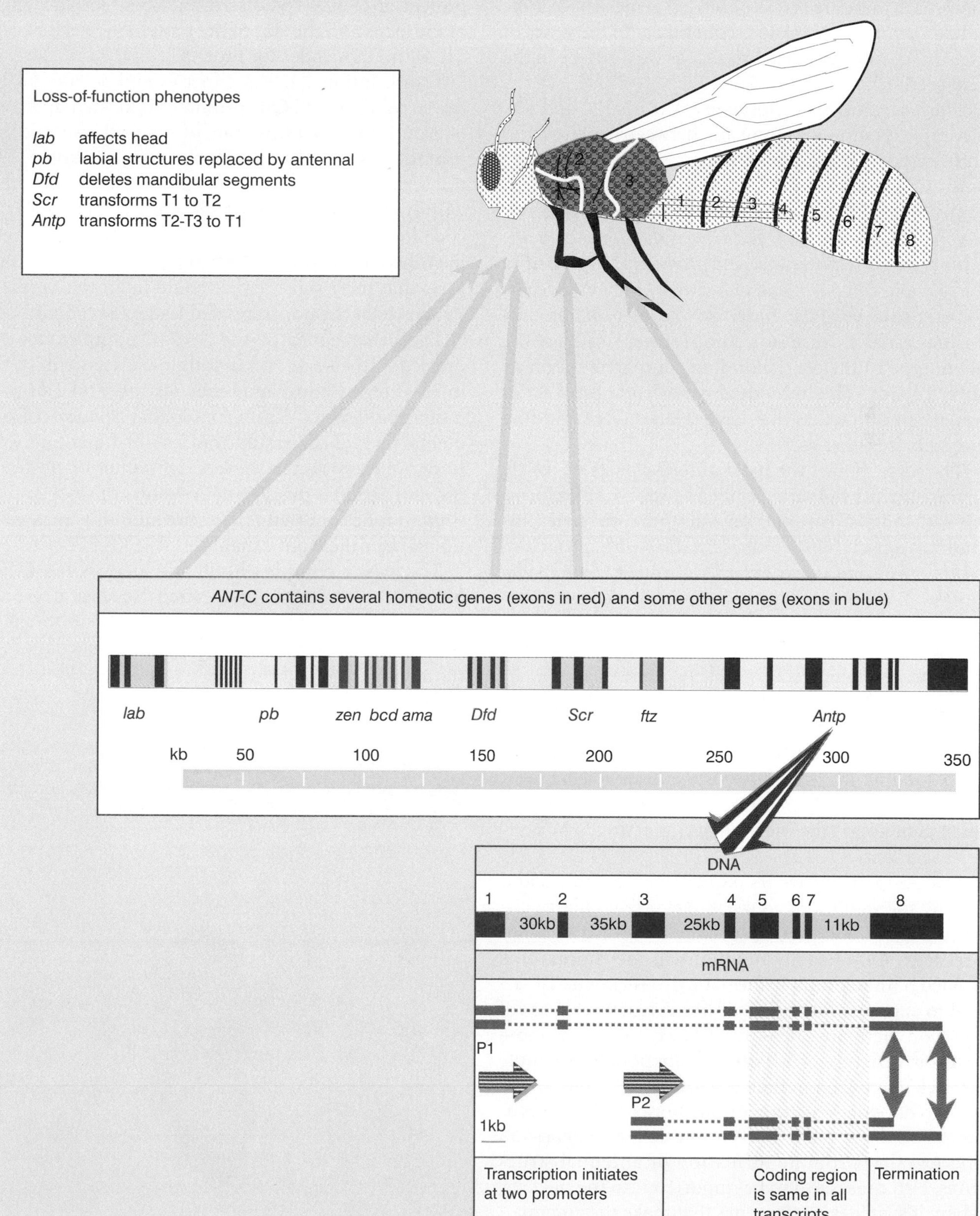

Tribolium, the *ANT-C* and *BX-C* complexes are found together at a single chromosomal location. The individual genes may have been derived from duplications and mutations of an original ancestral gene. And in mammals, there are arrays of related genes whose individual members are related sequentially to the genes of the *ANT-C* and *BX-C* complexes (as we describe in the next section).

The homeotic genes clustered at the *ANT-C* and *BX-C* complexes show a relationship between genetic order and the position in which they are expressed in the body of the fly. *Proceeding from left to right, each homeotic gene in the complex acts upon a more posterior region of the fly. The basic principle is that formation of a compartment requires the gene product(s) expressed in the previous compartment, plus a new function coded by the next gene along the cluster.* So loss-of-function mutations usually cause one compartment to have the phenotype of the corresponding compartment on its anterior side. The individual genes code for a set of transcription factors that have related DNA-binding domains (see next section).

The identities of the most anterior parts of the fly (parasegments 1–4) are specified by the complex locus, *ANT-C*. *ANT-C* contains several homeotic genes, including *labial (lab)*, *proboscipedia (pb)*, *Deformed (Dfd)*, *Sex combs reduced (Scr)*, and *Antennapedia (Antp)*. The homeotic genes lie in a cluster over a region of ~350 kb, but several other genes are interspersed; most of these genes are regulators that function at different stages of development.

Figure 29.30 correlates the organization of *ANT-C* with its effects upon body parts. Adjacent genes are expressed in successively more posterior parts of the embryo, ranging from the leftmost gene *labial* (the most anterior acting, which affects the head) to the rightmost gene *Antp* (the most posterior acting, which affects segments T2–T3).

The *Antp* gene gave its name to the complex, and among the mutations in it are alleles that change antennae into second legs, or second and third legs into first legs. *Antp* usually functions in the thorax; it is needed both to promote formation of segments T2–T3 and to suppress formation of head structures. Loss of function therefore causes T2–T3 to resemble the more anterior structure of T1; gain of function, for example by over-expression in the head, causes the anterior region to develop structures of the thorax. (The molecular action of *Antp* is to prevent the action of genes *hth* and *exd* that promote formation of antennal structures. Hth causes exd to be imported into the nucleus, where it switches on the genes that make the antenna.)

The figure summarizes the organization of the gene. It has 8 exons, separated by very large introns, and altogether spans ~103 kb. The single open reading frame begins only in exon 5, and apparently gives rise to a protein of 43 kD. The discrepancy between the length of the locus and the size of the protein means that only 1% of its DNA codes for protein.

Transcription starts at either of two promoters, located ~70 kb apart! One promoter is located upstream of exon 1, the other upstream of exon 3. Use of the first promoter is associated with omission of exon 3. The transcripts generated from either promoter end either within or after exon 8. All the transcripts appear to code for the same protein. No difference has yet been identified in the use of the two promoters, which suggests that their significance could lie in the different structures of the nontranslated leaders of the mRNAs.

The other genes of the *ANT-C* complex are expressed in the head and first thoracic segment. In the most anterior compartments, *lab*, *pb*, and *Dfd* have unique patterns of expression, so that deletions of segmental regions can result from loss-of-function mutations. An exception to the left-right/anterior-posterior order of action is that loss of *Scr* allows the overlapping *Antp* to function; that is, the direction of transformation is opposite from usual.

The classic complex homeotic locus is *BX-C*, the **bithorax complex,** characterized by several groups of homeotic mutations that affect development of the thorax, causing major morphological changes in the abdomen. When the whole complex is deleted, the

Figure 29.31 A four-winged fly is produced by a triple mutation in *abx*, *bx*, and *pbx* at the *BX-C* complex. Photograph kindly provided by Ed Lewis.

Figure 29.32 The *bithorax* (*BX-C*) locus has 3 coding units. A series of regulatory mutations affects successive segments of the fly. The sites of the regulatory mutations show the regions within which deletions, insertions, and translocations confer a given phenotype.

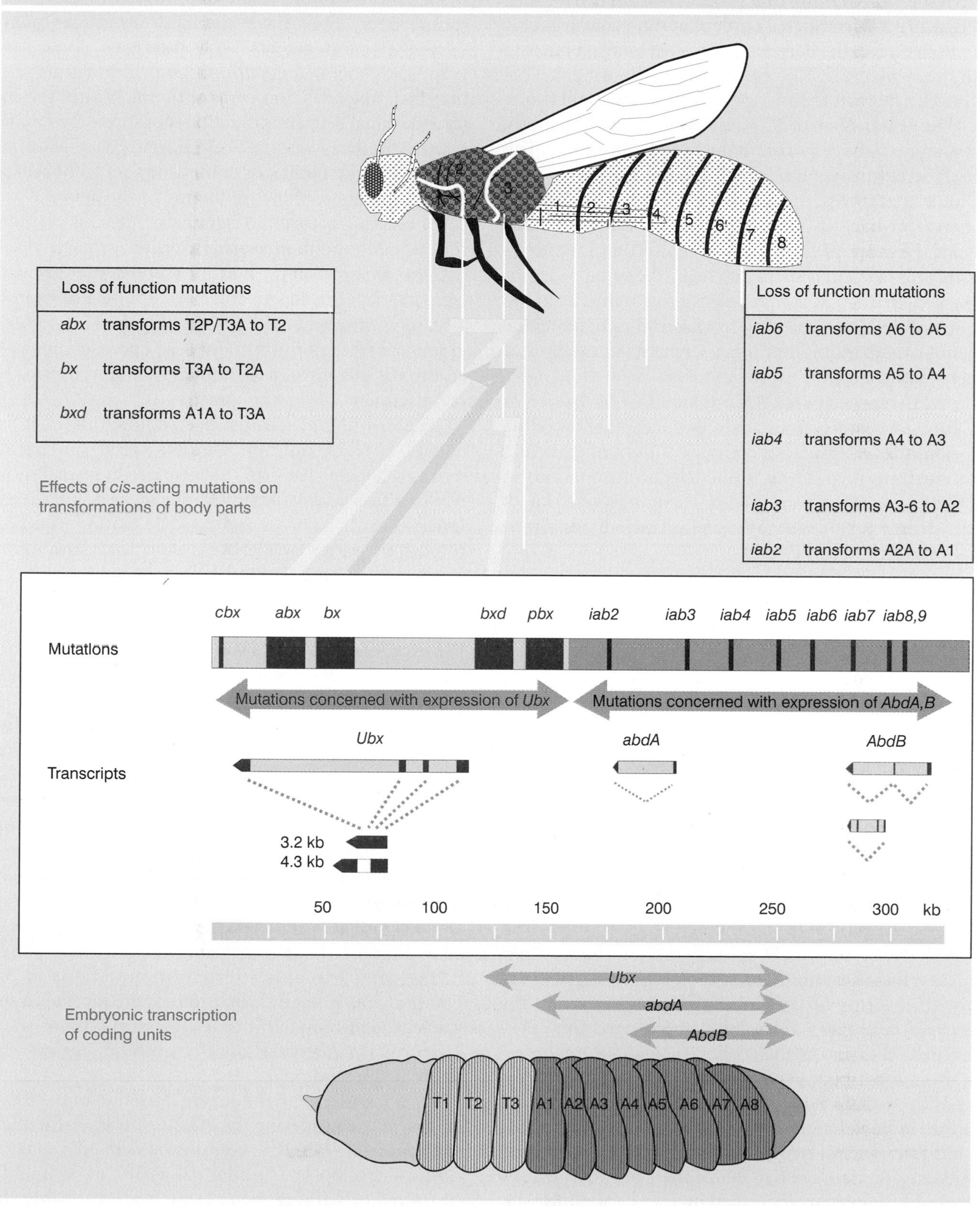

insect dies late in embryonic development. Within the complex, however, are mutations that are viable, but which change the phenotype of certain segments. An extreme case of homeotic transformation is shown in **Figure 29.31,** in which a triple mutation converts T3A (which carries the halteres [truncated wings]) into the tissue type of the T2 (which carries the wings). This creates a fly with four wings instead of the usual two.

The genetic map of *BX-C* is correlated with the body structures that it controls in the fly in **Figure 29.32.** The body structures extend from T2 to A8. The *BX-C* complex is therefore concerned with the development of the major part of the body of the fly. Like *ANT-C*, a crucial feature of this complex is also that mutations affecting particular segments lie in the same order on the genetic map as the corresponding segments in the body of the fly. Proceeding from left to right along the genetic map, mutations affect segments in the fly that become successively more posterior.

A difference between *ANT-C* and *BX-C* is that *ANT-C* functions largely or exclusively via its protein-coding loci, but *BX-C* displays a complex pattern of *cis*-acting interactions in addition to the effects of mutations in protein-coding regions. The *BX-C* occupies 315 kb, of which only 1.4% codes for protein. The individual mutations fall into two classes:

- Three transcription units (*Ubx, abdA, AbdB*) produce mRNAs that code for proteins. The transcription units are large (>75 kb for *Ubx*, and >20 kb each for *abdA* and *AbdB*). Each contains several large introns. (The *bxd* and *iab4* regions produce RNAs that do not code for proteins; again, the transcription units are large, and the RNA products are spliced. Their functions are unknown.)
- There are *cis*-acting mutations at intervals throughout the entire cluster. They control expression of the transcription units. *Cis*-acting mutations of any particular type may occur in a large region. The locations shown on the map are only approximate, and the boundaries within which mutations of each type may occur are not well defined.

As a historical note, the complex was originally defined in terms of two "domains." Mutations in the *Ultrabithorax* domain were characterized first; they have the thoracic segments T2P–T3P and the abdominal compartment A1A as their targets (this corresponds to parasegments 5–6). These mutations lie either in the *Ubx* transcription unit or in the *cis*-acting sites that control it. The mutations within the ultrabithorax domain are named for their phenotypes. The *bx* and *bxd* types are identified by a series of mutations, in each case dispersed over ~10 kb. The *abx* and *pbx* mutations are caused by deletions, which vary from 1 to 10 kb.

Mutations in the *Infraabdominal* domain were found later; they have the abdominal segments A1P–A8P (parasegments 7–14) as targets.) These mutations lie either in the *AbdA,B* transcription units or in the *cis*-acting sites that control them. Within the infraabdominal domain, *cis*-acting mutations are named systematically as *iab2–9*. These mutations affect individual compartments, or sometimes adjacent sets of compartments, as shown at the top of the figure.

Proceeding from left to right along the cluster, transcripts are found in increasingly posterior parts of the embryo, as shown at the bottom of the figure. The patterns overlap. *Ubx* has an anterior boundary of expression in compartment T2P (parasegment 5), *abdA* is expressed from compartment A1P (parasegment 7), and *AbdB* is expressed in compartments posterior to compartment A4P (parasegment 10).

Transcription of *Ubx* has been studied in the most detail. The *Ubx* transcription unit is ~75 kb, and has alternative splicing patterns that give rise to several short RNAs. A transient 4.7 kb RNA appears first, and then is replaced by RNAs of 3.2 and 4.3 kb. A feature common to both the latter two RNAs is their inclusion of sequences from both ends of the primary transcript. Of course, there may be other RNAs that have not yet been identified.

We do not yet have a good idea of whether there are significant differences in the coding functions of these RNAs. The first and last exons are quite lengthy, but the interior exons are rather small. Small exons from within the long transcription unit may enter mRNA products by means of alternative splicing patterns. So far, however, we do not know of any functional differences in the Ubx proteins produced by the various modes of expression.

Ubx proteins are found in the compartments that correspond to the sites of transcription, that is in T2P–A1 and at lower levels in A2–A8. So the Ubx unit codes for a set of related proteins that are concentrated in the compartments affected by mutations in the *Ultrabithorax* domain. Ubx proteins are located in the nucleus, and they fall into the general type of transcriptional regulators whose DNA-binding region consists of a homeodomain.

We can understand the general function of the *BX-C* complex by considering the effects of loss-of-function mutations. If the entire complex is deleted, the larva cannot develop the individual types of segments. In terms of parasegments (which are probably the affected units), all the parasegments differentiate in the same way as parasegment 4; the embryo has 10 repetitions of the repeating structure T1P/T2A all along its length, in place of the usual compartments between parasegments 5 and 14. In effect, the absence of *BX-C* functions allows *Antp* to be expressed throughout the abdomen, so that all the segments take on the characteristic of a segment determined by *Antp*; *BX-C* functions are needed to add more posterior-type information.

Each of the transcription units affects successive segments, according to its pattern of expression. So if *Ubx* alone is present, the larva has parasegment 4 (T1P/T2A), parasegment 5 (T2P/T3A), and then 8 copies of parasegment 6 (T3P/A1A). This suggests that the expression of *Ubx* is needed for the compartments anterior to A1A. *Ubx* is also expressed in the more posterior segments, but in the wild type, *abdA* and *AbdB* are also present. If they are removed, the expression of *Ubx* alone in all the posterior segments has the same effect that it usually has in parasegment 6 (T3P/A1A).

The addition of *abdA* to *Ubx* adds the wild-type pattern to parasegments 7, 8, and 9. In other words, *Ubx* plus *abdA* can specify up to compartments A3P/A4A, and in the absence of *AbdB*, this continues to be the default pattern for all the more posterior compartments. The addition of *AbdB* is needed to specify parasegments 10–14.

The general model for the function of the *ANT-C* and *BX-C* complexes is to suppose that additional functions are added to define successive segments proceeding in the posterior direction. It functions by reliance on a *combinatorial pattern* in which the addition of successive gene products confers new specificities. This explains the rule that a loss-of-function mutation in one of the genes of the *ANT-C/BX-C* complexes generally allows the gene on the more anterior side of the mutated gene to determine phenotype, that is, loss-of-function results in homeotic transformation of posterior regions into more anterior phenotypes.

Expression of *Ubx* in a more anterior segment than usual should have the opposite effect to a loss-of-function; the segment develops a more anterior phenotype. When this is tested by arranging for *Ubx* to be expressed in the head, the anterior segments are converted to the phenotype of parasegment 6. So lack of expression of *Ubx* causes a homeotic transformation in which posterior segments acquire more anterior phenotypes; and over-expression of *Ubx* causes a homeotic transformation in which anterior segments acquire more posterior phenotypes.

This type of relationship is true generally for the cluster as a whole, and explains the properties of *cis*-acting mutations as well as those in the transcription units. These regulatory mutations cause loss of the protein in part of its domain of expression or cause additional expression in new domains. So they may have either loss-of-function or gain-of-function phenotypes (or sometimes both). The most common is loss-of-function in an individual compartment. For example, *bx* specifically controls expression of *Ubx* in compartment T3A; a *bx* mutation loses expression of *Ubx* in that compartment, which is therefore transformed to the more anterior type of T2A. This example is typical of the general rule for individual *cis*-acting mutations in the complex; each converts a target compartment *so that it develops as though it were located at the corresponding position in the previous segment.* The order of the *cis*-acting sites of mutation on the chromosomes reflects the order of the compartments in which they function. So the expression of *Ubx* in parasegments 4, 5, 6 is controlled sequentially by *abx* (affects parasegment 5), *bx* (affects T3A), etc.

The presence of only 3 genes within the *BX-C* complex poses two major questions. First, how do the combinations of 3 proteins specify the identity of 10 parasegments? One possibility is that there are quantitative differences in the various regions, allowing for the same sort of varying responses in target genes that we described previously for the combinatorial functioning of the segmentation genes. Second, how do the proteins function in different tissue types? The pattern of expression described above refers generally to the epidermis; the development of other tissues is controlled in a way that is parallel, but not identical. For example, although *Ubx* is expressed in all posterior segments up to A8 in the epidermis, in mesoderm, it is repressed posterior of segment A7. The posterior boundary reflects repression by *abdA*, since in *abdA* mutants, *Ubx* expression extends posterior in the mesoderm.

Why are loci involved in regulating development of the adult insect from the embryonic larva different from genes coding for the everyday proteins of the organism? Is their enormous length necessary to generate the alternative products? Could it be connected with some timing mechanism, determined by how long it takes to transcribe the unit? At a typical rate of transcription, it would take ~100 minutes to transcribe *Antp*, which is a significant proportion of the 22 hour duration of *D. melanogaster* embryogenesis.

Proceeding from anterior to posterior along the embryo, we encounter the changing patterns of expression of the genes of the *ANT-C* and *BC-C* loci. What controls their transcription? As in the case of the segmentation loci, the homeotic loci are controlled *partially by the genes that were expressed at the previous stage of development, and partially by interactions among themselves.* For example, the expression of *Ubx* is changed by mutations in *bicoid, hunchback,* or *Kruppel.* The anterior boundary of expression respects the parasegment border defined by *ftz* and *eve.* The general principle is that all of these regulatory genes function by controlling transcription, either by activating it or by repressing it, and that the gene products may exert specific effects by both qualitative and quantitative combinations.

The homeobox is a common coding motif in homeotic genes

THE three groups of genes that control *D. melanogaster* development—maternal genes, segmentation genes, and homeotic genes—regulate one another and (presumably) target genes that code for structural proteins. Interactions between the regulator genes have been defined by analyses that show defects in expression of one gene in mutants of another. However, we have identified rather few of the structural targets on which these groups of genes act to cause differentiation of individual body parts.

Consistent with the idea that the segmentation genes code for proteins that regulate transcription, the genes of 3 gap loci (*hb, Kr, kni*) contain zinc finger motifs. As first identified in the transcription factors TFIIIA and Sp1 (see Figure 21.3), these motifs are responsible for making contacts with DNA. The products of other loci in the gap class also have DNA-binding motifs; *giant* encodes a protein with a basic zipper motif, and *tailless* encodes a protein that resembles the steroid receptors. This suggests that the general function of gap genes is to function as transcriptional regulators.

Conserved motifs are found in many of the homeotic and segmentation genes. The most common of the conserved motifs is the **homeobox,** a 180 bp region located near the 3′ end in several segmentation and homeotic genes. There are ~40 genes that contain a homeobox, and almost all are known to be involved in developmental regulation. (The homeobox was first identified by its predominance in the homeotic genes, from which it took its name.) The protein sequence coded by the homeobox is called the homeodomain; we described in Chapter 21 how it has been characterized in detail as a DNA-binding motif in transcription factors (see Figure 21.11).

The fly homeodomains fall into several groups. A major group in *Drosophila* consists of the homeotic genes in the *BX-C/ANT-C* complexes; they are called the *Antennapedia* group. Their homeodomains are 70–80% conserved, and usually occur at the C-terminal end of the protein (see Figure 21.9). A distinct homeodomain sequence is found in the related genes *engrailed* and *invected;* it has only 45% sequence conservation with the *Antennapedia* group (see Figure 21.10). Other types of homeodomain sequences are represented in 2–4 genes each.

Many of the *Drosophila* genes that contain homeoboxes are organized into clusters. Three of the homeotic genes in the *BX-C* cluster have homeoboxes, the *ANT-C* complex contains a group of 5 homeotic genes with homeoboxes, and 4 other genes at *ANT-C* also contain homeoboxes. The homeotic genes at *BX-C* and *ANT-C* are sometimes described under the general heading of *HOM-C* genes.

What is the basic function of the *HOM-C* genes in determining identity on the anterior-posterior axis? We assume that homeodomains with different amino acid sequences recognize different target sequences in DNA. Experiments in which regions have been swapped between different proteins suggest that a major part of the specificity of these proteins rests with the homeodomain. However, the ability to bind to a particular DNA target site may not account entirely for their properties. For example, some of these proteins can either activate or repress transcription in response to the context, that is, their actions depend on the set of other proteins that are bound, not just on recognition of the DNA-binding site.

The similarities between the homeodomains of the more closely related members of the group suggest that they could recognize overlapping patterns of target sites. This would open the way for combinatorial effects that could be based on quantitative as well as qualitative differences, that is, there could be competition between proteins with related homeodomains for the same sites. In some cases, different homeoproteins recognize the same target sites on DNA, which poses a puzzle with regard to defining their specificity of action; we assume that there are subtle differences in DNA-binding yet to be discovered, or there are other interactions, such as protein-protein interactions, that play a role.

The homeobox motif is extensively represented in evolution. A striking extension of the significance of homeoboxes is provided by the discovery that a DNA probe representing the homeobox hybridizes with the genomes of many eukaryotes. Genes containing homeoboxes have been characterized in particular in frog, mouse, and human DNA. The frog and mammalian genes are expressed in early embryogenesis, which strengthens the parallel with the fly genes, and suggests the possibility that genes containing homeoboxes are involved in regulation of embryogenesis in a variety of species.

Genes in mammals (and possibly all animals) that are related to the *HOM-C* group have a striking property: like those of the *BX-C/ANT-C* complexes, they are organized in clusters. The individual mammalian genes are called *Hox* genes. A cluster of *Hox* genes may extend 20–100 kb and contain up to 10 genes. Four *Hox* clusters of genes containing homeoboxes have been characterized in the mouse and human genomes. Their organization is compared with the two large fly clusters in **Figure 29.33.**

By comparing the sequences of the homeoboxes (and sometimes other short regions), the mammalian genes can be placed into groups that correspond with the fly genes. This is shown by vertical alignment in the figure. For example, *HoxA4* and *HoxB4* are best related to *Dfd*. When these relationships are defined for the cluster as a whole, it appears that within each cluster we can recognize a series of genes that are related to the genes in the *ANT-C* and *BX-C* clusters. Groups 1–9 in the mammalian loci are defined as corresponding to the genes of the *ANT-C* and *BX-C* loci organized end to end in anterior-posterior orientation. Groups 10–13 appear to have arisen by tandem duplications and divergence of group 9 (the *AbdB* homolog). The corresponding loci in each cluster are sometimes called **paralogs** (for example, *HoxA4* and *HoxB4* are paralogous).

This situation could have arisen if the fly and mammalian loci diverged at a point when there was only a single complex, containing all of the genes that define anterior-posterior polarity. The organism *Amphioxus*, which corresponds to a line of evolution parallel to the vertebrates, has a single *Hox* cluster containing one member of each paralogous group; this appears to be a direct representative of the original cluster. During evolution, the *Drosophila* genes broke into two separate clusters, while the entire group of mammalian genes became duplicated, some individual members being lost from each complex after the duplication.

The parallel between the mouse and fly genes extends to their pattern of spatial expression. The genes within a *Hox* cluster are expressed in the embryo in a manner that matches their organization in the genome. Progressing from the left toward the right end of the cluster drawn in Figure 29.33, genes are expressed in the embryo in locations progressively more restricted to the posterior end. The patterns of expression for fly and mouse are compared schematically in **Figure 29.34.** The domain of expression extends strongly to the posterior boundary shown in the figure, and then tails off into more posterior segments.

These results raise the extraordinary possibility that the clusters of genes not only share a common evolution, but also have maintained a common general function in which genome organization is related to spatial expression in fly and mouse, and there is some correspondence between the homologous genes. The idea of such a relationship is strengthened by the observation that ectopic expression of mouse *HoxD4* or *HoxB6* in *Drosophila* causes homeotic transformations virtually identical to those caused by homeotic expression of *Dfd* or *Antp*, respectively! Since the homology between these mouse and fly proteins rests almost exclusively with their homeodomains, this reinforces the view that these domains determine specificity.

There are some differences in the apparent behavior of the vertebrate Hox clusters and the *ANT-C/BX-C* fly clusters:

- The *Hox* genes are small, and there is a greater number of protein-coding units. The mouse *HoxB* cluster is ~120 kb and contains 9 genes. The connection between genomic position and embryonic expression is analogous to that in *Drosophila*, but describes only the genes themselves; we have no information about *cis*-acting sites. Of course, this may be a consequence of the much greater difficulty in generating mutations in verte-

Figure 29.33 Mouse and human genomes each contain 4 clusters of genes that have homeoboxes. The order of genes reflects the regions in which they are expressed on the anterior-posterior axis. The *Hox* genes are aligned with the fly genes according to homology, which is strong for groups 1, 2, 4, and 9. The genes are named according to the group and the cluster, e.g., *HoxA1* is the most anterior gene in the *HoxA* group. All *Hox*genes are present in both man and mice except for some mouse genes missing from cluster C (indicated by half boxes).

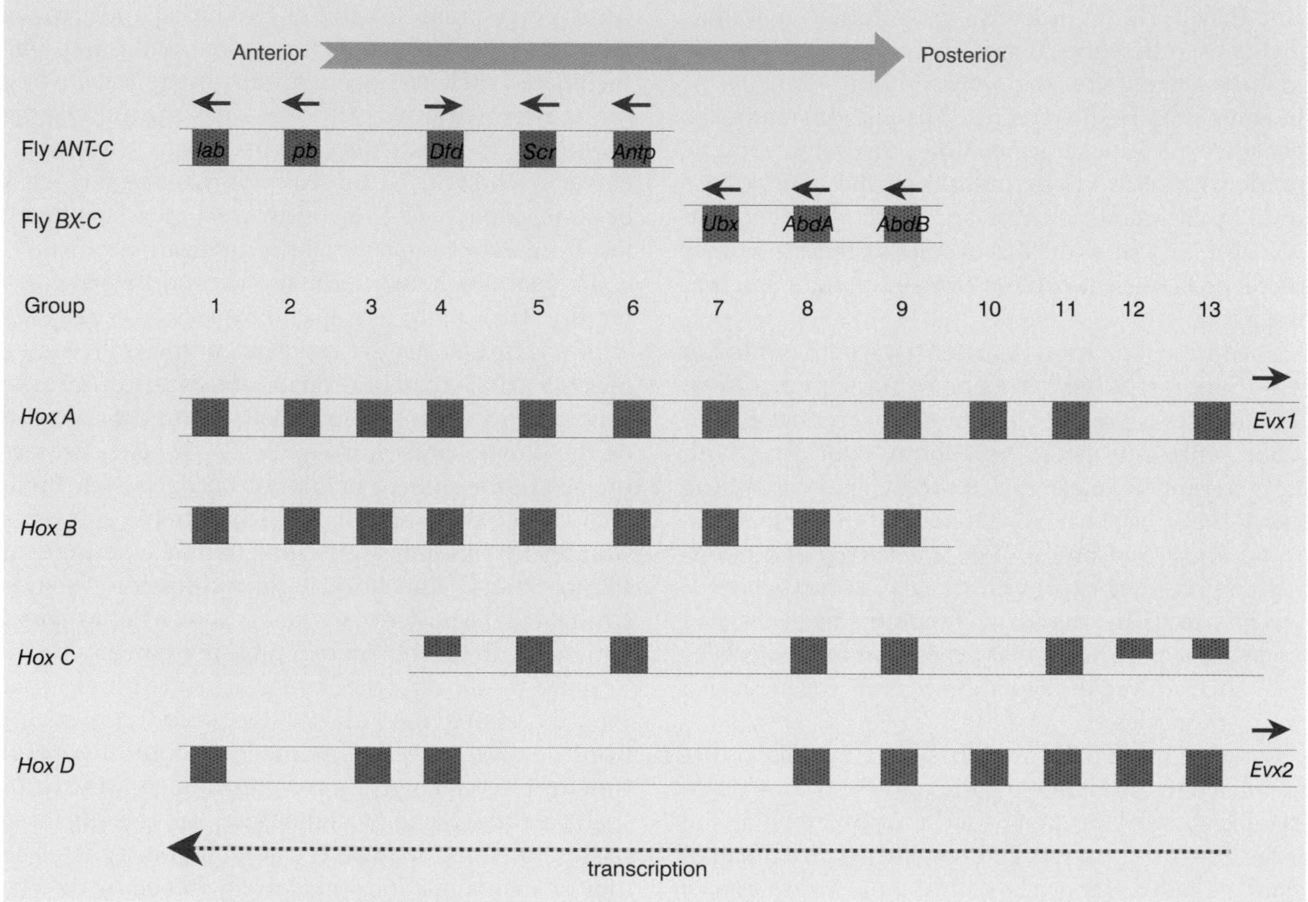

brates. However, our present information identifies the control of *Hox* genes only by promoters and enhancers in the region upstream of the startpoint. It remains to be seen whether there is any counterpart to the very extensive and complex regulatory regions of the *Drosophila* homeotic genes.

■ In *Drosophila*, each gene is unique; but in vertebrates, the duplication of the clusters enables multiple genes (paralogs) to have the same or very similar patterns of expression. If the paralogs have redundant or partially redundant functions, so that the absence of one product may be at least partially substituted by the corresponding protein of another cluster, the effects of mutations will be minimized.

Disruptions of *Hox* genes in mice often generate recessive lethals. In the examples of *HoxA1* and *HoxA3* various structures of the head and thorax are absent. Not all of the structures that usually express the mutant genes are missing, suggesting that there is indeed some functional redundancy, that is, other *Hox* genes of group 1 or group 3 can substitute in some but not all other tissues for the absence of the *HoxA* gene.

Homeotic transformations are less common with mutants in mice than in *Drosophila*, but sometimes occur. Loss of *HoxC8,* for example, causes some skeletal segments to show more anterior phenotypes. It remains to be seen whether this is a general rule.

Ectopic expression of *Hox* genes has been used successfully to demonstrate that gain-of-function can transform the identity of a segment towards the identity usually conferred by the gene. The most common type of effect in *Drosophila* is to transform a segment into a phenotype that is usually more posterior; in ef-

Figure 29.34 A comparison of *ANT-C/BX-C* and *HoxB* expression patterns shows that the individual gene products share a progressive localization of expression towards the more posterior of the animal proceeding along the gene cluster from left to right. Expression patterns show the regions of transcription in the fly epidermis at 10 hours, and in the central nervous system of the mouse embryo at 12 days.

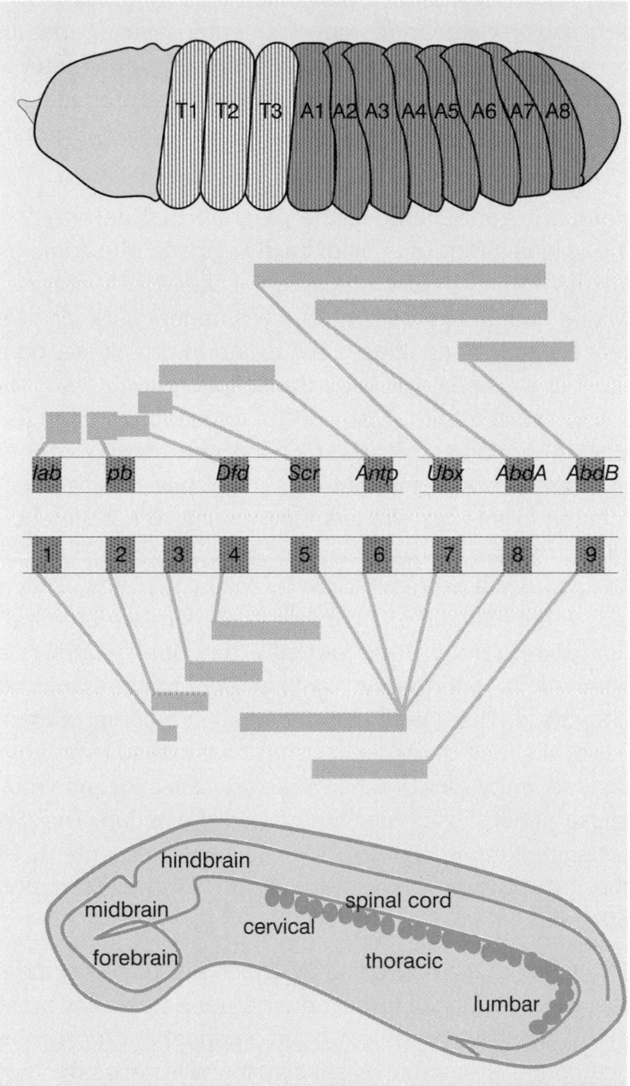

fect, the expression of the homeotic gene has added additional information that confers a more posterior identity on the segment. Similar effects are observed in some cases in the mouse. However, the pattern is not completely consistent.

Taken together, these results make it clear that the *Hox* genes resemble their counterparts in *Drosophila* in determining patterning along the anterior-posterior axis. It may be the case that there is a combinatorial code of *Hox* gene expression, or there may be differences in degree of functional redundancy between paralogs, but we cannot yet provide a systematic model for their role in determining pattern.

The most striking feature of organization of the *Hox* loci still defies explanation: why has the organization of the cluster, in which genomic position correlates with embryonic expression, been maintained in evolution? The obvious explanation is that there is some overall control of gene expression to ensure that it proceeds through the cluster, with the result that a gene could be properly expressed *only* when it is within the cluster. But this does not appear to be true, at least for those individual cases in which genes have been removed from the cluster. Analysis of promoter regions suggests that a *Hox* gene may be controlled by a series of promoter or enhancer elements that together ensure its overall pattern of expression. Usually these elements are in the region upstream of the startpoint. For example, *HoxB4* expression can be reconstructed as the sum of the properties of a series of such elements, tested by introducing appropriate constructs to make transgenic mice. But then why should there have been evolutionary pressure to retain genes in an ordered cluster? One possibility is that an enhancer for one gene might be embedded within another gene, in such a way that, even if an individual gene could function when translocated elsewhere, its removal would impede the expression of other gene(s).

Summary

The development of segments in *Drosophila* occurs by the actions of segmentation genes that delineate successively smaller regions of the embryo. Asymmetry in the distribution of maternal gene products is established by interactions between the oocyte and surrounding cells. This leads to the expression of the gap genes, in 4 broad regions of the embryo. The gap genes in turn control the pair-rule genes, each of which is distributed in 7 stripes; and the pair-rule genes define the pattern of expression of the segment polarity genes, which delineate individual compartments. At each stage of expression, the relevant genes are controlled both by the products of genes that were expressed at the previous stage, and by interactions among themselves. The segmentation genes act upon the homeotic genes, which determine the identities of the individual compartments.

Each of the 4 maternal systems consists of a cascade which generates a locally distributed or locally active morphogen. The morphogen either is a transcription factor or causes the activation of a transcription factor. The transcription factor is the last component in each pathway.

The major anterior-posterior axis is determined by two systems: the anterior system establishes a gradient of bicoid from the anterior pole; and the posterior system produces nanos protein in the posterior half of the egg. These systems function to define a gradient of hunchback protein from the anterior end, with broad bands of knirps and giant in the posterior half. The terminal system acts to produce localized events at both termini. The dorsal-ventral system produces a gradient of nuclear localization of dorsal protein on the ventral side, which represses expression of *dpp* and *zen*; this leads to the ventral activation of *twist* and *snail*, and the dorsal-side activation of *dpp* and *zen*.

The early embryo consists of a syncytium, in which nuclei are exposed to common cytoplasm. It is this feature that allows all 4 maternal systems to control the function of a nucleus according to the coordinates of its position on the anterior-posterior and dorsal-ventral axes. At cellular blastoderm, zygotic RNAs are transcribed, and the developing embryo becomes dependent upon its own genes. Cells form at the blastoderm stage, after which successive interactions involve a cascade of transcriptional regulators.

Three gap genes are zinc-finger proteins, and one is a basic zipper protein. Their concentrations control expression of the pair-rule genes, which are also transcription factors. In particular, the expression of *eve* and *ftz* controls the boundaries of compartments, functioning in every other segment. The segment polarity genes represent the first step in the developmental cascade that involves functions other than transcription factors. Interactions between the segmentation gene products define unique combinations of gene expression for each segment.

Homeotic genes impose the program that determines the unique differentiation of each segment. The complex loci *ANT-C* and *BX-C* each contain a cluster of functions, whose spatial expression on the anterior-posterior axis reflects genetic position in the cluster. Each cluster contains one exceedingly large transcription unit as well as other, shorter units. Many of the transcription units (including the largest genes, *Ubx* and *Antp*) have patterns of alternative splicing, but no significance has been attributed to this yet. Proceeding from left to right in each cluster, genes are expressed in more posterior tissues. The genes are expressed in overlapping patterns in such a way that addition of a function confers new features of posterior identity; thus loss of a function results in a homeotic transformation from posterior to more anterior phenotype. The genes are controlled in a complex manner by a series of regulatory sites that extend over large regions; mutations in these sites are *cis*-acting, and may cause either loss-of-function or gain-of-function. The *cis*-acting mutations tend to act on successive segments of the fly, by controlling expression of the homeotic proteins.

The genes of the *ANT-C* and *BX-C* loci, and many segmentation genes (including the maternal gene *bicoid* and most of the pair-rule genes), contain a conserved motif, the homeobox. Homeoboxes are also found in genes of other eukaryotes, including worms, frogs, and mammals. In each case, these genes are expressed during early embryogenesis. In mammals, the *Hox* genes (which specify homeodomains in the *Antennapedia* class) are organized in clusters. There are 4 *Hox* clusters in both man and mouse. These clusters can be aligned with the *ANT-C/BX-C* clusters in such a way as to recognize homologies between genes at corresponding positions. Proceeding towards the right in a *Hox* cluster, a gene is expressed more towards the posterior of the embryo. The *Hox* genes have roles in conferring identity on segments of the brain and skeleton (and other tissues). The analogous clusters represent regulators of embryogenesis in mammals and flies. *Hox* clusters may be a characteristic of all animals.

Drosophila genes containing homeoboxes form an intricate regulatory network, in which one gene may activate or repress another. The relationship between the sequence of the homeodomain, the DNA target it recognizes, and the regulatory consequences remains to be fully elucidated. Specificity in target choice appears to reside largely in the homeodomain; we have yet to explain the abilities of a particular homeoprotein to activate or to repress gene transcription at its various targets. The general principle is that segmentation and homeotic genes act in a transcriptional cascade, in which a series of hierarchical interactions between the regulatory proteins is succeeded by the activation of structural genes coding for body parts.

Further reading

Historical

Nusslein-Vollhard, C. and Wieschaus, E. (1980). Mutations affecting segment number and polarity in *Drosophila*. *Nature* 287, 795–801.

Reviews

Mahowald, A. P. and Hardy, P. A. (1985). Genetics of *Drosophila* embryogenesis. *Ann. Rev. Genet.* **19,** 149–177.

Regulski, M., Harding, K., Kostriken, R., Karch, F., Levine, M., and McGinnis, W. (1985). Homeo box genes of the Antennapedia and bithorax complexes of *Drosophila*. *Cell* **43,** 71–80.

Scott, M. P. (1987). Complex loci of *Drosophila*. *Ann. Rev. Biochem.* **56,** 195–227.

Scott, M. P. and Carroll, S. B. (1987). The segmentation and homeotic gene network in early *Drosophila* development. *Cell* **51,** 689–698.

Ingham, P. W. (1988). The molecular genetics of embryonic pattern formation in *Drosophila*. *Nature* **335,** 25–34.

Scott, M. P. *et al.* (1989). The structure and function of the homeodomain. *Biochim. Biophys. Acta* **989,** 25–48.

Hunt, P. and Krumlauf, R. (1992). Hox codes and positional specification in vertebrate embryonic axes. *Ann. Rev. Cell Biol.* **8,** 227–256.

Ingham, P. W. and Martinez-Arias, A. (1992). Boundaries and fields in early embryos. *Cell* **68,** 221–235.

Lawrence, P. (1992). *The Making of a Fly.* (Blackwell Scientific, Oxford).

McGinnis, W. and Krumlauf, R. (1992). Homeobox genes and axial patterning. *Cell* **68,** 283–302.

St. Johnston, D. and Nusslein-Volhard, C. (1992). The origin of pattern and polarity in the *Drosophila* embryo. *Cell* **68,** 201–219.

Krumlauf, R. (1994). Hox genes in vertebrate development. *Cell* **78,** 191–201.

Morisato, D. and Anderson, K. V. (1995). Signaling pathways that establish the dorsal-ventral pattern of the *Drosophila* embryo. *Ann. Rev. Genet.* **29,** 371–399.

Discoveries

Bender, W. *et al.* (1983). Molecular genetics of the bithorax complex in *D. melanogaster*. *Science* **221,** 23–29.

Scott, M. P. *et al.* (1983). The molecular organization of the *Antennapedia* locus of *Drosophila*. *Cell* **35,** 763–766.

McGinnis, W. *et al.* (1984). A homologous protein-coding sequence in *Drosophila* homeotic genes and its conservation in other metazoans. *Cell* **37,** 403–408.

Anderson, K. V., Bokla, L., and Nusslein-Volhard, C. (1985). Establishment of dorsal-ventral polarity in the *Drosophila* embryo: the induction of polarity by the *Toll* gene product. *Cell* **42,** 791–798.

Anderson, K. V., Jurgens, G., and Nusslein-Volhard, C. (1985). Establishment of dorsal-ventral polarity in the *Drosophila* embryo: genetic studies on the role of the *Toll* gene product. *Cell* **42,** 779–789.

Beachy, P. A., Helfand, S. L., and Hogness, D. S. (1985). Segmental distribution of bithorax complex proteins during *Drosophila* development. *Nature* **313,** 545–551.

Karch, F. *et al.* (1985). The abdominal region of the bithorax complex. *Cell* **43,** 81–96.

Schupbach, T. (1987). Germ line and soma cooperate during oogenesis to establish the dorsoventral pattern of egg shell and embryo in *Drosophila melanogaster*. *Cell* **49**, 699–707.

Driever, W. and Nusslein-Volhard, C. (1988). A gradient of bicoid protein in *Drosophila* embryos. *Cell* **54**, 83–93.

Driever, W. and Nusslein-Volhard, C. (1988). The bicoid protein determines position in the *Drosophila* embryo in a concentration dependent manner. *Cell* **54**, 95–104.

Graham, A., Papalopulu, N., and Krumlauf, R. (1989). The murine and *Drosophila* homeobox gene complexes have common features of organization and expression. *Cell* **57**, 367–378.

Roth, S., Stein, D., and Nusslein-Volhard, C. (1989). A gradient of nuclear localization of the dorsal protein determines dorsoventral pattern in the *Drosophila* embryo. *Cell* **59**, 1189–1202.

Malicki, J., Schughart, K., and McGinnis, W. (1990). Mouse *hox-2.2* specifies thoracic segmental identity in *Drosophila* embryos and larvae. *Cell* **63**, 961–967.

Sprenger, F. and Nusslein-Volhard, C. (1992). Torso receptor activity is regulated by a diffusible ligand produced at the extracellular terminal regions of the *Drosophila* egg. *Cell* **71**, 987–1001.

St. Johnston, D. and Nusslein-Volhard, C. (1992). The origin of pattern and polarity in the *Drosophila* embryo. *Cell* **68**, 201–219.

Struhl, G., Johnston, P., and Lawrence, P. A. (1992). Control of *Drosophila* body pattern by the hunchback morphogen gradient. *Cell* **69**, 237–249.

Ferrandon, D., Elphick, L., Nusslein-Volhard, C., and St. Johnston, D. (1994). Staufen protein associates with the 3' UTR of bicoid mRNA to form particles that move in a microtubule-dependent manner. *Cell* **79**, 1221–1232.

Garcia-Fernandez, J. and Holland, P. W. H. (1994). Archetypal organization of the amphioxus *Hox* gene cluster. *Nature* **370**, 563–566.

Martin, C. H. *et al.* (1995). Complete sequence of the bithorax complex of *Drosophila. Proc. Nat. Acad. Sci. USA* **92**, 8398–8402.

Glossary

Abundance of an mRNA is the average number of molecules per cell.

Abundant mRNAs consist of a small number of individual species, each present in a large number of copies per cell.

Acceptor splicing site—*see* right splicing junction.

Acentric fragment of a chromosome (generated by breakage) lacks a centromere and is lost at cell division.

Acrocentric chromosome has the centromere located nearer one end than the other.

Active site is the restricted part of a protein to which a substrate binds.

Allele is one of several alternative forms of a gene occupying a given locus on a chromosome.

Allelic exclusion describes the expression in any particular lymphocyte of only one allele coding for the expressed immunoglobulin.

Allosteric control refers to the ability of an interaction at one site of a protein to influence the activity of another site.

Alu-equivalent family is the set of sequences in a mammalian genome that is related to the human Alu family.

Alu family is a set of dispersed, related sequences, each ~300 bp long, in the human genome. The individual members have Alu cleavage sites at each end (hence the name).

α-Amanitin is a bicyclic octapeptide derived from the poisonous mushroom *Amanita phalloides*; it inhibits transcription by certain eukaryotic RNA polymerases, especially RNA polymerase II.

Amber codon is the nucleotide triplet UAG, one of three codons that cause termination of protein synthesis.

Amber mutation describes any change in DNA that creates an amber codon at a site previously occupied by a codon representing an amino acid in a protein.

Amber suppressors are mutant genes that code for tRNAs whose anticodons have been altered so that they can respond to UAG codons as well as or instead of to their previous codons.

Aminoacyl-tRNA is transfer RNA carrying an amino acid; the covalent linkage is between the NH_2 group of the amino acid and either the 3′- or 2′-OH group of the terminal base of the tRNA.

Aminoacyl-tRNA synthetases are enzymes responsible for covalently linking amino acids to the 2′- or 3′-OH position of tRNA.

Amphipathic structures have two surfaces, one hydrophilic and one hydrophobic. Lipids are amphipathic; and some protein regions may form amphipathic helices, with one charged face and one neutral face.

Amplification refers to the production of additional copies of a chromosomal sequence, found as intrachromosomal or extrachromosomal DNA.

Anchorage dependence describes the need of normal eukaryotic cells for a surface to attach to in order to grow in culture.

Aneuploid chromosome constitution differs from the usual diploid constitution by loss or duplication of chromosomes or chromosomal segments.

Annealing is the pairing of complementary single strands of DNA to form a double helix.
Anterograde transport describes the forward movement of proteins from the endoplasmic reticulum through the Golgi to the plasma membrane.
Antibody is a protein (immunoglobulin) produced by B lymphocyte cells that recognizes a particular foreign "antigen," and thus triggers the immune response.
Anticoding strand of duplex DNA is used as a template to direct the synthesis of RNA that is complementary to it.
Antigen is any molecule whose entry into an organism provokes synthesis of an antibody (immunoglobulin).
Antiparallel strands of the double helix are organized in opposite orientation, so that the 5′ end of one strand is aligned with the 3′ end of the other strand.
Antitermination proteins allow RNA polymerase to transcribe through certain terminator sites.
AP endonucleases make incisions in DNA on the 5′ side of either apurinic or apyrimidinic sites.
Apoinducer is a protein that binds to DNA to switch on transcription by RNA polymerase.
Apoptosis is the capacity of a cell to undergo programmed cell death; in response to a stimulus, a pathway is triggered that leads to destruction of the cell by a characteristic set of reactions.
Archeae are a branch of evolution distinct from prokaryotes and eukaryotes.
Ascus of a fungus contains a tetrad or an octad of the (haploid) spores, representing the products of a single meiosis.
att sites are the loci on a phage and the bacterial chromosome at which recombination integrates the phage into, or excises it from, the bacterial chromosome.
Attenuation describes the regulation of termination of transcription that is involved in controlling the expression of some bacterial operons.
Attenuator is the terminator sequence at which attenuation occurs.
Autogenous control describes the action of a gene product that either inhibits (negative autogenous control) or activates (positive autogenous control) expression of the gene coding for it.
Autonomous controlling element in maize is an active transposon with the ability to transpose (*cf.* nonautonomous controlling elements).
Autoradiography detects radioactively labeled molecules by their effect in creating an image on photographic film.
Autosomes are all the chromosomes except the sex chromosomes; a diploid cell has two copies of each autosome.
B lymphocytes (or **B cells**) are the cells responsible for synthesizing antibodies.
Backcross is another (earlier) term for a **testcross.**
Back mutation reverses the effect of a mutation that had inactivated a gene; thus it restores wild type.
Bacteriophages are viruses that infect bacteria; often abbreviated as **phages.**
Balbiani ring is an extremely large puff at a band of a polytene chromosome.
Bands of polytene chromosomes are visible as dense regions that contain the majority of DNA; **bands of normal chromosomes** are relatively much larger and are generated in the form of regions that retain a stain on certain chemical treatments.
Base pair (bp) is a partnership of A with T or of C with G in a DNA double helix; other pairs can be formed in RNA under certain circumstances.
Bidirectional replication is accomplished when two replication forks move away from the same origin in different directions.
Bivalent is the structure containing all four chromatids (two representing each homolog) at the start of meiosis.
Blastoderm is a stage of insect embryogenesis in which a layer of nuclei or cells around the embryo surround an internal mass of yolk.
Blocked reading frame cannot be translated into protein because it is interrupted by termination codons.
Blunt-end ligation is a reaction that joins two DNA duplex molecules directly at their ends.
bp is an abbreviation for base pair(s); distance along DNA is measured in bp.
Branch migration describes the ability of a DNA strand partially paired with its complement in a duplex to extend its pairing by displacing the resident strand with which it is homologous.
Breakage and reunion describes the mode of genetic recombination in which two DNA duplex molecules are broken at corresponding points and then rejoined crosswise (involving formation of a length of heteroduplex DNA around the site of joining).
Buoyant density measures the ability of a substance to float in some standard fluid, for example CsCl.

C banding is a technique for generating stained regions around centromeres.
C genes code for the constant regions of immunoglobulin protein chains.
C value is the total amount of DNA in a haploid genome.
CAAT box is part of a conserved sequence located upstream of the startpoints of eukaryotic transcription

units; it is recognized by a large group of transcription factors.

Cap is the structure at the 5′ end of eukaryotic mRNA, introduced after transcription by linking the terminal phosphate of 5′ GTP to the terminal base of the mRNA. The added G (and sometimes some other bases) are methylated, giving a structure of the form $^{7Me}G^{5'}ppp^{5'}Np\ldots$.

CAP (CRP) is a positive regulator protein activated by cyclic AMP. It is needed for RNA polymerase to initiate transcription of certain (catabolite-sensitive) operons of *E. coli.*

Capsid is the external protein coat of a virus particle.

Caspases comprise a family of protease some of whose members are involved in apoptosis (programmed cell death).

Catabolite repression describes the decreased expression of many bacterial operons that results from addition of glucose. It is caused by a decrease in the level of cyclic AMP, which in turn inactivates the CAP regulator.

cDNA is a single-stranded DNA complementary to an RNA, synthesized from it by reverse transcription *in vitro.*

cDNA clone is a duplex DNA sequence representing an RNA, carried in a cloning vector.

Cell cycle is the period from one division to the next.

Cell hybrid is a somatic cell containing chromosomes derived from parental cells of different species (e.g. a man-mouse somatic cell hybrid), generating by fusing the cells to form a heterokaryon in which the nuclei subsequently fused.

Centrioles are small hollow cylinders consisting of microtubules that become located near the poles during mitosis. They reside within the centrosomes.

Centromere is a constricted region of a chromosome that includes the site of attachment to the mitotic or meiotic spindle. (*See also* kinetochore.)

Centrosomes are the regions from which microtubules are organized at the poles of a mitotic cell. In animal cells, each centrosome contains a pair of centrioles surrounded by a dense amorphous region to which the microtubules attach. (*See also* microtubule organizing center.)

Chaperone—*see* molecular chaperone.

Chemical complexity is the amount of a DNA component measured by chemical assay.

***Chi* sequence** is an octamer that provides a hotspot for RecA-mediated genetic recombination in *E. coli.*

Chi structure is a joint between two duplex molecules of DNA revealed by cleaving an intermediate of two joined circles to generate linear ends in each circle. It resembles a Greek chi in outline, hence the name.

Chiasma (*pl.* chiasmata) is a site at which two homologous chromosomes appear to have exchanged material during meiosis.

Chromatids are the copies of a chromosome produced by replication. The name is usually used to describe them in the period before they separate at the subsequent cell division.

Chromatin is the complex of DNA and protein in the nucleus of the interphase cell. Individual chromosomes cannot be distinguished in it. It was originally recognized by its reaction with stains specific for DNA.

Chromatin remodeling describes the energy-dependent displacement or reorganization of nucleosomes that occurs in conjunction with activation of genes for transcription.

Chromocenter is an aggregate of heterochromatin from different chromosomes.

Chromomeres are densely staining granules visible in chromosomes under certain conditions, especially early in meiosis, when a chromosome may appear to consist of a series of chromomeres.

Chromosome is a discrete unit of the genome carrying many genes. Each chromosome consists of a very long molecule of duplex DNA and an approximately equal mass of proteins. It is visible as a morphological entity only during cell division.

Chromosome walking describes the sequential isolation of clones carrying overlapping sequences of DNA, allowing large regions of the chromosome to be spanned. Walking is often performed in order to reach a particular locus of interest.

***cis*-acting locus** affects the activity only of DNA sequences on its own molecule of DNA; this property usually implies that the locus does not code for protein.

***cis*-acting protein** has the exceptional property of acting only on the molecule of DNA from which it was expressed.

***cis* configuration** describes two sites on the same molecule of DNA.

***cis/trans* test** assays the effect of relative configuration on expression of two mutations. In a double heterozygote, two mutations in the same gene show mutant phenotype in *trans* configuration, wild-type in *cis* configuration.

Cistron is the genetic unit defined by the *cis/trans* test; equivalent to **gene** in comprising a unit of DNA representing a protein.

Class switching is a change in the expression of the C region of an immunoglobulin heavy chain during lymphocyte differentiation.

Clone describes a large number of cells or molecules identical with a single ancestral cell or molecule.

Cloning vector is a plasmid or phage that is used to "carry" inserted foreign DNA for the purposes of producing more material or a protein product.
Closed reading frame contains termination codons that prevent its translation into protein.
Coated vesicles are vesicles whose membrane has on its surface a layer of a protein such as clathrin, COP-I, or COP-II.
Coconversion is the simultaneous correction of two sites during gene conversion.
Coding strand of DNA has the same sequence as mRNA.
Codominant alleles both contribute to the phenotype; neither is dominant over the other.
Codon is a triplet of nucleotides that represents an amino acid or a termination signal.
Coevolution—*see* concerted evolution.
Cognate tRNAs are those recognized by a particular aminoacyl-tRNA synthetase.
Coincidental evolution—*see* concerted evolution.
Cointegrate structure is produced by fusion of two replicons, one originally possessing a transposon, the other lacking it; the cointegrate has copies of the transposon present at both junctions of the replicons, oriented as direct repeats.
Cold-sensitive mutant is defective at low temperature but functional at normal temperature.
Colony hybridization is a technique for using *in situ* hybridization to identify bacteria carrying chimeric vectors whose inserted DNA is homologous with some particular sequence.
Compatibility group of plasmids contains members unable to coexist in the same bacterial cell.
Complementation refers to the ability of independent (nonallelic) genes to provide diffusible products that produce wild phenotype when two mutants are tested in *trans* configuration in a heterozygote.
Complementation assay— *see in vitro* complementation assay.
Complementation group is a series of mutations unable to complement when tested in pairwise combinations in *trans;* defines a genetic unit (the cistron).
Complex locus (of *D. melanogaster*) has genetic properties inconsistent with the function of a gene representing a single protein. Complex loci are usually very large (>100 kb) at the molecular level.
Complexity is the total length of different sequences of DNA present in a given preparation.
Composite transposons have a central region flanked on each side by insertion sequences, either or both of which may enable the entire element to transpose.
Concatemer of DNA consists of a series of unit genomes repeated in tandem.
(Con)catenated circles of DNA are interlocked like rings on a chain.
Concerted evolution describes the ability of two related genes to evolve together as though constituting a single locus.
Condensation reaction is one in which a covalent bond is formed with loss of a water molecule, as in the addition of an amino acid to a polypeptide chain.
Conditional lethal mutations kill a cell or virus under certain (nonpermissive) conditions, but allow it to survive under other (permissive) conditions.
Conjugation describes "mating" between two bacterial cells, when (part of) the chromosome is transferred from one to the other.
Consensus sequence is an idealized sequence in which each position represents the base most often found when many actual sequences are compared.
Conservative recombination involves breakage and reunion of pre-existing strands of DNA without any synthesis of new stretches of DNA.
Conservative transposition refers to the movement of large elements, originally classified as transposons, but now considered to be episomes. The mechanism of movement resembles that of phage lambda.
Constant regions of immunoglobulins are coded by C genes and are the parts of the chain that vary least. Those of heavy chains identify the type of immunoglobulin.
Constitutive genes are expressed as a function of the interaction of RNA polymerase with the promoter, without additional regulation; sometimes also called housekeeping genes in the context of describing functions expressed in all cells at a low level.
Constitutive heterochromatin describes the inert state of permanently nonexpressed sequences, usually satellite DNA.
Constitutive mutations cause genes that usually are regulated to be expressed without regulation.
Contractile ring is a ring of actin filaments that forms around the equator at the end of mitosis and is responsible for pinching the daughter cells apart.
Controlling elements of maize are transposable units originally identified solely by their genetic properties. They may be autonomous (able to transpose independently) or nonautonomous (able to transpose only in the presence of an autonomous element).
Coordinate regulation refers to the common control of a group of genes.
Cordycepin is 3′ deoxyadenosine, an inhibitor of polyadenylation of RNA.
Core DNA is the 146 bp of DNA contained on a core particle.

Core particle is a digestion product of the nucleosome that retains the histone octamer and has 146 bp of DNA; its structure appears similar to that of the nucleosome itself.
Corepressor is a small molecule that triggers repression of transcription by binding to a regulator protein.
Cosmids are plasmids into which phage lambda *cos* sites have been inserted; as a result, the plasmid DNA can be packaged *in vitro* in the phage coat.
Cot is the product of DNA concentration and time of incubation in a reassociation reaction.
$Cot_{1/2}$ is the Cot required to proceed to half completion of the reaction; it is directly proportional to the unique length of reassociating DNA.
Cotransfection is the simultaneous transfection of two markers.
Crossing-over describes the reciprocal exchange of material between chromosomes that occurs during meiosis and is responsible for genetic recombination.
Crossover fixation refers to a possible consequence of unequal crossing-over that allows a mutation in one member of a tandem cluster to spread through the whole cluster (or to be eliminated).
Cruciform is the structure produced at inverted repeats of DNA if the repeated sequence pairs with its complement on the same strand (instead of with its regular partner in the other strand of the duplex).
Cryptic satellite is a satellite DNA sequence not identified as such by a separate peak on a density gradient; that is, it remains present in main-band DNA.
ctDNA is chloroplast DNA.
Cyclic AMP (cAMP) is a molecule of AMP in which the phosphate group is joined to both the 3′ and 5′ positions of the ribose; its binding activates the CAP, a positive regulator of prokaryotic transcription.
Cyclins are proteins that accumulate continuously throughout the cell cycle and are then destroyed by proteolysis during mitosis. (*See also* MPF.)
Cytokinesis is the final process involved in separation and movement apart of daughter cells at the end of mitosis.
Cytological hybridization—*see in situ* hybridization.
Cytoplasm describes the material between the plasma membrane and the nucleus.
Cytoplasmic inheritance is a property of genes located in mitochondria or chloroplasts (or possibly other extranuclear organelles).
Cytoplasmic protein synthesis is the translation of mRNAs representing nuclear genes; it occurs via ribosomes attached to the cytoskeleton.
Cytoskeleton consists of networks of fibers in the cytoplasm of the eukaryotic cell.
Cytosol describes the general volume of cytoplasm in which organelles (such as the mitochondria) are located.

D loop is a region within mitochondrial DNA in which a short stretch of RNA is paired with one strand of DNA, displacing the original partner DNA strand in this region. The same term is used also to describe the displacement of a region of one strand of duplex DNA by a single-stranded invader in the reaction catalyzed by RecA protein.
Degeneracy in the genetic code refers to the lack of an effect of many changes in the third base of the codon on the amino acid that is represented.
Deletions are generated by removal of a sequence of DNA, the regions on either side being joined together.
Denaturation of DNA or RNA describes its conversion from the double-stranded to the single-stranded state; separation of the strands is most often accomplished by heating.
Denaturation of protein describes its conversion from the physiological conformation to some other (inactive) conformation.
Derepressed state describes a gene that is turned on. It is synonymous with *induced* when describing the normal state of a gene; it has the same meaning as *constitutive* in describing the effect of mutation.
Dicentric chromosome is the product of fusing two chromosome fragments, each of which has a centromere. It is unstable and may be broken when the two centromeres are pulled to opposite poles in mitosis.
Diploid set of chromosomes contains two copies of each autosome and two sex chromosomes.
Direct repeats are identical (or related) sequences present in two or more copies in the same orientation in the same molecule of DNA; they are not necessarily adjacent.
Discontinuous replication refers to the synthesis of DNA in short (Okazaki) fragments that are later joined into a continuous strand.
Disjunction describes the movement of members of a chromosome pair to opposite poles during cell division. At mitosis and the second meiotic division, disjunction applies to sister chromatids; at first meiotic division it applies to sister chromatid pairs.
Divergence is the percent difference in nucleotide sequence between two related DNA sequences or in amino acid sequences between two proteins.
Divergent transcription refers to the initiation of transcription at two promoters facing in the opposite direction, so that transcription proceeds away in both directions from a central region.

dna mutants of bacteria are temperature-sensitive; they cannot synthesize DNA at 42°C, but can do so at 37°C.

DNAase is an enzyme that attacks bonds in DNA.

DNAase I hypersensitive site is a short region of chromatin detected by its extreme sensitivity to cleavage by DNAase I and other nucleases; probably comprises an area from which nucleosomes are excluded.

DNA-driven hybridization involves the reaction of an excess of DNA with RNA.

DNA polymerase is an enzyme that synthesizes a daughter strand(s) of DNA (under direction from a DNA template). May be involved in repair or replication.

DNA replicase is a DNA-synthesizing enzyme required specifically for replication.

Domain of a chromosome may refer *either* to a discrete structural entity defined as a region within which supercoiling is independent of other domains; *or* to an extensive region including an expressed gene that has heightened sensitivity to degradation by the enzyme DNAase I.

Domain of a protein is a discrete continuous part of the amino acid sequence that can be equated with a particular function.

Dominant allele determines the phenotype displayed in a heterozygote with another (recessive) allele.

Donor splicing site—*see* left splicing junction.

Dosage compensation describes mechanisms employed to compensate for the discrepancy between the presence of two X chromosomes in one sex but only one X chromosome in the other sex.

Down promoter mutations decrease the frequency of initiation of transcription.

Downstream identifies sequences proceeding farther in the direction of expression; for example, the coding region is downstream of the initiation codon.

Early development refers to the period of a phage infection before the start of DNA replication.

Ectopic expression describes the expression of a gene in a tissue in which it is not usually expressed; for example, in a transgenic animal or as the result of injection into an unusual location in an embryo.

Elongation factors (EF in prokaryotes, eEF in eukaryotes) are proteins that associate with ribosomes cyclically, during addition of each amino acid to the polypeptide chain.

End labeling describes the addition of a radioactively labeled group to one end (5′ or 3′) of a DNA strand.

End-product inhibition describes the ability of a product of a metabolic pathway to inhibit the activity of an enzyme that catalyzes an early step in the pathway.

Endocytic vesicles are membranous particles that transport proteins through endocytosis; also known as clathrin-coated vesicles.

Endocytosis is a process by which proteins at the surface of the cell are internalized, being transported into the cell within membranous vesicles.

Endonucleases cleave bonds within a nucleic acid chain; they may be specific for RNA or for single-stranded or double-stranded DNA.

Endoplasmic reticulum is a highly convoluted sheet of membranes, extending from the outer layer of the nuclear envelope into the cytoplasm.

Enhancer element is a *cis*-acting sequence that increases the utilization of (some) eukaryotic promoters, and can function in either orientation and in any location (upstream or downstream) relative to the promoter.

Envelopes surround some organelles (for example, nucleus or mitochondrion) and consist of concentric membranes, each membrane consisting of the usual lipid bilayer.

Epigenetic changes influence the phenotype without altering the genotype. They consist of changes in the properties of a cell that are inherited but that do not represent a change in genetic information.

Episome is a plasmid able to integrate into bacterial DNA.

Epistasis describes a situation in which expression of one gene obscures the phenotypic effects of another gene.

Essential gene is one whose deletion is lethal to the organism. (*See also* lethal locus.)

Established cell lines consist of eukaryotic cells that have been adapted to indefinite growth in culture (they are said to be immortalized).

Eubacteria comprise the major line of prokaryotes.

Euchromatin comprises all of the genome in the interphase nucleus except for the heterochromatin.

Evolutionary clock is defined by the rate at which mutations accumulate in a given gene.

Excision of phage or episome or other sequence describes its release from the host chromosome as an autonomous DNA molecule.

Excision-repair systems remove a single-stranded sequence of DNA containing damaged or mispaired bases and replace it in the duplex by synthesizing a sequence complementary to the remaining strand.

Exocytosis is the process of secreting proteins from a cell into the medium, by transport in membranous vesicles from the endoplasmic reticulum, through the Golgi, to storage vesicles, and finally (upon a regulatory signal) through the plasma membrane.

Exon is any segment of an interrupted gene that is represented in the mature RNA product.
Exonucleases cleave nucleotides one at a time from the end of a polynucleotide chain; they may be specific for either the 5′ or 3′ end of DNA or RNA.
Expression vector is a cloning vector designed so that a coding sequence inserted at a particular site will be transcribed and translated into protein.
Extranuclear genes reside outside the nucleus in organelles such as mitochondria and chloroplasts.

F factor is a bacterial sex or fertility plasmid.
F1 generation is the first generation produced by crossing two parental (homozygous) lines.
Facultative heterochromatin describes the inert state of sequences that also exist in active copies—for example, one mammalian X chromosome in females.
Fast component of a reassociation reaction is the first to renature and contains highly repetitive DNA.
Fate map is a map of an embryo showing the adult tissues that will develop from the descendants of cells that occupy particular regions of the embryo.
Figure eight describes two circles of DNA linked together by a recombination event that has not yet been completed.
Filter hybridization is performed by incubating a denatured DNA preparation immobilized on a nitrocellulose filter with a solution of radioactively labeled RNA or DNA.
Fingerprint of DNA is a pattern of polymorphic restriction fragments that differ between individual genomes.
Fingerprint of a protein is the pattern of fragments (usually resolved on a two-dimensional electrophoretic gel) generated by cleavage with an enzyme such as trypsin.
Fluidity is a property of membranes; it indicates the ability of lipids to move laterally within their particular monolayer.
Focus formation describes the ability of transformed eukaryotic cells to grow in dense clusters, piled up on one another.
Focus forming unit (ffu) is a quantitative measure of focus formation.
Foldback DNA consists of inverted repeats that have renatured by intrastrand reassociation of denatured DNA.
Footprinting is a technique for identifying the site on DNA bound by some protein by virtue of the protection of bonds in this region against attack by nucleases.
Forward mutations inactivate a wild-type gene.
Founder effect refers to the presence in a population of many individuals all with the same chromosome (or region of a chromosome) derived from a single ancestor.
Frameshift mutations arise by deletions or insertions that are not a multiple of 3 bp; they change the frame in which triplets are translated into protein.

G banding is a technique that generates a striated pattern in metaphase chromosomes that distinguishes the members of a haploid set.
G1 is the period of the eukaryotic cell cycle between the last mitosis and the start of DNA replication.
G2 is the period of the eukaryotic cell cycle between the end of DNA replication and the start of the next mitosis.
Gamete is either type of reproductive (germ) cell—sperm or egg—with haploid chromosome content.
Gap in DNA is the absence of one or more nucleotides in one strand of the duplex.
Gene (cistron) is the segment of DNA involved in producing a polypeptide chain; it includes regions preceding and following the coding region (leader and trailer) as well as intervening sequences (introns) between individual coding segments (exons).
Gene cluster is a group of adjacent genes that are identical or related.
Gene conversion is the alteration of one strand of a heteroduplex DNA to make it complementary with the other strand at any position(s) where there were mispaired bases.
Gene dosage gives the number of copies of a particular gene in the genome.
Gene family consists of a set of genes whose exons are related; the members were derived by duplication and variation from some ancestral gene.
Genetic code is the correspondence between triplets in DNA (or RNA) and amino acids in protein.
Genetic marker—*see* marker.
Genomic (chromosomal) DNA clones are sequences of the genome carried by a cloning vector.
Genotype is the genetic constitution of an organism.
Golgi apparatus consists of individual stacks of membranes near the endoplasmic reticulum; involved in glycosylating proteins and sorting them for transport to different cellular locations.
G proteins are guanine nucleotide-binding trimeric proteins that reside in the plasma membrane. When bound by GDP the trimer remains intact and is inert. When the GDP bound to the α subunit is replaced by GTP, the α subunit is released from the $\beta\gamma$ dimer. One of the separated units (either the α monomer or the $\beta\gamma$ dimer) then activates or represses a target protein.

Gratuitous inducers resemble authentic inducers of transcription but are not substrates for the induced enzymes.
GT-AG rule describes the presence of these constant dinucleotides at the first two and last two positions of introns of nuclear genes.
Gyrase is a type II topoisomerase of *E. coli* with the ability to introduce negative supercoils into DNA.

Hairpin describes a double-helical region formed by base pairing between adjacent (inverted) complementary sequences in a single strand of RNA or DNA.
Haploid set of chromosomes contains one copy of each autosome and one sex chromosome; the haploid number *n* is characteristic of gametes of diploid organisms.
Haplotype is the particular combination of alleles in a defined region of some chromosome, in effect the genotype in miniature. Originally used to described combinations of MHC alleles, it now may be used to describe particular combinations of RFLPs.
Hapten is a small molecule that acts as an antigen when conjugated to a protein.
HAT (histone acetyltransferase) enzymes modify histones by addition of acetyl groups; some transcriptional coactivators have HAT activity.
HDAC (histone deacetyltransferase) enzymes remove acetyl groups from histones; they may be associated with repressors of transcription.
Helper virus provides functions absent from a defective virus, enabling the latter to complete the infective cycle during a mixed infection.
Hemizygote is a diploid individual that has lost its copy of a particular gene (for example, because a chromosome has been lost) and which therefore has only a single copy.
Heterochromatin describes regions of the genome that are permanently in a highly condensed condition, are not transcribed, and are late-replicating. May be constitutive or facultative.
Heteroduplex (hybrid) DNA is generated by base pairing between complementary single strands derived from the different parental duplex molecules; it occurs during genetic recombination.
Heterogametic sex has the diploid chromosome constitution 2A + XY.
Heterogeneous nuclear (hn) RNA comprises transcripts of nuclear genes made by RNA polymerase II; it has a wide size distribution and low stability.
Heterokaryon is a cell containing two (or more) nuclei in a common cytoplasm, generated by fusing somatic cells.
Heteromultimeric proteins consist of nonidentical subunits (coded by different genes).
Heterozygote is an individual with different alleles at some particular locus.
Highly repetitive DNA is the first component to reassociate and is equated with satellite DNA.
Histones are conserved DNA-binding proteins of eukaryotes that form the nucleosome, the basic subunit of chromatin.
Homeobox describes the conserved sequence that is part of the coding region of *D. melanogaster* homeotic genes; it is also found in amphibian and mammalian genes expressed in early embryonic development.
Homeotic genes are defined by mutations that convert one body part into another; for example, an insect leg may replace an antenna.
Homogametic sex has the diploid chromosome constitution 2A + XX.
Homologs are chromosomes carrying the same genetic loci; a diploid cell has two copies of each homolog, one derived from each parent.
Homomultimeric protein consists of identical subunits.
Homozygote is an individual with the same allele at corresponding loci on the homologous chromosomes.
Hotspot is a site at which the frequency of mutation (or recombination) is very much increased.
Housekeeping (constitutive) genes are those (theoretically) expressed in all cells because they provide basic functions needed for sustenance of all cell types.
Hox genes are clusters of mammalian genes containing homeoboxes; the individual members are related to the genes of the complex loci *ANT-C* and *BX-C* in *D. melanogaster.*
Hybrid-arrested translation is a technique that identifies the cDNA corresponding to an mRNA by relying on the ability to base pair with the RNA *in vitro* to inhibit translation.
Hybrid DNA—*see* heteroduplex (hybrid) DNA.
Hybrid dysgenesis describes the inability of certain strains of *D. melanogaster* to interbreed, because the hybrids are sterile (although otherwise they may be phenotypically normal).
Hybridization is the pairing of complementary RNA and DNA strands to give an RNA-DNA hybrid.
Hybridoma is a cell line produced by fusing a myeloma with a lymphocyte; it continues indefinitely to express the immunoglobulins of both parents.
Hydrolytic reaction is one in which a covalent bond is broken with the incorporation of a water molecule.
Hydropathy plot is a measure of the hydrophobicity of

a protein region and therefore of the likelihood that it will reside in a membrane.

Hydrophilic groups interact with water, so that hydrophilic regions of protein or the faces of a lipid bilayer reside in an aqueous environment.

Hydrophobic groups repel water, so that they interact with one another to generate a nonaqueous environment.

Hyperchromicity is the increase in optical density that occurs when DNA is denatured.

Hypervariable regions of an immunoglobulin are the parts of the variable region that show maximum alteration when different antibodies are compared.

Ideogram is a diagrammatic representation of the G-banding pattern of a chromosome.

Idling reaction is the production of pppGpp and ppGpp by ribosomes when an uncharged tRNA is present in the A site; triggers the stringent response.

Immortalization describes the acquisition by a eukaryotic cell line of the ability to grow through an indefinite number of divisions in culture.

Immunity in phages refers to the ability of a prophage to prevent another phage of the same type from infecting a cell. It results from the synthesis of phage repressor by the prophage genome.

Immunity in plasmids describes the ability of a plasmid to prevent another of the same type from becoming established in a cell. It results usually from interference with the ability to replicate.

Immunity in transposons refers to the ability of certain transposons to prevent others of the same type from transposing to the same DNA molecule. It results from a variety of mechanisms.

Imprinting describes a change in a gene that occurs during passage through the sperm or egg with the result that the paternal and maternal alleles have different properties in the very early embryo. May be caused by methylation of DNA.

***In situ* hybridization** is performed by denaturing the DNA of cells squashed on a microscope slide so that reaction is possible with an added single-stranded RNA or DNA; the added preparation is radioactively labeled and its hybridization is followed by autoradiography.

***In vitro* complementation assay** consists of identifying a component of a wild-type cell that can confer activity on an extract prepared from a mutant cell. The assay identifies the component rendered inactive by the mutation.

Incompatibility is the inability of certain bacterial plasmids to coexist in the same cell. It is a cause of plasmid immunity.

Indirect end-labeling is a technique for examining the organization of DNA by making a cut at a specific site and isolating all fragments containing the sequence adjacent to one side of the cut; it reveals the distance from the cut to the next break(s) in DNA.

Induced mutations result from the addition of a mutagen.

Inducer is a small molecule that triggers gene transcription by binding to a regulator protein.

Induction refers to the ability of bacteria (or yeast) to synthesize certain enzymes only when their substrates are present; applied to gene expression, refers to switching on transcription as a result of interaction of the inducer with the regulator protein.

Induction of prophage describes its excision from the host genome and entry into the lytic (infective) cycle as a result of destruction of the lysogenic repressor.

Initiation factors (IF in prokaryotes, eIF in eukaryotes) are proteins that associate with the small subunit of the ribosome specifically at the stage of initiation of protein synthesis.

Insertion sequence (IS) is a small bacterial transposon that carries only the genes needed for its own transposition.

Insertions are identified by the presence of an additional stretch of base pairs in DNA.

Integral membrane protein is a protein (non-covalently) inserted into a membrane; it retains its membranous association by means of a stretch of ~25 amino acids that are uncharged and/or hydrophobic.

Integration of viral or another DNA sequence is its insertion into a host genome as a region covalently linked on either side to the host sequences.

Interallelic complementation describes the change in the properties of a heteromultimeric protein brought about by the interaction of subunits coded by two different mutant alleles; the mixed protein may be more or less active than the protein consisting of subunits only of one or the other type.

Interbands are the relatively dispersed regions of polytene chromosomes that lie between the bands.

Intercistronic region is the distance between the termination codon of one gene and the initiation codon of the next gene.

Intermediate component(s) of a reassociation reaction are those reacting between the fast (satellite DNA) and slow (nonrepetitive DNA) components; contain moderately repetitive DNA.

Interphase is the period between mitotic cell divisions; divided into G1, S, and G2.

Intervening sequence is an intron.

Intron is a segment of DNA that is transcribed, but removed from within the transcript by splicing together the sequences (exons) on either side of it.
Inversion is a chromosomal change in which a segment has been rotated by 180° relative to the regions on either side and reinserted.
Inverted repeats comprise two copies of the same sequence of DNA repeated in opposite orientation on the same molecule. Adjacent inverted repeats constitute a palindrome.
Inverted terminal repeats are the short related or identical sequences present in reverse orientation at the ends of some transposons.
IS is an abbreviation for **insertion sequence,** a small bacterial transposon carrying only the genetic functions involved in transposition.
Isoaccepting tRNAs represent the same amino acid.
Isotype is a group of closely related immunoglobulin chains.

Karyotype is the entire chromosomal complement of a cell or species (as visualized during mitosis).
kb (kilobase) is an abbreviation for 1000 base pairs of DNA or 1000 bases of RNA.
Kinase is an enzyme that phosphorylates (adds a phosphate group) to a substrate; the substrates for **protein kinases** are amino acids in other proteins, and they are divided into those specific for tyrosine and those specific for threonine/serine (and histidine in prokaryotes).
Kinetic complexity is the complexity of a DNA component measured by the kinetics of DNA reassociation.
Kinetochore is the structural feature of the chromosome to which microtubules of the mitotic spindle attach. (*See also* centromere.)

Lagging strand of DNA must grow overall in the 3′–5′ direction and is synthesized discontinuously in the form of short fragments (5′–3′) that are later connected covalently.
Lampbrush chromosomes are the large meiotic chromosomes found in amphibian oocytes.
Lariat is an intermediate in RNA splicing in which a circular structure with a tail is created by a 5′–2′ bond.
Late period of phage development is the part of infection following the start of DNA replication.
Late-replicating material does not replicate until the end of S phase. Typically this consists of the heterochromatin.
Leader is the nontranslated sequence at the 5′ end of mRNA that precedes the initiation codon.
Leader sequence of a protein is a short N-terminal sequence responsible for passage into or through a membrane.
Leading strand of DNA is synthesized continuously in the 5′–3′ direction.
Leaky mutations allow some residual level of gene expression.
Left splicing junction is the boundary between the right end of an exon and the left end of an intron.
Lethal locus is any gene in which a lethal mutation can be obtained (usually by deletion of the gene).
Library is a set of cloned fragments together representing the entire genome.
Ligation is the formation of a phosphodiester bond to link two adjacent bases separated by a nick in one strand of a double helix of DNA. (The term can also be applied to blunt-end ligation and to joining of RNA.)
LINES are long-period interspersed sequences in mammalian genomes that are retroposons generated from RNA polymerase II transcripts.
Linkage describes the tendency of genes to be inherited together as a result of their location on the same chromosome; measured by percent recombination between loci.
Linkage disequilibrium describes a situation in which some combinations of genetic markers occur more or less frequently in the population than would be expected from their distance apart. It implies that a group of markers has been inherited coordinately. It can result from reduced recombination in the region or from a founder effect, in which there has been insufficient time to reach equilibrium since one of the markers was introduced into the population.
Linkage group includes all loci that can be connected (directly or indirectly) by linkage relationships; equivalent to a chromosome.
Linker DNA is all DNA contained on a nucleosome in excess of the 146 bp core DNA.
Linker fragment is a short synthetic duplex oligonucleotide containing the target site for some restriction enzyme; may be added to ends of a DNA fragment prepared by cleavage with some other enzyme during reconstruction of recombinant DNA.
Linker scanner mutations are introduced by recombining two DNA molecules *in vitro* at a restriction fragment added to the end of each; the result is to insert the linker sequence at the site of recombination.
Linking number is the number of times the two strands of a closed DNA duplex cross over each other.
Linking number paradox describes the discrepancy between the existence of –2 supercoils in the path of DNA on the nucleosome compared with the measurement of –1 supercoil released when histones are removed.

Lipids have polar heads, containing phosphate (phospholipid), sterol (such as cholesterol), or saccharide (glycolipid) connected to a hydrophobic tail consisting of fatty acid(s).
Lipid bilayer is the form taken by concentration of lipids in which the hydrophobic fatty acids occupy the interior and the polar heads face the exterior.
Liquid (solution) hybridization is a reaction between complementary nucleic acid strands performed in solution.
Locus is the position on a chromosome at which the gene for a particular trait resides; locus may be occupied by any one of the alleles for the gene.
LOD score is a measure of genetic linkage, defined as the $\log_{10}$ ratio of the probability that the data would have arisen if the loci are linked to the probability that the data could have arisen from unlinked loci. The conventional threshold for declaring linkage is a LOD score of 3.0, that is, a 1000:1 ratio (which must be compared with the 50:1 probability that any random pair of loci will be unlinked).
Long-period interspersion is a pattern in the genome in which long stretches of moderately repetitive and nonrepetitive DNA alternate.
Loop is a single-stranded region at the end of a hairpin in RNA (or single-stranded DNA); corresponds to the sequence between inverted repeats in duplex DNA.
LTR is an abbreviation for **long-terminal repeat,** a sequence directly repeated at both ends of a retroviral DNA.
Lumen describes the interior of a compartment bounded by membranes, usually the endoplasmic reticulum or the mitochondrion.
Luxury genes are those coding for specialized functions synthesized (usually) in large amounts in particular cell types.
Lysis describes the death of bacteria at the end of a phage infective cycle when they burst open to release the progeny of an infecting phage. Also applies to eukaryotic cells, for example, infected cells that are attacked by the immune system.
Lysogen is a bacterium that possesses a repressed prophage as part of its genome.
Lysogenic immunity is the ability of a prophage to prevent another phage genome of the same type from becoming established in the bacterium.
Lysogenic repressor is the protein responsible for preventing a prophage from re-entering the lytic cycle.
Lysogeny describes the ability of a phage to survive in a bacterium as a stable prophage component of the bacterial genome.
Lysosomes are small bodies, enclosed by membranes, that contain hydrolytic enzymes in eukaryotic cells.
Lytic infection of bacteria by a phage ends in destruction of bacteria and release of progeny phage.

Main band of genomic DNA consists of a broad peak on a density gradient, excluding any visible satellite DNAs that form separate bands.
Major histocompatibility locus is a large chromosomal region containing a giant cluster of genes that code for transplantation antigens and other proteins found on the surfaces of lymphocytes.
Map distance is measured as cM (centiMorgans) = percent recombination (sometimes subject to adjustments).
MAR (matrix attachment site; also known as SAR for scaffold attachment site) is a region of DNA that attaches to the nuclear matrix.
Marker (DNA) is a fragment of known size used to calibrate an electrophoretic gel.
Marker (genetic) is any allele of interest in an experiment.
Maternal inheritance describes the preferential survival in the progeny of genetic markers provided by one parent.
Mb (megabase) is an abbreviation for 10^6 bp of DNA.
Meiosis occurs by two successive divisions (meiosis I and II) that reduce the starting number of $4n$ chromosomes to $1n$ in each of four product cells. Products may mature to germ cells (sperm or eggs).
Melting of DNA means its denaturation.
Melting temperature (T_m) is the midpoint of the temperature range over which DNA is denatured.
Membranes consist of an asymmetrical lipid bilayer that has lateral fluidity and contains proteins.
Membrane proteins have hydrophobic regions that allow part or all of the protein structure to reside within the membrane; the bonds involved in this association are usually noncovalent.
Metastasis describes the ability of tumor cells to leave their site of origin and migrate to other locations in the body, where a new colony is established.
Micrococcal nuclease is an endonuclease that cleaves DNA; in chromatin, DNA is cleaved preferentially between nucleosomes.
Microsomes are fragmented pieces of endoplasmic reticulum associated with ribosomes.
Microtubules are filaments consisting of dimers of tubulin; interphase microtubules are reorganized into spindle fibers at mitosis, when they are responsible for chromosome movement.
Microtubule associated proteins (MAPs) are proteins

associated with microtubules and responsible for influencing their stability and organization.
Microtubule organizing center (MTOC) is a structure from which microtubules may be extended. The major MTOCs in a mitotic cell are the centrosomes.
Minicell is an anucleate bacterial (*E. coli*) cell produced by a division that generates a cytoplasm without a nucleus.
Minichromosome of SV40 or polyoma is the nucleosomal form of the viral circular DNA.
Mitosis is the division of a eukaryotic somatic cell.
Modification of DNA or RNA includes all changes made to the nucleotides after their initial incorporation into the polynucleotide chain.
Modified bases are all those except the usual four from which DNA (T, C, A, G) and RNA (U, C, A, G) are synthesized; they result from postsynthetic changes in the nucleic acid.
Molecular chaperone is a protein that is needed for the assembly or proper folding of some other protein, but which is not itself a component of the target complex.
Monocistronic mRNA codes for one protein.
Monolayer describes the growth of eukaryotic cells in culture as a layer only one cell deep.
Morphogen is a factor that induces development of particular cell types in a manner that depends on its concentration.
MPF (maturation- or M phase-promoting factor) is a dimeric kinase, containing the p34 catalytic subunit and a cyclin regulatory subunit, whose activation triggers the onset of mitosis.
mtDNA is mitochondrial DNA.
MTOC—*see* microtubule organizing center.
Multicopy plasmids are present in bacteria at amounts greater than one per chromosome.
Multiforked chromosome (in bacterium) has more than one replication fork, because a second initiation has occurred before the first cycle of replication has been completed.
Multimeric proteins consist of more than one subunit.
Mutagens increase the rate of mutation by inducing changes in DNA.
Mutation describes any change in the sequence of genomic DNA.
Mutation frequency is the frequency at which a particular mutant is found in the population.
Mutation rate is the rate at which a particular mutation occurs, usually given as the number of events per gene per generation.
Myeloma is a tumor cell line derived from a lymphocyte; usually produces a single type of immunoglobulin.
Negative complementation occurs when interallelic complementation allows a mutant subunit to suppress the activity of a wild-type subunit in a multimeric protein.
Negative regulators function by switching off transcription or translation.
Negative supercoiling comprises the twisting of a duplex of DNA in space in the opposite sense to the turns of the strands in the double helix.
Neutral substitutions in a protein are those changes of amino acids that do not affect activity.
Nick in duplex DNA is the absence of a phosphodiester bond between two adjacent nucleotides on one strand.
Nick translation describes the ability of *E. coli* DNA polymerase I to use a nick as a starting point from which one strand of a duplex DNA can be degraded and replaced by resynthesis of new material; is used to introduce radioactively labeled nucleotides into DNA *in vitro*.
Nonautonomous controlling elements are defective transposons that can transpose only when assisted by an autonomous controlling element of the same type.
Nondisjunction describes failure of chromatids (duplicate chromosomes) to move to opposite poles during mitosis or meiosis.
Nonpermissive conditions do not allow conditional lethal mutants to survive.
Nonrepetitive DNA shows reassociation kinetics expected of unique sequences.
Nonreplicative transposition describes the movement of a transposon that leaves a donor site (usually generating a double-strand break) and moves to a new site.
Nonsense codon is any one of three triplets (UAG, UAA, UGA) that cause termination of protein synthesis. (UAG is known as amber; UAA as ochre.)
Nonsense mutation is any change in DNA that causes a (termination) codon to replace a codon representing an amino acid.
Nonsense suppressor is a gene coding for a mutant tRNA able to respond to one or more of the termination codons.
Nontranscribed spacer is the region between transcription units in a tandem gene cluster.
Northern blotting is a technique for transferring RNA from an agarose gel to a nitrocellulose filter on which it can be hybridized to a complementary DNA.
Nuclear envelope is a layer of two membranes surrounding the nucleus. It is penetrated by nuclear pores. The inner membrane is bound on the interior by the nuclear laminin. The outer membrane extends in the cytosol into the network of the endoplasmic reticulum.
Nuclear lamina consists of a proteinaceous layer on the

inside of the nuclear envelope. It consists of (up to) three lamin proteins.
Nuclear matrix is a network of fibers surrounding and penetrating the nucleus.
Nuclear pores are large structures that extend across the nuclear envelope and are to transport macromolecules both into and out of the nucleus.
Nucleoid is the compact body that contains the genome in a bacterium.
Nucleolar organizer is the region of a chromosome carrying genes coding for rRNA.
Nucleolus is a discrete region of the nucleus created by the transcription of rRNA genes.
Nucleolytic reactions involve the hydrolysis of a phosphodiester bond in a nucleic acid.
Nucleosome is the basic structural subunit of chromatin, consisting of ~200 bp of DNA and an octamer of histone proteins.
Null mutation completely eliminates the function of a gene, usually because it has been physically deleted.

Ochre codon is the triplet UAA, one of three codons that cause termination of protein synthesis.
Ochre mutation is any change in DNA that creates a UAA codon at a site previously occupied by another codon.
Ochre suppressor is a gene coding for a mutant tRNA able to respond to the UAA codon to allow continuation of protein synthesis; ochre suppressors also suppress amber codons.
Okazaki fragments are the short stretches of 1000–2000 bases produced during discontinuous replication; they are later joined into a covalently intact strand.
Oncogenes are genes whose products have the ability to transform eukaryotic cells so that they grow in a manner analogous to tumor cells. Oncogenes carried by retroviruses have names of the form *v-onc.* (*See also* proto-oncogenes.)
Open reading frame (ORF) contains a series of triplets coding for amino acids without any termination codons; sequence is (potentially) translatable into protein.
Operator is the site on DNA at which a repressor protein binds to prevent transcription from initiating at the adjacent promoter.
Operon is a unit of bacterial gene expression and regulation, including structural genes and control elements in DNA recognized by regulator gene product(s).
Organelles are compartments located in the cytoplasm and surrounded by a membrane.
Origin (*ori*) is a sequence of DNA at which replication is initiated.
Orphans are isolated individual genes found in isolated locations, but related to members of a gene cluster.
Overwinding of DNA is caused by positive supercoiling (which applies further tension in the direction of winding of the two strands about each other in the duplex).

Packing ratio is the ratio of the length of DNA to the unit length of the fiber containing it.
Pairing of chromosomes—*see* synapsis.
Palindrome is a sequence of DNA that is the same when one strand is read left to right or the other is read right to left; consists of adjacent inverted repeats.
Papovaviruses are a class of animal viruses with small genomes, including SV40 and polyoma.
Paranemic joint describes a region in which two complementary sequences of DNA are associated side by side instead of being intertwined in a double helical structure.
pBR322 is one of the standard plasmid cloning vectors.
PCR (polymerase chain reaction) describes a technique in which cycles of denaturation, annealing with primer, and extension with DNA polymerase, are used to amplify the number of copies of a target DNA sequence by $>10^6$ times.
Perinuclear space lies between the inner and outer membranes of the nuclear envelope.
Periodicity of DNA is the number of base pairs per turn of the double helix.
Permissive conditions allow conditional lethal mutants to survive.
Petite strains of yeast lack mitochondrial function.
Phage (bacteriophage) is a bacterial virus.
Phase variation describes an alternation in the type of flagella produced by a bacterium.
Phenotype is the appearance or other characteristics of an organism, resulting from the interaction of its genetic constitution with the environment.
Phosphatase is an enzyme that removes phosphate groups from substrates.
Plasma membrane is the continuous membrane defining the boundary of every cell.
Plasmid is an autonomous self-replicating extrachromosomal circular DNA.
Playback experiment describes the retrieval of DNA that has hybridized with RNA to check that it is nonrepetitive by a further reassociation reaction.
Plectonemic winding describes the intertwining of the two strands in the classical double helix of DNA.

Pleiotropic gene affects more than one (apparently unrelated) characteristic of the phenotype.

Ploidy refers to the number of copies of the chromosome set present in a cell; a haploid has one copy, a diploid has two copies, etc.

Point mutations are changes involving single base pairs.

Polarity refers to the effect of a mutation in one gene in influencing the expression (at transcription or translation) of subsequent genes in the same transcription unit.

Polyadenylation is the addition of a sequence of polyadenylic acid to the 3′ end of a eukaryotic RNA after its transcription.

Polycistronic mRNA includes coding regions representing more than one gene.

Polymorphism refers to the simultaneous occurrence in the population of genomes showing allelic variations (as seen either in alleles producing different phenotypes or—for example—in changes in DNA affecting the restriction pattern).

Polyploid cell has more than two sets of the haploid genome.

Polyprotein is a gene product that is cleaved into several independent proteins.

Polysome (polyribosome) is an mRNA associated with a series of ribosomes engaged in translation.

Polytene chromosomes are generated by successive replications of a chromosome set without separation of the replicas.

Position effect refers to a change in the expression of a gene brought about by its translocation to a new site in the genome; for example, a previously active gene may become inactive if placed near heterochromatin.

Position effect variegation describes the consequences when a gene is inactivated in some cells but not in others as the result of variable spread of inactivity from heterochromatin.

Positive regulator proteins are required for the activation of a transcription unit.

Positive supercoiling describes the coiling of the double helix in space in the same direction as the winding of the two strands of the double helix itself.

Postmeiotic segregation describes the segregation of two strands of a duplex DNA that bear different information (created by heteroduplex formation during meiosis) when a subsequent replication allows the strands to separate.

Primary cells are eukaryotic cells taken into culture directly from the animal.

Primary transcript is the original unmodified RNA product corresponding to a transcription unit.

Primer is a short sequence (often of RNA) that is paired with one strand of DNA and provides a free 3′-OH end at which a DNA polymerase starts synthesis of a deoxyribonucleotide chain.

Primosome describes the complex of proteins involved in the priming action that initiates synthesis of each Okazaki fragment during discontinuous DNA replication; the primosome may move along DNA to engage in successive priming events.

Prion is a proteinaceous infectious agent, which behaves as an inheritable trait, although it contains no nucleic acid. Examples are PrP^{Sc}, the agent of scrapie in sheep and bovine spongiform encephalopathy, and Psi, which confers an inherited state in yeast.

Procentriole is an immature centriole, formed in the vicinity of a mature centriole.

Processed pseudogene is an inactive gene copy that lacks introns, contrasted with the interrupted structure of the active gene. Such genes presumably originate by reverse transcription of mRNA and insertion of a duplex copy into the genome.

Processive enzymes continue to act on a particular substrate, that is, do not dissociate between repetitions of the catalytic event.

Prokaryotic organisms (bacteria) lack nuclei.

Promoter is a region of DNA involved in binding of RNA polymerase to initiate transcription.

–10 sequence is the consensus sequence TATAATG centered about 10 bp before the startpoint of a bacterial gene. It is involved in the initial melting of DNA by RNA polymerase.

–35 sequence is the consensus sequence centered about 35 bp before the startpoint of a bacterial gene. It is involved in initial recognition by RNA polymerase.

Proofreading refers to any mechanism for correcting errors in protein or nucleic acid synthesis that involves scrutiny of individual units *after* they have been added to the chain.

Prophage is a phage genome covalently integrated as a linear part of the bacterial chromosome.

Proteolytic reactions comprise the hydrolysis of peptide bonds in protein.

Proto-oncogenes are the normal counterparts in the eukaryotic genome to the oncogenes carried by some retroviruses. They are given names of the form *c-onc*.

Provirus is a duplex DNA sequence in the eukaryotic chromosome corresponding to the genome of an RNA retrovirus.

Pseudogenes are inactive but stable components of the genome derived by mutation of an ancestral active gene.

Puff is an expansion of a band of a polytene chromosome associated with the synthesis of RNA at some locus in the band.

Pulse-chase experiments are performed by incubating cells very briefly with a radioactively labeled precursor (of some pathway or macromolecule); then the fate of the label is followed during a subsequent incubation with a nonlabeled precursor.

Quaternary structure of a protein refers to its multimeric constitution.

Quick-stop *dna* mutants of *E. coli* cease replication immediately when the temperature is increased to 42°C.

R loop is the structure formed when an RNA strand hybridizes with its complementary strand in a DNA duplex, thereby displacing the original strand of DNA in the form of a loop extending over the region of hybridization.

Rapid lysis (*r*) **mutants** display a change in the pattern of lysis of *E. coli* at the end of an infection by a T-even phage.

Reading frame is one of three possible ways of reading a nucleotide sequence as a series of triplets.

Reassociation of DNA describes the pairing of complementary single strands to form a double helix.

RecA is the product of the *recA* locus of *E. coli;* a protein with dual activities, activating proteases and also able to exchange single strands of DNA molecules. The protease-activating activity controls the SOS response; the nucleic acid handling facility is involved in recombination-repair pathways.

Receptor is a transmembrane protein, located in the plasma membrane, that binds a ligand in a domain on the extracellular side, and as a result has a change in activity of the cytoplasmic domain. (The same term is sometimes used also for the steroid receptors, which are transcription factors that are activated by binding ligands that are steroids or other small molecules.)

Recessive allele is obscured in the phenotype of a heterozygote by the dominant allele, often due to inactivity or absence of the product of the recessive allele.

Recessive lethal is an allele that is lethal when the cell is homozygous for it.

Reciprocal recombination is the production of new genotypes with the reverse arrangements of alleles according to maternal and paternal origin.

Reciprocal translocation exchanges part of one chromosome with part of another chromosome.

Recombinant progeny have a different genotype from that of either parent.

Recombinant joint is the point at which two recombining molecules of duplex DNA are connected (the edge of the heteroduplex region).

Recombination nodules (nodes) are dense objects present on the synaptonemal complex; could be involved in crossing-over.

Recombination-repair is a mode of filling a gap in one strand of duplex DNA by retrieving a homologous single strand from another duplex.

Regulatory gene codes for an RNA or a protein product whose function is to control the expression of other genes.

Relaxed mutants of *E. coli* do not display the stringent response to starvation for amino acids (or other nutritional deprivation).

Relaxed replication control refers to the ability of some plasmids to continue replicating after bacteria cease dividing.

Release (termination) factors respond to termination codons to cause release of the completed polypeptide chain and the ribosome from mRNA.

Renaturation is the reassociation of denatured complementary single strands of a DNA double helix.

Repeating unit in a tandem cluster is the length of the sequence that is repeated; appears circular on a restriction map.

Repetition frequency is the (integral) number of copies of a given sequence present in the haploid genome; equals 1 for nonrepetitive DNA, >2 for repetitive DNA.

Repetitive DNA behaves in a reassociation reaction as though many (related or identical) sequences are present in a component, allowing any pair of complementary sequences to reassociate.

Replacement sites in a gene are those at which mutations alter the amino acid that is coded.

Replication-defective virus has lost one or more genes essential for completing the infective cycle.

Replication eye is a region in which DNA has been replicated within a longer, unreplicated region.

Replication fork is the point at which strands of parental duplex DNA are separated so that replication can proceed.

Replicative transposition describes the movement of a transposon by a mechanism in which first it is replicated, and then one copy is transferred to a new site.

Replicon is a unit of the genome in which DNA is replicated; contains an origin for initiation of replication.

Replisome is the multiprotein structure that assembles at the bacterial replicating fork to undertake synthesis of DNA. Contains DNA polymerase and other enzymes.

Reporter gene is a coding unit whose product is easily assayed (such as chloramphenicol transacetylase); it may be connected to any promoter of interest so that expression of the gene can be used to assay promoter function.
Repression is the ability of bacteria to prevent synthesis of certain enzymes when their products are present; more generally, refers to inhibition of transcription (or translation) by binding of repressor protein to a specific site on DNA (or mRNA).
Repressor protein binds to operator on DNA or RNA to prevent transcription or translation, respectively.
Resolvase is enzyme activity involved in site-specific recombination between two transposons present as direct repeats in a cointegrate structure.
Restriction enzymes recognize specific short sequences of DNA and cleave the duplex (sometimes at target site, sometimes elsewhere, depending on type).
Restriction fragment length polymorphism (RFLP) refers to inherited differences in sites for restriction enzymes (for example, caused by base changes in the target site) that result in differences in the lengths of the fragments produced by cleavage with the relevant restriction enzyme. RFLPs are used for genetic mapping to link the genome directly to a conventional genetic marker.
Restriction map is a linear array of sites on DNA cleaved by various restriction enzymes.
Retrograde transport describes movement of proteins in the reverse direction in the reticuloendothelial system, typically from Golgi to endoplasmic reticulum.
Retroposon is a transposon that mobilizes via an RNA form; the DNA element is transcribed into RNA, and then reverse-transcribed into DNA, which is inserted at a new site in the genome.
Retrovirus is an RNA virus that propagates via conversion into duplex DNA.
Reverse transcription is synthesis of DNA on a template of RNA; accomplished by reverse transcriptase enzyme.
Reverse translation is a technique for isolating genes (or mRNAs) by their ability to hybridize with a short oligonucleotide sequence prepared by predicting the nucleic acid sequence from the known protein sequence.
Reversion of mutation is a change in DNA that either reverses the original alteration (true reversion) or compensates for it (second site reversion in the same gene).
Revertants are derived by reversion of a mutant cell or organism.
RFLP—*see* restriction fragment length polymorphism.
Rho factor is a protein involved in assisting *E. coli* RNA polymerase to terminate transcription at certain (rho-dependent) sites.
Rho-independent terminators are sequences of DNA that cause *E. coli* RNA polymerase to terminate *in vitro* in the absence of rho factor.
Rifamycins (including rifampicin) inhibit transcription in bacteria.
Right splicing junction is the boundary between the right end of an intron and the left end of the adjacent exon.
RNAase is an enzyme whose substrate is RNA.
RNA-driven hybridization reactions use an excess of RNA to react with all complementary sequences in a single-stranded preparation of DNA.
RNA polymerase is an enzyme that synthesizes RNA using a DNA template (formally described as DNA-dependent RNA polymerase).
RNA replicase is an enzyme that synthesizes RNA using an RNA template (used for replication by RNA viruses).
Rolling circle is a mode of replication in which a replication fork proceeds around a circular template for an indefinite number of revolutions; the DNA strand newly synthesized in each revolution displaces the strand synthesized in the previous revolution, giving a tail containing a linear series of sequences complementary to the circular template strand.
Rot is the product of RNA concentration and time of incubation in an RNA-driven hybridization reaction.
Rough ER consists of endoplasmic reticulum associated with ribosomes.

S phase is the restricted part of the eukaryotic cell cycle during which synthesis of DNA occurs.
S1 nuclease is an enzyme that specifically degrades unpaired (single-stranded) sequences of DNA.
Saltatory replication is a sudden lateral amplification to produce a large number of copies of some sequence.
Satellite DNA consists of many tandem repeats (identical or related) of a short basic repeating unit.
Saturation density is the density to which cultured eukaryotic cells grow *in vitro* before division is inhibited by cell-cell contacts.
Saturation hybridization experiment has a large excess of one component, causing all complementary sequences in the other component to enter a duplex form.
Scaffold of a chromosome is a proteinaceous structure in the shape of a sister chromatid pair, generated when chromosomes are depleted of histones.
Scarce (complex) mRNA consists of a large number of

individual mRNA species, each present in very few copies per cell.
scRNA is any one of several small cytoplasmic RNA molecules present in the cytoplasm and (sometimes) nucleus.
scRNPs are small cytoplasmic ribonucleoproteins (scRNAs associated with proteins).
Segmentation genes are concerned with controlling the number or polarity of body segments in insects.
Selection describes the use of particular conditions to allow survival only of cells with a particular phenotype.
Semiconservative replication is accomplished by separation of the strands of a parental duplex, each then acting as a template for synthesis of a complementary strand.
Semidiscontinuous replication is the mode in which one new strand is synthesized continuously while the other is synthesized discontinuously.
Septum constitutes the material that forms in the center of a bacterium to divide it into two daughter cells at the end of a division cycle.
Serum dependence describes the need of eukaryotic cells for factors contained in serum in order to grow in culture.
Sex chromosomes are those whose contents are different in the two sexes; usually labeled X and Y (or W and Z), one sex has XX (or WW), the other sex has XY (or WZ).
Sex linkage is the pattern of inheritance shown by genes carried on a sex chromosome (usually the X).
Sex plasmid is actually an episome; it is able to initiate the process of conjugation, by which chromosomal material is transferred from one bacterium to another.
Shine-Dalgarno sequence is part or all of the polypurine sequence AGGAGG located on bacterial mRNA just prior to an AUG initiation codon; is complementary to the sequence at the 3′ end of 16S rRNA; involved in binding of ribosome to mRNA.
Short-period interspersion is a pattern in a genome in which moderately repetitive DNA sequences of ~300 bp alternate with nonrepetitive sequences of ~1000 bp.
Shotgun experiment is cloning of an entire genome in the form of randomly generated fragments.
Shuttle vector is a plasmid constructed to have origins for replication for two hosts (for example, *E. coli* and *S. cerevisiae*) so that it can be used to carry a foreign sequence in either prokaryotes or eukaryotes.
Sigma factor is the subunit of bacterial RNA polymerase needed for initiation; is the major influence on selection of binding sites (promoters).
Signal hypothesis describes the role of the N-terminal sequence of a secreted protein in attaching nascent polypeptide to membrane; that is, mRNA and ribosome are attached to membrane via the N-terminal end of the protein under synthesis.
Signal sequence is the region of a protein (usually N-terminal) responsible for co-translational insertion into membranes of the endoplasmic reticulum.
Signal transduction describes the process by which a receptor interacts with a ligand at the surface of the cell and then transmits a signal to trigger a pathway within the cell.
Silent mutations do not change the product of a gene.
Silent sites in a gene describe those positions at which mutations do not alter the product.
Simple-sequence DNA equals satellite DNA.
SINES are a class of retroposons found as short interspersed repeats in mammalian genomes; derived from transcripts of RNA polymerase III.
Single-copy plasmids are maintained in bacteria at a ratio of one plasmid for every host chromosome.
Single-strand assimilation describes the ability of RecA protein to cause a single strand of DNA to displace its homologous strand in a duplex; that is, the single strand is assimilated into the duplex.
Single-strand exchange is a reaction in which one of the strands of a duplex of DNA leaves its former partner and instead pairs with the complementary strand in another molecule, displacing its homolog in the second duplex.
Single X hypothesis describes the inactivation of one X chromosome in female mammals.
Sister chromatids are the copies of a chromosome produced by its replication.
Site-specific recombination occurs between two specific (not necessarily homologous) sequences, as in phage integration/excision or resolution of cointegrate structures during transposition.
Slow component of a reassociation reaction is the last to reassociate; usually consists of nonrepetitive DNA.
Slow-stop *dna* mutants of *E. coli* complete the current round of bacterial replication but cannot initiate another at 42°C.
Smooth ER consists of a region of endoplasmic reticulum devoid of ribosomes.
snRNA (small nuclear RNA) is any one of many small RNA species confined to the nucleus; several of the snRNAs are involved in splicing or other RNA processing reactions.
snRNPs are small nuclear ribonucleoproteins (snRNAs associated with proteins).
Solution hybridization—*see* liquid hybridization.
Somatic cells are all the cells of an organism except those of the germline.

Somatic mutation is a mutation occurring in a somatic cell, and therefore affecting only its daughter cells; it is not inherited by descendants of the organism.
SOS box is the DNA sequence (operator) of ~20 bp recognized by LexA repressor protein.
SOS response in *E. coli* describes the coordinate induction of many enzymes, including repair activities, in response to irradiation or other damage to DNA; results from activation of protease activity by RecA to cleave LexA repressor.
Southern blotting describes the procedure for transferring denatured DNA from an agarose gel to a nitrocellulose filter where it can be hybridized with a complementary nucleic acid.
Spheroplast is a bacterial or yeast cell whose wall has been largely or entirely removed.
Spindle describes the reorganized structure of a eukaryotic cell passing through division; the nucleus has been dissolved and chromosomes are attached to the spindle by microtubules.
Splice sites are the sequences immediately surrounding the exon-intron boundaries.
Splicing describes the removal of introns and joining of exons in RNA; thus introns are spliced out, while exons are spliced together.
Spontaneous mutations are those that occur in the absence of any added reagent to increase the mutation rate.
Sporulation is the generation of a spore by a bacterium (by morphological conversion) or by a yeast (as the product of meiosis).
SSB is the single-strand protein of *E. coli*, a protein that binds to single-stranded DNA.
Staggered cuts in duplex DNA are made when two strands are cleaved at different points near each other.
Startpoint (startsite) refers to the position on DNA corresponding to the first base incorporated into RNA.
Stem is the base-paired segment of a hairpin.
Sticky ends are complementary single strands of DNA that protrude from opposite ends of a duplex or from ends of different duplex molecules; can be generated by staggered cuts in duplex DNA.
Stop codons are the three triplets (UAA, UAG, UGA) which terminate protein synthesis.
Strand displacement is a mode of replication of some viruses in which a new DNA strand grows by displacing the previous (homologous) strand of the duplex.
Streptolydigins inhibit the elongation of transcription by bacterial RNA polymerase.
Stringent replication describes the limitation of single-copy plasmids to replication *pari passu* with the bacterial chromosome.
Stringent response refers to the ability of a bacterium to shut down synthesis of tRNA and ribosomes in a poor-growth medium.
Structural gene codes for any RNA or protein product other than a regulator.
Supercoiling describes the coiling of a closed duplex DNA in space so that it crosses over its own axis.
Superrepressed means the same as uninducible.
Suppression describes the occurrence of changes that eliminate the effects of a mutation without reversing the original change in DNA.
Suppressor (extragenic) is usually a gene coding a mutant tRNA that reads the mutated codon either in the sense of the original codon or to give an acceptable substitute for the original meaning.
Suppressor (intragenic) is a compensating mutation that restores the original reading frame after a frameshift.
SWI/SNF is a chromatin remodeling complex; it uses hydrolysis of ATP to change the organization of nucleosomes.
Synapsis describes the association of the two pairs of sister chromatids representing homologous chromosomes that occurs at the start of meiosis; the resulting structure is called a bivalent.
Synaptonemal complex describes the morphological structure of synapsed chromosomes.
Syntenic genetic loci lie on the same chromosome.

T cells are lymphocytes of the T (thymic) lineage; may be subdivided into several functional types. They carry a TCR (T-cell receptor) and are involved in the cell-mediated immune response.
T_m is the abbreviation for melting temperature.
Tandem repeats are multiple copies of the same sequence lying in series.
TATA box is a conserved A•T-rich septamer found about 25 bp before the startpoint of each eukaryotic RNA polymerase II transcription unit; may be involved in positioning the enzyme for correct initiation.
Telomerase is the ribonucleoprotein enzyme that creates repeating units of one strand at the telomere, by adding individual bases.
Telomere is the natural end of a chromosome; the DNA sequence consists of a simple repeating unit with a protruding single-stranded end that may fold into a hairpin.
Temperature-sensitive mutation creates a gene product that is functional at low temperature but inactive at higher temperature (the reverse relationship is usually called cold-sensitive).
Terminal redundancy describes the repetition of the

same sequence at both ends of (for example) a phage genome.

Termination codon is one of three triplet sequences, UAG (amber), UAA (ochre), or UGA, that cause termination of protein synthesis; they are also called "nonsense" codons.

Terminator is a sequence of DNA, represented at the end of the transcript, that causes RNA polymerase to terminate transcription.

Tertiary structure of a protein describes the organization in space of its polypeptide chain.

Testcross involves crossing an unknown genotype to a recessive homozygote so that the phenotypes of the progeny correspond directly to the chromosomes carried by the parent of unknown genotype.

Thalassemia is disease of red blood cells resulting from lack of either α or β globin.

Thymine dimer comprises a chemically cross-linked pair of adjacent thymine residues in DNA, a result of damage induced by ultraviolet irradiation.

Topoisomerase is an enzyme that can change the linking number of DNA (in steps of 1 by type I; in steps of 2 by type II).

Topological isomers are molecules of DNA that are identical except for a difference in linking number.

Tracer is a radioactively labeled nucleic acid component included in a reassociation reaction in amounts too small to influence the progress of reaction.

Trailer is a nontranslated sequence at the 3′ end of an mRNA following the termination codon.

Trans configuration of two sites refers to their presence on two different molecules of DNA (chromosomes).

Transcribed spacer is the part of an rRNA transcription unit that is transcribed but discarded during maturation; that is, it does not give rise to part of rRNA.

Transcription is synthesis of RNA on a DNA template.

Transcription unit is the distance between sites of initiation and termination by RNA polymerase; may include more than one gene.

Transduction refers to the transfer of a bacterial gene from one bacterium to another by a phage; a phage carrying host as well as its own genes is called transducing phage. Also describes the acquisition and transfer of eukaryotic cellular sequences by retroviruses.

Transfection of eukaryotic cells is the acquisition of new genetic markers by incorporation of added DNA.

Transformation of bacteria describes the acquisition of new genetic markers by incorporation of added DNA.

Transformation of eukaryotic cells refers to their conversion to a state of unrestrained growth in culture, resembling or identical with the tumorigenic condition.

Transgenic animals are created by introducing new DNA sequences into the germline via addition to the egg.

Transit peptide is the short leader sequence cleaved from proteins that are imported into cellular organelles by post-translational passage of the membrane.

Transition is a mutation in which one pyrimidine is substituted by the other or in which one purine is substituted for the other.

Translation is synthesis of protein on the mRNA template.

Translocation of a chromosome describes a rearrangement in which part of a chromosome is detached by breakage and then becomes attached to some other chromosome.

Translocation of a gene refers to the appearance of a new copy at location in the genome elsewhere from the original copy.

Translocation of a protein refers to its movement across a membrane.

Translocation of the ribosome is its movement one codon along mRNA after the addition of each amino acid to the polypeptide chain.

Transmembrane protein is a component of a membrane; a hydrophobic region or regions of the protein resides in the membrane, and hydrophilic regions are exposed on one or both sides of the membrane.

Transplantation antigen is protein coded by a major histocompatibility locus, present on all mammalian cells, involved in interactions between lymphocytes.

Transposase is the enzyme activity involved in insertion of transposon at a new site.

Transposition immunity refers to the ability of certain transposons to prevent others of the same type from transposing to the same DNA molecule.

Transposon is a DNA sequence able to insert itself at a new location in the genome (without any sequence relationship with the target locus).

Transposition refers to the movement of a transposon to a new site in the genome. (*See also* nonreplicative transposition, replicative transposition, and conservative transposition.)

Transvection describes the ability of a locus to influence activity of an allele on the other homolog only when two chromosomes are synapsed.

Transversion is a mutation in which a purine is replaced by a pyrimidine or vice versa.

True-breeding organisms are homozygous for the trait under consideration.

Twisting number of a DNA is the number of base pairs divided by the number of base pairs per turn of the double helix.

Underwinding of DNA is produced by negative supercoiling (because the double helix is itself coiled in the opposite sense from the intertwining of the strands).
Unequal crossing-over describes a recombination event in which the two recombining sites lie at nonidentical locations in the two parental DNA molecules.
Unidirectional replication refers to the movement of a single replication fork from a given origin.
Uninducible mutants cannot be induced.
Unscheduled DNA synthesis is any DNA synthesis occurring outside the S phase of the eukaryotic cell.
Up promoter mutations increase the frequency of initiation of transcription.
Upstream identifies sequences proceeding in the opposite direction from expression; for example, the bacterial promoter is upstream from the transcription unit, the initiation codon is upstream of the coding region.
URF is an open (unidentified) reading frame, presumed to code for protein, but for which no product has been found.

V gene is sequence coding for the major part of the variable (N-terminal) region of an immunoglobulin chain.
Variable region of an immunoglobulin chain is coded by the V gene and varies extensively when different chains are compared, as the result of multiple (different) genomic copies and changes introduced during construction of an active immunoglobulin.
Variegation of phenotype is produced by a change in genotype during somatic development.
Vector—*see* cloning vector.
Vesicles are small bodies bounded by membrane, derived by budding from one membrane, often able to fuse with another membrane.
Virion is the physical virus particle (irrespective of its ability to infect cells and reproduce).
Virulent phage mutants are unable to establish lysogeny.

Wobble hypothesis accounts for the ability of a tRNA to recognize more than one codon by unusual (non-G•C, non-A•T) pairing with the third base of a codon.
Writhing number is the number of times a duplex axis crosses over itself in space.

Zero time-binding DNA enters the duplex form at the start of a reassociation reaction; results from intramolecular reassociation of inverted repeats.
Zinc finger protein has a repeated motif of amino acids with characteristic spacing of cysteines that may be involved in binding zinc; is characteristic of some proteins that bind DNA and/or RNA.
Zoo blot describes the use of Southern blotting to test the ability of a DNA probe from one species to hybridize with the DNA from the genomes of a variety of other species.
Zygote is produced by fusion of two gametes—that is, it is a fertilized egg.

Index

Page numbers in italic refer to illustrations.